P9-DIA-543

Student Study Guide and Solutions Manual, 2e

for
Organic Chemistry, 2e

David Klein
Johns Hopkins University

WILEY

This book is printed on acid free paper. ∞

Founded in 1807, John Wiley & Sons, Inc. has been a valued source of knowledge and understanding for more than 200 years, helping people around the world meet their needs and fulfill their aspirations. Our company is built on a foundation of principles that include responsibility to the communities we serve and where we live and work. In 2008, we launched a Corporate Citizenship Initiative, a global effort to address the environmental, social, economic, and ethical challenges we face in our business. Among the issues we are addressing are carbon impact, paper specifications and procurement, ethical conduct within our business and among our vendors, and community and charitable support. For more information, please visit our website: www.wiley.com/go/citizenship.

Copyright © 2015, 2012   John Wiley & Sons, Inc. All rights reserved. No part of this publication may be reproduced, stored in a retrieval system or transmitted in any form or by any means, electronic, mechanical, photocopying, recording, scanning or otherwise, except as permitted under Sections 107 or 108 of the 1976 United States Copyright Act, without either the prior written permission of the Publisher, or authorization through payment of the appropriate per-copy fee to the Copyright Clearance Center, Inc. 222 Rosewood Drive, Danvers, MA 01923, website www.copyright.com. Requests to the Publisher for permission should be addressed to the Permissions Department, John Wiley & Sons, Inc., 111 River Street, Hoboken, NJ 07030-5774, (201)748-6011, fax (201)748-6008, website http://www.wiley.com/go/permissions.

Evaluation copies are provided to qualified academics and professionals for review purposes only, for use in their courses during the next academic year. These copies are licensed and may not be sold or transferred to a third party. Upon completion of the review period, please return the evaluation copy to Wiley. Return instructions and a free of charge return shipping label are available at www.wiley.com/go/return label. Outside of the United States, please contact your local representative.

CONTENTS

HOW TO USE THIS BOOK

Organic chemistry is much like bicycle riding. You cannot learn how to ride a bike by watching other people ride bikes. Some people might fool themselves into believing that it's possible to become an expert bike rider without ever getting on a bike. But you know that to be incorrect (and very naïve). In order to learn how to ride a bike, you must be willing to get on the bike, and you must be willing to fall. With time (and dedication), you can quickly train yourself to avoid falling, and to ride the bike with ease and confidence. The same is true of organic chemistry. In order to become proficient at solving problems, you must "ride the bike". You must try to solve the problems yourself (without the solutions manual open in front of you). Once you have solved the problems, this book will allow you to check your solutions. If, however, you don't attempt to solve each problem on your own, and instead, you read the problem statement and then immediately read the solution, you are only hurting yourself. You are not learning how to avoid falling. Many students make this mistake every year. They use the solutions manual as a crutch, and then they never really attempt to solve the problems on their own. It really is like believing that you can become an expert bike rider by watching hundreds of people riding bikes. The world doesn't work that way!

The textbook has thousands of problems to solve. Each of these problems should be viewed as an opportunity to develop your problem-solving skills. By reading a problem statement and then reading the solution immediately (without trying to solve the problem yourself), you are robbing yourself of the opportunity provided by the problem. If you repeat that poor study habit too many times, you will not learn how to solve problems on your own, and you will not get the grade that you want.

Why do so many students adopt this bad habit (of using the solutions manual too liberally)? The answer is simple. Students often wait until a day or two before the exam, and then they spend all night cramming. Sound familiar? Unfortunately, organic chemistry is the type of course where cramming is insufficient, because you need time in order to ride the bike yourself. You need time to think about each problem until you have developed a solution *on your own*. For some problems, it might take days before you think of a solution. This process is critical for learning this subject. Make sure to allot time every day for studying organic chemistry, and use this book to check your solutions. This book has also been designed to serve as a study guide, as described below.

WHAT'S IN THIS BOOK

This book contains more than just solutions to all of the problems in the textbook. Each chapter of this book also contains a series of exercises that will help you review the concepts, skills and reactions presented in the corresponding chapter of the textbook. These exercises are designed to serve as study tools that can help you identify your weak areas. Each chapter of this solutions manual/study guide has the following parts:

- **Review of Concepts**. These exercises are designed to help you identify which concepts are the least familiar to you. Each section contains sentences with missing words (blanks). Your job is to fill in the blanks, demonstrating mastery of the concepts. To verify that your answers are correct, you can open your textbook to the end of the corresponding chapter, where you will find a section entitled *Review of Concepts and Vocabulary*. In that section, you will find each of the sentences, verbatim.

- **Review of Skills**. These exercises are designed to help you identify which skills are the least familiar to you. Each section contains exercises in which you must demonstrate mastery of the skills developed in the *SkillBuilders* of the corresponding textbook chapter. To verify that your answers are correct, you can open your textbook to the end of the corresponding chapter, where you will find a section entitled *SkillBuilder Review*. In that section, you will find the answers to each of these exercises.

- **Review of Reactions**. These exercises are designed to help you identify which reagents are not at your fingertips. Each section contains exercises in which you must demonstrate familiarity with the reactions covered in the textbook. Your job is to fill in the reagents necessary to achieve each reaction. To verify that your answers are correct, you can open your textbook to the end of the corresponding chapter, where you will find a section entitled *Review of Reactions*. In that section, you will find the answers to each of these exercises.

- **Common Mistakes to Avoid**. This is a new feature to this edition. The most common student mistakes are described, so that you can avoid them when solving problems.

- **A List of Useful Reagents**. This is a new feature to this edition. This list provides a review of the reagents that appear in each chapter, as well as a description of how each reagent is used.

- **Solutions**. At the end of each chapter, you'll find detailed solutions to all problems in the textbook, including all SkillBuilders, conceptual checkpoints, additional problems, integrated problems, and challenge problems.

The sections described above have been designed to serve as useful tools as you study and learn organic chemistry. Good luck!

David Klein
Senior Lecturer, Department of Chemistry
Johns Hopkins University

Chapter 1
A Review of General Chemistry:
Electrons, Bonds and Molecular Properties

Review of Concepts

Fill in the blanks below. To verify that your answers are correct, look in your textbook at the end of Chapter 1. Each of the sentences below appears verbatim in the section entitled *Review of Concepts and Vocabulary*.

- _constitutional_ **isomers** share the same molecular formula but have different connectivity of atoms and different physical properties.
- Second-row elements generally obey the _octet_ **rule**, bonding to achieve noble gas electron configuration.
- A pair of unshared electrons is called a _lone pair_ .
- A **formal charge** occurs when an atom does not exhibit the appropriate number of _valence electrons_ .
- An **atomic orbital** is a region of space associated with _an atom_ , while a **molecular orbital** is a region of space associated with _molecule_ .
- Methane's tetrahedral geometry can be explained using four degenerate _sp^3_ **-hybridized orbitals** to achieve its four single bonds.
- Ethylene's planar geometry can be explained using three degenerate _sp^2_ **-hybridized orbitals**.
- Acetylene's linear geometry is achieved via _sp_ **-hybridized** carbon atoms.
- The geometry of small compounds can be predicted using valence shell electron pair repulsion (**VSEPR**) theory, which focuses on the number of _σ_ bonds and _lone pairs_ exhibited by each atom.
- The physical properties of compounds are determined by _IMF_ forces, the attractive forces between molecules.
- **London dispersion forces** result from the interaction between transient _dipole moments_ and are stronger for larger alkanes due to their larger surface area and ability to accommodate more interactions.

Review of Skills

Fill in the blanks and empty boxes below. To verify that your answers are correct, look in your textbook at the end of Chapter 1. The answers appear in the section entitled *SkillBuilder Review*.

SkillBuilder 1.1   Determining the Constitution of Small Molecules

STEP 1 - DETERMINE THE VALENCY (NUMBER OF EXPECTED BONDS) FOR EACH ATOM IN C_2H_5Cl	STEP 2 - DRAW THE STRUCTURE OF C_2H_5Cl BY PLACING ATOMS WITH THE HIGHEST VALENCY AT THE CENTER, AND PLACING MONOVALENT ATOMS AT THE PERIPHERY
Each carbon atom is expected to form _4_ bonds. Each hydrogen atom is expected to form _1_ bonds. The chlorine atom is expected to form _1_ bonds.	$$\begin{array}{cccc} & H & H & \\ & | & | & \\ H - & C - & C - & Cl \\ & | & | & \\ & H & H & \end{array}$$

SkillBuilder 1.2   Drawing the Lewis Dot Structure of an Atom

STEP 1 - DETERMINE THE NUMBER OF VALENCE ELECTRONS	STEP 2 - PLACE ONE ELECTRON BY ITSELF ON EACH SIDE OF THE ATOM	STEP 3 - IF THE ATOM HAS MORE THAN FOUR VALENCE ELECTRONS, PAIR THE REMAINING ELECTRONS WITH THE ELECTRONS ALREADY DRAWN
Nitrogen is in Group 5 of the periodic table, and is expected to have 5 valence electrons.		

SkillBuilder 1.3   Drawing the Lewis Structure of a Small Molecule

STEP 1 - DRAW THE LEWIS DOT STRUCTURE OF EACH ATOM IN CH_2O	STEP 2 - FIRST CONNECT ATOMS THAT FORM MORE THAN ONE BOND	STEP 3 - CONNECT THE HYDROGEN ATOMS	STEP 4 - PAIR ANY UNPAIRED ELECTRONS, SO THAT EACH ATOM ACHIEVES AN OCTET
	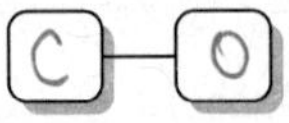	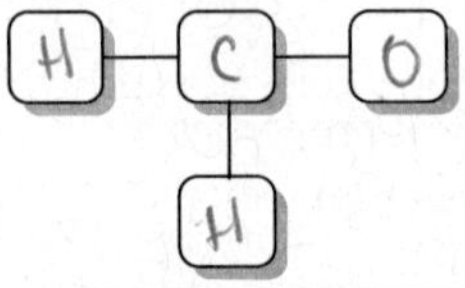	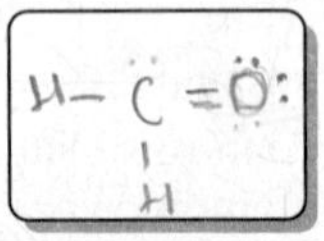

SkillBuilder 1.4   Calculating Formal Charge

STEP 1 - DETERMINE THE APPROPRIATE NUMBER OF VALENCE ELECTRONS	STEP 2 - DETERMINE THE NUMBER OF VALENCE ELECTRONS IN THIS CASE	STEP 3 - ASSIGN A FORMAL CHARGE TO THE NITROGEN ATOM IN THIS CASE
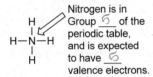 Nitrogen is in Group 5 of the periodic table, and is expected to have 5 valence electrons.	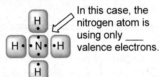 In this case, the nitrogen atom is using only ___ valence electrons.	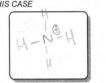

SkillBuilder 1.5   Locating Partial Charges Resulting from Induction

STEP 1 - CIRCLE THE BONDS BELOW THAT ARE POLAR COVALENT	STEP 2 - FOR EACH POLAR COVALENT BOND, DRAW AN ARROW THAT SHOWS THE DIRECTION OF THE DIPOLE MOMENT	STEP 3 - INDICATE THE LOCATION OF ALL PARTIAL CHARGES ($\delta+$ and $\delta-$)
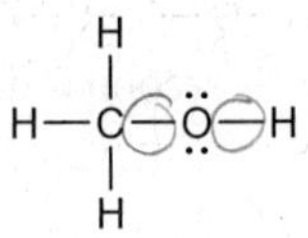	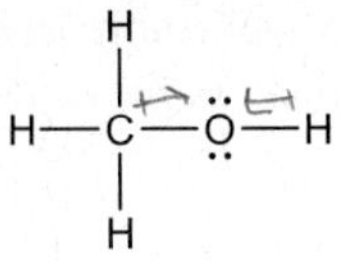 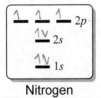	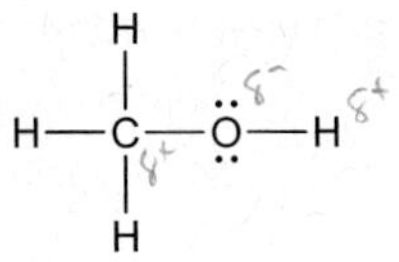

SkillBuilder 1.6   Identifying Electron Configurations

STEP 1 - IN THE ENERGY DIAGRAM SHOWN HERE, DRAW THE ELECTRON CONFIGURATION OF NITROGEN (USING ARROWS TO REPRESENT ELECTRONS).	STEP 2 - FILL IN THE BOXES BELOW WITH THE NUMBERS THAT CORRECTLY DESCRIBE THE ELECTRON CONFIGURATION OF NITROGEN
___↑_ ___↑_ ___↑_ $2p$ _↑↓_ $2s$ _↑↓_ $1s$ Nitrogen	$1s$ ② $2s$ ② $2p$ ③

SkillBuilder 1.7   Identifying Hybridization States

A CARBON ATOM WITH FOUR SINGLE BONDS WILL BE ____ HYBRIDIZED	A CARBON ATOM WITH ONE DOUBLE BOND WILL BE ____ HYBRIDIZED	A CARBON ATOM WITH A TRIPLE BOND WILL BE ____ HYBRIDIZED
sp^3 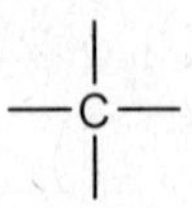 	sp^2 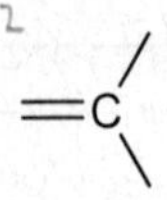 	sp 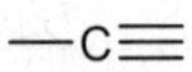

SkillBuilder 1.8 Predicting Geometry

STEP 1 *- DETERMINE THE STERIC NUMBER OF THE NITROGEN ATOM BELOW BY ADDING THE NUMBER OF SINGLE BONDS AND LONE PAIRS*

\# of single bonds = [3]

\# of lone pairs = [1]

Steric Number = [4]

STEP 2 *- THE STERIC NUMBER DETERMINES THE HYBRIDIZATION STATE AND ELECTRONIC GEOMETRY. FILL IN THE CHART BELOW:*

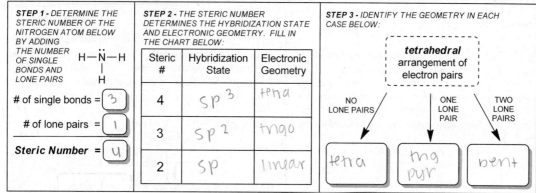

Steric #	Hybridization State	Electronic Geometry
4	[sp³]	[tetra]
3	[sp²]	[trigo]
2	[sp]	[linear]

STEP 3 *- IDENTIFY THE GEOMETRY IN EACH CASE BELOW:*

tetrahedral arrangement of electron pairs

- NO LONE PAIRS → [tetra]
- ONE LONE PAIR → [trig pyr]
- TWO LONE PAIRS → [bent]

SkillBuilder 1.9 Identifying the Presence of Molecular Dipole Moments

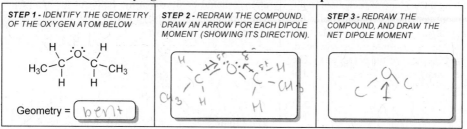

STEP 1 *- IDENTIFY THE GEOMETRY OF THE OXYGEN ATOM BELOW*

Geometry = [bent]

STEP 2 *- REDRAW THE COMPOUND. DRAW AN ARROW FOR EACH DIPOLE MOMENT (SHOWING ITS DIRECTION).*

STEP 3 *- REDRAW THE COMPOUND, AND DRAW THE NET DIPOLE MOMENT*

SkillBuilder 1.10 Predicting Physical Properties

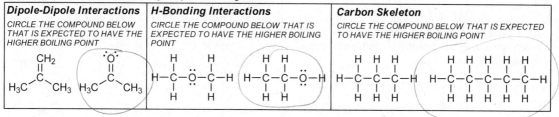

Dipole-Dipole Interactions

CIRCLE THE COMPOUND BELOW THAT IS EXPECTED TO HAVE THE HIGHER BOILING POINT

H-Bonding Interactions

CIRCLE THE COMPOUND BELOW THAT IS EXPECTED TO HAVE THE HIGHER BOILING POINT

Carbon Skeleton

CIRCLE THE COMPOUND BELOW THAT IS EXPECTED TO HAVE THE HIGHER BOILING POINT

A Common Mistake to Avoid

When drawing a structure, don't forget to draw formal charges, as forgetting to do so is a common error. If a formal charge is present, it MUST be drawn. For example, in the following case, the nitrogen atom bears a positive charge, so the charge must be drawn:

INCORRECT **CORRECT**

As we progress though the course, we will see structures of increasing complexity. If formal charges are present, failure to draw them constitutes an error, and must be scrupulously avoided. If you have trouble drawing formal charges, go back and master that skill. You can't go on without it. Don't make the mistake of underestimating the importance of being able to draw formal charges with confidence.

Solutions

1.1.
(a) Begin by determining the valency of each atom in the compound. The carbon atom is tetravalent, the oxygen atom is divalent, and the hydrogen atoms are all monovalent. The atoms with more than one bond (in this case, C and O) should be drawn in the center of the compound. The hydrogen atoms are then placed at the periphery, as shown.

$$
\begin{array}{c}
\quad\; H \\
\quad\; | \\
H-C-O-H \\
\quad\; | \\
\quad\; H
\end{array}
$$

(b) Begin by determining the valency of each atom in the compound. The carbon atom is tetravalent, while the chlorine and hydrogen atoms are all monovalent. The carbon atom is the only atom with more than one bond, so it must be drawn in the center of the compound. The chlorine atom and the hydrogen atoms are then placed at the periphery, as shown.

$$
\begin{array}{c}
\quad\; H \\
\quad\; | \\
H-C-Cl \\
\quad\; | \\
\quad\; H
\end{array}
$$

The chlorine atom can be placed in any one of the four available positions. The following four drawings all represent the same compound, in which the central carbon atom is connected to the Cl.

$$
\begin{array}{cccc}
H & Cl & H & H \\
| & | & | & | \\
H-C-Cl & H-C-H & Cl-C-H & H-C-H \\
| & | & | & | \\
H & H & H & Cl
\end{array}
$$

(c) The carbon atoms are tetravalent, while the hydrogen atoms are all monovalent. The atoms with more than one bond (in this case, the two carbon atoms) should be drawn in the center of the compound. The hydrogen atoms are then placed at the periphery, as shown.

$$
\begin{array}{c}
H\;\; H \\
|\;\;\; | \\
H-C-C-H \\
|\;\;\; | \\
H\;\; H
\end{array}
$$

(d) The carbon atom is tetravalent, the nitrogen atom is trivalent, and the hydrogen atoms are all monovalent. The atoms with more than one bond (in this case, C and N) should be drawn in the center of the compound. The hydrogen atoms are then placed at the periphery, as shown.

$$
\begin{array}{c}
H\;\;\;\; H \\
|\;\;\;\;\; / \\
H-C-N \\
|\;\;\;\;\; \backslash \\
H\;\;\;\; H
\end{array}
$$

(e) The carbon atoms are tetravalent, while the fluorine atoms are all monovalent. The atoms with more than one bond (in this case, the two carbon atoms) should be drawn in the center of the compound. The fluorine atoms are then placed at the periphery, as shown.

$$
\begin{array}{c}
F\;\;\; F \\
|\;\;\;\; | \\
F-C-C-F \\
|\;\;\;\; | \\
F\;\;\; F
\end{array}
$$

(f) The carbon atoms are tetravalent, while the bromine atom and hydrogen atoms are all monovalent. The atoms with more than one bond (in this case, the two carbon atoms) should be drawn in the center of the compound. The bromine atom and hydrogen atoms are then placed at the periphery, as shown.

$$
\begin{array}{c}
H\;\; H \\
|\;\;\; | \\
H-C-C-Br \\
|\;\;\; | \\
H\;\; H
\end{array}
$$

The bromine atom can be placed in any one of the six available positions. The following six drawings all represent the same compound, in which the two carbon atoms are connected to each other, and the bromine atom is connected to one of the carbon atoms.

$$
\begin{array}{ccc}
H\;\; H & H\;\; Br & Br\;\; H \\
|\;\;\; | & |\;\;\; | & |\;\;\; | \\
H-C-C-Br & H-C-C-H & H-C-C-H \\
|\;\;\; | & |\;\;\; | & |\;\;\; | \\
H\;\; H & H\;\; H & H\;\; H
\end{array}
$$

$$
\begin{array}{ccc}
H\;\; H & H\;\; H & H\;\; H \\
|\;\;\; | & |\;\;\; | & |\;\;\; | \\
Br-C-C-H & H-C-C-H & H-C-C-H \\
|\;\;\; | & |\;\;\; | & |\;\;\; | \\
H\;\; H & Br\;\; H & H\;\; Br
\end{array}
$$

(g) The carbon atoms are tetravalent, while the hydrogen atoms are all monovalent. The atoms with more than one bond (in this case, the three carbon atoms) should be drawn in the center of the compound. The hydrogen atoms are then placed at the periphery, as shown.

$$
\begin{array}{c}
H\;\; H\;\; H \\
|\;\;\; |\;\;\; | \\
H-C-C-C-H \\
|\;\;\; |\;\;\; | \\
H\;\; H\;\; H
\end{array}
$$

1.2. Begin by determining the valency of each atom that appears in the molecular formula. The carbon atoms are tetravalent, while the chlorine atom and hydrogen atoms are all monovalent. The atoms with more than one bond (in this case, the three carbon atoms) should be drawn in the center of the compound. Then, as indicated in the problem statement, the chlorine atom can be placed in either of two locations: i) connected to the central carbon atom, or ii) connected to one of the other two (equivalent) carbon atoms. The hydrogen atoms are then placed at the periphery.

```
  H  Cl  H            H   H   H
  |   |  |            |   |   |
H-C - C -C-H      H-C - C - C-Cl
  |   |  |            |   |   |
  H   H  H            H   H   H
```

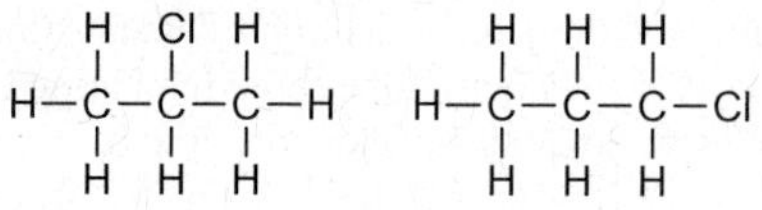

1.3. The carbon atoms are tetravalent, the oxygen atom is divalent, and the hydrogen atoms are all monovalent. The atoms with more than one bond (in this case, C, C, C and O) should be drawn in the center of the compound. There are several different ways to connect these four atoms. We can connect the three carbon atoms in a linear chain (C-C-C), which gives two different locations where the oxygen atom can be placed (see the solution to Problem 1.2). Specifically, the oxygen atom can either be connected to the central carbon atom (C2) or connected to one of the other two carbon atoms (C1 or C3).

```
    O                O
    |                |
 C - C - C        C - C - C
 1   2   3        1   2   3
```

Connecting the oxygen atom to C3 is the same as connecting it to C1 (if we just assign numbers in the other direction, right-to-left).

There is another way in which the three carbon atoms and the one oxygen atom can be connected. Instead of connecting the carbon atoms to each other in a chain of three carbon atoms, we can insert the oxygen atom between two of the carbon atoms, like this:

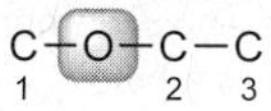

```
 C - O - C - C
 1       2   3
```

Inserting the oxygen atom between C2 and C3 is the same as inserting it between C1 and C2 (if we just assign numbers in the other direction, right-to-left).

In summary, we have seen exactly three different ways to connect three carbon atoms and one oxygen atom. For each of these possibilities, shown below, the hydrogen atoms are placed at the periphery.

```
      H                     H
      |                     |
  H   O   H             O   H   H
  |   |   |             |   |   |
H-C - C - C-H       H-C - C - C-H
  |       |             |   |   |
  H       H             H   H   H

              H       H  H
              |       |  |
          H-C - O - C - C-H
              |       |  |
              H       H  H
```

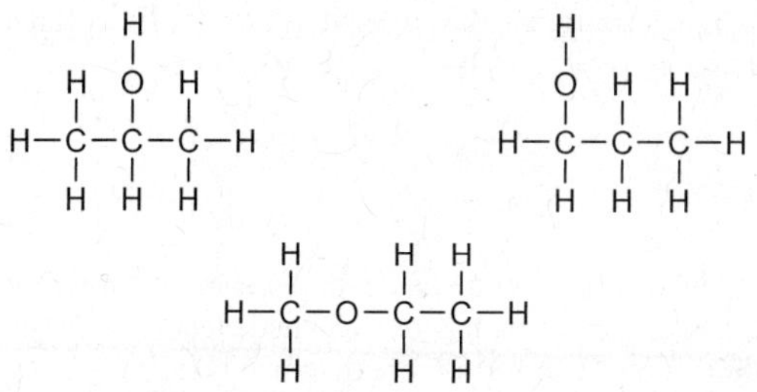

1.4. Begin by determining the valency of each atom that appears in the molecular formula. The carbon atoms are tetravalent, the oxygen atom is divalent, and the hydrogen atoms are all monovalent. Any atoms with more than one bond (in this case, the four carbon atoms and the one oxygen atom) should be drawn in the center of the compound, with the hydrogen atoms at the periphery. There are several different ways to connect

four carbon atoms and one oxygen atom. Let's begin with the four carbon atoms. There are two different ways to connect four carbon atoms. They can either be arranged in a linear fashion or in a branched fashion.

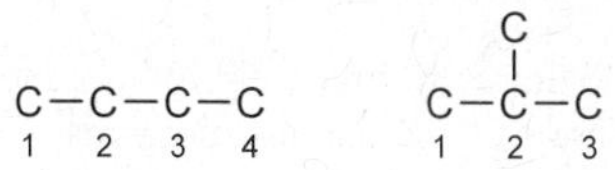

```
                          C
                          |
 C - C - C - C        C - C - C
 1   2   3   4        1   2   3
    Linear              Branched
```

Next, the oxygen atom must be inserted. For each of the two skeletons above (linear or branched), there are several different locations to insert the oxygen atom. The linear skeleton has four possibilities, shown here:

```
 O                        O
 |                        |
 C - C - C - C        C - C - C - C
 1   2   3   4        1   2   3   4

 C - O - C - C - C    C - C - O - C - C
 1       2   3   4    1   2       3   4
```

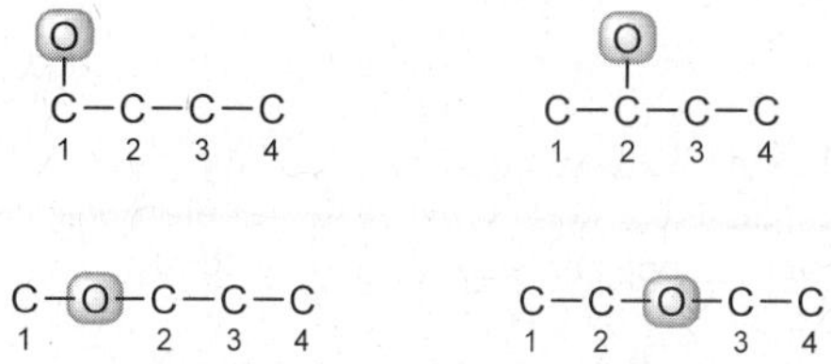

and the branched skeleton has three possibilities shown here:

```
    C            C                C
    |            |                |
  C - C - C    C - C - C      C - O - C - C
  |   2   3    1   2   3      1       2   3
  O            |
               O
```

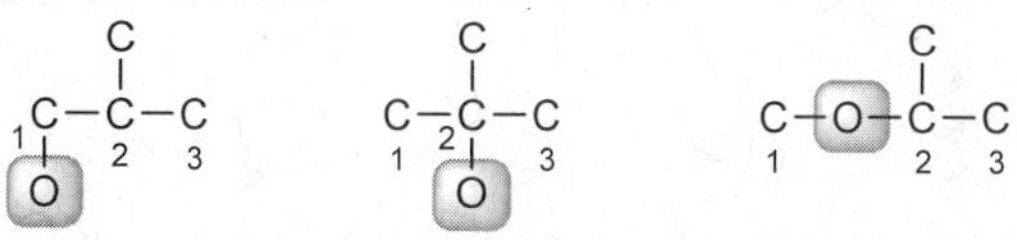

Finally, we complete all of the structures by drawing the bonds to hydrogen atoms.

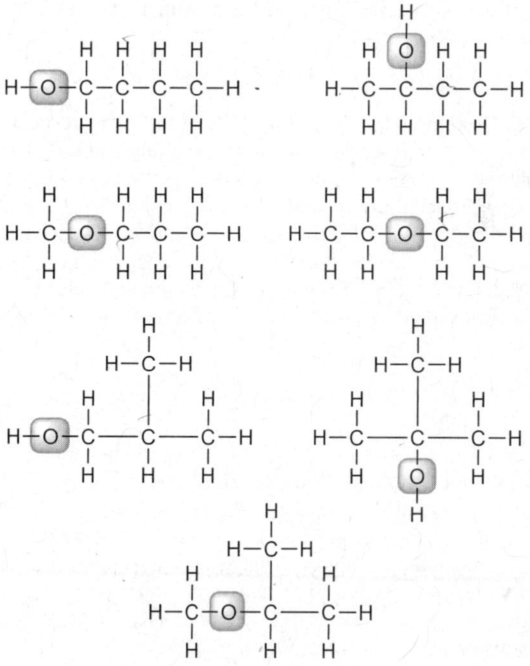

```
                                         H
                                         |
    H   H   H   H            H   O   H   H
    |   |   |   |            |   |   |   |
H-O - C - C - C - C-H  ·  H-C - C - C - C-H
    |   |   |   |            |   |   |   |
    H   H   H   H            H   H   H   H


    H       H   H   H        H   H       H   H
    |       |   |   |        |   |       |   |
H-C - O - C - C - C-H    H-C - C - O - C - C-H
    |       |   |   |        |   |       |   |
    H       H   H   H        H   H       H   H


            H                        H
            |                        |
          H-C-H                    H-C-H
    H       |       H        H       |       H
    |       |       |        |       |       |
H-O - C -   C   -  C-H    H-C -     C    -  C-H
    |       |       |        |       |       |
    H       H       H        H       O       H
                                     |
                                     H


                    H
                    |
                  H-C-H
          H         |        H
          |         |        |
      H-C - O - C   -   C-H
          |         |        |
          H         H        H
```

1.5.
(a) Carbon belongs to group 4A of the periodic table, and it therefore has four valence electrons. The periodic symbol for carbon (C) is drawn, and each valence

electron is placed by itself (unpaired), around the C, like this:

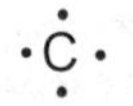

·Ċ·

(b) Oxygen belongs to group 6A of the periodic table, and it therefore has six valence electrons. The periodic symbol for oxygen (O) is drawn, and each valence electron is placed by itself (unpaired) on a side of the O, until all four sides are occupied. That takes care of four of the six electrons, leaving just two more electrons to draw. Each of the two remaining electrons is then paired up with an electron already drawn, like this:

:Ö:

(c) Fluorine belongs to group 7A of the periodic table, and it therefore has seven valence electrons. The periodic symbol for fluorine (F) is drawn, and each valence electron is placed by itself (unpaired) on a side of the F, until all four sides are occupied. That takes care of four of the seven electrons, leaving three more electrons to draw. Each of the three remaining electrons is then paired up with an electron already drawn, like this:

:F̈·

(d) Hydrogen belongs to group 1A of the periodic table, and it therefore has one valence electron. The periodic symbol for hydrogen (H) is drawn, and the one and only valence electron is placed on a side of the H, like this:

H·

(e) Bromine belongs to group 7A of the periodic table, and it therefore has seven valence electrons. The periodic symbol for bromine (Br) is drawn, and each valence electron is placed by itself (unpaired) on a side of the Br, until all four sides are occupied. That takes care of four of the seven electrons, leaving three more electrons to draw. Each of the three remaining electrons is then paired up with an electron already drawn, like this:

:B̈r·

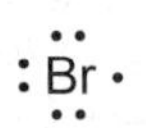

(f) Sulfur belongs to group 6A of the periodic table, and it therefore has six valence electrons. The periodic symbol for sulfur (S) is drawn, and each valence electron is placed by itself (unpaired) on a side of the S, until all four sides are occupied. That takes care of four of the six electrons, leaving just two more electrons to draw. Each of the two remaining electrons is then paired up with an electron already drawn, like this:

:S̈:

(g) Chlorine belongs to group 7A of the periodic table, and it therefore has seven valence electrons. The

periodic symbol for chlorine (Cl) is drawn, and each valence electron is placed by itself (unpaired) on a side of the Cl, until all four sides are occupied. That takes care of four of the seven electrons, leaving three more electrons to draw. Each of the three remaining electrons is then paired up with an electron already drawn, like this:

:C̈l·

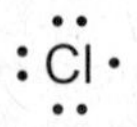

(h) Iodine belongs to group 7A of the periodic table, and it therefore has seven valence electrons. The periodic symbol for iodine (I) is drawn, and each valence electron is placed by itself (unpaired) on a side of the I, until all four sides are occupied. That takes care of four of the seven electrons, leaving three more electrons to draw. Each of the three remaining electrons is then paired up with an electron already drawn, like this:

:Ï·

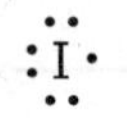

1.6. Both nitrogen and phosphorus belong to group **5A** of the periodic table, and therefore, each of these atoms has five valence electrons. In order to achieve an octet, we expect each of these elements to form three bonds.

1.7. Aluminum is directly beneath boron on the periodic table (group **3A**), and therefore both elements exhibit three valence electrons.

1.8. The Lewis dot structure for a carbon atom is shown in the solution to Problem **1.5a**. That drawing must be modified by removing one electron, resulting in a formal positive charge, as shown below. This resembles boron because it exhibits three valence electrons.

·Ċ⊕

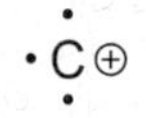

1.9. The Lewis dot structure for a carbon atom is shown in the solution to Problem **1.5a**. That drawing must be modified by drawing one additional electron, resulting in a formal negative charge, as shown below. This resembles nitrogen because it exhibits five valence electrons.

·Ċ:⊖

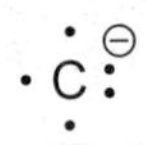

1.10.
(a) Each carbon atom has four valence electrons, and each hydrogen atom has one valence electron. Only the carbon atoms can form more than one bond, so we begin by connecting the carbon atoms to each other. Then, we connect all of the hydrogen atoms, as shown.

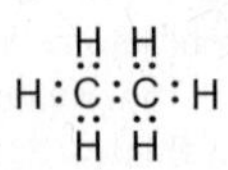

H H
H:C:C:H
H H

(b) Each carbon atom has four valence electrons, and each hydrogen atom has one valence electron. Only the carbon atoms can form more than one bond, so we begin by connecting the carbon atoms to each other. Then, we connect all of the hydrogen atoms, and the unpaired electrons are shared to give a double bond. In this way, each of the carbon atoms achieves an octet.

$$H:C::C:H$$
$$\overset{..}{H} \quad \overset{..}{H}$$

(c) Each carbon atom has four valence electrons, and each hydrogen atom has one valence electron. Only the carbon atoms can form more than one bond, so we begin by connecting the carbon atoms to each other. Then, we connect all of the hydrogen atoms, and the unpaired electrons are shared to give a triple bond. In this way, each of the carbon atoms achieves an octet.

$$H:C:::C:H$$

(d) Each carbon atom has four valence electrons, and each hydrogen atom has one valence electron. Only the carbon atoms can form more than one bond, so we begin by connecting the carbon atoms to each other. Then, we connect all of the hydrogen atoms, as shown.

$$\overset{H \; H \; H}{H:C:C:C:H}$$
$$\overset{}{H \; H \; H}$$

(e) Each carbon atom has four valence electrons, and each hydrogen atom has one valence electron. Only the carbon atoms can form more than one bond, so we begin by connecting the carbon atoms to each other. Then, we connect all of the hydrogen atoms, and the unpaired electrons are shared to give a double bond. In this way, each of the carbon atoms achieves an octet.

$$\overset{H}{H:C:C::C:H}$$
$$\overset{}{H \; H \; H}$$

(f) The carbon atom has four valence electrons, the oxygen atom has six valence electrons, and each hydrogen atom has one valence electron. Only the carbon atom and the oxygen atom can form more than one bond, so we begin by connecting them to each other. Then, we connect all of the hydrogen atoms, as shown.

$$\overset{H}{H:C:\overset{..}{\underset{..}{O}}:H}$$
$$\overset{}{H}$$

1.11. Boron has three valence electrons, each of which is shared with a hydrogen atom, shown below. The central boron atom lacks an octet of electrons, and it is therefore very unstable and reactive.

$$\overset{H}{H:B}$$
$$\overset{}{H}$$

1.12. Each of the carbon atoms has four valence electrons; the nitrogen atom has five valence electrons; and each of the hydrogen atoms has one valence electron. We begin by connecting the atoms that have more than one bond (in this case, the three carbon atoms and the nitrogen atom). There are four different ways that these four atoms can be connected to each other, shown below.

For each of these possible arrangements, we connect the hydrogen atoms, giving the following four constitutional isomers.

In each of these four structures, the nitrogen atom has one lone pair.

1.13.
(a) Aluminum is in group 3A of the periodic table, and it should therefore have three valence electrons. In this case, the aluminum atom exhibits four valence electrons (one for each bond). With one extra electron, this aluminum atom will bear a negative charge.

(b) Oxygen is in group 6A of the periodic table, and it should therefore have six valence electrons. In this case, the oxygen atom exhibits only five valence electrons (one for each bond, and two for the lone pair). This oxygen atom is missing an electron, and it therefore bears a positive charge.

(c) Nitrogen is in group 5A of the periodic table, and it should therefore have five valence electrons. In this case, the nitrogen atom exhibits six valence electrons (one for each bond and two for each lone pair). With one extra electron, this nitrogen atom will bear a negative charge.

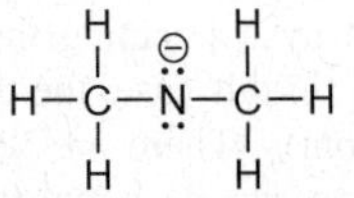

(d) Oxygen is in group 6A of the periodic table, and it should therefore have six valence electrons. In this case, the oxygen atom exhibits only five valence electrons (one for each bond, and two for the lone pair). This oxygen atom is missing an electron, and it therefore bears a positive charge.

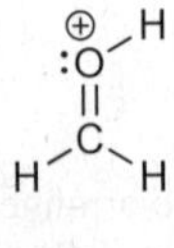

(e) Carbon is in group 4A of the periodic table, and it should therefore have four valence electrons. In this case, the carbon atom exhibits five valence electrons (one for each bond and two for the lone pair). With one extra electron, this carbon atom will bear a negative charge.

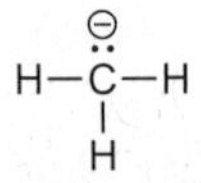

(f) Carbon is in group 4A of the periodic table, and it should therefore have four valence electrons. In this case, the carbon atom exhibits only three valence electrons (one for each bond). This carbon atom is missing an electron, and it therefore bears a positive charge.

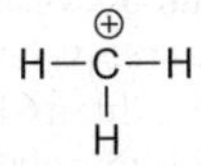

(g) Oxygen is in group 6A of the periodic table, and it should therefore have six valence electrons. In this case, the oxygen atom exhibits only five valence electrons (one for each bond, and two for the lone pair). This oxygen atom is missing an electron, and it therefore bears a positive charge.

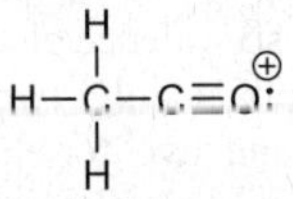

(h) Two of the atoms in this structure exhibit a formal charge because each of these atoms does not exhibit the appropriate number of valence electrons. The aluminum atom (group 3A) should have three valence electrons, but it exhibits four (one for each bond). With one extra electron, this aluminum atom will bear a negative charge. The neighboring chlorine atom (to the right) should have seven valence electrons, but it exhibits only six (one for each bond and two for each lone pair). It is missing one

electron, so this chlorine atom will bear a positive charge.

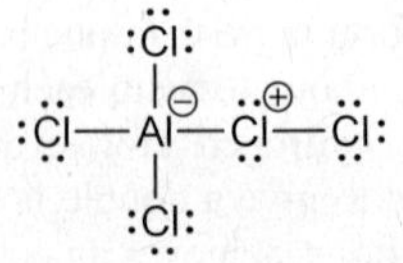

(i) Two of the atoms in this structure exhibit a formal charge because each of these atoms does not exhibit the appropriate number of valence electrons. The nitrogen atom (group 5A) should have five valence electrons, but it exhibits four (one for each bond). It is missing one electron, so this nitrogen atom will bear a positive charge. One of the two oxygen atoms (the one on the right) exhibits seven valence electrons (one for the bond, and two for each lone pair), although it should have only six. With one extra electron, this oxygen atom will bear a negative charge.

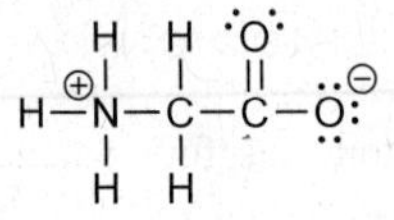

1.14.
(a) The boron atom in this case exhibits four valence electrons (one for each bond), although boron (group 3A) should only have three valence electrons. With one extra electron, this boron atom bears a negative charge.

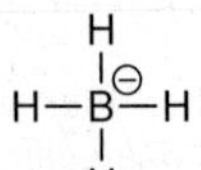

(b) The nitrogen atom in this case exhibits six valence electrons (one for each bond and two for each lone pair). But nitrogen (group 5A) should only have five valence electrons. With one extra electron, this nitrogen atom bears a negative charge.

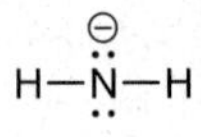

(c) One of the carbon atoms (below right) exhibits three valence electrons (one for each bond), but carbon (group 4A) is supposed to have four valence electrons. It is missing one electron, so this carbon atom therefore bears a positive charge.

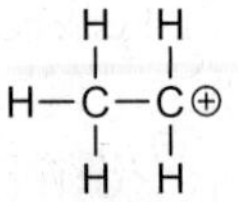

1.15.
(a) Oxygen is more electronegative than carbon, and a C–O bond is polar covalent. For each C–O bond, the O will be electron rich (δ–), and the C will be electron-poor (δ+), as shown below.

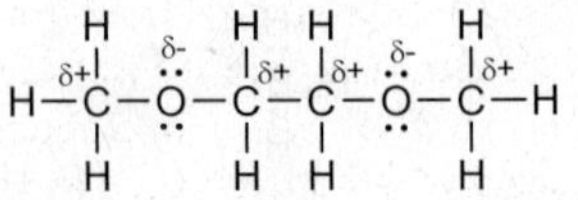

(b) Fluorine is more electronegative than carbon, and a C–F bond is polar covalent. For a C–F bond, the F will be electron-rich (δ–), and the C will be electron-poor (δ+). Chlorine is also more electronegative than carbon, so a C–Cl bond is also polar covalent. For a C–Cl bond, the Cl will be electron-rich (δ–), and the C will be electron-poor (δ+), as shown below.

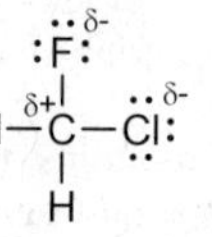

(c) Carbon is more electronegative than magnesium, so the C will be electron-rich (δ–) in a C–Mg bond, and the Mg will be electron-poor (δ+). Also, bromine is more electronegative than magnesium. So in a Mg–Br bond, the Br will be electron-rich (δ–), and the Mg will be electron-poor (δ+), as shown below.

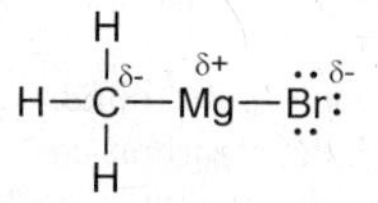

(d) Oxygen is more electronegative than carbon or hydrogen, so all C–O bonds and all O–H bond are polar covalent. For each C–O bond and each O–H bond, the O will be electron-rich (δ–), and the C or H will be electron-poor (δ+), as shown below.

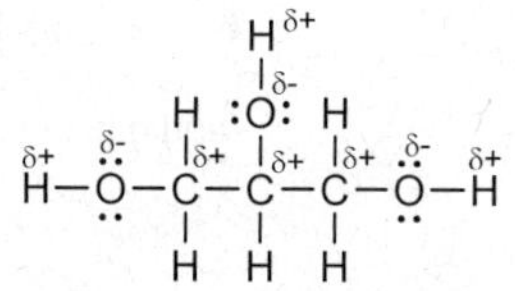

(e) Oxygen is more electronegative than carbon. As such, the O will be electron-rich (δ–) and the C will be electron-poor (δ+) in a C=O bond, as shown below.

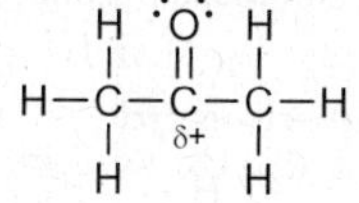

(f) Chlorine is more electronegative than carbon. As such, for each C–Cl bond, the Cl will be electron-rich (δ–) and the C will be electron-poor (δ+), as shown below.

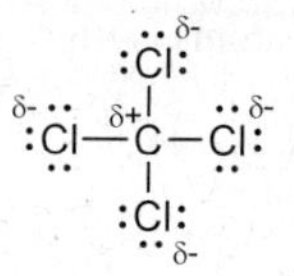

1.16. Oxygen is more electronegative than carbon. As such, the O will be electron-rich (δ–) and the C will be electron-poor (δ+) in a C=O bond. In addition, chlorine is more electronegative than carbon. So for a C–Cl

bond, the Cl will be electron-rich (δ–) and the C will be electron-poor (δ+), as shown below.

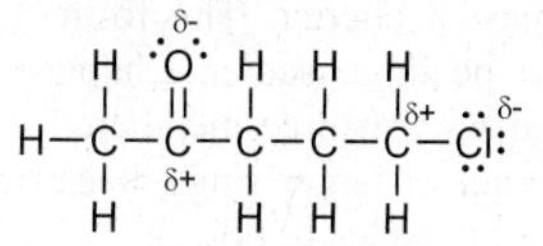

Notice that two carbon atoms are electron-poor (δ+). These are the positions that are most likely to be attacked by an anion, such as hydroxide.

1.17.
(a) As indicated in Figure 1.10, carbon has two *1s* electrons, two *2s* electrons, and two *2p* electrons. This information is represented by the following electron configuration: $1s^2 2s^2 2p^2$

(b) As indicated in Figure 1.10, oxygen has two *1s* electrons, two *2s* electrons, and four *2p* electrons. This information is represented by the following electron configuration: $1s^2 2s^2 2p^4$

(c) As indicated in Figure 1.10, boron has two *1s* electrons, two *2s* electrons, and one *2p* electron. This information is represented by the following electron configuration: $1s^2 2s^2 2p^1$

(d) As indicated in Figure 1.10, fluorine has two *1s* electrons, two *2s* electrons, and five *2p* electrons. This information is represented by the following electron configuration: $1s^2 2s^2 2p^5$

(e) Sodium has two *1s* electrons, two *2s* electrons, six *2p* electrons, and one *3s* electron. This information is represented by the following electron configuration: $1s^2 2s^2 2p^6 3s^1$

(f) Aluminum has two *1s* electrons, two *2s* electrons, six *2p* electrons, two *3s* electrons, and one *3p* electron. This information is represented by the following electron configuration: $1s^2 2s^2 2p^6 3s^2 3p^1$

1.18.
(a) The electron configuration of a carbon atom is $1s^2 2s^2 2p^2$ (see the solution to Problem 1.17a). However, if a carbon atom bears a negative charge, then it must have one extra electron, so the electron configuration should be as follows: $1s^2 2s^2 2p^3$

(b) The electron configuration of a carbon atom is $1s^2 2s^2 2p^2$ (see the solution to Problem 1.17a). However, if a carbon atom bears a positive charge, then it must be missing an electron, so the electron configuration should be as follows: $1s^2 2s^2 2p^1$

(c) As seen in Skillbuilder 1.6, the electron configuration of a nitrogen atom is $1s^2 2s^2 2p^3$. However, if a nitrogen atom bears a positive charge, then it must be missing an electron, so the electron configuration should be as follows: $1s^2 2s^2 2p^2$

(d) The electron configuration of an oxygen atom is $1s^2 2s^2 2p^4$ (see the solution to Problem 1.17b). However, if an oxygen atom bears a negative charge, then it must have one extra electron, so the electron configuration should be as follows: $1s^2 2s^2 2p^5$

1.19. The bond angles of an equilateral triangle are 60°, but each bond angle of cyclopropane is supposed to be 109.5°. Therefore, each bond angle is severely strained, causing an increase in energy. This form of strain, called ring strain, will be discussed in Chapter 4. The ring strain associated with a three-membered ring is greater than the ring strain of larger rings, because larger rings do not require bond angles of 60°.

1.20.
(a) The C=O bond of formaldehyde is comprised of one σ bond and one π bond.
(b) Each C–H bond is formed from the interaction between an sp^2 hybridized orbital from carbon and an s orbital from hydrogen.
(c) The oxygen atom is sp^2 hybridized, so the lone pairs occupy sp^2 hybridized orbitals.

1.21. Rotation of a single bond does not cause a reduction in the extent of orbital overlap, because the orbital overlap occurs on the bond axis. In contrast, rotation of a π bond results in a reduction in the extent of orbital overlap, because the orbital overlap is NOT on the bond axis.

1.22.
(a) The highlighted carbon atom (below) has four σ bonds, and is therefore sp^3 hybridized. The other carbon atoms in this structure are all sp^2 hybridized, because each of them has three σ bonds and one π bond.

(b) Each of the highlighted carbon atoms (below) has four σ bonds, and is therefore sp^3 hybridized. The other carbon atoms in this structure are all sp^2 hybridized, because each of them has three σ bonds and one π bond.

1.23.
(a) Each of the two central carbon atoms has two σ bonds and two π bonds, and as such, each of these carbon atoms is sp hybridized. The other two carbon atoms (the outer ones) are sp^2 hybridized because each has three σ bonds and one π bond.

(b) One of the carbon atoms (the one connected to oxygen) has two σ bonds and two π bonds, and as such, it is sp hybridized. The other carbon atom is sp^2 hybridized because it has three σ bonds and one π bond.

1.24. Carbon-carbon triple bonds generally have a shorter bond length than carbon-carbon double bonds, which are generally shorter than carbon-carbon single bonds (see Table 1.2).

c < b < a
a is the longest bond
and c is the shortest bond

1.25.
(a) The nitrogen atom has three σ bonds and one lone pair, so the steric number is 4 (sp^3 hybridization), which means that the electronic arrangement will be tetrahedral. One corner of the tetrahedron is occupied by a lone pair, so the arrangement of atoms is trigonal pyramidal.

(b) The oxygen atom has three σ bonds and one lone pair, so the steric number is 4 (sp^3 hybridization), which means that the electronic arrangement will be tetrahedral. One corner of the tetrahedron is occupied by a lone pair, so the arrangement of atoms is trigonal pyramidal.

(c) The boron atom has four σ bonds and no lone pairs, so the steric number is 4 (sp^3 hybridization), which means that the electronic arrangement will be tetrahedral. With no lone pairs, the arrangement of atoms (geometry) is the same as the electronic arrangement.   It is tetrahedral.

$$H-\overset{\overset{\displaystyle H}{|}}{\underset{\underset{\displaystyle H}{|}}{\overset{\ominus}{B}}}-H$$

(d) The boron atom has three σ bonds and no lone pairs, so the steric number is 3 (sp^2 hybridization), which means that the electronic arrangement will be trigonal planar. With no lone pairs, the arrangement of atoms (geometry) is the same as the electronic arrangement. It is trigonal planar.

$$\overset{\overset{\displaystyle Cl}{|}}{Cl}\overset{\displaystyle B}{\diagdown}Cl$$

(e) The boron atom has four σ bonds and no lone pairs, so the steric number is 4 (sp^3 hybridization), which means that the electronic arrangement will be tetrahedral. With no lone pairs, the arrangement of atoms (geometry) is the same as the electronic arrangement.   It is tetrahedral.

$$Cl-\overset{\overset{\displaystyle Cl}{|}}{\underset{\underset{\displaystyle Cl}{|}}{\overset{\ominus}{B}}}-Cl$$

(f) The carbon atom has four σ bonds and no lone pairs, so the steric number is 4 (sp^3 hybridization), which means that the electronic arrangement will be tetrahedral. With no lone pairs, the arrangement of atoms (geometry) is the same as the electronic arrangement.   It is tetrahedral.

$$Cl-\overset{\overset{\displaystyle Cl}{|}}{\underset{\underset{\displaystyle Cl}{|}}{C}}-Cl$$

(g) The carbon atom has four σ bonds and no lone pairs, so the steric number is 4 (sp^3 hybridization), which means that the electronic arrangement will be tetrahedral. With no lone pairs, the arrangement of atoms (geometry) is the same as the electronic arrangement.   It is tetrahedral.

$$Cl-\overset{\overset{\displaystyle Cl}{|}}{\underset{\underset{\displaystyle Cl}{|}}{C}}-H$$

(h) The carbon atom has four σ bonds and no lone pairs, so the steric number is 4 (sp^3 hybridization), which means that the electronic arrangement will be tetrahedral. With no lone pairs, the arrangement of atoms (geometry) is the same as the electronic arrangement.   It is tetrahedral.

$$Cl-\overset{\overset{\displaystyle H}{|}}{\underset{\underset{\displaystyle Cl}{|}}{C}}-H$$

1.26.

(a)   The carbon atom highlighted (below) has three σ bonds and no lone pairs (steric number = 3) and is therefore sp^2 hybridized and trigonal planar.

Each of the other carbon atoms has four σ bonds (steric number = 4) and is therefore tetrahedral.   The nitrogen atom has three σ bonds and one lone pair (steric number = 4), so the electronic arrangement is tetrahedral. But one corner of the tetrahedron is occupied by a lone pair, so the arrangement of atoms is trigonal pyramidal. The oxygen atom (of the OH group) has two σ bonds and two lone pairs (steric number = 4), so the electronic arrangement is tetrahedral. But two corners of the tetrahedron are occupied by lone pairs, so the arrangement of atoms is bent. The oxygen atom of the C=O group has one σ bond and two lone pairs (steric number = 3), so it is sp^2 hybridized.

(b)   Each of the highlighted carbon atoms has three σ bonds and no lone pairs (steric number = 3) and is therefore trigonal planar. Each of the other carbon atoms (not highlighted) has four σ bonds (steric number = 4), with tetrahedral geometry.   The highlighted nitrogen atom has two σ bonds and one lone pair (steric number = 3), so the electronic arrangement is trigonal planar. But there is one lone pair, so the arrangement of atoms is bent. The other nitrogen atom (not highlighted) has three σ bonds and one lone pair (steric number = 4), so the electronic arrangement will be tetrahedral.   One corner of the tetrahedron is occupied by a lone pair, so the arrangement of atoms is trigonal pyramidal. The oxygen atom has two σ bonds and two lone pairs (steric number = 4), so the electronic arrangement will be tetrahedral. Two corners of the tetrahedron are occupied by lone pairs, so the arrangement of atoms is bent.

(c) Each of the carbon atoms has three σ bonds and no lone pairs (steric number = 3) and is therefore trigonal planar:

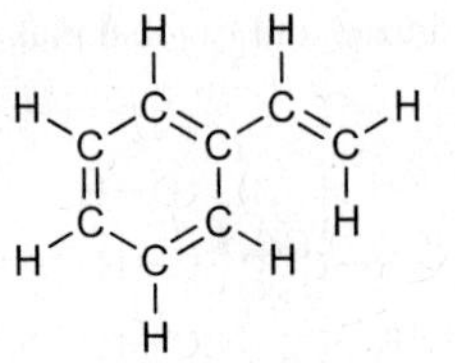

1.27. The carbon atom of the carbocation has three σ bonds and no lone pairs (steric number = 3), and is therefore trigonal planar. The carbon atom of the carbanion has three σ bonds and one lone pair (steric number = 4), and is therefore trigonal pyramidal (the electronic arrangement is tetrahedral, but one corner of the tetrahedron is occupied by a lone pair, giving a trigonal pyramidal arrangement of atoms).

1.28. Every carbon atom in benzene has three σ bonds and no lone pairs. With a steric number of 3, each of these carbon atoms is sp^2 hybridized and trigonal planar. Therefore, the entire molecule is planar (all of the atoms in this molecule occupy the same plane).

1.29.
(a) This compound has three C–Cl bonds, each of which exhibits a dipole moment. To determine if these dipole moments cancel each other, we must identify the molecular geometry. The central carbon atom has four σ bonds so it is sp^3 hybridized, with tetrahedral geometry. As such, the three C–Cl bonds do not lie in the same plane, and they do not completely cancel each other out. There is a net molecular dipole moment, as shown:

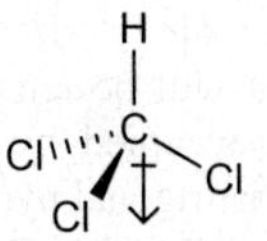

(b) The oxygen atom has two σ bonds and two lone pairs (steric number = 4), so it is sp^3 hybridized (electronically tetrahedral), with bent geometry (because two corners of the tetrahedron are occupied by the lone pairs). As such, the dipole moments associated with the C–O bonds do not fully cancel each other. There is a net molecular dipole moment, as shown:

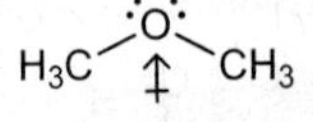

(c) The nitrogen atom has three σ bonds and one lone pair (steric number = 4), so it is sp^3 hybridized (electronically tetrahedral), with trigonal pyramidal geometry (because one corner of the tetrahedron is occupied by a lone pair). As such, the dipole moments associated with the N–H bonds do not fully cancel each

other. There is a net molecular dipole moment, as shown:

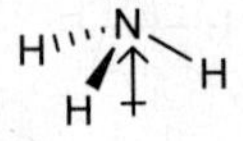

(d) The central carbon atom has four σ bonds (steric number = 4) and is therefore sp^3 hybridized, with tetrahedral geometry. There are individual dipole moments associated with each of the C–Cl bonds and each of the C–Br bonds. If all four dipole moments had the same magnitude, then we would expect them to completely cancel each other to give no molecular dipole moment (as in the case of CCl₄). However, the dipole moments for the C–Cl bonds are larger than the dipole moments of the C–Br bonds, and as such, there is a net molecular dipole moment, shown here:

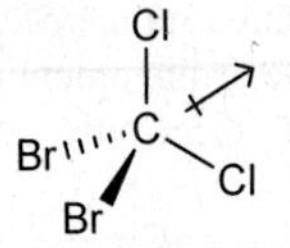

(e) The oxygen atom has two σ bonds and two lone pairs (steric number = 4), so it is sp^3 hybridized (electronically tetrahedral), with bent geometry (because two corners of the tetrahedron are occupied by lone pairs). As such, the dipole moments associated with the C–O bonds do not fully cancel each other. There is a net molecular dipole moment, as shown:

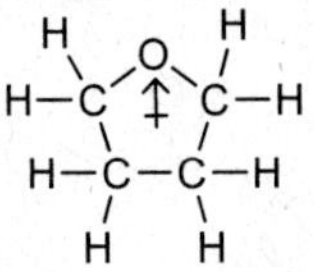

(f) There are individual dipole moments associated with each C–O bond (just as we saw in the solution to **1.29e**), but in this case, they fully cancel each other to give no net molecular dipole moment.

(g) Each C=O bond has a strong dipole moment, and they do not fully cancel each other because they are not pointing in opposite directions. As such, there will be a net molecular dipole moment, as shown here:

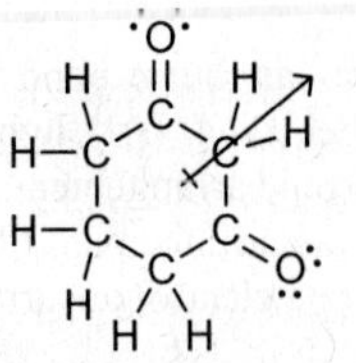

(h) Each C=O bond has a strong dipole moment, and in this case, they are pointing in opposite directions. As such, they fully cancel each other, giving no net molecular dipole moment.

(i) Each C–Cl bond has a dipole moment, and they do not fully cancel each other because they are not pointing in opposite directions. As such, there will be a net molecular dipole moment, as shown here:

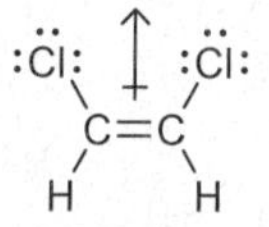

(j) Each C–Cl bond has a dipole moment, and in this case, they are pointing in opposite directions. As such, they fully cancel each other, giving no net molecular dipole moment.

(k) Each C–Cl bond has a dipole moment, and they do not fully cancel each other because they are not pointing in opposite directions. As such, there will be a net molecular dipole moment, as shown here:

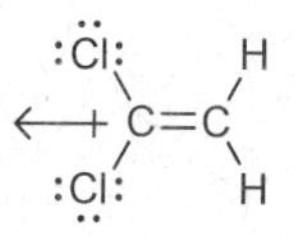

(l) Each C–Cl bond has a dipole moment, but in this case, they fully cancel each other to give no net molecular dipole moment.

1.30. $CHCl_3$ is expected to have a larger molecular dipole moment than $CBrCl_3$, because the bromine atom in the latter compound serves to nearly cancel out the effects of the other three chlorine atoms (as is the case for CCl_4).

1.31. The carbon atom of O=C=O has two σ bonds and no lone pairs (steric number = 2) and is therefore *sp* hybridized, with linear geometry. As a result, the individual dipole moments of each C=O bond cancel each other completely to give no overall molecular dipole moment. In contrast, the sulfur atom in SO_2 has a steric number of three (because it also has a lone pair, in addition to the two S=O bonds), which means that it has bent geometry. As a result, the individual dipole moments of each S=O bond do NOT cancel each other completely, and the molecule does have a molecular dipole moment.

1.32.
(a) The latter compound is expected to have a higher boiling point, because it is less branched.
(b) The latter compound is expected to have a higher boiling point, because it has more carbon atoms.
(c) The latter compound is expected to have a higher boiling point, because it has an OH bond, which will lead to hydrogen bonding interactions.
(d) The first compound is expected to have a higher boiling point, because it is less branched.

1.33. Two compounds possess OH groups, and these compounds will have the highest boiling points. Among these two compounds, the one with more carbon atoms

(six) will be higher boiling than the one with fewer carbon atoms (four). The remaining three compounds all have five carbon atoms and lack an OH group. The difference between these three compounds is the extent of branching. Among these three compounds, the compound with the greatest extent of branching has the lowest boiling point, and the one with the least branching has the highest boiling point.

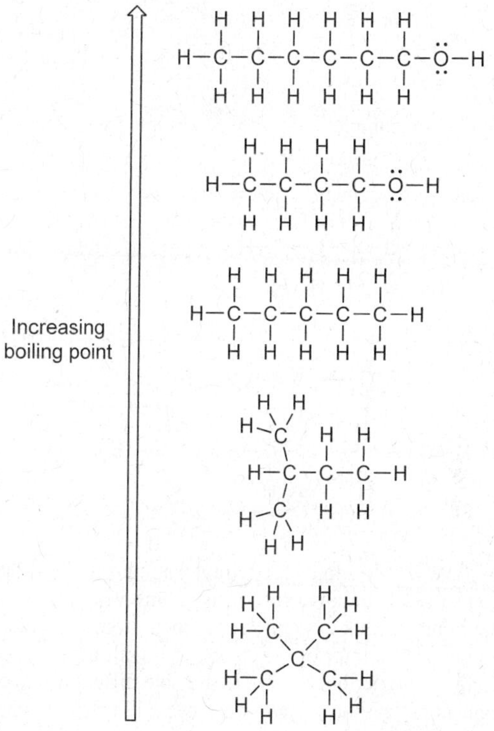

1.34.
(a) Begin by determining the valency of each atom in the compound. The carbon atoms are tetravalent, and the hydrogen atoms are all monovalent. Any atoms with more than one bond (in this case, the four carbon atoms) should be drawn in the center of the compound, with the hydrogen atoms at the periphery. There are two different ways to connect four carbon atoms. They can either be arranged in a linear fashion or in a branched fashion.

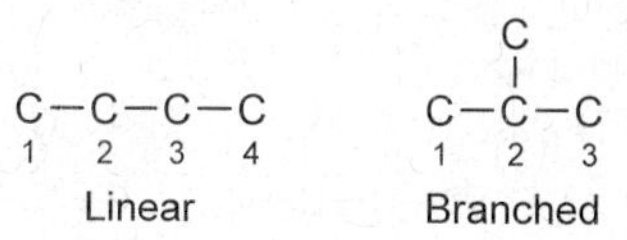

Finally, we complete both of the structures by drawing the bonds to hydrogen atoms.

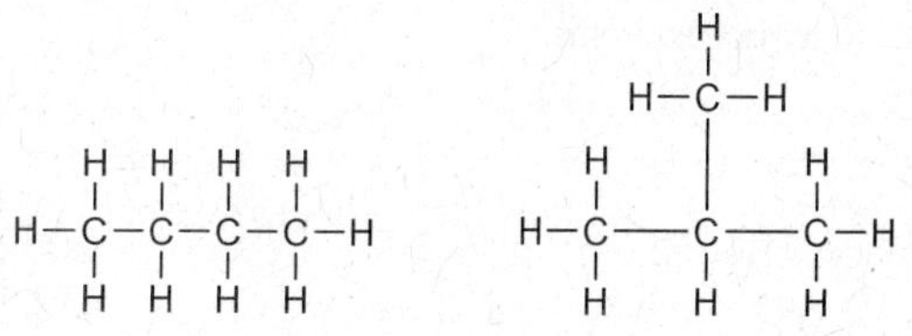

(b) The carbon atoms are tetravalent, and the hydrogen atoms are all monovalent. Any atoms with more than one bond (in this case, the five carbon atoms) should be drawn in the center of the compound, with the hydrogen atoms at the periphery. There are three different ways to connect five carbon atoms, as shown here:

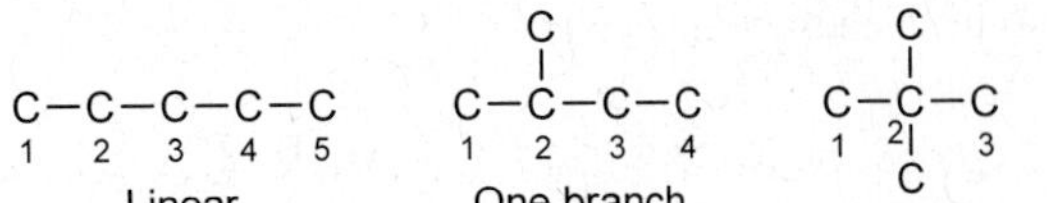

```
              C                    C
              |                    |
C—C—C—C—C   C—C—C—C         C—C—C
1  2  3 4 5   1  2  3 4          1  2| 3
   Linear      One branch           C
                              Two branches
```

Finally, we complete all of the structures by drawing the bonds to hydrogen atoms.

```
    H  H  H  H  H
    |  |  |  |  |
H—C—C—C—C—C—H
    |  |  |  |  |
    H  H  H  H  H

                    H
                    |
                 H—C—H
         H          H
         |          |
      H—C—— C ——C—H
         |     |    |
         H          H
                 H—C—H
                    H

        H
        |
     H—C—H
   H  |   H  H
   |  |   |  |
H—C— C — C—C—H
   |  |   |  |
   H  H   H  H
```

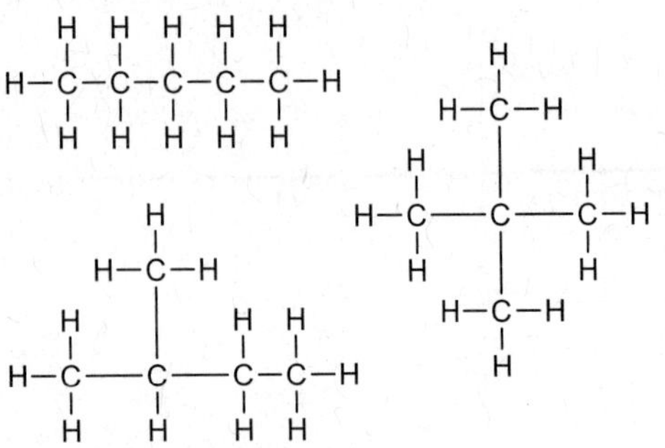

(c) The carbon atoms are tetravalent, and the hydrogen atoms are all monovalent. Any atoms with more than one bond (in this case, the six carbon atoms) should be drawn in the center of the compound, with the hydrogen atoms at the periphery. There are five different ways to connect six carbon atoms, which we will organize based on the length of the longest chain.

In a 6-carbon chain:
```
C—C—C—C—C—C
1  2  3  4  5  6
```

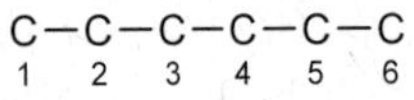

In a 5-carbon chain:
```
      C                       C
      |                       |
C—C—C—C—C            C—C—C—C—C
1  2  3  4  5          1  2  3  4  5
```

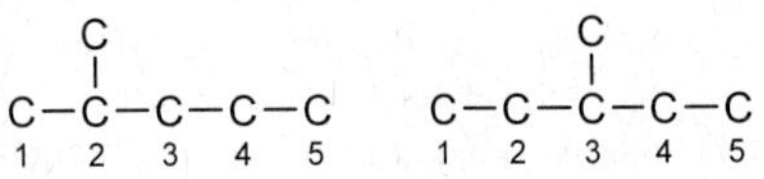

In a 4-carbon chain:
```
      C                    C   C
      |                    |   |
C—C—C—C            C—C—C—C
1  2| 3  4            1  2  3  4
    C
```

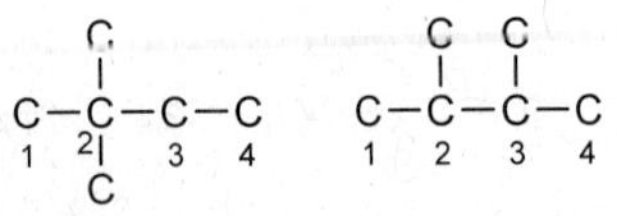

Finally, we complete all of the structures by drawing the bonds to hydrogen atoms.

```
    H  H  H  H  H  H
    |  |  |  |  |  |
H—C—C—C—C—C—C—H
    |  |  |  |  |  |
    H  H  H  H  H  H

                    H
                    |
                 H—C—H
    H  H          H  H
    |  |          |  |
 H—C—C — C —C—C—H
    |  |     |    |  |
    H  H          H  H

    H
    |
 H—C—H
    H          H  H  H
    |          |  |  |
H—C — C —C—C—C—H
    |  |       |  |  |
    H  H       H  H  H

    H
    |
 H—C—H
    H          H  H
    |          |  |
H—C— C —C—C—H
    |  |       |  |
    H          H  H
 H—C—H
    H

                 H
                 |
              H—C—H
    H          H     H
    |          |     |
H—C — C —C—C—H
    |  |     |    |
    H  H          H
              H—C—H
                 H
```

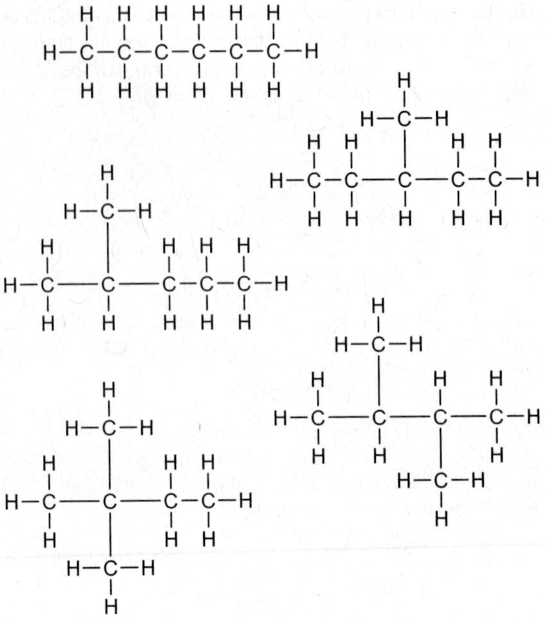

(d) The carbon atoms are tetravalent, while the chlorine atom and hydrogen atoms are all monovalent. The atoms with more than one bond (in this case, the two carbon atoms) should be drawn in the center of the compound. The chlorine atom and hydrogen atoms are then placed at the periphery, as shown.

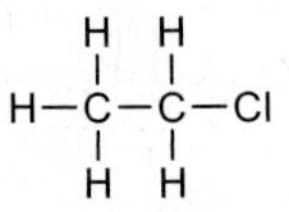

```
    H  H
    |  |
H—C—C—Cl
    |  |
    H  H
```

The chlorine atom can be placed in any one of the six available positions. The following six drawings all represent the same compound, in which the two carbon atoms are connected to each other, and the chlorine atom is connected to one of the carbon atoms.

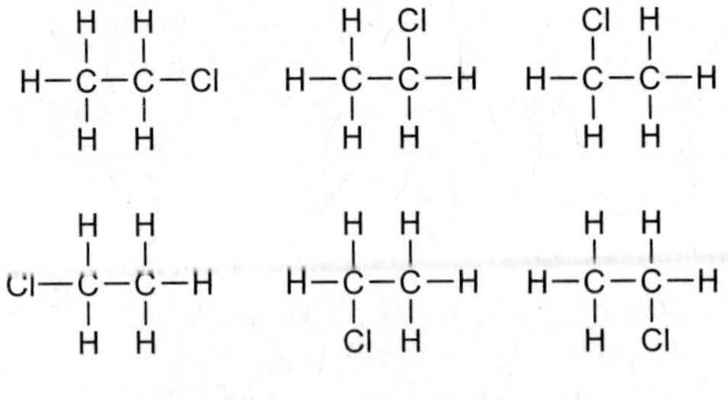

```
   H  H          H  Cl         Cl  H
   |  |          |  |          |   |
H—C—C—Cl     H—C—C—H     H—C—C—H
   |  |          |  |          |   |
   H  H          H  H          H   H

   H  H          H  H          H  H
   |  |          |  |          |  |
Cl—C—C—H     H—C—C—H     H—C—C—H
   |  |          |  |          |  |
   H  H          Cl H          H  Cl
```

(e) The carbon atoms are tetravalent, while the chlorine atoms and hydrogen atoms are all monovalent. The atoms with more than one bond (in this case, the two carbon atoms) should be drawn in the center of the compound. The chlorine atoms and hydrogen atoms are then placed at the periphery, and there are two different ways to do this. The two chlorine atoms can either be connected to the same carbon atom or to different carbon atoms, as shown.

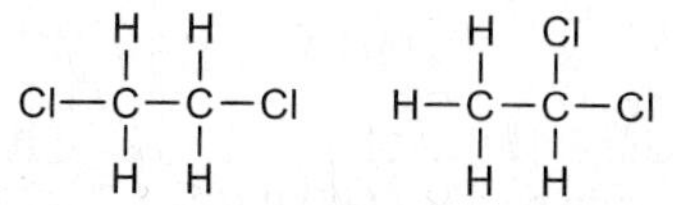

(f) The carbon atoms are tetravalent, while the chlorine atoms and chlorine atoms and hydrogen atoms are all monovalent. The atoms with more than one bond (in this case, the two carbon atoms) should be drawn in the center of the compound. The chlorine atoms and hydrogen atoms are then placed at the periphery, and there are two different ways to do this. One way is to connect all three chlorine atoms to the same carbon atom. Alternatively, we can connect two chlorine atoms to one carbon atom, and then connect the third chlorine atom to the other carbon atom, as shown here:

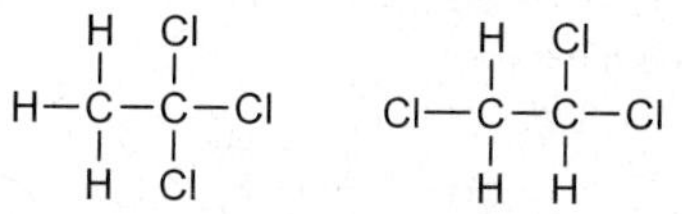

1.35.

(a) The molecular formula (C_4H_8) indicates that we must draw structures with four carbon atoms and eight hydrogen atoms. The carbon atoms are tetravalent, while the hydrogen atoms are all monovalent. The atoms with more than one bond (in this case, the four carbon atoms) should be drawn in the center of the compound, with the hydrogen atoms at the periphery. When we connect four carbon atoms, either in a linear fashion or in a branched fashion (see solution to 1.34a), we find that ten hydrogen atoms are required in order for all four carbons atom to achieve an octet (to have four bonds).

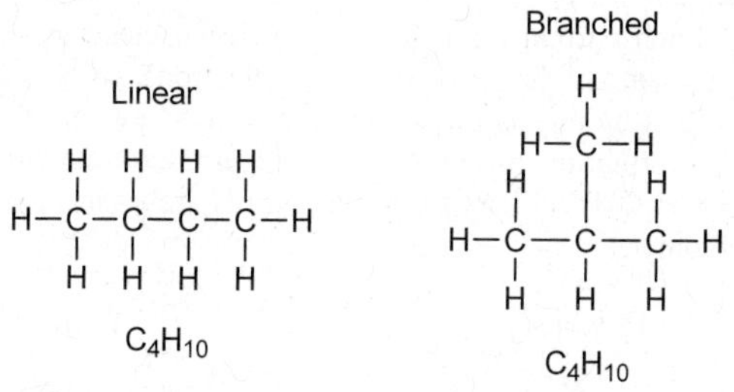

But the molecular formula (C_4H_8) indicates only eight hydrogen atoms, so we must remove two hydrogen atoms. This gives two carbon atoms that lack an octet, because each of them has an unpaired electron.

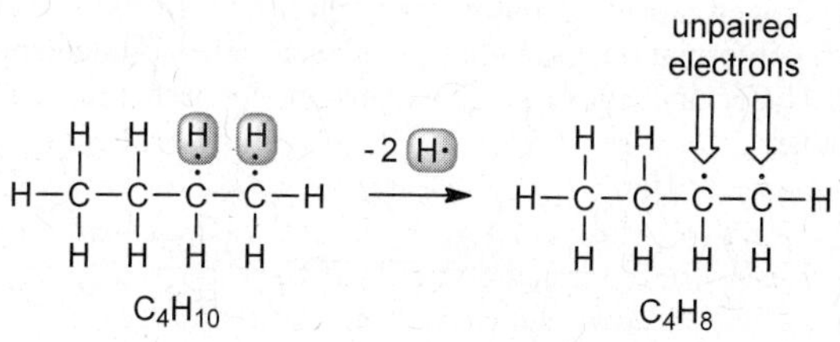

These electrons can be paired as a double bond:

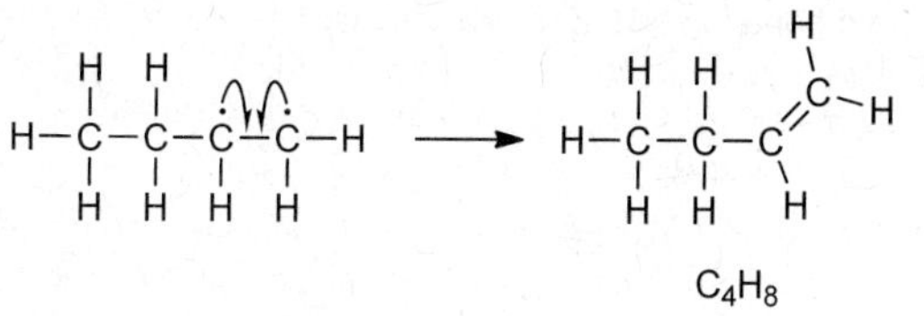

but the problem statement directs us to draw only those constitutional isomers in which all of the bonds are single bonds. So we must think of another way to pair up the unpaired electrons. It is difficult to see how this can be accomplished if the unpaired electrons are on adjacent carbon atoms. But suppose the unpaired electrons are on distant carbon atoms:

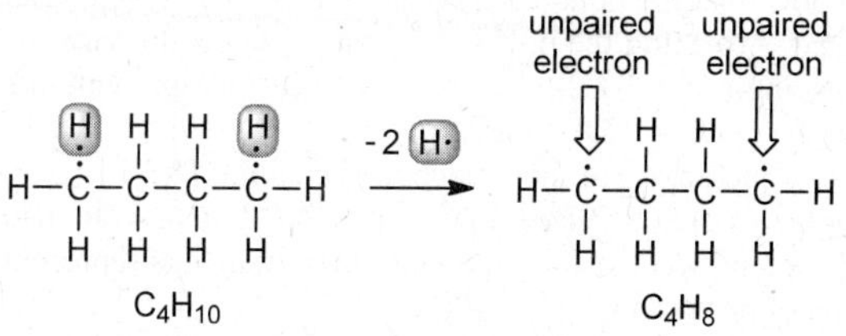

When drawn like this, it becomes apparent that we can pair the unpaired electrons by forming a C – C bond, giving a ring:

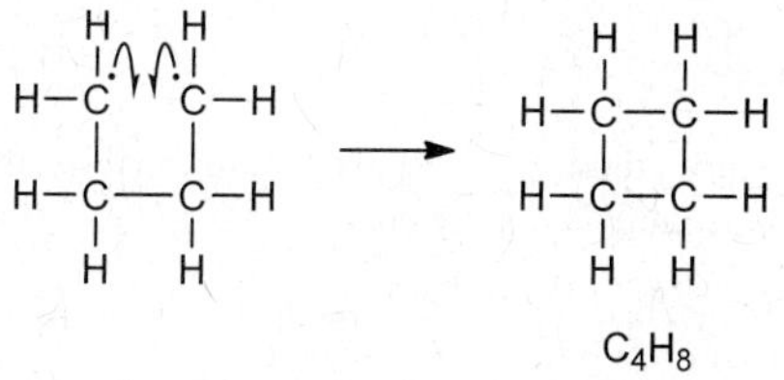

When the structure contains a ring, then eight hydrogen atoms are sufficient to provide all four carbon atoms with an octet of electrons. The ring can either be a 3-membered ring or a 4-membered ring, giving the following two constitutional isomers:

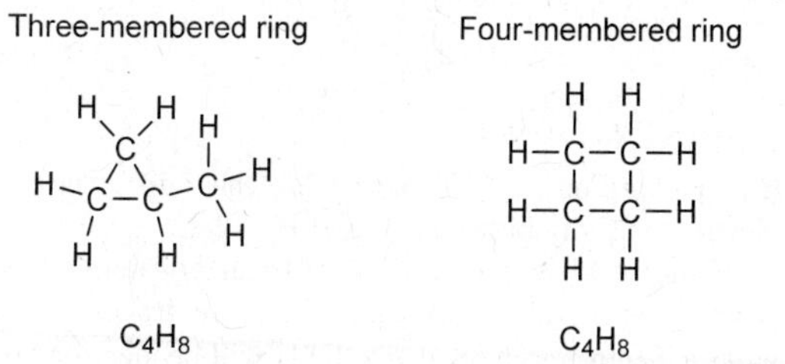

(b) See the solution to **1.35a** as an introduction to the following solution.

Since the unpaired electrons were paired as a double bond (rather than as a ring), we are looking for compounds that contain one double bond and do NOT have a ring. Since the structure does not contain a ring, we can imagine arranging the carbon atoms either in a linear fashion or in a branched fashion:

Linear   Branched

In the linear skeleton, there are two locations where we can place the double bond:

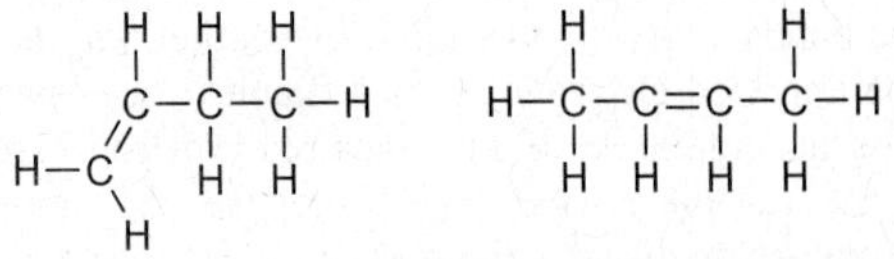

Notice that the double bond can be placed at C1-C2 or at C2-C3 (placing the double bond at C3-C4 is the same as placing it at C1-C2, because we can just assign numbers in the opposite direction).

Now let's explore the branched skeleton. There is only one location to place the double bond in a branched skeleton, because the following three drawings represent the same compound:

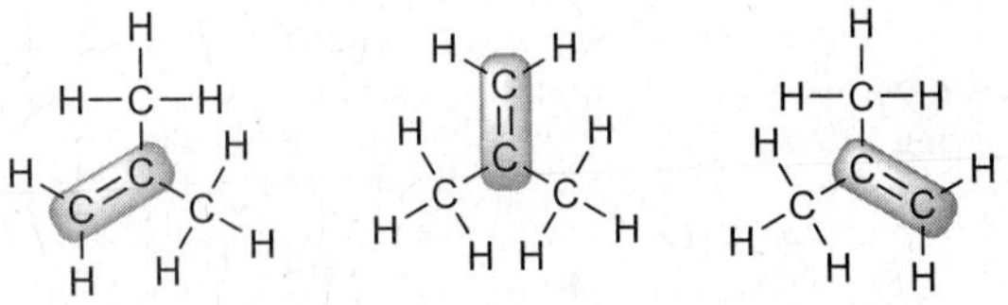

In summary, there are three constitutional isomers of C_4H_8 that contain a double bond:

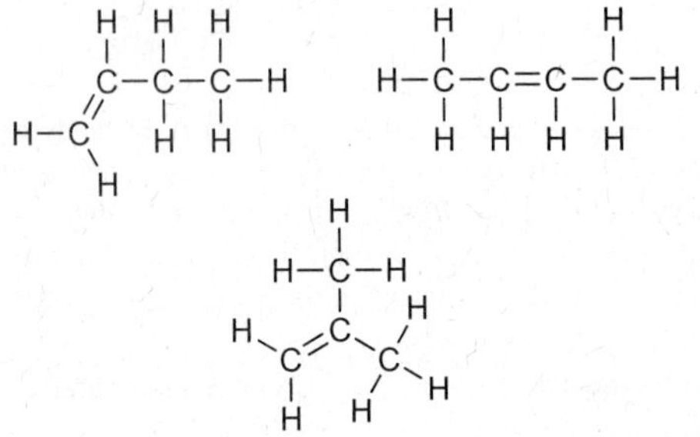

1.36.
(a) According to Table 1.1, the difference in electronegativity between Br and H is 2.8 – 2.1 = 0.7, so an H–Br bond is expected to be polar covalent. Since bromine is more electronegative than hydrogen, the Br will be electron rich (δ–), and the H will be electron-poor (δ+), as shown below:

$$\overset{\delta+}{H}\!\!-\!\!\overset{\delta-}{Br}$$

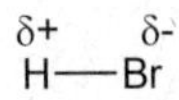

(b) According to Table 1.1, the difference in electronegativity between Cl and H is 3.0 – 2.1 = 0.9, so an H–Cl bond is expected to be polar covalent. Since chlorine is more electronegative than hydrogen, the Cl will be electron rich (δ–), and the H will be electron-poor (δ+), as shown below:

$$\overset{\delta+}{H}\!\!-\!\!\overset{\delta-}{Cl}$$

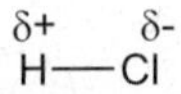

(c) According to Table 1.1, the difference in electronegativity between O and H is 3.5 – 2.1 = 1.4, so an O–H bond is expected to be polar covalent. Oxygen is more electronegative than hydrogen, so for each O–H bond, the O will be electron rich (δ–) and the H will be electron-poor (δ+), as shown below:

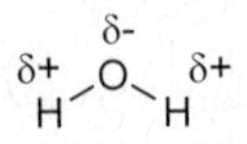

(d) Oxygen (3.5) is more electronegative than carbon (2.5) or hydrogen (2.1), and a C–O or H–O bond is polar covalent. For each C–O or H–O bond, the O will be electron rich (δ–), and the C or H will be electron-poor (δ+), as shown below:

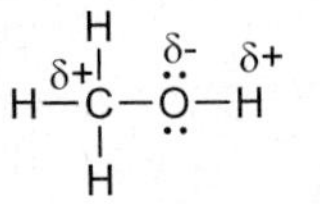

1.37.
(a) The difference in electronegativity between Na and Br is greater than the difference in electronegativity between H and Br. Therefore, NaBr is expected to have more ionic character than HBr.
(b) The difference in electronegativity between F (4.0) and Cl (3.0) is greater than the difference in electronegativity between Br (2.8) and Cl (3.0). Therefore, FCl is expected to have more ionic character than BrCl.

1.38.
(a) Each carbon atom has four valence electrons, the oxygen atom has six valence electrons, and each hydrogen atom has one valence electron. In this case, the information provided in the problem statement (CH₃CH₂OH) indicates how the atoms are connected to each other:

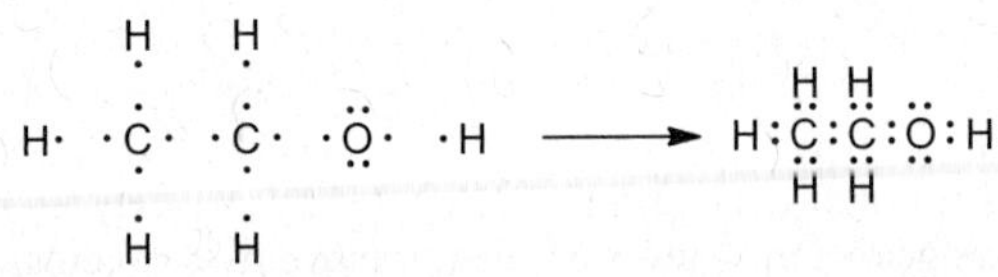

(b) Each carbon atom has four valence electrons, the nitrogen atom has five valence electrons, and each hydrogen atom has one valence electron. In this case, the information provided in the problem statement (CH₃CN) indicates how the atoms are connected to each other:

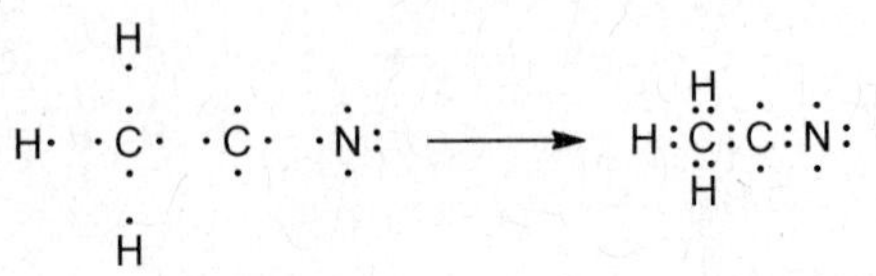

The unpaired electrons are then paired up to give a triple bond. In this way, each of the atoms achieves an octet.

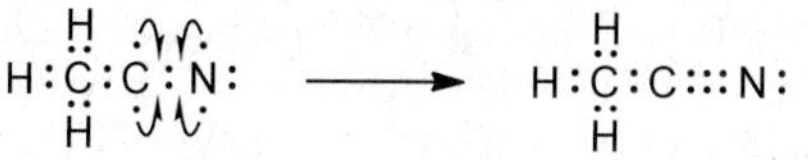

1.39.

(a) The carbon atom bearing a negative charge (highlighted) has three σ bonds and one lone pair, so the steric number is 4 (sp^3 hybridization), which means that the electronic arrangement will be tetrahedral. One corner of the tetrahedron is occupied by a lone pair, so the geometry of that carbon atom (the arrangement of atoms around that carbon atom) is trigonal pyramidal.
The structure has three other carbon atoms. Each of them has four σ bonds (steric number = 4) and is therefore tetrahedral.

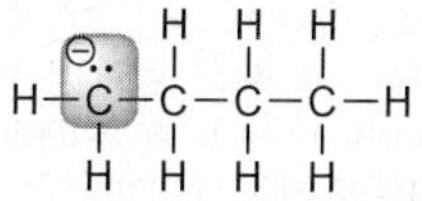

(b) The highlighted carbon atom has four σ bonds, so the steric number is 4 (sp^3 hybridization), which means that this carbon atom will be tetrahedral.   Each of the other two carbon atoms has three σ bonds and no lone pairs, so the steric number is 3 (sp^2 hybridization), which means that these carbon atoms will be trigonal planar. The oxygen atom has three σ bonds and one lone pair, so the steric number is 4 (sp^3 hybridization), which means that the electronic arrangement will be tetrahedral. One corner of the tetrahedron is occupied by a lone pair, so the geometry of that oxygen atom (the arrangement of atoms around that oxygen atom) is trigonal pyramidal.

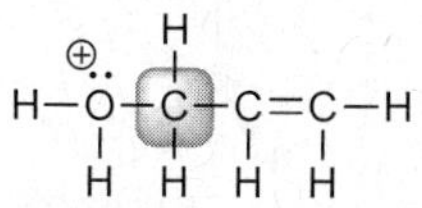

(c) The nitrogen atom and each of the two carbon atoms has four σ bonds, so the steric number is 4 (sp^3 hybridization), which means that each of these atoms will have tetrahedral geometry. The oxygen atom has two σ bonds and two lone pairs (steric number = 4), so the electronic arrangement is tetrahedral. But two corners of the tetrahedron are occupied by lone pairs, so the geometry of the oxygen atom (the arrangement of atoms around the oxygen atom) is bent.

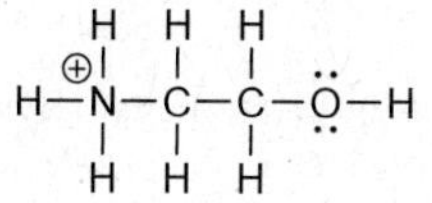

(d) Each of the three carbon atoms has four σ bonds, so the steric number is 4 (sp^3 hybridization), which means that each of these carbon atoms will have tetrahedral

geometry. The oxygen atom is only connected to one group so its geometry is not relevant.

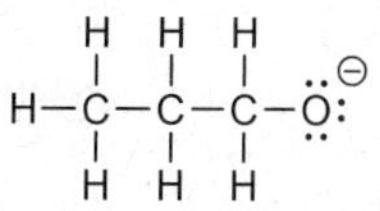

1.40. Each of the carbon atoms has four valence electrons; the nitrogen atom has five valence electrons; and each of the hydrogen atoms has one valence electron. We begin by connecting the atoms that have more than one bond (in this case, the four carbon atoms and the nitrogen atom). The problem statement indicates how we should connect them:

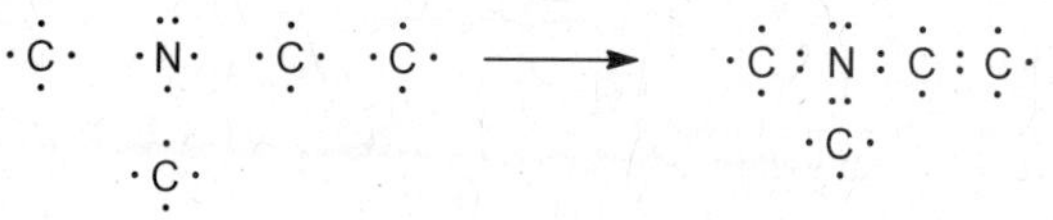

Then, we connect all of the hydrogen atoms, as shown.

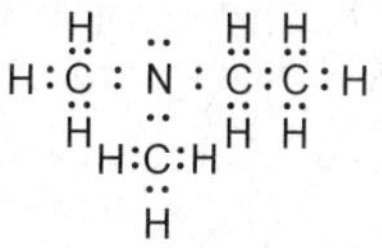

The nitrogen atom has three σ bonds and one lone pair, so the steric number is 4 (sp^3 hybridization), which means that the electronic arrangement will be tetrahedral. One corner of the tetrahedron is occupied by a lone pair, so the geometry of the nitrogen atom (the arrangement of atoms around that nitrogen atom) is trigonal pyramidal. As such, the individual dipole moments associated with the C–N bonds do not fully cancel each other. There is a net molecular dipole moment, as shown:

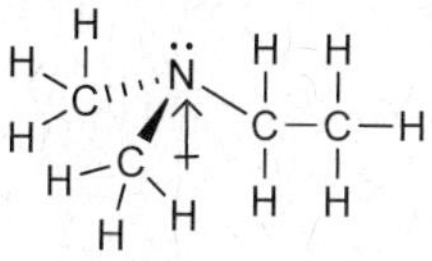

1.41. Bromine is in group 7A of the periodic table, so each bromine atom has seven valence electrons. Aluminum is in group 3A of the periodic table, so aluminum is supposed to have three valence electrons, but the structure bears a negative charge, which means that there is one extra electron. That is, the aluminum atom has four valence electrons, rather than three, which is why it has a formal negative charge. This gives the following Lewis structure:

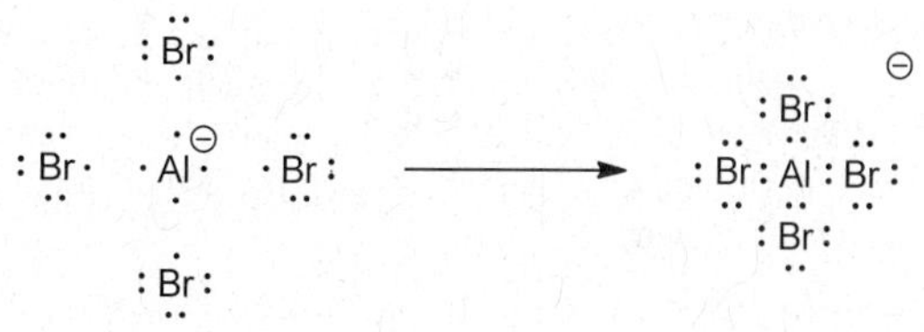

The aluminum atom has four bonds and no lone pairs, so the steric number is 4 (sp^3 hybridization), which means that this aluminum atom will have tetrahedral geometry.

1.42. The molecular formula of cyclopropane is C_3H_6, so we are looking for a different compound that has the same molecular formula, C_3H_6. That is, we need to find another way to connect the carbon atoms, other than in a ring (there is only one way to connect three carbon atoms in a ring, so we must be looking for something other than a ring). If we connect the three carbon atoms in a linear fashion and then draw four bonds for each carbon atom, we notice that the molecular formula (C_3H_8) is not correct:

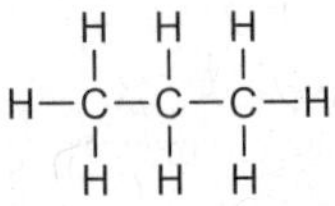

C_3H_8

We are looking for a structure with molecular formula C_3H_6. If we remove two hydrogen atoms from our drawing, we are left with two unpaired electrons, indicating that we should consider drawing a double bond:

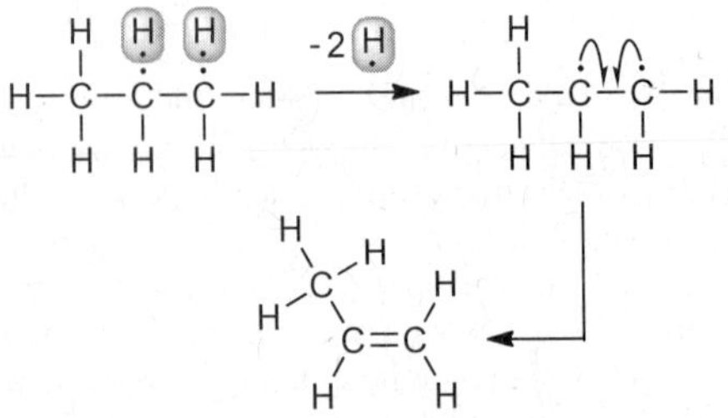

C_3H_6

The structure of this compound (called propylene) is different from the structure of cyclopropane, but both compounds share the same molecular formula, so they are constitutional isomers.

1.43.
(a) C–H bonds are considered to be covalent, although they do have a very small dipole moment, because there is a small difference in electronegativity between carbon (2.5) and hydrogen (2.1). Despite the very small dipole moments associated with the C–H bonds, the compound has no net dipole moment. The carbon atom has tetrahedral geometry (because it has four σ bonds), so the small effects from each C-H bond completely cancel each other.

(b) The nitrogen atom has trigonal pyramidal geometry (see the solution to Problem **1.25a**). As such, the dipole moments associated with the N–H bonds do not fully cancel each other. There is a net molecular dipole moment, as shown:

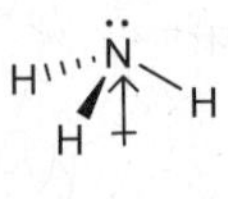

(c) The oxygen atom has two σ bonds and two lone pairs (steric number = 4), so it is sp^3 hybridized (electronically tetrahedral), with bent geometry (because two corners of the tetrahedron are occupied by lone pairs). As such, the dipole moments associated with the O–H bonds do not cancel each other. There is a net molecular dipole moment, as shown:

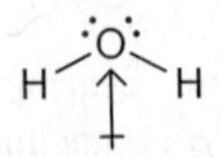

(d) The central carbon atom of carbon dioxide (CO_2) has two σ bonds and no lone pairs, so it is sp hybridized and has linear geometry. Each C=O bond has a strong dipole moment, but in this case, they are pointing in opposite directions. As such, they fully cancel each other, giving no net molecular dipole moment.

(e) Carbon tetrachloride (CCl_4) has four C–Cl bonds, each of which exhibits a dipole moment. However, the central carbon atom has four σ bonds so it is sp^3 hybridized, with tetrahedral geometry. As such, the four dipole moments completely cancel each other out, and there is no net molecular dipole moment.

(f) This compound has two C–Br bonds, each of which exhibits a dipole moment. To determine if these dipole moments cancel each other, we must identify the molecular geometry. The central carbon atom has four σ bonds so it is sp^3 hybridized, with tetrahedral geometry. As such, the C–Br bonds are separated by a bond angle of approximately 109.5°, and they do not completely cancel each other out. There is a net molecular dipole moment, as shown:

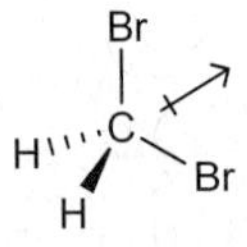

1.44.
(a) As indicated in Figure 1.10, oxygen has two $1s$ electrons, two $2s$ electrons, and four $2p$ electrons.
(b) As indicated in Figure 1.10, fluorine has two $1s$ electrons, two $2s$ electrons, and five $2p$ electrons.
(c) As indicated in Figure 1.10, carbon has two $1s$ electrons, two $2s$ electrons, and two $2p$ electrons.
(d) As seen in SkillBuilder 1.6, the electron configuration of a nitrogen atom is $1s^2 2s^2 2p^3$
(e) This is the electron configuration of chlorine.

1.45.
(a) The difference in electronegativity between sodium (0.9) and bromine (2.8) is 2.8 − 0.9 = 1.9. Since this difference is greater than 1.7, the bond is classified as ionic.

(b) The difference in electronegativity between sodium (0.9) and oxygen (3.5) is 3.5 − 0.9 = 2.6. Since this difference is greater than 1.7, the Na–O bond is classified as ionic. In contrast, the O–H bond is polar covalent, because the difference in electronegativity between oxygen (3.5) and hydrogen (2.1) is less than 1.7 but more than 0.5.

(c) Each C–H bond is considered to be covalent, because the difference in electronegativity between carbon (2.5) and hydrogen (2.1) is less than 0.5.
The C–O bond is polar covalent, because the difference in electronegativity between oxygen (3.5) and carbon (2.5) is less than 1.7 but more than 0.5. The Na–O bond is classified as ionic, because the difference in electronegativity between oxygen (3.5) and sodium (0.9) is greater than 1.7.

(d) Each C–H bond is considered to be covalent, because the difference in electronegativity between carbon (2.5) and hydrogen (2.1) is less than 0.5.
The C–O bond is polar covalent, because the difference in electronegativity between oxygen (3.5) and carbon (2.5) is less than 1.7 but more than 0.5. The O–H bond is polar covalent, because the difference in electronegativity between oxygen (3.5) and hydrogen (2.1) is less than 1.7 but more than 0.5.

(e) Each C–H bond is considered to be covalent, because the difference in electronegativity between carbon (2.5) and hydrogen (2.1) is less than 0.5.
The C=O bond is polar covalent, because the difference in electronegativity between oxygen (3.5) and carbon (2.5) is less than 1.7 but more than 0.5.

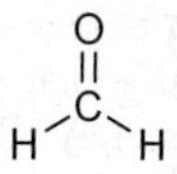

1.46.
(a) Begin by determining the valency of each atom in the compound. The carbon atoms are tetravalent, the oxygen atom is divalent, and the hydrogen atoms are all monovalent. Any atoms with more than one bond (in this case, the two carbon atoms and the oxygen atom) should be drawn in the center of the compound, with the hydrogen atoms at the periphery. There are two different ways to connect two carbon atoms and an oxygen atom, shown here:

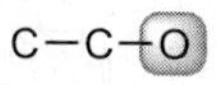

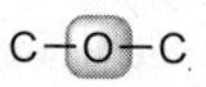

We then complete both structures by drawing the remaining bonds to hydrogen atoms:

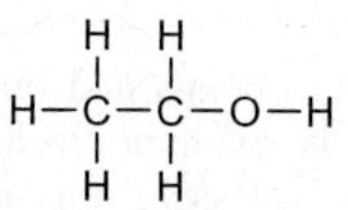

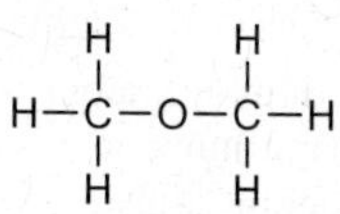

(b) Begin by determining the valency of each atom in the compound. The carbon atoms are tetravalent, the oxygen atoms are divalent, and the hydrogen atoms are all monovalent. Any atoms with more than one bond (in this case, the two carbon atoms and the two oxygen atoms) should be drawn in the center of the compound,

with the hydrogen atoms at the periphery. There are several different ways to connect two carbon atoms and two oxygen atoms (highlighted, for clarity of comparison), shown here:

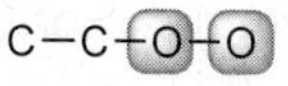

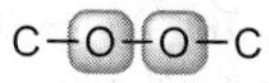

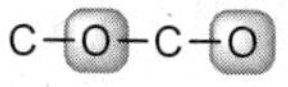

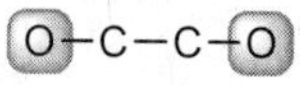

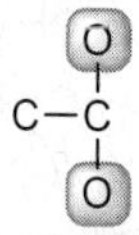

We then complete all of these structures by drawing the remaining bonds to hydrogen atoms:

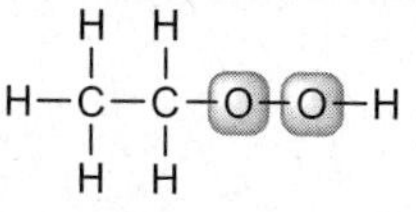

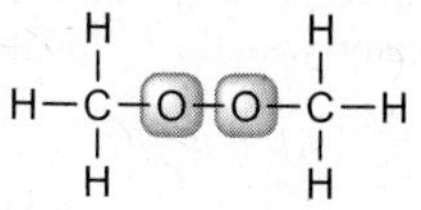

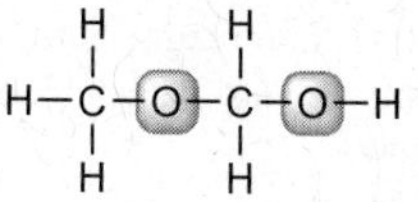

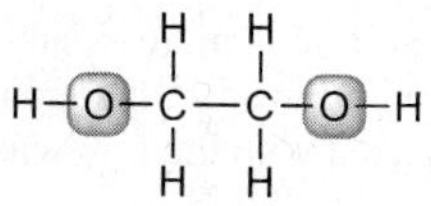

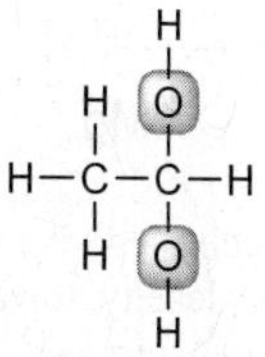

(c) The carbon atoms are tetravalent, while the bromine atoms and hydrogen atoms are all monovalent. The atoms with more than one bond (in this case, the two carbon atoms) should be drawn in the center of the compound. The bromine atoms and hydrogen atoms are then placed at the periphery, and there are two different ways to do this. The two bromine atoms can either be connected to the same carbon atom or to different carbon atoms, as shown.

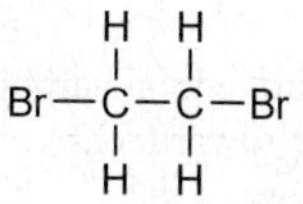

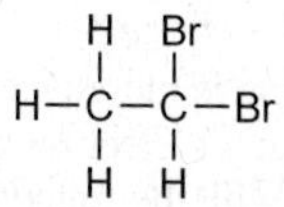

1.47. Begin by determining the valency of each atom in the compound. The carbon atoms are tetravalent, the oxygen atoms are divalent, and the hydrogen atoms are all monovalent. Any atoms with more than one bond (in this case, the two carbon atoms and the three oxygen atoms) should be drawn in the center of the compound, with the hydrogen atoms at the periphery. There are many different ways to connect two carbon atoms and three oxygen atoms (see the solution to Problem 1.46b for comparison). Five such ways are shown below, although there are certainly others:

H–C–C–OH with H, OH (top left) ; H–C–C–OH with OH, OH / H, H

H–C–O–C–H ; H–C–O–O–C–H

H–C–O–C–H

1.48.
(a) Oxygen is more electronegative than carbon, and the withdrawal of electron density toward oxygen can be indicated with the following arrow:

$$\longmapsto$$
C–O

(b) Carbon is more electronegative than magnesium, and the withdrawal of electron density toward carbon can be indicated with the following arrow:

$$\longleftarrow$$
C–Mg

(c) Nitrogen is more electronegative than carbon, and the withdrawal of electron density toward nitrogen can be indicated with the following arrow:

$$\longmapsto$$
C–N

(d) Carbon is more electronegative than lithium, and the withdrawal of electron density toward carbon can be indicated with the following arrow:

$$\longleftarrow$$
C–Li

(e) Chlorine is more electronegative than carbon, and the withdrawal of electron density toward chlorine can be indicated with the following arrow:

$$\longmapsto$$
C–Cl

(f) Carbon is more electronegative than hydrogen, and the withdrawal of electron density toward carbon can be indicated with the following arrow:

$$\longleftarrow$$
C–H

(g) Oxygen is more electronegative than hydrogen, and the withdrawal of electron density toward oxygen can be indicated with the following arrow:

$$\longleftarrow$$
O–H

(h) Nitrogen is more electronegative than hydrogen, and the withdrawal of electron density toward nitrogen can be indicated with the following arrow:

$$\longleftarrow$$
N–H

1.49.
(a) The oxygen atom has two σ bonds and two lone pairs (steric number = 4), so it is sp^3 hybridized. Each of the carbon atoms is also sp^3 hybridized (because each has four σ bonds). Therefore, all bond angles are expected to be approximately 109.5°.

(b) The central carbon atom has three σ bonds and no lone pairs (steric number = 3), so it is sp^2 hybridized, with trigonal planar geometry. As such, all bond angles are approximately 120°.

(c) Each of the carbon atoms has three σ bonds and no lone pairs (steric number = 3), so each carbon atom is sp^2 hybridized, with trigonal planar geometry. As such, all bond angles are approximately 120°.

(d) Each of the carbon atoms has two σ bonds and no lone pairs (steric number = 2), so each carbon atom is sp hybridized, with linear geometry. As such, all bond angles are approximately 180°.

H–C≡C–H

(e) The oxygen atom has two σ bonds and two lone pairs (steric number = 4), so it is sp^3 hybridized. Each of the carbon atoms is also sp^3 hybridized (because each has four σ bonds). Therefore, all bond angles are expected to be approximately 109.5°.

(f) The nitrogen atom has three σ bonds and one lone pair (steric number = 4), so it is sp^3 hybridized. The carbon atom is also sp^3 hybridized (because it has four σ bonds). Therefore, all bond angles are expected to be approximately 109.5°.

H–C–N:

(g) Each of the carbon atoms has four σ bonds (steric number = 4), so each of these carbon atoms is sp^3 hybridized and has tetrahedral geometry. Therefore, all bond angles are expected to be approximately 109.5°.

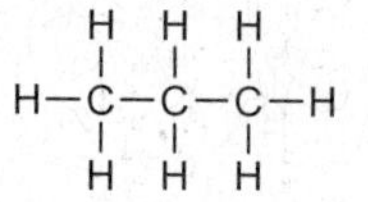

$$
\begin{array}{ccccc}
 & H & H & H & \\
 & | & | & | & \\
H- & C- & C- & C & -H \\
 & | & | & | & \\
 & H & H & H &
\end{array}
$$

(h) The structure of acetonitrile (CH_3CN) is shown below (see the solution to Problem **1.38b**).

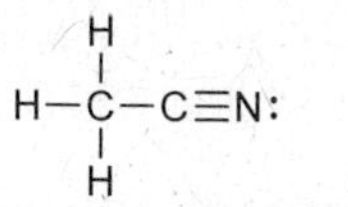

$$
\begin{array}{ccc}
 & H & \\
 & | & \\
H- & C- & C\equiv N: \\
 & | & \\
 & H &
\end{array}
$$

One of the carbon atoms has four σ bonds (steric number = 4), so it is sp^3 hybridized and has tetrahedral geometry. The other carbon atom (connected to nitrogen) has two σ bonds and no lone pairs (steric number = 2), so it is sp hybridized with linear geometry.
As such, the C–C≡N bond angle is 180°, and all other bond angles are approximately 109.5°.

1.50.
(a) The nitrogen atom has three σ bonds and one lone pair (steric number = 4), so it is sp^3 hybridized (electronically tetrahedral), with trigonal pyramidal geometry (because one corner of the tetrahedron is occupied by a lone pair).

(b) The boron atom has three σ bonds and no lone pairs (steric number = 3), so it is sp^2 hybridized, with trigonal planar geometry.

(c) This carbon atom has three σ bonds and no lone pairs (steric number = 3), so it is sp^2 hybridized, with trigonal planar geometry.

(d) This carbon atom has three σ bonds and one lone pair (steric number = 4), so it is sp^3 hybridized (electronically tetrahedral), with trigonal pyramidal geometry (because one corner of the tetrahedron is occupied by a lone pair).

(e) This oxygen atom has three σ bonds and one lone pair (steric number = 4), so it is sp^3 hybridized (electronically tetrahedral), with trigonal pyramidal geometry (because one corner of the tetrahedron is occupied by a lone pair).

1.51. The double bond represents one σ bond and one π bond, while the triple bond represents one σ bond and two π bonds. All single bonds are σ bonds. Therefore, this compound has sixteen σ bonds and three π bonds.

1.52.
(a) The latter compound is expected to have a higher boiling point, because it has an O–H bond, which will lead to hydrogen bonding interactions.

(b) The latter compound is expected to have a higher boiling point, because it has more carbon atoms, and thus more opportunity for London interactions.
(c) Both compounds have the same number of carbon atoms, but the first compound has a C=O bond, which has a strong dipole moment. The first compound is therefore expected to exhibit strong dipole-dipole interactions and to have a higher boiling point than the second compound.

1.53.
(a) This compound possesses an O–H bond, so it is expected to exhibit hydrogen bonding interactions.

$$
\begin{array}{cccc}
 & H & H & \\
 & | & | & \\
H- & C- & C- & O-H \\
 & | & | & \\
 & H & H &
\end{array}
$$

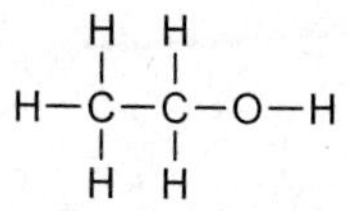

(b) This compound lacks a hydrogen atom that is connected to an electronegative element. Therefore, this compound cannot serve as a hydrogen bond donor (although the lone pairs can serve as hydrogen bond acceptors). In the absence of another hydrogen bond donor, we do not expect there to be any hydrogen bonding interactions.

$$
\begin{array}{c}
\ddot{O} \\
\| \\
C \\
H \quad H
\end{array}
$$

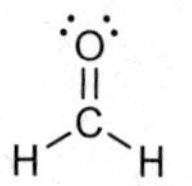

(c) This compound lacks a hydrogen atom that is connected to an electronegative element. Therefore, this compound will not exhibit hydrogen bonding interactions.

$$
\begin{array}{ccc}
H & & H \\
\diagdown & & \diagup \\
 & C=C & \\
\diagup & & \diagdown \\
H & & H
\end{array}
$$

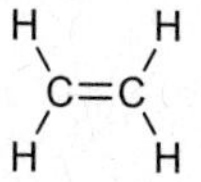

(d) This compound lacks a hydrogen atom that is connected to an electronegative element. Therefore, this compound will not exhibit hydrogen bonding interactions.

$$H-C\equiv C-H$$

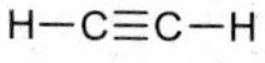

(e) This compound lacks a hydrogen atom that is connected to an electronegative element. Therefore, this compound cannot serve as a hydrogen bond donor (although lone pairs can serve as hydrogen bond acceptors). In the absence of another hydrogen bond donor, we do not expect there to be any hydrogen bonding interactions.

$$
\begin{array}{ccccc}
 & H & & H & \\
 & | & & | & \\
H- & C- & \ddot{O}- & C & -H \\
 & | & & | & \\
 & H & & H &
\end{array}
$$

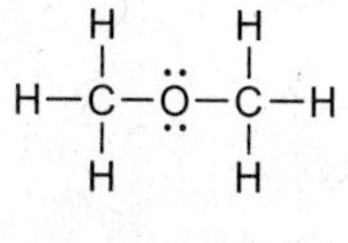

(f) This compound possesses an N–H bond, so it is expected to exhibit hydrogen bonding interactions.

$$
\begin{array}{ccc}
 & H & H \\
 & | & | \\
H- & C- & N: \\
 & | & | \\
 & H & H
\end{array}
$$

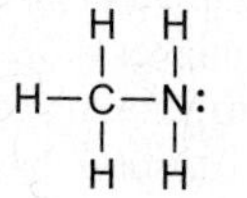

(g) This compound lacks a hydrogen atom that is connected to an electronegative element. Therefore, this compound will not exhibit hydrogen bonding interactions.

$$H-\overset{\overset{\displaystyle H}{|}}{\underset{\underset{\displaystyle H}{|}}{C}}-\overset{\overset{\displaystyle H}{|}}{\underset{\underset{\displaystyle H}{|}}{C}}-\overset{\overset{\displaystyle H}{|}}{\underset{\underset{\displaystyle H}{|}}{C}}-H$$

(h) This compound possesses an N–H bond, so it is expected to exhibit hydrogen bonding interactions.

$$H-\overset{\displaystyle ..}{\underset{\underset{\displaystyle H}{|}}{N}}-H$$

1.54.
(a) Boron is in group 3A of the periodic table, and therefore has three valence electrons. It can use each of its valence electrons to form a bond, so we expect the molecular formula to be BH_3.
(b) Carbon is in group 4A of the periodic table, and therefore has four valence electrons. It can use each of its valence electrons to form a bond, so we expect the molecular formula to be CH_4.

(c) Nitrogen is in group 5A of the periodic table, and therefore has five valence electrons. But it cannot form five bonds, because it only has four orbitals with which to form bonds. One of those orbitals must be occupied by a lone pair (two electrons), and each of the remaining three electrons is available to form a bond. Nitrogen is therefore trivalent, and we expect the molecular formula to be NH_3.

(d) Carbon is in group 4A of the periodic table, and therefore has four valence electrons. It can use each of its valence electrons to form a bond, and indeed, we expect the carbon atom to have four bonds. Two of the bonds are with hydrogen atoms, so the other two bonds must be with chlorine atoms. The molecular formula is CH_2Cl_2.

1.55.
(a) Each of the highlighted carbon atoms has three σ bonds and no lone pairs (steric number = 3), so each of these carbon atoms is sp^2 hybridized, with trigonal planar geometry. Each of the other four carbon atoms has two σ bonds and no lone pairs (steric number = 2), and therefore, those four carbon atoms are all sp hybridized, with linear geometry.

(b) The highlighted carbon atom has three σ bonds and no lone pairs (steric number = 3), so this carbon atom is sp^2 hybridized, with trigonal planar geometry. Each of the other three carbon atoms has four σ bonds (steric

number = 4), and therefore, those three carbon atoms are all sp^3 hybridized, with tetrahedral geometry.

(c) Each of the carbon atoms in this compound has four σ bonds (steric number = 4). Therefore, all of these carbon atoms are sp^3 hybridized, with tetrahedral geometry, or as close to it as this "triangle" will allow. We will discuss the geometry of this three-membered ring in more detail in Chapter 4.

1.56. Each of the highlighted carbon atoms has four σ bonds (steric number = 4), and is therefore sp^3 hybridized, with tetrahedral geometry. Each of the other fourteen carbon atoms in this structure has three σ bonds and no lone pairs (steric number = 3), so each of these fourteen carbon atoms is sp^2 hybridized, with trigonal planar geometry.

1.57.
(a) Oxygen is the most electronegative atom in this compound. See Table 1.1 for electronegativity values.

(b) Fluorine is the most electronegative atom. See Table 1.1 for electronegativity values.
(c) Carbon is the most electronegative atom in this compound. See Table 1.1 for electronegativity values.

1.58. The highlighted nitrogen atom (below) has two σ bonds and one lone pair (steric number = 3), so this nitrogen atom is sp^2 hybridized. It is electronically trigonal planar, but one of the sp^2 hybridized orbitals is occupied by a lone pair, so the geometry (arrangement of atoms) is bent. The other nitrogen atom (not highlighted) has three σ bonds and a lone pair (steric number = 4), so that nitrogen atom is sp^3 hybridized and electronically tetrahedral. One corner of the tetrahedron is occupied by a lone pair, so the geometry (arrangement of atoms) is trigonal pyramidal.

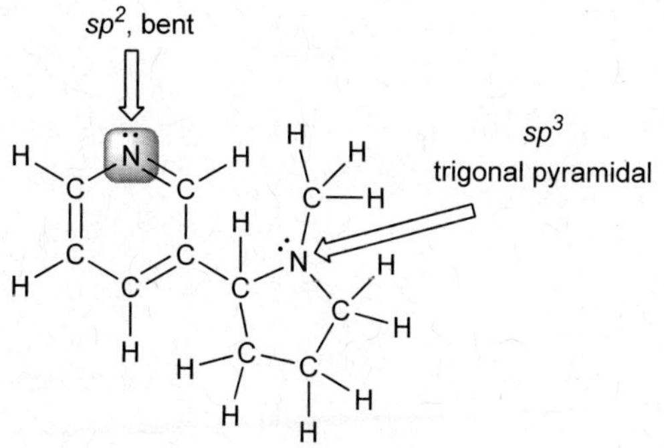

sp², bent

sp³
trigonal pyramidal

1.59. Each of the nitrogen atoms in this structure achieves an octet with three bonds and one lone pair, while each oxygen atom in this structure achieves an octet with two bonds and two lone pairs, as shown:

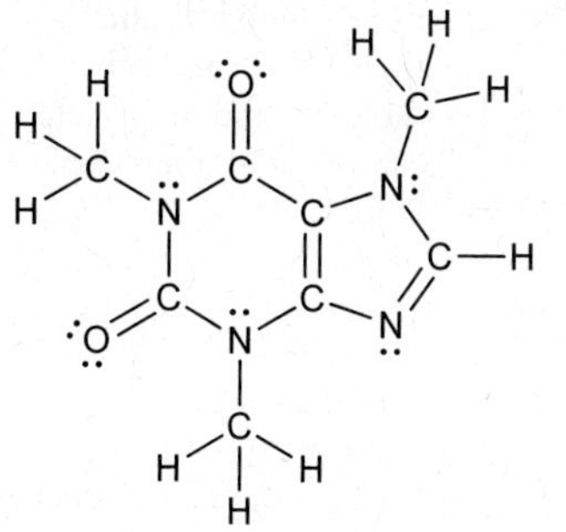

1.60. In the solution to Problem 1.46a, we saw that the following two compounds have the molecular formula C_2H_6O.

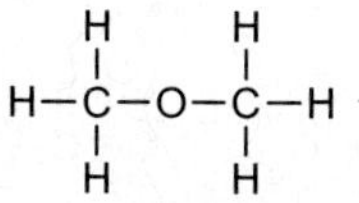

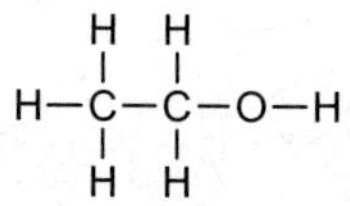

The second compound will have a higher boiling point because it possesses an OH group which can form hydrogen bonding interactions.

1.61.
(a) Each C–Cl bond has a dipole moment, and the two dipole moments do not fully cancel each other because they are not pointing in opposite directions. As such, there will be a net molecular dipole moment, as shown here:

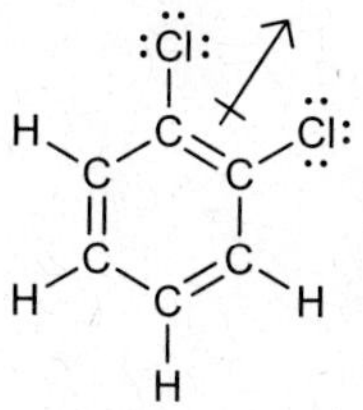

(b) Each C–Cl bond has a dipole moment, and the two dipole moments do not fully cancel each other because they are not pointing in opposite directions. As such, there will be a net molecular dipole moment, as shown here:

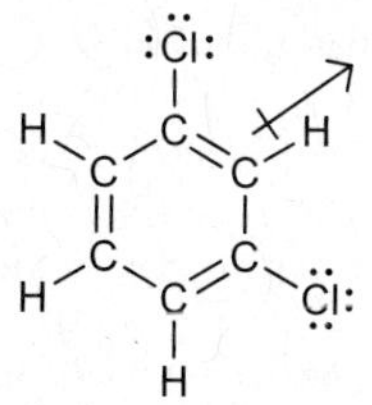

(c) Each C–Cl bond has a dipole moment, and in this case, the two dipole moments are pointing in opposite directions. As such, they fully cancel each other, giving no net molecular dipole moment.

(d) The C–Cl bond has a dipole moment, and the C–Br bond also has a dipole moment. These two dipole moments are in opposite directions, but they do not have the same magnitude. The C–Cl bond has a larger dipole moment than the C–Br bond, because chlorine is more electronegative than bromine. Therefore, there will be a net molecular dipole moment, as shown here:

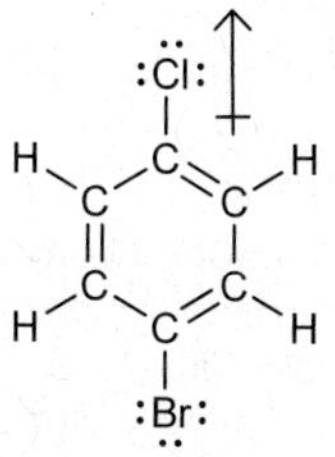

1.62. The third chlorine atom in chloroform partially cancels the effects of the other two chlorine atoms, thereby reducing the molecular dipole moment relative to methylene chloride.

1.63.
(a) Compounds A and B share the same molecular formula (C_4H_9N) but differ in their constitution (connectivity of atoms), and they are therefore constitutional isomers.
(b) The nitrogen atom in compound B has three σ bonds and one lone pair (steric number = 4), so it is *sp³* hybridized (electronically tetrahedral), with trigonal pyramidal geometry (because one corner of the tetrahedron is occupied by a lone pair).
(c) A double bond represents one σ bond and one π bond, while a triple bond represents one σ bond and two π bonds. A single bond represents a σ bond. With this in mind, compound B has 14 σ bonds, as compared with compounds A and C, which have 13 and 11 σ bonds, respectively.
(d) As explained in the solution to Problem **1.63c**, compound C has the fewest σ bonds.
(e) A double bond represents one σ bond and one π bond, while a triple bond represents one σ bond and two π bonds. As such, compound C exhibits two π bonds.

(f) Compound A has a C=N bond, in which the carbon atom has three σ bonds and no lone pairs (steric number = 3), and is therefore sp^2 hybridized.

(g) Each of the carbon atoms in compound B is sp^3 hybridized, because each has four σ bonds (steric number = 4). Similarly, the nitrogen atom in compound B has three σ bonds and one lone pair (steric number = 4), so this nitrogen atom is also sp^3 hybridized.

(h) Compound A has an N–H bond, and is therefore expected to form hydrogen bonding interactions. Compounds B and C do not contain an N–H bond, so compound A is expected to have the highest boiling point.

1.64.

(a) In each of the following two compounds, all of the carbon atoms are sp^2 hybridized (each carbon atom has three σ bonds and one π bond). There are certainly many other possible compounds for which all of the carbon atoms are sp^2 hybridized.

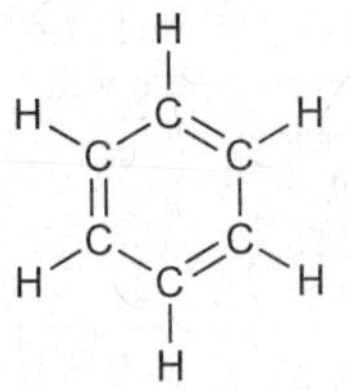

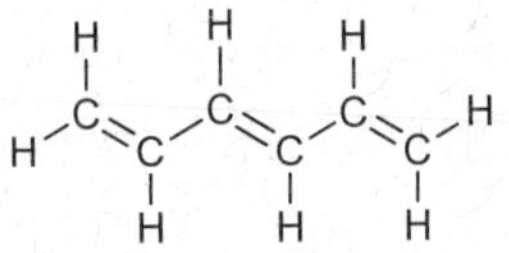

(b) In each of the following two compounds, all of the carbon atoms are sp^3 hybridized (because each carbon atom has four σ bonds) with the exception of the carbon atom connected to the nitrogen atom. That carbon atom has two σ bonds and is therefore sp hybridized. There are certainly many other acceptable answers.

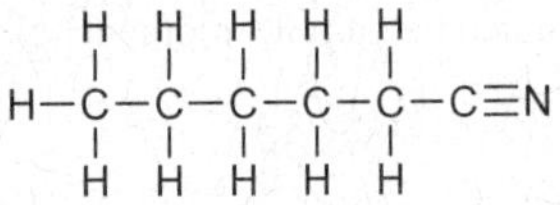

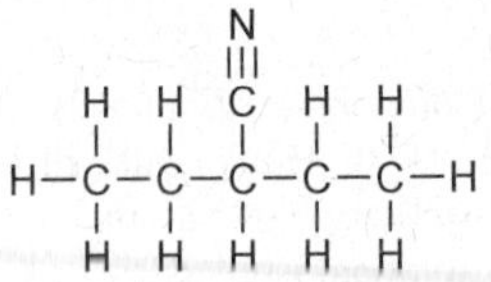

(c) In each of the following two compounds, there is a ring, and all of the carbon atoms are sp^3 hybridized (because each carbon atom has four σ bonds). There are certainly many other acceptable answers.

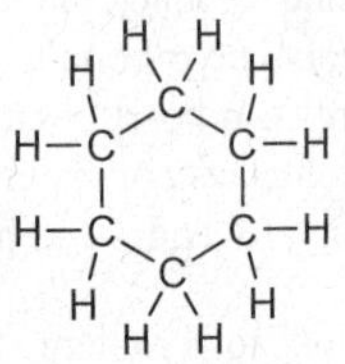

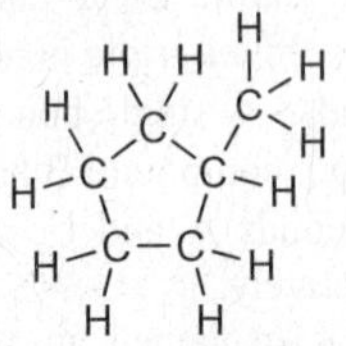

(d) In each of the following two compounds, all of the carbon atoms are sp hybridized (because each carbon

atom has two σ bonds). There are certainly many other acceptable answers.

$$N≡C-C≡C-C≡C-C≡N$$

$$F-C≡C-C≡C-C≡C-F$$

1.65. In the solution to Problem **1.34b**, we saw that there are three ways to arrange five carbon atoms:

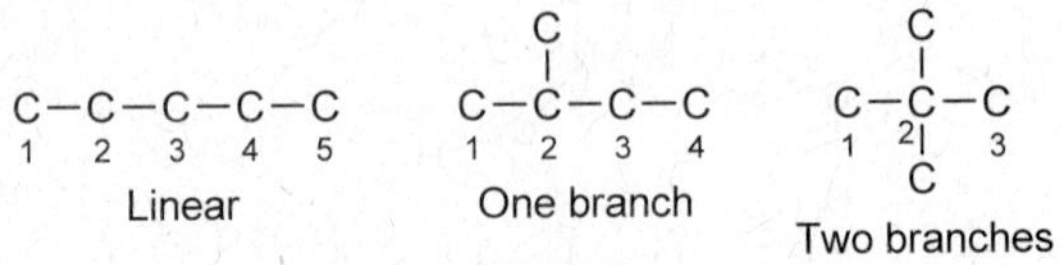

For each of these three skeletons, we must consider each possible location where a double bond can be placed. The skeleton with two branches cannot support a double bond, because the central carbon atom already has four bonds to carbon atoms, and it cannot accommodate a fifth bond (it cannot form another bond with any one of the four carbon atoms to which it is already connected).

So we only have to consider the other two skeletons above (the linear skeleton and the skeleton with one branch). In the linear skeleton, the double bond can be placed at C1-C2 or at C2-C3.

Placing the double bond at C3-C4 is the same as placing the double bond at C2-C3. Similarly, placing the double bond at C4-C5 is the same as placing the double bond at C1-C2.

For the skeleton with one branch, there are three different locations where the double bond can be placed, shown here:

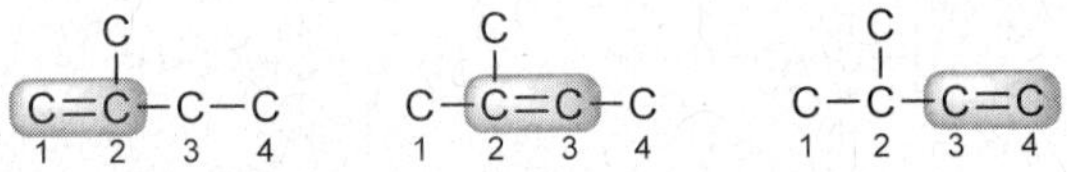

Be careful, the following two locations are the same:

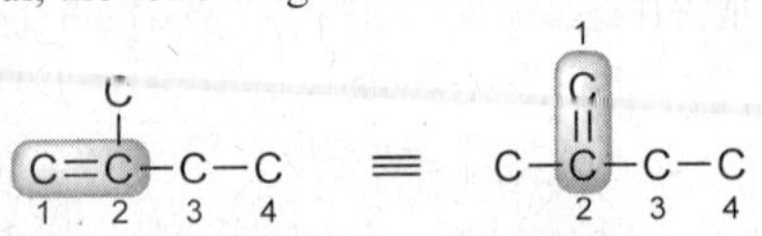

Finally, we complete all five possible structures by drawing the remaining bonds to the hydrogen atoms (see next page):

Linear skeleton

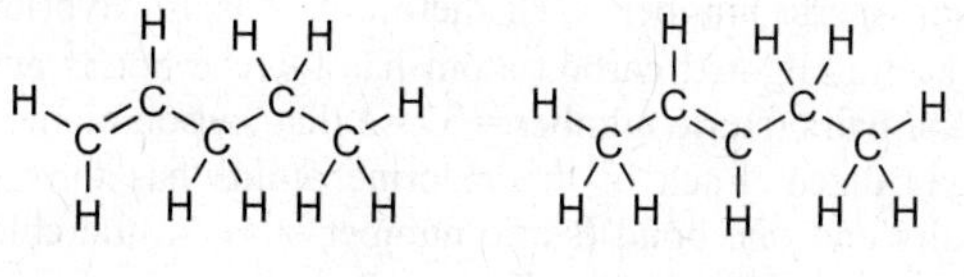

$C_5H_{13}N$

Branched skeleton

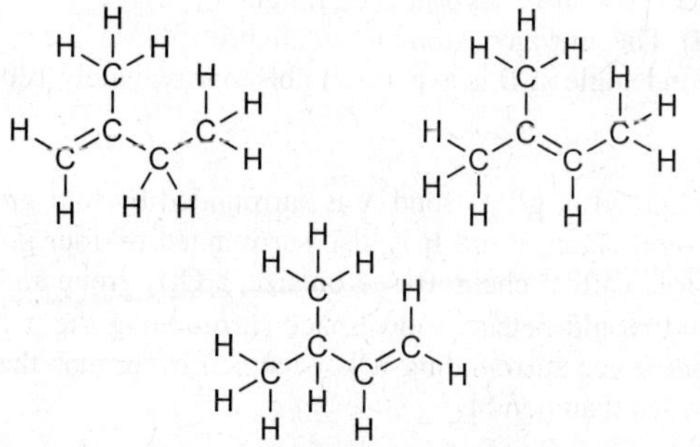

1.66. In each of the following two compounds, the molecular formula is $C_4H_{10}N_2$, there is a ring (as suggested in the hint given in the problem statement), there are no π bonds, there is no net dipole moment, and there is an N-H bond, which enables hydrogen bonding interactions. There are certainly other acceptable answers.

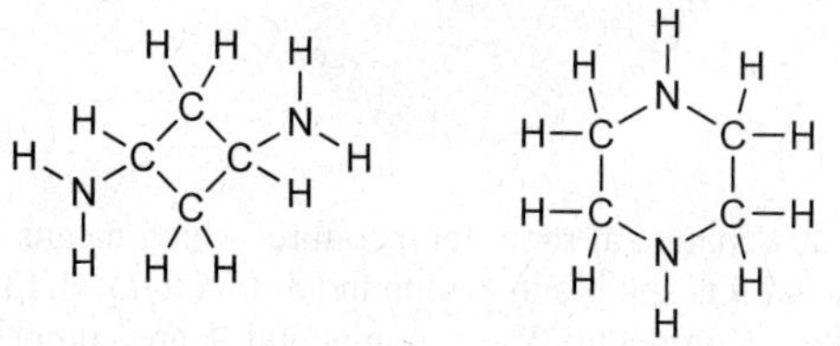

1.67. If we try to draw a linear skeleton with five carbon atoms and one nitrogen atom, we find that the number of hydrogen atoms is not correct (there are thirteen, rather than eleven):

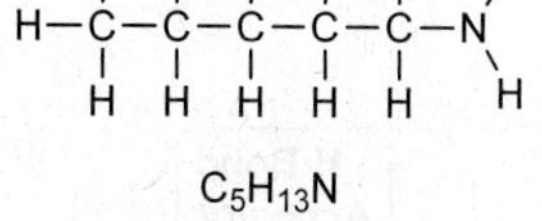

$C_5H_{13}N$

This will be the case even if try to draw a branched skeleton:

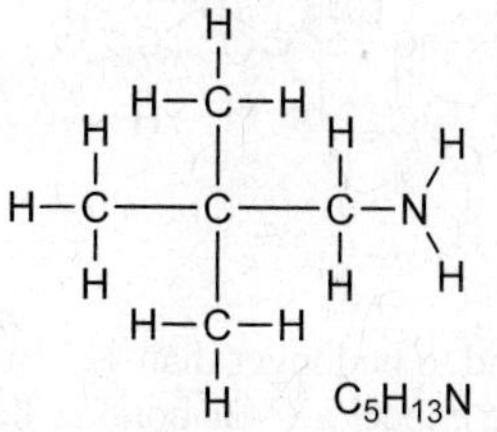

$C_5H_{13}N$

In fact, regardless of how the skeleton is branched, it will still have 13 hydrogen atoms. But we need to draw a structure with only 11 hydrogen atoms ($C_5H_{11}N$). So we

must remove two hydrogen atoms, which gives two unpaired electrons:

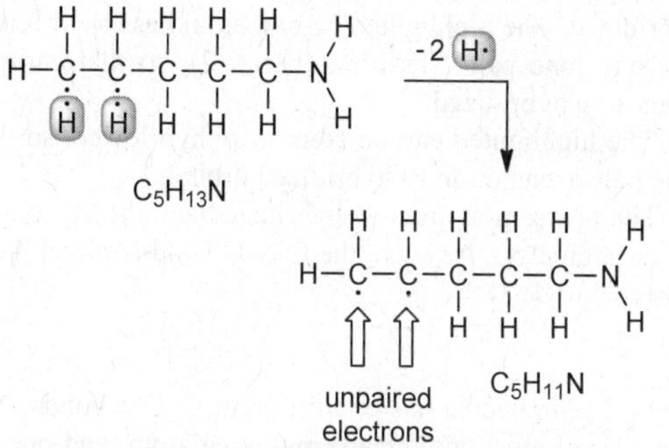

$C_5H_{13}N$

$-2\ \text{H·}$

unpaired electrons $C_5H_{11}N$

This indicates that we should consider pairing these electrons as a double bond. However, the problem statement specifically indicates that the structure cannot contain a double bond. So, we must find another way to pair the unpaired electrons. We encountered a similar issue in the solution to problem **1.35a**, in which we paired the electrons by forming a ring. We can do something similar here:

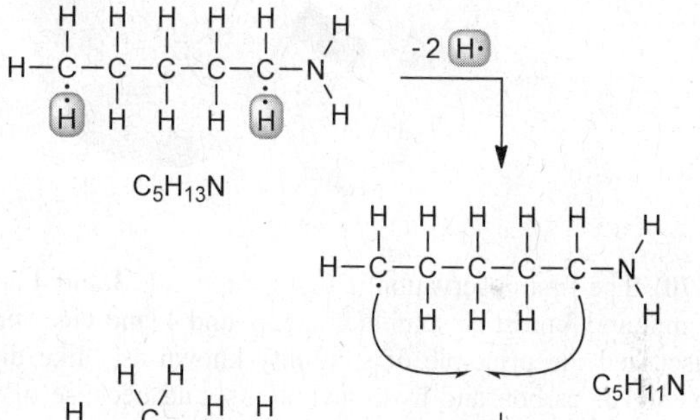

$C_5H_{13}N$

$-2\ \text{H·}$

$C_5H_{11}N$

$C_5H_{11}N$

Now we have the correct number of hydrogen atoms (eleven), which means that our structure must indeed contain a ring. But this particular cyclic structure (cyclic = containing a ring) does not meet all of the criteria described in the problem statement. Specifically, each carbon atom must be connected to exactly two hydrogen atoms. This is not the case in the structure above. This issue can be remedied in the following structure, which has a ring, and each of the carbon atoms is connected to exactly two hydrogen atoms, as required by the problem statement.

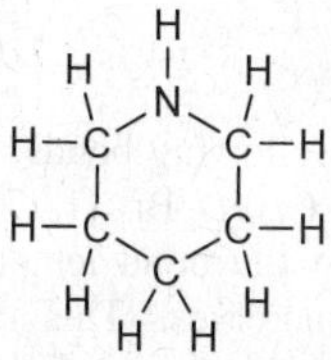

1.68.

(a) In compound **A**, the nitrogen atom has two σ bonds and no lone pairs (steric number = 2), so it is *sp* hybridized. The highlighted carbon atom has one σ bond and one lone pair (steric number = 2), so that carbon atom is *sp* hybridized.

(b) The highlighted carbon atom is *sp* hybridized, so the lone pair occupies an *sp* hybridized orbital.

(c) The nitrogen atom is *sp* hybridized and therefore has linear geometry. As such, the C-N-C bond angle in **A** is expected to be 180°.

(d) The nitrogen atom in **B** has two σ bonds and one lone pair (steric number = 3); therefore, it is *sp²* hybridized. The highlighted carbon atom has three σ bonds and no lone pairs (steric number = 3), so that carbon atom is *sp²* hybridized. Each of the chlorine atoms has three lone pairs and one bond (steric number = 4), so the chlorine atoms are *sp³* hybridized.

(e) The nitrogen atom is *sp²* hybridized, so the lone pair occupies an *sp²* hybridized orbital.

(f) The nitrogen atom is *sp²* hybridized so the C-N-C bond angle in **B** is expected to be approximately 120°.

1.69. The molecule has two different C=C π bonds, both highlighted in grey. Bond **A** is surrounded by four groups: two CH₃ groups, one (small) hydrogen atom, and one large carbon chain. Bond **B** is also surrounded by four groups, but they are different: three (small) hydrogen atoms, and one large carbon chain. Based on size, a CH₃ group and/or a carbon chain are much larger than a single hydrogen atom, and would occupy more space surrounding the π bond. Since there is more branching on bond **A** (and as a result more space surrounding it is occupied by groups that are larger than H), we would say that bond **A** is more sterically crowded than bond **B**.

Steric Crowding of π-bond A

Steric Crowding of π-bond B

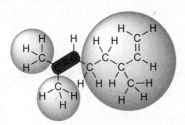
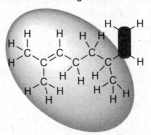

1.70. The first observation is that compounds **3** and **4** are difficult to separate at room temperature, which means that compound **3** must be soluble in compound **4**, and vice versa; compound **4** is soluble in compound **3**. In Section 1.13 we discussed the principle of solubility known as, "like dissolves like". Compound **3** and compound **4** are comprised mainly of carbon and hydrogen atoms, and because of the oxygen atoms present in each, both will have molecular dipole moments. Also of note, compound **3** possesses an O-H group; it has the ability to form hydrogen-bonding interactions with itself, *or with* the oxygen atoms of compound **4**.

When a mixture of compound **3** and **4** is heated, they can be separated from each other based on a difference in their boiling points. Compound **3** will have a much higher boiling point than compound **4**, because the former has an O-H group and the latter does not. Compound **4** does not have the ability to form hydrogen-bonding interactions with itself, so it will have a lower boiling point. When this mixture is heated, the compound that boils first (**4**) can be collected, leaving behind compound **3**.

1.71.

(a) We compare the following bonds: C(*sp³*)–F (1.40Å), C(*sp³*)–Cl (1.79Å), C(*sp³*)–Br (1.97Å), and C(*sp³*)–I (2.16Å). Notice that the bond length increases as the size of the halogen increases. This should make sense, since the valence electrons in iodine are farther away from the nucleus than the valence electrons in Br, so we

expect a C-I bond to be longer than a C-Br bond. For the same reason, we expect a C-Br bond to be longer than a C-Cl bond, which in turn is longer than a C-F bond.

(b) We compare the following bonds: C$_{sp}$3–F (1.40Å), C(*sp²*)–F (1.34Å), and C(*sp*)–F (1.27Å). Notice that the bond length decreases as the hybridization state goes from sp^3 to sp^2 to sp. This should make sense, because sp hybridized atoms hold their valence electrons closer to the nucleus (see Table 1.2 in the textbook), and therefore form shorter bonds.

(c) According to the trend that we discovered in part (a), we would expect C(sp^2)–Cl to be shorter than C(sp)–I, because the Cl atom holds its valence electrons closer to the nucleus (when compared with iodine). However, according to the trend that we discovered in part (b), we would expect C(sp^2)–Cl to be longer than C(sp)–I, because the carbon atom is sp^2 hybridized, so its valence electrons are farther from the nucleus than in the case of an sp hybridized carbon atom. These two trends are in conflict, and we would not be able to determine which factor dominates without the data that is provided. By analyzing the data, we can see that C(sp^2)–Cl must be shorter than 1.79Å [compare with C(sp^3)–Cl], while C(sp)–I must be longer than 1.79Å [compare with C(sp)–Br]. Therefore, C(sp)–I must be longer than C(sp^2)–Cl.

1.72.
(a) In the first compound, the fluorine isotope (^{18}F) has no formal charge. Therefore, it must have three lone pairs (see Section 1.4 for a review of how formal charges are calculated). Since it has one σ bond and three lone pairs, it must have a steric number of 4, so it is sp^3 hybridized. The bromine atom also has no formal charge. So, it too, like the fluorine isotope, must have three lone pairs. Once again, one σ bond and three lone pairs give a steric number of 4, so the bromine atom is sp^3 hybridized.
In the second compound, the nitrogen atom has no formal charge. Therefore, it must have one lone pair. Since the nitrogen atom has three σ bonds and one lone pair, it must have a steric number of 4, so it is sp^3 hybridized. The oxygen atom also has no formal charge. Therefore, it must have two lone pairs. Since the oxygen atom has two σ bonds and two lone pairs, it must have a steric number of 4, so it is sp^3 hybridized.
In the product, the fluorine isotope (^{18}F) has no formal charge. Therefore, it must have three lone pairs. Since it has one σ bond and three lone pairs, it must have a steric number of 4, so it is sp^3 hybridized. The nitrogen atom does have a positive formal charge. Therefore, it must have no lone pairs. Since it has four σ bonds and no lone pairs, it must have a steric number of 4, so it is sp^3 hybridized. The oxygen atom also has no formal charge. Therefore, it must have two lone pairs. Since the oxygen atom has two σ bonds and two lone pairs, it must have a steric number of 4, so it is sp^3 hybridized. Finally, the bromine atom has a negative charge and no bonds. So it must have four lone pairs. With four lone pairs and no bonds, it will have a steric number of 4, so it is sp^3 hybridized.
In summary, all of the heteroatoms (in all of the structures) are sp^3 hybridized.

(b) The nitrogen atom is sp^3 hybridized. With four bonds, we expect the geometry around the nitrogen atom to be tetrahedral. So, the bond angle for each C-N-C bond is expected to be approximately 109.5°.

(c) The oxygen atom is sp^3 hybridized, so each of the lone pairs occupies an sp^3 hybridized orbital.

1.73.
(a) Boron is in group 3A of the periodic table and is therefore expected to be trivalent. That is, it has three valence electrons, and it uses each one of those valence electrons to form a bond, giving rise to three bonds. It does not have any electrons left over for a lone pair (as in the case of nitrogen). With three σ bonds and no lone pairs, the boron atom has a steric number of three, so it is sp^2 hybridized.

(b) Since the boron atom is sp^2 hybridized, we expect the bond angle to be approximately 120°. However, in this case, the O-B-O system is part of a five-membered ring. That is, there are five different bond angles (of which the O-B-O angle is one of them) that together must form a closed loop. That requirement could conceivably force some of the bond angles (including the O-B-O bond angle) to deviate from the predicted value. In fact, we will explore this very phenomenon, called ring strain, in Chapter 4, and we will see that five-membered rings actually possess very little ring strain.

(c) Each of the oxygen atoms has no formal charge, and must therefore have two bonds and two lone pairs. The boron atom has no lone pairs, as explained in the solution to Problem **1.73a**.

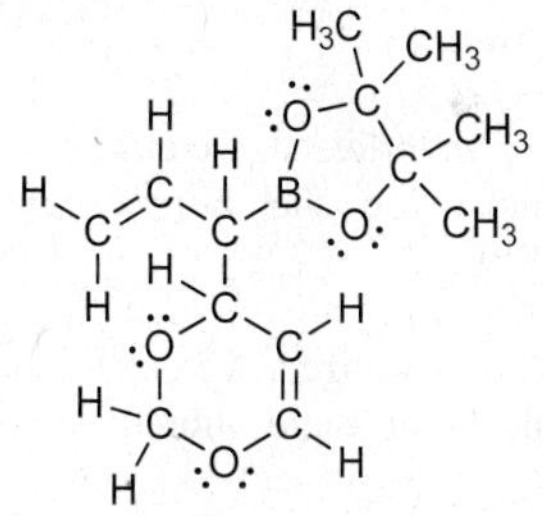

1.74.
(a) If we analyze each atom (in both **1** and **2**) using the procedure outlined in Section 1.4, we find that none of the atoms in compound **1** have a formal charge, while compound **2** possesses two formal charges:

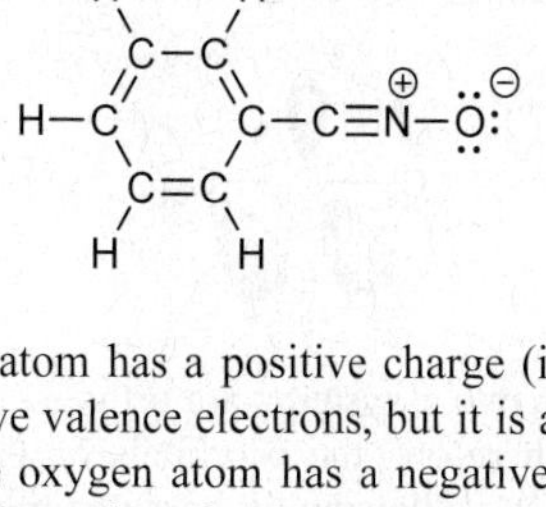

The nitrogen atom has a positive charge (it is supposed to be using five valence electrons, but it is actually using four), and the oxygen atom has a negative charge (it is supposed to be using six valence electrons, but it is actually using seven).

(b) Compound **1** possesses polar bonds, as a result of the presence of partial charges (δ+ and δ-). The associated dipole moments can form favorable interactions with the dipole moments present in the polar solvent molecules (dipole-dipole interactions). However, compound **2** has formal charges (negative on O and positive on N), so the dipole moment of the N-O bond is expected to be much more significant than the dipole moments in compound **1**. The dipole moment of the N-O bond in compound **2** is the result of full charges, rather than partial charges. As such, compound **2** is expected to experience much stronger interactions with the solvent molecules, and therefore, **2** should be more soluble than **1** in a polar solvent.

(c) In compound **1**, the carbon atom (attached to nitrogen) has three σ bonds and no lone pairs (steric number = 3), so that carbon atom is sp^2 hybridized, with trigonal planar geometry. As such, the C-C-N bond angle in compound **1** is expected to be approximately 120°. However, in compound **2**, the same carbon atom has two σ bonds and no lone pairs (steric number = 2), so now this carbon atom is sp hybridized, with linear geometry. As such, the C-C-N bond angle in **2** is expected to be 180°. The conversion of **1** to **2** therefore involves an increase in the C-C-N bond angle of approximately 60°.

1.75.
(a) C_a has three σ bonds and no lone pairs, so it has a steric number of 3, and is therefore sp^2 hybridized. The same is true for C_c. In contrast, C_b has two σ bonds and no lone pairs, so it has a steric number of 2, and is therefore sp hybridized.

(b) Since C_a is sp^2 hybridized, we expect its geometry to be trigonal planar, so the bond angle should be approximately 120°.

(c) Since C_b is sp hybridized, we expect its geometry to be linear, so the bond angle should be approximately 180°.

(d) The central carbon atom (C_b) is sp hybridized, so it is using two sp hybridized orbitals to form its two σ bonds, which will be arranged in a linear fashion. The remaining two p orbitals of C_b (used for π bonding) need to be as far apart from each other as possible, so they will be 90° apart from one another (just as we saw for the carbon atoms of a triple bond; see Figure 1.33).

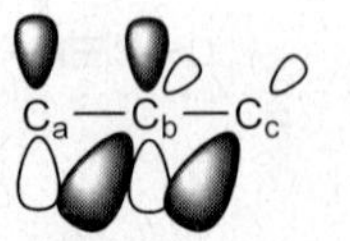

As a result, the two π systems are orthogonal (or 90°) to each other. Therefore, the p orbitals on C_a and C_c are orthogonal. The following is another drawing from a

different perspective (looking down the axis of the linear C_a-C_b-C_c system.

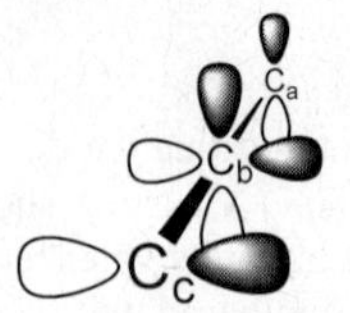

1.76.
(a) The following highlighted regions represent the two different N-C-N units in the structure:

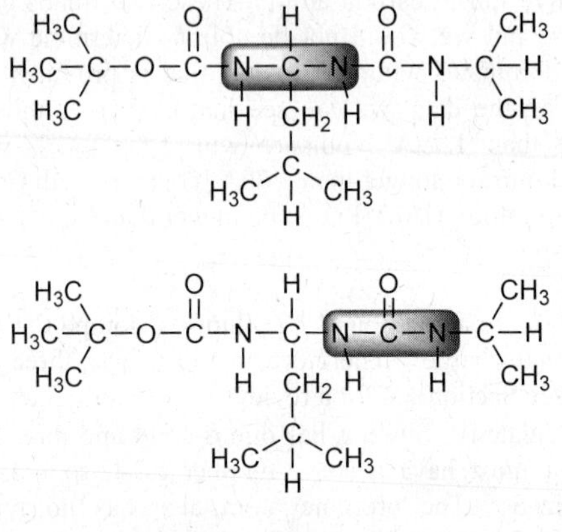

The first N-C-N unit (shown above) exhibits a central carbon atom that is sp^3 hybridized and is therefore expected to have tetrahedral geometry. Accordingly, the bond angles about that carbon atom are expected to be approximately 109.5°.
The other N-C-N unit exhibits a central carbon atom that is sp^2 hybridized and is therefore expected to have trigonal planar geometry. Accordingly, the bond angles about that carbon atom are expected to be approximately 120°.

(b) The non-covalent interaction is an intramolecular, hydrogen bonding interaction between the H (connected to the highlighted nitrogen atom) and the lone pair of the oxygen atom:

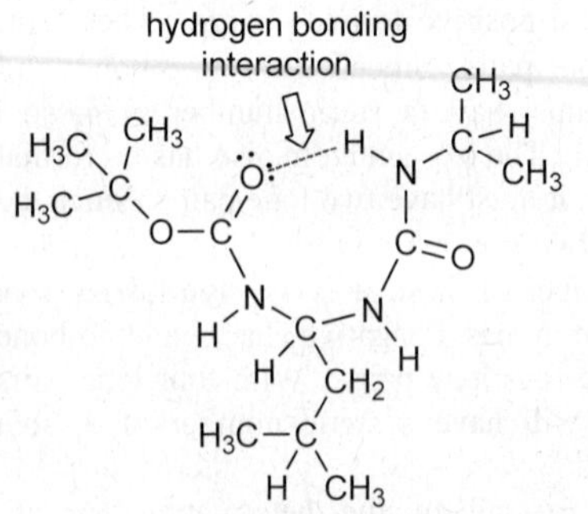

1.77. First, let's compare the functional groups at the peripheries of the binding sites in each of the two polymers. Each carboxylic acid group (COOH) in polymer **A** can serve as both a hydrogen-bond *donor* and *acceptor*, while each of the nitrogen atoms in polymer **B** can only serve as a hydrogen-bond *acceptor*.

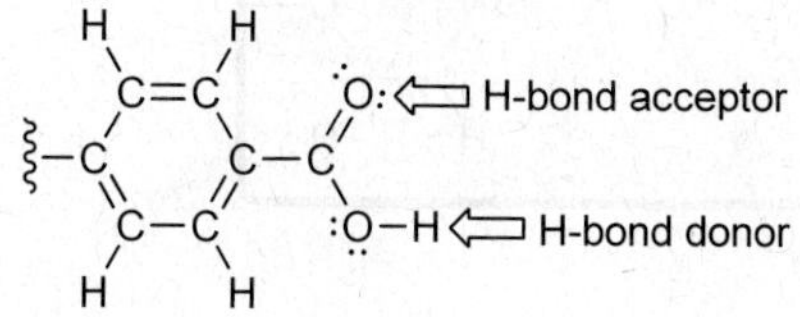

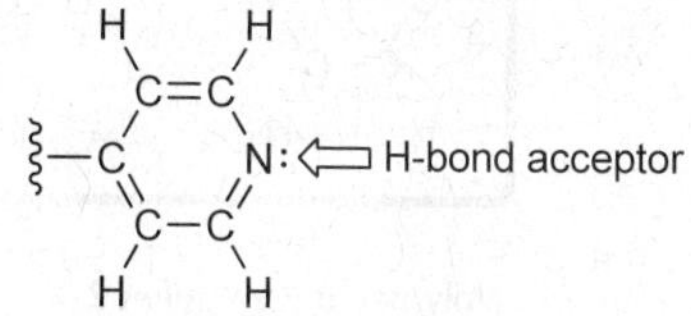

Now, let's compare the functional groups at the termini of the two guest molecules. The OH and NH_2 groups can each serve as a donor or an acceptor. The OH group is a stronger hydrogen-bond donor than the NH_2 group, because oxygen is more electronegative than nitrogen, thus resulting in a greater partial positive charge on the hydrogen atom attached to oxygen.

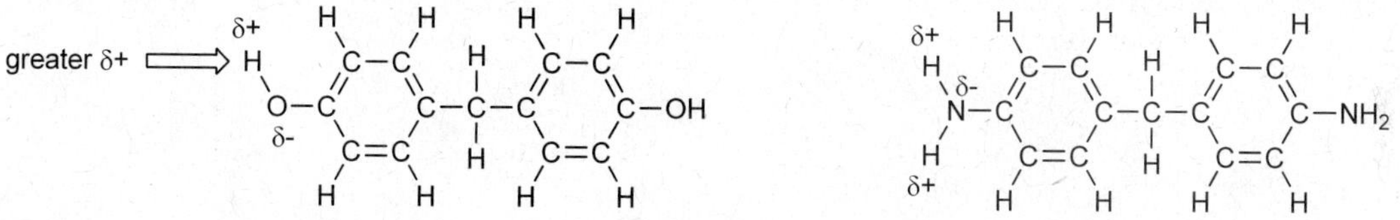

On the other hand, the NH_2 group is a better hydrogen-bond acceptor. Nitrogen is less electronegative than oxygen, so the lone pair on nitrogen is held less strongly to the nitrogen, and thus it is more available to form a hydrogen bond.

With this in mind, we can draw pictures proposing the mode of binding of each guest within each polymer. Each carboxylic acid group (COOH) in the binding site of polymer **A** can form two H-bonds to the functional group on each side of the guest molecule (OH in guest **1**, NH_2 in guest **2**), as shown below.

Polymer **A** with guest **1**

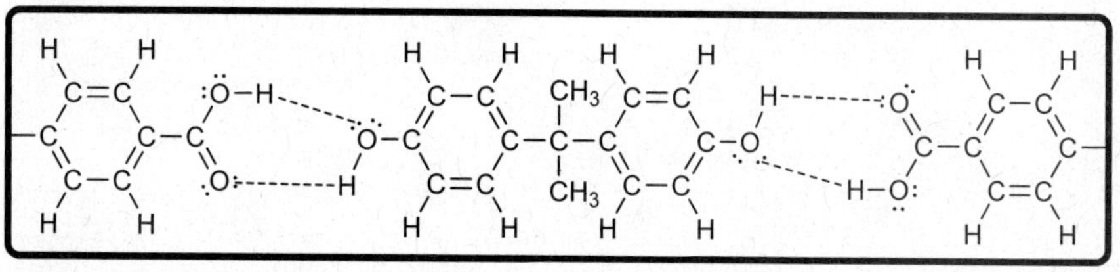

Polymer **A** with guest **2**

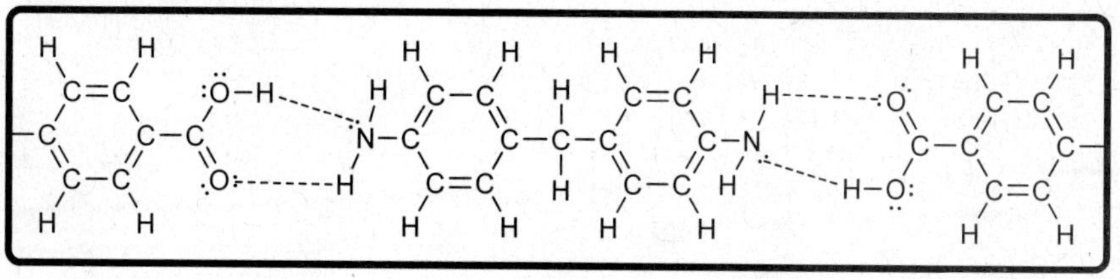

Each guest molecule binds similarly because: (a) the two guest molecules have similar size and shape, so they can both fit into the binding site, and (b) the increased H-bond *donating* ability of the OH group (which leads to stronger binding of guest **1**) is balanced by the increased H-bond *accepting* ability of the NH_2 group (which leads to stronger binding of guest **2**).

In contrast, the groups on the periphery of polymer **B** can only accept (but not donate) hydrogen bonds as explained above. Guest **1** binds strongly with polymer **B** (due to the formation of strong hydrogen-bonds). The binding of guest **2** is significantly weaker (due to the formation of relatively weaker hydrogen-bonds).

Polymer **B** with guest **1**

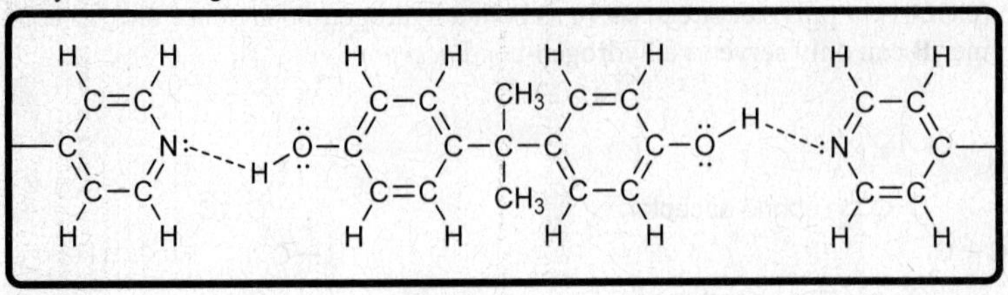

Polymer **B** with guest **2**

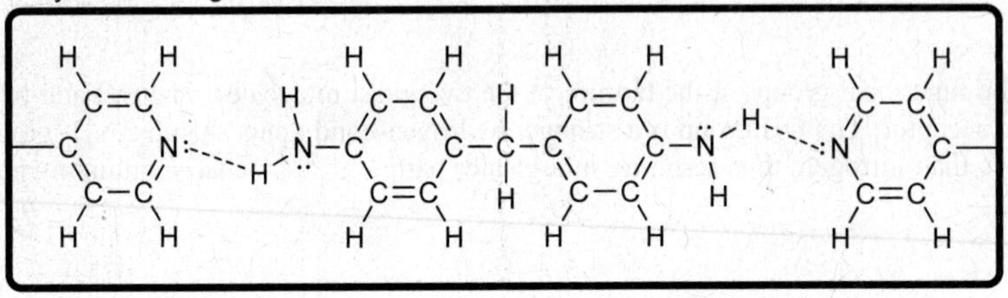

Chapter 2
Molecular Representations

Review of Concepts

Fill in the blanks below. To verify that your answers are correct, look in your textbook at the end of Chapter 2. Each of the sentences below appears verbatim in the section entitled *Review of Concepts and Vocabulary*.

- In **bond-line structures**, ___C___ atoms and most ___H___ atoms are not drawn.
- A _functional grp_ is a characteristic group of atoms/bonds that show a predictable behavior.
- When a carbon atom bears either a positive charge or a negative charge, it will have _____, rather than four, bonds.
- In bond-line structures, a **wedge** represents a group coming _out_ the page, while a **dash** represents a group _into_ the page.
- _curved_ **arrows** are tools for drawing resonance structures.
- When drawing curved arrows for resonance structures, avoid breaking a __σ__ bond and never exceed _octet_ for second-row elements.
- There are three rules for identifying significant resonance structures:
    1. Minimize _charge_.
    2. Electronegative atoms can bear a positive charge, but only if they possess an _octet_ of electrons.
    3. Avoid drawing a resonance structure in which two carbon atoms bear _opposite_ charges.
- A _decolaiued_ lone pair participates in resonance and is said to occupy a ____ orbital.
- A _iocaiued_ lone pair does not participate in resonance.

Review of Skills

Fill in the blanks and empty boxes below. To verify that your answers are correct, look in your textbook at the end of Chapter 2. The answers appear in the section entitled *SkillBuilder Review*.

SkillBuilder 2.1   Converting Between Different Drawing Styles

DRAW THE LEWIS STRUCTURE OF THE FOLLOWING COMPOUND

$(CH_3)_3COCH_3$ ⟶

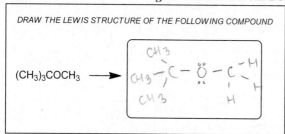

SkillBuilder 2.2   Reading Bond-Line Structures

CIRCLE ALL CARBON ATOMS IN THE COMPOUND BELOW	DRAW ALL HYDROGEN ATOMS IN THE COMPOUND BELOW

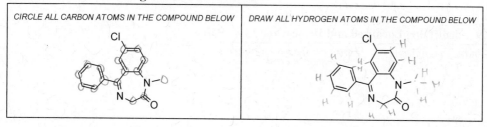

SkillBuilder 2.3   Drawing Bond-Line Structures

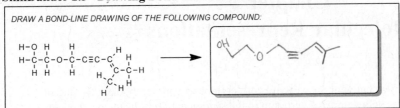

DRAW A BOND-LINE DRAWING OF THE FOLLOWING COMPOUND:

SkillBuilder 2.4   Identifying Lone Pairs on Oxygen Atoms

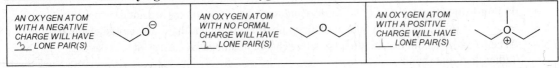

AN OXYGEN ATOM WITH A NEGATIVE CHARGE WILL HAVE _3_ LONE PAIR(S)

AN OXYGEN ATOM WITH NO FORMAL CHARGE WILL HAVE _2_ LONE PAIR(S)

AN OXYGEN ATOM WITH A POSITIVE CHARGE WILL HAVE _1_ LONE PAIR(S)

SkillBuilder 2.5   Identifying Lone Pairs on Nitrogen Atoms

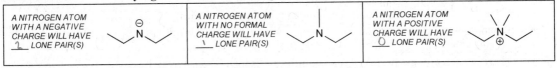

A NITROGEN ATOM WITH A NEGATIVE CHARGE WILL HAVE _2_ LONE PAIR(S)

A NITROGEN ATOM WITH NO FORMAL CHARGE WILL HAVE _1_ LONE PAIR(S)

A NITROGEN ATOM WITH A POSITIVE CHARGE WILL HAVE _0_ LONE PAIR(S)

SkillBuilder 2.6   Identifying Valid Resonance Arrows

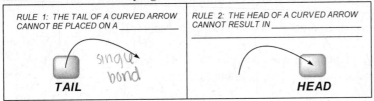

RULE 1: THE TAIL OF A CURVED ARROW CANNOT BE PLACED ON A _____

single bond

TAIL

RULE 2: THE HEAD OF A CURVED ARROW CANNOT RESULT IN _____

HEAD

SkillBuilder 2.7   Assigning Formal Charges in Resonance Structures

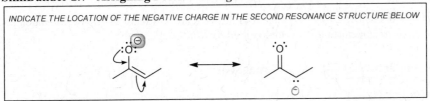

INDICATE THE LOCATION OF THE NEGATIVE CHARGE IN THE SECOND RESONANCE STRUCTURE BELOW

SkillBuilder 2.8   Drawing Significant Resonance Structures

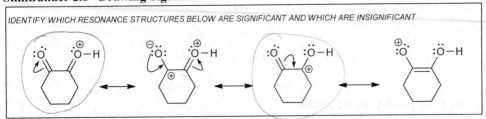

IDENTIFY WHICH RESONANCE STRUCTURES BELOW ARE SIGNIFICANT AND WHICH ARE INSIGNIFICANT

SkillBuilder 2.9   Identifying Localized and Delocalized Lone Pairs

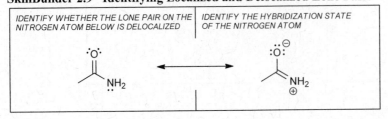

IDENTIFY WHETHER THE LONE PAIR ON THE NITROGEN ATOM BELOW IS DELOCALIZED

IDENTIFY THE HYBRIDIZATION STATE OF THE NITROGEN ATOM

Common Mistakes to Avoid

When drawing a structure, make sure to avoid drawing a pentavalent carbon atom, or even a hexavalent or heptavalent carbon atom:

INCORRECT   **INCORRECT**   **INCORRECT**

Carbon cannot   Carbon cannot   Carbon cannot
have five bonds   have six bonds   have seven bonds

Carbon cannot have more than four bonds.   Avoid drawing a carbon atom with more than four bonds, as that is one of the worst mistakes you can make as a student of organic chemistry.

Also, when drawing a structure, either draw all carbon atom labels (C) and all hydrogen atom labels (H), like this:

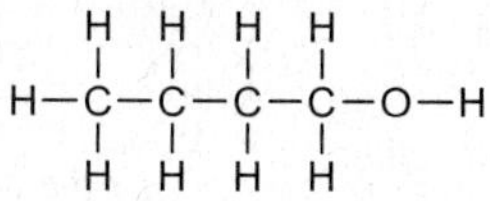

$$\begin{array}{ccccc} & H & H & H & H \\ & | & | & | & | \\ H- & C- & C- & C- & C-O-H \\ & | & | & | & | \\ & H & H & H & H \end{array}$$

or don't draw any labels (except H attached to a heteroatom), like this:

That is, if you draw all C labels, then you should really draw all H labels also. Avoid drawings in which the C labels are drawn and the H labels are not, as shown here:

INCORRECT   **INCORRECT**

C—C—C—C—OH   C—C—C—OH

These types of drawings (where C labels are shown and H labels are not shown) should only be used when you are working on a scratch piece of paper and trying to draw constitutional isomers. For example, if you are considering all constitutional isomers with molecular formula C_3H_8O, you might find it helpful to use drawing like these as a form of "short-hand" so that you can identify all of the different ways of connecting three carbon atoms and one oxygen atom:

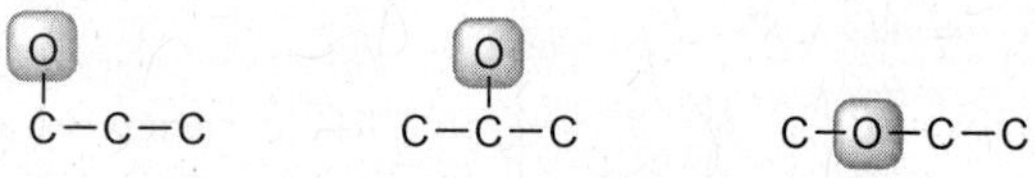

C—C—C   C—C—C   C—O—C—C

But your final structures should either show all C and H labels, or no labels at all. The latter is the more commonly used method:

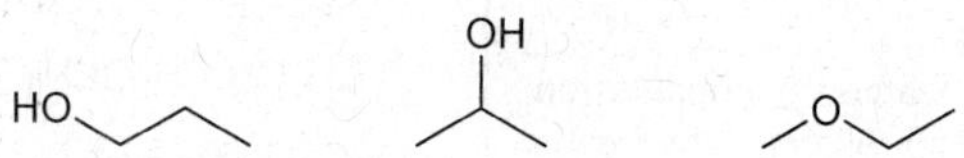

34   CHAPTER 2

Solutions

2.1.

(a) The condensed structure indicates the constitution (how the atoms are connected to each other). The Lewis structure shows these connections more clearly, because every bond is drawn:

$CH_2=CHOCH_2CH(CH_3)_2$
Condensed structure

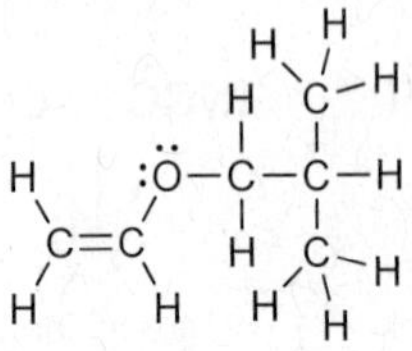

Lewis structure

(b) The condensed structure indicates the constitution (how the atoms are connected to each other). The Lewis structure shows these connections more clearly, because every bond is drawn:

$(CH_3CH_2)_2CHCH_2CH_2OH$
Condensed structure

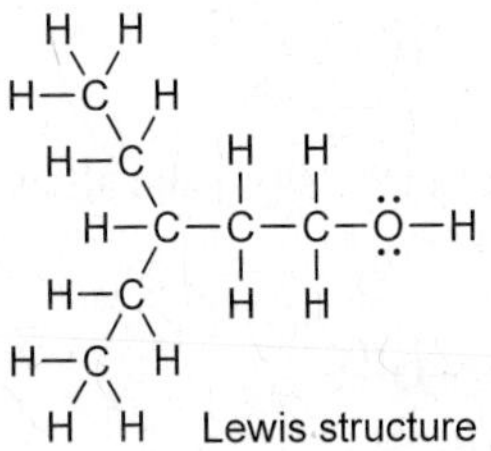

Lewis structure

(c) The condensed structure indicates the constitution (how the atoms are connected to each other). The Lewis structure shows these connections more clearly, because every bond is drawn:

$(CH_3CH_2)_3COH$
Condensed structure

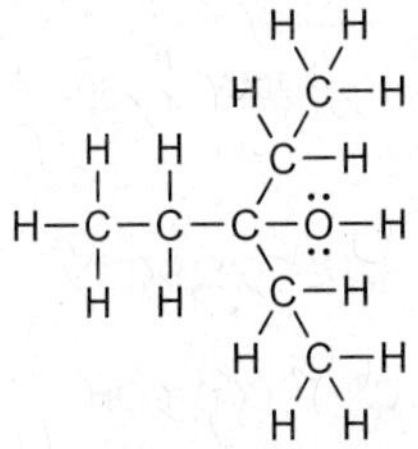

Lewis structure

(d) The condensed structure indicates the constitution (how the atoms are connected to each other). The Lewis structure shows these connections more clearly, because every bond is drawn:

$(CH_3)_2C=CHCH_2CH_3$
Condensed structure

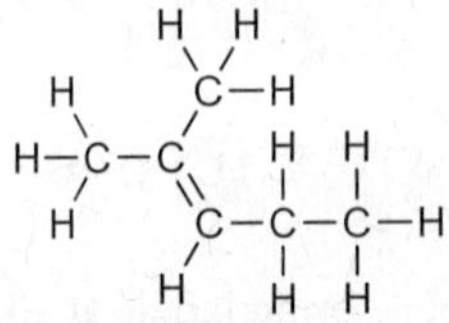

Lewis structure

(e) The condensed structure indicates the constitution (how the atoms are connected to each other). The Lewis structure shows these connections more clearly, because every bond is drawn:

$CH_2=CHCH_2OCH_2CH(CH_3)_2$
Condensed structure

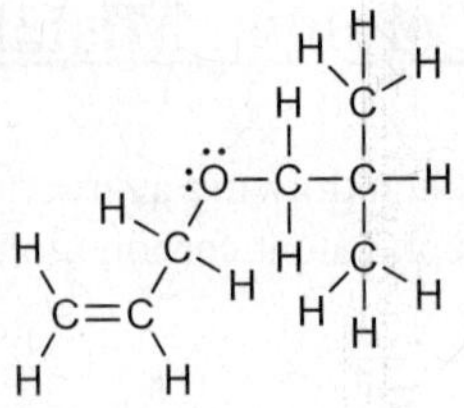

Lewis structure

(f) The condensed structure indicates the constitution (how the atoms are connected to each other). The Lewis structure shows these connections more clearly, because every bond is drawn:

$(CH_3CH_2)_2C=CH_2$
Condensed structure

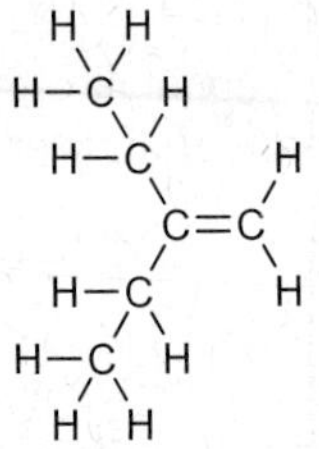

Lewis structure

(g) The condensed structure indicates the constitution (how the atoms are connected to each other). The Lewis structure shows these connections more clearly, because every bond is drawn:

$(CH_3)_3CCH_2CH_2OH$
Condensed structure

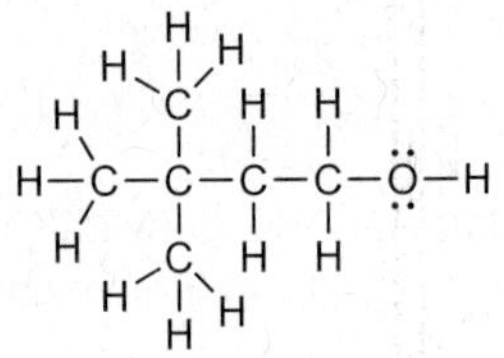

Lewis structure

(h) The condensed structure indicates the constitution (how the atoms are connected to each other). The Lewis structure shows these connections more clearly, because every bond is drawn:

$CH_3CH_2CH_2CH_2CH_2CH_3$
Condensed structure

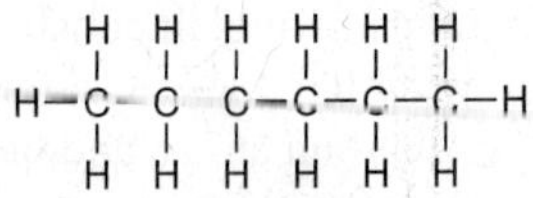

Lewis structure

(i) The condensed structure indicates the constitution (how the atoms are connected to each other). The Lewis structure shows these connections more clearly, because every bond is drawn:

$CH_3CH_2CH_2OCH_3$
Condensed structure

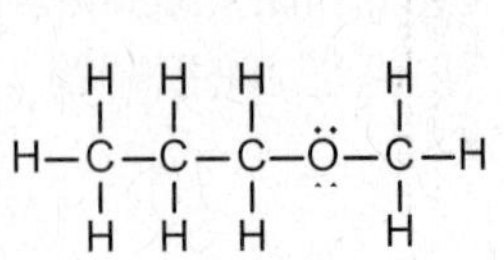

Lewis structure

(j) The condensed structure indicates the constitution (how the atoms are connected to each other). The Lewis structure shows these connections more clearly, because every bond is drawn:

(CH₃CH₂CH₂)₂CHOH
Condensed structure

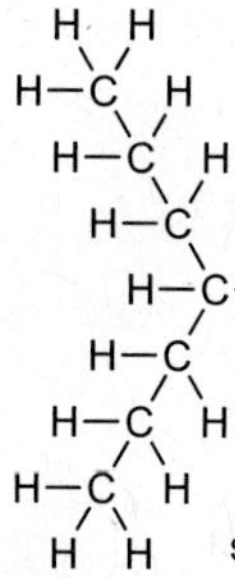

Lewis structure

(k) The condensed structure indicates the constitution (how the atoms are connected to each other). The Lewis structure shows these connections more clearly, because every bond is drawn:

(CH₃CH₂)₂CHCH₂OCH₃
Condensed structure

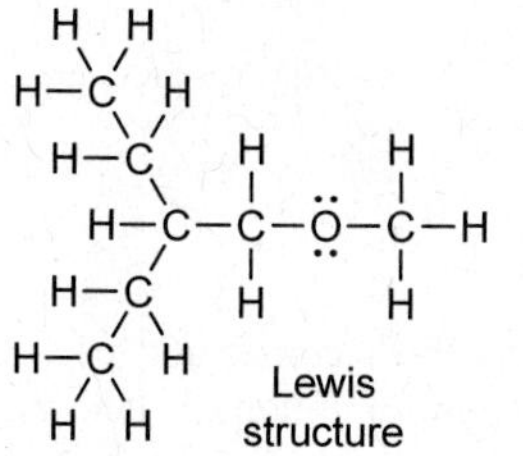

Lewis structure

(l) The condensed structure indicates the constitution (how the atoms are connected to each other). The Lewis structure shows these connections more clearly, because every bond is drawn:

(CH₃)₂CHCH₂OH
Condensed structure

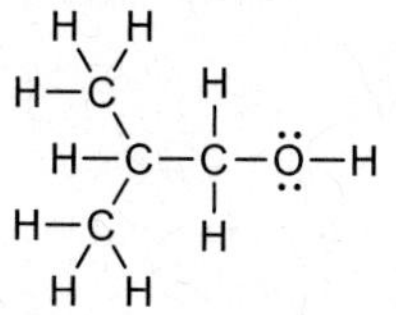

Lewis structure

2.2 The following two compounds are constitutional isomers because they share the same molecular formula (C₅H₁₂O). The third compound (not shown here) has a different molecular formula (C₄H₁₀O).

(CH₃)₃CÖCH₃   (CH₃)₂CHÖCH₂CH₃

2.3   Begin by drawing a Lewis structure, so that the bonding of each carbon atom is shown more clearly:

(CH₃)₂C=CHC(CH₃)₃
Condensed structure

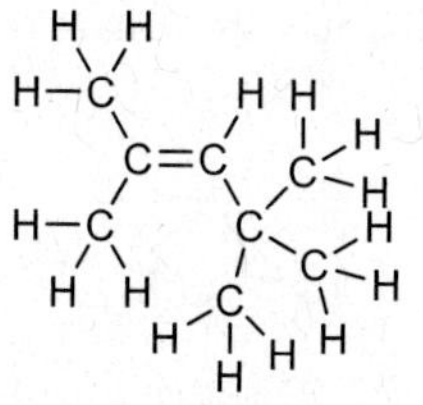

Lewis structure

Notice that two of the carbon atoms are sharing a double bond. These two atoms are sp^2 hybridized. Each of the other six carbon atoms exhibits four single bonds, and as

such, each of these six carbon atoms (highlighted) is sp^3 hybridized.

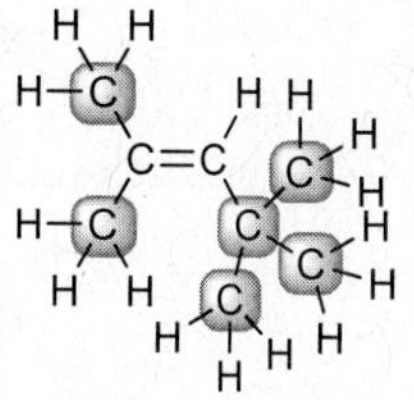

2.4   Propylene and cyclopropane are constitutional isomers because they are different compounds that share the same molecular formula, C₃H₆ (see the solution for problem **1.42** for an explanation):

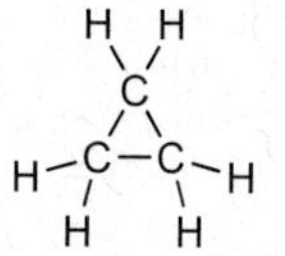

Cyclopropane
C₃H₆

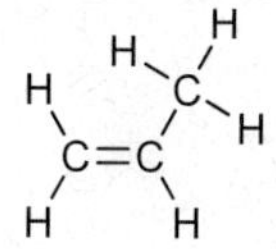

Propylene
C₃H₆

Propylene has the following condensed structure:

CH₂=CHCH₃

2.5.
(a)   Each corner and each endpoint represents a carbon atom (highlighted below), so this compound has eight carbon atoms. Each carbon atom will have enough hydrogen atoms to have exactly four bonds, as shown:

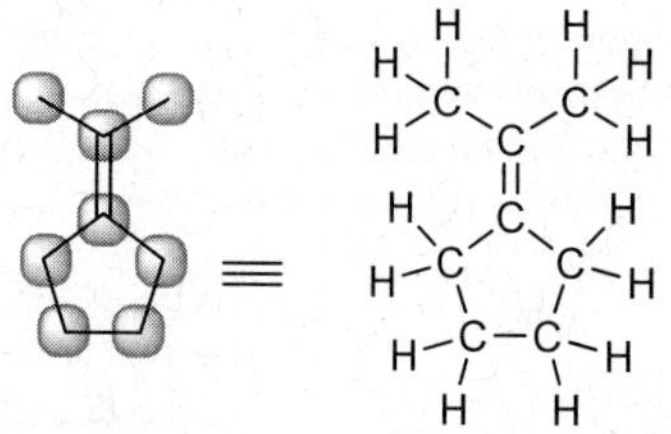

(b)   Each corner and each endpoint represents a carbon atom (highlighted below), so this compound has five carbon atoms. Each carbon atom will have enough hydrogen atoms to have exactly four bonds, as shown:

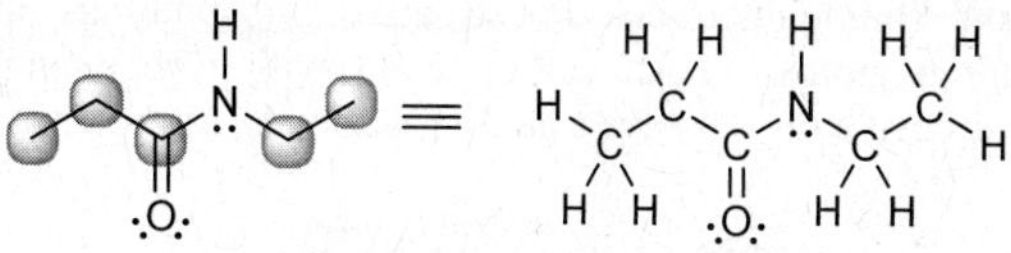

(c)   Each corner and each endpoint represents a carbon atom (highlighted below), so this compound has six carbon atoms. Each carbon atom will have enough hydrogen atoms to have exactly four bonds, as shown:

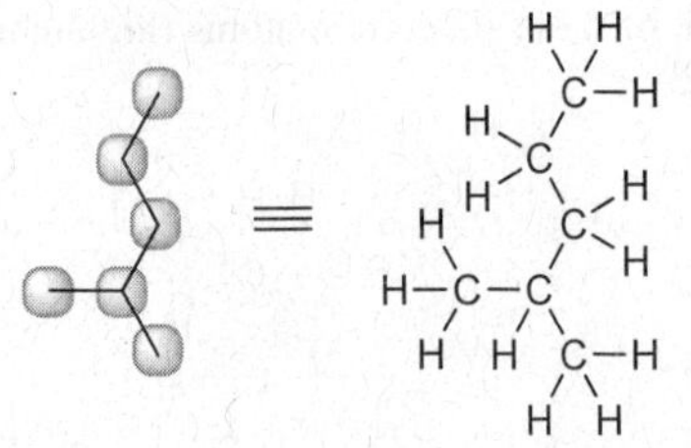

(d) Each corner and each endpoint represents a carbon atom (highlighted below), so this compound has twelve carbon atoms. Each carbon atom will have enough hydrogen atoms to have exactly four bonds, as shown:

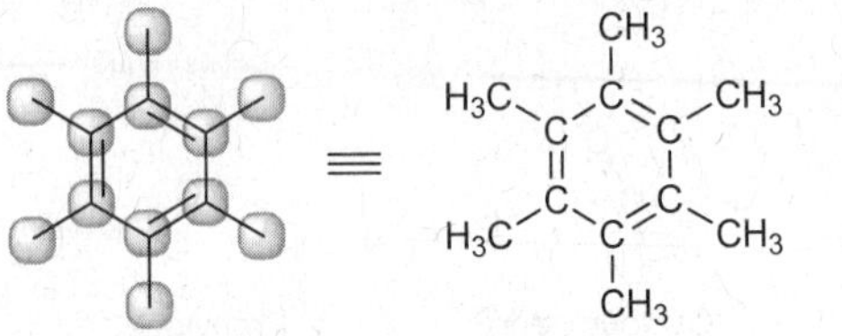

(e) Each corner and each endpoint represents a carbon atom (highlighted below), so this compound has six carbon atoms. Each carbon atom will have enough hydrogen atoms to have exactly four bonds, as shown:

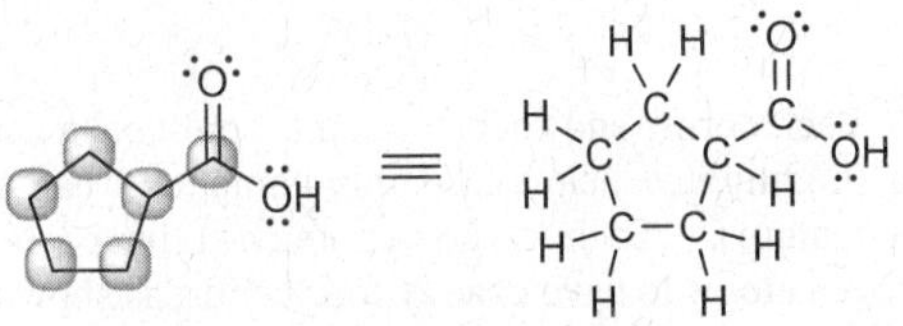

(f) Each corner and each endpoint represents a carbon atom (highlighted below), so this compound has six carbon atoms. Each carbon atom will have enough hydrogen atoms to have exactly four bonds, as shown:

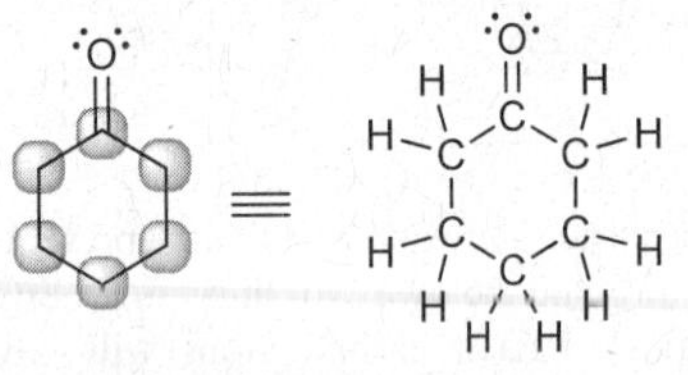

(g) Each corner and each endpoint represents a carbon atom (highlighted below), so this compound has six carbon atoms. Each carbon atom will have enough hydrogen atoms to have exactly four bonds, as shown:

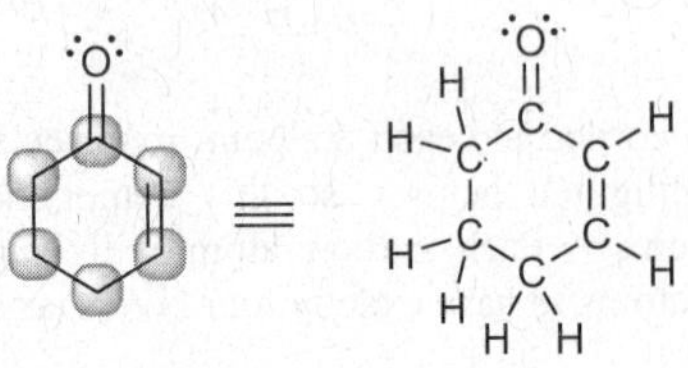

(h) Each corner and each endpoint represents a carbon atom (highlighted below), so this compound has six

carbon atoms. Each carbon atom will have enough hydrogen atoms to have exactly four bonds, as shown:

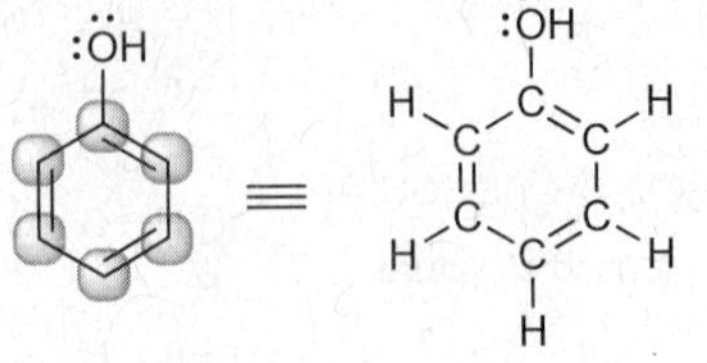

(i) Each corner and each endpoint represents a carbon atom (highlighted below), so this compound has seven carbon atoms. Each carbon atom will have enough hydrogen atoms to have exactly four bonds, as shown:

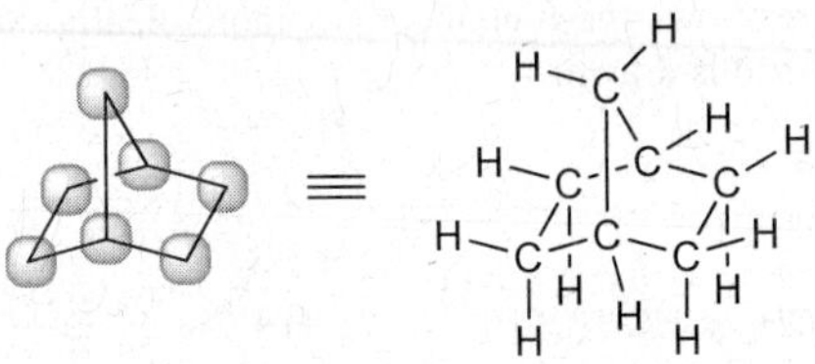

(j) Each corner and each endpoint represents a carbon atom (highlighted below), so this compound has six carbon atoms. Each carbon atom will have enough hydrogen atoms to have exactly four bonds, as shown:

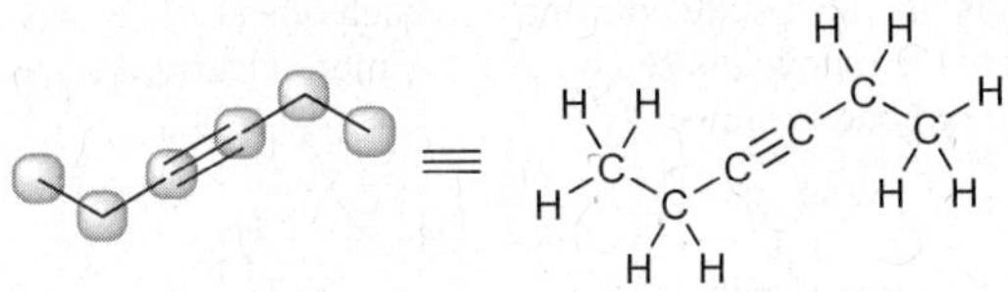

(k) Each corner and each endpoint represents a carbon atom (highlighted below), so this compound has seven carbon atoms. Each carbon atom will have enough hydrogen atoms to have exactly four bonds, as shown:

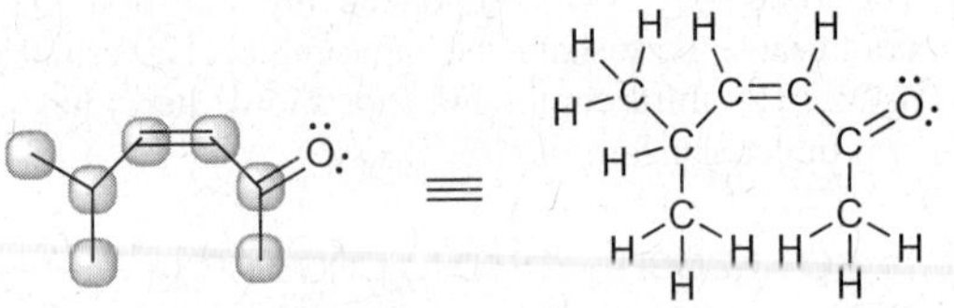

(l) Each corner and each endpoint represents a carbon atom (highlighted below), so this compound has seven carbon atoms. Each carbon atom will have enough hydrogen atoms to have exactly four bonds, as shown:

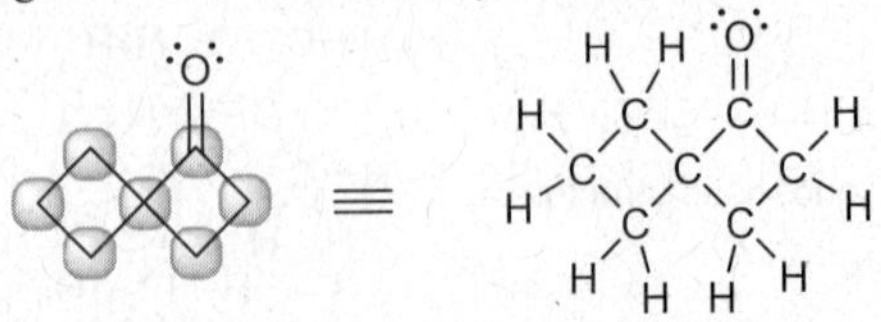

2.6
(a) The starting material has seven carbon atoms (highlighted), while the product has six carbon atoms

(highlighted), representing a decrease in the number of carbon atoms.

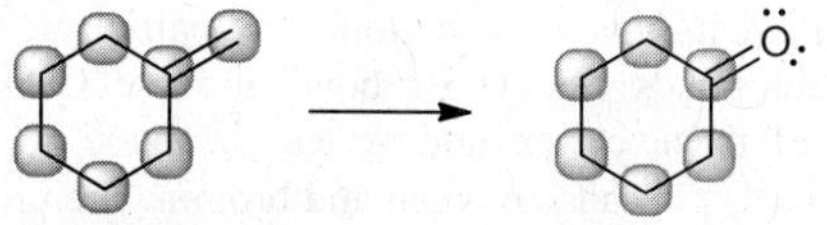

seven carbon atoms → six carbon atoms

(b) The starting material has eight carbon atoms (highlighted), and the product also has eight carbon atoms (highlighted). This transformation does not involve a change in the number of carbon atoms.

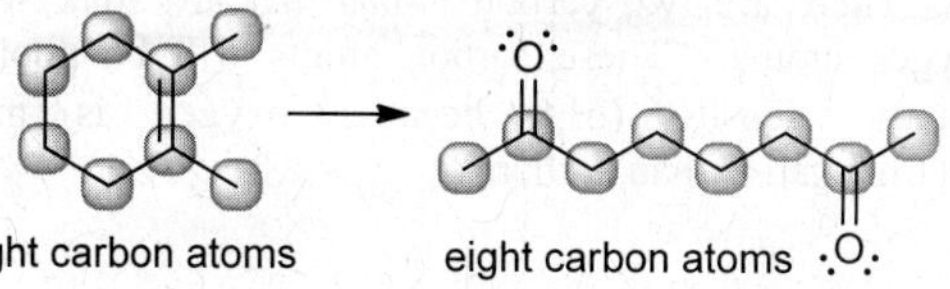

eight carbon atoms → eight carbon atoms

(c) The starting material has eight carbon atoms (highlighted), and the product also has eight carbon atoms (highlighted). This transformation does not involve a change in the number of carbon atoms.

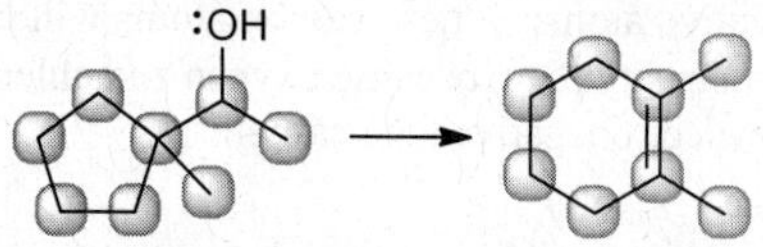

eight carbon atoms → eight carbon atoms

(d) The starting material has five carbon atoms (highlighted), while the product has seven carbon atoms (highlighted), representing an increase in the number of carbon atoms.

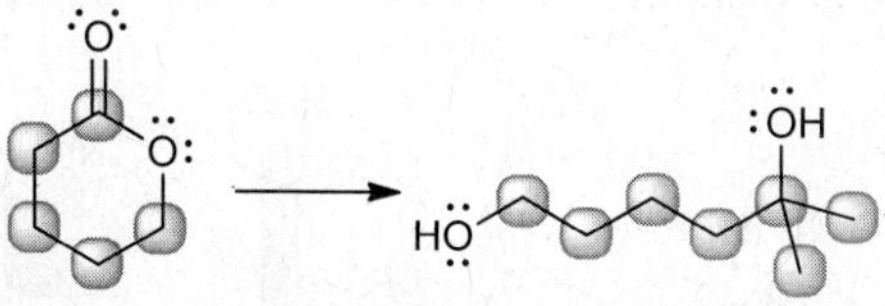

five carbon atoms → seven carbon atoms

2.7
(a) The starting material has twelve hydrogen atoms, while the product has fourteen hydrogen atoms. As such, this transformation involves an increase in the number of hydrogen atoms.

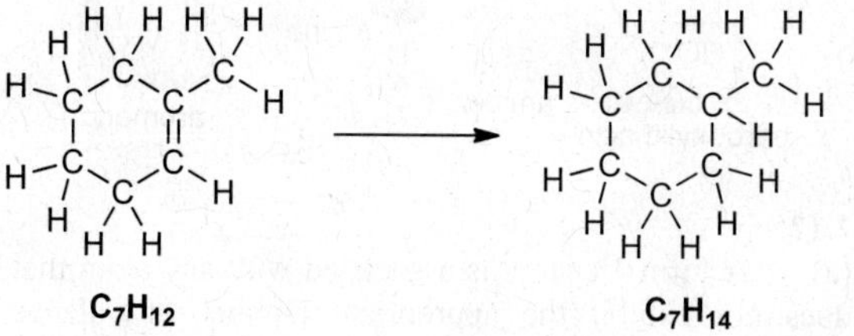

C_7H_{12} → C_7H_{14}

(b) The starting material has eight hydrogen atoms, while the product has six hydrogen atoms. As such, this

transformation involves a decrease in the number of hydrogen atoms.

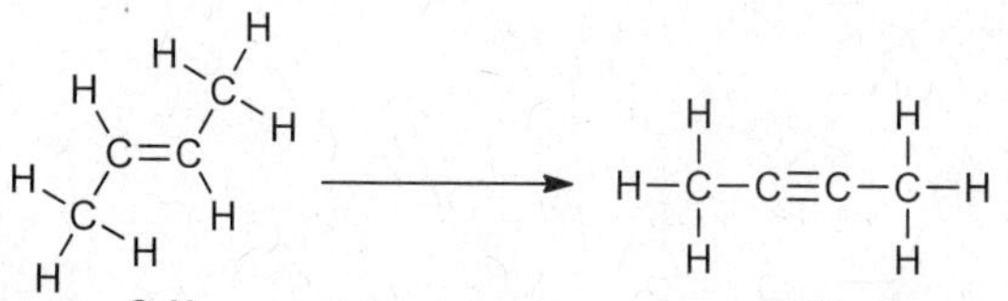

C_4H_8 → C_4H_6

2.8. In each of the following structures, the carbon skeleton is drawn in a zig-zag format, in which carbon atoms represent each corner and endpoint. Hydrogen atoms are only drawn if they are connected to heteroatoms:

(a)   **(b)**

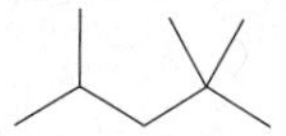

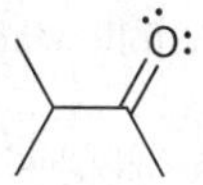

(c) 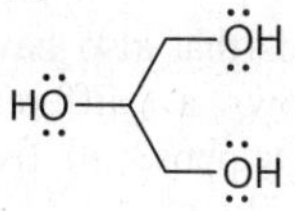   **(d)**

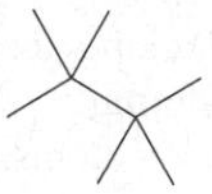

(e) 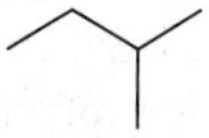   **(f)**

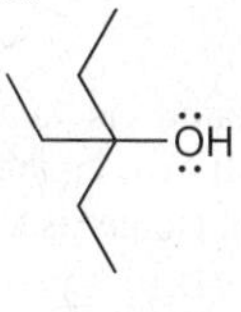

(g) 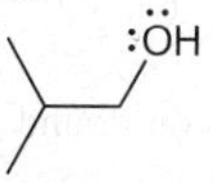   **(h)**

(i) 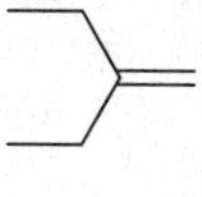   **(j)**

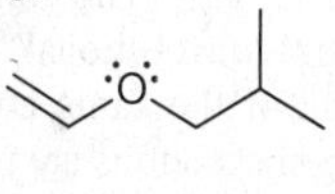

(k) 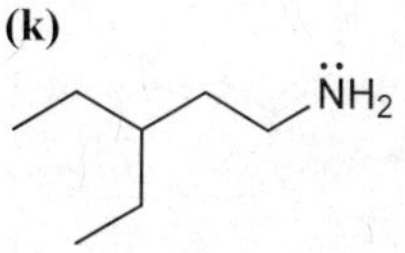   **(l)**

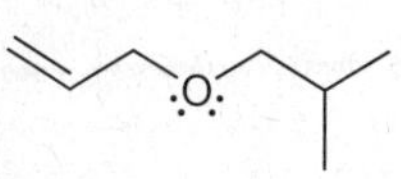

(m)    **(n)**

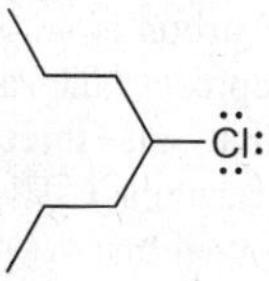

(o) 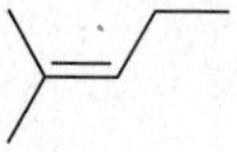   **(p)**

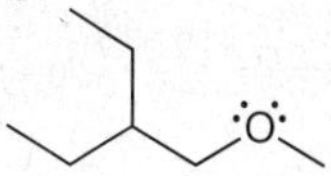

(q)

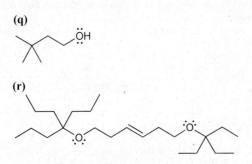

(r)

2.9. Recall that constitutional isomers are compounds that share the same molecular formula, but differ in constitution (the connectivity of atoms). The problem statement shows a compound with molecular formula C_5H_{12} and the following structure:

$$CH_3CH_2CH(CH_3)_2 \equiv$$

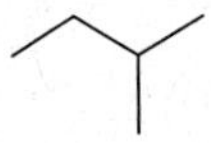

So we are looking for other compounds that also have molecular formula C_5H_{12} but show a different connectivity of atoms. For example, suppose all five carbon atoms are connected in a linear fashion:

This compound is a constitutional isomer of the compound shown in the problem statement. Another constitutional isomer is shown here:

Once again, the molecular formula of this compound is C_5H_{12}, but the connectivity of atoms is different.
There are no other constitutional isomers with molecular formula C_5H_{12}. The following two structures do NOT represent constitutional isomers, but are in fact two drawings of the same compound, as can be seen when the carbon skeletons are numbered, as shown:.

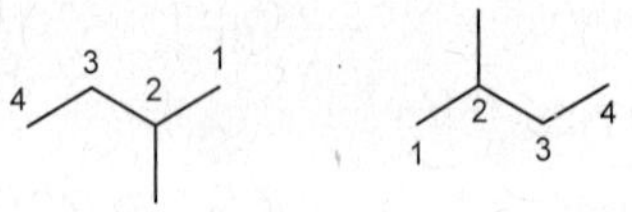

Notice that in both drawings, the longest linear chain is four carbon atoms, and there is a CH_3 group attached to the second carbon atom of the chain. As such, these two drawings represent the same compound. In contrast, we can see that all three constitutional isomers with molecular formula C_5H_{12} exhibit different connectivity of the carbon atoms:

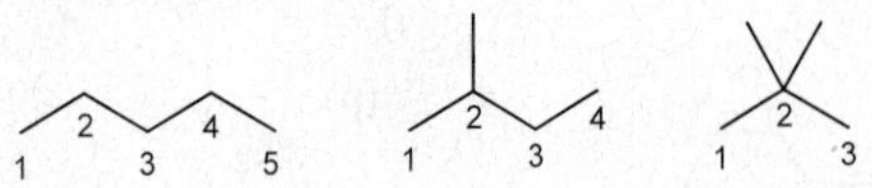

2.10.
(a) In Section 1.5, we discussed inductive effects and we learned how to identify polar covalent bonds. In this case, there are two carbon atoms that participate in polar covalent bonds (the C–Br bond and the C–O bond). Each of these carbon atoms will be poor in electron density ($\delta+$) because oxygen and bromine are each more electronegative than carbon:

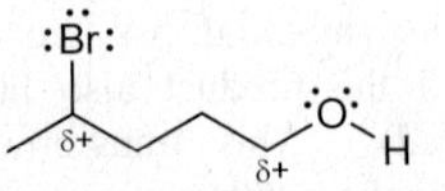

(b) There are two carbon atoms that are adjacent to oxygen atoms. These carbon atoms will be poor in electron density ($\delta+$), because oxygen is more electronegative than carbon:

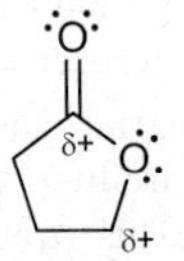

(c) There are two carbon atoms that are adjacent to electronegative atoms. These carbon atoms will be poor in electron density ($\delta+$), because oxygen and chlorine are each more electronegative than carbon:

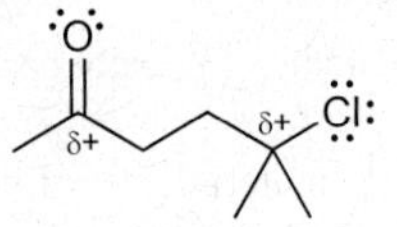

2.11. The functional groups in the following compounds are highlighted and identified, using the terminology found in Table 2.1.

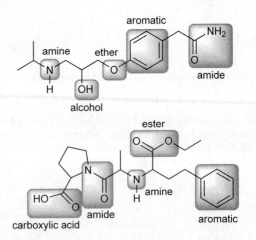

2.12.
(a) A formal charge is associated with any atom that does not exhibit the appropriate number of valence electrons. Nitrogen is in group **5A** of the periodic table, and it should therefore have five valence electrons. Next, we count how many valence electrons are

exhibited by the nitrogen atom in this case. The nitrogen atom has two bonds and two lone pairs. Each of the bonds requires one valence electron, and each of the lone pairs represents two valence electrons for a total of 1+1+2+2 = 6 valence electrons. So in this case, the nitrogen atom has an extra valence electron, and therefore, it must bear a negative charge:

(b) Nitrogen is in group **5A** of the periodic table, and it should therefore have five valence electrons. In this case, the nitrogen atom has three bonds and one lone pair. Each of the bonds requires one valence electron, and the lone pair represents two valence electrons, for a total of 1+1+1+2 = 5 valence electrons. So in this case, the nitrogen atom has the appropriate number of valence electrons, and therefore, it does not bear a formal charge:

(c) Nitrogen is in group **5A** of the periodic table, and it should therefore have five valence electrons. In this case, the nitrogen atom has four bonds and no lone pairs. Each of the bonds requires one valence electron, for a total of 1+1+1+1 = 4 valence electrons. So in this case, the nitrogen atom is missing a valence electron, and therefore, it must bear a positive charge:

(d) Nitrogen is in group **5A** of the periodic table, and it should therefore have five valence electrons. In this case, the nitrogen atom has three bonds and one lone pair. Each of the bonds requires one valence electron, and the lone pair represents two valence electrons, for a total of 1+1+1+2 = 5 valence electrons. So in this case, the nitrogen atom has the appropriate number of valence electrons, and therefore, it does not bear a formal charge:

2.13.
(a) A formal charge is associated with any atom that does not exhibit the appropriate number of valence electrons. Oxygen is in group **6A** of the periodic table, and it should therefore have six valence electrons. Next, we count how many valence electrons are exhibited by the oxygen atom in this case. The oxygen atom has one bond and three lone pairs. The bond requires one valence electron, and each of the lone pairs represents

two valence electrons, for a total of 1+2+2+2 = 7 valence electrons. So in this case, the oxygen atom has an extra valence electron, and therefore, it must bear a negative charge.

(b) Oxygen is in group **6A** of the periodic table, and it should therefore have six valence electrons. The oxygen atom in this case has three bonds and one lone pair. Each of the bonds requires one valence electron, and the lone pair represents two valence electrons, for a total of 1+1+1+2 = 5 valence electrons. So in this case, the oxygen atom is missing a valence electron, and therefore, it must bear a positive charge.

(c) Oxygen is in group **6A** of the periodic table, and it should therefore have six valence electrons. The oxygen atom in this case has two bonds and two lone pairs. Each of the bonds requires one valence electron, and each of the lone pairs represents two valence electrons, for a total of 1+1+2+2 = 6 valence electrons. So in this case, the oxygen atom has the appropriate number of valence electrons, and therefore, it does not bear a formal charge.

(d) Oxygen is in group **6A** of the periodic table, and it should therefore have six valence electrons. The oxygen atom in this case has three bonds and one lone pair. Each of the bonds requires one valence electron, and the lone pair represents two valence electrons, for a total of 1+1+1+2 = 5 valence electrons. So in this case, the oxygen atom is missing a valence electron, and therefore, it must bear a positive charge.

2.14.
(a) In this case, the oxygen atom has two bonds and no formal charge, so it must have two lone pairs (see Table 2.2).

(b) In this case, each of the oxygen atoms has two bonds and no formal charge, so each oxygen atom must have two lone pairs (see Table 2.2).

(c) In this case, each of the oxygen atoms has two bonds and no formal charge, so each oxygen atom must have two lone pairs (see Table 2.2).

(d) One of the oxygen atoms has two bonds and no formal charge, so that oxygen atom must have two lone pairs (see Table 2.2). The other oxygen atom has one bond and a negative charge, so that oxygen atom must have three lone pairs.

(e) In this case, the oxygen atom has one bond and a negative charge, so it must have three lone pairs (see Table 2.2).

(f) In this case, the oxygen atom has two bonds and no formal charge, so it must have two lone pairs (see Table 2.2).

(g) In this case, the oxygen atom has three bonds and a positive charge, so it must have one lone pair (see Table 2.2).

(h) In this case, the oxygen atom has three bonds and a positive charge, so it must have one lone pair (see Table 2.2).

(i) From left to right, the first oxygen atom has two bonds and no formal charge, so it must have two lone pairs (see Table 2.2). The second oxygen atom has three

bonds and a positive charge, so it must have one lone pair. Finally, the third oxygen atom has one bond and a negative charge, so it has three lone pairs.

(j) One of the oxygen atoms has two bonds and no formal charge, so that oxygen atom must have two lone pairs (see Table 2.2). The other oxygen atom has three bonds and a positive charge, so that oxygen atom must have one lone pair.

2.15. Carbon is in group **4A** of the periodic table, and it should therefore have four valence electrons. We are told that, in this case, the central carbon atom does not bear a formal charge.

Therefore, it must exhibit the appropriate number of valence electrons (four). This carbon atom already has two bonds (each of which requires one valence electron) and a lone pair (which represents two electrons), for a total of 1+1+2=4 valence electrons. This is the appropriate number of valence electrons, which means that this carbon atom does not have any bonds to hydrogen.
Notice that the carbon atom lacks an octet, so it should not be surprising that this structure is highly reactive and very short-lived.

2.16.
(a) In this case, the nitrogen atom has three bonds and no formal charge, so it must have one lone pair (see Table 2.3).

(b) In this case, the nitrogen atom has three bonds and no formal charge, so it must have one lone pair (see Table 2.3).

(c) In this case, the nitrogen atom has three bonds and no formal charge, so it must have one lone pair (see Table 2.3).

(d) In this case, the nitrogen atom has four bonds and a positive charge, so it must have no lone pairs (see Table 2.3).

(e) In this case, the nitrogen atom has two bonds and a negative charge, so it must have two lone pairs (see Table 2.3).

(f) In this case, the nitrogen atom has three bonds and no formal charge, so it must have one lone pair (see Table 2.3).

(g) In this case, the nitrogen atom has four bonds and a positive charge, so it must have no lone pairs (see Table 2.3).

(h) One of the nitrogen atoms has four bonds and a positive charge, so it must have no lone pairs (see Table 2.3). The other nitrogen atom has three bonds and no formal charge, so it must have one lone pair.

2.17.
(a) In this case, the oxygen atom has three bonds and a positive charge, so it must have one lone pair (see Table 2.2). The nitrogen atom has three bonds and no formal charge, so it must have one lone pair (see Table 2.3).

(b) In this case, the oxygen atom has two bonds and no formal charge, so it must have two lone pairs (see Table 2.2). The nitrogen atom has two bonds and a negative charge, so it must have two lone pairs (see Table 2.3).

$$:\ddot{O}=C=\overset{\ominus}{\ddot{N}}:$$

(c) Each of the oxygen atoms has two bonds and no formal charge, so each oxygen atom must have two lone

pairs (see Table 2.2). The nitrogen atom has four bonds and a positive charge, so it must have no lone pairs (see Table 2.3).

(d) One of the oxygen atoms (on the right) has two bonds and no formal charge, so that oxygen atom must have two lone pairs (see Table 2.2). The other oxygen atom has one bond and a negative charge, so that oxygen atom must have three lone pairs. The nitrogen atom has four bonds and a positive charge, so it must have no lone pairs (scc Table 2.3).

(e) The oxygen atom has two bonds and no formal charge, so it must have two lone pairs (see Table 2.2). The nitrogen atom has three bonds and no formal charge, so it must have one lone pair (see Table 2.3).

(f) The oxygen atom has one bond and a negative charge, so it must have three lone pairs (see Table 2.2). The nitrogen atom has three bonds and no formal charge, so it must have one lone pair (see Table 2.3).

$$\overset{\ominus}{:}\ddot{O}-C\equiv N:$$

2.18.
(a) The negative charge resides on a carbon atom that has three bonds, so that carbon atom must have one lone pair. None of the other carbon atoms exhibits a lone pair, so there is only one lone pair in this structure.

(b) The positive charge resides on a carbon atom that has three bonds, so that carbon atom must have no lone pairs. None of the other carbon atoms exhibits a lone pair, so there are no lone pairs in this structure.

(c) The negative charge resides on a carbon atom that has three bonds, so that carbon atom must have one lone pair. None of the other carbon atoms exhibits a lone pair, so there is only one lone pair in this structure.

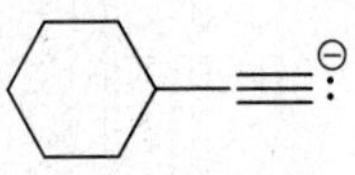

(d)   The negative charge is drawn on a carbon atom that has three bonds (one bond to hydrogen is not shown), so that carbon atom must have one lone pair. None of the other carbon atoms exhibits a lone pair, but each of the oxygen atoms has two lone pairs. So there are a total of five lone pairs in this structure.

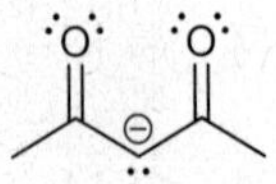

2.19 One of the oxygen atoms has two bonds and no formal charge, so that oxygen atom must have two lone pairs (see Table 2.2). The other oxygen atom has one bond and a negative charge, so that oxygen atom must have three lone pairs. The nitrogen atom has four bonds and a positive charge, so it must have no lone pairs (see Table 2.3). Therefore, there are a total of five lone pairs in this structure.

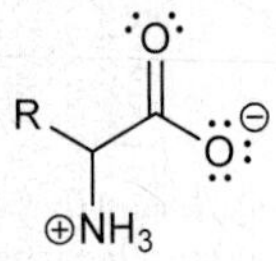

2.20
(a)   The highlighted regions (below) are similar, so the pharmacophore is likely to be all or part of these highlighted regions.

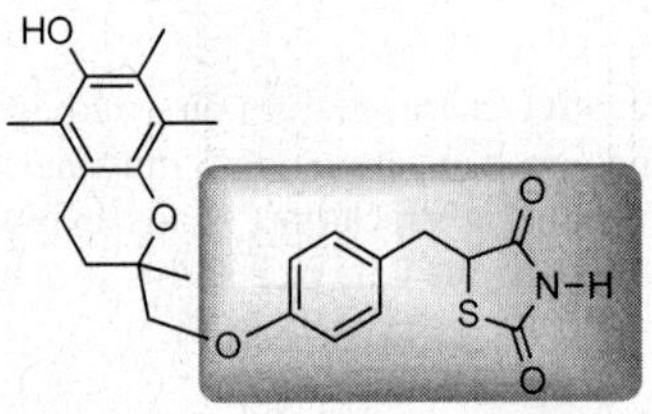

Troglitazone

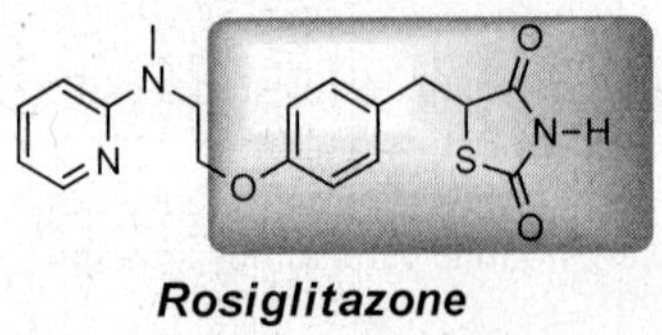

Rosiglitazone

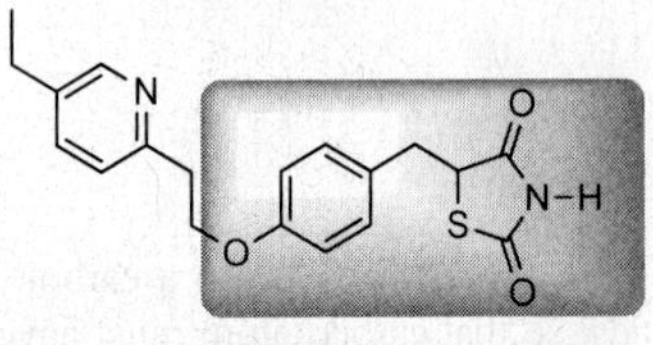

Pioglitazone

(b)   Yes, because the structure contains the likely pharmacophore highlighted above.

2.21
(a) This curved arrow violates the second rule by giving a fifth bond to a nitrogen atom.
(b) This curved arrow does not violate either rule.
(c) This curved arrow violates the second rule by giving five bonds to a carbon atom.
(d) This curved arrow violates the second rule by giving three bonds and two lone pairs to an oxygen atom.
(e) This curved arrow violates the second rule by giving five bonds to a carbon atom.
(f) This curved arrow violates the second rule by giving five bonds to a carbon atom.
(g) This curved arrow violates the first rule by breaking a single bond, and violates the second rule by giving five bonds to a carbon atom.
(h) This curved arrow violates the first rule by breaking a single bond, and violates the second rule by giving five bonds to a carbon atom.
(i) This curved arrow does not violate either rule.
(j) This curved arrow does not violate either rule.
(k) This curved arrow violates the second rule by giving five bonds to a carbon atom.
(l) This curved arrow violates the second rule by giving five bonds to a carbon atom.

2.22. The tail of the curved arrow must be placed on the double bond in order to avoid violating the first rule (avoid breaking a single bond).

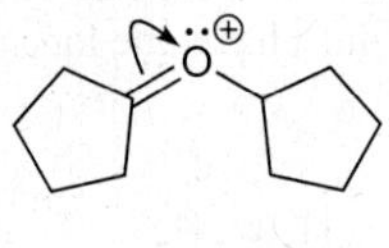

2.23.
(a)   The curved arrow indicates that we should draw a resonance structure in which the π bond has been pushed over. We then complete the resonance structure by assigning any formal charges.   Notice that both resonance structures show a positive charge, but in different locations:

(b)   The curved arrows indicate that we should draw a resonance structure in which the lone pair has been pushed to become a π bond, and the π bond has been pushed to become a lone pair. We then complete the resonance structure by assigning any formal charges. Notice that both resonance structures show a negative charge, but in different locations:

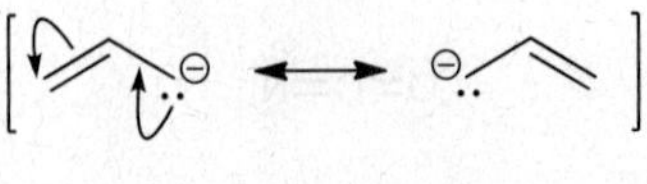

(c)   The curved arrows indicate that we should draw a resonance structure in which a lone pair has been pushed to become a π bond, and the π bond has been pushed to become a lone pair. We then complete the resonance structure by assigning any formal charges. Notice that both resonance structures have zero net charge:

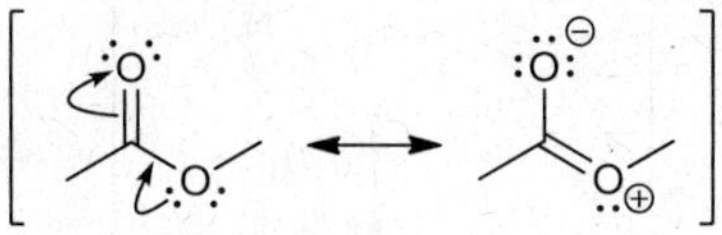

(d) The curved arrows indicate that we should draw a resonance structure in which a lone pair has been pushed to become a π bond, and the π bond has been pushed to become a lone pair. We then complete the resonance structure by assigning any formal charges. Notice that both resonance structures show a negative charge, but in different locations:

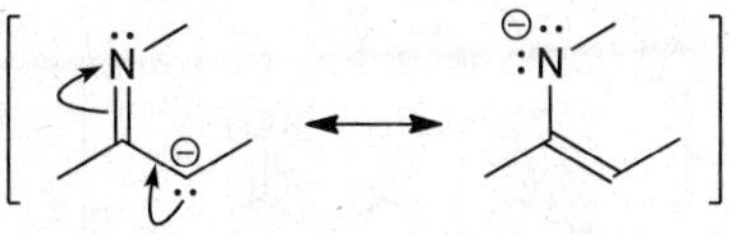

(e) The curved arrows indicate that we should draw the following resonance structure. Notice that both resonance structures have zero net charge:

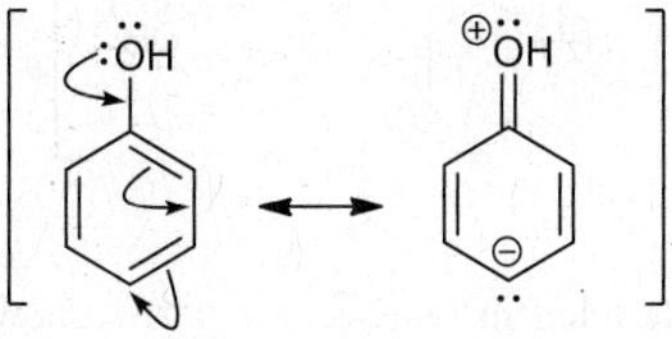

(f) The curved arrows indicate that we should draw the following resonance structure. Notice that both resonance structures have zero net charge:

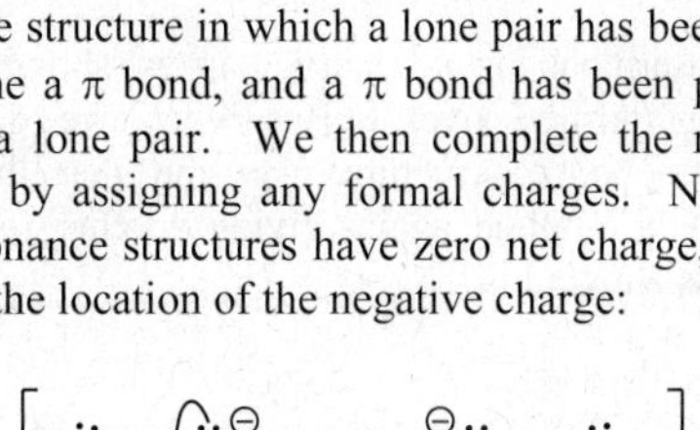

(g) The curved arrows indicate that we should draw a resonance structure in which a lone pair has been pushed to become a π bond, and a π bond has been pushed to become a lone pair. We then complete the resonance structure by assigning any formal charges. Notice that both resonance structures have zero net charge, but they differ in the location of the negative charge:

(h) The curved arrows indicate that we should draw the following resonance structure. Notice that both resonance structures have zero net charge:

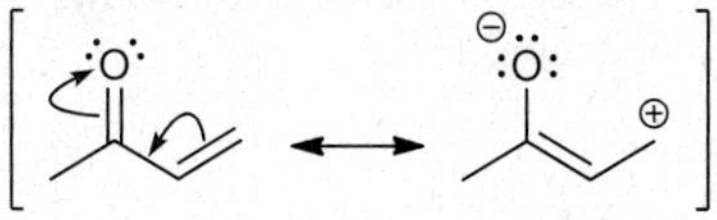

2.24.
(a) One curved arrow is required, showing the π bond being pushed to become a lone pair:

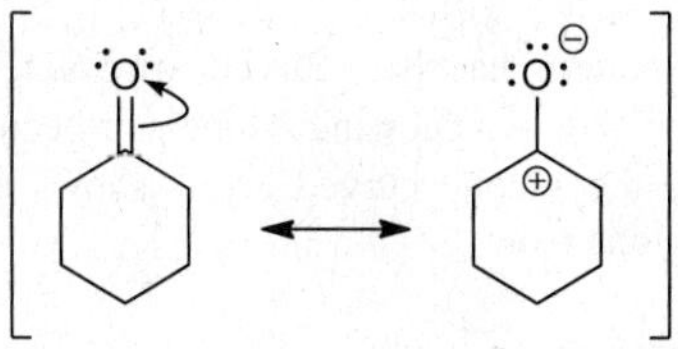

(b) Two curved arrows are required. One curved arrow shows the carbon-carbon π bond being pushed up, and the other curved arrow shows the carbon-oxygen π bond becoming a lone pair:

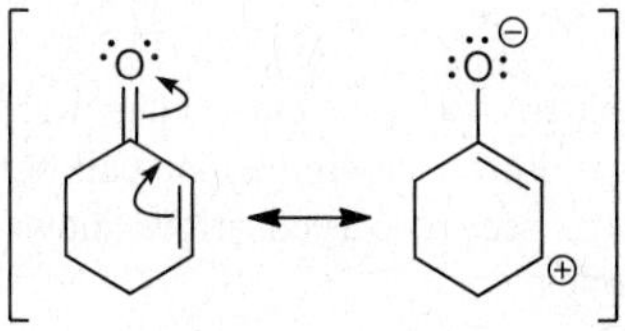

(c) Two curved arrows are required. One curved arrow shows a lone pair from the nitrogen atom becoming a π bond, and the other curved arrow shows the carbon-oxygen π bond becoming a lone pair:

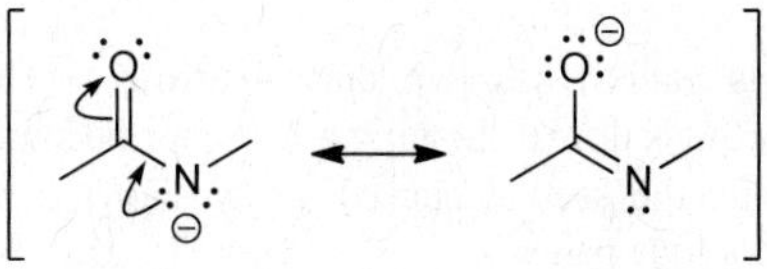

(d) One curved arrow is required, showing the π bond being pushed over:

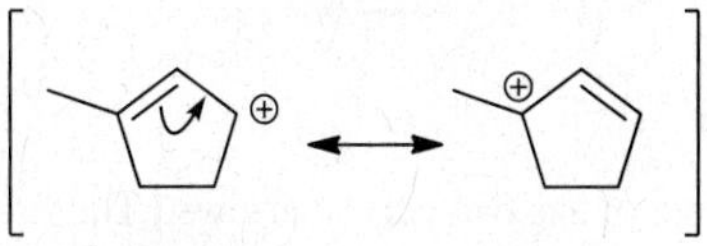

2.25.
(a) This pattern has two curved arrows. The first curved arrow is drawn showing a lone pair becoming a π bond, while the second curved arrow shows a π bond becoming a lone pair:

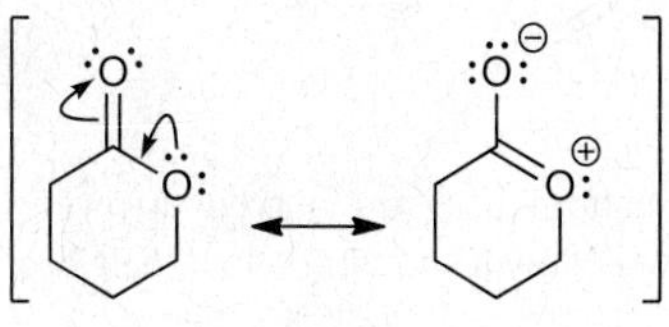

(b) This pattern has two curved arrows. The first curved arrow is drawn showing a lone pair becoming a π bond, while the second curved arrow shows a π bond becoming a lone pair:

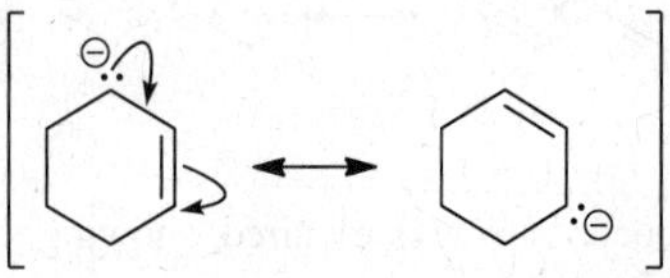

(c) This pattern has two curved arrows. The first curved arrow is drawn showing a lone pair becoming a π bond, while the second curved arrow shows a π bond becoming a lone pair:

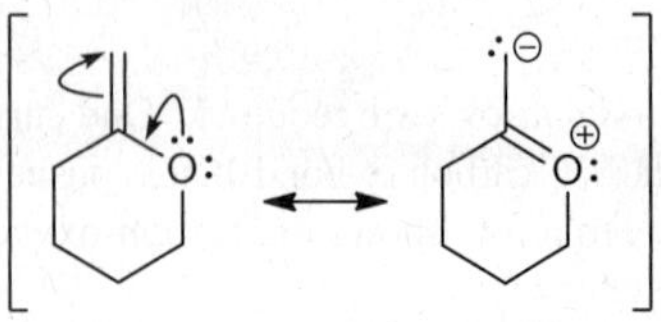

(d) This pattern has two curved arrows. The first curved arrow is drawn showing a lone pair becoming a π bond, while the second curved arrow shows a π bond becoming a lone pair:

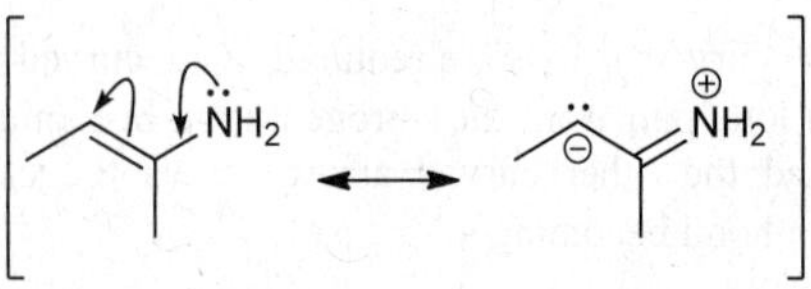

(e) This pattern has two curved arrows. The first curved arrow is drawn showing a lone pair becoming a π bond, while the second curved arrow shows a π bond becoming a lone pair:

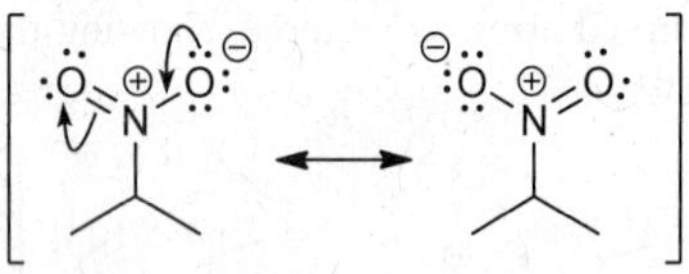

(f) This pattern has two curved arrows. The first curved arrow is drawn showing a lone pair becoming a π bond, while the second curved arrow shows a π bond becoming a lone pair:

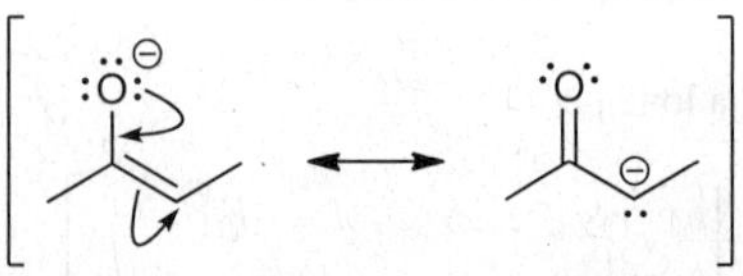

(g) This pattern has two curved arrows. The first curved arrow is drawn showing a lone pair becoming a π

bond, while the second curved arrow shows a π bond becoming a lone pair:

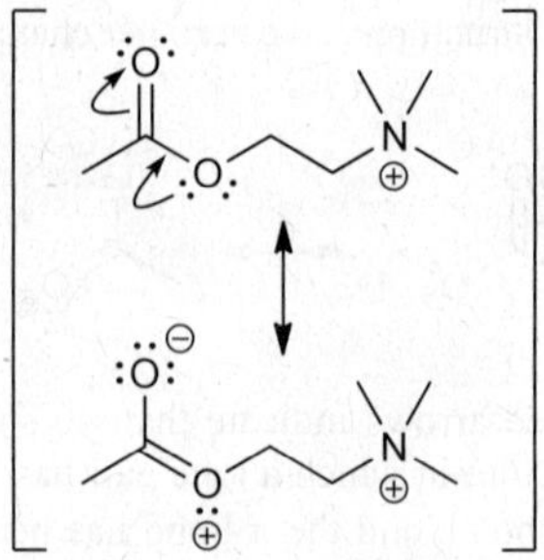

(h) This pattern has two curved arrows. The first curved arrow is drawn showing a lone pair becoming a π bond, while the second curved arrow shows a π bond becoming a lone pair:

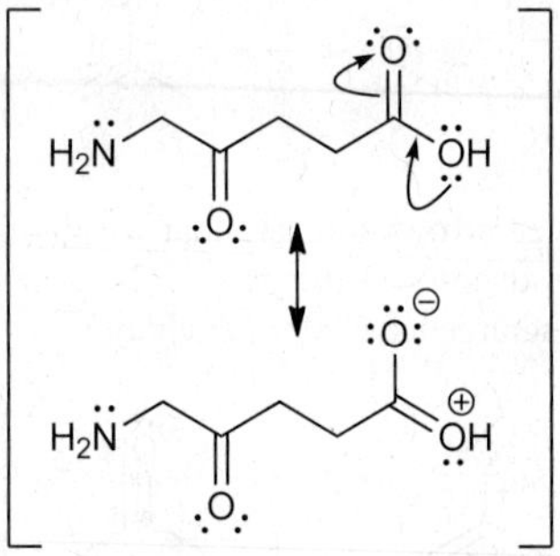

2.26.
(a) This pattern has just one curved arrow, showing the π bond being pushed over:

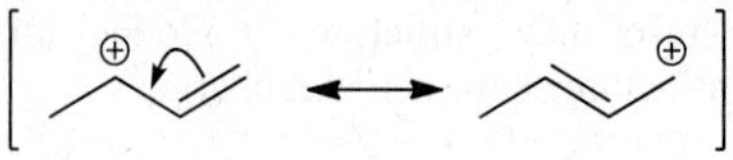

(b) This pattern has just one curved arrow, showing the π bond being pushed over:

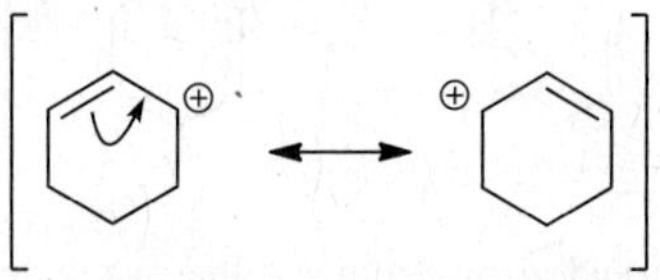

(c) This pattern has just one curved arrow, showing the π bond being pushed over. But when we draw the resulting resonance structure, we find that the same pattern can be applied again, giving another resonance structure, as shown:

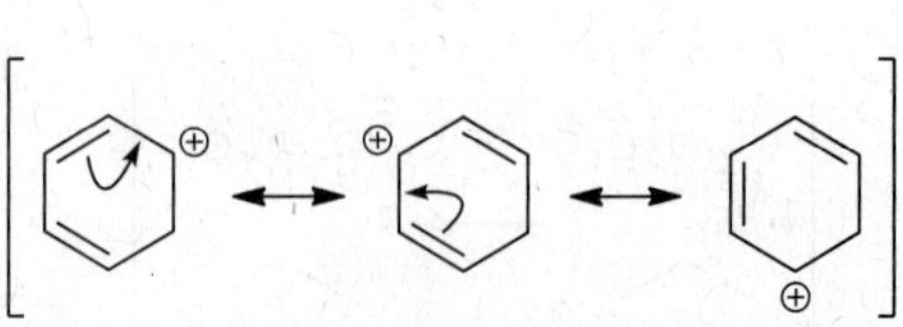

(d) This pattern has just one curved arrow, showing the π bond being pushed over. But when we draw the resulting resonance structure, we find that the same pattern can be applied again, giving another resonance structure. This process continues several more times, and we can see that the positive charge is spread (via resonance) over all seven carbon atoms of the ring:

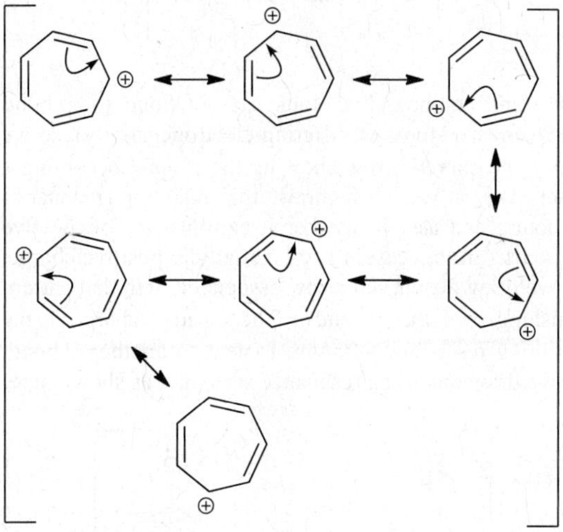

2.27.
(a) This pattern has just one curved arrow, showing the lone pair becoming a π bond:

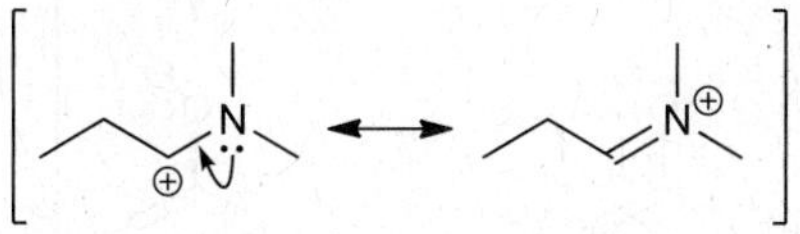

(b) This pattern has just one curved arrow, showing the lone pair becoming a π bond:

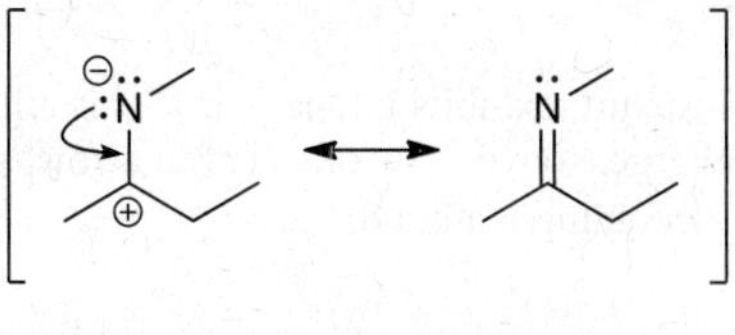

(c) This pattern has just one curved arrow, showing the lone pair becoming a π bond:

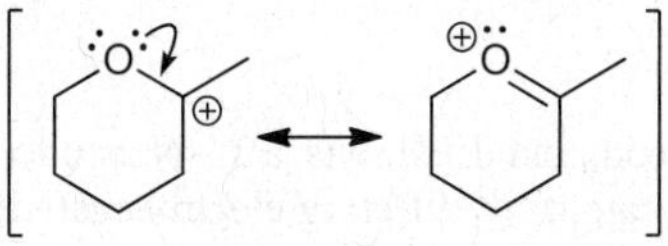

2.28.
(a) This pattern has just one curved arrow, showing the π bond becoming a lone pair:

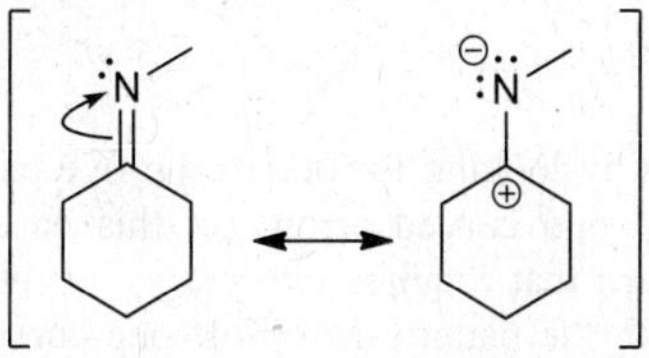

(b) This pattern has just one curved arrow, showing the π bond becoming a lone pair:

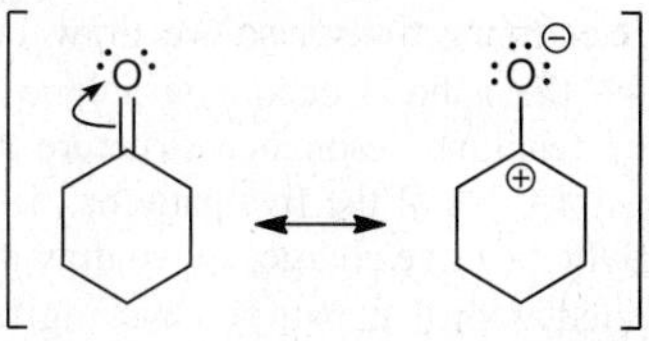

(c) This pattern has just one curved arrow, showing the π bond becoming a lone pair:

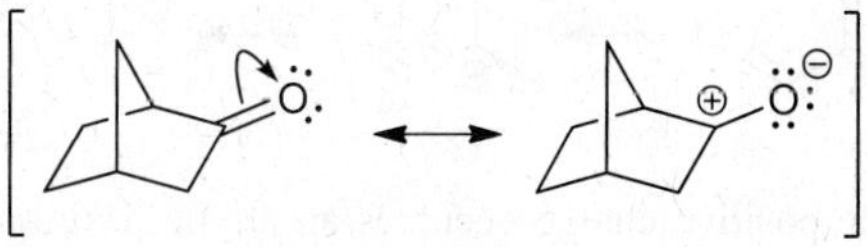

2.29. This pattern has just one curved arrow, showing the π bond becoming a lone pair:

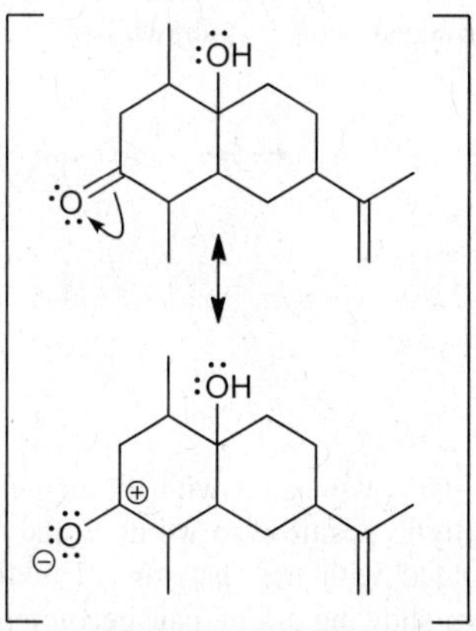

2.30. This pattern has just one curved arrow, showing the π bond becoming a lone pair:

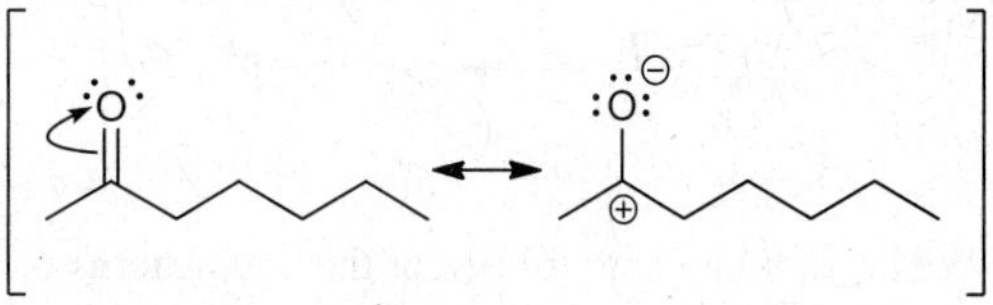

2.31. This pattern has three curved arrows, showing the π bonds moving in a circle.

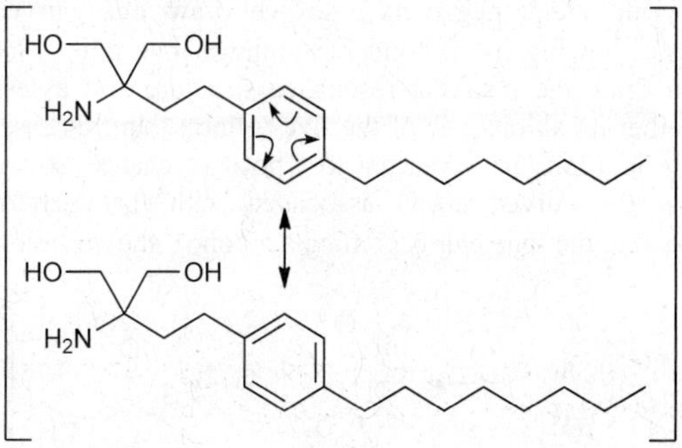

2.32.

(a)   We begin by looking for the five patterns. In this case, there is a C=O bond (a π bond between two atoms of differing electronegativity), so we draw one curved arrow showing the π bond becoming a lone pair. We then draw the resulting resonance structure and assess whether it exhibits one of the five patterns. In this case, there is an allylic positive charge, so we draw the curved arrow associated with that pattern (pushing over the π bond), shown here:

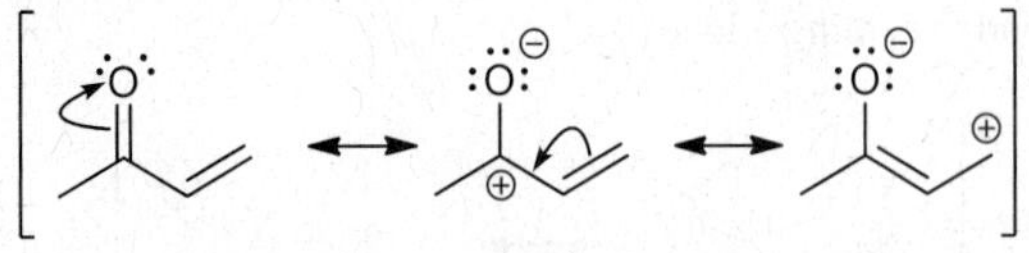

(b) The positive charge occupies an allylic position, so we draw the one curved arrow associated with that pattern (pushing over the π bond). The positive charge in the resulting resonance structure is again next to another π bond, so we draw one curved arrow and another resonance structure, as shown here:

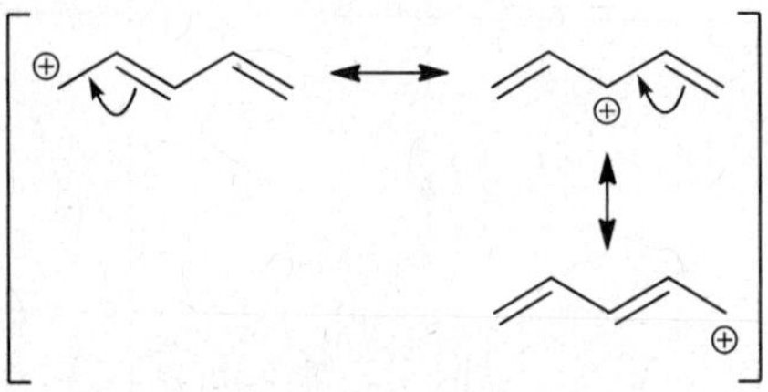

(c) The lone pair (associated with the negative charge) occupies an allylic position, so we draw the two curved arrows associated with that pattern. The first curved arrow is drawn showing a lone pair becoming a π bond, while the second curved arrow shows a π bond becoming a lone pair:

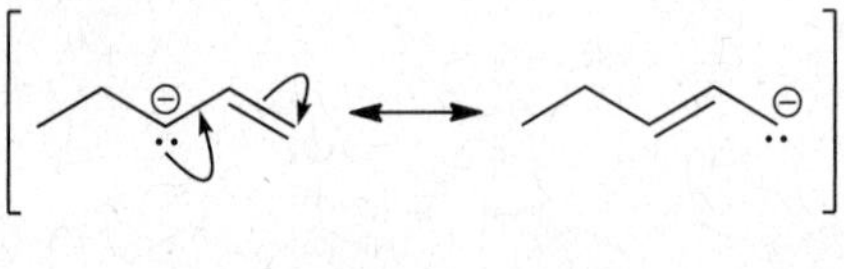

(d) We begin by looking for one of the five patterns that employs just one curved arrow (in this case, there is another pattern that requires two curved arrows, but we will start with the pattern using just one curved arrow). There is a C=O bond (a π bond between two atoms of differing electronegativity), so we draw one curved arrow showing the π bond becoming a lone pair. We then draw the resulting resonance structure and assess whether it exhibits one of the five patterns. In this case, there is a lone pair adjacent to a positive charge, so we draw the curved arrow associated with that pattern (showing the lone pair becoming a π bond), shown here:

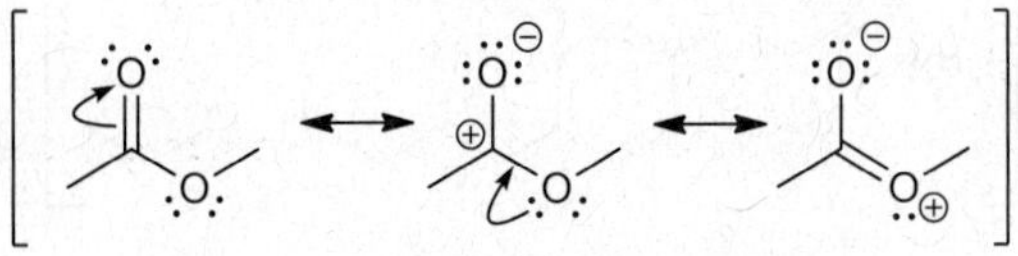

(e) This structure exhibits a lone pair that is adjacent to a positive charge, so we draw one curved arrow, showing a lone pair becoming a π bond:

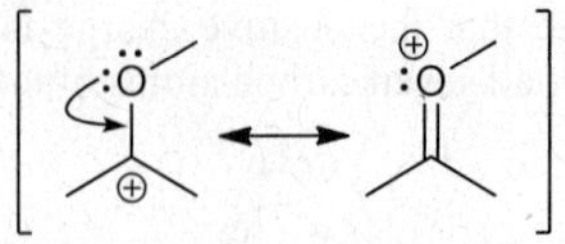

(f)   This compound exhibits a C=O bond (a π bond between two atoms of differing electronegativity), so we draw one curved arrow showing the π bond becoming a lone pair. We then draw the resulting resonance structure and assess whether it exhibits one of the five patterns. In this case, there is an allylic positive charge, so we draw the curved arrow associated with that pattern (pushing over the π bond). The positive charge in the resulting resonance structure is next to another π bond, so we draw one more resonance structure, as shown here:

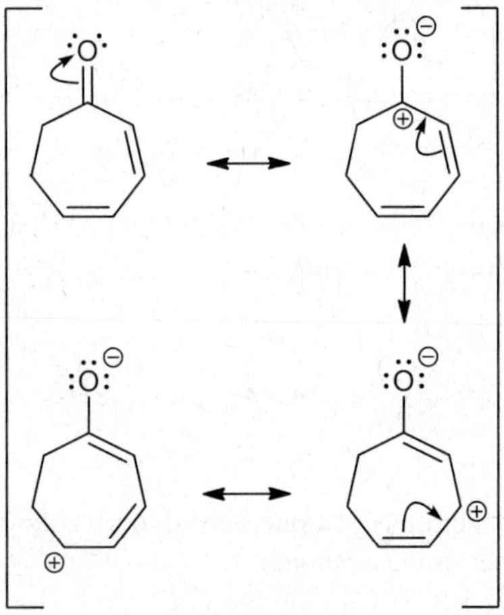

(g)   This structure exhibits a lone pair that is adjacent to a positive charge, so we draw one curved arrow, showing a lone pair becoming a π bond:

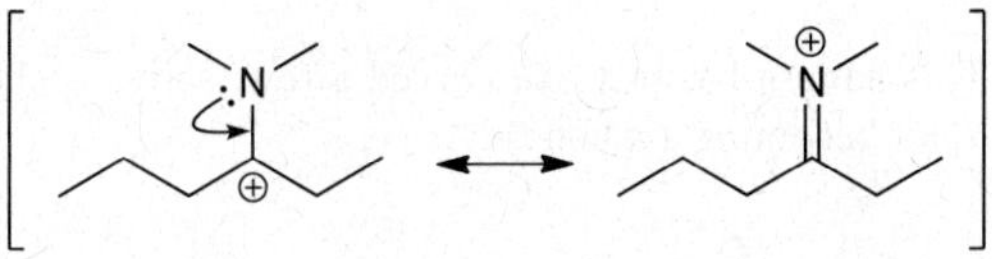

(h)   This compound exhibits a C=N bond (a π bond between two atoms of differing electronegativity), so we draw one curved arrow showing the π bond becoming a lone pair:

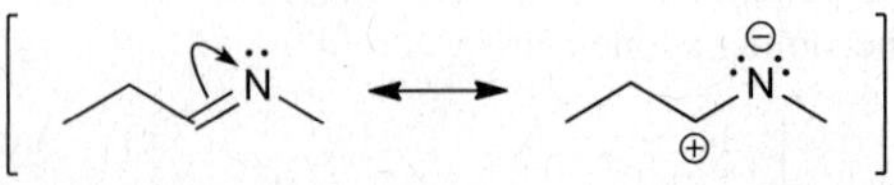

(i) We begin by looking for one of the five patterns that employs just one curved arrow (in this case, there is another pattern that requires two curved arrows, but we will start with the pattern using just one curved arrow).

There is a C=O bond (a π bond between two atoms of differing electronegativity), so we draw one curved arrow showing the π bond becoming a lone pair. We then draw the resulting resonance structure and assess whether it exhibits one of the five patterns. In this case, there is a lone pair adjacent to a positive charge, so we draw the curved arrow associated with that pattern (showing the lone pair becoming a π bond), shown here:

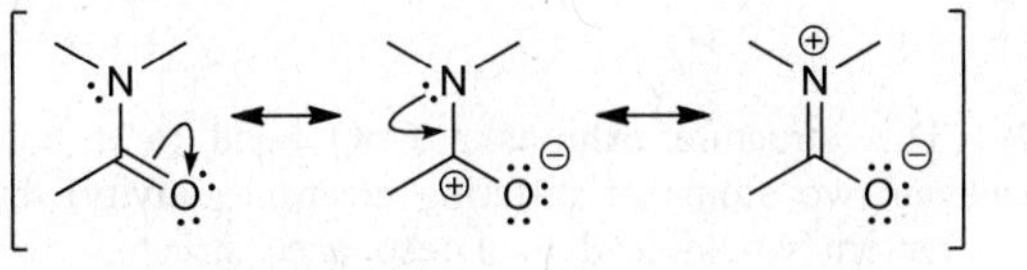

(j) We begin by looking for one of the five patterns that employs just one curved arrow (in this case, there is another pattern that requires two curved arrows, but we will start with the pattern using just one curved arrow). There is a C=O bond (a π bond between two atoms of differing electronegativity), so we draw one curved arrow showing the π bond becoming a lone pair. We then draw the resulting resonance structure and assess whether it exhibits one of the five patterns. In this case, there is a lone pair adjacent to a positive charge, so we draw the curved arrow associated with that pattern (showing the lone pair becoming a π bond), shown here:

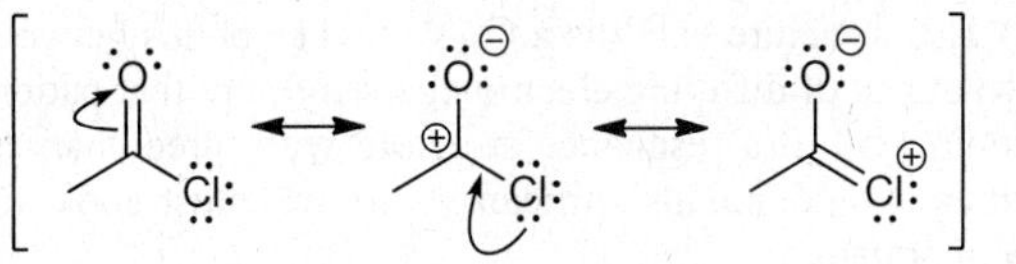

2.33.
(a) This compound exhibits a C=N bond (a π bond between two atoms of differing electronegativity), so we draw one curved arrow showing the π bond becoming a lone pair. We then draw the resulting resonance structure and assess whether it exhibits one of the five patterns. In this case, there is an allylic positive charge, so we draw the curved arrow associated with that pattern (pushing over the π bond), shown here:

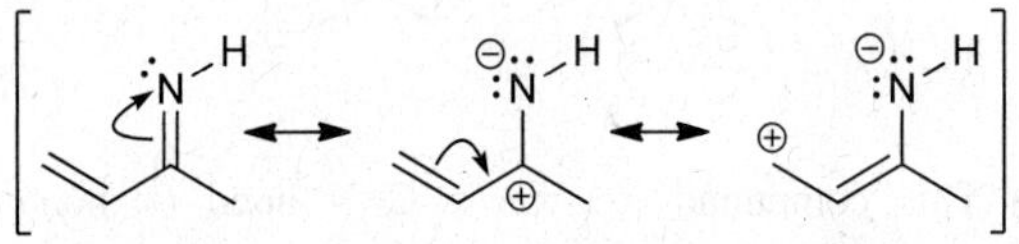

These are all of the significant resonance structures. The following resonance structure is not significant, because it has two carbon atoms with opposite charges:

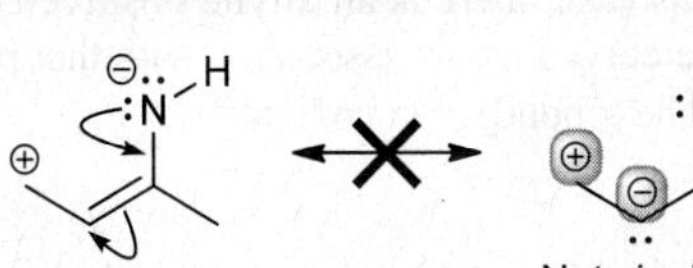

Not significant

(b) The lone pair on the nitrogen atom is next to a π bond, so we draw the two curved arrows associated with that pattern. The first curved arrow is drawn showing

the lone pair becoming a π bond, while the second curved arrow shows a π bond becoming a lone pair. We then draw the resulting resonance structure and assess whether it exhibits one of the five patterns. In this case, the lone pair is next to another π bond, so again, we draw the two curved arrows associated with that pattern. The first curved arrow is drawn showing the lone pair becoming a π bond, while the second curved arrow shows a π bond becoming a lone pair.

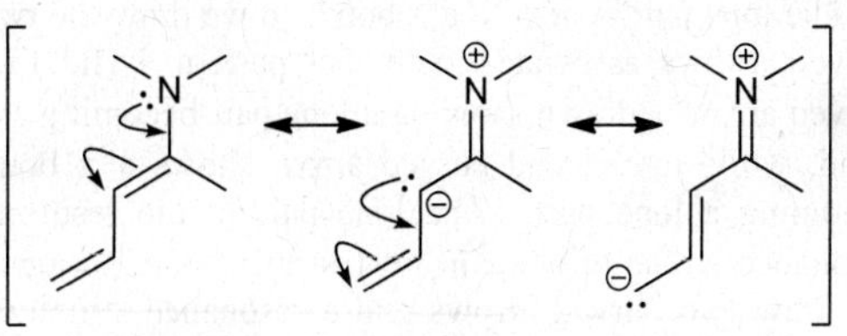

These are all of the significant resonance structures. The following resonance structure is not significant, because it has two carbon atoms with opposite charges:

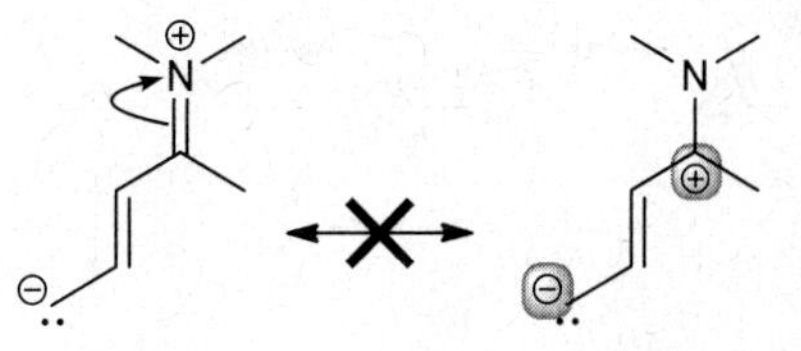

Not significant

(c) This structure exhibits a lone pair next to a π bond, so we draw the two curved arrows associated with that pattern. The first curved arrow is drawn showing a lone pair becoming a π bond, while the second curved arrow shows a π bond becoming a lone pair:

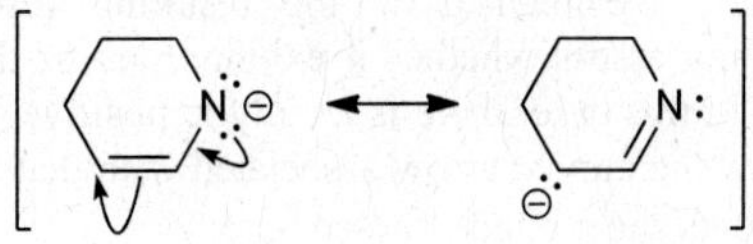

(d) We begin by looking for one of the five patterns that employs just one curved arrow (in this case, there is another pattern that requires two curved arrows, but we will start with the pattern using just one curved arrow). There is a C≡N bond (π bonds between two atoms of differing electronegativity), so we draw one curved arrow showing a π bond becoming a lone pair. We then draw the resulting resonance structure and assess whether it exhibits one of the five patterns. In this case, there is a lone pair adjacent to a positive charge, so we draw the curved arrow associated with that pattern (showing the lone pair becoming a π bond), shown here:

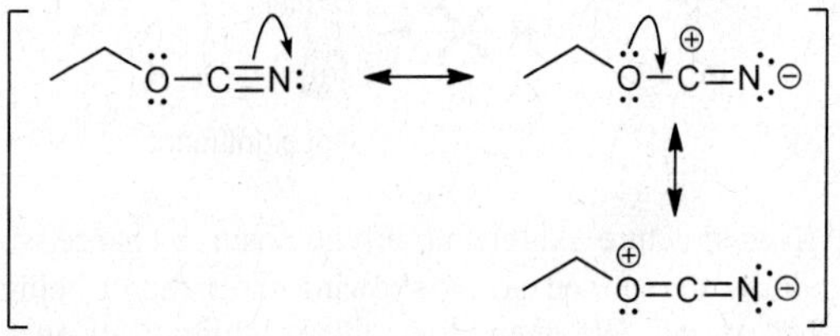

(e) This compound exhibits an S=O bond (a π bond between two atoms of differing electronegativity), so we draw one curved arrow showing the π bond becoming a lone pair.

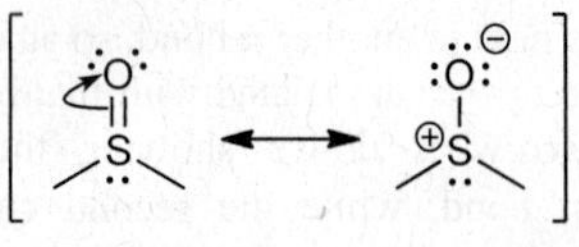

(f) The lone pair is next to a π bond, so we draw the two curved arrows associated with that pattern. The first curved arrow is drawn showing a lone pair becoming a π bond, while the second curved arrow shows a π bond becoming a lone pair. The lone pair in the resulting resonance structure is again next to a π bond, so again we draw two curved arrows and a resonance structure. This continues, until the negative charge has been "pushed" onto all five carbon atoms of the ring:

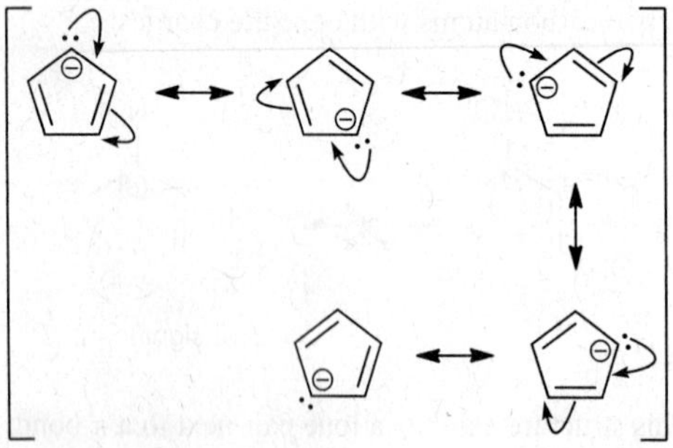

(g) This compound exhibits a C=O bond (a π bond between two atoms of differing electronegativity), so we draw one curved arrow showing the π bond becoming a lone pair. We then draw the resulting resonance structure and assess whether it exhibits one of the five patterns. In this case, there is an allylic positive charge, so we draw the curved arrow associated with that pattern (pushing over the π bond), shown here:

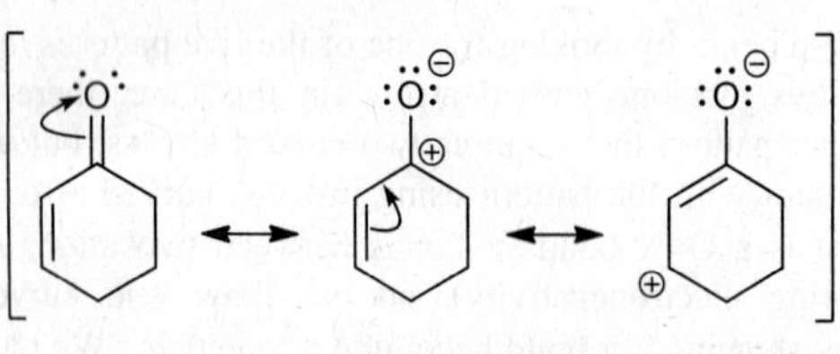

The following resonance structure is not significant, because it has two carbon atoms with opposite charges:

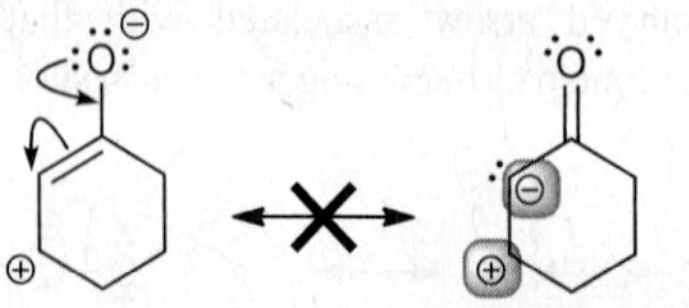

Not significant

(h) This structure exhibits an allylic positive charge, so we draw one curved arrow showing the π bond being pushed over. We then draw the resulting resonance

structure and assess whether it exhibits one of the five patterns. In this case, the positive charge is again next to a π bond, so again, we draw the curved arrow associated with that pattern (pushing over the π bond), shown here:

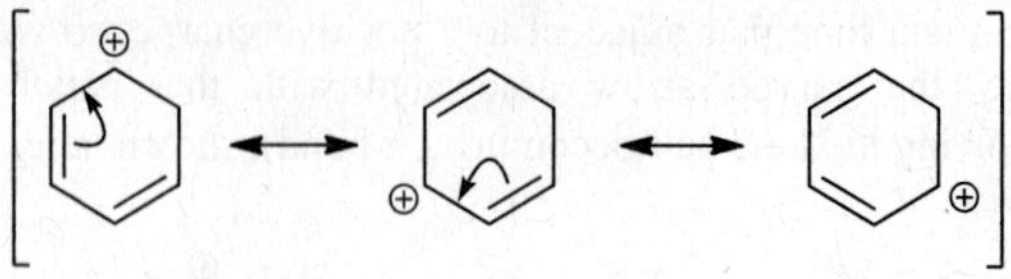

(i) This structure exhibits a C=O bond (a π bond between two atoms of differing electronegativity), but that pattern would lead to a resonance structure with three charges, which would be insignificant. So we must look for another pattern.

This structure does also exhibit a lone pair next to a π bond, so we draw the two curved arrows associated with that pattern. The first curved arrow is drawn showing a lone pair becoming a π bond, while the second curved arrow shows a π bond becoming a lone pair:

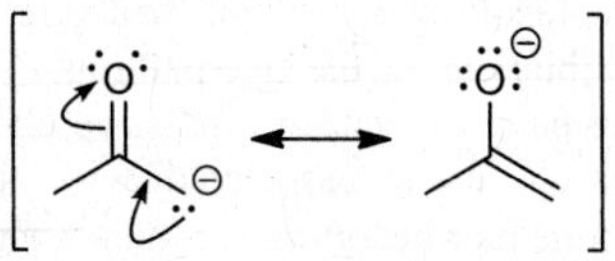

(j) This structure exhibits a C≡N bond (π bonds between two atoms of differing electronegativity), but that pattern would lead to a resonance structure with three charges, which would be insignificant. So we must look for another pattern.

This structure does also exhibit a lone pair next to a π bond, so we draw the two curved arrows associated with that pattern. The first curved arrow is drawn showing a lone pair becoming a π bond, while the second curved arrow shows a π bond becoming a lone pair:

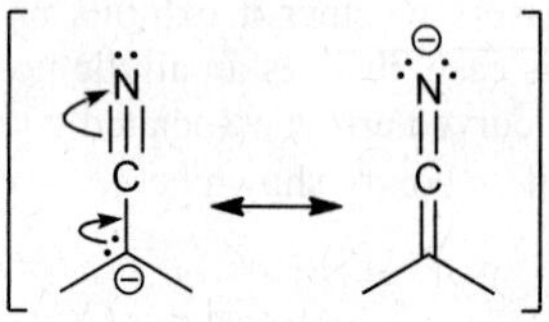

(k) This compound exhibits a C≡N bond (π bonds between two atoms of differing electronegativity), so we draw one curved arrow showing the π bond becoming a lone pair. We then draw the resulting resonance structure and assess whether it exhibits one of the five patterns. In this case, there is an allylic positive charge, so we draw the curved arrow associated with that pattern (pushing over the π bond), shown here:

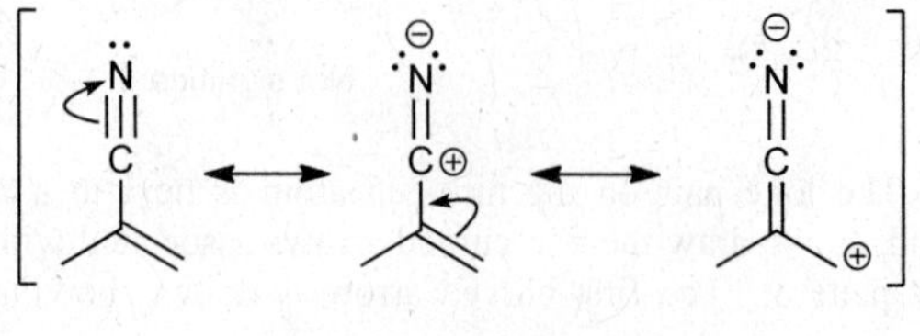

The following resonance structure is not significant, because it has two carbon atoms with opposite charges:

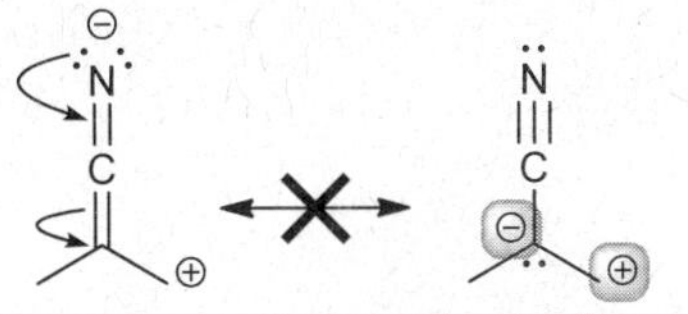

Not significant

(l) This structure exhibits a lone pair that is adjacent to a positive charge, so we draw one curved arrow, showing a lone pair becoming a π bond:

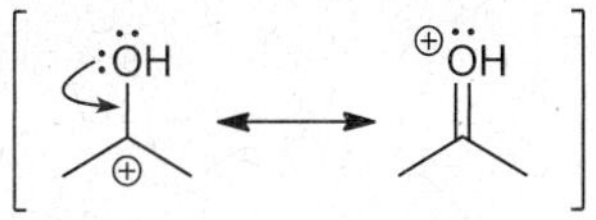

2.34. This compound exhibits a C=O bond (a π bond between two atoms of differing electronegativity), so we draw one curved arrow showing the π bond becoming a lone pair. We then draw the resulting resonance structure and assess whether it exhibits one of the five patterns. In this case, there is an allylic positive charge, so we draw the curved arrow associated with that pattern (pushing over the π bond), shown here. This pattern continues, many more times, spreading the positive charge over many locations:

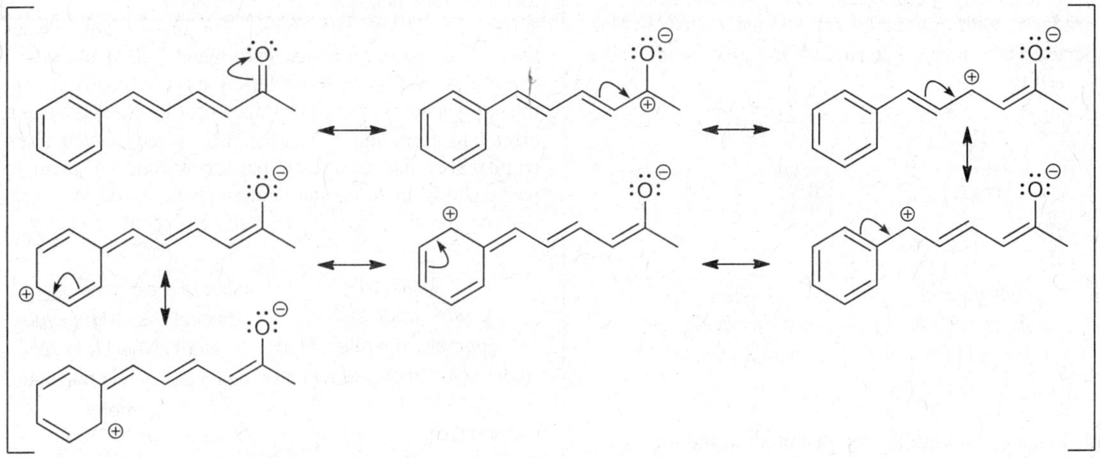

By considering the significant resonance structures (drawn above), we can determine the positions that are electron deficient (δ+). This information is summarized here.

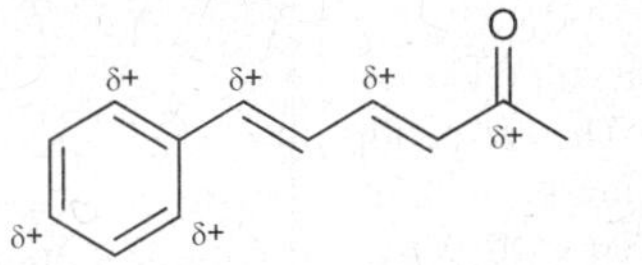

2.35. This compound exhibits a lone pair next to a π bond, so we draw two curved arrows. The first curved arrow is drawn showing a lone pair becoming a π bond, while the second curved arrow shows a π bond becoming a lone pair. We then draw the resulting resonance structure and assess whether it exhibits one of the five patterns. In this case, there is a lone pair next to a π bond, so once again, we draw the two curved arrows associated with that pattern. The resulting resonance structure again exhibits a lone pair next to a π bond. This pattern continues, many more times, spreading a negative charge over many locations (see next page):

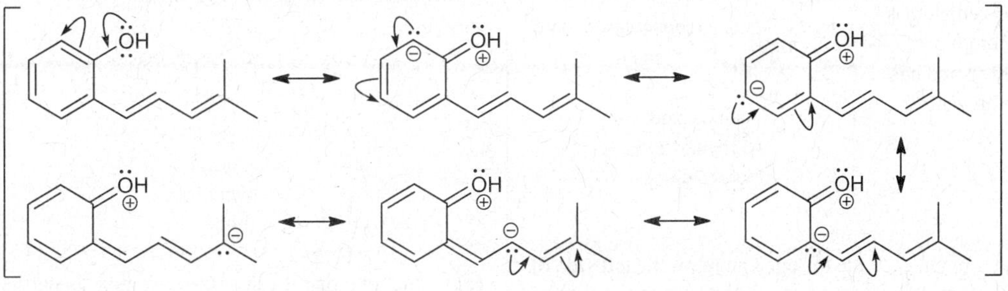

By considering the significant resonance structures (drawn above), we can determine the positions that are electron rich (δ–). This information is summarized here.

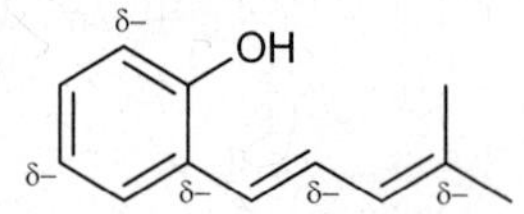

2.36.

(a) Let's begin with the nitrogen atom on the left side of the structure. The lone pair on this nitrogen atom is delocalized by resonance (because it is next to a π bond). Therefore, this lone pair occupies a *p* orbital, which means that the nitrogen atom is *sp*² hybridized. As a result, the geometry is trigonal planar.

On the right side of the structure, there is a nitrogen atom with a localized lone pair (it does not participate in resonance). This nitrogen atom is therefore *sp*³ hybridized, with trigonal pyramidal geometry, just as expected for a nitrogen atom with σ sigma bonds and a localized lone pair.

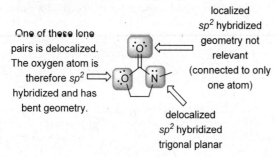

delocalized
*sp*² hybridized
trigonal planar

localized
*sp*³ hybridized
trigonal pyramidal

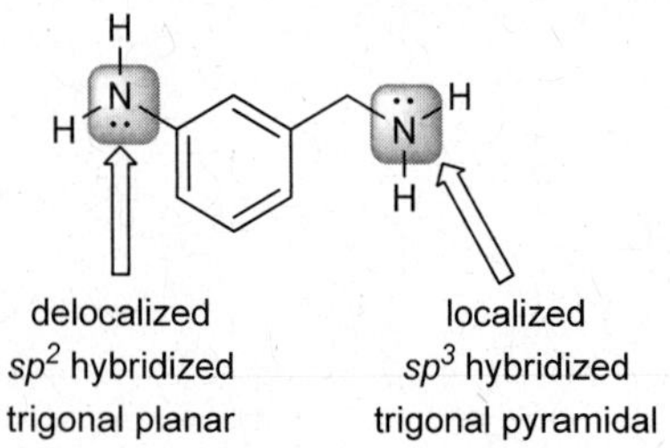

(b) Let's begin with the oxygen atom at the top of the structure. This oxygen atom has two lone pairs, both of which are localized (they do not participate in resonance). This oxygen atom is *sp*² hybridized, because it is using a *p* orbital to form the π bond.

The other oxygen atom also has two lone pairs, but one of them is delocalized via resonance. Therefore, this lone pair occupies a *p* orbital, which means that the oxygen atom is *sp*² hybridized (with bent geometry).

The lone pair on the nitrogen atom is also delocalized via resonance. Therefore, this lone pair occupies a *p* orbital, which means that the nitrogen atom is *sp*² hybridized. As a result, the geometry is trigonal planar.

One of these lone pairs is delocalized. The oxygen atom is therefore *sp*² hybridized and has bent geometry.

localized
*sp*² hybridized
geometry not relevant
(connected to only one atom)

delocalized
*sp*² hybridized
trigonal planar

(c) Let's begin with the nitrogen atom on the left side of the structure. The lone pair on this nitrogen atom is

delocalized by resonance (because it is next to a π bond). Therefore, this lone pair occupies a *p* orbital, which means that the nitrogen atom is *sp*² hybridized. As a result, the geometry is trigonal planar.

On the right side of the structure, there is a nitrogen atom with a localized lone pair (it does not participate in resonance). This nitrogen atom is therefore *sp*³ hybridized, with trigonal pyramidal geometry, just as expected for a nitrogen atom with three σ bonds and a localized lone pair.

There are two oxygen atoms. The one on the left has two lone pairs, neither of which participates in resonance. As such, those lone pairs are localized, and the oxygen atom is *sp*² hybridized (because it is using a *p* orbital to form the π bond). The other oxygen atom (right) also has two lone pairs, but one of them is participating in resonance and is delocalized. As such, that oxygen atom is *sp*² hybridized with bent geometry.

localized
*sp*² hybridized
geometry not relevant
(connected to only one atom)

One of these lone pairs is delocalized. The oxygen atom is therefore *sp*² hybridized and has bent geometry.

delocalized
*sp*² hybridized
trigonal planar

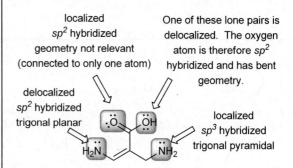

localized
*sp*³ hybridized
trigonal pyramidal

(d) As we saw with pyridine, the lone pair on this nitrogen atom is not participating in resonance, because the nitrogen atom is already using a *p* orbital for the π bond. As a result, the lone pair cannot join in the conduit of overlapping *p* orbitals, and therefore, it cannot participate in resonance. In this case, the lone pair occupies an *sp*² hybridized orbital, which is in the plane of the ring. Since this lone pair is not participating in resonance, it is localized. The nitrogen atom is *sp*² hybridized, and the geometry is bent.

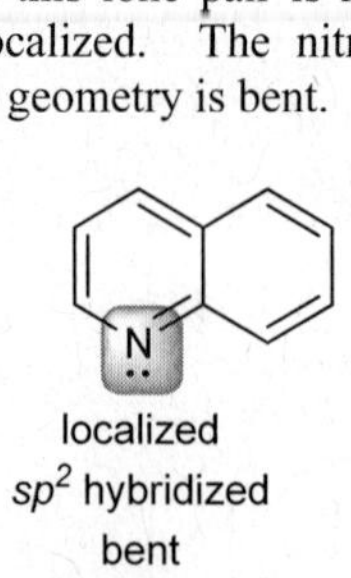

localized
*sp*² hybridized
bent

(e) The lone pair on this nitrogen atom is participating in resonance (it is next to a π bond), so it is delocalized via resonance. As such, the nitrogen atom is sp^2 hybridized, with trigonal planar geometry.

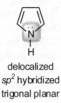

delocalized
sp^2 hybridized
trigonal planar

(f) This compound has three oxygen atoms. The lone pairs on two of these oxygen atoms are not participating in resonance, so those lone pairs are localized. But one of the oxygen atoms has a lone pair that is participating in resonance, so that lone pair is delocalized.

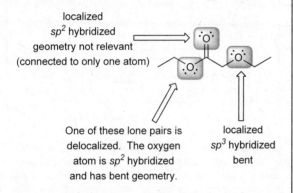

localized
sp^2 hybridized
geometry not relevant
(connected to only one atom)

One of these lone pairs is delocalized. The oxygen atom is sp^2 hybridized and has bent geometry.

localized
sp^3 hybridized
bent

2.37. Each of these lone pairs is not participating in resonance. So each of these lone pairs is localized. Therefore, both lone pairs are expected to be reactive.

2.38. Lone pairs that participate in resonance are delocalized, while those that do not participate in resonance are localized:

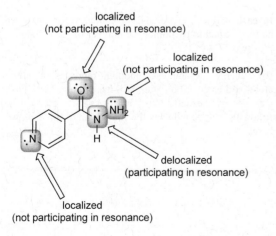

localized
(not participating in resonance)

localized
(not participating in resonance)

delocalized
(participating in resonance)

localized
(not participating in resonance)

2.39.
(a) Each corner and each endpoint represents a carbon atom (highlighted below), so this compound has nine

carbon atoms. Each carbon atom will have enough hydrogen atoms to have exactly four bonds, as shown. Each of the oxygen atoms has two bonds and no formal charge, so each oxygen atom will have two lone pairs.

(b) Each corner and each endpoint represents a carbon atom (highlighted below), so this compound has eight carbon atoms. Each carbon atom will have enough hydrogen atoms to have exactly four bonds, as shown. Each of the oxygen atoms has two bonds and no formal charge, so each oxygen atom will have two lone pairs. The nitrogen atom has three bonds and no formal charge, so it must have one lone pair of electrons.

(c) Each corner and each endpoint represents a carbon atom (highlighted below), so this compound has eight carbon atoms. Each carbon atom will have enough hydrogen atoms to have exactly four bonds, as shown. Each of the oxygen atoms has two bonds and no formal charge, so each oxygen atom will have two lone pairs. Each nitrogen atom has three bonds and no formal charge, so each nitrogen atom must have one lone pair of electrons.

2.40. The molecular formula indicates that there are four carbon atoms. Recall that constitutional isomers are compounds that share the same molecular formula, but differ in constitution (the connectivity of atoms). So we are looking for different ways that four carbon atoms can be connected together. They can be connected in a linear fashion (below left), or they can be connected with a branch (below right).

C_4H_{10} C_4H_{10}

These two compounds are the only constitutional isomers that have the molecular formula C_4H_{10}, because there are no other ways to connect four carbon atoms without changing the number of hydrogen atoms. For example, if we try to connect the carbon atoms into a ring, we find that the number of hydrogen atoms is reduced:

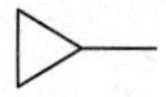

C_4H_8 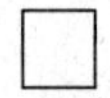 C_4H_8

2.41. As described in the solution to Problem 2.9, there are only three constitutional isomers with molecular formula C_5H_{12}, shown here again.

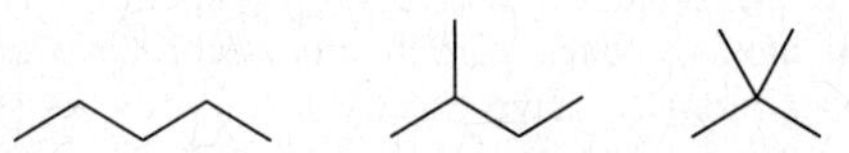

2.42. In each of the following structures, each corner and endpoint represents a carbon atom. Hydrogen atoms are only drawn if they are connected to heteroatoms (such as oxygen).

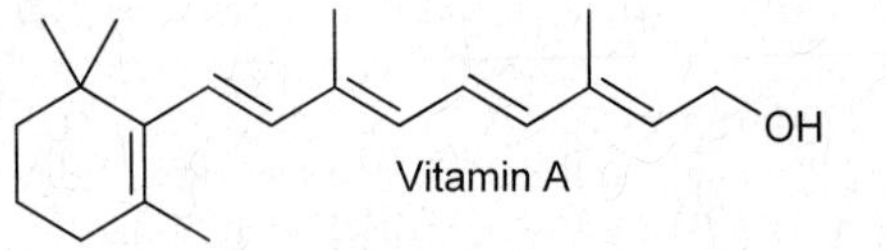

Vitamin A

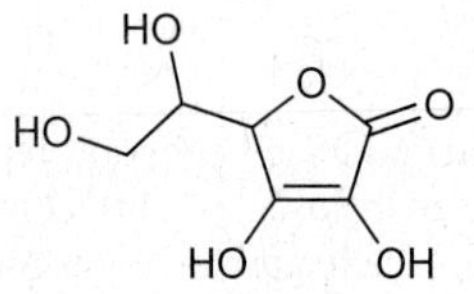

Vitamin C

2.43. Each oxygen atom has two bonds and no formal charge. Therefore, each oxygen atom has two lone pairs, for a total of twelve lone pairs.

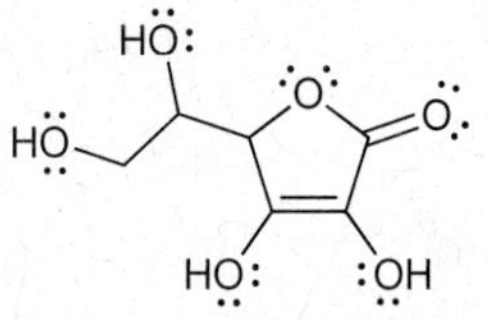

2.44. An oxygen atom will bear a negative charge if it has one bond and three lone pairs, and it will bear a positive charge if it has three bonds and one lone pair (see Table 2.2). A nitrogen atom will bear a negative charge if it has two bonds and two lone pairs, and it will bear a positive charge if it has four bonds and no lone pairs (see Table 2.3).

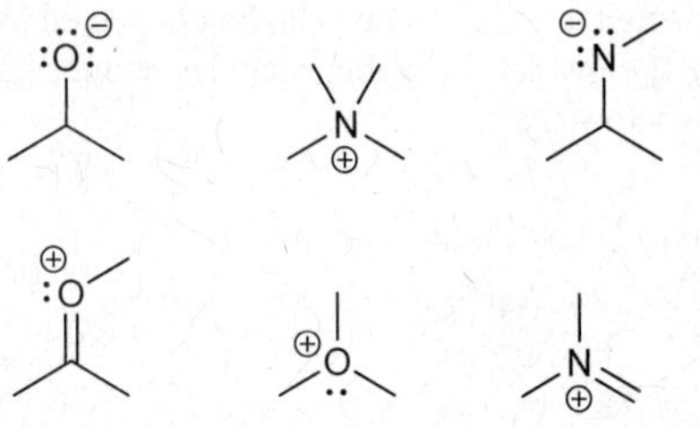

2.45. This compound exhibits a C=O bond (a π bond between two atoms of differing electronegativity), so we draw one curved arrow showing the π bond becoming a lone pair. We then draw the resulting resonance structure and assess whether it exhibits one of the five patterns. In this case, there is an allylic positive charge, so we draw the curved arrow associated with that pattern (pushing over the π bond). This pattern continues, many more times, spreading the positive charge over many locations, as shown here:

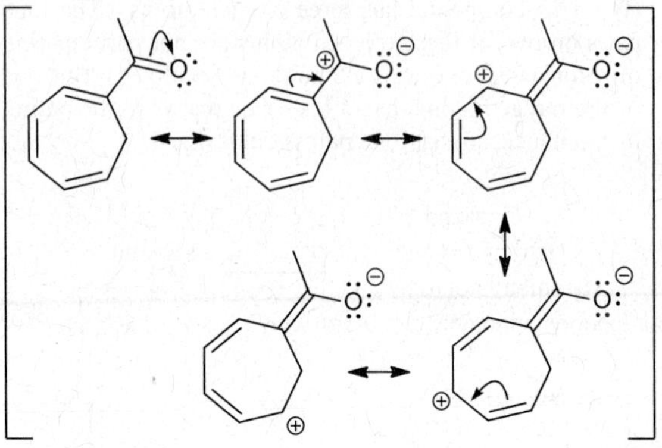

2.46.
(a) Each corner and each endpoint represents a carbon atom, and each carbon atom will have enough hydrogen atoms to have exactly four bonds. As such, each of the following compounds has the molecular formula shown here:

C_4H_{10}  C_6H_{14}

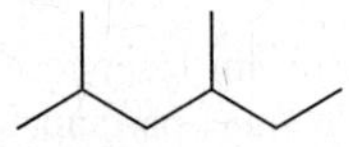

C_8H_{18} 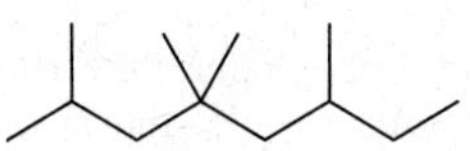 $C_{12}H_{26}$

In each of these compounds, the number of hydrogen atoms is equal to <u>two</u> times the number of carbon atoms, plus <u>two</u>.

(b) Each corner and each endpoint represents a carbon atom, and each carbon atom will have enough hydrogen atoms to have exactly four bonds. As such, each of the following compounds has the molecular formula shown here:

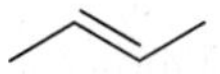

C_4H_8 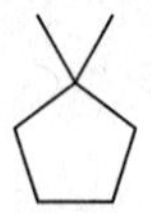 C_7H_{14}

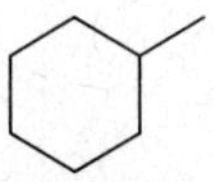

C_7H_{14} 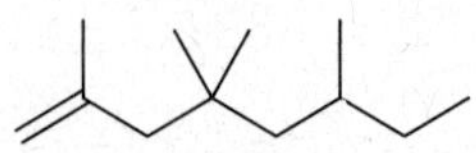 $C_{12}H_{24}$

In each of these compounds, the number of hydrogen atoms is <u>two</u> times the number of carbon atoms.

(c) Each corner and each endpoint represents a carbon atom, and each carbon atom will have enough hydrogen atoms to have exactly four bonds. As such, each of the following compounds has the molecular formula shown here:

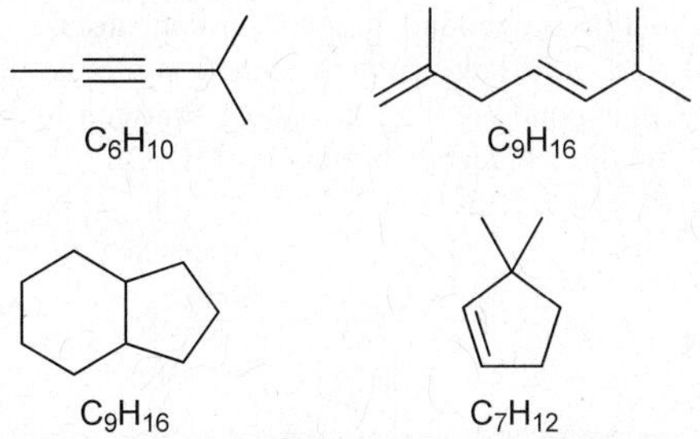

C_6H_{10} C_9H_{16}

C_9H_{16} C_7H_{12}

In each of these compounds, the number of hydrogen atoms is <u>two</u> times the number carbon atoms, minus <u>two</u>.

(d) The number of hydrogen atoms is exactly twice the number of carbon atoms. As such, we expect that the compound will contain either one ring or one double bond, but not both. A compound with the molecular formula $C_{24}H_{48}$ cannot have a triple bond, but it may have a double bond.

(e) The molecular formula C_4H_8 contains exactly twice as many hydrogen atoms as carbon atoms. As such, every constitutional isomer of C_4H_8 must contain either one ring or one double bond. Below are the structures that fit this description. Each of the first three constitutional isomers (shown here) exhibits a double bond, while each of the last two constitutional isomers exhibits a ring.

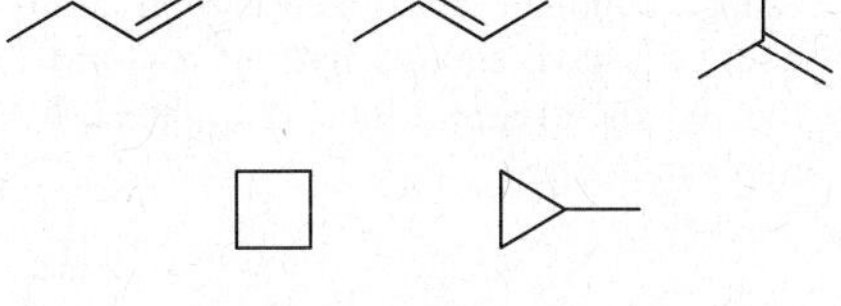

2.47.
(a) In order for the lone pair to participate in resonance, it must occupy a p orbital, which would render the nitrogen atom sp hybridizxed. With sp hybridization, the geometry of the nitrogen atom should be linear, which cannot be accommodated in a six-membered ring.

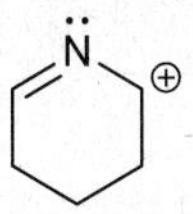

(b) There is a lone pair associated with the negative charge, and this lone pair is delocalized via resonance (the lone pair is allylic):

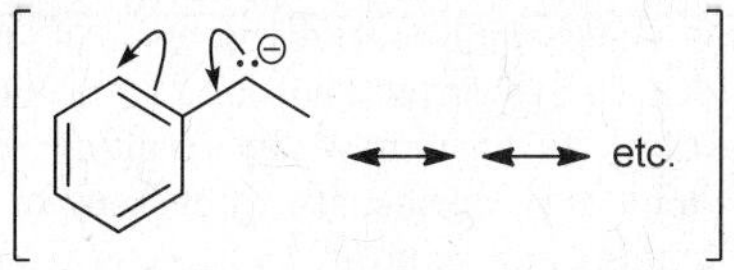

As such, the lone pair must occupy a p orbital.

(c) The nitrogen atom has a lone pair, which is delocalized via resonance (there is an adjacent positive charge):

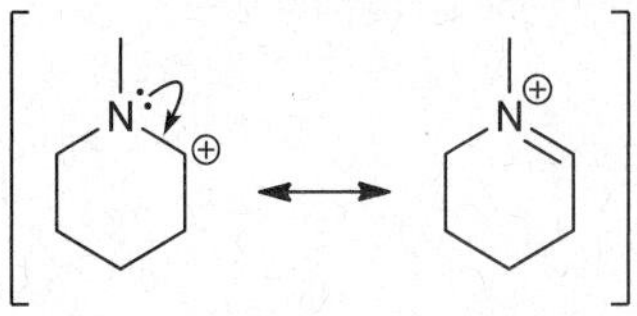

As such, the lone pair must occupy a p orbital.

2.48.
(a) This structure exhibits a lone pair that is adjacent to a positive charge. In fact, there are two such lone pairs (on the nitrogen and oxygen atoms). We will begin with the lone pair on the oxygen atom, although we would have arrived at the same solution either way (we will draw a total of three resonance structures, below, and it is just a matter of the order in which we draw them). We draw one curved arrow, showing a lone pair on the oxygen atom becoming a π bond. We then draw the resulting resonance structure and assess whether it exhibits one of the five patterns. In this case, there is a lone pair next to a π bond, so we draw the two curved arrows associated with that pattern. The first curved arrow is drawn showing a lone pair on the nitrogen atom becoming a π bond, while the second curved arrow shows a π bond becoming a lone pair on the oxygen atom:

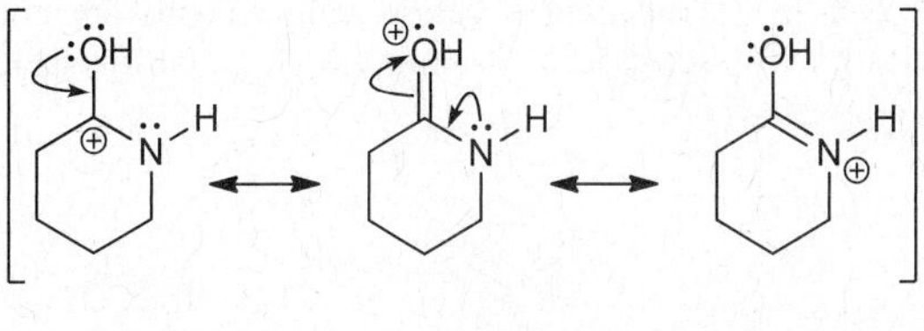

(b) This structure exhibits an allylic positive charge, so we draw one curved arrow showing the π bond being pushed over. We then draw the resulting resonance structure and assess whether it exhibits one of the five patterns. In this case, the positive charge is adjacent to a lone pair, so we draw the curved arrow associated with that pattern (the lone pair is shown becoming a π bond):

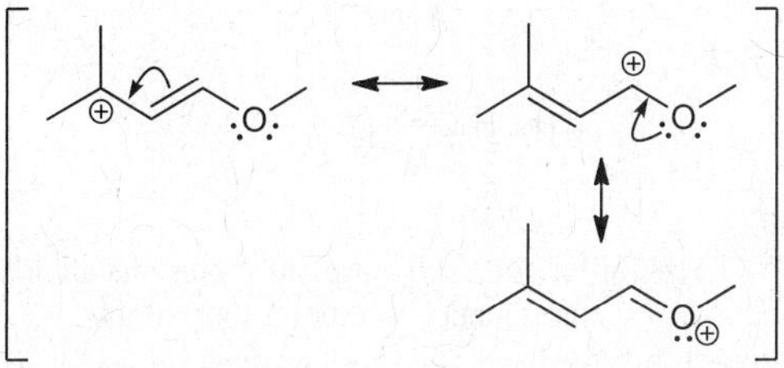

(c) This structure exhibits an allylic positive charge, so we draw one curved arrow showing the π bond being pushed over. We then draw the resulting resonance structure and assess whether it exhibits one of the five patterns. In this case, the positive charge is again next to a π bond, so again, we draw the curved arrow associated with that pattern (pushing over the π bond again). The resulting resonance structure has the positive charge next to yet another π bond, so we draw a curved arrow showing the π bond being pushed over one more time to give our final resonance structure:

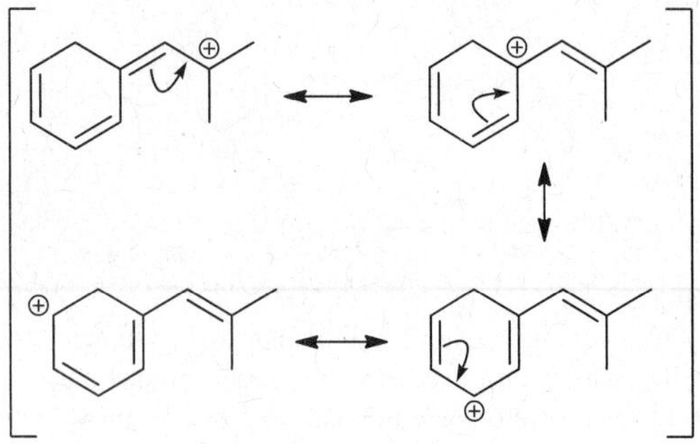

2.49.
(a) In a condensed structure, single bonds are not drawn. Instead, groups of atoms are clustered together, as shown here:

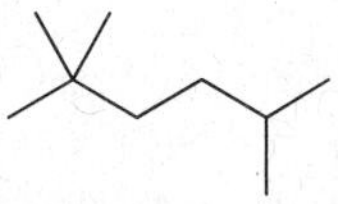

(CH$_3$)$_3$CCH$_2$CH$_2$CH(CH$_3$)$_2$

(b) In a condensed structure, single bonds are not drawn. Instead, groups of atoms are clustered together, as shown here:

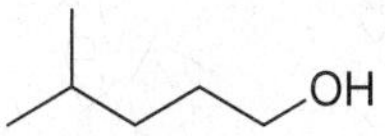

(CH$_3$)$_2$CHCH$_2$CH$_2$CH$_2$OH

(c) In a condensed structure, single bonds are not drawn. Instead, groups of atoms are clustered together, as shown here:

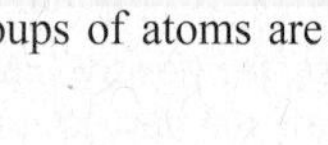

CH$_3$CH$_2$CH=C(CH$_2$CH$_3$)$_2$

2.50.
(a) Each corner and each endpoint represents a carbon atom, so this compound has nine carbon atoms. Each carbon atom will have enough hydrogen atoms to have

exactly four bonds, giving a total of twenty hydrogen atoms. So the molecular formula is C$_9$H$_{20}$.

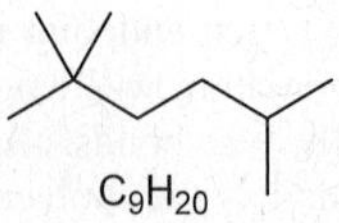

C$_9$H$_{20}$

(b) Each corner and each endpoint represents a carbon atom, so this compound has six carbon atoms. Each carbon atom will have enough hydrogen atoms to have exactly four bonds, giving a total of fourteen hydrogen atoms. So the molecular formula is C$_6$H$_{14}$O.

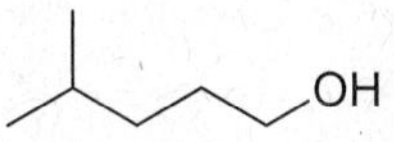

C$_6$H$_{14}$O

(c) Each corner and each endpoint represents a carbon atom, so this compound has eight carbon atoms. Each carbon atom will have enough hydrogen atoms to have exactly four bonds, giving a total of sixteen hydrogen atoms. So the molecular formula is C$_8$H$_{16}$.

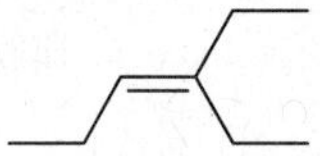

C$_8$H$_{16}$

2.51. The final structure **(d)** is not a valid resonance structure, because it violates the octet rule. The nitrogen atom has five bonds in this drawing, which is not possible, because the nitrogen atom only has four orbitals with which it can form bonds.

2.52. Each corner and each endpoint represents a carbon atom, so this compound has fifteen carbon atoms. Each carbon atom will have enough hydrogen atoms to have exactly four bonds, giving a total of eighteen hydrogen atoms, as shown here:

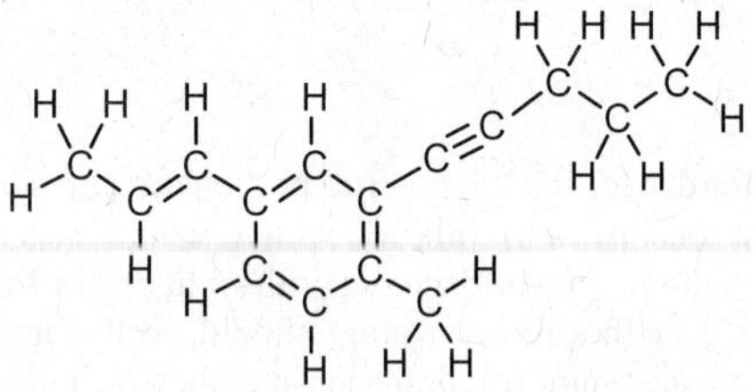

C$_{15}$H$_{18}$

2.53.
(a) A negative charge is assigned to an oxygen atom with one bond and three lone pairs (see Table 2.2).

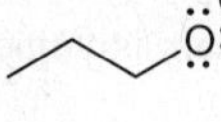

(b) The nitrogen atom on the left has three bonds and one lone pair, so it does not have a formal charge. The other nitrogen atom has two bonds and one lone pair. Each bond requires one valence electron, and the lone pair represents two valence electrons. So, this nitrogen atom is using 1+1+2 = 4 valence electrons. The appropriate number of valence electrons for a nitrogen atom is five, which means that this nitrogen is missing a valence electron. As such, the nitrogen atom is assigned a positive formal charge.

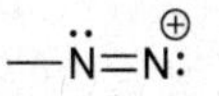

$$—\ddot{N}{=}\overset{\oplus}{N}{:}$$

(c) A positive charge is assigned to an oxygen atom with three bonds and one lone pair (see Table 2.2).

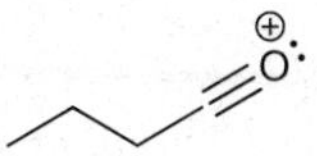

(d) A negative charge is assigned to a nitrogen atom with two bonds and two lone pairs (see Table 2.3).

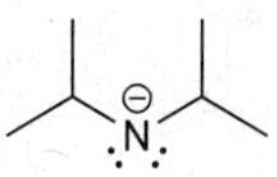

2.54. As seen in the solution to Problem 2.40, there are only two ways to connect four carbon atoms in a compound with molecular formula C_4H_{10}:

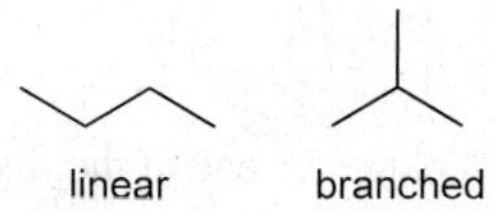

linear   branched

In our case, the molecular formula is C_4H_9Cl, which is similar to C_4H_{10}, but one H has been replaced with a chlorine atom. So, we must explore all of the different locations where a chlorine atom can be placed on each of the carbon skeletons above (the linear skeleton and the branched skeleton).

Let's begin with linear skeleton. There are two distinctly different locations where a chlorine atom can be placed on this skeleton: either at position 1 or position 2, shown here.

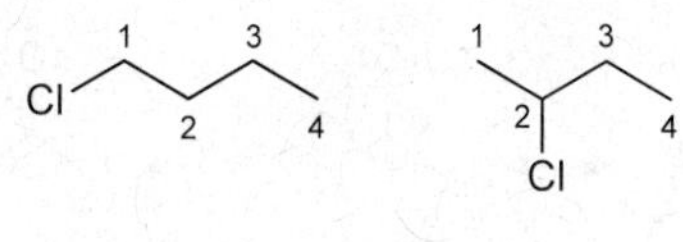

Placing the chlorine atom at position 3 would be the same as placing it at position 2; and placing the chlorine atom at position 4 would be the same as it as position 1:

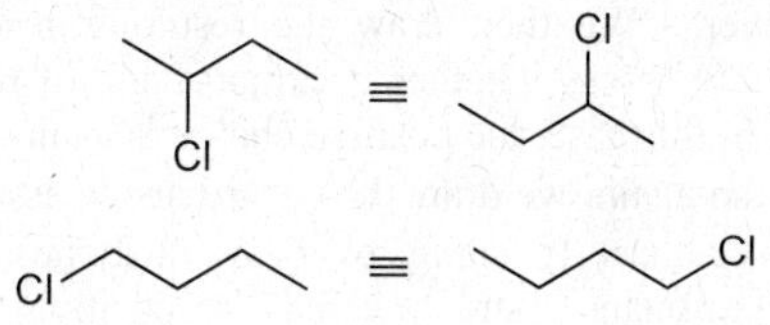

Next, we move on to the other carbon skeleton, containing a branch. Once again, there are two distinctly different locations where a chlorine atom can be placed: either at position 1 or position 2, shown here.

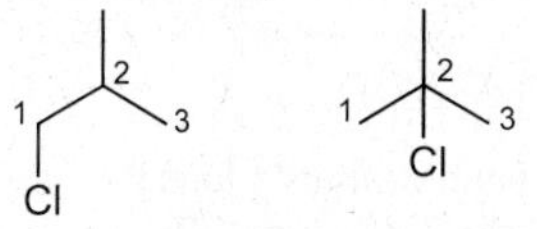

Placing the chlorine atom on any of the peripheral carbon atoms will lead to the same compound:

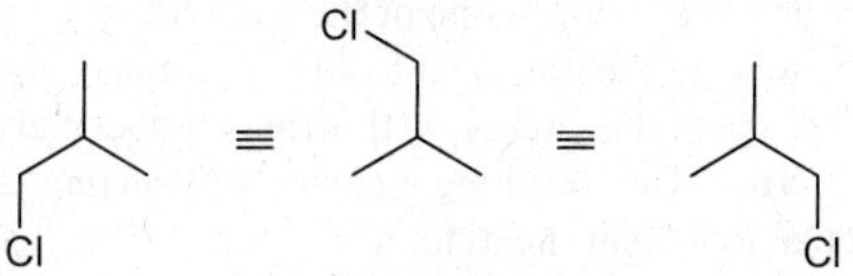

In summary, there are a total of four constitutional isomers with molecular formula C_4H_9Cl:

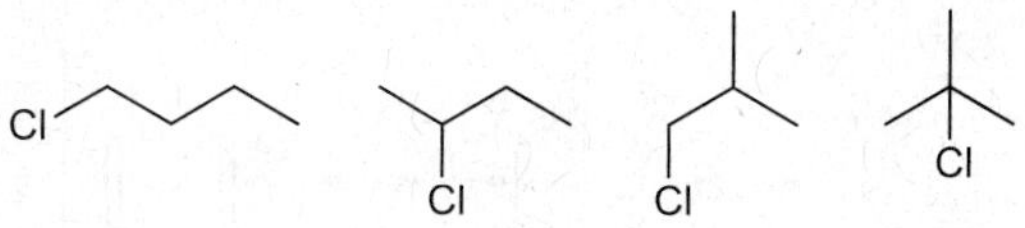

2.55.
(a) This compound exhibits a lone pair next to a π bond, so we draw the two curved arrows associated with that pattern. The first curved arrow is drawn showing a lone pair becoming a π bond, while the second curved arrow shows a π bond becoming a lone pair:

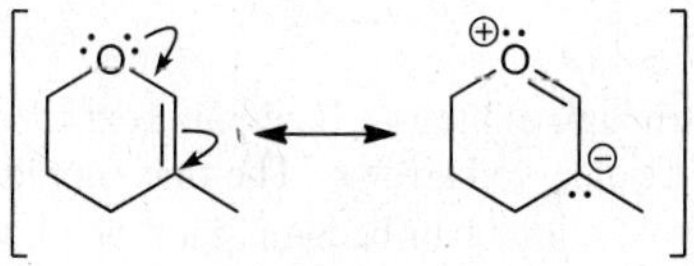

(b) This structure exhibits a lone pair that is adjacent to a positive charge, so we draw one curved arrow, showing a lone pair becoming a π bond:

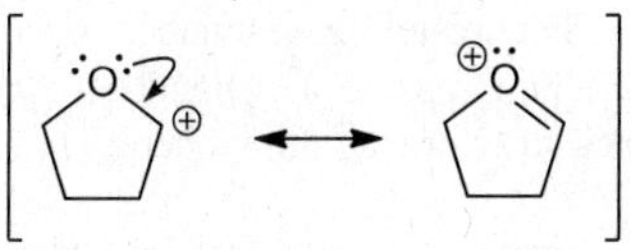

(c) This structure exhibits an allylic positive charge, so we draw one curved arrow showing the π bond being pushed over.

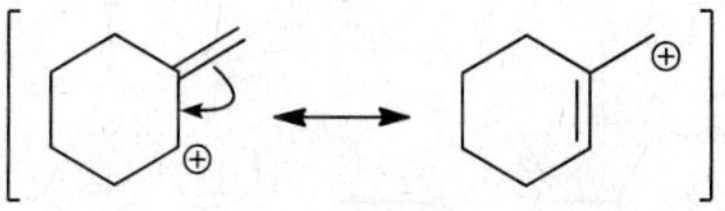

(d) This compound exhibits a C=N bond (a π bond between two atoms of differing electronegativity), so we draw one curved arrow showing the π bond becoming a lone pair.

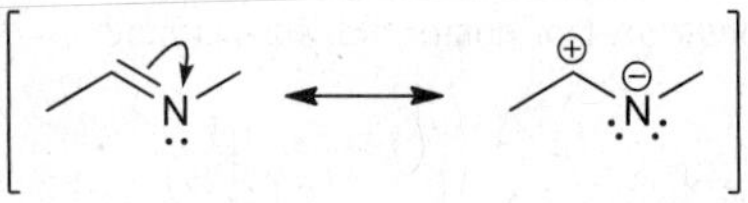

(e) This compound exhibits a lone pair next to a π bond, so we draw two curved arrows. The first curved arrow is drawn showing a lone pair becoming a π bond, while the second curved arrow shows a π bond becoming a lone pair. We then draw the resulting resonance structure and assess whether it exhibits one of the five patterns. In this case, the lone pair is now next to another π bond, so once again, we draw the two curved arrows associated with that pattern. The resulting resonance structure again exhibits a lone pair next to a π bond. This pattern continues again, thereby spreading a negative charge over many locations, as shown here:

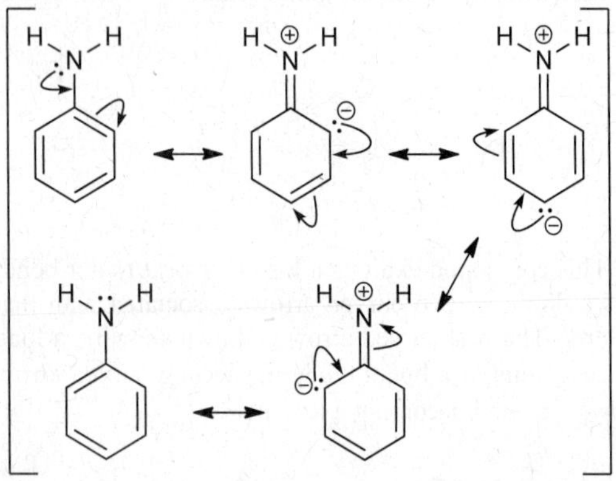

(f) This structure exhibits a lone pair next to a π bond, so we draw two curved arrows. The first curved arrow is drawn showing a lone pair becoming a π bond, while the second curved arrow shows a π bond becoming a lone pair. We then draw the resulting resonance structure and assess whether it exhibits one of the five patterns. In this case, the lone pair is now next to another π bond, so once again, we draw the two curved arrows associated with that pattern. The resulting resonance structure again exhibits a lone pair next to a π bond, so we draw one more resonance structure, as shown here:

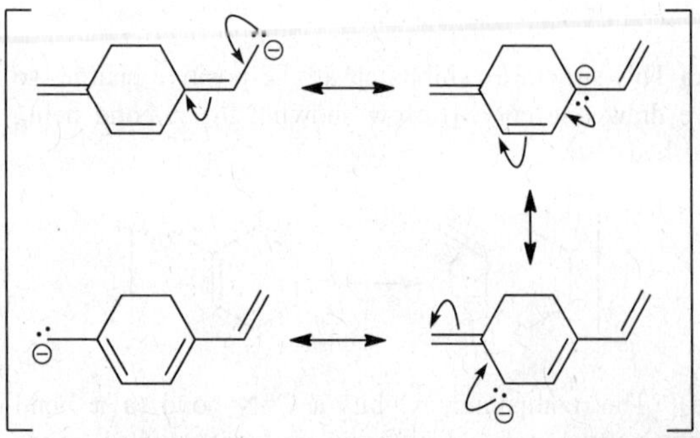

(g) This compound exhibits a C=O bond (a π bond between two atoms of differing electronegativity), so we draw one curved arrow showing the π bond becoming a lone pair. We then draw the resulting resonance structure and assess whether it exhibits one of the five patterns. In this case, there is an allylic positive charge, so we draw the curved arrow associated with that pattern (pushing over the π bond). This pattern continues, many more times, spreading the positive charge over many locations, as shown here:

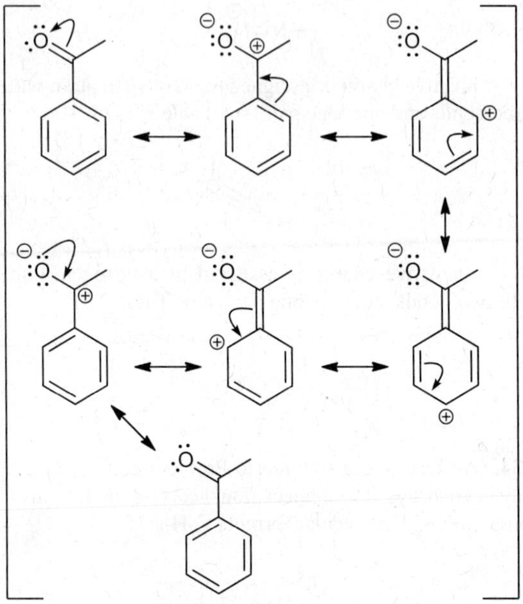

(h) We begin by looking for one of the five patterns that employs just one curved arrow (in this case, there is another pattern that requires two curved arrows, but we will start with the pattern using just one curved arrow). There is a C=O bond (a π bond between two atoms of differing electronegativity), so we draw one curved arrow showing the π bond becoming a lone pair. We then draw the resulting resonance structure and assess whether it exhibits one of the five patterns. In this case, there is a lone pair adjacent to a positive charge, so we draw the curved arrow associated with that pattern (showing the lone pair becoming a π bond), shown here:

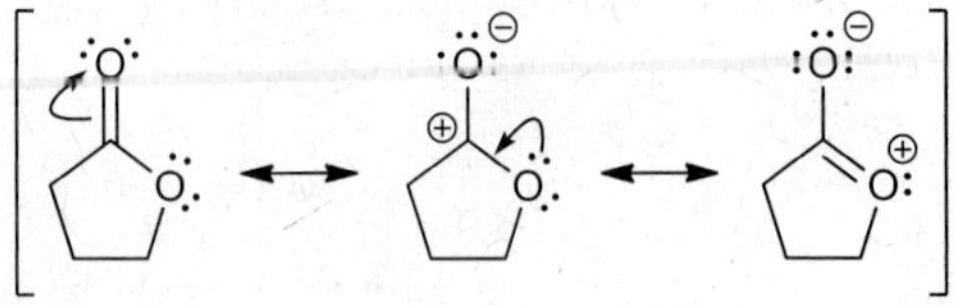

(i) This structure exhibits an allylic positive charge, so we draw one curved arrow showing the π bond being pushed over. We then draw the resulting resonance structure and assess whether it exhibits one of the five patterns. In this case, the positive charge is again next to a π bond, so again, we draw the curved arrow associated with that pattern (pushing over the π bond). The resulting resonance structure has a positive charge adjacent to a lone pair, so we draw the one curved arrow associated with that pattern (showing the lone pair becoming a π bond), shown here:

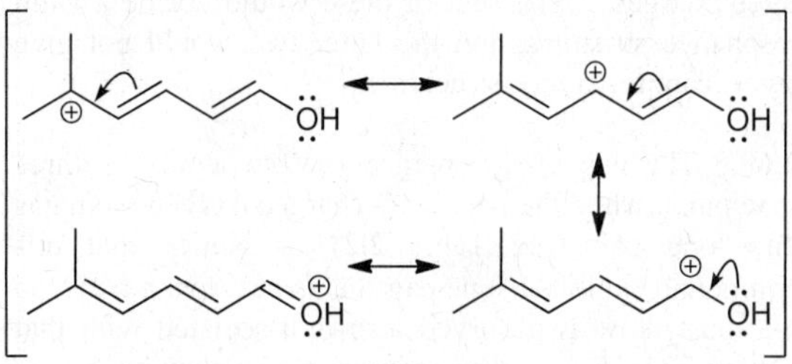

(j) This structure exhibits an allylic positive charge, so we draw one curved arrow showing the π bond being pushed over. We then draw the resulting resonance structure and assess whether it exhibits one of the five patterns. In this case, the positive charge is again next to a π bond, so again, we draw the curved arrow associated with that pattern (pushing over the π bond). The resulting resonance structure again has a positive charge next to a π bond, so again, we draw the curved arrow associated with that pattern (pushing over the π bond). The resulting resonance structure has a positive charge adjacent to a lone pair, so we draw the one curved arrow associated with that pattern (showing the lone pair becoming a π bond), shown here:

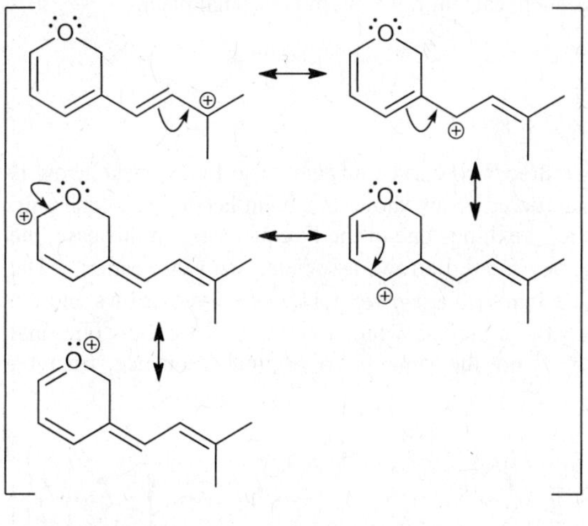

2.56. These structures do not differ in their connectivity of atoms. They differ only in the placement of electrons. Therefore, these structures are resonance structures, as shown here:

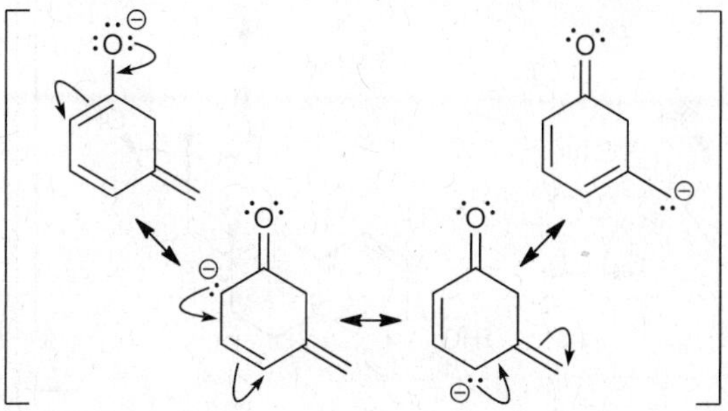

2.57.
(a) These compounds both have the same molecular formula (C_7H_{12}), but they differ in their connectivity of atoms, or constitution. Therefore, they are constitutional isomers.
(b) These structures have the same molecular formula (C_7H_{16}), AND they have the same constitution (connectivity of atoms), so they represent the same compound.
(c) The first compound has molecular formula C_5H_{10}, while the second compound has molecular formula C_5H_8. As such, they are different compounds that are not isomeric.
(d) These compounds both have the same molecular formula (C_5H_8), but they differ in their connectivity of atoms, or constitution. Therefore, they are constitutional isomers.

2.58.
(a) The condensed structure (shown in the problem statement) indicates the constitution (how the atoms are connected to each other). In the bond-line structure, hydrogen atoms are not drawn (they are implied). Each corner and each endpoint represents a carbon atom, so the carbon skeleton is shown more clearly.

CH_2=$CHCH_2C(CH_3)_3$ 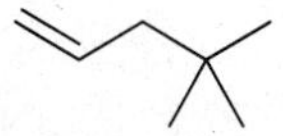
Condensed structure   Bond-line structure

(b) The condensed structure indicates how the atoms are connected to each other. In the bond-line structure, hydrogen atoms are not drawn (they are implied), except for the hydrogen atom attached to the oxygen atom (hydrogen atoms must be drawn if they are connected to a heteroatom, such as oxygen). Each corner and each endpoint represents a carbon atom, so the carbon skeleton is shown more clearly.

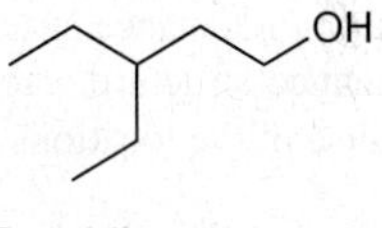

$(CH_3CH_2)_2CHCH_2CH_2OH$
Condensed structure   Bond-line structure

(c) The condensed structure indicates how the atoms are connected to each other. In the bond-line structure, hydrogen atoms are not drawn (they are implied). Each corner and each endpoint represents a carbon atom, so the carbon skeleton is shown more clearly.

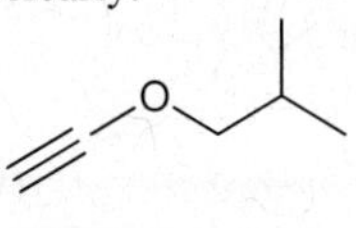

CH≡$COCH_2CH(CH_3)_2$
Condensed structure   Bond-line structure

(d) The condensed structure indicates how the atoms are connected to each other. In the bond-line structure, hydrogen atoms are not drawn (they are implied). Each corner and each endpoint represents a carbon atom, so the carbon skeleton is shown more clearly.

CH₃CH₂OCH₂CH₂OCH₂CH₃

Condensed structure   Bond-line structure

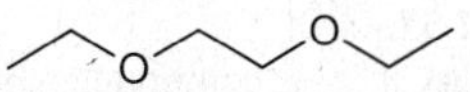

(e) The condensed structure indicates how the atoms are connected to each other. In the bond-line structure, hydrogen atoms are not drawn (they are implied). Each corner and each endpoint represents a carbon atom, so the carbon skeleton is shown more clearly.

(CH₃CH₂)₃CBr

Condensed structure

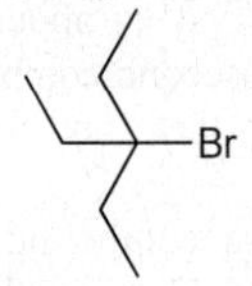

Bond-line structure

(f) The condensed structure indicates how the atoms are connected to each other. In the bond-line structure, hydrogen atoms are not drawn (they are implied). Each corner and each endpoint represents a carbon atom, so the carbon skeleton is shown more clearly.

(CH₃)₂C=CHCH₃

Condensed structure

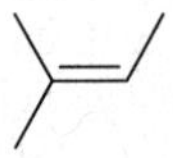

Bond-line structure

2.59. The nitronium ion does *not* have any significant resonance structures because any attempts to draw a

resonance structure will either 1) exceed an octet for the nitrogen atom or 2) generate a nitrogen atom with less than an octet of electrons, or 3) generate a structure with three charges. The first of these would not be a valid resonance structure, and the latter two would not give significant resonance structures.

2.60. The negatively charged oxygen atom has three lone pairs, while the positively charged oxygen atom has one lone pair (see Table 2.2). Notice that this compound exhibits a lone pair that is next to a π bond, so we must draw two curved arrows associated with that pattern. The first curved arrow is drawn showing a lone pair becoming a π bond, while the second curved arrow shows a π bond becoming a lone pair:

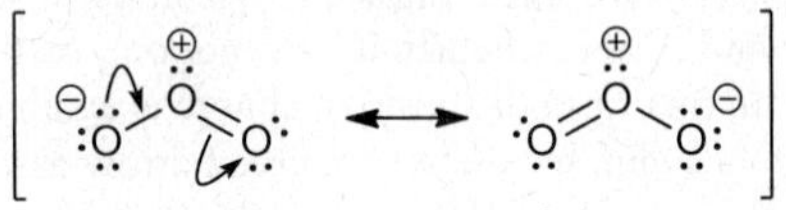

There are no other valid resonance structures that are significant.

2.61. Each nitrogen atom has a lone pair that is delocalized via resonance. In order to be delocalized via resonance, the lone pair must occupy a *p* orbital, and therefore, each nitrogen atom must be *sp²* hybridized. As such, each nitrogen atom is trigonal planar.

2.62.
(a) This compound exhibits a lone pair next to a π bond, so we draw two curved arrows. The first curved arrow is drawn showing a lone pair becoming a π bond, while the second curved arrow shows a π bond becoming a lone pair. We then draw the resulting resonance structure and assess whether it exhibits one of the five patterns. In this case, the lone pair is next to another π bond, so once again, we draw the two curved arrows associated with that pattern. The resulting resonance structure again exhibits a lone pair next to a π bond, so again we draw two curved arrows and the resulting resonance structure. Once again, there is a lone pair next to a π bond, which requires that we draw one final resonance structure, shown below. This last resonance structure is not the same as the original resonance structure, because of the locations in which the π bonds are drawn.

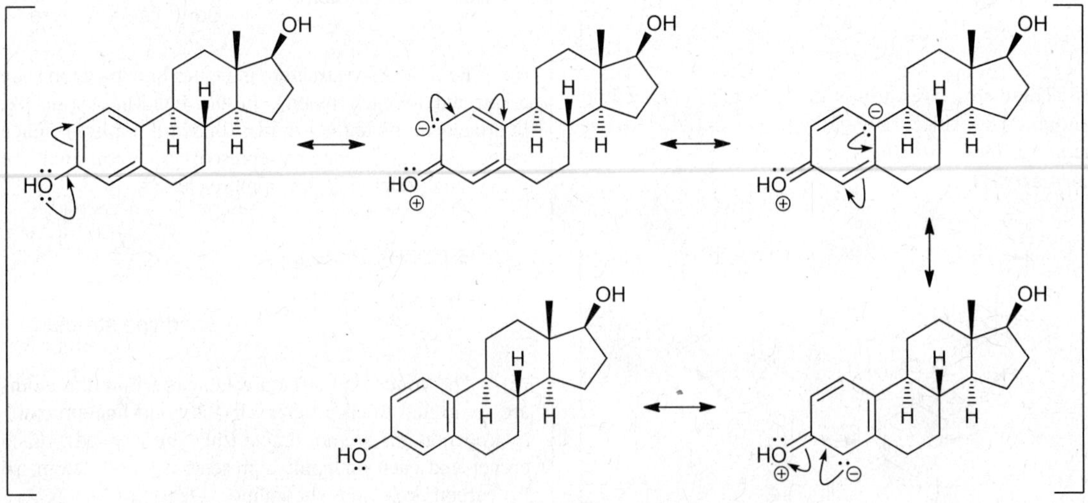

(b) The following compound exhibits a C=O bond (a π bond between two atoms of differing electronegativity), so we draw one curved arrow showing the π bond becoming a lone pair. We then draw the resulting resonance structure and assess whether it exhibits one of the five patterns. In this case, there is an allylic positive charge, so we draw the curved arrow associated with that pattern (pushing over the π bond), shown here:

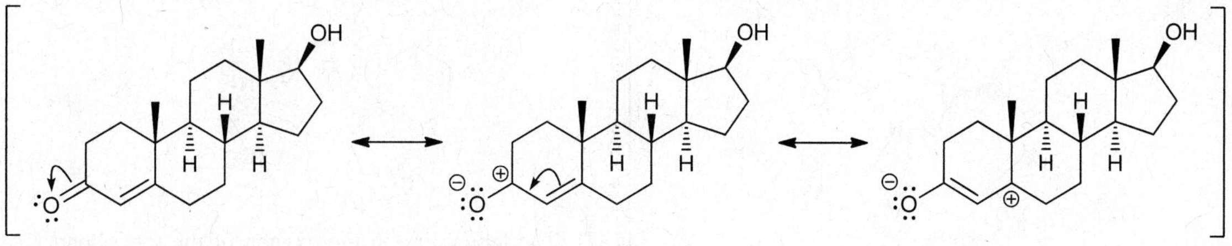

2.63.
(a) The molecular formula is $C_3H_6N_2O_2$.
(b) Each of the highlighted carbon atoms (below) has four sigma bonds (the bonds to hydrogen are not shown). As such, these two carbon atoms are sp^3 hybridized.

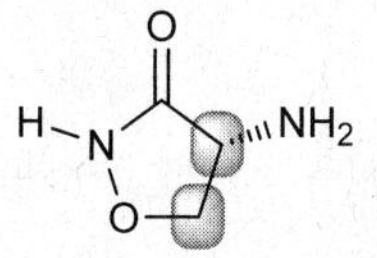

(c) There is one carbon atom that is using a p orbital to form a π bond. As such, this carbon atom (highlighted) is sp^2 hybridized.

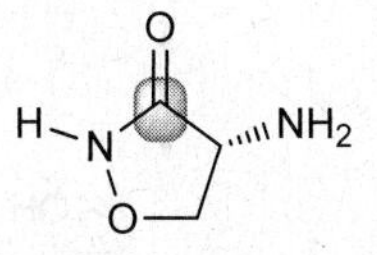

(d) There are no sp hybridized carbon atoms in this structure.
(e) There are six lone pairs (each nitrogen atom has one lone pair and each oxygen atom has two lone pairs):

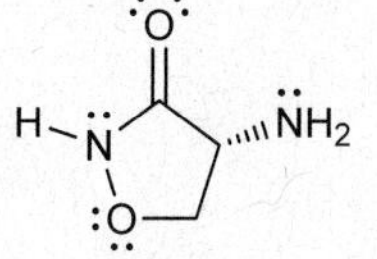

(f) Only the lone pair on one of the nitrogen atoms is delocalized via resonance (to see why it is delocalized, see the solution to 2.63h). The other lone pairs are all localized.

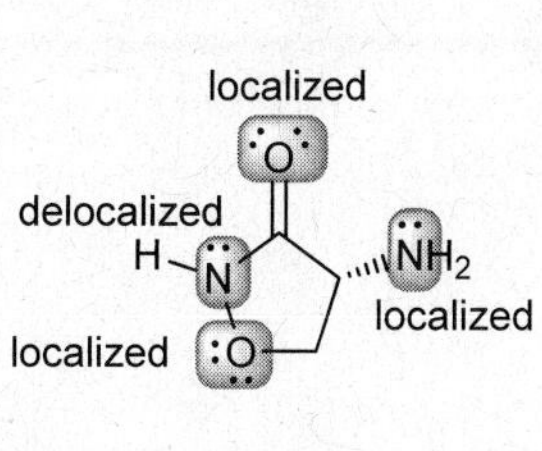

(g) The geometry of each atom is shown below (see SkillBuilder 1.8):

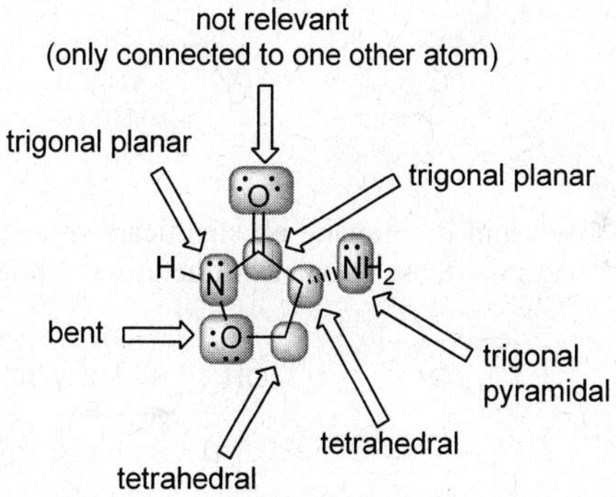

not relevant
(only connected to one other atom)

trigonal planar

trigonal planar

bent

trigonal pyramidal

tetrahedral

tetrahedral

(h) We begin by looking for one of the five patterns that employs just one curved arrow (in this case, there is another pattern that requires two curved arrows, but we will start with the pattern using just one curved arrow). There is a C=O bond (a π bond between two atoms of differing electronegativity), so we draw one curved arrow showing the π bond becoming a lone pair. We then draw the resulting resonance structure and assess whether it exhibits one of the five patterns. In this case, there is a lone pair adjacent to a positive charge, so we draw the curved arrow associated with that pattern (showing the lone pair becoming a π bond), shown here:

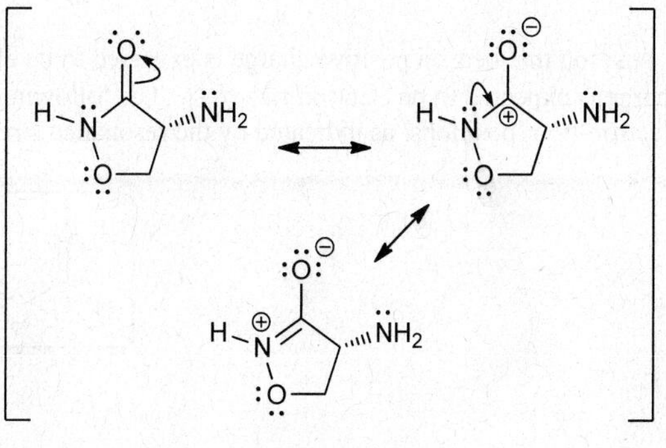

2.64.
(a) The molecular formula is $C_{16}H_{21}NO_2$.
(b) Each of the highlighted carbon atoms (below) has four sigma bonds (the bonds to hydrogen are not shown). As such, these nine carbon atoms are sp^3 hybridized.

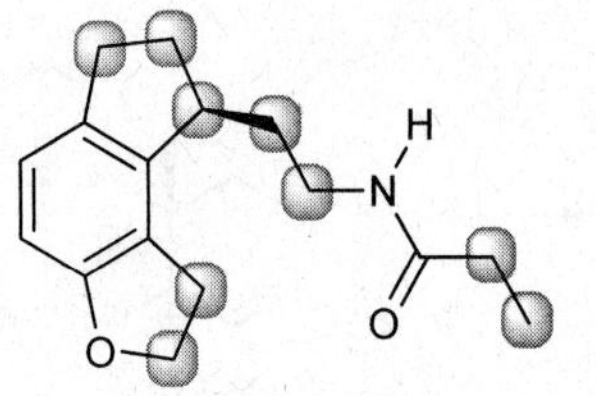

(c) There are seven carbon atoms that are each using a p orbital to form a π bond. As such, these seven carbon atoms (highlighted) are sp^2 hybridized.

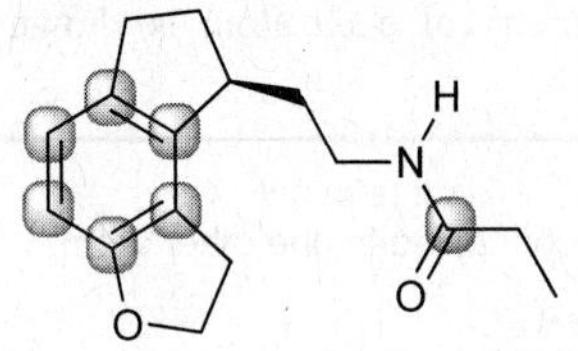

(d) There are no sp hybridized carbon atoms in this structure.
(e) There are five lone pairs (the nitrogen atom has one lone pair and each oxygen atom has two lone pairs):

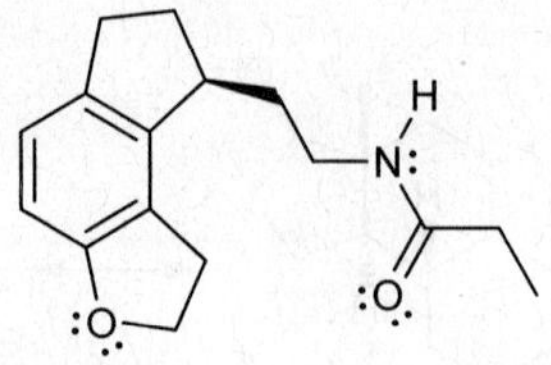

(f) The lone pairs on the oxygen of the C=O bond are localized. One of the lone pairs on the other oxygen atom is delocalized via resonance. The lone pair on the nitrogen atom is delocalized via resonance.

(g) All sp^2 hybridized carbon atoms are trigonal planar. All sp^3 hybridized carbon atoms are tetrahedral. The nitrogen atom is trigonal planar. The oxygen atom of the C=O bond does not have a geometry because it is connected to only one other atom, and the other oxygen atom has bent geometry (see SkillBuilder 1.8).

2.65. We begin by drawing all significant resonance structures, and then considering the placement of the formal charges in each of those resonance structures (highlighted below)

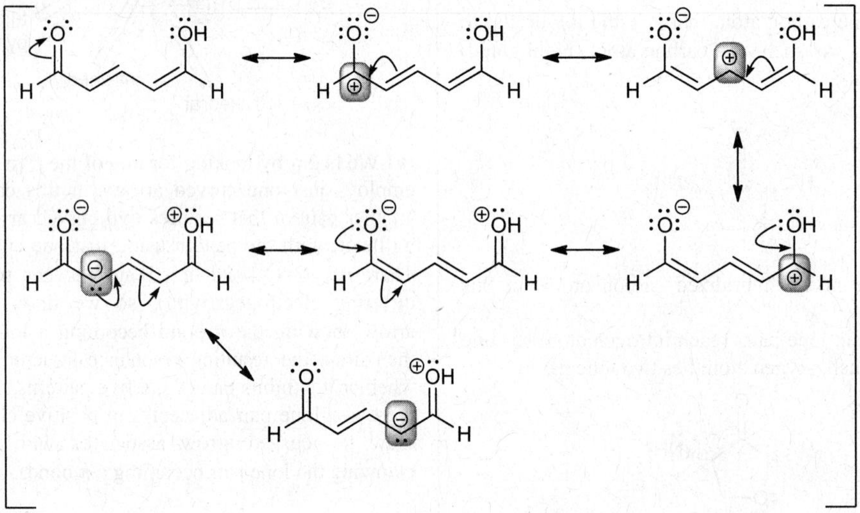

A position that bears a positive charge is expected to be electron deficient ($\delta+$), while a position that bears a negative charge is expected to be electron rich ($\delta-$). The following is a summary of the electron-deficient positions and the electron-poor positions, as indicated by the resonance structures above.

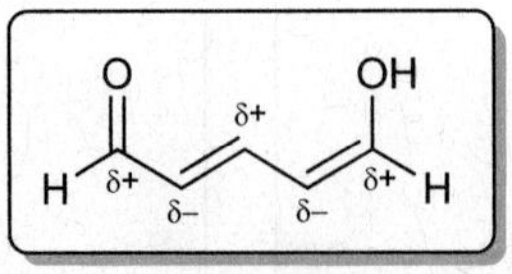

2.66.

(a) Compound **B** has one additional resonance structure that compound **A** lacks, because of the relative positions of the two groups on the aromatic ring. Specifically, compound **B** has a resonance structure in which one oxygen atom has a negative charge and the other oxygen atom has a positive charge:

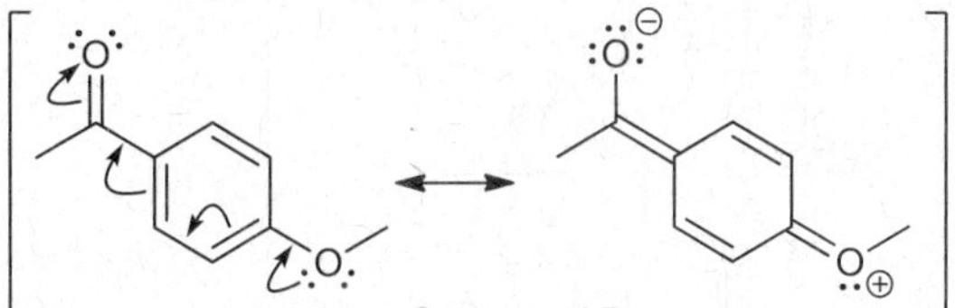

Compound B

Compound **A** does *not* have a significant resonance structure in which one oxygen atom has a negative charge and the other oxygen atom has a positive charge. That is, compound **A** has fewer resonance structures than compound **B**. Accordingly, compound **B** has greater resonance stabilization.

(b) Compound **C** is expected to have resonance stabilization similar to that of compound **B**, because compound **C** also has a resonance structure in which one oxygen atom has a negative charge and the other oxygen atom has a positive charge:

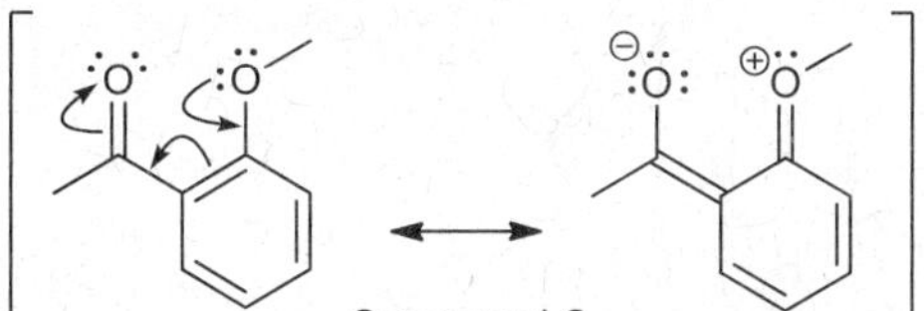

Compound C

2.67.

The single bond mentioned in this problem has some double bond character, as a result of resonance:

Each of the carbon atoms of this single bond uses an atomic *p* orbital to form a conduit (as described in Section 2.7):

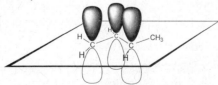

Rotation about this single bond will destroy the overlap of the *p* orbitals, thereby destroying the resonance stabilization. This single bond therefore exhibits a large barrier to rotation.

2.68. Polymer 2 contains only ester groups, so the IR spectrum of polymer **2** is expected to exhibit a signal near 1740 cm^{-1} (typical for esters), associated with vibrational excitation (stretching) of the C=O bond. Polymer **4** lacks any ester groups, so the signal near 1740 cm^{-1} is expected to be absent in the IR spectrum of polymer **4**. Instead, polymer **4** has OH groups, which are expected to produce a broad signal in the range 3200-3600 cm^{-1}. Polymer **3** has both functional groups (alcohol group and ester group), so an IR spectrum of polymer **3** is expected to exhibit both characteristic signals. When polymer **3** is converted to polymer **4**, the signal near 1740 cm^{-1} is expected to vanish, which would indicate complete hydrolysis of polymer **3**.

In practice, the signal for the C=O stretch in polymer 2 appears at 1733 cm^{-1}, which is very close to our estimated value of 1740 cm^{-1}.

2.69. Compound 1 has an OH group, which is absent in compound 2. Therefore, the IR spectrum of **1** should exhibit a broad signal in the range 3200-3600 cm^{-1} (associated with O-H stretching), while the IR spectrum of **2** would be expected to lack such a signal. The conversion of **1** to **2** could therefore be confirmed with the disappearance of the signal corresponding with excitation of the O-H bond.

Another way to monitor the conversion of **1** to **2** is to focus on the C-H bond of the aldehyde group in compound **1**, which is expected to produce a signal in the range 2750-2850 cm^{-1}. Since the aldehyde group is not present in compound **2**, we expect this signal to vanish when **1** is converted to **2**.

There is yet another way to monitor this reaction with IR spectroscopy. Compound **1** possesses only one C=O bond, while compound **2** has two C=O bonds. As such, the latter should exhibit two C=O signals. One signal is expected to be near 1680 cm^{-1} (for the conjugated ketone), and the other signal should be near 1700 cm^{-1} (corresponding to the conjugated ester). In contrast compound **1** has only one C=O bond, which is expected to produce a signal near 1680 cm^{-1} (for the conjugated aldehyde). Therefore, the conversion of **1** to **2** can be monitored by the appearance of a signal near 1700 cm^{-1}.

2.70.
(a) The molecular formula for CL-20 is $C_6H_6N_{12}O_{12}$. The molecular formula for HMX is $C_4H_8N_8O_8$.
(b) The lone pair is delocalized (see resonance structures below).

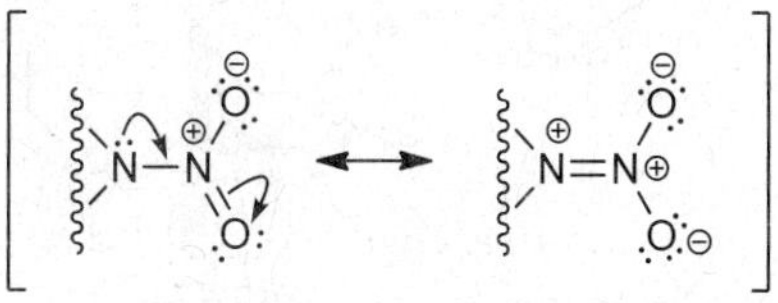

2.71.
This intermediate is highly stabilized by resonance. The positive charge is spread over one carbon atom and three oxygen atoms.

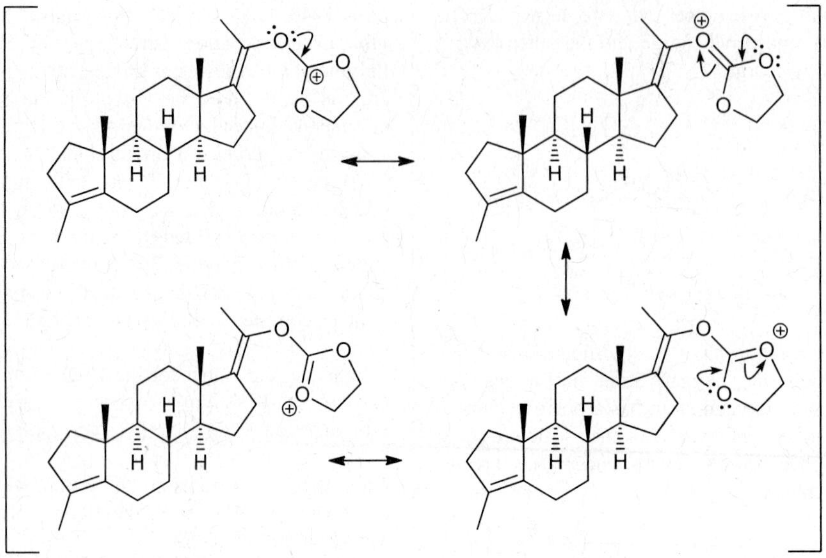

2.72.
(a) Both molecules have identical functional groups (alcohol + alkene).

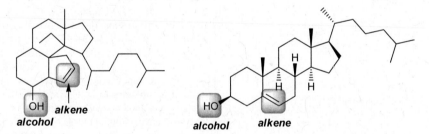

The structure on the left exhibits two six-membered rings and two five-membered rings, while the structure on the right has three six-membered rings and only one five-membered ring.
The incorrect structure suffers from having a five-membered ring in place of the six-membered ring. The long alkane group is apparently located in the wrong position on the five-membered ring of the incorrect structure.

(b) Both structures contain an alkene group, an aromatic ring, an amide group, and two ether functional groups. But the incorrect structure has a third ether functional group (in the eight-membered ring), while the correct structure has an alcohol functional group. The incorrect structure has an eight-membered ring, while the correct structure has a five-membered ring. The two carbon atoms and oxygen atom in the ring of the incorrect structure are not part of the ring for the correct structure.

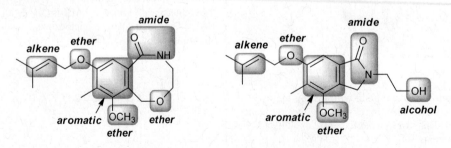

2.73.

(a) This compound contains the following functional groups:

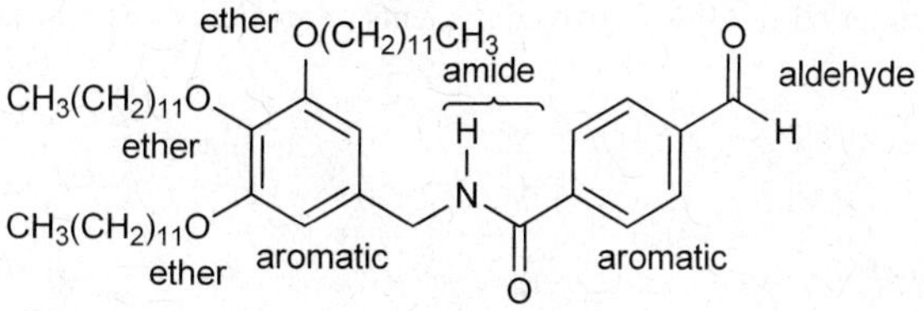

(b) The two bonds on the right (indicated below) have restricted rotation.

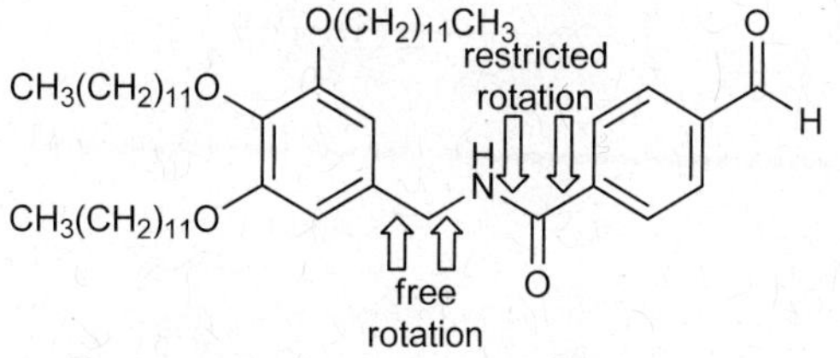

In each of the resonance structures shown below, one of these bonds can be drawn as a double bond, indicating their partial double-bond character. Each bond is thus a hybrid between a single and double bond; the partial double bond character results in partially restricted rotation around these bonds. As a result of the sp^3 hybridized carbon atom (just to the left of the nitrogen) atom, the two bonds on the left (shown above) have only single-bond character and can freely rotate.

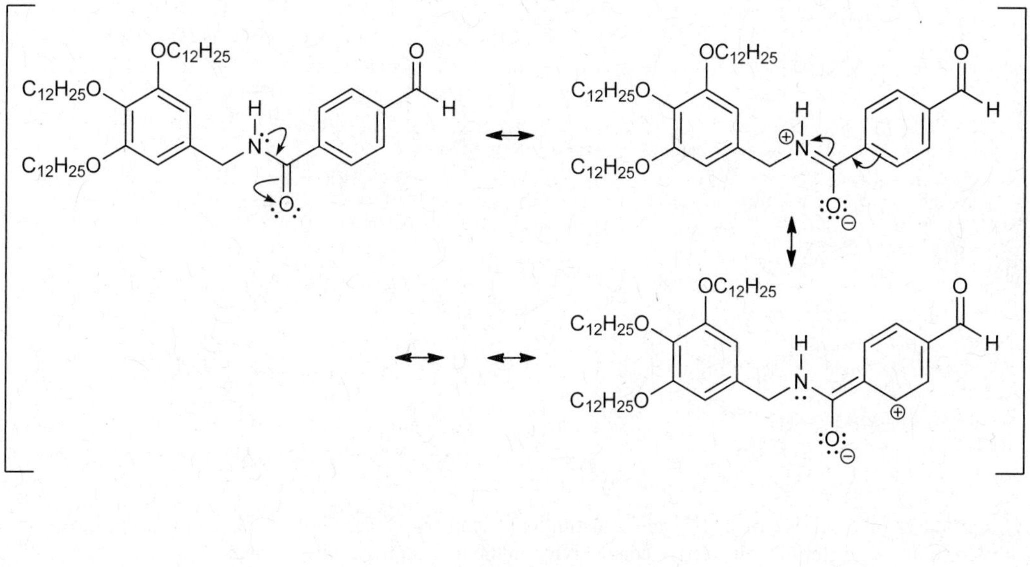

2.74.

(a) For each of the four reactions (i–iv), the product should have two imine groups, resulting from the reaction between a compound with two amino groups (**B** or **C**) with two equivalents of an aldehyde (**A** or **D**).

Product of reaction i

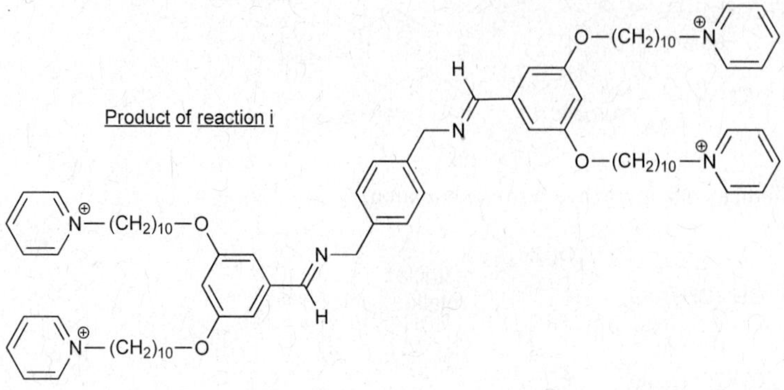

Product of reaction ii

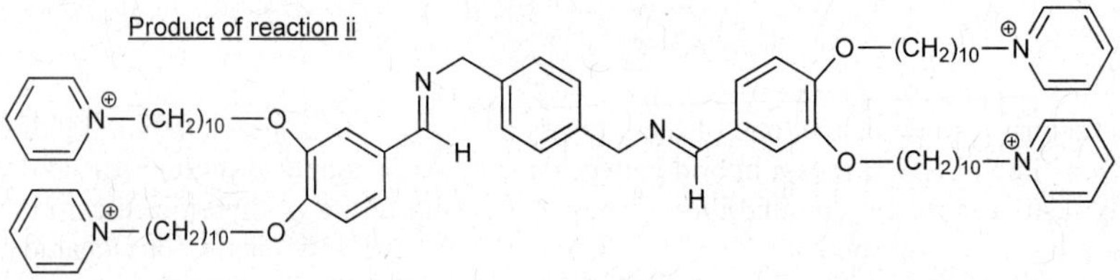

Product of reaction iii

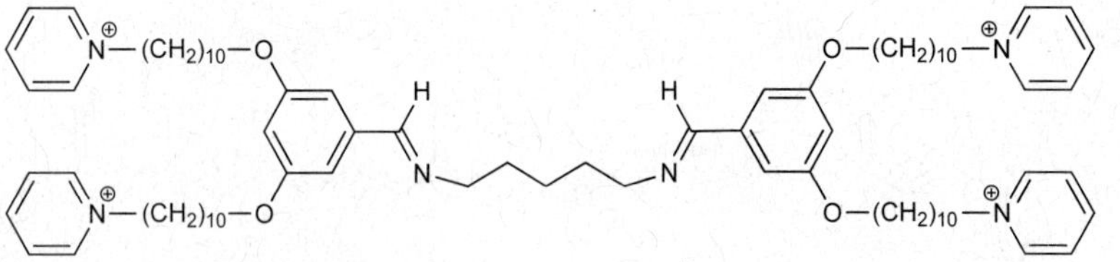

Product of reaction iv

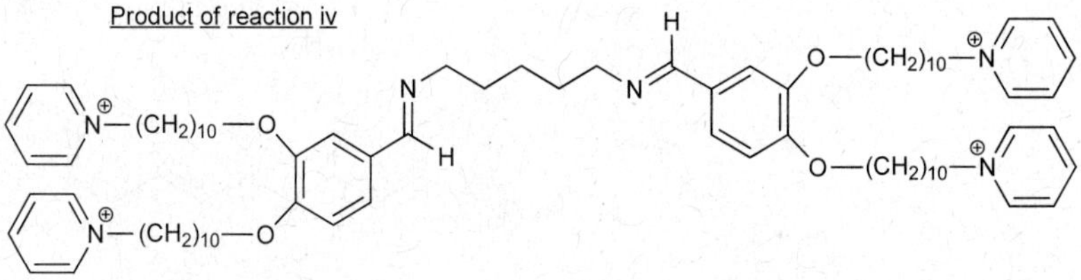

(b) The products of reactions iii and iv are constitutional isomers of each other. These products have the same molecular formula, but differ in their relative connectivity on the two central aromatic rings.

2.75.

(a) Each of the four amides can be represented as a resonance hybrid (one example shown below). The charge-separated resonance structure indicates that there is a δ+ on the amide nitrogen, which thus pulls the electrons in the N-H bond closer to the nitrogen atom, leaving the hydrogen atom with a greater δ+. This resonance effect is not present in the N-H bond of the amines. Thus, the δ+ on an amide H is greater than that on an amine H, leading to a stronger hydrogen bond.

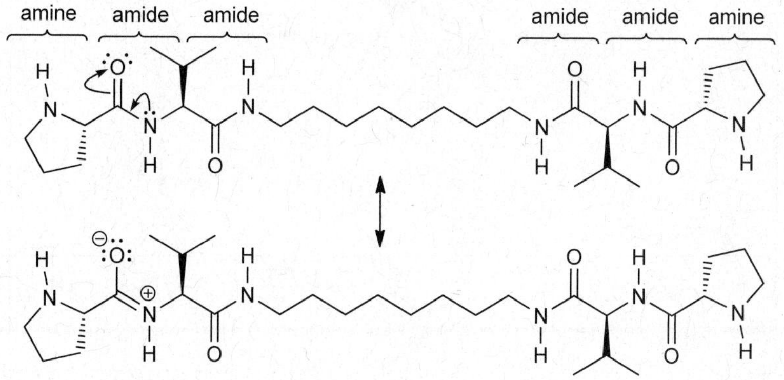

(b) The following intermolecular hydrogen bonds are formed during self-assembly:

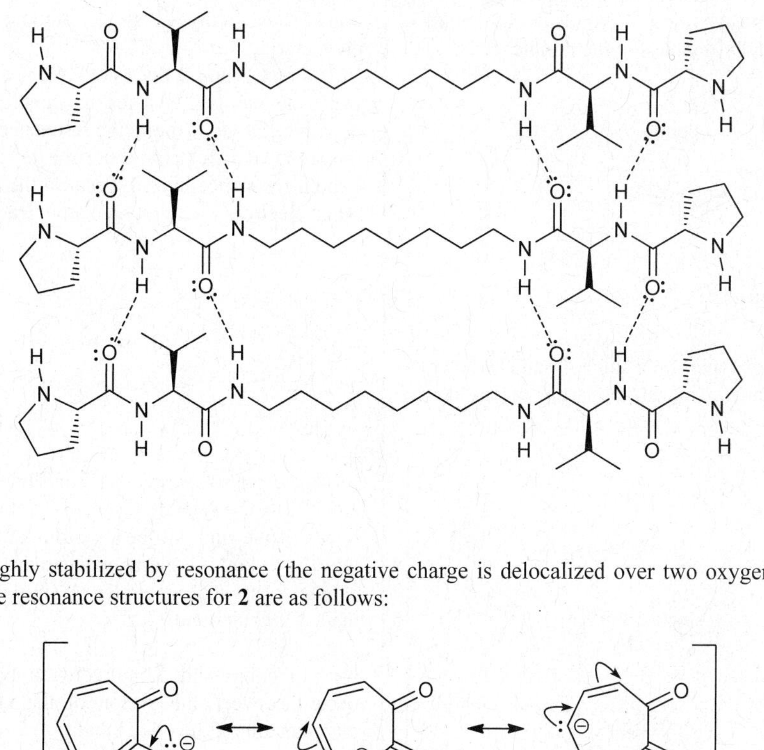

2.76.

(a) Anion **2** is highly stabilized by resonance (the negative charge is delocalized over two oxygen atoms and three carbon atoms). The resonance structures for **2** are as follows:

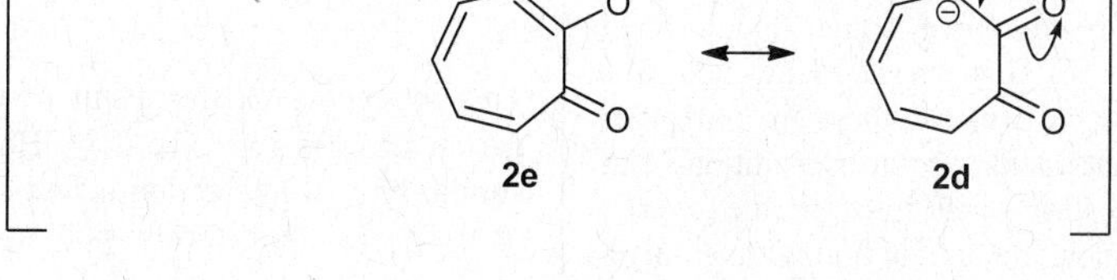

Cation **3** is highly stabilized by resonance (the positive charge is delocalized over two oxygen atoms and four carbon atoms). The resonance structures of **3** are as follows:

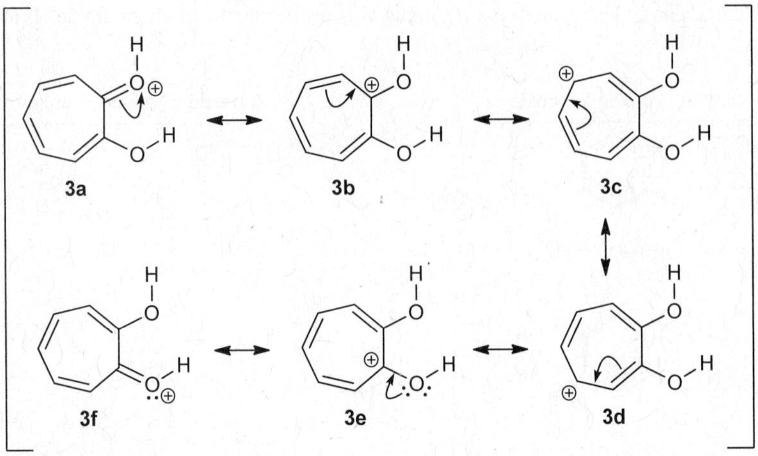

(b) Double bonds are shorter in length than single bonds (see Table 1.2). As such, the C-C bonds in compound **1** will alternate in length (double, single, double, etc.):

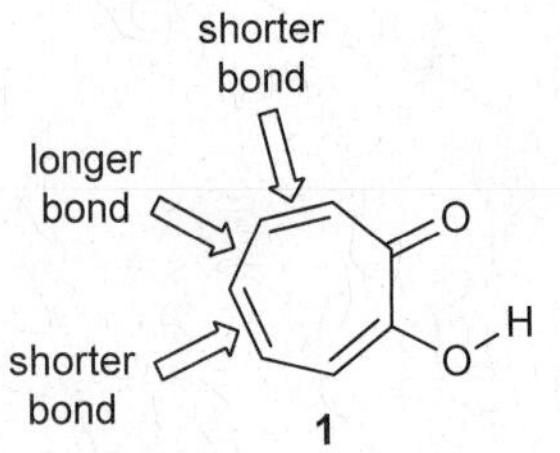

The double bonds do have some single-bond character as a result of resonance, as can be seen in resonance structure **1c**:

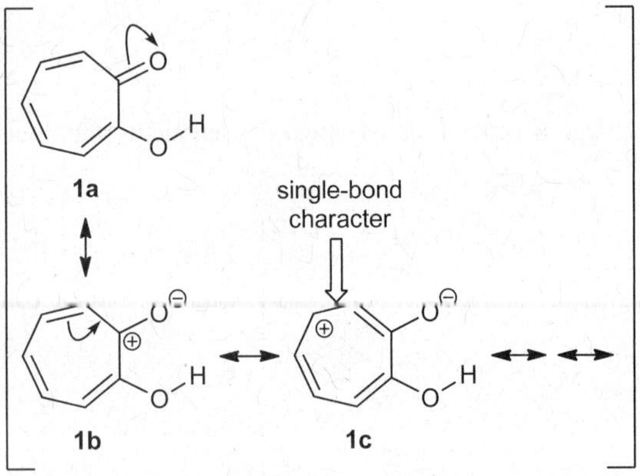

Similarly, the single bonds have some double-bond character, also because of resonance. However, this effect is relatively small, because there is only one resonance structure that lacks charge separation, and therefore, it is the greatest contributor to the overall resonance hybrid. As such, the double bonds have only a small amount of single-bond character, and the single bonds have only a small amount of double-bond character.

In contrast, anion **2** does not have a resonance structure that lacks charges. All resonance structures of **2** bear a negative charge. Among the resonance structures, two of them (**2a** and **2e**) contribute the most character to the overall resonance hybrid, because the negative charge is on an electronegative oxygen atom (rather than carbon).

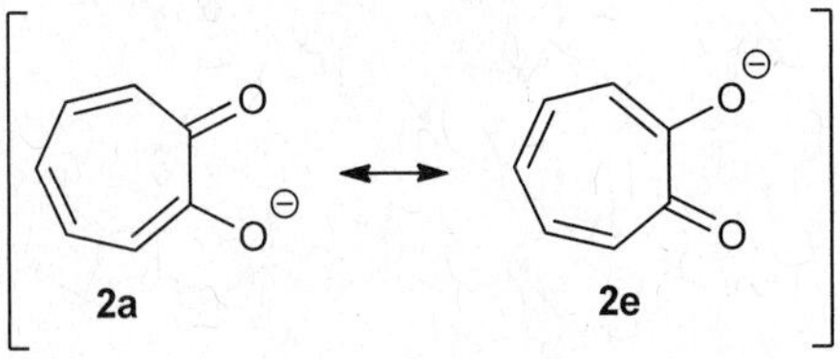

In fact, these two resonance contributors will contribute equally to the overall resonance hybrid. As such, the bonds of the ring will be very similar in length, because they have both single-bond character and double-bond character in equal amounts. A similar argument can be made for compound **3**.

(c) In compound **1**, a hydrogen bonding interaction occurs between the proton of the OH group and the oxygen atom of the C=O bond:

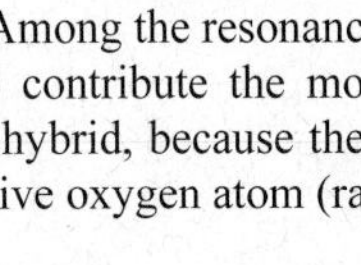

This interaction is the result of the attraction between partial charges ($\delta+$ and $\delta-$). However, in cation **3**, a similar type of interaction is less effective because the O of the C=O bond is now poor in electron density, and therefore less capable of forming a hydrogen bonding interaction, as can be seen in resonance structure **3a**.

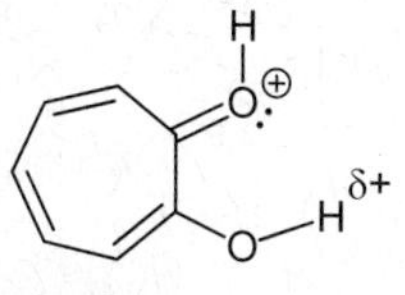

3a

The other oxygen atom is also ineffective at forming an intramolecular hydrogen bond because it too is poor in electron density, as can be seen in resonance structure **3f**:

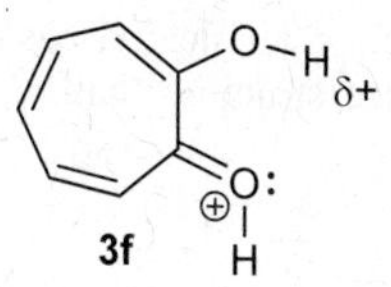

3f

2.77.

(a) The positive charge in basic green 4 is resonance-stabilized (delocalized) over twelve positions (two nitrogen atoms and ten carbon atoms), as seen in the following resonance structures.

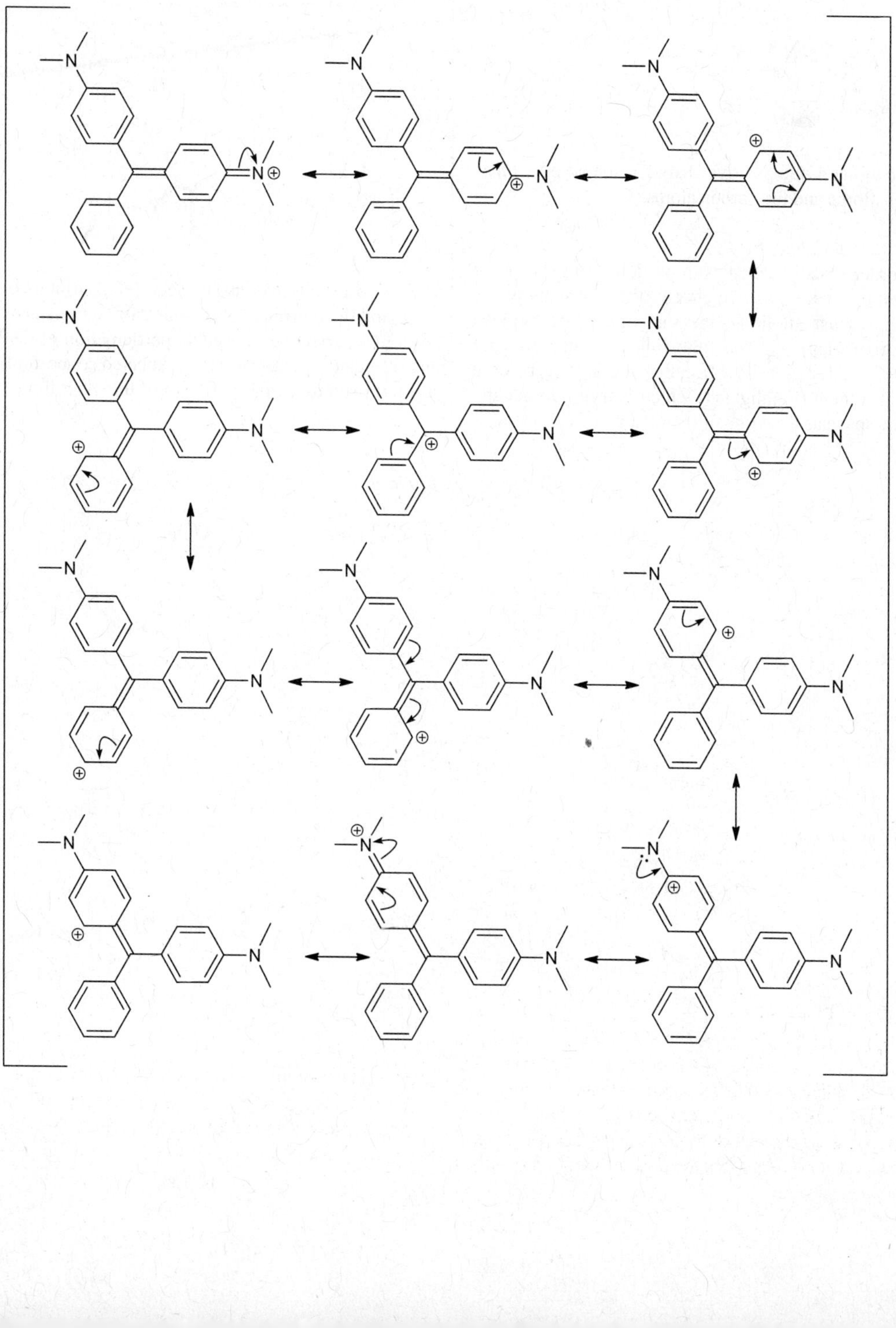

68 **CHAPTER 2**

(b) The positive charge in basic violet 4 is expected to be more stabilized than the positive charge in basic green 4, because the former is delocalized over thirteen positions, rather than twelve. Specifically, basic violet 4 has an additional resonance structure that basic green 4 lacks, shown below:

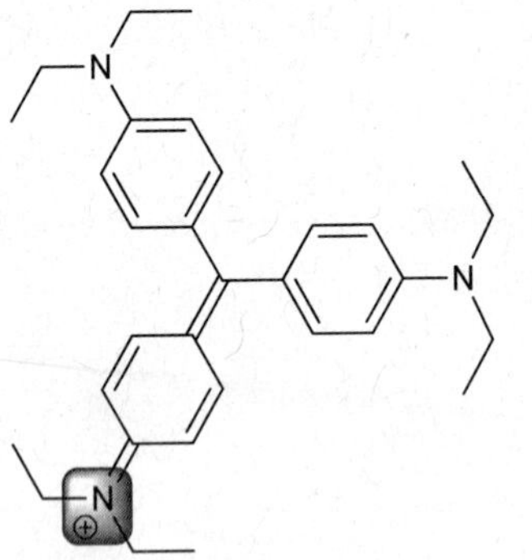

In basic violet 4, the positive charge is spread over *three* nitrogen atoms and ten carbon atoms.

2.78.
(a) In order for all four rings to participate in resonance stabilization of the positive charge, the *p* orbitals in the four rings must all lie in the same plane (to achieve effective overlap). In the following drawing, the four rings are labeled A-D. Notice that the D ring bears a large substituent (highlighted) which is trying to occupy the same space as a portion of the C ring:

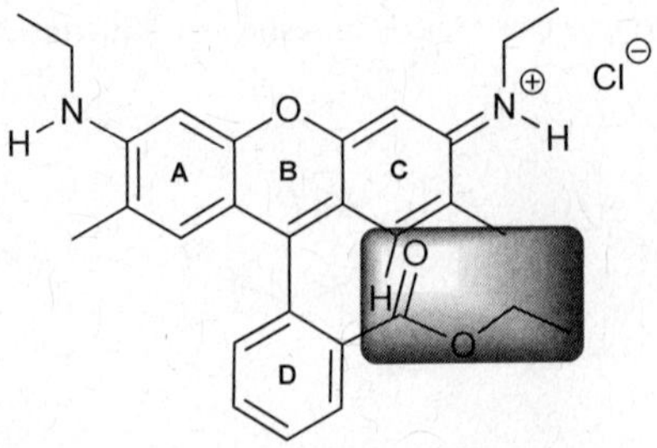

This type of interaction, called a steric interaction, forces the D ring to twist out of plane with respect to the other three rings, like this:

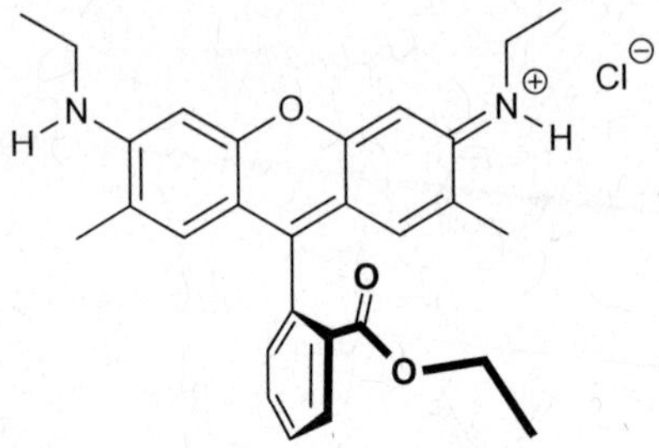

In this way, the overlap between the *p* orbitals of the D ring and the *p* orbitals of the other three rings is expected to be less effective. As such, participation of the D ring in resonance stabilization is expected to be diminished with respect to the participation of the other three rings.

Chapter 3
Acids and Bases

Review of Concepts

Fill in the blanks below. To verify that your answers are correct, look in your textbook at the end of Chapter 3. Each of the sentences below appears verbatim in the section entitled *Review of Concepts and Vocabulary*.

- A **Brønsted-Lowry acid** is a proton _____, while a **Brønsted-Lowry base** is a proton _____.

- The mechanism of **proton transfer** always involves at least _____ curved arrows.
- A strong acid has a _____ pK_a, while a weak acid has a _____ pK_a.
- There are four factors to consider when comparing the _____ of conjugate bases.
- The equilibrium of an acid-base reaction always favors the more _____ negative charge.
- A **Lewis acid** is an electron _____, while a **Lewis base** is an electron _____.

Review of Skills

Fill in the blanks and empty boxes below. To verify that your answers are correct, look in your textbook at the end of Chapter 3. The answers appear in the section entitled *SkillBuilder Review*.

SkillBuilder 3.1   Drawing the Mechanism of a Proton Transfer

DRAW THE CURVED ARROWS FOR THE FOLLOWING ACID-BASE REACTION

$$H\text{—}\overset{..}{\underset{..}{O}}\text{—}H \ + \ CH_3\overset{..}{\underset{..}{O}}{:}^{\ominus} \ \rightleftharpoons \ HO{:}^{\ominus} \ + \ CH_3\overset{..}{O}H$$

SkillBuilder 3.2   Using pK_a Values to Compare Acids

CIRCLE THE COMPOUND BELOW THAT IS MORE ACIDIC

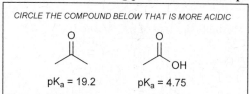

pK_a = 19.2   pK_a = 4.75

SkillBuilder 3.3   Using pK_a Values to Compare Basicity

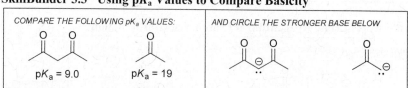

COMPARE THE FOLLOWING pK_a VALUES:

pK_a = 9.0   pK_a = 19

AND CIRCLE THE STRONGER BASE BELOW

SkillBuilder 3.4   Using pK_a Values to Predict the Position of Equilibrium

CIRCLE THE SIDE OF THE EQUILIBRIUM THAT IS FAVORED:

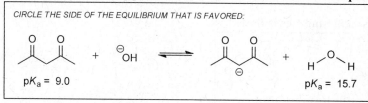

pK_a = 9.0   pK_a = 15.7

SkillBuilder 3.5   Assessing Relative Stability. Factor #1: Atom

COMPARE THE TWO HIGHLIGHTED PROTONS, AND CIRCLE THE ONE THAT IS MORE ACIDIC. USE THE EXTRA SPACE TO DRAW THE TWO POSSIBLE CONJUGATE BASES.

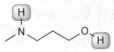

SkillBuilder 3.6   Assessing Relative Stability. Factor #2: Resonance

COMPARE THE TWO HIGHLIGHTED PROTONS, AND CIRCLE THE ONE THAT IS MORE ACIDIC. USE THE EXTRA SPACE TO DRAW THE TWO POSSIBLE CONJUGATE BASES.

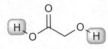

SkillBuilder 3.7   Assessing Relative Stability. Factor #3: Induction

COMPARE THE TWO HIGHLIGHTED PROTONS, AND CIRCLE THE ONE THAT IS MORE ACIDIC. USE THE EXTRA SPACE TO DRAW THE TWO POSSIBLE CONJUGATE BASES.

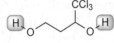

SkillBuilder 3.8   Assessing Relative Stability. Factor #4: Orbital

COMPARE THE TWO HIGHLIGHTED PROTONS, AND CIRCLE THE ONE THAT IS MORE ACIDIC. USE THE EXTRA SPACE TO DRAW THE TWO POSSIBLE CONJUGATE BASES.

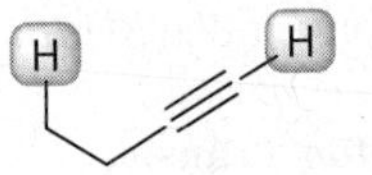

SkillBuilder 3.9   Assessing Relative Stability. Using All Four Factors

COMPARE THE TWO HIGHLIGHTED PROTONS, AND CIRCLE THE ONE THAT IS MORE ACIDIC. USE THE EXTRA SPACE TO DRAW THE TWO POSSIBLE CONJUGATE BASES.

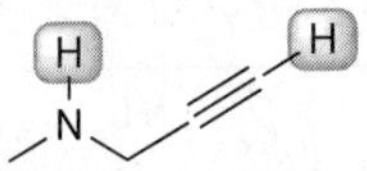

SkillBuilder 3.10   Predicting the Position of Equilibrium Without the Use of pK_a Values

CIRCLE THE SIDE OF THE EQUILIBRIUM THAT IS FAVORED:

SkillBuilder 3.11   Choosing the Appropriate Reagent for a Proton Transfer Reaction

DETERMINE WHETHER WATER IS A SUITABLE PROTON SOURCE TO PROTONATE THE ACETATE ION, AS SHOWN BELOW:

SkillBuilder 3.12   Identifying Lewis Acids and Lewis Bases

IDENTIFY THE LEWIS ACID AND THE LEWIS BASE IN THE FOLLOWING REACTION:

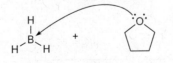

Common Mistakes to Avoid

When drawing the mechanism of a proton transfer, two curved arrows are required. The first curved arrow shows the base attacking the proton, and the second curved arrow shows the bond to H being broken.

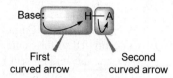

First
curved arrow

Second
curved arrow

It is a common mistake to draw only the first curved arrow and not the second, so make sure to draw both curved arrows. When drawing the second curved arrow, make sure that the tail is placed on the middle of the bond to the H, as shown:

Correct

Base: H—A

NOT correct

Base: HA

When the acid is H_3O^+, you must draw at least one of the O–H bonds in order to draw the second curved arrow properly.

Correct

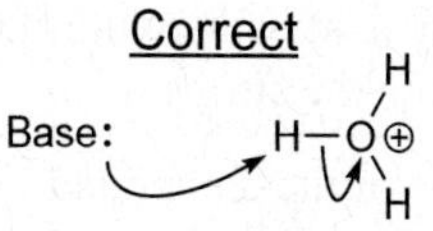

Base: H—O⊕

NOT correct

Base: H_3O^+

This is also the case when the acid is H_2SO_4:

Correct

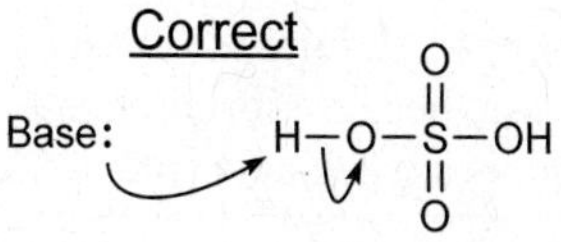

Base: H—O—S—OH

NOT correct

Base: H_2SO_4

Useful reagents

In Chapter 3, we explored the behavior of acids and bases. Throughout the remainder of the textbook, many acids and bases will be frequently encountered. It would be wise to become familiar with the following reagents (and their uses), as they will appear many times in the upcoming chapters:

Structure	Name	Use
$H-O-\overset{\overset{O}{\|\|}}{\underset{\overset{\|\|}{O}}{S}}-OH$	Sulfuric acid	A very strong acid. Commonly used as a source of protons. Concentrated sulfuric acid is an aqueous mixture of H_2SO_4 and H_2O, and the acid present in solution is actually H_3O^+, because of the leveling effect. That is, H_3O^+ is a weaker acid than H_2SO_4, so the protons are transferred from H_2SO_4 to water, giving a high concentration of H_3O^+.
$H-Cl$	Hydrochloric acid	A very strong acid. Similar in function to H_2SO_4. In an aqueous solution of HCl, the acid that is present is H_3O^+, because of the leveling effect, as described above for H_2SO_4.
(acetic acid structure)	Acetic acid	A weak acid. Mild source of protons.
$H \overset{O}{\frown} H$	Water	A weak acid and a weak base. It can function as either, depending on the conditions. When treated with a strong base, water will function as an acid (a source of protons). When treated with a strong acid, water will function as a base and remove a proton from the strong acid.
$R \overset{O}{\frown} H$	An alcohol	R represents the rest of the compound. Alcohols are compounds that possess an O-H group, and will be the subject of Chapter 13. Alcohols can function very much like water (either as weak acids or as weak bases).
$H \overset{\overset{H}{\|}}{\underset{H}{N}} H$	Ammonia	A fairly strong base, despite the absence of a negative charge. It is a strong base, because its conjugate acid (NH_4^+), called an ammonium ion, is a weak acid ($pK_a = 9.2$).
CH_3CH_2ONa	Sodium ethoxide	The ethoxide ion ($CH_3CH_2O^-$) is a strong base, and Na^+ is the counterion. Other alkoxide ions (RO^-), such as methoxide (CH_3O^-), are also strong bases.
$NaNH_2$	Sodium amide	H_2N^- is a very strong base, and Na^+ is the counterion.
$H-\overset{\overset{H}{\|}}{\underset{\overset{\|}{H}}{C}}-\overset{\overset{H}{\|}}{\underset{\overset{\|}{H}}{C}}-\overset{\overset{H}{\|}}{\underset{\overset{\|}{H}}{C}}-\overset{\overset{H}{\|}}{\underset{\overset{\|}{H}}{C}}:^{\ominus} \overset{\oplus}{Li}$	Butyllithium	An extremely strong base. This is one of the strongest bases that you will encounter.

Solutions

3.1.

(a) Phenol (C_6H_5OH) loses a proton and is therefore functioning as an acid. Hydroxide (HO^-) functions as the base that removes the proton. Two curved arrows must be drawn. The first curved arrow shows a lone pair of the base attacking the proton, and the second curved arrow comes from the O–H bond (being broken) and goes to the oxygen atom, as shown:

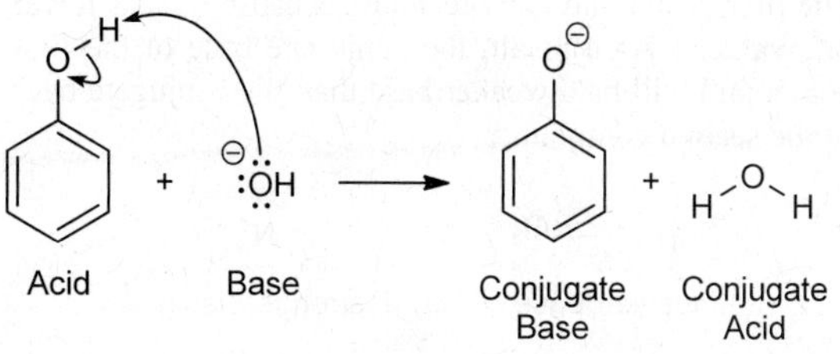

(b) H_3O^+ loses a proton and is therefore functioning as an acid. The ketone functions as the base that removes the proton. Two curved arrows must be drawn. The first curved arrow shows a lone pair of the base attacking the proton, and the second curved arrow comes from the O–H bond (being broken) and goes to the oxygen atom, as shown:

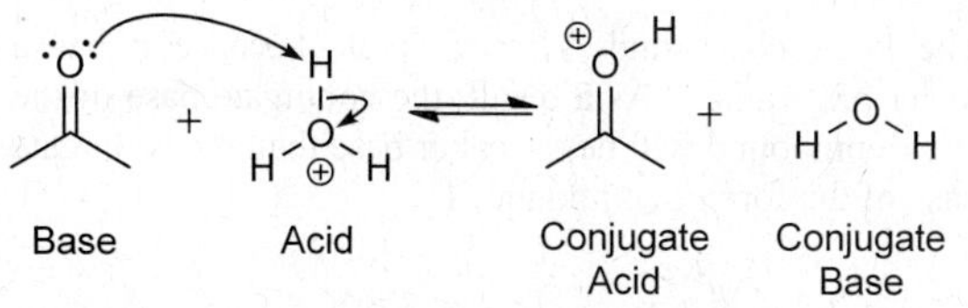

(c) The ketone loses a proton and is therefore functioning as an acid. The other reagent functions as the base that removes the proton. Two curved arrows must be drawn. The first curved arrow shows a lone pair of the base attacking the proton, and the second curved arrow comes from the C–H bond (being broken) and goes to the carbon atom, as shown:

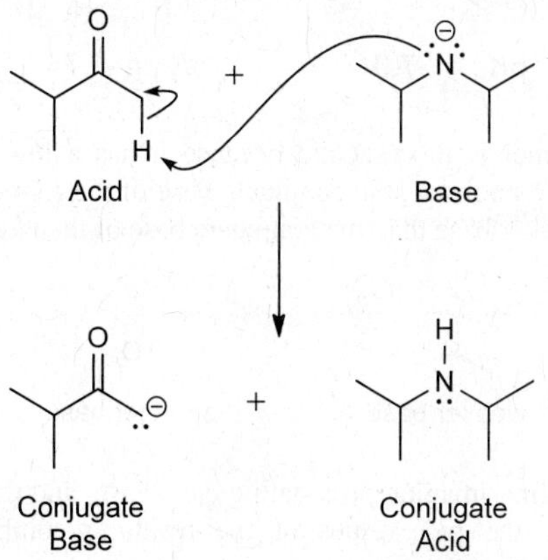

(d) Benzoic acid ($C_6H_5CO_2H$) loses a proton and is therefore functioning as an acid. Hydroxide (HO^-) functions as the base that removes the proton. Two curved arrows must be drawn. The first curved arrow

shows a lone pair of the base attacking the proton, and the second curved arrow comes from the O–H bond (being broken) and goes to the oxygen atom, as shown:

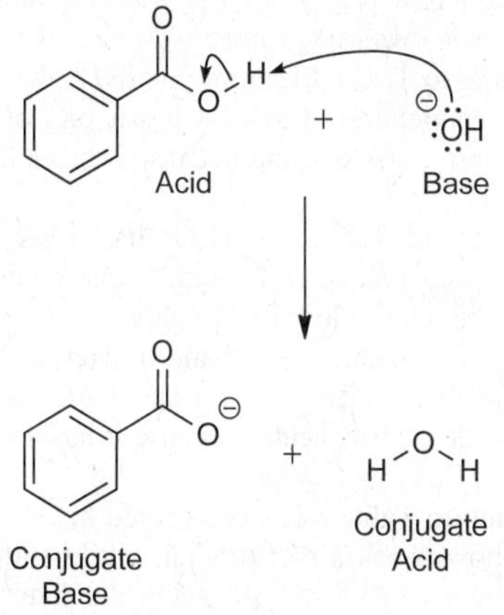

3.2.

(a) There is only one curved arrow, and it is going in the wrong direction. The tail has been placed on the hydrogen atom, and this is incorrect. Curved arrows do not show the motion of atoms, but the motion of electrons. The tail of this curved arrow should be on the lone pair of the nitrogen atom, and the head of the curved arrow should be on the proton. In addition, a second curved arrow is also required. It should look like this:

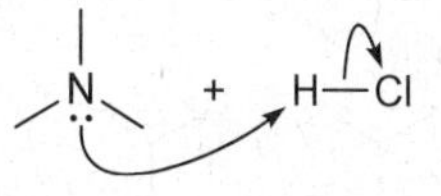

(b) The first arrow (from the lone pair of nitrogen to the proton) is correct, but the second curved arrow is not correct. Specifically, the tail is placed on the proton, and instead should be placed on the bond between the proton and the oxygen atom. This bond must be drawn in order to properly place the second curved arrow:

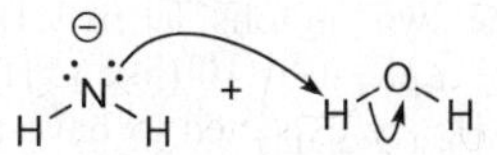

(c) The second curved arrow is missing:

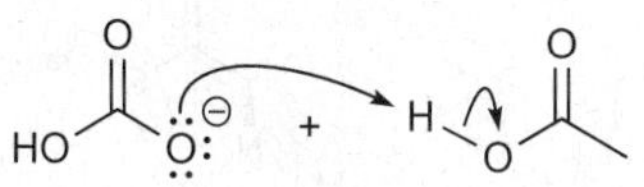

3.3. Two curved arrows must be drawn. The first curved arrow shows a lone pair of the base attacking the proton, and the second curved arrow comes from the O–H bond (being broken) and goes to the oxygen atom, as shown:

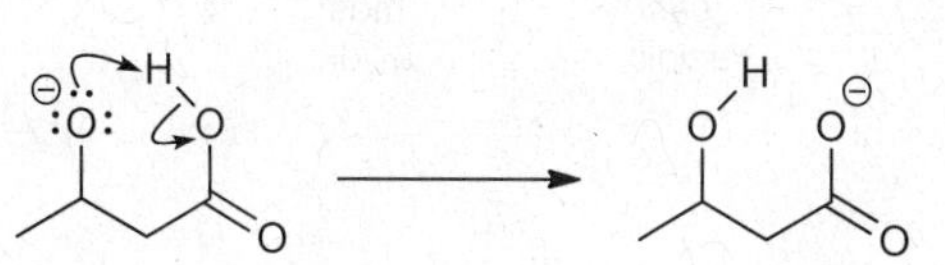

3.4.

(a) According to Table 3.1, phenol (C_6H_5OH) has a pK_a of 9.9, while water has a pK_a of 15.7. The former is more acidic because it has a lower pK_a value.

(b) According to Table 3.1, $(CH_3)_3COH$ has a pK_a of 18, while water has a pK_a of 15.7. Water is more acidic because it has a lower pK_a value.

(c) According to Table 3.1, ammonia (NH_3) has a pK_a of 38, while acetylene ($H–C\equiv C–H$) has a pK_a of 25. As such, the latter is more acidic because it has a lower pK_a value.

(d) According to Table 3.1, H_3O^+ has a pK_a of -1.7, while HCl has a pK_a of -7. As such, the latter is more acidic because it has a lower pK_a value.

(e) According to Table 3.1, ethane (C_2H_6) has a pK_a of 50, while acetylene ($H–C\equiv C–H$) has a pK_a of 25. As such, the latter is more acidic because it has a lower pK_a value.

(f) According to Table 3.1, a protonated ketone (the first structure shown) has a pK_a of -7.3, while sulfuric acid (H_2SO_4) has a pK_a of -9. As such, the latter is more acidic because it has a lower pK_a value.

3.5. According to Table 3.1, the following two highlighted protons are expected to be the two most acidic protons in the compound. The proton connected to the oxygen atom is expected to have a pK_a value near 16 (similar to CH_3CH_2OH), while the proton connected to the nitrogen atom is expected to have a pK_a value near 38 (similar to NH_3).

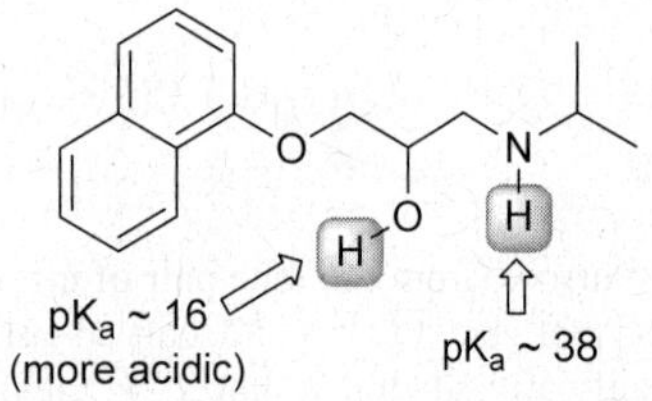

$pK_a \sim 16$
(more acidic)

$pK_a \sim 38$

3.6. According to Table 3.1, the proton of the carboxylic acid group (H_a below) is expected to be the most acidic ($pK_a \sim 5$). The two protons labeled H_b and H_c are expected to have a pK_a near 10 (like C_6H_5OH), and the protons labeled H_d are expected to have a pK_a near 38 (like NH_3). The order of acidity is shown below:

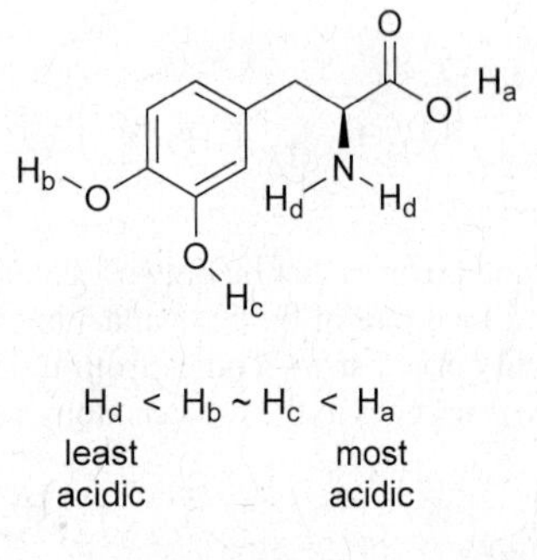

$$H_d < H_b \sim H_c < H_a$$
least   most
acidic   acidic

3.7.

(a) We first imagine protonating each base, and then we compare the pK_a values of the resulting compounds (using Table 3.1):

$H–C\equiv C–H$
$pK_a = 25$

$pK_a = 38$

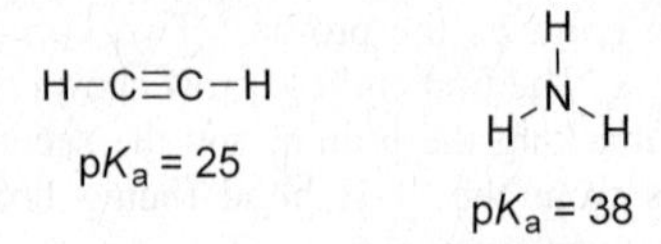

The first compound is more acidic because it has a lower pK_a value. As a result, the conjugate base of the first compound will be a weaker base than the conjugate base of the second compound:

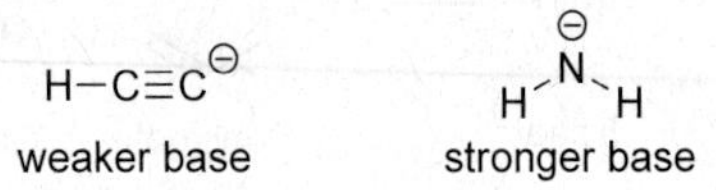

weaker base   stronger base

(b) We first imagine protonating each base, and then we compare the pK_a values of the resulting compounds (using Table 3.1):

$pK_a = 18$ $pK_a = 16$

The latter compound is more acidic because it has a lower pK_a value. As a result, the conjugate base of the latter compound will be a weaker base than the conjugate base of the former compound:

stronger base   weaker base

(c) We first imagine protonating each base, and then we compare the pK_a values of the resulting ions (using Table 3.1):

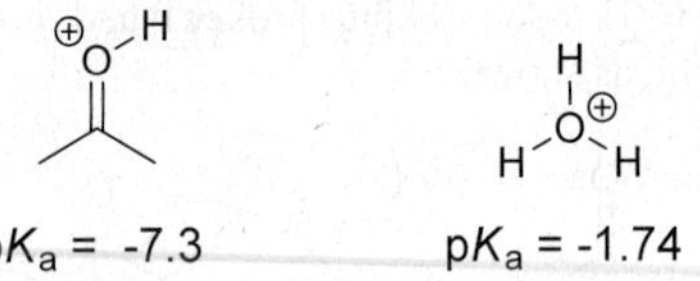

$pK_a = -7.3$ $pK_a = -1.74$

The former is more acidic because it has a lower pK_a value. As a result, the conjugate base of the former will be a weaker base than the conjugate base of the latter:

weaker base   stronger base

(d) We first imagine protonating each base, and then we compare the pK_a values of the resulting compounds (using Table 3.1):

$pK_a = 15.7$ $pK_a = 16$

The former compound (water) is more acidic because it has a lower pK_a value. As a result, the conjugate base of the latter compound will be a stronger base than the conjugate base of water:

H–O$^{\ominus}$ weaker base

$^{\ominus}$O–CH$_2$CH$_3$ stronger base

(e) We first imagine protonating each base, and then we compare the pK_a values of the resulting compounds (using Table 3.1):

H–C–C–H (with H's) $pK_a = 50$

H–C≡C–H $pK_a = 25$

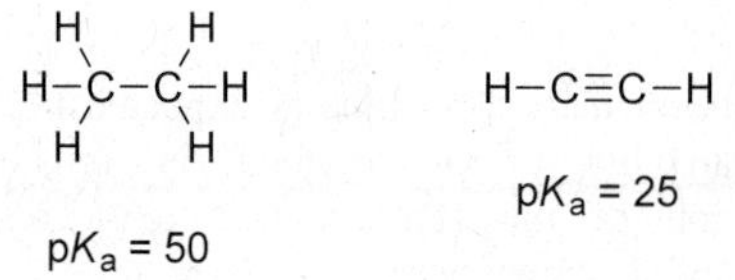

The latter compound is more acidic because it has a lower pK_a value. As a result, the conjugate base of the latter compound will be a weaker base than the conjugate base of the former compound:

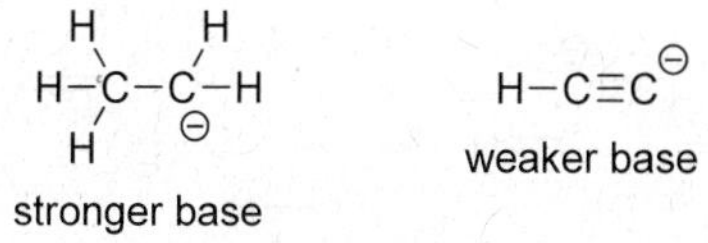

H–C–C–H$^{\ominus}$ stronger base

H–C≡C$^{\ominus}$ weaker base

(f) We first imagine protonating each base, and then we compare the pK_a values of the resulting compounds (using Table 3.1). HCl has a pK_a of -7, while H$_2$O has a pK_a of 15.7. Since HCl has a lower pK_a value, it is more acidic than water (significantly). As a result, the conjugate base of HCl will be a much weaker base than the conjugate base of H$_2$O.

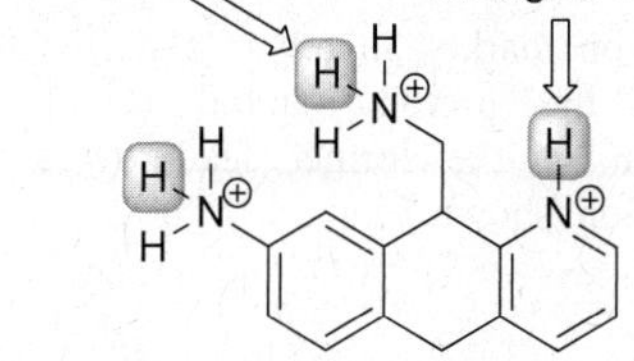

Cl$^{\ominus}$ weaker base

HO$^{\ominus}$ stronger base

3.8. The problem statement provides data that support the following conclusions regarding relative acidity:

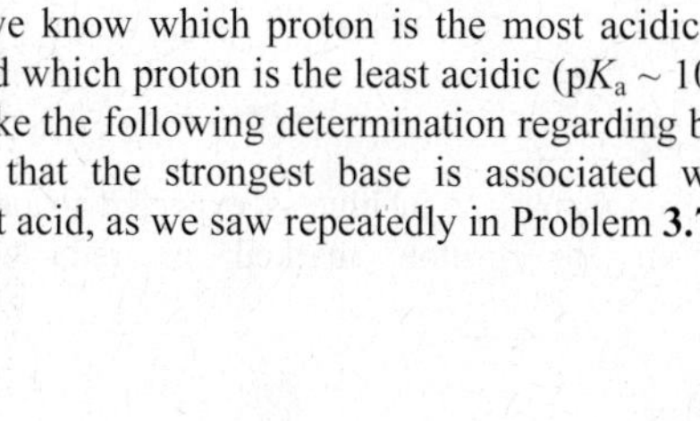

Once we know which proton is the most acidic ($pK_a \sim$ 3.4) and which proton is the least acidic ($pK_a \sim$ 10.5), we can make the following determination regarding basicity. Notice that the strongest base is associated with the weakest acid, as we saw repeatedly in Problem **3.7**.

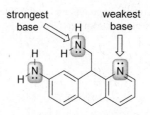

3.9.
(a) The pK_a values indicate that a proton is more acidic when it is connected to a positively charged oxygen atom ($pK_a \sim$ -2.2) than when it is connected to a positively charged nitrogen atom ($pK_a \sim$ 10.5). The weaker acid (CH$_3$NH$_3^+$) gives the stronger base upon deprotonation, and therefore, a lone pair on a nitrogen atom will be more basic than a lone pair on an oxygen atom.
(b) The difference in the pK_a values (-2.2 and 10.5) is approximately 13 pK_a units, so the lone pair on the nitrogen atom is 13 orders of magnitude more basic than the lone pair on the oxygen atom.

3.10.
(a) We begin by identifying the acid on each side of the equilibrium. In this case, the acid on the left side is ethanol (CH$_3$CH$_2$OH) and the acid on the right side is water (H$_2$O). We compare their pK_a values (Table 3.1), and we find that ethanol (pK_a = 16) is less acidic than water (pK_a = 15.7). The equilibrium will favor the weaker acid (ethanol).
(b) Identify the acid on each side of the equilibrium. In this case, the acid on the left side is phenol (C$_6$H$_5$OH) and the acid on the right side is water (H$_2$O). We compare their pK_a values (Table 3.1), and we find that water (pK_a = 15.7) is less acidic than phenol (pK_a = 9.9). The equilibrium will favor the weaker acid (water).
(c) Identify the acid on each side of the equilibrium. In this case, the acid on the left side is HCl and the acid on the right side is H$_3$O$^+$. We compare their pK_a values (Table 3.1), and we find that H$_3$O$^+$ (pK_a = -1.74) is less acidic than HCl (pK_a = -7). The equilibrium will favor the weaker acid (H$_3$O$^+$).
(d) Identify the acid on each side of the equilibrium. In this case, the acid on the left side is acetylene (H–C≡C–H) and the acid on the right side is ammonia (NH$_3$). We compare their pK_a values (Table 3.1), and we find that ammonia (pK_a = 38) is less acidic than acetylene (pK_a = 25). The equilibrium will favor the weaker acid (ammonia).

3.11. The equilibrium does not favor deprotonation of acetylene by hydroxide, because water is more acidic than acetylene. The equilibrium will favor the weaker acid (acetylene). A suitable base would be one whose conjugate acid is less acidic than acetylene. For example, H$_2$N$^-$ would be a suitable base, because ammonia (NH$_3$) is less acidic than acetylene.

3.12. At a pH of 7.4, the carboxylic acid group (RCO$_2$H) will exist primarily as its conjugate base (RCO$_2^-$), called a carboxylate ion. At the same pH, the ammonium group

(RNH$_3$$^+$) will retain its proton, and will primarily exist in the charged form, as shown here:

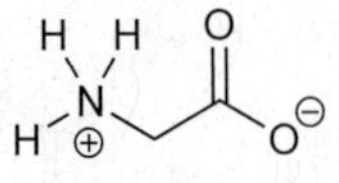

glycine

3.13.
(a)   Carbon and oxygen are in the same row of the periodic table, so we must compare their electronegativity values.   Oxygen is more electronegative than carbon and can better stabilize the negative charge that will be generated upon deprotonation. Therefore, a proton connected to an oxygen atom is expected to be more acidic than a proton connected to a carbon atom:

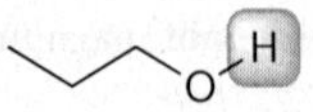

(b) Carbon and nitrogen are in the same row of the periodic table, so we must compare their electronegativity values.   Nitrogen is more electronegative than carbon and can better stabilize the negative charge that will be generated upon deprotonation. Therefore, a proton connected to a nitrogen atom is expected to be more acidic than a proton connected to a carbon atom:

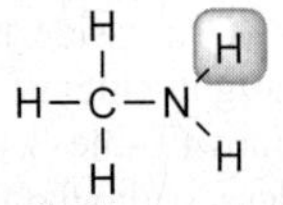

(c) Sulfur and oxygen are in the same column of the periodic table, so we must compare their size. Sulfur is larger than oxygen and can better stabilize the negative charge that will be generated upon deprotonation. Therefore, a proton connected to a sulfur atom is expected to be more acidic than a proton connected to an oxygen atom:

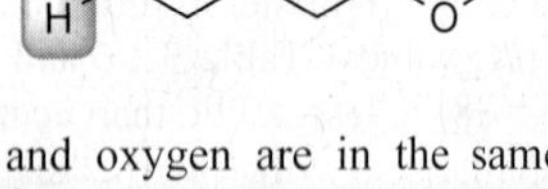

(d) Nitrogen and oxygen are in the same row of the periodic table, so we must compare their electronegativity values.   Oxygen is more electronegative than nitrogen and can better stabilize the negative charge that will be generated upon deprotonation. Therefore, a proton connected to an oxygen atom is expected to be more acidic than a proton connected to a nitrogen atom:

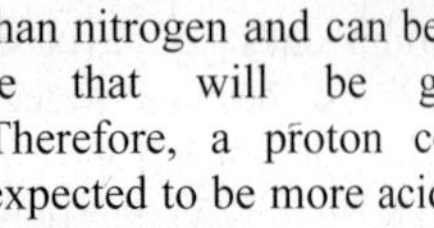

3.14. A proton connected to a sulfur atom will be more acidic than a proton connected to an oxygen atom, which

will be more acidic than a proton connected to a nitrogen atom.   Therefore, the proton on the sulfur atom will definitely be more acidic than the proton on the nitrogen atom.

3.15.
(a) The proton marked in blue is expected to be more acidic than the proton marked in red, because deprotonation of the former leads to a resonance-stabilized anion, shown here:

(b) The proton marked in blue is expected to be more acidic than the proton marked in red, because deprotonation of the former leads to a resonance-stabilized anion, shown here:

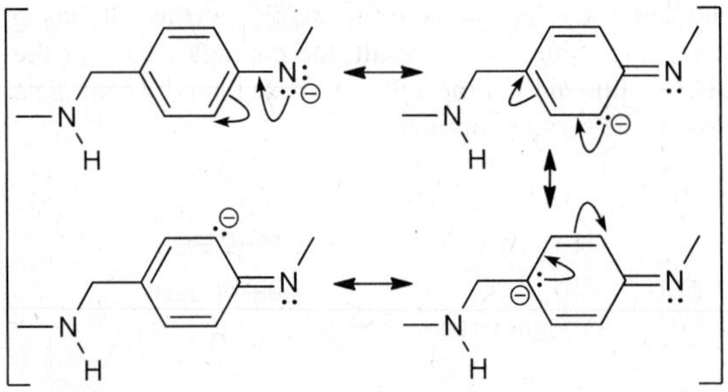

(c) The proton marked in red is expected to be more acidic than the proton marked in blue, because deprotonation of the former leads to a resonance-stabilized anion, shown here:

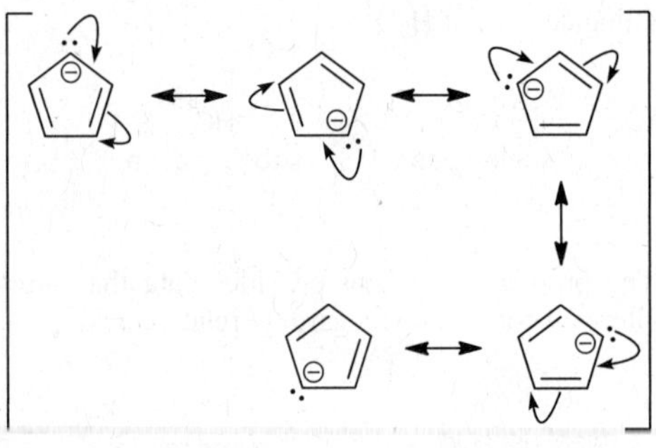

(d) The proton marked in red is expected to be more acidic than the proton marked in blue, because deprotonation of the former leads to a resonance-stabilized anion, shown here:

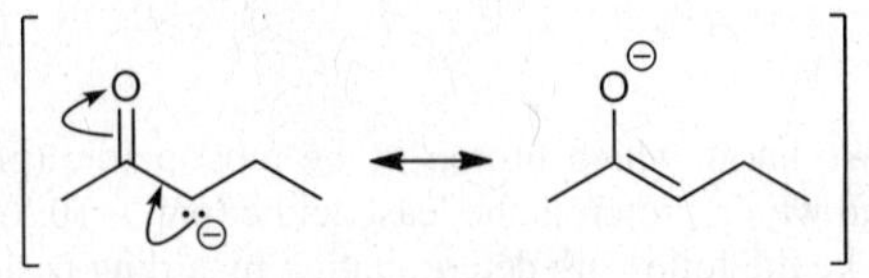

(e) The proton marked in blue is expected to be more acidic than the proton marked in red, because deprotonation of the former leads to a resonance-stabilized anion in which the negative charge is spread over two oxygen atoms.

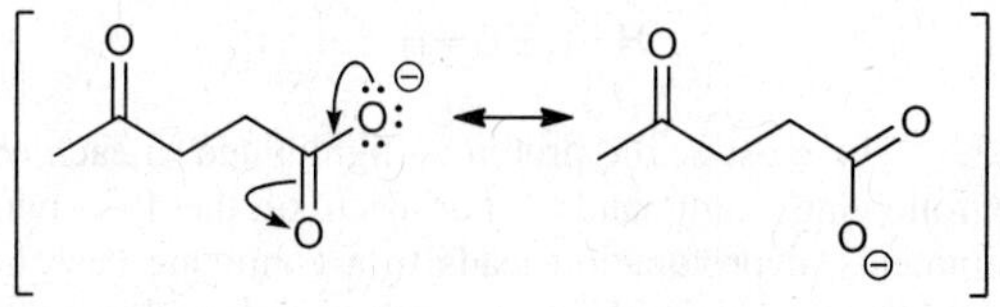

Deprotonation at the location of the red proton also leads to a resonance-stabilized anion, but the negative charge is spread over an oxygen atom and a carbon atom, which is less stable than spreading the charge over two oxygen atoms.

(f) The proton marked in red is expected to be more acidic than the proton marked in blue, because deprotonation of the former leads to a resonance-stabilized anion in which the negative charge is spread over one oxygen atom and three carbon atoms.

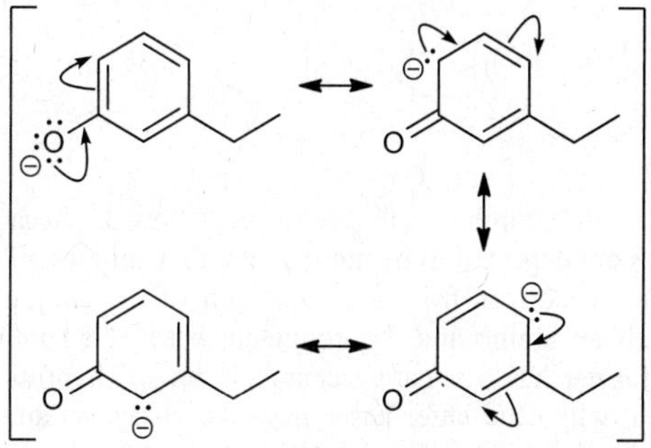

Deprotonation at the location of the blue proton also leads to a resonance-stabilized anion, but the negative charge is spread over four carbon atoms, which is less stable than spreading the charge over one oxygen atom and three carbon atoms.

3.16.

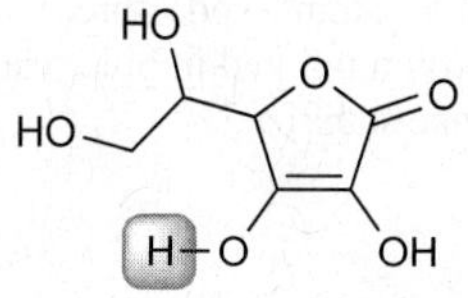

The proton highlighted above is the most acidic proton in the structure, because deprotonation at that location generates a resonance-stabilized anion, in which the negative charge is spread over two oxygen atoms and one carbon atom:

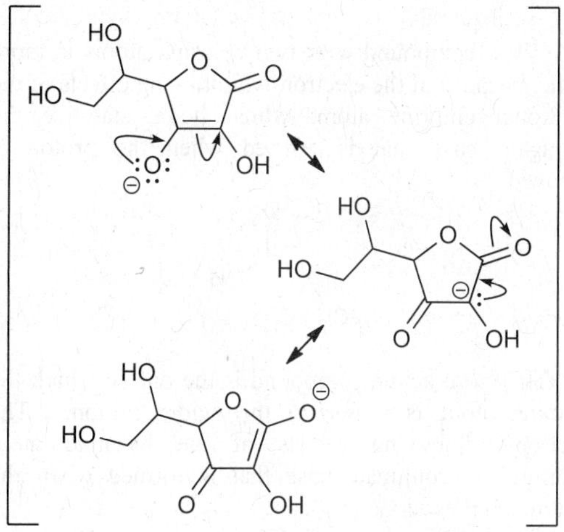

3.17. Deprotonation at the location marked by the red proton leads to a resonance-stabilized anion in which the negative charge is spread over one oxygen atom and three carbon atoms, just as we saw in Problem **3.15f**. Deprotonation at the location marked by the blue proton leads to a resonance-stabilized anion in which the negative charge is spread over two oxygen atoms, just as we saw in Problem **3.15e**. A negative charge will be more stabilized when spread over two oxygen atoms, rather than being spread over one oxygen atom and three carbon atoms (oxygen is more electronegative than carbon).

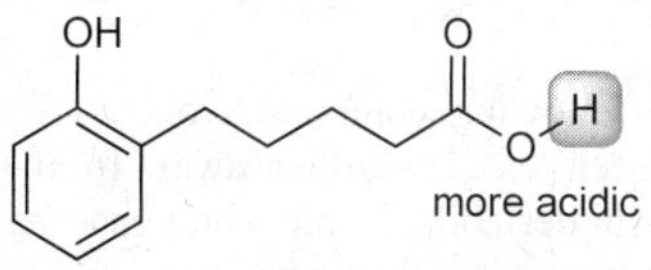

more acidic

3.18.
(a)   The highlighted proton is the most acidic. When this location is deprotonated, the resulting conjugate base is stabilized by the electron-withdrawing effects of the electronegative fluorine atoms:

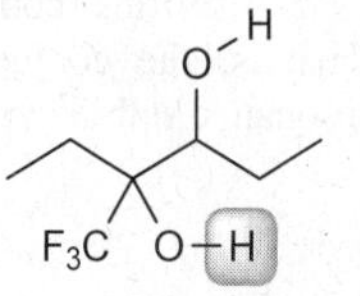

more acidic

(b) The highlighted proton is more acidic. When this location is deprotonated, the resulting conjugate base is stabilized by the electron-withdrawing effects of the electronegative chlorine atoms, which are closer to this proton than the other acidic proton (left):

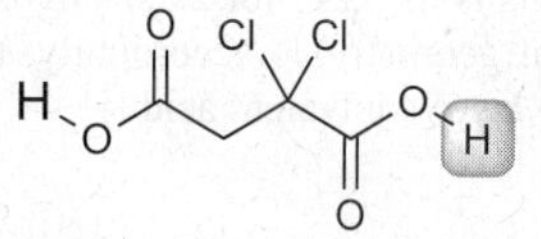

3.19.
(a) The compound with two chlorine atoms is more acidic, because of the electron-withdrawing effects of the additional chlorine atom, which helps stabilize the conjugate base that is formed when the proton is removed.

(b) The more acidic compound is the one in which the bromine atom is closer to the acidic proton. The electron-withdrawing effects of the bromine atom stabilize the conjugate base that is formed when the proton is removed.

3.20.
(a) In the following compound, one of the chlorine atoms has been moved closer to the acidic proton, which further stabilizes the conjugate base that is formed when the proton is removed.

(b) In the following compound, one of the chlorine atoms has been moved farther away from the acidic proton, which destabilizes the conjugate base that is formed when the proton is removed.

(c) The following compound is less acidic than the compounds above, because this compound is not a carboxylic acid. That is, the conjugate base of this compound is NOT resonance stabilized:

3.21. Both protons are the same distance from the fluorine atom, and both protons are the same distance from the chlorine atom (if you have trouble seeing this, you should build a molecular model and convince yourself that this is the case for an sp^3 hybridized center with tetrahedral geometry). Accordingly, these protons are expected to be of equivalent acidity.

3.22. The following compound (acetylene) is more acidic. The conjugate base of this compound has a negative charge associated with a lone pair in an sp hybridized orbital, which is more stable than a negative charge associated with a lone pair in an sp^2 hybridized orbital.

$$H-C\equiv C-H$$

3.23. The most acidic proton is highlighted in each of the following compounds. For each of the first two compounds, deprotonation leads to a conjugate base in which the negative charge is associated with an sp hybridized orbital (which is more stable than being associated with an sp^2 or sp^3 hybridized orbital). In the final compound, deprotonation leads to a conjugate base in which the negative charge is associated with an sp^2 hybridized orbital (which is more stable than being associated with an sp^3 hybridized orbital).

3.24. Most imines will have a pK_a below 35, because imines are expected to be more acidic than amines. This prediction derives from a comparison of the conjugate base of an amine and the conjugate base of an imine. The former has a negative charge in an sp^3 hybridized orbital, while the latter has a negative charge in an sp^2 hybridized orbital. The latter is expected to be more stable, and therefore, imines are expected to be more acidic.

3.25.
(a) The proton marked in red is expected to be more acidic than the proton marked in blue, because deprotonation of the former leads to a resonance-stabilized anion in which the negative charge is spread over one oxygen atom and three carbon atoms. Removal of the proton marked in blue results in an anion that is not resonance stabilized.

(b) The proton marked in red is expected to be more acidic than the proton marked in blue, because deprotonation of the former leads to a resonance-stabilized anion in which the negative charge is spread over two oxygen atoms, rather than just one.

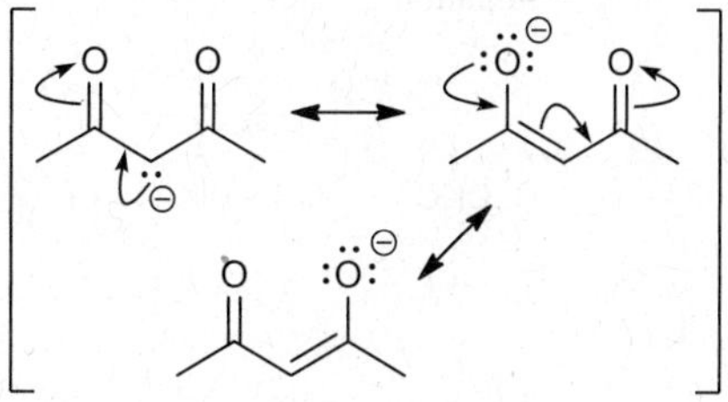

(c) The proton marked in red is expected to be more acidic than the proton marked in blue, because deprotonation of the former leads to a resonance-stabilized anion in which the negative charge is spread over two oxygen atoms, rather than being spread over two nitrogen atoms:

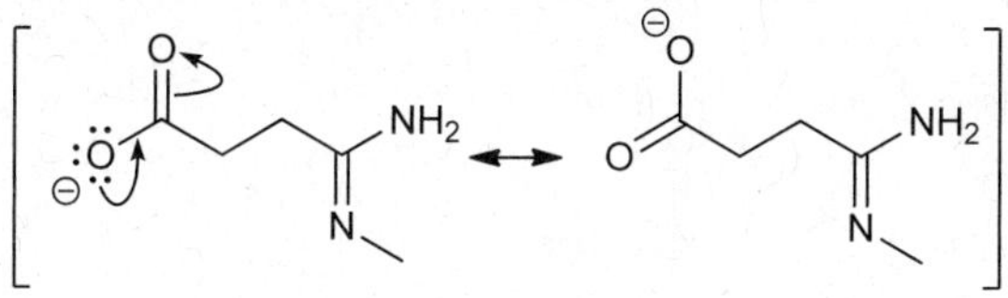

(d) The proton marked in blue is expected to be more acidic than the proton marked in red, because deprotonation of the former leads to a resonance-stabilized anion in which the negative charge is spread over a nitrogen atom and a carbon atom.

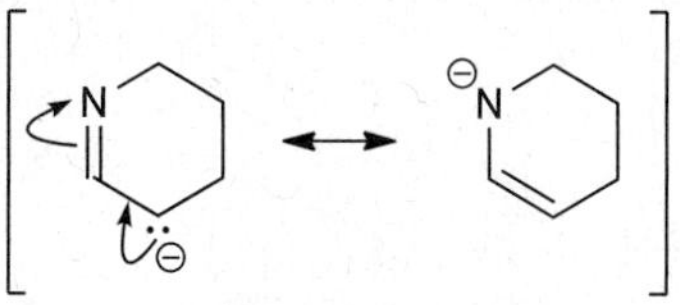

(e) The proton marked in red is expected to be more acidic than the proton marked in blue, because deprotonation of the former leads to a resonance-stabilized anion in which the negative charge is spread over two nitrogen atoms, rather than being spread over two carbon atoms.

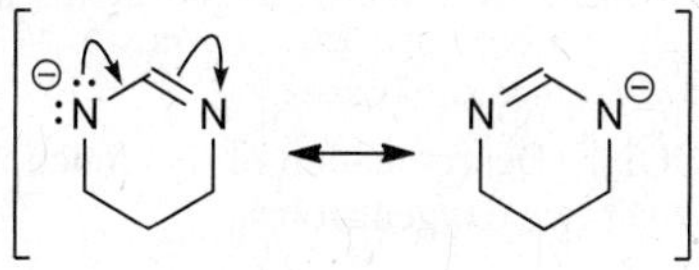

(f) The proton marked in blue is expected to be more acidic than the proton marked in red, because deprotonation of the former leads to a resonance-stabilized anion in which the negative charge is spread over two sulfur atoms, rather than two oxygen atoms.

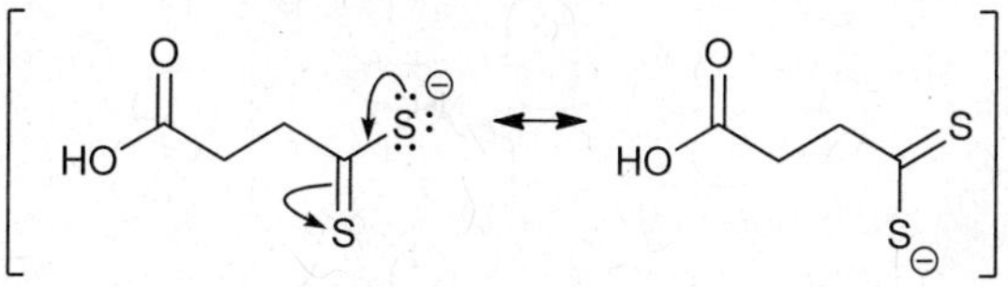

(g) The proton marked in red is expected to be more acidic than the proton marked in blue, because deprotonation of the former leads to a conjugate base in which the negative charge is associated with an *sp* hybridized orbital. This case represents an exception to the ARIO priority scheme, because factor 4 (orbital) trumps factor 1 (atom).

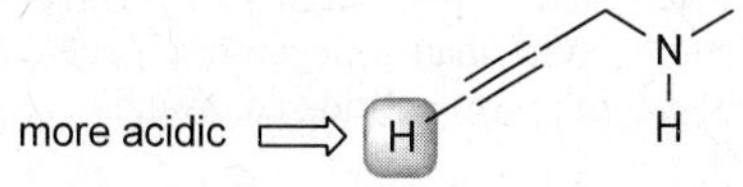

(h) The proton marked in blue is expected to be more acidic than the proton marked in red, because deprotonation of the former leads to a conjugate base in which the negative charge is spread over two oxygen atoms, rather than being spread over one oxygen atom and three carbon atoms.

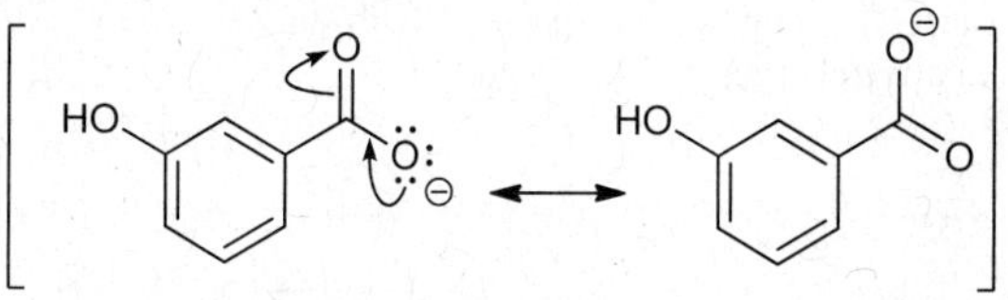

(i) The proton marked in blue is expected to be more acidic than the proton marked in red, because deprotonation of the former leads to a conjugate base in which the negative charge is spread over three oxygen atoms, rather than being spread over one oxygen atom and one nitrogen atom.

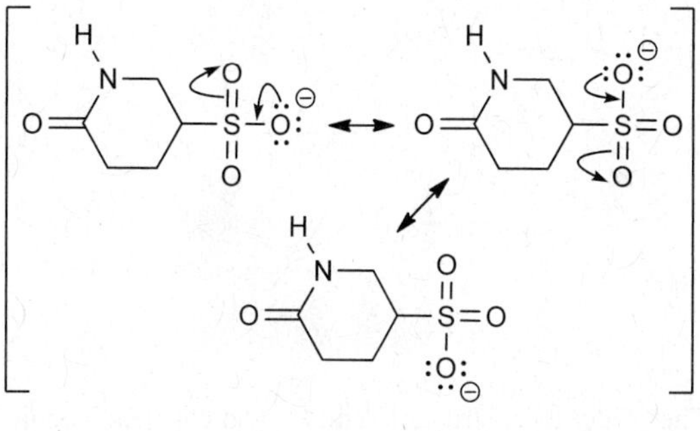

3.26.
(a) Bromine and chlorine are in the same column of the periodic table (group 7A), so we must compare their size. Bromine is larger than chlorine and can better stabilize the negative charge that will be generated upon deprotonation. HBr is expected to be more acidic than HCl.

(b) Sulfur and oxygen are in the same column of the periodic table (group 6A), so we must compare their size. Sulfur is larger than oxygen and can better stabilize the negative charge that will be generated upon deprotonation. Therefore, H_2S is expected to be more acidic than H_2O.

(c) Carbon and nitrogen are in the same row of the periodic table, so we must compare their electronegativity values. Nitrogen is more electronegative than carbon and can better stabilize the negative charge that will be generated upon deprotonation. Therefore, NH$_3$ is expected to be more acidic than CH$_4$.

(d) Acetylene (H–C≡C–H) is more acidic. The conjugate base of acetylene has a negative charge associated with a lone pair in an *sp* hybridized orbital, which is more stable than a negative charge associated with a lone pair in an *sp*^{2} hybridized orbital.

(e) The first compound (shown below) is more acidic, because of the electron-withdrawing effects of the nearby chlorine atoms that stabilize the conjugate base formed upon deprotonation.

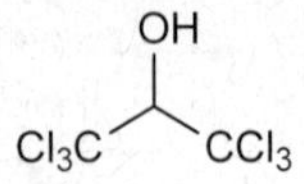

3.27.
(a) When the proton is removed, the resulting conjugate base is highly resonance-stabilized because the negative charge is spread over four nitrogen atoms and seven oxygen atoms. In addition, the inductive effects of the trifluoromethyl groups (-CF$_3$) further stabilize the negative charge.
(b) There are certainly many, many acceptable answers to this problem. Below are two separate modifications that would render the compound even more acidic:
The OH group can be replaced with an SH group. Sulfur is larger than oxygen and more capable of stabilizing a negative charge:

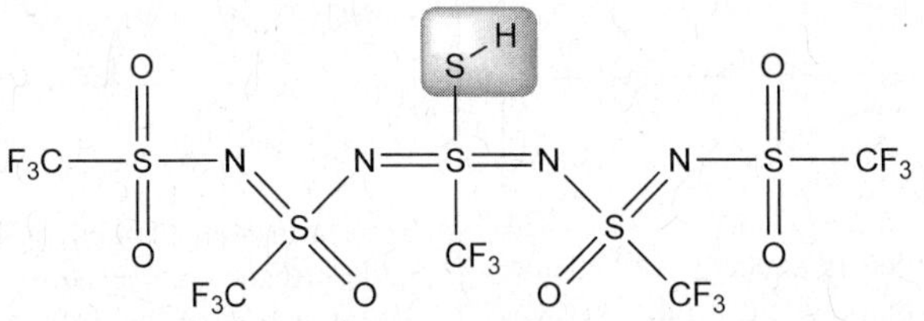

Alternatively, the conjugate base could be further stabilized by spreading the charge over an even larger number of nitrogen and oxygen atoms. For example, consider the structural changes, highlighted here:

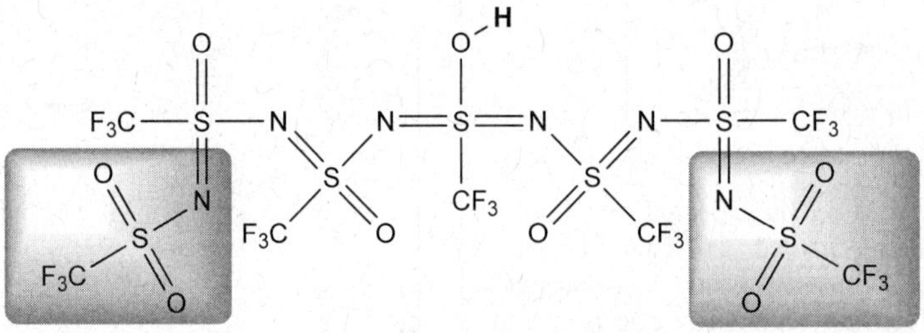

These additional structural units would enable the conjugate base to spread its negative charge over six nitrogen atoms and nine oxygen atoms, which should be even more stable than being spread over four nitrogen atoms and seven oxygen atoms.

3.28. The most acidic proton belongs to the carboxylic acid group (COOH). Deprotonation of this functional group gives a resonance-stabilized anion in which the negative charge is spread over two oxygen atoms.

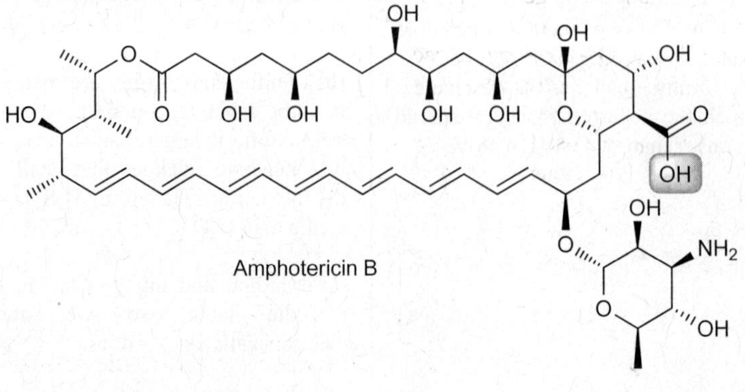

Amphotericin B

3.29.

(a) We compare the bases on either side of the equilibrium. The first (left side) has a negative charge on a carbon atom, while the second (right side) has a negative charge on an oxygen atom, so we turn to factor #1 (atom). Carbon and oxygen are in the same row of the periodic table, so we compare their electronegativity values. Oxygen is more electronegative than carbon, so a negative charge will be more stable on an oxygen atom. As such, the reaction favors the products.

(b) We compare the bases on either side of the equilibrium. The base on the left side is a resonance-stabilized anion, in which the negative charge is spread over two oxygen atoms. The base on the right side has the negative charge localized on a nitrogen atom. Both factor 1 (atom) and factor 2 (resonance) indicate that the reaction does not favor the products.

(c) We compare the bases on either side of the equilibrium. The base on the left side is a hydroxide ion, while the base on the right side is a resonance-stabilized anion, in which the negative charge is spread over one oxygen atom and three carbon atoms. The latter is more stable because of factor 2 (resonance). Therefore, the reaction favors the products.

3.30. In an intramolecular acid-base reaction, a proton is transferred from one region of the molecule to another region within the same molecule, because the acid and base are tethered together, as in this case.

The equilibrium favors the product shown because the negative charge in the product is resonance-stabilized (factor 2), and the reaction will favor the formation of a weak base from a stronger one.

3.31.

(a) Yes, because a negative charge on an oxygen atom will be more stable than a negative charge on a nitrogen atom.

(b) Yes, because a negative charge on a nitrogen atom will be more stable than a negative charge on an sp^3 hybridized carbon atom.

(c) No, because a negative charge on an sp^2 hybridized carbon atom will be less stable than a negative charge on a nitrogen atom.

(d) No, because this base is resonance-stabilized, with the negative charge spread over two oxygen atoms and one carbon atom. Protonating this base with water would result in the formation of a hydroxide ion, which is less stable because the negative charge is localized on one oxygen atom.

(e) Yes, because a negative charge on an oxygen atom will be more stable than a negative charge on a carbon atom.

(f) Yes, because a negative charge on an sp hybridized carbon atom will be more stable than a negative charge on a nitrogen atom.

3.32.

(a) No, water will not be a suitable proton source to protonate this anion, because the anion is resonance-stabilized and is more stable than hydroxide.

(b) No, water will not be a suitable proton source to protonate this anion, because the anion is resonance-stabilized and is more stable than hydroxide.

3.33. The CH_3CH_2 group in ethanol provides a small amount of steric bulk that is absent in water. As such, water is more acidic than ethanol (for the same reason that ethanol is more acidic than *tert*-butanol). Indeed, the pK_a of water (15.7) is slightly lower than the pK_a of ethanol (16).

3.34.

(a) A lone pair of the oxygen atom attacks the aluminum atom. $AlCl_3$ functions as the Lewis acid by accepting the electrons, and the ketone functions as the Lewis base by serving as an electron donor.

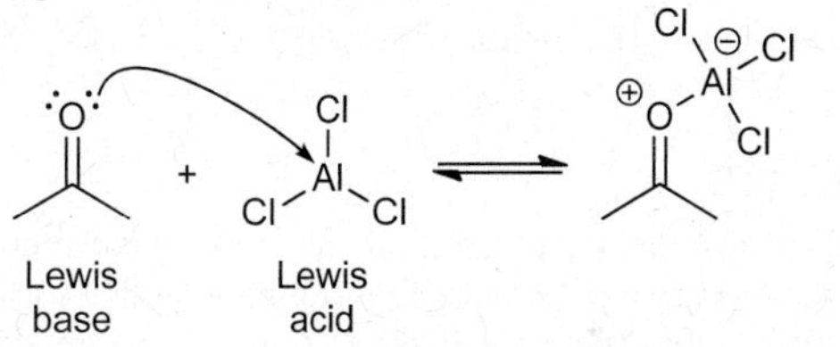

(b) A lone pair of the oxygen atom attacks a proton, as shown below. H_3O^+ functions as the Lewis acid by accepting the electrons, and the ketone functions as the Lewis base by serving as an electron donor.

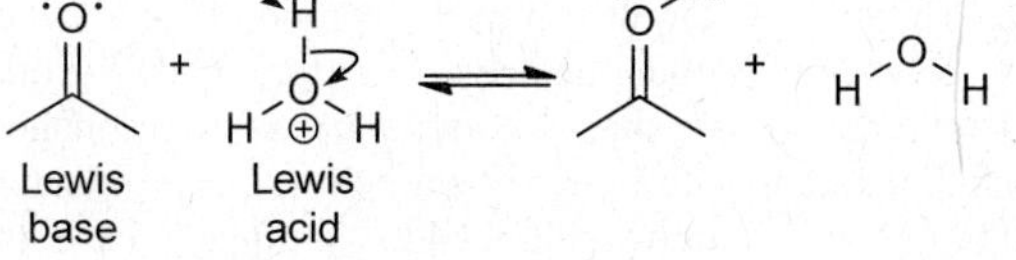

(c) A lone pair of a bromine atom attacks the aluminum atom. $AlBr_3$ functions as the Lewis acid by accepting the electrons, and molecular bromine (Br_2) functions as the Lewis base by serving as an electron donor.

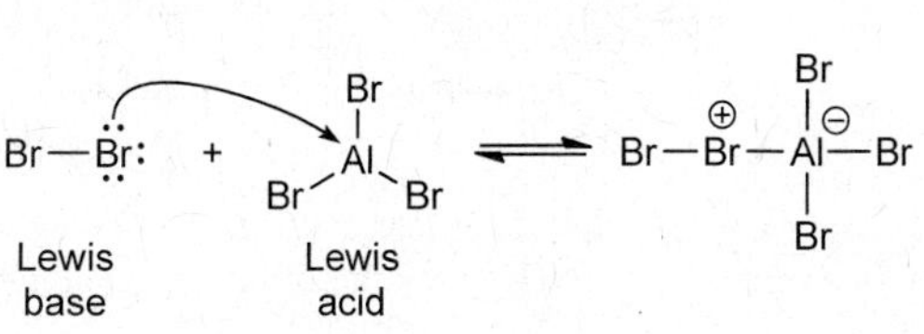

(d) A lone pair of the oxygen atom attacks a proton, as shown below. H_3O^+ functions as the Lewis acid by accepting the electrons, and the ester functions as the Lewis base by serving as an electron donor.

Lewis base   Lewis acid

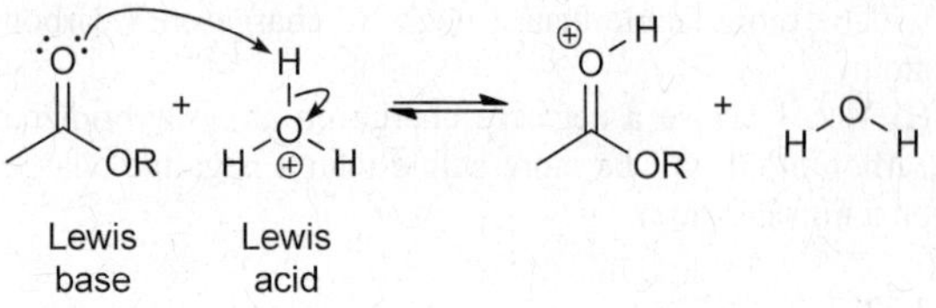

(e) A lone pair of the oxygen atom attacks the boron atom. BF_3 functions as the Lewis acid by accepting the electrons, and the ketone functions as the Lewis base by serving as an electron donor.

Lewis base   Lewis acid

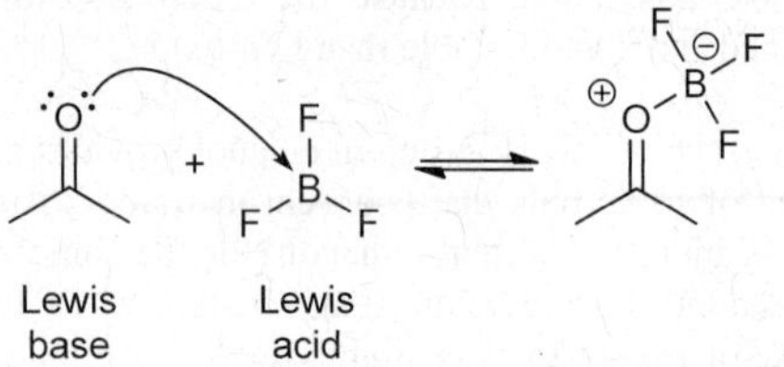

3.35. Each of the following two compounds can serve as a Lewis base, because each of these compounds has a lone pair that can serve as an electron donor.

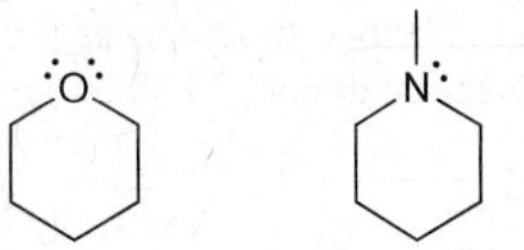

3.36.
(a) The most acidic proton in the compound is the proton of the O-H group, because deprotonation at that location gives a conjugate base with a negative charge on an oxygen atom, shown below. Deprotonation at any other location would lead to a conjugate base with a negative charge on carbon (which is MUCH less stable).

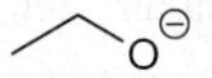

(b) The most acidic proton in the compound is attached to a carbon atom adjacent to the C=O group. Deprotonation at that location gives a resonance-stabilized conjugate base, shown below. Deprotonation at any other location would lead to a conjugate base that is not resonance-stabilized, and therefore much less stable.

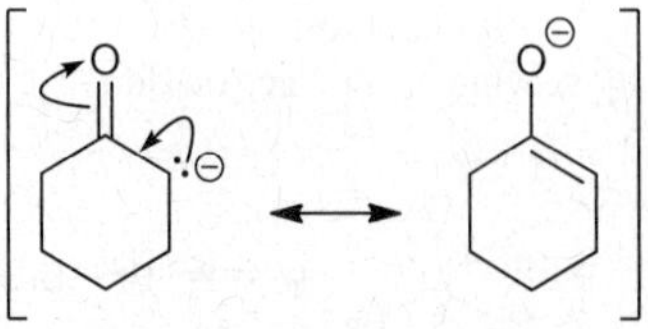

(c) Deprotonation of NH_3 gives the following conjugate base.

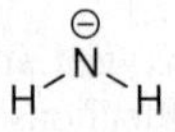

(d) Deprotonation of H_3O^+ gives the following conjugate base.

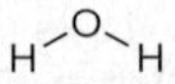

(e) The most acidic proton in the compound is the one attached to oxygen. Deprotonation at that location gives a resonance-stabilized conjugate base, in which the charge is spread over two oxygen atoms.

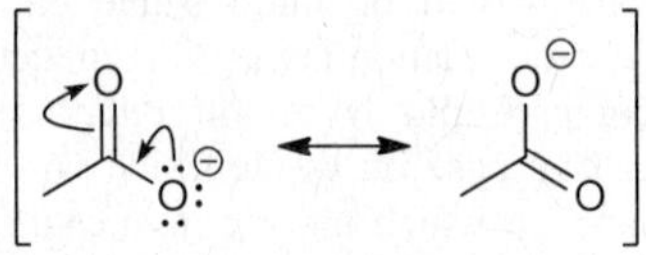

(f) The most acidic proton in the compound is the proton connected to nitrogen, because deprotonation at that location gives a conjugate base with a negative charge on a nitrogen atom, shown below. Deprotonation at any other location would lead to a conjugate base with a negative charge on carbon (which is less stable).

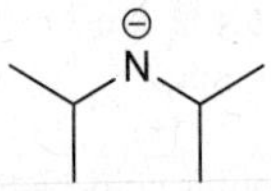

(g) Deprotonation of NH_4^+ gives the following conjugate base.

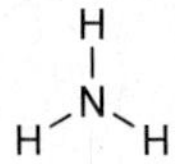

3.37.
(a) Protonation occurs at the site bearing the negative charge, giving the following compound:

(b) Protonation gives the following ketone:

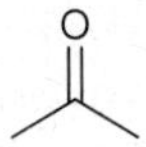

(c) Protonation of H_2N^- gives NH_3:

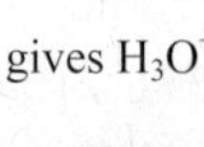

(d) Protonation of H_2O gives H_3O^+:

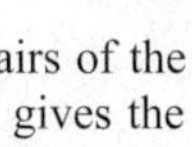

(e) One of the lone pairs of the oxygen atom can serve as a base. Protonation gives the following oxonium ion (a cation in which the positive charge is located on an oxygen atom).

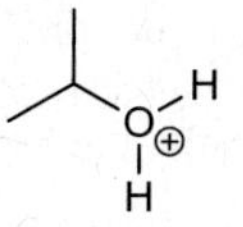

(f) The lone pair of the nitrogen atom can serve as a base. Protonation gives the following ammonium ion (a cation in which the positive charge is located on a nitrogen atom).

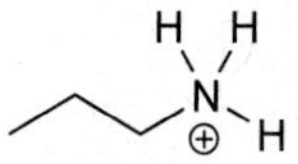

(g) One of the lone pairs of the oxygen atom can serve as a base. Protonation gives the following resonance-stabilized cation.

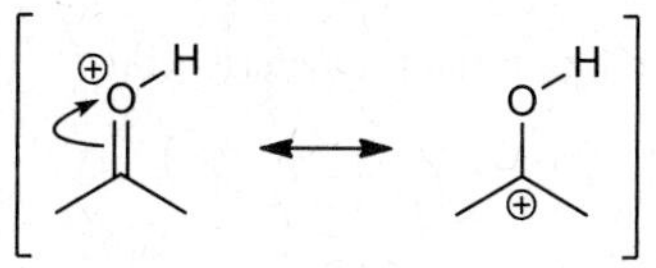

(h) Protonation of HO⁻ gives H₂O:

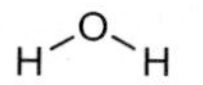

3.38. The difference in acidity between compounds A and B is $10 - 7 = 3$ pK_a units. Each pK_a unit represents an order of magnitude, so compound A is 1000 times more acidic than compound B.

3.39.
(a) A lone pair of the oxygen atom attacks the carbocation (C+), as shown below. The carbocation functions as the Lewis acid by accepting the electrons, and ethanol (CH_3CH_2OH) functions as the Lewis base by serving as an electron donor.

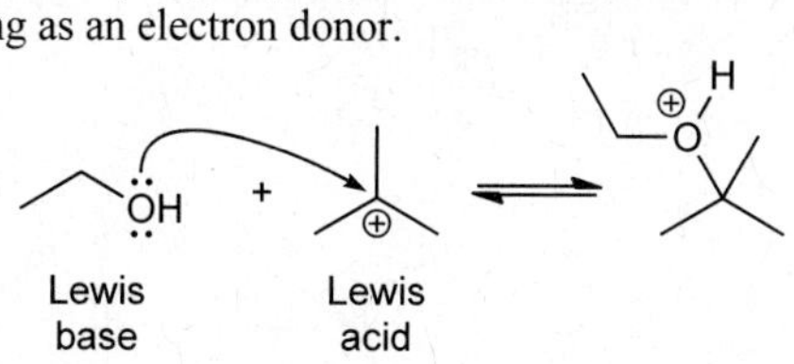

(b) A lone pair of the oxygen atom attacks the boron atom. BF_3 functions as the Lewis acid by accepting the electrons, and ethanol (CH_3CH_2OH) functions as the Lewis base by serving as an electron donor.

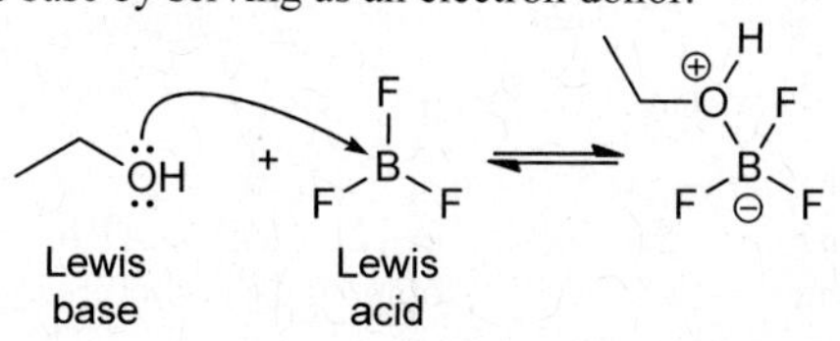

(c) A lone pair of a chlorine atom attacks the aluminum atom. $AlCl_3$ functions as the Lewis acid by accepting the

electrons, and ethyl chloride (CH_3CH_2Cl) functions as the Lewis base by serving as an electron donor.

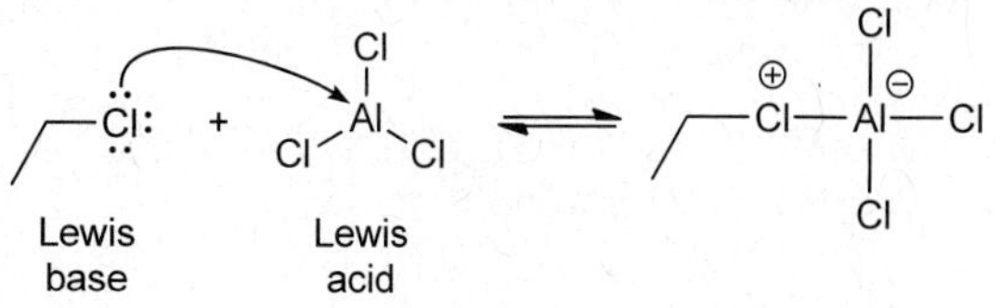

3.40. The amide ion (H_2N^-) is a strong base, and in the presence of H_2O, a proton transfer reaction will occur, generating hydroxide, which is a more stable base than the amide ion (Factor #1: oxygen is more electronegative than nitrogen). The reaction will greatly favor products (ammonia and hydroxide).

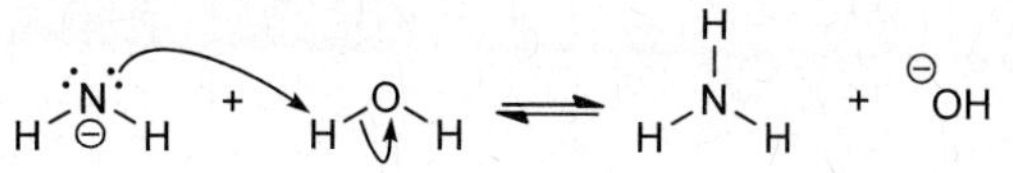

3.41. No, the reaction cannot be performed in the presence of ethanol, because the leveling effect would cause deprotonation of ethanol to form ethoxide ions, and the desired anion would not be formed under these conditions.

3.42. No, water would not be a suitable proton source in this case. This anion is the conjugate base of a carboxylic acid. The negative charge is resonance stabilized and is more stable than hydroxide.

3.43.
(a) Water functions as a base and deprotonates HBr. Two curved arrows are required. The first curved arrow shows a lone pair of the base attacking the proton, and the second curved arrow comes from the H–Br bond (being broken) and goes to the bromine atom, as shown:

(b) Water functions as a base and deprotonates sulfuric acid (H_2SO_4). Two curved arrows are required. The first curved arrow shows a lone pair of the base attacking the proton, and the second curved arrow comes from the O–H bond (being broken) and goes to the oxygen atom, as shown:

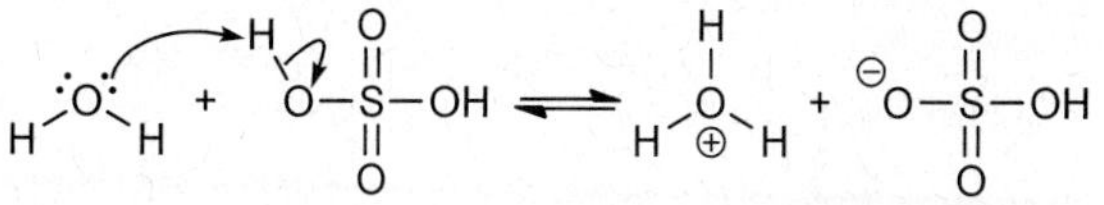

(c) Water functions as a base and deprotonates this strong acid (See Table 3.1). Two curved arrows are required. The first curved arrow shows a lone pair of the base attacking the proton, and the second curved arrow comes from the O–H bond (being broken) and goes to the oxygen atom, as shown:

3.44.

(a) Water functions as an acid in this case, by giving a proton to the strong base, as shown below. Two curved arrows are required. The first curved arrow shows a lone pair of the base attacking the proton, and the second curved arrow comes from the O–H bond (being broken) and goes to the oxygen atom, as shown:

(b) Water functions as an acid in this case, by giving a proton to the base, as shown below. Two curved arrows are required. The first curved arrow shows a lone pair of the base attacking the proton, and the second curved arrow comes from the O–H bond (being broken) and goes to the oxygen atom, as shown:

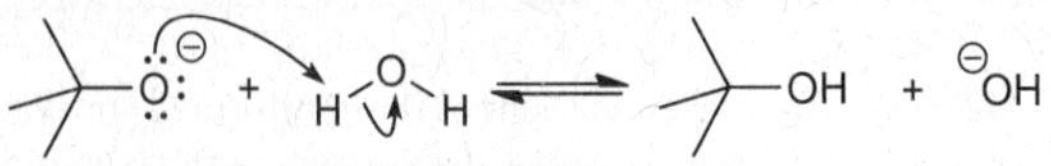

(c) Water functions as an acid in this case, by giving a proton to the strong base, as shown below. Two curved arrows are required. The first curved arrow shows a lone pair of the base attacking the proton, and the second curved arrow comes from the O–H bond (being broken) and goes to the oxygen atom, as shown:

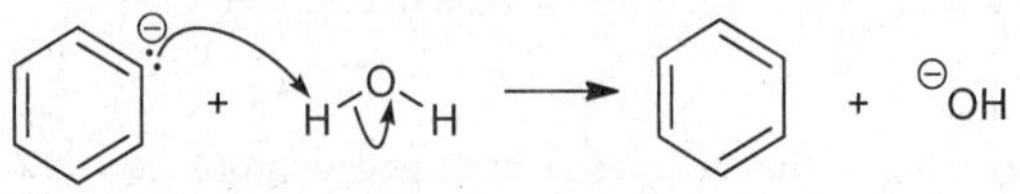

(d) Water functions as an acid in this case, by giving a proton to the strong base, as shown below. Two curved arrows are required. The first curved arrow shows a lone pair of the base attacking the proton, and the second curved arrow comes from the O–H bond (being broken) and goes to the oxygen atom, as shown:

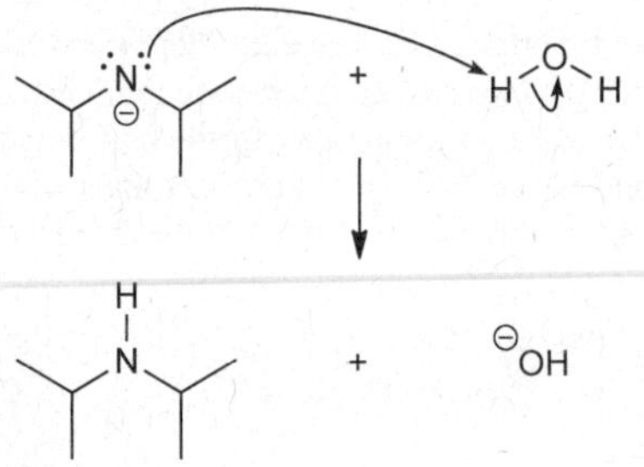

3.45.
(a) The second anion is more stable because it is resonance stabilized.
(b) The second anion is more stable because the negative charge is on a nitrogen atom (factor #1 of ARIO), rather than an sp^3 hybridized carbon atom.
(c) The second anion is more stable because the negative charge is on an sp hybridized carbon atom, rather than an sp^3 hybridized carbon atom.

3.46.

(a) The last compound in the group is the most acidic, because deprotonation of this compound gives a resonance-stabilized conjugate base (shown below) in which the negative charge is spread between a nitrogen atom and a carbon atom:

(b) The third compound can be immediately ruled out, because deprotonation of this compound does not give a resonance-stabilized conjugate base, whereas deprotonation of each of the other three compounds does give a resonance-stabilized conjugate base. The fourth compound also has the inductive effects of the electron-withdrawing chlorine atoms, which further stabilize its conjugate base, shown below. Therefore, the fourth compound in this group is the most acidic.

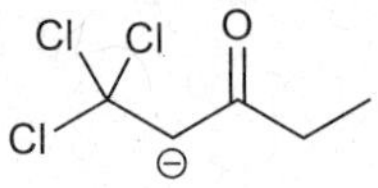

3.47.
(a) The second compound is more acidic, because its conjugate base (shown below) has a negative charge on a sulfur atom, which is more stable than a negative charge on an oxygen atom (factor #1 of ARIO):

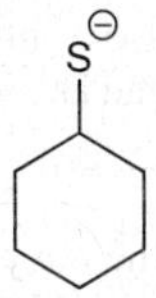

(b) The first compound (called phenol) is more acidic, because its conjugate base (shown below) is resonance stabilized (factor #2 of ARIO):

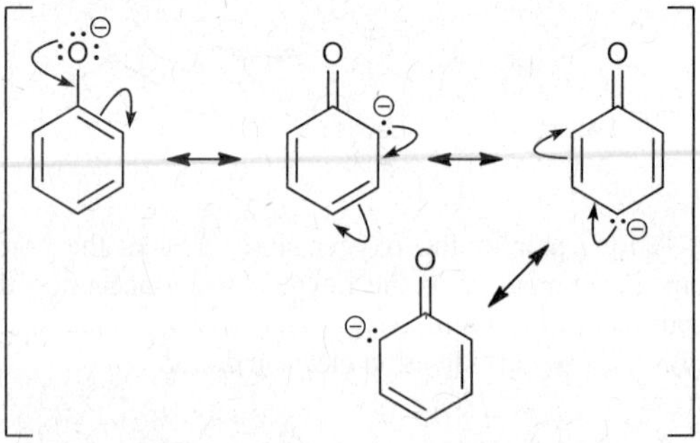

(c) The conjugate base for each of these compounds is resonance-stabilized (as in Problem **3.47b**). The difference between these compounds is the presence of electron-withdrawing chlorine atoms. The first compound is more acidic as a result of the combined inductive effects of the five chlorine atoms which stabilize the conjugate base shown below (factor #3 of ARIO).

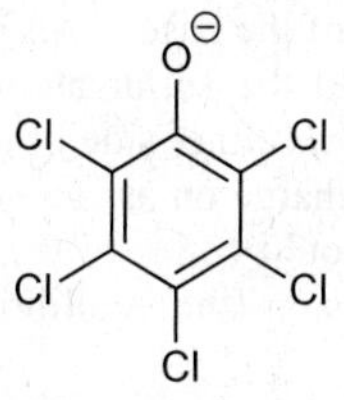

(d) The second compound is more acidic, because its conjugate base (shown below) has a negative charge associated with an *sp* hybridized orbital, while the first compound is less acidic because its conjugate base has a negative charge on a nitrogen atom. We learned that this example constitutes an exception to the order of priorities, ARIO. In this case, factor #4 (orbital) trumps factor #1 (atom). This exception applies whenever we compare a negative charge on an *sp* hybridized carbon atom and a negative charge on an sp^3 hybridized nitrogen atom.

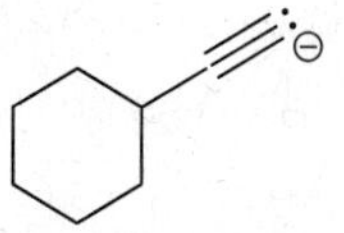

(e) The first compound is more acidic because its conjugate base (shown below) is resonance stabilized. The conjugate base of the second compound is not resonance stabilized.

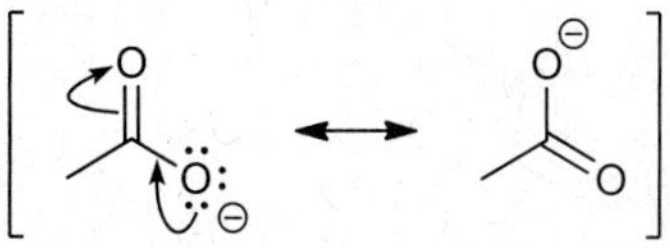

(f) The first compound is more acidic because its conjugate base (shown below) is resonance stabilized, and one of the resonance structures has the negative charge on an oxygen atom. The conjugate base of the second compound is not resonance stabilized (the negative charge would be localized on a carbon atom).

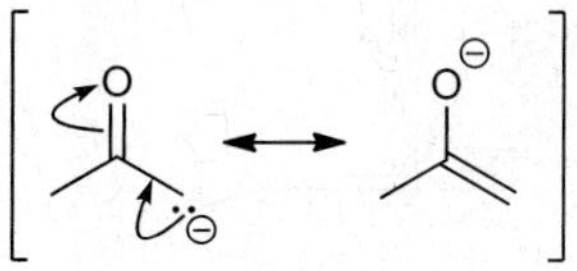

(g) The first compound is more acidic because its conjugate base (shown below) is resonance stabilized, and one of the resonance structures has the negative charge on an oxygen atom. The conjugate base of the second compound is not resonance stabilized (the negative charge would be localized on a carbon atom).

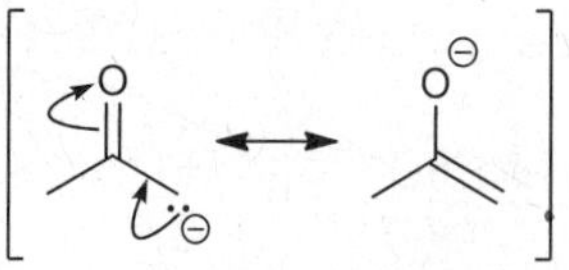

(h) The second compound is more acidic because its conjugate base is resonance-stabilized, with the negative charge being spread over two oxygen atoms (shown below). The conjugate base of the first compound is also resonance stabilized, but the negative charge would be spread over an oxygen atom and a nitrogen atom, which is less stable than being spread over two oxygen atoms (because oxygen is more electronegative than nitrogen, as described in factor #1 of ARIO):

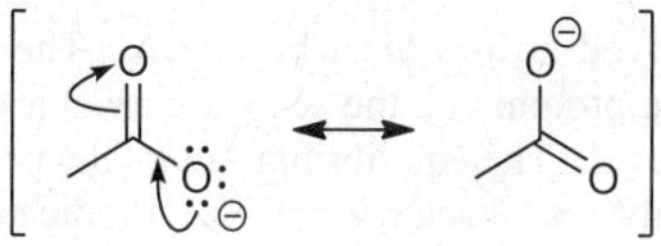

3.48. NaA represents an ionic compound, comprised of cations (Na^+) and anions (A^-). When H–B is treated with A^-, a proton can be transferred from H–B to A^-, as shown in the following equilibrium:

The equilibrium will favor the weaker acid (the acid with the higher pK_a value). In this case, the equilibrium favors formation of HA.

3.49.

(a) Water (H_2O) loses a proton and is therefore functioning as an acid. Two curved arrows must be drawn. The first curved arrow shows a lone pair of the base attacking the proton of water, and the second curved arrow comes from the O–H bond (being broken) and goes to the oxygen atom, as shown. The equilibrium favors the products, which can be determined by comparing the bases on either side of the equilibrium. Hydroxide is more stable, and the equilibrium favors the more stable base. Alternatively, we could compare the pK_a values of the acids on either side of the equilibrium [H_2O and $(CH_3)_2CHOH$] and we would arrive at the same conclusion (the equilibrium will favor the products in this case because the equilibrium favors the weaker acid).

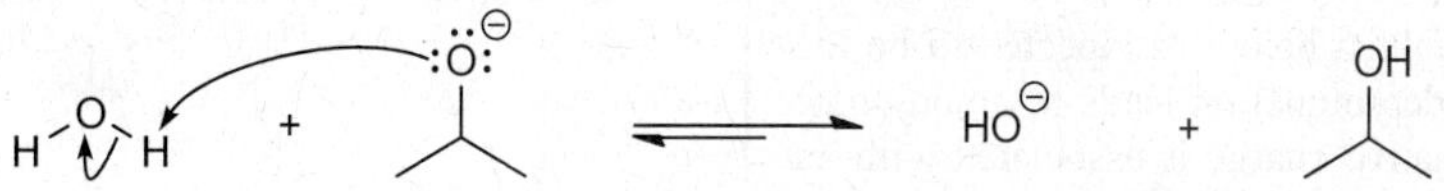

(b) Two curved arrows must be drawn. The first curved arrow shows a lone pair of the base attacking the proton, and the second curved arrow comes from the S–H bond (being broken) and goes to the sulfur atom, as shown. The equilibrium favors the products, which can be determined by comparing the bases on either side of the equilibrium. A negative charge on a sulfur atom is expected to be more stable than the negative charge on an oxygen atom (factor #1 of ARIO), and the equilibrium favors the more stable base. Alternatively, we could compare the pK_a values of the acids on either side of the equilibrium and we would arrive at the same conclusion (the equilibrium will favor the products in this case because the equilibrium favors the weaker acid).

(c) Two curved arrows must be drawn. The first curved arrow shows a lone pair of the base (HS⁻ in this case) attacking the proton, and the second curved arrow comes from the S–H bond (being broken) and goes to the sulfur atom, as shown. The equilibrium favors the products, which can be determined by comparing the bases on either side of the equilibrium. The base on the right side of the equilibrium is resonance stabilized, with the negative charge being spread over two sulfur atoms. The base on the left side of the equilibrium is not resonance stabilized, and the negative charge is localized on one sulfur atom. The equilibrium favors the more stable, resonance-stabilized base.

(d) Two curved arrows must be drawn. The first curved arrow shows a lone pair of the base (the nitrogen atom) attacking the proton, and the second curved arrow comes from the O–H bond (being broken) and goes to the oxygen atom, as shown. The equilibrium favors the products, which can be determined by comparing the bases on either side of the equilibrium. Factor #1 of ARIO indicates that the product is favored in this case, because a negative charge is more stable on an oxygen atom than a nitrogen atom.

3.50. One of the anions is resonance stabilized, with the negative charge spread over two oxygen atoms. That anion is the weakest (most stable) base. Among the remaining three anions, they do not differ from each other in any of the four factors (ARIO), but they are expected to differ from each other in terms of solvent effects. That is, an anion will be less stable (stronger base) if it has steric bulk in close proximity with the negative charge. The steric bulk reduces the stability of the anion by limiting its ability to interact with solvent molecules, as described in Section 3.7.

3.51.
(a) The proton highlighted below is expected to be the most acidic, because deprotonation leads to a conjugate base in which the negative charge is associated with an *sp* hybridized orbital. This case represents an exception to the ARIO priority scheme, because factor #4 (orbital) trumps factor #1 (atom). This exception applies whenever we compare a negative charge on an *sp* hybridized carbon atom and a negative charge on an sp^3 hybridized nitrogen atom.

(b) The proton highlighted below is expected to be the most acidic, because deprotonation leads to a conjugate base in which the negative charge is on a sulfur atom (more stable than being on an oxygen atom or a nitrogen atom, according to factor #1 of ARIO).

(c) The proton highlighted below is expected to be the most acidic, because deprotonation leads to a resonance-stabilized conjugate base.

(d) The proton highlighted below is expected to be the most acidic, because deprotonation leads to a resonance-stabilized conjugate base in which the negative charge is spread over two carbon atoms and one oxygen atom.

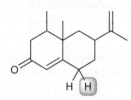

(e) The proton highlighted below is expected to be the most acidic, because deprotonation leads to a resonance-stabilized conjugate base in which the negative charge is spread over two oxygen atoms (factor #2 of ARIO).

(f) The proton highlighted below is expected to be the most acidic, because deprotonation leads to a resonance-stabilized conjugate base in which the negative charge is

spread over one carbon atom and one oxygen atom (factor #2 of ARIO).

(g) There are three COOH groups, each of which bears an acidic proton. Removing any one of these protons will result in a resonance-stabilized conjugate base. Among these three protons, the highlighted proton is the most acidic because of the electron-withdrawing effects of the nearby chlorine atoms. When this proton is removed, the conjugate base is stabilized not only by resonance, but also by induction (factor #3 of ARIO).

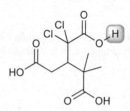

(h) The proton highlighted below is expected to be the most acidic, because deprotonation leads to a conjugate base in which the negative charge is on a sulfur atom, which is more stable than being on an oxygen atom (factor #1 of ARIO).

3.52.
(a) Acetic acid (CH$_3$CO$_2$H) loses a proton and is therefore functioning as an acid. Hydroxide (HO$^-$) functions as the base that removes the proton. Two curved arrows must be drawn. The first curved arrow shows a lone pair of the base attacking the proton, and the second curved arrow comes from the O–H bond (being broken) and goes to the oxygen atom, as shown. The equilibrium favors the products, because the base on the right side of the equilibrium is resonance stabilized and therefore more stable than a hydroxide ion. The equilibrium favors the more stable base. Alternatively, we could compare the pK_a values of the acids on either side of the equilibrium and we would arrive at the same conclusion (the equilibrium will favor the products in this case because the equilibrium favors the weaker acid).

(b) Water (H$_2$O) loses a proton and is therefore functioning as an acid. The carbanion (an anion in which the negative charge is on a carbon atom) functions as the base that removes the proton. Two curved arrows must be drawn. The first curved arrow shows a lone pair of the base attacking the proton, and the second curved arrow comes from the O–H bond (being broken) and goes to the oxygen atom, as shown. The reaction favors the products, because hydroxide is more stable than a carbanion (factor #1 of ARIO). Alternatively, we could compare the pK_a values of the acids on either side of the equilibrium and we would arrive at the same conclusion (the reaction favors the products). In fact, the pK_a values of H_2O (15.7) and C_4H_{10} (~50) are so vastly different that the reaction is essentially irreversible.

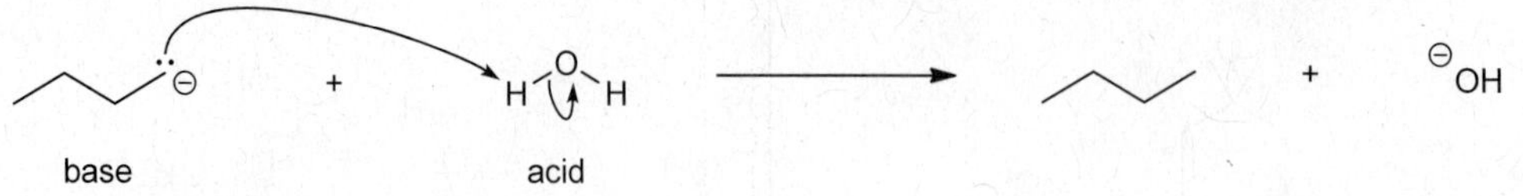

base   acid

(c) Hydroxide (HO⁻) functions as a base and removes a proton from the acid. Two curved arrows must be drawn. The first curved arrow shows a lone pair of the base attacking the proton, and the second curved arrow comes from the C–H bond (being broken) and goes to the carbon atom, as shown.

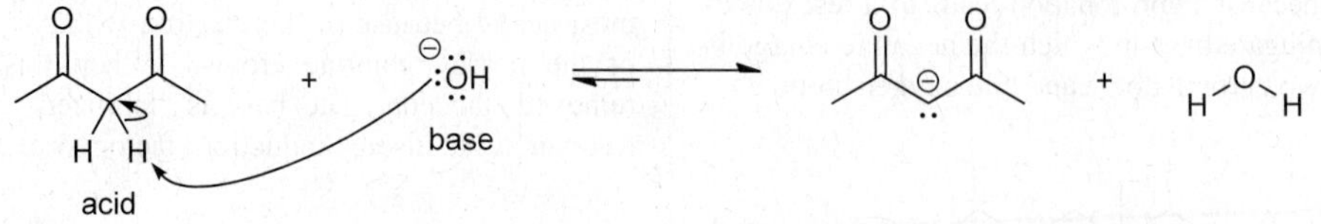

Notice that the conjugate base is resonance stabilized, so its formation can alternatively be shown with more than two curved arrows (leading directly to the resonance structure that contributes the most character to the overall resonance hybrid).

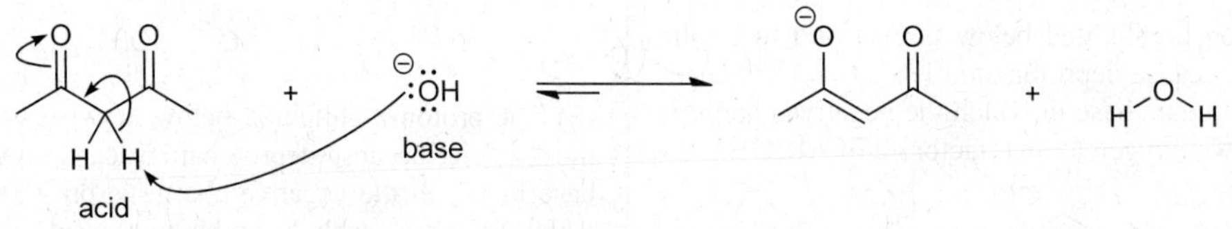

The equilibrium favors the products, because the base on the right side of the equilibrium is resonance stabilized, with the negative charge spread over two oxygen atoms. This is more stable than the negative charge being localized on one oxygen atom, as in hydroxide. The equilibrium favors the more stable base. Alternatively, we could compare the pK_a values of the acids on either side of the equilibrium and we would arrive at the same conclusion. That is, the equilibrium will favor the products in this case because the equilibrium favors formation of the weaker acid (H_2O).

3.53. Each of the carbon atoms is tetravalent; the sulfur atom is divalent; and each of the hydrogen atoms is monovalent. We begin by connecting the atoms that have more than one bond (in this case, the two carbon atoms and the sulfur atom). There are only two different ways that these three atoms can be connected to each other, shown below:

C—S—C   C—C—S

For each of these arrangements, we connect the hydrogen atoms, giving the following two constitutional isomers:

The second isomer is more acidic because deprotonation of that isomer gives a conjugate base with a negative charge on a sulfur atom. The first compound above is less acidic, because its conjugate base would have a negative charge on a carbon atom, which is much less stable (factor #1 of ARIO).

3.54. As seen in the solution to Problem **1.3**, there are three constitutional isomers with molecular formula C_3H_8O, shown here again:

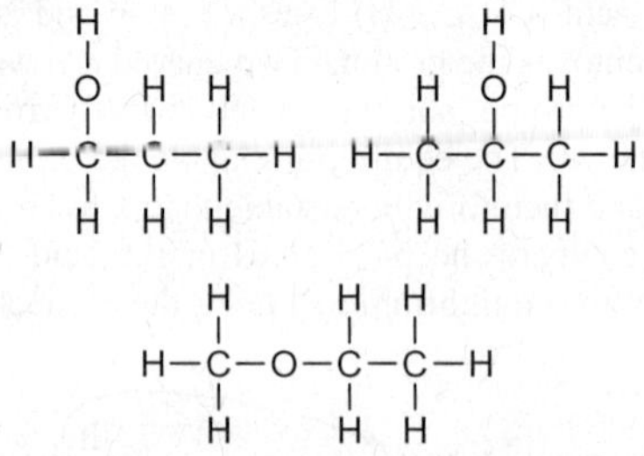

One of these compounds lacks an O-H group, so that compound will be the least acidic (its conjugate base will have a negative charge on a carbon atom). Of the two remaining compounds, the compound with the least branching will be the most acidic, because its conjugate base is the most stable (due to steric effects, discussed in Section 3.7).

In summary, there are three constitutional isomers with molecular formula C_3H_8O, arranged below in order of increasing acidity:

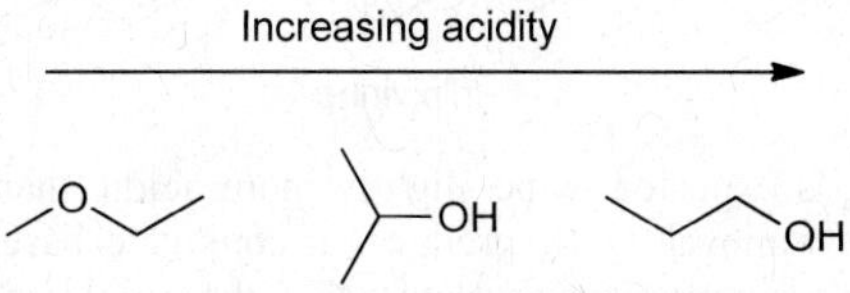

Increasing acidity

3.55.
(a) A carbon atom must have four sigma bonds in order to be sp^3 hybridized. There is only one such carbon atom in cyclopentadiene, highlighted below.

(b) The most acidic proton in cyclopentadiene is highlighted below:

The corresponding conjugate base is highly resonance stabilized (see the solution to Problem **3.55c**). In addition, the conjugate base is further stabilized by yet another factor that we will discuss in Chapter 18.

(c) Deprotonation of cyclopentadiene gives a conjugate base that is highly stabilized by resonance, as shown here:

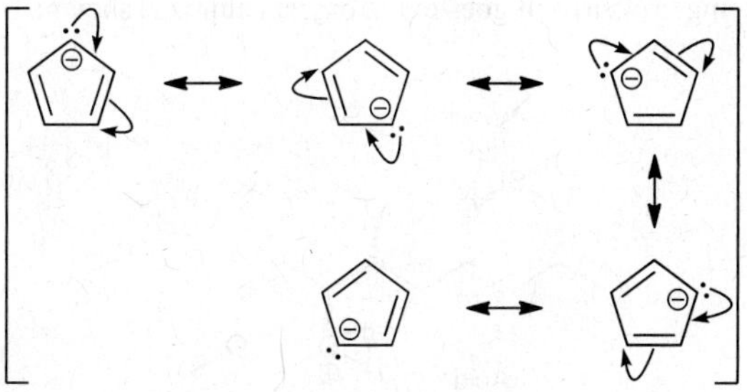

(d) There are no sp^3 hybridized carbon atoms in the conjugate base. All five carbon atoms are sp^2 hybridized.

(e) All carbon atoms are sp^2 hybridized and trigonal planar. Therefore, the entire compound has planar geometry.

(f) There are five hydrogen atoms in the conjugate base.

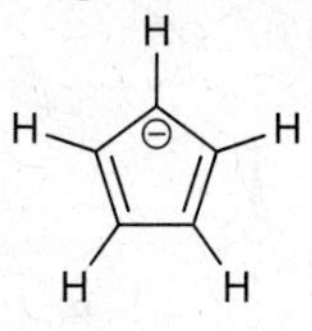

(g) As seen in the resonance structures (see the solution to Problem **3.55c**), there is one lone pair in the conjugate base, and it is highly delocalized.

3.56. We begin by drawing the conjugate base of each compound and comparing them:

The first conjugate base is stabilized by resonance, while the second conjugate base is not. However, the second conjugate base exhibits a negative charge on a sulfur atom, which is larger than an oxygen atom. Therefore, there is a competition between two factors. Using the ARIO order of priority that is generally applied ("atom" is more important than "resonance"), we would expect that the second conjugate base should be more stable than the first. Yet, when we compare the pK_a values, we find that our prediction is not correct. Therefore, this is an exception, in which "resonance" is more important than "atom."

3.57. Two possible explanations can be given:
1) In salicylic acid, the inductive effect (electron-withdrawal) of the OH group is expected to be more pronounced because of its proximity to the site where deprotonation will occur.
2) When salicylic acid is deprotonated, the resulting conjugate base is significantly stabilized by intramolecular hydrogen bonding (this explanation is likely more significant than the first explanation):

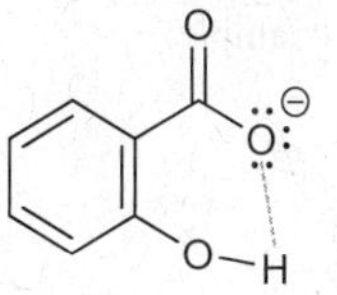

3.58.
(a) We are looking for a constitutional isomer with an acidic proton. The following two compounds are both carboxylic acids, because in each case, the conjugate base is resonance stabilized, with the negative charge being spread over two oxygen atoms. Carboxylic acids typically have a pK_a in the range of 4-5, while the structure shown in the problem statement is expected to have a pK_a in the range of 16-18. Therefore, the structures below satisfy the criteria described in the problem statement (they are both expected to be approximately one trillion times more acidic that the structure shown in the problem statement).

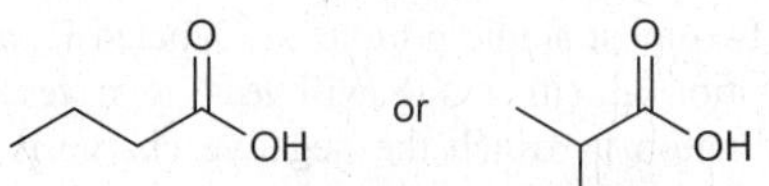

(b) The structure shown in the problem statement has an O-H group, and that proton is the most acidic proton in the compound. If we draw a constitutional isomer that lacks an O-H group, we would expect that isomer to be significantly less acidic. The following two compounds fit this criterion. There are certainly many other constitutional isomers that also lack an O-H group, so there are many correct answers to this problem.

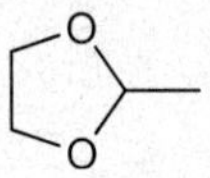

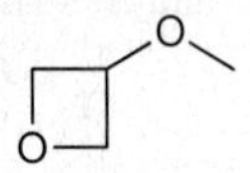

(c) Each of the following two isomers are expected to have a pK_a value that is similar to the pK_a of the structure shown in the problem statement, because each of these compounds possesses an O-H group. There are certainly many other constitutional isomers that also contain an O-H group, so there are many correct answers to this problem.

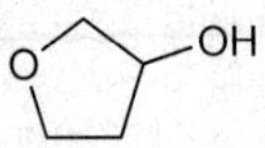

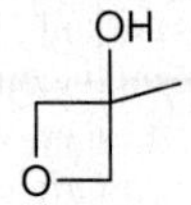

3.59. The four constitutional isomers are shown below.

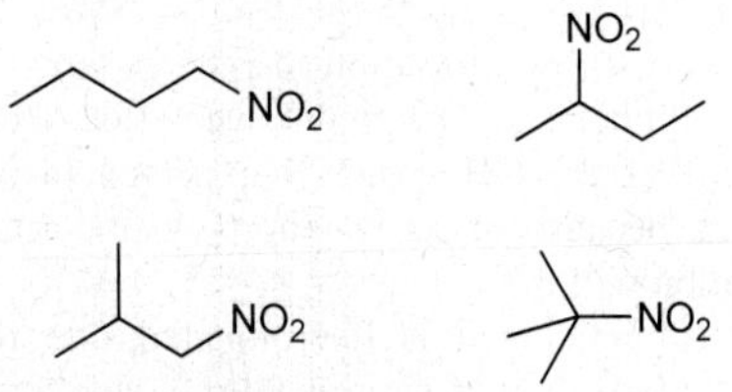

The last compound is expected to have the highest pK_a because its conjugate base is not resonance stabilized. The other three compounds have resonance-stabilized conjugate bases, for example:

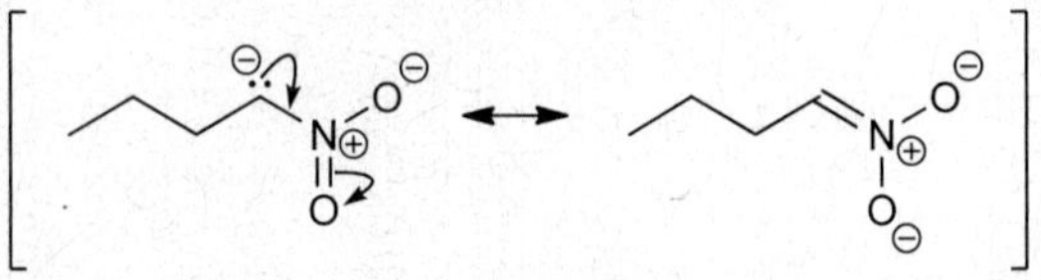

3.60. Compare the conjugate bases. Both are resonance stabilized. But the conjugate base of the first compound has a negative charge spread over two nitrogen atoms and two carbon atoms, while the conjugate base of the second compound has a negative charge spread over one nitrogen atom and three carbon atoms. Since nitrogen is more electronegative than carbon, nitrogen is more capable of stabilizing a negative charge. Therefore, the conjugate base of the first compound is more stable than the conjugate base of the second compound. As a result, the first compound will be more acidic.

3.61.
(a) The two most acidic protons are labeled H_a and H_b. Deprotonation at either site will lead to a resonance-stabilized anion in which the negative charge is highly delocalized (spread over many nitrogen atoms).

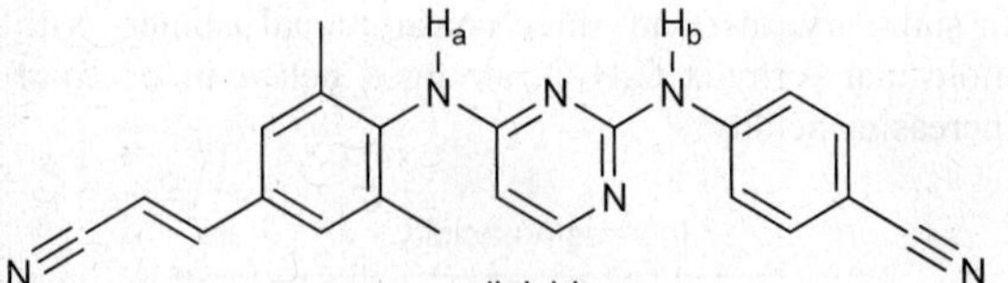

(b) H_a is expected to be slightly more acidic than H_b, because removal of H_a produces a conjugate base that has one more resonance structure than the conjugate base formed from removal of H_b. The former has the negative charge spread over four nitrogen atoms and *five* carbon atoms, while the latter has the negative charge spread over four nitrogen atoms and *four* carbon atoms.

3.62.
(a) When R is a cyano group, the conjugate base is resonance stabilized:

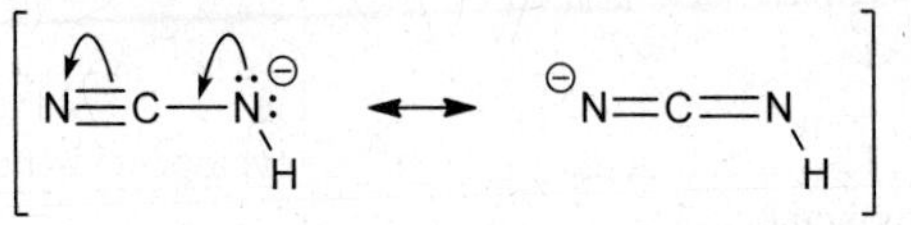

(b) There are many possible answers. Here is one example, for which the conjugate base has the negative charge spread over three nitrogen atoms, rather than just two nitrogen atoms:

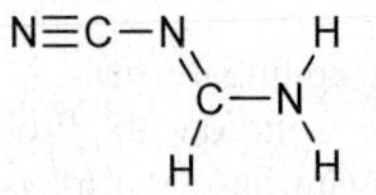

3.63.
(a) The positively charged structure serves as the acid, while the lone pair on the nitrogen atom serves as the base. Two curved arrows are required. The first curved arrow shows a lone pair of the base attacking the proton, and the second curved arrow comes from the O–H bond (being broken) and goes to the oxygen atom, as shown:

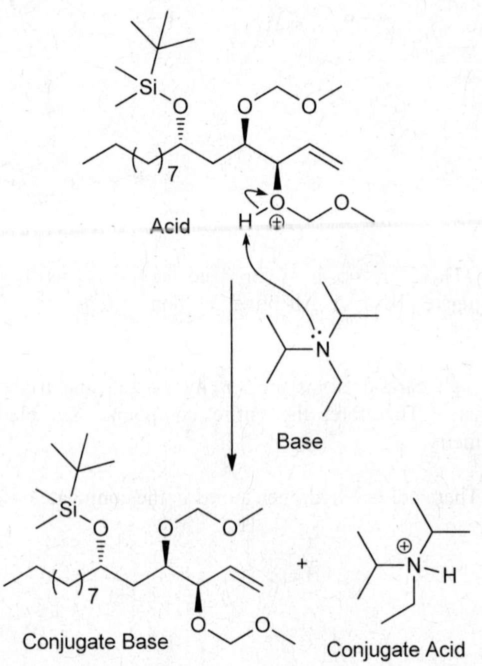

(b) In Problem **3.9**, we saw that an oxonium ion has a $pK_a \sim -2.2$, while an ammonium ion has a $pK_a \sim 10.5$. Therefore, we expect the former to be 13 orders of magnitude more acidic than the latter, and as such, the acid-base reaction above will significantly favor the products.

3.64.
(a) As seen in Table 1.1, the electronegativity value for carbon is 2.5, while the electronegativity value for magnesium is only 1.2. The difference (1.3) is significant, and as explained in Section 1.5, this bond can be drawn as either covalent or ionic:

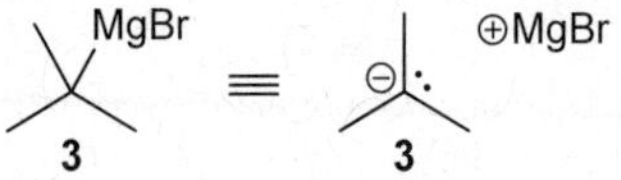

When drawn in this way, the anion exhibits a negative charge on a carbon atom (called a carbanion), which is a very strong base (because it is the conjugate base of a very, very weak acid). If compound **3** were treated with H_2O, we would expect the following proton-transfer reaction:

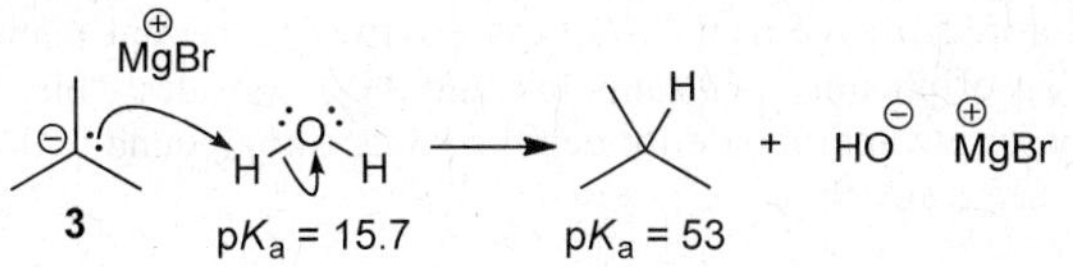

This reaction would be irreversible as a result of the enormous difference in pK_a values (see inside cover of the textbook for a table of pK_a values). In a similar way, if compound **3** is treated with D_2O, rather than H_2O, the following irreversible reaction occurs, in which a deuteron (rather than a proton) is transferred:

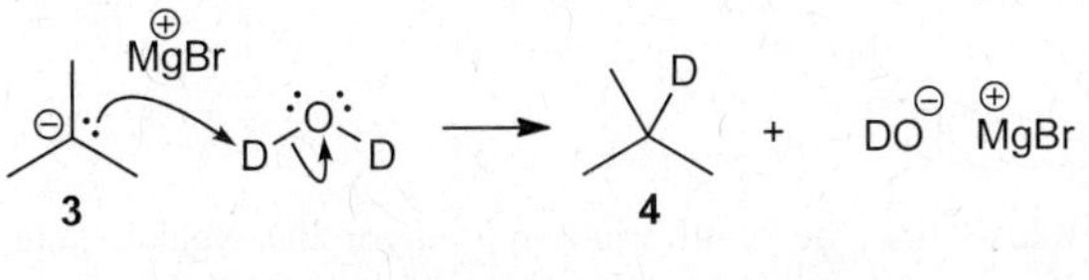

(b) As explained in Section 15.3, signals associated with the stretching of single bonds generally appear in the fingerprint region of an IR spectrum (400 – 1500 cm^{-1}), while signals associated with the stretching of double bonds and triple bonds appear in the diagnostic region (1500 – 4000 cm^{-1}). Notable exceptions are C-H bonds, which produce high-energy signals in the diagnostic region (2800 – 3000 cm^{-1}). This was explained in Section 15.3, by exploring the following equation, derived from Hooke's law:

$$\tilde{\nu} = \left(\frac{1}{2\pi c}\right)\left(\frac{f}{m_{red}}\right)^{\frac{1}{2}}$$

$$reduced\ mass = \left(\frac{m_1 m_2}{m_1 + m_2}\right)$$

Specifically, mass is found in the denominator, rather than the numerator, and as such, atoms with lower mass produce higher energy signals. Hydrogen has the smallest mass of all atoms, which explains why C-H bonds appear in the range 2800 – 3000 cm^{-1}. Extending this logic, a C-D bond is also expected to produce a higher energy signal, although not quite as high as C-H, because D has greater mass than H. So, we expect the C-D signal to appear somewhere below 2800 cm^{-1}, but still within the diagnostic region (greater than 1500 cm^{-1}). There is only one signal in the IR spectrum of compound **4** that fits this description, and that is the signal at 2180 cm^{-1}.

Also, 2180 cm^{-1} is in the region of the IR spectrum where signals for triple bonds are expected to appear (2100 – 2300 cm^{-1}). But we know that compound **4** lacks a triple bond. So this signal must be attributed to something else, and the C-D bond is the only candidate, because the C-H bonds appear in the range 2800 – 3000 cm^{-1} and the C-C bonds appear in the range 1250 and 1500 cm^{-1}.

3.65.
(a) The lone pair in compound **4** functions as a base and deprotonates intermediate **3**. This requires two curved arrows, as shown:

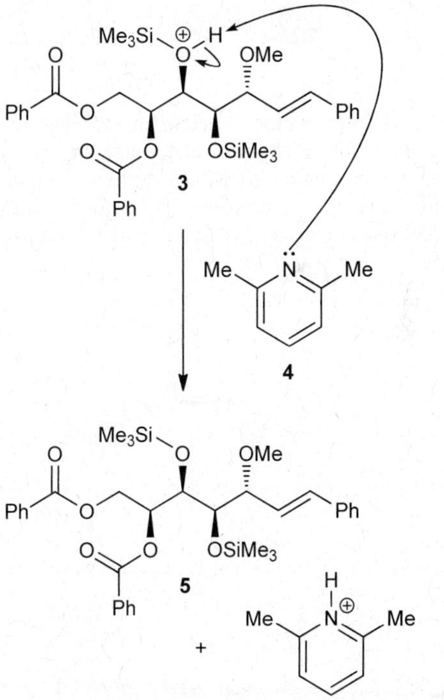

(b) If we consult the pK_a table found in the inside cover of the textbook, we find the following entries:

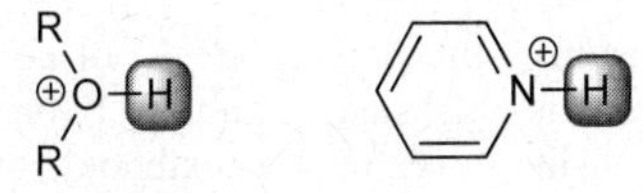

Notice that the first cation (an oxonium ion) has a pK_a value of approximately -3.6, while the second cation (a pyridinium ion) has a pK_a value of 3.4. The difference between them is seven pK_a units. In other words, an oxonium ion is approximately 10^7 (or 10,000,000) times more acidic than a pyridinium ion. Since a proton transfer step will proceed in the direction that favors the weaker acid, we expect compound **4** to be successful in deprotonating intermediate **3** to give compound **5**.

(c) The conversion of **1** to **5** involves the loss of an O-H bond. That is, the IR spectrum of compound **1** should exhibit a broad signal between 3200 and 3600 cm^{-1}, due to the O-H stretching vibration. This signal is expected to be absent in the IR spectrum of compound **5**, which lacks an O-H bond. This can be used to verify that the desired reaction has occurred. Specifically, the disappearance of the O-H signal indicates the conversion of **1** to **5**.

3.66.
(a) The proton highlighted below is the most acidic proton.

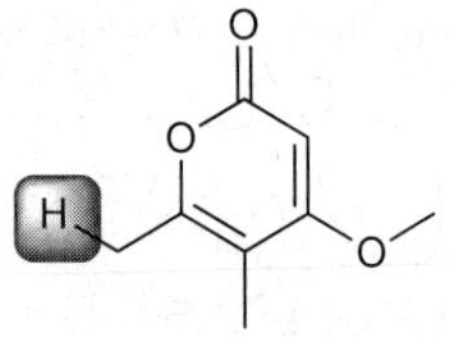

Removal of this proton generates a conjugate base in which the negative charge is delocalized by resonance. Notice that the charge is spread over four positions; and in the final resonance structure, the negative charge is on an electronegative oxygen atom, which stabilizes the charge.

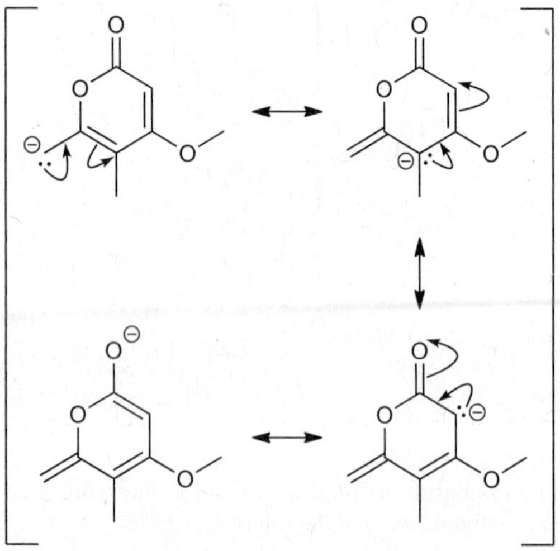

(b) Let's begin by estimating the pK_a of compound **1**. This compound is not listed in Table 3.1 (or in the pK_a table in the inside cover of the textbook), but we can make a rough estimation in the following way. The conjugate base of compound **1** has four resonance structures (above), in which the negative charge is spread

over three carbon atoms and an oxygen atom. If we look for a compound whose conjugate base has similar features, we will find that the conjugate base of phenol (listed in Table 3.1) also has the negative charge spread over three carbon atoms and an oxygen atom:

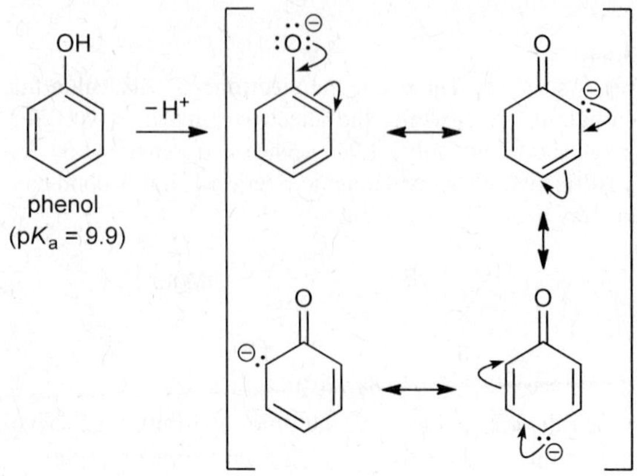

So, we might expect compound **1** to have a similar pK_a as phenol ($pK_a \sim 10$). Certainly, compound **1** and phenol are not identical, so we don't expect the pK_a of compound **1** to be exactly 9.9. But we would be surprised if the actual pK_a was extremely different than that of phenol. With this in mind, let's consider which bases are appropriate for deprotonating a compound with a pK_a near 10:

(i) NaOH – The conjugate acid of NaOH is H_2O (water). According to Table 3.1, H_2O has a pK_a of 15.7:

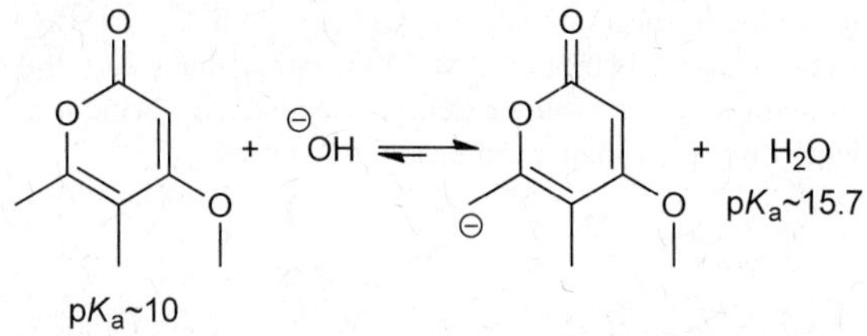

Recall that the equilibrium will favor the weaker acid (H_2O). Accordingly, NaOH should be a sufficient base for deprotonating compound **1**. This is true even if our estimate for the pK_a of compound **1** was incorrect by several pK_a units.

(ii) NaNH$_2$ – The conjugate acid of NaNH$_2$ is NH$_3$ (ammonia), which has a pK_a of 38. As such, NaNH$_2$ is most certainly a sufficiently strong base to deprotonate compound **1**. In fact, the difference in pK_a values is so vast that the reaction is essentially irreversible:

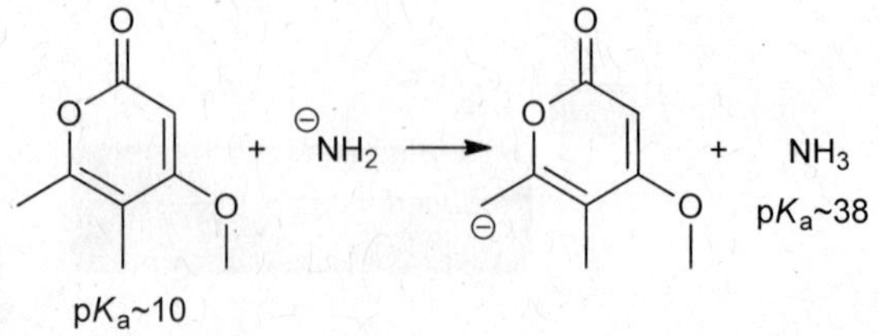

(iii) CH$_3$CO$_2$Na – The conjugate acid of CH$_3$CO$_2$Na is acetic acid (CH$_3$CO$_2$H), which has a pK_a of 4.8. Since the pK_a of acetic acid is less than 10, the equilibrium will not favor deprotonation of compound **1**. So, CH$_3$CO$_2$Na is not a sufficiently strong base to deprotonate compound **1**.

3.67. The second structure is an oxonium ion (O+) and is expected to function as an acid. The first structure has nitrogen atoms with lone pairs, so that compound will function as a base. First we must determine which nitrogen atom will function as a stronger base.

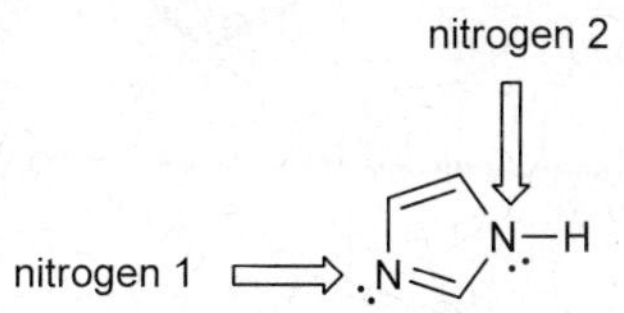

As we saw in SkillBuilder 2.9, the lone pair on nitrogen 1 will more readily function as a base than the lone pair on nitrogen 2, because the latter occupies a p-orbital and is participating in resonance. Thus, the lone pair on nitrogen 2 is not available to act as a base. In contrast, the lone pair on nitrogen 1 occupies an sp^2 hybridized orbital and is not participating in resonance. Therefore, it is readily available to function as a base, as seen in the following mechanism:

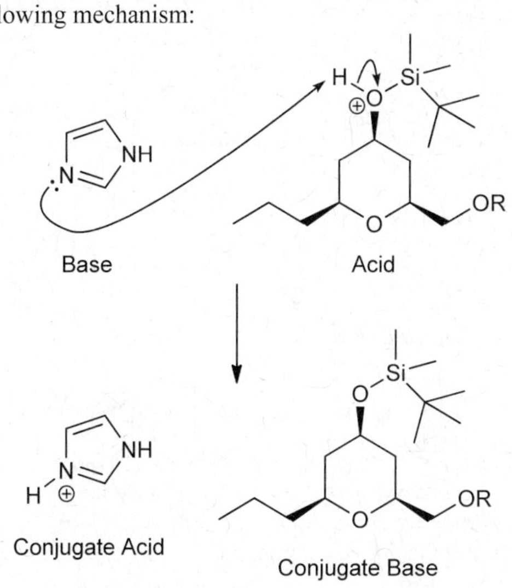

3.68.
(a) One of the lone pairs on the ketone group functions as a base and abstracts a proton from p-TsOH. Two curved arrows are required, as shown:

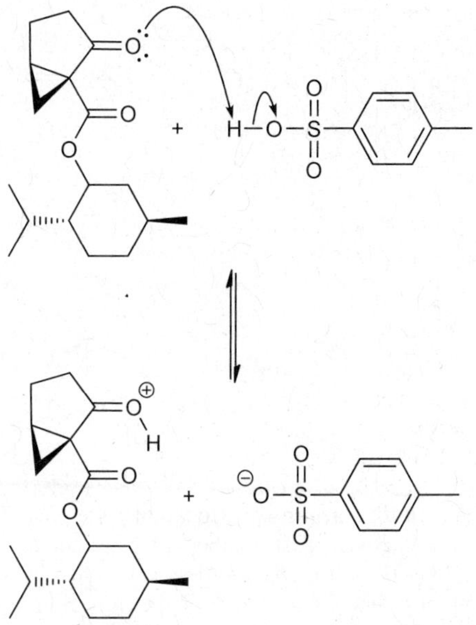

(b) Consult the pK_a table in the front cover of the book. We expect that p-TsOH will have a pK_a that is approximately -0.6, while the protonated ketone will have a pK_a that is approximately -7.3. As such, the protonated ketone is more acidic, which means that the equilibrium will not favor the protonated ketone. That is, there will be very little protonated ketone present at any moment in time. In Chapter 20, we will see that the presence of even a catalytic amount of protonated ketone is sufficient to achieve this type of transformation.

3.69.
(a) If we take the structure of **1**, as drawn, and rotate it 180 degrees, the same image is obtained. As such, there are only four different locations (rather than eight) where deprotonation can occur.

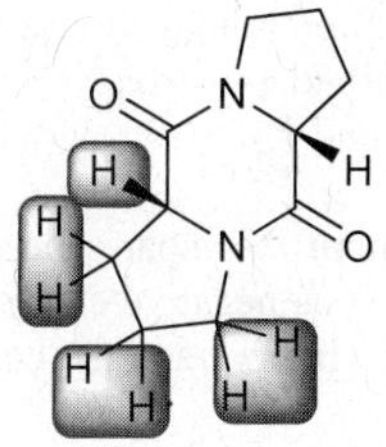

The most acidic proton is the one whose removal generates a resonance-stabilized conjugate base.

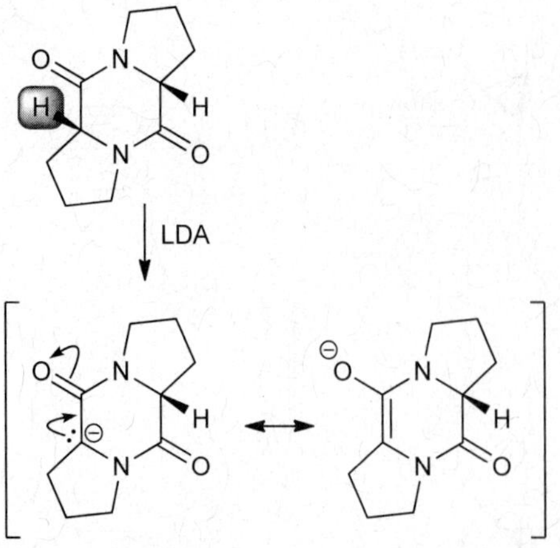

(b) Quantitative argument: Comparing the pK_a values on the inside cover of the textbook, we see that amines (pK_a ~ 38–40) are *significantly* less acidic than ketones (16–19), or esters (24–25), both of which bear an acidic proton on the carbon atom connected to the C=O group, allowing for a resonance stabilized conjugate base.

pK_a ~ 16-19   pK_a ~ 24-25

With this in mind, it is reasonable to expect the following proton of an amide to exhibit a lower pK_a than an amine:

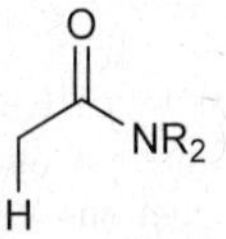

As such, the conjugate base of this compound should be more stable than the conjugate base of an amine. Therefore, LDA should be strong enough to remove a proton from compound **1**.

Qualitative argument: Compare the structures of the anions. LDA exhibits a negative charge that is localized on a nitrogen atom. In contrast, the conjugate base of **1**

is resonance-stabilized, with a major resonance contributor that places the negative charge on the more electronegative oxygen atom. As such, the conjugate base of **1** is more stable than LDA, so LDA should be a suitable base.

(c) LDA functions as a base and deprotonates compound **1**. This proton transfer step can either be drawn with two curved arrows, like this,

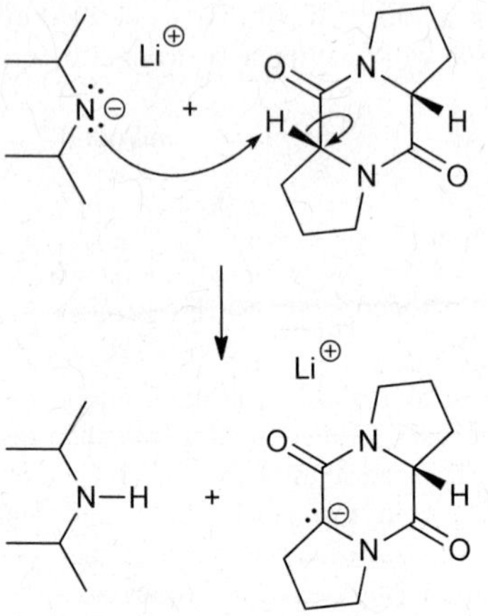

or with three curved arrows, like this:

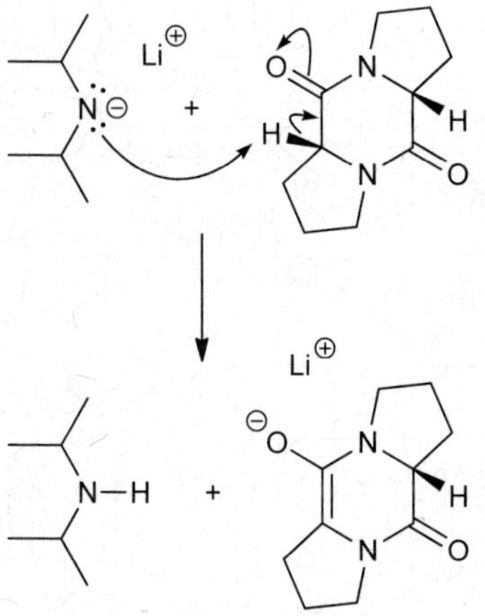

3.70. AlCl$_3$ is a Lewis acid. That is, it is capable of accepting electron density. So we must inspect compound **1** and determine the locations of high electron density. There are two oxygen atoms that bear lone pairs, and either location could certainly function as a Lewis base to donate an electron pair. But let's compare the structure of the complex that is obtained in each scenario. If the oxygen atom of the C=O bond interacts with AlCl$_3$, then the following resonance-stabilized complex is obtained:

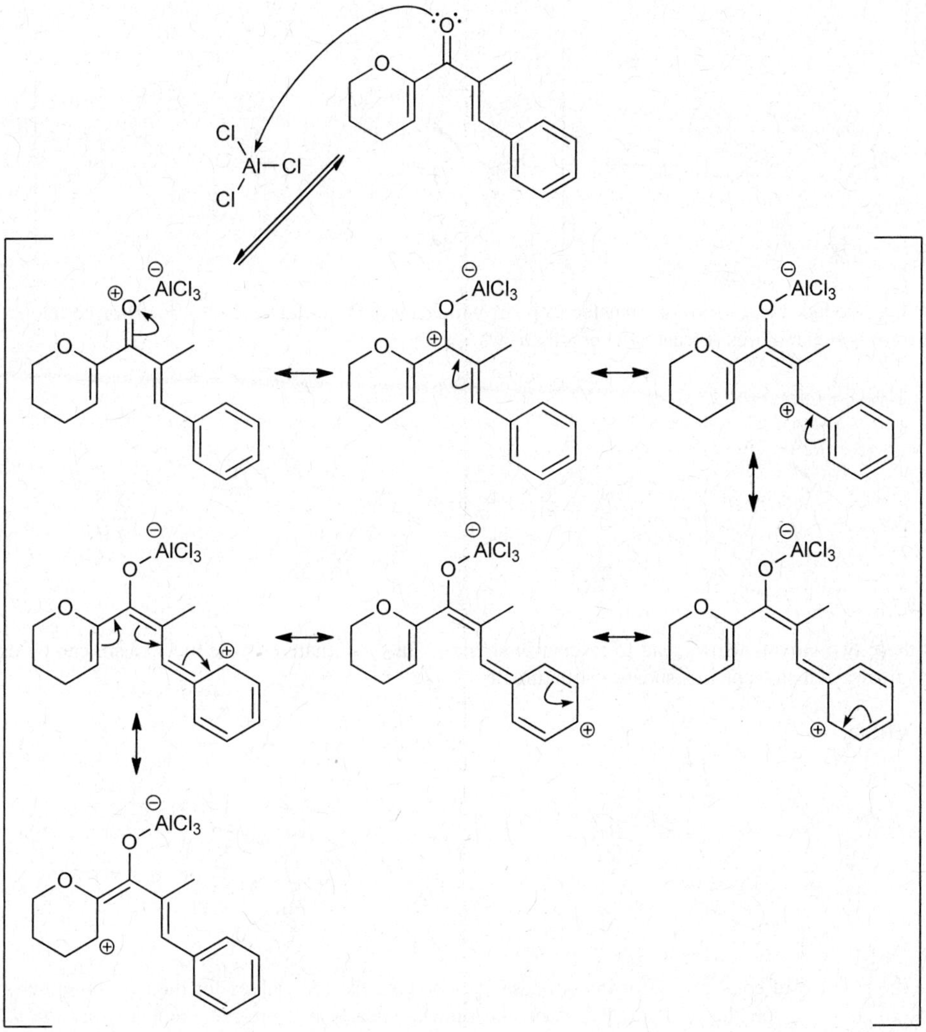

However, if the other oxygen atom interacts with AlCl₃, the following complex is obtained:

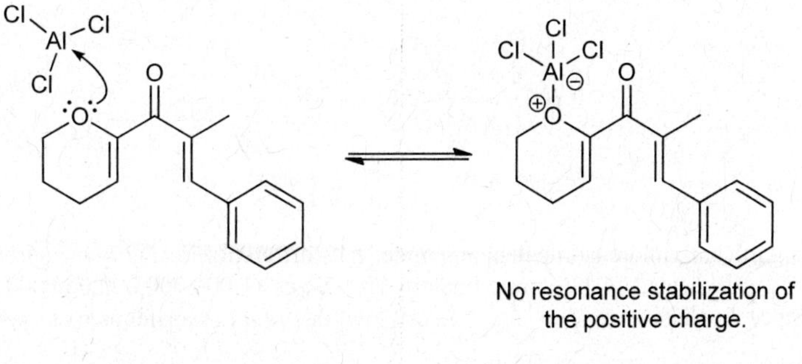

No resonance stabilization of
the positive charge.

Notice that the positive charge in this complex is not resonance stabilized. As such, we expect the oxygen atom of the C=O group to interact with the AlCl₃ to form the lower-energy, resonance-stabilized complex.

3.71.
(a) As seen in Problem 3.69, LDA (compound **2**) is a very strong base because it has a negative charge on a nitrogen atom (much like NaNH₂). So it will deprotonate compound **1** to give an anion. The most acidic proton in compound **1**

is connected to the carbon atom adjacent to the C=O bond, since deprotonation at that location generates a resonance-stabilized conjugate base:

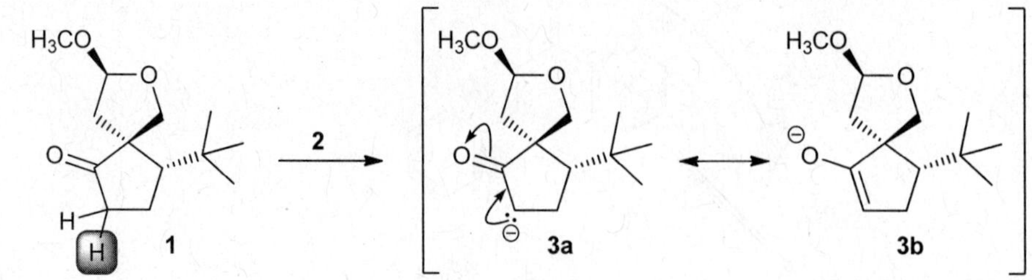

(b) Anion **3** is produced via a proton transfer step, in which the base removes a proton from compound **1**. The mechanism for this step requires at least two curved arrows:

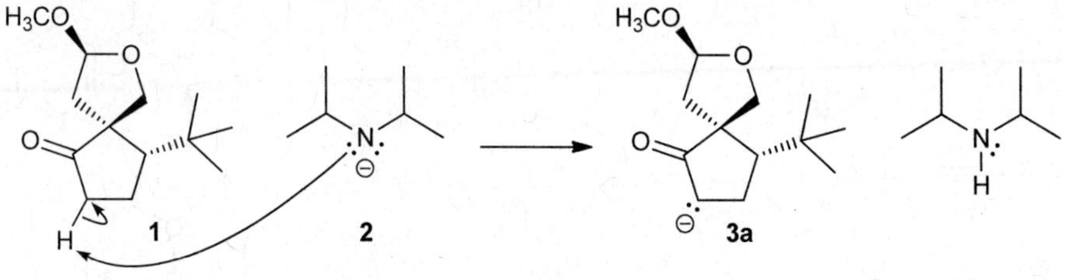

Notice that these two curved arrows lead to resonance structure **3a**. Alternatively, the mechanism can be drawn with three curved arrows, which leads to resonance structure **3b**:

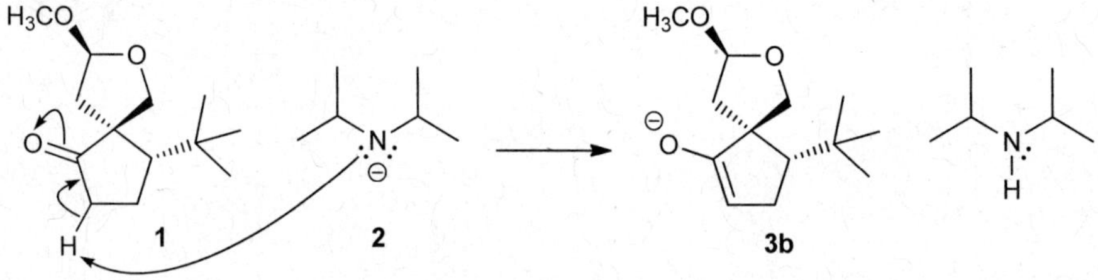

To evaluate the position of equilibrium for this process, we compare the pK_a values for the acid on either side of the equilibrium (compound **1** on the left, and R_2NH on the right). Clearly, we are not going to find the structure of compound **1** in Table 3.1. However, we can see from Table 3.1 that a proton connected to a carbon atom adjacent to the C=O bond of a ketone typically has a pK_a of approximately 19. In contrast, an amine is expected to have a pK_a near 38:

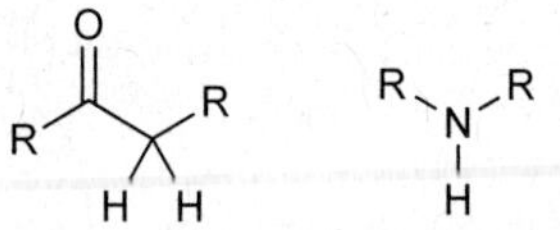

Notice that a ketone is significantly more acidic than an amine. The difference in acidity is approximately 21 pK_a units. In other words, a ketone is expected to be approximately 10^{21} times (1,000,000,000,000,000,000,000 times) more acidic than an amine. Since the difference is so large, we can treat this proton transfer step as irreversible.

(c) Compound **6** bears a negative charge on a nitrogen atom, which is generally fairly unstable. However, this anion is stabilized by several factors. The charge is delocalized into the neighboring triflate (Tf) group via resonance:

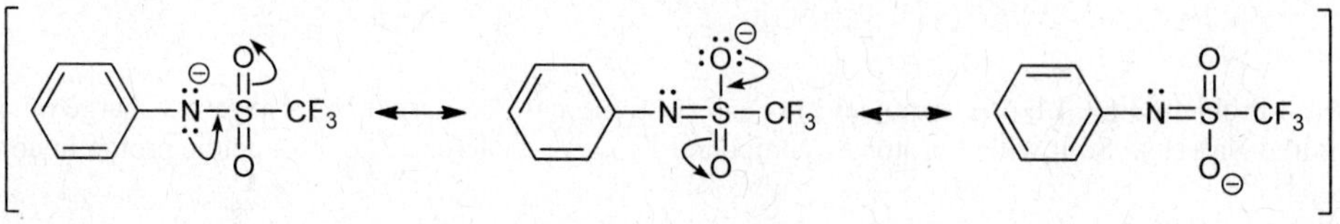

And the charge is further delocalized into the ring via resonance:

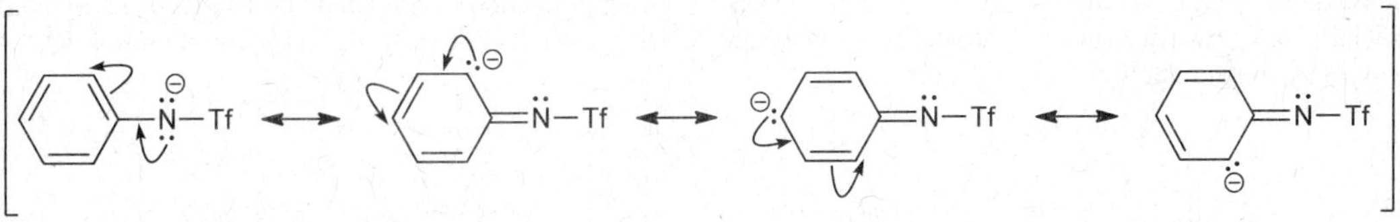

In total, the negative charge is spread over three carbon atoms, one nitrogen atom, and two oxygen atoms. The charge is therefore highly delocalized.

In addition, the electronegative fluorine atoms in the triflate group withdraw electron density via induction, thereby stabilizing the negative charge even further. As such the negative charge in this case is stabilized by resonance as well as induction, and is therefore highly stabilized.

3.72. The conjugate base of compound **5** possesses many resonance contributors. The negative charge can be delocalized into both aromatic rings and so the conjugate base is highly stabilized. Notice, however, that the placement of the negative charge is exclusively on carbon atoms in the ring and never on an electronegative atom such as an oxygen atom that can better stabilize the charge.

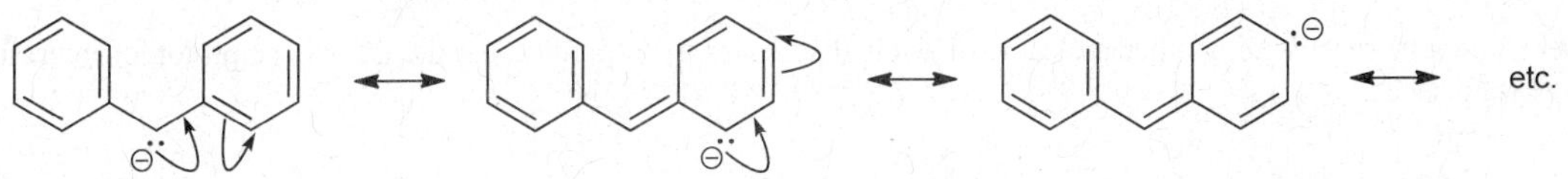

In contrast, in the conjugate base of **6**, the negative charge is delocalized into the carbonyl oxygen atom:

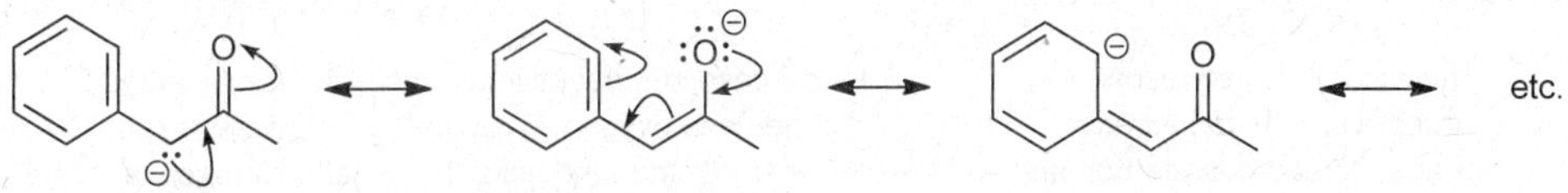

We now have to contend with two competing arguments: on the one hand, the conjugate base of **6** should be more stable than the conjugate base of **5** because the former has a resonance structure in which the negative charge is on an oxygen atom, while the latter lacks such a resonance structure. On the other hand, the conjugate base of **5** has more resonance structures (seven) than the conjugate base of **6** (which has only five resonance structures). In other words, is it more stable to spread a negative charge over seven carbon atoms or to spread the negative charge over four carbon atoms and an oxygen atom? The table of pK_a values will give us guidance on this question. It is apparent that compounds **1** and **3** represent the comparison we are asked to make in this problem. The conjugate base of compound **1** involves delocalization of the negative charge into the keto group while that of compound **3** will involve the aromatic ring. From the given pK_a values, it is clear that the keto group is more acidifying than the phenyl group. So, the same trend must hold when comparing the acidities of **5** and **6**. Therefore, the keto group of the conjugate base of **6** is better able to stabilize the negative charge than the phenyl group of the conjugate base of **5**, despite the fact that there are overall more resonance contributors in the latter than the former.

3.73. Let's estimate the relative acidity of the protons in the compound by comparing them to similar acids in Table 3.1. We are considering reactions with 1-4 molar equivalents of base, so we should aim to determine the four most acidic protons in the compound. We focus on the protons directly bound to electronegative heteroatoms (O, N, etc) and protons bound to non-sp^3 carbons. Three of the four most acidic protons are directly comparable to acids in Table 3.1. Note that the pK_a values shown are not meant to be precise (due to other structural factors that can affect these values), but rather, they should be considered as rough approximations.

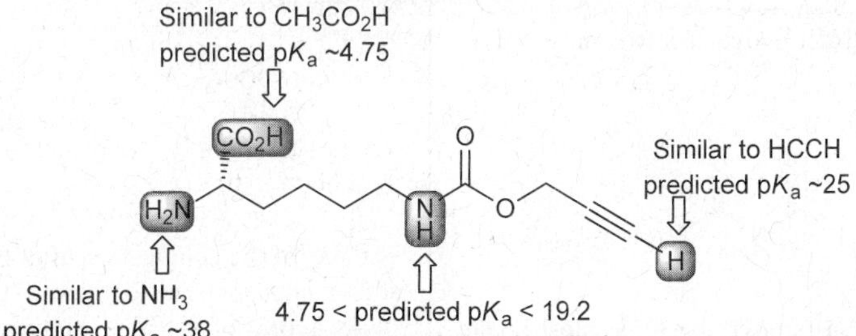

Similar to CH_3CO_2H
predicted pK_a ~4.75

Similar to HCCH
predicted pK_a ~25

Similar to NH_3
predicted pK_a ~38

4.75 < predicted pK_a < 19.2
(see text below for explanation)

The proton on the nitrogen atom adjacent to the C=O group is also a good candidate for one of our four most acidic protons, because it is bound to an electronegative atom (nitrogen). In this case there is no obvious molecule of comparison in Table 3.1. However, we can still make a reasonable prediction of pK_a range by using our knowledge of the relationship between structure and acidity. First of all, we can predict that the pK_a of the indicated proton is **greater than 4.75** by making the following comparison:

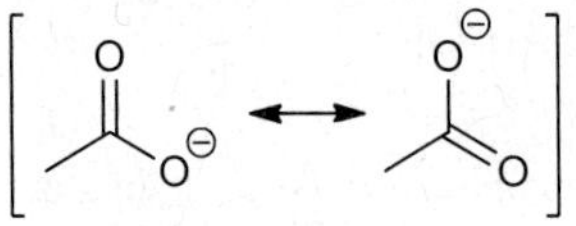

This anion has two resonance structures, each with a negative charge on oxygen.

[The pK_a of the conjugate acid of this anion is 4.75]

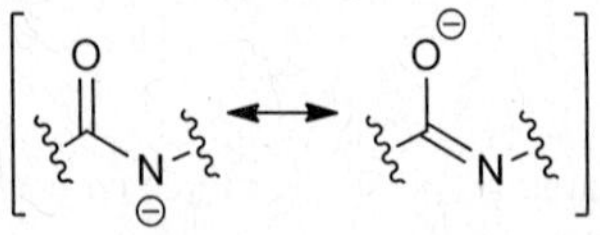

In this case, the negative charge is shared between nitrogen and oxygen. Therefore, this anion is predicted to be *less stable* than the anion to the left. (Nitrogen is less electronegative than oxygen, so it is not as stable with a negative charge.)

[The pK_a of the conjugate acid of this anion is predicted to be >4.75]

A second comparison with another molecule in Table 3.1 allows us to predict that the pK_a of the proton of interest is **lower than 19.2.**

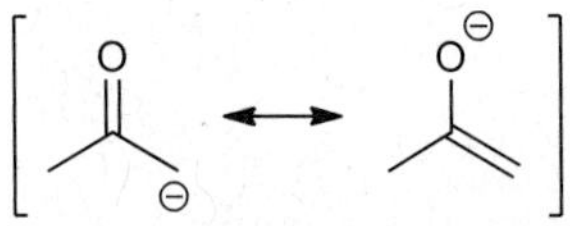

This anion has two resonance structures, with the negative charge spread over carbon and oxygen.

[The pK_a of the conjugate acid of this anion is 19.2]

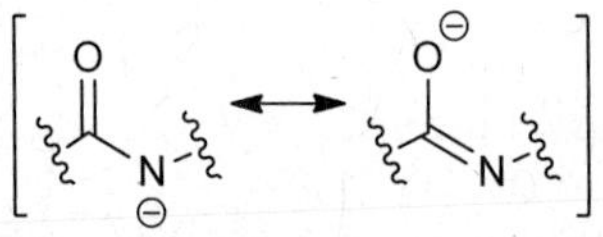

In this case, the negative charge is shared between nitrogen and oxygen. This anion is predicted to be *more stable* than the anion to the left. (Nitrogen is more electronegative than carbon and therefore more stable with a negative charge.)

[The pK_a of the conjugate acid of this anion is predicted to be <19.2]

The other piece of information we need to solve this problem is the relative basicities of the bases EtNa and EtONa. Both of these anions are listed in Table 3.1. EtNa is a strong enough base to remove any of the indicated protons in our compound; the pK_a of its conjugate acid is 50 – higher than the pK_a values of all four indicated protons in our compound. With this information in hand, we can make the following predictions:

(a) One mole of EtNa removes the most acidic proton, producing:

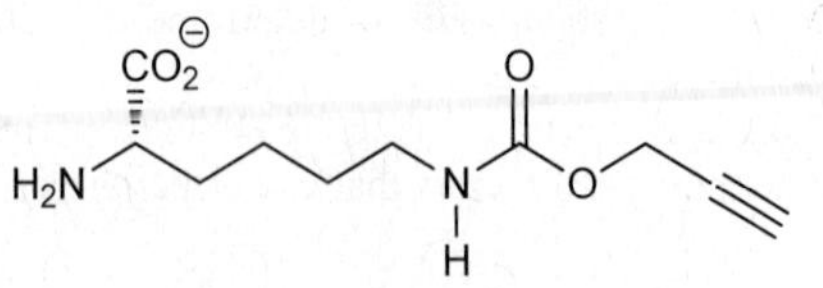

(b) Two moles of EtNa remove the two most acidic protons, producing:

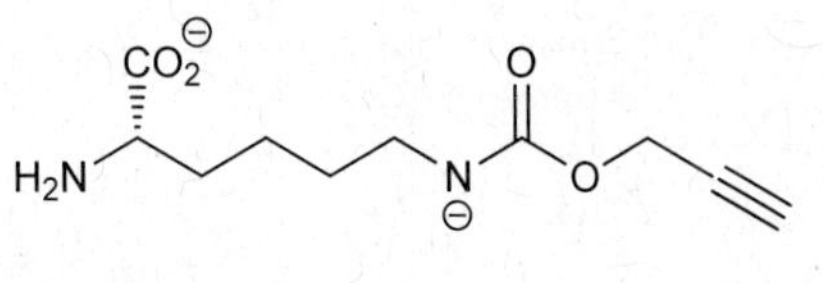

(c) Three moles of EtNa remove the three most acidic protons, producing:

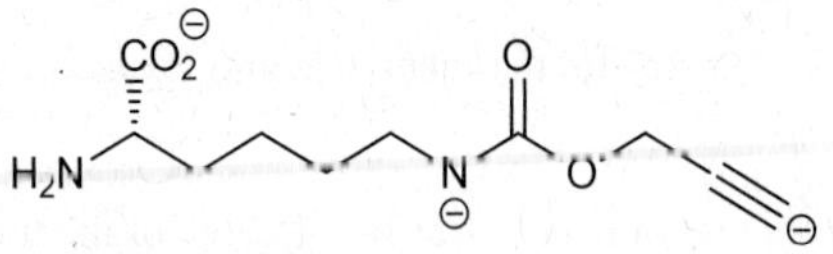

(d) Four moles of EtNa remove the four most acidic protons, producing:

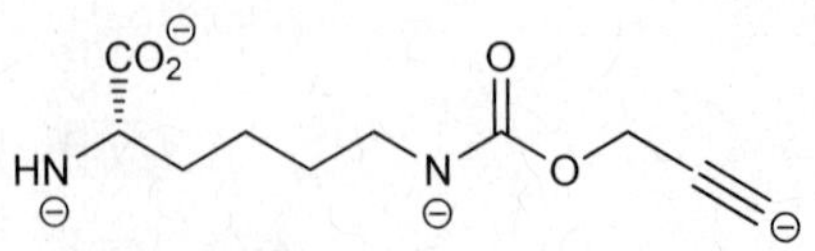

The pK_a of the conjugate acid of EtONa is 16. It is thus a strong enough base to remove the most acidic proton (with a pK_a of ~4.75 – significantly below 16). The second most acidic proton is estimated to have a pK_a value between 4.75 and 19, so EtONa should be able to deprotonate this position to a certain degree. The specific extent of deprotonation is dependent on the precise pK_a value within this range (below 16 or above 16). All other protons in this molecule have pK_a values significantly higher than 16, and thus will not be deprotonated by EtONa. This allows us to make the following predictions.

(e) One mole of EtONa removes the most acidic proton, producing:

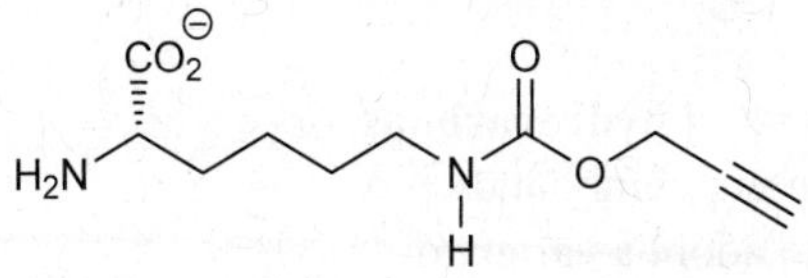

(f) Two moles of EtONa remove the two most acidic protons, producing:

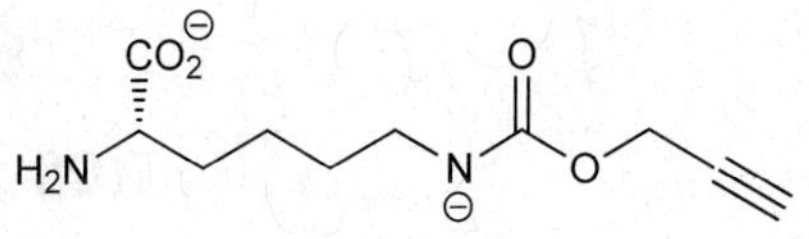

(g) Three moles of EtONa remove only the two most acidic protons, producing:

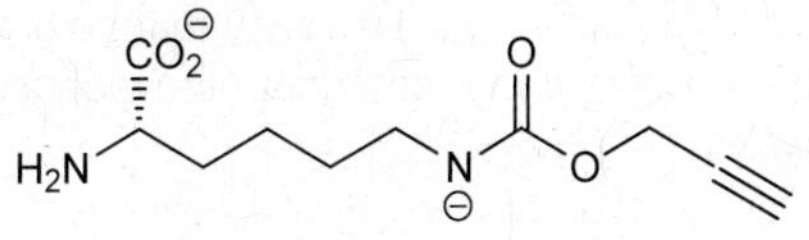

(h) Four moles of EtONa remove only the two most acidic protons, producing:

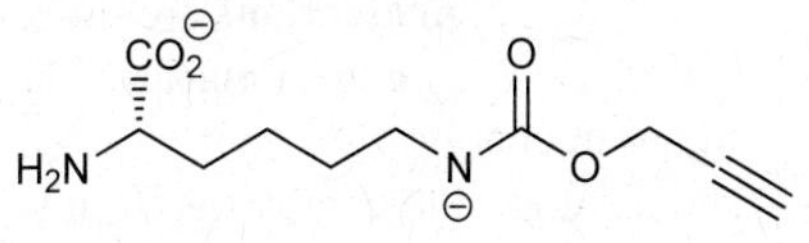

Chapter 4
Alkanes and Cycloalkanes

Review of Concepts

Fill in the blanks below. To verify that your answers are correct, look in your textbook at the end of Chapter 4. Each of the sentences below appears verbatim in the section entitled *Review of Concepts and Vocabulary*.

- Hydrocarbons that lack _____ are called **saturated hydrocarbons**, or _____.
- _____ provide a systematic way for naming compounds.
- Rotation about C-C single bonds allows a compound to adopt a variety of _____.
- _____ **projections** are often used to draw the various conformations of a compound.
- _____ **conformations** are lower in energy, while _____ **conformations** are higher in energy.
- The difference in energy between staggered and eclipsed conformations of ethane is referred to as _____ **strain**.
- _____ **strain** occurs in cycloalkanes when bond angles deviate from the preferred _____°.
- The _____ conformation of cyclohexane has no torsional strain and very little angle strain.
- The term **ring flip** is used to describe the conversion of one _____ conformation into the other. When a ring has one substituent…the equilibrium will favor the chair conformation with the substituent in the _____ position.

Review of Skills

Fill in the blanks and empty boxes below. To verify that your answers are correct, look in your textbook at the end of Chapter 4. The answers appear in the section entitled *SkillBuilder Review*.

SkillBuilder 4.1 Identifying the Parent

IDENTIFY THE PARENT IN EACH OF THE FOLLOWING COMPOUNDS.

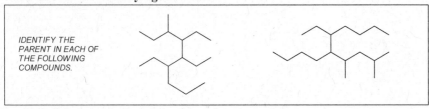

SkillBuilder 4.2 Identifying and Naming Substituents

STEP 1 - IDENTIFY THE PARENT IN THE FOLLOWING COMPOUND

STEPS 2 AND 3 - CIRCLE AND NAME ALL ALKYL SUBSTITUENTS CONNECTED TO THE PARENT

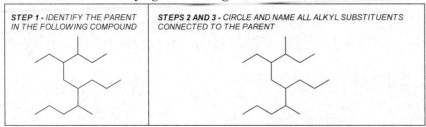

SkillBuilder 4.3 Identifying and Naming Complex Substituents

PROVIDE A NAME FOR THE FOLLOWING COMPLEX SUBSTITUENT (HIGHLIGHTED)

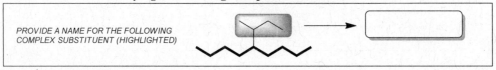

SkillBuilder 4.4 Assembling the Systematic Name of an Alkane

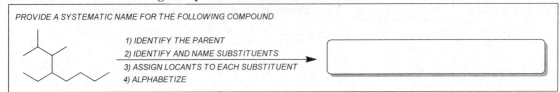

PROVIDE A SYSTEMATIC NAME FOR THE FOLLOWING COMPOUND

1) IDENTIFY THE PARENT
2) IDENTIFY AND NAME SUBSTITUENTS
3) ASSIGN LOCANTS TO EACH SUBSTITUENT
4) ALPHABETIZE

SkillBuilder 4.5 Assembling the Name of a Bicyclic Compound

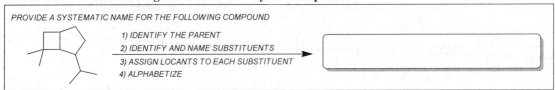

PROVIDE A SYSTEMATIC NAME FOR THE FOLLOWING COMPOUND

1) IDENTIFY THE PARENT
2) IDENTIFY AND NAME SUBSTITUENTS
3) ASSIGN LOCANTS TO EACH SUBSTITUENT
4) ALPHABETIZE

SkillBuilder 4.6 Identifying Constitutional Isomers

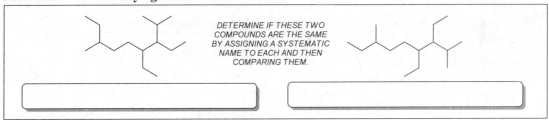

DETERMINE IF THESE TWO COMPOUNDS ARE THE SAME BY ASSIGNING A SYSTEMATIC NAME TO EACH AND THEN COMPARING THEM.

SkillBuilder 4.7 Drawing Newman Projections

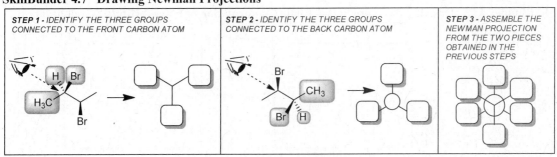

STEP 1 - IDENTIFY THE THREE GROUPS CONNECTED TO THE FRONT CARBON ATOM

STEP 2 - IDENTIFY THE THREE GROUPS CONNECTED TO THE BACK CARBON ATOM

STEP 3 - ASSEMBLE THE NEWMAN PROJECTION FROM THE TWO PIECES OBTAINED IN THE PREVIOUS STEPS

SkillBuilder 4.8 Identifying Relative Energy of Conformations

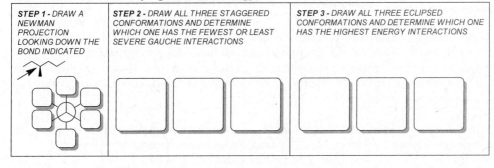

STEP 1 - DRAW A NEWMAN PROJECTION LOOKING DOWN THE BOND INDICATED

STEP 2 - DRAW ALL THREE STAGGERED CONFORMATIONS AND DETERMINE WHICH ONE HAS THE FEWEST OR LEAST SEVERE GAUCHE INTERACTIONS

STEP 3 - DRAW ALL THREE ECLIPSED CONFORMATIONS AND DETERMINE WHICH ONE HAS THE HIGHEST ENERGY INTERACTIONS

SkillBuilder 4.9
Drawing a Chair Conformation

DRAW A CHAIR CONFORMATION

SkillBuilder 4.10
Drawing Axial and Equatorial Positions

DRAW A CHAIR CONFORMATION SHOWING ALL SIX AXIAL POSITIONS AND ALL SIX EQUATORIAL POSITIONS

SkillBuilder 4.11 Drawing Both Chair Conformations of a Monosubstituted Cyclohexane

DRAW BOTH CHAIR CONFORMATIONS OF BROMOCYCLOHEXANE

SkillBuilder 4.12 Drawing Both Chair Conformations of Disubstituted Cyclohexanes

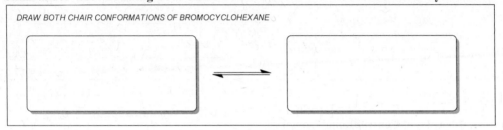

DRAW BOTH CHAIR CONFORMATIONS OF THE FOLLOWING COMPOUND

SkillBuilder 4.13 Drawing the More Stable Chair Conformation of Polysubstituted Cyclohexanes

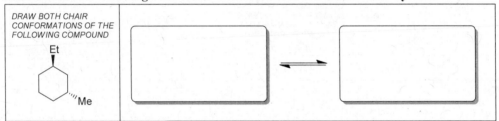

DRAW BOTH CHAIR CONFORMATIONS OF THE FOLLOWING COMPOUND AND DETERMINE WHICH ONE IS MORE STABLE

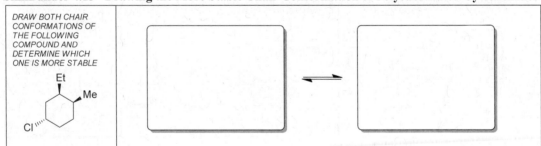

Common Mistakes to Avoid

You might find that you struggle with problems that ask you to draw constitutional isomers. Don't be discouraged. Many students struggle with drawing constitutional isomers. In particular, it is sometimes difficult to find ALL of the constitutional isomers with a particular molecular formula, and it is also difficult to avoid drawing the same compound more than once. However, these skills will be critical as we progress through the upcoming chapters. So, in order to get more proficient with constitutional isomers, do the following:

1) Skip to Section 15.16 of your textbook, and read that entire section (it is only a few pages, and you do not need any background to understand that section in its entirety). Then, do SkillBuilder 15.4, including all of the problems in that SkillBuilder.

2) Review the methodical approach for drawing constitutional isomers that is presented in the solution to problem 4.3.

3) Try to do problem 4.15, using the approach outlined in the solution to problem 4.3, and then check the solution to problem 4.15 to make sure that you applied the approach correctly.

If you complete the three tasks above, you should gain confidence in your ability to draw constitutional isomers and to avoid drawing the same isomer twice.

Solutions

4.1.

(a) The longest chain is six carbon atoms (shown below), so the parent is hexane:

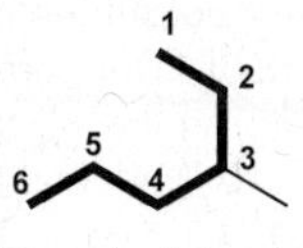

Parent = hexane

(b) The longest chain is seven carbon atoms (shown below), so the parent is heptane:

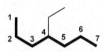

Parent = heptane

(c) The longest chain is seven carbon atoms (shown below), so the parent is heptane:

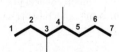

Parent = heptane

(d) The longest chain is nine carbon atoms (shown below), so the parent is nonane:

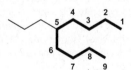

Parent = nonane

(e) The longest chain is eight carbon atoms (shown below), so the parent is octane:

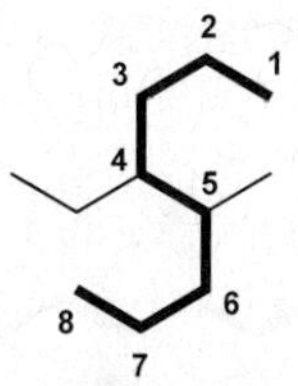

Parent = octane

(f) The longest chain is seven carbon atoms, so the parent is heptane. In this case, there is more than one seven-carbon chain. While the parent will be heptane

regardless of which chain we choose, the correct parent chain is the one with the most substituents, shown below (this will be important later when we must use the numbering scheme to identify the locations of the substituents connected to the chain):

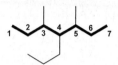

Parent = heptane

(g) This compound has a five-membered ring (shown below), so the parent is cyclopentane:

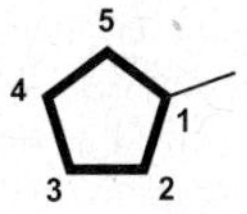

Parent = cyclopentane

(h) This compound has a seven-membered ring (shown below), so the parent is cycloheptane:

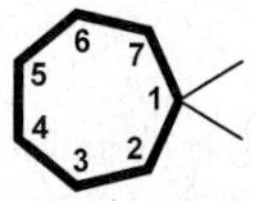

Parent = cycloheptane

(i) This compound has a three-membered ring (shown below), so the parent is cyclopropane:

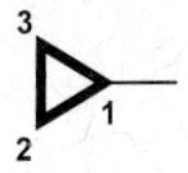

Parent = cyclopropane

4.2. Each of the following two compounds has a parent chain of eight carbon atoms (octane). The other two compounds that appear in the problem statement (not shown here) have parent chains of seven and nine carbon atoms, respectively.

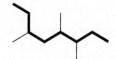   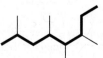

Parent = octane   Parent = octane

4.3. Recall that constitutional isomers are compounds that have the same molecular formula but differ in their constitution (connectivity of atoms). We must draw all of the constitutional isomers with molecular formula C_6H_{14}, so we are looking for all of the different ways of connecting six carbon atoms. We begin with a linear chain (parent = hexane):

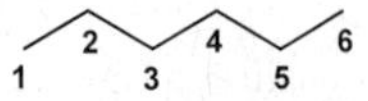

Next, we look for any constitutional isomers where the parent is pentane (five carbon atoms). There are only two such isomers. Specifically, we can either connect the extra CH_3 group to positions C2 or C3 of the pentane chain:

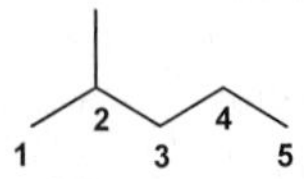

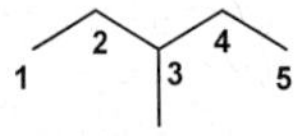

We cannot connect the CH_3 group to positions C1 or C5, as that would simply give us the linear chain (hexane), which we already drew (above). We also cannot connect the CH_3 group to position C4 as that would generate the same structure as placing the CH_3 group at the C2 position:

Next, we look for any constitutional isomers where the parent is butane (four carbon atoms). There are only two such isomers. Specifically, we can either connect two CH_3 groups to adjacent positions (C2 and C3) or the same position:

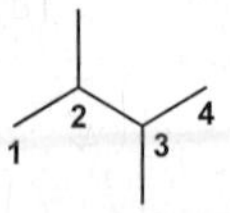

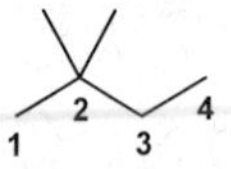

If we try to connect a CH_3CH_2 group to a butane chain, we end up with a pentane chain (an isomer already drawn above):

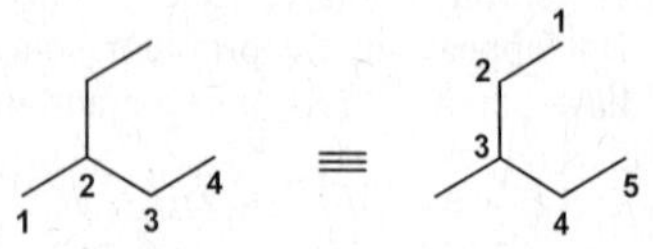

In summary, there are five constitutional isomers with molecular formula C_6H_{14}:

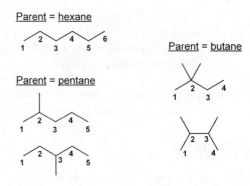

Parent = hexane

Parent = butane

Parent = pentane

4.4. We are looking for constitutional isomers of C_8H_{18} that have a parent chain of seven carbon atoms (heptane). The extra CH_3 group can be placed in any of three positions (C2, C3, or C4) giving the following three constitutional isomers:

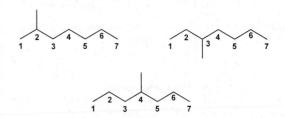

We cannot connect the CH_3 group to positions C1 or C7, as that would give a parent chain of eight carbon atoms (octane). We also cannot connect the CH_3 group to position C5 as that would generate the same structure as placing the CH_3 group at the C3 position:

Similarly, we cannot connect the CH_3 group to position C6 as that would generate the same structure as placing the CH_3 group at the C2 position. In summary, there are only three constitutional isomers of C_8H_{18} that have a parent name of heptane (shown above).

4.5.
(a) First we identify the parent (octane), and then we identify any alkyl substituents connected to the parent (highlighted here). In this case, all of the alkyl groups are methyl groups.

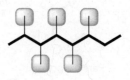

(b) First we identify the parent (nonane), and then we identify any alkyl substituents (highlighted) that are connected to the parent.

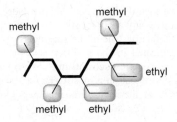

(c) First we identify the parent (cyclohexane), and then we identify any alkyl substituents (highlighted) that are connected to the parent.

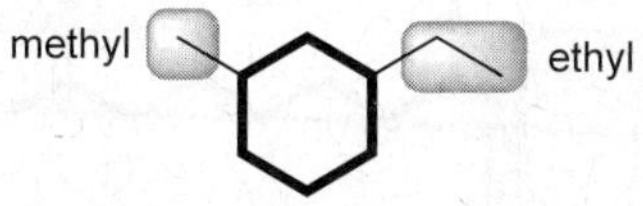

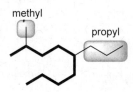

(d) First we identify the parent (cyclopropane), and then we identify any alkyl substituents (highlighted) that are connected to the parent.

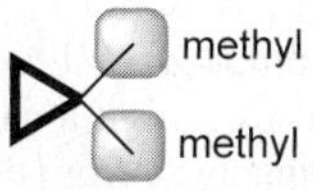

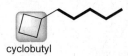

(e) First we identify the parent (nonane), and then we identify any alkyl substituents (highlighted) that are connected to the parent.

(f) First we identify the parent. In this case, there is a ring (of four carbon atoms) connected to an alkyl group of five carbon atoms. There are more carbon atoms in the alkyl chain than in the ring, so the former becomes the parent (pentane), and the latter is listed as a substituent (cyclobutyl group).

cyclobutyl

(g) First we identify the parent (undecane), and then we identify any alkyl substituents (highlighted) that are connected to the parent.

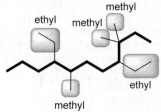

4.6.

(a) If the parent must be pentane, then we must draw constitutional isomers for which the longest chain is five carbon atoms. The problem statement also indicates that there must be two methyl groups. We cannot place a methyl group at C1 or C5 of the parent chain, as that would make the parent chain longer. If we place the first methyl group at C2, then the second methyl group can be placed at either C2, at C3 or at C4:

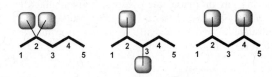

If, however, the first methyl group is placed at C3, there is only one location (C3) where we can place the second methyl group to obtain a new and unique isomer that we have not already drawn above:

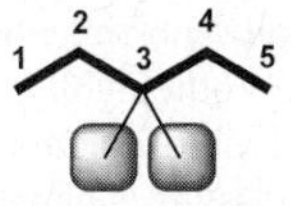

In summary, there are a total of four constitutional isomers that fit the criteria described in the problem statement.

(b) The C3 position is the only position on a pentane chain where an ethyl group can be placed without creating a longer chain. If the ethyl group is placed at C2, the longest chain would have six carbon atoms (hexane).

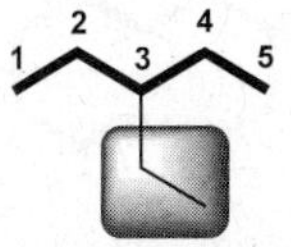

4.7.

(a) First we identify the parent chain (heptane), and then we identify any alkyl substituents connected to the parent. In this case, the substituent (highlighted) is complex, so we treat it as a "substituent on a substituent," and we assign a name based on numbers going away from the parent:

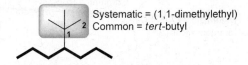

Systematic = (1,1-dimethylethyl)
Common = *tert*-butyl

(b) First we identify the parent chain (nonane), and then we identify any alkyl substituents connected to the parent. In this case, one of the highlighted substituents is complex, so we treat it as a "substituent on a substituent," and we assign a name based on numbers going away from the parent:

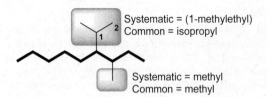

Systematic = (1-methylethyl)
Common = isopropyl

Systematic = methyl
Common = methyl

(c) First we identify the parent chain (nonane), and then we identify any alkyl substituents connected to the parent. In this case, the substituent (highlighted) is complex, so we treat it as a "substituent on a substituent," and we assign a name based on numbers going away from the parent:

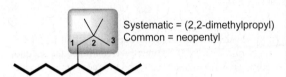

Systematic = (2,2-dimethylpropyl)
Common = neopentyl

(d) First we identify the parent chain (decane), and then we identify any alkyl substituents connected to the parent. In this case, both highlighted substituents are complex, so we treat each of them as a "substituent on a substituent," and we assign a name based on numbers going away from the parent:

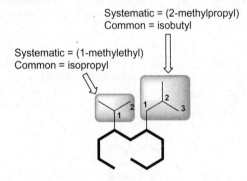

Systematic = (2-methylpropyl)
Common = isobutyl

Systematic = (1-methylethyl)
Common = isopropyl

(e) First we identify the parent chain (cyclohexane), and then we identify any alkyl substituents connected to the parent. In this case, all four substituents (highlighted) are complex, so we treat each of them as a "substituent on a substituent," and we assign a name based on numbers going away from the parent:

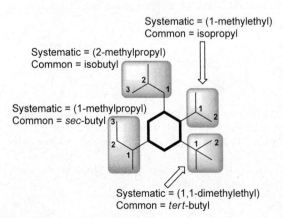

Systematic = (1-methylethyl)
Common = isopropyl

Systematic = (2-methylpropyl)
Common = isobutyl

Systematic = (1-methylpropyl)
Common = sec-butyl

Systematic = (1,1-dimethylethyl)
Common = tert-butyl

4.8. First we identify the parent chain (dodecane), and then we identify any alkyl substituents connected to the parent (highlighted). One of the substituents is a phenyl group, as indicated in the problem statement, and the other two are complex substituents. We treat each of them as a "substituent on a substituent," and we assign a name to each of them based on numbers going away from the parent, as seen here:

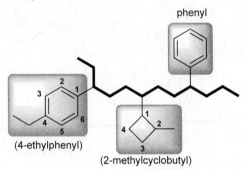

phenyl

(4-ethylphenyl)

(2-methylcyclobutyl)

4.9. We treat each complex substituent as a "substituent on a substituent," and we assign a name to each of them based on numbers going away from the parent, as shown:

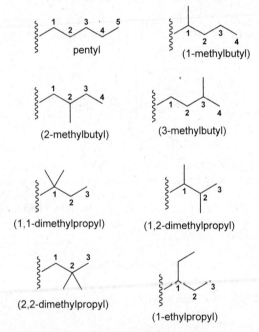

pentyl

(1-methylbutyl)

(2-methylbutyl)

(3-methylbutyl)

(1,1-dimethylpropyl)

(1,2-dimethylpropyl)

(2,2-dimethylpropyl)

(1-ethylpropyl)

4.10. For each of the following compounds, we assign its name via a four-step process: First identify the parent, then the substituents, then assign locants, and finally, arrange the substituents alphabetically. In each case, use commas to separate numbers from each other, and use hyphens to separate letters from numbers.

(a) 3,4,6-trimethyloctane
(b) sec-butylcyclohexane
(c) 3-ethyl-2-methylheptane
(d) 3-isopropyl-2,4-dimethylpentane
(e) 3-ethyl-2,2-dimethylhexane

(f) 2-cyclohexyl-4-ethyl-5,6-dimethyloctane
(g) 3-ethyl-2,5-dimethyl-4-propylheptane
(h) 5-*sec*-butyl-4-ethyl-2-methyldecane
(i) 2,2,6,6,7,7-hexamethylnonane
(j) 4,5-dimethylnonane
(k) 2,4,4,6-tetramethylheptane
(l) 2,2,4-trimethylpentane
(m) 4-*tert*-butylheptane
(n) 3-ethyl-6-isopropyl-2,4-dimethyldecane
(o) 3,5-diethyl-2-methyloctane
(p) 1,3-diisopropylcyclopentane
(q) 3-ethyl-2,5-dimethylheptane

4.11.

(a) The name indicates that the parent is a five-carbon chain and there are three substituents (a methyl group at C2, an isopropyl group at C3 and another methyl group at C4):

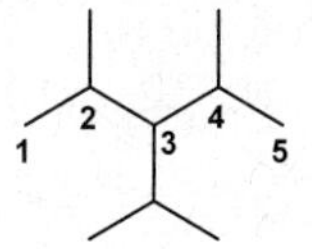

(b) The name indicates that the parent is a six-carbon chain and there are two substituents (a methyl group at C2 and an ethyl group at C4:

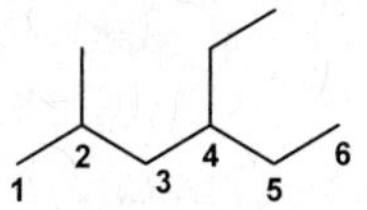

(c) The name indicates that the parent is a three-membered ring and there are four substituents (all methyl groups), as shown:

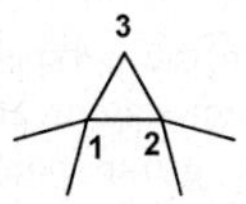

4.12.   For each of the following compounds, we assign its name via a four-step process: First identify the parent, then the substituents, then assign locants, and finally, arrange the substituents alphabetically. When assigning locants, make sure to start at a bridgehead and continue numbering along the longest path to the second bridgehead. Then continue assigning locants along the second longest path, and then finally, along the shortest path that connects the two bridgehead positions.

(a) 4-ethyl-1-methylbicyclo[3.2.1]octane
(b) 2,2,5,7-tetramethylbicyclo[4.2.0]octane
(c) 2,7,7-trimethylbicyclo[4.2.2]decane
(d) 3-*sec*-butyl-2-methylbicyclo[3.1.0]hexane
(e) 2,2-dimethylbicyclo[2.2.2]octane
(f) 2,7-dimethylbicyclo[3.3.0]octane
(g) bicyclo[1.1.0]butane
(h) 5,5-dimethylbicyclo[2.1.1]hexane
(i) 3-(3-methylbutyl)bicyclo[4.4.0]decane

4.13.

(a) The name indicates a bicyclic parent with two methyl groups at C2 and two methyl groups at C3, as shown:

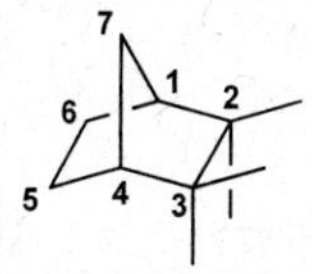

(b) The name indicates a bicyclic parent with two ethyl groups at C8, as shown:

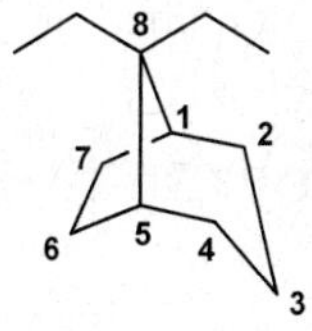

(c) The name indicates a bicyclic parent with an isopropyl group at C3, as shown:

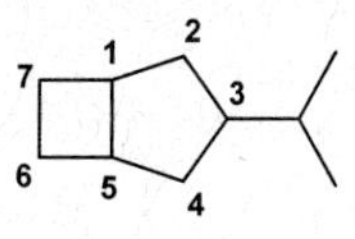

4.14.

(a) If we assign a systematic name for each of these structures, we find that they share the same name (2,3-dimethylpentane). Therefore, these drawings are simply different representations of the same compound.

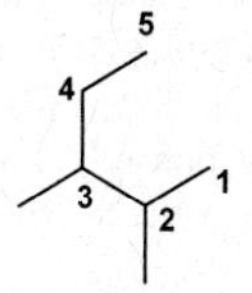

2,3-dimethylpentane 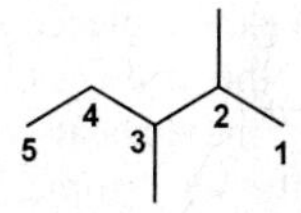   2,3-dimethylpentane

(b) If we assign a systematic name for each of these structures, we find that they share the same name (3-ethyl-2,4-dimethylpentane). Therefore, these drawings are simply different representations of the same compound.

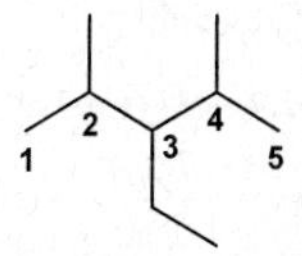

3-ethyl-2,4-dimethyl   3-ethyl-2,4-dimethyl
pentane   pentane

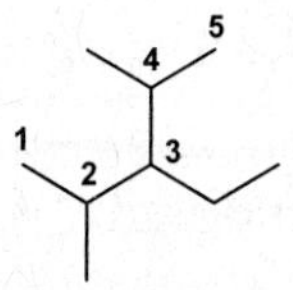

(c) If we assign a systematic name for each of these structures, we find that they share the same name (4-isobutyl-2,8-dimethylnonane). Therefore, these drawings are simply different representations of the same compound.

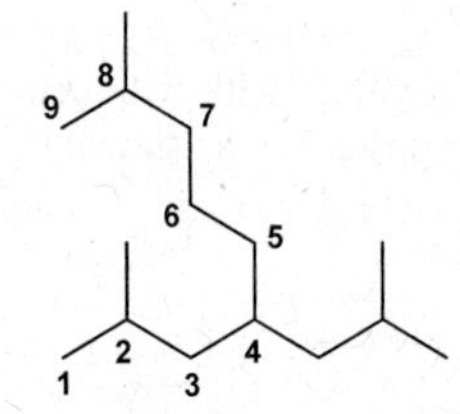

4-isobutyl-2,8-dimethyl
nonane

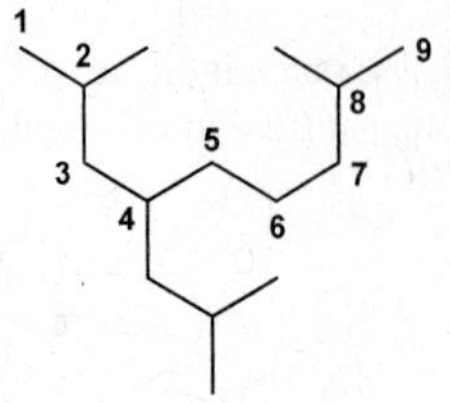

4-isobutyl-2,8-dimethyl
nonane

(d) We assign a systematic name to each structure, and find that they have different names (shown below). Therefore, they must differ in their constitution (connectivity of atoms). These compounds are therefore different from each other, but they share the same molecular formula ($C_{11}H_{24}$), so they are constitutional isomers.

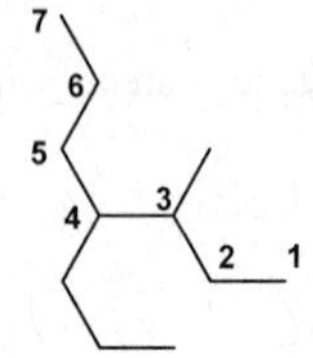

3-methyl-4-propyl
heptane

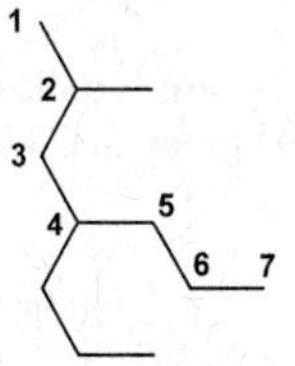

2-methyl-4-propyl
heptane

4.15. To draw all of the constitutional isomers of C_7H_{16}, we will follow the same methodical approach that we used in Problem **4.3**. We start with a linear chain of seven carbon atoms (heptane). Then, we draw all possible substituted hexanes with molecular formula C_7H_{16}. There are only two possibilities - the methyl group can be placed at either C2 or C3. Then, we move on to the pentanes, and finally any possible butanes. This methodical analysis gives the following constitutional isomers:

Parent = heptane

Parent = butane

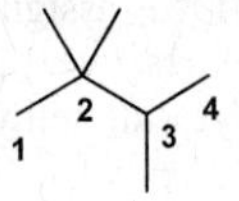

Parent = hexane

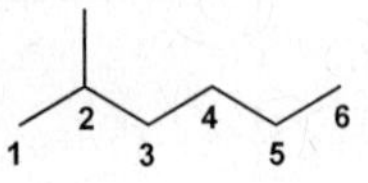

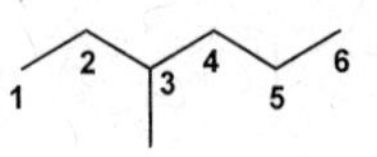

Parent = pentane

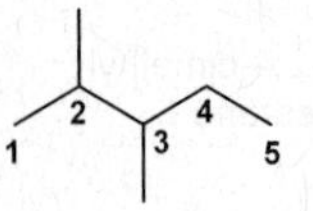

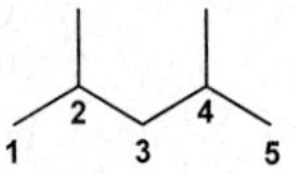

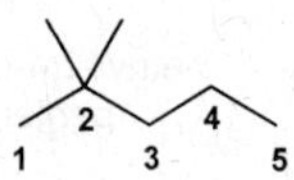

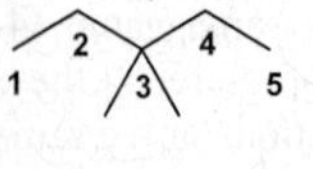

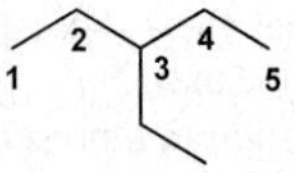

4.16.
(a) When looking from the perspective of the observer (as shown in the problem statement), the front carbon is connected to three methyl groups and the back carbon is connected to one methyl group pointing up.

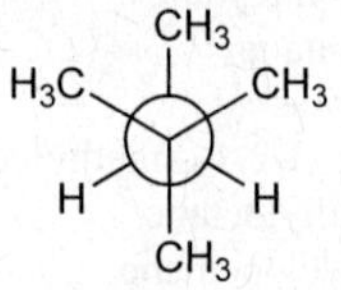

(b) When looking from the perspective of the observer (as shown in the problem statement), the front carbon is connected to a chlorine atom pointing up and to the right, as well as a methyl group pointing down. The back carbon atom is connected to a chlorine atom pointing down and to the left, as well as a methyl group pointing up.

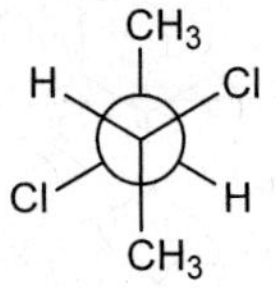

(c) When looking from the perspective of the observer (as shown in the problem statement), the front carbon is connected to a methyl group pointing up and to the right, as well as an ethyl group pointing down. The back carbon atom is connected to an ethyl group pointing up.

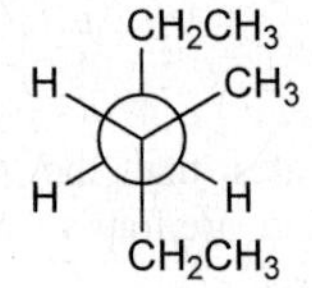

(d) When looking from the perspective of the observer (as shown in the problem statement), the front carbon is connected to a methyl group pointing up. The back carbon atom is connected to a methyl group pointing down and two chlorine atoms, as shown.

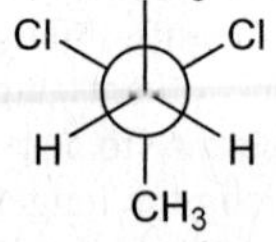

(e) When looking from the perspective of the observer (as shown in the problem statement), the front carbon is connected to a methyl group pointing up and a chlorine atom pointing down and to the right. The back carbon atom is connected to a methyl group pointing down and a chlorine atom pointing up and to the right, as shown.

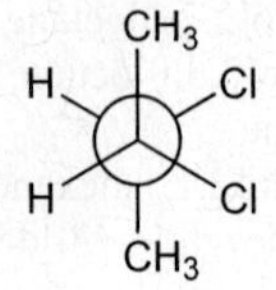

(f) When looking from the perspective of the observer (as shown in the problem statement), the front carbon is connected to a chlorine atom pointing up and to the right, as well as a methyl group pointing down. The back carbon atom is connected to a methyl group pointing up and a bromine atom pointing down and to the left.

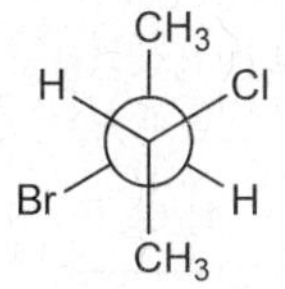

4.17.

(a) The Newman projection indicates that the front carbon atom is connected to two methyl groups, while the back carbon atom is connected to two ethyl groups and a methyl group. This corresponds with the following bond-line drawing:

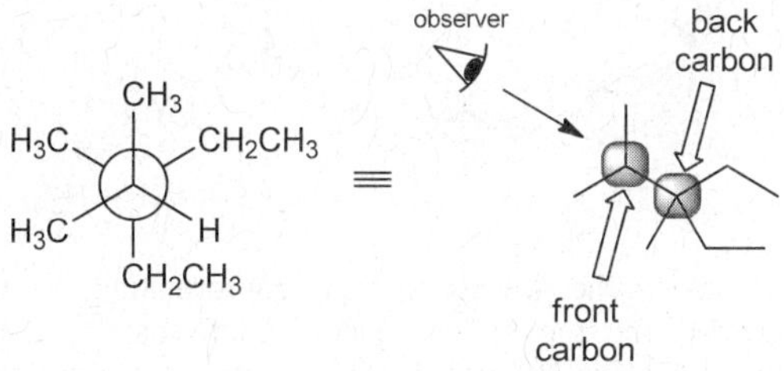

(b) The Newman projection indicates that the front carbon atom and the back carbon atom are part of a five-membered ring:

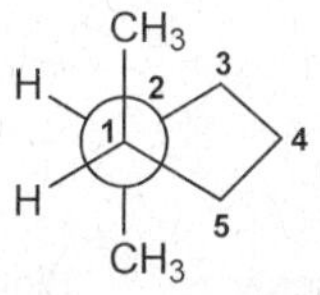

The front carbon atom is connected to a methyl group pointing above the ring, while the back carbon atom is connected to a methyl group pointing below the ring. This corresponds with the following bond-line drawing:

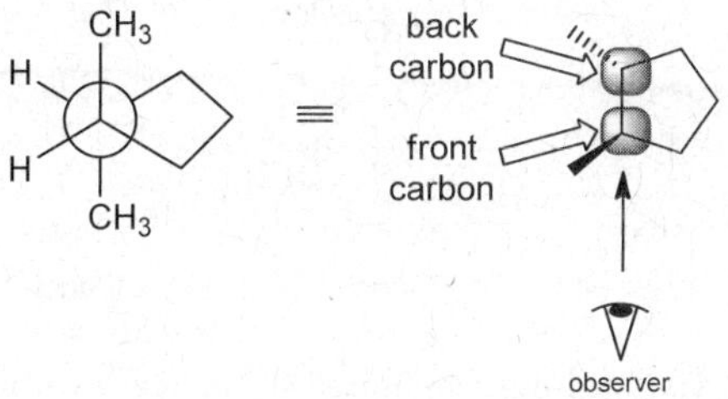

(c) In this case, there are two Newman projections connected to each other, indicating a ring. If we count

the carbon atoms, we can see that it is a six-membered ring (cyclohexane):

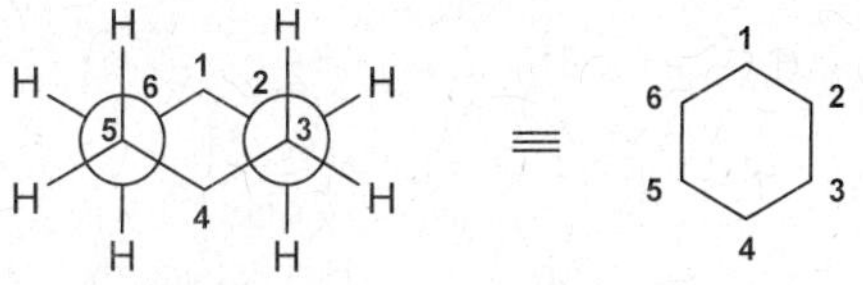

4.18. If we convert each Newman projection into a bond-line structure, we see that these structures are representations of the same compound (2,3-dimethylbutane). They are not constitutional isomers.

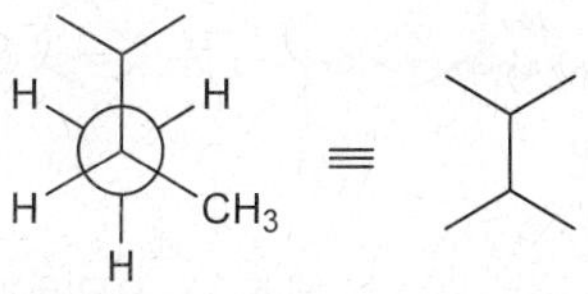

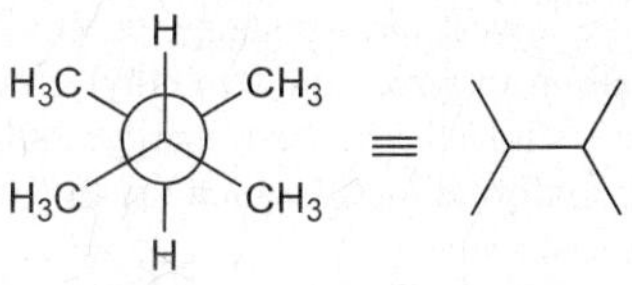

4.19.

(a) The energy barrier is expected to be approximately 18 kJ/mol (calculation below):

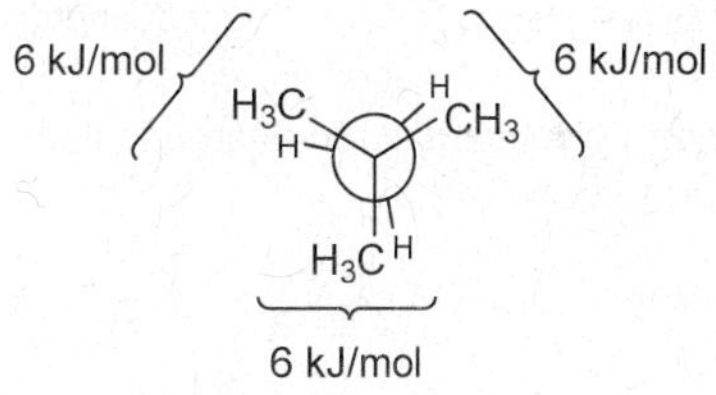

(b) The energy barrier is expected to be approximately 16 kJ/mol (calculation below):

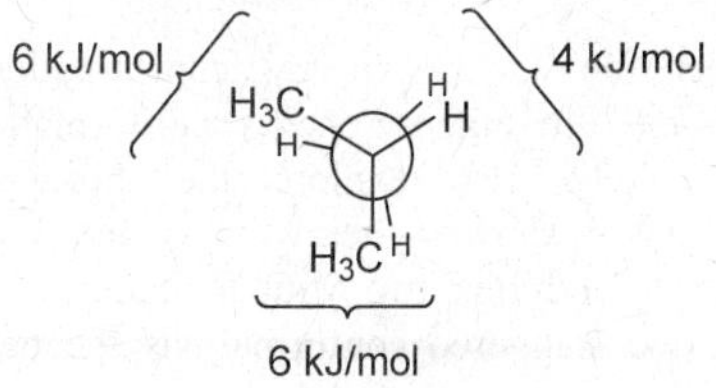

4.20.

(a) In the Newman projection, the front carbon atom has three methyl groups, and the back carbon has one methyl group. Since the front carbon atom has three identical groups, we expect all staggered conformations to be degenerate. Similarly, we expect all eclipsed conformations to be degenerate as well. The lowest energy conformation is the staggered conformation, and the highest energy conformation is the eclipsed conformation.

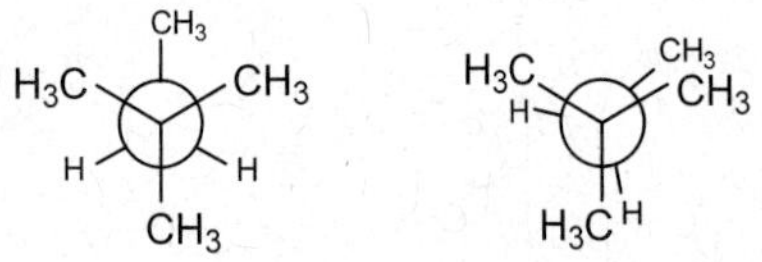

CH₃ / H₃C / CH₃ / H / H / CH₃

Lowest energy

H₃C / H / CH₃ / CH₃ / H₃C / H

Highest energy

(b) We begin by converting the bond-line drawing into a Newman projection.

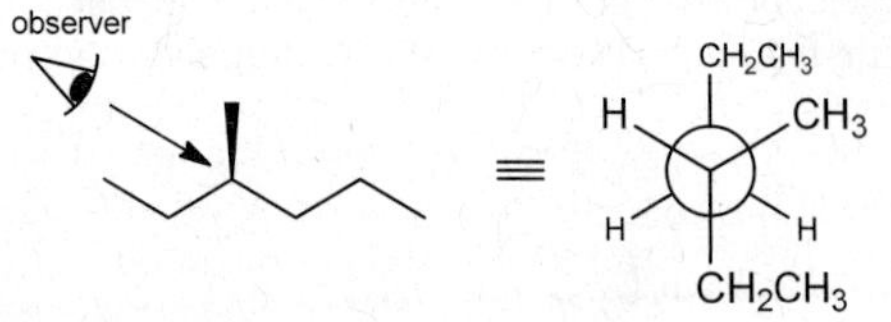

observer

≡

CH₂CH₃ / H / CH₃ / H / H / CH₂CH₃

We expect the lowest energy conformation to be staggered and the highest energy conformation to be eclipsed. The lowest energy staggered conformation is the conformation in which the two ethyl groups are *anti* to each other, while the highest energy conformation is the eclipsed conformation in which the two ethyl groups are eclipsing each other:

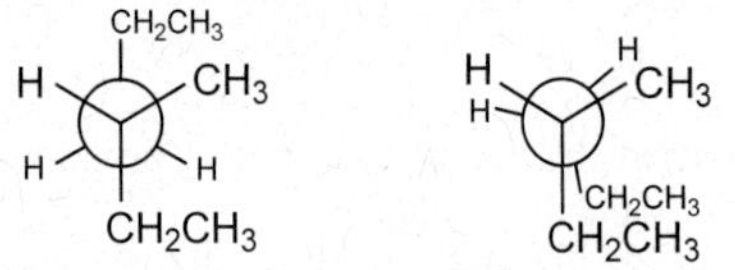

CH₂CH₃ / H / CH₃ / H / H / CH₂CH₃

Lowest energy

H / H / CH₃ / H / CH₂CH₃ / CH₂CH₃

Highest energy

(c) We begin by converting the bond-line drawing into a Newman projection.

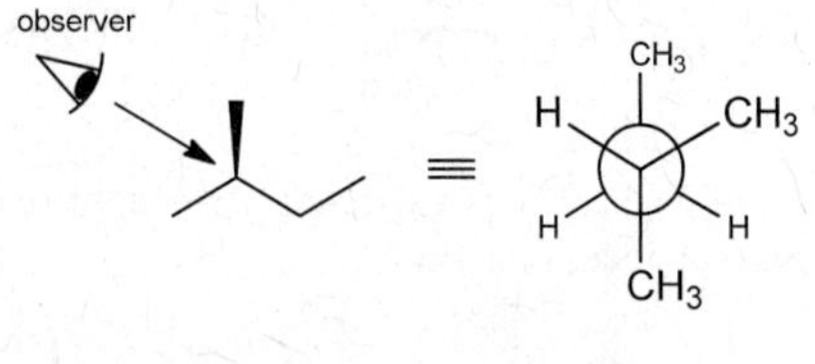

observer

≡

CH₃ / H / CH₃ / H / H / CH₃

We expect the lowest energy conformation to be staggered and the highest energy conformation to be eclipsed. Let's first explore the three staggered conformations, shown below. One of these conformations exhibits two gauche interactions, while the other two staggered conformations are degenerate, with each having only one gauche interaction:

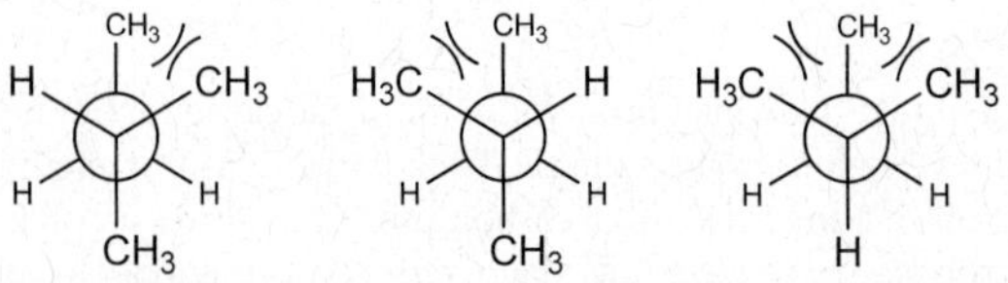

CH₃ / H / CH₃ / H / H / CH₃

One gauche interaction

CH₃ / H₃C / H / H / H / CH₃

One gauche interaction

CH₃ / H₃C / CH₃ / H / H / H

Two gauche interactions

degenerate

The two degenerate conformations are lowest in energy among all available conformations.

In order to determine which conformation is highest in energy, we must examine the eclipsed conformation. There are three eclipsed conformations, shown below. Two of these conformations are degenerate because they exhibit a methyl-methyl eclipsing interaction:

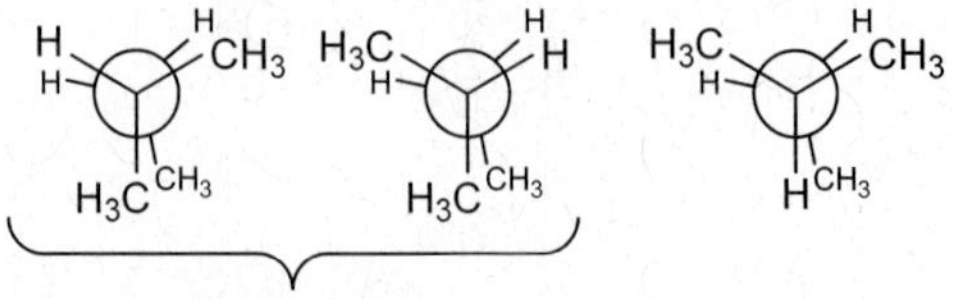

H / H / CH₃ / H₃C / CH₃ / H₃C / H / H / CH₃ / H₃C / CH₃ / H / H / CH₃ / HCH₃

degenerate

These two degenerate conformations are highest in energy among all available conformations.

(d) We begin by converting the bond-line drawing into a Newman projection.

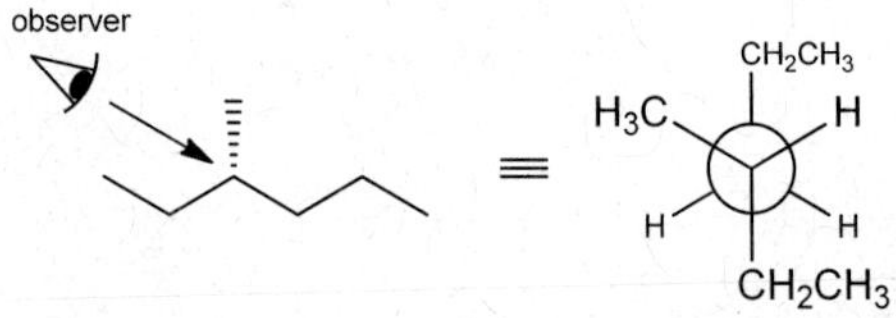

observer

≡

CH₂CH₃ / H₃C / H / H / H / CH₂CH₃

We expect the lowest energy conformation to be staggered and the highest energy conformation to be eclipsed. The lowest energy conformation is the staggered conformation in which the two ethyl groups are *anti* to each other, while the highest energy conformation is the eclipsed conformation in which the two ethyl groups are eclipsing each other:

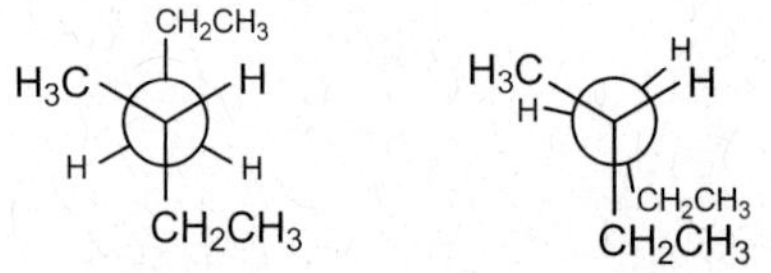

CH₂CH₃ / H₃C / H / H / H / CH₂CH₃

Lowest energy

H₃C / H / H / CH₂CH₃ / CH₂CH₃

Highest energy

4.21. The gauche conformations are capable of intramolecular hydrogen bonding, as shown below. The *anti* conformation lacks this stabilizing effect.

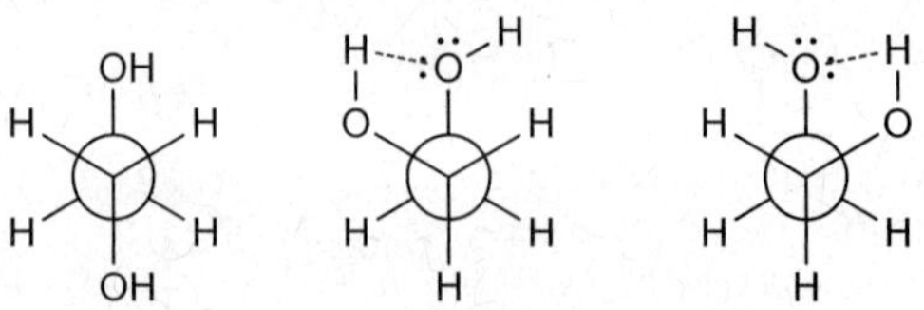

OH / H / H / H / H / OH

Anti

H⋯Ö—H / O / H / H / H / H

Gauche

H⋯Ö⋯H / O / H / H / H / H

Gauche

4.22. The step-by-step procedure in the SkillBuilder should provide the following drawing:

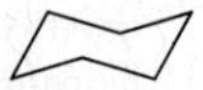

4.23.

(a) The chair conformation for this compound is similar to the chair conformation of cyclohexane, but one of the carbon atoms has been replaced with a nitrogen atom, and the hydrogen atom can adopt either a pseudo-equatorial or a pseudo-axial position:

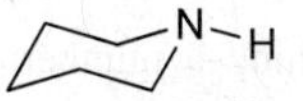

(b) The chair conformation for this compound is similar to the chair conformation of cyclohexane, but two of the carbon atoms have been replaced with oxygen atoms:

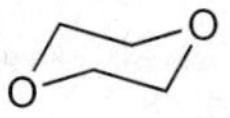

4.24. The step-by-step procedure in the SkillBuilder should provide the following drawing:

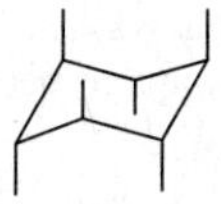

4.25. The step-by-step procedure in the SkillBuilder should provide the following drawing:

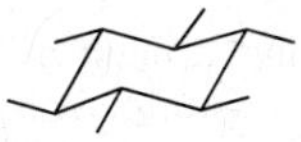

4.26. The step-by-step procedure in the SkillBuilder should provide the following drawing:

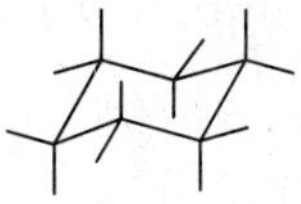

4.27. There are eight hydrogen atoms in axial positions and seven hydrogen atoms in equatorial positions.

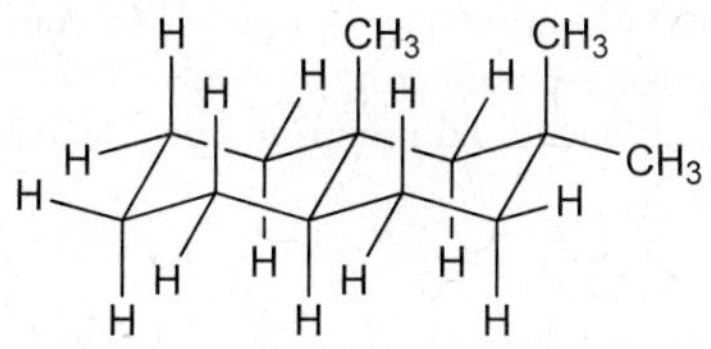

4.28.

(a) In one chair conformation, the substituent (OH) occupies an axial position. In the other chair conformation, the substituent occupies an equatorial position:

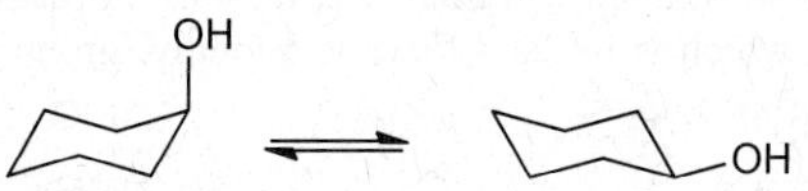

(b) In one chair conformation, the substituent (NH_2) occupies an axial position. In the other chair

conformation, the substituent occupies an equatorial position:

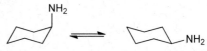

(c) In one chair conformation, the substituent (Cl) occupies an axial position. In the other chair conformation, the substituent occupies an equatorial position:

(d) In one chair conformation, the methyl group occupies an axial position. In the other chair conformation, the methyl group occupies an equatorial position:

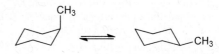

(e) In one chair conformation, the *tert*-butyl group occupies an axial position. In the other chair conformation, the *tert*-butyl group occupies an equatorial position:

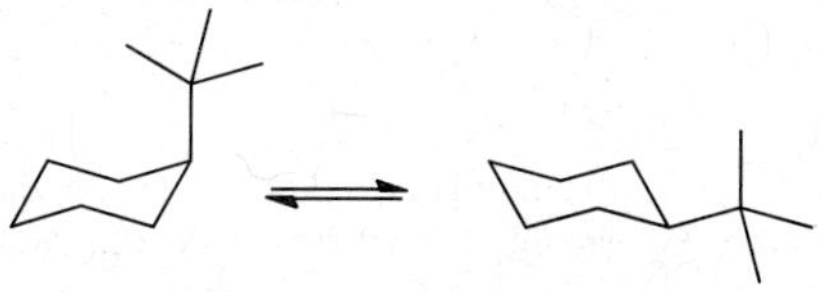

4.29.

(a) In this double Newman projection, the axial hydrogen atoms point straight up and down, while the equatorial hydrogen atoms go out to the sides. The bromine atom occupies an equatorial position.

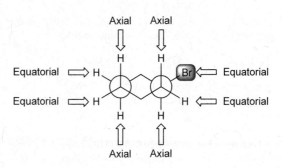

(b) In this chair conformation, the bromine atom occupies an equatorial position, so we draw a chair conformation with an equatorial bromo group:

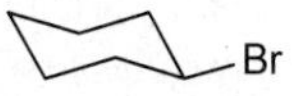

(c) As the result of the ring flip, the equatorial bromo group becomes axial:

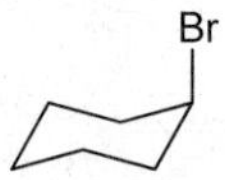

4.30. Although the OH group is in an axial position, this conformation is capable of intramolecular hydrogen bonding, which is a stabilizing effect:

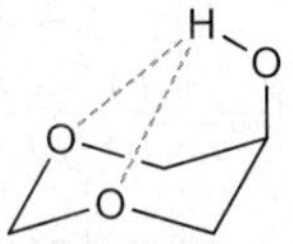

4.31.
(a) Begin by assigning a numbering system (which does not need to adhere to IUPAC rules), and then determine the location and three-dimensional orientation of each substituent. In this case, there is a methyl group at C-1, which is up, and there is an ethyl group at C-2, which is down:

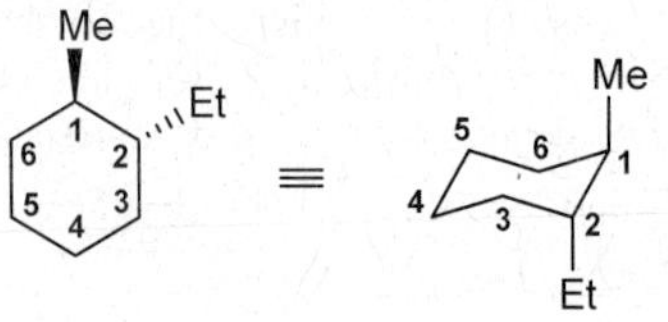

When assigning the numbers to the chair drawing, the first number can be placed anywhere on the ring. For example, it would be perfectly acceptable to draw the chair like this:

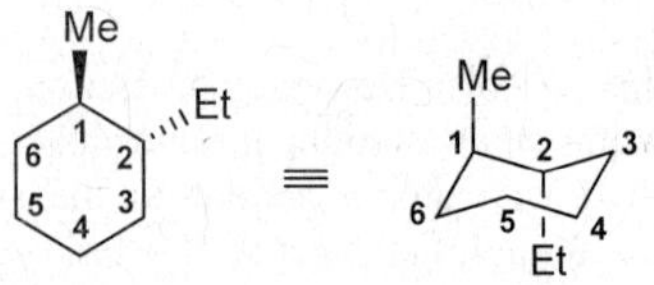

At first it might look like this is a different chair than the one previously drawn. But let's compare them:

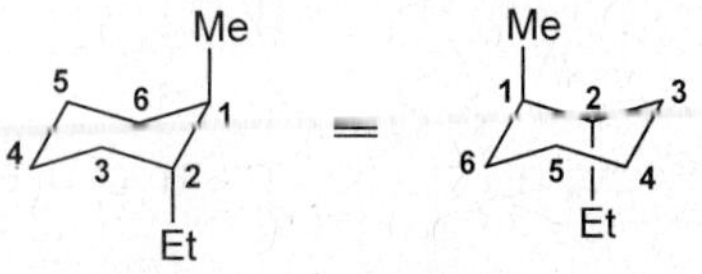

In both drawings, if we travel clockwise around the ring, we will encounter the methyl group first and the ethyl group second. Also, in both chair drawings, the methyl group is up and the ethyl group is down. Therefore, either of these drawings is an acceptable chair representation. Neither one is "more correct" than the other (to see this more clearly, you may find it helpful to build a molecular model and view it from different angles).

Once we have drawn the first chair conformation, we then draw the second chair conformation, once again using a numbering system. Notice that a ring flip causes

all equatorial groups to become axial groups, and vice versa.

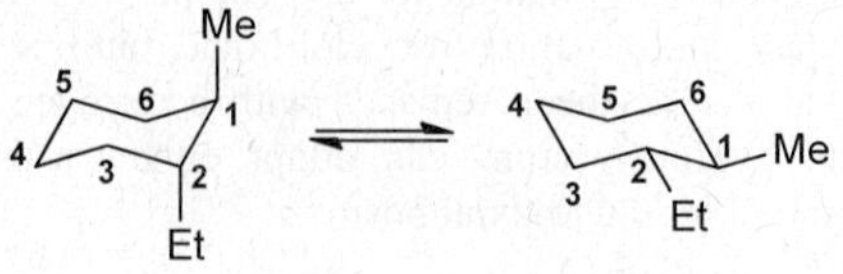

(b) Begin by assigning a numbering system, and then determine the location and three-dimensional orientation of each substituent. In this case, there is a methyl group at C-1, which is up, and there is an ethyl group at C-2, which is also up:

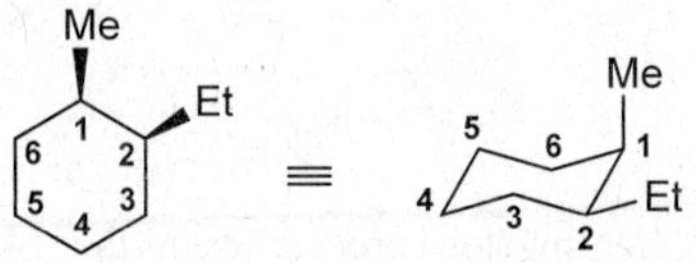

Finally, we draw the second chair conformation, once again using a numbering system. Notice that a ring flip causes all equatorial groups to become axial groups, and vice versa.

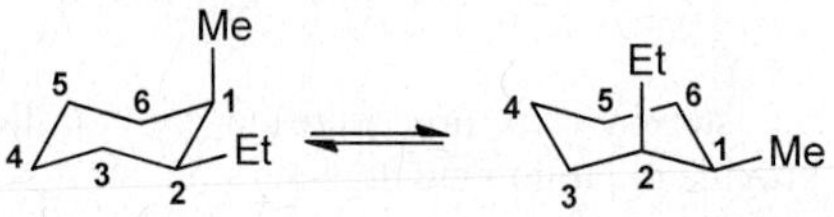

(c) Begin by assigning a numbering system, and then determine the location and three-dimensional orientation of each substituent. In this case, there is a methyl group at C-1, which is up, and there is a bromo group at C-3, which is also up:

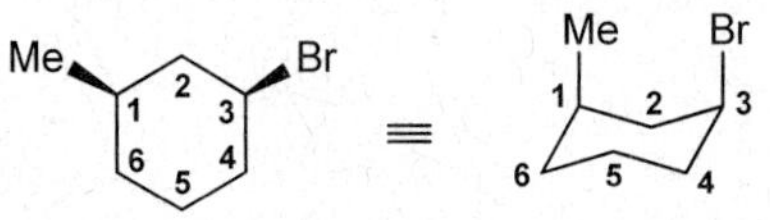

When assigning the numbers to the chair drawing, the first number can be placed anywhere on the ring (as long as the numbers go clockwise).

Finally, we draw the second chair conformation, once again using a numbering system. Notice that a ring flip causes all equatorial groups to become axial groups, and vice versa.

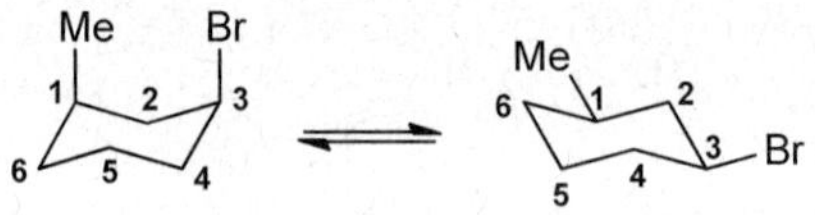

(d) Begin by assigning a numbering system, and then determine the location and three-dimensional orientation of each substituent. In this case, there is a bromo group at C-1, which is up, and there is a methyl group at C-3, which is also up:

When assigning the numbers to the chair drawing, the first number can be placed anywhere on the ring (as long as the numbers go clockwise), as explained in the solution to Problem **4.31a**.

Finally, we draw the second chair conformation, once again using a numbering system. Notice that a ring flip causes all equatorial groups to become axial groups, and vice versa.

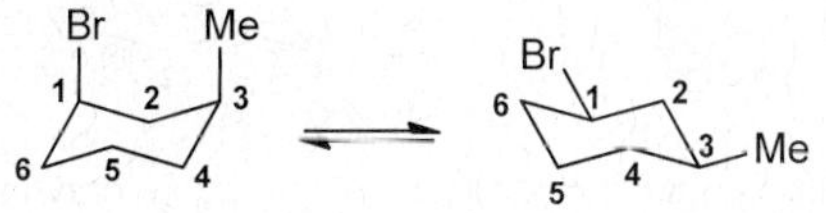

(e) Begin by assigning a numbering system, and then determine the location and three-dimensional orientation of each substituent. In this case, there is a methyl group at C-1, which is up, and a *tert*-butyl group at C-4, which is down:

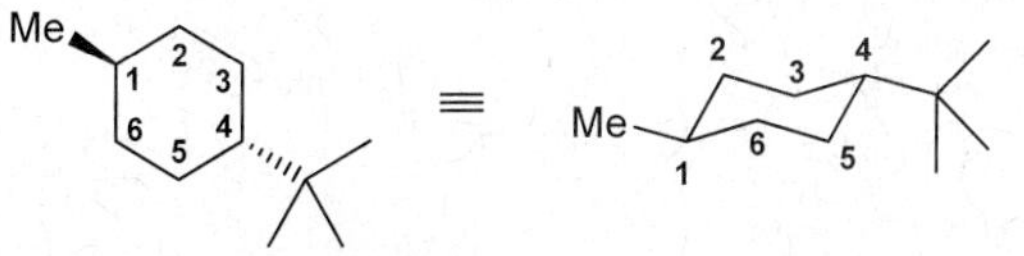

Notice that the numbering system need not adhere to IUPAC rules – the numbering system is simply a tool that we are using to guide us. When assigning the numbers to the chair drawing, the first number can be placed anywhere on the ring (as long as the numbers go clockwise), as seen in the solution to Problem **4.31a**.

Finally, we draw the second chair conformation, once again using a numbering system. Notice that a ring flip causes all equatorial groups to become axial groups, and vice versa.

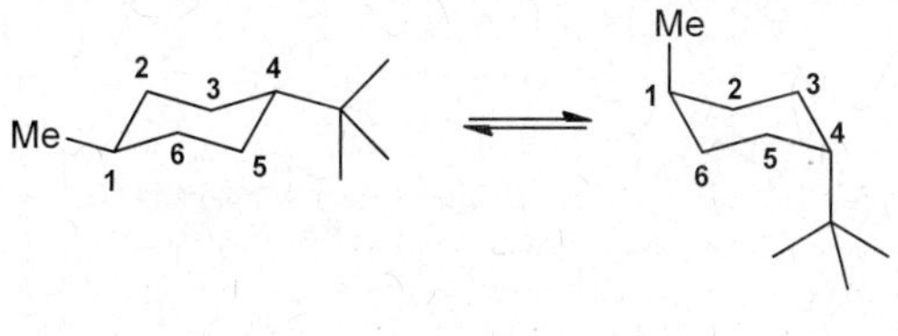

(f) Begin by assigning a numbering system, and then determine the location and three-dimensional orientation of each substituent. In this case, there is a methyl group at C-1, which is up, and another methyl group at C-3, which is down:

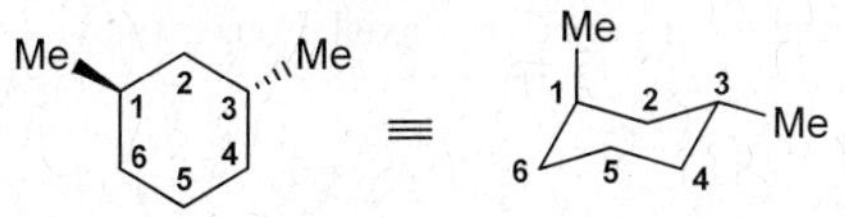

When assigning the numbers to the chair drawing, the first number can be placed anywhere on the ring (as long as the numbers go clockwise), as explained in the solution to Problem **4.31a**.

Finally, we draw the second chair conformation, once again using a numbering system. Notice that a ring

flip causes all equatorial groups to become axial groups, and vice versa.

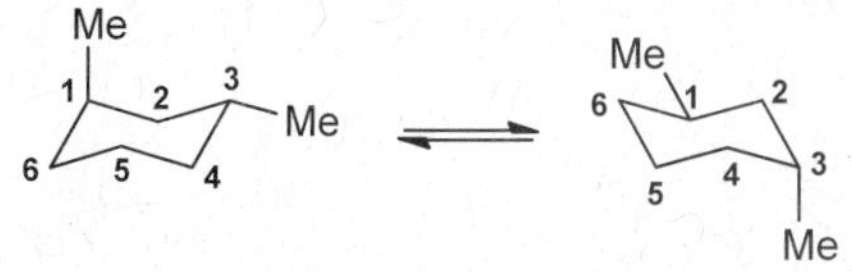

(g) Begin by assigning a numbering system, and then determine the location and three-dimensional orientation of each substituent. In this case, there is an isopropyl group at C-1, which is up, and another isopropyl group at C-3, which is also up:

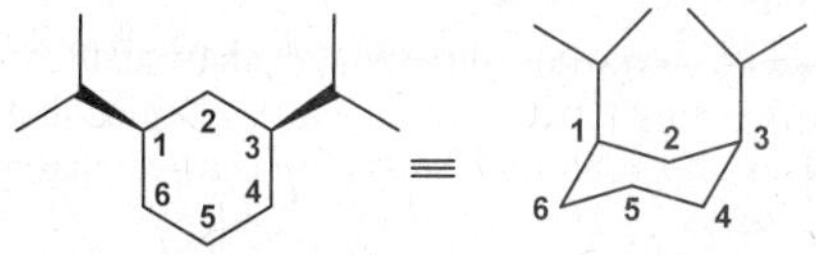

When assigning the numbers to the chair drawing, the first number can be placed anywhere on the ring (as long as the numbers go clockwise), as explained in the solution to Problem **4.31a**.

Finally, we draw the second chair conformation, once again using a numbering system. Notice that a ring flip causes all equatorial groups to become axial groups, and vice versa.

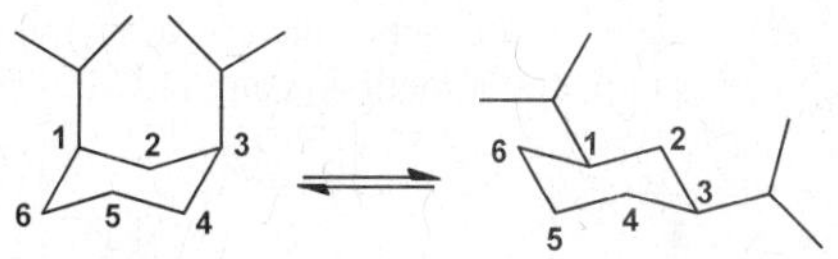

(h) Begin by assigning a numbering system, and then determine the location and three-dimensional orientation of each substituent. In this case, there is a methyl group at C-1, which is up, and another methyl group at C-4, which is also up:

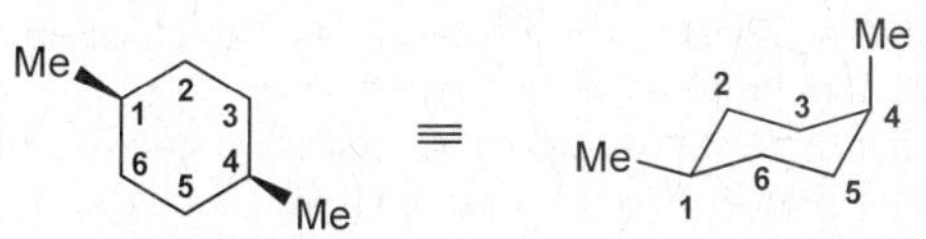

When assigning the numbers to the chair drawing, the first number can be placed anywhere on the ring (as long as the numbers go clockwise), as explained in the solution to Problem **4.31a**.

Finally, we draw the second chair conformation, once again using a numbering system. Notice that a ring flip causes all equatorial groups to become axial groups, and vice versa.

4.32. Begin by assigning a numbering system, and then determine the location and three-dimensional orientation of each substituent:

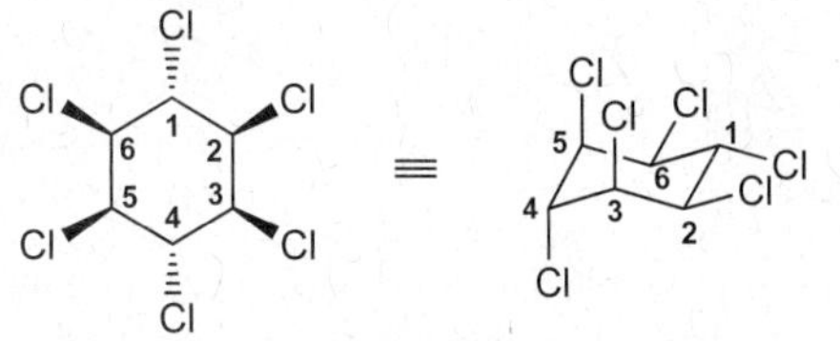

When assigning the numbers to the chair drawing, the first number can be placed anywhere on the ring (as long as the numbers go clockwise), as explained in the solution to Problem **4.31a**.

Finally, we draw the second chair conformation, once again using a numbering system. Notice that a ring flip causes all equatorial groups to become axial groups, and vice versa.

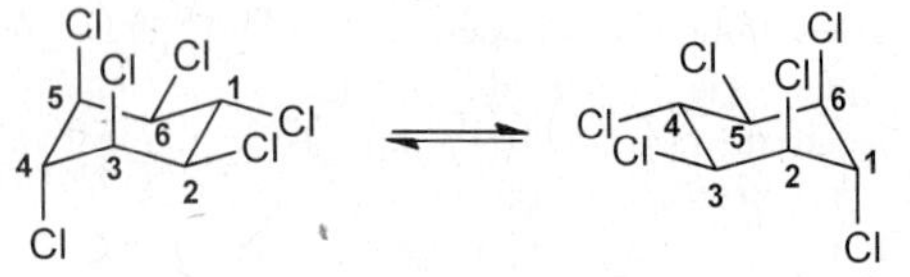

4.33.
(a) Begin by assigning a numbering system, and then determine the location and three-dimensional orientation of each substituent. In this case, there is a methyl group at C-1, which is up, and a methyl group at C-2, which is down:

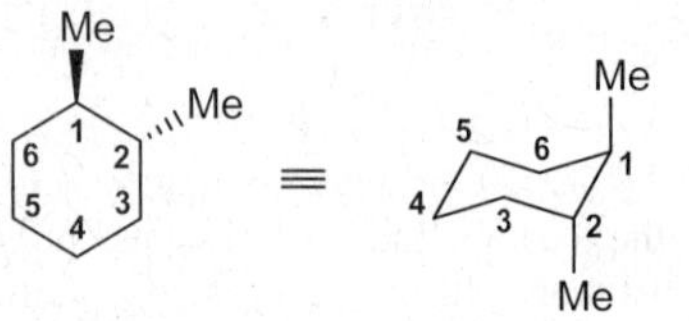

When assigning the numbers to the chair drawing, the first number can be placed anywhere on the ring (as long as the numbers go clockwise), as explained in the solution to Problem **4.31a**.

Once we have drawn the first chair conformation, we then draw the second chair conformation, again using a numbering system. Notice that a ring flip causes all equatorial groups to become axial groups, and vice versa.

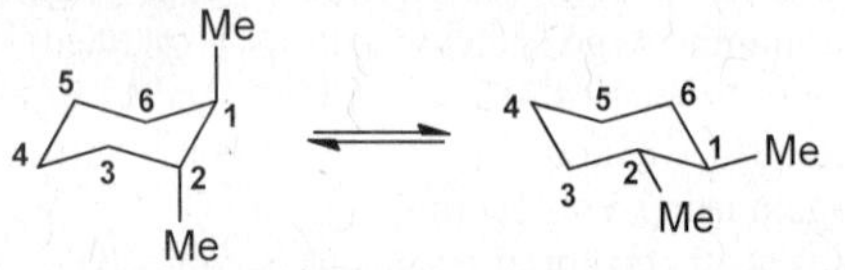

Finally, we compare both chair drawings and determine which one has fewer or less severe 1,3-diaxial interactions. In the first drawing, both substituents occupy axial positions, but in the second drawing, both substituents occupy equatorial positions. As such, the latter is more stable, since it lacks 1,3-diaxial interactions.

(b) Begin by assigning a numbering system, and then determine the location and three-dimensional orientation of each substituent. In this case, there is a methyl group at C-1, which is up, and an isopropyl group at C-2, which is also up:

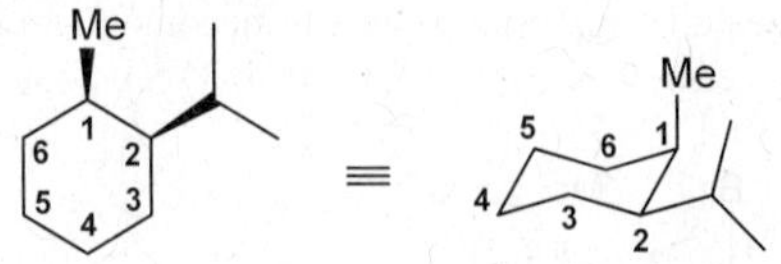

When assigning the numbers to the chair drawing, the first number can be placed anywhere on the ring (as long as the numbers go clockwise), as explained in the solution to Problem **4.31a**.

Once we have drawn the first chair conformation, we then draw the second chair conformation, again using a numbering system. Notice that a ring flip causes all equatorial groups to become axial groups, and vice versa.

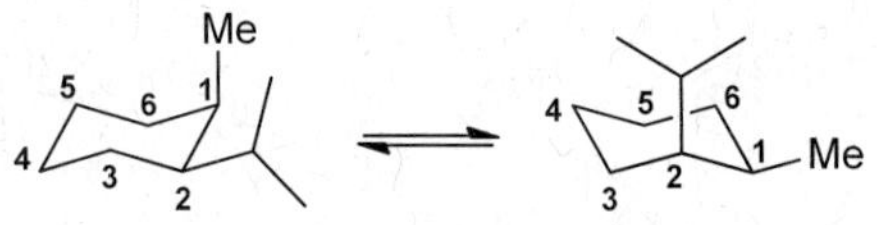

Finally, we compare both chair drawings and determine which one has fewer or less severe 1,3-diaxial interactions. In the first drawing, the methyl group occupies an axial position, but in the second drawing, the isopropyl group occupies an axial position. As such, the former is more stable, since it is expected to have less severe 1,3-diaxial interactions.

(c) Begin by assigning a numbering system, and then determine the location and three-dimensional orientation of each substituent:

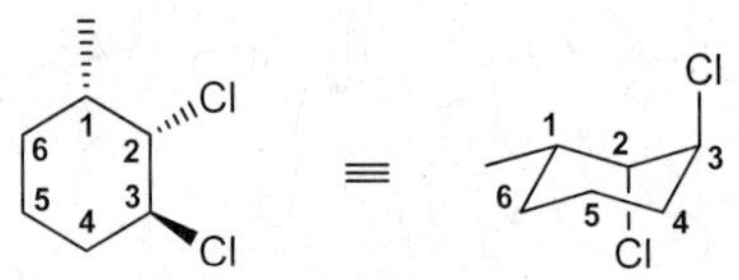

When assigning the numbers to the chair drawing, the first number can be placed anywhere on the ring (as long as the numbers go clockwise), as explained in the solution to Problem **4.31a**.

Once we have drawn the first chair conformation, we then draw the second chair conformation, again using a numbering system. Notice that a ring flip causes all equatorial groups to become axial groups, and vice versa.

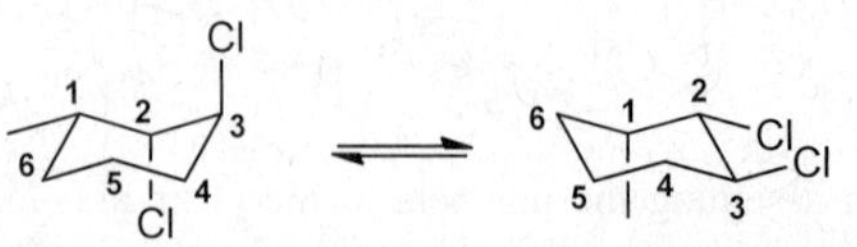

Finally, we compare both chair drawings and determine which one has fewer or less severe 1,3-diaxial interactions. In the first drawing, the methyl group occupies an equatorial position, while both chloro groups occupy axial positions. In the second drawing, the methyl group occupies an axial position, while both chloro groups occupy equatorial positions. According to the data presented in Table 4.8, each methyl group experiences 1,3-diaxial interactions of 7.6 kJ/mol, while each chloro group experiences 1,3-diaxial interactions of 2.0 kJ/mol. As such, the 1,3-diaxial interactions from the methyl group are more severe than the combined 1,3-diaxial interactions of the two chloro groups (4.0 kJ/mol). As such, the more stable conformation is the one in which the methyl group occupies an equatorial position.

(d) Begin by assigning a numbering system, and then determine the location and three-dimensional orientation of each substituent:

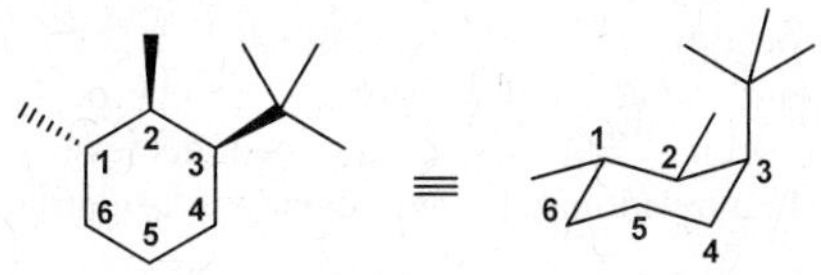

When assigning the numbers to the chair drawing, the first number can be placed anywhere on the ring (as long as the numbers go clockwise), as explained in the solution to Problem **4.31a**.

Once we have drawn the first chair conformation, we then draw the second chair conformation, again using a numbering system. Notice that a ring flip causes all equatorial groups to become axial groups, and vice versa.

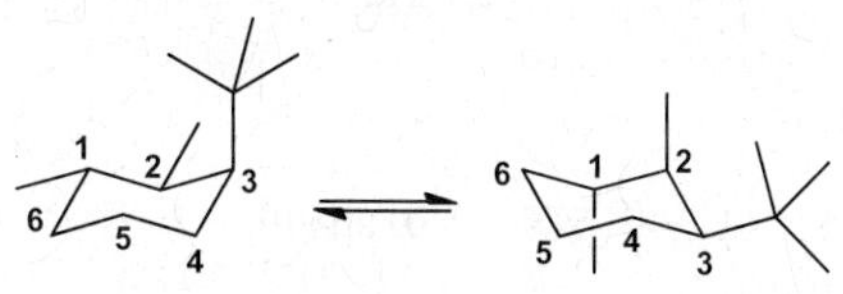

Finally, we compare both chair drawings and determine which one has fewer or less severe 1,3-diaxial interactions. In the first drawing, the *tert*-butyl group occupies an axial position, while both methyl groups occupy equatorial positions. In the second drawing, the *tert*-butyl group occupies an equatorial position, while both methyl groups occupy equatorial positions. According to the data presented in Table 4.8, each methyl group experiences 1,3-diaxial interactions of 7.6 kJ/mol, while a *tert*-butyl group experiences 1,3-diaxial interactions of 22.8 kJ/mol. As such, the 1,3-diaxial interactions from the one *tert*-butyl group are more severe than the combined 1,3-diaxial interactions of the methyl groups (15.2 kJ/mol). As such, the more stable conformation is the one in which the *tert*-butyl group occupies an equatorial position.

(e) Begin by assigning a numbering system, and then determine the location and three-dimensional orientation of each substituent:

When assigning the numbers to the chair drawing, the first number can be placed anywhere on the ring (as long as the numbers go clockwise), as explained in the solution to Problem **4.31a**.

Once we have drawn the first chair conformation, we then draw the second chair conformation, again using a numbering system. Notice that a ring flip causes all equatorial groups to become axial groups, and vice versa.

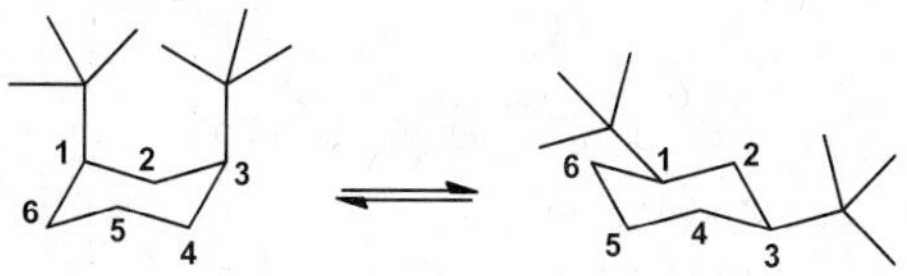

Finally, we compare both chair drawings and determine which one has fewer or less severe 1,3-diaxial interactions. In the first drawing, both *tert*-butyl groups occupy axial positions, but in the second drawing, both occupy equatorial positions. As such, the latter is more stable, since it lacks 1,3-diaxial interactions.

4.34. Each chair conformation has three substituents occupying equatorial positions and three substituents occupying axial positions:

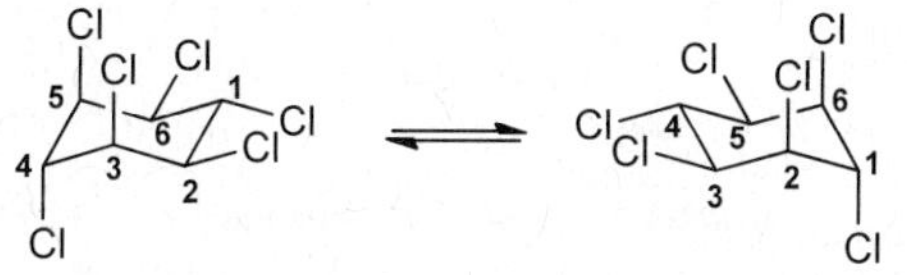

As such, the two chair conformations of lindane are degenerate. There is no difference in energy between them.

4.35. *trans*-1,4-di-*tert*-butylcyclohexane exists predominantly in a chair conformation, because both substituents can occupy equatorial positions. In contrast, *cis*-1,4-di-*tert*-butylcyclohexane cannot have both of its substituents in equatorial positions. Each chair conformation has one of the substituents in an axial position, which is too high in energy. The compound can achieve a lower energy state by adopting a twist boat conformation.

4.36. *cis*-1,3-dimethylcyclohexane is expected to be more stable than *trans*-1,3-dimethylcyclohexane because the former can adopt a chair conformation in which both substituents are in equatorial positions (highlighted below):

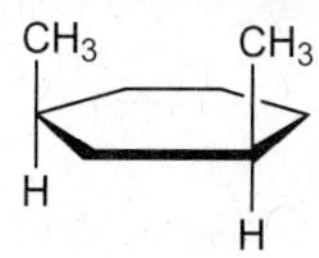

cis-1,3-dimethylcyclohexane

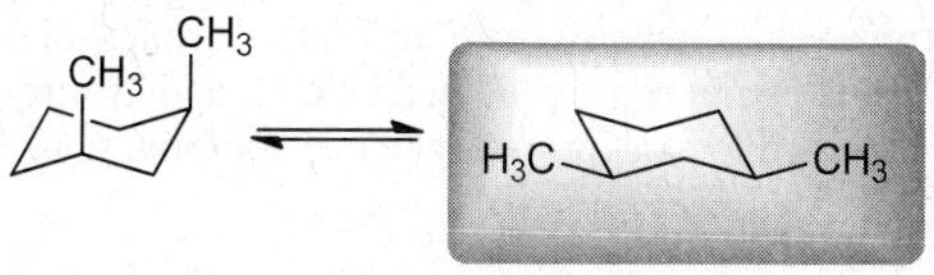

In contrast, *trans*-1,3-dimethylcyclohexane cannot adopt a chair conformation in which both substituents are in equatorial positions. Each chair conformation has one methyl group in an axial position:

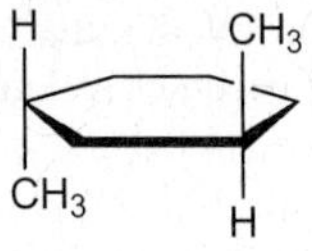

trans-1,3-dimethylcyclohexane

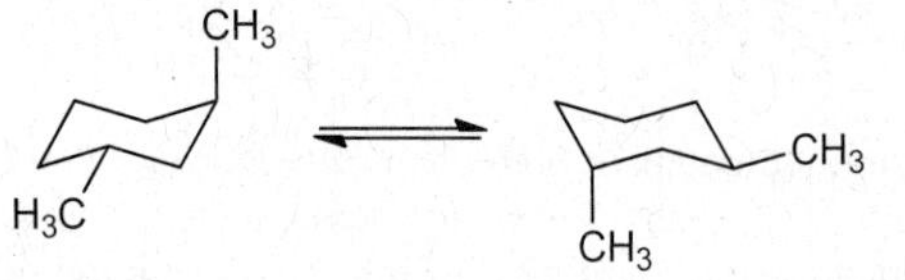

4.37. *trans*-1,4-dimethylcyclohexane is expected to be more stable than *cis*-1,4-dimethylcyclohexane because the former can adopt a chair conformation in which both substituents are in equatorial positions (highlighted below):

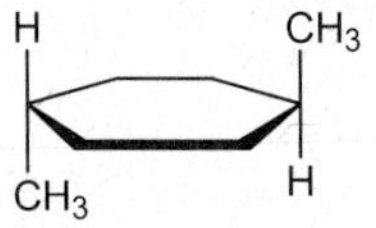

trans-1,4-dimethylcyclohexane

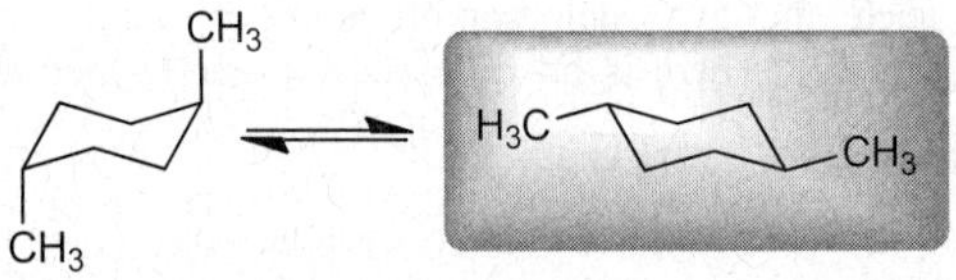

In contrast, *cis*-1,4-dimethylcyclohexane cannot adopt a chair conformation in which both substituents are in equatorial positions. Each chair conformation has one methyl group in an axial position:

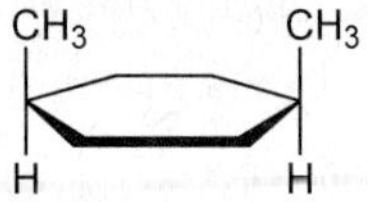

cis-1,4-dimethylcyclohexane

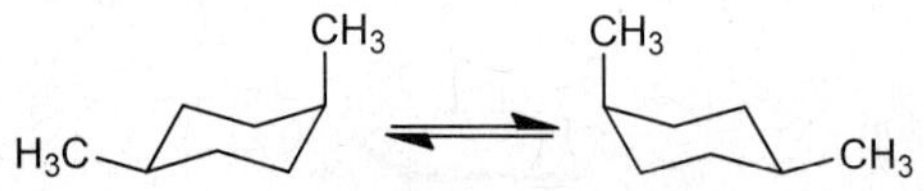

4.38. *cis*-1,3-di-*tert*-butylcyclohexane can adopt a chair conformation in which both *tert*-butyl groups occupy equatorial positions (highlighted below), and as a result, it is expected to exist primarily in that conformation.

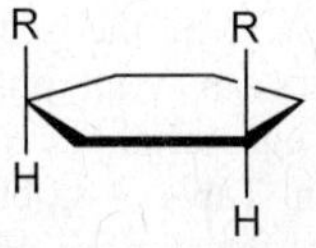

cis-1,3-di-*tert*-butylcyclohexane

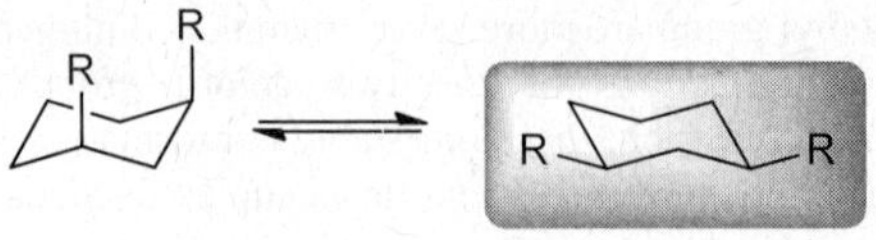

(R = *tert*-butyl group)

In contrast, *trans*-1,3-di-*tert*-butylcyclohexane cannot adopt a chair conformation in which both *tert*-butyl groups occupy equatorial positions. In either chair conformation, one of the *tert*-butyl groups occupies an axial position. This compound can achieve a lower energy state by adopting a twist-boat conformation.

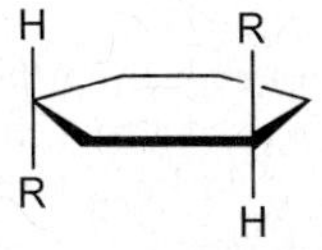

trans-1,3-di-*tert*-butylcyclohexane

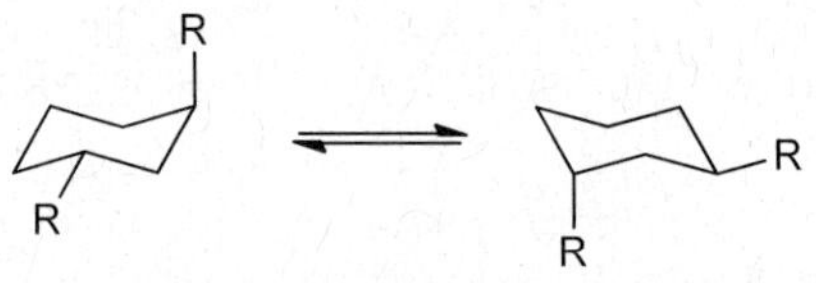

(R = *tert*-butyl group)

4.39.
(a) The longest chain is eight carbon atoms (shown below), so the parent is octane:

Parent = octane

(b) The longest chain is nine carbon atoms (shown below), so the parent is nonane:

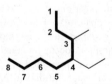

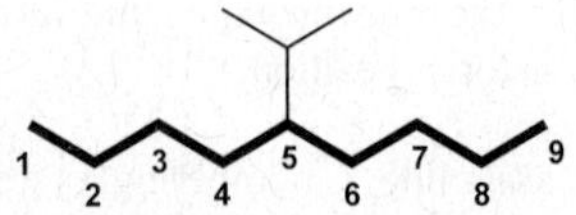

Parent = nonane

(c) The longest chain is eight carbon atoms, so the parent is octane. In this case, there is more than one eight-carbon chain, and the correct parent chain is the one with the most substituents, shown below:

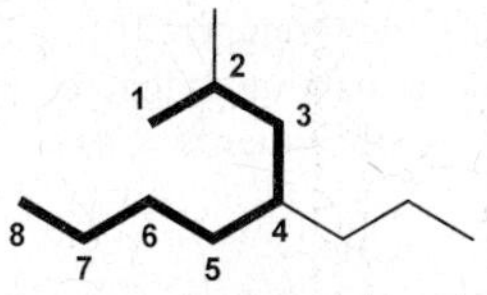

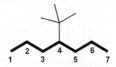

Parent = octane

(d) The longest chain is seven carbon atoms (shown below), so the parent is heptane:

Parent = heptane

4.40.

(a) First we identify the parent (octane), and then we identify any alkyl substituents (highlighted) that are connected to the parent.

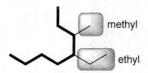

methyl

ethyl

(b) First we identify the parent (nonane), and then we identify any alkyl substituents (highlighted) that are connected to the parent. In this case, the substituent is an isopropyl group, which can also be called (1-methylethyl) because it is a complex substituent.

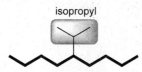

isopropyl

(c) First we identify the parent (octane), and then we identify any alkyl substituents (highlighted) that are connected to the parent.

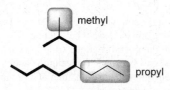

methyl

propyl

(d) First we identify the parent (heptane), and then we identify any alkyl substituents (highlighted) that are connected to the parent. In this case, the substituent is a *tert*-butyl group, which can also be called (1,1-dimethylethyl) because it is a complex substituent.

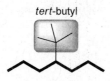

tert-butyl

4.41. For each of the following compounds, we assign its name via a four-step process: First identify the parent, then the substituents, then assign locants, and finally, arrange the substituents alphabetically. In each case, use commas to separate numbers from each other, and use hyphens to separate letters from numbers.
(a) 2,3,5-trimethyl-4-propylheptane
(b) 1,2,4,5-tetramethyl-3-propylcyclohexane
(c) 2,3,5,9-tetramethylbicyclo[4.4.0]decane
(d) 1,4-dimethylbicyclo[2.2.2]octane

4.42.
(a) If we assign a systematic name for each of these structures, we find that they share the same name (2-methylpentane). Therefore, these drawings are simply different representations of the same compound.

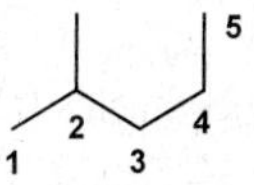

2-methylpentane 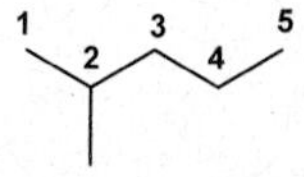   2-methylpentane

(b) We assign a systematic name to each structure, and find that they have different names (shown below). Therefore, they must differ in their constitution (connectivity of atoms). These compounds are therefore different from each other, but they share the same molecular formula (C_7H_{16}), so they are constitutional isomers.

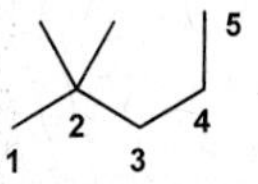

2,2-dimethylpentane 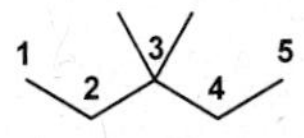   3,3-dimethylpentane

(c) If we assign a systematic name for each of these structures, we find that they share the same name (3-ethyl-2,4-dimethylheptane). Therefore, these drawings are simply different representations of the same compound.

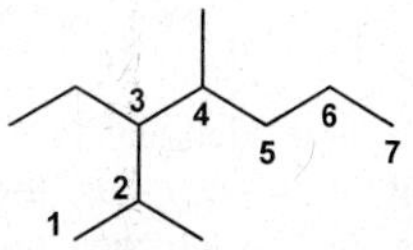

3-ethyl-2,4-dimethyl   3-ethyl-2,4-dimethyl
heptane   heptane

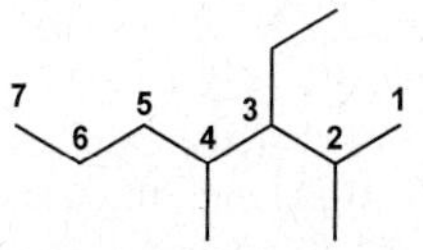

4.43. When looking down the C2-C3 bond, the front carbon atom has one methyl group and two H's, while the back carbon atom has an ethyl group, a methyl group, and a hydrogen atom. The lowest energy conformation is the staggered conformation with the fewest and least severe gauche interactions, shown below. In this conformation, there is only one Me-Me gauche interaction.

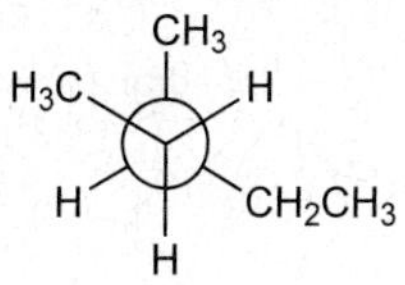

4.44.   Both compounds share the same molecular formula (C_6H_{14}). That is, they are constitutional isomers, and the unbranched isomer is expected to have the larger heat of combustion:

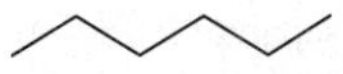

4.45.

(a) The name indicates that the parent is a five-carbon chain and there are three substituents (two methyl groups at C2, and one methyl group at C4):

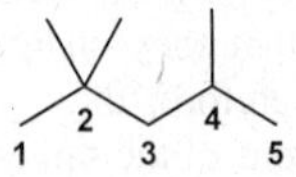

(b) The name indicates that the parent is a seven-membered ring and there are four substituents (all methyl groups), as shown:

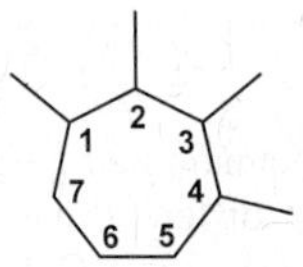

(c) The name indicates a bicyclic parent with two ethyl groups at C2 and two ethyl groups at C4, as shown:

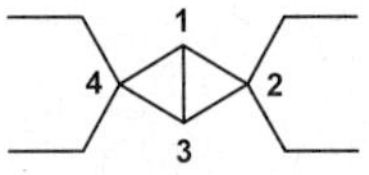

4.46. We begin by drawing a Newman projection of 2,2-dimethylpropane:

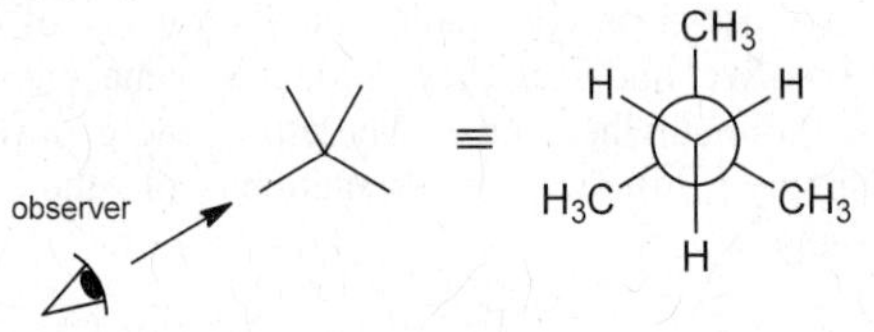

Notice that the front carbon has three identical groups (all H's), and the back carbon atom also has three identical groups (all methyl groups). As such, we expect all staggered conformations to be degenerate, and we expect all eclipsed conformations to be degenerate as well. Therefore, the energy diagram will more closely resemble the shape of the energy diagram for the conformational analysis of ethane.

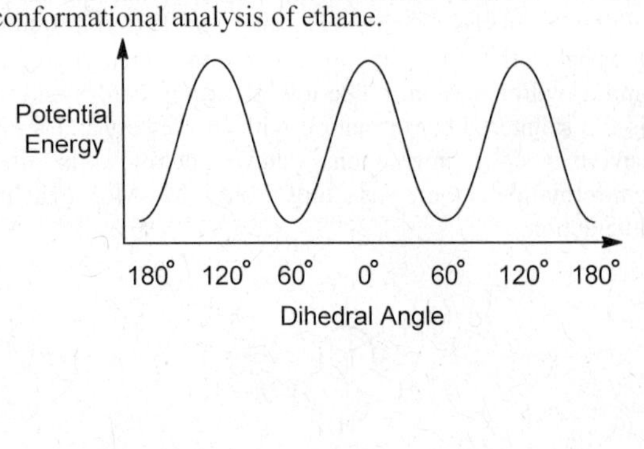

4.47. Two of the staggered conformations are degenerate. The remaining staggered conformation is lower in energy than the other two, as shown:

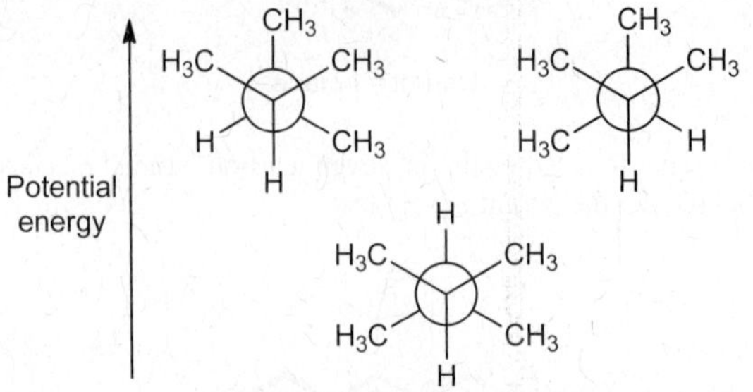

4.48. For each of the following cases, we draw the second chair conformation using a numbering system to ensure the substituents are placed correctly. The numbering system does NOT need to adhere to IUPAC rules, as it is just a tool that we are using to draw both chair conformations correctly. Notice that a ring flip causes all equatorial groups to become axial groups, and vice versa.

(a)

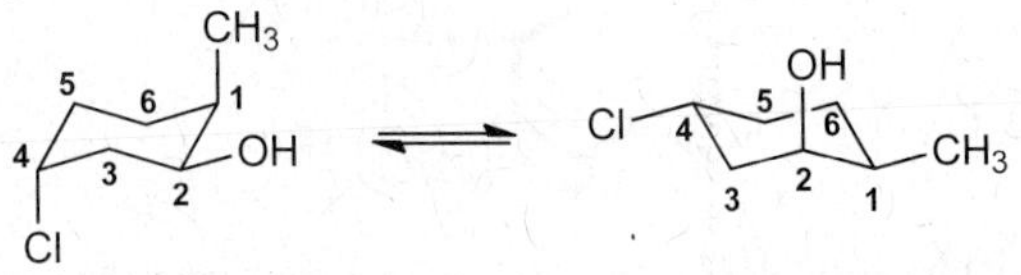

(b)

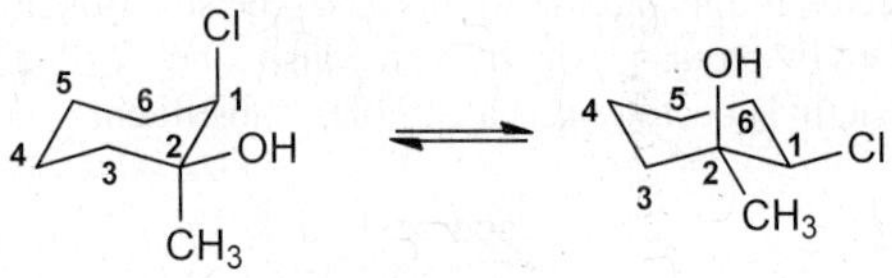

(c)

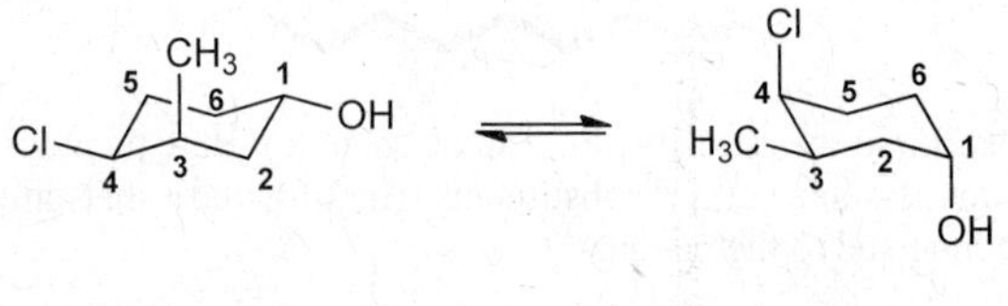

4.49.

(a) The second compound is expected to have a higher heat of combustion because it has more carbon atoms.

(b) The first compound is expected to have a higher heat of combustion because it cannot adopt a chair conformation in which both methyl groups occupy equatorial positions.

(c) The second compound is expected to have a higher heat of combustion because it cannot adopt a chair conformation in which both methyl groups occupy equatorial positions.

(d) The first compound is expected to have a higher heat of combustion because it cannot adopt a chair conformation in which both methyl groups occupy equatorial positions.

4.50. The energy diagram of 1,2-dichloroethane is similar to the energy diagram of butane (see Figure 4.11). The CH₃ groups have simply been replaced with chloro groups.

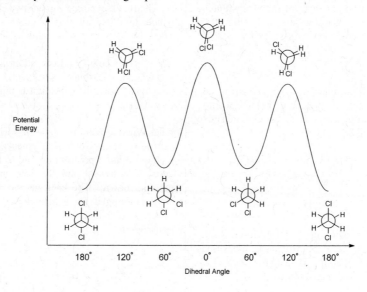

4.51.
(a) We first convert the Newman projection into a bond-line drawing, because it is easier to assign a systematic name to a bond-line drawing. In this case, the compound is hexane:

(b) We first convert the Newman projection into a bond-line drawing, because it is easier to assign a systematic name to a bond-line drawing. In this case, the compound is methylcyclohexane:

(c) We first convert the Newman projection into a bond-line drawing, because it is easier to assign a systematic name to a bond-line drawing. In this case, the compound is methylcyclopentane:

(d) As seen in the solution to Problem **4.17b**, this compound is *trans*-1,2-dimethylcyclopentane:

4.52. Each H-H eclipsing interaction is 4 kJ/mol, and there are two of them (for a total of 8 kJ/mol). The remaining energy cost is associated with the Br-H eclipsing interaction: 15 − 8 = 7 kJ/mol.

4.53. In order to draw the first chair conformation, begin by assigning a numbering system, and then determine the location and three-dimensional orientation of each substituent:

When assigning the numbers to the chair drawing, the first number can be placed anywhere on the ring (as long as the numbers go clockwise), as explained in the solution to Problem **4.31a**.

Once we have drawn the first chair conformation, we then draw the second chair conformation, again using a numbering system. Notice that a ring flip causes all equatorial groups to become axial groups, and vice versa.

Finally, we compare both chair drawings and determine which one has fewer or less severe 1,3-diaxial interactions. In the first drawing, all three substituents occupy axial positions, but in the second drawing, they all occupy equatorial positions. As such, the latter is more stable.

4.54.
(a) The methyl group occupies an axial position in one chair conformation, and occupies an equatorial position in the other chair conformation. The latter is more stable because it lacks 1,3-diaxial interactions.

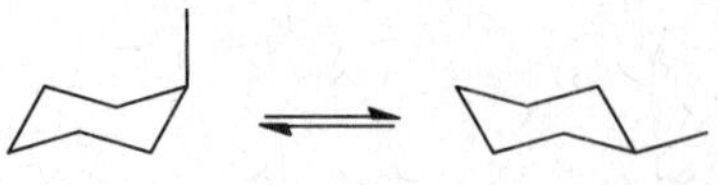

more stable

(b) Both isopropyl groups occupy axial positions in one chair conformation, and both occupy equatorial positions in the other chair conformation. The latter is more stable because it lacks 1,3-diaxial interactions.

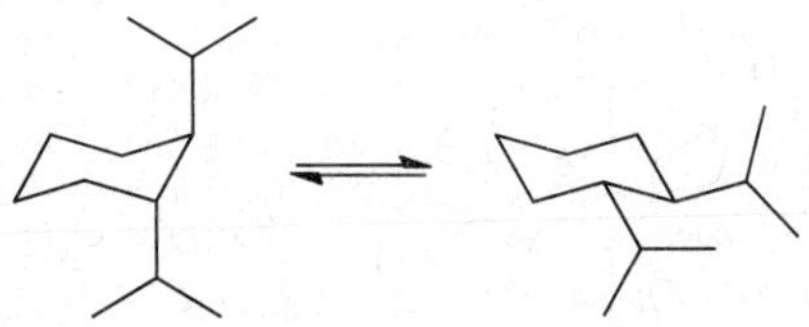

more stable

(c) Both isopropyl groups occupy axial positions in one chair conformation, and both occupy equatorial positions in the other chair conformation. The latter is more stable because it lacks 1,3-diaxial interactions.

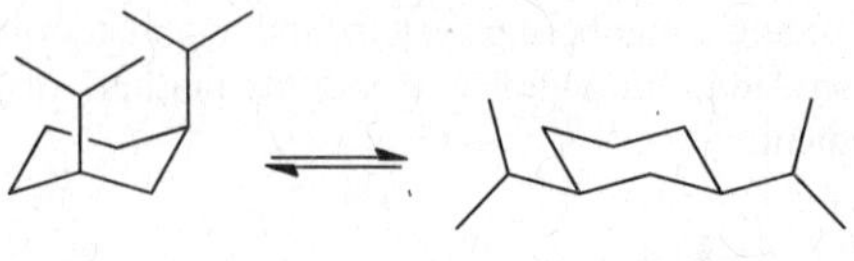

more stable

(d) Both isopropyl groups occupy axial positions in one chair conformation, and both occupy equatorial positions in the other chair conformation. The latter is more stable because it lacks 1,3-diaxial interactions.

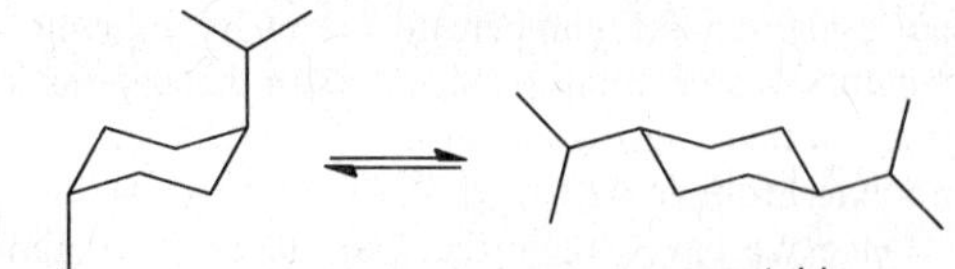

more stable

4.55.
(a) The second compound can adopt a chair conformation in which all three substituents occupy equatorial positions. Therefore, the second compound is expected to be more stable.
(b) The first compound can adopt a chair conformation in which all three substituents occupy equatorial

positions. Therefore, the first compound is expected to be more stable.
(c) The first compound can adopt a chair conformation in which both substituents occupy equatorial positions. Therefore, the first compound is expected to be more stable.
(d) The first compound can adopt a chair conformation in which all four substituents occupy equatorial positions. Therefore, the first compound is expected to be more stable.

4.56. When looking from the perspective of the observer (as shown in the problem statement), the front carbon has a chlorine atom pointing up and to the right, a bromine atom pointing up and the left, and a methyl group pointing down. The back carbon atom has a chlorine atom pointing down and to the right, a bromine atom pointing down and to the left, as well as a methyl group pointing up.

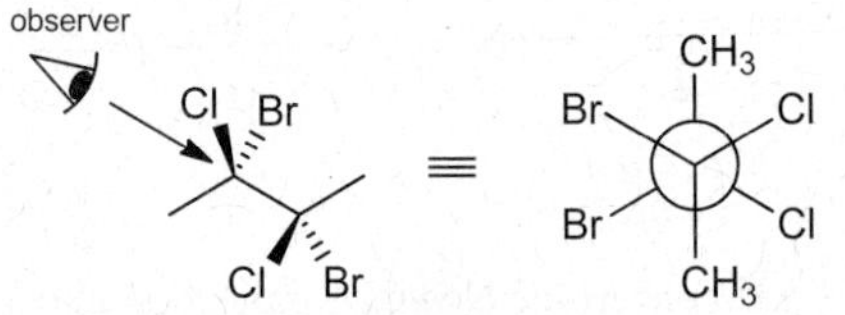

4.57. Two chair conformations can be drawn. In one of these conformations, all substituents occupy equatorial positions. In the other conformation, all substituents occupy axial positions. The former (shown below) lacks 1,3-diaxial interactions, and is therefore the most stable conformation of glucose:

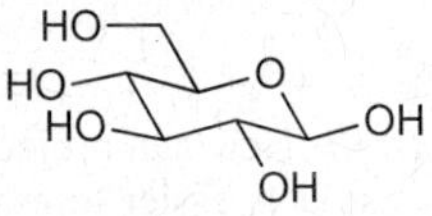

4.58. Begin by drawing a Newman projection:

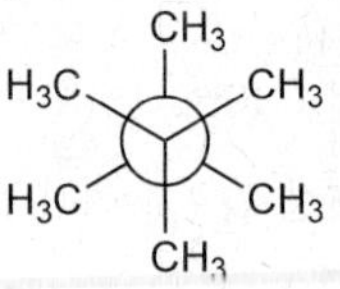

2,2,4-tetramethylbutane

All staggered conformations are degenerate, and the same is true for all eclipsed conformations. As such, the energy diagram has a shape that is similar to the energy diagram for the conformational analysis of ethane:

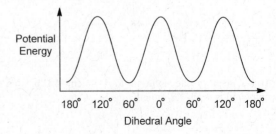

The staggered conformations have six gauche interactions, each of which has an energy cost of 3.8 kJ/mol. Therefore, each staggered conformation has an energy cost of 22.8 kJ/mol. The eclipsed conformations have three methyl-methyl eclipsing interactions, each of which has an energy cost of 11 kJ/mol. Therefore, each eclipsed conformation has an energy cost of 33 kJ/mol. The difference in energy between staggered and eclipsed conformations is therefore expected to be approximately 10.2 kJ/mol.

4.59. The two staggered conformations are lower in energy than the two eclipsed conformations. Among the staggered conformations, the *anti* conformation is the lowest in energy. Among the eclipsed conformations, the highest energy conformation is the one in which the bromine atoms are eclipsing each other. This information is summarized below:

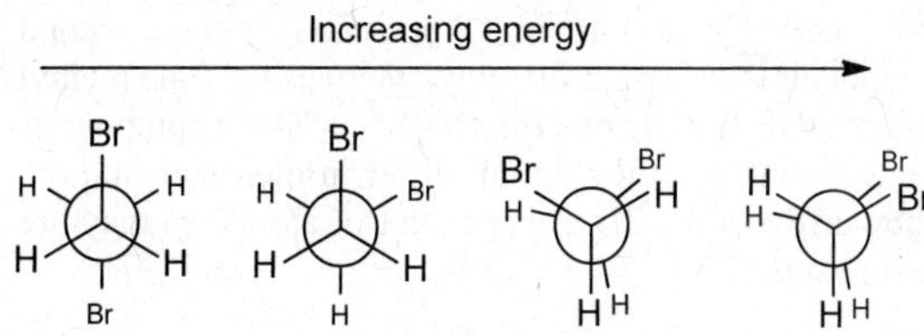

Increasing energy →

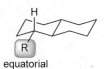

4.60.
(a) This conformation has three gauche interactions, each of which has an energy cost of 3.8 kJ/mol. Therefore, this conformation has a total energy cost of 11.4 kJ/mol associated with steric strain.
(b) This conformation has two methyl-H eclipsing interactions, each of which has an energy cost of 6 kJ/mol. In addition, it also has one methyl-methyl eclipsing interaction, which has an energy cost of 11 kJ/mol. Therefore, this conformation has a total energy cost of 23 kJ/mol associated with torsional strain and steric strain.

4.61. There are two chair conformations that can be drawn. In one chair conformation, all groups are equatorial except for one. In the other chair conformation, all groups are axial except for one. The former conformation is more stable because it has fewer 1,3-diaxial interactions.

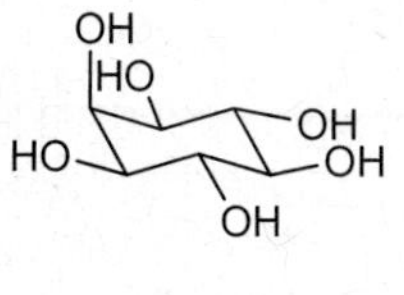

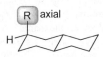

4.62.
(a) A group at C-2, pointing up, will occupy an equatorial position, as seen here:

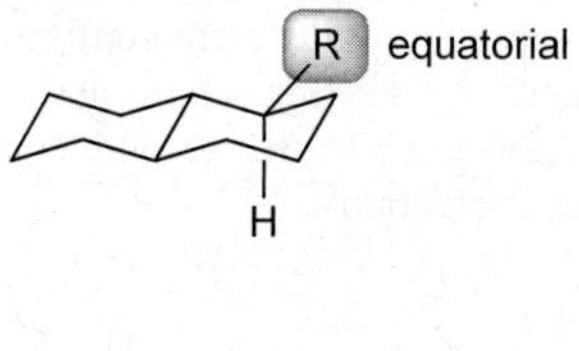

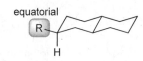

(b) A group at C-3, pointing down, will occupy an equatorial position, as seen here:

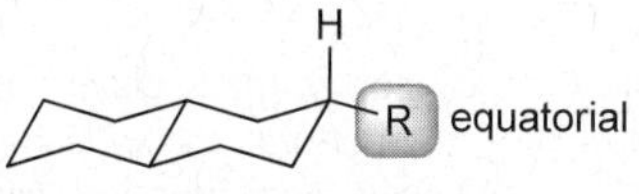

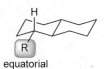

(c) A group at C-4, pointing down, will occupy an axial position, as seen here:

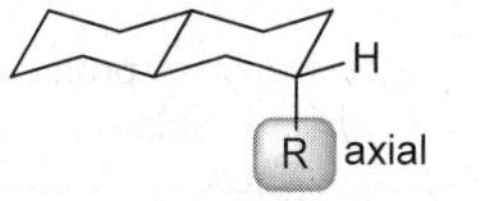

(d) A group at C-7, pointing down, will occupy an equatorial position, as seen here:

(e) A group at C-8, pointing up, will occupy an equatorial position, as seen here:

(f) A group at C-9, pointing up, will occupy an axial position, as seen here:

4.63. Propylene (C_3H_6) has one constitutional isomer (see Problem **1.42**). This isomer has a three-membered ring, so it is called cyclopropane.

cyclopropane

4.64. As mentioned in Section 4.9, cyclobutene adopts a slightly puckered conformation in order to alleviate some of the torsional strain associated with the eclipsing hydrogen atoms:

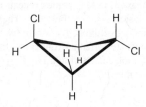

In this non-planar conformation, the individual dipole moments of the C-Cl bonds in *trans*-1,3-dichlorocyclobutane do not fully cancel each other, giving rise to a small molecular dipole moment.

4.65. Cyclohexene cannot adopt a chair conformation because two of the carbon atoms are sp^2 hybridized and trigonal planar. A chair conformation can only be achieved when all six carbon atoms are sp^3 hybridized and tetrahedral (with bond angles of 109.5°).

4.66.

(a) If we convert each Newman projection into a bond-line structure, we will be able to compare the two structures more easily. Then, if we assign a systematic name to each bond-line structure, we find that they have the same name. Therefore, these two structures represent the same compound.

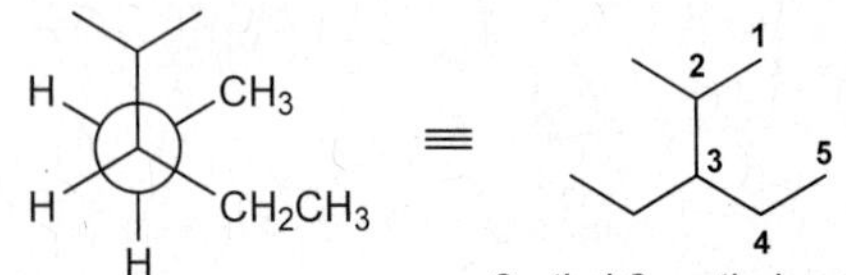

3-ethyl-2-methylpentane

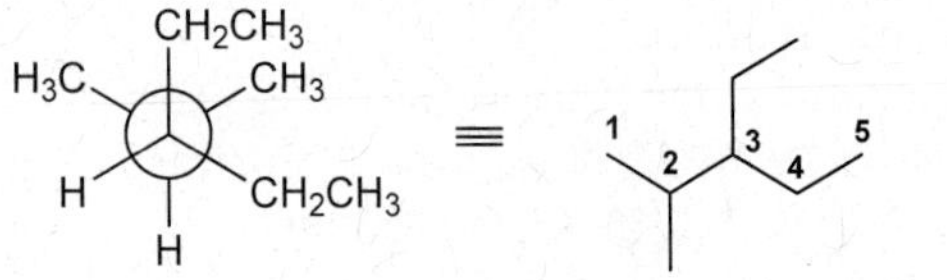

3-ethyl-2-methylpentane

(b) In the first compound, the two methyl groups are attached to C-1 and C-2, but in the second compound, the two methyl groups are attached to C-1 and C-3. These compounds share the same molecular formula (C_8H_{16}), but they have different constitution (connectivity of atoms). Therefore, these two compounds are constitutional isomers.

(c) These two structures are both representations of bicyclo[2.2.1]heptane, as can be seen when we apply the numbering system below. These two compounds are the same.

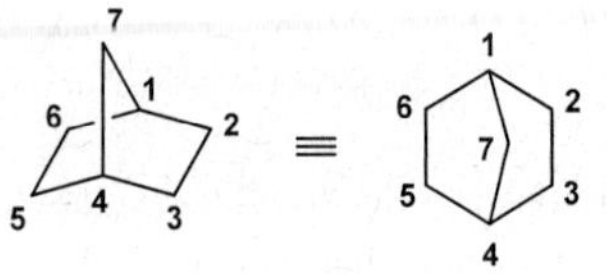

(d) We assign a systematic name to each structure, and find that they have different names (because they have different ring fusions, shown below). Therefore, they differ in their constitution (connectivity of atoms). These compounds are different from each other, but they share the same molecular formula ($C_{12}H_{22}$), so they are constitutional isomers.

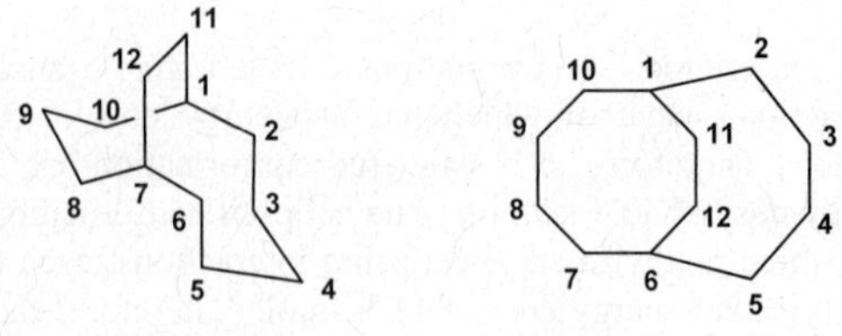

Bicyclo[5.3.2]dodecane   Bicyclo[4.4.2]dodecane

(e) Both of these structures are *cis*-1,4-dimethylcyclohexane. They are representations of the same compound.

(f) The first compound is *cis*-1,4-dimethylcyclohexane, and the second compound is *trans*-1,4-dimethylcyclohexane. These compounds are different, but they do not differ in their constitution. They differ in the 3D arrangement of atoms, so they are stereoisomers.

(g) The first compound is *cis*-1,2-dimethylcyclohexane (both methyl groups are in UP positions), and the second compound is *trans*-1,2-dimethylcyclohexane (one methyl group is UP and the other is DOWN). These compounds are different, but they do not differ in their constitution. They differ in the 3D arrangement of atoms, so they are stereoisomers.

(h) The first compound is *cis*-1,2-dimethylcyclohexane (both methyl groups are in UP positions), and the second compound is *trans*-1,2-dimethylcyclohexane (one methyl group is UP and the other is DOWN). These compounds are different, but they do not differ in their constitution. They differ in the 3D arrangement of atoms, so they are stereoisomers.

(i) If we convert each Newman projection into a bond-line structure, we will be able to compare the two structures more easily. Then, if we assign a systematic name to each bond-line structure, we find that they have different names. Since they share the same molecular formula, they are constitutional isomers.

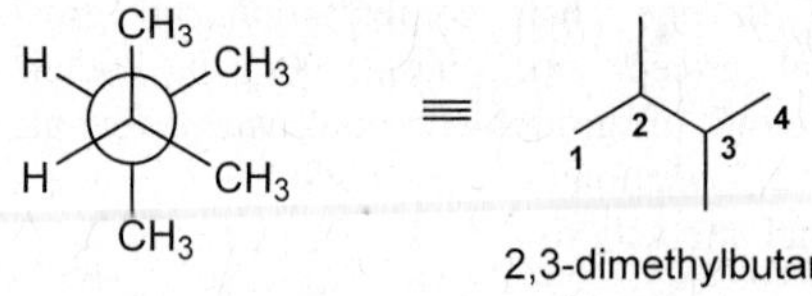

2,3-dimethylbutane

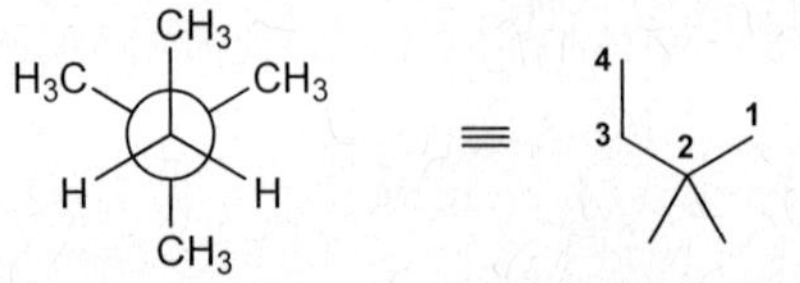

2,2-dimethylbutane

(j) In both cases, the front carbon atom has two methyl groups and the back carbon atom has two methyl groups. These structures represent different conformations of the same compound. The first conformation has three gauche interactions, while the second compound only has two gauche interactions.

(k) The first compound is *cis*-1,3-dimethylcyclohexane, and the second compound is *trans*-1,3-dimethylcyclohexane. These compounds are different, but they do not differ in their constitution. They differ in the 3D arrangement of atoms, so they are stereoisomers.

(l) In the first compound, the two methyl groups are attached to C-1 and C-3, but in the second compound, the two methyl groups are attached to C-1 and C-2. These compounds share the same molecular formula (C_8H_{16}), but they have different constitution (connectivity of atoms). Therefore, these two compounds are constitutional isomer.

4.67.

(a) The *trans* isomer is expected to be more stable, because the *cis* isomer has a very high-energy methyl-methyl eclipsing interaction (11 kJ/mol). See calculation below.

(b) We calculate the energy cost associated with all eclipsing interactions in both compounds. Let's begin with the *trans* isomer. It has the following eclipsing interactions, below the ring and above the ring, giving a total of 32 kJ/mol:

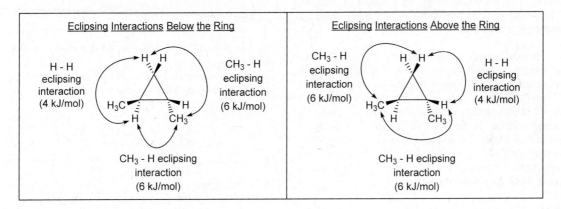

Now let's focus on the *cis* isomer. It has the following eclipsing interactions, below the ring and above the ring, giving a total of 35 kJ/mol:

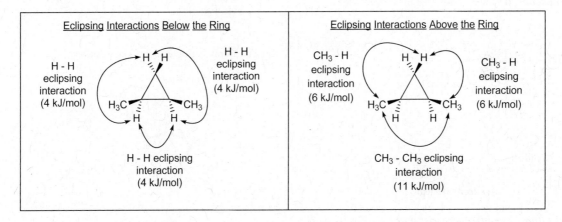

The difference between these two isomers is therefore predicted to be (35 kJ/mol) – (32 kJ/mol) = 3 kJ/mol.

4.68.

(a) Begin by assigning a numbering system, and then determine the location and three-dimensional orientation of each substituent:

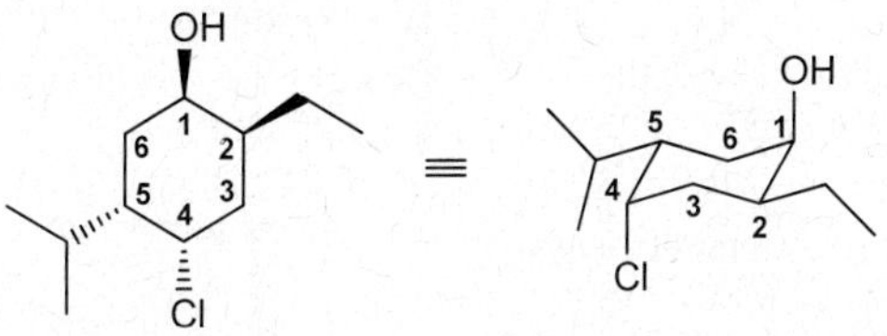

When assigning the numbers to the chair drawing, the first number can be placed anywhere on the ring (as long as the numbers go clockwise), as explained in the solution to Problem **4.31a**.

Once we have drawn the first chair conformation, we then draw the second chair conformation, again using a numbering system. Notice that a ring flip causes all equatorial groups to become axial groups, and vice versa.

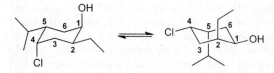

(b) Comparison of these chair conformations requires a comparison of the energy costs associated with all axial substituents (see Table 4.8). The first chair conformation has two axial substituents: an OH group (energy cost = 4.2 kJ/mol) and a Cl group (energy cost = 2.0 kJ/mol), giving a total of 6.2 kJ/mol. The second chair conformation has two axial substituents: an isopropyl group (energy cost = 9.2 kJ/mol) and an ethyl group (energy cost = 8.0 kJ/mol), giving a total of 17.2 kJ/mol. The first chair conformation has a lower energy cost, and is therefore more stable.

(c) Using the numbers calculated in part b, the difference in energy between these two chair conformations is expected to be (17.2 kJ/mol) – (6.2 kJ/mol) = 11 kJ/mol. Using the numbers in Table 4.8, we see that a difference of 9.2 kJ/mol corresponds with a ratio of 97:3 for the two conformations. In this case, the difference in energy is more than 9.2 kJ/mol, so the ratio should be even higher (more than 97%). Therefore, we do expect the compound to spend more than 95% of its time in the more stable chair conformation.

4.69.

(a) *cis*-Decalin has three gauche interactions, while *trans*-decalin has only two gauche interactions. Therefore, the latter is expected to be more stable.

(b) *trans*-Decalin is incapable of ring flipping, because a ring flip of one ring would cause its two alkyl substituents (which comprise the second ring) to be too far apart to accommodate the second ring.

4.70.

(a) The two substituents at the bridgehead carbons are *cis* to each other, analogous to the two bridgehead hydrogen atoms on *cis*-decalin.

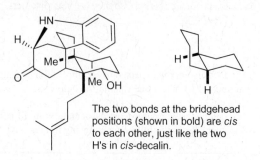

The two bonds at the bridgehead positions (shown in bold) are *cis* to each other, just like the two H's in *cis*-decalin.

(b) The dihedral angle between the two methyl groups should be approximately 60° (From the perspective of the chair on the right, one is equatorial down, and the other is axial down. They are gauche to each other.)

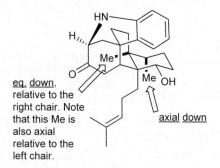

eq. down, relative to the right chair. Note that this Me is also axial relative to the left chair.

axial down

(c) The aromatic ring is in an axial position:

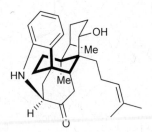

(d) The branch is *equatorial* to the left chair,

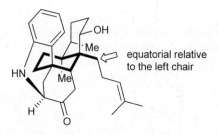

equatorial relative to the left chair

but *axial* to the right chair:

axial relative to
the right chair

4.71. The three staggered conformations are as follows:

A   B   C

Several types of gauche interactions are present in these conformers (Me-Me, Me-OH, and/or OH-OH). The strain of a methyl-methyl gauche interaction is 3.8 kJ/mol (see Table 4.6). The destabilization due to a Me-OH gauche interaction can be estimated as being roughly half of the value for the 1,3-diaxial interaction associated with an OH group (an OH group in an axial position experiences two gauche interactions, each of which might be expected to be somewhat similar to a Me-OH gauche interaction). Therefore, a Me-OH gauche interaction is expected to be approximately 4.2 / 2 = 2.1 kJ/mol (see Table 4.8).

Before exploring the OH-OH gauche interaction, our analysis thus far gives the following calculations for each conformer

Conformer **A**: 2 x 2.1 kJ/mol = 4.2 kJ/mol
Conformer **B**: 3.8 kJ/mol
Conformer **C**: 2 x 2.1 kJ/mol + 3.8 kJ/mol = 8.0 kJ/mol

This calculation must be modified when we take into account the effect of two OH groups that are gauche to each other, as seen in conformers **A** and **B**. We should expect an OH-OH gauche interaction to be less than a Me-Me gauche interaction (less than 3.8 kJ/mol), because an OH group appears to be less sterically encumbering than a methyl group (compare CH_3 and OH in Table 4.8). Therefore, the destabilizing effect associated with an OH-OH gauche interaction (less than

3.8kJ/mol) should be overshadowed by the *stabilizing* effect that results from the hydrogen bonding interactions between the two OH groups, which is expected to be approximately 20 kJ/mol (see section 1.12). As a result, we expect extra stabilization to be associated with any conformer in which two OH groups are gauche to each other. This occurs in conformers **A** and **B**, but the hydroxyl groups in **C** are too far to form this type of interaction. If we assume that the stabilization achieved through hydrogen bonding is the same for conformers **A** and **B**, then conformer **B** should be the lowest-energy staggered conformation for this isomer of 2,3-butanediol.

4.72.
(a) The staggered and eclipsed conformations of methanol, **A** and **B**, respectively, are as follows:

A   B

In conformation **B**, there are three eclipsing interactions. One of those interactions is between the O-H bond and a C-H bond, while the other two eclipsing interactions are between a lone pair and a C-H bond. The corresponding conformations for ethane are structures **C** and **D**:

C   D

(b) We know that each eclipsing interaction between two C-H bonds in ethane is 4 kJ/mol. This value is very close to the total destabilization energy for the eclipsed conformation of methanol. And since we can identify exactly one eclipsing interaction between an O-H bond and a C-H bond in methanol, this interaction is likely responsible for nearly all of the 4.2 kJ/mol of destabilization. We can conclude, then, that the eclipsing interaction between a lone pair and a C-H bond in methanol contributes a negligible amount of destabilization in this molecule.

4.73. The following is a summary of the changes in bonding, conformation, and intramolecular hydrogen bonding during the following process:

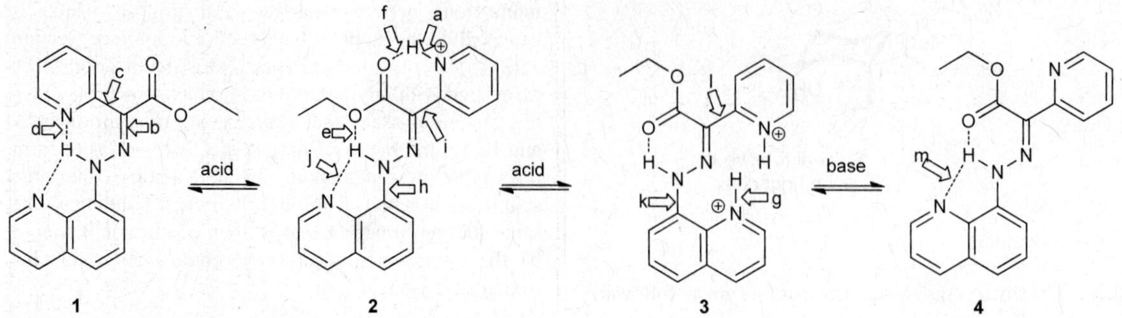

1 → 2

Bonding change:

    -new N-H bond (a), as a result of protonation of a nitrogen atom.

Conformation change:

    - rotation around C-C single bond (c)

    - there is also a change in orientation around the N=C double bond (b), although we will see in the next chapter that this change is not considered to be a change in *conformation*, but rather, it is a change in *configuration.*

Intramolecular hydrogen bonding changes:

    - N---H interaction broken (d) and replaced with O---H interaction (e)

    - O---H interaction (f) is formed

2 → 3

Bonding change:

    -new N-H bond (g) as a result of protonation of a nitrogen atom

Conformation changes:

    - rotation around N-C single bond (h)

    - rotation around C-C single bond (i)

Intramolecular hydrogen bonding change:

    - N---H interaction broken (j)

    - O---H interaction broken (f)

3 → 4

Bonding changes:

    -two N-H bonds broken (deprotonation) (a, g)

Conformation changes:

    - rotation around N-C single bond (k)

    - rotation around C-C single bond (l)

Intramolecular hydrogen bonding change:

    - N---H interaction forms (m)

4.74.

(a) The following two σ bonds must be rotated by 180° to convert from conformation A to conformation B.

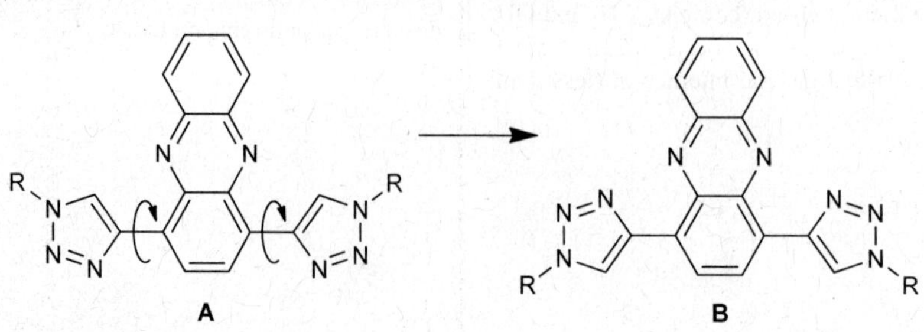

rotate around these
bonds by 180 degrees

In conformation **A**, the molecule is planar because all of atoms are sp^2-hybridized (except for the R groups). The same is true for conformation **B**. But during the conformational change from **A** to **B**, this planarity is temporarily broken. A molecular model might be helpful in order to see how the planarity of the entire compound is temporarily disrupted.

(b) The decreased stability of **B** can be explained by the fact that the indicated lone pairs (regions of high electron density) in **B** are oriented toward each other, leading to mutual electrostatic repulsion in this conformation, and thus a higher relative energy. In conformation **A**, these lone pairs are oriented away from each other, thus avoiding this significant electrostatic repulsion.

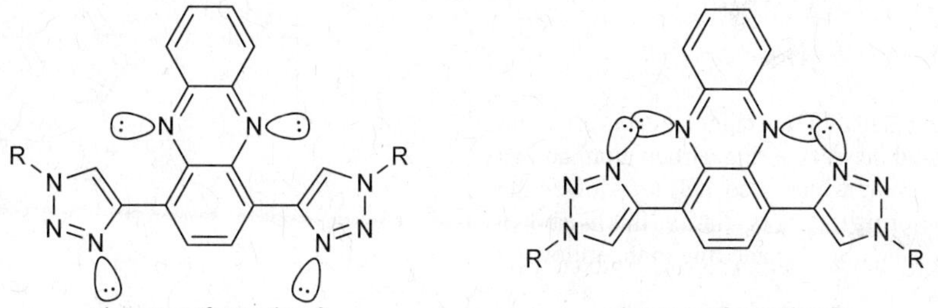

planar conformation A

planar conformation B

4.75.
(a) The three possible staggered conformations of **1**, viewed along the C_a-C_b bond, are as follows:

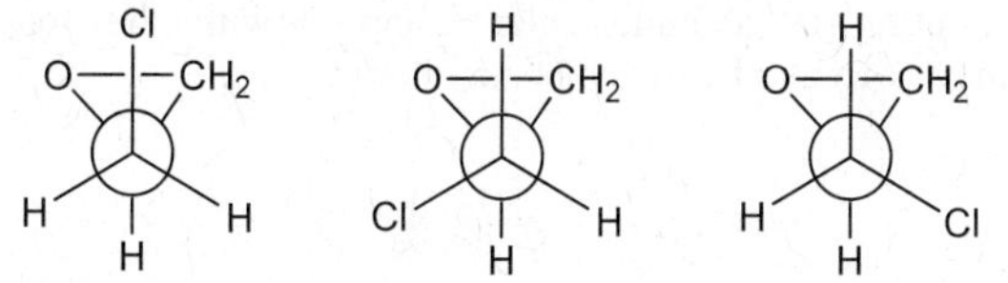

A   **B**   **C**

(b) In conformation **A**, there are two gauche interactions: one between the chlorine atom and the CH_2 group, and another between the chlorine atom and the oxygen atom. This conformation is the highest in energy (least stable) due to these two interactions. Conformations **B** and **C** each exhibit only one of these two gauche interactions.

(c) In conformation **B**, the only gauche interaction is that between the oxygen atom and chlorine atom, while in **C**, there is one gauche interaction between the chlorine atom and the CH_2 group. Based on the information provided for 1,2-dichloropropane, we can conclude that the energy cost of a gauche interaction between lone pairs and lone pairs is lower than the energy cost of a gauche interaction between a chlorine atom and a CH_2 group:

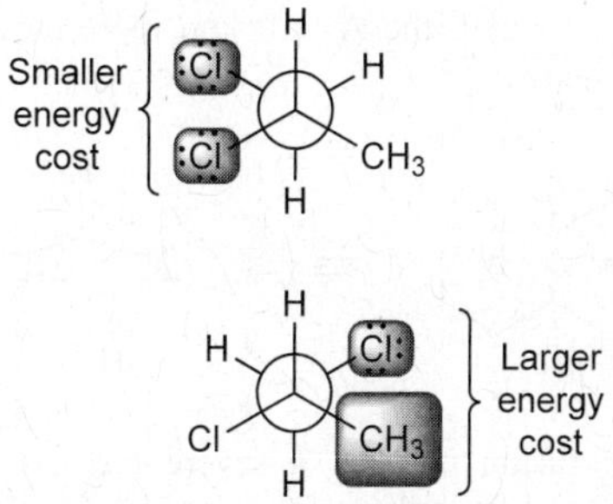

Smaller energy cost

Larger energy cost

Applying this concept to compound **1**, we should expect the gauche interaction in **C** (between the Cl and the CH_2 group) to be greater than the gauche interaction in **B** (between the oxygen atom and the chlorine atom, which is similar to the interaction between the two chlorine atoms in 1,2-dichloropropane). Therefore, **B** is expected to be the most stable conformation.

4.76.
(a) The all-equatorial chair conformation of compound **2** experiences severe steric repulsion due to the nearby isopropyl groups as evident in the following drawing, in which we focus on the interactions between a pair of neighboring isopropyl groups. These interactions, highlighted below, occur for each pair of neighboring isopropyl groups.

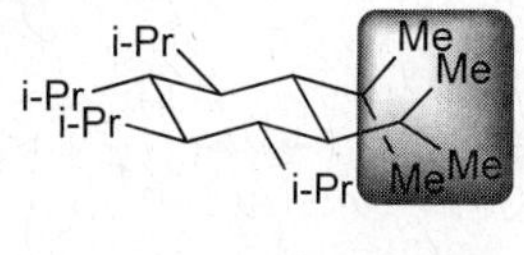

A Newman projection of **2**, viewed down one of the C-C bonds connecting an equatorial isopropyl group to the cyclohexane ring, shows how crowded the equatorial groups are as the methyl groups are in constant contact with each other. These interactions, highlighted below, occur for each pair of neighboring isopropyl groups.

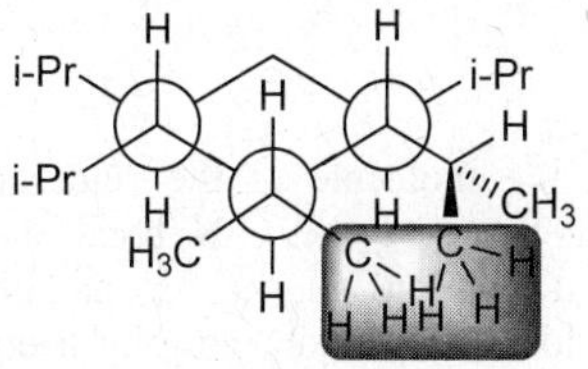

(b) In the all axial chair conformation for **2**, the C-H bond of each isopropyl group can all point towards each other (highlighted below) such that steric repulsion can be minimized. For clarity, this is shown only for the top face of the cyclohexane ring in the following structure.

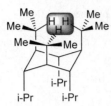

Furthermore, this results in a staggered conformation along each C-C bond between a ring carbon atom and an isopropyl carbon atom, which also helps to lower the energy of this conformation. For clarity, this is shown only on one face of the cyclohexane ring in the following structure.

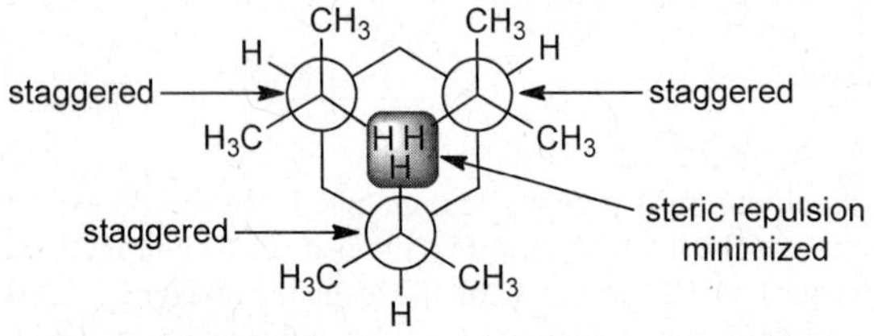

4.77.
In conformation **1a**, *p* orbitals on N, C, C, and B can overlap in one continuous system, allowing for resonance delocalization between the lone pair on nitrogen and the empty 2*p* orbital of the boron atom,

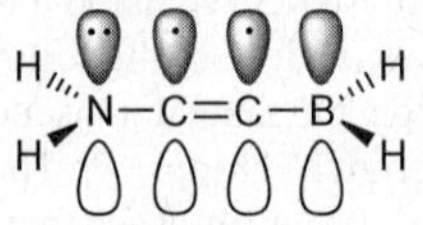

giving the following resonance structures:

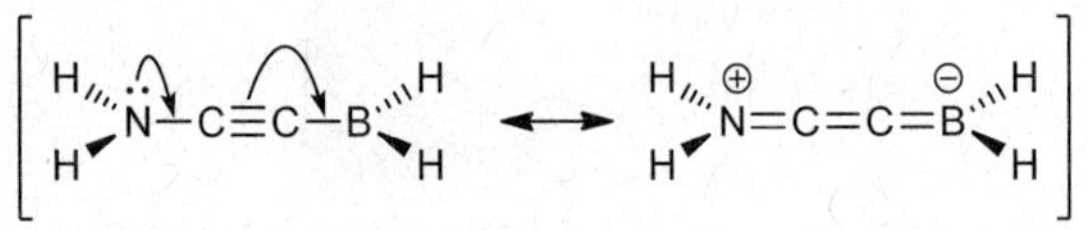

However, in conformation **1b**, the 2*p* orbital of the nitrogen atom is perpendicular to the empty 2*p* orbital of the boron, so delocalization is not possible. Therefore, conformation **1a** is expected be lower in energy (more favorable) than conformation **1b**, because of the additional resonance stabilization that is present in **1a**.

4.78.
First, to get the molecule in the right conformation, rotate by 180° around the C-N bond indicated. The resulting conformation allows for an intramolecular hydrogen bond as indicated below by a dotted line. In this conformation, there are two hydrogen-bond acceptors ("A") and two hydrogen bond donors ("D") along the top edge of the molecule as drawn.

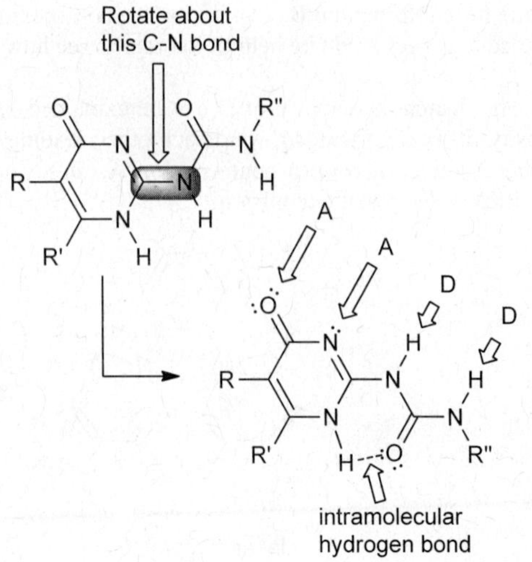

Next, to show the four intermolecular hydrogen bonds, we draw a second molecule rotated by 180° relative to the first. In this orientation the donor/donor/acceptor/acceptor pattern of the bottom molecule is complementary to the acceptor/acceptor/donor/donor motif on the top molecule resulting in a bimolecular complex with the four intermolecular hydrogen bonds shown below.

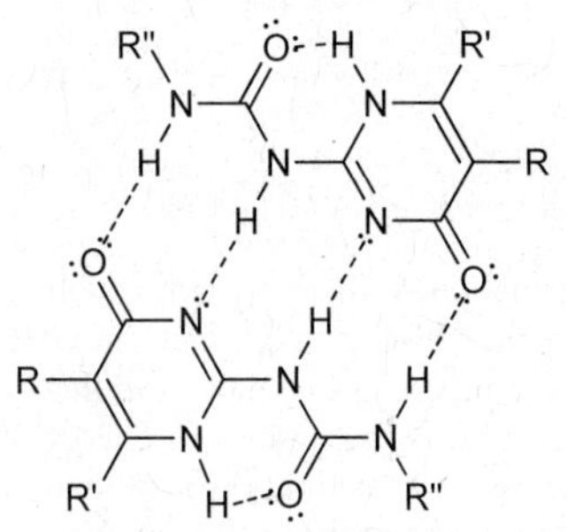

4.79.
Each compound has three rings, labeled A, B, and C.

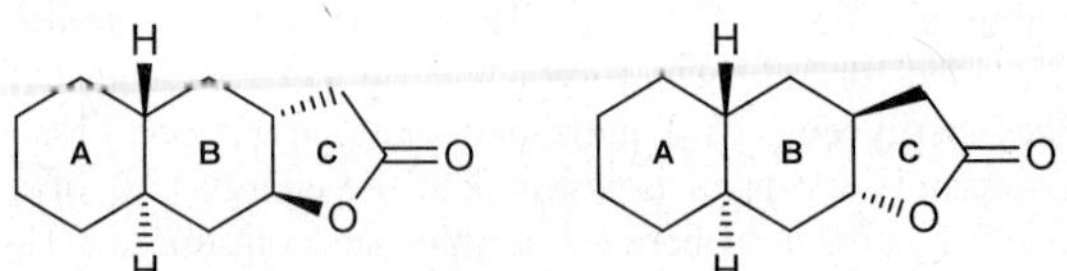

In each compound, the A-B fusion represents a *trans*-decalin system:

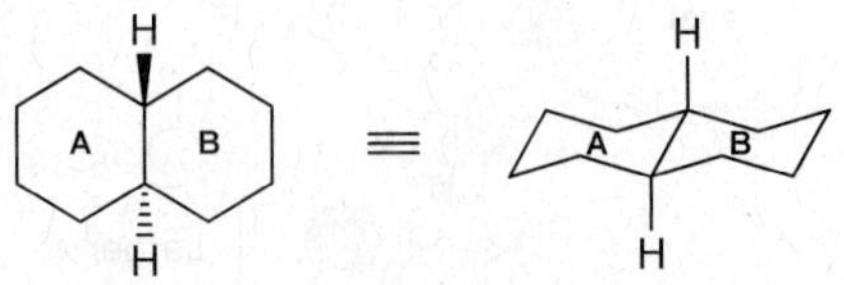

The *trans* fusion imposes a severe restriction on the conformational flexibility of the system. Specifically, in a *trans*-decalin system, neither ring can undergo a ring-flip to give a different chair conformation. For each ring, the chair conformation shown above is the only chair conformation that is achievable. However, both rings are still free to adopt a higher energy boat conformation. For example, the B ring of *trans*-decalin can adopt the following boat conformation:

In compound **1**, the C ring imposes a further conformational restriction, by locking the B ring into a chair conformation:

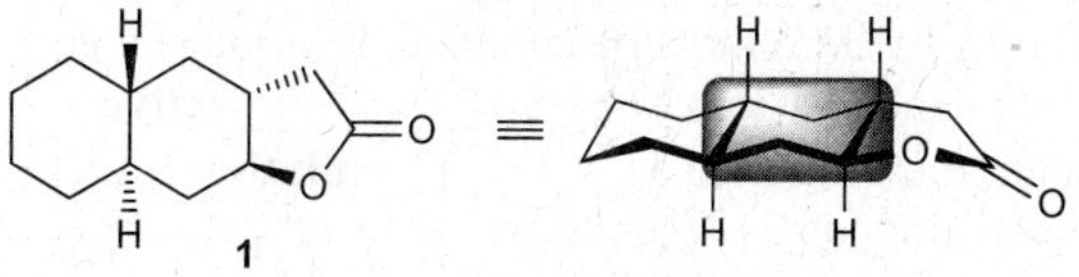

In contrast, the C ring in compound **2** imposes the restriction of locking the B ring in a boat conformation:

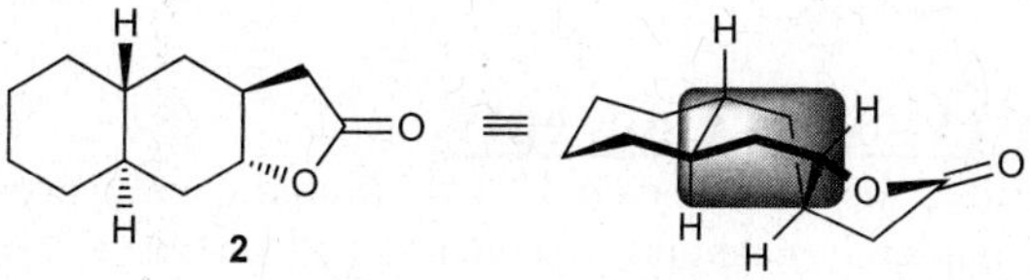

The boat conformation of compound **2** is expected to be higher in energy than the chair conformation of compound **1**. Therefore, compound **2** is expected to have the higher heat of combustion. In fact, the investigators prepared compounds **1** and **2** for the purpose of measuring the difference in energy between chair and boat conformations.

Chapter 5
Stereoisomerism

Review of Concepts

Fill in the blanks below. To verify that your answers are correct, look in your textbook at the end of Chapter 5. Each of the sentences below appears verbatim in the section entitled *Review of Concepts and Vocabulary*.

- _____isomers have the same connectivity of atoms but differ in their spatial arrangement.
- **Chiral** objects are not **superimposable** on their _____. The most common source of molecular chirality is the presence of a _____, a carbon atom bearing _____ different groups.
- A compound with one chirality center will have one non-superimposable mirror image, called its _____.
- The Cahn-Ingold-Prelog system is used to assign the _____ of a chirality center.
- A **polarimeter** is a device used to measure the ability of chiral organic compounds to rotate the plane of _____ light. Such compounds are said to be _____ **active**.
- A solution containing equal amounts of both enantiomers is called a _____ **mixture**. A solution containing a pair of enantiomers in unequal amounts is described in terms of **enantiomeric** _____ (*ee*).
- For a compound with multiple chirality centers, a family of stereoisomers exists. Each stereoisomer will have at most one enantiomer, with the remaining members of the family being _____.
- A _____ **compound** contains multiple chirality centers but is nevertheless achiral because it possesses reflectional symmetry.
- _____ **projections** are drawings that convey the configuration of chirality centers, without the use of wedges and dashes.

Review of Skills

Fill in the blanks and empty boxes below. To verify that your answers are correct, look in your textbook at the end of Chapter 5. The answers appear in the section entitled *SkillBuilder Review*.

SkillBuilder 5.1   Identifying *cis-trans* Stereoisomerism

ASSIGN THE CONFIGURATION OF THE FOLLOWING DOUBLE BOND AS CIS OR TRANS

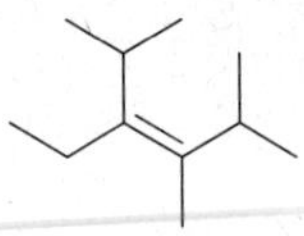

SkillBuilder 5.2   Locating Chirality Centers

CIRCLE THE CHIRALITY CENTER IN THE FOLLOWING COMPOUND

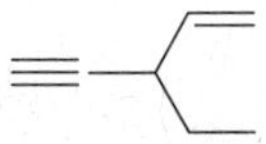

SkillBuilder 5.3   Drawing an Enantiomer

SHOW THREE WAYS TO DRAW THE ENANTIOMER OF THE FOLLOWING COMPOUND. PLACE YOUR ANSWERS IN THE BOXES SHOWN.

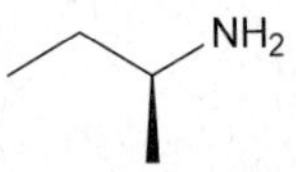

SkillBuilder 5.4   Assigning Configuration

ASSIGN THE CONFIGURATION
OF THE CHIRALITY CENTER IN
THE FOLLOWING COMPOUND

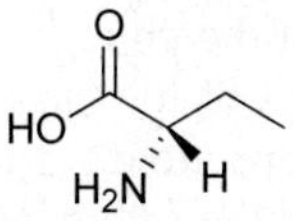

SkillBuilder 5.5   Calculating specific rotation

CALCULATE THE
SPECIFIC ROTATION
GIVEN THE FOLLOWING
INFORMATION:

0.300 grams sucrose
dissolved in 10.0 mL of water
sample cell = 10.0 cm
observed rotation = +1.99°

$$\text{specific rotation} = [\,\alpha\,] = \frac{\alpha}{c \times l} = \frac{\boxed{}}{\boxed{} \times \boxed{}} = \boxed{}$$

SkillBuilder 5.6   Calculating % *ee*

CALCULATE THE
ENANTIOMERIC EXCESS
GIVEN THE FOLLOWING
INFORMATION:

The specific rotation of optically
pure adrenaline is –53 .
A mixture of (*R*)- and (*S*)-
adrenaline was found to have a
specific rotation of –45.
Calculate the % *ee* of the
mixture

$$\% \, ee \;=\; \frac{\text{observed } [\alpha]}{[\alpha] \text{ of pure enantiomer}} \;\times\; 100\,\%$$

$$=\; \frac{\boxed{}}{\boxed{}} \;\times\; 100\,\% \;=\; \boxed{}$$

SkillBuilder 5.7   Determining Stereoisomeric Relationship

IDENTIFY THE STEREOISOMERIC
RELATIONSHIP BETWEEN THE
FOLLOWING TWO COMPOUNDS

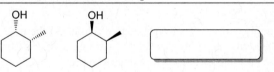

SkillBuilder 5.8   Identifying *Meso* Compounds

DRAW ALL POSSIBLE STEREOISOMERS OF 1,2-CYCLOHEXANEDIOL (SHOWN LEFT), AND THEN LOOK FOR A
PLANE OF SYMMETRY IN ANY OF THE DRAWINGS. THE PRESENCE OF A PLANE OF SYMMETRY INDICATES A
MESO COMPOUND

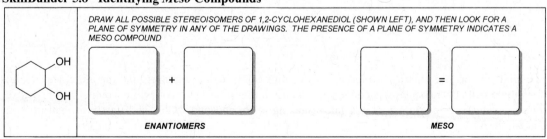

ENANTIOMERS   **MESO**

SkillBuilder 5.9   Assigning configuration from a Fischer projection

ASSIGN THE CONFIGURATION
OF THE CHIRALITY CENTER IN
THE FOLLOWING COMPOUND

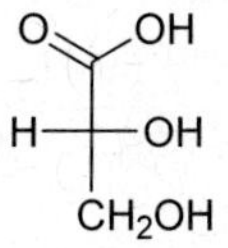

Common Mistakes to Avoid

When drawing a chirality center, the four groups connected to the chirality center must be drawn so that one group is on a wedge (which indicates that it is coming out of the page), and one group is on a dash (which indicates that it is going behind the page), and two groups are on straight lines (which indicate that these two groups are in the plane of the page), as seen in each of the following drawings:

Notice that in all of these cases, the two straight lines form a V, and neither the dash nor the wedge is placed inside that V. This is very important. If either the dash or the wedge is placed inside the V, the drawing becomes ambiguous and inaccurate. Don't make this mistake, as it is a common mistake:

WRONG

This dash cannot be placed here, inside the V

WRONG

This wedge cannot be placed here, inside the V

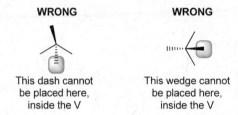

The drawings above do not make any sense, and if a chirality center is drawn like either of the drawings above, it would be impossible to assign a configuration to the chirality center. Never draw a chirality center that way. For the same reason, never draw a chirality center like this:

WRONG

WRONG

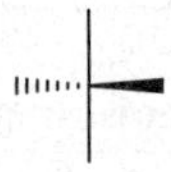

These two drawings imply square planar geometry, which is not the case for an sp^3 hybridized carbon atom (the geometry is tetrahedral). In some rare cases, you might find a chirality center for which three of the lines are drawn as straight lines, as in the following example:

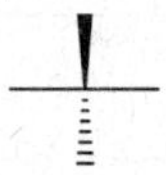

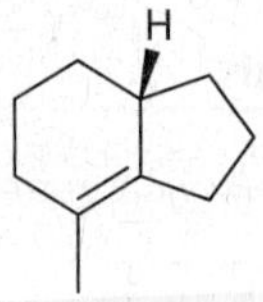

This compound has one chirality center, and its configuration is unambiguous (and therefore acceptable), although you will not encounter this convention often. In most cases that you will encounter in this course, a chirality center will be drawn as two lines (making a V), and one wedge and one dash that are both *outside* of the V:

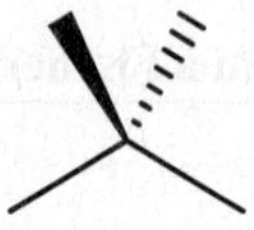

Solutions

5.1.

(a) The methyl groups are on opposite sides of the double bond, so this compound has a *trans* configuration.

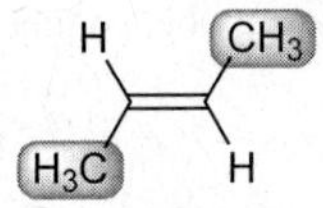

trans

(b) Two identical groups (fluorine atoms) are connected to the same position. Therefore, this compound is not stereoisomeric.

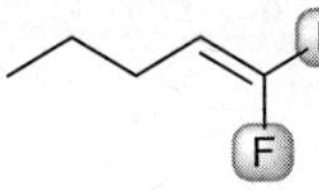

(c) The hydrogen atoms are on opposite sides of the double bond, so this compound has a *trans* configuration.

trans

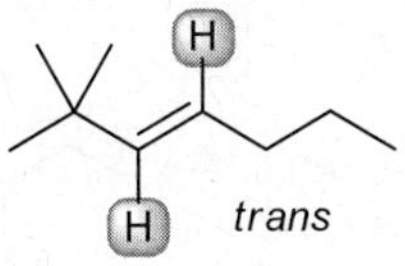

(d) The hydrogen atoms are on opposite sides of the double bond, so this compound has a *trans* configuration.

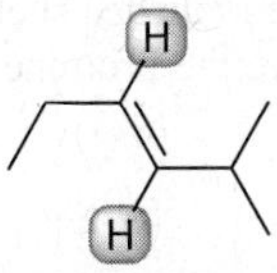

trans

(e) The phenyl groups (highlighted below) are on opposite sides of the double bond, so this compound has a *trans* configuration.

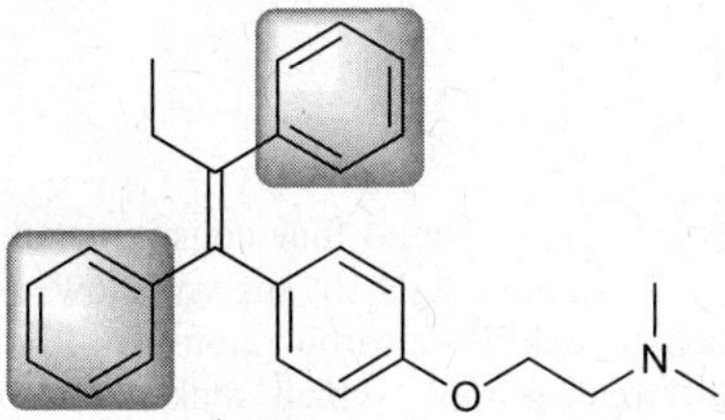

trans

(f) There are three butyl groups in this case, and two of them (highlighted below) are connected to the same position. Therefore, this compound is not stereoisomeric.

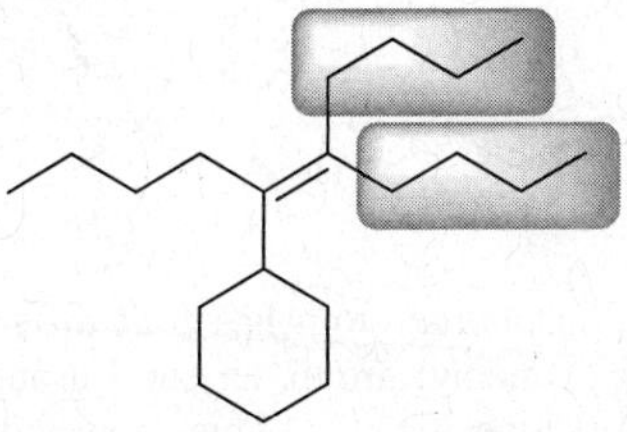

(g) The carboxylic acid groups (highlighted below) are on the same side of the double bond, so this compound has a *cis* configuration.

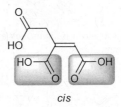

cis

5.2. We first draw a bond-line structure, which makes it easier to see the groups that are connected to each of the double bonds.

Each of the double bonds has two identical groups (hydrogen atoms) connected to the same position.

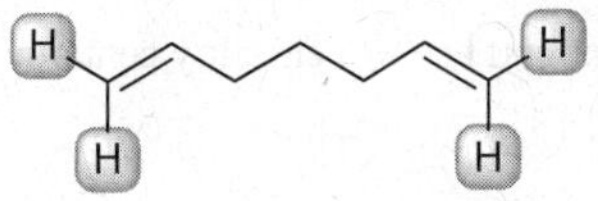

As such, neither double bond exhibits stereoisomerism, so this compound does not have any stereoisomers.

5.3.

(a) Compound X must contain a carbon-carbon double bond in the *trans* configuration, which accounts for four of the five carbon atoms:

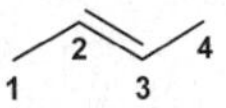

Now we must decide where to place the fifth carbon atom. We cannot attach this carbon atom to a vinylic position (C2 or C3), as that would give a double bond that is not stereoisomeric, and compound X is supposed to have the *trans* configuration.

same
group

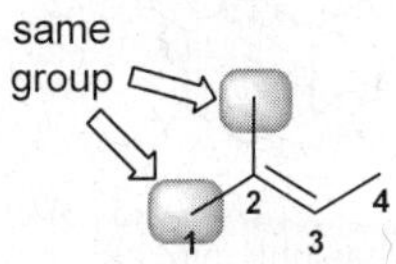

not stereoisomeric

Therefore, we must attach the fifth carbon atom to an allylic position, giving the following compound:

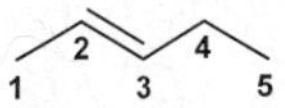

(b) Compound Y possesses a carbon-carbon double bond that is not stereoisomeric, which means that it must contain two identical groups connected to the same vinylic position. Those identical groups can be methyl groups, as in the following compound,

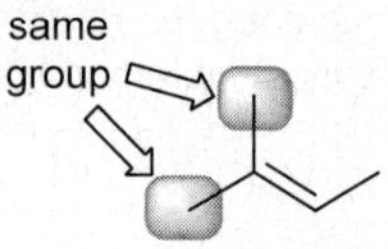

or the identical groups can be hydrogen atoms, as in the following three compounds:

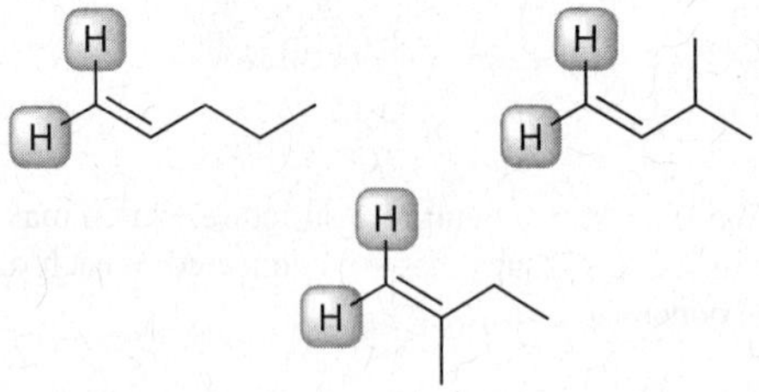

5.4. In each of the following cases, we ignore all sp^2 hybridized carbon atoms, all sp hybridized carbon atoms, and all CH_2 and CH_3 groups. We identify those carbon atoms (highlighted below) bearing four different groups:

(a) This compound has two chirality centers:

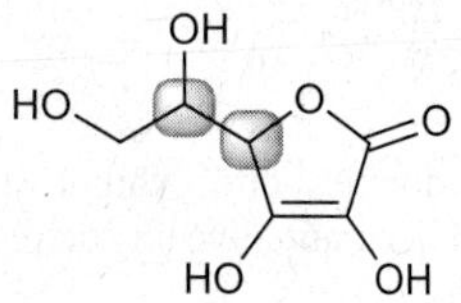

(b) This compound has five chirality centers:

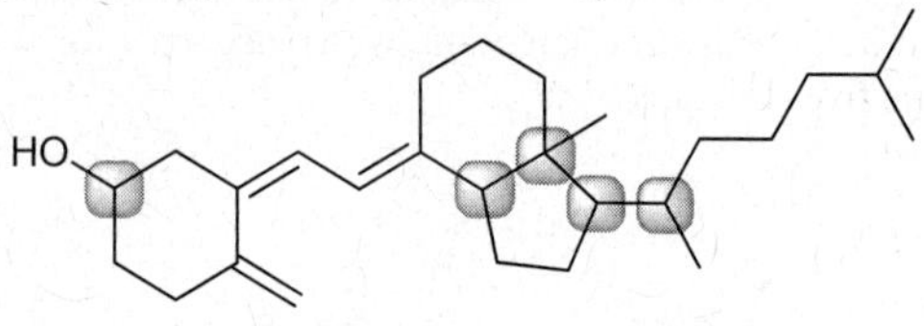

(c) This compound has five chirality centers:

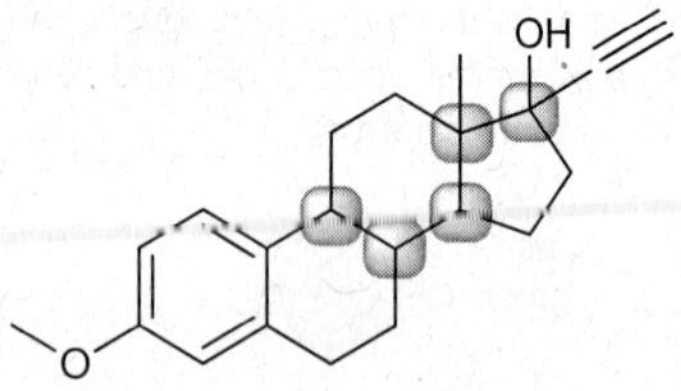

(d) This compound has only one chirality center:

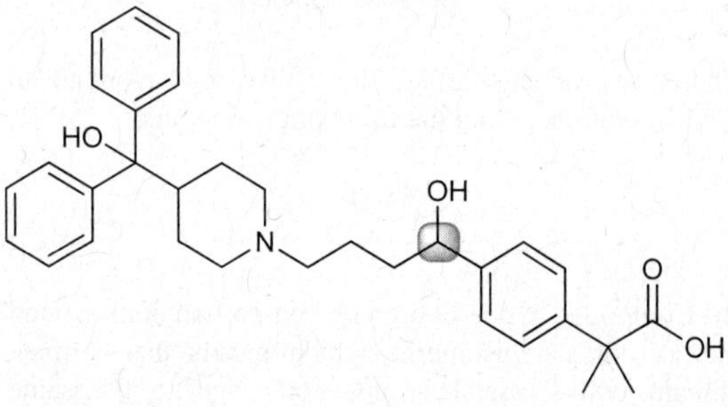

5.5. Recall that constitutional isomers are compounds that share the same molecular formula, but differ in constitution (the connectivity of atoms). There are two different ways that four carbon atoms can be connected together. They can be connected in a linear fashion (below left), or they can be connected with a branch (below right).

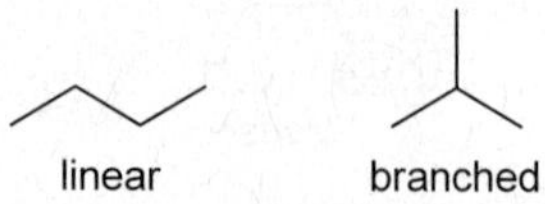

For each of these skeletons, we must consider the different locations where a bromine atom can be placed. In the linear skeleton, the bromine atom can either be placed at C1 or at C2.

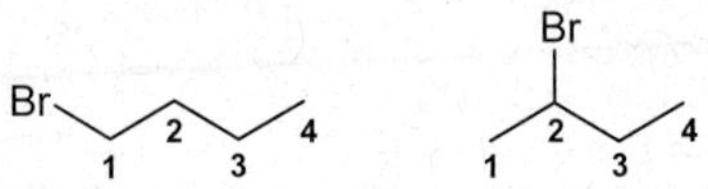

Placing the bromine atom at C3 is the same as placing it at C2:

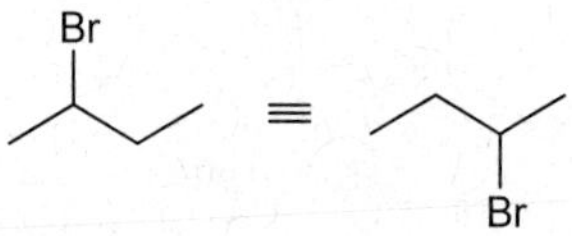

Similarly, placing the bromine atom at C4 is the same as placing it at C1.
Now let's consider the branched skeleton. There are two unique locations where the bromine atom can be placed (either at C1 or at C2).

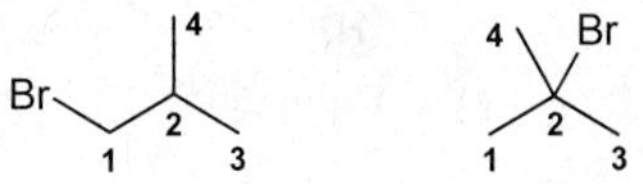

Placing the bromine atom at C3 or C4 is the same as placing it at C1:

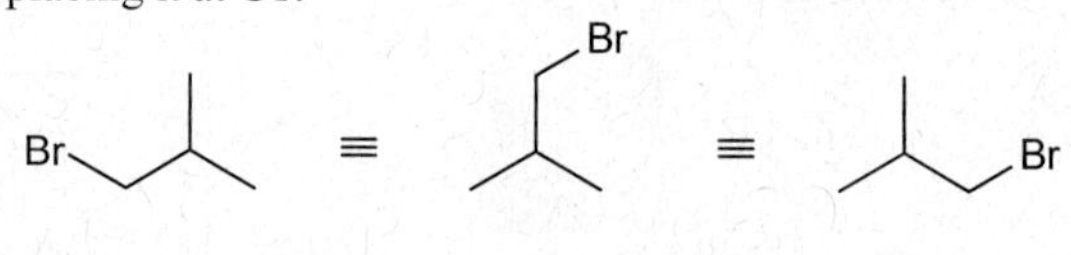

In summary, we have found four constitutional isomers with molecular formula C_4H_9Br, shown below. Only one of these isomers exhibits a carbon atom that is connected to four different groups, which makes it a chirality center.

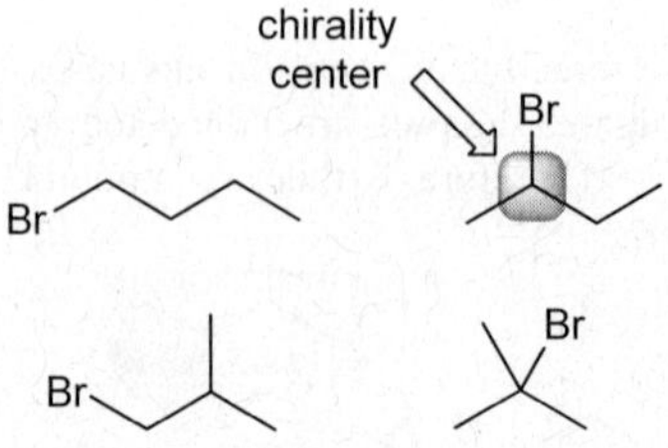

5.6. The phosphorus atom has four different groups attached to it (a methyl group, an ethyl group, a phenyl group, and a lone pair). This phosphorous atom therefore represents a chirality center. This compound is not superimposable on its mirror image (this can be seen more clearly by building and comparing molecular models).

5.7.
(a) To draw the enantiomer, we simply redraw the structure in the problem statement, except that we replace the wedge with a dash, as shown:

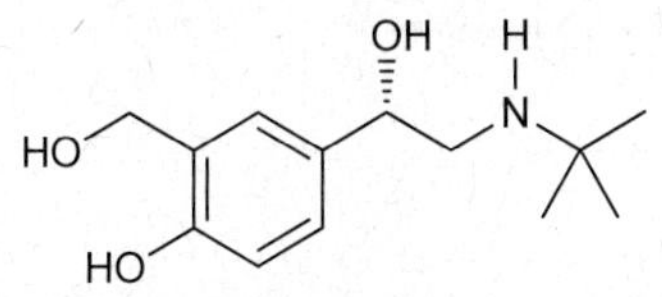

(b) To draw the enantiomer, we simply redraw the structure in the problem statement, except that we replace the wedge with a dash, as shown:

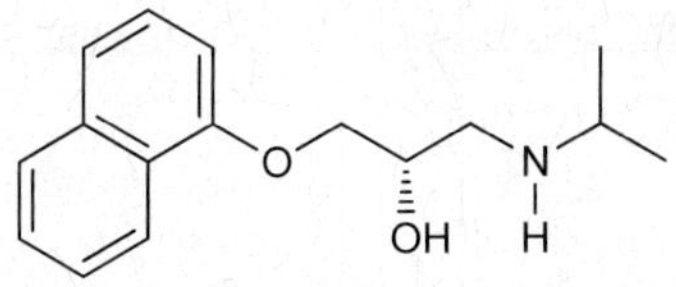

(c) To draw the enantiomer, we simply redraw the structure in the problem statement, except that we replace the wedge with a dash, and we replace the dash with a wedge, as shown:

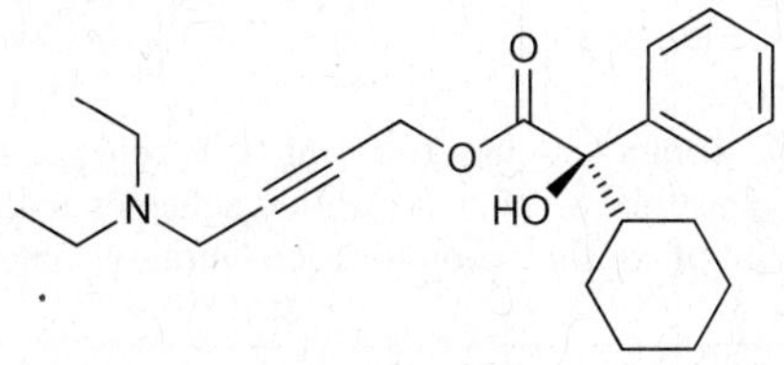

(d) To draw the enantiomer, we simply redraw the structure in the problem statement, except that we replace the wedge with a dash, as shown:

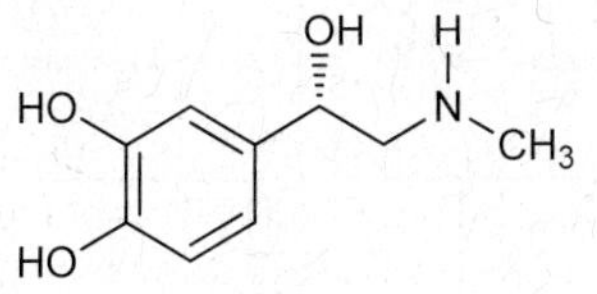

(e) To draw the enantiomer, we simply redraw the structure in the problem statement, except that we replace the wedge with a dash, and we replace the dash with a wedge, as shown:

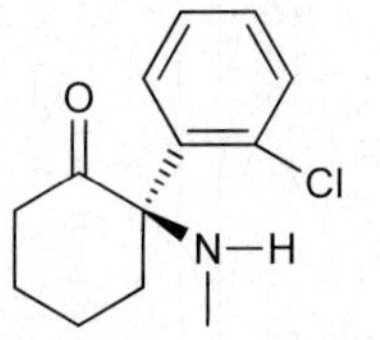

(f) To draw the enantiomer, we simply redraw the structure in the problem statement, except that we replace the wedge with a dash, and we replace the dash with a wedge, as shown:

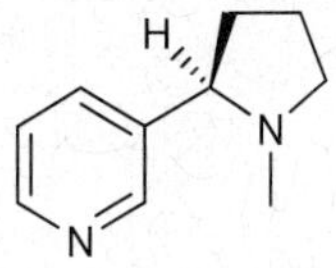

(g) Wedges and dashes are not drawn in the structure in the problem statement, because the three-dimensional geometry is implied by the drawing. In this case, it will be easier to place the mirror on the side of the molecule, giving the following structure for its enantiomer:

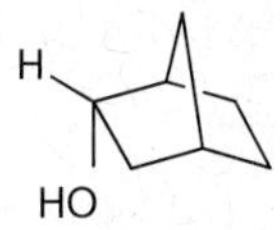

5.8. To draw the enantiomer, we simply redraw the structure in the problem statement, except that we replace the wedges with dashes, and we replace the dashes with wedges, as shown:

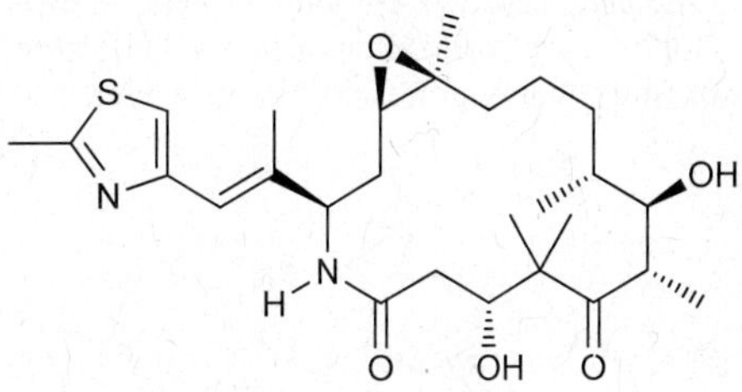

5.9.
(a) This compound has two chirality centers, shown below. The following prioritization schemes led to the assignment of configuration for each chirality center.

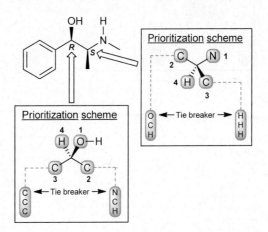

(b) This compound has two chirality centers, shown below. The following prioritization schemes led to the assignment of configuration for each chirality center.

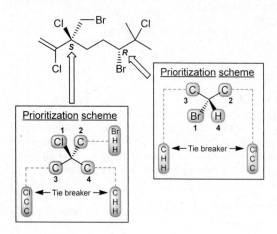

(c) This compound has two chirality centers, shown below. The following prioritization schemes led to the assignment of configuration for each chirality center.

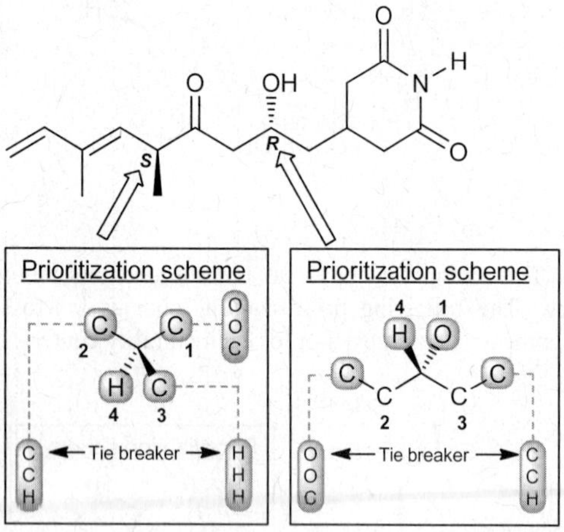

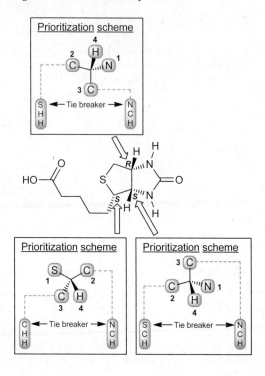

(d) This compound has three chirality centers. The following prioritization schemes led to the assignment of configuration for each chirality center.

(e) This compound has four chirality centers, shown below. The following prioritization schemes led to the assignment of configuration for each chirality center.

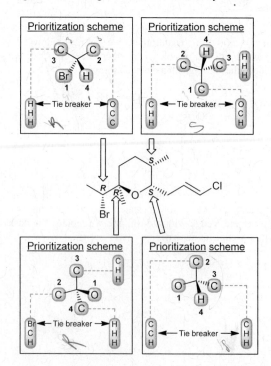

(f) This compound has two chirality centers, shown below. The following prioritization schemes led to the assignment of configuration for each chirality center.

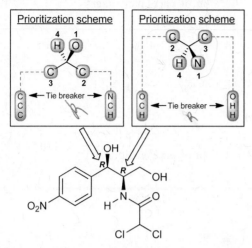

5.10. To assign the configuration, we must first assign a prioritization scheme to the four atoms connected to the chirality center:

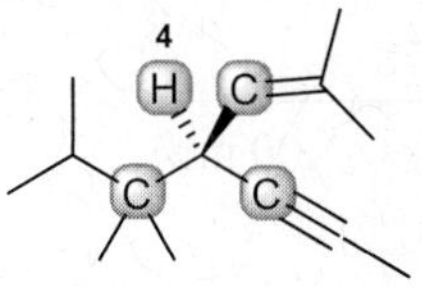

One of these atoms is H, so that atom is immediately assigned the fourth priority. The remaining three atoms are all carbon atoms, so for each of them, we prepare a list of the atoms attached to them. Let's begin with the double bond. Recall that a double bond is comprised of one σ bond and one π bond. For purposes of assigning configurations, we treat the π bond as if it were a σ bond to another carbon atom, like this:

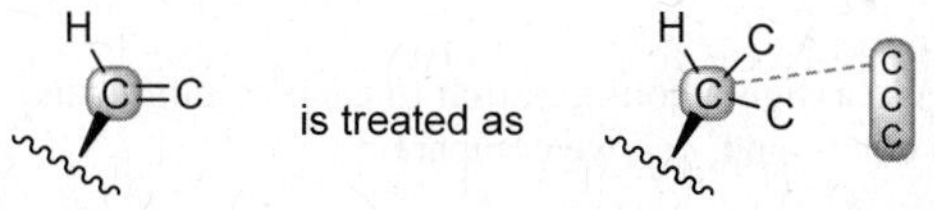

is treated as

We treat a triple bond similarly. Recall that triple bonds are comprised of one σ bond and two π bonds. As such, each of the two π bonds is treated as if it were a σ bond to another carbon atom:

is treated as

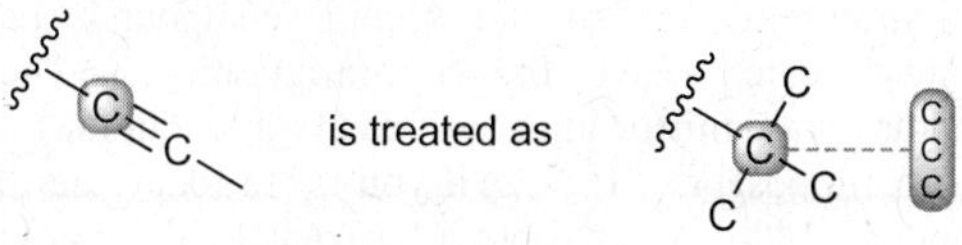

This gives the following competition:

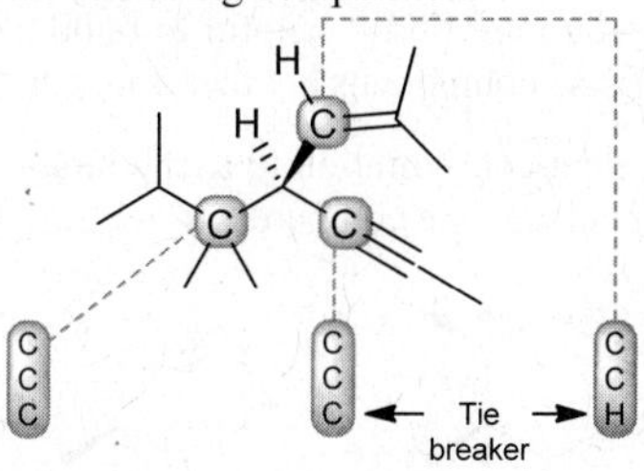

Among these groups, the double bond is assigned the third priority in our prioritization scheme, and we must continue to assign the first and second priorities.

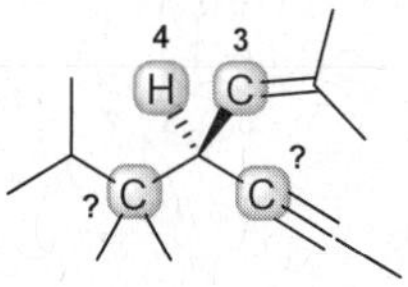

We move one carbon atom away from the chirality center, and we compare the following two positions:

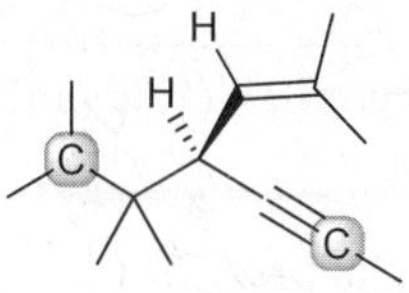

In each case, we must construct a list of the three atoms that are connected to each position. This is easy to do for the left side, which is connected to two carbon atoms and one hydrogen atom:

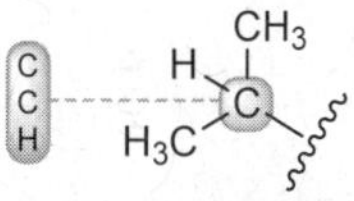

For the right side, the highlighted carbon atom is part of a triple bond, so we treat each π bond as if it were a sigma bond to another carbon atom:

is treated as

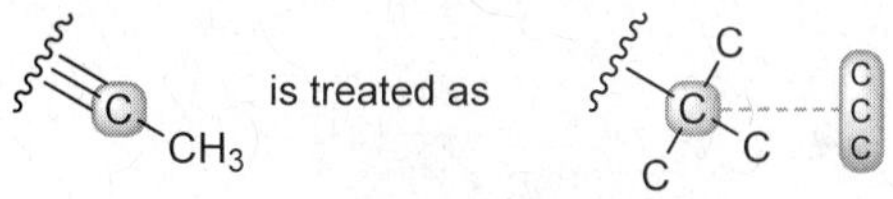

As such, the right side wins the tie-breaker:

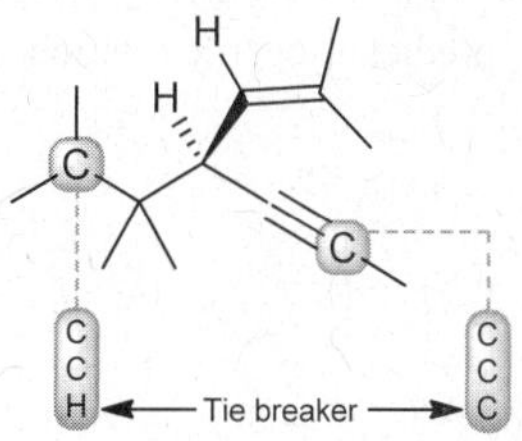

Our prioritization scheme indicates that the chirality center has the *R* configuration:

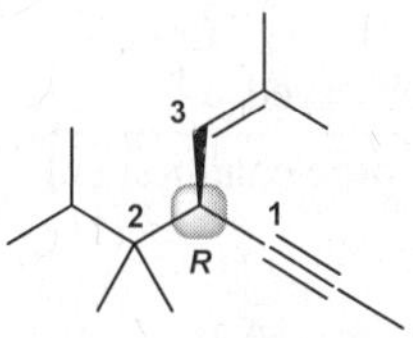

5.11. The chirality center has the *S* configuration, determined by the prioritization scheme shown here.

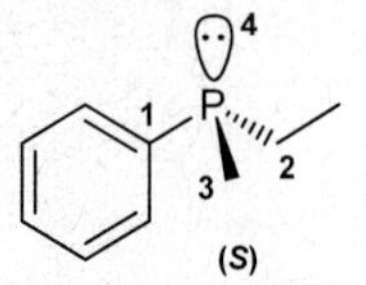

(S)

5.12. specific rotation $= [\alpha] = \dfrac{\alpha}{c \times l}$

$$= \dfrac{(+1.47^\circ)}{(0.0575 \text{ g/mL}) \times (1.00 \text{ dm})} = \mathbf{+25.6}$$

5.13. specific rotation $= [\alpha] = \dfrac{\alpha}{c \times l}$

$$= \dfrac{(-2.99^\circ)}{(0.095 \text{ g/mL}) \times (1.00 \text{ dm})} = \mathbf{-31.5}$$

5.14. specific rotation $= [\alpha] = \dfrac{\alpha}{c \times l}$

$$= \dfrac{(+0.57^\circ)}{(0.260 \text{ g/mL}) \times (1.00 \text{ dm})} = \mathbf{+2.2}$$

5.15. This compound does not have a chirality center, because two of the groups are identical:

Accordingly, the compound is achiral and is not optically active. We thus predict a specific rotation of zero.

5.16.

$$[\alpha] = \dfrac{\alpha}{c \times l}$$

$$\alpha = [\alpha] \times c \times l$$

$$= (-13.5)(0.100 \text{ g/mL})(1.00 \text{ dm}) = -1.35^\circ$$

5.17.

$$\% \, ee = \dfrac{|\text{ observed } \alpha \,|}{|\alpha \text{ of pure enantiomer}|} \times 100 \, \%$$

$$= \dfrac{37}{39.5} \times 100 \, \%$$

$$= 94 \, \%$$

5.18.

$$\% \, ee = \dfrac{|\text{ observed } \alpha \,|}{|\alpha \text{ of pure enantiomer}|} \times 100 \, \%$$

$$= \dfrac{6.0}{6.3} \times 100 \, \%$$

$$= 95 \, \%$$

5.19.

$$\% \, ee = \dfrac{|\text{ observed } \alpha \,|}{|\alpha \text{ of pure enantiomer}|} \times 100 \, \%$$

$$= \dfrac{85}{92} \times 100 \, \%$$

$$= 92 \, \%$$

5.20. Observed $[\alpha] = \dfrac{\alpha}{c \times l}$

$$= \dfrac{(+0.78^\circ)}{(0.350 \text{ g/mL}) \times (1.00 \text{ dm})} = \mathbf{+2.2}$$

$$\% \, ee = \dfrac{|\text{ observed } \alpha \,|}{|\alpha \text{ of pure enantiomer}|} \times 100 \, \%$$

$$= \dfrac{2.2}{2.8} \times 100 \, \%$$

$$= 79 \, \%$$

5.21.
(a) We assign a configuration to each chirality center in each compound, and we compare:

In the first compound, both chirality centers have the *R* configuration, while in the second compound, both chirality centers have the *S* configuration. These compounds are mirror images of each other, but they are nonsuperimposable. That is, if you try to rotate the first compound 180 degrees about a horizontal axis, you will not generate the second compound (if you have trouble seeing this, you may find it helpful to build a molecular model). These compounds are therefore enantiomers.

(b) We assign a configuration to each chirality center in each compound, and we compare:

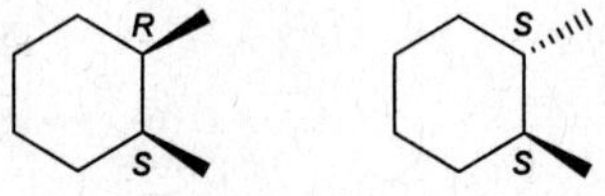

In the first compound, the configurations of the chirality centers are *R* and *S*, while in the second compound, they are *S* and *S*. These compounds are stereoisomers, but they are not mirror images of each other. Therefore, they are diastereomers.

(c) We assign a configuration to each chirality center in each compound, and we compare:

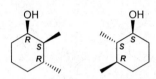

In the first compound, the configurations of the chirality centers are *R* and *R*, while in the second compound, they are *R* and *S*. These compounds are stereoisomers, but they are not mirror images of each other. Therefore, they are diastereomers.

(d) We assign a configuration to each chirality center in each compound, and we compare:

In the first compound, the configurations of the chirality centers are *R*, *S*, and *R*, respectively. In the second compound, they are *S*, *S*, and *R*. These compounds are stereoisomers, but they are not mirror images of each other. Therefore, they are diastereomers.

(e) We assign a configuration to each chirality center in each compound, and we compare:

In the first compound, the configurations of the chirality centers are *R* and *S*, while in the second compound, they are *R* and *R*. These compounds are stereoisomers, but they are not mirror images of each other. Therefore, they are diastereomers.

5.22. There are three chirality centers, and only one of these chirality centers has a different configuration in these two compounds. The other two chirality centers have the same configuration in both compounds. Therefore, these compounds are diastereomers.

5.23.
(a) Yes, there is a plane of symmetry that chops the goggles in half, with the right side reflecting the left side.
(b) Yes, there is a plane of symmetry that goes through the handle of the cup.
(c) No, there is no plane of symmetry.
(d) Yes, there is a plane of symmetry that chops the whistle in half.
(e) Yes, there are three planes of symmetry in this cinder block.
(f) No, there are no planes of symmetry in a hand.

5.24. The cinder block (5.23e) has three planes of symmetry, each of which chops the block in half.

5.25. Each of the following compounds has a plane of symmetry, as shown:

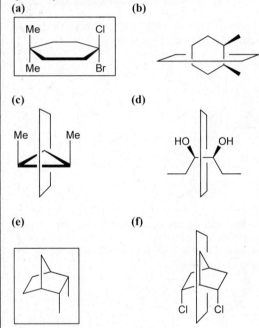

5.26.
(a) With two chirality centers, we would expect four stereoisomers. However, there are only three stereoisomers in this case, because the first one shown below is a *meso* compound.

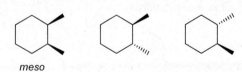

(b) With two chirality centers, we would expect four stereoisomers. However, there are only three stereoisomers in this case, because the first one shown below is a *meso* compound.

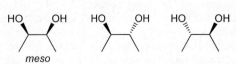

(c) With two chirality centers, we would expect four stereoisomers. However, there are only three stereoisomers in this case, because the first one shown below is a *meso* compound.

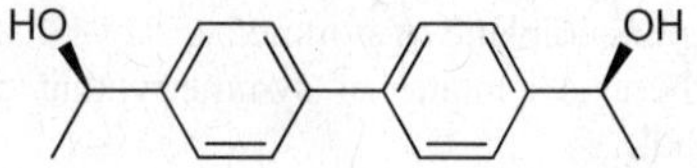

meso

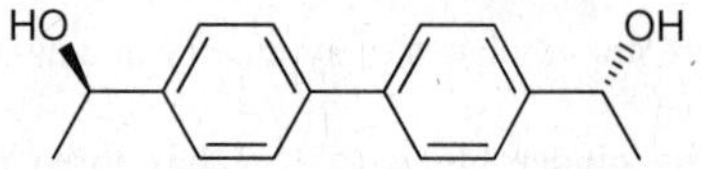

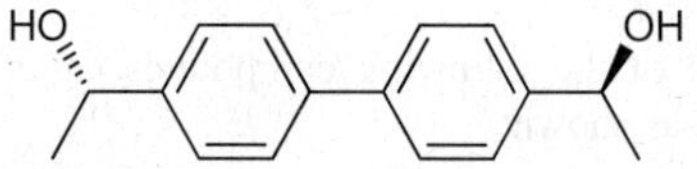

(d) With two chirality centers, there are four stereoisomers (no *meso* compounds).

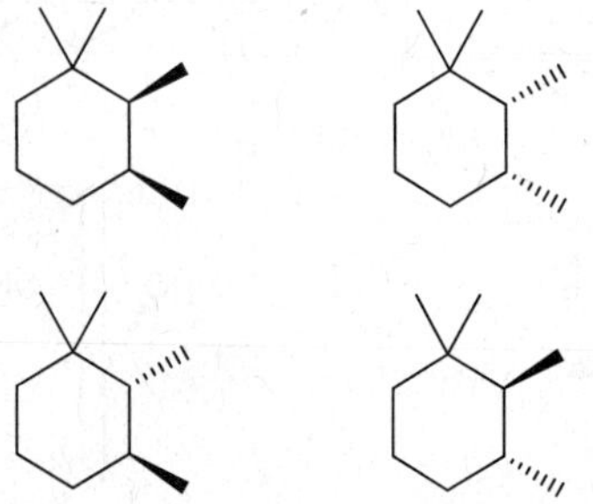

(e) With two chirality centers, we would expect four stereoisomers. However, there are only three stereoisomers in this case, because the first one shown below is a *meso* compound.

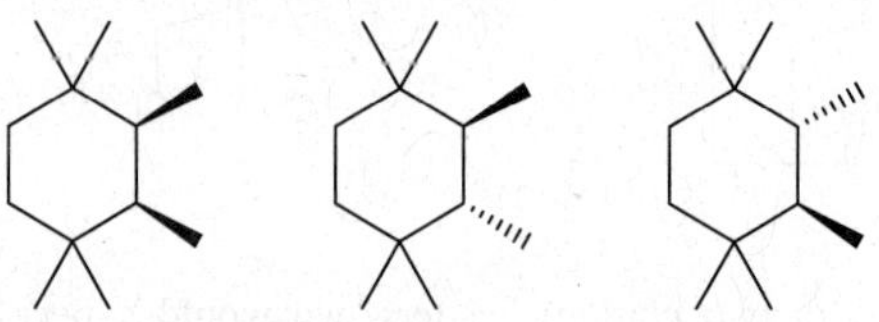

meso

5.27. Each of these compounds is a *meso* compound and does not have an enantiomer.

5.28 With three chirality centers, we would expect eight stereoisomers ($2^3 = 8$), shown below (labeled **1–8**). However, structures **1** and **2** represent one compound (a *meso* compound), while structures **3** and **4** also represent one compound (a *meso* compound). In addition, structure **5** is the same as structure **8**, while structure **6** is the same as structure **7** (you might find it helpful to build molecular models to see this). Structures **5-8** are not *meso* structures. The reason for the equivalence of

structures **5** and **8** (and also for the equivalence of **6** and **7**) is that the central carbon atom in each of these four structures is actually not a chirality center. For structures **5-8**, changing the "configuration" at the central carbon atom does not produce a stereoisomer, which proves that the central carbon atom is not a chirality center in these cases (see the solution to Problem **5.48** for more on this). In summary, there are only four stereoisomers (structures **1**, **3**, **5**, and **6**).

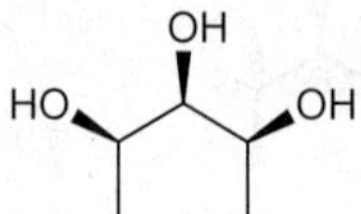

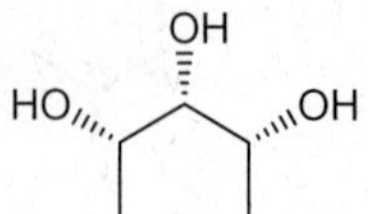

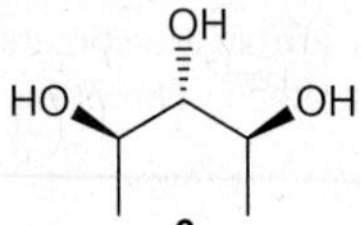

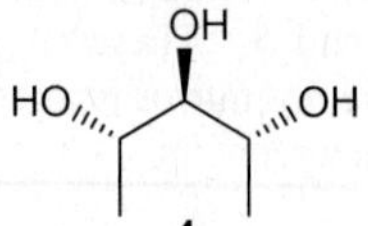

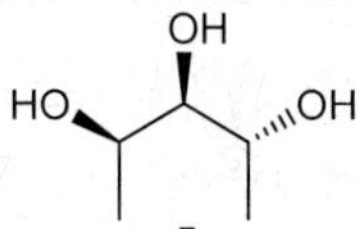

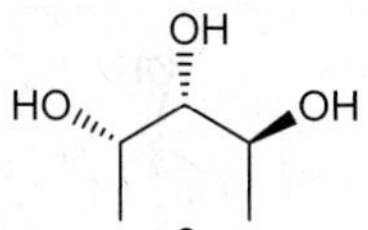

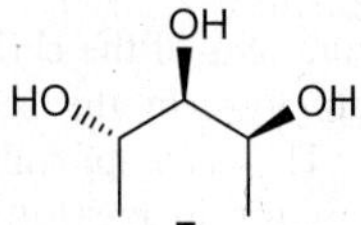

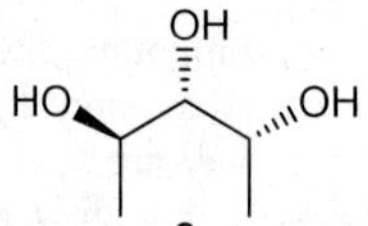

5.29.
(a) This chirality center has the R configuration, as shown below:

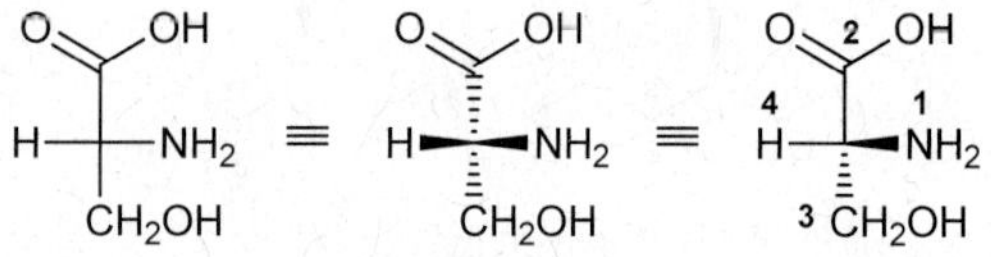

R

(b) This chirality center has the S configuration, as shown below:

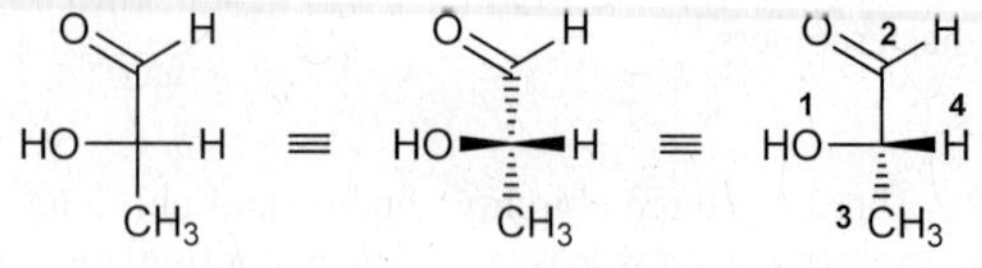

S

(c) This chirality center has the S configuration, as shown below:

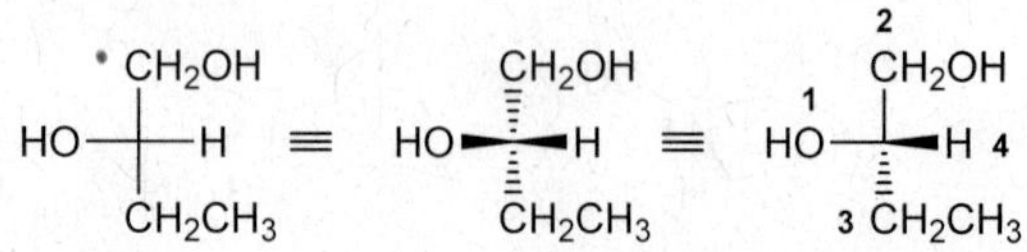

S

(d) This chirality center has the *S* configuration, as shown below:

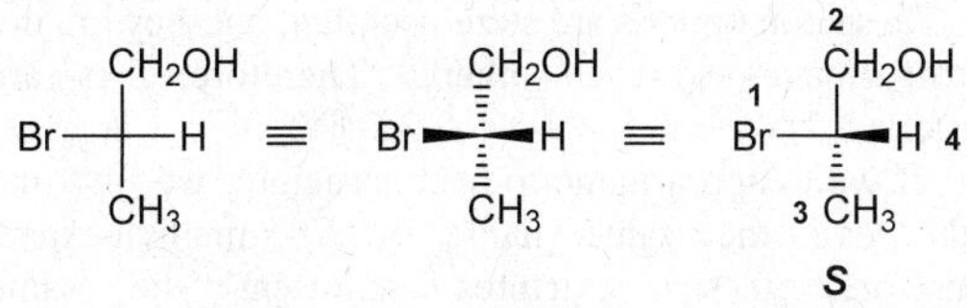

Br—CH₂OH—H ≡ Br▬CH₂OH▬H ≡ Br—CH₂OH—H 4

CH₃   CH₃   3 CH₃

S

5.30. For each chirality center, we follow the same procedure that we used in the previous problem. We first redraw the chirality center so that there are two lines, one wedge and one dash:

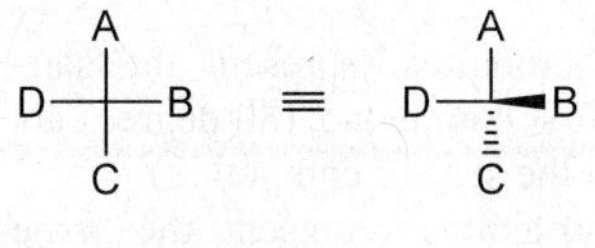

D—C—B   ≡   D—C—B
  A   A
  C   C

Then, we assign priorities, and determine the configuration. This process gives the following configurations:

(a)

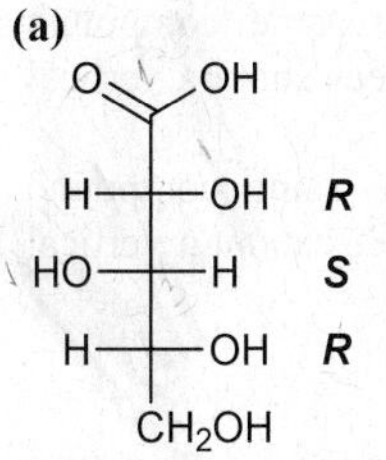

O   OH

H——OH   *R*

HO——H   *S*

H——OH   *R*

CH₂OH

(b)

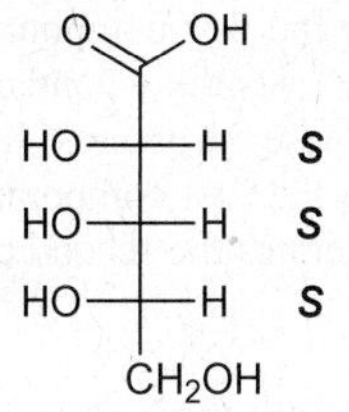

O   OH

HO——H   *S*

HO——H   *S*

HO——H   *S*

CH₂OH

(c)

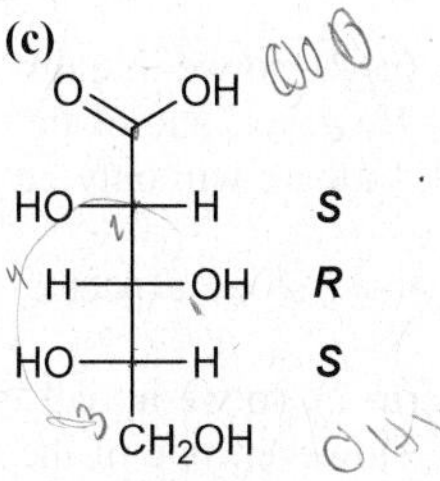

O   OH

HO——H   *S*

H——OH   *R*

HO——H   *S*

CH₂OH

5.31. In order to draw the enantiomer for each compound, we simply change the configuration at every chirality center by switching the groups on the left with the groups on the right.

(a)

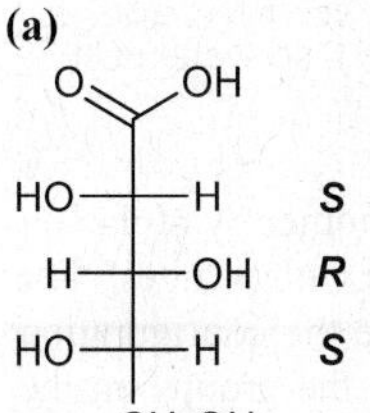

O   OH

HO——H   *S*

H——OH   *R*

HO——H   *S*

CH₂OH

(b)

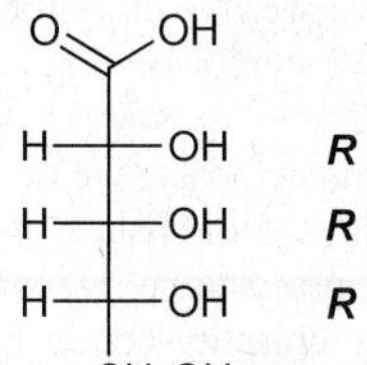

O   OH

H——OH   *R*

H——OH   *R*

H——OH   *R*

CH₂OH

(c)

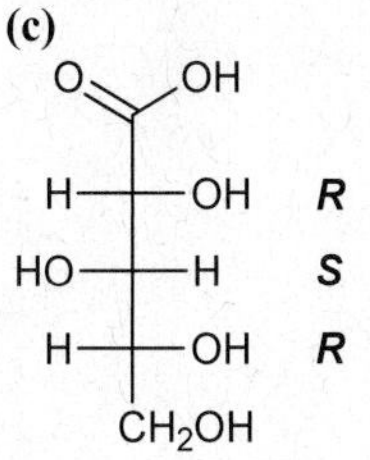

O   OH

H——OH   *R*

HO——H   *S*

H——OH   *R*

CH₂OH

5.32. This compound has two chirality centers, shown below. The following prioritization schemes led to the assignment of configuration for each chirality center.

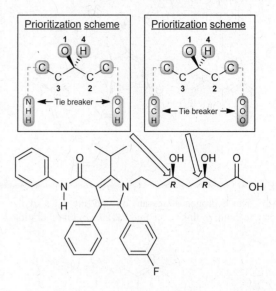

5.33. In this case, it will be easiest to place the mirror behind the molecule, giving the following structure for its enantiomer:

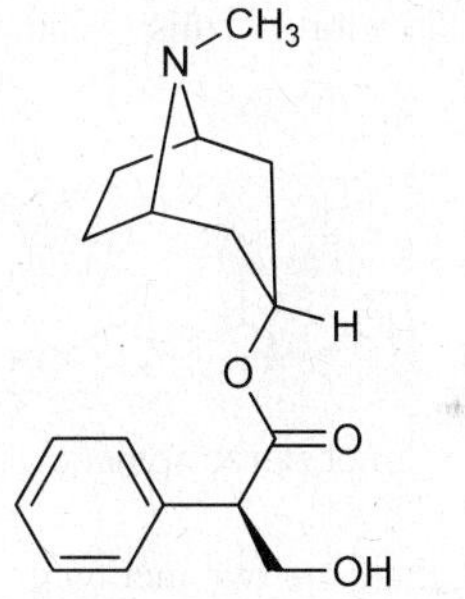

5.34.
(a) In order to draw the enantiomer of paclitaxel, we convert every wedge into a dash, and we convert every dash into a wedge, as shown:

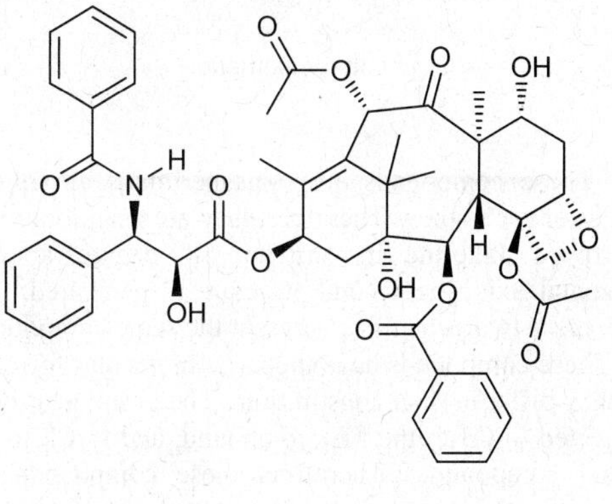

(b) Paclitaxel has eleven chirality centers (highlighted below).

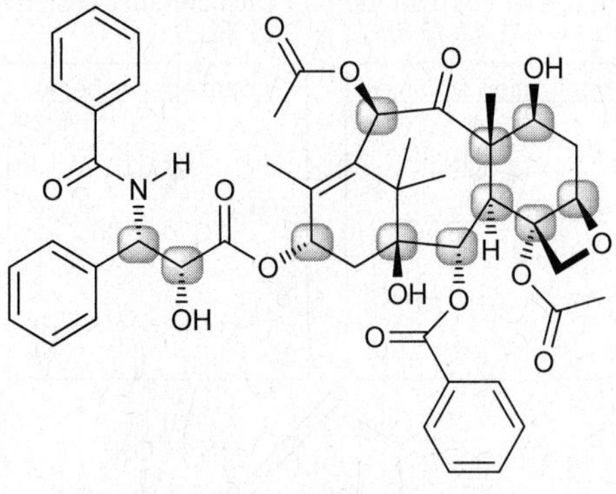

5.35.
(a) The hydrogen atoms are on opposite sides of the double bond, so this compound has a *trans* configuration.

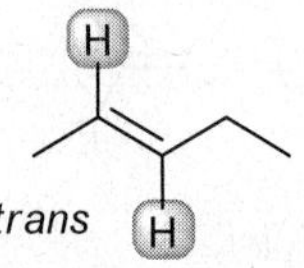

trans

(b) In this case, there are two identical groups (highlighted below) connected to the same vinylic position. Therefore, this compound is not stereoisomeric.

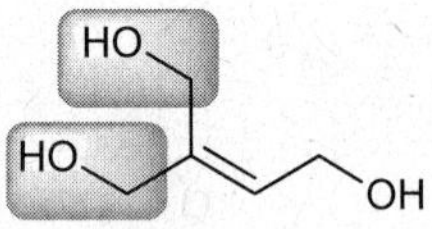

not stereoisomeric

(c) In this case, there are two methyl groups (highlighted below) connected to the same vinylic position. Therefore, this compound is not stereoisomeric.

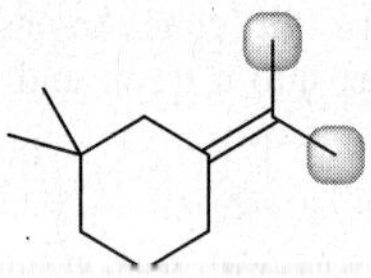

not stereoisomeric

5.36.
(a) These compounds are nonsuperimposable mirror images of each other. Therefore, they are enantiomers.
(b) If we rotate the first structure 180 degrees about a horizontal axis, the second structure is generated. As such, these two structures represent the same compound.
(c) These compounds have the same molecular formula, but they differ in their constitution. The bromine atom is connected to C3 in the first compound, and to C2 in the second compound. Therefore, these compounds are constitutional isomers.

(d) These compounds have the same molecular formula (C_6H_{12}), but they differ in their constitution. Therefore, these compounds are constitutional isomers.
(e) These compounds are stereoisomers, but they are not mirror images of each other. Therefore, they are diastereomers.
(f) If we assign a name to each structure, we find that both have the same name: (S)-3-methylhexane. Therefore, these structures represent the same compound.
(g) These compounds are nonsuperimposable mirror images of each other. Therefore, they are enantiomers.
(h) These compounds are stereoisomers, but they are not mirror images of each other. Therefore, they are diastereomers.
(i) These structures represent the same compound (rotating the first compound 180 degrees about a vertical axis generates the second compound).
(j) These structures represent the same compound, which does not contain a chirality center, because there are two ethyl groups connected to the central carbon atom.
(k) These structures represent the same compound (rotating the first compound 180 degrees about a vertical axis generates the second compound).
(l) These structures represent the same compound (rotating the first compound 180 degrees about a vertical axis generates the second compound).

5.37.
(a) There are three chirality centers (n=3), so we expect $2^n = 2^3 = 8$ stereoisomers.
(b) There are two chirality centers (n=2), so we initially expect $2^n = 2^2 = 4$ stereoisomers. However, one of the stereoisomers is a *meso* compound, so there will only be 3 stereoisomers.
(c) There are four chirality centers (n=4), so we expect $2^n = 2^4 = 16$ stereoisomers.
(d) There are two chirality centers (n=2), so we initially expect $2^n = 2^2 = 4$ stereoisomers. However, one of the stereoisomers is a *meso* compound, so there will only be 3 stereoisomers.
(e) There are two chirality centers (n=2), so we initially expect $2^n = 2^2 = 4$ stereoisomers. However, one of the stereoisomers is a *meso* compound, so there will only be 3 stereoisomers.
(f) There are five chirality centers (n=5), so we expect $2^n = 2^5 = 32$ stereoisomers.

5.38. In each case, we draw the enantiomer by replacing all wedges with dashes, and all dashes with wedges. For Fischer projections, we simply change the configuration at every chirality center by switching the groups on the left with the groups on the right:

(a)   **(b)**

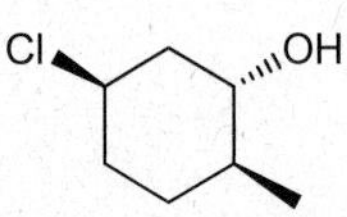

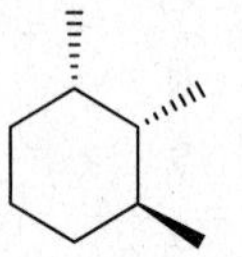

(c)

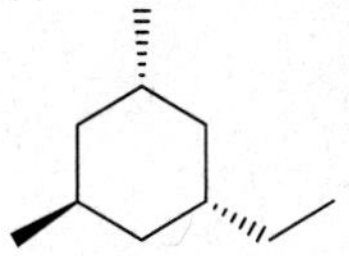

(d)

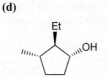

(e)

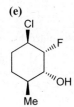

(f)

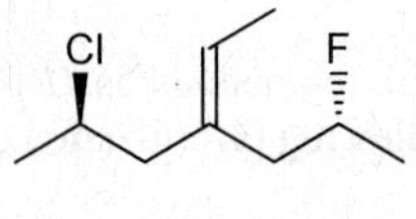

Cl, F (structure with Cl and F)

(g)

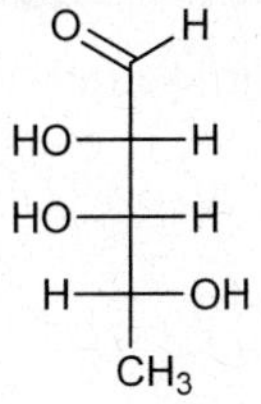

(h)

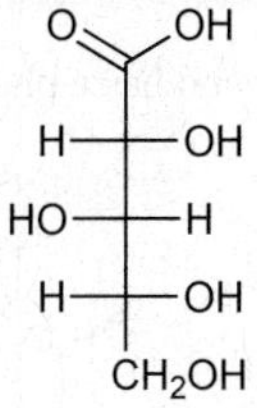

(Fischer projection: O, OH / H, OH / HO, H / H, OH / CH₂OH)

(i)

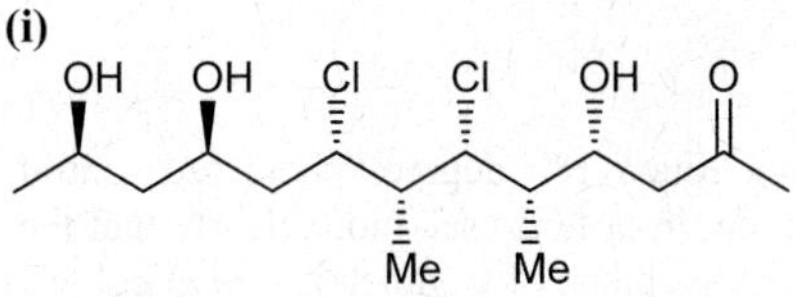

(long chain structure with OH OH Cl Cl OH O and Me Me)

(j)

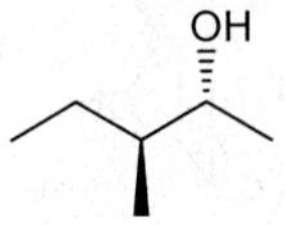

(k)

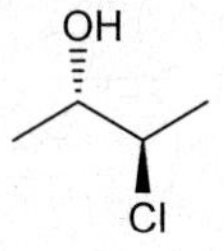

(structure with OH and Cl)

(l)

(structure with OH)

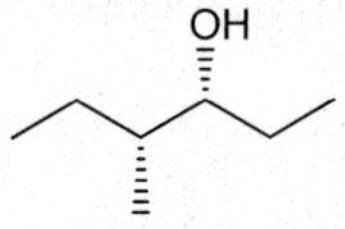

5.39. The configuration of each chirality center is shown below:

(a)

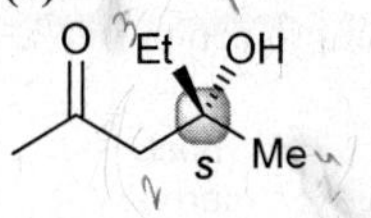

(b)

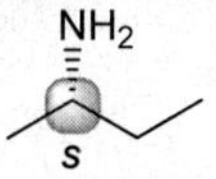

(structure with NH₂, S)

(c)

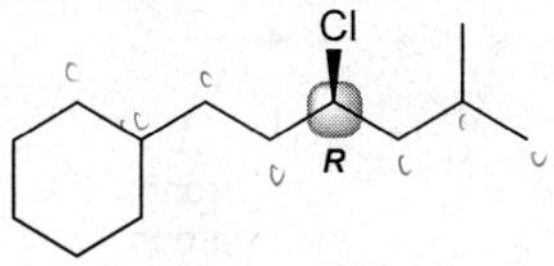

(d)

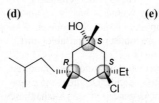

(e)

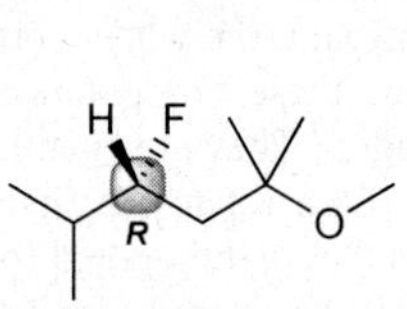

(structure with H, F, R)

(f)

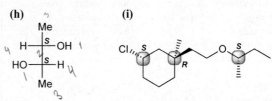

(g)

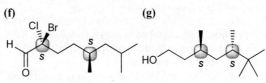

(h)

Me / H—S—OH / HO—S—H / Me

(i)

(structure with Cl, S, R, O, S)

5.40.   In this case, there is a 96% excess of A. The remainder of the solution is a racemic mixture of both enantiomers (2% A and 2% B).   Therefore, the enantiomeric excess (*ee*) is 96%.

5.41.
(a) One of the chirality centers has a different configuration in each compound, while the other chirality center has the same configuration in each compound.   As such, these compounds are stereoisomers, but they are not mirror images of each other. They are diastereomers.
(b) One of the chirality centers has a different configuration in each compound, while the other chirality center has the same configuration in each compound.   As such, these compounds are stereoisomers, but they are not mirror images of each other. They are diastereomers.
(c) Each of the chirality centers has a different configuration when comparing these compounds. As such, these compounds are nonsuperimposable mirror images. They are enantiomers.
(d) These structures represent the same compound (rotating the first compound 180 degrees about a horizontal axis generates the second compound).
(e) These compounds are nonsuperimposable mirror images. Therefore, they are enantiomers.

5.42.
(a) One of the chirality centers has a different configuration in each compound, while the other chirality center has the same configuration in each compound.   As such, these compounds are stereoisomers, but they are not mirror images of each other. They are diastereomers.
(b) Each of the chirality centers has a different configuration when comparing these compounds. As such, these compounds are nonsuperimposable mirror images. They are enantiomers.

(c) These compounds are stereoisomers, but they are not mirror images of each other. They are diastereomers.
(d) Each of the chirality centers has a different configuration when comparing these compounds. As such, these compounds are nonsuperimposable mirror images. They are enantiomers.
(e) If we rotate the first structure 180 degrees about an axis that is orthogonal (perpendicular) to the plane of the page, the second structure is generated. As such, these two structures represent the same compound.
(f) Each of the chirality centers has a different configuration when comparing these compounds. As such, these compounds are nonsuperimposable mirror images. They are enantiomers.
(g) One of the chirality centers has a different configuration in each compound, while the other chirality center has the same configuration in each compound. As such, these compounds are stereoisomers, but they are not mirror images of each other. They are diastereomers.

5.43.
(a) True.
(b) False. A *meso* compound cannot have an enantiomer. Its mirror image IS superimposable.
(c) True. In a Fischer projection, all horizontal lines represent wedges, and all vertical lines represent dashes. If we rotate the structure by 90 degrees, we are changing all wedges into dashes and all dashes into wedges. Therefore, the new structure will be an enantiomer of the original structure. Be careful though – you can only rotate a Fischer projection by 90 degrees (to draw the enantiomer) if there is one chirality center. If there is more than one chirality center, then you cannot rotate the Fischer projection by 90 degrees in order to draw its enantiomer. Instead, you must change the configuration at every chirality center by switching the groups on the left with the groups on the right.

5.44. specific rotation $= [\alpha] = \dfrac{\alpha}{c \times l}$

$$= \dfrac{(-0.47°)}{(0.0075 \text{ g / mL}) \times (1.00 \text{ dm})} = \textbf{-63}$$

5.45.
(a) This compound is (*S*)-limonene, as determined by the following prioritization scheme:

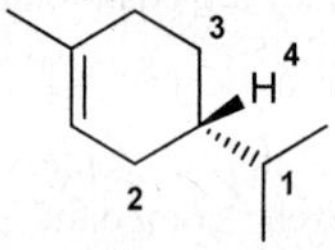

(b) This compound is (*R*)-limonene, as determined by the following prioritization scheme:

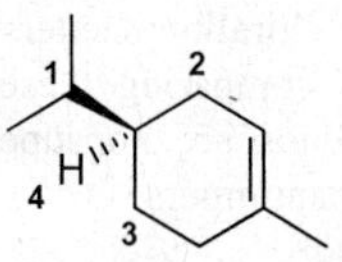

(c) This compound is (*S*)-limonene, as determined by the following prioritization scheme:

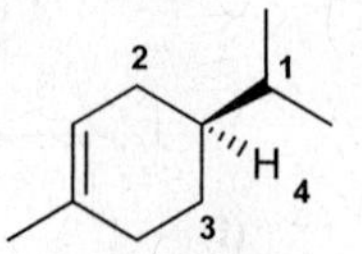

(d) This compound is (*R*)-limonene, as determined by the following prioritization scheme:

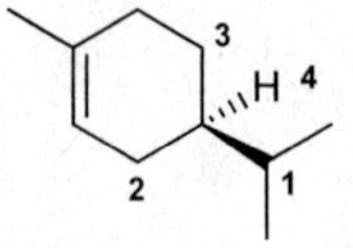

5.46.
(a) This compound has a plane of symmetry as shown:

(b) We must first rotate 180 degrees about the central carbon-carbon bond, in order to see more clearly that the compound possesses a plane of symmetry, shown below:

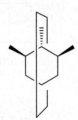

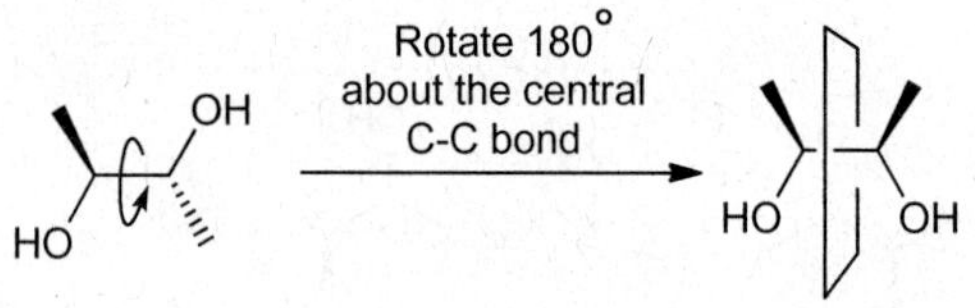

(c) We must first rotate 180 degrees about the central carbon-carbon bond, in order to see more clearly that the compound possesses a plane of symmetry, shown below:

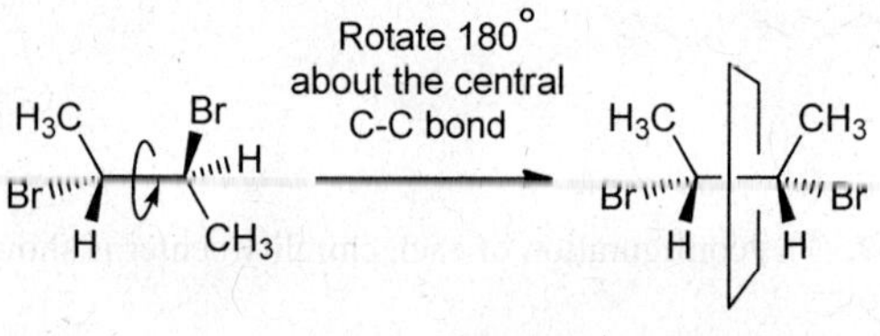

(d) We must first convert the Newman projection into a bond-line drawing:

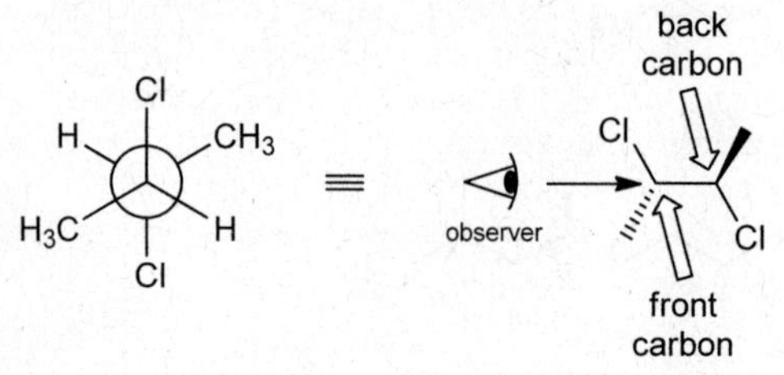

Then, we rotate 180 degrees about the central carbon-carbon bond, in order to see more clearly that the compound possesses a plane of symmetry, shown below:

Rotate 180°
about the central
C-C bond

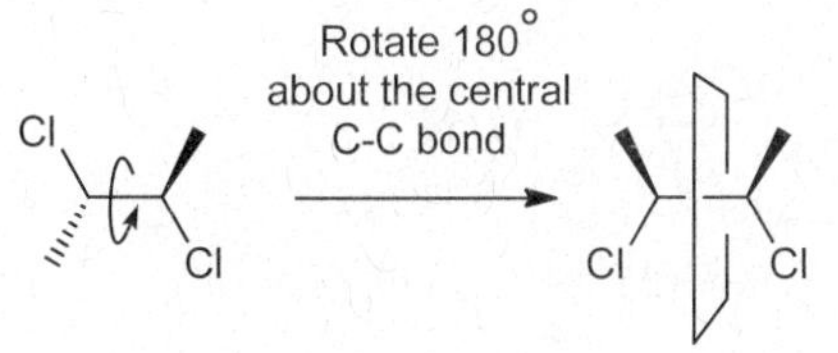

5.47. As seen in Section 4.9, cyclobutane adopts a slightly puckered conformation. It has two planes of symmetry, shown below:

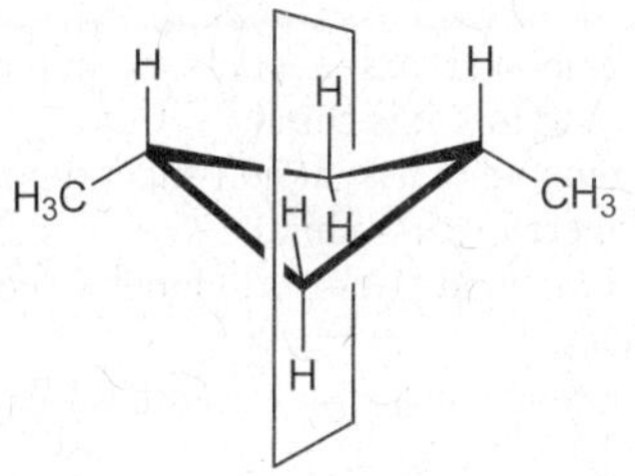

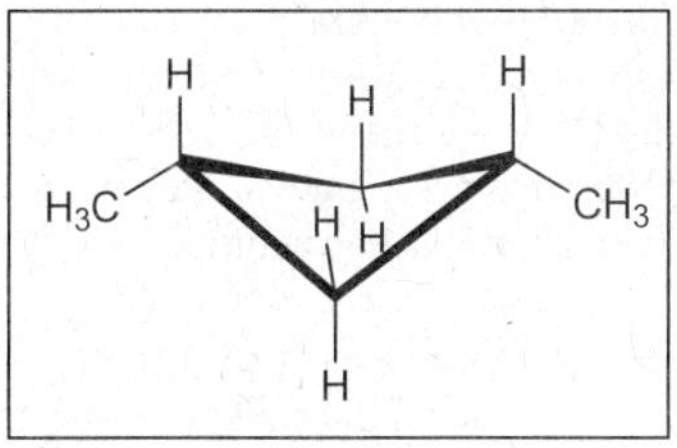

5.48. The first compound has three chirality centers:

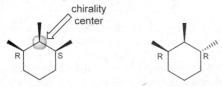

three chirality centers   two chirality centers

This is apparent if we assign the configuration at C1 and C3 of the cyclohexane ring. In the first compound, the configuration at C1 is different than the configuration at C3. As a result, there are four different groups attached to the C2 position. That is, C1 and C3 represent two different groups: one with the R configuration and the other with the S configuration. In contrast, consider the configuration at C1 and C3 in the second compound. Both of these positions have the same configuration, and therefore, the C2 position in that compound does not have four different groups. Two of the groups are identical, so C2 is not a chirality center.

There is an alternative argument that can be provided to explain why C2 is not a chirality center in the second compound. To understand this alternative argument, let's treat the C2 position "as if" it is a chirality center

and let's draw both possible configurations for that location:

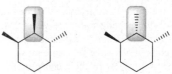

If we compare these two structures, we find that rotating the first structure 180 degrees about a vertical axis generates the second structure. As such, these two structures represent the same compound. Therefore, C2 cannot possibly be a chirality center, because it does NOT have two different configurations. In contrast, the C2 position in the first compound DOES have two different configurations. The following two compounds are indeed different, which proves that C2 is a chirality center in the first compound.

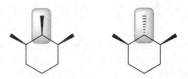

5.49.
(a) Each of the chirality centers has a different configuration when comparing these compounds. As such, these compounds are nonsuperimposable mirror images. They are enantiomers. Be careful – rotating one of these structures by 180 degrees does NOT generate the other structure. You may find it helpful to build a molecular model to prove this to yourself.
(b) One of the chirality centers has a different configuration in each compound, while the other chirality center has the same configuration in each compound. As such, these compounds are stereoisomers, but they are not mirror images of each other. They are diastereomers.
(c) Each of the chirality centers has a different configuration when comparing these compounds. As such, these compounds are nonsuperimposable mirror images. They are enantiomers. Be careful – rotating one of these structures by 180 degrees does NOT generate the other structure. You may find it helpful to build a molecular model to prove this to yourself.
(d) If we rotate the first structure 180 degrees about a vertical axis, the second structure is generated. As such, these two structures represent the same compound (which is *meso*).
(e) Each of the chirality centers has a different configuration when comparing these compounds. As such, these compounds are nonsuperimposable mirror images. They are enantiomers. Be careful – rotating one of these structures by 180 degrees does NOT generate the other structure. You may find it helpful to build a molecular model to prove this to yourself.
(f) Two of the chirality centers have a different configuration in each compound, while the other chirality center has the same configuration in each compound. As such, these compounds are

stereoisomers, but they are not mirror images of each other. They are diastereomers.

(g) If we rotate the first structure 180 degrees about a vertical axis, the second structure is generated. As such, these two structures represent the same compound.

(h) These compounds have the same molecular formula (C_8H_{16}), but they differ in their constitution. The first compound is 1,2-disubstituted, while the second compound is 1,3-disubstituted. Therefore, they are constitutional isomers.

(i) Let's redraw the compounds in a way that shows the configuration of each chirality center without showing the conformation (chair). This will make it easier for us to determine the stereoisomeric relationship between the two compounds:

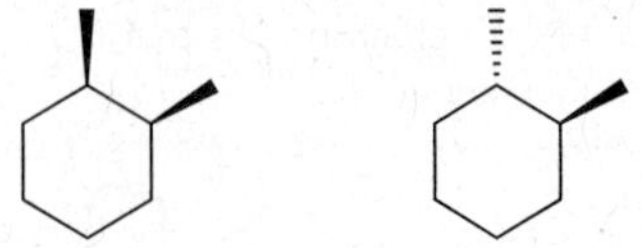

When drawn in this way, we can see clearly that one of the chirality centers has a different configuration in each compound, while the other chirality center has the same configuration in each compound. As such, these compounds are stereoisomers, but they are not mirror images of each other. They are diastereomers.

(j) Let's redraw the compounds in a way that shows the configuration of each chirality center without showing the conformation (chair). This will make it easier for us to determine the stereoisomeric relationship between the two compounds:

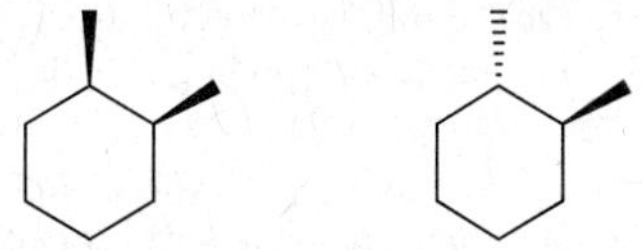

When drawn in this way, we can see clearly that one of the chirality centers has a different configuration in each compound, while the other chirality center has the same configuration in each compound. As such, these compounds are stereoisomers, but they are not mirror images of each other. They are diastereomers.

(k) It might appear as if each of the chirality centers has a different configuration when comparing these compounds. However, each of these compounds has an internal plane of symmetry (horizontal plane). As such, both structures represent the same *meso* compound. These two structures are the same.

(l) Each of the chirality centers has a different configuration when comparing these compounds. As such, these compounds are nonsuperimposable mirror images. They are enantiomers.

5.50.

(a) The specific rotation of (R)-carvone should be the same magnitude but opposite sign as the specific rotation of (S)-carvone (assuming both are measured at the same temperature). Therefore, we expect the specific rotation of (R)-carvone at 20 °C to be –61.

(b)

$$\% \; ee = \frac{|\text{ observed } \alpha \; |}{|\alpha \text{ of pure enantiomer}|} \times 100\,\%$$

$$= \frac{55}{61} \times 100\,\%$$

$$= 90\,\%$$

(c) Since the *ee* is 90%, the mixture must be comprised of 95% (R)-carvone and 5% (S)-carvone (95 – 5 = 90).

5.51.

(a) This compound has a non-superimposable mirror image, and therefore it is chiral.

(b) This compound has a non-superimposable mirror image, and therefore it is chiral.

(c) This compound lacks a chirality center and is therefore achiral.

(d) This compound lacks a chirality center and is therefore achiral.

(e) This compound has a non-superimposable mirror image, and therefore it is chiral.

(f) This compound is a *meso* compound, which we can see more clearly if we rotate the central carbon-carbon bond by 180 degrees, shown below. Since the compound is *meso*, it must be achiral.

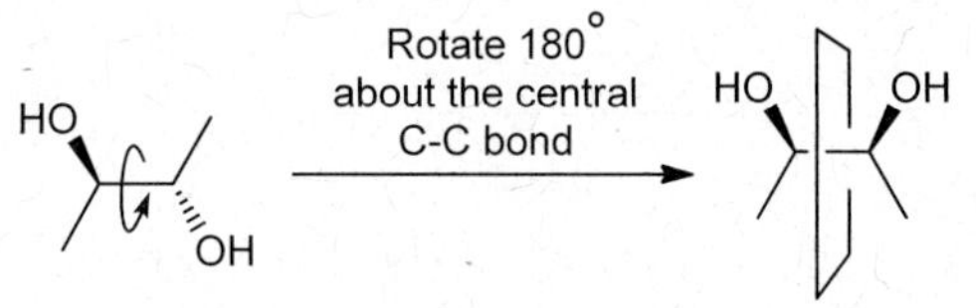

(g) This compound has an internal plane of symmetry and is therefore a *meso* compound. As such, it must be achiral.

(h) This compound has a non-superimposable mirror image, and therefore it is chiral.

(i) This compound has a non-superimposable mirror image, and therefore it is chiral.

(j) This compound has an internal plane of symmetry (chopping the OH group in half and chopping the methyl group in half) and is therefore a *meso* compound. As such, it must be achiral.

(k) This compound has a non-superimposable mirror image, and therefore it is chiral.

(l) This compound has a non-superimposable mirror image, and therefore it is chiral.

(m) This compound lacks a chirality center and is therefore achiral.

(n) This compound has a non-superimposable mirror image, and therefore it is chiral.

(o) This compound has an internal plane of symmetry and is therefore a *meso* compound. As such, it must be achiral.

(p) This compound has an internal plane of symmetry (shown below), and is therefore a *meso* compound. As such, it must be achiral.

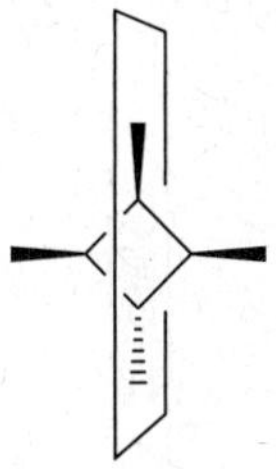

5.52.

$$[\alpha] = \frac{\alpha}{c \times l}$$

$$\alpha = [\alpha] \times c \times l$$

$$= (+24)(0.0100 \text{ g} / \text{mL})(1.00 \text{ dm}) = +0.24 \degree$$

5.53.
(a) If we rotate the central carbon-carbon bond of the Newman projection 180°, we arrive at a conformation in which the two OH groups are eclipsing each other, the two methyl groups are eclipsing each other, and the two H's are eclipsing each other.

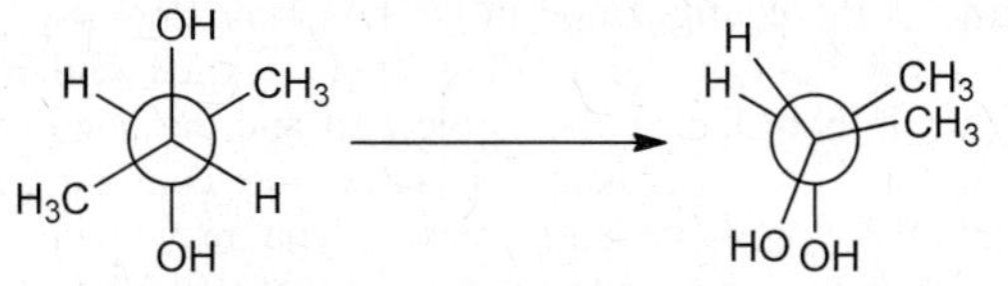

In this eclipsed conformation, we can clearly see that the molecule has an internal plane of symmetry (it is a *meso* compound). Therefore, the compound is optically inactive.

(b) We first convert the Newman projection into a bond-line drawing:

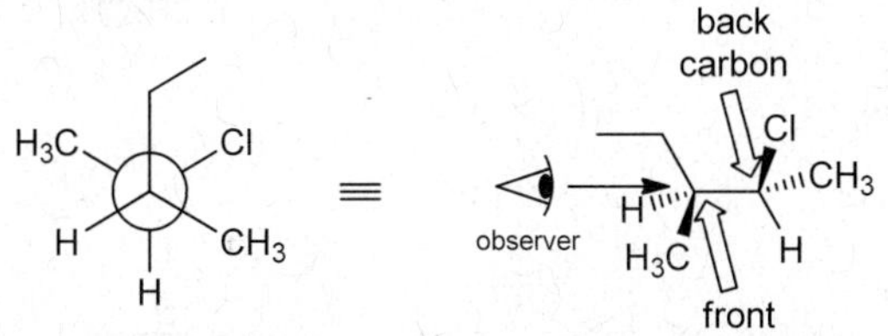

When drawn in bond-line format, we can see that the compound has two chirality centers (highlighted):

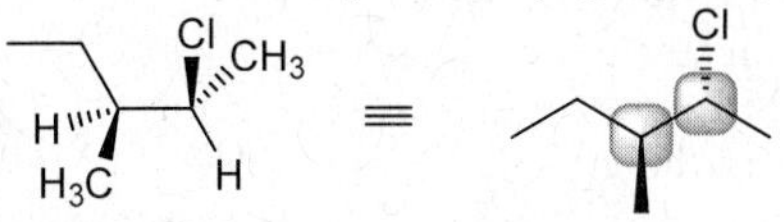

This compound is chiral and therefore optically active.

(c) This compound has a non-superimposable mirror image, so it is chiral. Therefore, it is optically active.

(d) Let's redraw the compound in a way that shows the configuration of each chirality center without showing the conformation (chair). This will make it easier for us to evaluate:

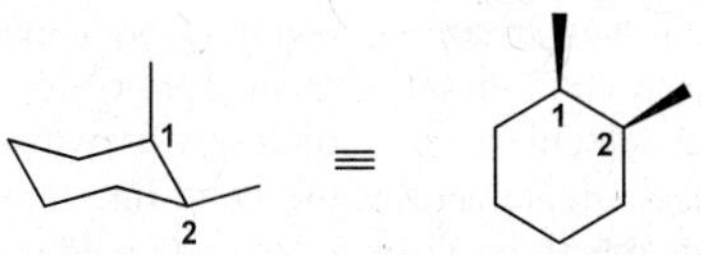

This compound has an internal plane of symmetry, so it is a *meso* compound. As such, it is achiral and optically inactive.

(e) This compound has a non-superimposable mirror image, so it is chiral. Therefore, it is optically active.

(f) We first convert the Newman projection into a bond-line drawing:

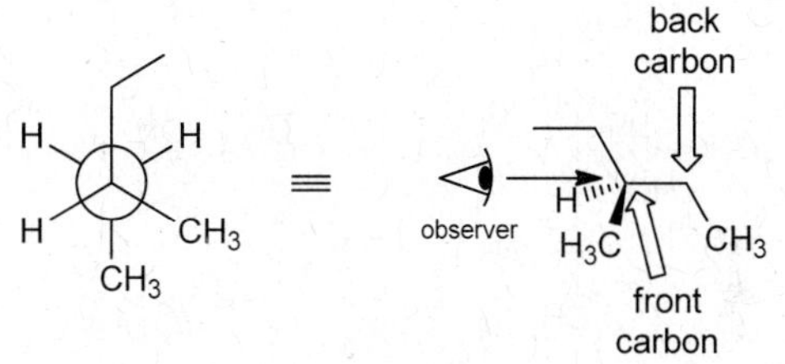

This compound is 3-methylpentane, which does not have a chirality center. Therefore, it is optically inactive.

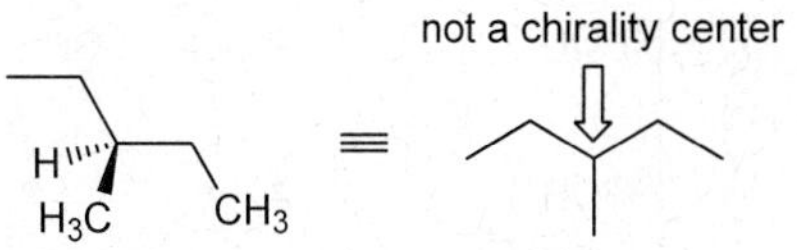

(g) This compound has an internal plane of symmetry (a vertical plane that chops one of the methyl groups in half), so it is a *meso* compound. As such, it is achiral and optically inactive.

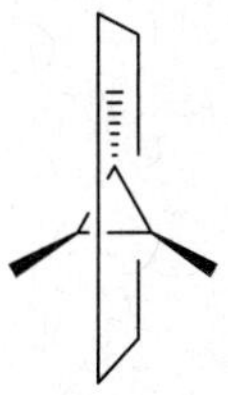

(h) This compound has an internal plane of symmetry, so it is a *meso* compound. As such, it is achiral and optically inactive.

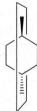

5.54. In each case, begin by numbering the carbon atoms in the Fischer projection (from top to bottom) and then draw the skeleton of a bond-line drawing with the same number of carbon atoms. Then, place the substituents in their correct locations (by comparing the numbering system in the Fischer projection with the numbering system in the bond-line drawing). When drawing each substituent in the bond-line drawing, you must decide whether it is on a dash or a wedge. For each chirality center, make sure that the configuration is the same as the configuration in the Fischer projection. If necessary, assign the configuration of each chirality center in both the Fischer projection and the bond-line drawing to ensure that you drew the configuration correctly. With enough practice, you may begin to notice some trends (rules of thumb) that will allow you to draw the configurations more quickly.

(a)

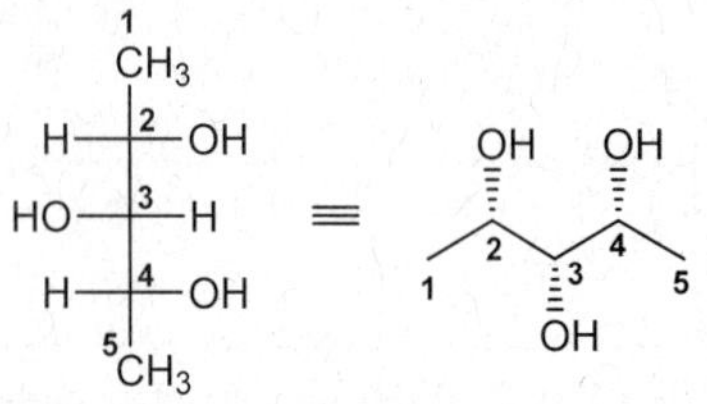

(b)

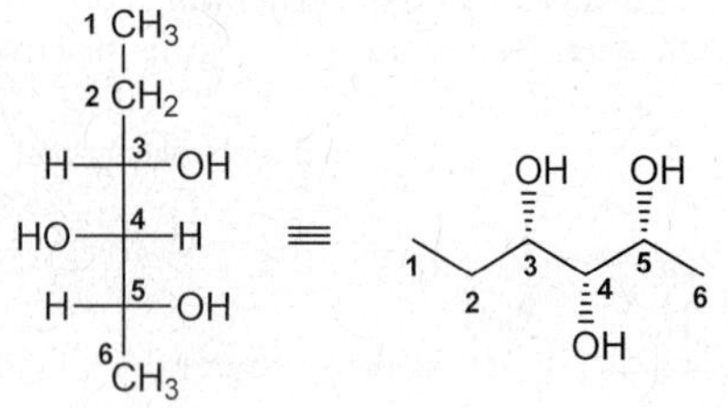

(c)

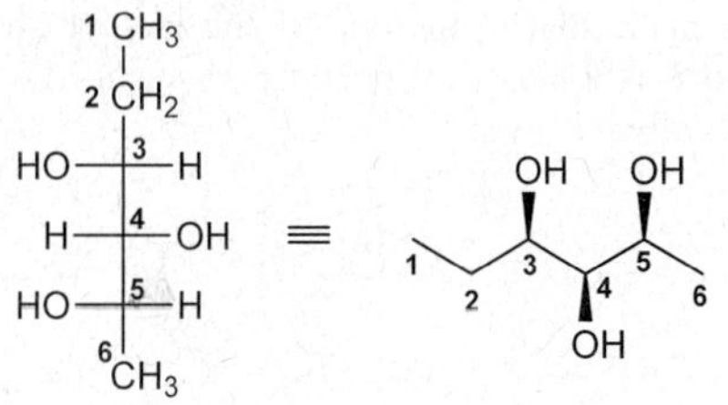

(d)

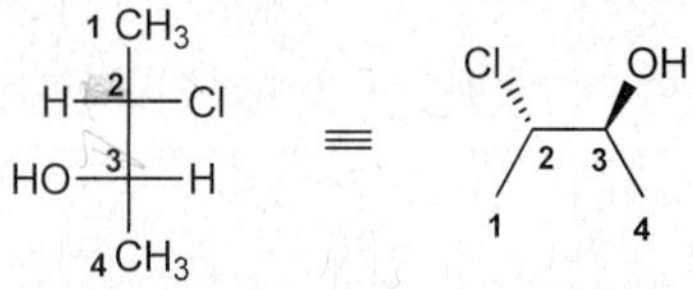

(e)

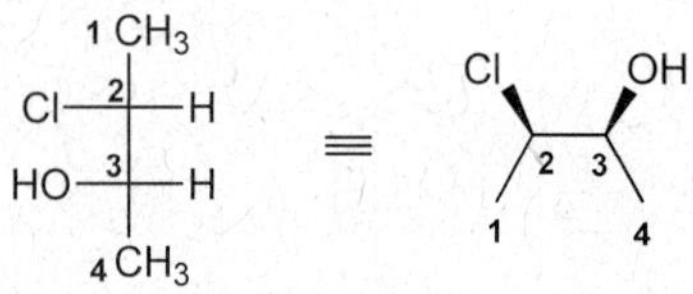

5.55.
(a) The compound in problem **5.54a** has an internal plane of symmetry, and is therefore a *meso* compound:

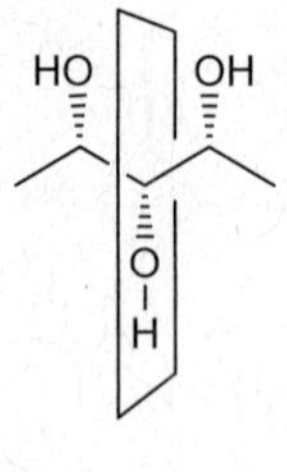

(b) The structures shown in **5.54b** and **5.54c** are enantiomers. An equal mixture of these two compounds is a racemic mixture, which will be optically inactive.

(c) Yes, this mixture is expected to be optically active, because the structures shown in **5.54d** and **5.54e** are not enantiomers. They are diastereomers, which are not expected to exhibit equal and opposite rotations.

5.56. As we saw in problem **5.54**, it is helpful to use a numbering system when converting one type of drawing into another. When drawing each substituent in the Fischer projection, you must decide whether it is on the right or left side of the Fischer projection. For each chirality center, make sure that the configuration is the same as the configuration in the bond-line drawing. If necessary, assign the configuration of each chirality center in both the Fischer projection and the bond-line drawing to ensure that you drew the configuration correctly. With enough practice, you may begin to notice some trends (rules of thumb) that will allow you to draw the configurations more quickly.
(a)

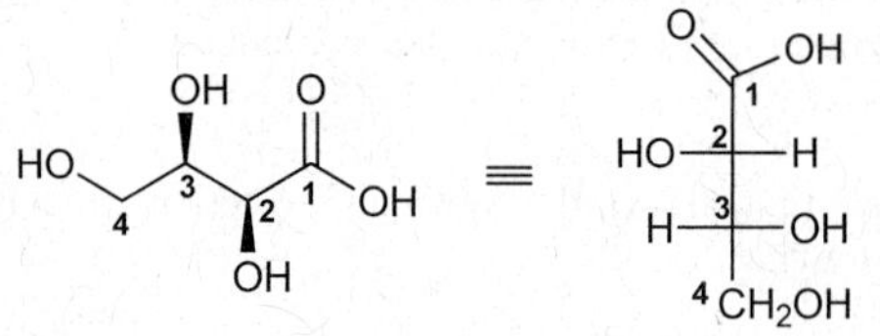

(b)

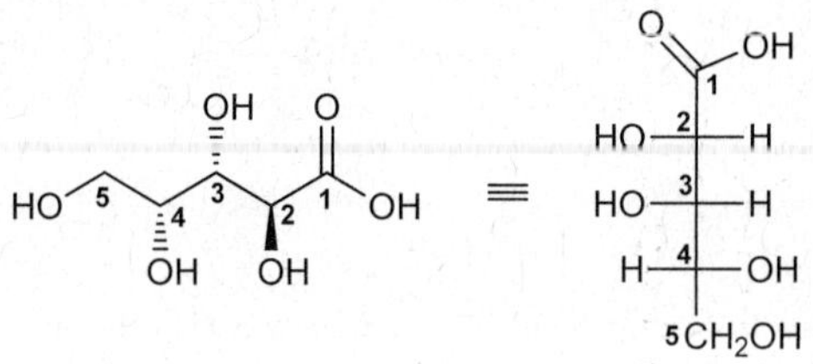

(c)

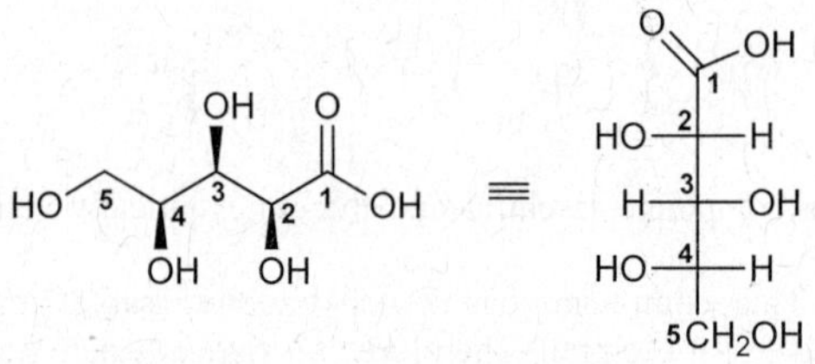

5.57.

(a) The second compound is 2-methylpentane. If we redraw the first compound as a bond-line drawing, rather than a Newman projection, we see that the first compound is 3-methylpentane.

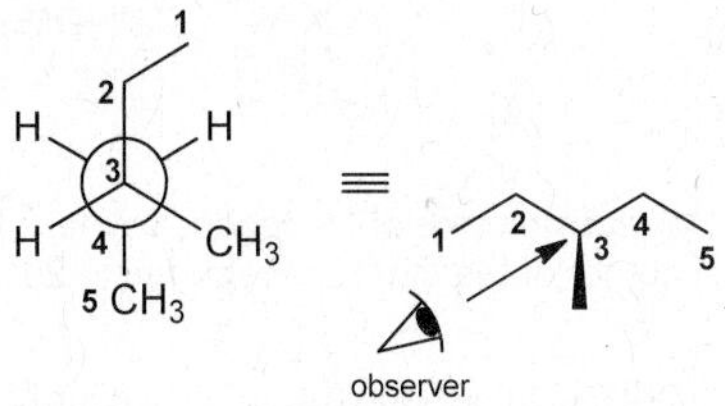

These two compounds, 2-methylpentane and 3-methylpentane, have the same molecular formula (C_6H_{14}) but different constitution, so they are constitutional isomers.

(b) The first compound is *trans*-1,2-dimethylcyclohexane:

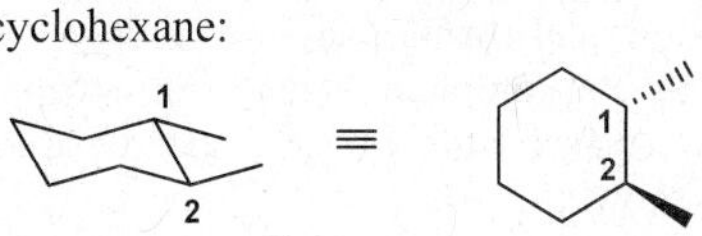

In contrast, the second compound is *cis*-1,2-dimethylcyclohexane. These compounds are stereoisomers, but they are not mirror images of each other. Therefore, they are diastereomers.

5.58. The following two compounds are enantiomers because they are nonsuperimposable mirror images. You may find it helpful to construct molecular models to help visualize the mirror image relationship between these two compounds.

5.59. This compound has two identical groups (methyl groups) attached to the same position (shown below). For this reason, the compound will be superimposable on its mirror image (once again, molecular models may help visualize this). Therefore, the compound will be achiral.

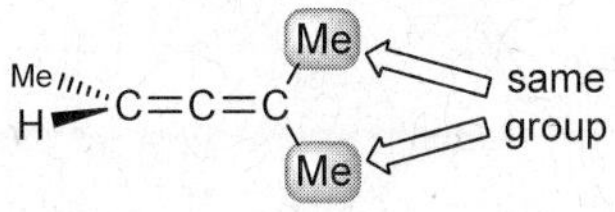

5.60.

(a) This compound cannot be completely planar because steric hindrance prevents the two ring systems from rotating with respect to each other. The compound is locked in a particular conformation that is chiral.

(b) This ring system cannot be planar because of steric hindrance, and must therefore adopt a spiral shape (like a spiral staircase). The spiral can be right handed or left handed, and the relationship between these two forms is enantiomeric.

5.61. The compound is chiral because it is not superimposable on its mirror image, shown below.

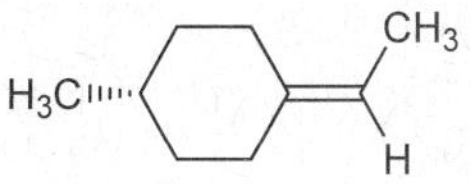

5.62. This compound has a center of inversion, which is a form of reflection symmetry. As a result, this compound is superimposable on its mirror image and is therefore optically inactive.

5.63. The compound contains three chirality centers, with the following assignments:

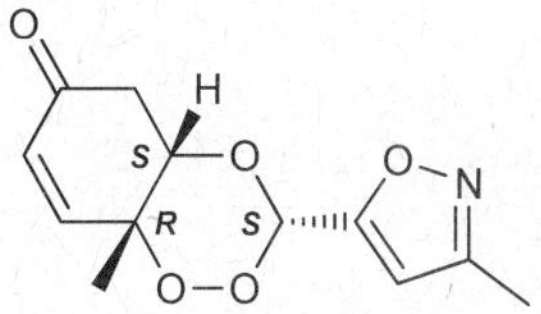

With three chirality centers, there should be a total of 2^3 = 8 stereoisomers, shown below. Pairs of enantiomers are highlighted together. All other relationships are diastereomeric.

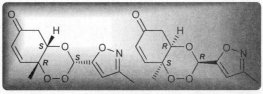

enantiomers

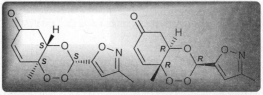

enantiomers

enantiomers

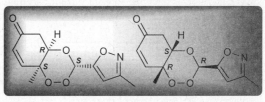

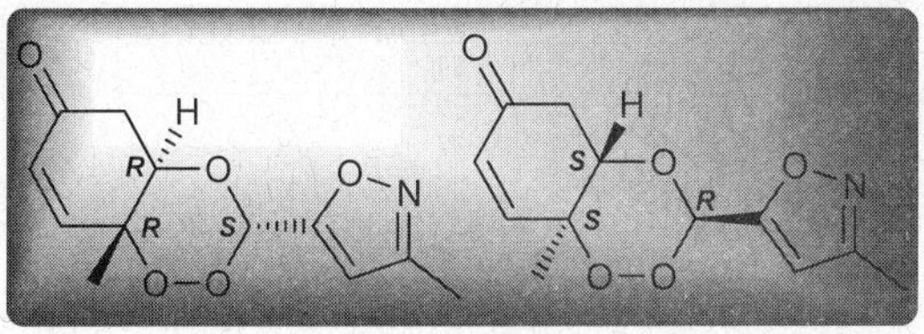

enantiomers

5.64.

(a) The following are the prioritization schemes that give rise to the correct assignment of configuration for each chirality center.

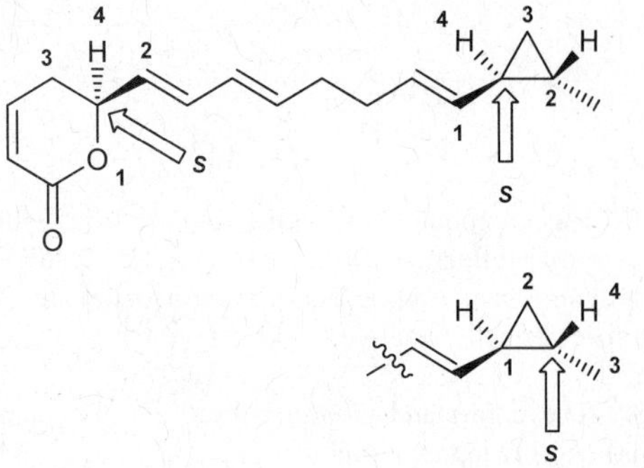

(b) The total number of possible stereoisomers is 2^n (where n = the number of chirality centers). With three chirality centers, we expect $2^3 = 8$ possible stereoisomers, one of which is the natural product coibacin B.

5.65.

(a) The product has one chirality center, which can either have the R configuration or the S configuration, as shown here.

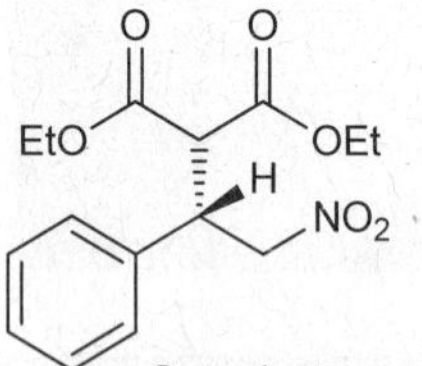

S-product
(major using Cu^I catalyst)

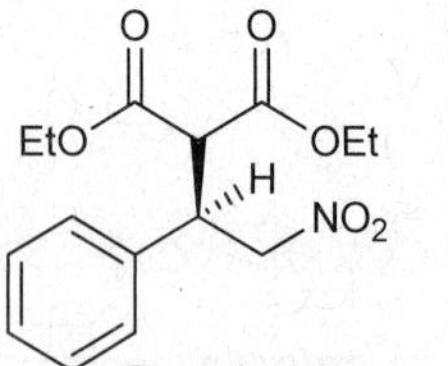

R-product
(major using Cu^{II} catalyst)

(b) Acetonitrile (CH_3CN) is the best choice of solvent because it results in the highest combination of enantioselectivity (72% *ee)* and percent yield (55%). While toluene gives the same % yield, enantioselectivity in this solvent is significantly lower.

Solvent	%ee	%S	%R
toluene	24	62	38
tetrahydrofuran	48	74	26
CH_3CN	72	86	14
CHCl$_3$	30	65	35
CH_2Cl_2	46	73	27
hexane	51	75.5	24.5

5.66.

(a) There is one chirality center, which was incorrectly assigned. So, it must have the R configuration, as shown

below, rather than the S configuration (as originally thought).

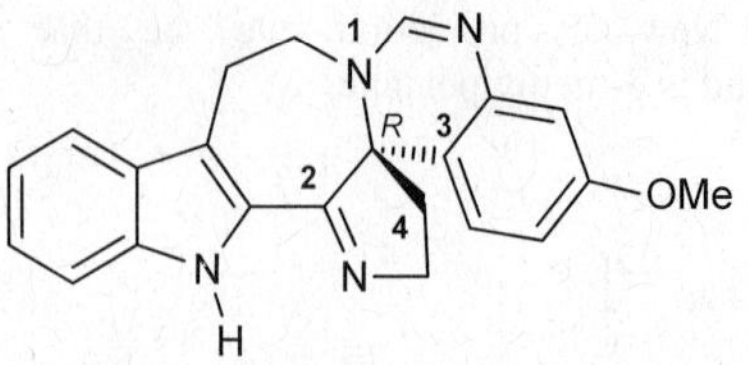

Correct structure of (+)-trigonoliimine A
(enantiomer of the structure reported in 2010).

(b) The synthesized compound [(−)-trigonoliimine A] was found to rotate plane polarized light in the opposite direction of the natural compound [(+)-trigonoliimine A]. The investigators therefore knew that the structure they had prepared was the enantiomer of the natural product.

5.67.

(a) To draw the enantiomer, we simply redraw the structure in the problem statement, except that we replace all dashes with wedges, and all wedges with dashes, as shown:

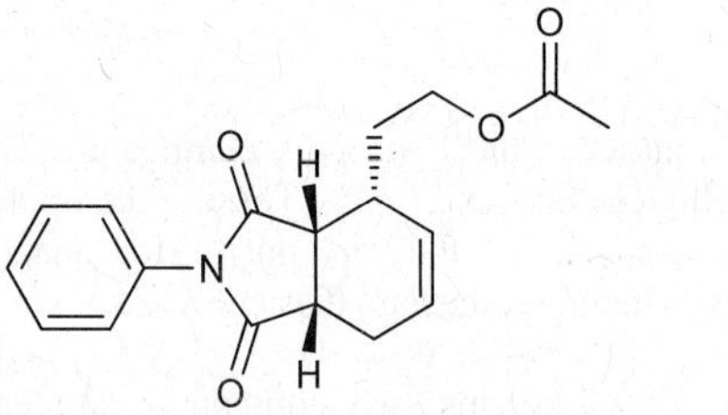

(2nd major product)

(b) The following compounds are the minor products, as described in the problem statement.

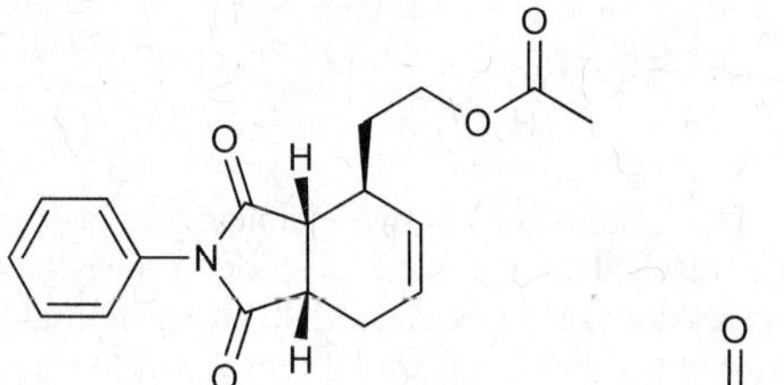

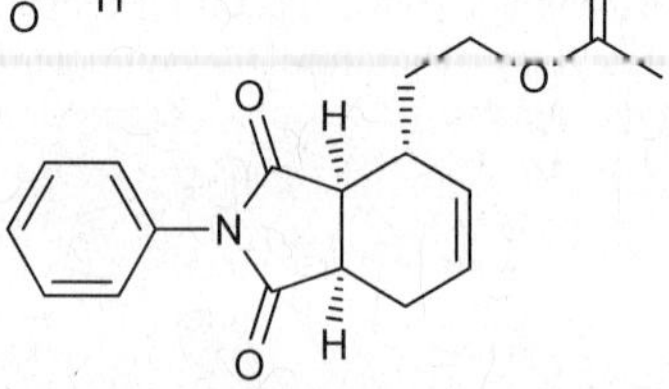

(c) The minor products are nonsuperimposable mirror images of each other. Therefore, they are enantiomers.

(d) They are stereoisomers, but they are not mirror images of each other. Therefore, they are diastereomers.

5.68.

The following are two examples of correct answers, where the molecule is viewed from different perspectives.

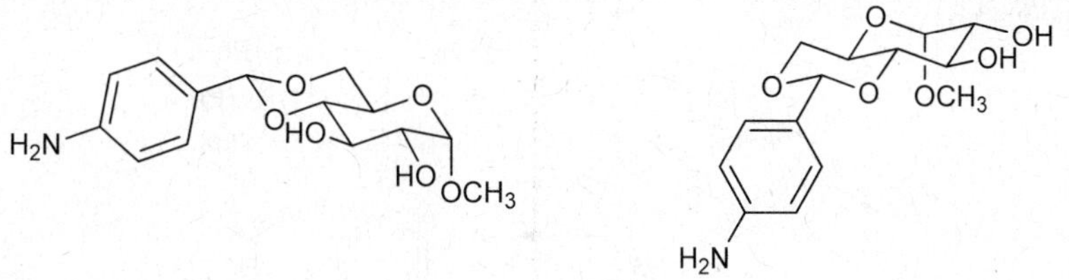

A suggested approach to this problem:

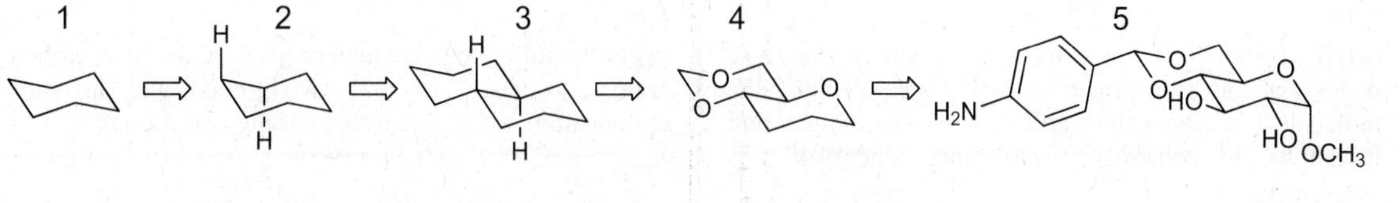

1   2   3   4   5

1) Draw a chair structure of the ring on the right side of the compound.
2) Now, to find an appropriate place to connect the second ring, find two axial positions on adjacent carbons so that they are down and up when going counterclockwise around the ring, as they are in the wedge and dash drawing. These are the two bridgehead positions.
3) Draw the second ring (with connecting bonds equatorial to the first ring).
4) Replace appropriate methylene groups in the rings with oxygen atoms.
5) Draw all substituents.

5.69.

We will need to determine the configuration at each of the two chirality centers in **A**. Let's begin with the chirality center connected to the oxygen atom. If we assign priorities to the four groups attached to the chirality center, the first priority will be O, and the last priority will be H. Since there are two carbon atoms directly attached to the chirality center, we will need to find a tie breaker. Notice that one of the carbon atoms (of the phenyl ring) is attached to C, C, and C, while the other carbon atom is attached to P, C, and H. The latter takes priority, because P has a higher atomic number than C. If we were to place the H on a dash and the phenyl ring on a wedge, the sequence 1-2-3 would be clockwise, corresponding to the *R* configuration:

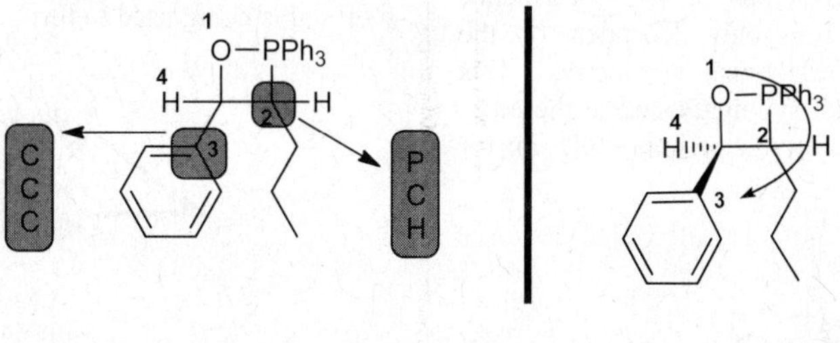

If we were to assign priorities for the second chirality center, phosphorus will receive the first priority, and hydrogen will receive the fourth priority. Since there are two carbon atoms attached at this chirality center, we will need a tie breaker for this side as well. The carbon atom that is connected to the propyl chain is directly attached to C, H, and H; while the other carbon atom is connected to the O, C, and H. Since oxygen has the highest atomic number, this side will win. For this chirality center, when H is placed on the wedge, and the propyl on a dash, the configuration is *R*.

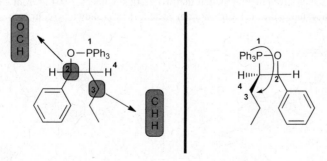

Finally, since the phenyl group is on a wedge and the propyl is on a dash, the resulting alkene must have *trans* geometry. This is because formation of the alkene occurs simultaneously with the breakdown of the four-membered ring, so the substituents do not have a chance to rotate; thus the relative *trans* configuration on the ring translates to a *trans* alkene in compound B.

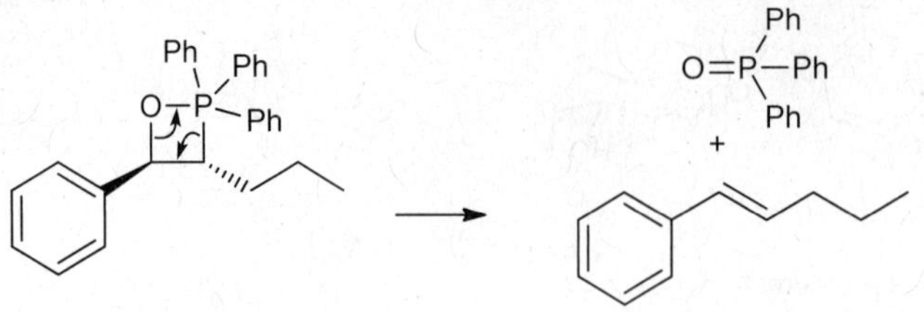

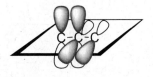

5.70.
(a) The compound exhibits rotational symmetry, because it possesses an axis of symmetry (consider rotating the molecule 180° about this vertical axis). You might find it helpful to construct a molecular model of this compound.

Axis of rotation

(b) The compound lacks reflectional symmetry; it does not have a plane of symmetry.
(c) Chirality is not dependent on the presence or absence of rotational symmetry. It is only dependent on the presence or absence of reflectional symmetry. This compound lacks reflectional symmetry and is therefore chiral. That is, it has a non-superimposable mirror image, drawn here:

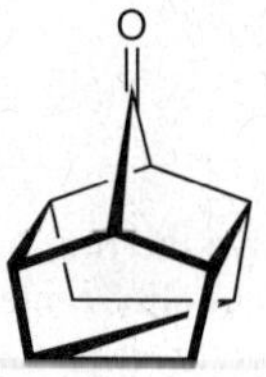

5.71.
(a) In the following Newman projection, the front carbon atom is connected to only two groups, and the back carbon atom is also connected to only two groups:

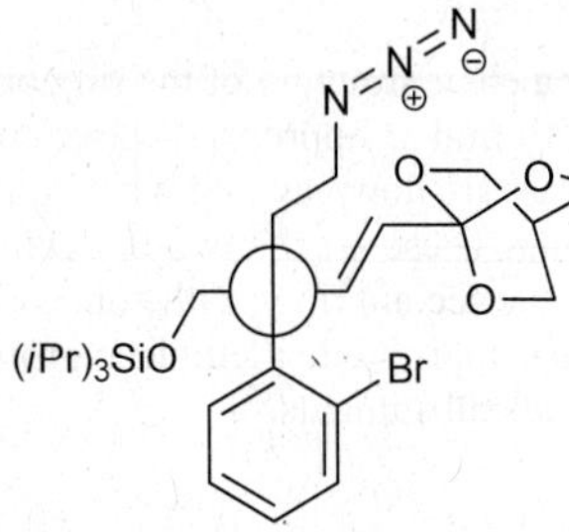

Notice that the two groups connected to the front carbon atom are twisted 90° with respect to the two groups connected to the back carbon atom. This is because the central carbon atom (in between the front carbon atom and the back carbon atom) is *sp* hybridized – it has two *p* orbitals, which are 90° apart from each other. One *p* orbital is being used to form one π bond, while the other *p* orbital is being used to form the other π bond.

(b) To draw the enantiomer, we could either switch the two groups connected to the front carbon atom, or we could switch the two groups connected to the back carbon atom. The former is shown here, in both bond-line format and in a Newman projection.

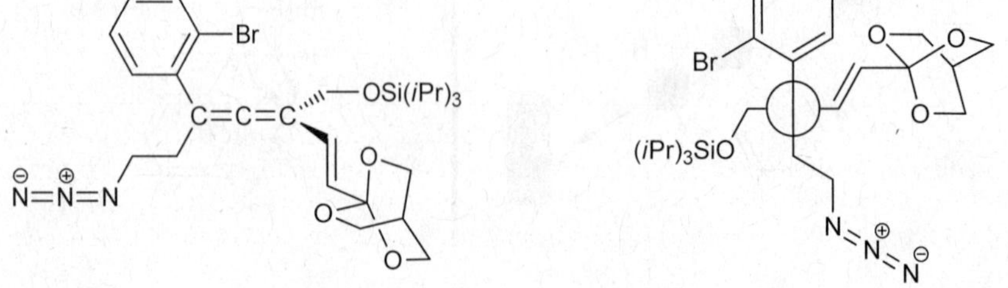

(c) To draw a diastereomer of the original compound, simply convert the *trans* configuration of the alkene to a *cis* configuration. The two diastereomers (both *cis* alkenes) are enantiomers of each other.

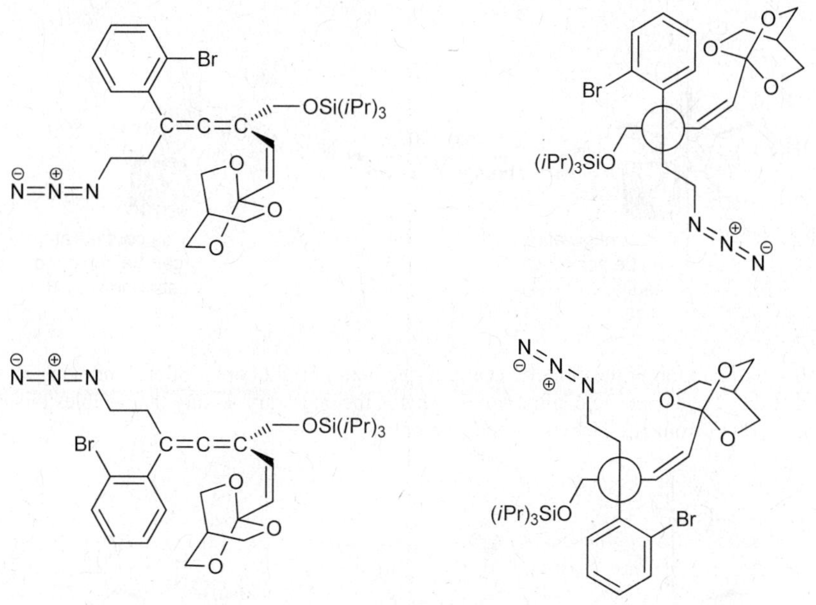

5.72.
(a) Glucuronolactone 1L is the enantiomer of 1D, which is shown in the problem statement. To draw the enantiomer of 1D, we simply redraw it, except that we replace all dashes with wedges, and all wedges with dashes, as shown:

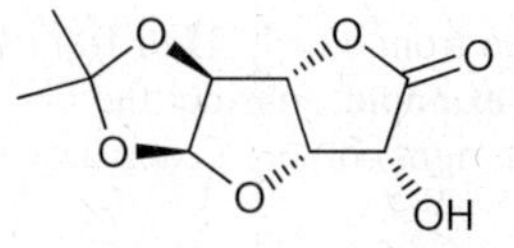

(b) There are 5 chirality centers, so there are 32 (or 2^5) possible stereoisomers.

(c)

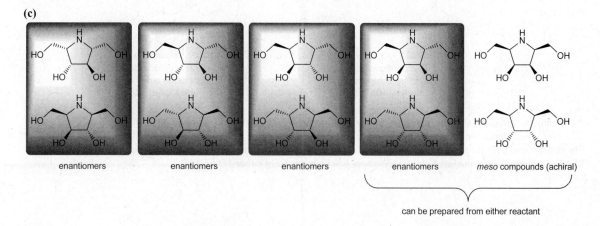

| enantiomers | enantiomers | enantiomers | enantiomers | *meso* compounds (achiral) |

can be prepared from either reactant

(d) The four products that are accessible from either of the reactants are the four products shown on the right in the solution to part c, as indicated above. Recall that the synthetic protocol allows for control of configurations at C2, C3 and C5, but not at C4. Therefore, in order for a specific stereoisomer to be accessible from either 1D or from 1L, that stereoisomer must display a specific feature. To understand this feature, we must draw one of the ten stereoisomers and then redraw it again after rotating it 180 degrees about a vertical axis. For example, let's do this for one of the *meso* compounds:

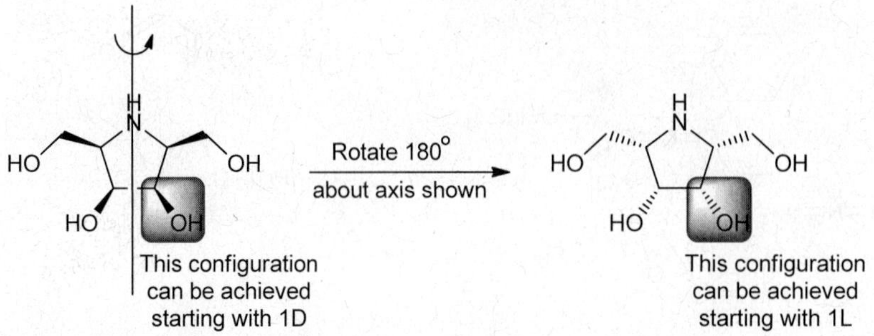

This configuration
can be achieved
starting with 1D

Rotate 180°
about axis shown

This configuration
can be achieved
starting with 1L

Now we look at the configuration of the chirality center in the bottom right corner of each drawing above (highlighted in gray). Notice that they have opposite configuration. This is the necessary feature that enables this compound to be accessible from either 1D or from 1L. Here is another example:

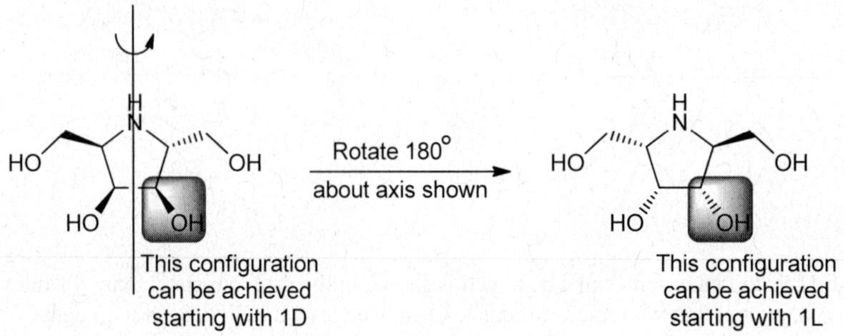

This configuration
can be achieved
starting with 1D

Rotate 180°
about axis shown

This configuration
can be achieved
starting with 1L

Once again, this stereoisomer will be accessible from either 1D or from 1L. In contrast, the first six structures (in the answer to part c) do not have this feature. For example, consider the first structure: let's draw it, rotate it 180 degrees, and then inspect the configuration in the bottom right corner of each drawing:

This configuration
can be achieved
starting with 1D

Rotate 180°
about axis shown

This configuration
cannot be achieved
starting with 1L

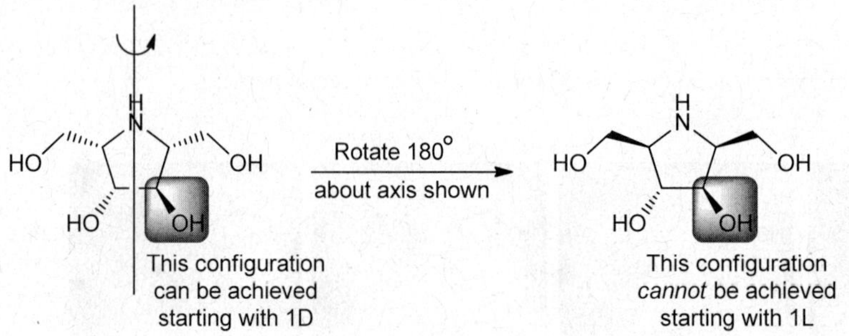

Note that in this case, the configuration in the bottom right corner of each drawing of this structure is the same. Therefore, this stereoisomer can ONLY be made from 1D. It cannot be made from 1L. A similar analysis for the first six stereoisomers (in the answer to part c) shows that all six of these stereoisomers require a specific enantiomer for the starting material. Only the last four stereoisomers can be made from either 1D or from 1L.

SkillBuilder 6.5   Drawing Curved Arrows

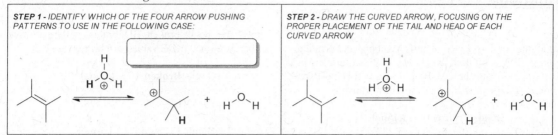

SkillBuilder 6.6   Predicting Carbocation Rearrangements

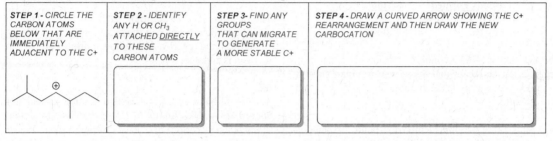

Common Mistakes to Avoid

In this chapter, we learned many skills that are necessary for drawing reaction mechanisms (identifying nucleophilic and electrophilic centers, drawing curved arrows, identifying arrow-pushing patterns, etc.). We will use these skills frequently in the upcoming chapters. In particular, it is important to become proficient with curved arrows, as they represent the language of reaction mechanisms, and you will have to become fluent in that language as we progress through the chapters. There are many common mistakes that students make when drawing curved arrows, and most of those mistakes can be avoided if you always remember that curved represent the motion of electrons. That is, the tail of every curved arrow must identify which electrons are moving, and the head of every curved arrow must show where those electrons are going. Let's first focus on the tail. The tail must always be placed on electrons, which means that it must be placed either on a lone pair or on a bond. If we examine each of the four characteristic arrow-pushing patterns below, we see this clearly:

Nucleophilic attack	Loss of a leaving group
Proton transfer	Rearrangement

Notice that the tail of each curved arrow is placed on a lone pair or on a bond. Similarly, the head of every curved arrow must either show the formation of a lone pair or the formation of a bond. Look at all of the curved arrows above and convince yourself that this is correct. If you keep this in mind when drawing curved arrows, you can avoid many silly errors.

Solutions

6.1.
(a) Using Table 6.1, we identify the bond dissociation energy (BDE) of each bond that is either broken or formed. For bonds broken, BDE values will be positive. For bonds formed, BDE values will be negative.

Bonds Broken	kJ/mol
H—CH(CH$_3$)$_2$	+ 397
Br—Br	+ 193

Bonds Formed	kJ/mol
(CH$_3$)$_2$CH—Br	− 285
H—Br	− 368

The net sum is **− 63 kJ/mol**. $\Delta H°$ for this reaction is negative, which means that the system is losing energy. It is giving off energy to the environment, so the reaction is exothermic.

(b) Using Table 6.1, we identify the bond dissociation energy (BDE) of each bond that is either broken or formed. For bonds broken, BDE values will be positive. For bonds formed, BDE values will be negative.

Bonds Broken	kJ/mol
(CH$_3$)$_3$C—Cl	+ 331
H—OH	+ 498

Bonds Formed	kJ/mol
(CH$_3$)$_3$C—OH	− 381
H—Cl	− 431

The net sum is **+ 17 kJ/mol**. $\Delta H°$ for this reaction is positive, which means that the system is gaining energy. It is receiving energy from the environment, so the reaction is endothermic.

(c) Using Table 6.1, we identify the bond dissociation energy (BDE) of each bond that is either broken or formed. For bonds broken, BDE values will be positive. For bonds formed, BDE values will be negative.

Bonds Broken	kJ/mol
(CH$_3$)$_3$C—Br	+ 272
H—OH	+ 498

Bonds Formed	kJ/mol
(CH$_3$)$_3$C—OH	− 381
H—Br	− 368

The net sum is **+ 21 kJ/mol**. $\Delta H°$ for this reaction is positive, which means that the system is gaining energy. It is receiving energy from the environment, so the reaction is endothermic.

(d) Using Table 6.1, we identify the bond dissociation energy (BDE) of each bond that is either broken or

formed. For bonds broken, BDE values will be positive. For bonds formed, BDE values will be negative.

Bonds Broken	kJ/mol
(CH$_3$)$_3$C—I	+ 209
H—OH	+ 498

Bonds Formed	kJ/mol
(CH$_3$)$_3$C—OH	− 381
H—I	− 297

The net sum is **+ 29 kJ/mol**. $\Delta H°$ for this reaction is positive, which means that the system is gaining energy. It is receiving energy from the environment, so the reaction is endothermic.

6.2.
The C—C bond of CH$_3$—CH$_3$ has a bond dissociation energy of = +368 kJ/mol. If a C=C bond has a total bond dissociation energy of +632 kJ/mol, then the π component of the double bond can be estimated to be (632 kJ/mol) − (368 kJ/mol) = 264 kJ/mol. In other words, the π component of the C=C bond is not as strong as the σ component of the C=C bond. In the reaction shown in this problem, the π component of the C=C bond is broken but the σ component remains intact. Accordingly, the calculation is as follows:

Bonds Broken	kJ/mol
C=C (just the π component)	+ 264
H—OH	+ 498

Bonds Formed	kJ/mol
CH$_3$CH$_2$—OH	− 381
H—CH$_2$R	~ − 410

The net sum is **− 29 kJ/mol**. $\Delta H°$ for this reaction is predicted to be negative, which means that the system is losing energy. It is giving off energy to the environment, so the reaction is exothermic.

6.3.
(a) ΔS_{sys} is expected to be negative (a decrease in entropy) because two molecules are converted into one molecule.
(b) ΔS_{sys} is expected to be negative (a decrease in entropy) because an acylic compound is converted into a cyclic compound.
(c) ΔS_{sys} is expected to be positive (an increase in entropy) because one molecule is converted into two molecules.
(d) ΔS_{sys} is expected to be positive (an increase in entropy) because one molecule is converted into two ions.
(e) ΔS_{sys} is expected to be negative (a decrease in entropy) because two chemical entities are converted into one.

(f) ΔS_{sys} is expected to be positive (an increase in entropy) because a cyclic compound is converted into an acyclic compound.

6.4.
(a) There is a competition between the two terms contributing to ΔG. In this case, the reaction is endothermic, which contributes to a positive value for ΔG, but the second term contributes to a negative value for ΔG:

$$\Delta G = \underset{\oplus}{\Delta H} + \underset{\ominus}{(-T\Delta S)}$$

The sign of ΔG will therefore depend on the competition between these two terms, which is affected by temperature. A high temperature will cause the second term to dominate, giving rise to a negative value of ΔG. A low temperature will render the second term insignificant, and the first term will dominate, giving rise to a positive value of ΔG.

(b) In this case, both terms contribute to a negative value for ΔG, so ΔG will definitely be negative (the process will be spontaneous).

(c) In this case, both terms contribute to a positive value for ΔG, so ΔG will definitely be positive (the process will not be spontaneous).

(d) There is a competition between the two terms contributing to ΔG. In this case, the reaction is exothermic, which contributes to a negative value for ΔG, but the second term contributes to a positive value for ΔG:

$$\Delta G = \underset{\ominus}{\Delta H} + \underset{\oplus}{(-T\Delta S)}$$

The sign of ΔG will therefore depend on the competition between these two terms, which is affected by temperature. A high temperature will cause the second term to dominate, giving rise to a positive value of ΔG. A low temperature will render the second term insignificant, and the first term will dominate, giving rise to a negative value of ΔG.

6.5. A system can only achieve a lower energy state by transferring energy to its surroundings (conservation of energy). This increases the entropy of the surroundings, which more than offsets the decrease in entropy of the system. As a result, ΔS_{tot} increases.

6.6.
(a) A positive value of ΔG favors reactants.
(b) A reaction for which $K_{eq} < 1$ will favor reactants.
(c) $\Delta G = \Delta H - T\Delta S = (33 \text{ kJ/mol}) - (298 \text{ K})(0.150 \text{ kJ/mol} \cdot \text{K}) = -11.7$ kJ/mol
A negative value of ΔG favors products.
(d) Both terms contribute to a negative value of ΔG, which favors products.

(e) Both terms contribute to a positive value of ΔG, which favors reactants.

6.7.
(a) Process D will occur more rapidly because it has a lower energy of activation than process A.
(b) Process A will more greatly favor products than process B, because the former is exergonic (the products are lower in energy than the reactants) while the latter is not exergonic.
(c) None of these processes exhibits an intermediate, because none of the energy diagrams has a local minimum (a valley). But all of the processes proceed via a transition state, because all of the energy diagrams have a local maximum (a peak).
(d) In process A, the transition state resembles the reactants more than products because the transition state is closer in energy to the reactant than the products (the Hammond postulate).
(e) Process A will occur more rapidly because it has a lower energy of activation than process B.
(f) Process D will more greatly favor products at equilibrium than process B, because the former is exergonic (the products are lower in energy than the reactants) while the latter is not exergonic.
(g) In process C, the transition state resembles the products more than reactants because the transition state is closer in energy to the products than the reactants (the Hammond postulate).

6.8.
(a) Carbon is significantly more electronegative than lithium. As such, the carbon atom of the C-Li bond withdraws electron density (via induction) from the lithium atom, rendering that carbon atom highly nucleophilic.

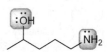

(b) Lone pairs are regions of high electron density. As such, this compound has nucleophilic centers, highlighted below:

(c) The lone pair on the nitrogen atom constitutes a nucleophilic center:

(d) The π bond (highlighted) is a nucleophilic center:

6.9.

(a) The carbon atom of the carboxylic acid group (COOH) is electrophilic. This is the only electrophilic center in the compound.

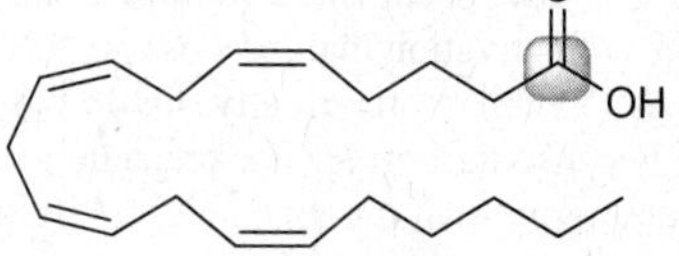

(b) The carbon atom of the carbonyl group (C=O) is electrophilic. This is the only electrophilic center in the compound.

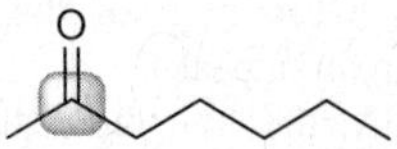

(c) The following six positions are all electrophilic centers.

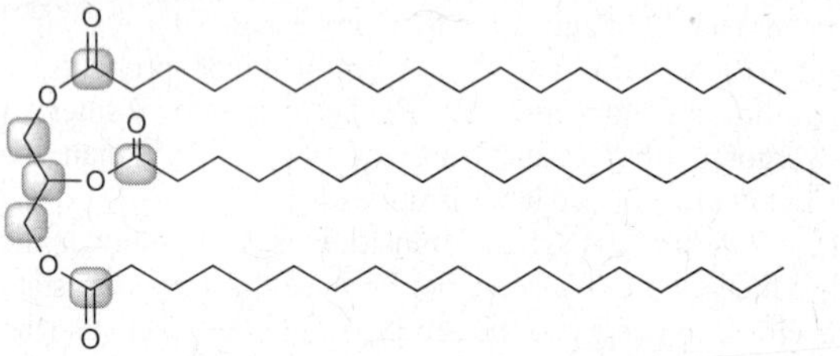

6.10. In this hypothetical compound, the boron atom would have an empty p orbital, and would therefore be an electrophilic center. The carbon atom of the C-Li bond withdraws electron density (via induction) from the lithium atom, rendering that carbon atom highly nucleophilic.

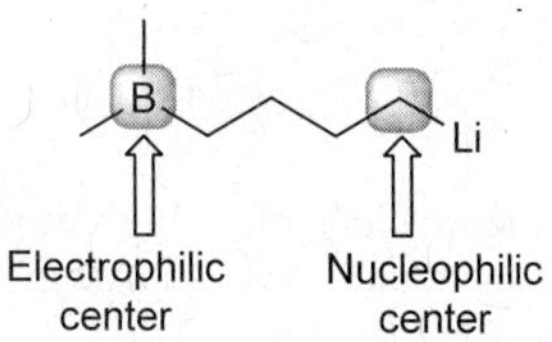

Electrophilic   Nucleophilic
center   center

6.11.

(a) When we draw all significant resonance structures, we find that there are two positions (highlighted) that are deficient in electron density:

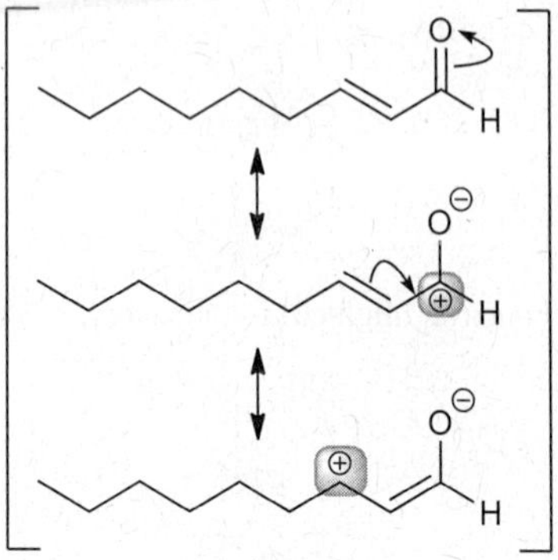

Therefore, these two positions are electrophilic:

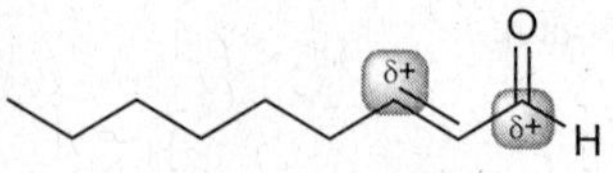

(b) When we draw all significant resonance structures, we find that there are two positions (highlighted) that are deficient in electron density:

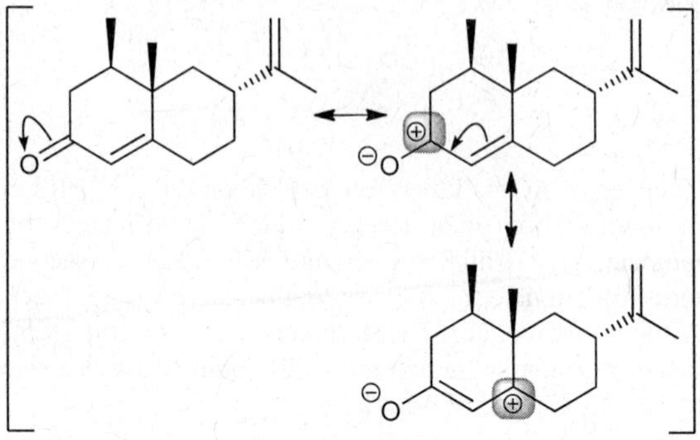

Therefore, these two positions are electrophilic:

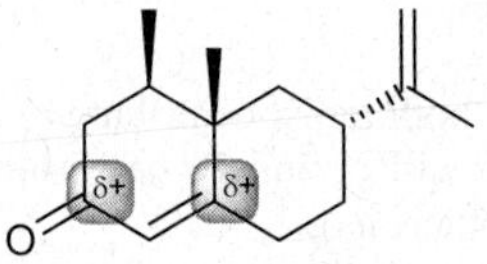

(c) When we draw all significant resonance structures, we find that there are two positions (highlighted) that are deficient in electron density:

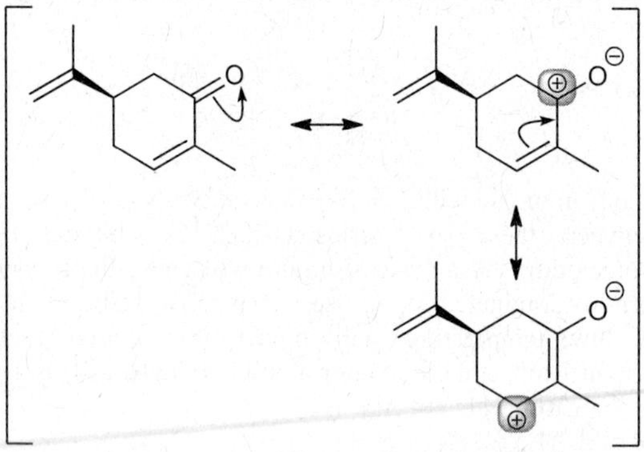

Therefore, these two positions are electrophilic:

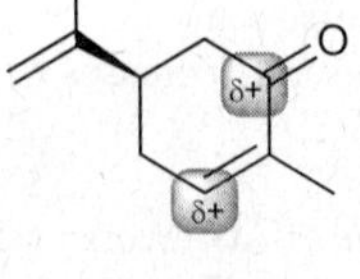

6.12.

(a) The curved arrow indicates loss of a leaving group (Cl$^-$).

(b) The curved arrows indicate a proton transfer. In this case, H_2SO_4 is the acid and HNO_3 functions as the base that removes the proton from H_2SO_4.

(c) The curved arrow indicates a hydride shift, which is a type of carbocation rearrangement.

(d) The curved arrow indicates a nucleophilic attack. In this case, water functions as a nucleophile and attacks the carbocation.

(e) The curved arrows indicate a proton transfer. In this case, water functions as the base that removes the proton.

(f) The curved arrows indicate a nucleophilic attack. In this case, the lone pair on the nitrogen atom functions as the nucleophilic center that attacks the electrophile.

(g) The curved arrow indicates a methyl shift, which is a type of carbocation rearrangement.

(h) The curved arrow indicates loss of a leaving group $(CH_3SO_3^-)$.

(i) The curved arrows indicate a nucleophilic attack. In this case, one of the lone pairs on the oxygen atom functions as the nucleophilic center that attacks the electrophilic center.

6.13. The π bond functions as a nucleophile and attacks the electrophilic carbocation. This step is therefore a nucleophilic attack.

6.14.
(a)   The sequence of arrow-pushing patterns is as follows:
    (i) proton transfer
    (ii) nucleophilic attack
    (iii) proton transfer

(b) The sequence of arrow-pushing patterns is as follows:
    (i) nucleophilic attack
    (ii) proton transfer
    (iii) proton transfer

(c) The sequence of arrow-pushing patterns is as follows:
    (i) proton transfer
    (ii) nucleophilic attack
    (iii) loss of a leaving group

(d) The sequence of arrow-pushing patterns is as follows:
    (i) proton transfer
    (ii) loss of a leaving group
    (iii) nucleophilic attack
    (iv) proton transfer

(e) The sequence of arrow-pushing patterns is as follows:
    (i) proton transfer
    (ii) nucleophilic attack
    (iii) proton transfer

6.15.   Both reactions have the same sequence: (i) nucleophilic attack, followed by (ii) loss of a leaving group. In both cases, a hydroxide ion functions as a nucleophile and attacks a compound that can accept the negative charge and store it temporarily. The charge is then expelled as a chloride ion in both cases.

6.16.
(a) In this case, a C–O bond is formed, indicating a nucleophilic attack.   Water (H_2O) functions as a nucleophilic center and attacks the carbocation. This is shown with one curved arrow. The tail of this curved arrow is placed on the lone pair of the oxygen atom, and the head is placed on the electrophilic center (the empty p orbital of the carbocation), as shown here:

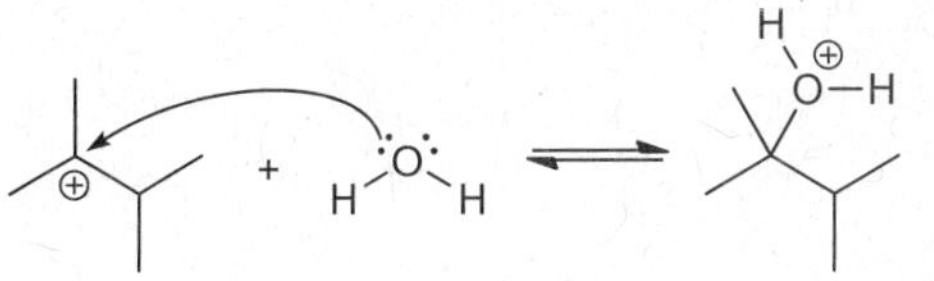

(b)   This is a proton transfer step, in which water functions as a base and removes a proton, thereby generating H_3O^+. A proton transfer step requires two curved arrows. The tail of the first curved arrow is placed on a lone pair of H_2O, and the head is placed on the proton that is being transferred. Don't forget the second curved arrow. The tail is placed on the O–H bond (that is being broken) and the head is placed on the oxygen atom, as shown:

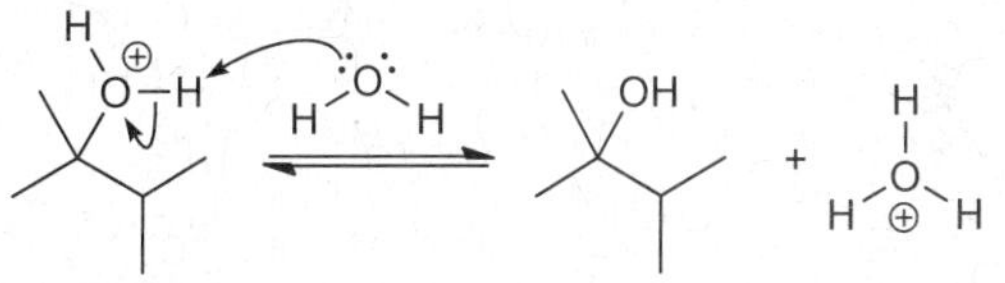

(c) This step represents the loss of a leaving group (where the leaving group is H_2O). One curved arrow is required. The tail is placed on the C–O bond that is broken, and the head is placed on the oxygen atom.

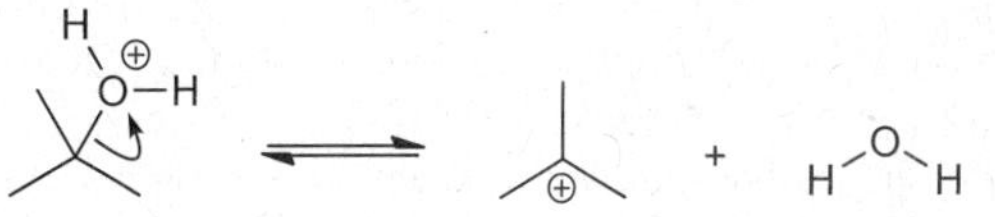

6.17.
(a) The C–Br bond is broken, indicating the loss of a leaving group (Br^-), while the C–Cl bond is formed, indicating a nucleophilic attack. This is, in fact, a concerted process in which nucleophilic attack and loss of the leaving group occur in a simultaneous fashion. One curved arrow is required to show the nucleophilic attack, and another curved arrow is required to show loss of the leaving group:

(b) The first step represents the loss of a leaving group (Br^-), while the second step is a nucleophilic attack (in which Cl^- functions as the nucleophile). Each of these steps requires one curved arrow, as shown.

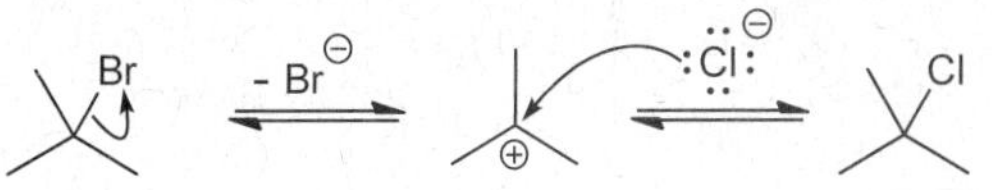

(c) The C–Cl bond is broken, indicating the loss of a leaving group (Cl⁻). In addition, a proton transfer step occurs, with RO⁻ functioning as a base). This is, in fact, a concerted process in which loss of the leaving group occurs at the same time as the proton transfer. In the process, a π bond is formed. This requires three curved arrows, as shown:

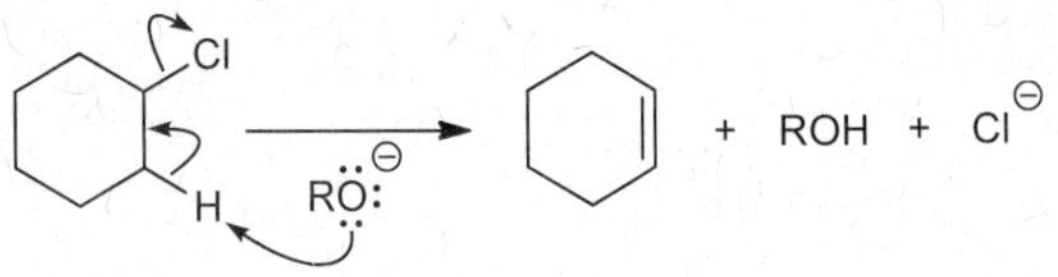

(d) The first step represents the loss of a leaving group (Cl⁻), while the second step is a proton transfer (in which water functions as the base). One curved arrow is required to show the loss of the leaving group, and two curved arrows are required to show the proton transfer, as shown:

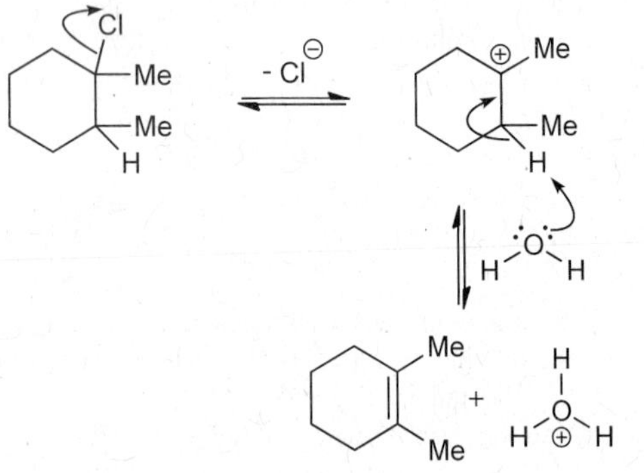

6.18.

(a) This carbocation is secondary, and it can rearrange via a hydride shift (shown below) to give a more stable, tertiary carbocation:

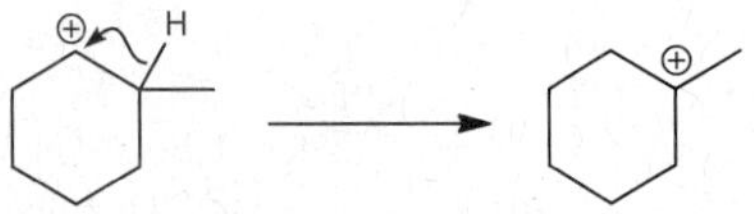

(b) This carbocation is tertiary, and it cannot become more stable via a rearrangement.

(c) This carbocation is tertiary. Yet, in this case, rearrangement via a methyl shift will generate a more stable, tertiary allylic carbocation, which is resonance stabilized, as shown:

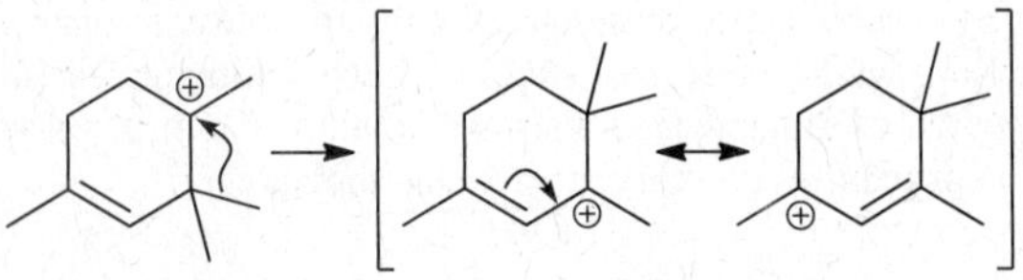

tertiary   tertiary allylic

(d) This carbocation is secondary, but there is no way for it to rearrange to form a tertiary carbocation.

(e) This carbocation is secondary, and it can rearrange via a hydride shift (shown below) to give a more stable, tertiary carbocation:

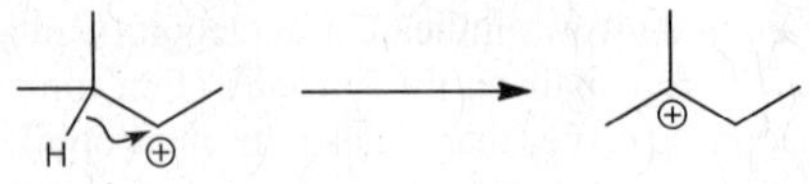

(f) This carbocation is secondary, and it can rearrange via a methyl shift (shown below) to give a more stable, tertiary carbocation:

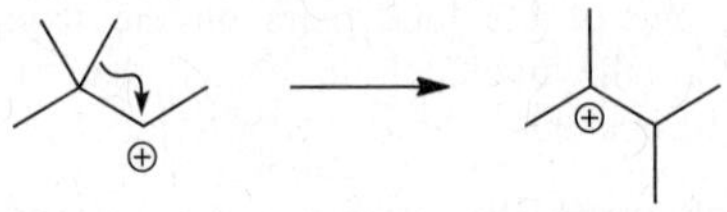

(g) This carbocation is primary, and it can rearrange via a hydride shift (shown below) to give a resonance stabilized carbocation (we will see in Chapter 7 that this carbocation is called a benzylic carbocation):

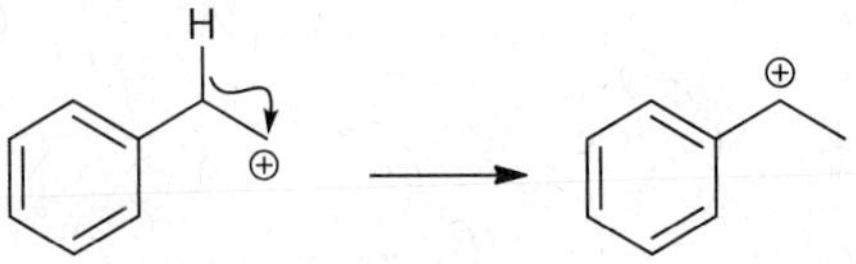

(h) This carbocation is tertiary and it is resonance-stabilized (we will see in Chapter 7 that this carbocation is called a benzylic carbocation). It will not rearrange.

6.19. In this case, a five-membered ring is being converted into a six-membered ring (a process called ring expansion). In order for this to occur, the migrating carbon atom must be part of the five-membered ring. The migrating group must be connected to the position that is adjacent to C+. The migrating carbon atom is highlighted below:

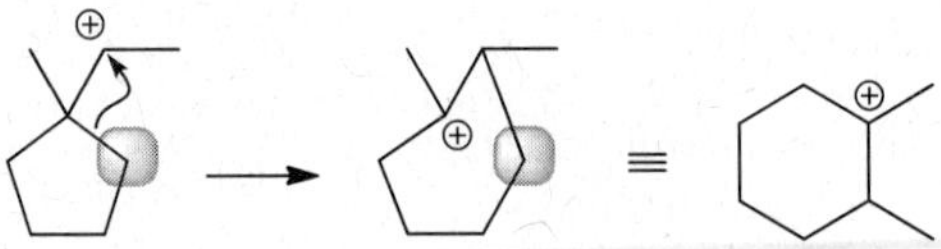

6.20.
(a) A carbon-carbon triple bond is comprised of one σ bond and two π bonds, and is therefore stronger than a carbon-carbon double bond (one σ and one π bond) or a carbon-carbon single bond (only one σ bond).
(b) The data in Table 6.1 indicate that the C-F bond will have the largest bond dissociation energy.

6.21.
(a) Using Table 6.1, we identify the bond dissociation energy (BDE) of each bond that is either broken or

formed. For bonds broken, BDE values will be positive. For bonds formed, BDE values will be negative.

Bonds Broken	kJ/mol
RCH_2—Br	+ 285
RCH_2O—H	+ 435

Bonds Formed	kJ/mol
RCH_2—OR	– 381
H—Br	– 368

The net sum is – **29 kJ/mol**. $\Delta H°$ for this reaction is negative, which means that the system is losing energy. It is giving off energy to the environment, so the reaction is exothermic.

(b) ΔS of this reaction is positive because one mole of reactant is converted into two moles of product.

(c) Both terms (ΔH) and ($-T\Delta S$) contribute to a negative value of ΔG.

(d) Since both terms (ΔH and $-T\Delta S$) have negative values, the value of ΔG will be negative at all temperatures.

(e) Yes. At high temperatures, the value of $-T\Delta S$ is large and negative (while at low temperatures, the value of $-T\Delta S$ is small and negative).

Since $\Delta G = \Delta H + (-T\Delta S)$, the magnitude of ΔG will be dependent on temperature.

6.22.
(a) A reaction for which $K_{eq} > 1$ will favor products.
(b) A reaction for which $K_{eq} < 1$ will favor reactants.
(c) A positive value of ΔG favors reactants.
(d) Both terms contribute to a negative value of ΔG, which favors products.
(e) Both terms contribute to a positive value of ΔG, which favors reactants.

6.23. $K_{eq} = 1$ when $\Delta G = 0$ kJ/mol (See Table 6.2).

6.24. $K_{eq} < 1$ when ΔG has a positive value. The answer is therefore "a" (+1 kJ/mol)

6.25.
(a) ΔS_{sys} is expected to be negative (a decrease in entropy) because two moles of reactant are converted into one mole of product.
(b) ΔS_{sys} is expected to be positive (an increase in entropy) because one mole of reactant is converted into two moles of product.
(c) ΔS_{sys} is expected to be approximately zero, because two moles of reactant are converted into two moles of product.
(d) ΔS_{sys} is expected to be negative (a decrease in entropy) because an acylic compound is converted into a cyclic compound.
(e) ΔS_{sys} is expected to be approximately zero, because one mole of reactant is converted into one mole of product, and both the reactant and the product are acyclic.

6.26.
(a) If the reaction has only one step, then the energy diagram will have only one hump. Since ΔG for this reaction is negative, the product will be lower in free energy than the reactant, as shown here.

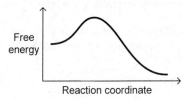

(b) If the reaction has only one step, then the energy diagram will have only one hump. Since ΔG for this reaction is positive, the product will be higher in free energy than the reactant, as shown here.

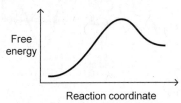

(c) If the reaction has two steps, then the energy diagram will have two humps. Since ΔG for this reaction is negative, the product will be lower in free energy than the reactant. And the problem statement indicates that the transition state for the first step is higher in energy than the transition state for the second step, as shown here.

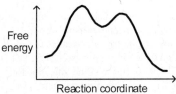

6.27.
(a) Energy diagrams B and D each exhibit two humps, characteristic of a two-step process.
(b) Energy diagrams A and C each exhibit only one hump, characteristic of a one-step process.
(c) The energy of activation (E_a) is determined by the difference in energy between the reactant and the transition state (the top of the hump in the energy diagram). This energy difference is greater in C than it is in A.
(d) Energy diagram A has a negative ΔG, because the product is lower in free energy than the reactant. This is not the case in energy diagram C.
(e) Energy diagram D has a positive ΔG, because the product is higher in free energy than the reactant. This is not the case in energy diagram A.
(f) The energy of activation (E_a) is determined by the difference in energy between the reactant and the transition state (the top of the hump in the energy diagram). This energy difference is greatest in D.
(g) $K_{eq} > 1$ when ΔG has a negative value. This is the case in energy diagrams A and B, because in each of these energy diagrams, the product is lower in free energy than the reactant.

(h) $K_{eq} = 1$ when $\Delta G = 0$ kJ/mol. This is the case in energy diagram C, in which the reactant and product have approximately the same free energy.

6.28. All local minima (valleys) represent intermediates, while all local maxima (peaks) represent transition states:

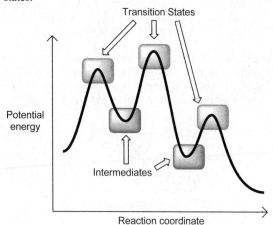

6.29.
(a) The reaction is second order, so the rate should be dependent on the concentrations of the reactants, as shown in the following rate equation:

$$\text{Rate} = k \left[\overset{\ominus}{:} \overset{..}{\underset{..}{O}} H \right] \left[CH_3CH_2Cl \right]$$

(b) The rate will be tripled, because the rate is linearly dependent on the concentration of hydroxide.
(c) The rate will be tripled, because the rate is linearly dependent on the concentration of CH_3CH_2Cl.
(d) As a rule of thumb, the rate doubles for every increase of 10° C. Therefore an increase of 40° C will correspond to increase in rate of approximately 16-fold $(2 \times 2 \times 2 \times 2)$.

6.30.
(a) The curved arrow indicates loss of a leaving group (Cl⁻).
(b) The curved arrow indicates a methyl shift, which is a type of carbocation rearrangement.
(c) The curved arrows indicate a nucleophilic attack.
(d) The curved arrows indicate a proton transfer. In this case, the proton transfer step occurs in an intramolecular fashion (because the acidic proton and the base are tethered together in one structure).

6.31.
(a) A tertiary carbocation is more stable than a secondary carbocation, which is more stable than a primary carbocation.

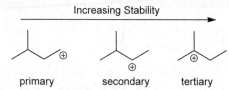

(b) The most stable carbocation is the one that is resonance-stabilized. Among the other two carbocations, the secondary carbocation is more stable than the primary carbocation, as shown here.

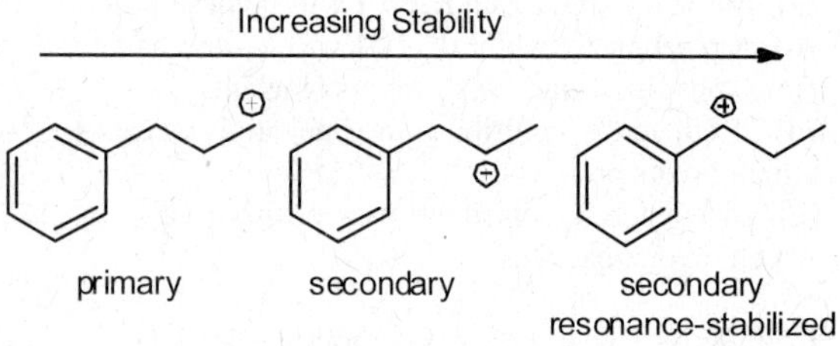

6.32. The sequence of arrow-pushing patterns is as follows:

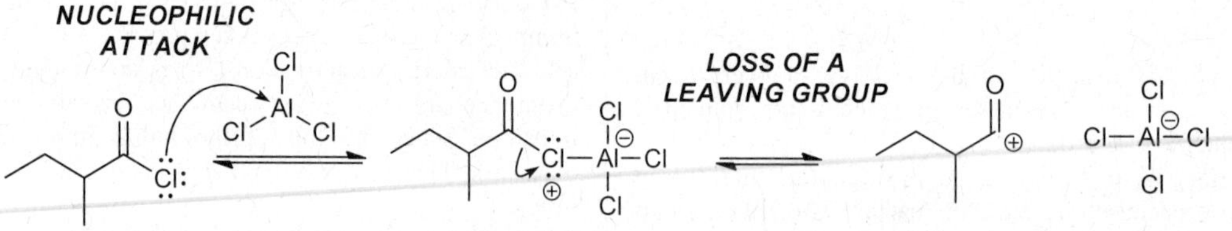

6.33. The sequence of arrow-pushing patterns is as follows:

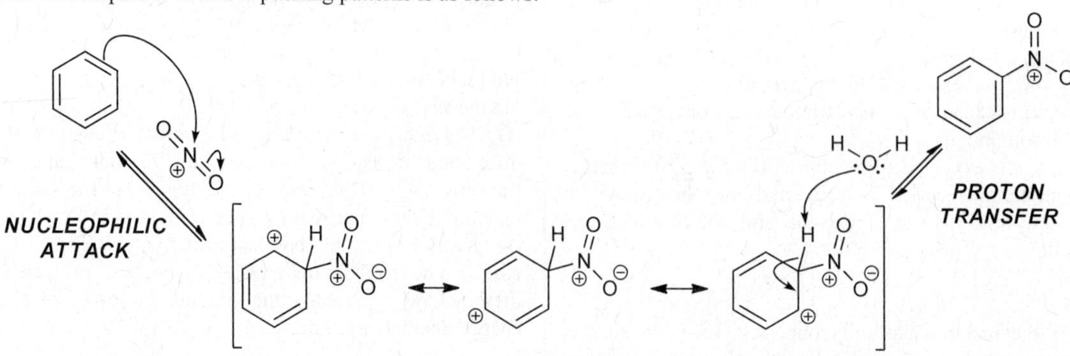

6.34. The sequence of arrow-pushing patterns is as follows:

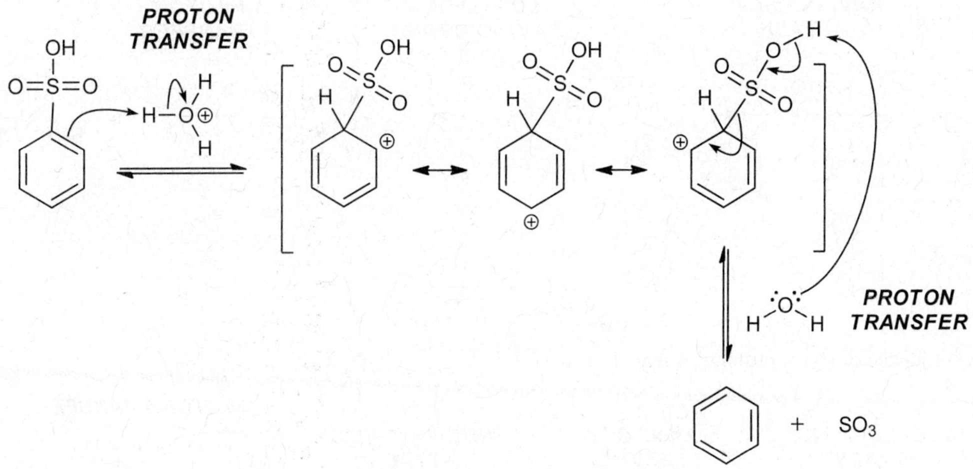

6.35. The sequence of arrow-pushing patterns is as follows:

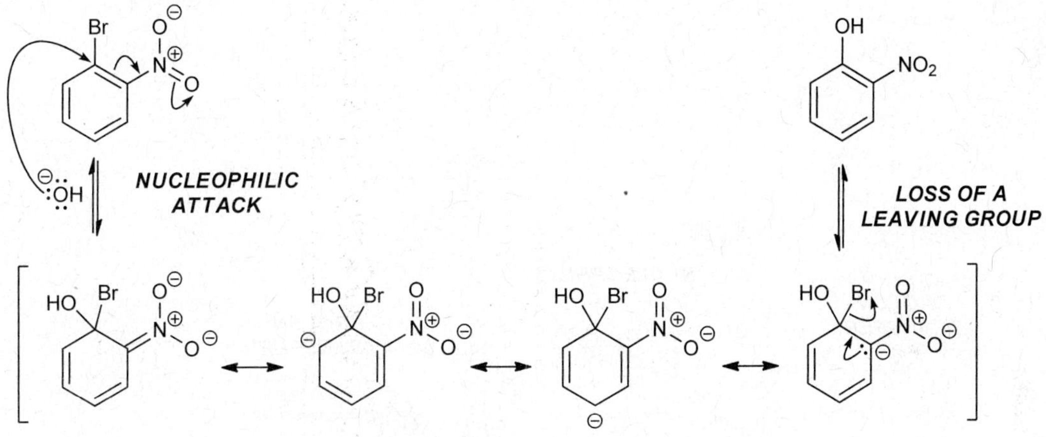

6.36. The sequence of arrow-pushing patterns is as follows:

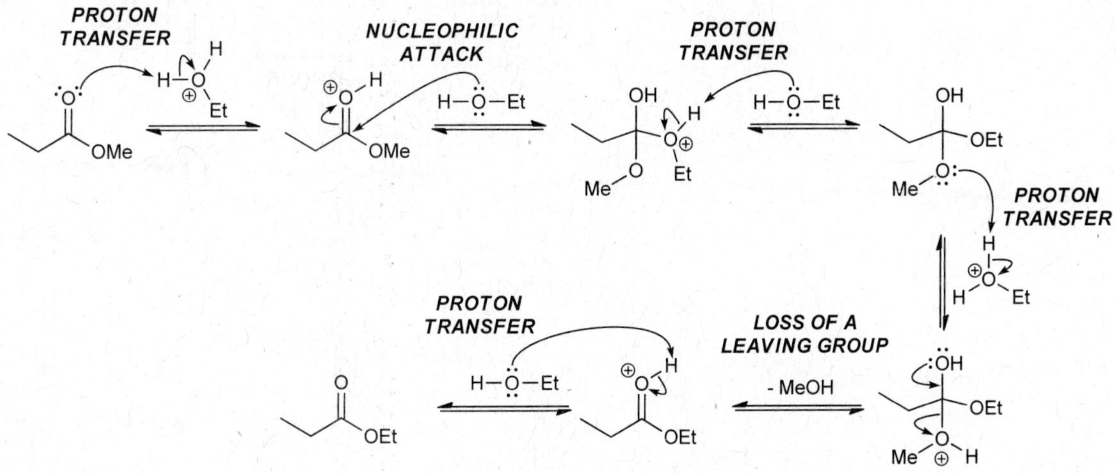

6.37. The sequence of arrow-pushing patterns is as follows:

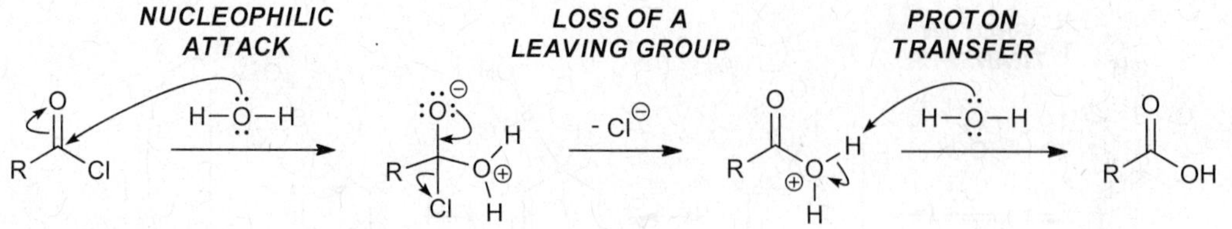

6.38. The sequence of arrow-pushing patterns is as follows:

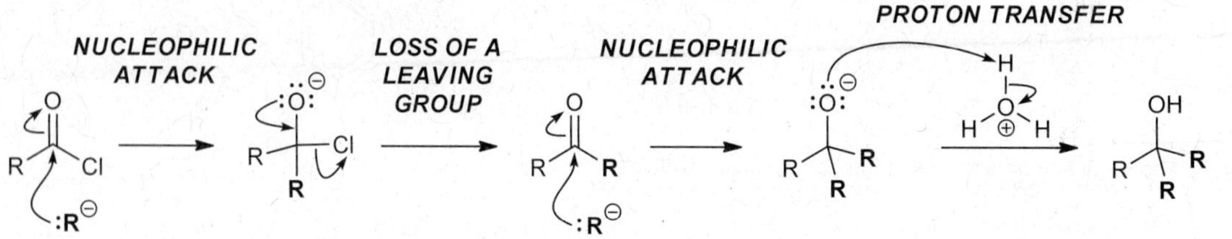

6.39. The sequence of arrow-pushing patterns is as follows:

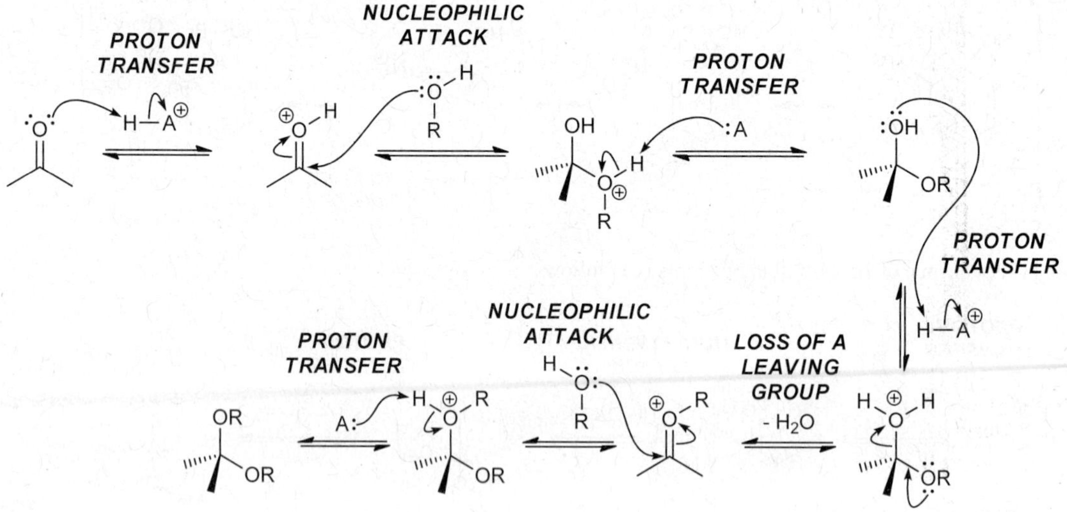

6.40. The sequence of arrow-pushing patterns is as follows:

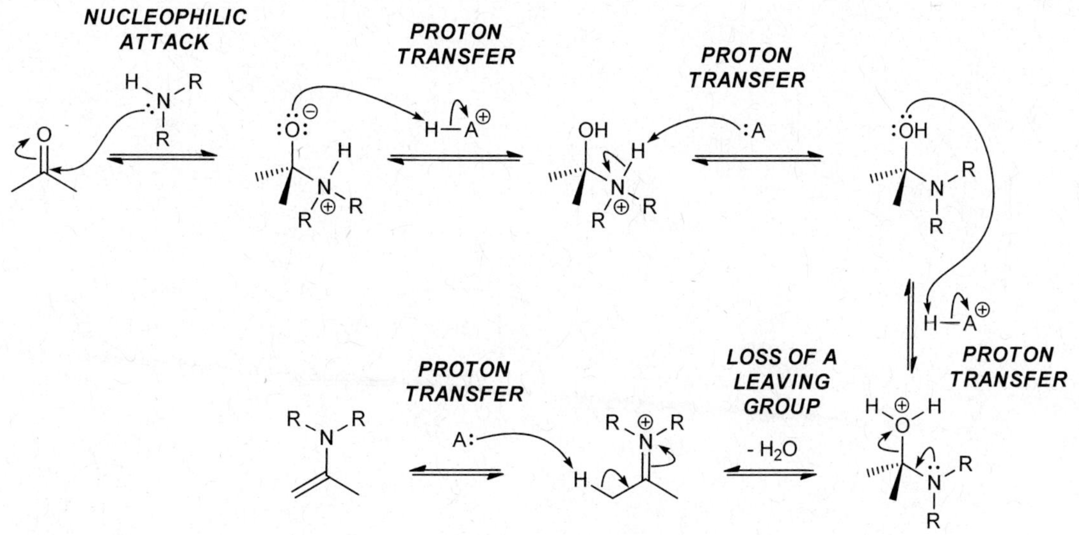

6.41. The sequence of arrow-pushing patterns is as follows:

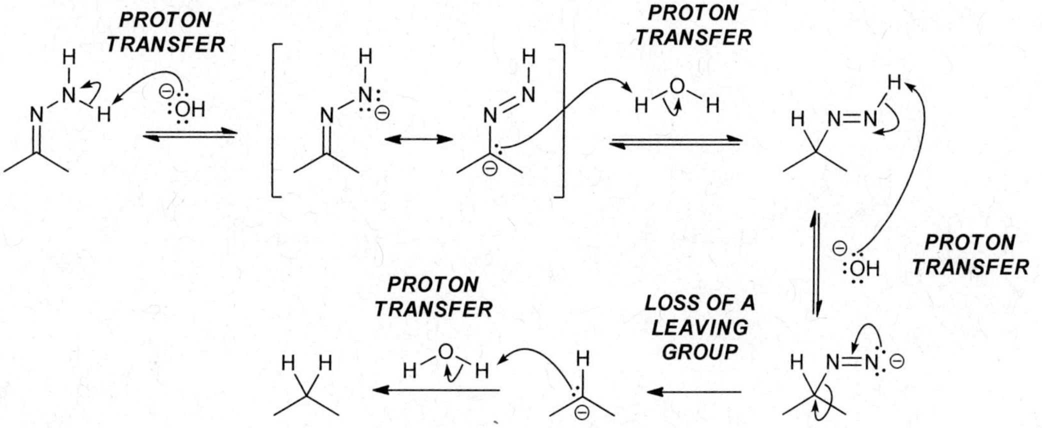

6.42. The first step is a nucleophilic attack, and the second step is loss of a leaving group. In this case, each of these steps requires several curved arrows, as shown:

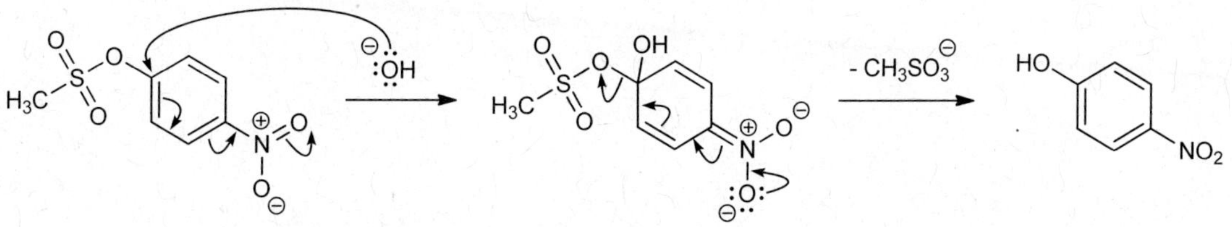

6.43. The following curved arrows show the flow of electrons that achieve the transformation as shown:

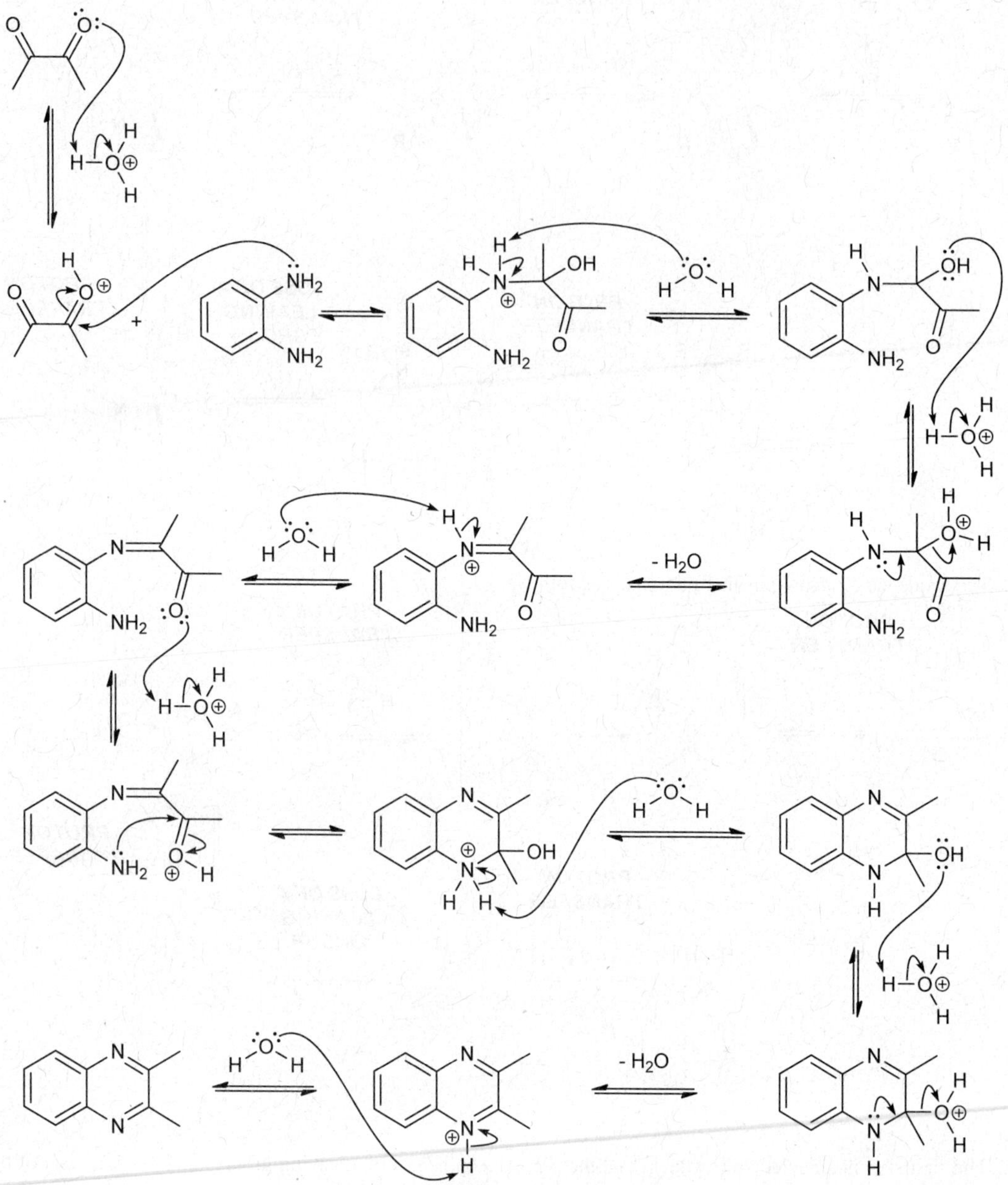

6.44. The following curved arrows show the flow of electrons that achieve the transformation as shown:

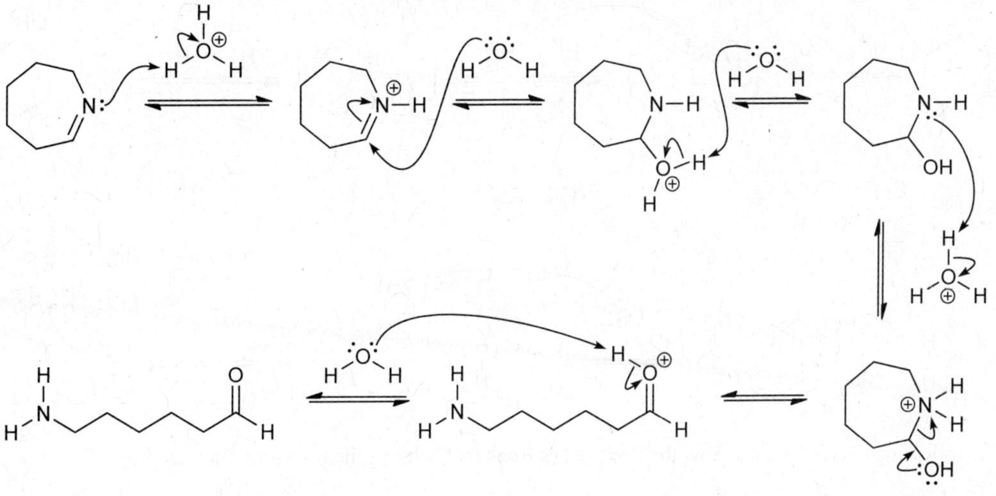

6.45. The following curved arrows show the flow of electrons that achieve the transformation as shown:

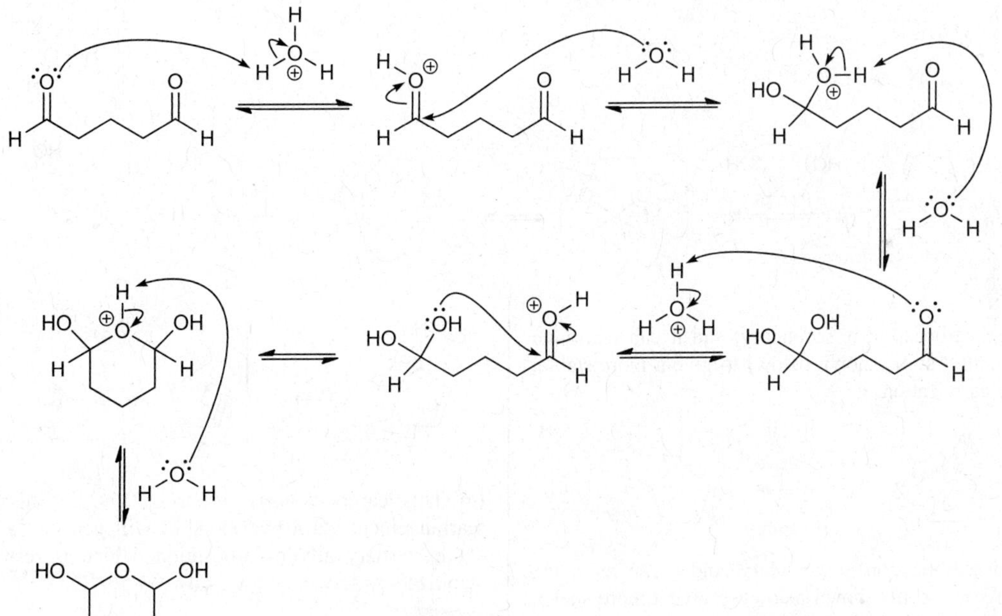

6.46. The following curved arrows show the flow of electrons that achieve the transformation as shown:

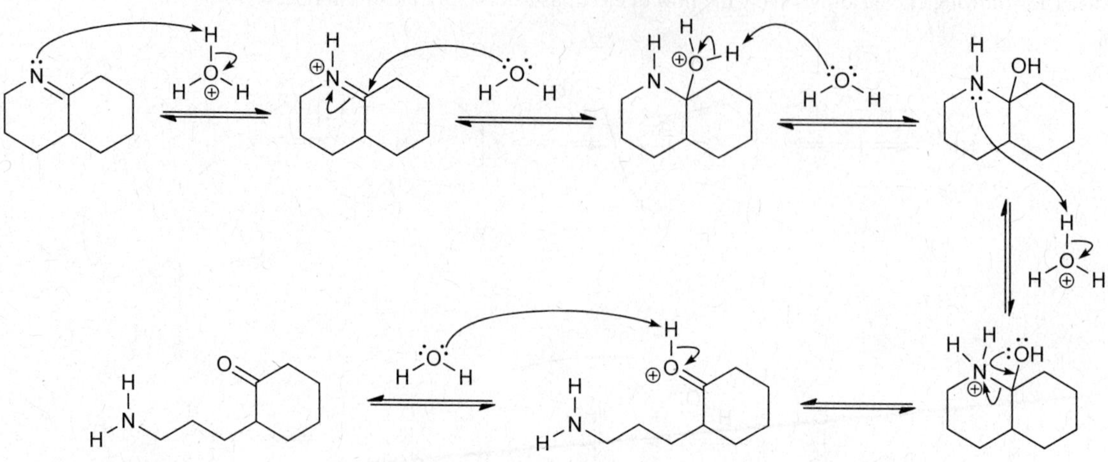

6.47. The following curved arrows show the flow of electrons that achieve the transformation as shown:

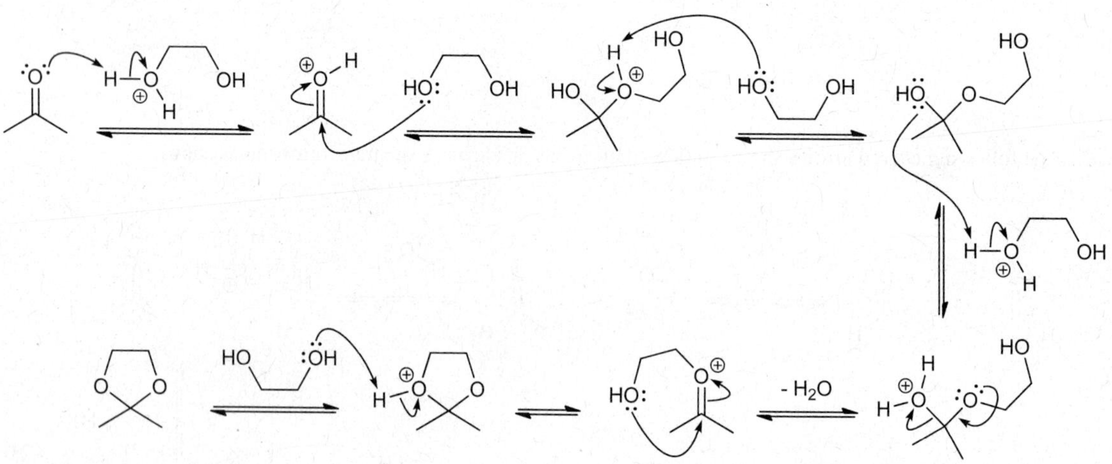

6.48.
(a) This carbocation is secondary, and it can rearrange via a methyl shift (shown below) to give a more stable, tertiary carbocation:

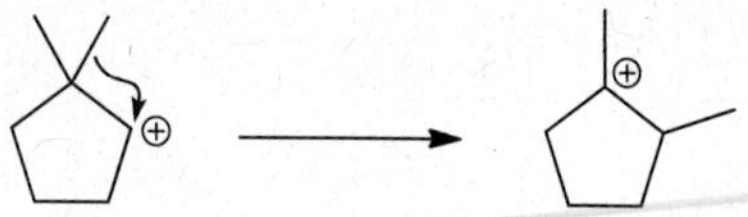

(b) This carbocation is secondary, and it can rearrange via a hydride shift (shown below) to give a more stable, tertiary carbocation:

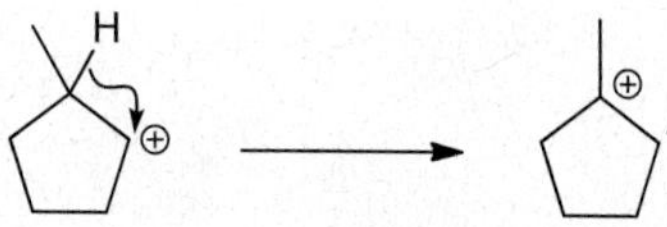

(c) This carbocation is secondary, but it cannot rearrange to generate a tertiary carbocation.
(d) This carbocation is secondary, and it can rearrange via a hydride shift (shown below) to generate a more stable, secondary allylic carbocation, which is resonance stabilized:

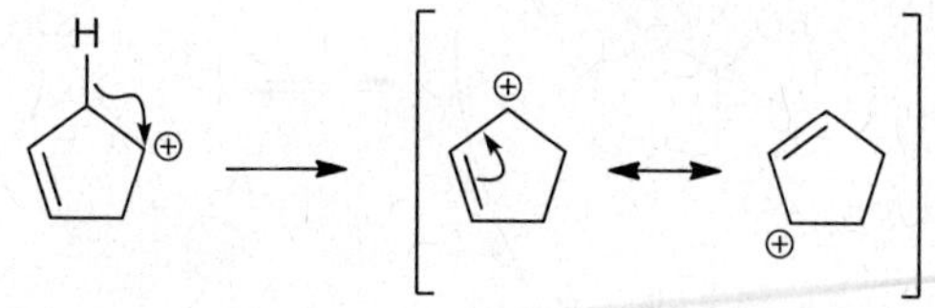

(e) This carbocation is tertiary. Yet, in this case, rearrangement via a hydride shift will generate a more stable, tertiary allylic carbocation, which is resonance stabilized, as shown below.

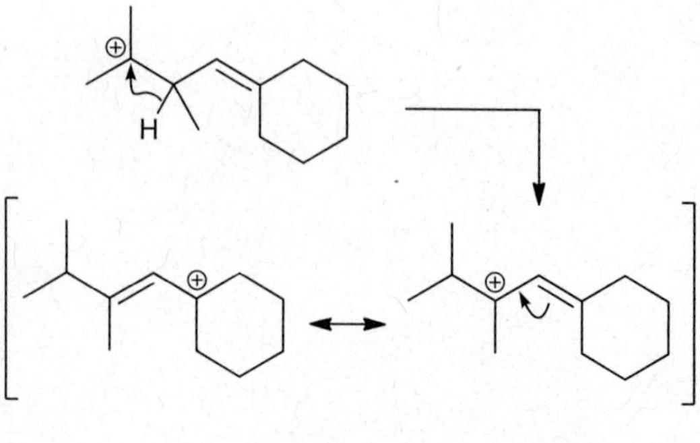

(f) This carbocation is secondary, and it can rearrange via a hydride shift (shown below) to give a more stable, tertiary carbocation:

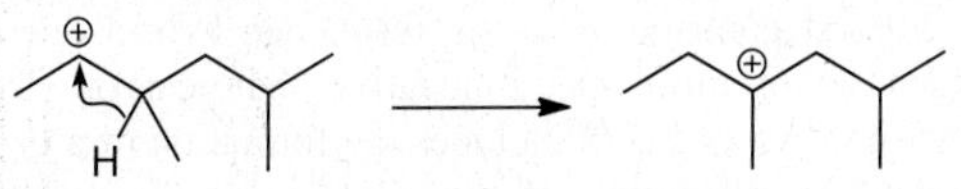

(g) This carbocation is tertiary and will not rearrange.

6.49.
(a) The C–Br bond is broken, indicating the loss of a leaving group (Br⁻), while the C–O bond is formed, indicating a nucleophilic attack. This is, in fact, a concerted process in which nucleophilic attack and loss of the leaving group occur in a simultaneous fashion. One curved arrow is required to show the nucleophilic attack, and another curved arrow is required to show loss of the leaving group:

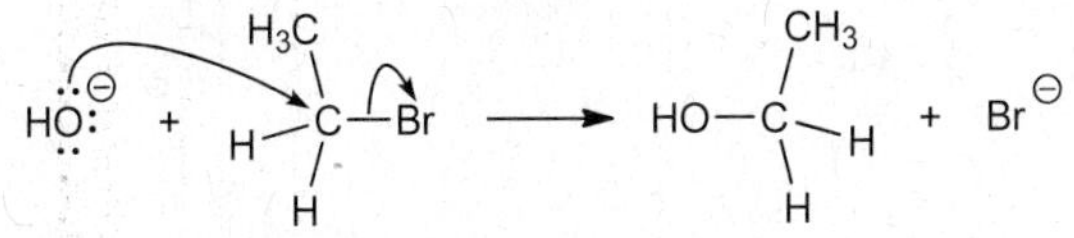

(b) As seen in the solution to the previous problem (6.49a), there are two characteristic arrow-pushing patterns in this case: nucleophilic attack and loss of a leaving group.
(c) We identify the bond broken (CH_3CH_2—Br), and the bond formed (CH_3CH_2—OH). Using the data in Table 6.1, ΔH for this reaction is expected to be approximately (285 kJ/mol) – (381 kJ/mol). The sign of ΔH is therefore predicted to be negative, which means that the reaction should be exothermic.
(d) Two chemical entities are converted into two chemical entities. Both the reactants and products are acyclic. Therefore, ΔS for this process is expected to be approximately zero.
(e) ΔG has two components: (ΔH) and ($-T\Delta S$). Based on the answers to the previous questions, the first term has a negative value and the second term is insignificant. Therefore, ΔG is expected to have a negative value. This is confirmed by the energy diagram, which shows the products having lower free energy than the reactants.
(f) The position of equilibrium is dependent on the sign and value of ΔG. As mentioned in part e, ΔG is comprised of two terms. The effect of temperature appears in the second term ($-T\Delta S$), which is insignificant because ΔS is approximately zero. Therefore, an increase (or decrease) in temperature is not expected to have a significant impact on the position of equilibrium.
(g) This transition state corresponds with the peak of the curve, and has the following structure:

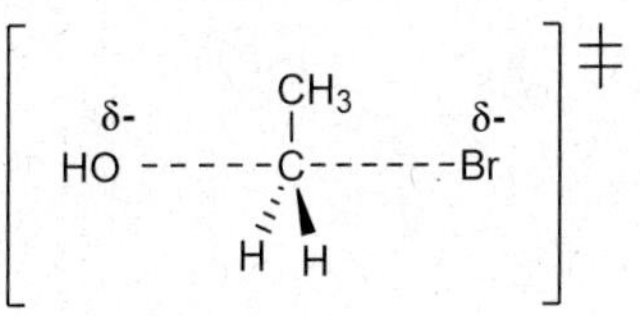

(h) The transition state in this case is closer in energy to the reactants than the products, and therefore, it is closer in structure to the reactants than the products (the Hammond postulate).
(i) The reaction is second order.
(j) According to the rate equation, the rate is linearly dependent on the concentration of hydroxide. Therefore, the rate will be doubled if the concentration of hydroxide is doubled.
(k) Yes, the rate will increase with increasing temperature.

6.50.
(a) K_{eq} does not affect the rate of the reaction. It only affects the equilibrium concentrations.
(b) ΔG does not affect the rate of the reaction. It only affects the equilibrium concentrations.
(c) Temperature does affect the rate of the reaction, by increasing the number of collisions that result in a reaction.
(d) ΔH does not affect the rate of the reaction. It only affects the equilibrium concentrations.
(e) E_a greatly affects the rate of the reaction. Lowering the E_a will increase the rate of reaction.
(f) ΔS does not affect the rate of the reaction. It only affects the equilibrium concentrations.

6.51. In order to determine if reactants or products are favored at high temperature, we must consider the effect of temperature on the sign of ΔG. Recall that ΔG has two components: (ΔH) and ($-T\Delta S$). The reaction is exothermic, so the first term (ΔH) has a negative value, which contributes to a negative value of ΔG. This favors products. At low temperature, the second term will be insignificant and the first term will dominate. Therefore, the process will be thermodynamically favorable, and the reaction will favor the formation of products. However, at high temperature, the second term becomes more significant. In this case, two moles of reactants are converted into one mole of product. Therefore, ΔS for this process is negative, which means that ($-T\Delta S$) is positive. At high enough temperature, the second term ($-T\Delta S$) should dominate over the first term (ΔH), generating a positive value for ΔG. Therefore, the reaction will favor reactants at high temperature.

6.52. Recall that ΔG has two components:

$$(\Delta H) \text{ and } (-T\Delta S)$$

We must analyze each term separately. The first term is expected to have a negative value, because three π bonds are being converted into one π bond and two σ bonds. A σ bond is stronger (lower in energy) than the π component of a double bond (see problems 6.2 and 6.20). Therefore, reaction is expected to release energy to the environment, which means the reaction should be exothermic. In other words, the first term (ΔH) has a negative value, which contributes to a negative value of ΔG. This favors products.
Now let's consider the second term ($-T\Delta S$) contributing to ΔG. In this case, two moles of reactants are converted

into one mole of product. Therefore, ΔS for this process is negative, which means that $(-T\Delta S)$ is positive. At low temperature, the second term will be insignificant and the first term will dominate. Therefore, the process will be thermodynamically favorable, and the reaction will favor the formation of products. However, at high temperature, the second term becomes more significant. At high enough temperature, the second term $(-T\Delta S)$ should dominate over the first term (ΔH), generating a positive value for ΔG. Therefore, the reaction will favor reactants at high temperature.

6.53.   The nitrogen atom of an ammonium ion is positively charged, but that does not render it electrophilic. In order to be electrophilic, it must have an empty orbital that can be attacked by a nucleophile. The nitrogen atom in this case does not have an empty orbital, because nitrogen is a second row element and therefore only has four orbitals with which to form bonds. All four orbitals are being used for bonding, leaving none of the orbitals vacant. As a result, the nitrogen atom is not electrophilic, despite the fact that is positively charged.

In contrast, an iminium ion is resonance stabilized:

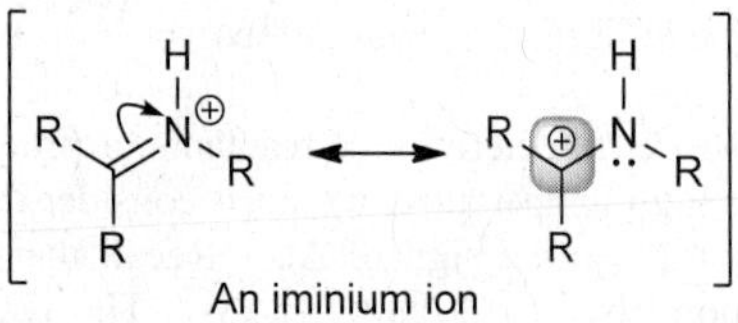

An iminium ion

The second resonance structure exhibits a positive charge on a carbon atom, which serves as an electrophilic center (a carbocation is an empty p orbital). Therefore, an iminium ion is an electrophile and is subject to attack by a nucleophile:

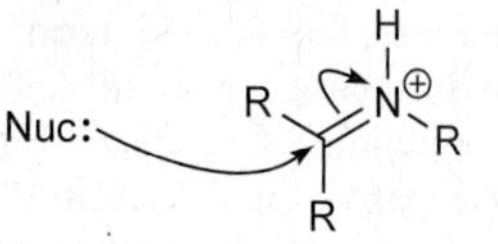

6.54.   The following curved arrows show the flow of electrons that achieve the transformation as shown:

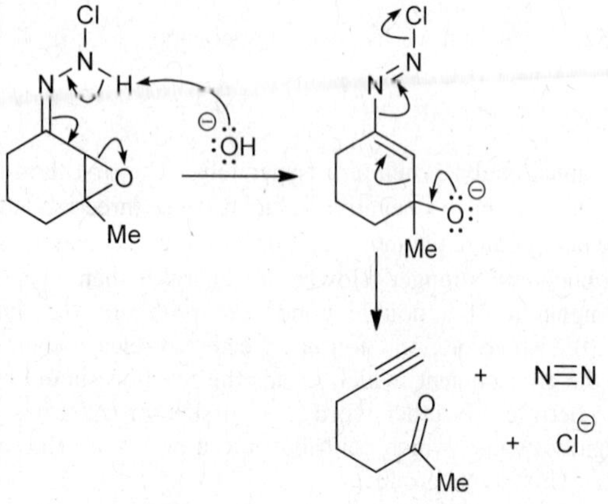

6.55.
The first step ($1 \rightarrow 2$) is a proton transfer, in which MeO⁻ deprotonates **1** to produce **2**, the conjugate base of **1**. The resonance structure of **2** with an anionic oxygen is the greatest contributor to the resonance hybrid due to oxygen being more electronegative than carbon. The transformation of **2** to **3** includes the formation of a C=O π bond, the relocation of two C=C π bonds, and the breaking of a C–O σ bond. The resulting anionic oxygen of **3** then serves as a nucleophile, attacking the indicated carbon, pushing electrons up to the oxygen of the C=O bond, as shown. In the final step (**4 → 5**), the C=O π bond is reformed and a proton is transferred from MeOH to the cyclic structure, regenerating MeO⁻.

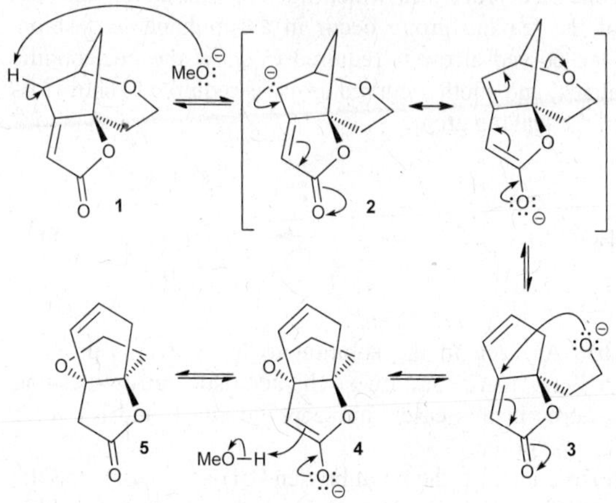

6.56.
In step 1, the hydroxide ion functions as a nucleophile and attacks the carbon atom of the ester group, pushing the π electrons up to the oxygen atom. In step 2, the anionic oxygen atom serves as a base and removes a proton in an intramolecular fashion (proton transfer), as shown. The electrons in the C–H bond form a C=C π bond; the C–O σ bond breaks thus converting the adjacent C–O bond to a double bond, expelling HO⁻ as a leaving group. Step 3 is a proton transfer to create a resonance-stabilized anion.

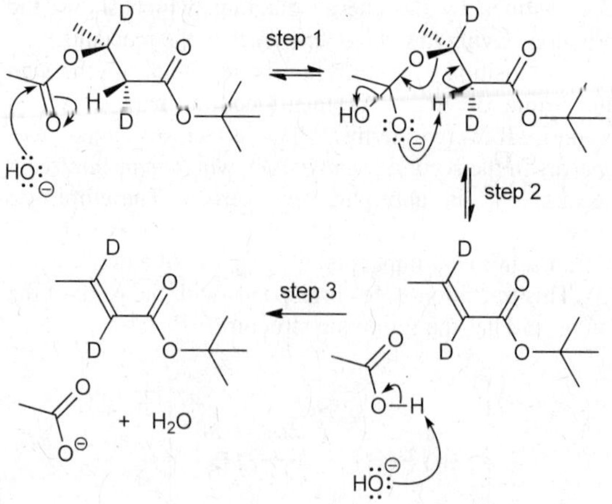

6.57.
The conversion of **1** to **2** involves two steps: First the lone pair on the phosphorus atom functions as a nucleophile and attacks the azide (RN$_3$), followed by loss of a leaving group to give compound **2**. Note that nitrogen gas (N$_2$) is liberated, which renders the conversion from **1** to **2** irreversible. When **2** is treated with **3**, the nitrogen atom of **2** functions as a nucleophile and attacks one of the C=O bonds in **3** to give **5**. Loss of a leaving group gives **6**, which undergoes an intramolecular nucleophilic attack to give **7**. Another intramolecular nucleophilic attack gives **8**, which then undergoes loss of a leaving group to give the product (**4**). The curved arrows are drawn here:

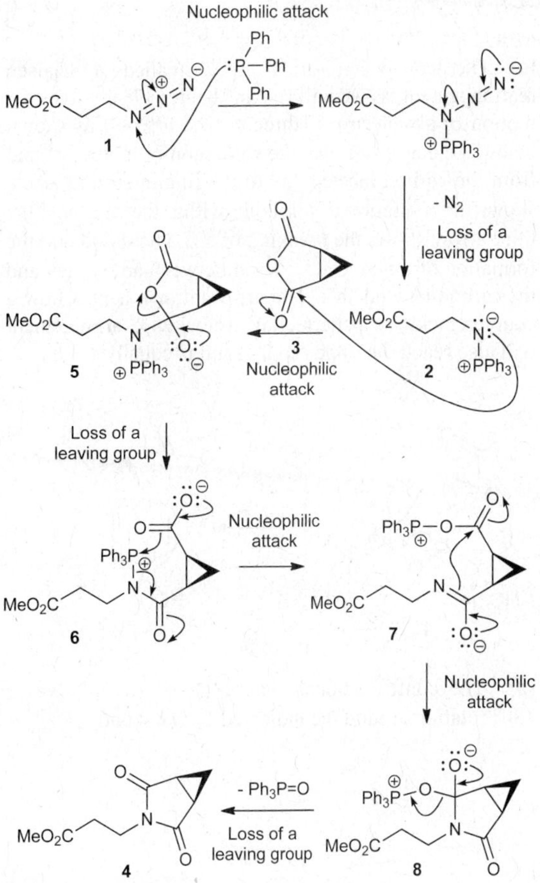

6.58.
(a) The following curved arrows show the flow of electrons that achieve the transformation as shown:

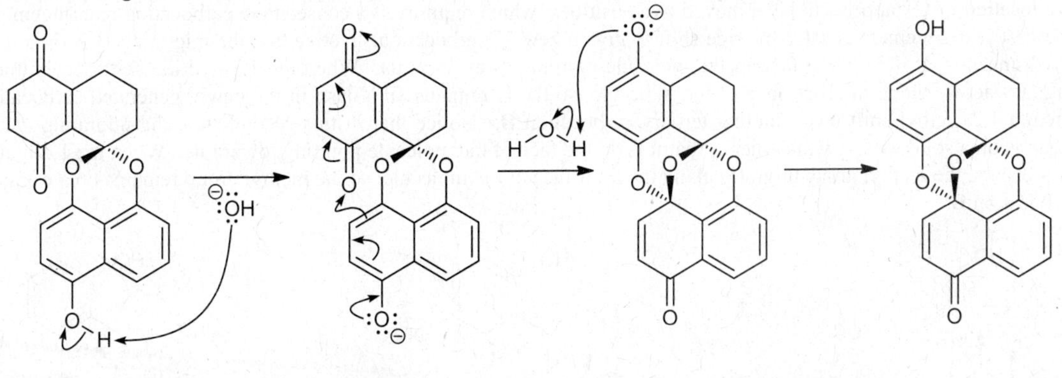

(b) When the second intermediate is redrawn from the following perspective, it becomes clear that the electrons in the carbon-carbon π-bond come from above the carbon-oxygen π bond, allowing for the chirality center to be generated as shown. If you have trouble seeing this, you might find it helpful to build a molecular model.

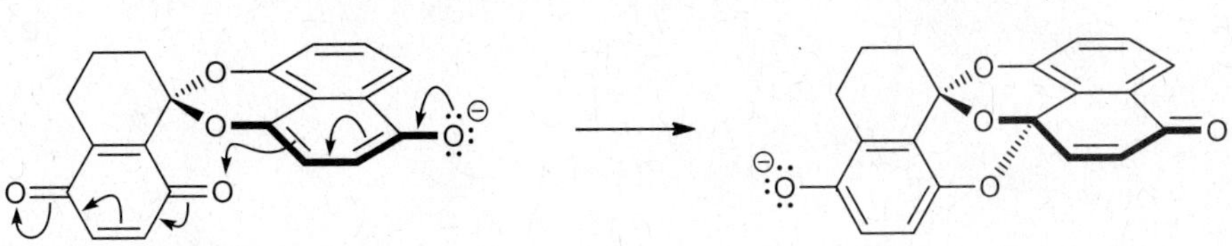

6.59.

(a) The conversion of **1** to **2** (called a Claisen rearrangement, as we will see in Chapter 17) involves the motion of six electrons (three curved arrows) as shown below. One arrow shows the formation of a new σ bond from the carbon labeled "a" to the ring, a second arrow shows the migration of a π bond within the ring, and the third arrow shows the breaking of a C–O σ-bond and the formation of a new C=O π bond between an oxygen and the carbon labeled "b". The arrows drawn here follow a counterclockwise pattern, but a clockwise arrangement will also reach the same product and is equally valid.

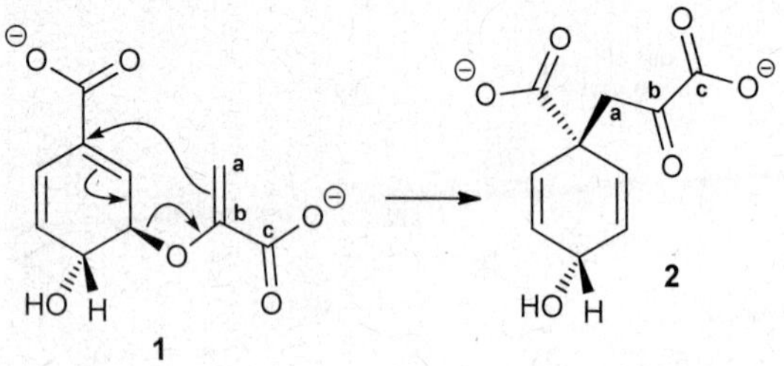

(b) The conformational change (**1** → **1a**) involves a 180° rotation around the indicated C–O σ-bond.

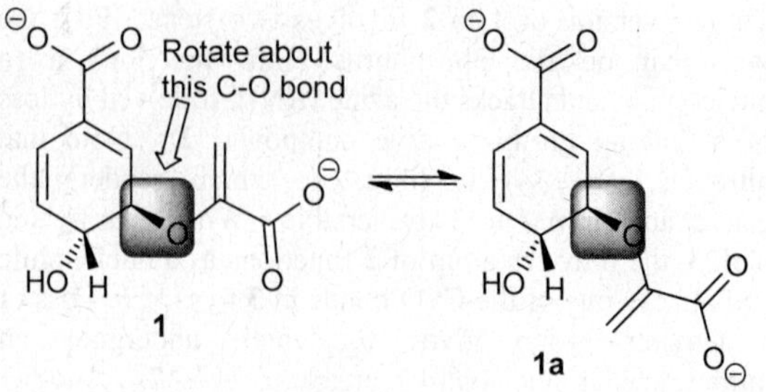

(c) The conversion of **1a** to **3 + 4** is accomplished via a pericyclic reaction called a [1,5] sigmatropic rearrangement (as we will see in Chapter 17), in which six electrons move as shown below. It is similar to the rearrangement shown in part (a), except that this process yields two products. Note that the highlighted H on the ring is transferred to the terminal carbon that ends up being part of molecule **4**.

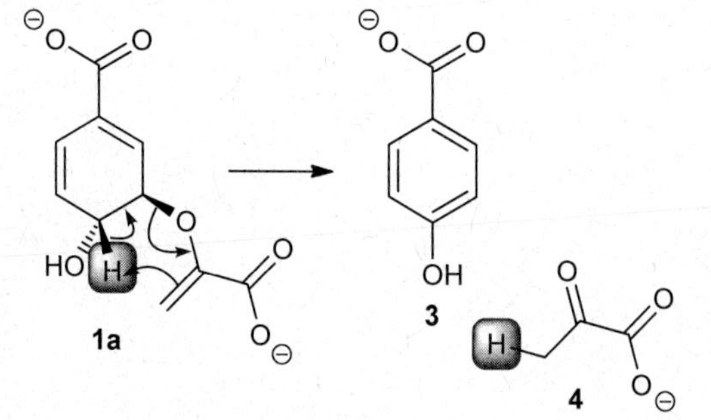

6.60.

The location of C+ appears to have moved two positions, which requires two consecutive carbocation rearrangements. The first rearrangement is a 1,2-hydride shift to give a new 3° carbocation. Notice that the migrating H is on a dash, which means that it is on the face of the molecule pointing away from us. When the 1,2-hydride shift occurs, the H migrates across along the back face of the molecule, so the H remains on a dash in the newly generated carbocation. Then, a 1,2-methyl shift gives another tertiary carbocation **B**. Notice that in this second step, the migrating methyl group is on a solid wedge, which means that it is on the face of the molecule pointing toward us. When the 1,2-methyl shift occurs, the methyl group migrates along the front face of the molecule, so the methyl group remains on a wedge in carbocation **B**.

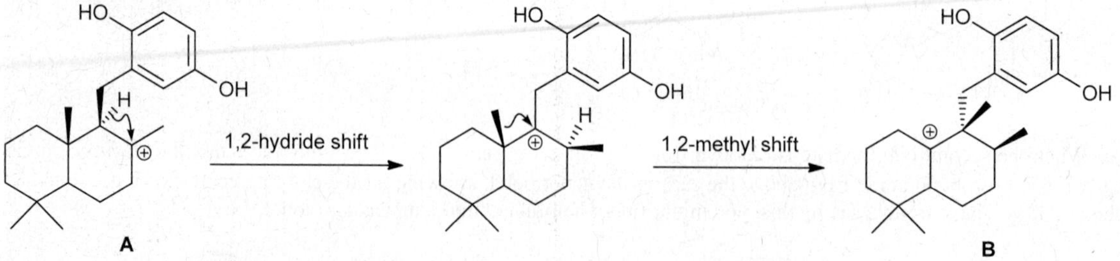

6.61.

(a) The sequence of arrow pushing patterns is shown below, together with all curved arrows:

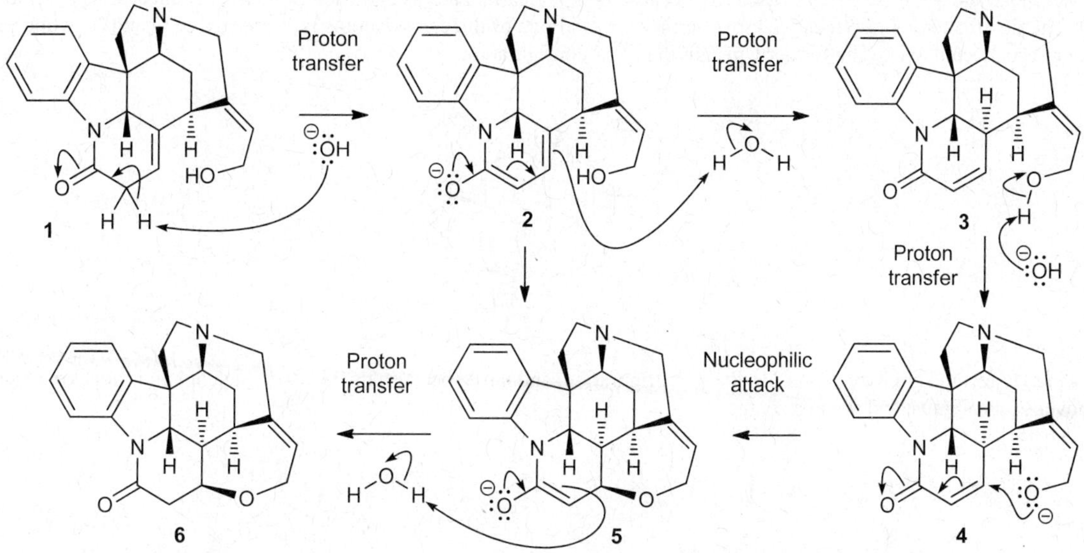

(b) Consider where the nucleophilic attack is occurring. The lone pairs on the oxygen atom (highlighted) are attacking the highlighted carbon atom.

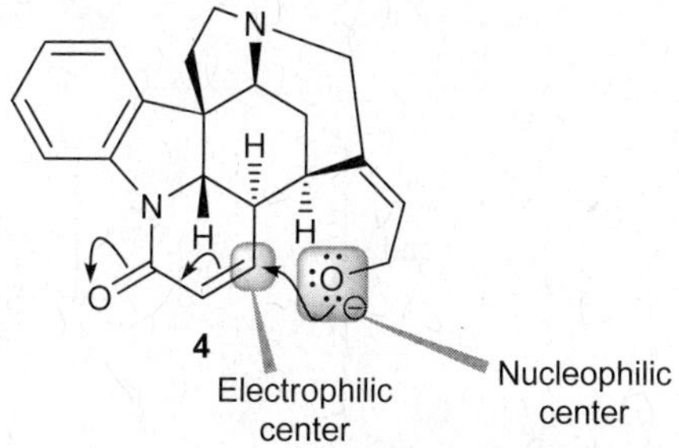

The oxygen atom is the nucleophilic center, and the carbon atom is the electrophilic center. In order to understand why this carbon atom is electron-poor, we draw resonance structures of **4**:

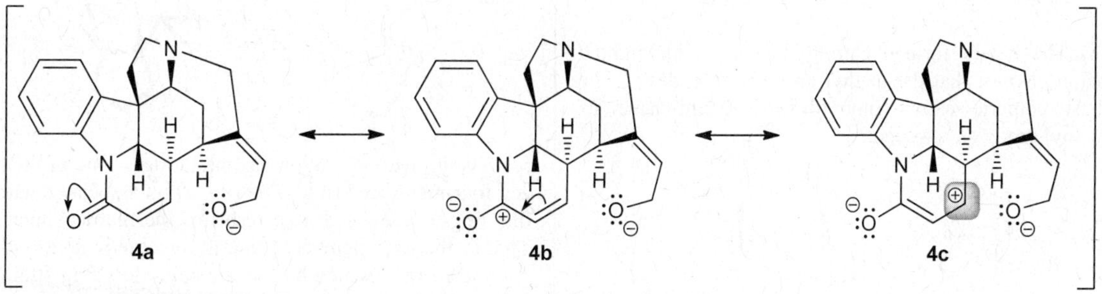

If we inspect resonance structure 4c, we see why the highlighted carbon atom is electrophilic. In fact, when we draw this resonance structure, we can see the nucleophilic attack more clearly:

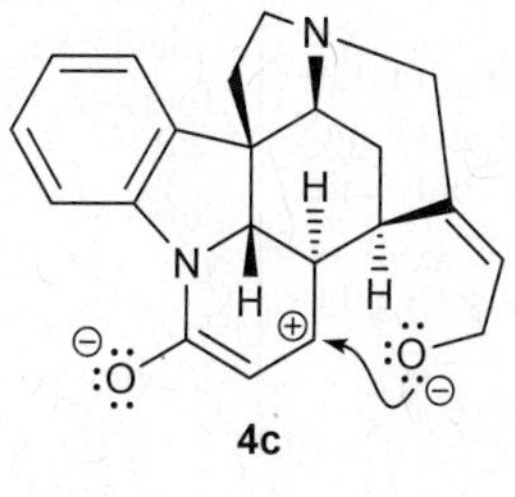

4c

(c) The new chirality center is highlighted below. The oxygen atom takes priority #1, while the H has priority #4. Between #2 and #3, it is difficult to choose because both are carbon atoms and each of them is connected to C, C, and H. The tie breaker comes when we move farther out, and one of the carbon atoms is connected to O and N, while the other is connected to N. The former wins, giving the *R* configuration:

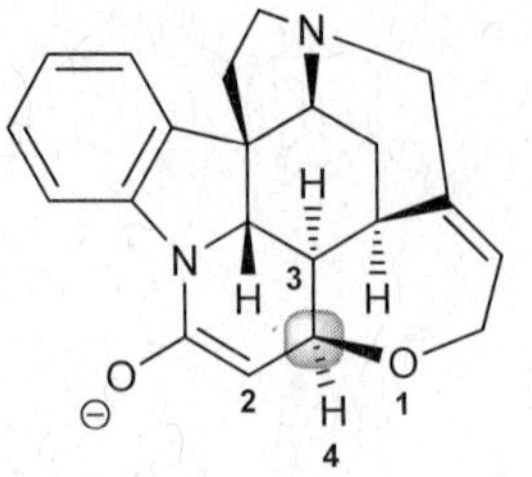

The newly formed chirality center has the *R* configuration, and not *S*, because the pendant nucleophile is attacking from above, so the O must end up on a wedge:

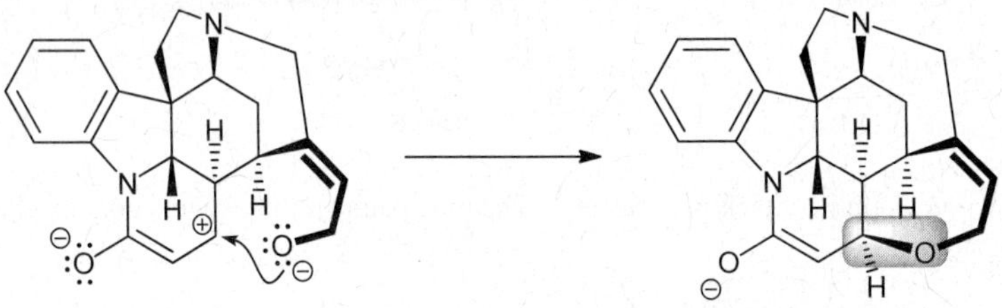

6.62.
(a) The carbon atom highlighted below is migrating, and the following curved arrow shows the migration:

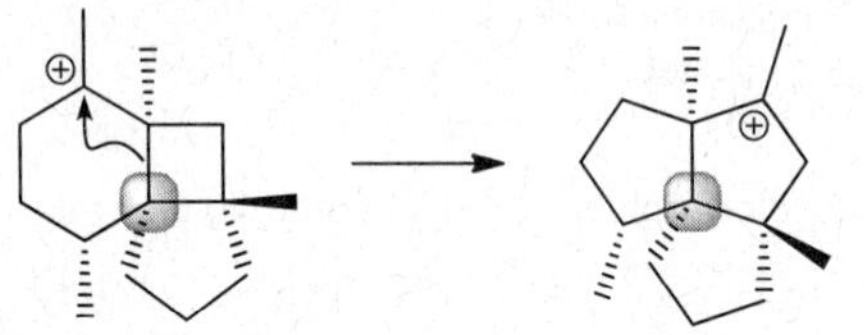

(b) The newly formed chirality center is highlighted below. Notice that the methyl group is on a dash. The diastereomeric cation (not formed) would have its methyl group on a wedge:

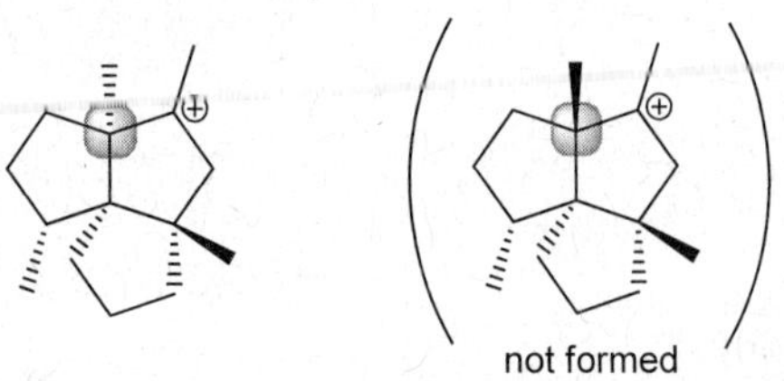

not formed

The other configuration is not formed because of the structural rigidity (and lack of conformational freedom) imposed by the tricyclic system. Specifically, only one face of the empty *p* orbital (associated with C+) is accessible to the migrating carbon atom, as seen in the following scheme:

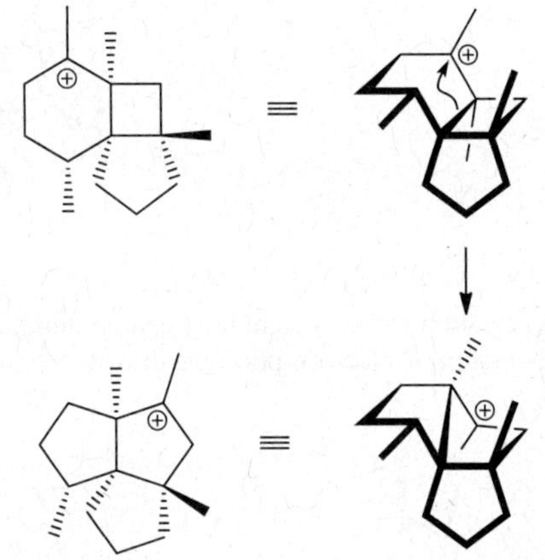

(c) The initial carbocation has three rings. One of them is a four-membered ring. The ring strain associated with this ring is alleviated as a result of the rearrangement. That is, the four-membered ring is converted into a five-membered ring, which has considerably less ring strain. It is true that a six-membered ring (which generally has very little, if any, ring strain) is converted into a five-membered ring, which does possess some ring strain. However, this energy cost is more than off-set by the alleviation of ring strain resulting from enlarging the four-membered ring.

6.63.
(a) The following curved arrows show an intramolecular nucleophilic attack, with the simultaneous loss of a leaving group:

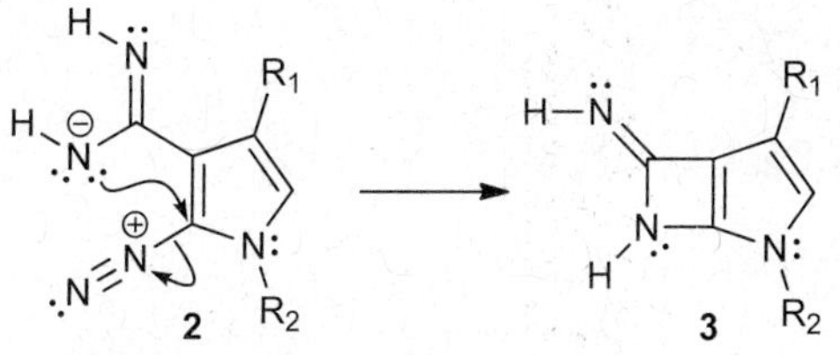

(b) Resonance structures (**2a** and **2b**) can be drawn for intermediate **2**:

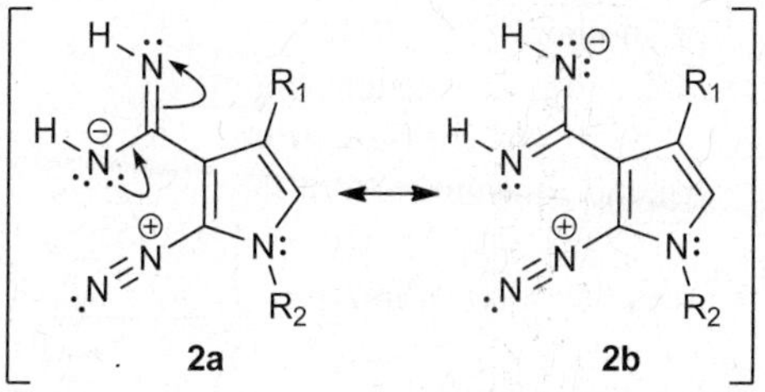

Notice that the negative charge is spread over two nitrogen atoms via resonance. As such, either of these nitrogen atoms can function as the nucleophilic center during the nucleophilic attack. Either the isotopically labeled nitrogen atom can attack, like this:

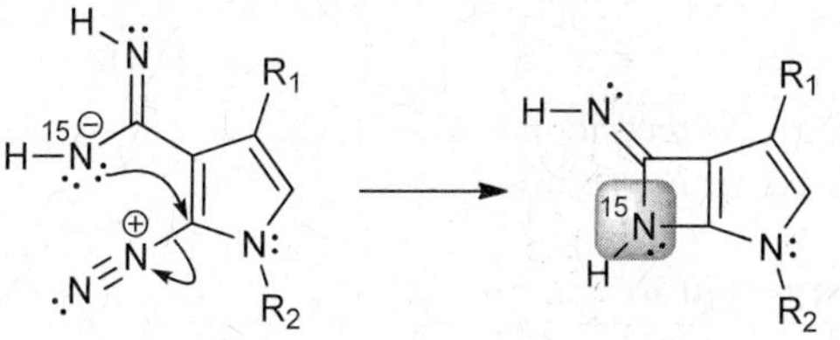

or the other nitrogen atom can attack, like this:

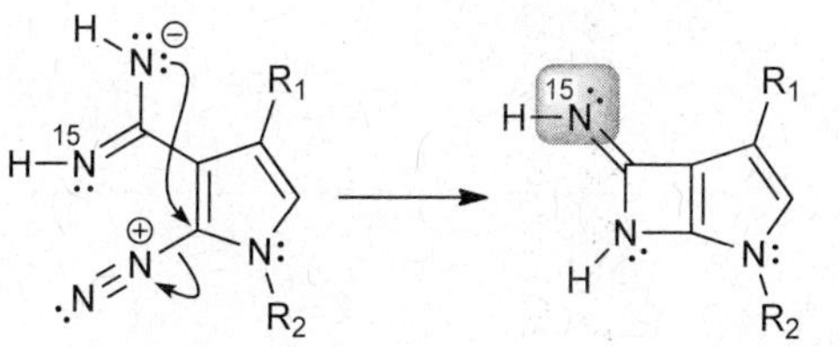

So, if the proposed mechanism were truly operating, we would expect that either nitrogen atom would have an equal probability of forming the four-membered ring, and so only 50% of the ^{15}N atom would be incorporated into the nitrile. That is, both **4a** and **4b** should both be formed. If that had been the case, we would have said that the ^{15}N atom was "scrambled" during the reaction. Since that did not occur, the proposed mechanism was refuted.

6.64.
(a) Compound **1** is converted to intermediate **2** upon loss of a leaving group.

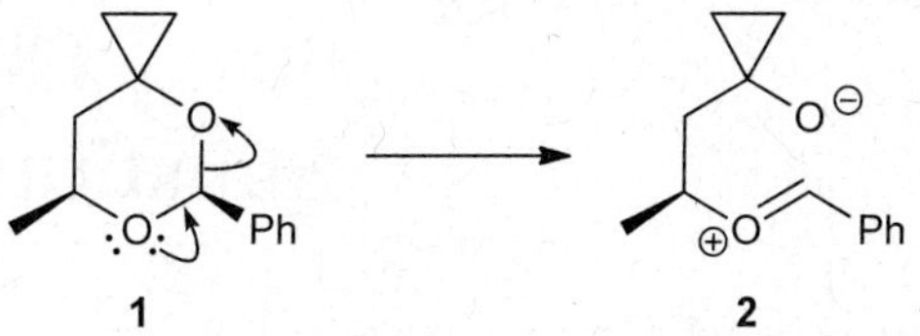

Notice that two curved arrows are employed in this case. The arrow with its tail on a C–O bond represents loss of the leaving group. The other curved arrow (with its tail on a lone pair) can be viewed in two ways: It can be viewed as the electrons coming up from the O to push out the leaving group, as shown above, or it can be viewed as a resonance arrow that allows us to draw resonance structure **2**, thereby bypassing resonance structure **2a**:

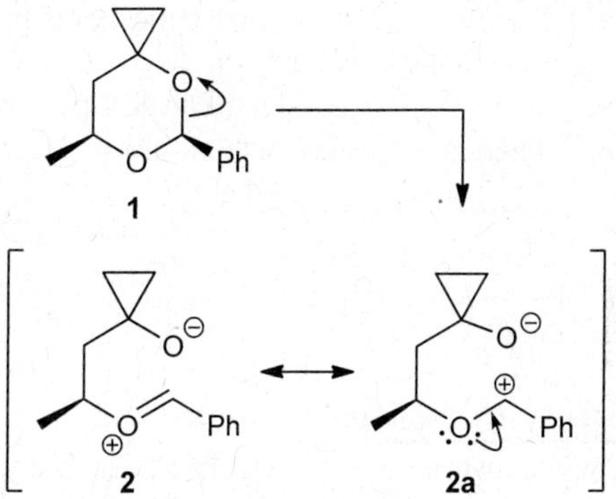

Structure **2** is a greater contributor than **2a** to the overall resonance hybrid because all atoms possess an octet of electrons. That is not the case for **2a** or any of the other resonance structures (not shown) resulting from conjugation with the benzene ring.

(b) In order to draw the curved arrows in this case, it will be helpful if we first rotate about the C–C bond indicated below.

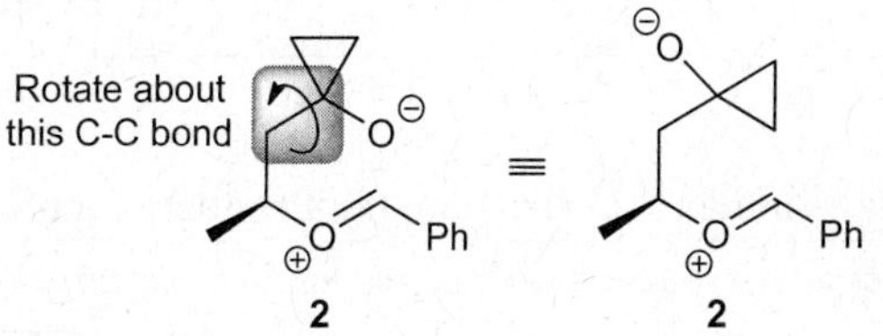

This gives a conformation in which the atoms are all arranged in the proper orientation necessary to show the conversion of **2** into **3**. Three curved arrows are required to show the flow of electrons that correspond with the transformation of **2** into **3**:

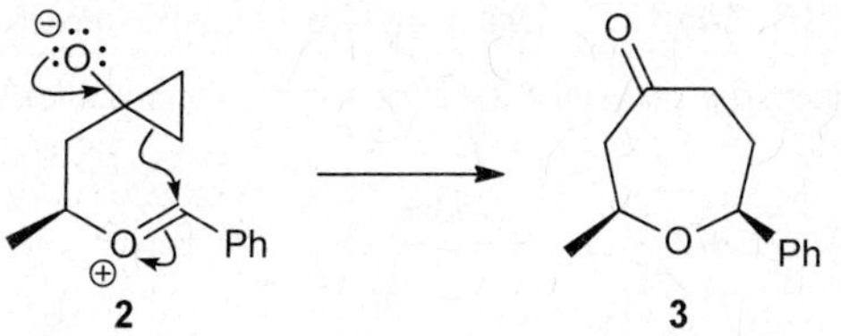

Chapter 7
Substitution Reactions

Review of Concepts

Fill in the blanks below. To verify that your answers are correct, look in your textbook at the end of Chapter 7. Each of the sentences below appears verbatim in the section entitled *Review of Concepts and Vocabulary.*

- Substitution reactions exchange one _____ for another.
- Evidence for the concerted mechanism, called **S_N2**, includes the observation of a _____-**order** rate equation. The reaction proceeds with _____ **of configuration**.
- S_N2 reactions are said to be _____ because the configuration of the product is determined by the configuration of the substrate.
- Evidence for the stepwise mechanism, called **S_N1**, includes the observation of a _____-**order** rate equation.
- The _____ step of an S_N1 process is the **rate-determining step**.
- There are four factors that impact the competition between the S_N2 mechanism and S_N1: 1) the _____, 2) the _____, 3) the _____ _____, and 4) the _____.
- _____ solvents favor S_N2.

Review of Skills

Follow the instructions below. To verify that your answers are correct, look in your textbook at the end of Chapter 7. The answers appear in the section entitled *SkillBuilder Review.*

SkillBuilder 7.1   Drawing the Curved Arrows of a Substitution Reaction

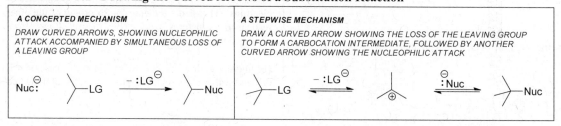

A CONCERTED MECHANISM

DRAW CURVED ARROWS, SHOWING NUCLEOPHILIC ATTACK ACCOMPANIED BY SIMULTANEOUS LOSS OF A LEAVING GROUP

A STEPWISE MECHANISM

DRAW A CURVED ARROW SHOWING THE LOSS OF THE LEAVING GROUP TO FORM A CARBOCATION INTERMEDIATE, FOLLOWED BY ANOTHER CURVED ARROW SHOWING THE NUCLEOPHILIC ATTACK

SkillBuilder 7.2   Drawing the Product of an S_N2 Process

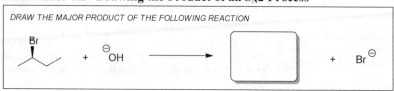

DRAW THE MAJOR PRODUCT OF THE FOLLOWING REACTION

SkillBuilder 7.3   Drawing the Transition State of an S_N2 Process

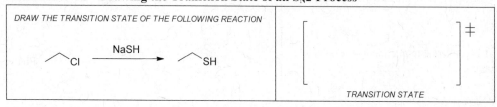

DRAW THE TRANSITION STATE OF THE FOLLOWING REACTION

TRANSITION STATE

SkillBuilder 7.4   Drawing the Carbocation Intermediate of an S$_N$1 Process

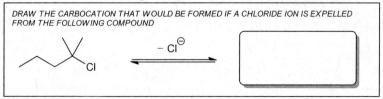

DRAW THE CARBOCATION THAT WOULD BE FORMED IF A CHLORIDE ION IS EXPELLED FROM THE FOLLOWING COMPOUND

$- Cl^{\ominus}$

SkillBuilder 7.5   Drawing the Products of an S$_N$1 Process

PREDICT THE PRODUCTS OF THE FOLLOWING S$_N$1 REACTION

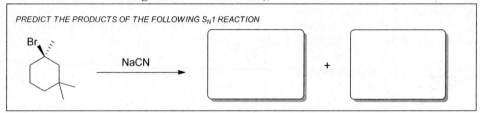

NaCN

+

SkillBuilder 7.6   Drawing the Complete Mechanism of an S$_N$1 Process

IDENTIFY THE TWO CORE STEPS AND THREE POSSIBLE ADDITIONAL STEPS OF AN S$_N$1 PROCESS

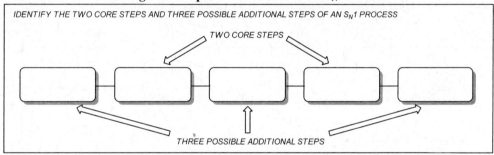

TWO CORE STEPS

THREE POSSIBLE ADDITIONAL STEPS

SkillBuilder 7.7   Drawing the Complete Mechanism of an S$_N$2 Process

IDENTIFY THE ONE CORE STEP (CONCERTED) AND TWO POSSIBLE ADDITIONAL STEPS OF AN S$_N$2 PROCESS

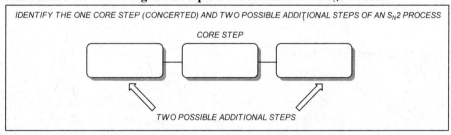

CORE STEP

TWO POSSIBLE ADDITIONAL STEPS

SkillBuilder 7.8   Determining Whether a Reaction Proceeds via an S$_N$1 or S$_N$2 Mechanism

FILL IN THE TABLE BELOW, SHOWING THE FEATURES THAT FAVOR S$_N$2 OR S$_N$1 REACTIONS

	S$_N$2	S$_N$1
SUBSTRATE		
NUC		
LG		
SOLVENT		

SkillBuilder 7.9 Identifying the Reagents Necessary for a Substitution Reaction

IDENTIFY THE REAGENTS NECESSARY TO ACHIEVE THE FOLLOWING TRANSFORMATION

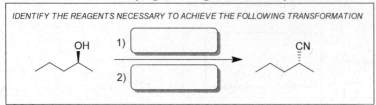

Review of Reactions

Follow the instructions below. To verify that your answers are correct, look in your textbook at the end of Chapter 7. The answers appear in the section entitled *Review of Reactions*.

S_N2

DRAW THE CURVED ARROWS THAT SHOW THE FLOW OF ELECTRON DENSITY DURING THE FOLLOWING S_N2 REACTION

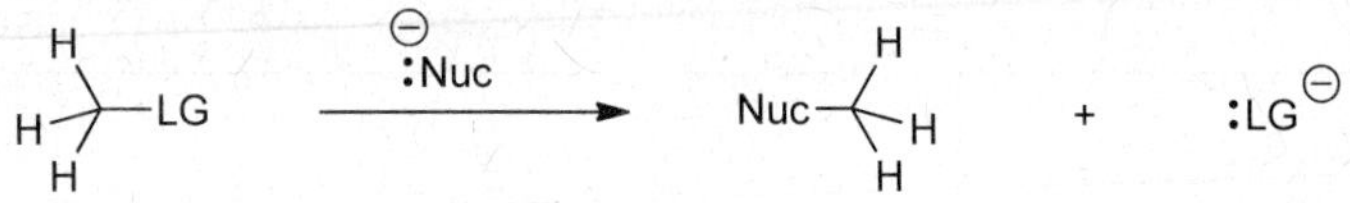

S_N1

DRAW THE CURVED ARROWS THAT SHOW THE FLOW OF ELECTRON DENSITY DURING THE FOLLOWING S_N1 REACTION

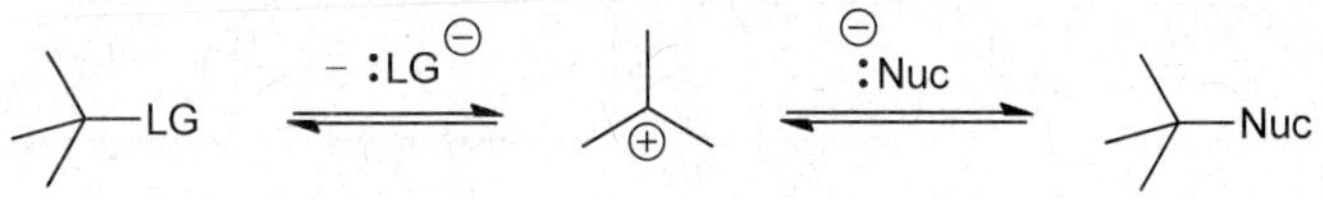

Useful reagents

The following is a list of commonly encountered nucleophilic agents:

Nucleophilic agent	Name	Function
NaCl	Sodium chloride	An ionic salt consisting of Na^+ and Cl^- ions. The former (Na^+) can be ignored in most cases, while the latter (chloride) is a strong nucleophile. NaCl is a source of chloride ions.
NaBr	Sodium bromide	An ionic salt consisting of Na^+ and Br^- ions. The former (Na^+) can be ignored in most cases, while the latter (bromide) is a strong nucleophile. NaBr is a source of bromide ions.
NaI	Sodium iodide	An ionic salt consisting of Na^+ and I^- ions. The former (Na^+) can be ignored in most cases, while the latter (iodide) is a strong nucleophile. NaI is a source of iodide ions.
NaOH	Sodium hydroxide	Hydroxide (HO^-) is a strong nucleophile, used in S_N2 reactions. In Chapter 8, we will see that hydroxide is also a strong base, used in E2 reactions. In many cases, S_N2 and E2 reactions compete with each other, as will be explored in Chapter 8.

NaOR	Sodium alkoxide	R is an alkyl group. Examples include sodium methoxide (NaOMe) and sodium ethoxide (NaOEt). Alkoxide ions are strong nucleophiles. In Chapter 8, we will see that alkoxide ions are also strong bases, used in E2 reactions. In many cases, S_N2 and E2 reactions compete with each other, as will be explored in Chapter 8.
NaSH	Sodium hydrosulfide	HS^- is a strong nucleophile, used in S_N2 reactions.
H_2O	Water	Water is a weak nucleophile, used in S_N1 reactions. In Chapter 8, we will see that water is also a weak base, used in E1 reactions. In most cases, S_N1 and E1 reactions compete with each other, as will be explored in Chapter 8.
ROH	An alcohol	Examples include methanol (CH_3OH) and ethanol (CH_3CH_2OH)
HX (X = Cl, Br, or I)	Hydrogen halides	A strong acid that serves as both a source of H^+ and nucleophilic X^- where X = Cl, Br, or I.

Common Mistakes to Avoid

When drawing the mechanism of a reaction, you must always consider what reagents are being used, and your mechanism must be consistent with the conditions employed. As an example, consider the following S_N1 reaction:

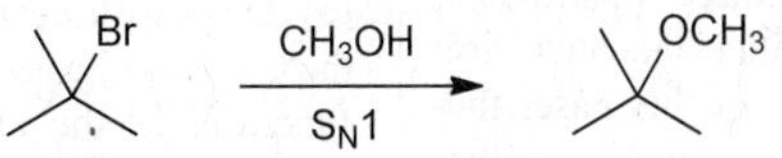

The following proposed mechanism is unacceptable, because the reagent employed in the second step is not present:

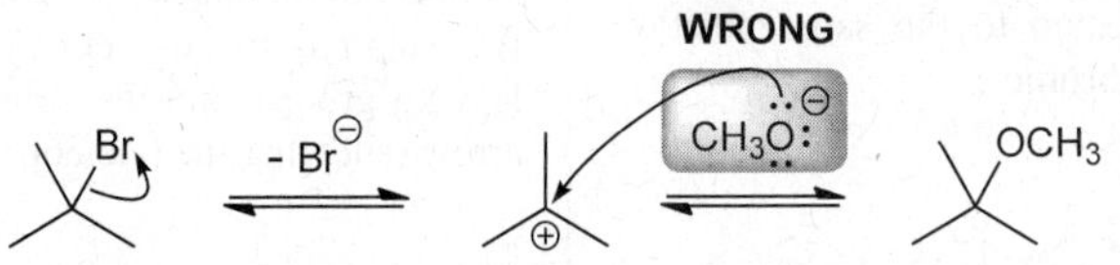

This is a common student error. To see what's wrong, let's look closely at the reagent. Methanol (CH_3OH) is not a strong acid. Rather, it is a weak acid, because its conjugate base, methoxide (CH_3O^-), is a strong base. Therefore, methoxide is not present in substantial quantities, so the mechanism in this case should not employ methoxide. Below is the correct mechanism:

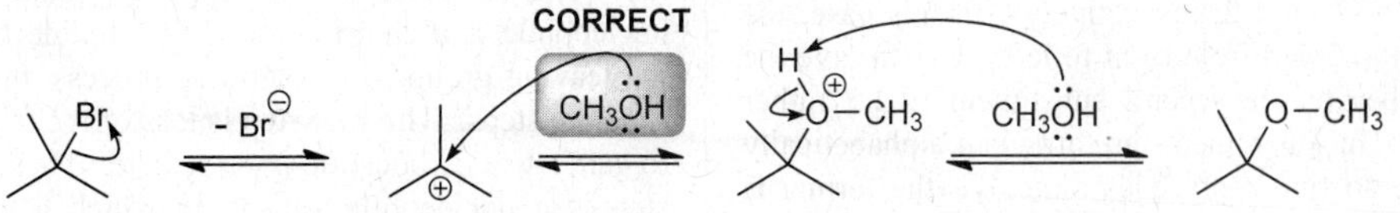

Methanol (rather than methoxide) functions as the nucleophile in the second step, because methoxide was not indicated as a reagent, and it is not expected to be present. The result of the nucleophilic attack is an oxonium ion (an intermediate with a positive charge on an oxygen atom), which is then deprotonated by another molecule of methanol. Once again, in this final step of the mechanism, methanol functions as the base, rather than methoxide, because the latter is not present.

This example is just one illustration of the importance of analyzing the reagent and considering what entities can be used in your mechanism. This will become increasingly important in upcoming chapters.

Solutions

7.1.

(a) The parent is the longest chain, which is seven carbon atoms in this case (heptane). There are two substituents (ethyl and chloro), both of which are located at the C4 position. This will be the case whether we number the parent from left to right or from right to left. Substituents are alphabetized in the name (chloro precedes ethyl).

4-chloro-4-ethylheptane

(b) The parent is a six-membered ring (cyclohexane). There are two substituents (methyl and bromo), both of which are located at the C1 position. Substituents are alphabetized in the name (bromo precedes methyl).

1-bromo-1-methylcyclohexane

(c) The parent is the longest chain, which is five carbon atoms in this case (pentane). There are three substituents (bromo, bromo, and chloro), and their locants are assigned as 4, 4, and 1, respectively. In this case, the parent was numbered from right to left, so as to give the lowest number to the first substituent (**1**,4,4 rather than **2**,2,5). Notice that two locants are necessary (rather than one) to indicate the locations of the two bromine atoms, even though they are connected to the same position (4,4-dibromo rather than 4-dibromo).

4,4-dibromo-1-chloropentane

(d) The parent is the longest chain, which is six carbon atoms in this case (hexane). There are three substituents (fluoro, methyl, and methyl), and their locants are assigned as 5, 2, and 2, respectively. In this case, the parent was numbered from right to left, so as to give the lowest number to the second substituent (2,**2**,5 rather than 2,**5**,5). The substituents are arranged alphabetically in the name, so fluoro precedes dimethyl (the former is "f" and the latter is "m"). In this case, there is also a chirality center, so we must assign the configuration (*S*), which must be indicated at the beginning of the name.

(*S*)-5-fluoro-2,2-dimethylhexane

7.2.

(a) The substrate is 2-bromopropane, and the nucleophile is HS⁻. Bromide functions as the leaving group. In a concerted process, nucleophilic attack and loss of the leaving group occur in a simultaneous fashion (in one step). Two curved arrows are required. One curved arrow shows the nucleophilic attack, and the other curved arrow shows loss of the leaving group.

(b) The substrate is 1-iodopropane, and the nucleophile is methoxide (MeO⁻). Iodide functions as the leaving group. In a concerted process, nucleophilic attack and loss of the leaving group occur in a simultaneous fashion (in one step). Two curved arrows are required. One curved arrow shows the nucleophilic attack, and the other curved arrow shows loss of the leaving group.

7.3.

(a) The substrate is 1-bromo-1-methylcyclohexane, and the nucleophile is an acetate ion (CH₃CO₂⁻). Bromide functions as the leaving group. In a stepwise process, there are two separate steps. The first step is loss of a leaving group to generate a carbocation intermediate. Then, the second step is a nucleophilic attack, in which an acetate ion attacks the carbocation intermediate. The first step requires one curved arrow (showing loss of the leaving group), and the second step requires one curved arrow (showing the nucleophilic attack).

(b) The substrate is 2-iodo-2-methylbutane, and the nucleophile is a chloride ion (Cl⁻). Iodide functions as the leaving group. In a stepwise process, there are two separate steps. The first step is loss of the leaving group to generate a carbocation intermediate. Then, the second step is a nucleophilic attack, in which a chloride ion attacks the carbocation intermediate. The first step requires one curved arrow (showing loss of the leaving group), and the second step requires one curved arrow (showing the nucleophilic attack).

7.4. In a concerted process, nucleophilic attack and loss of the leaving group occur in a simultaneous fashion (in one step). Since the nucleophilic center and the electrophilic center are tethered to each other, the reaction occurs in an intramolecular fashion, as shown below. Two curved arrows are required. One curved arrow shows the nucleophilic attack, and the other curved arrow shows loss of the leaving group (bromide).

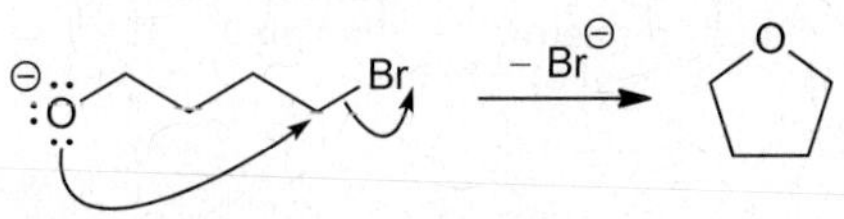

7.5. In a stepwise process, the first step is loss of a leaving group to generate a carbocation intermediate. In this case, the carbocation intermediate is resonance-stabilized, so we draw both resonance structures. Nucleophilic attack (at the electrophilic position indicated in the second resonance structure) affords the observed product.

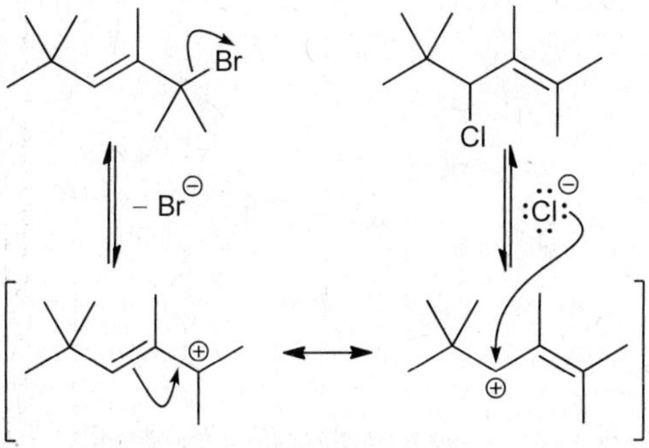

7.6.
(a) The reaction has a second-order rate equation, which means that the rate should be linearly dependent on the concentrations of two compounds (the nucleophile AND the substrate). If the concentration of the substrate is tripled, the rate should also be tripled.
(b) As described above, the rate is linearly dependent on the concentrations of both the nucleophile and the substrate. If the concentration of the nucleophile is doubled, the rate of the reaction is doubled.
(c) As described above, the rate is linearly dependent on the concentrations of both the nucleophile and the substrate. If the concentration of the substrate is doubled and the concentration of the nucleophile is tripled, then the rate of the reaction will be six times faster ($\times 2 \times 3$).

7.7.
(a) The substrate is (S)-2-chloropentane, and the nucleophile is HS⁻. Chloride is ejected as a leaving group, with inversion of configuration.

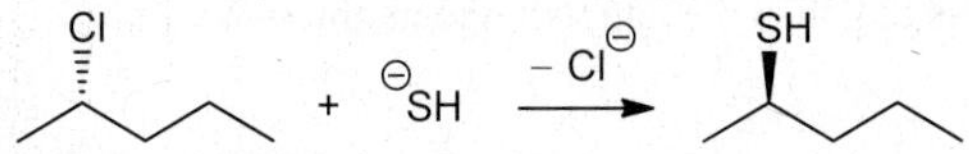

(S)-2-chloropentane

(b) The substrate is (R)-3-iodohexane, and the nucleophile is chloride (Cl⁻). Iodide is ejected as a leaving group, with inversion of configuration.

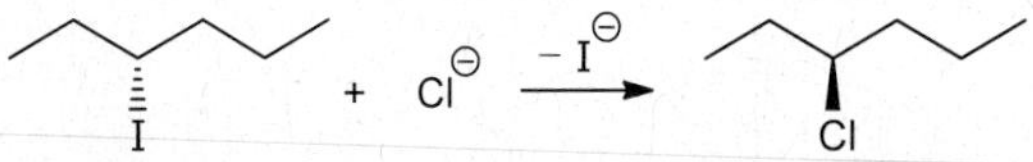

(R)-3-iodohexane

(c) The substrate is (R)-2-bromohexane, and the nucleophile is hydroxide (HO⁻). Bromide is ejected as a leaving group, with inversion of configuration.

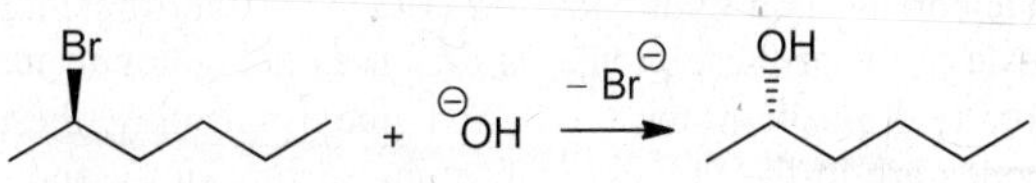

(R)-2-bromohexane

7.8. The reaction does proceed with inversion of configuration. However, the Cahn-Ingold-Prelog system for assigning a stereodescriptor (R or S) is based on a prioritization scheme. Specifically, the four groups connected to a chirality center are ranked (one through four). In the reactant, the highest priority group is the leaving group (bromide) which is then replaced by a group that does not receive the highest priority. In the product, the fluorine atom has been promoted to the highest priority as a result of the reaction, and as such, the prioritization scheme has changed. In this way, the stereodescriptor (S) remains unchanged, despite the fact that the chirality center undergoes inversion.

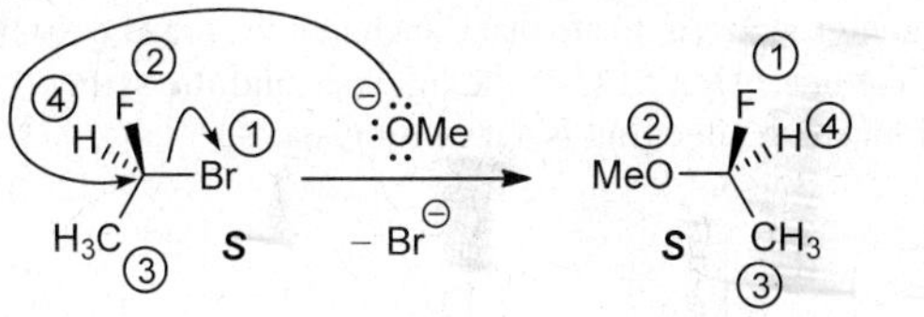

7.9.
(a) The leaving group is a bromide ion (Br⁻) and the nucleophile is a hydroxide ion (HO⁻). In the transition state, each of these groups is drawn as being connected to the α position with a dotted line (indicating these bonds are in the process of forming or breaking), and a δ– is placed on each group to indicate that the charge is spread over both locations. Don't forget the brackets and the symbol that indicate the drawing is a transition state.

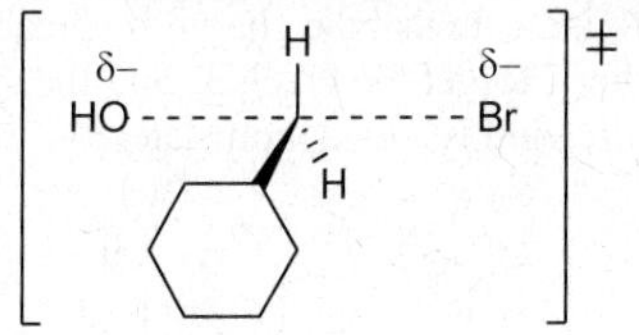

(b) The leaving group is an iodide ion (I⁻) and the nucleophile is an acetate ion (CH₃CO₂⁻). In the transition state, each of these groups is drawn as being connected to the α position with a dotted line (indicating these bonds are in the process of forming or breaking), and a δ– is placed on each group to indicate that the charge is spread over both locations. Don't forget the brackets and the symbol that indicate the drawing is a transition state.

(c) The leaving group is a chloride ion (Cl⁻) and the nucleophile is a hydroxide ion (HO⁻). In the transition state, each of these groups is drawn as being connected to the α position with a dotted line (indicating these bonds are in the process of forming or breaking), and a δ– is placed on each group to indicate that the charge is spread over both locations. Don't forget the brackets and the symbol that indicate the drawing is a transition state.

(d) The leaving group is a bromide ion (Br⁻) and the nucleophile is HS⁻. In the transition state, each of these groups is drawn as being connected to the α position with a dotted line (indicating these bonds are in the process of forming or breaking), and a δ– is placed on each group to indicate that the charge is spread over both locations. Don't forget the brackets and the symbol that indicate the drawing is a transition state.

7.10. The leaving group is a bromide ion (Br⁻) and the nucleophilic center is the oxygen atom bearing the negative charge. In the transition state, each of these groups is drawn as being connected to the α position with a dotted line (indicating these bonds are in the process of forming or breaking), and a δ– is placed on each group to indicate that the charge is spread over both locations. Don't forget the brackets and the symbol that indicate the drawing is a transition state.

7.11. This step is favorable (downhill in energy) because ring strain is alleviated when the three-membered ring is opened.

7.12.
(a) The lone pair on the sp^3 hybridized nitrogen atom functions as a nucleophilic center and attacks the electrophilic methyl group in SAM, forming an ammonium ion which loses a proton to give the product.

(b) The nitrogen atom functions as a nucleophilic center and attacks the electrophilic methyl group in SAM, forming an ammonium ion.

7.13.
(a) The reaction occurs via an S_N1 pathway, which means that the rate is not dependent on the concentration of the nucleophile. The rate is only dependent on the concentration of the substrate. If the concentration of substrate is doubled, the rate will be doubled (the change in concentration of the nucleophile will not affect the rate).
(b) The reaction occurs via an S_N1 pathway, which means that the rate is not dependent on the concentration of the nucleophile. If that is the only factor that is changed, then the rate will remain the same.

7.14. In each case, the bond between the α position and the leaving group is broken, and the carbon atom obtains a positive charge.

(a)   (b)

(c)   (d)

7.15. The first compound will generate a tertiary carbocation, while the second compound will generate a tertiary benzylic carbocation that is resonance stabilized. The second compound leads to a more stable carbocation, so the second compound will lose its leaving group more rapidly than the first compound.

7.16.
(a) The leaving group is iodide, and the nucleophile is chloride. The former is replaced by the latter, giving the following products.

+ NaI

(b) The leaving group is bromide, and the nucleophile is HS⁻. The former is replaced by the latter. In this case, the reaction is taking place at a chirality center, so we expect that both enantiomers will be produced (as expected for an S_N1 process).

+

+ Br⁻

(c) The leaving group is chloride, and the nucleophile is an acetate ion ($CH_3CO_2^-$). The former is replaced by the latter. In this case, the reaction is taking place at a chirality center, so we expect that both enantiomers will be produced (as expected for an S_N1 process).

+

+ Cl⁻

7.17. The leaving group is bromide, and the nucleophile is HS⁻. The former is replaced by the latter. In this case, the reaction is taking place at one of the two chirality centers that are present in the compound. The other

chirality center remains unchanged. This gives the following two stereoisomers.

+

Notice that in each stereoisomer, the chirality center on the right side has the *S* configuration, as it did in the starting material. The difference between these two stereoisomers is the configuration of the other chirality center (where the reaction took place). These compounds are stereoisomers that are not mirror images of each other, so they are diastereomers.

7.18.
(a) The substrate is an alkyl iodide, which has an excellent leaving group (iodide), so an S_N1 process will not require a proton transfer at the beginning of the mechanism.
(b) The substrate is an alcohol (ROH), which does not have a good leaving group. The OH group must be protonated to function as a leaving group. So, an S_N1 process will require a proton transfer at the beginning of the mechanism.
(c) The substrate is an alkyl bromide, which has an excellent leaving group (bromide), so an S_N1 process will not require a proton transfer at the beginning of the mechanism.
(d) The substrate is an alcohol (ROH), which does not have a good leaving group. The OH group must be protonated to function as a leaving group. So, an S_N1 process will require a proton transfer at the beginning of the mechanism.
(e) The substrate is an alcohol (ROH), which does not have a good leaving group. The OH group must be protonated to function as a leaving group. So, an S_N1 process will require a proton transfer at the beginning of the mechanism.
(f) The substrate is an alkyl chloride, which has an excellent leaving group (chloride), so an S_N1 process will not require a proton transfer at the beginning of the mechanism.

7.19.
(a) The nucleophile (HS⁻) is negatively charged, so a proton transfer will not be necessary at the end of the mechanism.
(b) The nucleophile (H_2S) is neutral, so nucleophilic attack will produce a positively charged species. Removal of the positive charge will require a proton transfer at the end of the mechanism.
(c) The nucleophile (H_2O) is neutral, so nucleophilic attack will produce a positively charged species. Removal of the positive charge will require a proton transfer at the end of the mechanism.
(d) The nucleophile (EtOH) is neutral, so nucleophilic attack will produce a positively charged species. Removal of the positive charge will require a proton transfer at the end of the mechanism.

(e) The nucleophile (N≡C⁻) is negatively charged, so a proton transfer will not be necessary at the end of the mechanism.

(f) The nucleophile (Cl⁻) is negatively charged, so a proton transfer will not be necessary at the end of the mechanism.

(g) The nucleophile (H₂N⁻) is negatively charged, so a proton transfer will not be necessary at the end of the mechanism.

(h) The nucleophile (NH₃) is neutral, so nucleophilic attack will produce a positively charged species. Removal of the positive charge will require a proton transfer at the end of the mechanism.

(i) The nucleophile (MeO⁻) is negatively charged, so a proton transfer will not be necessary at the end of the mechanism.

(j) The nucleophile (EtO⁻) is negatively charged, so a proton transfer will not be necessary at the end of the mechanism.

(k) The nucleophile (MeOH) is neutral, so nucleophilic attack will produce a positively charged species. Removal of the positive charge will require a proton transfer at the end of the mechanism.

(l) The nucleophile (Br⁻) is negatively charged, so a proton transfer will not be necessary at the end of the mechanism.

7.20.
(a) Loss of a leaving group results in a tertiary carbocation, which cannot rearrange to become more stable. So a carbocation rearrangement will not occur in this case.

(b) In order for this compound to function as a substrate in an S$_N$1 reaction, acidic conditions will be necessary (to protonate the OH group, rendering it a better leaving group). Then, loss of the leaving group will result in a secondary carbocation. This carbocation can rearrange via a methyl shift to produce a more stable, tertiary carbocation. So we do expect that a rearrangement can and will occur.

(c) In order for this compound to function as a substrate in an S$_N$1 reaction, acidic conditions will be necessary (to protonate the OH group, rendering it a better leaving group). Then, loss of the leaving group will result in a secondary carbocation. This carbocation can rearrange via a hydride shift to produce a more stable, tertiary carbocation. So we do expect that a rearrangement can and will occur.

(d) In order for this compound to function as a substrate in an S$_N$1 reaction, acidic conditions will be necessary (to protonate the OH group, rendering it a better leaving group). Then, loss of the leaving group will result in a secondary carbocation. In this case, there is no possible rearrangement that will lead to a more stable, tertiary carbocation. As such, we expect no rearrangement will occur.

(e) Loss of a leaving group results in a tertiary carbocation, which cannot rearrange to become more stable. So a carbocation rearrangement will not occur in this case.

(f) Loss of a leaving group results in a secondary carbocation. In this case, there is no possible rearrangement that will lead to a more stable, tertiary carbocation. As such, we expect no rearrangement will occur.

7.21.
(a) The substrate is an alcohol, so acidic conditions are employed so that the OH group can be protonated, rendering it a better leaving group. Then, loss of a leaving group generates a carbocation, which is then captured by a bromide ion to give the product. Notice that the mechanism is comprised of a proton transfer, followed by the two core steps of an S$_N$1 process (loss of a leaving group and nucleophilic attack).

(b) The substrate is an alcohol, so acidic conditions are employed so that the OH group can be protonated, rendering it a better leaving group. Then, loss of a leaving group generates a secondary carbocation, which undergoes a hydride shift to give a more stable, tertiary carbocation. This carbocation is then captured by a bromide ion to give the product.

(c) The leaving group is bromide. Loss of the leaving group generates a tertiary carbocation, which is captured by a water molecule to generate an oxonium ion. Deprotonation of the oxonium ion gives the product. Notice that the mechanism is comprised of the two core steps of an S_N1 process (loss of a leaving group and nucleophilic attack), followed by a proton transfer step.

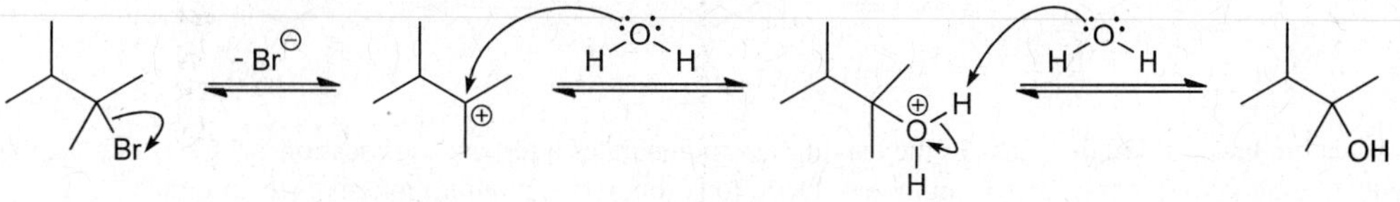

(d) The leaving group is iodide. Loss of the leaving group generates a secondary carbocation, which then undergoes a methyl shift to give a more stable, tertiary carbocation. This carbocation is then captured by a molecule of ethanol to generate an oxonium ion. Deprotonation of the oxonium ion gives the product.

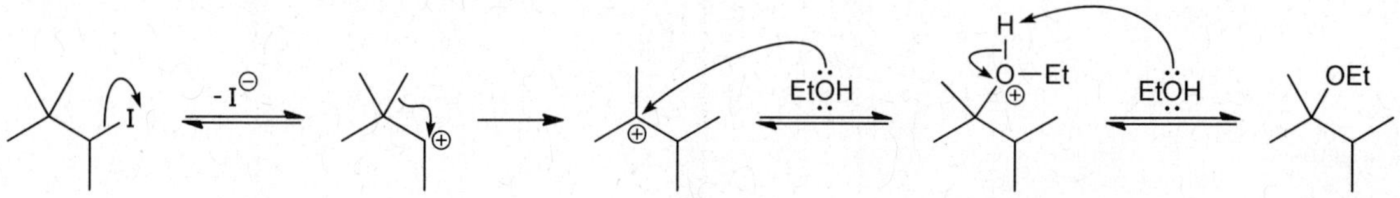

(e) The substrate is an alcohol, so acidic conditions are employed so that the OH group can be protonated, rendering it a better leaving group. The likely source of the proton is $MeOH_2^+$, which received its proton from sulfuric acid (a comparison of the pK_a values for $MeOH_2^+$ and H_2SO_4 indicates that there is likely very little H_2SO_4 present at equilibrium). Loss of a leaving group generates a tertiary carbocation, which is captured by a molecule of methanol to give an oxonium ion. Deprotonation of the oxonium ion gives the product.

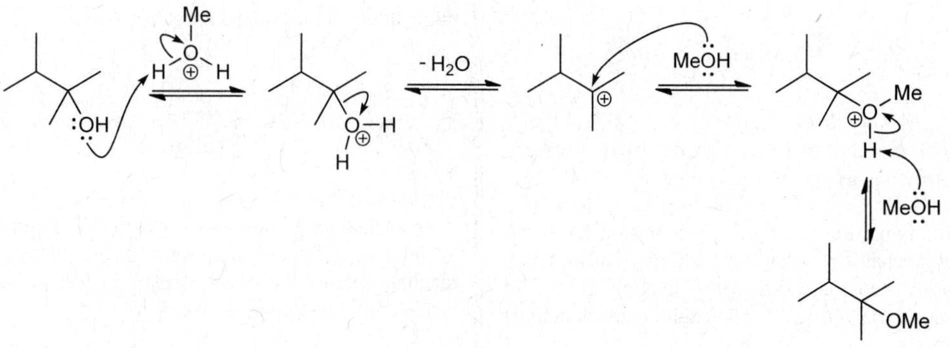

(f) The substrate is an alcohol, so acidic conditions are employed so that the OH group can be protonated, rendering it a better leaving group (as in the previous problem, the likely source of the proton is $MeOH_2^+$, which received its proton from sulfuric acid). Then, loss of a leaving group generates a secondary carbocation, which undergoes a methyl shift to give a more stable, tertiary carbocation. This carbocation is then captured by a molecule of methanol to give an oxonium ion. Deprotonation of the oxonium ion gives the product.

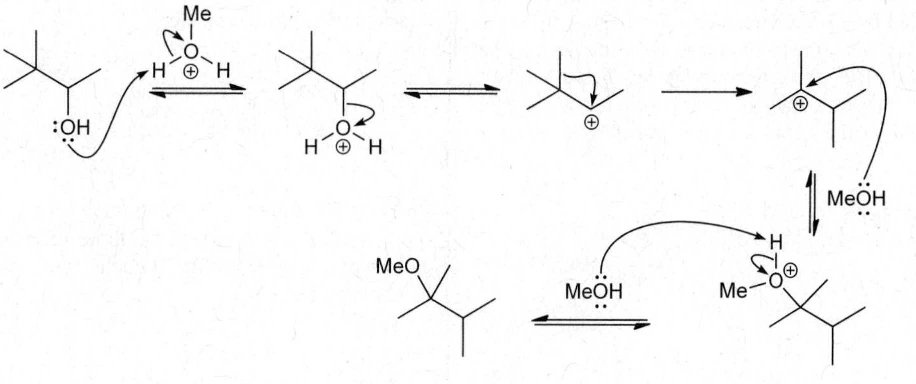

(g) The leaving group is bromide. Loss of the leaving group generates a tertiary carbocation, which is then captured by the nucleophile (HS⁻) to give the product.

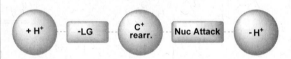

(h) The leaving group is iodide. Loss of the leaving group generates a tertiary carbocation, which is then captured by a molecule of ethanol to generate an oxonium ion. Deprotonation of the oxonium ion gives the product.

7.22.

(c) In problem 7.21c, we saw that the two core steps (loss of leaving group and nucleophilic attack) are accompanied by a proton transfer at the end of the mechanism. Therefore, the mechanism has a total of three steps: 1) loss of a leaving group, 2) nucleophilic attack, and 3) proton transfer.

(d) In problem 7.21d, we saw that the two core steps (loss of leaving group and nucleophilic attack) are accompanied by a carbocation rearrangement (in between the two core steps) and a proton transfer at the end of the mechanism. Therefore, the mechanism has a total of four steps: 1) loss of a leaving group, 2) carbocation rearrangement, 3) nucleophilic attack, and 4) proton transfer.

(e) In problem 7.21e, we saw that the two core steps (loss of leaving group and nucleophilic attack) are accompanied by a proton transfer at the beginning of the mechanism and another proton transfer at the end of the mechanism. Therefore, the mechanism has a total of four steps: 1) proton transfer, 2) loss of a leaving group, 3) nucleophilic attack, and 4) proton transfer.

(f) In problem 7.21f, we saw that the two core steps (loss of leaving group and nucleophilic attack) are accompanied by three additional steps (a proton transfer at the beginning of the mechanism, a carbocation rearrangement between the two core steps, and finally, a proton transfer at the end of the mechanism). Therefore, the mechanism has a total of five steps: 1) proton transfer, 2) loss of a leaving group, 3) rearrangement, 4) nucleophilic attack, and 5) proton transfer.

(g) In problem 7.21g, we saw that the two core steps (loss of leaving group and nucleophilic attack) are not accompanied by any additional steps. Therefore, the mechanism has only two steps: 1) loss of a leaving group and 2) nucleophilic attack.

(h) In problem 7.21h, we saw that the two core steps (loss of leaving group and nucleophilic attack) are accompanied by a proton transfer at the end of the mechanism. Therefore, the mechanism has a total of three steps: 1) loss of a leaving group, 2) nucleophilic attack, and 3) proton transfer.

Problem 7.21c and 7.21h exhibit the same pattern. Both problems are characterized by three mechanistic steps: 1) loss of a leaving group, 2) nucleophilic attack, and 3) proton transfer.

7.23. The mechanism has a total of five steps (two core steps, and three additional steps).

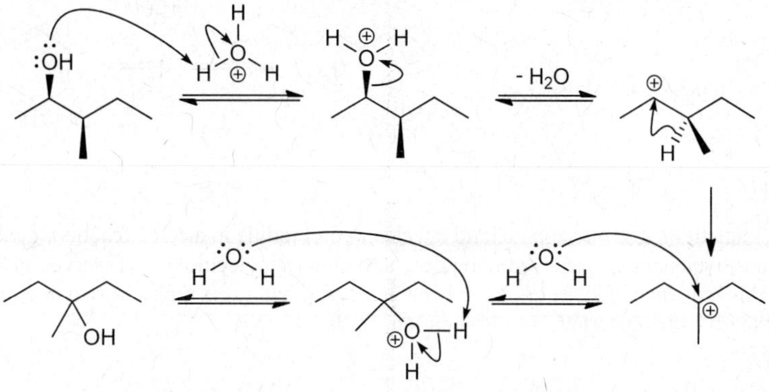

In the first step, the OH group is protonated to give a better leaving group. The chirality center at C2 is lost when the leaving group leaves to form a carbocation with trigonal planar geometry. This carbocation undergoes rearrangement via a hydride shift, during which the chirality center at C3 is also lost (once again via the trigonal planar sp^2 hybridized center). The tertiary carbocation is then captured by water to give an oxonium ion, which undergoes deprotonation to generate the product.

7.24.

(a) An S_N2 process is a concerted process in which the nucleophile attacks and the leaving group leaves simultaneously. The nucleophile (methanol) is uncharged, so a proton transfer is required at the end of the mechanism in order to remove the positive charge. A molecule of methanol serves as the base for this deprotonation step.

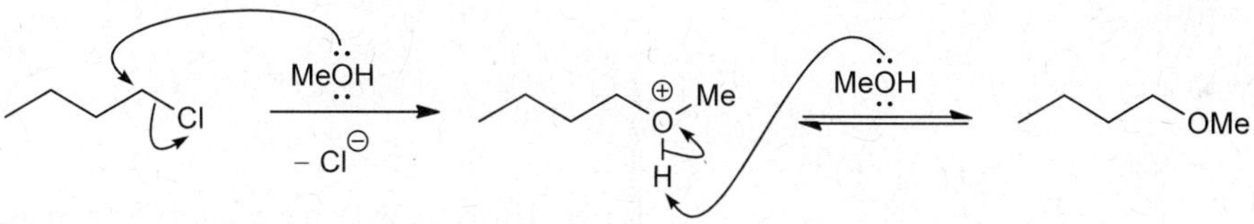

(b) An S_N2 process is a concerted process in which the nucleophile attacks and the leaving group leaves simultaneously. The nucleophile (ethanol) is uncharged, so a proton transfer is required at the end of the mechanism in order to remove the positive charge. A molecule of ethanol serves as the base for this deprotonation step.

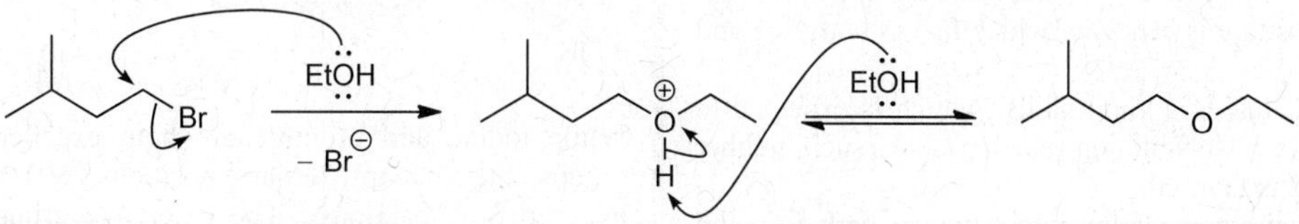

(c) An S_N2 process is a concerted process in which the nucleophile attacks and the leaving group leaves simultaneously. The nucleophile (water) is uncharged, so a proton transfer is required at the end of the mechanism in order to remove the positive charge. A molecule of water serves as the base for this deprotonation step.

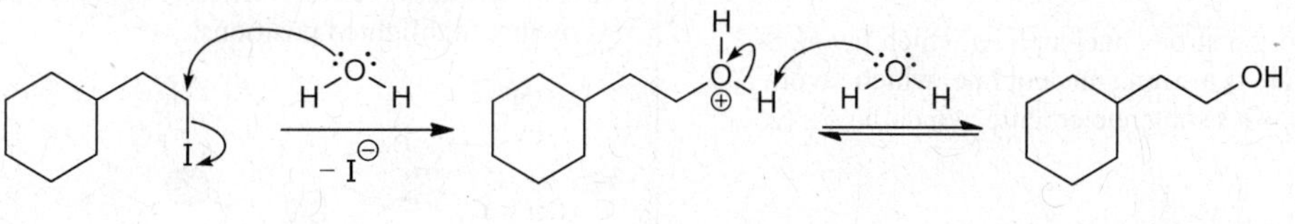

(d) An S_N2 process is a concerted process in which the nucleophile attacks and the leaving group leaves simultaneously. The nucleophile (1-propanol) is uncharged, so a proton transfer is required at the end of the mechanism in order to remove the positive charge. A molecule of 1-propanol serves as the base for this deprotonation step.

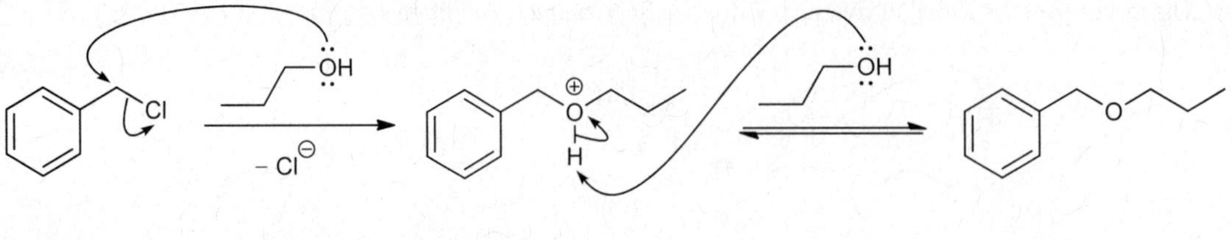

7.25. Ammonia (NH_3) functions as a nucleophile and attacks methyl iodide in an S_N2 reaction, generating an ammonium ion. The positive charge is removed upon deprotonation (a molecule of NH_3 serves as the base for this deprotonation step). This sequence of steps (S_N2 followed by deprotonation) is repeated two more times, followed by one final S_N2 reaction to give the quaternary ammonium ion, as shown here.

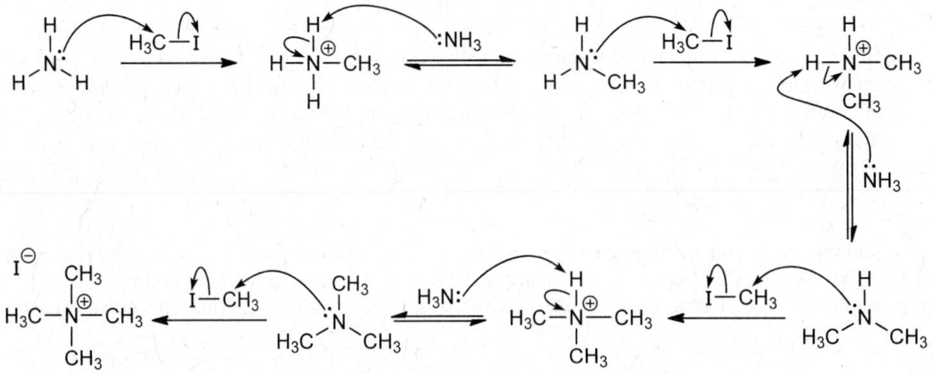

7.26.
(a) The substrate is tertiary, which favors S_N1.
(b) The substrate is primary, which favors S_N2.
(c) The substrate is an aryl halide, which is expected to be unreactive in substitution reactions. As such, neither S_N1 nor S_N2 is favored.
(d) The substrate is tertiary, which favors S_N1.
(e) The substrate is benzylic, which favors both S_N1 and S_N2.
(f) The substrate is a vinyl halide, which is expected to be unreactive in substitution reactions. As such, neither S_N1 nor S_N2 is favored.
(g) The substrate is allylic, which favors both S_N1 and S_N2.
7.27.
(a) Ethanol is a weak nucleophile, which disfavors S_N2 (and thereby allows S_N1 to compete successfully).
(b) Ethanethiol is a strong nucleophile, which favors S_N2.
(c) Ethoxide is a strong nucleophile, which favors S_N2.
(d) Hydroxide is a strong nucleophile, which favors S_N2.
(e) Cyanide is a strong nucleophile, which favors S_N2.

7.28.
(a) In order for an S_N2 reaction to occur, the position must be unhindered (primary, or perhaps secondary, but not tertiary), AND there must be a suitable leaving group. These criteria are satisfied in the following, highlighted positions.

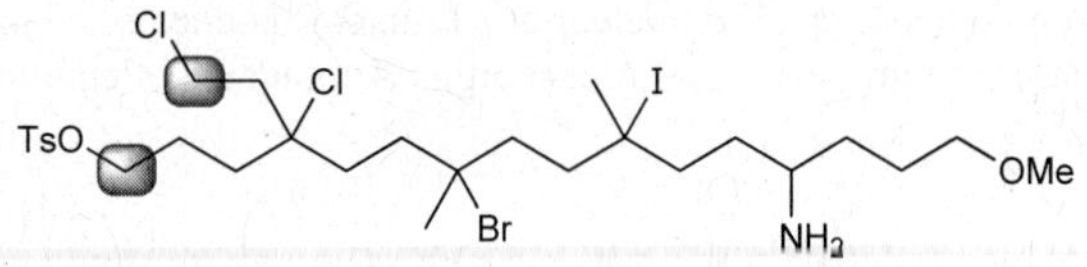

While iodide and bromide are both excellent leaving groups, they occupy tertiary positions, which will not undergo S_N2 reactions. The other two groups are not good leaving groups (MeO^- and NH_2^-).

(b) In order for an S_N1 reaction to occur, the position must be capable of stabilizing a carbocation (i.e. tertiary rather than primary), AND there must be a suitable leaving group. These criteria are satisfied in the following, highlighted positions.

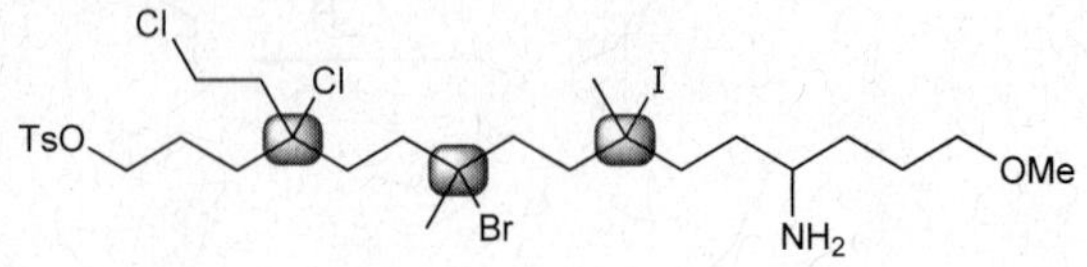

7.29.
(a) Ethanol is a protic solvent, which favors S_N1 (see Table 7.2).
(b) DMSO is a polar aprotic solvent, which favors S_N2 (see Table 7.2).
(c) Acetic acid is a protic solvent, which favors S_N1 (see Table 7.2).
(d) DMF is a polar aprotic solvent, which favors S_N2 (see Table 7.2).
(e) Methanol is a protic solvent, which favors S_N1 (see Table 7.2).
(f) Acetonitrile is a polar aprotic solvent, which favors S_N2 (see Table 7.2).
(g) HMPA is a polar aprotic solvent, which favors S_N2 (see Table 7.2).
(h) Ammonia is a protic solvent, which favors S_N1 (see Table 7.2).

7.30.
Acetone is a polar aprotic solvent and will favor S_N2 by raising the energy of the nucleophile, giving a smaller E_a.

7.31.
(a) The substrate is tertiary, and the nucleophile (MeOH) is weak, both of which favor an S_N1 process, giving the following product.

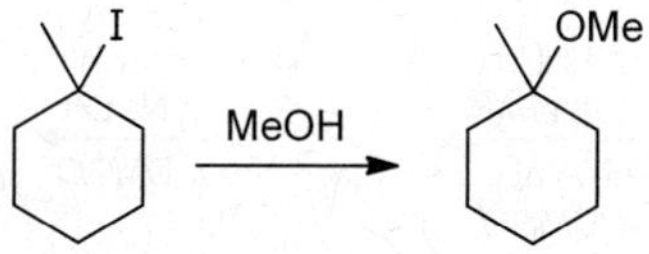

(b) The substrate is secondary, which is not very helpful in the determination between S_N1 and S_N2. But the nucleophile is strong, and the solvent is polar aprotic, and these two factors favor an S_N2 process. The product is shown below.

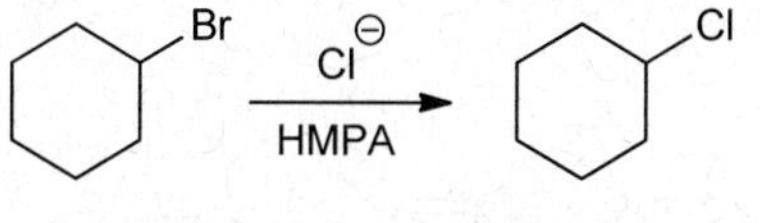

(c) The substrate is tertiary, which favors S_N1. The OH group is protonated under these conditions, giving an excellent leaving group (water). The nucleophile (bromide) is strong, but an S_N2 reaction will not occur at a tertiary substrate, so we expect an S_N1 process will occur, giving a pair of enantiomers in this case.

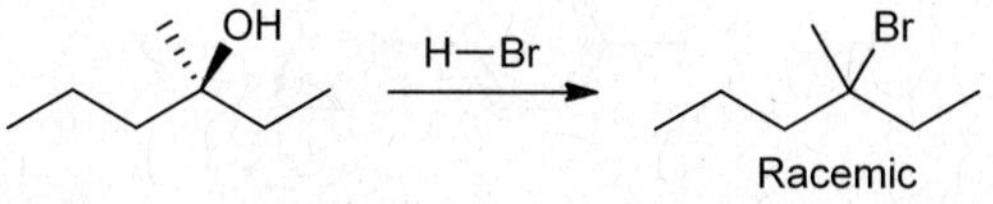

(d) The substrate is primary, which favors S_N2. The nucleophile is strong, and the solvent is polar aprotic, and these two factors also favor an S_N2 process. The product is shown below.

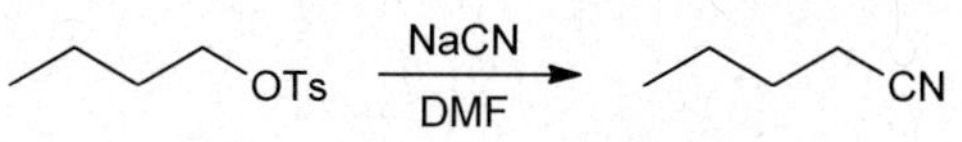

(e) The substrate is tertiary, and the nucleophile (H_2O) is weak, both of which favor an S_N1 process, giving a pair of enantiomers in this case.

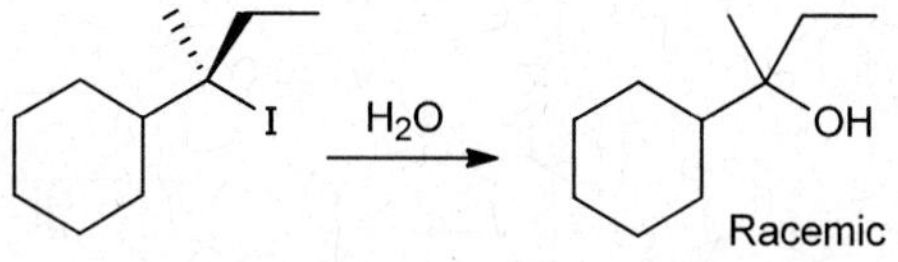

(f) The substrate is secondary, which is not very helpful in the determination between S_N1 and S_N2. But the nucleophile is strong, and the solvent is polar aprotic, and these two factors favor an S_N2 process. We therefore expect inversion of configuration, as shown below.

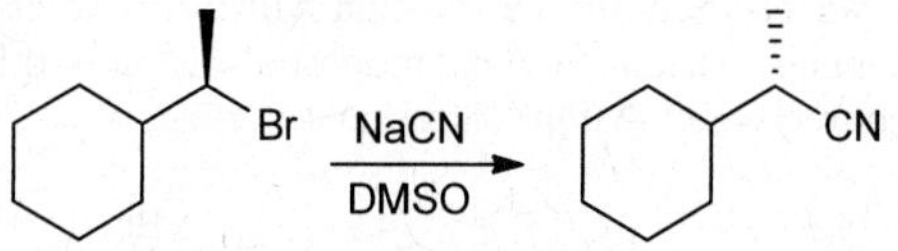

7.32. No. Preparation of this amine via the Gabriel synthesis would require the use of a tertiary alkyl halide, which will not undergo an S_N2 process.

7.33.
(a) The substrate is primary, so we will need to use an S_N2 process. We must therefore use a strong nucleophile (hydroxide), rather than a weak nucleophile (water).

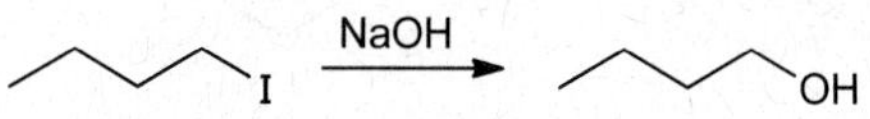

(b) The substrate is tertiary, so we will need to use an S_N1 process. The OH group cannot function as a leaving group without protonation, so we must use strongly acidic conditions. HBr supplies both the proton and the nucleophile:

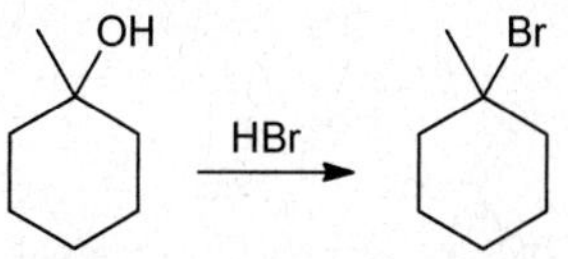

(c) The substrate is primary, so we will need to use an S_N2 process. The OH group cannot function as a leaving group, so we must convert it to a better leaving group, which can be accomplished by first converting the alcohol to a tosylate, followed by an S_N2 reaction with iodide.

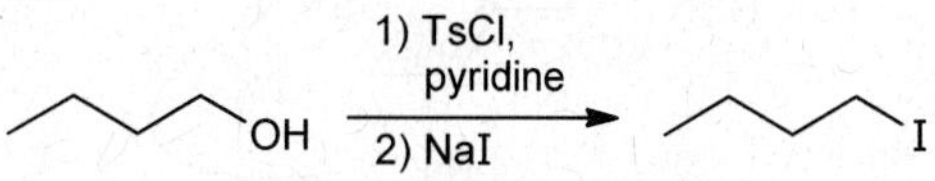

(d) The desired transformation involves inversion of configuration, so we will need to use an S_N2 process. We use a strong nucleophile (HS^-) and a polar aprotic solvent.

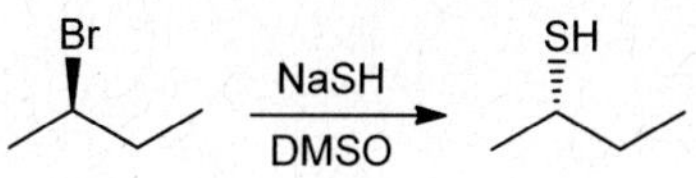

(e) The desired transformation involves inversion of configuration, so we will need to use an S_N2 process. The use of a polar aprotic solvent will help overcome the fact that the substrate is secondary.

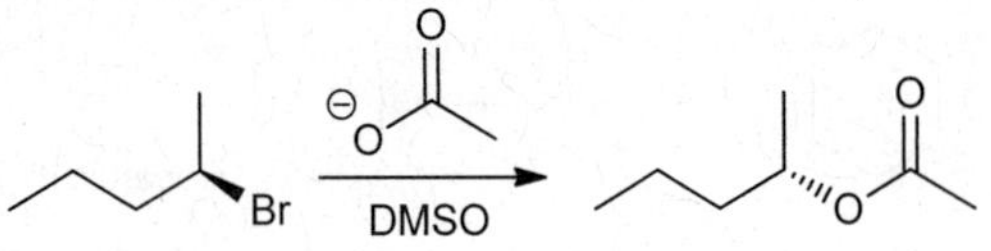

(f) The desired transformation involves inversion of configuration, so we will need to use an S_N2 process. The OH group cannot function as a leaving group, so we must convert it to a better leaving group, which can be accomplished by converting the alcohol to a tosylate. Then, we perform an S_N2 reaction with bromide as the nucleophile. The use of a polar aprotic solvent will help overcome the fact that the substrate is secondary.

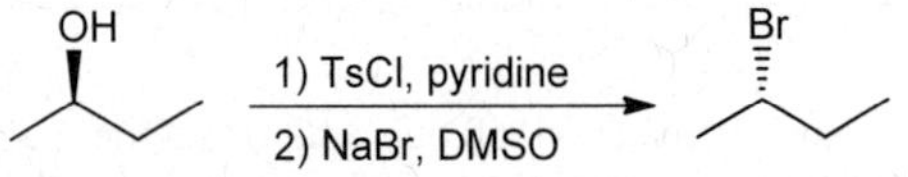

(g) The substrate is primary, so we will need to use an S_N2 process. We must therefore use a strong nucleophile (ethoxide), rather than a weak nucleophile (ethanol).

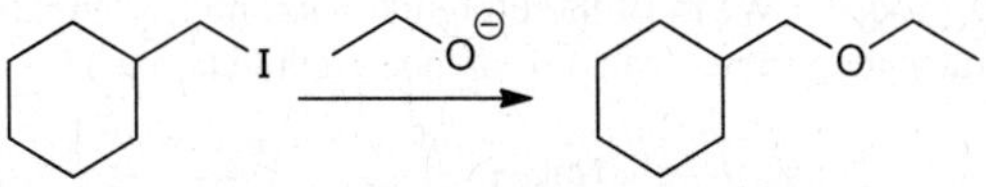

(h) The substrate is tertiary, so we will need to use an S_N1 process. We therefore use a weak nucleophile (H_2O) rather than a strong nucleophile (HO^-).

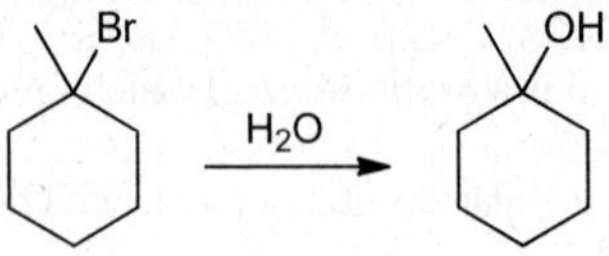

7.34. We have not learned a direct way to perform a substitution reaction with retention of configuration. However, if we perform two successive S_N2 reactions (each of which proceeds via inversion of configuration), the net result of the overall process will be retention of configuration. The first S_N2 reaction cannot be performed with hydroxide as a leaving group, so the alcohol must first be converted into a tosylate (by treatment with TsCl and pyridine). After converting the OH group into a better leaving group, we are now ready for the first S_N2 reaction. Iodide makes an excellent choice for the nucleophile, because it is both a strong nucleophile AND an excellent leaving group (the former is important for the first S_N2 reaction, while the latter will be important for the second S_N2 reaction). Since the substrate is secondary, polar aprotic solvents are used to enhance the rate of each S_N2 process.

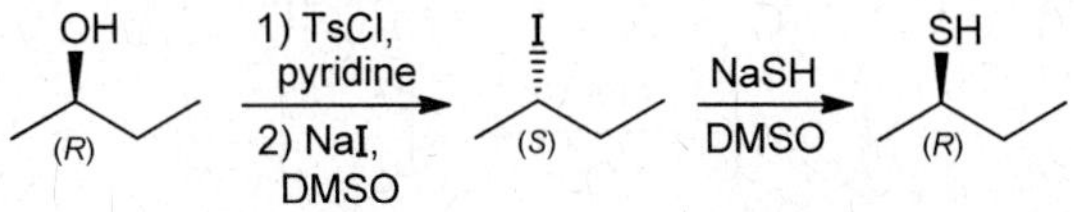

7.35. The lone pair of the nitrogen atom (connected to the aromatic ring) provides anchimeric assistance, ejecting the chloride ion in an intramolecular S_N2-type reaction, generating a high-energy intermediate that exhibits a three-membered ring. The ring is opened upon attack of a nucleophile in an S_N2 process. These two steps are then repeated, as shown here.

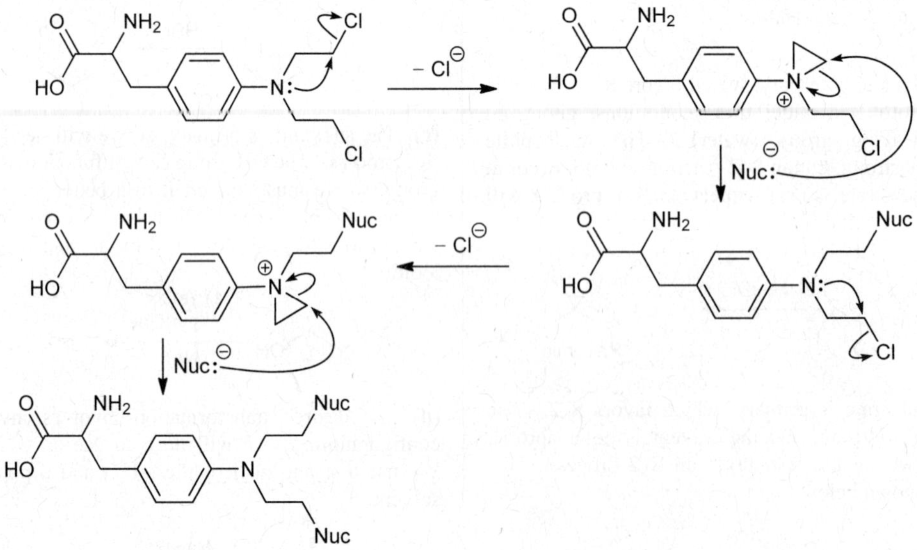

7.36.

(a) The parent is the longest chain, which is three carbon atoms in this case (propane).   There is only one substituent (chloro), and its locant is assigned as 2 (as shown below), so the systematic name for this compound is 2-chloropropane.   The common name is isopropyl chloride.

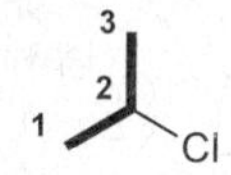

2-chloropropane

(b) The parent is the longest chain, which is three carbon atoms in this case (propane). There are two substituents (bromo and methyl), and their locants are assigned as 2 and 2, as shown below. Substituents are alphabetized in the name (bromo precedes methyl), so the systematic name is 2-bromo-2-methylpropane. The common name is *tert*-butyl bromide.

2-bromo-2-methylpropane

(c) The parent is the longest chain, which is three carbon atoms in this case (propane).   There is only one substituent (iodo), and its locant is assigned as 1 (as shown below), so the systematic name for this compound is 1-iodopropane. The common name is propyl iodide.

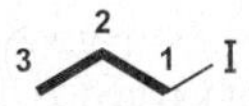

1-iodopropane

(d) The parent is the longest chain, which is four carbon atoms in this case (butane). There is only one substituent (bromo), and its locant is assigned as 2 (as shown below). The compound has a chirality center, so the configuration must be indicated at the beginning of the name: (*R*)-2-bromobutane. The common name is (*R*)-*sec*-butyl bromide.

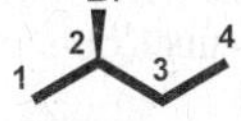

(*R*)-2-bromobutane

(e) The parent is the longest chain, which is three carbon atoms in this case (propane). There are three substituents (chloro, methyl, and methyl), and their locants are assigned as 1, 2, and 2, respectively, as shown below. Substituents are alphabetized in the name (chloro precedes methyl). Make sure that each methyl group receives a locant (2,2-dimethyl rather than 2-dimethyl). The systematic name is therefore 1-chloro-2,2-dimethylpropane.   The common name is neopentyl chloride.

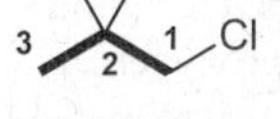

1-chloro-2,2-dimethylpropane

7.37.   The constitutional isomers of C_4H_9I are shown below, arranged in order of increasing reactivity toward S_N2.   Notice that the tertiary substrate is the least reactive because it is the most hindered. Among the two primary substrates, butyl iodide is the least sterically hindered (and therefore the most reactive toward S_N2), because it does not contain a substituent at the beta position.

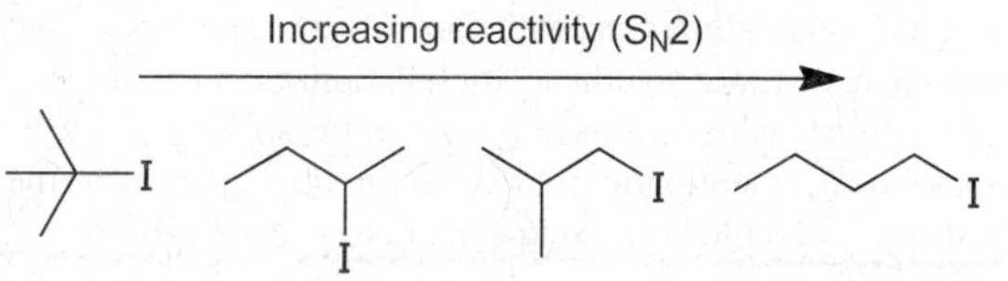

Increasing reactivity (S_N2)

7.38.
(a) The secondary substrate is more hindered than the primary substrate, and therefore, the latter reacts more rapidly in an S_N2 reaction.

secondary   primary

(b)   Both substrates are primary, but one is more hindered than the other.   The less hindered substrate reacts more rapidly in an S_N2 reaction.

more sterically   less sterically
hindered   hindered

(c)   The secondary substrate is less hindered than the tertiary substrate, and therefore, the former reacts more rapidly in an S_N2 reaction.

secondary   tertiary

(d) Both substrates are primary, but they have different leaving groups. The compound with the better leaving group (iodide) will react more rapidly in an S_N2 reaction.

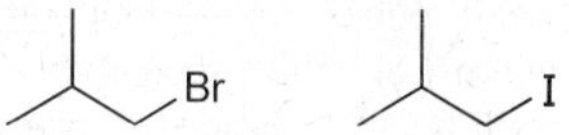

better
leaving group

7.39. No. Preparation of this compound via an acetylide ion would require the use of the following tertiary alkyl halide, which will not participate in an S_N2 reaction because of steric crowding.

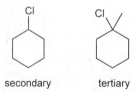

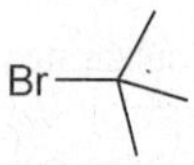

7.40.

(a) HS⁻ bears a negative charge, so it is a stronger nucleophile than H_2S.

(b) HO⁻ bears a negative charge, so it is a stronger nucleophile than H_2O.

(c) In a polar aprotic solvent (such as DMSO), cations are solvated and surrounded by a solvent shell, but anions are not (see Table 7.2). This greatly enhances the nucleophilicity of the methoxide ions because they are not strongly interacting with the solvent molecules, and are instead more available to function as nucleophiles. In contrast, when dissolved in a protic solvent (such as methanol), methoxide ions interact strongly with the solvent molecules and are therefore less nucleophilic.

7.41.

(a) The tertiary substrate will react more rapidly in an S_N1 reaction, because upon loss of a leaving group, the tertiary substrate gives a tertiary carbocation, while the primary substrate gives a primary carbocation. A tertiary carbocation is more stable than a primary carbocation because of hyperconjugation, so the energy of activation (E_a) for formation of the tertiary carbocation is expected to be lower than the E_a for formation of the primary carbocation.   Therefore, the tertiary carbocation is expected to form more rapidly, which explains why the tertiary substrate is more reactive towards S_N1 than the primary substrate.

tertiary   primary

(b) The tertiary substrate will react more rapidly in an S_N1 reaction (see explanation in Problem **7.41a**).

primary   tertiary

(c) The allylic substrate will react more rapidly in an S_N1 reaction, because upon loss of a leaving group, the allylic substrate gives a resonance-stabilized allylic carbocation. This carbocation is more stable than the carbocation that would form from chlorocyclohexane (which would be secondary and not resonance-stabilized). The energy of activation (E_a) for formation of the allylic carbocation is expected to be lower than the E_a for formation of the secondary carbocation. Therefore, the allylic carbocation is expected to form more rapidly.

allylic

(d) In this case, both substrates are tertiary, and both substrates lead to the same carbocation intermediate. The difference between these compounds is the identity of the leaving group. TsO⁻ is a better leaving group than chloride (Cl⁻), so we expect the tosylate to undergo S_N1 at a faster rate.

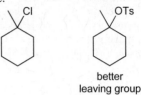

better
leaving group

7.42.

(a)   There are many indications that this reaction proceeds via an S_N2 pathway (a strong nucleophile in a polar aprotic solvent, and inversion of configuration). As such, the reaction has a second-order rate equation, which means that the rate should be linearly dependent on the concentrations of two compounds (the nucleophile AND the substrate). If the concentration of the substrate is doubled, the rate should also be doubled.

(b)   The reaction proceeds via an S_N2 pathway (see solution to Problem **7.42a**). As such, the reaction has a second-order rate equation, which means that the rate should be linearly dependent on the concentrations of two compounds (the nucleophile AND the substrate). If the concentration of the nucleophile is doubled, the rate should also be doubled.

7.43.

(a) The substrate is a tertiary alcohol, and the reaction proceeds via an S_N1 process (under acidic conditions, the OH group is protonated to give an excellent leaving group).   As such, the reaction has a first-order rate equation, which means that the rate should be linearly dependent only on the concentration of the substrate (not the nucleophile). If the concentration of the substrate is doubled, the rate should also be doubled.

(b)   The reaction proceeds via an S_N1 pathway (see solution to Problem **7.43a**). As such, the reaction has a first-order rate equation, which means that the rate should be linearly dependent only on the concentration of the substrate (not the nucleophile).   If the concentration of the nucleophile is doubled, the rate should not be affected.

7.44.

(a) DMF is a polar aprotic solvent, because it does not contain any hydrogen atoms that are connected directly to an electronegative atom (see Table 7.2).

(b) Ethanol (CH_3CH_2OH) is a protic solvent because it contains a hydrogen atom connected to an electronegative atom (oxygen).

(c) DMSO is a polar aprotic solvent, because it does not contain any hydrogen atoms that are connected directly to an electronegative atom (see Table 7.2).

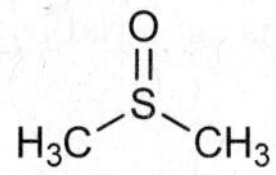

(d) Water (H_2O) is a protic solvent because it contains hydrogen atoms connected to an electronegative atom (oxygen).

(e) Ammonia (NH_3) is a protic solvent because it has hydrogen atoms that are connected to an electronegative atom (nitrogen).

7.45.
(a) The chirality center in the substrate has the *R* configuration, as shown below.

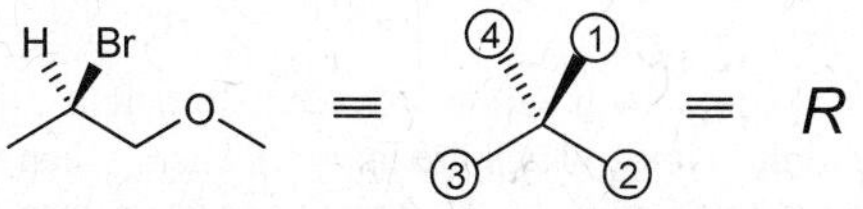

(b) The chirality center in the product has the *R* configuration, as shown below.

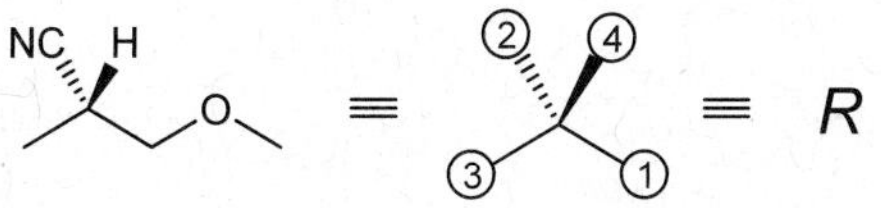

(c) The reaction is an S_N2 process, and it does proceed with inversion of configuration. However, the prioritization scheme changes when the bromo group (#1) is replaced with a cyano group (#2). As a result, the Cahn-Ingold-Prelog system assigns the same configuration to the reactant and the product.

7.46. The leaving group is an iodide ion (I^-) and the nucleophile is an acetate ion ($CH_3CO_2^-$). In the transition state, each of these groups is drawn as being connected to the α position with a dotted line (indicating these bonds are in the process of forming or breaking), and a δ– is placed on each group to indicate that the charge is spread over both locations. Don't forget the brackets and the symbol that indicate the drawing is a transition state.

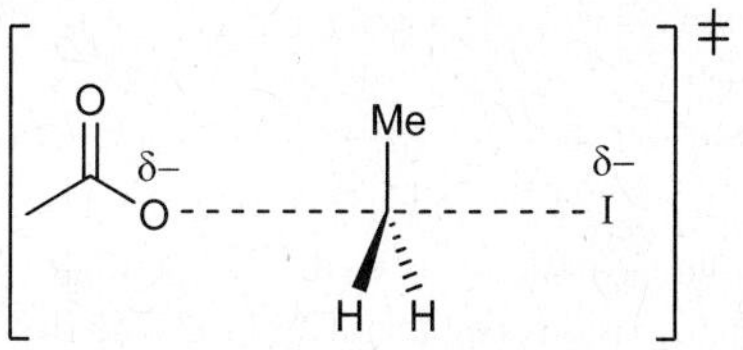

7.47. Iodide functions as a nucleophile and attacks (*S*)-2-iodopentane, displacing iodide as a leaving group. The reaction is an S_N2 process, and therefore proceeds via inversion of configuration. The product is (*R*)-2-iodopentane. The reaction continues repeatedly until a racemic mixture is eventually obtained.

7.48. The substrate is an alcohol, so acidic conditions are employed so that the OH group can be protonated, rendering it a better leaving group. Then, loss of a leaving group generates a carbocation, which is then captured by a bromide ion to give the product. Notice that the mechanism is comprised of a proton transfer, followed by the two core steps of an S_N1 process (loss of a leaving group and nucleophilic attack).

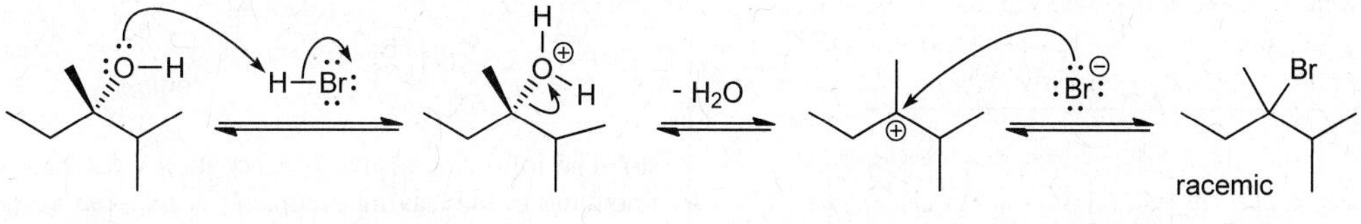

The chirality center is lost when the leaving group leaves to form a carbocation with trigonal planar geometry. The nucleophile can then attack either face of the planar carbocation, leading to a racemic mixture.

7.49. The substrate is an alcohol, so acidic conditions are employed so that the OH group can be protonated, rendering it a better leaving group. In dilute sulfuric acid, the source of the proton is H_3O^+, which received its proton from sulfuric acid (a comparison of the pK_a values for H_3O^+ and H_2SO_4 indicates that there is likely very little H_2SO_4 present at equilibrium). Loss of a leaving group generates a secondary carbocation, which is captured by a molecule of water to give an oxonium ion. Deprotonation of the oxonium ion gives the product.

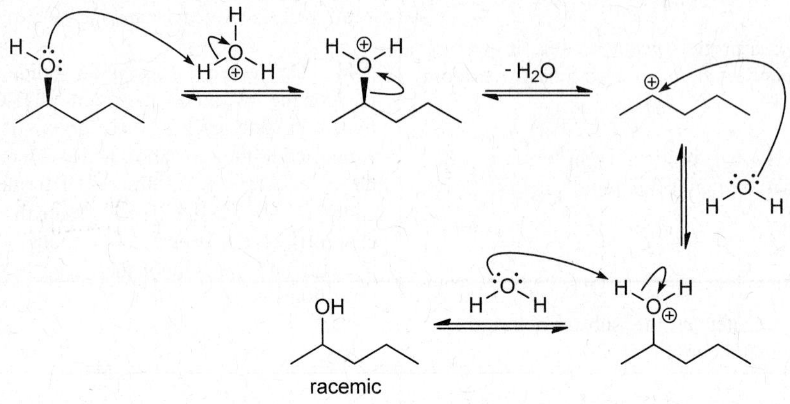

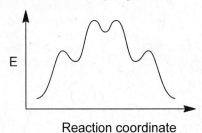

racemic

The reaction has four steps, so the energy diagram should have four steps, as shown below. Notice that there are three intermediates (valleys on the diagram). The first and last represent oxonium ions, which are lower in energy than the carbocation intermediate. Of the three intermediates, the carbocation intermediate is the highest in energy, as can be seen in the middle of the energy diagram. Also note that the starting material and product are represented at the same energy level, because in this case, they have the same identity (2-pentanol).

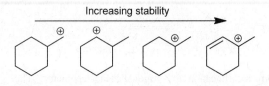

Reaction coordinate

7.50. The tertiary allylic carbocation is the most stable, because it is stabilized by hyperconjugation as well as resonance. The next most stable is the tertiary carbocation, followed by the secondary carbocation. The primary carbocation is the least stable.

Increasing stability

7.51.
(a) The following secondary carbocation will be formed upon loss of the leaving group.

secondary

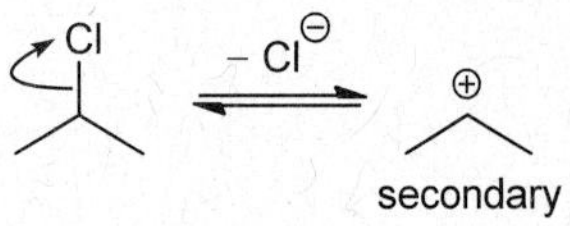

(b) The following tertiary carbocation will be formed upon loss of the leaving group.

tertiary

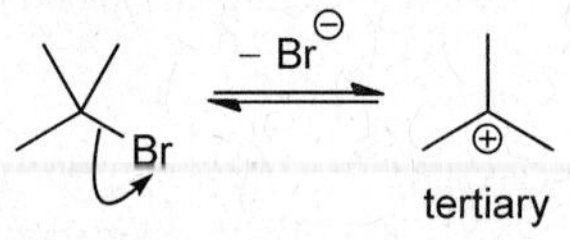

(c) The following primary carbocation would be formed upon loss of the leaving group.

primary

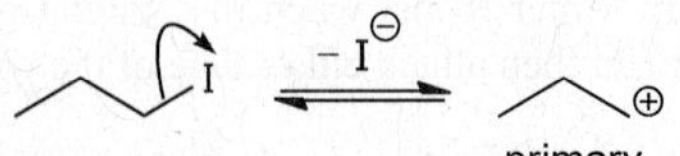

(d) The following secondary carbocation will be formed upon loss of the leaving group.

secondary

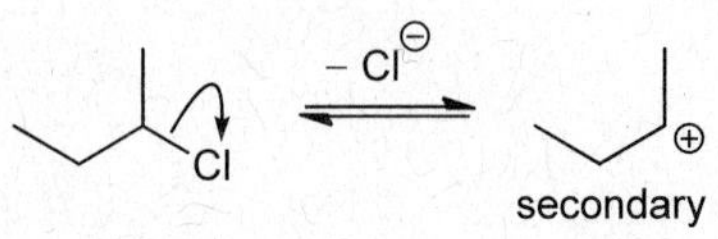

7.52. The substrate is an alcohol, which does not possess a good leaving group. However, acidic conditions are employed, and under these conditions, the OH group is protonated, thereby forming an excellent leaving group. Loss of the leaving group (H$_2$O) gives a secondary carbocation, which can undergo a hydride shift to form a tertiary carbocation. This carbocation is then captured by chloride to give the product.

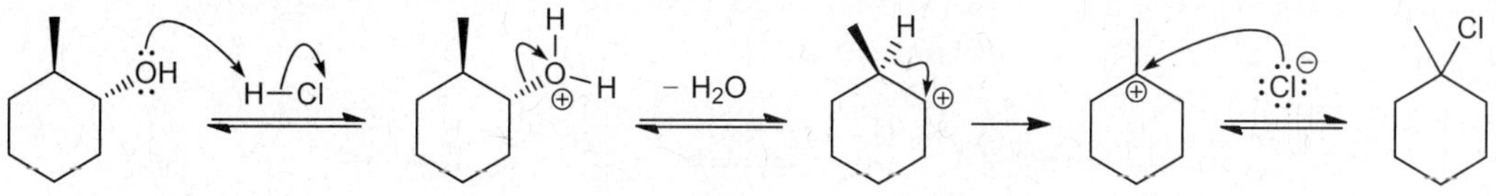

7.53. This is a substitution reaction in which an acetate ion (CH$_3$CO$_2^-$) functions as a nucleophile. The reaction occurs with inversion of configuration, which indicates that an S$_N$2 process must be operating. As such, we draw a concerted process in which nucleophilic attack and loss of the leaving group occur simultaneously, as shown below.

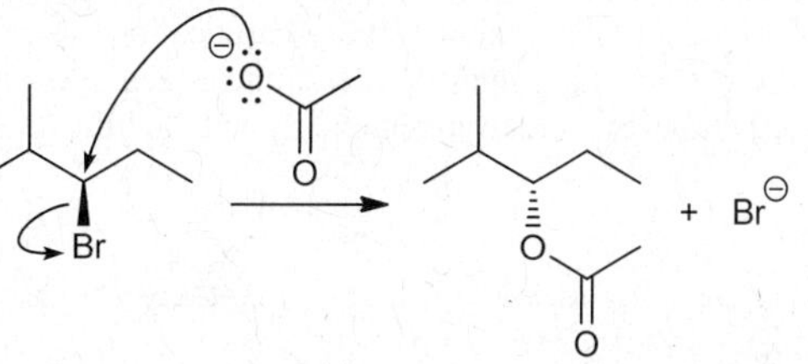

7.54.
(a) In an S$_N$1 process, there must be at least two core steps: loss of the leaving group, and nucleophilic attack. In addition, there can be other steps that accompany the S$_N$1 process, including a proton transfer step at the beginning of the mechanism, a carbocation rearrangement in between the two core steps, and a proton transfer step at the end of the mechanism.

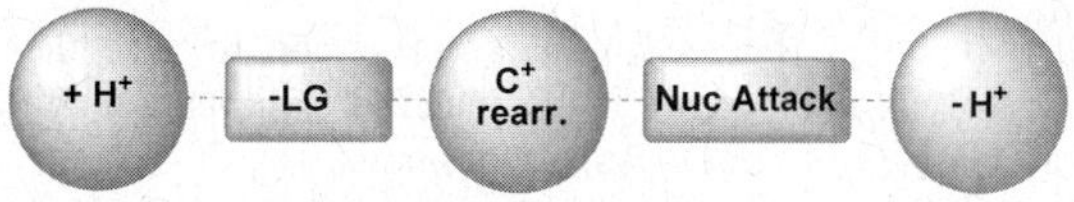

In this case, the first proton transfer step does not occur because chloride is an excellent leaving group (and acidic conditions are not employed). The intermediate carbocation is tertiary and cannot rearrange to become more stable, so the mechanism will not involve a carbocation rearrangement. A proton transfer step at the end of the mechanism will be required because the attacking nucleophile is uncharged. As such, our mechanism will have three steps: 1) loss of a leaving group, 2) nucleophilic attack, and 3) proton transfer.

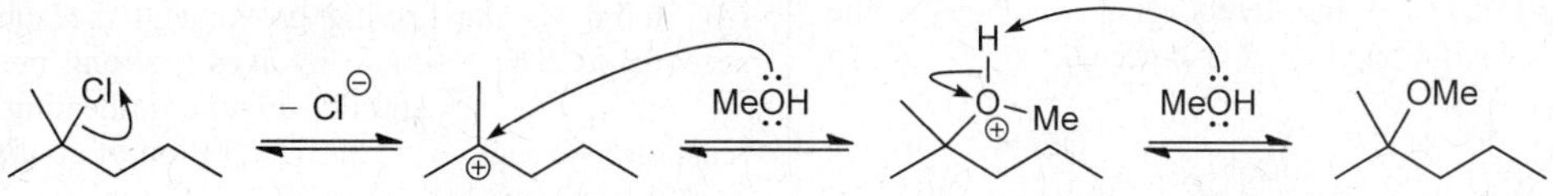

(b) As in the solution to **7.54a**, we consider the three possible steps that can accompany the two core steps of an S$_N$1 process, shown in circles below.

In this case, the first proton transfer step does not occur because chloride is an excellent leaving group (and acidic conditions are not employed). The intermediate carbocation is tertiary and cannot rearrange to become more stable, so the mechanism will not involve a carbocation rearrangement. A proton transfer step at the end of the mechanism will not be required because the attacking nucleophile is an anion. As such, our mechanism will have only the two core steps: 1) loss of a leaving group, and 2) nucleophilic attack.

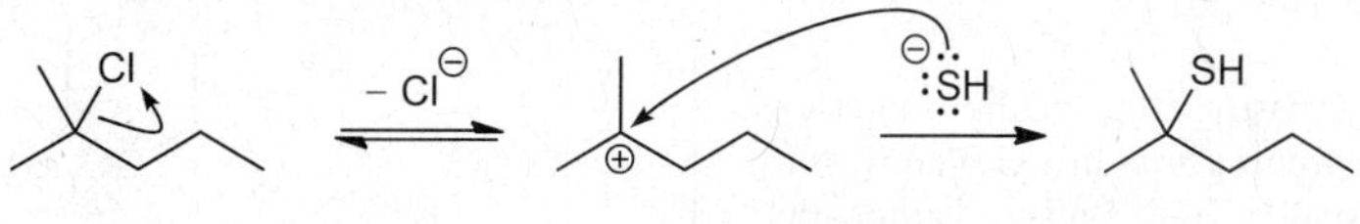

(c) As in the solution to **7.54a**, we consider the three possible steps that can accompany the two core steps of an S$_N$1 process. In this case, a proton transfer step must occur at the beginning of the mechanism, because the OH group is not a good leaving group, and acidic conditions are employed (which cause protonation of the OH group, thereby converting it into a good leaving group). The intermediate carbocation is tertiary and cannot rearrange to become more stable, so the mechanism will not involve a carbocation rearrangement. A proton transfer step at the end of the mechanism will not be required because the attacking nucleophile is an anion. As such, our mechanism will have three steps: 1) proton transfer to convert the OH group into a better leaving group, 2) loss of the leaving group, and 3) nucleophilic attack.

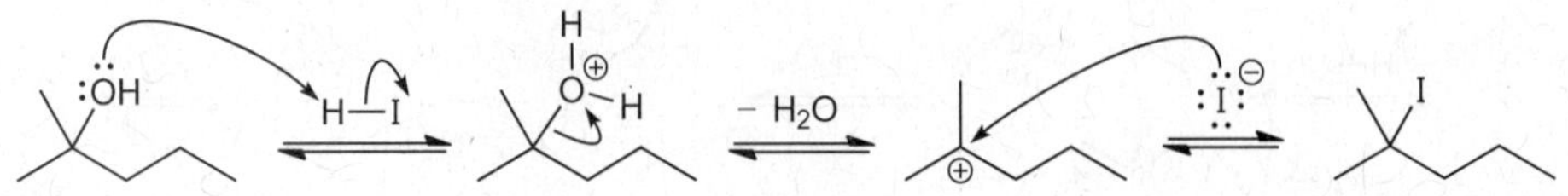

(d) As in the solution to **7.54a**, we consider the three possible steps that can accompany the two core steps of an S_N1 process. In this case, a proton transfer step does not occur at the beginning of the mechanism, because the OTs group is an excellent leaving group (and acidic conditions are not employed). The intermediate carbocation is secondary and can rearrange (via a hydride shift) to give a more stable, tertiary carbocation. A proton transfer step at the end of the mechanism will be required because the attacking nucleophile is uncharged. As such, our mechanism will have four steps: 1) loss of the leaving group, 2) carbocation rearrangement, 3) nucleophilic attack, and 4) proton transfer.

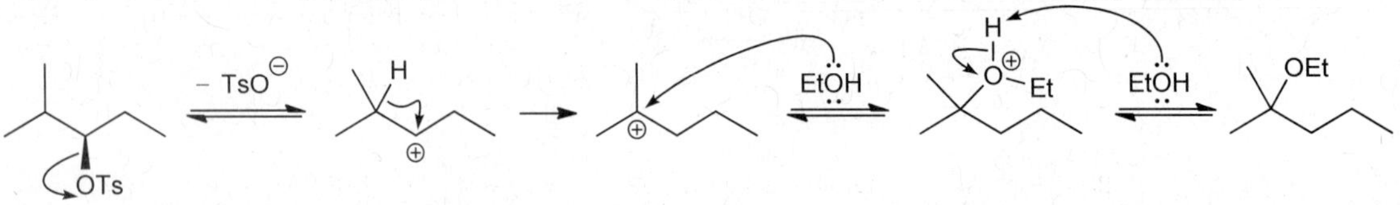

7.55.
(a) The substrate is tertiary, and a weak nucleophile is employed, both of which indicate an S_N1 process. As such, we must consider the possibility of a carbocation rearrangement. In this case, a tertiary carbocation will be formed, which cannot rearrange to form a more stable carbocation. Also, we must consider the stereochemical outcome whenever the reaction occurs at a chirality center, but in this case, there is no chirality center, so we don't need to consider the stereochemical outcome. The following substitution product is expected.

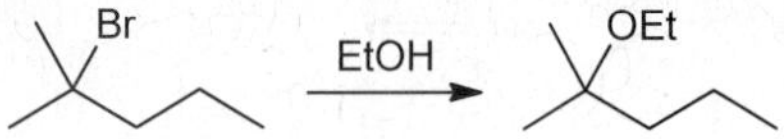

(b) The substrate is primary, indicating an S_N2 process. In this case, there is no chirality center, so we don't need to consider the stereochemical outcome. The following substitution product is expected.

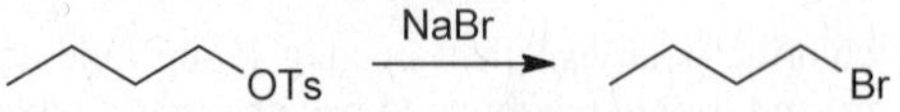

(c) The substrate is a secondary alcohol, and the OH group cannot function as a leaving group without being converted into a good leaving group.

That is, in fact, the purpose of the acidic conditions. Protonation of the OH group gives an excellent leaving group, which leaves to give a secondary carbocation. This carbocation can rearrange via a hydride shift to give a tertiary carbocation, which is then captured by the

chloride ion to give the product. Notice the starting material exhibits two chirality centers, while the product has none.

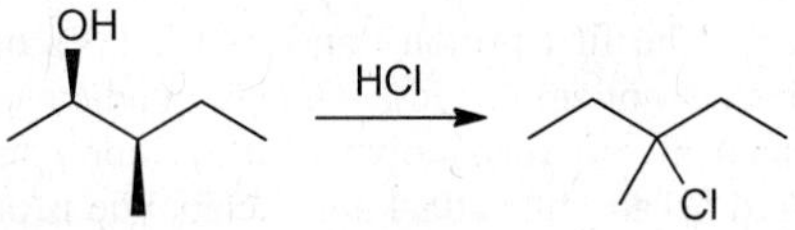

(d) Iodide is the leaving group, and the substrate is secondary. The reaction involves a strong nucleophile (N≡C⁻) in a polar aprotic solvent, indicating an S_N2 reaction. As such, we expect inversion of configuration, as shown below.

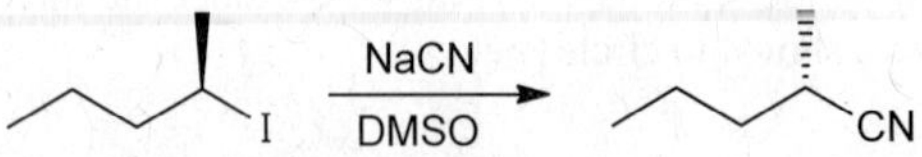

7.56. The dianion has two nucleophilic centers, and the electrophile has two electrophilic centers. As such, these compounds can react with each other via two successive S_N2 reactions, as shown below, giving a six-membered ring with molecular formula $C_4H_8O_2$.

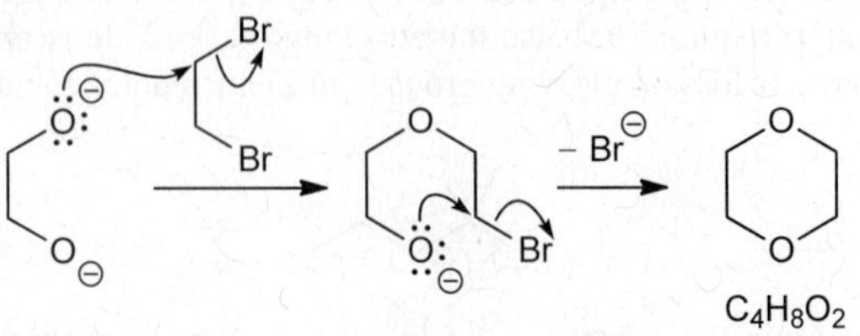

$C_4H_8O_2$

7.57. The strong nucleophile (HO⁻) and the primary substrate indicate an S$_N$2 reaction, as shown below.

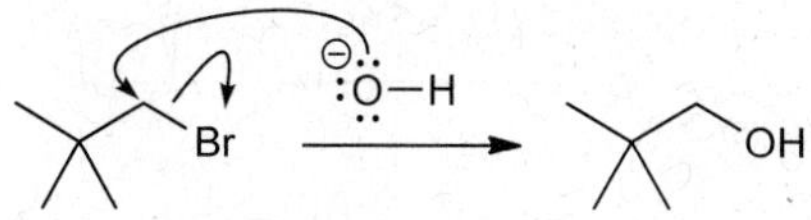

Although the compound is a primary substrate, nonetheless, it has three methyl groups attached to the beta position. These methyl groups provide steric hindrance that causes the energy of the transition state to be very high (see Figure 7.11). For this reason, neopentyl halides do not undergo S$_N$2 reactions at an appreciable rate.

7.58.
(a) The substrate is primary, and therefore, the reaction must proceed via an S$_N$2 process. Water functions as the nucleophile, attacking the substrate (propyl bromide) and ejecting bromide as a leaving group in one concerted step. Since the nucleophile is uncharged, a proton transfer step will be required at the end of the mechanism, in order to remove the positive charge. Water can function as the base for this proton transfer step.

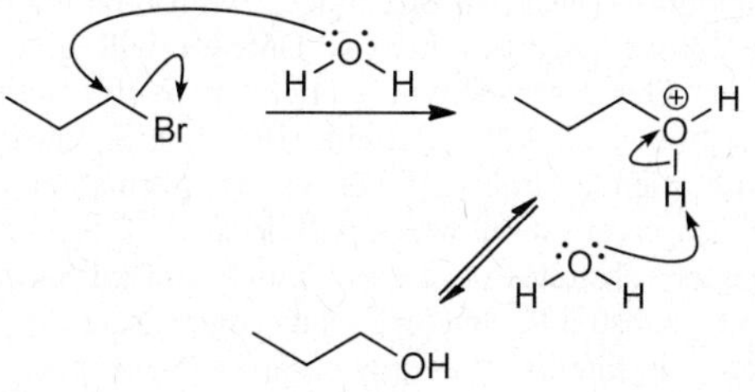

(b) S$_N$2 reactions are highly sensitive to the strength of the nucleophile, and water is a weak nucleophile. As a result, the reaction occurs slowly.

(c) S$_N$2 reactions are highly sensitive to the strength of the nucleophile, and hydroxide is a strong nucleophile. As a result, the reaction occurs rapidly. The reaction occurs in one concerted step, and a proton transfer step is not required at the end of the mechanism, because the nucleophile was an anion.

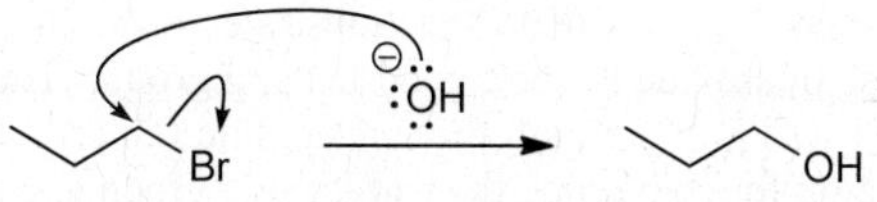

7.59.
(a) The substrate is primary, so we will need to perform an S$_N$2 reaction. We must therefore use a strong nucleophile (hydroxide, rather than water).

(b) The substrate is a primary alcohol, and the OH group is not a good leaving group. So we first convert the OH group into a better leaving by treating the alcohol with tosyl chloride and pyridine. Then, an S$_N$2 reaction can be performed (since the substrate is primary) with cyanide as the nucleophile, giving the desired product.

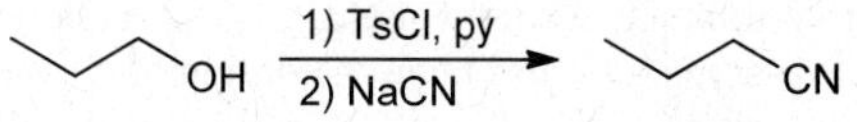

(c) The substrate is tertiary, so we will need to perform an S$_N$1 reaction. The nucleophile must be bromide, but we cannot simply treat the substrate with bromide, because hydroxide is not a good leaving group. The use of HBr will provide both the nucleophile (bromide) and the proton for converting the bad leaving group to a good leaving group (water).

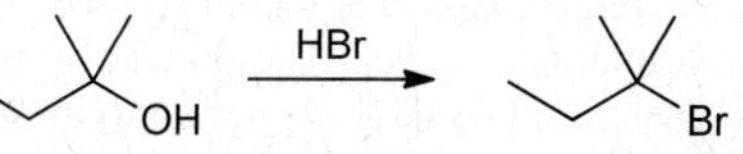

(d) The desired transformation involves inversion of configuration, so we must use an S$_N$2 process. As such, we want to use a strong nucleophile (HS⁻).

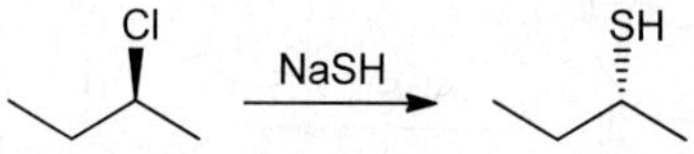

Since the substrate is secondary, the use of a polar aprotic solvent, such as DMSO, might be helpful.

(e) The desired transformation involves inversion of configuration, so we must use an S$_N$2 process. The nucleophile must be an acetate ion (CH$_3$CO$_2$⁻).

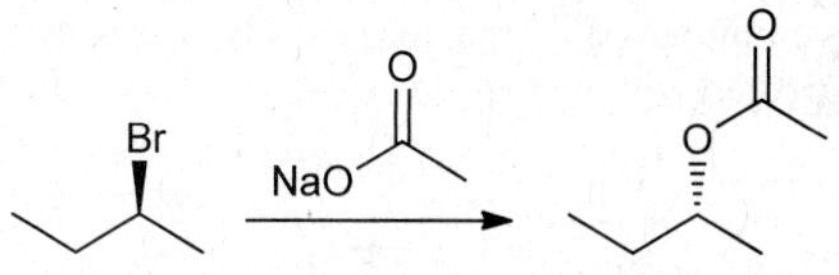

Since the substrate is secondary, the use of a polar aprotic solvent, such as DMSO, might be helpful.

7.60.
(a) The desired compound is a primary alcohol, so we will need to start with the corresponding primary alkyl iodide. The substrate is primary, which dictates that we employ an S$_N$2 process. Therefore, we must use a strong nucleophile (hydroxide, rather than water).

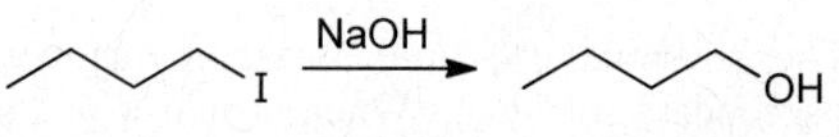

(b) The desired compound can be prepared if we use acetate (CH$_3$CO$_2$⁻) as a nucleophile, and perform an S$_N$2 reaction with the primary alkyl iodide, as shown below.

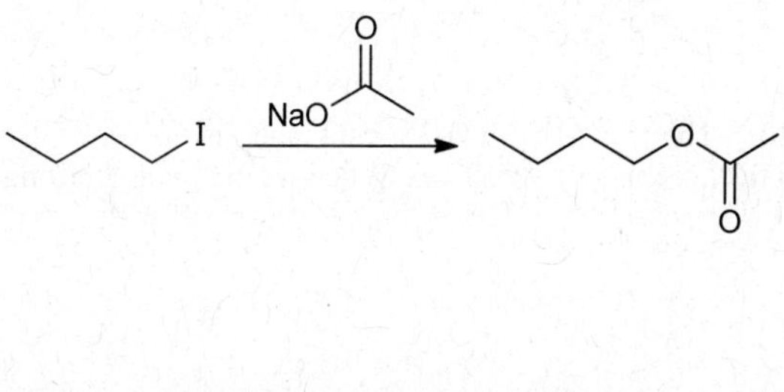

(c) The desired compound is a nitrile (R–C≡N), so we must use cyanide (N≡C⁻) as the nucleophile. The product is chiral, and only one enantiomer is desired (not a racemic mixture). Since an S_N1 process would produce a racemic mixture, we must use an S_N2 process. Since S_N2 processes exhibit inversion of configuration, we will need to start with an alkyl iodide with the *R* configuration in order to produce a nitrile with the *S* configuration.

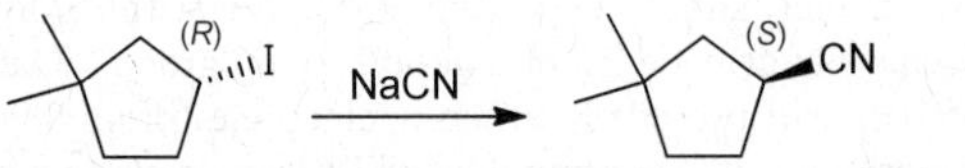

(d) The desired compound is a thiol (RSH), so we must use HS⁻ as the nucleophile. The product is chiral, and only one enantiomer is desired (not a racemic mixture). Since an S_N1 process would produce a racemic mixture, we must use an S_N2 process. Since S_N2 processes exhibit inversion of configuration, we will need to start with an alkyl iodide with the *S* configuration in order to produce a thiol with the *R* configuration.

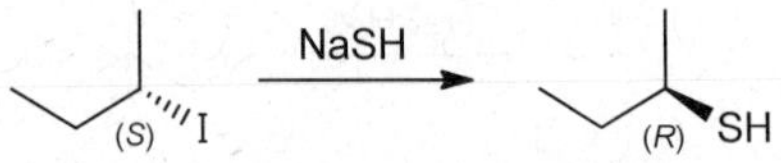

(e) The desired compound is a tertiary alcohol, so we will need to start with the corresponding tertiary alkyl iodide. The substrate is tertiary, which dictates that we employ an S_N1 process. Therefore, we must use a weak nucleophile (water, rather than hydroxide). In Chapter 8, we will learn about a different reaction that would occur (called elimination) if the tertiary alkyl iodide shown here is treated with hydroxide.

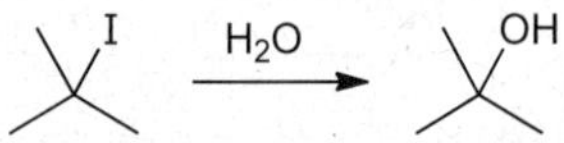

(f) The desired compound is a tertiary thiol (RSH), so we will need to start with the corresponding tertiary alkyl iodide. The substrate is tertiary, which dictates that we employ an S_N1 process, and use H_2S as the nucleophile.

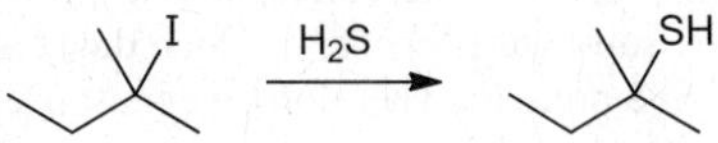

7.61.
(a) The substrate, (*S*)-2-iodobutane, is an optically active, secondary substrate. When treated with a strong nucleophile, such as HS⁻, an S_N2 reaction will occur, with the expected inversion of configuration.

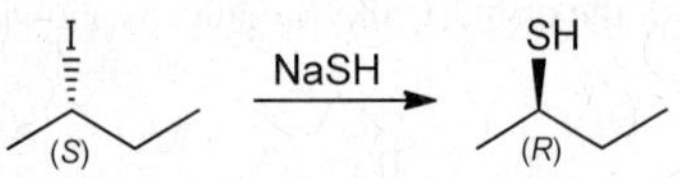

(b) The substrate, (*S*)-2-iodobutane, is an optically active, secondary substrate. When treated with a strong

nucleophile, such as EtS⁻, an S_N2 reaction will occur, with the expected inversion of configuration.

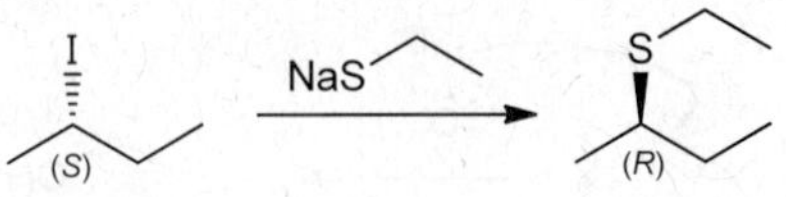

(c) The substrate, (*S*)-2-iodobutane, is an optically active, secondary substrate. When treated with a strong nucleophile, such as cyanide (N≡C⁻), an S_N2 reaction will occur, with the expected inversion of configuration.

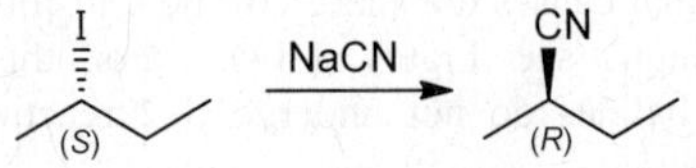

7.62. The second method is more efficient because the alkyl halide (methyl iodide) is not sterically hindered. The first method is not efficient because it employs a tertiary alkyl halide, and S_N2 reactions do not occur at tertiary substrates.

7.63.
(a) The substrate is a secondary alcohol, so we must first convert the OH group (bad leaving group) into a better leaving group, by treating the alcohol with tosyl chloride and pyridine. Then, an S_N2 process with bromide (in a polar aprotic solvent, such as DMSO) will give the product. This approach will likely be more efficient than treating the alcohol directly with HBr in an S_N1 process, because such a process relies on the formation of a secondary carbocation, which will be slow. It is true that secondary substrates react more slowly toward S_N2 than primary substrates, but we can compensate for the reduction in rate by using a polar aprotic solvent (which significantly improves the rate of reaction of an S_N2 process).

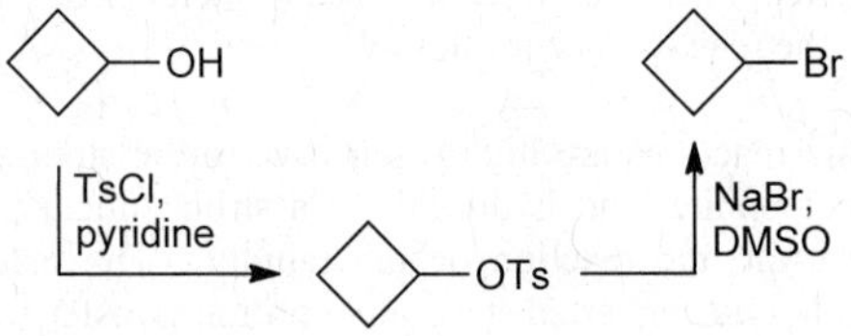

(b) The substrate is tertiary, so we will need to perform an S_N1 reaction. The nucleophile must be chloride, but we cannot simply treat the substrate with chloride, because hydroxide is not a good leaving group. The use of HCl will provide both the nucleophile (chloride) and the proton for converting the bad leaving group to a good leaving group (water).

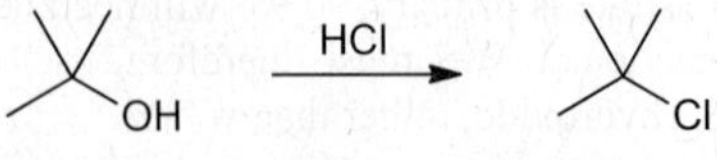

(c) The substrate is primary, so we will need to perform an S_N2 reaction. We must therefore use a strong nucleophile (hydroxide, rather than water).

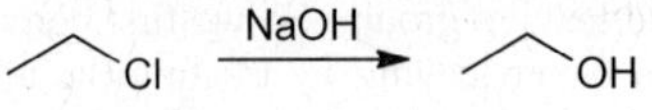

7.64.

(a) The substrate is primary, and a strong nucleophile is used in a polar aprotic solvent. All of these factors indicate an S_N2 reaction, in which nucleophilic attack and loss of the leaving group occur in a concerted fashion (in one step), as shown below.

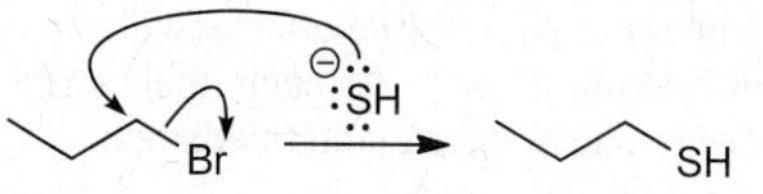

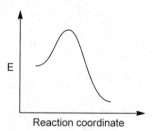

(b) Since the reaction is an S_N2 process, we expect a second-order rate equation that is linearly dependent on both the concentration of the substrate and the concentration of the nucleophile.

$$\text{Rate} = k \left[\text{\raisebox{0pt}{\,}} \begin{array}{c} \text{Br} \end{array} \right] \left[\text{NaSH} \right]$$

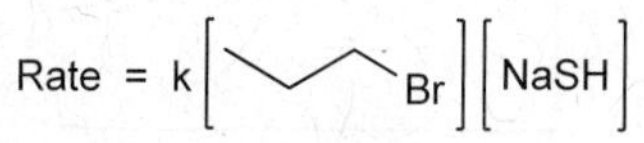

(c) DMSO is a polar aprotic solvent, which enhances the rate of an S_N2 process. By using a protic solvent, rather than a polar aprotic solvent, we expect the reaction to occur at a slower rate (because the nucleophile is surrounded by a solvent shell).

(d) As seen in the solution to **7.64a**, the reaction occurs via an S_N2 reaction, which is comprised of one concerted step (in which the nucleophile attacks with simultaneous loss of the leaving group). As such, the energy diagram will have only one hump (no intermediates).

(e) The leaving group is a bromide ion (Br^-) and the nucleophile is HS^-. In the transition state, each of these groups is drawn as being connected to the α position with a dotted line (indicating these bonds are in the process of forming or breaking), and a δ− is placed on each group to indicate that the charge is spread over both locations. Don't forget the brackets and the symbol that indicate the drawing is a transition state.

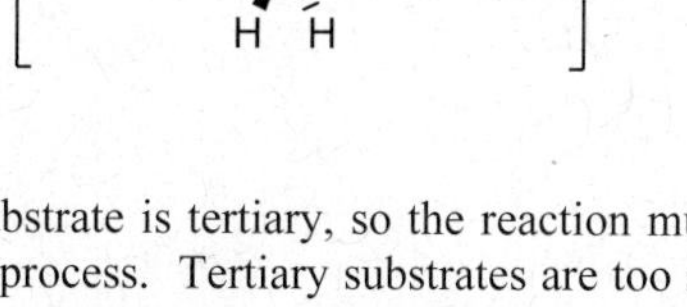

$$\left[\begin{array}{c} \overset{CH_2CH_3}{\underset{\underset{H\quad H}{|}}{\overset{\delta-}{HS} - - - - \overset{}{C} - - - - \overset{\delta-}{Br}}} \end{array} \right]^{\ddagger}$$

7.65.

(a) The substrate is tertiary, so the reaction must occur via an S_N1 process. Tertiary substrates are too sterically hindered to undergo S_N2 reactions.

(b) Hydroxide is not a good leaving group, but in acidic conditions, a proton is transferred to the OH group, thereby converting it into a better leaving group. Loss of the leaving group generates a tertiary carbocation, which is then captured by a bromide ion to give the product.

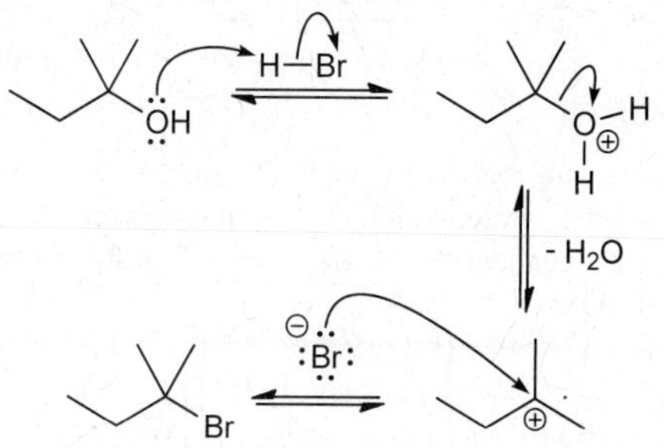

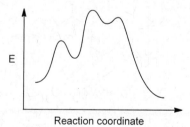

(c) Since the reaction is an S_N1 process, we expect a first-order rate equation that is linearly dependent on the concentration of the substrate alone (and not dependent on the concentration of the nucleophile).

$$\text{Rate} = k \left[\begin{array}{c} \text{\raisebox{0pt}{\,}} \text{OH} \end{array} \right]$$

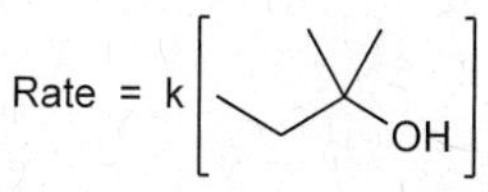

(d) No, the reaction would not occur at a faster rate. In an S_N1 process, the rate is not dependent on the concentration of the nucleophile (as seen in the rate equation for this reaction). Increasing the concentration of the nucleophile will not affect the rate of the rate-determining step (loss of the leaving group), and will therefore not affect the rate.

(e) As seen in the solution to **7.65b**, the reaction occurs via an S_N1 reaction, in which there are three steps (proton transfer, followed by loss of a leaving group, and then nucleophilic attack). As such, the energy diagram must show three steps (three humps), as shown here.

Notice that there are two intermediates (valleys) in the energy diagram. The first intermediate is an oxonium ion, while the second intermediate is a carbocation. We have seen that oxonium ions are lower in energy than carbocations, and this is shown in the energy diagram (the first intermediate is lower in energy than the second).

7.66.

(a) A polar aprotic solvent is used, and the reaction occurs with inversion of configuration. These factors indicate an S_N2 process.

(b) In an S_N2 process, nucleophilic attack and loss of the leaving group occur in a concerted fashion (in one step), as shown below.

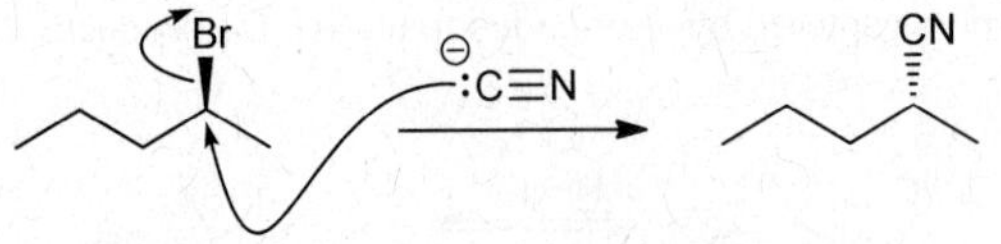

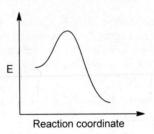

(c) Since the reaction is an S_N2 process, we expect a second-order rate equation that is linearly dependent on both the concentration of the substrate and the concentration of the nucleophile.

$$\text{Rate} = k \left[\begin{array}{c} \text{Br} \\ \end{array} \right] \left[\text{NaCN} \right]$$

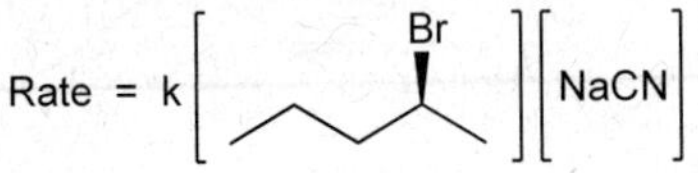

(d) The rate of an S_N2 reaction is linearly dependent on the concentration of the nucleophile. As such, if the concentration of the nucleophile (cyanide) is doubled, the reaction rate is expected to double.

(e) As seen in the solution to **7.66b**, the reaction occurs via an S_N2 reaction, which is comprised of one concerted step (in which the nucleophile attacks with simultaneous loss of the leaving group). As such, the energy diagram will have only one hump (no intermediates).

E

Reaction coordinate

7.67. The substrate is tertiary and the nucleophile (H_2O) is weak, indicating S_N1 conditions. Loss of the leaving group (iodide) generates a tertiary carbocation, which undergoes a hydride shift to give another tertiary carbocation that is also resonance-stabilized (the initially formed, tertiary carbocation was not resonance-stabilized, so this carbocation rearrangement is favorable). The resonance-stabilized carbocation is then captured by the nucleophile (H_2O) to give an oxonium ion, which loses a proton to generate the product. Water serves as the base for this proton transfer step, as shown here.

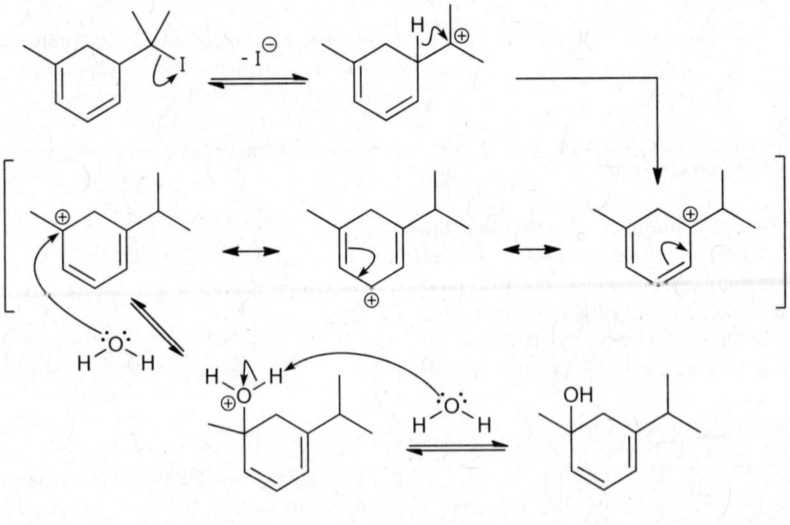

7.68.

(a) The nucleophile is iodide and the solvent is a polar aprotic solvent (DMF), indicating an S_N2 reaction. The substrate (which is primary) has an electrophilic center shown here.

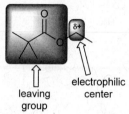

leaving
group

electrophilic
center

Iodide attacks this position in an S_N2 process, ejecting the leaving group, as shown.

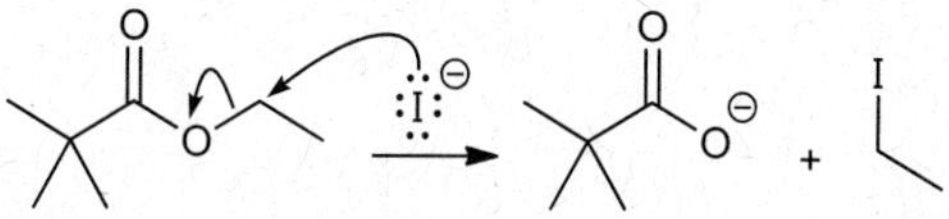

(b) This reaction occurs via an S_N2 process. As such, the rate of the reaction is highly sensitive to the nature of the substrate. The reaction will be faster in this case, because the methyl ester is less sterically hindered than the ethyl ester.

7.69. A strong base will remove the most acidic proton in the starting alcohol (the proton of the OH group), giving an anion that contains both a nucleophilic center and an electrophilic center, allowing for an intramolecular S_N2-type process (bromide is ejected as a leaving group), as shown here.

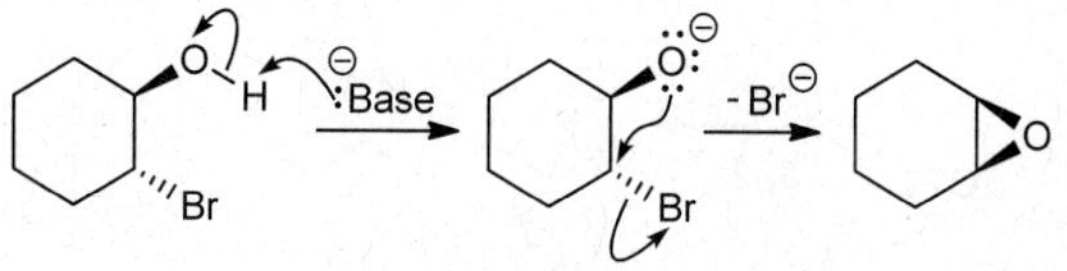

7.70. Iodide is a much stronger nucleophile than ethanol, and the former attacks butyl bromide (a primary substrate) in an S_N2 reaction to give butyl iodide.

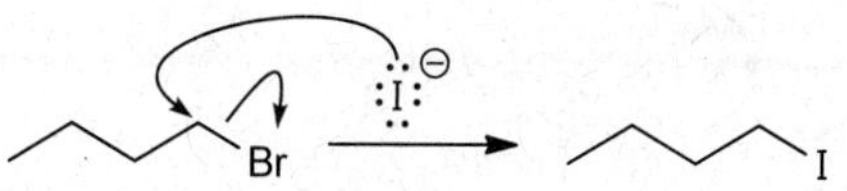

As a result of this initial rapid reaction, the concentration of iodide quickly decreases. Then, in the presence of ethanol (a weak nucleophile), a slow S_N2 process occurs

in which butyl iodide functions as the substrate (iodide is an excellent leaving group) and ethanol functions as the nucleophile. The resulting oxonium ion is then deprotonated (by ethanol, this time functioning as a base), giving the ether shown here.

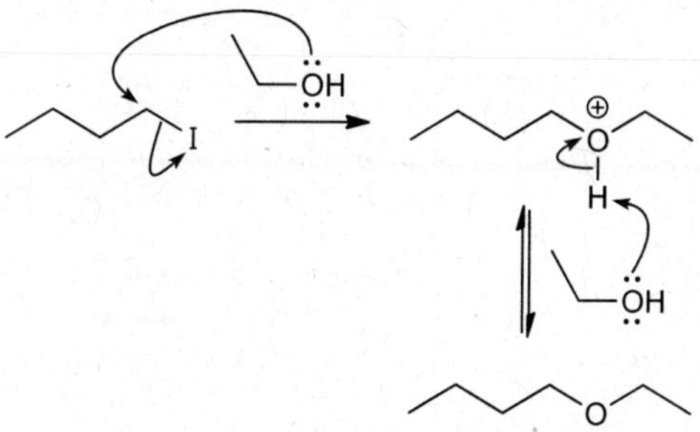

7.71.
Primary substrates generally do not readily undergo S_N1 reactions, because the intermediate primary carbocation is too high in energy to form at an appreciable rate. However, in this case, loss of the leaving group generates a resonance-stabilized cation. Because this intermediate is stabilized, it can form at an appreciable rate, allowing an S_N1 process to successfully compete, despite the fact that the substrate is primary.

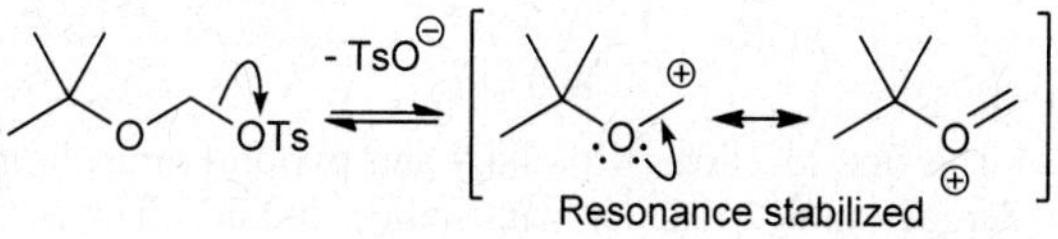

Resonance stabilized

7.72.
Iodide is a very good nucleophile (because it is polarizable), and it is also a very good leaving group (because it is a very weak base). As such, iodide will function as a nucleophile to displace the chloride ion. Once installed, the iodide group is a better leaving group than chloride, thereby increasing the rate of the reaction.

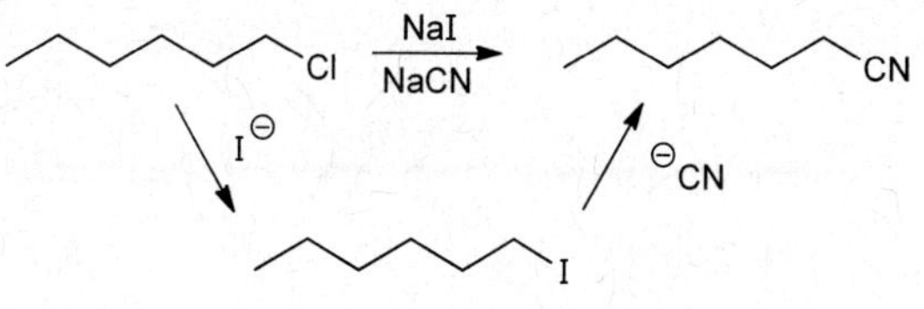

7.73. In the presence of aqueous acid, the OH group of the starting alcohol is protonated, giving an oxonium ion. Loss of a leaving group (H$_2$O) then gives a resonance-stabilized carbocation that can be captured by water to give another oxonium ion. Deprotonation (with water functioning as the base) generates the product, as shown here.

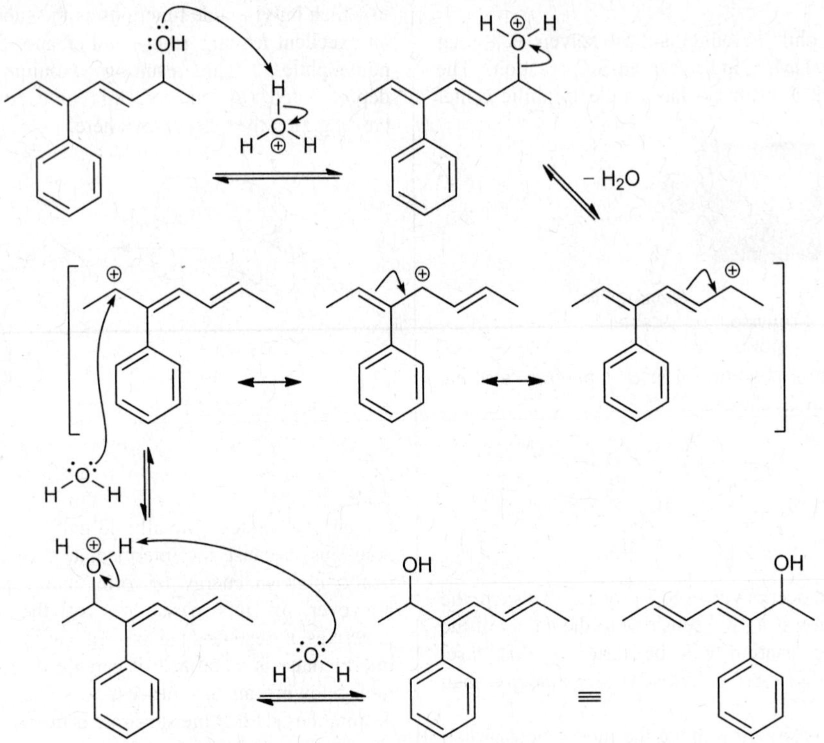

7.74.

(a) The first reaction (with TsCl and pyridine) transforms the OH group into a good leaving group (a tosylate) that undergoes an S$_N$2 reaction with sodium iodide. The net result of these two steps is the conversion of an alcohol to an alkyl iodide.

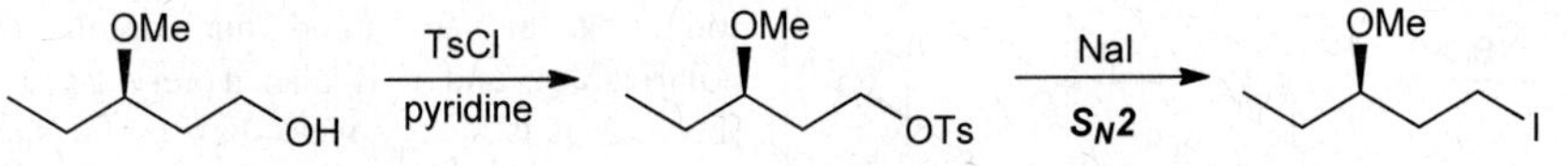

In the final step of the process, the primary alkyl iodide is then treated with triphenylphosphine (PPh$_3$), which functions as a nucleophile, giving another S$_N$2 reaction, to afford a phosphonium salt.

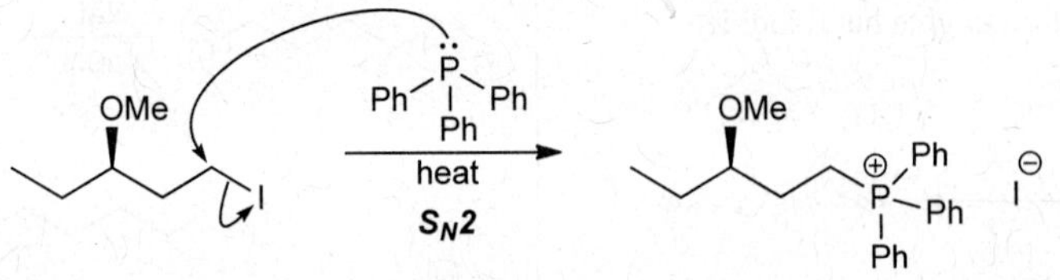

The factors favoring this step are the leaving group and the nucleophile. Let's explore each separately. The leaving group is one of the best leaving groups that could be used (because HI is one of the strongest acids, pK_a = -10). The nucleophile is PPh$_3$. Why is it such a powerful nucleophile? Phosphorus is in the same column of the periodic table as nitrogen (5A), but it is in the third row, rather than the second row. As such, phosphorus is larger and more polarizable than nitrogen, and therefore more strongly nucleophilic. This argument is similar to the argument we saw in the text when we compared sulfur and oxygen. Recall that sulfur is larger and more polarizable than oxygen, and therefore, sulfur is very strongly nucleophilic, even if it lacks a negative charge. Similarly, PPh$_3$ is a powerful nucleophile, even though it lacks a negative charge.

(b) In the second step, one leaving group (tosylate) is replaced with another (iodide). If we compare the pK_a of HI (-10) with the pK_a of sulfonic acids (approximately -0.6), we see a difference of approximately 9 pK_a units. In other words, HI is approximately 10^9 times (1 billion times) more acidic than sulfonic acids. As a result, we expect iodide to be a *significantly* better leaving group than a tosylate (a billion times better). A tosylate group is certainly a good leaving group; nevertheless, iodide is a *much* better leaving group. This is useful, because it renders step 3 more favorable.

7.75.
(a) The proton connected to the oxygen atom is the most acidic proton in compound **1**, so it is removed upon treatment with a strong base.

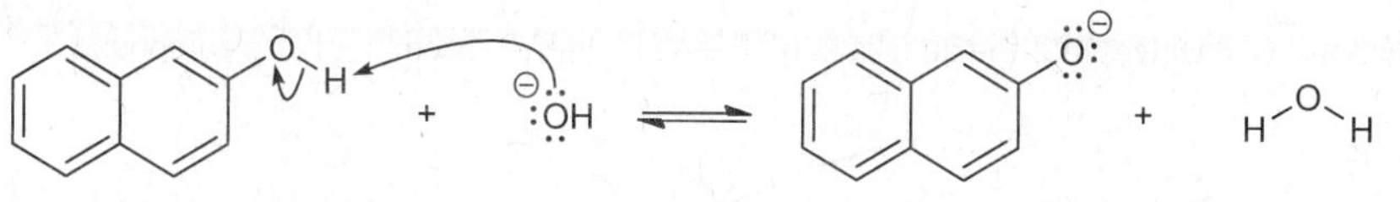

We can justify that hydroxide is a suitable base to achieve the conversion of **1** to **2**, with either a qualitative argument (based on structural comparisons) or with a quantitative argument (based on pK_a values). Let's start with the qualitative argument. Compare the structures of the anions on either side of the reaction.

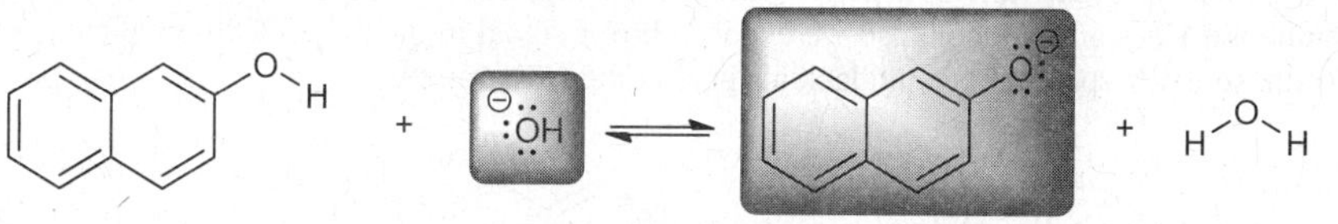

The negative charge in a hydroxide ion is localized on one oxygen atom, while the negative charge in the other anion is delocalized over one oxygen atom and five carbon atoms. As such, we expect the latter anion to be more stabilized (via resonance delocalization).

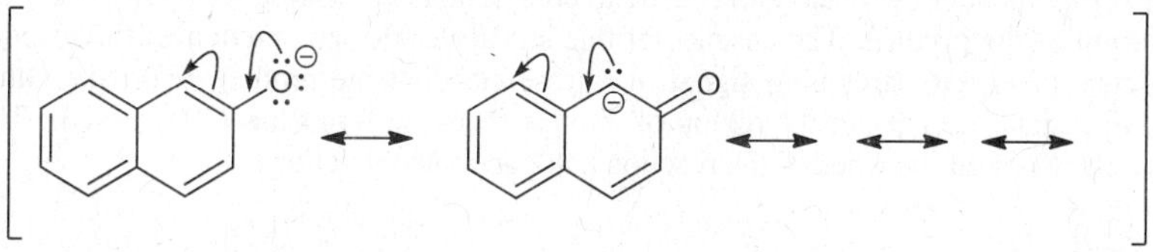

The equilibrium will favor formation of the more stable anion. That is, hydroxide is a sufficiently strong base, because it is stronger (less stable) than anion **2**.

Alternatively, we can use a quantitative argument to justify why hydroxide is an appropriate base to use in this case. Specifically, we compare the pK_a values of the acids on either side of the equilibrium.

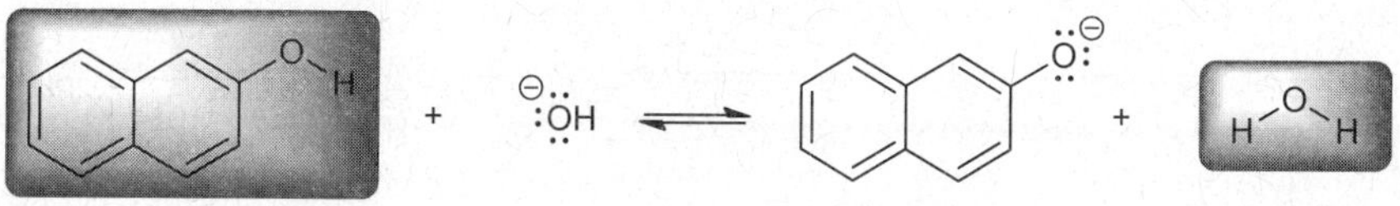

We know that the pK_a of water is 15.7, but we need a way to assess the pK_a of compound **1**. When we explore Table 3.1, we see that phenol is similar in structure, and has a pK_a of 9.9.

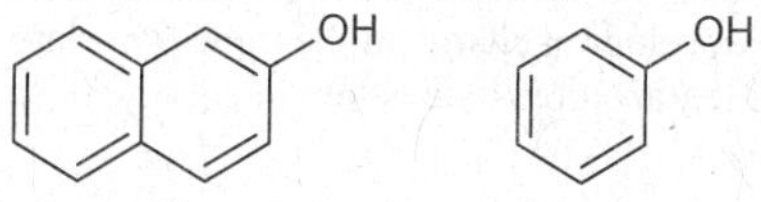

2-naphthol   phenol

We expect the pK_a of compound **1** (2-naphthol) to be more similar to the pK_a of phenol (than to the pK_a of water).

Therefore, we expect the pK_a of compound **1** to be lower than the pK_a of water, and the equilibrium will favor formation of the weaker acid.

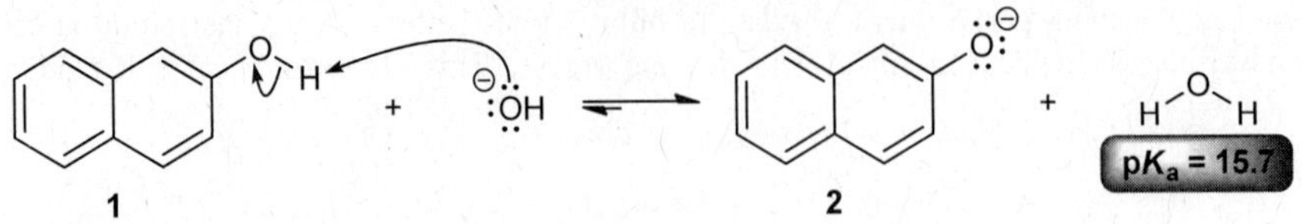

1
$pK_a \sim 10$

2

$pK_a = 15.7$

As such, hydroxide is a suitable base to favor deprotonation of compound **1**.

(b) Anion **2** functions as a nucleophile and attacks butyl iodide in an S_N2 reaction, giving compound **3**.

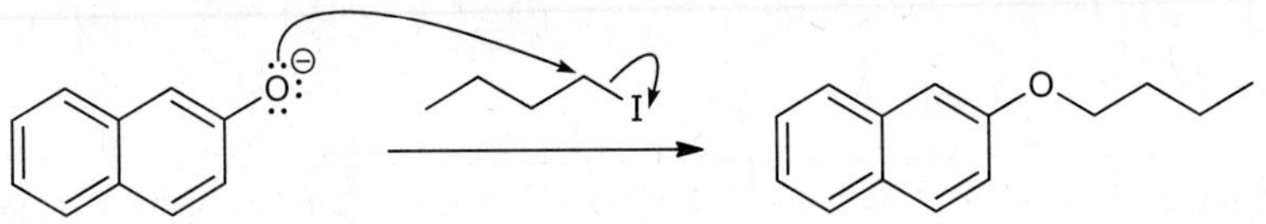

2

3

(c) If compound **3** was properly dried, then it could be distinguished from compound **1** with IR spectroscopy. Specifically, compound **1** has an O-H bond, so we expect a broad signal in the range 3200-3600 cm^{-1}, while compound **3** lacks such a bond, so its IR spectrum should lack a signal in the same range.

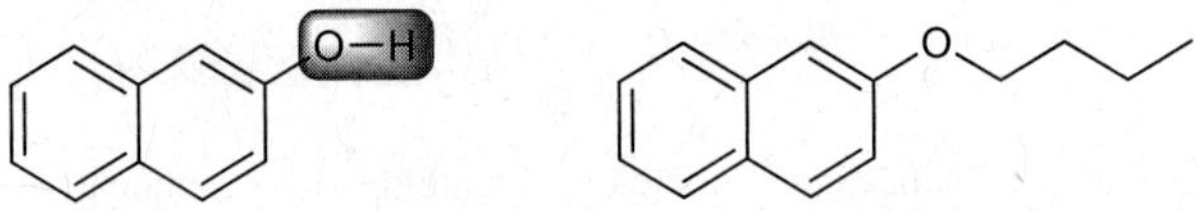

1

3

In this way, we can verify whether the reaction has gone to completion, by looking for a signal in the range 3200-3600 cm^{-1} in the IR spectrum of the product. The absence of this signal verifies completion of the reaction. Since we are looking for the absence of an OH stretching signal, it is essential that the product is dried. Otherwise, the water molecules would give a signal exactly in the region of interest (because water has O-H bonds). This would prevent from us from being able to determine whether the reaction had gone to completion.

7.76.
(a) The substrate is tertiary, so we expect the reaction to proceed exclusively through an S_N1 pathway (steric crowding prevents S_N2 from competing). The first step involves loss of a bromide to give a carbocation, which is then captured by the nucleophile to produce an oxonium ion. Deprotonation of the oxonium ion gives the product, **2b**.

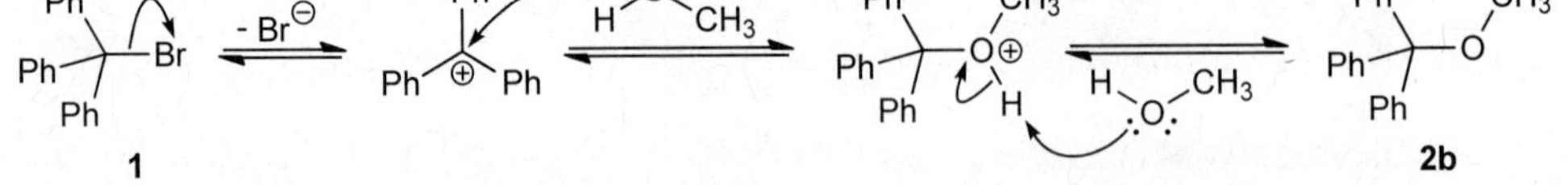

1

2b

(b) When bromide leaves, the resulting carbocation is benzylic to three different aromatic rings. As such, it is highly stabilized because the positive charge is delocalized over 10 carbon atoms via resonance (see below). Since this intermediate is so stabilized (low in energy), we can infer that the transition state for formation of the carbocation will also be very low in energy (because any developing charge in the transition state is stabilized by resonance, just as seen in the intermediate carbocation). Since the transition state is low in energy, this step will occur very rapidly.

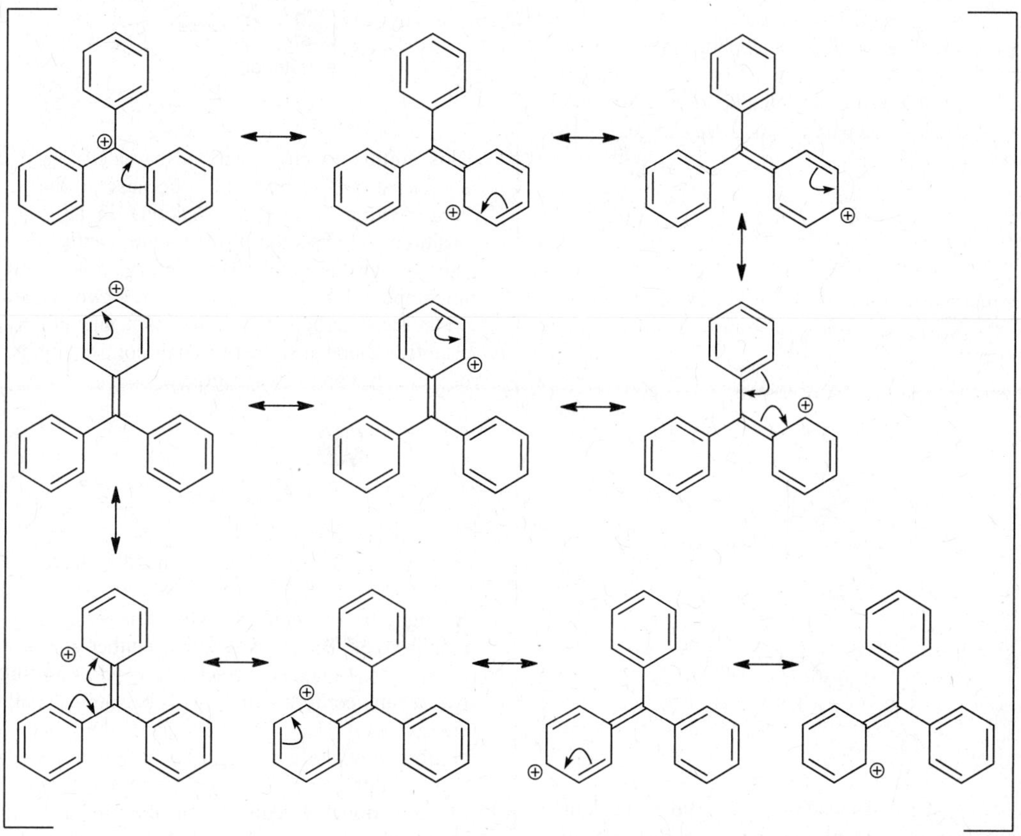

(c) The conversion of **1** to **2a** involves introduction of an OH group, which should produce a broad, easily detectable signal in the range 3200-3600 cm⁻¹. Therefore, by taking an IR spectrum of the product, we can verify formation of **2a** by looking for a broad signal in the range 3200-3600 cm⁻¹. In contrast, **2b** and **2c** do not have an OH group. As such, it will be difficult to distinguish the diagnostic regions of the IR spectra of compounds **1**, **2b**, and **2c**. The utility of IR in spectroscopy in these cases must rely on analysis of the fingerprint region (C-Br stretch vs. C-O stretch), although fingerprint regions are often more difficult to interpret (except in the hands of a trained expert). Therefore, IR spectroscopy is not the best tool for verifying the conversion of **1** to either **2b** or **2c**. NMR spectroscopy would be a better tool for confirming completion of those reactions.

7.77. The substrate is a primary alkyl halide, so we expect an S_N2 reaction to occur if the substrate is treated with a suitable nucleophile. The product obtained is not the product that would be expected if *tert*-butoxide had functioned as a nucleophile and attacked the alkyl halide directly. Something else is happening here. Careful inspection of the product indicates that a bond is formed between two carbon atoms. This indicates that the nucleophilic center must have been a carbon atom within the substrate. That is, the nucleophile and electrophile are tethered to each other (rather than being separate compounds), and the reaction occurs in an intramolecular fashion.

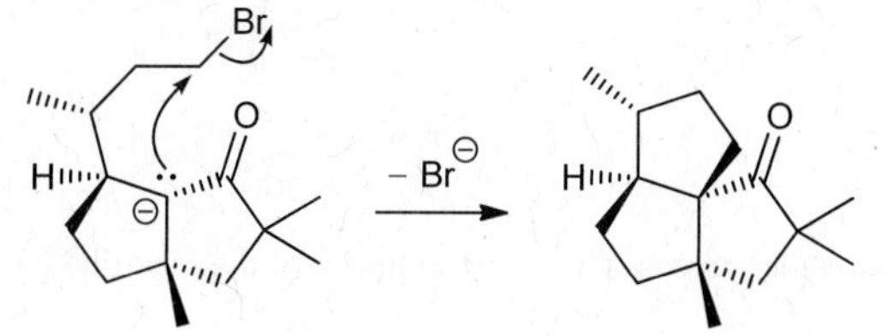

How is this nucleophilic center formed in the first place? That explains the role of *tert*-butoxide – to function as a base and remove a proton from the starting material to form a resonance stabilized anion, which then undergoes an intramolecular S_N2-type reaction.

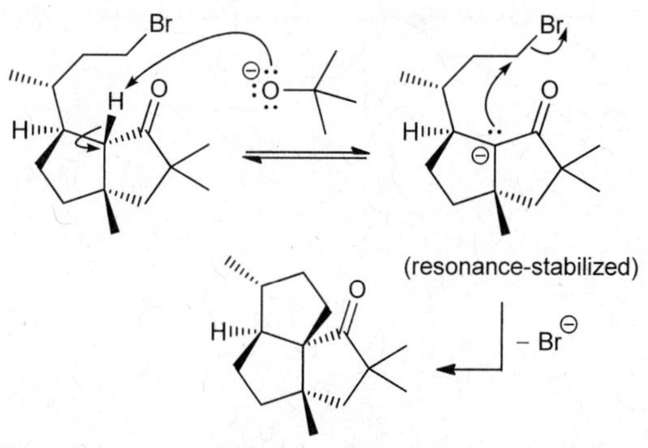

7.78.

(a) Compound **2** is the nucleophile in this S_N2 reaction. To see why, recall from Chapter 1 that a C-Li bond can be viewed as an ionic bond, in which the carbon atom has a lone pair and negative charge.

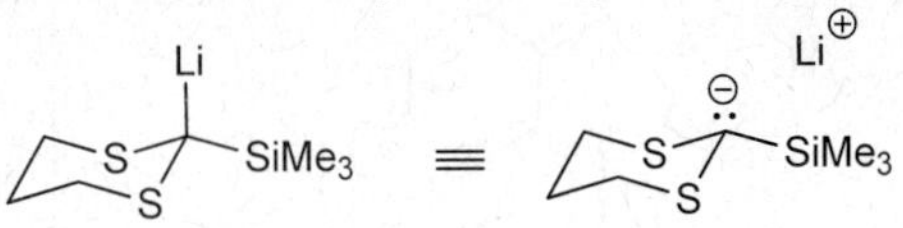

This compound is indeed a very strong nucleophile, and it attacks the alkyl halide in an S_N2 process, as shown here:

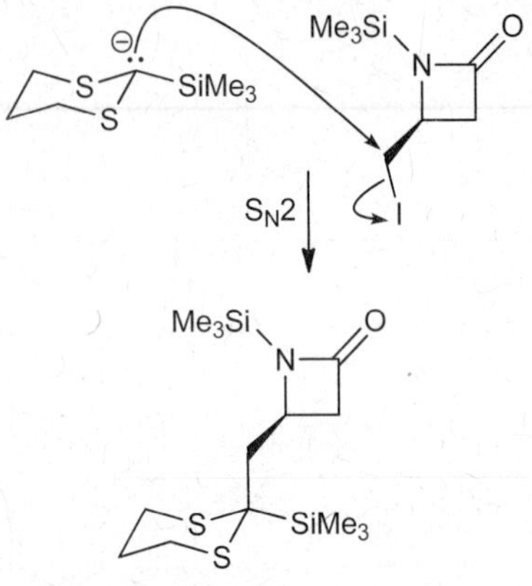

(b) Compound **2** will spend most of its time in a chair conformation in which the bulky SiMe₃ group occupies an equatorial position.

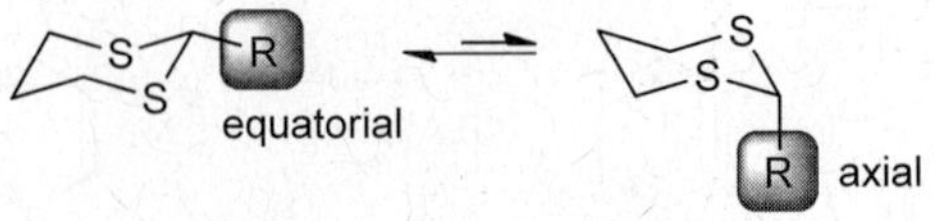

This dithiane ring will experience very little conformational freedom because one chair conformation (with the substituent in an equatorial position) is expected to be significantly lower in energy than the other chair conformation. However, in compound **3**, the dithiane ring has two substituents on the same carbon atom. Therefore, an equilibrium is established between the two chair conformations, each of which has an axial substituent.

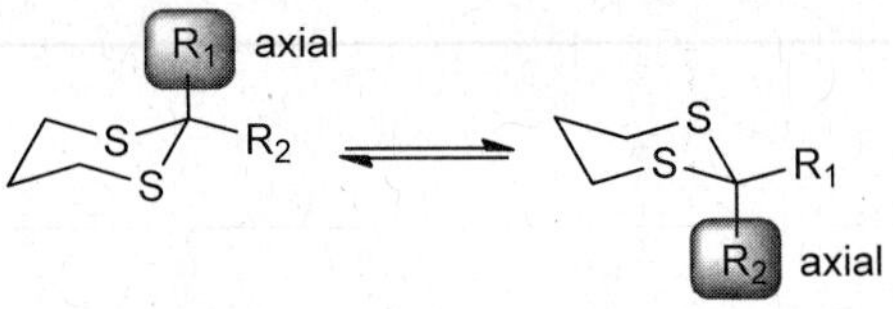

It might be difficult to predict the position of this equilibrium. But we know that neither conformation is free of 1,3-diaxial interactions. And as a result, the two chair conformations will be more similar in energy (to each other) than the two chair conformations of compound **2**. Therefore, we expect that the dithiane ring in compound **3** will have more conformational freedom than the dithiane ring in compound **2**.

7.79.

In the absence of dioxane, the only nucleophile is water (the solvent) which reacts with the optically pure 2-octyl sulfonate through an S_N2 pathway to form optically pure 2-octanol, with inverted stereochemistry.

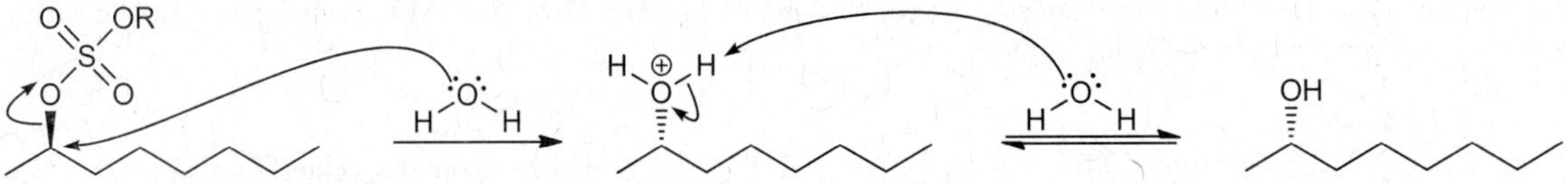

However, if dioxane is also present, it can function as a nucleophile and attack 2-octyl sulfonate in an S_N2 reaction, to form an inverted intermediate that can then undergo another S_N2 reaction with water. The product is 2-octanol with an overall retention of stereochemistry, due to two successive S_N2 steps taking place.

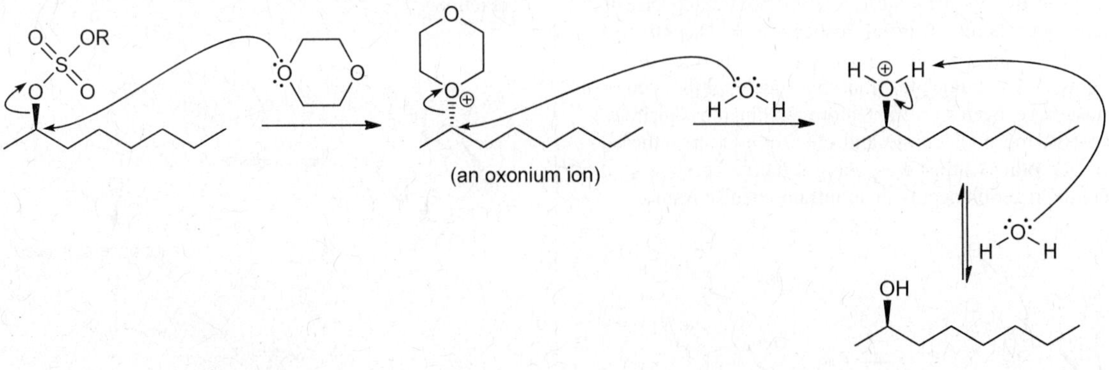

(an oxonium ion)

Since this process increases in frequency as the concentration of dioxane increases, the optical purity of the resulting 2-octanol decreases as dioxane's concentration is increased.

7.80.

(a) Compound **2** functions as a nucleophile, which means that the lone pair on the carbon atom will attack the substrate. Based on the structure of the product, we can deduce that the oxygen atom (next to the *tert*-butyl group) is attacked by the nucleophile. The leaving group is a resonance-stabilized anion (an acetate ion).

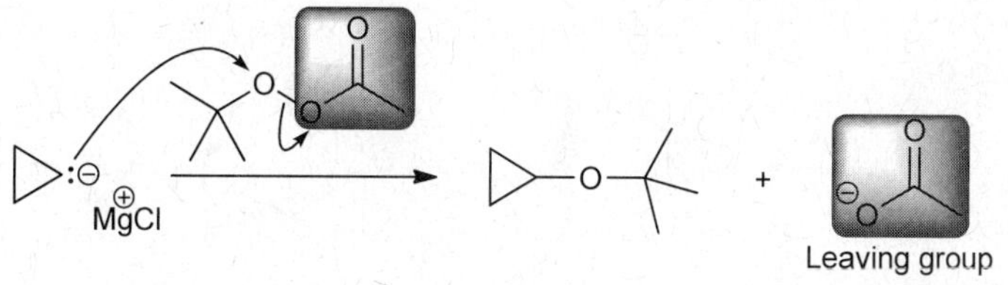

Leaving group

(b) The reverse process would involve an acetate ion functioning as a nucleophile and the expulsion of **2** as a leaving group. That is extremely unlikely to occur, because **2** is not a good leaving group. It is a very strong base, because its conjugate acid is an alkane, which is an extremely weak acid (compare pK_a values of alkanes with other organic compounds). Since **2** is not a weak base, it cannot function as a leaving group. And as a result, the reaction is irreversible.

(c) Since an alkoxide group (RO⁻) cannot function as a leaving group, it must be protonated first, in either pathway. The S_N2 pathway involves a *simultaneous* nucleophilic attack and loss of a leaving group. This step is then followed by deprotonation to give the product. As expected for an S_N2 process, the nucleophilic attack occurs at the secondary position, rather than the sterically crowded tertiary position.

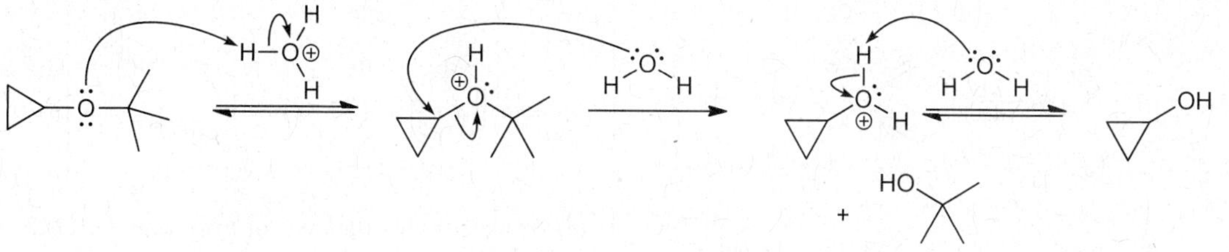

The S_N1 pathway also begins with protonation of the oxygen atom to produce a better leaving group. But in this pathway, the leaving group first leaves to generate a carbocation, and only then does the nucleophile attack. As expected for an S_N1 process, loss of the leaving group generates a tertiary carbocation, rather than a secondary carbocation. In this case, the leaving group is the product (cyclopropanol).

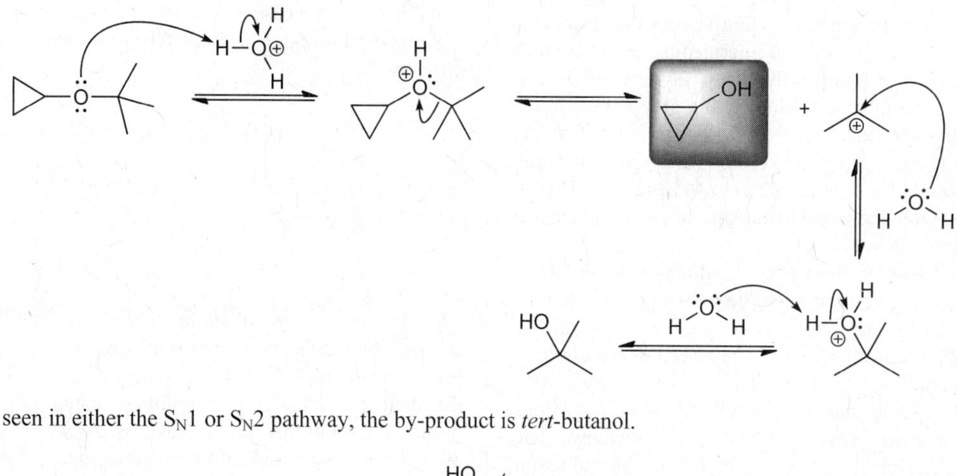

(d) As seen in either the S_N1 or S_N2 pathway, the by-product is *tert*-butanol.

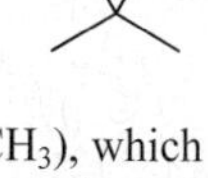

7.81. Compound **1** exhibits a mesylate group (OSO_2CH_3), which is listed in Figure 7.28 (Section 7.8) as a good leaving group. Upon its formation, compound **1** undergoes an <u>intramolecular</u> S_N2-type reaction, in which the nitrogen atom functions as the nucleophile. This results in the formation of a three-membered nitrogen-containing ring fused to a six-membered ring (**3**). Notice the stereochemistry: the nitrogen atom displaced the mesylate from the back face of the molecule, resulting in inversion at the original chirality center. Next, an <u>intermolecular</u> S_N2 reaction occurs: the chloride ion attacks the three-membered ring at the least hindered site, resulting in the formation of the desired product (**2**). Notice that the configuration of the chirality center (that underwent inversion in the first S_N2-type reaction) does not change during this second S_N2 reaction (in this step, chloride attacks a carbon atom that is not a chirality center).

7.82.

(a) Figure 7.27 indicates that bromide is expected to be a better leaving group than chloride, because bromide is a more stable base (HBr is a stronger acid than HCl). This is supported by the hydrolysis data, when we compare the rate of hydrolysis for PhCHCl$_2$ and PhCHBrCl.

In both cases, the first step involves loss of a leaving group, and in both cases, the intermediate carbocation is the same. The only difference between these two reactions is the identity of the leaving group. According to the data provided, hydrolysis occurs more rapidly when the leaving group is bromide ($k = 31.1 \times 10^4$ /min) rather than chloride ($k = 2.21 \times 10^4$ /min). This is consistent with the expectation that bromide is a better leaving group than chloride.

(b) If we compare the rates of hydrolysis for PhCH$_2$Cl and PhCHCl$_2$, we find that the presence of a chloro group (attached to C+) causes an increased rate of hydrolysis. A similar trend is observed if we compare PhCHCl$_2$ and PhCCl$_3$. Therefore, we can conclude that a chloro group will stabilize a carbocation (if the chloro group is attached directly to C+ of the carbocation). This stabilizing effect is unlikely to be caused by induction, because we expect the chloro group to be electron-

withdrawing via induction, which would destabilize the carbocation (rather than stabilize it). Instead, the effect must be explained with resonance, which overwhelms the inductive effect. Specifically, the presence of a chloro group stabilizes the carbocation intermediate by spreading the charge via resonance.

(c) If we compare the rates of hydrolysis for PhCHBr$_2$ and PhCBr$_3$, we find that the presence of a bromo group (attached to C+) causes an increased rate of hydrolysis. This is likely explained as a resonance effect, just as we saw in the previous part to this problem.

(d) Compare hydrolysis of PhCHBrCl with hydrolysis of PhCHBr$_2$. In both cases, the identity of the leaving group is the same (bromide). But the resulting carbocations are different.

Comparing the rates of hydrolysis indicates that an adjacent chloro group more effectively stabilizes a carbocation than an adjacent bromo group.

(e) Comparison of PhCHCl$_2$ versus PhCHBr$_2$ suggests that the better leaving group ability of the bromide ion is more important than the greater carbocation stability afforded by the chlorine atom via resonance.

7.83.

(a) Treating compound **1** with TsCl and pyridine results in conversion of the OH group to a tosylate group (OTs), which is a better leaving group. Treating **2** with a nucleophile then results in an S_N2 reaction in which the nucleophile is an amine, and the tosylate group functions as a leaving group. This S_N2 process gives an initial ammonium ion (an intermediate with a positively charged nitrogen atom), which loses a proton to give compound **3**.

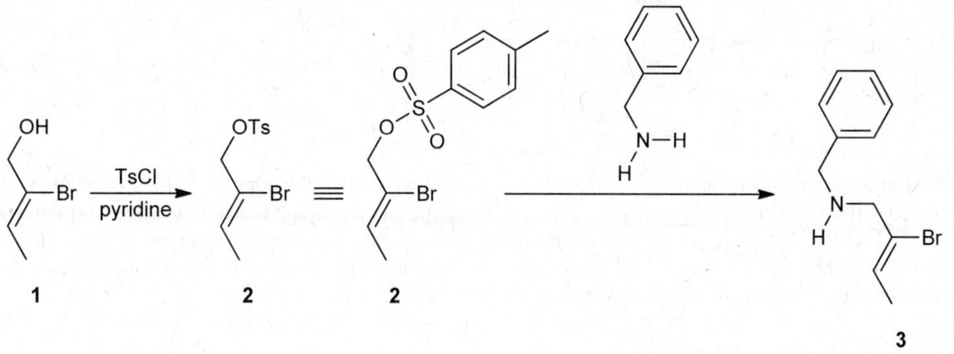

Notice that compound **2** has two leaving groups, but only one is ejected. The other leaving group (bromide) is connected to a vinyl position, and as indicated in the problem statement, an S_N2 reaction will not occur at this position because it is sp^2 hybridized.

(b) Sodium amide functions as a base and removes the proton connected to the nitrogen atom, generating an intermediate that contains a strong nucleophilic center and a strong electrophilic center. The resulting intramolecular, S_N2-type reaction gives compound **4**, via the transition state shown below.

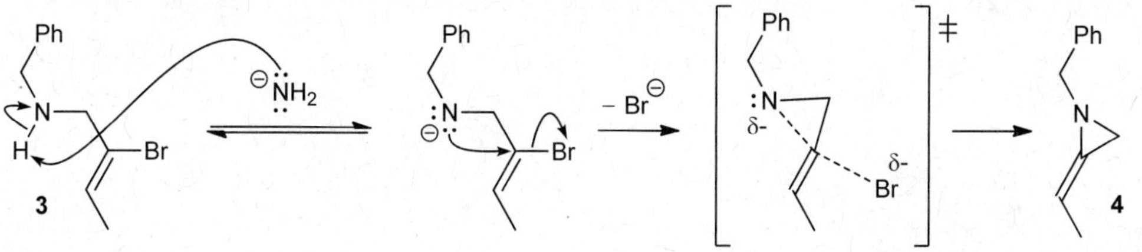

(c) In compound **3**, the methyl group is *trans* to the methylene (CH$_2$) group; but in the product, the methyl group is *cis* to the methylene group.

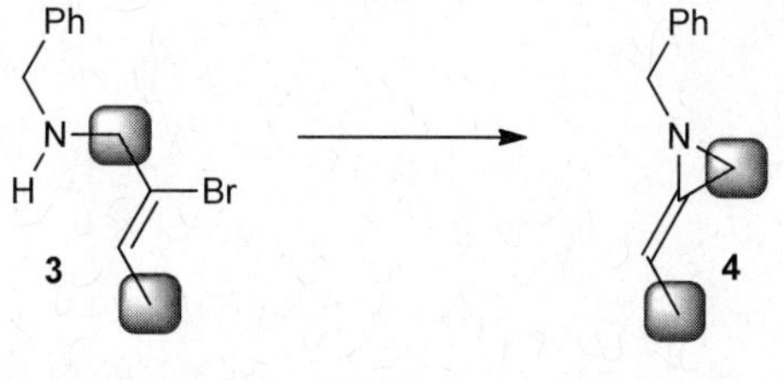

In other words, this transformation proceeds with inversion of configuration at the vinylic center. Inversion of configuration is consistent with an S_N2 mechanism.

7.84. Acetate is a good nucleophile, and CH$_3$CN is a polar aprotic solvent, indicating an S_N2 process. The minor product can form via a simple S_N2 displacement of the bromide by the acetate anion.

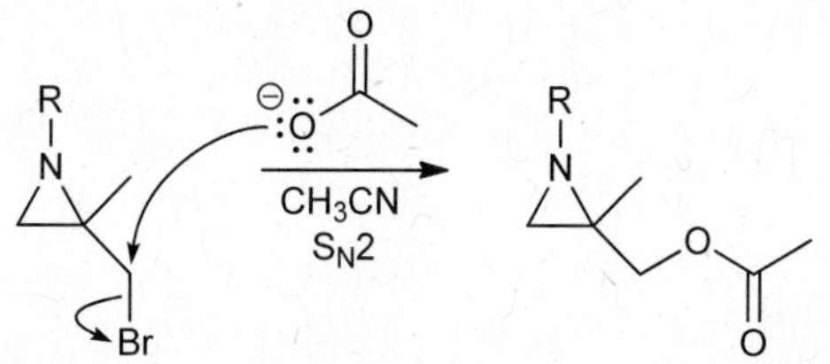

The formation of the major product, however, is not a simple S_N2 displacement. In the first step, the lone pair on the nitrogen of the three-membered ring can function as a nucleophilic center (just as we saw in the medically speaking application at the end of the chapter) and displace the primary bromide in an S_N2 fashion, forming a four-membered intermediate. This intermediate then rapidly opens to a carbocation to alleviate ring strain. Next, the acetate ion traps the carbocation in an S_N1 process to form the major product.

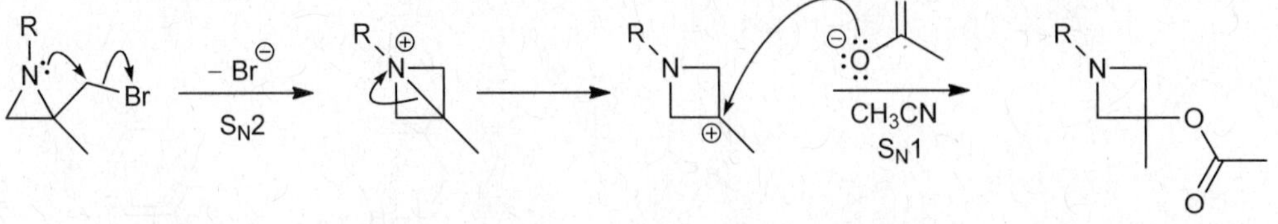

7.85. The sulfur atom provides anchimeric assistance via an intramolecular nucleophilic S_N2-type reaction. That is, a lone pair on the sulfur atom functions as a nucleophile, ejecting the leaving group (causing the liberation of SO_2 gas, as described in the problem statement) to form an intermediate with a positively charged sulfur atom. This intermediate is then attacked by a chloride ion to give **3**. Each of these two steps proceeds with inversion of configuration, as expected, which gives a net overall retention of configuration.

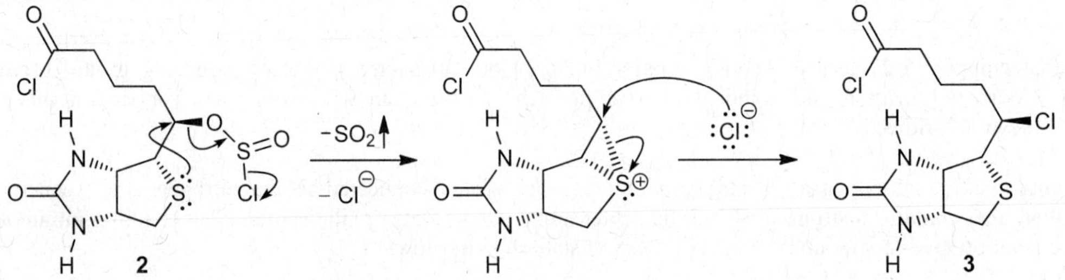

Chapter 8
Alkenes: Structure and
Preparation via Elimination Reactions

Review of Concepts

Fill in the blanks below. To verify that your answers are correct, look in your textbook at the end of Chapter 8. Each of the sentences below appears in the section entitled *Review of Concepts and Vocabulary*.

- Alkene stability increases with increasing degree of _____.
- E2 reactions are said to be **regioselective**, because the more substituted alkene, called the _____ **product**, is generally the major product.
- When both the substrate and the base are sterically hindered, the less substituted alkene, called the _____ **product,** is the major product.
- E2 reactions are **stereospecific** because they generally occur via the _____ conformation.
- Substituted cyclohexanes only undergo E2 reactions from the chair conformation in which the leaving group and the proton both occupy _____ positions.
- E1 reactions exhibit a regiochemical preference for the _____ product.
- E1 reactions are not stereospecific, but they are stereo_____.
- Strong nucleophiles are compounds that contain a _____ and/or are _____.
- Strong bases are compounds whose conjugate acids are _____.

Review of Skills

Fill in the blanks and empty boxes below. To verify that your answers are correct, look in your textbook at the end of Chapter 8. The answers appear in the section entitled *SkillBuilder Review*.

8.1 Assembling the Systematic Name of an Alkene

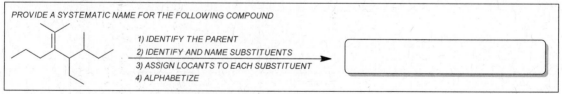

8.2 Assigning the Configuration of a double bond

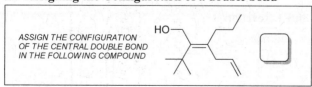

8.3 Comparing the Stability of Isomeric Alkenes

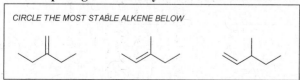

8.4 Drawing the Curved Arrows of an Elimination Reaction

A CONCERTED MECHANISM

DRAW THREE CURVED ARROWS, SHOWING A PROTON TRANSFER ACCOMPANIED BY SIMULTANEOUS LOSS OF A LEAVING GROUP

A STEPWISE MECHANISM

DRAW A CURVED ARROW SHOWING THE LOSS OF THE LEAVING GROUP TO FORM A CARBOCATION INTERMEDIATE, FOLLOWED BY ANOTHER TWO CURVED ARROWS SHOWING A PROTON TRANSFER

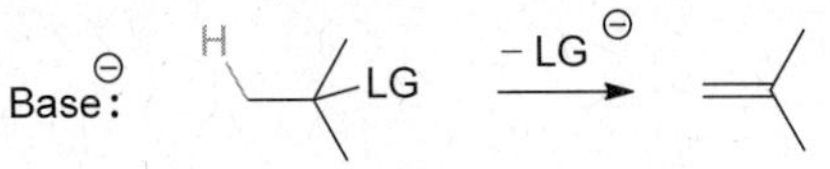

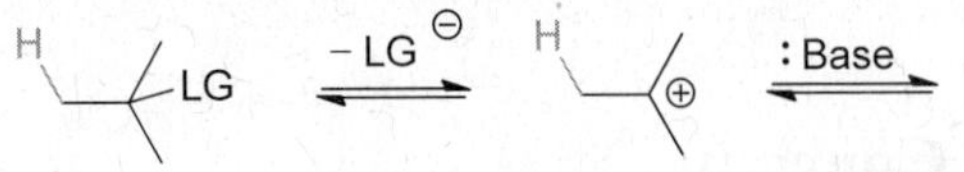

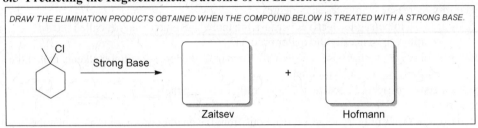

8.5 Predicting the Regiochemical Outcome of an E2 Reaction

DRAW THE ELIMINATION PRODUCTS OBTAINED WHEN THE COMPOUND BELOW IS TREATED WITH A STRONG BASE.

Strong Base

Zaitsev   +   Hofmann

8.6 Predicting the Stereochemical Outcome of an E2 Reaction

PREDICT THE STEREOCHEMICAL OUTCOME OF THE FOLLOWING REACTION, AND DRAW THE PRODUCT.

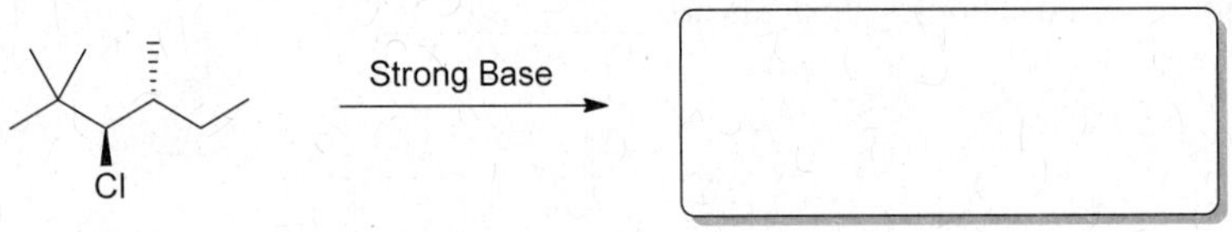

Strong Base

8.7 Drawing the Products of an E2 Reaction

PREDICT THE MAJOR AND MINOR PRODUCTS OF THE FOLLOWING REACTION.

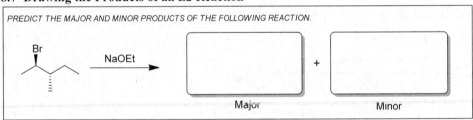

NaOEt

Major   +   Minor

8.8 Predicting the Regiochemical Outcome of an E1 Reaction

PREDICT THE MAJOR AND MINOR PRODUCTS OF THE FOLLOWING REACTION.

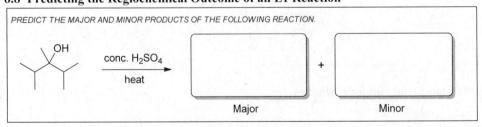

conc. H_2SO_4

heat

Major   +   Minor

8.9 Drawing the Complete Mechanism of an E1 Reaction

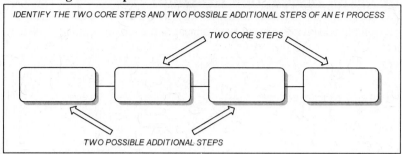

8.10 Determining the Function of a Reagent

IDENTIFY REAGENTS THAT FALL INTO EACH OF THE FOUR CATEGORIES BELOW:

NUCLEOPHILE (ONLY)	BASE (ONLY)	STRONG NUC / STRONG BASE	WEAK NUC / WEAK BASE

8.11 Identifying the Expected Mechanism(s)

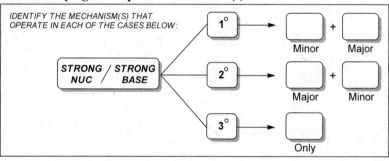

8.12 Predicting the Products of Substitution and Elimination Reactions

FILL IN THE BLANKS BELOW:

STEP 1	STEP 2	STEP 3
DETERMINE THE FUNCTION OF THE _____	ANALYZE THE _____ AND DETERMINE THE EXPECTED MECHANISM(S).	CONSIDER ANY RELEVANT REGIOCHEMICAL AND _____ REQUIREMENTS

Review of Synthetically Useful Elimination Reactions

Identify reagents that will achieve each of the transformations below. To verify that your answers are correct, look in your textbook at the end of Chapter 8. The answers appear in the section entitled *Review of Synthetically Useful Elimination Reactions.*

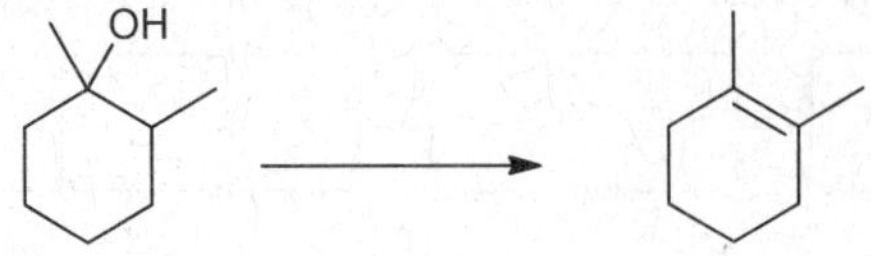

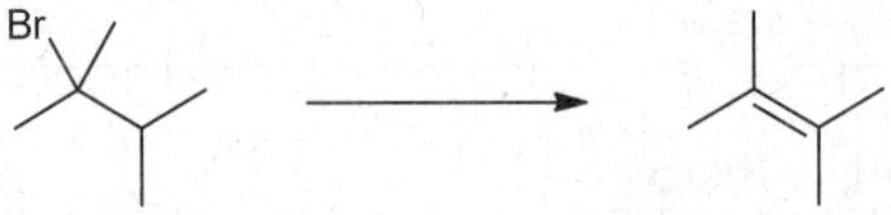

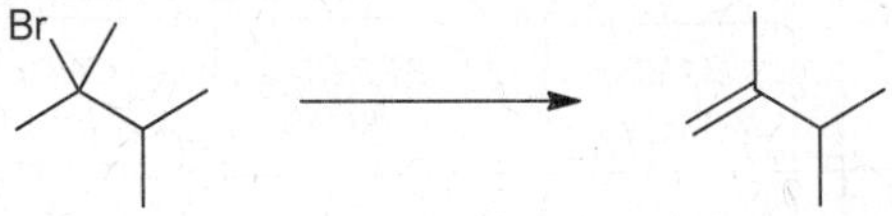

Useful reagents

The following is a list of commonly encountered reagents for substitution and elimination reactions:

Reagent	Name	Function
NaX (X = Cl, Br, or I)	Sodium halide	An ionic salt consisting of Na^+ and X^- ions (where X = Cl, Br, or I). The former (Na^+) can be ignored in most cases, while the latter (chloride, bromide, or iodide) is a strong nucleophile.
NaH	Sodium hydride	Hydride is a strong base that can be used in E2 reactions.
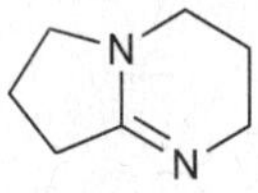	DBN	A strong base that can be used in E2 reactions.
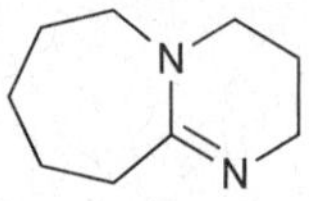	DBU	A strong base that can be used in E2 reactions.
NaOH	Sodium hydroxide	Hydroxide (HO^-) is both a strong nucleophile AND a strong base, and can therefore be used for either E2 or S_N2 reactions, depending on the substrate (S_N2 is favored for primary substrates, E2 is favored for secondary substrates, and E2 is the exclusive pathway for tertiary substrates).
NaOR	Sodium alkoxide	R is an alkyl group. Examples include sodium methoxide (NaOMe) and sodium ethoxide (NaOEt). Alkoxide ions are both strong nucleophiles and strong bases. They can therefore be used for either E2 or S_N2 reactions, depending on the substrate (S_N2 is favored for primary substrates, E2 is favored for secondary substrates, and E2 is the exclusive pathway for tertiary substrates).
t-BuOK	Potassium *tert*-butoxide	*tert*-Butoxide is both a strong nucleophile and a strong base. But it is sterically hindered, which favors E2 over S_N2 even for primary substrates. For E2 reactions, when more than one regiochemical outcome is possible, *tert*-butoxide will favor formation of the less substituted alkene.

NaSH	Sodium hydrosulfide	HS⁻ is a strong nucleophile, used in S_N2 reactions.
H_2O	Water	Water is a weak nucleophile and a weak base, used in S_N1 and E1 reactions. Heat will often favor E1 over S_N1.
ROH	An alcohol	Examples include methanol (CH_3OH) and ethanol (CH_3CH_2OH). Alcohols are weak nucleophiles and weak bases, used in S_N1 and E1 reactions. Heat will often favor E1 over S_N1. Alcohols can also serve as substrates for S_N1 or E1 reactions under acidic conditions.
HX (X = Cl, Br, or I)	Hydrogen halides	A strong acid that serves as both a source of H^+ and nucleophilic X^- where X = Cl, Br, or I.
conc. H_2SO_4	Concentrated sulfuric acid	A strong acid, used to convert alcohols into alkenes via an E1 process.

Common Mistakes to Avoid

We have seen that an OH group is a bad leaving group (because hydroxide is a strong base). Therefore, in order for an alcohol (ROH) to serve as a substrate in a substitution or elimination reaction, the OH group must first be converted into a better leaving group. We have seen two ways to do this. One method involves converting the alcohol into a tosylate:

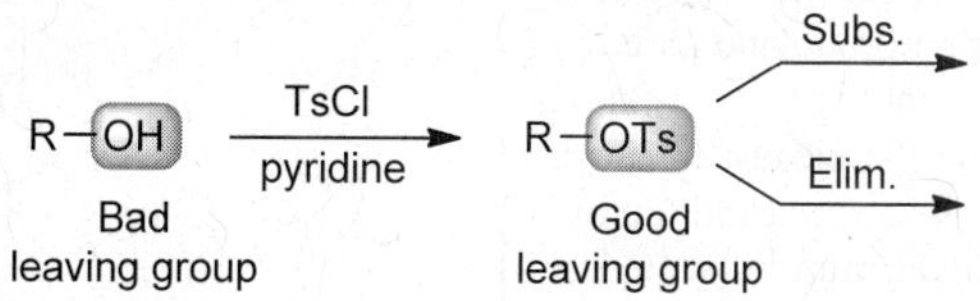

The other method involves protonation of the OH group, as seen in the following example:

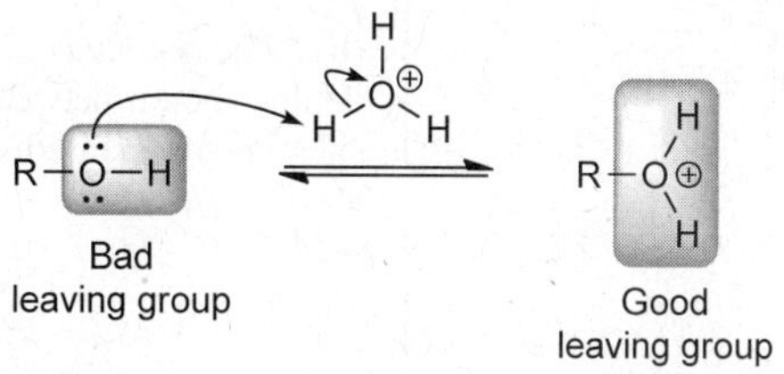

This latter approach (protonation) has a serious limitation. Specifically, it cannot be used if the reagent is a strong base. For example, the following reaction sequence does not work:

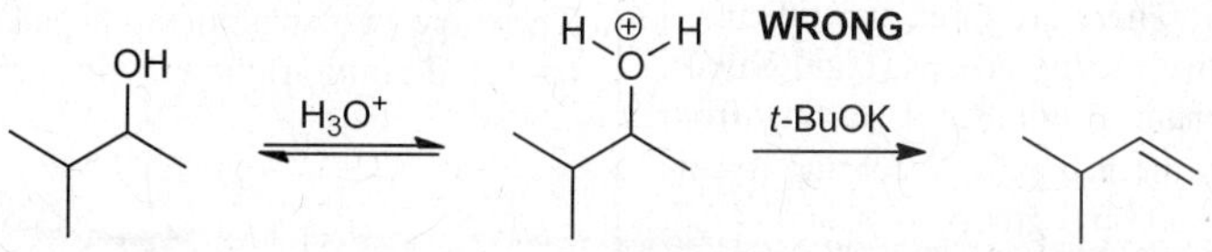

It doesn't work because it is not possible to have a strong acid (H_3O^+) and a strong base (*t*-BuOK) present in the same reaction flask at the same time (they would simply neutralize each other). In order to achieve the desired transformation, the OH group must first be converted to a tosylate (rather than simply being protonated), and then the desired reaction can be performed, as shown here:

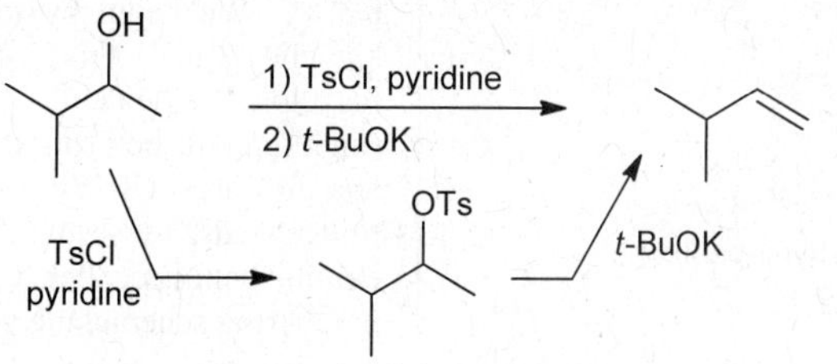

Solutions

8.1.

(a) We begin by identifying the parent. The longest chain is seven carbon atoms, so the parent is heptene. There are three substituents (highlighted), all of which are methyl groups. Notice that the parent chain is numbered starting from the side that is closest to the π bond. According to this numbering scheme, the methyl groups are located at C2, C3, and C5. Finally, we use the prefix "tri" to indicate the presence of three methyl groups, and we include a locant that identifies the position of the double bond ("2" indicates that the double bond is located between C2 and C3):

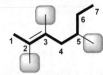

2,3,5-trimethyl-2-heptene

(b) We begin by identifying the parent. The longest chain is seven carbon atoms, so the parent is heptene. There are two substituents – a methyl group and an ethyl group (highlighted). Notice that the parent chain is numbered starting from the side that is closest to the π bond. According to this numbering scheme, the methyl group is located at C2, and the ethyl group is located at C3. Finally, we arrange the substituents alphabetically, and we include a locant that identifies the position of the double bond:

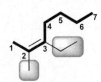

3-ethyl-2-methyl-2-heptene

(c) We begin by identifying the parent. The longest chain (containing the double bond) is five carbon atoms, so the parent is pentene. There are three substituents – an isopropyl group and two methyl groups (highlighted). Notice that the parent chain is numbered starting from the side that is closest to the π bond. According to this numbering scheme, the isopropyl group is located at C3, and the methyl groups are located at C2 and C4. Finally, we arrange the substituents alphabetically, and we include a locant that identifies the position of the double bond:

3-isopropyl-2,4-dimethyl-1-pentene

(d) We begin by identifying the parent. The longest chain is seven carbon atoms, so the parent is heptene. There is one substituent – a *tert*-butyl group (highlighted). Notice that the parent chain is numbered starting from the side that is closest to the π bond. According to this numbering scheme, the *tert*-butyl group is located at C4. Finally, we include a locant that identifies the position of the double bond:

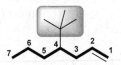

4-*tert*-butyl-1-heptene

8.2.

(a) The parent is five carbon atoms (pentene), with the double bond between C2 and C3. There are three substituents – an isopropyl group at C3, and two methyl groups at C2 and C4:

3-isopropyl-2,4-dimethyl-2-pentene

(b) The parent is six carbon atoms (hexene), with the double bond between C2 and C3. There are two substituents – an ethyl group at C4, and a methyl group at C2:

(c) The parent is a four-membered ring (cyclobutene). There are two substituents located at C1 and C2, which are (by definition) the vinylic positions:

8.3. We begin by identifying the parent, which is bicyclic in this case. The parent is bicyclo[2.2.1]heptene. There are two substituents (highlighted), both of which are methyl groups. Notice that the parent chain is numbered starting from one of the bridgeheads, as seen in Section 4.2, which places the double bond between C2 and C3. According to this numbering scheme, the methyl groups are also located at

C2 and C3. Finally, we include a locant that identifies the position of the double bond (C2).

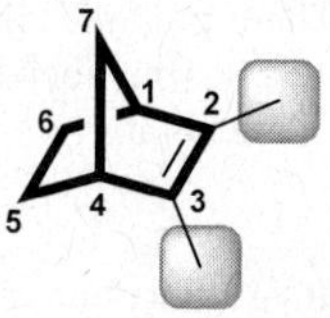

2,3-dimethylbicyclo[2.2.1]hept-2-ene

8.4.
(a) This alkene is trisubstituted because there are three groups (highlighted) connected to the double bond:

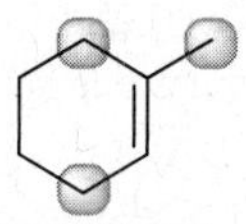

trisubstituted

(b) This alkene is disubstituted because there are two groups (highlighted) connected to the double bond:

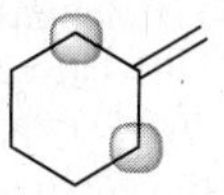

disubstituted

(c) This alkene is trisubstituted because there are three groups (highlighted) connected to the double bond. Notice that one of the groups counts twice because it is connected to both vinylic positions:

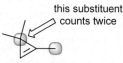

this substituent counts twice

trisubstituted

(d) This alkene is trisubstituted because there are three groups (highlighted) connected to the double bond:

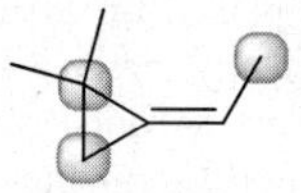

trisubstituted

(e) This alkene is monosubstituted because there is only one group (highlighted) connected to the double bond:

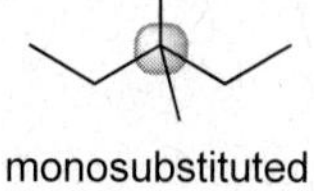

monosubstituted

8.5.
(a) The priorities (as shown below) are on opposite sides of the double bond, so this alkene has the *E* configuration.

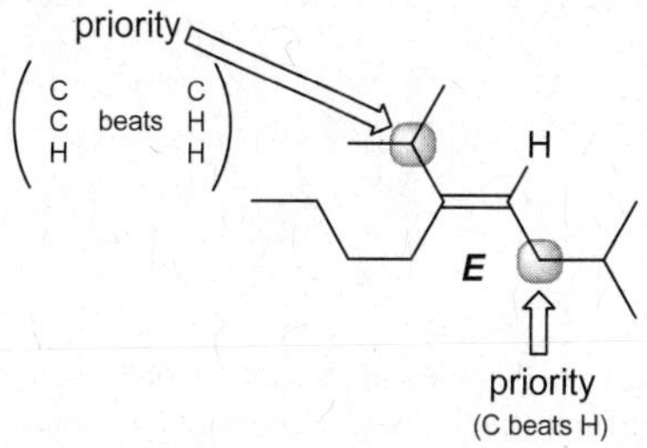

$$\left(\begin{matrix} C \\ C \\ H \end{matrix}\right) \text{ beats } \left(\begin{matrix} C \\ H \\ H \end{matrix}\right)$$

priority
(C beats H)

E

(b) The priorities (as shown below) are on the same sides of the double bond, so this alkene has the *Z* configuration.

priority
$$\left(\begin{matrix} C \\ C \\ H \end{matrix}\right) \text{ beats } \left(\begin{matrix} H \\ H \\ H \end{matrix}\right)$$

priority
$$\left(\begin{matrix} C \\ C \\ C \end{matrix}\right) \text{ beats } \left(\begin{matrix} C \\ C \\ H \end{matrix}\right)$$

Z

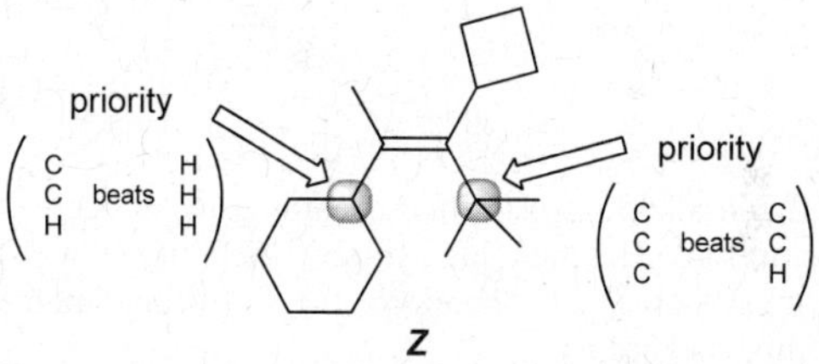

(c) The priorities (as shown below) are on the same sides of the double bond, so this alkene has the *Z* configuration.

priority
(F beats C)

priority
$$\left(\begin{matrix} O \\ H \\ H \end{matrix} \longleftrightarrow \begin{matrix} C \\ C \\ C \end{matrix}\right) \text{ beats }$$

Z

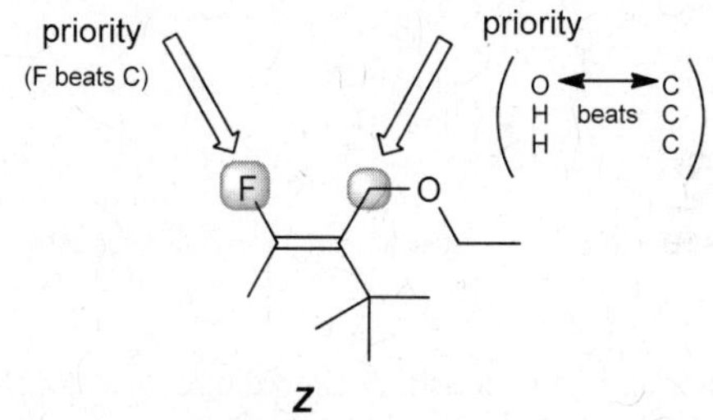

(d) The priorities (as shown below) are on the same sides of the double bond, so this alkene has the *Z* configuration.

priority
(Cl beats C)

priority
(C beats H)

Z

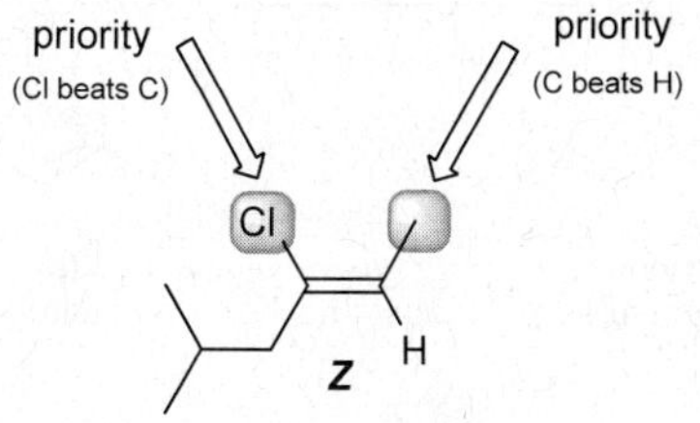

8.6. When using *cis-trans* terminology, we look for two identical groups. In this case, there are two ethyl groups that are in the *trans* configuration:

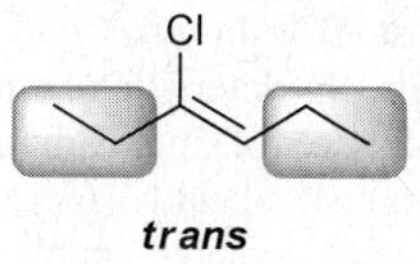

trans

However, when using *E-Z* terminology, we look for the highest priority at each vinylic position. Chlorine receives a higher priority than ethyl, so in this case, the highest priority groups are on the same side of the π bond:

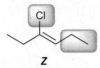

Z

Below are two other examples of alkenes that have the *trans* configuration, but nevertheless have the *Z* configuration. There are certainly other acceptable answers.

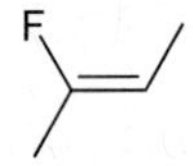

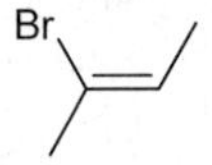

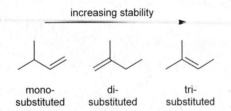

8.7.
(a) Each alkene is classified according to its degree of substitution. The most highly substituted alkene will be the most stable, and therefore, the following order of stability is expected.

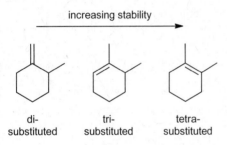

increasing stability

| di- substituted | tri- substituted | tetra- substituted |

(b) Each alkene is classified according to its degree of substitution. The most highly substituted alkene will be the most stable, and therefore, the following order of stability is expected.

increasing stability

| mono- substituted | di- substituted | tri- substituted |

8.8.   In the first compound, all of the carbon atoms of the ring are sp^3 hybridized and tetrahedral. As a result, they are supposed to have bond angles of approximately 109.5°, but their bond angles are compressed due to the ring (and are almost 90°). In other words, the compound exhibits angle strain characteristic of small rings. In the second compound, two of the carbon atoms are sp^2 hybridized and trigonal planar. As a result, they are supposed to have bond angles of approximately 120°, but

their bond angles are compressed due to the ring (and are almost 90°). The resulting angle strain (120° → 90°) is greater than the angle strain in the first compound (109.5° → 90°). Therefore, the second compound is higher in energy, despite the fact that it has a more highly substituted double bond.

8.9.
(a)   Hydroxide (HO⁻) functions as the base. In a concerted mechanism, the base abstracts a proton with simultaneous loss of the leaving group (Cl⁻). This requires a total of three curved arrows. The tail of the first curved arrow is placed on a lone pair of the base, and the head of that curved arrow is placed on the proton being removed. The second curved arrow shows formation of the π bond, and the third curved arrow shows loss of the leaving group, as shown here.

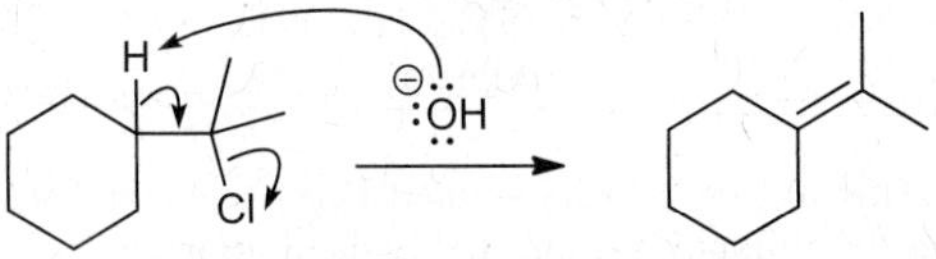

(b)   Ethoxide (EtO⁻) functions as the base. In a concerted mechanism, the base abstracts a proton with simultaneous loss of the leaving group (TsO⁻). This requires a total of three curved arrows. The tail of the first curved arrow is placed on a lone pair of the base, and the head of that curved arrow is placed on the proton being removed. The second curved arrow shows formation of the π bond, and the third curved arrow shows loss of the leaving group, as shown here.

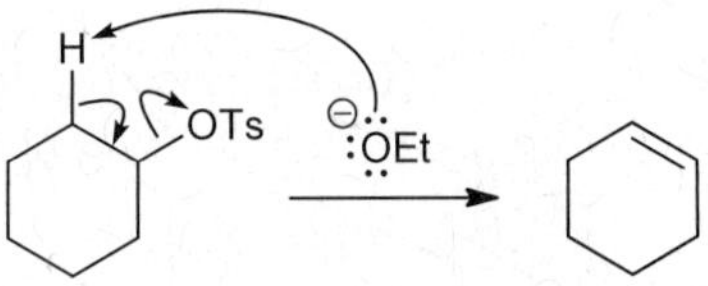

(c) Methoxide (MeO⁻) functions as the base. In a concerted mechanism, the base abstracts a proton with simultaneous loss of the leaving group (Br⁻). This requires a total of three curved arrows. The tail of the first curved arrow is placed on a lone pair of the base, and the head of that curved arrow is placed on the proton being removed. The second curved arrow shows formation of the π bond, and the third curved arrow shows loss of the leaving group, as shown here.

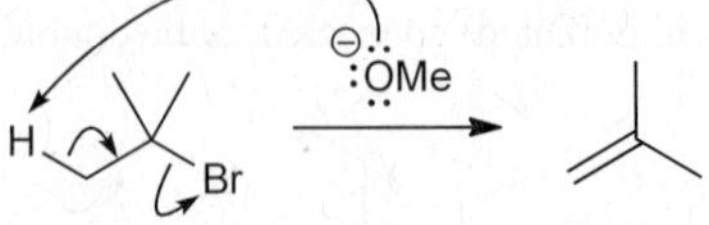

8.10.
(a) In a stepwise process, the leaving group first leaves, and then (in a separate step) the base removes a proton. The first step of the mechanism (loss of the leaving group) requires one curved arrow. The resulting carbocation intermediate is the deprotonated by the base

to give the alkene. This step requires two curved arrows, as shown:

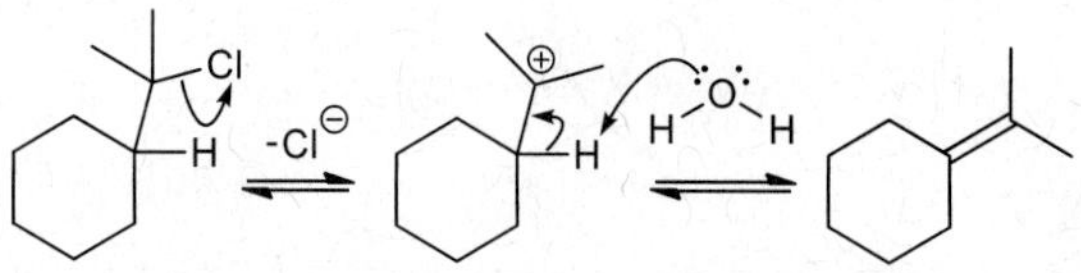

(b) In a stepwise process, the leaving group first leaves, and then (in a separate step) the base removes a proton. The first step of the mechanism (loss of the leaving group) requires one curved arrow. The resulting carbocation intermediate is the deprotonated by the base to give the alkene. This step requires two curved arrows, as shown:

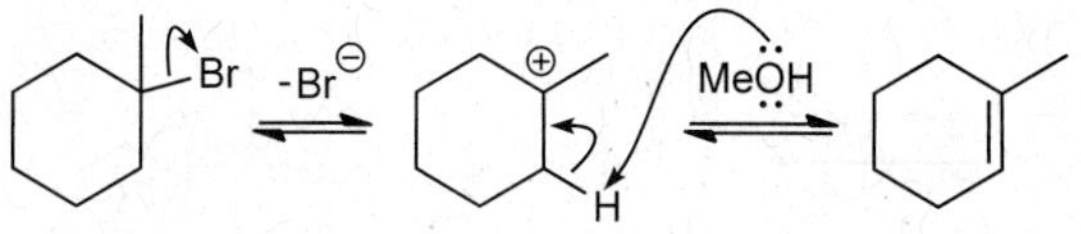

(c) In a stepwise process, the leaving group first leaves, and then (in a separate step) the base removes a proton. The first step of the mechanism (loss of the leaving group) requires one curved arrow. The resulting carbocation intermediate is the deprotonated by the base to give the alkene. This step requires two curved arrows, as shown:

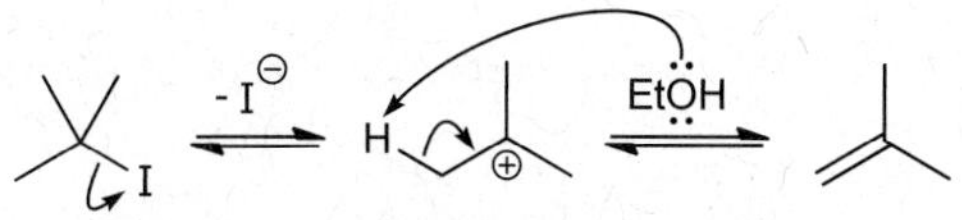

8.11. In this case, the base deprotonates the substrate with *simultaneous* loss of the leaving group (TsO⁻). Therefore, this is a concerted process, generating the following alkene:

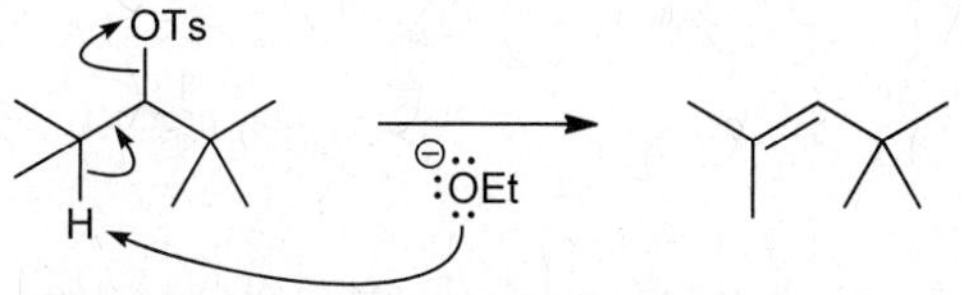

8.12. In the first step of this stepwise process, loss of the leaving group (Br⁻) gives the carbocation intermediate shown here:

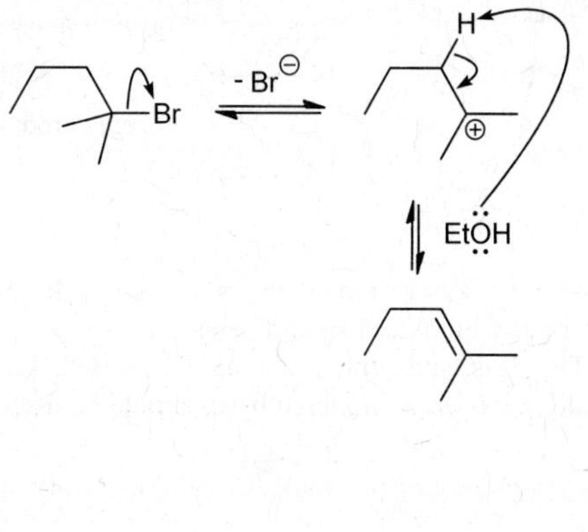

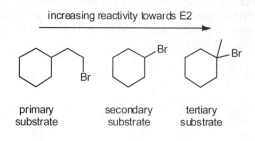

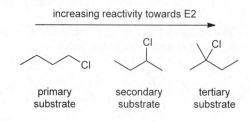

8.13.
(a) With a second-order rate equation, the rate is expected to be linearly dependent on the concentrations of the substrate and the base. If the concentration of the substrate is tripled, then the rate is expected to be three times faster.
(b) With a second-order rate equation, the rate is expected to be linearly dependent on the concentrations of the substrate and the base. If the concentration of the base is doubled, then the rate is expected to be two times faster.
(c) With a second-order rate equation, the rate is expected to be linearly dependent on the concentrations of the substrate and the base. If the concentration of the substrate is doubled and the concentration of the base is tripled, then the rate is expected to be six times faster (×2×3).

8.14.
(a) A tertiary substrate will be more reactive toward E2 than a secondary substrate, which is more reactive than a primary substrate.

increasing reactivity towards E2

primary substrate   secondary substrate   tertiary substrate

(b) A tertiary substrate will be more reactive toward E2 than a secondary substrate, which is more reactive than a primary substrate.

increasing reactivity towards E2

primary substrate   secondary substrate   tertiary substrate

8.15.
(a) This compound has three β positions, but one of them (highlighted) does not bear protons:

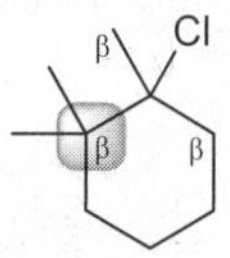

Since there are two β positions bearing protons, there are two possible elimination products, shown below. Since the base (ethoxide) is not sterically hindered, we expect that the major product will be the more-substituted

alkene, and the minor product will be the less-substituted alkene.

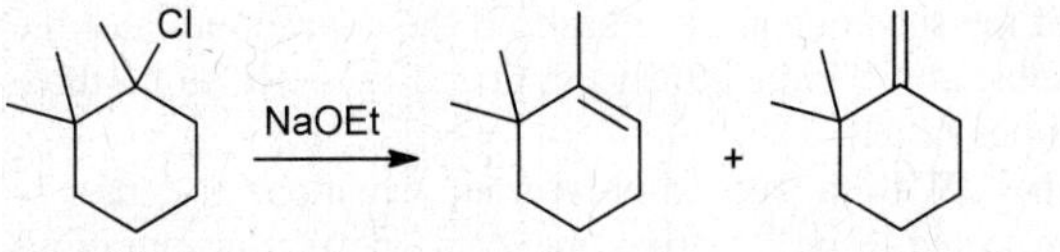

major (more substituted) minor (less substituted)

(b) This compound has three β positions that bear protons, but two of them (highlighted) are identical:

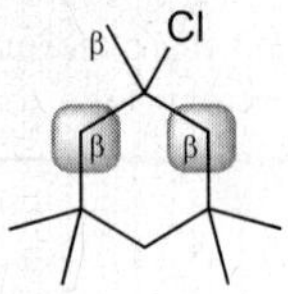

Thus, there are only two unique β positions, giving rise to two possible elimination products, shown below. Since the base (*tert*-butoxide) is sterically hindered, we expect that the major product will be the less-substituted alkene, and the minor product will be the more-substituted alkene.

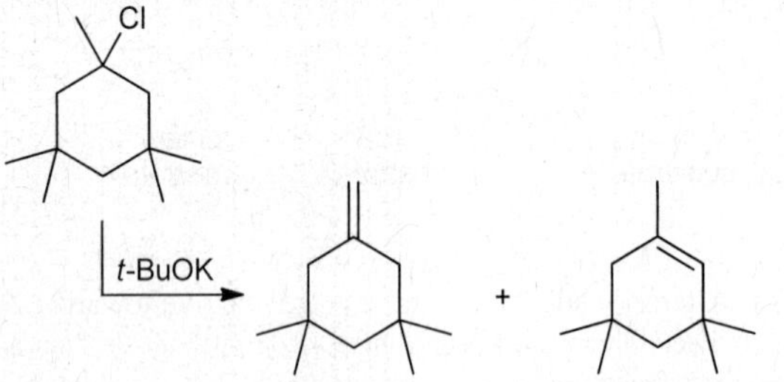

major (less substituted) minor (more substituted)

(c) This compound has three β positions that bear protons, but two of them (highlighted) are identical:

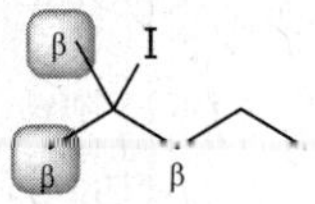

Thus, there are only two unique β positions, giving rise to two possible elimination products, shown below. Since the base (hydroxide) is not sterically hindered, we expect that the major product will be the more-substituted alkene, and the minor product will be the less-substituted alkene.

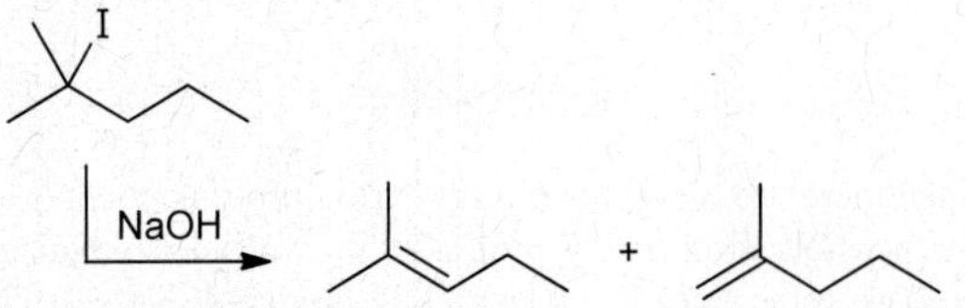

major (more substituted) minor (less substituted)

(d) This compound has three β positions that bear protons, but two of them (highlighted) are identical:

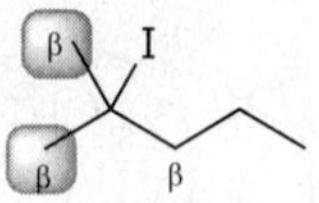

Thus, there are only two unique β positions, giving rise to two possible elimination products, shown below. Since the base (*tert*-butoxide) is sterically hindered, we expect that the major product will be the less-substituted alkene, and the minor product will be the more-substituted alkene.

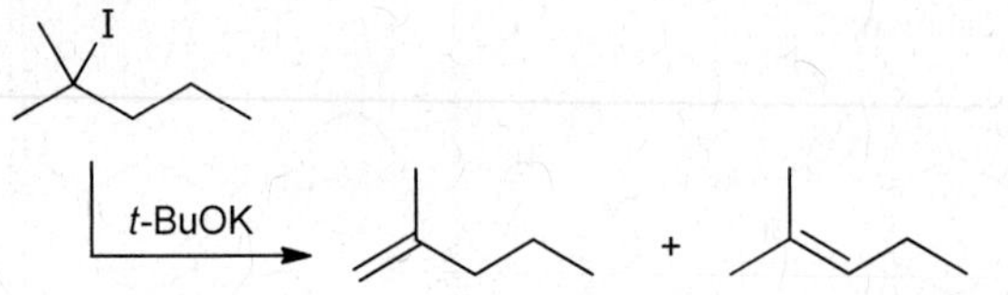

major (less substituted) minor (more substituted)

(e) This compound has three β positions that bear protons, but all of them are identical, because removing a proton from any one of these positions will lead to the same product:

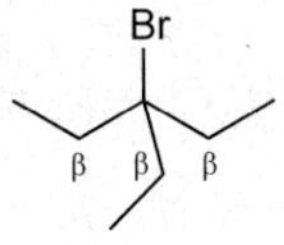

As such, there is only one possible elimination product:

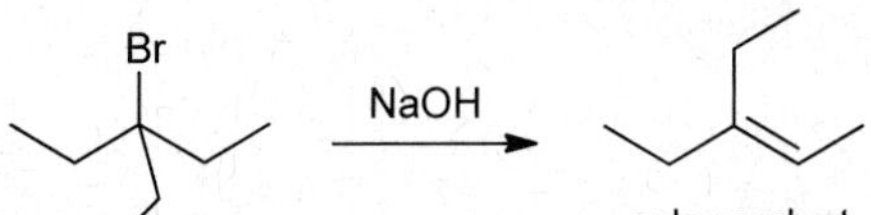

only product

(f) As seen in the previous solution (problem **8.15e**), all three β positions are identical, so only one elimination product is possible.

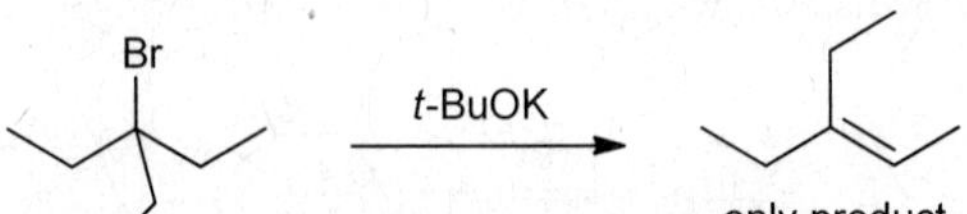

only product

8.16.
(a) The more substituted alkene is desired, so hydroxide (not sterically hindered) should be used.
(b) The less substituted alkene is desired, so *tert*-butoxide (a sterically hindered base) should be used.

8.17.

(a) Each of the following substrates is expected to afford the desired alkene when treated with a sterically hindered base:

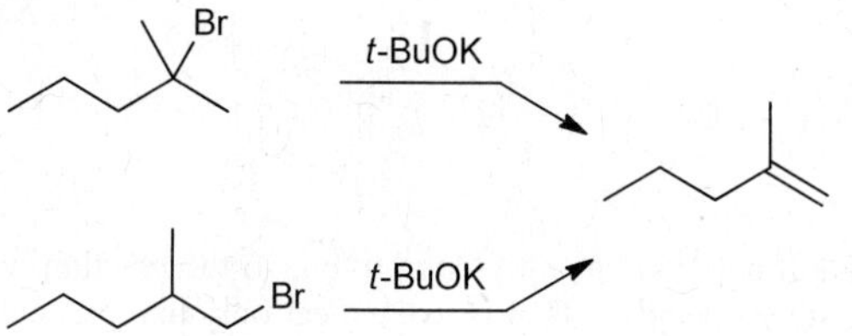

(b) Each of the following substrates is expected to afford the desired alkene when treated with a sterically hindered base:

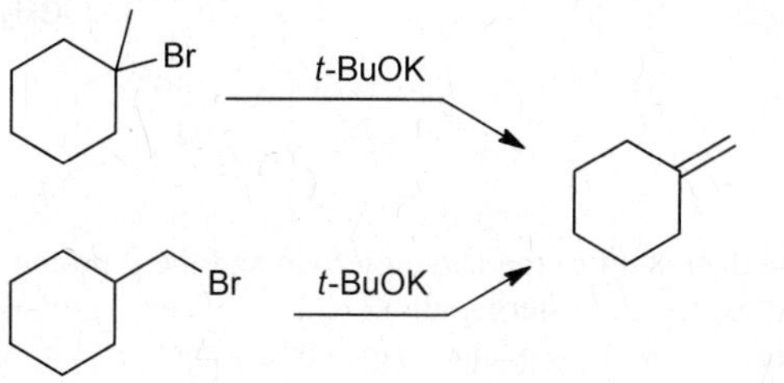

8.18.

(a) The substrate has two β positions, but only one of these positions (highlighted) bears protons.

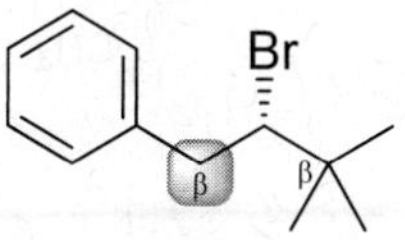

This β position has two protons, so the reaction will be stereoselective. That is, we expect both *cis* and *trans* isomers, with a preference for the *trans* isomer.

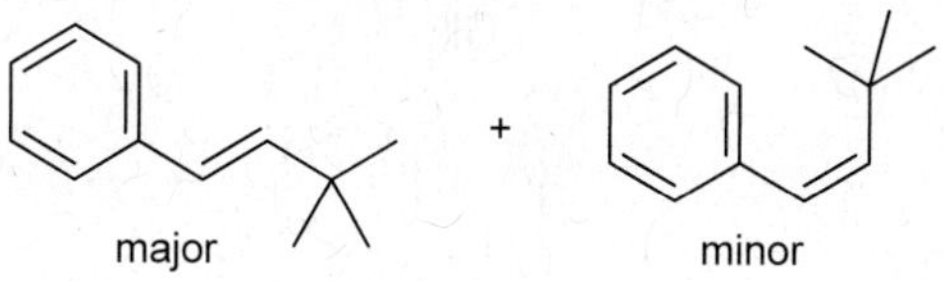

(b) The substrate has two β positions, but only one of these positions (highlighted) bears a proton.

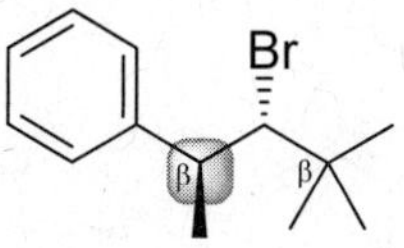

This β position has only one proton, so the reaction will be stereospecific. That is, we expect only one particular

stereoisomeric product. To determine which product to expect, we draw a Newman projection:

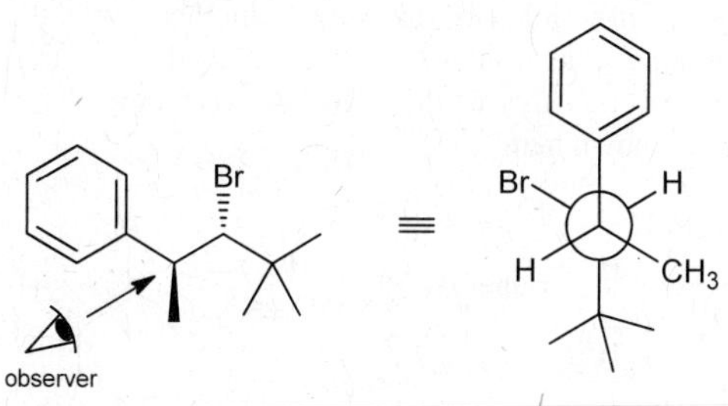

We then identify the leaving group (highlighted below) and the β proton (highlighted below) that will be removed during an E2 reaction. We rotate the central C-C bond, so that the β proton and the leaving group are *anti*-periplanar to one another:

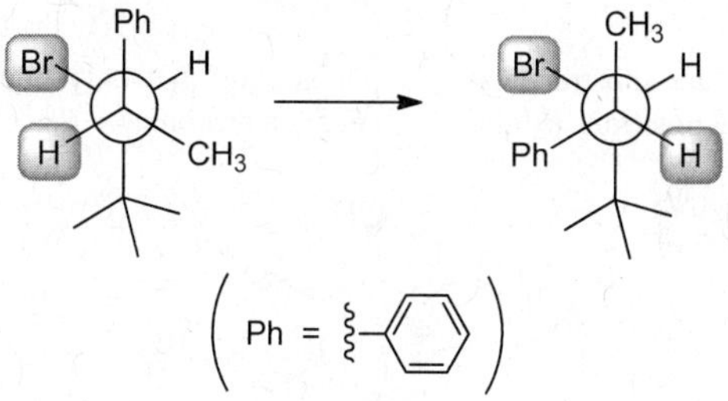

The reaction proceeds through this conformation, so we expect the *Z* isomer, as shown here:

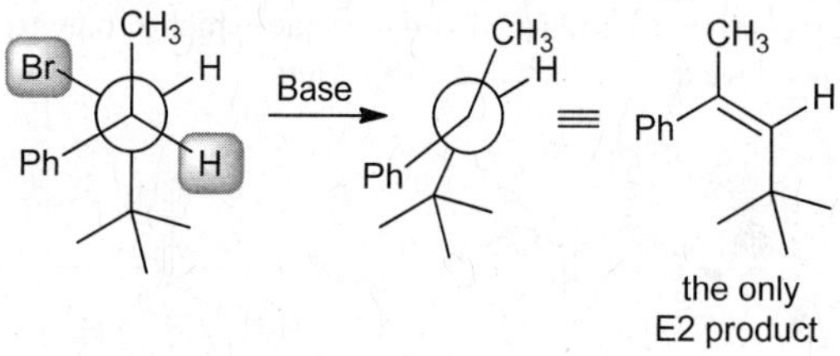

(c) The substrate has two β positions, but only one of these positions (highlighted) bears a proton.

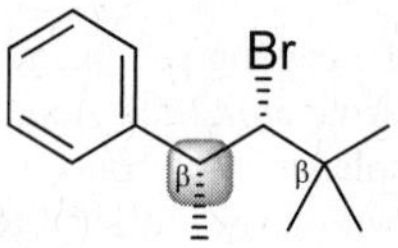

This β position has only one proton, so the reaction will be stereospecific. That is, we expect only one particular stereoisomeric product. To determine which product to expect, we draw a Newman projection:

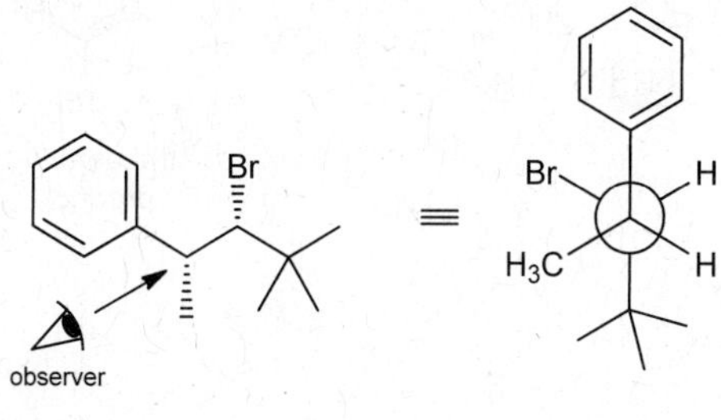

We then identify the leaving group and the β proton that will be removed during an E2 reaction. These groups are already *anti*-periplanar to one another in the conformation drawn above. Therefore, the reaction proceeds through this conformation to give only the *E* isomer, as shown here:

the only
E2 product

$$\left(Ph = \right)$$

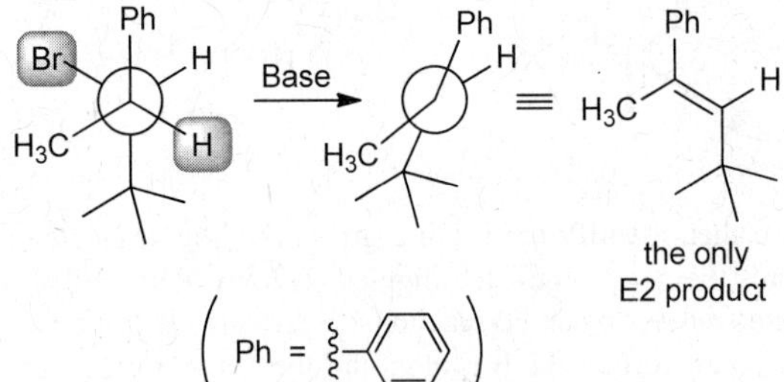

(d) The substrate has two β positions, but only one of these positions (highlighted) bears a proton.

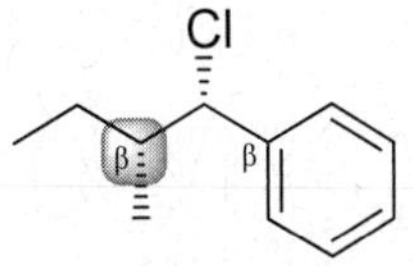

This β position has only one proton, so the reaction will be stereospecific. That is, we expect only one particular stereoisomeric product. To determine which product to expect, we draw a Newman projection:

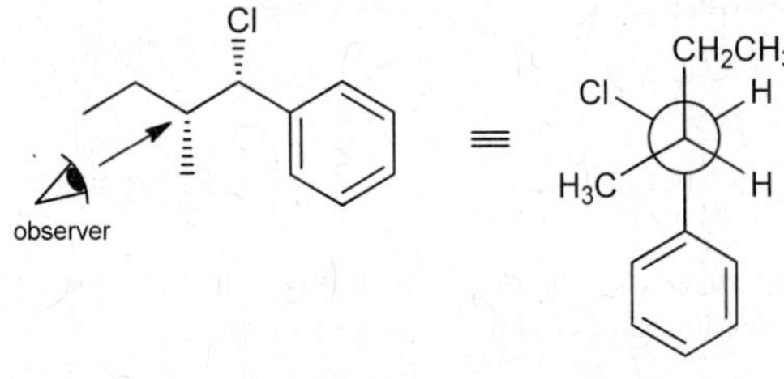

observer

We then identify the leaving group and the β proton that will be removed during an E2 reaction. These groups are already *anti*-periplanar to one another in the conformation drawn above. Therefore, the reaction proceeds through this conformation to give only the *E* isomer, as shown here:

the only
E2 product

$$\left(Ph = \right)$$

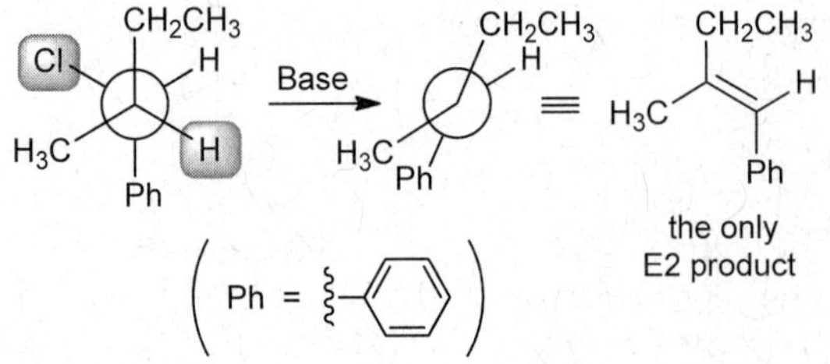

(e) The substrate has two β positions, but only one of these positions (highlighted) bears a proton.

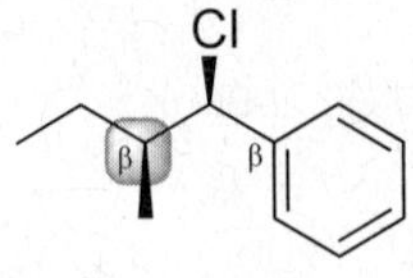

This β position has only one proton, so the reaction will be stereospecific. That is, we expect only one particular stereoisomeric product. To determine which product to expect, we draw a Newman projection:

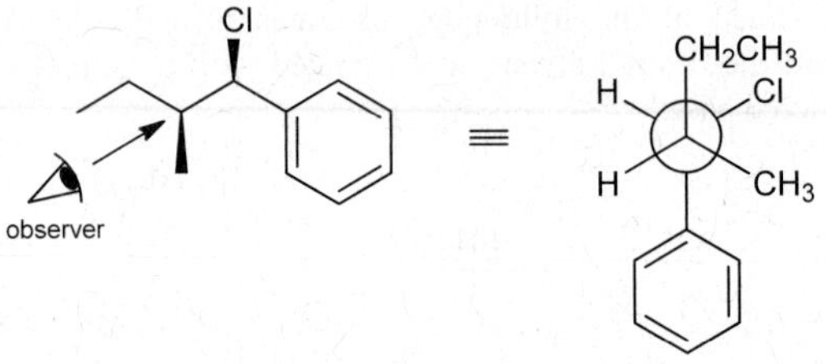

observer

We then identify the leaving group and the β proton that will be removed during an E2 reaction. These groups are already *anti*-periplanar to one another in the conformation drawn above. Therefore, the reaction proceeds through this conformation to give only the *E* isomer, as shown here:

the only
E2 product

$$\left(Ph = \right)$$

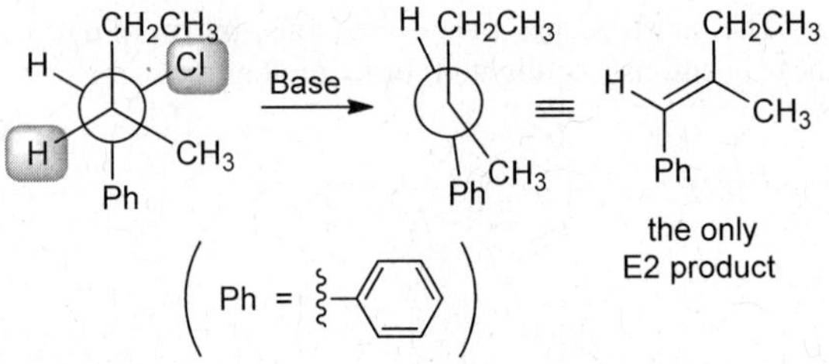

(f) The substrate has two β positions, but only one of these positions (highlighted) bears a proton.

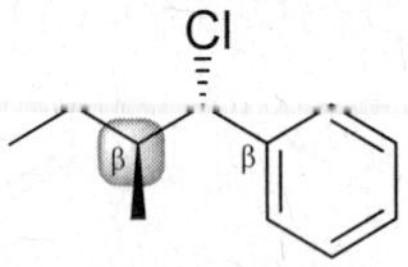

This β position has only one proton, so the reaction will be stereospecific. That is, we expect only one particular stereoisomeric product. To determine which product to expect, we draw a Newman projection:

observer

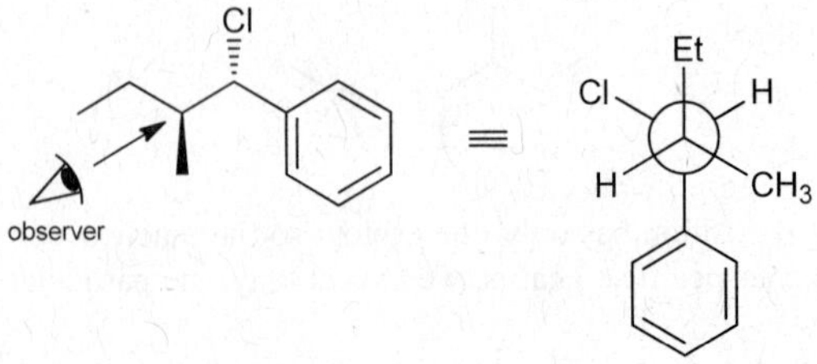

We then identify the leaving group (highlighted below) and the β proton (highlighted below) that will be removed during an E2 reaction. We rotate the central C-C bond, so that the β proton and the leaving group are *anti*-periplanar to one another:

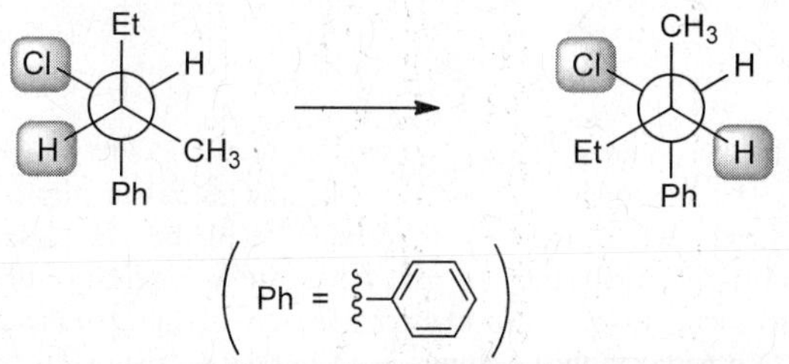

The reaction proceeds through this conformation to give only the *Z* isomer, as shown here:

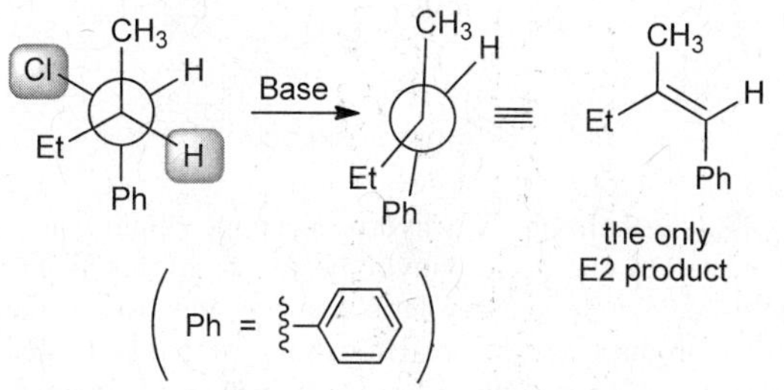

(g) The substrate has two β positions, but only one of these positions (highlighted) bears protons.

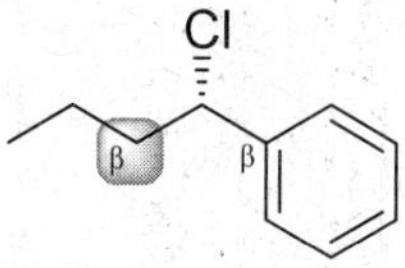

This β position has two protons, so the reaction will be stereoselective. That is, we expect both *cis* and *trans* isomers, with a preference for the *trans* isomer.

major   minor

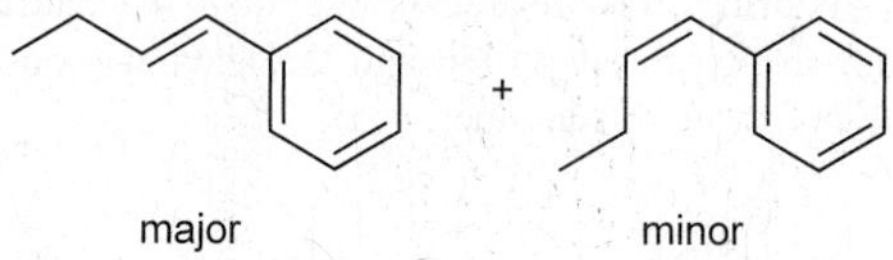

(h) The substrate has two β positions, but only one of these positions (highlighted) bears protons.

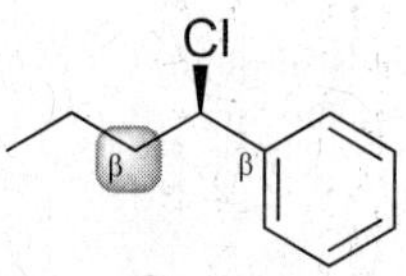

This β position has two protons, so the reaction will be stereoselective. That is, we expect both *cis* and *trans* isomers, with a preference for the *trans* isomer.

major   minor

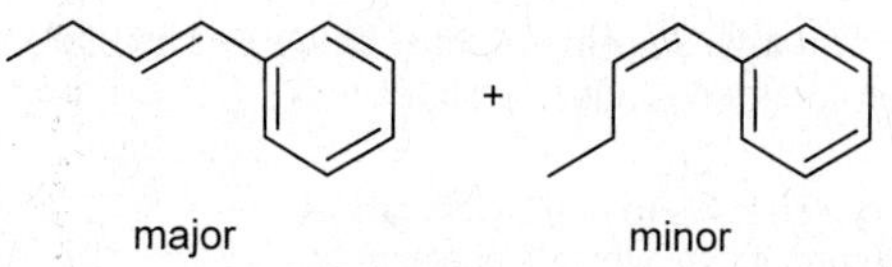

8.19. The following alkyl halide will give the desired alkene, as shown here:

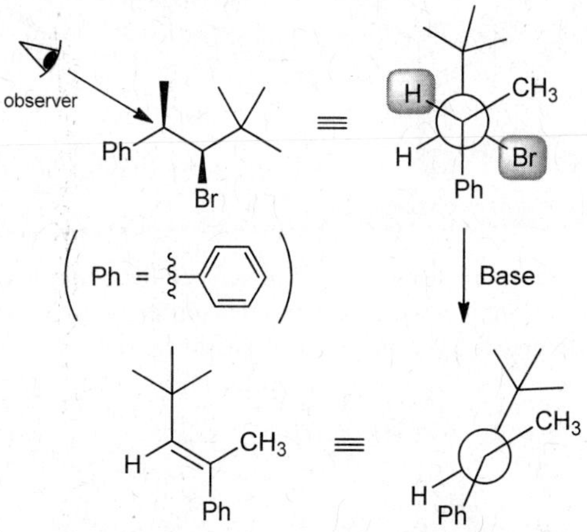

The alkyl halide above is not the only alkyl halide that will generate the desired alkene. The enantiomer of the alkyl halide above can also be used, as shown:

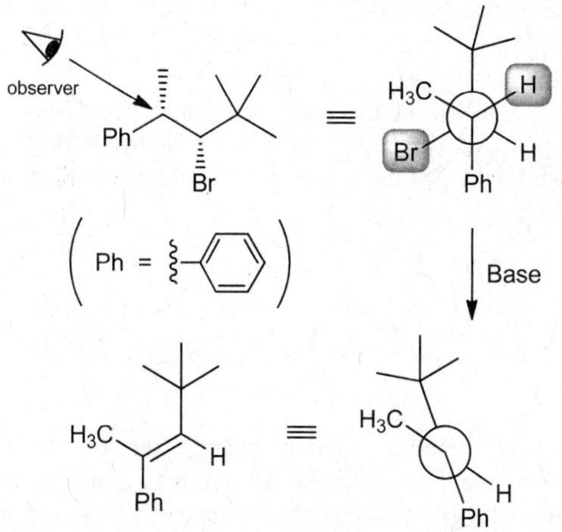

Alternatively, either enantiomer of the following alkyl halide (or a racemic mixture of the alkyl halide) may be converted into the desired product, albeit via a stereoselective (rather than stereospecific) E2 process.

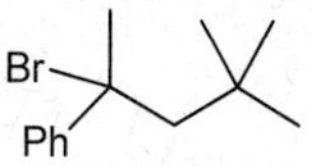

8.20. In the structure of menthyl chloride (shown below), the leaving group (Cl⁻) is on a dash. Therefore, we are looking for a β proton that is on a wedge, in order that it should be able to achieve antiperiplanarity with the leaving group. In this case, there is only one β proton on a wedge (highlighted below). Therefore, only one elimination product is observed:

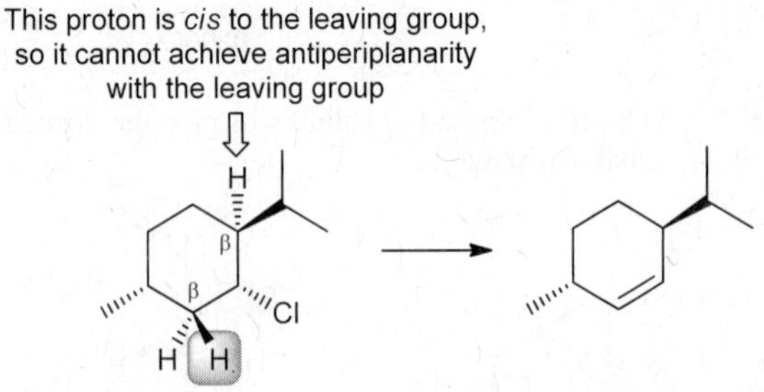

This proton is *cis* to the leaving group, so it cannot achieve antiperiplanarity with the leaving group

menthyl chloride

In contrast, the leaving group in neomenthyl chloride can achieve antiperiplanarity with two different β protons (highlighted), giving rise to two possible products:

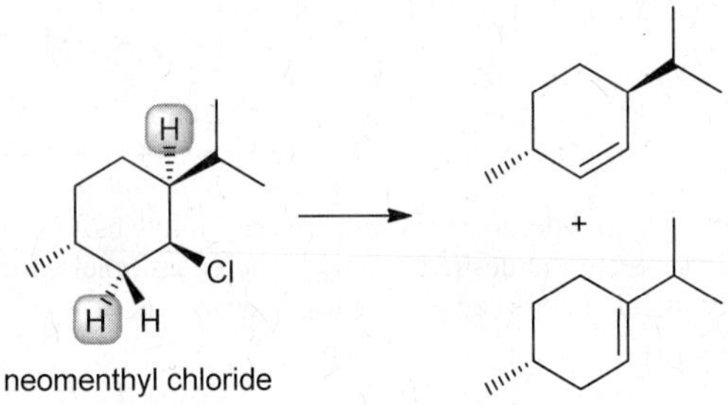

neomenthyl chloride

8.21. Because of the bulky *tert*-butyl group, the first compound is essentially locked in a chair conformation in which the leaving group (Cl⁻) occupies an equatorial position.

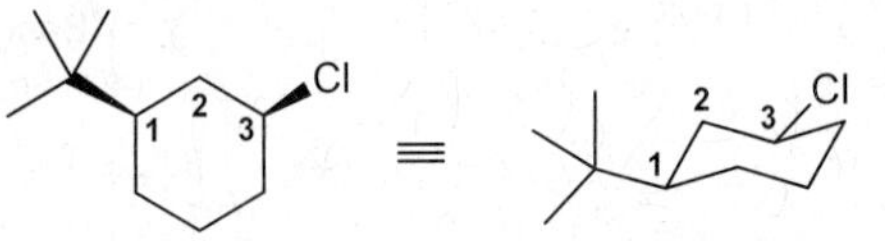

This conformation cannot undergo an E2 reaction because the leaving group is not antiperiplanar to a proton. However, the second compound is locked in a chair conformation in which the chlorine occupies an axial position.

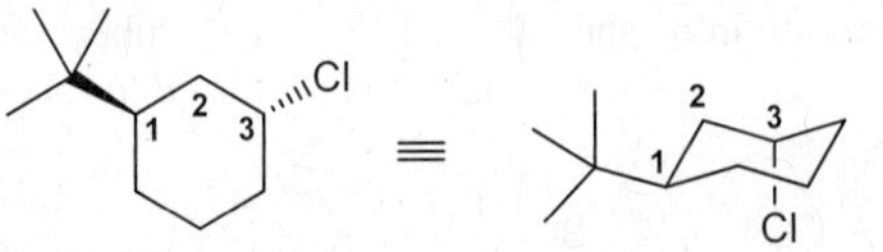

This conformation rapidly undergoes an E2 reaction. Therefore, the second compound is expected to be more reactive towards an E2 process than the first compound.

8.22.
(a) We must determine both the regiochemical outcome and the stereochemical outcome. Let's begin with

regiochemistry. There are two β positions in this case, so there are two possible regiochemical outcomes.

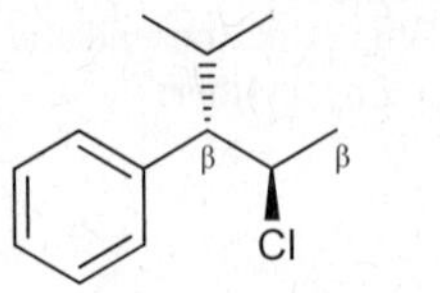

The base (ethoxide) is not sterically hindered, so we expect the major product will be the more-substituted alkene, while the minor product will be the less-substituted alkene. Next, we must identify the stereochemistry of formation of each of the products. Let's begin with the minor product (the less substituted alkene), because its double bond does not exhibit stereoisomerism:

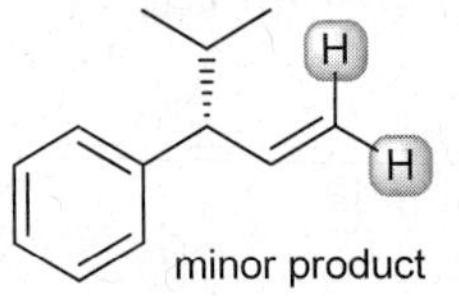

minor product

As shown, there are two hydrogen atoms (highlighted) connected to one of the vinylic positions, so this alkene is neither *E* nor *Z*. Now let's turn our attention to the major product of the reaction (the more-substituted alkene). To determine which stereoisomer is obtained, we must draw a Newman projection:

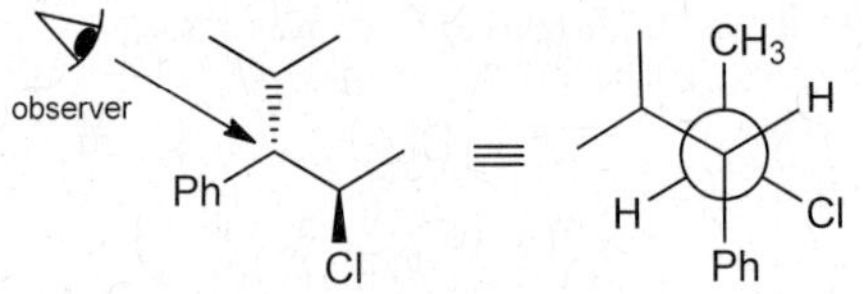

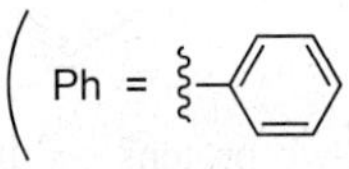

$$\left(Ph = \text{—}\bigcirc \right)$$

We then identify the leaving group (highlighted below) and the β proton (highlighted below) that will be removed during an E2 reaction. We rotate the central C-C bond, so that the β proton and the leaving group are *anti*-periplanar to one another:

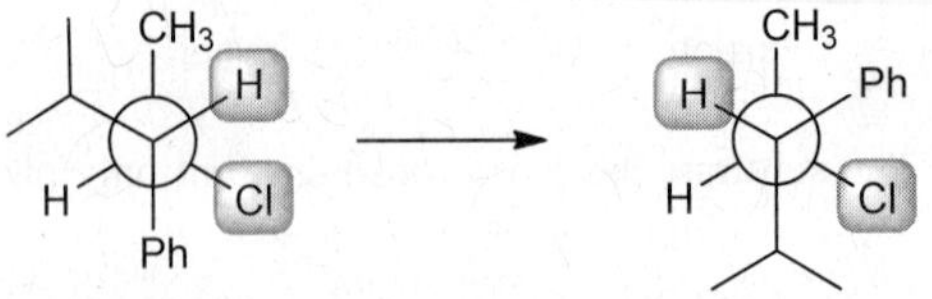

The reaction proceeds through this conformation to give only the *Z* isomer, as shown here:

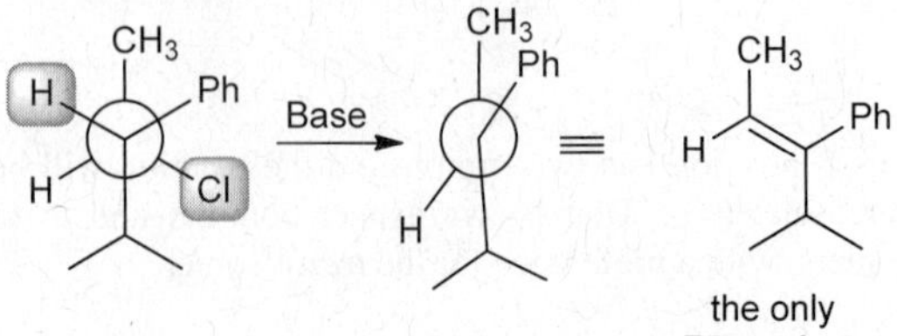

the only E2 product

In summary, we expect the following two products:

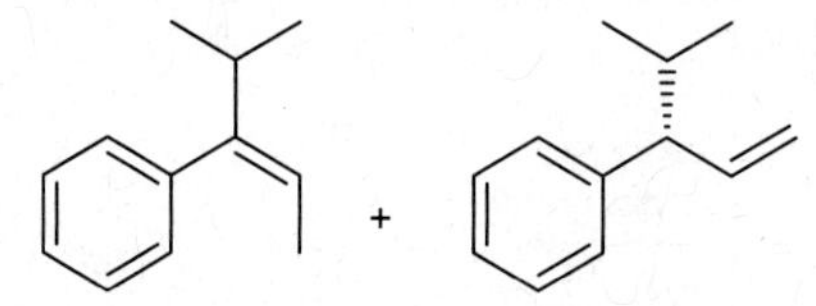

major   +   minor

(b) For substituted cyclohexanes, an E2 reaction will occur if the leaving group and the β proton can achieve antiperiplanarity. In order to achieve this, one must be on a wedge and the other must be on a dash. The leaving group (Br⁻) is on a wedge. Therefore, we are looking for a β proton that is on a dash. In this case, there are two different β protons on a dash (highlighted below), giving rise to two products. Since the base (ethoxide) is not sterically hindered, we expect the more-substituted alkene to be the major product:

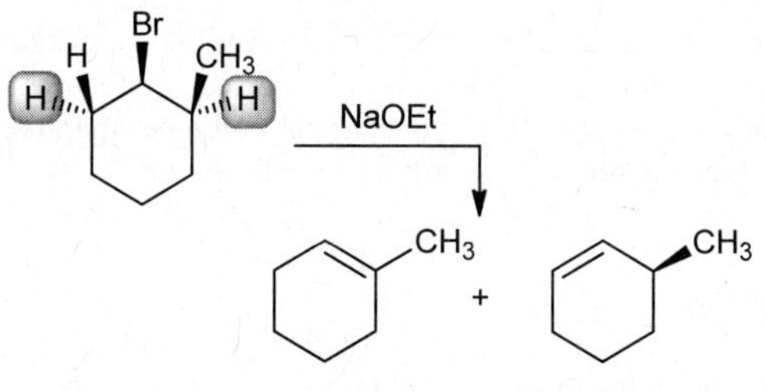

major   +   minor

(c) For substituted cyclohexanes, an E2 reaction will occur if the leaving group and the β proton can achieve antiperiplanarity. In order to achieve this, one must be on a wedge and the other must be on a dash. The leaving group (Br⁻) is on a wedge. Therefore, we are looking for a β proton that is on a dash. In this case, there is only one β proton on a dash (highlighted below), giving rise to only one product:

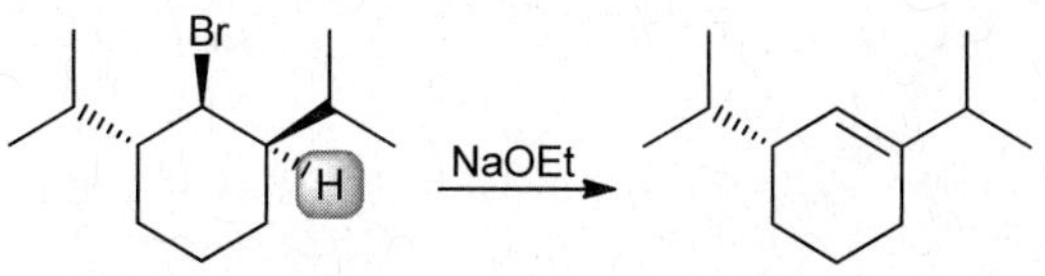

only E2 product

(d) We must determine both the regiochemical outcome and the stereochemical outcome. Let's begin with regiochemistry. This compound has three β positions, but two of them (highlighted) are identical because deprotonation at either of these locations will result in the same alkene:

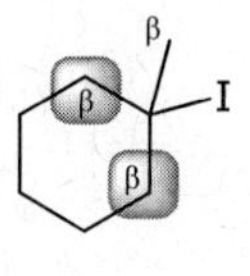

As such, we expect two possible regiochemical outcomes. Since the base (ethoxide) is not sterically

hindered, we expect the more-substituted alkene as the major product.

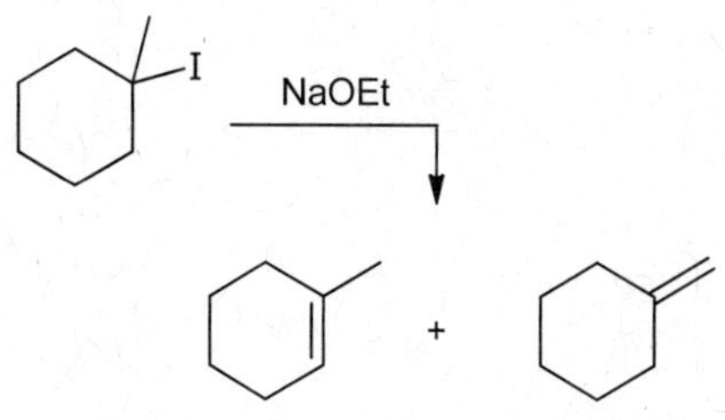

major   +   minor

Stereochemisty is not a consideration for either product. The minor product is not stereoisomeric, and the major product cannot exist as an *E* isomer (because the ring makes that impossible).

(e) For substituted cyclohexanes, an E2 reaction will occur if the leaving group and the β proton can achieve antiperiplanarity. In order to achieve this, one must be on a wedge and the other must be on a dash. The leaving group (Br⁻) is on a wedge. Therefore, we are looking for a β proton that is on a dash. In this case, there are two different β protons on a dash (highlighted below), giving rise to two products. Since the base (*tert*-butoxide) is sterically hindered, we expect the less-substituted alkene to be the major product:

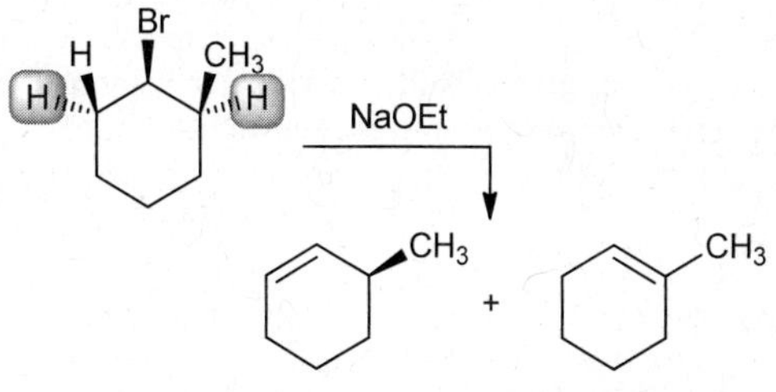

major   +   minor

(f) For substituted cyclohexanes, an E2 reaction will occur if the leaving group and the β proton can achieve antiperiplanarity. In order to achieve this, one must be on a wedge and the other must be on a dash. The leaving group (Br⁻) is on a wedge. Therefore, we are looking for a β proton that is on a dash. In this case, there is only one β proton on a dash (highlighted below), giving rise to only one product:

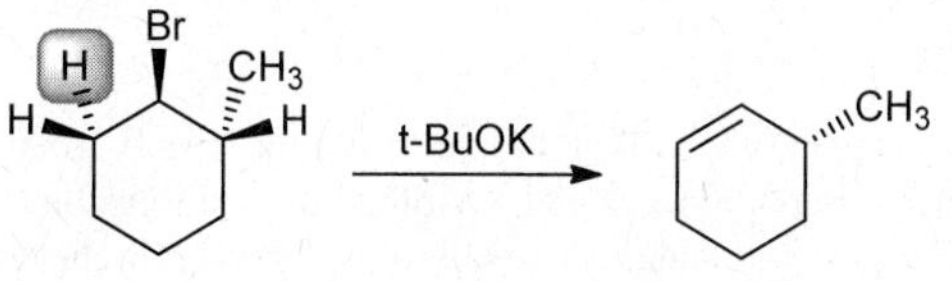

only E2 product

8.23. There are certainly many acceptable answers. For example, in the following structure, if all R groups are identical (either all H's or all the same alkyl group), then the two β positions are identical, because deprotonation at either position leads to the same product.

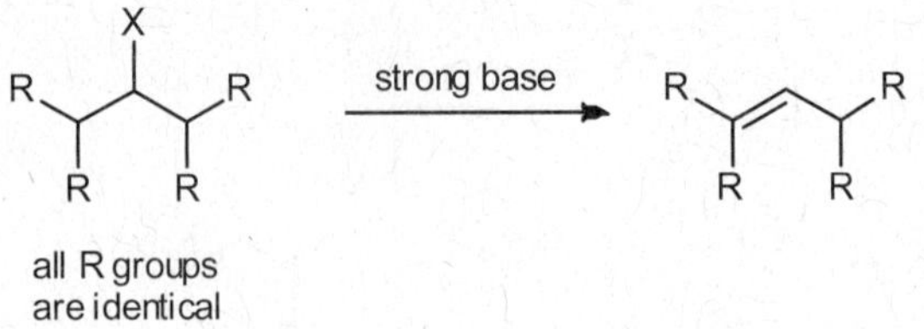

all R groups are identical

Once again, there are many acceptable answers to this problem, and not all of them will conform to the general structure above. For example, here is another acceptable answer to the problem:

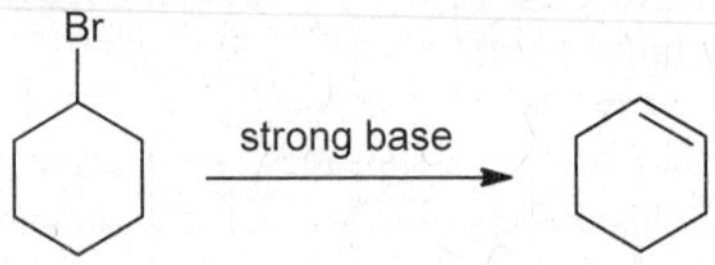

Just as we saw in the previous example, the two β positions are identical, because deprotonation at either position leads to the same compound.

8.24. There are certainly many acceptable answers. We are looking for an example in which the alkyl halide that contains one β position bearing two protons, and no other β positions bearing protons. As such, there is only one regiochemical outcome, and the reaction will be stereoselective, giving both possible stereoisomers (*cis* and *trans*). One such example is shown below:

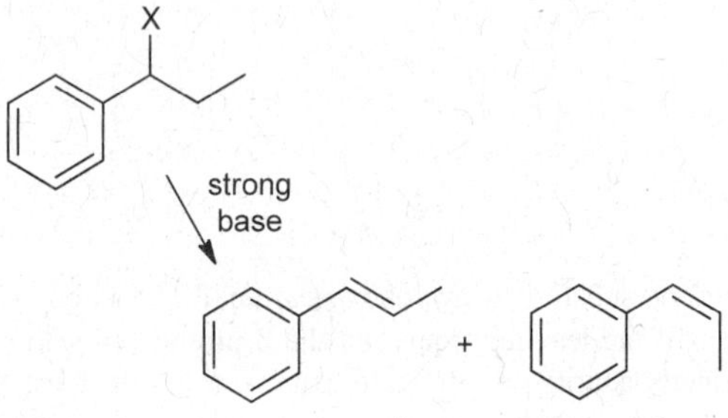

Here is another example of an acceptable answer:

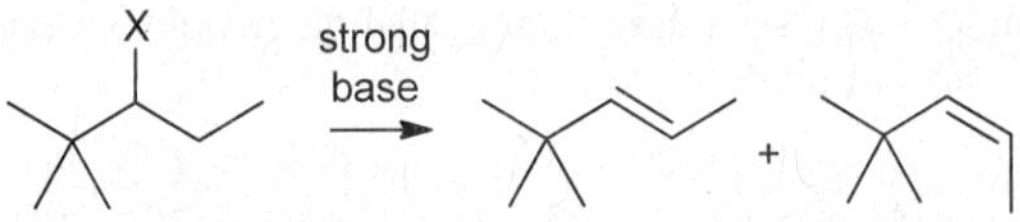

Once again, this alkyl halide contains one β position bearing two protons, and no other β positions bearing protons. As such, there is only one regiochemical outcome, and the reaction will be stereoselective, giving both possible stereoisomers (*cis* and *trans*).

8.25. There are certainly many acceptable answers. We are looking for an example in which the alkyl halide contains two β positions bearing protons, giving two possible regiochemical outcomes. For each

regiochemical outcome, the alkene produced must not be stereoisomeric (neither *E* nor *Z*). One such example is shown below:

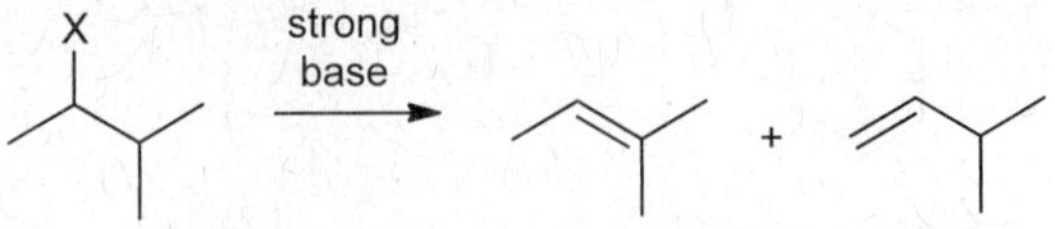

Here is another example of an acceptable answer:

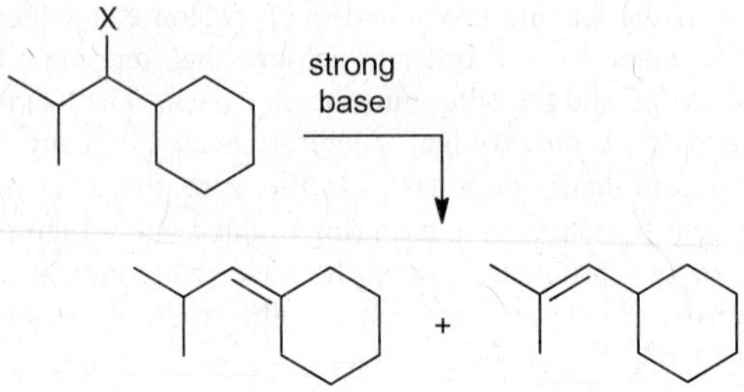

Once again, this alkyl halide contains two β positions bearing protons, giving two possible regiochemical outcomes. For each regiochemical outcome, the alkene produced is not stereoisomeric (neither *E* nor *Z*).

8.26.
(a) Only the concentration of *tert*-butyl iodide affects the rate, so the rate will double.
(b) Only the concentration of *tert*-butyl iodide affects the rate, so the rate will remain the same.

8.27.
(a) Loss of the leaving group (bromide) results in the following carbocation:

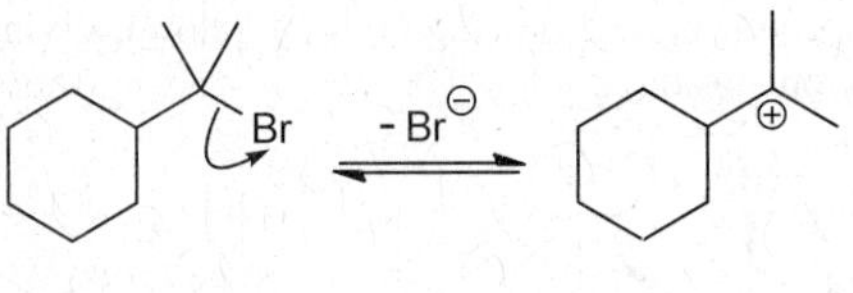

(b) Loss of the leaving group (chloride) results in the following carbocation:

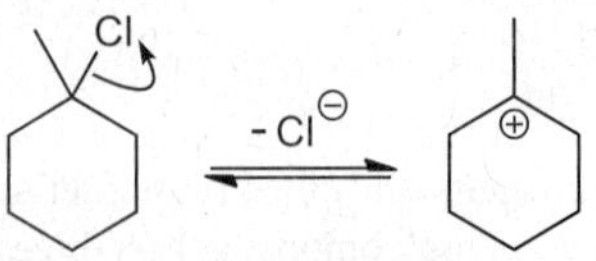

(c) Loss of the leaving group (iodide) results in the following carbocation:

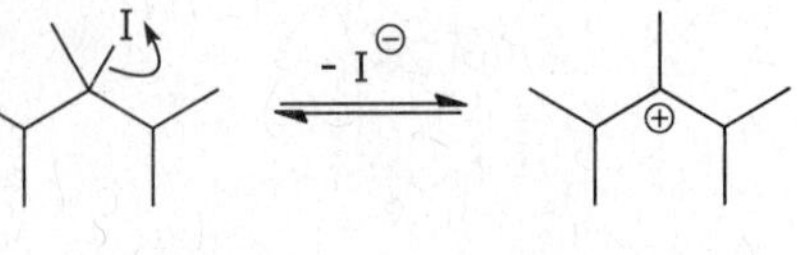

(d) Loss of the leaving group (bromide) results in the following carbocation:

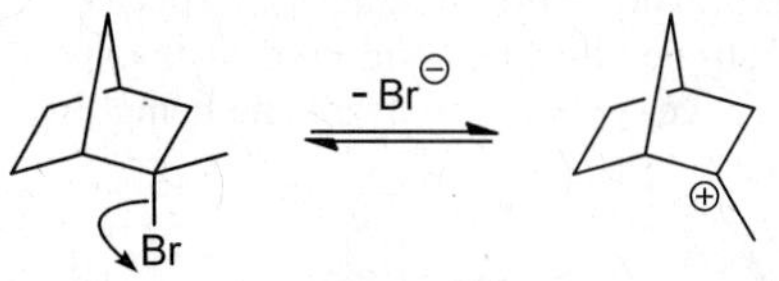

8.28.

(a) The OH group is first protonated, giving an excellent leaving group (H_2O), which leaves to give the following carbocation:

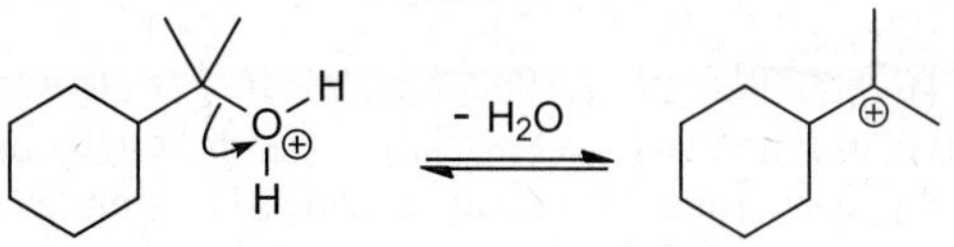

(b) The OH group is first protonated, giving an excellent leaving group (H_2O), which leaves to give the following carbocation:

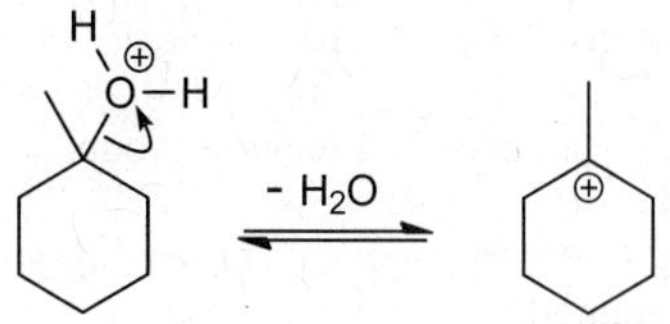

(c) The OH group is first protonated, giving an excellent leaving group (H_2O), which leaves to give the following carbocation:

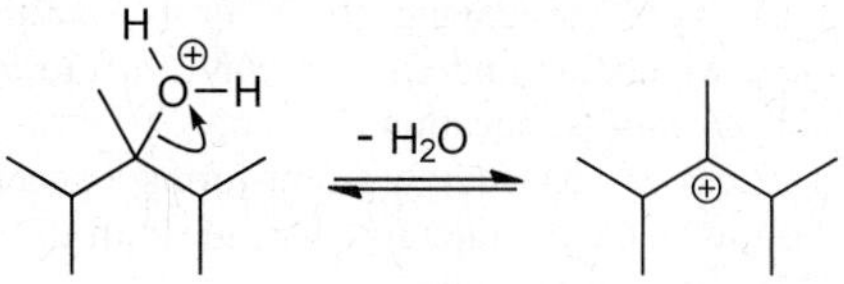

(d) The OH group is first protonated, giving an excellent leaving group (H_2O), which leaves to give the following carbocation:

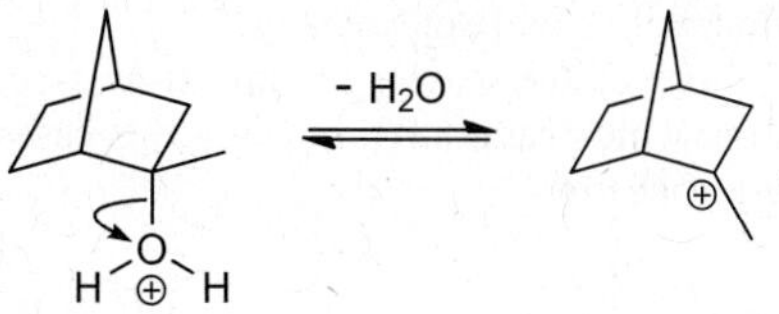

8.29.

(a) There are three β positions, and all three of them bear protons. But two of them (highlighted) are identical, because deprotonation at either of these locations results in the same product:

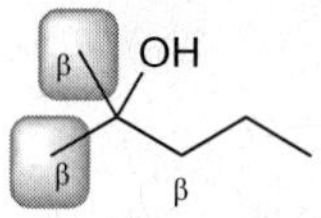

Therefore, there are two possible products, and we expect a mixture of both products, although the more-substituted alkene will be favored.

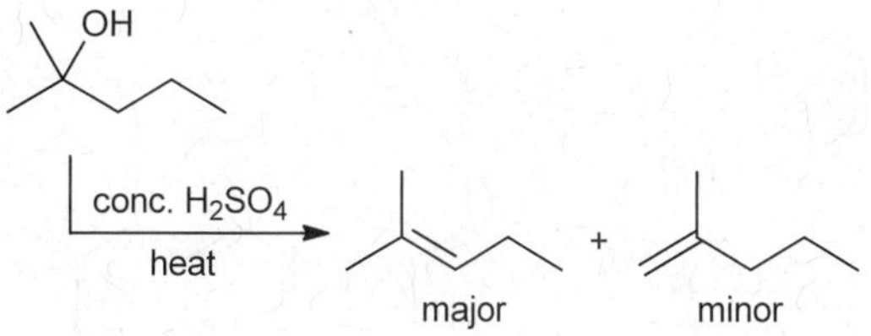

(b) There are three β positions, and all three of them bear protons. But two of them (highlighted) are identical, because deprotonation at either of these locations results in the same product:

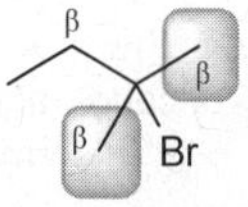

Therefore, there are two possible products, and we expect a mixture of both products, although the more-substituted alkene will be favored.

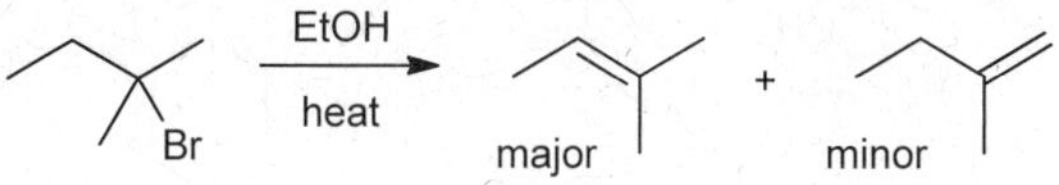

(c) There are two β positions, but only one of them (highlighted) bears protons.

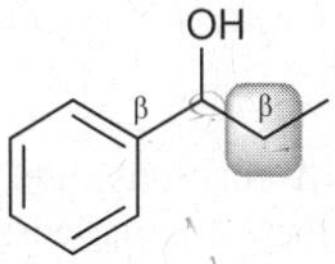

Therefore, there is only one possible regiochemical outcome. Since this β position bears two protons, the reaction is expected to be stereoselective. That is, we expect both stereoisomeric products (*cis* and *trans*), although the *trans* alkene will be favored.

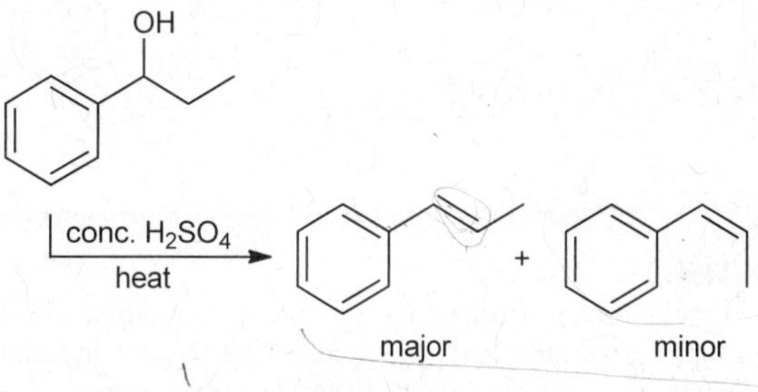

(d) There are three β positions, but only two of them (highlighted) bear protons.

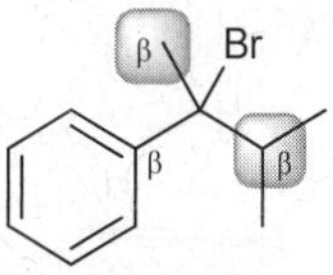

Therefore, there are two possible regiochemical outcomes, giving the following two products:

major + minor

Stereochemistry is not a consideration for either product, as neither product is stereoisomeric.

8.30. Both alcohols below can be used to form the product. The tertiary alcohol below will react more rapidly because the rate determining step involves formation of a tertiary carbocation rather than a secondary carbocation.

8.31.
(a) There are three β positions, and two of them bear protons. Therefore, there are two possible regiochemical outcomes. The major product is expected to be the more-substituted alkene. The process is expected to be stereoselective, with the *trans* alkene being favored over the *cis* alkene.

(b) There are three β positions, and two of them bear protons. Therefore, there are two possible regiochemical outcomes. The major product is expected to be the more-substituted alkene. The process is expected to be stereoselective, with the *trans* alkene being favored over the *cis* alkene.

8.32.
(a) Iodide is an excellent leaving group, so an acid is not required.
(b) Hydroxide (HO⁻) is not a good leaving group, so an acid is required in order to protonate the OH group, thereby converting it into an excellent leaving group (H_2O).
(c) Tosylate (TsO⁻) is an excellent leaving group, so an acid is not required.
(d) Hydroxide (HO⁻) is not a good leaving group, so an acid is required in order to protonate the OH group, thereby converting it into an excellent leaving group (H_2O).
(e) Bromide is an excellent leaving group, so an acid is not required.
(f) Tosylate (TsO⁻) is an excellent leaving group, so an acid is not required.

8.33.
(a) No. Loss of the leaving group forms a tertiary carbocation, which will not rearrange.
(b) Yes. Loss of the leaving group forms a secondary carbocation, which can undergo a methyl shift to form a more stable tertiary carbocation.
(c) Yes. Loss of the leaving group forms a secondary carbocation, which can undergo a hydride shift to form a more stable tertiary carbocation.
(d) No. Loss of the leaving group forms a secondary carbocation, which cannot rearrange in this case to form a tertiary carbocation.
(e) No. Loss of the leaving group forms a tertiary carbocation, which will not rearrange.
(f) No. Loss of the leaving group forms a secondary carbocation, which cannot rearrange in this case to form a tertiary carbocation.

8.34.
(a) In acidic conditions, the OH group is protonated, which converts it from a bad leaving group to a good leaving group. In sulfuric acid, the acid that is present in solution is H_3O^+ (because of the leveling effect, as explained in Section 3.6), so we use H_3O^+ as the proton source in the first step of the mechanism. The next two steps of the mechanism constitute the core steps of an E1 process: (i) loss of a leaving group (H_2O), which requires one curved arrow, and (ii) proton transfer, which requires two curved arrows, as shown:

(b) The leaving group is bromide, which is not protonated prior to its departure. Loss of the leaving group gives a tertiary carbocation that cannot rearrange to become more stable. Finally, ethanol serves as the base that removes the proton to generate the product.

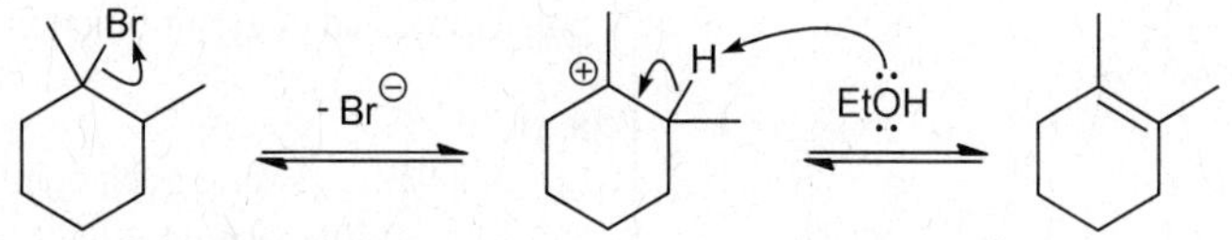

(c) The leaving group is chloride, which is not protonated prior to its departure. Loss of the leaving group gives a tertiary, resonance-stabilized carbocation that cannot rearrange to become more stable. Finally, ethanol serves as the base that removes the proton to generate the product.

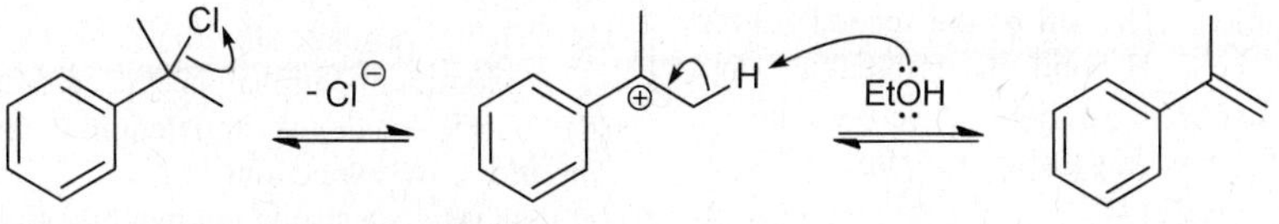

(d) In acidic conditions, the OH group is protonated, which converts it from a bad leaving group to a good leaving group. In aqueous sulfuric acid, the acid that is present in solution is H_3O^+ (because of the leveling effect, as explained in Section 3.6), so we use H_3O^+ as the proton source in the first step of the mechanism. Loss of the leaving group gives a secondary carbocation that can undergo a hydride shift, giving a more stable, tertiary carbocation. Finally, water serves as the base that removes the proton to generate the product.

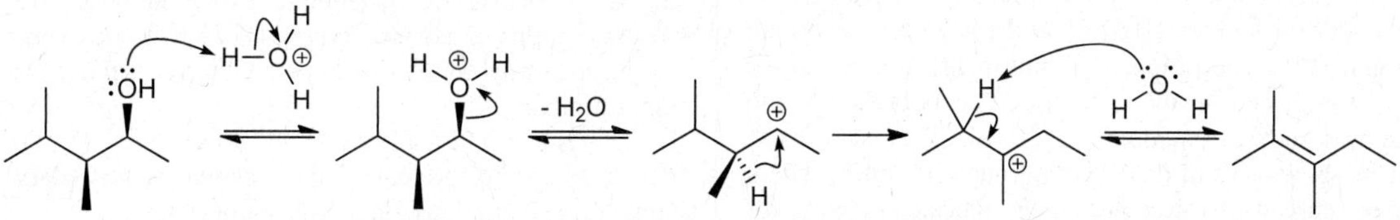

8.35. The patterns for 8.34a-d are shown below. Problem 8.34b and 8.34c exhibit the same pattern because both have leaving groups that can leave without being protonated, and both do not exhibit a carbocation rearrangement. As a result both mechanisms involve only two steps: 1) loss of a leaving group and 2) proton transfer.

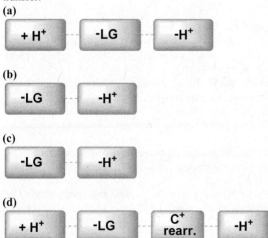

8.36. The first method is more efficient because it employs a strong base to promote an E2 process for a secondary substrate bearing a good leaving group. The second method relies on an E1 process occurring at a secondary substrate, which will be slow and will involve a carbocation rearrangement to produce a different product.

8.37.
(a) For an E2 process, three curved arrows are required. The tail of the first curved arrow is placed on a lone pair of the base (methoxide) and the head is placed on a proton at one of the three (identical) β positions. The tail of the second curved arrow is placed on the C–H bond that is breaking, and the head shows formation of a π bond. The third curved arrow shows loss of the leaving group (chloride).

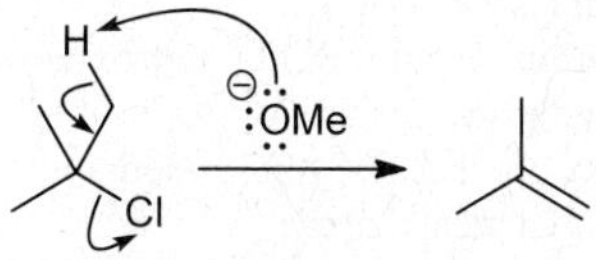

(b) For an E2 process, three curved arrows are required. The tail of the first curved arrow is placed on a lone pair of the base (*tert*-butoxide) and the head is placed on a proton at the β position. The tail of the second curved arrow is placed on the C–H bond that is breaking, and the head shows formation of a π bond. The third curved arrow shows loss of the leaving group (bromide).

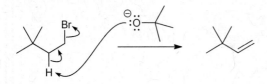

(c) For an E2 process, three curved arrows are required. The tail of the first curved arrow is placed on a lone pair of the base (hydroxide) and the head is placed on the proton at the β position. The tail of the second curved arrow is placed on the C–H bond that is breaking, and the head shows formation of a π bond. The third curved arrow shows loss of the leaving group (bromide).

8.38. For an E2 process, three curved arrows are required. The tail of the first curved arrow is placed on a lone pair of the base (H$_2$N$^-$) and the head is placed on a proton at the β position. The tail of the second curved arrow is placed on the C–H bond that is breaking, and the head shows formation of a π bond. The third curved arrow shows loss of the leaving group (chloride). These three curved arrows are then repeated again for conversion of the vinyl chloride into an alkyne.

8.39.
(a)   As seen in Figure 8.25, ethanol is a weak nucleophile and a weak base.
(b) As seen in Figure 8.25, EtSH is a strong nucleophile and a weak base.
(c)   As seen in Figure 8.25, ethoxide is a strong nucleophile and a strong base.
(d)   As seen in Figure 8.25, bromide is a strong nucleophile and a weak base.
(e)   As seen in Figure 8.25, hydroxide is a strong nucleophile and a strong base.
(f)   As seen in Figure 8.25, methanol is a weak nucleophile and a weak base.
(g)   As seen in Figure 8.25, methoxide is a strong nucleophile and a strong base.
(h) As seen in Figure 8.25, DBN is a strong base and a weak nucleophile.

8.40.   Aluminum is a larger atom and is polarizable. Therefore, the entire complex can function as a strong nucleophile, and can serve as a delivery agent of a hydride ion. In contrast, the hydride ion by itself is not polarizable and does not function as a nucleophile.

8.41.
(a) NaOH is a strong nucleophile and strong base. The substrate in this case is primary. Therefore, we expect S$_N$2 (giving the major product) and E2 (giving the minor product).
(b) NaSH is a strong nucleophile and weak base. The substrate in this case is primary. Therefore, we expect only S$_N$2.
(c) When a primary alkyl halide is treated with *t*-BuOK, the predominant pathway is expected to be E2.
(d)   DBN is a weak nucleophile and a strong base. Therefore, we expect only E2.
(e) NaOMe is a strong nucleophile and strong base. The substrate in this case is primary. Therefore, we expect S$_N$2 (giving the major product) and E2 (giving the minor product).

8.42.
(a) NaOEt is a strong nucleophile and strong base. The substrate in this case is secondary. Therefore, we expect the E2 pathway to predominate, with S$_N$2 giving the minor product.
(b) NaI is a strong nucleophile and weak base. DMSO is a polar aprotic solvent. The substrate is secondary. Under these conditions, only S$_N$2 can occur.
(c) DBU is a weak nucleophile and a strong base. Therefore, we expect only E2.
(d) NaOH is a strong nucleophile and strong base. The substrate in this case is secondary. Therefore, we expect the E2 pathway to predominate, with S$_N$2 giving the minor product.
(e)   *t*-BuOK is a strong, sterically hindered base. Therefore, we expect only E2.

8.43.
(a) EtOH is a weak nucleophile and weak base. The substrate in this case is tertiary. Therefore, we expect both S$_N$1 and E1.
(b)   *t*-BuOK is a strong, sterically hindered base. Therefore, we expect only E2.
(c) NaI is a strong nucleophile and weak base. The substrate in this case is tertiary. Therefore, we expect only S$_N$1.
(d) NaOEt is a strong nucleophile and strong base. The substrate in this case is tertiary. Therefore, we expect only E2.
(e) NaOH is a strong nucleophile and strong base. The substrate in this case is tertiary. Therefore, we expect only E2.

8.44.
(a) An E2 reaction does not readily occur because the base is weak.

(b) An E1 reaction does not readily occur because the substrate is primary.
(c) Replacing the weak base (EtOH) with a strong base would greatly enhance the rate of an E2 process. Since the substrate is primary, we use a sterically hindered base, such as *tert*-butoxide, to suppress the competing S_N2 process.
(d) Replacing the primary substrate with a tertiary substrate (such as 2-chloro-2-methylpentane) would greatly enhance the rate of an E1 process.

8.45. The substrate is sterically hindered (severely), so S_N2 cannot occur at a reasonable rate. There are no β protons, so E2 also cannot occur.

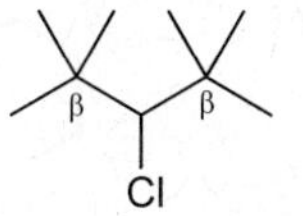

8.46.
(a) The reagent is chloride, which functions as a nucleophile, so we expect a substitution reaction. The substrate is secondary and the solvent is polar aprotic, indicating an S_N2 process. As such, we expect inversion of configuration, as shown:

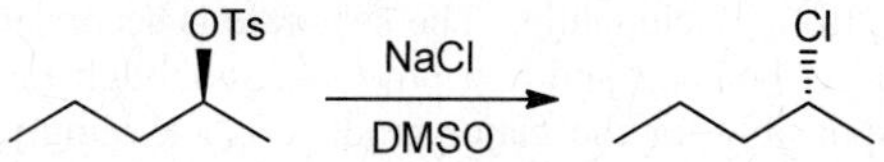

(b) The reagent is hydroxide, which is both a strong base and a strong nucleophile. The substrate is tertiary, so we expect an E2 process. There are three β positions, but two of them are identical, so there are two possible regiochemical outcomes. The more-substituted alkene is the major product, as shown. The products are not stereoisomeric, so stereochemistry is not a consideration.

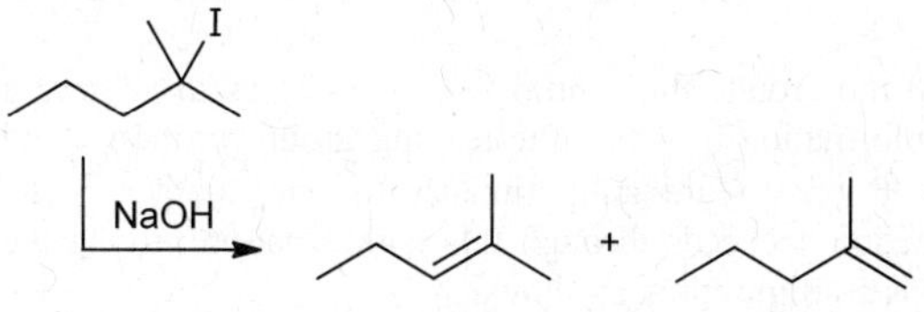

major (more substituted) minor (less substituted)

(c) The reagent is *tert*-butoxide, which is a strong, sterically hindered base. The substrate is tertiary, so we expect an E2 process. There are three β positions, but two of them are identical, so there are two possible regiochemical outcomes. Since the base is sterically hindered, we expect that the less-substituted alkene will be the major product, as shown. The products are not stereoisomeric, so stereochemistry is not a consideration.

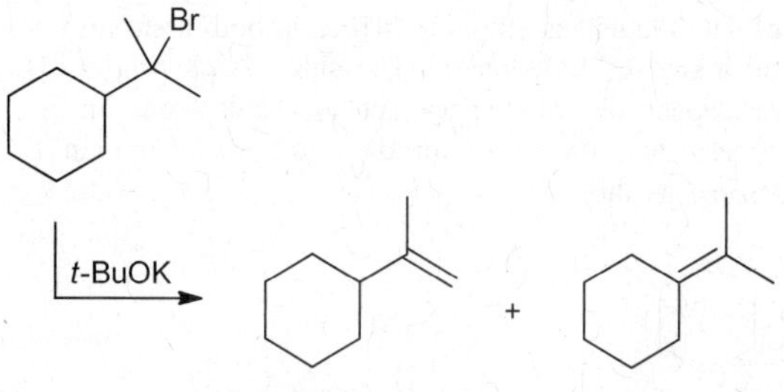

major (less substituted) minor (more substituted)

(d) The reagent is DBN, which is a strong base. The substrate is tertiary, so we expect an E2 process. There are two β positions bearing protons, and both of these positions are identical, so there is only one possible regiochemical outcome. The product is not stereoisomeric, so stereochemistry is not a consideration.

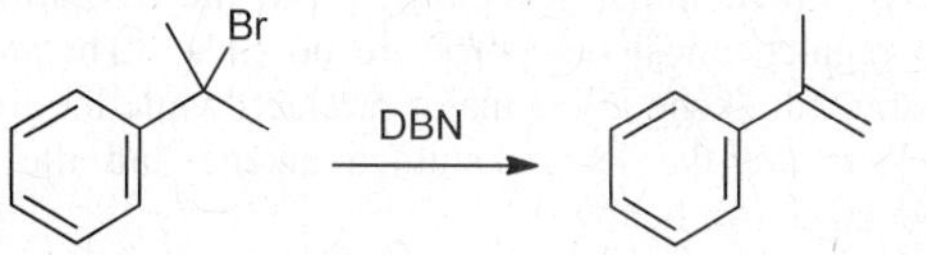

(e) The reagent is *tert*-butoxide, which is a strong base and strong nucleophile. For most reagents in this category, treatment with a primary alkyl halide will give S_N2 as the major pathway and E2 as the minor pathway. However, *tert*-butoxide is sterically hindered, which reduces the rate of the S_N2 process, such that E2 now prevails. Therefore, the major product will result from an E2 process, and the minor product will result from an S_N2 process.

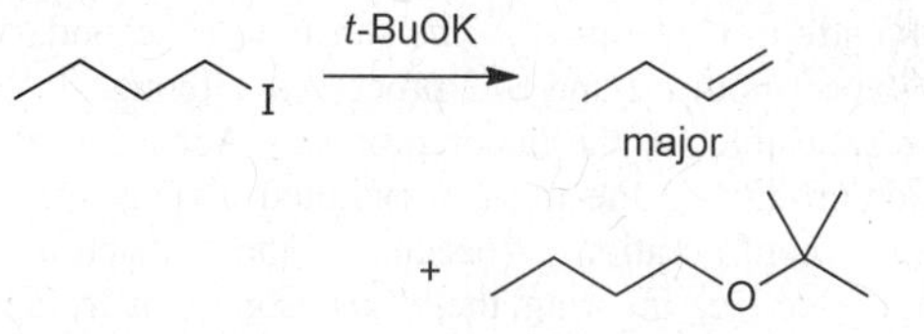

minor

(f) The reagent is HS^-, which is a strong nucleophile and a weak base. The substrate is primary, so we expect an S_N2 process, giving the following product:

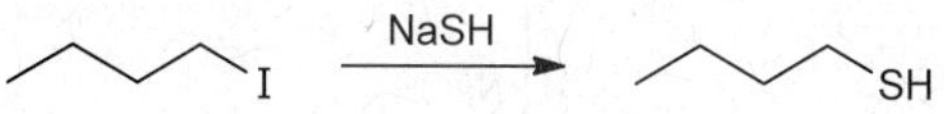

(g) The reagent is hydroxide, which is both a strong base and a strong nucleophile. The substrate is primary, so we expect the major product to result from an S_N2 process, and the minor product to result from an E2 process, as shown:

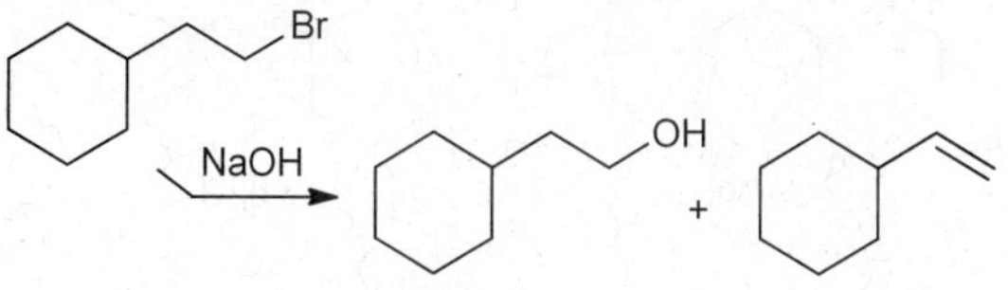

major minor

(h) The reagent is ethoxide, which is both a strong base and a strong nucleophile. The substrate is primary, so we expect the major product to result from an S_N2 process, and the minor product to result from an E2 process, as shown:

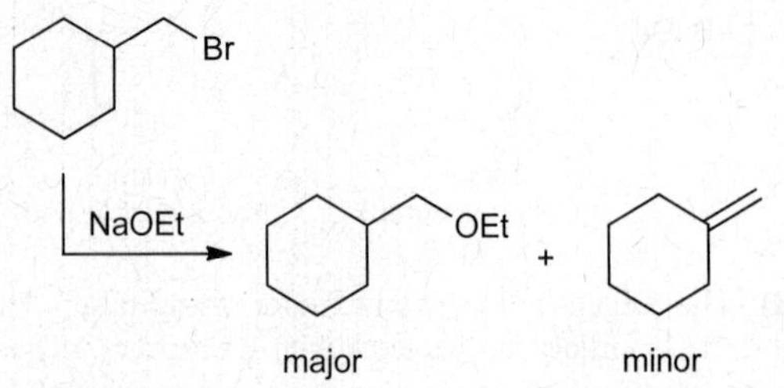

(i) The reagent is ethanol, which is both a weak base and a weak nucleophile. The substrate is tertiary, so we expect E1 and S_N1 processes. Heat is indicated, so E1 is likely to predominate over S_N1. For the E1 pathway, two regiochemical outcomes are possible. The more-substituted alkene is the major product, while the minor products are the less-substituted alkene and the S_N1 product, shown below:

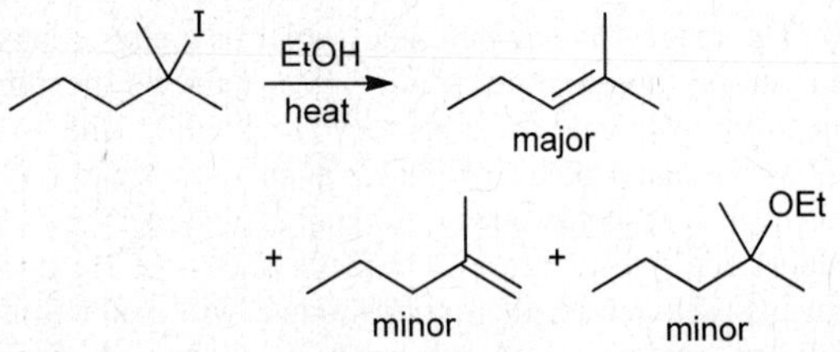

(j) The reagent is hydroxide, which is both a strong base and a strong nucleophile. The substrate is secondary so we expect both E2 and S_N2 processes, although E2 will be responsible for the major product. Accordingly, the major product is the more substituted alkene, with the *trans* configuration (because the reaction is stereoselective, favoring the *trans* isomer over the *cis* isomer). The minor products include the *cis* isomer, as well as the less-substituted alkene and the S_N2 product:

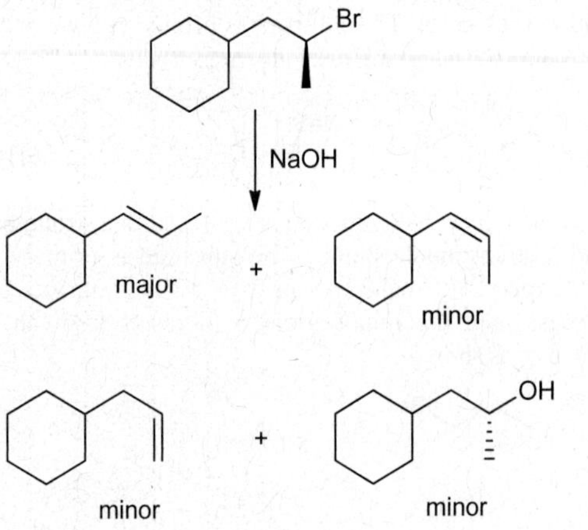

(k) The reagent is methoxide, which is both a strong base and a strong nucleophile. The substrate is secondary so we expect both E2 and S_N2 processes, although E2 will be responsible for the major product. Accordingly, the major product is the more substituted alkene, with the *trans* configuration (because the reaction is stereoselective, favoring the *trans* isomer over the *cis* isomer). The minor products include the *cis* isomer, as well as the less-substituted alkene and the S_N2 product:

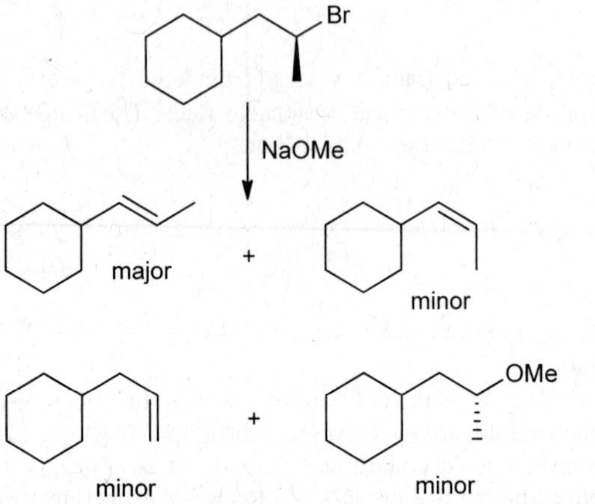

(l) The reagent is methoxide, which is both a strong base and a strong nucleophile. The substrate is secondary so we expect both E2 and S_N2 processes, although E2 will be responsible for the major product. Accordingly, the major product is the more substituted alkene. To determine the stereochemistry of formation of the major product, we must draw a Newman projection,

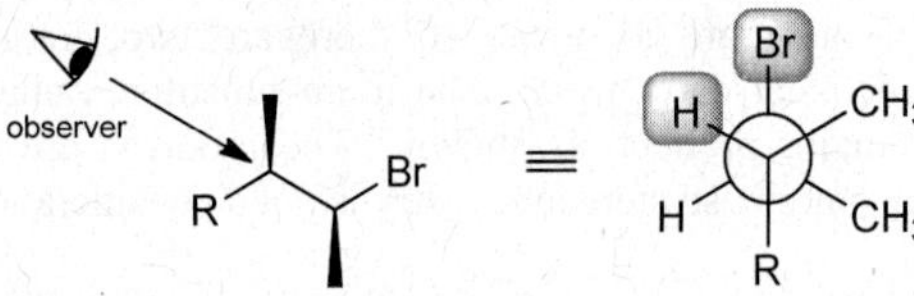

and then rotate the central C-C bond so as to achieve a conformation in which the leaving group (bromide) and the β proton are antiperiplanar to one another. The reaction proceeds through this conformation, giving the trans configuration, as shown:

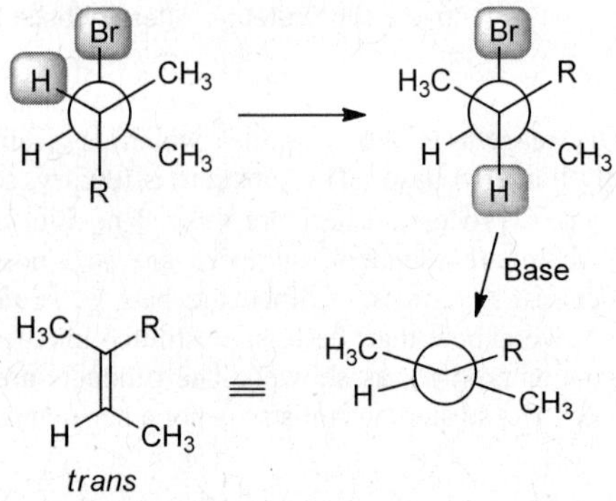

The minor products include the less-substituted alkene, as well as the S_N2 product (via inversion of configuration), a shown:

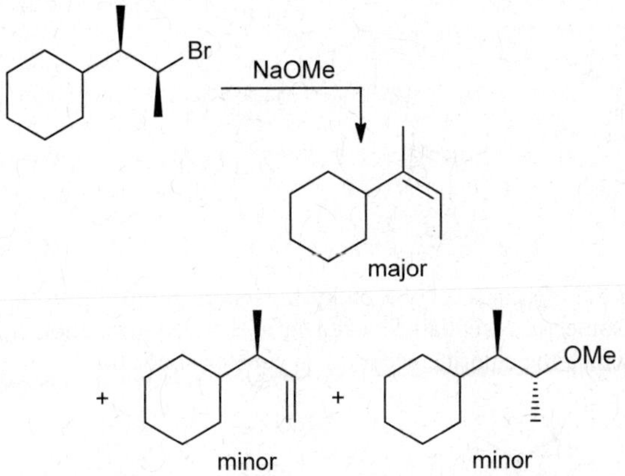

major

+ minor + minor

(m) The reagent is hydroxide, which is both a strong base and a strong nucleophile. The substrate is secondary so we expect both E2 and S_N2 processes, although E2 will be responsible for the major product. For substituted cyclohexanes, an E2 reaction occurs via a conformation in which the leaving group and the β proton are antiperiplanar to one another (one must be on a wedge and the other must be on a dash). The leaving group (Br^-) is on a wedge. Therefore, we are looking for a β proton that is on a dash. In this case, there is only one β proton that is on a dash (highlighted below), giving rise to only one elimination product, as shown:

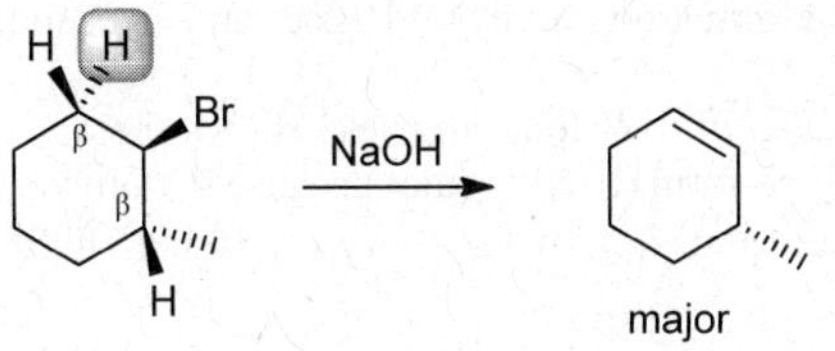

major

The minor product is generated via an S_N2 pathway (with inversion of configuration, as expected):

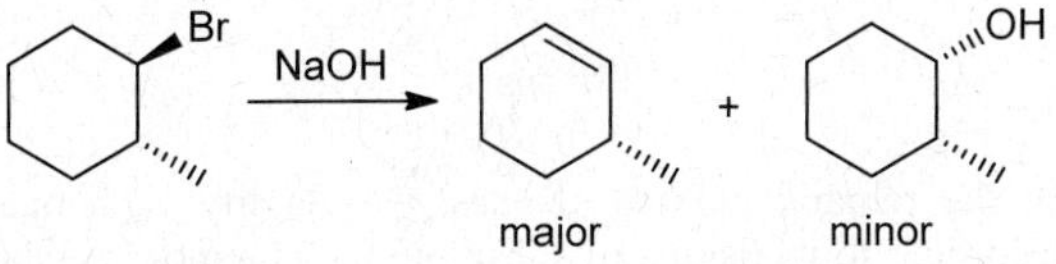

major   minor

(n) The reagent is hydroxide, which is both a strong base and a strong nucleophile. The substrate is secondary so we expect both E2 and S_N2 processes, although E2 will be responsible for the major product. For substituted cyclohexanes, an E2 reaction occurs via a conformation in which the leaving group and the β proton are antiperiplanar to one another (one must be on a wedge and the other must be on a dash). The leaving group (Br^-) is on a wedge. Therefore, we are looking for a β proton that is on a dash. In this case, there are two such protons (highlighted below), giving rise to two elimination products, as shown. The major product is the more-substituted alkene (because the base is not

sterically hindered), while the minor products include the less-substituted alkene and the S_N2 product (with the expected inversion of configuration):

major

+ minor + minor

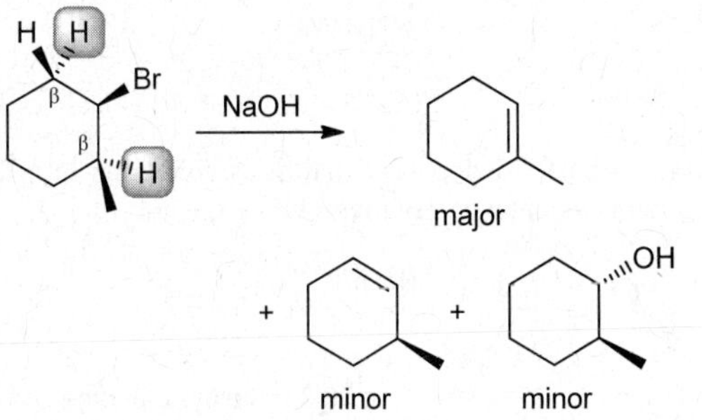

8.47. There are only two constitutional isomers with molecular formula C_3H_7Cl:

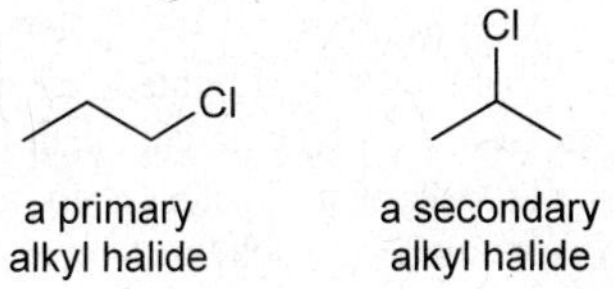

a primary   a secondary
alkyl halide   alkyl halide

Sodium methoxide is both a strong nucleophile and a strong base. When compound **A** is treated with sodium methoxide, a substitution reaction predominates. Therefore, compound **A** must be the primary alkyl chloride above. When compound **B** is treated with sodium methoxide, an elimination reaction predominates. Therefore, compound **B** must be the secondary alkyl chloride:

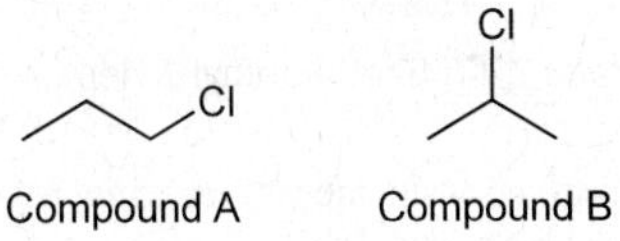

Compound A   Compound B

8.48.
(a) There are four constitutional isomers with molecular formula C_4H_9Cl, shown below. Only one of them affords the desired product when treated with methoxide:

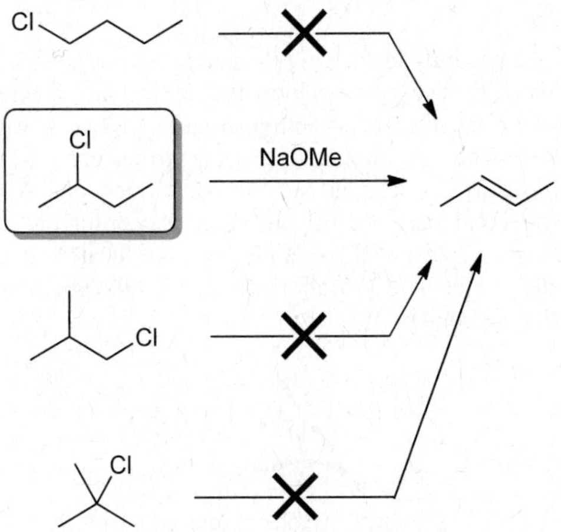

(b) Among the four constitutional isomers with molecular formula C_4H_9Cl, only one isomer is converted into a disubstituted alkene upon treatment with methoxide:

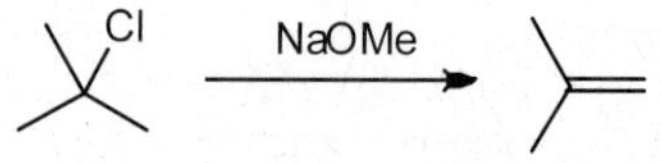

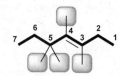

8.49. The following alkyl halide is converted into the desired alkene, upon treatment with ethoxide.

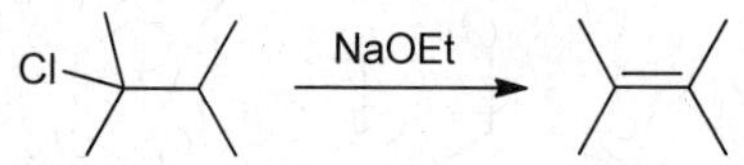

2,3-dimethyl-2-butene

8.50.
(a) We begin by identifying the parent. The longest chain is seven carbon atoms, so the parent is heptene. There are four substituents (highlighted), all of which are methyl groups. Notice that the parent chain is numbered starting from the side that is closest to the π bond. According to this numbering scheme, the methyl groups are located at C3, C4, C5 and C5. Finally, we use the prefix "tetra" to indicate the presence of four methyl groups, and we include a locant that identifies the position of the double bond ("3" indicates that the double bond is located between C3 and C4):

trans-3,4,5,5-tetramethyl-3-heptene

(b) The parent is a six-membered ring (cyclohexene). There is only substituent (located at C1:

5 6 1
4 2
3

1-ethylcyclohexene

(c) We begin by identifying the parent, which is bicyclic in this case. The parent is bicyclo[2.2.2]octene. There is only one substituent (a methyl group). Notice that the parent chain is numbered starting from one of the bridgeheads, as seen in Section 4.2, which places the double bond between C2 and C3. According to this numbering scheme, the methyl group is located at C2. Finally, we include a locant that identifies the position of the double bond (C2).

8 7 1
6   CH3
        2
5 4 3

2-methylbicyclo[2.2.2]oct-2-ene

8.51. The configuration of each π bond is shown below, together with the priorities (highlighted) that were used to determine the configuration in each case.

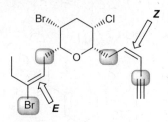

8.52. Because of the bulky *tert*-butyl group, the *trans* isomer is essentially locked in a chair conformation in which the chlorine occupies an equatorial position.

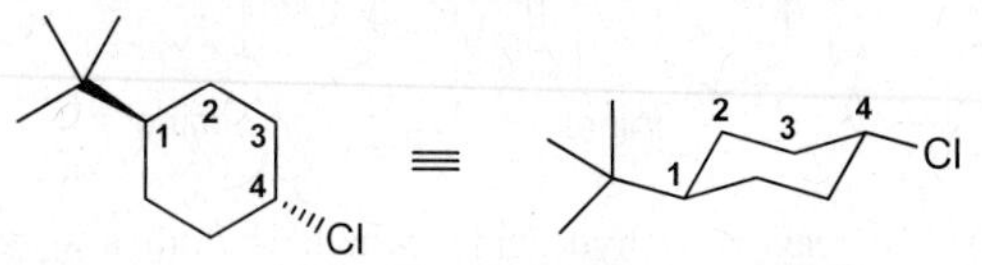

1 2 3 ≡ 2 3 4
      4 Cl   1   Cl

This conformation cannot readily undergo an E2 reaction because the leaving group is not antiperiplanar to a proton. However, the *cis* isomer is locked in a chair conformation in which the chlorine occupies an axial position:

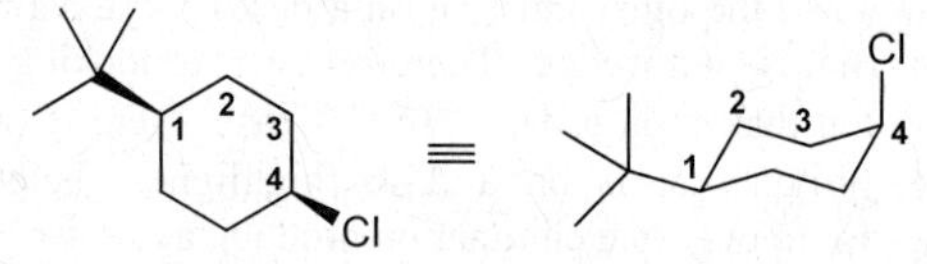

1 2 3 ≡ 2 3 Cl
      4 Cl   1   4

This conformation rapidly undergoes an E2 reaction.

8.53. One of the four alkenes, shown below, is too unstable to form (Bredt's rule) because it would exhibit a six-membered ring with a *trans* π bond, highlighted below:

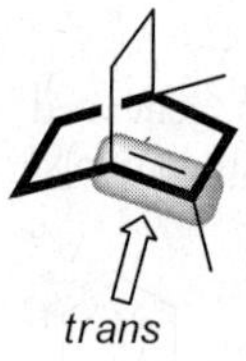

trans

Of the remaining three alkenes, we classify each one according to its degree of substitution. The most highly substituted alkene will be the most stable, and therefore, the following order of stability is expected:

Increasing Stability

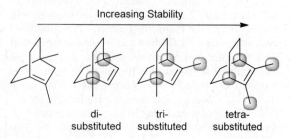

di-   tri-   tetrasubstituted   substituted   substituted

8.54.

(a) A tertiary carbocation is more stable than a primary carbocation, and the former will form more readily than the latter. Therefore, a tertiary substrate will be more reactive toward E1 than a primary substrate.

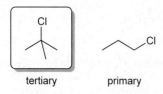

(b) The first compound is a secondary alkyl halide, so an E1 process would have to proceed via a secondary carbocation. But the other compound is secondary allylic, so an E1 process proceeds via a more stable, allylic (resonance-stabilized) carbocation. The allylic halide is therefore more reactive toward E1:

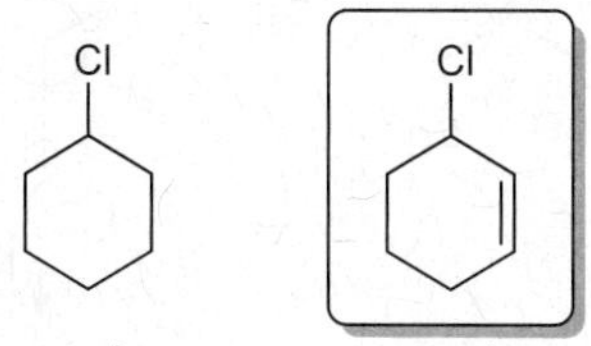

8.55.

(a) Hydroxide is a stronger base than water, because the former bears a negative charge.

(b) Ethoxide is a stronger base than ethanol, because the former bears a negative charge.

8.56. We first draw a bond-line drawing of the reactant, and then convert it into a Newman projection:

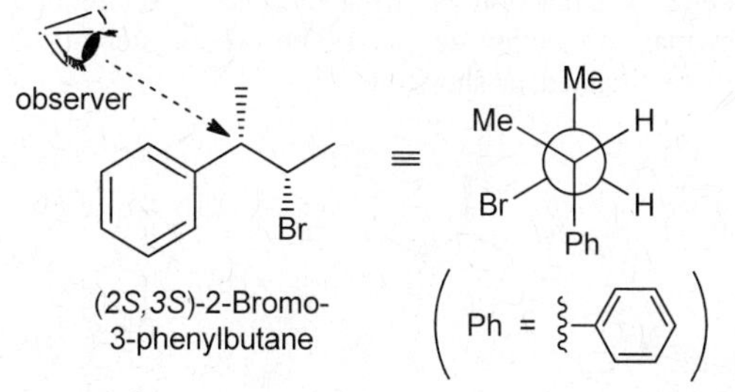

(2S,3S)-2-Bromo-3-phenylbutane

In this conformation, the β proton and the leaving group (highlighted below) are antiperiplanar to one another, so the reaction proceeds through this conformation, giving the product, as shown:

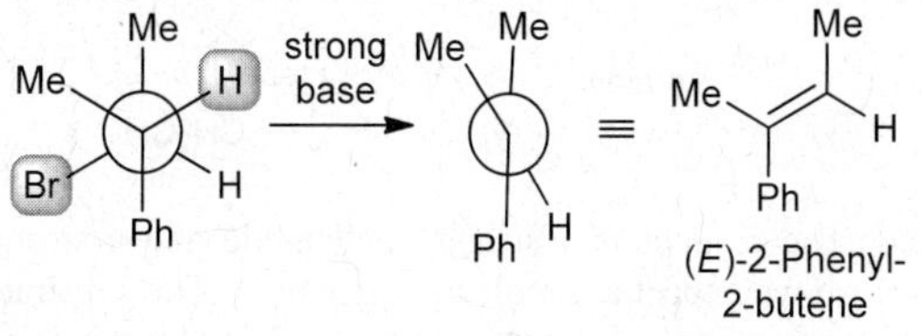

(E)-2-Phenyl-2-butene

8.57.

(a) The rate of an E2 process is dependent on the concentrations of the substrate and the base. Therefore, the rate will be doubled if the concentration of *tert*-butyl bromide is doubled.

(b) The rate of an E2 process is dependent on the concentrations of the substrate and the base. Therefore, the rate will be doubled if the concentration of sodium ethoxide is doubled.

8.58.

(a) The rate of an E1 process is dependent only on the concentration of the substrate (not the base). Therefore, the rate will be doubled if the concentration of *tert*-butyl bromide is doubled.

(b) The rate of an E1 process is dependent only on the concentration of the substrate (not the base). Therefore, the rate will remain the same if the concentration of ethanol is doubled.

8.59. We begin by drawing the substrate and identifying the β positions (highlighted):

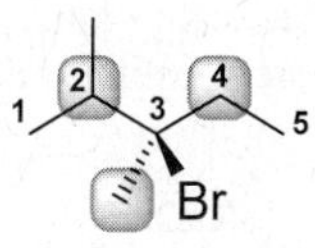

(R)-3-Bromo-2,3-dimethylpentane

There are three β positions, each of which contains protons, so there are three possible regiochemical outcomes. If the double bond is formed between C3 and C4, then both *cis* and *trans* stereoisomers are possible, giving a total of four alkenes, shown below:

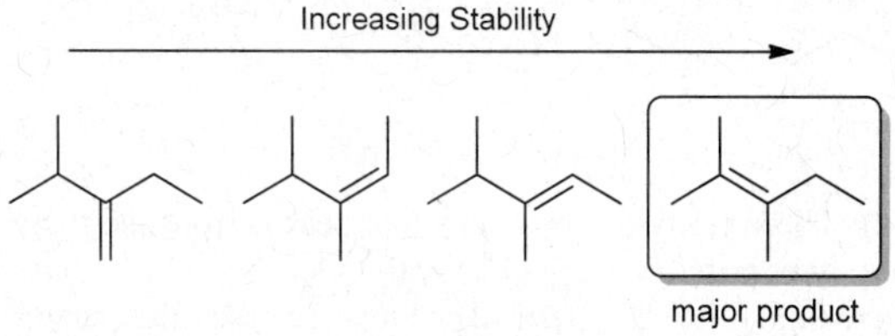

The tetrasubstituted alkene is the most stable, while the disubstituted alkene is the least stable. Among the two trisubstituted alkenes, the *E* isomer is more stable, because it exhibits fewer steric interactions than the *Z* isomer.

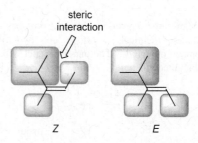

8.60. There are only two β protons to abstract: one at C2 and the other at C4. Abstraction of either proton leads to the same product.

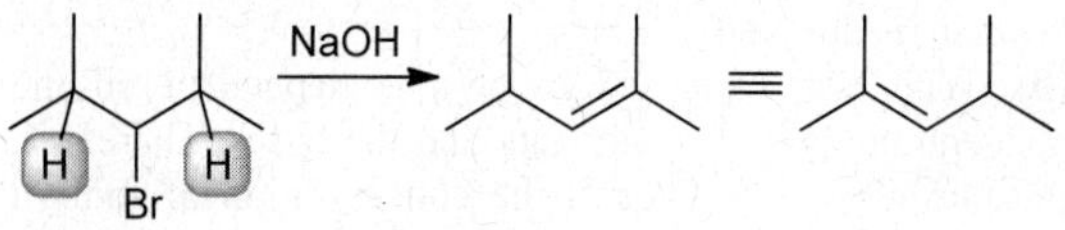

8.61.

(a) There are two β positions that bear protons, so there are two possible regiochemical outcomes. The base (hydroxide) is not sterically hindered, so the major product will be the more-substituted alkene. Two stereoisomers are possible (*cis* and *trans*), and the trans isomer is favored:

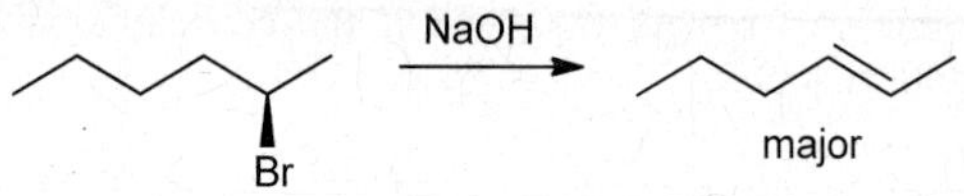

(b) There are two β positions that bear protons, so there are two possible regiochemical outcomes. The base (hydroxide) is not sterically hindered, so the major product will be the more-substituted alkene:

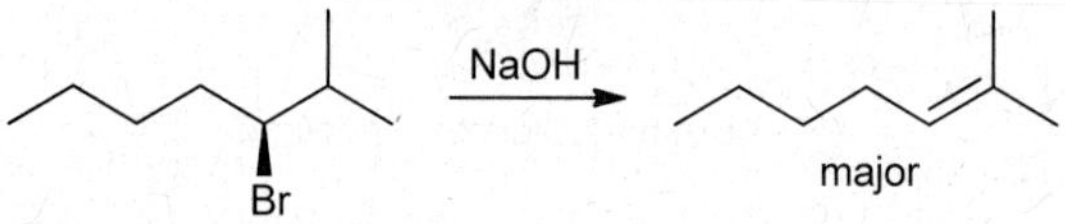

(c) There are two β positions that bear protons, so there are two possible regiochemical outcomes. The base (*tert*-butoxide) is sterically hindered, so the major product will be the less-substituted alkene:

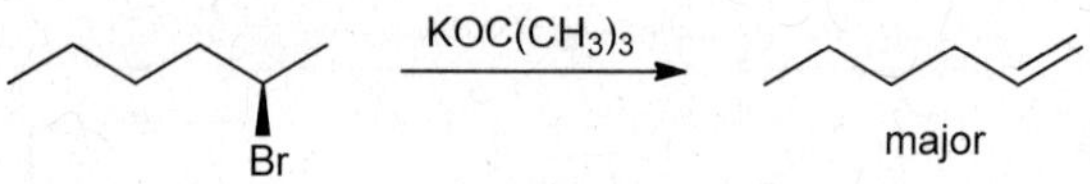

(d) There are two β positions that bear protons, so there are two possible regiochemical outcomes. The base (*tert*-butoxide) is sterically hindered, so the major product will be the less-substituted alkene:

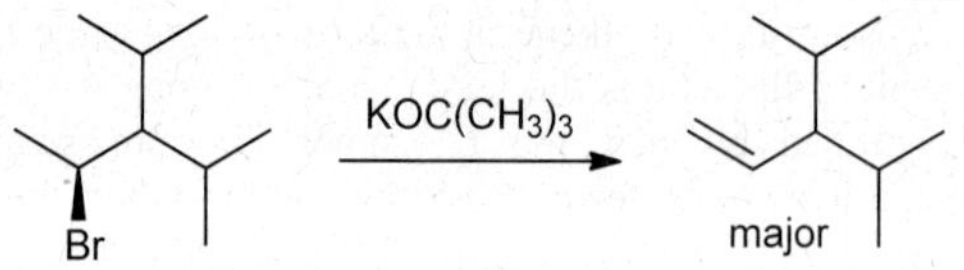

8.62.

(a) There are three β positions that bear protons, but two of them are identical, so there are two possible regiochemical outcomes. The major product will be the more-substituted alkene:

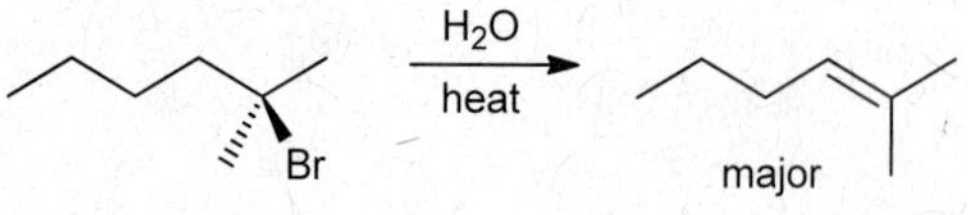

(b) There are two β positions that bear protons, so there are two possible regiochemical outcomes. The major product will be the more-substituted alkene:

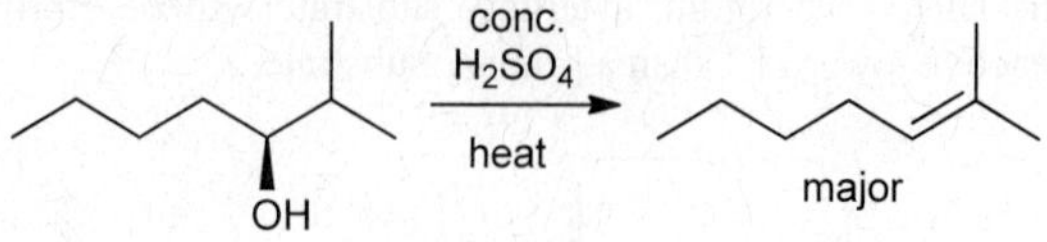

8.63.

(a) There are two β positions, so there are two possible regiochemical outcomes. Since the base (hydroxide) is not sterically hindered, the major product will be the more-substituted alkene. The reaction will be stereospecific, so we draw a Newman projection to determine which stereoisomer will be obtained, as shown:

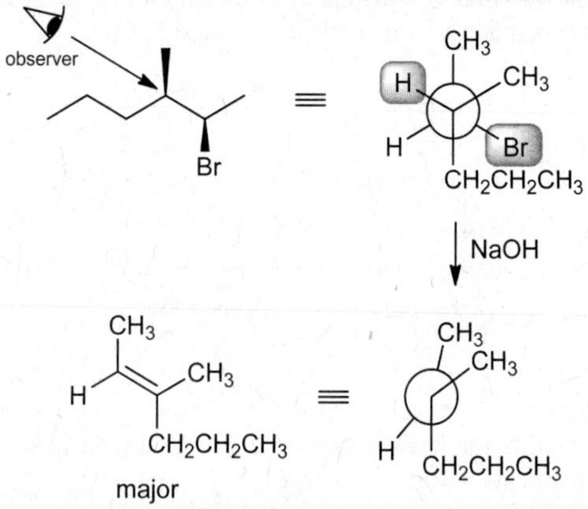

(b) There are three β positions, but only two of them bear protons, so there are two possible regiochemical outcomes. Since the base (hydroxide) is not sterically hindered, the major product will be the more-substituted alkene. The reaction will be stereospecific, so we draw a Newman projection to determine which stereoisomer will be obtained, as shown:

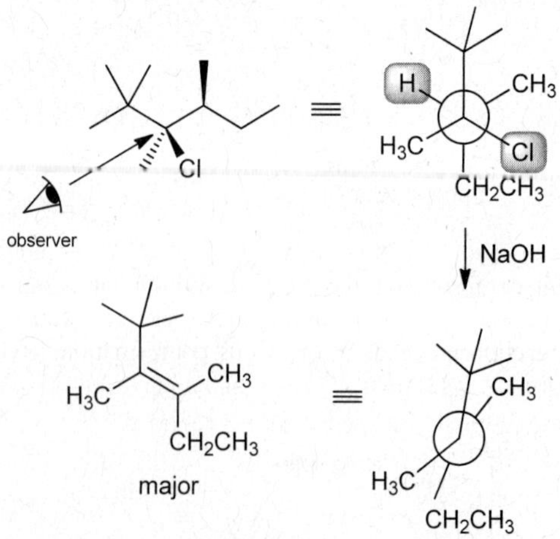

8.64. The reagent is a strong nucleophile and a strong base, so we expect a bimolecular reaction. The substrate is tertiary so only E2 can operate (S$_N$2 is too sterically hindered to occur). There is only one possible regiochemical outcome for the E2 process, because the other β positions lack protons.

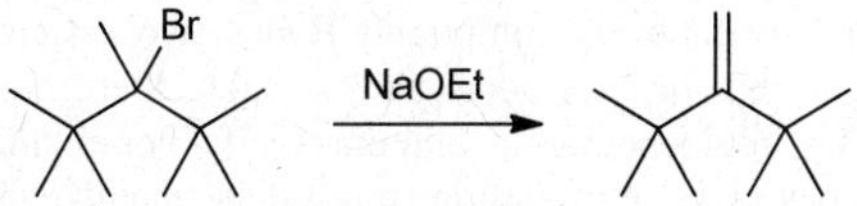

8.65.
(a) There are certainly a very large number of alkyl halides that will give four different alkenes when treated with a strong base. Below is just one example:

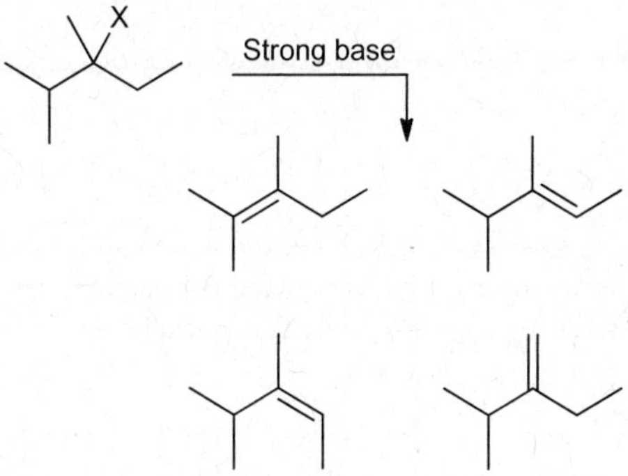

(b) There are certainly a very large number of alkyl halides that will give three different alkenes when treated with a strong base. Below is just one example:

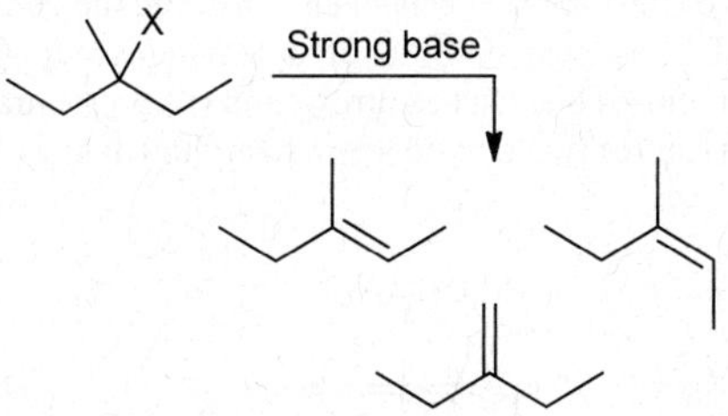

(c) There are certainly a very large number of alkyl halides that will give two different alkenes when treated with a strong base. Below is just one example:

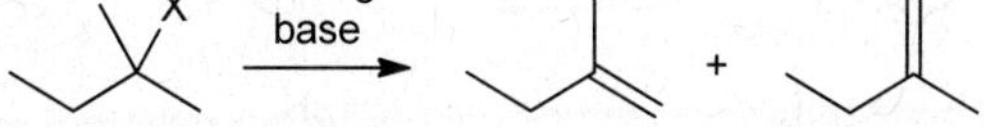

(d) There are certainly a large number of alkyl halides that will give only one alkene when treated with a strong base. Below is just one example:

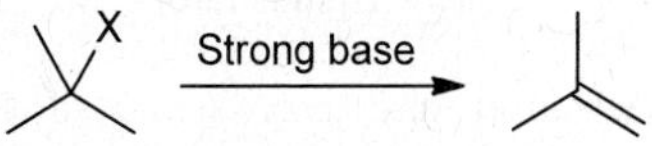

8.66.
(a) There is only one β position, and the resulting alkene is not stereoisomeric, so only one alkene will be produced, as shown:

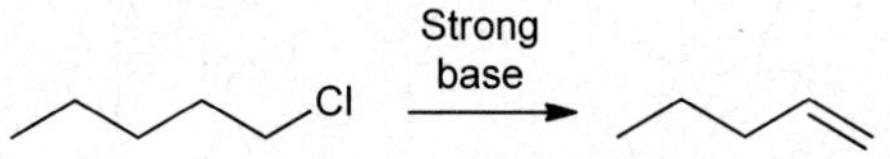

(the only alkene produced)

(b) The alkyl halide has two β positions, so there are two possible regiochemical outcomes. The more-substituted alkene can be formed as the E or Z isomer, giving a total of three alkenes, as shown:

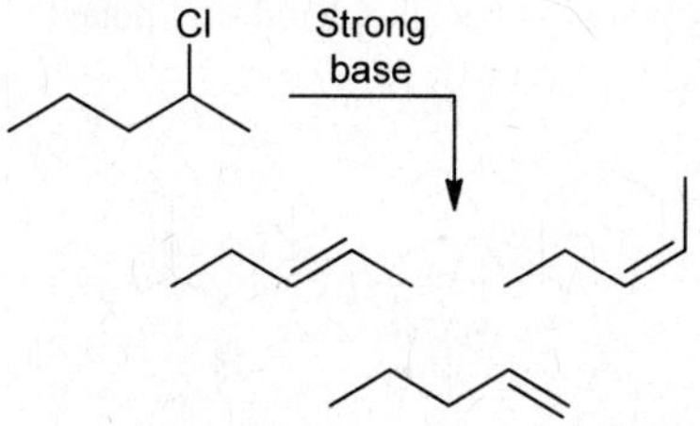

(c) The alkyl halide has two β positions, but they are identical, so there is only one possible regiochemical outcome. Two stereoisomers are possible (*cis* and *trans*), giving a total of two alkenes, as shown:

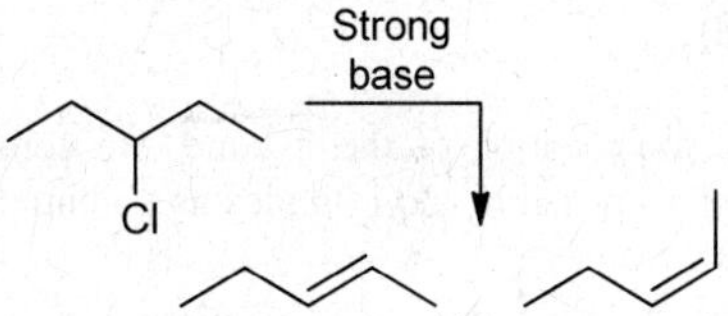

(d) The alkyl halide has three β positions, but two of them are identical. As such, there are two possible regiochemical outcomes, giving the following two alkenes:

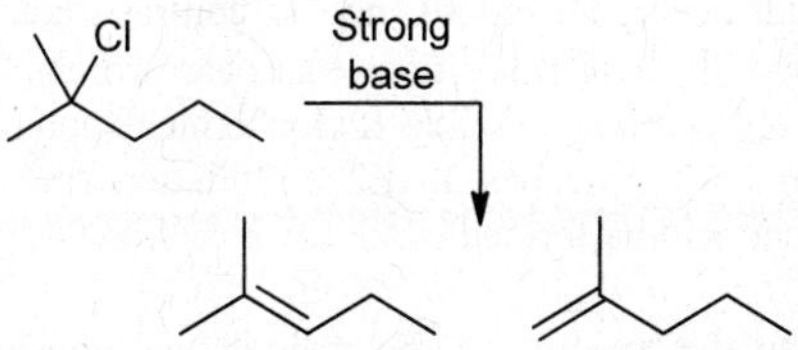

(e) There are three different β positions, and each of them has protons, giving rise to three different regiochemical outcomes. For two of these outcomes, both *cis* and *trans* isomers are possible, giving a total of five alkenes, as shown:

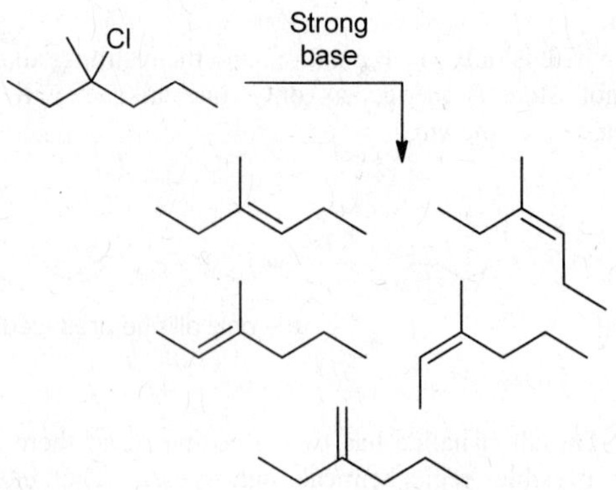

8.67.

(a) Given the location of the π bond, we consider the following two possible alkyl halides as potential starting materials.

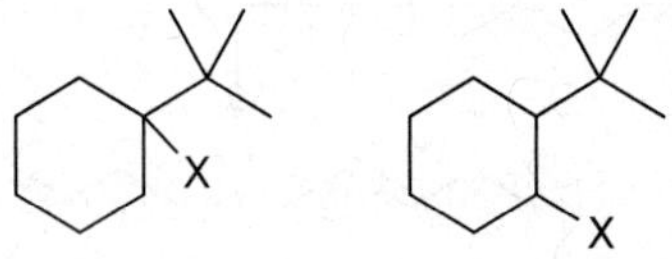

A   **B**

Compound **A** has three β positions, but only two of them bear protons, and those two positions are identical. Deprotonation at either location will result in the desired alkene. In contrast, compound **B** has two different β positions that bear protons, and as such, there are two possible regiochemical outcomes if compound **B** undergoes an E2 elimination (more than one alkene will be formed).

(b) Given the location of the π bond, we consider the following two possible alkyl halides as potential starting materials.

A   **B**

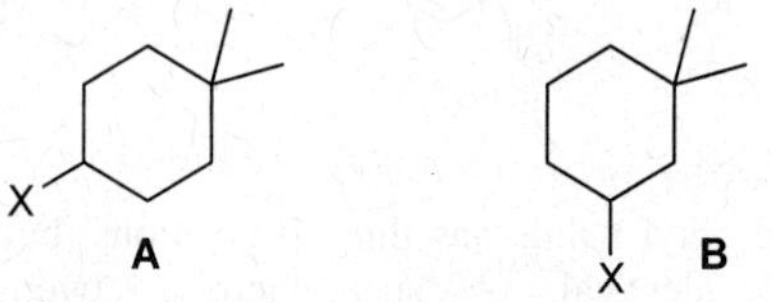

Compound **A** has two β positions, and those two positions are identical. Deprotonation at either location will result in the desired alkene. In contrast, compound **B** has two different β positions that bear protons, and as such, there are two possible regiochemical outcomes if compound **B** undergoes an E2 elimination (more than one alkene will be formed).

(c) Given the location of the π bond, we consider the following two possible alkyl halides as potential starting materials.

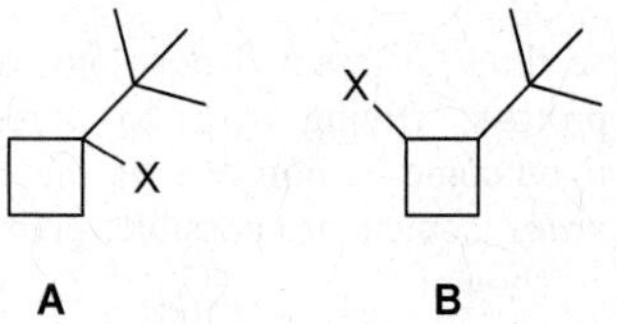

A   **B**

Compound **A** has three β positions, but only two of them bear protons, and those two positions are identical. Deprotonation at either location will result in the desired alkene. In contrast, compound **B** has two different β positions that bear protons, and as such, there are two possible regiochemical outcomes if compound **B** undergoes an E2 elimination (more than one alkene will be formed).

(d) Given the location of the π bond, we consider the following two possible alkyl halides as potential starting materials.

A   **B**

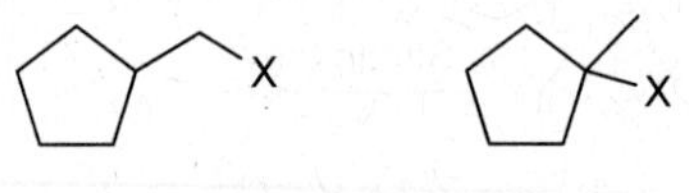

Compound **A** has only one β position, giving rise to only one alkene. In contrast, compound **B** has more than one β position, giving rise to more than one alkene.

8.68.

(a) In acidic conditions, the OH group is protonated, which converts it from a bad leaving group to a good leaving group. In aqueous sulfuric acid, the acid that is present in solution is H_3O^+ (because of the leveling effect, as explained in Section 3.6), so we use H_3O^+ as the proton source in the first step of the mechanism. The next two steps of the mechanism constitute the core steps of an E1 process: (i) loss of a leaving group (H_2O), which requires one curved arrow, and (ii) proton transfer, which requires two curved arrows, as shown:

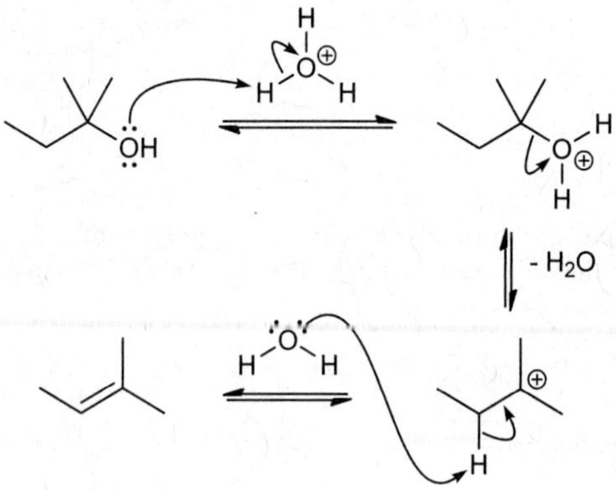

(b) This is an E1 process, so the rate is dependent only on the substrate:

$$Rate = k \text{ [substrate]}$$

(c) The mechanism has three steps, so the energy diagram must have three humps, as shown below. There are two intermediates: an oxonium ion and a carbocation. Notice that the former is lower in energy than the latter (because all atoms in the oxonium ion have a full octet, which is not the case for the carbocation).

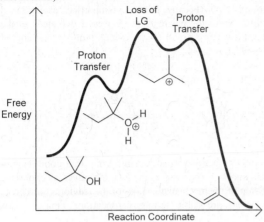

8.69.

(a) Loss of the leaving group (chloride) results in the following secondary carbocation:

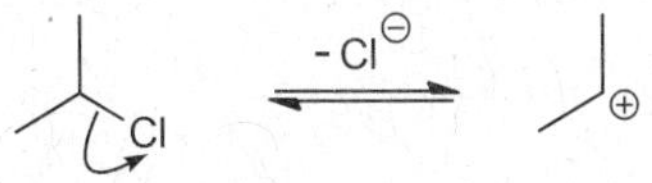

(b) Loss of the leaving group (bromide) results in the following tertiary carbocation:

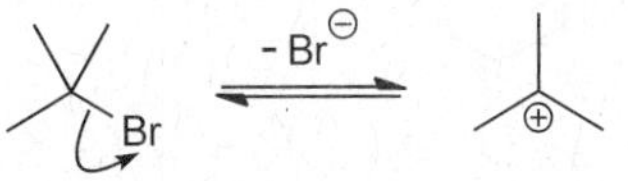

(c) Loss of the leaving group (iodide) would result in a primary carbocation, shown below, which is too high in energy to form readily. This alkyl halide will not undergo an E1 reaction.

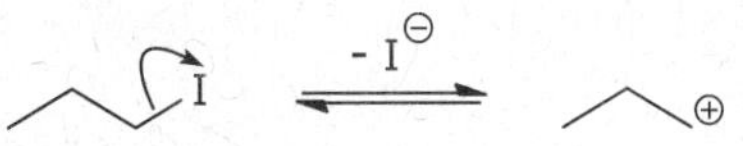

(d) Loss of the leaving group (chloride) results in the following secondary carbocation:

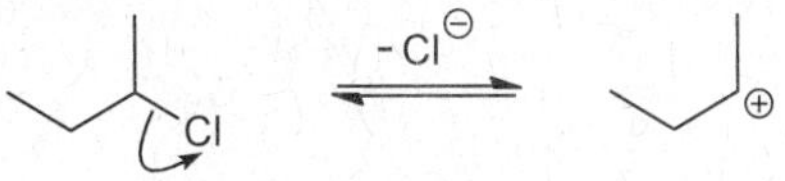

8.70. In the transition state, the hydroxide ion is in the process of removing the proton, the double bond is in the process of forming, and the leaving group is in the process of leaving. We use dotted lines to indicate the bonds that are in the process of being formed or broken:

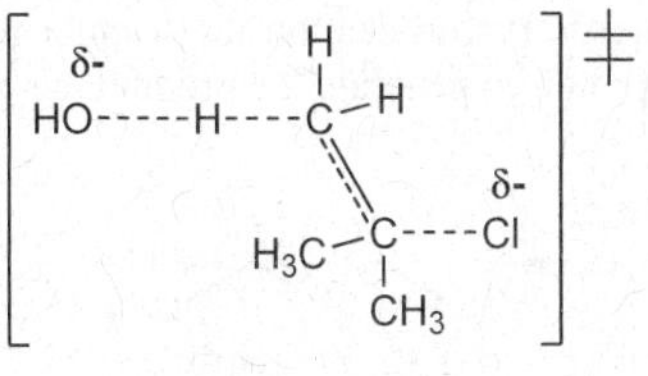

8.71. The first reaction is very slow, because the *tert*-butyl group effectively locks the ring in a chair conformation in which the leaving group (Br) occupies an equatorial position. In this conformation, the leaving group cannot be antiperiplanar to a β proton. So the reaction can only occur from the other chair conformation, which the compound does not readily adopt. The second reaction is very rapid, because the *tert*-butyl group effectively locks the ring in a chair conformation in which the leaving group (Br) occupies an axial position. In this conformation, the leaving group is antiperiplanar to a β proton. The third reaction does not occur at all because there are no β protons that are antiperiplanar to the leaving group (on a cyclohexane ring, there must be at least one β proton that is *trans* to the leaving group in order to be able to adopt an antiperiplanar conformation).

8.72. Pi bonds cannot be formed at the bridgehead of a bicyclic compound, unless one of the rings is large (at least eight carbon atoms). This rule is known as Bredt's rule.

8.73.
(a) The first compound will react more rapidly because it is tertiary.
(b) The second compound will react more rapidly in an E2 reaction because the first compound does not have any β protons (and therefore cannot undergo E2 at all).

8.74.
(a) The Zaitsev product is desired, so sodium ethoxide should be used.
(b) The Hofmann product is desired, so potassium *tert*-butoxide should be used.
(c) The Zaitsev product is desired, so sodium ethoxide should be used.
(d) The Hofmann product is desired, so potassium *tert*-butoxide should be used.

8.75. For substituted cyclohexanes, an E2 reaction will occur if the leaving group and the β proton can achieve antiperiplanarity. In order to achieve this, one must be on a wedge and the other must be on a dash. The leaving group (Br⁻) is on a wedge. Therefore, we are looking for a β proton that is on a dash. In this case, there is only one such proton, highlighted below, so there is only one possible regiochemical outcome. In this case, the Hofmann product is formed regardless of the choice of base.

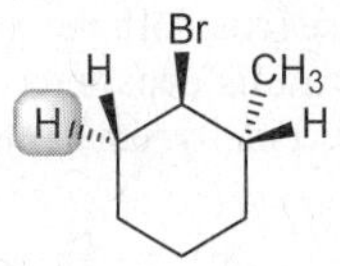

8.76.

(a) In acidic conditions, the OH group is protonated, which converts it from a bad leaving group to a good leaving group. In aqueous sulfuric acid, the acid that is present in solution is H_3O^+ (because of the leveling effect, as explained in Section 3.6), so we use H_3O^+ as the proton source in the first step of the mechanism. The next two steps of the mechanism constitute the core steps of an E1 process: (i) loss of a leaving group (H_2O), which requires one curved arrow, and (ii) proton transfer, which requires two curved arrows, as shown:

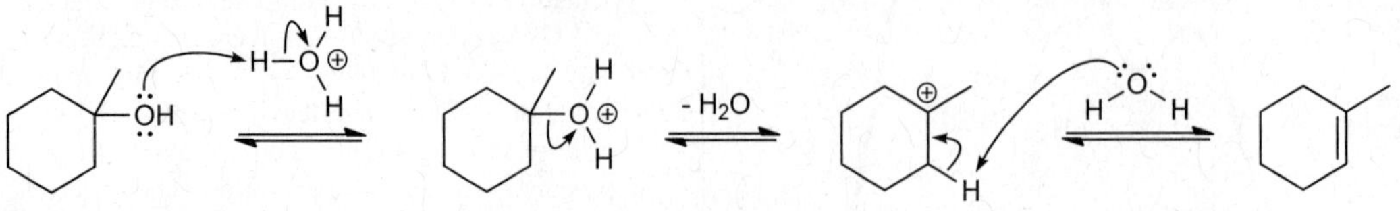

(b) In acidic conditions, the OH group is protonated, which converts it from a bad leaving group to a good leaving group. In aqueous sulfuric acid, the acid that is present in solution is H_3O^+ (because of the leveling effect, as explained in Section 3.6), so we use H_3O^+ as the proton source in the first step of the mechanism. Loss of the leaving group gives a secondary carbocation that can undergo a hydride shift, giving a more stable, tertiary carbocation. Finally, water serves as the base that removes the proton to generate the product:

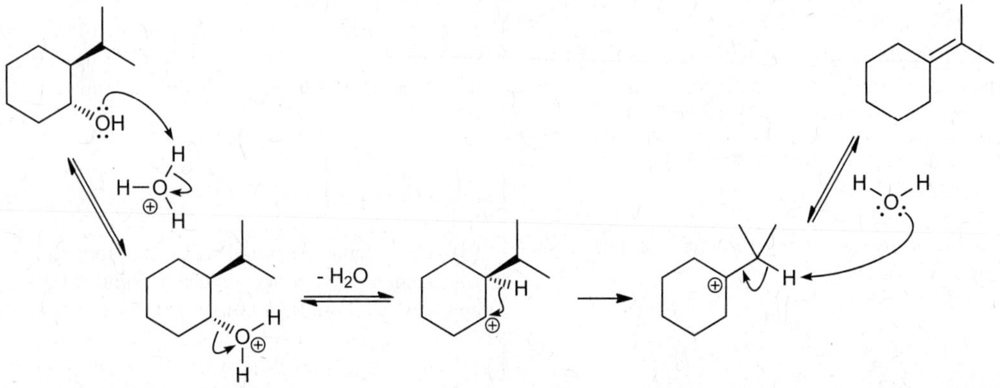

(c) In acidic conditions, the OH group is protonated, which converts it from a bad leaving group to a good leaving group. In aqueous sulfuric acid, the acid that is present in solution is H_3O^+ (because of the leveling effect, as explained in Section 3.6), so we use H_3O^+ as the proton source in the first step of the mechanism. Loss of the leaving group gives a secondary carbocation that can undergo a methyl shift, giving a more stable, tertiary carbocation. Finally, water serves as the base that removes the proton to generate the product:

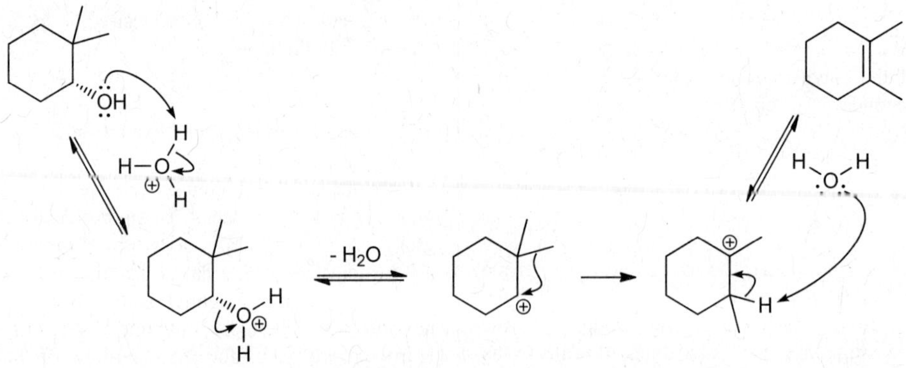

(d) With a weak base (ethanol), the reaction must proceed via an E1 mechanism. The leaving group is iodide, which is not protonated prior to its departure. Loss of the leaving group gives a tertiary carbocation that cannot rearrange to become more stable. Finally, ethanol serves as the base that removes the proton to generate the product.

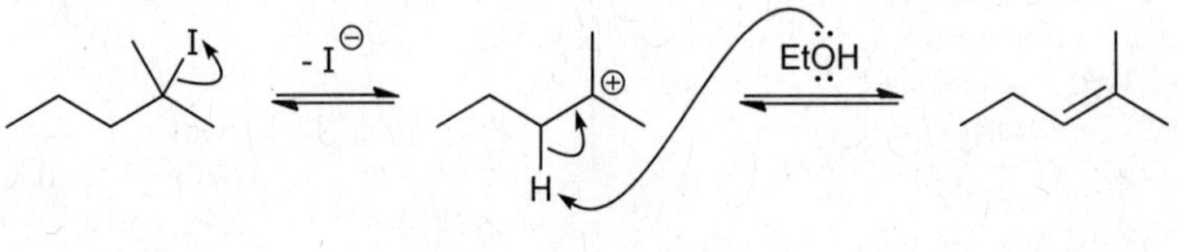

(e) The reagent (ethoxide) is a strong base, and the substrate is tertiary, so the reaction must proceed via an E2 process. Three curved arrows are required. The tail of the first curved arrow is placed on a lone pair of the base (ethoxide) and the head is placed on the proton that is removed. The tail of the second curved arrow is placed on the C–H bond that is breaking, and the head shows formation of the π bond. The third curved arrow shows loss of the leaving group (iodide), as shown here.

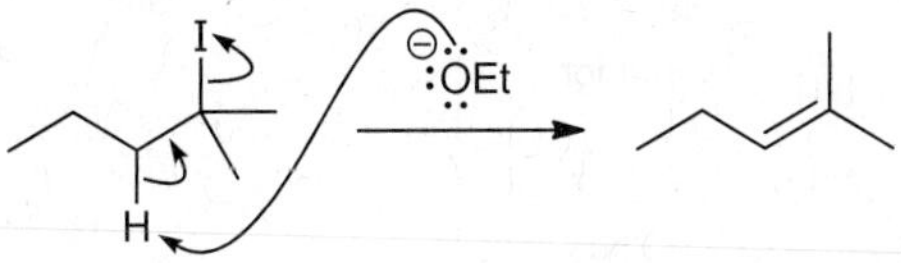

8.77.

(a) The reagent is *tert*-butoxide, which is a strong, sterically hindered base. The substrate is secondary so we expect E2 processes to predominate (S$_N$2 is highly disfavored because of steric interactions). The major product is the less-substituted alkene. The more-substituted alkene can be formed as either of two stereoisomers (*cis* and *trans*), giving the following three products:

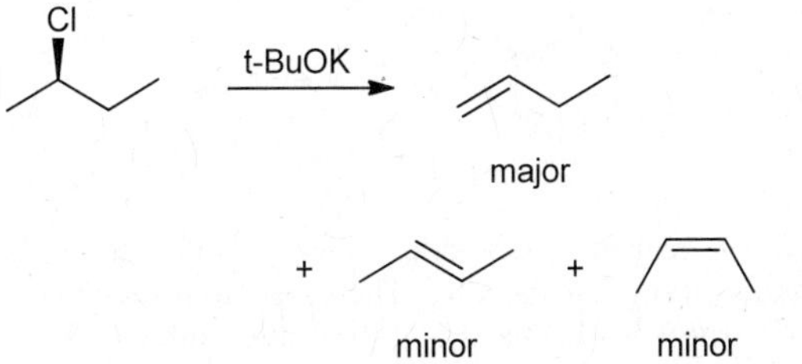

(b) The reagent is hydroxide, which is both a strong base and a strong nucleophile. The substrate is primary, so we expect both E2 and S$_N$2 processes, although S$_N$2 will be responsible for the major product, as shown here.

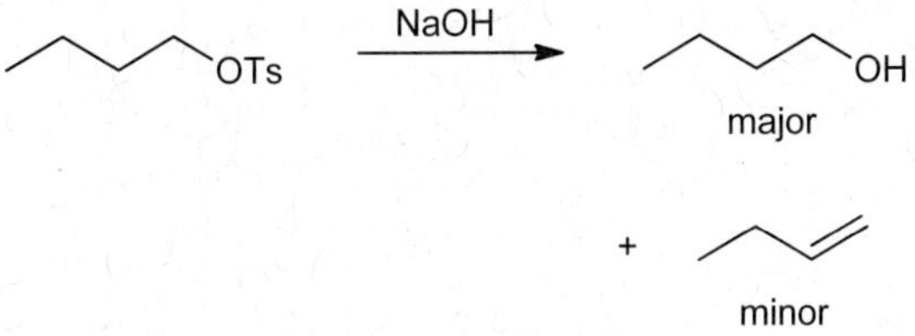

(c) The conditions (treatment an alcohol with sulfuric acid at elevated temperature) favor an E1 process. There are two different β positions that bear protons, so there are two possible regiochemical outcomes. The more-substituted alkene is the major product, and the less-substituted alkene is a minor product. Another minor product can result if the initially formed secondary carbocation undergoes a rearrangement to give a tertiary carbocation, followed by deprotonation to give a disubstituted alkene, shown below:

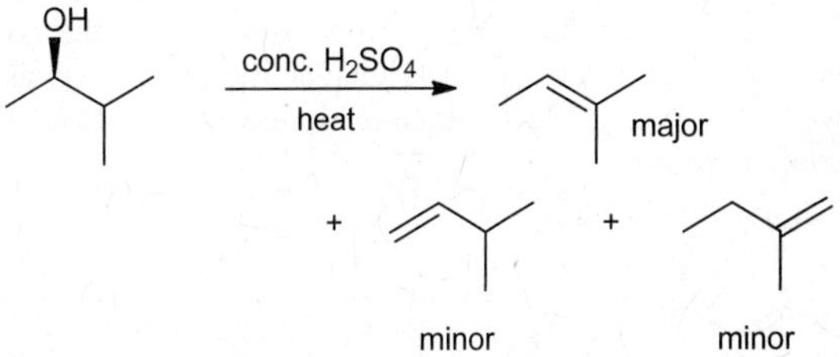

(d) The reagent is chloride, which functions as a nucleophile, so we expect a substitution reaction. The substrate is secondary and the solvent is polar aprotic, indicating an S$_N$2 process. As such, we expect inversion of configuration, as shown:

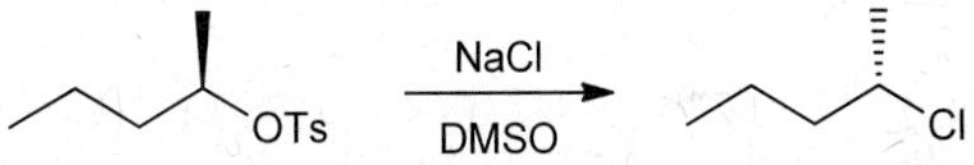

(e) The reagent is ethoxide, which is both a strong base and a strong nucleophile. The substrate is secondary so we expect both E2 and S$_N$2 processes, although E2 will be responsible for the major product. Accordingly, the major product is the more substituted alkene, with the *trans* configuration (because the reaction is stereoselective, favoring the *trans* isomer over the *cis* isomer). The minor products include the *cis* isomer, as well as the less-substituted alkene and the S$_N$2 product:

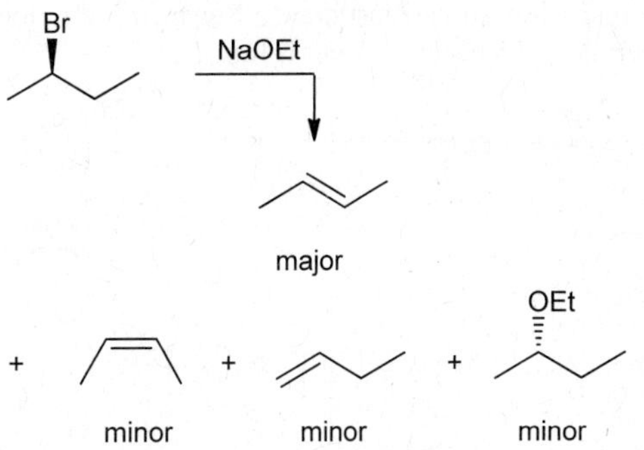

(f) The reagent is ethoxide, which is both a strong base and a strong nucleophile. The substrate is tertiary, so we expect an E2 process. There are three β positions, but two of them are identical, so there are two possible regiochemical outcomes. Since the base is not sterically hindered, we expect that the more-substituted alkene will be the major product, as shown. The less-substituted alkene is the minor product.

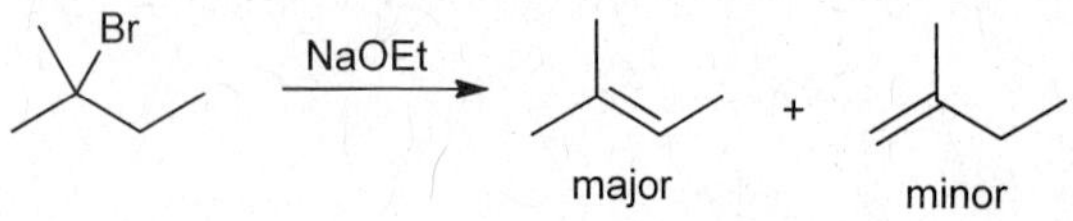

(g) The reagent is ethoxide, which is both a strong base and a strong nucleophile. The substrate is secondary so we expect both E2 and S_N2 processes, although E2 will be responsible for the major product. In this case, there is only one β position that bears protons, so there is only one possible regiochemical outcome. The *trans* isomer is expected to be the major product (because the reaction is stereoselective, favoring the *trans* isomer over the *cis* isomer). The minor products include the *cis* isomer, as well as the S_N2 product (which is formed via inversion of configuration):

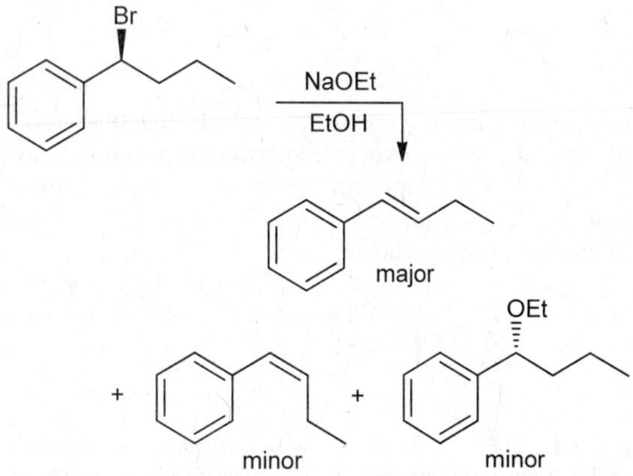

(h) The reagent is ethoxide, which is both a strong base and a strong nucleophile. The substrate is secondary so we expect both E2 and S_N2 processes, although E2 will be responsible for the major product. In this case, there is only one β position that bears a proton, so there is only one possible regiochemical outcome. Since there is only one β proton, the reaction is expected to be stereospecific, so we must draw a Newman projection to determine which stereoisomer is formed:

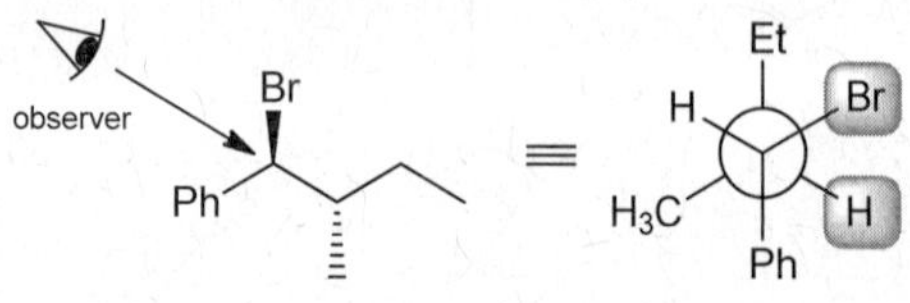

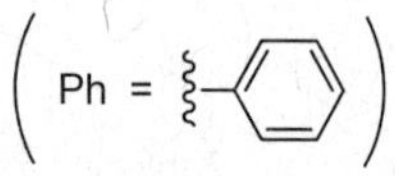

We then identify the leaving group (highlighted below) and the β proton (highlighted below) that will be removed during an E2 reaction. We rotate the central C-C bond, so that the β proton and the leaving group are

anti-periplanar to one another. The reaction proceeds through this conformation, so we expect the *Z* isomer, as shown here:

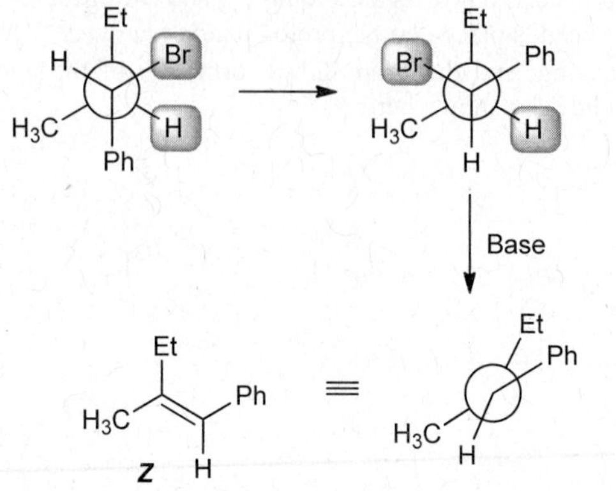

The minor product is formed via an S_N2 process (with inversion of configuration). In summary, the following products are expected:

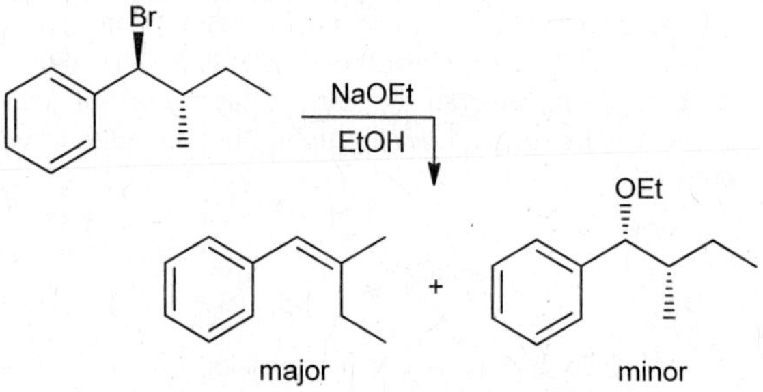

(i) The reagent is ethoxide, which is both a strong base and a strong nucleophile. The substrate is secondary so we expect both E2 and S_N2 processes, although E2 will be responsible for the major product. In this case, there is only one β position that bears a proton, so there is only one possible regiochemical outcome. Since there is only one β proton, the reaction is expected to be stereospecific, so we must draw a Newman projection to determine which stereoisomer is formed:

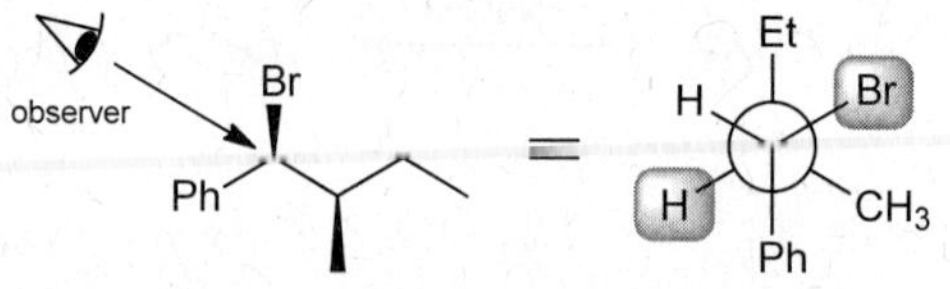

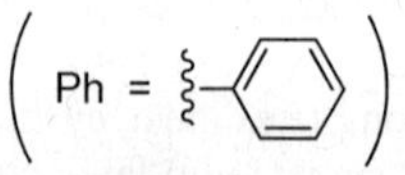

In the conformation drawn above, the β proton and the leaving group are *anti*-periplanar to one another. The reaction proceeds through this conformation, so we expect the *E* isomer, as shown here:

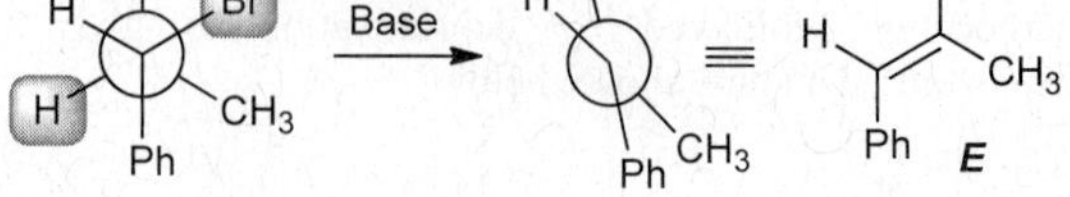

The minor product is formed via an S$_N$2 process (with inversion of configuration). In summary, the following products are expected:

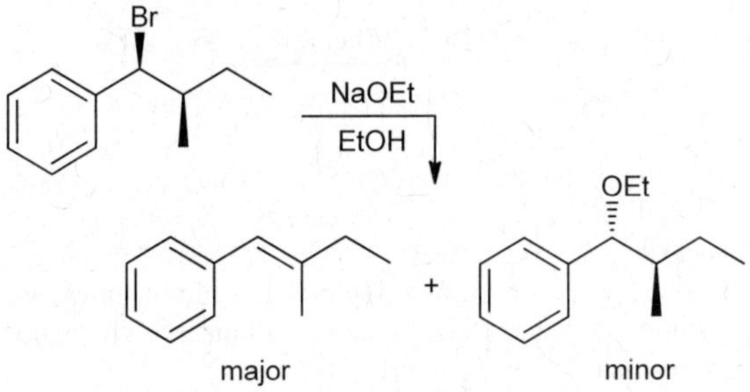

major   +   minor

(j) This compound has three β positions that bear protons, but two of them are identical, giving rise to two possible regiochemical outcomes. Since the base (*tert*-butoxide) is sterically hindered, we expect that the major product will be the less-substituted alkene, and the minor product will be the more-substituted alkene. The latter is formed as a mixture of *cis* and *trans* stereoisomers, giving a total of three products, shown here:

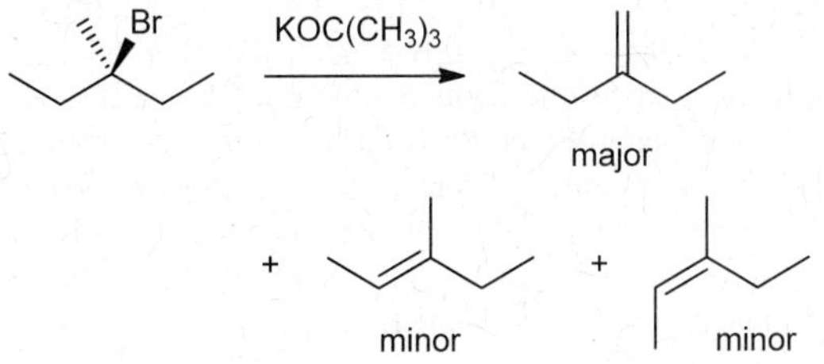

major

+   minor   +   minor

(k) This compound has three β positions that bear protons, but two of them are identical, giving rise to two possible regiochemical outcomes. Since the base (methoxide) is not sterically hindered, we expect that the major product will be the more-substituted alkene (specifically, the *E* isomer, because the process is stereoselective). The minor products include the *Z* isomer, as well as the less-substituted alkene:

major

+   minor   +   minor

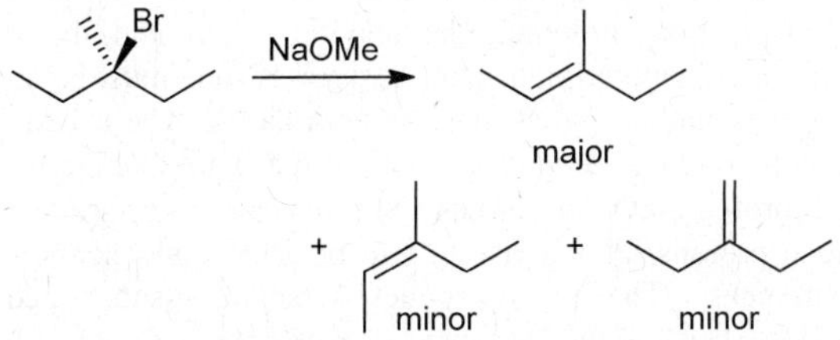

(l) The reagent is hydroxide, which is both a strong base and a strong nucleophile. The substrate is secondary so we expect both E2 and S$_N$2 processes, although E2 will be responsible for the major product. Accordingly, the major product is the more substituted alkene, with the *trans* configuration (because the reaction is stereoselective, favoring the *trans* isomer over the *cis* isomer). The minor products include the *cis* isomer, as well as the less-substituted alkene and the S$_N$2 product:

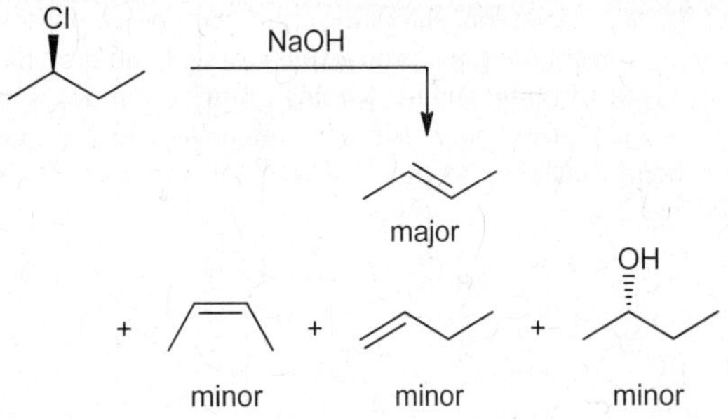

major

+   minor   +   minor   +   minor

8.78.
(a) The reagent (ethoxide) is a strong base, and the substrate is tertiary, so the reaction must proceed via an E2 process. Three curved arrows are required. The tail of the first curved arrow is placed on a lone pair of the base (ethoxide) and the head is placed on the proton that is removed. The tail of the second curved arrow is placed on the C–H bond that is breaking, and the head shows formation of the π bond. The third curved arrow shows loss of the leaving group (bromide), as shown here.

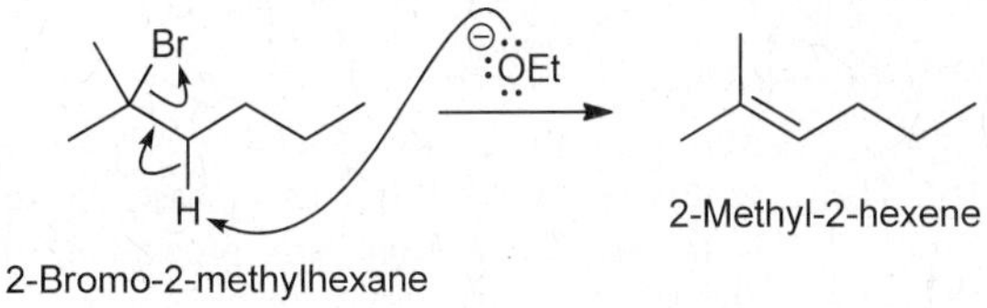

2-Bromo-2-methylhexane   2-Methyl-2-hexene

(b) For an E2 process, the rate is dependent on the concentrations of the substrate and the base:

$$\text{Rate} = k\,[\text{substrate}]\,[\text{base}]$$

(c) If the concentration of base is doubled, the rate will be doubled.
(d) The mechanism has one step, so the energy diagram must have only one hump. The products are lower in energy than the reactants, because bromide is more stable than ethoxide.

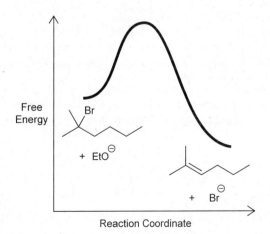

Reaction Coordinate

(e) In the transition state, the ethoxide ion is in the process of removing the proton, the double bond is in the process of forming, and the leaving group (bromide) is in the process of leaving. We use dotted lines to indicate the bonds that are in the process of being formed or broken:

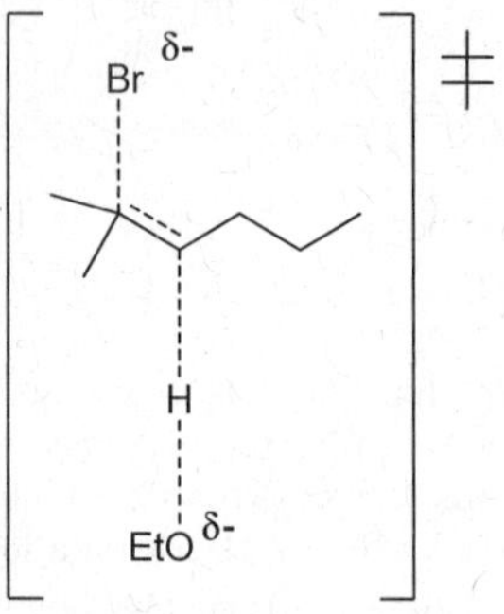

8.79.
(a) The reagent is HS⁻, which is a strong nucleophile, and the substrate is secondary, so we expect an S_N2 process, with inversion of configuration:

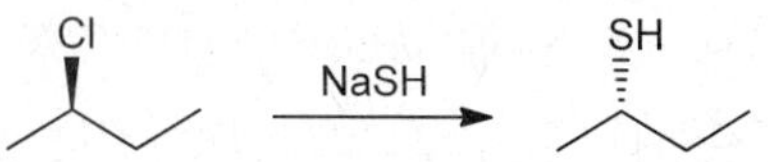

(b) The reagent is DBN, which is a strong base, so we expect an E2 process. There is only one β position, so only one regiochemical outcome is possible.

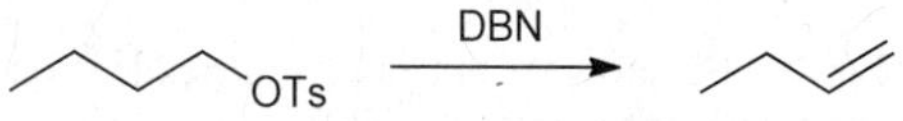

(c) The reagent is hydroxide, which is both a strong base and a strong nucleophile. The substrate is tertiary, so we expect an E2 process. There are three β positions, but two of them are identical, so there are two possible regiochemical outcomes. The more-substituted alkene is the major product, as shown.

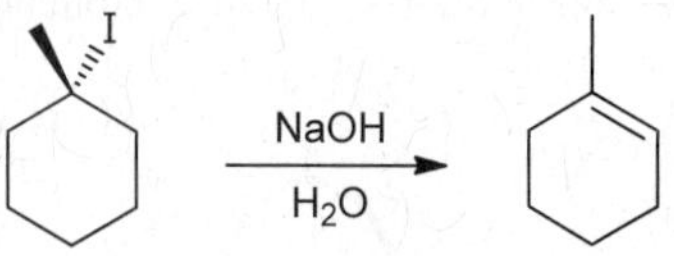

(d) The reagent is water, which is both a weak base and a weak nucleophile. The substrate is tertiary, so we expect E1 and S_N1 processes. Heat is indicated, so E1 is likely to predominate over S_N1. For the E1 pathway, two regiochemical outcomes are possible. The more-substituted alkene is the major product, as shown:

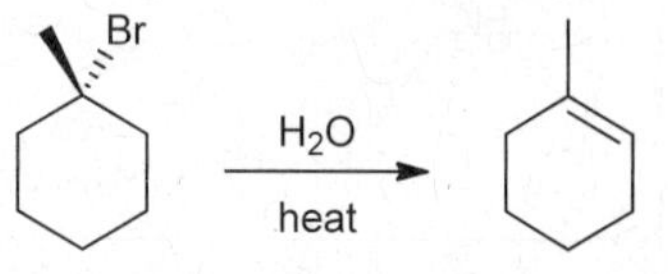

(e) The reagent is *tert*-butoxide, which is a strong, sterically hindered base. The substrate is primary, but

with *tert*-butoxide as the base, we expect an E2 process. There is only one β position, so only one regiochemical outcome is possible.

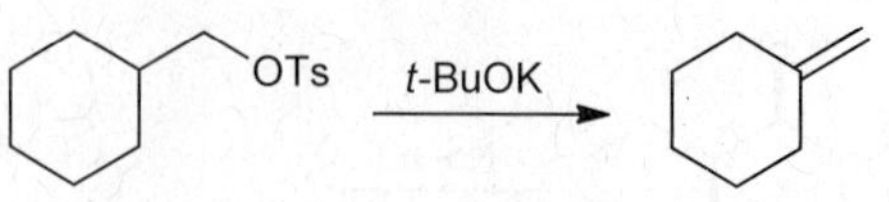

(f) The reagent is *tert*-butoxide, which is a strong, sterically hindered base. The substrate is secondary, so we expect an E2 process. There are two β positions bearing protons, so two regiochemical outcomes are possible. Since the base is sterically hindered, the major product is the less-substituted alkene:

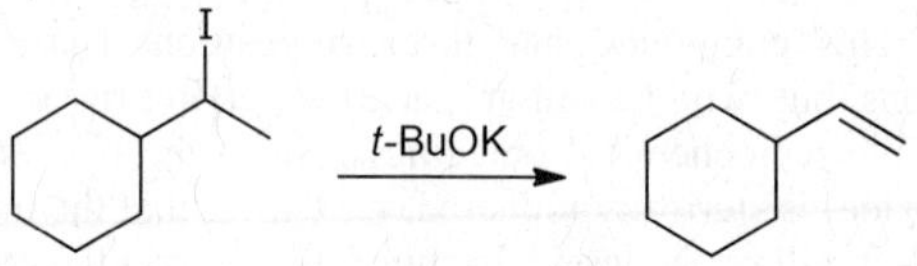

(g) The reagent is hydroxide, which is both a strong base and a strong nucleophile. The substrate is secondary so we expect an E2 process to dominate. For substituted cyclohexanes, an E2 reaction occurs via a conformation in which the leaving group and the β proton are antiperiplanar to one another (one must be on a wedge and the other must be on a dash). The leaving group (TsO⁻) is on a dash. Therefore, we are looking for a β proton that is on a wedge. In this case, there is only one β proton that is on a wedge, giving rise to only one elimination product, as shown:

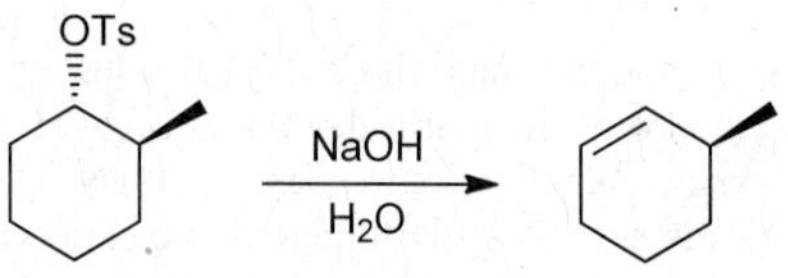

(h) The reagent is hydroxide, which is both a strong base and a strong nucleophile. The substrate is secondary so we expect an E2 process to dominate. For substituted cyclohexanes, an E2 reaction occurs via a conformation in which the leaving group and the β proton are antiperiplanar to one another (one must be on a wedge and the other must be on a dash). The leaving group (TsO⁻) is on a dash. Therefore, we are looking for a β proton that is on a wedge. In this case, there are two such protons, giving rise to two possible regiochemical outcomes. The major product is the more-substituted alkene (because the base is not sterically hindered):

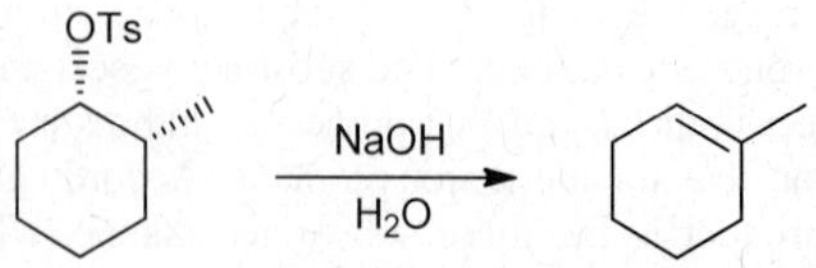

(i) The reagent is methoxide, which is both a strong base and a strong nucleophile. The substrate is tertiary, so we expect an E2 process. There are three β positions, but two of them are identical, so there are two possible regiochemical outcomes.   The more-substituted alkene is the major product, as shown.

(j) The reagent is *tert*-butoxide, which is a strong, sterically hindered base. The substrate is tertiary, so we expect an E2 process. There are three β positions, but two of them are identical, so there are two possible regiochemical outcomes.   Since the base is sterically hindered, we expect that the less-substituted alkene will be the major product, as shown.

8.80.   There are four constitutional isomers with molecular formula C_4H_9Br (as seen in the solution to Problem **2.54**).   The tertiary alkyl halide is the most reactive towards E2, followed by the secondary alkyl halide.   Among the two primary alkyl halides, the one leading to a disubstituted alkene will be more reactive towards E2 than the one leading to a monosubstituted alkene.

Increasing reactivity towards E2

8.81.   In acidic conditions, the OH group is protonated, which converts it from a bad leaving group to a good leaving group. In aqueous sulfuric acid, the acid that is present in solution is H_3O^+ (because of the leveling effect, as explained in Section 3.6), so we use H_3O^+ as the proton source in the first step of the mechanism. Loss of the leaving group gives a secondary carbocation that can undergo a methyl shift, giving a more stable, tertiary carbocation.   Another carbocation rearrangement (this time, a hydride shift) gives an even more stable, tertiary allylic carbocation (stabilized by resonance).   Finally, water serves as the base that removes the proton to generate the product.

8.82.
(a) We begin by drawing a Newman projection, and we find that the front carbon atom bears the leaving group (bromide), while the back carbon atom bears two β protons, either of which can be removed. The following two Newman projections represent the two conformations in which a β proton is antiperiplanar to the leaving group. In the first Newman projection, the phenyl groups (highlighted) are anti to each other, so the transition state is not expected to exhibit a steric interaction between the phenyl groups. In contrast, the second Newman projection exhibits a gauche interaction between the two phenyl groups (highlighted), so the transition state is expected to exhibit a steric interaction.   Therefore, *trans*-stilbene is formed at a faster rate than *cis*-stilbene (because formation of the latter involves a higher energy transition state).

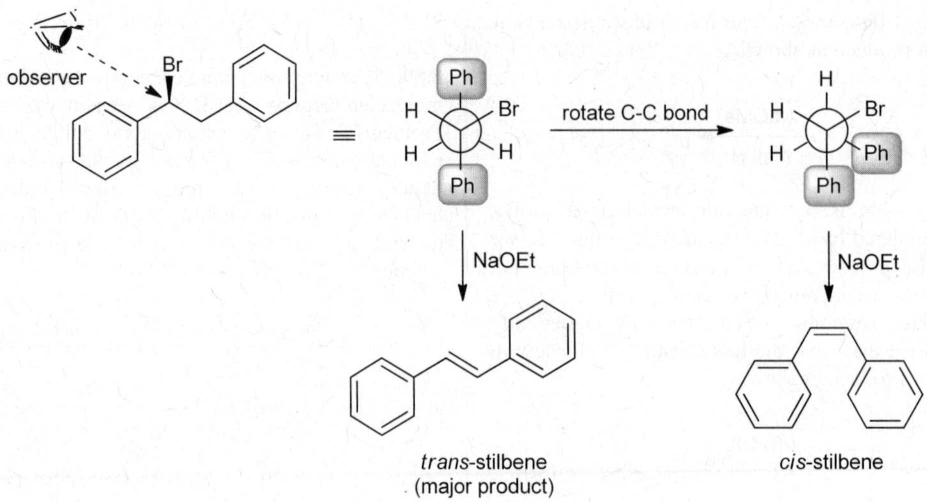

trans-stilbene
(major product)

cis-stilbene

(b) The same argument (as seen in 8.82a) can be applied again in this case. That is, there are still two β protons that can be abstracted in a β elimination, so both products are still possible, as shown below.

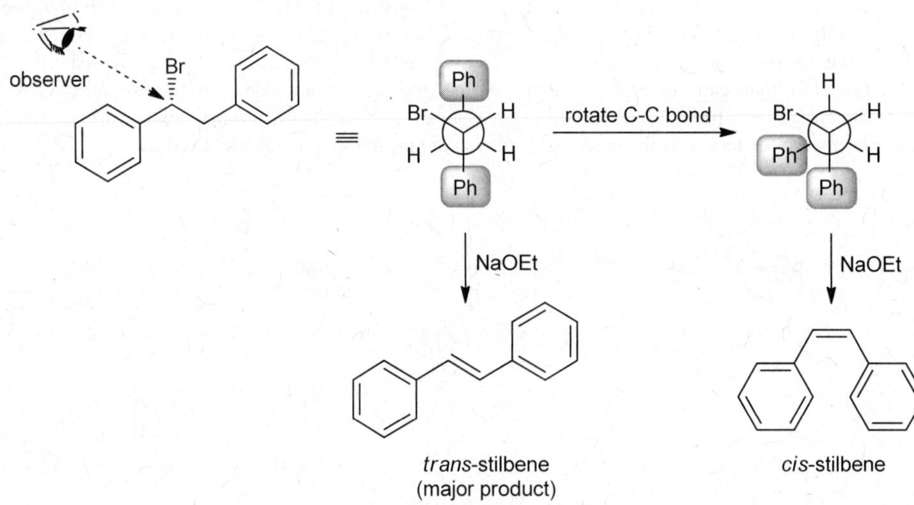

trans-stilbene
(major product)

cis-stilbene

8.83. In acidic conditions, the OH group is protonated, thereby being converted from a bad leaving group to a good leaving group. In aqueous sulfuric acid, the acid that is present in solution is H_3O^+ (because of the leveling effect, as explained in Section 3.6), so we use H_3O^+ as the proton source in the first step of the mechanism. Loss of the leaving group gives a secondary carbocation that can undergo a carbocation rearrangement. In this case, it is not a hydride shift or a methyl shift. Rather, the rearrangement takes place when one of the carbon atoms of the ring migrates (ring expansion), generating a more stable, tertiary carbocation. Finally, water serves as the base that removes the proton to generate the product.

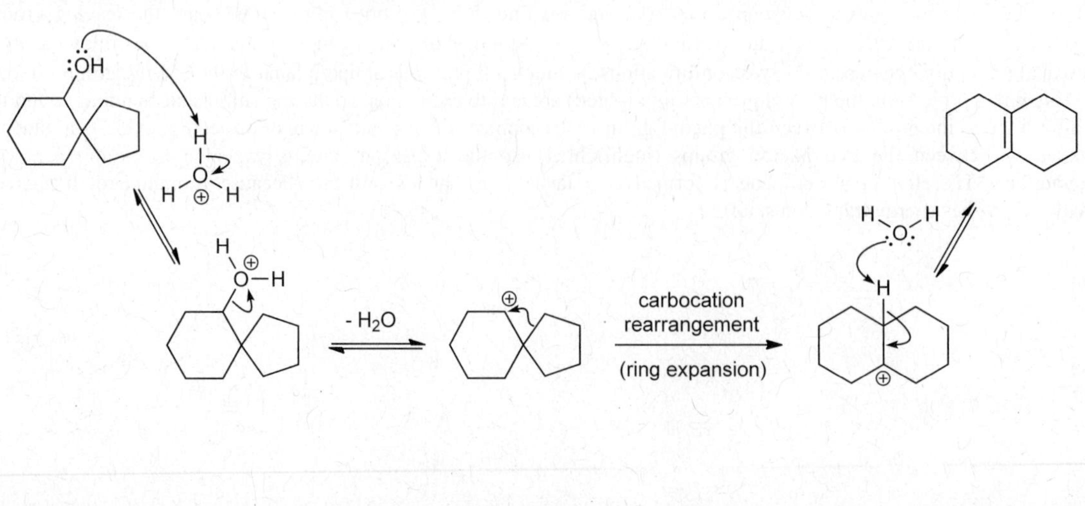

carbocation
rearrangement

(ring expansion)

8.84. The conditions (no strong nucleophile or strong base; polar protic solvent) favor unimolecular processes (E1 and S$_N$1), so we must explain the formation of the products with those mechanisms.

Formation of the first product can be explained with the following S$_N$1 mechanism, in which the first step is loss of the leaving group to generate a resonance-stabilized cation (resonance structures not shown). This carbocation is then captured by ethanol (which functions as a nucleophile), and the resulting oxonium ion is then deprotonated by another molecule of ethanol (which functions as a base):

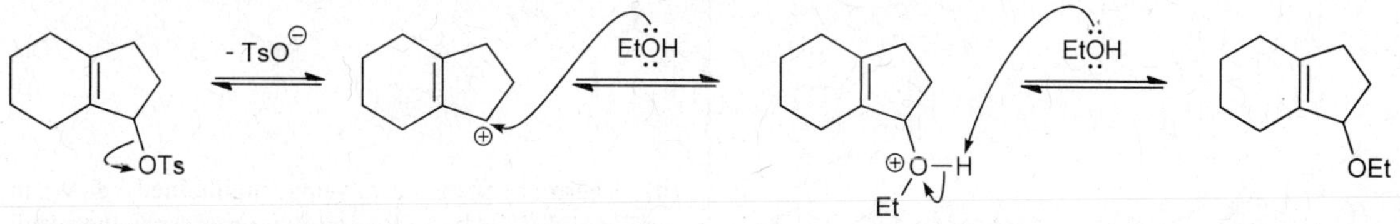

Formation of the second product can be explained via a similar mechanism (also S$_N$1). The first step is loss of the leaving group to generate a resonance-stabilized cation (this time, the resonance structures are drawn). This carbocation is then captured by ethanol (which functions as a nucleophile), and the resulting oxonium ion is then deprotonated by another molecule of ethanol (which functions as a base):

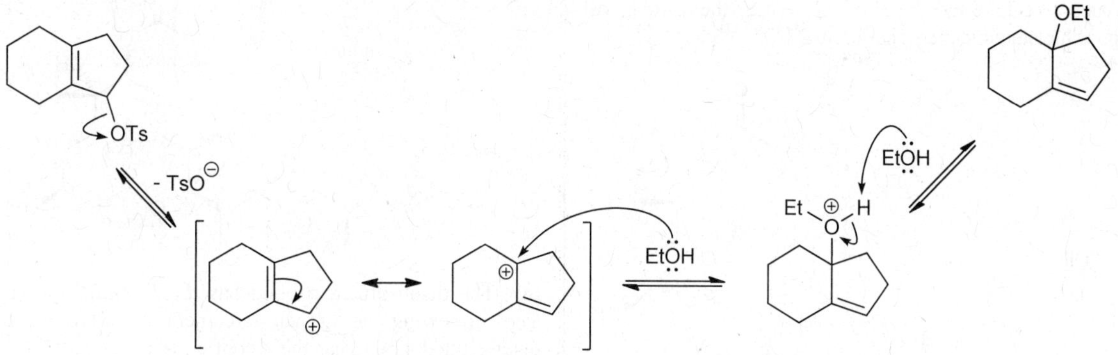

Finally, the third product is formed via an E1 process. In the first step of the mechanism, loss of the leaving group generates a resonance-stabilized cation. Then, in the second (and final) step, ethanol functions as a base and removes a proton, giving the product.

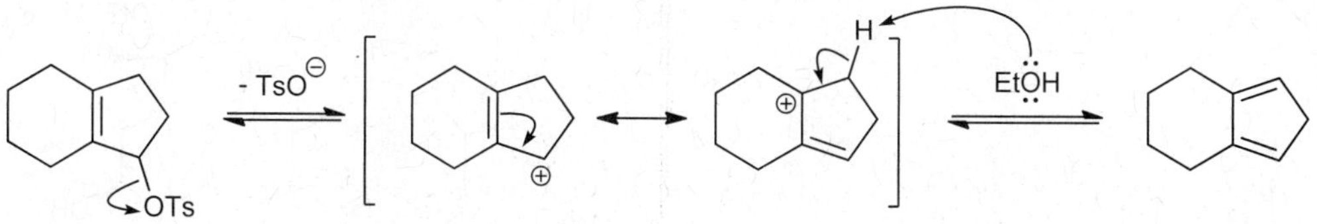

8.85. The stereoisomer shown below does not readily undergo E2 elimination because none of the chlorine atoms can be antiperiplanar to a β proton in a chair conformation. Recall that for substituted cyclohexanes, the leaving group must be *trans* to a β proton in order to achieve antiperiplanarity. In the isomer below, none of the chlorine atoms are *trans* to a β proton.

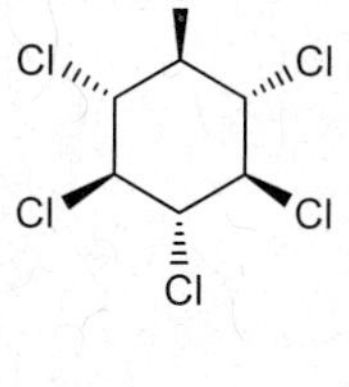

8.86. The first compound is a tertiary substrate. The second compound is a tertiary allylic substrate. The latter will undergo E1 more rapidly because a tertiary allylic carbocation is more highly stabilized than a tertiary carbocation. The rate-determining step (loss of the leaving group) will therefore occur more rapidly for the second compound.

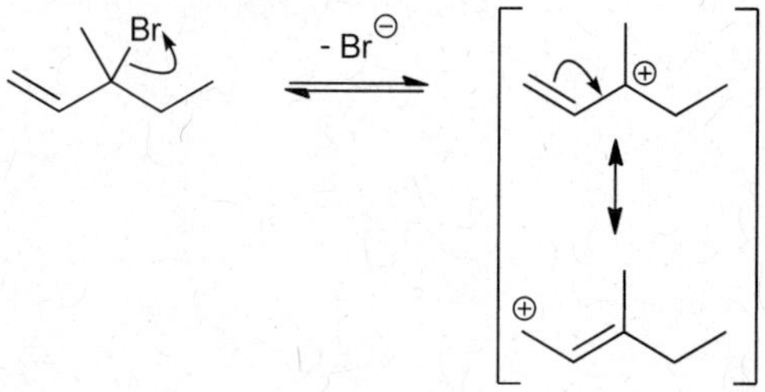

8.87.
(a) The trisubstituted π bond has the *E* configuration. The higher priority groups (highlighted below) are on opposite sides (*E*).

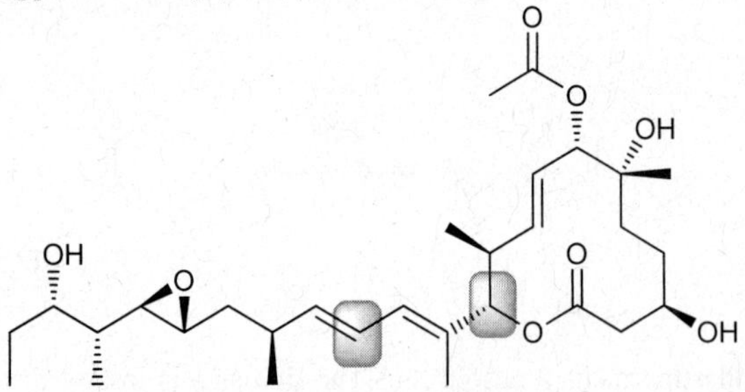

(b) There are 10 chirality centers (indicated below, each of which can be *R* or *S*), and there are three C=C π bonds (each of which can be *E* or *Z*). Thus, the number of possible stereoisomers should be $2^{13} = 8192$.

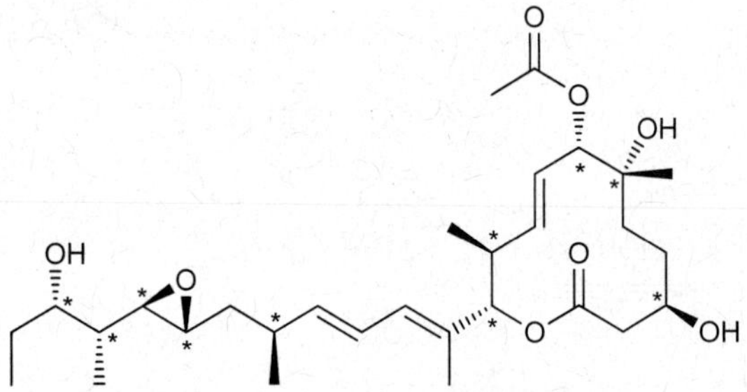

(c) The enantiomer of pladienolide B has the opposite configuration at each and every chirality center, but the configuration of each C=C π bond remains the same as in pladienolide B.

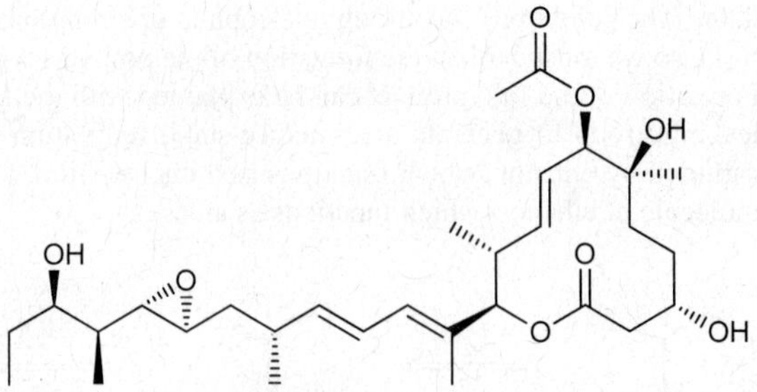

(d) There are two ester groups (highlighted below) in pladienolide B. The following diastereomer is the result of inverting the configuration of the two chirality centers adjacent to the ester groups.

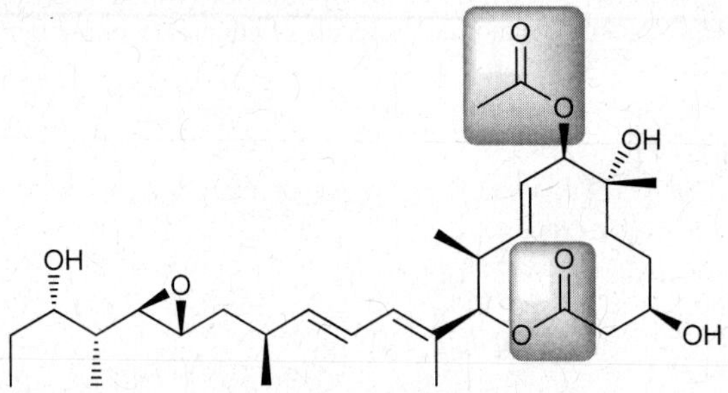

(e) The disubstituted π bond has the *E* configuration. The following is a diastereomer in which the disubstituted π bond has the *Z* configuration.

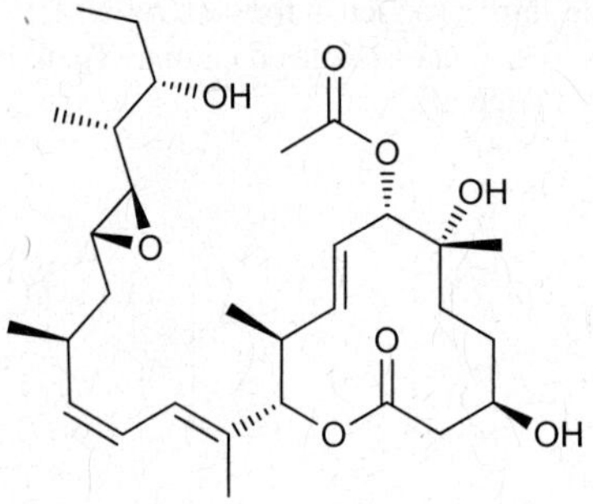

8.88.

(a) When compound **1a** is treated with TsCl and pyridine, the OH group is converted to OTs (a better leaving group). Then sodium acetate functions as a base and removes a proton to give alkene **2a**. Notice that the axial proton (highlighted) is removed, since that proton is antiperiplanar with the leaving group. The equatorial proton is <u>not</u> antiperiplanar with the leaving group.

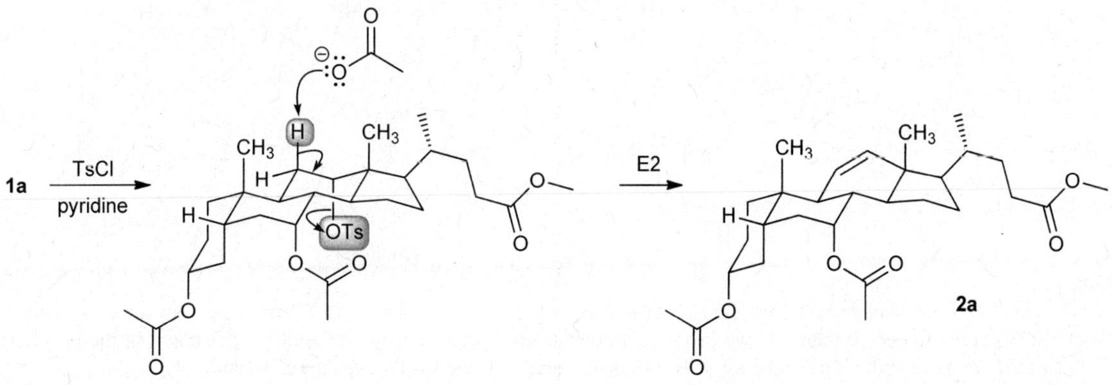

(b) We know that the axial proton (or deuteron) is removed in the elimination step. In compound **1b**, the axial position is occupied by a deuteron, so the deuteron is removed, and the product (**2b**) will not have a deuteron (and thus, **2b** is the same as **2a**). In compound **1c**, the axial position is occupied by a proton, which is removed during the elimination step. The deuteron in **1c** is in an equatorial position, so it survives the reaction. Compound **2c** will be deuterated, as seen below:

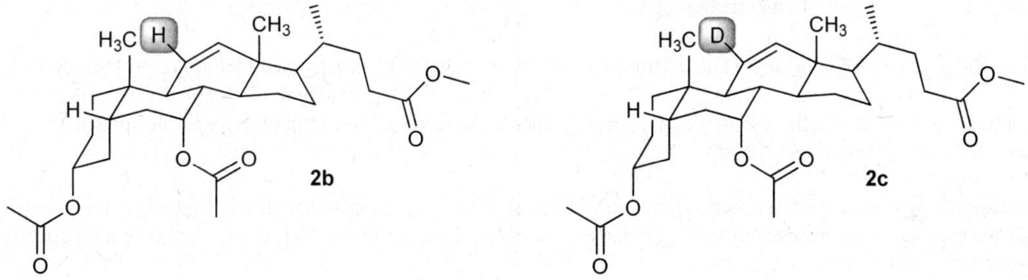

8.89.

When compound **1** is treated with TsCl and pyridine, the OH group is converted to a better leaving group (OTs). Then, upon treatment with a strong base, an E2 elimination reaction occurs to give the more substituted alkene (despite the use of a sterically hindered base, as indicated in the problem statement). Evaluation of the Newman projection shows that the proton on the oxetane and the tosylate are *anti*-periplanar which leads to the Z-alkene.

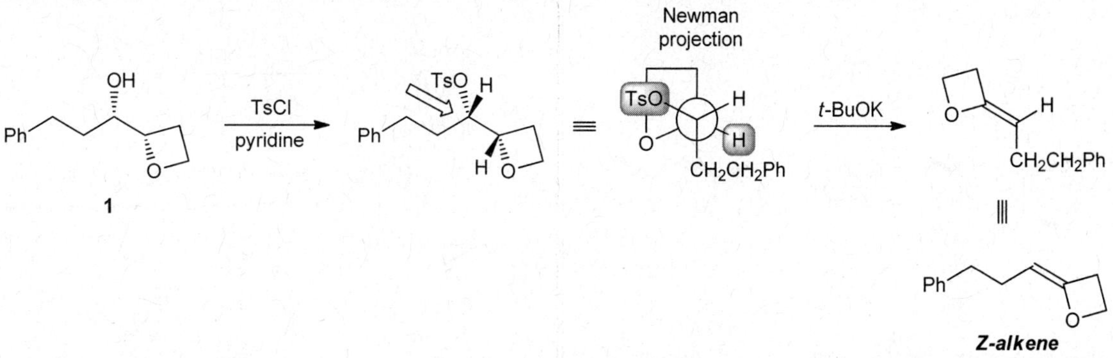

The reaction of compound **2** begins the same way - with tosylation of the alcohol to make it a better leaving group. Evaluation of a Newman projection shows that the H and the leaving group are not *anti*-periplanar, so we must rotate the front carbon atom to achieve *anti*-periplanarity. Elimination will occur via this conformation, and the resulting E2 process will lead to the *E*-alkene:

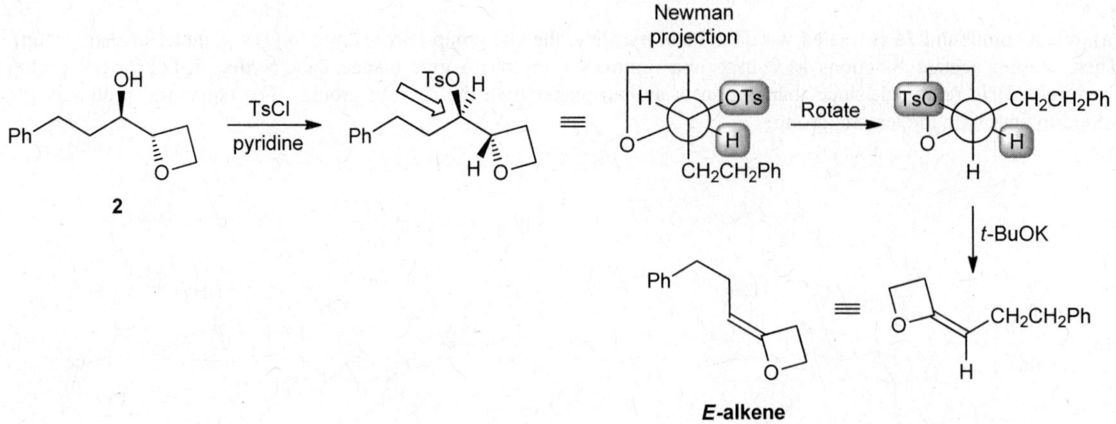

E-alkene

8.90.

(a) The proposed mechanism is a concerted process that very closely resembles an E2 process. As such, we would expect this process to occur when the two bromine atoms are *anti*-periplanar to one another, as shown in the problem statement. If we perform the same reaction with a diastereomer of **1**, we would expect a *Z*-alkene:

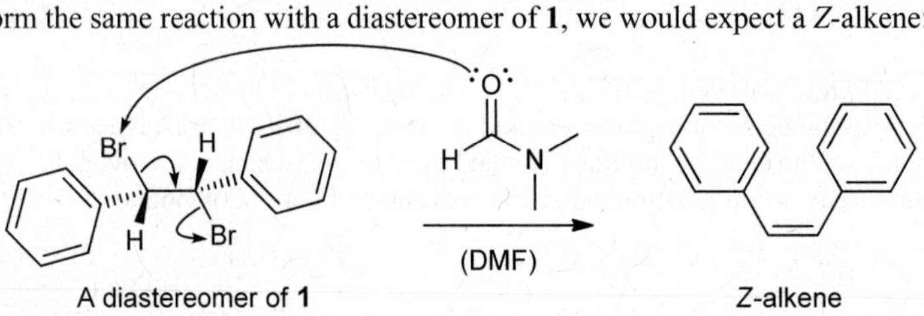

A diastereomer of **1**   *Z*-alkene

But none of the *Z*-alkene is formed. The *E* isomer is obtained exclusively, which means that the preference for the *E*-isomer is not dependent on the configuration of the starting dibromide. The *E*-alkene is obtained from either **1** or from a diastereomer of **1**. As such, the reaction does not appear to have a requirement for *anti*-periplanarity. This is evidence against a concerted mechanism.

(b) The reaction is stereoselective, because one configuration of the alkene product (the *E* isomer) is favored over the other. The reaction is not considered to be stereospecific, because the preference for the *E* isomer is not dependent on the configuration of the starting dibromide.

8.91.

In the first reaction, the OH group is converted into a better leaving group:

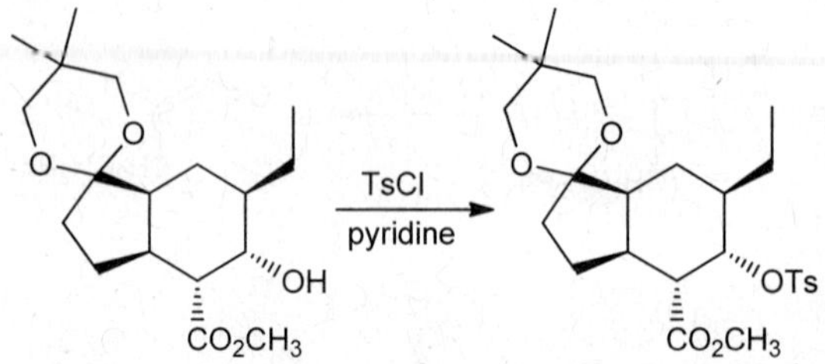

Then, the second reaction employs DBU which is a strong base that generally does not function as a nucleophile. We therefore expect an E2 elimination process to occur. During an E2 process, the base removes a proton that is antiperiplanar to the leaving group, as shown, giving the following E2 product.

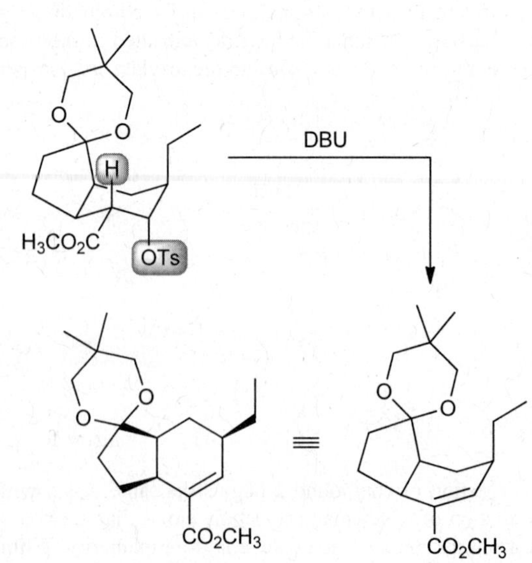

8.92.
The bond to the N_2^+ group and the *anti*-periplanar C-C bond are both shown in bold in the drawing of structure **2** below. The 1,2-alkyl shift displaces the N_2 leaving group, as shown (x). The resulting resonance-stabilized carbocation (resonance structures not shown) is then redrawn to more easily show the next two steps: (y) the alcohol oxygen attacks the carbocation and (z) deprotonation to produce compound **3**.

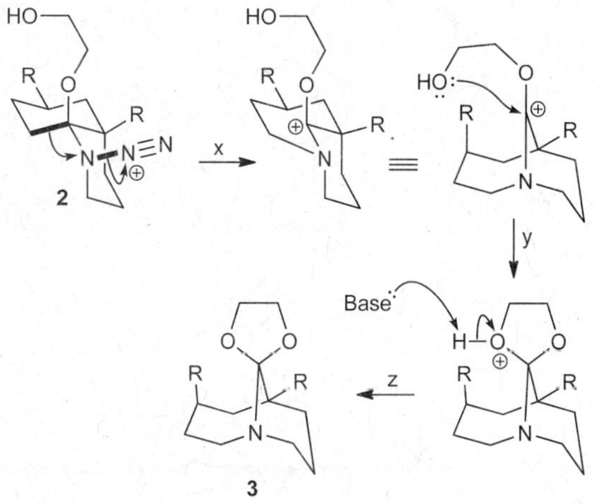

8.93.
(a) In order to compare the strength of these four bases, we can compare the pK_a values of their conjugate acids (see the pK_a table on the inside cover of the textbook). Benzoic acid ($pK_a = 4.8$) is more acidic than phenol ($pK_a = 9.9$), which is in turn more acidic than trifluoroethanol ($pK_a = 12.5$), which is more acidic than ethanol ($pK_a = 16.0$):

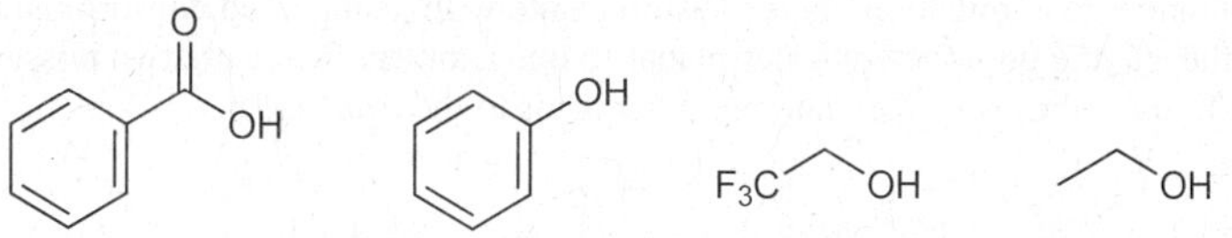

Therefore, basicity is expected to increase in the order that the bases were presented. That is, potassium benzoate is the weakest base (among the four bases listed), while ethoxide is the strongest base in the group.
The data indicates that the percentage of 1-butene increases as the basicity of the base increases. This observation can also be stated in the following way: the preference for formation of the more-substituted alkene (2-butene) decreases as the base strength increases. A stronger base is a more reactive base. So we see that there is an inverse relationship between reactivity and selectivity. Specifically, a more-reactive reagent results in lower selectivity, while a less-reactive reagent results in higher selectivity. This is a trend that we will encounter several times throughout the remaining chapters of the textbook, so it would be wise to remember this trend.
(b) Based on the pK_a value of 4-nitrophenol, we can conclude that it is more acidic than phenol, but not quite as acidic as benzoic acid. Based on our answer for part (a), we would expect that the conjugate base of 4-nitrophenol will be a stronger base than potassium benzoate, but a weaker base than potassium phenoxide. As such, we expect the selectivity to be somewhere in between the selectivity of potassium benzoate and potassium phenoxide. So we would expect that the percentage of 1-butene should be somewhere between 7.2% and 11.4%.

8.94.
In the first step, a lone pair of the alcohol attacks the Lewis acid, thereby converting the hydroxyl group into a good leaving group. Loss of the leaving group forms a stable, 3° carbocation. Next, a 1,2-hydride shift occurs to generate a different 3° carbocation. Note the stereochemistry of the newly formed bond. The migrating hydride is on the back face of the molecule. When it breaks to form a new bond with the neighboring carbocation, which is flat, it reattaches itself on the same face of the molecule (back face). At this stage, a 1,2-methyl shift occurs to generate another 3° carbocation. Also take note of the stereochemistry for the newly formed bond. The migrating methyl group is on the top face of the molecule. When it breaks to form a new bond with the neighboring carbocation, it reattaches itself on the same face of the molecule (top face). In the final step, loss of a proton will produce the alkene, thereby completing an E1 process.

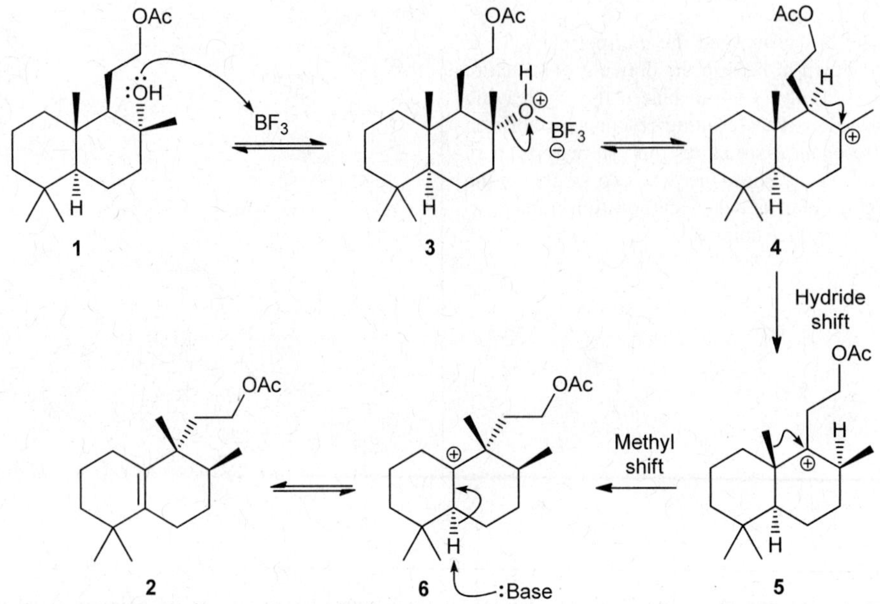

8.95.

The 2-butyl halides will undergo elimination via a transition state with a staggered conformation, in which the H (that is being removed) and the leaving group are *anti*-periplanar to one another. There are two possible transition states that fit this description; one leads to the *cis* isomer, and the other leads to the *trans* isomer.

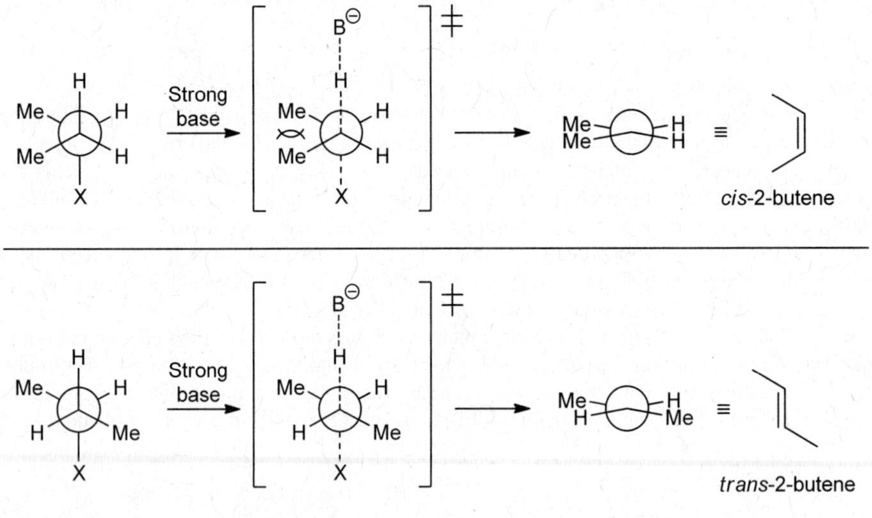

Formation of the *cis* isomer occurs via a transition state that exhibits a gauche interaction, while formation of the *trans* isomer occurs via a transition state that lacks this steric interaction. As such, the latter transition state will be lower in energy, and as a result, the *trans* isomer will form at a faster rate than the *cis* isomer.

On the other hand, the 2-butyl tosylate possesses additional conformational influences due to the steric repulsion between the methyl group and the tosyl group attached to the same carbon atom (the back carbon atom in the Newman projection). This steric repulsion can be alleviated via rotation about the C-O bond:

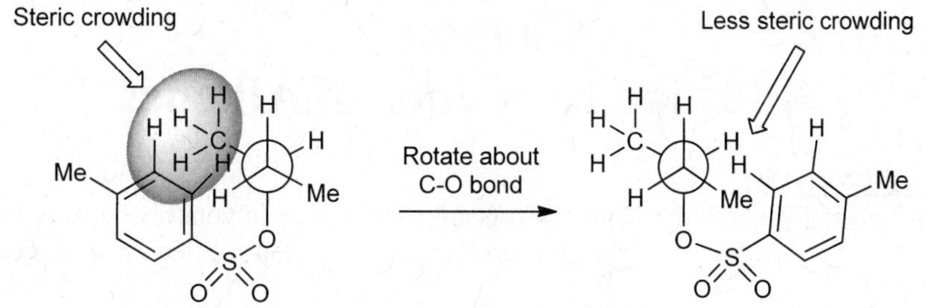

Steric crowding   Less steric crowding   Rotate about C-O bond

In other words, the C-O bond does not experience unhindered rotation, but rather, the molecule will adopt a conformation in which the tosyl group is oriented away from the methyl group (that is attached to the back carbon atom in the Newman projection), as seen in the figure above (right). Although this conformation does alleviate the steric crowding (shown above left), there is a trade-off here, because this causes an increase in the gauche interaction between the tosyl group and the methyl group connected to the front carbon atom of the Newman projection:

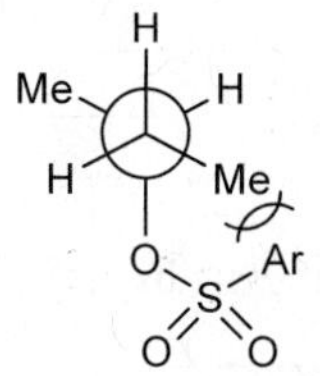

This raises the activation energy of the transition state that leads to the *trans* alkene.   In fact, in the case of 2-butyl tosylate, this effect is so strong that it causes the transition state for formation of the *trans* alkene to be even higher in energy than the transition state for formation of the *cis* alkene. Both transition states are shown below.

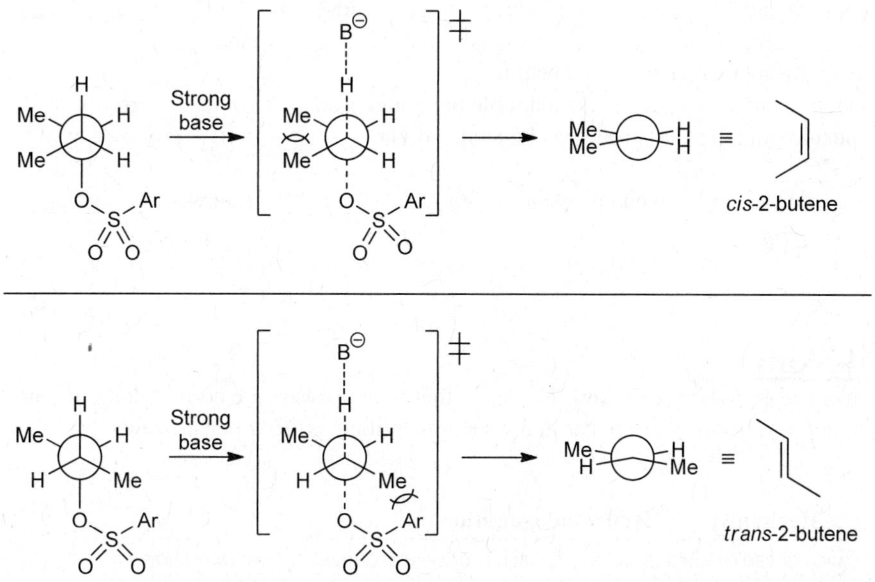

Strong base   cis-2-butene

Strong base   trans-2-butene

Both transition states exhibit a gauche interaction, but the latter is more significant than the former. As such, the *cis* alkene forms at a faster rate than the *trans* alkene.

Chapter 9
Addition Reactions of Alkenes

Review of Concepts

Fill in the blanks below. To verify that your answers are correct, look in your textbook at the end of Chapter 9. Each of the sentences below appears verbatim in the section entitled *Review of Concepts and Vocabulary*.

- Addition reactions are thermodynamically favorable at _____ temperature and disfavored at _____ temperature.

- Hydrohalogenation reactions are **regioselective**, because the halogen is generally placed at the _____ substituted position, called _____ **addition**.

- In the presence of _____, addition of HBr proceeds via an ***anti*-Markovnikov addition**.

- The regioselectivity of an ionic addition reaction is determined by the preference for the reaction to proceed through _____.

- Acid-catalyzed hydration is inefficient when _____ are possible. Dilute acid favors formation of the _____, while concentrated acid favors the _____.

- **Oxymercuration-demercuration** achieves hydration of an alkene without _____.

- _____ - _____ can be used to achieve an *anti*-Markovnikov addition of water across an alkene. The reaction is stereospecific and proceeds via a _____ **addition**.

- **Asymmetric hydrogenation** can be achieved with a _____ catalyst.

- Bromination proceeds through a bridged intermediate, called a _____ _____, which is opened by an S_N2 process that produces an _____ **addition**.

- A two-step procedure for *anti* dihydroxylation involves conversion of an alkene to an _____, followed by acid-catalyzed ring opening.

- Ozonolysis can be used to cleave a double bond and produce two _____ groups.

- The position of a leaving group can be changed via _____ followed by _____.

- The position of a π bond can be changed via _____ followed by _____.

Review of Skills

Fill in the blanks and empty boxes below. To verify that your answers are correct, look in your textbook at the end of Chapter 9. The answers appear in the section entitled *SkillBuilder Review*.

9.1 Drawing a Mechanism for Hydrohalogenation

STEP 1 - DRAW TWO CURVED ARROWS SHOWING PROTONATION OF THE ALKENE, AND DRAW THE CARBOCATION THAT IS PREFERENTIALLY FORMED.	STEP 2 - DRAW ONE CURVED ARROW THAT SHOWS THE HALIDE ION ATTACKING THE CARBOCATION, AND DRAW THE PRODUCT.

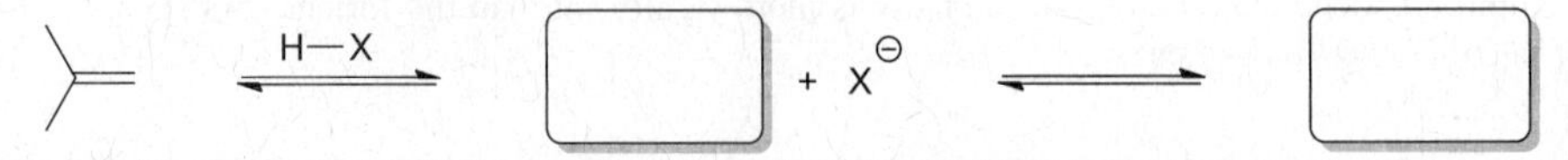

9.2 Drawing a Mechanism for Hydrohalogenation with a Carbocation Rearrangement

STEP 1 - *DRAW TWO CURVED ARROWS SHOWING PROTONATION OF THE ALKENE AND DRAW THE CARBOCATION THAT IS INITIALLY FORMED.*

STEP 2 - *DRAW ONE CURVED ARROW SHOWING A CARBOCATION REARRANGEMENT AND DRAW THE RESULTING, MORE STABLE CARBOCATION.*

STEP 3 - *DRAW ONE CURVED ARROW SHOWING THE HALIDE ION ATTACKING THE CARBOCATION, AND DRAW THE PRODUCT.*

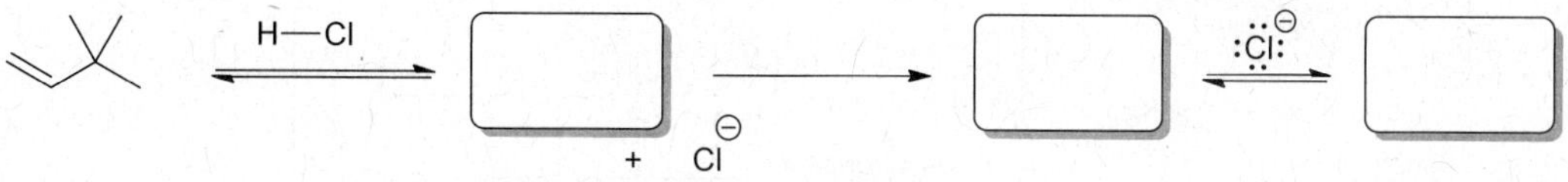

H—Cl

$+$ $\overset{\ominus}{Cl}$

$:\overset{..}{\underset{..}{Cl}}:^{\ominus}$

9.3 Drawing a Mechanism for an Acid-Catalyzed Hydration

STEP 1 - *DRAW TWO CURVED ARROWS SHOWING PROTONATION OF THE ALKENE, AND DRAW THE RESULTING CARBOCATION.*

STEP 2 - *DRAW ONE CURVED ARROW SHOWING WATER ATTACKING THE CARBOCATION, AND DRAW THE RESULTING OXONIUM ION.*

STEP 3 - *DRAW TWO CURVED ARROWS SHOWING DEPROTONATION OF THE OXONIUM ION, AND DRAW THE RESULTING PRODUCT.*

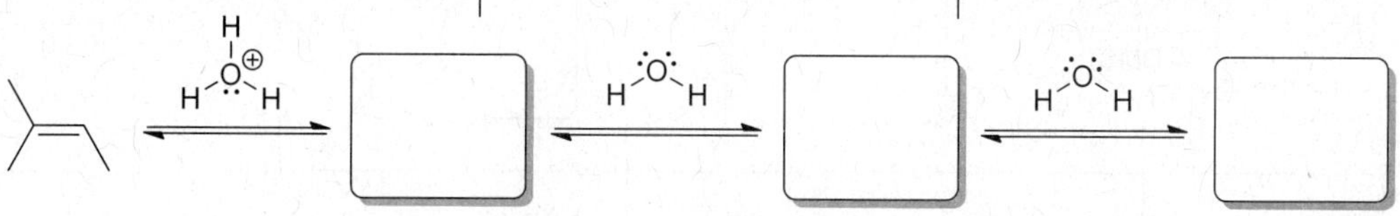

9.4 Predicting the Products of Hydroboration-Oxidation

DRAW THE EXPECTED PRODUCTS OF THE FOLLOWING REACTION SEQUENCE, AND DETERMINE THEIR RELATIONSHIP

1) $BH_3 \cdot THF$

2) H_2O_2, NaOH

$+$

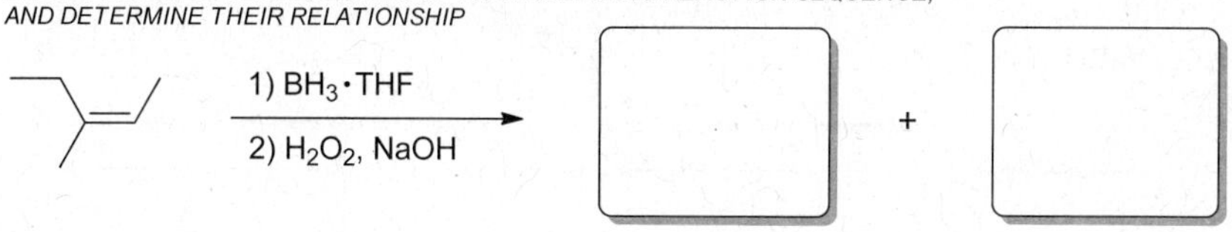

RELATIONSHIP = _____

9.5 Predicting the Products of Catalytic Hydrogenation

DRAW THE EXPECTED PRODUCTS OF THE FOLLOWING REACTION, AND DETERMINE THEIR RELATIONSHIP.

$\dfrac{H_2}{Pt}$

$+$

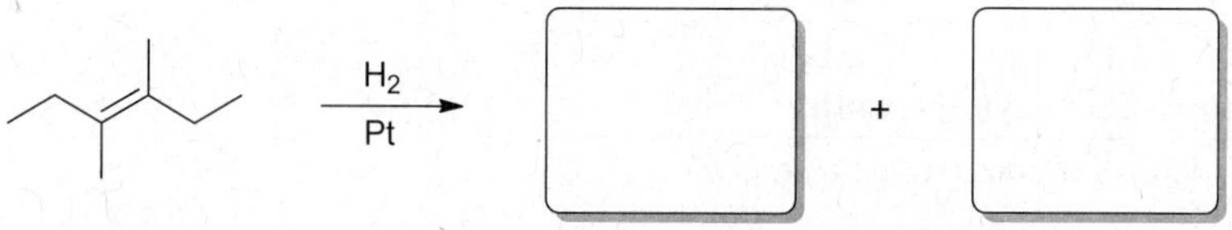

RELATIONSHIP = _____

9.6 Predicting the Products of Halohydrin Formation

DRAW THE EXPECTED PRODUCTS OF THE FOLLOWING REACTION, AND DETERMINE THEIR RELATIONSHIP.

$\dfrac{Br_2}{H_2O}$

$+$

RELATIONSHIP = _____

9.7   Drawing the Products of *Anti* Dihydroxylation

DRAW THE EXPECTED PRODUCTS OF THE FOLLOWING REACTION, AND DETERMINE THEIR RELATIONSHIP.

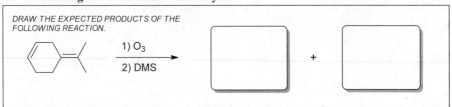

1) MCPBA

2) H_3O^+

+

RELATIONSHIP = _____

9.8   Predicting the Products of Ozonolysis

DRAW THE EXPECTED PRODUCTS OF THE FOLLOWING REACTION.

1) O_3

2) DMS

+

9.9   Predicting the Products of an Addition Reaction

DRAW THE EXPECTED PRODUCTS OF THE FOLLOWING REACTION.

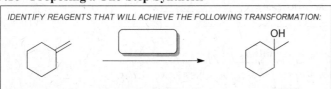

1) $BH_3 \cdot THF$

2) H_2O_2, NaOH

+

9.10   Proposing a One-Step Synthesis

IDENTIFY REAGENTS THAT WILL ACHIEVE THE FOLLOWING TRANSFORMATION:

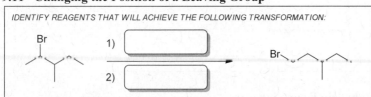

9.11   Changing the Position of a Leaving Group

IDENTIFY REAGENTS THAT WILL ACHIEVE THE FOLLOWING TRANSFORMATION:

1)

2)

9.12   Changing the Position of a π Bond

IDENTIFY REAGENTS THAT WILL ACHIEVE THE FOLLOWING TRANSFORMATION:

1)

2)

Review of Reactions

Identify the reagents necessary to achieve each of the following transformations. To verify that your answers are correct, look in your textbook at the end of Chapter 9. The answers appear in the section entitled *Review of Reactions*.

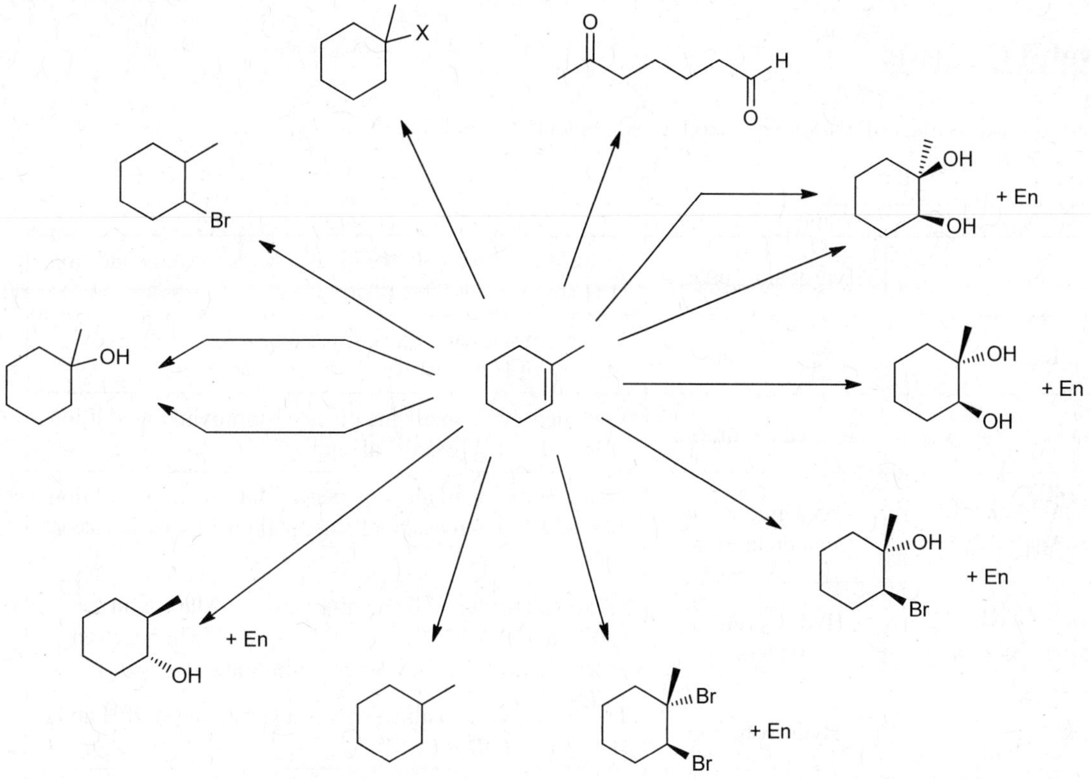

Common Mistakes to Avoid

This chapter introduces several stereospecific addition reactions. Some of them occur exclusively via a *syn* addition (such as hydrogenation or hydroboration-oxidation), while others occur exclusively via an *anti* addition (such as bromination or halohydrin formation). When drawing the products of a stereoscpecific addition reaction, be careful to avoid drawing a wedge or a dash on a location that is not a chirality center. For example, consider the following *syn* dihydroxylation. In such a case, it is tempting for students to draw the products as if they have two chirality centers, like this:

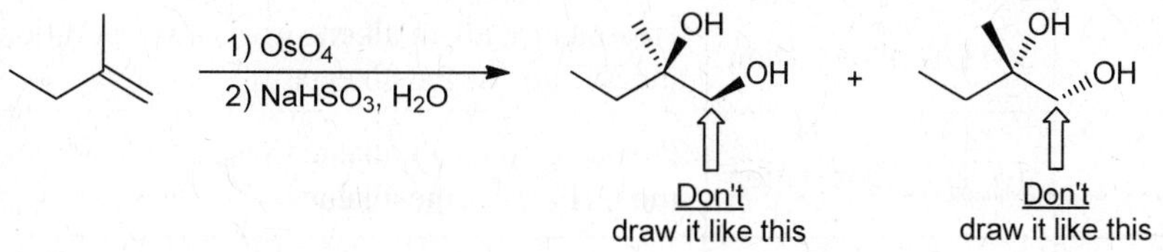

This mistake is understandable – after all, the two OH groups are indeed added in a *syn* fashion. But the product does not contain two chirality centers. It has only one chirality center. As such, the products should be drawn like this:

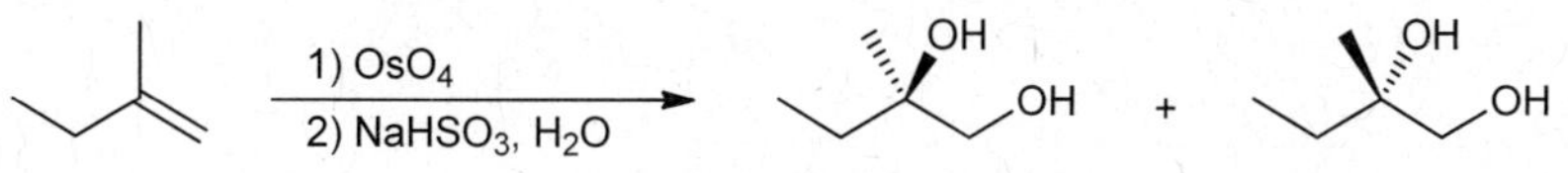

Notice that the stereochemical requirement for *syn* addition is not relevant in this case, because only one chirality center is formed. As such, both enantiomers are produced, because *syn* addition can occur on either face of the alkene to give either enantiomer.

Useful reagents

The following is a list of commonly encountered reagents for addition reactions:

Reagents	Name of Reaction	Description
HX	Hydrohalogenation	Treatment with an alkene gives a Markovnikov addition of H and X across the alkene.
HBr, ROOR	Hydrobromination	Treatment with an alkene gives an *anti*-Markovnikov addition of H and Br across the alkene.
H_3O^+	Acid-cat. hydration	Treatment with an alkene gives a Markovnikov addition of H and OH across the alkene.
1) $Hg(OAc)_2$, H_2O 2) $NaBH_4$	Oxymercuration-demercuration	Treatment with an alkene gives a Markovnikov addition of H and OH across the alkene, without any carbocation rearrangements.
1) $BH_3 \cdot THF$ 2) H_2O_2, NaOH	Hydroboration-oxidation	Treatment with an alkene gives an *anti*-Markovnikov addition of H and OH across the alkene. The reaction proceeds exclusively via a *syn* addition.
H_2, Pt	Hydrogenation	Treatment with an alkene gives a *syn* addition of H and H across the alkene.
Br_2	Bromination	Treatment with an alkene gives an *anti* addition of Br and Br across the alkene.
Br_2, H_2O	Halohydrin formation	Treatment with an alkene gives an *anti* addition of Br and OH across the alkene, with the OH group being installed at the more substituted position.
1) RCO_3H 2) H_3O^+	*Anti* Dihydroxylation	Treatment of an alkene with a peroxy acid (RCO_3H) converts the alkene into an epoxide, which is then opened upon treatment with aqueous acid to give a *trans*-diol.
$KMnO_4$, NaOH, cold	*Syn* Dihydroxylation	Treatment with an alkene gives a *syn* addition of OH and OH across the alkene.
1) OsO_4 2) $NaHSO_3$, H_2O	*Syn* Dihydroxylation	Treatment with an alkene gives a *syn* addition of OH and OH across the alkene.
1) O_3 2) DMS	Ozonolysis	Ozonolysis of an alkene causes cleavage of the C=C bond, giving two compounds, each of which possesses a C=O bond.

Solutions

9.1.

(a) An alkene is treated with HBr (in the absence of peroxides), so we expect a Markovnikov addition of H and Br across the π bond. That is, Br is placed at the more-substituted position:

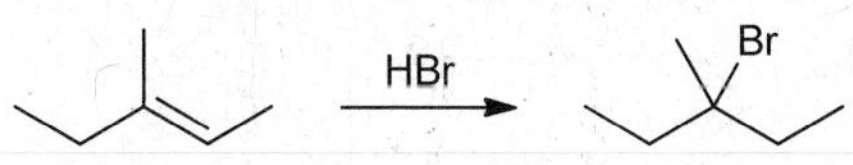

(b) An alkene is treated with HBr in the presence of peroxides, so we expect an *anti*-Markovnikov addition of H and Br across the π bond. That is, Br is placed at the less-substituted position:

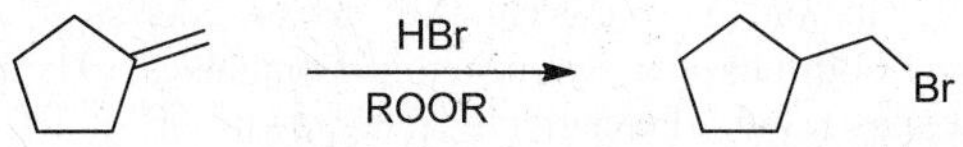

(c) An alkene is treated with HBr (in the absence of peroxides), so we expect a Markovnikov addition of H and Br across the π bond. That is, Br is placed at the more-substituted position:

(d) An alkene is treated with HCl, so we expect a Markovnikov addition of H and Cl across the π bond. That is, Cl is placed at the more-substituted position:

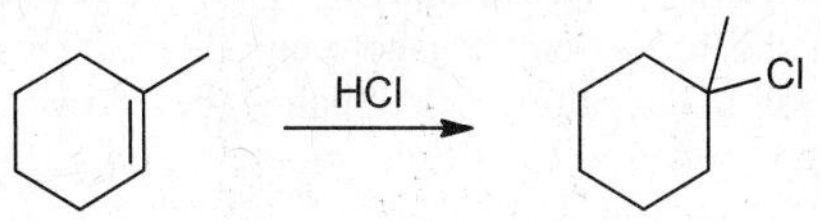

(e) An alkene is treated with HI, so we expect a Markovnikov addition of H and I across the π bond. That is, the iodo group is placed at the more-substituted position:

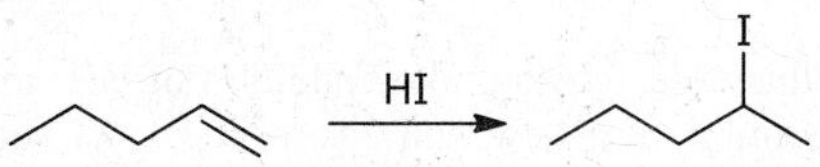

(f) An alkene is treated with HBr in the presence of peroxides, so we expect an *anti*-Markovnikov addition of H and Br across the π bond. That is, Br is placed at the less-substituted position:

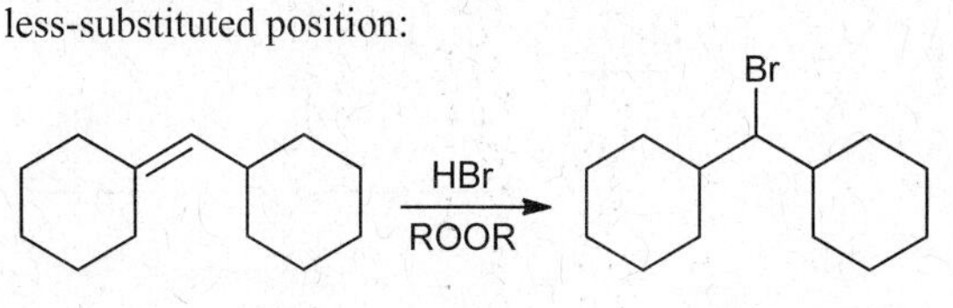

9.2.

(a) The desired transformation is a Markovnikov addition of H and Br across the π bond. This can be achieved by treating the alkene with HBr (in the absence of peroxides).

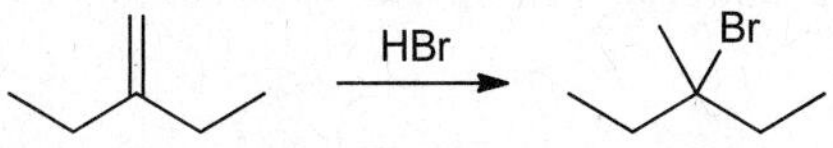

(b) The desired transformation is an *anti*-Markovnikov addition of H and Br across the π bond. This can be achieved by treating the alkene with HBr in the presence of peroxides.

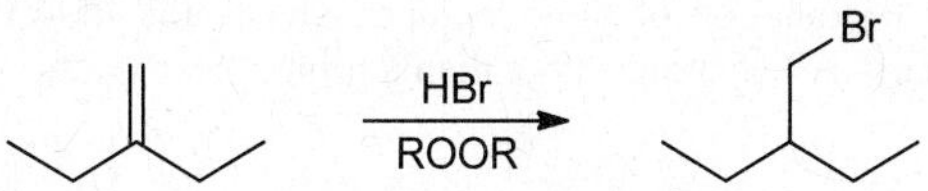

9.3.

(a) In this reaction, H and Br are added across the alkene in a Markovnikov addition, which indicates an ionic process. There are two mechanistic steps in the ionic addition of HBr across an alkene: 1) proton transfer, followed by 2) nucleophilic attack. In the first step, a proton is transferred from HBr to the alkene, which requires two curved arrows, as shown below. The resulting, tertiary carbocation is then captured by a bromide ion in the second step of the mechanism. This step requires one curved arrow, going from the nucleophile to the electrophile, as shown:

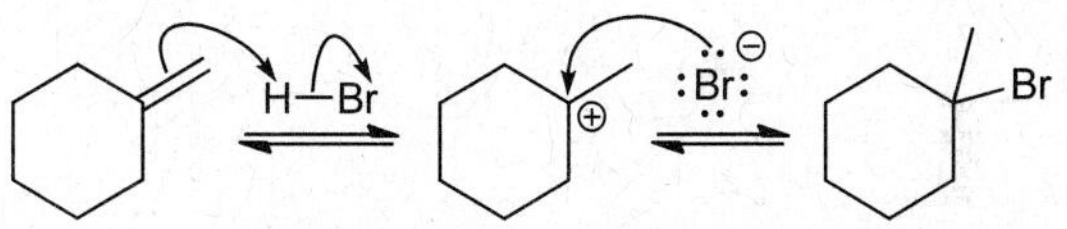

(b) In this reaction, H and Cl are added across the alkene in a Markovnikov addition. There are two mechanistic steps in the ionic addition of HCl across an alkene: 1) proton transfer, followed by 2) nucleophilic attack. In the first step, a proton is transferred from HCl to the alkene, which requires two curved arrows, as shown below. The resulting, tertiary carbocation is then captured by a chloride ion in the second step of the mechanism. This step requires one curved arrow, going from the nucleophile to the electrophile, as shown:

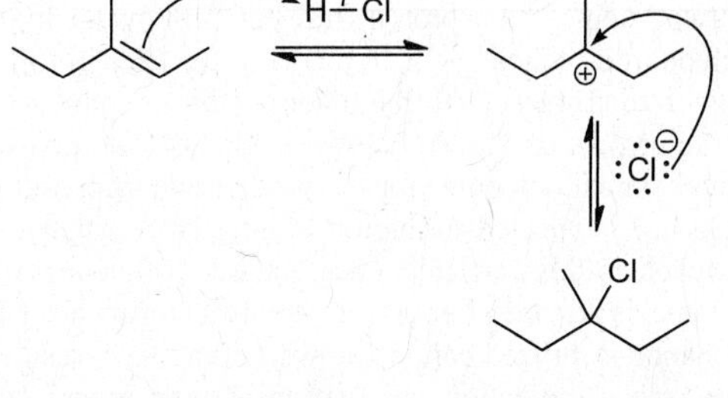

(c) In this reaction, H and Cl are added across the alkene in a Markovnikov addition. There are two mechanistic steps in the ionic addition of HCl across an alkene: 1) proton transfer, followed by 2) nucleophilic attack. In the first step, a proton is transferred from HCl to the alkene, which requires two curved arrows, as shown below.   The resulting, tertiary carbocation is then captured by a chloride ion in the second step of the mechanism. This step requires one curved arrow, going from the nucleophile to the electrophile, as shown:

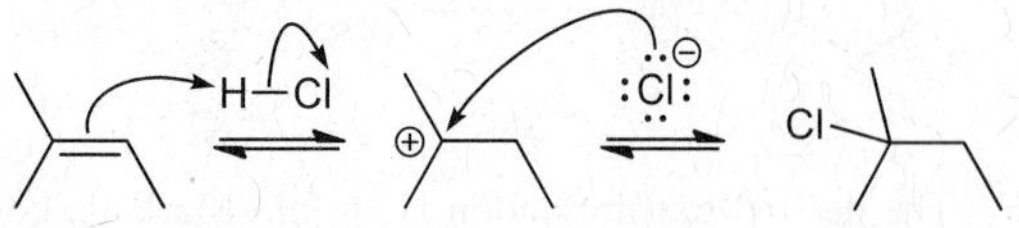

9.4.
(a) Protonation of the π bond results in the following tertiary carbocation (rather than a primary carbocation):

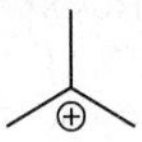

(b)   Protonation of the π bond results in the following tertiary carbocation (rather than a secondary carbocation):

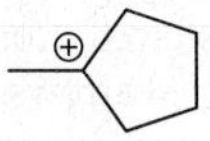

(c) Protonation of the π bond results in the following tertiary carbocation:

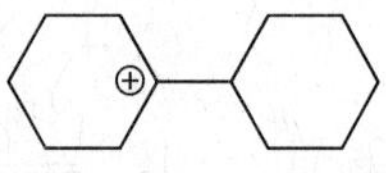

(d) Protonation of the π bond results in the following tertiary carbocation (rather than a secondary carbocation):

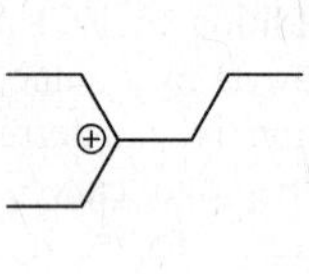

9.5. Since this reaction proceeds through an ionic mechanism, we expect the mechanism to be comprised of two steps: 1) proton transfer, followed by 2) nucleophilic attack.   In the first step, a proton is transferred from HCl to the alkene, which requires two curved arrows, as shown below. There are two possible regiochemical outcomes for the protonation step, and we might have expected formation of a tertiary carbocation. However, in this particular case, the other regiochemical outcome is favored because it involves formation of a resonance-stabilized cation, shown below. As a result of resonance stabilization, this cation is even more stable than a tertiary carbocation, and the reaction proceeds via the more stable intermediate.   This cation is then

captured by a chloride ion in the second step of the mechanism, which requires two curved arrows, as shown:

resonance-stabilized

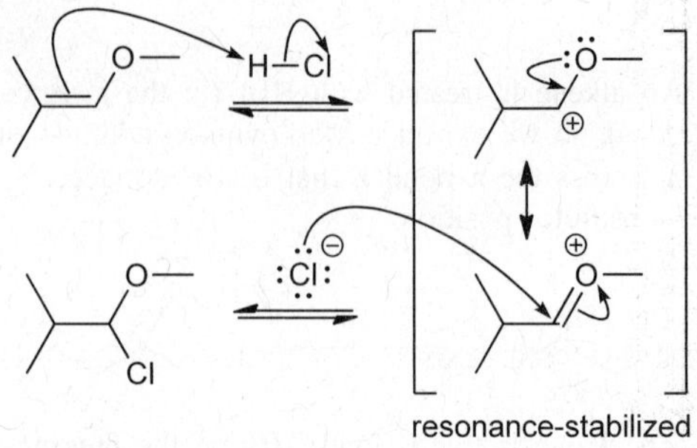

9.6.
(a) In this case, Markovnikov addition of HBr involves the formation of a new chirality center. As such, we expect both possible stereochemical outcomes. That is, we expect a pair of enantiomers, as shown:

enantiomers

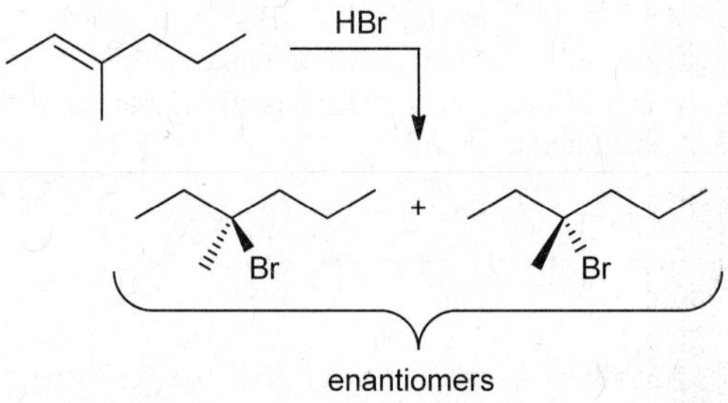

(b) In this case, Markovnikov addition of HCl does not involve the formation of a new chirality center (the α carbon of the resulting alkyl halide bears two propyl groups):

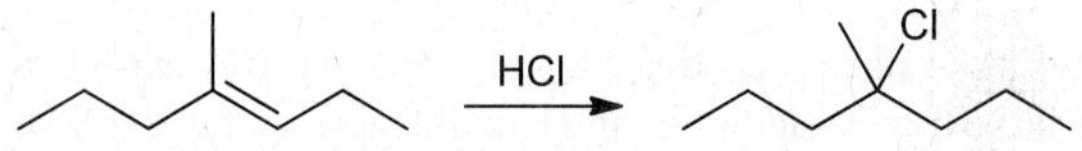

(c) In this case, Markovnikov addition of HBr involves the formation of a new chirality center. As such, we expect both possible stereochemical outcomes. That is, we expect a pair of enantiomers, as shown:

enantiomers

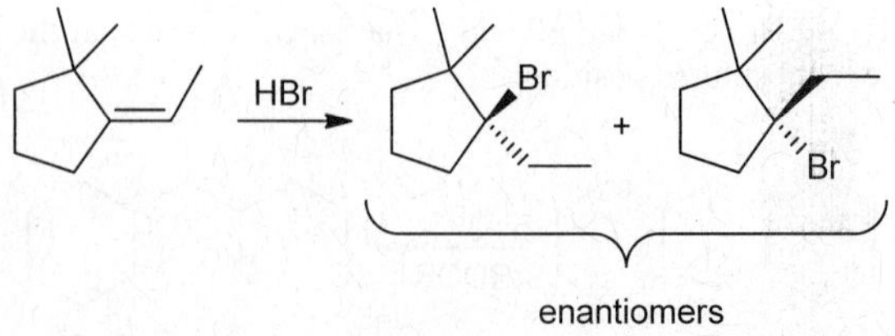

(d) In this case, Markovnikov addition of HI involves the formation of a new chirality center. As such, we expect both possible stereochemical outcomes. That is, we expect a pair of enantiomers, as shown:

enantiomers

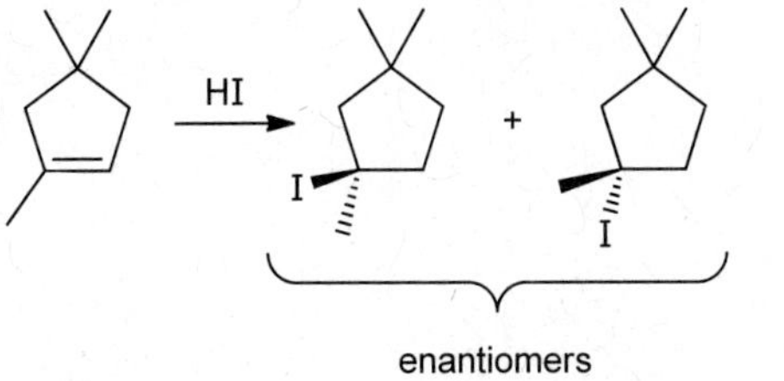

(e) In this case, Markovnikov addition of HCl does not involve the formation of a new chirality center (the α carbon of the resulting alkyl halide bears two methyl groups):

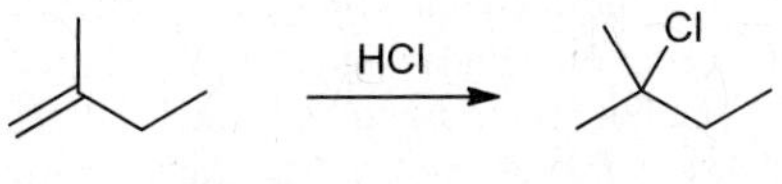

(f) In this case, Markovnikov addition of HCl involves the formation of a new chirality center. As such, we expect both possible stereochemical outcomes. That is, we expect a pair of enantiomers, as shown:

enantiomers

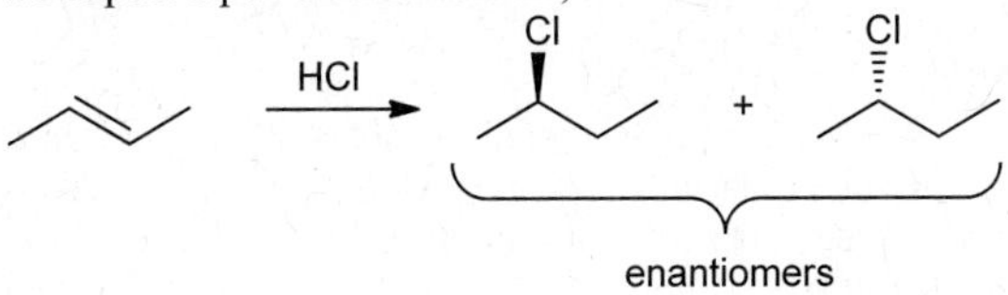

9.7.
(a) Protonation of the alkene requires two curved arrows, as shown, and leads to the secondary carbocation (rather than a primary carbocation). This secondary carbocation then undergoes a hydride shift, shown with one curved arrow, generating a more stable, tertiary carbocation. In the final step of the mechanism (nucleophilic attack), the carbocation is captured by a bromide ion. This step requires one curved arrow, going from the nucleophile (bromide) to the electrophile (the carbocation), as shown:

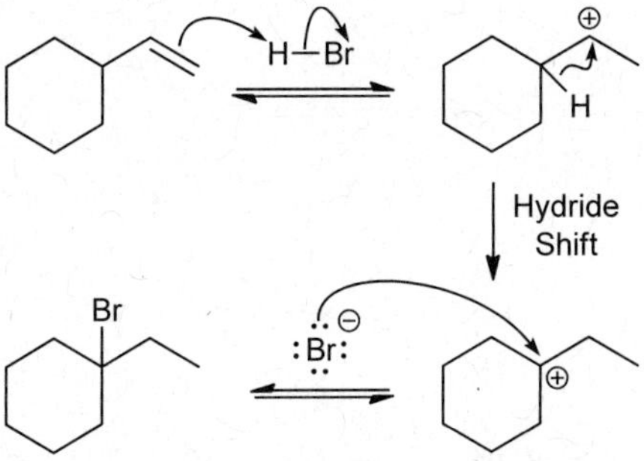

(b) Protonation of the alkene requires two curved arrows, as shown, and leads to the more stable, secondary carbocation (rather than a primary carbocation). This secondary carbocation then undergoes a hydride shift, shown with one curved arrow, generating a more stable, tertiary carbocation. In the final step of the mechanism (nucleophilic attack), the carbocation is captured by a bromide ion. This step requires one curved arrow, going

from the nucleophile (bromide) to the electrophile (the carbocation), as shown:

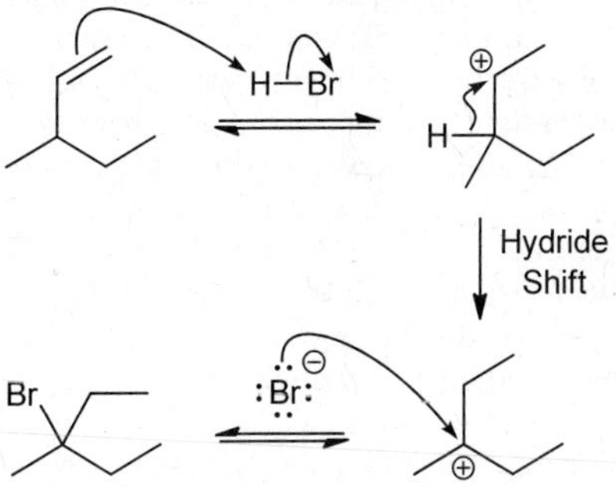

(c) Protonation of the alkene requires two curved arrows, generating a secondary carbocation. This secondary carbocation then undergoes a methyl shift, shown with one curved arrow, generating a more stable, tertiary carbocation. In the final step of the mechanism (nucleophilic attack), the carbocation is captured by a chloride ion. This step requires one curved arrow, going from the nucleophile (chloride) to the electrophile (the carbocation), as shown:

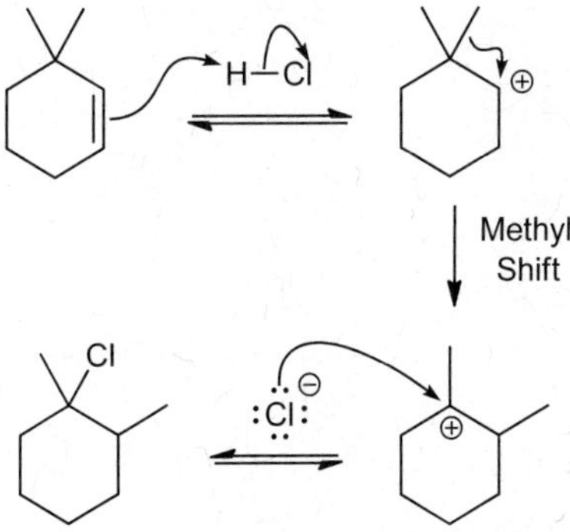

9.8. Protonation of the alkene requires two curved arrows, as shown in the first step of the following mechanism. This leads to the more stable, secondary carbocation (rather than a primary carbocation). This secondary carbocation then undergoes a rearrangement, in which one of the carbon atoms of the ring migrates (as described in the problem statement). This is represented with one curved arrow that shows the formation of a more stable, tertiary carbocation. In the final step of the mechanism (nucleophilic attack), the carbocation is captured by a bromide ion. This step requires one curved arrow, going from the nucleophile (bromide) to the electrophile (the carbocation), as shown:

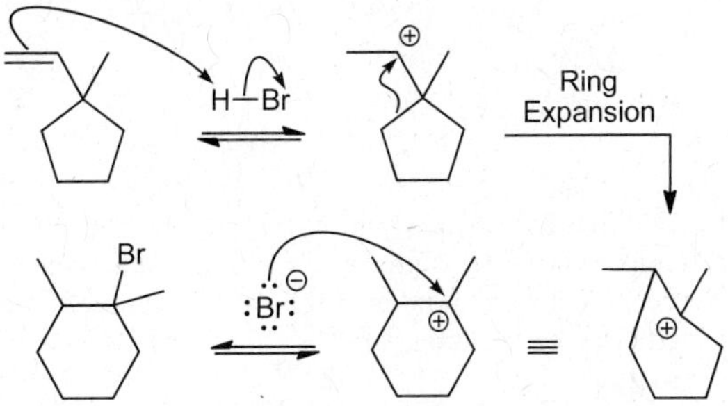

9.9. Protonation of the alkene requires two curved arrows, as shown in the following mechanism, and leads to the more stable, secondary carbocation (rather than a primary carbocation). This secondary carbocation then undergoes a methyl shift, shown with one curved arrow, generating a tertiary carbocation. This rearrangement is favorable because tertiary carbocations are more stable than secondary carbocations. Then, another rearrangement can occur (this time, a hydride shift). This rearrangement is favorable because the resulting carbocation is resonance-stabilized (and even more stable than a tertiary carbocation). In the final step of the mechanism (nucleophilic attack), the carbocation is captured by a bromide ion. This step requires one curved arrow, going from the nucleophile (bromide) to the electrophile (the carbocation):

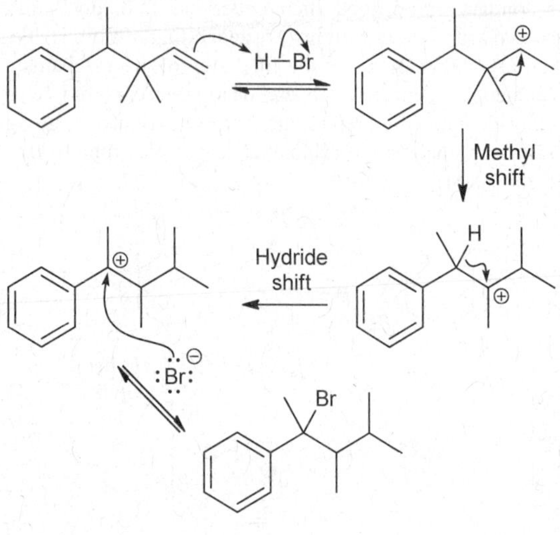

9.10.

(a) The second compound (highlighted) is expected to be more reactive toward acid-catalyzed hydration than the first compound, because the reaction proceeds via a tertiary carbocation, rather than via a secondary carbocation, as shown.

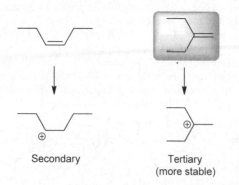

Secondary   Tertiary
                          (more stable)

(b) The first compound (2-methyl-2-butene) is expected to be more reactive toward acid-catalyzed hydration than the second compound, because the reaction proceeds via a tertiary carbocation, rather than a secondary carbocation.

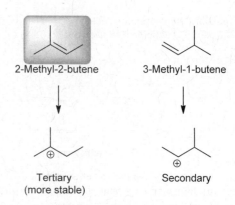

2-Methyl-2-butene   3-Methyl-1-butene

Tertiary   Secondary
(more stable)

9.11.
(a) To favor the alcohol, dilute sulfuric acid (mostly water) is used. Having a high concentration of water favors the alcohol according to Le Châtelier's principle.
(b) To favor the alkene, concentrated sulfuric acid (which has very little water) is used. Having a low concentration of water favors the alkene according to Le Châtelier's principle.

9.12.
(a) Water (H and OH) is added across the alkene in a Markovnikov fashion. The mechanism is expected to have three steps: 1) proton transfer, 2) nucleophilic attack, and 3) proton transfer. In the first step, a proton is transferred from H_3O^+ to the alkene, which requires two curved arrows, as shown below. The resulting tertiary carbocation is then captured by a water molecule in the second step of the mechanism. This step requires one curved arrow, going from the nucleophile (water) to the electrophile (the carbocation). Then, in the final step of the mechanism, a molecule of water functions as a base and removes a proton, thereby generating the product. This final step is a proton transfer step, and therefore requires two curved arrows, as shown:

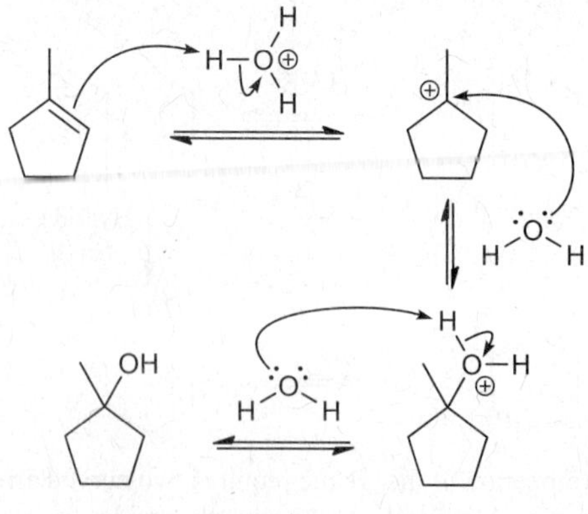

(b) Water (H and OH) is added across the alkene in a Markovnikov fashion. The mechanism is expected to have three steps: 1) proton transfer, 2) nucleophilic attack, and 3) proton transfer. In the first step, a proton is transferred from H_3O^+ to the alkene, which requires two curved arrows, as shown below. The resulting tertiary carbocation is then captured by a water molecule in the second step of the mechanism. This step requires one curved arrow, going from the nucleophile (water) to the electrophile (the carbocation). Then, in the final step of the mechanism, a molecule of water functions as a base and removes a proton, thereby generating the product. This final step is a proton transfer step, and therefore requires two curved arrows, as shown:

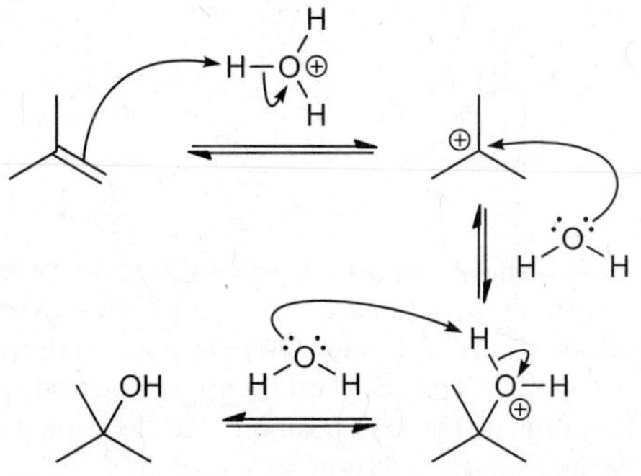

(c) Water (H and OH) is added across the alkene in a Markovnikov fashion. The mechanism is expected to have three steps: 1) proton transfer, 2) nucleophilic attack, and 3) proton transfer. In the first step, a proton is transferred from H_3O^+ to the alkene, which requires two curved arrows, as shown below. The resulting tertiary carbocation is then captured by a water molecule in the second step of the mechanism. This step requires one curved arrow, going from the nucleophile (water) to the electrophile (the carbocation). Then, in the final step of the mechanism, a molecule of water functions as a base and removes a proton, thereby generating the product. This final step is a proton transfer step, and therefore requires two curved arrows, as shown:

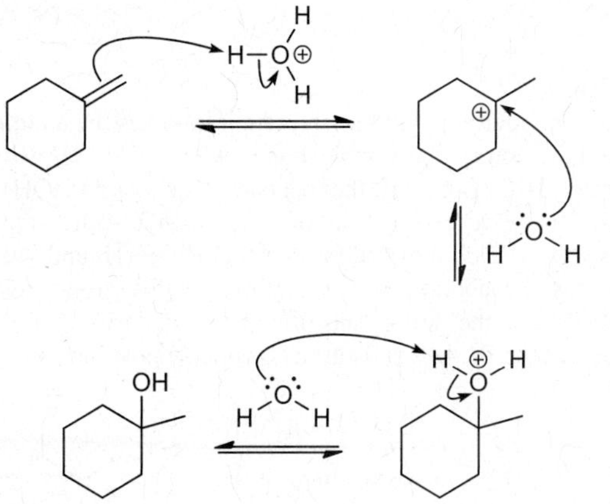

9.13. Methanol (H and OCH_3) is added across the alkene in a Markovnikov fashion. The reaction is extremely similar to the addition of water across an alkene under acid-catalyzed conditions, so we expect the mechanism to have three steps: 1) proton transfer, 2) nucleophilic attack, and 3) proton transfer. In the first step, a proton is transferred from $CH_3OH_2^+$ to the alkene, which requires two curved arrows, as shown below. The resulting tertiary carbocation is then captured by a

molecule of methanol in the second step of the mechanism. This step requires one curved arrow, going from the nucleophile (methanol) to the electrophile (the carbocation). Then, in the final step of the mechanism, a molecule of methanol functions as a base and removes a proton, thereby generating the product. This final step is a proton transfer step, and therefore requires two curved arrows, as shown:

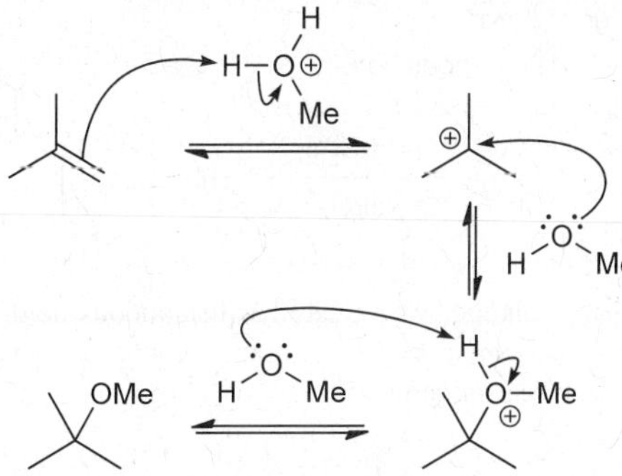

9.14. The reactant is acyclic (it does not have a ring), and the product is cyclic, indicating an intramolecular reaction. We can justify an intramolecular reaction if we inspect the cation that is obtained upon protonation of the alkene:

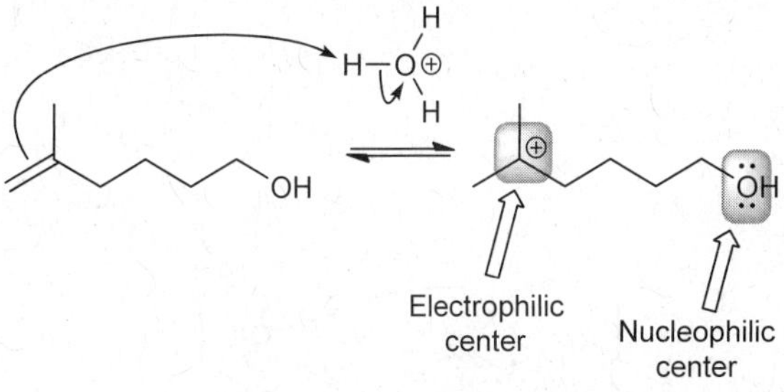

Notice that this intermediate exhibits both an electrophilic center and a nucleophilic center. In other words, the reactive centers are tethered together via a chain of methylene (CH_2) groups. As such, a ring is formed in the following intramolecular nucleophilic attack, which is shown with one curved arrow:

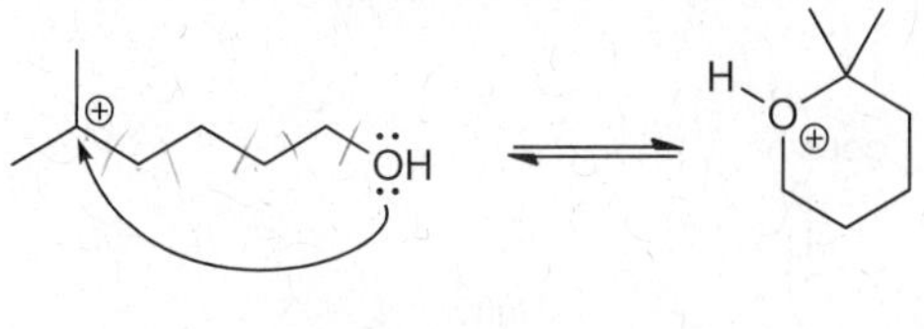

Finally, water functions as a base and removes a proton, thereby generating the product. This final step is a proton transfer step, and therefore requires two curved arrows, as shown:

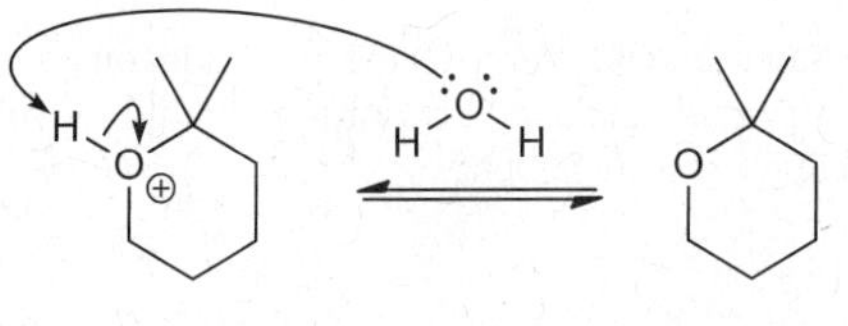

9.15.

(a) Oxymercuration-demercuration gives Markov-nikov addition of water (H and OH) without carbocation rearrangements. That is, the OH group ends up at the more substituted (secondary) position, and the proton ends up at the less substituted (primary) position:

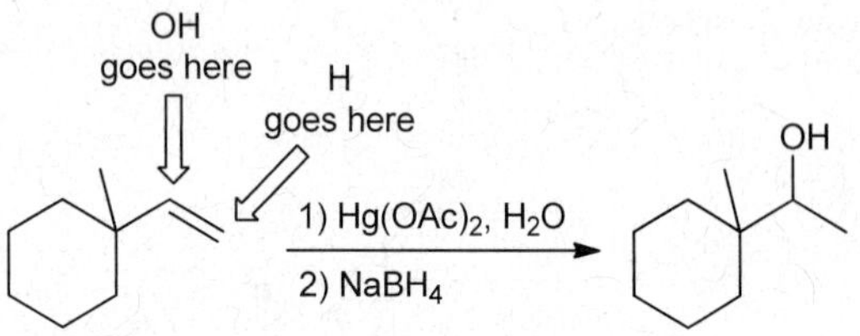

If the same alkene were treated with aqueous acid, the resulting acid-catalyzed hydration would involve a carbocation rearrangement:

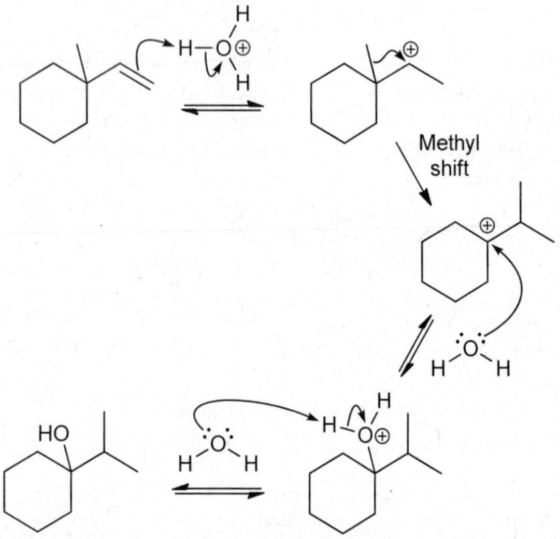

(b) Oxymercuration-demercuration gives Markov-nikov addition of water (H and OH) without carbocation rearrangements. That is, the OH group ends up at the more substituted (secondary) position, and the proton ends up at the less substituted (primary) position:

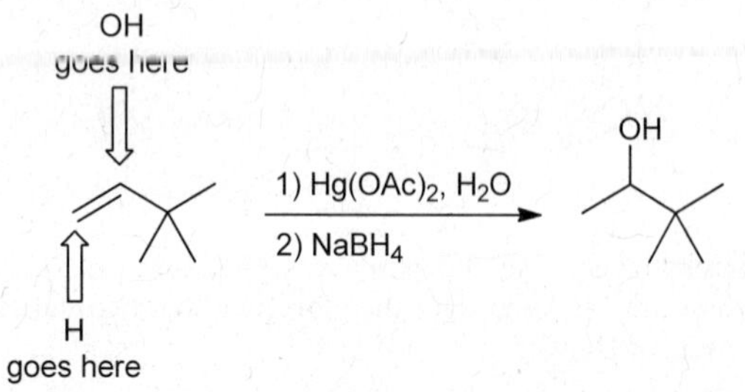

If the same alkene were treated with aqueous acid, the resulting acid-catalyzed hydration would involve a carbocation rearrangement:

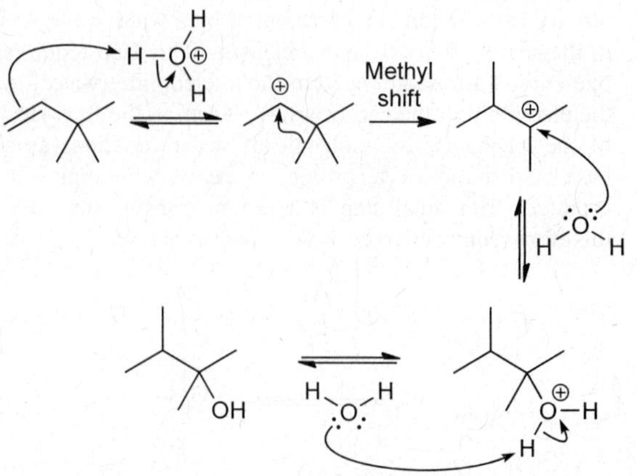

(c) Oxymercuration-demercuration gives Markov-nikov addition of water (H and OH) without carbocation rearrangements. That is, the OH group ends up at the more substituted (tertiary) position, and the proton ends up at the less substituted (primary) position.

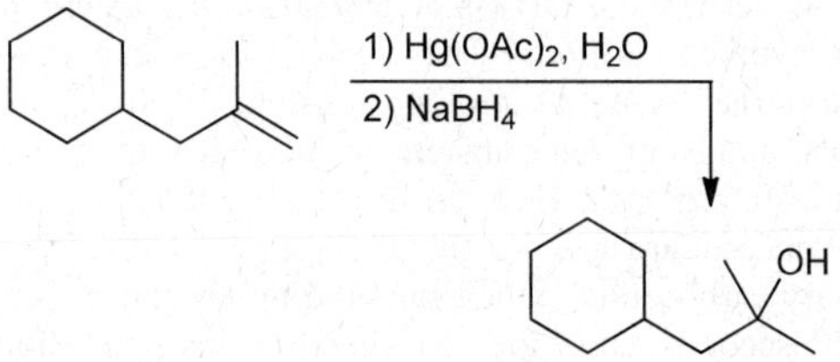

In this case, acid-catalyzed hydration gives the same product, because the intermediate tertiary carbocation does not undergo rearrangement:

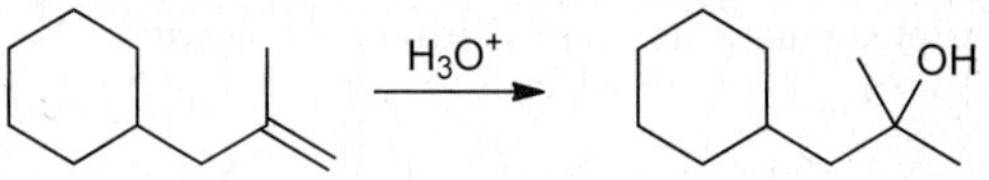

9.16.

(a) Oxymercuration-demercuration involves the addition of H-Z across the double bond (where Z = OH when water, H_2O, is used as the reagent). If ethanol (EtOH) is used as the reagent instead of water, then Z = OEt, so we expect Markvnikov addition of ethanol (H and OEt) across the alkene, with the ethoxy (OEt) group being placed at the more substituted (secondary) position, rather than the less substituted (primary) position.

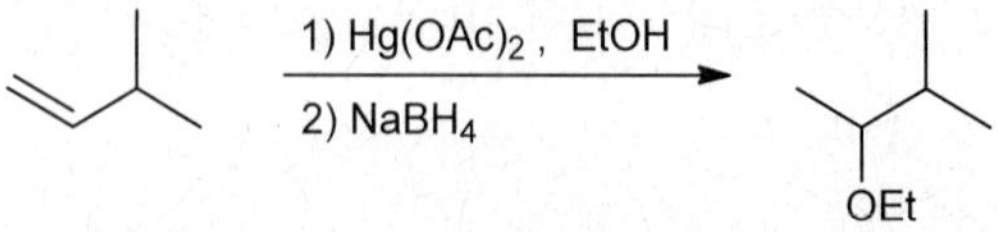

(b) Oxymercuration-demercuration involves the addition of H-Z across the double bond (where Z = OH when water, H_2O, is used as the reagent). If ethylamine ($EtNH_2$) is used as the reagent instead of water, then Z = NHEt, so we expect Markvnikov addition of H and NHEt across the alkene, with the ethylamino group (NHEt) being placed at the more substituted (secondary) position, rather than the less substituted (primary) position.

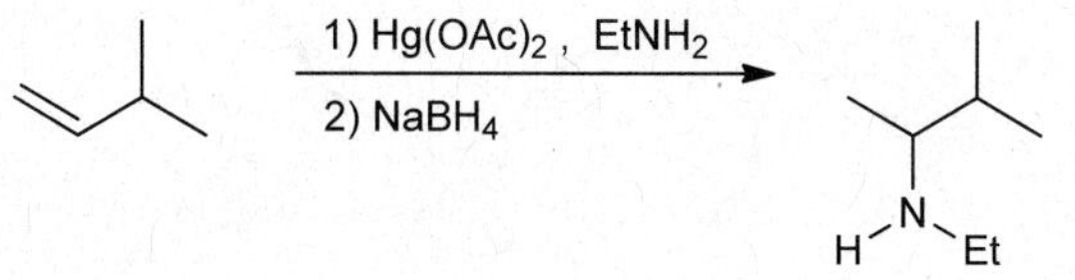

9.17.

(a) Hydroboration-oxidation results in the *anti*-Markovnikov addition of water (H and OH) across the π bond. That is, the OH group is placed at the less-substituted (primary) position, rather than the more substituted (tertiary) position:

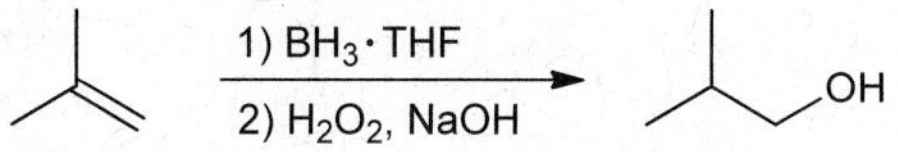

(b) Hydroboration-oxidation results in the *anti*-Markovnikov addition of water (H and OH) across the π bond. That is, the OH group is placed at the less-substituted (primary) position, rather than the more substituted (tertiary) position:

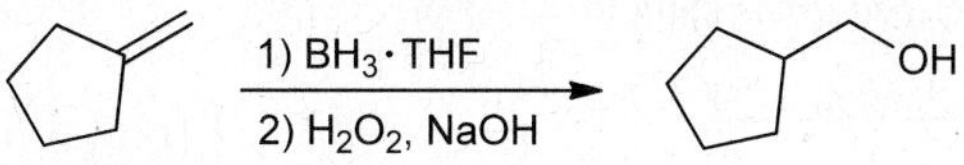

(c) Hydroboration-oxidation results in the *anti*-Markovnikov addition of water (H and OH) across the π bond. That is, the OH group is placed at the less-substituted (primary) position, rather than the more substituted (secondary) position:

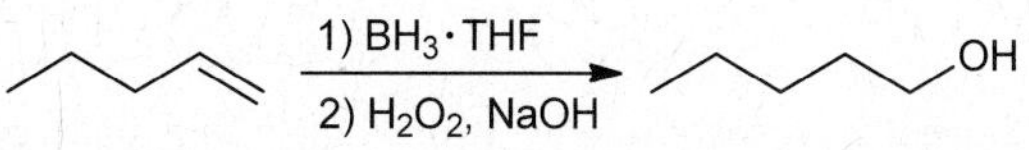

9.18. There is only one alkene, shown below, that will afford the desired product upon hydroboration-oxidation.

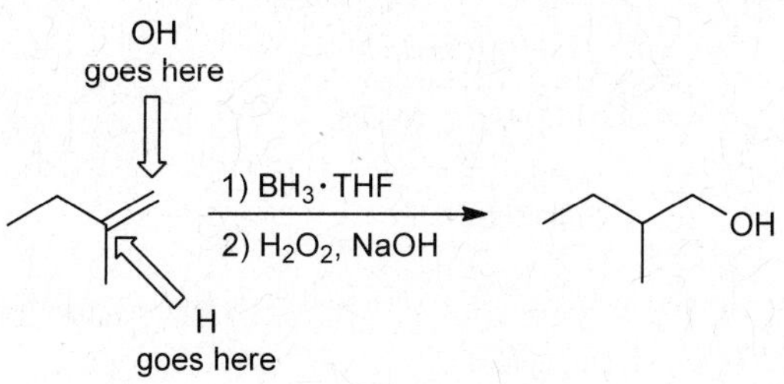

9.19.

(a) The reagents indicate a hydroboration-oxidation. The net result of this two-step process is the *anti*-Markovnikov addition of H and OH across the π bond. That is, the OH group is placed at the less-substituted position, while the H is placed at the more substituted position. In this case, two chirality centers are created. Therefore, the stereochemical requirement for *syn*

addition determines that the H and OH are added on the same face of the alkene, giving the following products:

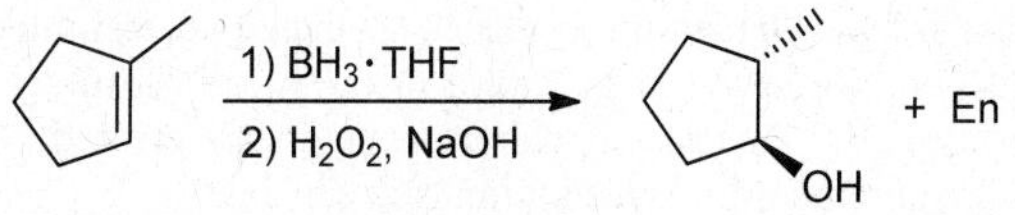

In this case, it might seem as if there was an *anti* addition, rather than a *syn* addition, because we see that the product has one wedge and one dash. But this is an optical illusion. Recall, that most hydrogen atoms are not drawn in bond-line drawings, so the H that was added during the process has not been drawn. However, if we draw that hydrogen atom, we will see that the H and OH were indeed added in a *syn* fashion:

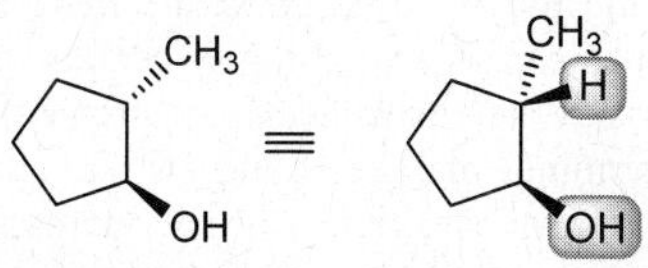

(b) The reagents indicate a hydroboration-oxidation. The net result of this two-step process is the *anti*-Markovnikov addition of H and OH across the π bond. That is, the OH group is placed at the less-substituted position, while the H is placed at the more substituted position. In this case, only one chirality center is created. Since *syn* addition can take place from either face of the alkene with equal likelihood, we expect a pair of enantiomers, as shown:

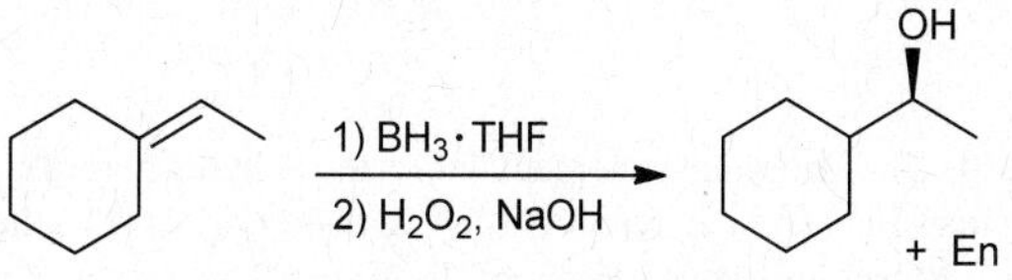

(c) The reagents indicate a hydroboration-oxidation. The net result of this two-step process is the *anti*-Markovnikov addition of H and OH across the π bond. That is, the OH group is placed at the less-substituted position, while the H is placed at the more substituted position. In this case, no chirality centers are created, so the requirement for *syn* addition is irrelevant.

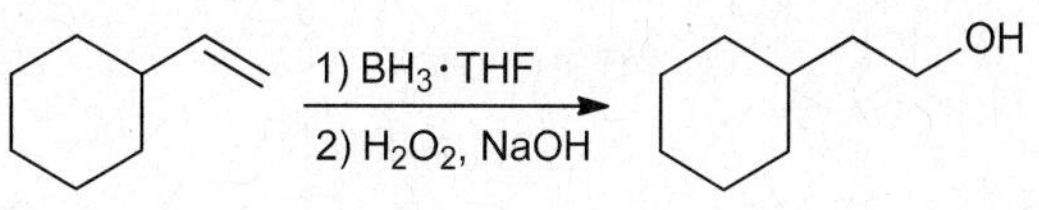

(d) The reagents indicate a hydroboration-oxidation. The net result of this two-step process is the *anti*-Markovnikov addition of H and OH across the π bond. That is, the OH group is placed at the less-substituted position, while the H is placed at the more substituted position. In this case, only one chirality center is created. Since *syn* addition can take place from either face of the alkene with equal likelihood, we expect a pair of enantiomers, as shown:

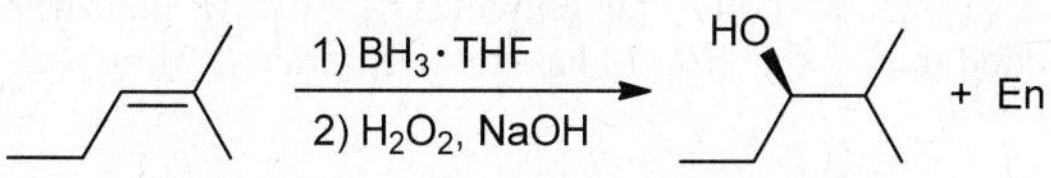

(e) The reagents indicate a hydroboration-oxidation. The net result of this two-step process is the *anti*-Markovnikov addition of H and OH across the π bond. That is, the OH group is placed at the less-substituted position, while the H is placed at the more substituted position. In this case, no chirality centers are created, so the requirement for *syn* addition is irrelevant.

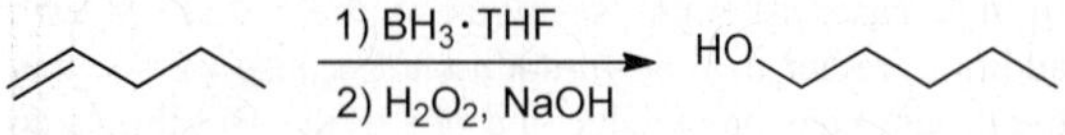

$$\text{1) BH}_3 \cdot \text{THF}$$
$$\text{2) H}_2\text{O}_2, \text{NaOH}$$

(f) The reagents indicate a hydroboration-oxidation. The net result of this two-step process is the *anti*-Markovnikov addition of H and OH across the π bond. That is, the OH group is placed at the less-substituted position, while the H is placed at the more substituted position. In this case, two chirality centers are created. Therefore, the stereochemical requirement for *syn* addition determines that the H and OH are added on the same face of the alkene, giving the following products:

$$\text{1) BH}_3 \cdot \text{THF}$$
$$\text{2) H}_2\text{O}_2, \text{NaOH}$$

+ En

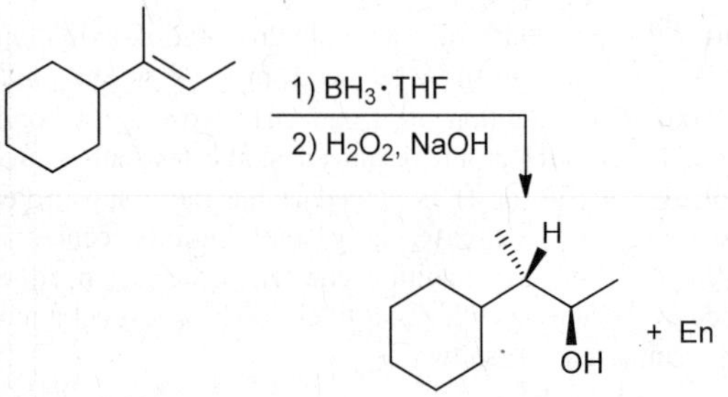

9.20. The problem statement indicates a hydroboration-oxidation. The net result of this two-step process is the *anti*-Markovnikov addition of H and OH across the π bond. That is, the OH group is placed at the less-substituted position, while the H is placed at the more substituted position. In this case, two chirality centers are created. Therefore, the stereochemical requirement for *syn* addition determines that the H and OH are added on the same face of the alkene, giving the following products:

$$\text{1) BH}_3 \cdot \text{THF}$$
$$\text{2) H}_2\text{O}_2, \text{NaOH}$$

+ En

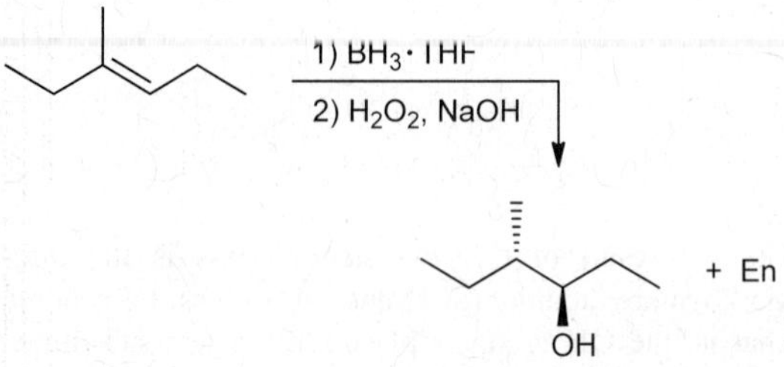

In this case, it might seem as if there was an *anti* addition, rather than a *syn* addition, because we see that the product has one wedge and one dash. But this is an optical illusion. Recall, that most hydrogen atoms are not drawn in bond-line drawings, so the H that was added during the process has not been drawn. However,

if we draw that hydrogen atom, we will see that the H and OH were indeed added in a *syn* fashion:

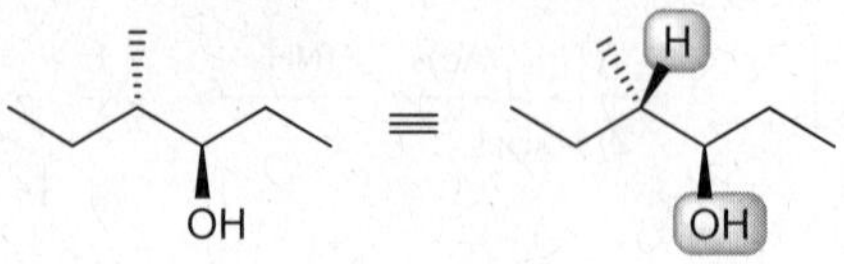

$$\equiv$$

9.21. Only one chirality center is formed, so both possible stereoisomers (enantiomers) are obtained, regardless of the configuration of the starting alkene:

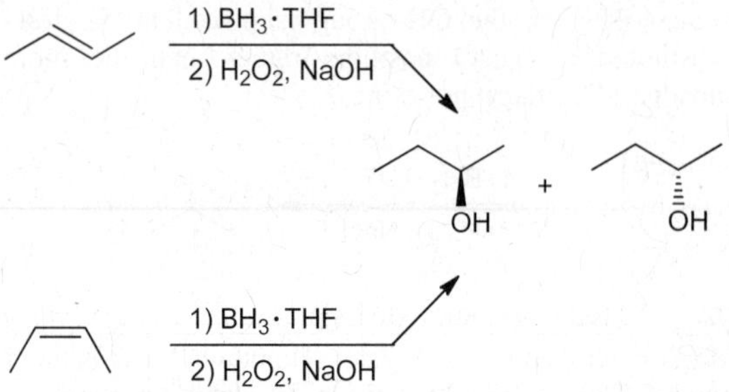

$$\text{1) BH}_3 \cdot \text{THF}$$
$$\text{2) H}_2\text{O}_2, \text{NaOH}$$

+

$$\text{1) BH}_3 \cdot \text{THF}$$
$$\text{2) H}_2\text{O}_2, \text{NaOH}$$

9.22. We begin by drawing all possible alkenes with molecular formula C_5H_{10} (using the methodical approach described in the solution to Problem 4.3):

Five-carbon chain

Four-carbon chain

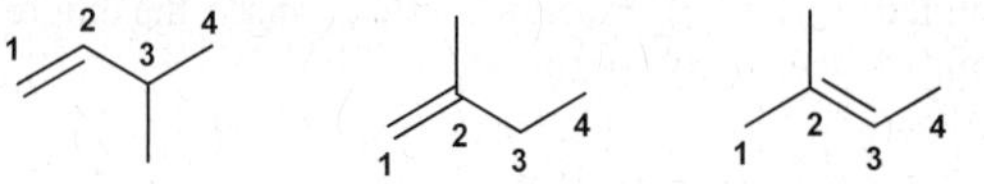

Among these isomers, only two of them will undergo hydroboration-oxidation to afford an alcohol with no chirality centers, shown below. The remaining four isomers can undergo hydroboration-oxidation to produce alcohols that do possess a chirality center.

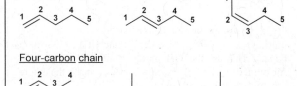

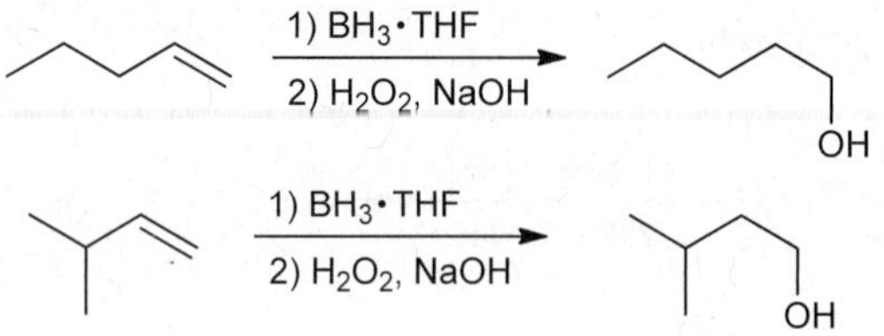

$$\text{1) BH}_3 \cdot \text{THF}$$
$$\text{2) H}_2\text{O}_2, \text{NaOH}$$

$$\text{1) BH}_3 \cdot \text{THF}$$
$$\text{2) H}_2\text{O}_2, \text{NaOH}$$

9.23.
(a) The reagents indicate a catalytic hydrogenation process, so we expect the addition of H and H across the alkene. In this case, the product does not have a chirality center, so stereochemistry is not a relevant consideration.

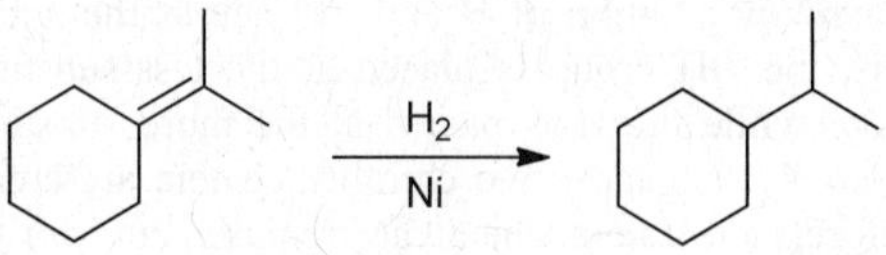

$$\xrightarrow[\text{Ni}]{\text{H}_2}$$

(b) The reagents indicate a catalytic hydrogenation process, so we expect the addition of H and H across the alkene. In this case, the product does not have a chirality center, so stereochemistry is not a relevant consideration.

(c) The reagents indicate a catalytic hydrogenation process, so we expect the addition of H and H across the alkene. In this case, the product has one chirality center, so we expect both possible enantiomers (*syn* addition can occur from either face of the π bond).

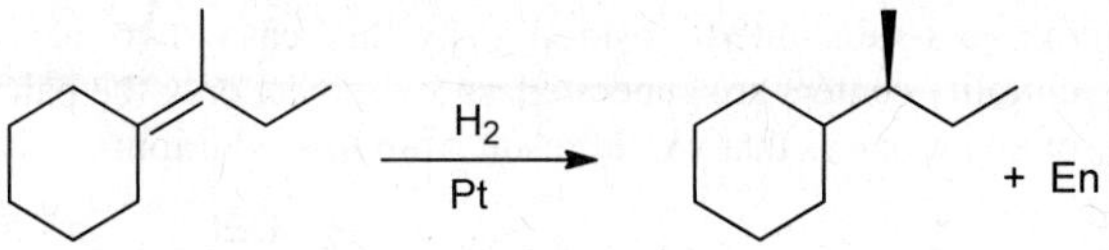

(d) The reagents indicate a catalytic hydrogenation process, so we expect the addition of H and H across the alkene. In this case, the product has one chirality center, so we expect both possible enantiomers (*syn* addition can occur from either face of the π bond).

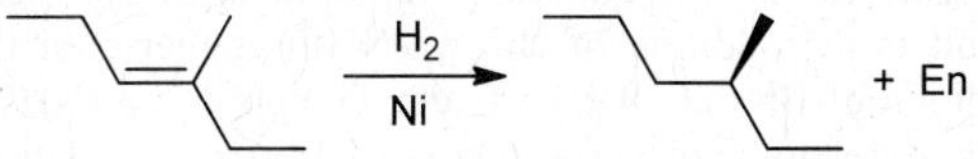

(e) The reagents indicate a catalytic hydrogenation process, so we expect the addition of H and H across the alkene. In this case, the product does not have a chirality center, so stereochemistry is not a relevant consideration.

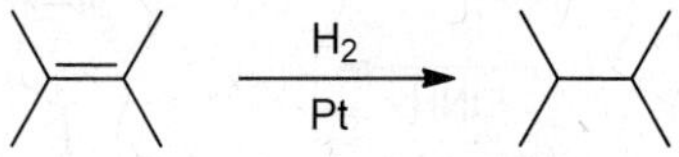

(f) The reagents indicate a catalytic hydrogenation process, so we expect the addition of H and H across the alkene. In this case, the reaction generates two chirality centers. The requirement for *syn* addition results in the formation of a *meso* compound, so there is only one product.

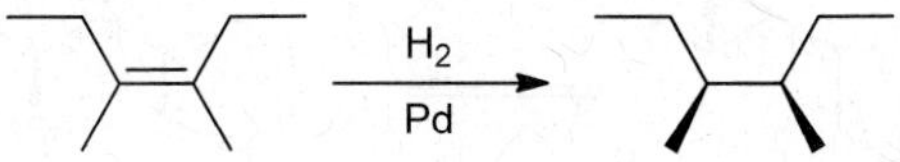

(meso)

9.24. We expect a *syn* addition of D and D across the alkene, giving a pair of enantiomers:

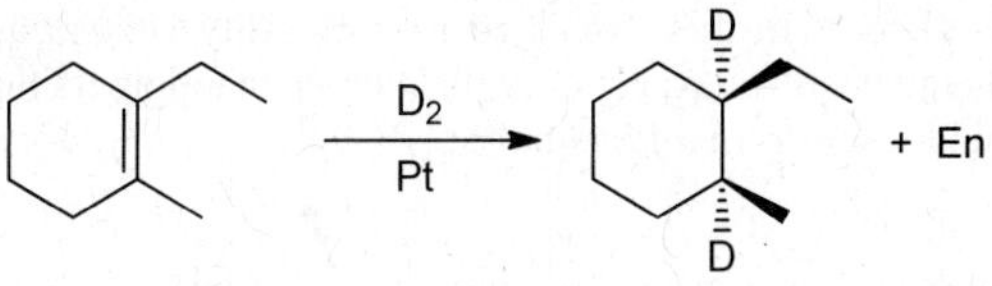

9.25.
(a) Compound X reacts with H_2 in the presence of a catalyst, so compound X is an alkene. The product of hydrogenation is 2-methylbutane, so compound X must have the same carbon skeleton as 2-methylbutane:

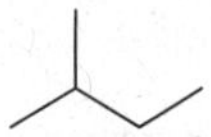

2-methylbutane

We just have to decide where to place the double bond in compound X. Keep in mind that the following two positions are identical:

2-methyl-1-butene   2-methyl-1-butene

So, there are only three possible locations where we can place the double bond:

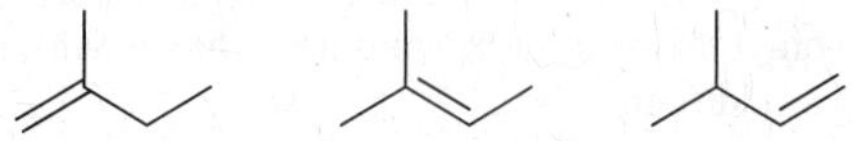

(b) Upon hydroboration-oxidation, only one of the three proposed alkenes will be converted to an alcohol without any chirality centers, shown below. Each of the other two compounds will be converted into an alcohol with a chirality center.

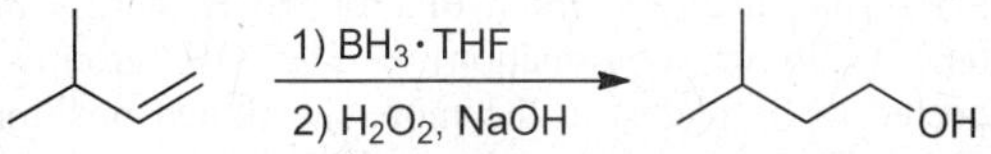

1) $BH_3 \cdot THF$
2) H_2O_2, NaOH

9.26.
(a) When an alkene is treated with molecular bromine (Br_2), we expect an *anti* addition of Br and Br across the alkene, giving the following pair of enantiomers:

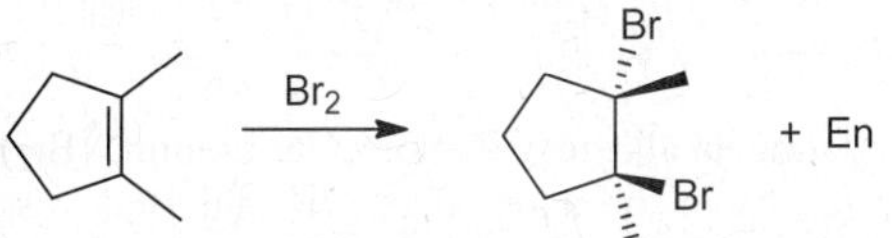

(b) When an alkene is treated with molecular bromine (Br_2), we expect an *anti* addition of Br and Br across the alkene, giving the following pair of enantiomers:

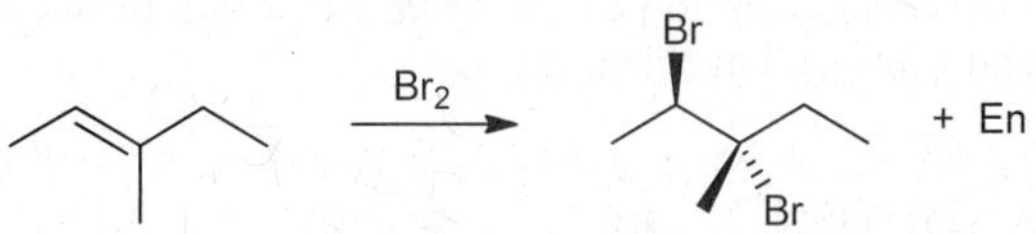

(c) When an alkene is treated with molecular bromine (Br_2), we expect an *anti* addition of Br and Br across the alkene. In this case, only one chirality center is created, so we expect both possible enantiomers (formation of the

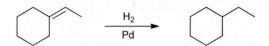

initial bromonium ion can occur on either face of the π bond with equal likelihood):

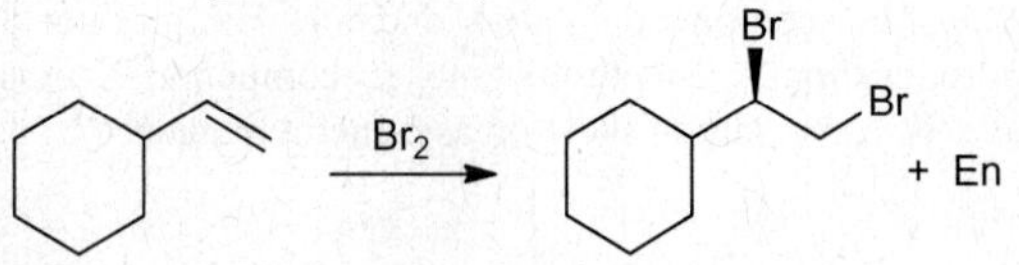

(d) When an alkene is treated with molecular bromine (Br₂), we expect an *anti* addition of Br and Br across the alkene, giving the following pair of enantiomers:

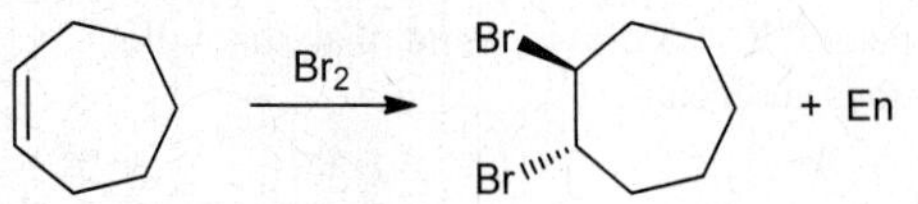

9.27.
(a) Treating an alkene with molecular bromine (Br₂) and water results in the addition of OH and Br across the alkene (halohydrin formation). The OH group is expected to be placed at the more-substituted position, while the Br is placed at the less-substituted position. In this case, two new chirality centers are generated, so we expect only the pair of enantiomers that would result from *anti* addition.

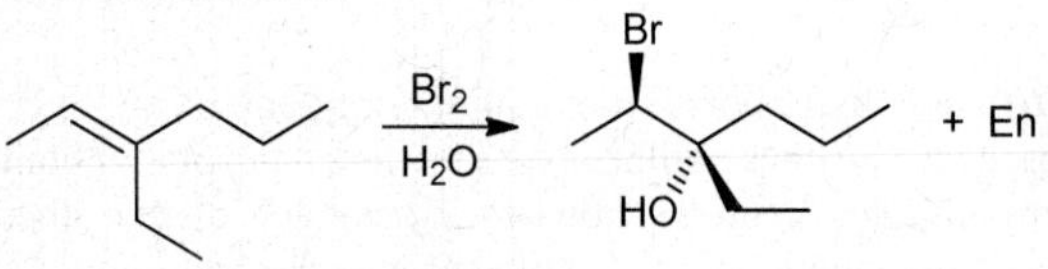

(b) Treating an alkene with molecular bromine (Br₂) and water results in the addition of OH and Br across the alkene (halohydrin formation). The OH group is expected to be placed at the more-substituted position, while the Br is placed at the less-substituted position. In this case, two new chirality centers are generated, so we expect only the pair of enantiomers that would result from *anti* addition.

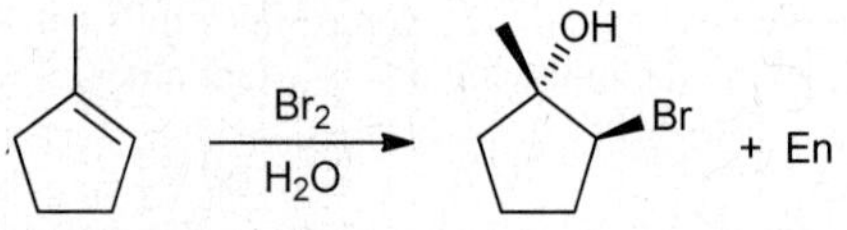

(c) Treating an alkene with molecular bromine (Br₂) and water results in the addition of OH and Br across the alkene (halohydrin formation). The OH group is expected to be placed at the more-substituted position, while the Br is placed at the less-substituted position. In this case, only one new chirality center is generated, so we expect both possible enantiomers (formation of the initial bromonium ion can occur on either face of the π bond with equal likelihood).

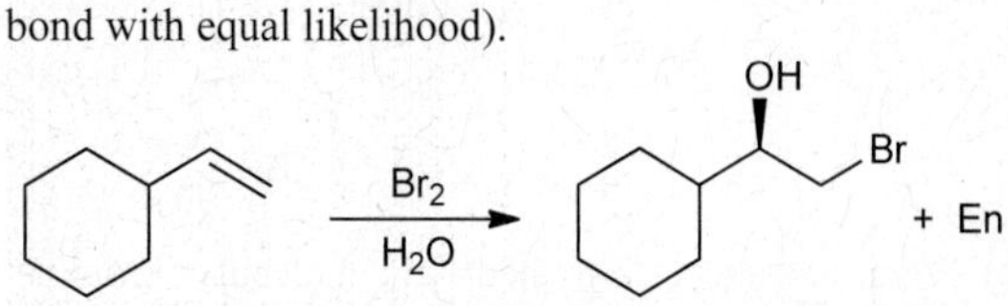

(d) Treating an alkene with molecular bromine (Br₂) and water results in the addition of OH and Br across the

alkene (halohydrin formation). In this case, two new chirality centers are generated, so we expect only the pair of enantiomers that would result from *anti* addition.

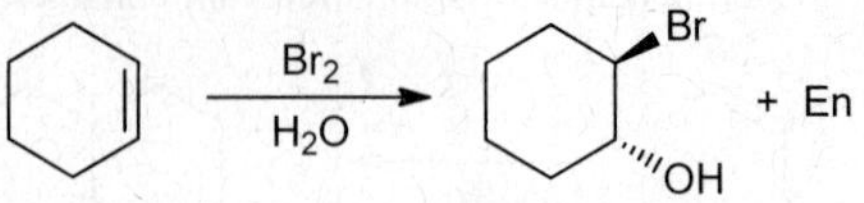

9.28.
(a) The alkene reacts with molecular bromine to give a bromonium ion, which is then attacked by a molecule of solvent (EtOH, in this case, rather than H₂O). The result is the addition of Br and OEt (rather than the addition of Br and OH). The OEt group is expected to be placed at the more-substituted position, while the Br is placed at the less-substituted position. In this case, two new chirality centers are generated, so we expect only the pair of enantiomers that would result from *anti* addition:

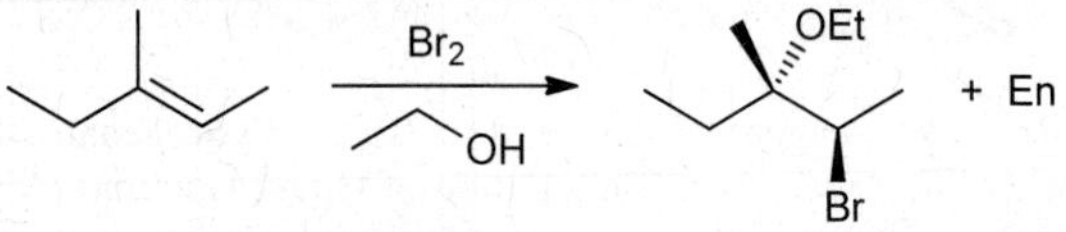

(b) The alkene reacts with molecular bromine to give a bromonium ion, which is then captured by a molecule of solvent (EtNH₂, in this case, rather than H₂O). The result is the addition of Br and NHEt (rather than the addition of Br and OH). The ethylamino group (NHEt) is expected to be placed at the more-substituted position, while the Br is placed at the less-substituted position. In this case, two new chirality centers are generated, so we expect only the pair of enantiomers that would result from *anti* addition:

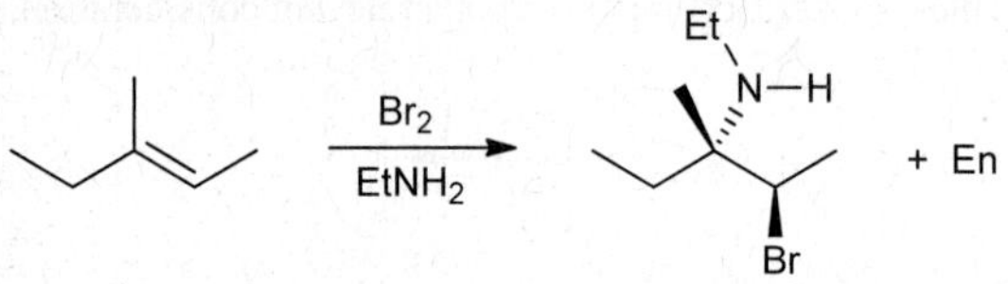

9.29. The bromonium ion can open (before a bromide ion attacks), forming a resonance-stabilized carbocation. This carbocation is trigonal planar and can be attacked from either side:

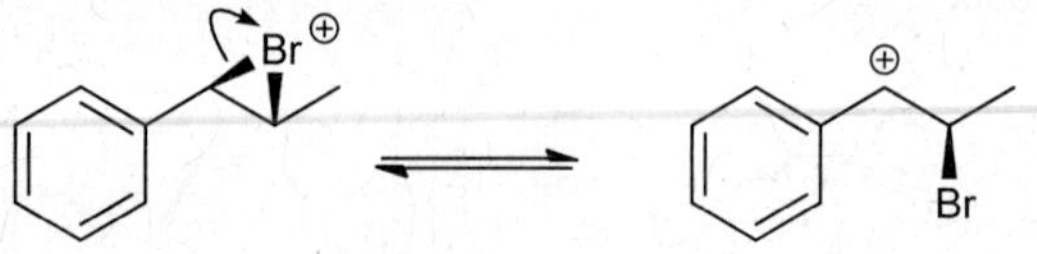

resonance-stabilized

9.30.
(a) Treating an alkene with MCPBA followed by aqueous acid results in the addition of OH and OH across the alkene. In this case, two new chirality centers are generated, so we expect only the pair of enantiomers that would result from *anti* addition:

(b) Treating an alkene with MCPBA followed by aqueous acid results in the addition of OH and OH across the alkene. In this case, two new chirality centers are generated, so we expect only the pair of enantiomers that would result from *anti* addition:

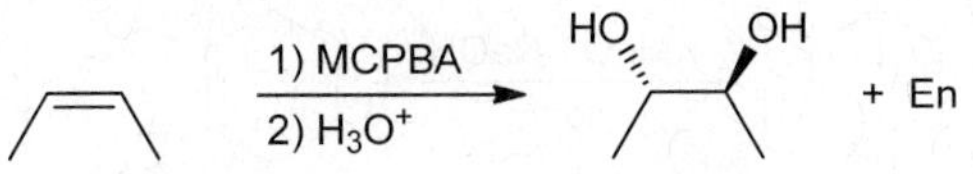

(c) Treating an alkene with MCPBA followed by aqueous acid results in the addition of OH and OH across the alkene. In this case, two new chirality centers are generated, so we expect only the pair of enantiomers that would result from *anti* addition:

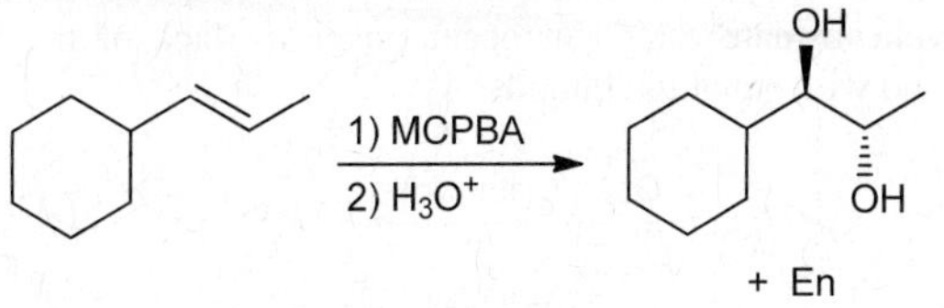

(d) Treating an alkene with MCPBA followed by aqueous acid results in the addition of OH and OH across the alkene. In this case, the product has no chirality centers, so stereochemistry is not a relevant consideration.

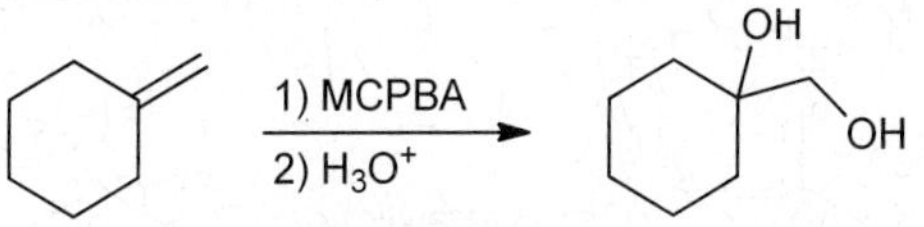

(e) Treating an alkene with MCPBA followed by aqueous acid results in the addition of OH and OH across the alkene. In this case, two new chirality centers are generated. The requirement for *anti* addition results in the formation of a *meso* compound:

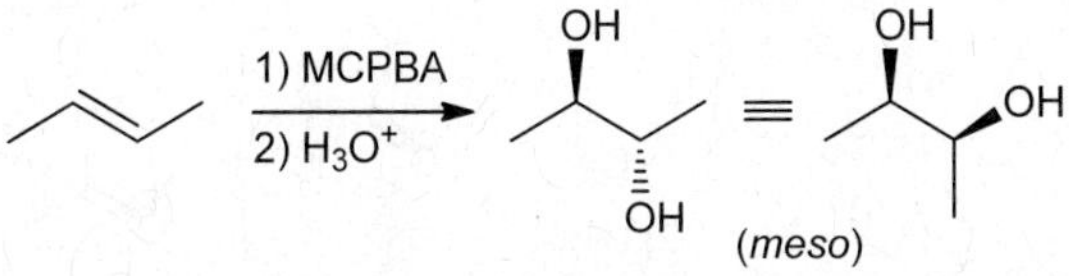

(f) Treating an alkene with MCPBA followed by aqueous acid results in the addition of OH and OH across the alkene. In this case, two new chirality centers are generated, so we expect only the pair of enantiomers that would result from *anti* addition:

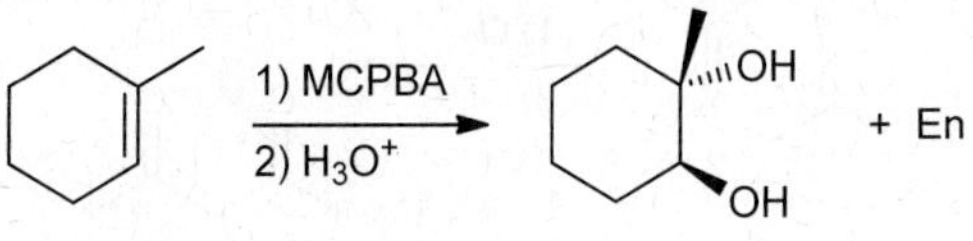

9.31.
(a) Treating an alkene with MCPBA results in an epoxide. Further treatment of the epoxide with ethanol under acid conditions results in a ring opening reaction in which ethanol serves as the nucleophile. Nucleophilic attack occurs at the more-substituted (tertiary) position,

so the net result is the addition of OH and OEt across the alkene, with the latter being placed at the more-substituted position, as shown:

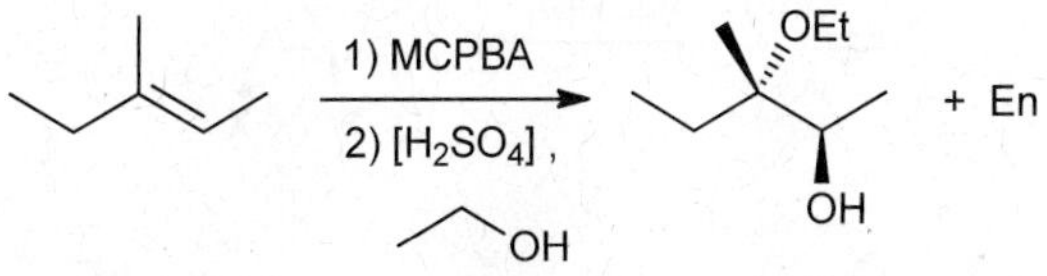

(b) Treatment of the epoxide with phenol (C_6H_5OH) under acid conditions results in a ring opening reaction in which the oxygen atom of phenol serves as the nucleophilic center. Nucleophilic attack occurs at the more-substituted (tertiary) position, so the net result is the addition of OH and OR (where R is C_6H_5) across the alkene, with the latter being placed at the more-substituted position. Since the starting epoxide is enantiomerically pure (we are starting only with the enantiomer shown), we expect an enantiomerically pure product (not a mixture of enantiomers), as shown.

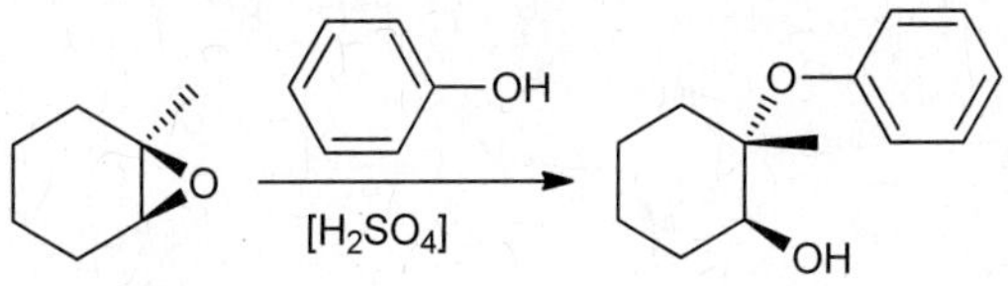

9.32.
(a) Compound A is converted to an epoxide upon treatment with MCPBA, so compound A must be an alkene. There are many alkenes with molecular formula C_6H_{12}, and it would be time-consuming to try to draw them all. Instead, we notice the following: in order for the product to have no chirality centers, each of the vinylic positions must already contain two identical groups, like this:

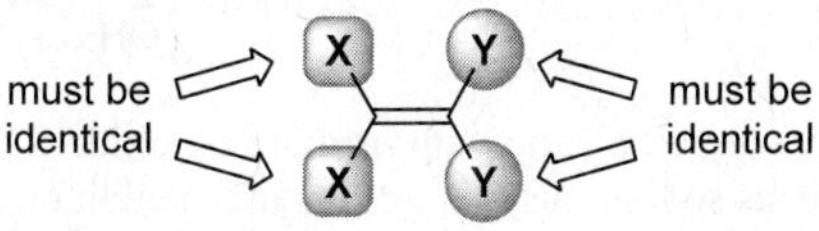

There are only two alkenes with molecular formula C_6H_{12} that fit this criterion:

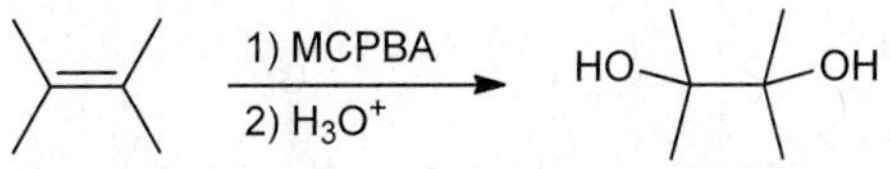

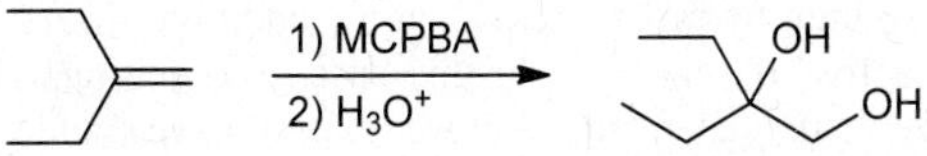

(b) In order to be a *meso* compound, the resulting diol must contain two chirality centers, as well as reflectional symmetry (such as an internal plane of symmetry). In order to achieve this result, the starting alkene must have the following structural features:

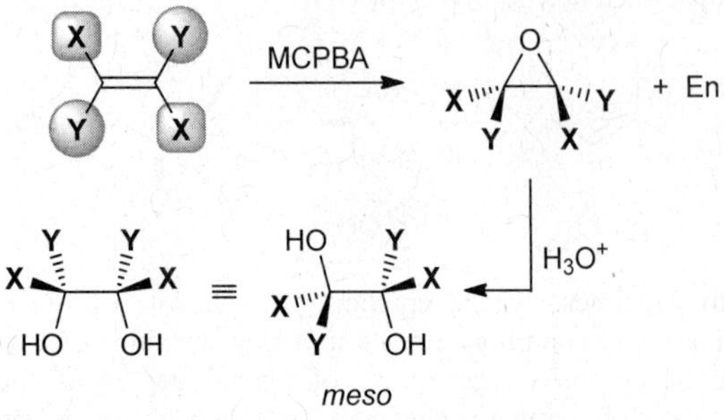

The identity of X and Y must be different, or the resulting diol would have no chirality centers. There is only one alkene with molecular formula C_6H_{12} that fits this criterion:

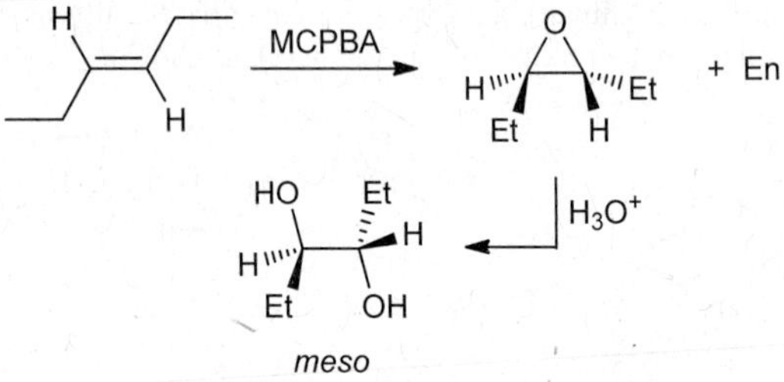

9.33.
(a) Treating an alkene with catalytic osmium tetroxide and NMO results in the addition of OH and OH across the alkene. In this case, two new chirality centers are generated, so we expect only the pair of enantiomers that would result from *syn* addition:

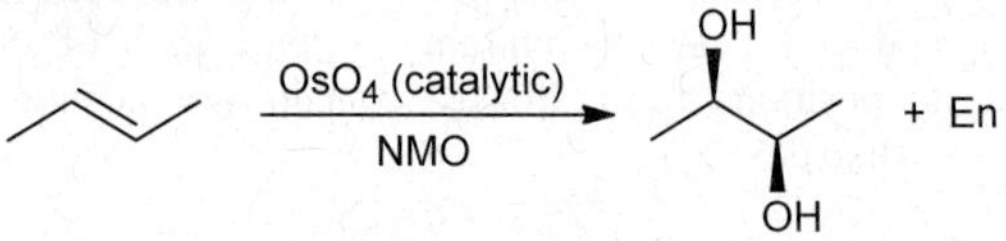

(b) Treating an alkene with osmium tetroxide followed by aqueous sodium bisulfite results in the addition of OH and OH across the alkene. In this case, only one chirality center is created, so we expect both possible enantiomers (formation of the initial cyclic osmate ester can occur on either face of the π bond with equal likelihood):

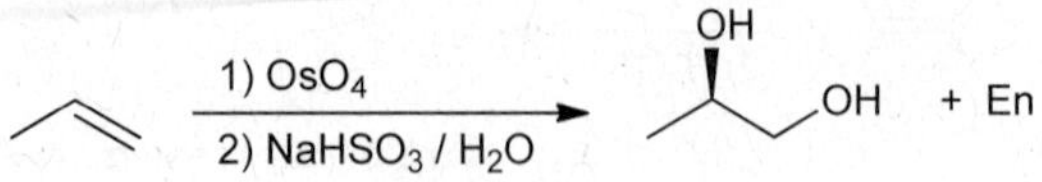

(c) Treating an alkene with cold potassium permanganate and sodium hydroxide results in the addition of OH and OH across the alkene. In this case, two new chirality centers are generated, and we expect a *syn* addition, giving the following *meso* compound:

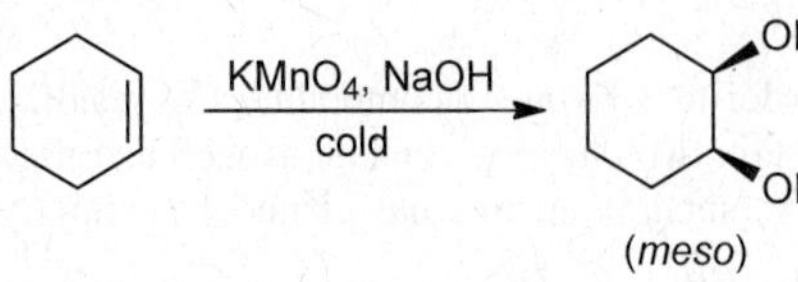

(d) Treating an alkene with cold potassium permanganate and sodium hydroxide results in the addition of OH and OH across the alkene. In this case, the product has no chirality centers, so stereochemistry is not a relevant consideration.

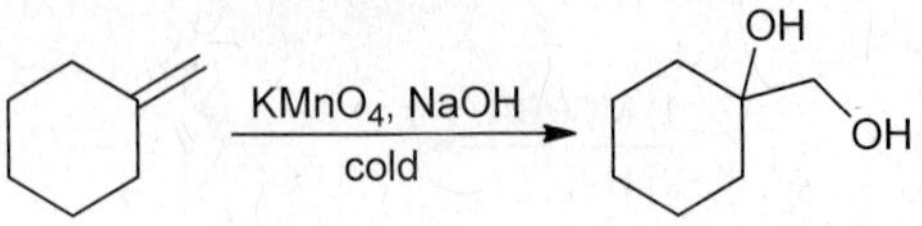

(e) Treating an alkene with catalytic osmium tetroxide and a suitable co-oxidant (*tert*-butyl hydroperoxide) results in the addition of OH and OH across the alkene. In this case, only one chirality center is created, so we expect both possible enantiomers (formation of the initial cyclic osmate ester can occur on either face of the π bond with equal likelihood):

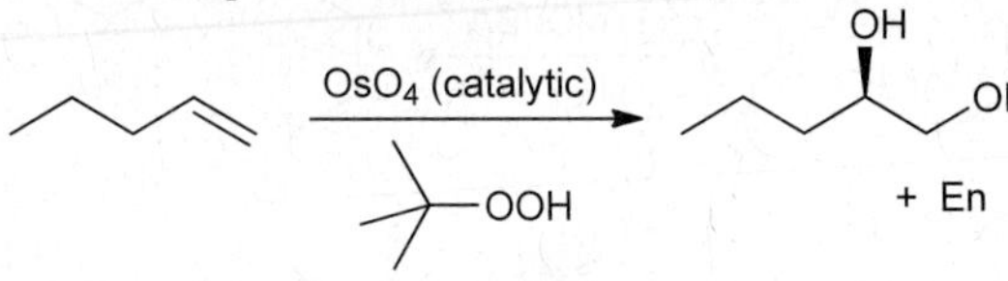

(f) Treating an alkene with catalytic osmium tetroxide and NMO results in the addition of OH and OH across the alkene. In this case, two new chirality centers are generated, so we expect only the pair of enantiomers that would result from *syn* addition:

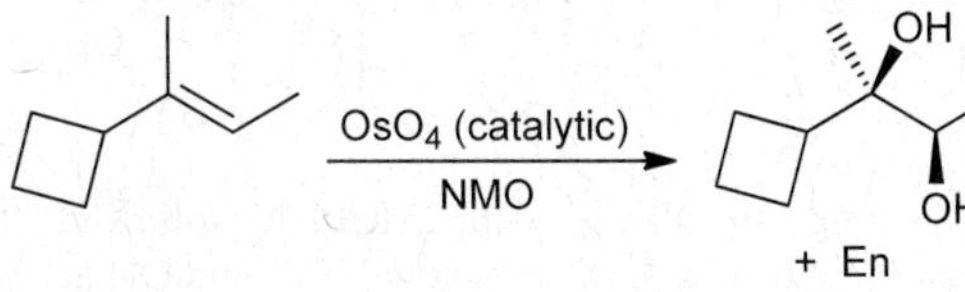

9.34.
(a) Each C=C bond is split apart and redrawn as two C=O bonds, giving the following two products:

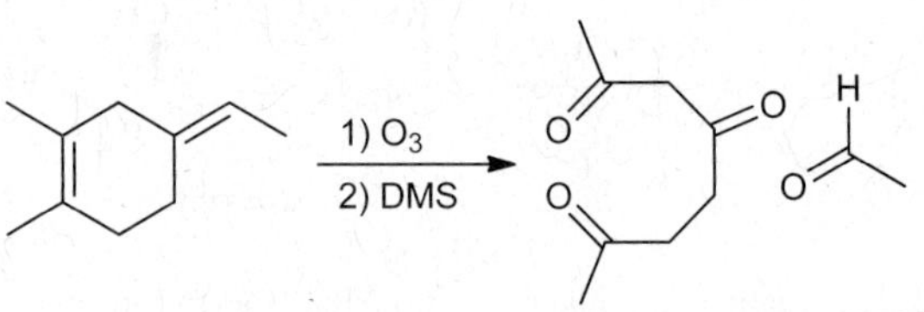

(b) Each C=C bond is split apart and redrawn as two C=O bonds, giving two equivalents of one product:

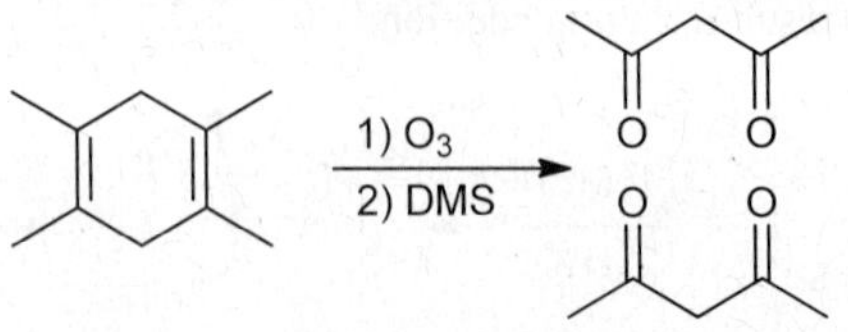

(c) The C=C bond is split apart and redrawn as two C=O bonds, giving two equivalents of one product:

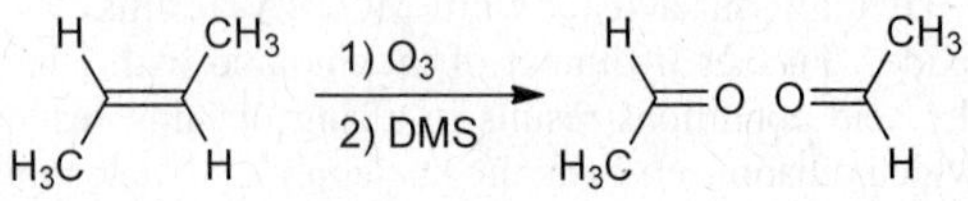

(d) The C=C bond is split apart and redrawn as two C=O bonds, giving the following product:

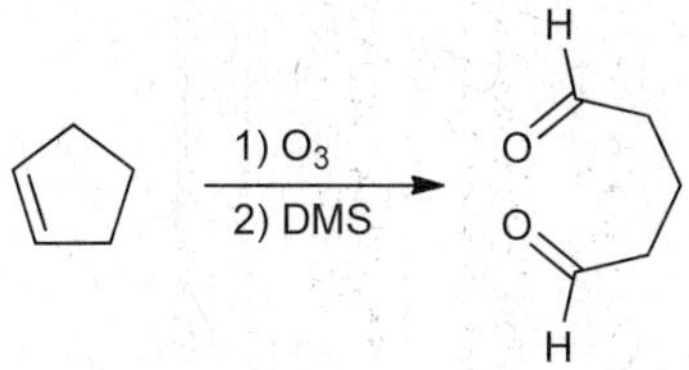

(e) The C=C bond is split apart and redrawn as two C=O bonds, giving the following *meso* compound:

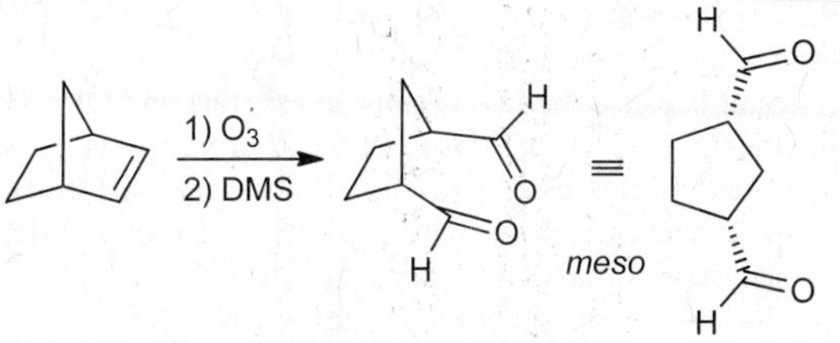

(f) The C=C bond is split apart and redrawn as two C=O bonds, giving two equivalents of one product:

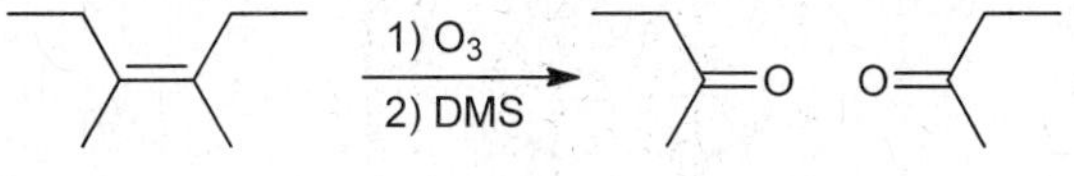

9.35.
(a) We can draw the starting alkene by removing the two oxygen atoms from the product, and connecting the sp^2 hybridized carbon atoms as a C=C bond:

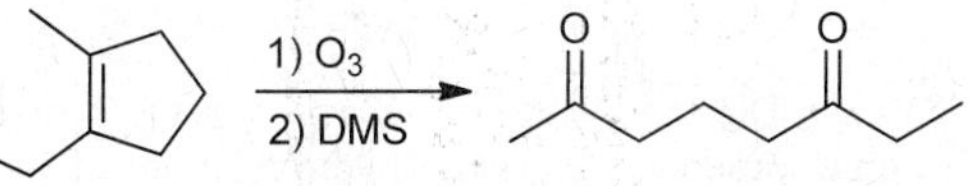

(b) In this case, the starting alkene has ten carbon atoms while the product has only five carbon atoms. Therefore, one equivalent of the starting alkene must produce two equivalents of the product:

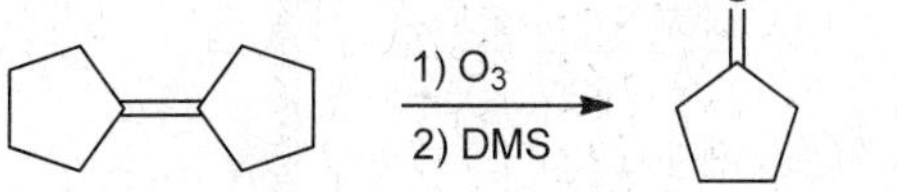

(two equivalents)

(c) In this case, the starting alkene has ten carbon atoms while the product has only five carbon atoms. Therefore, one equivalent of the starting alkene must produce two equivalents of the product:

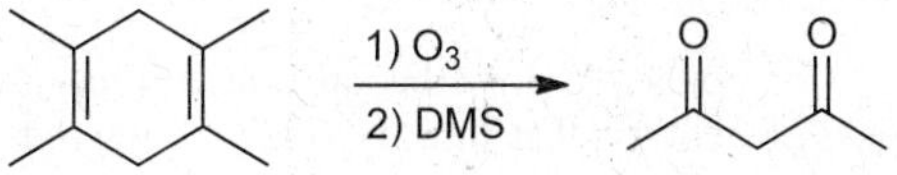

(two equivalents)

9.36.
(a) The reagents indicate a hydroboration-oxidation, so the net result will be the addition of H and OH across the alkene. For the regiochemical outcome, we expect an *anti*-Markovnikov addition, so the OH group is placed at the less-substituted position. The stereochemical outcome (*syn* addition) is not relevant in this case, because the product has no chirality centers:

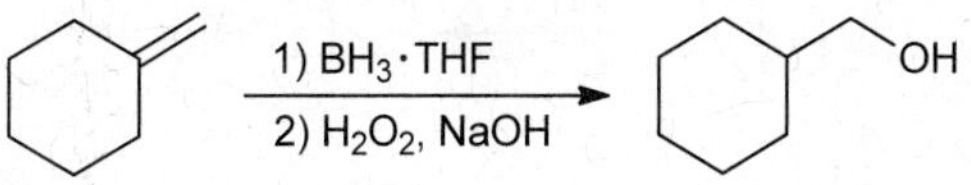

(b) The reagents indicate a hydrogenation reaction, so the net result will be the addition of H and H across the alkene. The regiochemical outcome is not relevant because the two groups added (H and H) are identical. We expect the reaction to proceed via a *syn* addition, but only one chirality center is formed. Therefore, both enantiomers are obtained because *syn* addition can occur from either face of the starting alkene:

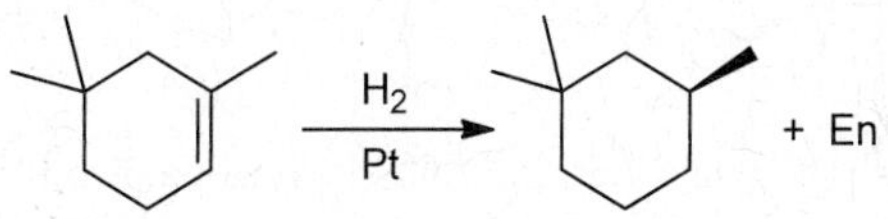

(c) The first reagent is a peroxy acid, indicating formation of an epoxide, which is then opened under aqueous acidic conditions. The net result is expected to be the addition of OH and OH across the alkene. The regiochemical outcome is not relevant because the two groups added (OH and OH) are identical. For the stereochemical outcome, we notice that two chirality centers are formed, and we expect only the pair of enantiomers resulting from an *anti* addition:

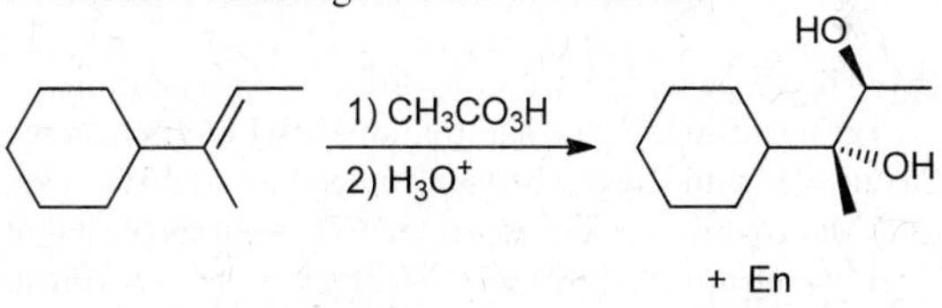

(d) The reagents indicate a dihydroxylation reaction, so the net result will be the addition of OH and OH across the alkene. The regiochemical outcome is not relevant because the two groups added (OH and OH) are identical. We expect the reaction to proceed via a *syn* addition. In this case, two chirality centers are formed, so we expect only the pair of enantiomers resulting from a *syn* addition:

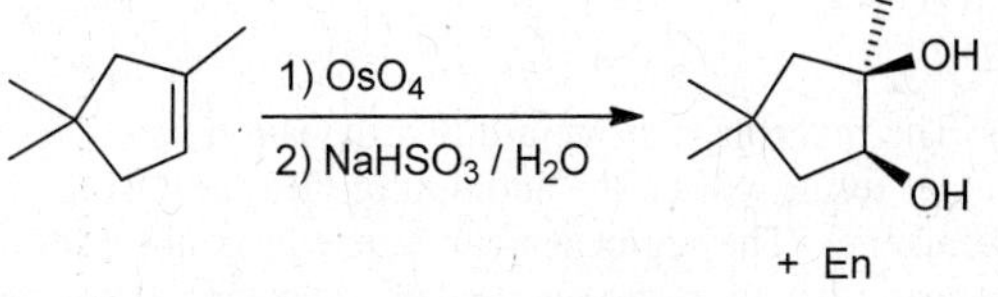

(e) The reagent indicates an acid-catalyzed hydration, so the net result will be the addition of H and OH across the alkene. We expect a Markovnikov addition, so the OH group will be placed at the more-substituted position. Only one chirality center is formed, so we expect the following pair of enantiomers:

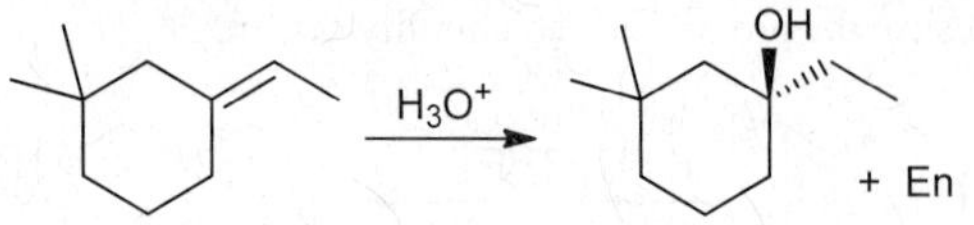

(f) The reagent indicates a hydrobromination reaction, so the net result will be the addition of H and Br across the alkene. We expect a Markovnikov addition, so the Br group will be placed at the more-substituted position. No chirality centers are formed in this case, so stereochemistry is irrelevant:

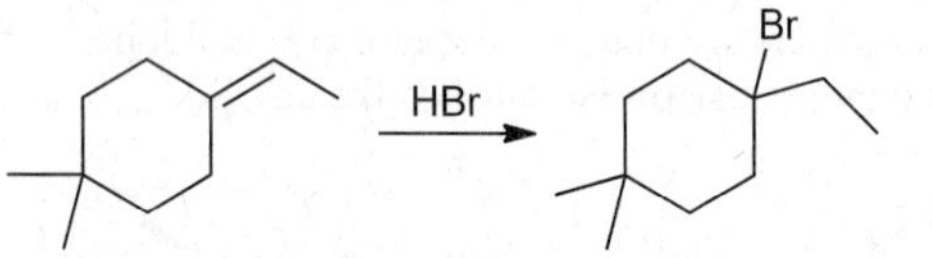

(g) The reagents indicate a dihydroxylation process (via an epoxide), so the net result will be the addition of OH and OH across the alkene. The regiochemical outcome is not relevant because the two groups added (OH and OH) are identical. We expect the reaction to proceed via an *anti* addition. In this case, two chirality centers are formed, so we expect only the pair of enantiomers resulting from an *anti* addition:

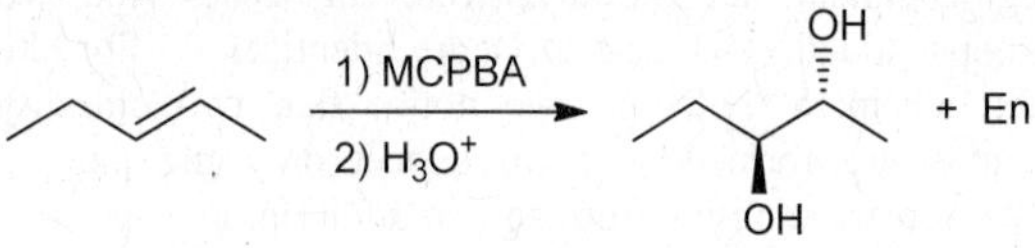

(h) The reagents indicate a hydroboration-oxidation, so the net result will be the addition of H and OH across the alkene. For the regiochemical outcome, we expect an *anti*-Markovnikov addition, so the OH group is placed at the less-substituted position. We expect the reaction to proceed via a *syn* addition, but only one chirality center is formed, so we expect both enantiomers (*syn* addition can occur on either face of the starting alkene):

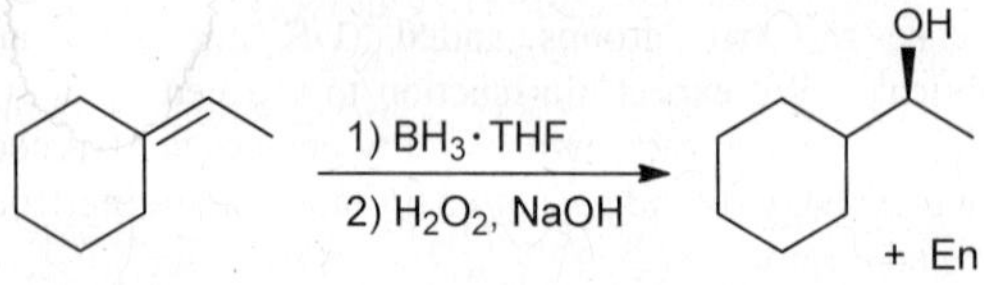

(i) The reagents indicate a dihydroxylation reaction, so the net result will be the addition of OH and OH across the alkene. The regiochemical outcome is not relevant because the two groups added (OH and OH) are identical. We expect the reaction to proceed via a *syn* addition. In this case, two chirality centers are formed,

so we expect only the pair of enantiomers resulting from a *syn* addition:

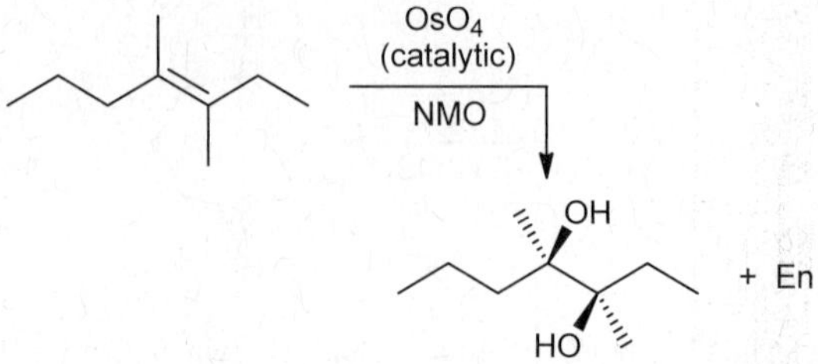

9.37. The net result will be the addition of OH and OH across the alkene. The regiochemical outcome is not relevant because the two groups added (OH and OH) are identical. We expect the reaction to proceed via a *syn* addition. In this case, two chirality centers are formed, so we expect the following two products. Because of the presence of a third chirality center, these two products are diastereomers, rather than enantiomers.

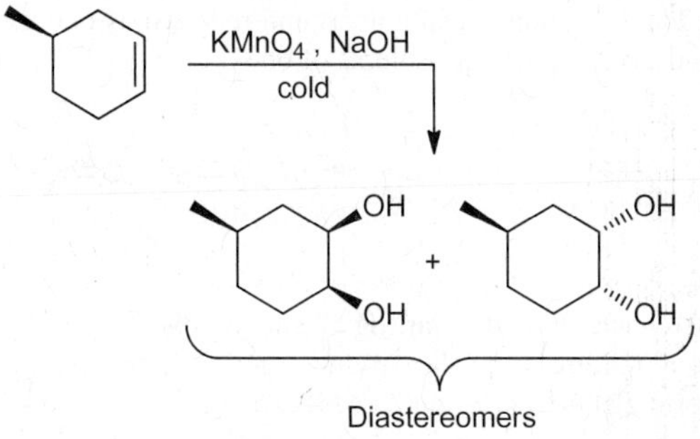

Diastereomers

9.38. *syn*-Dihydroxylation of a *trans* alkene results in the same products as *anti*-dihydroxylation of a *cis* alkene, as shown below. The configuration of each chirality center has been assigned to demonstrate that the products are indeed the same for these two reaction sequences:

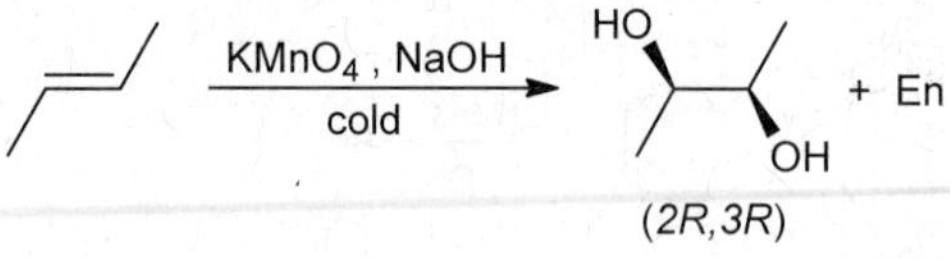

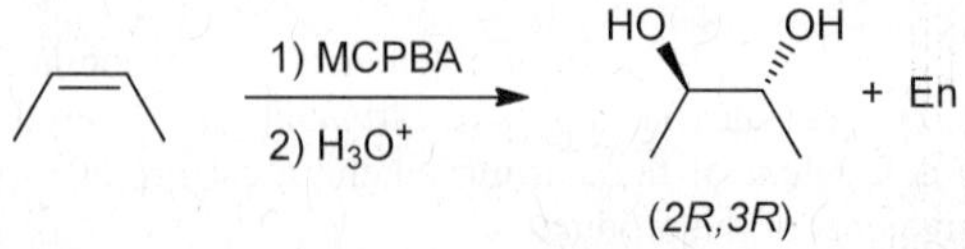

9.39. Compound A must be an alkene (because it undergoes reactions that are typically observed for alkenes, such as hydroboration-oxidation, hydrobromination and ozonolysis). So, we begin by drawing all possible alkenes with molecular formula

C_5H_{10} (using the methodical approach described in the solution to Problem 4.3):

Five-carbon chain

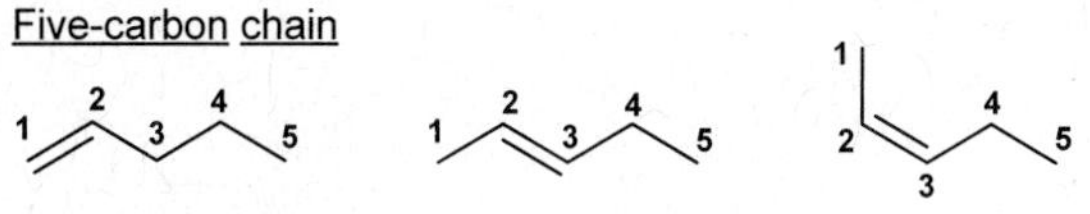

Four-carbon chain

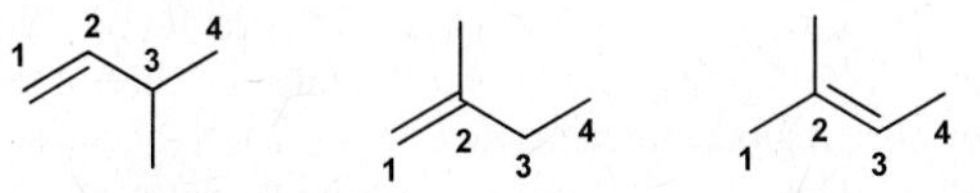

Among these isomers, only the last two will afford a tertiary alkyl halide upon treatment with HBr. And among these two alkenes, only the latter will undergo ozonolysis to produce a compound with three carbon atoms and another compound with two carbon atoms. Now that we have identified the starting alkene, we can draw the products B-F, as shown here:

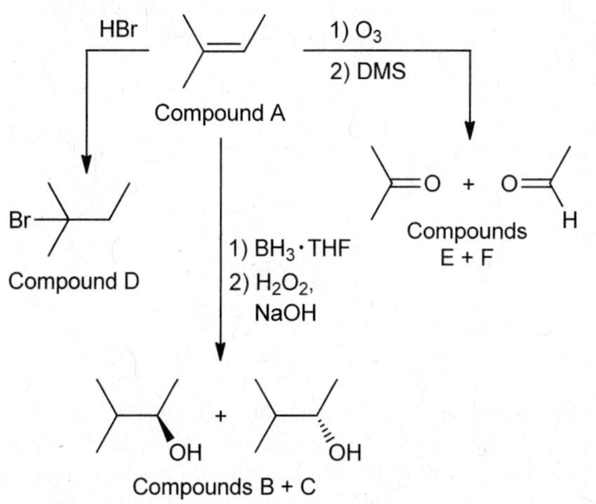

9.40.
(a) The two groups being added across the alkene are H and OH. The OH group is installed at the less-substituted carbon atom, so we must use conditions that give an *anti*-Markovnikov addition of H and OH. This can be accomplished via hydroboration-oxidation. The reaction proceeds via a *syn* addition, which can occur on either face of the alkene, giving a pair of enantiomers:

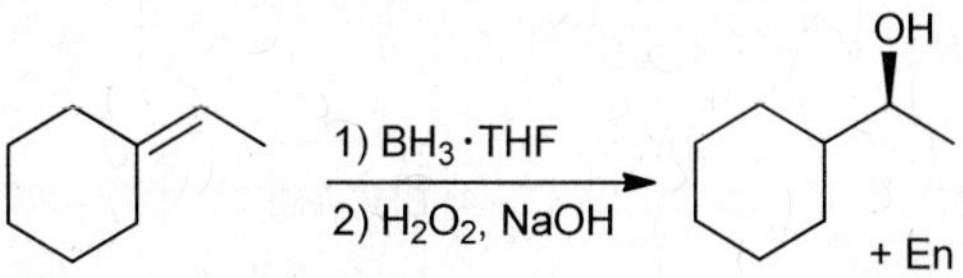

(b) This reaction involves elimination of H and Br to give the less-substituted alkene, so a sterically hindered base is required:

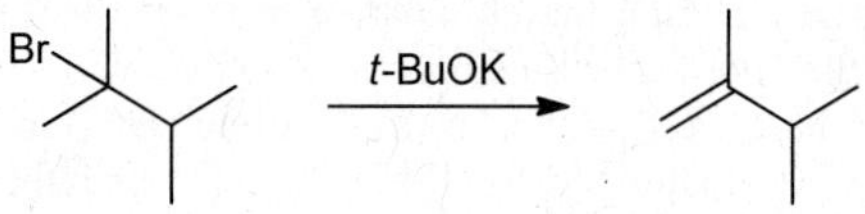

(c) The two groups being added across the alkene are H and Br. The Br group is installed at the less-substituted carbon atom, so we must use conditions that give an *anti*-Markovnikov addition of H and Br. This can be accomplished by treating the alkene with HBr in the presence of peroxides.

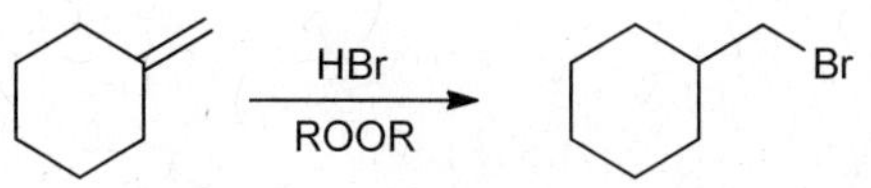

(d) The two groups being added across the alkene are H and H, which can be accomplished by treating the alkene with molecular hydrogen (H_2) in the presence of a suitable catalyst.

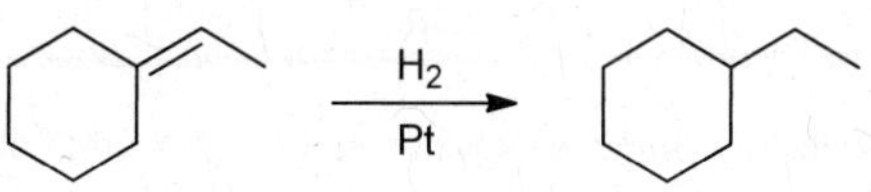

(e) The two groups being added across the alkene are H and Cl. The Cl group is installed at the more-substituted carbon atom, so we must use conditions that give a Markovnikov addition of H and Cl. This can be accomplished by treating the alkene with HCl.

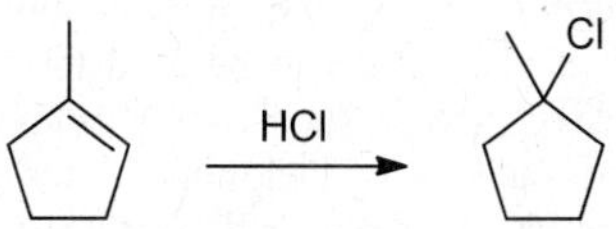

(f) The two groups being added across the alkene are H and OH. The OH group is installed at the less-substituted carbon atom, so we must use conditions that give an *anti*-Markovnikov addition of H and OH. Also, the H and OH are added in a *syn* fashion (this can be seen more clearly if you draw the H that was installed, as shown below). This can be accomplished via hydroboration-oxidation:

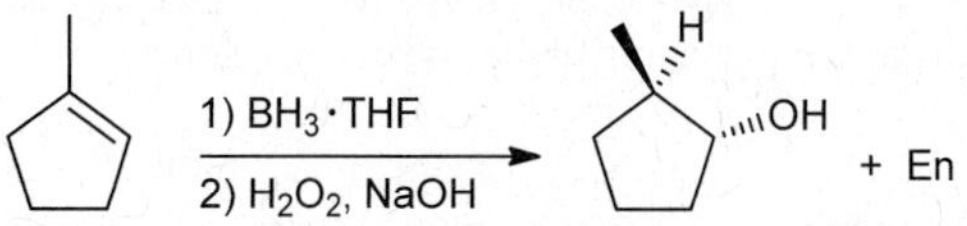

(g) This reaction involves elimination of H and Br to give the more-substituted alkene, so we must use a strong base that is not sterically hindered. We can use hydroxide, methoxide or ethoxide as the base. All of these bases are suitable, as the substrate is tertiary so S_N2 reactions will not compete.

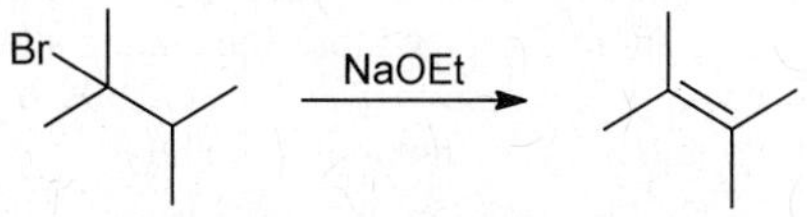

(h) The two groups being added across the alkene are H and Br. The Br group is installed at the more-substituted carbon atom, so we must use conditions that give a

Markovnikov addition of H and Br. This can be accomplished by treating the alkene with HBr.

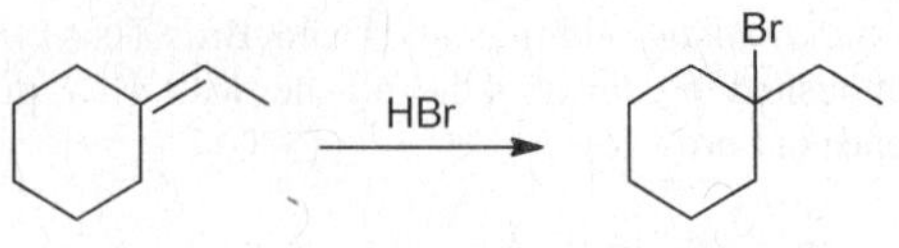

9.41.
(a) First draw the starting alkene. Treating this alkene with HBr will result in a tertiary alkyl halide. But if peroxides are present, a radical process will occur, resulting in the formation of a secondary alkyl halide, as shown:

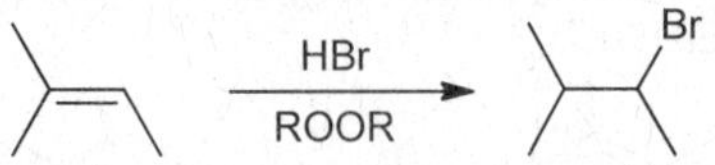

(b) First draw the starting alkene. Treating this alkene with HBr will result in a tertiary alkyl halide, as shown:

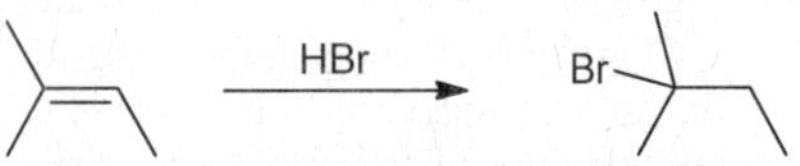

(c) First draw the starting alkene. A *syn* dihydroxylation is required in order to produce a *meso* diol. We have seen several reagents that can be used to accomplish a *syn* dihydroxylation, such as cold potassium permanganate (as shown below). Alternatively, we could achieve the same result with osmium tetroxide and a suitable co-oxidant.

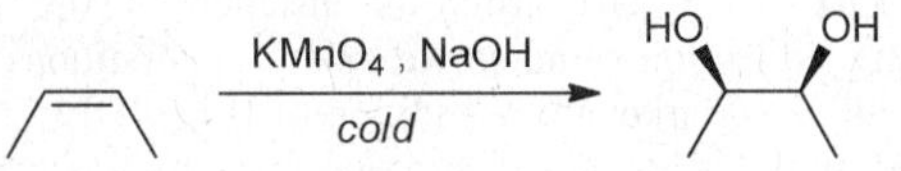

(d) First draw the starting alkene. An *anti* dihydroxylation is required in order to prepare enantiomeric diols. This can be accomplished by converting the alkene into an epoxide, followed by acid-catalyzed ring opening of the epoxide, as shown.

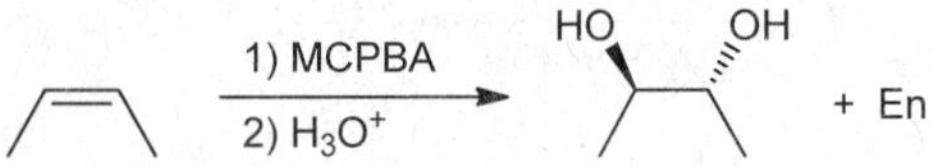

9.42.
(a) The desired transformation can be achieved via a two-step process (elimination, followed by addition). We must be careful to control the regiochemical outcome of each of these processes. During the elimination process, we want to form the more-substituted alkene, so we must use a strong base that is not sterically hindered (such as hydroxide, methoxide, or ethoxide). Then, during the addition process, we want to add HCl in a Markovnikov fashion (with the Cl being installed at the more-substituted position). This can be accomplished by treating the alkene with HCl, as shown here:

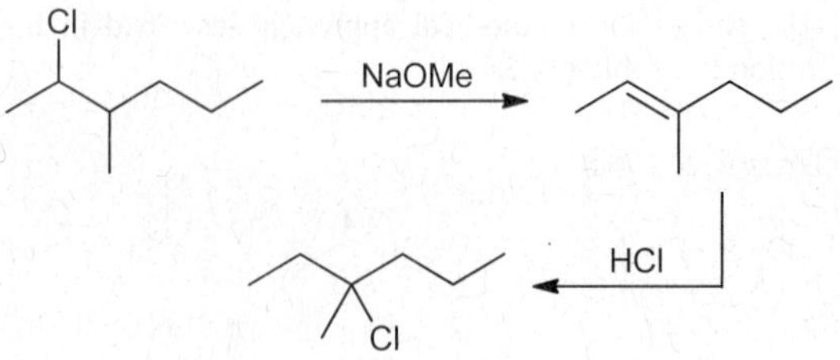

(b) The desired transformation can be achieved via elimination, followed by addition. We must be careful to control the regiochemical outcome of each of these processes. During the elimination process, we want to form the less-substituted alkene, so we must use a strong, sterically hindered base, such as potassium *tert*-butoxide. Notice that the substrate is an alcohol, so we must first convert the OH group (bad leaving group) into a tosylate group (good leaving group) before performing the elimination process. Then, during the addition process, we want to add H and OH in an *anti*-Markovnikov fashion (with the OH being installed at the less-substituted position). This can be accomplished via hydroboration-oxidation, as shown here:

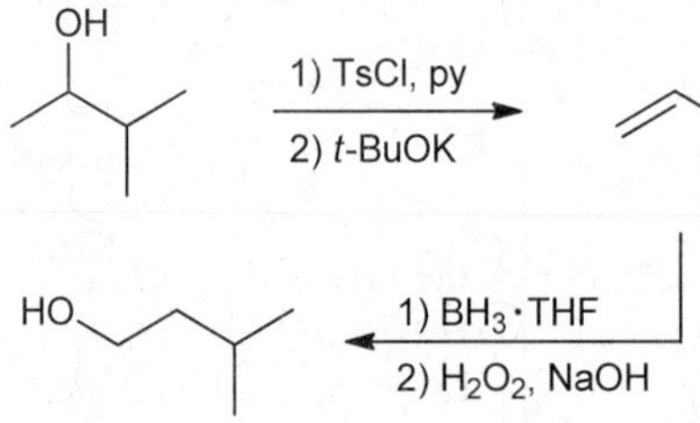

(c) The desired transformation can be achieved via elimination, followed by addition. We must be careful to control the regiochemical outcome of each of these processes. During the elimination process, we want to form the less-substituted alkene, so we must use a strong, sterically hindered base, such as potassium *tert*-butoxide. Then, during the addition process, we want to add H and OH in an *anti*-Markovnikov fashion (with the OH being installed at the less-substituted position). This can be accomplished via hydroboration-oxidation, as shown here:

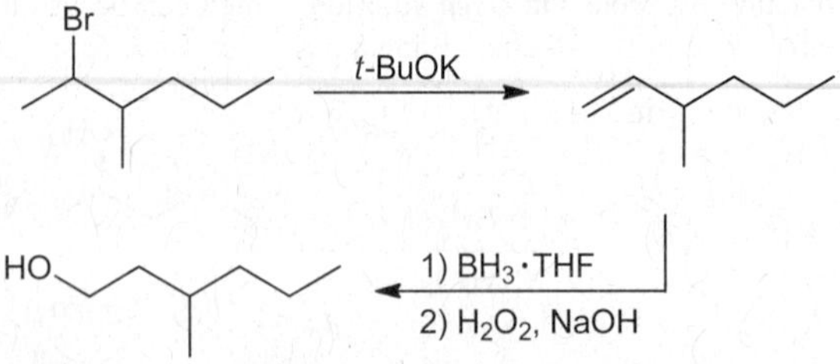

(d) The desired transformation can be achieved via elimination, followed by addition. We must be careful to control the regiochemical outcome of each of these processes. During the elimination process, we want to form the more-substituted alkene, so we will need a strong base that is not sterically hindered (such as hydroxide, methoxide, or ethoxide). Notice that the substrate is an alcohol, so we must first convert the OH group (bad leaving group) into a tosylate group (good leaving group) before performing the elimination process. Alternatively, we can simply perform the elimination process in one step by treating the alcohol with concentrated aqueous sulfuric acid (via an E1 process). Then, during the addition process, we want to add H and OH in an *anti*-Markovnikov fashion (with the OH being installed at the less-substituted position) via a *syn* addition (this can be seen more clearly if you draw the H that is installed, as shown). This can be accomplished via hydroboration-oxidation:

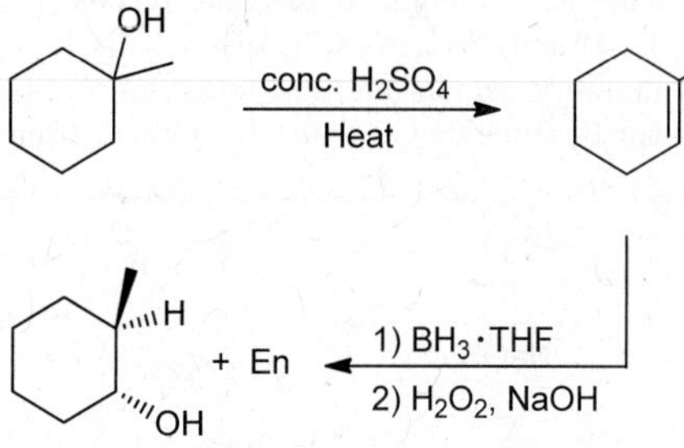

9.43.

(a) Begin by drawing the starting alkyl halide. This tertiary alkyl halide can be converted into a primary alkyl halide via a two-step process (elimination followed by addition). In each case, we must carefully consider the regiochemical outcome. During the elimination process, there is only one regiochemical outcome, so any strong base will work (even if it is sterically hindered, although that is not necessary). In the addition process, we want to install Br at the less-substituted position, so we will need an *anti*-Markovnikov addition of HBr (using peroxides):

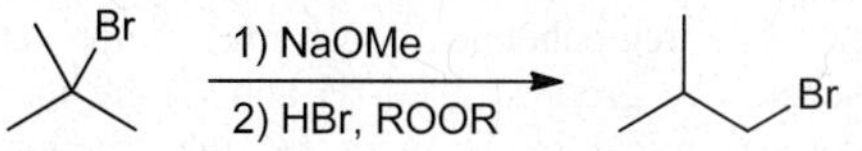

(b) Begin by drawing the starting alkyl halide. This secondary alkyl halide can be converted into a primary alkyl halide via a two-step process (elimination followed by addition). In each case, we must carefully consider the regiochemical outcome. During the elimination process, there is only one regiochemical outcome, so any strong base will work (even if it is sterically hindered). In fact, in this case, there is a distinct advantage to using a sterically hindered base. Specifically, it will suppress the competing S_N2 process (the substrate is secondary, so S_N2 should be a minor product, unless a sterically hindered base is used). During the addition process, we want to install Br at the less-substituted position, so we will need an *anti*-Markovnikov addition of HBr (using peroxides):

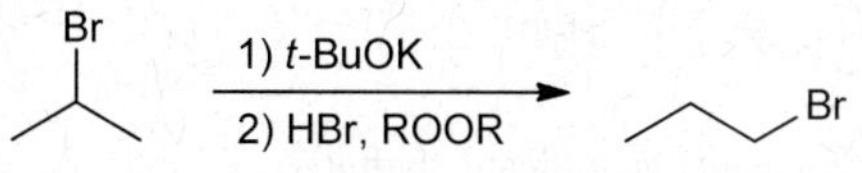

9.44.
The two-step process (elimination followed by addition) must be used twice in this case. First, we use a strong base to give an elimination reaction, followed by

Markovnikov addition of HBr. Then, we repeat the two-step procedure, but this time, the elimination process must be performed with a sterically hindered base (to give the less-substituted alkene), and the addition process must be performed in the presence of peroxides to give an *anti*-Markovnikov addition of HBr, as shown.

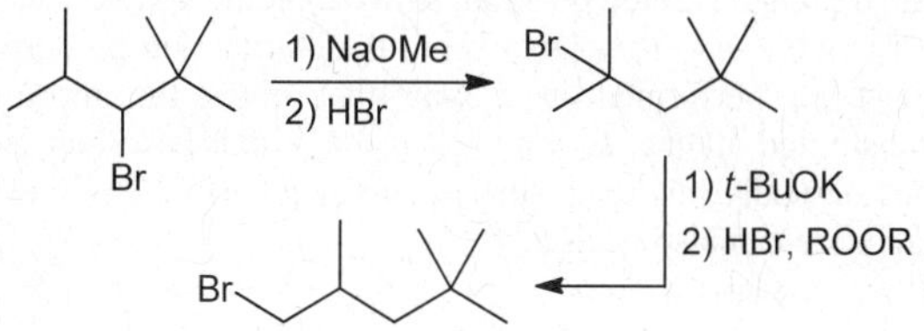

9.45.

(a) The desired transformation can be achieved via a two-step process (addition, followed by elimination). We must be careful to control the regiochemical outcome of each step of the process. During the addition reaction, we want to install the Br at the more-substituted position, so we treat the alkene with HBr (without peroxides present). Then, the elimination process must be performed in a way that gives the more-substituted alkene, so we must use a strong base that is not sterically hindered, such as methoxide (hydroxide or ethoxide can also be used).

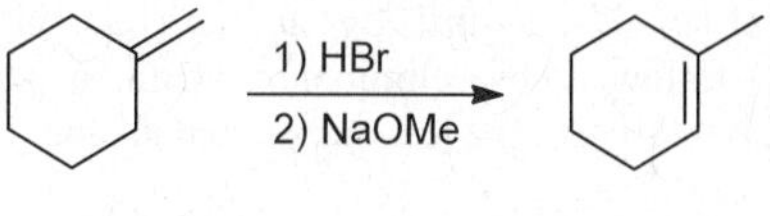

(b) These two alkenes can be interconverted via a two-step process (addition, followed by elimination). We must be careful to control the regiochemical outcome of each step of the process. In one case, a sterically hindered base is required, while in the other case, we must use a base that is not sterically hindered, as shown.

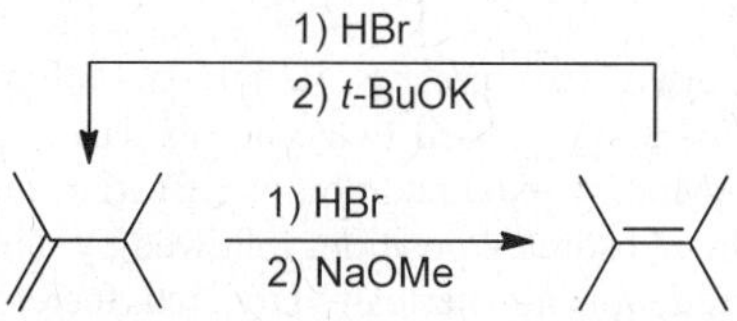

9.46.

(a) Begin by drawing the starting alkene. This trisubstituted alkene can be converted into a monosubstituted alkene via a two-step process (addition, followed by elimination). We must be careful to control the regiochemical outcome of each step of the process. During the addition reaction, we want to install the Br at the less-substituted position, so we treat the alkene with HBr in the presence of peroxides. Then, the elimination reaction must be performed in a way that gives the less-substituted alkene, so we must use a strong, sterically hindered base (such as *tert*-butoxide).

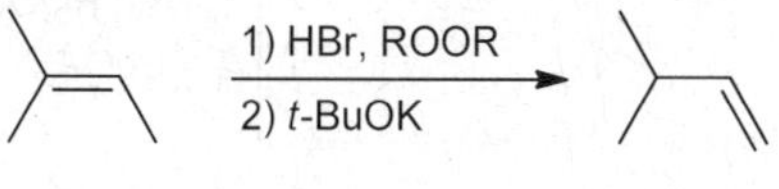

(b) Begin by drawing the starting alkene. This disubstituted alkene can be converted into a tetrasubstituted alkene via a two-step process (addition, followed by elimination). We must be careful to control the regiochemical outcome of each step of the process. During the addition reaction, we want to install the Br at the more-substituted position, so we treat the alkene with HBr (without peroxides). Then, the elimination process must be performed in a way that gives the more-substituted alkene, so we must use a strong base that is not sterically hindered, such as methoxide (hydroxide or ethoxide can also be used).

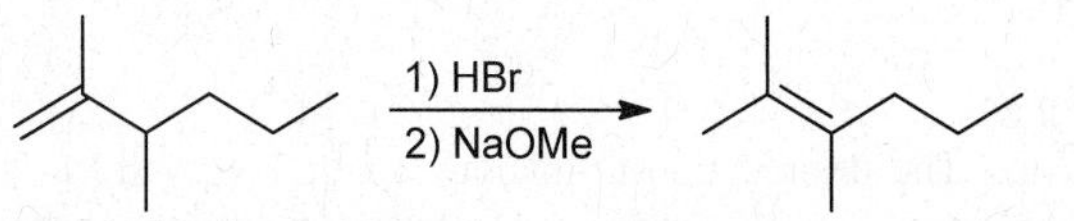

9.47.

(a) The two-step process (addition followed by elimination) must be used twice in this case. First, we perform a Markovnikov addition of HBr to install the Br at the more-substituted position, followed by elimination with a base that is not sterically hindered, thereby giving the more-substituted alkene. Then, we perform the two-step process again. But this time, we begin with an *anti*-Markovnikov addition of HBr (in the presence of peroxides) to install the Br at the less-substituted position, followed by elimination with a sterically hindered base to give the less substituted alkene:

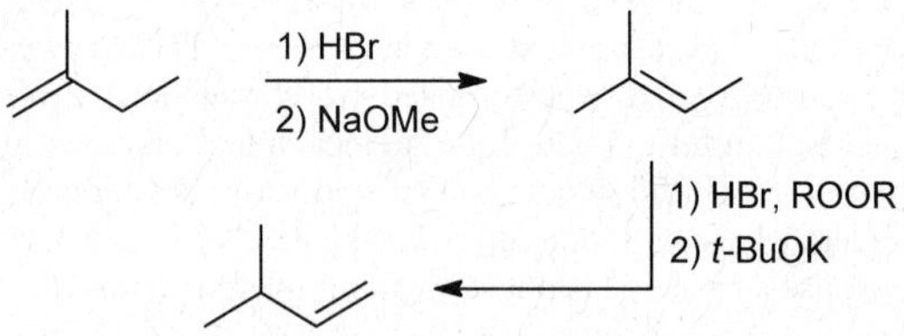

(b) The two-step process (addition followed by elimination) must be used twice in this case. First, we perform a Markovnikov addition of HBr to install the Br at the more-substituted position, followed by elimination with a base that is not sterically hindered, thereby giving the more-substituted alkene. Then, we perform the two-step process again. But this time, we begin with an *anti*-Markovnikov addition of HBr (in the presence of peroxides) to install the Br at the less-substituted position, followed by elimination with a sterically hindered base to give the less substituted alkene.

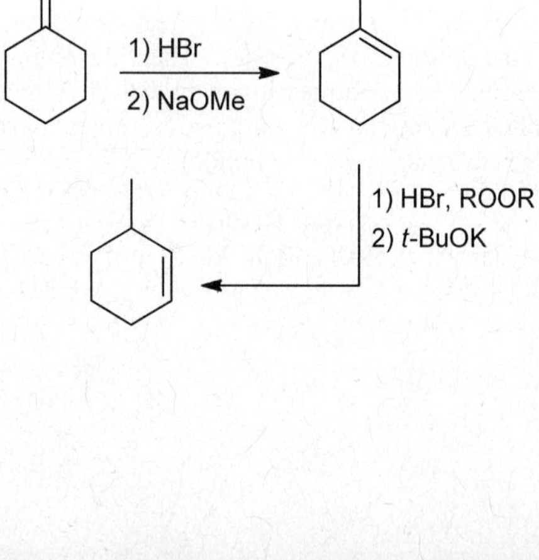

9.48. A reaction is only favorable if ΔG is negative. Recall that ΔG has two components: (ΔH) and (-$T\Delta S$). The first term (ΔH) is positive for this reaction (two σ bonds are converted into one σ bond and one π bond). The second term (-$T\Delta S$) is negative because ΔS is positive (one molecule is converted into two molecules). Therefore, the reaction is only favorable if the second term is greater in magnitude than the first term. This only occurs at high temperature.

9.49.

(a) Treating an alkene with cold potassium permanganate and sodium hydroxide results in the addition of OH and OH across the alkene. In this case, two new chirality centers are generated, so we expect only the pair of enantiomers that would result from *syn* addition:

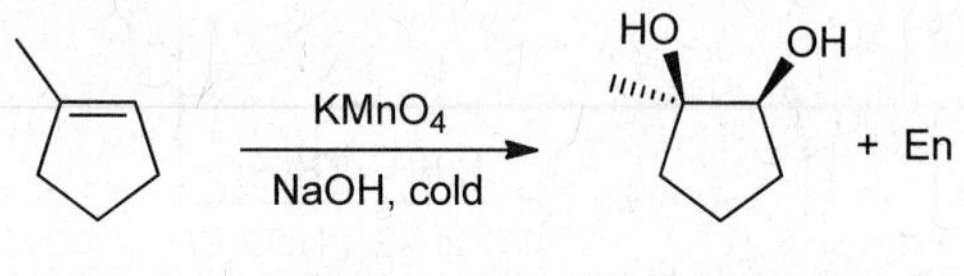

(b) The reagent indicates a hydrochlorination reaction, so the net result will be the addition of H and Cl across the alkene. We expect a Markovnikov addition, so the Cl group will be placed at the more-substituted position. No chirality centers are formed in this case, so stereochemistry is irrelevant:

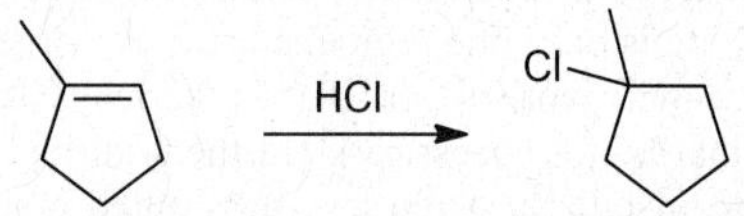

(c) The reagents indicate a hydrogenation reaction, so the net result will be the addition of H and H across the alkene. The regiochemical outcome is not relevant because the two groups added (H and H) are identical. The stereochemical requirement for the reaction (*syn* addition) is not relevant in this case, as no chirality centers are formed:

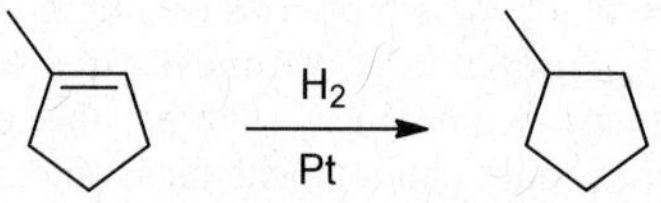

(d) The reagents indicate bromohydrin formation, so the net result will be the addition of Br and OH across the alkene. The OH group is expected to be installed at the more-substituted position. The reaction proceeds via an *anti* addition, giving the following pair of enantiomers:

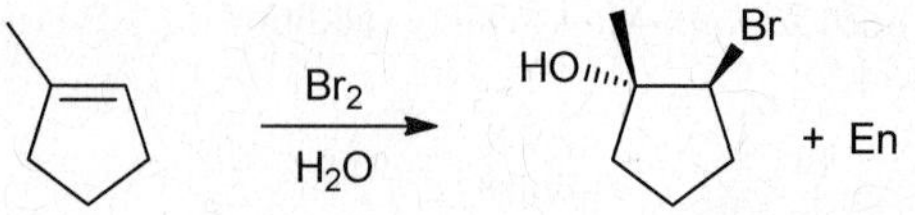

(e) The reagents indicate hydration of the alkene via oxymercuration-demercuration. The net result will be the addition of H and OH across the alkene, with the OH group being installed at the more-substituted position.

The product has no chirality centers, so stereochemistry is not a consideration:

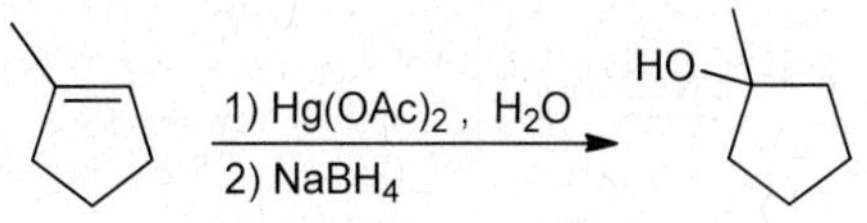

9.50.

(a) The reagents indicate the addition of OH and OH across the alkene. The regiochemical outcome is not relevant because the two groups added (OH and OH) are identical. We expect the reaction to proceed via an *anti* addition, but only one chirality center is formed, so we expect both enantiomers:

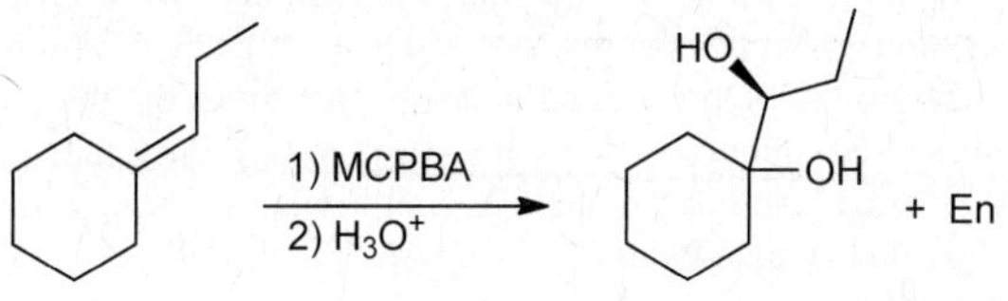

(b) The reagent indicates a hydrobromination reaction, so the net result will be the addition of H and Br across the alkene. We expect a Markovnikov addition, so the Br group will be placed at the more-substituted position. No chirality centers are formed in this case, so stereochemistry is irrelevant:

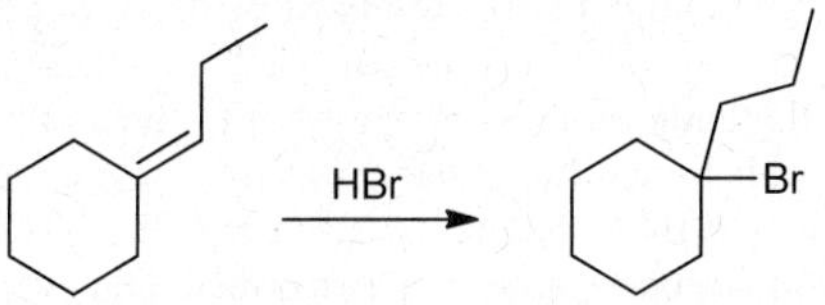

(c) The reagents indicate a hydrogenation reaction, so the net result will be the addition of H and H across the alkene. The regiochemical outcome is not relevant because the two groups added (H and H) are identical. No chirality centers are formed in this case, so stereochemistry is also irrelevant.

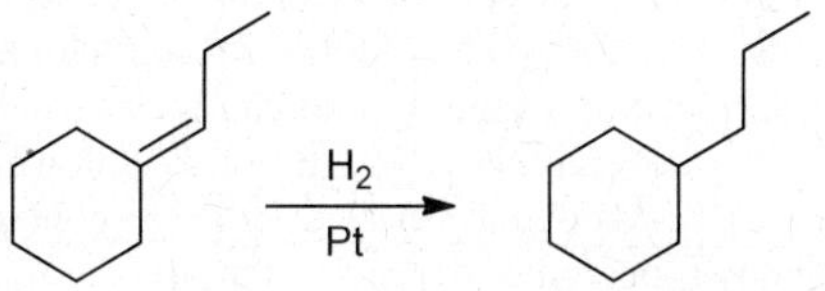

(d) The reagent indicates a bromination reaction, so the net result will be the addition of Br and Br across the alkene. The regiochemical outcome is not relevant because the two groups added (Br and Br) are identical. We expect the reaction to proceed via an *anti* addition, but only one chirality center is formed, so we expect both enantiomers:

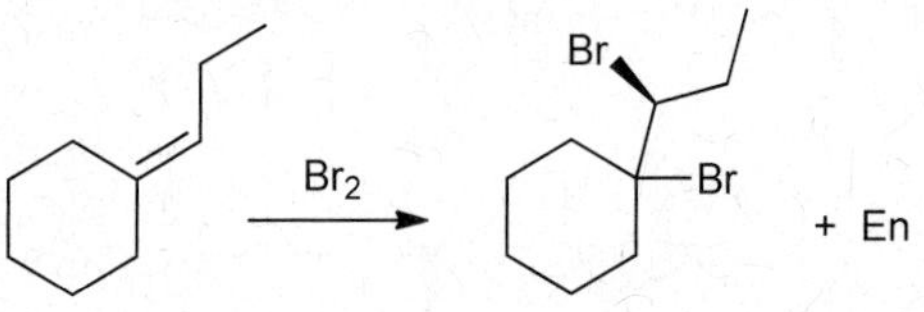

(e) The reagents indicate a hydroboration-oxidation, so the net result will be the addition of H and OH across the alkene. For the regiochemical outcome, we expect an *anti*-Markovnikov addition, so the OH group is placed at the less-substituted (secondary) position. We expect the reaction to proceed via a *syn* addition, but only one chirality center is formed, so we expect both enantiomers (*syn* addition can occur on either face of the alkene):

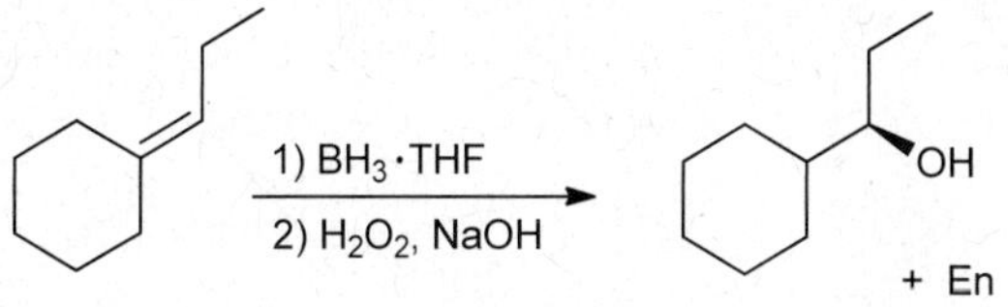

9.51.

(a) Water (H and OH) is added across the alkene in a Markovnikov fashion. The mechanism is expected to have three steps: 1) proton transfer, 2) nucleophilic attack, and 3) proton transfer. In the first step, a proton is transferred from H_3O^+ to the alkene, which requires two curved arrows, as shown below. The resulting tertiary carbocation is then captured by a water molecule in the second step of the mechanism. This step requires one curved arrow, going from the nucleophile (water) to the electrophile (the carbocation). Then, in the final step of the mechanism, a molecule of water functions as a base and removes a proton, thereby generating the product. This final step is a proton transfer step, and therefore requires two curved arrows, as shown:

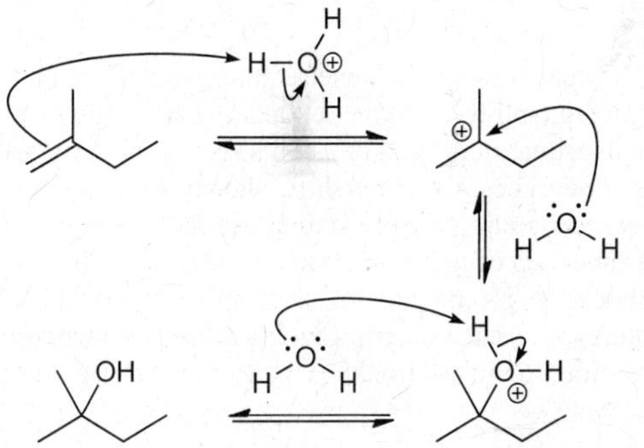

(b) In the first step of the mechanism, a proton is transferred from H_3O^+ to the alkene, which requires two curved arrows, as shown below. The resulting secondary carbocation then rearranges via a hydride shift, giving a more stable, tertiary carbocation. That step is shown with one curved arrow. The tertiary carbocation is then captured by a water molecule, which is shown with one curved arrow, going from the nucleophile (water) to the electrophile (the carbocation). Then, in the final step of the mechanism, a molecule of water functions as a base and removes a proton, thereby generating the product.

This final step is a proton transfer step, and therefore requires two curved arrows, as shown:

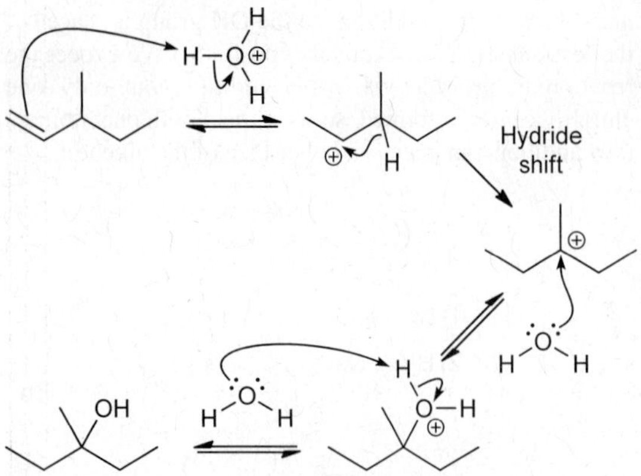

(c) In this reaction, H and Br are added across the alkene in a Markovnikov addition, which indicates an ionic process. There are two mechanistic steps in the ionic addition of HBr across an alkene: 1) proton transfer, followed by 2) nucleophilic attack. In the first step, a proton is transferred from HBr to the alkene, which requires two curved arrows, as shown below. The resulting tertiary carbocation is then captured by a bromide ion in the second step of the mechanism. This step requires one curved arrow, going from the nucleophile to the electrophile, as shown:

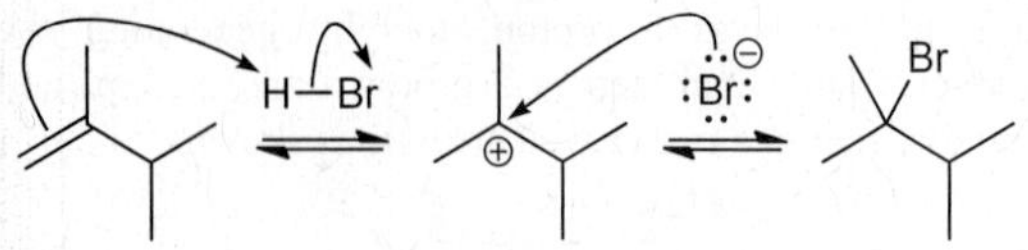

(d) Protonation of the alkene requires two curved arrows, as shown, and leads to the secondary carbocation (rather than a primary carbocation). This secondary carbocation then undergoes a methyl shift, shown with one curved arrow, generating a more stable, tertiary carbocation. In the final step of the mechanism (nucleophilic attack), the carbocation is captured by a bromide ion. This step requires one curved arrow, going from the nucleophile (bromide) to the electrophile (the carbocation), as shown:

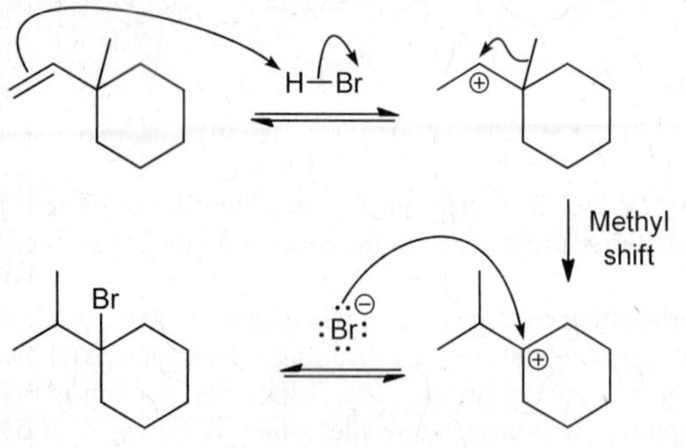

9.52. The starting material (1-bromo-1-methylcyclohexane) is a tertiary alkyl halide, and will undergo an E2 reaction when treated with a strong base such as methoxide, to give the more substituted alkene

(compound A). Hydrogenation of compound A gives methylcyclohexane:

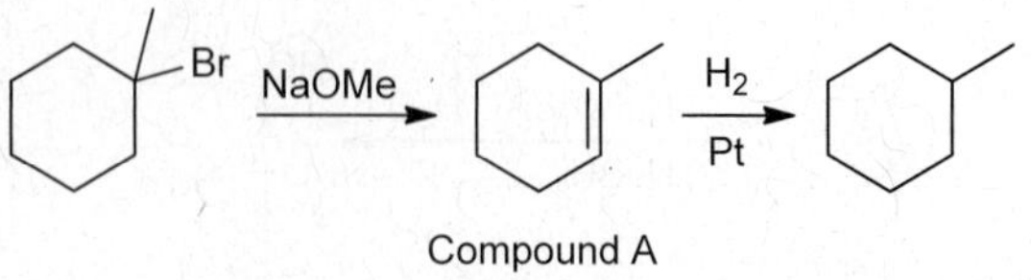

Compound A

9.53.
(a) The desired transformation can be achieved via a two-step process (addition, followed by elimination). We must be careful to control the regiochemical outcome of each step of the process. During the addition reaction, we want to install the Br at the more-substituted (tertiary) position, so we treat the alkene with HBr (without peroxides present). Then, the elimination process must be performed in a way that gives the more-substituted alkene, so we must use a strong base that is not sterically hindered, such as methoxide (hydroxide or ethoxide can also be used).

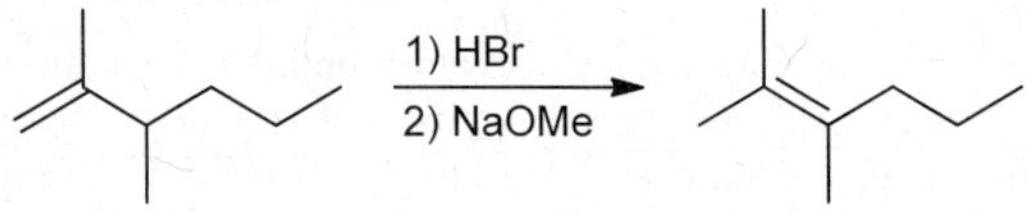

(b) This trisubstituted alkene can be converted into the monosubstituted alkene via a two-step process (addition, followed by elimination). We must be careful to control the regiochemical outcome of each of these steps. During the addition process, we want to install the Br at the less-substituted (secondary) position, so we treat the alkene with HBr in the presence of peroxides. Then, the elimination process must be performed in a way that gives the less-substituted alkene, so we must use a strong, sterically hindered base (such as *tert*-butoxide).

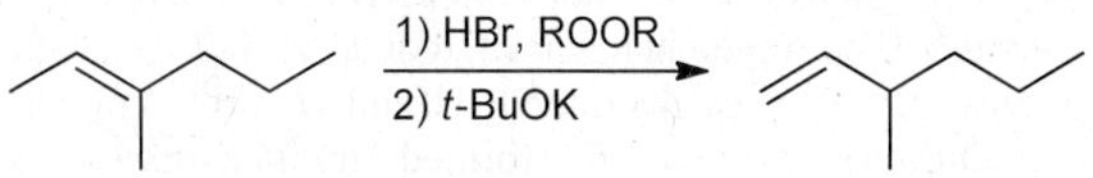

9.54. Treatment of the starting alcohol with concentrated sulfuric acid affords the more substituted alkene. Moving the position of the π bond can then be achieved via a two step-process (addition, followed by elimination). We must be careful to control the regiochemical outcome of each step of the process. During the addition reaction, we want to install the Br at the less-substituted (secondary) position, so we treat the alkene with HBr in the presence of peroxides. Then, the elimination reaction must be performed in a way that gives the less-substituted alkene, so we must use a strong, sterically hindered base (such as *tert*-butoxide).

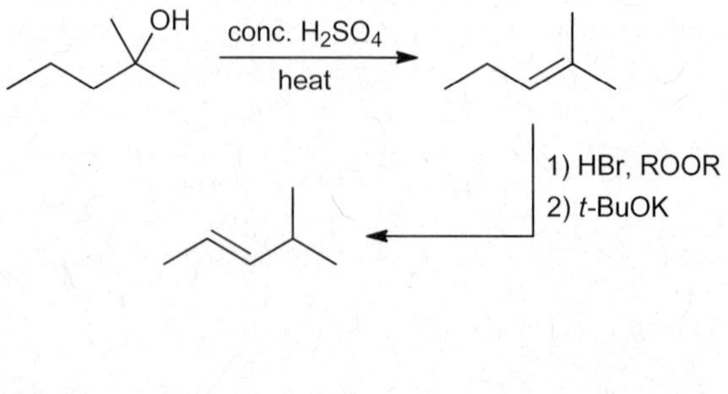

9.55. Two different alkenes will produce 2,4-dimethylpentane upon hydrogenation:

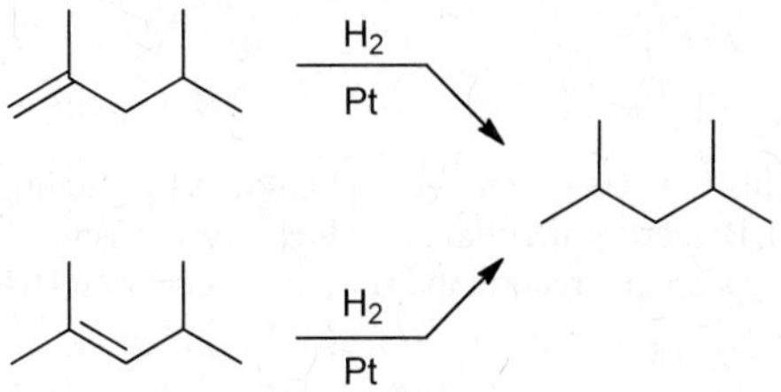

Note that the following four drawings all represent the same compound:

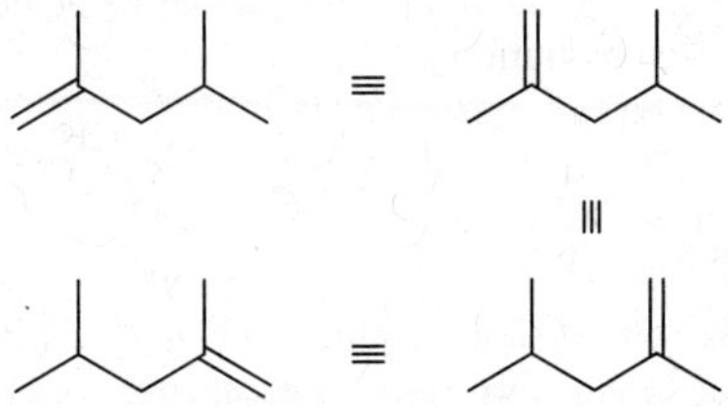

9.56. We must first determine the structure of compound **A**. The necessary information has been provided. Specifically, ozonolysis of compound **A** gives only one product, which has only one C=O bond. Therefore, the starting alkene must be symmetrical, leading to two equivalents of the product:

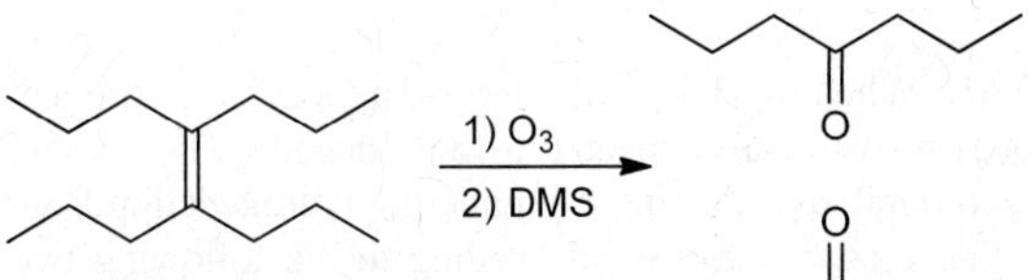

Treatment of compound **A** with a peroxy acid (such as MCPBA), followed by aqueous acid, affords a diol. No chirality centers are formed, so stereochemistry is not a relevant consideration.

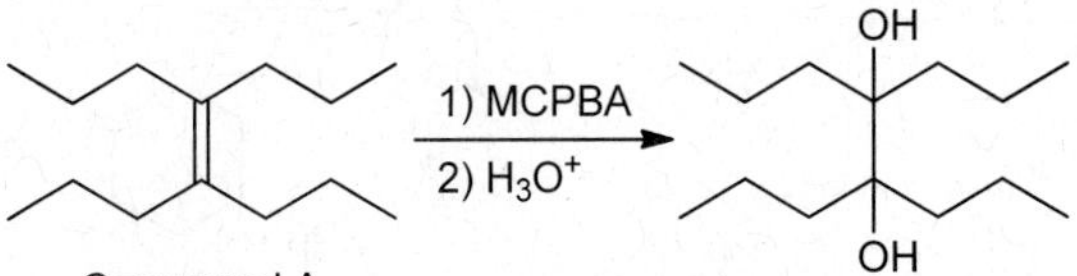

9.57.
(a) Interconversion between the two alcohols requires moving the position of the OH group. In each case, this can be accomplished via a two step-process (elimination followed by addition). In each case, the elimination step can be achieved by treating the alcohol with concentrated sulfuric acid. For the addition step, the regiochemical outcome must be carefully considered. In the first case below, dilute aqueous acid is used to give a Markovnikov addition, while in the second case below,

hydroboration-oxidation is employed to give an *anti*-Markovnikov addition.

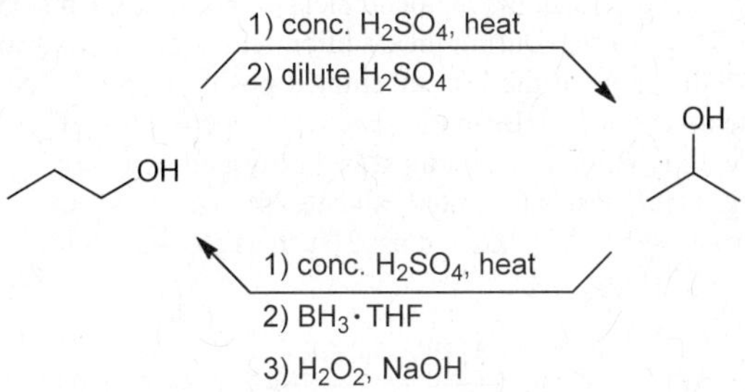

(b) Interconversion between the two alkyl halides requires moving the position of the Br group. In each case, this can be accomplished via a two step-process (elimination followed by addition). In each case, the elimination step can be achieved via an E2 reaction, using a strong base (such as hydroxide, or methoxide or ethoxide) to give the more substituted alkene. For the addition step, the regiochemical outcome must be carefully considered. In the first case below, HBr and peroxides are used to give an *anti*-Markovnikov addition, while in the second case below, HBr is used to give a Markovnikov addition.

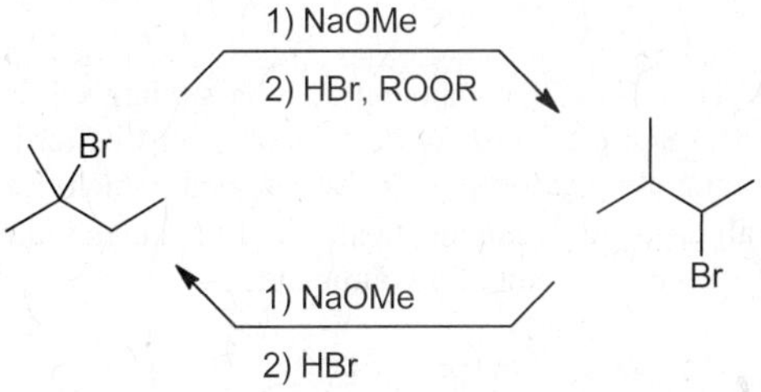

(c) Treating the starting material with a strong base (such as hydroxide, methoxide or ethoxide) gives the more substituted (tetrasubstituted) alkene, which can then be converted to the desired *meso* compound upon hydrogenation.

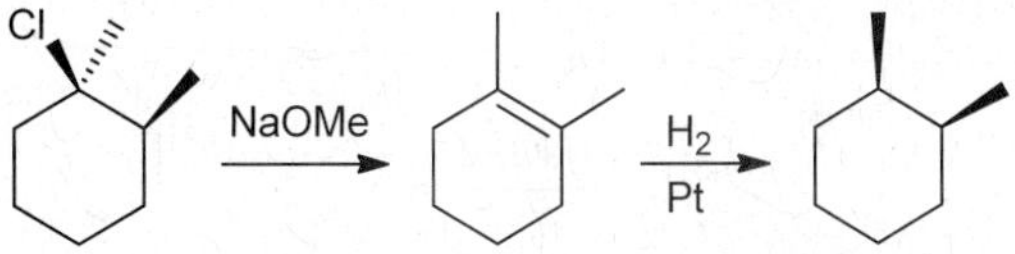

(d) The product is a *cis*-diol which can be prepared via a *syn* dihydroxylation. The necessary alkene (cyclohexene) can be made in one step from the starting alcohol, upon treatment with concentrated sulfuric acid (an E1 reaction):

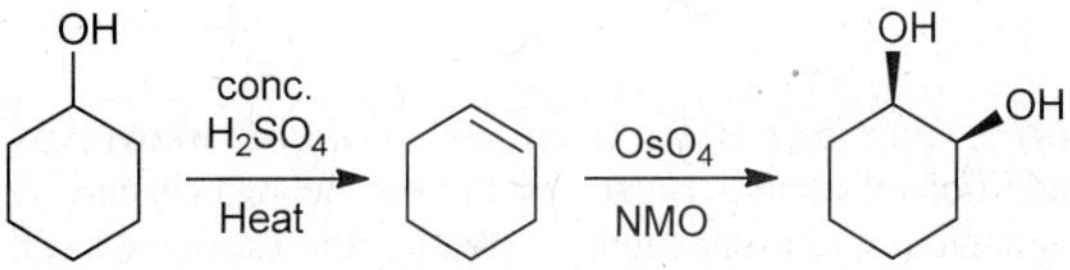

9.58.

(a) This conversion can be achieved via a two-step process (addition, followed by elimination). We must be careful to control the regiochemical outcome of each step of the process. During the addition reaction, we want to install the Br at the less-substituted position, so we treat the alkene with HBr in the presence of peroxides. Then, the elimination reaction must be performed in a way that gives the less-substituted alkene, so we must use a strong, sterically hindered base (such as *tert*-butoxide).

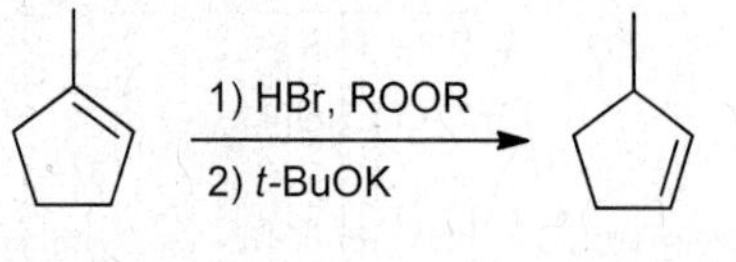

(b) This conversion can be accomplished via a two step-process (elimination followed by addition). The elimination step can be achieved by treating the alcohol with concentrated sulfuric acid. For the addition step, dilute aqueous acid can be used to give a Markovnikov addition.

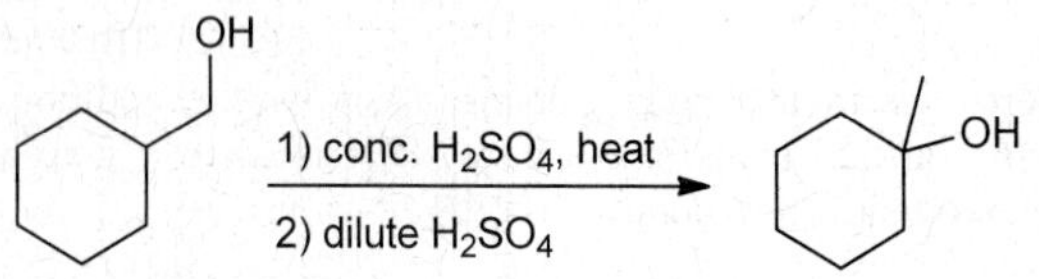

9.59. Treatment of compound **A** with sodium ethoxide gives no S$_N$2 products, so the substrate must be tertiary. Only one elimination product is obtained, which means that all β positions are identical. These features indicate the following structure for compound A:

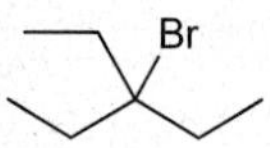

Compound A

Treatment of compound A with ethoxide gives alkene B, shown below, which undergoes acid-catalyzed hydration (Markovnikov addition of water) to give alcohol C:

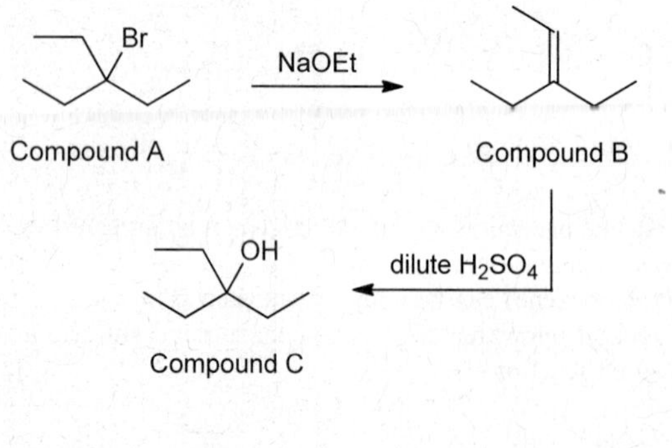

Compound A   Compound B

Compound C

9.60.

(a) This conversion requires an *anti*-Markovnikov addition of H and Br across the alkene, which can be achieved in just one step, by treating the starting alkene with HBr in the presence of peroxides:

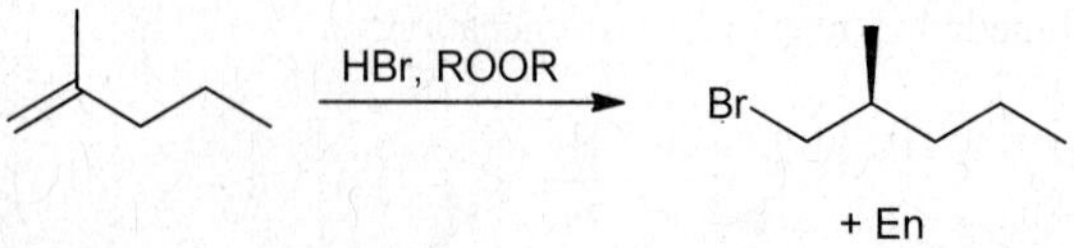

(b) This conversion requires a Markovnikov addition of H and Br across the alkene, which can be achieved in just one step, by treating the starting alkene with HBr:

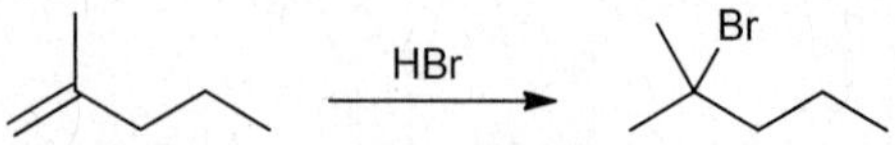

(c) This conversion requires a Markovnikov addition of H and OH across the alkene, which can be achieved via acid-catalyzed hydration:

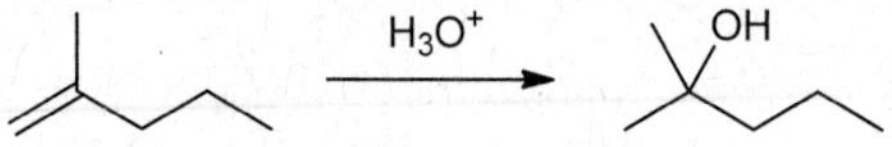

(d) This conversion requires an *anti*-Markovnikov addition of H and OH across the alkene, which can be achieved via hydroboration-oxidation. The process does proceed via *syn* addition, but only one chirality center is formed, and the *syn* addition can take place on either face of the alkene, giving a pair of enantiomers.

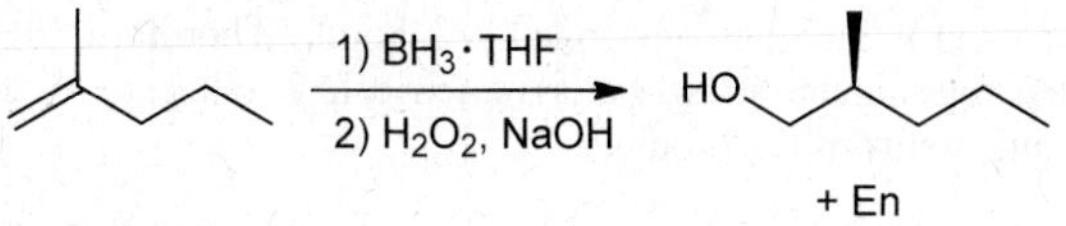

9.61. When treated with excess molecular hydrogen, both π bonds are expected to be reduced. The π bond incorporated in the ring can undergo hydrogenation from either face of the π bond, leading to the following two compounds. These disubstituted cyclohexanes are diastereomers because they are stereoisomers that are not mirror images of each other.

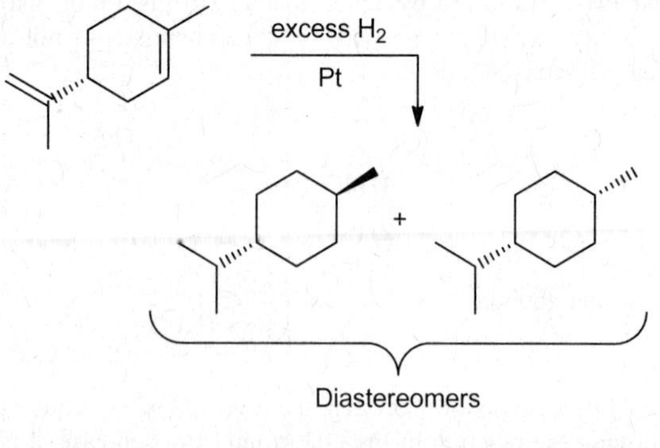

Diastereomers

9.62. This conversion requires the Markovnikov addition of water *without* carbocation rearrangement. This can be achieved via oxymercuration-demercuration:

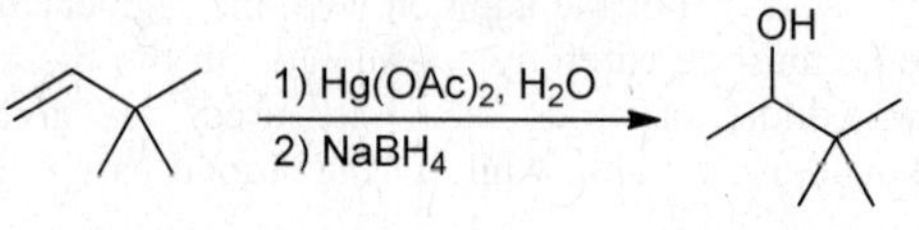

racemic

9.63. In the presence of acid, the epoxide is first protonated, which requires two curved arrows, as shown below. The resulting intermediate is then attacked by a molecule of methanol, which functions as a nucleophile. This step requires two curved arrows. Then, in the final step of the mechanism, a molecule of methanol functions as a base and removes a proton, thereby generating the product. This final step is a proton transfer step, and therefore requires two curved arrows:

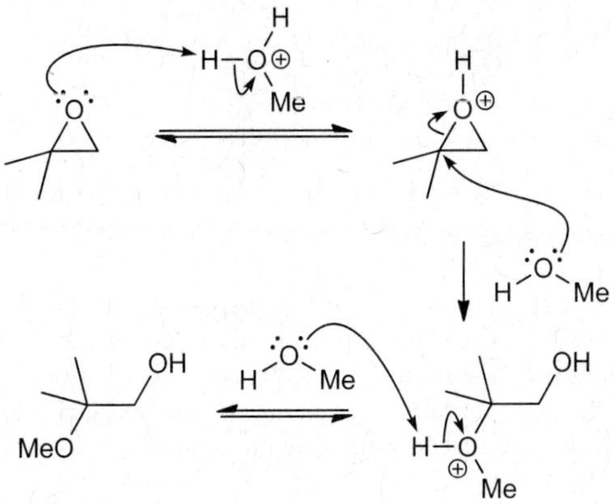

9.64.
(a) In the first step of the mechanism, a proton is transferred from H_3O^+ to the alkene, which requires two curved arrows, as shown below. The resulting secondary carbocation then rearranges via a methyl shift, giving a more stable, tertiary carbocation. That step is shown with one curved arrow. The tertiary carbocation is then captured by a water molecule, which is shown with one curved arrow, going from the nucleophile (water) to the electrophile (the carbocation). Then, in the final step of the mechanism, a molecule of water functions as a base and removes a proton, thereby generating the product. This final step is a proton transfer step, and therefore requires two curved arrows, as shown:

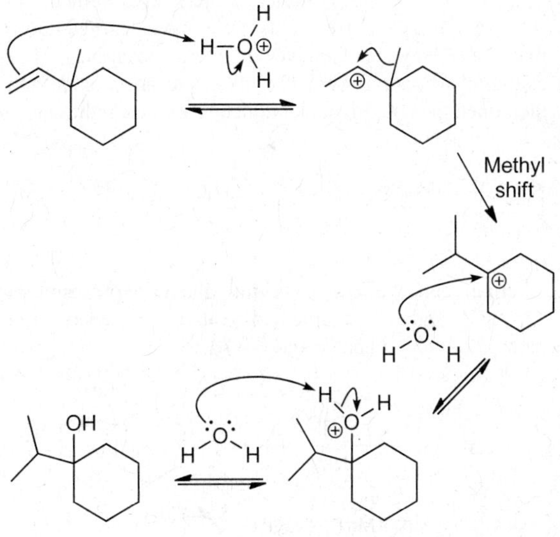

(b) In the first step, the alkene is protonated, which requires two curved arrows. The resulting, tertiary

carbocation is then captured by a bromide ion in the second step of the mechanism. This step requires one curved arrow, going from the nucleophile to the electrophile, as shown:

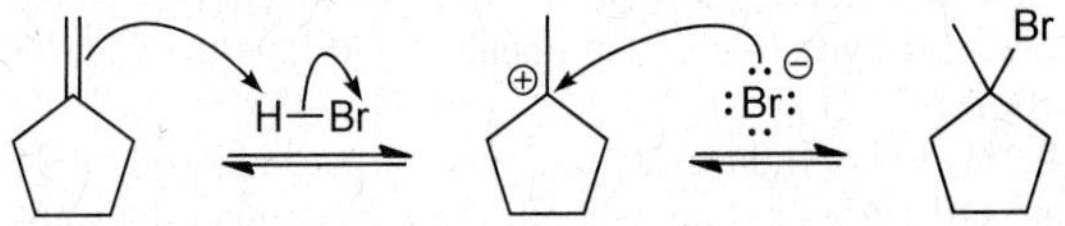

9.65.
(a) The reagents indicate a hydrogenation reaction, so the net result will be the addition of H and H across the alkene. The regiochemical outcome is not relevant because the two groups added (H and H) are identical. We expect the reaction to proceed via a *syn* addition, giving the following *meso* compound:

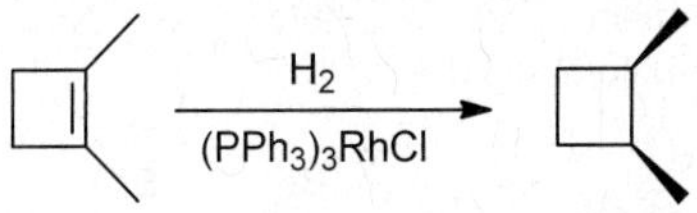

(b) The reagents indicate an acid-catalyzed hydration, so the net result will be the addition of H and OH across the alkene. We expect a Markovnikov addition, so the OH group will be placed at the more-substituted position. No chirality centers are formed in the process, so stereochemistry is not a relevant consideration:

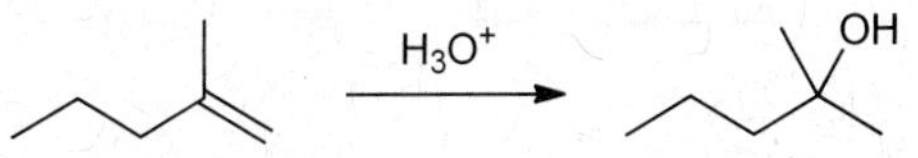

(c) The reagents indicate a hydroboration-oxidation, so the net result will be the addition of H and OH across the alkene. For the regiochemical outcome, we expect an *anti*-Markovnikov addition, so the OH group is placed at the less-substituted position. The stereochemical outcome (*syn* addition) is not relevant in this case, because the product has no chirality centers:

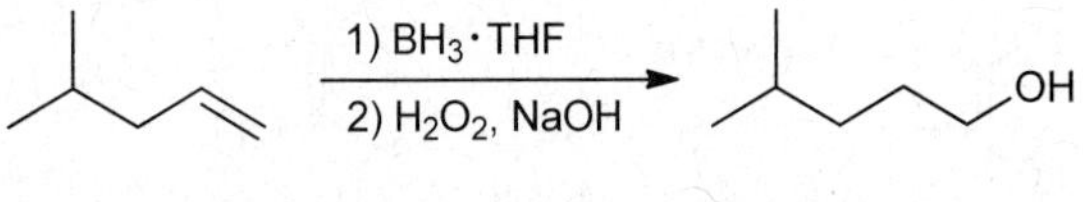

(d) The reagents indicate a dihydroxylation process (via an epoxide), so the net result will be the addition of OH and OH across the alkene. The regiochemical outcome is not relevant because the two groups added (OH and OH) are identical. We expect the reaction to proceed via an *anti* addition. In this case, two chirality centers are formed, so we expect the pair of enantiomers resulting from an *anti* addition:

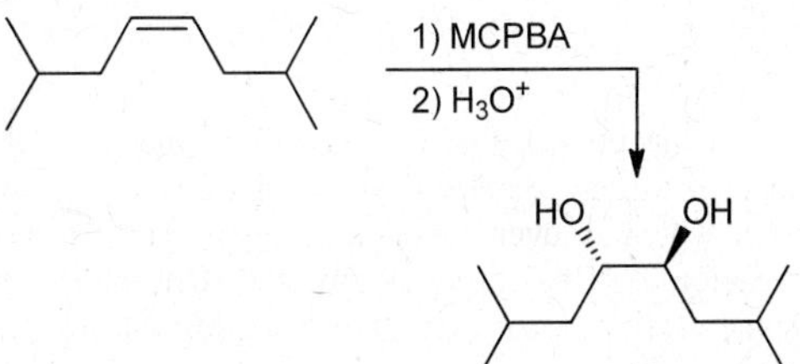

9.66.

(a) Hydroboration-oxidation gives an *anti*-Markovnikov addition. If 1-propene is the starting material, the OH group will not be installed in the correct location. Acid-catalyzed hydration of 1-propene would give the desired product.

(b) Hydroboration-oxidation gives a *syn* addition of H and OH across a double bond. This compound does not have a proton that is *cis* to the OH group, and therefore, hydroboration-oxidation cannot be used to make this compound.

(c) Hydroboration-oxidation gives an *anti*-Markovnikov addition. There is no starting alkene that would yield the desired product via an *anti*-Markovnikov addition.

9.67. Bromination of *cis*-2-butene does NOT give the desired *meso* compound:

NOT *meso*

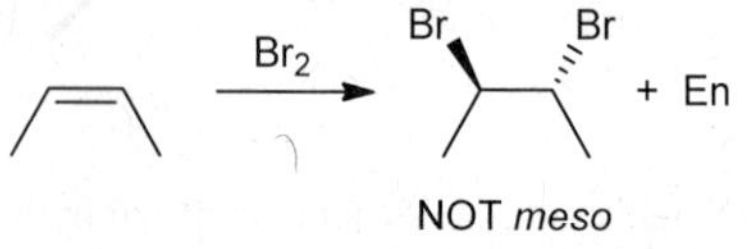

In contrast, *trans*-2-butene gives the desired *meso* compound, as shown:

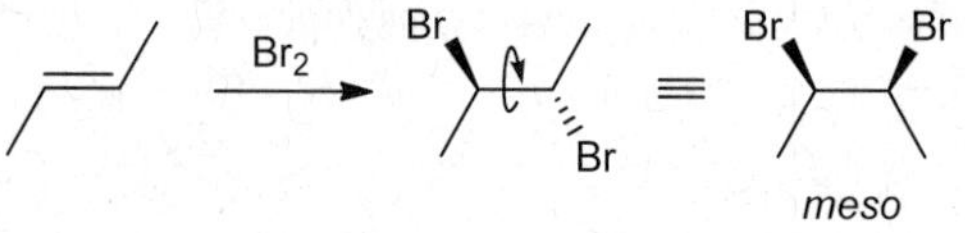

meso

9.68. In each of the following cases, we draw the necessary alkene by removing the oxygen atoms from the product and connecting the *sp*² hybridized carbon atoms to form a C=C bond:

(a)

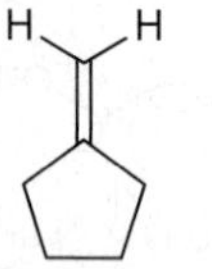

b)

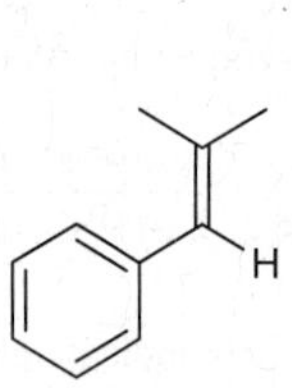

c)

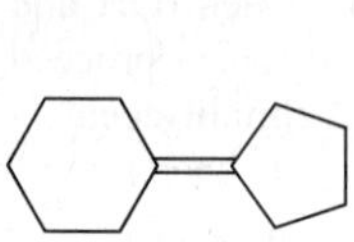

d)

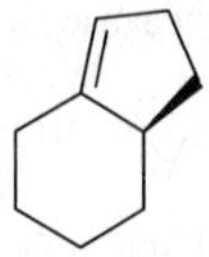

9.69. In the presence of a strong acid, the π bond is protonated to give a resonance-stabilized cation (shown below), which is even lower in energy than a tertiary carbocation. This protonation step determines the regiochemical outcome of the reaction, because the

resonance-stabilized cation is captured by a bromide ion to give the product, as shown.

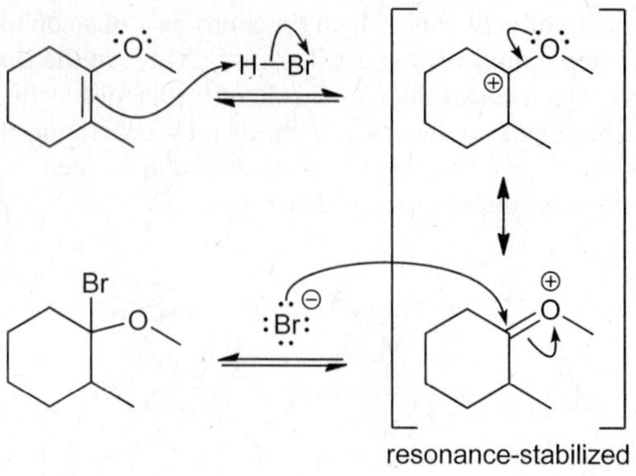

resonance-stabilized

9.70.

(a) The two groups being added across the alkene are H and OH. The OH group is installed at the less-substituted carbon atom, so we must use conditions that give an *anti*-Markovnikov addition of H and OH. This can be accomplished via hydroboration-oxidation.

1) BH₃·THF
2) H₂O₂, NaOH

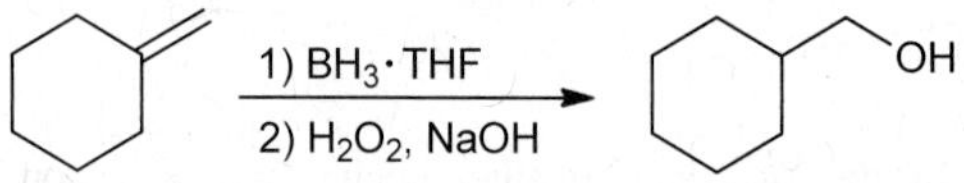

(b) The two groups being added across the alkene are H and Br. The Br group is installed at the less-substituted carbon atom, so we must use conditions that give an *anti*-Markovnikov addition of H and Br. This can be accomplished by treating the alkene with HBr in the presence of peroxides.

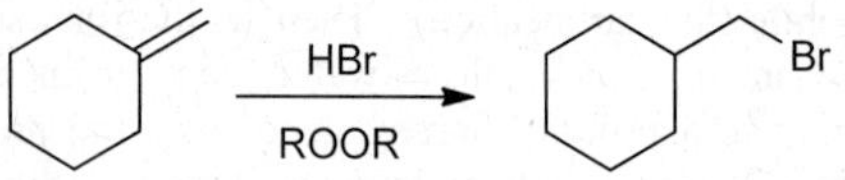

HBr
ROOR

(c) The two groups being added across the alkene are OH and OH. No chirality centers are formed, so stereochemistry is irrelevant. We have learned more than one way to achieve a dihydroxylation. For example, we can convert the alkene to an epoxide and then open the epoxide under aqueous acidic conditions.

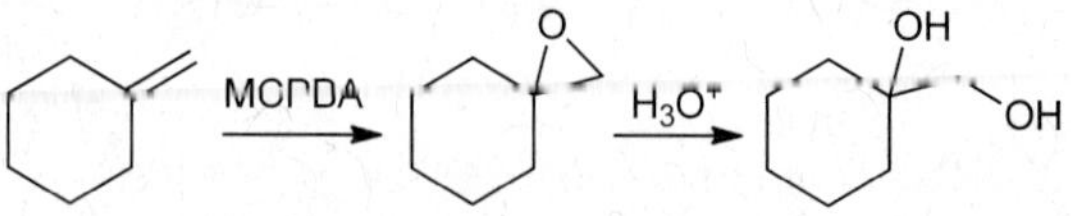

MCPBA   H₃O⁺

Alternatively, we can treat the alkene with osmium tetroxide and a suitable co-oxidant, or even with potassium permanganate and NaOH.

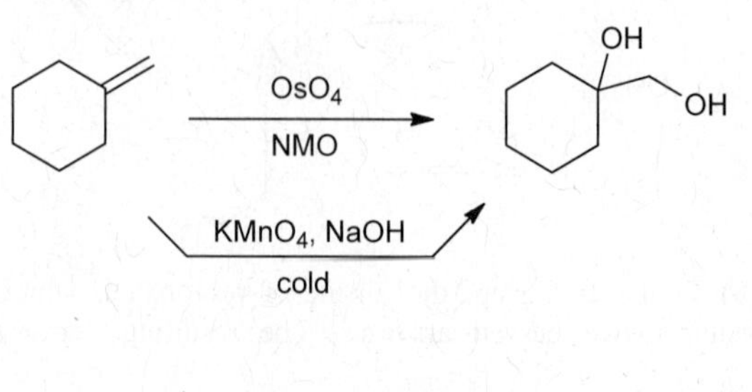

OsO₄
NMO

KMnO₄, NaOH
cold

(d) The two groups being added across the alkene are Cl and Cl. This can be achieved by treating the alkene with Cl$_2$:

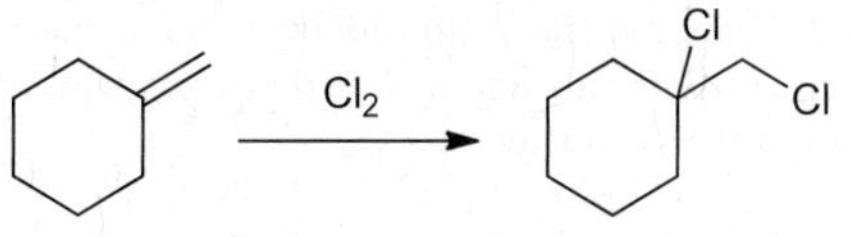

(e) The two groups being added across the alkene are Br and OH, with the latter being placed at the more substituted position. This can be achieved by treating the alkene with Br$_2$ in the presence of water (halohydrin formation):

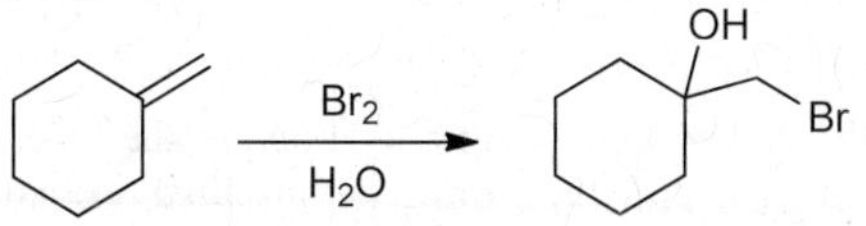

(f) The two groups being added across the alkene are H and Br. The Br group is installed at the more-substituted, tertiary position, so we must use conditions that give a Markovnikov addition of H and Br. This can be accomplished by treating the alkene with HBr.

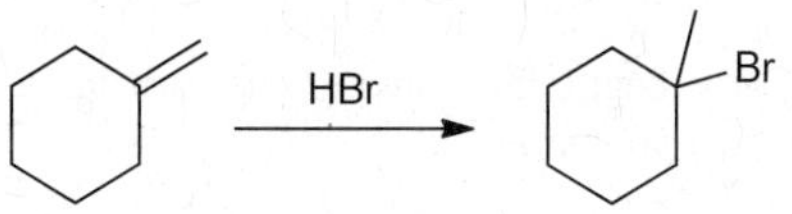

(g) This transformation requires the elimination of H and Br to give the more-substituted alkene, so a strong base is required (such as hydroxide, methoxide, or ethoxide):

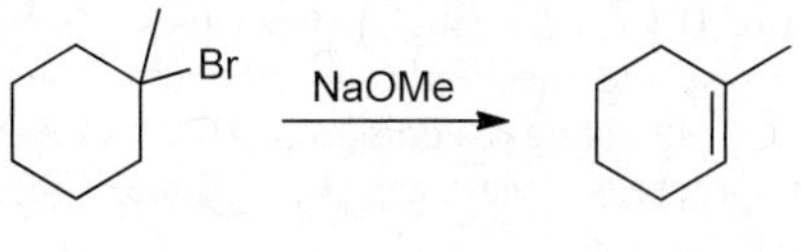

(h) The two groups being added across the alkene are H and H, which can be accomplished by treating the alkene with molecular hydrogen (H$_2$) in the presence of a suitable catalyst.

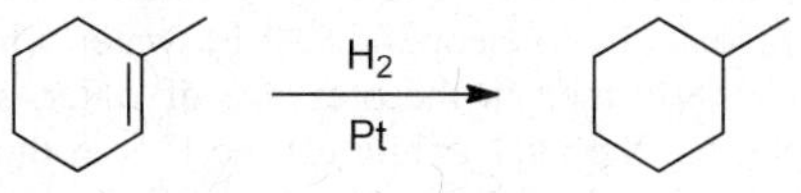

(i) The two groups being added across the alkene are Br and OH in an *anti* fashion, with the latter being placed at the more substituted position. This can be achieved by treating the alkene with Br$_2$ in the presence of water (halohydrin formation):

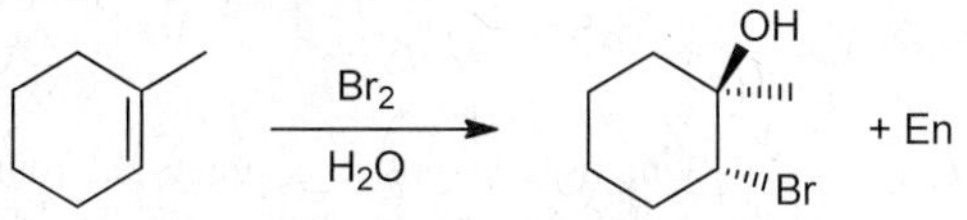

(j) Cleavage of the C=C double bond can be achieved via ozonolysis:

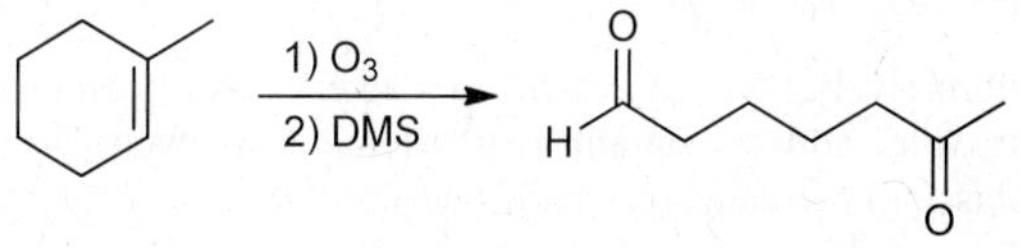

(k) The two groups being added across the alkene are H and OH. The OH group is installed at the less-substituted carbon atom, so we must use conditions that give an *anti*-Markovnikov addition of H and OH. This can be accomplished via hydroboration-oxidation, which proceeds via a *syn* addition:

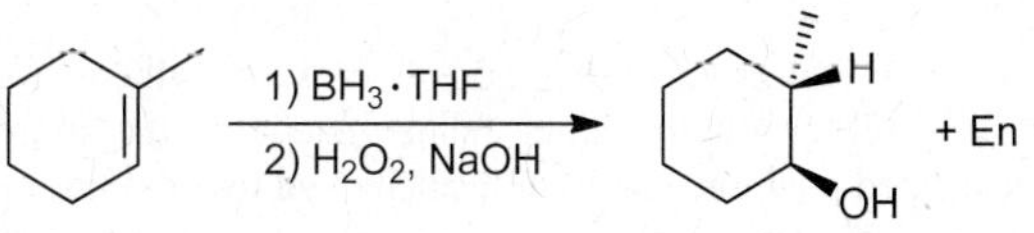

9.71.
(a) Cleavage of the C=C double bond can be achieved via ozonolysis:

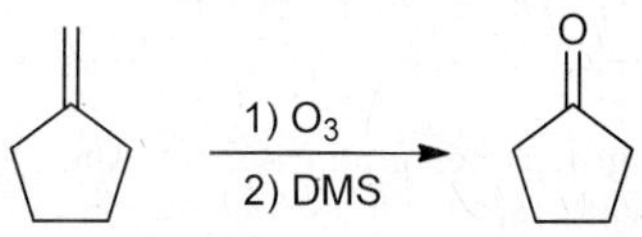

(b) The two groups being added across the alkene are H and Br. The Br group is installed at the less-substituted carbon atom, so we must use conditions that give an *anti*-Markovnikov addition of H and Br. This can be accomplished by treating the alkene with HBr in the presence of peroxides.

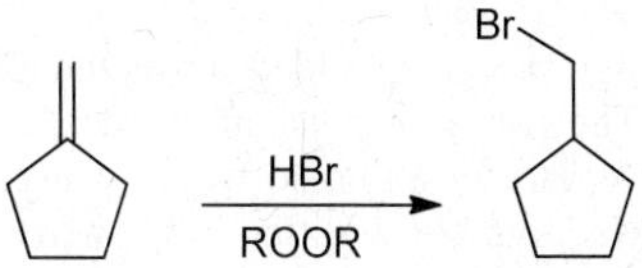

(c) The two groups being added across the alkene are H and OH. The OH group is installed at the less-substituted carbon atom, so we must use conditions that give an *anti*-Markovnikov addition of H and OH. This can be accomplished via hydroboration-oxidation.

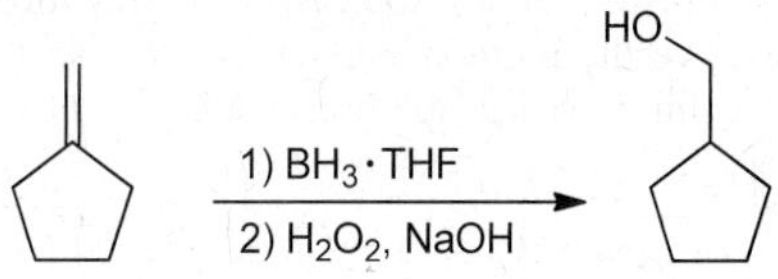

(d) The two groups being added across the alkene are OH and OH. No chirality centers are formed, so stereochemistry is irrelevant. We have learned more than one way to achieve a dihydroxylation. For example, we can convert the alkene to an epoxide and then open the epoxide under aqueous acidic conditions.

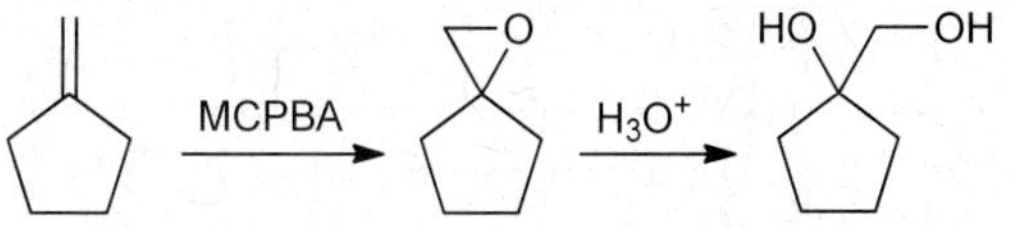

Alternatively, we can treat the alkene with osmium tetroxide and a suitable co-oxidant, or even with potassium permanganate and NaOH.

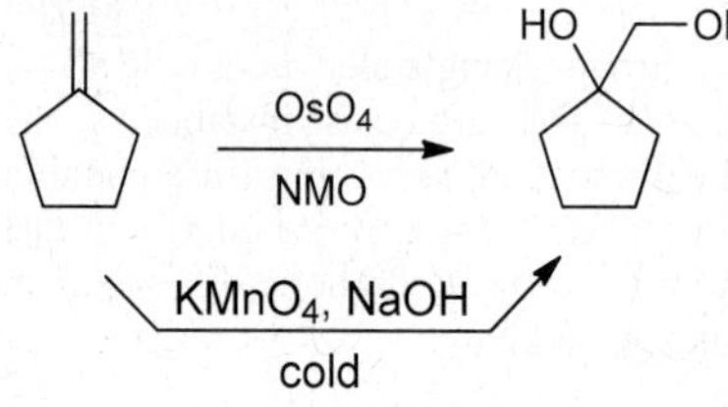

(e) The two groups being added across the alkene are Br and OH, with the latter being placed at the more substituted position. This can be achieved by treating the alkene with Br_2 in the presence of water (halohydrin formation):

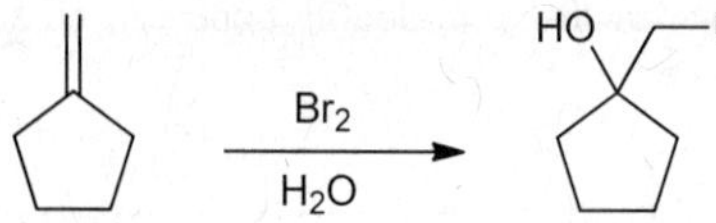

(f) The two groups being added across the alkene are H and OH, with the latter being placed at the more substituted position (Markovnikov addition). This can be achieved via acid-catalyzed hydration:

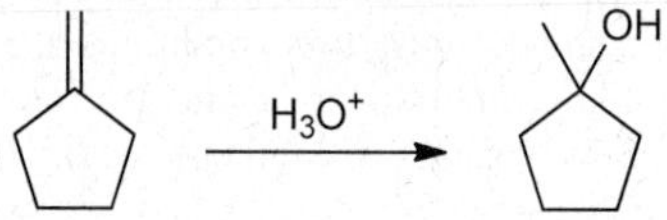

(g) The two groups being added across the alkene are H and Br. The Br group is installed at the more-substituted, tertiary position, so we must use conditions that give a Markovnikov addition of H and Br. This can be accomplished by treating the alkene with HBr.

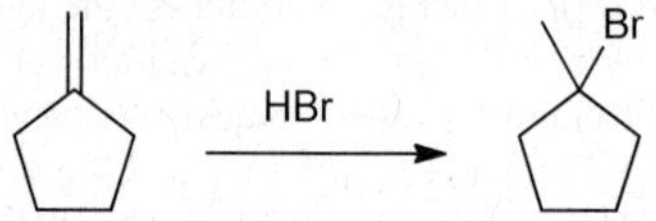

(h) This transformation requires the elimination of H and Br to give the more-substituted alkene, so a strong base is required (such as hydroxide, methoxide, or ethoxide):

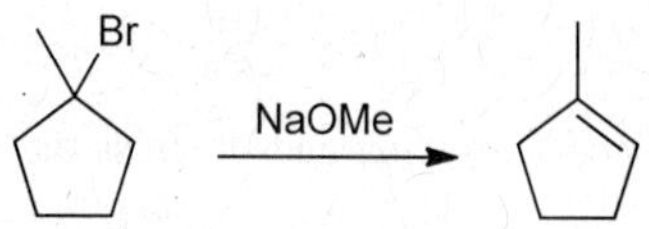

(i) The two groups being added across the alkene are H and H, which can be accomplished by treating the alkene with molecular hydrogen (H_2) in the presence of a suitable catalyst.

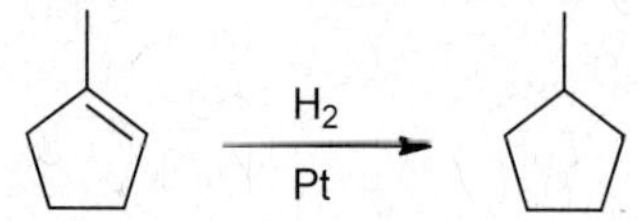

(j) The two groups being added across the alkene are OH and OH, and they must be installed via a *syn* addition. This can be achieved by treating the alkene with osmium tetroxide and a suitable co-oxidant, or with potassium permanganate:

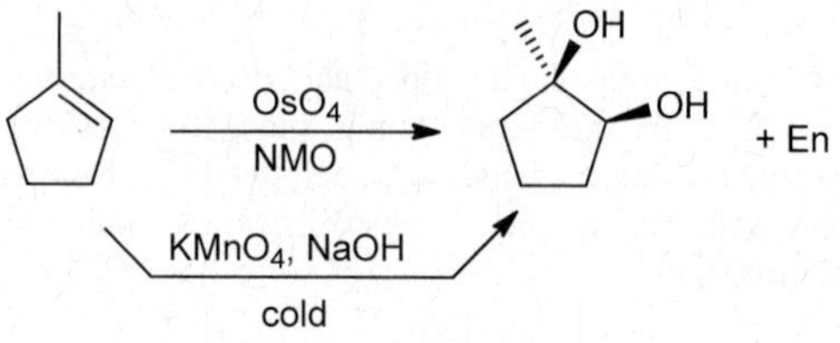

(k) The two groups being added across the alkene are OH and OH, and they must be installed via an *anti* addition. This can be achieved by treating the alkene with MCPBA, followed by aqueous acid:

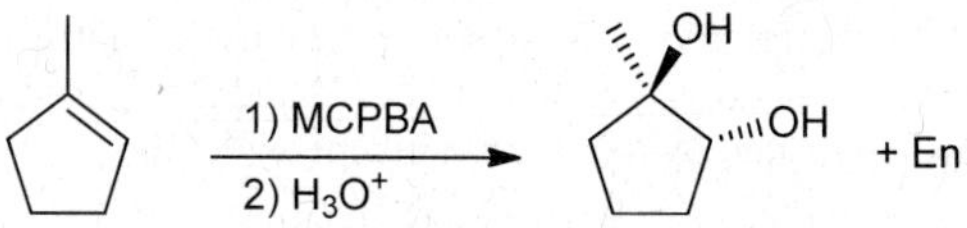

9.72. Let's begin by drawing the structures of the alkenes under comparison:

Addition of HBr to 2-methyl-2-pentene should be more rapid because the reaction can proceed via a tertiary carbocation. In contrast, addition of HBr to 4-methyl-1-pentene proceeds via a less stable, secondary carbocation.

9.73. When treated with molecular bromine (Br_2), the alkene is converted to an intermediate bromonium ion, which is then subject to attack by a nucleophile. We have seen that the nucleophile can be water when the reaction is performed in the presence of water, so it is reasonable that the nucleophile can be H_2S in this case. This should give the installation of an SH group (rather than an OH group) at the more substituted position:

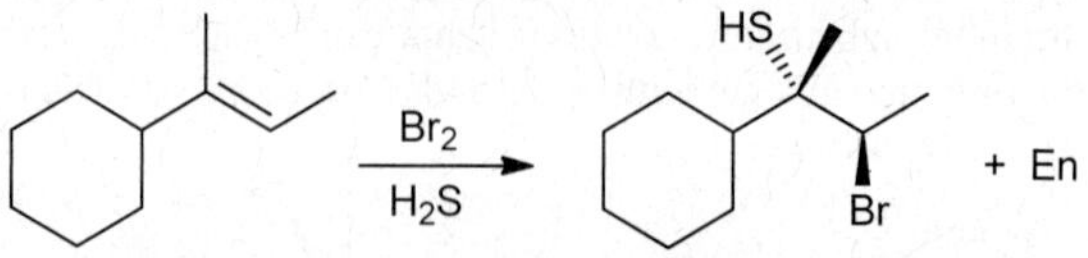

9.74. The following is one possible suggested route. Other acceptable solutions are certainly possible. For example, after the first step (elimination with *tert*-butoxide), the next two steps (addition of HBr, followed by elimination) could be replaced with acid-catalyzed hydration, followed by elimination with conc. H_2SO_4.

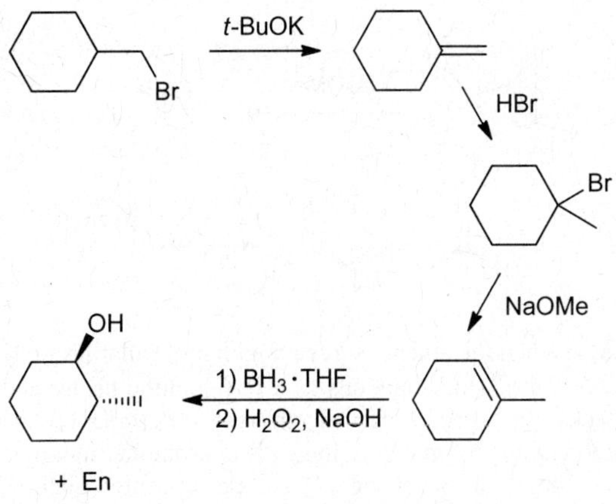

9.75. There is only one alkene (compound X, shown below) that can be converted to 2,4-dimethyl-1-pentanol via hydroboration-oxidation. Treatment of that alkene with aqueous acid affords an alcohol (via Markvonikov addition):

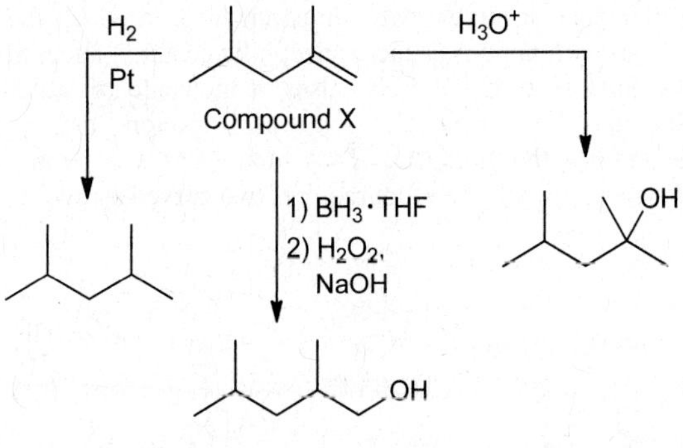

2,4-dimethyl-1-pentanol

9.76. The substrate is a secondary alkyl halide, and treatment with *tert*-butoxide gives the less substituted alkene. When that alkene is treated with HBr, the π bond is protonated to give a secondary carbocation (rather than a primary carbocation). This carbocation can either be captured by a bromide ion, giving products **A** and **B** below, or the carbocation can undergo a rearrangement (hydride shift) to give a tertiary carbocation, which is then captured by a bromide ion, affording product **C**.

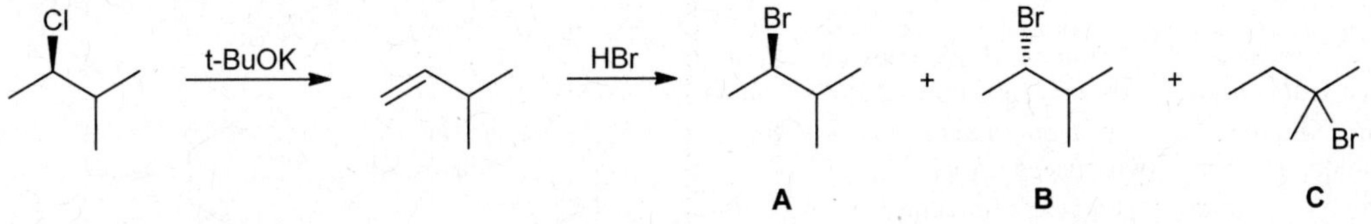

9.77. There is only one alkene (compound Y, shown below) that is consistent with the information provided in the problem statement. Ozonolysis of that alkene results in cleavage of the C=C bond to give two separate compounds, each of which has a C=O bond:

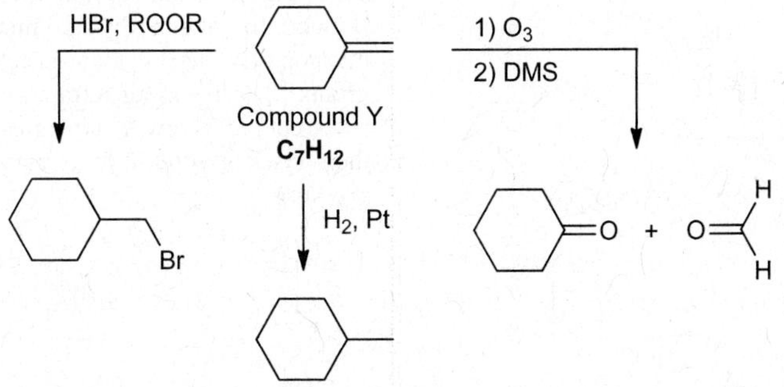

Compound Y
C₇H₁₂

9.78. Each of the products is an aldehyde, and their *sp²* hybridized carbon atoms were once connected to each other as a C=C bond in the original alkene. That gives the following two possibilities (stereoisomers) for the structure of the original alkene:

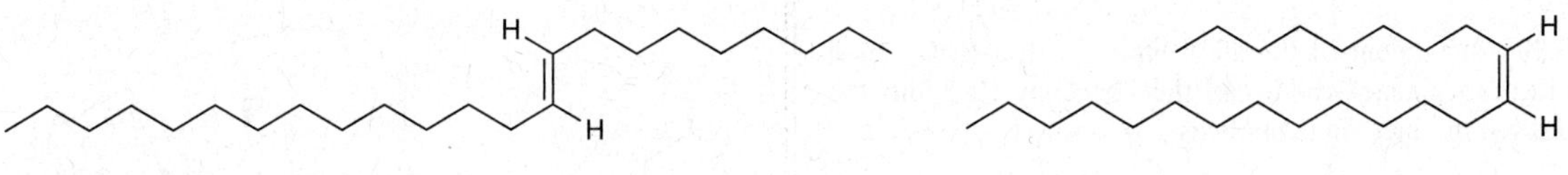

9.79.

(a) In the presence of aqueous acid, the epoxide is first protonated (two curved arrows), as shown below. The resulting intermediate can then undergo an S$_N$2-like, intramolecular attack (two curved arrows), in which the OH group functions as the nucleophilic center. Then, in the final step of the mechanism, a molecule of water functions as a base and removes a proton, thereby generating the product. This final step is a proton transfer step, and therefore requires two curved arrows:

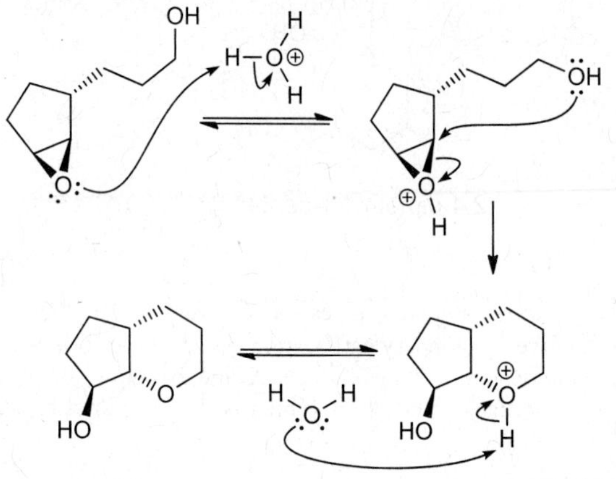

(b) In the presence of aqueous acid, the epoxide is first protonated (two curved arrows), as shown below. The resulting intermediate can then undergo an S$_N$2-like, intramolecular attack (two curved arrows), in which the π bond functions as the nucleophilic center. Then, in the final step of the mechanism, a molecule of water functions as a base and removes a proton, thereby generating the product. This final step is a proton transfer step, and therefore requires two curved arrows:

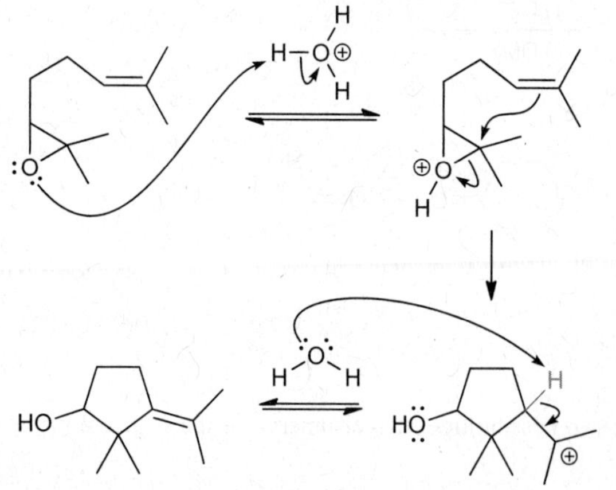

9.80. Treatment of the alkyl halide with a strong base gives an alkene which can then be converted into the desired product via ozonolysis:

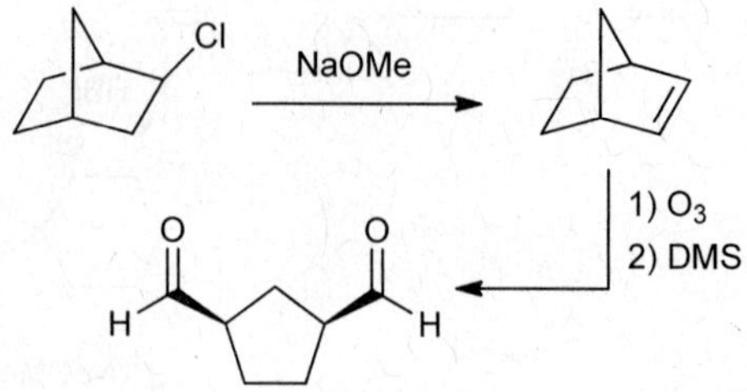

9.81. When the alkene is treated with molecular bromine (Br$_2$), the π bond functions as a nucleophilic center and attacks Br$_2$ (three curved arrows), resulting in an intermediate bromonium ion. The bromonium ion is then subject to attack by a nucleophilic center, such as the OH group that is tethered to the bromonium group. The resulting intramolecular nucleophilic attack (two curved arrows) generates an oxonium ion, which then loses a proton (two curved arrows) to give the product.

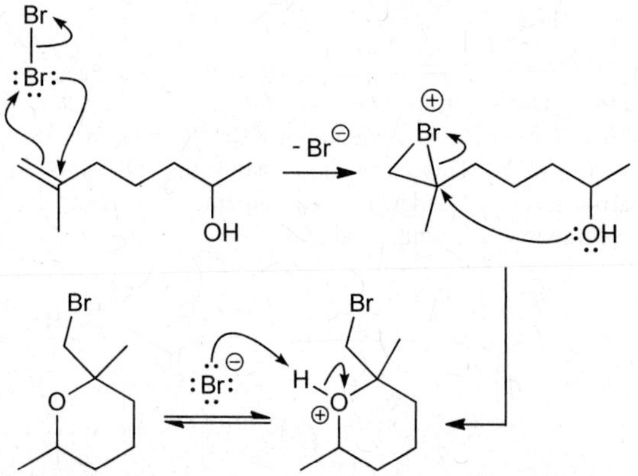

9.82. When the alkene is treated with molecular iodine (I$_2$), the π bond functions as a nucleophilic center and attacks I$_2$ (three curved arrows), resulting in an intermediate iodonium ion. The iodonium ion is then subject to attack by a nucleophile, such as the nucleophilic center that is tethered to the iodonium group. The resulting intramolecular nucleophilic attack (two curved arrows) generates an intermediate which then loses a proton (two curved arrows) to give the product.

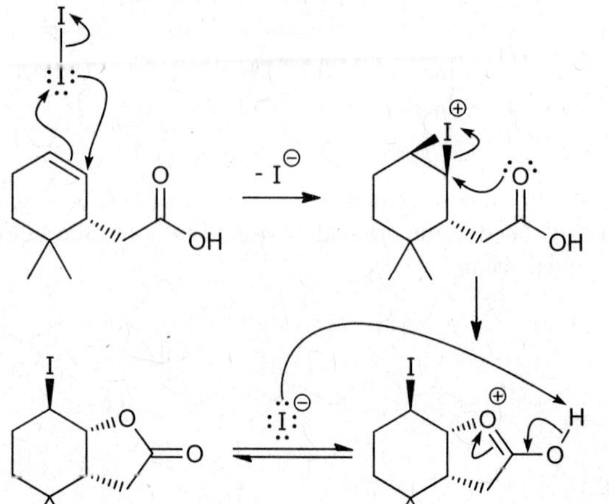

9.83. The *cis*-dibromide is not obtained, suggesting that the reaction proceeds via an *anti* addition process. This can be explained if we argue that the carbocation (formed upon protonation of the π bond) is converted into a bromonium ion, as shown here. The incoming nucleophile (bromide) would have to attack from the backside of the bromonium bridge, giving a *trans* dibromide:

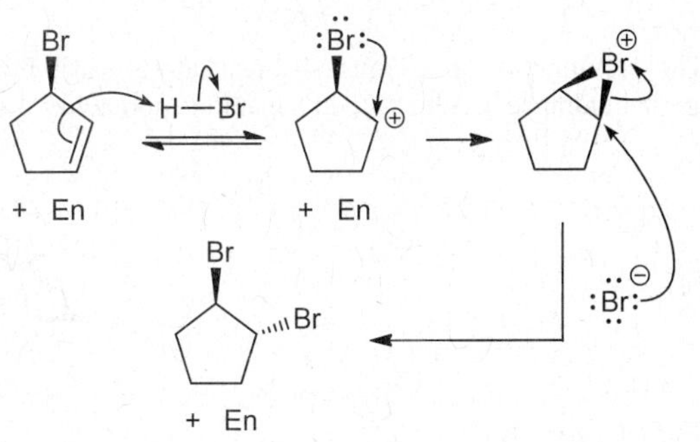

9.84.

(a) The reagents indicate a hydroboration-oxidation, in which an alkene is converted to an alcohol. In compound **1**, there are two alkene groups, so we must choose which one is more likely to react with BH₃. One alkene group is disubstituted, while the other is tetrasubstituted:

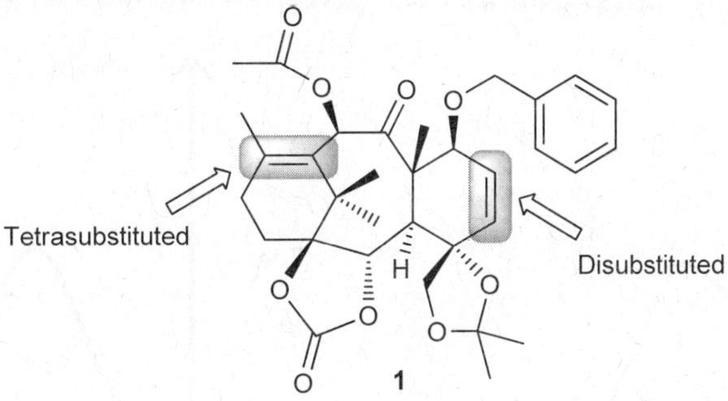

Since the rate of hydroboration is particularly sensitive to steric factors, we expect the disubstituted alkene group to undergo hydroboration more readily.

(b) As mentioned in part (a), hydroboration is sensitive to steric considerations. When we inspect both vinylic positions, we find that both are equally substituted. The tie-breaker will likely be the nearby presence of a six-membered, aromatic ring, which provides significant steric crowding that favors the following regiochemical outcome:

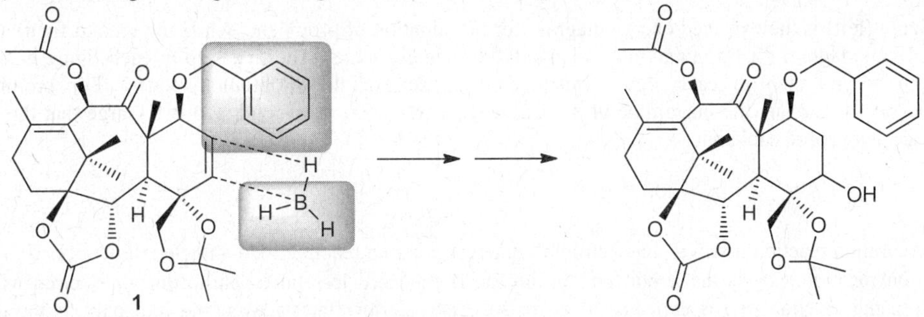

In predicting the stereochemical outcome, we once again invoke the steric bulk of the six-membered, aromatic ring. Specifically, the front face of the alkene group is blocked by the large six-membered, aromatic ring. As a result, the back face of the alkene is more accessible, so the reaction occurs more readily on the back face, giving the following expected product:

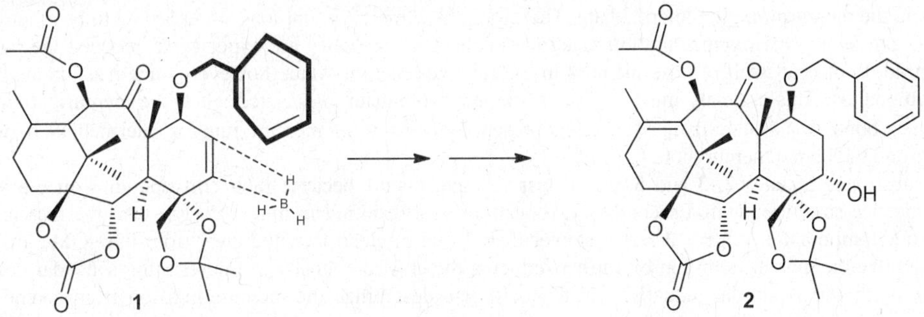

9.85.

The bromonium ion is unusually resistant towards nucleophilic attack by the bromide anion because of significant steric hindrance involved when the anion approaches the electrophilic carbon atoms of the bromonium ion.

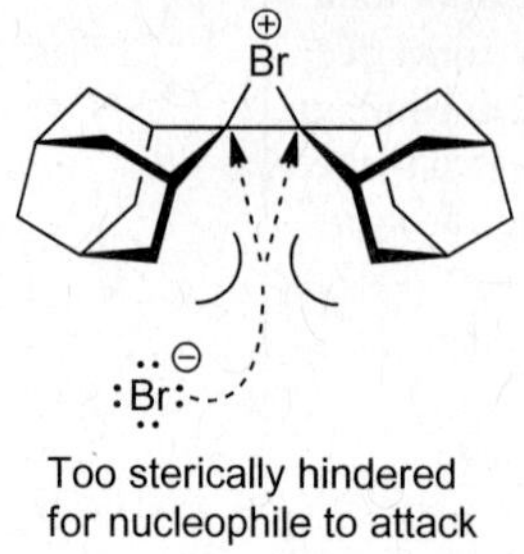

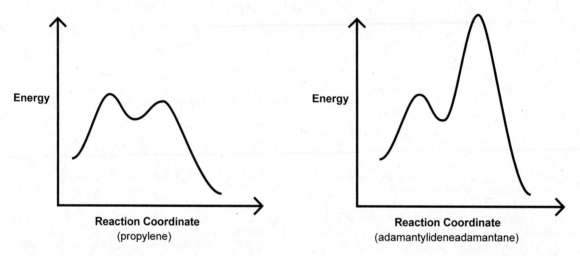

Too sterically hindered
for nucleophile to attack

Thus, there is a very large activation energy associated with this bromonium ion being attacked by a nucleophile. This can be illustrated by comparing the following reaction coordinate diagrams.

The first figure (left) is the expected energy diagram for bromination of propylene, while the second figure (right) is a proposed energy diagram for bromination of adamantylideneadamantane. The first step in each figure is similar, but compare the second step in each figure (nucleophilic attack of the bromonium ion). For bromination of adamantylideneadamantane, the magnitude of the activation energy for the second step is so large that this step does not take place at an appreciable rate.

9.86.

The oxymercuration reaction involves an electrophilic mercuric cation reacting with a nucleophilic π bond of an alkene in an addition reaction. So, as the π bond of the alkene is rendered less nucleophilic due to electron-withdrawing substituent(s), the reaction rate is expected to decrease. Also, steric effects may come into play as the number of substituents around the π bond increases.

Among the alkenes listed, alkene **1** is disubstituted while **4** is trisubstituted. All of the rest are monosubstituted alkenes. Given that alkyl substituents are generally electron-donating groups, we would expect **1** and **4** to be the most reactive. More specifically, compounds **1** and **4** are the only ones capable of having a tertiary carbocation as a resonance contributor in the mercurinium ion intermediate. Therefore, these mercurinium ions are expected to be among the most stable ones, and hence, the oxymercuration reactions of these two alkenes are expected to proceed the fastest. This expectation does not bear itself out for alkene **4** in the relative reactivity data, however, since it is among the slower reacting compounds. This anomaly must be due to the steric repulsion associated when the mercuric cation tries to approach the π bond, or a destabilizing steric effect present in the resulting mercurinium ion intermediate between these substituents and the bound mercury ion.

The monosubstituted alkenes **2**, **3** and **5** are all less reactive than **1** because their corresponding mercurinium ions involve resonance structures with a secondary carbocation, thus resulting in higher energy than the mercurinium ion obtained from compound **1**. Alkene **3** reacts slower than **2** due to electron-withdrawal from the –OMe group, which would destabilize the mercurinium ion by further reducing the electron density of the resulting secondary carbocation resonance contributor. A similar inductive effect would also destabilize the mercurinium ion from alkene **5**, but an additional steric destabilization due to the large chlorine atom may also be in effect to make this alkene the slowest reacting compound among the series.

9.87.
Hydroboration of an alkene typically occurs with equal probability from both faces of the π-bond. However, in the case of α-pinene, the top face of the π-bond is blocked by one of the methyl substituents, so the approach of borane on this face of the π-bond is severely hindered. Therefore, hydroboration cannot occur at an appreciable rate on the top face of the π-bond. However, the bottom face of the π-bond is unhindered; it is much more accessible to hydroboration. As a result, boron and hydrogen add in a *syn*-fashion to the bottom face of α-pinene resulting in the IpcBH₂ enantiomer shown; as always, boron is added to the less hindered position on the alkene.

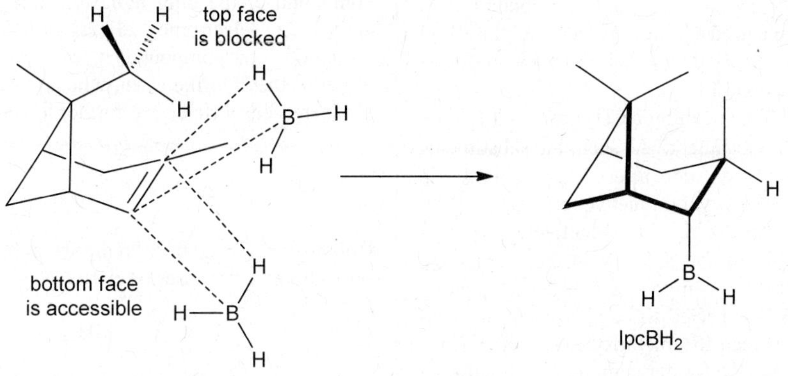

Notice that the methyl group of the alkene is now occupying an axial position in the product.
    Next, the π-bond of a second molecule of α-pinene will react with IpcBH₂, also from the same face, to produce the observed enantiomer of Ipc₂BH.

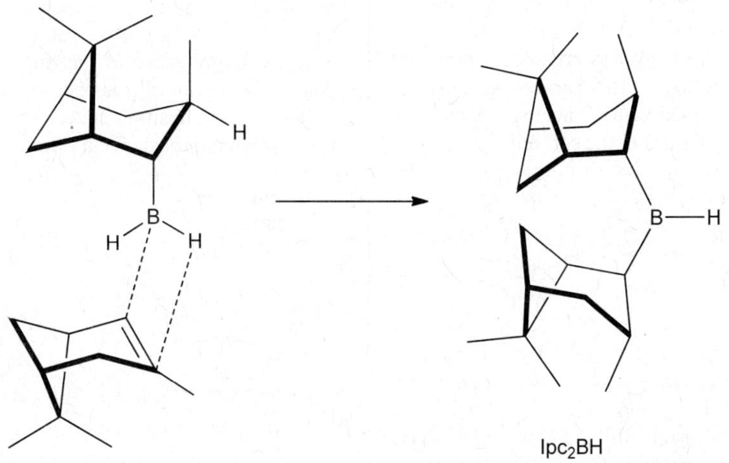

9.88.
The hydroboration reaction involves an electrophilic borane (or, organoborane such as 9-BBN) reacting with a nucleophilic alkene in an addition reaction. So, as the π bond of the alkene is rendered more nucleophilic due to electron-donating substituent(s), the reaction rate is expected to increase. Steric effects (that arise because of the bulky reagent) are also expected to play an important role in determining the relative rates of reactivity.

(a) Alkene **1** possesses an alkoxy substituent (OR) in an allylic position. An alkoxy group is expected to be inductively electron-withdrawing, because oxygen is an

electronegative atom and will therefore withdraw electron density away from the π bond. This effect should render the π bond less reactive (less nucleophilic). However, the alkoxy group is expected to be electron-donating via resonance, as seen when we draw the resonance structures:

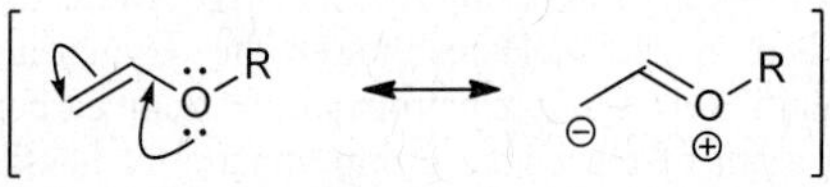

So there are two effects in competition with each other. The alkoxy group is expected to be electron-withdrawing via induction, but it is expected to be electron-donating via resonance. Which effect is stronger? We have seen that, in general, resonance is a stronger effect than induction. As such, we would expect the alkoxy group to be electron-donating, which would render the alkene more nucleophilic (more reactive). This prediction is verified by the high rate of reactivity of compound **1**.

The π bond in compound **2** is adjacent to an alkyl group, rather than an alkoxy group, so there is no resonance effect. The only effect is induction (we have seen that alkyl groups are generally electron donating). As such, the nucleophilicity of the π bond in compound **2** is expected to be enhanced by the presence of the alkyl group, but it is not expected to be quite as nucleophilic as the π bond in compound **1**.

Compounds **3** and **5** both exhibit a CH_2 group in between the π bond and the substituent. As such, the substituents in these compounds do not affect the π bond via resonance effects; only via inductive effects. Both substituents are expected to be inductively electron-withdrawing. Therefore, the π bonds in these cases are less electron-rich than in compounds **1** or **2**. The acetoxy group of alkene **5** is expected to be a stronger electron-withdrawing group than the methoxy group of **3** since the carbonyl group in the former compound will enhance the electron-withdrawing ability of the oxygen atom via resonance. Compound **4** is interesting in that the acetoxy oxygen atom is expected to be an electron-donating group with respect to the alkene π bond (via resonance), much like the oxygen atom in compound **1**. From the relative reactivity data, however, the lone pairs on the acetoxy oxygen are apparently less efficient at donating to the carbon-carbon π bond. This can be explained by recognizing that the lone pairs on the acetoxy oxygen are already partially delocalized into the carbonyl group:

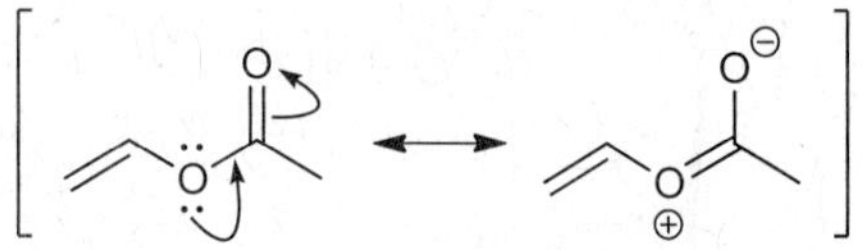

That is, the lone pairs on the acetoxy oxygen are less available to donate electron density to the π bond. Therefore, the acetoxy group of **4** influences the reactivity primarily through its inductive and steric effects, which will be much like that of the substituents in compounds **3** and **5**.

(b) Compound **6** is apparently able to impose the electron-withdrawing effect due to induction of the –CN group on the alkene π bond via the shorter σ bond (due to the sp^3-sp orbital overlap) between the CH_2 and the CN groups, and thus the closer proximity of the partial positive charge to the π electrons of the alkene. This is apparent when comparing the relative rates between compounds **5** and **6**. Here, though the oxygen atom in compound **5** is more electronegative than carbon, the carbon atom of the cyano group apparently possesses a very large partial positive change that effectively renders this carbon atom more electronegative than that of the acetoxy oxygen atom of **5**. The low reactivity associated

with compound **7** may be due to both induction and steric effects, since the inductive withdrawal of electron density by the chlorine atom is not expected to exceed the substituent in compound **5**.

(c) Compounds **2**, **8** and **9** illustrate the consequences of steric effects on the hydroboration reaction. Compounds **8** and **9** are the only disubstituted alkenes among the series, and it is not surprising that they have the lowest relative reactivity, due to increased steric effects. They are more than 100 times lower in reactivity than compound **2**, the only monosubstituted alkene with no significant electronic effects (i.e. resonance and induction). In compound **9**, the electron-withdrawing effect of the chlorine atom, which is superimposed upon the increased steric effect, further lowers the reactivity.

9.89.
Compound **1** has been redrawn below in a Haworth projection for purposes of clarity:

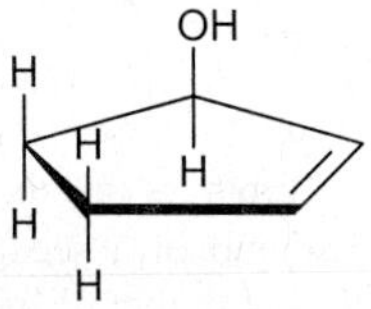

Osmium tetroxide can approach the alkene either on the top face of the ring (on the same side as the OH group) or on the bottom face of the ring (opposite side as OH group). When the former occurs, a steric interaction is present in the transition state. This steric interaction is absent when approach occurs from the bottom face:

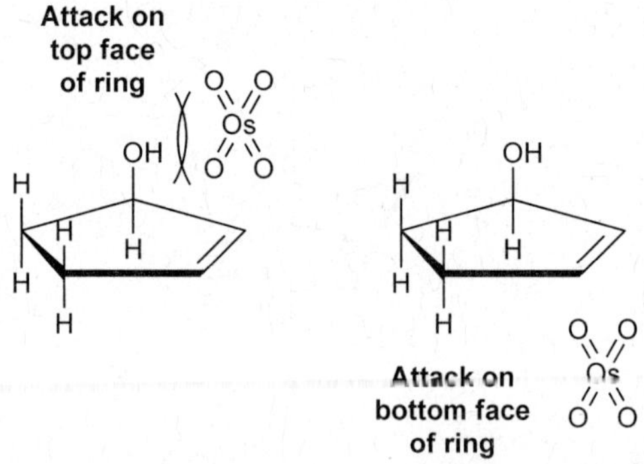

As such, the transition state for approach on the bottom face will be lower in energy than the transition state for approach on the top face. So diastereomer **3** is favored over diastereomer **2**. Any hydrogen bonding interactions between the OH group and the oxygen atoms of OsO_4 are clearly not strong enough to overcome the energy cost associated with the steric interaction.

However, when TMEDA is employed, a complex is formed. In this complex, the oxygen atoms are electron rich (δ-) at the expense of the electron-donating nitrogen atoms (which are δ+). Because the oxygen atoms are electron-rich, they are capable of forming very strong hydrogen bonding interactions with the proton of the OH group.

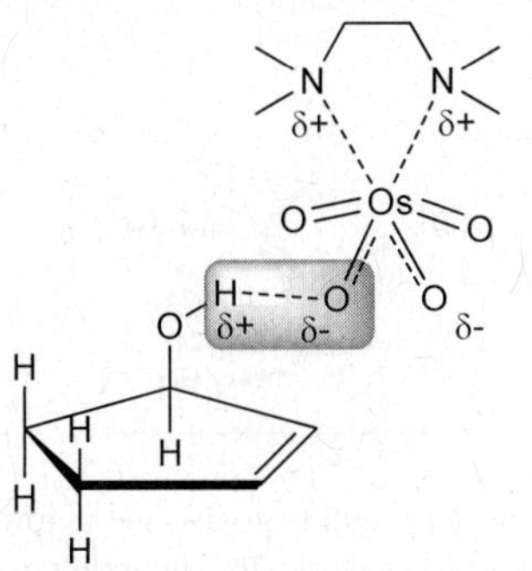

In this case, the hydrogen bonding interactions (which stabilize the transition state) are sufficiently strong to overcome the energy cost associated with the steric interaction. As such, the transition state for approach on the top face will be lower in energy than the transition state for approach on the bottom face. And consequently, diastereomer **2** is favored over diastereomer **3**.

9.90.

When approaching this problem it would be advisable to first label the carbon side chain coming off the benzene ring so that you can determine what new connections have been made. Your numbering system does not need to conform to IUPAC rules for assigning locants. Rather, it is OK to use an arbitrary numbering system, because the goal of the numbering system is to track the fate of all atoms during the transformation:

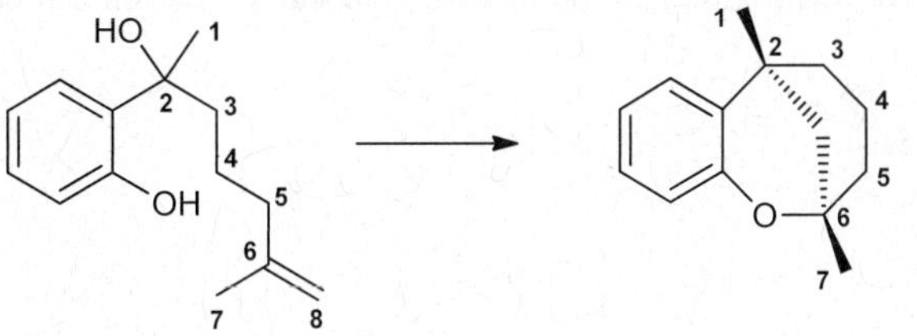

With this numbering system, the benzene ring is attached to C2 of the chain and the phenolic oxygen is attached to C6 of the chain.

Based on this, a possible mechanism is illustrated below. Protonation of the tertiary alcohol, followed by loss of water, gives the stable tertiary benzylic carbocation (at C2). Markovnikov attack by the terminal alkene (C6 and C8) at the tertiary benzylic carbocation affords a tertiary carbocation at C6. This carbocation is then attacked by the phenolic oxygen to afford the final product after removal of the acidic proton.

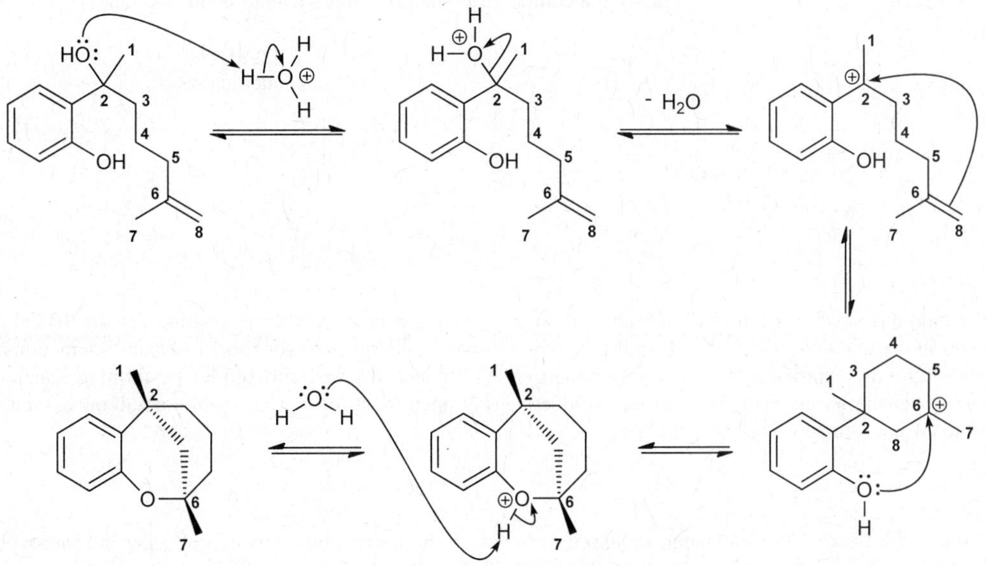

9.91.
Building a molecular model is perhaps the best way to see that the bottom face of the π bond is more hindered than the top face. Alternatively, this can be seen if we redraw the compound in a Haworth projection:

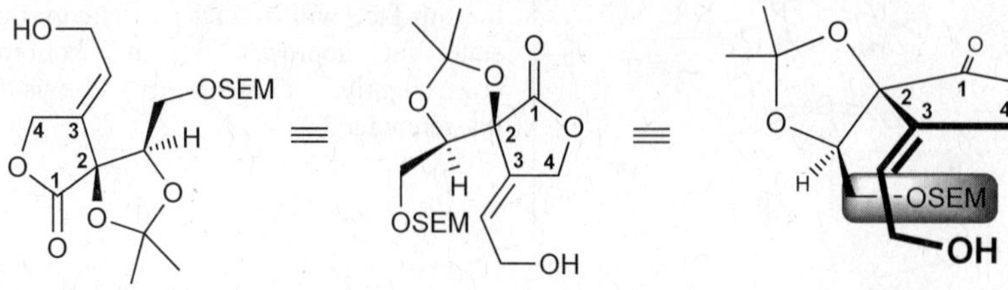

Notice that the π bond is in the plane of the ring, and the large and bulky OSEM group is positioned below the plane of the ring, directly underneath the π bond. As such, the bottom face of the π bond is sterically encumbered, so approach of the oxidizing agent (OsO$_4$) from that face is blocked (it would involve a transition state that is too high in energy). Attack on the top face is unencumbered, so it involves a lower energy transition state, and as a result, the reaction occurs more readily on this face.

9.92.
(a) In the first step, the mercuric cation is formed, which rapidly reacts with the π bond in intermediate A to produce mercurinium ion **4**. Next, an intramolecular reaction will occur – one of the lone pairs on the hydroxyl group will add to the more substituted side of the mercurinium ion to form compound **5**, which can be deprotonated by the acetate anion to form compound **2**.

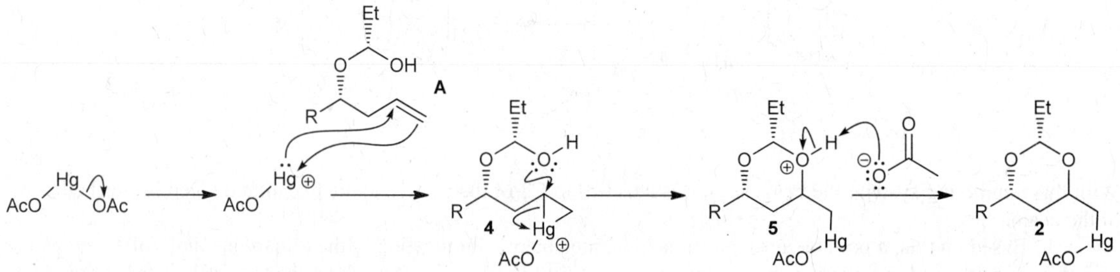

(b) In order to better understand why only intermediate **A** reacts to from the cyclic product (while intermediate **B** does not), we must redraw intermediates **4A** and **4B** in a conformation that resembles a chair conformation:

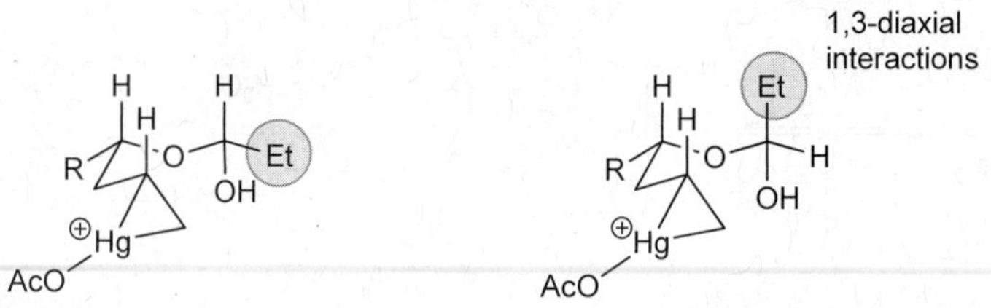

When depicted this way, notice that **4A** has all of its substituents in pseudo-equatorial positions; while **4B** exhibits an ethyl group in a pseudo-axial position. Recall (Chapter 4) that the energy cost for the 1,3-diaxial interactions for an axial ethyl group are approximately 8 kJ/mol in energy. As a result, **4A** will undergo a ring-forming reaction via a lower energy transition state than **4B**. As such, productive ring-opening of the mercurinium ion will only occur via the more stable **4A** conformation.

9.93.
(a) As indicated in the problem statement, sodium bicarbonate functions as a base and deprotonates the carboxylic acid group to give a carboxylate ion. Then, the π bond reacts with I$_2$ to give an iodonium ion (similar to a bromonium ion), which is then opened via an intramolecular nucleophilic attack to give the product: This process is called iodolactonization, because the product features a newly installed cyclic ester group (a lactone) as well as an iodo group:

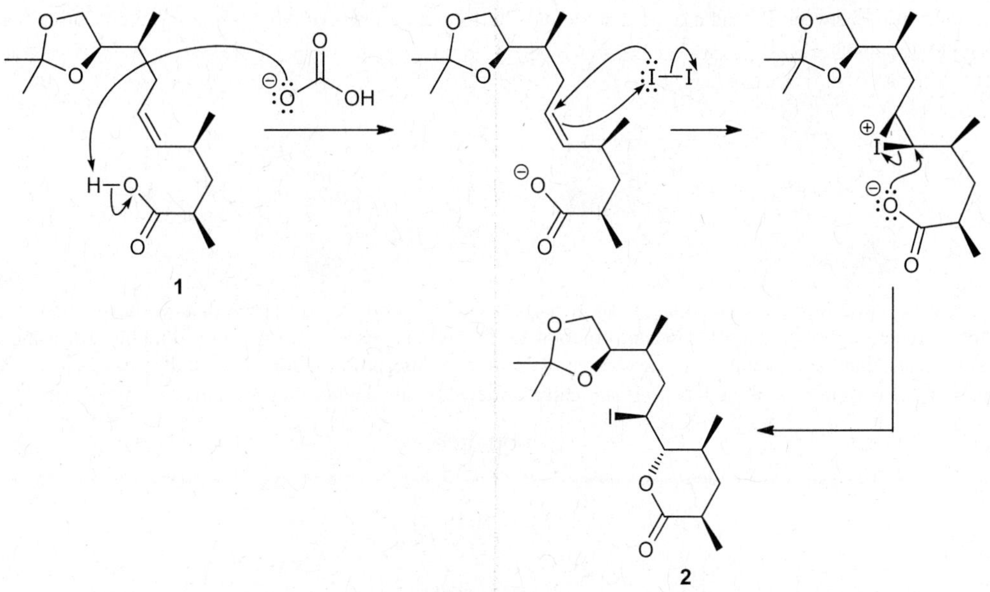

1

2

(b) Let's simplify our drawings by referring to the following large groups as R and R':

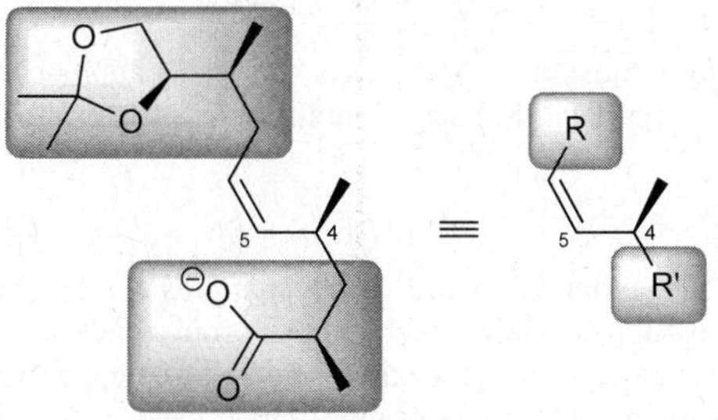

Now we are ready to look down the C4–C5 bond, like this:

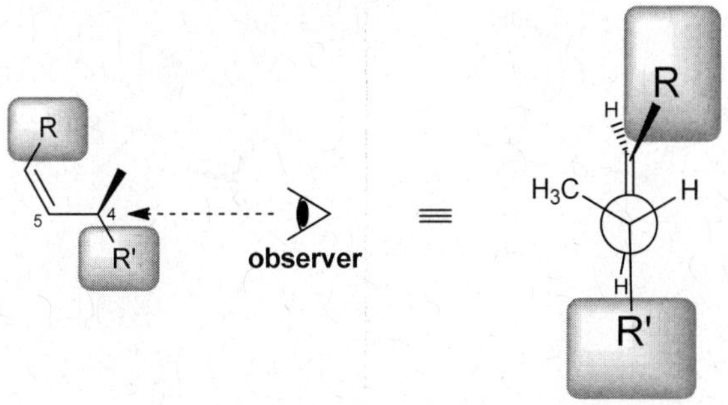

Let's rotate this entire Newman projection by 90° (which does not change the conformation at all) so that we can clearly see the top face and bottom face of the π bond:

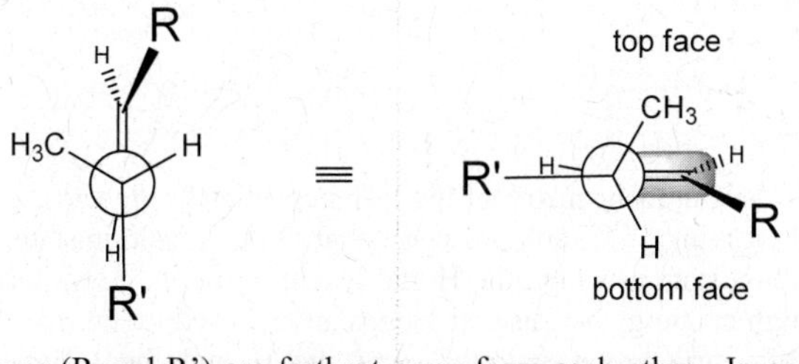

Notice that the two largest groups (R and R') are farthest away from each other. In our search for the lowest energy conformation, this conformation should be the first one that we examine, because significant steric interactions will be present if R and R' are near each other in space. When we analyze this conformation, we see two additional factors contributing to its overall energy: 1) gauche interactions between the methyl group and the R group, and 2) an eclipsing interaction between R' and a hydrogen atom. The former can be avoided by rotating the front carbon atom counterclockwise, like this:

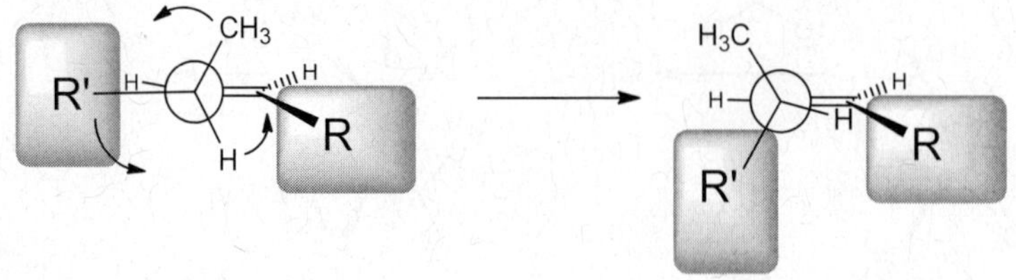

We have traded one eclipsing interaction for another (presumably similar), but notice that we have lost the gauche interaction. Accordingly, we expect this conformation to be the most stable conformation, looking down the C4-C5 bond. This means that the molecule will spend most of its time in this conformation. Notice that, in this lowest energy conformation, the bottom face of the π bond is sterically hindered, while the top face is relatively unhindered:

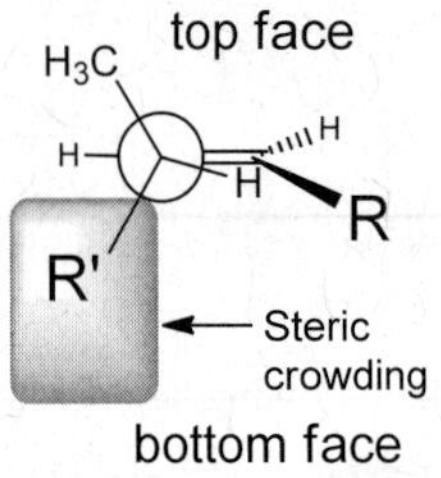

That is, the top face is more accessible (most of the time). As such, if I_2 approaches from the top face of the π bond, the transition state will be lower in energy then if attack occurs from the bottom face. So attack of the top face occurs more readily.

9.94.
The first step involves formation of a bromonium ion, which requires three curved arrows (See Mechanism 9.5), followed by an intramolecular nucleophilic attack in which the OH group functions as a nucleophilic center and attacks the bromonium ion. Deprotonation then affords a cyclic ether. In the last step, a methoxide ion functions as a base and removes a proton, which leads to expulsion of bromide in an E2 process.

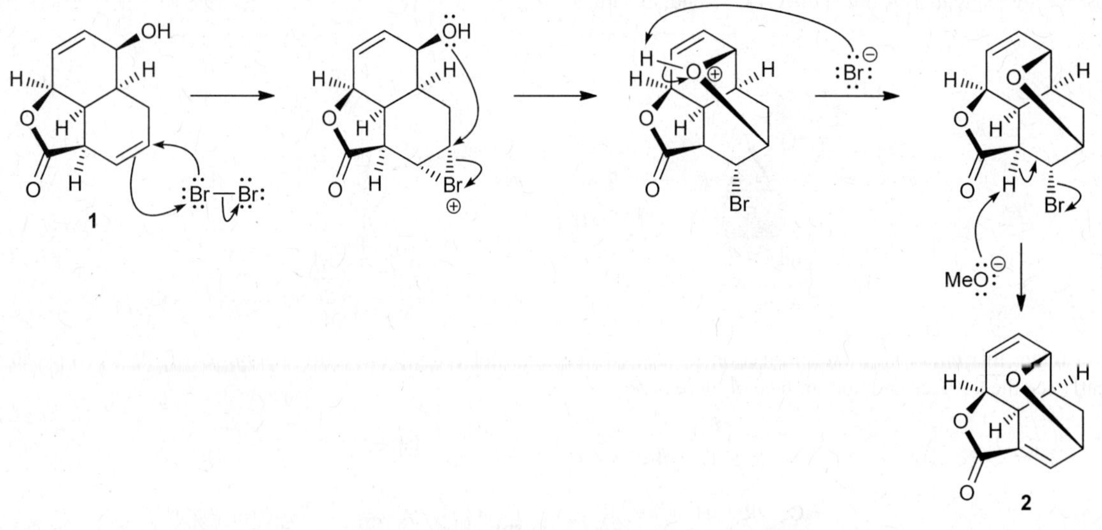

You might be wondering about the stereochemistry of the last step (the E2 process). In general, E2 processes occur more rapidly when the H and the leaving group are *anti*-periplanar in the transition state. However, it is possible for an E2 process to occur via a transition state in which the H and leaving group are *syn*-periplanar. In general, this is not favored (the transition state is high in energy because all groups are eclipsed rather than staggered), but in this case, the rigid geometry of the polycyclic structure essentially locks the H and the leaving group into a *syn*-periplanar arrangement, where the H and the leaving group are eclipsing each other. As such, the reaction can occur, because the minimum requirement of periplanarity is still met.

Chapter 10
Alkynes

Review of Concepts

Fill in the blanks below. To verify that your answers are correct, look in your textbook at the end of Chapter 10. Each of the sentences below appears verbatim in the section entitled *Review of Concepts and Vocabulary*.

- A triple bond is comprised of three separate bonds: one ____ bond and two ____ bonds.
- Alkynes exhibit _____ geometry and can function either as bases or as _____.
- Monosubstituted acetylenes are **terminal alkynes**, while disubstituted acetylenes are _____ **alkynes**.
- Catalytic hydrogenation of an alkyne yields an _____.
- A **dissolving metal reduction** will convert an alkyne into a _____ alkene.
- Acid-catalyzed hydration of alkynes is catalyzed by mercuric sulfate to produce an _____ that cannot be isolated because it is rapidly converted into a ketone.
- Enols and ketones are _____, which are constitutional isomers that rapidly interconvert via the migration of a proton.
- When treated with ozone, followed by water, internal alkynes undergo oxidative cleavage to produce _____.
- Alkynide ions undergo _____ when treated with an alkyl halide (methyl or primary).

Review of Skills

Fill in the blanks and empty boxes below. To verify that your answers are correct, look in your textbook at the end of Chapter 10. The answers appear in the section entitled *SkillBuilder Review*.

10.1 Assembling the Systematic Name of an Alkyne

PROVIDE A SYSTEMATIC NAME FOR THE FOLLOWING COMPOUND

1) IDENTIFY THE PARENT
2) IDENTIFY AND NAME SUBSTITUENTS
3) ASSIGN LOCANTS TO EACH SUBSTITUENT
4) ALPHABETIZE

10.2 Predicting the Position of Equilibrium for the Deprotonation of a Terminal Alkyne

CIRCLE THE SIDE OF THE EQUILIBRIUM THAT IS FAVORED IN THE FOLLOWING ACID-BASE REACTION

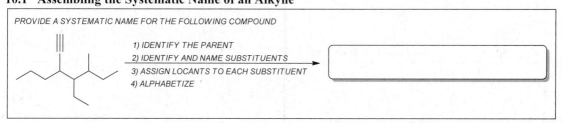

$$R-C\equiv C-H \quad + \quad {}^{\ominus}{:}\ddot{O}H \quad \rightleftharpoons \quad R-C\equiv C{:}^{\ominus} \quad + \quad H_2O$$

10.3 Drawing the Mechanism of Acid-Catalyzed Keto-Enol Tautomerization

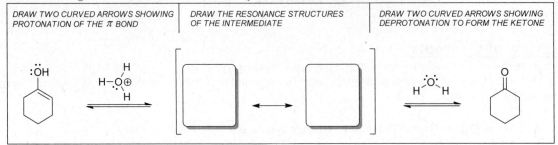

10.4 Choosing the Appropriate Reagents for the Hydration of an Alkyne

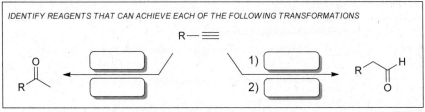

10.5 Alkylating Terminal Alkynes

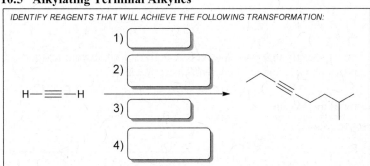

10.6 Interconverting Alkenes and Alkynes

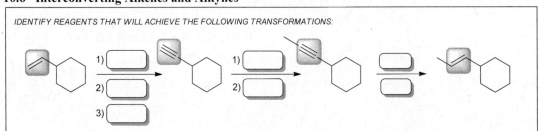

Review of Reactions

Identify the reagents necessary to achieve each of the following transformations. To verify that your answers are correct, look in your textbook at the end of Chapter 10. The answers appear in the section entitled *Review of Reactions*.

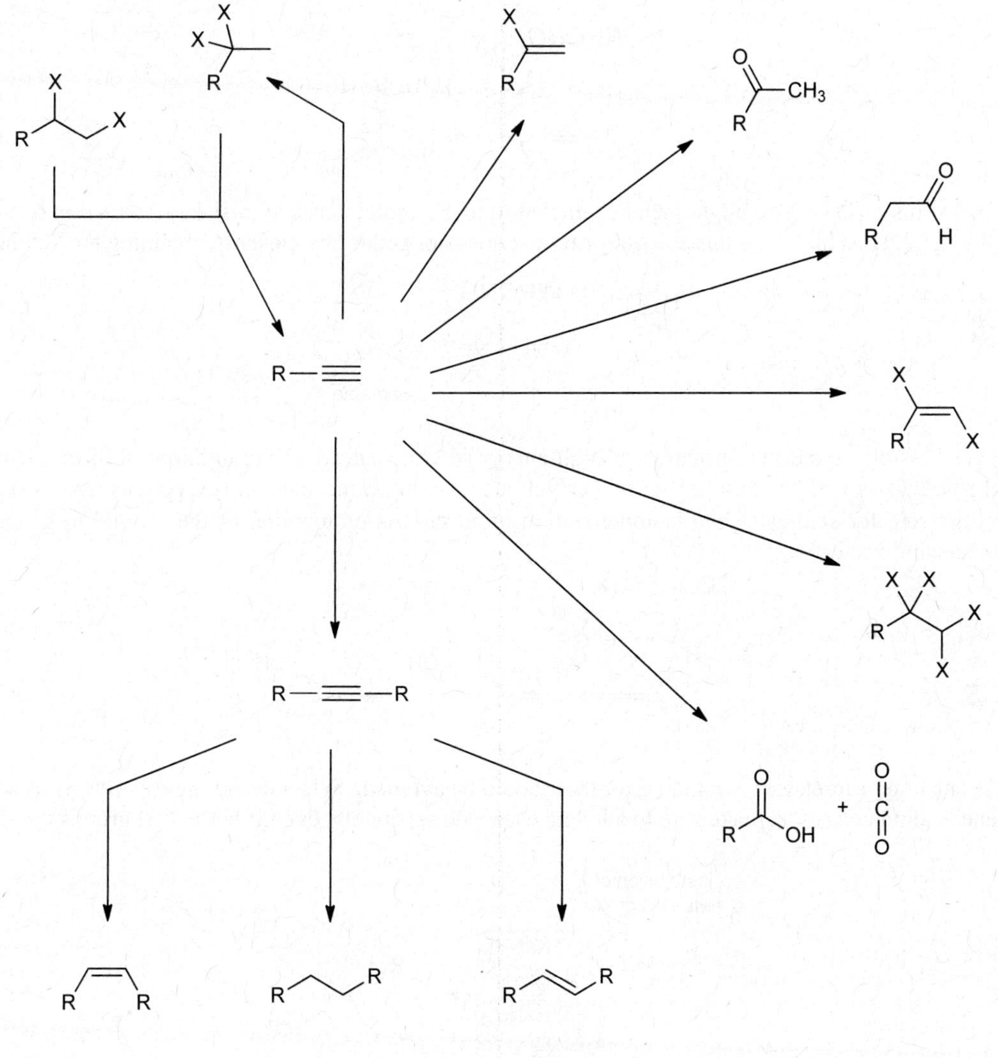

Common Mistakes to Avoid

When drawing a mechanism for the acid-catalyzed tautomerization of an enol, the first step is protonation. Students sometimes get confused about where to place the proton during this first step. Indeed, there are two possible locations where protonation could occur (a lone pair or the π bond):

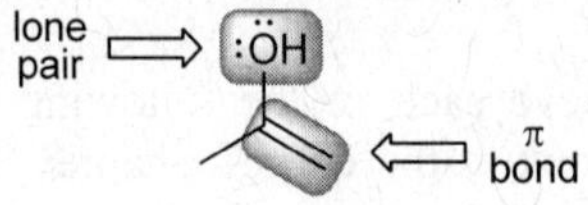

Don't forget that it is the π bond that is protonated during tautomerization of the enol. It is a common student error to protonate the OH group, like this:

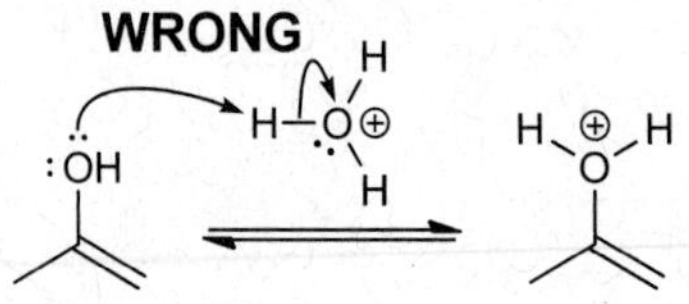

If you make this mistake, you might then be tempted to make another critical mistake - formation of a vinyl carbocation, which is likely too unstable to form (and does not get us any closer to obtaining the ketone).

Whenever possible, avoid the formation of high-energy intermediates that are unlikely to form. This is a general rule that should be followed whenever you are drawing a mechanism (exceptions are rare). The correct first step for acid-catalyzed tautomerization of an enol is protonation of the π bond to generate a resonance-stabilized cation:

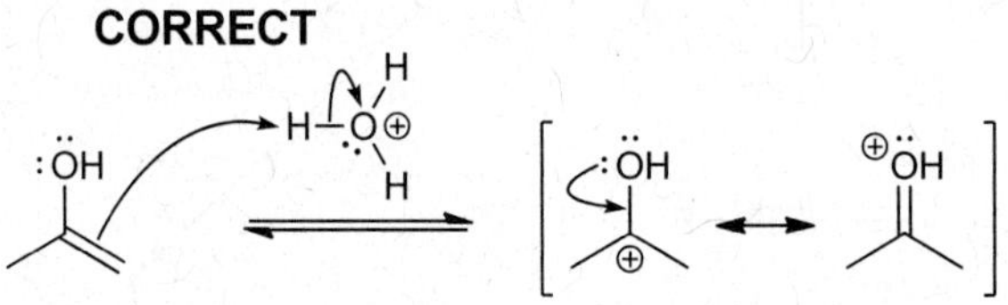

There is one other (unrelated) common error that should be avoided. When designing a synthesis in which acetylene is alkylated twice, make sure to alkylate each side separately, even if both alkyl groups are the same:

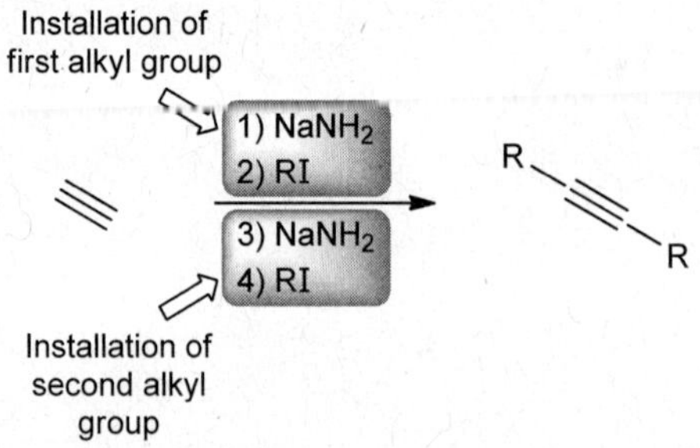

It is a common student error to show the reagents just one time, assuming that alkylation will occur on both sides of acetylene. If the reagents for alkylation are only shown once, then alkylation will only occur once:

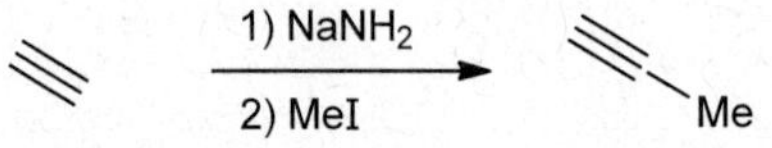

Useful Reagents

The following is a list of commonly encountered reagents for reactions of alkynes:

Reagents	Name of Reaction	Description of Reaction
1) excess $NaNH_2$ 2) H_2O	Elimination	When treated with these reagents, a vicinal or geminal dibromide is converted to an alkyne.
HX	Hydrohalogenation	When treated with HX, an alkyne undergoes Markovnikov addition (excess HX gives two addition reactions to afford a geminal dihalide).
H_2SO_4, H_2O, $HgSO_4$	Acid-cat. hydration	When treated with these reagents, a terminal alkyne undergoes Markovnikov addition of H and OH to give an enol, which quickly tautomerizes to give a ketone.
1) R_2BH 2) H_2O_2, NaOH	Hydroboration-oxidation	When treated with these reagents, a terminal alkyne undergoes *anti*-Markovnikov addition of H and OH to give an enol, which quickly tautomerizes to give an aldehyde.
X_2	Halogenation	When treated with this reagent, an alkyne undergoes addition of X and X (excess X_2 gives a tetrahalide).
1) O_3 2) H_2O	Ozonolysis	When treated with these reagents, an alkyne undergoes oxidative cleavage of the C≡C bond. Internal alkynes are converted into two carboxylic acids, while terminal alkynes are converted into a carboxylic acid and carbon dioxide.
H_2, Lindlar's catalyst	Hydrogenation	When treated with these reagents, an alkyne is converted to a *cis*-alkene.
H_2, Pt	Hydrogenation	When treated with these reagents, an alkyne is converted to an alkane.
Na, NH_3 (*l*)	Dissolving metal reduction	When treated with these reagents, an internal alkyne is converted to a *trans*-alkene.

Solutions

10.1.
(a) We begin by identifying the parent. The longest chain is six carbon atoms, so the parent is hexyne. There are no substituents. We must include a locant that identifies the position of the triple bond ("3" indicates that the triple bond is located between C3 and C4). This is determined by numbering the parent, which can be done in this case either from left to right or vice versa (either way gives the same result).

3-Hexyne

(b) We begin by identifying the parent. The longest chain is six carbon atoms, so the parent is hexyne. There is one substituent – a methyl group (highlighted). In this

case, the triple bond is at C3 regardless of which way we number the parent, so the parent chain is numbered starting from the side that gives the substituent the lowest possible number. According to this numbering scheme, the methyl group is located at C2:

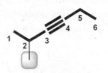

2-Methyl-3-hexyne

(c) We begin by identifying the parent. The longest chain is eight carbon atoms, so the parent is octyne. There are no substituents. We must include a locant that identifies the position of the triple bond. The parent chain is numbered so that the triple bond is assigned the lowest possible locant ("3" indicates that the triple bond is located between C3 and C4).

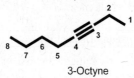

3-Octyne

(d) We begin by identifying the parent. The longest chain is four carbon atoms, so the parent is butyne. There are two substituents – both methyl groups (highlighted). We number the parent so that the triple bond receives the lowest possible locant (C1). According to this numbering scheme, the methyl groups are both located at C3:

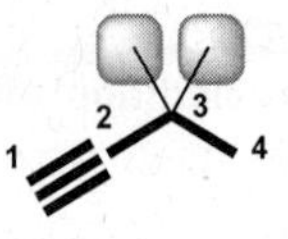

3,3-Dimethyl-1-butyne

10.2.
(a) The parent (pentyne) indicates a chain of five carbon atoms. The triple bond is between C2 and C3, and there are two methyl groups (highlighted), both located at C4.

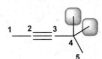

4,4-Dimethyl-2-pentyne

(b) The parent (heptyne) indicates a chain of seven carbon atoms. The triple bond is between C3 and C4, and there are three substituents (highlighted) – two methyl groups (at C2 and C5), as well as an ethyl group at C5.

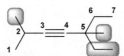

5-Ethyl-2,5-dimethyl-3-heptyne

10.3. The parent is cyclononyne, so we draw a nine-membered ring that incorporates a triple bond. The triple bond is (by definition) between C1 and C2, and a methyl group is located at C3. This position (C3) is a chirality center, with the *R* configuration, shown here:

10.4. Terminal alkynes have the structure R–C≡C–H. The molecular formula indicates six carbon atoms, so the

R group must be comprised of four carbon atoms. There are four different ways to connect these carbon atoms. They can be connected in a linear fashion, like this:

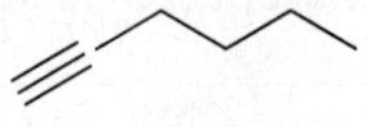

1-Hexyne

or there can be one methyl branch, which can be placed in either of two locations (C3 or C4), shown here:

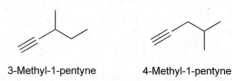

3-Methyl-1-pentyne   4-Methyl-1-pentyne

or there can be two methyl branches, as shown here:

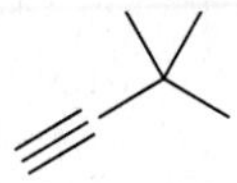

3,3-Dimethyl-1-butyne

10.5.
(a) Yes, as seen in Table 10.1, NaNH$_2$ is a sufficiently strong base to deprotonate a terminal alkyne.
(b) No, as seen in Table 10.1, NaOEt is not a sufficiently strong base to deprotonate a terminal alkyne.
(c) No, as seen in Table 10.1, NaOH is not a sufficiently strong base to deprotonate a terminal alkyne.

(d) Yes, as seen in Table 10.1, BuLi is a sufficiently strong base to deprotonate a terminal alkyne.
(e) Yes, as seen in Table 10.1, NaH is a sufficiently strong base to deprotonate a terminal alkyne.

(f) No, as seen in Table 10.1, *t*-BuOK is not a sufficiently strong base to deprotonate a terminal alkyne.
10.6.
(a) In the conjugate base of methyl amine (CH$_3$NH$_2$), the negative charge is associated with an sp^3 hybridized nitrogen atom. In the conjugate base of HCN, the negative charge is associated with an sp hybridized carbon atom. The latter is more stable, because the charge is closer to the positively charged nucleus. As a result, HCN is a stronger acid than methyl amine.
(b) The pK_a of HCN is lower than the pK_a of a terminal alkyne. Therefore, cyanide cannot be used as a base to deprotonate a terminal alkyne, as it would involve the formation of a stronger acid.

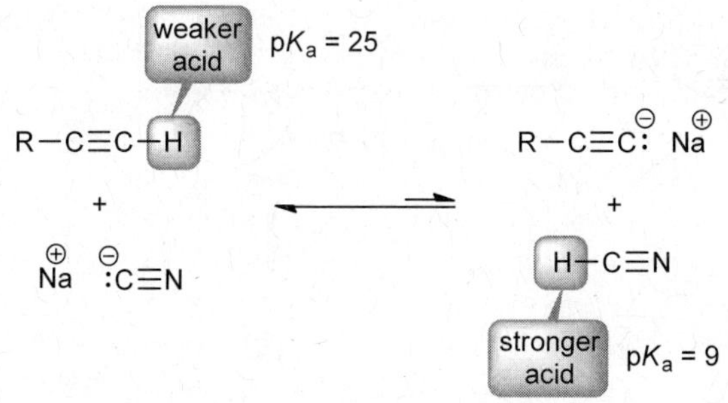

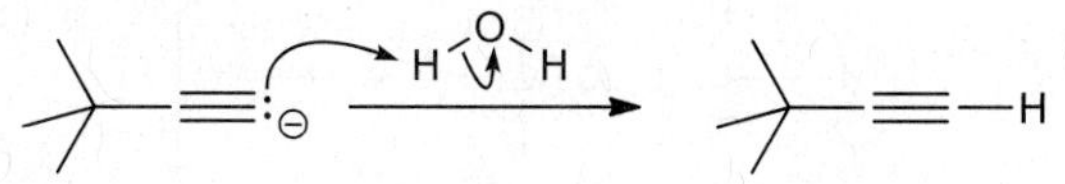

(b) The starting material is a vicinal dichloride. When treated with excess sodium amide (NaNH$_2$), two, successive E2 reactions occur (each of which requires three curved arrows, as shown below). The resulting terminal alkyne is then deprotonated to give an alkynide ion:

10.7.

(a) The starting material is a geminal dibromide. When treated with excess sodium amide (NaNH$_2$), two, successive E2 reactions occur (each of which requires three curved arrows, as shown below). The resulting terminal alkyne is then deprotonated to give an alkynide ion:

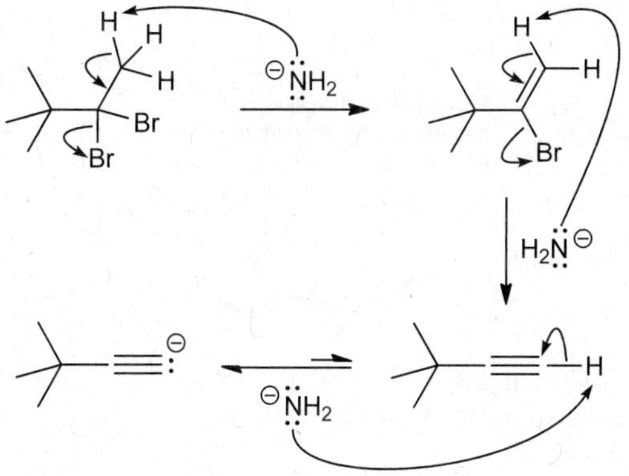

After the reaction is complete, water (H$_2$O) is introduced into the reaction flask to protonate the alkynide ion, thereby giving the terminal alkyne.

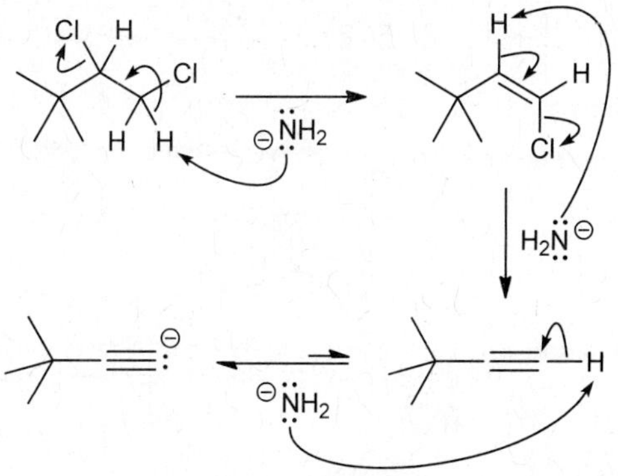

After the reaction is complete, water (H$_2$O) is introduced into the reaction flask to protonate the alkynide ion, thereby giving the terminal alkyne:

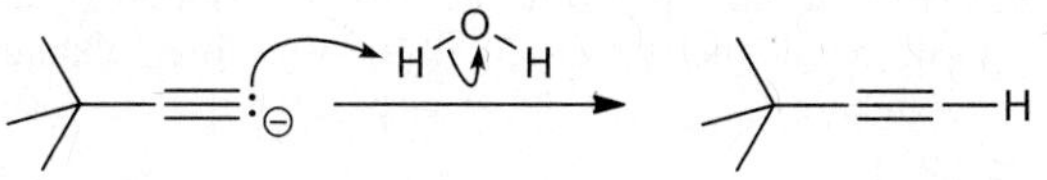

10.8. Deprotonation of 2-pentyne generates a resonance-stabilized anion, which is then protonated by NH$_3$ to give an allene (a compound with C=C=C). The allene is then deprotonated to give a resonance-stabilized anion, which is then protonated by NH$_3$ to give 1-pentyne. Deprotonation of this terminal alkyne gives an alkynide ion. Formation of this alkynide ion pushes the equilibrium to favor this isomerization process.

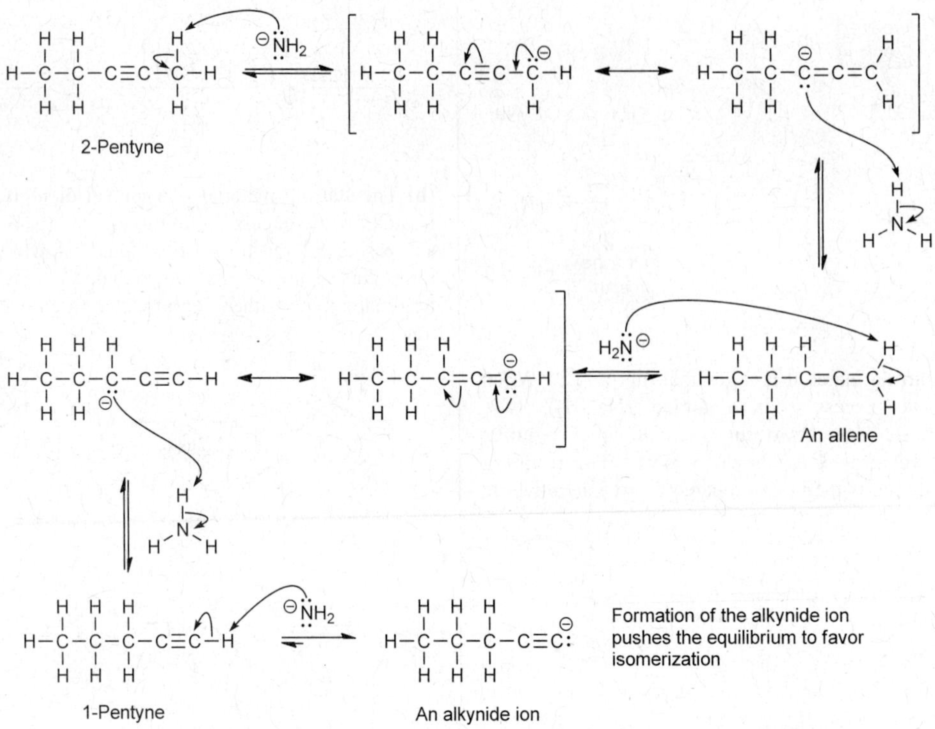

2-Pentyne

An allene

Formation of the alkynide ion pushes the equilibrium to favor isomerization

1-Pentyne   An alkynide ion

10.9.
(a) When hydrogenation is performed in the presence of a poisoned catalyst (such as Lindlar's catalyst), the alkyne is reduced to a *cis* alkene. When Pt is used as the catalyst, the alkyne is reduced all the way to an alkane, as shown here:

H_2 Lindlar's catalyst

H_2 Pt

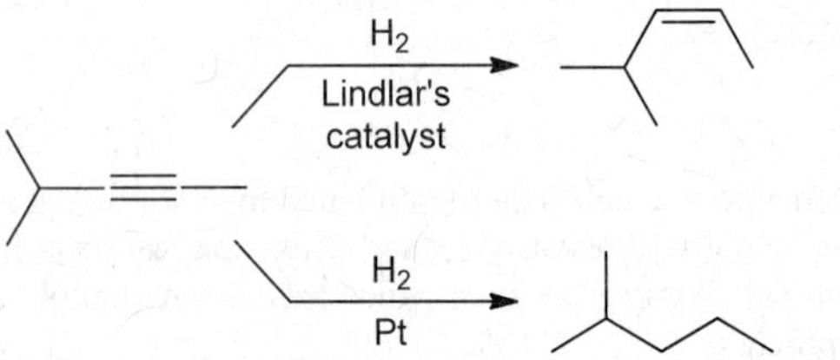

(b) When hydrogenation is performed in the presence of a poisoned catalyst (such as Ni_2B), the alkyne is reduced to a *cis* alkene. When nickel is used as the catalyst, the alkyne is reduced all the way to an alkane, as shown here:

H_2 Ni_2B

H_2 Ni

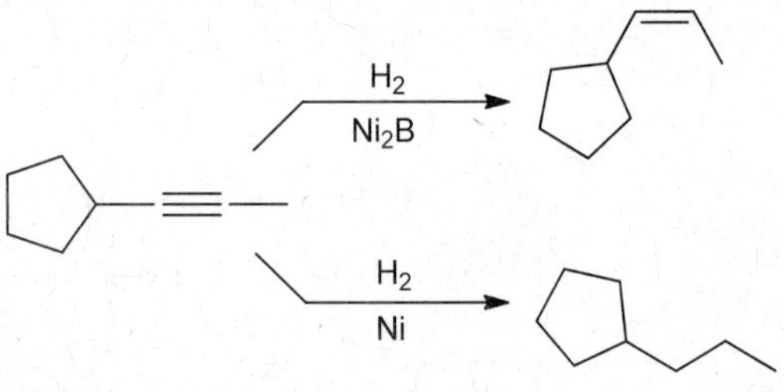

10.10.
(a) When treated with sodium in liquid ammonia, the alkyne is converted to a *trans* alkene:

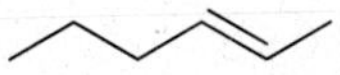

(b) When treated with sodium in liquid ammonia, the alkyne is converted to a *trans* alkene:

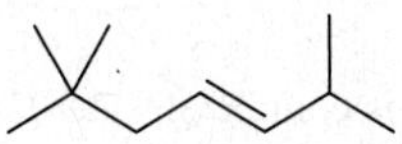

(c) When treated with sodium in liquid ammonia, the alkyne is converted to a *trans* alkene:

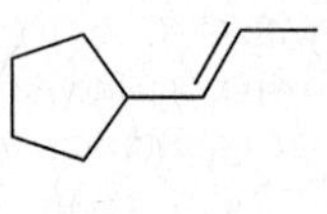

(d) When treated with sodium in liquid ammonia, the alkyne is converted to a *trans* alkene:

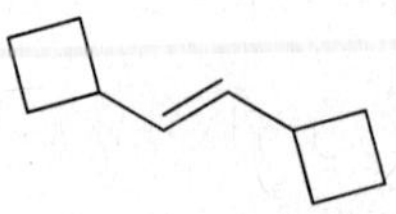

10.11.
(a) When the alkyne is treated with molecular hydrogen (H_2) in the presence of a poisoned catalyst (such as Lindlar's catalyst), the alkyne is reduced to a *cis* alkene. If instead, the alkyne is treated with sodium in liquid ammonia, a dissolving metal reduction occurs, giving a *trans* alkene, as shown:

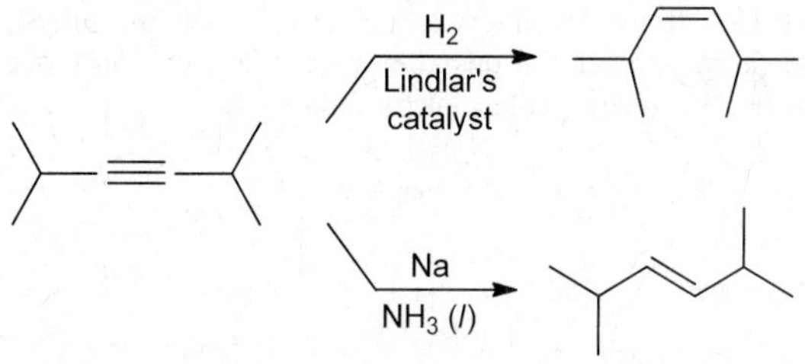

(b) When the alkyne is treated with sodium in liquid ammonia, a dissolving metal reduction occurs, giving a *trans* alkene. If instead, the alkyne is treated with molecular hydrogen (H_2) in the presence of a catalyst such as platinum (NOT a poisoned catalyst), the alkyne is reduced to an alkane, as shown:

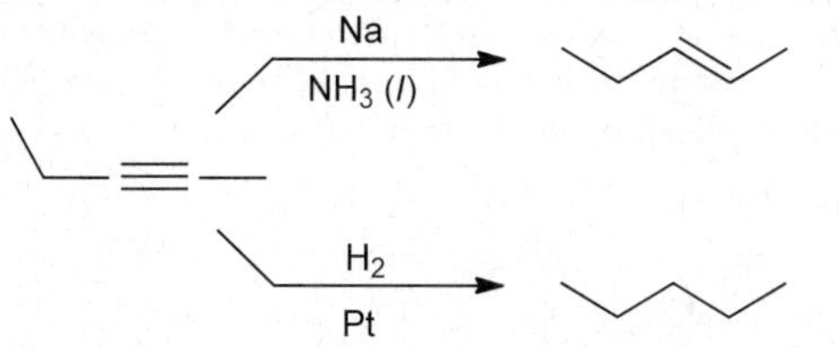

10.12. The product is a disubstituted alkene, so the starting alkyne must be an internal alkyne (rather than a terminal alkyne):

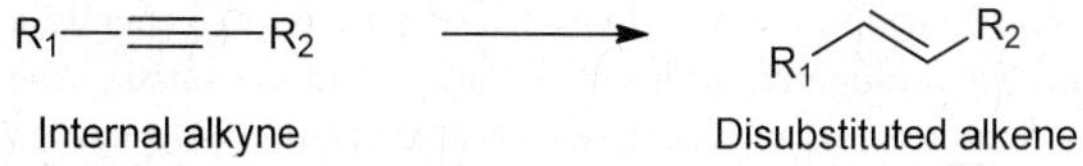

Internal alkyne   Disubstituted alkene

The molecular formula of the alkyne indicates five carbon atoms. Two of those atoms are the *sp* hybridized carbon atoms of the triple bond. The remaining three carbon atoms must be in the R groups. So, one R group must be a methyl group, and the other must be an ethyl group:

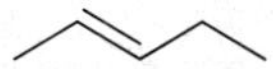

10.13.
(a) The starting alkyne is terminal, and when treated with excess HCl, two successive addition reactions occur, producing a geminal dihalide. The two chlorine atoms are installed at the more substituted, secondary position, rather than the less substituted, primary position:

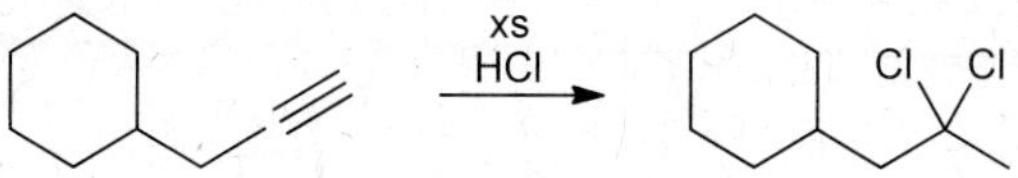

(b) The starting material is a geminal dichloride, and treatment with excess sodium amide (followed by work-up with water) gives a terminal alkyne.

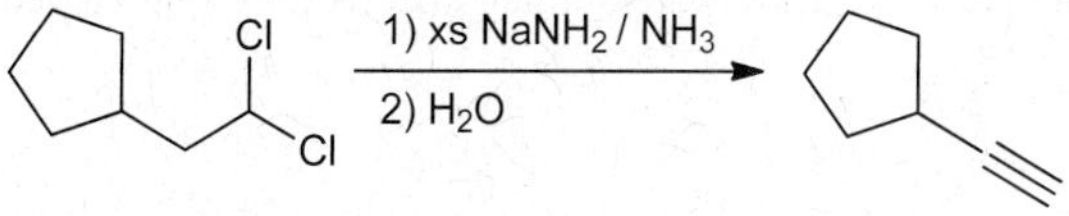

(c) The starting alkyne is terminal, and when treated with excess HBr, two successive addition reactions occur, producing a geminal dibromide. The two bromine atoms are installed at the more substituted, secondary position, rather than the less substituted, primary position:

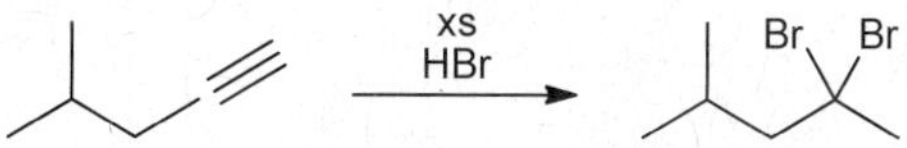

(d) The starting material is a geminal dibromide, and treatment with excess sodium amide (followed by work-up with water) gives a terminal alkyne.

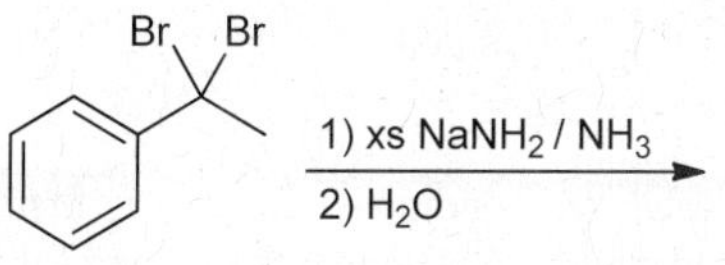

(e) The starting material is a geminal dichloride, and treatment with excess sodium amide (followed by work-up with water) gives a terminal alkyne. When this alkyne is treated with HBr, in the presence of peroxides, an *anti*-Markovnikov addition occurs, in which Br is installed at the less substituted position. This gives rise to two stereoisomers, as shown:

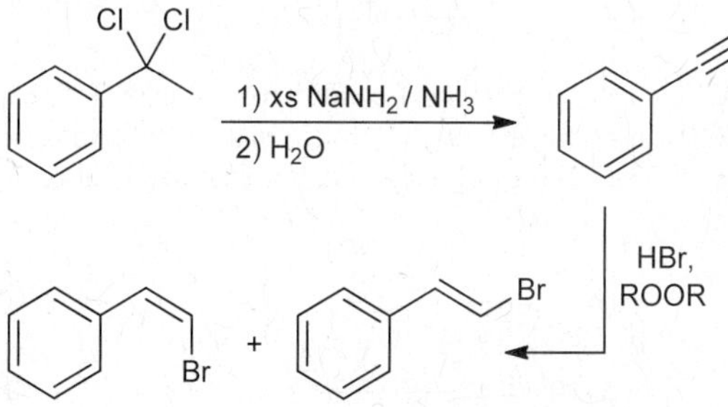

(f) The starting material is a geminal dichloride, and treatment with excess sodium amide (followed by work-up with water) gives a terminal alkyne. When this alkyne is treated with excess HBr, two addition reactions occur, installing two bromine atoms at the more substituted position (Markovnikov addition), as shown:

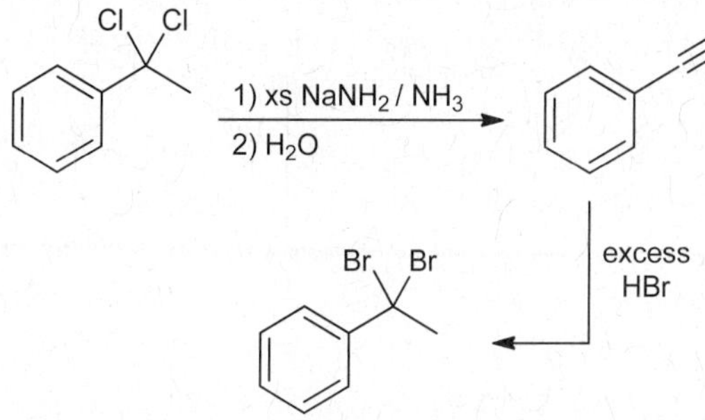

10.14. The starting material is a geminal dichloride, and the product is also a geminal dichloride. The difference between these compounds is the placement of the chlorine atoms. We did not learn a single reaction that will change the locations of the chlorine atoms. But the desired transformation can be achieved via an alkyne.

Specifically, the starting material is treated with excess sodium amide (followed by water work-up) to give the terminal alkyne, which is then treated with excess HCl to give the desired compound:

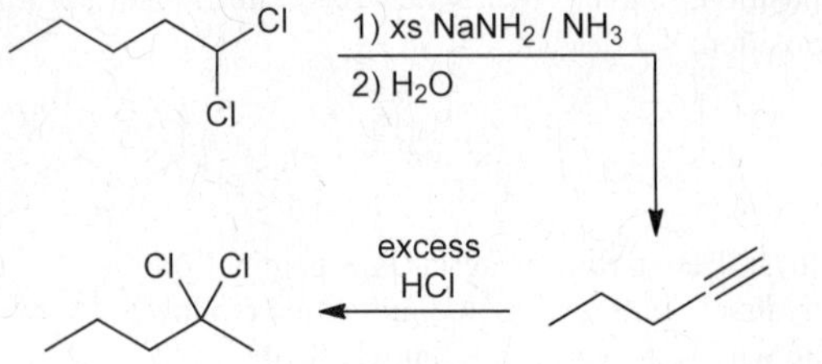

10.15. If two products are obtained, then the alkyne must be internal and unsymmetrical. There is only one such alkyne with molecular formula C_5H_8:

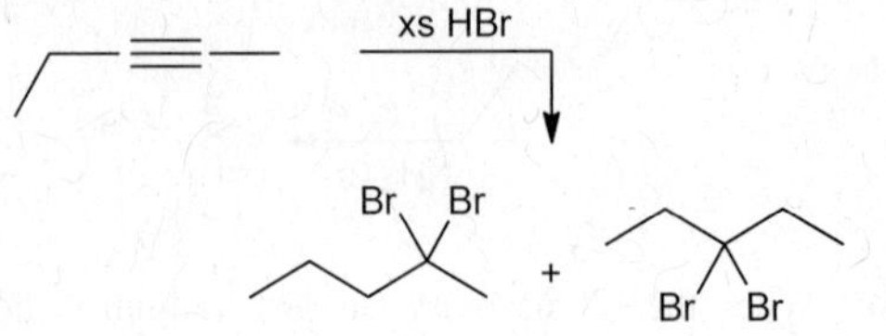

10.16.

(a) Under acid-catalyzed conditions, the enol is first protonated (which requires two curved arrows) to generate a resonance-stabilized cation. Notice that the π bond (of the enol) is protonated in this step (rather than protonating the OH group). The resulting resonance-stabilized cation is then deprotonated by water, which also requires two curved arrows, as shown:

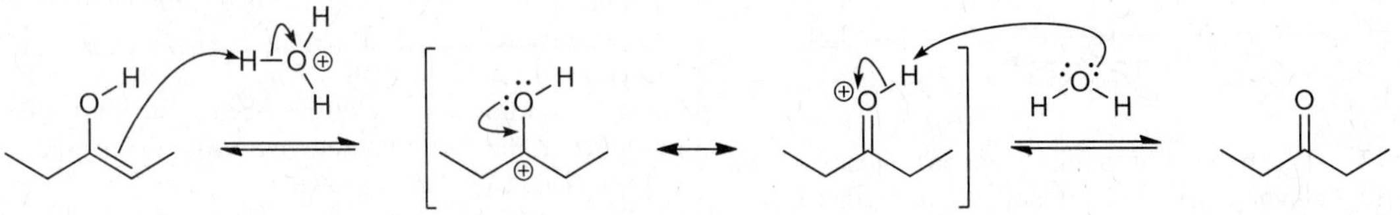

(b) Under acid-catalyzed conditions, the enol is first protonated (which requires two curved arrows) to generate a resonance-stabilized cation. Notice that the π bond (of the enol) is protonated in this step (rather than protonating the OH group). The resulting resonance-stabilized cation is then deprotonated by water, which also requires two curved arrows, as shown:

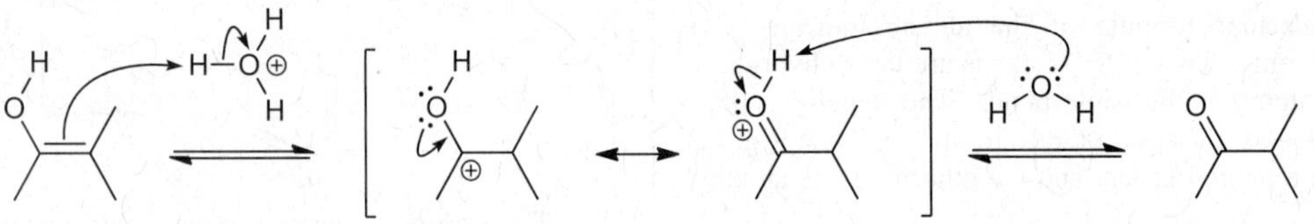

(c) Under acid-catalyzed conditions, the enol is first protonated (which requires two curved arrows) to generate a resonance-stabilized cation. Notice that the π bond (of the enol) is protonated in this step (rather than protonating the OH group). The resulting resonance-stabilized cation is then deprotonated by water, which also requires two curved arrows, as shown:

other
resonance
structures

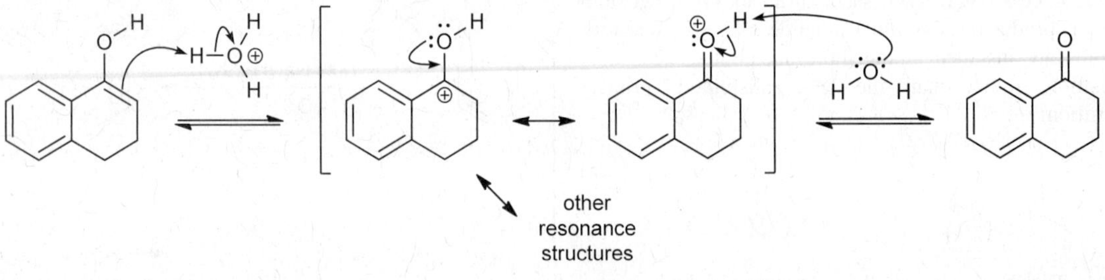

(d) Under acid-catalyzed conditions, the enol is first protonated (which requires two curved arrows) to generate a resonance-stabilized cation. Notice that the π bond (of the enol) is protonated in this step (rather than protonating the OH group). The resulting resonance-stabilized cation is then deprotonated by water, which also requires two curved arrows, as shown:

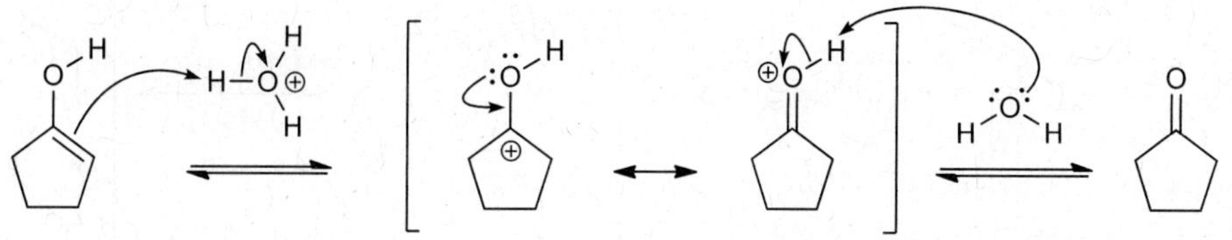

10.17. Under acid-catalyzed conditions, the starting material is first protonated (much the way an enol is protonated) to generate a resonance-stabilized cation. This step requires two curved arrows. Notice that the π bond is protonated (rather than protonating the nitrogen atoms), because protonation of the π bond results in a cation that is highly stabilized by resonance (notice that there are three resonance structures). This resonance-stabilized cation is then deprotonated by water, which also requires two curved arrows, as shown:

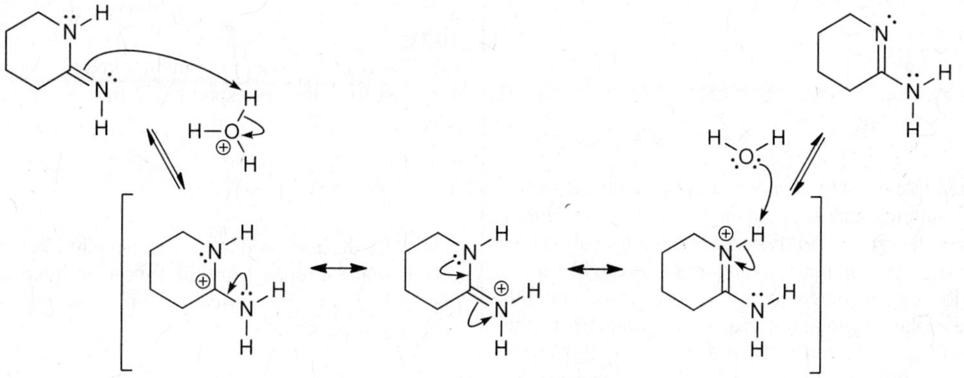

10.18.
(a) When treated with aqueous acid (in the presence of mercuric sulfate), the terminal alkyne undergoes Markovnikov addition of H and OH across the alkyne, giving an enol. The enol is not isolated, because upon its formation, it undergoes tautomerization to give a methyl ketone:

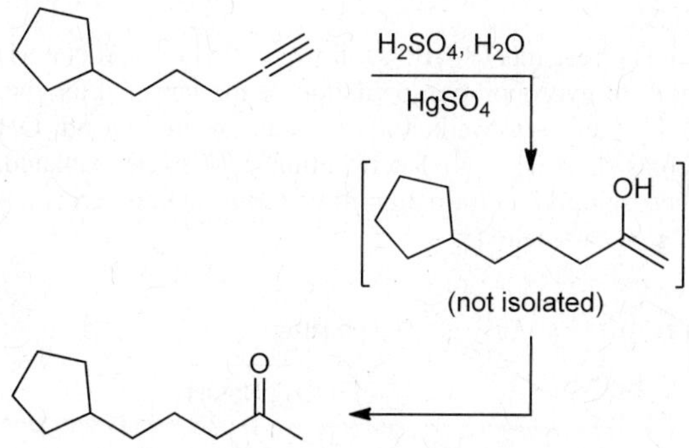

(b) When treated with aqueous acid (in the presence of mercuric sulfate), the terminal alkyne undergoes Markovnikov addition of H and OH across the alkyne, giving an enol. The enol is not isolated, because upon its formation, it undergoes tautomerization to give a methyl ketone:

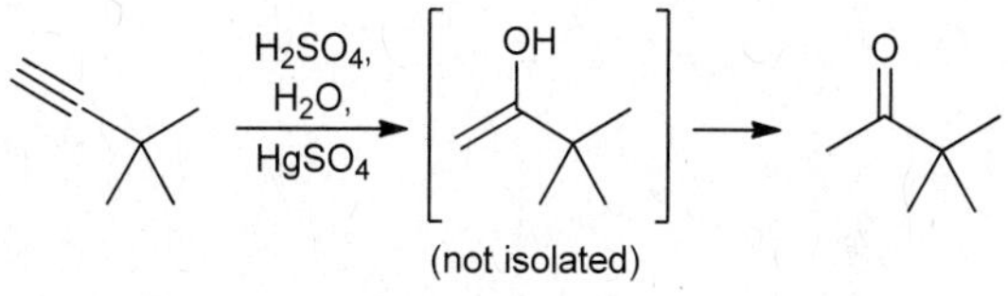

(c) When treated with aqueous acid (in the presence of mercuric sulfate), an alkyne undergoes addition of H and OH across the alkyne. In this case, the starting alkyne is not terminal – it is an internal alkyne. As such, there are two possible regiochemical outcomes, giving rise to two possible enols. Neither of these enols is isolated, because upon their formation, they each undergo tautomerization to give a ketone, as shown here:

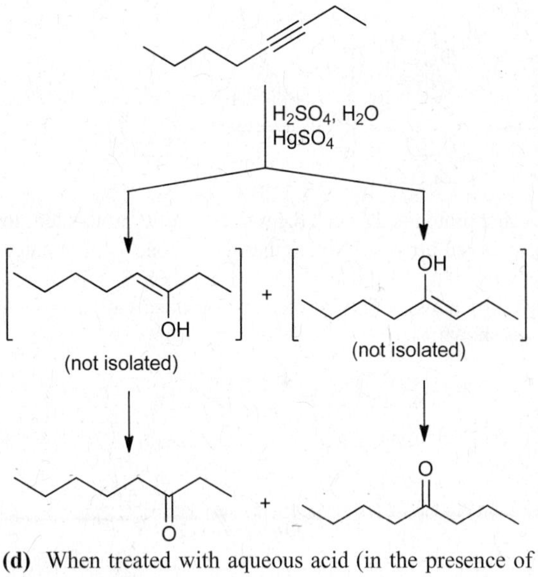

(d) When treated with aqueous acid (in the presence of mercuric sulfate), an alkyne undergoes addition of H and OH across the alkyne. In this case, the starting alkyne is not terminal – it is an internal alkyne, so we might expect two regiochemical outcomes. But look closely at the structure of the alkyne in this case. It is symmetrical (the triple bond is connected to two identical alkyl groups), and as such, there is only one possible enol that can be formed. This enol is not isolated, because upon its formation, it undergoes tautomerization to give a ketone:

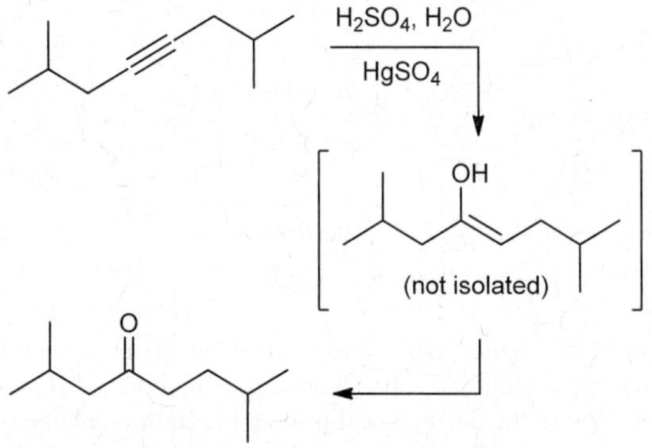

(e) When treated with aqueous acid (in the presence of mercuric sulfate), an alkyne undergoes addition of H and OH across the alkyne. In this case, the starting alkyne is a symmetrical internal alkyne, and as such, there is only one possible enol that can be formed. This enol is not isolated, because upon its formation, it undergoes tautomerization to give a ketone, as shown:

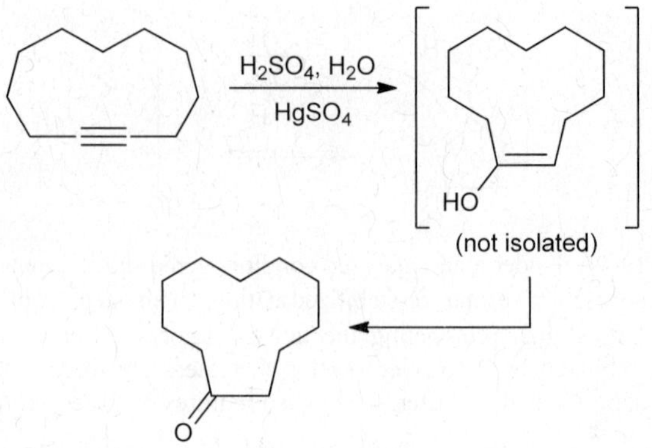

10.19.
(a) The desired product is a methyl ketone, which can be prepared from the corresponding terminal alkyne, shown here:

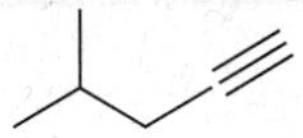

(b) The desired product is not a methyl ketone, but it can be made directly from the following internal alkyne. This alkyne is symmetrical, so only one regiochemical outcome is possible:

(c) The desired product is a methyl ketone, which can be prepared from the corresponding terminal alkyne, shown here:

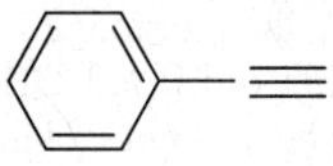

10.20.
(a) The reagents (9-BBN, followed by H_2O_2 and NaOH) indicate hydroboration-oxidation of the terminal alkyne, giving an *anti*-Markovnikov addition of H and OH across the alkyne. The resulting enol is not isolated, because upon its formation, it undergoes tautomerization to give an aldehyde:

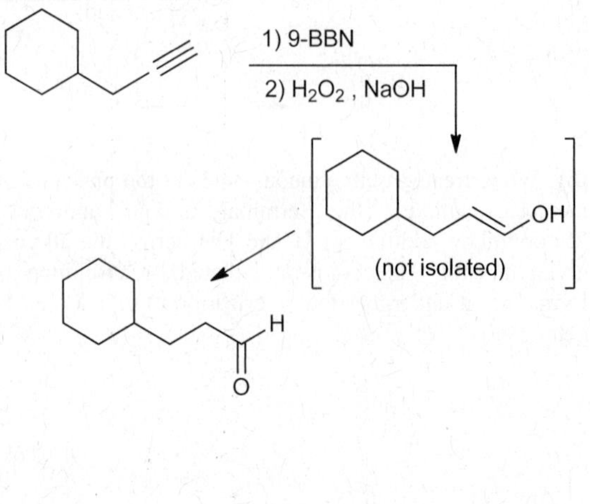

(b) The reagents (disiamylborane, followed by H_2O_2 and NaOH) indicate hydroboration-oxidation of the terminal alkyne, giving an *anti*-Markovnikov addition of H and OH across the alkyne. The resulting enol is not isolated, because upon its formation, it undergoes tautomerization to give an aldehyde:

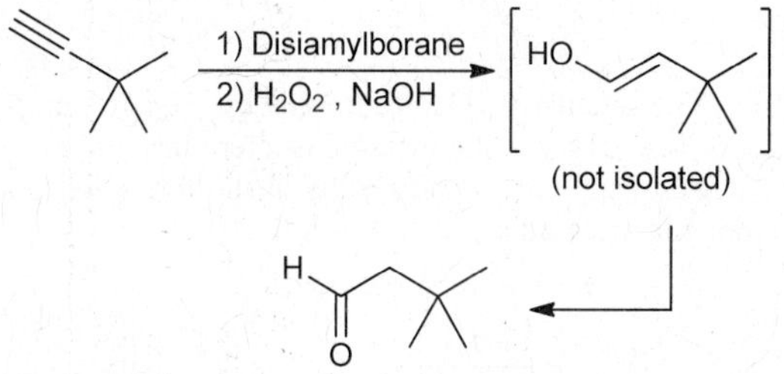

(c) The reagents (9-BBN, followed by H_2O_2 and NaOH) indicate hydroboration-oxidation of the terminal alkyne, giving addition of H and OH. Since the alkyne is symmetrical, regiochemistry is not relevant in this case. The resulting enol is not isolated, because upon its formation, it undergoes tautomerization to give a ketone:

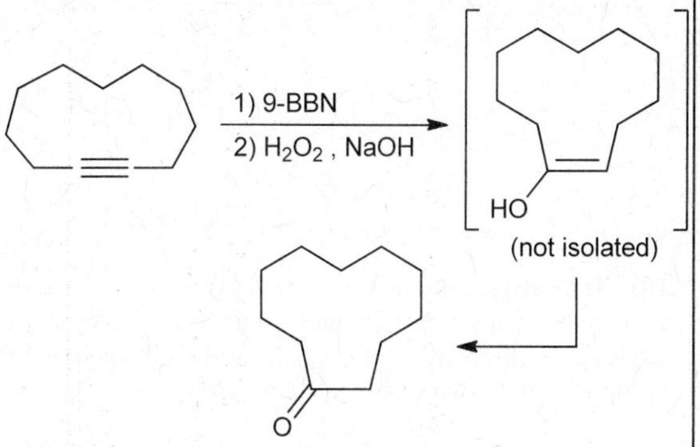

10.21.
(a) The desired product is an aldehyde, which can be prepared from the corresponding terminal alkyne, shown here:

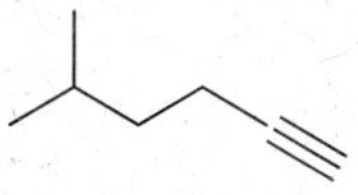

(b) The desired product is not an aldehyde, but it can be made directly from the following internal alkyne. This alkyne is symmetrical, so only one regiochemical outcome is possible:

(c) The desired product is an aldehyde, which can be prepared from the corresponding terminal alkyne, shown here:

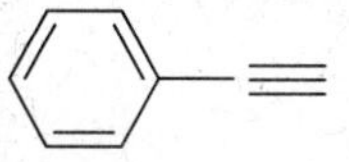

10.22.
(a) The starting material is a terminal alkyne and the product is a methyl ketone. This transformation requires a Markovnikov addition, which can be achieved via an acid-catalyzed hydration in the presence of mercuric sulfate.

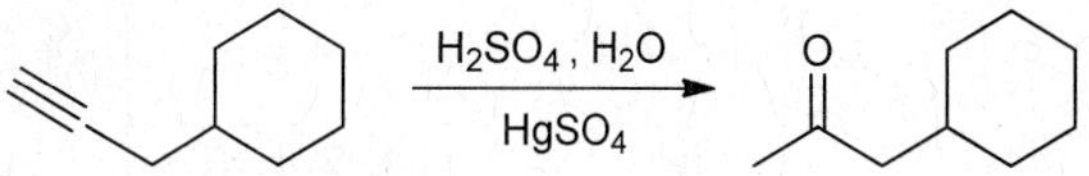

(b) The starting material is a terminal alkyne and the product is an aldehyde. This transformation requires an *anti*-Markovnikov addition, which can be achieved via hydroboration-oxidation.

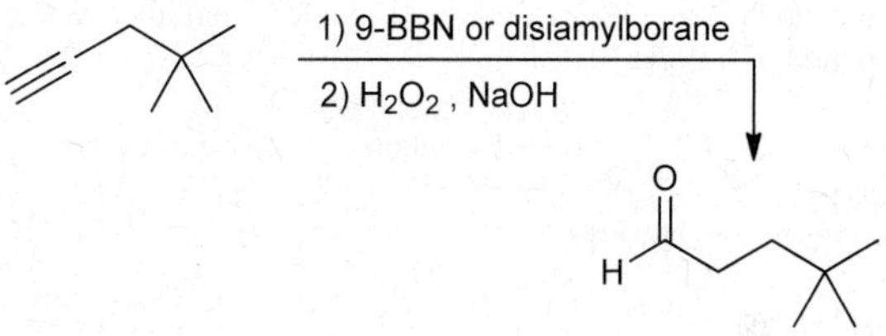

10.23.
(a) The desired product is an aldehyde, which can be made via hydroboration-oxidation of the appropriate terminal alkyne:

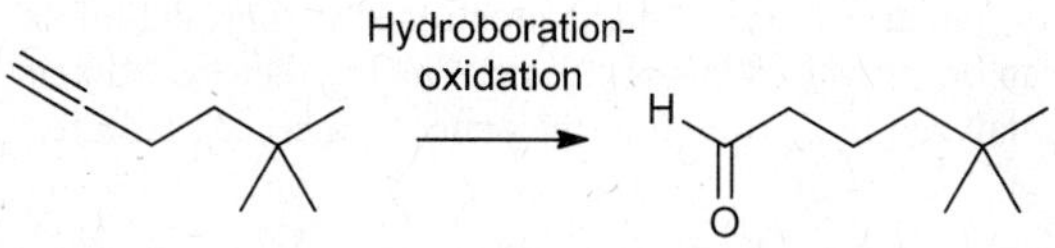

The necessary alkyne can be made from the starting material upon treatment with excess sodium amide, followed by work-up with water:

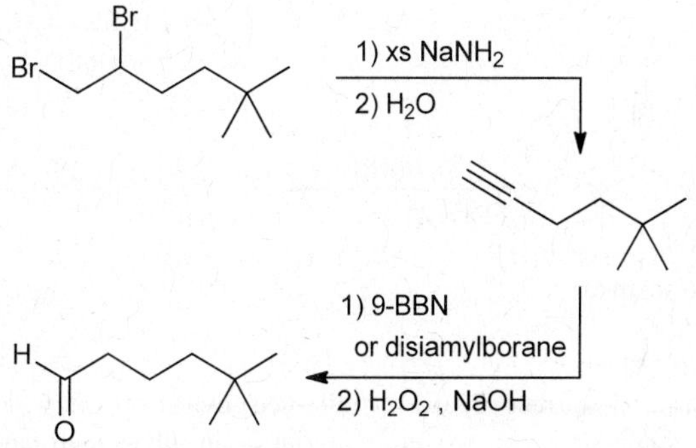

(b) The desired product is a methyl ketone, which can be made via acid-catalyzed hydration of the appropriate terminal alkyne:

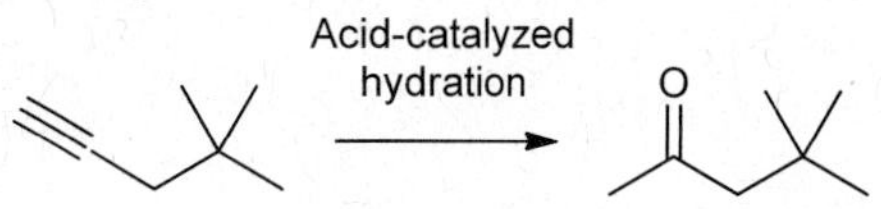

The necessary alkyne can be made from the starting material upon treatment with excess sodium amide, followed by work-up with water:

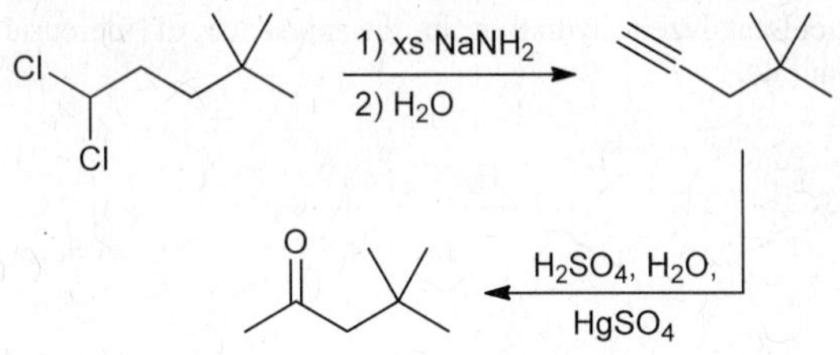

10.24. The product is a methyl ketone, and the starting material is an alkene. We have not seen a method for directly converting an alkene into a ketone. However, we have seen a way of converting a terminal alkyne into a methyl ketone.

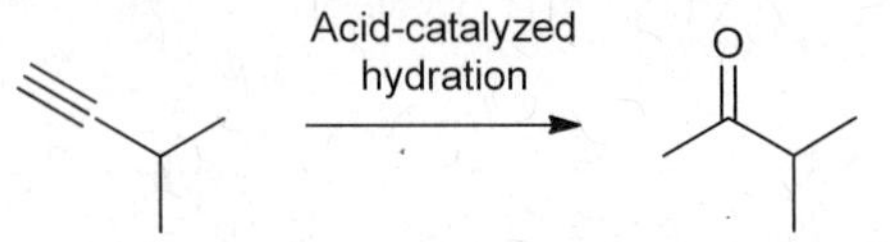

Completing the synthesis requires that we first prepare the alkyne above from the starting alkene, shown in the problem statement. This can be accomplished via a two-step procedure. The alkene is treated with molecular bromine (Br_2) to give a vicinal dibromide, which is then treated with excess $NaNH_2$ (followed by water work-up) to give an alkyne. And as mentioned earlier, the alkyne can be converted into the desired methyl ketone via acid-catalyzed hydration in the presence of mercuric sulfate:

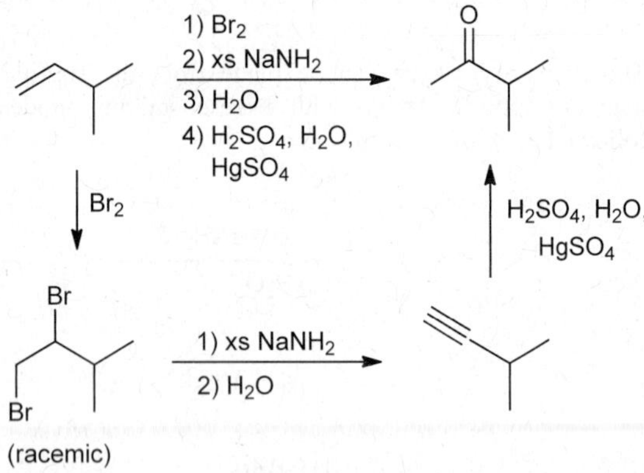

10.25.
(a) The reagents (O_3, followed by H_2O) indicate ozonolysis. The starting material is an unsymmetrical, internal alkyne, so cleavage of the C≡C bond results in the formation of two carboxylic acids:

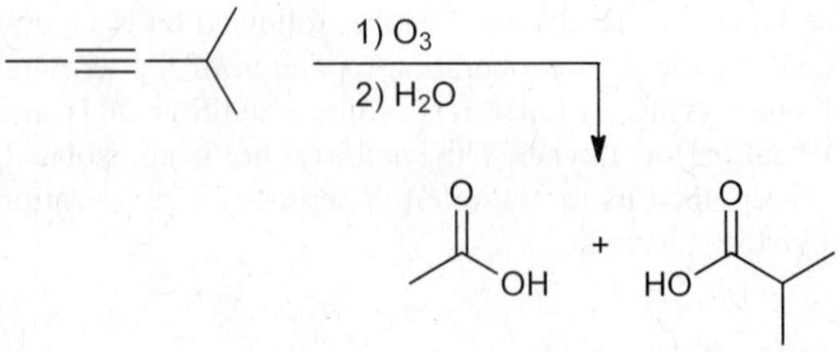

(b) The reagents (O_3, followed by H_2O) indicate ozonolysis. The starting material is a terminal alkyne, so cleavage of the C≡C bond results in the formation of a carboxylic acid and carbon dioxide (CO_2):

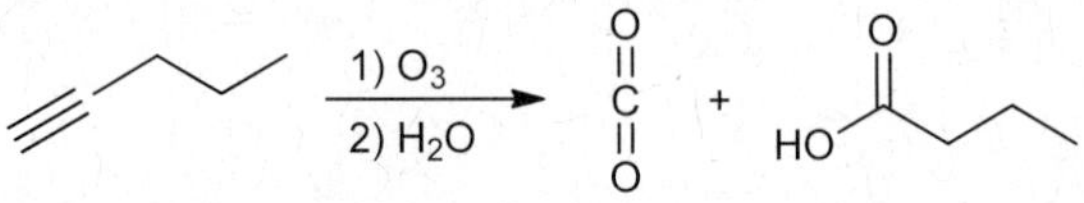

(c) The reagents (O_3, followed by H_2O) indicate ozonolysis. The starting material is an unsymmetrical, internal alkyne, so cleavage of the C≡C bond results in the formation of two carboxylic acids:

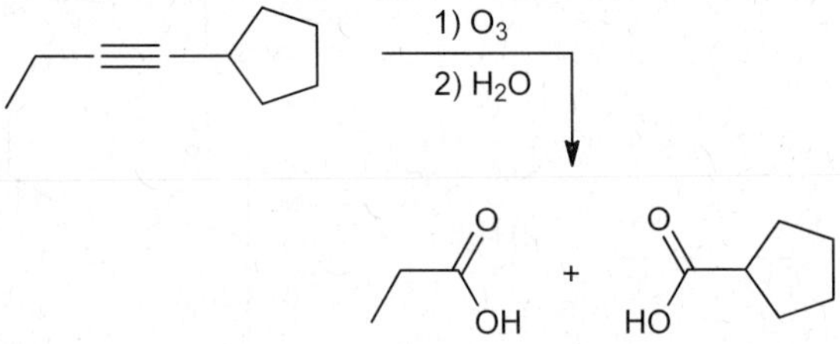

(d) The reagents (O_3, followed by H_2O) indicate ozonolysis. The starting material is a cycloalkyne, so cleavage of the C≡C bond results in the formation of a compound with two carboxylic acid groups:

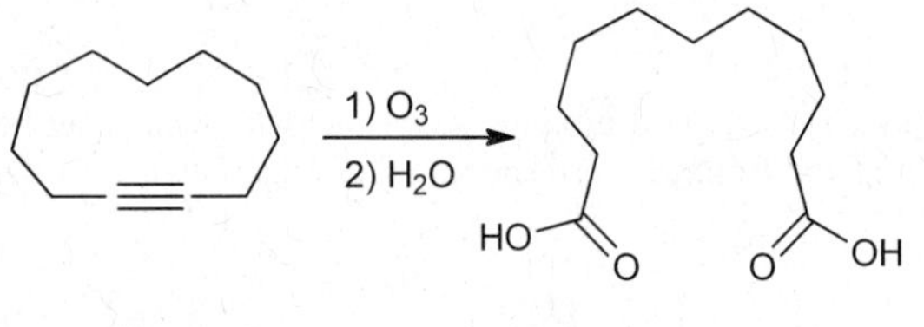

10.26. If ozonolysis produces only one product, then the starting alkyne must be symmetrical. There is only one symmetrical alkyne with molecular formula C_6H_{10}:

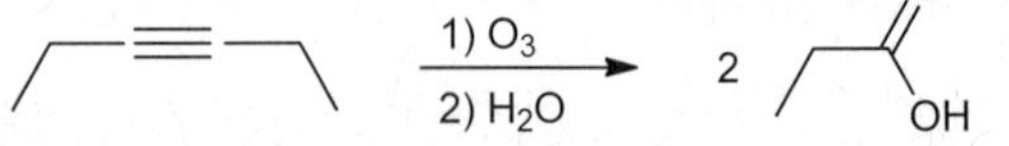

10.27. If ozonolysis produces a carboxylic acid and carbon dioxide, then the starting alkyne must be terminal. There is only one terminal alkyne with molecular formula C_4H_6:

C_4H_6

$$\xrightarrow[\text{2) } H_2O]{\text{1) } O_3}$$

When this alkyne is treated with aqueous acid in the presence of mercuric sulfate, the alkyne undergoes Markovnikov addition of H and OH, generating an enol. This enol is not isolated, because upon its formation, it undergoes tautomerization to give a methyl ketone:

$$\xrightarrow[\text{HgSO}_4]{\begin{array}{c}\text{H}_2\text{SO}_4\\ \text{H}_2\text{O}\end{array}}$$
(not isolated)

10.28.
(a) We begin by drawing the starting material (acetylene) and the product (1-butyne), which makes it clearer to see that the desired transformation involves a single alkylation process. This can be achieved by treating acetylene with sodium amide, followed by ethyl iodide:

$$\xrightarrow[\text{2) EtI}]{\text{1) NaNH}_2}$$

(b) We begin by drawing the starting material (acetylene) and the product (2-butyne), which makes it clearer to see that the desired transformation involves two, successive alkylation processes. The first alkylation is achieved by treating acetylene with sodium amide, followed by methyl iodide. The second alkylation process is then achieved upon further treatment with sodium amide, followed by methyl iodide. Notice that each alkylation process must be performed separately.

$$\xrightarrow[\begin{array}{c}\text{3) NaNH}_2\\ \text{4) MeI}\end{array}]{\begin{array}{c}\text{1) NaNH}_2\\ \text{2) MeI}\end{array}}$$

(c) We begin by drawing the starting material (acetylene) and the product (3-hexyne), which makes it clearer to see that the desired transformation involves two, successive alkylation processes. The first alkylation is achieved by treating acetylene with sodium amide, followed by ethyl iodide. The second alkylation process is then achieved upon further treatment with sodium amide, followed by ethyl iodide. Notice that each alkylation process must be performed separately.

$$\xrightarrow[\begin{array}{c}\text{3) NaNH}_2\\ \text{4) EtI}\end{array}]{\begin{array}{c}\text{1) NaNH}_2\\ \text{2) EtI}\end{array}}$$

(d) We begin by drawing the starting material (acetylene) and the product (2-hexyne), which makes it clearer to see that the desired transformation involves two, successive alkylation processes. That is, we must install a methyl group and a propyl group. The propyl group is installed by treating acetylene with sodium amide, followed by propyl iodide. And the methyl group is installed in a similar way (upon treatment with sodium amide, followed by methyl iodide). The propyl group can be installed first or last. Either way is acceptable. The following shows installation of the propyl group followed by installation of the methyl group:

$$\xrightarrow[\begin{array}{c}\text{3) NaNH}_2\\ \text{4) MeI}\end{array}]{\begin{array}{c}\text{1) NaNH}_2\\ \text{2)}\end{array}}$$

(e) We begin by drawing the starting material (acetylene) and the product (1-hexyne). This makes it clearer to see that the desired transformation involves a single alkylation process, which can be achieved with the following reagents:

$$\xrightarrow[\text{2)}]{\text{1) NaNH}_2}$$

(f) We begin by drawing the starting material (acetylene) and the product (2-heptyne), which makes it clearer to see that the desired transformation involves two, successive alkylation processes. That is, we must install a methyl group and a butyl group. These two groups can be installed in either order. The following shows installation of the butyl group followed by installation of the methyl group:

$$\xrightarrow[\begin{array}{c}\text{3) NaNH}_2\\ \text{4) MeI}\end{array}]{\begin{array}{c}\text{1) NaNH}_2\\ \text{2) I}\end{array}}$$

(g) We begin by drawing the starting material (acetylene) and the product (3-heptyne), which makes it clearer to see that the desired transformation involves two, successive alkylation processes. That is, we must install an ethyl group and a propyl group. These two groups can be installed in either order. The following shows installation of the propyl group followed by installation of the ethyl group:

$$\xrightarrow[\begin{array}{c}\text{3) NaNH}_2\\ \text{4) EtI}\end{array}]{\begin{array}{c}\text{1) NaNH}_2\\ \text{2) I}\end{array}}$$

(h) We begin by drawing the starting material (acetylene) and the product (2-octyne), which makes it clearer to see that the desired transformation involves two, successive alkylation processes. That is, we must install a methyl group and a pentyl group. These two groups can be installed in either order. The following shows installation of the pentyl group followed by installation of the methyl group:

1) NaNH₂
2) I⟋⟍⟋⟍⟋
3) NaNH₂
4) MeI

(i) We begin by drawing the starting material (acetylene) and the product (2-pentyne), which makes it clearer to see that the desired transformation involves two, successive alkylation processes. That is, we must install a methyl group and an ethyl group. These two groups can be installed in either order. The following shows installation of the ethyl group followed by installation of the methyl group:

1) NaNH₂
2) EtI
3) NaNH₂
4) MeI

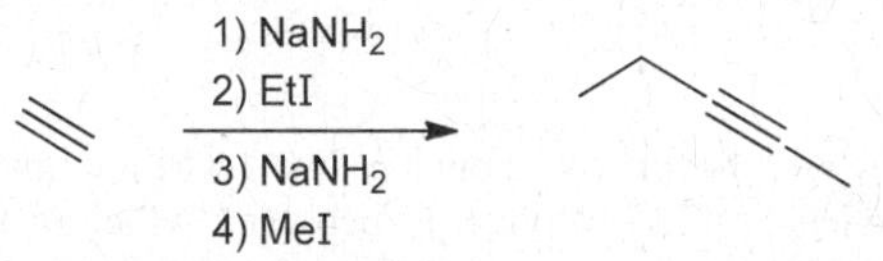

(j) The desired transformation involves two, successive alkylation processes. That is, we must install an ethyl group and a benzyl group. These two groups can be installed in either order. The following shows installation of the ethyl group followed by installation of the benzyl group:

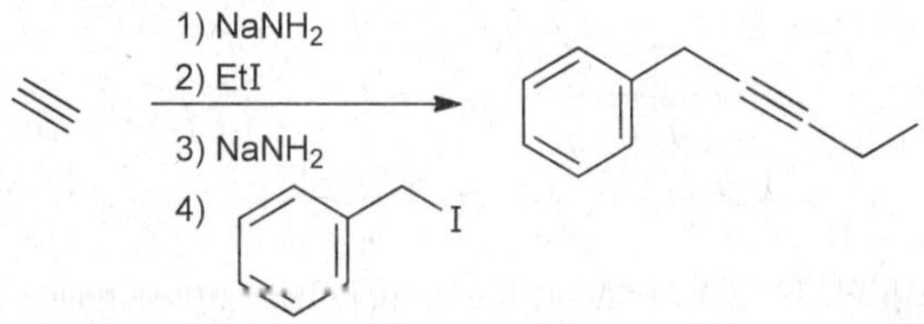

1) NaNH₂
2) EtI
3) NaNH₂
4)

(k) The desired transformation involves two, successive alkylation processes. Each alkylation is achieved by treating the alkyne with sodium amide, followed by benzyl iodide. Notice that each alkylation process must be performed separately.

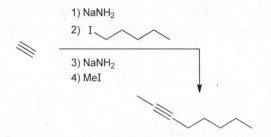

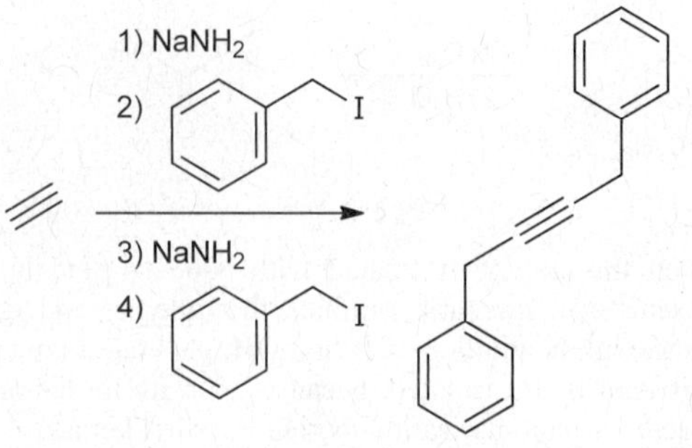

1) NaNH₂
2)
3) NaNH₂
4)

10.29. Formation of 2,2-dimethyl-3-octyne (shown below) from acetylene would require two alkylation processes. That is, we must install a butyl group and a *tert*-butyl group. The former can be readily achieved, but the latter cannot, because installation of a *tert*-butyl group would require the use of a tertiary substrate, which will not undergo an S_N2 process:

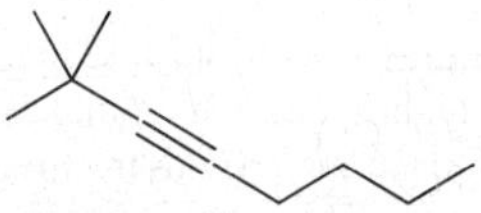

2,2-Dimethyl-3-octyne

10.30. Since ozonolysis of the internal alkyne leads to only one carboxylic acid (rather than two), we can deduce that the internal alkyne must be symmetrical. If it has to be symmetrical, then it must be 4-octyne, since we installed a propyl group ourselves (that is, the starting alkyne already had one propyl group).

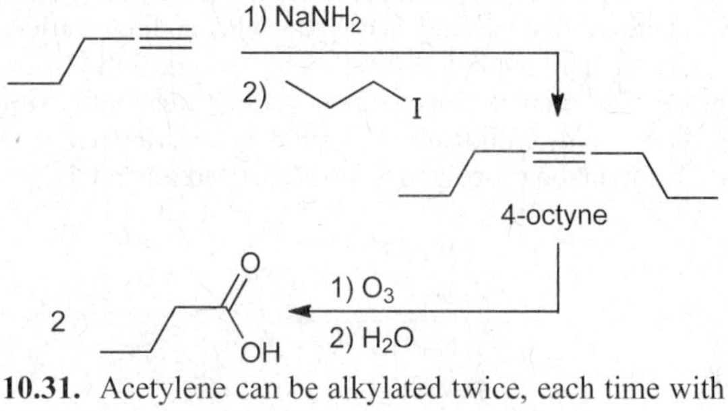

1) NaNH₂
2) ⟍⟋⟍I

4-octyne

1) O₃
2) H₂O

10.31. Acetylene can be alkylated twice, each time with ethyl bromide.

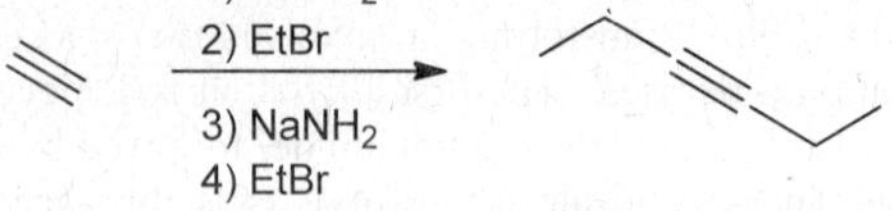

1) NaNH₂
2) EtBr
3) NaNH₂
4) EtBr

The problem statement indicates that acetylene must be the only source of carbon atoms, so we must prepare ethyl bromide from acetylene. This can be achieved via hydrogenation of acetylene (with a poisoned catalyst), followed by treatment with HBr:

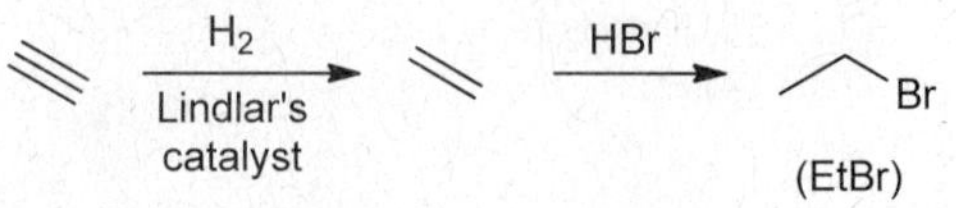

H₂
Lindlar's catalyst

HBr

Br
(EtBr)

10.32.

(a) The desired transformation appears to involve installation of an ethyl group. If the starting compound were a terminal alkyne, we could simply perform an alkylation reaction, but unfortunately, the starting compound is an alkene (not an alkyne). Alkenes do not undergo alkylation reactions under the same conditions used for the alkylation of alkynes, and we have not covered the conditions necessary for alkylating an alkene. So, in order to perform the necessary alkylation reaction, we must first convert the alkene into an alkyne. This is accomplished by treating the starting alkene with Br$_2$, giving a dibromide, followed by treatment with excess sodium amide (and then water work-up). At this point, the alkylation step can be performed with ease, and the resulting alkyne can then be reduced with molecular hydrogen (H$_2$) in the presence of Lindlar's catalyst to generate the *cis* alkene.

1) Br$_2$
2) xs NaNH$_2$
3) H$_2$O
4) NaNH$_2$
5) EtI
6) H$_2$, Lindlar's catalyst

H$_2$, Lindlar's catalyst

Br$_2$

Br
Br (racemic)

1) xs NaNH$_2$
2) H$_2$O

1) NaNH$_2$
2) EtI

Note: The alkyne produced after step 3 does not need to be isolated and purified, and therefore, steps 3 and 4 can be omitted. That is, the synthesis can be presented like this:

1) Br$_2$
2) xs NaNH$_2$
3) EtI
4) H$_2$, Lindlar's catalyst

H$_2$, Lindlar's catalyst

Br$_2$

Br
Br (racemic)

1) xs NaNH$_2$
2) EtI

(b) The desired product is an aldehyde, which can be made from the following alkyne (1-butyne) via hydroboration-oxidation:

Hydroboration-oxidation

H
O

This alkyne can be prepared from the starting alkene by brominating the alkene to give a dibromide, followed by elimination with excess sodium amide:

1) Br$_2$
2) xs NaNH$_2$
3) H$_2$O
4) 9-BBN
5) H$_2$O$_2$, NaOH

H
O

Br$_2$

Br
Br (racemic)

1) xs NaNH$_2$
2) H$_2$O

1) 9-BBN
2) H$_2$O$_2$, NaOH

(c) We see an OH group at the more substituted position, indicating a Markovnikov addition. But there is a problem. If we try to perform Markovnikov addition of H and OH across the alkyne, the resulting enol will immediately tautomerize to give a ketone:

H$_2$SO$_4$
H$_2$O
HgSO$_4$

OH

O

(not isolated)

And we have not yet learned a way to convert the ketone into the desired product, although we will learn how to achieve this transformation in Chapter 13:

O

(in Chapter 13)

OH

So, we must find another route. Instead of performing acid-catalyzed hydration first, we could first reduce the alkyne (in the presence of a poisoned catalyst), giving an alkene. This alkene can then be treated with dilute aqueous acid to give Markovnikov addition of H and OH, resulting in the desired product:

1) H$_2$, Lindlar's catalyst
2) dilute H$_2$SO$_4$

OH

(d) Much as we saw in the solution to Problem 10.32c, the desired product can be made by first reducing the alkyne in the presence of a poisoned catalyst (to give an alkene), followed by addition of H and OH across the alkene. In this case, we need an *anti*-Markovnikov addition, so we employ a hydroboration-oxidation procedure:

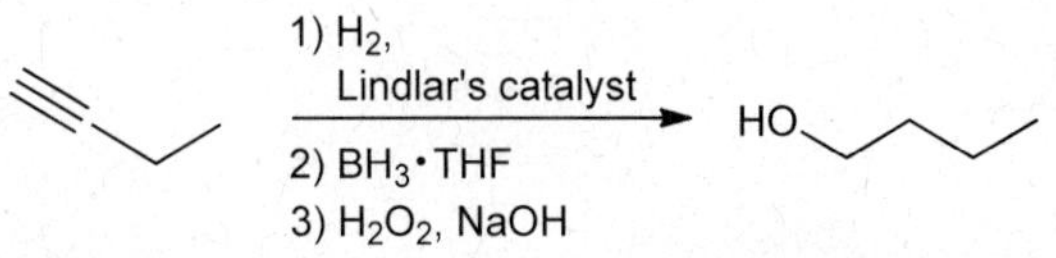

(e) The starting alkyne has only four carbon atoms, while the desired product has six carbon atoms. So we must install an ethyl group. This can be achieved by alkylating the starting alkyne. After the alkylation, the resulting, symmetrical alkyne can be reduced to an alkene, followed by bromination. Since the last step (bromination) proceeds via an *anti* addition, the desired stereoisomer can only be obtained if the previous step (reduction of the alkyne) is performed in an *anti* fashion. That is, we must use a dissolving metal reduction, rather than hydrogenation with a poisoned catalyst.

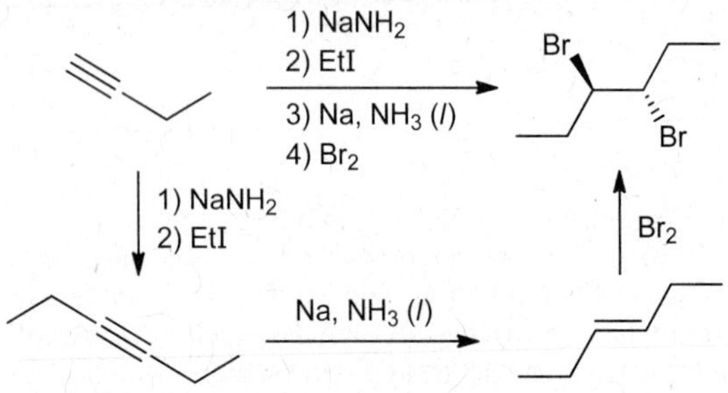

Notice that the product is a *meso* compound, which can be seen more clearly if we rotate about the central C-C bond, like this:

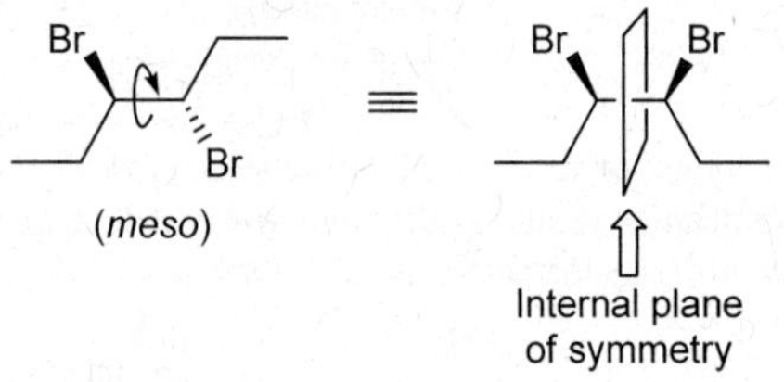

(f) The starting alkyne has only four carbon atoms, while the desired product has six carbon atoms. So we must install an ethyl group. This can be achieved by alkylating the starting alkyne. After the alkylation, the resulting, symmetrical alkyne can be reduced to an alkene, followed by bromination. Since the last step (bromination) proceeds via an *anti* addition, the desired stereoisomer can only be obtained if the previous step (reduction of the alkyne) is performed in a *syn* fashion. That is, we must perform a hydrogenation reaction with a poisoned catalyst, rather than using a dissolving metal reduction.

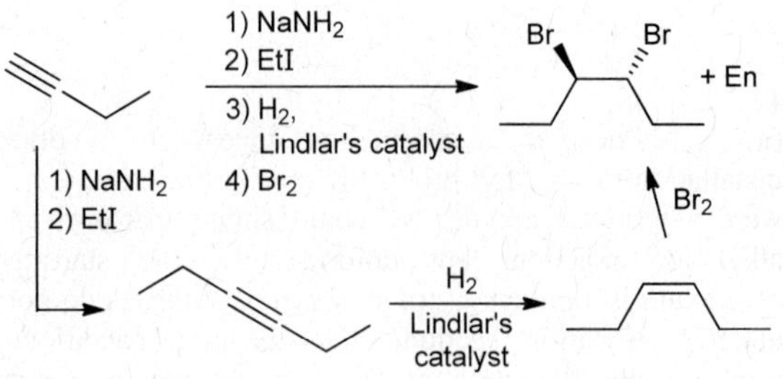

10.33.
(a) If we can convert bromoethane into acetylene, then ozonolysis will convert all of the carbon atoms into CO_2, as shown here:

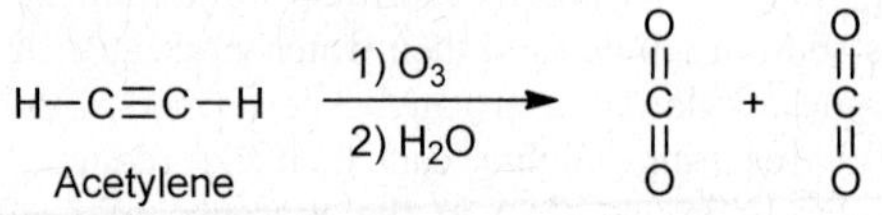

So, our goal is to make acetylene from bromoethane. This can be achieved in three steps. First an elimination reaction converts the alkyl halide into an alkene (notice that *tert*-butoxide is used to favor E2 over S_N2; in contrast, an unhindered base such as NaOEt or NaOH would favor S_N2 over E2, as seen in Chapter 8). Then, bromination gives a dibromide, followed by treatment with excess sodium amide (and water work-up) to give acetylene. And as mentioned above, ozonolysis of acetylene gives CO_2.

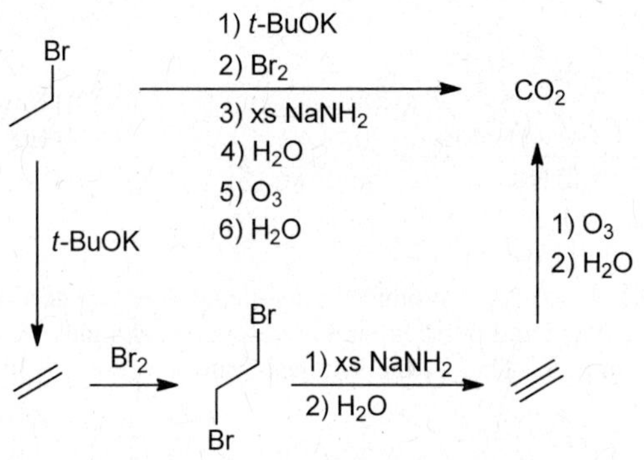

(b) We can take a similar approach that we took in solving the previous problem (**10.33a**) by realizing that acetic acid is the sole product that would be produced when 2-butyne undergoes ozonolysis:

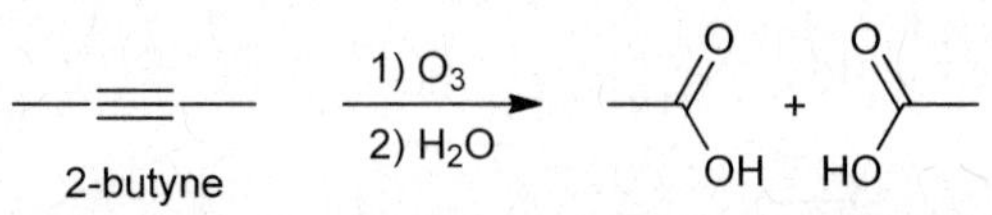

So, our goal is to make 2-butyne from 2-bromopropane. This requires several steps. First an elimination reaction converts the alkyl halide into an alkene (the alkyl halide is secondary, so E2 is favored over S_N2, as seen in Chapter 8). Then, bromination gives a dibromide, followed by treatment with excess sodium amide (and water work-up) to give 1-propyne. Alkylation then gives 2-butyne, which (as mentioned above) undergoes ozonolysis to give acetic acid.

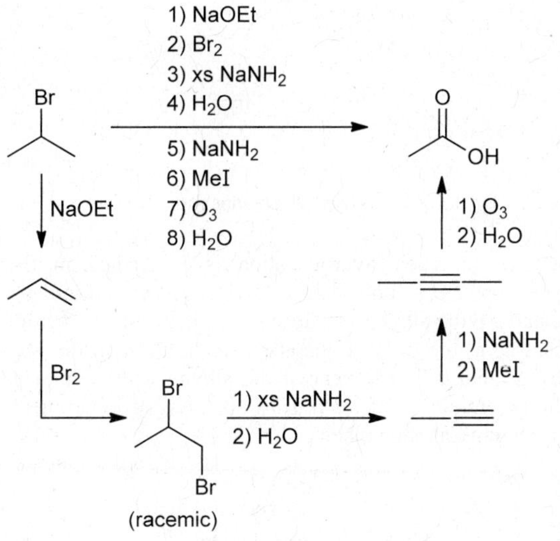

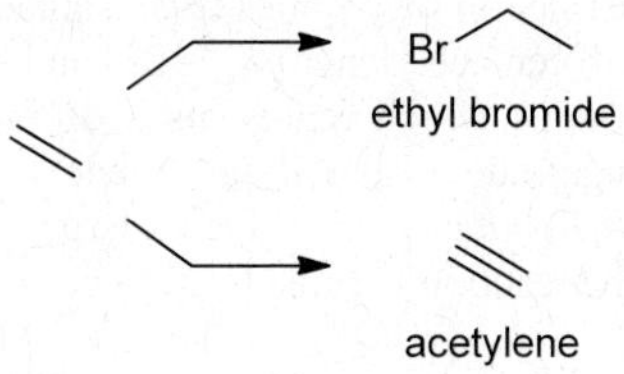

The former can be achieved in just one step, by treating ethylene with HBr:

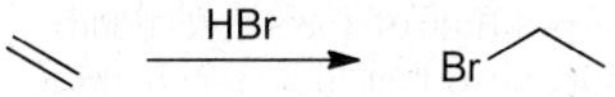

Acetylene can be made from ethylene via the following two step process:

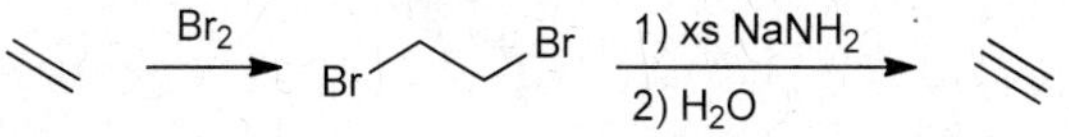

In summary, the following synthetic route converts ethylene into 3-hexanone:

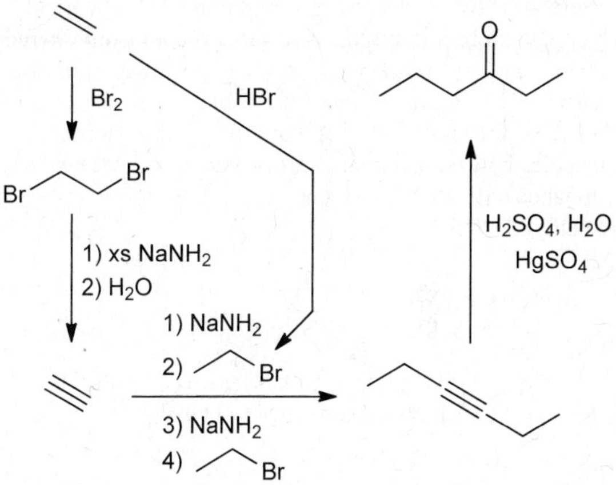

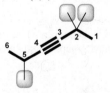

Note: The alkyne produced after step 4 does not need to be isolated and purified, and therefore, steps 4 and 5 can be omitted, like this:

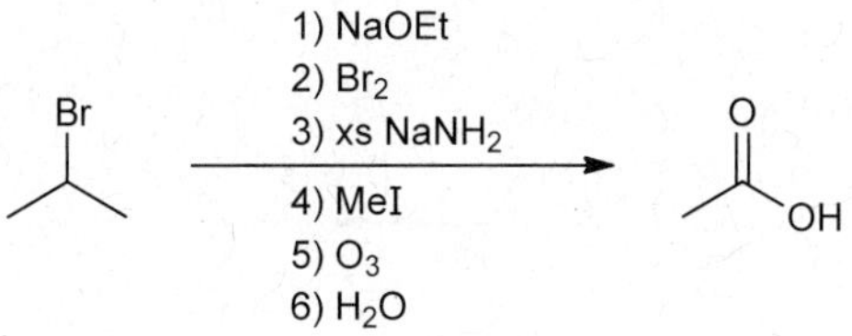

(racemic)

10.34. The starting material has two carbon atoms, and the desired product has six carbon atoms:

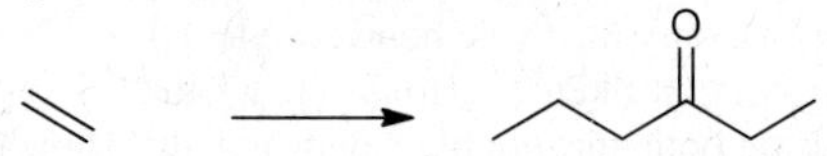

Two carbon atoms   Six carbon atoms

So, we must form C-C bonds. This can be achieved with acetylene and two equivalents of ethyl bromide:

Ethyl bromide   Acetylene   Ethyl bromide

And the resulting alkyne can be converted into the product in just one step:

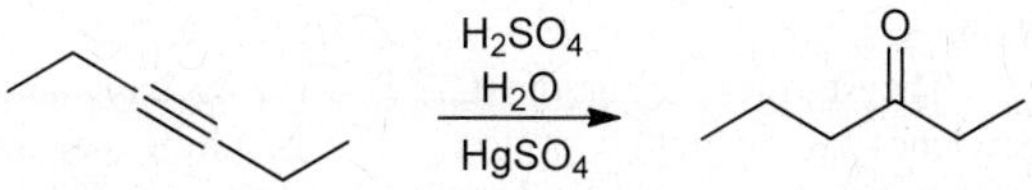

So, our goal is to convert ethylene into acetylene and ethyl bromide:

ethyl bromide

acetylene

10.35.

(a) We begin by identifying the parent. The longest chain is six carbon atoms, so the parent is hexyne. There are three substituents – all methyl groups (highlighted). Numbering the parent chain from either direction will place the triple bond at C3 (between C3 and C4), so we number in the direction that gives the lower number to the second substituent (2,2,5 rather than 2,5,5):

2,2,5-Trimethyl-3-hexyne

(b) We begin by identifying the parent. The longest chain is six carbon atoms, so the parent is hexyne. There are two substituents – both chloro groups. The parent chain is numbered to give the triple bond the lower possible number, C2 (because it is between C2 and C3). According to this numbering scheme, the chlorine atoms are both at C4:

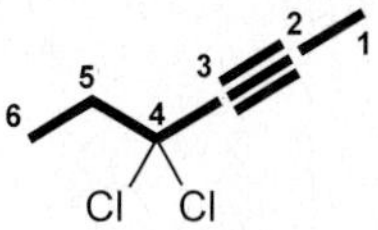

4,4-Dichloro-2-hexyne

(c) We begin by identifying the parent. The longest chain is six carbon atoms, so the parent is hexyne. There are no substituents. We must include a locant that identifies the position of the triple bond. The parent chain is numbered so that the triple bond is assigned the lowest possible locant ("1" indicates that the triple bond is located between C1 and C2).

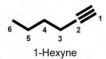

1-Hexyne

(d) We begin by identifying the parent. The longest chain is four carbon atoms, so the parent is butyne. There are two substituents (highlighted) – a methyl group and a bromo group. The parent chain is numbered so that the triple bond is assigned the lowest possible locant ("1" indicates that the triple bond is located between C1 and C2). According to this numbering scheme, both substituents are located at C3. They are alphabetized in the name, so "bromo" precedes "methyl."

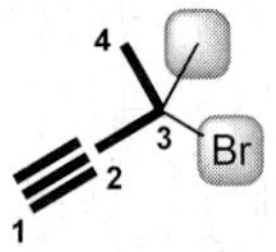

3-Bromo-3-methyl-1-butyne

10.36.
(a) The parent (heptyne) indicates a chain of seven carbon atoms. The triple bond is between C2 and C3, and there are no substituents.

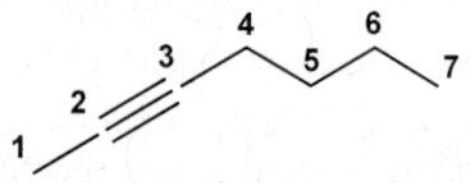

2-Heptyne

(b) The parent (octyne) indicates a chain of eight carbon atoms, with the triple bond located between C4 and C5. There are two methyl groups (highlighted), both located at position C2.

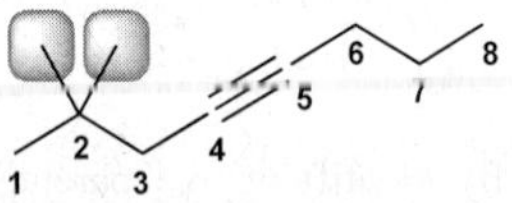

2,2-Dimethyl-4-octyne

(c) The parent (cyclodecyne) indicates a ring of ten carbon atoms. The numbering system is assigned such that the triple bond is between C1 and C2. There are two substituents (ethyl groups), each of which is located at position C3:

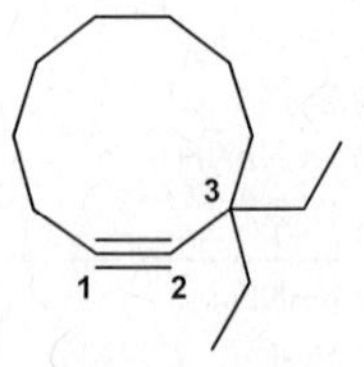

3,3-Diethylcyclodecyne

10.37. When hydrogenation is performed in the presence of a poisoned catalyst (such as Lindlar's catalyst), the alkyne is reduced to a *cis* alkene. When Pt is used as the catalyst, the alkyne is reduced all the way to an alkane. Treatment of the alkyne with sodium in liquid ammonia affords a *trans* alkene (dissolving metal reduction), as shown here:

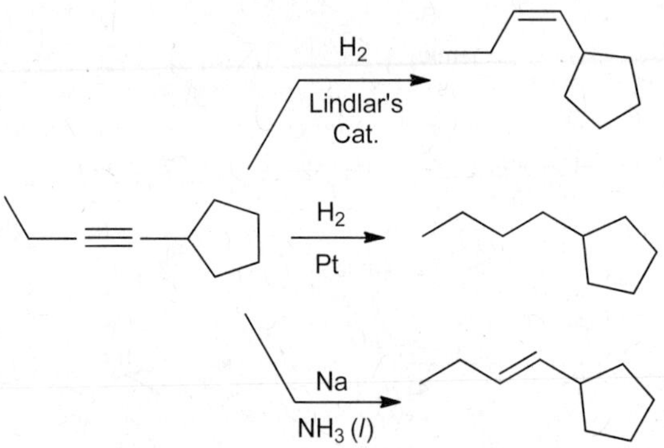

10.38. The starting material has both a double bond and a triple bond. When hydrogenation is performed in the presence of a poisoned catalyst (such as Lindlar's catalyst), only the triple bond is reduced (via *syn* addition to give a *cis* alkene). However, when Pt is used as the catalyst, both the double bond and the triple bond are reduced, giving an alkane:

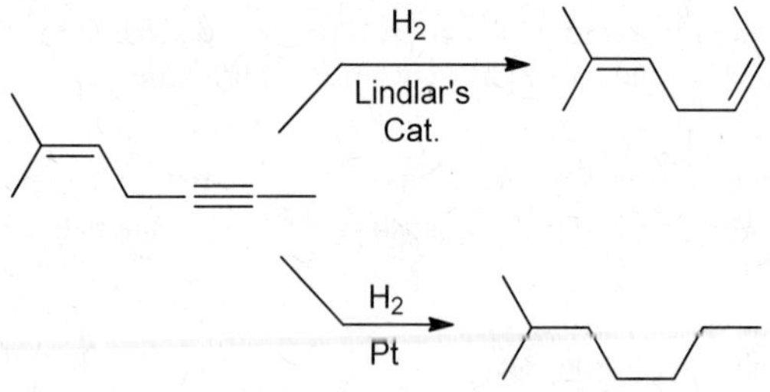

10.39.
(a) The starting carbanion (C^-) is a strong base, and acetylene has an acidic proton. The resulting proton transfer step gives an acetylide ion and butane. When we compare the acids (highlighted below), we find a massive difference in pK_a values (See Table 10.1). The difference between acetylene (pK_a = 25) and butane (pK_a = 50) is 25 units, which represents a difference of 25 orders of magnitude. That is, acetylene is 10^{25} times more acidic than butane. With such a large difference in pK_a values, the reaction is considered to be irreversible.

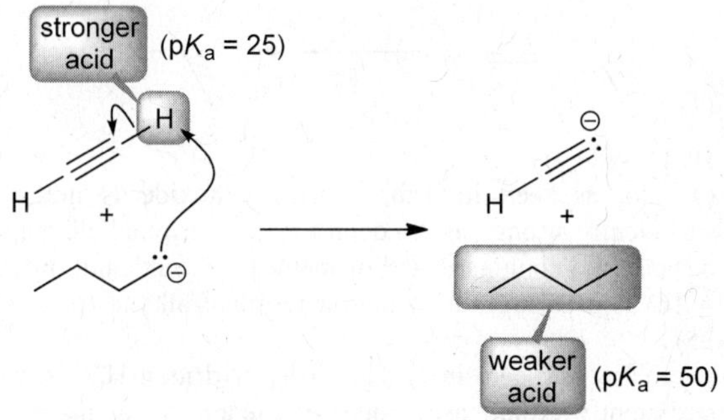

(b) Hydride (H⁻) is a strong base, and the terminal alkyne has an acidic proton. The resulting proton transfer step gives an alkynide ion and hydrogen gas (H_2). We compare the pK_a values of the acids (highlighted below) using the values given in Table 10.1. Acetylene (pK_a = 25) is more acidic than H_2 (pK_a = 35). As such, the equilibrium favors the weaker acid (H_2).

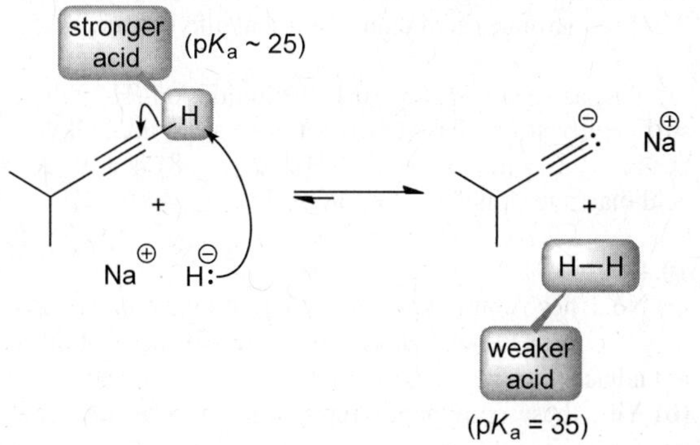

In practice, H_2 bubbles out of solution as a gas, and as a result, the reaction proceeds to completion (Le Chatelier's principle).

10.40.
(a) When treated with aqueous acid (in the presence of mercuric sulfate), the terminal alkyne undergoes Markovnikov addition of H and OH across the alkyne, giving an enol. The enol is not isolated, because upon its formation, it undergoes tautomerization to give a methyl ketone:

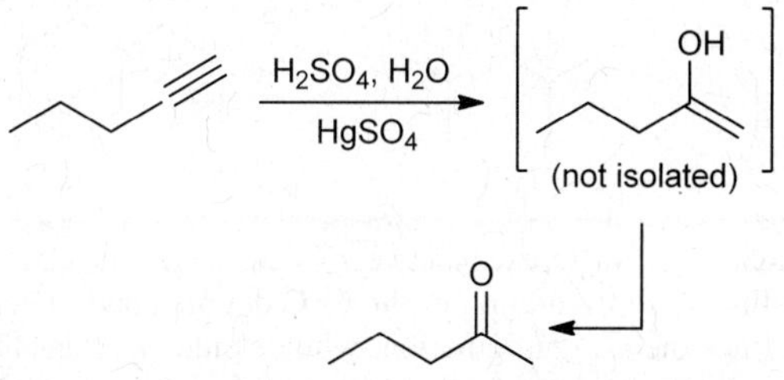

(b) The reagents (9-BBN, followed by H_2O_2 and NaOH) indicate hydroboration-oxidation of the terminal alkyne, giving an *anti*-Markovnikov addition of H and OH across the alkyne. The resulting enol is not isolated, because upon its formation, it undergoes tautomerization to give an aldehyde:

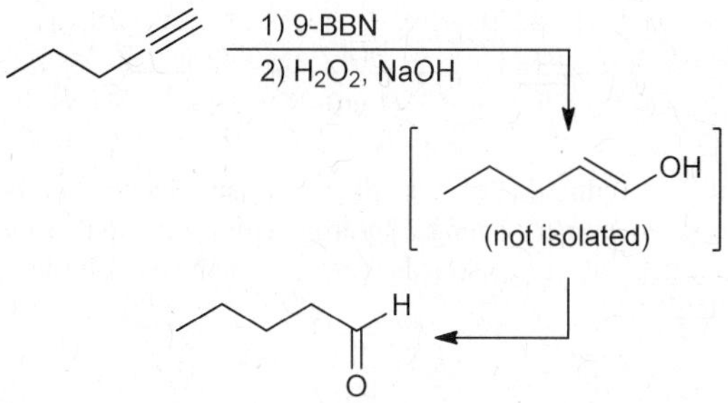

(c) When treated with two equivalents of HBr, a terminal alkyne undergoes two, successive addition reactions, each of which proceeds in a Markovnikov fashion, giving the following geminal dibromide:

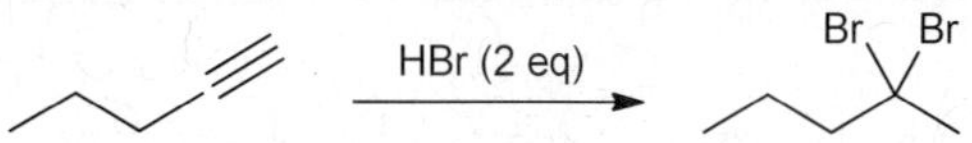

(d) When treated with one equivalent of HCl, a terminal alkyne undergoes an addition reaction that proceeds in a Markovnikov fashion, giving the following vinyl chloride:

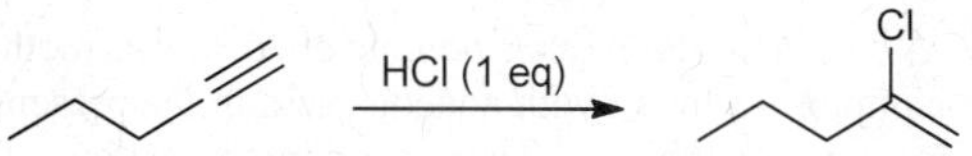

(e) When treated with two equivalents of Br_2, the alkyne undergoes two, successive addition reactions, giving the following tetrabromide:

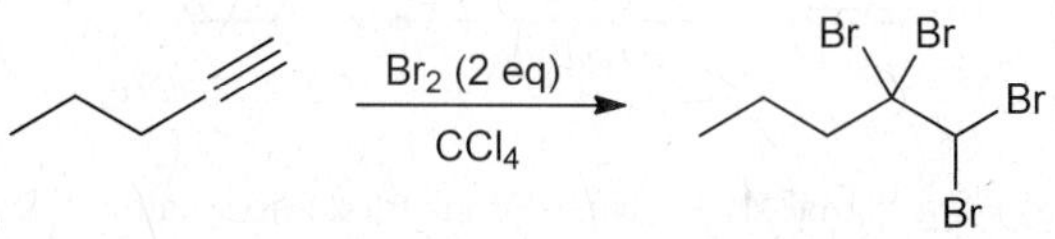

(f) These reagents indicate an alkylation process. In step 1, the alkyne is deprotonated by the strong base (H_2N^-) to give an alkynide ion. In step 2, this alkynide ion is used as a nucleophile to attack methyl iodide in an S_N2 reaction), thereby installing a methyl group:

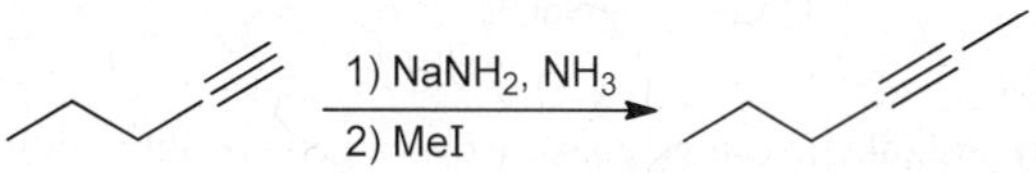

(g) Since platinum is used as the catalyst (rather than a poisoned catalyst), hydrogenation of the alkyne gives an alkane:

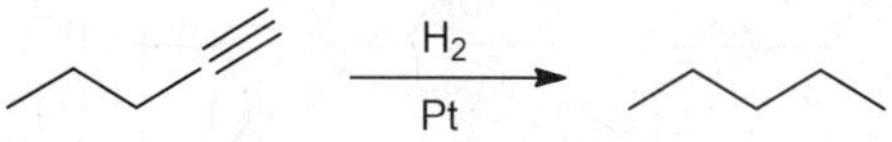

10.41.
(a) Markovnikov addition of addition of H and Cl can be achieved by treating the alkyne with one equivalent of HCl:

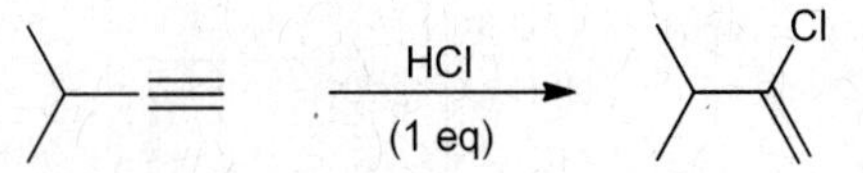

318 CHAPTER 10

(b) Reduction of the alkyne to an alkene can be achieved by performing a hydrogenation reaction in the presence of a poisoned catalyst, such as Lindlar's catalyst:

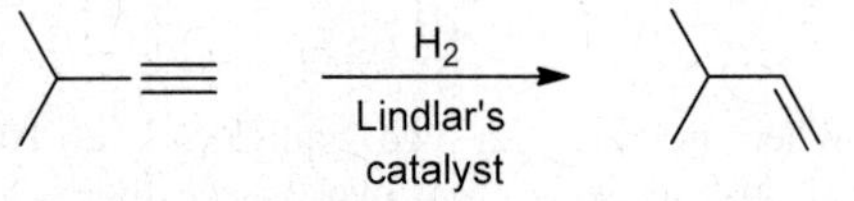

(c) Ozonolysis of the alkyne achieves cleavage of the C≡C bond to give a carboxylic acid (and carbon dioxide as a by-product):

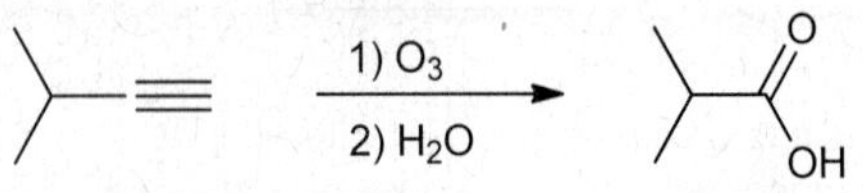

(+CO₂)

(d) A terminal alkyne can be converted into a methyl ketone upon treatment with aqueous acid in the presence of mercuric sulfate. These conditions allow for Markovnikov addition of H and OH, giving an enol, which tautomerizes to the methyl ketone:

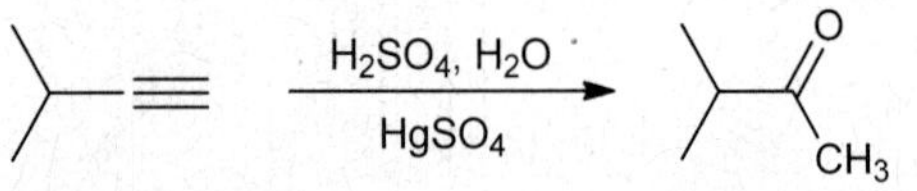

(e) A terminal alkyne can be converted into an aldehyde via hydroboration-oxidation. These conditions allow for an *anti*-Markovnikov addition of H and OH, giving an enol, which tautomerizes to the aldehyde:

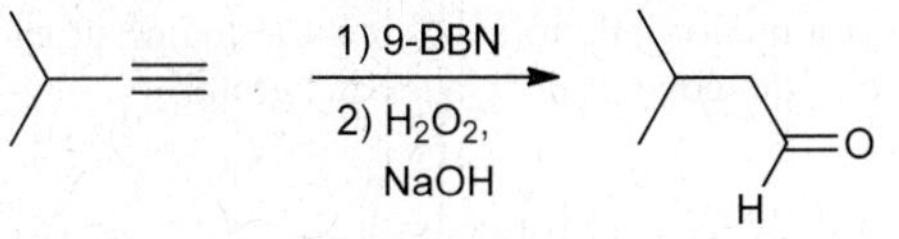

(f) An alkyne can be converted to a geminal dibromide via two successive addition reactions with HBr. Markovnikov addition is required, so we use excess HBr without peroxides.

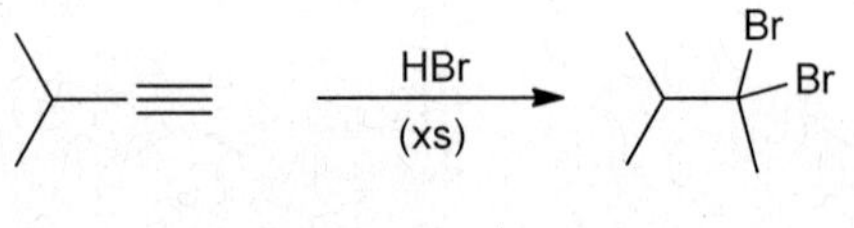

(g) Reduction of the alkyne to an alkane can be achieved by performing hydrogenation in the presence of a catalyst, such as platinum (not a poisoned catalyst):

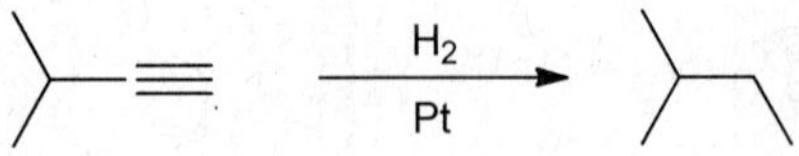

10.42.
(a) No, as seen in Table 10.1, methoxide is not a sufficiently strong base to deprotonate a terminal alkyne, because the conjugate acid of methoxide (methanol, $pK_a = 16$) is a stronger acid than a terminal alkyne ($pK_a \sim 25$).
(b) Yes, as seen in Table 10.1, hydride (H⁻) is a sufficiently strong base to deprotonate a terminal alkyne, because the conjugate acid of hydride (H₂, $pK_a = 35$) is a weaker acid than a terminal alkyne ($pK_a \sim 25$).
(c) Yes, as seen in Table 10.1, butyllithium is a sufficiently strong base to deprotonate a terminal alkyne, because the conjugate acid (butane, $pK_a = 50$) is a weaker acid than a terminal alkyne ($pK_a \sim 25$).
(d) No, as seen in Table 10.1, hydroxide is not a sufficiently strong base to deprotonate a terminal alkyne, because the conjugate acid of hydroxide (water, $pK_a = 15.7$) is a stronger acid than a terminal alkyne ($pK_a \sim 25$).

(e) Yes, as seen in Table 10.1, the amide ion (H₂N⁻) is a sufficiently strong base to deprotonate a terminal alkyne, because the conjugate acid (NH₃, $pK_a = 38$) is a weaker acid than a terminal alkyne ($pK_a \sim 25$).

10.43.
(a) No. These compounds are constitutional isomers, but they are not keto-enol tautomers because the π bond is not adjacent to the OH group in the first compound.
(b) Yes, these compounds represent a pair of keto-enol tautomers.
(c) Yes, these compounds represent a pair of keto-enol tautomers.
(d) Yes, these compounds represent a pair of keto-enol tautomers.

10.44. In each of the following cases, the enol is drawn by changing the C=O double bond to a C-O single bond, placing a proton on the oxygen atom, and drawing a C=C double bond adjacent to the OH group:

(a)   **(b)** 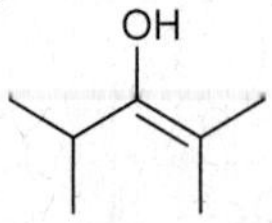   **(c)**

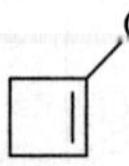

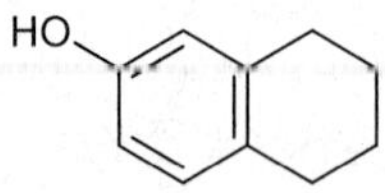

10.45. The difference between oleic acid and elaidic acid is the configuration of the C=C double bond. Oleic acid has the *cis* configuration, while elaidic acid has the *trans* configuration. Each of these stereoisomers can be obtained via reduction of the corresponding alkyne. Hydrogenation in the presence of a poisoned catalyst affords the *cis* alkene, while a dissolving metal reduction gives the *trans* alkene:

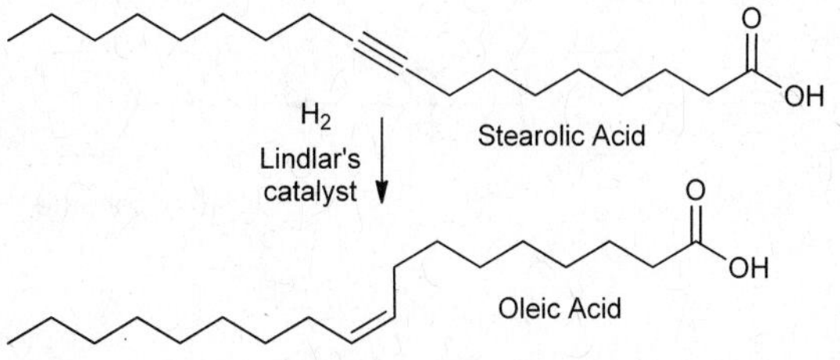

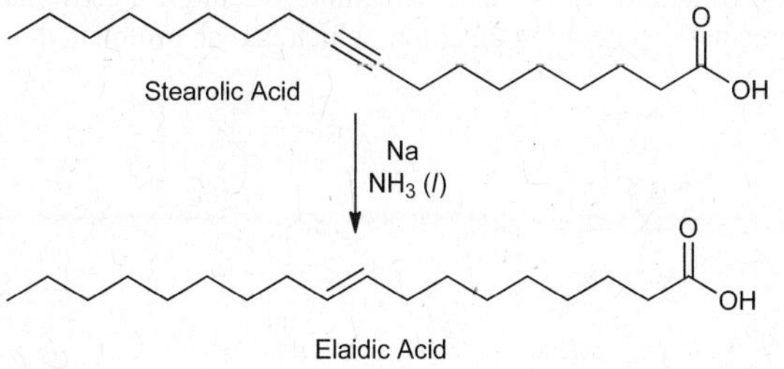

10.46.

(a) Upon treatment with excess sodium amide, the geminal dibromide undergoes elimination (twice) followed by deprotonation, to give an alkynide ion. This alkynide ion then undergoes alkylation when treated with ethyl chloride. Finally, hydrogenation in the presence of a poisoned catalyst affords the *cis* alkene:

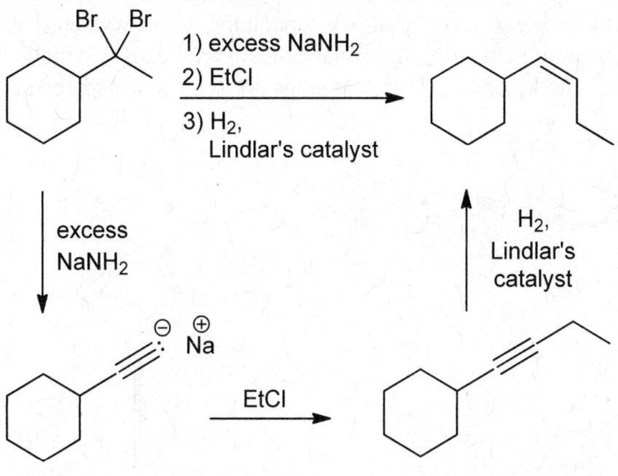

(b) Acetylene undergoes alkylation upon treatment with sodium amide, followed by methyl iodide. The resulting alkyne (1-propyne) then undergoes hydroboration-oxidation when treated with 9-BBN followed by H_2O_2 and NaOH, giving an aldehyde:

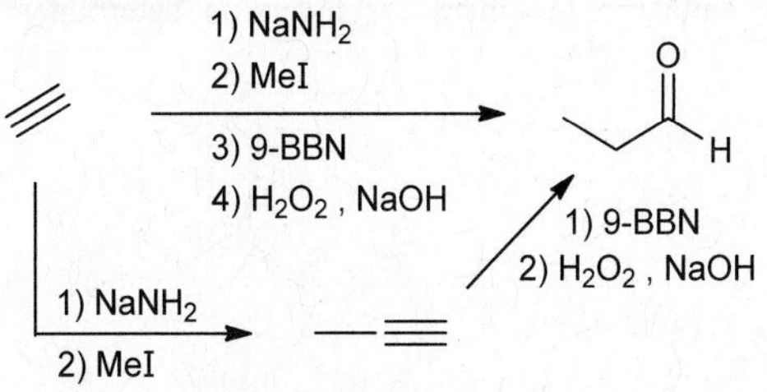

(c) Acetylene undergoes alkylation upon treatment with sodium amide, followed by ethyl iodide. The resulting

alkyne (1-butyne) then undergoes acid-catalyzed hydration when treated with aqueous acid in the presence of mercuric sulfate, giving a methyl ketone:

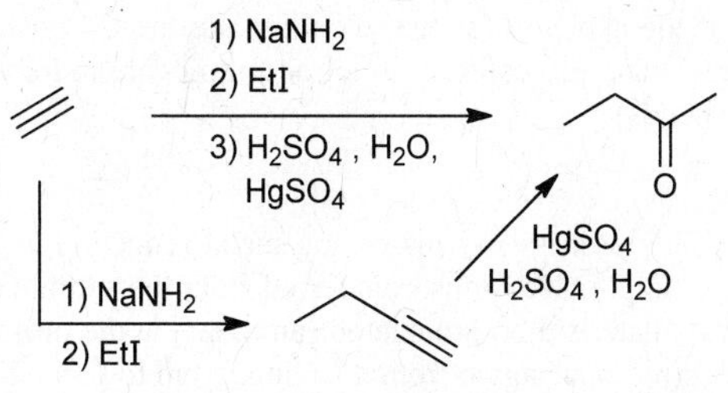

(d) Acetylene undergoes alkylation upon treatment with sodium amide, followed by methyl iodide. The resulting alkyne (1-propyne) then undergoes alkylation, once again, upon treatment with sodium amide, followed by ethyl iodide. The first alkylation process installs a methyl group, while the second process installs an ethyl group. Finally, a dissolving metal reduction converts the alkyne to a *trans* alkene:

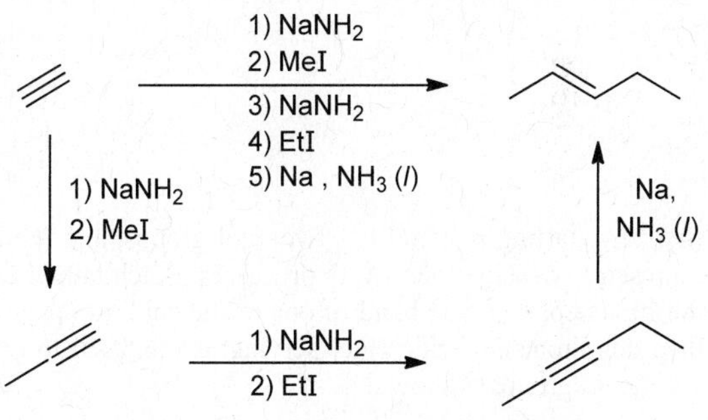

10.47. When *(R)*-4-bromohept-2-yne is treated with H_2 in the presence of Pt, the asymmetry is destroyed and C4 is no longer a chirality center:

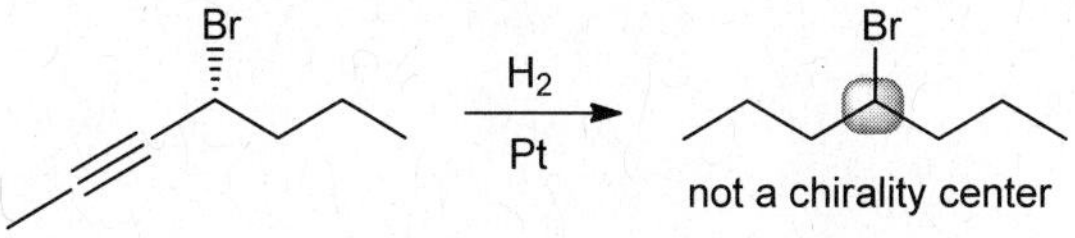

This is not the case for *(R)*-4-bromohex-2-yne:

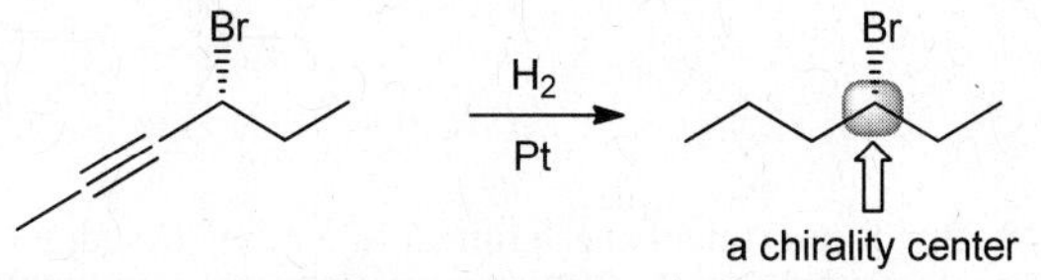

10.48. We are looking for an alkyne that will undergo hydrogenation to give the following alkane:

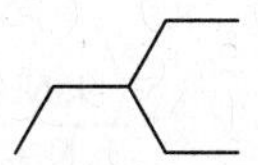

3-Ethylpentane

In order to produce this alkane, the starting alkyne must have the same carbon skeleton as this alkane. There is only one such alkyne (3-ethyl-1-pentyne, shown below), because the triple bond cannot be placed between C2 and C3 of the skeleton (as that would make the C3 position pentavalent, and carbon cannot accommodate more than four bonds):

3-Ethyl-1-pentyne   3-Ethylpentane

10.49.
(a)   This process is a dissolving metal reduction, so it follows the four steps shown in Mechanism 10.1. In the first step, a single electron is transferred from the sodium atom to the alkyne, generating a radical anion intermediate. This intermediate is then protonated (ammonia is the proton source), generating a radical intermediate. A single electron is transferred once again from a sodium atom to the radical intermediate, generating an anion, which is then protonated in the final step of the mechanism:

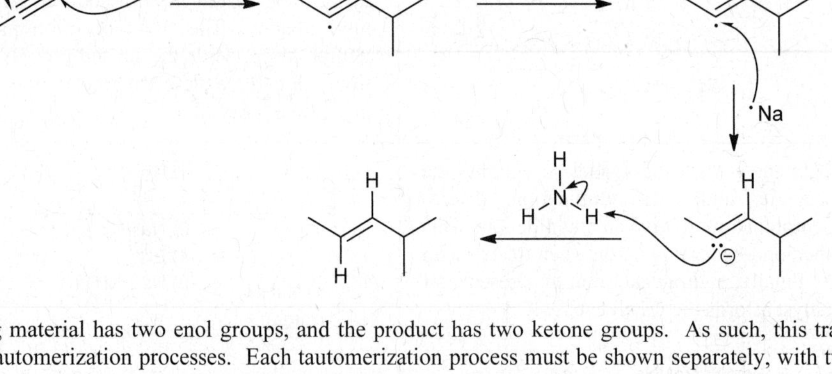

(b)   The starting material has two enol groups, and the product has two ketone groups. As such, this transformation represents two tautomerization processes. Each tautomerization process must be shown separately, with two steps. In the first step, a double bond of one of the enol groups is protonated, generating a resonance-stabilized cation, which is then deprotonated (with water serving as the base) to generate the ketone. These two steps are then repeated for the other enol group, as shown:

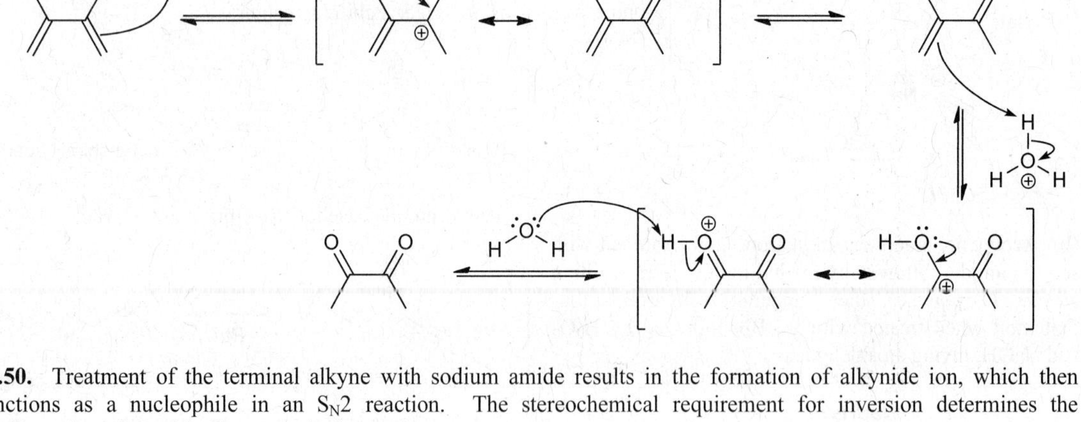

10.50.   Treatment of the terminal alkyne with sodium amide results in the formation of alkynide ion, which then functions as a nucleophile in an S_N2 reaction. The stereochemical requirement for inversion determines the configuration of the chirality center in the product.

10.51.

(a) In order to produce 2,4,6-trimethyloctane, the starting alkyne must have the same carbon skeleton as this alkane. There is only one such alkyne (shown below), because the triple bond cannot be placed in any other location (as that would give a pentavalent carbon atom, which is not possible):

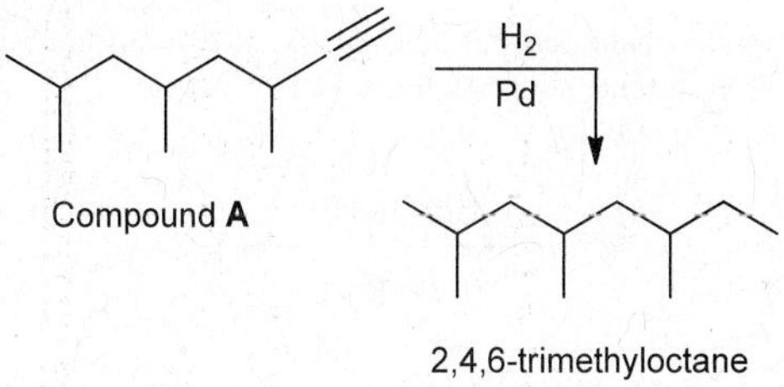

Compound **A**

2,4,6-trimethyloctane

(b) Compound A has two chirality centers, highlighted below:

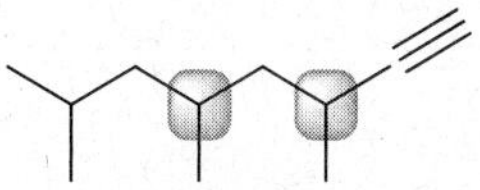

(c) The locants for the methyl groups in compound **A** are 3, 5, and 7, because locants are assigned in a way that gives the triple bond the lower possible number (1 rather than 7). In the alkane, the numbering scheme goes in the other direction, so as to give the first substituent the lower possible number (2 rather than 3)

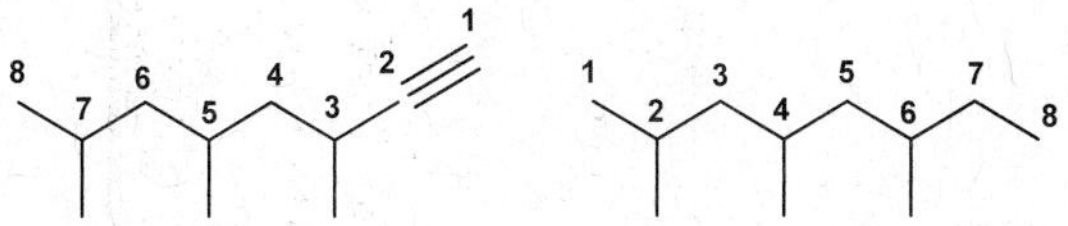

10.52. Hydrogenation of compound A produces 2-methylhexane:

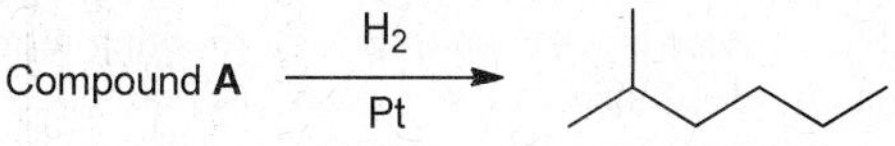

Compound **A**

There are several alkynes that can undergo hydrogenation to yield 2-methylhexane.

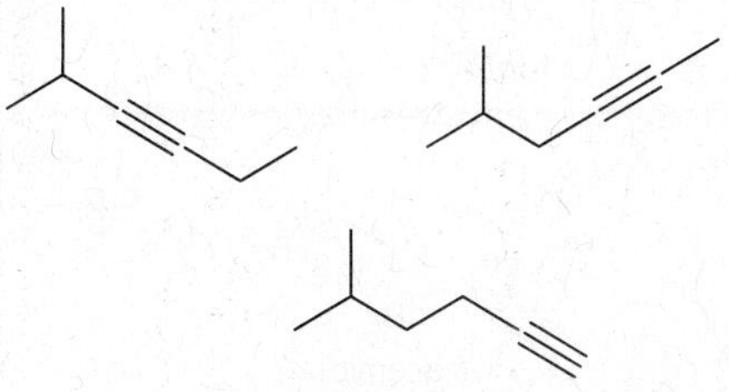

However, only one of these three possibilities will undergo hydroboration-oxidation to give an aldehyde. Specifically, the alkyne must be terminal in order to generate an aldehyde upon hydroboration-oxidation:

Compound **A**

1) 9-BBN

2) H$_2$O$_2$, NaOH

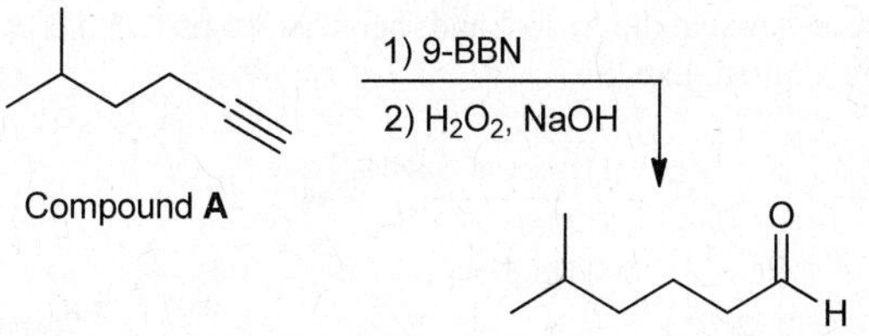

10.53.

(a) The starting material has four carbon atoms, while the product has six carbon atoms. Therefore, two carbon atoms must be introduced. This can be achieved via alkylation of the terminal alkyne, as seen in the first two steps of the following synthesis. The resulting, symmetrical alkyne can then undergo acid-catalyzed hydration to give the desired ketone. Alternatively, the alkyne can be converted into the desired ketone via hydroboration-oxidation.

1) NaNH$_2$

2) EtI

3) H$_2$SO$_4$, H$_2$O, HgSO$_4$

1) NaNH$_2$

2) EtI

HgSO$_4$

H$_2$SO$_4$, H$_2$O

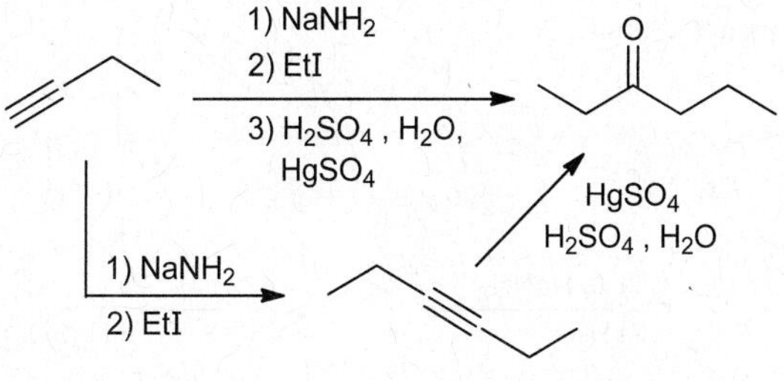

(b) The starting material is a geminal dihalide, and we have not learned a way to convert a geminal dihalide directly into an alkene. However, we have learned how to convert a geminal dihalide into a terminal alkyne, upon treatment with excess sodium amide (followed by water workup). The alkyne can then be reduced by hydrogenation with a poisoned catalyst, such as Lindlar's catalyst.

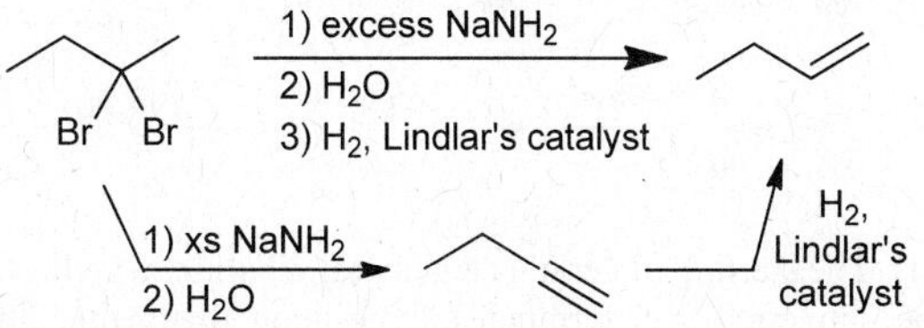

1) excess NaNH$_2$

2) H$_2$O

3) H$_2$, Lindlar's catalyst

1) xs NaNH$_2$

2) H$_2$O

H$_2$, Lindlar's catalyst

(c) This problem is similar to the previous problem, but an additional methyl group must be installed. We have not seen a way to alkylate an alkene (only an alkyne). So, the extra alkylation process must be performed before the alkyne is reduced to an alkene in the last step of the synthesis:

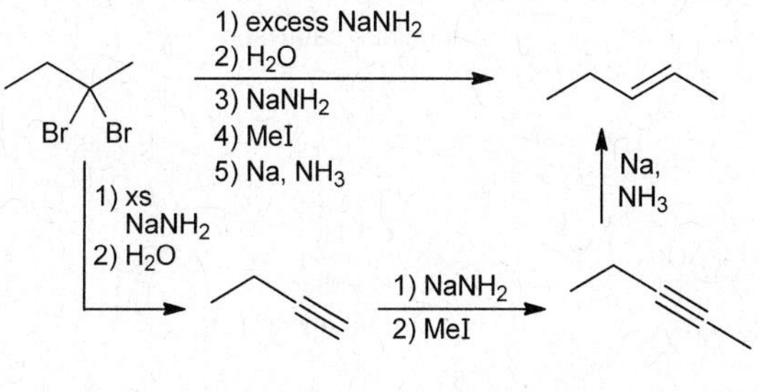

1) excess NaNH$_2$

2) H$_2$O

3) NaNH$_2$

4) MeI

5) Na, NH$_3$

1) xs NaNH$_2$

2) H$_2$O

1) NaNH$_2$

2) MeI

Na, NH$_3$

Note: The alkyne produced after step 2 does not need to be isolated and purified, and therefore, steps 2 and 3 can be omitted, like this:

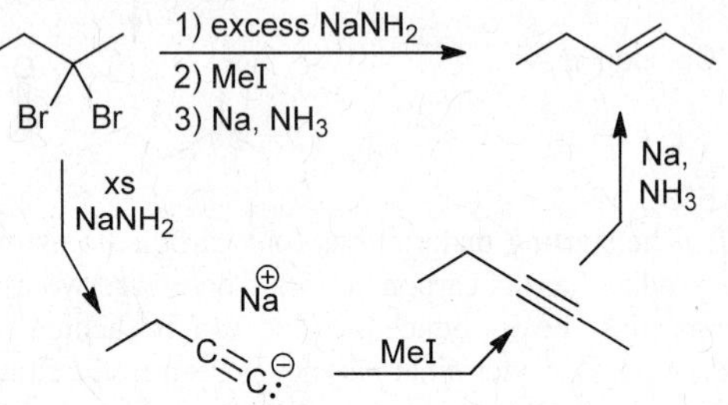

(d) The starting material is a geminal dihalide, which can be converted to a terminal alkyne upon treatment with excess sodium amide (followed by water workup). The terminal alkyne can then undergo acid-catalyzed hydration in the presence of mercuric sulfate to give a methyl ketone:

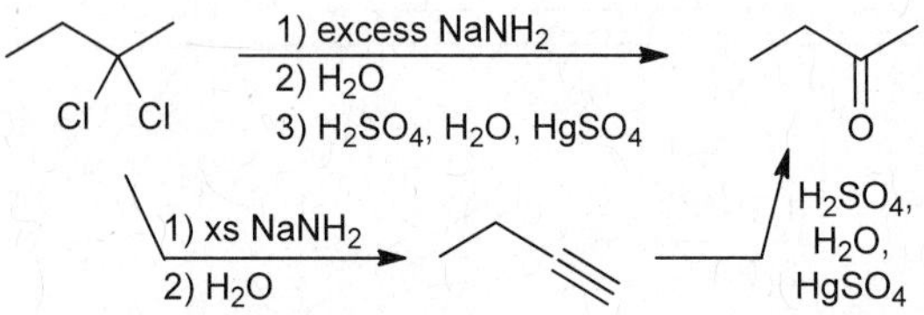

(e) The starting material is a geminal dihalide, which can be converted to a terminal alkyne upon treatment with excess sodium amide (followed by water workup). The terminal alkyne can then undergo bromination to give the dibromide, as shown:

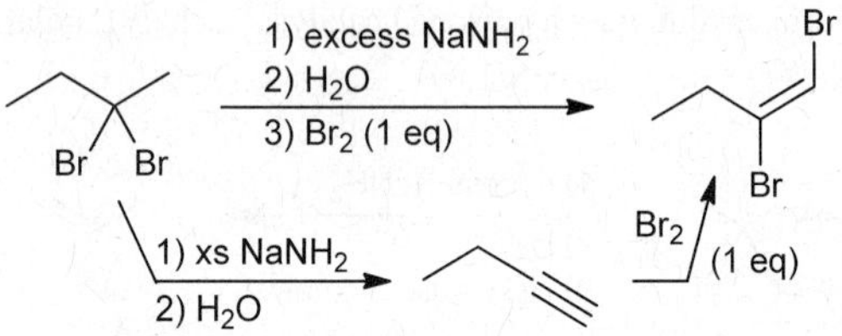

(f) The starting material is a geminal dihalide, which can be converted to a terminal alkyne upon treatment with excess sodium amide (followed by water workup). The terminal alkyne can then be reduced to give an alkene, which can then be treated with aqueous acid to give a Markvonikov addition of water, affording the desired product:

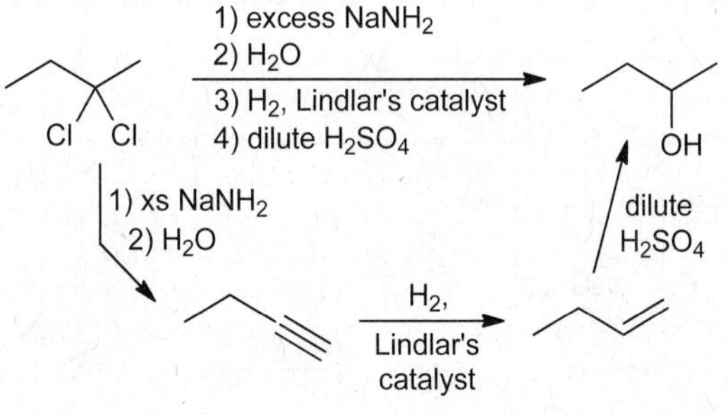

10.54. Treatment of 1,2-dichloropentane with excess sodium amide (followed by water-workup) gives 1-pentyne (compound X).

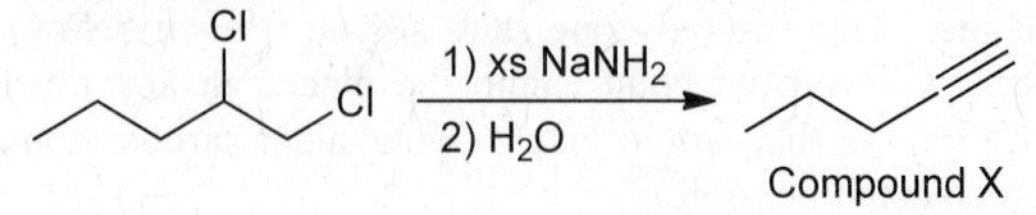

Compound X undergoes acid-catalyzed hydration to give a methyl ketone, shown below:

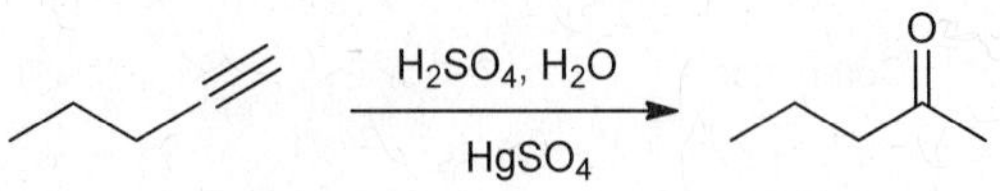

10.55. Acetic acid has two carbon atoms, and carbon dioxide has one, so our starting alkyne has three carbon atoms:

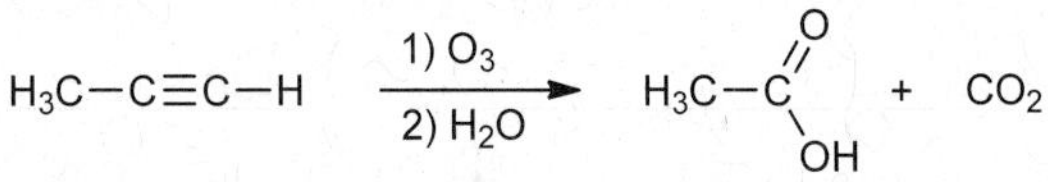

10.56. If two products are obtained, then the alkyne must be internal and unsymmetrical. There is only one such alkyne with molecular formula C_5H_8:

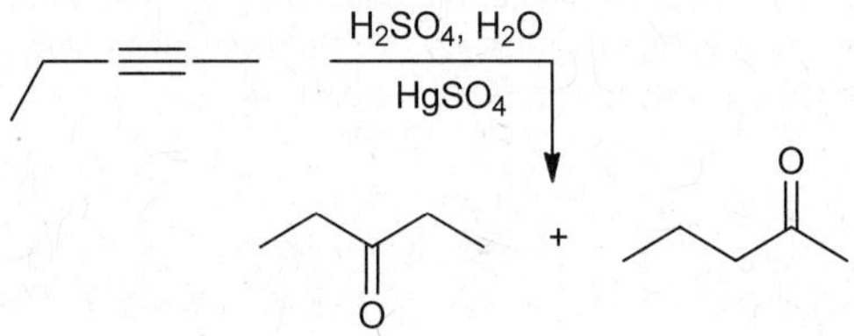

10.57.
(a) As seen in Figure 10.7, a double bond can be converted into a triple bond via bromination, followed by two successive elimination reactions (with excess NaNH$_2$). Under these conditions, the resulting terminal alkyne is deprotonated, so water is introduced to protonate the alkynide ion, generating the alkyne:

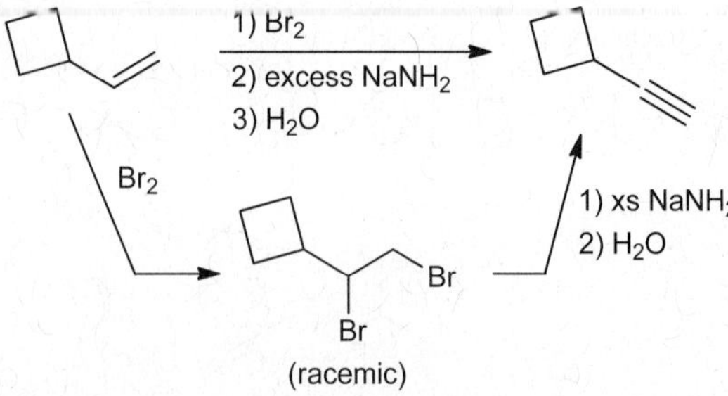

(b) The product is a methyl ketone, which can be prepared from the corresponding alkyne via acid-catalyzed hydration. To convert the starting material into the necessary alkyne, we simply perform the process shown in the previous problem (10.57a):

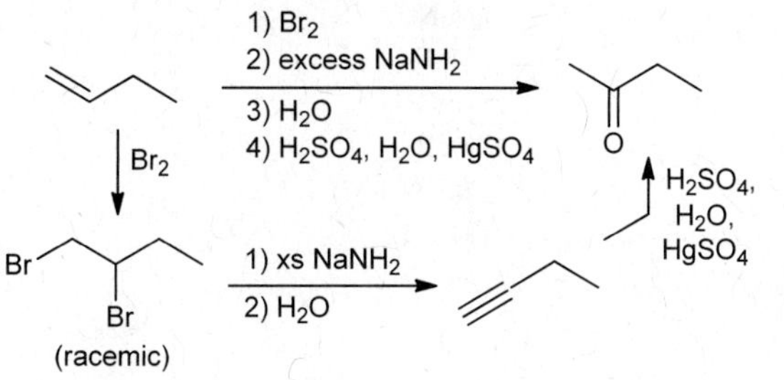

(c) The starting material has four carbon atoms, and the product has six carbon atoms, so two carbon atoms must be installed. This can be accomplished by alkylating the terminal alkyne. After the alkylation process, a dissolving metal reduction will convert the alkyne into the desired *trans* alkene:

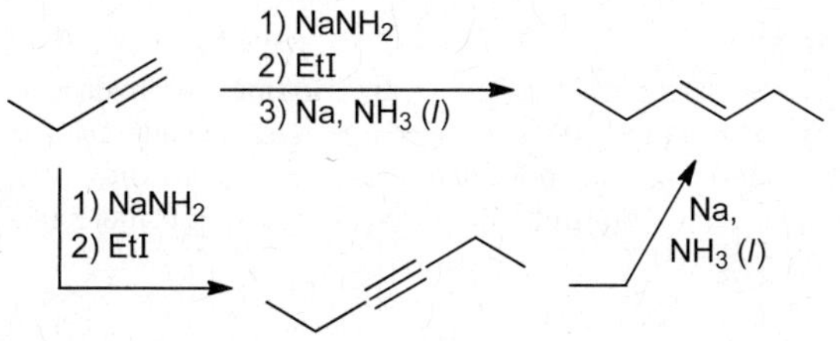

(d) The starting material has eight carbon atoms, and the product has eleven carbon atoms, so three carbon atoms must be installed. This can be accomplished by alkylating the terminal alkyne. After the alkylation process, the resulting internal alkyne can be converted to the desired alkane upon hydrogenation:

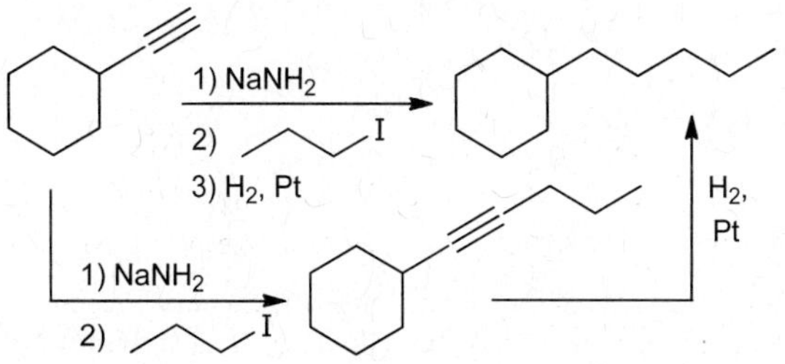

10.58. The starting material can either be a geminal dichloride or a vicinal dichloride:

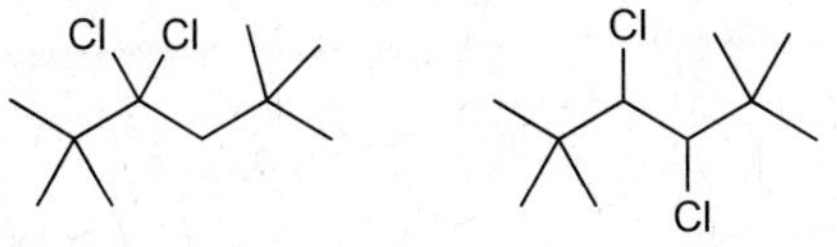

10.59. In order to determine the structures of compounds A-D, we must work backwards. The last step is an alkylation process that installs an ethyl group, so compound D must be an alkynide ion, which is prepared from the dibromide (compound B) formed in

the first step when an alkene (compound A) is treated with Br$_2$:

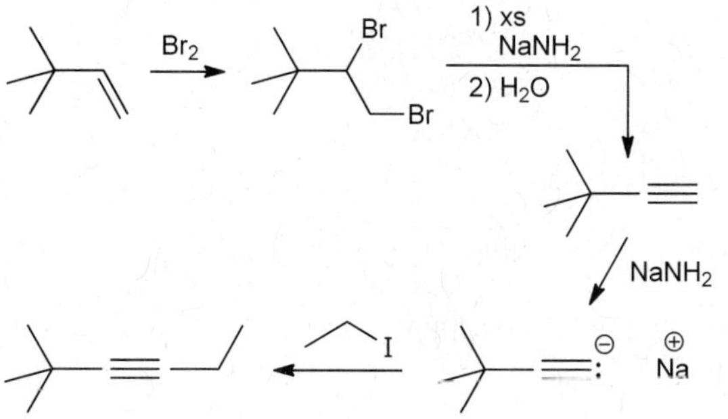

10.60.
(a) The alkyne can be reduced to an alkene, followed by bromination. Since the last step (bromination) proceeds via an *anti* addition, the desired stereoisomer can only be obtained if the previous step (reduction of the alkyne) is performed in an *anti* fashion as well. That is, we must perform a dissolving metal reduction:

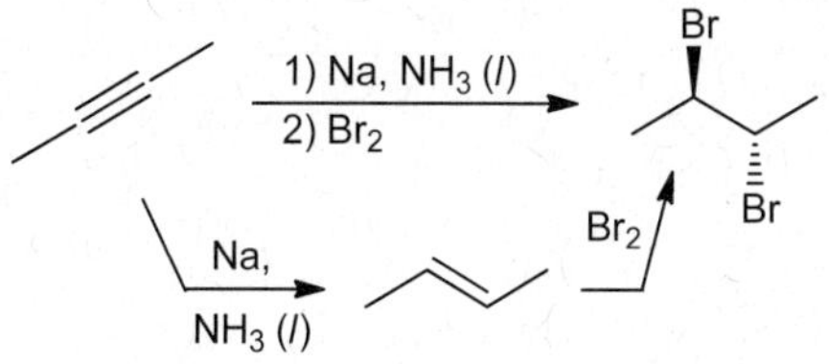

(b) The alkyne can be reduced to an alkene, followed by bromination. Since the last step (bromination) proceeds via an *anti* addition, the desired stereoisomer can only be obtained if the previous step (reduction of the alkyne) is performed in a *syn* fashion. That is, we must perform a hydrogenation reaction with a poisoned catalyst, rather than using a dissolving metal reduction.

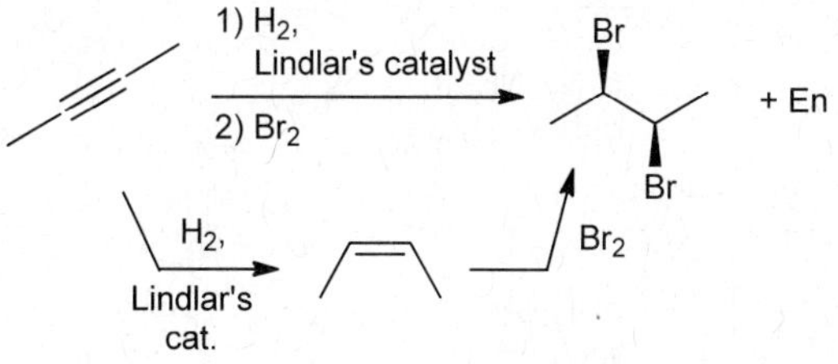

(c) This transformation requires two processes: 1) reduction of the alkyne to give an alkene, and 2) dihydroxylation to give a diol. In order to achieve the desired stereochemical outcome, we must perform one process in an *anti* fashion, and the other in a *syn* fashion. In the first answer below, the reduction is performed in an *anti* fashion, while the dihydroxylation process is performed in a *syn* fashion. In the second answer below, the reduction is performed in a *syn* fashion, while the

324   **CHAPTER 10**

dihydroxylation process is performed in an *anti* fashion. Both answers are acceptable.

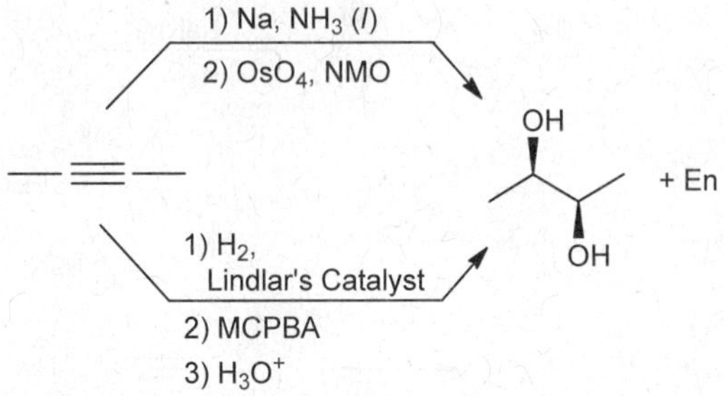

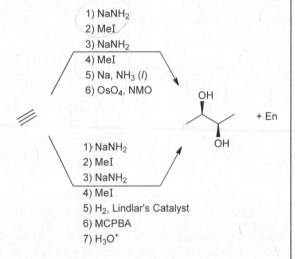

(d) This transformation requires two processes: 1) reduction of the alkyne to give an alkene, and 2) dihydroxylation to give a diol. In order to achieve the desired stereochemical outcome, we must either perform each process in an *anti* fashion (as shown in the first answer below), or we must perform each process in a *syn* fashion (as shown in the second answer below). Both answers are acceptable.

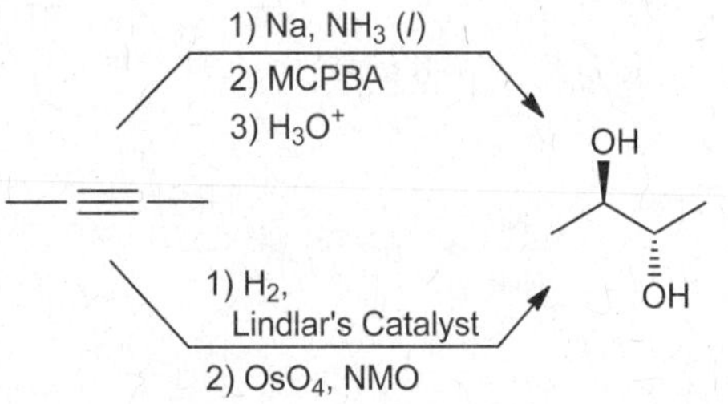

(e) This problem is similar to problem 10.60c, but the starting alkyne is acetylene, which must first be alkylated twice:

(f) This problem is similar to problem 10.60d, but the starting alkyne is acetylene, which must first be alkylated twice:

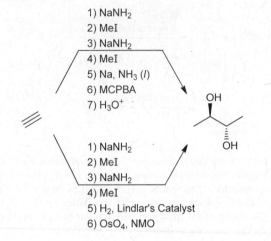

10.61.
(a) Deuterium (D) is an isotope of hydrogen. Therefore, this process is similar to hydrogenation (in the presence of a poisoned catalyst), but molecular deuterium (D_2) must be used in place of molecular hydrogen (H_2):

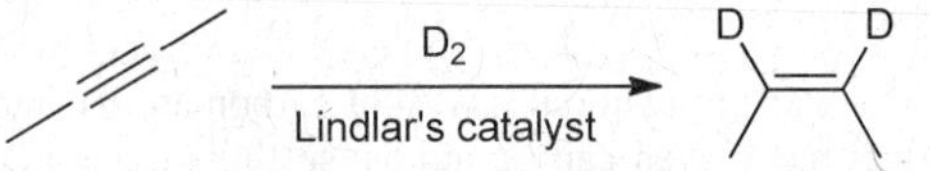

(b) Deuterium (D) is an isotope of hydrogen. Therefore, this process is similar to a dissolving metal reduction, but the source of protons (NH_3) must be replaced with a source of deuterons (ND_3):

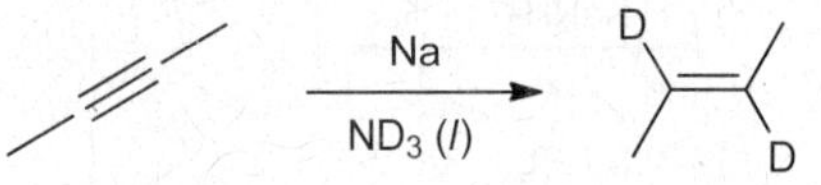

(c) Treatment of the alkyne with sodium amide results in an alkynide ion. If the alkynide ion is treated with water (H_2O), the alkynide ion will be protonated again to regenerate the starting alkyne. If, however, the alkynide ion is treated with a source of deuterons (D_2O), then the desired deuterated compound will be produced:

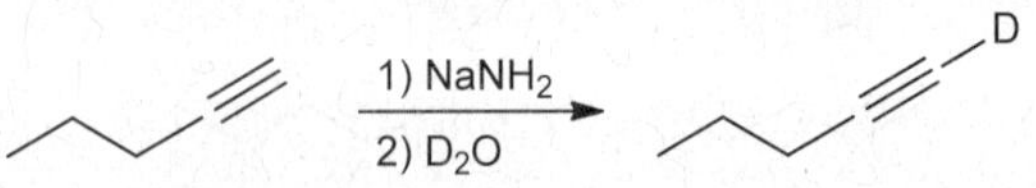

10.62. The acetylenic proton (pK_a ~ 25) is not the most acidic proton in the compound. The OH group bears a more acidic proton (pK_a ~ 18), so treatment with a strong base (such as sodium amide) will result in deprotonation of the OH group. The resulting alkoxide ion then serves as a nucleophile when treated with methyl iodide, giving an S$_N$2 reaction to generate the following product:

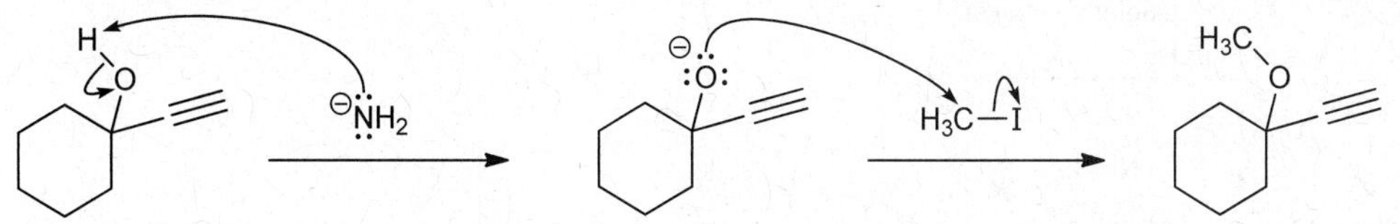

10.63. The key to solving this problem is to recognize that the methyl groups are not actually migrating. We can see this more clearly, if we flip the product horizontally, and then compare it to the starting material:

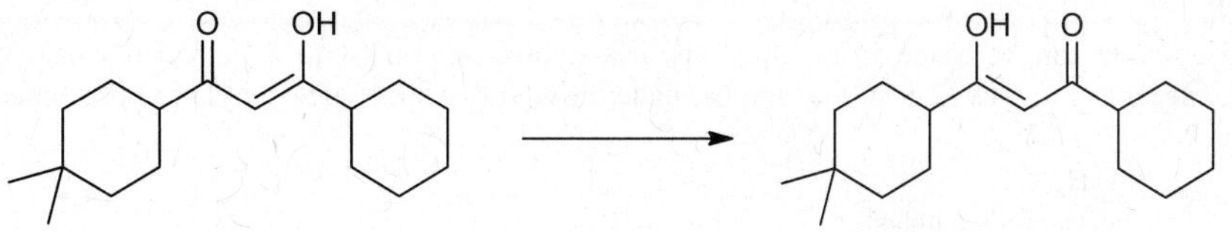

When drawn in this way, we can see more clearly that it is just a tautomerization process. Protonation of the C=O bond gives an intermediate that is highly resonance stabilized. Deprotonation then gives the product:

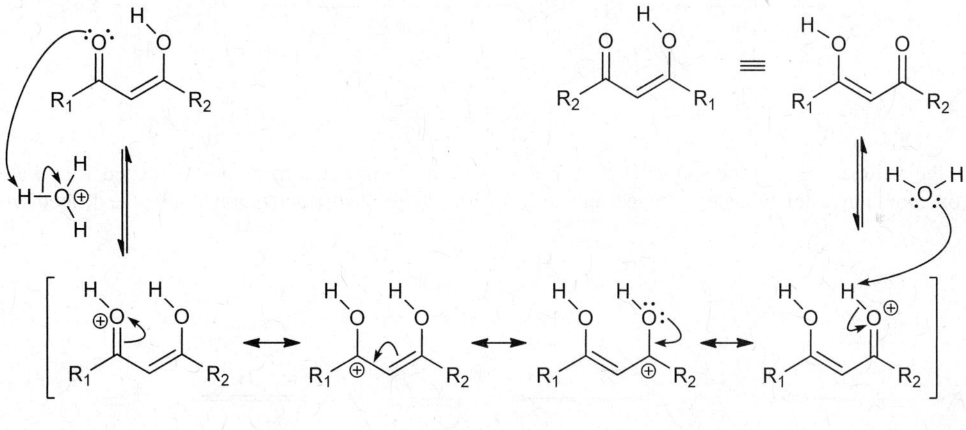

10.64. Much like an enol, the starting material in this case (called an enamine) also undergoes tautomerization via a similar mechanism. That is, the double bond is first protonated to generate a resonance-stabilized cation, which is then deprotonated. Notice that, in the first step, the double bond is protonated, rather than the nitrogen atom. Protonation of the nitrogen atom does not result in a resonance-stabilized cation.

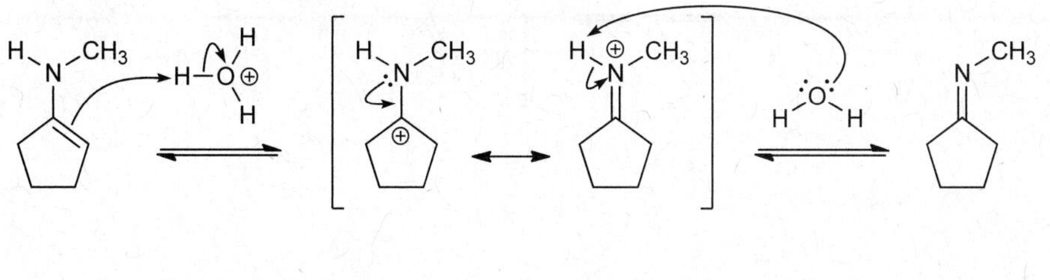

10.65.
(a) The desired epoxide can be made from 2-pentyne via a dissolving metal reduction, followed by epoxidation. The necessary alkyne (2-pentyne) can be made through two successive alkylation processes, shown here:

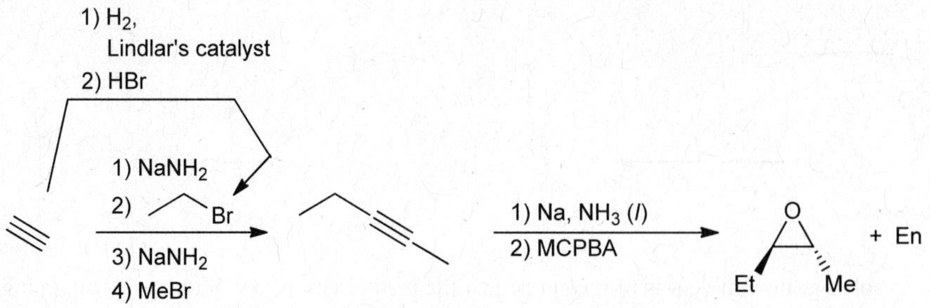

(b) The desired epoxide can be made from 2-pentyne via hydrogenation (with a poisoned catalyst), followed by epoxidation. The necessary alkyne (2-pentyne) can be made through two successive alkylation processes, shown here:

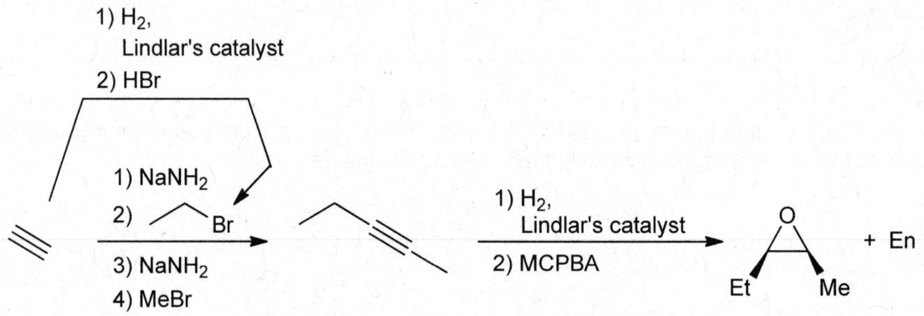

10.66. In the presence of Br_2, the alkyne is converted to a bromonium ion, which can be opened with water. The resulting oxonium ion is deprotonated to reveal an enol, which undergoes tautomerization to give the desired product:

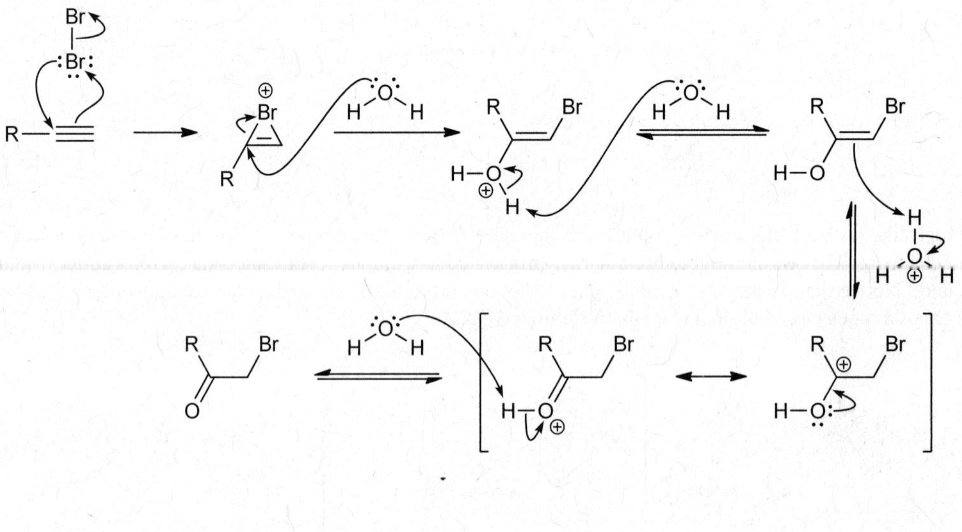

10.67. D_3O^+ is directly analogous to H_3O^+, but the protons have been replaced with deuterons. In the presence of D_3O^+, tautomerization processes can occur. Below are two, successive tautomerization processes that can successfully explain formation of the deuterated product:

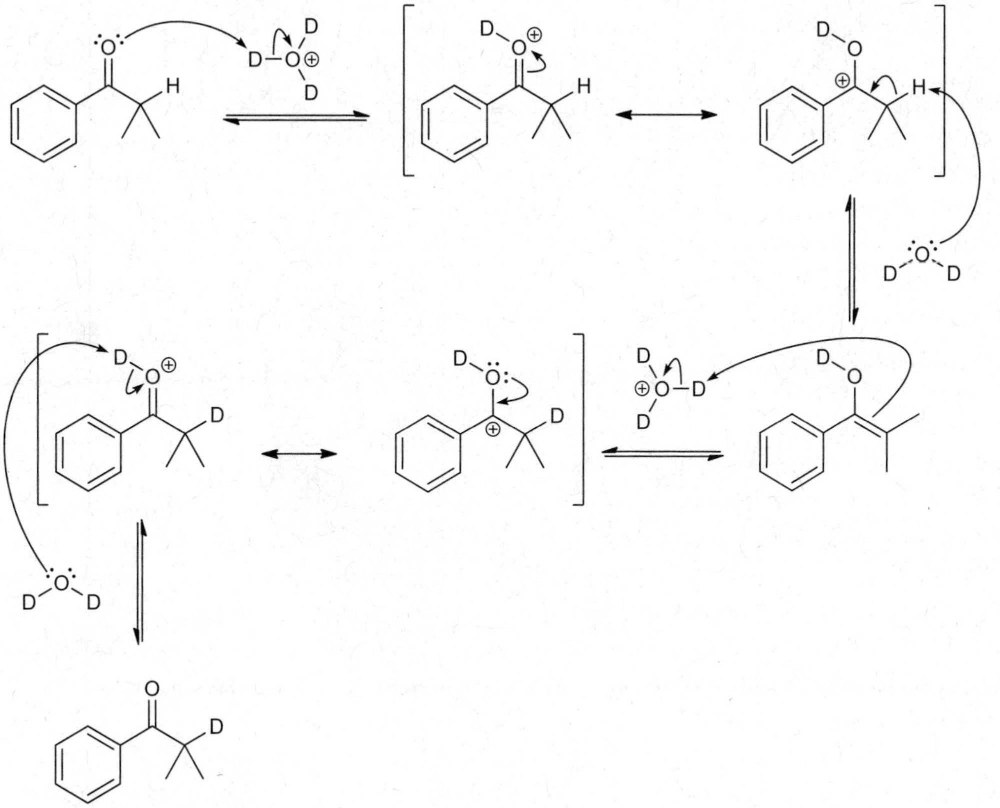

10.68.

(a) If the tautomerization process is base-catalyzed, then the first step will be deprotonation to give a resonance stabilized anion, which is then protonated to give the tautomer.

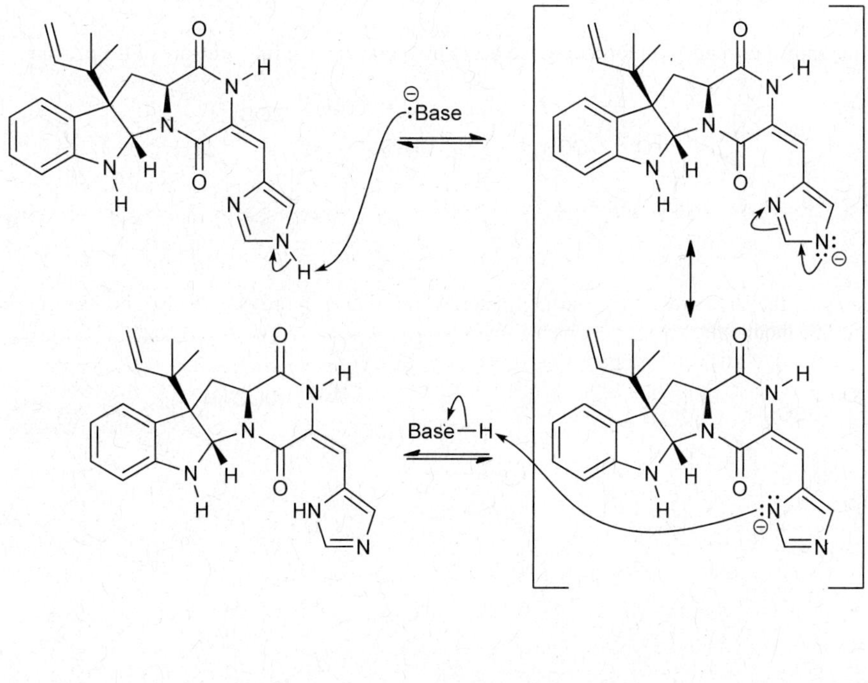

(b) If the tautomerization process is acid-catalyzed, then the first step will be protonation to give a resonance stabilized cation, which is then deprotonated to give the tautomer.

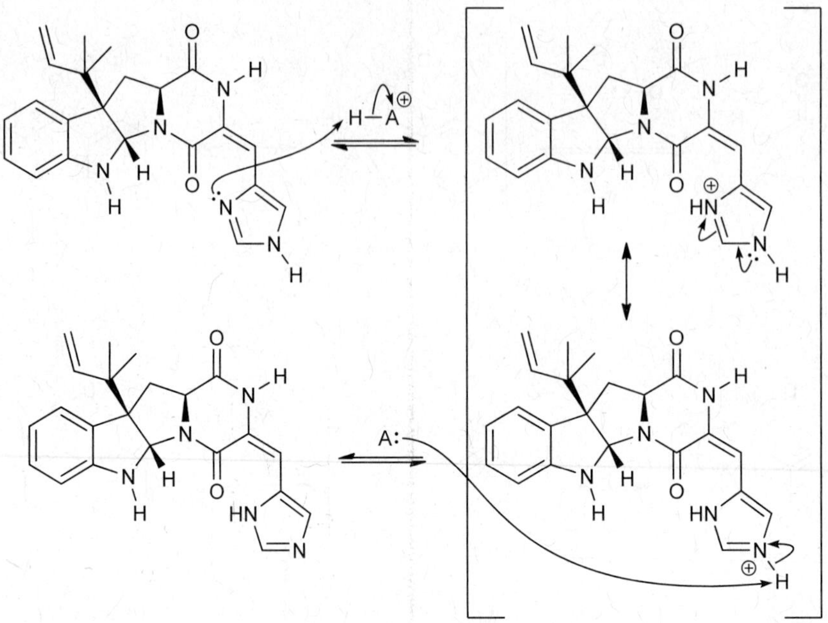

(c) The more stable tautomer is likely the one that can form an intramolecular hydrogen bonding interaction:

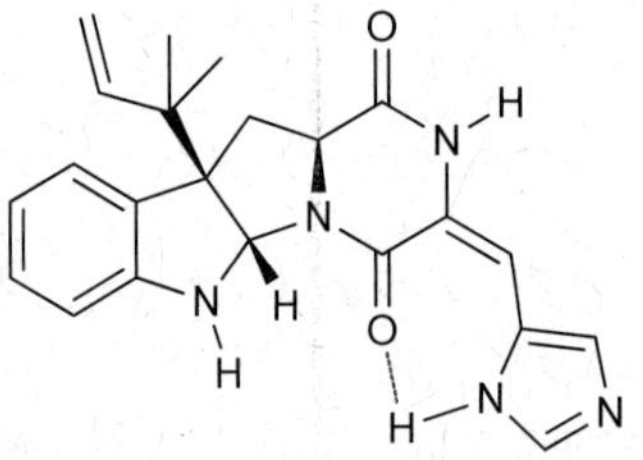

10.69.
(a) The product is shown in condensed format, so we begin by redrawing the full structure of the product:

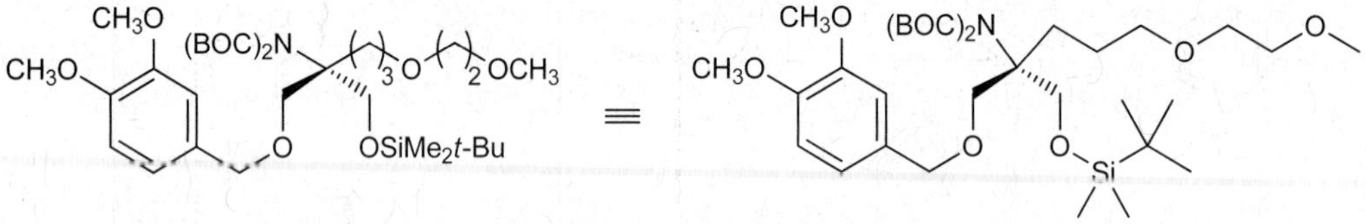

Now we can compare the structures of the starting material and product more effectively, and identify the part of the structure that must be modified:

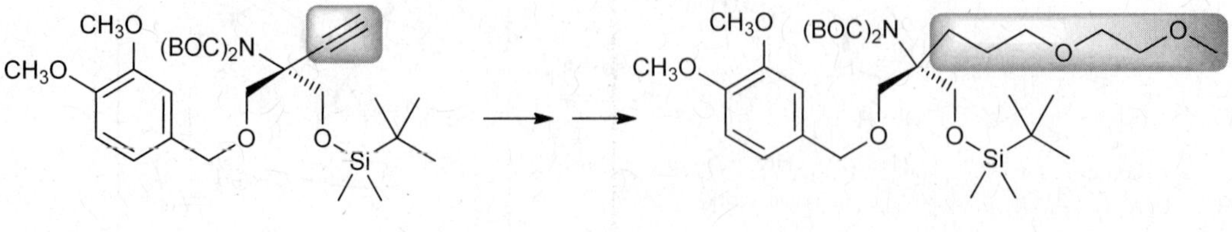

The rest of the structure remains unchanged. If we represent that part of the structure with an R group, the desired transformation can be shown as follows:

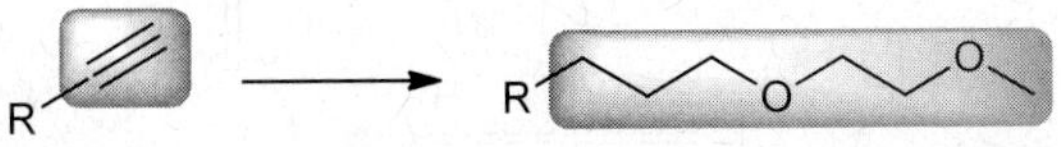

This transformation can be achieved by deprotonation of the terminal alkyne (with $NaNH_2$) and subsequent treatment with an appropriate electrophile (in this case the electrophile is $ClCH_2OCH_2CH_2OCH_3$, called methoxy ethoxymethyl chloride or MEM chloride), followed by reduction of the alkyne group via hydrogenation.

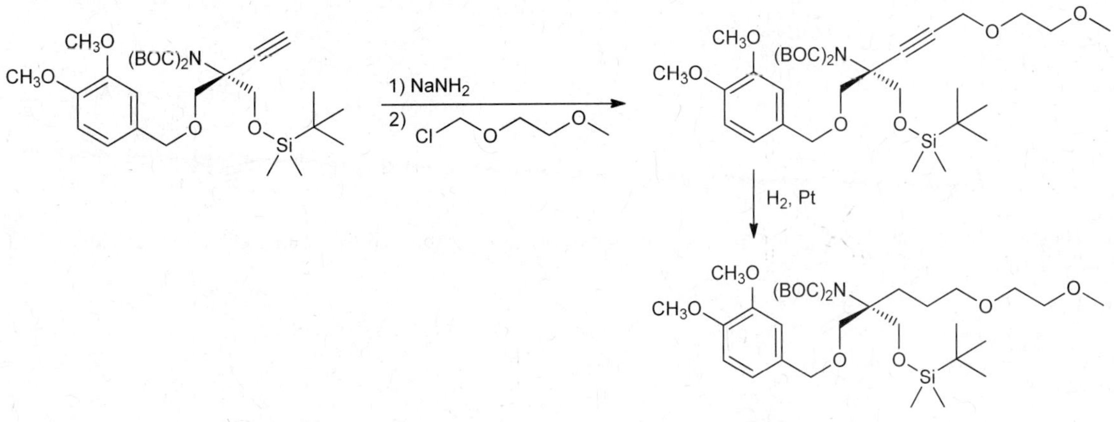

(b) The chirality center has the *S* configuration in both the starting material and the final product (the priorities, shown below, do not change as a result of the transformation).

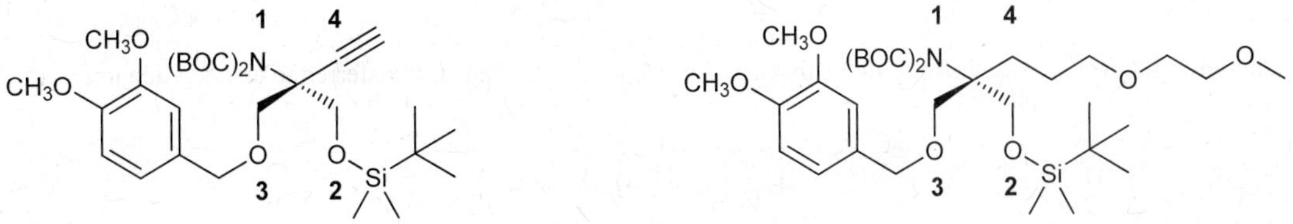

10.70.
(a) The molecular formula of compound **2** indicates that five carbon atoms have been installed (compound **1** contains only eight carbon atoms, while compound **2** contains thirteen carbon atoms). The lack of oxygen atoms in the molecular formula of compound **2** also indicates that the acetate group (OAc) has been completely removed, but the nitrogen atom is still present. These observations are consistent with the following structure, which can be formed via a substitution reaction in which the nitrogen atom of the indoline ring functions as the nucleophilic center, and the acetate group functions as a leaving group.

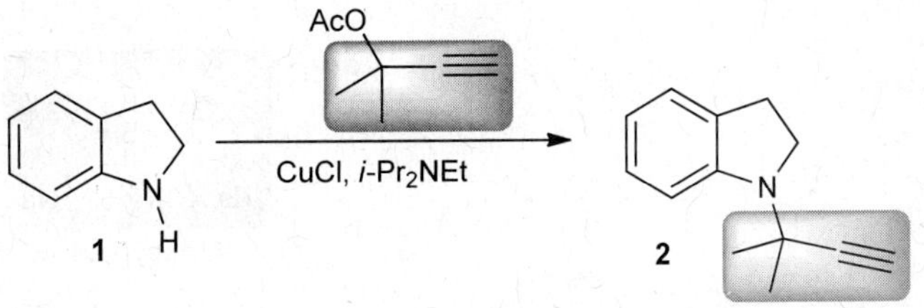

Since the substrate is tertiary, an S_N2 pathway is too slow to be viable, as the result of steric hindrance. As such, the reaction must proceed via an S_N1 pathway:

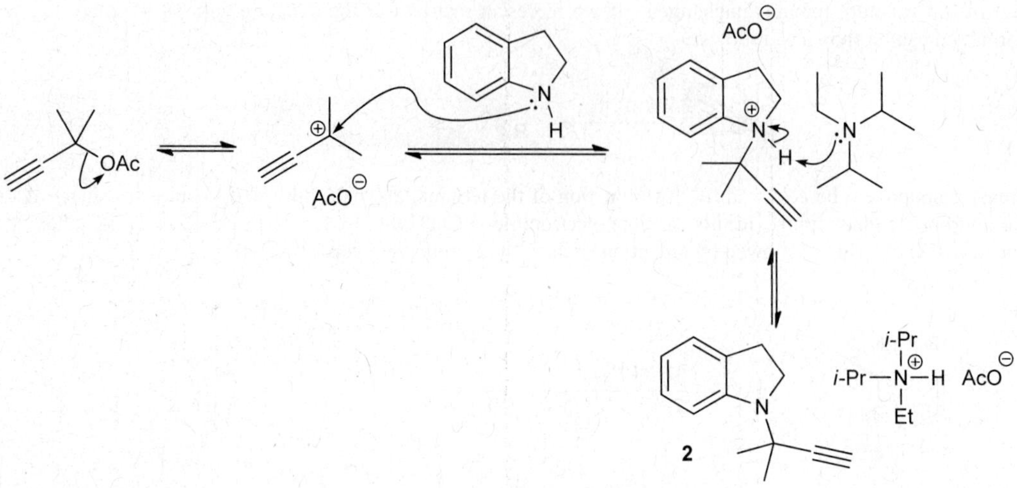

(b) The reduction of the alkyne group to an alkene group can be accomplished by treating compound **2** with H_2 and Lindlar's catalyst.

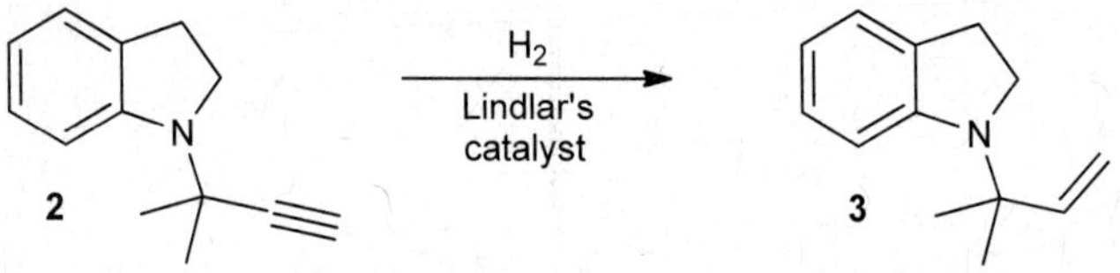

10.71.
(a) Dimethyl sulfate has two electrophilic methyl groups, each of which can be transferred to an acetylide ion:

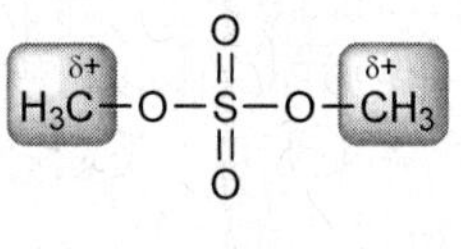

A mechanism is shown below. First, an acetylide ion functions as a nucleophile and attacks one of the methyl groups of dimethyl sulfate, thereby methylating the acetylide ion, giving propyne. Then, another acetylide ion attacks the remaining methyl group, to give a second equivalent of propyne. The ionic byproduct is Na_2SO_4 (sodium sulfate), as shown.

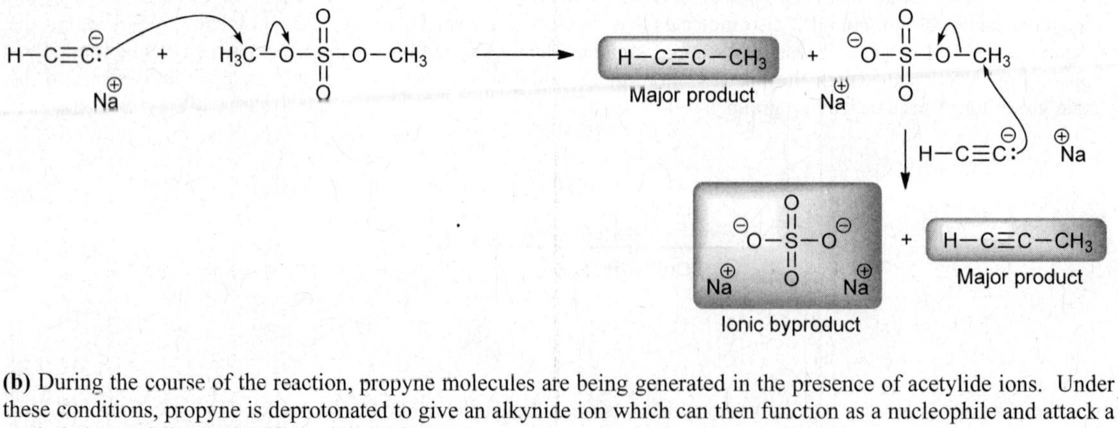

(b) During the course of the reaction, propyne molecules are being generated in the presence of acetylide ions. Under these conditions, propyne is deprotonated to give an alkynide ion which can then function as a nucleophile and attack a methyl group of dimethyl sulfate, giving 2-butyne:

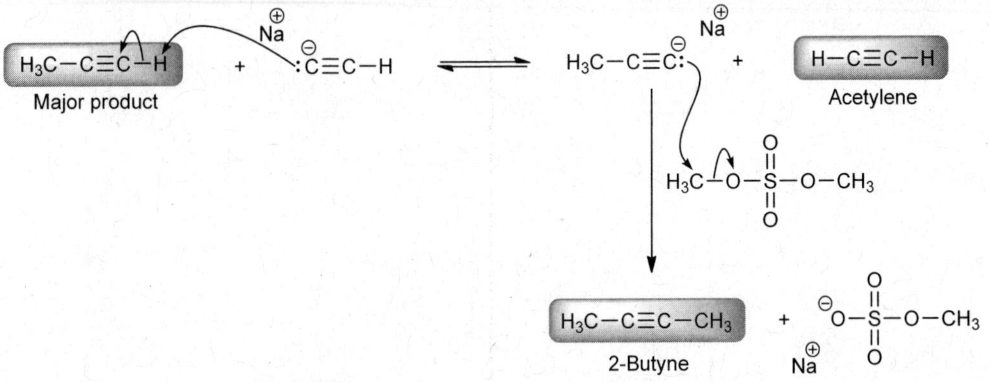

Notice that in the first step of this process, an acetylide ion is converted to acetylene. This mechanism is therefore consistent with the observation that acetylene is present among the reaction products.

(c) With diethyl sulfate, an ethyl group is transferred (rather than a methyl group). Therefore, the major product would be 1-butyne, as shown below:

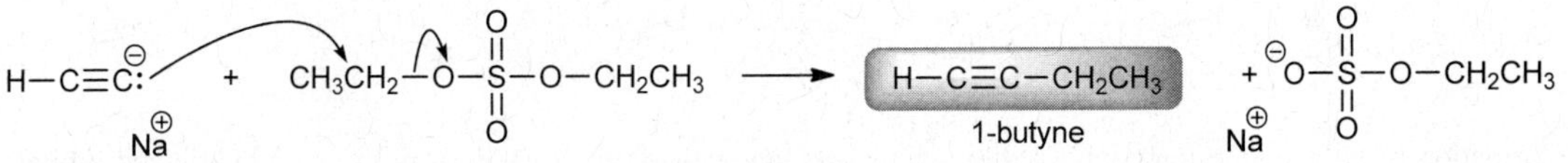

The minor product would be 3-hexyne, as shown below:

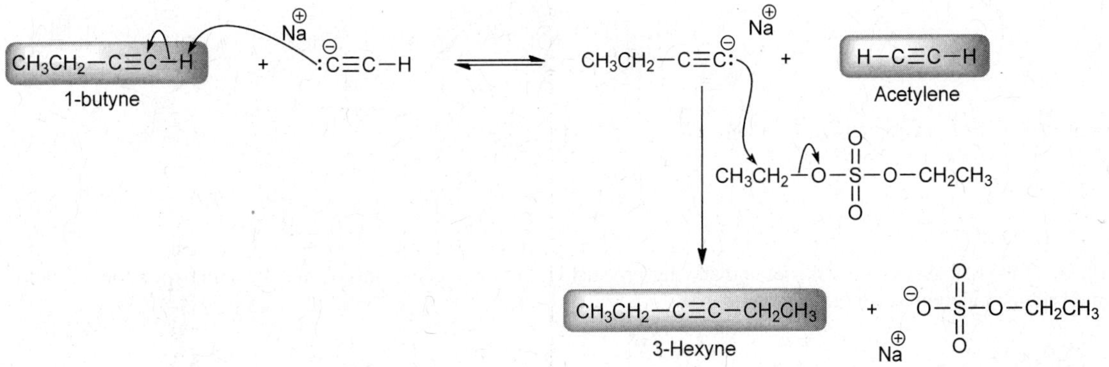

10.72. Sodium amide is a strong base; before the alkyne isomerization occurs, the alcohol group in **1** will be deprotonated to form an alkoxide ion (a negative charge on an oxygen atom) which is carried through until the final reprotonation event (when H_2O is introduced, which takes place *after* the transformation below is complete).

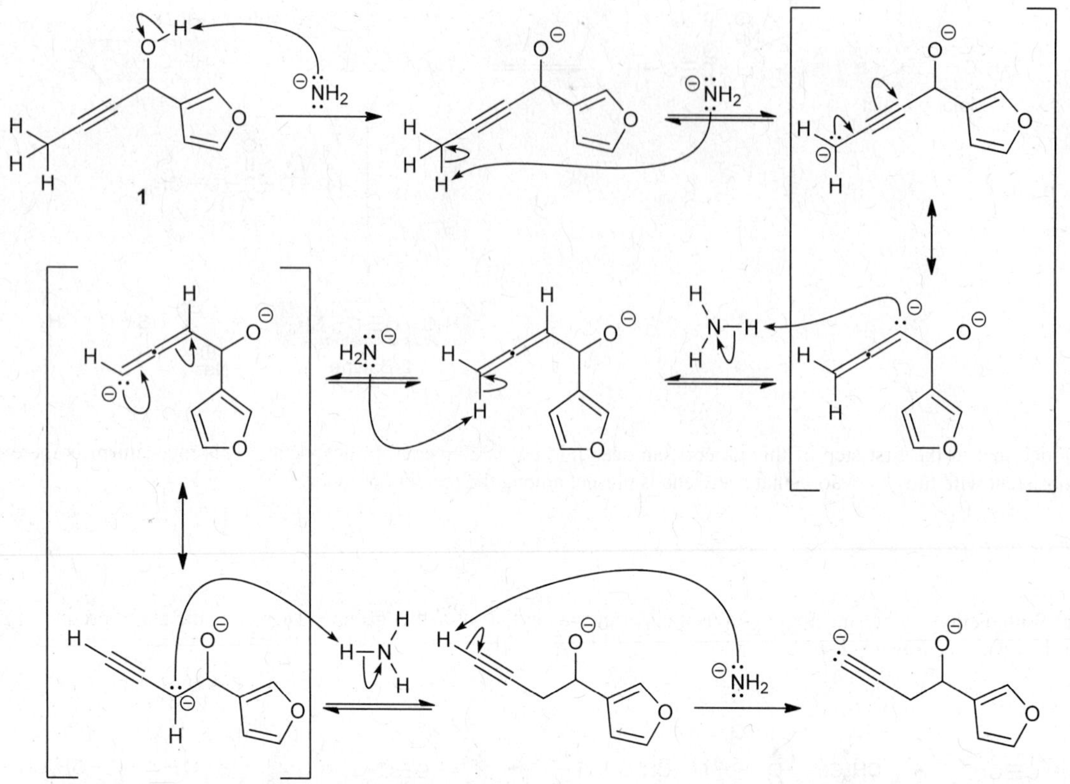

The formation of an alkynide ion (stabilized by having the lone pair on the carbon atom in an *sp* hybridized orbital) serves as a driving force for the reaction (See solution to problem 10.8). After the transformation above is complete, water is introduced as a proton source, giving **2**. The more strongly basic position is likely protonated first, followed by protonation of the alkoxide:

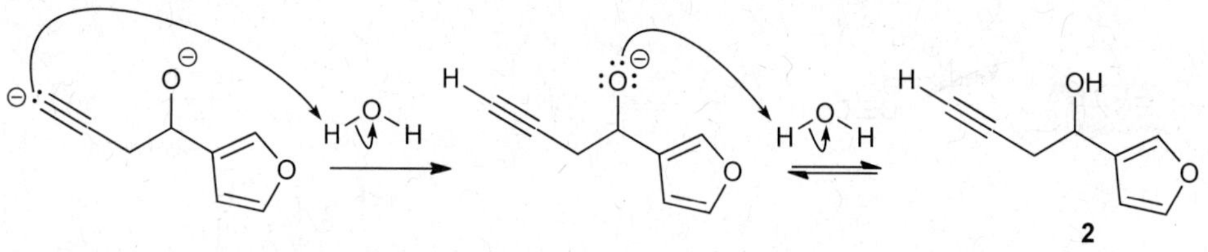

10.73. If the halogenation of alkenes and alkynes proceed via similar mechanisms, then we would expect the formation of a bridged iodonium ion intermediate:

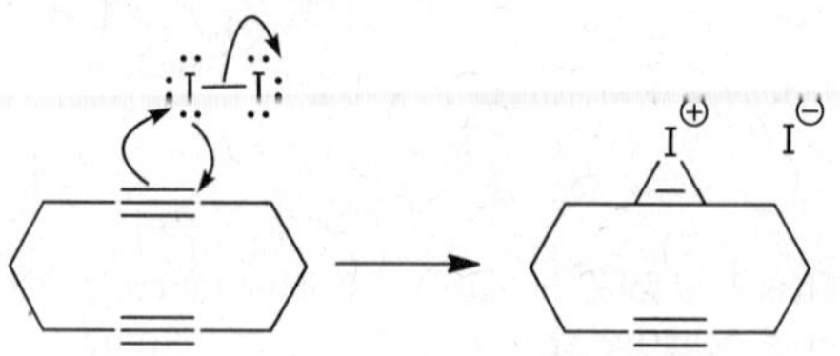

Iodonium ion
intermediate

From this point there are two possible mechanistic paths that could be envisioned. The first follows **Path A** where the second alkyne group attacks the iodonium ion, giving a vinylic carbocation, which then is attacked by the iodide anion to afford the product. However, this does not seem likely because the vinylic carbocation is a high energy intermediate. **Path B** is more likely because it avoids the formation of a vinylic carbocation. After the initial formation of the iodonium ion, the iodide anion can attack the second alkyne group, causing the π-electrons to attack the iodonium ion, giving the final product.

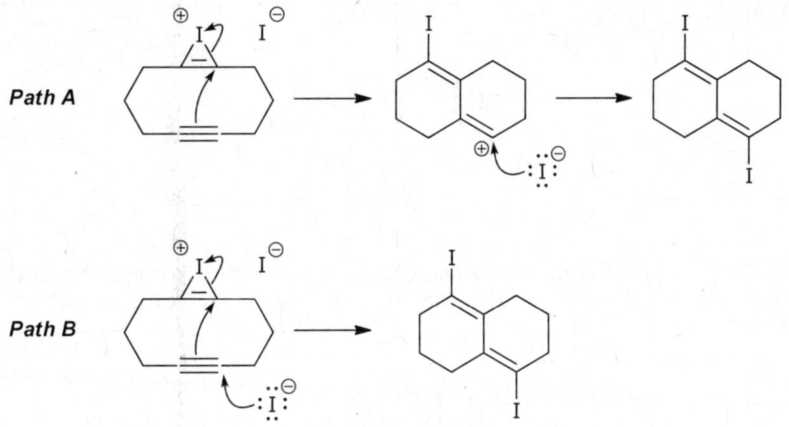

Path A

Path B

10.74.
(a) The following equilibrium is established between the aldehyde and the enol:

aldehyde   enol

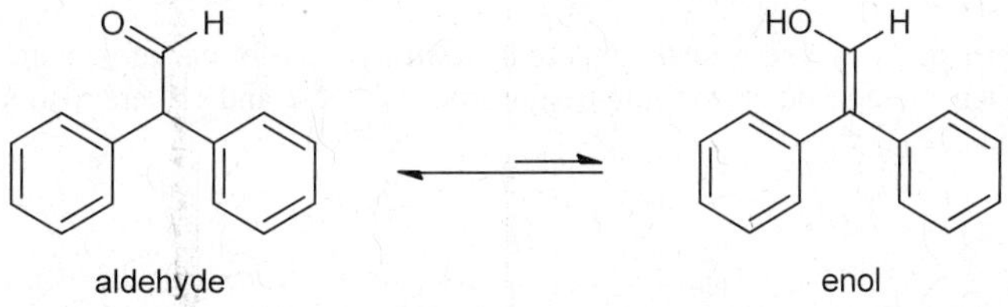

Notice that there is a change in hybridization state for the carbon atom that is connected to both aromatic rings. That carbon atom is sp^3 hybridized in the aldehyde, yet it is sp^2 hybridized in the enol. Recall that sp^3 hybridized carbon atoms are expected to exhibit bond angles of approximately 109.5°, while sp^2 hybridized carbon atoms are expected to exhibit bond angles of approximately 120°.

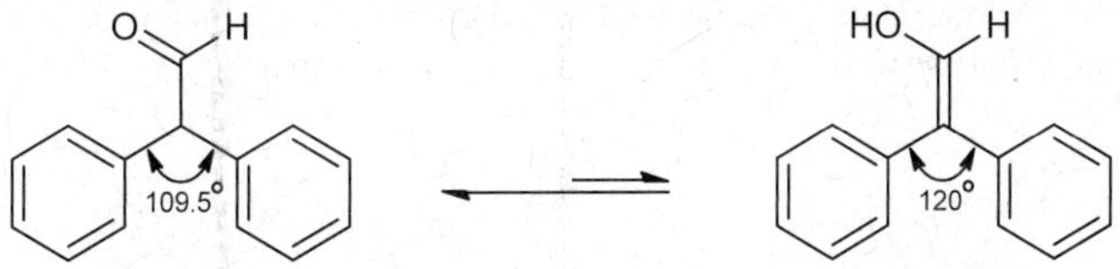

As such, the sterically demanding (bulky) aromatic rings are able to alleviate some steric strain in the enol form, relative to the aldehyde. In the aldehyde form, the aromatic rings are forced to be closer together in space. This steric effect causes the enol form to be particularly stable, and its concentration is significant (9.1%).

(b) The following equilibrium is established between the aldehyde and the enol:

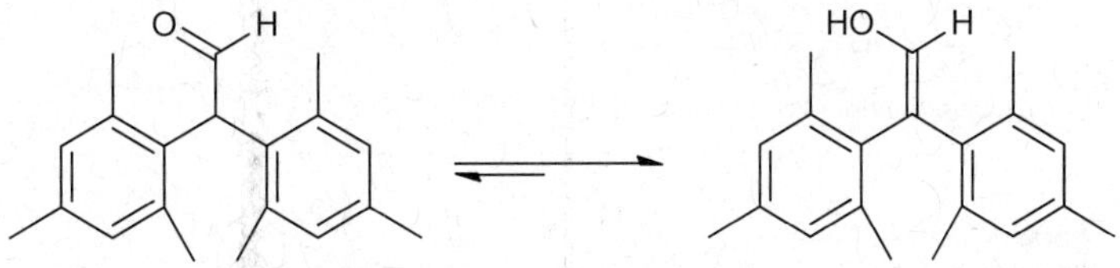

Once again, there is a change in hybridization state for the central carbon atom, and once again there is a steric effect. But in this case, the steric effect is more pronounced. The presence of the methyl groups causes the steric effect to be much greater, and as a result, the enol form is actually more stable than the aldehyde, because the enol form alleviates much of the significant steric strain present in the aldehyde form.

10.75.
(a) The following is a wedge-and-dash structure for the *anti-anti* conformation of pentane. Notice that C1 and C4 are *anti* to each other (when looking down the C2-C3 bond), while C2 and C5 are also *anti* to each other (when looking down the C3-C4 bond).

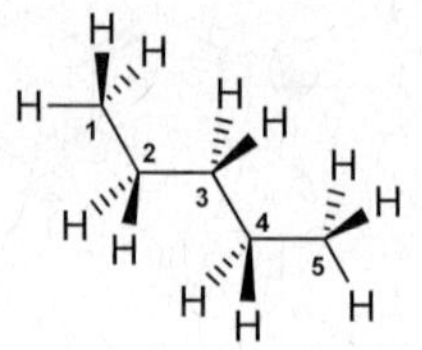

The following is a wedge-and-dash structure for the *anti-gauche* conformation of pentane. Notice that C1 and C4 are *anti* to each other (when looking down the C2-C3 bond), while C2 and C5 are *gauche* to each other (when looking down the C3-C4 bond).

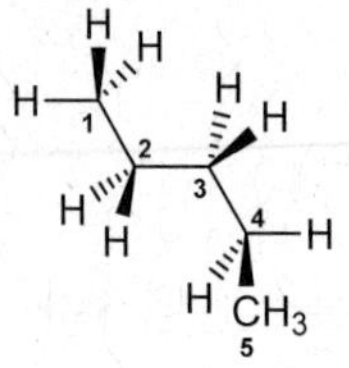

(b) The following conformer of 3-heptyne is analogous to the *anti-anti* conformer of pentane. Notice that C1 and C6 are *anti* to each other (when looking down the alkyne group), while C4 and C7 are also *anti* to each other (when looking down the C5-C6 bond).

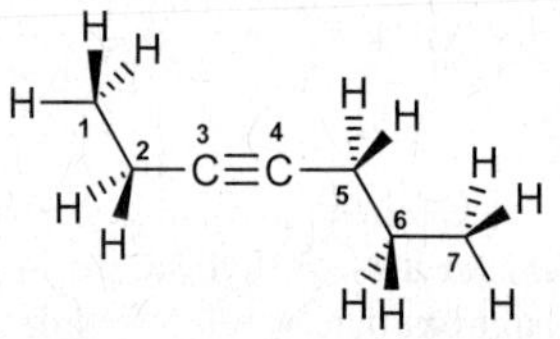

(c) In each of the lowest energy conformations of 3-heptyne, C1 and C6 are eclipsing each other (when looking down the alkyne group), as shown below. Note that the wedge-and-dash structure below is one of the two low-energy conformations, where C4 and C7 are *anti*:

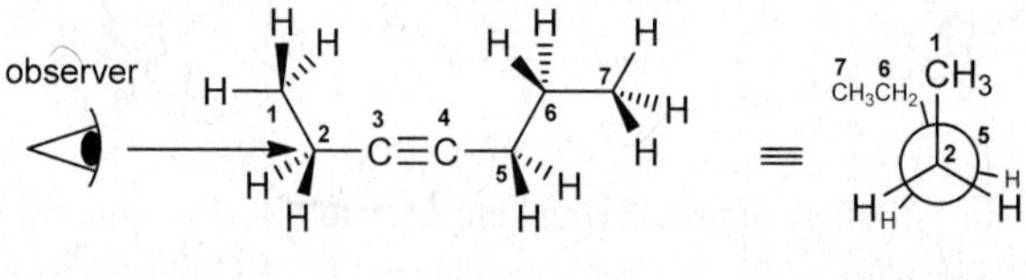

(d) The difference between the two lowest energy conformations of 3-heptyne can be seen when looking down the C5-C6 bond. In one conformation, C4 and C7 are *anti* to each other, as shown below.

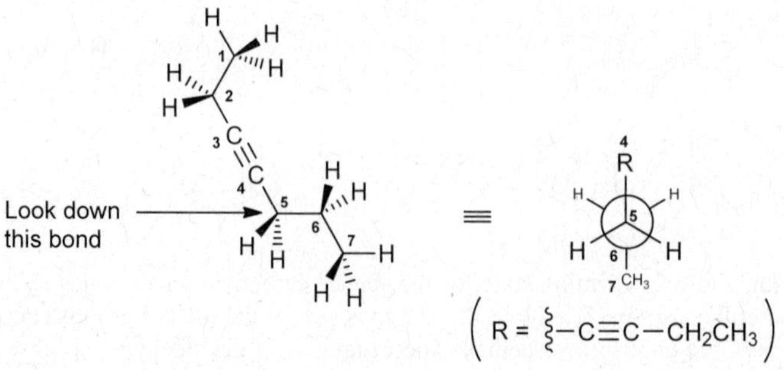

In the other conformation, C4 and C7 are *gauche*, as shown here:

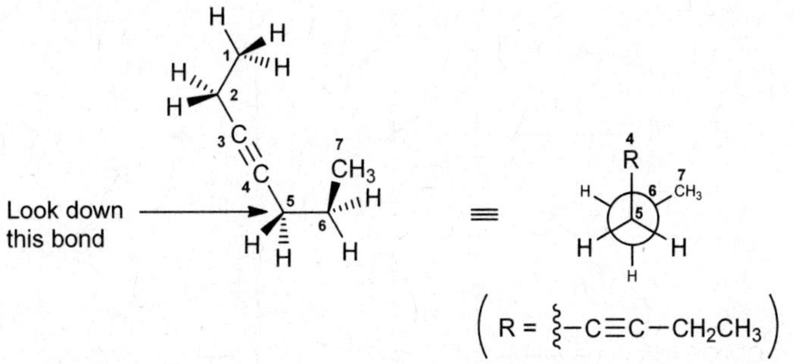

$$\left(R = \text{{}}-C\equiv C-CH_2CH_3 \right)$$

10.76.
(a) The following are the two possible vinyl cations (**A** and **B**) that can be produced when phenyl-substituted acetylenes (**1a-d**) are treated with HCl:

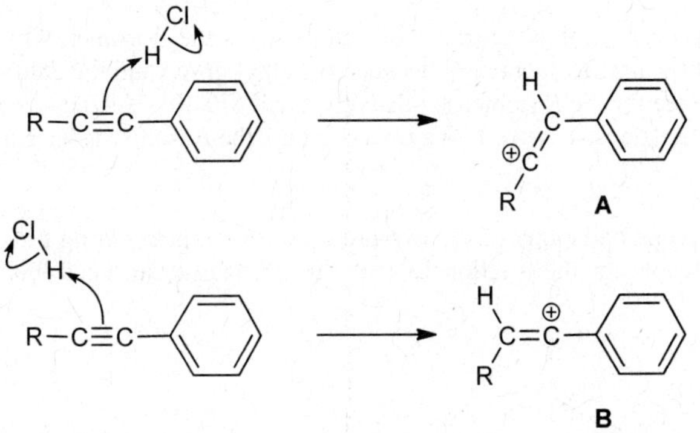

The empty $2p$ orbital (of the C+ atom) in cation **A** cannot be stabilized via resonance interaction with the aromatic ring, so we expect this cation to be highly unstable (as is the case for most vinyl carbocations). In contrast, the empty p-orbital of cation **B** *can* overlap effectively with the π system of the aromatic ring:

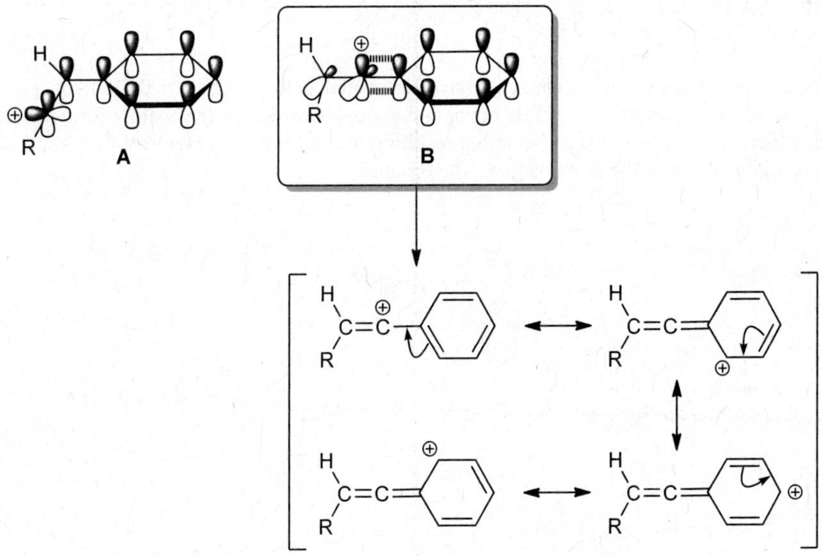

Since cation **B** is resonance-stabilized, it is much more stable than a typical vinyl carbocation. That is, cation **B** is more stable than cation **A**, explaining the regioselectivity observed in this series of hydrohalogenation reactions. This explains why Cl is installed at the benzylic position (the position next to the aromatic ring), because that is the location of the most stable carbocation.

(b) The stereoselectivity can be explained by considering the steric effects involved in the two competing transition states:

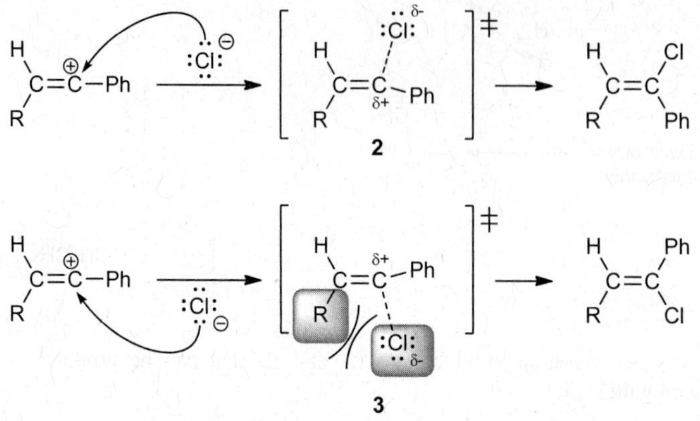

Attack of chloride on the vinyl carbocation via transition state **2** gives the *E* isomer, while transition state **3** results in the *Z* isomer. The latter involves a steric interaction between the alkyl group and the chloro group. For a small R group (such as R = Me), the preference for the *E* isomer is relatively small (70:30). As the size of the R group increases, the preference for the *E* isomer is enhanced. When R is a *t*-butyl group, the *E* isomer is the exclusive product.

10.77. An analysis of the reactant and either of the two products reveals that during the course of the reaction, a new C-C bond forms, as shown below for the reaction in acid. It is important to keep this in mind when developing a mechanism.

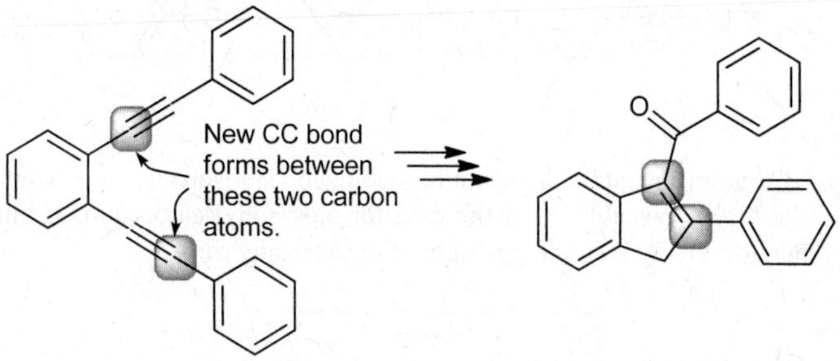

A reasonable mechanism for the bromination reaction is presented below. One of the alkyne groups reacts with bromine, resulting in the formation of a new C-Br bond, and a vinyl carbocation (see below for details on this). The π electrons from the other alkyne attack this carbocation, resulting in the formation of a new C-C bond, and a new vinyl carbocation. Nucleophilic attack of the bromide gives the product.

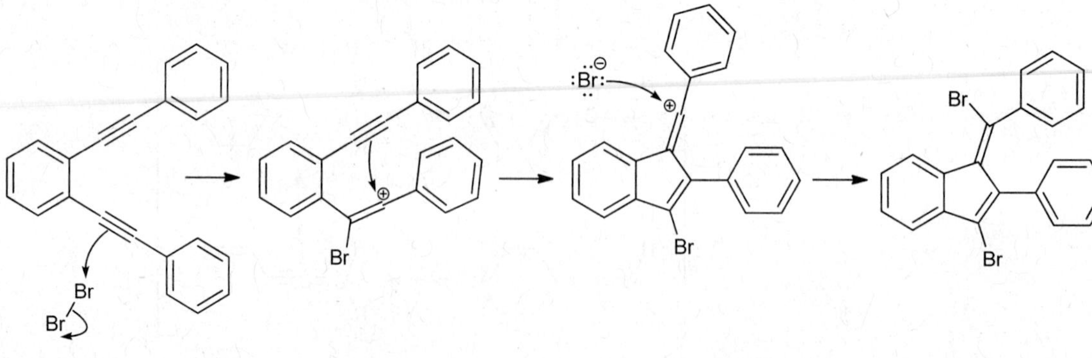

In this chapter, we noted that vinyl carbocations are not very stable, and thus we should consider whether or not this is a reasonable intermediate. In this case, the vinyl carbocation is not an unreasonable intermediate, because it is stabilized by the adjacent aromatic ring, as demonstrated by the following resonance structures.

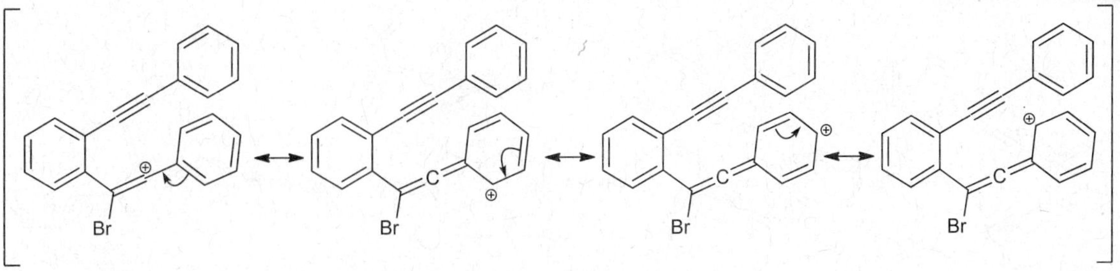

An alternate, but also reasonable mechanism is shown below, where the first two steps from our original mechanism are combined in a single, concerted step.

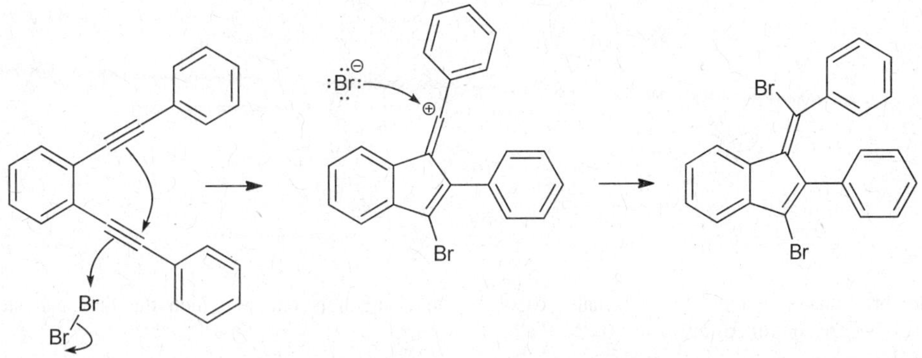

The stereochemistry of the exocyclic alkene can be explained by recognizing that bromide attack from the left side (distal to the indicated phenyl ring) is likely a more sterically accessible approach, as shown below.

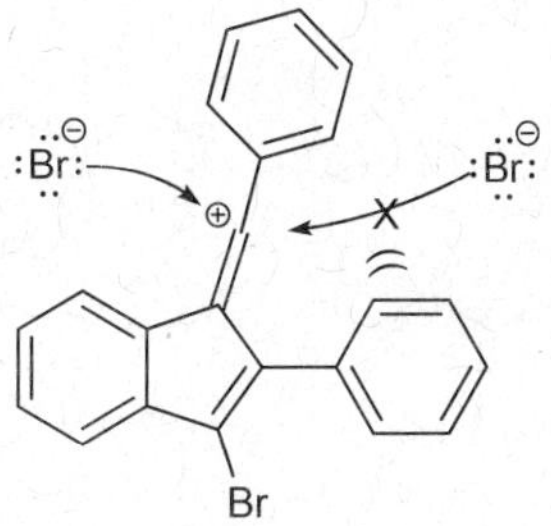

A reasonable mechanism for the reaction in sulfuric acid and water is presented below. One of the alkyne groups is protonated resulting in the formation of a new C-H bond, and a resonance-stabilized vinyl carbocation, analogous to the one described for the bromination reaction above. The π electrons from the other alkyne attack this carbocation, resulting in the formation of a new C-C bond, and a new vinyl carbocation. Nucleophilic attack of water, followed by deprotonation gives the enol, which tautomerizes to form the ketone, as shown (next page):

As with the bromination reaction, an alternate reasonable mechanism is one in which the first two steps of the mechanism above are combined, as shown:

Chapter 11
Radical Reactions

Review of Concepts

Fill in the blanks below. To verify that your answers are correct, look in your textbook at the end of Chapter 11. Each of the sentences below appears verbatim in the section entitled *Review of Concepts and Vocabulary*.

- Radical mechanisms utilize **fishhook arrows**, each of which represents the flow of _____.

- Every step in a radical mechanism can be classified as **initiation**, _____, or **termination**.

- A **radical initiator** is a compound with a weak bond that readily undergoes _____.

- A _____, also called a radical scavenger, is a compound that prevents a chain process from either getting started or continuing.

- _____ is more selective than chlorination.

- When a new chirality center is created during a radical halogenation process, a _____ mixture is obtained.

- _____ can undergo **allylic bromination**, in which bromination occurs at the allylic position.

- Organic compounds undergo oxidation in the presence of atmospheric oxygen to produce **hydroperoxides**. This process, called _____, is believed to proceed via a _____ mechanism.

- **Antioxidants**, such as BHT and BHA, are used as food preservatives to prevent autooxidation of _____ oils.

- When vinyl chloride is polymerized, _____ is obtained.

- Radical halogenation provides a method for introducing _____ into an alkane.

Review of Skills

Fill in the blanks and empty boxes below. To verify that your answers are correct, look in your textbook at the end of Chapter 11. The answers appear in the section entitled *SkillBuilder Review*.

11.1 Drawing Resonance Structures of Radicals

DRAW A RESONANCE STRUCTURE OF THE RADICAL BELOW, AS WELL AS THE FISHHOOK ARROWS THAT SHOW ITS FORMATION:

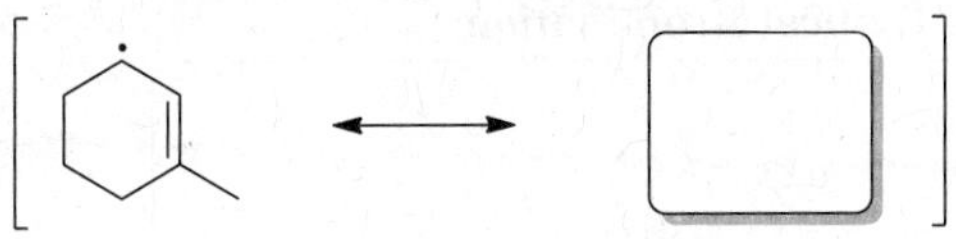

11.2 Identifying the Weakest C-H Bond in a Compound

IDENTIFY THE WEAKEST C-H BOND IN THE FOLLOWING COMPOUND:

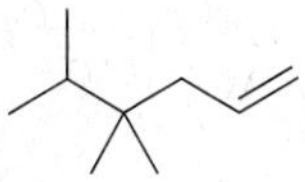

11.3   Identifying a Radical Pattern and Drawing Fishhook Arrows

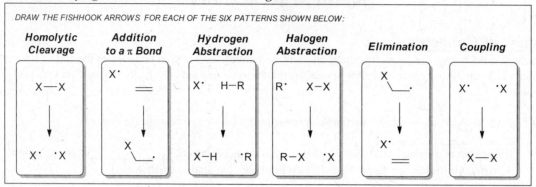

11.4   Drawing a Mechanism for Radical Halogenation

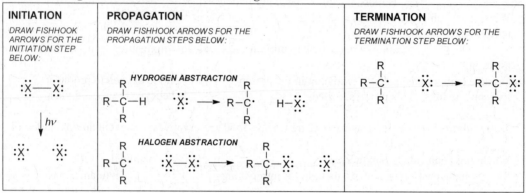

11.5   Predicting the Selectivity of Radical Bromination

DRAW THE EXPECTED MAJOR PRODUCT OF THE FOLLOWING MONOBROMINATION REACTION:

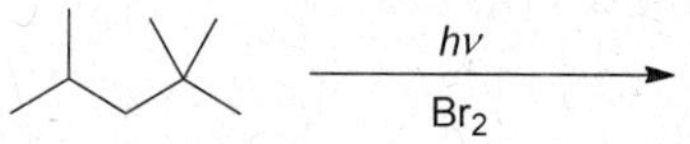

$$\xrightarrow[Br_2]{hv}$$

11.6   Predicting the Stereochemical Outcome of Radical Bromination

DRAW THE EXPECTED PRODUCTS OF THE FOLLOWING REACTION:

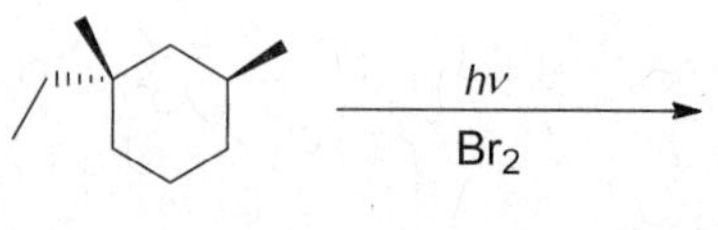

$$\xrightarrow[Br_2]{hv}$$

11.7   Predicting the Products of Allylic Bromination

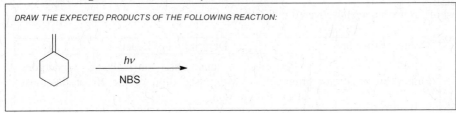

DRAW THE EXPECTED PRODUCTS OF THE FOLLOWING REACTION:

$$\xrightarrow[\text{NBS}]{h\nu}$$

11.8   Predicting the Products for Radical Addition of HBr

DRAW THE EXPECTED PRODUCTS OF THE FOLLOWING REACTION:

$$\xrightarrow[\text{ROOR}]{\text{HBr}}$$

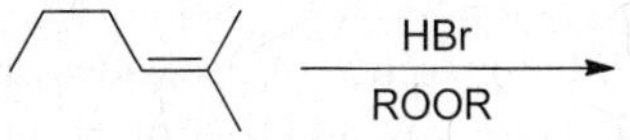

Review of Reactions

Identify the reagents necessary to achieve each of the following transformations. To verify that your answers are correct, look in your textbook at the end of Chapter 11. The answers appear in the section entitled *Review of Synthetically Useful Radical Reactions*.

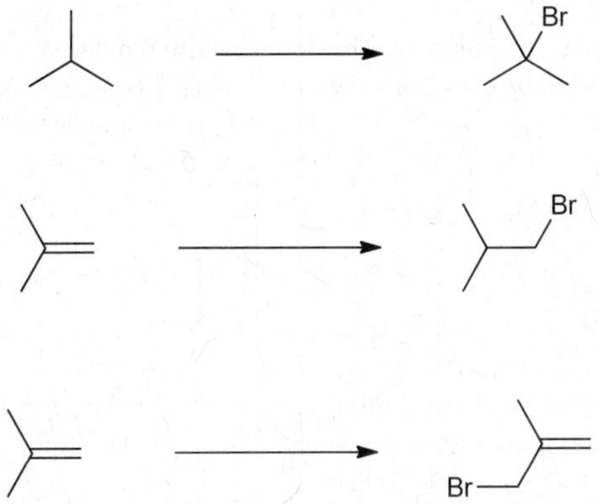

Common Mistakes to Avoid

When drawing a mechanism for a radical process, make sure that all curved arrows are single-barbed (called fishhook arrows), rather than double-barbed. For example, look closely at the head of each of the following fishhook arrows:

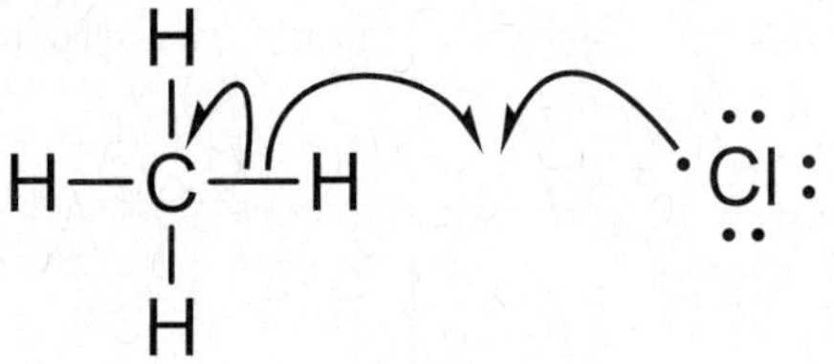

Each of these single-barbed arrows indicates the motion of one electron, while double-barbed arrows indicate the motion of two electrons. All of the mechanisms presented in this chapter utilize single-barbed arrows. Make sure to draw them properly.

Useful Reagents

The following is a list of common reagents covered in this chapter:

Reagents	Name of Reaction	Description of Reaction
Br$_2$, $h\nu$	Radical bromination	Under these conditions, an alkane undergoes bromination, with installation of the Br at the most substituted position.
Cl$_2$, $h\nu$	Radical chlorination	Under these conditions, an alkane undergoes chlorination. This reaction is less selective than bromination (but faster), and as such, it is generally most useful in situations where only one regiochemical outcome is possible (such as chlorination of cyclohexane or chlorination of 2,2-dimethylpropane).
HBr, ROOR	Hydrobromination	When treated with HBr in the presence of peroxides, an alkene undergoes *anti*-Markovnikov addition of H and Br.
NBS, $h\nu$	Allylic bromination	NBS, or *N*-bromosuccinimide, is a reagent that can be used to install a bromine atom at the allylic position of an alkene.

Solutions

11.1.

(a) The tertiary radical is the most stable, because alkyl groups stabilize the unpaired electron via a delocalization effect, called hyperconjugation. The primary radical is the least stable, because it lacks the stabilizing effect provided by multiple alkyl groups.

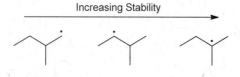

(b) The tertiary radical is the most stable, because alkyl groups stabilize the unpaired electron via a delocalization effect, called hyperconjugation. The primary radical is the least stable, because it lacks the stabilizing effect provided by multiple alkyl groups.

Increasing Stability

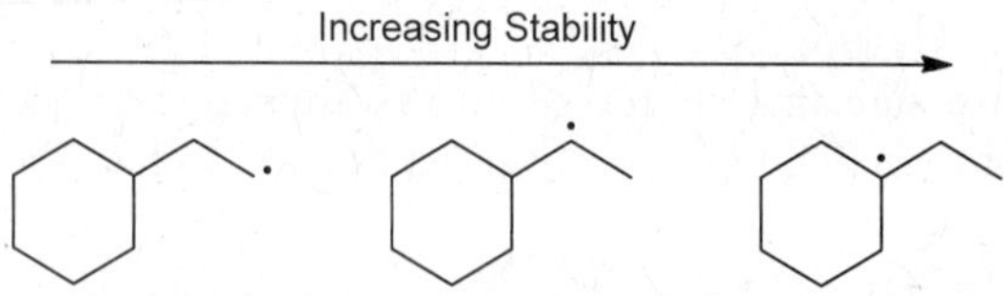

11.2.

(a) The unpaired electron occupies an allylic position, so it is resonance-stabilized. Three fishhook arrows are required, as shown:

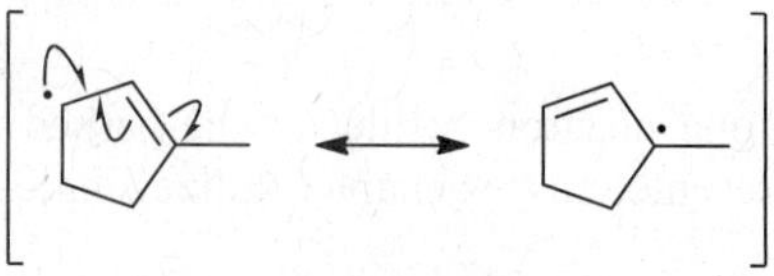

(b) The unpaired electron occupies an allylic position, so it is resonance-stabilized. Three fishhook arrows are required, as shown below. The resulting resonance structure exhibits an unpaired electron that is allylic to another π bond, so again we draw three fishhook arrows to arrive at another resonance structure:

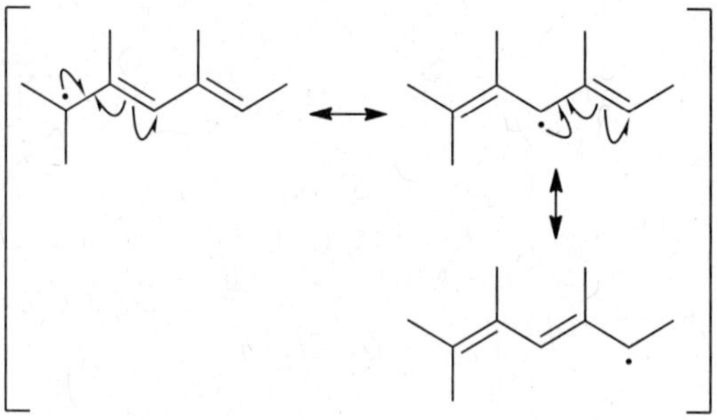

(c) The unpaired electron occupies an allylic position, so it is resonance-stabilized. Three fishhook arrows are required, as shown below. The resulting resonance structure exhibits an unpaired electron that is allylic to another π bond, so again we draw three fishhook arrows to arrive at another resonance structure:

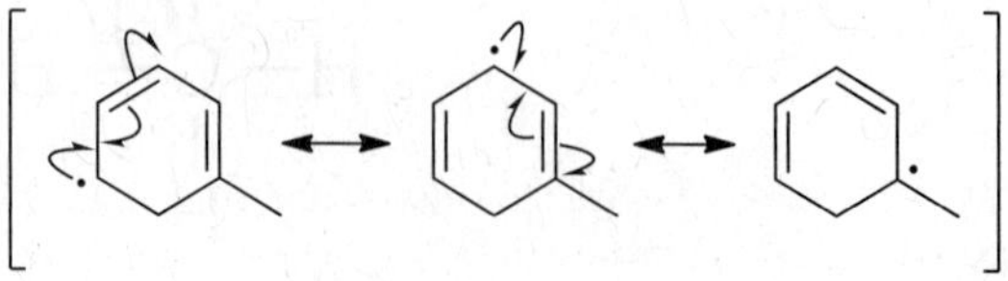

(d) The unpaired electron occupies an allylic position, so it is resonance-stabilized. Three fishhook arrows are required, as shown:

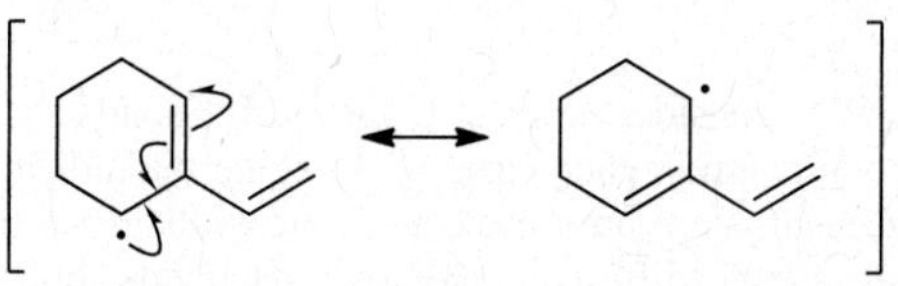

Note that the other π bond does not participate in resonance, because the unpaired electron is not allylic to that π bond.

11.3. This radical is unusually stable, because it has a large number of resonance structures, so the unpaired electron is highly delocalized. As shown below, three fishhook arrows are required in order to draw each resonance structure:

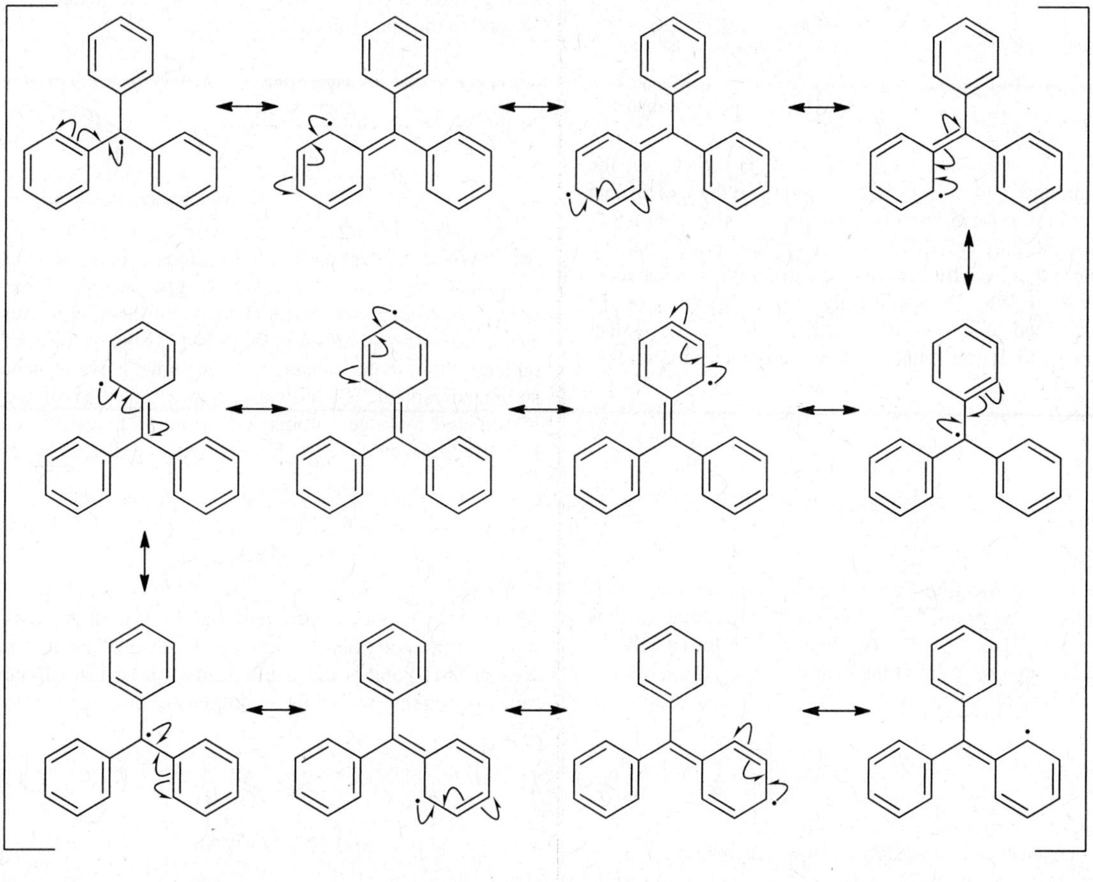

11.4. The following hydrogen atom (highlighted) is connected to an allylic position, and its removal will generate a resonance-stabilized radical:

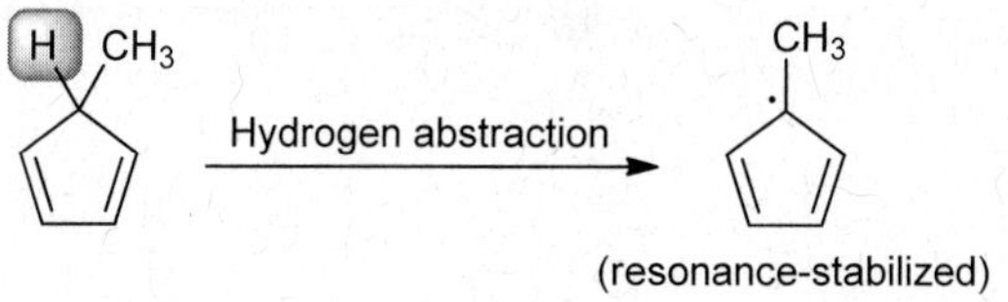

structure. This process continues until we have drawn all five resonance structures, shown here:

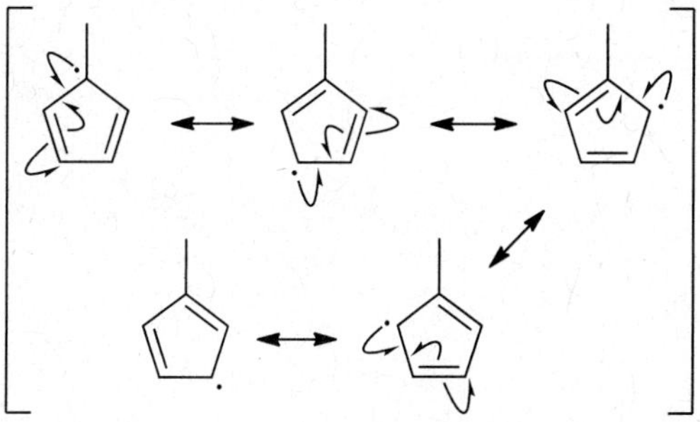

To draw the resonance structures for this radical, we begin by drawing three fishhook arrows. The resulting resonance structure also exhibits an unpaired electron that is allylic to a π bond, so again we draw three fishhook arrows as well as the resulting resonance

11.5.
(a)   We consider each of the C-H bonds in the compound, and in each case, we imagine the radical that would result from homolytic cleavage of that C-H bond. Among all the C-H bonds, only one of them can undergo homolytic cleavage to generate a resonance-stabilized radical.   Specifically, removal of the highlighted hydrogen atom will result in an allylic radical.   And therefore, this C-H bond is the weakest C-H bond in the compound:

(c)   We consider each of the C-H bonds in the compound, and in each case, we imagine the radical that would result from homolytic cleavage of that C-H bond. Among all the different types of C-H bonds, only one location can undergo homolytic cleavage to generate a resonance-stabilized radical. Specifically, removal of the highlighted hydrogen atom will result in an allylic radical.   And therefore, this C-H bond is the weakest C-H bond in the compound:

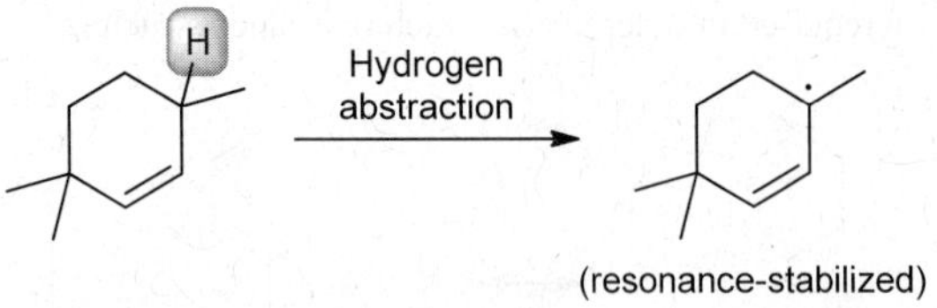

(b)   We consider each of the C-H bonds in the compound, and in each case, we imagine the radical that would result from homolytic cleavage of that C-H bond. There are only two different kinds of C-H bonds that can undergo homolytic cleavage to generate a resonance-stabilized radical.   Specifically, removing either of the highlighted hydrogen atoms will result in a resonance-stabilized intermediate (with three resonance structures):

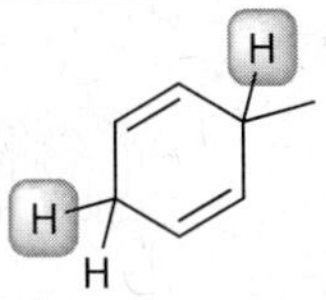

These two locations represent the two weakest C-H bonds in the compound.   Between the two of them, the weaker C-H bond is the one that gives a tertiary allylic radical upon removal of the hydrogen atom (rather than a secondary allylic radical):

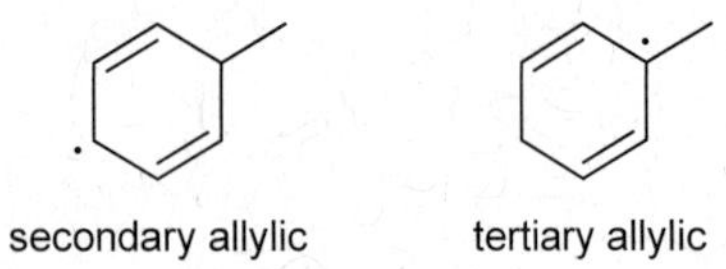

secondary allylic   tertiary allylic

A tertiary allylic radical is more stable than a secondary allylic radical, so the following C-H bond is the weakest C-H bond in the compound:

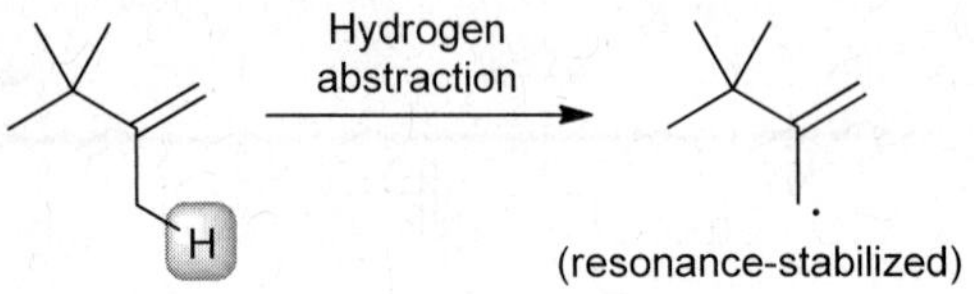

(d)   We consider each of the C-H bonds in the compound, and in each case, we imagine the radical that would result from homolytic cleavage of that C-H bond. There are only two different kinds of C-H bonds that can undergo homolytic cleavage to generate a resonance-stabilized radical.   Specifically, removing either of the highlighted hydrogen atoms will result in a resonance-stabilized intermediate (with two resonance structures):

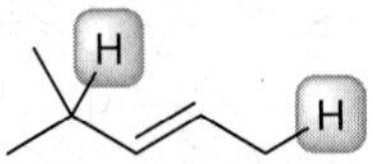

These two locations represent the two weakest C-H bonds in the compound.   Between the two of them, the weaker C-H bond is the one that gives a tertiary allylic radical upon removal of the hydrogen atom:

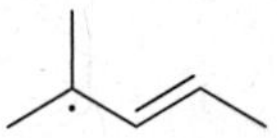

tertiary allylic

Therefore, the following C-H bond is the weakest C-H bond in the compound:

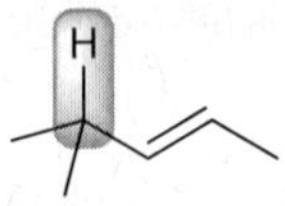

11.6. We begin by drawing the resonance structures of the radical that is formed when H$_a$ is abstracted, as well as the resonance structures of the radical that is formed when H$_b$ is abstracted:

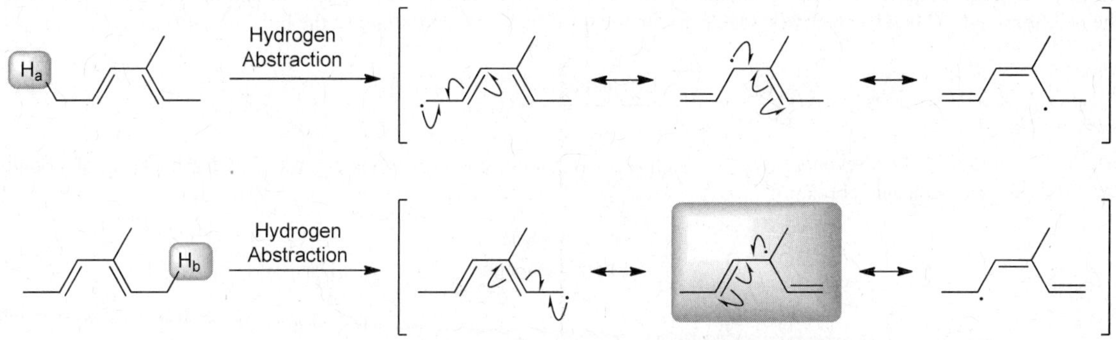

Compare the resonance structures in each case. Specifically, look at the middle resonance structure in each case. When H$_b$ is abstracted, the middle resonance structure is tertiary, and the effect of the methyl group is to stabilize the radical. This stabilizing factor is not present when H$_a$ is abstracted. Therefore, we expect the C-H$_b$ bond to be slightly weaker than the C-H$_a$ bond.

11.7.

(a) This step is a *hydrogen abstraction*, which requires a total of three fishhook arrows. In order to place the fishhook arrows properly, we must identify any bonds being formed (C–H) and any bonds being broken (H–Br). Formation of the C–H bond is shown with two fishhook arrows: one coming from the carbon radical and the other coming from the H–Br bond. The latter, together with the third fishhook arrow, shows the breaking of the H–Br bond:

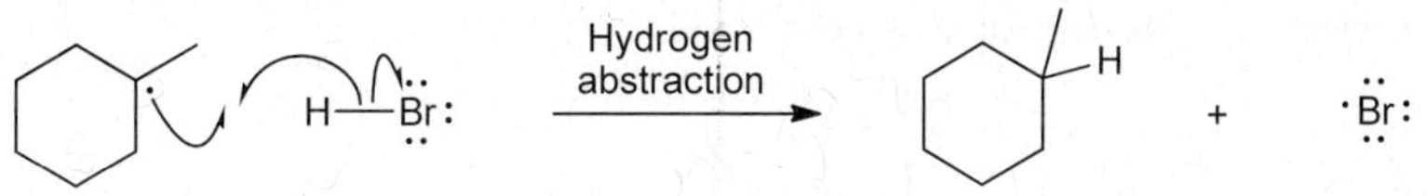

(b) This step is *addition to a π bond*, which requires a total of three fishhook arrows. In order to place the fishhook arrows properly, we must identify any bonds being formed (C–Br) and any bonds being broken (C=C π bond). Formation of the C–Br bond is shown with two fishhook arrows: one coming from the π bond and the other coming from the bromine radical. The former, together with the third fishhook arrow, shows the breaking of the C=C π bond:

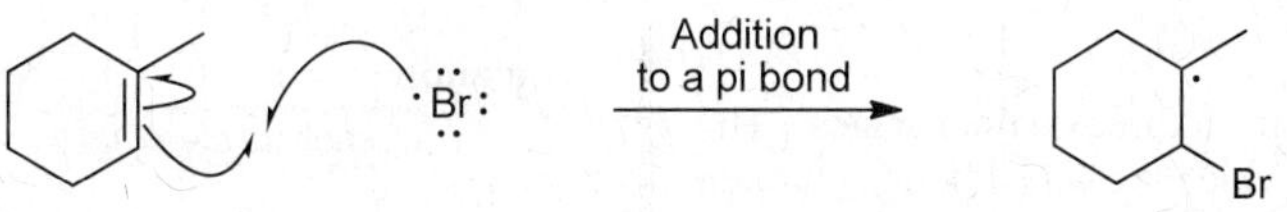

(c) This step is a *coupling* process, which requires a total of two fishhook arrows, showing formation of a bond (in this case, a C–C bond is formed):

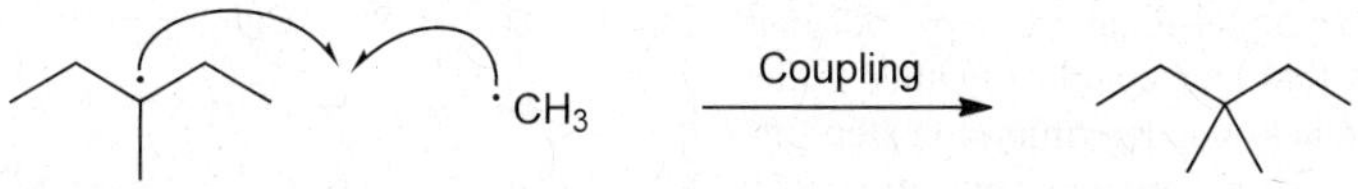

(d) This step is a *hydrogen abstraction*, which requires a total of three fishhook arrows. In order to place the fishhook arrows properly, we must identify any bonds being formed (H–Br) and any bonds being broken (C–H). Formation of the H–Br bond is shown with two fishhook arrows: one coming from the bromine radical and the other coming from the C–H bond. The latter, together with the third fishhook arrow, shows the breaking of the C–H bond:

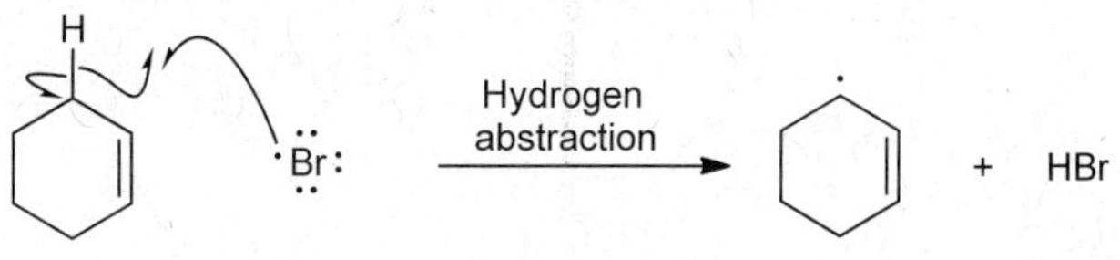

(e) This step is an *elimination*, which requires a total of three fishhook arrows. In order to place the fishhook arrows properly, we must identify any bonds being formed (a C=C π bond) and any bonds being broken (C–C). Formation of the C=C π bond is shown with two fishhook arrows: one coming from the carbon radical and the other coming from the neighboring C–C bond. The latter, together with the third fishhook arrow, shows the breaking of the C–C bond:

(f) This step is a *homolytic bond cleavage*, which requires a total of two fishhook arrows, showing breaking of a bond (in this case, an O–O bond is broken):

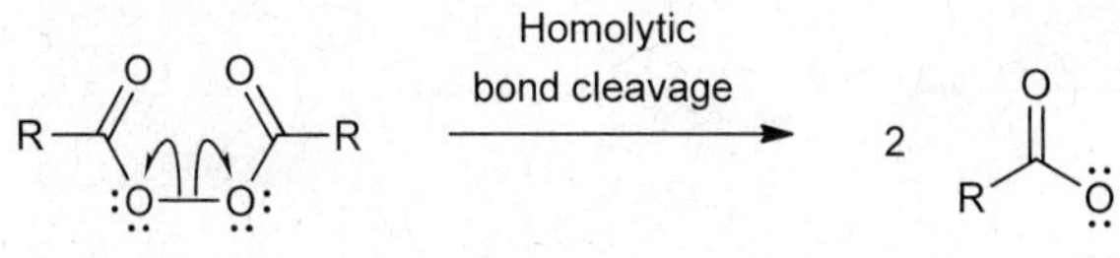

11.8. The reaction is intramolecular, which means that the reaction is taking place between two different regions of the same radical intermediate. Specifically, the unpaired electron can react with the π bond in a step called *addition to a π bond*. The resulting allylic radical is resonance-stabilized:

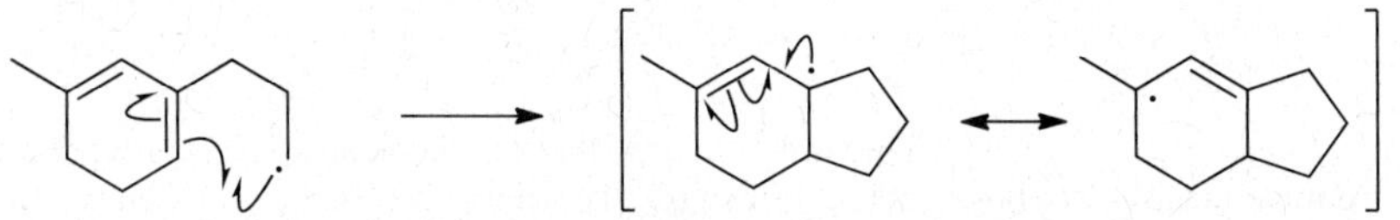

11.9. This process shows two radicals (see Solution to Problem 11.3 for resonance structures of these radicals) joining together to form a compound with no unpaired electrons. This step is therefore called a coupling step, and it requires two fishhook arrows, which show the formation of a C–C bond:

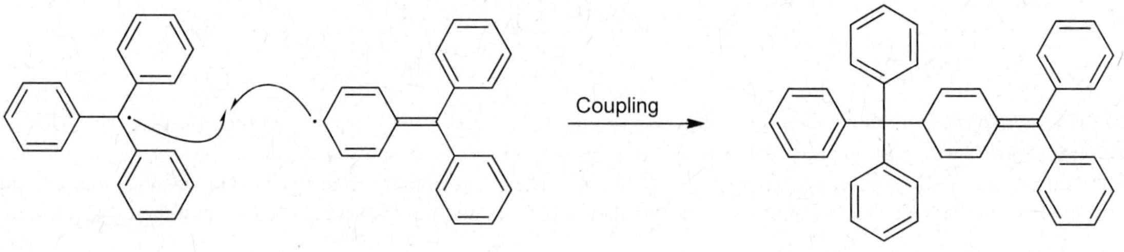

11.10.
(a) The mechanism will have three distinct stages. The first stage is initiation, in which the Cl–Cl bond is broken to generate chlorine radicals. This step requires two fishhook arrows, as shown. There are two propagation steps (hydrogen abstraction and halogen abstraction), each of which requires three fishhook arrows (to show the bonds being broken and formed). Finally, there are many steps that can serve as termination steps, because there are many radicals that can couple together under these conditions. Shown below is a termination step that actually generates the desired product, but it is still considered to be a termination step because this step reduces the number of radicals present in the reaction flask (two radicals are destroyed, without generating new radicals):

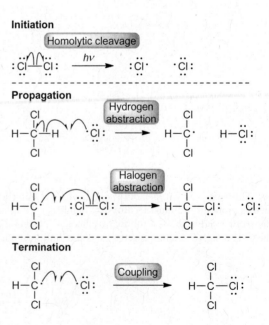

(b) The mechanism will have three distinct stages. The first stage is initiation, in which the Cl–Cl bond is broken to generate chlorine radicals. This step requires two fishhook arrows, as shown. There are two propagation steps (hydrogen abstraction and halogen abstraction), each of which requires three fishhook arrows (to show the bonds being broken and formed). Finally, there are many steps that can serve as termination steps, because there are many radicals that can couple together under these conditions. Shown below is a termination step that generates the desired product, but it is still considered to be a termination step because this step reduces the number of radicals present in the reaction flask (two radicals are destroyed, without generating new radicals):

Initiation

Propagation

Termination

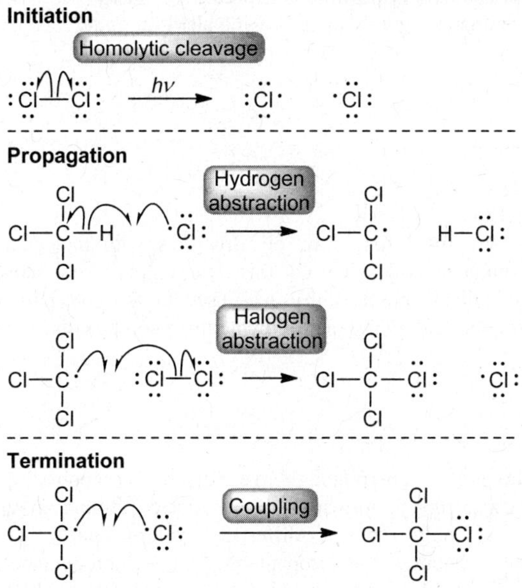

(c) The mechanism will have three distinct stages. The first stage is initiation, in which the Cl–Cl bond is broken to generate chlorine radicals. This step requires two fishhook arrows, as shown. There are two propagation steps (hydrogen abstraction and halogen abstraction), each of which requires three fishhook arrows (to show the bonds being broken and formed). Finally, there are many steps that can serve as termination steps, because there are many radicals that can couple together under these conditions. Shown below is a termination step that actually generates the desired product, but it is still considered to be a termination step because this step reduces the number of radicals present in the reaction flask (two radicals are destroyed, without generating new radicals):

Initiation

Propagation

Termination

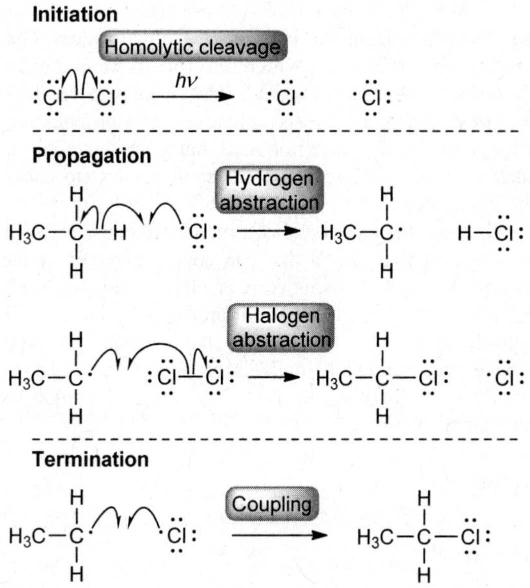

(d) The mechanism will have three distinct stages. The first stage is initiation, in which the Cl–Cl bond is broken to generate chlorine radicals. This step requires two fishhook arrows, as shown. There are two propagation steps (hydrogen abstraction and halogen abstraction), each of which requires three fishhook arrows (to show the bonds being broken and formed). Finally, there are many steps that can serve as termination steps, because there are many radicals that can couple together under these conditions. Shown below is a termination step that actually generates the desired product, but it is still considered to be a termination step because this step reduces the number of radicals present in the reaction flask (two radicals are destroyed, without generating new radicals):

Initiation

Propagation

Termination

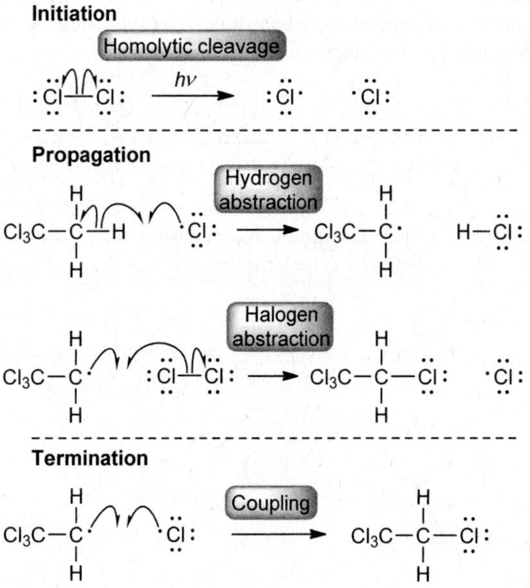

(e) The mechanism will have three distinct stages. The first stage is initiation, in which the Cl–Cl bond is broken to generate chlorine radicals. This step requires two fishhook arrows, as shown. There are two propagation steps (hydrogen abstraction and halogen abstraction), each of which requires three fishhook arrows (to show the bonds being broken and formed). Finally, there are many steps that can serve as termination steps, because there are many radicals that can couple together under these conditions. Shown below is a termination step that actually generates the desired product, but it is still considered to be a termination step because this step reduces the number of radicals present in the reaction flask (two radicals are destroyed, without generating new radicals):

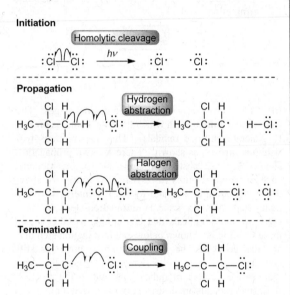

11.11. During the chlorination of methane, methyl radicals are generated. Two of these methyl radicals can couple together to form ethane:

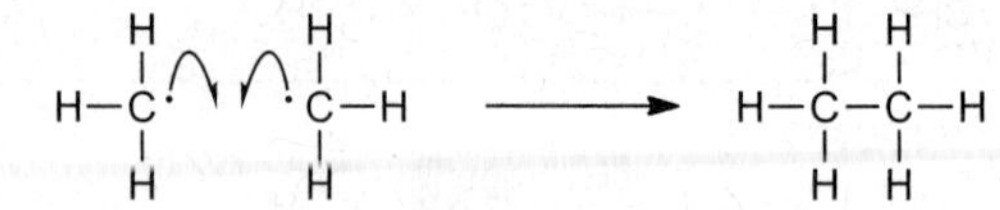

Ethane can then undergo hydrogen abstraction, followed by halogen abstraction to generate ethyl chloride:

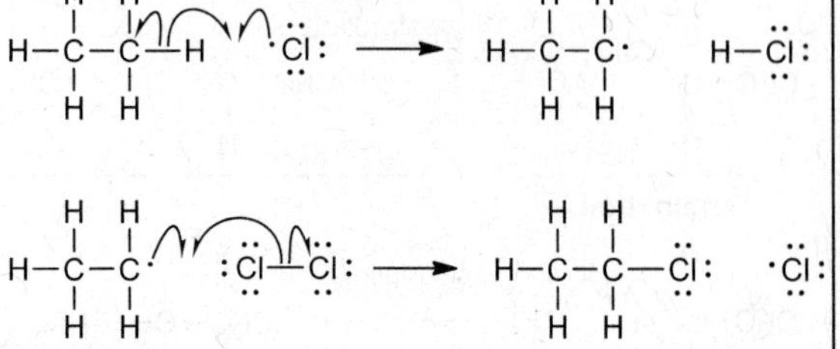

11.12.
(a) The tertiary position is expected to undergo selective bromination, giving the following alkyl bromide:

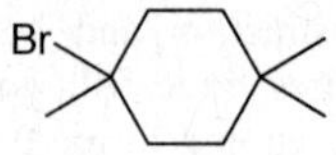

(b) The tertiary position is expected to undergo selective bromination, giving the following alkyl bromide:

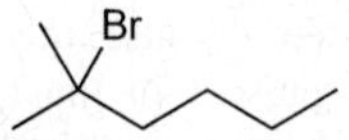

(c) The tertiary position is expected to undergo selective bromination, giving the following alkyl bromide:

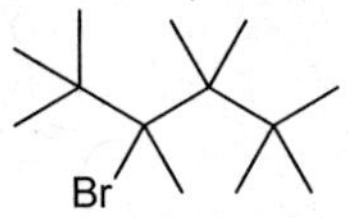

11.13.
(a) We must first draw all structures with molecular formula C_5H_{12}. To do this, we employ the same methodical approach that was used in the solution to Problem 4.3, giving the following three compounds:

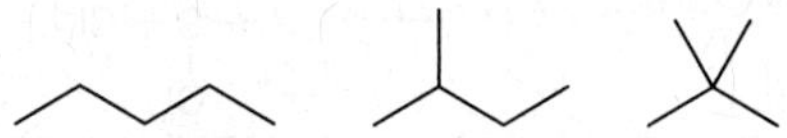

The middle compound above must be compound A, because monochlorination of the first compound above gives only three different chloroalkanes (1-chloropentane, 2-chloropentane, or 3-chloropentane), while monochlorination of the last compound above gives only one regiochemical outcome. The middle compound is 2-methylbutane.
(b) Among the possibilities that we explored in part (a), we found that compound A is 2-methylbutane. This compound undergoes monochlorination to produce four constitutionally isomeric alkyl chlorides, shown here:

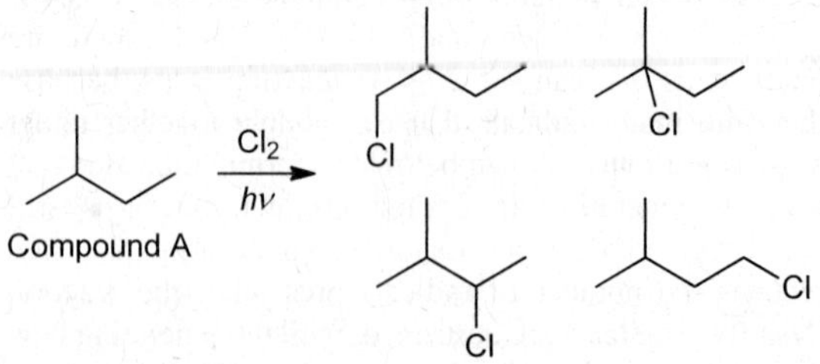

Compound A $\xrightarrow[hv]{Cl_2}$

(c) The tertiary position undergoes selective bromination, giving the following tertiary alkyl bromide:

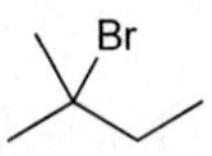

11.14.

(a) The starting compound contains only one tertiary position, so bromination occurs at this position. The product does not contain a chirality center:

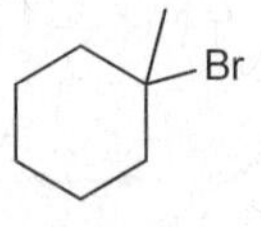

no chirality center

(b) The starting compound contains only one tertiary position, so bromination occurs at this position. This position is an existing chirality center, so we expect loss of configuration to produce both possible enantiomers:

+

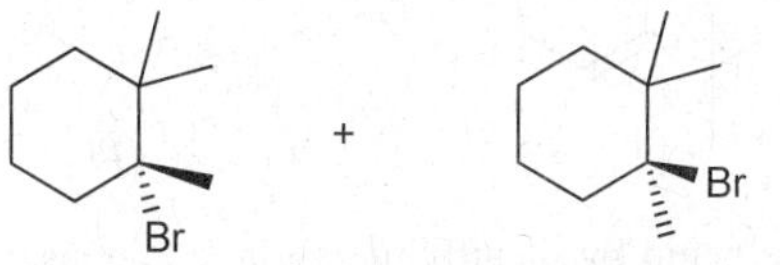

(c) The starting compound contains only one tertiary position, so bromination occurs at this position. This position is an existing chirality center, so we expect loss of configuration to produce both possible enantiomers:

+

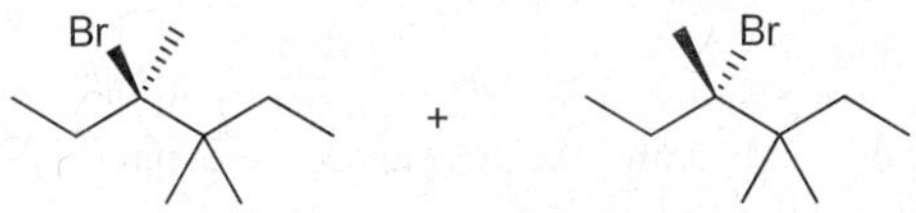

(d) The starting compound contains only one tertiary position, so bromination occurs at this position. This position is an existing chirality center, so we expect loss of configuration to produce both possible stereoisomers:

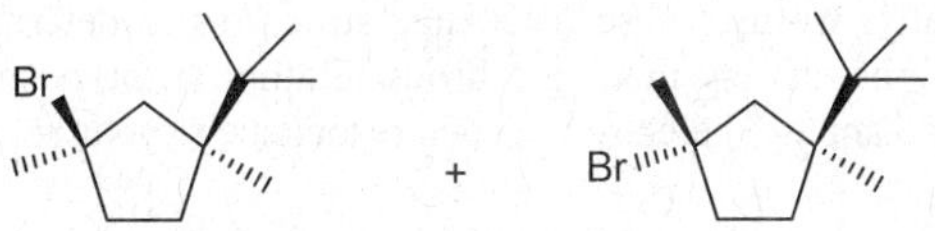

+

Notice that the other chirality center was unaffected by the reaction. The products are diastereomers.

11.15. Bromination of compound A gives 2,2-dibromopentane:

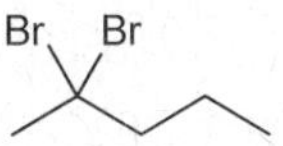

2,2-dibromopentane

So compound A must be 2-bromopentane:

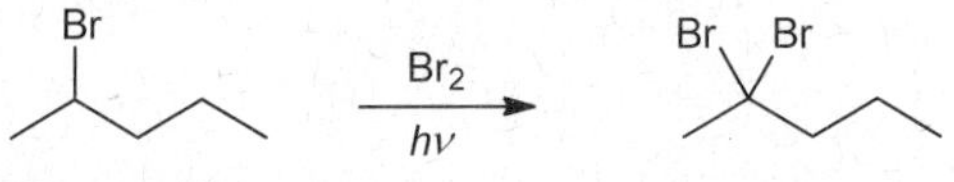

Compound A

Compound A exhibits one chirality center. The problem statement provides the information that is necessary to determine the configuration of this chirality center. Specifically, we are told that treating compound A with a strong nucleophile (S$_N$2 conditions) results in a product with the *R* configuration. Since S$_N$2 reactions proceed via inversion of configuration, the chirality center must have the *S* configuration in compound A:

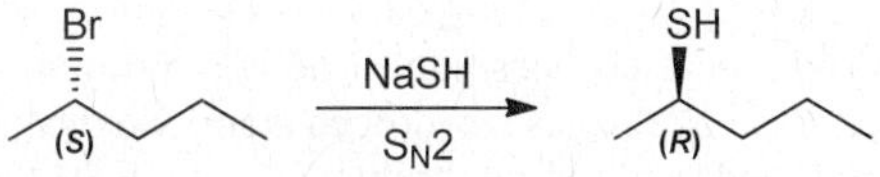

Compound **A**
(*S*)-2-bromopentane

11.16.

(a) We begin by identifying allylic positions. There are four:

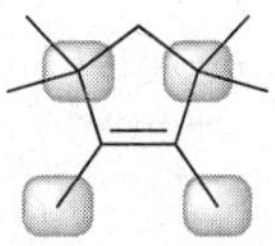

allylic positions

But two of these positions (the top two positions in the structure above) lack hydrogen atoms. So those positions cannot undergo radical bromination (a C–H bond is necessary because a key step in the mechanism is a hydrogen abstraction step). The remaining two allylic positions are identical (the molecule has a plane of symmetry that renders these two positions identical):

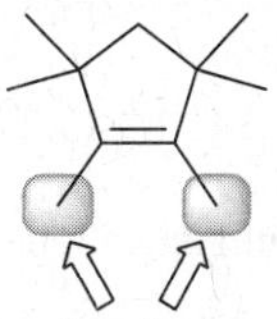

identical

So, we only need to consider allylic bromination occurring at one of these positions. To do that, we remove a hydrogen atom from the allylic position and draw the resonance structures of the resulting allylic radical:

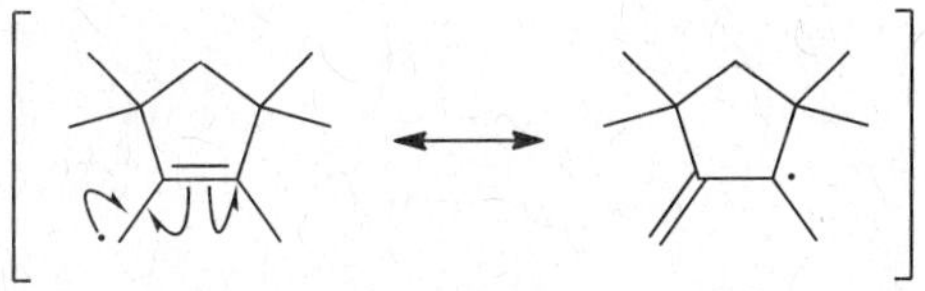

Finally, we use these resonance structures to determine the products, by placing a bromine atom at the position of the unpaired electron in each resonance structure:

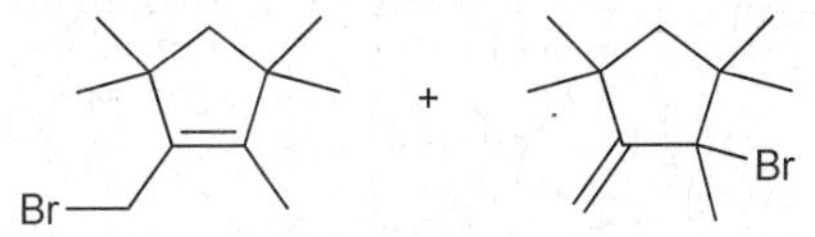

+

(b) We begin by identifying the allylic positions. There are two:

allylic
positions

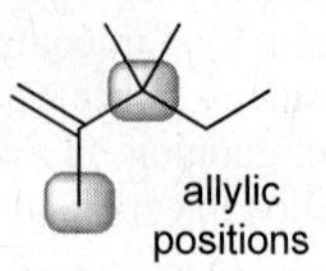

But one of these positions (the top position in the structure above) lacks a hydrogen atom. So this position cannot undergo radical bromination (a C–H bond is necessary because a key step in the mechanism is a hydrogen abstraction step). So, we only need to consider allylic bromination occurring at one position. To do that, we remove a hydrogen atom from the allylic position and draw the resonance structures of the resulting allylic radical:

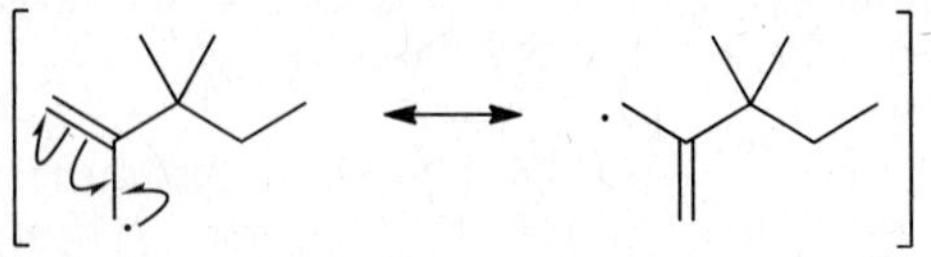

Finally, we use these resonance structures to determine the products, by placing a bromine atom at the position of the unpaired electron in each resonance structure. In this case, our methodical approach has produced two structures that are identical compounds, so this reaction has only one product:

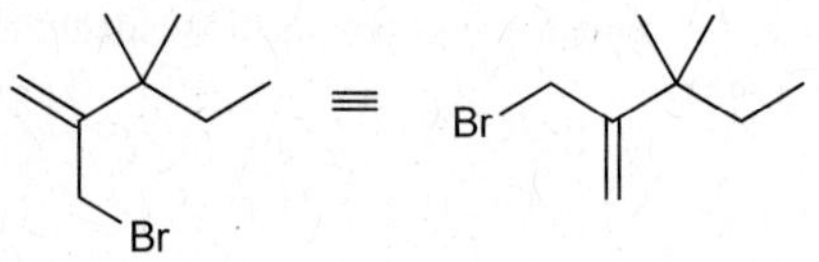

(c) We begin by identifying allylic positions. There are two:

allylic positions

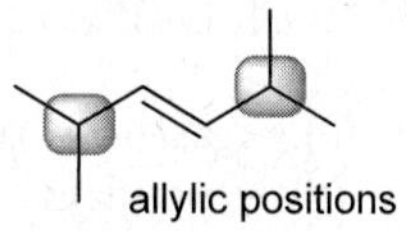

But these two positions are identical (symmetry). So, we only need to consider allylic bromination occurring at one of these positions. To do that, we remove a hydrogen atom from the allylic position and draw the resonance structures of the resulting allylic radical:

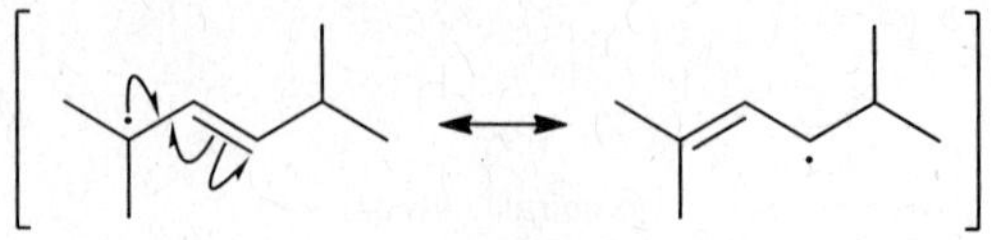

Finally, we use these resonance structures to determine the products, by placing a bromine atom at the position of the unpaired electron in each resonance structure:

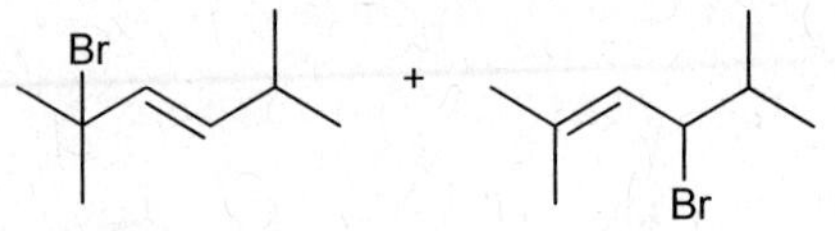

(d) We begin by identifying allylic positions. There is only one in this case:

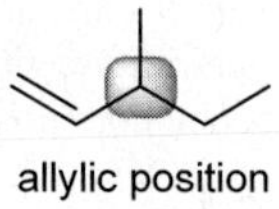

allylic position

Next, we remove a hydrogen atom from the allylic position and draw the resonance structures of the resulting allylic radical:

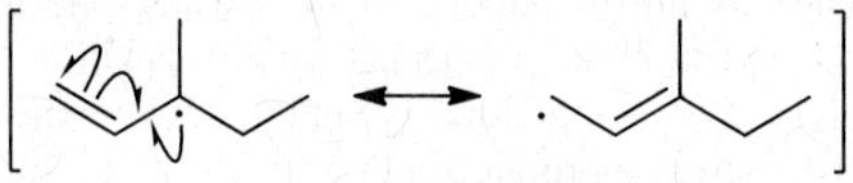

Finally, we use these resonance structures to determine the products, by placing a bromine atom at the position of the unpaired electron in each resonance structure:

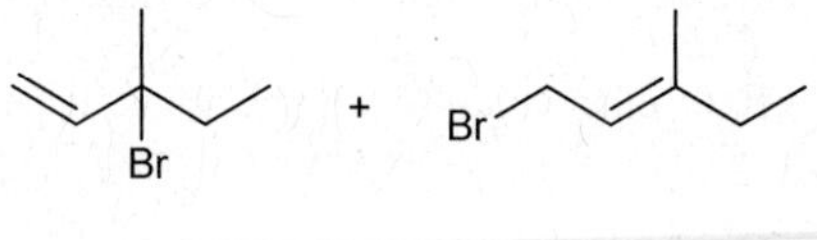

11.17. We begin by drawing the starting compound (2-methyl-2-butene) and identifying the allylic positions (highlighted below). Each of these three positions can undergo hydrogen abstraction to give a resonance-stabilized radical, shown below. The resulting products are also shown. Notice that there are a total of five products, labeled **1 – 5**. Notice that compound **4** is the only product that exhibits a chirality center. As such, a racemic mixture of compound **4** is expected.

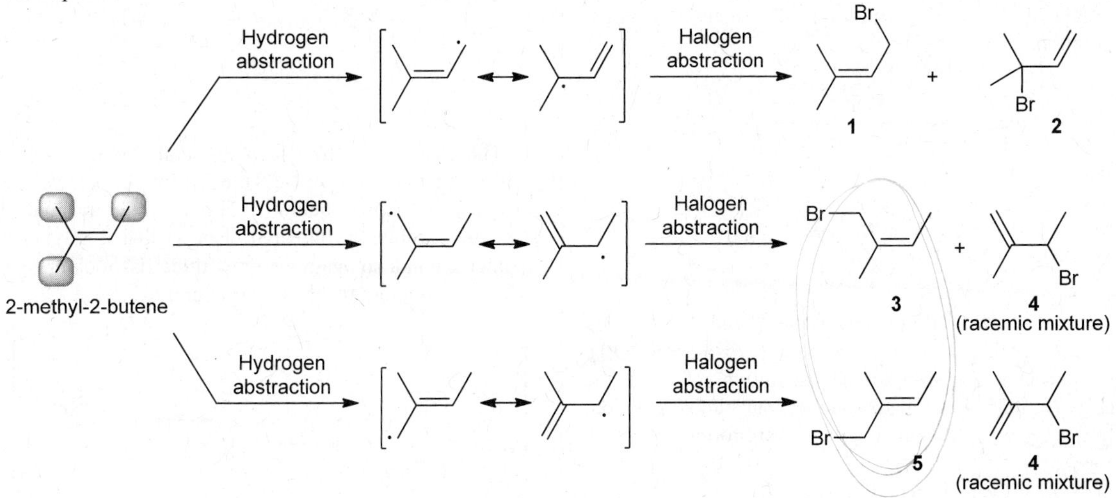

11.18. As seen in Section 11.8, destruction of ozone in the atmosphere occurs via the following two propagation steps (where R• represents a radical, such as Cl•, that is responsible for destroying ozone):

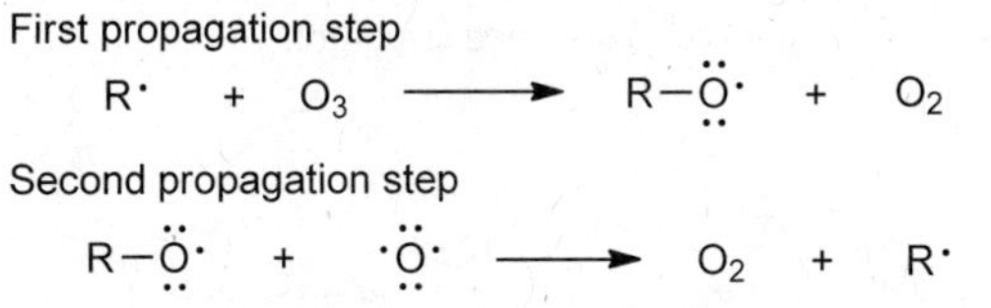

First propagation step

$$R^{\cdot} + O_3 \longrightarrow R-\ddot{O}^{\cdot} + O_2$$

Second propagation step

$$R-\ddot{O}^{\cdot} + {}^{\cdot}\ddot{O}^{\cdot} \longrightarrow O_2 + R^{\cdot}$$

If we redraw this mechanism with nitric oxide serving as the radical that destroys ozone, we get the following:

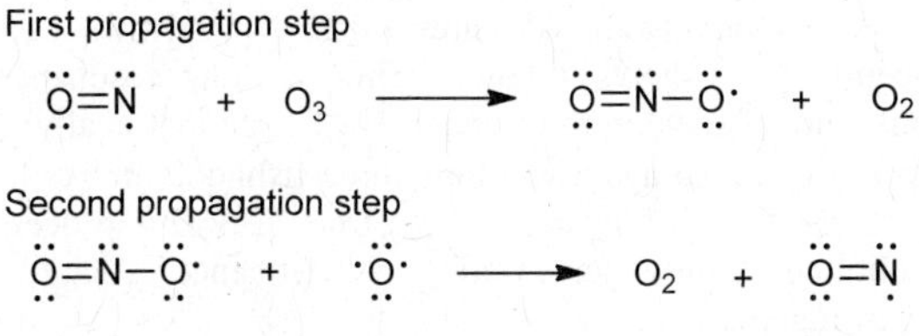

First propagation step

$$\ddot{O}=\dot{N} + O_3 \longrightarrow \ddot{O}=N-\ddot{O}^{\cdot} + O_2$$

Second propagation step

$$\ddot{O}=N-\ddot{O}^{\cdot} + {}^{\cdot}\ddot{O}^{\cdot} \longrightarrow O_2 + \ddot{O}=\ddot{N}$$

11.19. Radicals will react with BHT and BHA because each of these compounds has a hydrogen atom that can be readily abstracted, thereby generating a resonance-stabilized radical. In each case, the phenolic hydrogen atom is abstracted (the hydrogen atom of the OH group connected to the aromatic ring). Similarly, vitamin E also exhibits a phenolic hydrogen atom. Abstraction of this hydrogen atom (highlighted below) gives a resonance-stabilized radical.

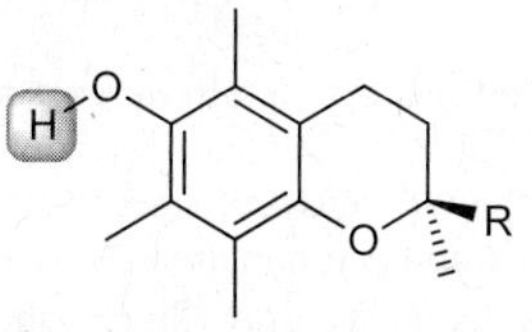

Vitamin E

11.20.

(a) The reagent (HBr) indicates that H and Br are added across the π bond. In the presence of peroxides (ROOR), we expect an *anti*-Markovnikov addition. That is, the bromine atom is installed at the less substituted position. In this case, one new chirality center is created, which results in a racemic mixture of the two possible enantiomers.

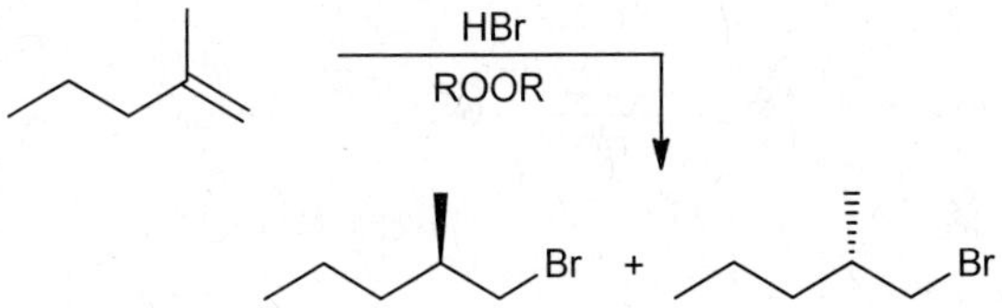

(b) The reagent (HBr) indicates that H and Br are added across the π bond. In the presence of peroxides (ROOR), we expect an *anti*-Markovnikov addition. That is, the bromine atom is installed at the less substituted position. In this case, no new chirality centers are formed:

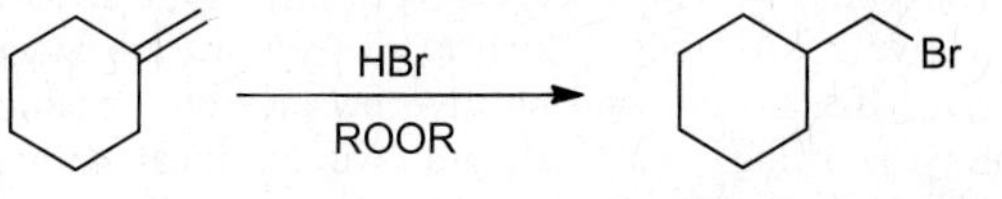

(no chirality centers)

(c) The reagent (HBr) indicates that H and Br are added across the π bond. In the presence of peroxides (ROOR), we expect an *anti*-Markovnikov addition. That is, the bromine atom is installed at the less substituted position. In this case, one new chirality center is created, which results in a racemic mixture of the two possible enantiomers.

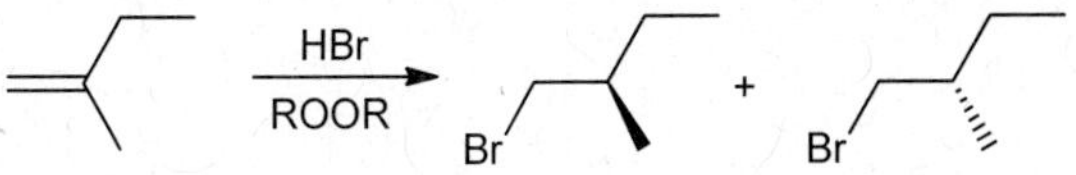

(d) The reagent (HBr) indicates that H and Br are added across the π bond. In the presence of peroxides (ROOR), we expect an *anti*-Markovnikov addition. That is, the bromine atom is installed at the less substituted position. In this case, one new chirality center is created, which results in a racemic mixture of the two possible enantiomers.

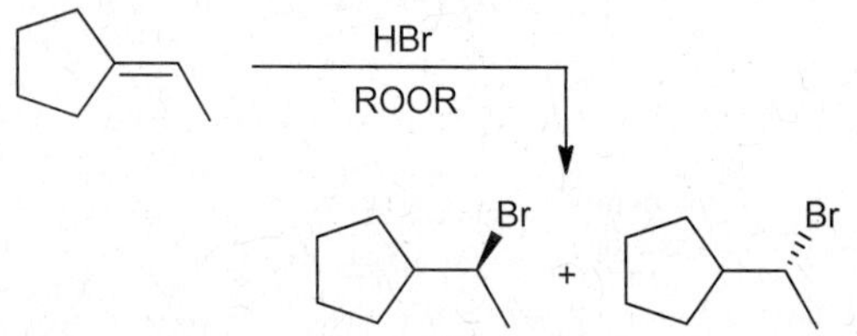

(e) The reagent (HBr) indicates that H and Br are added across the π bond. In the presence of peroxides (ROOR), we expect an *anti*-Markovnikov addition. That is, the bromine atom is installed at the less substituted position. In this case, no new chirality centers are formed:

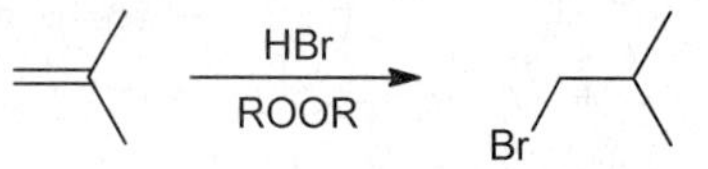

(no chirality centers)

(f) The reagent (HBr) indicates that H and Br are added across the π bond. In the presence of peroxides (ROOR), we expect an *anti*-Markovnikov addition, although that is irrelevant in this case, because the alkene is symmetrical. No new chirality centers are formed:

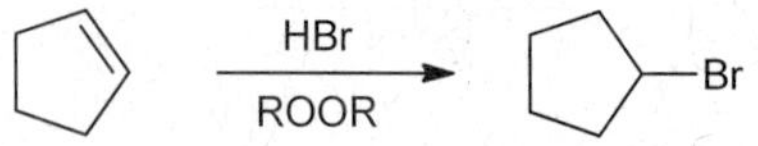

(no chirality centers)

11.21.

(a) Recall that ΔG has two components: (ΔH) and (-TΔS). The first term is positive, so the second term must have a large negative value in order for ΔG to be negative (which is necessary in order for the process to be thermodynamically favorable). This will be the case if ΔS and T are both large and positive (T cannot be negative). At high T, both of these terms are indeed large. ΔS is large and positive because one chemical entity is being converted into two chemical entities, which significantly increases the entropy of the system.

(b) Recall that ΔG has two components: (ΔH) and (-TΔS). The magnitude of the latter term is dependent on the temperature. At high temperature, the latter term dominates over the former, and the reaction is thermodynamically favorable. However, at low temperature, the first term (enthalpy) dominates, and the reaction is no longer thermodynamically favored.

11.22.

(a) The unpaired electron occupies an allylic position, so it is resonance-stabilized. Three fishhook arrows are required, as shown:

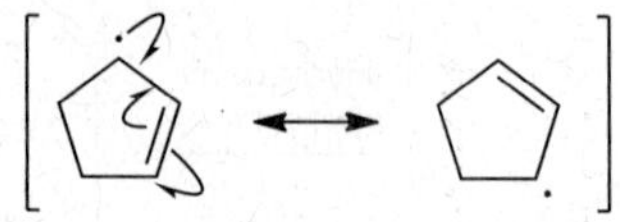

(b) The unpaired electron occupies an allylic position, so it is resonance-stabilized. Three fishhook arrows are required, as shown below. The resulting resonance structure exhibits an unpaired electron that is allylic to another π bond, so again we draw three fishhook arrows to arrive at another resonance structure:

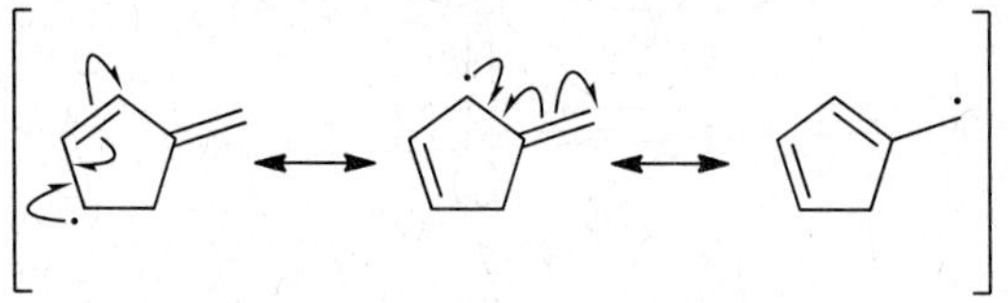

(c) The unpaired electron occupies an allylic position, so it is resonance-stabilized. Three fishhook arrows are required, as shown below. The resulting resonance structure exhibits an unpaired electron that is allylic to another π bond, so again we draw three fishhook arrows to arrive at another resonance structure:

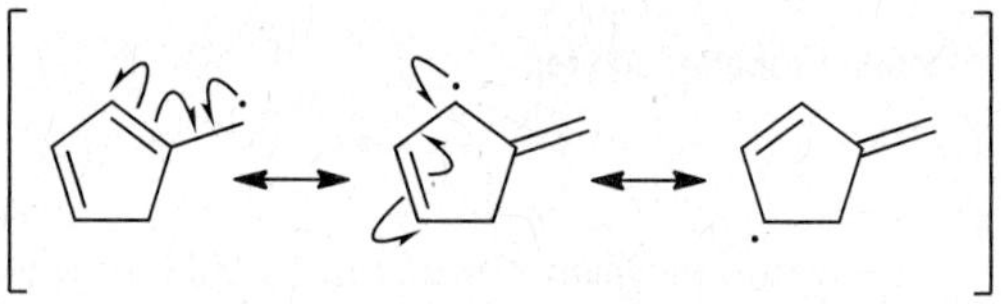

(d) The unpaired electron is allylic to a π bond, so it is resonance-stabilized. Three fishhook arrows are required, as shown below. The resulting resonance structure also exhibits an unpaired electron that is allylic to a π bond, so again we draw three fishhook arrows as well as the resulting resonance structure. This process continues until we have drawn five resonance structures, shown here:

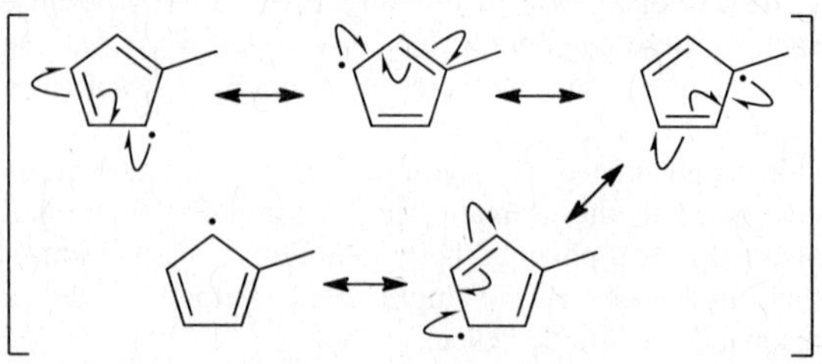

(e) The unpaired electron is allylic to a π bond, so it is resonance-stabilized. Three fishhook arrows are required, as shown below. The resulting resonance structure also exhibits an unpaired electron that is allylic to a π bond, so again we draw three fishhook arrows as well as the resulting resonance structure. This process continues until we have drawn five resonance structures, shown here:

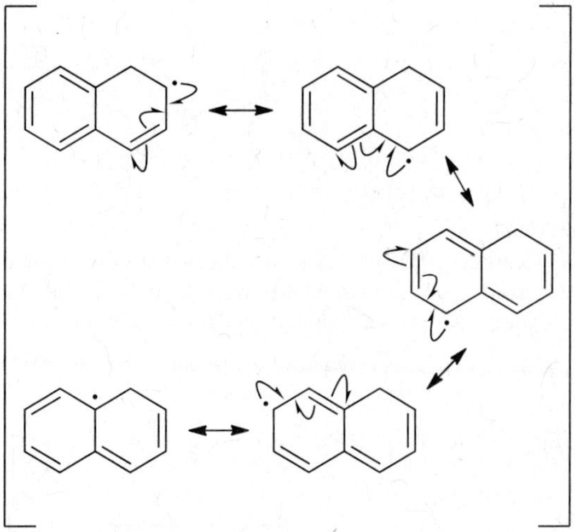

11.23. There are three different hydrogen atoms to consider, labeled H_a, H_b and H_c:

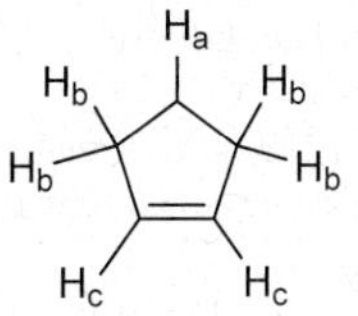

The weakest C–H bond is C–H_b, because abstraction of H_b generates a radical that is stabilized by resonance:

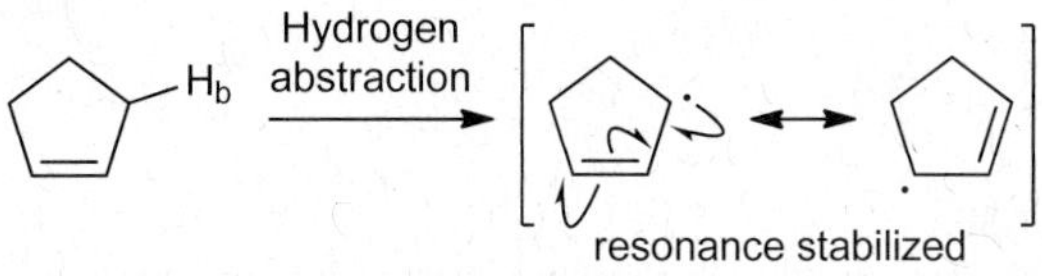

The strongest C–H bond is C–H_c, because abstraction of H_c generates an unstable vinyl radical:

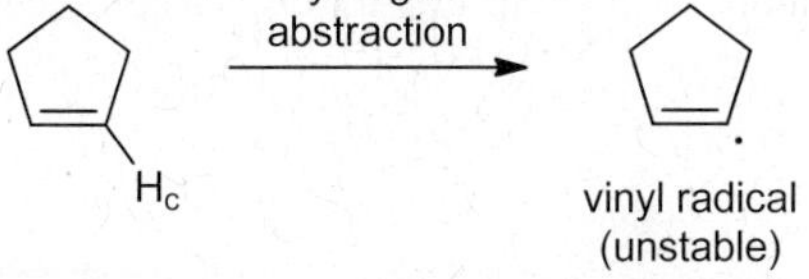

We therefore expect the C–H bonds of cyclopentene to exhibit the following order of increasing bond strength:

C–H_b $<$ C–H_a $<$ C–H_c

(weakest bond)   (strongest bond)

11.24.
(a) The tertiary allylic radical is the most stable (it is resonance-stabilized). The second most stable radical is the tertiary radical, and the least stable radical is the primary radical:

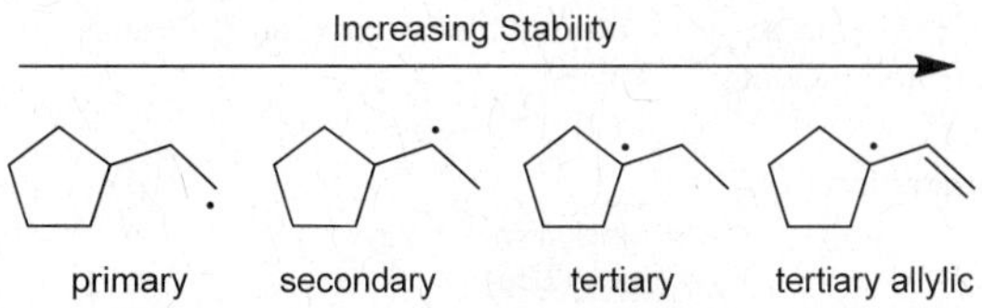

Increasing Stability

primary   secondary   tertiary   tertiary allylic

(b) The tertiary radical is the most stable, because alkyl groups stabilize the unpaired electron via a delocalization effect, called hyperconjugation. The primary radical is the least stable, because it lacks the stabilizing effect provided by multiple alkyl groups.

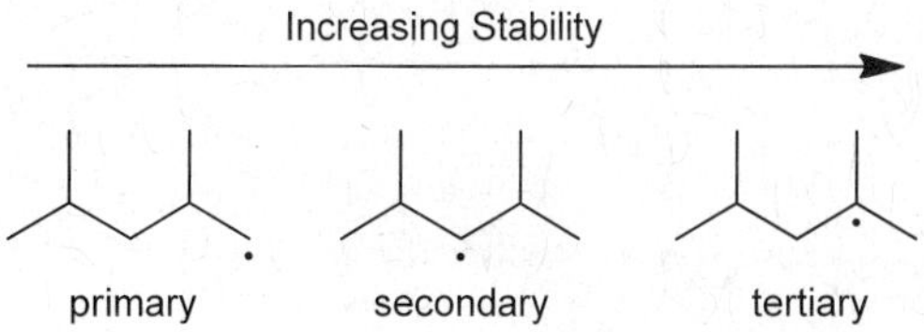

Increasing Stability

primary   secondary   tertiary

11.25. The unpaired electron is allylic to a π bond, so it is resonance-stabilized. Three fishhook arrows are required, as shown below. The resulting resonance structure exhibits an unpaired electron that is allylic to another π bond, so again we draw three fishhook arrows to arrive at another resonance structure. This pattern appears one more time, giving a fourth resonance structure. This radical is particularly stable, because it has many resonance structures, so the unpaired electron is highly delocalized.

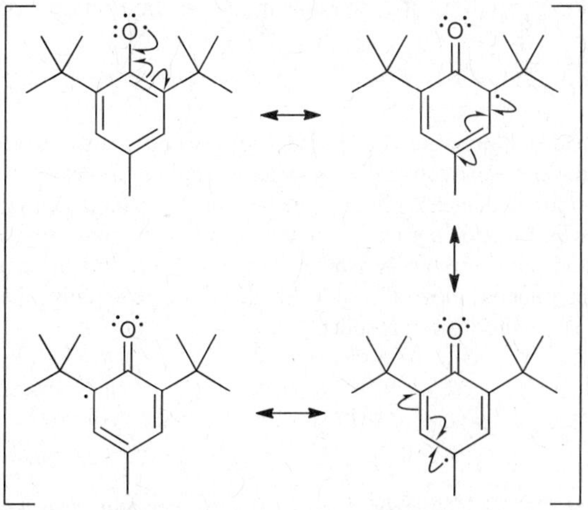

11.26. The benzylic hydrogen atom is the only hydrogen atom that can be abstracted to generate a resonance-stabilized radical. As such, the benzylic position is selectively brominated. The mechanism will have three distinct stages. The first stage is initiation, in which the N–Br bond (of NBS) is broken to generate a bromine radical. This step requires two fishhook arrows, as shown. There are two propagation steps (hydrogen abstraction and halogen abstraction), each of which requires three fishhook arrows (to show the bonds being broken and formed). Finally, there are many steps that can serve as termination steps, because there are many radicals that can couple together under these conditions. Shown below is a termination step that generates the desired product:

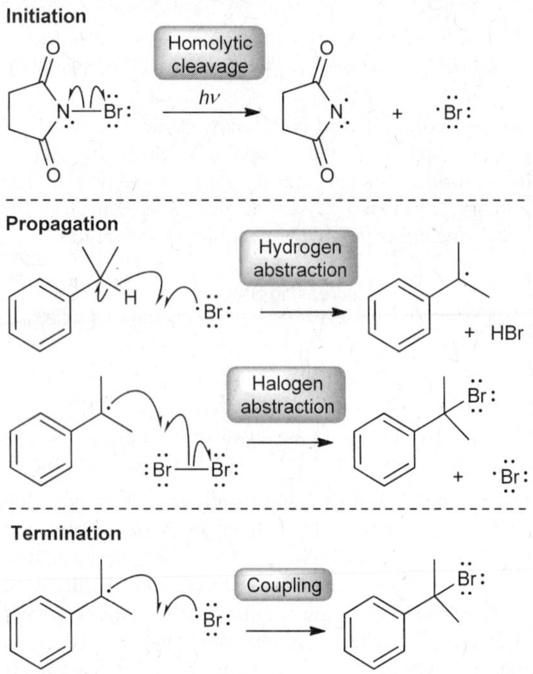

Initiation

Propagation

Termination

11.27. We must first draw all structures with molecular formula C_5H_{12}. To do this, we employ the same methodical approach that was used in the solution to Problem 4.3, giving the following three compounds:

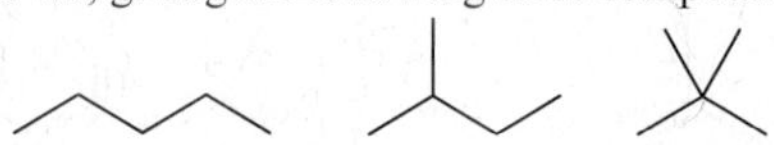

Monochlorination of the first compound above gives three constitutionally isomeric chloroalkanes (1-chloropentane, 2-chloropentane, or 3-chloropentane), while monochlorination of the second compound gives four constitutionally isomeric chloroalkanes. The last compound above (2,2-dimethylpropane) gives only one monochlorination product:

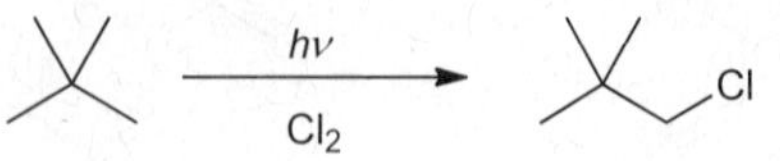

11.28. Selective bromination at the benzylic position generates a new chirality center. The intermediate benzylic radical is expected to be attacked from either face of the planar radical with equal likelihood, giving rise to a racemic mixture of enantiomers:

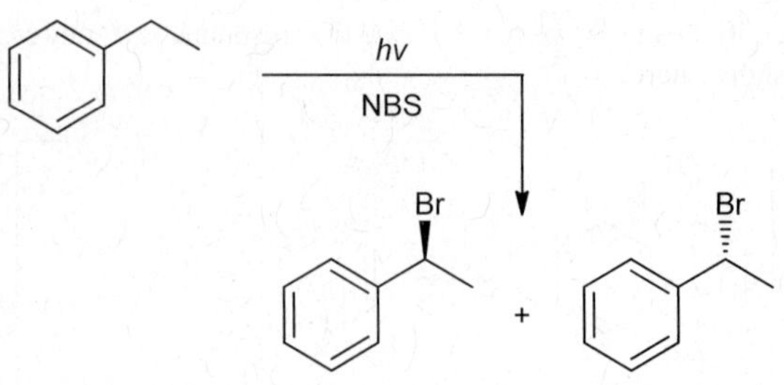

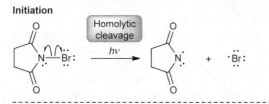

11.29.
(a) Heating AIBN generates a radical that is resonance-stabilized (the unpaired electron is delocalized via resonance, which is a stabilizing effect):

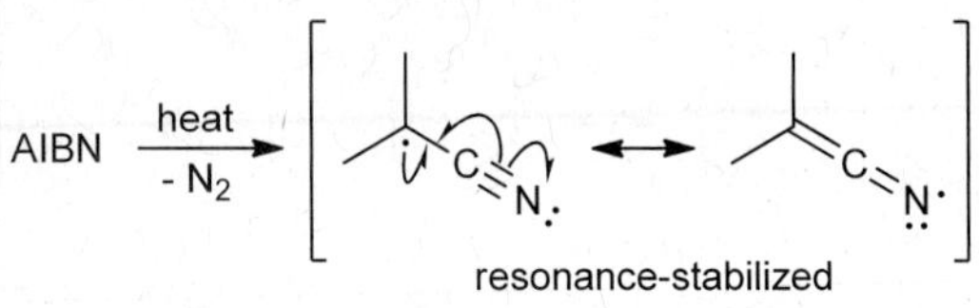

resonance-stabilized

In addition, the radical is further stabilized by the presence of methyl groups, which are capable of stabilizing the unpaired electron via a delocalization effect, called hyperconjugation.

(b) Loss of nitrogen gas would result in the formation of vinyl radicals, which are too unstable to form under normal conditions:

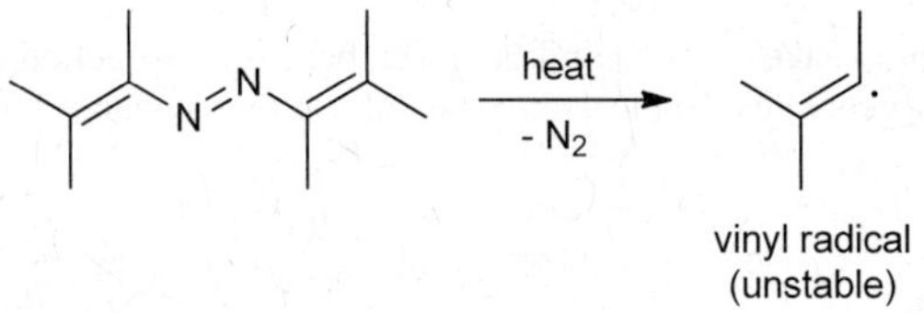

vinyl radical
(unstable)

11.30.
(a) The central carbon atom is benzylic to three aromatic rings. As such, that C–H bond is expected to be extremely weak. Hydrogen abstraction (which initiates the autooxidation process) occurs at this location, generating a radical that is highly stabilized by resonance (see solution to Problem 11.3 for resonance structures):

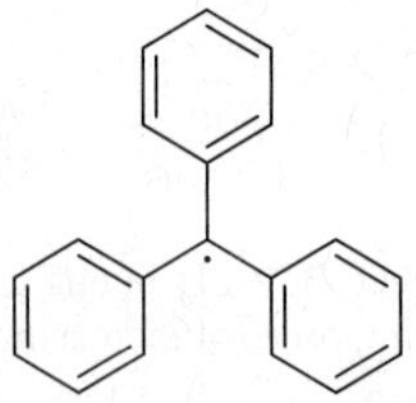

Autooxidation at this location gives the following hydroperoxide:

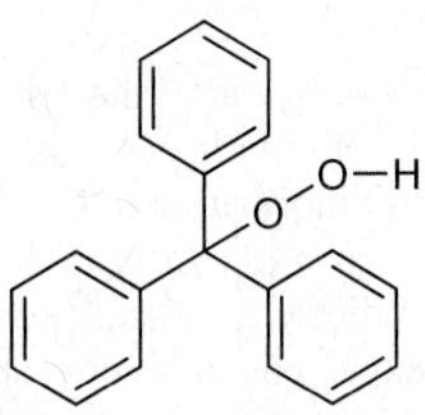

(b) As explained in the solution to part (a) of this problem, hydrogen abstraction leads to an exceptionally stable radical, with many, many resonance structures (see solution to Problem 11.3).

(c) Phenol acts as a radical scavenger (much like BHA and BHT), thereby preventing the chain process from beginning.

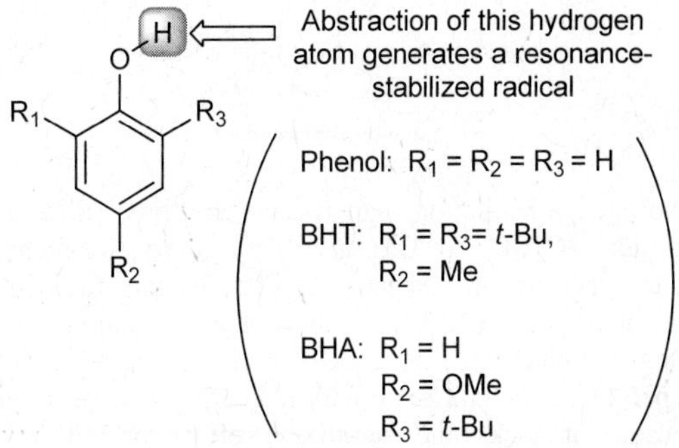

Abstraction of this hydrogen atom generates a resonance-stabilized radical

Phenol: $R_1 = R_2 = R_3 = H$

BHT: $R_1 = R_3 = t\text{-}Bu$, $R_2 = Me$

BHA: $R_1 = H$ $R_2 = OMe$ $R_3 = t\text{-}Bu$

11.31. The mechanism will have three distinct stages. The first stage is initiation, in which the N–Br bond (of NBS) is broken to generate a bromine radical. This step requires two fishhook arrows, as shown. The next stage is propagation. There are two propagation steps. The first is hydrogen abstraction, which requires three fishhook arrows and generates a resonance-stabilized radical. The second propagation step (halogen abstraction) also requires three fishhook arrows. Notice that this step can occur in either of two locations, as shown. The final stage is termination, and there are many steps that can serve as termination steps, because there are many radicals that can couple together under these conditions. Shown below are termination steps that generate the desired products:

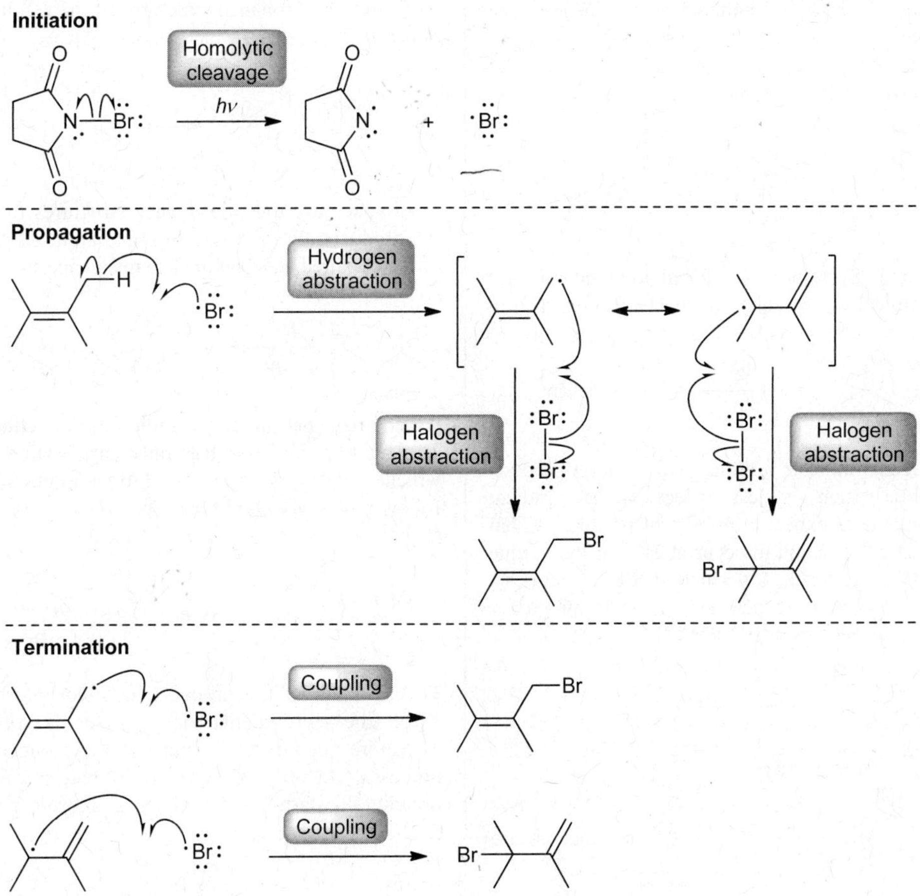

11.32. We begin by drawing the structure of 3-ethylpentane:

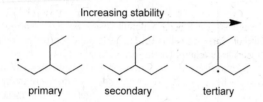

3-ethylpentane

Hydrogen abstraction can occur at three different positions (C1, C2, or C3), giving a primary, secondary, or tertiary radical, respectively. The tertiary radical is the most stable, because alkyl groups stabilize the unpaired electron via a delocalization effect, called hyperconjugation. The primary radical is the least stable, because it lacks the stabilizing effect provided by multiple alkyl groups.

Increasing stability →

primary   secondary   tertiary

11.33.
(a) The reagents indicate a radical bromination process. The starting material has only one tertiary position, which undergoes selective bromination, giving the following tertiary alkyl bromide.

$$\xrightarrow[h\nu]{Br_2}$$

(b) As seen in Section 11.4, radical iodination is not thermodynamically favorable. We expect no reaction in this case:

$$\xrightarrow[h\nu]{I_2}\ \text{no reaction}$$

(c) The reagents indicate a radical chlorination process. Radical chlorination is less selective than radical bromination, so we expect a mixture of products. That is, we expect chlorination to occur at each of the unique locations: C1, C2, or C3. Chlorination at C2 generates a product with a chirality center, so a racemic mixture is expected:

$$\xrightarrow[h\nu]{Cl_2}$$

(racemic mixture)

Chlorination at C4 yields the same product as chlorination at C2. Similarly, chlorination at C5 yields the same product as chlorination at C1.

(d) The reagents indicate a radical bromination process. Bromination is expected to occur selectively at the benzylic position (because hydrogen abstraction occurs at that position to generate a resonance-stabilized radical):

$$\xrightarrow[h\nu]{NBS}$$

(e) The reagents indicate a radical bromination process. Bromination is expected to occur selectively at the allylic position (because hydrogen abstraction occurs at that position to generate a resonance-stabilized radical). In this case, there is only one allylic position:

allylic position

We remove a hydrogen atom from the allylic position and draw the resonance structures of the resulting allylic radical:

$$\left[\quad \longleftrightarrow \quad \right]$$

Finally, we use these resonance structures to determine the products, by placing a bromine atom at the position of the unpaired electron in each resonance structure:

$$\xrightarrow[h\nu]{NBS} \quad + $$

(f) The reagents indicate a radical bromination process. The starting material has only one tertiary position, which undergoes selective bromination, giving the following tertiary alkyl bromide.

$$\xrightarrow[h\nu]{Br_2}$$

11.34. Each of the products has two chlorine atoms, while the starting material has one chlorine atom. Therefore, this reaction must be a monochlorination process. In order to draw the products, we begin by drawing the starting material, (S)-2-chloropentane:

(S)-2-chloropentane

This compound has five carbon atoms, each of which represents a unique location (no symmetry). Therefore, we must consider the products from chlorination at each position. As shown below, there are a total of seven products, which are labeled **A** – **G** below. Let's begin with chlorination at position C1, which results in a single product:

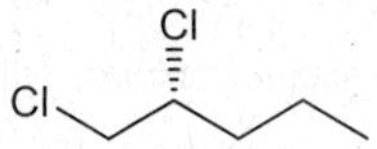

Compound **A**

Notice that the existing chirality center (C2) retains its configuration because formation of compound **A** does not involve the chirality center.
Next, we consider chlorination at C2, which also gives only one product. This product (compound **B**) lacks a chirality center:

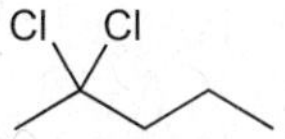

Compound **B**

Now let's consider chlorination at C3. In this case, a new chirality is being created, and we expect both possible outcomes (either the *R* configuration at C3 or the *S* configuration at C3):

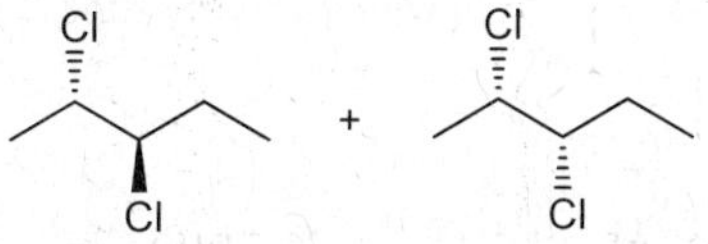

Compound **C**   Compound **D**

Notice that the existing chirality center (C2) is not involved in the formation of compounds **C** and **D**. As such, the configuration at C2 is retained. Compounds **C** and **D** are diastereomers.
Now let's consider chlorination at C4. Once again, a new chirality is being created, and we expect both possible outcomes (either the *R* configuration at C4 or the *S* configuration at C4):

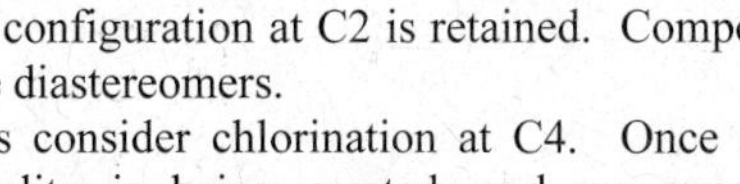

Compound **E**   Compound **F**

Notice that the existing chirality center (C2) is not involved in the formation of compounds **E** and **F**. As such, the configuration at C2 is retained in each of these compounds. Compounds **E** and **F** are diastereomers.
Finally, we consider chlorination at C5, which results in a single product:

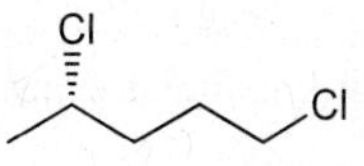

Compound **G**

Notice that the existing chirality center (at C2) retains its configuration because formation of compound **G** does not involve the chirality center.
In summary, monochlorination of (*S*)-2-chloropentane gives a total of seven products, labeled **A** – **G** above.

11.35. We begin by drawing the starting material, (*S*)-3-methylhexane:

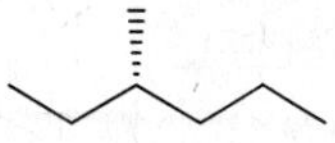

(*S*)-3-methylhexane

This compound has only one tertiary position, so bromination will occur selectively at that site. In this case, the reaction is occurring at a chirality center, so we expect a racemic mixture:

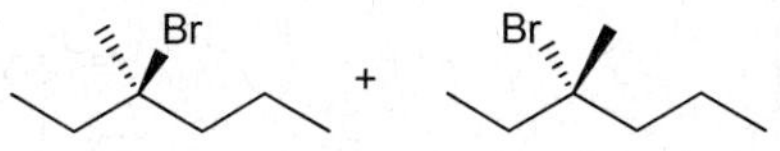

11.36.
(a) We begin by identifying allylic positions. There are two:

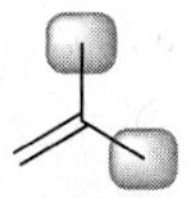

allylic positions

But these two positions are identical (symmetry). So, we only need to consider allylic bromination occurring at one of these positions. To do that, we remove a hydrogen atom from the allylic position and draw the resonance structures of the resulting allylic radical:

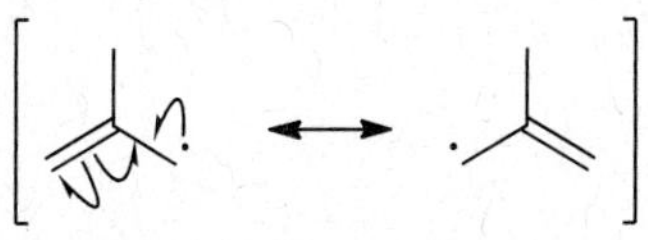

Finally, we use these resonance structures to determine the products, by placing a bromine atom at the position of the unpaired electron in each resonance structure. In this case, the two resulting structures represent the same compound, so only one product is obtained:

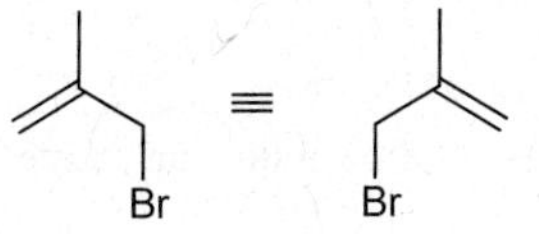

(b) We begin by identifying the allylic positions. There are three of them (highlighted below), each of which can undergo hydrogen abstraction to give a resonance-stabilized radical, shown below. The resulting products are also shown. Notice that there are a total of five products, labeled **1 – 5**. Notice that many of the products exhibit a chirality center, and a racemic mixture is expected in each case.

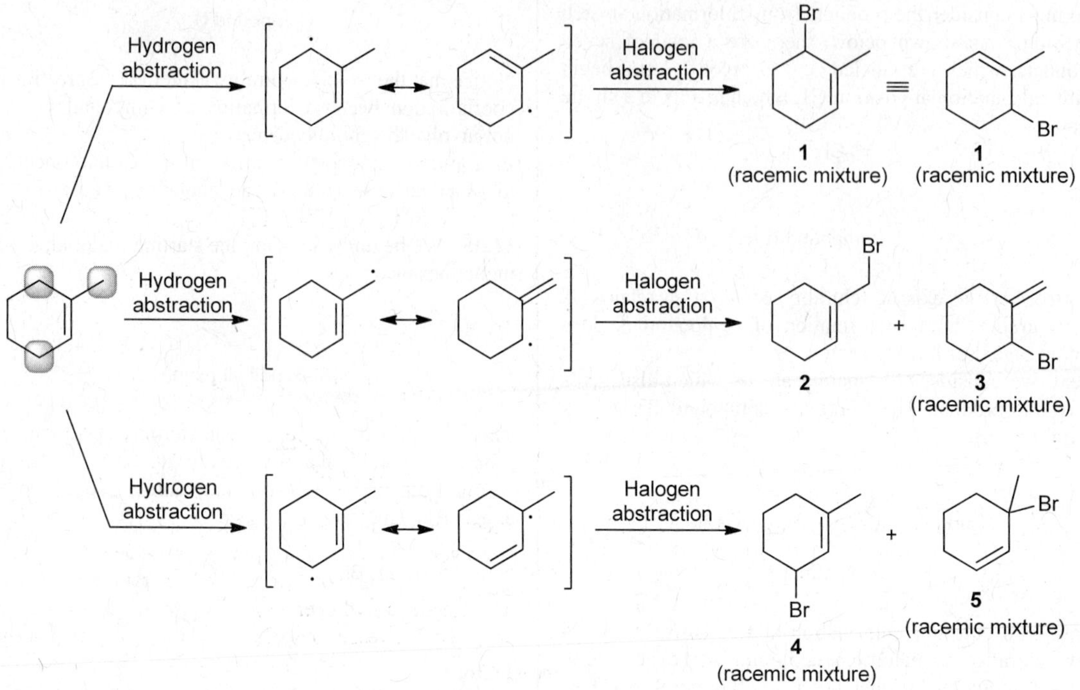

11.37. As seen in Section 11.9, diethyl ether undergoes autooxidation to give the following hydroperoxide:

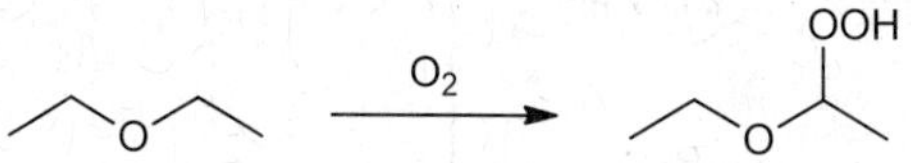

As seen in Mechanism 11.2, autooxidation is believed to occur via two propagation steps. The first is a coupling step, and the second is a hydrogen abstraction, as shown here:

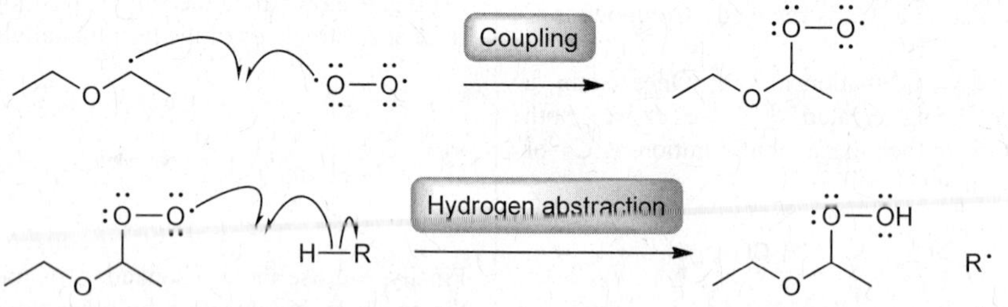

11.38.
(a) Compound A has molecular formula C_5H_{12}, so it must be one of the following three constitutional isomers:

The first compound has no tertiary positions, so we would expect bromination to occur at one of the secondary positions. While there are three such positions in pentane (C2, C3, and C4), two of these positions are identical (C2 = C4). So there are two unique positions that are likely to be brominated: C2 and C3. That is, monobromination of pentane should produce a mixture of 2-bromopentane and 3-bromopentane. So, compound A cannot be pentane, because the problem statement indicates that monobromination would result in only one product. The other two constitutional isomers above are candidates, because each of them would give one alkyl halide as the product. The second compound above (2-methylbutane) will undergo bromination selectively at the tertiary position. For the last compound above (2,2-dimethylpropane), all of the methyl groups are identical, so only one regiochemical outcome is possible.

To determine which isomer is compound A, we must interpret the other piece of information provided in the problem statement. When compound B is treated with a strong base, two products are obtained. This would not be true if compound A were 2,2-dimethylpropane, as then, compound B would not undergo elimination at all (it would have no β protons):

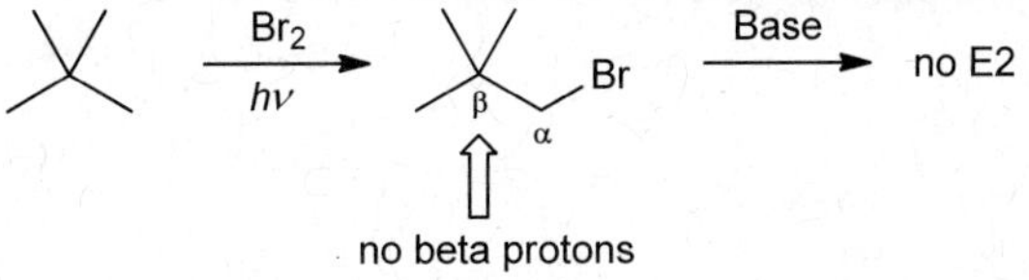

Therefore, compound A must be 2-methylbutane:

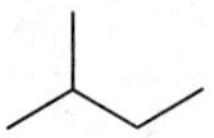

Compound A

(b) As mentioned in the solution to part (a), compound A undergoes monobromination selectively at the tertiary position, giving the corresponding tertiary alkyl halide (compound B):

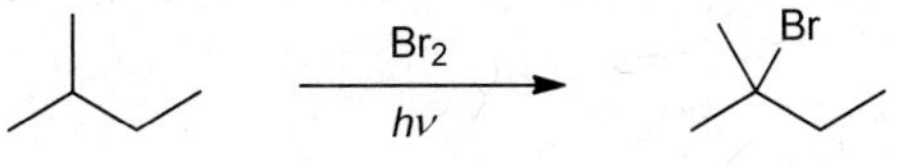

(c) When compound B is treated with a strong base, an E2 reaction is expected. Two regiochemical outcomes are possible, so we expect a mixture of both products (the major product is determined by the choice of base, as seen in the remaining parts of this problem):

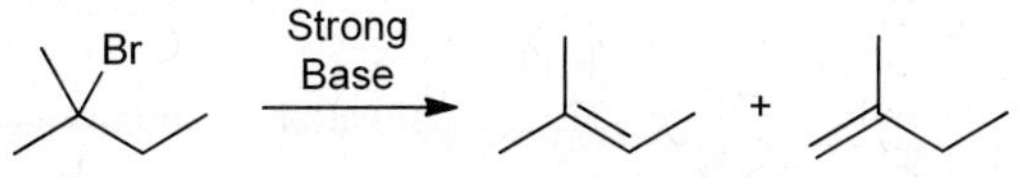

(d) When compound B is treated with *tert*-butoxide (a sterically hindered base), the Hofmann product is favored.

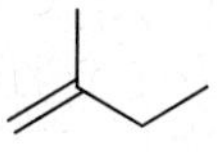

(e) When compound B is treated with sodium ethoxide, the Zaitsev product is favored:

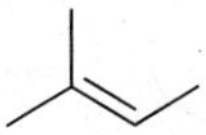

11.39. We begin by identifying the allylic positions. There are two:

allylic
positions

But one of these positions (the lower position in the structure above) lacks a hydrogen atom. So this position cannot undergo radical bromination (a C–H bond is necessary because a key step in the mechanism is a hydrogen abstraction step). So, we only need to consider allylic bromination occurring at one position. To do that, we remove a hydrogen atom from that allylic position and draw the resonance structures of the resulting allylic radical:

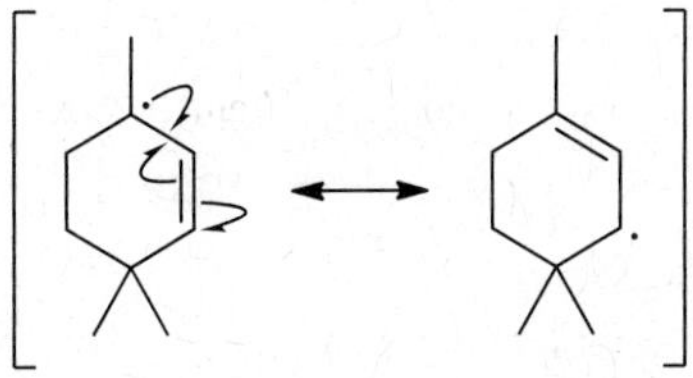

Finally, we use these resonance structures to determine the products, by placing a bromine atom at the position of the unpaired electron in each resonance structure. Notice that each product has a chirality center and is therefore expected to be produced as a racemic mixture:

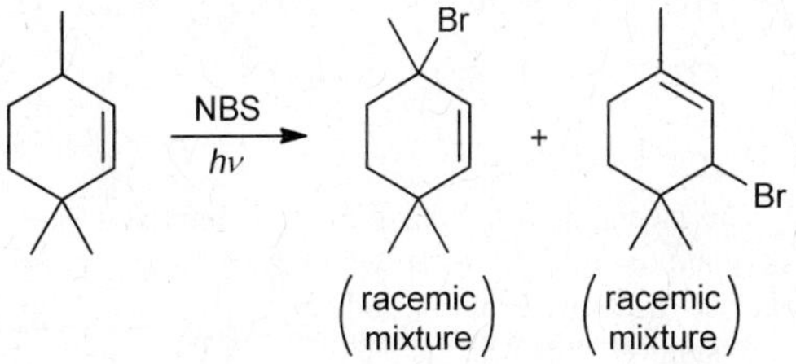

11.40.
(a) Bromination occurs selectively at the tertiary position, giving the following tert-butyl bromide:

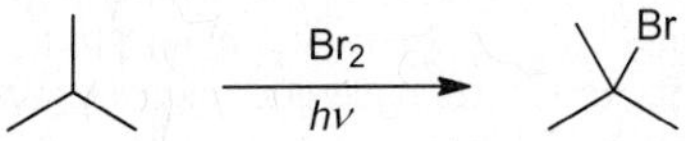

(b) The minor product occurs via bromination at the primary position:

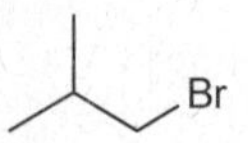

(c) The mechanism will have three distinct stages. The first stage is initiation, in which the Br–Br bond is broken to generate bromine radicals. This step requires two fishhook arrows, as shown. There are two propagation steps (hydrogen abstraction and halogen abstraction), each of which requires three fishhook arrows (to show the bonds being broken and formed). In the first propagation step, a hydrogen atom is abstracted to give the more stable tertiary radical. In the second step, this radical undergoes halogen abstraction to give the product. There are many steps that can serve as termination steps, because there are many radicals that can couple together under these conditions. Shown below is a termination step that generates the desired product:

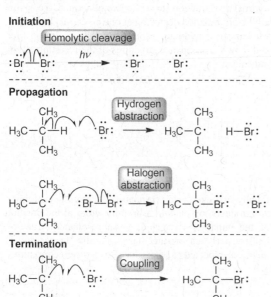

(d) The mechanism will have three distinct stages. The first stage is initiation, in which the Br–Br bond is broken to generate bromine radicals. This step requires two fishhook arrows, as shown. There are two propagation steps (hydrogen abstraction and halogen abstraction), each of which requires three fishhook arrows (to show the bonds being broken and formed). In the first propagation step, a hydrogen atom is abstracted to give the less stable primary radical (we are drawing the mechanism that leads to the minor product). In the second step, this radical undergoes halogen abstraction to give the product. There are many steps that can serve as termination steps, because there are many radicals that can couple together under these conditions. Shown

below is a termination step that generates the desired product:

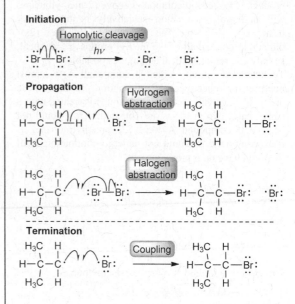

(e) The minor product is only formed via a primary radical, which does not readily form under bromination conditions. The tertiary radical is selectively formed, which leads to the tertiary alkyl bromide as the major product.

11.41.
(a) There are two tertiary positions in this case, highlighted here:

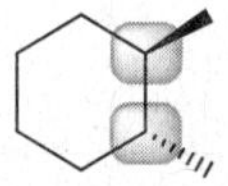

But these positions are identical, because they can be interchanged by an axis of symmetry, shown here (when we rotate 180 degrees about this axis, the same image is regenerated – you might want to build a molecular model to prove this to yourself):

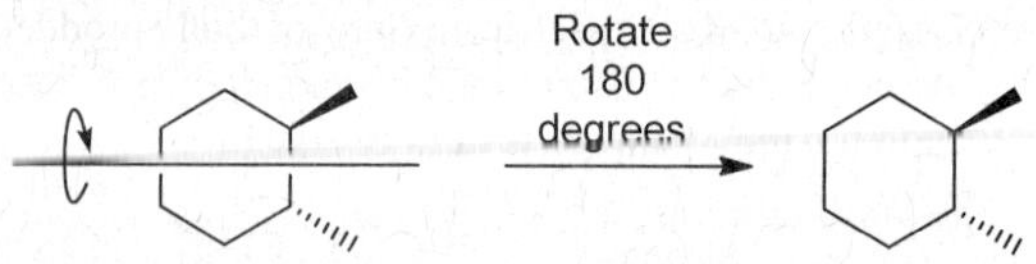

So we only need to consider bromination at one of these positions (either one will lead to the same products). Since the reaction occurs at a chirality center, we expect monobromination to give both possible configurations (*R* and *S*) for that chirality center center. The configuration of the other chirality center (not involved in the reaction) is retained. Therefore, the products are diastereomers:

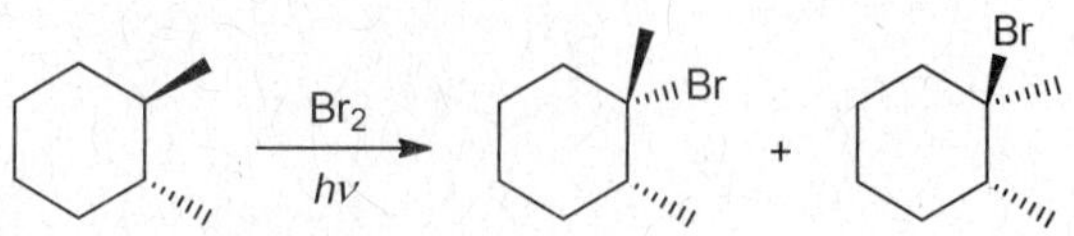

(b) The starting compound has two tertiary positions, so dibromination will install one bromine atom at each of the tertiary positions:

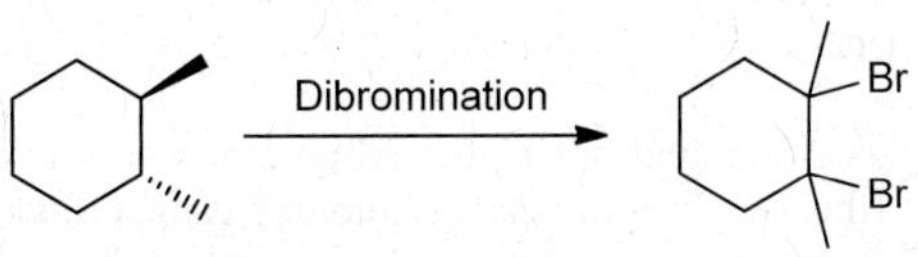

Each chirality center can be produced with either the *R* or *S* configuration, so all possible stereoisomers are expected. With two chirality centers, we might expect four stereoisomers (2^n, where n = # of chirality centers = 2). However, in this case, there are only three stereoisomers, because one of them is a *meso* compound, shown here. (For a review of *meso* compounds, see Section 5.6)

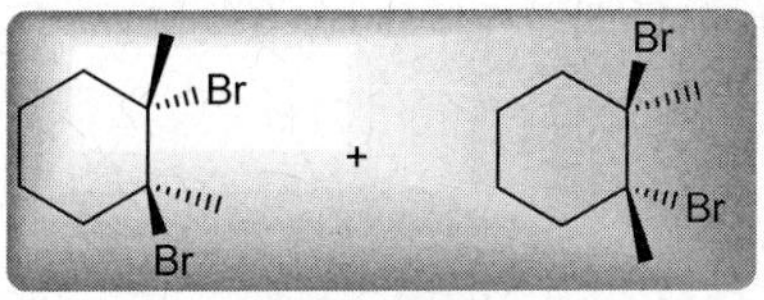

enantiomers

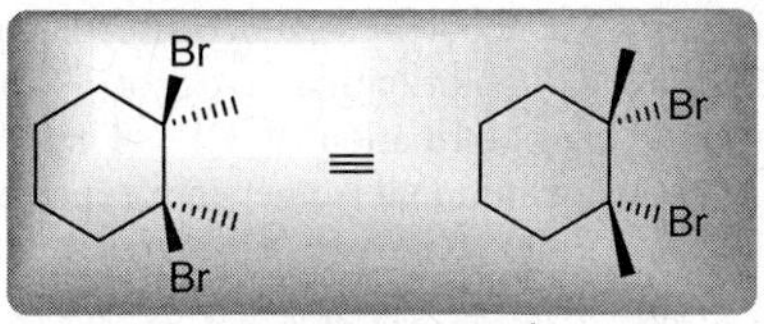

a *meso* compound

11.42. Methyl radicals are less stable than *tert*-butyl radicals so the former react with each other more rapidly than the latter. Also, methyl radicals are less hindered than *tert*-butyl radicals, so the former react with each other more rapidly than the latter.

11.43.
(a) Monochlorination of cyclopentane gives only one product, because a reaction at any position generates the same product as a reaction at any other position:

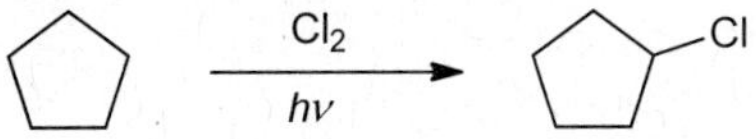

(b) This compound has six carbon atoms, but there are only four unique positions (highlighted) where chlorination can occur:

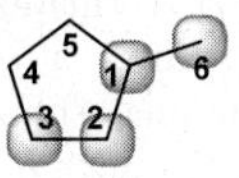

Chlorination at C4 produces the same result as chlorination at C3, and chlorination at C5 produces the same result as chlorination at C2. Therefore, the following four constitutional isomers are expected:

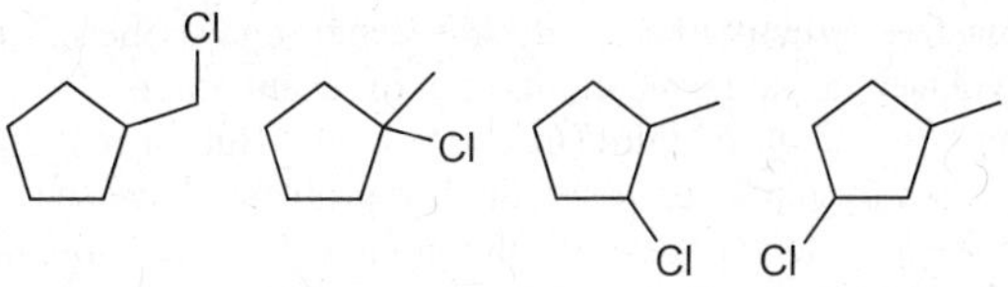

The last two structures have chirality centers and are therefore produced as mixtures of stereoisomers. But this can be ignored for purposes of solving this problem, because the problem statement asks for the number of *constitutional isomers* that are obtained.

(c) This compound has seven carbon atoms, but there are only three unique positions (highlighted) where chlorination can occur:

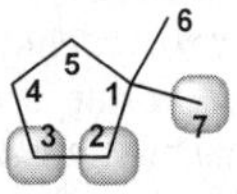

Chlorination cannot occur at C1, because that position does not have a C–H bond (which is necessary, because hydrogen abstraction is the first step of the chlorination process). Chlorination at C4 produces the same result as chlorination at C3; chlorination at C5 produces the same result as chlorination at C2; and chlorination at C6 produces the same result as chlorination at C7. Therefore, the following three constitutional isomers are expected:

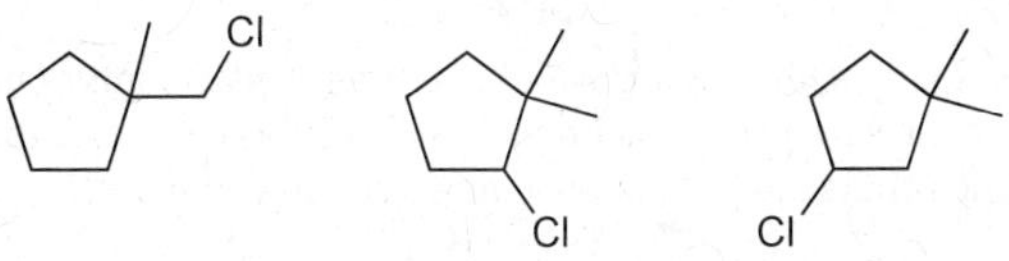

Each of the last two structures has a chirality center and is therefore produced as a mixture of enantiomers. But this can be ignored for purposes of solving this problem, because the problem statement asks for the number of *constitutional isomers* that are obtained.

(d) This compound has five carbon atoms, but there are only four unique positions (highlighted) where chlorination can occur:

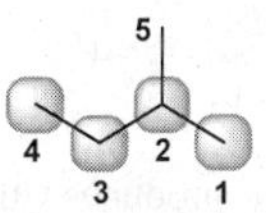

Chlorination at C5 produces the same result as chlorination at C1. Therefore, the following four constitutional isomers are expected:

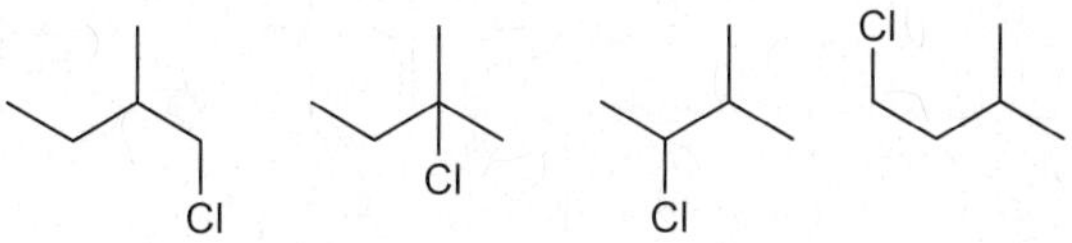

The first structure has a chirality center and is therefore produced as a racemic mixture of enantiomers. The same is true of the third structure above. But this can be ignored for purposes of solving this problem, because the problem statement asks for the number of *constitutional isomers* that are obtained.

(e) This compound has six carbon atoms, but there are only two unique positions (highlighted) where chlorination can occur:

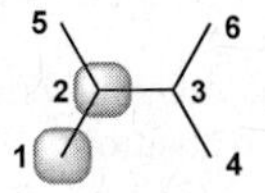

Chlorination at C3 produces the same result as chlorination at C2. Similarly, chlorination at C4, C5, or C6 produces the same result as chlorination at C1. Therefore, the following two constitutional isomers are expected:

The first structure has a chirality center and is therefore produced as a racemic mixture of enantiomers. But this can be ignored for purposes of solving this problem, because the problem statement asks for the number of *constitutional isomers* that are obtained.

(f) Monochlorination of cyclohexane gives only one product, because a reaction at any position generates the same product as a reaction at any other position:

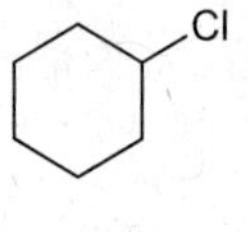

(g) This compound has seven carbon atoms, but there are only five unique positions (highlighted) where chlorination can occur:

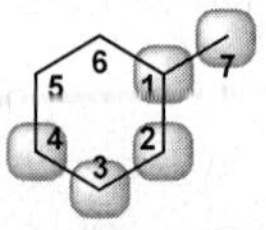

Chlorination at C5 produces the same result as chlorination at C3, and chlorination at C6 produces the same result as chlorination at C2. Therefore, the following five constitutional isomers are expected:

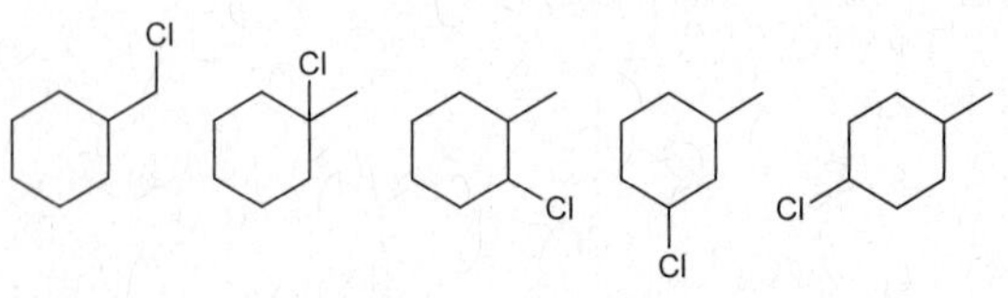

Each of the last three structures is produced as a mixture of stereoisomers. But this can be ignored for purposes of solving this problem, because the problem statement asks for the number of *constitutional isomers* that are obtained.

(h) This compound has eight carbon atoms, but two of them (the two central carbon atoms) cannot undergo chlorination because they lack a C–H bond. The remaining six carbon atoms (the methyl groups) all are identical, because a reaction at any position generates the same product as a reaction at any other position.

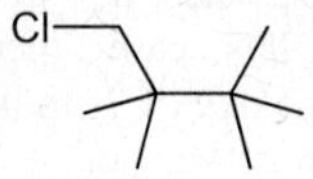

(i) This compound has six carbon atoms, but there are only three unique positions (highlighted) where chlorination can occur:

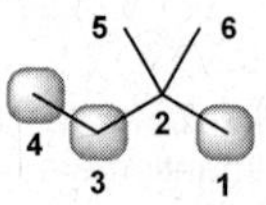

Chlorination cannot occur at C2, because that position lacks a C–H bond. Chlorination at C5 or C6 produces the same result as chlorination at C1. Therefore, the following three constitutional isomers are expected:

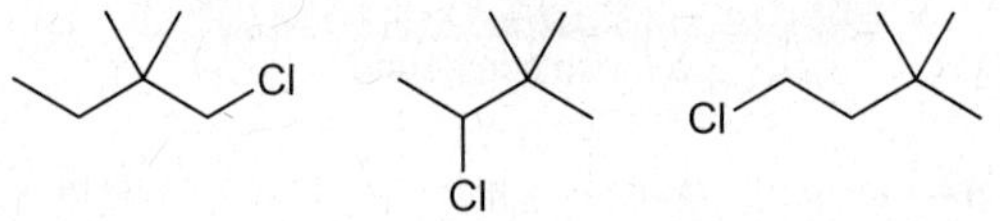

The middle structure has a chirality center and is therefore produced as a racemic mixture of enantiomers. But this can be ignored for purposes of solving this problem, because the problem statement asks for the number of *constitutional isomers* that are obtained.

(j) This compound has five carbon atoms, but there are only three unique positions (highlighted) where chlorination can occur:

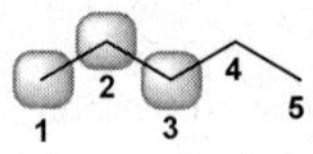

Chlorination at C4 produces the same result as chlorination at C2, and chlorination at C5 produces the same result as chlorination at C1. Therefore, the following three constitutional isomers are expected:

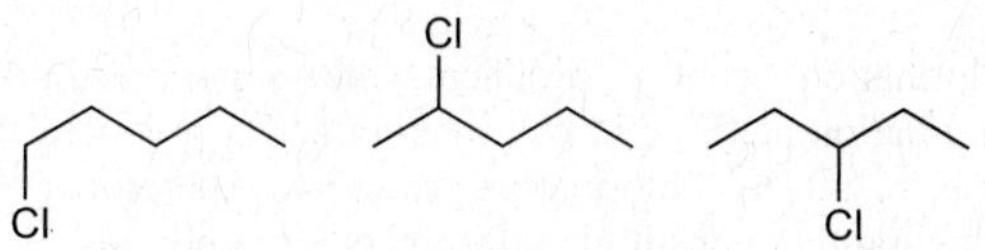

The middle structure has a chirality center and is therefore produced as a racemic mixture of enantiomers. But this can be ignored for purposes of solving this problem, because the problem statement asks for the number of *constitutional isomers* that are obtained.

11.44. Let's begin by drawing our starting materials, so that we can see them more clearly:

Acetylene   2-Methylpropane

These starting materials have two carbon atoms and four carbon atoms, respectively, while the product has six carbon atoms. Therefore, we must find a way to join the starting materials. So far, we have only seen one way to make a carbon-carbon bond. Specifically, acetylene can undergo alkylation, as shown here:

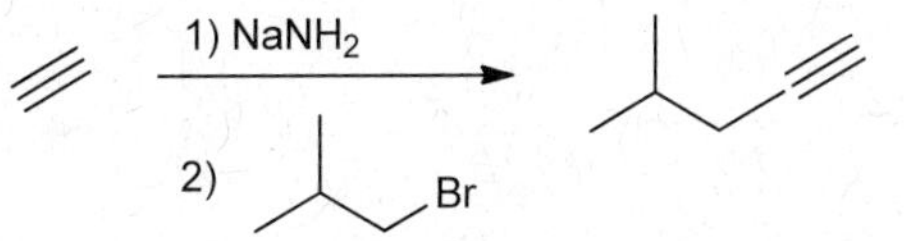

The resulting terminal alkyne can then be converted into the desired product via acid-catalyzed hydration in the presence of mercuric sulfate:

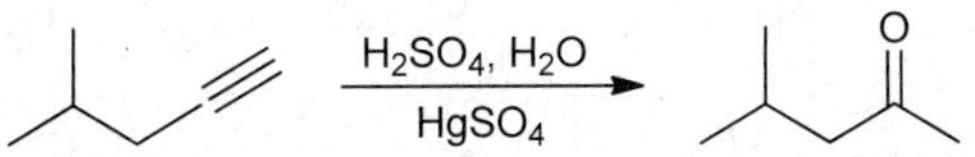

All that remains is to show how the starting alkane (2-methylpropane) can be converted into the necessary primary alkyl halide:

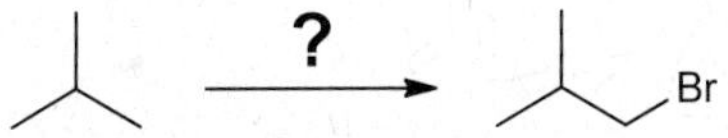

Radical bromination will indeed install a bromine atom, but this occurs selectively at the tertiary position:

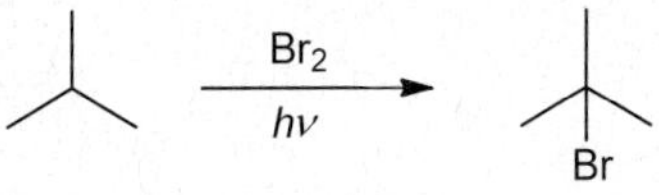

This tertiary alkyl bromide must now be converted into the desired primary alkyl bromide. That is, we must move the position of the bromine atom, which can be accomplished via elimination, followed by *anti*-Markovnikov addition of HBr:

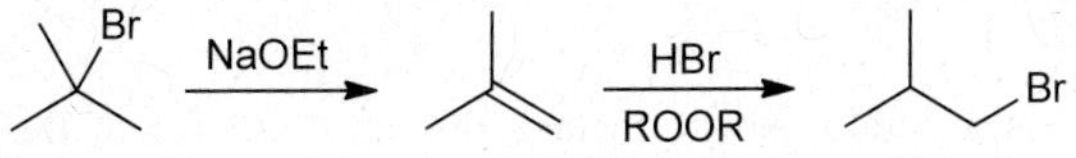

In summary, the entire synthesis is shown below.

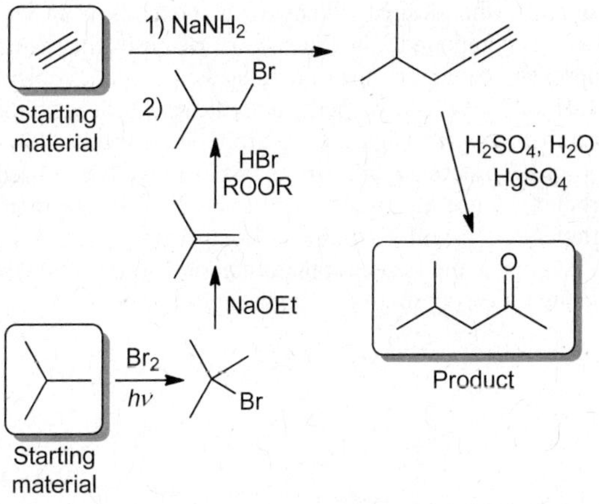

11.45.
(a) This transformation can be achieved via radical chlorination (under conditions that favor monochlorination):

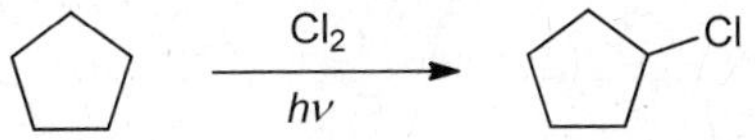

(b) Radical iodination is not a feasible process (it is not thermodynamically favorable), so we cannot directly iodinate the starting cycloalkane. However, radical bromination can be performed, followed by an S$_N$2 reaction in which iodide replaces bromide:

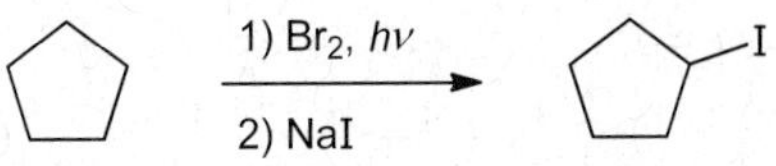

(c) We have not seen a way to convert an alkane into an alkene in one step. However, it can be achieved in two steps. First, radical chlorination or bromination can be performed, thereby installing a leaving group, which then allows for an E2 process with a strong base to give the desired product. Any strong base can be used (there is only one regiochemical outcome, so a sterically hindered base is not required). It is acceptable to use hydroxide, methoxide or ethoxide for the E2 process. That being said, in this particular case, *tert*-butoxide will likely be more efficient, as it will suppress the competing S$_N$2 process (we would expect S$_N$2 to give a minor product if the base is not sterically hindered).

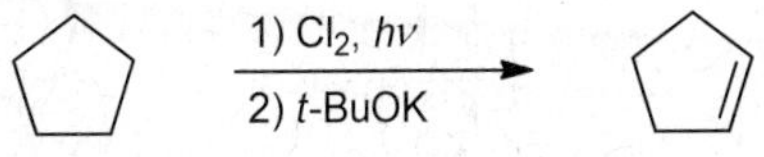

(d) The product is a *trans*-dibromide, so the last step of the synthesis is likely to be addition of Br$_2$ across a π bond:

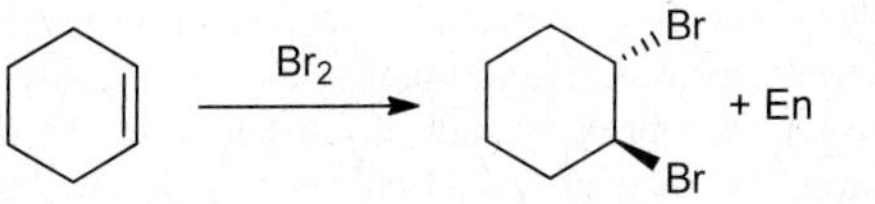

364 CHAPTER 11

To make the desired alkene (cyclohexene) from the starting compound (cyclohexane), we use the same approach taken in the previous solution (Problem 11.45c). That is, we perform radical chlorination or bromination to install a leaving group, which then allows for an E2 process with a strong base to give the desired product. Once again, *tert*-butoxide will likely be more efficient, as it will suppress the competing S_N2 process. In summary, the desired transformation can be achieved via the following three-step synthesis:

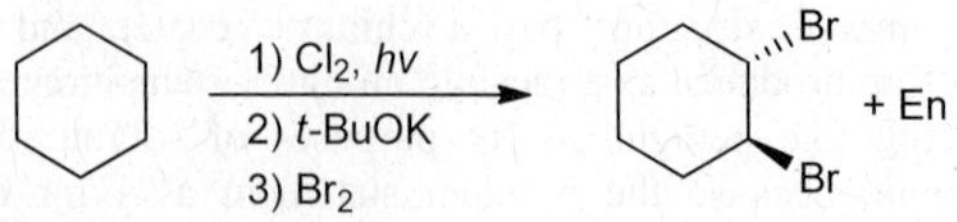

(e) Radical bromination installs a bromine atom selectively at the tertiary position. Then, a strong base is used to give an E2 reaction (Zaitsev product, since the base employed is not sterically hindered):

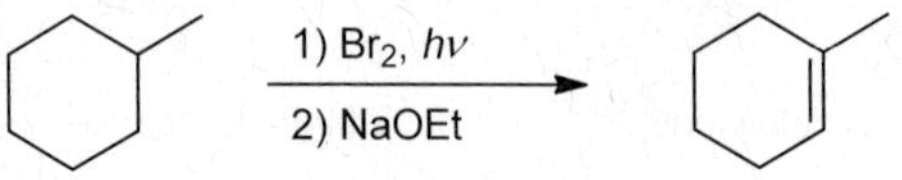

11.46. *cis*-1,2-Dimethylcyclopentane produces six pairs of compounds, where each pair have a diastereomeric relationship (except for one pair, which is enantiomeric, indicated below). In contrast, *trans*-dimethylcyclopentane produces only six different compounds, as shown below:

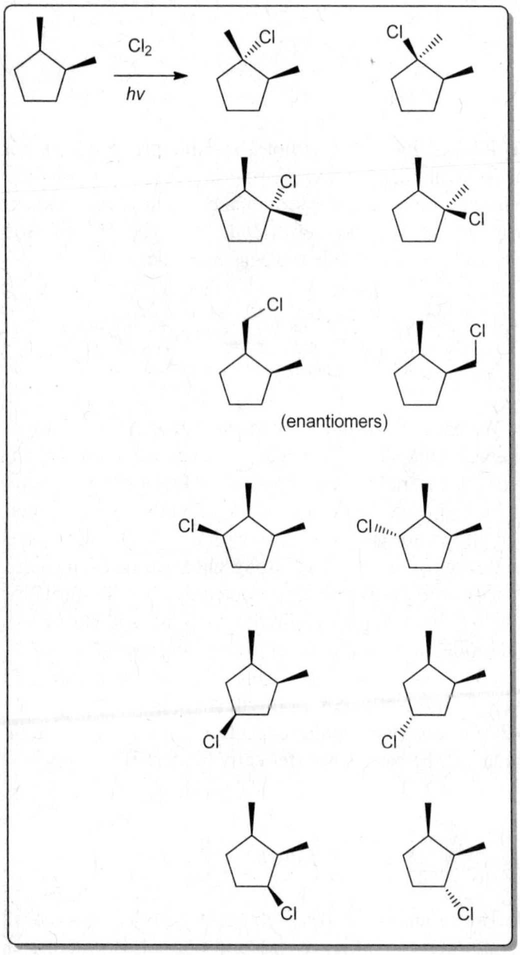

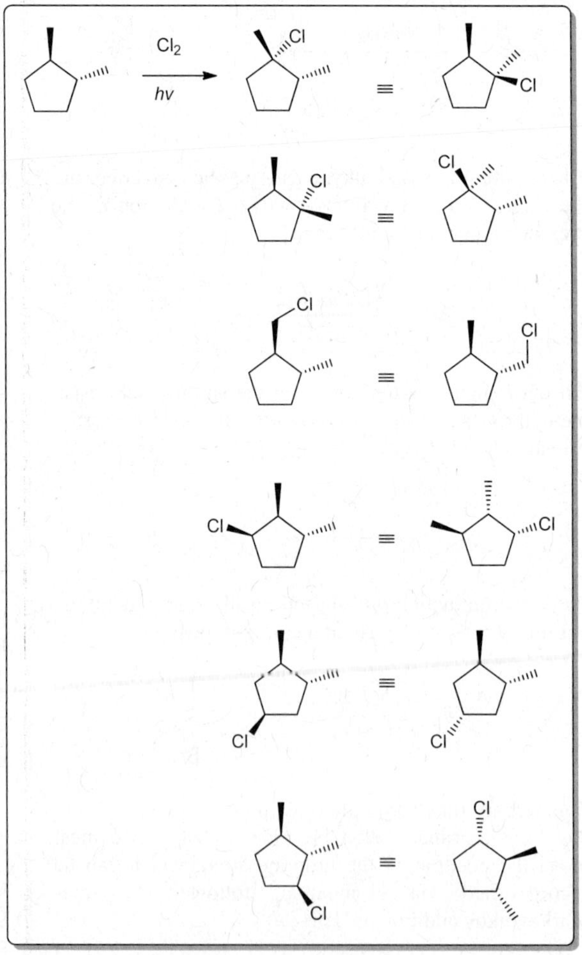

(enantiomers)

11.47. The first propagation step in a bromination process is generally slow. In fact, this is the source of the selectivity for bromination processes. A pathway via a secondary radical will be significantly lower in energy than a pathway via a primary radical. As a result, bromination occurs predominantly at the more substituted (secondary) position. However, when chlorine is present, chlorine radicals can perform the first propagation step (hydrogen abstraction) very rapidly, and with little selectivity. Under these conditions, primary radicals are formed almost as easily as secondary radicals. The resulting radicals then react with bromine in the second propagation step to yield monobrominated products. Therefore, in the presence of chlorine, the selectivity normally observed for bromination is lost.

11.48. As shown in the problem statement, an acyl peroxide will undergo cleavage to give a radical that can then lose carbon dioxide:

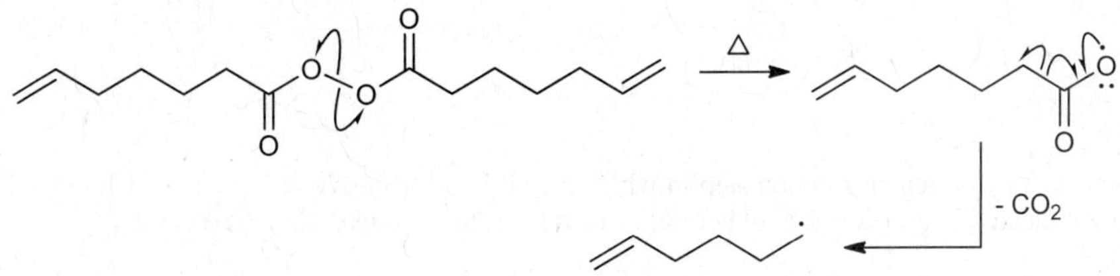

This radical is responsible for the formation of each of the reported products. The first product is formed when the radical undergoes hydrogen abstraction:

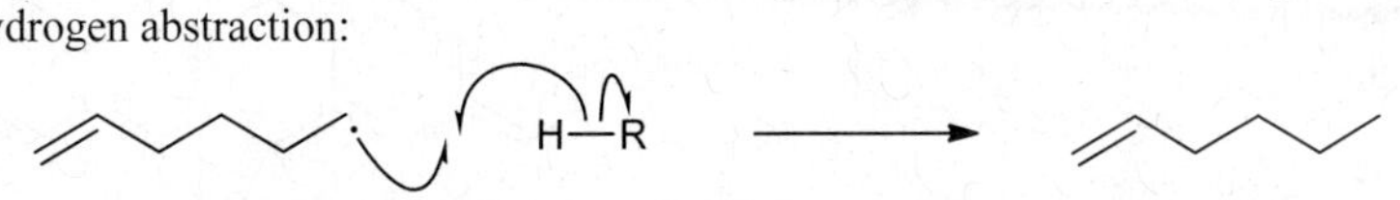

The second product is formed when two of the radicals couple with each other, as shown:

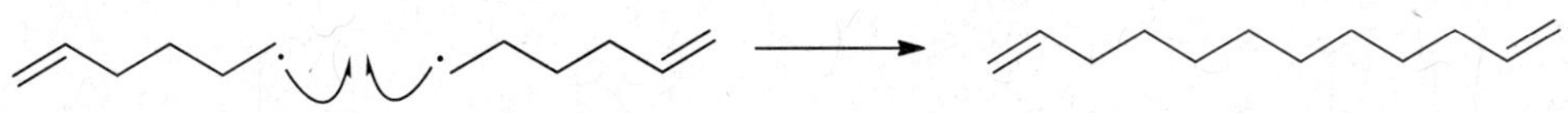

And finally, each of the cyclic products is formed via addition to a π bond (in an intramolecular fashion), followed by hydrogen abstraction:

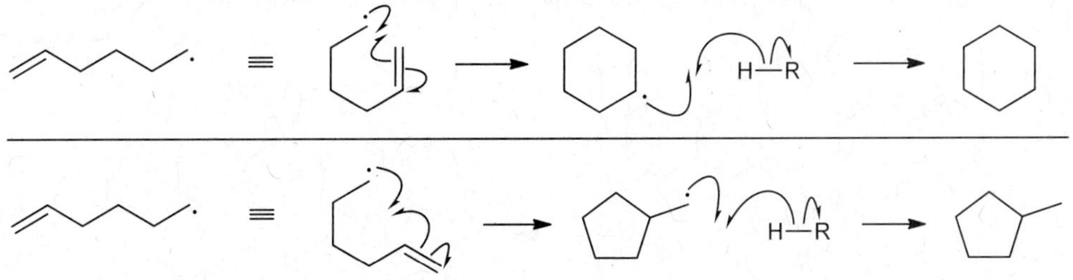

11.49.
(a) In this reaction, two groups (R′S and H) are being added across the π bond in an *anti*-Markovnikov fashion. This reaction resembles the *anti*-Markovnikov addition of HBr across a π bond (Section 11.10). Using that reaction as a guide, we can draw the following initiation and propagation steps. In the initiation step, an R′S radical is formed. The first propagation step is an addition to a π bond, and the second propagation step is a hydrogen abstraction to regenerate an R′S radical.

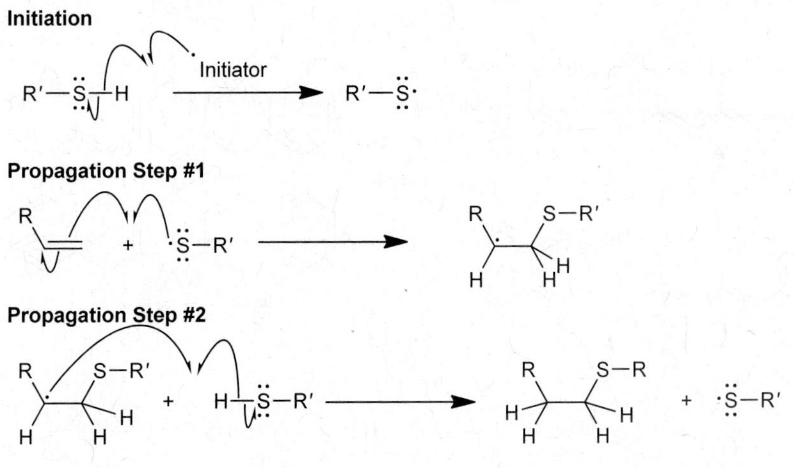

Initiation

Propagation Step #1

Propagation Step #2

(b) We expect an *anti*-Markovnikov addition of R′S and H across the π bond, as follows:

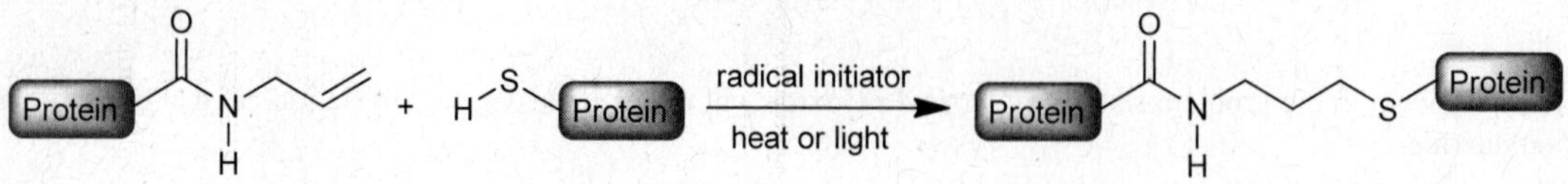

11.50.

A + B → 2 C. This is a hydrogen abstraction step in which a C-H bond from **B** undergoes homolytic cleavage, leading to a transfer of the indicated hydrogen (in bold below) from **B** to **A**, to form two radicals (**C** and **C**).

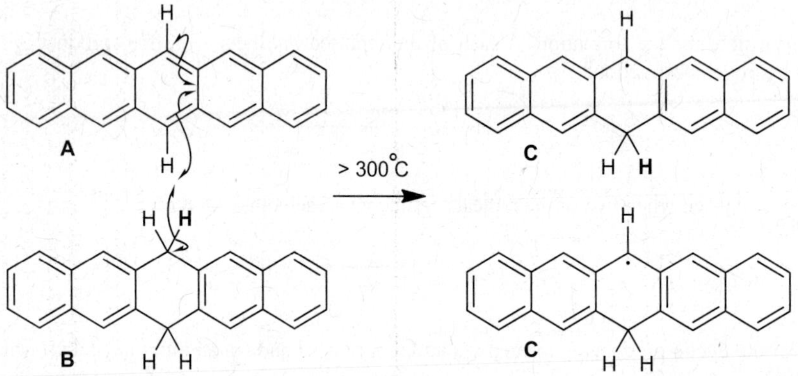

C + C → D. The two radicals produced above (**C** and **C**) can undergo a coupling reaction to give **D**.

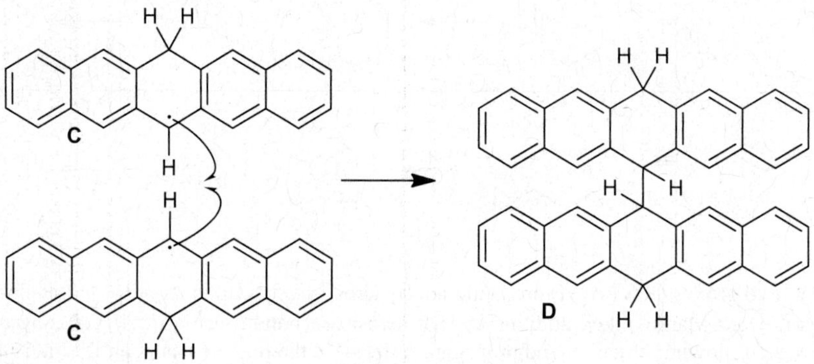

C + A → E. This step can be described as "addition to a π bond." The additional fishhook arrows are consistent with the resulting resonance structure shown.

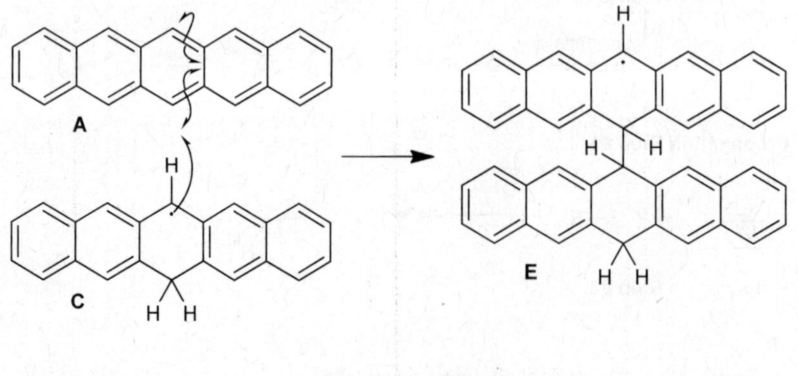

11.51. Overall, in this oxidation reaction, the two C-H bonds on the central carbon atom are replaced with a C=O double bond, as shown below. It is useful to keep this in mind as we develop our mechanism.

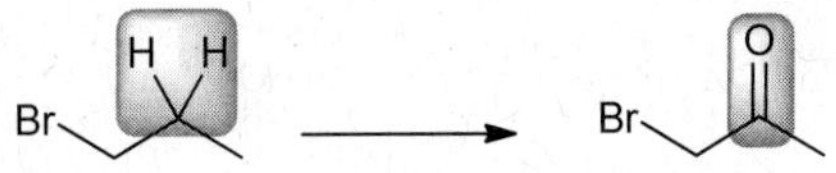

The following mechanism is described in the problem statement.

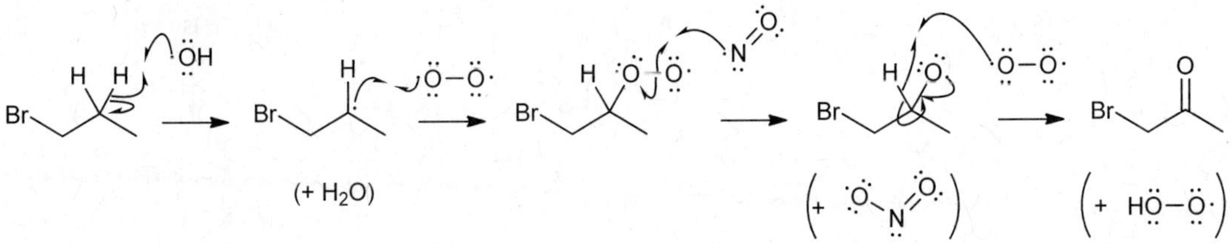

(+ H₂O)

11.52. The first tricyclic radical can be formed with the following two steps. First, the initial radical undergoes an intramolecular addition to a π bond (see Figure 11.9), generating a 5-membered ring and a new radical. This radical once again undergoes addition to a π bond, destroying the second π bond and resulting in the formation of two 5-membered rings and a new radical.

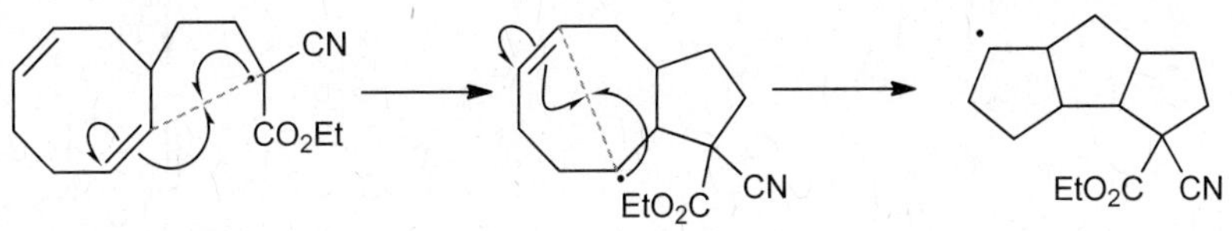

The second tricyclic radical can be formed if the initial radical undergoes addition with the π bond that is farther away, followed by another addition with the remaining π bond.

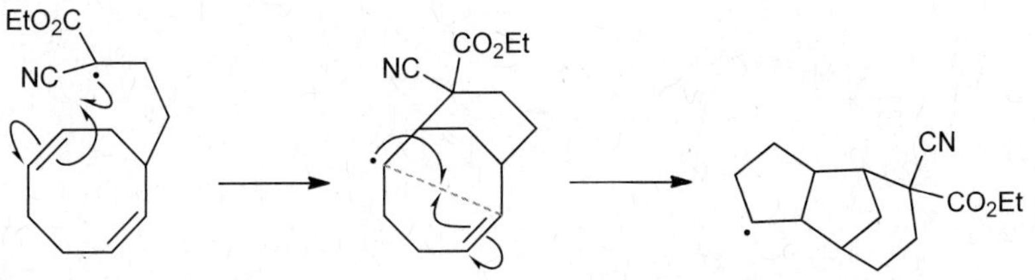

11.53. A comparison of the first product with the starting material reveals which bond in the starting material undergoes homolytic cleavage:

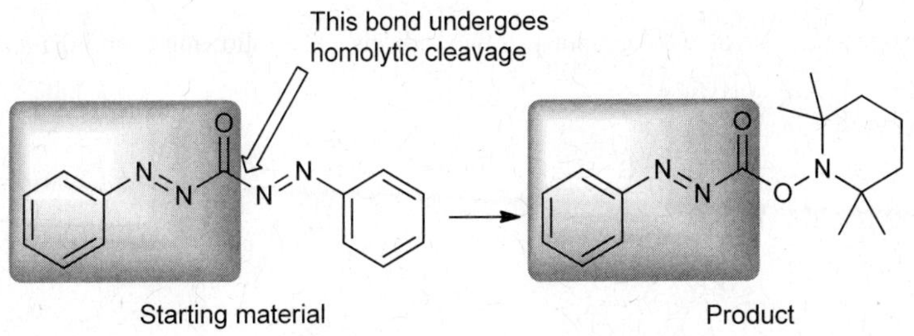

So we begin our mechanism by drawing homolytic cleavage of that bond, giving two radicals (**A** and **B**):

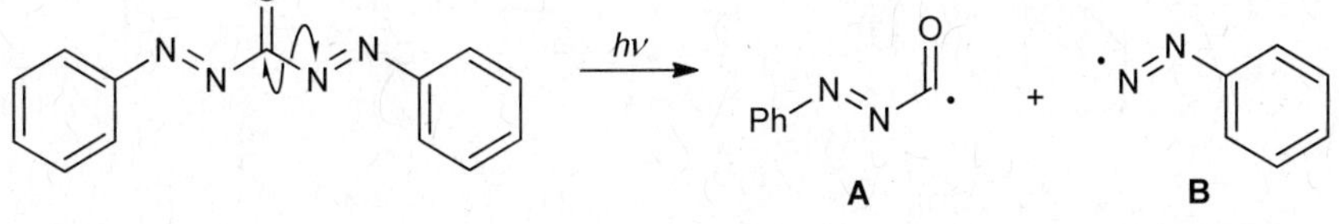

Radical **A** then reacts with TEMPO to give the first product, while radical **B** undergoes further homolytic cleavage to give nitrogen gas and a phenyl radical, which reacts with TEMPO to give the second product.

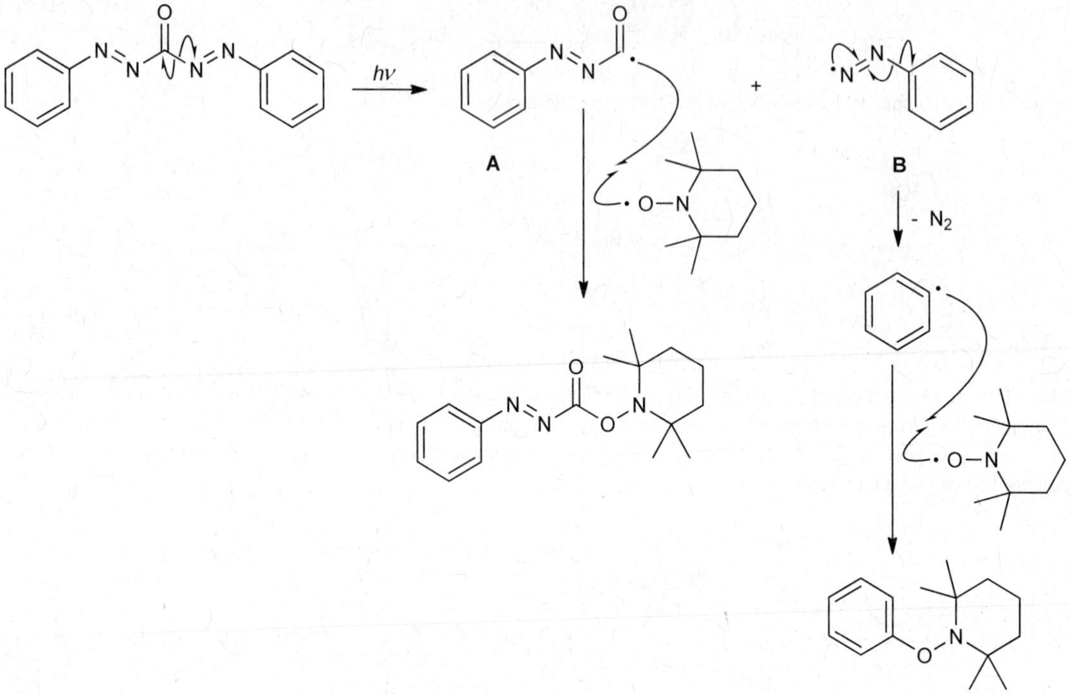

11.54.

(a) The first step of the propagation cycle is abstraction of a proton from cubane, generating a cubyl radical. In the second step of the propagation cycle, halogen abstraction gives the product (iodocubane) and also regenerates the triiodomethyl radical.

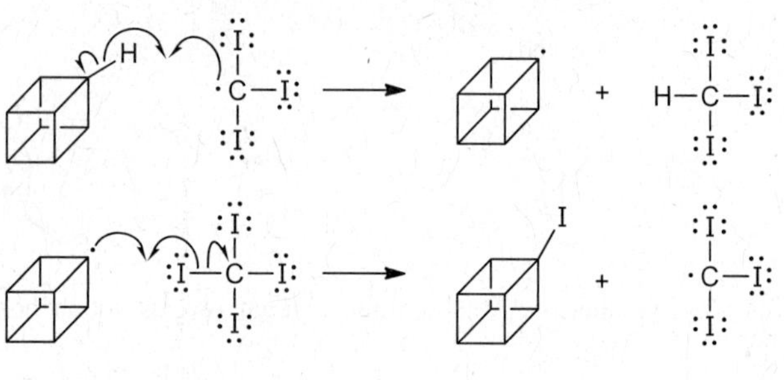

(b) Termination steps generally involve the coupling of two radicals. The following coupling reactions are all possible termination steps.

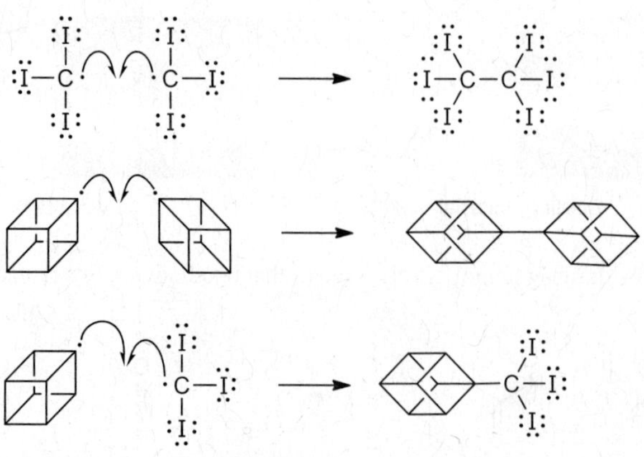

(c) Recall (Section 11.4) that for radical halogenation reactions, $\Delta G \approx \Delta H$. For the iodination of an alkane with I_2 (compound **2a**), ΔH will have a positive value:

$$(H_3C)C-H \;+\; I-I \longrightarrow (H_3C)C-I \;+\; H-I \qquad \Delta H = +26 \text{ kJ/mol}$$

~ 381 kJ/mol **2a** ~ 151 kJ/mol ~ 209 kJ/mol ~ 297 kJ/mol

Since ΔH for the reaction is positive, ΔG for the reaction will also be positive. Therefore, the reaction is thermodynamically unfavorable. This argument is expected to hold true for cubane as well, assuming that the BDE for the C–H bond in cubane is not too different from the BDE for the C–H bond in $(CH_3)_3CH$. That is, iodination of cubane with molecular iodine (**2a**) is unfavorable.

In contrast, when **2b** is used as the reagent for iodination, ΔH for the reaction will be negative, which gives a negative value for ΔG. This can be rationalized by comparing the BDE values for **4a** and **4b**. The BDE of **4a** is 297 kJ/mol, while the BDE of **4b** is 423 kJ/mol. This significant difference in BDE values causes ΔH for the reaction to be negative when **2b** is used rather than **2a**. Recall that bonds broken require an input of energy, contributing to a positive value of ΔH, while bonds formed release energy when they are formed, contributing to a negative value of ΔH. Since the bonds in **4a** and **4b** are being formed during the process, a large BDE (such as that of **4b**) will contribute to a negative value of ΔH, while a small BDE (such as that of **4a**) will contribute to a positive value of ΔH.

11.55.
Abstraction of any of the hydrogen atoms attached to carbon leads to a vinyl radical that is not resonance stabilized. Abstraction of any of the hydrogen atoms attached to oxygen leads to a resonance-stabilized radical. Thus, we should focus our attention on the three hydroxyl groups.

Abstraction of a hydrogen atom from the hydroxyl group on the right side of the molecule leads to a radical with eight reasonable resonance structures, as shown below (**A-H**).

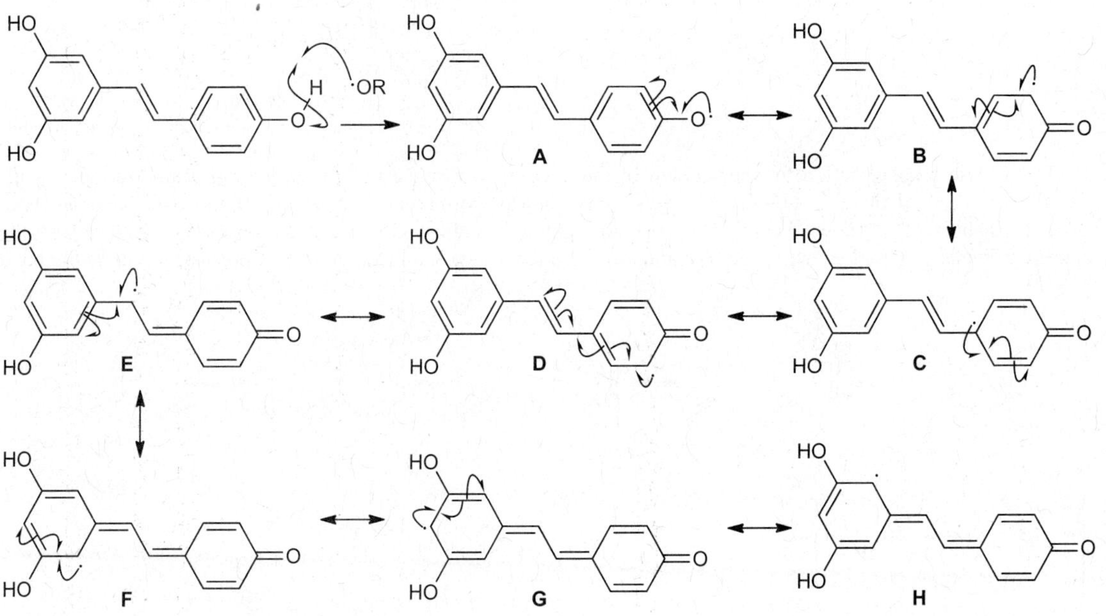

Abstraction of a hydrogen atom from either of the hydroxyl groups on the left side of the molecule leads to a radical with four reasonable resonance structures, as shown here (**I-L**).

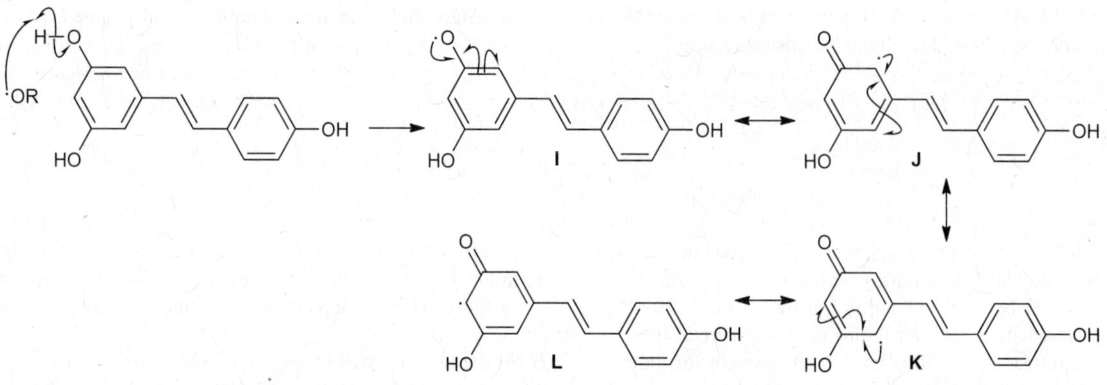

This suggests that the hydrogen atom of the hydroxyl group on the right side of the molecule is more susceptible to abstraction because it leads to the formation of a more stable radical.

11.56.

(a) In the first step, the C-O bond undergoes homolytic cleavage, yielding the two radicals shown. A resonance structure of the sulfoxide radical demonstrates that the unpaired electron is delocalized over the oxygen and sulfur atoms, as shown. Recombination of the radicals then provides the product, as shown.

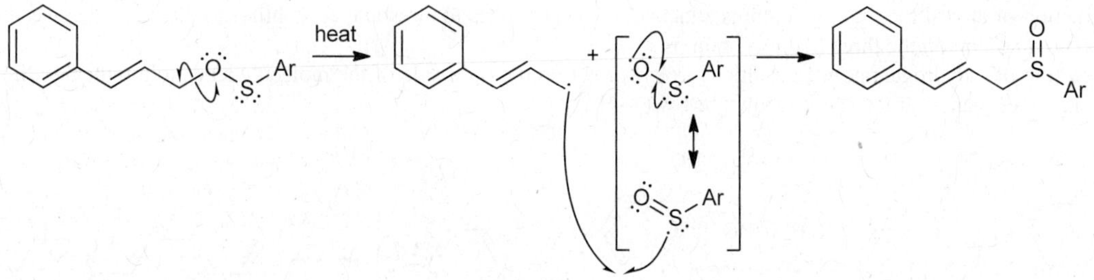

(b) To propose an explanation for the scission of the C-O bond over the O-S bond, we must analyze the radicals formed from each of these homolytic cleavages. The radicals formed from cleavage of the C-O bond are both resonance stabilized, as shown below. Note that the sulfoxide radical is also further resonance-stabilized by the adjacent aromatic ring (not shown). The formation of these resonance-stabilized radicals is consistent with facile bond cleavage.

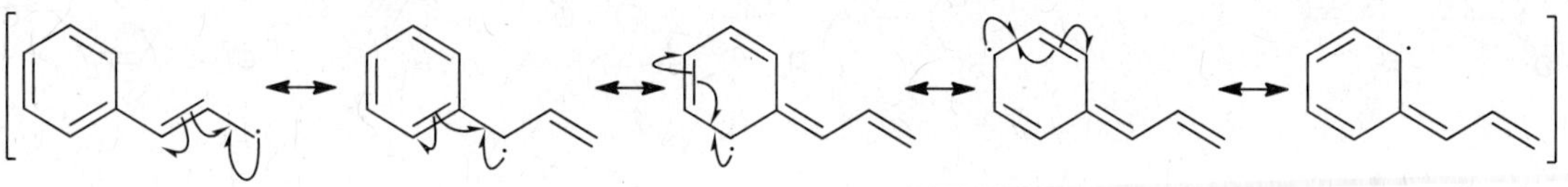

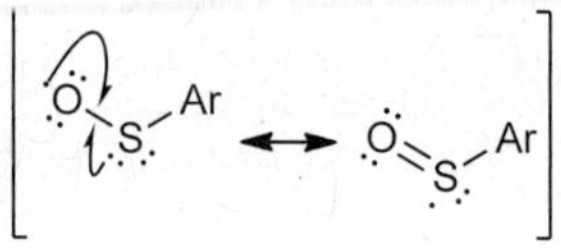

The radicals formed from homolytic cleavage of the O-S bond are shown below. The oxygen radical is not resonance-stabilized. The sulfur radical is resonance-stabilized due to being adjacent to an aromatic ring. However, analogous resonance stabilization is also present in the sulfoxide radical shown above, so this does not provide any additional relative stabilization.

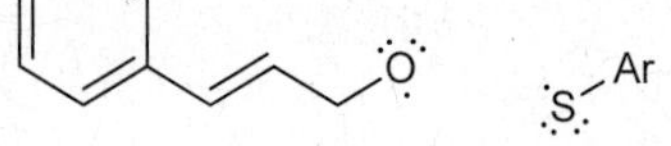

11.57.

In the first step, *tert*-butoxide is a strong, sterically hindered base, so we expect an E2 process, generating an alkene. Then, when treated with NBS under radical conditions, the alkene undergoes hydrogen abstraction to give a resonance stabilized radical, followed by halogen abstraction to generate the product.

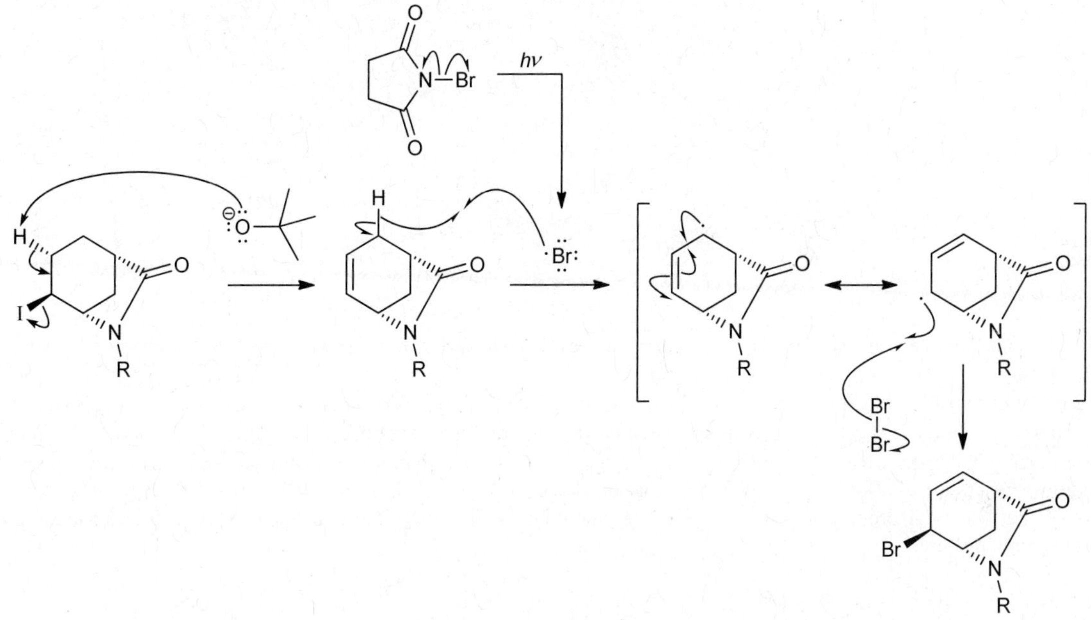

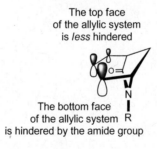

To rationalize the observed stereochemical outcome, consider the two faces of the allylic system (which are essentially the two faces of the six-membered ring). Notice that the bottom face is blocked by the amide. As a result, it would be difficult for bromine to react on this face of the molecule. This steric consideration can successfully explain why bromine traps the carbon radical on the top face of the molecule, leading to the observed stereochemical outcome.

The top face
of the allylic system
is *less* hindered

The bottom face
of the allylic system   R
is hindered by the amide group

11.58.

(a) Benzoyl peroxide is a radical initiator, and upon heating, it readily undergoes homolytic bond cleavage, giving resonance-stabilized radicals. These radicals can then abstract a halogen from CCl_4 in a halogen abstraction step to give a trichloromethyl radical ($Cl_3C\cdot$), shown below.

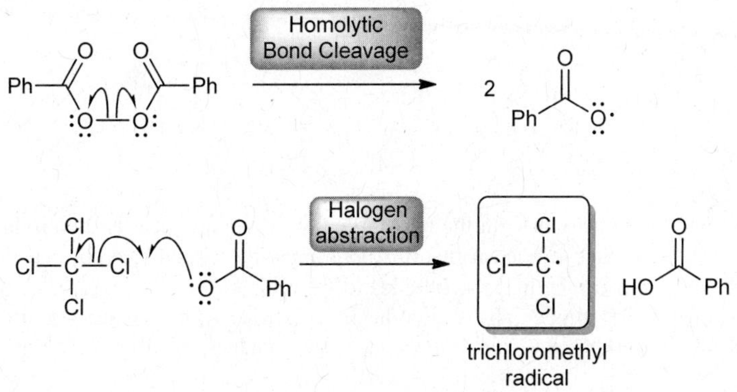

Homolytic
Bond Cleavage

Halogen
abstraction

trichloromethyl
radical

There are two propagation steps, shown below. In the first propagation step, the π bond reacts with a trichloromethyl radical to give the more stable secondary radical (rather than the less stable primary radical). This step explains the regiochemical outcome of the process. Then, in the second propagation step, a halogen abstraction gives the product and regenerates the reactive intermediate (Cl₃C•), as expected for a propagation cycle.

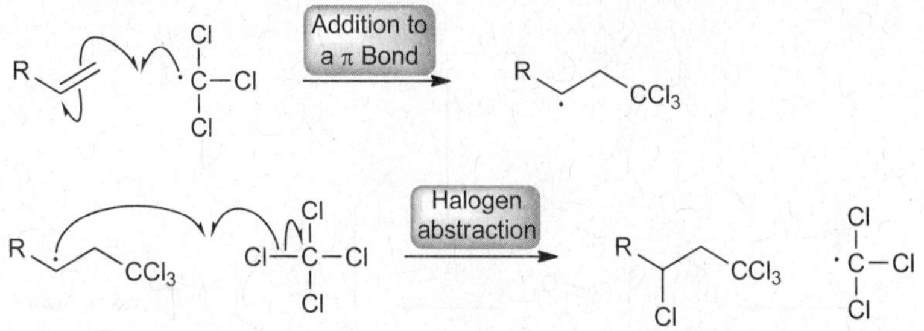

(b) The initiation steps are the same as those for a standard Kharasch reaction [see the initiation steps in part (a) of this problem]

Based on our answer to part (a), we expect the first propagation step to involve addition of Cl₃C• across a π bond. In this case, there are two π bonds. The Cl₃C• radical can react with either π bond, but reaction with the more substituted alkene (di-substituted) affords the more stable 3° radical. This intermediate radical can then add to the other π bond (mono-substituted) to afford a primary radical that quickly reacts with CCl₄ to give the desired product. Once again, notice that the final propagation step involves regeneration of the reactive intermediate, Cl₃C•, as expected for a propagation cycle.

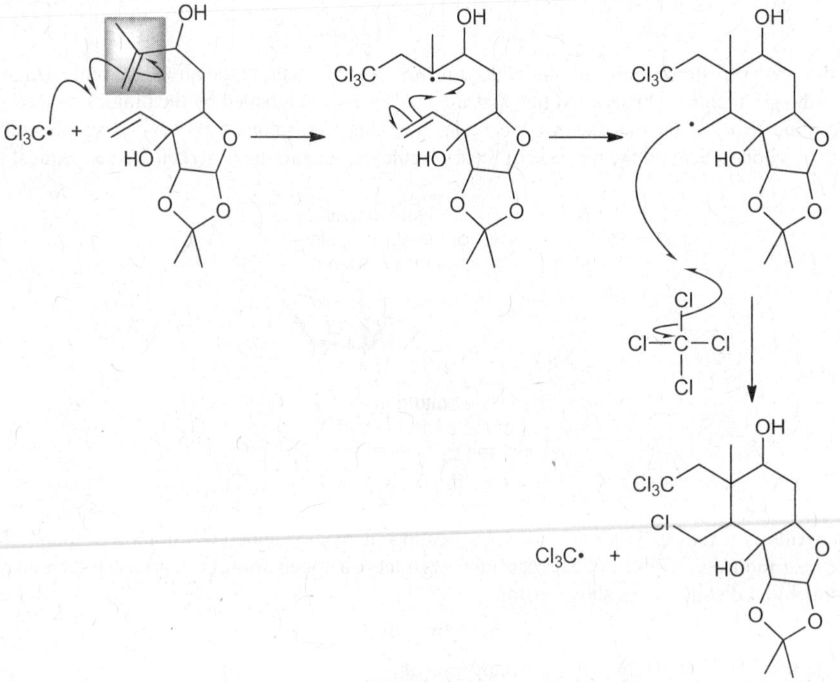

11.59. A reasonable mechanism consistent with the transformation of the diradical to the product is presented below. Based on the structure of the product, it is clear that a σ bond must be formed (thereby forming the five-membered ring). This σ bond can be formed if the vinyl radical adds to the aromatic carbon atom, leading to a new resonance-stabilized radical, shown below. Hydrogen abstraction then occurs when this resonance-stabilized radical encounters another radical (*i.e.*, R₂NO• = another equivalent of the same intermediate), thereby enabling the reformation of the aromatic ring. The remaining unpaired electron (on the oxygen atom) then abstracts a hydrogen atom from another entity (*i.e.*, R'H = another equivalent of the resonance-stabilized radical above) producing the product.

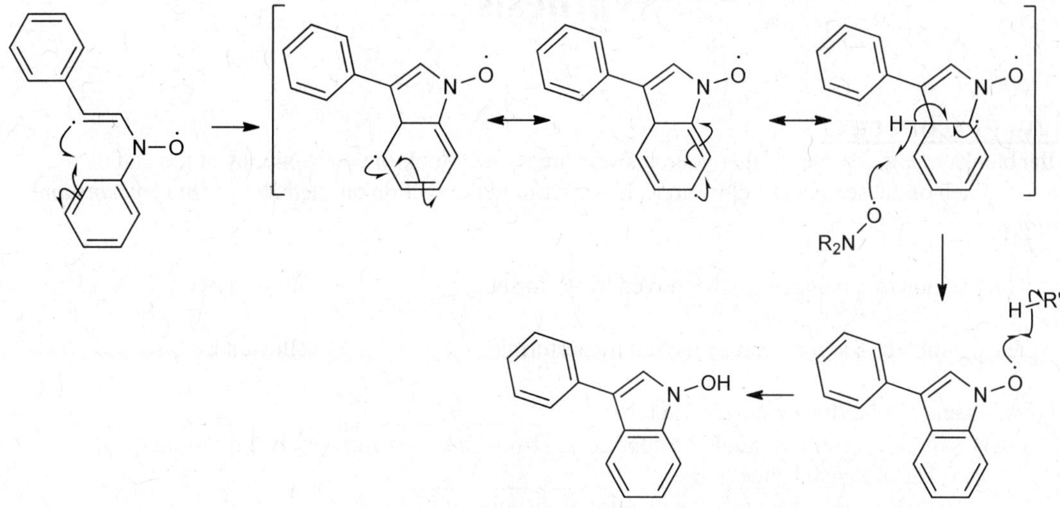

Chapter 12
Synthesis

Review of Concepts

Fill in the blanks below. To verify that your answers are correct, look in your textbook at the end of Chapter 12. Each of the sentences below appears verbatim in the section entitled *Review of Concepts and Vocabulary*.

- The position of a halogen can be moved by performing _____ followed by _____.

- The position of a π bond can be moved by performing _____ followed by _____.

- An alkane can be functionalized via radical _____.
- Every synthesis problem should be approached by asking the following two questions:
    1. Is there any change in the _____?
    2. Is there any change in the identity or location of the _____?
- In a _____ **analysis**, the last step of the synthetic route is first established, and the remaining steps are determined, working backwards from the product.

Review of Skills

Fill in the blanks and empty boxes below. To verify that your answers are correct, look in your textbook at the end of Chapter 12. The answers appear in the section entitled *SkillBuilder Review*.

12.1 Changing the Identity or Position of a Functional Group

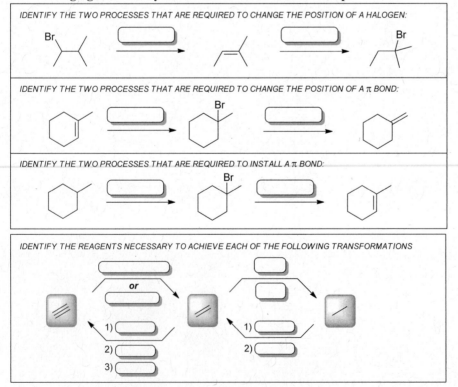

12.2   Changing the Carbon Skeleton

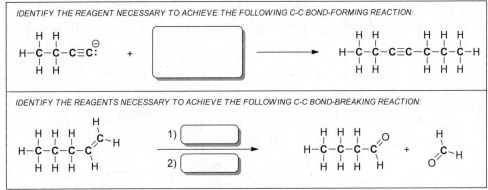

12.3   Approaching a Synthesis Problem by Asking Two Questions

IDENTIFY THE TWO QUESTIONS TO ASK WHEN APPROACHING A SYNTHESIS PROBLEM:

1) _____*?*

2) _____*?*

12.4   Retrosynthetic Analysis

COMPLETE THE FOLLOWING RETROSYNTHETIC ANALYSIS BY DRAWING THE APPROPRIATE STRUCTURES IN THE BOXES PROVIDED:

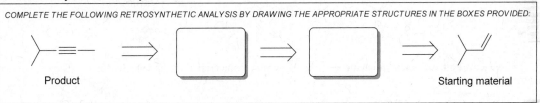

Useful Reagents:

This chapter does not cover any new reactions. As such, there are no new reagents in this chapter. The problems in this chapter require the use of the reagents covered in previous chapters (specifically, Chapter 3 and Chapters 7-11). For each of those chapters, a summary of reagents can be found in the corresponding chapters of this solutions manual.

Common Mistakes to Avoid

When proposing a synthesis, avoid drawing curved arrows (unless the problem statement asks you to draw a mechanism). So often, students will begin drawing mechanism, rather than a synthesis, when asked to propose a synthesis. If you look through all of the solutions to the problems in this chapter, you will find that none of the solutions exhibit curved arrows.

Also, avoid using steps for which you have no control over the regiochemical outcome or the stereochemical outcome. For example, acid-catalyzed hydration of the following alkyne will produce two different ketones.

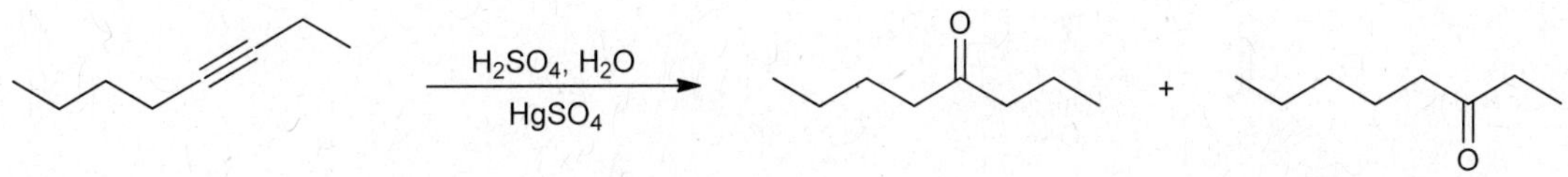

This process should not be used if only one of these ketones is desired.

Solutions

12.1. The reagents for these reactions can be found in the summary material at the end of Chapter 9. They are shown again here:

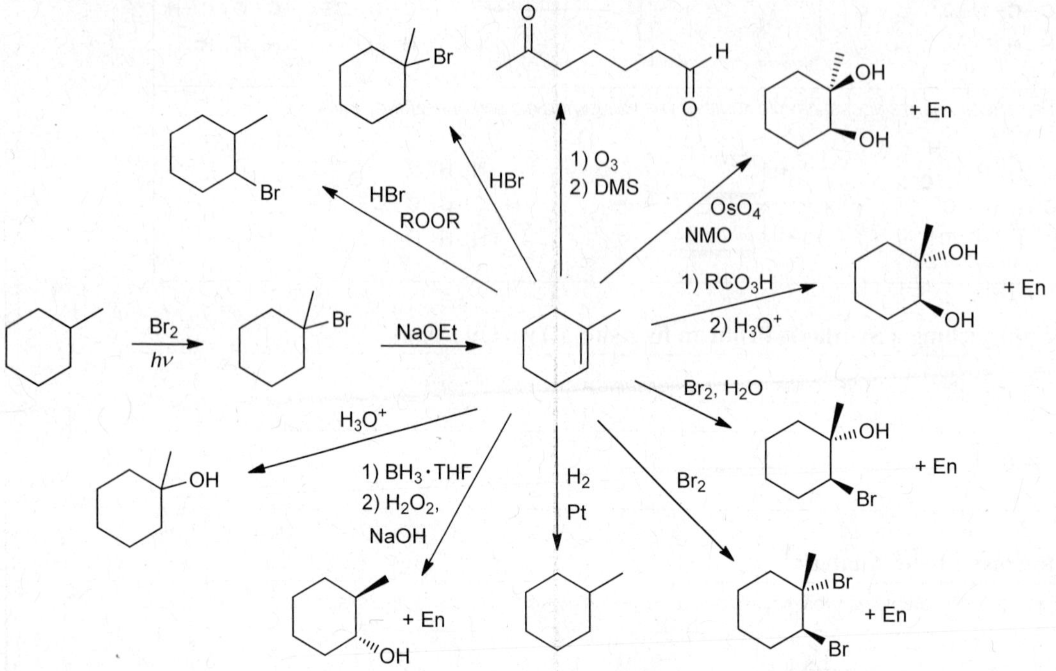

12.2. The reagents for these reactions can be found in the summary material at the end of Chapter 10. They are shown again here:

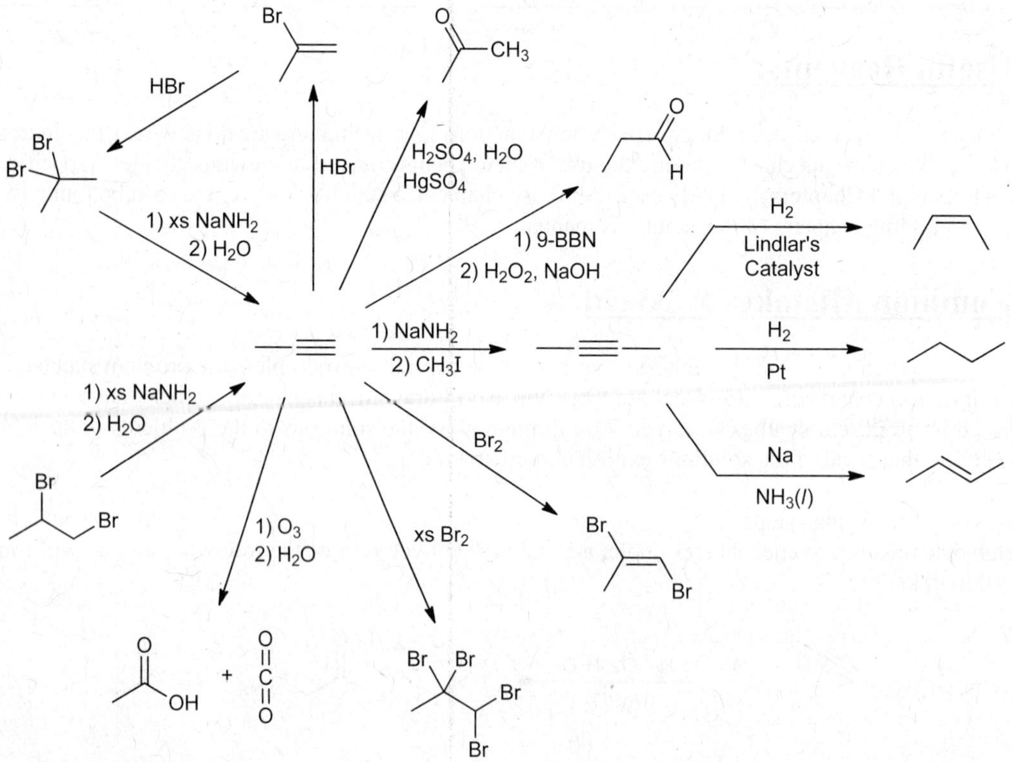

12.3.

(a) We begin by analyzing the identity and location of the functional group in both the starting material and the product:

Starting material   Product

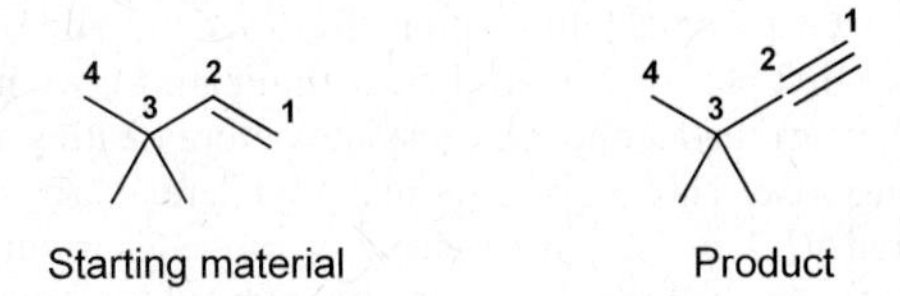

The identity of the functional group has changed (double bond → triple bond), but its location has not changed (it remains between C1 and C2). The conversion of a double bond to a triple bond can be accomplished via the following two-step process (bromination, followed by elimination):

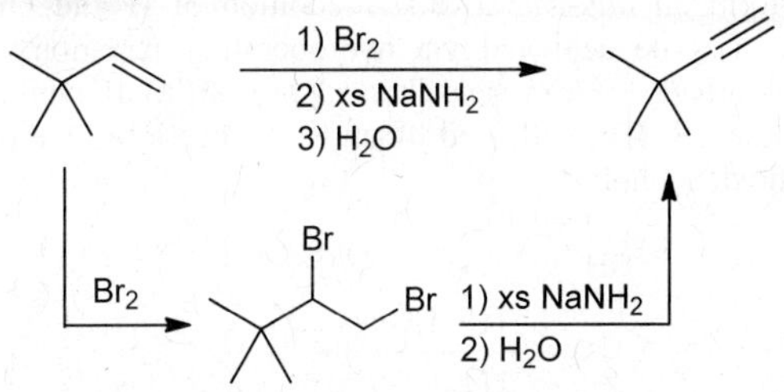

(b) We begin by analyzing the identity and location of the functional group in both the starting material and the product. The identity of the functional group (Br) has not changed, but its location has changed. We have seen that this transformation can be accomplished via a two-step process (elimination, followed by addition). For the elimination process, there is only one possible regiochemical outcome; nevertheless, *tert*-butoxide is used as the base (rather than hydroxide, methoxide, or ethoxide) in order to suppress the S_N2 reaction, which will otherwise dominate for a primary bromide. For the next step of our synthesis, Markovnikov addition is required, so we use HBr without peroxides.

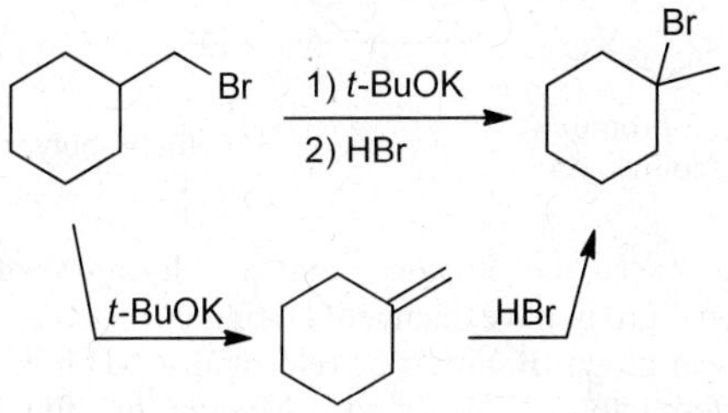

(c) We begin by analyzing the identity and location of the functional group in both the starting material and the product. The identity of the functional group (a π bond) has not changed, but its location has changed. We have seen that this transformation can be accomplished via a two-step process (addition, followed by elimination). For the first step of our synthesis, Markovnikov addition is required, so we use HBr without peroxides. For the next step of our synthesis, the more substituted alkene is desired (trisubstituted rather than disubstituted), so the

base cannot be sterically hindered. Appropriate choices include hydroxide, methoxide, and ethoxide.

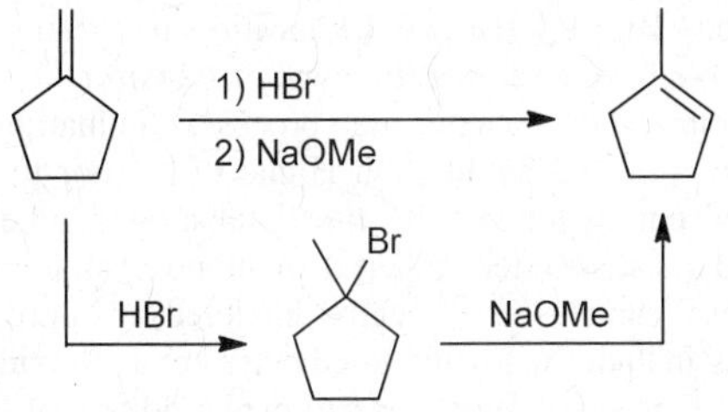

(d) We begin by analyzing the identity and location of the functional group in both the starting material and the product. In doing so, we immediately realize that the starting material has no functional group (it is an alkane), so the first step of our synthesis must be the installation of a functional group. This can be accomplished via radical bromination, which selectively installs a bromine atom at the tertiary position. The resulting tertiary alkyl halide can then be converted into the product via an elimination process. The more substituted alkene is desired (trisubstituted rather than disubstituted), so the base cannot be sterically hindered. Appropriate choices include hydroxide, methoxide, and ethoxide.

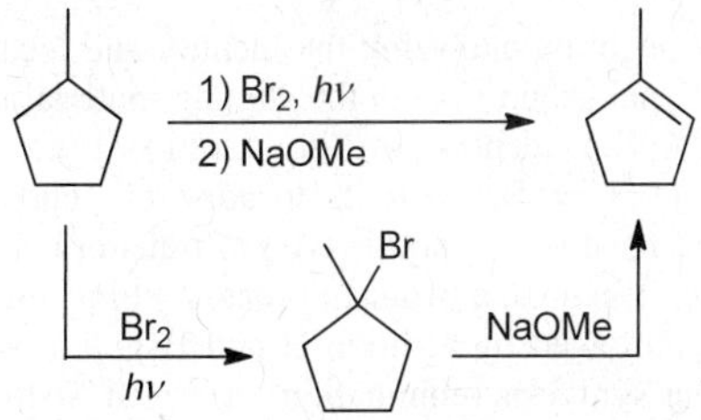

(e) The identity of the functional group (OH) has not changed, but its location has changed. We have seen that this transformation can be accomplished via a two-step process (elimination, followed by addition). Elimination can be achieved either by heating the starting alcohol with concentrated sulfuric acid (E1), or by converting the alcohol to a tosylate followed by treatment with a strong base (E2). The resulting alkene can then be converted to the desired product via an addition process. Specifically, a hydration reaction must be performed (addition of H and OH) in a Markovnikov fashion. This can be achieved by treating the alkene with dilute acid (acid-catalyzed hydration):

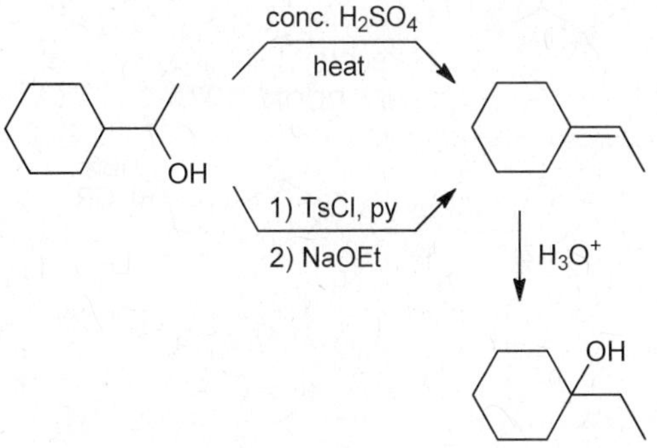

(f) We begin by analyzing the identity and location of the functional group in both the starting material and the product. The identity of the functional group has changed (Br → OH), and its location has changed as well. We have seen that this type of transformation can be accomplished via a two-step process (elimination of H and Br, followed by addition H and OH). For the first step of our synthesis, the more substituted alkene is desired (trisubstituted rather than monosubstituted), so the base cannot be sterically hindered. Appropriate choices include hydroxide, methoxide, and ethoxide. A hydration reaction must be performed (addition of H and OH) in a Markovnikov fashion. This can be achieved by treating the alkene with dilute acid (acid-catalyzed hydration):

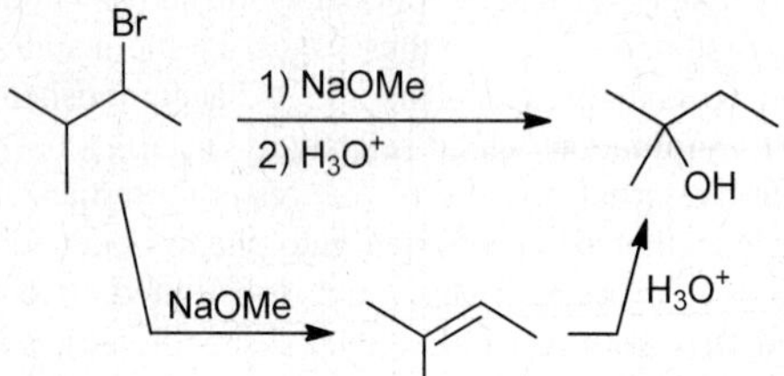

(g) We begin by analyzing the identity and location of the functional group in both the starting material and the product. The identity of the functional group has changed (OH → Br), and its location has changed as well. We have seen that this type of transformation can be accomplished via a two-step process (elimination of H and OH, followed by addition H and Br). For the first part of our synthesis (elimination), we must first convert the alcohol to a tosylate before treating with a strong base. The use of a sterically hindered base is not required (E2 = major product; while S_N2 = minor product). Nevertheless, a sterically hindered base will be helpful here as it will suppress the competing S_N2 reaction. So *tert*-butoxide is used in the synthesis below. Note that an E1 process should be avoided, because heating the alcohol with concentrated sulfuric acid will likely involve a carbocation rearrangement (methyl shift). For the second part of our synthesis, we must perform an *anti*-Markovnikov addition of H and Br, so we use HBr in the presence of peroxides.

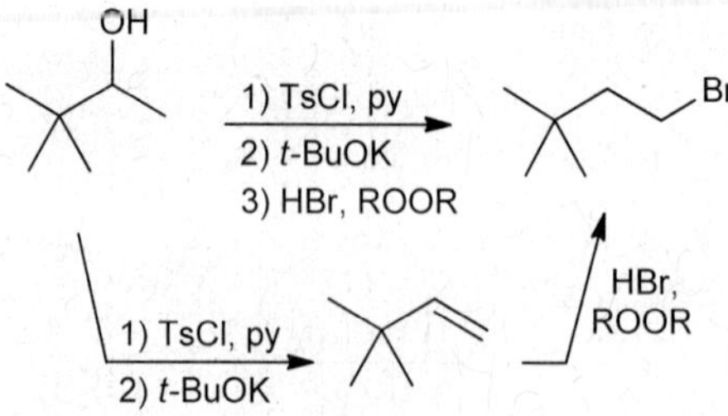

(h) We begin by analyzing the identity and location of the functional group in both the starting material and the product. The identity of the functional group (OH) has not changed, but its location has changed. We have seen that this type of transformation can be accomplished via a two-step process (elimination, followed by addition). For the first part of our synthesis (elimination), we must first convert the alcohol to a tosylate before treating with a strong base. The use of a sterically hindered base is not required (E2 = major product; while S_N2 = minor product). Nevertheless, a sterically hindered base will be helpful here as it will suppress the competing S_N2 reaction. So *tert*-butoxide is used in the synthesis below. Note that an E1 process should be avoided, because heating the alcohol with concentrated sulfuric acid will likely involve a carbocation rearrangement (methyl shift). For the second part of our synthesis, we must perform an *anti*-Markovnikov addition of H and OH. This can be achieved via hydroboration-oxidation, or alternatively, it can be achieved by *anti*-Markovnikov addition of HBr followed by an S_N2 process to give the desired alcohol:

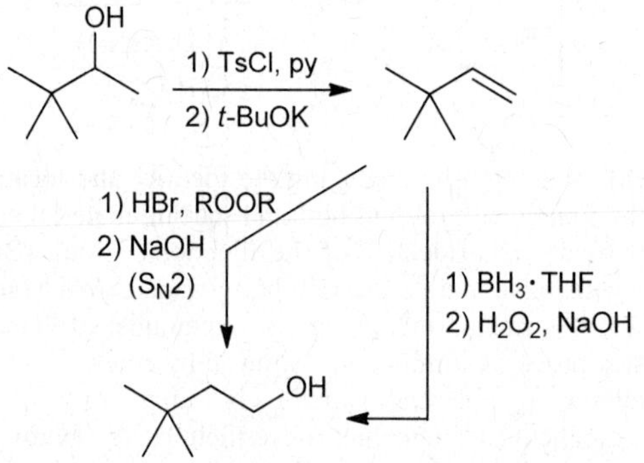

12.4. Let's begin by drawing the starting material and the product, so that we can see the desired transformation more clearly:

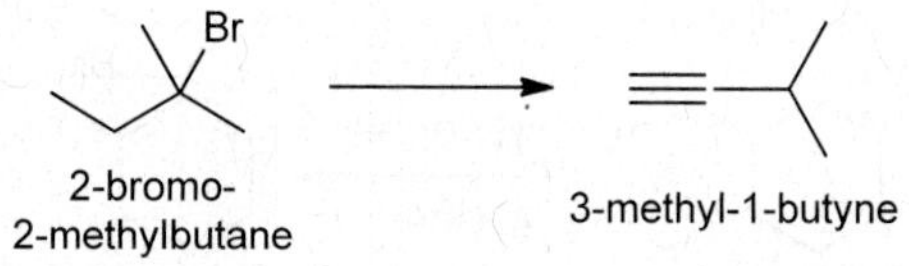

In this case, the identity and the location of the functional group have changed. During our synthesis, the functional group (Br) must be relocated, AND it must be converted into a triple bond. Moving the functional group can be achieved via elimination (to give the Zaitsev product) followed by addition (in an *anti*-Markovnikov fashion).

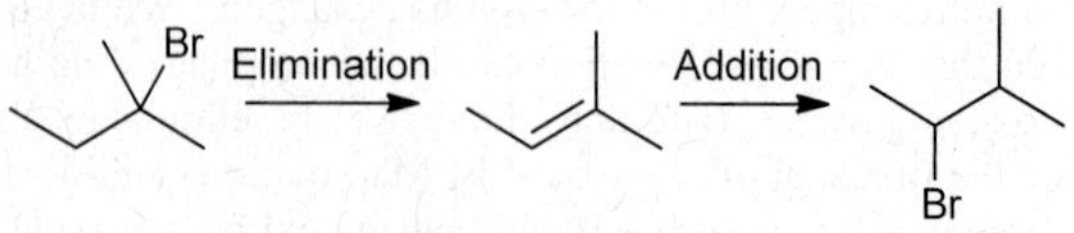

Then, changing the identity of the functional group can be achieved via the following three-step process:

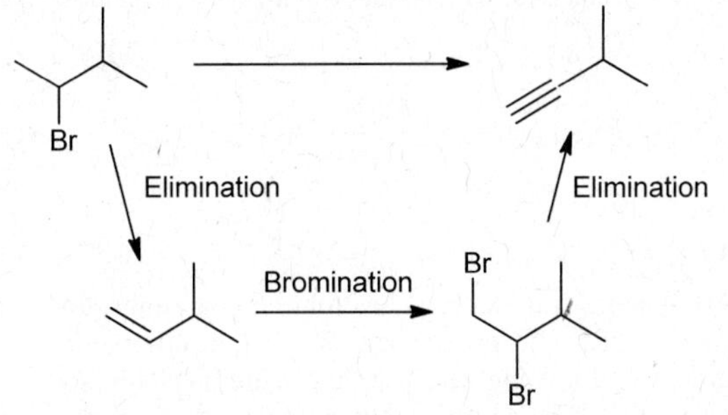

The necessary reagents are shown here:

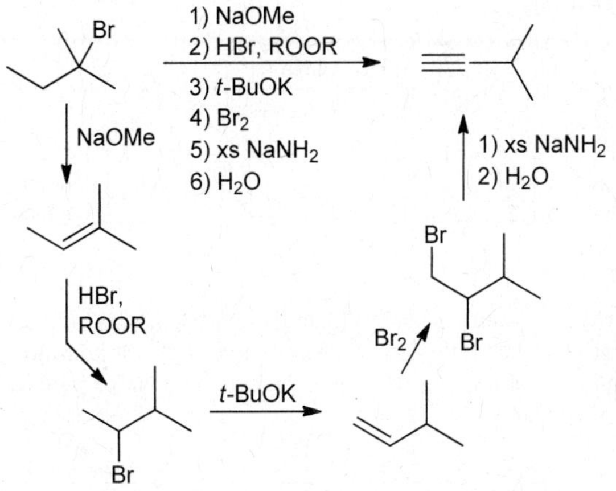

12.5. We have not seen a way to convert an *alkene* directly into a geminal dihalide. But we have seen a way to convert an *alkyne* into a geminal dihalide. So, we must convert the starting alkene into an alkyne. This can be accomplished via bromination, followed by elimination with excess sodium amide. The resulting alkyne then undergoes two successive addition reactions upon treatment with excess HBr to give a geminal dihalide.

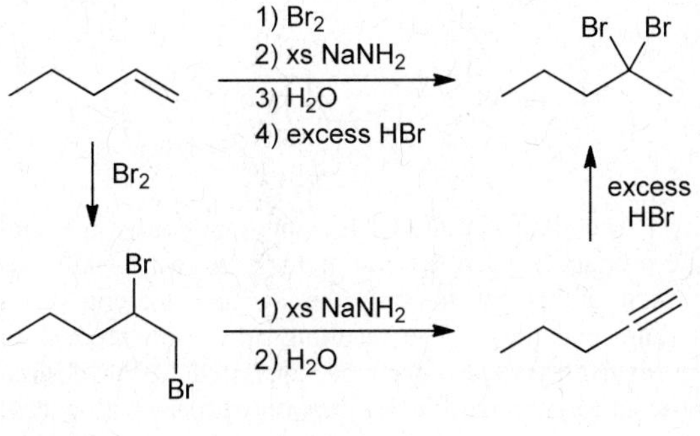

12.6.
(a) Radical bromination of 1-methylcyclohexane will result in the following tertiary alkyl bromide:

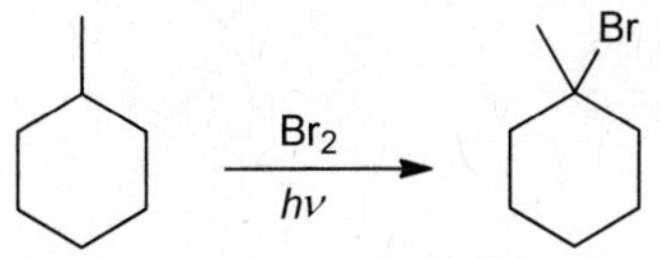

(b) Alkenes can be formed via elimination reactions, so we will need a compound with a leaving group, such as an alkyl halide, in order to make the desired product. If we use the tertiary alkyl halide from the previous exercise (**12.6a**), we can make the desired cycloalkene via an E2 process. A strong, non-hindered base should be used, such as hydroxide, methoxide, or ethoxide.

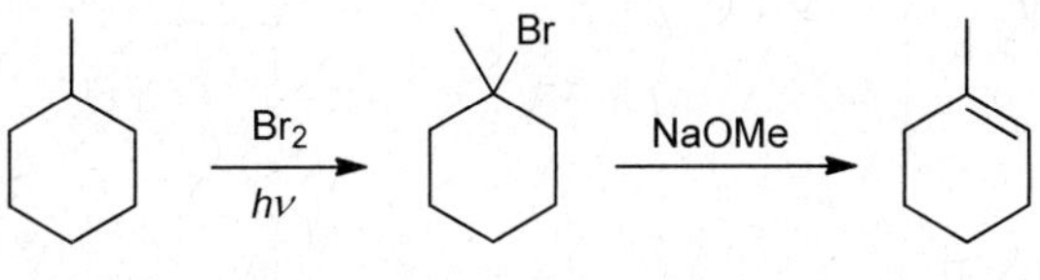

(c) In exercise **12.6b**, we saw that methylcyclohexane can be converted to 1-methyl-cyclohexene via a two step process:

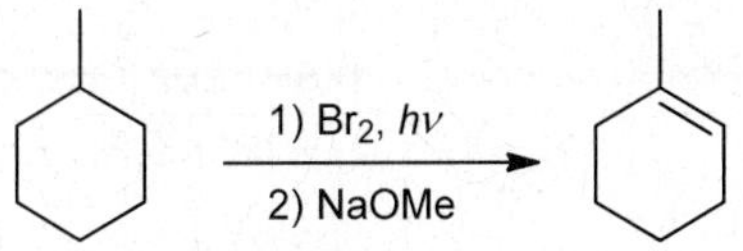

This cycloalkene can be converted into the product by moving the position of the π bond. We have seen that this can be accomplished via a two-step process (addition, followed by elimination):

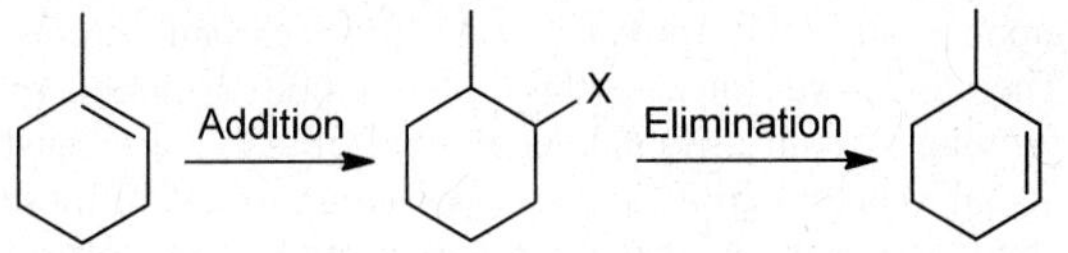

During the addition step, we must use a reaction that gives an *anti*-Markovnikov addition (such as hydrobromination in the presence of peroxides). During the elimination step, we need the Hofmann product (less substituted alkene), so we use *tert*-butoxide as the base. The entire synthesis is shown here:

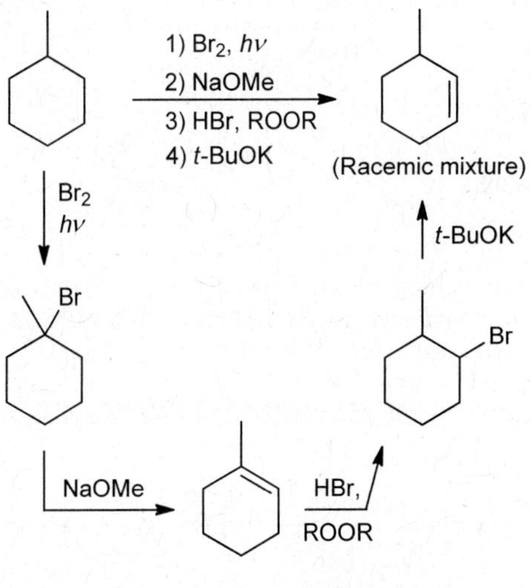

Alternatively, the end of our synthesis (moving the position of the π bond) can be accomplished via hydroboration-oxidation, followed by conversion of the resulting alcohol to a tosylate, followed by an elimination reaction:

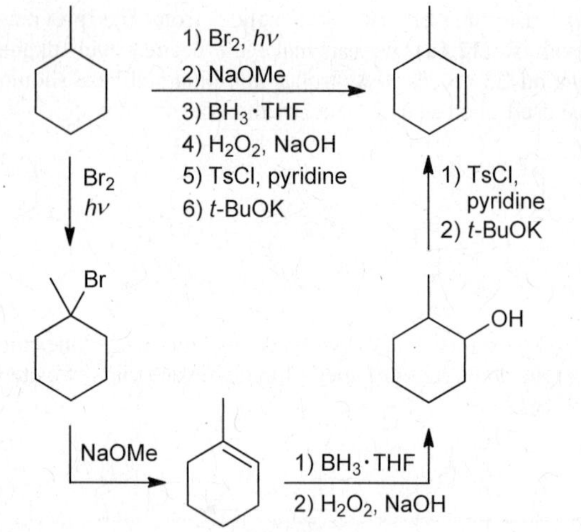

This alternate answer illustrates a very important point. *With synthesis problems, there are often multiple acceptable answers.*

12.7.
(a) The starting material (acetylene) has two carbon atoms, and the product has five carbon atoms. Therefore, we must change the carbon skeleton by forming carbon-carbon bonds. Specifically, we must install a methyl group and an ethyl group. Each of these alkylation reactions can be achieved upon treatment of the alkyne with sodium amide followed by the appropriate alkyl halide. Notice that each alkyl group must be installed separately, and it does not matter which alkyl group is installed first and which is installed second.

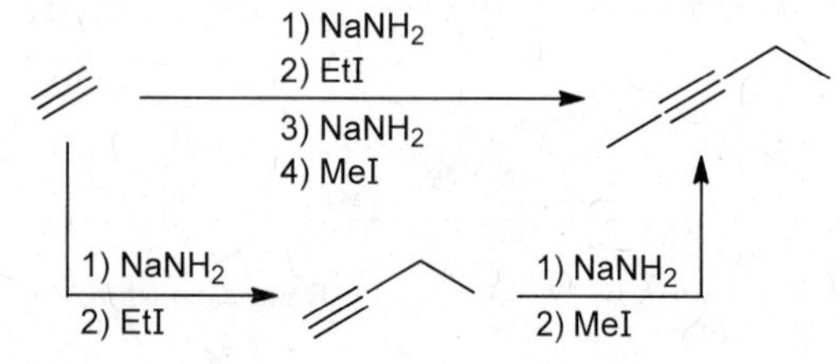

(b) The starting material (benzyl bromide) has seven carbon atoms, and the product has nine carbon atoms. Therefore, we must install two carbon atoms. This alkylation process can be achieved in one step, by treating the starting material with sodium acetylide:

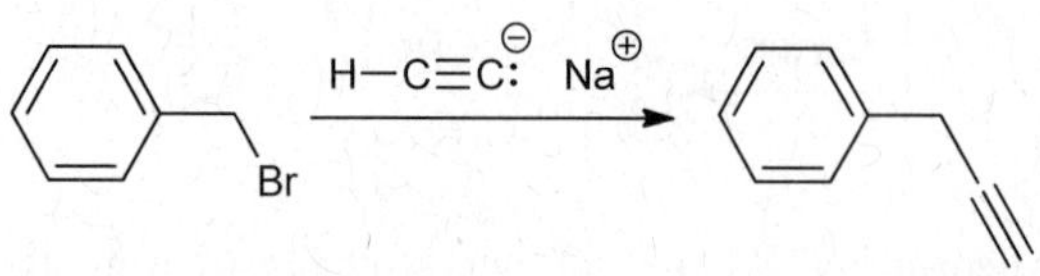

(c) The starting material has eight carbon atoms, and the product has only seven carbon atoms. One carbon atom must be removed, which can be achieved via ozonolysis:

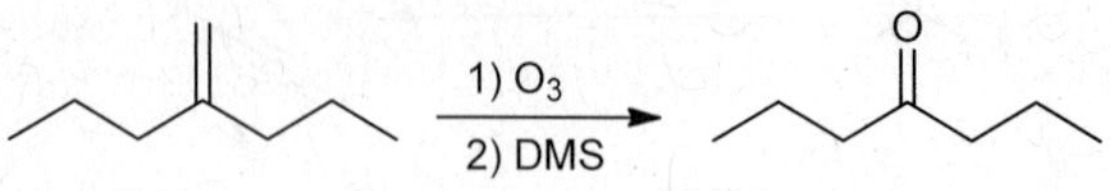

12.8.
(a) The starting material has four carbon atoms, and the product has six carbon atoms. We can install two carbon atoms by treating the starting material with sodium acetylide. The resulting alkylation product (a terminal alkyne) can then be reduced to an alkene via hydrogenation with a poisoned catalyst, such as Lindlar's catalyst:

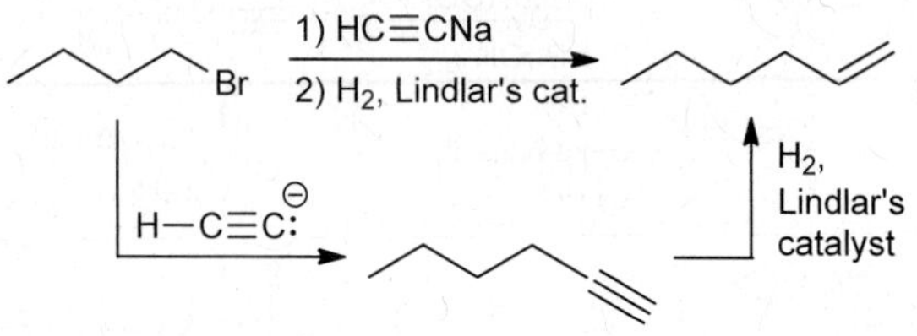

(b) The starting material has five carbon atoms, and the product has only four carbon atoms. One carbon atom must be removed, which can be achieved via ozonolysis of an alkene:

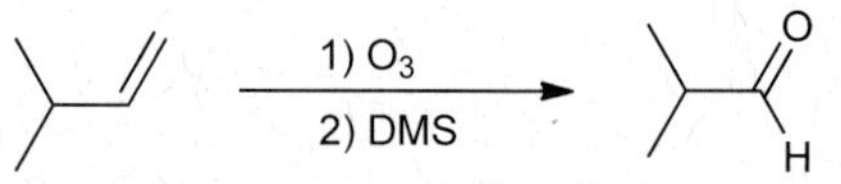

The alkene can be prepared from the starting material via an elimination reaction with a sterically hindered base (giving the less substituted alkene):

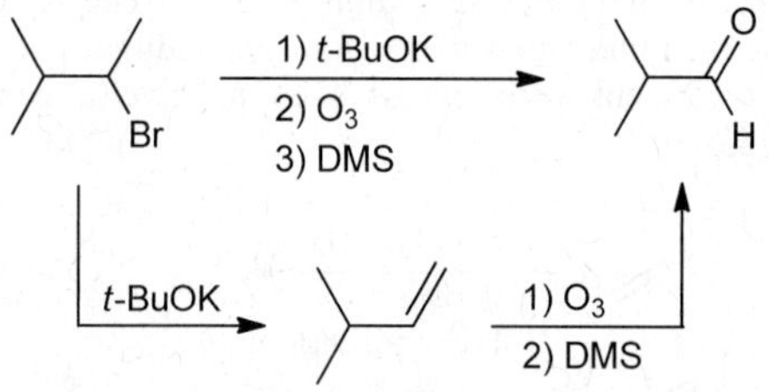

(c) The starting material has three carbon atoms, and the product has five carbon atoms. We can install two carbon atoms by treating the starting material with sodium acetylide. The resulting alkylation product (a terminal alkyne) can then be converted to the desired product (a geminal dihalide) upon treatment with excess HBr:

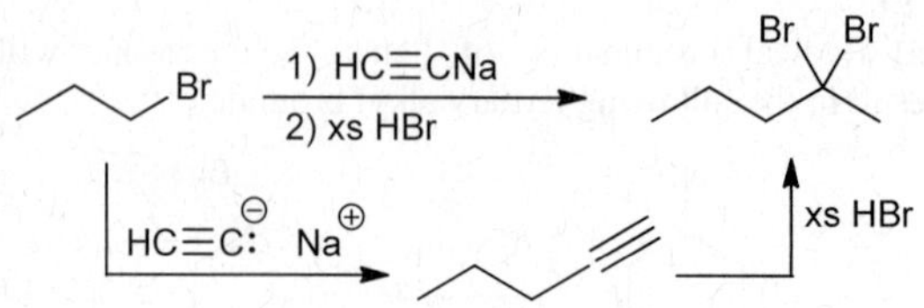

12.9. The starting material (3-bromo-3-ethylpentane) is a tertiary substrate and will not readily undergo an S_N2 reaction. Under these conditions, the acetylide ion functions as a base, rather than a nucleophile, giving an E2 reaction, instead of S_N2. The product is an alkene, while the base (acetylide) is protonated to give acetylene as a byproduct:

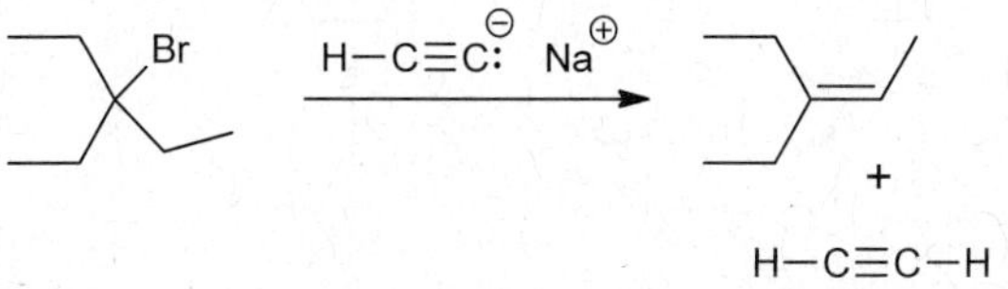

12.10.

(a) The starting material has five carbon atoms, and the product has eight carbon atoms. So our synthesis must involve the installation of three carbon atoms. In addition, the identity of the functional group must be changed (from a triple bond to a double bond). Reduction of the alkyne (to give an alkene) must be the last step of our synthesis, because if we first reduce the triple bond, then we would not be able to perform the alkylation step. The alkylation step must be performed first, followed by reduction of the alkyne (with a poisoned catalyst), as shown here:

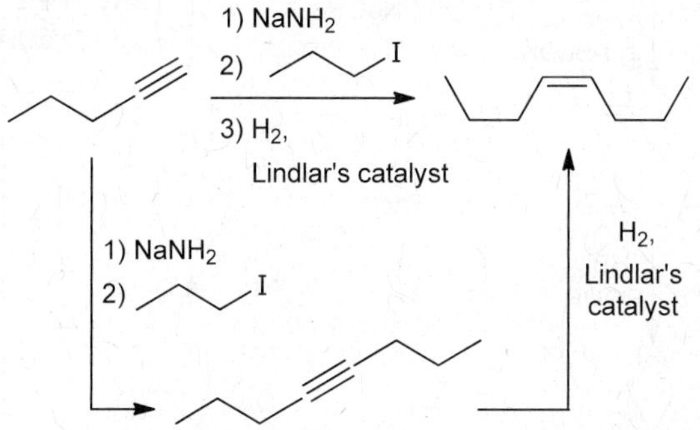

(b) The starting material has five carbon atoms, and the product has seven carbon atoms. So our synthesis must involve the installation of two carbon atoms. If we simply alkylate the starting alkyne (by treating with sodium amide followed by ethyl iodide), then we will need to move the location of the triple bond. Instead, we can convert the starting alkyne into an alkyl halide (via hydrogenation with a poisoned catalyst followed by an *anti*-Markovnikov addition of HBr). If this alkyl halide is treated with acetylide, the desired product is formed:

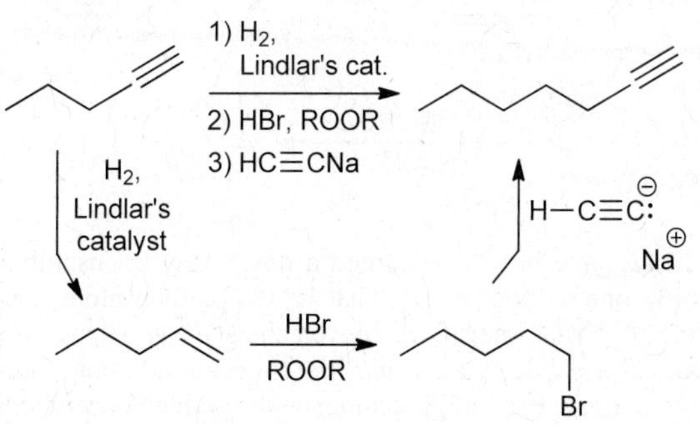

(c) The starting material has one more carbon atom than the product. Therefore, our synthesis must employ an ozonolysis process. Since the product is a carboxylic acid (rather than an aldehyde or ketone), we can conclude that the last step of our process must involve ozonolysis of an *alkyne*, rather than ozonolysis of an *alkene*:

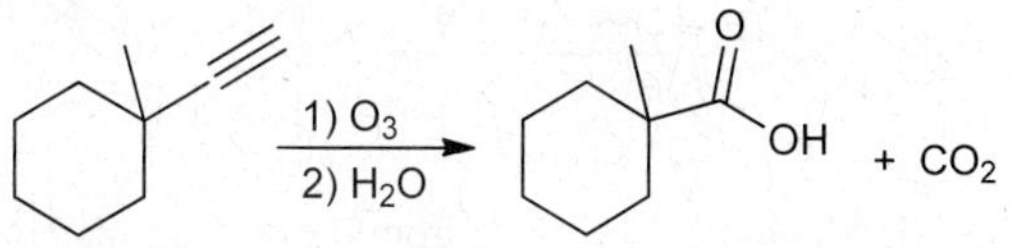

This alkyne can be prepared directly from the starting material upon treatment with excess sodium amide (followed by water workup). The complete synthesis is shown here:

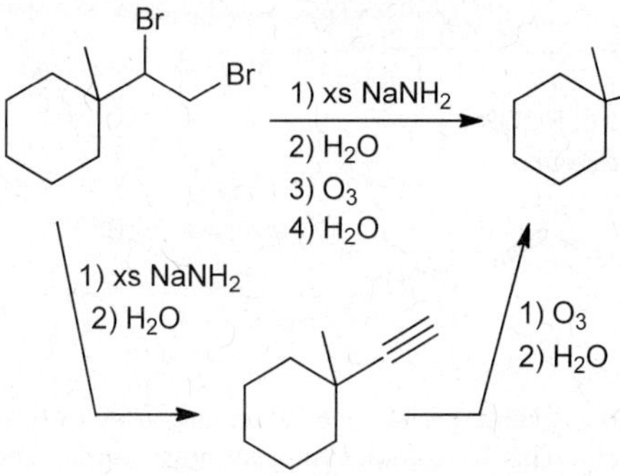

(d) The starting material has six carbon atoms, and the product has nine carbon atoms. So our synthesis must involve the installation of three carbon atoms. Also, the location of the functional group has been changed. The product is a *trans* alkene, which can be made if the last step of our synthesis is a dissolving metal reduction:

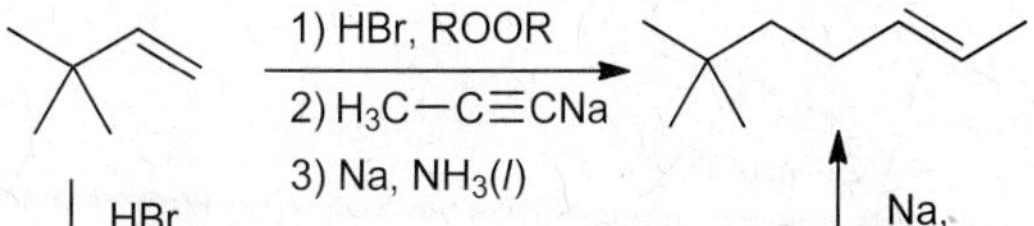

This alkyne can be made from the starting alkene via an *anti*-Markovnikov addition of HBr, followed by treatment with the appropriate alkynide ion, as shown here:

(e) The product has two more carbon atoms than the starting material. The installation of two carbon atoms can be achieved via an alkylation process. In fact, the desired product is a terminal alkyne, which can be made from the following alkyl halide in just one step:

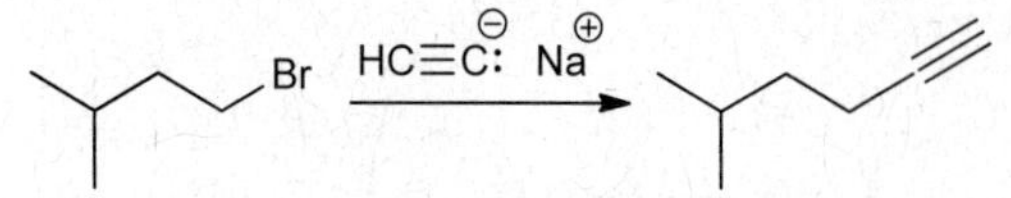

This alkyl halide can be made from the starting material by first moving the location of the π bond, followed by *anti*-Markovnikov addition:

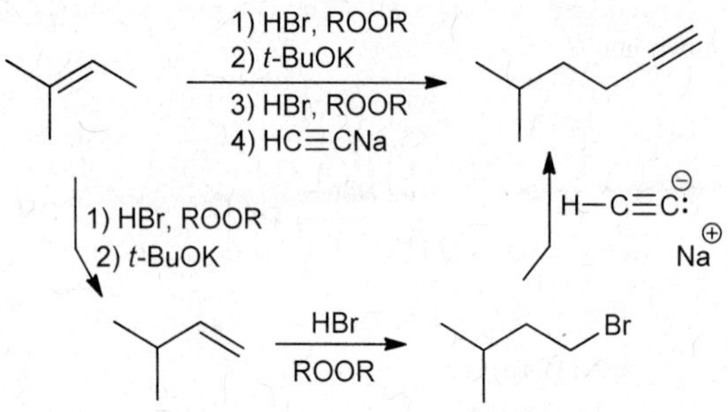

(f) The starting material has one more carbon atom than the product. Therefore, our synthesis must employ an ozonolysis process. Since the product is an aldehyde, we can conclude that the last step of our process must involve ozonolysis of an alkene:

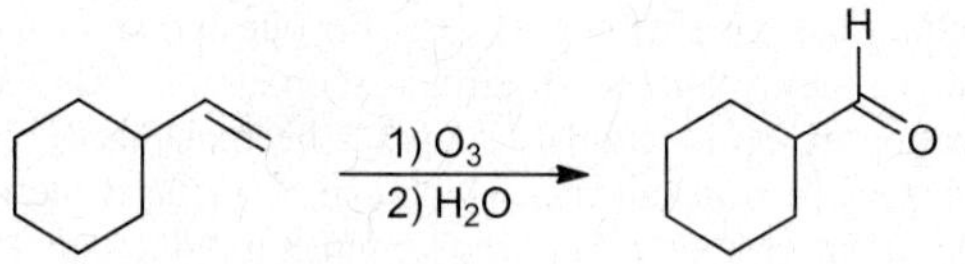

This alkene can be prepared from the starting material by converting the alcohol to a tosylate, and then performing an E2 reaction with a sterically hindered base (giving the less substituted alkene):

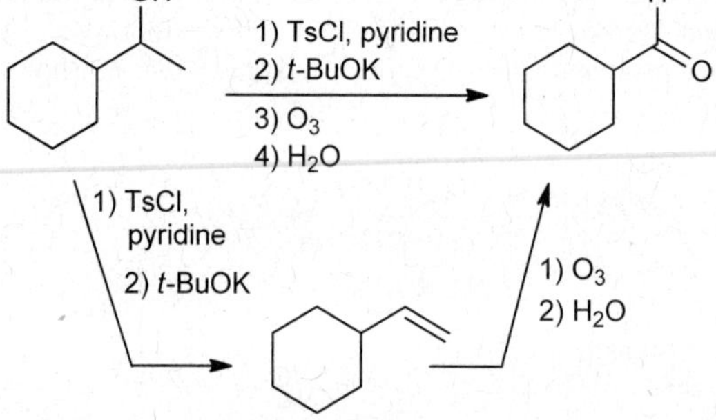

bromide, performing an alkylation process and then converting the triple bond into the desired alcohol:

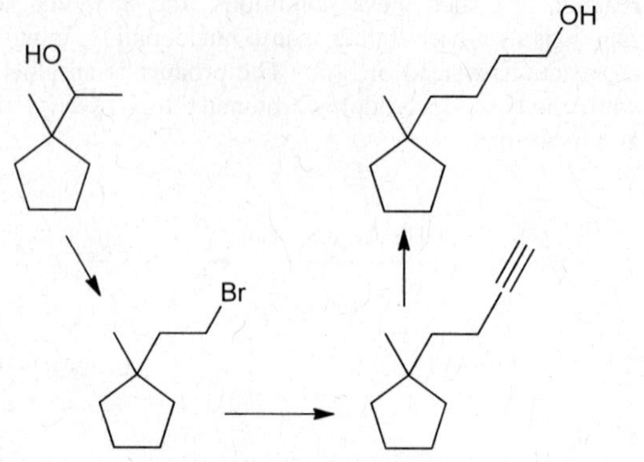

The following reagents can be used to achieve the desired transformations:

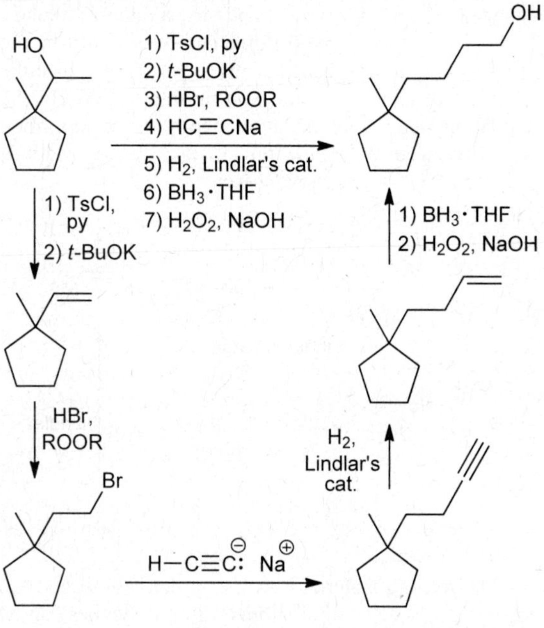

It should be noted that there are other acceptable answers. As one example, the last part of our synthesis (hydroboration-oxidation) could be replaced with *anti*-Markovnikov addition of HBr to give a primary alkyl bromide, followed by an S$_N$2 process with hydroxide as the nucleophile:

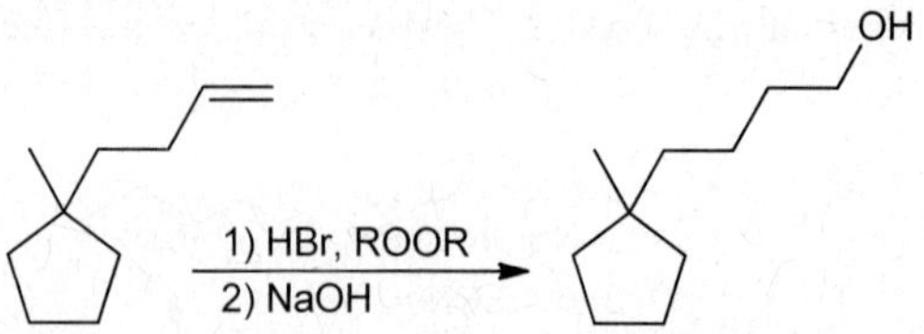

12.11. The desired transformation involves the installation of two carbon atoms, as well as a change in the location of the functional group. This can be achieved by converting the alcohol into a primary alkyl

12.12. We have not learned a direct way of installing only one carbon atom. That is, two carbon atoms are installed (not one) if we convert the starting alkene into an alkyl halide (via an *anti*-Markovnikov addition), and then treat the alkyl halide with sodium acetylide.

However, after installing two carbon atoms, we can remove one of them with an ozonolysis procedure. In order to obtain an aldehyde, the last step must be ozonolysis of an alkene, so the second-to-last step (called the penultimate step) must be reduction of the alkyne to an alkene in the presence of a poisoned catalyst:

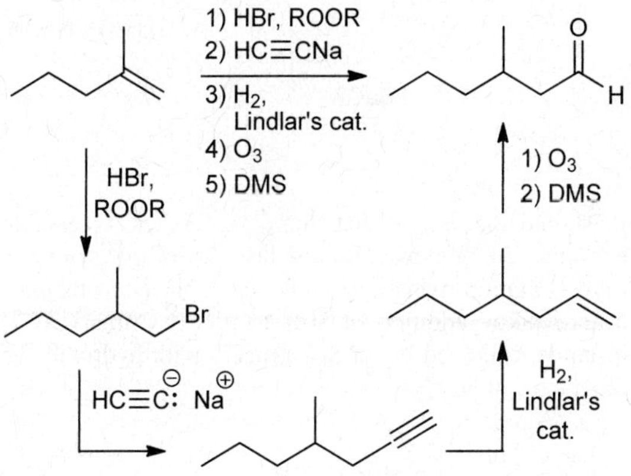

12.13.
(a) The product is a halohydrin, which can be made from an alkene:

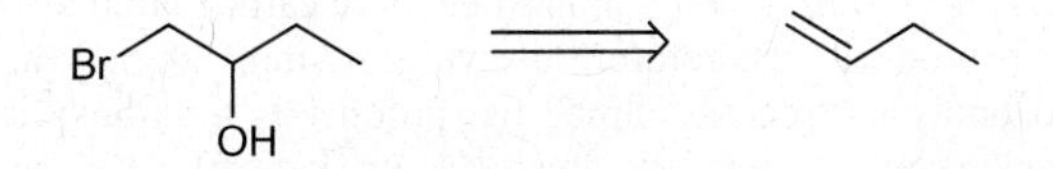

So the last step of our synthesis will likely be conversion of the alkene into the halohydrin.

Now let's work forward from the starting material. The starting material has only two carbon atoms, while the product has four carbon atoms, so two carbon atoms must be installed. This can be achieved via an alkylation process:

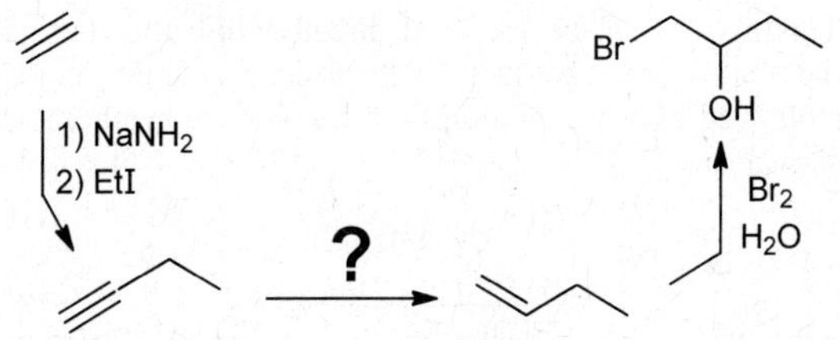

Now we must bridge the gap (between the terminal alkyne and the alkene). This can be achieved in one step, via hydrogenation with a poisoned catalyst. In summary, the desired transformation can be achieved with the following synthesis:

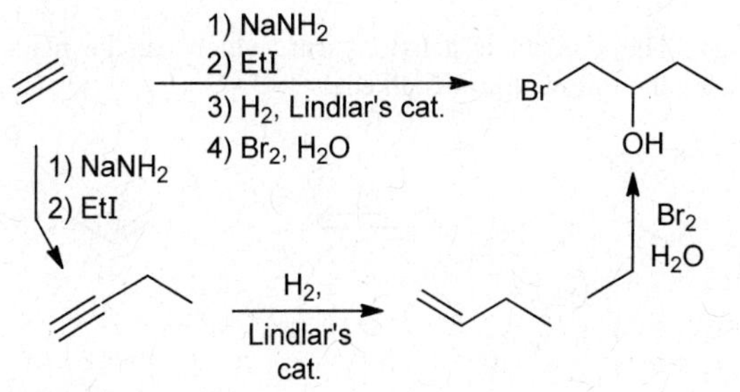

(b) The product is an epoxide, which can be made from an alkene:

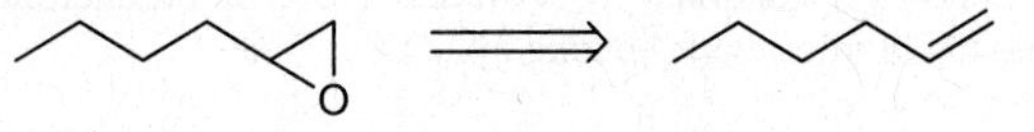

So the last step of our synthesis will likely be conversion of the alkene into the epoxide.

Now let's work forward from the starting material. The starting material has only four carbon atoms, while the product has six carbon atoms, so two carbon atoms must be installed. This can be achieved via an alkylation process:

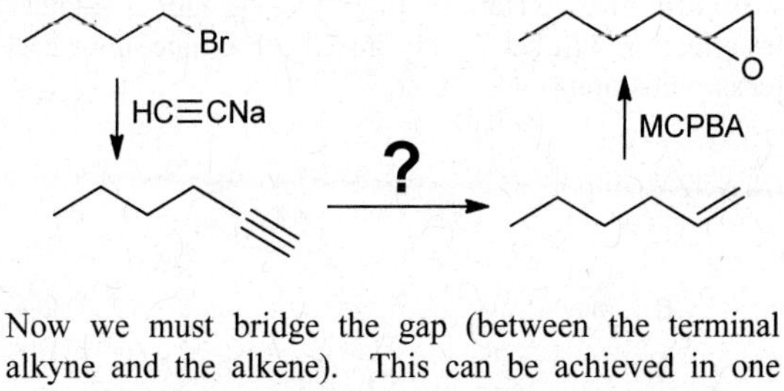

Now we must bridge the gap (between the terminal alkyne and the alkene). This can be achieved in one step, via hydrogenation with a poisoned catalyst. In summary, the desired transformation can be achieved with the following synthesis:

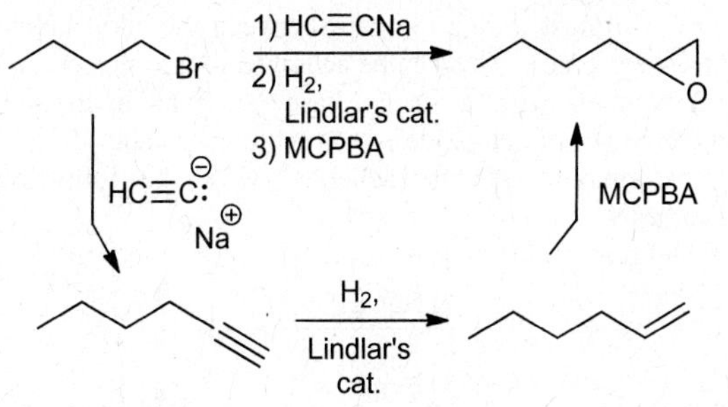

(c) The product is a methyl ketone, and we have seen that a methyl ketone can be prepared from an alkyne (via acid-catalyzed hydration):

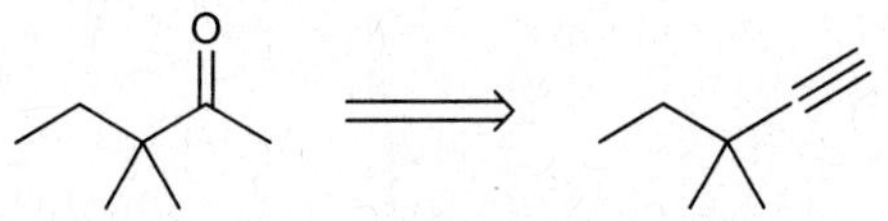

This alkyne can be prepared directly from the starting material upon treatment with excess sodium amide (followed by water workup). In summary, the desired transformation can be achieved with the following synthesis:

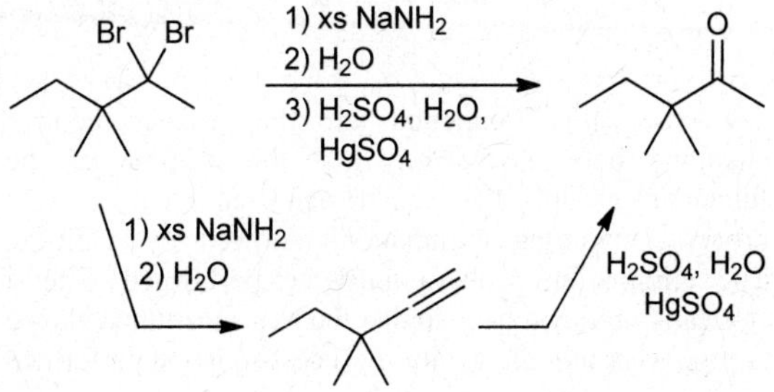

(d) The starting material has six carbon atoms and the product has three carbon atoms, indicating that an ozonolysis reaction will be necessary. The product can be made from the following alkene:

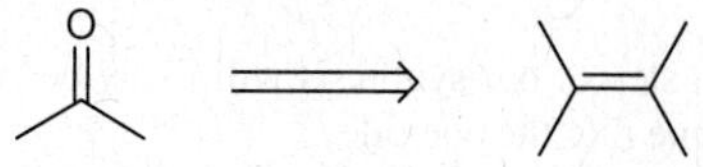

So the last step of our synthesis will likely be ozonolysis of this alkene.

Now let's work forward from the starting material. The starting material is an alkane (no functional group), so we must first install a functional group. Radical bromination will selectively install a bromine atom at a tertiary position:

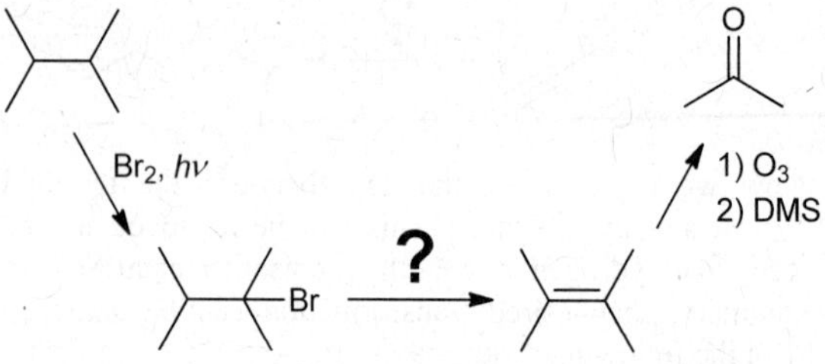

Now we must bridge the gap (between the alkyl halide and the alkene). This can be achieved in one step, via an E2 reaction with a strong base (such as hydroxide, methoxide, or ethoxide). In summary, the desired transformation can be achieved with the following synthesis:

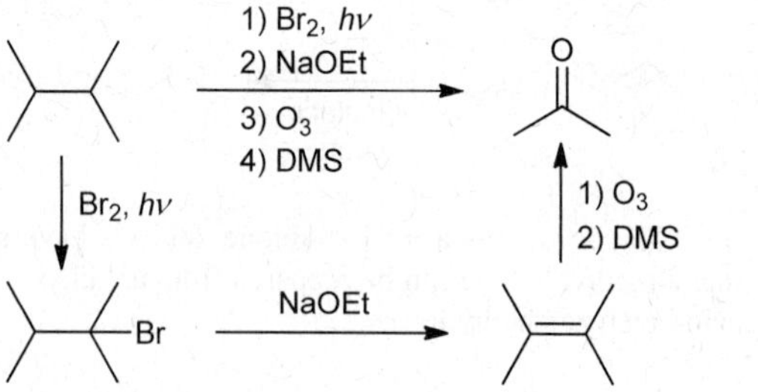

(e) The starting material is an alkane (no functional group), so we must first install a functional group. Radical bromination will selectively install a bromine atom at a tertiary position:

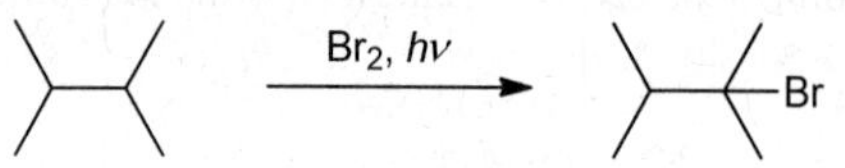

Converting this compound into the product requires changing both the identity and the location of the functional group. This can be achieved via a two-step process, involving elimination followed by addition. The elimination process must be performed with a sterically hindered base so that the less substituted alkene is produced, and the addition process must be performed

in an *anti*-Markovnikov fashion (via hydroboration-oxidation):

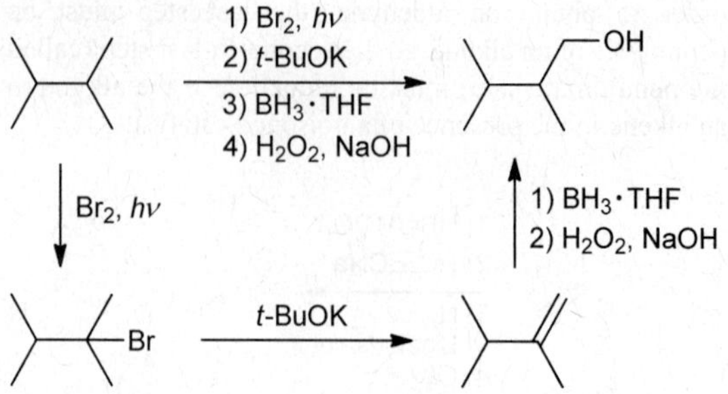

It should be noted that there are other acceptable answers. As one example, the last part of our synthesis (hydroboration-oxidation) could be replaced with *anti*-Markovnikov addition of HBr to give a primary alkyl bromide, followed by an S_N2 process with hydroxide as the nucleophile:

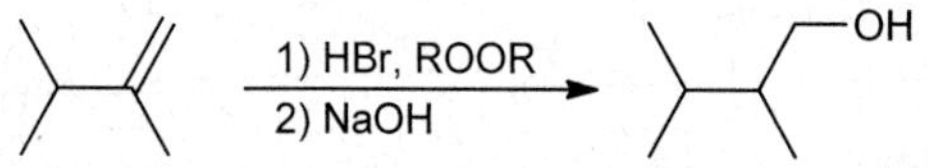

(f) The starting material has one more carbon atom than the product. Therefore, our synthesis must employ an ozonolysis process. Since the product is a carboxylic acid (rather than an aldehyde or ketone), we can conclude that the last step of our process must involve ozonolysis of an *alkyne*, rather than ozonolysis of an *alkene*:

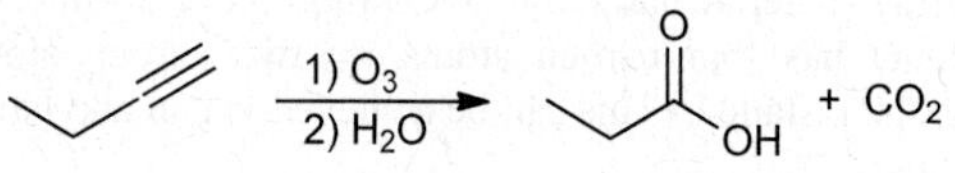

This alkyne can be prepared directly from the starting material upon treatment with excess sodium amide (followed by water workup). The complete synthesis is shown here:

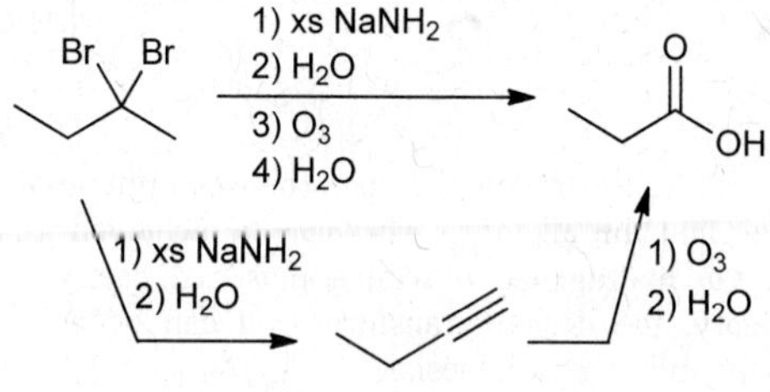

(g) The product is a halohydrin, which can be made from the following *trans* alkene:

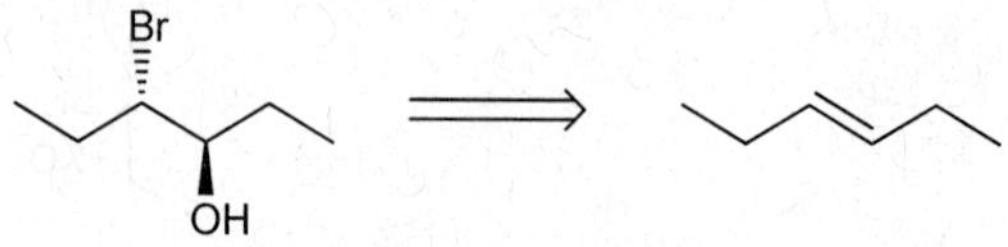

So the last step of our synthesis will likely be conversion of this alkene into the halohydrin.

Now let's work forward from the starting material. The starting material has only two carbon atoms, while the product has six carbon atoms. Two alkylation processes are required:

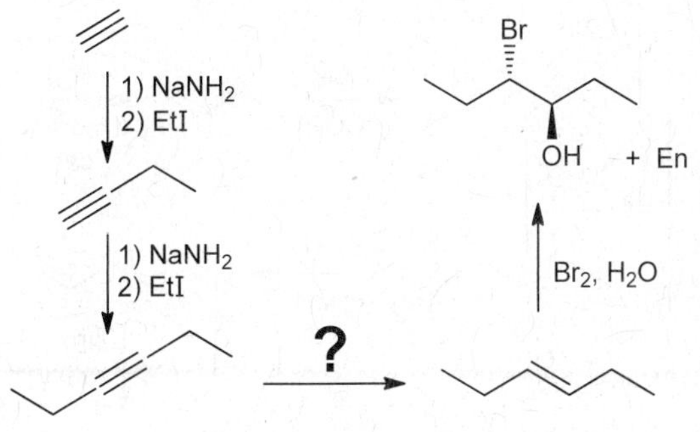

Now we must bridge the gap (between the internal alkyne and the *trans* alkene). This can be achieved in one step, via a dissolving metal reduction. In summary, the desired transformation can be achieved with the following synthesis:

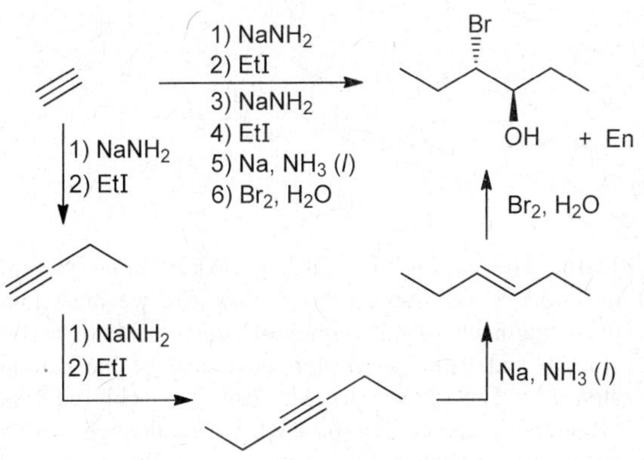

(h) In this case, the carbon skeleton remains the same. The starting material is an alkane (no functional group) so we must begin our synthesis by installing a functional group. Radical bromination will selectively install a bromine atom at the tertiary position, giving the following tertiary alkyl halide:

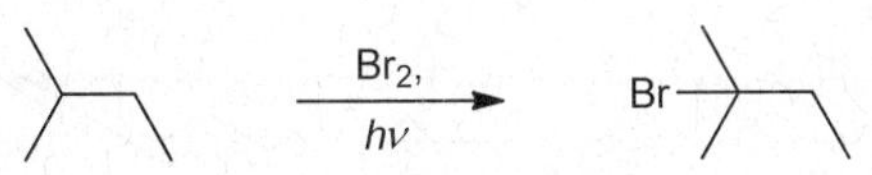

Now we must change both the location and the identity of the functional group. We can move the functional group into the right location through the following series of reactions:

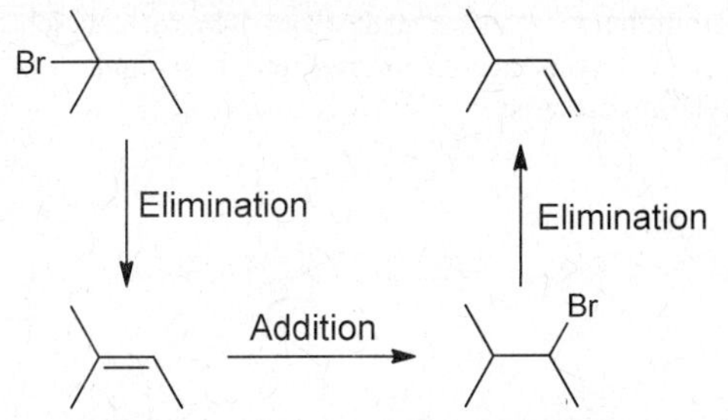

And then finally, the double bond can be converted to the triple bond via the following two-step process:

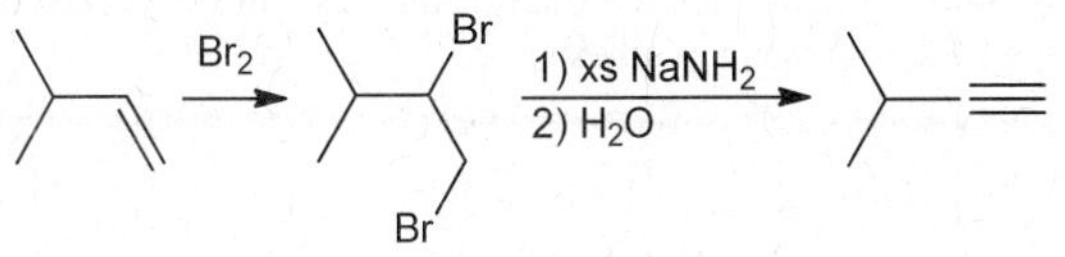

In summary, the desired transformation can be achieved with the following synthesis:

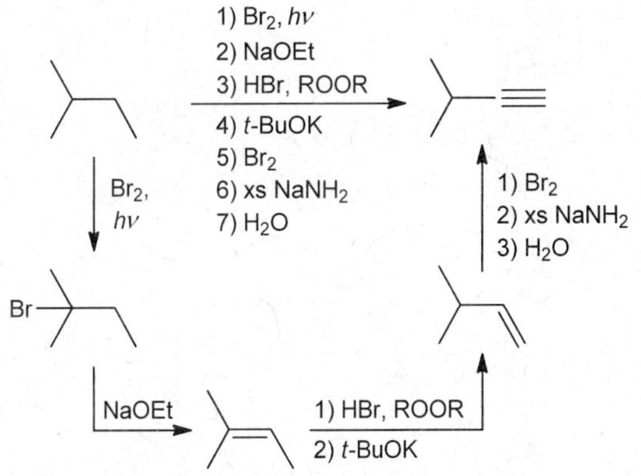

12.14. The product is a *trans* alkene, which can be made from an alkyne. So the last step of our synthesis might be a dissolving metal reduction to convert the alkyne below into the product. This alkyne can be made from acetylene and 1-bromobutane via alkylation processes:

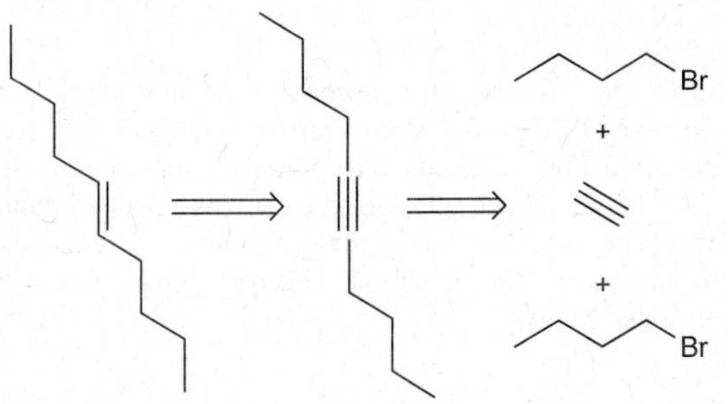

1-Bromobutane can be made from 1-butyne, which can be made from acetylene and ethyl bromide via an alkylation process:

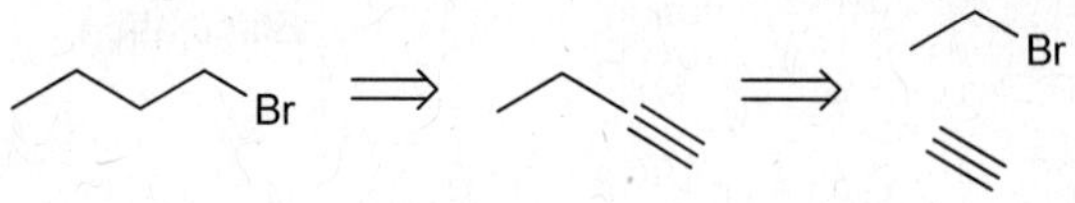

And ethyl bromide can be made from acetylene:

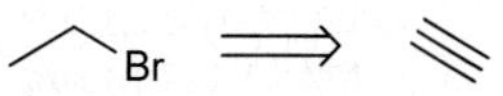

In summary, the desired transformation can be achieved with the following synthesis:

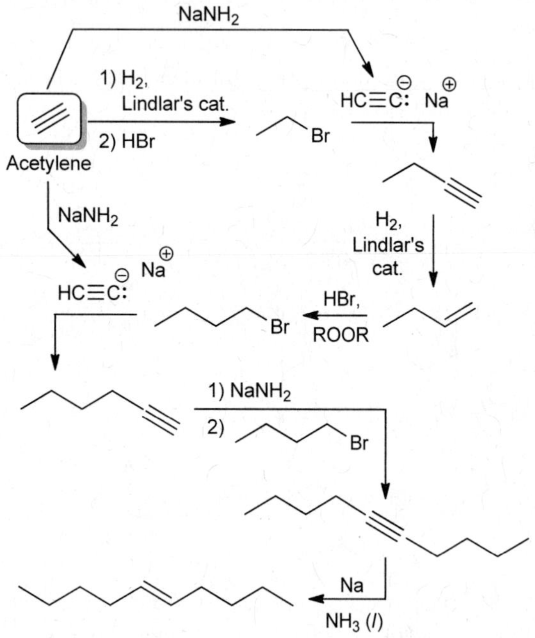

12.15. We use the same approach taken in the previous problem (**12.14**). All carbon-carbon bonds are prepared via alkylation of an alkynide ion with the appropriate alkyl halide. Each alkyl halide must be prepared from acetylene. The last step of the synthesis is the reduction of an alkyne to a *cis* alkene via hydrogenation with a poisoned catalyst:

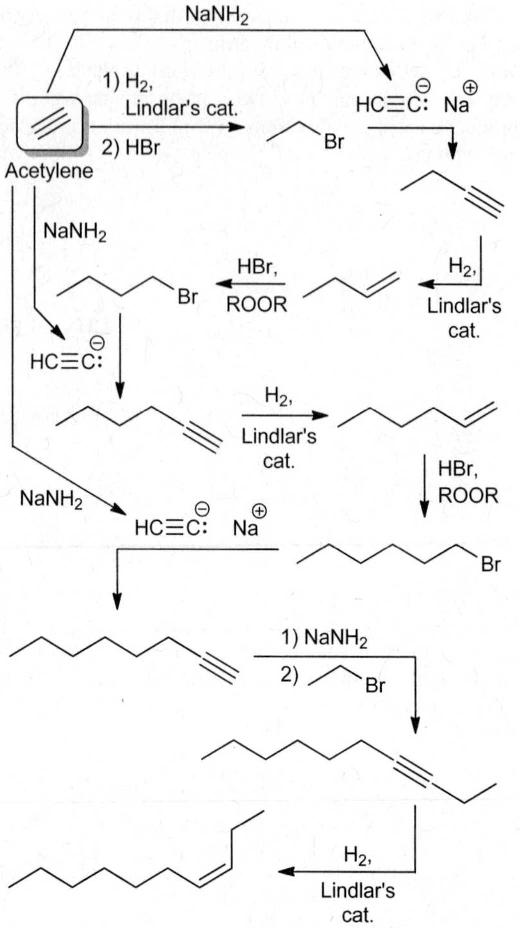

12.16. The starting material has two carbon atoms, and the product has five carbon atoms. So, we must join three fragments together (each of which has two carbon atoms), and then we must remove one of the carbon atoms. The latter process can be achieved via ozonolysis. Since the product is an aldehyde, it is reasonable to explore using ozonolysis as the last step of our synthesis:

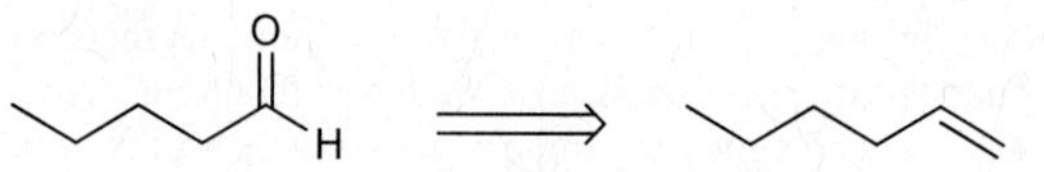

This alkene can be prepared from an alkyne, which can be prepared from acetylene and 1-bromobutane:

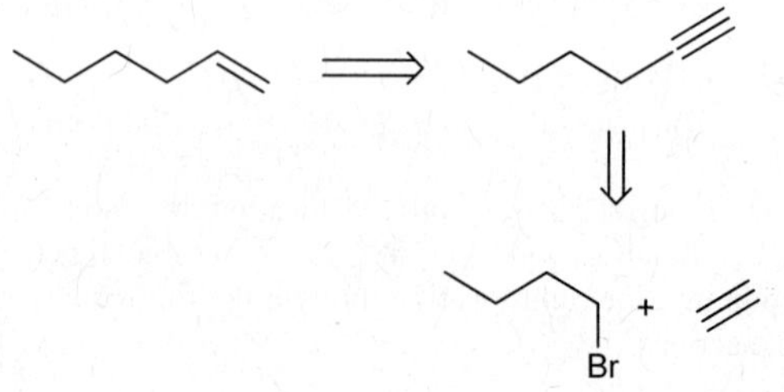

1-Bromobutane can be made from 1-butyne, which can be made from acetylene and ethyl bromide via an alkylation process:

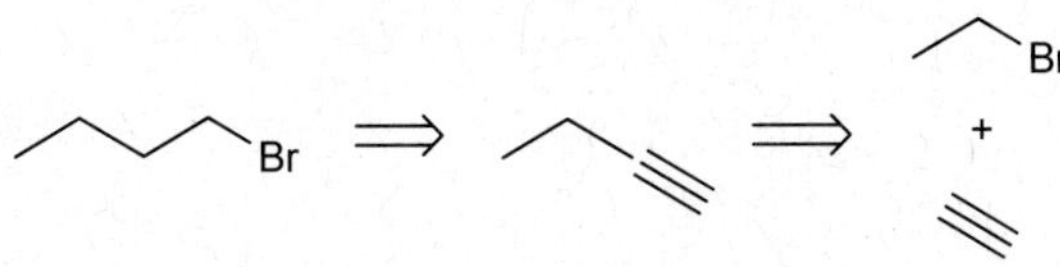

And ethyl bromide can be made from acetylene:

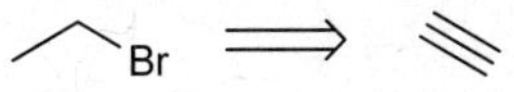

In summary, the desired transformation can be achieved with the following synthesis:

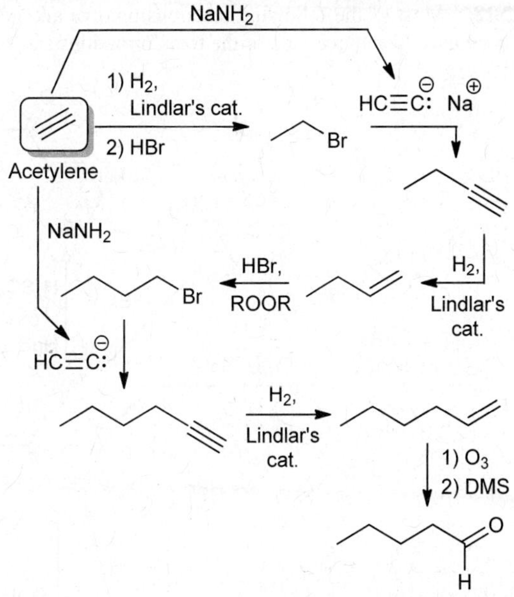

This synthesis represents just one correct answer to the problem. There are certainly other acceptable answers to this problem.

12.17. Most of the following reactions are addition reactions, which can be found in Chapter 10. The following reagents can be used to achieve each of the transformations shown:

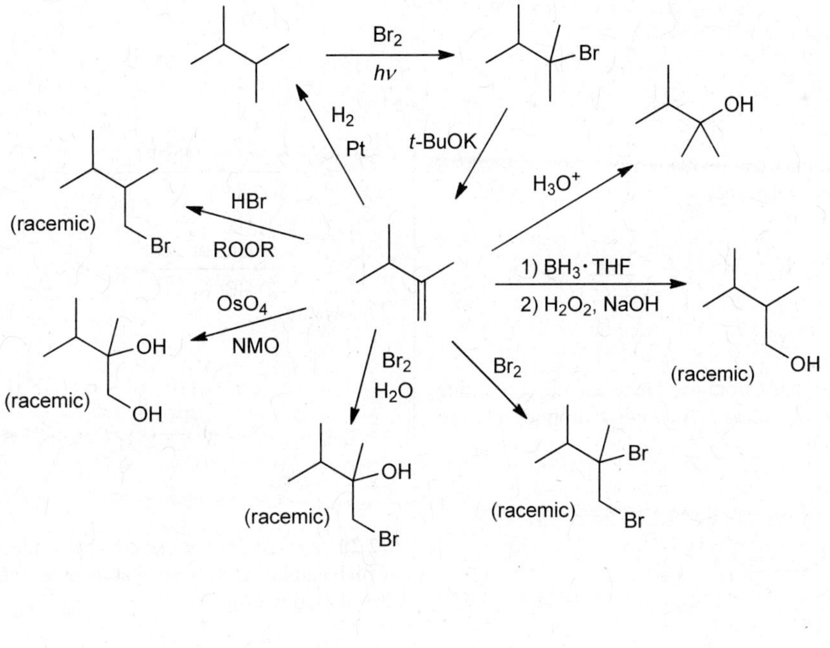

12.18. Most of the following reactions involve alkynes, which can be found in Chapter 11. The following reagents can be used to achieve each of the transformations shown:

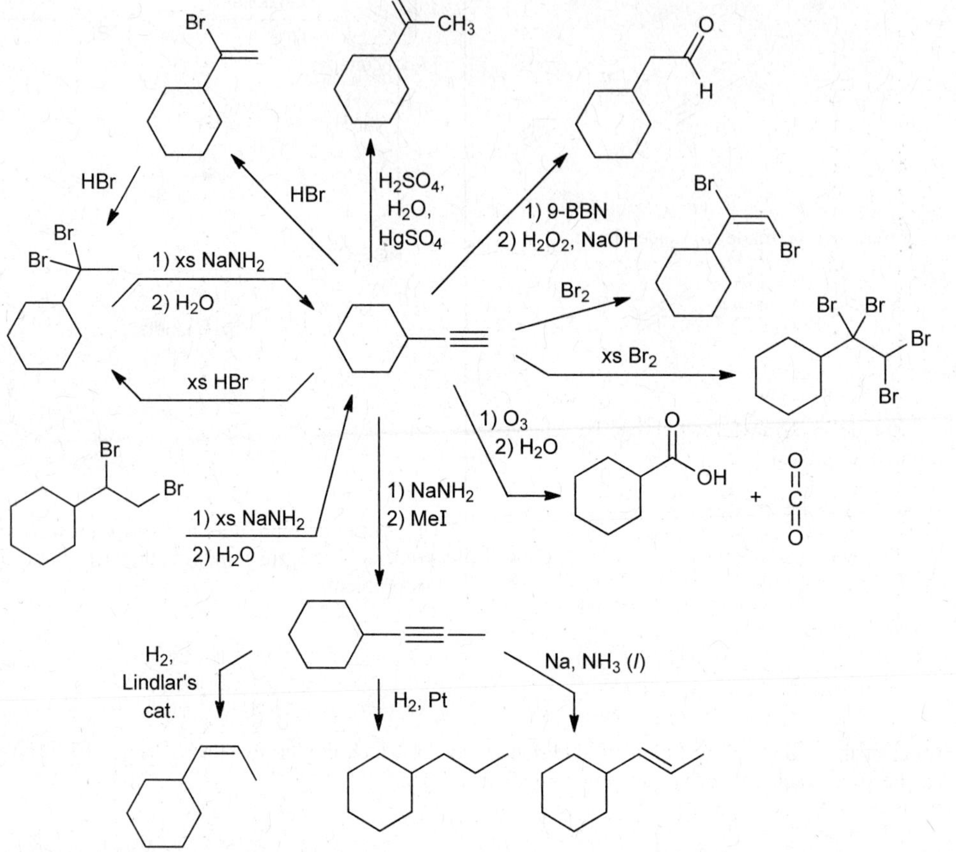

HBr

HBr

H₂SO₄,
H₂O,
HgSO₄

1) 9-BBN
2) H₂O₂, NaOH

Br

Br

1) xs NaNH₂

2) H₂O

Br₂

xs Br₂

xs HBr

Br

Br

1) O₃
2) H₂O

1) NaNH₂
2) MeI

1) xs NaNH₂

2) H₂O

H₂,
Lindlar's
cat.

H₂, Pt

Na, NH₃ (*l*)

12.19. The product can be made from 1-butene, which can be made from 1-butyne:

Br

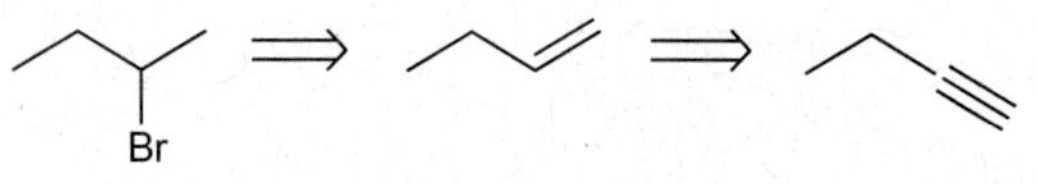

1-Butyne can be made from acetylene and ethyl bromide via an alkylation process. And ethyl bromide can be made from acetylene:

Br

+

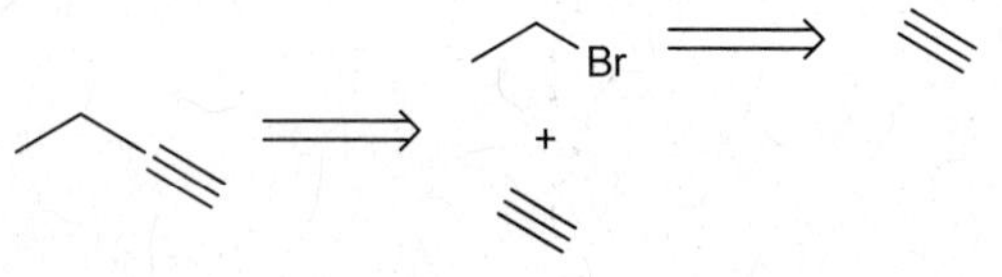

In summary, the desired transformation can be achieved with the following synthesis:

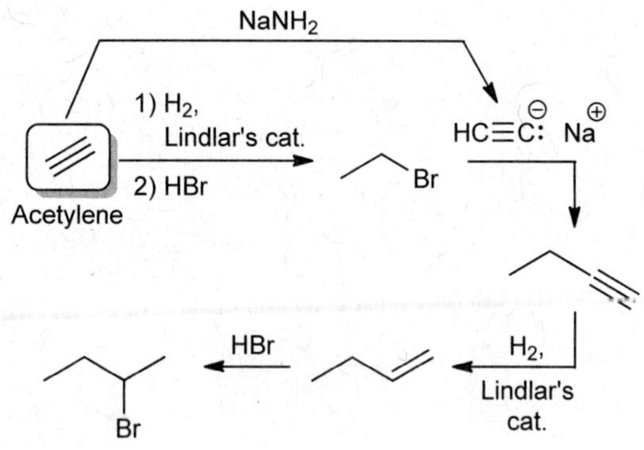

NaNH₂

1) H₂,
Lindlar's cat.

2) HBr

Br

Acetylene

HC≡C: ⁻ Na⁺

HBr

H₂,
Lindlar's
cat.

Br

12.20. 1-Bromobutane can be made from 1-butyne, which can be made from acetylene and ethyl bromide via an alkylation process:

Br

Br

+

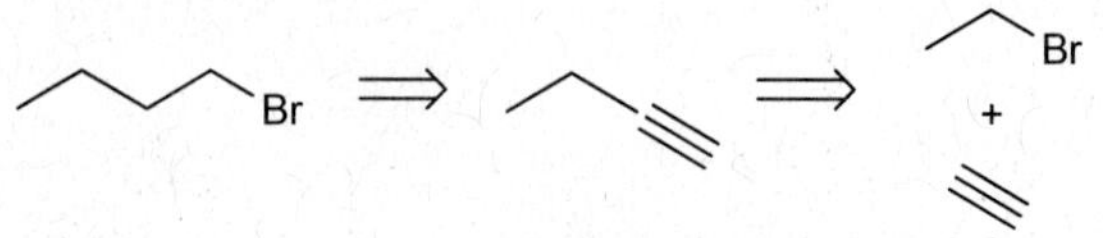

And ethyl bromide can be made from acetylene:

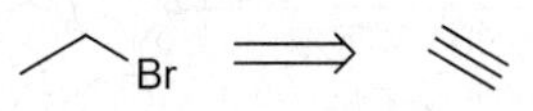

In summary, the desired transformation can be achieved with the following synthesis:

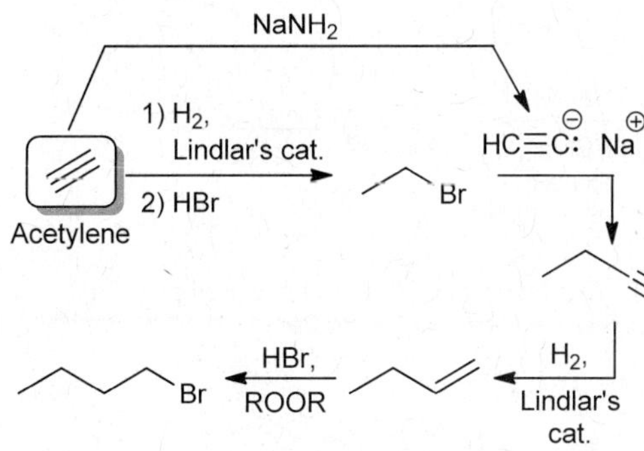

12.21.

(a) The identity of the functional group (OH) has not changed, but its location has changed. We have seen that this type of transformation can be accomplished via a two-step process (elimination, followed by addition). For the first part of our synthesis (elimination), we must first convert the alcohol to a tosylate before performing the elimination process. Then, for the elimination, we must use a sterically hindered base, *tert*-butoxide, in order to obtain the less-substituted alkene. For the second part of our synthesis, we must perform an *anti*-Markovnikov addition of H and OH. This can be achieved via hydroboration-oxidation, or alternatively, it can be achieved by *anti*-Markovnikov addition of HBr followed by an S_N2 process (with hydroxide as a nucleophile) to give the desired alcohol:

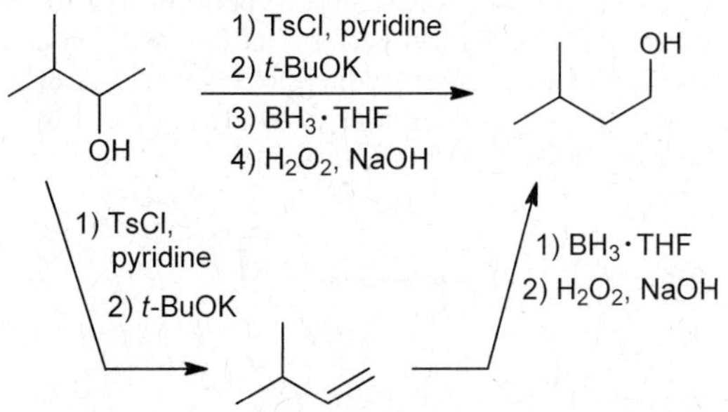

(b) The product is a methyl ketone, which can be made from an alkyne (via acid-catalyzed hydration):

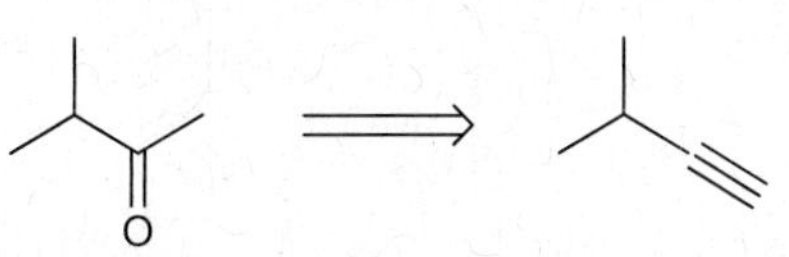

This alkyne can be made from the starting alkene via a two-step process (bromination, followed by elimination with excess sodium amide):

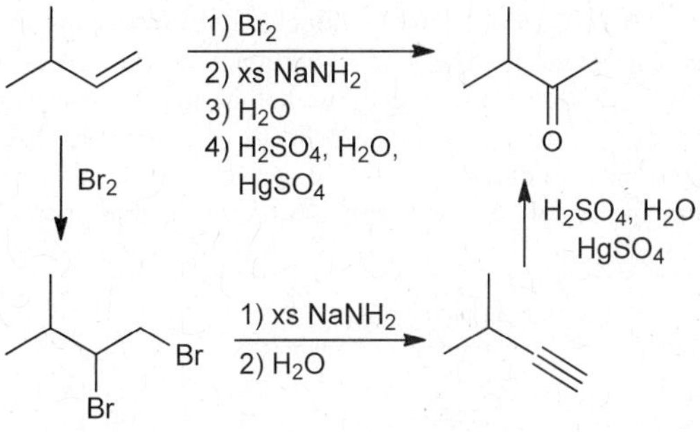

12.22.
We must move the location of the π bond, and we have seen that this can be achieved via a two-step process (addition, followed by elimination). The addition step must occur in an *anti*-Markovnikov fashion, which can be achieved by treating the starting material with HBr in the presence of peroxides. The elimination process must give the less substituted alkene, so a sterically hindered base is required:

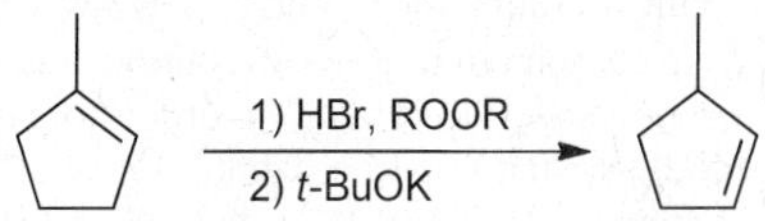

Alternatively, the addition of HBr can be replaced with hydroboration-oxidation (addition of H and OH), which is also an *anti*-Markovnikov addition. In that scenario, the resulting alcohol must first be converted to a tosylate before the elimination step can be performed.

12.23.
(a) The starting material has one more carbon atom than the product. Therefore, our synthesis must employ an ozonolysis process. Since the product is a carboxylic acid (rather than an aldehyde or ketone), we can conclude that the last step of our process must involve ozonolysis of an *alkyne*, rather than ozonolysis of an *alkene*:

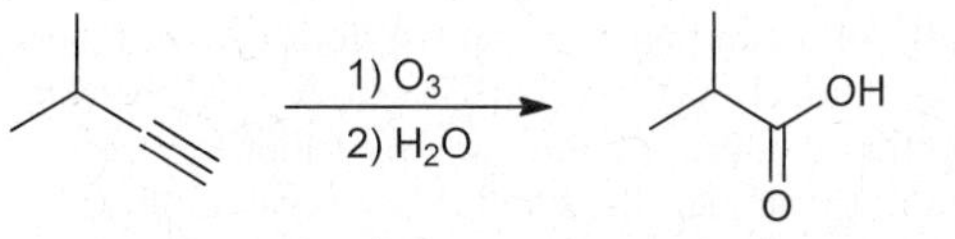

This alkyne can be made from the starting alkene via a two-step process (bromination, followed by elimination with excess sodium amide):

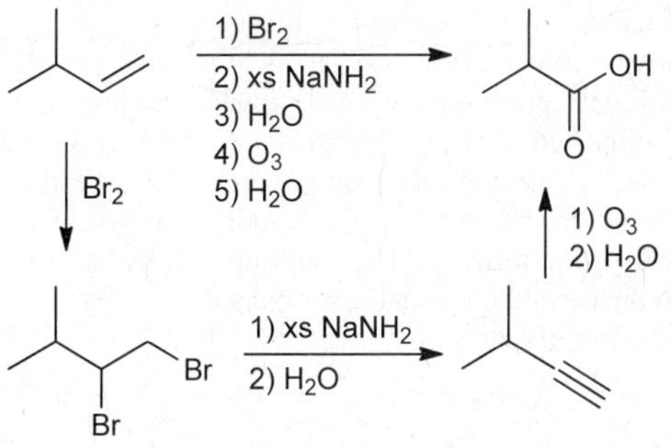

(b) We have not learned a direct way of installing only one carbon atom. That is, two carbon atoms are installed (not one) if we use the alkyl halide in an alkylation process (with sodium acetylide). However, after installing two carbon atoms, we can remove one of them with an ozonolysis procedure, giving the desired product:

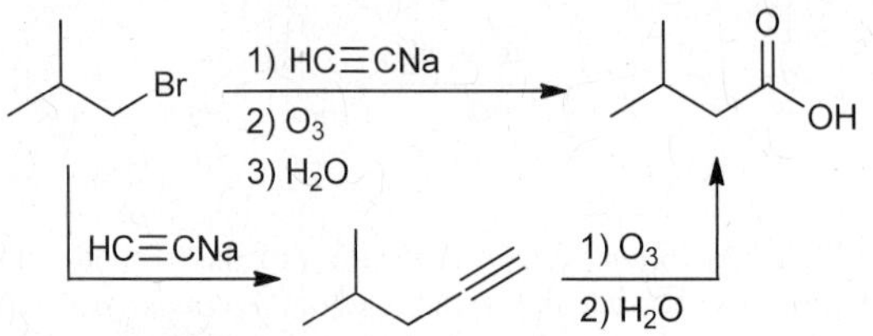

(c) We have not learned a direct way of installing only one carbon atom. That is, two carbon atoms are installed (not one) if we use the alkyl halide in an alkylation process (with sodium acetylide). However, after installing two carbon atoms, we can remove one of them with an ozonolysis procedure. In order to obtain an aldehyde, the last step must be ozonolysis of an alkene, so the penultimate step must be reduction of the alkyne to an alkene in the presence of a poisoned catalyst:

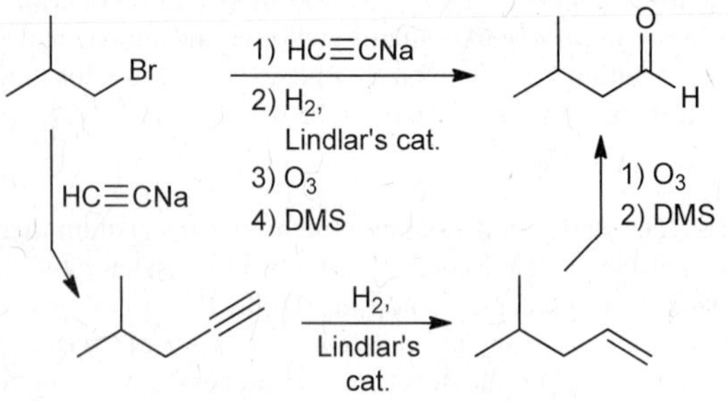

(d) The starting material has four carbon atoms, and the product has six carbon atoms. Therefore, our synthesis must employ an alkylation process. The starting material (2-methylpropane) cannot be converted into an alkyne without giving five bonds to the central carbon atom (which is impossible). Therefore, the starting material must be converted into an alkyl halide (so that it can be treated with sodium acetylide to give an alkylation process). Radical bromination provides a tertiary alkyl halide, which must be converted to a primary alkyl halide. That is, the position of the functional group must be moved, which can be accomplished via a two-step process (elimination followed by *anti*-Markovnikov addition). The primary alkyl halide is then treated with sodium acetylide to give an alkylation process. The resulting terminal alkyne can be reduced in the presence of a poisoned catalyst to give the desired alkene:

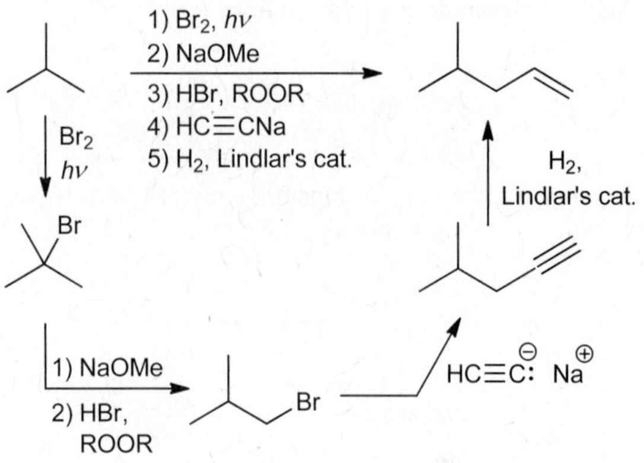

12.24. We begin by drawing the desired products:

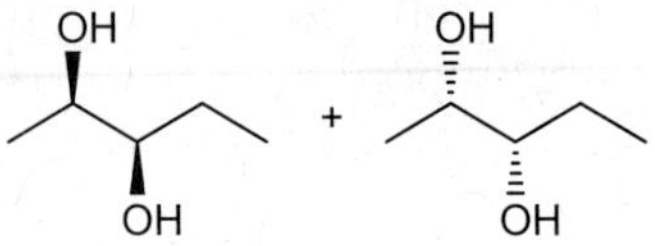

These compounds have five carbon atoms, but our starting materials can contain no more than two carbon atoms. So our synthesis must involve the formation of carbon-carbon bonds. This can be accomplished via the alkylation of acetylene (a compound with two carbon atoms). The location of the functional groups (C2 and C3) indicates that we need two alkylation processes (one to install a methyl group and the other to install an ethyl group). This places the triple bond between C2 and C3, which enables the installation of the functional groups at those locations. Conversion of the internal alkyne into the desired product requires the addition of H and H to give an alkene, followed by the addition of OH and OH. In order to achieve the correct stereochemical outcome, one of these addition processes must be performed in a *syn* fashion, while the other must be performed in an *anti* fashion. That is, we can perform an *anti* addition of H and H, followed by a *syn* addition of OH and OH, or we can perform a *syn* addition of H and H, followed by an *anti* addition of OH and OH, as shown:

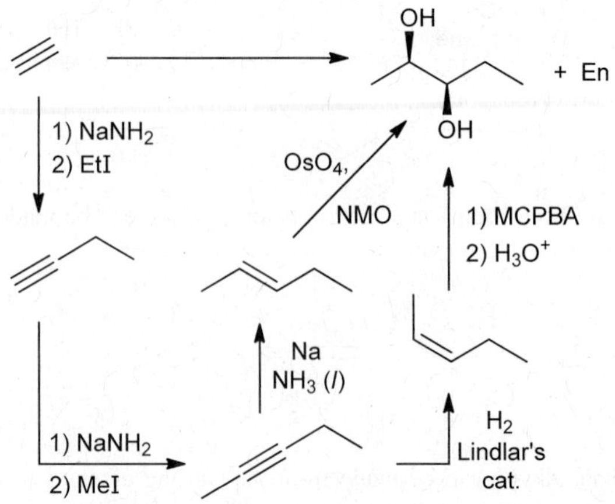

12.25.   We begin by drawing the desired products:

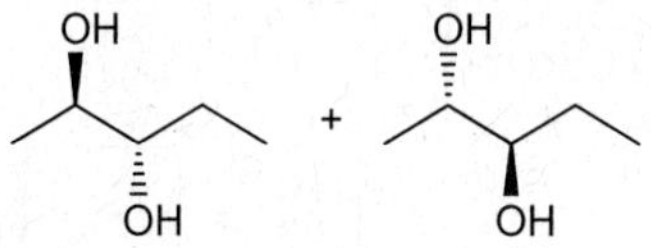

These compounds have five carbon atoms, but our starting materials can contain no more than two carbon atoms. So our synthesis must involve the formation of carbon-carbon bonds. This can be accomplished via the alkylation of acetylene (a compound with two carbon atoms). The location of the functional groups (C2 and C3) indicates that we need two alkylation processes (one to install a methyl group and the other to install an ethyl group). This places the triple bond between C2 and C3, which enables the installation of the functional groups at those locations. Conversion of the internal alkyne into the desired product requires the addition of H and H to give an alkene, followed by the addition of OH and OH. In order to achieve the correct stereochemical outcome, both of these addition processes must be performed in an *anti* fashion, or both must be performed in a *syn* fashion. That is, we can perform an *anti* addition of H and H, followed by an *anti* addition of OH and OH, or we can perform a *syn* addition of H and H, followed by a *syn* addition of OH and OH, as shown:

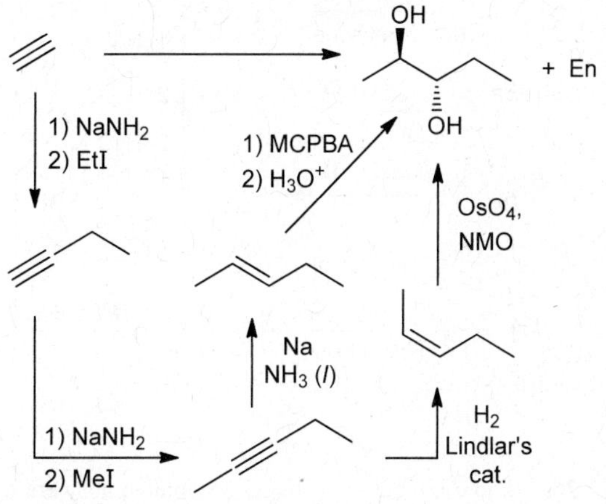

12.26.
(a)   The starting material has four carbon atoms, and the product has six carbon atoms. So our synthesis must involve the installation of two carbon atoms. Also, the location of the functional group has been changed. The product is a methyl ketone, which can be made from a terminal alkyne (via acid catalyzed hydration):

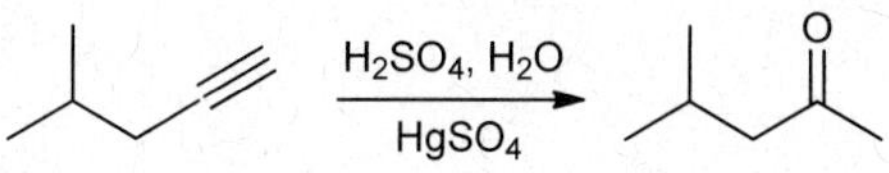

This alkyne can be made from the starting alkene via an *anti*-Markovnikov addition of HBr, followed by treatment with sodium acetylide, as shown here:

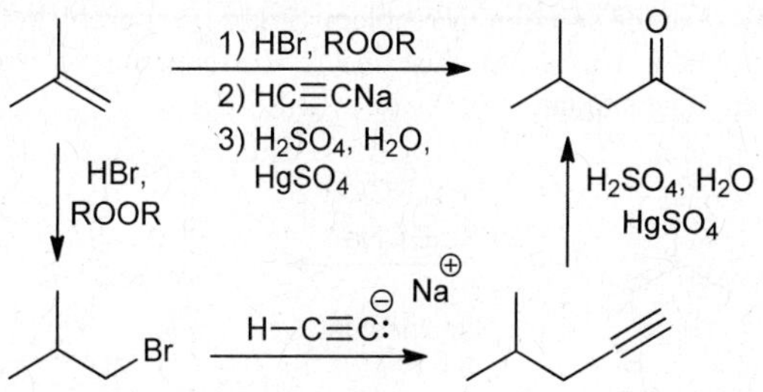

(b)   The starting material has four carbon atoms, and the product has six carbon atoms. So our synthesis must involve the installation of two carbon atoms. Also, the location of the functional group has been changed. The product is an aldehyde, which can be made from a terminal alkyne (via hydroboration-oxidation):

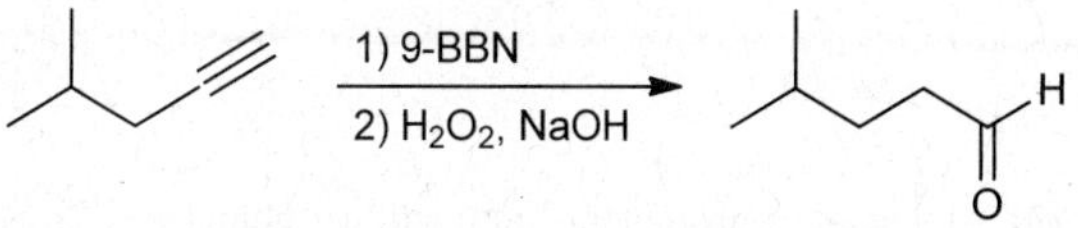

As seen in the previous problem (**12.26a**), this alkyne can be made from the starting alkene via an *anti*-Markovnikov addition of HBr, followed by treatment with sodium acetylide, as shown here:

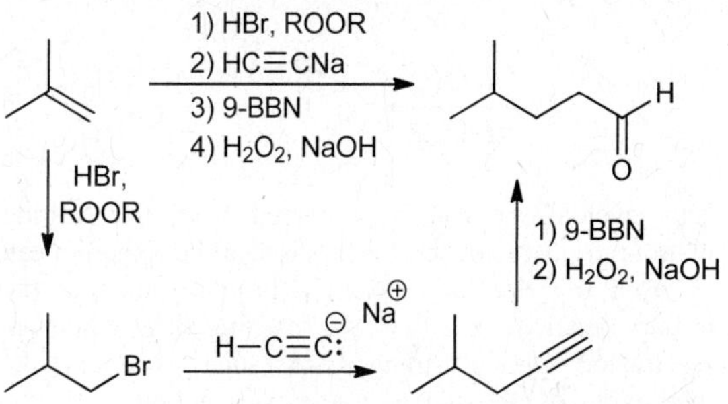

(c)   The starting material has four carbon atoms, and the product has five carbon atoms. We have not learned a direct way of installing only one carbon atom. That is, two carbon atoms are installed (not one) if we convert the starting alkene into an alkyl halide (via an *anti*-Markovnikov addition), and then treat the alkyl halide with sodium acetylide. However, after installing two carbon atoms, we can remove one of them with ozonolysis, giving the product:

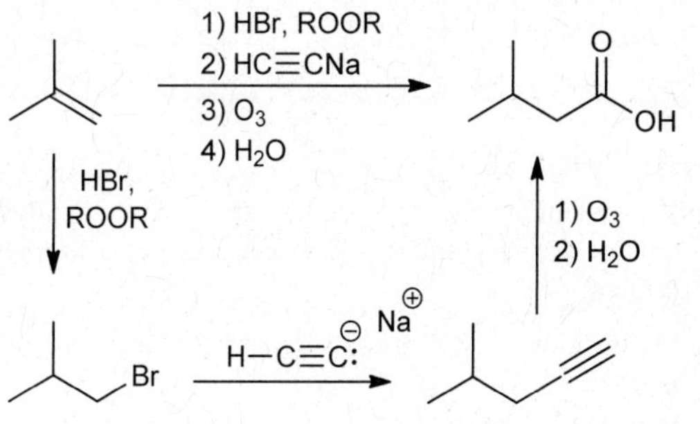

(d) See the solution to Problem 12.12. That problem is extremely similar to this one. The solution is also extremely similar:

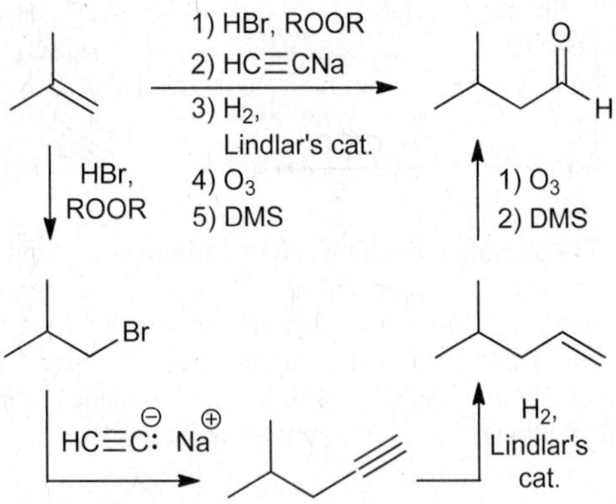

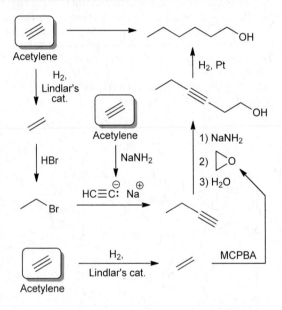

(e) The starting material is cyclic (it contains a ring) and the product lacks a ring. So, we must break one of the carbon-carbon bonds of the ring. We have only learned one way (ozonolysis) to break a carbon-carbon bond. So, the last step of our synthesis is likely the following reaction:

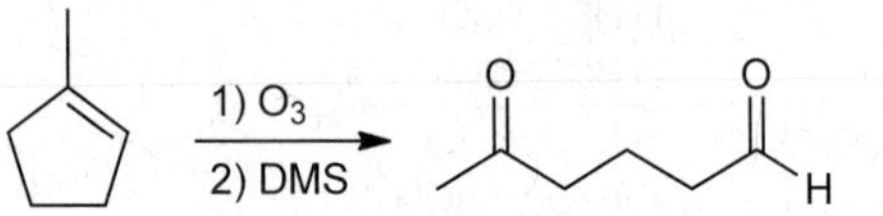

This cycloalkene can be prepared from the starting material in just two steps. First, radical bromination can be used to selectively install a bromine atom at the tertiary position. And then, the resulting alkyl halide can be treated with a strong base (such as hydroxide, methoxide, or ethoxide) to give an E2 reaction:

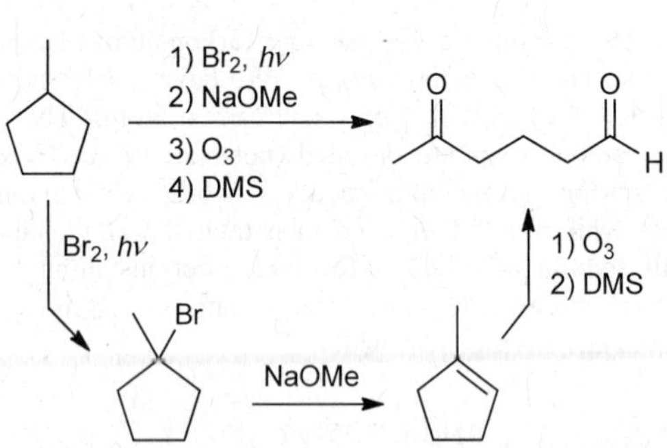

12.27. There are certainly many acceptable answers to this problem. The following retrosynthetic analysis employs the technique described in the problem statement:

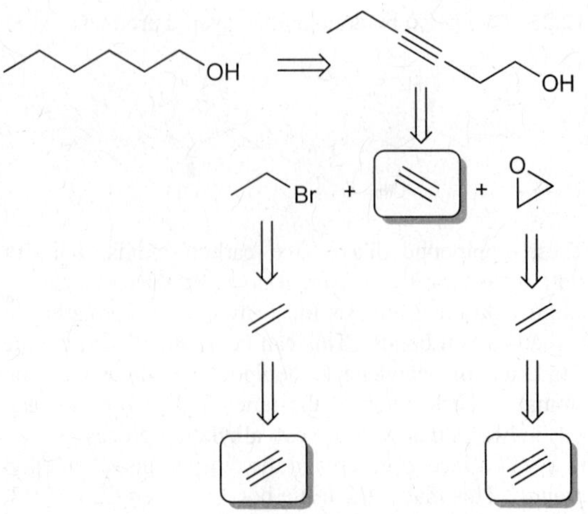

This retrosynthetic analysis gives the following synthesis:

12.28. There are certainly many acceptable answers to this problem. The following retrosynthetic analysis employs the technique described in the problem statement:

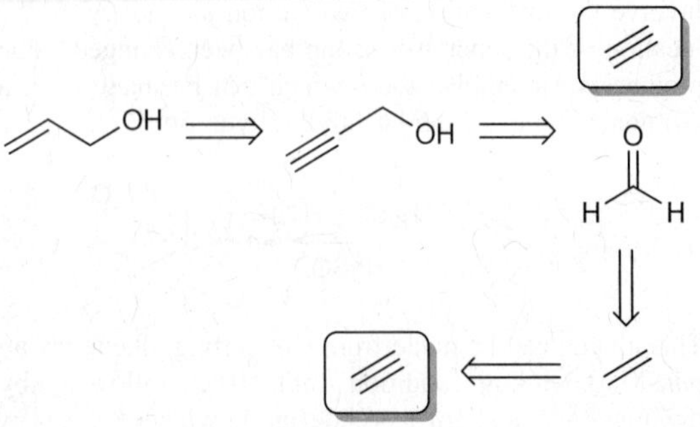

This retrosynthetic analysis gives the following synthesis:

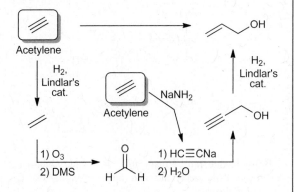

12.29. The key to solving this problem is recognizing that the cyclic product can be made from the following acyclic compound (which can be prepared from the starting material in just one step):

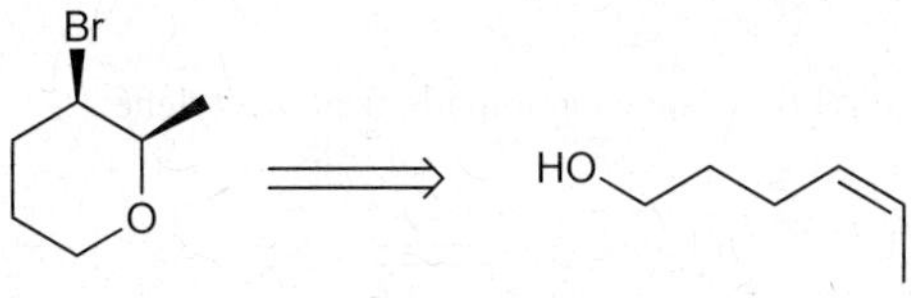

This reaction is similar to halohydrin formation:

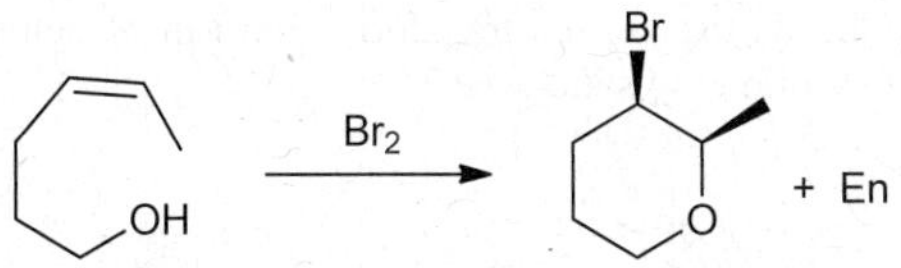

The π bond reacts with molecular bromine to give a bromonium ion, which is then attacked by the OH group in an intramolecular process. You may find it helpful to build molecular models to help visualize the stereochemistry of the ring-closing step.

According to the retrosynthetic analysis above, the desired transformation can be achieved in just two steps, shown here:

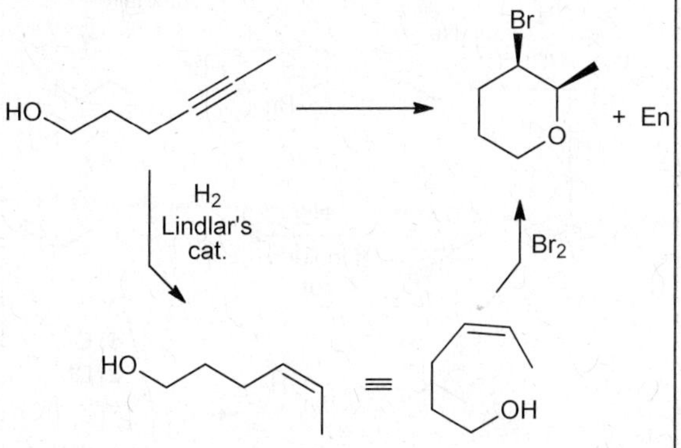

12.30.
(a) The desired compound can be prepared from acetylene in just one step (via acid-catalyzed hydration):

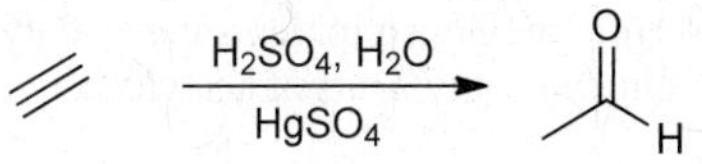

Alternatively, this transformation can also be achieved via hydroboration-oxidation of acetylene.

(b) The following synthesis represents just one correct answer to the problem. There are certainly other acceptable answers to this problem.

We have seen in previous problems that 1-butyne can be prepared from two equivalents of acetylene:

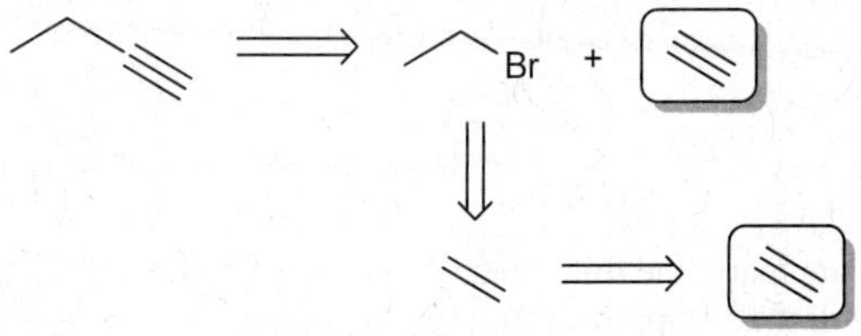

The product can be made from 1-butyne in just two steps (hydrogenation, followed by ozonolysis):

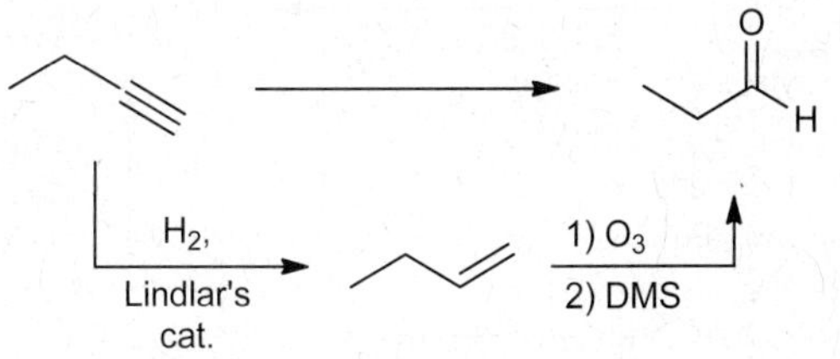

In summary, the following synthesis can be used to make the desired compound from acetylene:

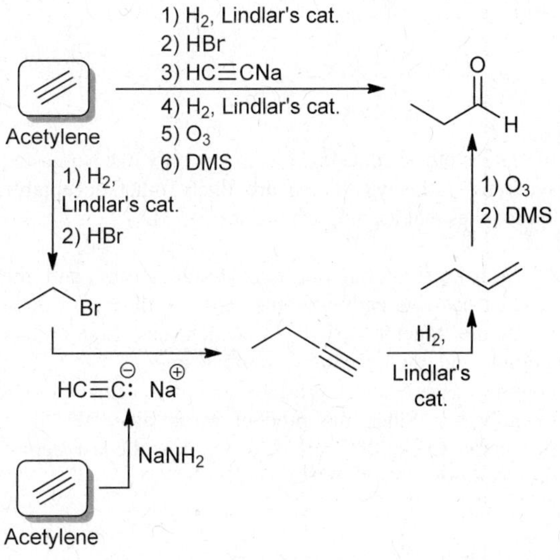

(c) The synthesis developed below is only one suggested synthetic pathway. There are likely other acceptable approaches that accomplish the same goal.

We have seen in previous problems that 1-butyne can be prepared from two equivalents of acetylene:

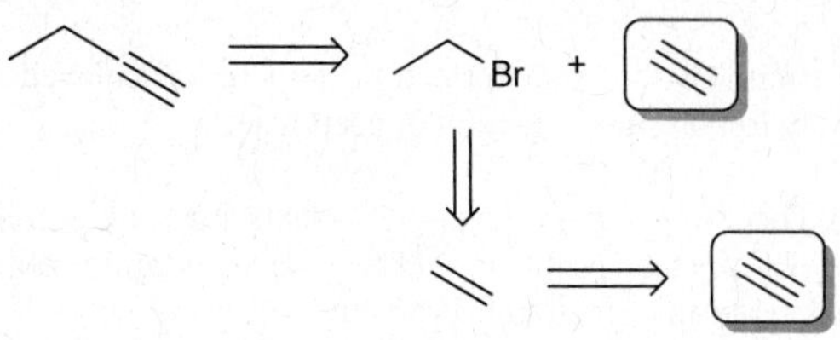

And the product can be made from 1-butyne via hydroboration-oxidation:

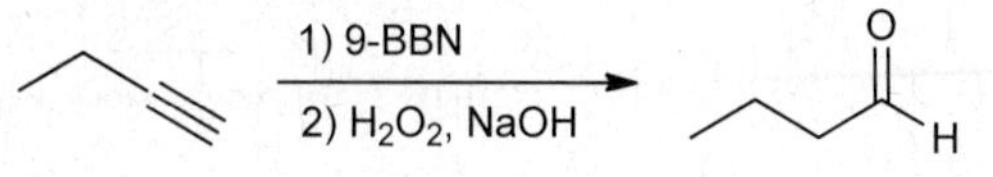

In summary, the following synthesis can be used to make the desired compound from acetylene:

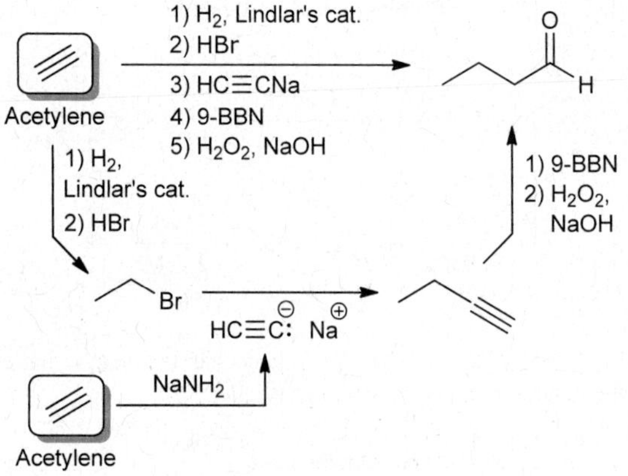

(d) The synthesis developed below is only one suggested synthetic pathway. There are likely other acceptable approaches that accomplish the same goal.

The starting material has two carbon atoms, and the product has five carbon atoms. So, we must join three fragments together (each of which has two carbon atoms), and then we must remove one of the carbon atoms. The latter process can be achieved via ozonolysis. Since the product is an aldehyde, it is reasonable to explore using ozonolysis as the last step of our synthesis:

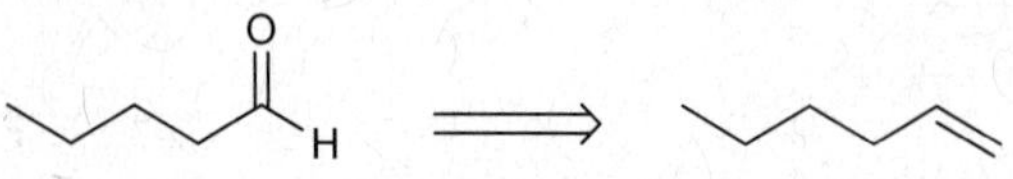

This alkene can be prepared from an alkyne, which can be prepared from acetylene and 1-bromobutane:

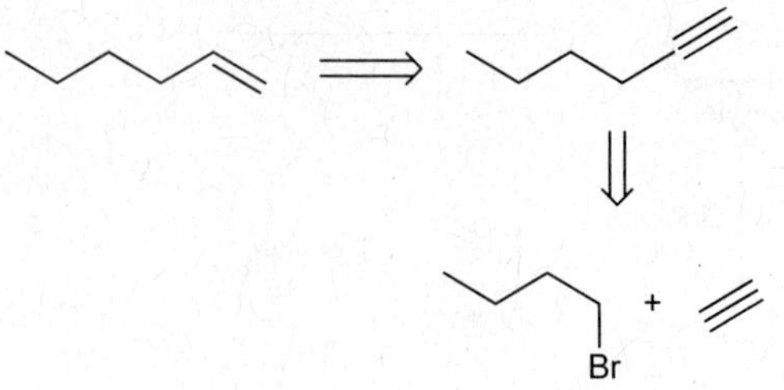

1-Bromobutane can be made from 1-butyne, which can be made from acetylene and ethyl bromide via an alkylation process:

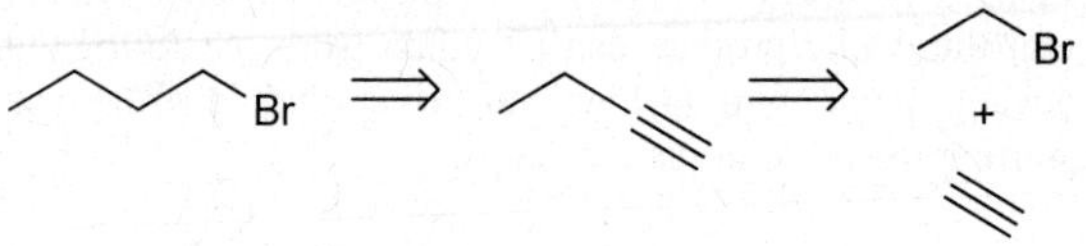

And ethyl bromide can be made from acetylene:

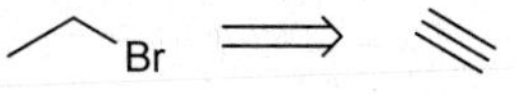

In summary, the desired transformation can be achieved with the following synthesis:

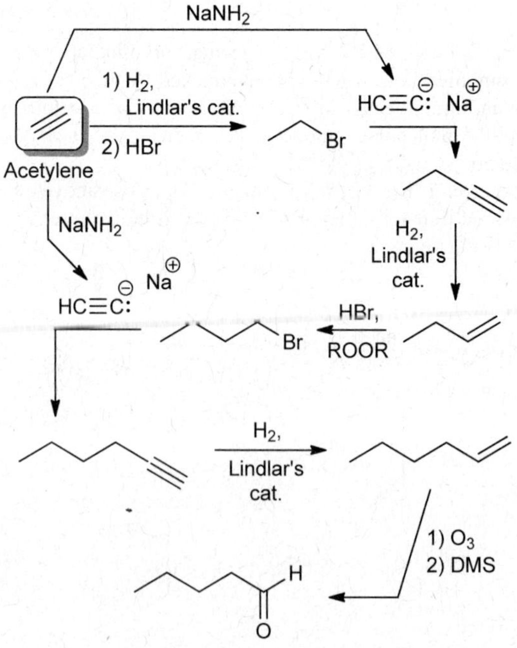

12.31. The key to solving this problem is recognizing that the cyclic product can be made from the following acyclic compounds via two S_N2 reactions:

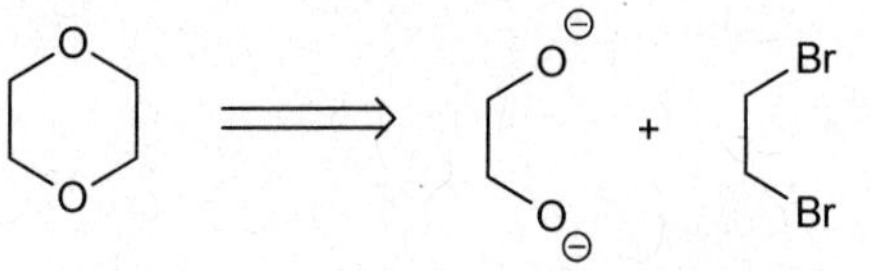

Each of these starting materials can be made from acetylene, as seen in the following synthesis:

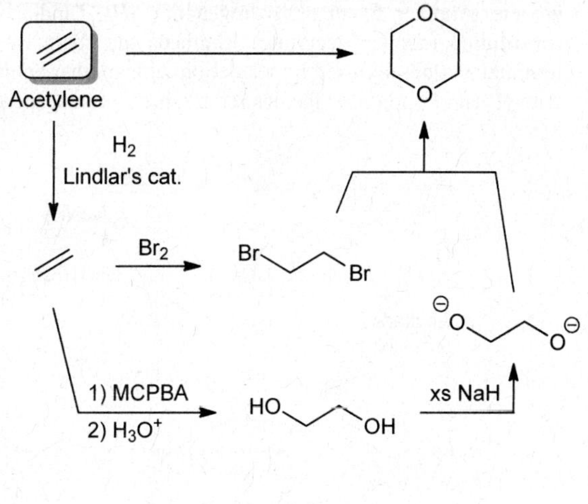

12.32.
The synthesis developed below is only one suggested synthetic pathway. There are likely other acceptable approaches that accomplish the same goal.

As seen in Table 4.1, tetradecane is a saturated hydrocarbon with 14 carbon atoms (with no branching), and our source of carbon (acetylene) has two carbon atoms, so we will likely use 7 equivalents of acetylene in this synthesis. There are a number of different approaches to complete this synthesis, including connecting two carbon atoms at a time sequentially from one end, or disconnecting it symmetrically from the center. The retrosynthesis below takes the latter of these two tactics.

The figure below outlines a retrosynthetic analysis for our target molecule, employing squiggly lines to indicate C-C bonds that are disconnected in the retrosynthetic direction. An explanation of each of the steps (*a-j*) follows.

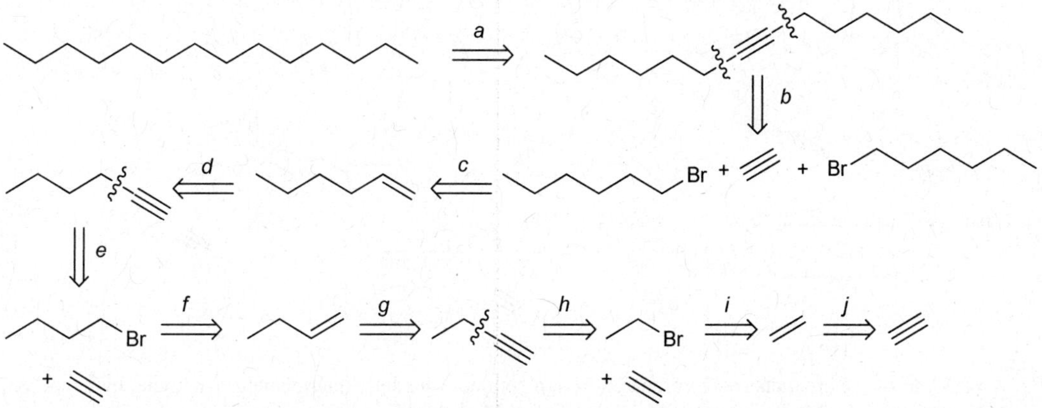

  a.   Tetradecane can be made via hydrogenation of 7-tetradecyne.
  b.   7-Tetradecyne can be made by sequentially alkylating both sides of acetylene with 1-bromohexane.
  c.   1-Bromohexane is made via an *anti*-Markovnikov addition of HBr across 1-hexene.
  d.   1-Hexene is made by reduction of 1-hexyne using H_2 and Lindlar's catalyst.
  e.   1-Hexyne can be produced from acetylene (after deprotonation to make a nucleoph
  f.   1-Bromobutane is made via an *anti*-Markovnikov addition of HBr across 1-butene.
  g.   1-Butene is made by reduction of 1-butyne using H_2 and Lindlar's catalyst.
  h.   1-Butyne is made from acetylene (after deprotonation) and 1-bromoethane.
  i.   1-Bromoethane is made by addition of HBr across ethylene.
  j.   Ethylene is made by reduction of acetylene using H_2 and Lindlar's catalyst.

Now, let's draw out the forward scheme. Acetylene is reduced to ethylene using H_2 an addition, followed by S_N2 substitution with an acetylide nucleophile (made by deprotonation amide) gives 1-butyne. Reduction to 1-butene with H_2 and Lindlar's catalyst followed by *anti*-HBr in the presence of peroxide produces 1-bromobutane. A substitution reaction with s hexyne. Another round of hydrogenation (H$_2$, Lindlar's catalyst), *anti*-Markovnikov addition (HBr, peroxide) and substitution (sodium acetylide) lengthens the chain by two more carbons, giving 1-octyne. Deprotonation of this terminal alkyne, followed by alkylation with another equivalent of 1-bromohexane yields 7-tetradecyne. Hydrogenation with H$_2$ and Pt produces the desired product, tetradecane.

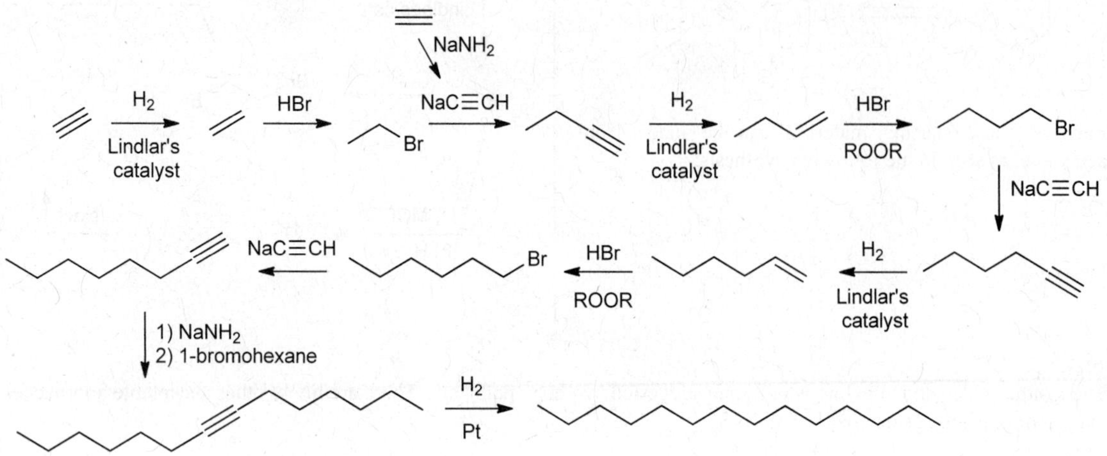

12.33. The following synthesis is one suggested synthetic pathway. There are likely many other acceptable approaches that accomplish the same goal.

Each of the target compounds has a nine carbon linear chain, and our reagents must be alkenes with fewer than six carbon atoms. Thus, it is clear that we will be forming new C-C bonds in the course of this synthesis. The figure below outlines a retrosynthetic analysis for our target compound. An explanation of each of the steps (*a-h*) follows.

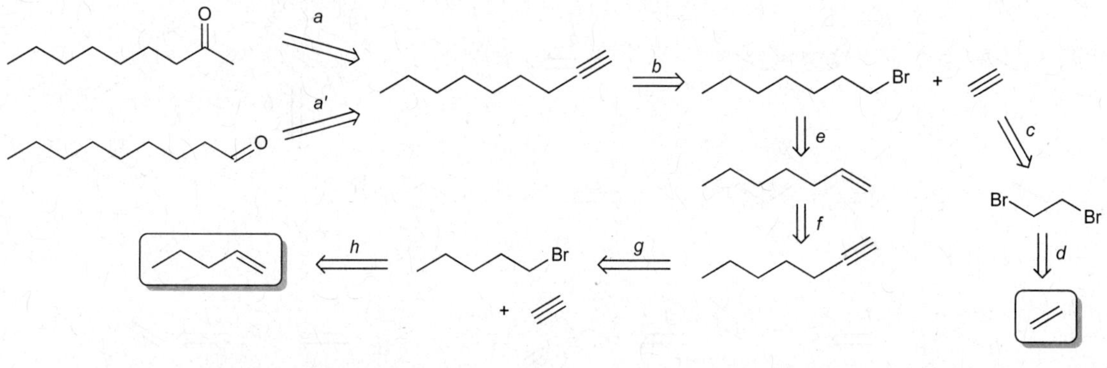

  a.   Either of the products can be prepared from a common synthetic intermediate, 1-nonyne, by hydration of the alkyne via (*a*) Markovnikov addition, or (a') *anti*-Markovnikov addition.
  b.   The terminal alkyne can be prepared via reaction of 1-bromoheptane with an acetylide anion (formed by deprotonating acetylene).
  c.   Acetylene is prepared via a double elimination from 1,2-dibromoethane.
  d.   1,2-Dibromoethane is prepared via bromination of ethylene.
  e.   1-Bromoheptane is prepared via *anti*-Markovnikov addition of HBr across 1-heptene.
  f.   1-Heptene is prepared via hydrogenation of 1-heptyne in the presence of a poisoned catalyst.
  g.   1-Heptyne is prepared via reaction of 1-bromopentane with an acetylide anion.
  h.   1-Bromopentane is prepared via *anti*-Markovnikov addition of HBr across 1-pentene.

Now, let's draw the forward scheme. In the presence of peroxides, the reaction of 1-pentene with HBr produces 1-bromopentane (via *anti*-Markovnikov addition). Subsequent reaction with acetylide [produced from ethylene as shown by bromination (Br$_2$), double elimination and deprotonation (excess NaNH$_2$)] provides 1-heptyne. Reduction to the alkene (H$_2$ / Lindlar's catalyst) followed by *anti*-Markovnikov addition (HBr / peroxide) yields 1-bromoheptane. This primary alkyl bromide can then undergo an S$_N$2 reaction when treated with acetylide (prepared above), giving the common intermediate, 1-nonyne. A hydroboration / oxidation protocol (9-BBN then H_2O_2, NaOH) produces the target aldehyde. Acid and mercury catalyzed hydration gives the target ketone.

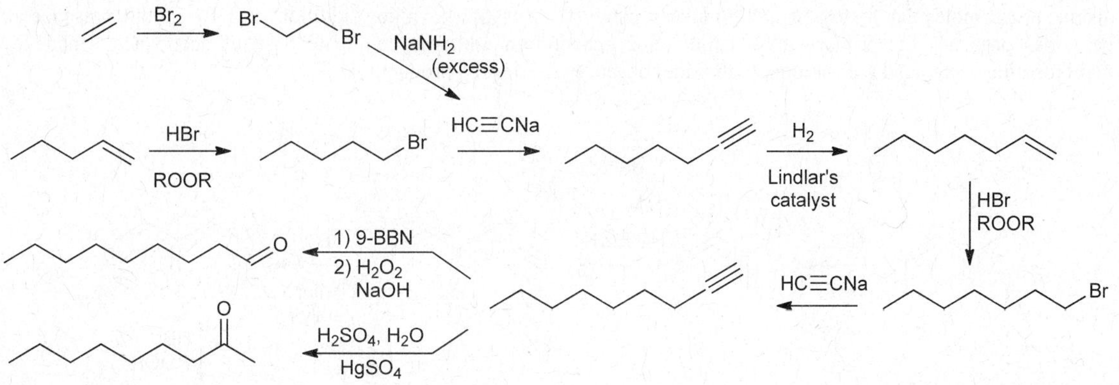

12.34.
The synthesis developed below is only one suggested synthetic pathway. There are likely other acceptable approaches that accomplish the same goal.

An analysis of the structure of the product suggests the following origins of each of the carbon atoms in the product.

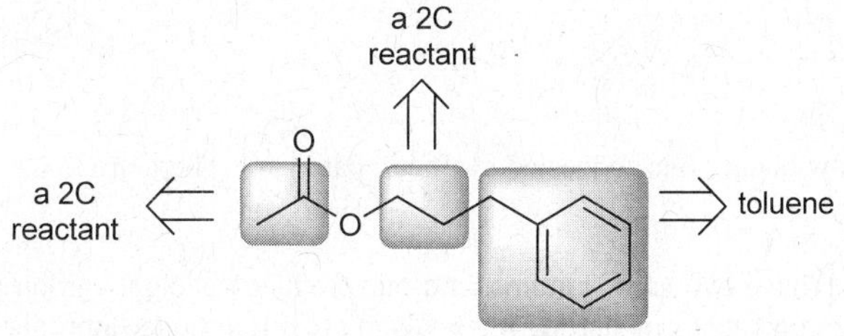

The following figure outlines a retrosynthetic analysis for our target molecule. An explanation of each of the steps (*a-f*) follows.

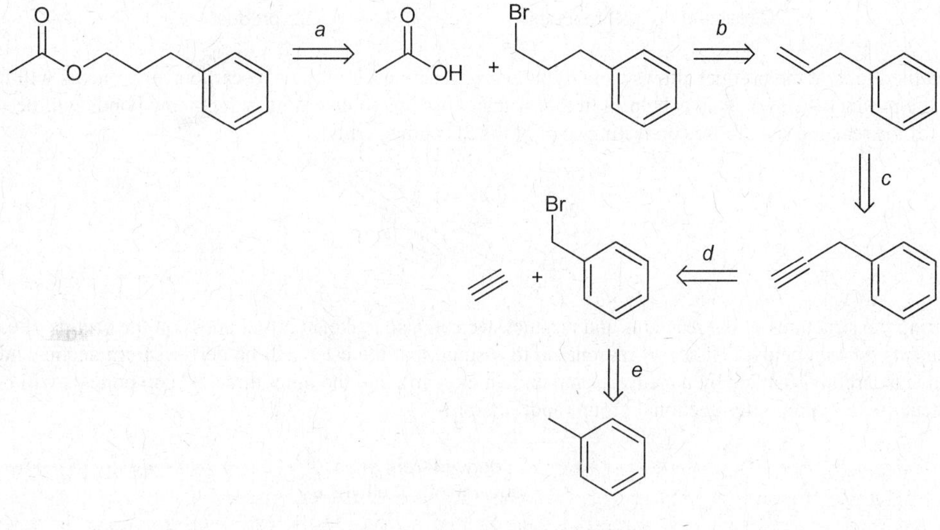

a. The product 3-phenylpropyl acetate can be made via an S_N2 reaction between the carboxylic acid (after deprotonation to make a competent nucleophile) and the primary alkyl bromide.
b. The primary alkyl bromide can be made by *anti*-Markovnikov addition of HBr to the monosubstituted alkene.
c. The alkene is made by reduction of the corresponding terminal alkyne.
d. The terminal alkyne is made by alkylating acetylene (using sodium amide to deprotonate) with benzyl bromide.
e. Benzyl bromide is made via radical bromination of toluene. (The carbon adjacent to the aromatic ring is activated toward bromination due to the resonance-stabilized radical intermediate that forms.)

Now, let's draw the forward scheme. Toluene is brominated using NBS and heat. Reaction with sodium acetylide (made by deprotonating acetylene with sodium amide) produces the terminal alkyne. The alkyne is reduced to the alkene using molecular hydrogen and Lindlar's catalyst. *Anti*-Markovnikov addition of HBr in the presence of peroxides produces the primary alkyl halide. S_N2 substitution with the conjugate base of acetic acid (made by deprotonating acetic acid with sodium hydroxide) produces the desired product.

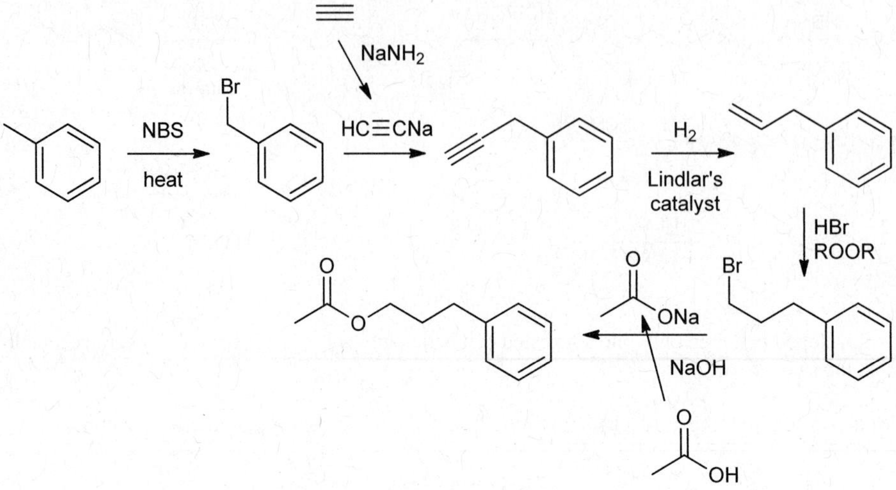

12.35.
The synthesis developed below is only one suggested synthetic pathway. There are likely other acceptable approaches that accomplish the same goal.

Acetic acid and ethylene each have two carbon atoms, and our product has eight carbon atoms. So our synthesis will need to involve a total of *four* equivalents of starting materials in order to produce a product with eight carbon atoms.

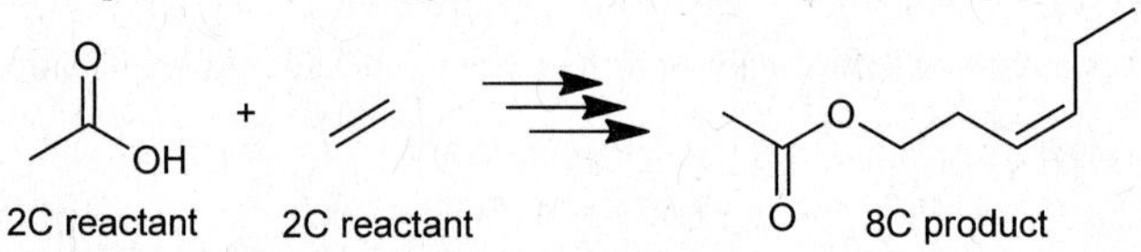

A more detailed look at the product allows us to hypothesize where each of the two-carbon components will ultimately end up in the product (below). This is helpful in that it may allow us to determine which new bonds will be formed in the course of the reaction (*i.e.*, those connecting each of the 2C components).

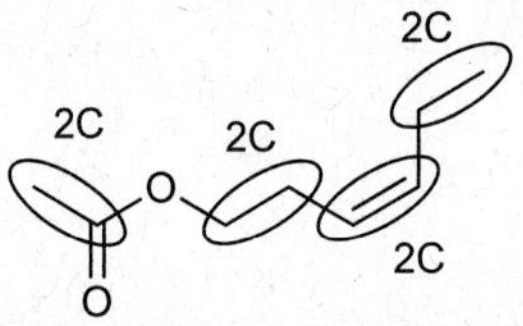

By comparing the structures of the reactants and product, we can also make an initial guess on the origins of each of the 2C components (shown below). It seems reasonable to assume that the ester will be derived from acetic acid (as both of these have a carbonyl flanked by a methyl group and an oxygen), and the other three 2C components will be derived from ethylene (with appropriate functional group modification).

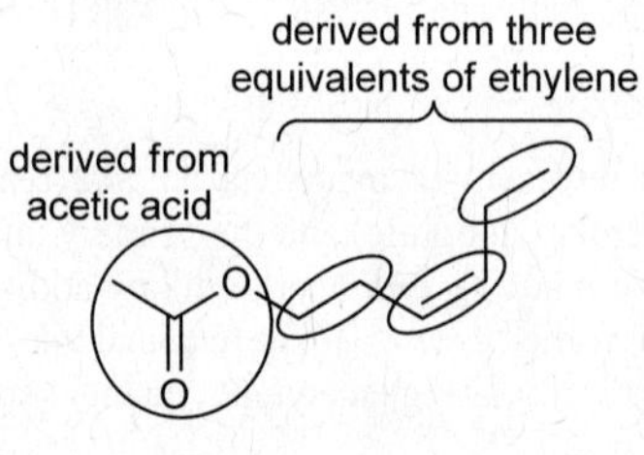

When considering which types of reactions to use, we will connect these pieces using a number of substitution reactions. Also, the only way we have learned to produce a *cis* alkene is via hydrogenation of an alkyne using H_2 and Lindlar's catalyst, so this will clearly be one of our steps.

The figure below outlines a retrosynthetic analysis for our target molecule. An explanation of each of the steps (*a-j*) follows.

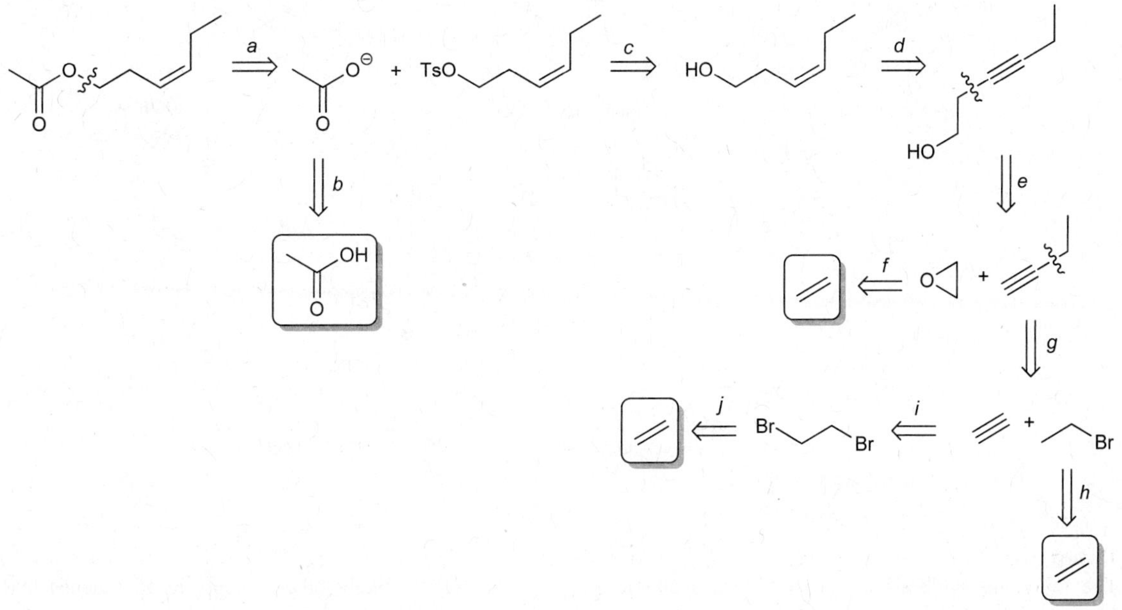

a. The indicated C-O bond (wavy line) can be made via an S_N2 reaction between a carboxylate (the conjugate base of a carboxylic acid) and a substrate with an appropriate leaving group (*e.g.*, tosylate).
b. The carboxylate can be prepared from acetic acid (one of our given reactants) by treatment with a suitable base, such as NaOH.
c. The tosylate can be prepared from the corresponding alcohol.
d. The *cis* alkene can be produced from the corresponding alkyne (H_2 / Lindlar's catalyst).
e. This retrosynthetic step is the key disconnection that utilizes the reaction described in the problem statement. We can make this internal alkyne/alcohol by the reaction of an alkynide ion (formed by deprotonating 1-butyne) and an epoxide.
f. The epoxide is prepared from ethylene via epoxidation.
g. 1-Butyne is prepared by alkylating the conjugate base of acetylene using bromoethane.
h. Bromoethane is prepared via HBr addition to ethylene.
i. Acetylene is prepared via a double elimination from 1,2-dibromoethane.
j. 1,2-Dibromoethane is prepared via bromination of ethylene.

Now, let's draw out the forward scheme. This multi-step synthesis uses three equivalents of ethylene (labeled **A**, **B**, **C** in the scheme below) and one equivalent of acetic acid (labeled **D**). Ethylene (**A**) is converted to 1,2-dibromoethane upon treatment with bromine. Subsequent reaction with excess sodium amide produces an acetylide anion which is then treated with bromoethane [made from ethylene (**B**) and HBr] to produce 1-butyne. Deprotonation with sodium amide, followed by reaction with an epoxide [prepared by epoxidation of ethylene (**C**)] and water workup, produces a compound with an alkyne group and an alcohol group. Reduction of the alkyne to the *cis* alkene is accomplished with H_2 and Lindlar's catalyst, after which the alcohol is converted to a tosylate with tosyl chloride. Reaction with the conjugate base of acetic acid [produced by treating acetic acid (**D**) with NaOH] allows for an S_N2 reaction, thus yielding the desired product, *Z*-hexenyl acetate.

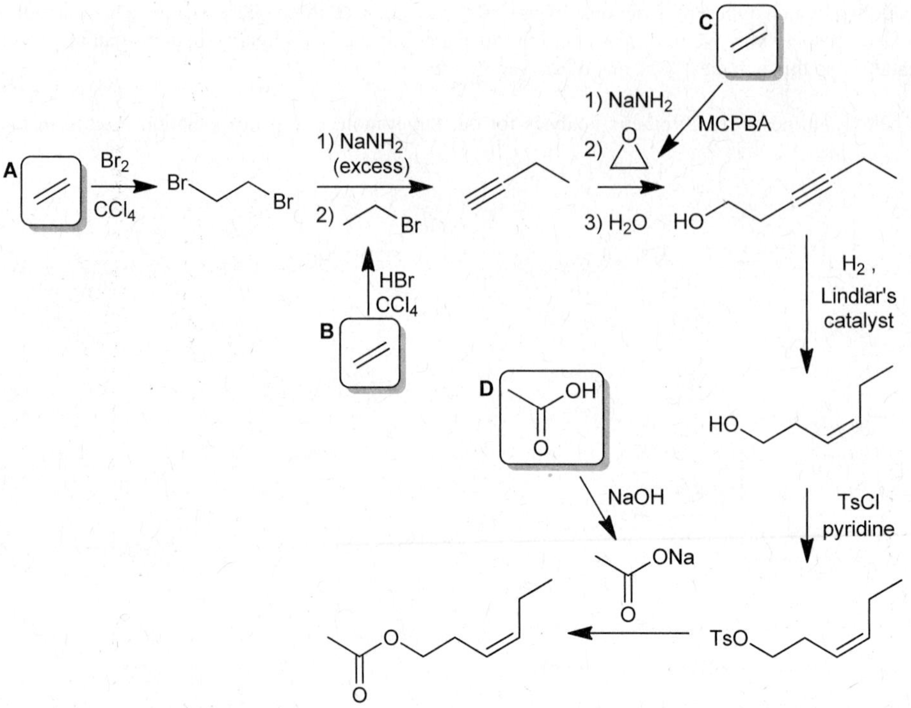

12.36.

The following synthesis is one suggested synthetic pathway. There are likely other acceptable approaches that accomplish the same goal.

Take note that the reactant and product each have six carbon atoms. This suggests that our synthetic plan will not necessarily involve any C-C bond-forming reactions. However, there is a change in the carbon skeleton, and we will need a C-C bond-*breaking* reaction to convert the cyclic starting material into an acyclic product.

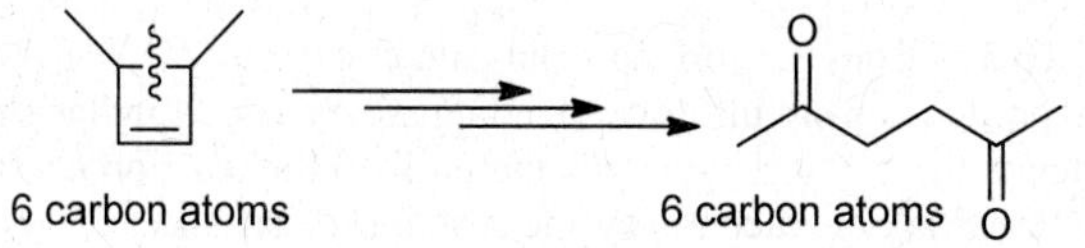

6 carbon atoms   6 carbon atoms

The product contains two ketone groups, which is suggestive of an ozonolysis. The figure below outlines a retrosynthetic analysis for our target molecule. An explanation of each of the steps (*a-e*) follows.

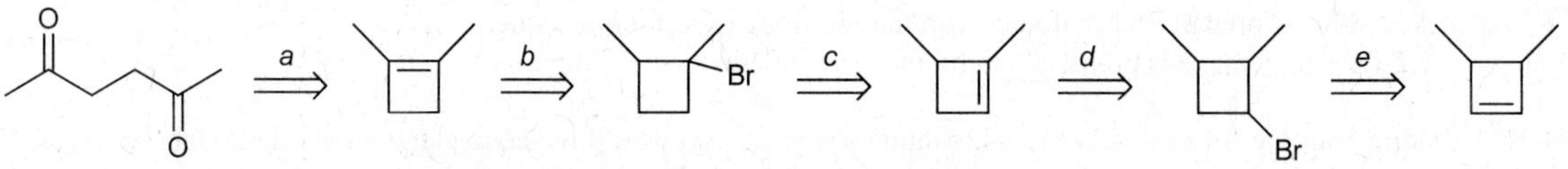

a. The two ketone groups can be prepared via ozonolysis of 1,2 dimethylcyclobutene. This is a key disconnection as it brings us to a synthetic intermediate with the same basic connectivity as the provided reactant (3,4-dimethylcyclobutene). The remaining steps involve reactions to move the position of the double bond.

b. 1,2-dimethylcyclobutene is prepared via elimination from a suitable alkyl halide (*e.g.*, 1-bromo-1,2-dimethylcyclobutane).

c. The tertiary alkyl bromide is prepared via Markovnikov addition of HBr to 1,4-dimethylcyclobutene.

d. This alkene can be prepared via Zaitsev elimination of 1-bromo-2,3-dimethylcyclobutane.

e. This secondary alkyl bromide can be made via addition of HBr to 3,4-dimethylcyclobutene (our provided reactant).

Now, let's draw out the forward scheme. HBr converts 3,4-dimethylcyclobutene to 1-bromo-2,3-methylcyclobutane. Zaitsev elimination using sodium ethoxide affords 1,4-dimethylcyclobutene, which is subsequently converted to 1-bromo-1,2-dimethylcyclobutane using HBr (Markovnikov addition). Elimination, followed by azonolysis of the resulting alkene gives the product, 2,5-hexanedione.

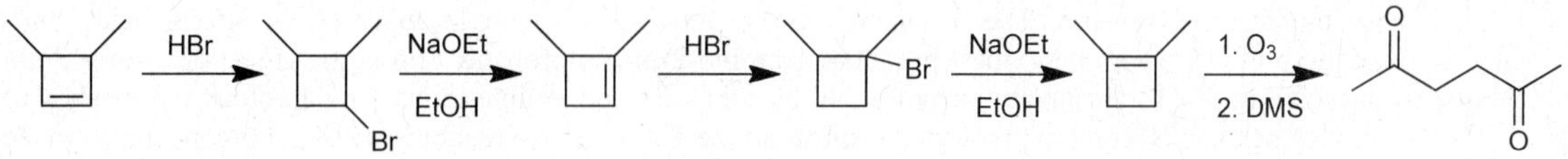

12.37.
The following synthesis is one suggested synthetic pathway. There are likely other acceptable approaches that accomplish the same goal.

Thus far, we have learned two ways to make aldehydes: (i) *anti*-Markovnikov hydration of a terminal alkyne or (ii) ozonolysis of an alkene, either of which is potentially reasonable here. However, in order to produce both of these compounds from a single synthetic protocol, a key recognition is that they can be produced from ozonolysis of the following disubstituted alkene.

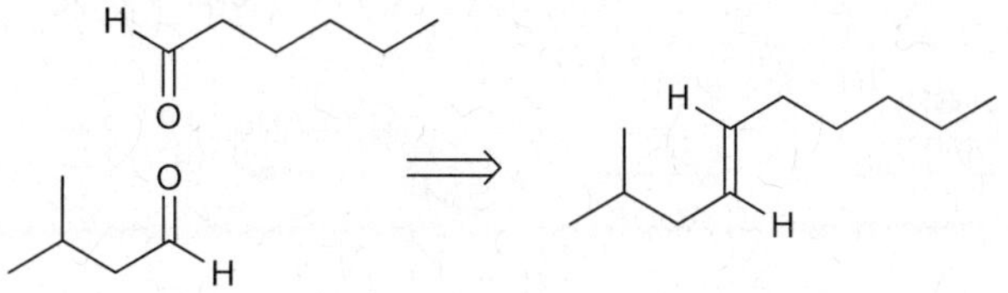

The following figure outlines a retrosynthetic analysis for our target molecule. An explanation of each of the steps (*a-k*) follows.

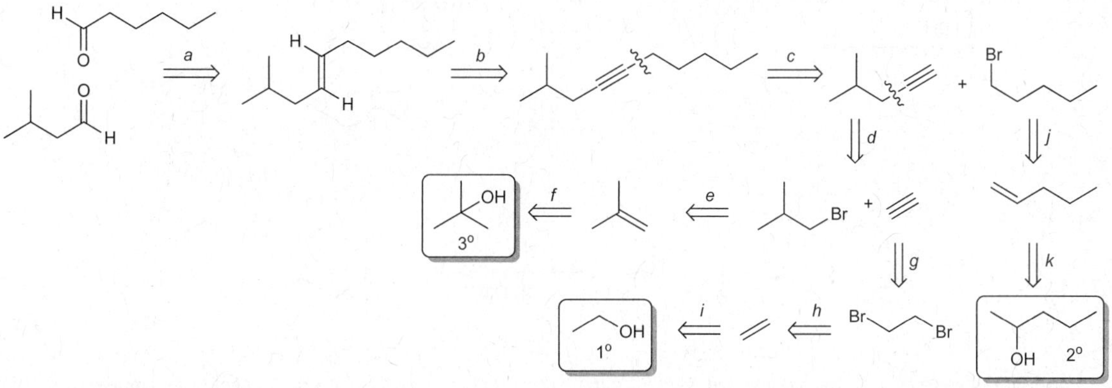

a. The mixture of aldehydes can be made by ozonolysis of this disubstituted alkene (*trans*-2-methyl-4-decene), as described above. (Note that the *E* alkene is shown here, but ozonolysis of the *Z* alkene would also produce the same two aldehydes.)

b. Thus far, we have learned two methods to make an alkene: 1) reduction of an alkyne or 2) elimination. In this case, the better choice to make this alkene is by reduction of the corresponding internal alkyne. The reason for this is explained in the next step.

c. We have to start our synthesis with compounds with fewer than six carbons, so this alkyne is a useful intermediate because we know how to make bonds between *sp* and *sp³* hybridized carbon atoms. This internal alkyne can thus be made from 4-methyl-1-pentyne (which must be deprotonated to produce a nucleophile) and 1-bromopentane.

d. The terminal alkyne can be made from acetylene (which must be deprotonated to produce a nucleophile) and 1-bromo-2-methylpropane.

e. Recall that we need to start with one 1°, one 2° and one 3° alcohol. The synthetic intermediate 1-bromo-2-methylpropane is the only one with a 3° carbon, so it follows that this compound is the one produced from a 3° alcohol. With this in mind, the alkyl halide can be produced from *anti*-Markovnikov addition of HBr to an alkene.

f. The alkene can be made from acid-catalyzed dehydration of the 3° alcohol.

g. Acetylene has only two carbons, so the only type of alcohol that can be used to make it is a 1° alcohol. Acetylene is thus made from double elimination of 1,2-dibromoethane.

h. 1,2-Dibromoethane is produced by bromination of ethylene.

i. Ethylene is produced by acid-catalyzed elimination from ethanol, a 1° alcohol.

j. 1-Bromopentane is made by *anti*-Markovnikov addition of HBr to 1-pentene.

k. 1-Pentene is made from the 2° alcohol by tosylation followed by reaction with a bulky base to give the less substituted product.

Now let's draw the forward scheme. The 3° alcohol is converted to 2-methylpropene using strong acid. *Anti*-Markovnikov addition of HBr (with peroxides) produces 1-bromo-2-methylpropane. Subsequent reaction with sodium acetylide (produced from the 1° alcohol by dehydration, bromination and double elimation/deprotonation as shown) produces 4-methyl-1-pentyne. Deprotonation with sodium amide followed by reaction with 1-bromopentane (made from the 2° alcohol by tosylation, elimination and *anti*-Markovnikov addition) yields 2-methyl-4-decyne. Reduction using sodium in liquid ammonia produces the *E* alkene. Ozonolysis followed by treatment with dimethylsulfide produces an equimolar ratio of the two products, 3-methylbutanal and hexanal.

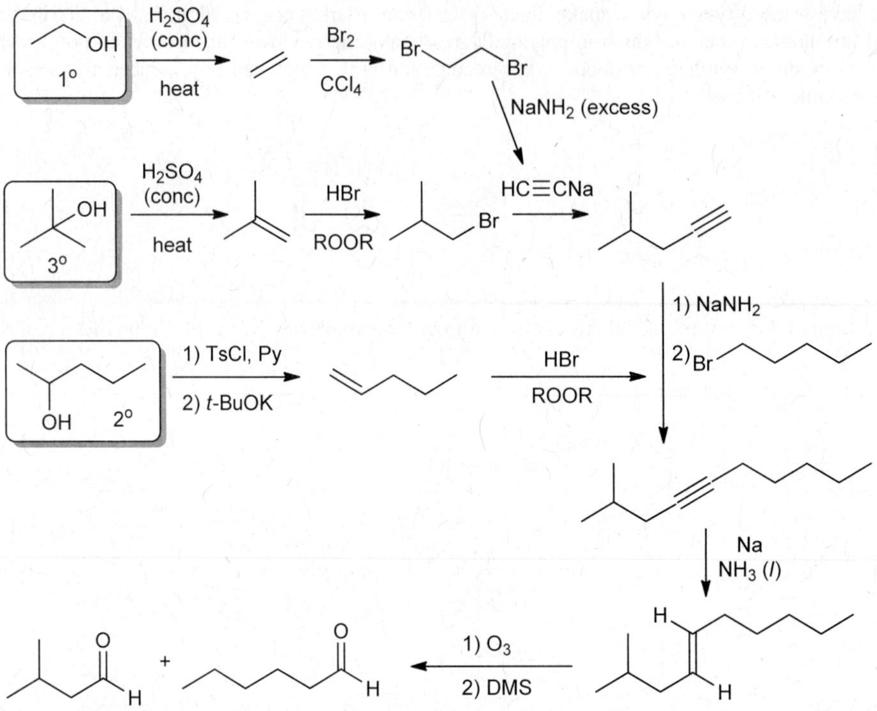

12.38.

The following synthesis is one suggested synthetic pathway. There are likely other acceptable approaches that accomplish the same goal.

An analysis of the structure of the product reveals that the five-carbon alkyl group (highlighted below) matches the skeletal structure of 2-methylbutane, the given starting material. This indicates which C-C bond (arrow) must be made in the course of the synthesis.

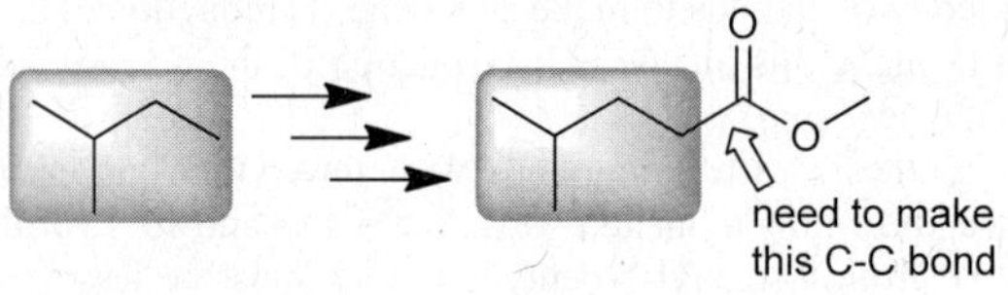

The figure below outlines a retrosynthetic analysis for our target compound. An explanation of each of the steps (*a-h*) follows.

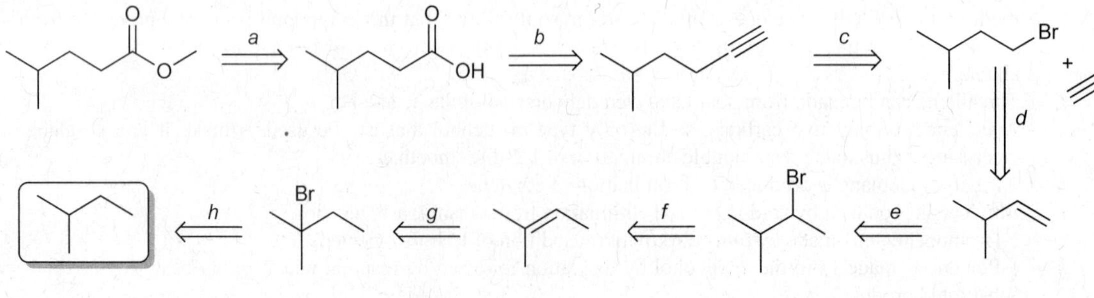

a.   The only way we have learned to make an ester (so far) is via a reaction of a carboxylate nucleophile (the conjugate base of a carboxylic acid) and an alkyl halide (in this case, bromomethane).

b.   The only way we have learned to make a carboxylic acid (so far) is by ozonolysis of an alkyne. An alternative intermediate to the terminal alkyne shown would be the symmetric internal alkyne, 2,9-dimethyl-5-decyne (not shown), which would produce two equivalents of the target carboxylic acid upon ozonolysis.

c.   The alkyne is prepared from the reaction of 1-bromo-3-methylbutane with an acetylide anion (formed by deprotonating acetylene). This alkyl halide has the same carbon skeleton as our given starting material (2-methylbutane), so our remaining steps involve primarily functional group manipulation.

d.   Knowing that the first synthetic step must be radical halogenation of 2-methylbutane to produce the tertiary alkyl halide (the only useful reaction of alkanes), we need to migrate the functionality back toward the tertiary carbon in this retrosynthetic analysis. Thus, 1-bromo-3-methylbutane can be prepared via *anti*-Markovnikov addition of HBr to 3-methyl-1-butene.

e.   3-Methyl-1-butene is prepared via elimination (with a sterically hindered base) from 2-bromo-3-methylbutane.

f.   2-Bromo-3-methylbutane is prepared via *anti*-Markovnikov addition of HBr to 2-methyl-2-butene.

g.   2-Methyl-2-butene is prepared via Zaitsev elimination from 2-bromo-2-methylbutane.

h.   2-bromo-2-methylbutane is made from our given starting material, 2-methylbutane, via radical bromination.

Now, let's draw the forward scheme. Radical bromination of 2-methylbutane produces the tertiary alkyl halide, selectively. Then, elimination with NaOEt, followed by *anti*-Markovnikov addition (HBr / peroxides), and then elimination with *tert*-butoxide, followed by another *anti*-Markovnikov addition (HBr / peroxides) produces 1-bromo-3-methylbutane. This alkyl halide will then undergo an S_N2 reaction when treated with an acetylide ion to give 5-methyl-1-hexyne. Ozonolysis of this terminal alkyne cleaves the CC triple bond, producing the carboxylic acid. Deprotonation (with NaOH) produces a carboxylate nucleophile that subsequently reacts with bromomethane in an S_N2 reaction to give the desired ester.

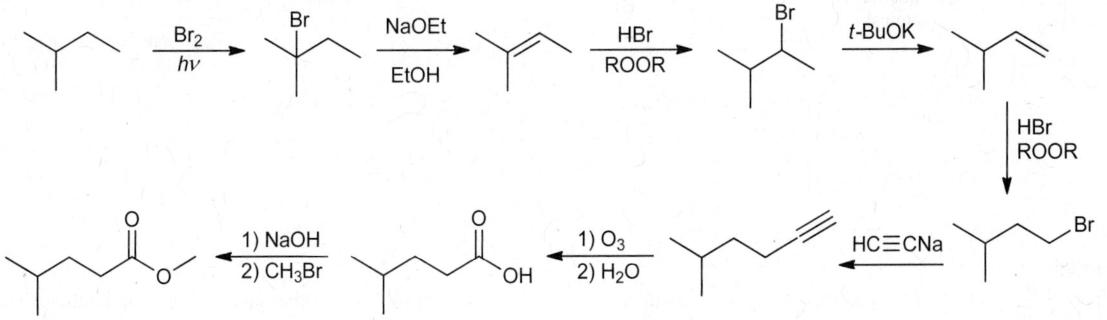

12.39.
The synthesis developed below is only one suggested synthetic pathway. There are likely other acceptable approaches that accomplish the same goal.

The following figure outlines a retrosynthetic analysis for our target molecule. An explanation of each of the steps (*a-h*) follows.

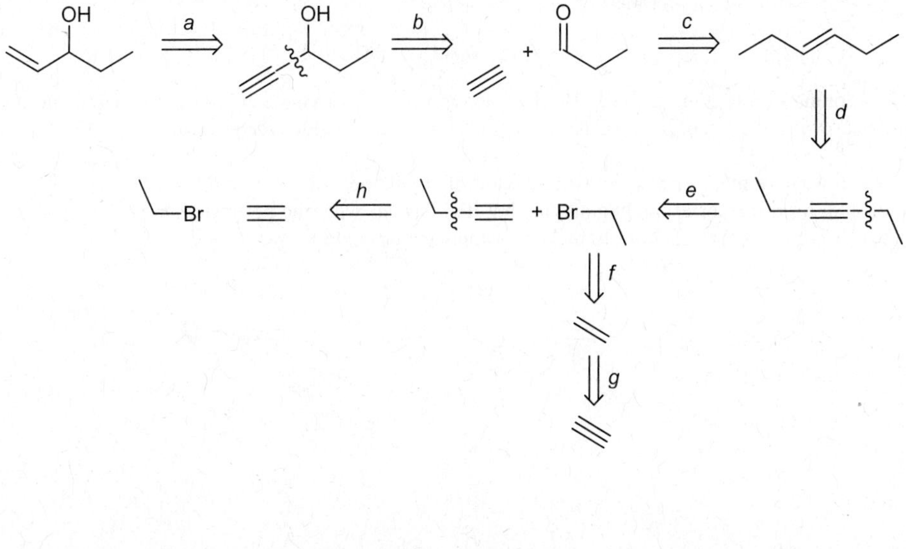

a. 1-Penten-3-ol can be made by reduction of the corresponding terminal alkyne.
b. The terminal alkyne can be made from acetylene (after deprotonation to form a nucleophile) and the aldehyde shown.
c. The aldehyde can be made by ozonolysis of (*E*)-3-hexene.
d. (*E*)-3-Hexene is prepared via reduction of 3-hexyne.
e. 3-Hexyne is made by alkylating 1-butyne with bromoethane.
f. Bromoethane is made by HBr addition to ethylene.
g. Ethylene is made by reduction of acetylene.
h. 1-Butyne is made by alkylation of acetylene using bromoethane (made as described above in steps *f-g*).

Now, let's draw the forward scheme. Acetylene is reduced to ethylene using molecular hydrogen and Lindlar's catalyst. Addition of HBr affords bromoethane. Reaction with sodium acetylide (prepared from acetylene and sodium amide) gives 1-butyne. Deprotonation with sodium amide followed by reaction with bromoethane produces 3-hexyne, which is subsequently reduced to (*E*)-3-hexene using a dissolving metal reduction. Alternatively, hydrogenation in the presence of Lindlar's catalyst will provide the *Z* alkene, which will also lead to the product via the same steps. Ozonolysis produces two equivalents of the desired aldehyde. Reaction with sodium acetylide gives the alkyne/alcohol which is reduced to the product using molecular hydrogen and Lindlar's catalyst.

12.40.
The following synthesis is one suggested synthetic pathway. There are likely other acceptable approaches that accomplish the same goal.
The target molecule, (*E*)-2-hexenal, is bifunctional - containing both an alkene group and an aldehyde group. We have learned two ways to make aldehydes: (i) ozonolysis of alkenes and (ii) *anti*-Markovnikov hydration of terminal alkynes. In this case, it is not immediately apparent which of these methods we should use. For example, ozonolysis of the compound below would not be a good approach, as both C=C bonds are susceptible to cleavage, yielding the following three products.

Anti-Markovnikov hydration of a terminal alkyne also does not appear to be a viable approach, as there is no obvious precursor that would allow installation of the C=C double bond adjacent to the aldehyde group.

Likewise, we know two ways to make an alkene: (i) reduction of an alkyne and (ii) elimination. We could potentially make the target molecule from the corresponding alkyne, but it is unclear what the next retrosynthetic step should be. (Note that in Chapter 13, we will learn reactions to make this approach possible.)

Installing the alkene group by elimination turns out to be a viable approach in this case, as described below.
The figure below outlines a retrosynthetic analysis for our target molecule. An explanation of each of the steps (*a-f*) follows.

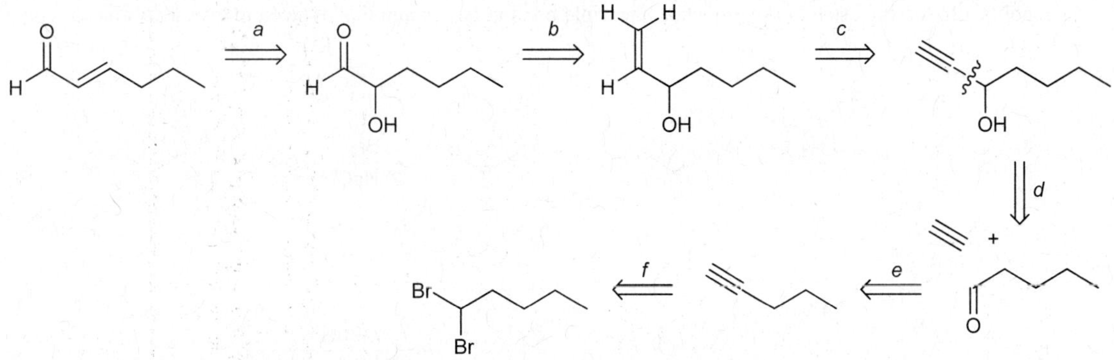

a. The target compound can be made via elimination of the alcohol under acidic conditions.
b. The aldehyde can be made by ozonolysis of the alkene shown here. Note that this compound only has one C=C double bond, so we avoid the problem described above with ozonolysis of a diene.
c. The alkene can be made by partial reduction of the corresponding alkyne.
d. This compound can be made from the reaction between an acetylide ion and the aldehyde shown (see problem 12.28).
e. The aldehyde is made from *anti*-Markovnikov addition of water to 1-pentyne.
f. 1-Pentyne is made from 1,1-dibromopentane by double elimination.

Now let's draw the forward scheme. 1,1-Dibromopentane is converted to 1-pentyne by reaction with excess sodium amide (to afford double elimination followed by deprotonation of the resulting alkyne), followed by aqueous workup to protonate the terminal alkynide. 1-Pentyne is converted to the aldehyde via hydroboration/oxidation. Subsequent reaction with sodium acetylide, followed by aqueous workup, produces an alcohol. Reduction with H_2 and Lindlar's catalyst converts the alkyne group to an alkene group. Ozonolysis converts the alkene to an aldehyde. Reaction with concentrated acid allows for elimination of the alcohol, producing the target compound.

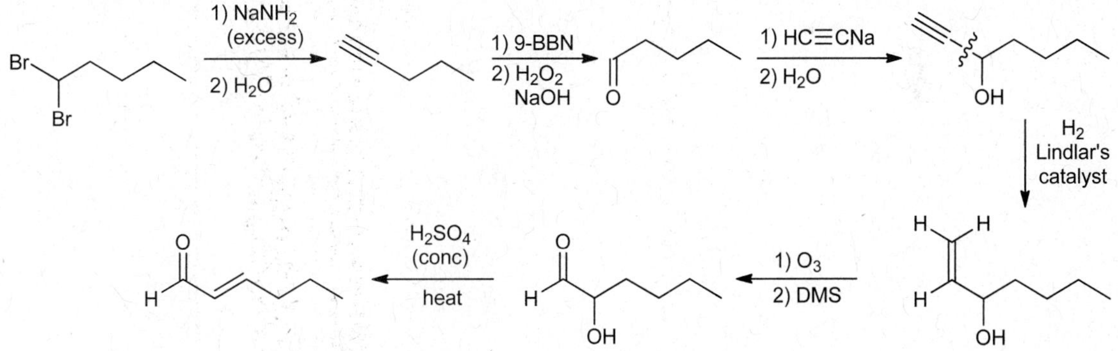

12.41. Treatment of compound **1** with BH_3 affords a chiral organoborane (**2**). Compound **3** is an alkyl halide, which is expected to react with an acetylide ion in an S_N2 reaction, to give compound **4**. Upon treatment with the strong base BuLi, compound **4** is deprotonated to give an alkynide ion, which then serves as a nucleophile in an S_N2 reaction to give compound **5**. As described in the problem statement, treatment of compound **5** with TsOH in methanol gives an alcohol (**6**). In the final step of the sequence, the triple bond in compound **6** is reduced to give a *cis* alkene (compound **7**).

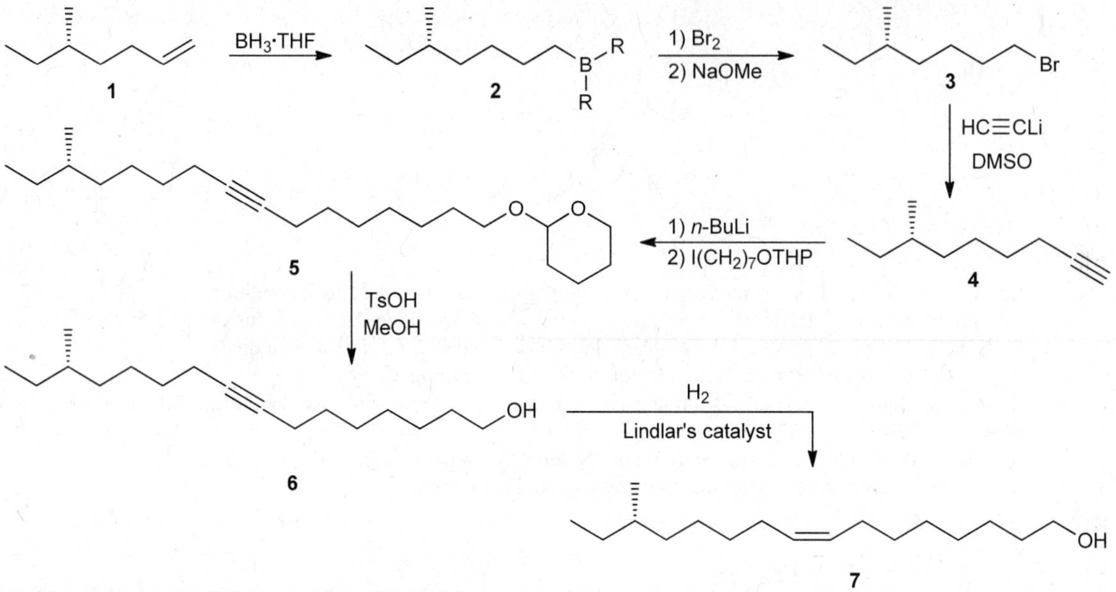

Chapter 13
Alcohols

Review of Concepts

Fill in the blanks below. To verify that your answers are correct, look in your textbook at the end of Chapter 13. Each of the sentences below appears verbatim in the section entitled *Review of Concepts and Vocabulary*.

- When naming an alcohol, the parent is the longest chain containing the _____ group.
- The conjugate base of an alcohol is called an _____ ion.
- Several factors determine the relative acidity of alcohols, including _____, _____, and _____.
- The conjugate base of phenol is called a _____, or _____ ion.

- When preparing an alcohol via a substitution reaction, primary substrates will require S_N___ conditions, while tertiary substrates will require S_N___ conditions.
- Alcohols can be formed by treating a _____ **group** (C=O bond) with a _____ **agent**.
- **Grignard reagents** are carbon nucleophiles that are capable of attacking a wide range of _____, including the carbonyl group of ketones or aldehydes, to produce an alcohol.
- _____ **groups**, such as the trimethylsilyl group, can be used to circumvent the problem of Grignard incompatibility and can be easily removed after the desired Grignard reaction has been performed.
- Tertiary alcohols will undergo an S_N___ reaction when treated with a hydrogen halide.
- Primary and secondary alcohols will undergo an S_N___ process when treated with either HX, $SOCl_2$, PBr_3, or when the hydroxyl group is converted into a tosylate group followed by nucleophilic attack.
- Tertiary alcohols undergo E1 elimination when treated with _____.
- Primary alcohols undergo **oxidation** twice to give a _____:
- Secondary alcohols are oxidized only once to give a _____
- PCC is used to convert a primary alcohol into an _____.
- NADH is a biological reducing agent that functions as a _____ delivery agent (very much like $NaBH_4$ or LAH), while NAD^+ is an _____ agent.
- There are two key issues to consider when proposing a synthesis:
    1. A change in the _____.
    2. A change in the _____.

Review of Skills

Fill in the blanks and empty boxes below. To verify that your answers are correct, look in your textbook at the end of Chapter 13. The answers appear in the section entitled *SkillBuilder Review*.

13.1 Naming an Alcohol

PROVIDE A SYSTEMATIC NAME FOR THE FOLLOWING COMPOUND

1) IDENTIFY THE PARENT
2) IDENTIFY AND NAME SUBSTITUENTS
3) ASSIGN LOCANTS TO EACH SUBSTITUENT
4) ALPHABETIZE
5) ASSIGN CONFIGURATION

13.2 Comparing the Acidity of Alcohols

FOR EACH PAIR OF COMPOUNDS BELOW, CIRCLE THE COMPOUND THAT IS MORE ACIDIC:

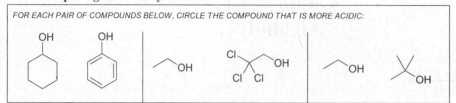

13.3 Identifying Oxidation and Reduction Reactions

IN THE FOLLOWING REACTION, DETERMINE WHETHER THE STARTING MATERIAL HAS BEEN OXIDIZED, REDUCED, OR NEITHER:

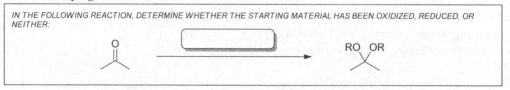

13.4 Drawing a Mechanism, and Predicting the Products of Hydride Reductions

COMPLETE THE MECHANISM BELOW BY DRAWING ALL CURVED ARROWS, INTERMEDIATES AND PRODUCTS.

13.5 Preparing an Alcohol via a Grignard Reaction

IDENTIFY REAGENTS THAT CAN ACHIEVE EACH OF THE FOLLOWING TRANSFORMATIONS

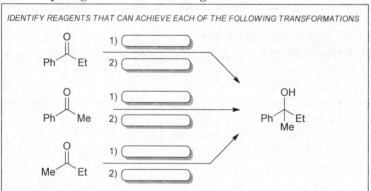

13.6 Proposing Reagents for the Conversion of an Alcohol into an Alkyl Halide

IDENTIFY REAGENTS THAT CAN ACHIEVE EACH OF THE FOLLOWING TRANSFORMATIONS

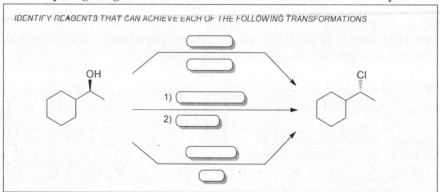

13.7 Predicting the Products of an Oxidation Reaction

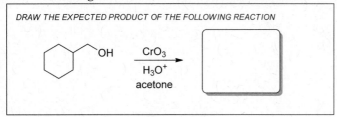

DRAW THE EXPECTED PRODUCT OF THE FOLLOWING REACTION

13.8 Converting Functional Groups

IDENTIFY REAGENTS THAT CAN ACHIEVE EACH OF THE FOLLOWING FUNCTIONAL GROUP TRANSFORMATIONS

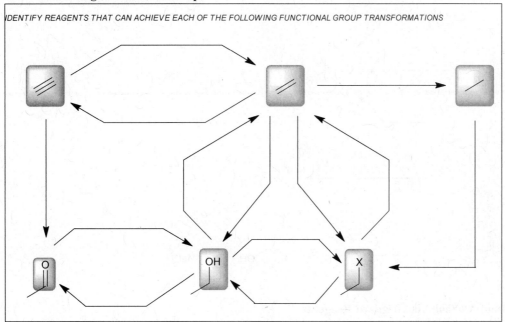

13.9 Proposing a Synthesis

AS A GUIDE FOR PROPOSING A SYNTHESIS, ASK THE FOLLOWING TWO QUESTIONS:

1) IS THERE A CHANGE IN THE _____ SKELETON?

2) IS THERE A CHANGE IN THE LOCATION OR IDENTITY OF THE _____?

AFTER PROPOSING A SYNTHESIS, USE THE FOLLOWING TWO QUESTIONS TO ANALYZE YOUR ANSWER:

1) IS THE _____ OUTCOME OF EACH STEP CORRECT?

2) IS THE _____ OUTCOME OF EACH STEP CORRECT?

Review of Reactions

Identify the reagents necessary to achieve each of the following transformations. To verify that your answers are correct, look in your textbook at the end of Chapter 13. The answers appear in the section entitled *Review of Reactions*.

Preparation of Alkoxides

$ROH \longrightarrow RO^{\ominus} \ Na^{\oplus}$

Preparation of Alcohols via Reduction

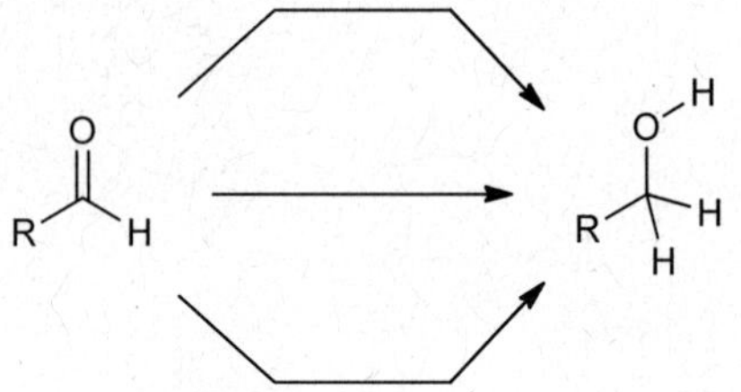

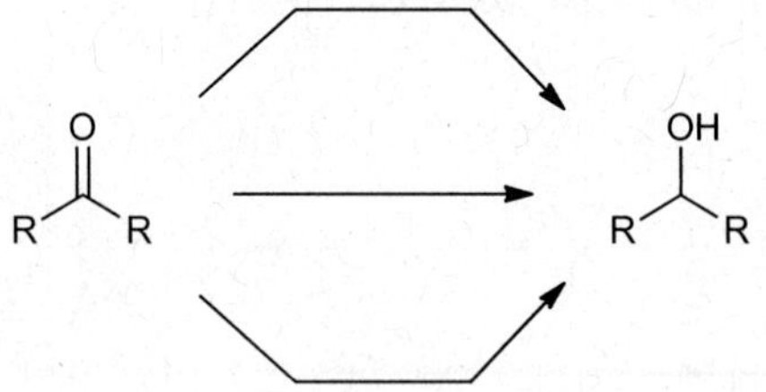

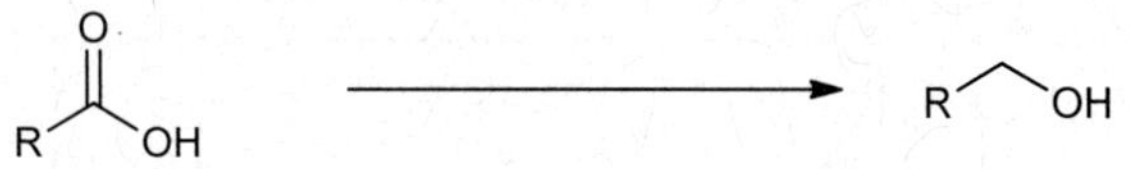

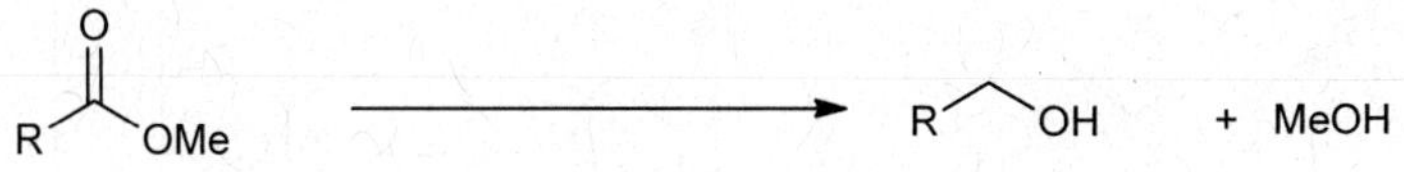

Preparation of Alcohols via Grignard Reagents

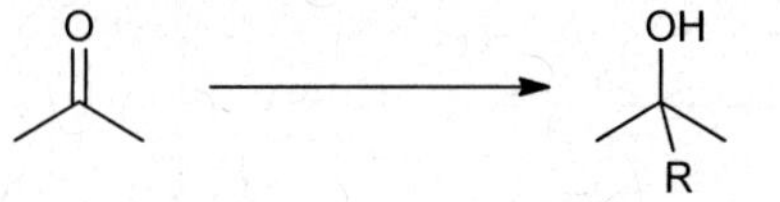

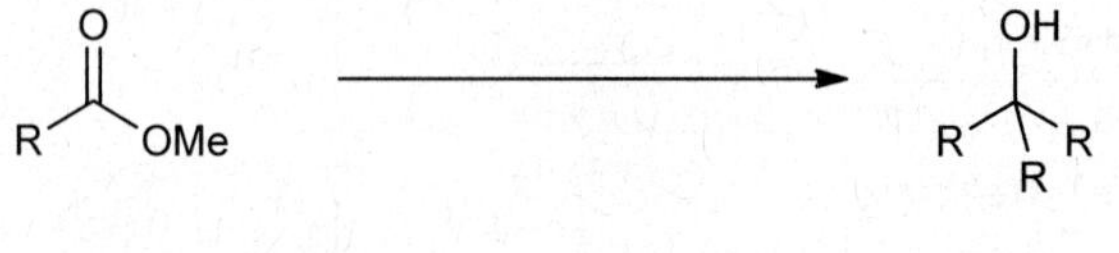

Protection and Deprotection of Alcohols

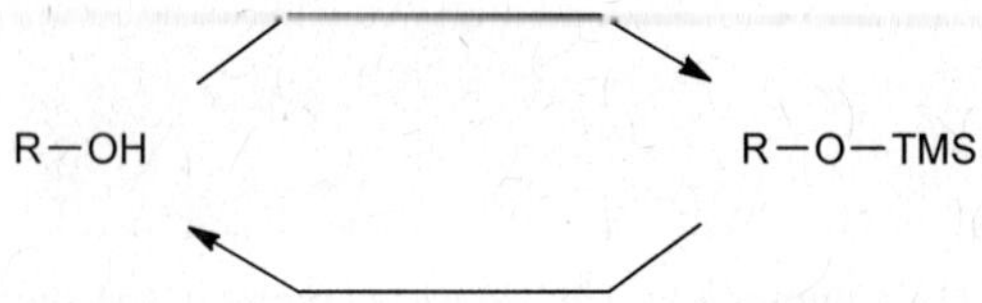

S$_N$1 Reactions with Alcohols

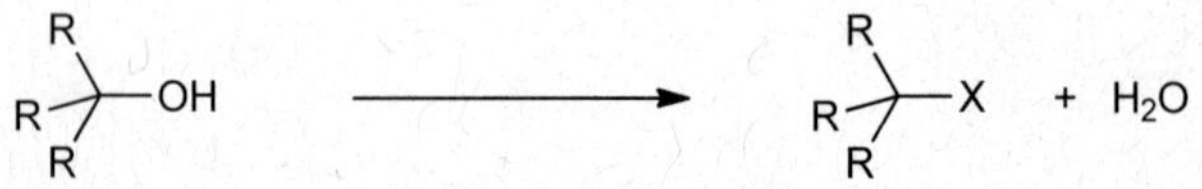

S_N2 Reactions with Alcohols

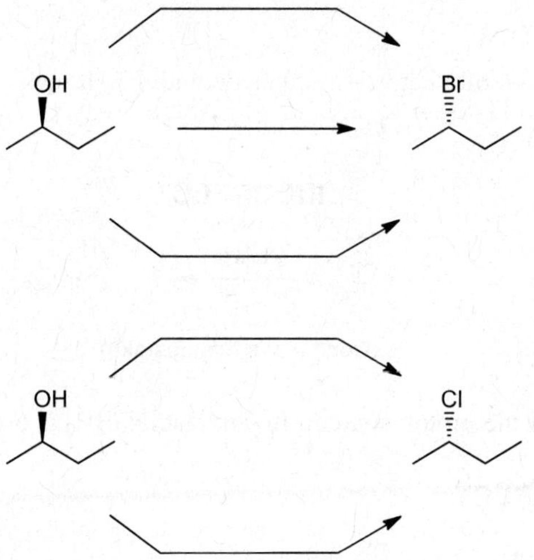

E1 and E2 Reactions with Alcohols

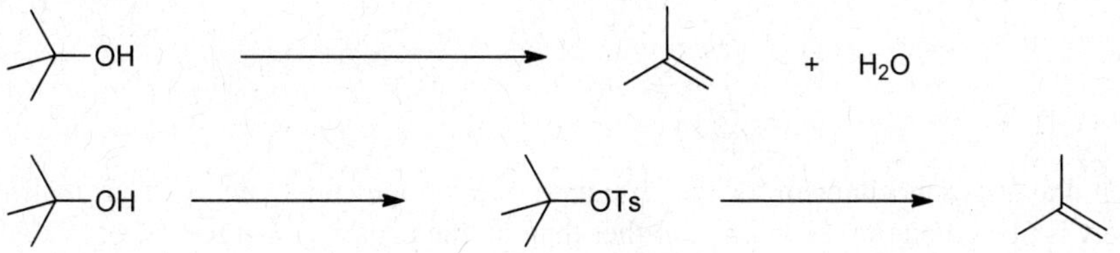

Oxidation of Alcohols and Phenols

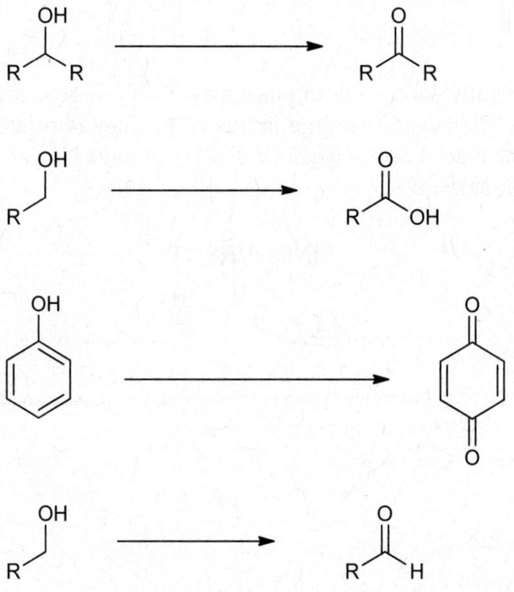

Common Mistakes to Avoid

If you are proposing a synthesis that involves reduction of a ketone or aldehyde with LAH, make sure that the water workup is shown as a separate step:

CORRECT
1) LAH
2) H_2O
OH
INCORRECT
LAH,
H_2O
OH
(Don't make this mistake)

This is important because LAH is incompatible with the proton source. In contrast, $NaBH_4$ is used in the presence of a proton source:

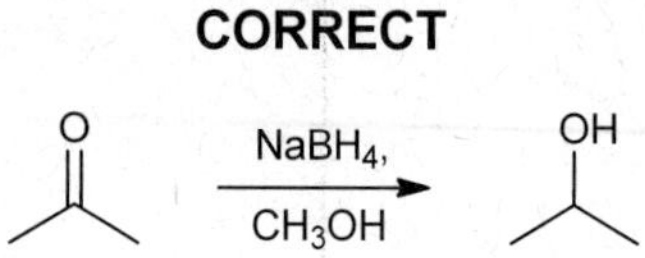
CORRECT
$NaBH_4$,
CH_3OH
OH

So, when using $NaBH_4$ as a reducing agent, do not show the proton source (MeOH) as a separate step.

Also, when drawing a mechanism for the reduction of a ketone, aldehyde, or ester, make sure that the first curved arrow is placed on the Al–H bond, rather than on the negative charge:

CORRECT
INCORRECT

In previous chapters, we have seen that it is generally acceptable to place the tail of a curved arrow on a negative charge, but this is an exceptional case. The negative charge in this case is not associated with a lone pair, so the tail of the curved arrow cannot be placed on the negative charge. It must be placed on the bond. This is true for reductions involving $NaBH_4$ as well:

CORRECT
INCORRECT

Useful reagents

The following is a list of reagents that were new in this chapter:

Reagents	Function
NaH	A very strong base, used to deprotonate an alcohol to give an alkoxide ion.
Na	Will react with an alcohol to liberate hydrogen gas, giving an alkoxide ion.
NaBH$_4$, MeOH	A reducing agent. Can be used to reduce ketones or aldehydes to alcohols. Will not reduce esters or carboxylic acids.
1) LAH 2) H$_2$O	A strong reducing agent. Can be used to reduce ketones, aldehydes, esters, or carboxylic acids to give an alcohol.
H$_2$, Pt	Reducing agent. Generally used to reduce alkenes to alkanes, but in some cases, it can also be used to reduce ketones to alcohols.
Mg	Can be used to convert an alkyl halide (RX) into a Grignard reagent (RMgX).
RMgX	A Grignard reagent. Examples include MeMgBr, EtMgBr and PhMgBr. These reagents are very strong nucleophiles (and very strong bases as well), and they will react with aldehydes or ketones. Aldehydes are converted into secondary alcohols (except for formaldehyde which is converted to a primary alcohol), while ketones are converted to tertiary alcohols. Esters are converted to tertiary alcohols when treated with excess Grignard.
TMSCl, Et$_3$N	Trimethylsilyl chloride, in the presence of a base (such as triethylamine), will protect an alcohol.
TBAF	Tetrabutyl ammonium fluoride. Used for deprotection of alcohols with silyl protecting groups.
HX	HBr and HCl are strong acids that also provide a source of a strong nucleophile. Can be used to convert an alcohol into an alkyl halide.
TsCl, pyridine	Will convert an alcohol into a tosylate. This is important because it converts a bad leaving group (HO$^-$) into a good leaving group (TsO$^-$).
PBr$_3$	Can be used to convert a primary or secondary alcohol into an alkyl bromide. If the OH group is connected to a chirality center, we expect inversion of configuration (typical for an S$_N$2 process).
SOCl$_2$, pyridine	Can be used to convert a primary or secondary alcohol into an alkyl chloride. If the OH group is connected to a chirality center, we expect inversion of configuration (typical for an S$_N$2 process).
HCl, ZnCl$_2$	Can be used to convert an alcohol into an alkyl chloride.
Na$_2$Cr$_2$O$_7$, H$_2$SO$_4$, H$_2$O	A mixture of sodium dichromate and sulfuric acid gives chromic acid, which is a strong oxidizing agent. Primary alcohols are oxidized to give carboxylic acids, while secondary alcohols are oxidized to give ketones. Tertiary alcohols are generally unreactive.
PCC, CH$_2$Cl$_2$	A mild oxidizing agent that will oxidize a primary alcohol to give an aldehyde, rather than a carboxylic acid. Secondary alcohols are oxidized to give ketones.

Solutions

13.1.

(a) We begin by identifying the parent. The carbon atom connected to the OH group must be included in the parent. The longest chain that includes this carbon atom is six carbon atoms in length, so the parent is hexanol. There are three substituents (highlighted). Notice that the parent chain is numbered starting from the side that is closest to the OH group (the OH group is at C2 rather than C5). According to this numbering scheme, the methyl group is located at C2, and the bromine atoms are both at C5. Finally, we assemble the substituents alphabetically. The compound does not contain any chirality centers.

5,5-dibromo-2-methylhexan-2-ol

(b) We begin by identifying the parent. The carbon atom connected to the OH group must be included in the parent. The longest chain that includes this carbon atom is five carbon atoms in length, so the parent is pentanol. There are three substituents (highlighted), all of which are methyl groups. Notice that the parent chain is numbered starting from the side that is closest to the OH group (thereby placing the OH group at C1). According to this numbering scheme, the methyl groups are at C2, C3 and C4. We use the prefix "tri" to indicate three methyl groups. Finally, we assign a configuration to each chirality center.

(2S,3R)-2,3,4-trimethylpentan-1-ol

(c) We begin by identifying the parent. The carbon atom connected to the OH group must be included in the parent. That carbon atom is part of a five-membered ring, so the parent is cyclopentanol. There are four substituents (highlighted), all of which are methyl groups. Notice that the parent chain is numbered starting from the carbon atom bearing the OH group (it is not necessary to indicate a locant for the OH group, because in a ring, it is assumed to be at C1, by definition). According to this numbering scheme, the methyl groups are at C2, C2, C5 and C5. We use the prefix "tetra" to indicate four methyl groups. The compound does not contain any chirality centers.

2,2,5,5-tetramethylcyclopentanol

(d) We begin by identifying the parent (phenol). There are two substituents (highlighted), both of which are ethyl groups. Notice that the ring is numbered starting from the carbon atom bearing the OH group (it is not necessary to indicate a locant for the OH group, because in a ring, it is assumed to be at C1, by definition). According to this numbering scheme, the ethyl groups are at C2 and C6. We use the prefix "di" to indicate two ethyl groups. The compound does not contain any chirality centers.

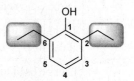

2,6-diethylphenol

(e) We begin by identifying the parent. The carbon atom connected to the OH group must be included in the parent. That carbon atom is part of a six-membered ring, so the parent is cyclohexanol. There are four substituents (highlighted), all of which are methyl groups. Notice that the parent chain is numbered starting from the carbon atom bearing the OH group (it is not necessary to indicate a locant for the OH group, because in a ring, it is assumed to be at C1, by definition). The numbers go counterclockwise, so as to give the lowest number to the first substituents (C2). According to this numbering scheme, the methyl groups are at C2, C2, C4 and C4. We use the prefix "tetra" to indicate four methyl groups. Finally, we assign a configuration to each chirality center.

(S)-2,2,4,4-tetramethylcyclohexanol

13.2.

(a) The parent (cyclohexanol) is a six-membered ring containing an OH group. The name indicates that there are two substituents (both Br) at the C3 position. The C1 position is a chirality center, and it has the *R* configuration.

(b) The parent (3-pentanol) is a chain of five carbon atoms with an OH group connected to the C3 position. The name indicates that there are two substituents (both methyl groups) at C2 and C3.   The C3 position is a chirality center, and it has the *S* configuration.

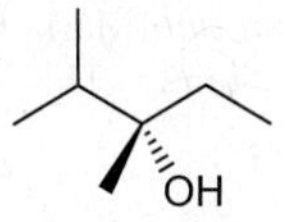

13.3.   Nonyl mandelate has a longer alkyl chain than octyl mandelate and is therefore more effective at penetrating cell membranes, rendering it a more potent agent.   Nonyl mandelate has a shorter alkyl chain than decyl mandelate and is therefore more water-soluble, enabling it to be transported through aqueous media and to reach its target destination more effectively.

13.4.
(a) When an alcohol is treated with elemental sodium (Na), the OH group is deprotonated, giving the following alkoxide ion.

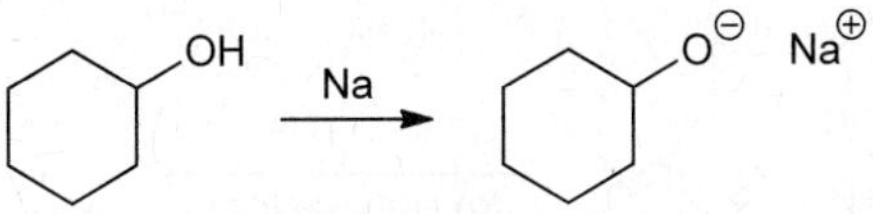

(b) When an alcohol is treated with sodium hydride (NaH), the OH group is deprotonated, giving the following alkoxide ion.

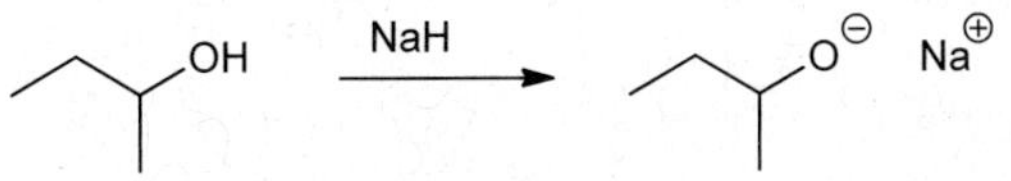

(c) When an alcohol is treated with elemental lithium (Li), the OH group is deprotonated, giving the following alkoxide ion.

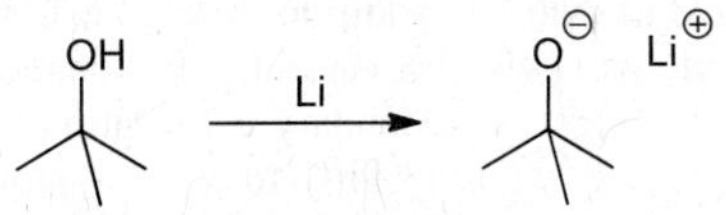

(d) When an alcohol is treated with sodium hydride (NaH), the OH group is deprotonated, giving the following alkoxide ion.

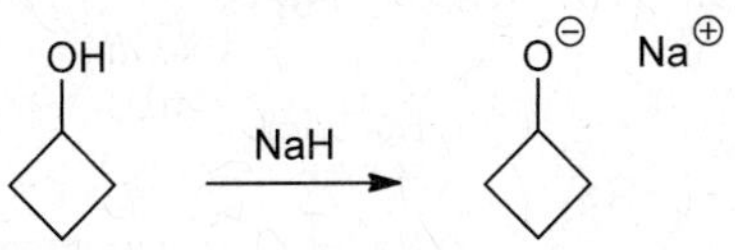

13.5.
(a) The second compound is more acidic because the electron-withdrawing effects of the fluorine atoms stabilize the conjugate base.

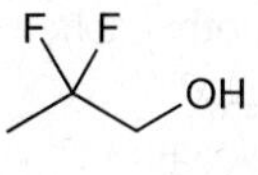

(b) The first compound is more acidic because the conjugate base of a primary alcohol will be more easily solvated than the conjugate base of a tertiary alcohol.

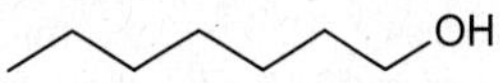

(c) The first compound is more acidic because the electron-withdrawing effects of the chlorine atoms stabilize the conjugate base.

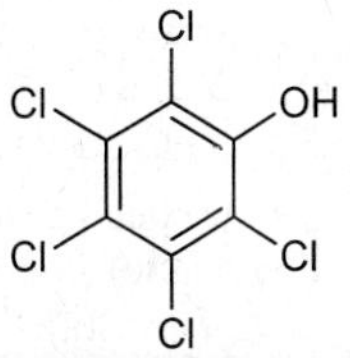

(d) The second compound is more acidic because its conjugate base is more stabilized by resonance, with the negative charge spread over two oxygen atoms, rather than just one oxygen atom.

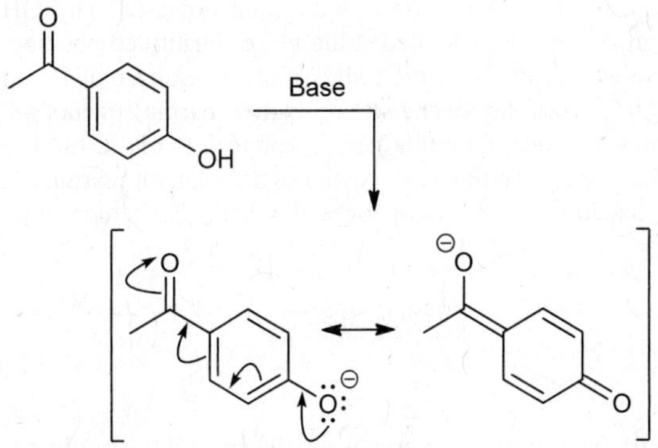

(e) The second compound is more acidic because its conjugate base is stabilized by resonance.   In contrast, the conjugate base of the first compound is not resonance-stabilized.

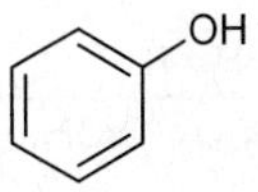

13.6.   2-Nitrophenol is expected to be more acidic (lower pK_a) because its conjugate base has a resonance structure in which the negative charge is spread onto an oxygen atom of the nitro group, shown below.   In contrast, 3-nitrophenol does not have such a resonance structure.

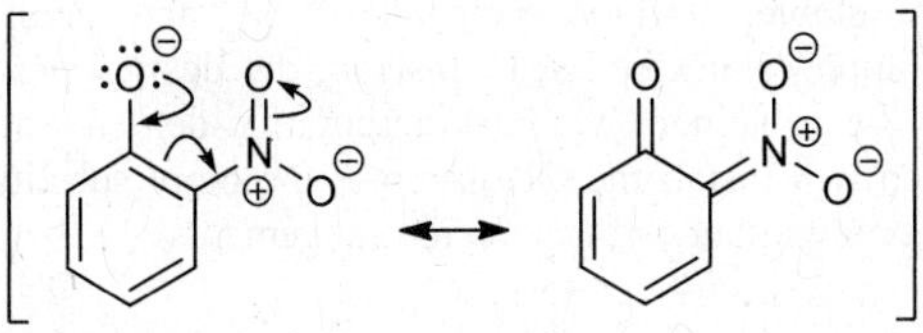

13.7.

(a)   The starting material is an alkyl halide, and the product is an alcohol, so we need a substitution reaction. The substrate (the alkyl halide) is tertiary, so we must use an S_N1 process. That is, we must use a weak nucleophile (water rather than hydroxide, as the latter would give E2).

(b)   The starting material is an alkyl halide, and the product is an alcohol, so we need a substitution reaction. The substrate (the alkyl halide) is primary, so we must use an S_N2 process. Therefore, we use a strong nucleophile (hydroxide).

(c) The starting material is an alkene, and the product is an alcohol, so we need an addition process. The OH group must be installed at the more substituted position, so we need to perform a Markovnikov addition of H and OH across the alkene. Carbocation rearrangements are not a concern in this case (protonation of the alkene generates a tertiary carbocation which cannot rearrange), so acid-catalyzed hydration will give the desired product.

(d)   The starting material is an alkene, and the product is an alcohol, so we need an addition process. The OH group must be installed at the less substituted position, so we need to perform an *anti*-Markovnikov addition of H and OH across the alkene. This can be achieved via hydroboration-oxidation.

(e) The starting material is an alkene, and the product is an alcohol, so we need an addition process. The OH group must be installed at the more substituted position, so we need to perform a Markovnikov addition of H and OH across the alkene. Carbocation rearrangements are a concern in this case (protonation of the alkene generates a secondary carbocation which can rearrange to give a more stable, tertiary carbocation), so acid-catalyzed hydration cannot be used. Instead, the desired product can be obtained via oxymercuration-demercuration, which will install the OH group at the more substituted position without carbocation rearrangements.

(f) The starting material is an alkene, and the product is an alcohol, so we need an addition process. The OH group must be installed at the less substituted position, so we need to perform an *anti*-Markovnikov addition of H and OH across the alkene. This can be achieved via hydroboration-oxidation.

13.8

(a) Let's begin by drawing the starting material.

1-hexene

Addition of H and OH across this alkene will provide an alcohol. Markovnikov addition will give a secondary alcohol, so we must perform an *anti*-Markovnikov addition in order to obtain a primary alcohol. This can be achieved via hydroboration-oxidation.

(b) Let's begin by drawing the starting material.

3,3-dimethyl-1-hexene

Addition of H and OH across this alkene will provide an alcohol. Markovnikov addition will give a secondary alcohol, but we must be careful. Protonation of the alkene will generate a secondary carbocation which can rearrange (via a methyl shift) to give a more stable, tertiary carbocation. Therefore, acid-catalyzed hydration cannot be used. Instead, the desired product can be obtained via oxymercuration-demercuration, which will install the OH group at the more substituted position without carbocation rearrangements.

(c) Let's begin by drawing the starting material.

2-methyl-1-hexene

Addition of H and OH across this alkene will provide an alcohol. Specifically, Markovnikov addition will give a tertiary alcohol. Carbocation rearrangements are not a concern in this case (protonation of the alkene generates a tertiary carbocation which cannot rearrange), so acid-catalyzed hydration will give the desired product.

13.9.
(a) We focus on the carbon atom that undergoes a change in bonding as a result of the transformation. In the starting material, that carbon atom (the central carbon atom) has an oxidation state of +2. In the product, the same carbon atom has an oxidation state of +2. Since the oxidation state does not change, the starting material is neither oxidized nor reduced.

(b) We focus on the carbon atom that undergoes a change in bonding as a result of the transformation. In the starting material, that carbon atom has an oxidation state of +1. In the product, the same carbon atom has an oxidation state of +3. Since the oxidation state increases as a result of the transformation, the starting material is oxidized.

(c) We focus on the carbon atom that undergoes a change in bonding as a result of the transformation. In the starting material, that carbon atom has an oxidation state of +3. In the product, the same carbon atom has an oxidation state of -1. Since the oxidation state decreases as a result of the transformation, the starting material is reduced.

(d) We focus on the carbon atom that undergoes a change in bonding as a result of the transformation. In the starting material, that carbon atom has an oxidation state of +3. In the product, the same carbon atom has an oxidation state of +3. Since the oxidation state does not change, the starting material is neither oxidized nor reduced.

(e) We focus on the carbon atom that undergoes a change in bonding as a result of the transformation. In the starting material, that carbon atom has an oxidation state of 0. In the product, the same carbon atom has an oxidation state of +2. Since the oxidation state increases as a result of the transformation, the starting material is oxidized.

(f) We focus on the carbon atom that undergoes a change in bonding as a result of the transformation. In the starting material, that carbon atom has an oxidation state of +2. In the product, the same carbon atom has an oxidation state of +3. Since the oxidation state increases as a result of the transformation, the starting material is oxidized.

13.10. One carbon atom is reduced from an oxidation state of 0 to an oxidation state of -1, while the other carbon atom is oxidized from an oxidation state of 0 to an oxidation state of +1. Overall, the starting material does not undergo a net change in oxidation state and is, therefore, neither reduced nor oxidized.

13.11. One carbon atom is reduced from an oxidation state of 0 to an oxidation state of -2, while the other carbon atom is oxidized from an oxidation state of 0 to an oxidation state of +2. Overall, the starting material does not undergo a net change in oxidation state and is, therefore, neither reduced nor oxidized.

13.12.
(a) Two curved arrows are used to show hydride delivery. Note that the tail of the first curved arrow is placed on the bond between Al and H (it is NOT placed on the negative charge). The resulting alkoxide ion is then protonated upon treatment with water. This protonation step requires two curved arrows, as shown.

(b) Two curved arrows are used to show hydride delivery. Note that the tail of the first curved arrow is placed on the bond between Al and H (it is NOT placed on the negative charge). The resulting alkoxide ion is then protonated upon treatment with water. This protonation step requires two curved arrows, as shown.

(c) Two curved arrows are used to show hydride delivery. Note that the tail of the first curved arrow is placed on the bond between B and H (it is NOT placed on the negative charge). The resulting alkoxide ion is then protonated, which requires two curved arrows, as shown.

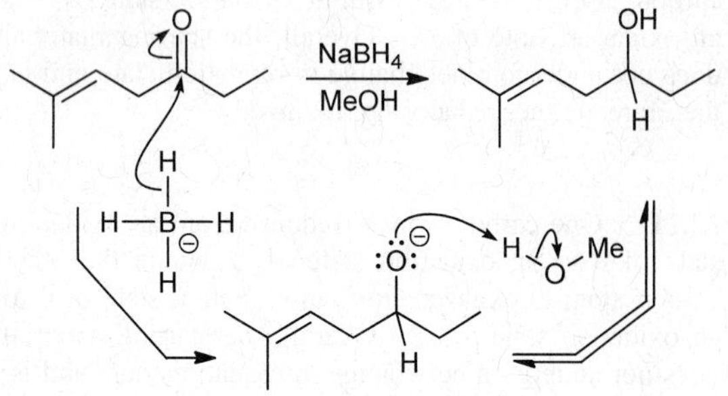

(d) Two curved arrows are used to show hydride delivery. Note that the tail of the first curved arrow is placed on the bond between Al and H (it is NOT placed on the negative charge). The resulting alkoxide ion is then protonated upon treatment with water. This protonation step requires two curved arrows, as shown.

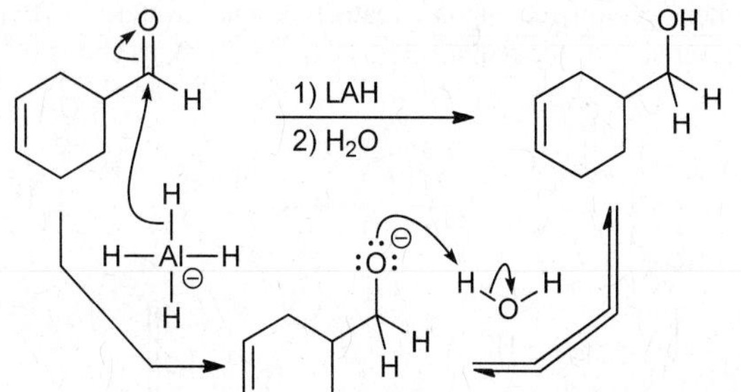

(e) Two curved arrows are used to show hydride delivery to the ester. Note that the tail of the first curved arrow is placed on the bond between Al and H (it is NOT placed on the negative charge). The resulting intermediate then ejects ethoxide as a leaving group, which requires two curved arrows. The resulting aldehyde is then further reduced by another equivalent of LAH. Once again, two curved arrows are used to show hydride delivery. The resulting alkoxide ion is then protonated upon treatment with water. This protonation step requires two curved arrows, as shown.

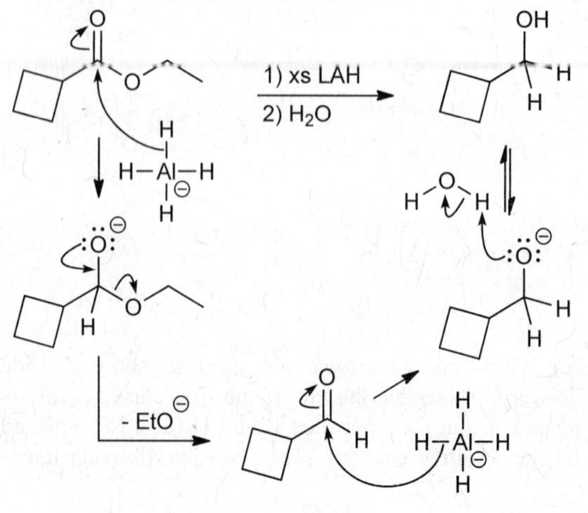

(f) Two curved arrows are used to show hydride delivery. Note that the tail of the first curved arrow is placed on the bond between B and H (it is NOT placed on the negative charge). The resulting alkoxide ion is then protonated, which requires two curved arrows, as shown.

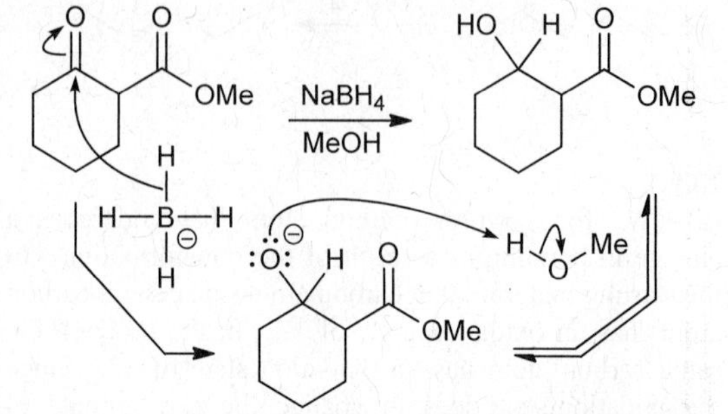

13.13. Two curved arrows are used to show delivery of hydride to the ester. Note that the tail of the first curved arrow is placed on the bond between Al and H (it is NOT placed on the negative charge). The resulting intermediate then ejects a leaving group, which requires two curved arrows. The resulting aldehyde is then further reduced by another equivalent of LAH. Once again, two curved arrows are used to show hydride delivery, resulting in a dianion.

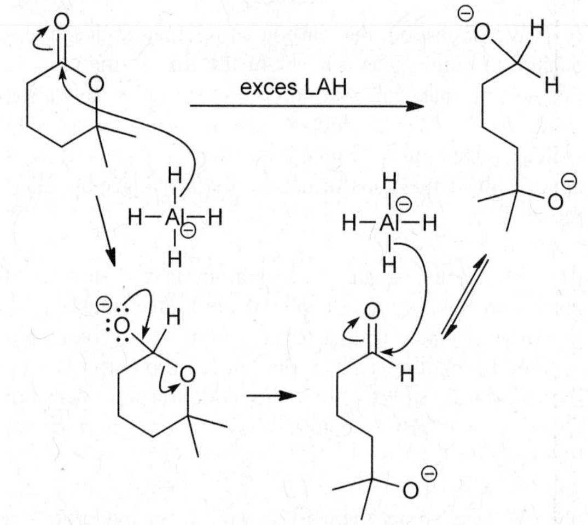

The resulting dianion is then protonated upon treatment with aqueous acid. There are two locations that are protonated, each of which requires two curved arrows, as shown. Notice that each protonation step is drawn separately. The less stable negative charge (the stronger base) is protonated first (see Section 3.7 to determine which negative charge is less stable).

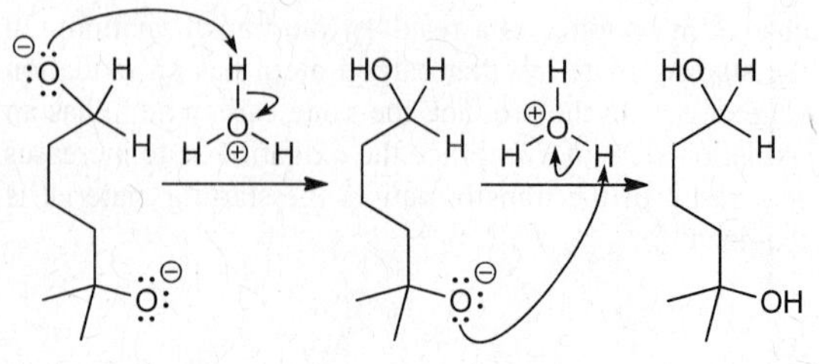

13.14.

(a) The desired product has only two groups connected to the α position. Either one of these groups could have been installed via a Grignard reaction with the appropriate aldehyde, as shown.

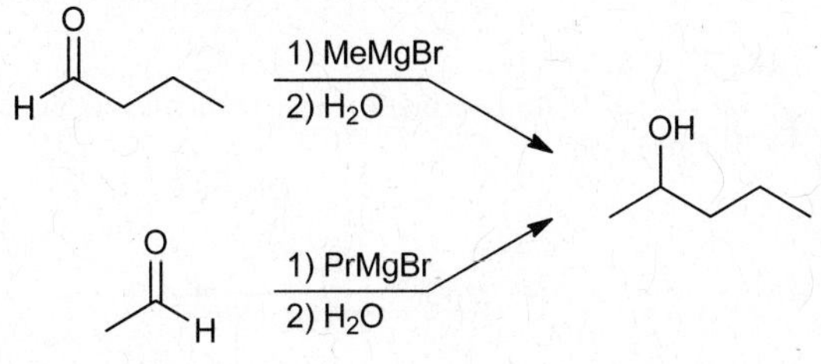

(b) The desired product has three groups connected to the α position. Any one of these groups could have been installed via a Grignard reaction with the appropriate ketone, as shown.

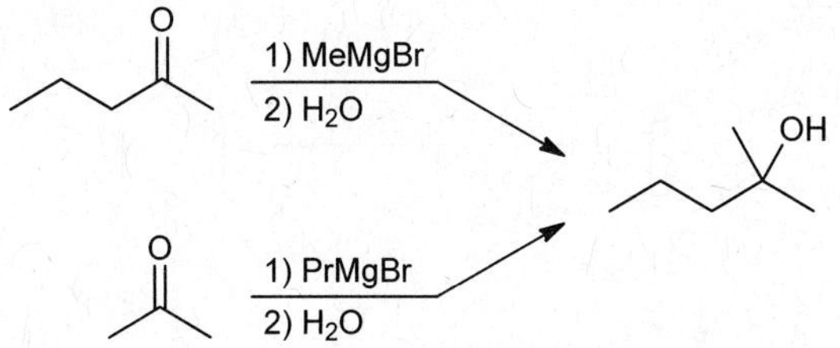

(c) The desired product has only one group connected to the α position. That group could have been installed via a Grignard reaction with formaldehyde, as shown.

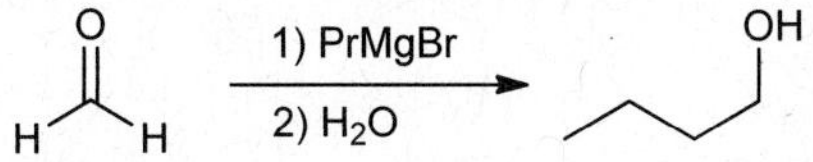

(d) The desired product has only two groups connected to the α position. Either one of these groups could have been installed via a Grignard reaction with the appropriate aldehyde, as shown.

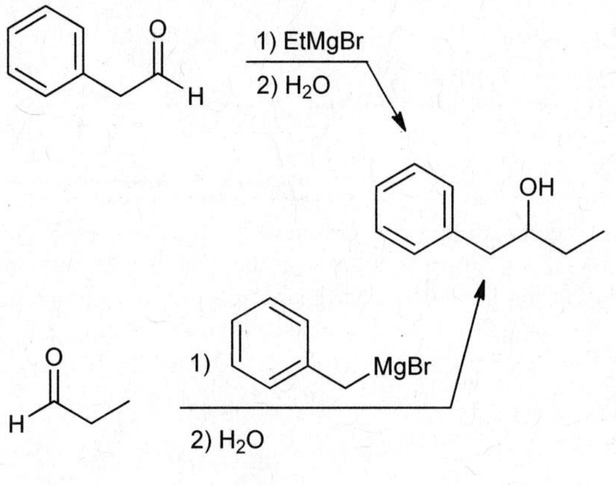

(e) The desired product has three groups connected to the α position. Any one of these groups could have been installed via a Grignard reaction with the appropriate ketone, as shown.

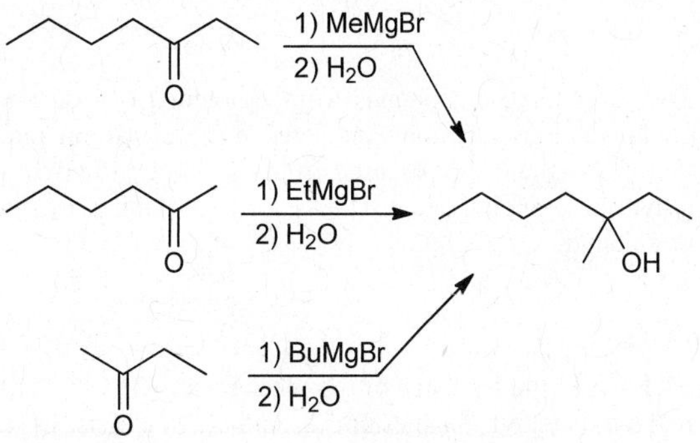

(f) The desired product has three groups connected to the α position, although two of them are identical. So there are only two different groups that could have been installed via a Grignard reaction, shown here.

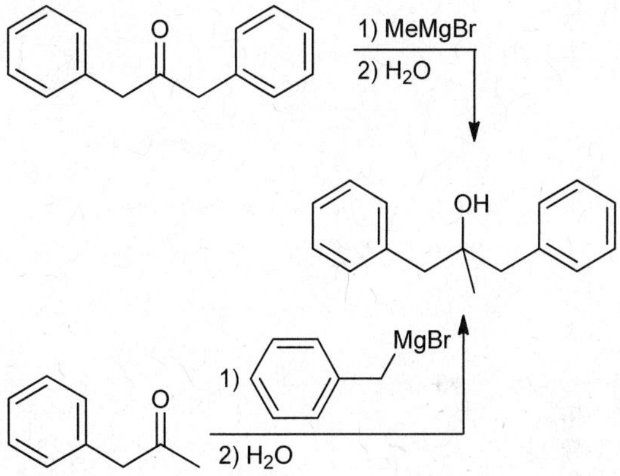

13.15 Each of the following two compounds can be prepared from the reaction between a Grignard reagent and an ester, because each of these compounds has two identical groups connected to the α position:

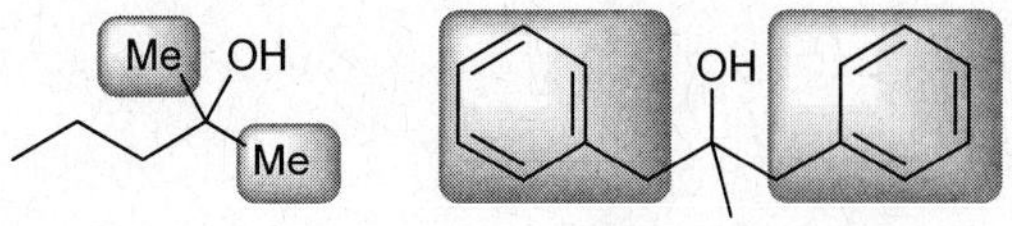

The other four compounds from Problem 13.14 do not contain two identical groups connected to the α position, and cannot be prepared from the reaction between an ester and a Grignard reagent.

13.16 Each of the following three compounds can be prepared from the reaction between a hydride reducing agent (NaBH$_4$ or LAH) and a ketone or aldehyde, because each of these compounds has a hydrogen atom connected to the α position:

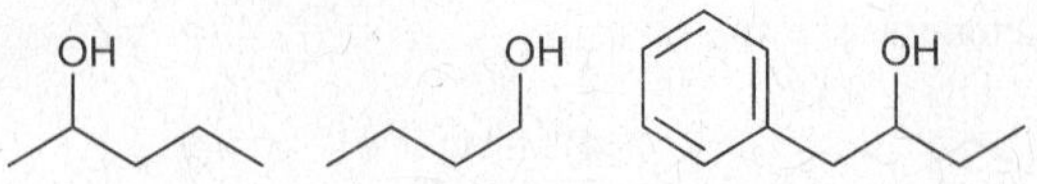

The other three compounds from Problem 13.14 do not contain a hydrogen atom connected to the α position and, therefore, cannot be prepared from the reaction between a hydride reducing agent (NaBH$_4$ or LAH) and a ketone or aldehyde.

13.17. In the first step of the mechanism, the Grignard reagent (methyl magnesium bromide) functions as a nucleophile and attacks the C=O bond of the ester. This step requires two curved arrows. The resulting intermediate then ejects a leaving group to give a ketone, which requires two curved arrows. The ketone is then further attacked by another equivalent of the Grignard reagent. Once again, two curved arrows are used to show the nucleophilic attack, resulting in a dianion.

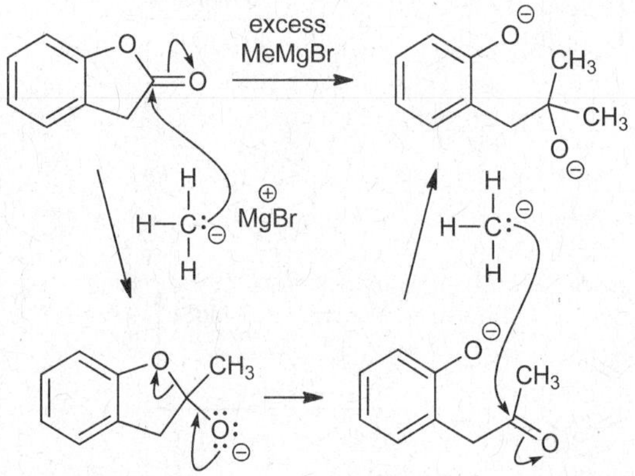

The resulting dianion is then protonated upon treatment with aqueous acid. There are two locations that are protonated, each of which requires two curved arrows, as shown.

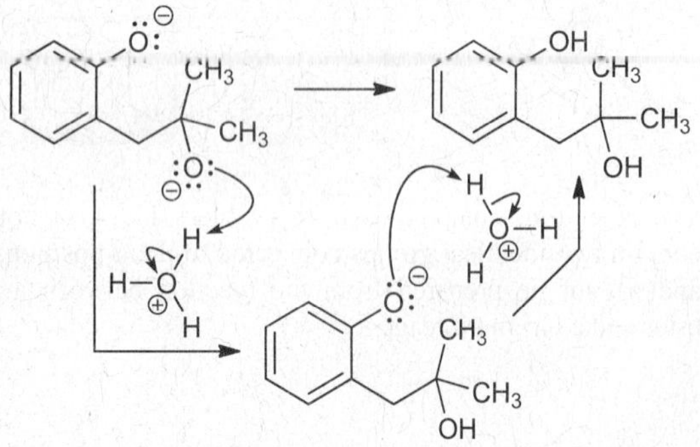

Notice that each protonation step is drawn separately. The less stable negative charge (the alkoxide) is protonated first (rather than phenolate, which is more

stable because it is resonance-stabilized). In fact, the phenolate anion is even more stable than a hydroxide ion, which explains why the proton source must be H$_3$O$^+$ rather than water. Water is not sufficiently acidic to protonate a phenolate ion (see Section 13.2, Acidity of Alcohols and Phenols).

13.18.
(a) This type of transformation can be achieved via a Grignard reaction.

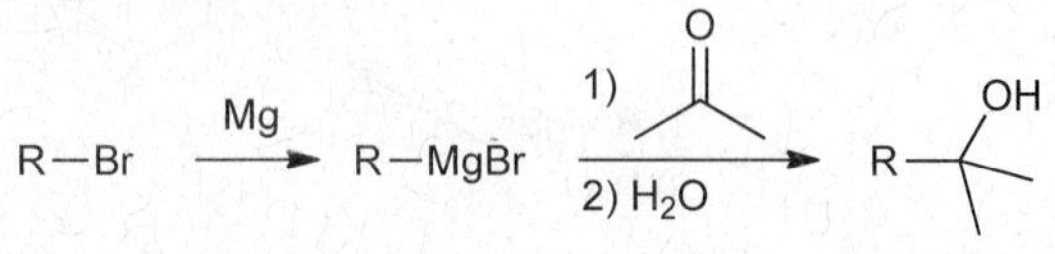

However, the starting material has an OH group, which is incompatible with a Grignard reaction. To resolve this issue, we must first protect the OH group and then perform the desired Grignard reaction. Deprotection then gives the desired product, as shown.

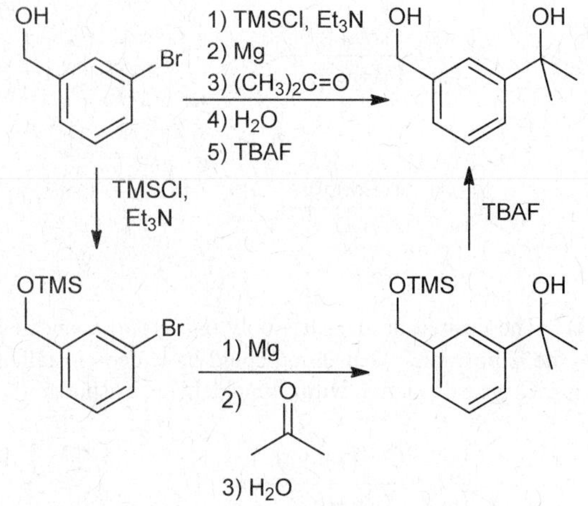

(b) This type of transformation can be achieved via a Grignard reaction in which the Grignard reagent is treated with 0.5 equivalents of an ester.

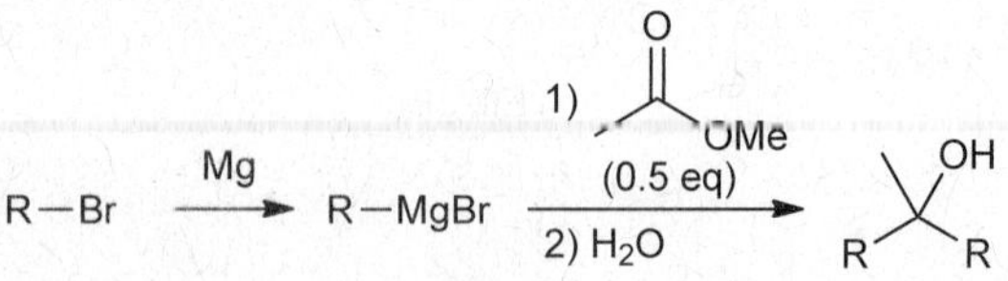

However, the starting material has an OH group, which is incompatible with a Grignard reaction. To resolve this issue, we must first protect the OH group and then perform the desired Grignard reaction. Deprotection then gives the desired product, as shown (next page).

13.19.

(a) We must convert a secondary alcohol into a secondary alkyl bromide, with inversion of configuration, so an S_N2 process is required. However, the OH group is a bad leaving group. One way around this issue is to convert the OH group into a tosylate (by treating the alcohol with tosyl chloride and pyridine). The tosylate can then be treated with bromide to give the desired product. Alternatively, the alcohol can be treated with PBr_3 to give the desired product.

(b) We must convert a tertiary alcohol into a tertiary alkyl bromide. This can be achieved upon treatment with HBr.

(c) We must convert a secondary alcohol into a secondary alkyl chloride, with inversion of configuration, so an S_N2 process is required. However, the OH group is a bad leaving group. One way around this issue is to convert the OH group into a tosylate (by treating the alcohol with tosyl chloride and pyridine). The tosylate can then be treated with chloride to give the desired product. Alternatively, the alcohol can be treated with thionyl chloride and pyridine, or with HCl in the presence of $ZnCl_2$.

(d) We must convert a secondary alcohol into a secondary alkyl bromide, with inversion of configuration, so an S_N2 process is required. However, the OH group is a bad leaving group. One way around this issue is to convert the OH group into a tosylate (by treating the alcohol with tosyl chloride and pyridine). The tosylate can then be treated with bromide to give the desired product. Alternatively, the alcohol can be treated with PBr_3 to give the desired product.

(e) We must convert a secondary alcohol into a secondary alkyl chloride, so a substitution process is required. However, the OH group is a bad leaving group. One way around this issue is to convert the OH group into a tosylate (by treating the alcohol with tosyl chloride and pyridine). The tosylate can then be treated with chloride to give the desired product. Alternatively, the alcohol can be treated with thionyl chloride and pyridine.

(f) We must convert a primary alcohol into a primary alkyl chloride, so an S_N2 process is required. However, the OH group is a bad leaving group. One way around this issue is to convert the OH group into a tosylate (by treating the alcohol with tosyl chloride and pyridine). The tosylate can then be treated with bromide to give the desired product. Alternatively, the alcohol can be treated with HBr or with PBr_3.

13.20. We must convert a secondary alcohol into a secondary alkyl chloride, so a substitution reaction is required. However, the OH group is a bad leaving group, so we must first convert the OH group into a tosylate (by treating the alcohol with tosyl chloride and pyridine). Notice that during the conversion of the OH group into a tosylate group, the configuration of the chirality center remains unchanged.

The desired transformation must now be performed without changing the configuration of the chirality center (called retention of configuration).

However, we have not yet encountered a substitution reaction that gives retention of configuration (S_N2 gives inversion, and S_N1 gives racemization). The key to solving this problem is to recognize that two successive S_N2 reactions can achieve the desired retention of configuration.

This strategy requires that Z is both a good nucleophile (in order for OTs to be replaced with Z) and a good leaving group (in order for Z to be replaced with Cl). Iodide is an excellent candidate to fill this function, because it is both a good nucleophile and a good leaving group.
In summary, the following synthesis can achieve the desired transformation.

13.21.
(a) An alcohol is converted to an alkene upon treatment with concentrated sulfuric acid. In this case, there are two possible regiochemical outcomes, and we expect that the more substituted alkene will be the major product.

(b) An alcohol is converted to a tosylate upon treatment with tosyl chloride and pyridine. This tosylate is a secondary substrate, and ethoxide is both a strong nucleophile and a strong base. Recall from Section 8.13 that a secondary substrate is expected to react with ethoxide via an E2 process to give the major product (while S_N2 gives the minor product).

13.22.
(a) The alcohol in this case is primary. PCC will oxidize the alcohol to give an aldehyde, and the other aldehyde group (already present) is unaffected.

(b) The alcohol in this case is primary. Chromic acid will oxidize the alcohol to give a carboxylic acid.

(c) The alcohol in this case is primary. This alcohol is oxidized upon treatment with chromic acid to give a carboxylic acid group, and the aldehyde group (already present) also undergoes oxidation to give a carboxylic acid group.

(d) The alcohol in this case is primary. PCC will oxidize the alcohol to give an aldehyde.

(e) The alcohol in this case is secondary. PCC will oxidize the alcohol to give a ketone.

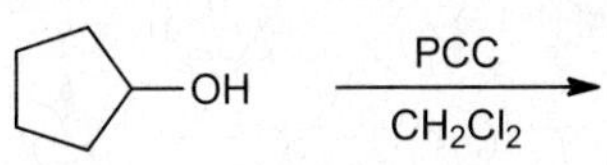

(f) The alcohol in this case is secondary. Chromic acid will oxidize the alcohol to give a ketone.

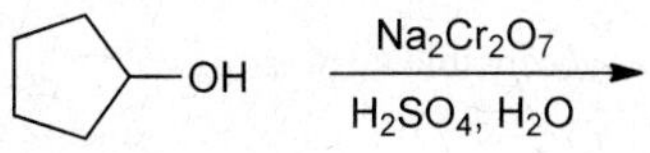

13.23.
(a) The desired aldehyde can be made from the corresponding primary alcohol via an oxidation process.

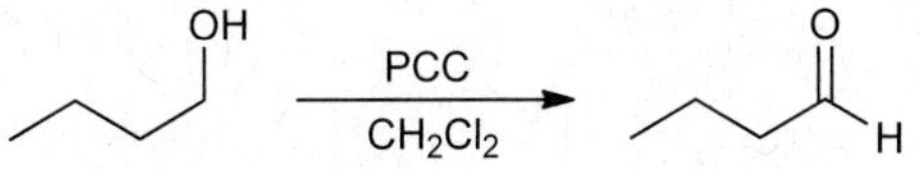

This primary alcohol (1-butanol) can be made from the starting alkyl bromide via an S_N2 process (upon treatment with hydroxide).

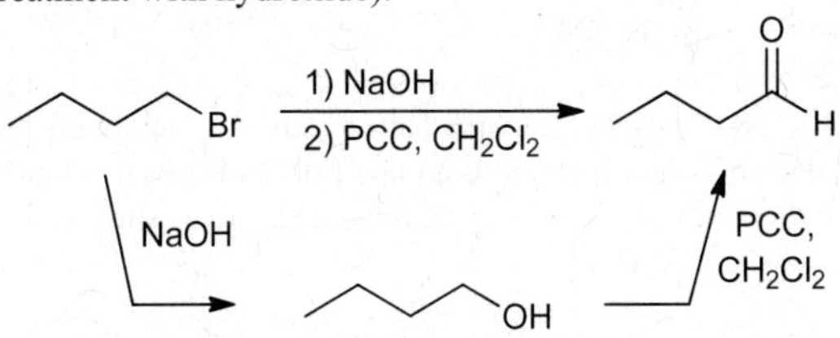

(b) The desired aldehyde can be made from the corresponding primary alcohol via an oxidation process.

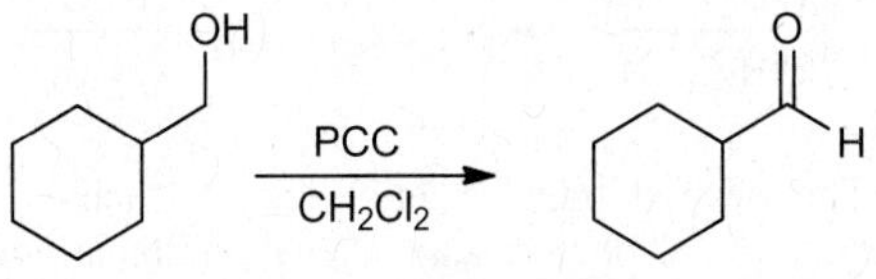

This primary alcohol can be made from the starting alkene via an *anti*-Markovnikov addition of H and OH, which can be achieved via hydroboration-oxidation.

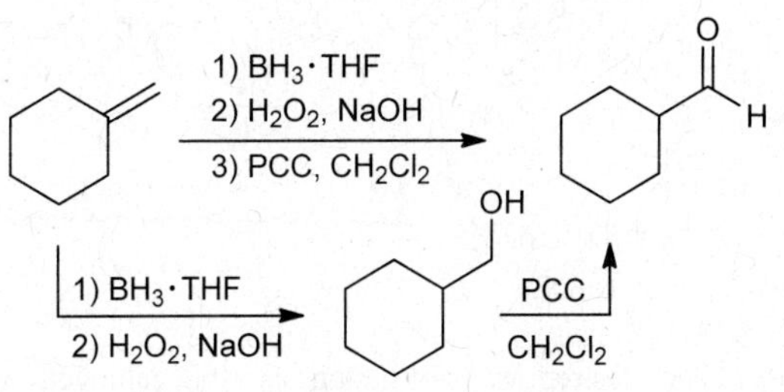

Alternatively, the primary alcohol can be made via *anti*-Markovnikov addition of H and Br, followed by an S_N2 reaction with hydroxide as the nucleophile.

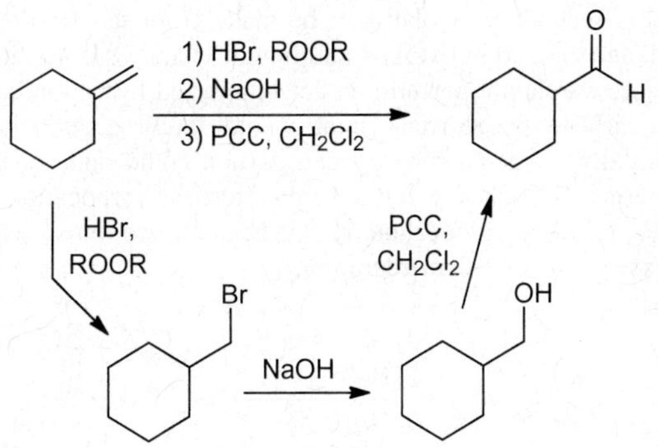

(c) The desired ketone can be made from the corresponding secondary alcohol via an oxidation process.

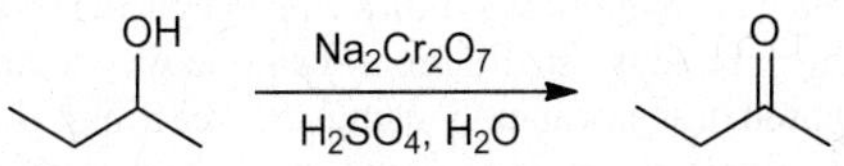

This secondary alcohol (2-butanol) can be made from the starting alkene via a Markovnikov addition of H and OH, which can be achieved via acid-catalyzed hydration.

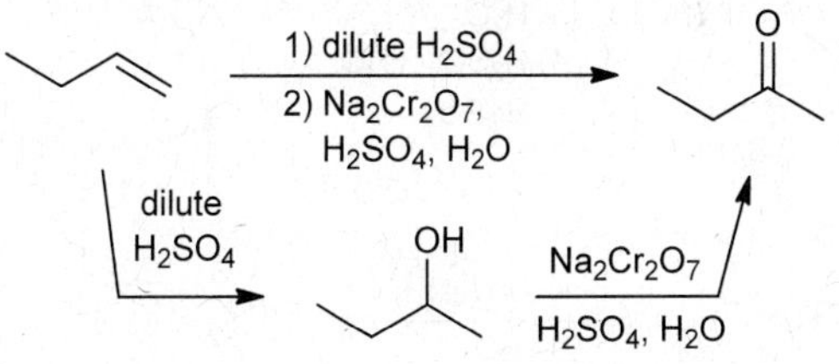

As is the case with most synthesis problems that you will encounter, there is usually more than one way to achieve the desired transformation using the reactions that we have learned in previous chapters. For example, the following alternative synthesis is perfectly acceptable.

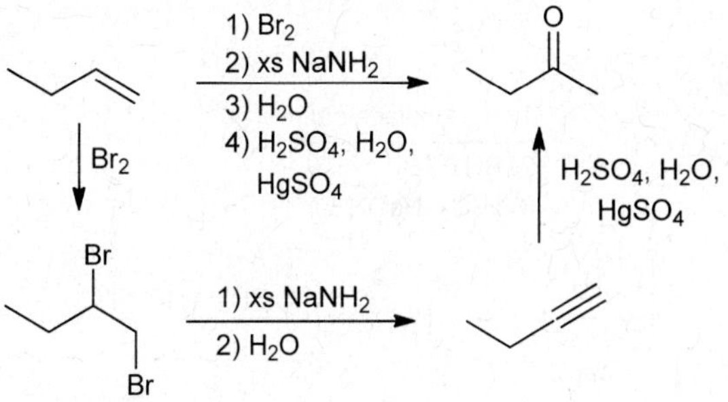

(d) The desired ketone can be made from the corresponding secondary alcohol via an oxidation process.

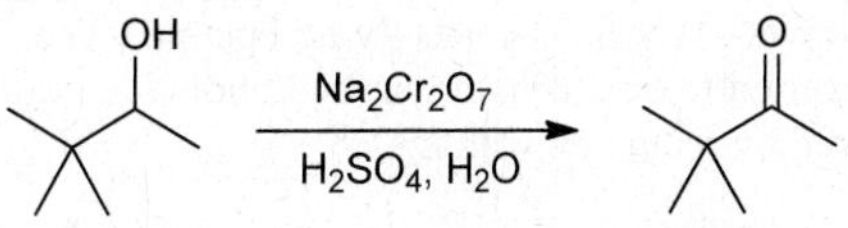

This secondary alcohol can be made from the starting alkene via a Markovnikov addition of H and OH. In this case, we cannot perform an acid-catalyzed hydration, as a carbocation rearrangement would likely occur (the initially formed secondary carbocation could undergo a methyl shift to give a more stable tertiary carbocation). To avoid a carbocation rearrangement, we must use oxymercuration-demercuration.

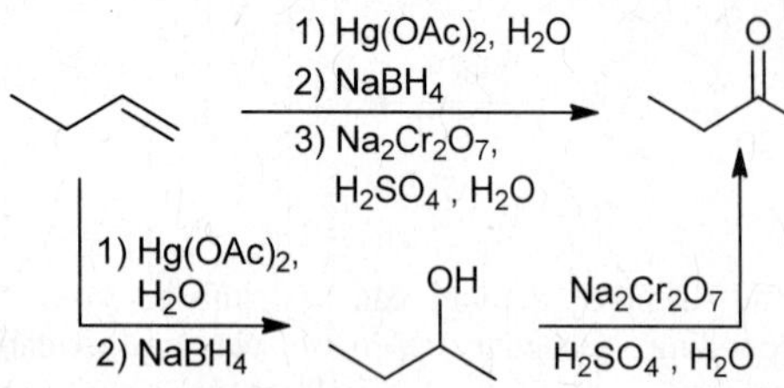

As is the case with most synthesis problems that you will encounter, there is usually more than one way to achieve the desired transformation using the reactions that we have learned in previous chapters. For example, the following alternative synthesis is perfectly acceptable.

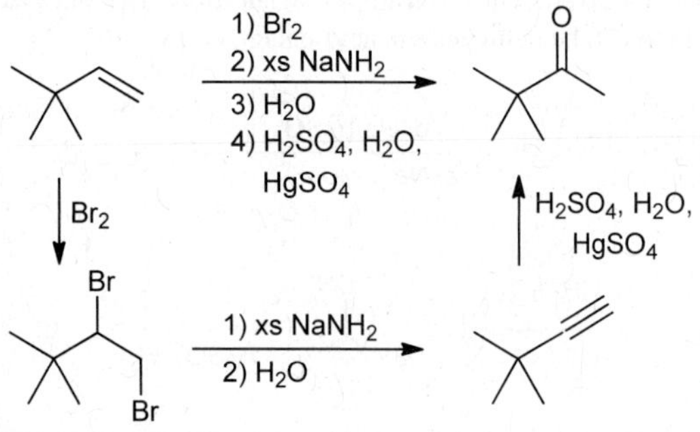

13.24.
(a) The desired transformation can be achieved by converting the alkyne into an alkene, followed by *anti*-Markovnikov addition of H and OH.

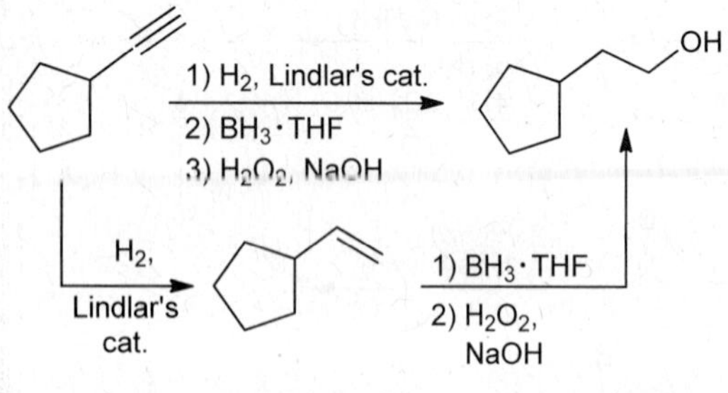

As is the case with most synthesis problems that you will encounter, there is usually more than one way to achieve the desired transformation using the reactions that we have learned so far. For example, the following alternative synthesis is perfectly acceptable. The alkyne is converted to an aldehyde via hydroboration-oxidation, followed by reduction with LAH.

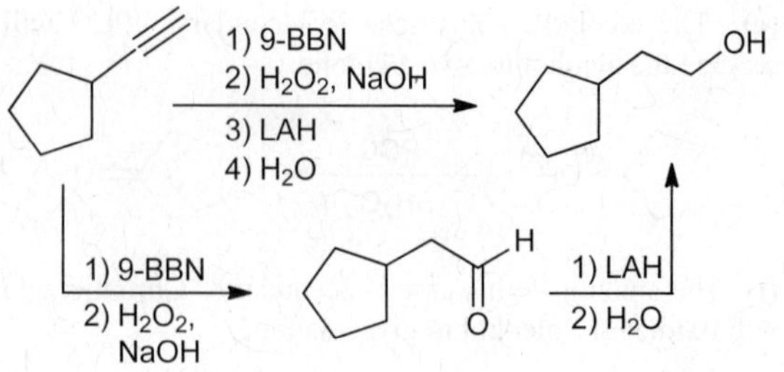

There are likely other acceptable answers as well.

(b) The desired transformation can be achieved by converting the alcohol into an alkene, and then converting the alkene into an alkyne, as shown.

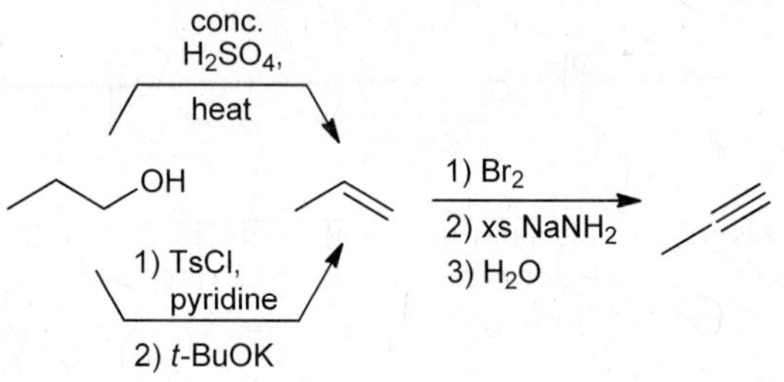

(c) The desired transformation can be achieved by converting the alkene into an alcohol, and then oxidizing the alcohol with PCC to give the aldehyde, as shown.

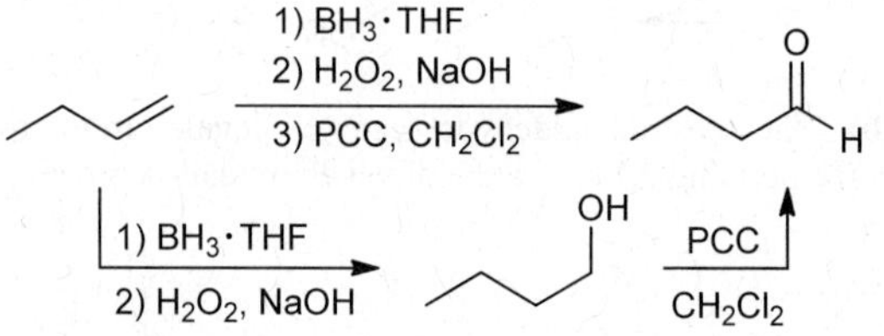

(d) The desired transformation can be achieved by converting the alcohol into an alkene, and then reducing the alkene to an alkane, as shown.

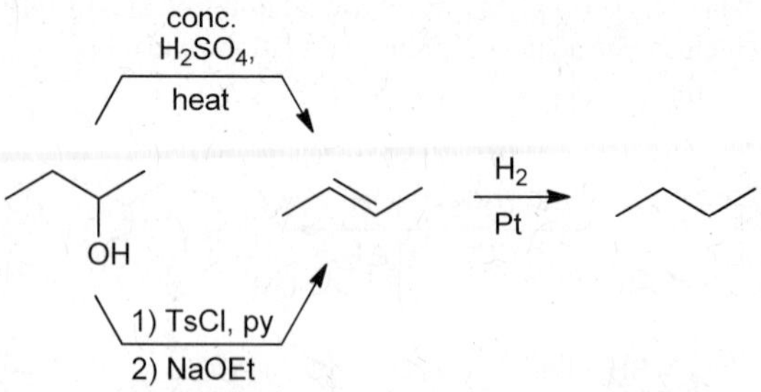

(e) The desired transformation can be achieved by reducing the aldehyde, and then converting the resulting alcohol into an alkene, as shown. If the second process is performed by converting the alcohol to a tosylate, followed by treatment with a strong base, then it is important that the base is sterically hindered (*tert*-butoxide). With a primary substrate, S_N2 will likely predominate over E2 if the base is not sterically hindered.

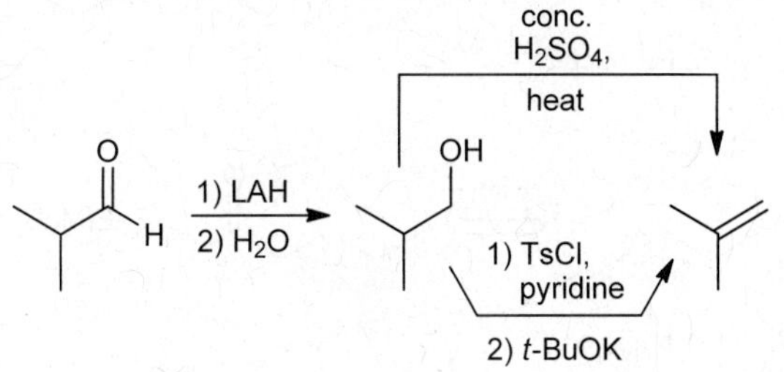

(f) The desired transformation can be achieved by reducing the ketone, and then converting the resulting secondary alcohol into an alkene, as shown. If the second process is performed by converting the alcohol to a tosylate, followed by treatment with a strong base, then it is important that the base is not sterically hindered. A sterically hindered base would give the less substituted alkene as the major product, and we need the more substituted alkene. Appropriate bases include hydroxide, methoxide and ethoxide (ethoxide is shown in the following synthesis).

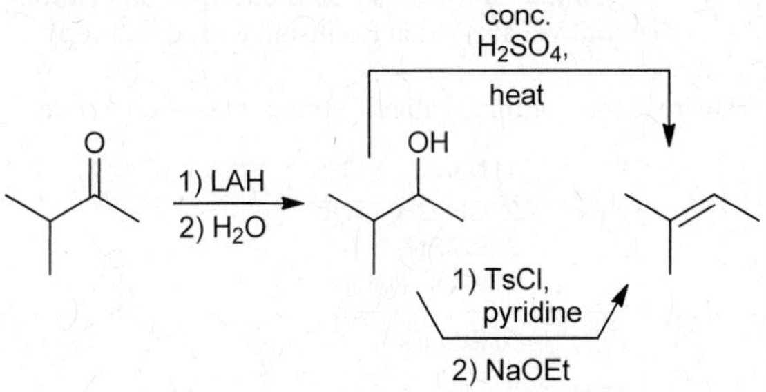

13.25. The alcohol can be converted to an alkene via either of the two processes shown here. Notice that the second method involves converting the alcohol to a tosylate, followed by treatment with a strong base. If this method is used, it is important that the base is not sterically hindered. A sterically hindered base might give the less substituted alkene as the major product, and we need the more substituted alkene. Appropriate bases include hydroxide, methoxide and ethoxide (ethoxide is shown in the following synthesis).

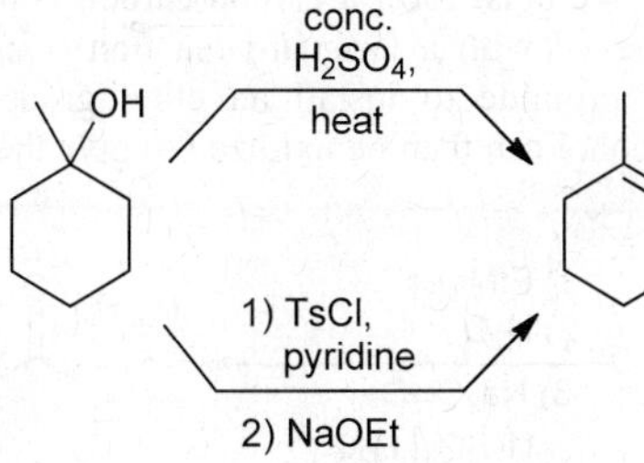

13.26. The desired transformation can be achieved by converting the alcohol to an alkene (via either of the two methods discussed in the previous problem), followed by

anti-Markovnikov addition of H and OH to give the product.

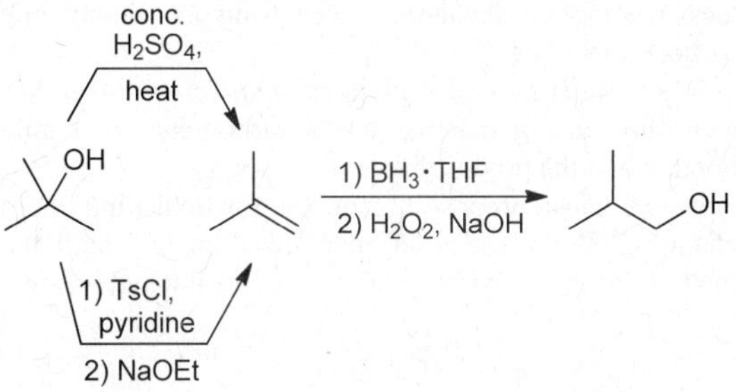

There likely other acceptable answers. For example, conversion of the alkene to the desired alcohol can be achieved via *anti*-Markovnikov addition of HBr (in the presence of peroxides) followed by an S$_N$2 process in which hydroxide functions a nucleophile.

13.27.
(a) The product has more carbon atoms than the starting material, so we must form a carbon-carbon bond. This can be achieved with a Grignard reaction (using ethyl magnesium bromide to install an ethyl group). The resulting alcohol can then be oxidized to give the desired product.

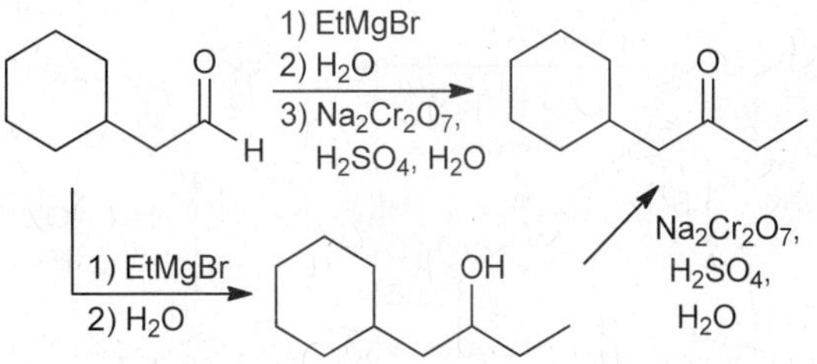

(b) The product has one more carbon atom than the starting material, so we must form a carbon-carbon bond. This can be achieved with a Grignard reaction (using methyl magnesium bromide to install a methyl group). The resulting alcohol can then be oxidized to give the desired product.

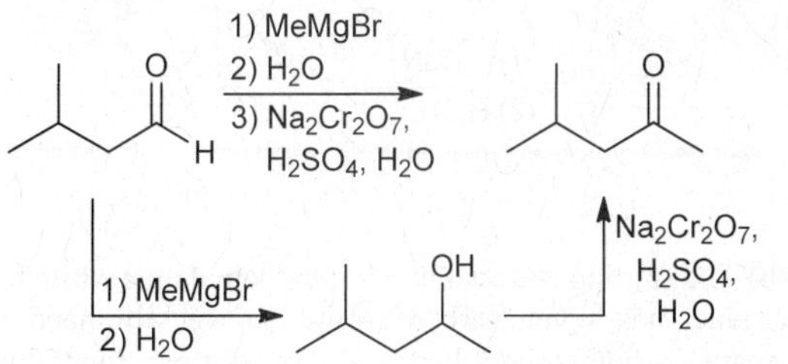

13.28.

(a) We begin by asking the following two questions:

1) *Is there a change in the carbon skeleton?* Yes, the carbon skeleton is increasing in size by one carbon atom.

2) *Is there a change in the functional groups?* Yes, the starting material has a carbon-carbon double bond, while the product has an OH group.

Now we must propose a strategy for achieving these changes. If we use a Grignard reaction to install the methyl group, we would need to use the aldehyde shown here:

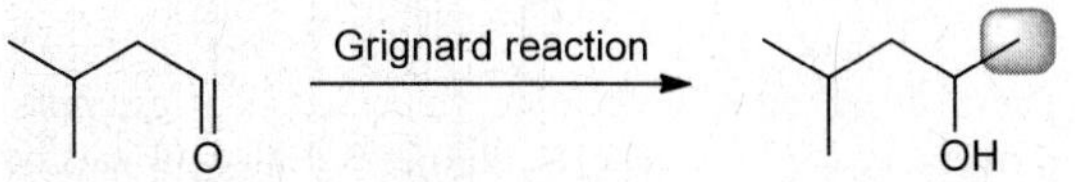

The resulting alcohol has the correct carbon skeleton, and it can be converted directly into the product via oxidation. So we only need to determine if the aldehyde above can be made from the starting material. Indeed, there are at least two methods for converting the starting alkene into the necessary aldehyde. One method is to convert the alkene into an alkyne (via bromination followed by elimination) and then to perform hydroboration-oxidation to obtain the aldehyde. Another method is to perform hydroboration-oxidation with the starting alkene, followed by oxidation with PCC to give the aldehyde.

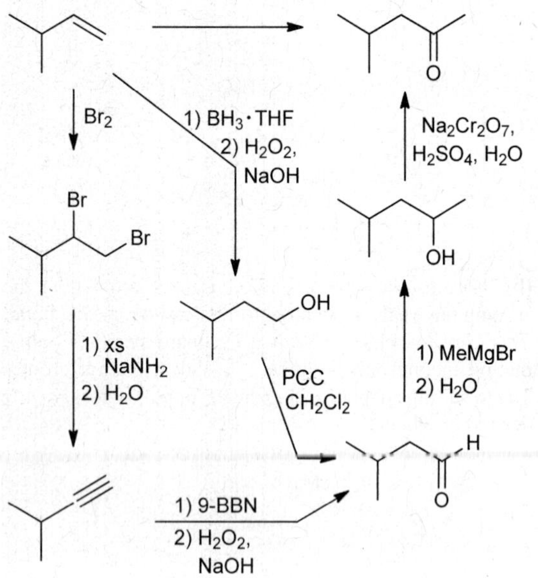

(b) The starting material has two carbon atoms, while the product has seven carbon atoms, so we will need to create carbon-carbon bonds. The starting functional group is a triple bond, and the product is a ketone. There are certainly many acceptable answers to this problem. One such answer can be rationalized with the following retrosynthetic analysis.

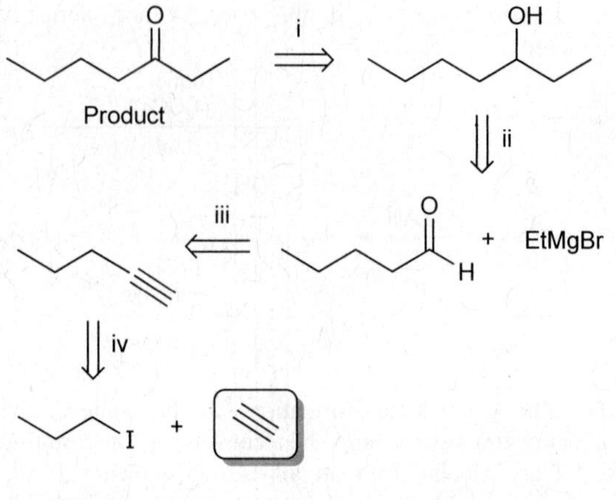

i. The product is a ketone, which can be made via oxidation of the corresponding secondary alcohol.
ii. The secondary alcohol can be made by treating the appropriate aldehyde with ethyl magnesium bromide.
iii. The aldehyde can be made from the corresponding terminal alkyne (via hydroboration-oxidation).
iv. The terminal alkyne can be made from the starting material via alkylation (to install a propyl group).

The reagents for this synthetic strategy are shown here.

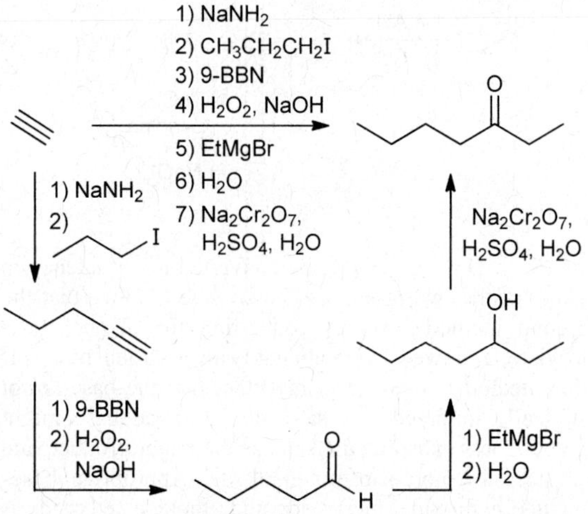

(c) The product has more carbon atoms than the starting material, so we must form a carbon-carbon bond. This can be achieved with a Grignard reaction (using ethyl magnesium bromide to install an ethyl group). The resulting alcohol can then be oxidized to give the desired product.

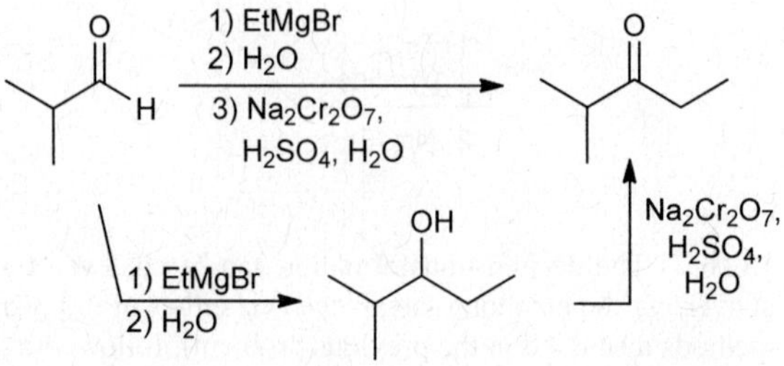

(d) The starting material has five carbon atoms, while the product has six carbon atoms, so we will need to install a methyl group. The starting functional group is an alcohol, and the product is a ketone. There are certainly many acceptable answers to this problem. One such answer can be rationalized with the following retrosynthetic analysis.

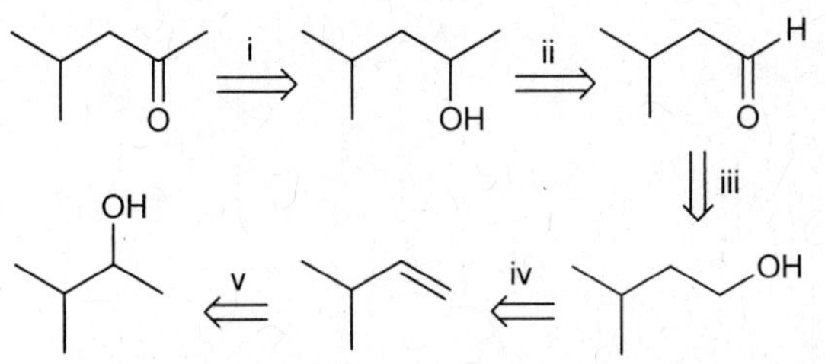

i. The ketone can be made via oxidation of the corresponding secondary alcohol.
ii. The alcohol can be made by treating the appropriate aldehyde with methyl magnesium bromide.
iii. The aldehyde can be made from the corresponding primary alcohol.
iv. The alcohol can be made from an alkene via hydroboration-oxidation (*anti*-Markovnikov addition of H and OH)
v. The alkene can be made from the starting alcohol via an elimination process. Since OH is a bad leaving group, it must first be converted into a tosylate in order to perform an E2 reaction. A sterically hindered base is then used to favor the less-substituted alkene.

The reagents for this synthetic strategy are shown here.

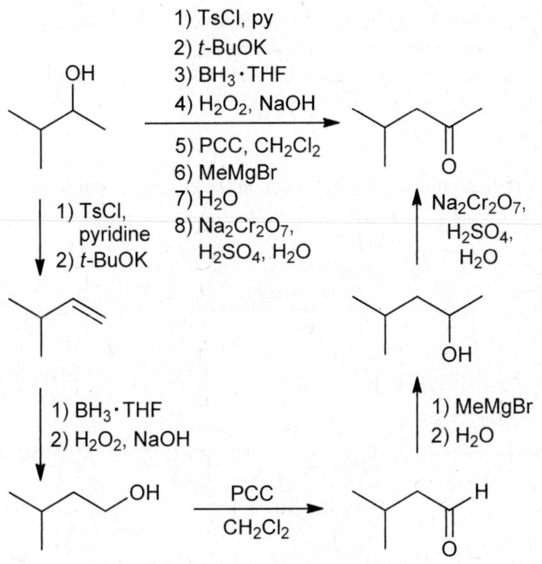

(e) The product has one more carbon atom than the starting material, so we will need to install a methyl group. The starting functional group is a triple bond, and the product is a ketone. There are certainly many acceptable answers to this problem. One such answer

can be rationalized with the following retrosynthetic analysis.

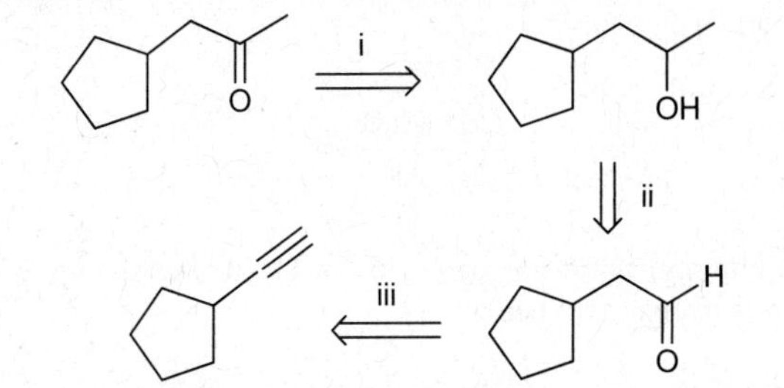

i. The product is a ketone, which can be made via oxidation of the corresponding secondary alcohol.
ii. The secondary alcohol can be made by treating the appropriate aldehyde with methyl magnesium bromide.
iii. The aldehyde can be made from the corresponding terminal alkyne (via hydroboration-oxidation).

The reagents for this synthetic strategy are shown here.

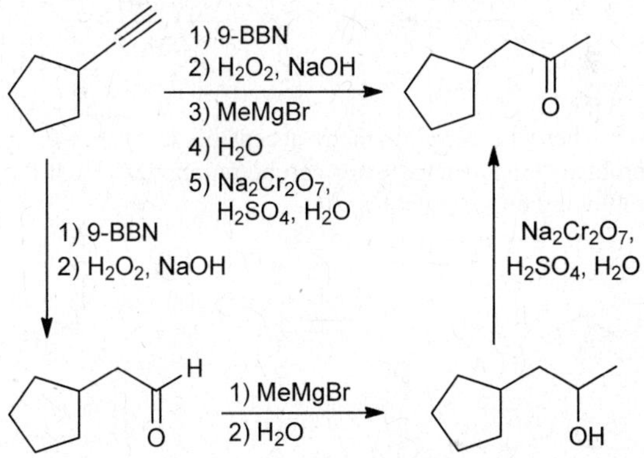

(f) The functional group has not changed, but the product has an additional methyl group. This methyl group can be installed with a Grignard reaction, if the starting alcohol is first converted to a ketone (oxidation). Then, a Grignard reaction with methyl magnesium bromide will give the desired product.

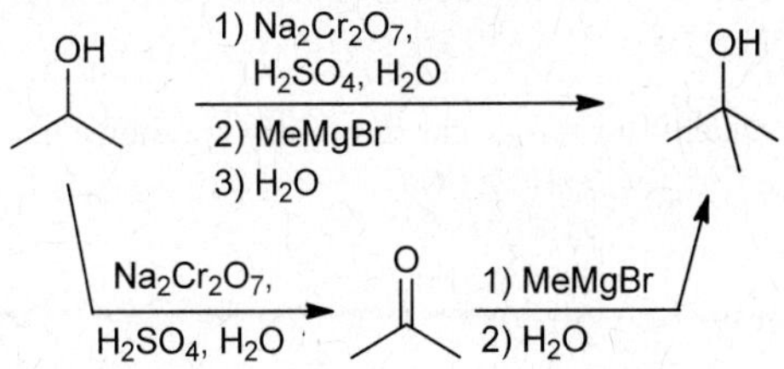

13.29.
(a) The desired product can be made in just one step (a Grignard reaction).

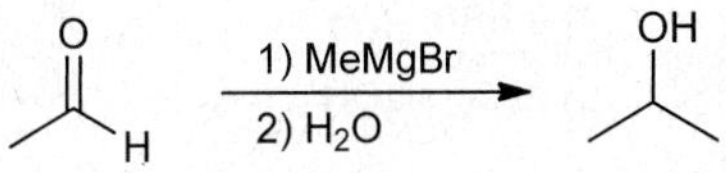

(b) The desired transformation can be achieved via a Grignard reaction in which acetaldehyde (two carbon atoms) is treated with ethyl magnesium bromide (two carbon atoms):

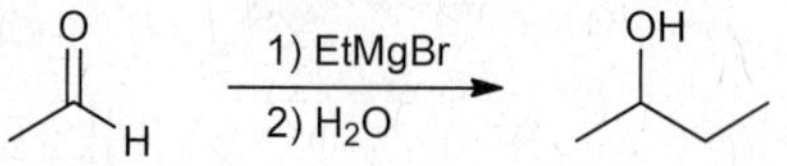

Certainly, there are other acceptable answers, two of which are shown below:

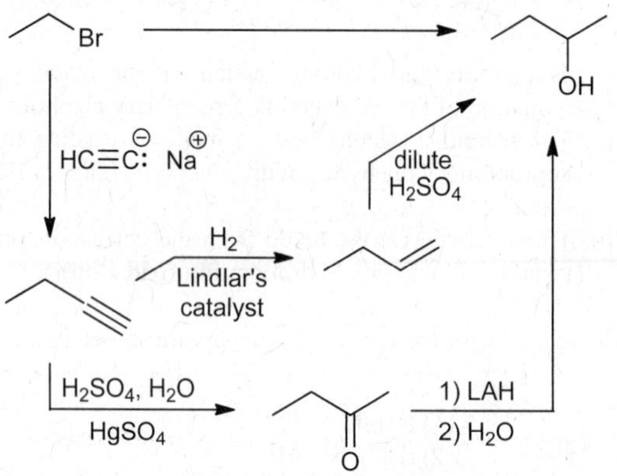

(c) There are certainly many acceptable answers to this problem. One such answer can be rationalized with the following retrosynthetic analysis.

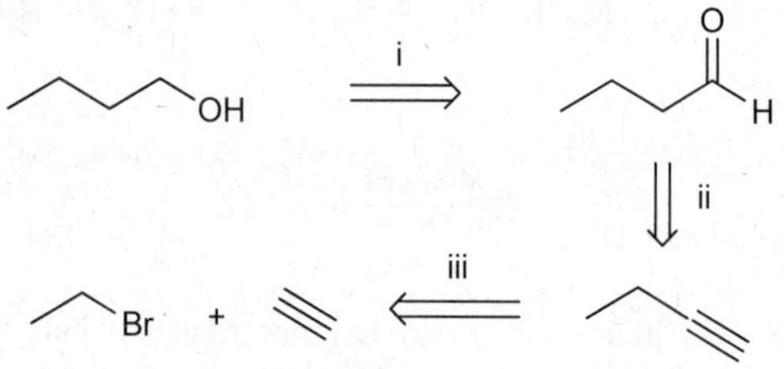

i.   The product is an alcohol, which can be made from the corresponding aldehyde via reduction.
ii. The aldehyde can be made from a terminal alkyne via hydroboration-oxidation.
iii. The terminal alkyne can be made via an alkylation process.

The reagents for this synthetic strategy are shown here.

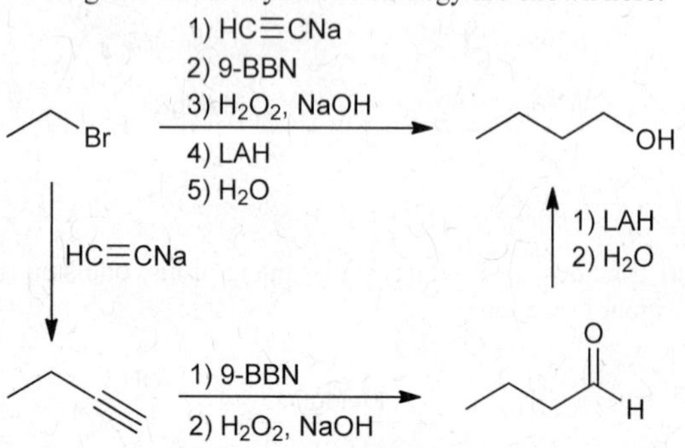

The following are alternative synthetic pathways that also achieve the desired transformation:

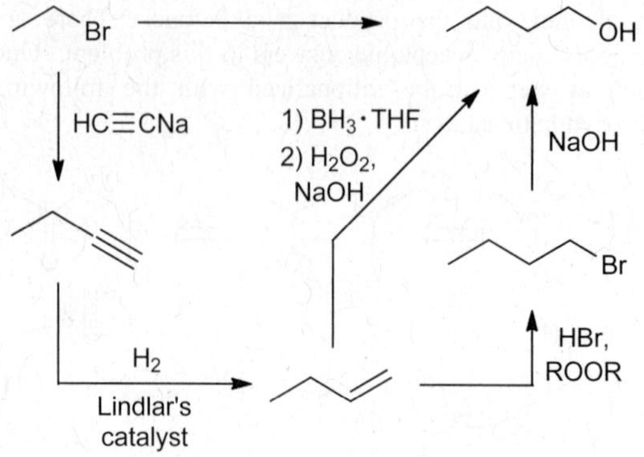

(d) There are certainly many acceptable answers to this problem. One such answer can be rationalized with the following retrosynthetic analysis.

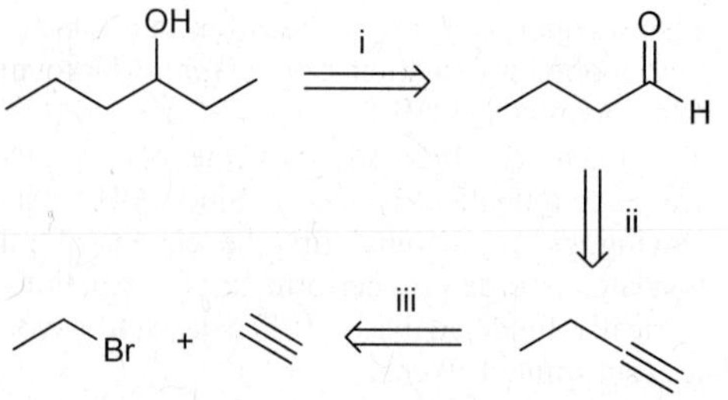

i.   The alcohol can be made by treating the appropriate aldehyde with ethyl magnesium bromide.
ii. The aldehyde can be made from a terminal alkyne (via hydroboration-oxidation).
iii. The terminal alkyne can be made via an alkylation process.

The reagents for this synthetic strategy are shown here.

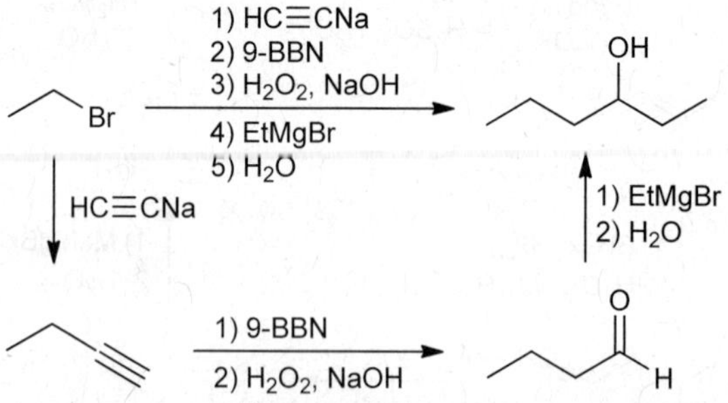

The following is an alternative synthetic pathway that also achieves the desired transformation:

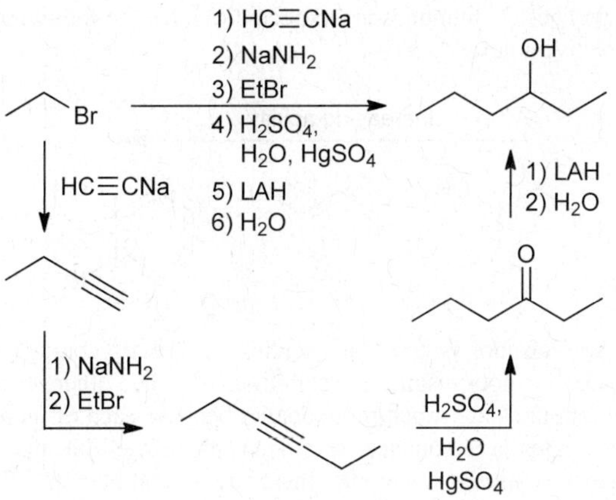

The reaction sequence at the top left shows:

Starting material (Br compound) with reagents:
1) HC≡CNa
2) NaNH₂
3) EtBr
4) H₂SO₄, H₂O, HgSO₄
5) LAH
6) H₂O

leading to OH product.

HC≡CNa leads downward to an alkyne.

1) NaNH₂
2) EtBr

leading to an internal alkyne, then H₂SO₄, H₂O, HgSO₄

1) LAH
2) H₂O

13.30.

(a) We begin by identifying the parent. The carbon atom connected to the OH group must be included in the parent. The longest chain that includes this carbon atom is five carbon atoms in length. So the parent is pentanol. There is only one substituent (highlighted) – a propyl group, located at C2). A locant is included to indicate the location of the OH group.

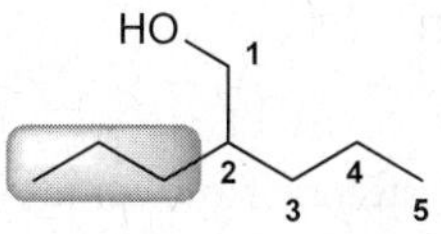

2-propyl-1-pentanol

(b) We begin by identifying the parent. The carbon atom connected to the OH group must be included in the parent. The longest chain that includes this carbon atom is five carbon atoms in length. So the parent is pentanol. There is only one substituent (highlighted) – a methyl group, located at C4. Finally, we assign a configuration to the chirality center.

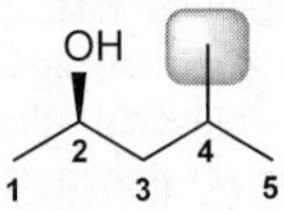

(R)-4-methyl-2-pentanol

(c) We begin by identifying the parent (phenol). There are two substituents (highlighted) – a bromo group and a methyl group. Notice that the ring is numbered starting from the carbon atom bearing the OH group (it is not necessary to indicate a locant for the OH group, because in a ring, it is assumed to be at C1, by definition). We then number in the direction that gives the second substituent the lowest possible number (C2 rather than C4). According to this numbering scheme, the bromo group is at C2 and the methyl group is at C4. The substituents are arranged alphabetically in the name.

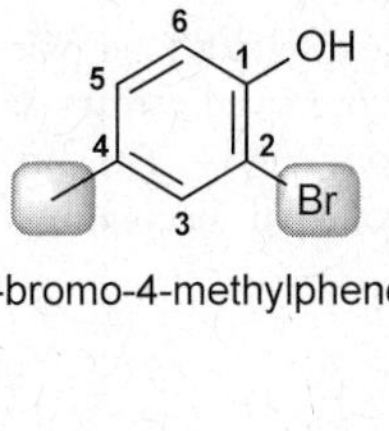

2-bromo-4-methylphenol

(d) We begin by identifying the parent. The carbon atom connected to the OH group must be included in the parent. That carbon atom is part of a six-membered ring, so the parent is cyclohexanol. There is only one substituent (highlighted) – a methyl group. Notice that the parent chain is numbered starting from the carbon atom bearing the OH group (it is not necessary to indicate a locant for the OH group, because in a ring, it is assumed to be at C1, by definition). The numbers go clockwise, so as to give the lowest number to the substituent (C2 rather than C6). Finally, we assign a configuration to each chirality center.

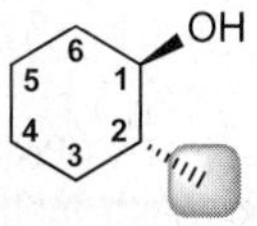

(1R,2R)-2-methylcyclohexanol

13.31.

(a) The parent is cyclohexanediol, which is a six-membered ring containing two OH groups. The locants for the OH groups (C1 and C2) indicate that they are on adjacent carbon atoms, and the name of the compound indicates a *cis* configuration. That is, the OH groups are on the same side of the ring, giving the following *meso* compound.

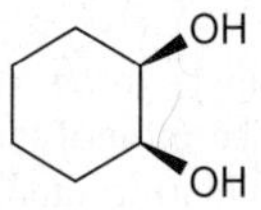

(b) Isobutanol is the common name for 2-butanol. The parent is a chain of four carbon atoms with the OH group connected to the C2 position.

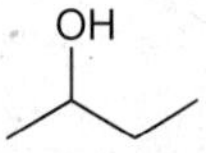

(c) The parent is phenol and there are three substituents (all of which are nitro groups) located at C2, C4 and C6.

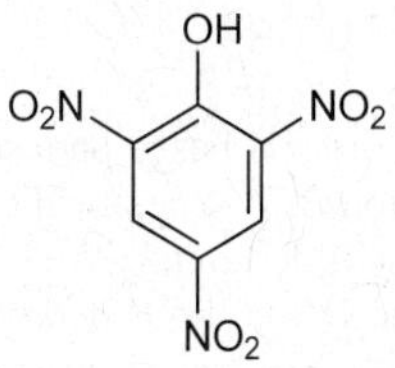

(d) The parent (3-heptanol) is a chain of seven carbon atoms containing an OH group at the C3 position. The name indicates that there are two substituents (both methyl groups) at the C2 position. The C3 position is a chirality center, and it has the *R* configuration.

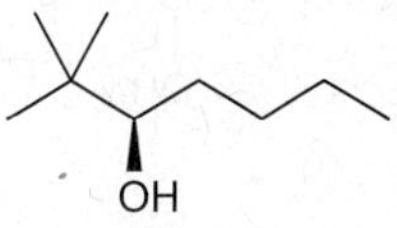

(e) Ethylene glycol is the common name for the following compound.

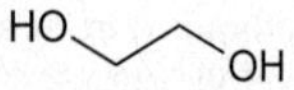

(f) The parent (1-butanol) is a chain of four carbon atoms with an OH group connected to the C1 position. The name indicates that there is one substituent (a methyl group) at C2. The C2 position is a chirality center, and it has the *S* configuration.

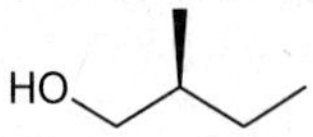

13.32. The solution to Problem 1.4 shows all constitutional isomers with molecular formula $C_4H_{10}O$. Four of these isomers are alcohols, and have the following systematic names:

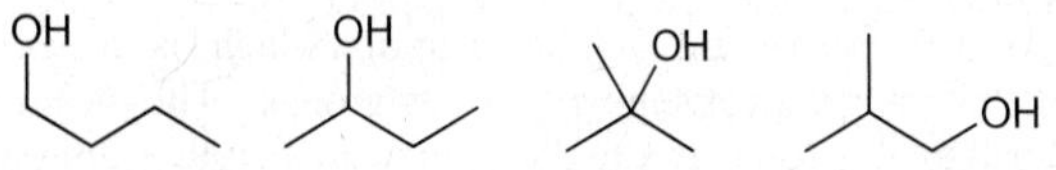

1-butanol   2-butanol   2-methyl-   2-methyl-
                            2-propanol   1-propanol

13.33.
(a) The trichloromethyl group (CCl_3) is powerfully electron-withdrawing because of the combined inductive effects of the three chlorine atoms. As a result, the presence of a trichloromethyl group stabilizes the conjugate base (alkoxide ion) that is formed when the OH group is deprotonated. So the presence of the trichloromethyl group (in close proximity to the OH group) renders the alcohol more acidic. The compound with two such groups is the most acidic.

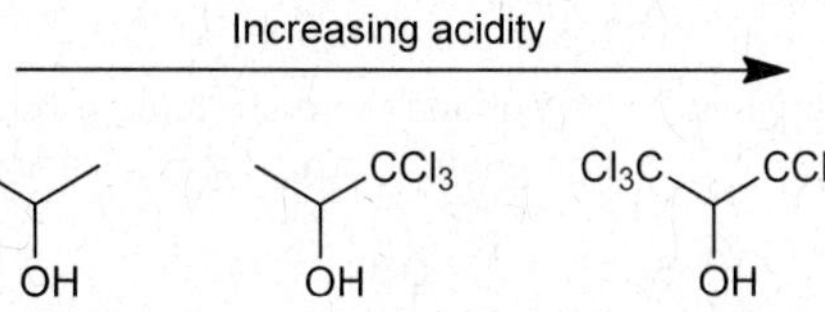

Increasing acidity

(b) The following order is based on a solvating effect, as described in Section 3.7. Specifically, the presence of *tert*-butyl groups will destabilize the conjugate base that is formed when the OH group is deprotonated. As such, the compound with two *tert*-butyl groups will be the least acidic.

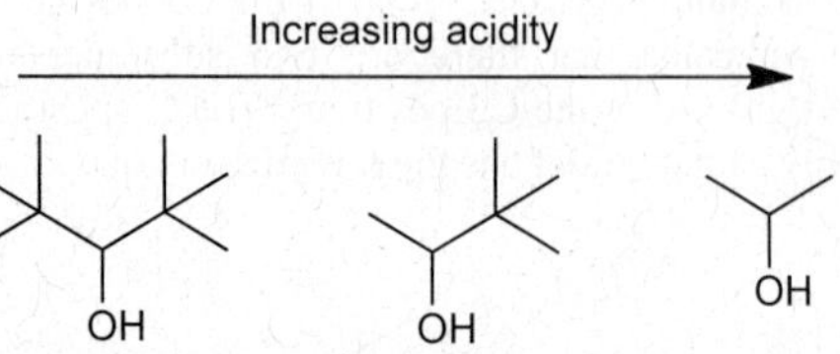

Increasing acidity

(c) These compounds are expected to have the following relative acidity.

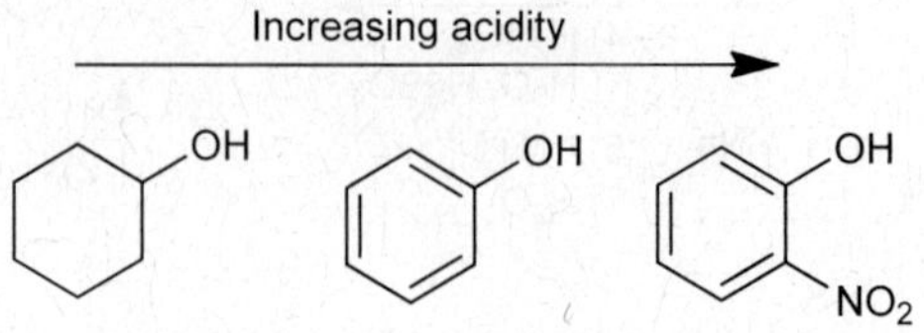

Increasing acidity

Cyclohexanol is the least acidic because its conjugate base is not resonance stabilized. The other two compounds are much more acidic, because each of them generates a resonance stabilized phenolate ion upon deprotonation. Among these two compounds, 2-nitrophenol is more acidic, because its conjugate base has an additional resonance structure in which the negative charge is placed on an oxygen atom of the nitro group.

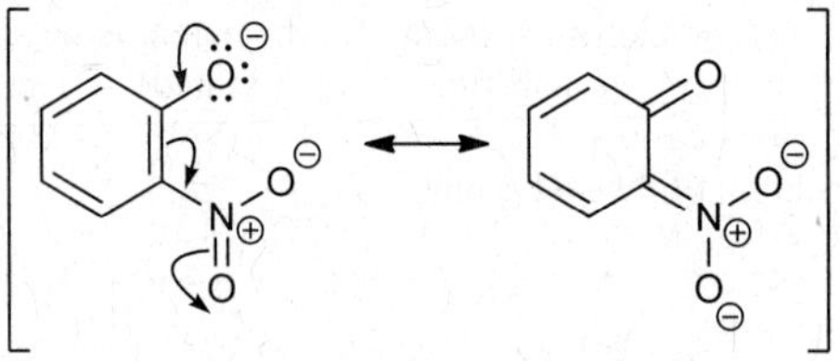

Note: This conjugate base has other resonance structures not shown here.

13.34.
(a) This structure exhibits a lone pair next to a π bond, so we draw the two curved arrows associated with that pattern (see Section 2.10). The first curved arrow is drawn showing a lone pair becoming a π bond, while the second curved arrow shows a π bond becoming a lone pair. The resulting resonance structure also exhibits a lone pair next to a π bond, so again we draw the two curved arrows associated with that pattern. This is continued until we have drawn all of the resonance structures, shown here.

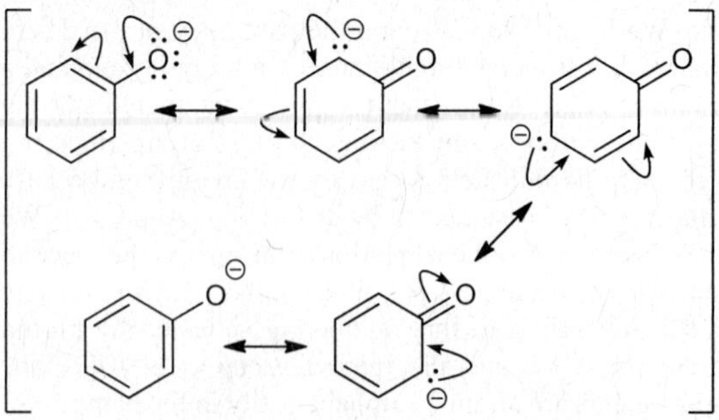

(b) This structure exhibits a lone pair next to a π bond, so we draw the two curved arrows associated with that pattern (see Section 2.10). The first curved arrow is drawn showing a lone pair becoming a π bond, while the second curved arrow shows a π bond becoming a lone pair.

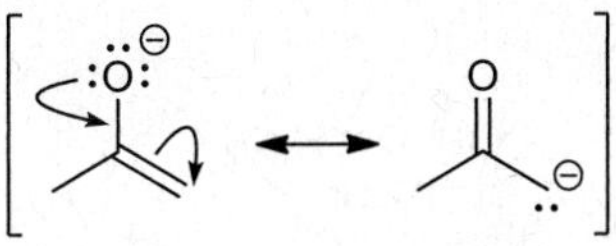

(c) This structure exhibits a lone pair next to a π bond, so we draw the two curved arrows associated with that pattern (see Section 2.10). The first curved arrow is drawn showing a lone pair becoming a π bond, while the second curved arrow shows a π bond becoming a lone pair. The resulting resonance structure also exhibits a lone pair next to a π bond, so again we draw the two curved arrows associated with that pattern, giving the third and final resonance structure.

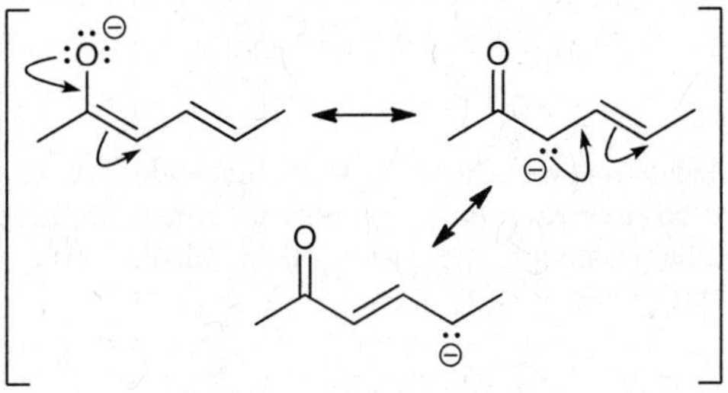

13.35.
(a) 1-Butanol is a primary alcohol. When treated with PBr₃, the OH group is replaced with Br, so the product is 1-bromobutane.

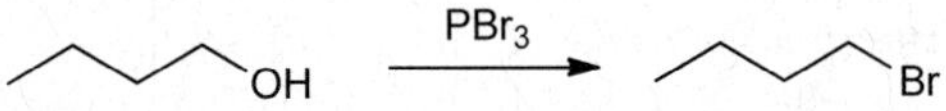

(b) 1-Butanol is a primary alcohol. When treated with SOCl₂ and pyridine, the OH group is replaced with Cl, so the product is 1-chlorobutane.

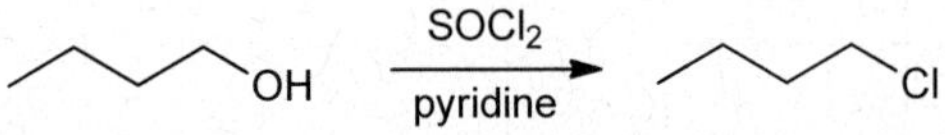

(c) 1-Butanol is a primary alcohol. When treated with HCl and ZnCl₂, the OH group is replaced with Cl, so the product is 1-chlorobutane.

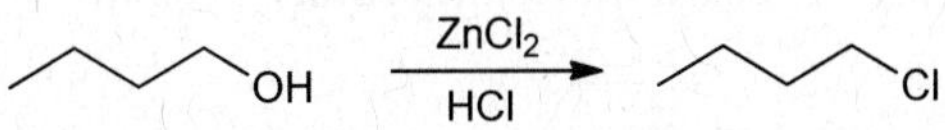

(d) When treated with concentrated sulfuric acid, the alcohol is converted into an alkene via an E1 process. After the OH group is protonated (thereby forming a better leaving group), loss of the leaving group results in a primary carbocation, which will rearrange to give a secondary carbocation (or more likely, loss of the leaving group occurs simultaneously with the rearrangement, so that the primary carbocation is avoided altogether). As such, the major product is likely to be *trans*-2-butene, rather than 1-butene. Notice that

the *trans* isomer is expected, because it is lower in energy than the *cis* isomer.

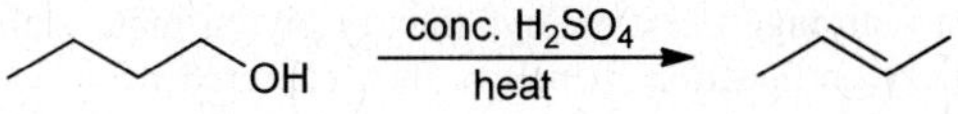

(e) When treated with PCC, a primary alcohol is oxidized to give an aldehyde (which is not further oxidized).

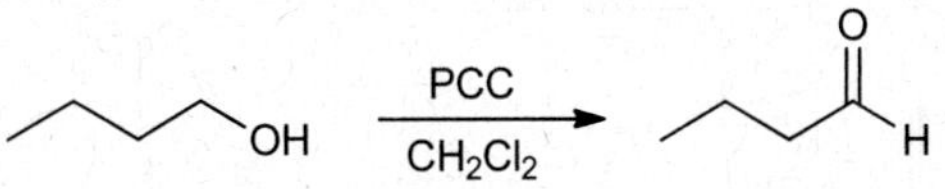

(f) When treated with chromic acid, a primary alcohol is oxidized to give a carboxylic acid.

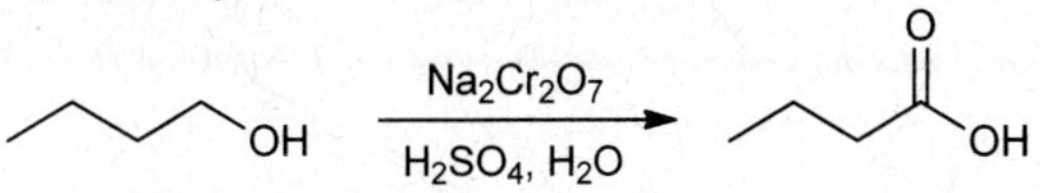

(g) When treated with lithium, an alcohol is deprotonated to give an alkoxide ion.

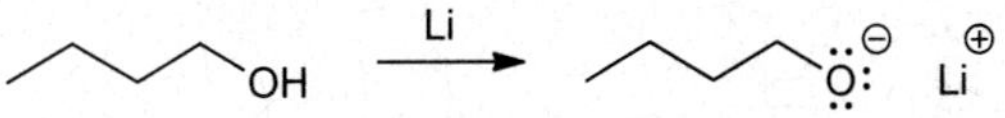

(h) When treated with sodium hydride (a strong base), an alcohol is deprotonated to give an alkoxide ion.

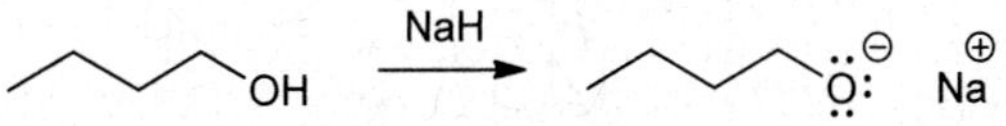

(i) When treated with TMSCl and a base (Et₃N), the OH group is protected (it is converted to OTMS).

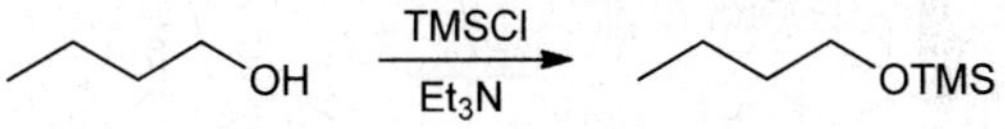

(j) When treated with tosyl chloride and a base (pyridine), the OH group is converted to a tosylate group.

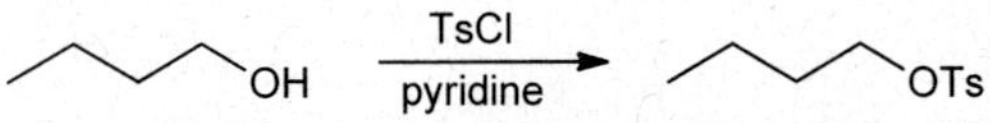

(k) When treated with sodium, an alcohol is deprotonated to give an alkoxide ion.

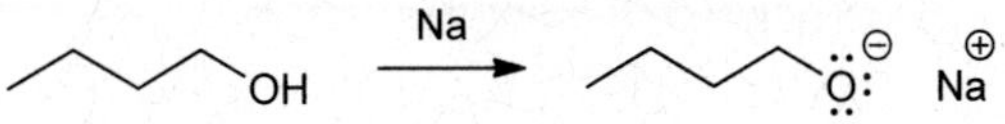

(l) When treated with *tert*-butoxide (a strong base), an alcohol is deprotonated to give an alkoxide ion.

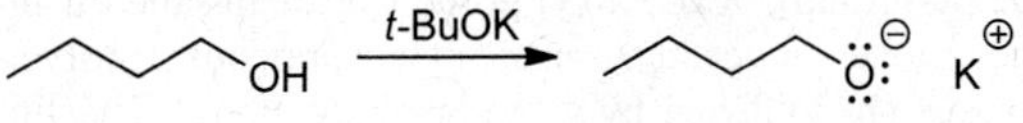

13.36. When treated with aqueous acid, the π bond is protonated, giving a secondary carbocation (rather than a primary carbocation). This secondary carbocation can then rearrange via a methyl shift to give a more stable, tertiary carbocation, which is then captured by a water molecule. The resulting oxonium ion is then deprotonated by a molecule of water to give the product:

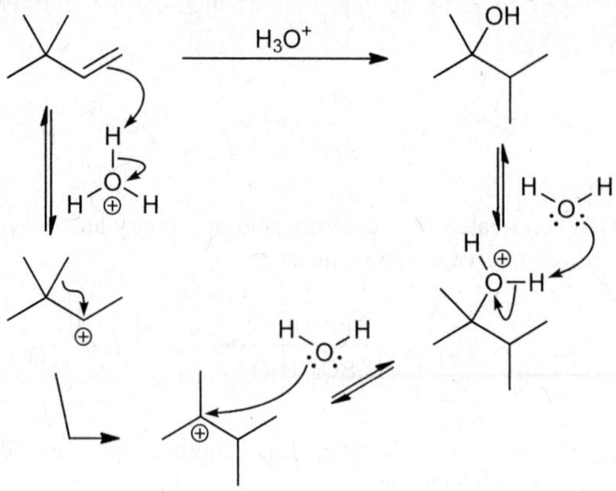

13.37.
(a) The desired aldehyde can be prepared in one step, using PCC as the oxidizing agent.

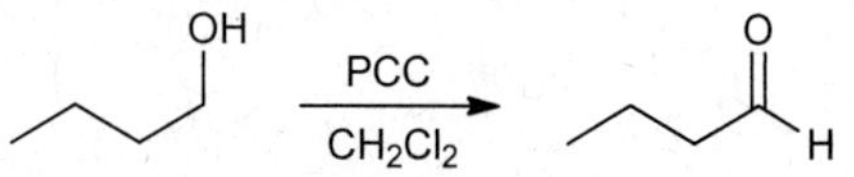

(b) The desired carboxylic acid can be prepared in one step, using chromic acid as the oxidizing agent.

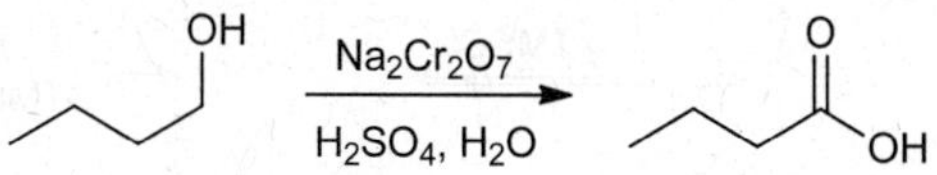

(c) The desired alcohol can be prepared via oxidation (with PCC to give the aldehyde) followed by a Grignard reaction with ethyl magnesium bromide.

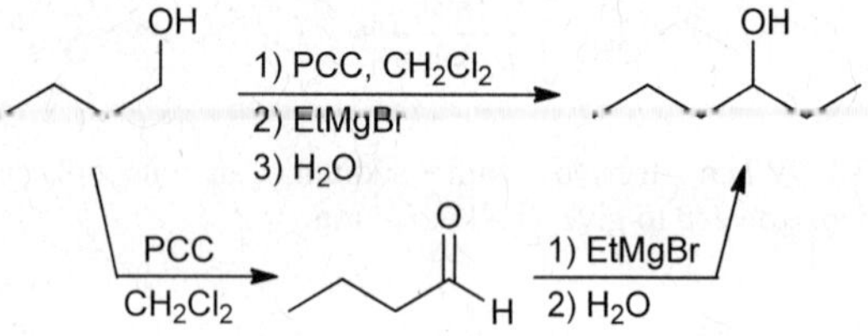

(d) This problem is similar to the previous problem, but two alkyl groups must be installed (an ethyl group and a methyl group). Each alkyl group can be installed using the same procedure from the previous problem (oxidation, followed by a Grignard reaction). The first oxidation procedure must be performed with PCC to give

the aldehyde. The second oxidation procedure can be achieved with either PCC or chromic acid (to give a ketone).

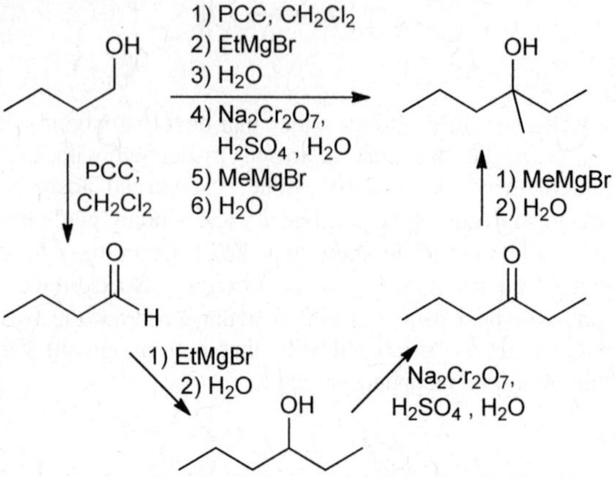

(e) Oxidation with PCC gives an aldehyde, which can then be treated with a Grignard reagent to give a secondary alcohol. Oxidation of this alcohol gives the desired ketone.

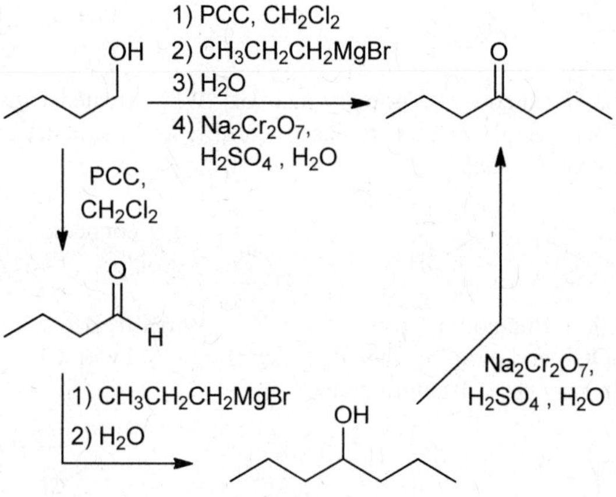

13.38.
(a) The desired product has only one group connected to the α position. That group could have been installed via a Grignard reaction with formaldehyde, as shown.

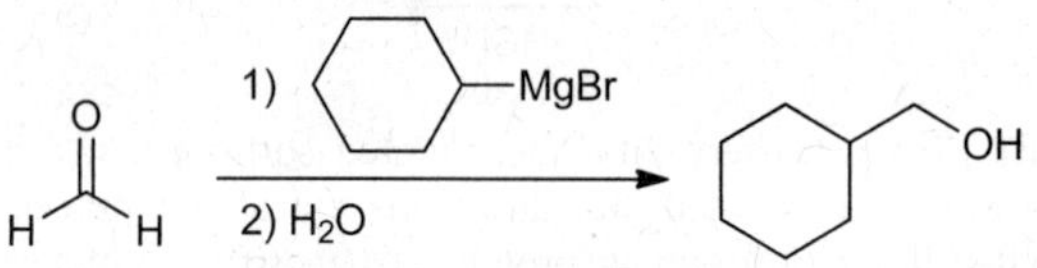

(b) The desired product has three groups connected to the α position. Any one of these groups could have been installed via a Grignard reaction with the appropriate ketone, as shown.

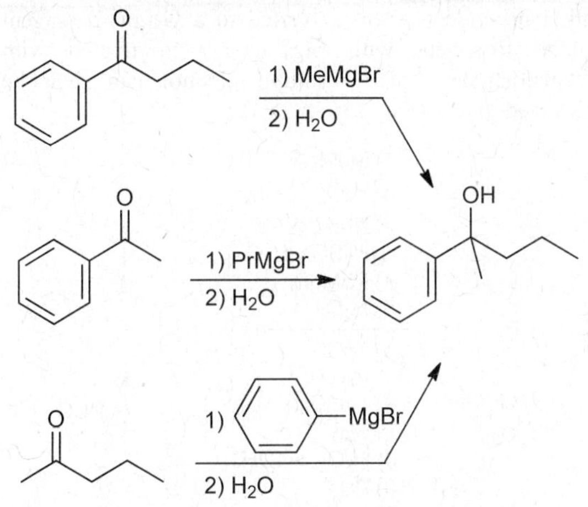

(c) The desired product has only two groups connected to the α position. Either one of these groups could have been installed via a Grignard reaction with the appropriate aldehyde, as shown.

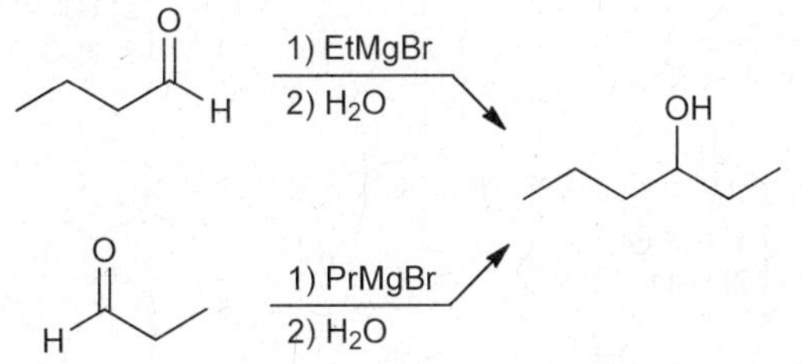

(d) The desired product has three groups connected to the α position. Any one of these groups could have been installed via a Grignard reaction with the appropriate ketone, as shown.

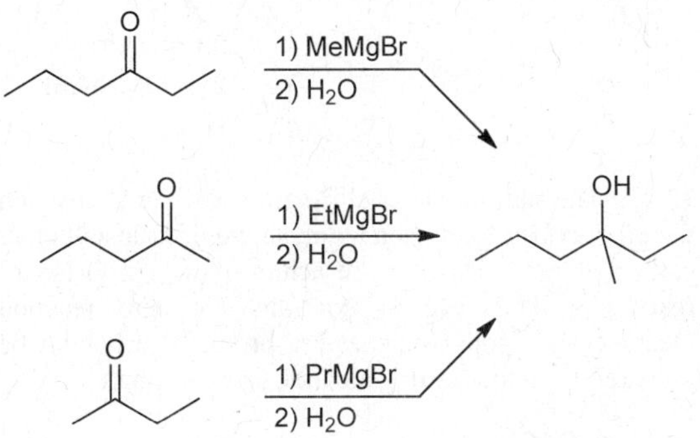

13.39.
(a) Reduction of the following aldehyde will afford the desired product.

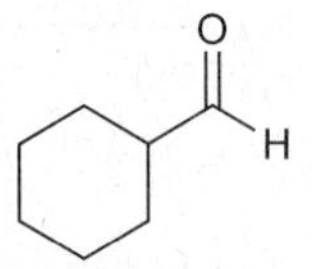

(b) Reduction of the following ketone will afford the desired product.

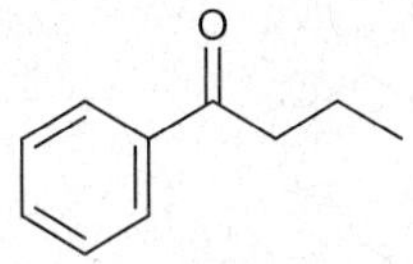

(c) Reduction of the following ketone will afford the desired product.

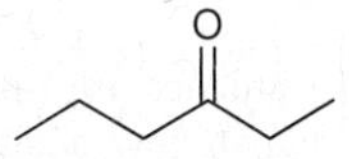

(d) Reduction of the following ketone will afford the desired product.

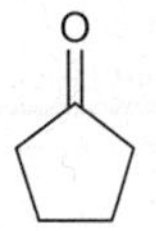

13.40.
(a) The product has more carbon atoms than the starting material, so we must form a carbon-carbon bond. This can be achieved with a Grignard reaction (using ethyl magnesium bromide to install an ethyl group). The resulting alcohol can then be oxidized to give the desired product.

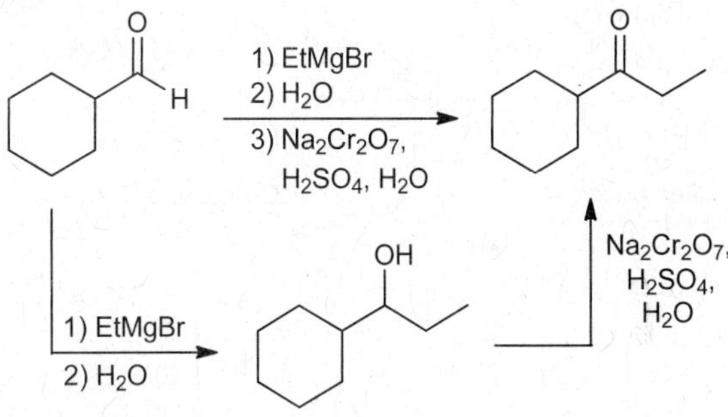

(b) Reduction of the aldehyde can be achieved with either LAH or NaBH₄. This transformation cannot be achieved via catalytic hydrogenation, as that process would also reduce the carbon-carbon π bond.

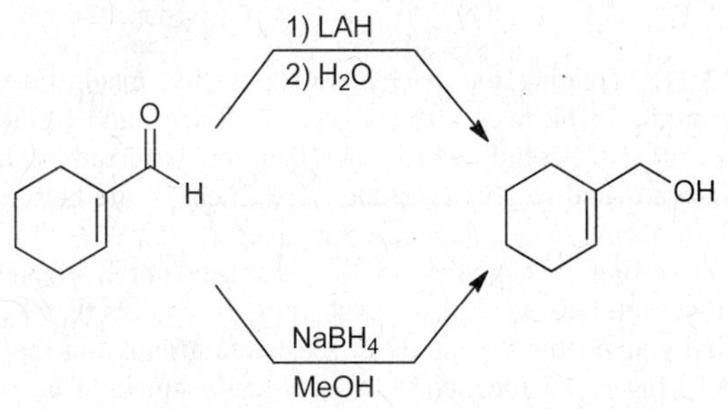

13.41. Hydride functions as a base and removes a proton from the alcohol, giving an alkoxide ion. This intermediate has both a nucleophilic region (the negatively charged oxygen atom) and an electrophilic region (the position that is α to the bromine atom). As such, an intramolecular, S$_N$2-type process can occur, giving the cyclic product.

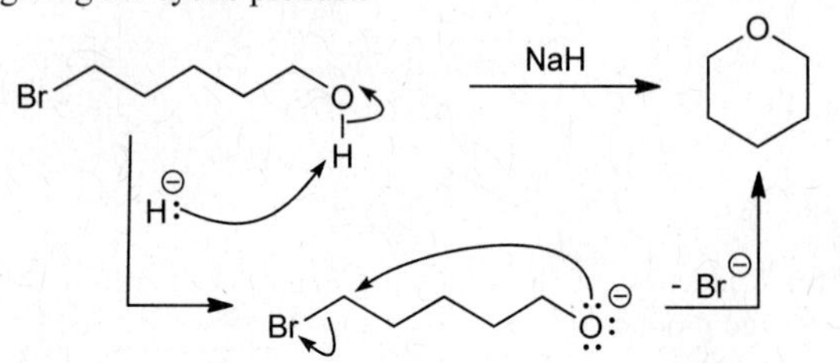

13.42. The major product is 1-methylcyclohexanol (resulting from Markovnikov addition), which is a tertiary alcohol.

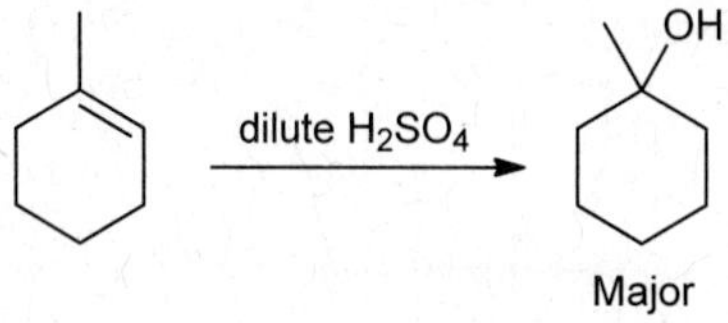

Major

Tertiary alcohols do not generally undergo oxidation. In contrast, the minor product (2-methylcyclohexanol) is a secondary alcohol and can undergo oxidation to yield a ketone.

13.43. The conversion of compound **B** to compound **C** is achieved via a Grignard reaction that employs acetone as the electrophile. Therefore, the Grignard reagent (compound **B**) must be cyclohexyl magnesium bromide, as shown.

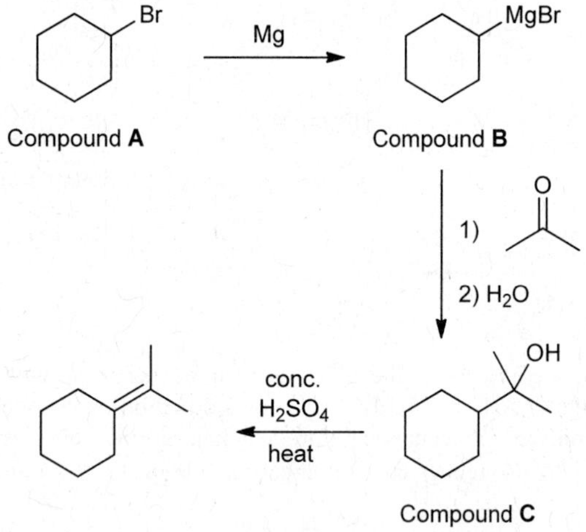

Compound **A**   Compound **B**

Compound **C**

13.44. Treating the aldehyde with methyl magnesium bromide (followed by water work-up) gives the secondary alcohol, which can then be oxidized with chromic acid to give a ketone. Treatment of the ketone with phenyl magnesium bromide gives a tertiary alcohol. Converting this alcohol to the less substituted alkene (disubstituted, rather than trisubstituted) requires that we first convert the OH group to a tosylate group, and then perform an E2 reaction with a sterically hindered base, such as *tert*-butoxide. Conversion of this alkene into a primary alcohol requires an *anti*-Markovnikov addition of H and OH across the alkene, which can be achieved via hydroboration-oxidation. Treatment with PBr$_3$ then converts the primary alcohol to a primary bromide. This

alkyl bromide is then converted to a Grignard reagent (upon treatment with Mg) and then treated with formaldehyde. The resulting alcohol can then be oxidized to an aldehyde with PCC.

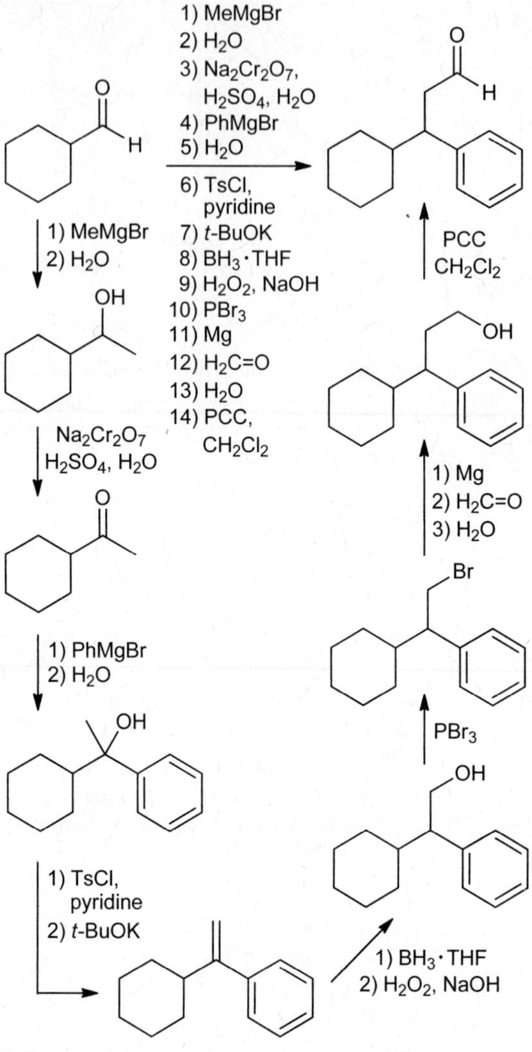

13.45. The starting material has three carbon atoms, and the product has six carbon atom, so we must install three carbon atoms. This can be achieved with a Grignard reaction. The reagents for this Grignard reaction (acetone and propyl magnesium bromide) can both be prepared from the starting alcohol, as shown here.

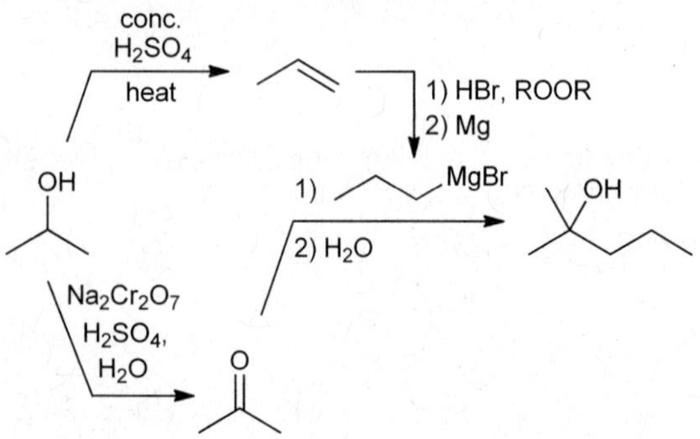

13.46.

(a) Two curved arrows are used to show hydride delivery. Note that the tail of the first curved arrow is placed on the bond between Al and H (it is NOT placed on the negative charge). The resulting alkoxide ion is then protonated upon treatment with water. This protonation step requires two curved arrows, as shown.

1) LAH
2) H$_2$O

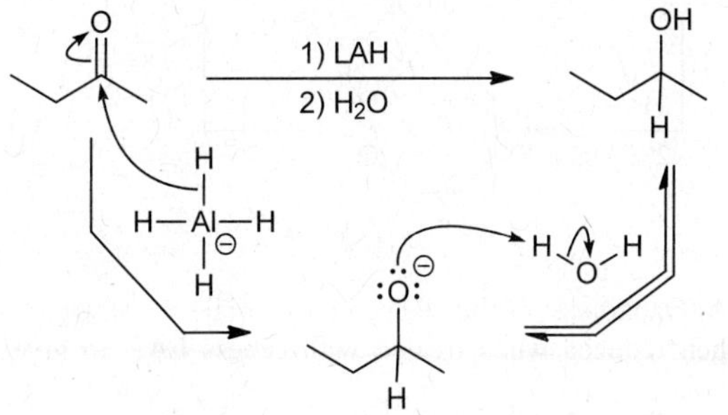

(b) Two curved arrows are used to show hydride delivery. Note that the tail of the first curved arrow is placed on the bond between Al and H (it is NOT placed on the negative charge). The resulting alkoxide ion is then protonated upon treatment with water. This protonation step requires two curved arrows, as shown.

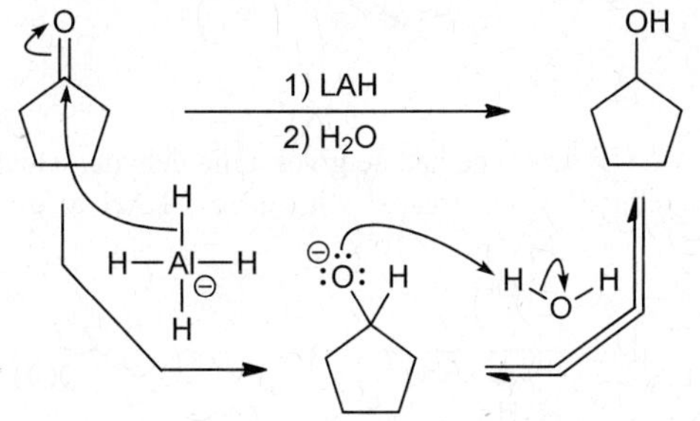

1) LAH
2) H$_2$O

(c) Two curved arrows are used to show hydride delivery. Note that the tail of the first curved arrow is placed on the bond between B and H (it is NOT placed on the negative charge). The resulting alkoxide ion is then protonated, which requires two curved arrows, as shown.

NaBH$_4$, MeOH

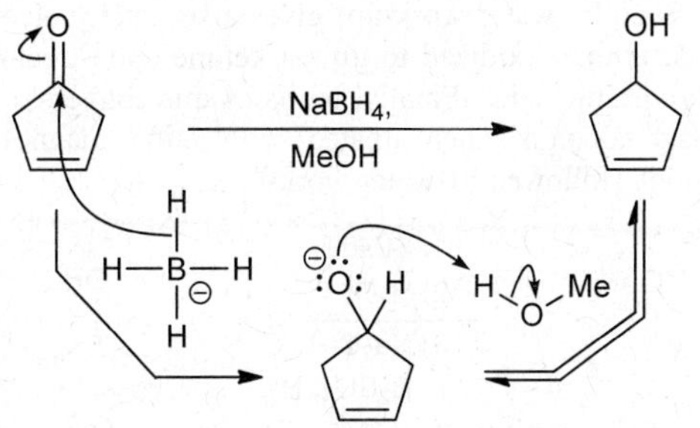

13.47.

(a) As seen in Mechanism 13.6, the OH group is transformed into a better leaving group (through a series of steps shown below), and then chloride attacks as a nucleophile and expels the leaving group.

SOCl$_2$
pyridine

Cl + SO$_2$ + Cl$^{\ominus}$

- Cl$^{\ominus}$

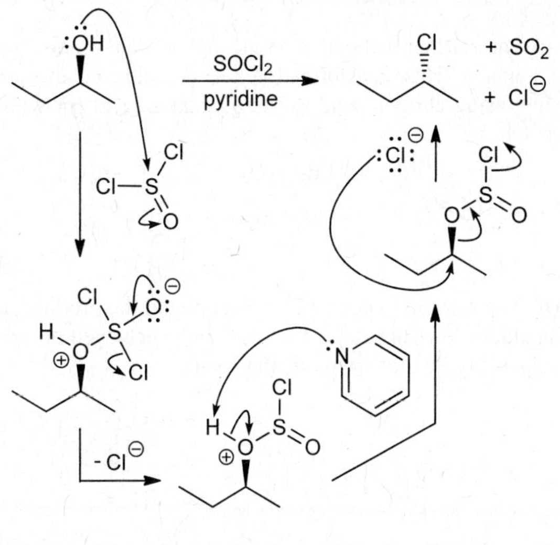

(b) As seen in Mechanism 13.7, the OH group is transformed into a better leaving group, and then bromide attacks as a nucleophile and expels the leaving group.

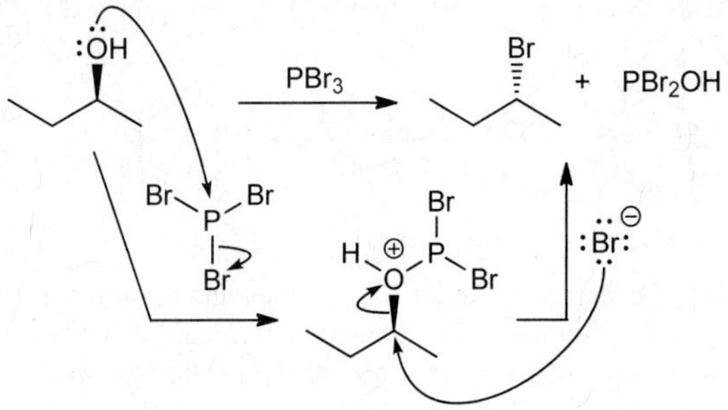

PBr$_3$

Br + PBr$_2$OH

(c) Two curved arrows are used to show hydride delivery to the ester. Note that the tail of the first curved arrow is placed on the bond between Al and H (it is NOT placed on the negative charge). The resulting intermediate then ejects ethoxide as a leaving group, which requires two curved arrows. The resulting aldehyde is then further reduced by another equivalent of LAH. Once again, two curved arrows are used to show hydride delivery. The resulting alkoxide ion is then protonated upon treatment with water. This protonation step requires two curved arrows.

1) LAH
2) H$_2$O

- EtO$^{\ominus}$

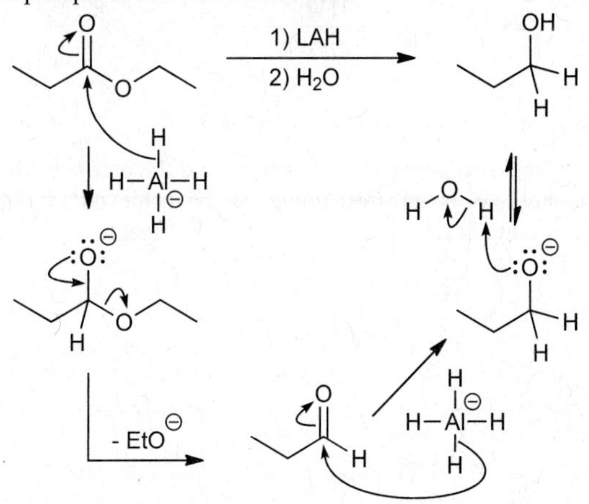

13.48.

(a) The starting material is an alcohol and the product is a ketone. This transformation can be achieved in one step, using chromic acid as the oxidizing agent (or with PCC).

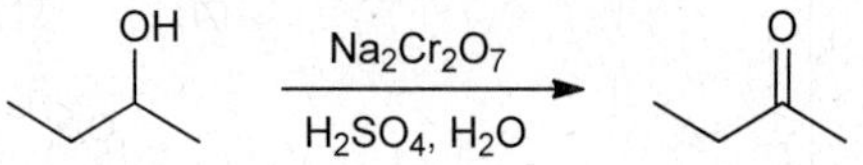

(b) The starting material is an alcohol and the product is an aldehyde. This transformation can be achieved in one step, using PCC as the oxidizing agent.

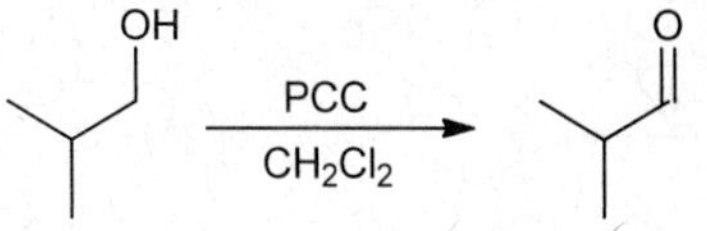

(c) The starting material is an alcohol and the product is an aldehyde. This transformation can be achieved in one step, using PCC as the oxidizing agent.

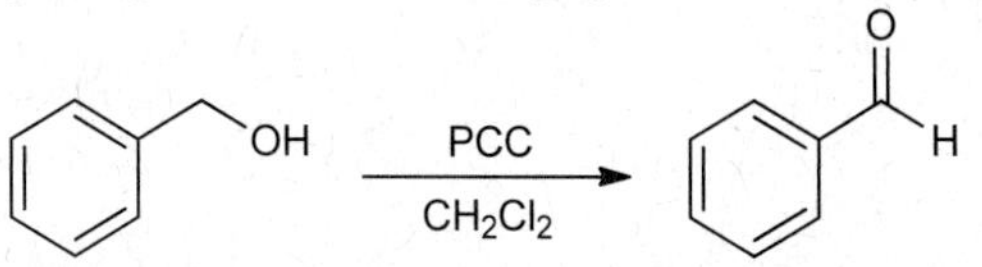

(d) The starting material is an alcohol and the product is a carboxylic acid. This transformation can be achieved in one step, using chromic acid as the oxidizing agent.

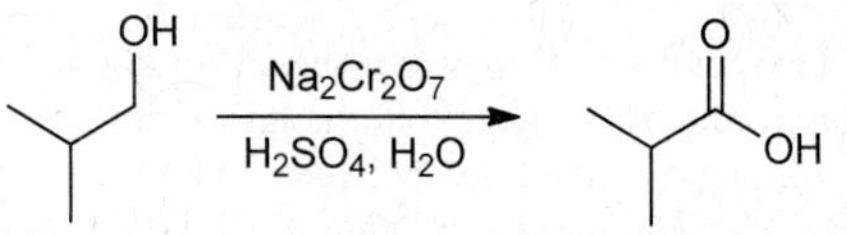

(e) The starting material is an aldehyde and the product is an alcohol. This transformation can be achieved in one step, using LAH or NaBH$_4$ as a reducing agent.

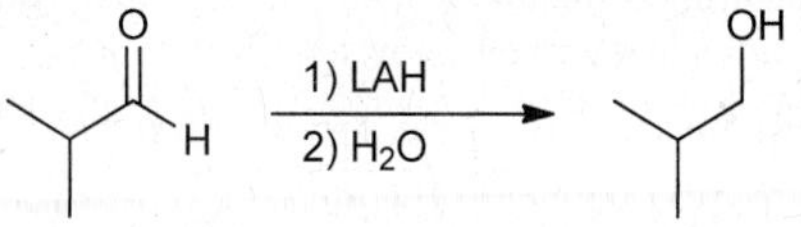

(f) The starting material is a ketone and the product is an alcohol. This transformation can be achieved in one step, using LAH or NaBH$_4$ as a reducing agent.

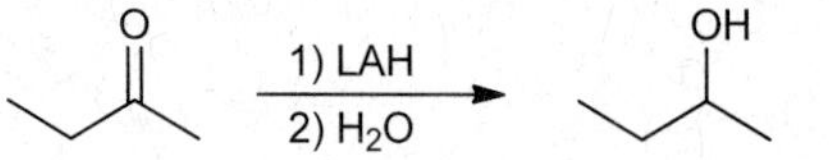

13.49

(a) Ozonolysis of the alkene gives a dialdehyde, which is then reduced when treated with excess LAH to give a diol.

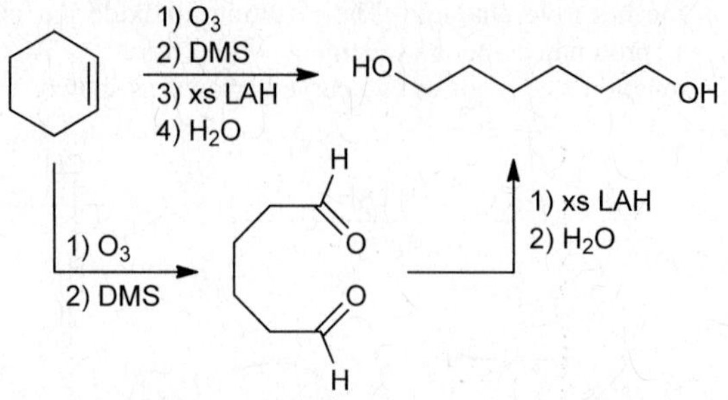

(b) Ozonolysis of the alkene gives a dialdehyde, which is then reduced when treated with excess LAH to give a diol.

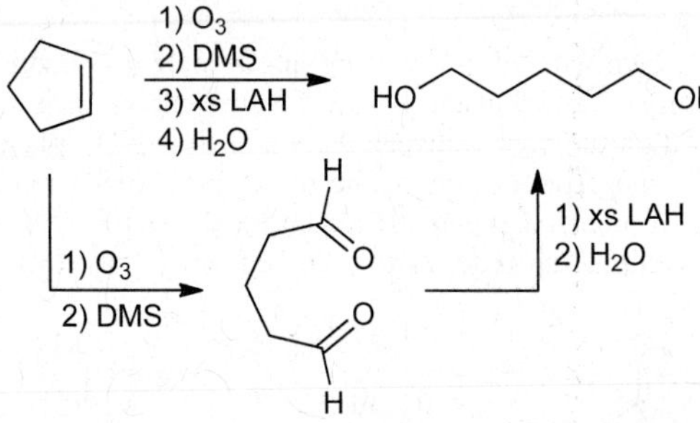

(c) Ozonolysis of the alkene gives a dialdehyde, which is then reduced when treated with excess LAH to give a diol.

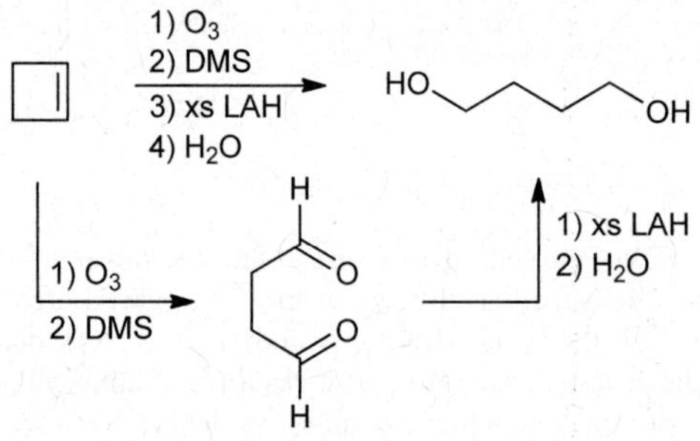

(d) Treating the aldehyde with ethyl magnesium bromide (followed by water work-up) gives a secondary alcohol, which is then oxidized to give a ketone upon treatment with chromic acid. Finally, the ketone is converted to a tertiary alcohol when treated with ethyl magnesium bromide (followed by water work-up).

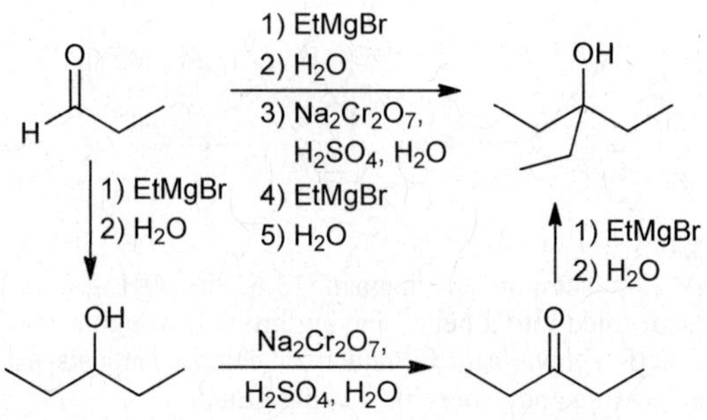

(e) The aldehyde is reduced upon treatment with LAH to give an alcohol. Treating the alcohol with tosyl chloride and pyridine converts the alcohol into a tosylate.

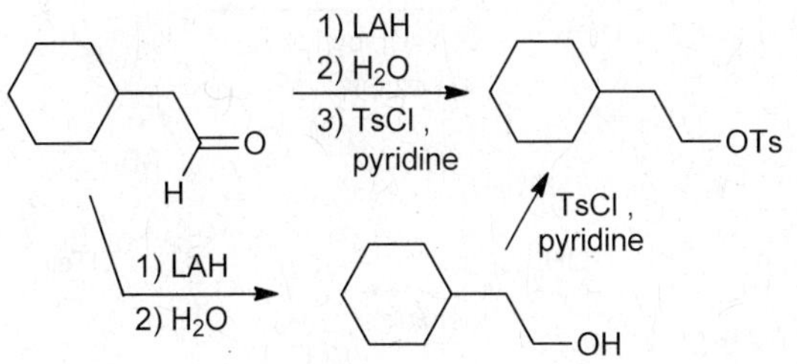

(f) Acid-catalyzed hydration of the alkene gives a secondary alcohol, which is then oxidized to a ketone upon treatment with chromic acid. Finally, the ketone is converted to a tertiary alcohol when treated with a Grignard reagent (followed by water work-up).

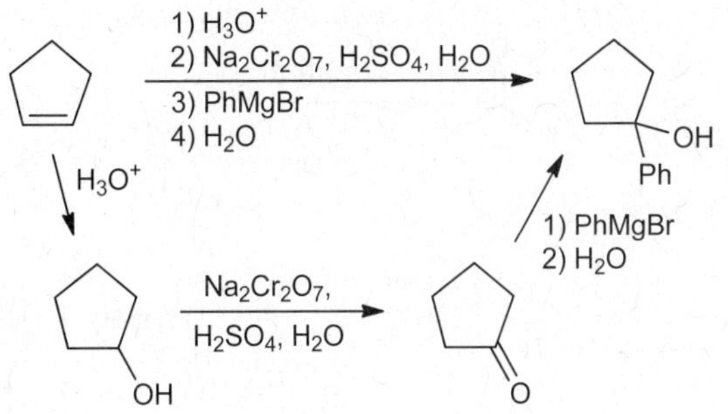

13.50.

(a) In the first step of the mechanism, the Grignard reagent (methyl magnesium bromide) functions as a nucleophile and attacks the C=O bond of the ketone. This step requires two curved arrows. The resulting alkoxide ion is then protonated upon treatment with water. This proton transfer step also requires two curved arrows, as shown.

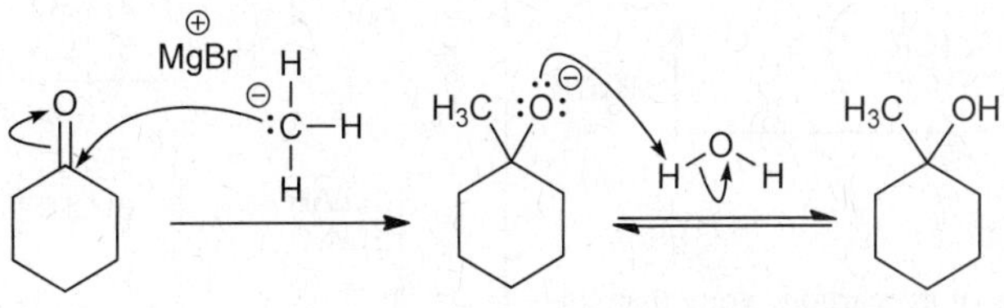

(b) In the first step of the mechanism, the Grignard reagent (methyl magnesium bromide) functions as a nucleophile and attacks the C=O bond of the ester. This step requires two curved arrows. The resulting intermediate then ejects a leaving group to give a ketone, which requires two curved arrows. The ketone is then further attacked by another equivalent of the Grignard reagent. Once again, two curved arrows are used to show the nucleophilic attack, resulting in a dianion.

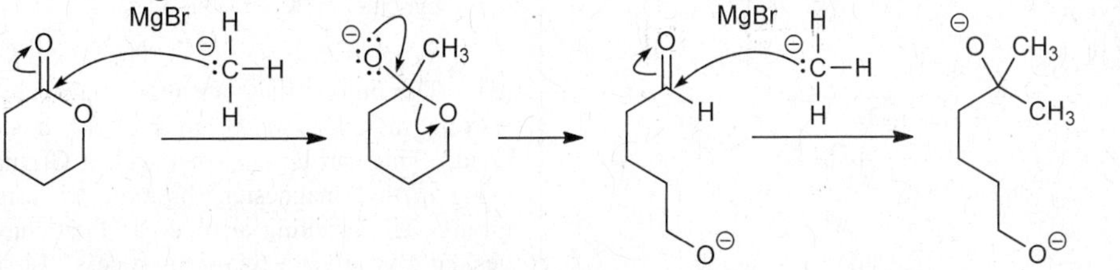

The resulting dianion is then protonated upon treatment with water. There are two locations that are protonated, each of which requires two curved arrows, as shown. Notice that each anion is protonated in a separate step (this should not be drawn as one step with four curved arrows, because there are two distinct processes occurring, and it is unlikely that they occur precisely at the same moment).

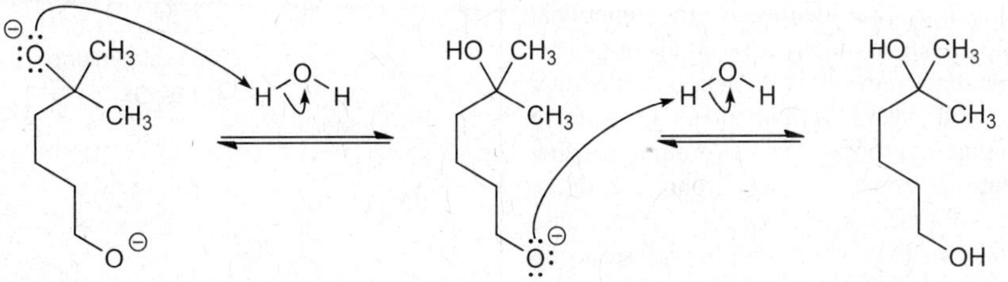

13.51. Most of the reagents for these transformations can be found in Figure 13.11.

1) 9-BBN
2) H₂O₂, NaOH

PCC
CH₂Cl₂

1) LAH
2) H₂O

1) Br₂
2) xs NaNH₂
3) H₂O

H₂, Lindlar's cat.

1) TsCl
2) t-BuOK

NaOH   PBr₃

1) BH₃·THF
2) H₂O₂, NaOH

HBr
ROOR

Br

t-BuOK

13.52.

(a) The product has one more carbon atom than the starting material, so we must form a carbon-carbon bond. This can be achieved with a Grignard reaction (using methyl magnesium bromide to install a methyl group). The resulting alcohol can then be oxidized to give the desired product.

1) MeMgBr
2) H₂O
3) Na₂Cr₂O₇, H₂SO₄, H₂O

1) MeMgBr
2) H₂O

OH

Na₂Cr₂O₇, H₂SO₄, H₂O

(b) The starting material and the product have the same carbon skeleton, so we do not need to form or break any carbon-carbon bonds. The identity of the functional group has changed from an aldehyde to an alkene. This transformation can be achieved via reduction (with either LAH or NaBH₄) followed by elimination. In order to perform the elimination process, the OH group must first be converted into a better leaving group (toyslate). Then, treating the tosylate with a sterically hindered base will give an alkene (the base should be sterically hindered to suppress the competing S_N2 process).

1) LAH
2) H₂O
3) TsCl, pyridine
4) t-BuOK

1) LAH
2) H₂O

OH

TsCl
pyridine

OTs

t-BuOK

(c) The product has one less carbon atom then the starting material, so we must break a carbon-carbon bond. Therefore, the last step of our synthesis must be ozonolysis of an alkene to give the desired ketone. With this in mind, preparation of the alkene can be achieved by reducing the starting material (with LAH or NaBH₄), followed by converting the OH group to a tosylate and then treating the tosylate with a sterically hindered base.

1) LAH
2) H₂O
3) TsCl, pyridine
4) t-BuOK
5) O₃
6) DMS

1) LAH
2) H₂O

OH

TsCl
pyridine

OTs

t-BuOK

1) O₃
2) DMS

(d) The product has one more carbon atom than the starting material, so we must make a carbon-carbon bond. This can be achieved with a Grignard reaction, using methyl magnesium bromide to install a methyl group. The resulting alcohol can then converted to the desired alkene via a two-step process. First the alcohol is converted to a tosylate (because OH is a bad leaving group), and then the tosylate is treated with a sterically hindered base to give the less substituted alkene.

1) MeMgBr
2) H₂O
3) TsCl, pyridine
4) t-BuOK

1) MeMgBr
2) H₂O

OH

TsCl
pyridine

OTs

t-BuOK

(e) The starting material and the product have the same carbon skeleton. The identity of the functional group must be changed. This can be achieved in two steps. The alkyl chloride is first converted to a primary alcohol via an S_N2 process, and then the alcohol can be oxidized with PCC to give the desired product.

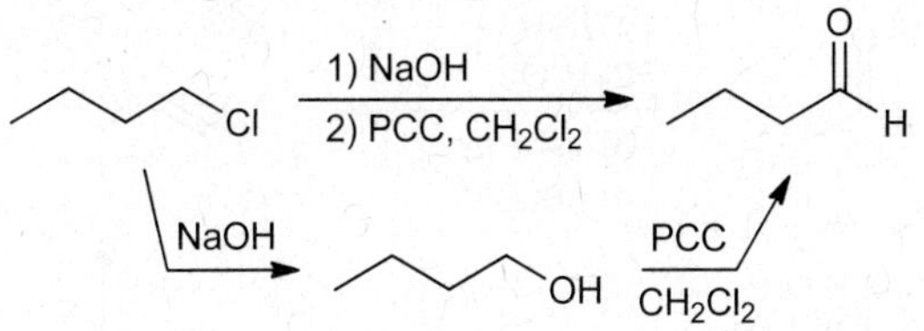

(f) In the previous problem, we saw a two-step procedure for converting the starting material into an aldehyde.

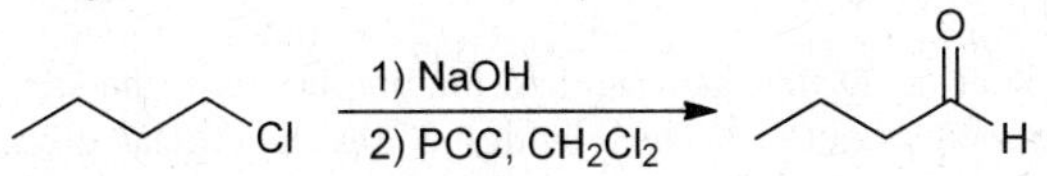

This aldehyde can then be converted into the desired product via a two-step process (a Grignard reaction to install a methyl group, followed by oxidation).

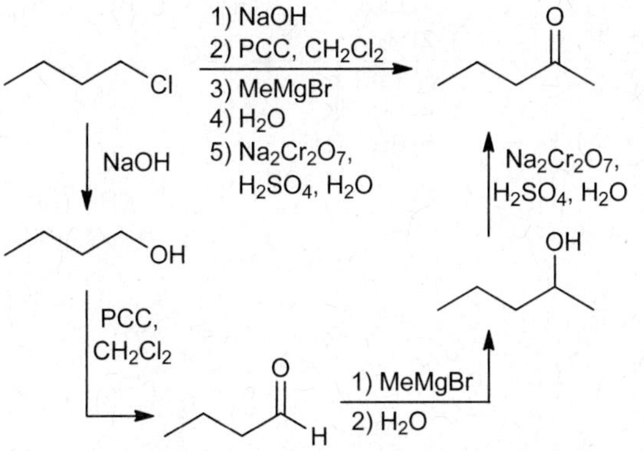

(g) The starting material and the product have the same carbon skeleton. The identity of the functional group must be changed. This can be achieved in two steps. The alkene is first converted to an alcohol via acid-catalyzed hydration, and then the alcohol can be oxidized with chromic acid to give the desired product.

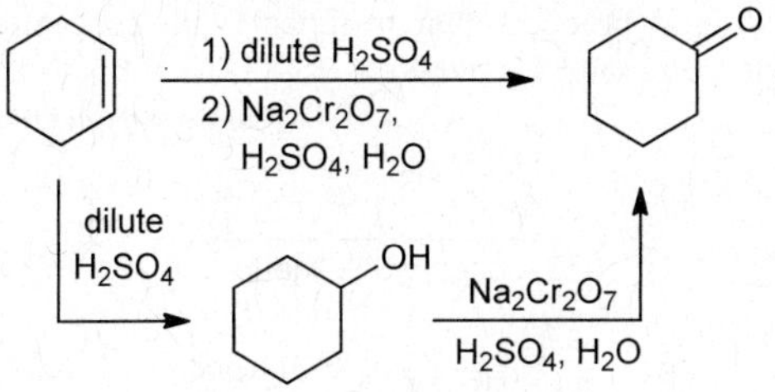

(h) In the previous problem, we saw that the starting material can be converted into a ketone in just two steps.

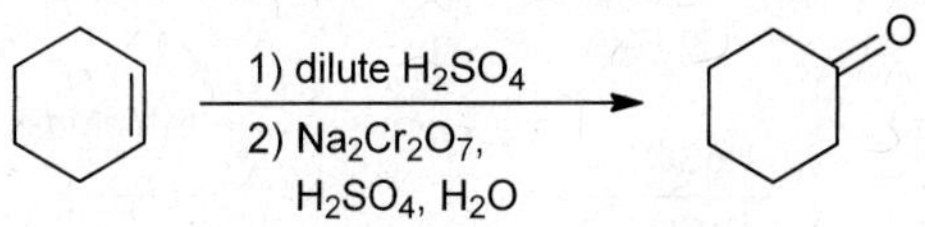

This ketone can be converted into the desired product with a Grignard reaction.

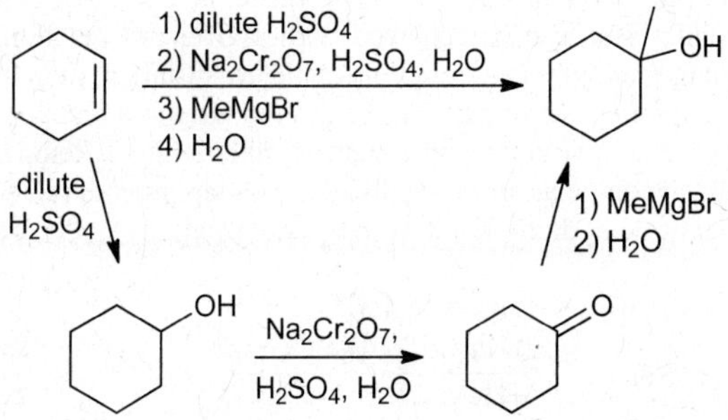

(i) In the previous problem, we saw that the starting material can be converted into a tertiary alcohol.

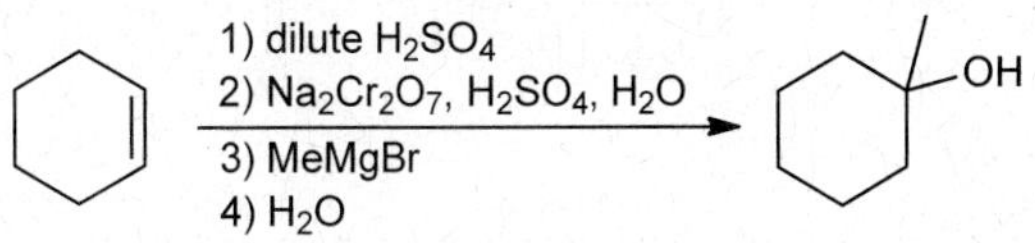

This alcohol can be converted into the desired product in just one step (upon treatment with concentrated sulfuric acid).

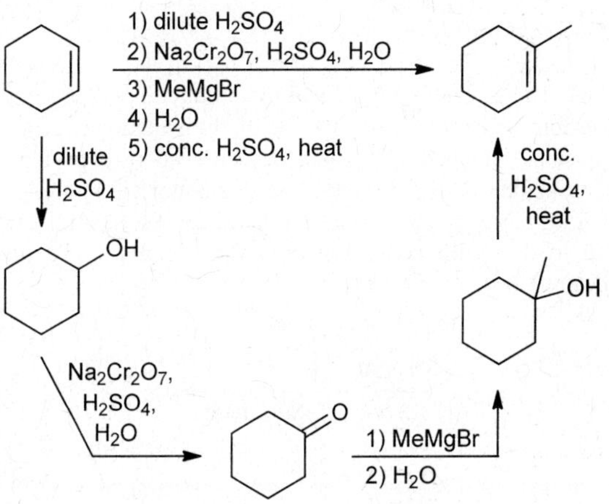

(j) The product has one more carbon atom than the starting material, so we must make a carbon-carbon bond. This can be achieved with a Grignard reaction, using methyl magnesium bromide to install a methyl group. In order to perform the desired Grignard reaction, we must first convert the starting alkyne into a ketone, which can be accomplished via acid-catalyzed hydration in the presence of mercuric sulfate.

1) HgSO$_4$, H$_2$SO$_4$, H$_2$O
2) MeMgBr
3) H$_2$O

HgSO$_4$, H$_2$SO$_4$, H$_2$O

1) MeMgBr
2) H$_2$O

(k) The product has one more carbon atom than the starting material, so we must make a carbon-carbon bond. This can be achieved with a Grignard reaction, using methyl magnesium bromide to install a methyl group. In order to perform the desired Grignard reaction, we must first convert the starting alkene into a ketone, which can be accomplished via a two-step process (acid-catalyzed hydration, followed by oxidation).

1) dilute H$_2$SO$_4$
2) Na$_2$Cr$_2$O$_7$, H$_2$SO$_4$, H$_2$O
3) MeMgBr
4) H$_2$O

dilute H$_2$SO$_4$

1) MeMgBr
2) H$_2$O

Na$_2$Cr$_2$O$_7$, H$_2$SO$_4$, H$_2$O

(l) The desired transformation can be achieved in one step, using a Grignard reaction.

1) EtMgBr
2) H$_2$O

(m) The starting material and the product have the same carbon skeleton. The identity of the functional group must be changed. This can be achieved in two steps. The ketone is first converted to an alcohol via reduction (using either LAH or NaBH$_4$), and then the alcohol can be treated with concentrated sulfuric acid to give the desired alkene.

1) LAH
2) conc. H$_2$SO$_4$, heat

LAH

conc. H$_2$SO$_4$ heat

(n) The starting material and the product have the same carbon skeleton. The identity of the functional group must be changed. The ketone is first converted to an alcohol via reduction (using either LAH or NaBH$_4$), and then we must perform an elimination process. Since the less substituted alkene is desired, we must use a

sterically hindered base. Since OH is a bad leaving group, it must first be converted to a good leaving group. This can be accomplished by treating the alcohol with tosyl chloride and the resulting tosylate can then be treated with *tert*-butoxide to give the desired product.

1) LAH
2) H$_2$O
3) TsCl, py
4) *t*-BuOK

1) LAH
2) H$_2$O

TsCl, pyridine

t-BuOK

(o) The starting material and the product have the same carbon skeleton. The identity of the functional group must be changed. The answer to the previous problem allows us to convert the starting material into alkene, which can be then converted into the product via hydroboration-oxidation.

1) LAH
2) H$_2$O
3) TsCl, py
4) *t*-BuOK
5) BH$_3$·THF
6) H$_2$O$_2$, NaOH

1) LAH
2) H$_2$O

1) BH$_3$·THF
2) H$_2$O$_2$, NaOH

TsCl, pyridine

t-BuOK

(p) The product has one more carbon atom than the starting material, so we must make a carbon-carbon bond. This can be achieved with a Grignard reaction, using methyl magnesium bromide to install a methyl group. The resulting alcohol can then be converted to the desired product upon treatment with concentrated sulfuric acid (an E1 process).

1) MeMgBr
2) H$_2$O
3) conc. H$_2$SO$_4$, heat

1) MeMgBr
2) H$_2$O

conc. H$_2$SO$_4$, heat

(q) The desired transformation can be achieved in one step, using a Grignard reaction.

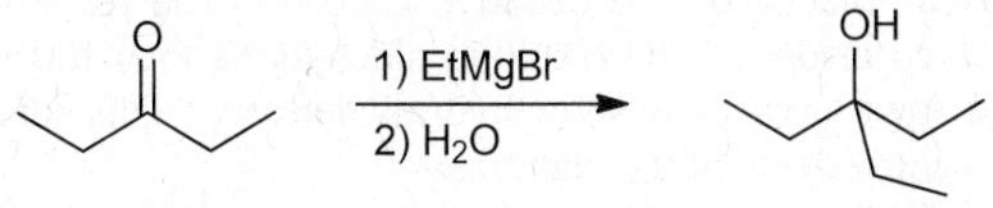

(r) The starting material and the product have the same carbon skeleton. The identity of the functional group must be changed. This can be achieved by reducing the aldehyde (with either LAH or NaBH$_4$) followed by treatment with PBr$_3$ to give the desired alkyl bromide.

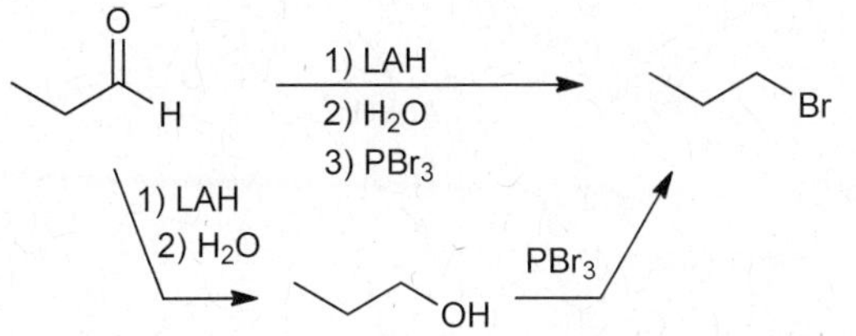

(s) The product has one more carbon atom than the starting material, so we must make a carbon-carbon bond. This can be achieved with a Grignard reaction, using methyl magnesium bromide to install a methyl group. In order to perform the desired Grignard reaction, we must first convert the starting alkene into an aldehyde, which can be accomplished via hydroboration-oxidation, followed by oxidation with PCC.

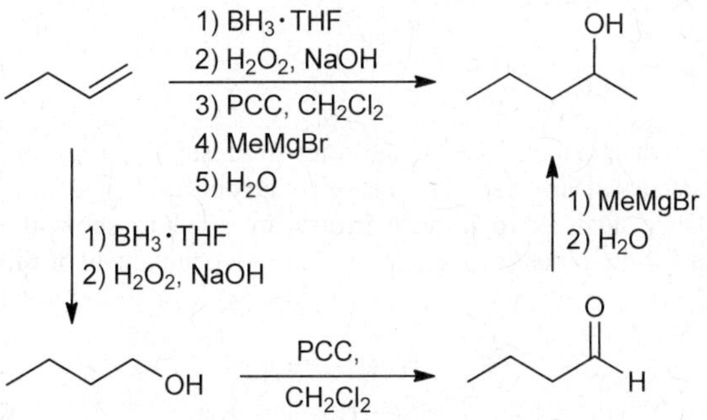

13.53. The singlet with an integration of 9 is characteristic of a *tert*-butyl group. The signals near 7 ppm (with a total integration of 4) indicate a disubstituted aromatic ring. The splitting pattern of these signals indicates that the compound is *para*-disubstituted (because of symmetry). The signal at 5 ppm with an integration of 1 is likely an OH group. Our analysis produces the following three fragments:

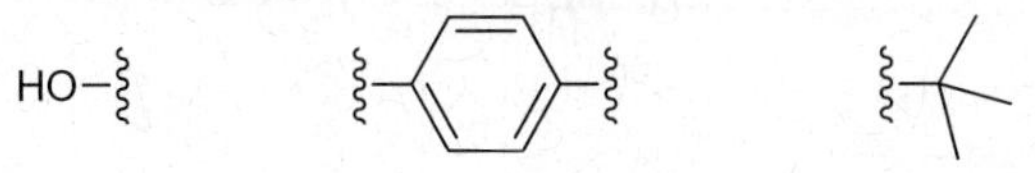

These three fragments can only be assembled in one way:

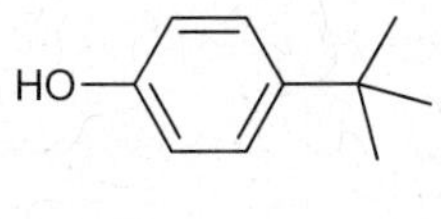

13.54. The ^{13}C NMR spectrum indicates that all three carbon atoms are in different environments. One of these signals appears between 50 and 100 ppm, indicating that one carbon atom is connected to an oxygen atom (the molecular formula indicates the presence of an oxygen atom). Now we turn to the ^{1}H NMR spectrum. The signal at 1 ppm has an integration of 3, indicating a CH$_3$ group. Since this signal is a triplet, the CH$_3$ group must be adjacent to a CH$_2$ group. The signal at 3.6 ppm indicates a CH$_2$ group (integration = 2) that is neighboring an oxygen atom (thus it is shifted downfield, as expected for a protons that are α to an OH group). The singlet at 2.4 ppm is likely an OH group, and the signal at 1.6 ppm results from the CH$_2$ group that is being split by two sets of neighbors. Using all of this information, we can arrive at the following structure.

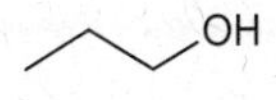

13.55. In the IR spectrum, the broad signal between 3200 and 3600 cm^{-1} indicates an OH group. The NMR spectrum indicates that there are only three different kinds of carbon atoms, yet the molecular formula indicates that the compound has five carbon atoms. Therefore, we must draw structures that possess enough symmetry such that there are only three unique kinds of carbon atoms. The following two structures are consistent with this analysis.

13.56. There is a multiplet just above 7 ppm, indicating an aromatic ring. Since the integration of this signal is 5, we expect the aromatic ring to be monosubstituted. The two triplets (at 2.8 ppm and 3.8 ppm) indicate two CH$_2$ groups that are neighboring each other, and the singlet at 2 ppm (with an integration of 1) is likely an OH group. Our analysis produces the following three fragments:

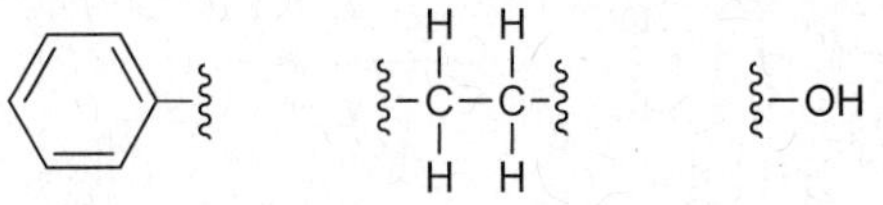

These three fragments can only be assembled in one way:

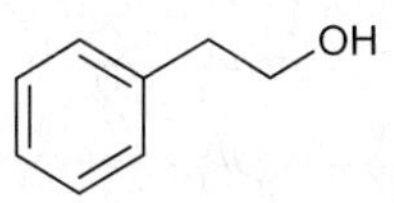

Notice that the signals for the CH$_2$ groups (the triplets) are shifted downfield. The signal at 3.8 ppm is the CH$_2$ group next to the oxygen atom, and the signal at 2.8 ppm is the CH$_2$ group next to the aromatic ring.

13.57. Two curved arrows are used to show hydride delivery to one of the ester groups. Note that the tail of the first curved arrow is placed on the bond between Al and H (it is NOT placed on the negative charge). The resulting intermediate then ejects a leaving group, which requires two curved arrows. The resulting aldehyde is then further reduced by another equivalent of LAH. Once again, two curved arrows are used to show hydride delivery. This entire process is then repeated again for the other ester group, giving two equivalents of the dianion.

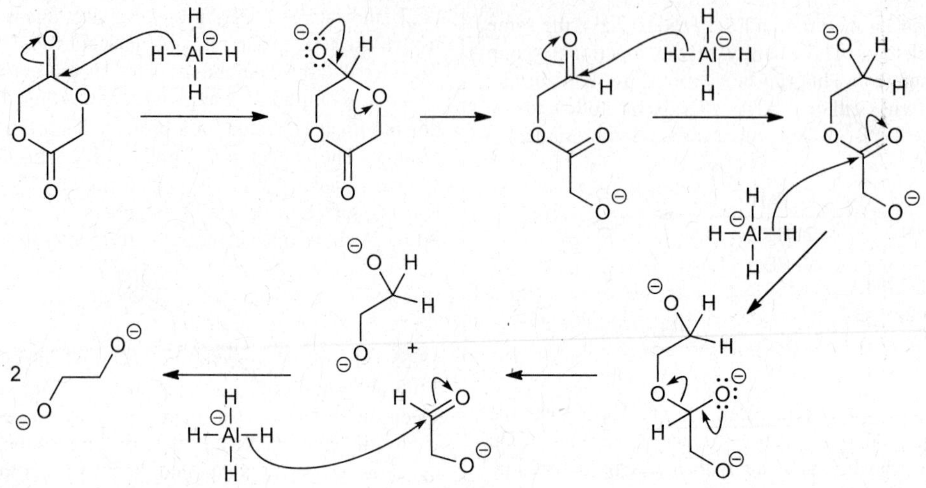

The resulting dianion is then protonated upon treatment with water. Notice that each protonation step is drawn separately.

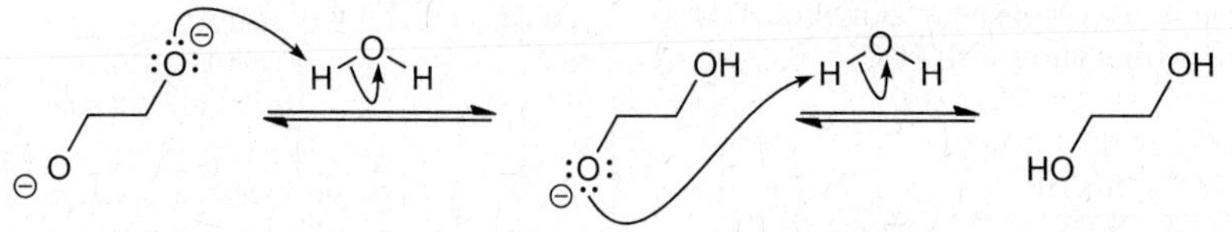

13.58. Two curved arrows are used to show a Grignard reagent attacking one of the ester groups. The resulting intermediate then ejects a leaving group, which requires two curved arrows. The resulting ketone is then further attacked by another equivalent of the Grignard reagent. Once again, two curved arrows are used to show the nucleophilic attack. This entire process is then repeated again for the other ester group, giving two equivalents of the dianion.

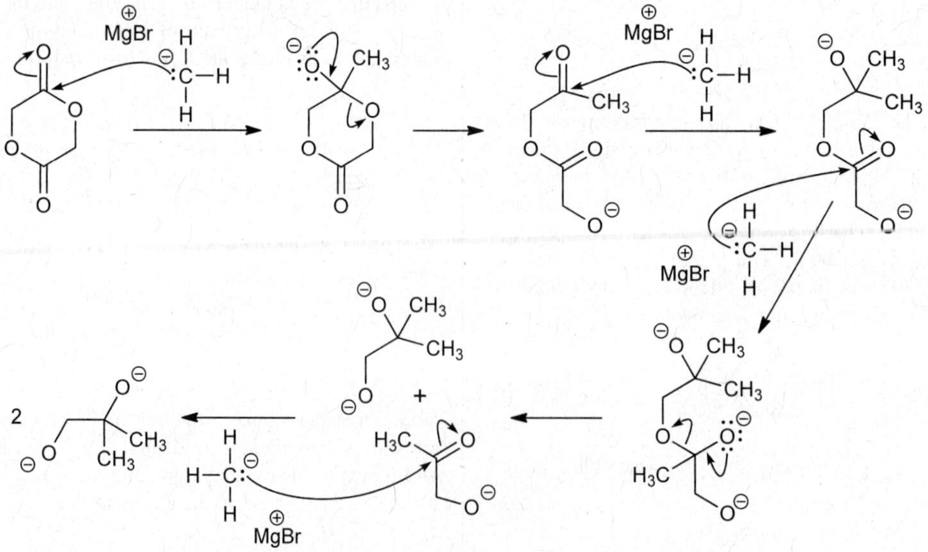

13.64. We have seen that tetrabutylammonium fluoride (TBAF) can be used to remove silyl protecting groups. In this case, there are two such groups:

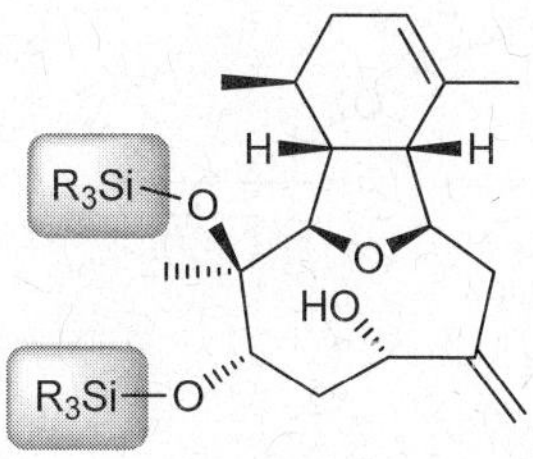

We learned about the trimethylsilyl protecting group (TMS), and the protecting groups employed in this case are very similar (in each case, one of the methyl (R) groups has been replaced with a *tert*-butyl group). This protecting group is called the *tert*-butyldimethyl silyl group (TBDMS, or just TBS for short), and it is removed upon treatment with TBAF, in much the same way that the TMS group is removed under similar conditions. The rest of the compound is expected to remain unchanged.

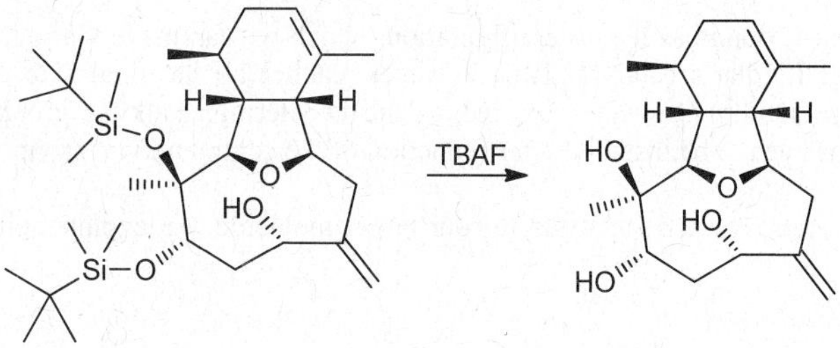

13.65.
(a) When diol **1** is treated with PBr$_3$, each of the two OH groups can separately react with the reagent to produce dibromide **2**. Notice that the stereochemistry is now inverted, since the mechanism for bromide displacement involves an S$_N$2 reaction.

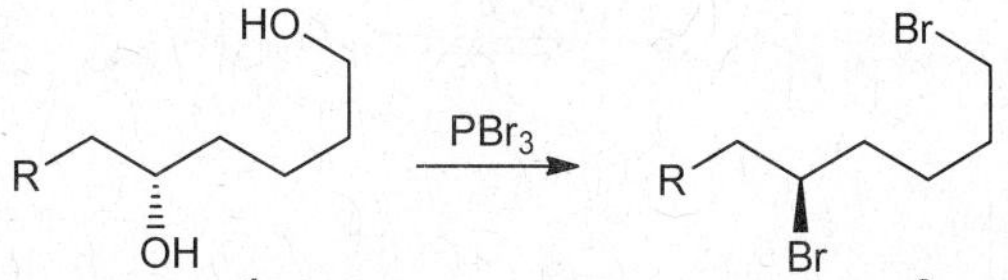

(b) The byproduct can be formed via the following process. First, the terminal alcohol is converted into a good leaving group upon treatment with PBr$_3$. However, before a bromide ion can attack this intermediate, an intramolecular S$_N$2-type reaction occurs – the internal OH group can displace the good leaving group of **4** to form cyclic ether **5** which can be deprotonated to produce byproduct **3**.

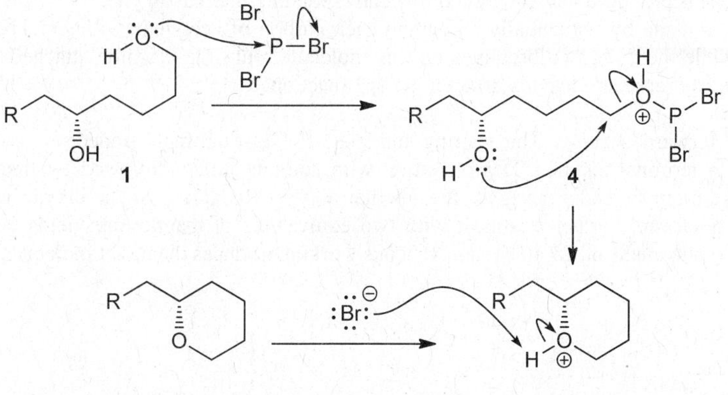

Notice that the primary OH group is converted into a good leaving group, and the secondary OH group functions as a nucleophile. If instead the secondary OH group had reacted with PBr$_3$ (**6**), and the primary OH group had functioned as a nucleophile, then compound **7** would have been produced, which is not the byproduct (it is the enantiomer of the byproduct).

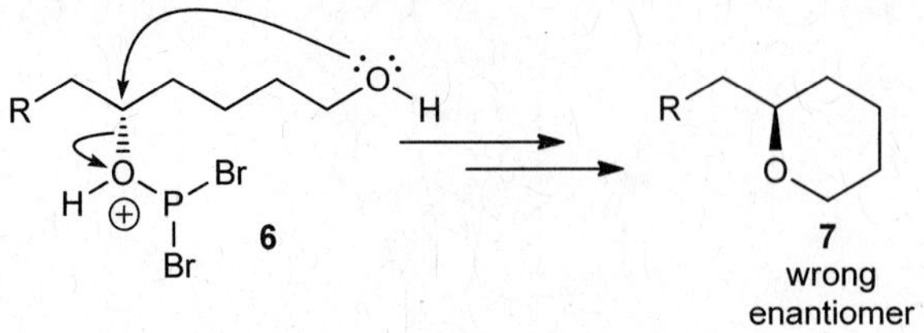

13.66. The following synthesis is one suggested synthetic pathway. There are likely other acceptable approaches that accomplish the same goal.

Let's start with a few general observations:

1. Our starting materials cannot have more than eleven carbon atoms, and the target compound has thirty carbon atoms, so we will need to form more than one C-C bond.
2. The molecule is symmetric, so it makes sense to propose equivalent reactions on each side of the molecule.
3. The central C=C bond has the *cis* configuration, which we can make via reduction of an internal alkyne using H_2 and Lindlar's catalyst. This, however, cannot be the final step of our proposed synthesis, because these conditions would also reduce the two terminal alkyne groups. Thus, the two terminal alkyne groups need to be installed *after* reduction of the central alkyne group.

The figure below outlines a retrosynthetic analysis for our target molecule. An explanation of each of the steps (*a-d*) follows.

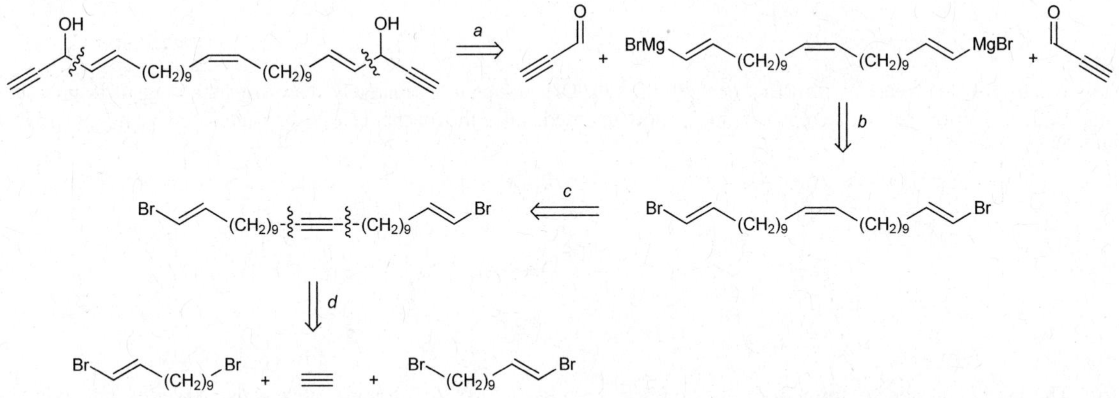

a. Duryne can be made from the reaction between the bis-vinyl Grigard reagent (a dianion) and two equivalents of the aldehyde shown.
b. The bis-vinyl Grignard is made from the bis-vinyl bromide and two equivalents of magnesium.
c. The *cis* alkene is produced via reduction of the corresponding internal alkyne.
d. The alkyne is made by sequentially alkylating each carbon of acetylene with *E*-1,11-dibromo-1-undecene. Note that while there are two bromines on this molecule, only the bromide attached to the sp^3 hybridized atom carbon can serve as a leaving group in an S_N2 reaction.

Now let's draw the forward scheme. The starting material, *E*-1,11-dibromo-1-undecene, is treated with sodium acetylide to produce a terminal alkyne. Deprotonation with sodium amide, followed by treatment with a second equivalent of *E*-1,11-dibromo-1-undecene gives the internal alkyne. Reduction of the alkyne with H_2 and Lindlar's catalyst affords the *cis* alkene. Further treatment with two equivalents of magnesium yields the bis-vinyl Grignard, which reacts with two equivalents of the aldehyde. Aqueous workup produces the target molecule, duryne.

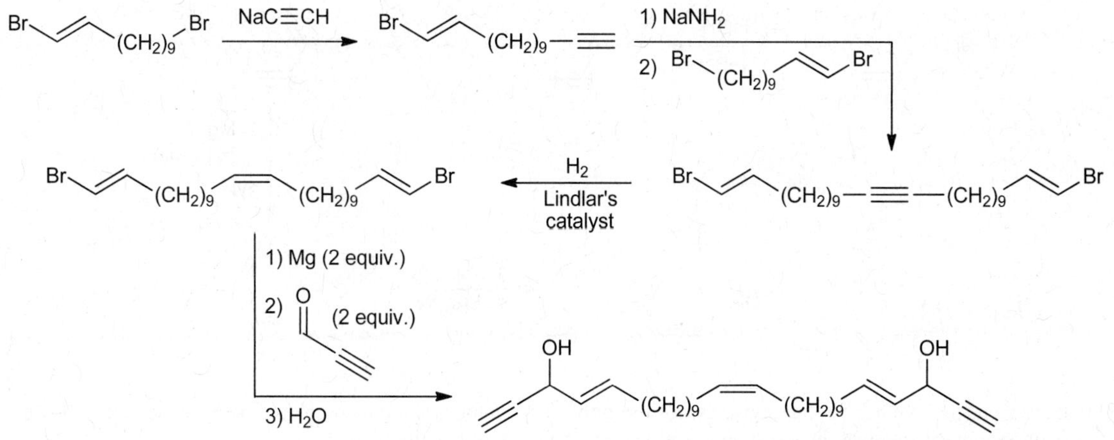

13.67. The following synthesis is one suggested synthetic pathway. There are likely other acceptable approaches that accomplish the same goal.

By comparing the structures of the starting material (4-methylphenol) and the product, it is clear that the following bonds (indicated by wavy lines below) need to be made in this synthesis.

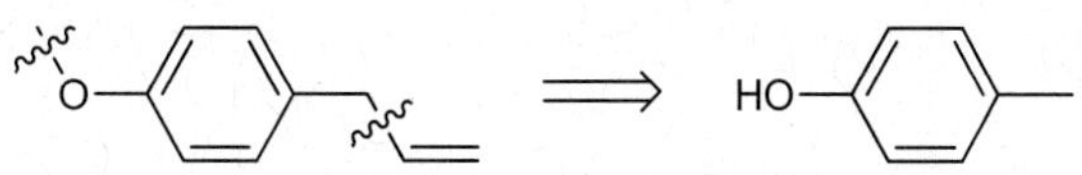

The left bond (C-O) can be made via an S_N2 process, while the right bond (C-C) can be made by either a Grignard reaction or by using an acetylide ion as a nucleophile. It is important that we make the ether bond early in our scheme to avoid an acid/base reaction between the phenolic proton ($pK_a \approx 10$) and the Grignard reagent or acetylide (both of which are strong bases). If the phenolic proton is subjected to a Grignard reagent, the latter would be destroyed via protonation. The same fate would occur for an acetylide ion that is treated with a compound bearing a phenolic proton. The following is a retrosynthetic analysis for our target compound. An explanation of each of the steps (a-e) follows.

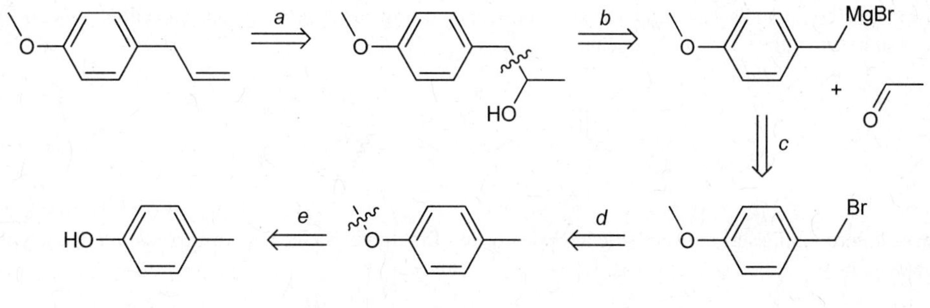

  a.   We can make the monosubstituted alkene by converting the OH group into a tosylate group and then performing an elimination reaction with a sterically hindered base.
  b.   The alcohol can be made via a Grignard reaction between the benzylic Grignard reagent and the aldehyde shown.
  c.   The Grignard reagent is made from the corresponding benzylic bromide.
  d.   The bromine atom can be installed via radical bromination at the benzylic position.
  e.   The methyl ether can be produced via an S_N2 reaction from the starting material.

Now let's draw the forward scheme. The starting material, 4-methylphenol, is deprotonated with sodium hydroxide and the resulting phenoxide serves as a nucleophile in an S_N2 reaction with bromomethane. The bromine atom is installed using N-bromosuccinimide and light. This compound is then converted into a Grignard reagent using magnesium. The Grignard reagent reacts with the appropriate aldehyde (CH_3CHO), followed by a water workup to produce the alcohol. Conversion to the tosylate and subsequent reaction with *tert*-butoxide produces the less-substituted elimination product, estragole.

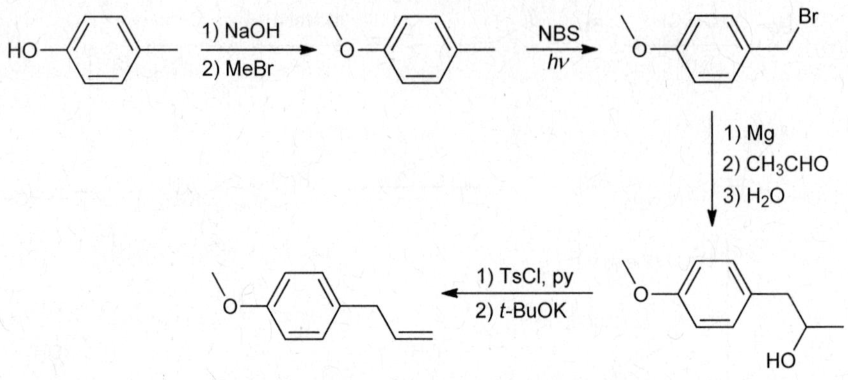

As mentioned, the synthetic route above is not the only method for making estragole. For example, the following alternative synthesis involves an acetylide ion, rather than a Grignard reaction:

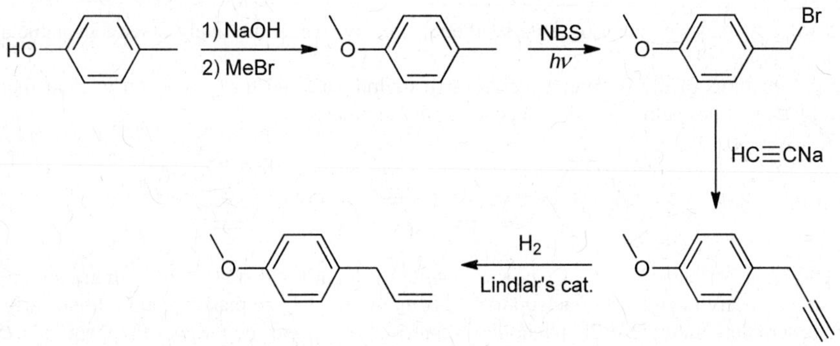

13.68. The following two syntheses are suggested synthetic pathways. There are likely other acceptable approaches that accomplish the same goal.

Each of the target compounds has eight carbon atoms, suggesting the following disconnections, which break each carbon skeleton into two four-carbon fragments.

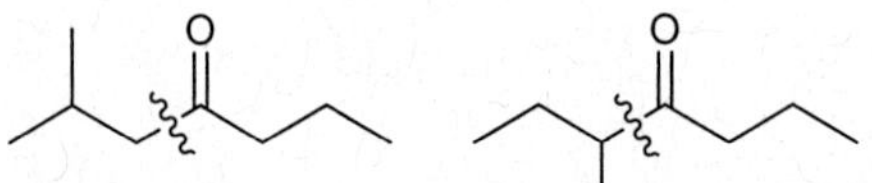

The following figure outlines a retrosynthetic analysis for our first target molecule. An explanation of each of the steps (*a-h*) follows.

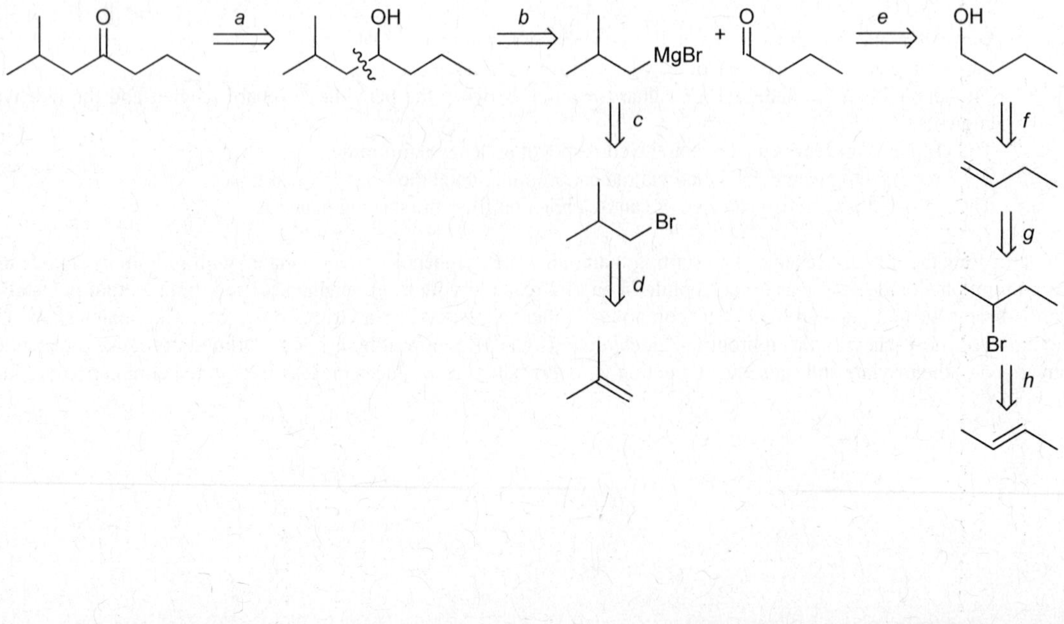

a.   The target ketone is made by oxidation of the corresponding alcohol, 2-methyl-4-heptanol.
b.   The alcohol is produced via a Grignard reaction between the Grignard reagent and aldehyde shown.
c.   The Grignard reagent is made from the corresponding alkyl halide, 1-bromo-2-methylpropane.
d.   1-Bromo-2-methylpropane is produced via an *anti*-Markovnikov addition of HBr to 2-methylpropene.
e.   The aldehyde is made from oxidation of 1-butanol.
f.   1-Butanol is made from 1-butene via hydroboration/oxidation.
g.   1-Butene is made from 2-bromobutane using a sterically hindered base to produce the less substituted alkene.
h.   2-Bromobutane is made from HBr addition to *trans*-2-butene. (Note: *Cis*-2-butene also produces the same product.)

Now let's draw the forward scheme. Addition of HBr to *trans*-2-butene produces 2-bromobutane, which is subsequently treated with *tert*-butoxide to give 1-butene. *Anti*-Markovnikov addition of water (via hydroboration/oxidation) followed by PCC oxidation gives the aldehyde. Reaction with the Grignard reagent (produced by *anti*-Markovnikov addition of HBr to 2-methylpropene, then magnesium, as shown) gives 2-methyl-4-heptanol after water workup. Oxidation with PCC produces the target ketone.

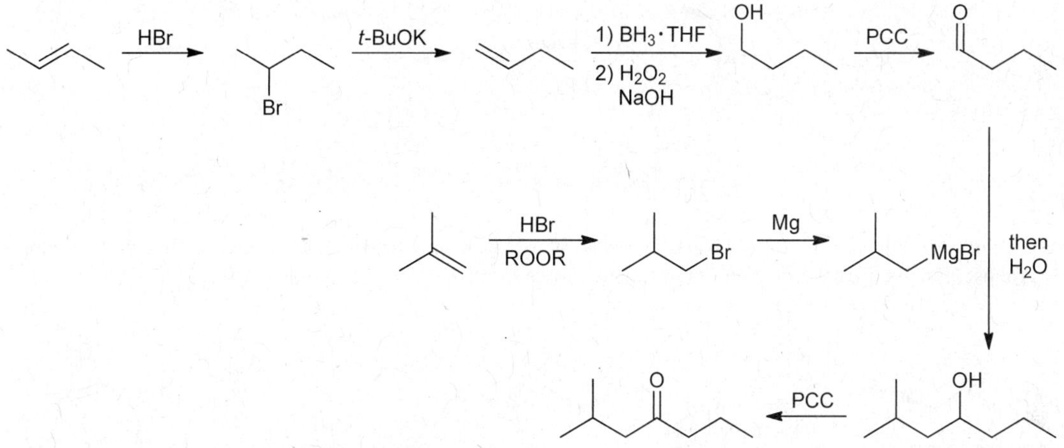

The following figure outlines a retrosynthetic analysis for our second target molecule. An explanation of each of the steps (*a-e*) follows.

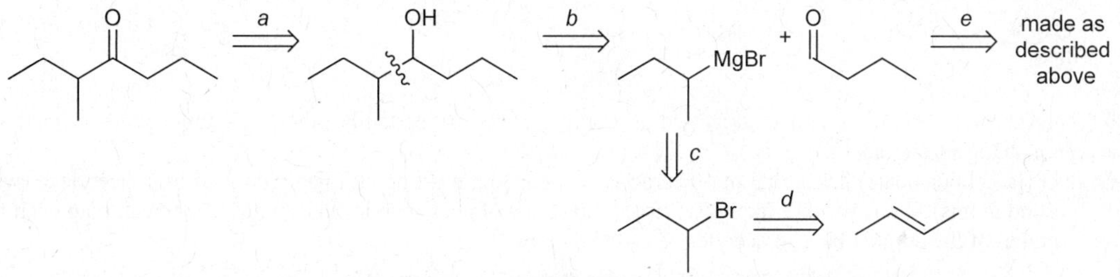

a.   The target ketone is made by oxidation of the corresponding alcohol, 3-methyl-4-heptanol.
b.   The alcohol is produced via a Grignard reaction between the Grignard reagent and aldehyde shown.
c.   The Grignard reagent is made from the corresponding alkyl halide, 2-bromobutane.
d.   2-bromobutane is produced via addition of HBr to *trans*-2-butene. (Note: *Cis*-2-butene also produces the same product.)
e.   The aldehyde is made as described in the synthesis of the first target ketone.

Now let's draw the forward scheme. Addition of HBr to *trans*-2-butene produces 2-bromobutane. 2-Bromobutane is converted to a Grignard reagent which is subsequently treated with the aldehyde (made as described above), followed by water workup, to produce 3-methyl-4-heptanol. Oxidation with PCC gives the target ketone.

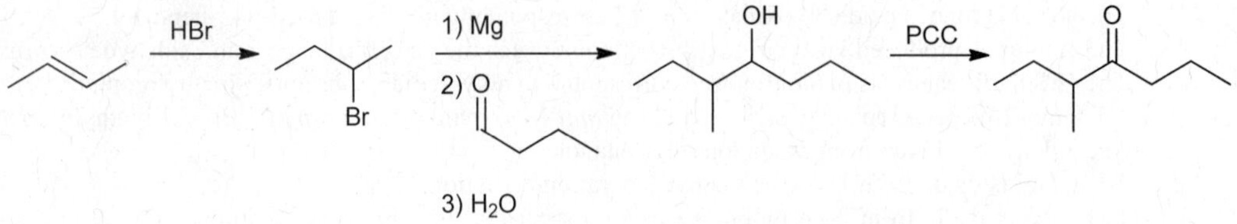

13.69. Compound **1** contains both an ester group and an amide group. As described in the problem statement, treatment of compound **1** with LiBH$_4$ is expected to result in reduction of the ester group, while the amide group will remain unchanged.

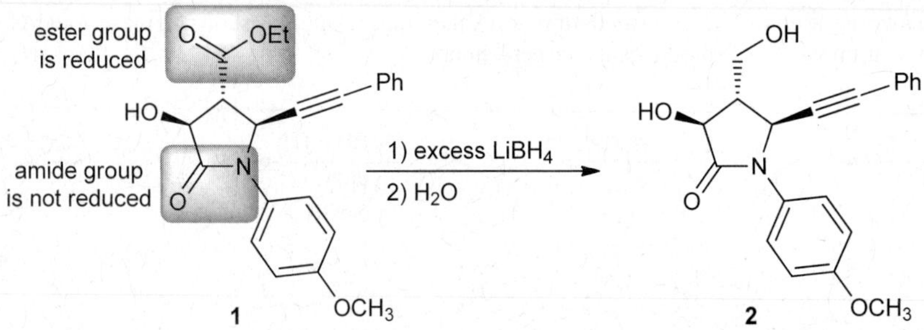

Treatment of compound **2** with excess NaH, followed by excess benzyl bromide, converts both hydroxyl groups into ether groups. Finally, partial reduction of the alkyne affords the *cis*-alkene.

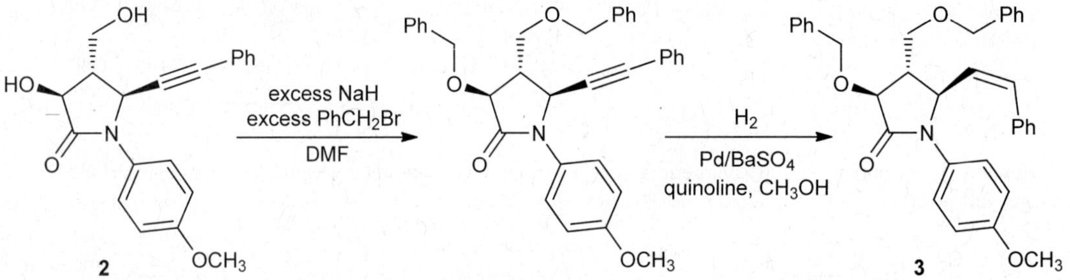

13.70. The following synthesis is one suggested approach. There are certainly other acceptable synthetic pathways that accomplish the same goal.

The product (hexyl butanoate) has a total of 10 carbons, all of which must be ultimately derived from acetylene (which has two carbon atoms). Thus, we will likely use five equivalents of acetylene in this synthesis. We can map each two-carbon fragment of the product back to acetylene as shown below.

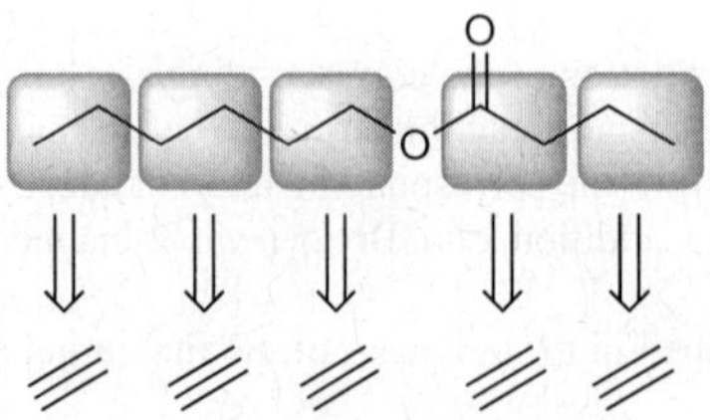

The figure below outlines a retrosynthetic analysis for our target molecule. An explanation of each of the steps (*a-l*) follows.

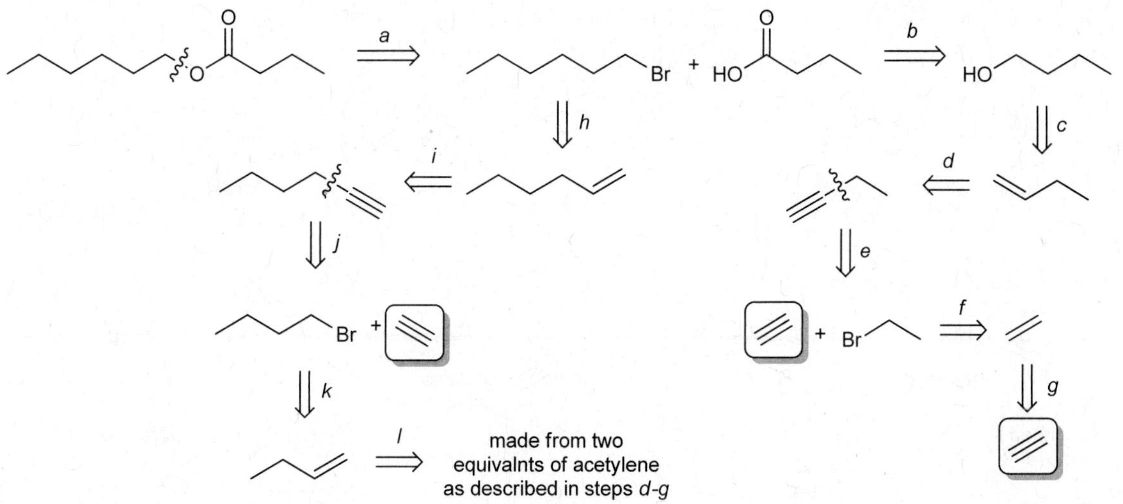

a.   Hexyl butanoate is made from an S_N2 reaction between 1-bromohexane and the carboxylic acid shown (after deprotonation of the carboxylic acid to make a good nucleophile).

b.   The carboxylic acid is made by oxidation of 1-butanol.

c.   1-Butanol is made by *anti*-Markovnikov hydration of 1-butene.

d.   1-Butene is made by partial reduction of 1-butyne.

e.   1-Butyne is made from an S_N2 reaction between acetylene (which must be deprotonated to form an acetylide ion) and bromoethane.

f.   Bromoethane is made by addition of HBr to ethylene.

g.   Ethylene is made by partial reduction of acetylene.

h.   1-Bromohexane is made by *anti*-Markovnikov addition of HBr to 1-hexene.

i.   1-Hexene is made by partial reduction of 1-hexyne.

j.   1-Hexyne is made from an S_N2 reaction between acetylene (which must be deprotonated to form an acetylide ion) and 1-bromobutane.

k.   1-Bromobutane is made by *anti*-Markovnikov addition of HBr to 1-butene.

l.   1-Butene is made as described in steps *d-g*.

Now let's draw the forward scheme. Acetylene is converted to bromoethane in two steps by hydrogenation with Lindlar's catalyst followed by addition of HBr to the resulting alkene. Reaction with sodium acetylide (made from acetylene and sodium amide) produces 1-butyne, which is then treated with H_2 and Lindlar's catalyst to furnish 1-butene. *Anti*-Markovnikov addition of water (via hydroboration / oxidation) gives 1-butanol, which is subsequently oxidized to the carboxylic acid using chromic acid. Deprotonation with sodium hydroxide followed by reaction with 1-bromohexane produces the product, hexyl butanoate. (1-Bromohexane is made in four steps from 1-butene, as shown. *Anti*-Markovnikov addition using HBr and peroxides gives 1-bromobutane. Reaction with sodium acetylide gives 1-hexyne, which is subsequently reduced to 1-hexene with H_2 and Lindlar's catalyst. *Anti*-Markovnikov addition using HBr and peroxides gives 1-bromohexane.)

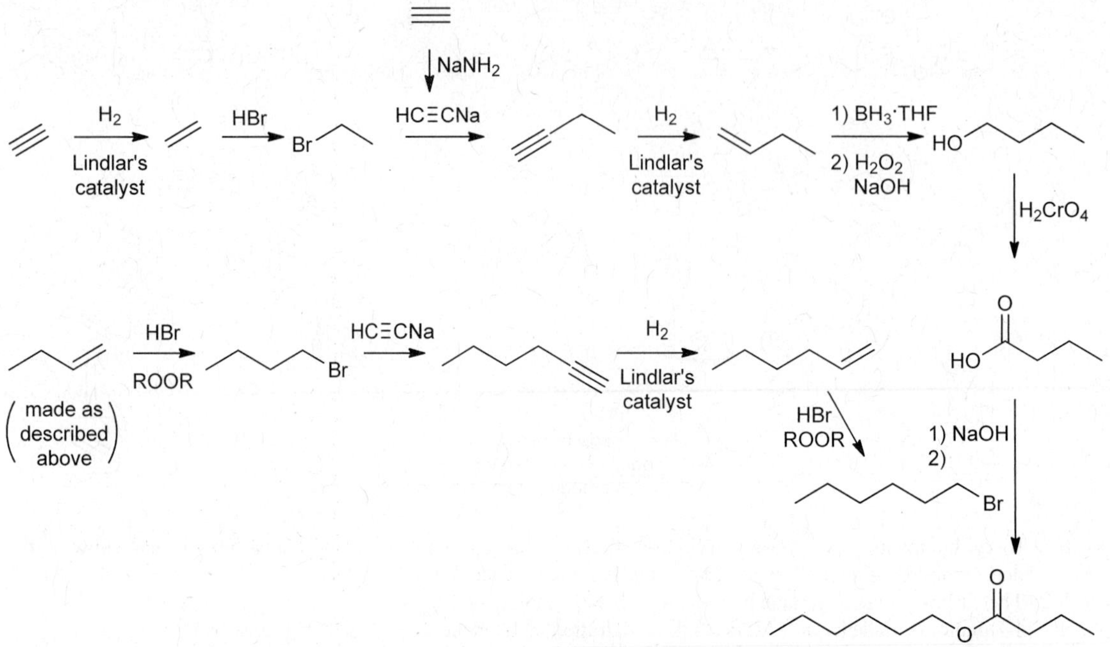

H₂
Lindlar's catalyst

HBr
Br

NaNH₂

HC≡CNa

H₂
Lindlar's catalyst

1) BH₃·THF
2) H₂O₂ NaOH
HO

H₂CrO₄

HBr
ROOR
Br

HC≡CNa

H₂
Lindlar's catalyst

HBr
ROOR

1) NaOH
2)
Br

O
HO
(butanoic acid)

O
O
(ester product)

(made as described above)

13.71.

(a) The alcohol is protonated by HBr, converting it into a good leaving group. Water leaves, producing a tertiary carbocation intermediate. Bromide attacks the indicated carbon atom of the cyclopropyl group, thus opening up the ring to produce the product. The last step of the mechanism is aided by the relief of ring strain present in the three-membered ring.

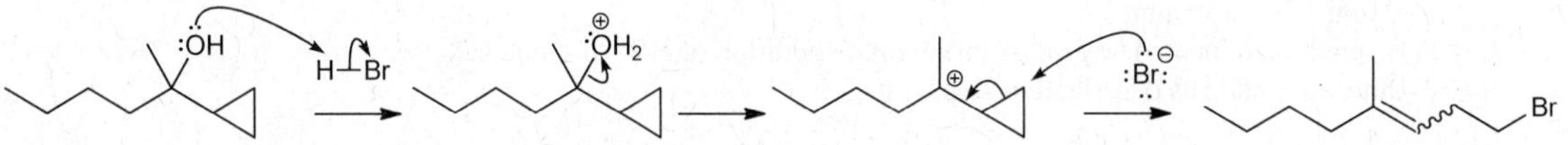

:OH   H—Br   ⁺OH₂   ⊕   :Br:⁻   Br

(b) The following synthesis is one suggested synthetic pathway. There are likely other acceptable approaches that accomplish the same goal.

Let's start by looking at the product and attempting to map out the destinations of the carbon atoms in each of our given starting materials. The product is symmetric, which suggests that the groups from the left and right halves will have analogous origins. Compounds **A**, **B**, and **C** have four, five, and six carbon atoms, respectively. Both of the four-carbon linear termini of the product likely originate from the two equivalents of 1-bromobutane, which also has a linear four-carbon chain.

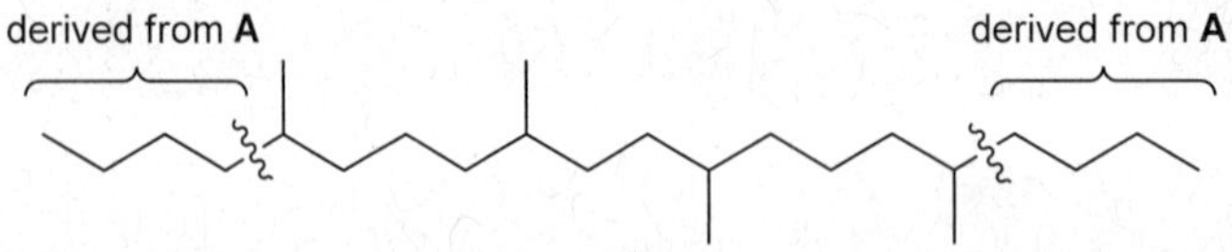

derived from **A**   derived from **A**

This leaves 16 carbon atoms in the central portion of the target structure (between the two wavy lines in the figure above). A careful analysis of this fragment (and keeping in mind the symmetry of the product) suggests that the six central carbon atoms are from **C**, while the two five-carbon units flanking this portion are derived from two equivalents of **B**.

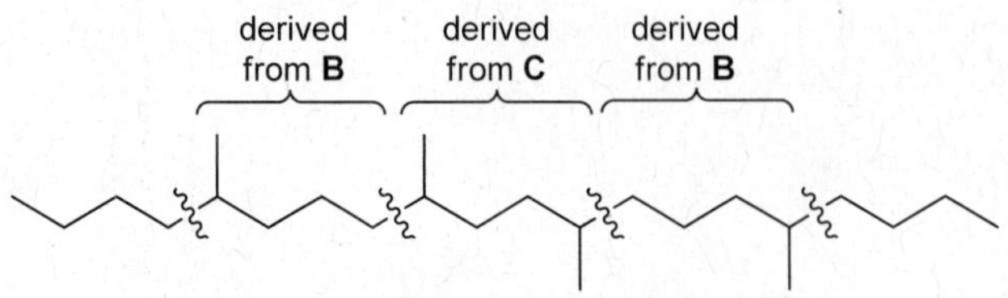

derived   derived   derived
from **B**   from **C**   from **B**

Now let's consider the key step (described in the problem statement) and how it will fit into our synthesis. We need to fill in the gap between the starting materials and the key step, as well as the gap between the key step and the final product.

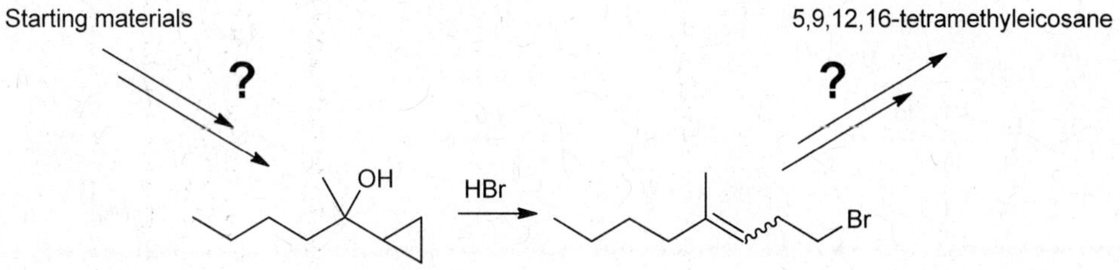

The tertiary alcohol above can be prepared via a Grignard reaction between butyl Grignard (prepared from **A**) and ketone **B**.

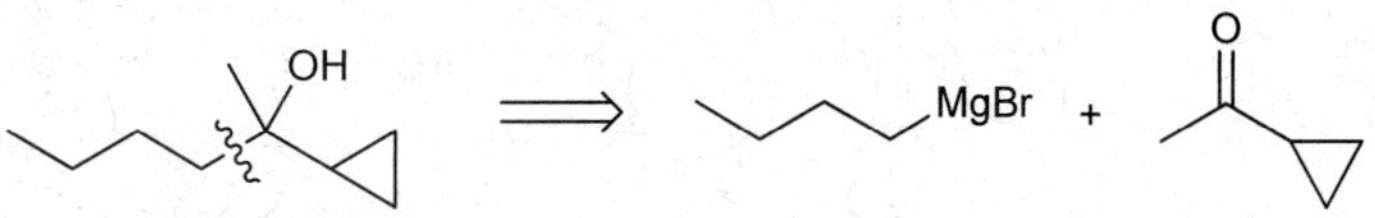

Now let's focus on the steps following the key step. The carbon atoms in the product can be mapped on to two equivalents of the alkenyl bromide and one equivalent of the diketone as shown in the figure below.

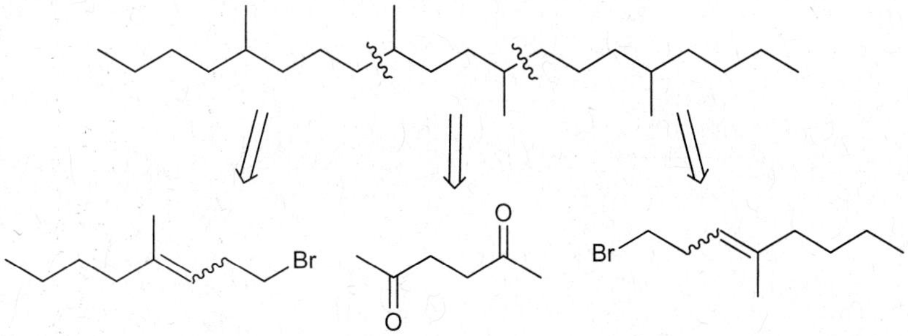

This suggests the following retrosynthetic approach.

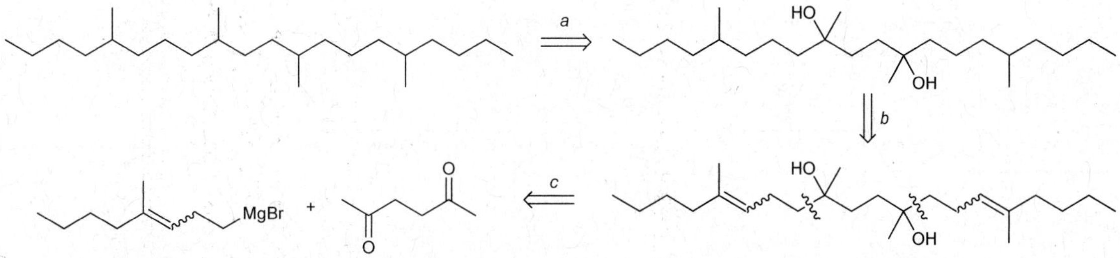

a. The product can be prepared in two steps from the diol: acid-catalyzed elimination followed by hydrogenation of the resulting diene.
b. The two double bonds can be removed via hydrogenation.
c. A double Grignard reaction of two equivalents of the Grignard reagent and one equivalent of the diketone allows for the formation of the indicated bonds.

Now, let's draw out the forward scheme. 1-Bromobutane (**A**) is treated with magnesium to give a Grignard reagent, and subsequently reacted with ketone **B**. Aqueous workup produces the tertiary alcohol. Reaction with HBr drives the ring-opening reaction to give the alkenyl bromide. Conversion to the Grignard (Mg), followed by addition of diketone **C** (2:1 ratio) and a water workup, yields the dialkenyl diol, which is converted to the diol via hydrogenation. Acid-catalyzed elimination followed by hydrogenation affords the product, 5,9,12,16-tetramethyleicosane.

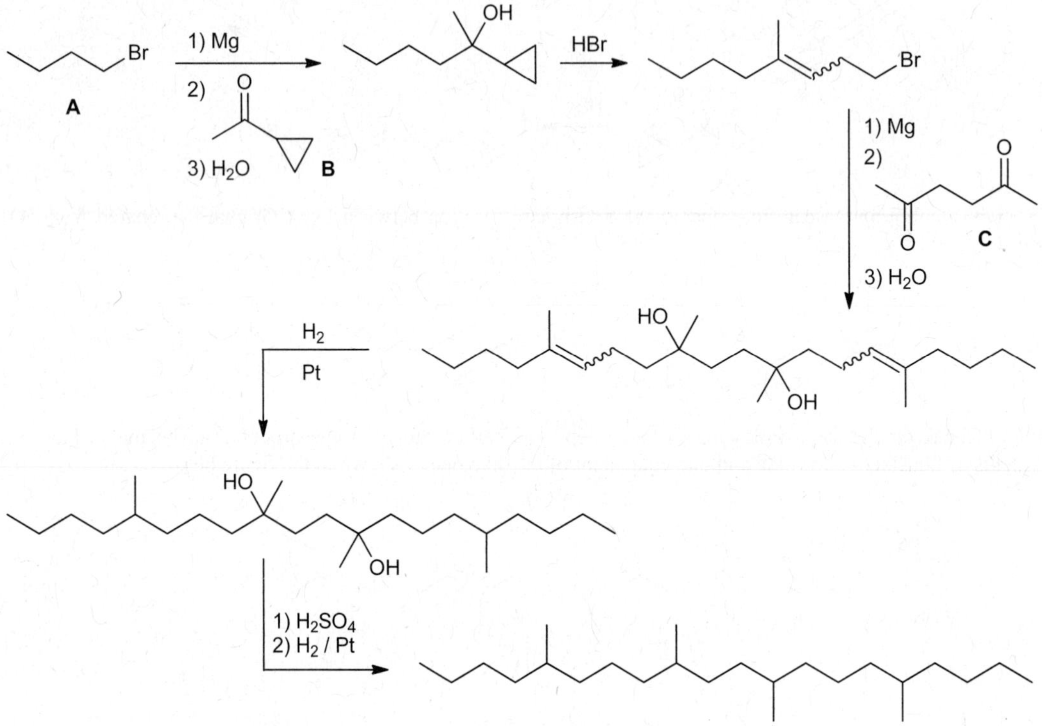

13.72.

(a) TBAF removes the silyl protecting group to give compound **3**, which then undergoes selective tosylation of the primary hydroxyl group (as mentioned in the problem statement) to give compound **4**:

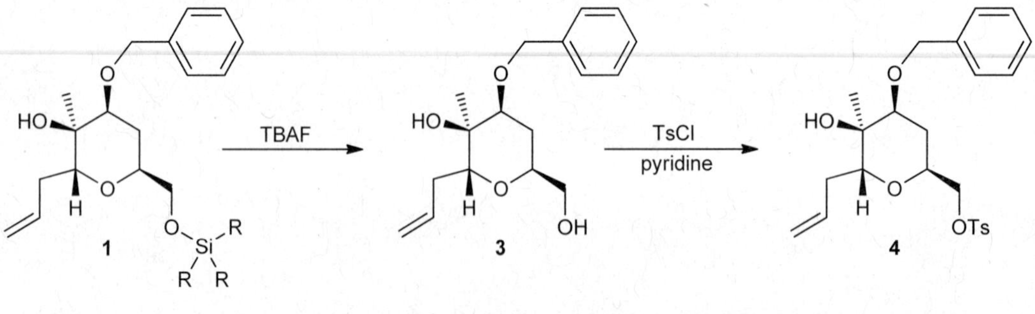

(b) Methoxide functions as a base and deprotonates the tertiary hydroxyl group. The resulting alkoxide functions as a nucleophile and attacks the primary tosylate in an intramolecular, S$_N$2-type process. Locants have been assigned to help redraw the alkoxide ion.

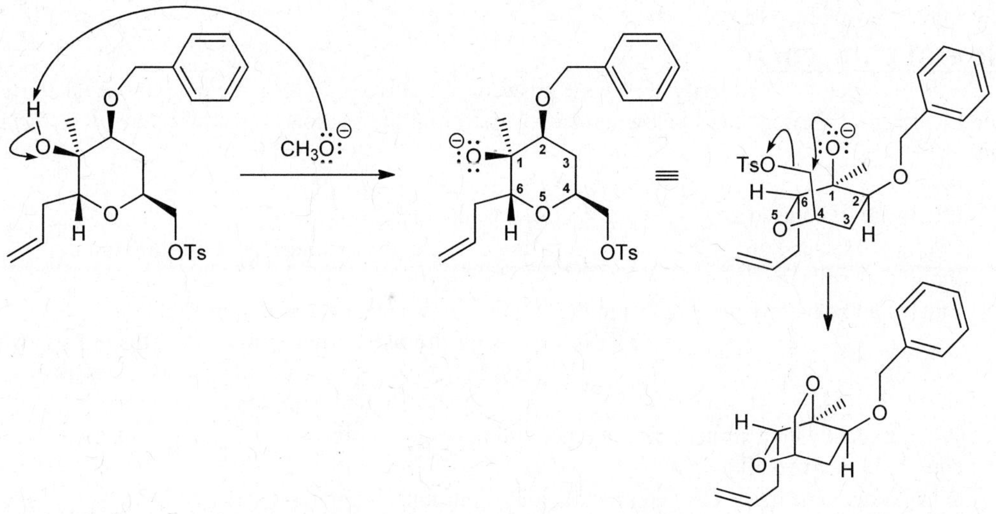

Note that these locants do not necessarily adhere to IUPAC guidelines for assigning locants, but rather, they are simply tools that are used to verify that the alkoxide ion has been drawn correctly (it enables you to compare the configurations in each drawing, and prove to yourself that these drawings are the same). The use of locants can be especially helpful in situations like this.

(c) The *syn* relationship places the negatively charged oxygen atom in close proximity with the carbon atom bearing the tosylate group, thereby enabling an intramolecular attack. The *syn* relationsip is necessary in order for the reaction to occur. The following structure lacks the *syn* relationship, so the reactive centers are too far apart for an intramolecular reaction to occur:

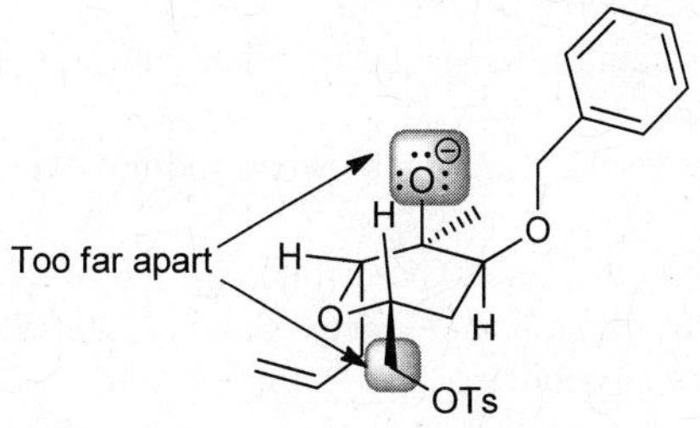

Chapter 14
Ethers and Epoxides; Thiols and Sulfides

Review of Concepts

Fill in the blanks below. To verify that your answers are correct, look in your textbook at the end of Chapter 14. Each of the sentences below appears verbatim in the section entitled *Review of Concepts and Vocabulary*.

- Ethers are often used as _____ for organic reactions.
- Cyclic polyethers, or _____ **ethers**, are capable of solvating metal ions in organic (nonpolar) solvents.
- Ethers can be readily prepared from the reaction between an alkoxide ion and an _____, a process called a **Williamson ether synthesis**. This process works best for _____ or _____ alkyl halides. _____ alkyl halides are significantly less efficient, and _____ alkyl halides cannot be used.
- When treated with a strong acid, an ether will undergo **acidic** _____ in which it is converted into two alkyl halides.
- When a phenyl ether is cleaved under acidic conditions, the products are _____ and an alkyl halide.
- Ethers undergo autooxidation in the presence of atmospheric oxygen to form _____.
- Substituted oxiranes are also called _____.
- _____ can be converted into epoxides by treatment with peroxy acids or via halohydrin formation and subsequent epoxidation.
- _____ catalysts can be used to achieve the enantioselective epoxidation of allylic alcohols.
- Epoxides will undergo **ring-opening reactions** in: 1) conditions involving a strong nucleophile, or under 2) _____-catalyzed conditions. When a strong nucleophile is used, the nucleophile attacks at the _____-substituted position.
- Sulfur analogs of alcohols contain an SH group rather than an OH group, and are called _____.
- Thiols can be prepared via an S_N2 reaction between sodium hydrosulfide (NaSH) and a suitable _____.
- The sulfur analogs of ethers (thioethers) are called _____.
- Sulfides can be prepared from thiols in a process that is essentially the sulfur analog of the Williamson ether synthesis, involving a _____ ion, rather than an alkoxide.

Review of Skills

Fill in the blanks and empty boxes below. To verify that your answers are correct, look in your textbook at the end of Chapter 14. The answers appear in the section entitled *SkillBuilder Review*.

14.1 Naming an Ether

PROVIDE A SYSTEMATIC NAME FOR THE FOLLOWING COMPOUND

1) IDENTIFY THE PARENT
2) IDENTIFY AND NAME SUBSTITUENTS
3) ASSIGN LOCANTS TO EACH SUBSTITUENT
4) ALPHABETIZE
5) ASSIGN CONFIGURATION

14.2 Preparing an Ether via a Williamson Ether Synthesis

IDENTIFY REAGENTS THAT WILL ACHIEVE THE FOLLOWING TRANSFORMATION:

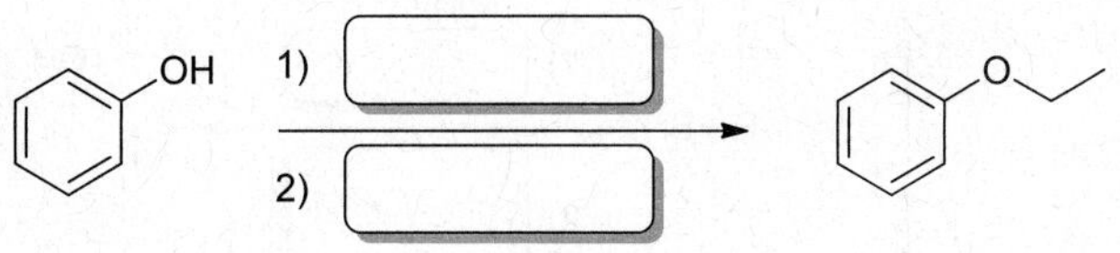

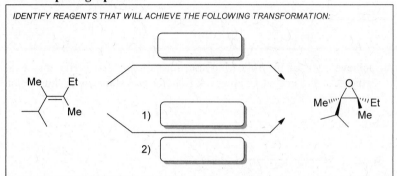

14.3 Preparing Epoxides

IDENTIFY REAGENTS THAT WILL ACHIEVE THE FOLLOWING TRANSFORMATION:

14.4 Drawing the Mechanism and Predicting the Product of the Reaction between a Strong Nucleophile and an Epoxide

COMPLETE THE MECHANISM BELOW BY DRAWING ALL CURVED ARROWS, INTERMEDIATES AND PRODUCTS.

NaCN   H$_2$O$^-$

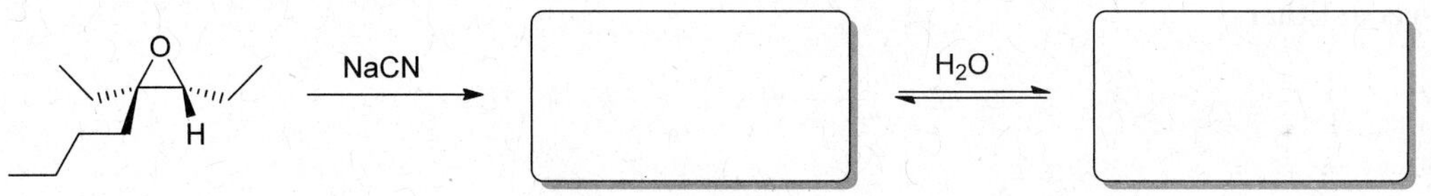

14.5 Drawing the Mechanism and Predicting the Product of Acid-Catalyzed Ring-Opening

COMPLETE THE MECHANISM BELOW BY DRAWING ALL CURVED ARROWS, INTERMEDIATES AND PRODUCTS.

EtOH   EtOH

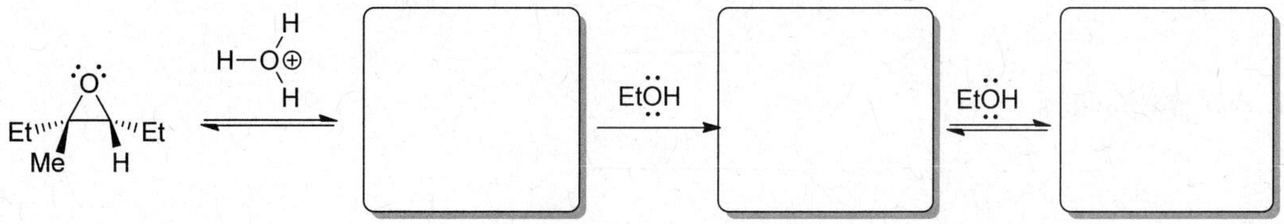

14.6 Installing Two Adjacent Functional Groups

IDENTIFY WHETHER EACH RING-OPENING REACTION BELOW REQUIRES ACIDIC CONDITIONS OR BASIC CONDITIONS:

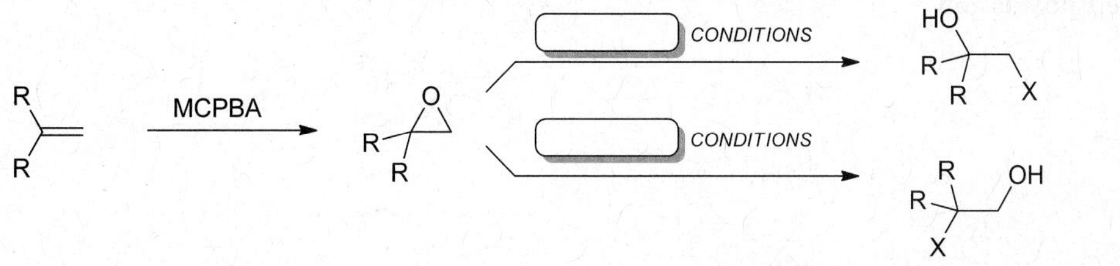

MCPBA

CONDITIONS

CONDITIONS

14.7   Choosing the Appropriate Grignard Reaction

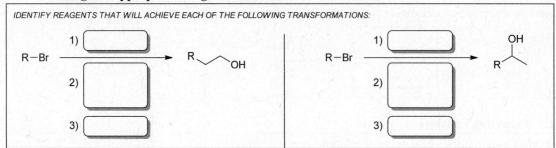

IDENTIFY REAGENTS THAT WILL ACHIEVE EACH OF THE FOLLOWING TRANSFORMATIONS:

Review of Reactions

Identify the reagents necessary to achieve each of the following transformations. To verify that your answers are correct, look in your textbook at the end of Chapter 14. The answers appear in the section entitled *Review of Reactions*.

Preparation of Ethers

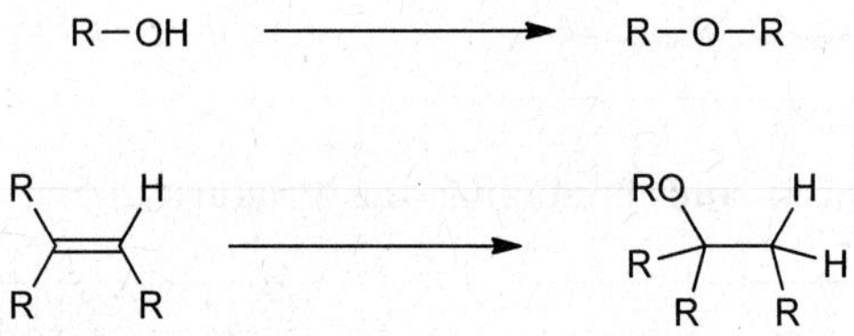

$$R{-}OH \longrightarrow R{-}O{-}R$$

Reactions of Ethers

$$R{-}O{-}R \longrightarrow 2\ R{-}X\ +\ H_2O$$

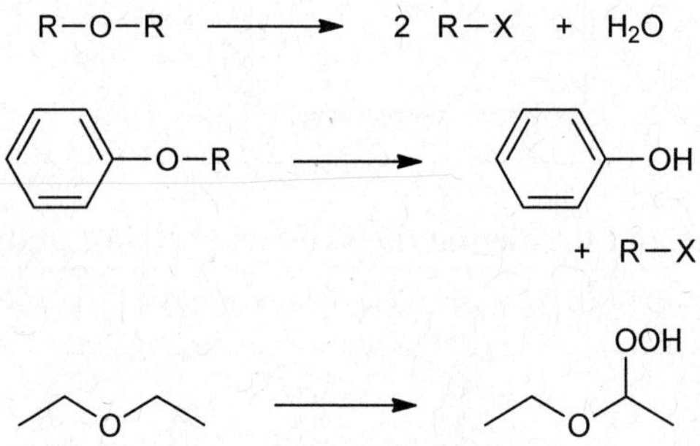

Preparation of Epoxides

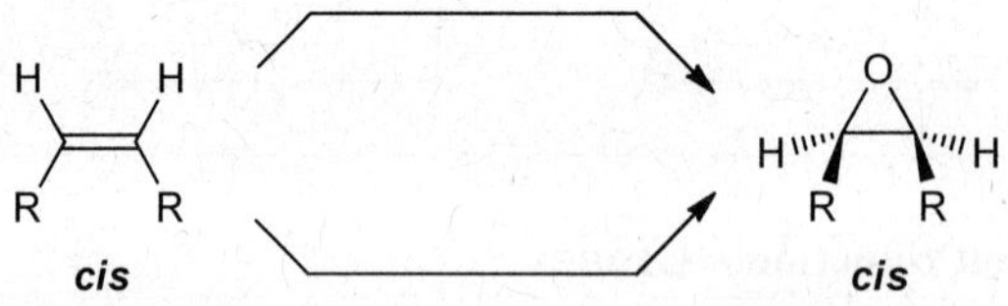

Enantioselective Epoxidation

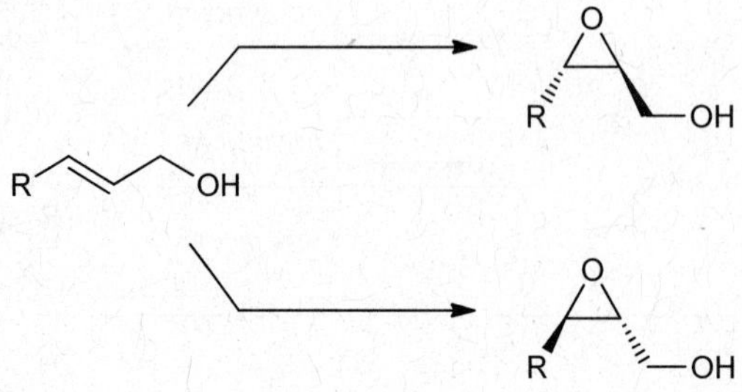

Ring-Opening Reactions of Epoxides

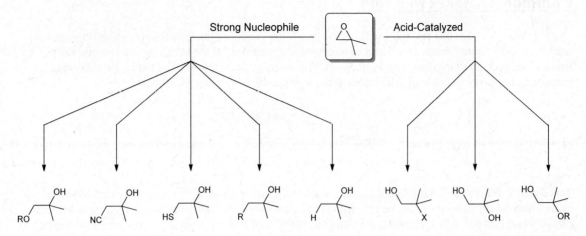

Thiols and Sulfides
Thiols

Sulfides

Common Mistakes to Avoid

We have seen that epoxides will react with a wide variety of nucleophiles (causing ring-opening reactions), either under acidic conditions or under basic conditions. If the attacking nucleophile is itself a strong base, such as a Grignard reagent, then acidic conditions cannot be used. That is, the following transformation is not possible to achieve, so avoid trying to do something like this:

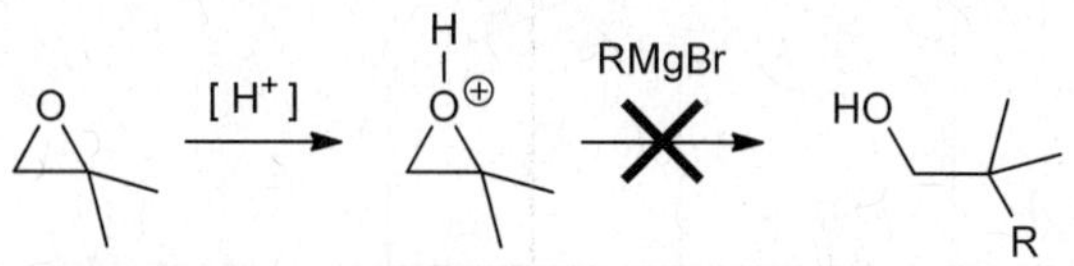

This doesn't work because Grignard reagents are not only strong nucleophiles, but they are also strong bases. And strong bases are incompatible with acidic conditions. If a Grignard reagent is subjected to a source of acid (even a relatively weak acid, such as H_2O), the Grignard reagent is irreversibly protonated to give an alkane, for example:

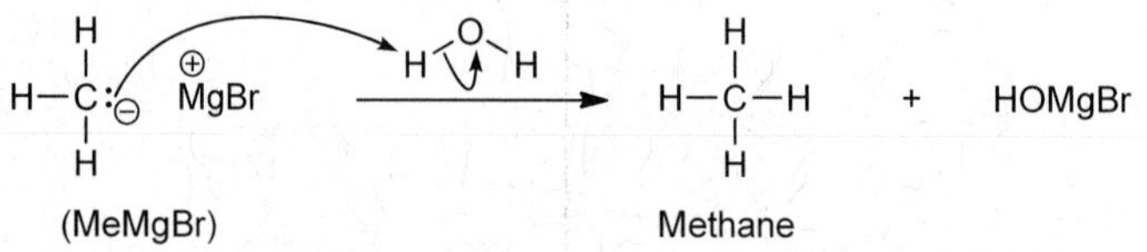

In summary, never use a Grignard reagent in the presence of an acid. The same rule applies to the use of LAH (lithium aluminum hydride), which is both a strong nucleophile and a strong base. Therefore, much like a Grignard reagent, LAH also cannot be used to open an epoxide under acidic conditions:

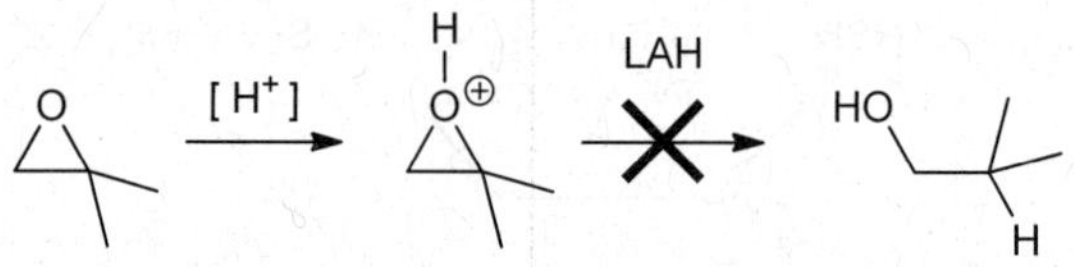

Once again, this doesn't work because LAH is incompatible with acidic conditions. Avoid making this mistake.

Useful reagents

The following is a list of commonly encountered reagents for reactions involving ethers, epoxides, thiols, and sulfides:

Reagents	Description
RX	An alkyl halide. Used for the alkylation of alcohols or thiols. First, the alcohol or thiol is deprotonated with a base, such as NaH or NaOH, and the resulting anion is then treated with the alkyl halide, thereby installing an alkyl group.
1) $Hg(OAc)_2$, ROH 2) $NaBH_4$	These reagents will achieve alkoxymercuration-demercuration of an alkene. This process adds H and OR in a Markovnikov fashion across the alkene.
HX	Will convert dialkyl ethers into alkyl halides via cleavage of the C–O bonds. Will also react with an epoxide, thereby opening the ring, and installing a halogen at the more substituted position.
MCPBA	*meta*-Chloro-peroxybenzoic acid. An oxidizing agent that will convert an alkene into an epoxide
1) Br_2, H_2O 2) NaOH	Alternative reagents for converting an alkene into an epoxide.
$(CH_3)_3COOH$, $Ti[OCH(CH_3)_2]_4$. (+)-DET or (−)-DET	Reagents for enantioselective (Sharpless) epoxidation
NaOR (or RONa)	An alkoxide ion is both a strong nucleophile and a strong base. It can be used to open an epoxide under basic conditions (the alkoxide ion attacks the less substituted position).
NaCN	A good nucleophile that will react with an epoxide in a ring-opening reaction.
NaSH	A very strong nucleophile that will react with an epoxide in a ring-opening reaction. NaSH can also be used to prepare thiols from alkyl halides.
RMgBr	A Grignard reagent. A strong base and a strong nucleophile. Will react with an epoxide in a ring-opening reaction, to attack the less substituted side (it is not possible to use acidic conditions and have the Grignard reagent attack the more substituted side – see the previous section on common mistakes to avoid).
LAH	Lithium aluminum hydride is a source of nucleophilic hydride ions. It will react with an epoxide in a ring-opening reaction, to attack the less substituted side (it is not possible to use acidic conditions and have a hydride ion attack the more substituted side – see the previous section on common mistakes to avoid).
$[H^+]$, H_2O	Aqueous acidic conditions. Under these conditions, an epoxide is opened to give a diol.
$[H^+]$, ROH	Under these conditions, an epoxide is opened, with a molecule of the alcohol attacking a protonated epoxide at the more substituted position.
$NaOH/H_2O$, Br_2	Reagents for converting thiols into disulfides.
HCl, Zn	Reagents for converting thiols into disulfides.
H_2O_2	Strong oxidizing agent, used to oxidize sulfides to sulfoxides, and then further to sulfones.
$NaIO_4$	Oxidizing agent, used to oxidize sulfides to sulfoxides.

Solutions

14.1.

(a)   The oxygen atom has two groups attached to it. One group has two carbon atoms and the other group has three carbon atoms. The parent is named after the larger group, so the parent is propane. The smaller group (together with the oxygen atom) is treated as an ethoxy substituent, and a locant is included to identify the location (C2) of the substituent on the parent chain.

2-Ethoxypropane

(b) The oxygen atom has two groups attached to it. One group has three carbon atoms and the other group has two carbon atoms. The parent is named after the larger group, so the parent is propane. The smaller group (together with the oxygen atom) is treated as an ethoxy substituent, and a locant is included to identify the location (C1) of the substituent on the parent chain. The parent chain also has a chloro substituent, located at C2 of the parent. That position is a chirality center, so a stereodescriptor is required to identify the configuration (*S*). When assembling the name, the substituents are arranged alphabetically (chloro precedes ethoxy).

(*S*)-2-Chloro-1-ethoxypropane

(c) The oxygen atom has two groups attached to it. One group has two carbon atoms and the other group has six carbon atoms. The parent is named after the larger group, so the parent is benzene. The smaller group (together with the oxygen atom) is treated as an ethoxy substituent, and a locant (C1) is included to identify the location of the substituent on the parent chain. The parent chain also has two chloro substituents, located at C2 and C4 of the parent. The parent is numbered to give the substituents the lowest possible numbers (1,2,4 instead of 1,3,4). When assembling the name, the substituents are arranged alphabetically (chloro precedes ethoxy).

2,4-Dichloro-1-ethoxybenzene

(d) The parent is cyclohexanol, and the ethoxy group is listed as a substituent. A locant (C2) is included to indicate the location of the ethoxy group on the cyclohexanol parent, and stereodescriptors are included to indicate the configuration of each chirality center:

(*1R,2R*)-2-Ethoxycyclohexanol

(e) The parent is cyclohexene, and the ethoxy group is listed as a substituent. A locant (C1) is included to indicate the location of the ethoxy group on the cyclohexene parent:

1-Ethoxycyclohexene

14.2.

(a) The parent (cyclobutane) is a four-membered ring, and there is an ethoxy group connected to the ring (which is defined as position C2 of the ring). This position is a chirality center, and it has the *R* configuration. There are two methyl groups, both located at C1.

(*R*)-2-Ethoxy-1,1-dimethylcyclobutane

(b) This name has the format of a common name. It is an ether in which the oxygen atom is connected to a cyclopropyl group (a three-membered ring) and an isopropyl group (a three-membered chain, connected at the middle carbon atom).

Cyclopropyl isopropyl ether

14.3.

(a) As seen in the solution to Problem 2.9, there are only three different ways to connect five carbon atoms:

Pentane   2-Methylbutane   2,3-Dimethyl
                                            propane

For each of these three skeletons, we will consider all of the possible locations where an oxygen atom can be inserted. Let's begin with the linear skeleton (pentane). We can place the oxygen atom between C1 and C2, or we can place it between C2 and C3, giving the following two ethers:

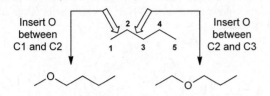

Notice that placing the oxygen atom between C3 and C4 gives the same ether as placing the oxygen atom between C2 and C3.

Similarly, placing the oxygen atom between C4 and C5 gives the same ether as placing the oxygen atom between C1 and C2.

Now let's move on to our second carbon skeleton (2-methylbutane). In this skeleton, we can insert the oxygen atom in one of three places (between C1 and C2, or between C2 and C3, or between C3 and C4), giving the following three ethers:

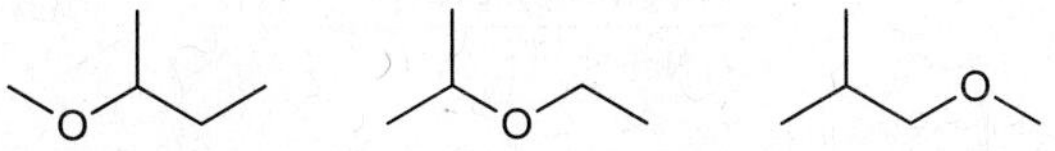

Notice that inserting the oxygen atom between C2 and C5 gives the same ether as inserting the oxygen atom between C1 and C2.

Finally, we explore our last skeleton (2,2-dimethylpropane). Inserting the oxygen atom between any one of the C-C bonds will result in the same ether:

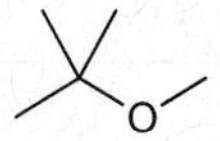

In summary, there are six constitutionally isomeric ethers with molecular formula $C_5H_{12}O$, shown here:

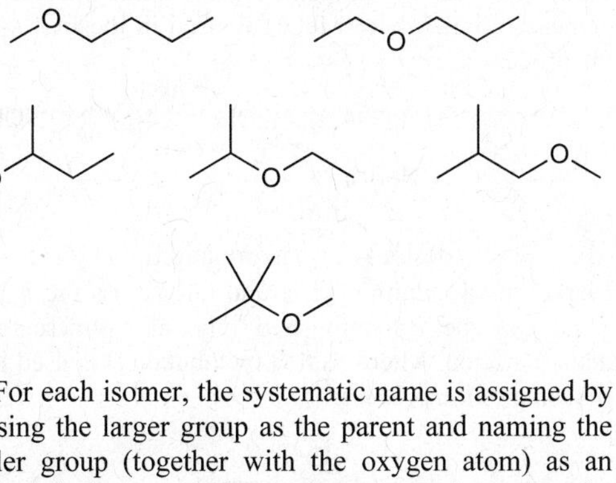

(b) For each isomer, the systematic name is assigned by choosing the larger group as the parent and naming the smaller group (together with the oxygen atom) as an alkoxy substituent. Each substituent must be assigned a locant, and substituents are assembled alphabetically in the name. Common names are assigned by identifying

the alkyl group on either side of the oxygen atom (in alphabetical order). Systematic and common names for all six constitutional isomers are shown here (common names are shown in parentheses).

1-methoxybutane
(butyl methyl ether)

1-ethoxypropane
(ethyl propyl ether)

2-methoxybutane
(*sec*-butyl methyl ether)

2-ethoxypropane
(ethyl isopropyl ether)

1-methoxy-
2-methylpropane
(isobutyl methyl ether)

2-methoxy-
2-methylpropane
(*tert*-butyl methyl ether)

(c) See solution to **14.3b.**

(d) Only one of the six isomers, shown here, contains a carbon atom that is connected to four different groups (a methoxy group, a methyl group, an ethyl group, and a hydrogen atom).

Chirality
center

2-methoxybutane
(*sec*-butyl methyl ether)

14.4.
(a) The cation is potassium, so we must use 18-crown-6, which solvates potassium ions.

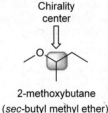

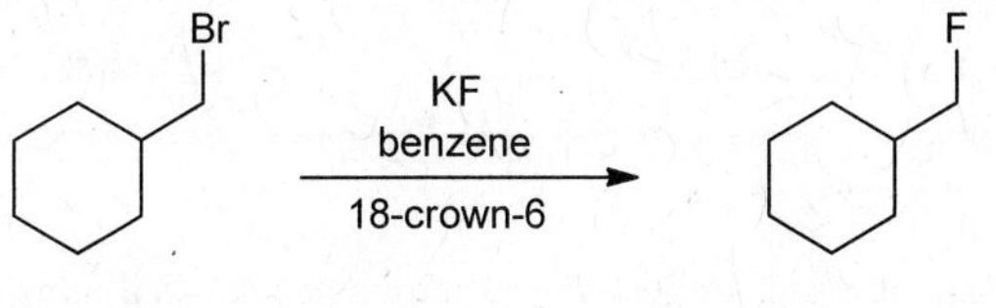

(b) The cation is sodium, so we must use 15-crown-5, which solvates sodium ions.

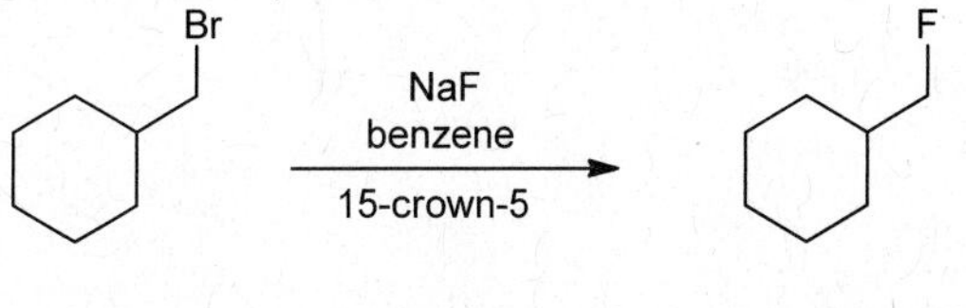

(c) The cation is lithium, so we must use 12-crown-4, which solvates lithium ions.

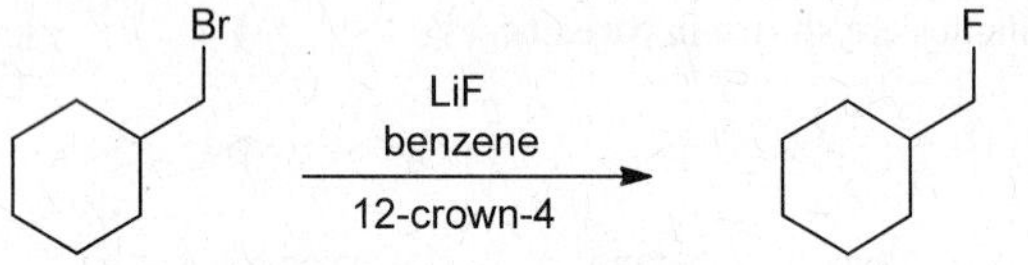

(d) The cation is potassium, so we must use 18-crown-6, which solvates potassium ions.

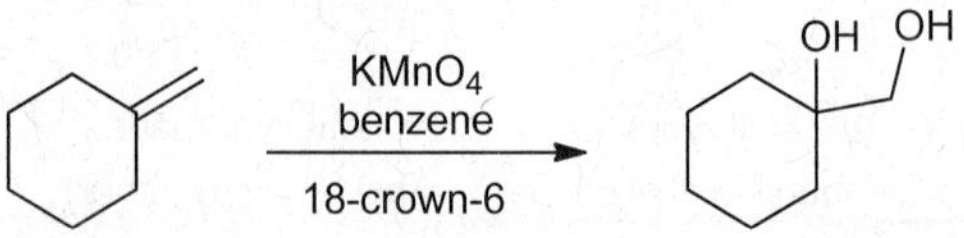

14.5.

(a) A Williamson ether synthesis will be more efficient with a less sterically hindered substrate, since the process involves an S_N2 reaction. Therefore, in this case, it is better to start with a secondary alcohol and a primary alkyl halide, rather than a primary alcohol and a secondary alkyl halide:

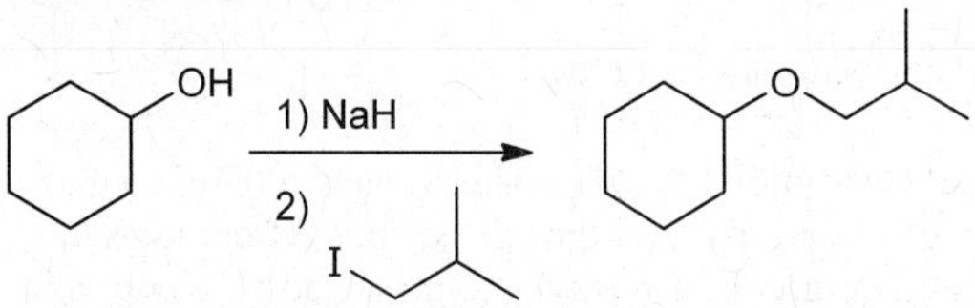

(b) In this case, it is better to start with a secondary alcohol and a primary alkyl halide, rather than a primary alcohol and a secondary alkyl halide:

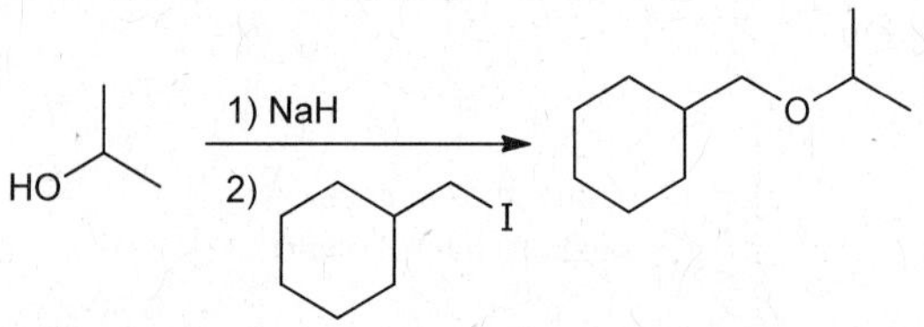

(c) In this case, it is better to start with a tertiary alcohol and a methyl halide, rather than methanol and a tertiary alkyl halide:

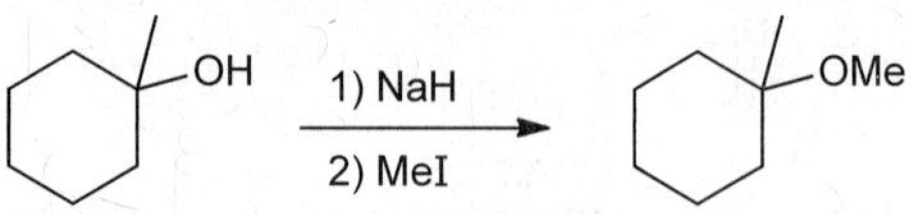

14.6. In order to perform an intramolecular Williamson ether synthesis, we must choose a starting compound that contains both an OH group and a halogen, as shown here:

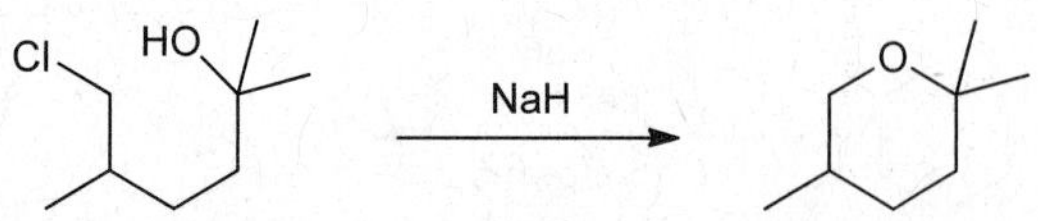

The OH group is deprotonated upon treatment with NaH (a strong base). The resulting alkoxide ion can then function as a nucleophile in an intramolecular, S_N2-type reaction, expelling chloride as a leaving group, and giving a six-membered ring.

The alternative starting compound, shown below, cannot be used, because the leaving group is attached to a tertiary position, so an S_N2-type process cannot occur at that location.

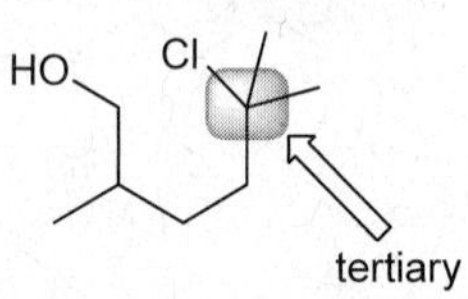

14.7. Making this compound via a Williamson ether synthesis would require an S_N2 process at either a tertiary position or a vinylic position. An S_N2 process will not occur readily at either of these positions, so a Williamson ether synthesis cannot be used to make the desired compound.

14.8.

(a) The desired transformation involves the Markovnikov addition of H and OEt across the π bond. This can be accomplished via alkoxymercuration-demercuration, where EtOH is used during the oxymercuration process.

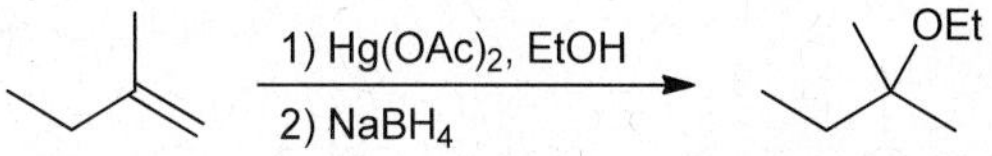

(b) The desired transformation involves the Markovnikov addition of H and an isopropoxy group across the π bond. This can be accomplished via alkoxymercuration-demercuration, where isopropanol is used in the first step of the process.

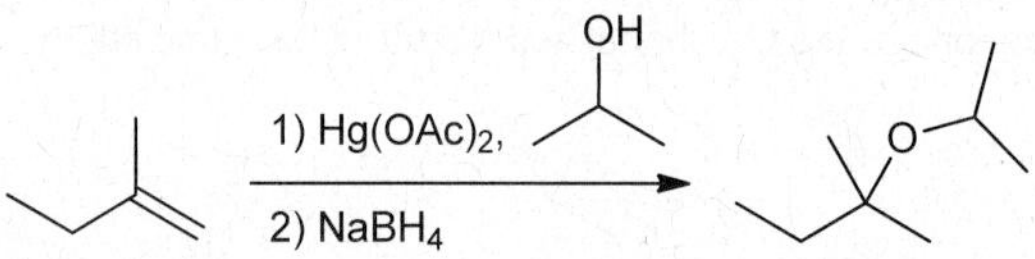

(c) The desired transformation involves the Markovnikov addition of H and a methoxy group across the π bond. This can be accomplished via alkoxymercuration-demercuration, where MeOH is used in the first step of the process.

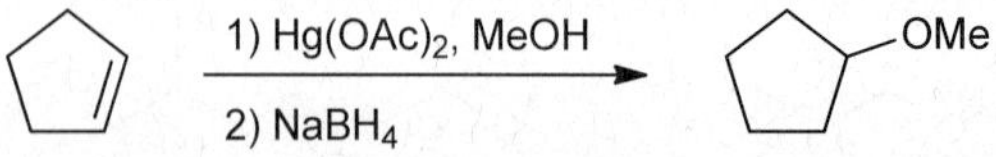

(d) The desired transformation involves the Markovnikov addition of H and OR across the π bond. This can be accomplished via alkoxymercuration-demercuration, where ROH (cyclobutanol) is used in the first step of the process.

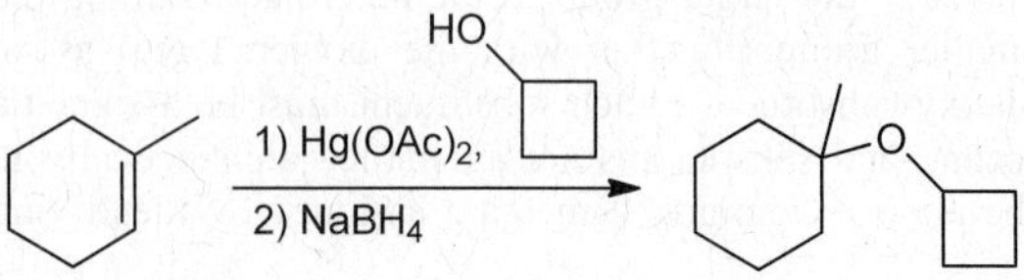

14.9. Cyclopentene can be converted to cyclopentanol via acid catalyzed hydration (upon treatment with dilute aqueous H_2SO_4). Cyclopentanol can then be used for the alkoxymercuration of cyclopentene, giving the desired product.

14.10. Propene can be converted to 1-propanol via an *anti*-Markovnikov addition of H and OH across the π bond, which can be achieved with hydroboration-oxidation. This alcohol can then be used for the alkoxymercuration of propene, giving the desired product.

14.11.
(a) The oxygen atom is connected to two carbon atoms, each of which is sp^3 hybridized. As such, treatment with HBr is expected to cleave each of the C–O bonds and replace them with C–Br bonds. Because the starting ether is symmetrical, the resulting two alkyl bromides are identical. Therefore, two equivalents of this alkyl bromide are expected.

(b) The oxygen atom is connected to two carbon atoms, each of which is sp^3 hybridized. As such, treatment with HI is expected to cleave each of the C–O bonds and replace them with C–I bonds, giving the following diiodide.

(c) The oxygen atom is connected to two carbon atoms. One of these carbon atoms (left) is sp^2 hybridized, while the other (right) is sp^3 hybridized. As such, treatment with HBr is expected to cleave only the C–O bond involving the sp^3 hybridized carbon atom. The other C–O bond is not cleaved. This gives phenol and ethyl bromide as products.

(d) The oxygen atom is connected to two carbon atoms. One of these carbon atoms (left) is sp^2 hybridized, while the other (right) is sp^3 hybridized. As such, treatment with HI is expected to cleave only the C–O bond involving the sp^3 hybridized carbon atom. The other C–O bond is not cleaved, giving the following product.

(e) The oxygen atom is connected to two carbon atoms, each of which is sp^3 hybridized. As such, treatment with HI is expected to cleave each of the C–O bonds and replace them with C–I bonds. One of these carbon atoms is a chirality center. Since this position is tertiary, cleavage will occur via an S_N1 process, so we expect racemization.

(racemic mixture)

(f) The oxygen atom is connected to two carbon atoms, each of which is sp^3 hybridized. As such, treatment with HBr is expected to cleave each of the C–O bonds and replace them with C–Br bonds, giving cyclohexyl bromide and ethyl bromide as products.

14.12.
(a) There are two methods for naming epoxides. In one method, the parent will be propane, and the oxygen atom is considered to be an epoxy substituent connected to the parent at C1 and C2. In addition, there is a methyl substituent at C2.

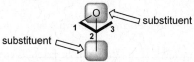

2-Methyl-1,2-epoxypropane

According to the second method, the parent is considered to be the oxirane ring, which has two methyl groups attached to it, both located at the 2 position.

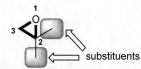

2,2-Dimethyloxirane

(b) There are two methods for naming epoxides. In one method, the parent will be ethane, and the oxygen atom is considered to be an epoxy substituent connected to the parent at C1 and C2. In addition, there are two phenyl substituents at C1.

1,1-Diphenyl-1,2-epoxyethane

According to the second method, the parent is considered to be the oxirane ring, which has two phenyl groups attached to it, both located at the 2 position.

2,2-Diphenyloxirane

(c) There are two methods for naming epoxides, although one of these methods will be less helpful because the two substituents (connected to the oxirane ring) are actually closed in a ring. This makes it difficult to name the compound as an oxirane. According to the first method for naming ethers, the parent is cyclohexane, and the oxygen atom is considered to be an epoxy substituent connected to the parent at C1 and C2.

1,2-Epoxycyclohexane

14.13.
(a) There are two methods for naming epoxides. In one method, the parent will be propane, and the oxygen atom is considered to be an epoxy substituent connected to the parent at C1 and C2. In addition, there is a phenyl substituent at C2, and the configuration of the chirality center is indicated.

(*R*-2-Phenyl-1,2-epoxypropane)

According to the second method, the parent is considered to be the oxirane ring. The methyl group and the phenyl group are both considered to be substituents, and their locations are identified with locants. Finally, the configuration of the chirality center is indicated (at the beginning of the name).

(*R*)-2-Methyl-2-phenyloxirane

In this case, the chirality center has the *R* configuration, as a result of the following prioritization scheme.

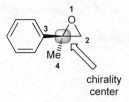

chirality center

(b) There are two methods for naming epoxides. In one method, the parent will be heptane, and the oxygen atom is considered to be an epoxy substituent connected to the parent at C3 and C4. The configuration of each chirality center is indicated.

(*3R,4R*)-3,4-Epoxyheptane

According to the second method, the parent is considered to be the oxirane ring, which is connected to two substituents (a propyl group and an ethyl group). Their locations are identified with locants, and the configuration of each chirality center is indicated (at the beginning of the name).

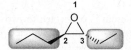

(*2R,3R*)-3-Ethyl-2-propyloxirane

(c) There are two methods for naming epoxides. In one method, the parent will be pentane, and the oxygen atom is considered to be an epoxy substituent connected to the parent at C2 and C3. There is also a methyl substituent located at C4. The configuration of each chirality center is indicated.

(*2R,3S*)-4-Methyl-2,3-epoxypentane

According to the second method, the parent is considered to be the oxirane ring, which is connected to two substituents (a methyl group and an isopropyl group). Their locations are identified with locants, and the configuration of each chirality center is indicated (at the beginning of the name).

(*2S,3R*)-2-Isopropyl-3-methyloxirane

14.14.

(a) We begin by identifying the four groups attached to the epoxide ring. On the left, there is a methyl group and a phenyl group. On the right, there is a phenyl group and a hydrogen atom. Notice that the two phenyl groups are *trans* to each other:

These two groups must be *trans* to each other in the starting alkene.

(b) We begin by identifying the four groups attached to the epoxide ring. On the left, there is a cyclohexyl group and a hydrogen atom. On the right, there is a methyl group and a hydrogen atom. Notice that the two hydrogen atoms are *trans* to each other.

These two hydrogen atoms must be *trans* to each other in the starting alkene.

(c) The starting alkene must have the *E* configuration in order to obtain the desired epoxide, as shown.

(d) The starting alkene must have the *E* configuration in order to obtain the desired epoxide, as shown.

14.15. This process for epoxide formation involves deprotonation of the hydroxyl group, followed by an intramolecular S_N2-type attack. Recall that S_N2 processes occur via back-side attack, which can only be achieved when both the hydroxyl group and the bromine occupy axial positions on the ring. Due to the steric bulk of a *tert*-butyl group, compound A spends most of its time in a chair conformation that has the *tert*-butyl group in an equatorial position. In this conformation, the OH and Br are indeed in axial positions, so the reaction can occur quite rapidly. In contrast, compound B spends most of its time in a chair conformation in which the OH and Br occupy equatorial positions. The S_N2 process cannot occur from this conformation.

Most stable conformation of compound A   Most stable conformation of compound B

14.16.

(a) The allylic hydroxyl group appears in the upper right corner, so a Sharpless epoxidation with (+)-DET will generate an epoxide ring above the plane of the π bond, giving the following enantiomer.

(b) We begin by redrawing the compound so that the allylic hydroxyl group appears in the upper right corner. A Sharpless epoxidation with (−)-DET will generate an epoxide ring below the plane of the π bond, giving the following enantiomer.

(c) We begin by redrawing the compound so that the allylic hydroxyl group appears in the upper right corner. A Sharpless epoxidation with (+)-DET will generate an epoxide ring above the plane of the π bond, giving the following enantiomer.

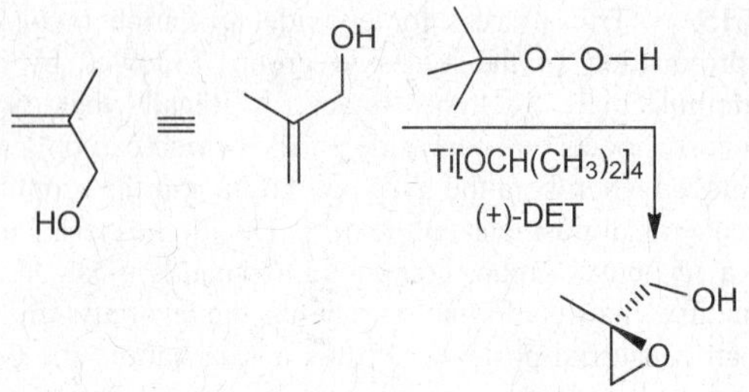

(d) We begin by redrawing the compound so that the allylic hydroxyl group appears in the upper right corner. A Sharpless epoxidation with (−)-DET will generate an epoxide ring below the plane of the π bond, giving the following enantiomer.

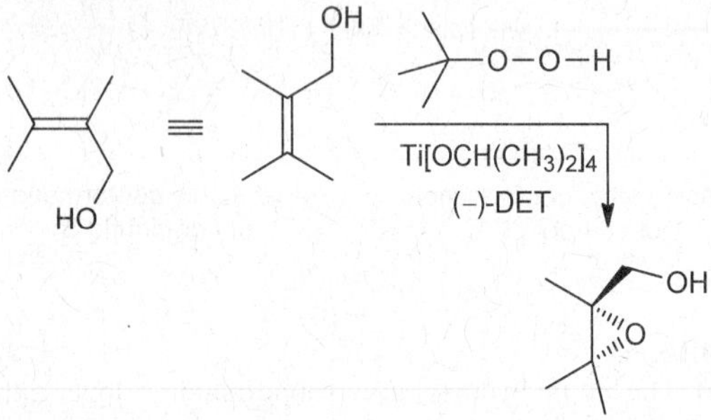

14.17.
(a) The Grignard reagent (PhMgBr) is a strong nucleophile, and it attacks the epoxide at the less substituted position. The epoxide is opened, resulting in an alkoxide ion. This alkoxide is then protonated upon treatment with water.

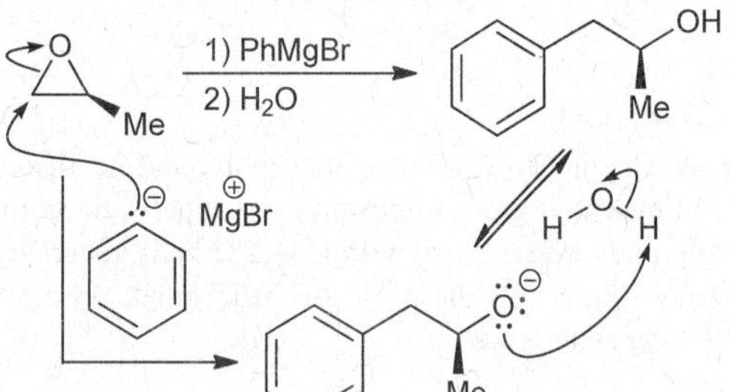

(b) Cyanide (NC⁻) is a good nucleophile, and it attacks the epoxide at the less substituted position. This opens the epoxide, resulting in an alkoxide ion, which is then protonated upon treatment with water.

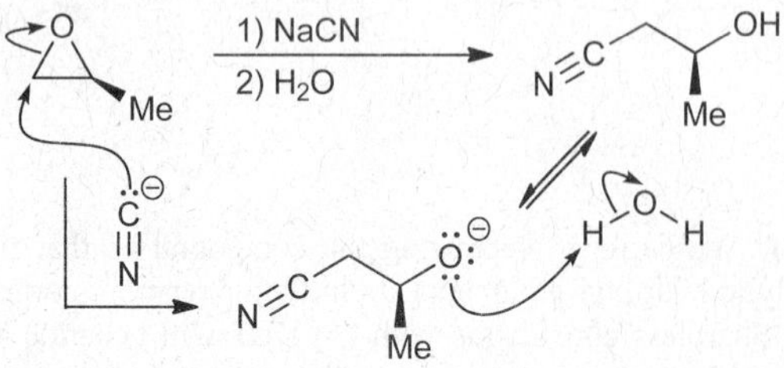

(c) HS⁻ is a very strong nucleophile, and it attacks the epoxide at the less substituted position. This opens the epoxide, resulting in an alkoxide ion, which is then protonated upon treatment with water.

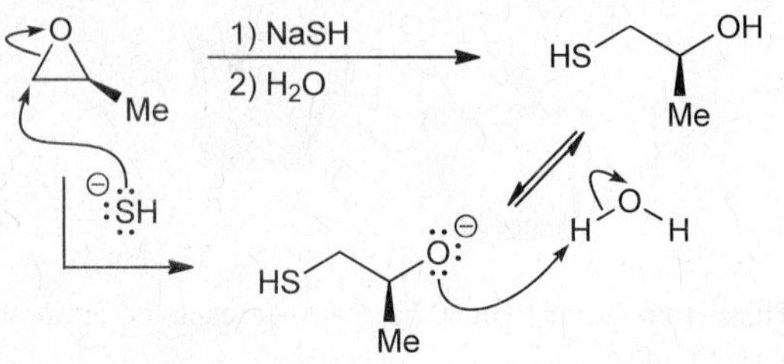

(d) LAH is a source of nucleophilic hydride (H⁻), and it attacks the epoxide at the less substituted position. This opens the epoxide, resulting in an alkoxide ion, which is then protonated upon treatment with water.

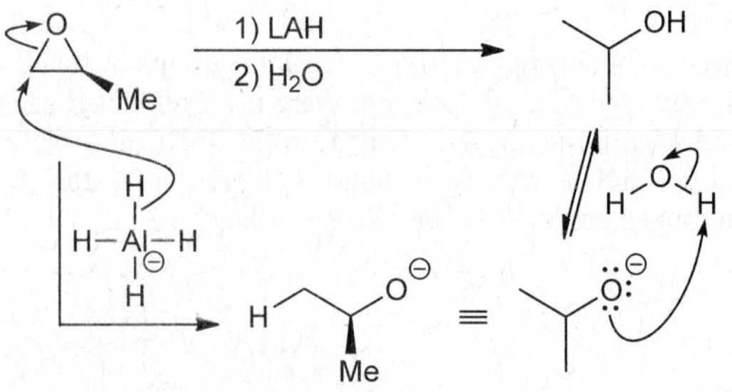

(e) HS⁻ is a very strong nucleophile, and it attacks the epoxide at the less substituted position. That position is a chirality center, so we expect an inversion of configuration at that center. The resulting alkoxide ion is then protonated upon treatment with water.

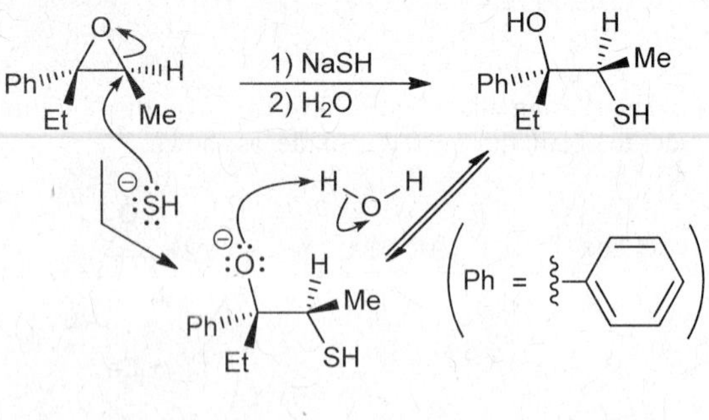

(f) LAH is a source of nucleophilic hydride (H⁻), and it attacks the epoxide at the less substituted position. That position is a chirality center, so we expect an inversion of configuration at that center. The resulting alkoxide ion is then protonated upon treatment with water.

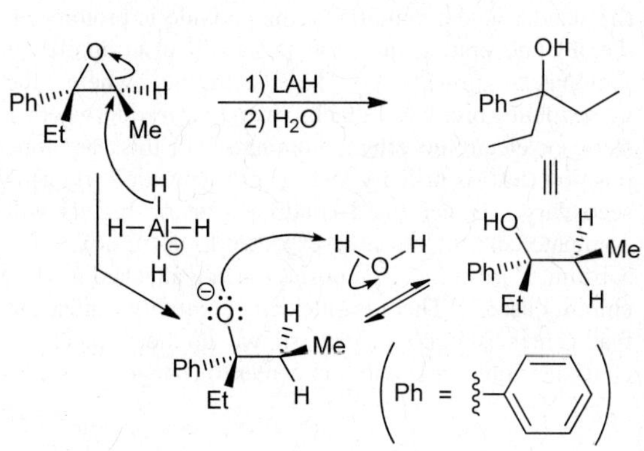

14.18. Hydroxide is a strong nucleophile, and it can attack the epoxide at either of two locations, highlighted here:

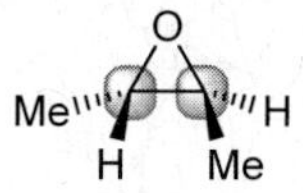

Let's first consider the ring-opening reaction occurring at the position on the left. The resulting alkoxide ion is protonated by water to give a diol:

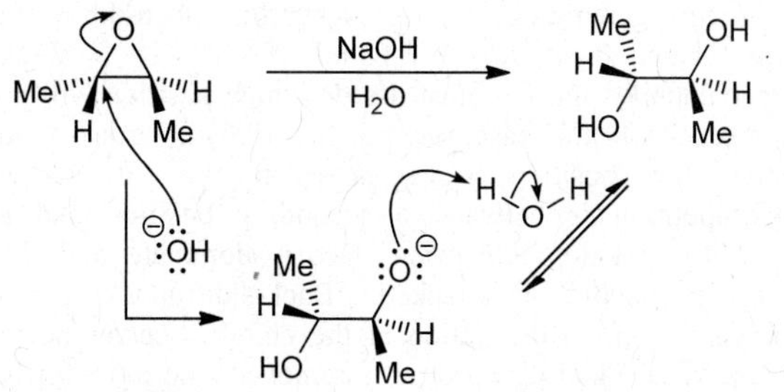

This diol is a *meso* compound, which can be seen more clearly if we draw a Newman projection and then rotate about the central C–C bond:

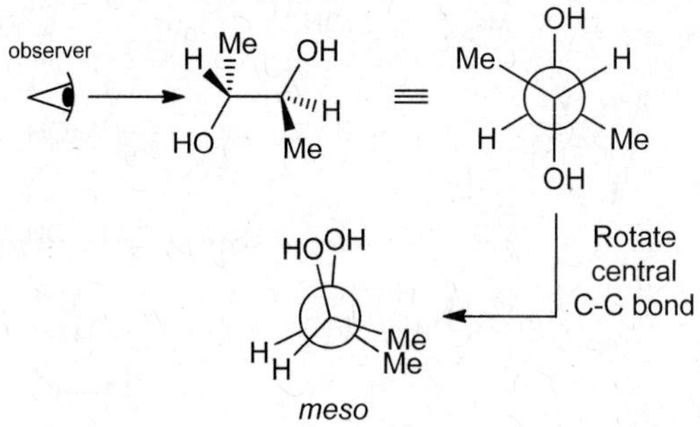

Now let's consider the product that is obtained if the ring-opening reaction occurs at the position on the right:

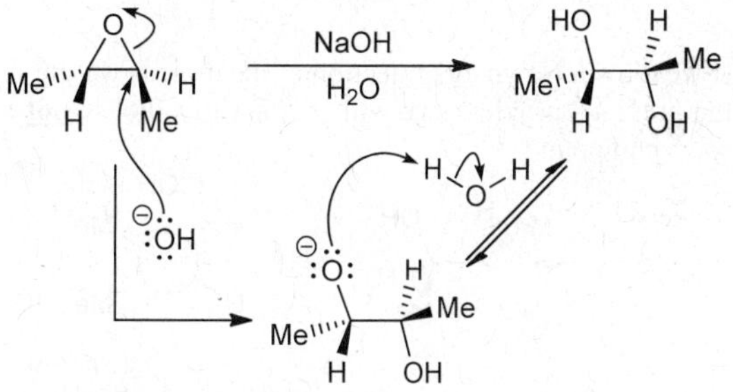

Once again, notice that the resulting diol is a *meso* compound:

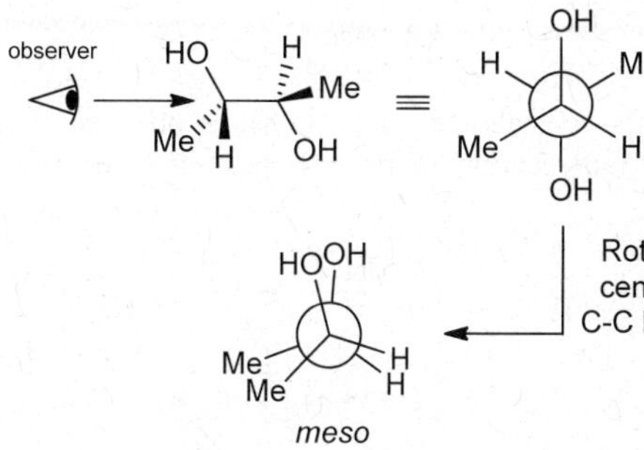

In fact, this *meso* compound is the same *meso* compound that was obtained earlier (when the attack occurred at the position on the left).

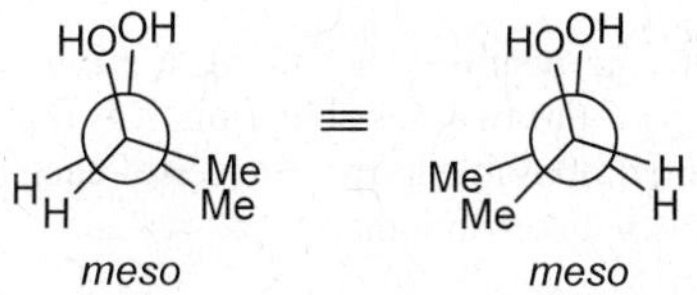

That is, the same product is obtained, regardless of which electrophilic position is attacked by hydroxide.

14.19. We begin by drawing the structure of *meso*-2,3-epoxybutane.

Hydroxide is a strong nucleophile, and it can attack the epoxide at either of two locations, just as we saw in the previous problem (14.18). Let's first consider the ring-opening reaction occurring at the position on the left. The resulting alkoxide ion is protonated by water to give a diol:

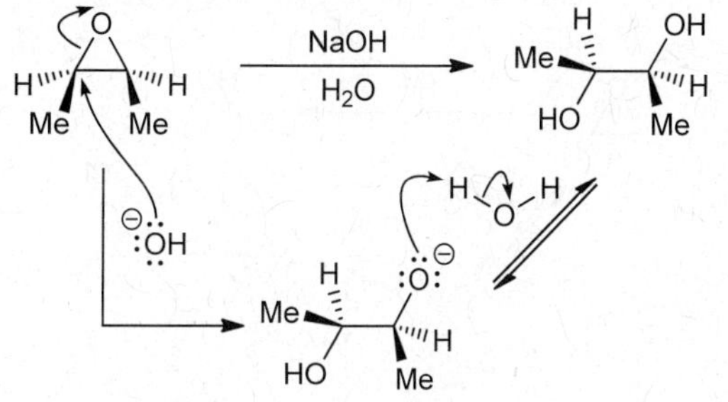

If we draw a Newman projection of the diol, as we did in the previous problem, we will see that this diol is **not** a *meso* compound.

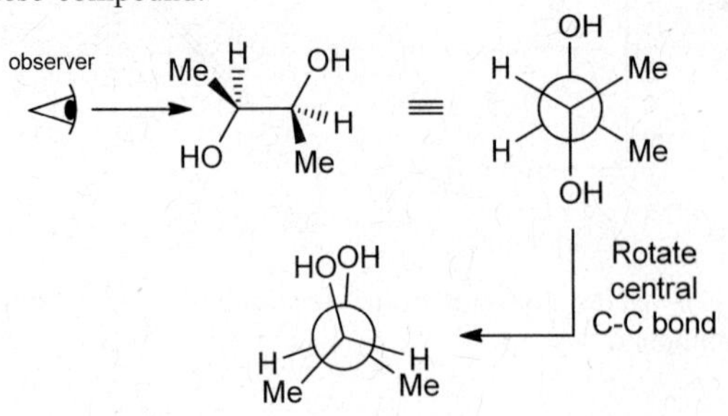

not *meso*

Now let's consider the product that is obtained if the ring-opening reaction occurs at the position on the right:

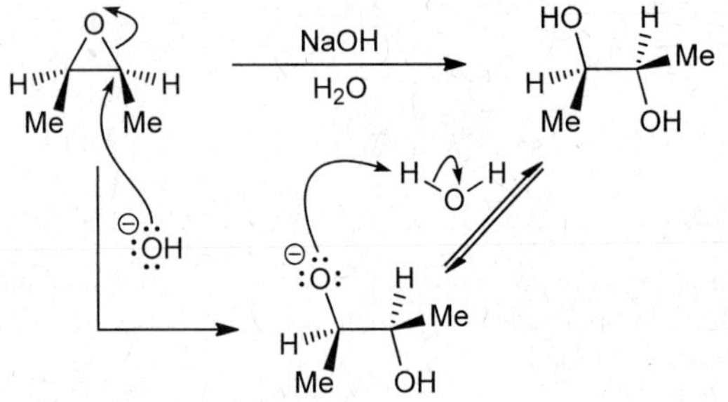

Once again, the resulting diol is **not** a *meso* compound. If we compare the two possible diols, we find that they are non-superimposable mirror images of each other, and therefore, they are enantiomers:

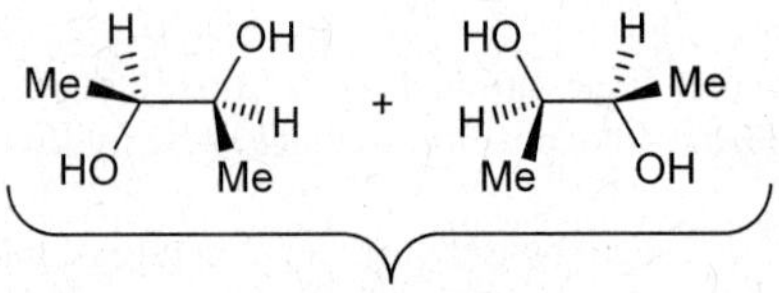

Enantiomers

14.20.

(a) Under acidic conditions, the epoxide is protonated, thereby generating a very powerful electrophile (a protonated epoxide). Since the starting epoxide is symmetrical, regiochemistry is not a concern in this case. That is, the nucleophile can attack the epoxide at either position, giving the same product either way. Stereochemistry is also not an issue in this case, because the product does not contain any chirality centers.

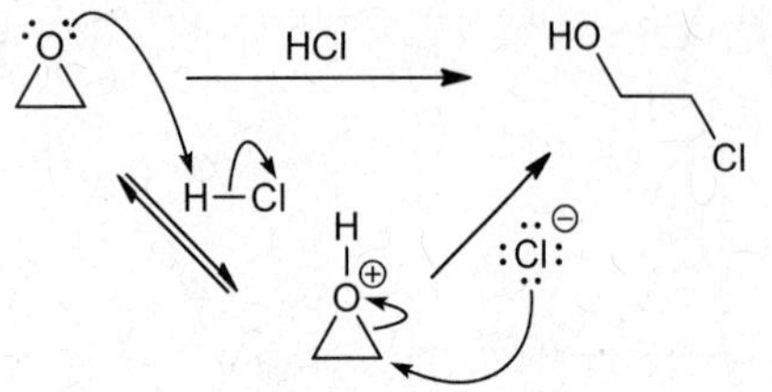

(b) Under acidic conditions, the epoxide is protonated, thereby generating a very powerful electrophile (a protonated epoxide). To determine where the nucleophile (bromide) attacks, we must decide whether steric or electronic effects dominate. In this case, one position (left) is primary and the other position (right) is secondary. Under these conditions, steric effects will dominate, and the attack is expected to occur at the less substituted position. The position being attacked is not a chirality center. There is an existing chirality center, but that center is not attacked, so we do not expect the configuration of that chirality center to change.

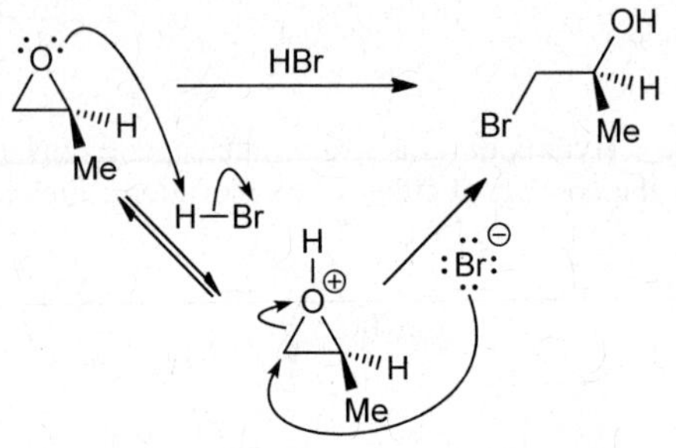

(c) Under acidic conditions, the epoxide is protonated, thereby generating a very powerful electrophile (a protonated epoxide). Ethanol (EtOH) is a weak nucleophile, and we must decide which position will be attacked. In this case, one position (left) is tertiary, and the other position (right) is secondary. When the competition is between a secondary position and a tertiary position, electronic factors dominate and the tertiary position is attacked. Back-side attack causes inversion of configuration at the chirality center being attacked. Finally, a proton is removed (the most likely base is the solvent, ethanol).

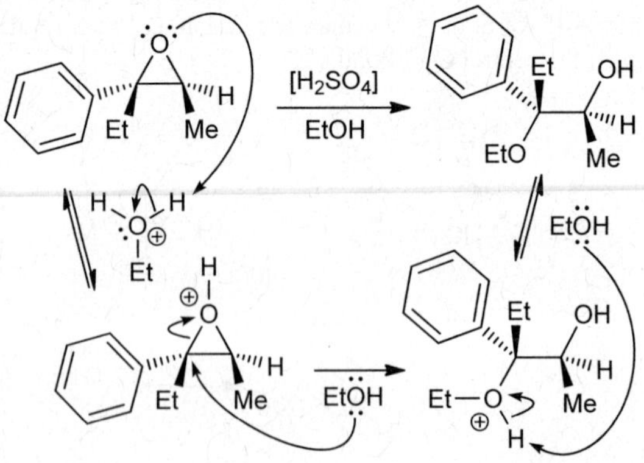

(d) Under acidic conditions, the epoxide is protonated, thereby generating a very powerful electrophile (a protonated epoxide). Water (H_2O) is a weak nucleophile, and we must decide which position will be attacked. In this case, one position (left) is tertiary, and the other position (right) is secondary. When the competition is between a secondary position and a tertiary position, electronic factors dominate and the tertiary position is attacked. Back-side attack causes inversion of configuration at the chirality center being attacked. Finally, a proton is removed (the most likely base is the solvent, water).

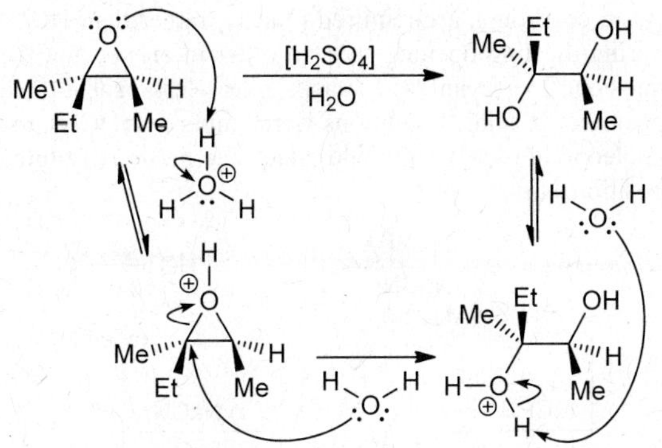

Back-side attack causes inversion of configuration at the chirality center being attacked.

(e) Under acidic conditions, the epoxide is protonated, thereby generating a very powerful electrophile (a protonated epoxide). Methanol (CH_3OH) is a weak nucleophile, and we must decide which position will be attacked. In this case, one position (left) is tertiary, and the other position (right) is secondary. When the competition is between a secondary position and a tertiary position, electronic factors dominate and the tertiary position is attacked. Back-side attack causes inversion of configuration at the chirality center being attacked. Finally, a proton is removed (the most likely base is the solvent, methanol).

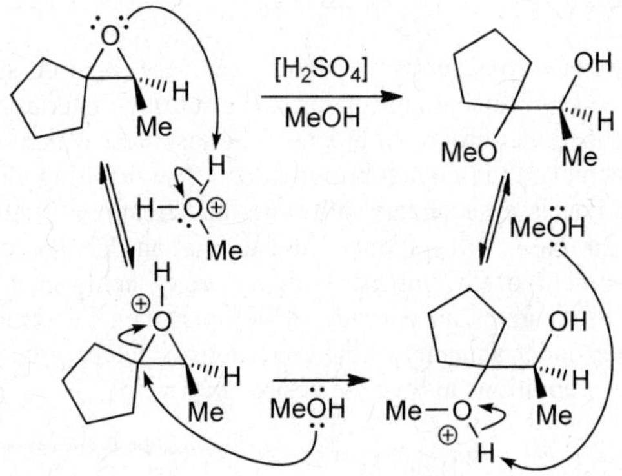

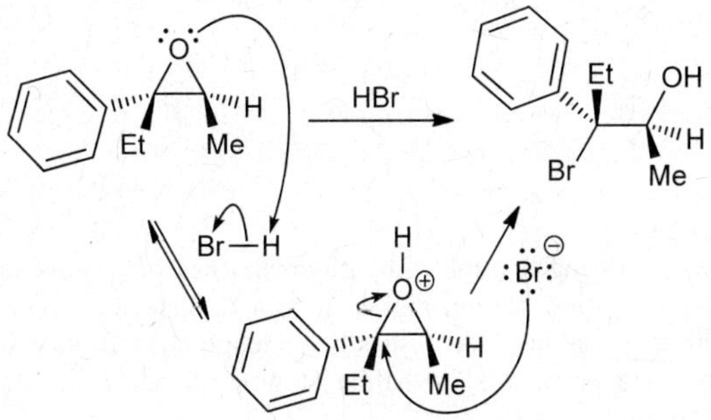

14.21. Under acidic conditions, the epoxide is protonated, thereby generating a very powerful electrophile (a protonated epoxide). This intermediate has an OH group that can function as a nucleophilic center, attacking the epoxide and opening the ring. In this case, one position (left) is tertiary, and the other position (right) is secondary. When the competition is between a secondary position and a tertiary position, electronic factors dominate and the tertiary position is attacked. Back-side attack causes inversion of configuration at the chirality center being attacked. Finally, a proton is removed (the most likely base is water, which is present in a solution of H_2SO_4).

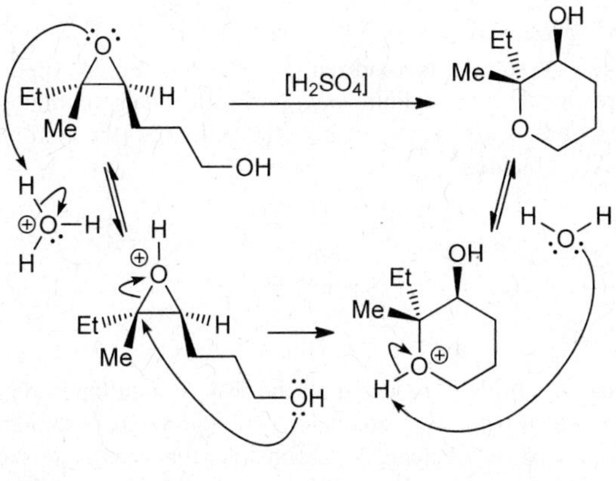

14.22.
(a) The desired thiol can be prepared by treating an appropriate alkyl halide (shown below) with sodium hydrosulfide (NaSH).

(f) Under acidic conditions, the epoxide is protonated, thereby generating a very powerful electrophile (a protonated epoxide). Bromide is a nucleophile, and we must decide which position will be attacked. In this case, one position (left) is tertiary, and the other position (right) is secondary. When the competition is between a secondary position and a tertiary position, electronic factors dominate and the tertiary position is attacked.

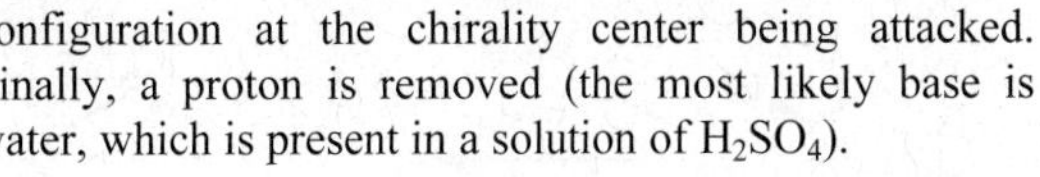

(b) The desired thiol can be prepared by treating an appropriate alkyl halide (shown below) with sodium hydrosulfide (NaSH).

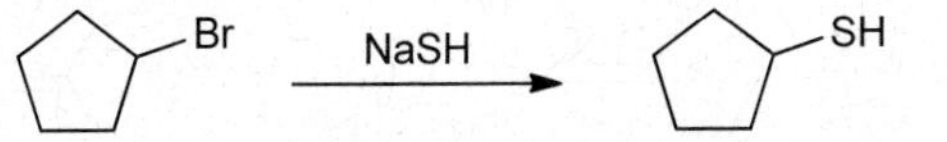

(c) The desired thiol can be prepared by treating an appropriate alkyl halide (shown below) with sodium hydrosulfide (NaSH).   Notice that inversion of configuration is expected, so the starting alkyl halide must have a different configuration than the desired product.

14.23.

(a) Treating a thiol with sodium hydroxide results in deprotonation of the thiol to give a thiolate ion. This thiolate ion is a very strong nucleophile, and it will attack a primary alkyl halide to give an S_N2 reaction. The product is a sulfide.

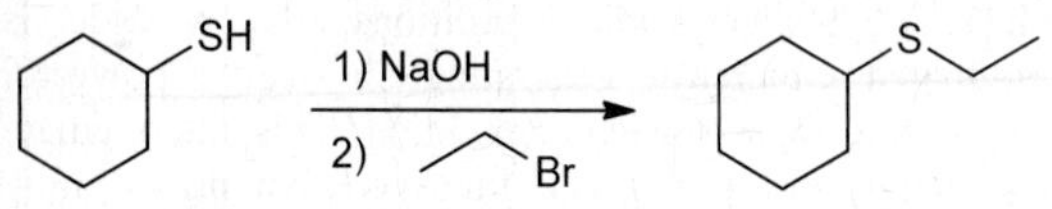

(b) A secondary alkyl bromide will serve as a substrate (electrophile) in an S_N2 reaction, upon treatment with a thiolate ion. The reaction occurs at a chirality center, so we expect inversion of configuration, which is characteristic of S_N2 processes.

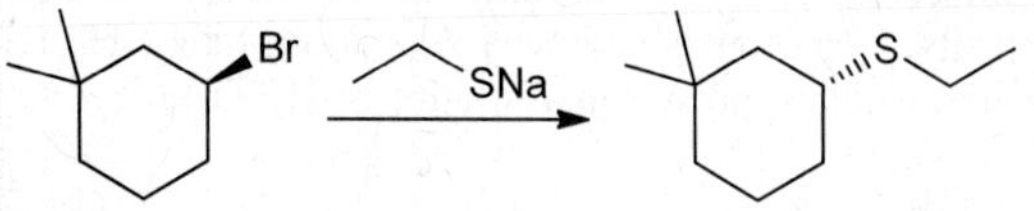

(c) A sulfide is oxidized to give a sulfoxide upon treatment with sodium *meta*-periodate. This oxidizing agent does not further oxidize the sulfoxide (the sulfone is not obtained).

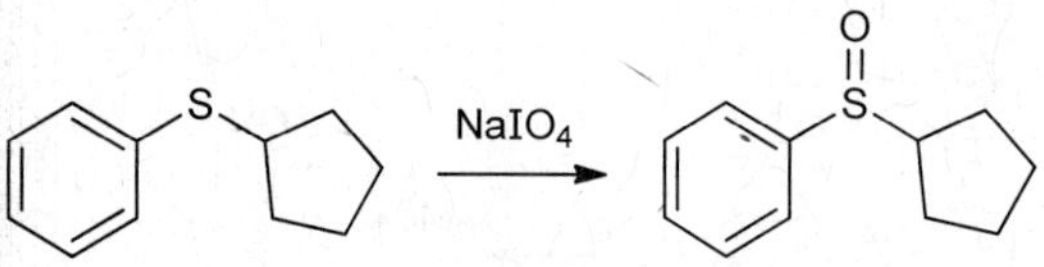

(d) A sulfide is oxidized all the way to a sulfone upon treatment with two equivalents of hydrogen peroxide. The first equivalent is responsible for oxidizing the sulfide to a sulfoxide, and the second equivalent oxidizes the sulfoxide to a sulfone.

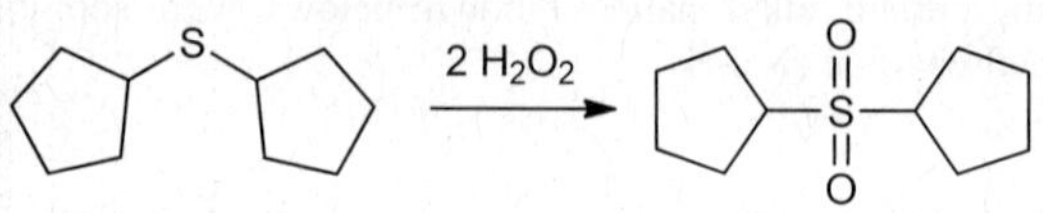

14.24.
(a) The desired product can be made from an epoxide, which can be made from the starting alkene, as illustrated in the following retrosynthetic analysis:

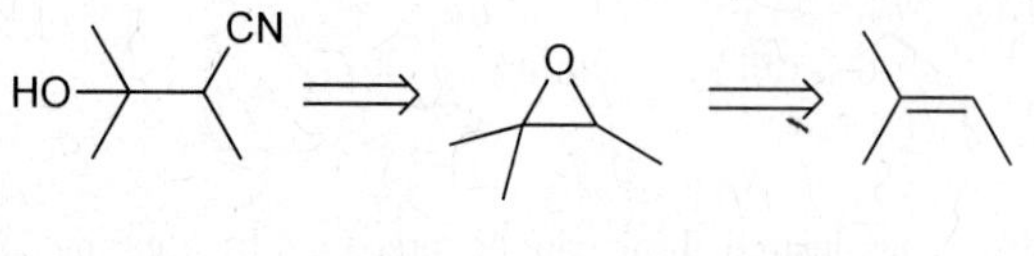

Now let's consider the reagents necessary for each step of the synthesis. First, the alkene must be converted into an epoxide, which can be achieved upon treatment with a peroxy acid, such as MCPBA. Notice that the resulting epoxide contains one chirality center. Since the epoxide can form on either face of the π bond with equal likelihood, we expect a racemic mixture of the epoxide. Basic conditions are required (NaCN, rather than HCN) during the ring opening step, in order to ensure that the nucleophile (cyanide) attacks the less substituted position. If acidic conditions were employed (HCN), the nucleophile (cyanide) would attack the more substituted position.

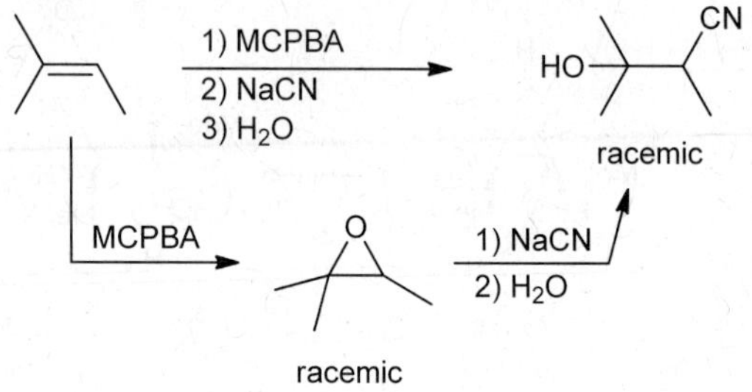

(b) The desired *trans*-diol can be made from an epoxide, which can be made from the corresponding alkene.

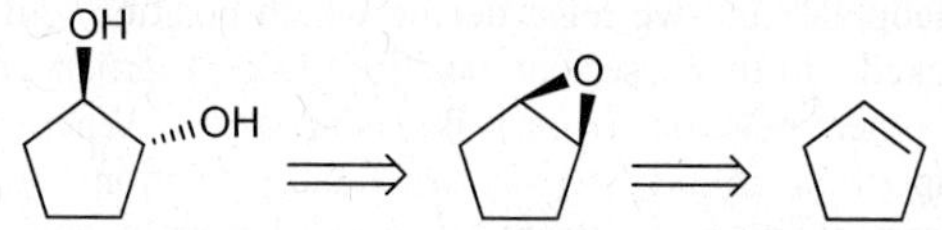

And the alkene can be prepared in two steps from the starting material.

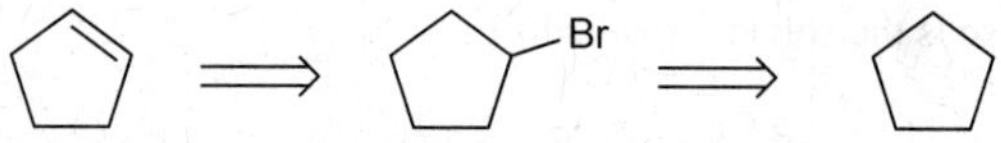

Now let's consider the reagents necessary for each step of the synthesis, going forward. The starting material has no functional group, so one must be installed, which can be achieved via radical bromination. The resulting alkyl bromide is a secondary substrate, and treatment with a strong base will give an alkene via an E2 process. Treatment of the alkene with a peroxy acid, such as MCPBA, gives an epoxide, which can then be opened either under aqueous acidic conditions or under aqueous basic conditions to give the desired *trans*-diol.

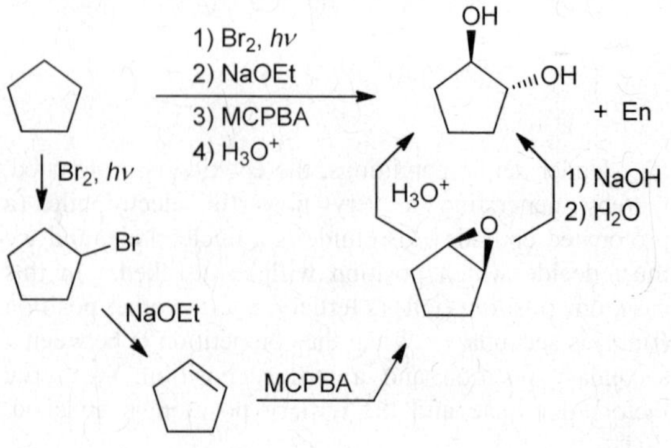

(c) The desired product can be made from an epoxide, which can be made from the starting alkene, as illustrated in the following retrosynthetic analysis:

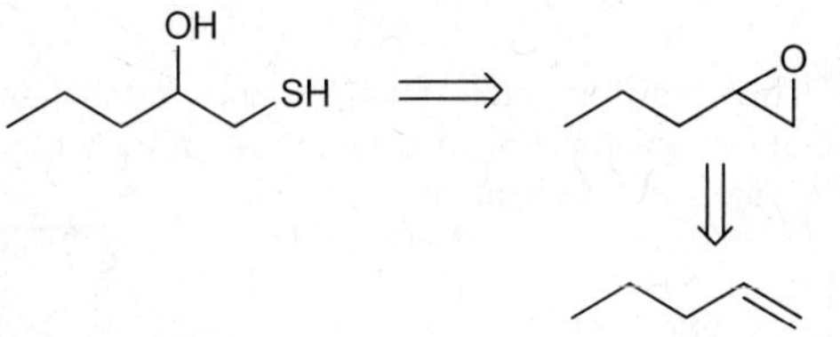

Now let's consider the reagents necessary for each step of the synthesis. First, the alkene must be converted into an epoxide, which can be achieved upon treatment with a peroxy acid, such as MCPBA. Notice that the resulting epoxide contains one chirality center. Since the epoxide can form on either face of the π bond with equal likelihood, we expect a racemic mixture of the epoxide. Basic conditions are required (NaSH, rather than H_2S and sulfuric acid) during the ring opening step, in order to ensure that the nucleophile (HS⁻) attacks the less substituted position.

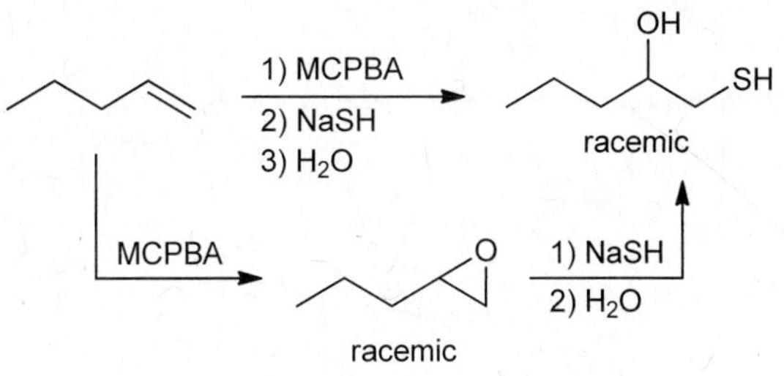

(d) The desired product can be made from an epoxide, which can be made from the starting alkene, as illustrated in the following retrosynthetic analysis:

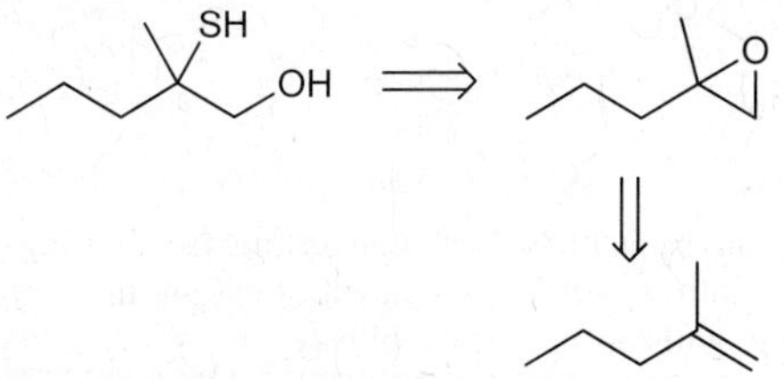

Now let's consider the reagents necessary for each step of the synthesis. First, the alkene must be converted into an epoxide, which can be achieved upon treatment with a peroxy acid, such as MCPBA. Notice that the resulting epoxide contains one chirality center. Since the epoxide can form on either face of the π bond with equal likelihood, we expect a racemic mixture of the epoxide. Acidic conditions are required (H_2S and sulfuric acid, rather than NaSH) during the ring opening step, in order to ensure that the nucleophile (H_2S) attacks the more substituted position.

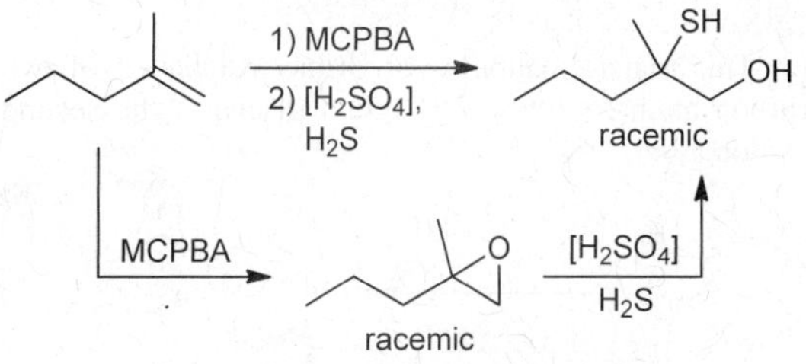

(e) The desired product (a ketone) can be made via oxidation of a secondary alcohol, which can be made from the starting epoxide, as illustrated in the following retrosynthetic analysis:

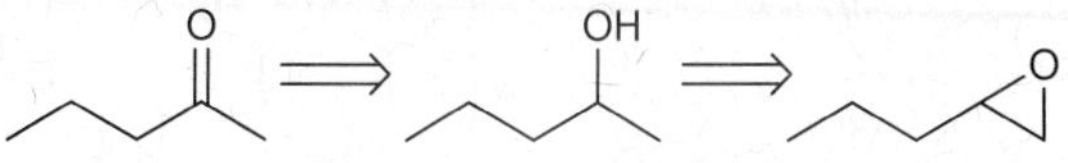

Now let's consider the reagents necessary for each step of the synthesis. First, the epoxide must be converted into an alcohol, which can be achieved upon treatment with LAH, followed by water work-up. The resulting alcohol is then oxidized to give the desired ketone.

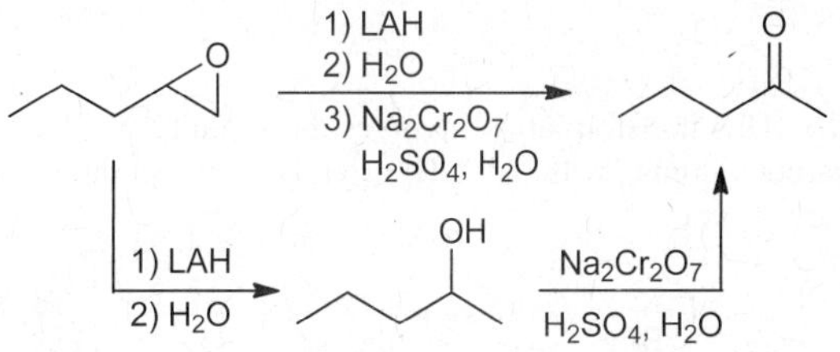

14.25. The desired product can be made from an epoxide, which can be made from an alkene, which can be made from the starting alkyl iodide, as illustrated in the following retrosynthetic analysis:

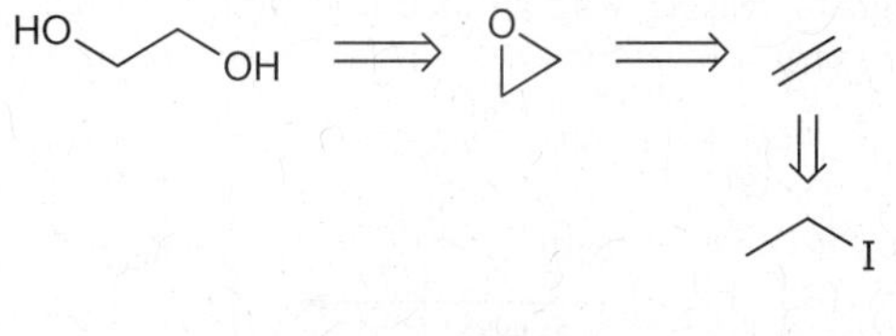

Now let's consider the reagents necessary for each step of the synthesis. First, the alkyl iodide is converted into an alkene upon treatment with a sterically hindered base (via an E2 process). The alkene is then converted into an epoxide upon treatment with a peroxy acid, such as MCPBA. The epoxide can then be opened, either under aqueous acidic conditions or under aqueous basic conditions, to give the desired diol.

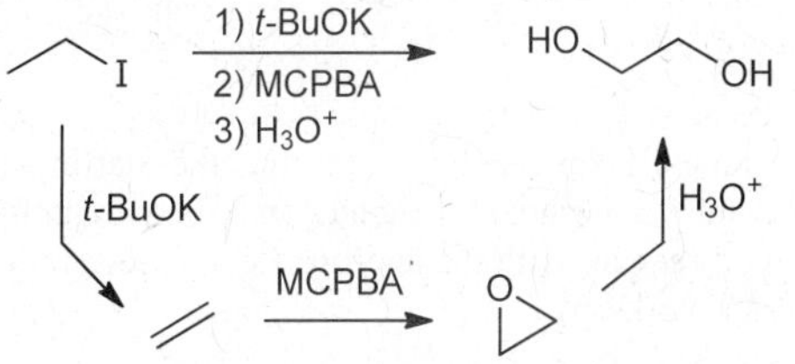

14.26.

(a) This transformation involves the installation of two carbon atoms, with an OH group placed at the second carbon atom.

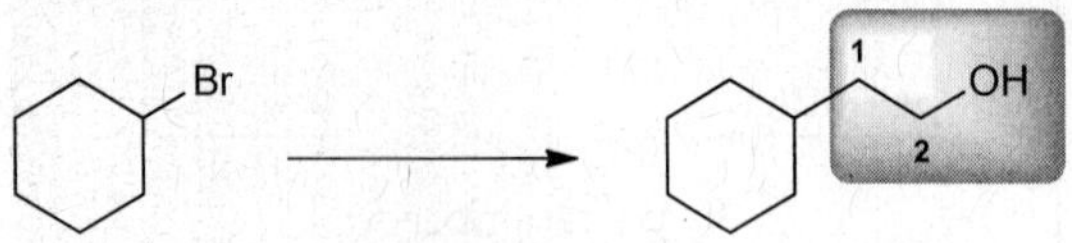

This can be achieved by converting the starting alkyl halide into a Grignard reagent, and then treating the Grignard reagent with an epoxide (followed by water work-up).

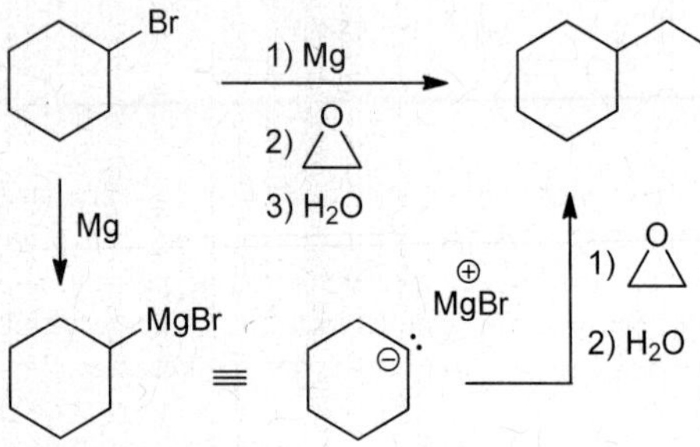

(b) This transformation involves the installation of two carbon atoms, with an OH group placed at the first carbon atom.

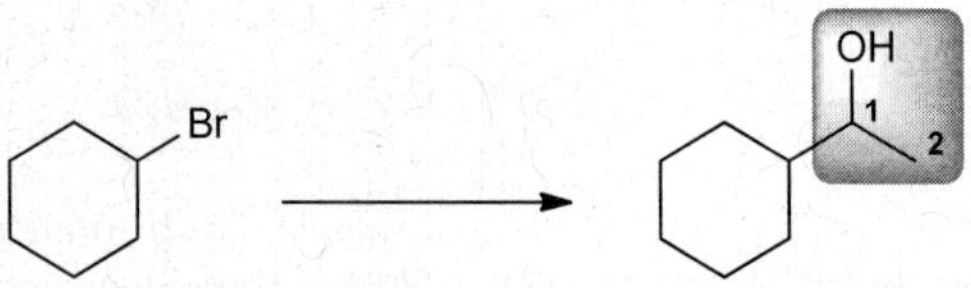

This can be achieved by converting the starting alkyl halide into a Grignard reagent, and then treating the Grignard reagent with an aldehyde (followed by water work-up).

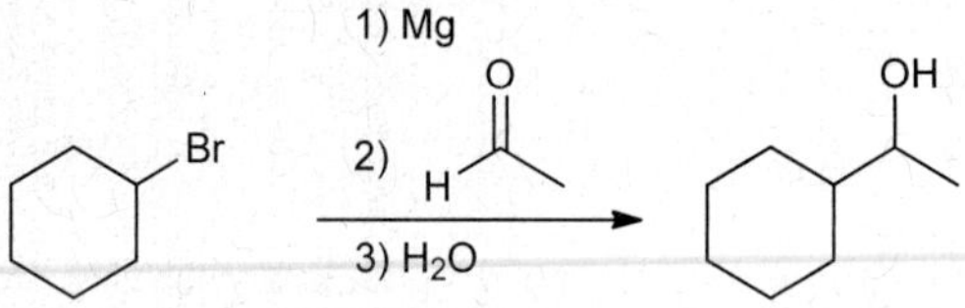

(c) This transformation involves the installation of an alkyl chain, with an OH group placed at the second carbon atom of the chain.

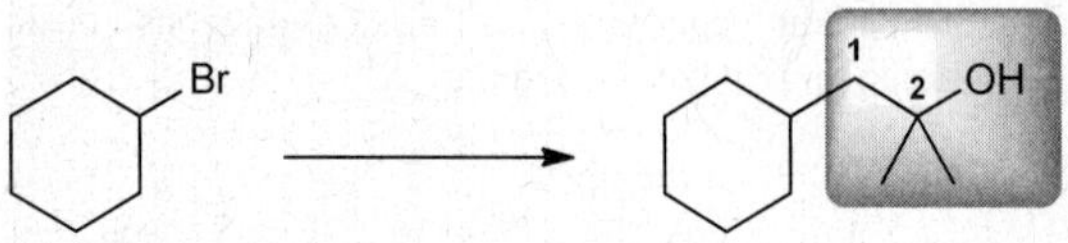

This can be achieved by converting the starting alkyl halide into a Grignard reagent, and then treating the Grignard reagent with the appropriate epoxide (followed by water work-up).

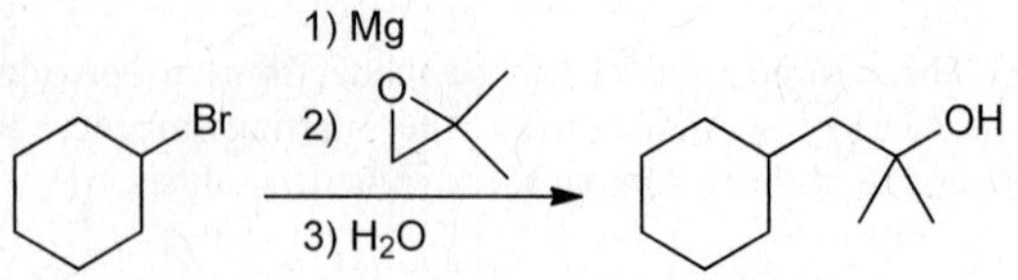

(d) This transformation involves the installation of several carbon atoms, with a functional group at the first carbon atom of the chain.

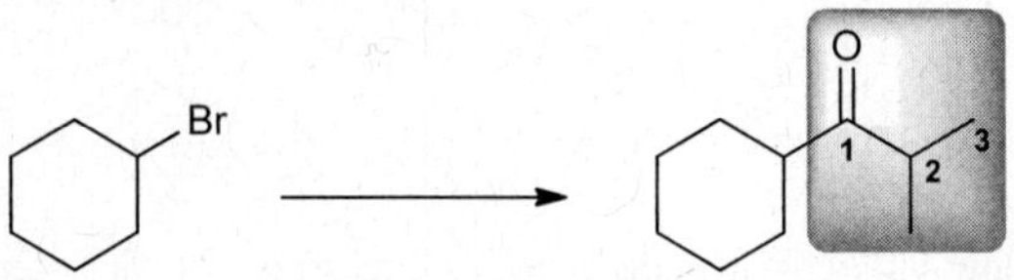

This can be achieved by converting the starting alkyl halide into a Grignard reagent, treating the Grignard reagent with an aldehyde (followed by water work-up) to give an alcohol, and then oxidizing the alcohol to a ketone.

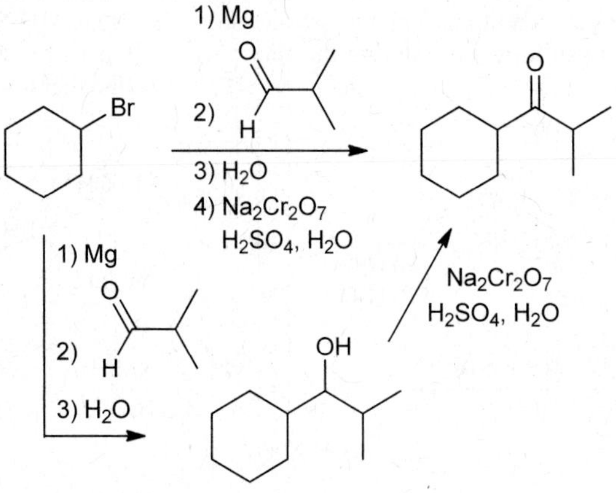

(e) This transformation involves the installation of two carbon atoms, with a functional group placed at the second carbon atom.

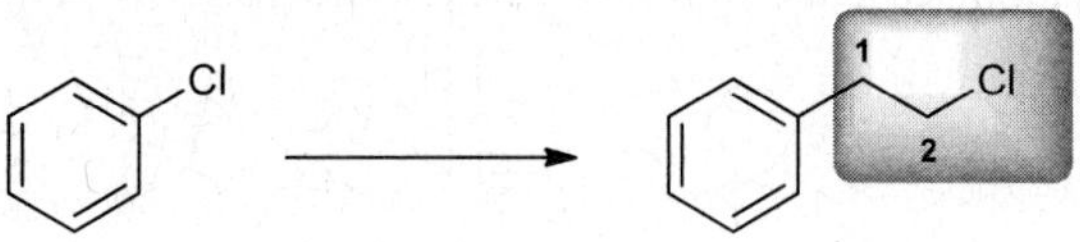

This can be achieved by converting the starting alkyl halide into a Grignard reagent, treating the Grignard reagent with an epoxide (followed by water work-up), and then converting the resulting alcohol into an alkyl chloride (via an S_N2 process).

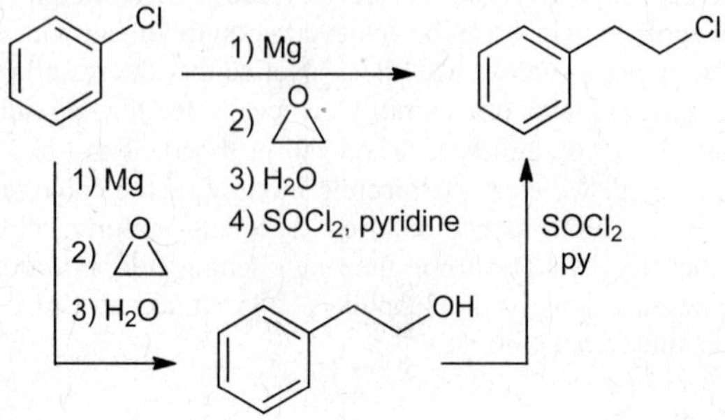

(f) This transformation involves the installation of two carbon atoms, with a functional group placed at the first carbon atom.

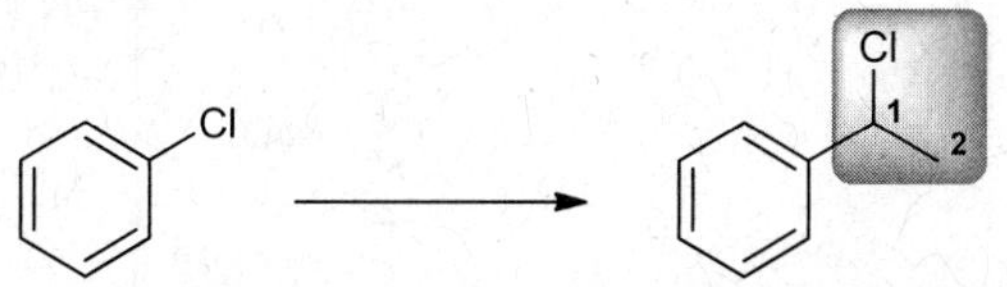

This can be achieved by converting the starting alkyl halide into a Grignard reagent, treating the Grignard reagent with an aldehyde (followed by water work-up), and then converting the resulting alcohol into an alkyl chloride.

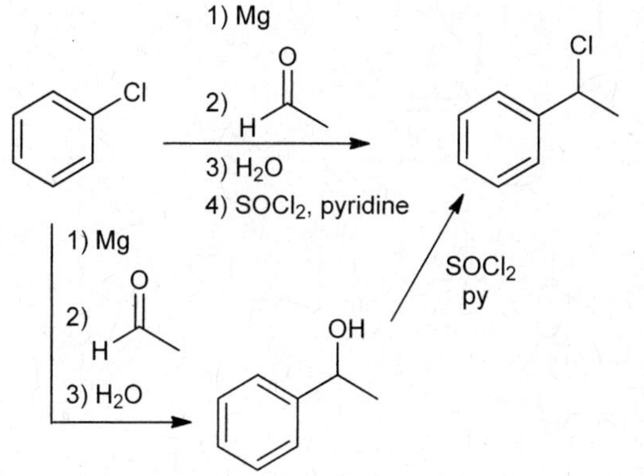

(g) This transformation involves the installation of an alkyl chain, with an OH group placed at the second carbon atom of the chain.

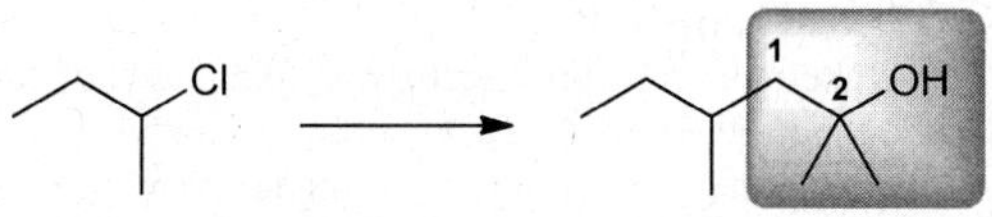

This can be achieved by converting the starting alkyl halide into a Grignard reagent, and then treating the Grignard reagent with the appropriate epoxide (followed by water work-up).

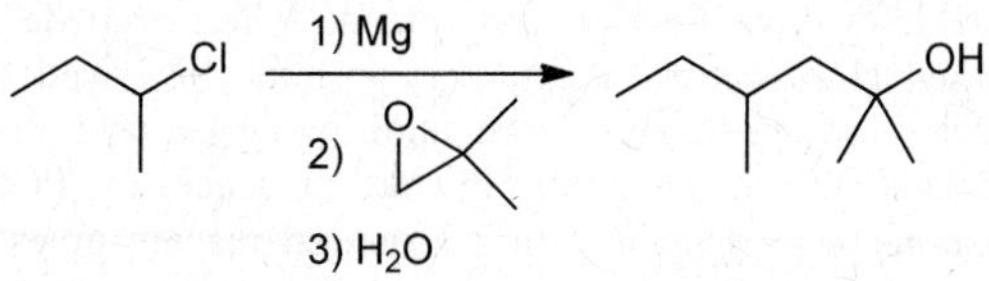

(h) This transformation involves the installation of an alkyl chain, with an OH group placed at the second carbon atom of the chain.

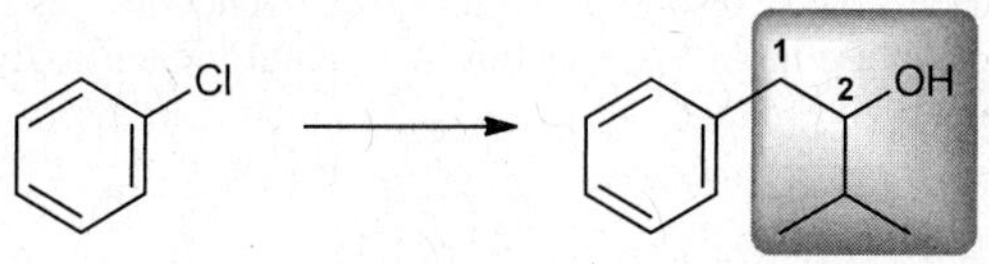

This can be achieved by converting the starting alkyl halide into a Grignard reagent, and then treating the Grignard reagent with the appropriate epoxide (followed by water work-up).

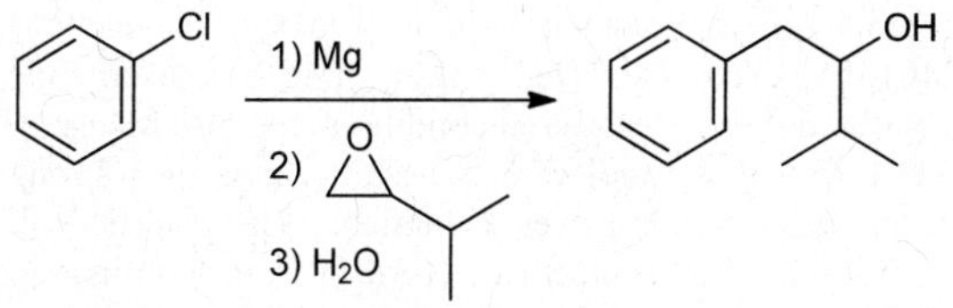

(i) The desired transformation can be achieved by converting the alkene into an epoxide, and then attacking the epoxide with a Grignard reagent, as shown here:

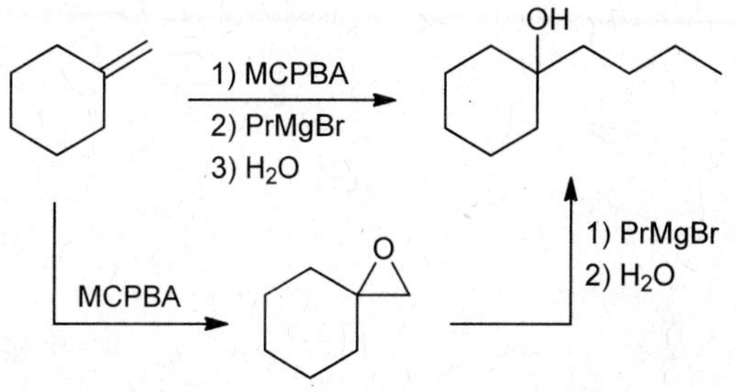

14.27. There are certainly many acceptable methods for achieving the desired transformation. The following retrosynthetic analysis represents one such method. An explanation of each of the steps (*a-f*) follows.

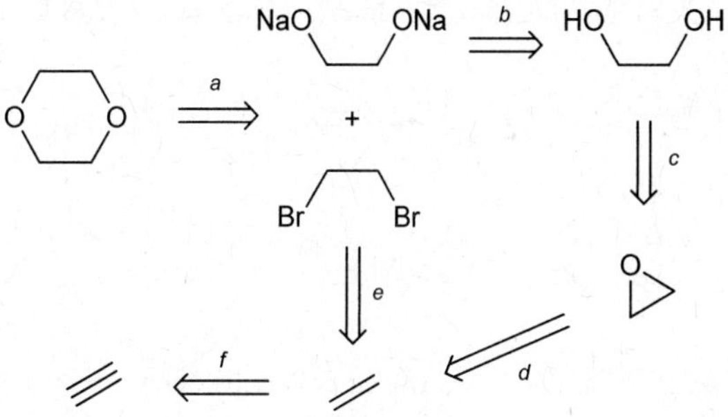

a. The cyclic product can be made by treating a dianion with a dihalide (via two successive S_N2 reactions).
b. The dianion can be made by treating the corresponding diol with two equivalents of a strong base (such as NaH).
c. The diol can be made from an epoxide, via a ring-opening reaction (either under acidic conditions or under basic conditions).
d. The epoxide can be made by treating the corresponding alkene with a peroxy acid, such as MCPBA.
e. The dibromide can be made via bromination of the corresponding alkene.
f. The alkene can be made via hydrogenation of the corresponding alkyne, in the presence of Lindlar's catalyst.

Now let's draw the forward scheme. Acetylene undergoes hydrogenation in the presence of Lindlar's catalyst to afford ethylene, which can be converted to an epoxide upon treatment with a peroxy acid, such as MCPBA. Acid-catalyzed ring-opening of the epoxide gives a diol (base-catalyzed conditions can also be used). Treatment of the diol with two equivalents of a strong base, such as NaH, gives a dianion. The dianion will react with 1,2-dibromoethane (formed from bromination of ethylene) to give the desired cyclic product via two successive S_N2 reactions.

14.28. There are certainly many acceptable methods for achieving the desired transformation. The following retrosynthetic analysis represents one such method. An explanation of each of the steps (a-d) follows.

a. The product has two ether groups, each of which can be formed via a Williamson ether synthesis, from the dianion shown.
b. The dianion can be made by treating the corresponding diol with two equivalents of a strong base (such as NaH).
c. The diol can be made from an alkene, via a dihydroxylation process.
d. The alkene can be made via hydrogenation of the corresponding alkyne, in the presence of Lindlar's catalyst.

Now let's draw the forward scheme. Acetylene undergoes hydrogenation in the presence of Lindlar's catalyst to afford ethylene, which can be converted to a diol via a dihydroxylation process. Treatment of the diol with two equivalents of a strong base, such as NaH, gives a dianion. The dianion will react with two equivalents of methyl iodide giving the product (via a Williamson ether synthesis, twice).

14.29. There are certainly many acceptable methods for achieving the desired transformation. The following retrosynthetic analysis represents one such method. An explanation of each of the steps (a-f) follows.

a. The product can be made via a Williamson ether synthesis, by treating the alcohol shown with base followed by ethyl iodide.
b. The alcohol can be made by treating a ketone with a Grignard reagent.
c. The ketone can be made via oxidation of the corresponding secondary alcohol.
d. The secondary alcohol can be made by treating an aldehyde with a Grignard reagent.
e. The aldehyde can be made via oxidation of the primary corresponding primary alcohol (with PCC).
f. The alcohol can be made by treating an epoxide with a Grignard reagent.

Now let's draw the forward scheme. The epoxide is opened with methyl magnesium bromide, followed by water work-up, to give 1-propanol. 1-Propanol is then oxidized to an aldehyde in the presence of PCC. Treating the aldehyde with a Grignard reagent gives a secondary alcohol. Oxidation of the alcohol gives a ketone, which can be treated with ethyl magnesium bromide to give a tertiary alcohol. This alcohol is then deprotonated upon treatment with a strong base, such as NaH. The resulting anion then functions as a nucleophile in an S_N2 reaction with ethyl iodide to give the desired product.

14.30.

(a) The oxygen atom has two groups attached to it. The parent is named after the larger group (cyclohexane). The smaller group (together with the oxygen atom) is treated as an ethoxy substituent. Locants are used to identify the positions of the ethoxy group and of the methyl group. The compound has two chirality centers, each of which is assigned a configuration in the beginning of the name, placed in parentheses.

(1S,2S)-1-Ethoxy-2-methylcyclohexane

(b) The oxygen atom has two groups attached to it. One group has two carbon atoms and the other group has four carbon atoms. The parent is named after the larger group, so the parent is butane. The smaller group (together with the oxygen atom) is treated as an ethoxy substituent, and a locant (2) is included to identify the location of the substituent on the parent chain. The configuration of the chirality center is indicated at the beginning of the name, placed in parentheses.

(R)-2-Ethoxybutane

(c) The compound is named in the same way that we would name an alcohol (if the SH group were an OH group), except that the term "thiol" is used in the suffix of the name. In this case, the parent is hexane, and the location of the SH group is indicated with a locant (3).

The configuration of the chirality center is indicated at the beginning of the name, placed in parentheses.

(S)-3-Hexanethiol

(d) This compound is a sulfoxide (S=O group) with an ethyl group and a propyl group. The groups are alphabetized in the name, so ethyl precedes propyl.

Ethyl propyl sulfoxide

(e) The oxygen atom has two groups attached to it. The parent is named after the larger group. In this case, the parent is pentene (an alkene), and the smaller group (together with the oxygen atom) is treated as an ethoxy substituent.

Locants are used to identify the positions of the ethoxy group and of the methyl group. The configuration of the alkene is indicated in the beginning of the name, placed in parentheses.

(E)-2-Ethoxy-3-methyl-2-pentene

(f) The parent is benzene, and the two methoxy groups are treated as substituents. Locants are used to indicate the relative placement of the methoxy groups on the ring.

1,2-Dimethoxybenzene

(g) This compound is a sulfide (R–S–R), and it is named very much like an ether. The sulfur atom is connected to two groups, ethyl and propyl, which are arranged alphabetically in the name.

Ethyl propyl sulfide

14.31.
(a) The oxygen atom is connected to two carbon atoms, each of which is sp^3 hybridized. Therefore, each of the C–O bonds is cleaved, giving two alkyl halides.

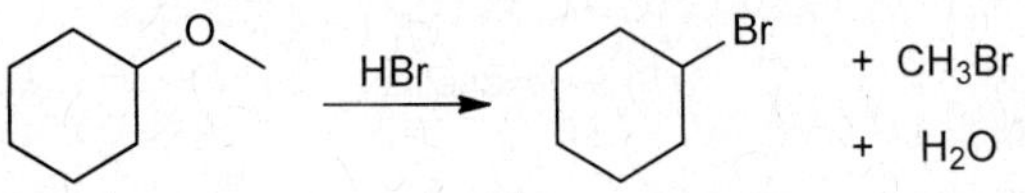

(b) The oxygen atom is connected to two carbon atoms, but only one of them is sp^3 hybridized. Therefore, only one C–O bond is cleaved. The other C–O bond (where the C is sp^2 hybridized) is not cleaved. The products are phenol and methyl halide.

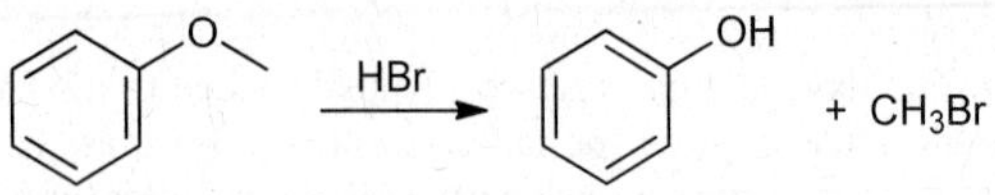

(c) The oxygen atom is connected to two carbon atoms, each of which is sp^3 hybridized. Therefore, each of the C–O bonds is cleaved, giving two equivalents of 2-bromopropane.

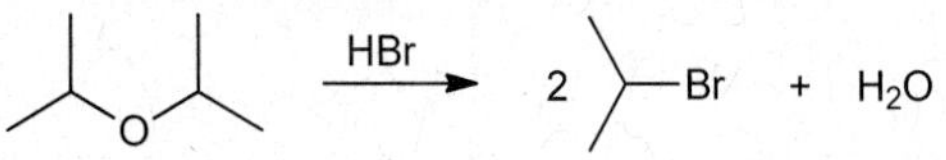

(d) The oxygen atom is connected to two carbon atoms, each of which is sp^3 hybridized. Therefore, each of the C–O bonds is cleaved. As a result, the ring is opened, giving a dihalide.

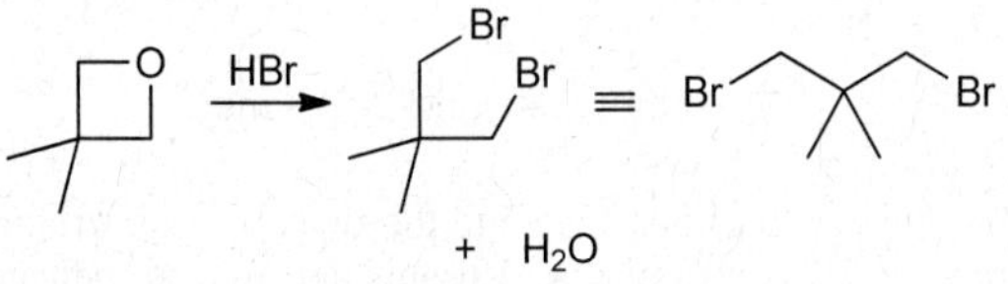

14.32. Ethers have the following structure:

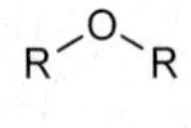

We are looking for ethers with the molecular formula $C_4H_{10}O$, which means that the four carbon atoms must be contained in the two R groups of the ether. One possibility is that each R group has two carbon atoms.

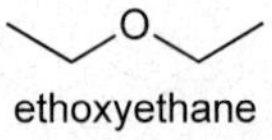

ethoxyethane
(diethyl ether)

Alternatively, one R group can have three carbon atoms and the other R group can have one carbon atom. The R

group containing three carbon atoms can either be a propyl group or an isopropyl group:

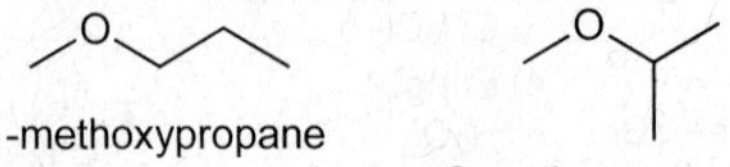

1-methoxypropane
(methyl propyl ether)

2-methoxypropane
(isopropyl methyl ether)

In total, there are three constitutionally isomeric ethers with the molecular formula $C_4H_{10}O$.

14.33.
(a) The desired transformation involves addition of H and OEt across a π bond. This can be achieved via alkoxymercuration-demercuration, where EtOH is used as the alcohol during the alkoxymercuration step.

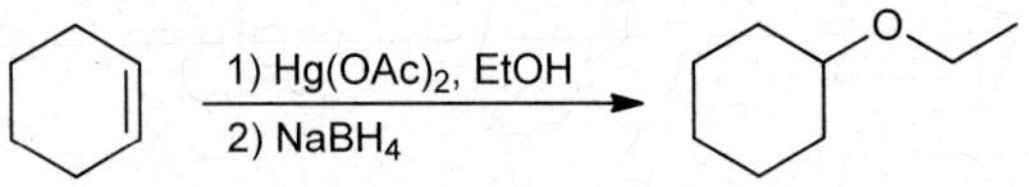

(b) The desired transformation involves an *anti* addition of OH and OMe across a π bond. This can be achieved by converting cyclohexene into an epoxide (upon treatment with a peroxy acid, such as MCPBA), followed by a ring opening reaction. The second step of this process (opening the ring) can be performed under either acidic conditions or basic conditions.

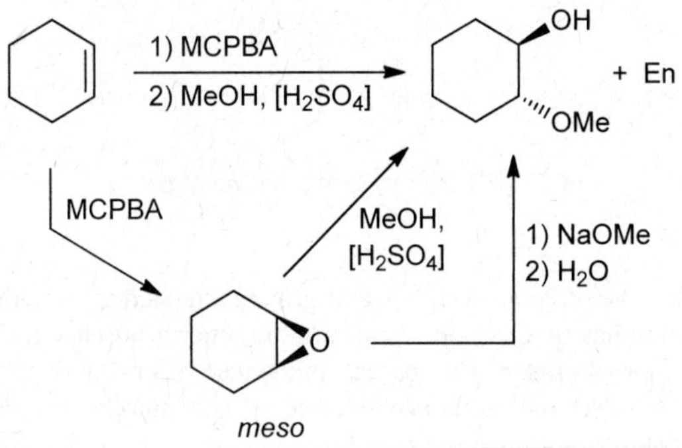

(c) The desired transformation involves addition of H and OR across a π bond. This can be achieved via alkoxymercuration-demercuration, where $(CH_3)_3COH$ is used as the alcohol during the alkoxymercuration step.

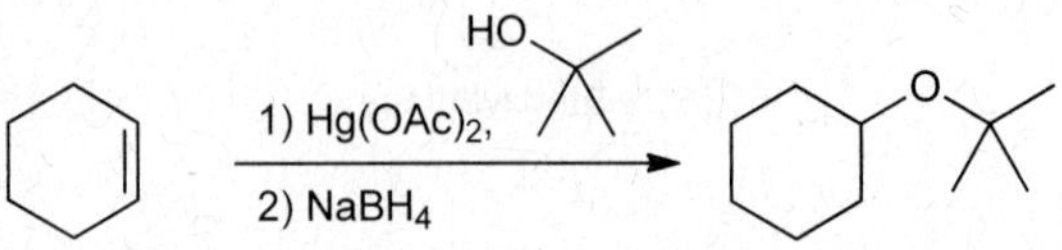

14.34.
(a) There are four C–O bonds, and each of them is cleaved under acidic conditions and replaced with a C–I

bond, giving two moles of compound **A** and two moles of water.

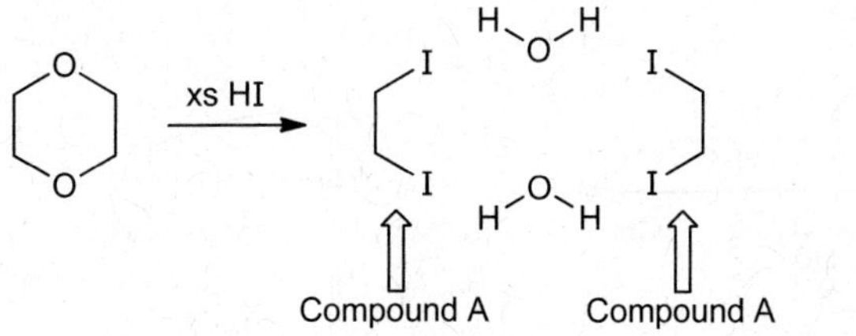

(b) As seen in the previous solution (14.34a), two moles of compound **A** are produced for every one mole of 1,4-dioxane.

(c) Each of the C–O bonds is cleaved via a two-step process: (i) protonation of the oxygen atom to give an oxonium ion, and (ii) an S_N2 reaction, in which iodide functions as a nucleophile and attacks the substrate. Each of these two steps requires two curved arrows, as shown. Since there are four C–O bonds that undergo cleavage, our mechanism will have a total of eight steps, where each step utilizes two curved arrows, as shown.

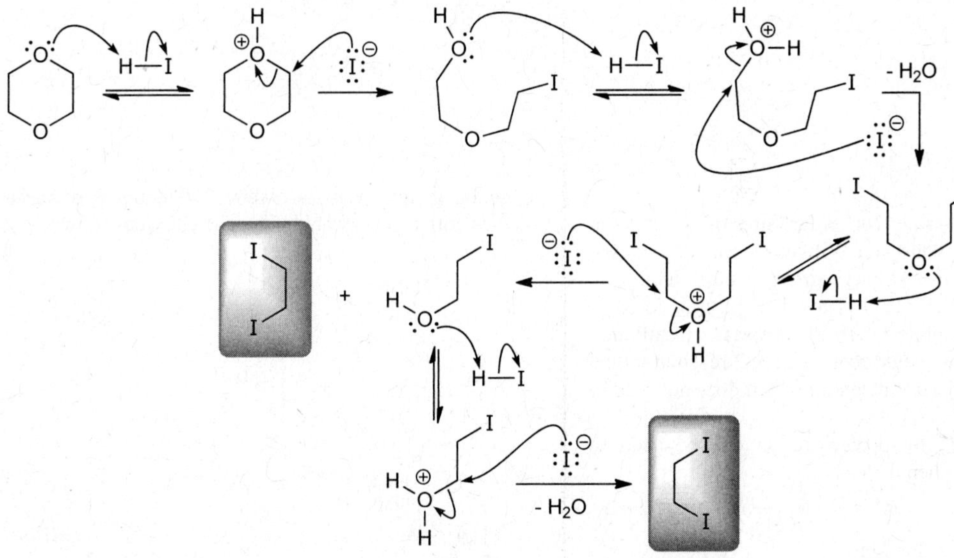

14.35. In the presence of an acid catalyst, an OH group is protonated to give an oxonium ion, thereby converting a bad leaving group into a good leaving group. Then, the other OH group (that was not protonated) functions as a nucleophilic center in an S_N2-type process, forming a ring. Finally, deprotonation gives the product. Notice that water (not hydroxide) functions as the base in the deprotonation step (because there is virtually no hydroxide present in acidic conditions).

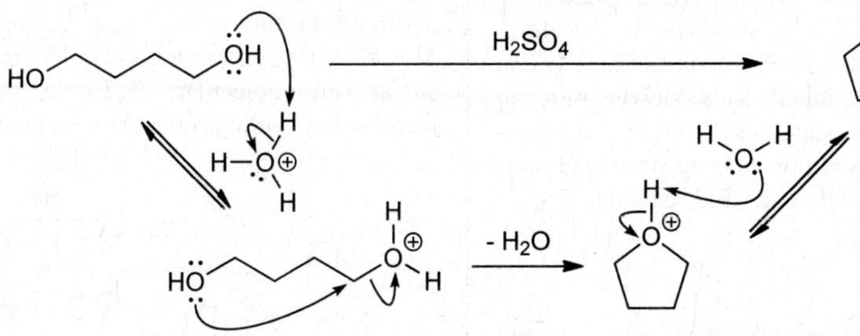

14.36. In the presence of an acid catalyst, an OH group is protonated to give an oxonium ion, thereby converting a bad leaving group into a good leaving group. Then, one molecule of the diol (that was not yet protonated) can function as a nucleophile and attack the oxonium ion in an S_N2 process, expelling water as a leaving group. The resulting oxonium ion is then deprotonated. The previous three steps are then repeated (proton transfer, S_N2, and then proton transfer), giving the cyclic product. Notice that each of the deprotonation steps is shown with water functioning as the base (not hydroxide), because there is virtually no hydroxide present in acidic conditions.

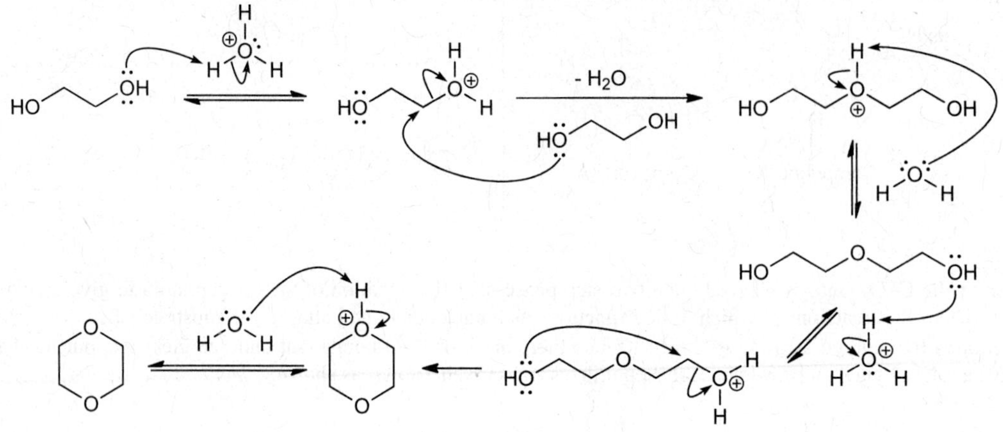

14.37.
(a) Neither alkyl group (on either side of the oxygen atom) can be installed via a Williamson ether synthesis. Installation of the *tert*-butyl group would require a tertiary alkyl halide, which is too sterically hindered to serve as an electrophile for an S_N2 process. Installation of the phenyl group would require an S_N2 reaction taking place at an sp^2 hybridized center, which does not readily occur.

(b) Alkoxymercuration-demercuration can be used to prepare *tert*-butyl phenyl ether:

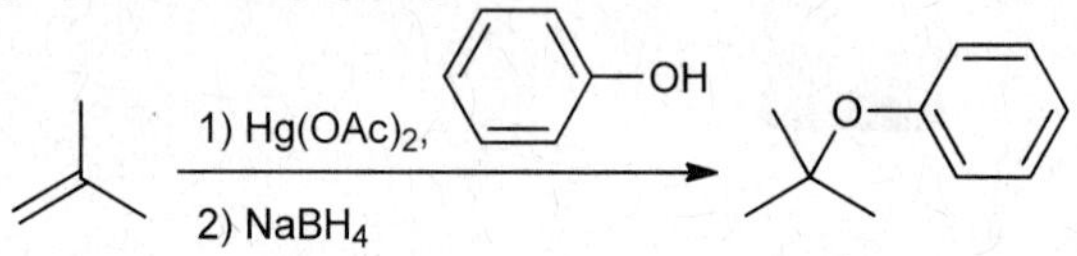

14.38. Ethylene oxide has a high degree of ring strain, and readily functions as an electrophile in an S_N2 reaction. The reaction opens the ring and alleviates the ring strain. Oxetane has less ring strain and is, therefore, less reactive as an electrophile towards S_N2. The reaction can still occur, albeit at a slower rate, to alleviate the ring strain associated with the four-membered ring. THF has almost no ring strain and does not function as an electrophile in an S_N2 reaction.

14.39. Acetylene undergoes alkylation when treated with a strong base (such as $NaNH_2$) followed by an alkyl halide. This process is then repeated to install a second methyl group, giving 2-butyne. This alkyne can be reduced either via hydrogenation with a poisoned catalyst to give a *cis* alkene, or via a dissolving metal

reduction to give a *trans* alkene. Treatment of these alkenes with a peroxy acid gives the epoxides shown.

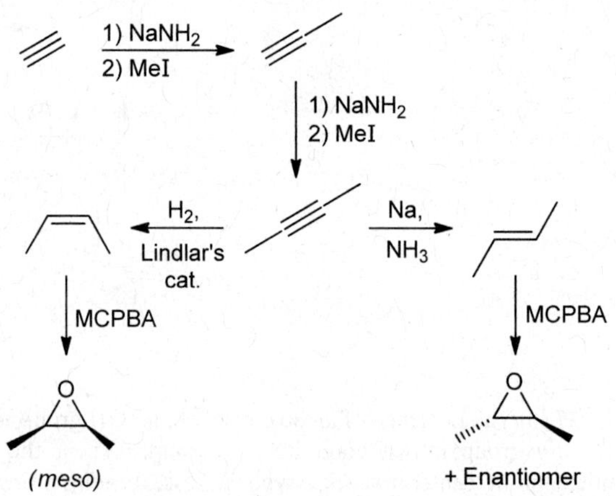

14.40. Upon treatment with NaH (a strong base), the hydroxyl proton is removed, giving an alkoxide ion. This alkoxide ion has a built-in leaving group (bromide), so an intramolecular, S_N2-type process can occur, forming the cyclic product shown here.

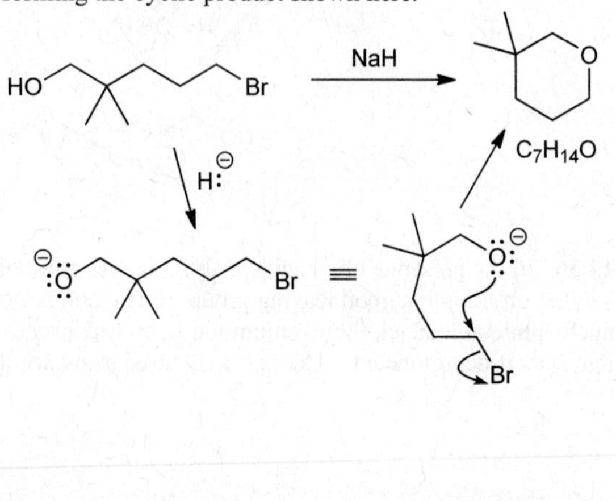

14.41.

(a) Acetylene can be treated with NaNH$_2$ followed by PhCH$_2$Br to install a benzyl group (PhCH$_2$). This process is then repeated (with methyl iodide as the alkyl halide) to install a methyl group. These two alkylation processes could have been performed in reverse order (with installation of the methyl group first, followed by installation of the benzyl group). The resulting alkyne can then be reduced via hydrogenation with a poisoned catalyst to give a *cis* alkene, which gives the desired epoxide upon treatment with a peroxy acid.

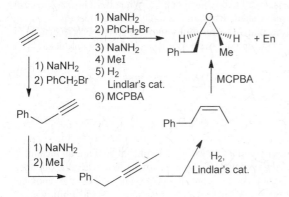

(b) Acetylene can be treated with NaNH$_2$ followed by PhCH$_2$Br to install a benzyl group (PhCH$_2$). This process is then repeated (with ethyl iodide as the alkyl halide) to install an ethyl group. These two alkylation processes could have been performed in reverse order (with installation of the ethyl group first, followed by installation of the benzyl group). The resulting alkyne can then be reduced via a dissolving metal reduction to give a *trans* alkene, which gives the desired epoxide upon treatment with a peroxy acid.

(c) Acetylene can be treated with NaNH$_2$ followed by ethyl iodide to install an ethyl group. This process is then repeated (with methyl iodide as the alkyl halide) to install a methyl group. These two alkylation processes could have been performed in reverse order (with installation of the methyl group first, followed by installation of the ethyl group). The resulting alkyne can then be reduced via hydrogenation with a poisoned

catalyst to give a *cis* alkene, which gives the desired epoxide upon treatment with a peroxy acid.

(d) Acetylene can be treated with NaNH$_2$ followed by ethyl iodide to install the first ethyl group. This process is then repeated (again with ethyl iodide as the alkyl halide) to install the second ethyl group. Notice that two alkylation processes are required, even though the same group is being installed on both sides of acetylene. The reagents (NaNH$_2$ followed by ethyl iodide) will install only one ethyl group. So these reagents must be repeated to install the second ethyl group (treating acetylene with two equivalents of NaNH$_2$ followed by two equivalents of ethyl iodide will NOT produce the desired internal alkyne). The resulting alkyne can then be reduced via a dissolving metal reduction to give a *trans* alkene, which gives the desired epoxide upon treatment with a peroxy acid.

14.42.

(a) Treating the alkene with a peroxy acid generates an epoxide. This epoxide has one chirality center, so we expect a racemic mixture of enantiomers (epoxide formation can occur on either face of the π bond with equal likelihood). Then, the epoxide is treated with methyl magnesium bromide, a strong nucleophile, resulting in a ring-opening reaction. Attack occurs at the less substituted position, giving an alkoxide ion, which is then protonated upon treatment with water to give a secondary alcohol as the product.

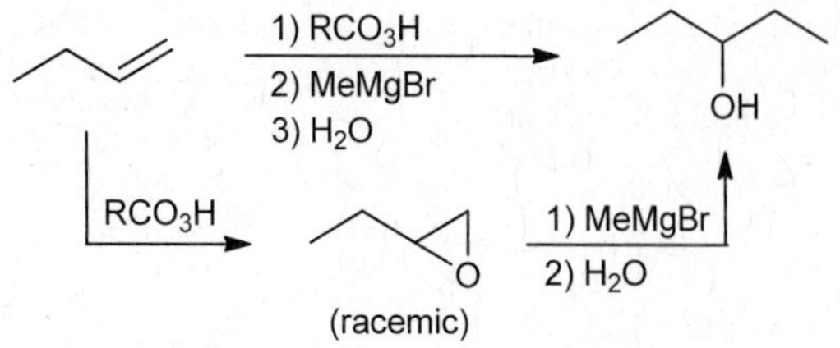

(b) The starting material is an alkene, and the reagents indicate an alkoxymercuration-demercuration, resulting in the Markovnikov addition of H and OMe across the π bond. One chirality center is generated in the process, so we expect a racemic mixture of enantiomers.

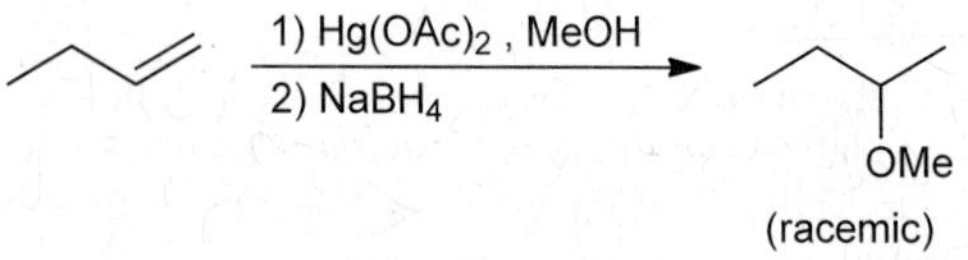

(c) Treating the alkene with a peroxy acid generates an epoxide. This epoxide has one chirality center, so we expect a racemic mixture of enantiomers (epoxide formation can occur on either face of the π bond with equal likelihood). Then, the epoxide is treated with a strong nucleophile (HS⁻), resulting in a ring-opening reaction. Nucleophilic attack occurs at the less substituted position, giving an alkoxide ion, which is then protonated upon treatment with water to give the product shown here.

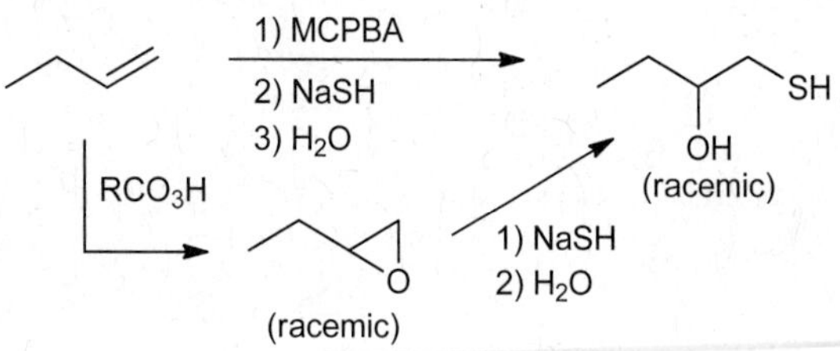

(d) Upon treatment with elemental sodium, an alcohol is deprotonated to give an alkoxide ion. The alkoxide then functions as a nucleophile when treated with ethyl chloride, giving an S_N2 reaction that affords the following ether:

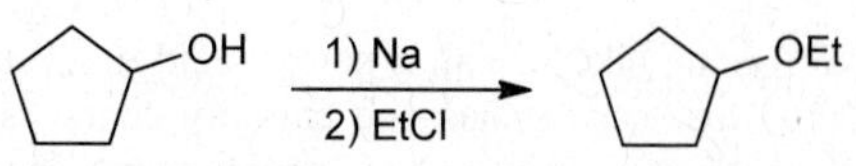

(e) Upon treatment with elemental sodium, an alcohol is deprotonated to give an alkoxide ion. The alkoxide then functions as a nucleophile and attacks the epoxide, giving a ring-opening reaction. The resulting alkoxide

ion is then protonated upon treatment with water to give the following product:

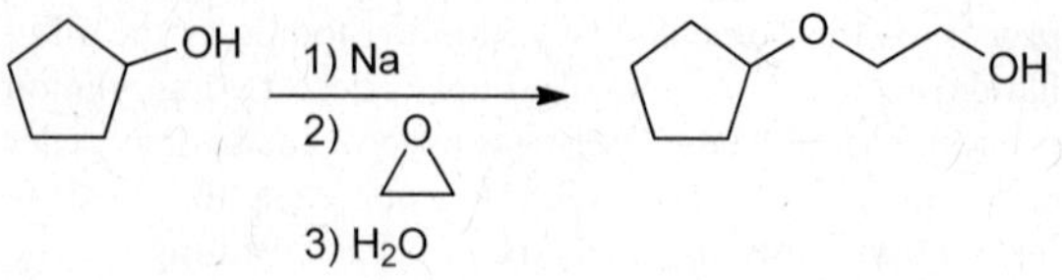

(f) Upon treatment with magnesium (Mg), the starting alkyl halide is converted into a Grignard reagent, which is a very strong nucleophile. When this Grignard reagent is treated with an epoxide, a ring-opening reaction occurs. The resulting alkoxide ion is protonated upon work-up with water to give the following product.

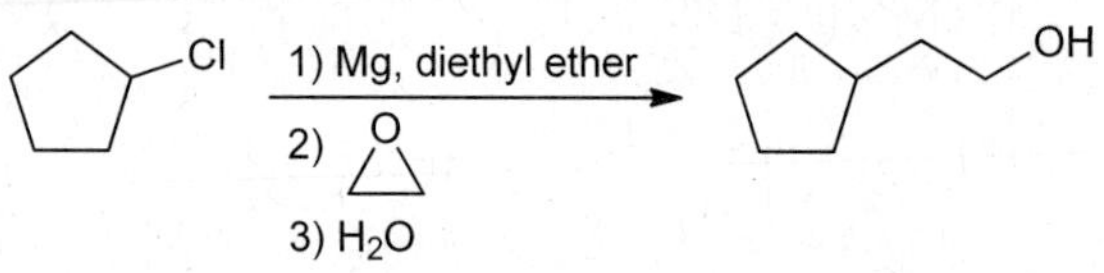

(g) Upon treatment with elemental sodium, an alcohol is deprotonated to give an alkoxide ion. The alkoxide then functions as a nucleophile and attacks the epoxide, giving a ring-opening reaction. Nucleophilic attack occurs at the less substituted position, giving another alkoxide ion, which is then protonated upon treatment with water to give the following product:

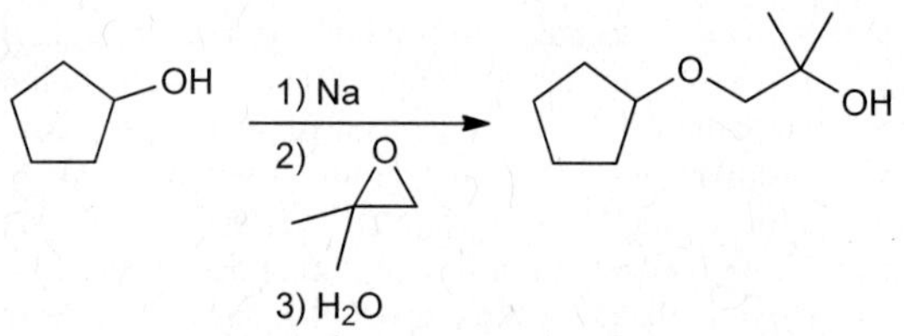

(h) Upon treatment with magnesium (Mg), the starting alkyl halide is converted into a Grignard reagent, which is a very strong nucleophile. When this Grignard reagent is treated with an epoxide, a ring-opening reaction occurs. Nucleophilic attack occurs at the less substituted position, giving an alkoxide ion, which is then protonated upon treatment with water to give the following product:

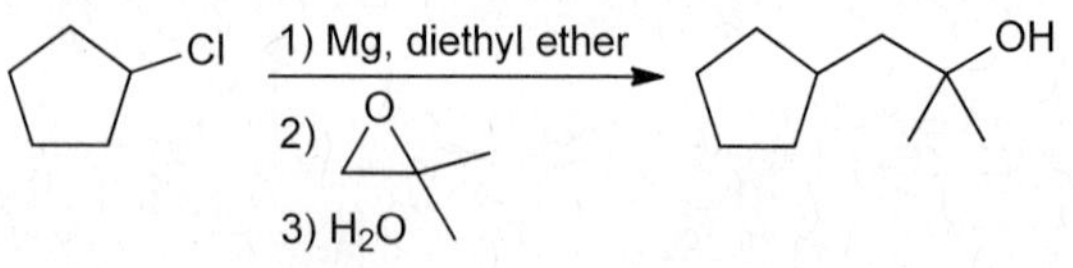

14.43
(a) Ethyl magnesium bromide is a strong nucleophile, and it will attack the epoxide at the less substituted position, giving a ring-opening reaction. The resulting alkoxide ion is then protonated upon treatment with water to give the tertiary alcohol as the product.

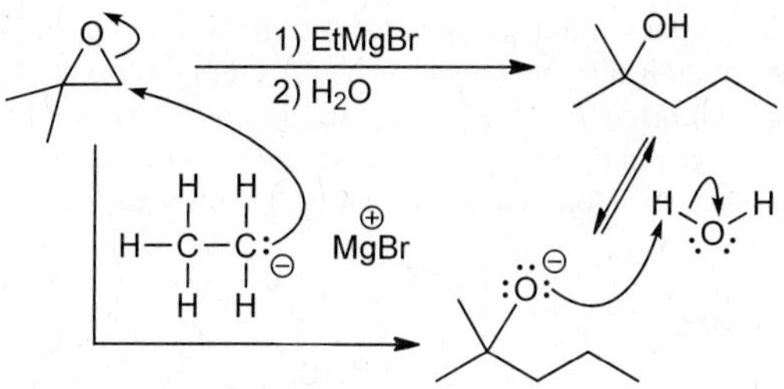

(b) Sodium hydride is a strong base, and it will deprotonate an alcohol to give an alkoxide ion. This alkoxide ion will then function as a nucleophile when treated with a primary alkyl halide, giving an S_N2 reaction, as shown here:

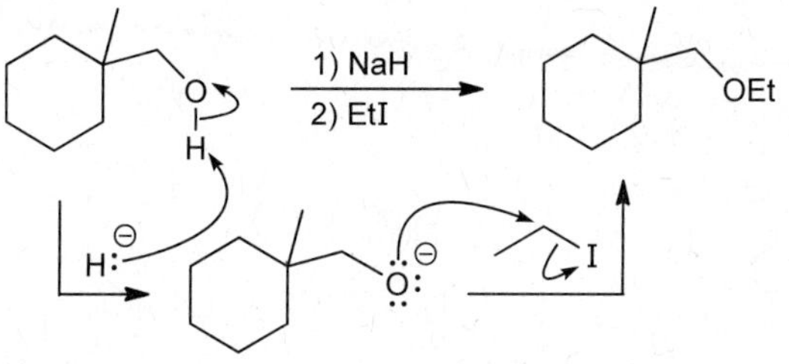

(c) The acetylide ion is a strong nucleophile and it will attack the epoxide at the less substituted position, giving a ring-opening reaction. The resulting alkoxide ion is then protonated upon treatment with water, as shown:

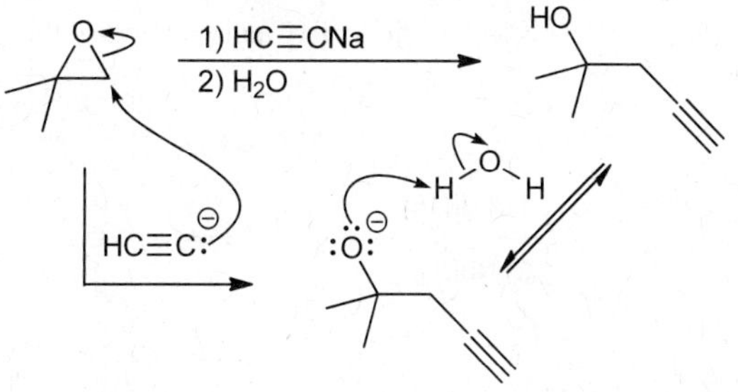

(d) Upon treatment with a strong acid, the epoxide is protonated. The resulting protonated epoxide is then attacked by MeSH (a good nucleophile). Nucleophilic attack occurs at the more substituted position, as a result of electronic effects (the tertiary position bears more partial carbocationic character than the primary position). In the final step, a proton is removed to give the product, shown here:

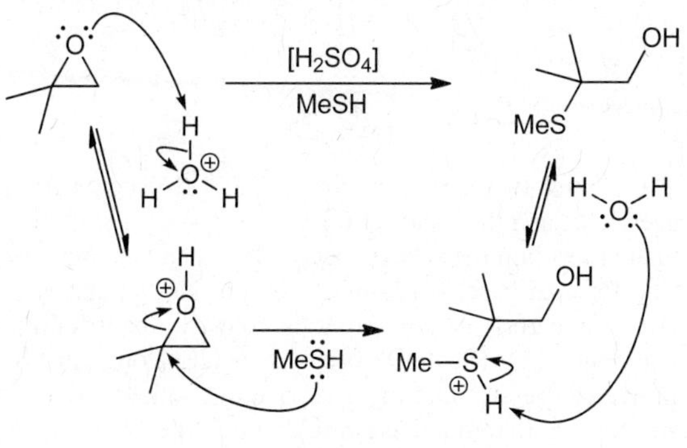

(e) Sodium hydride is a strong base, and it will deprotonate the alcohol to give an alkoxide ion. This alkoxide ion contains a built-in leaving group, so it can undergo an intramolecular S_N2-type process in which a chloride ion is ejected as a leaving group, giving the cyclic product shown:

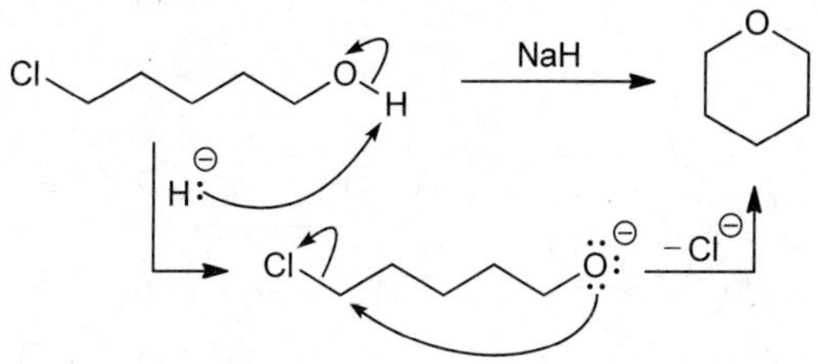

(f) Hydroxide functions as a nucleophile in an S_N2 reaction, ejecting chloride as a leaving group. The resulting alcohol is then deprotonated by hydroxide, giving an alkoxide ion. This alkoxide ion contains a built-in leaving group, so it can undergo an intramolecular S_N2-type process in which a chloride ion is ejected as a leaving group, giving the cyclic product shown:

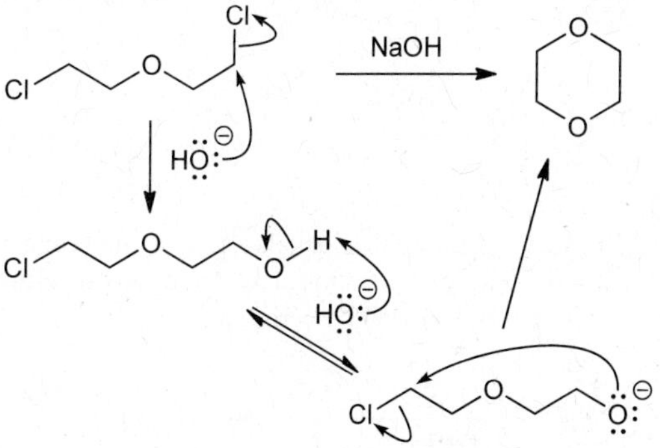

14.44. The starting material is a cyclic ether, in which the oxygen atom is connected to two carbon atoms, each of which is sp^3 hybridized. As such, treatment with HBr is expected to cleave each of the C–O bonds and replace them with C–Br bonds, giving the following dibromide.

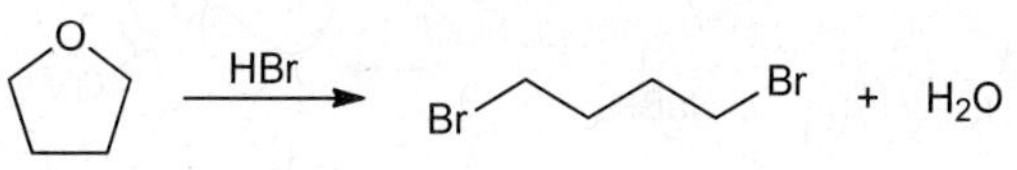

14.45. The starting material has six carbon atoms, and the product has six carbon atoms. So, the starting material must be some sort of cyclic ether, which opens to give a dibromide. By inspecting the dibromide product, we can determine which two carbon atoms must have been connected to the oxygen atom in the starting cyclic ether.

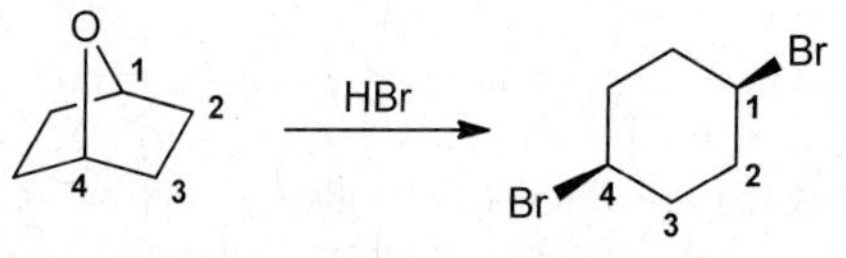

14.46. Ethyl magnesium bromide is a strong nucleophile, and it will attack the epoxide at the less substituted position, in a ring-opening process, with inversion of configuration at the position that is attacked. The resulting alkoxide ion can then undergo an intramolecular S_N2-type process, expelling a chloride ion and generating a new epoxide. This epoxide can be attacked once again by ethyl magnesium bromide, once again at the less substituted position, and once again with inversion of configuration. The resulting alkoxide ion is then protonated upon treatment with water, to give the observed product.

14.47.
(a) The starting material is an alkene, and the reagents indicate an alkoxymercuration-demercuration, resulting in the addition of H and OMe across the π bond.

(b) The starting material is an alkene, and the reagents indicate an alkoxymercuration-demercuration, resulting in the Markovnikov addition of H and OMe across the π bond.

(c) The starting material is an alkene, and the reagents indicate an alkoxymercuration-demercuration, resulting in the addition of H and OR across the π bond.

(d) The starting material is an alkene, and the reagents indicate an alkoxymercuration-demercuration, resulting in the Markovnikov addition of H and OR across the π bond.

14.48.
(a) There are certainly many ways to prepare the target compound. The following strategy represents just one possible synthetic approach.

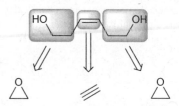

The central two carbon atoms (C3 and C4) come from acetylene, and the bonds at C2-C3 and C4-C5 are formed via the reaction between an acetylide ion and an epoxide. The forward process is shown below. Notice that after the first carbon-carbon bond is formed, the resulting compound is an alcohol. As such, the OH group must be protected before treating the terminal alkyne with a strong base (a strong base would simply deprotonate the alcohol, rather than the alkyne, if the OH group were not protected). The protecting group can be removed after the second carbon-carbon bond is formed. Finally, the desired product is obtained upon reduction of the alkyne via hydrogenation in the presence of a poisoned catalyst, to give the *cis* alkene.

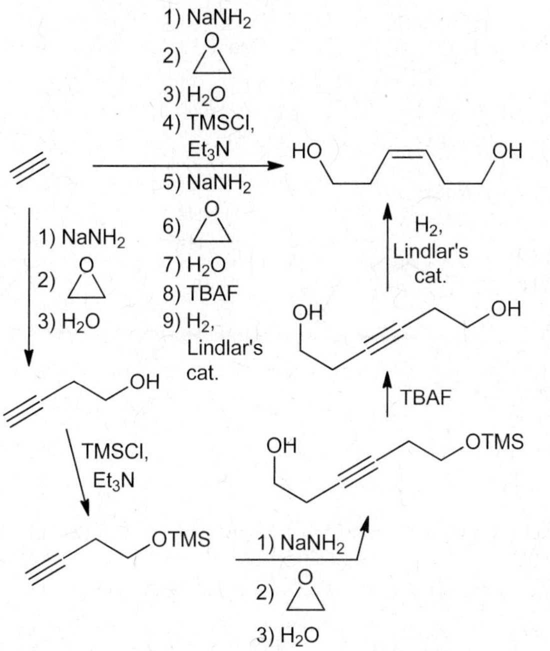

(b) There are certainly many ways to prepare the target compound. The following strategy represents just one possible synthetic approach.

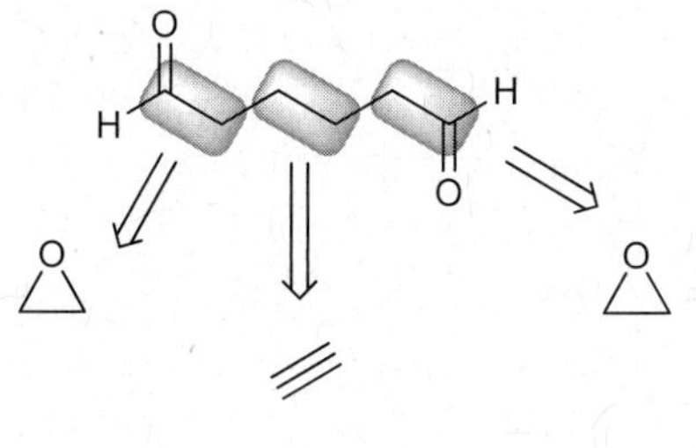

The central two carbon atoms (C3 and C4) come from acetylene, and the bonds at C2-C3 and C4-C5 are formed via the reaction between an acetylide ion and an epoxide. The forward process is shown below. Notice that after the first carbon-carbon bond is formed, the resulting compound is an alcohol. As such, the OH group must be protected before treating the terminal alkyne with a strong base (a strong base would simply deprotonate the alcohol, rather than the alkyne, if the OH group were not protected). The protecting group can be removed after the second carbon-carbon bond is formed. Finally, the desired product is obtained via hydrogenation of the alkyne followed by oxidation of the primary alcohol groups to aldehyde groups (using PCC as an oxidizing agent):

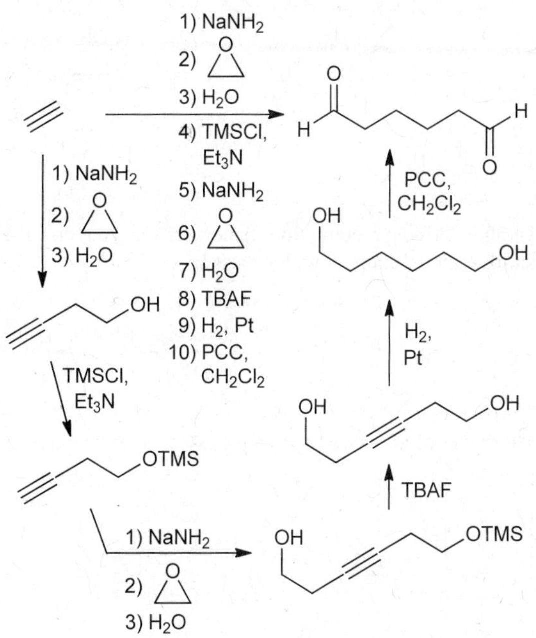

14.49. The following reagents can be used to achieve the desired transformations:

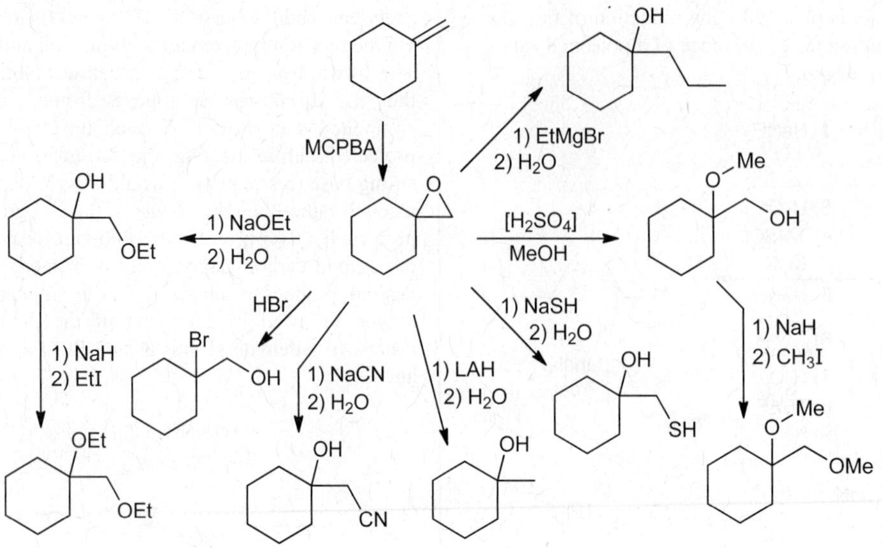

14.50. Alkoxymercuration-demercuration converts the alkene into an ether, which is then cleaved into two alkyl halides upon treatment with excess HI:

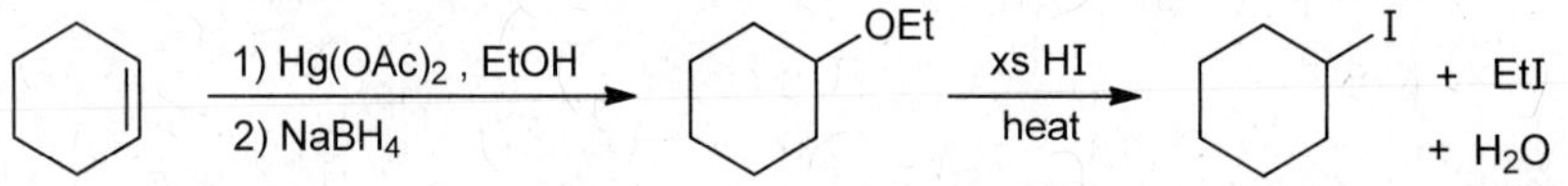

Treatment of the alkene with a peroxy acid results in an epoxide, which is a *meso* compound in this case:

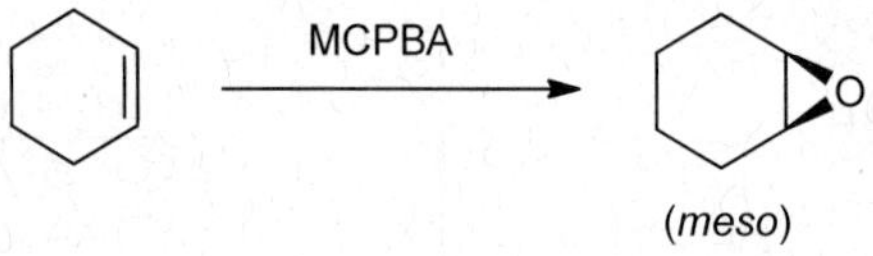

(*meso*)

The epoxide can be opened in the presence of a variety of nucleophiles, as shown:

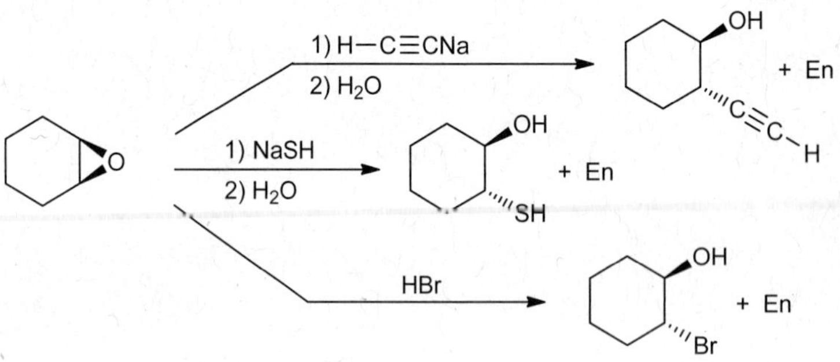

14.51.

(a) Treating the alkene with a peroxy acid gives an epoxide. This epoxide undergoes a ring-opening reaction when treated with methyl magnesium bromide (a strong nucleophile), to give an alkoxide ion which is protonated upon treatment with water to give the desired product.

1) MCPBA
2) MeMgBr
3) H_2O

MCPBA

(racemic)

1) MeMgBr
2) H_2O

(b) The following synthesis builds on the synthesis in the previous solution (**14.51a**). The product of that synthesis is treated with an oxidizing agent, such as chromic acid, to give the desired ketone:

1) MCPBA
2) MeMgBr
3) H_2O
4) $Na_2Cr_2O_7$
   H_2SO_4, H_2O

MCPBA

$Na_2Cr_2O_7$
H_2SO_4, H_2O

(racemic)

1) MeMgBr
2) H_2O

(c) Two carbon atoms are installed, with a functional group on the second carbon atom of the newly installed chain. This indicates a reaction involving a Grignard reagent and an epoxide. The starting alkyl halide is converted into a Grignard reagent, which is then treated with ethylene oxide, followed by water work-up. This installs the two carbon atoms, and simultaneously installs a functional group at the second carbon atom of the newly installed chain. This functional group (OH) is then converted into the desired functional group upon treatment with thionyl chloride and pyridine:

1) Mg
2) O (epoxide)
3) H_2O
4) $SOCl_2$, pyridine

1) Mg
2) O (epoxide)
3) H_2O

$SOCl_2$, py

(d) The solution to this problem is a slight modification of the solution to the previous problem (**14.51c**). The only change is the structure of the starting epoxide:

1) Mg
2) O (epoxide)
3) H_2O
4) $SOCl_2$, pyridine

1) Mg
2) O (epoxide)
3) H_2O

$SOCl_2$, py

(e) The starting material is an alcohol and the product is an ether. This transformation can be achieved via a Williamson ether synthesis. The alcohol is first deprotonated with a strong base (such as NaH) to give an alkoxide ion, which is then treated with ethyl iodide to give an S_N2 reaction (with iodide serving as the leaving group):

1) NaH
2) EtI

(f) Two carbon atoms are installed, with a functional group on the second carbon atom of the newly installed chain. This indicates a reaction involving an epoxide. The alcohol is first deprotonated with a strong base (such as NaH) to give an alkoxide ion, which is then treated with ethylene oxide to give a ring-opening reaction. The resulting alkoxide ion is protonated upon aqueous work-up to give the desired product:

1) NaH
2) O (epoxide)
3) H_2O

(g) The starting alkene will undergo hydroboration-oxidation (*anti*-Markovnikov addition of H and OH) to give the primary alcohol. A Williamson ether synthesis can then be used to convert the alcohol into the desired ether.

1) $BH_3 \cdot THF$
2) H_2O_2, NaOH
3) NaH
4) EtI

1) $BH_3 \cdot THF$
2) H_2O_2, NaOH

1) NaH
2) EtI

(h) The starting alkene will undergo acid-catalyzed hydration (Markovnikov addition of H and OH) to give the secondary alcohol. A Williamson ether synthesis can then be used to convert the alcohol into the desired ether.

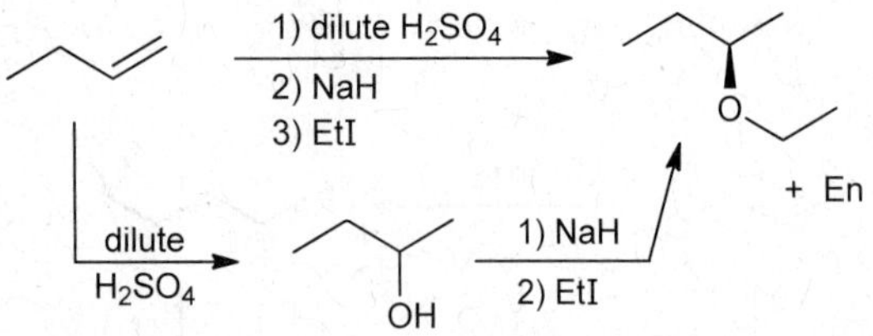

(i) Reduction of the alkyne in the presence of a poisoned catalyst affords an alkene, which is converted to the epoxide upon treatment with a peroxy acid.

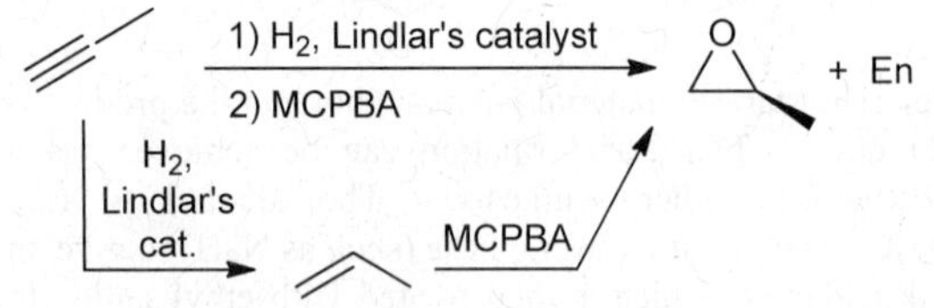

(j) Reduction of the alkyne in the presence of a poisoned catalyst affords a *cis*-alkene, which is converted to the desired epoxide upon treatment with a peroxy acid.

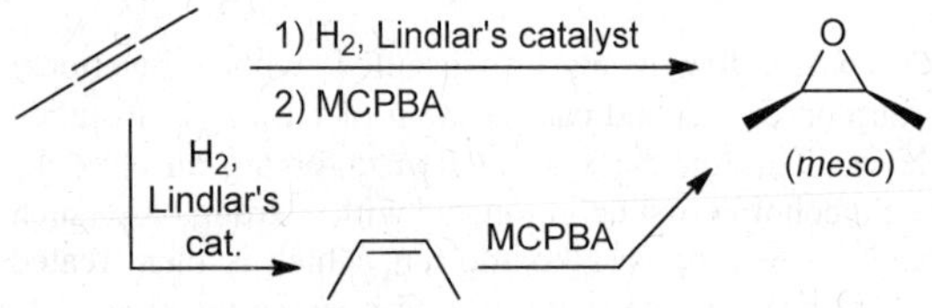

(k) Reduction of the alkyne via a dissolving metal reduction affords a *trans*-alkene, which is converted to the desired epoxide upon treatment with a peroxy acid.

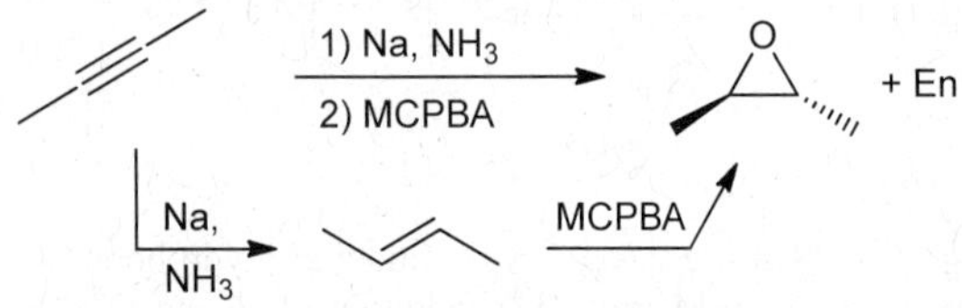

(l) The starting material has five carbon atoms and the product has seven carbon atoms. There are several ways to install two carbon atoms, but we must carefully consider where we want the functional group to be in the product. For example, if we try to alkylate the alkyne, we will obtain an unsymmetrical internal alkyne.

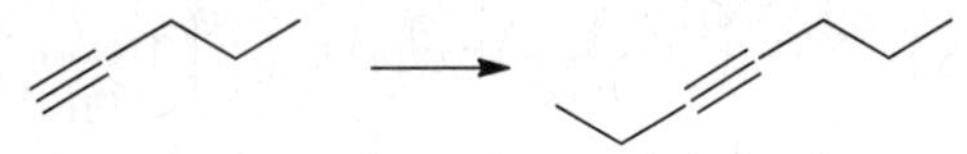

As such, we won't be able to control the regiochemical outcome of hydration of this internal alkyne (two products would be obtained, which is inefficient).
Instead, we can convert the alkyne into an epoxide, and then open the epoxide with a Grignard reagent. This strategy successfully installs the two carbon atoms while simultaneously installing a functional group in the desired location.

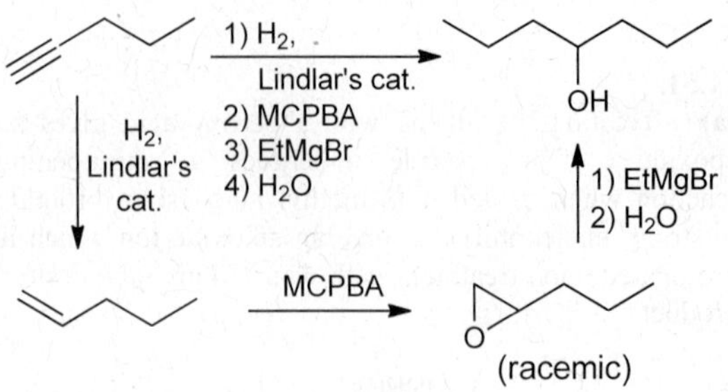

(m) We did not learn a way to alkylate an ether. However, we did learn a way to cleave a phenyl ether to give phenol. A Williamson ether synthesis can then be used to reinstall the alkyl group (this time an ethyl group, rather than a methyl group):

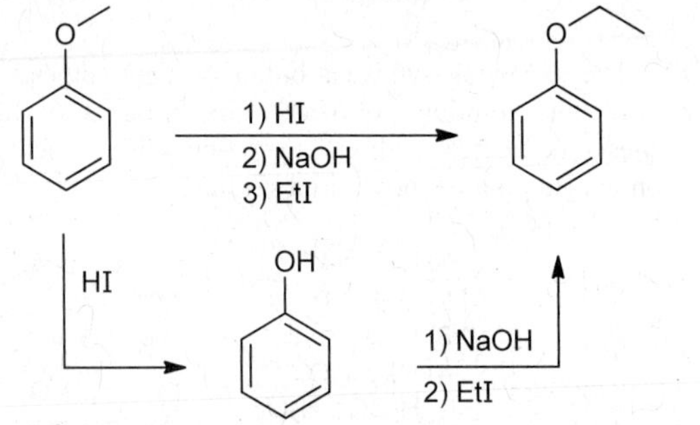

(n) The starting material is an ether, which will undergo cleavage when treated with HBr to give bromocyclohexane. Conversion of this alkyl halide into a Grignard reagent, followed by treatment with ethylene oxide (and aqueous work-up), gives the desired product.

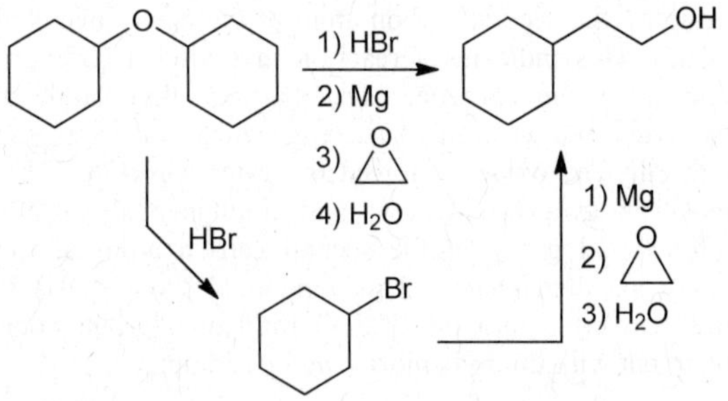

(o) Two carbon atoms are installed, with a functional group on the second carbon atom of the newly installed chain. This indicates a reaction involving an epoxide. The alcohol is first deprotonated with a strong base (such as NaH) to give an alkoxide ion, which is then treated with ethylene oxide to give a ring-opening reaction. The resulting alkoxide ion is protonated upon aqueous work-up to give the desired product:

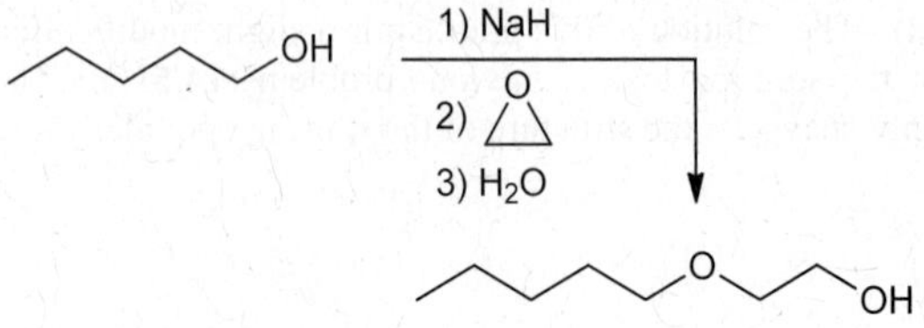

(p) Two carbon atoms are installed, with a functional group on the second carbon atom of the newly installed chain. This indicates a reaction involving an epoxide. The starting alkyl halide is converted into a Grignard reagent, which is then treated with ethylene oxide, followed by water work-up. This installs the two carbon atoms, and simultaneously installs the correct functional group in the correct location (at the second carbon atom of the newly installed chain):

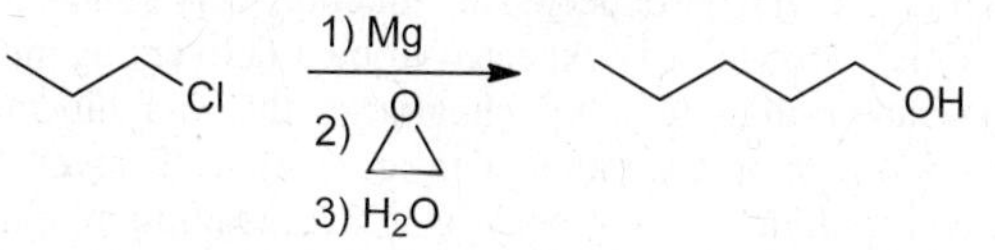

(q) Two carbon atoms are installed, with a functional group on the second carbon atom of the newly installed chain. This indicates a reaction involving an epoxide. The starting alkyl halide is converted into a Grignard reagent, which is then treated with ethylene oxide, followed by water work-up. This installs the two carbon atoms, and simultaneously installs a functional group in the correct location (at the second carbon atom of the newly installed chain). Oxidation of the primary alcohol (with PCC) gives the desired aldehyde:

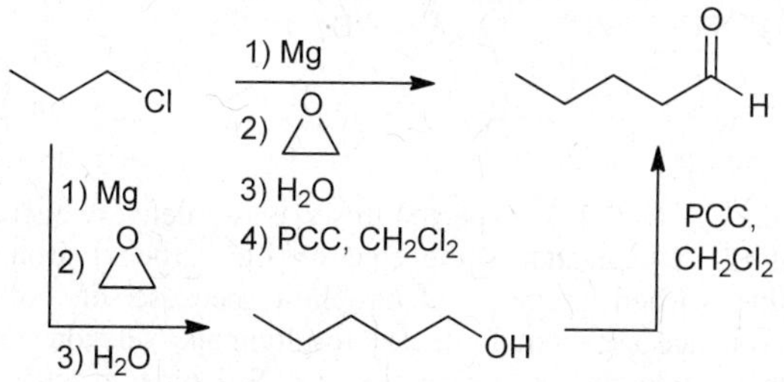

(r) This conversion can be achieved in one step, by treating the starting material with the epoxide shown, in the presence of acid catalysis. Under these conditions, the alcohol functions as a nucleophile and attacks a protonated epoxide to give a ring-opening reaction in which the nucleophile attacks the more substituted tertiary position (due to an electronic effect).

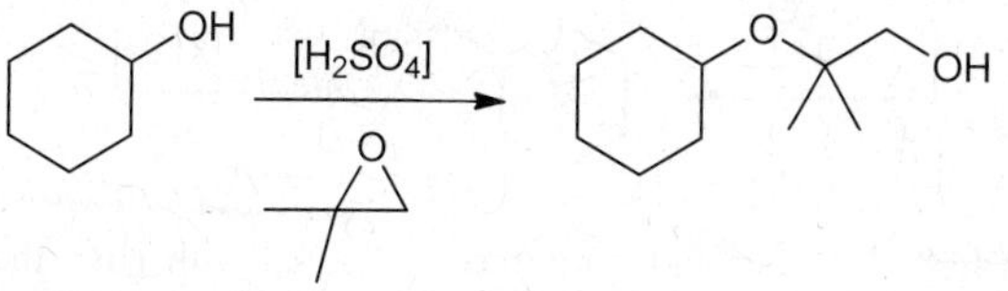

(s) This transformation is similar to the previous problem (**14.51r**), although in this case, the nucleophilic attack must occur at the less substituted position. This requires treating an epoxide with a strong nucleophile in basic conditions. The starting alcohol is first deprotonated with a strong base (such as NaH), and the resulting alkoxide ion is treated with the epoxide shown below.

The resulting ring-opening reaction, followed by aqueous work-up, gives the desired product.

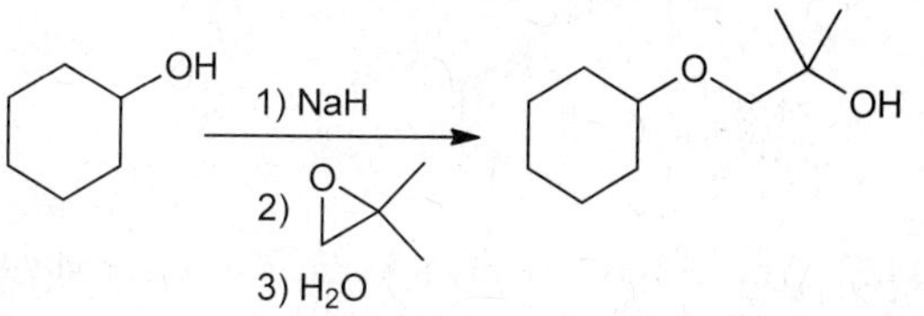

(t) The epoxide can be made from the following alkene, so we must find a way to make this alkene:

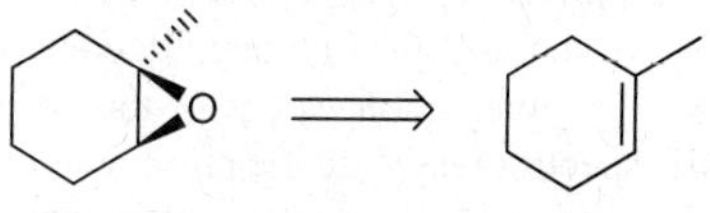

This alkene can be made from the starting alkane in two steps (radical bromination, followed by elimination), giving the following synthesis:

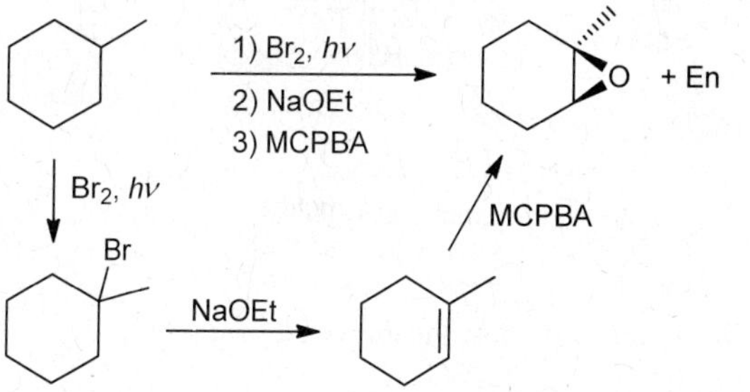

(u) The following synthesis builds on the synthesis in the previous solution (**14.51t**). The product of that synthesis is opened by a strong nucleophile (methoxide), which attacks the less substituted position. Aqueous work-up gives the desired product.

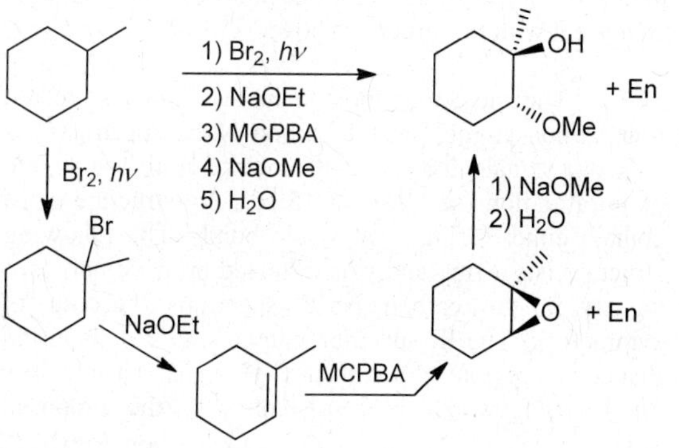

14.52. The molecular formula indicates that there are seven carbon atoms, but the spectrum has only five signals, indicating symmetry. With four degrees of unsaturation (see Section 15.16), we suspect an aromatic ring. There are four signals in the aromatic region of the spectrum, so we expect a monosubstituted ring, which explains the symmetry. The fifth signal appears above 50 ppm, indicating that it is next to an electronegative atom. We see in the molecular formula that there is an oxygen atom, so we propose the following structure, called methoxybenzene (also called anisole).

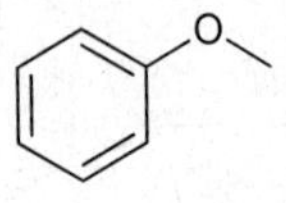

14.53. The ^{1}H NMR spectrum has four signals, and the total integration of those four signals is 2+2+2+3 = 9. However, the molecular formula indicates eighteen hydrogen atoms, so we expect a high degree of symmetry (one half of the molecule mirrors the other half). The compound has no degrees of unsaturation (see Section 15.16), so we expect an acyclic compound with no π bonds. The integration values of the signals in the proton NMR spectrum indicate the presence of three CH$_2$ groups and one CH$_3$ group, which appear to be connected to each other in a chain:

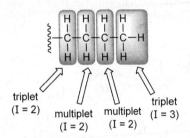

triplet (I = 2) multiplet (I = 2) multiplet (I = 2) triplet (I = 3)

We therefore propose the following structure:

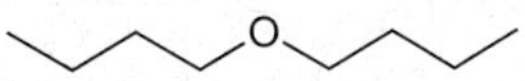

This structure is consistent with the carbon NMR spectrum as well. Notice that there are four different kinds of carbon atoms, thus giving rise to four signals. Only one of the four signals is above 50 ppm, indicating that it is next to an electronegative atom. This is consistent with the structure above.

14.54. The molecular formula indicates that there are four carbon atoms, but the carbon NMR spectrum has only two signals, indicating symmetry. With one degree of unsaturation (see Section 15.16), the structure must contain either a ring or a double bond. The following structure has a ring and would indeed produce only two signals in the carbon NMR spectrum (because of symmetry). The IR spectrum contains no signals in the diagnostic region (other than C-H signals just below 3000 cm^{-1}), which is consistent with the proposed structure.

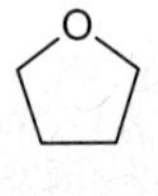

14.55. The spectrum has two signals, and the total integration of those two signals is 2+3 = 5. However, the molecular formula indicates ten hydrogen atoms, so we expect a high degree of symmetry (one half of the molecule mirrors the other half). The compound has no degrees of unsaturation (see Section 15.16), so we expect

an acyclic compound with no π bonds. The signals in the spectrum are consistent with an ethyl group (a quartet with an integration of 2, and a triplet with an integration of 3). The molecular formula indicates the presence of an oxygen atom, so we propose the following structure in which the two ethyl groups mirror each other:

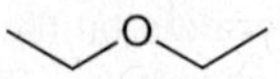

14.56. LiAlD$_4$ is expected to function very much like LiAlH$_4$. That is, it is expected to be a delivery agent of D$^-$ (rather than H$^-$), which attacks the less hindered position (secondary rather than tertiary), with inversion of configuration at that position. The resulting alkoxide ion is then protonated upon treatment with water, to give the product shown.

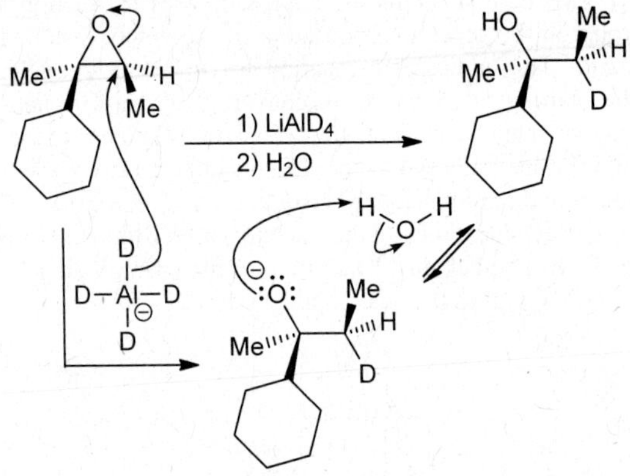

1) LiAlD$_4$
2) H$_2$O

14.57. NaBH$_4$ is expected to serve as a delivery agent of H$^-$, which attacks the electrophilic carbonyl group (that carbon atom is electrophilic because of both resonance and induction). The resulting alkoxide ion can then undergo an intramolecular S$_N$2-type reaction, expelling a halide as a leaving group, and generating the epoxide, as shown:

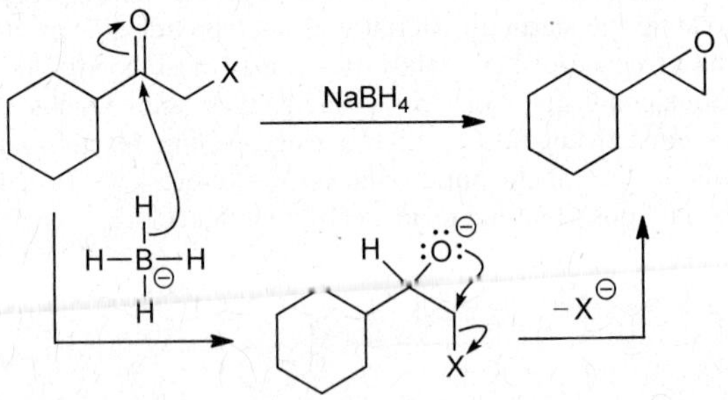

NaBH$_4$

14.58 When methyloxirane is treated with HBr, the regiochemical outcome is determined by a competition between steric and electronic factors, with steric factors prevailing – the Br is positioned at the less substituted position. However, when phenyloxirane is treated with HBr, electronic factors prevail in controlling the regiochemical outcome. Specifically, the position next to the phenyl group is a benzylic position and can stabilize a large partial positive charge. In such a case, electronic factors are more powerful than steric factors, and the Br is positioned at the more substituted position.

14.59. There are certainly many acceptable methods for achieving the desired transformation. The following retrosynthetic analysis represents one such method. An explanation of each of the steps (*a-e*) follows.

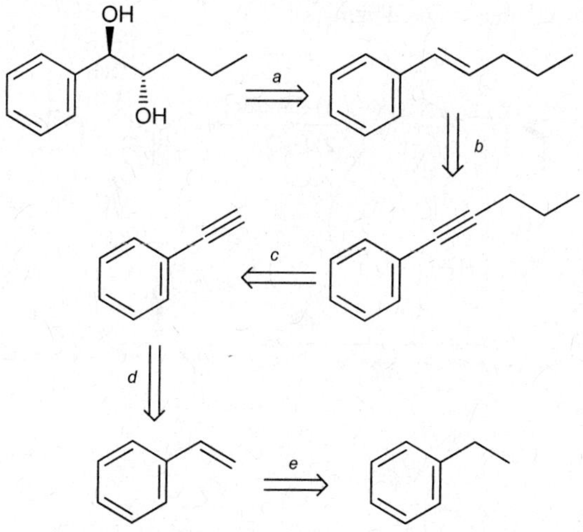

a. The diol can be made from the *trans* alkene, by converting the alkene into an epoxide and then opening under aqueous acidic conditions (or under basic conditions).

b. The alkene can be made from the corresponding alkyne via a dissolving metal reduction.

c. The alkyne can be made via alkylation of a terminal alkyne.

d. The terminal alkyne can be made from the corresponding alkene via bromination followed by elimination with $NaNH_2$.

e. The alkene can be made from the alkane via bromination, followed by elimination with a strong base, such as NaOEt.

Here is an alternative strategy. An explanation of each of the steps (*a-d*) follows.:

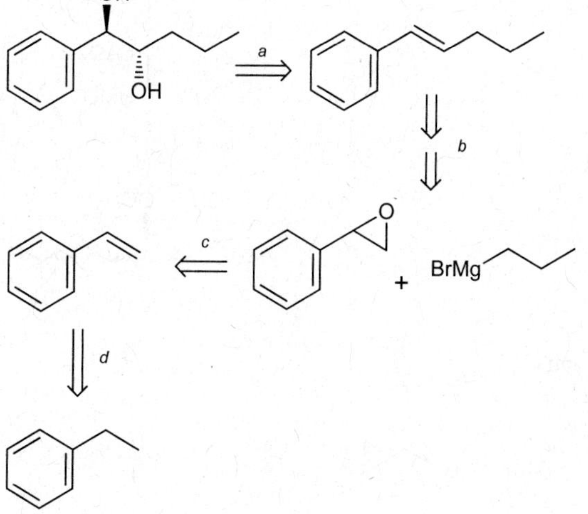

a. The diol can be made from the *trans* alkene, by converting the alkene into an epoxide and then opening under aqueous acidic conditions (or under basic conditions).

b. The alkene can be made from the corresponding alcohol, which can be made from the corresponding epoxide via a Grignard reaction.

c. The epoxide can be made from the corresponding alkene (called styrene)

d. The alkene can be made from the alkane via bromination, followed by elimination with a strong base, such as NaOEt.

The forward synthetic scheme for the second pathway is illustrated here:

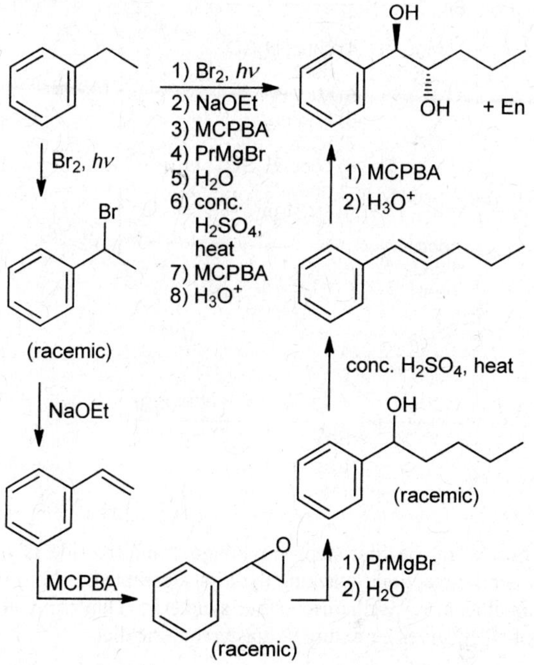

1) Br_2, *hv*
2) NaOEt
3) MCPBA
4) PrMgBr
5) H_2O
6) conc. H_2SO_4, heat
7) MCPBA
8) H_3O^+

Br_2, *hv*

(racemic)

NaOEt

MCPBA

(racemic)

1) PrMgBr
2) H_2O

(racemic)

conc. H_2SO_4, heat

1) MCPBA
2) H_3O^+

+ En

14.60. There are certainly many acceptable methods for achieving the desired transformation. The following retrosynthetic analysis represents one such method. An explanation of each of the steps (*a-e*) follows.

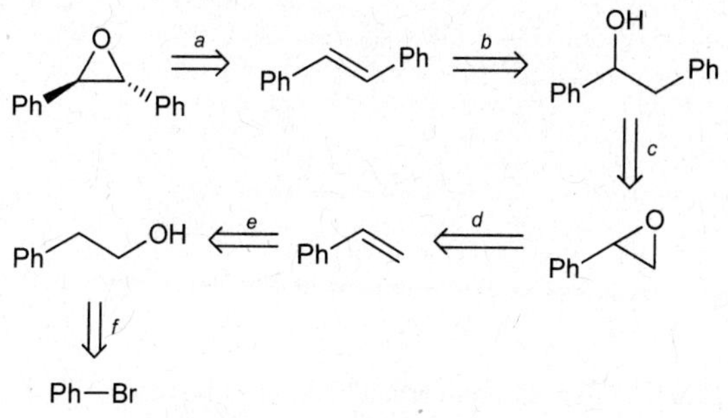

a. The epoxide can be made from the corresponding *trans* alkene.

b. The *trans* alkene can be made from the corresponding alcohol via a dehydration reaction (upon treatment with concentrated sulfuric acid).

c. The alcohol can be made from the reaction between a Grignard reagent (PhMgBr) and an epoxide, thereby installing a phenyl group.

d. The epoxide can be made from the corresponding alkene upon treatment with a peroxy acid.

e. The alkene can be made from the corresponding primary alcohol.

f. The primary alcohol can be made via Grignard reaction involving an epoxide.

The forward synthetic scheme is illustrated here:

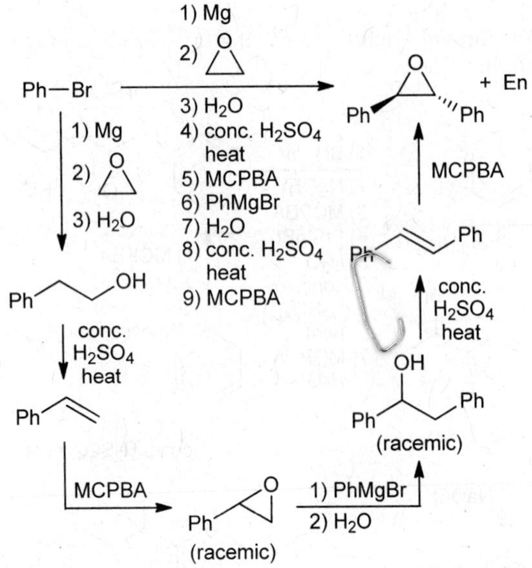

14.61   Since the Grignard reagent is both a strong base and a strong nucleophile, substitution and elimination can both occur. Indeed, they compete with each other. As we discussed in Chapter 8, elimination will be favored when the substrate is secondary. Both electrophilic positions in this epoxide are secondary, and so, elimination predominates:

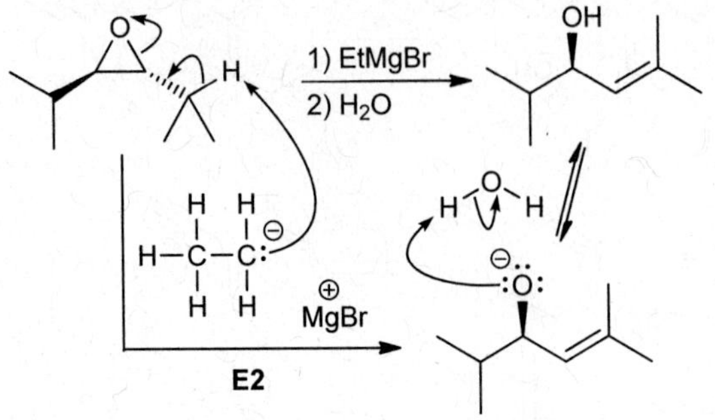

14.62. In the first step, vinylmagnesium bromide is a powerful nucleophile and can attack the epoxide in a ring-opening reaction (attacking the less substituted carbon) to afford an alcohol. This alcohol is subsequently converted to an ether via a Williamson ether synthesis. Dihydroxylation of the terminal alkene with catalytic OsO_4 in the presence of NMO gives a mixture of diastereomeric diols.

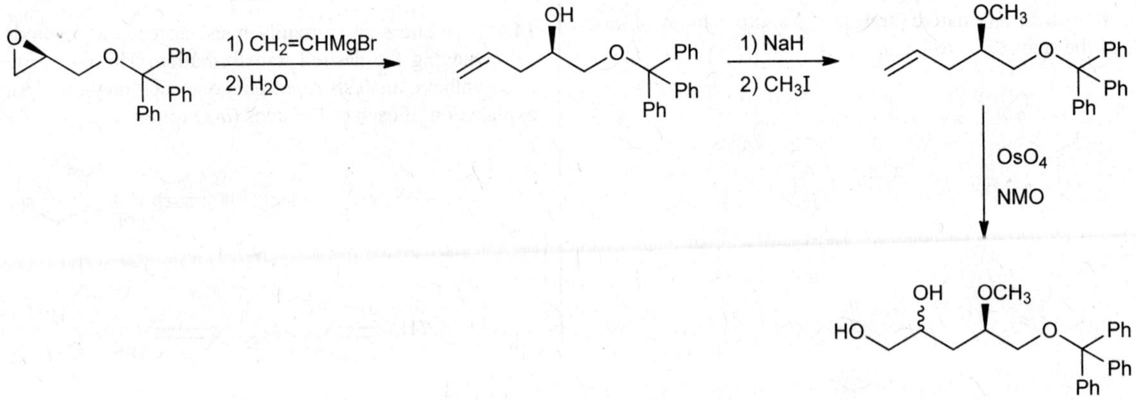

14.63. A new carbon-carbon bond must be made between C6 and C7:

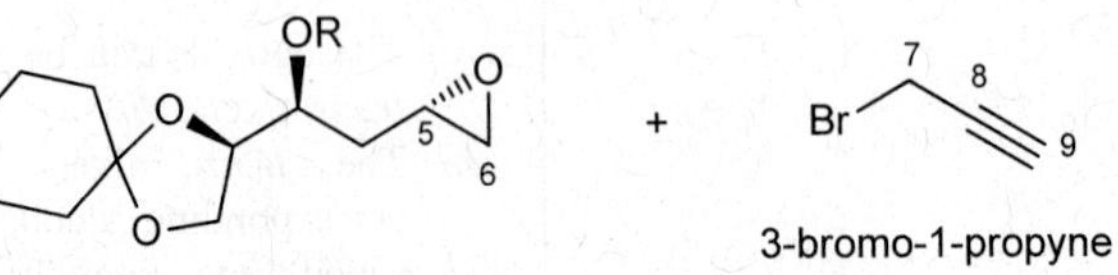

3-bromo-1-propyne

Epoxides are electrophilic functional groups and are subject to attack by a nucleophile. In order for the OH group to be ultimately positioned at C5, the nucleophile must attack the less substituted side of the epoxide (C6), which requires basic conditions (rather than acidic conditions, which often favors attack at the more substituted carbon).

The nucleophile for this reaction must be made from 3-bromo-1-propyne, which can be achieved by treatment with magnesium, thereby forming the Grignard reagent:

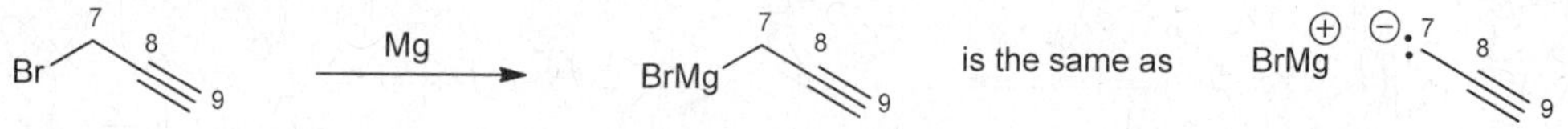

The desired transformation can therefore be achieved in the following way:

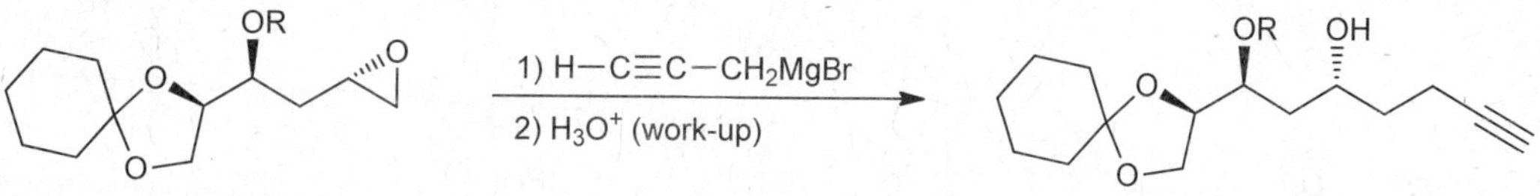

1) H—C≡C—CH₂MgBr
2) H₃O⁺ (work-up)

14.64.
(a) If we compare the starting material and the product, two changes become apparent: 1) the epoxide is converted to a *trans* diol, and 2) the OH group is converted to SPh:

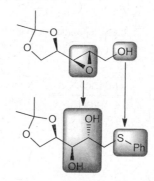

For the conversion of the epoxide, it might be tempting to suggest that the epoxide is directly opened by hydroxide to give a *trans* diol:

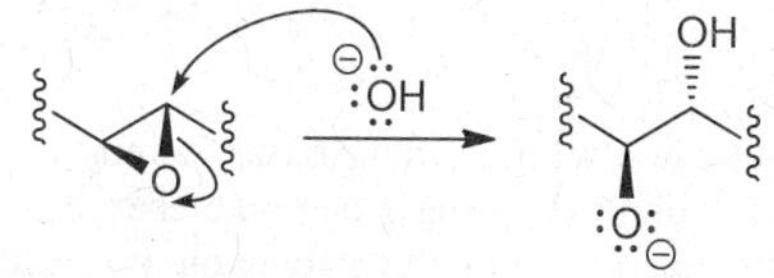

After all, this process would be expected to occur via back-side attack of the epoxide, ensuring the formation of a *trans* diol, rather than a *cis* diol. However, it does not explain why the other side of the epoxide is not attacked:

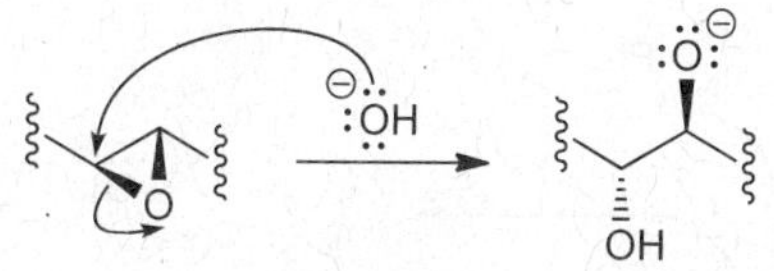

Our mechanism must somehow explain why only one side of the epoxide is attacked. Direct attack on the epoxide does not explain this observation. There is yet another curiosity that must also be explained by our mechanism. Specifically, the replacement of OH for SPh cannot occur via S$_N$2, since hydroxide is a bad leaving group:

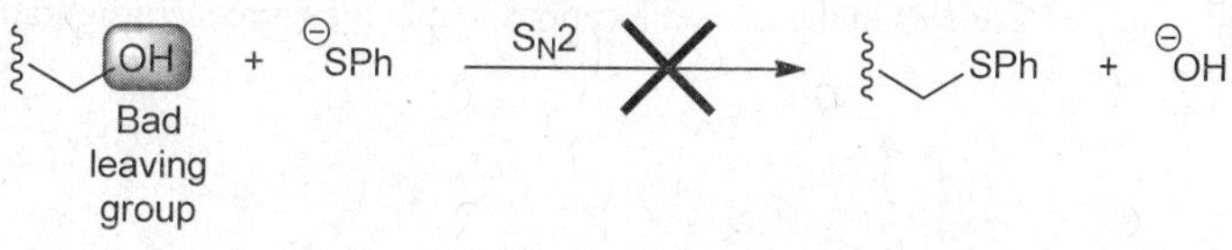

OH + ⁻SPh → [S$_N$2 ✗] → SPh + ⁻OH
Bad
leaving
group

Both of the issues described above can be resolved with the following mechanism, which explains the observed stereochemical outcome. The hydroxyl group is first deprotonated to give an alkoxide ion, which functions as a nucleophile and participates in an intramolecular S_N2-type process in which the epoxide is opened and a new epoxide is formed (notice that back-side attack places the oxygen atom of the new epoxide on a dash). After protonation of the resulting alkoxide, the epoxide group can then be opened with a thiophenolate ion (PhS⁻) to give another alkoxide, which is then protonated to give the observed product.

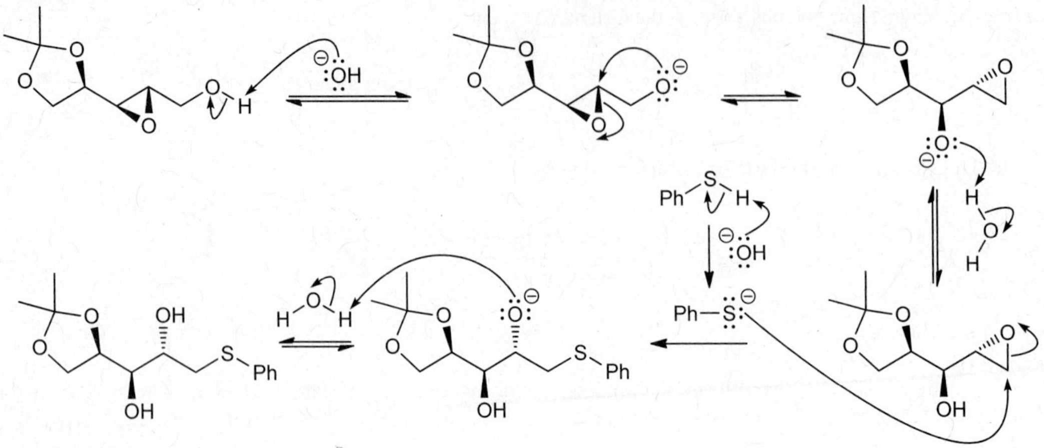

(b) The allylic alcohol undergoes a Sharpless asymmetric epoxidation to give the following epoxide, which then undergoes the reaction explored in part (a).

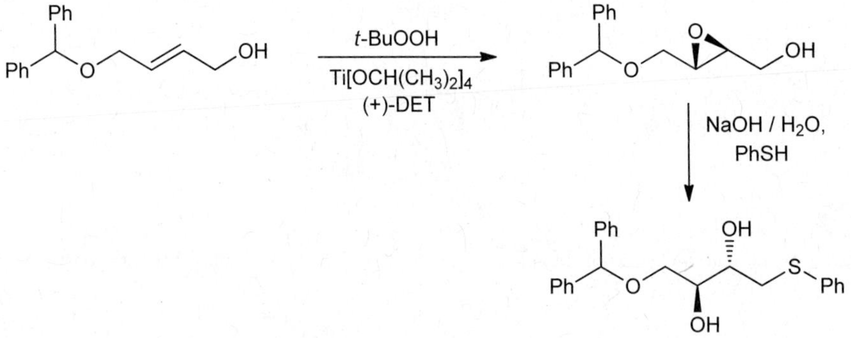

14.65. In the first step, compound **1** is treated with a strong base, thereby deprotonating one of the methyl groups, forming a sulfur ylide (a net-neutral compound exhibiting a negative charge on a carbon atom, connected to a sulfur atom with a positive charge). The carbanion of the ylide can function as a nucleophile and attack a carbonyl group, much like a Grignard reagent would. The resulting tetrahedral intermediate is not stable and will readily decompose to form the epoxide, via the loss of dimethyl sulfide. This is accomplished via an intramolecular S_N2-type reaction whereby the oxygen anion displaces the positively charged sulfur, expelling DMS as a leaving group.

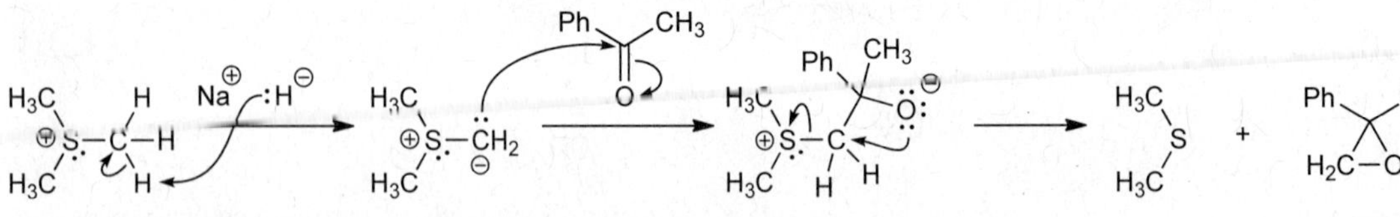

14.66. Let's begin by drawing the *trans*-decalin system in a way that illustrates the chair conformation of each six-membered ring (as instructed by the problem statement). Assigning locants can be helpful in this situation, since they make it easier to place all of the substituents in the correct locations and in the correct configurations:

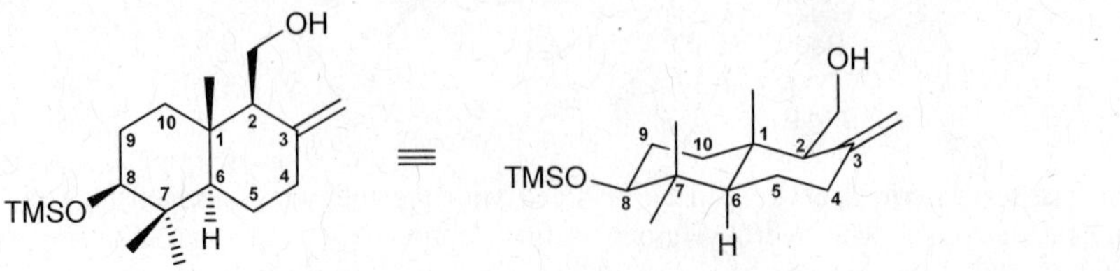

The *trans*-decalin structure imposes structural rigidity and limits the conformational freedom available to the compound.

Now let's consider the two faces of the π bond (the top face and the bottom face):

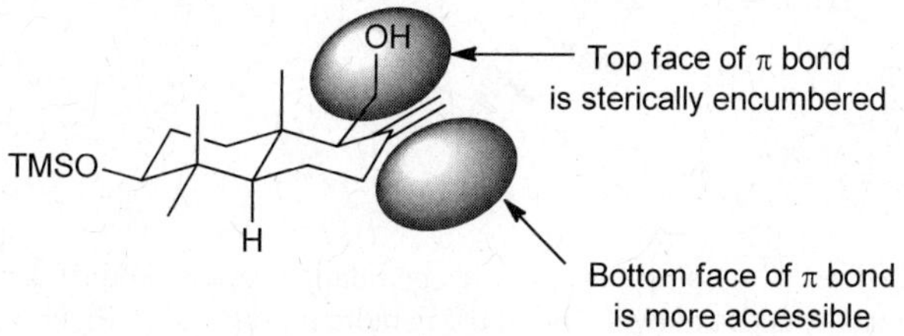

These two faces of the π bond are not equally accessible. That is, the compound is confined to a conformation in which one face of the π bond is more sterically encumbered than the other. This steric consideration is difficult to see in the drawing above, and can be visualized more clearly if we draw a Newman projection, looking down the C2-C3 bond.

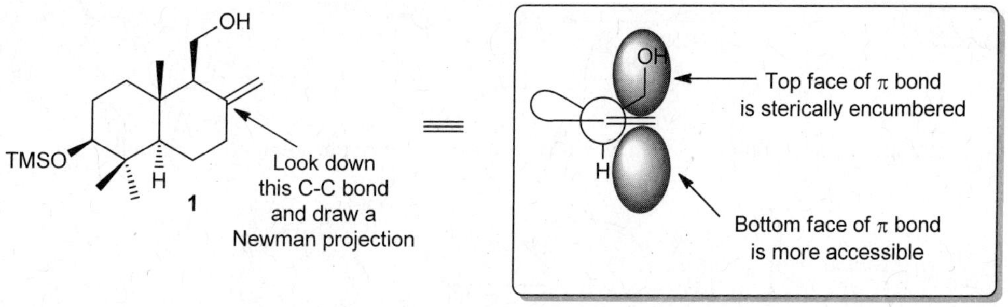

Since the bottom face of the π bond is more accessible, the epoxidation process will occur on that face, giving the following epoxide:

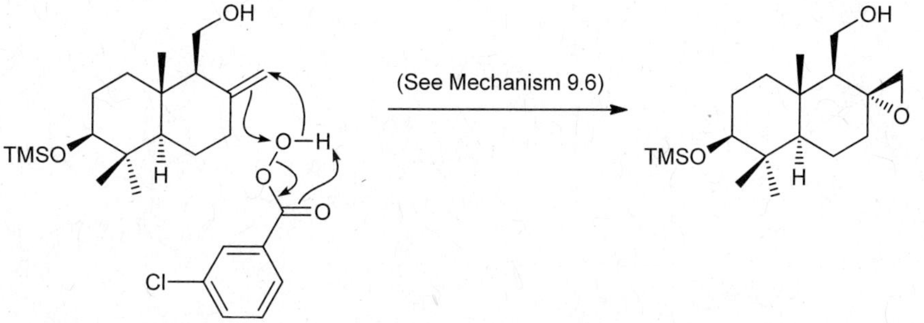

(See Mechanism 9.6)

This epoxide is then opened with LAH, and under these conditions, it is likely that the primary OH group will be deprotonated, giving a dianion:

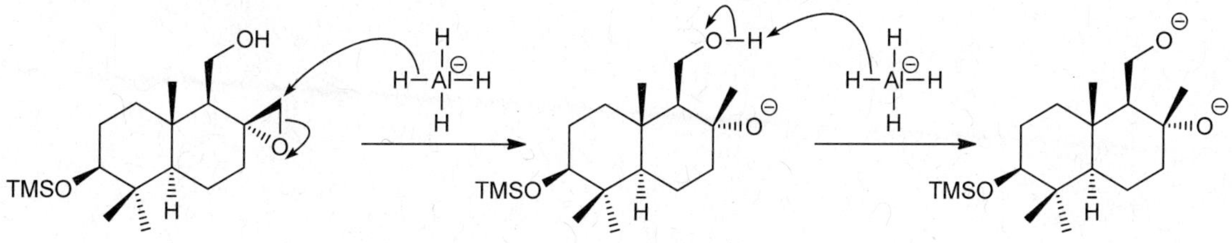

This dianion is then protonated upon aqueous work-up to give the following diol:

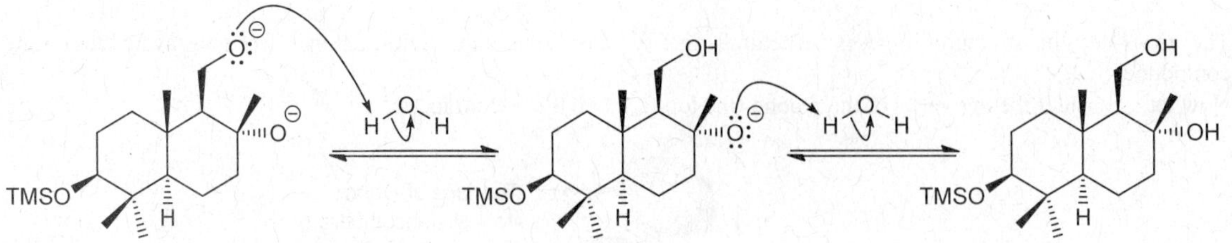

14.67.

(a) The conversion of **1** to **2** utilizes a Sharpless asymmetric epoxidation, so the product is expected to be an epoxide. The stereochemical outcome can be deduced by applying the paradigm provided in Figure 14.3. Since the chiral catalyst was formed using (–)-DIPT, which affords the same stereochemical outcome as (–)-DET, we can conclude that the epoxide forms "below the plane" of the alkene:

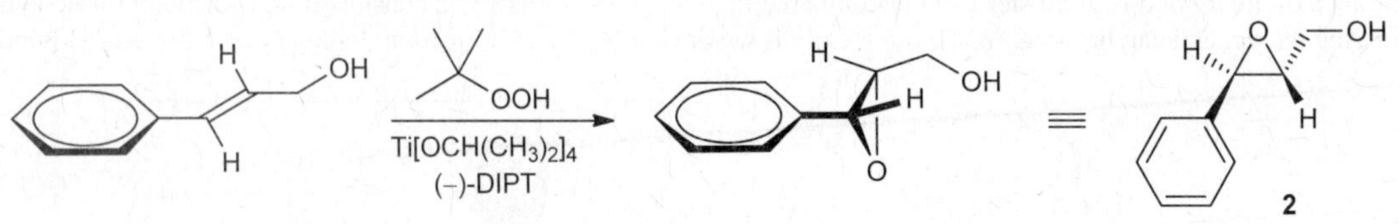

(b) First consider the reagents. Sodium hydroxide will deprotonate 2-ethoxyphenol to give a phenolate ion.

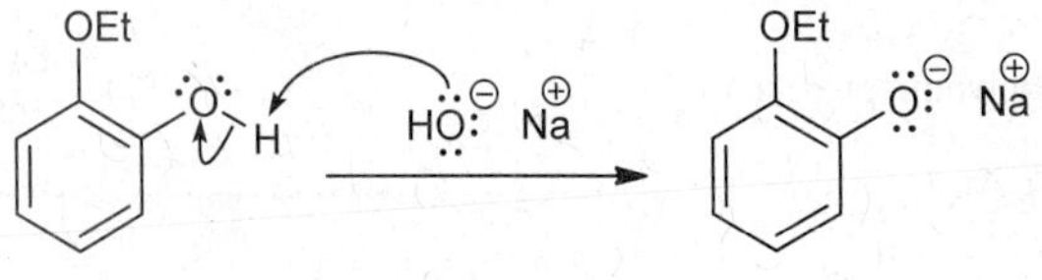

The phenolate ion can serve as a nucleophile and attack epoxide **2** in a ring-opening reaction. Nucleophilic attack occurs at the benzylic position (the carbon atom attached to the benzene ring) to form an alkoxide intermediate, that is protonated to give diol **3**.

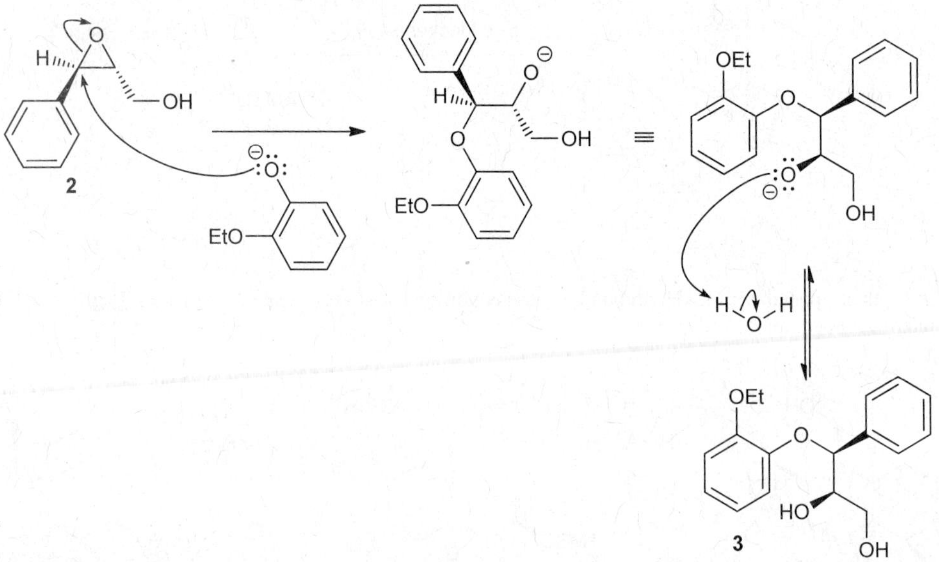

(c) Treatment of **3** with TMSCl in Et$_3$N results in the selective protection of the less sterically hindered primary alcohol (rather than the secondary alcohol) to give **6**. The secondary OH group of **6** is then converted to the mesylate with MsCl, thereby converting it into a good leaving group, as seen in compound **7**. Subjecting **7** to aqueous acid results in the removal of the silyl protecting group to form **4** in which the primary alcohol group is revealed. The net result of these three reactions is the selective mesylation of the secondary OH group.

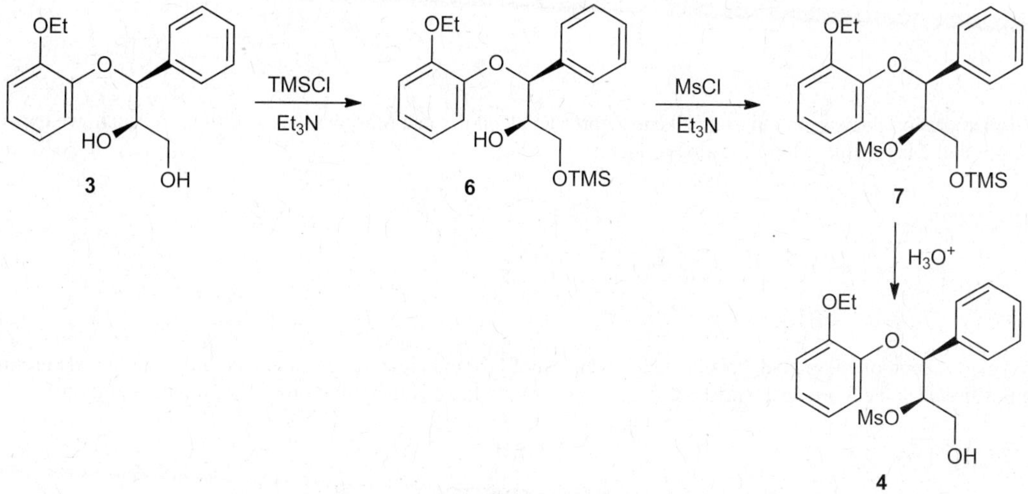

(d) Treatment of **4** with aqueous NaOH results in deprotonation of the primary alcohol to give alkoxide **8**, a strong nucleophile. This facilitates an intramolecular ring-forming reaction, with inversion of configuration at the carbon atom bearing the mesylate leaving group, to give compound **5**.

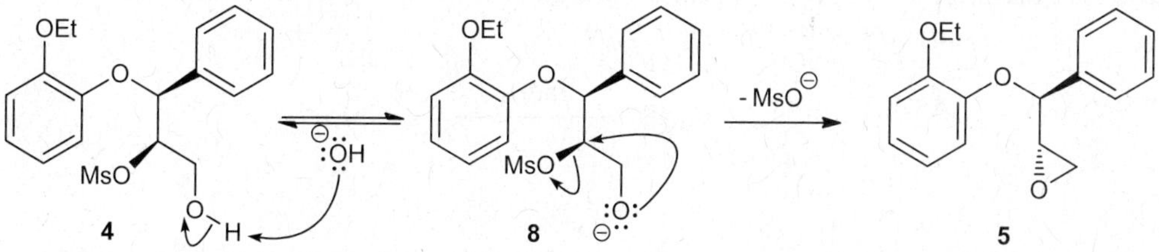

14.68. When compound **1** is treated with an electrophilic source of bromine, the alkene will readily react to form bromonium ion **3**. At this stage there are two possibilities for further reactivity. Typically, with an OH group present in the compound, the OH group can function as a nucleophile and attack the bromonium ion to form an ether. For example, consider path A. If this were to occur, an 8- or 9-membered ring would be formed, depending on which side of the bromonium ion is attacked by the alcohol. However, since we do not see the formation of a medium-sized ring in the product, a different reaction pathway must be occurring. Consider the lone pairs on the oxygen atom which is part of the 4-membered cyclic ether. If one of these lone pairs were to attack the bromonium ion (path B), the result would be a 5-membered ring fused to the 4-membered ring (compound **4**). However, this intermediate is not stable – the ring oxygen has 3 bonds and is positively charged. At this stage, if the free OH group reacts in an S_N2 fashion to open the 4-membered ring, followed by the loss of a proton, the product formed will have both the 5-membered cyclic ether and the epoxide:

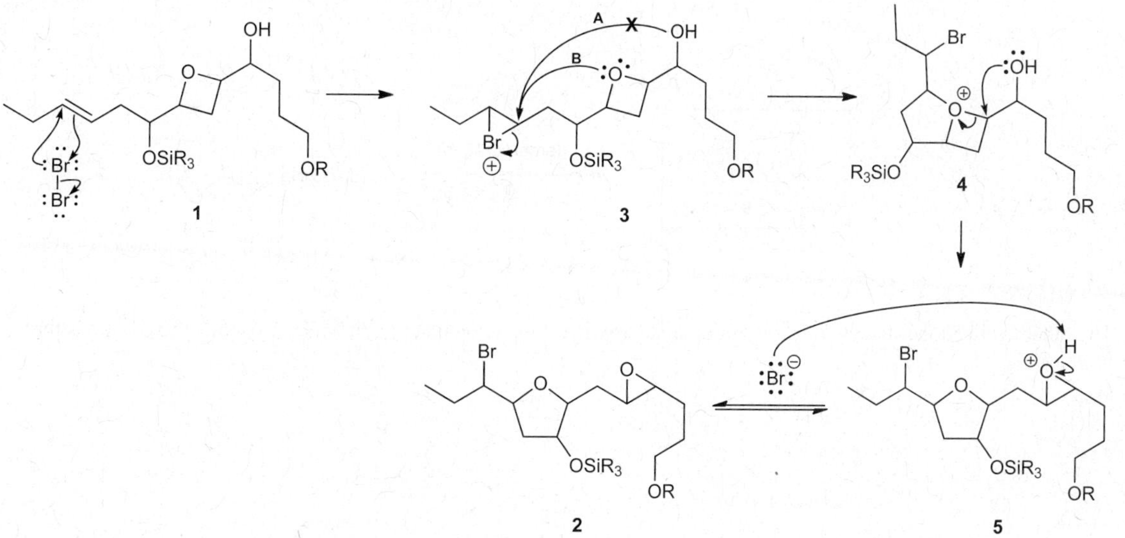

14.69.

(a) Oxymercuration-demercuration of an alkene affords an alcohol, via a Markovnikov addition. As such, we expect that compound **2** is the following hydration product:

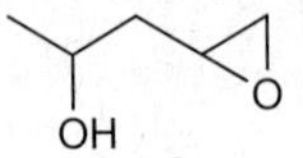

2

(b) Oxymercuration of compound **3** results in the addition of OH and HgOAc across the π bond, with the OH group being positioned at the more substituted position. Thus, compound **4** is the initial product of oxymercuration:

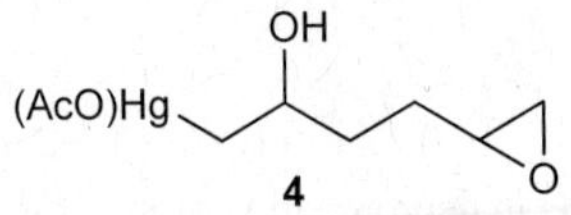

4

This intermediate can theoretically be converted into the Markovnikov hydration product via demercuration.

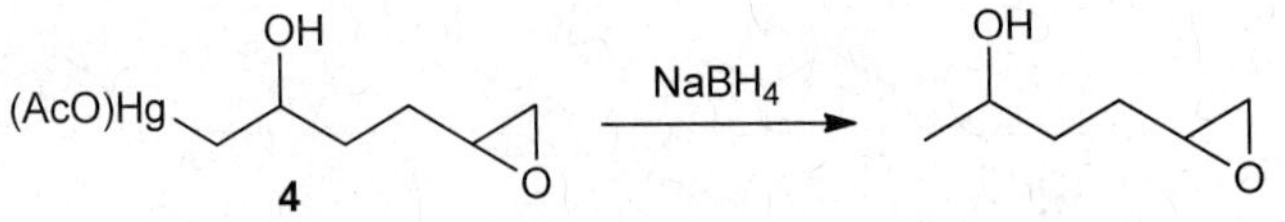

However, before NaBH₄ is introduced, **4** can also continue to react under the oxymercuration conditions, where Hg(OAc)₂ can behave as a Lewis acid. As shown, the Hg(OAc)₂ can interact with the oxygen atom of the epoxide group, much like an epoxide interacts with a proton, to facilitate the intramolecular nucleophilic attack by the nearby hydroxyl group. This intramolecular cyclization reaction forms a five-membered ring. A final proton transfer step, followed by the demercuration step, results in one of the major cyclization products, **8**.

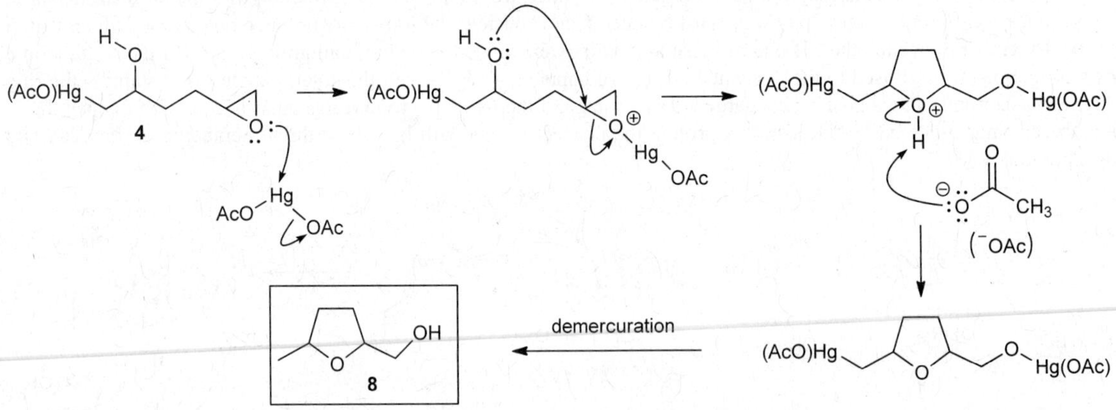

The other cyclization product, **7**, is a result of an alternative cyclization reaction in which a six-membered ring is the product:

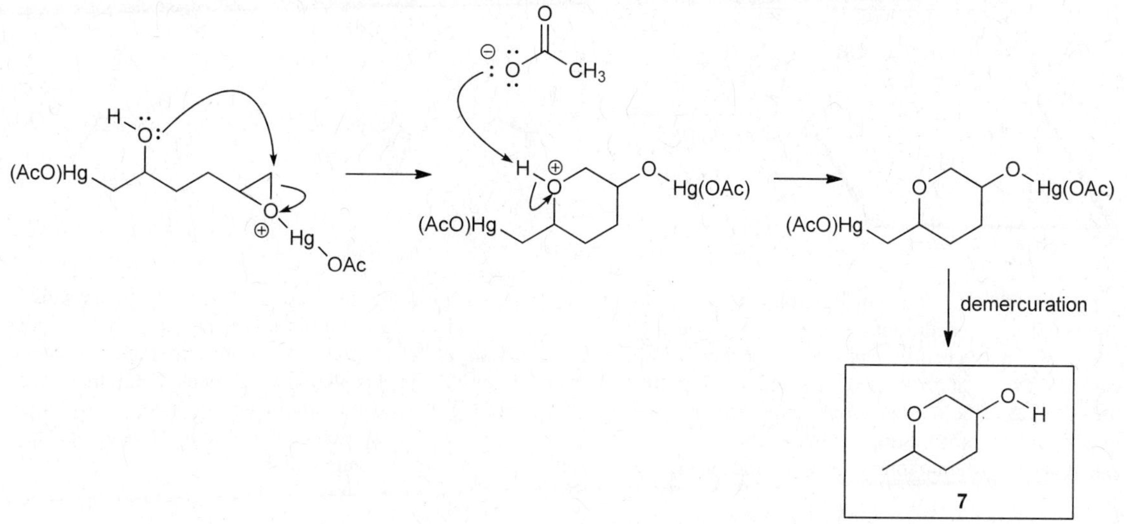

7

(c) Oxymercuration-demercuration of **1** can only result in a cyclic product if a four-membered ring is formed. Four-membered rings exhibit significant ring strain so their formation is slow under these conditions.

14.70. In the first step, iodine (the electrophile) adds to the terminal alkene to form an iodonium ion. Once formed, intramolecular attack of this intermediate by the pendant methoxy group occurs, resulting in the formation of a 6-membered ring. This intermediate, however, is not stable and is quickly trapped by the iodide ion, resulting in ring opening to form a stable product. In the second step, phenylsulfide displaces the iodide in an S_N2 fashion, followed by deprotonation of the resulting intermediate using triethylamine as a base.

Step 1

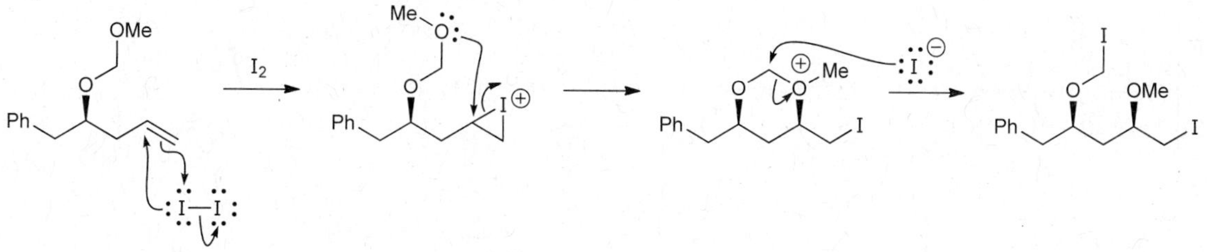

Step 2

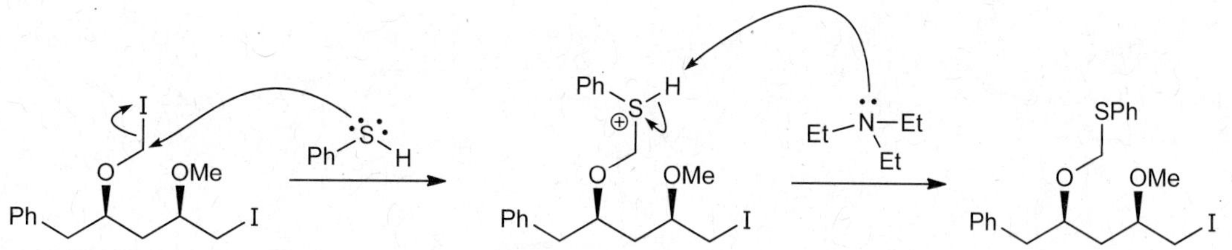

14.71.
(a) Epoxide **2** can be generated from allylic alcohol **1** via a Sharpless asymmetric epoxidation. The reagents for this reaction are: *tert*-butyl hydroperoxide, titanium tetraisopropoxide, and one enantiomer of diethyl tartrate (DET), depending on which epoxide enantiomer is required. In Figure 14.3, we saw a predictive tool for determining which enantiomer of DET would be required to afford epoxide **2**. Allylic alcohol **1** is orientated so that the allylic hydroxyl group appears in the upper right corner, and then we can see that (−)-DET is required in order for the epoxide ring to be formed on the bottom face of the molecule, corresponding to compound **2**.

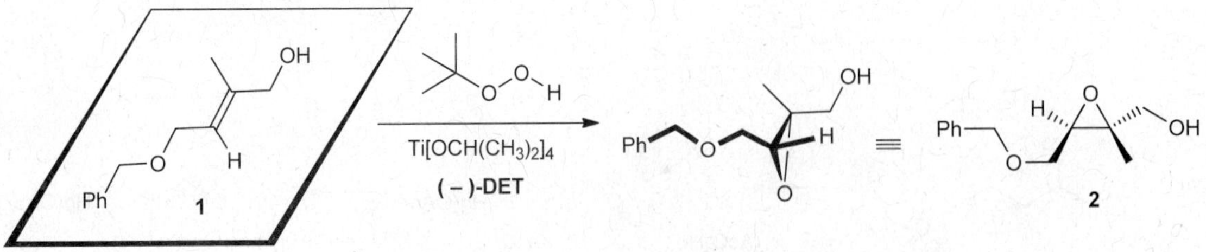

(b) In the first step, the epoxide group in compound **3** is protonated to form intermediate **5**. Notice that the epoxide is unsymmetrical; on the left side, the epoxide carbon is secondary, and on the right side, it is tertiary. Under acidic conditions, and when the epoxide has a tertiary carbon, the dominant effect is electronic (more important than steric considerations), and a nucleophile will attack at this site (Section 14.10). In this example, the nucleophile that opens the epoxide is the oxygen atom of the pendant ester group, generating a resonance-stabilized intermediate (**6**) which exhibits a 5-membered ring. At this stage, the alcohol on the 2° carbon, which was generated from the epoxide opening, can attack to generate intermediate **7**. And finally, deprotonation of **7** generates compound **4**.

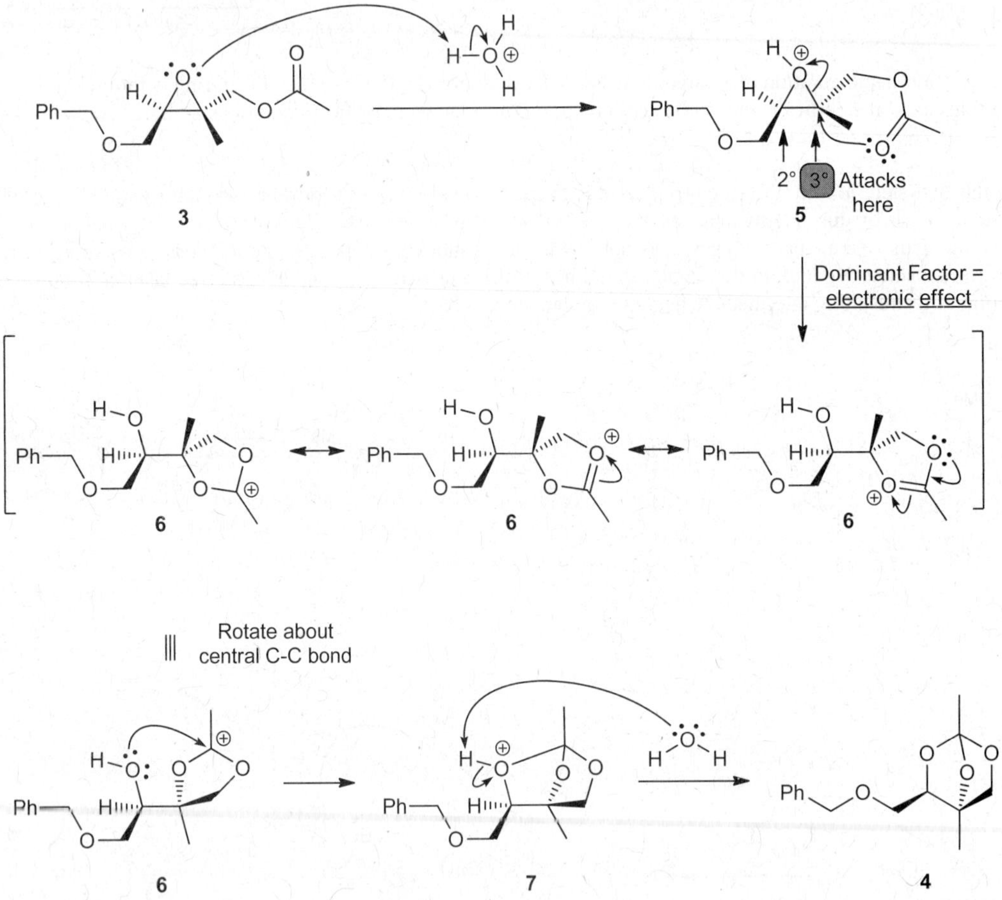

14.72. The epoxide in compound **2** has a tertiary carbon on the top side, and a primary carbon on the bottom side. Under acidic conditions, we saw that electronic effects override steric effects when there is a tertiary carbon on one side of the epoxide. If this line of reasoning were to govern product formation in this example, the alcohol should have attacked the top carbon. However, we know from looking at the structure of compound **2** that the alcohol must have attacked the primary carbon. Also of note is the stereochemistry at the ring opening site. After epoxide **1** is protonated, hexafluoro-2-propanol is expected to approach intermediate **3** from the bottom face of the molecule, resulting in compound **4**. Again, this is not what we observe.

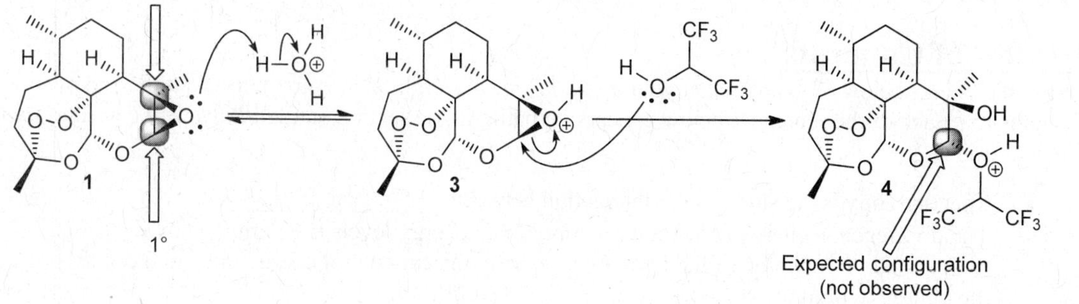

Expected configuration
(not observed)

As it turns out, when the epoxide is protonated, the neighboring oxygen atom is able to open the epoxide ring intramolecularly, resulting in a resonance-stabilized cation (**4a** and **4b**). This oxonium ion can then be trapped by hexafluoro-2-propanol, which approaches from the top face of the molecule (the bottom face is blocked by the nearby methyl group, and peroxide bridge), producing compound **5**. Finally, loss of a proton forms alcohol **2**.

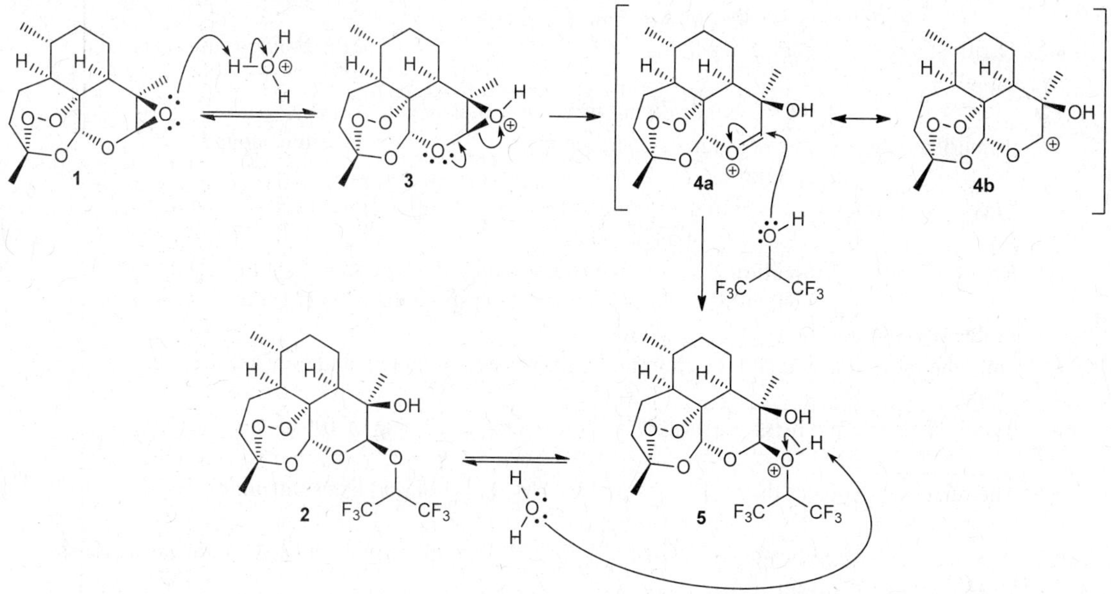

Notice that this example does not violate the principles underlying electronic trends. Specifically, when there is an oxygen atom attached directly to one side of the epoxide, under acidic conditions, ring opening can occur at this site, *because it results in a more stable (resonance-stabilized) cation.*

Chapter 15
Infrared Spectroscopy and Mass Spectrometry

Review of Concepts

Fill in the blanks below. To verify that your answers are correct, look in your textbook at the end of Chapter 15. Each of the sentences below appears verbatim in the section entitled *Review of Concepts and Vocabulary*.

- **Spectroscopy** is the study of the interaction between _____ and _____.
- The difference in energy (ΔE) between vibrational energy levels is determined by the nature of the bond. If a photon of light possesses exactly this amount of energy, the bond can absorb the photon to promote a _____ **excitation**.
- IR spectroscopy can be used to identify which _____ are present in a compound.
- The location of each signal in an IR spectrum is reported in terms of a frequency-related unit called _____.
- The wavenumber of each signal is determined primarily by bond _____ and the _____ of the atoms sharing the bond.
- The intensity of a signal is dependent on the _____ of the bond giving rise to the signal.
- _____ C=C bonds do not produce signals.
- Primary amines exhibit two signals resulting from _____ **stretching** and _____ **stretching**.
- **Mass spectrometry** is used to determine the _____ and _____ of a compound.
- **Electron impact ionization (EI)** involves bombarding the compound with high energy _____, generating a radical cation that is symbolized by $(M)^{+\bullet}$ and is called the **molecular ion**, or the _____ **ion**.
- Only the molecular ion and the cationic fragments are deflected, and they are then separated by their _____ (*m/z*).
- The tallest peak in a mass spectrum is assigned a relative value of 100% and is called the _____ **peak**.
- The relative heights of the $(M)^{+\bullet}$ peak and the $(M+1)^{+\bullet}$ peak indicates the number of _____.
- A signal at M−15 indicates the loss of a _____ group; a signal at M−29 indicates the loss of an _____ group.
- _____ alkanes have a molecular formula of the form C_nH_{2n+2}.
- Each double bond and each ring represents one **degree of** _____.

Review of Skills

Fill in the blanks and empty boxes below. To verify that your answers are correct, look in your textbook at the end of Chapter 15. The answers appear in the section entitled *SkillBuilder Review*.

15.1 Analyzing an IR Spectrum

STEP 1 - LOOK FOR _____ BONDS BETWEEN 1600 AND 1850 cm^{-1}	STEP 2 - LOOK FOR _____ BONDS BETWEEN 2100 AND 2300 cm^{-1}	STEP 3 - LOOK FOR _____ BONDS BETWEEN 2750 AND 4000 cm^{-1}
GUIDELINES: C=O BONDS PRODUCE _____ SIGNALS C=C BONDS GENERALLY PRODUCE _____ SIGNALS. SYMMETRICAL C=C BONDS DO NOT APPEAR AT ALL	**GUIDELINES:** _____ TRIPLE BONDS DO NOT PRODUCE SIGNALS	**GUIDELINES:** DRAW A LINE AT 3000 cm^{-1}, AND LOOK FOR _____ OR _____ C-H BONDS TO THE LEFT OF THE LINE THE SHAPE OF AN O-H SIGNAL IS AFFECTED BY _____ (DUE TO H-BONDING) PRIMARY AMINES EXHIBIT TWO N-H SIGNALS (_____ AND _____ STRETCHING)

15.2   Distinguishing Between Two Compounds Using IR Spectroscopy

STEP 1 - WORK METHODICALLY THROUGH THE EXPECTED _____ OF EACH COMPOUND	STEP 2 - DETERMINE IF ANY _____ WILL BE PRESENT FOR ONE COMPOUND BUT ABSENT FOR THE OTHER	STEP 3 - FOR EACH EXPECTED SIGNAL, COMPARE FOR ANY POSSIBLE DIFFERENCES IN _____, _____, OR _____.

15.3   Using the Relative Abundance of the (M+1)$^{+\bullet}$ Peak to Propose a Molecular Formula

STEP 1 - FILL IN THE BOXES BELOW TO COMPLETE THE FORMULA THAT CAN BE USED TO DETERMINE THE NUMBER OF CARBON ATOMS IN A COMPOUND USING MASS SPECTROMETRY:	STEP 2 - ANALYZE THE MASS OF THE MOLECULAR ION TO DETERMINE IF ANY _____ ARE PRESENT.

$$\left(\frac{\text{Abundance of } \boxed{\quad} \text{ peak}}{\text{Abundance of } \boxed{\quad} \text{ peak}} \right) \times 100\%$$

$$1.1 \%$$

15.4   Calculating HDI

STEP 1 - REWRITE THE MOLECULAR FORMULA "AS IF" THE COMPOUND HAD NO ELEMENTS OTHER THAN C AND H, USING THE FOLLOWING RULES:	STEP 2 - DETERMINE WHETHER ANY H'S ARE MISSING. EVERY TWO H'S REPRESENT ONE DEGREE OF UNSATURATION:
- ADD ONE H FOR EACH _____ - IGNORE ALL _____ ATOMS - SUBTRACT ONE H FOR EACH _____	$C_4H_9Cl \longrightarrow HDI = \_\_\_$ $C_4H_8O \longrightarrow HDI = \_\_\_$ $C_4H_9N \longrightarrow HDI = \_\_\_$

<u>Mistakes to Avoid</u>

We have seen that the IR spectrum of an alkene will exhibit a signal near 1650 cm^{-1} (the characteristic signal for a C=C bond) if there is a dipole moment associated with the C=C bond. In such a case, the dipole moment changes as the C=C bond vibrates, creating an oscillating electric field that serves as an antenna to absorb the appropriate frequency of IR radiation. If the C=C bond does not have a dipole moment, then it cannot efficiently absorb IR radiation, and the signal near 1650 cm^{-1} will be absent. For example, consider the following two compounds:

will produce a signal   will NOT produce a signal
near 1650 cm^{-1} near 1650 cm^{-1}

Don't be confused by the terms 'symmetrical alkene' and 'unsymmetrical alkene'. We might refer to the first compound as an unsymmetrical alkene (in reference to the dipole moment), but the truth is that this alkene still does possess some symmetry (an axis of symmetry, as well as two planes of symmetry).

Axis of symmetry

Plane of symmetry   Plane of symmetry

But these symmetry elements are not relevant for determining whether a C=C bond has a dipole moment, and therefore, they are not relevant for determining whether or not a C=C bond will produce a signal in an IR spectrum. When we refer to a symmetrical alkene, we are referring to the symmetry of the two vinylic positions:

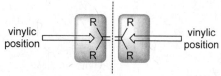

With this in mind, let's consider the following alkene:

While this molecule does possess some symmetry, you should avoid falling into the trap of calling it a symmetrical alkene and erroneously deciding that the C=C double bond will not produce a signal. In fact, the dipole moment for this C=C bond is expected to be quite large (because of the combined inductive effects of the chlorine atoms. One vinylic position (the one connected to the two chlorine atoms) is more electron-deficient ($\delta+$) than the other vinylic position. As a result, this C=C bond is expected to produce a rather strong signal in the IR spectrum.

Solutions

15.1.
(a) The C–H bond is expected to produce the signal with the largest wavenumber, because bonds to H typically produce high-energy signals (due to the low mass of the hydrogen atom). Among the remaining two bonds, the triple bond is stronger than the double bond, so we expect the double bond to produce the signal with lowest wavenumber.

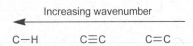

Increasing wavenumber

C—H   C≡C   C=C

(b) Each of the bonds in this case is a single bond. The C–H bond is expected to produce the signal with the largest wavenumber, because bonds to H typically produce high-energy signals (due to the low mass of the hydrogen atom).

Increasing wavenumber

C—H   C—C

15.2.
(a) This compound exhibits an sp^2 hybridized carbon atom that is connected to a hydrogen atom. As such, this C–H bond (highlighted) should produce a signal above 3000 cm^{-1} (at approximately 3100 cm^{-1}).

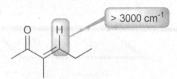

> 3000 cm^{-1}

(b) This compound has three sp^2 hybridized carbon atoms, but none of them are connected to hydrogen atoms. And there are no sp hybridized carbon atoms. Therefore, we do not expect a signal above 3000 cm^{-1}.

(c) This compound has two sp hybridized carbon atoms, but neither of them are connected to hydrogen atoms. And there are no sp^2 hybridized carbon atoms. Therefore, we do not expect a signal above 3000 cm^{-1}.

(d) This compound exhibits an sp^2 hybridized carbon atom that is connected to a hydrogen atom. As such, this C–H bond (highlighted) should produce a signal above 3000 cm^{-1} (at approximately 3100 cm^{-1}).

> 3000 cm^{-1}

(e) This compound has two sp^2 hybridized carbon atoms, but neither of them are connected to hydrogen atoms. And there are no sp hybridized carbon atoms. Therefore, we do not expect a signal above 3000 cm^{-1}.

(f) This compound exhibits an sp hybridized carbon atom that is connected to a hydrogen atom. As such, this C–H bond (highlighted) should produce a signal above 3000 cm^{-1} (at approximately 3300 cm^{-1}).

> 3000 cm^{-1}

15.3.
(a) One of the carbonyl groups (upper left) is not conjugated, so it is expected to produce a signal at approximately 1720 cm^{-1}. The other carbonyl group (bottom right) is conjugated to a C=C π bond, so it is expected to produce a signal at approximately 1680 cm^{-1}.

lower wavenumber
(carbonyl group is conjugated)

(b) One of the ester groups (bottom right) is not conjugated, so it is expected to produce a signal at approximately 1740 cm^{-1}. The other carbonyl group (upper left) is conjugated to a C=C π bond, so it is expected to produce a signal at approximately 1710 cm^{-1}.

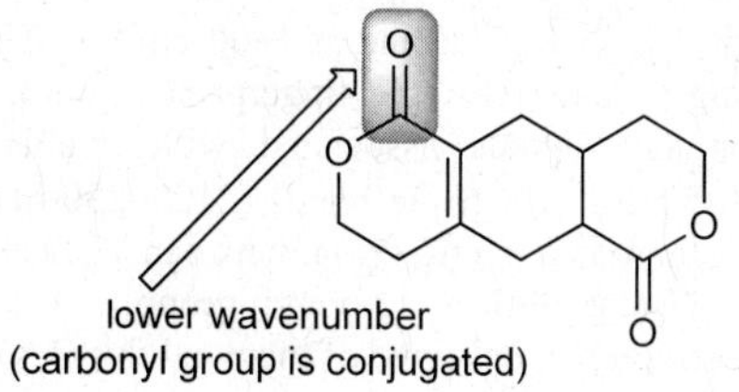

lower wavenumber
(carbonyl group is conjugated)

(c) The carbonyl group of a ketone is expected to produce a signal at approximately 1720 cm^{-1}, while the carbonyl group of an ester is expected to produce a signal at approximately 1740 cm^{-1}.

lower wavenumber
(a ketone)

higher wavenumber
(an ester)

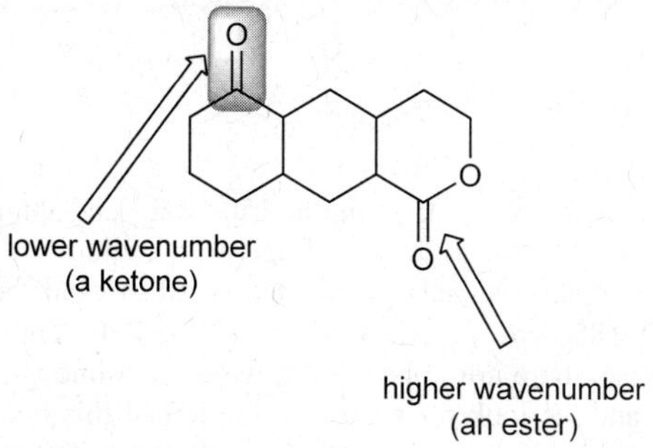

15.4. The C=C π bond in the conjugated compound produces a signal at lower wavenumber (1600 cm^{-1}) because it has some single bond character, as seen in the third resonance structure below. This additional single bond character renders the C=C π bond weaker (relative to the C=C π bond of the other compound, which does not exhibit any single bond character).

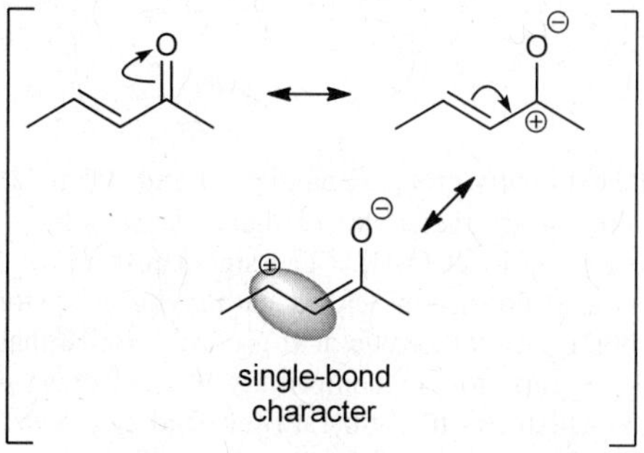

single-bond
character

15.5.
(a) The second compound has an electronegative chlorine atom, which withdraws electron density via induction.

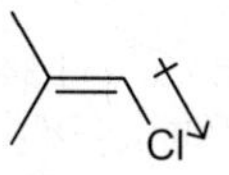

This causes the two vinylic positions to experience different electronic environments. The vinylic position connected directly to the chlorine atom is expected to be more electron-poor ($\delta+$) than the other vinylic position

(which is not directly connected to the chlorine atom). Therefore, the C=C bond in this compound has a larger dipole moment than the C=C bond in the other compound. As a result we expect this compound to be more efficient at absorbing IR radiation (thereby producing a stronger signal).

(b) The C=C bond in the following compound will have a larger dipole moment because one vinylic position is connected to two chlorine atoms while the other vinylic position is not directly connected to any chlorine atoms. As a result, the two vinylic positions are in very different electronic environments, giving rise to a large dipole moment. We therefore expect this C=C bond to be more efficient at absorbing IR radiation (and therefore produce a stronger signal).

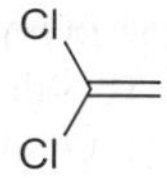

15.6. If we draw all significant resonance structures of 2-cyclohexenone, we see that one of the vinylic positions is electron-deficient (highlighted in the third resonance structure):

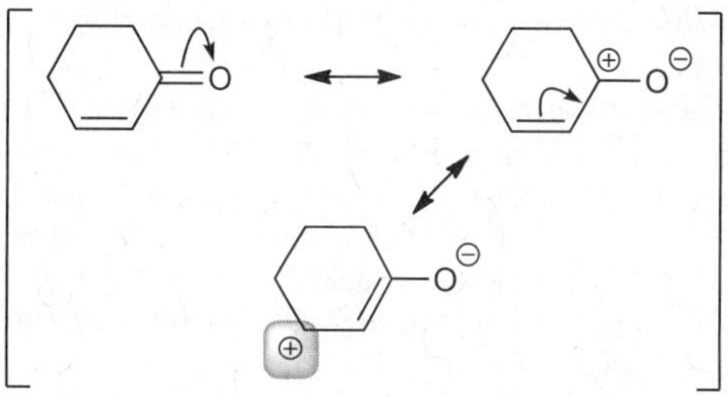

As a result, the two vinylic positions experience very different electronic environments, giving rise to a large dipole moment. With a large dipole moment, this C=C bond is expected to be very efficient at absorbing IR radiation, thereby producing a strong signal.

15.7. The vinylic C–H bond should produce a signal at approximately 3100 cm^{-1}.

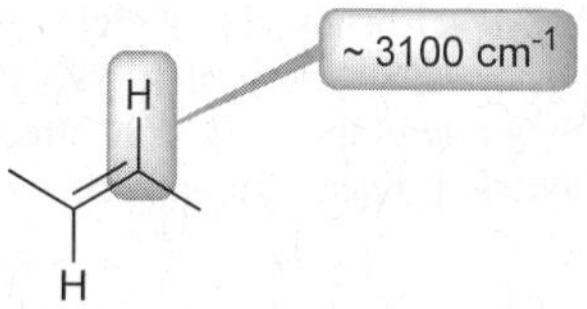

~ 3100 cm^{-1}

15.8. The narrow signal is produced by the O–H stretching in the absence of hydrogen bonding effect. The broad signal is produced by O–H stretching when hydrogen bonding is present. Hydrogen bonding effectively lowers the bond strength of the O–H bonds, because each hydrogen atom is slightly pulled away from the oxygen to which it is connected. A longer bond length (albeit temporary) corresponds with a weaker bond, which corresponds with a lower wavenumber.

15.9.
(a) The broad signal between 3200 and 3600 cm^{-1} is characteristic of an alcohol (ROH).
(b) This spectrum lacks broad signals above 3000 cm^{-1}, so the compound is neither an alcohol nor a carboxylic acid (both of which produce broad signals that reach as high as 3600 cm^{-1}).
(c) The extremely broad signal that extends from 2200 to 3600 cm^{-1} is characteristic of the O–H stretching of a carboxylic acid (RCO$_2$H). The signal just above 1700 cm^{-1} is also consistent with a carboxylic acid (for the C=O bond of the carboxylic acid).
(d) This spectrum lacks broad signals above 3000 cm^{-1}, so the compound is neither an alcohol nor a carboxylic acid (both of which produce broad signals that reach as high as 3600 cm^{-1}).
(e) The broad signal between 3100 and 3600 cm^{-1} is characteristic of an alcohol (ROH).
(f) The extremely broad signal that extends from 2200 to 3600 cm^{-1} is characteristic of the O–H stretching of a carboxylic acid (RCO$_2$H). The signal around 1700 cm^{-1} is also consistent with a carboxylic acid (for the C=O bond of the carboxylic acid).

15.10.
(a) The strong signal just above 1700 cm^{-1} is consistent with the stretching of the carbonyl group (C=O) of a ketone.
(b) The extremely broad signal that extends from 2200 to 3600 cm^{-1} is characteristic of the O–H stretching of a carboxylic acid (RCO$_2$H). The signal just above 1700 cm^{-1} is also consistent with a carboxylic acid (for the C=O bond of the carboxylic acid).
(c) The signal at approximately 3400 cm^{-1} is consistent with the stretching of the N–H bond of a secondary amine.
(d) The two signals at 3350 and 3450 cm^{-1} are consistent with the stretching of the N–H bonds (symmetric and asymmetric) of a primary amine.
(e) The strong signal just above 1700 cm^{-1} is consistent with the stretching of the carbonyl group (C=O) of a ketone.
(f) The broad signal between 3200 and 3600 cm^{-1} is characteristic of an alcohol (ROH).

15.11.   The C$_{sp^3}$–H bonds can stretch symmetrically, asymmetrically, or in a variety of ways with respect to each other. Each one of these possible stretching modes is associated with a different wavenumber of absorption.

15.12.
(a) Begin by drawing a line at 1500 cm^{-1} and ignoring everything to the right (the fingerprint region). Then, look for any signals associated with double bonds (1600–1850 cm^{-1}) or triple bonds (2100–2300 cm^{-1}). In this case, there is a weak signal between 1600 and 1700 cm^{-1}, consistent with an alkene. Finally, we draw a line at 3000 cm^{-1}, and we look for signals to the left of this line. In this case, there is a signal at approximately 3100 cm^{-1}, which is consistent with a C$_{sp^2}$–H bond of an alkene. Among the possible structures, the alkene is the structure that is consistent with the signals in the spectrum.

(b) Begin by drawing a line at 1500 cm^{-1} and ignoring everything to the right (the fingerprint region). Then, look for any signals associated with double bonds (1600–1850 cm^{-1}) or triple bonds (2100–2300 cm^{-1}). In this case, there is a strong signal between 1700 and 1800 cm^{-1}, consistent with a carbonyl group. Among the possible structures, only two of them exhibit C=O bonds. One of these structures has carboxylic acid groups, and the spectrum does not match that compound (because that compound is expected to give a broad signal from 2200-3600 cm^{-1}, which is absent in our spectrum). The following structure (an ester) is consistent with the IR spectrum.

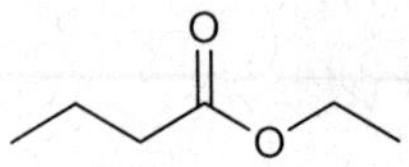

(c) Begin by drawing a line at 1500 cm^{-1} and ignoring everything to the right (the fingerprint region). Then, look for any signals associated with double bonds (1600–1850 cm^{-1}) or triple bonds (2100–2300 cm^{-1}). In this case, there are none. Next, we draw a line at 3000 cm^{-1}, and we look for signals to the left of this line. In this case, there are none. With no characteristic signals for any functional groups, this spectrum is consistent with an alkane.

(d) The broad signal between 3200 and 3600 cm^{-1} is characteristic of an alcohol (ROH). There is only one alcohol among the possible structures given.

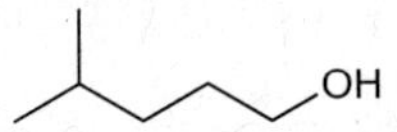

(e) The extremely broad signal that extends from 2200 to 3600 cm^{-1} is characteristic of the O–H stretching of a carboxylic acid (RCO$_2$H). The signal just above 1700 cm^{-1} is also consistent with a carboxylic acid (for the C=O bond of a carboxylic acid group). Notice that this signal appears to be comprised of two overlapping signals, which can likely be attributed to symmetric and asymmetric stretching of the two carboxylic acid groups.

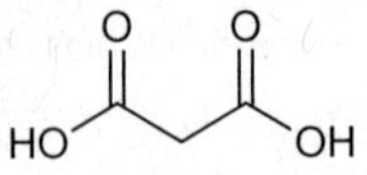

(f) The two signals at 3350 and 3450 cm^{-1} are consistent with the stretching of the N–H bonds (symmetric and asymmetric) of a primary amine. There is only one primary amine among the possible structures given.

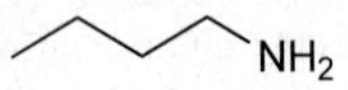

15.13. The following five signals are expected (presented in order of increasing wavenumber):
1) The C=C bond (expected to be ~ 1650 cm^{-1})
2) The C=O bond of the carboxylic acid group (expected to be ~ 1720 cm^{-1})
3) All C$_{sp^3}$–H bonds (expected to be <3000 cm^{-1})
4) The C$_{sp^2}$–H bond (expected to be ~ 3100 cm^{-1})
5) The O–H bond of the carboxylic acid group (expected to be 2200 - 3600 cm^{-1})

15.14.
(a) The starting material is an alcohol and is expected to produce a typical signal for an O–H stretch (a broad signal between 3200 and 3600 cm^{-1}). In contrast, the product is a carboxylic acid and is expected to produce an even broader O–H signal (2200-3600 cm^{-1}) as a result of more extensive hydrogen bonding. Alternatively, the product can be differentiated from the starting material by looking for a signal at around 1720 cm^{-1}. The product has a C=O bond and should exhibit this signal. The starting material lacks a C=O bond and will not show a signal at 1720 cm^{-1}.

(b) The starting material is a secondary amine and is expected to produce a typical signal for an N–H stretch at around 3400 cm^{-1}. In contrast, the product is a tertiary amine and is not expected to produce a similar signal.

(c) The starting material is an unsymmetrical alkyne and is expected to produce a signal at around 2200 cm^{-1}. In contrast, the product is an unsymmetrical alkene and is expected to produce a signal at around 1600 cm^{-1}. Also, the product has C$_{sp^2}$–H bonds that are absent in the starting material. The product is expected to have signals at around 3100 cm^{-1}, and the starting material will have no signal in that region.

(d) The C≡C triple bond in the starting material and the C=C double bond in the product are each symmetrical and will not produce signals. However the product has C$_{sp^2}$–H bonds that are absent in the starting material. The product is expected to have a signal at around 3100 cm^{-1}, and the starting material will have no signal in that region.

(e) The starting material will have two signals in the double-bond region: one for the C=O bond and one for the C=C bond. The product only has one signal in the double-bond region. It only has the signal for the C=O bond, which should be at a higher wavenumber (because the C=O bond is not conjugated in the product).

15.15. The starting material has a cyano group (C≡N) and is expected to produce a signal at approximately 2200 cm^{-1}. In contrast, the product is a carboxylic acid and is expected to produce a broad signal from 2200–3600 cm^{-1}, as well as a signal at 1720 cm^{-1} for the C=O bond.

15.16. The C≡C bond in the starting material (1-butyne) is unsymmetrical and produces a signal at 2200 cm^{-1}, corresponding with the C≡C stretch. In contrast, the C≡C bond in the product (3-hexyne) is symmetrical and does not produce a signal at 2200 cm^{-1}.

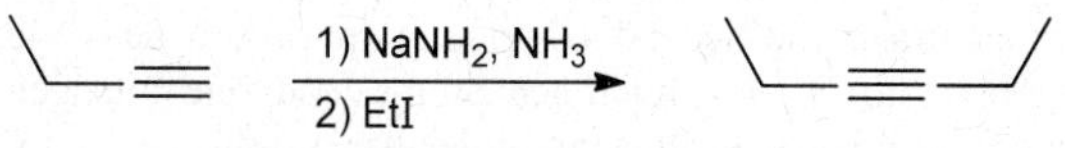

15.17. The starting material should have a C=O signal at 1720 cm^{-1}, while the product should have an O–H signal at 3200 – 3600 cm^{-1}.

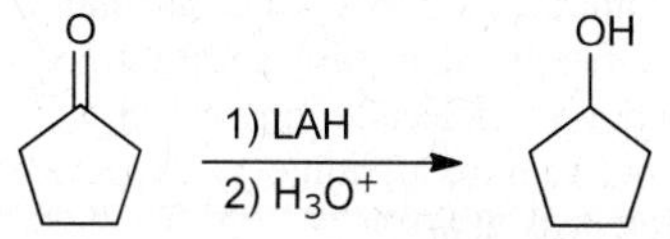

15.18. 1-Chlorobutane is a primary substrate. When treated with sodium hydroxide, substitution is expected to dominate over elimination (see Chapter 8), but both products are expected to be obtained:

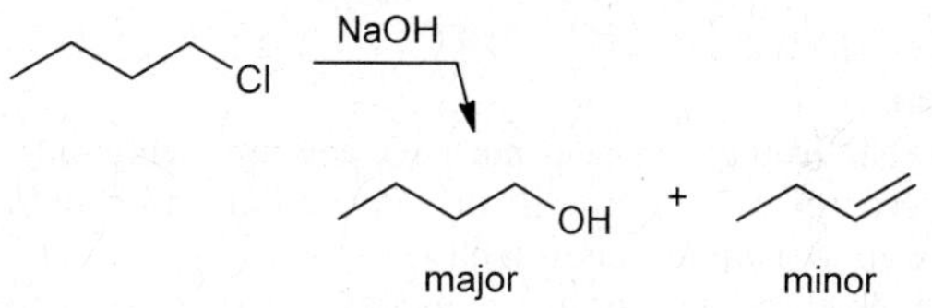

The substitution product is an alcohol and should have a broad signal from 3200 to 3600 cm^{-1}. The elimination product is an unsymmetrical alkene and is expected to give a C=C signal at approximately 1650 cm^{-1}, as well as a C$_{sp^2}$–H signal at 3100 cm^{-1}.

15.19.
(a) The first compound (called furan) has four carbon atoms, four hydrogen atoms and one oxygen atom, so it has the following molecular weight:

$$MW = (4 \times 12) + (4 \times 1) + (16) = 68$$

In contrast, the second compound (called cyclopentadiene) has five carbon atoms and six hydrogen atoms, so it has the following molecular weight:

$$MW = (5 \times 12) + (6 \times 1) = 66$$

These compounds have different molecular weights, so mass spectrometry can be used to distinguish between them. Specifically, the mass spectrum of furan is expected to have a molecular ion peak at $m/z = 68$, while the mass spectrum of cyclopentadiene is expected to have a molecular ion peak at $m/z = 66$.

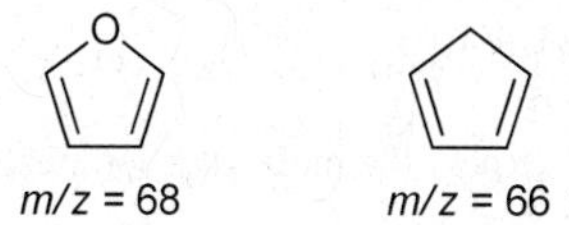

$m/z = 68$ $m/z = 66$

(b) The first compound (benzene) has six carbon atoms and six hydrogen atoms, so it has the following molecular weight:

$$MW = (6 \times 12) + (6 \times 1) = 78$$

In contrast, the second compound (pyridine) has five carbon atoms, five hydrogen atoms, and one nitrogen atom, so it has the following molecular weight:

$$MW = (5 \times 12) + (5 \times 1) + (14) = 79$$

These compounds have different molecular weights, so mass spectrometry can be used to distinguish between them. Specifically, the mass spectrum of benzene is expected to have a molecular ion peak at $m/z = 78$, while the mass spectrum of pyridine is expected to have a molecular ion peak at $m/z = 79$.

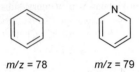

$m/z = 78$ $m/z = 79$

15.20.
(a) This compound does not have any nitrogen atoms. According to the nitrogen rule, this compound should have an even molecular weight.
With five carbon atoms, ten hydrogen atoms and one oxygen atom, the molecular weight is calculated as follows:

$$MW = (5 \times 12) + (10 \times 1) + (16) = 86$$

Therefore, we expect the molecular ion peak to appear at $m/z = 86$.

(b) This compound does not have any nitrogen atoms. According to the nitrogen rule, this compound should have an even molecular weight.
With five carbon atoms, eight hydrogen atoms and two oxygen atoms, the molecular weight is calculated as follows:

$$MW = (5 \times 12) + (8 \times 1) + (2 \times 16) = 100$$

Therefore, we expect the molecular ion peak to appear at $m/z = 100$.

(c) This compound has one nitrogen atom. According to the nitrogen rule, this compound should have an odd molecular weight.
With six carbon atoms, fifteen hydrogen atoms and one nitrogen atom, the molecular weight is calculated as follows:

$$MW = (6 \times 12) + (15 \times 1) + (14) = 101$$

Therefore, we expect the molecular ion peak to appear at $m/z = 101$.

(d) This compound has two nitrogen atoms. According to the nitrogen rule, this compound should have an even molecular weight.
With five carbon atoms, fourteen hydrogen atoms and two nitrogen atoms, the molecular weight is calculated as follows:

$$MW = (5 \times 12) + (14 \times 1) + (2 \times 14) = 102$$

Therefore, we expect the molecular ion peak to appear at $m/z = 102$.

15.21.
(a) In order to determine the number of carbon atoms in the compound, we compare the relative heights of the $(M+1)^{+\bullet}$ peak and the $(M)^{+\bullet}$ peak, like this:

$$\frac{1.7\%}{38.3\%} \times 100\% = 4.4\%$$

This means that the $(M+1)^{+\bullet}$ peak is 4.4% as tall as the $(M)^{+\bullet}$ peak. Recall that each carbon atom in the compound contributes 1.1% to the height of the $(M+1)^{+\bullet}$ peak, so we must divide by 1.1% to determine the number of carbon atoms in the compound:

$$\text{Number of C} = \frac{4.4\%}{1.1\%} = 4$$

The compound contains four carbon atoms, which account for $4 \times 12 = 48$ amu (atomic mass units), but the molecular weight is 72 amu. So we must account for the remaining $72 - 48 = 24$ amu. The molecular formula cannot be C_4H_{24}, because a compound with four carbon atoms cannot have that many hydrogen atoms.
Therefore, there must be another element present. It cannot be a nitrogen atom as that would give an odd molecular weight (the nitrogen rule). So we try oxygen (16 amu), leaving only 8 amu for hydrogen atoms. This gives the following proposed molecular formula: C_4H_8O.

(b) In order to determine the number of carbon atoms in the compound, we compare the relative heights of the $(M+1)^{+\bullet}$ peak and the $(M)^{+\bullet}$ peak, like this:

$$\frac{4.3\%}{100\%} \times 100\% = 4.3\%$$

This means that the $(M+1)^{+\bullet}$ peak is 4.3% as tall as the $(M)^{+\bullet}$ peak. Recall that each carbon atom in the compound contributes 1.1% to the height of the $(M+1)^{+\bullet}$ peak, so we must divide by 1.1% to determine the number of carbon atoms in the compound:

$$\text{Number of C} = \frac{4.3\%}{1.1\%} = 4$$

The compound contains four carbon atoms, which account for $4 \times 12 = 48$ amu (atomic mass units), but the molecular weight is 68 amu. So we must account for the remaining 68 − 48 = 20 amu. The molecular formula cannot be C_4H_{20}, because a compound with four carbon atoms cannot have that many hydrogen atoms.

Therefore, there must be another element present. It cannot be a nitrogen atom as that would give an odd molecular weight (the nitrogen rule). So we try oxygen (16 amu), leaving only 4 amu for hydrogen atoms. This gives the following proposed molecular formula: C_4H_4O.

(c) In order to determine the number of carbon atoms in the compound, we compare the relative heights of the $(M+1)^{+\bullet}$ peak and the $(M)^{+\bullet}$ peak, like this:

$$\frac{4.6\%}{100\%} \times 100\% = 4.6\%$$

This means that the $(M+1)^{+\bullet}$ peak is 4.6% as tall as the $(M)^{+\bullet}$ peak. Recall that each carbon atom in the compound contributes 1.1% to the height of the $(M+1)^{+\bullet}$ peak, so we must divide by 1.1% to determine the number of carbon atoms in the compound:

$$\text{Number of C} = \frac{4.6\%}{1.1\%} = 4.15 \sim 4$$

The compound cannot have 4.15 carbon atoms. It must be a whole number, so we round to the nearest whole number, which is 4. That is, the compound contains four carbon atoms, which account for 4 × 12 = 48 amu (atomic mass units), but the molecular weight is 54 amu. So we must account for the remaining 54 − 48 = 6 amu. This indicates six hydrogen atoms, giving the following proposed molecular formula: C_4H_6.

(d) In order to determine the number of carbon atoms in the compound, we compare the relative heights of the $(M+1)^{+\bullet}$ peak and the $(M)^{+\bullet}$ peak, like this:

$$\frac{1.5\%}{19.0\%} \times 100\% = 7.9\%$$

This means that the $(M+1)^{+\bullet}$ peak is 7.9% as tall as the $(M)^{+\bullet}$ peak. Recall that each carbon atom in the compound contributes 1.1% to the height of the $(M+1)^{+\bullet}$ peak, so we must divide by 1.1% to determine the number of carbon atoms in the compound:

$$\text{Number of C} = \frac{7.9\%}{1.1\%} = 7.2 \sim 7$$

The compound cannot have 7.2 carbon atoms. It must be a whole number, so we round to the nearest whole number, which is 7. That is, the compound contains seven carbon atoms, which account for 7 × 12 = 84 amu (atomic mass units), but the molecular weight is 96 amu. So we must account for the remaining 96 − 84 = 12 amu. This indicates twelve hydrogen atoms, giving the following proposed molecular formula: C_7H_{12}.

15.22. Each nitrogen atom in the molecular formula of a compound should contribute 0.37% to the $(M+1)^{+\bullet}$ peak. Three nitrogen atoms therefore contribute the same amount (1.1%) as one carbon atom. A compound with molecular formula $C_8H_{11}N_3$ should have an $(M+1)^{+\bullet}$ peak that is 9.9% as tall as the molecular ion peak. If the molecular ion peak is 24% of the base peak, then the $(M+1)^{+\bullet}$ peak must be 2.4% of the base peak.

15.23.
(a) This fragment is M − 79, which is formed by loss of a Br. So the fragment does not contain Br. Also, the fragment has a smaller mass than a single bromine atom (77 < 79).

(b) Loss of bromide generates a phenyl cation, as shown:

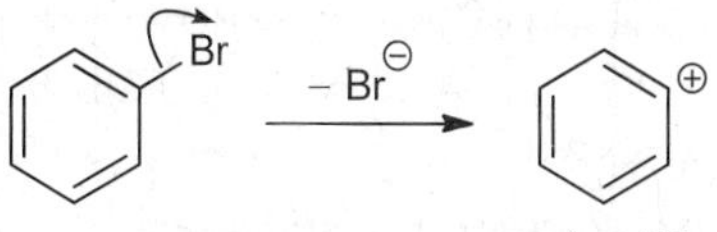

$m/z = 77$

Note: This type of process readily occurs under the high energy conditions in a mass spectrometer, but this process is generally not otherwise observed, because phenyl carbocations are generally too high in energy to form at an appreciable rate.

15.24.
(a) There is not a significant $(M+2)^{+\bullet}$ peak, so neither bromine nor chlorine is present.
(b) There is not a significant $(M+2)^{+\bullet}$ peak, so neither bromine nor chlorine is present.
(c) The $(M+2)^{+\bullet}$ peak is approximately equivalent in height to the molecular ion peak, indicating the presence of a bromine atom.
(d) The $(M+2)^{+\bullet}$ peak is approximately one-third as tall as the molecular ion peak, indicating the presence of a chlorine atom.

15.25.
(a) A peak at M−57 indicates the loss of a four-carbon radical fragment, which can result in the formation of a tertiary carbocation, as shown. This tertiary carbocation is the fragment responsible for the peak at M−57:

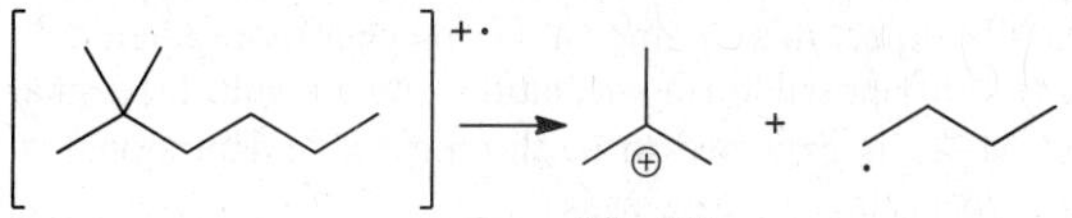

(M−57)

Remember that the radical fragment (the butyl radical) is not detected by the mass spectrometer.

(b) This carbocation is tertiary, and its formation is favored over the other possible secondary and primary carbocations.

(c) They readily fragment to produce tertiary carbocations.

(d) M−15 corresponds to loss of a methyl group. Indeed, loss of a methyl group would also produce a tertiary carbocation, but that pathway is less favorable because it involves formation of a methyl radical (which is less stable than a primary radical).

15.26. A fragment at M−29 should result from α cleavage:

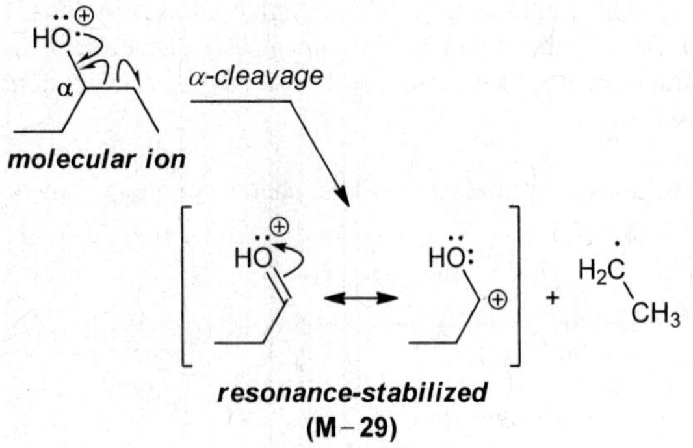

molecular ion

α-cleavage

resonance-stabilized (M−29)

and a fragment at M−18 should result from dehydration:

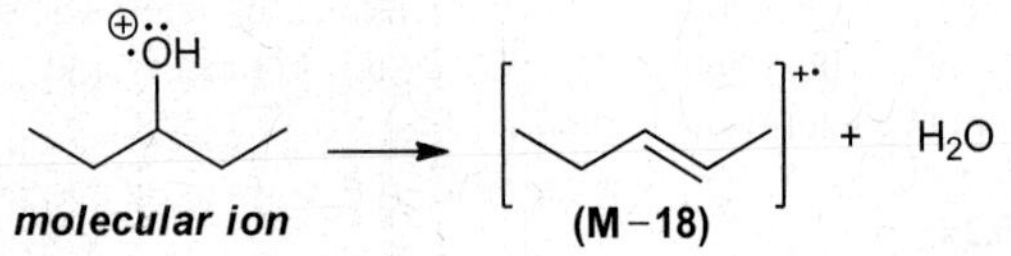

molecular ion   **(M−18)**   + H_2O

15.27. A peak at M−43 indicates the loss of a three-carbon radical fragment, which can result in the formation of a tertiary carbocation, as shown. This tertiary carbocation is the fragment responsible for the peak at M−43:

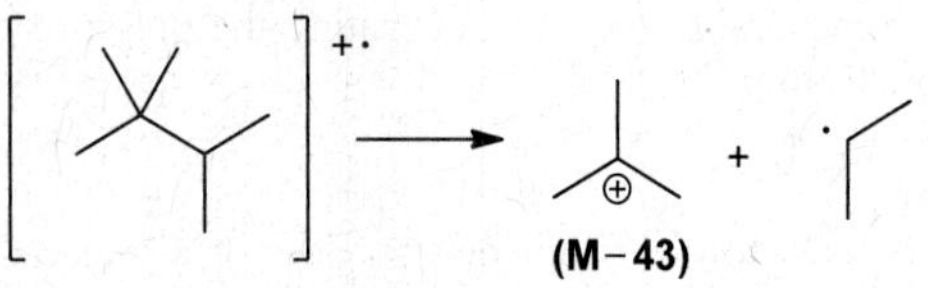

(M−43)

Notice that this fragmentation results in a tertiary carbocation as well as a secondary radical. No other possible fragmentation results in more stable fragments, so we expect this fragmentation to occur more often than any other possible fragmentation. As a result, the signal at M−43 is expected to be the most abundant signal in the spectrum (the base peak).

15.28. In the first spectrum, the base peak appears at M−29, signifying the loss of an ethyl group. This spectrum is likely the mass spectrum of ethylcyclohexane. The second spectrum has a peak at M−15, signifying the loss of a methyl group. The second spectrum is likely the mass spectrum of 1,1-dimethylcyclohexane.

15.29.
(a) If atomic masses are rounded to the nearest whole number, then both of these compounds have the same molecular weight (126 amu). However, when measured to four decimal places, these compounds do *not* have the same mass, and the difference is detectable via high-resolution mass spectrometry.

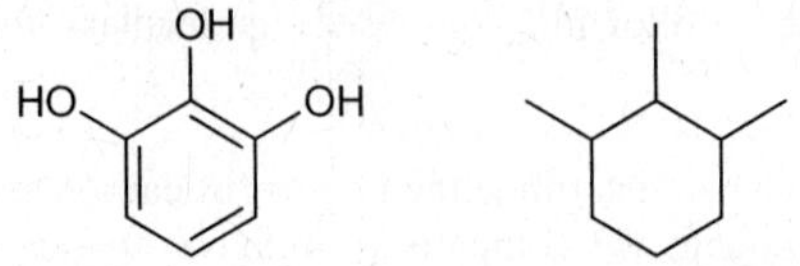

m/z = 126.0315   m/z = 126.1404

The calculation for each compound is shown here:

$C_6H_6O_3$ = (6×12.0000) + (6×1.0078) + (3×15.9949) = **126.0315** amu

C_9H_{18} = (9×12.0000) + (18×1.0078) = **126.1404** amu

(b) If atomic masses are rounded to the nearest whole number, then both of these compounds have the same molecular weight (112 amu). However, when measured to four decimal places, these compounds do *not* have the same mass, and the difference is detectable via high-resolution mass spectrometry:

m/z = 112.0522   m/z = 112.1248

The calculation for each compound is shown here:

$C_6H_8O_2$ = (6×12.0000) + (8×1.0078) + (2×15.9949) = **112.0522 amu**

C_8H_{16} = (8×12.0000) + (16×1.0078) = **112.1248 amu**

15.30.
(a) The first compound should have a broad signal between 3200 and 3600 cm^{-1}, corresponding with O−H stretching. The second compound will not have such a signal.
(b) The first compound should have a pair of strong signals around 1720 cm^{-1}, corresponding with symmetric and asymmetric stretching of the C=O bonds. In contrast, the second compound will have a weak signal at around 1650 cm^{-1}, corresponding to the C=C bond.

15.31.
(a) A compound with six carbon atoms would require (2×6) + 2 = 14 hydrogen atoms to be fully saturated. This compound has only ten hydrogen atoms, so four hydrogen atoms are missing. The compound therefore has two degrees of unsaturation (HDI = 2).

(b) A compound with five carbon atoms would require (2×5) + 2 = 12 hydrogen atoms to be fully saturated (the presence of the oxygen atom does not affect this calculation). This compound has only ten hydrogen atoms, so two hydrogen atoms are missing. Therefore, the compound has one degree of unsaturation (HDI = 1).

(c) A compound with the molecular formula C_5H_9N is expected to have the same HDI as a compound with the molecular formula C_5H_8. With five carbon atoms, the compound would need to have twelve hydrogen atoms to be fully saturated. With only eight hydrogen atoms, four are missing, representing two degrees of unsaturation (HDI = 2).

(d) A compound with the molecular formula C_3H_5ClO is expected to have the same HDI as a compound with the molecular formula C_3H_6 (Cl is treated like H, while O is ignored). With three carbon atoms, the compound would need to have eight hydrogen atoms to be fully saturated. With only six hydrogen atoms, two are missing, representing one degree of unsaturation (HDI = 1).

(e) A compound with ten carbon atoms would require $(2 \times 10) + 2 = 22$ hydrogen atoms to be fully saturated. This compound has only twenty hydrogen atoms, so two hydrogen atoms are missing. The compound therefore has one degree of unsaturation (HDI = 1).

(f) A compound with the molecular formula $C_4H_6Br_2$ is expected to have the same HDI as a compound with the molecular formula C_4H_8 (each Br is treated like an H). With four carbon atoms, the compound would need to have ten hydrogen atoms to be fully saturated. With only eight hydrogen atoms, two are missing, representing one degree of unsaturation (HDI = 1).

(g) A compound with six carbon atoms would require $(2 \times 6) + 2 = 14$ hydrogen atoms to be fully saturated. This compound has only six hydrogen atoms, so eight hydrogen atoms are missing. The compound therefore has four degrees of unsaturation (HDI = 4).

(h) A compound with the molecular formula C_2Cl_6 is expected to have the same HDI as a compound with the molecular formula C_2H_6 (each Cl is treated like an H). With two carbon atoms, the compound would need to have six hydrogen atoms to be fully saturated. No hydrogen atoms are missing, representing zero degrees of unsaturation (HDI = 0).

(i) A compound with the molecular formula $C_2H_4O_2$ is expected to have the same HDI as a compound with the molecular formula C_2H_4 (the oxygen atoms can be ignored for purposes of calculating HDI). With two carbon atoms, the compound would need to have six hydrogen atoms to be fully saturated. With only four hydrogen atoms, two are missing, representing one degree of unsaturation (HDI = 1).

(j) A compound with the molecular formula $C_{100}H_{200}Cl_2O_{16}$ is expected to have the same HDI as a compound with the molecular formula $C_{100}H_{202}$ (each Cl is treated like an H, and all of the oxygen atoms are ignored). No hydrogen atoms are missing, representing zero degrees of unsaturation (HDI = 0).

15.32. A compound with the molecular formula $C_3H_5ClO_2$ is expected to have the same HDI as a compound with the molecular formula C_3H_6 (the Cl is treated like an H, and both oxygen atoms are ignored). Both $C_3H_5ClO_2$ and C_3H_6 have one degree of unsaturation.

15.33. A compound with the molecular formula C_4H_8O is expected to have the same HDI as a compound with the molecular formula C_4H_8 (the oxygen atom can be ignored for purposes of calculating HDI). With four carbon atoms, the compound would need to have ten hydrogen atoms to be fully saturated. With only eight hydrogen atoms, two are missing, representing one degree of unsaturation (HDI = 1). A compound with one degree of unsaturation must contain either one double bond or one ring (but not both). The IR spectrum indicates the presence of a C=O bond, which accounts for the one degree of unsaturation (this means that the compound does NOT have a ring).

In the absence of a ring, there are only two ways to connect four carbon atoms (linear or branched):

Linear   Branched

Let's begin with the linear skeleton. There are two unique locations on the linear skeleton where a C=O bond can be placed (either at C1 or at C2):

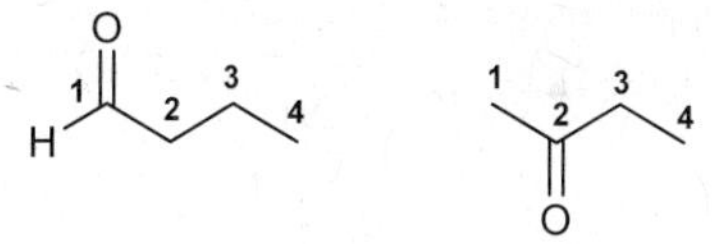

Placing the C=O bond at C3 would be the same as placing it at C2. Similarly, placing the C=O bond at C4 would be the same as placing it at C1.

Next we turn to the branched skeleton. We cannot place a C=O bond at the central position, because that would violate the octet rule:

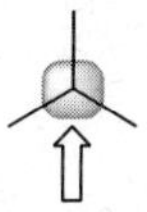

C=O bond can't be here,
because carbon can't have five bonds

And the remaining three positions are all identical. Placing the C=O bond at any of these three positions will give the same aldehyde:

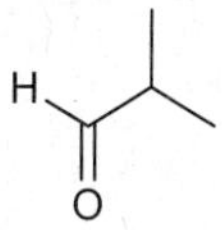

In summary, there are only three possible structures that have the molecular formula C_4H_8O and contain a C=O bond:

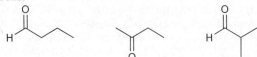

15.34. A compound with the molecular formula C_4H_8O is expected to have the same HDI as a compound with the molecular formula C_4H_8 (the oxygen atom can be ignored for purposes of calculating HDI). With four carbon atoms, the compound would need to have ten hydrogen atoms to be fully saturated. With only eight hydrogen atoms, two are missing, representing one degree of unsaturation (HDI = 1). A compound with one degree of unsaturation must contain either one double bond or one ring (but not both). The IR spectrum does not have signals between 1600 and 1850 cm⁻¹, which indicates the absence of either a C=O bond or an unsymmetrical C=C bond (and a symmetrical double bond is not possible because the compound has only one oxygen atom). Therefore, the structure likely contains a ring. With four carbon atoms, there are only two different types of rings that are possible (a four-membered ring or a three-membered ring):

The broad signal between 3200-3600 cm⁻¹ indicates an OH group, so our structure must include an OH group. There is only one unique location where an OH group can be placed on a four-membered ring:

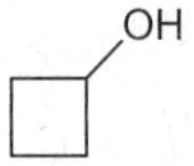

Placement of the OH group at any other location would give the same compound (cyclobutanol). However, there are three unique locations where the OH group can be placed on the other skeleton (containing the three-membered ring):

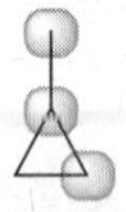

Our methodical analysis has revealed that there are four constitutional isomers with the molecular formula C_4H_8O that contain one ring and an OH group:

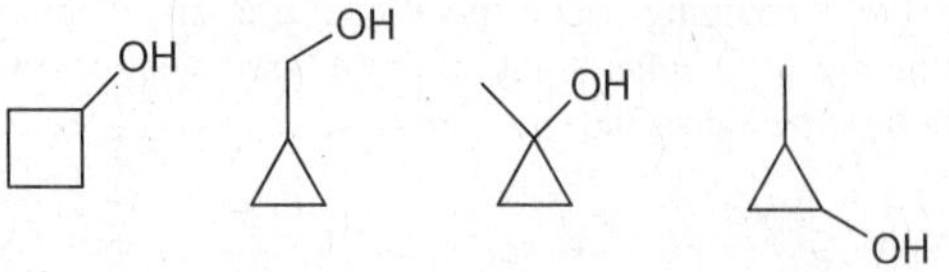

OH   OH   OH   OH

(Four possible stereoisomers)

15.35. A signal at 2200 cm⁻¹ signifies the presence of a C≡C bond. There are only two possible constitutional isomers: 1-butyne or 2-butyne. The latter is symmetrical and would not produce a signal at 2200 cm⁻¹. The compound must be 1-butyne.

15.36. A compound with the molecular formula $C_{10}H_{20}O$ is expected to have the same HDI as a compound with the molecular formula $C_{10}H_{20}$ (the oxygen atom can be ignored for purposes of calculating HDI). With ten carbon atoms, the compound would need to have twenty-two hydrogen atoms to be fully saturated. With only twenty hydrogen atoms, two are missing, representing one degree of unsaturation (HDI = 1). This compound must possess either either one ring or one double bond (not both).

15.37.
(a) The first compound has one degree of unsaturation (a ring) and the second compound also has one degree of unsaturation (a double bond). Since both compounds have the same number of carbon atoms and the same HDI, they should also have the same number of hydrogen atoms. Indeed, if we count the hydrogen atoms in each compound, we will see that both compounds share the molecular formula C_6H_{12}. These compounds are constitutional isomers.

HDI = 1   HDI = 1

(b) The first compound has two degrees of unsaturation (two rings) and the second compound also has two degrees of unsaturation (a ring and a double bond). Since both compounds have the same number of carbon atoms and the same HDI, they should also have the same number of hydrogen atoms. Indeed, if we count the hydrogen atoms in each compound, we will see that both compounds share the molecular formula C_8H_{14}. These compounds are constitutional isomers,

HDI = 2   HDI = 2

(c) The first compound has two degrees of unsaturation (two π bonds) while the second compound has only one degree of unsaturation (a ring). These compounds have the same number of carbon atoms, but they have a different HDI. Therefore, they must not have the same number of hydrogen atoms. Indeed, if we count the hydrogen atoms in each compound, we will see that the first compound has the molecular formula C_5H_8, while the second compound has the molecular formula C_5H_{10}. These compounds are not constitutional isomers.

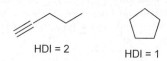

HDI = 2   HDI = 1

15.38. The range between 1600 and 1850 cm^{-1} is associated with the stretching of double bonds. Each of the compounds in this problem exhibits at least one double bond, if not two.
(a) The carbonyl group is conjugated so it is expected to produce a signal near 1680 cm^{-1}. The C=C double bond is conjugated so it is expected to produce a signal near 1600 cm^{-1}.

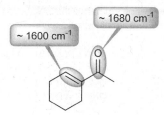

(b) The carbonyl group is isolated (not conjugated) so it is expected to produce a signal near 1720 cm^{-1}. The C=C double bond is also isolated so it is expected to produce a signal near 1650 cm^{-1}.

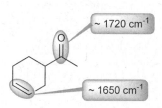

(c) The carbonyl group of an ester will typically produce a signal near 1740 cm^{-1}.

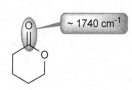

(d) The carbonyl group of the ester is conjugated so it is expected to produce a signal near 1710 cm^{-1}. The C=C double bond is conjugated so it is expected to produce a signal near 1600 cm^{-1}.

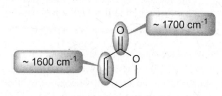

(e) The C=C double bond is isolated so it is expected to produce a signal near 1650 cm^{-1}.

15.39. The signals at highest wavenumber are expected to be associated with X–H bonds (because hydrogen has the lowest atomic mass). Among the list of bonds in the problem statement, there are three bonds that belong to this category (O–H, N–H, and C–H). These three bonds are expected to produce signals at approximately 3600, 3400, and 3100 cm^{-1}, respectively. Among the remaining bonds, the triple bond has the next highest wavenumber, followed by the double bonds (the C=O bond has a higher wavenumber than the C=C bond), followed by the C–O single bond.

Increasing Wavenumber

~ 3600 cm^{-1} ~ 3400 cm^{-1} ~ 3100 cm^{-1} ~ 2200 cm^{-1} ~ 1720 cm^{-1} ~ 1650 cm^{-1} ~ 1000 cm^{-1}

15.40.
(a) In addition to the C_{sp^3}–H stretching signals that are expected to appear just below 3000 cm^{-1}, the C=N bond and the C=O bond should each produce a signal in the double bond region (1600 – 1850 cm^{-1}).
(b) In addition to the C_{sp^3}–H stretching signals that are expected to appear just below 3000 cm^{-1}, the C_{sp^2}–H

bond should produce a signal near 3100 cm^{-1}, and the C=C bond should produce a signal in the double bond region (1600 – 1850 cm^{-1}).
(c) In addition to the C_{sp^3}–H stretching signals that are expected to appear just below 3000 cm^{-1}, the C_{sp^2}–H bond should produce a signal near 3100 cm^{-1}, and the C=C bond and the C=O bond should each produce a

signal in the double bond region (1600 – 1850 cm^{-1}). The two C≡C bonds should also produce two signals around 2200 cm^{-1}, and the C_{sp}–H bond should produce a signal around 3300 cm^{-1}.

(d) In addition to the C_{sp^3}–H stretching signals that are expected to appear just below 3000 cm^{-1}, the C=O bond should produce a signal in the double bond region around 1720 cm^{-1}, and the O–H of the carboxylic acid group should produce a very broad signal from 2200 – 3600 cm^{-1}.

15.41.

(a) The IR spectrum of the reactant should have a signal near 1650 cm^{-1} (for the C=C bond) and a signal near 3100 cm^{-1} (for the C_{sp^2}–H bond). In contrast, the IR spectrum of the product should not exhibit either of these signals.

(b) The IR spectrum of the reactant should have a broad signal from 3200 – 3600 cm^{-1} (for the O–H bond), while the IR spectrum of the product should not exhibit this signal, and instead should have a signal near 1720 cm^{-1} (for the C=O bond).

(c) The C=O bond of an ester should produce a signal at higher wavenumber (~1740 cm^{-1}) than the signal associated with the C=O bond of a ketone (~1720 cm^{-1}).

(d) The IR spectrum of the reactant should have a signal near 1650 cm^{-1} (for the C=C bond) and a signal near 3100 cm^{-1} (for the C_{sp^2}–H bonds). In contrast, the IR spectra of the products should not exhibit either of these signals. Instead, they should have signals near 1720 cm^{-1} (for the C=O bonds).

(e) The IR spectrum of the product should have a signal near 1650 cm^{-1} (for the C=C bond) and a signal near 3100 cm^{-1} (for the C_{sp^2}–H bond). In contrast, the IR spectrum of the reactant should not exhibit either of these signals.

15.42.

(a) The IR spectrum of this compound is expected to exhibit C_{sp^3}–H signals just below 3000 cm^{-1}, and a C=O signal near 1720 cm^{-1}.

(b) The IR spectrum of this compound is expected to exhibit C_{sp^3}–H signals just below 3000 cm^{-1}, and a C=O signal near 1680 cm^{-1} (conjugated), and a C=C signal near 1600 cm^{-1} (conjugated), and a C_{sp^2}–H signal near 3100 cm^{-1}.

(c) The IR spectrum of this compound is expected to exhibit C_{sp^3}–H signals just below 3000 cm^{-1}, and a C=O signal near 1720 cm^{-1} and a C=C signal near 1650 cm^{-1} and a C_{sp^2}–H signal near 3100 cm^{-1}.

(d) The IR spectrum of this compound is expected to exhibit C_{sp^3}–H signals just below 3000 cm^{-1}, and a C=C signal near 1650 cm^{-1} and a broad O–H signal in the range of 3200 – 3600 cm^{-1} and a C_{sp^2}–H signal near 3100 cm^{-1}.

(e) The IR spectrum of this compound is expected to exhibit C_{sp^3}–H signals just below 3000 cm^{-1}, and a C=O

signal near 1720 cm^{-1} and an extremely broad O-H signal in the range of 2200 – 3600 cm^{-1}.

(f) The IR spectrum of this compound is expected to exhibit C_{sp^3}–H signals just below 3000 cm^{-1}, and a C=O signal near 1720 cm^{-1} and a broad O-H signal in the range of 3200 – 3600 cm^{-1}.

15.43.

(a) The molecular formula for this compound is C_7H_8, so it has the following molecular weight:

$$MW = (7 \times 12) + (8 \times 1) = 92 \text{ amu}$$

Therefore, the mass spectrum of this compound is expected to have a molecular ion peak at $m/z = 92$.

(b) The molecular formula for this compound is C_6H_6O, so it has the following molecular weight:

$$MW = (6 \times 12) + (6 \times 1) + (1 \times 16) = 94 \text{ amu}$$

Therefore, the mass spectrum of this compound is expected to have a molecular ion peak at $m/z = 94$.

(c) The molecular formula for this compound is $C_6H_{10}O$, so it has the following molecular weight:

$$MW = (6 \times 12) + (10 \times 1) + (1 \times 16) = 98 \text{ amu}$$

Therefore, the mass spectrum of this compound is expected to have a molecular ion peak at $m/z = 98$.

(d) The molecular formula for this compound is C_6H_8O, so it has the following molecular weight:

$$MW = (6 \times 12) + (8 \times 1) + (1 \times 16) = 96 \text{ amu}$$

Therefore, the mass spectrum of this compound is expected to have a molecular ion peak at $m/z = 96$.

(e) The molecular formula for this compound is $C_6H_{15}N$, so it has the following molecular weight:

$$MW = (6 \times 12) + (15 \times 1) + (1 \times 14) = 101 \text{ amu}$$

Therefore, the mass spectrum of this compound is expected to have a molecular ion peak at $m/z = 101$.

15.44. The molecular ion peak appears at $m/z = 86$, so the compound must have a molecular weight of 86 amu. Each carbon atom contributes 12 amu, and each hydrogen atom contributes 1 amu. So, we begin by considering the following possibilities, each of which has one degree of unsaturation (as required by the problem statement).

$$C_4H_8 = (4 \times 12) + (8 \times 1) = 56 \text{ amu}$$
$$C_5H_{10} = (5 \times 12) + (10 \times 1) = 70 \text{ amu}$$
$$C_6H_{12} = (6 \times 12) + (12 \times 1) = 84 \text{ amu}$$

None of these possibilities exhibits the correct molecular weight. The closest is C_6H_{12}, but we are short by two amu (86 – 84 = 2 amu). If we add two hydrogen atoms (to give C_6H_{14}) then the compound will no longer have one degree of unsaturation (it would be fully saturated). So, we consider C_5H_{10}, which has a molecular weight of 70 amu. In order to get this number up to 86, we are missing 16 amu, which is exactly the weight of an oxygen atom. So, a compound with the molecular formula $C_5H_{10}O$ will have a molecular weight of 86 amu (and still has one degree of unsaturation, because the insertion of an oxygen atom does not affect the HDI).

There are certainly many acceptable answers to this problem. For example, a compound with the molecular formula $C_4H_{10}N_2$ will have a molecular weight of 86 amu and its structure will have one degree of unsaturation.

15.45. The problem statement indicates that the $(M+1)^{+\bullet}$ peak is 10% as tall as the $(M)^{+\bullet}$ peak. Recall that each carbon atom in the compound contributes 1.1% to the height of the $(M+1)^{+\bullet}$ peak, so we must divide by 1.1% to determine the number of carbon atoms in the compound:

$$\text{Number of C} = \frac{10\%}{1.1\%} = 9.1 \sim 9$$

The compound cannot have 9.1 carbon atoms. It must be a whole number, so we round to the nearest whole number, which is nine. That is, the compound contains nine carbon atoms.

15.46. We begin by determining the molecular formula and molecular weight of each compound:

Cl	NH_2	OH	Br
C_6H_5Cl	C_6H_7N	C_6H_6O	C_6H_5Br
(MW = 112)	(MW = 93)	(MW = 94)	(MW = 156)

Each of these compounds has a different molecular weight, so we should be able to use match each compound with its spectrum if we focus on the molecular ion peak in each case.

Chlorobenzene (C_6H_5Cl) is consistent with spectrum (b) which has an $(M)^{+\bullet}$ peak at $m/z = 112$, as well as an $(M+2)^{+\bullet}$ peak that is one-third the height of the $(M)^{+\bullet}$ peak.

Aniline (C_6H_7N) is consistent with spectrum (c) which has an $(M)^{+\bullet}$ peak at $m/z = 93$ (which is an odd number, consistent with the nitrogen rule).

Phenol (C_6H_6O) is consistent with spectrum (a) which has an $(M)^{+\bullet}$ peak at $m/z = 94$.

Bromobenzene (C_6H_5Br) is consistent with spectrum (d) which has an $(M)^{+\bullet}$ peak at $m/z = 156$, as well as an $(M+2)^{+\bullet}$ peak that is approximately the same height as the $(M)^{+\bullet}$ peak.

15.47.

(a) The broad signal between 3200 and 3600 cm^{-1} indicates the presence of an OH group. The signals between 1600 and 1700 cm^{-1} indicate the presence of double bonds.

(b) In order to determine the number of carbon atoms in the compound, we compare the relative heights of the $(M+1)^{+\bullet}$ peak and the $(M)^{+\bullet}$ peak, like this:

$$\frac{3.9\%}{27.2\%} \times 100\% = 14.3\%$$

This means that the $(M+1)^{+\bullet}$ peak is 14.3% as tall as the $(M)^{+\bullet}$ peak. Recall that each carbon atom in the compound contributes 1.1% to the height of the $(M+1)^{+\bullet}$ peak, so we must divide by 1.1% to determine the number of carbon atoms in the compound:

$$\text{Number of C} = \frac{14.3\%}{1.1\%} = 13$$

Our analysis reveals that the compound likely has thirteen carbon atoms.

(c) The molecular ion peak appears at $m/z = 196$, so the compound must have a molecular weight of 196 amu. Each carbon atom contributes 12 amu, and each hydrogen atom contributes 1 amu. So, we begin by considering the following possibilities, each of which has two degrees of unsaturation (as required by the problem statement).

$$C_{12}H_{22} = (12 \times 12) + (22 \times 1) = 166 \text{ amu}$$
$$C_{13}H_{24} = (13 \times 12) + (24 \times 1) = 180 \text{ amu}$$
$$C_{14}H_{26} = (14 \times 12) + (26 \times 1) = 194 \text{ amu}$$

None of these possibilities exhibits the correct molecular weight. The closest is $C_{14}H_{26}$, but we are short by two amu (196 – 194 = 2 amu). If we add two hydrogen atoms (to give $C_{14}H_{28}$), then the compound will no longer have two degrees of unsaturation (it would only have one degree of unsaturation). So, we consider $C_{13}H_{24}$, which has a molecular weight of 180 amu. In order to get this number up to 196, we are missing 16 amu, which is exactly the weight of an oxygen atom. So, a compound the molecular formula $C_{13}H_{24}O$ will have a molecular weight of 196 amu (and still has two degrees of unsaturation, because the insertion of an oxygen atom does not affect the HDI).

There are certainly many acceptable answers to this problem. For example, a compound with the molecular formula $C_{12}H_{24}N_2$ will have a molecular weight of 196 amu and its structure will have two degrees of unsaturation.

15.48.

(a) Each of these compounds has the molecular formula C_6H_{12}.

Cyclohexane
(C_6H_{12})

2-methyl-2-pentene
(C_6H_{12})

(b) A compound with six carbon atoms would require $(6×2) + 2 = 14$ hydrogen atoms to be fully saturated. Each of these compounds has only twelve hydrogen atoms, so each compound is missing two hydrogen atoms. Therefore, each of these compounds has one degree of unsaturation (HDI = 1).

(c) No. Both compounds have exactly six carbon atoms and twelve hydrogen atoms, so each of these compounds will produce an $(M)^{+\bullet}$ peak with the same m/z, even with high resolution mass spectrometry.

$$C_6H_{12} = (6×12.0000) + (12×1.0078) = \textbf{84.0936 amu}$$

(d) The IR spectrum of the alkene would have a signal near 1650 cm^{-1} for the C=C bond and another signal near 3100 cm^{-1} for the C_{sp^2}–H bond. The IR spectrum of cyclohexane lacks both of these signals.

15.49. The parent ion for 1-ethyl-1-methylcyclohexane is expected to appear at $m/z = 126$ (calculation below).

1-Ethyl-1-methylcyclohexane
(C_9H_{18})

MW of $C_9H_{18} = (9×12) + (18×1) = 126$ amu

If the parent ion appears at $m/z = 126$, then the signal that appears at $m/z = 111$ is at (M–15), which corresponds with the loss of a methyl group. The signal at $m/z = 97$ is at (M–29), which corresponds with the loss of an ethyl group. Both fragmentations lead to a tertiary carbocation:

$+ {}^{\bullet}CH_3$

(M–15)

$+\bullet$

$+ {}^{\bullet}CH_2CH_3$

(M–29)

15.50.

(a) This fragment appears at M–79, which corresponds with the loss a bromine atom. Therefore, the fragment no longer contains bromine. As such, we do not expect a signal at M–77 (which we would only observe if the fragment still contained the bromine atom).

(b) This fragment appears at (M–15), which corresponds with the loss of a methyl group. Therefore, this fragment still contains the bromine atom. As such, we do expect a signal at M–13 that is equal in height to the M–15 peak (because there are two isotopes of bromine in roughly equal abundance).

(c) This fragment appears at (M–29), which corresponds with the loss of an ethyl group. Therefore, this fragment still contains the bromine atom. As such, we do expect a signal at M–27 that is equal in height to the M–29 peak (because there are two isotopes of bromine in roughly equal abundance).

15.51. As seen in Chapter 8, the choice of base affects the regiochemical outcome. Ethoxide favors the more-substituted (Zaitsev) product, while *tert*-butoxide favors the less-substituted (Hofmann) product.

NaOEt

Br

Zaitsev product

t-BuOK

Hofmann product

The Zaitsev product is a symmetrical alkene, and as such, its IR spectrum will not exhibit a signal at 1650 cm^{-1} (where the signals for C=C bonds typically appear) or a signal at 3100 cm^{-1} (because this compound contains no C_{sp^2}–H bonds). In contrast, the Hofmann product will display both of these signals in its IR spectrum.

15.52.

(a)

The molecular ion peak appears at $m/z = 66$, so the compound must have a molecular weight of 66 amu. Each carbon atom contributes 12 amu, and each hydrogen atom contributes 1 amu. Five carbon atoms will account for 60 amu, leaving 6 amu for hydrogen atoms.

$$C_5H_6 = (5×12) + (6×1) = 66 \text{ amu}$$

The compound cannot have six carbon atoms, because that would exceed the known molecular weight ($6×12 = 72$ amu). And the compound cannot have four carbon atoms, because then we would not reach 66 amu even if the compound is fully saturated:

$C_4H_{10} = (4 \times 12) + (10 \times 1) = 58$ amu

And a hydrocarbon with four carbon atoms cannot have more than ten hydrogen atoms.

(b) The molecular ion peak appears at $m/z = 70$, so the compound must have a molecular weight of 70 amu. Each carbon atom contributes 12 amu, and each hydrogen atom contributes 1 amu. The signal in the IR spectrum indicates the presence of a carbonyl group (C=O), so we must also account for at least one oxygen atom, which contributes 16 amu. If we start with four carbon atoms ($4 \times 12 = 48$), and one oxygen atom (16), we get a total of $48 + 16 = 64$ amu, which is just 6 amu short of the known molecular weight (70 amu). Six hydrogen atoms will supply the remaining 6 amu, giving the molecular formula C_4H_6O.

15.53.
(a) Octane is a saturated alkane with the molecular C_8H_{18}. The molecular weight of octane is calculated here:

$$C_8H_{18} = (8 \times 12) + (18 \times 1) = 114 \text{ amu}$$

Therefore, the molecular ion peak is expected to appear at $m/z = 114$, and indeed, that peak can be seen in the spectrum.

(b) The base peak is (by definition) the tallest peak in the spectrum, which appears at $m/z = 43$.

(c) The base peak (at $m/z = 43$) corresponds with the loss of a radical fragment that is $114 - 43 = 71$ amu. That is, the base peak is at (M–71), which corresponds with the loss of a pentyl radical to give a propyl cation, as shown here.

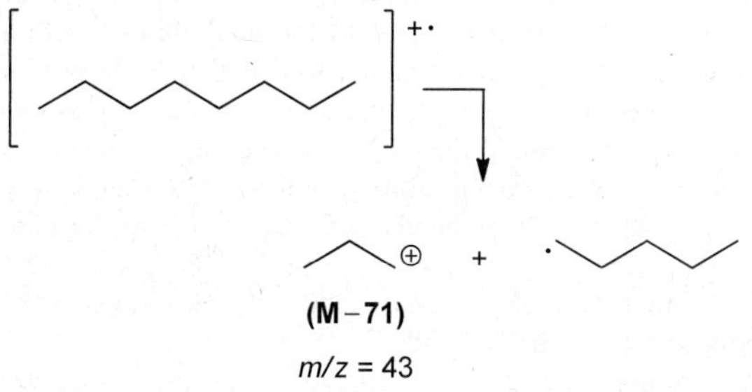

(M–71)

m/z = 43

15.54.
(a) A compound with four carbon atoms would require $(2 \times 4) + 2 = 10$ hydrogen atoms to be fully saturated. This compound has only six hydrogen atoms, so four hydrogen atoms are missing. The compound therefore has two degrees of unsaturation (HDI = 2).

(b) A compound with five carbon atoms would require $(2 \times 5) + 2 = 12$ hydrogen atoms to be fully saturated. This compound has only eight hydrogen atoms, so four hydrogen atoms are missing. The compound therefore has two degrees of unsaturation (HDI = 2).

(c) A compound with 40 carbon atoms would require $(2 \times 40) + 2 = 82$ hydrogen atoms to be fully saturated. This compound has only 78 hydrogen atoms, so four hydrogen atoms are missing. The compound therefore has two degrees of unsaturation (HDI = 2).

(d) A compound with 72 carbon atoms would require $(2 \times 72) + 2 = 146$ hydrogen atoms to be fully saturated. This compound has only 74 hydrogen atoms, so 72 hydrogen atoms are missing. The compound therefore has 36 degrees of unsaturation (HDI = 36).

(e) A compound with the molecular formula $C_6H_6O_2$ is expected to have the same HDI as a compound with the molecular formula C_6H_6 (the oxygen atoms can be ignored for purposes of calculating HDI). With six carbon atoms, the compound would need to have fourteen hydrogen atoms to be fully saturated. With only six hydrogen atoms, eight are missing, representing four degrees of unsaturation (HDI = 4).

(f) A compound with the molecular formula $C_7H_9NO_2$ is expected to have the same HDI as a compound with the molecular formula C_7H_8 (the oxygen atoms can be ignored, and we subtract one H because of the presence of an N). With seven carbon atoms, the compound would need to have sixteen hydrogen atoms to be fully saturated. With only eight hydrogen atoms (in C_7H_8), eight are missing, representing four degrees of unsaturation (HDI = 4).

(g) A compound with the molecular formula $C_8H_{10}N_2O$ is expected to have the same HDI as a compound with the molecular formula C_8H_8 (the oxygen atom can be ignored, and we subtract one H for each N). With eight carbon atoms, the compound would need to have eighteen hydrogen atoms to be fully saturated. With only eight hydrogen atoms (in C_8H_8), ten are missing, representing five degrees of unsaturation (HDI = 5).

(h) A compound with the molecular formula $C_5H_7Cl_3$ is expected to have the same HDI as a compound with the molecular formula C_5H_{10} (each Cl is treated like an H). With five carbon atoms, the compound would need to have twelve hydrogen atoms to be fully saturated. With only ten hydrogen atoms (in C_5H_{10}), two are missing, representing one degree of unsaturation (HDI = 1).

(i) A compound with the molecular formula C_6H_5Br is expected to have the same HDI as a compound with the molecular formula C_6H_6 (Br is treated like H). With six carbon atoms, the compound would need to have fourteen hydrogen atoms to be fully saturated. With only six hydrogen atoms, eight are missing, representing four degrees of unsaturation (HDI = 4).

(j) A compound with the molecular formula $C_6H_{12}O_6$ is expected to have the same HDI as a compound with the molecular formula C_6H_{12} (the oxygen atoms can be ignored for purposes of calculating HDI). With six carbon atoms, the compound would need to have fourteen hydrogen atoms to be fully saturated. With only twelve hydrogen atoms, two are missing, representing one degree of unsaturation (HDI = 1).

15.55. The compound contains only carbon atoms and hydrogen atoms, and there is a signal in the IR spectrum at 3300 cm^{-1}, indicating a C_{sp}–H bond. That is, the compound is a terminal alkyne. There are only two terminal alkynes with the molecular formula C_5H_8:

15.56. Limonene is a hydrocarbon, which means that it contains only carbon atoms and hydrogen atoms. That is, we must account for the entire molecular weight (136 amu) with only carbon atoms (12 amu) and hydrogen atoms (1 amu). With two double bonds and one ring, the compound has three degrees of unsaturation. So, we begin by considering the following possibilities, each of which has three degrees of unsaturation (as required by the problem statement).

$$C_9H_{14} = (9 \times 12) + (14 \times 1) = 122 \text{ amu}$$
$$C_{10}H_{16} = (10 \times 12) + (16 \times 1) = 136 \text{ amu}$$
$$C_{11}H_{18} = (11 \times 12) + (18 \times 1) = 150 \text{ amu}$$

The middle possibility ($C_{10}H_{16}$) matches exactly (136 amu).

15.57. Let's begin by drawing the structures of the compounds that must be distinguished:

Br—1
        2   3
            4

1-Bromo-
3-methyl-2-butene

1
  2   3
Br   4

2-Bromo-
3-methyl-2-butene

Both of these compounds contain an unsymmetrical C=C bond that is expected to produce a signal in the range between 1600 and 1700 cm^{-1}, so we cannot use that signal to distinguish the compounds. Next we consider any other signals that are expected to appear in the diagnostic region (1600 – 4000 cm^{-1}). Neither compound has a triple bond so the region between 2100 and 2300 cm^{-1} won't be helpful. So we look to the region of the spectrum that has signals from X–H bonds. The first compound does indeed have a C_{sp^2}–H bond that is absent in the second compound:

C_{sp^2}–H   no C_{sp^2}–H

As such, the IR spectrum of the first compound is expected to exhibit a signal near 3100 cm^{-1}, while the IR spectrum of the second compound is not expected to have any signals above 3000 cm^{-1}. These compounds

can therefore be distinguished by looking for a signal at 3100 cm^{-1} in the IR spectrum of each compound.

15.58. The compound can exhibit intramolecular hydrogen bonding even in dilute solutions.

intramolecular
hydrogen bonding

H—O⁀H—O
     1   3   5
       2   4

15.59. The IR spectrum has a strong signal just above 1700 cm^{-1}, which indicates that the compound has a C=O bond. The mass spectrum has a molecular ion peak at $m/z = 86$, which indicates a molecular weight of 86. The base peak is at M–43, indicating the loss of a propyl group. The compound likely has a three carbon chain (either a propyl group or isopropyl group) as shown in the following two structures.

15.60. The IR spectrum has a narrow signal near 2100 cm^{-1}, which indicates the presence of a triple bond. In addition, the signal at 3300 cm^{-1} is consistent with a C_{sp}–H bond, which indicates that the compound is a terminal alkyne. The mass spectrum has a molecular ion peak at $m/z = 68$, which indicates a molecular weight of 68 amu. So we are looking for terminal alkynes with a molecular weight of 68 amu. Recall that each carbon atom contributes 12 amu, so five carbon atoms contribute 60 amu to the molecular weight, leaving just 8 amu left for hydrogen atoms (each hydrogen atom is 1 amu). That is, compounds with the molecular formula C_5H_8 will have a molecular weight of 68 amu. This molecular formula is consistent with two degrees of unsaturation (a triple bond).

There are two terminal alkynes with the molecular formula C_5H_8, shown here:

15.61. All of the reactions in the following sequence were covered in previous chapters. The starting alkyl chloride is a tertiary substrate and will undergo an E2 reaction when treated with a strong base. With a base like ethoxide (which is not sterically hindered), the major product is the more-substituted alkene (the Zaitsev product), compound **A**. Hydroboration-oxidation of compound **A** gives an *anti*-Markovnikov addition of H and OH, affording alcohol **B**. When treated with tosyl chloride and pyridine, the alcohol is turned into the corresponding tosylate (compound **C**). Treatment of the tosylate with a sterically hindered base gives another E2 reaction, this time giving the less substituted alkene (the Hofmann product), compound **D**. Treating **D** with a peroxy acid, such as MCPBA, gives epoxide **E**. When the epoxide is treated with a Grignard reagent (such as MeMgBr), followed by an aqueous workup, a ring-opening reaction occurs in which the Grignard reagent attacks the less substituted side of the epoxide, to afford alcohol **F**. The last step of the sequence is a Williamson ether synthesis, in which a methyl group is installed to give compound **G**.

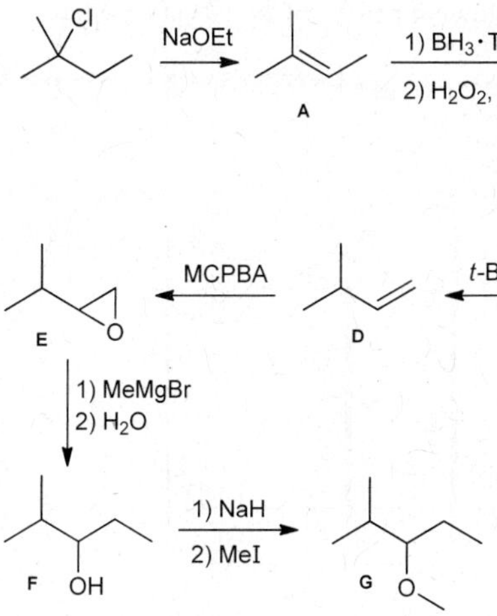

(a) Compound **F** is an alcohol and its IR spectrum will exhibit a broad signal between 3200 and 3600 cm^{-1}. Compound **G** is an ether and its IR spectrum will not exhibit the same signal.

(b) Compound **D** is an alkene and its IR spectrum will exhibit a signal near 1650 cm^{-1} (for the C=C bond), as well as a signal near 3100 cm^{-1} (for the C_{sp^2}–H bond). Compound **E** is an epoxide, and its IR spectrum will not have these two signals.

(c) IR spectroscopy would not be helpful to distinguish these two compounds because they are both alcohols. Mass spectrometry could be used to differentiate these two compounds because they have different molecular weights.

(d) No, they both have the same molecular formula, although a trained expert might be able to distinguish these compounds based on their fragmentation patterns.

15.62. The molecular formula C_4H_8 indicates an HDI of 1. As such, every constitutional isomer of C_4H_8 must contain either one ring or one double bond. The following structures are consistent with this description. Each of the first three constitutional isomers (shown here) exhibits a double bond, while each of the last two constitutional isomers exhibits a ring.

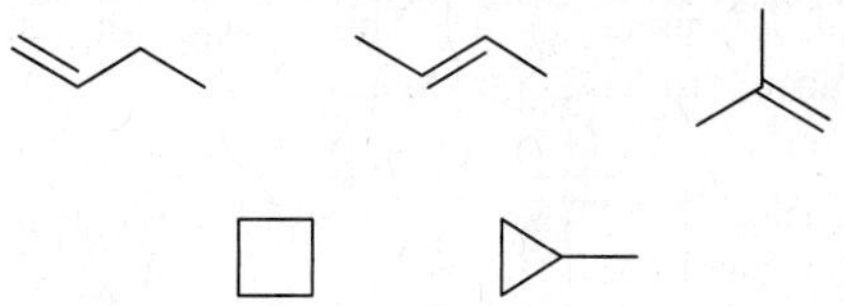

From among these isomers, only one of them (1-butene) can lose a methyl group to form a resonance stabilized carbocation at (M–15):

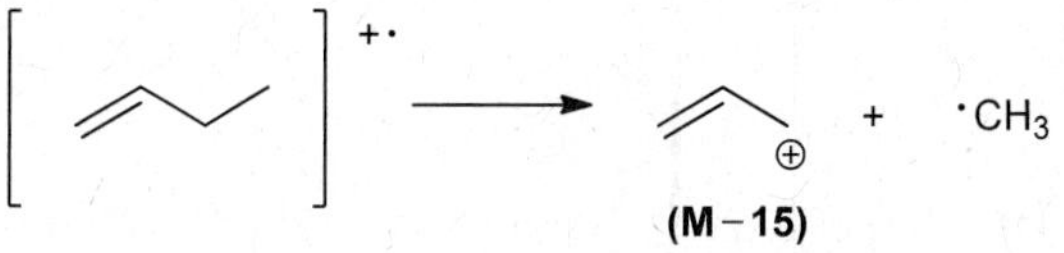

(M–15)
resonance-stabilized

15.63. As seen in the solution to Problem **2.54**, there are four constitutional isomers with the molecular formula C_4H_9Cl, shown here again:

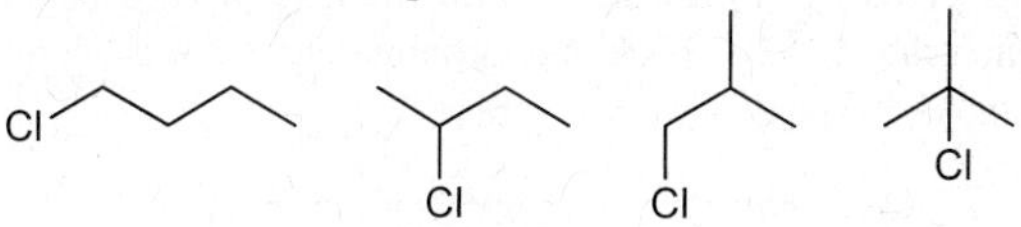

Compound **A**

Only one of these four isomers, compound **A**, exhibits a chirality center. When compound **A** is treated with a strong base, the following three alkenes are formed (**B**, **C**, and **D**):

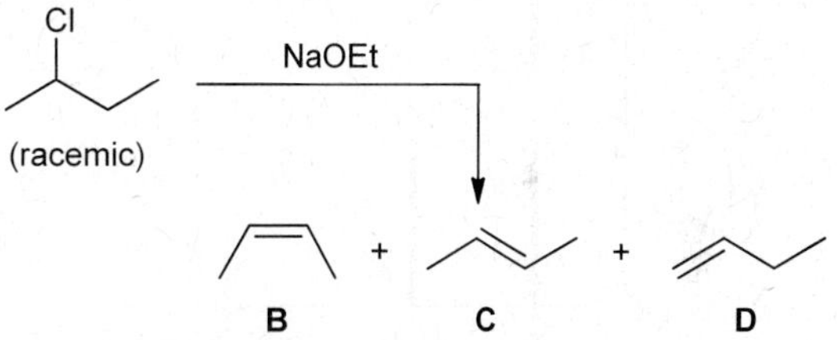

Compounds **B** and **C** are diastereomers (*cis* vs. *trans*), with the *trans* alkene being favored. That leaves the minor product **D**, which is an unsymmetrical alkene. Accordingly, we do expect that the C=C bond will have a small dipole moment, so there should be a signal near 1650 cm^{-1} in the IR spectrum of compound **D**.

15.64. In order to determine the pattern of signals that we expect in the mass spectrum of a compound with two chlorine atoms, we must quickly review the reason for the pattern of signals that are observed for a compound with only one chlorine atom. Let's first imagine that all chlorine atoms had the same mass (35 amu). In such a hypothetical world, we would see the $(M)^{+\bullet}$ peak, but we would not observe a significant $(M+2)^{+\bullet}$ peak. However, all chlorine atoms are not the same mass, because there are two naturally abundant isotopes: ^{35}Cl and ^{37}Cl. The latter represents approximately ¼ of all chlorine atoms in the world. As such, approximately ¼ of the parent ions (for a compound with one chlorine atom) will contain an atom of ^{37}Cl instead of ^{35}C. This causes ¼ of the signal to be moved from (M)$^{+\bullet}$ to (M+2)$^{+\bullet}$.

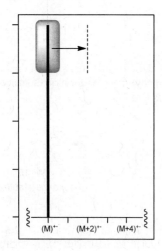

This gives the following familiar pattern of signals, in which the (M+2)$^{+\bullet}$ peak is approximately 1/3 the height of the (M)$^{+\bullet}$ peak:

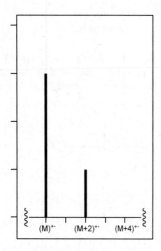

This is the characteristic pattern of signals for a compound with one chlorine atom.

Now let's consider a compound with *two* chlorine atoms. We will begin with the pattern of signals above, and determine how this pattern changes if the parent ion has two chlorine atoms, rather than just one. Once again, ¼ of each peak is moved up by two *m/z* units, because in ¼ of all of those ions, the second chlorine atom will be ^{37}Cl rather than ^{35}Cl.

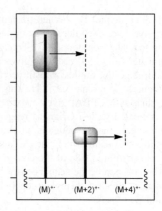

This gives the following pattern of signals at M, M+2, and M+4.

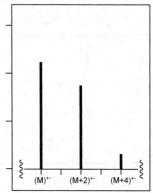

15.65. The OH group in ephedrine can engage in intramolecular hydrogen bonding, even in dilute solutions.

intramolecular hydrogen bonding

15.66. In order to determine the pattern of signals that we expect in the mass spectrum of a compound with two bromine atoms, we must quickly review the reason for the pattern of signals that are observed for a compound with only one bromine atom. Let's first imagine that all bromine atoms had the same mass (79 amu). In such a hypothetical world, we would see the (M)$^{+\bullet}$ peak, but we would not observe a significant (M+2)$^{+\bullet}$ peak. However, all bromine atoms do not have the same mass, because there are two naturally abundant isotopes: ^{79}Br and ^{81}Br. Each isotope represents approximately half of all bromine atoms in the world. As such, approximately half of the parent ions (for a compound with one bromine atom) will contain an atom of ^{81}Br instead of ^{79}Br. This causes approximately half of the signal to be moved from (M)$^{+\bullet}$ to (M+2)$^{+\bullet}$.

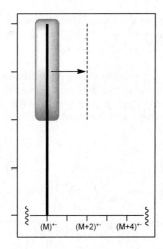

This gives the following familiar pattern of signals, in which the (M+2)$^{+\bullet}$ peak is approximately the same height as the (M)$^{+\bullet}$ peak:

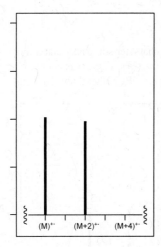

This is the characteristic pattern of signals for a compound with one bromine atom.

Now let's consider a compound with *two* bromine atoms. We will begin with the pattern of signals above, and determine how this pattern changes if the parent ion has two bromine atoms, rather than just one. Once again, half of each peak is moved up by two *m/z* units, because in half of all of those ions, the second bromine atom will be ^{81}Br rather than ^{79}Br.

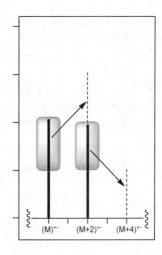

This gives the following pattern of signals at M, M+2, and M+4.

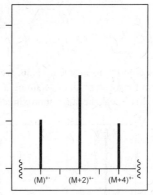

15.67.
Explanation #1) Outside of the carbonyl group, one of the C–O bonds of the ester has some double bond character, as can be seen in the third resonance structure below:

This C–O bond is a stronger bond than the other C–O single bond, which does not have any double bond character.   As a result, the stronger C–O bond (highlighted) produces a signal at higher wavenumber.

1300 cm^{-1} 1000 cm^{-1}

Explanation #2) The C–O bond at 1300 cm^{-1} involves an sp^2 hybridized carbon atom, rather than an sp^3 hybridized carbon atom. The former has more *s*-character and holds its electrons closer to the positively charged nucleus. A C_{sp^2}–O bond is therefore shorter and stronger than a C_{sp^3}–O bond, so the former should produce a signal at higher wavenumber.

15.68. A signal at 1720 cm^{-1} indicates the presence of a C=O bond. The following mechanism justifies the formation of a C=O bond. In the first step, one of the OH groups is protonated to give an oxonium ion, which then loses a leaving group (H$_2$O) to give a secondary carbocation. This carbocation then rearranges to give a more-stable, resonance-stabilized cation, which is deprotonated to give the product (a ketone), consistent with the signal at 1720 cm^{-1}.

15.69. In the solution to Problem **15.66**, we explained the source of the characteristic peaks that are observed in the mass spectrum for a compound containing a bromine atom. Specifically, the (M+2)$^{+\bullet}$ peak is approximately the same height as the (M)$^{+\bullet}$ peak, because approximately half of the bromine atoms are ^{81}Br, rather than ^{79}Br.

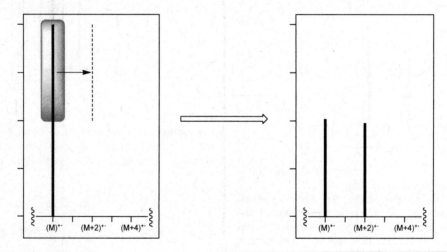

Now let's consider what happens if the parent ion also has a chlorine atom. In the solution to Problem **15.64**, we saw that the effect of a chlorine atom is to move ¼ of each peak by two *m/z* units, which gives the following pattern.

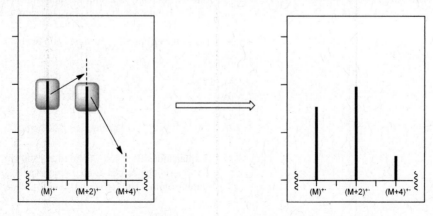

Incidentally, we would arrive at the same characteristic pattern if we had started our analysis by first considering the effect of the chlorine atom and only then considering the effect of the bromine atom, as illustrated here:

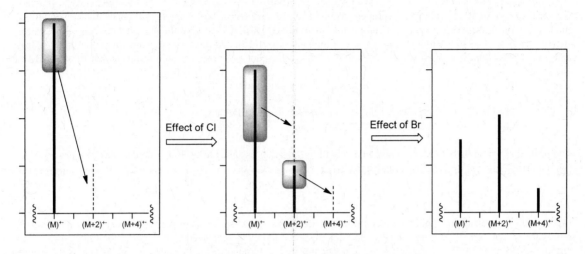

15.70.

(a) The molecular formula for the dye is $C_{21}H_{15}N_2O_6S^-$. Using the values in Table 15.5 for C, H, N and O, and the values given in the question for S, we can calculate the expected mass for the parent ion using the most abundant isotope of each element (^1H, ^{12}C, ^{14}N, ^{16}O, ^{32}S) as follows.

$$C_{21}H_{15}N_2O_6S^- =$$
$$(21 \times 12.000) + (15 \times 1.0078) + (2 \times 14.0031) + (6 \times 15.9949) + (1 \times 31.9721)$$
$$= 423.0647$$

This is consistent with reported base peak of the spectrum $[(M^-), 100\%$ relative intensity].

Now we must consider the other two peaks – the peak at 424.0681 is nominally 1 amu higher than the parent molecular ion, while the peak at 425.0605 is nominally two amu higher. Considering the relative natural isotopic abundances of C, H, N, O and S, the two elements that have the highest percentage of a second isotope are carbon ($^{12}C = 98.93\%$, $^{13}C = 1.07\%$) and sulfur ($^{32}S = 95.02\%$, $^{34}S = 4.21\%$). With this in mind, the peak at $m/z = 424.0681$ is consistent with an ion where one of the ^{12}C atoms is replaced with a ^{13}C.

$$^{12}C_{20}{}^{13}C_1H_{15}N_2O_6S^- =$$
$$(20 \times 12.000) + (1 \times 13.0034) + (15 \times 1.0078) + (2 \times 14.0031) + (6 \times 15.9949) + (1 \times 31.9721)$$
$$= 424.0681$$

The peak with a nominal mass of 425 is consistent with an ion with one ^{34}S, as demonstrated by the calculation below:

$$C_{21}H_{15}N_2O_6{}^{34}S^- =$$
$$(21 \times 12.000) + (15 \times 1.0078) + (2 \times 14.0031) + (6 \times 15.9949) + (1 \times 33.9679)$$
$$= 425.0605$$

Now, we must consider the relative abundances of each peak. The peak at $m/z = 423.0647$ is the base peak, because it has a relative abundance of 100% (it is, by definition, the most abundant ion). The peak at 424.0681 has an abundance of 22.5%, relative to the most abundant ion. This is consistent with an ion having 21 carbon atoms, because the natural abundance of ^{13}C is 1.07%. [21 $\times$ 1.07% = 22.5%]

The peak at 425.0605 has an abundance of 4.21%, relative to the most abundant ion. This is consistent with an ion having one sulfur atom, because the natural abundance of ^{34}S is 4.21%. [1 $\times$ 4.21% = 4.21%]

(b) The peak at $m/z = 425.0605$ is consistent with an ion in which two of the ^{12}C atoms are replaced with two ^{13}C atoms:

$$^{12}C_{19}{}^{13}C_2H_{15}N_2O_6S^- =$$
$$(19 \times 12.000) + (2 \times 13.0034) + (15 \times 1.0078) + (2 \times 14.0031) + (6 \times 15.9949) + (1 \times 31.9721)$$
$$= 425.0715$$

15.71. The molecular ion for cyclohexanone is expected at $m/z = 98$, so the two peaks at $m/z = 55$ must be due to ionic fragments. Considering the atoms present in cyclohexanone ($C_6H_{10}O$), we must consider all possible ionic formula with a nominal $m/z = 55$.

One approach to this is to start with the heavier elements (carbon, oxygen) to come up with a formula close to (but not greater than) 55 and then adding in hydrogens to make up the difference. For example, an ion with 4 carbon atoms (nominal mass = 48) would require 7 hydrogen atoms (nominal mass = 7) to give the correct total mass [48+7=55]. Alternatively, we can start with one oxygen atom (nominal mass = 16), add three carbon atoms (nominal mass = 36), and 3 hydrogen atoms (nominal mass = 3) [16+36+3=55]. Thus, our two possible ionic formulas that fit are $C_3H_3O^+$ and $C_4H_7^+$.

Using the values in Table 15.5, we can calculate the high resolution masses:

$$C_3H_3O^+ = (3 \times 12.000) + (3 \times 1.0078) + (1 \times 15.9949) = \textbf{55.0183 amu}$$
$$C_4H_7^+ = (4 \times 12.000) + (7 \times 1.0078) = \textbf{55.0546 amu}$$

The peak at $m/z = 55.0183$ (with a relative intensity of 86.7) is thus assigned to $C_3H_3O^+$, and the peak at $m/z = 55.0546$ (with a relative intensity of 13.3) is assigned to $C_4H_7^+$.

15.72.
(a) Tautomerization occurs rapidly (and is difficult to prevent) so an IR spectrum of **A** is essentially an IR spectrum of a mixture of **A** and **B**. The signal at 1740 cm^{-1} corresponds to the nonconjugated C=O bond of the ester group in **A**, and the signal at 1720 cm^{-1} corresponds to the nonconjugated C=O bond of the ketone group in **A**, just as we would expect.

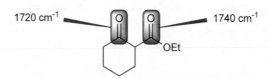

The C=O bond in compound **B** is part of a conjugated ester, so it appears at a lower wavenumber than a typical C=O bond of an ester. This must be the signal at 1660 cm^{-1}. This might be initially surprising, as we might have expected a conjugated ester to produce a signal around 1700 cm^{-1}. But on further inspection, we recognize that this conjugated system has an OH group, whose lone pairs are participating in resonance.

As such, there is one additional resonance structure (highlighted above); this additional resonance structure gives the C=O bond additional single bond character (relative to other conjugated esters).
Finally, the signal at 1620 cm^{-1} corresponds with the conjugated C=C bond in **B**.

(b) Compound **B** is capable of forming an intramolecular hydrogen bonding interaction, which lowers the energy of that tautomer. As such, it has a more significant presence at equilibrium.

There is one additional factor that favors the enol in this case, which we will discuss in Chapter 17. Specifically, we will see that conjugation is a source of stabilization. Compound **B** possesses this type of stabilization, while compound **A** does not.

(c) Compound **D**, as shown below, is the tautomer of **C**.

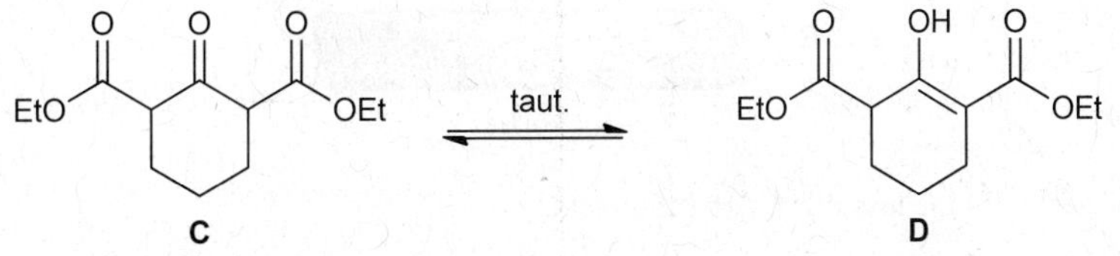

The absence of a signal at 1720 cm^{-1} indicates that the concentration of **C** is negligible, because **C** possesses an unconjugated ketone group (which should produce a signal at 1720 cm^{-1}). Therefore, the IR spectrum reveals that **D** is the dominant tautomer at equilibrium.

15.73.
Using the nitrogen rule, we can quickly determine that the molecular mass of the parent compound is an odd number since there is one nitrogen atom in the structure. Thus, the difference in mass between the expected molecular ion (an odd number) and the fragment of interest ($m/z = 113$, which is also an odd number) is an even number. This suggests the possibility of a McLafferty rearrangement. [This can be confirmed by calculating the molecular mass using the molecular formula ($C_{23}H_{43}NO$), resulting in a difference between the expected molecular ion ($m/z = 349$) and the fragment of interest ($m/z = 113$) of 236.] This amide has the structural feature required for such a rearrangement: a hydrogen atom that is gamma to a C=O group (analogous to the ketones and aldehydes discussed in section 15.12). The resulting ion ($m/z = 113$) and its mechanism of formation are presented below.

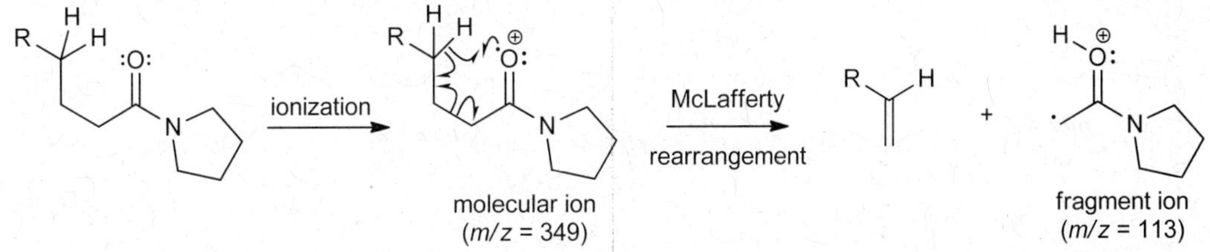

15.74.
(a) The mass of pinolenic acid (based on its given chemical formula) is 278 amu. The significant peak at 279 in positive ion mode is likely due to the addition of a proton, as this experiment is performed under conditions facilitating protonation. We can refer to this ion as the (M+1)$^{+}$ ion.

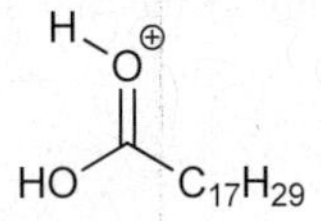

Next, we need to determine the positions of the three *cis* double bonds. The peaks at 211, 171 and 117 (which only appear when ozone is present) are likely due to ozonolysis products of each of the three alkenes. In each case, the C=C is replaced with C=O bonds. These three peaks represent losses of 68, 108, and 162 from the (M+1)$^{+}$ ion. The change in mass between the (M+1)$^{+}$ ion and each of the ozonolysis products corresponds to the loss of carbon atoms and hydrogen atoms (from the chain terminus to each sp^2 hybridized carbon atom) as well as the gain of an oxygen atom. Accounting for the additional mass of oxygen (16) in each ionic fragment, the portions of the molecule cleaved by ozonolysis have masses of **84** (=68+16), **124** (=108+16) and **178** (=162+16).

With this information, we can reconstruct the chain starting from the end distal to the carboxylic acid. A loss of 84 is consistent with the loss of a C_6H_{12} fragment.

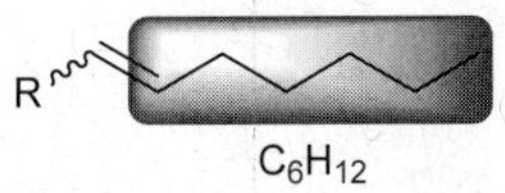

C_6H_{12}

Continuing up the chain, a loss of 124 is consistent with the loss of a C_9H_{16} fragment.

C_9H_{16}

A loss of 178 is consistent with the loss of a $C_{13}H_{22}$ fragment.

$C_{13}H_{22}$

With all of this information in hand, along with the knowledge that all of the double bonds have the *cis* configuration, we can propose the following structure for pinolenic acid:

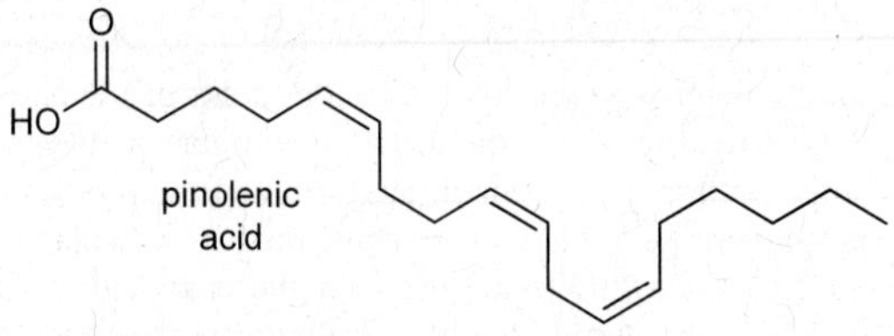

pinolenic
acid

And we can propose the following structures for the four indicated fragments in positive ion mode.

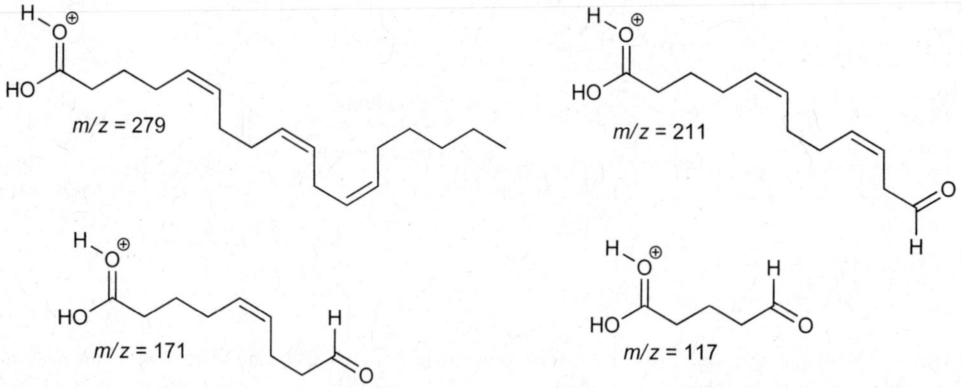

m/z = 279

m/z = 211

m/z = 171

m/z = 117

(b) In negative ion mode, under conditions facilitating deprotonation, we expect to form the conjugate base of the carboxylic acid instead of the conjugate acid, thus giving the following ions, each with two mass units lower than the corresponding ions in positive ion mode.

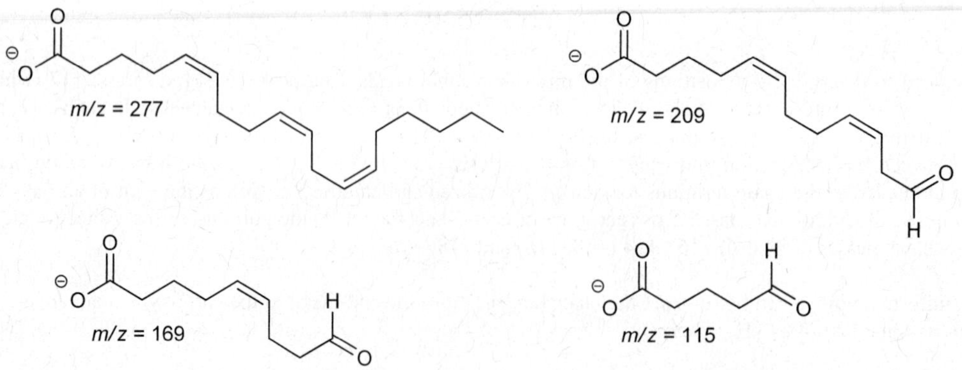

m/z = 277

m/z = 209

m/z = 169

m/z = 115

15.75.
The two compounds are isomers (both are C_8H_9BrO) so we expect the molecular ion from each of them to have the same m/z value. Recall that there are two isotopes of bromine with approximately equivalent natural abundance (^{81}Br and ^{79}Br), which results in two molecular ion peaks for each compound:

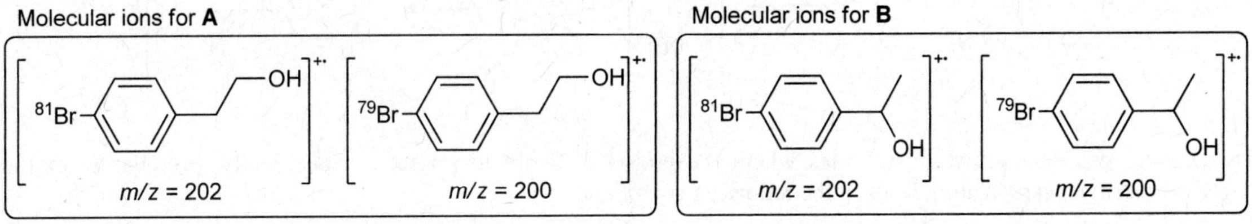

Molecular ions for **A**

Molecular ions for **B**

These molecular ions correspond to the two highest m/z peaks in each spectrum. Now we have to consider fragmentation patterns to match each spectrum to the correct isomer. Note that in each spectrum, there are several pairs of peaks separated by two mass units. This is consistent with the presence of bromine in these fragments, with the two peaks corresponding to fragments with ^{79}Br and ^{81}Br, respectively, just as observed for the molecular ion. In the peak list below, pairs separated by two mass units are shown in brackets.

Sample **X**, m/z = [202, 200], [187, 185], [159, 157], 121
Sample **Y**, m/z = [202, 200], [171, 169], 121

Focusing on sample **X**, the peaks at 187 and 185 represent a loss of 15 from the molecular ions 202 and 200, respectively. This is consistent with the loss of a •CH_3 radical fragment. Compound **B** has a methyl group, while compound **A** does not, suggesting that this spectrum is from compound **B**. The peaks at 159 and 157 represent a loss of 43 from the molecular ions 202 and 200, respectively. This is consistent with the loss of a •$CH(OH)CH_3$ radical fragment. The peak at 121 represents a loss of ^{81}Br• radical from the heavier molecular ion (202-81=121) or a loss of ^{79}Br• radical from the lighter molecular ion (200-79=121). These data are consistent with the structure of compound **B**, and the ions below.

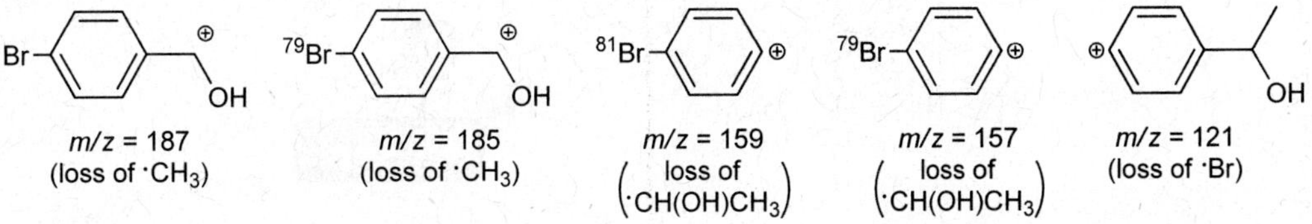

Focusing on sample **Y**, the peaks at 171 and 169 represent a loss of 31 from the molecular ions 202 and 200, respectively. This is consistent with the loss of a •CH_2OH radical fragment. The peak at 121 represents a loss of ^{81}Br• radical from the heavier molecular ion (202-81=121) or a loss of ^{79}Br• radical from the lighter molecular ion (200-79=121), analogous to the corresponding peak in sample **X**. These data are consistent with the structure of compound **A** and with the ions shown below.

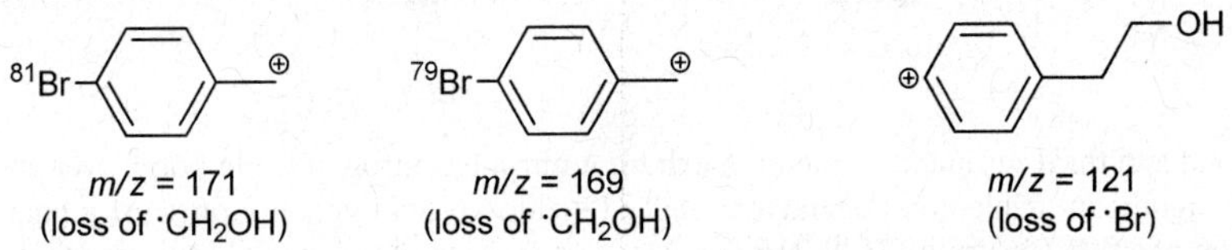

15.76.
(a) The structure contains an N-H bond, which is expected to produce a broad signal above 3300 cm^{-1}, but that signal is absent from the spectrum.

(b) Recall that tautomers are constitutional isomers that can rapidly interconvert via the migration of a proton. We know that the tautomer of the proposed structure must not exhibit an N-H bond, so it is clear that this is the proton that must migrate. To determine where it migrates, we remove the proton (this is how tautomerization occurs under base-catalyzed conditions) and draw the conjugate base, which is resonance stabilized:

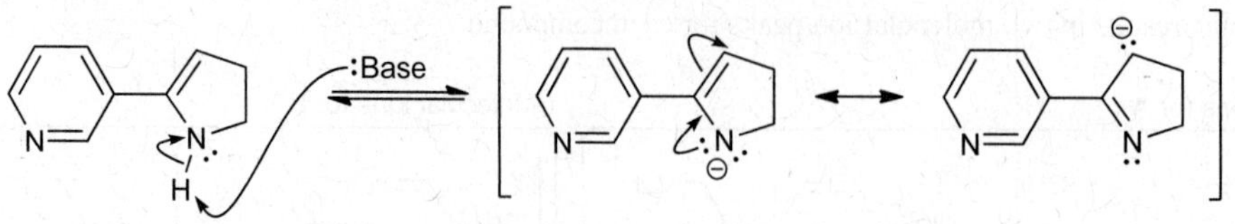

The second resonance structure indicates where the proton is likely to migrate. Specifically, protonation of the negatively charged carbon atom leads to the following compound:

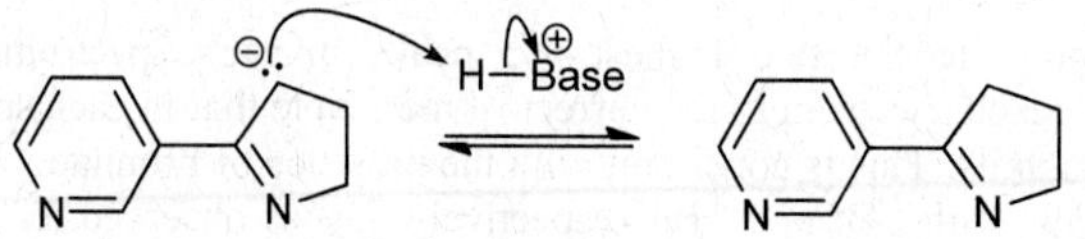

This compound lacks an N-H bond and is therefore consistent with the IR spectrum.

(c) The tautomer of the proposed structure exhibits a C=N bond, which is likely to be the source of the signal at 1621 cm^{-1} (this is the region of the spectrum where double bonds generally appear). An unconjugated C=N bond would likely appear at a higher wavenumber than 1621 cm^{-1}, but this C=N bond is conjugated, so the location of the signal is appropriate if the source of the signal is indeed the C=N bond.

15.77. Notice that the azido group in compound **2** exhibits charge separation. If we try to draw a resonance structure without charge separation, we find that we cannot do so without violating the octet rule:

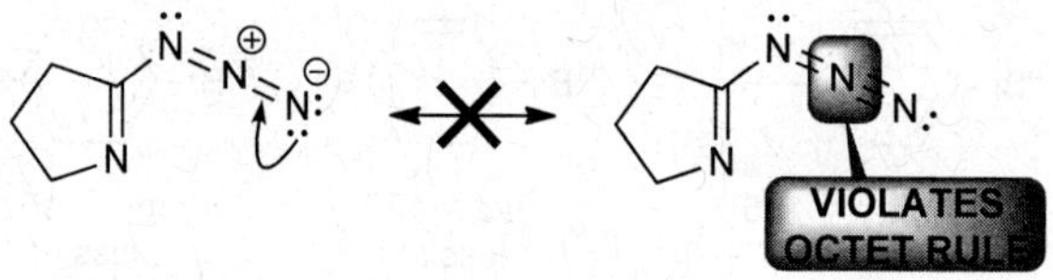

We see that the structure of the azido group cannot be drawn without charge separation (much like a nitro group, which also cannot be drawn without charge separation). Nevertheless, the azido group is still stabilized by resonance, as seen here:

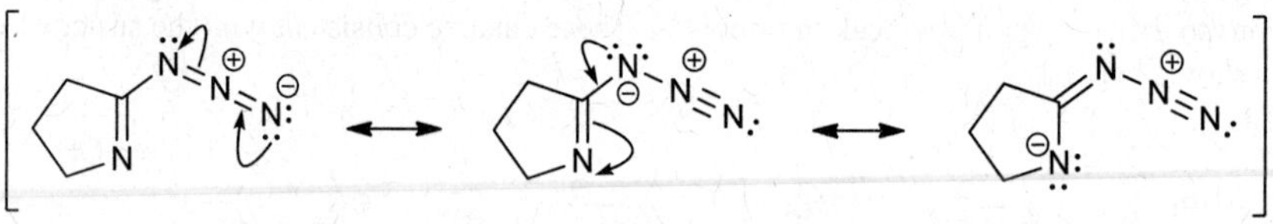

Notice that the second and third resonance structures exhibit a nitrogen-nitrogen triple bond. We therefore expect that this bond will have significant triple bond character. Indeed, azido groups typically produce a characteristic signal in the triple bond region of an IR spectrum (2120-2160 cm^{-1}). The absence of this signal indicates that compound **2** is not present in substantial quantities.

15.78.
(a) Each of these compounds has four rings, which we can label A-D:

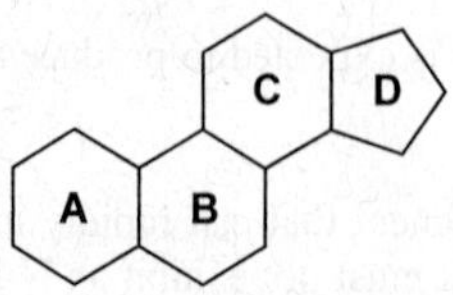

Notice that the A-B ring fusion resembles a *trans*-decalin system, which we explored at the end of Chapter 4:

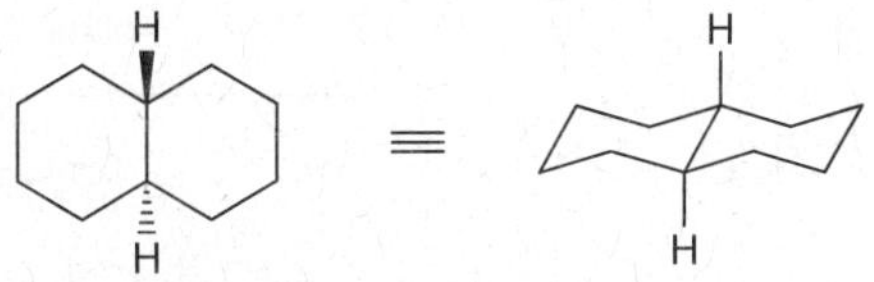

trans-decalin

But in our cases here, there four rings. Building on the decalin system as a foundation, can we redraw our compounds (starting with the A-B ring fusion), in a way that shows the conformation of each ring:

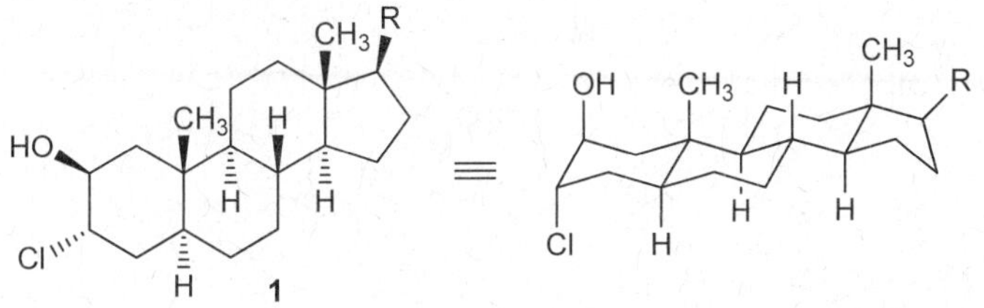

These polycyclic systems are conformationally rigid, because none of the six-membered rings can undergo a ring flip (you might find it helpful to build a model to prove to yourself that this is the case). Notice that in compound **2**, the OH group and the Cl group are in close proximity and can therefore participate in an intramolecular hydrogen-bonding interaction (which weakens the existing O-H bond in **2**, giving a lower wavenumber of absorption).   In contrast, the OH group and the Cl group are far apart from each other in compound **1**, and cannot participate in an intramolecular hydrogen-bonding interaction.

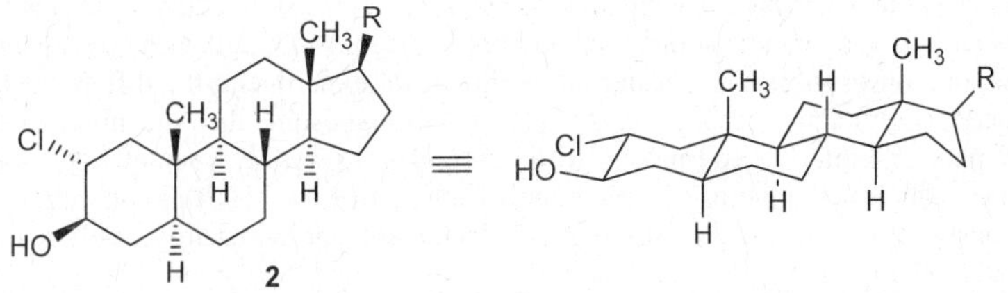

(b)   The problem statement specifies that dilute solutions were investigated. As such, the effects of *inter*molecular hydrogen bonds are negligible and can be ignored. Compound **2** is capable of forming an *intra*molecular hydrogen bonding interaction (while compound **1** is not), so the signal for the OH group in compound **2** is expected to be broader.

15.79.
(a) The problem statement indicates that one of the two hydrocarbon chains contains a *cis* alkene group. To determine which of these two chains contains the double bond, consider the relative numbers of carbon atoms and hydrogen atoms in each chain. One of these chains (-$C_{15}H_{31}$) is fully saturated, because there are two hydrogen atoms for every carbon atom in the chain, except for the terminal methyl group which has three hydrogen atoms. In contrast, the other chain (-$C_{17}H_{33}$) must be unsaturated, because a saturated chain with 17 carbon atoms would be expected to have 35 hydrogen atoms (16 C with two H's and one C with three H's). Since it has only 33 hydrogen atoms, this chain must contain the double bond.

Now that we know which chain contains the double bond, we can determine the position of the double bond by analyzing the mass spectrometry data. The introduction of ozone allows for ozonolysis of the alkene, breaking the C=C bond and producing two new C=O bonds. The peak at *m/z* = 563 provides the information needed to determine the position of the double bond. The change in mass between the parent ion and the product of ozonolysis (110 mass units) corresponds to the loss of carbon atoms and hydrogen atoms (from the chain terminus to the first sp^2 hybridized carbon) as well as the gain of an oxygen atom. Accounting for the additional mass of oxygen (16), the portion of the molecule that was cleaved by ozonolysis has a mass of 126. This is consistent with the loss of a fragment with the formula C_9H_{18}, as shown here:

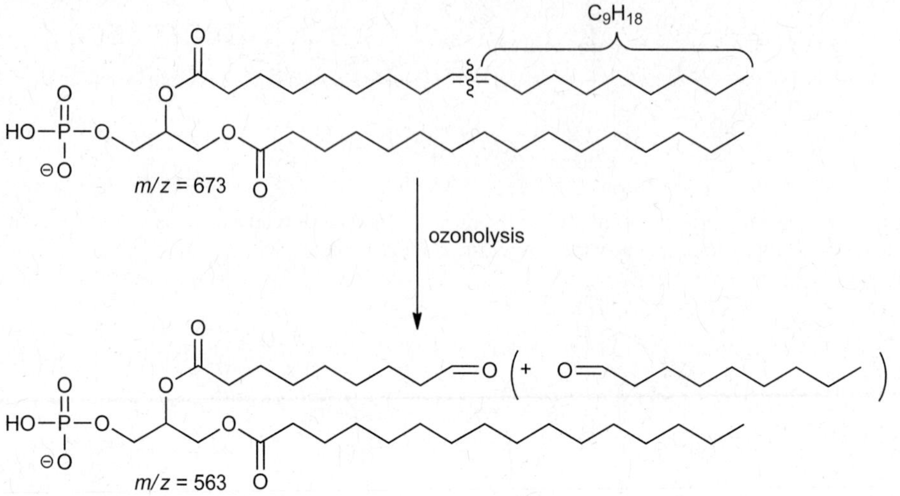

(b) The difference between the new peak in methanol (611) and the new peak in ethanol (625) is 14 mass units. This is equivalent to the difference in mass between the two alcohols (methanol = 32, ethanol = 46), suggesting that the new peak arises from incorporation of solvent. Focusing on the methanol experiment, the difference between the peak for the ozonolysis product (563) and the second peak (611) is 48, suggesting that the mystery product incorporates methanol (mass = 32) plus an extra oxygen (mass = 16) [32+16=48]. Likewise, when ethanol is used the new peak at 625 (which is 62 larger than 563) is due to incorporation of ethanol (mass = 46) plus an extra oxygen (mass = 16) [46+16=62]. This is consistent with the incorporation of the alcohol solvent by reacting with one of the intermediates of the ozonolysis reaction, as shown here.

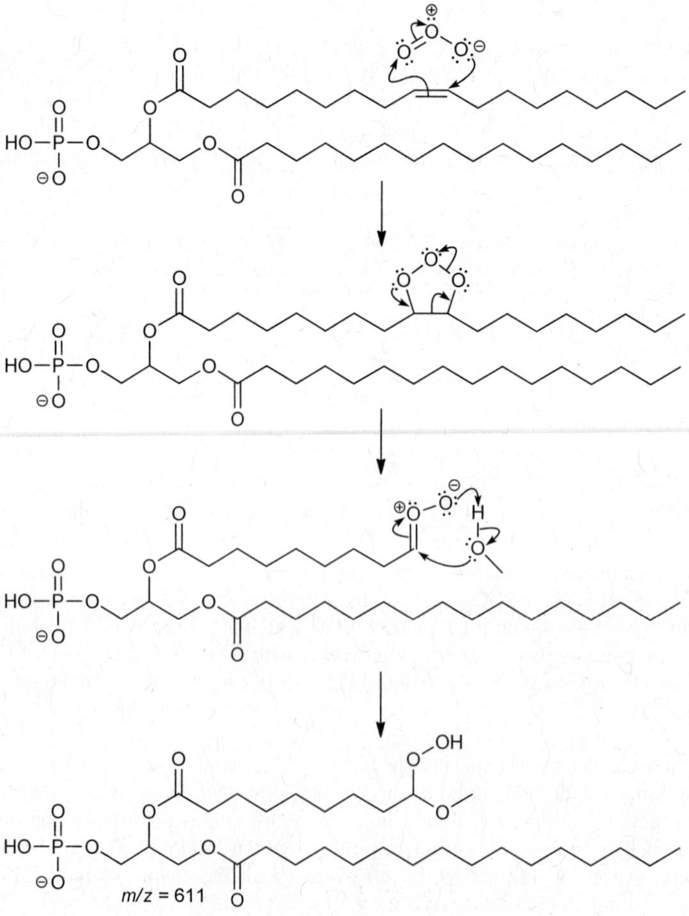

Chapter 16
Nuclear Magnetic Resonance Spectroscopy

Review of Concepts

Fill in the blanks below. To verify that your answers are correct, look in your textbook at the end of Chapter 16. Each of the sentences below appears verbatim in the section entitled *Review of Concepts and Vocabulary*.

- A spinning proton generates a **magnetic** _____, which must align either with or against an imposed external magnetic field.
- All protons do not absorb the same frequency because of _____, a weak magnetic effect due to the motion of surrounding electrons that either **shield** or **deshield** the proton.
- _____ solvents are generally used for acquiring NMR spectra.
- In a ^{1}H NMR spectrum, each signal has three important characteristics: _____, _____ and _____.
- When two protons are interchangeable by rotational symmetry, the protons are said to be _____.
- When two protons are interchangeable by reflectional symmetry, the protons are said to be _____.
- The left side of an NMR spectrum is described as _____**field**, and the right side is described as _____**field**.
- In the absence of inductive effects, a methyl group (CH_3) will produce a signal near _____ppm, a **methylene group** (CH_2) will produce a signal near _____, and a _____ **group** (CH) will produce a signal near _____. The presence of nearby groups increases these values somewhat predictably.
- The _____, or area under each signal, indicates the number of protons giving rise to the signal.
- _____ represents the number of peaks in a signal. A _____ has one peak, a _____ has two, a _____ has three, a _____ has four, and a _____ has five.
- Multiplicity is the result of **spin-spin splitting**, also called _____, which follows the **n+1 rule**.
- When signal splitting occurs, the distance between the individual peaks of a signal is called the **coupling constant**, or _____ **value**, and is measured in hertz.
- Complex splitting occurs when a proton has two different kinds of neighbors, often producing a _____.
- ^{13}C is an _____ of carbon, representing _____% of all carbon atoms.
- All ^{13}C-^{1}H splitting is suppressed with a technique called **broadband** _____, causing all of the ^{13}C signals to collapse to _____.

Review of Skills

Fill in the blanks and empty boxes below. To verify that your answers are correct, look in your textbook at the end of Chapter 16. The answers appear in the section entitled *SkillBuilder Review*.

16.1 Determining the Relationship between Two Protons in a Compound

FOR EACH OF THE FOLLOWING COMPOUNDS, IDENTIFY THE RELATIONSHIP BETWEEN THE TWO INDICATED PROTONS (ARE THEY HOMOTOPIC, ENANTIOTOPIC OR DIASTEREOTOPIC?) AND DETERMINE WHETHER THEY ARE CHEMICALLY EQUIVALENT.

RELATIONSHIP

CHEMICALLY EQUIVALENT?

16.2   Identifying the Number of Expected Signals in a ¹H NMR Spectrum

FOR EACH OF THE FOLLOWING COMPOUNDS, DETERMINE WHETHER THE INDICATED PROTONS ARE CHEMICALLY EQUIVALENT.

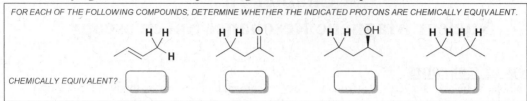

CHEMICALLY EQUIVALENT?

16.3   Predicting Chemical Shifts

FOR EACH OF THE FOLLOWING COMPOUNDS, PREDICT THE EXPECTED CHEMICAL SHIFT OF THE INDICATED PROTONS.

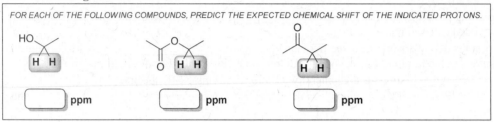

[   ] ppm   [   ] ppm   [   ] ppm

16.4   Determining the Number of Protons Giving Rise to a Signal

STEP 1 - COMPARE THE RELATIVE _____ VALUES, AND CHOOSE THE LOWEST NUMBER.	STEP 2 - DIVIDE ALL INTEGRATION VALUES BY THE NUMBER FROM STEP #1, WHICH GIVES THE RATIO OF _____	STEP 3 - IDENTIFY THE NUMBER OF PROTONS IN THE COMPOUND (FROM THE MOLECULAR FORMULA) AND THEN ADJUST THE RELATIVE INTEGRATION VALUES SO THAT THE SUM TOTAL EQUALS THE NUMBER OF _____.

16.5   Predicting the Multiplicity of a Signal

IDENTIFY THE EXPECTED MULTIPLICITY FOR EACH SIGNAL IN THE PROTON NMR SPECTRUM OF THE FOLLOWING COMPOUND.

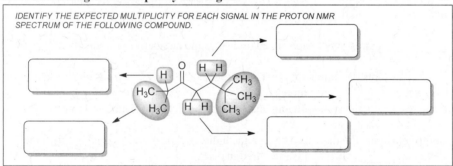

16.6   Drawing the Expected ¹H NMR Spectrum of a Compound

STEP 1 - IDENTIFY THE NUMBER OF _____.	STEP 2 - PREDICT THE _____ OF EACH SIGNAL.	STEP 3 - DETERMINE THE _____ OF EACH SIGNAL BY COUNTING THE NUMBER OF _____ GIVING RISE TO EACH SIGNAL.	STEP 4 - PREDICT THE _____ OF EACH SIGNAL	STEP 5 - DRAW EACH SIGNAL.

16.7   Using ¹H NMR Spectroscopy to Distinguish Between Compounds

STEP 1 - IDENTIFY THE NUMBER OF _____ THAT EACH COMPOUND WILL PRODUCE.	STEP 2 - IF EACH COMPOUND IS EXPECTED TO PRODUCE THE SAME NUMBER OF SIGNALS, THEN DETERMINE THE _____, _____, AND _____ OF EACH SIGNAL IN BOTH COMPOUNDS.	STEP 3 - LOOK FOR DIFFERENCES IN THE CHEMICAL SHIFTS, MULTIPLICITIES OR INTEGRATION VALUES OF THE EXPECTED SIGNALS.

16.8   Analyzing a ¹H NMR Spectrum and Proposing the Structure of a Compound

STEP 1 - USE THE _____ _____ TO DETERMINE THE HDI. AN HDI OF _____ INDICATES THE POSSIBILITY OF AN AROMATIC RING.	STEP 2 - CONSIDER THE NUMBER OF SIGNALS AND INTEGRATION OF EACH SIGNAL (GIVES CLUES ABOUT THE _____ OF THE COMPOUND).	STEP 3 - ANALYZE EACH SIGNAL (____, ____, AND ____), AND THEN DRAW FRAGMENTS CONSISTENT WITH EACH SIGNAL. THESE FRAGMENTS BECOME OUR PUZZLE PIECES THAT MUST BE ASSEMBLED TO PRODUCE A MOLECULAR STRUCTURE.	STEP 4 - ASSEMBLE THE FRAGMENTS.

16.9 Predicting the Number of Signals and Approximate Location of Each Signal in a ^{13}C NMR Spectrum

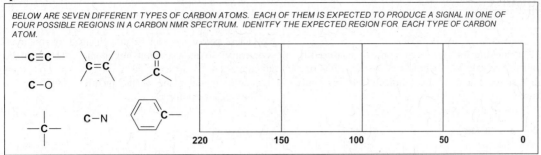

16.10 Determining Molecular Structure using DEPT ^{13}C NMR Spectroscopy

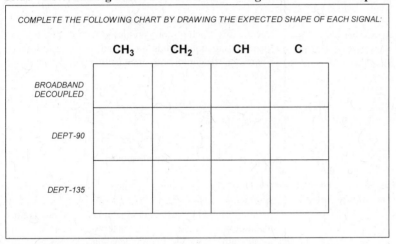

Mistakes to Avoid

In ^{1}H NMR spectroscopy, two neighboring CH_2 groups will only split each other if they are not chemically equivalent. For example compare the structures of butane and pentane, and in particular, compare the CH_2 group at the C2 position in each structure, highlighted below:

quartet   quartet of triplets (or triplet of quartets)

Butane   Pentane

In the case of butane, the signal for the highlighted CH_2 group is expected to be a quartet as a result of the neighboring methyl group. Notice that the signal is not further split by the neighboring CH_2 group (C3), because the CH_2 groups at C2 and C3 are chemically equivalent and therefore, they don't split each other. In contrast, the signal for the highlighted CH_2 group in pentane has a more complex multiplicity. In the case of pentane, the C2 and C3 positions are not identical. The C2 position is next to a methyl group, while the C3 position is not. They occupy different electronic environments, and they are not interchangeable by symmetry. As a result, the signal for the highlighted CH_2 group (C2) is expected to be split into a quartet of triplets (or into a triplet of quartets, depending on which J value is larger). When analyzing a ^{1}H NMR spectrum, make sure to take this into account, as students will often misinterpret a spectrum as a result of failing to take this into account.

Solutions

16.1.

(a) When looking for symmetry, don't be confused by the position of the double bonds in the aromatic ring. Recall that we can draw the following two resonance structures:

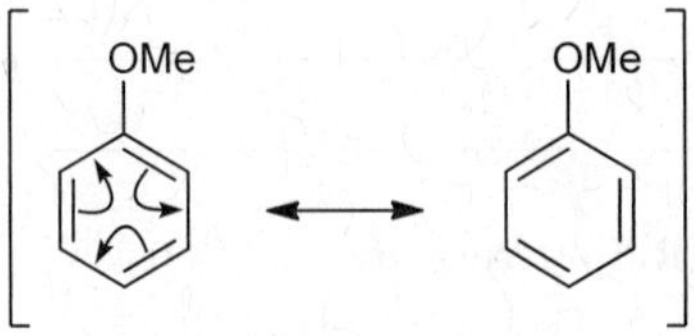

Neither resonance structure is more correct than the other. For purposes of looking for symmetry, it will be less confusing to draw the compound like this:

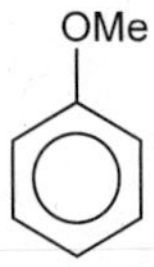

When drawn in this way, we can see that the two highlighted protons can be interchanged by rotational symmetry (the axis of symmetry is shown below). Therefore, the protons are homotopic.

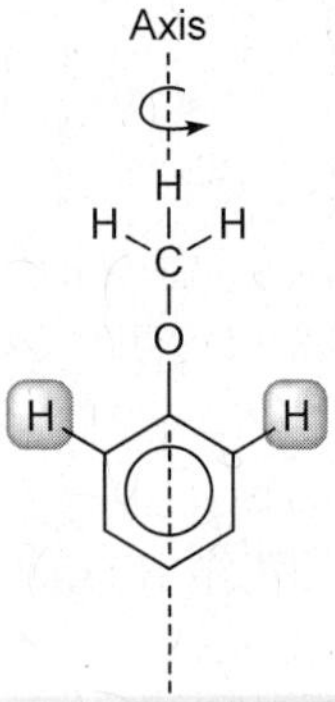

This conclusion can be verified by the replacement test. Specifically, each proton is replaced with deuterium, and the resulting compounds are found to be the same. Therefore, the compounds are homotopic:

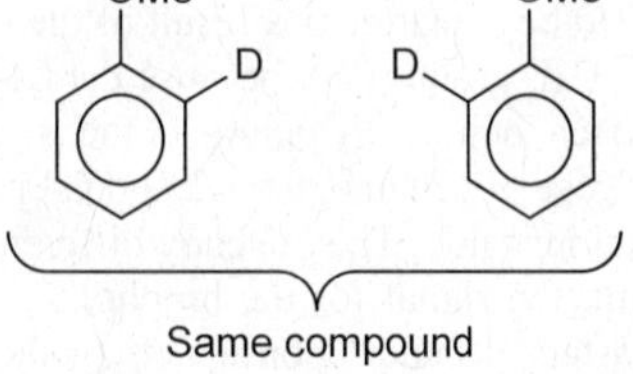

Same compound

(b) These protons cannot be interchanged by rotational symmetry, so they are not homotopic. They can be interchanged by reflectional symmetry (the plane of symmetry is the plane of the page). Therefore, the protons are enantiotopic.

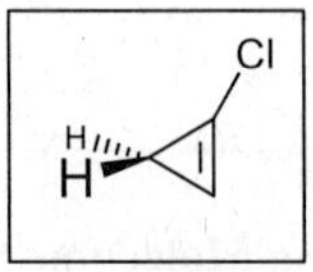

This conclusion can be verified by the replacement test. Specifically, each proton is replaced with deuterium, and the resulting compounds are found to be enantiomers. Therefore, the compounds are enantiotopic:

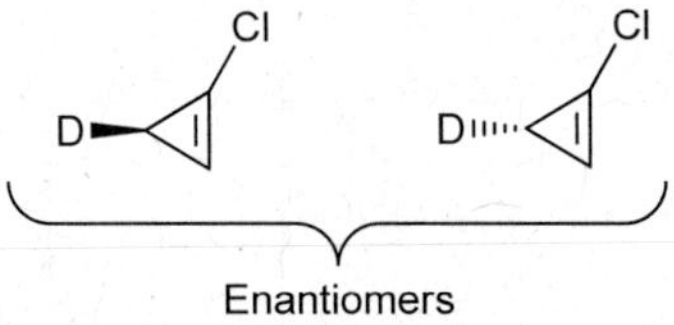

Enantiomers

(c) These protons cannot be interchanged by rotational symmetry, so they are not homotopic. They also cannot be interchanged by reflectional symmetry so they are not enantiotopic either. To determine if they are diastereotopic, we use the replacement test. Specifically, each proton is replaced with deuterium, and the resulting compounds are found to be diastereomers. Therefore, the compounds are diastereotopic:

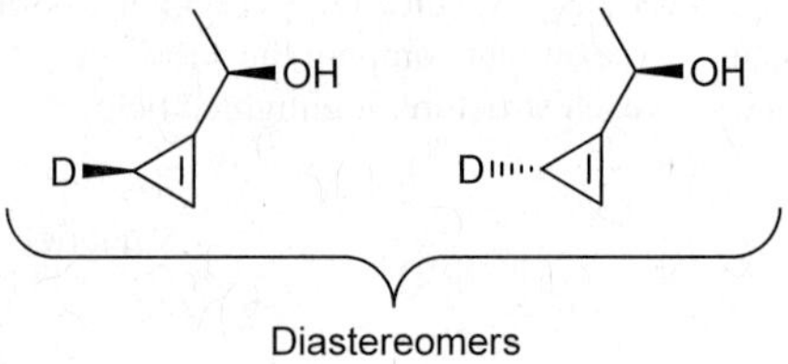

Diastereomers

(d) These protons cannot be interchanged by rotational symmetry, so they are not homotopic. They can be interchanged by reflectional symmetry (the plane of symmetry is the plane of the page). Therefore, the protons are enantiotopic.

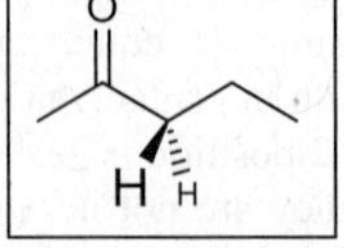

This conclusion can be verified by the replacement test. Specifically, each proton is replaced with deuterium, and the resulting compounds are found to be enantiomers. Therefore, the compounds are enantiotopic:

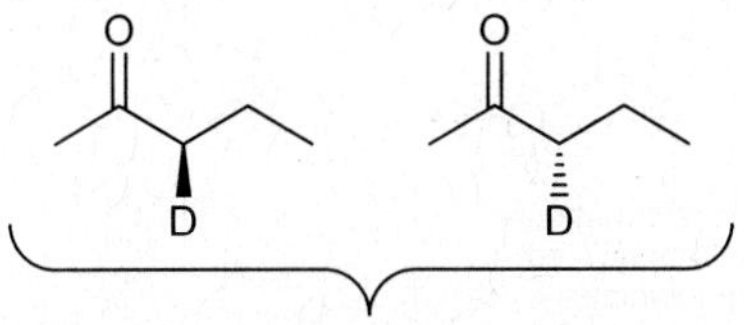

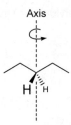

Enantiomers

(e) These protons can be interchanged by rotational symmetry (the axis of symmetry is shown below). Therefore, the protons are homotopic.

Axis

This conclusion can be verified by the replacement test. Specifically, each proton is replaced with deuterium, and the resulting compounds are found to be the same. Therefore, the compounds are homotopic:

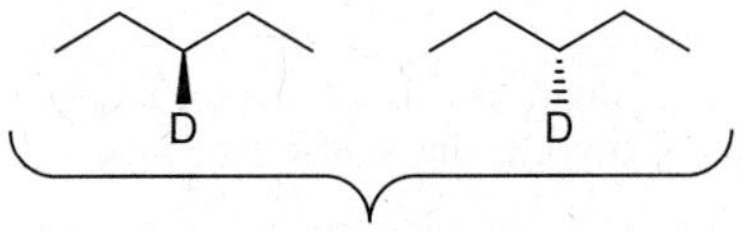

Same compound

16.2.

(a) All four protons shown in red can be interchanged either via rotation or reflection, so they are all chemically equivalent.

(b) The three protons of a methyl group are always equivalent (as we will soon see, immediately after the SkillBuilder), and in this case, the two methyl groups are equivalent to each other because they can be interchanged by rotation. Therefore, all six protons shown in blue are equivalent.

(c) Pentane has three different kinds of protons, shown here:

These two protons can be interchanged by rotational symmetry, so they are homotopic (chemically equivalent)

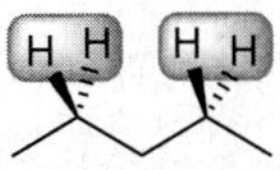

All four of these protons can be interchanged by either rotational or reflectional symmetry, so they are chemically equivalent

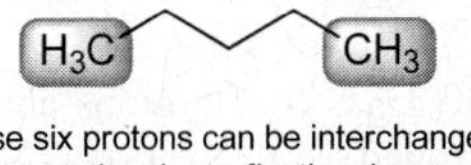

These six protons can be interchanged by either rotational or reflectional symmetry, so they are chemically equivalent

(d) Hexane has three different kinds of protons, shown here:

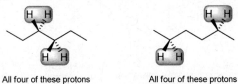

All four of these protons can be interchanged by either rotational or reflectional symmetry, so they are chemically equivalent

All four of these protons can be interchanged by either rotational or reflectional symmetry, so they are chemically equivalent

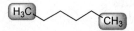

These six protons can be interchanged by either rotational or reflectional symmetry, so they are chemically equivalent

(e) The presence of a chlorine atom creates six different environments (in terms of proximity to the Cl), so there are six different kinds of protons, highlighted here:

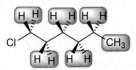

16.3.   The compound must have a high degree of symmetry in order to have only one kind of proton. The molecular formula indicates that there are twelve protons. The equivalence of twelve protons can be achieved by four methyl groups in identical environments. Four methyl groups account for four of the five carbon atoms in the compound. So, we can draw a structure in which the fifth carbon atom is connected to each of the methyl groups, providing the necessary symmetry, as shown here:

16.4.
(a) This compound has eight different kinds of protons (highlighted below), giving rise to eight signals.

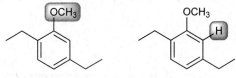

The three protons of a methyl group are always equivalent

These two protons can be interchanged by reflectional symmetry, so they are enantiotopic

The three protons of a methyl group are always equivalent

These two protons can be interchanged by reflectional symmetry, so they are enantiotopic

The three protons of a methyl group are always equivalent

(b) This compound has four different kinds of protons (highlighted below), giving rise to four signals.

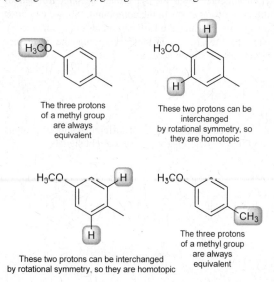

The three protons of a methyl group are always equivalent

These two protons can be interchanged by rotational symmetry, so they are homotopic

These two protons can be interchanged by rotational symmetry, so they are homotopic

The three protons of a methyl group are always equivalent

(c) This compound has two different kinds of protons (highlighted below), giving rise to two signals.

These two methyl groups can be interchanged by rotational symmetry, so all six protons are homotopic

These four protons can all be interchanged by rotational symmetry, so they are all equivalent

(d) This compound has three different kinds of protons (highlighted below), giving rise to three signals.

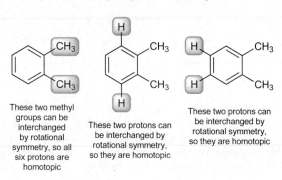

These two methyl groups can be interchanged by rotational symmetry, so all six protons are homotopic

These two protons can be interchanged by rotational symmetry, so they are homotopic

These two protons can be interchanged by rotational symmetry, so they are homotopic

(e) This compound has five different kinds of protons (highlighted below), giving rise to five signals.

The three protons of a methyl group are always equivalent

(f) This compound has three different kinds of protons (highlighted below), giving rise to three signals.

These two protons can be interchanged by rotational symmetry, so they are homotopic

(g) This compound has four protons and none of them can be interchanged by rotational or reflectional symmetry. Each of the four protons occupies a unique electronic environment, giving rise to four signals.

(h) This compound has two different kinds of protons (highlighted below), giving rise to two signals.

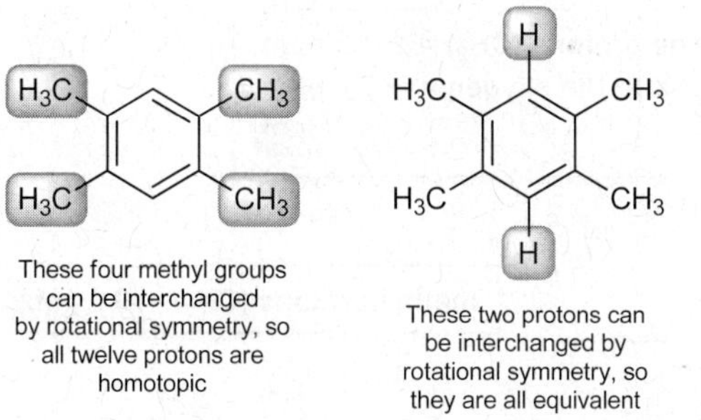

These four methyl groups can be interchanged by rotational symmetry, so all twelve protons are homotopic

These two protons can be interchanged by rotational symmetry, so they are all equivalent

(i) Due to the location of the two bromine atoms, each CH$_2$ group occupies a unique electron environment, giving rise to two separate signals (one for each CH$_2$ group). In addition, the two methyl groups are also in different electronic environments, giving two separate signals. In total, we expect four signals.

(j) Each of the protons in each CH$_2$ group is in a unique electronic environment, as a result of the presence of a chirality center. That is, each CH$_2$ group gives rise to two separate signals, so the two CH$_2$ groups collectively give rise to four different signals. The two methyl groups are also different from each other (because of their proximity to the bromine atom), giving two more signals. In addition, there is one signal from the proton attached to the carbon bearing the bromine atom. In total, we expect seven signals.

(k) As we saw in the solution to Problem **16.2c**, pentane is expected to produce three signals in its ^{1}H NMR spectrum. A similar analysis of heptane indicates that its ^{1}H NMR spectrum should contain four signals.

(l) Each of the three vinylic protons occupies a unique electronic environment, giving rise to three separate signals:

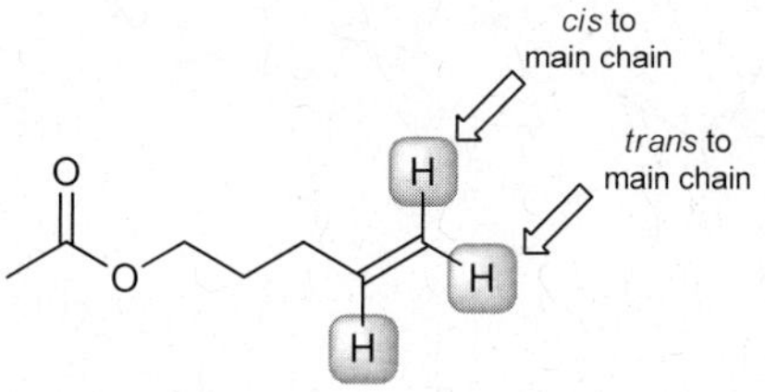

cis to main chain

trans to main chain

Each of these protons is unique. They are NOT chemically equivalent.

The two vinylic protons at the very end are different from each other because one is *cis* to the main chain and

the other is *trans* to the main chain, as shown. Each of the CH$_2$ groups provides one signal (because each CH$_2$ group occupies a unique electronic environment), and the CH$_3$ provides one more signal, giving a total of seven signals.

16.5. The presence of the bromine atom does not render C3 a chirality center because there are two ethyl groups connected to C3. Nevertheless, the presence of the bromine atom does prevent the two protons at C2 from being interchangeable by reflection. The replacement test gives a pair of diastereomers, so the protons are diastereotopic.

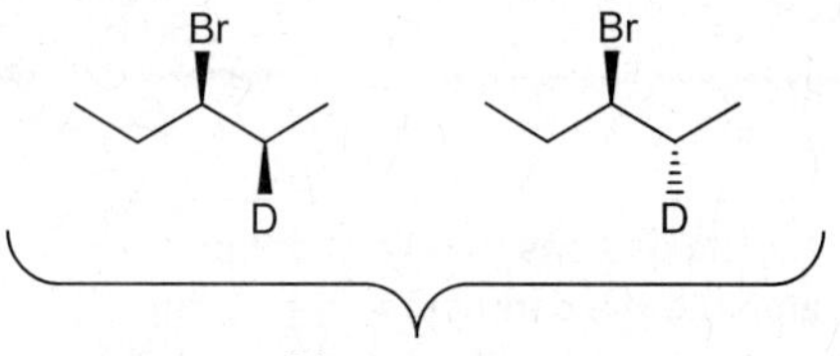

Diastereomers

16.6. Four CH$_2$ groups can be chemically equivalent if they are all attached to the same carbon atom (provided that the four CH$_2$ groups are all connected to identical groups:

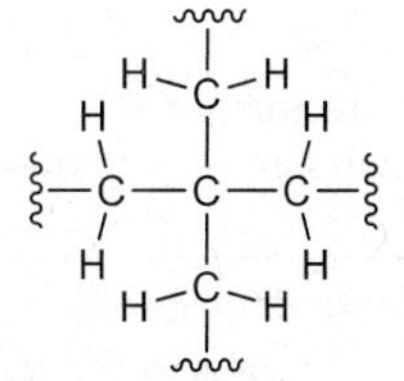

The molecular formula indicates a total of nine carbon atoms and twenty hydrogen atoms, but the structure above only accounts for five carbon atoms and eight hydrogen atoms. We must still account for another four carbon atoms and twelve hydrogen atoms. This can be accomplished if we simply connect a methyl group to each of the CH$_2$ groups, giving the following structure:

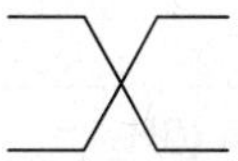

The four methyl groups are chemically equivalent, giving rise to only signal. As such, the ^{1}H NMR spectrum of this compound is expected to exhibit only two signals (one for the CH$_2$ groups and the other for the CH$_3$ groups).

16.7.
(a) The ^{1}H NMR spectrum of this compound is expected to exhibit five signals. The calculation for the estimated chemical shift of each signal is shown here:

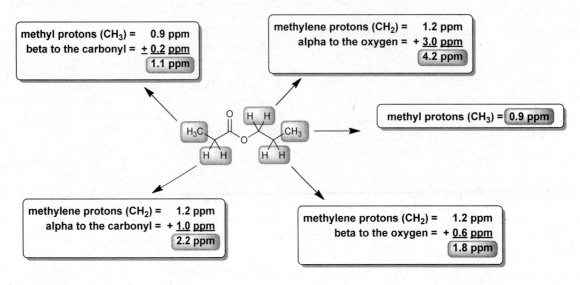

(b) The ^{1}H NMR spectrum of this compound is expected to exhibit three signals. The calculation for the estimated chemical shift of each signal is shown here:

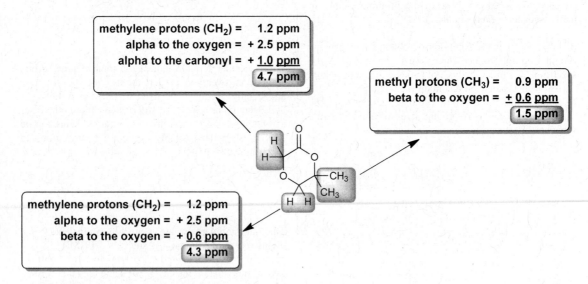

(c) The ^{1}H NMR spectrum of this compound is expected to exhibit four signals. The calculation for the estimated chemical shift of each signal is shown here:

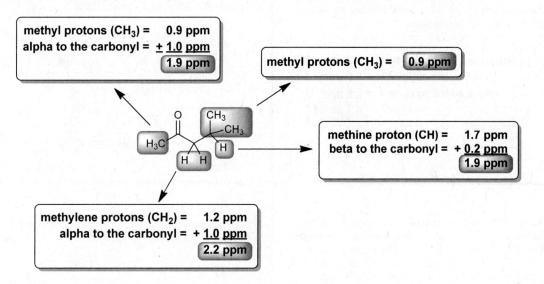

(d) The ^{1}H NMR spectrum of this compound is expected to exhibit four signals. The calculation for the estimated chemical shift of each signal is shown here:

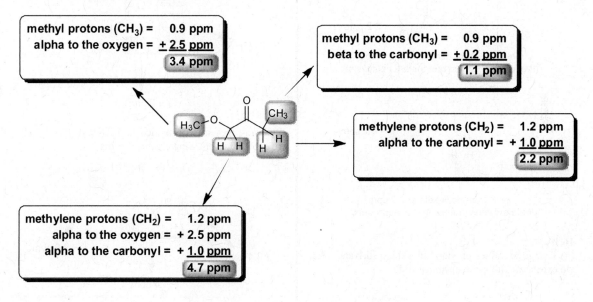

(e) All four methylene groups are equivalent, so the compound will have only one signal in its ^{1}H NMR spectrum. That signal is expected to appear at approximately $(1.2 + 2.5 + 0.5) = 4.2$ ppm.

16.8. The ¹H NMR spectrum of the compound shown here is expected to exhibit a signal between 6 ppm and 7 ppm (calculation shown) below.

methylene protons (CH₂) =	1.2 ppm
alpha to the oxygen =	+ 3.0 ppm
alpha to the oxygen =	+ 2.5 ppm
	6.7 ppm

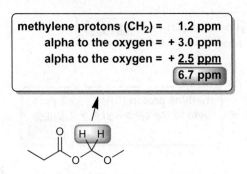

16.9. The first compound produces only one signal downfield of 2.2 ppm, because the highlighted methylene groups are equivalent (they can be interchanged by rotational symmetry). In contrast, the second compound exhibits two signals downfield of 2.2 ppm, because the highlighted methylene groups are not equivalent (they cannot be interchanged by rotational symmetry or reflectional symmetry).

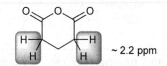

~ 2.2 ppm

Only one signal downfield of 2.0 ppm
(the four highlighted protons are equivalent)

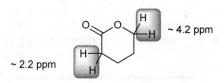

Two signals downfield of 2.0 ppm
(the methylene groups are not equivalent)

16.10.
(a) Using the values provided in Tables 16.1 and 16.2, we expect the following chemical shifts:

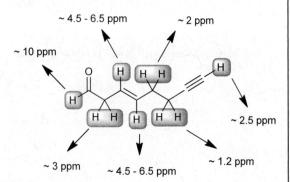

(b) Using the values provided in Tables 16.1 and 16.2, we expect the following chemical shifts:

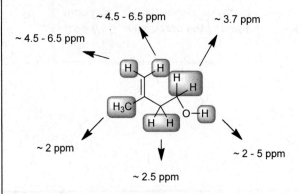

(c) Using the values provided in Tables 16.1 and 16.2, we expect the following chemical shifts:

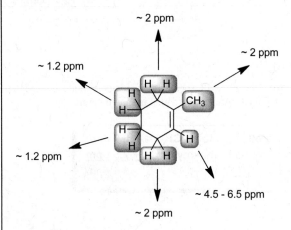

(d) Using the values provided in Tables 16.1 and 16.2, we expect the following chemical shifts:

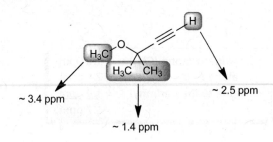

16.11. Among the integration values provided, the lowest number is 33.2, so we divide all integration values by 33.2, giving the following ratio:

$$1 : 1.5 : 1 : 1.5$$

Since there is no such thing as a half of a proton, these numbers must represent 2, 3, 2, and 3 protons, respectively. This is confirmed by the molecular formula, which indicates that the compound has ten hydrogen atoms.

Therefore,

> The signal at 4.0 ppm represents two protons.
> The signal at 2.0 ppm represents three protons.
> The signal at 1.6 ppm represents two protons.
> The signal at 0.9 ppm represents three protons.

16.12. Among the integration values provided, the lowest number is 17.1, so we divide all integration values by 17.1, giving the following ratio:

$$1 : 5 : 1 : 3$$

The molecular formula indicates that the compound has ten hydrogen atoms, so the numbers above are not only relative values, but they are also exact values.
Therefore,

> The signal at 9.6 ppm represents one proton.
> The signal at 7.5 ppm represents five protons.
> The signal at 7.3 ppm represents one proton.
> The signal at 2.1 ppm represents three protons.

16.13. Among the integration values provided, the lowest number is 18.92, so we divide all integration values by 18.92, giving the following ratio:

$$1 : 1 : 1$$

The molecular formula indicates that the compound has six hydrogen atoms (not just three), so the numbers above are only relative values. Each signal must actually represent two protons.

16.14. The relative integration values are 2 : 3, but the molecular formula indicates that the structure has fifteen protons. As such, the signals must represent six protons and nine protons, respectively. This indicates a high level of symmetry, which can be achieved if three ethyl groups are connected to a central carbon atom, as in the following structure.

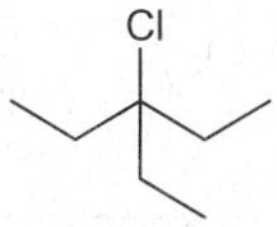

16.15.
(a) This compound has four different kinds of protons, highlighted here. In each case, we apply the $n+1$ rule, giving the multiplicities shown:

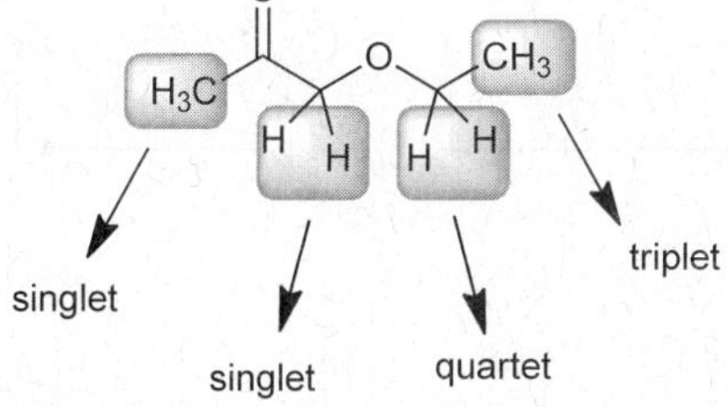

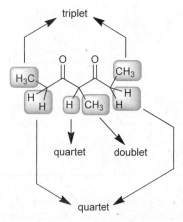

(b) This compound has four different kinds of protons, highlighted here. In each case, we apply the $n+1$ rule, giving the multiplicities shown:

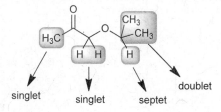

(c) This compound has four different kinds of protons, highlighted here. In each case, we apply the $n+1$ rule, giving the multiplicities shown:

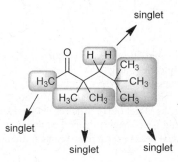

(d) This compound has six different kinds of protons, highlighted here. In each case, we apply the $n+1$ rule, giving the multiplicities shown:

16.16. A quartet indicates the presence of three neighboring protons (all of which are equivalent to each other). This is most commonly observed when a neighboring methyl group is present, however, it can also occur if the proton giving rise to the signal is connected to three chemically equivalent methine protons:

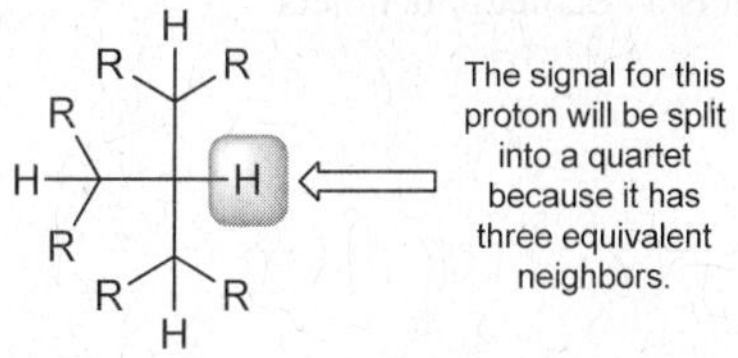

The signal for this proton will be split into a quartet because it has three equivalent neighbors.

There are certainly a large number of compounds that can be drawn that fit this description, bur remember that the problem statement indicates that our structure cannot contain any methyl groups. The following compounds are just two possible answers.

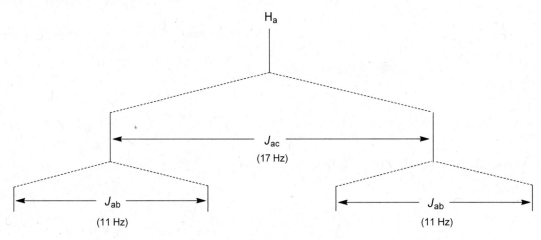

The signal for this proton is expected to be a quartet

The signal for this proton is expected to be a quartet

16.17.
(a) The spectrum exhibits the characteristic pattern of an isopropyl group (a septet with an integration of one, and a doublet with an integration of six).
(b) The spectrum exhibits the characteristic pattern of an isopropyl group (a septet with an integration of one, and a doublet with an integration of six) as well as the characteristic pattern of an ethyl group (a quartet with an integration of two, and a triplet with an integration of three).
(c) The spectrum exhibits a singlet with a relative integration of 9 (because $55.0 / 6.0 \approx 9$), which is the characteristic pattern of a *tert*-butyl group.
(d) The spectrum does not exhibit the characteristic pattern of an ethyl group, an isopropyl group, or a *tert*-butyl group.

16.18.
(a) To determine the expected splitting pattern for the signal corresponding to H_a, we must consider the effects of the two non-equivalent neighbors, H_b and H_c, The former is coupled to H_a with a coupling constant J_{ab} (11 Hz), and the latter is coupled to H_a with a coupling constant J_{ac} (17 Hz). We always begin with the larger coupling constant (J_{ac} in this case), which splits the signal into a doublet. Then, each peak of this doublet is then further split into a doublet because of the effect of H_b. The result is a doublet of doublets. It can be distinguished from a quartet because all four peaks should have roughly equivalent integrations values (if integrated individually), as opposed to a quartet, in which the individual peaks have a relative integration of $1 : 3 : 3 : 1$.

H_a

J_{ac}
(17 Hz)

J_{ab}
(11 Hz)

J_{ab}
(11 Hz)

(b) To determine the expected splitting pattern for the signal corresponding to H_b, we must consider the effects of the two non-equivalent neighbors, H_a and H_c, The former is coupled to H_b with a coupling constant J_{ab} (11 Hz), and the latter is coupled to H_b with a coupling constant J_{bc} (1 Hz). We always begin with the larger coupling constant (J_{ab} in this case), which splits the signal into a doublet. Then, each peak of this doublet is then further split into a doublet because of the effect of H_c. The result is a doublet of doublets.

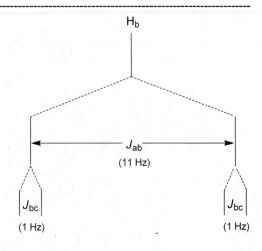

H_b

J_{ab}
(11 Hz)

J_{bc}
(1 Hz)

J_{bc}
(1 Hz)

(c) To determine the expected splitting pattern for the signal corresponding to H_c, we must consider the effects of the two non-equivalent neighbors, H_a and H_b. The former is coupled to H_c with a coupling constant J_{ac} (17 Hz), and the latter is coupled to H_c with a coupling constant J_{bc} (1 Hz). We always begin with the larger coupling constant (J_{ac} in this case), which splits the signal into a doublet. Then, each peak of this doublet is then further split into a doublet because of the effect of H_b. The result is a doublet of doublets.

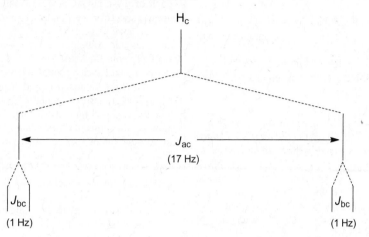

16.19.
(a) This compound is expected to produce four signals in its ^1H NMR spectrum. For each signal, its expected chemical shift, multiplicity, and integration are shown.

$\delta = 1.2 + 1 = 2.2$ ppm
m = triplet
I = 2H

$\delta = 1.2 + 0.2 + 0.1 = 1.5$ ppm
m = triplet of triplets
I = 2H

$\delta = 1.2 + 0.6 = 1.8$ ppm
m = triplet of triplets
I = 2H

$\delta = 1.2 + 3 = 4.2$ ppm
m = triplet
I = 2H

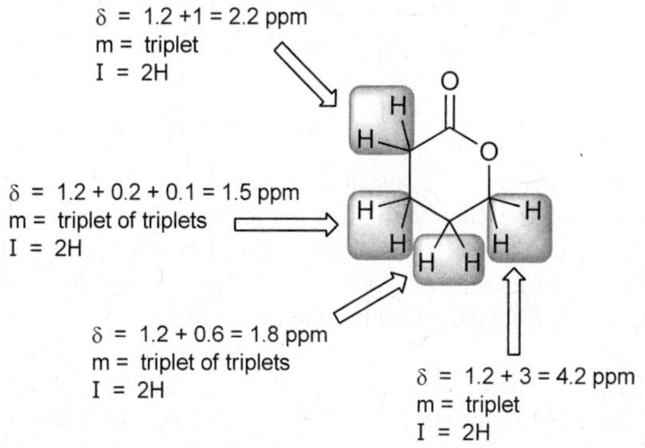

(b) This compound has rotational and reflectional symmetry:

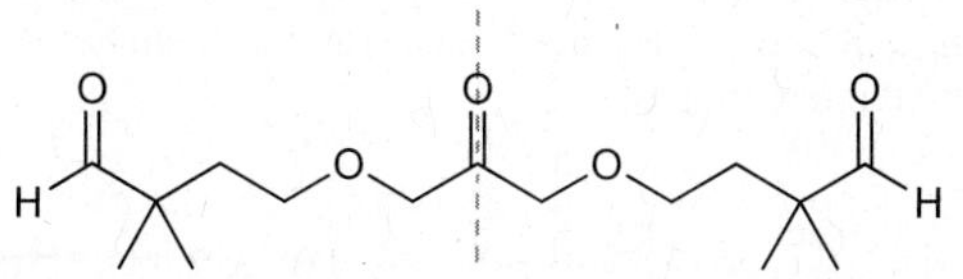

So we expect its ^1H NMR spectrum to exhibit only five signals, corresponding with the following highlighted protons. For each signal, its expected chemical shift, multiplicity, and integration are shown.

$\delta = 1.2 + 0.5 + 0.2 = 1.9$ ppm
m = triplet
I = 2H

$\delta = 1.2 + 2.5 = 3.7$ ppm
m = triplet
I = 2H

$\delta = 10$ ppm
m = singlet
I = 1H

$\delta = 0.9 + 0.2 = 1.1$ ppm
m = singlet
I = 6H

$\delta = 1.2 + 2.5 + 1.0 = 4.7$ ppm
m = singlet
I = 2H

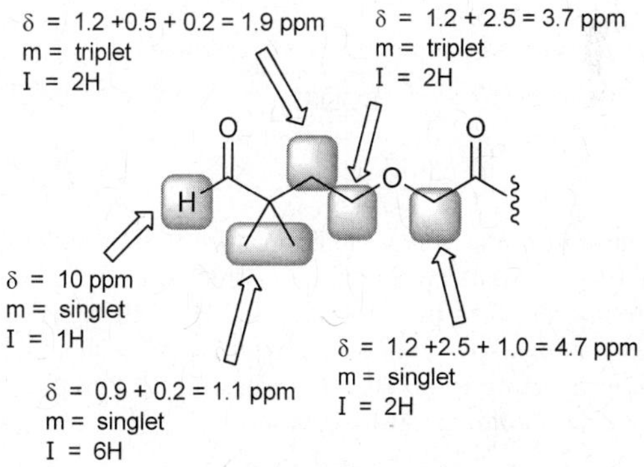

16.20. The compound has one degree of unsaturation (see Section 15.16). The compound has fourteen protons, but only three signals so we are looking for a compound with symmetry. Also, we must use the oxygen atoms carefully in order to ensure that all signals are downfield of 2 ppm. The following structure meets all of these requirements.

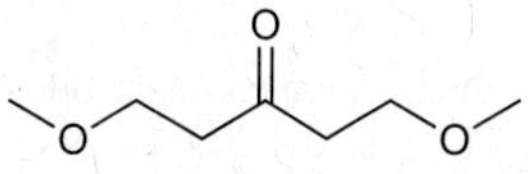

16.21.
(a) The first compound will have only three signals in its ^1H NMR spectrum, while the second compound will have six signals.
(b) Both compounds will exhibit ^1H NMR spectra with only two singlets. In each spectrum, the relative integration of the two singlets is 1:3. In the first compound, the singlet with the smaller integration value will be at approximately 2 ppm. In the second compound, the singlet with the smaller integration value will be at approximately 4 ppm.

(c) The first compound will have only two signals in its ^1H NMR spectrum, while the second compound will have three signals.

(d) The first compound will have five signals in its ^1H NMR spectrum, while the second compound will have only three signals.

(e) The first compound will have only two signals in its ^1H NMR spectrum, while the second compound will have four signals.

(f) The first compound will have only one signal in its ^1H NMR spectrum, while the second compound will have two signals.

16.22. As shown in the problem statement, Markovnikov addition of HBr will install the Br at the more substituted position, giving a compound with a high degree of symmetry (thus only two signals in its ^1H NMR spectrum). If the product has more than two signals, then the product must lack the symmetry possessed by the Markovnikov product. This can be explained if we consider the result of an *anti*-Markovnikov addition of HBr (for example, if peroxides are present):

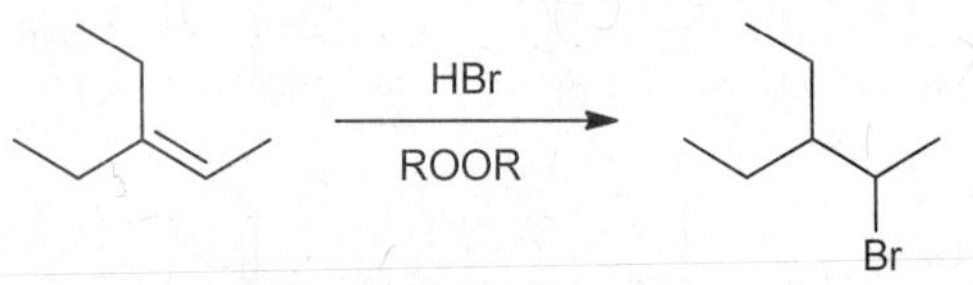

In an *anti*-Markovnikov addition, the Br is installed at the less substituted position. This product is expected to exhibit more than two signals in its ^1H NMR signal. If the Markovnikov addition is required, care should be taken to purify the HBr (to ensure the absence of peroxides or other radical initiators).

16.23.

(a) The molecular formula ($C_8H_{10}O$) indicates four degrees of unsaturation (see Section 15.16), which is highly suggestive of an aromatic ring. To determine the relative integration values, we divide each of the integration values by 9.1, giving a ratio of approximately 5 : 2 : 2 : 1. Since the compound has ten protons, the numbers above are not only relative values, but they are also exact values. Now let's analyze each of the signals individually.

The signal just above 7 ppm confirms our suspicion of an aromatic ring. This signal has an integration of 5H, indicating a monosubstituted aromatic ring:

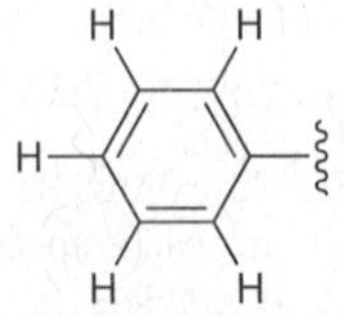

The spectrum also exhibits two triplets (just below 3 ppm and just below 4 ppm), indicating two methylene groups connected to each other:

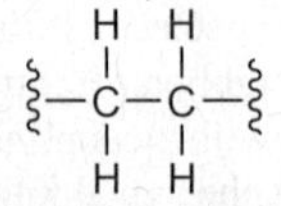

Each of these signals appears more downfield than we might expect for a methylene group (1.2 ppm), so each of these methylene groups must be connected to a group that causes a deshielding effect. This must be taken into account in our final structure.

If we inspect the two fragments that we have determined thus far (the monosubstituted aromatic ring and the methylene groups that neighbor each other), we will find that these two fragments account for nearly all of the atoms in the molecular formula ($C_8H_{10}O$). We only need to account for one more proton and one oxygen atom. The singlet at 2 ppm has an integration of 1, so this signal corresponds with only one proton (with no neighbors), so we conclude that the compound has an OH group.

There is only one way to assemble the three fragments:

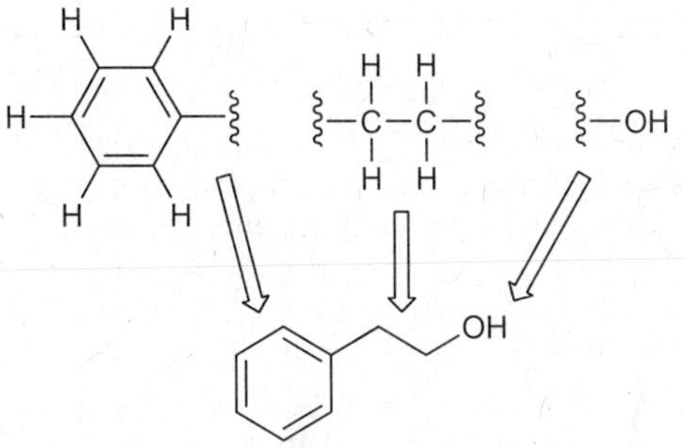

This structure is consistent with all of the signals, including the chemical shifts of the two triplets, which can be explained by the electron withdrawing effect of the oxygen atom as well as the local magnetic field established by the aromatic ring.

(b) The molecular formula ($C_7H_{14}O$) indicates one degree of unsaturation (see Section 15.16), which means that the compound must possess either a double bond or a ring. To determine the relative integration values, we divide each of the integration values by 10.8, giving a ratio of approximately 1 : 6. This spectrum has the characteristic pattern of an isopropyl group (a doublet with a relative integration of 6 and a septet with a relative integration of 1).

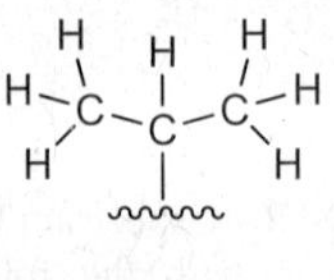

An isopropyl group only contains seven protons, but there are fourteen protons in the compound ($C_7H_{14}O$). We therefore conclude that the compound must contain two isopropyl groups, which are interchangeable by symmetry:

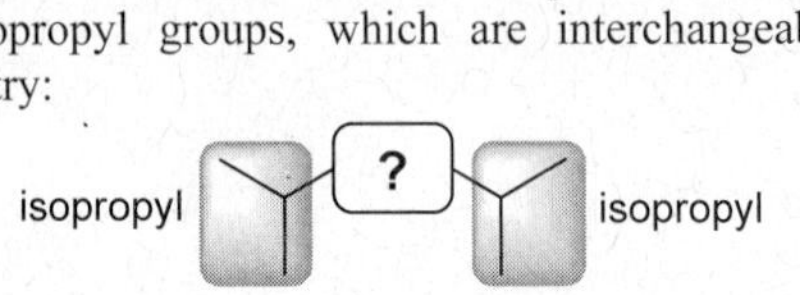

Notice that the two isopropyl groups account for all but one of the carbon atoms in the compound. Therefore, there can only be one carbon atom in between the two isopropyl groups. This carbon atom cannot have any protons, since we don't see any other signals in the ^1H NMR spectrum. Also, we must still account for one oxygen atom ($C_7H_{14}O$), and we said that the compound must contain one degree of unsaturation. This all points to a carbonyl group at the central position:

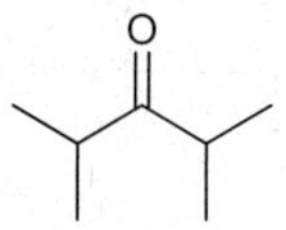

This structure is indeed consistent with the observed chemical shift at 2.7 ppm for the methine (CH) protons (1.7 + 1 = 2.7 ppm).

(c) The molecular formula ($C_{10}H_{14}O$) indicates four degrees of unsaturation (see Section 15.16), which is highly suggestive of an aromatic ring.

The signals near 7 ppm are likely a result of aromatic protons. Notice that the combined integration of these two signals is 4H. This, together with the distinctive splitting pattern (a pair of doublets), suggests a 1,4-disubstituted aromatic ring:

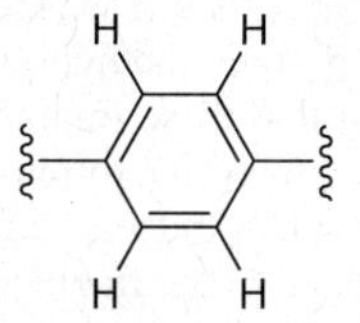

The spectrum also exhibits a singlet with an integration of 9H (at approximately 1.4 ppm) which is characteristic of a *tert*-butyl group.

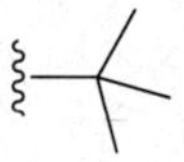

If we inspect the two fragments that we have determined thus far (the disubstituted aromatic ring and the *tert*-butyl group), we will find that these two fragments account for nearly all of the atoms in the molecular formula ($C_{10}H_{14}O$). We only need to account for one more proton and one oxygen atom. The peak just under 5 ppm has an integration of 1, so this signal corresponds with only one proton (with no neighbors), so we conclude that the compound has an OH group. This signal is broad, which is often (although not always) the case for signals arising from OH groups.

There is only one way to assemble the three fragments:

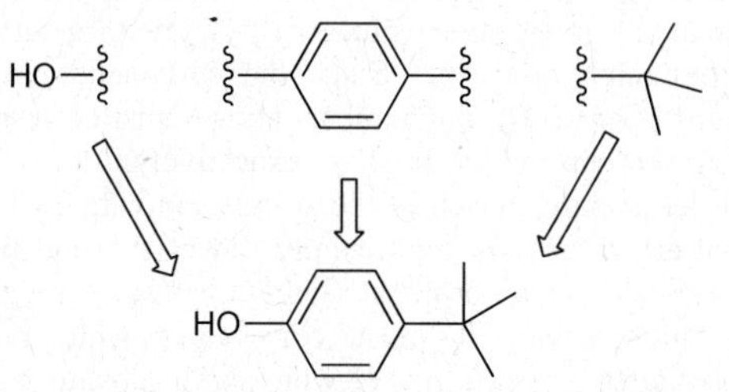

(d) The molecular formula ($C_4H_6O_2$) indicates two degrees of unsaturation (see Section 15.16), which means that the compound must possess either two double bonds, or two rings, or one ring and one double bond, or a triple bond.

To determine the relative integration values, we divide each of the integration values by 19.46, giving a ratio of approximately 1 : 1 : 1. The molecular formula indicates six protons (rather than three), so the relative integration values must correspond with two protons for each signal. That is, the spectrum indicates the presence of three different methylene groups. From the splitting patterns (a triplet, a triplet, and a triplet of triplets), we can conclude that the three methylene groups are connected to each other:

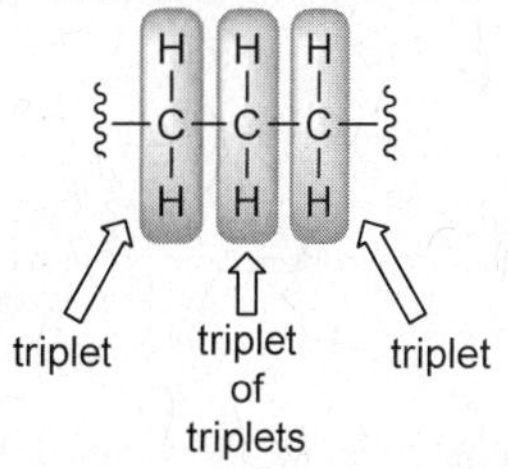

Now let's focus on the chemical shifts of the triplets (2.4 ppm and 4.3 ppm). Both signals are shifted downfield (relative to 1.2 ppm for a typical methylene group). One of these signals is significantly shifted downfield, perhaps because it is next to an oxygen atom (after all, the molecular formula indicates that there are two oxygen atoms in the compound):

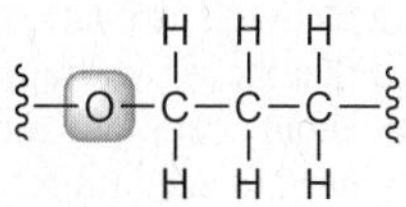

The other triplet is also shifted downfield, but the effect is weaker. This seems consistent with the effect of a carbonyl group:

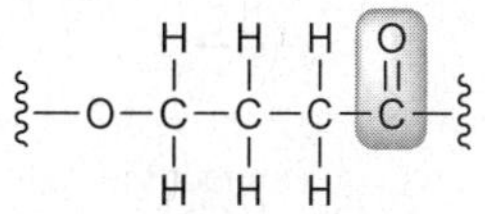

The central methylene group is beta to both the oxygen atom and the carbonyl group, and it feels the distant effects of both (explaining why that signal is shifted somewhat downfield itself).

The fragment above accounts for ALL of the atoms in the molecular formula, yet we are still missing one degree of unsaturation (the product must contain two degrees of unsaturation, but the fragment above has only one degree of unsaturation). Therefore, we close the ends together, giving the following structure:

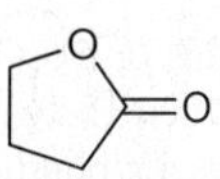

(e) The molecular formula ($C_9H_{10}O$) indicates five degrees of unsaturation (see Section 15.16), which is highly suggestive of an aromatic ring, in addition to either one double bond or one ring.

To determine the relative integration values, we divide each of the integration values by 13.9, giving a ratio of approximately 1 : 1.5 : 1 : 1.5. Since the compound has ten protons, the numbers above must correspond with:

$$2 : 3 : 2 : 3$$

Now let's analyze each of the signals individually.

The signals near 7 ppm are likely a result of aromatic protons. Notice that the combined integration of these signals is 5H, indicating a monosubstituted aromatic ring:

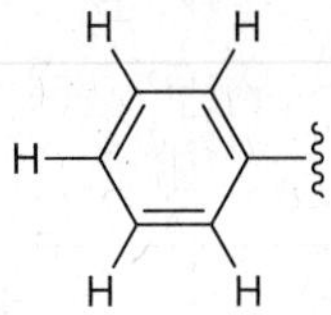

The spectrum also exhibits the characteristic pattern of an ethyl group (a quartet with an integration of 2 and a triplet with an integration of 3):

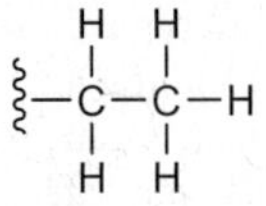

If we inspect the two fragments that we have determined thus far (the monosubstituted aromatic ring and the ethyl group), we will find that these two fragments account for nearly all of the atoms in the molecular formula ($C_9H_{10}O$). We only need to account for one more carbon atom and one oxygen atom. And let's not forget that our structure still needs one more degree of unsaturation, suggesting a carbonyl group:

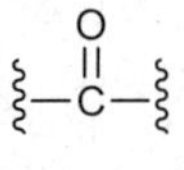

There is only one way to connect these three fragments.

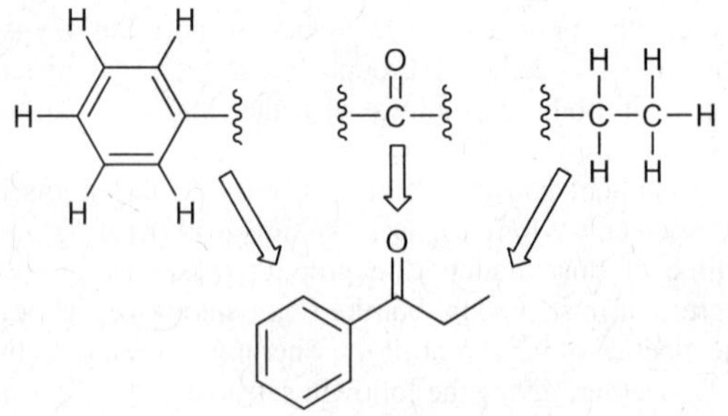

(f) The molecular formula ($C_5H_{12}O$) indicates no degrees of unsaturation (see Section 15.16), which means that the compound does not have a π bond or a ring.

To determine the relative integration values, we divide each of the integration values by 13.6, giving a ratio of approximately 1 : 2 : 6 : 3. Since the compound has twelve protons, the numbers above are not only relative values, but they are also exact values.

Now let's analyze each of the signals individually.

Let's begin with the two signals that represent the characteristic pattern for an ethyl group (a quartet with an integration of 2 and a triplet with an integration of 3).

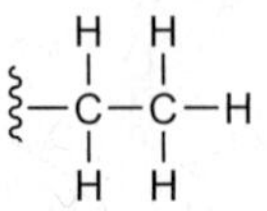

The singlet at 1.2 ppm has an integration of 6, indicating two methyl groups that can be interchanged via symmetry (and they cannot have any neighboring protons).

The singlet at 2.2 ppm has an integration of 1, so this signal corresponds with only one proton (with no neighbors), so we conclude that this likely represents an OH group (the molecular formula indicates the presence of an oxygen atom).

If we inspect the fragments that we have determined thus far (an ethyl group, two methyl groups and an OH group), we will find that these fragments account for all of the atoms in the molecular formula ($C_5H_{10}O$) except for one carbon atom. Indeed, this carbon atom is necessary to connect all of the fragments, as shown:

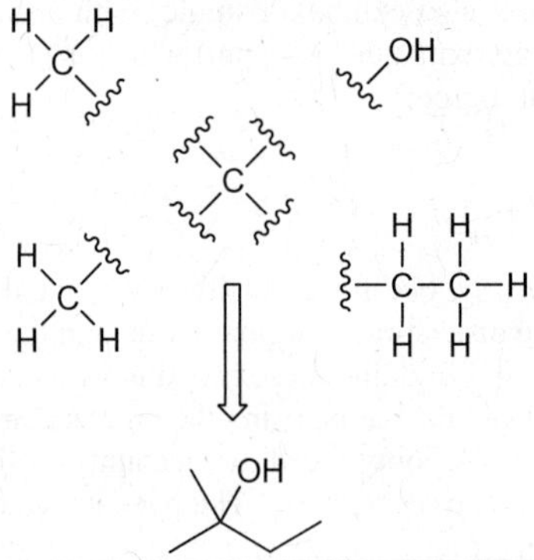

16.24. The molecular formula ($C_{10}H_{10}O_4$) indicates six degrees of unsaturation (see Section 15.16), which suggests an aromatic ring as well as two other degrees of unsaturation.

To determine the relative integration values, we divide each of the integration values by 52, giving a ratio of approximately 1 : 1.5. Since the compound has ten protons ($C_{10}H_{10}O_4$), the numbers above must correspond to four protons and six protons, respectively.

Now let's analyze each of the signals individually. The signal at 8.1 ppm is significantly downfield, and likely represents aromatic protons. Since it is a singlet with an integration of 4, it must correspond with a 1,4-disubstituted aromatic ring in which both substituents are identical (therefore rendering all four aromatic protons equivalent).

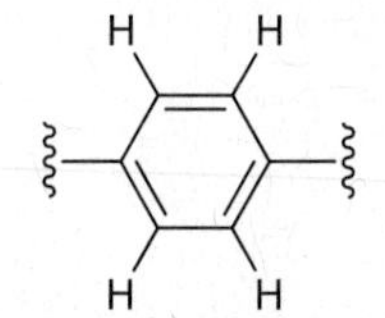

The other signal has an integration of 6, which likely represents two equivalent methyl groups (interchangeable by symmetry). Since the signal is a singlet, these methyl groups must not have any neighbors. The chemical shift is consistent with these methyl groups being connected to oxygen atoms:

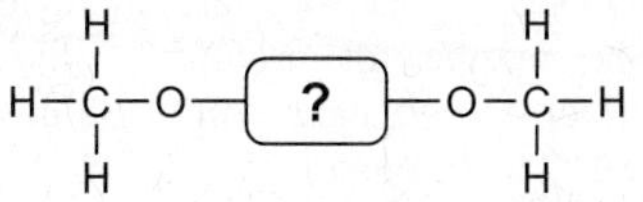

If we inspect the fragments that we have determined thus far (a 1,4-disubstituted aromatic ring, and two methoxy groups), we will find that these fragments account for all of the atoms in the molecular formula ($C_{10}H_{10}O_4$) except for two carbon atoms and two oxygen atoms. Recall that the compound must have six degrees of unsaturation, and the fragments above only account for four degrees of unsaturation. Therefore, the remaining two carbon atoms and two oxygen atoms are likely carbonyl groups:

There are certainly a few different ways to connect all of these fragments, but there is only one way to them without breaking the symmetry necessary to keep all four aromatic protons identical:

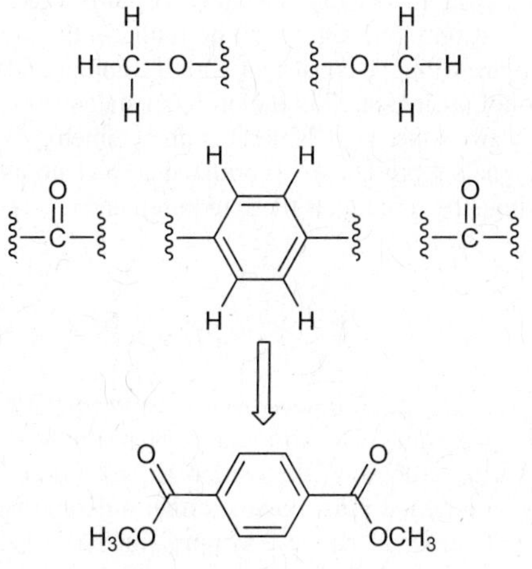

16.25.
(a) The two methyl groups occupy identical environments (there is a conformation of this molecule in which the two methyl groups are interchangeable by reflectional symmetry), so the methyl groups are chemically equivalent. All of the other carbon atoms are unique, giving rise to a total of four signals. Three of the signals are produced by sp^3 hybridized carbon atoms (not connected to electronegative atoms), and are therefore expected to appear in the region 0 – 50 ppm. The fourth signal is produced by the carbon atom of the carbonyl group (C=O), which is expected to appear in the region 150 – 220 ppm.

(b) This compound exhibits symmetry, rendering the two methyl groups equivalent. Similarly, the ring has only four unique signals, because of symmetry. In total, the ^{13}C NMR spectrum of this compound should exhibit five signals. All five are produced by sp^3 hybridized carbon atoms (not connected to electronegative atoms) and are therefore expected to appear in the region 0 – 50 ppm.

(c) Monosubstituted aromatic rings exhibit symmetry, giving only four unique carbon atoms in the aromatic ring. These four signals will appear in the region 100-150 ppm. Each of the carbon atoms of the ethyl group is unique, giving a total of six signals. The two signals of the ethyl group are produced by sp^3 hybridized carbon atoms (not connected to electronegative atoms) and are therefore expected to appear in the region 0 – 50 ppm.

(d) The compound has no symmetry. Each of the carbon atoms is unique, giving a total of nine signals. Two signals appear in the region 0 – 50 ppm (sp^3 hybridized), one signal appears in the region 50 – 100 ppm (the sp^3 hybridized carbon atom attached to an oxygen atom), and six signals appear in the region 100 – 150 ppm (sp^2 hybridized).

(e) The aromatic ring is 1,4-disubstituted so it exhibits symmetry, giving four unique carbon atoms in the aromatic ring (four signals). These four signals will appear in the region 100-150 ppm. In addition, there will be a signal for the carbon atom of the methoxy group, which will appear in the region 50 – 100 ppm (as expected for an sp^3 hybridized carbon atom attached to an oxygen atom). And finally there will be two signals in the region 0 – 50 ppm (for the sp^3 hybridized carbon atoms of the ethyl group). In total, there are seven signals.

(f) This compound has symmetry, giving only five unique positions (highlighted):

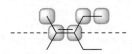

So the ^{13}C NMR spectrum should have five signals. Three of them should appear in the region 0 – 50 ppm (sp^3 hybridized), and two signals should appear in the region 100 – 150 ppm (sp^2 hybridized).

(g) None of the carbon atoms in this compound can be interchanged with any of the other carbon atoms in this compound via either reflection or rotation. All of the carbon atoms are unique, giving rise to seven signals.

Five of them should appear in the region 0 – 50 ppm (sp^3 hybridized), and two signals should appear in the region 100 – 150 ppm (sp^2 hybridized).

(h) This compound exhibits a high degree of symmetry. All four methyl groups are interchangeable by either rotation or reflection, and therefore, all four methyl groups give rise to one signal, appearing in the region 0 – 50 ppm (sp^3 hybridized). The two vinylic carbon atoms are also chemically equivalent, giving rise to one signal in the region 100 – 150 ppm (sp^2 hybridized). In total, there are only two signals.

(i) This compound exhibits a high degree of symmetry. All four carbon atoms are interchangeable by either rotation or reflection, and therefore, all four carbon atoms give rise to one signal, appearing in the region 50 – 100 ppm (as expected for an sp^3 hybridized carbon atom attached to an oxygen atom).

(j) None of the carbon atoms in this compound can be interchanged with any of the other carbon atoms in this compound via either reflection or rotation. All of the carbon atoms are unique, giving rise to five signals. Among the two sp^3 hybridized carbon atoms, one of them should produce a signal in the region 0 – 50 ppm, and the other sp^3 hybridized carbon atom (connected to an oxygen atom) should appear in the region 50 – 100 ppm. In addition, the two vinylic carbon atoms will each produce a signal in the region 100 – 150 ppm (sp^2 hybridized). And finally, the carbonyl group (C=O) will produce a signal in the region 150 – 220 ppm.

16.26. The first compound lacks a chirality center. The two methyl groups are enantiotopic and are therefore chemically equivalent. The second compound has a chirality center (the position bearing the OH group). As such, the two neighboring methyl groups are diastereotopic and are therefore not chemically equivalent. For this reason, the ^{13}C NMR spectrum of the second compound exhibits six signals, rather than five.

16.27. The molecular formula (C_8H_{10}) indicates four degrees of unsaturation (see Section 15.16), which is highly suggestive of an aromatic ring. The aromatic ring accounts for six of the eight carbon atoms, so we must account for the other two carbon atoms, which must be sp^3 hybridized. However, only one signal appears in the region 0 – 50 ppm, suggesting that the two sp^3 hybridized carbon atoms are interchangeable by symmetry. This can be achieved in any of these three compounds:

However, only one of these compounds, shown below, will exhibit four signals in the region 100 – 150 ppm.

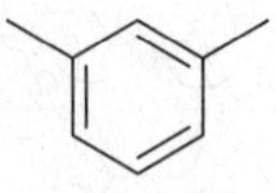

16.28. The molecular formula ($C_5H_{10}O$) indicates one degree of unsaturation (see Section 15.16), which means that the compound must either have a double bond or a ring. The signal above 200 ppm (in the broadband-decoupled spectrum) indicates the source of the degree of unsaturation (C=O). The four signals in the range 0 – 50 ppm represent four unique sp^3 hybridized carbon atoms. Two of them are upside-down in the DEPT-135 spectrum, indicating that they are methylene groups. Based on the number of protons in the compound ($C_5H_{10}O$), the other two signals must be methyl groups (to give a total of ten protons).

Now we must connect a carbonyl group, two methylene groups and two methyl groups. There are only two ways to do that:

The first possibility cannot be correct because it has too much symmetry, and would have fewer signals. The latter structure is the only structure that is consistent with the data.

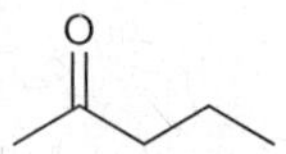

16.29. The molecular formula ($C_5H_{12}O$) indicates no degrees of unsaturation (see Section 15.16), which means that the compound cannot have either a double bond or a ring. The signal at 73.8 δ must be produced by the carbon atom that is connected to the oxygen atom. Notice that only one signal can be found in the range 50 – 100 ppm, which is consistent with the compound being an alcohol, as indicated in the problem statement. There are only two other signals, indicating symmetry. Both of those signals represent sp^3 hybridized carbon atoms. The following structure meets these requirements:

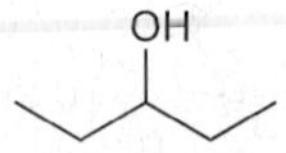

OH

16.30. The molecular formula ($C_7H_{14}O$) indicates one degree of unsaturation (see Section 15.16), which means that the compound must possess either a double bond or a ring. The signal above 200 ppm (in the broadband-decoupled spectrum) indicates the source of the degree of unsaturation (C=O). The five signals in the range 0 – 50 ppm represent five unique sp^3 hybridized carbon atoms. As such, the spectrum has a total of six signals for a compound with seven carbon atoms. That means that one of the signals in the range 0 – 50 ppm must represent two carbon atoms (for example, two equivalent methyl groups – the ^{1}H NMR data confirms the two equivalent methyl groups). Also notice that two of the signals are upside-down in the DEPT-135 spectrum, indicating that they are methylene groups. Based on the DEPT-90 spectrum, we can see that one of the signals is a CH group, while the last two signals must be methyl groups. There are several possible structures that are consistent with the information above, but only one of them is consistent with the ^{1}H NMR spectrum. Specifically, the singlet at 1.9 ppm (with an integration of 3) indicates a methyl ketone.

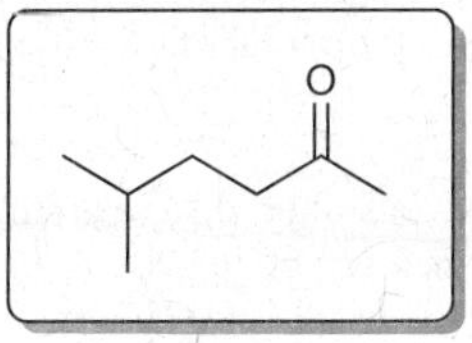

Consistent with
^{1}H NMR and ^{13}C NMR spectra

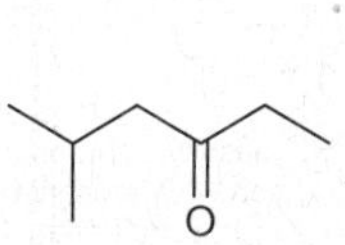

Consistent only with
^{13}C NMR spectrum

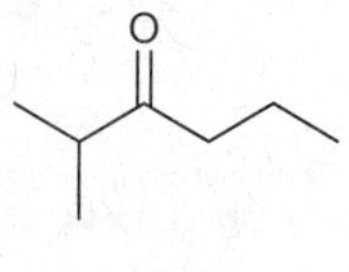

Consistent only with
^{13}C NMR spectrum

16.31.
(a) The molecular formula (C$_5$H$_{10}$) indicates one degree of unsaturation (see Section 15.16), which means that the compound must possess either a double bond or a ring. With only one signal in the spectrum, the structure must have a high degree of symmetry, such that all ten protons are equivalent. This is indeed the case for a five-membered ring (cyclopentane). There is no alkene with molecular formula C$_5$H$_{10}$ in which all ten protons occupy identical electronic environments. So cyclopentane is the only structure consistent with the spectral data:

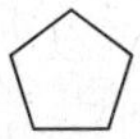

(b) The molecular formula (C$_5$H$_8$Cl$_4$) indicates no degrees of unsaturation (see Section 15.16), which means that the compound cannot have any double bonds, triple bonds, or rings. That is, the structure must be acyclic and cannot have any π bonds. With only one signal in the spectrum, the structure must have a high degree of symmetry, such that all eight protons are equivalent. This can be achieved with four equivalent methylene (CH$_2$) groups. Four CH$_2$ groups will be chemically equivalent if they are all attached to the same carbon atom (provided that all four CH$_2$ groups are connected to identical groups):

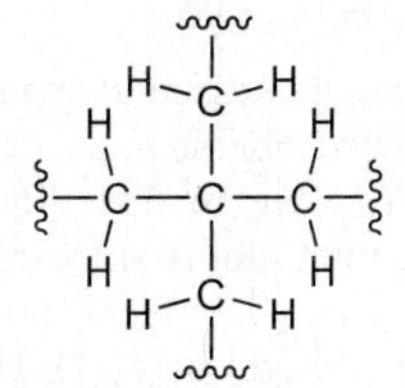

The molecular formula indicates five carbon atoms (which are now all accounted for) and four chlorine atoms, which can serve to cap the four loose ends shown above, like this:

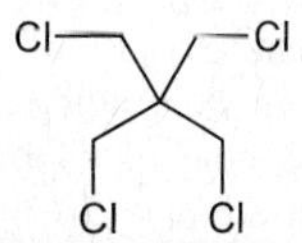

(c) The molecular formula (C$_{12}$H$_{18}$) indicates four degrees of unsaturation (see Section 15.16), which is highly suggestive of an aromatic ring. With only one signal in the spectrum, the structure must have a high degree of symmetry, such that all eighteen protons are equivalent. This can be achieved with six equivalent methyl groups, as seen in the following structure:

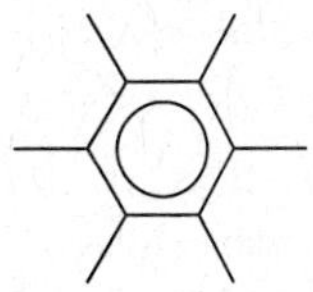

16.32. The molecular formula (C$_{12}$H$_{24}$) indicates one degree of unsaturation (see Section 15.16), which means that the compound must possess either a double bond or a ring. With only one signal in the ^{1}H NMR spectrum, the structure must have a high degree of symmetry, such that all twenty-four protons are equivalent. This can be accomplished with either twelve equivalent methylene (CH$_2$) groups or eight equivalent methyl (CH$_3$) groups. Since the former would use up all of the carbon atoms in the structure (all twelve), it is tempting to explore that possibility first. Indeed, a twelve-membered ring is comprised of twelve equivalent methylene groups, which should give rise to one signal in the ^{1}H NMR spectrum. And this compound (cyclododecane) also exhibits the correct degree of unsaturation (HDI = 1). However, this structure would give only one signal in its ^{13}C NMR spectrum, and the problem statement indicates that the ^{13}C NMR spectrum has two signals. So, we consider our other alternative (eight equivalent methyl groups). If the structure has eight equivalent methyl groups, then we still need to account for four more carbon atoms, as well as one degree of unsaturation. We can account for all four carbon atoms in a four-membered ring (HDI = 1). In the following structure, all eight methyl groups are equivalent.

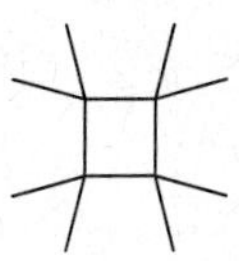

This structure has a high degree of symmetry, and it will indeed give rise to two signals in its ^{13}C NMR spectrum (one signal for all of the methyl groups, and another signal for all four carbon atoms of the ring).

16.33. The molecular formula ($C_{17}H_{36}$) indicates no degrees of unsaturation (see Section 15.16), which means that the compound cannot have any double bonds, triple bonds, or rings. That is, the structure must be acyclic and cannot have any π bonds. With only one signal in the ^{1}H NMR spectrum, the structure must have a high degree of symmetry, such that all thirty-six protons are equivalent. If all of these protons were methylene groups, then we would need at least eighteen carbon atoms, but the molecular formula indicates fewer than eighteen carbon atoms. So the thirty-six protons must be twelve equivalent methyl groups. Twelve methyl groups cannot all be attached to the same carbon atom, because the central carbon atom cannot have twelve bonds. However, twelve methyl groups will be equivalent if they comprise four equivalent *tert*-butyl groups, attached to one carbon atom.

This structure is consistent with all of the available data, and it is expected to give rise to three signals in its ^{13}C NMR spectrum (one signal for the central carbon atom, another signal for the four carbon atoms connected to the central carbon atom, and then one last signal for all twelve, equivalent methyl groups).

16.34.
(a) All six methyl groups are equivalent (giving rise to one signal), while all four aromatic protons are also equivalent (giving rise to another signal). In total, we expect two signals.
(b) All four aromatic protons are in unique environments (because of their proximity to the two substituents). Therefore, we expect four signals.
(c) All four aromatic protons are in unique environments (because of their proximity to the two substituents). Therefore, we expect four signals.
(d) This compound has two different kinds of protons (highlighted below), giving rise to two signals.

These two protons can be interchanged by rotational symmetry, so they are homotopic (they produce one signal)

These two protons can be interchanged by rotational symmetry, so they are homotopic (they produce one signal)

(e) The methine (CH) proton gives one signal, while the two methyl groups collectively give one signal (with an integration of 6). In total, we expect two signals.
(f) The replacement test indicates that the two protons of the methylene group are diastereotopic. Therefore, each of these protons will produce its own signal. That is, this methylene group will give rise to two signals (because of the presence of the chirality center). The structure also has two methyl groups which are not equivalent to each other (because of their proximity to the Cl) so they also produce two different signals. Finally, the methine (CH) proton gives a signal, for a total of five signals.

16.35.
(a) This compound has four different kinds of carbon atoms (highlighted below), giving rise to four signals.

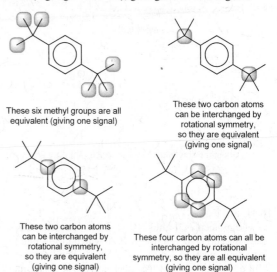

These six methyl groups are all equivalent (giving one signal)

These two carbon atoms can be interchanged by rotational symmetry, so they are equivalent (giving one signal)

These two carbon atoms can be interchanged by rotational symmetry, so they are equivalent (giving one signal)

These four carbon atoms can all be interchanged by rotational symmetry, so they are all equivalent (giving one signal)

(b) Each of the six carbon atoms is in a unique environment (because of its unique proximity to the two substituents). Therefore, we expect six signals.
(c) Each of the six carbon atoms is in a unique environment (because of its unique proximity to the two substituents). Therefore, we expect six signals.
(d) This compound has four different kinds of carbon atoms (highlighted below), giving rise to four signals.

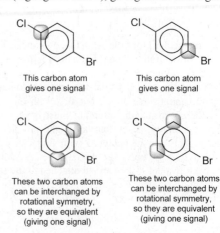

This carbon atom gives one signal

This carbon atom gives one signal

These two carbon atoms can be interchanged by rotational symmetry, so they are equivalent (giving one signal)

These two carbon atoms can be interchanged by rotational symmetry, so they are equivalent (giving one signal)

(e) The carbon atom of the methine (CH) group gives one signal, and the two methyl groups give one signal, for a total of two signals.

(f) Each of the four carbon atoms is in a unique environment, because of its proximity to the substituent. Therefore, we expect four signals.

16.36. The first compound exhibits symmetry that causes some of the carbon atoms to be equivalent (similar to the symmetry present in 16.35d). As such, the first compound will have five signals in its ^{13}C NMR spectrum. In contrast, the second compound lacks this symmetry. Each carbon atom occupies a unique environment, so the second compound is expected to produce seven signals in its ^{13}C NMR spectrum.

16.37. This compound has six different kinds of protons, highlighted here. In each case, we apply the *n*+1 rule, giving the multiplicities shown:

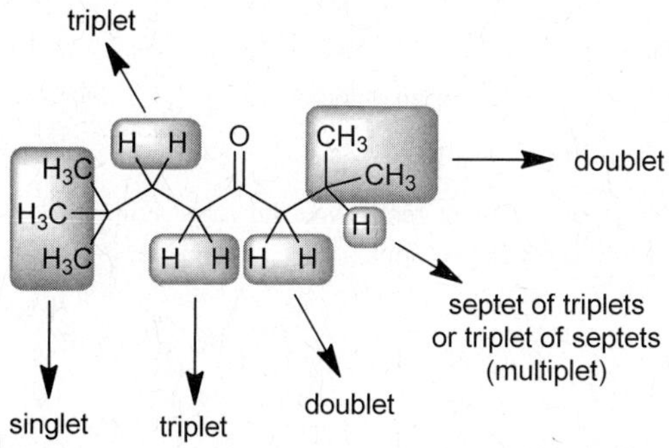

triplet

H₃C →

H₃C →

H₃C →

H H   O   CH₃

CH₃

H

H H H H

→ doublet

singlet

triplet

doublet

septet of triplets
or triplet of septets
(multiplet)

16.38.
(a) The first compound has a very high degree of symmetry, and will produce only four only signals in its ^{13}C NMR spectrum, while the second compound will produce twelve signals.
The first compound will produce only two signals in its ^{1}H NMR spectrum, while the second compound will produce eight signals.

(b) The first compound is a *meso* compound. Two of the protons are enantiotopic (the protons that are alpha to the chlorine atoms) and are therefore chemically equivalent. As such, the first compound will only have two signals in its ^{1}H NMR spectrum, while the second compound will have three signals. For a similar reason, the first compound will only have two signals in its ^{13}C NMR spectrum, while the second compound will have three signals.

(c) The ^{13}C NMR spectrum of the second compound will have one more signal than the ^{13}C NMR spectrum of the first compound. The ^{1}H NMR spectra will differ in the following way: the first compound will have a singlet somewhere between 2 and 5 ppm with an integration of 1, while the second compound will have a singlet at approximately 3.4 ppm with an integration of 3.

(d) The first compound has symmetry that is not present in the second compound. As such, the first compound will have three signals in its ^{13}C NMR spectrum, while the second compound will have five signals.
For similar reasons, the first compound will have two signals in its ^{1}H NMR spectrum, while the second compound will have four signals.

16.39. The molecular formula (C_8H_{18}) indicates no degrees of unsaturation (see Section 15.16), which means that the compound does not have a π bond or a ring. With only one signal in its ^{1}H NMR spectrum, the structure must have a high degree of symmetry, such that all eighteen protons are equivalent. This can be achieved with six equivalent methyl groups, which account for six of the eight carbon atoms. The remaining two carbon atoms can be placed at the center of the structure, rendering all six methyl groups equivalent, like this:

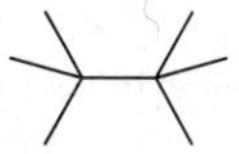

This compound will exhibit two signals in its ^{13}C NMR spectrum (one signal for the two central carbon atoms, and another signal for the six methyl groups).

16.40.
(a) The replacement test gives the same compound, so the protons are homotopic:

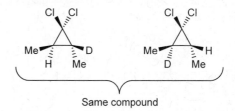

Same compound

(b) The replacement test gives enantiomers, so the protons are enantiotopic:

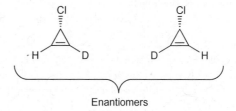

Enantiomers

(c) The replacement test gives enantiomers, so the protons are enantiotopic:

Enantiomers

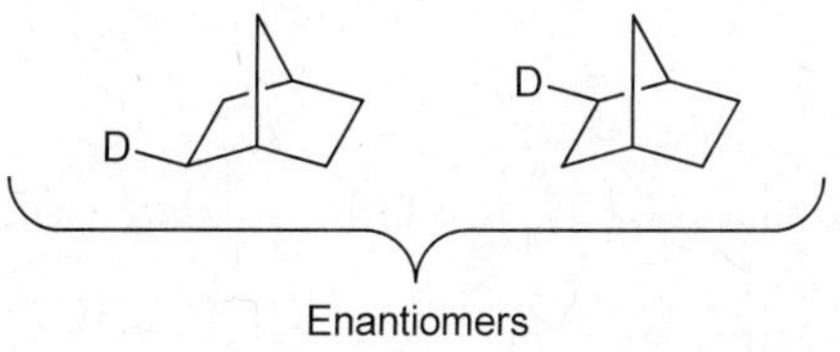

(d) The replacement test gives the same compound, so the protons are homotopic:

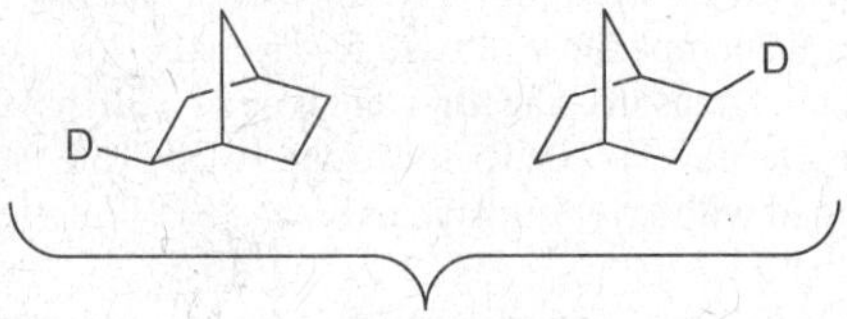

Same compound

(e) The replacement test gives diastereomers, so the protons are diastereotopic:

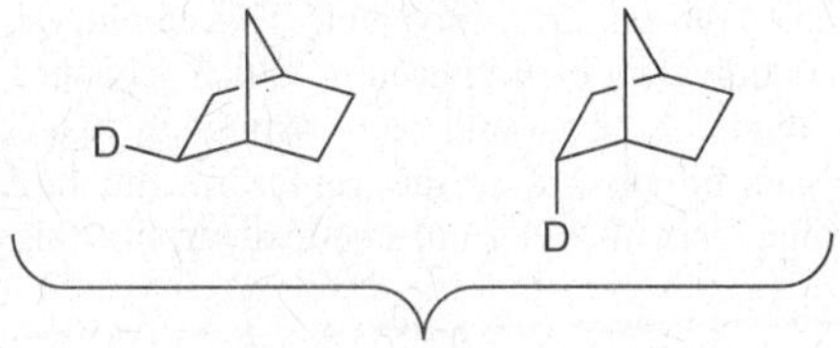

Diastereomers

These compounds are diastereomers because they are stereoisomers that are not mirror images of each other. If it seems like they should be enantiomers, keep in mind that each of these compounds has three chirality centers, and these compounds differ only in the configuration of one of the three chirality centers (each of the bridgehead positions is a chirality center).

(f) The replacement test gives the same compound, so the protons are homotopic:

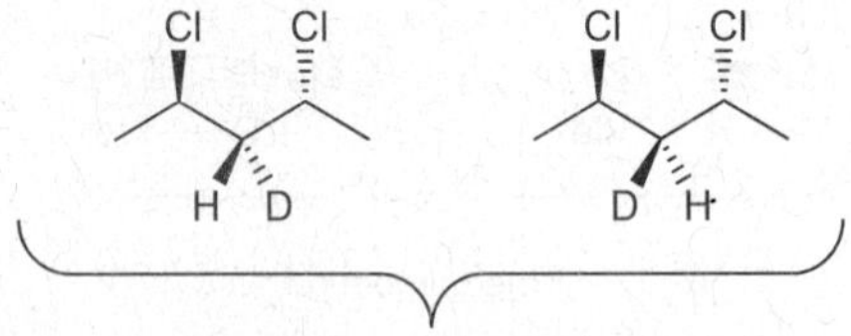

Same compound

(g) The replacement test gives diastereomers, so the protons are diastereotopic:

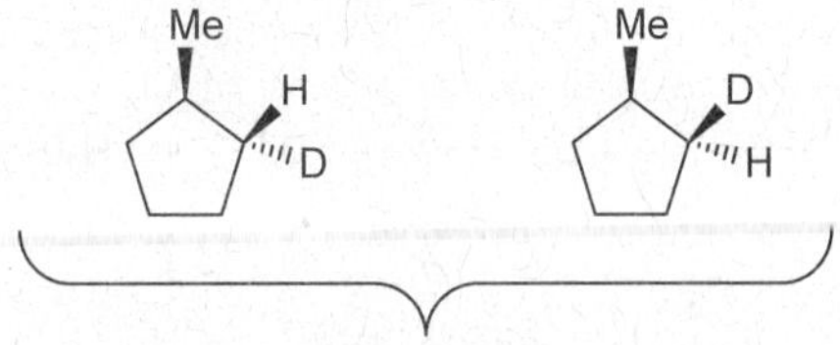

Diastereomers

(h) The replacement test gives diastereomers, so the protons are diastereotopic:

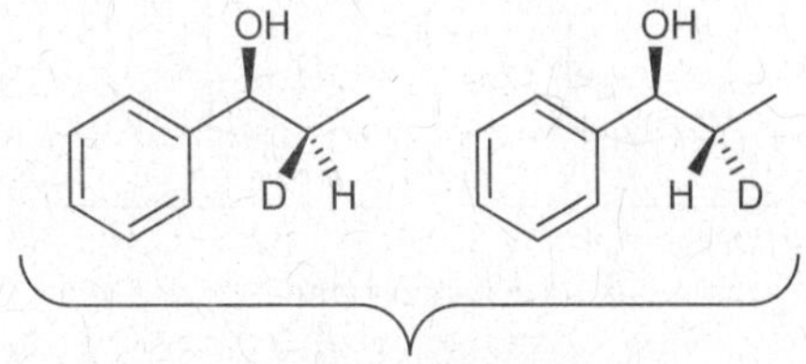

Diastereomers

(i) The replacement test gives the same compound, so the protons are homotopic:

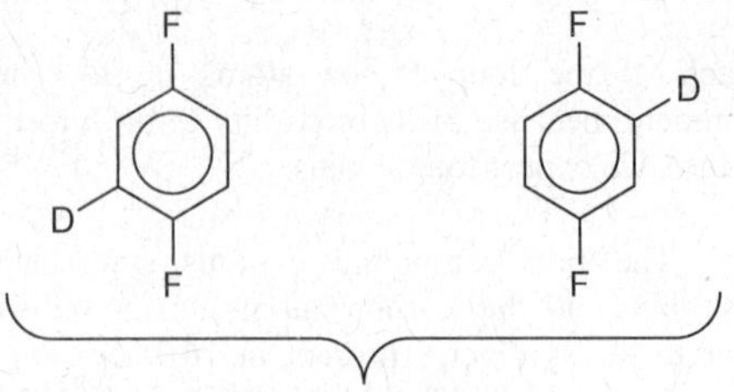

Same compound

(j) The replacement test gives the same compound, so the protons are homotopic:

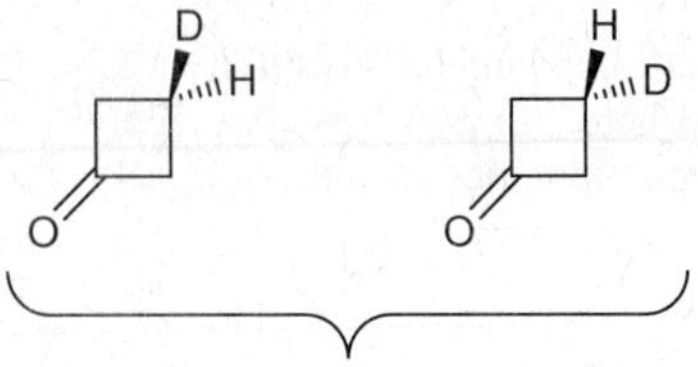

Same compound

(k) The replacement test gives the same compound, so the protons are homotopic:

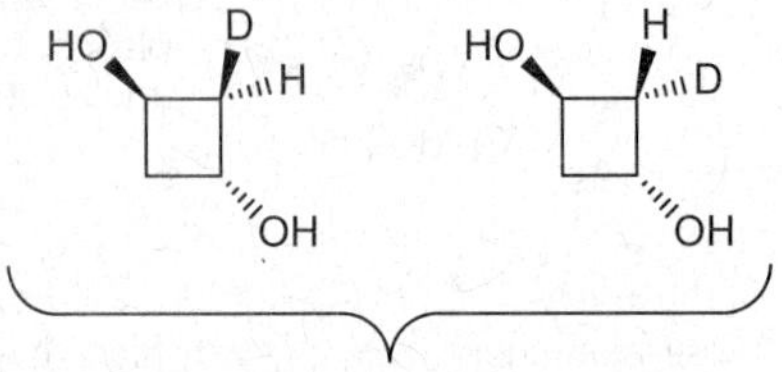

Same compound

(l) The replacement test gives diastereomers, so the protons are diastereotopic:

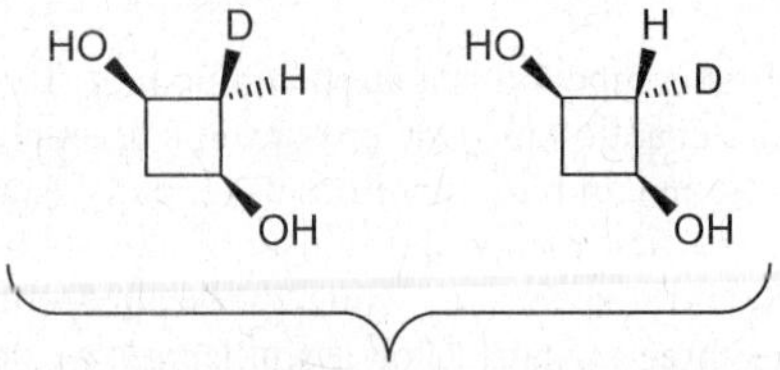

Diastereomers

(m) The replacement test gives enantiomers, so the protons are enantiotopic:

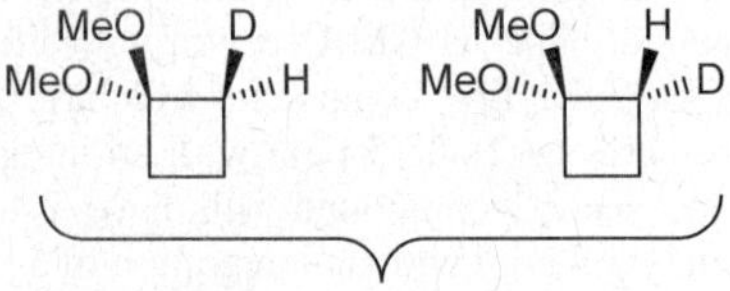

Enantiomers

(n) The replacement test gives diastereomers, so the protons are diastereotopic:

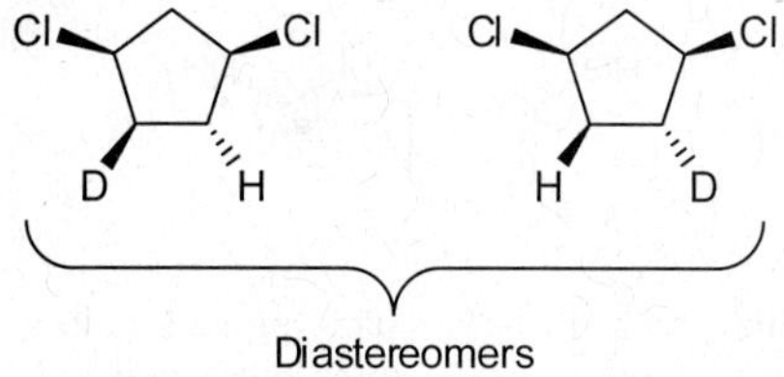

Diastereomers

(o) The replacement test gives the same compound, so the protons are homotopic:

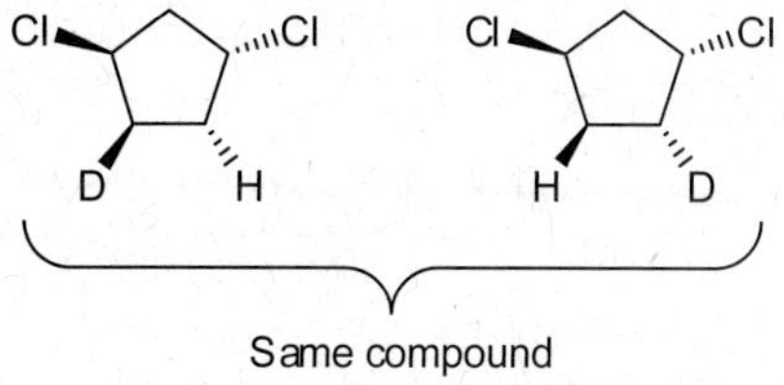

Same compound

16.41. This compound has four aromatic protons. Because of their relationship with the respect to the ring (symmetry), we expect two types of protons, giving rise to a pair of doublets between 7 and 8 ppm:

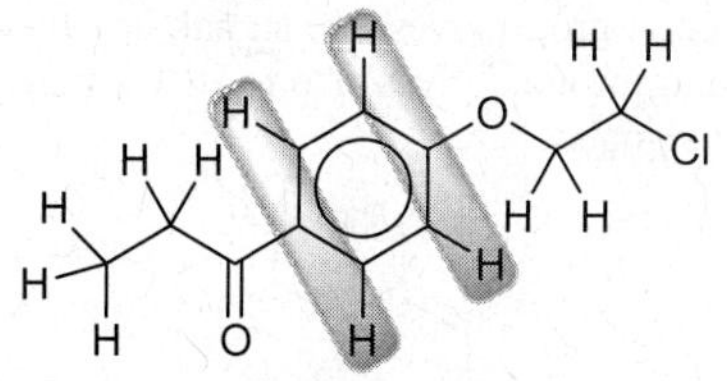

The structure also has an ethyl group, so we expect the characteristic pattern of signals for an ethyl group. Specifically, we expect a triplet with an integration of 3 (corresponding to the CH_3 of the ethyl group) and a quartet with an integration of 2 (corresponding to the CH_2 of the ethyl group).

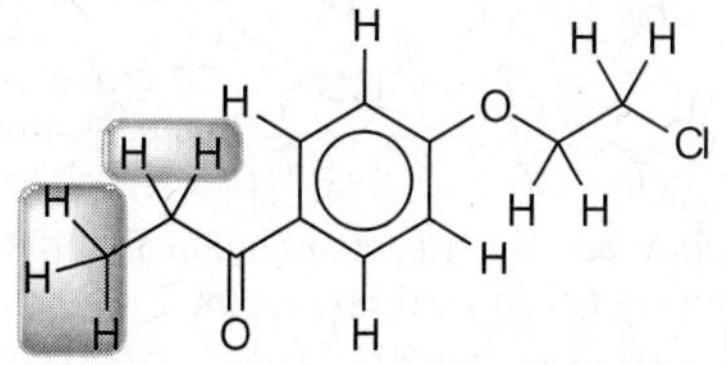

The signal for the CH_2 of the ethyl group is expected to appear at $1.2 + 1 = 2.2$ ppm, while the signal for the CH_3 of the ethyl group is expected to appear at $0.9 + 0.2 = 1.1$ ppm.

Finally, there are two neighboring methylene groups, shown here:

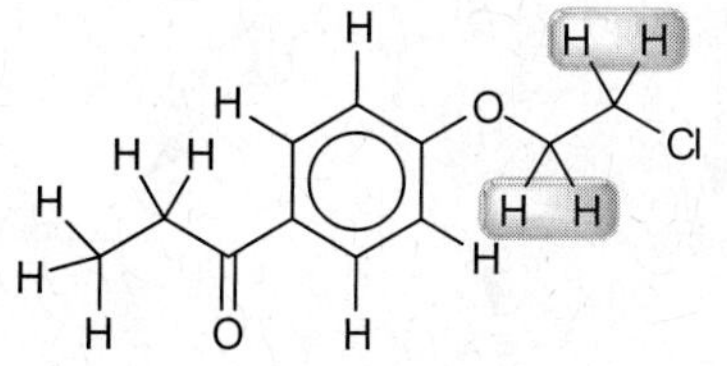

Each of these signals is expected to be a triplet with an integration of 2. And each of these signals is expected to be shifted downfield, because of the electron-withdrawing effects of the chlorine atom and oxygen atom. As seen in Section 16.5, each Cl adds approximately +2, and oxygen adds approximately +2.5 ppm. The CH_2 next to the oxygen atom should produce a signal near $1.2 + 2.5 + 0.4 = 4.1$ ppm. The last term (+0.4) was for the effect of the distant Cl (one-fifth of 2.0). The CH_2 next to the Cl should produce a signal near $1.2 + 2.0 + 0.5 = 3.7$ ppm.

The following hand-drawn spectrum shows all of the signals described above.

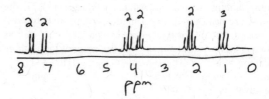

16.42.
(a) Each of the protons occupies a unique environment, and therefore, we expect four signals in the ^1H NMR spectrum of this compound.
(b) All of the halogens withdraw electron density from the neighboring proton, causing a downfield shift. But the effect will be strongest for fluorine (the most electronegative) and weakest for iodine (the least electronegative of the halogens in this compound).

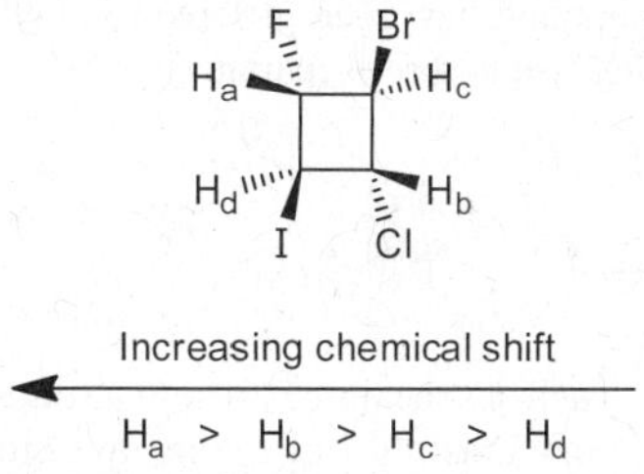

Increasing chemical shift

$H_a > H_b > H_c > H_d$

(c) Each of the carbon atoms occupies a unique environment, and therefore, we expect four signals in the ^{13}C NMR spectrum of this compound.
(d) The carbon atoms follow the same trend exhibited by the protons.

16.43. The molecular formula (C_9H_{18}) indicates one degree of unsaturation (see Section 15.16), which means that the compound must possess either a double bond or a ring. With only one signal in the ^1H NMR spectrum, the structure must have a high degree of symmetry, such that all eighteen protons are equivalent. This can be accomplished with either nine equivalent methylene (CH_2) groups or six equivalent methyl (CH_3) groups. Since the former would use up all of the carbon atoms in the structure (all nine), it is tempting to explore that possibility first. Indeed, a nine-membered ring is comprised of nine equivalent methylene groups, which should give rise to one signal in the ^{1}H NMR spectrum:

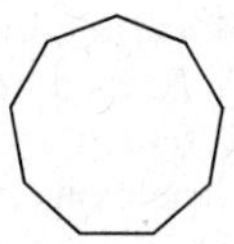

And this compound (cyclononane) also exhibits the correct degree of unsaturation (HDI = 1). However, this structure would give only one signal in its ^{13}C NMR spectrum, and the problem statement indicates that the ^{13}C NMR spectrum has two signals. So, we consider our other alternative (six equivalent methyl groups). If the structure has six equivalent methyl groups, then we still need to account for three more carbon atoms, as well as one degree of unsaturation. We can account for all three carbon atoms in a three-membered ring (HDI = 1). In the following structure, all six methyl groups are equivalent.

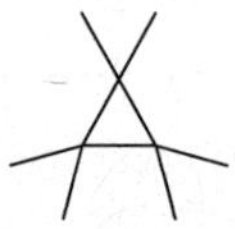

This structure has a high degree of symmetry, and it will indeed give rise to two signals in its ^{13}C NMR spectrum (one signal for all of the methyl groups, and another signal for all three carbon atoms of the ring).

16.44.
(a) This compound has three different methylene (CH$_2$) groups, giving rise to three separate signals.

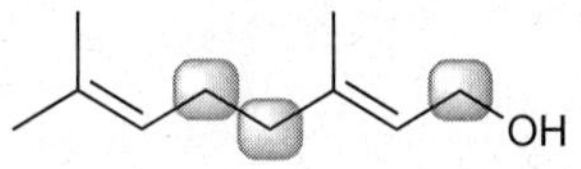

In addition, each of the three methyl groups gives its own unique signal, as none of the methyl groups are in identical electronic environments (the two methyl groups on the left side of the structure are not in identical environments, because one is *trans* to the main chain, while the other is *cis* to the main chain):

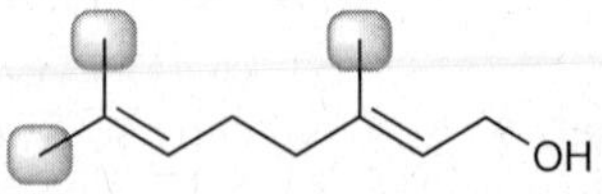

Each of the vinylic CH groups are different, giving rise to two more signals:

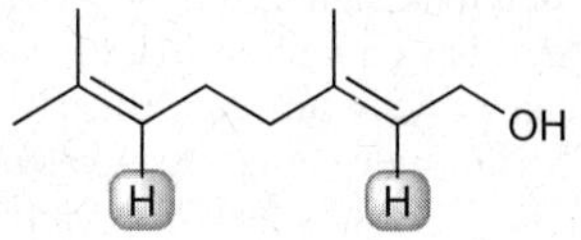

And finally, the proton of the OH group gives one last signal, for a total of 3 + 3 + 2 + 1 = 9 signals.

(b) This compound has two different methylene (CH$_2$) groups, giving rise to two separate signals.

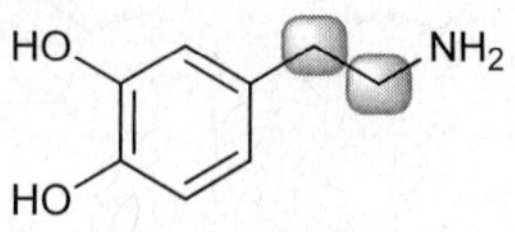

In addition, each of the aromatic protons is in a unique environment, giving rise to three more signals (some or all of these aromatic signals are likely to overlap with each other, giving the appearance of a multiplet):

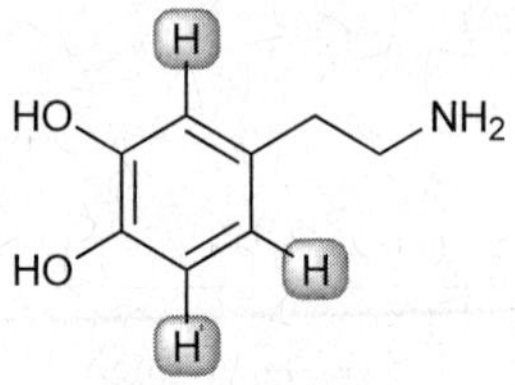

Each of the OH groups is in a unique environment (relative to the CH$_2$CH$_2$NH$_2$ substituent), giving rise to two signals. And finally, the NH$_2$ group gives one last signal, for a total of 2 + 3 + 2 + 1 = 8 signals.

(c) The methyl group gives one signal, and then each of the remaining protons gives rise to its own signal, for a total of six signals:

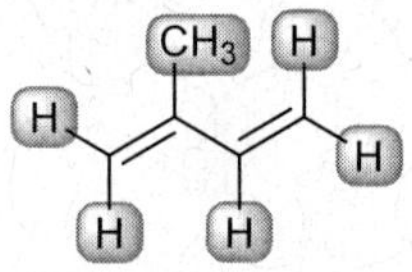

Note that the vinylic protons are different from each other:

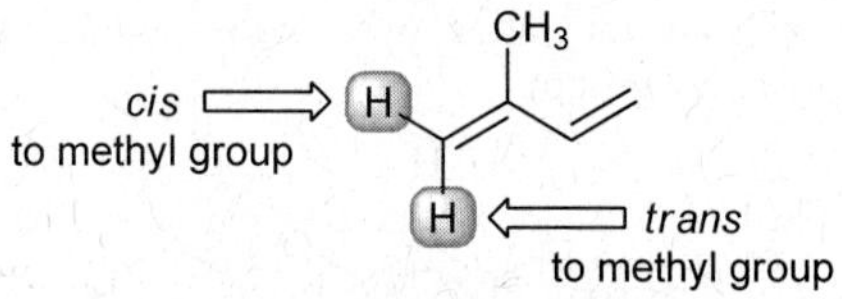

cis to methyl group

trans to methyl group

16.45. Below are the expected chemical shifts for each of the seven signals in this compound:

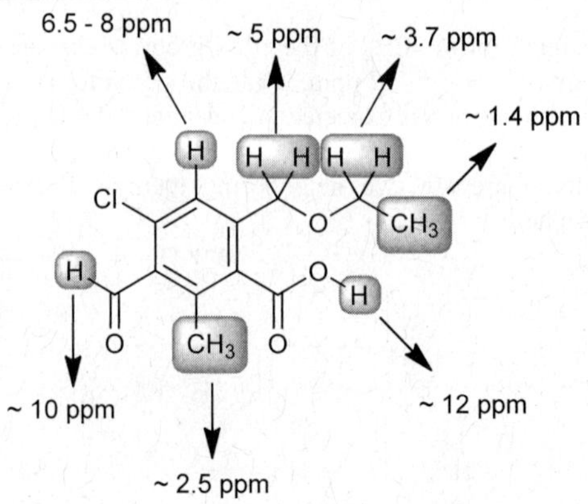

6.5 - 8 ppm   ~ 5 ppm   ~ 3.7 ppm

~ 1.4 ppm

~ 10 ppm

~ 12 ppm

~ 2.5 ppm

16.46.
(a) Symmetry in the ring gives four different signals for the carbon atoms of the ring, in addition to two signals for the vinylic carbon atoms. So in total, we expect six signals, all of which result from sp^2 hybridized carbon atoms, and therefore, we expect all six signals to appear in the region 100 – 150 ppm.
(b) Each of the carbon atoms of the ring occupies a unique environment, giving six signals. The two methyl groups occupy identical environments (they are interchangeable by reflectional symmetry), so they produce one signal. This can be seen more clearly if we draw wedges and dashes to illustrate the 3D orientation of the methyl groups:

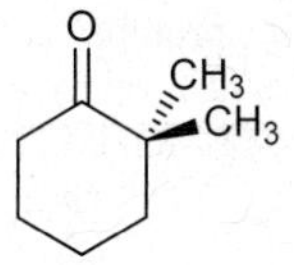

This gives a total of seven signals. The signal resulting from the carbon atom of the carbonyl group is expected to appear in the region 150 – 220 ppm, while the remaining six signals should appear in the region 0 – 50 ppm.
(c) The compound is symmetrical, so we only need to consider half of the structure. We expect a total of four signals, corresponding with the following unique positions:

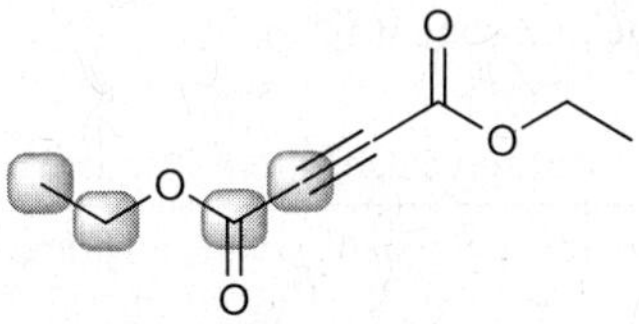

The signal from the carbon atom of the methyl group (sp^3 hybridized) will appear in the region 0 – 50 ppm. The carbon atom of the methylene (CH$_2$) group is also sp^3 hybridized, but it is next to an oxygen atom. So we expect that signal to appear in the region 50 – 100 ppm, together with the signal from the sp hybridized carbon. Finally, the signal from the carbonyl group is expected to appear in the region 150 – 220 ppm.

16.47. Let's begin by drawing the reaction described in the problem statement:

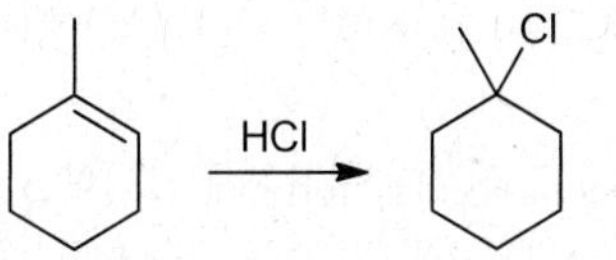

The Markovnikov product has symmetry that the *anti*-Markovnikov product lacks. As such, a ^{1}H NMR spectrum of the Markovnikov product should have fewer signals than a ^{1}H NMR spectrum of the *anti*-Markovnikov product.

16.48.
(a) The compound has a high degree of symmetry, and there are only two unique aromatic protons, highlighted below, giving rise to two signals.

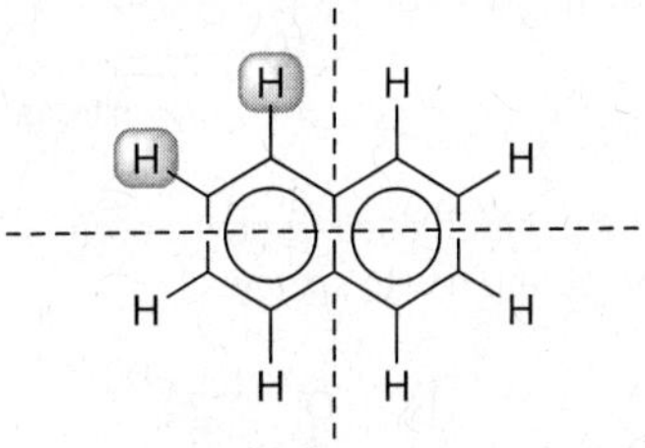

Each of the remaining aromatic protons can be interchanged with one of these positions (via either rotational or reflectional symmetry).

(b) The presence of the methyl group renders all of the aromatic protons different from each other (because of their proximity to the methyl group). As such, we expect eight signals:

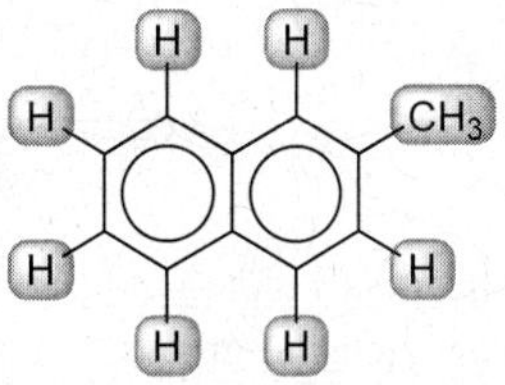

(c) The compound has symmetry that renders some positions identical to other positions. As such, there are only four unique types of protons, highlighted below, giving rise to four signals.

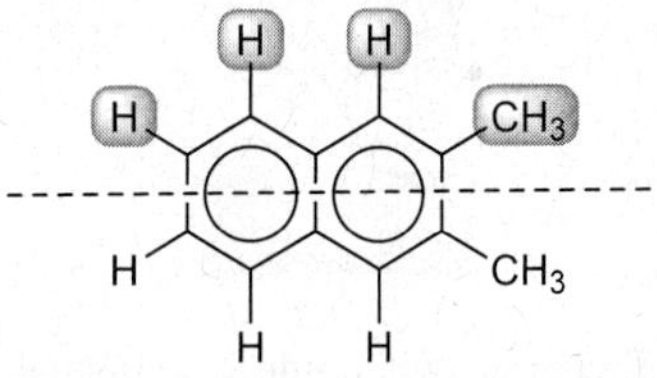

(d) The compound has a high degree of symmetry that renders some positions identical to other positions. As such, there are only two unique types of protons, highlighted below, giving rise to two signals.

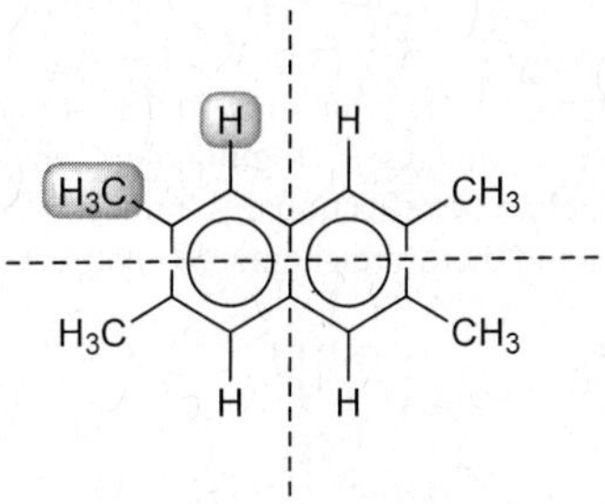

(e) All three vinylic protons are in unique environments, so we expect three signals.

trans to nitro group

cis to nitro group

(f) Each of the highlighted protons occupies a unique environment, giving rise to six signals:

Note that the following positions are different from each other:

trans to oxygen atom

cis to oxygen atom

(g) The compound has a high degree of symmetry. As such, the two methyl groups occupy identical environments and collectively give rise to one signal. Similarly, all four protons of the two methylene (CH_2) can be interchanged by either rotational or reflection symmetry, so these four protons will collectively give rise to one signal. In total, we expect only two signals:

(h) The methyl group will produce one signal.

Now let's consider the remaining four protons. The two protons on wedges (highlighted below) are interchangeable via reflectional symmetry, so they are enantiotopic and therefore chemically equivalent.

Similarly, the two protons on dashes (highlighted below) are also interchangeable via reflectional symmetry, so

they too are enantiotopic and therefore chemically equivalent.

In summary, we expect this compound to produce three signals in its ^1H NMR spectrum.

16.49. Among these three compounds, the first one (benzene) has aromatic protons, which are expected to produce a signal the farthest downfield (between 6.5 and 8 ppm). Acetylenic protons give signals that are relatively upfield (near 2.5 ppm) while vinylic protons are expected to produce a signal in the range of 4.5 – 6.5 ppm.

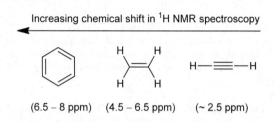

Increasing chemical shift in ^1H NMR spectroscopy

(6.5 – 8 ppm)   (4.5 – 6.5 ppm)   (~ 2.5 ppm)

16.50. In Section 16.5, the term "chemical shift" was defined in the following way:

$$\delta = \frac{\text{observed shift from TMS (in Hz)}}{\text{operating frequency of the instrument (in Hz)}}$$

The problem statement indicates that the chemical shift of the proton is 1.2 ppm and the operating frequency of the spectrometer is 300-MHz. We then plug these values into the equation above, as shown:

$$1.2 \times 10^{-6} = \frac{\text{observed shift from TMS (in Hz)}}{300 \times 10^6 \text{ Hz}}$$

which gives the following observed shift from TMS (in Hz):

$$\text{observed shift from TMS (in Hz)} = \left(1.2 \times 10^{-6}\right) \times \left(300 \times 10^6 \text{ Hz}\right) = 360 \text{ Hz}$$

16.51. The molecular formula ($C_{13}H_{28}$) indicates no degrees of unsaturation (see Section 15.16), which means that the compound does not have a π bond or a ring. The ^1H NMR spectrum exhibits the characteristic pattern of an isopropyl group (a septet with an integration of 1 and a doublet with an integration of 6):

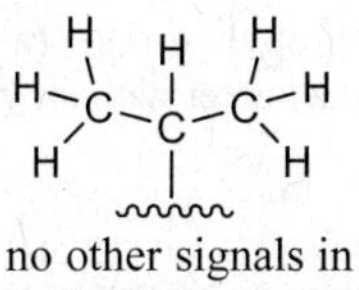

However, there are no other signals in this spectrum, and the molecular formula indicates that there are 28 protons (not just 7 protons, as we would expect for an isopropyl group). So the compound must be highly symmetrical, with four equivalent isopropyl groups (to account for all 28 protons). This also accounts for 12 of the 13 carbon atoms in this compound. The remaining carbon atom must be at the center, connected to all four isopropyl groups:

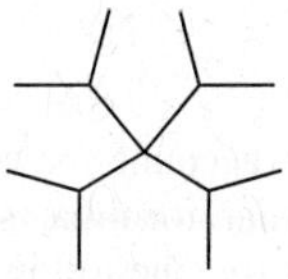

16.52. The molecular formula (C_8H_{10}) indicates four degrees of unsaturation (see Section 15.16), which is highly suggestive of an aromatic ring. This accounts for six of the eight carbon atoms in the structure. The other two carbon atoms must be connected to the ring, either as an ethyl group or as two methyl groups. Ethylbenzene would give an ^1H NMR spectrum with four signals and a ^{13}C NMR spectrum with six signals. The problem statement indicates fewer signals in each of these spectra, which means that the compound must have more symmetry than ethylbenzene. If we explore the three possible ways to connect two methyl groups to a ring (1,2 or 1,3 or 1,4), we will find that only 1,4-dimethylbenzene has the necessary symmetry to give only two signals in the ^1H NMR spectrum and three signals in the ^{13}C NMR spectrum.

16.53. The molecular formula (C_3H_8O) indicates no degrees of unsaturation (see Section 15.16), which means that the compound does not have a π bond or a ring. The broad signal between 3200 and 3600 cm^{-1} indicates the presence of an OH group. The molecular formula indicates that the structure has three carbon atoms, yet the ^{13}C NMR spectrum exhibits only two signals (not three), indicating the presence of symmetry. This is only true for 2-propanol (not for 1-propanol):

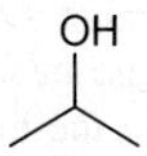

16.54. The molecular formula ($C_4H_6O_4$) indicates two degrees of unsaturation (see Section 15.16), which means that the compound must possess either two double bonds, or two rings, or one ring and one double bond, or a triple bond. The very broad signal (2500 – 3600 cm^{-1}) in the IR spectrum indicates the presence of a carboxylic acid group. The ^1H NMR spectrum has only two signals, with a total integration of 3, however the molecular formula indicates the presence of 6 protons. Therefore,

the actual integration values for the signals are 2H and 4H, respectively. The singlet at 12.1 ppm is characteristic of a carboxylic acid group (COOH), as suggested by the IR spectrum, and since this signal has an integration value of 2H, we conclude that the compound must have two carboxylic acid groups. This accounts for both degrees of unsaturation, which means that the compound does not possess a ring. In order for the remaining four protons to be identical, they must be interchangeable by symmetry, which is indeed the case when we place two methylene (CH$_2$) groups in between the two carboxylic acid groups, like this:

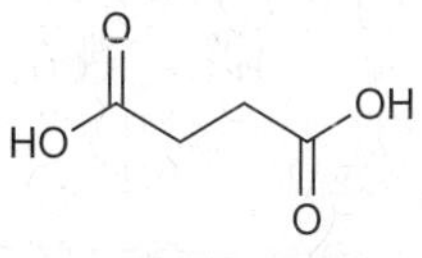

16.55.
(a)   The molecular formula ($C_5H_{10}O$) indicates one degree of unsaturation (see Section 15.16), which means that the compound must possess either a double bond or a ring. The ^1H NMR spectrum exhibits the characteristic pattern of an isopropyl group (a doublet with an integration of 6, and a septet with an integration of 1):

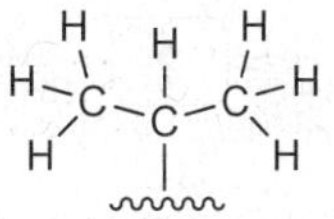

There is also a singlet with an integration of 3, indicating a methyl group. Notice that the signal for the methyl group appears at 2.12 ppm rather than 0.9 ppm, so it has been shifted downfield by just over 1 ppm, which is consistent with being adjacent to a C=O group (accounting for the one degree of unsaturation). The same downfield shift is true for the chemical shift of the CH of the isopropyl group. This gives the following structure:

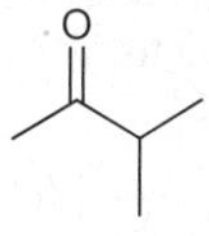

(b)   The molecular formula ($C_5H_{12}O$) indicates no degrees of unsaturation (see Section 15.16), which means that the compound does not have a π bond or a ring. The ^1H NMR spectrum exhibits the characteristic pattern of an ethyl group (a quartet with an integration of 2, and a triplet with an integration of 3):

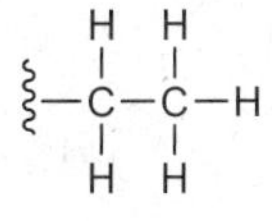

In addition, the singlet with an integration of 6 indicates two equivalent methyl groups:

methyl H_3C ? CH_3 methyl

And the singlet with an integration of 1 is suggestive of an OH group. These pieces account for all of the atoms in the compound except for one carbon atom, which we place in between the two methyl groups. The three fragments can then only be connected in one way, giving the following structure:

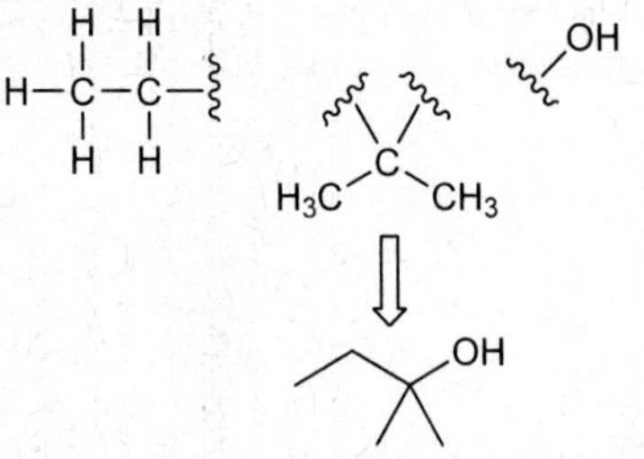

(c) The molecular formula ($C_4H_{10}O$) indicates no degrees of unsaturation (see Section 15.16), which means that the compound does not have a π bond or a ring. The ^1H NMR spectrum exhibits a signal with an integration of 6, which indicates two equivalent methyl groups:

methyl H_3C ? CH_3 methyl

The singlet with an integration of 1 is suggestive of an OH group. The doublet with an integration of 2 indicates a methylene group with only one neighboring proton:

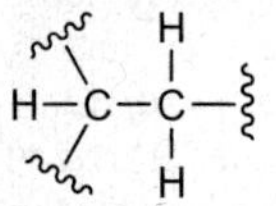

The CH (methine) proton is responsible for the last remaining signal, which is a multiplet, indicating that this CH group is adjacent to the CH_2 group as well as the methyl groups, like this:

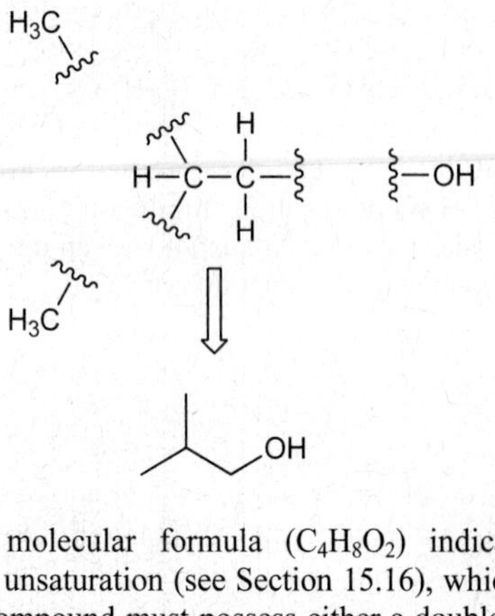

(d) The molecular formula ($C_4H_8O_2$) indicates one degree of unsaturation (see Section 15.16), which means that the compound must possess either a double bond or a ring. The ^1H NMR spectrum exhibits the characteristic

pattern of an isopropyl group (a doublet with an integration of 6, and a septet with an integration of 1):

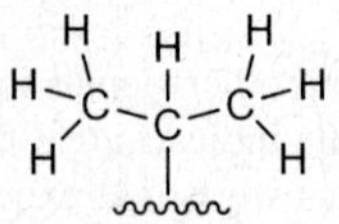

There is also a singlet with an integration of 1 at 11.38 ppm, indicating a carboxylic acid group (which accounts for the one degree of unsaturation).

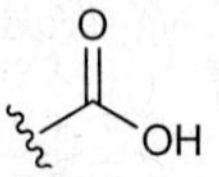

These two fragments account for all of the atoms that appear in the molecular formula, so we connect these fragments together to give the following structure:

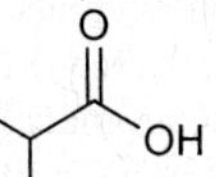

16.56. The molecular formula ($C_8H_{10}O$) indicates four degrees of unsaturation (see Section 15.16), which is highly suggestive of an aromatic ring. This accounts for six of the eight carbon atoms in the structure. The other two carbon atoms must be an ethyl group, as seen in the ^1H NMR spectrum. Specifically, the quartet with an integration of 2 and the triplet with an integration of 3 indicate the presence of an ethyl group:

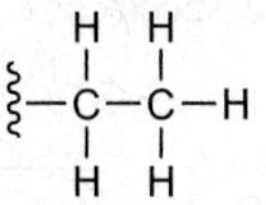

Note that the total integration of the aromatic region is 2H + 2H = 4H, which means that the ring is disubstituted. Furthermore, the aromatic signals are a pair of doublets, indicating symmetry, which is achieved with a 1,4-disubstituted aromatic ring.

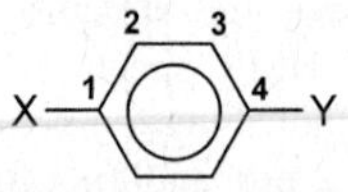

One of the substituents must be an ethyl group, so the remaining substituent must be an OH group (we have accounted for all atoms in the molecular formula except for an O and an H).

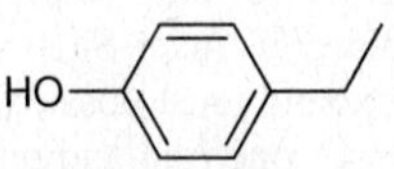

This structure is corroborated by the ^{13}C NMR spectrum, which exhibits six signals (four for the aromatic ring, because of symmetry, and two for the ethyl group).

16.57. The molecular formula (C_9H_{12}) indicates four degrees of unsaturation (see Section 15.16), which is highly suggestive of an aromatic ring. This is confirmed by the multiplet just above 7 ppm in the ^1H NMR spectrum. This signal has an integration of 5, indicating that the ring is monosubstituted:

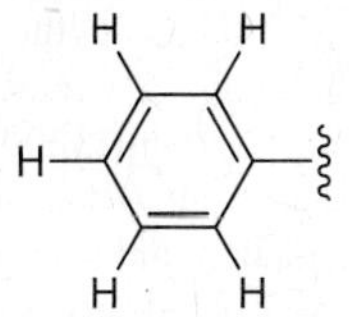

The ^1H NMR spectrum also shows the characteristic pattern of signals for an isopropyl group (a septet with an integration of 1 and a doublet with an integration of 6):

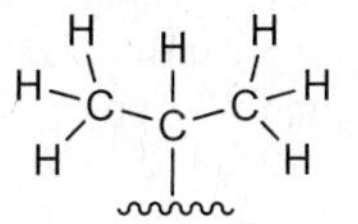

These two fragments (the monosubstituted aromatic ring and the isopropyl group) account for the entire structure:

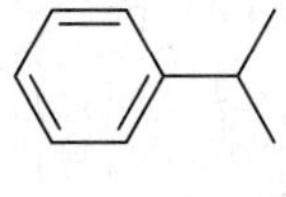

16.58. The molecular formula ($C_9H_{10}O_2$) indicates five degrees of unsaturation (see Section 15.16), which is highly suggestive of an aromatic ring, in addition to either one double bond or one ring. The multiplet just above 7 ppm in the ^1H NMR spectrum corresponds with aromatic protons, which confirms the presence of an aromatic ring. The integration of this signal is 5, which indicates that the ring is monosubstituted:

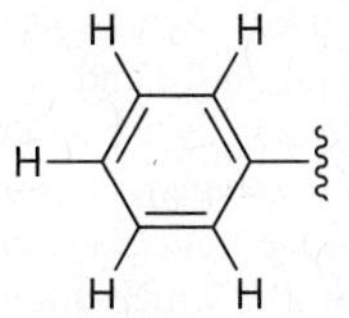

The presence of a monosubstituted ring is confirmed by the four signals between 100 and 150 ppm in the ^{13}C NMR spectrum (the region associated with sp^2 hybridized carbon atoms), just as expected for a monosubstituted aromatic ring:

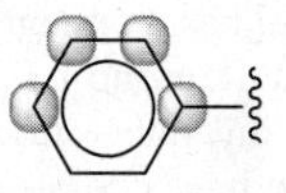

In the ^1H NMR spectrum, the signal near 12 ppm (with an integration of 1) indicates the presence of a carboxylic acid group, which is confirmed by the signal near 180 ppm in the ^{13}C NMR spectrum (the region associated with carbonyl groups):

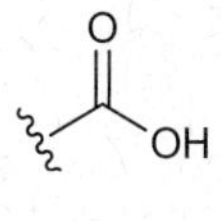

In the ^1H NMR spectrum, the pair of triplets (each with an integration of 2) indicates a pair of neighboring methylene groups:

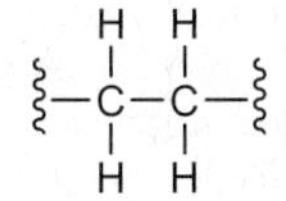

These methylene groups account for the two signals between 0 and 50 ppm in the ^{13}C NMR spectrum.
We have now analyzed all of the signals in both spectra, and we have uncovered three fragments, which can only be connected to each other in the following way:

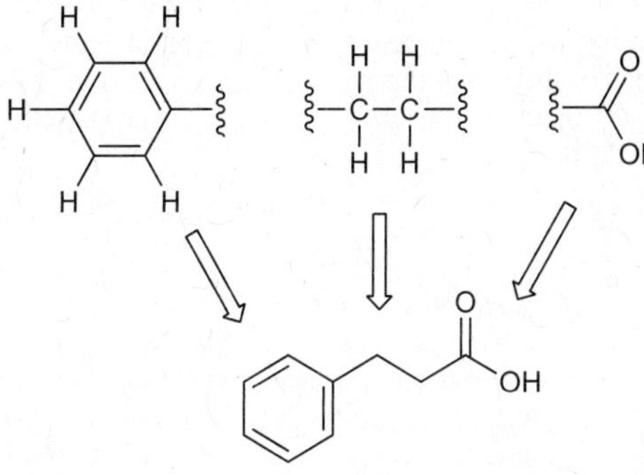

16.59.
(a) The molecular formula ($C_5H_{10}O$) indicates one degree of unsaturation (see Section 15.16), which means that the compound must possess either a double bond or a ring. One of the signals in the ^{13}C NMR spectrum appears above 200 ppm, indicating the presence of a carbonyl group (C=O), which accounts for the one degree of unsaturation.
In total, the ^{13}C NMR spectrum exhibits only three signals, while the molecular formula indicates the presence of five carbon atoms. Therefore, the structure must possess symmetry, giving only three different kinds of carbon atoms (one of which is a carbonyl group), as seen in the following structure:

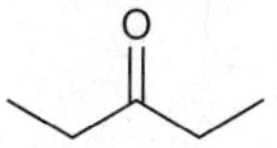

(b) The molecular formula ($C_6H_{10}O$) indicates two degrees of unsaturation (see Section 15.16), which means that the compound must possess either two double bonds, or two rings, or one double bond and one ring, or a triple bond. The ^{13}C NMR spectrum exhibits only three signals, while the molecular formula indicates the presence of six carbon atoms. Therefore, the structure must possess symmetry, giving only three different kinds of carbon atoms. Two of these types of carbon atoms must be sp^2 hybridized (because two signals appear between 100 and 150 ppm), while the third signal indicates a carbon atom attached to an oxygen atom.

The following structure accounts for all of the observations above:

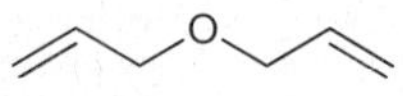

16.60. The problem statement indicates that the compound is an alcohol, so it must contain an OH group. The molecular formula ($C_4H_{10}O$) indicates no degrees of unsaturation (see Section 15.16), which means that the compound cannot have any double bonds, triple bonds, or rings. That is, the structure must be acyclic and cannot have any π bonds. The broadband decoupled spectrum has four signals, one of which appears above 50 ppm (this signal accounts for the carbon atom attached directly to the OH group). The other three signals in the broadband decoupled spectrum are all below 50 ppm, indicating that all of the carbon atoms are sp^3 hybridized (although we already knew that because HDI = 0). The DEPT-90 spectrum has only one signal, which means the compound has only one methine (CH) group. Furthermore, this signal is above 50 ppm, which indicates that this CH group is connected to directly to the OH group:

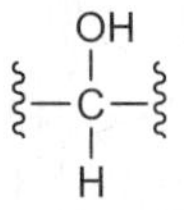

The DEPT-135 spectrum indicates that the other three signals (below 50 ppm) correspond with one methylene group (upside-down signal) and two methyl groups (right-side up signals that did not appear in the DEPT-90).

We have now analyzed all of the signals in all of the spectra, and we have uncovered four fragments, which can only be connected to each other in the following way:

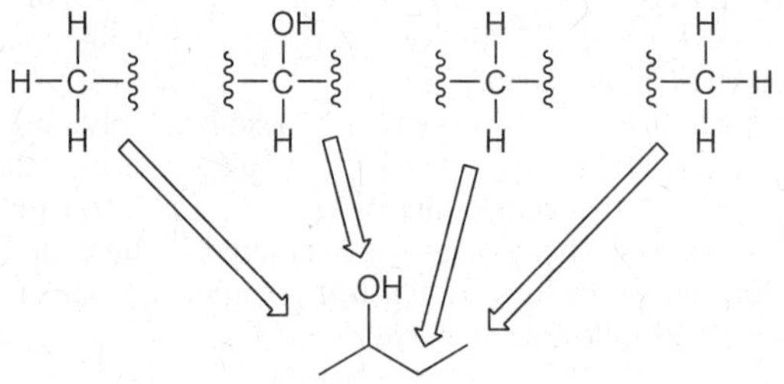

16.61. The problem statement indicates that the compound is an alcohol, so it must contain an OH group. The molecular formula ($C_6H_{14}O$) indicates no degrees of unsaturation (see Section 15.16), which means that the compound cannot have any double bonds, triple bonds, or rings. That is, the structure must be acyclic and cannot have any π bonds. The DEPT-135 spectrum exhibits five signals that are upside-down, indicating the presence of five methylene groups. One of these methylene groups must be connected to the OH group, because one of the upside-down signals appears above 50 ppm. The five methylene groups and the OH group account for 11 of the 14 protons in this compound. Therefore, the signal pointing up must correspond with a

methyl group (rather than a CH group). There is only one way to connect a methyl group, five methylene groups, and an OH group, as shown here:

16.62. The molecular formula ($C_6H_{14}O_2$) indicates no degrees of unsaturation (see Section 15.16), which means that the compound cannot have any double bonds, triple bonds, or rings. That is, the structure must be acyclic and cannot have any π bonds. The IR spectrum has a broad signal between 3200 and 3600 cm^{-1}, indicating the presence of an OH group. The ^{13}C NMR spectrum has six signals, and the molecular formula indicates there are six carbon atoms, which means that the compound lacks symmetry that would interchange any of the carbon atoms. Since the compound has no degrees of unsaturation, all of the signals must arise from sp^3 hybridized carbon atoms. Indeed, three of the signals appear between 0 and 50 ppm, as expected for sp^3 hybridized carbon atoms. But the other three signals appear between 50 and 100 ppm, indicating that three carbon atoms are connected to an oxygen atom. We can therefore draw the following fragments (since the compound has an OH group, and since the molecular formula indicates only two oxygen atoms):

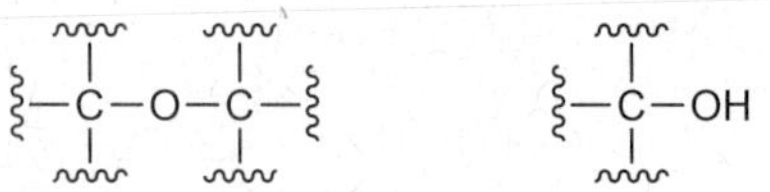

Now we explore the ^1H NMR spectrum. Let's begin with the signals downfield. There are three signals that appear between 3.5 and 4 ppm. One of these signals is clearly a triplet, but the other two signals are overlapping so it is difficult to determine their multiplicity (perhaps they are doublets that appear very close to each other, or perhaps they are overlapping triplets). We will revisit the multiplicity of these signals later. For now, let's focus on the chemical shifts and integration values for these signals. These signals are certainly from the three groups connected to oxygen atoms (because of their chemical shifts), and we notice that each of these signals has an integration of 2H, which allows us to modify the fragments above as follows:

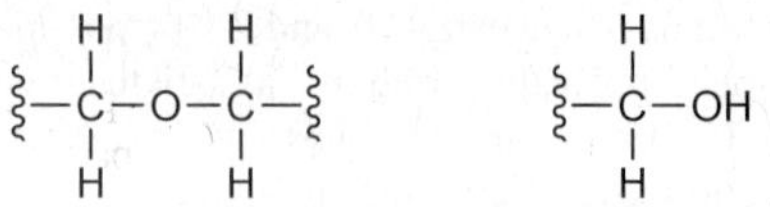

The singlet at 2.4 ppm has an integration of 1, which can be attributed to the OH group. Each of the multiplets near 1.5 ppm has an integration of 2H, indicating methylene groups that have complex splitting:

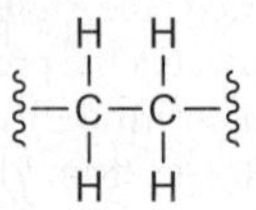

The triplet at 1.0 ppm has an integration of 3H, indicating a methyl group that is connected to a neighboring CH$_2$ group.

In summary, we have the following fragments:

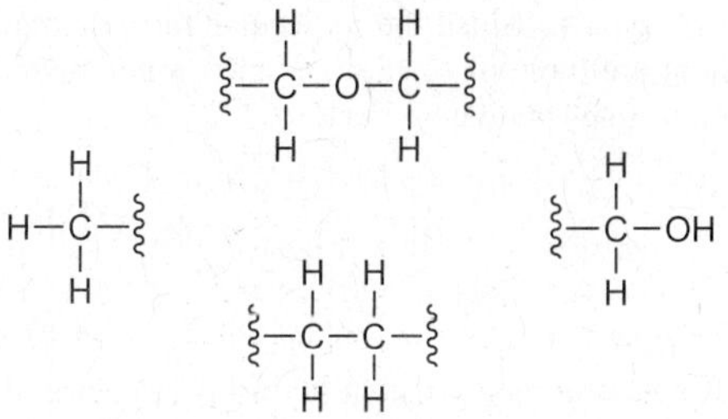

There are only two ways to connect these four fragments:

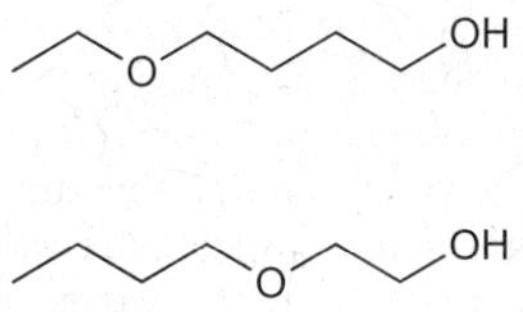

The first structure contains an ethyl group, which should produce a quartet with an integration of 2 (for the CH$_2$ portion of the ethyl group). If we inspect the three signals between 3.5 and 4 ppm, it is difficult to argue that any of these signals is a quartet. While it is difficult to be certain, because two of these signals overlap with each other, it looks more like each of these signals is a triplet, which would be consistent with the second structure:

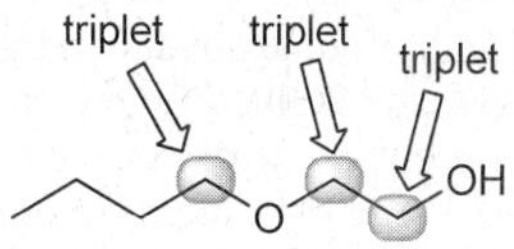

16.63. The molecular formula (C$_8$H$_{10}$O) indicates four degrees of unsaturation (see Section 15.16), which is highly suggestive of an aromatic ring. This is confirmed by the presence of signals just above 3000 cm^{-1} in the IR spectrum, and signals at approximately 1600 cm^{-1}. In the ^{1}H NMR spectrum, the signals near 7 ppm are likely a result of aromatic protons, which also confirms the presence of an aromatic ring. Notice that the combined integration of these two signals (near 7 ppm) is 4H. This, together with the distinctive splitting pattern (a pair of doublets), suggests a 1,4-disubstituted aromatic ring:

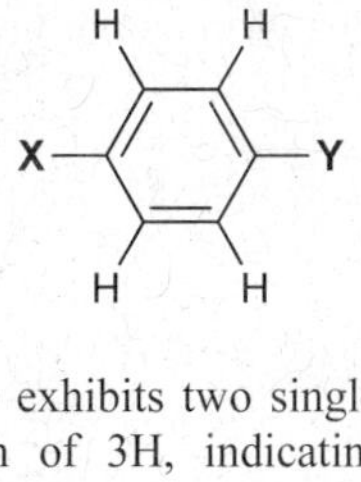

The spectrum also exhibits two singlets, each of which has an integration of 3H, indicating methyl groups.

Notice that both signals are shifted downfield (relative the benchmark value for a methyl group of 0.9 ppm). One of them is shifted much more than other, indicating that it is likely next to an oxygen atom. This gives the following structure:

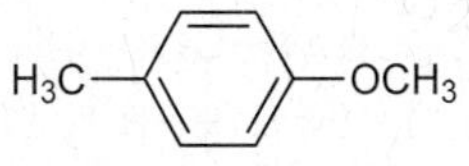

This structure is consistent with the ^{13}C NMR spectrum. Specifically, there are four signals for the aromatic ring, one of which is shifted downfield because it is next to an oxygen atom. And the other two signals are for the methyl groups, one of which appears above 50 ppm because the carbon atom giving rise to this signal is next to an oxygen atom.

16.64. The molecular formula (C$_5$H$_{10}$O) indicates one degree of unsaturation (see Section 15.16), which means that the compound must either have a double bond or a ring. The IR spectrum has a signal at approximately 3100 cm^{-1}, indicating the presence of a Csp^2–H bond:

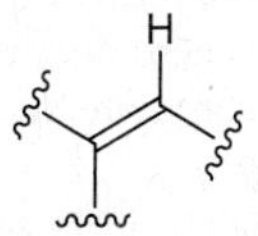

This is consistent with the weak signal at 1600 cm^{-1}, indicating the presence of a C=C double bond (which accounts for the one degree of unsaturation).
The ^{13}C NMR spectrum has two signals between 100 and 150 ppm, confirming the presence of a C=C double bond. In addition, there are three other signals, two of which appear between 50 and 100 ppm. These latter two signals are characteristic of carbon atoms connected to an oxygen atom. Since there is only one oxygen atom in the structure (C$_5$H$_{10}$O), these two carbon atoms must be connected to it:

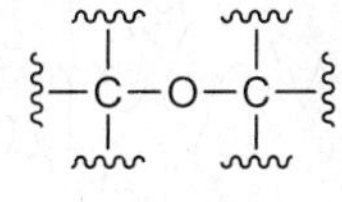

In the ^{1}H NMR spectrum, we see the characteristic pattern of an ethyl group (a triplet with an integration of 3 and a quartet with an integration of 2):

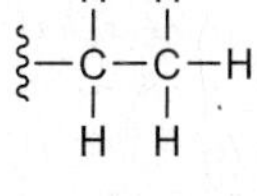

Notice that the quartet appears at 3.5 ppm, indicating that the ethyl group is one of the two groups that is connected to the oxygen atom:

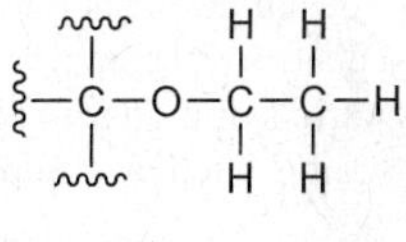

The signal at 4.0 ppm has an integration of 2, which represents a CH_2 group connected to the other side of the oxygen, like this:

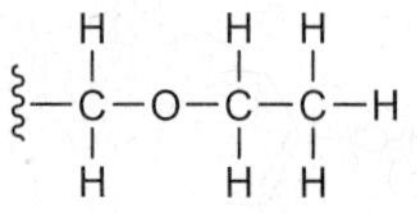

Remember that our structure must include a double bond, which completes the structure, and accounts for all of the atoms in the molecular formula ($C_5H_{10}O$):

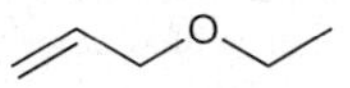

The signals above 5.0 correspond with the vinylic protons. There are three signals, and they are all splitting each other. Two of them are overlapping, to give an apparent integration of 2.

16.65. The molecular formula ($C_8H_{14}O_3$) indicates that the compound has two degrees of unsaturation, so the structure must either have two rings, or two double bonds, or a ring and a double bond, or a triple bond.
In the IR spectrum, there are two signals that are suggestive of C=O groups, which would account for both degrees of unsaturation.
In the ^1H NMR spectrum, there are three signals. By comparing the height of the S-curves, we can see that the *relative* integration values are 2:2:3. With a total of 14 protons (as seen in the molecular formula), these relative integration values must correspond to 4:4:6. This indicates a high level of symmetry in the compound.
The signal at approximately 1 ppm has an integration of 6, indicating two methyl groups that are identical because of symmetry. This signal is a triplet, which means that each of these CH_3 groups is neighboring a methylene (CH_2) group.

Based on the integration and multiplicities of the remaining two signals (at 1.7 ppm and 2.5 ppm), we can assemble the following two fragments, which must be identical by symmetry:

Each of the central methylene groups is being split by a neighboring methyl group and a neighboring methylene group. This could lead to a complex splitting pattern (either a triplet of quartets or a quartet of triplets). In this case, we are not seeing such a complex pattern. As

mentioned in the textbook, this can happen when the J-values are fortuitously similar. In such a case, the system behaves as if it has five neighbors, and the n+1 rule gives a sextet, which is what we see in this case.
Thus far, we have identified the presence of two, symmetrically positioned propyl groups, and there are two C=O bonds. Since the molecular formula indicates that there are three oxygen atoms in this compound, we can deduce the following structure:

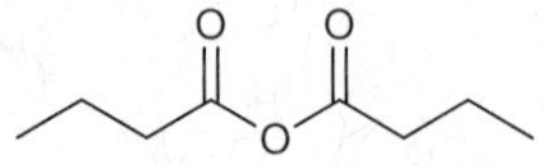

The IR spectrum is consistent with this structure, in that the two signals (at 1755 and 1820 cm^{-1}) represent symmetrical and unsymmetrical stretching of the anhydride unit.
The ^{13}C NMR spectrum is also consistent with this structure. We see a carbon atom (of a C=O bond) at 180 ppm - notice that there is only one signal at 180 ppm, because the two carbonyl groups are identical by symmetry. As expected, three signals appear between 0 and 50 ppm, corresponding to the three carbon atoms of the propyl group (once again, the two propyl groups are identical to each other, giving rise to three signals rather than six). We can see from the DEPT-135 spectrum that two of these signals are methylene groups (because the signals are upside down), which is also consistent with the structure that we deduced above.

16.66. A compound with molecular formula $C_6H_{10}O_4$ has two degrees of unsaturation, so any proposed structure must either have two rings, or two double bonds, or a ring and a double bond, or a triple bond.
The signal at 1747 cm^{-1} is likely a C=O bond of an ester, which accounts for one of the degrees of unsaturation.
In the proton NMR spectrum, there are four signals. The signal just above 5 ppm has an integration of 1, indicating a methine proton (CH), and it is a quartet, which indicates a neighboring methyl group. The signal for the methyl group should be a doublet (since it is next to the methine proton). That signal appears at 1.5ppm.

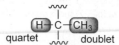

The quartet is significantly downfield (5 ppm), which indicates that it is neighbored by an oxygen atom, as well as some other group which can also shift the signal further downfield. The remaining two signals, just above 2 ppm and just below 4 ppm, are methyl groups (each has an integration of 3). Based on their chemical shifts, the former is likely to have a neighboring C=O group, while the latter is likely to be next to an oxygen atom:

The carbon NMR spectrum reveals that there are actually two carbonyl groups in this compound, and the molecular formula indicates four oxygen atoms, so it is likely that there are two ester groups that are overlapping in the IR spectrum (both at approximately 1747 cm⁻¹). This gives us three pieces that must be assembled:

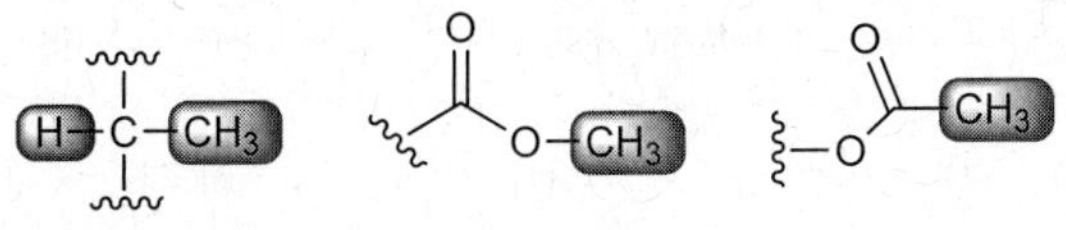

The following structure can be assembled from these three pieces:

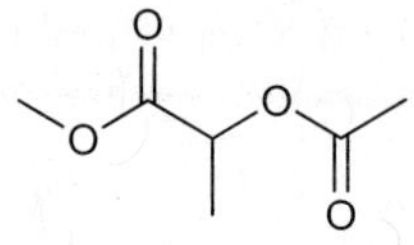

16.67. The molecular formula ($C_8H_{14}O_4$) indicates that the compound has two degrees of unsaturation, so the structure must either have two rings, or two double bonds, or a ring and a double bond, or a triple bond.

The signal at 1736 cm⁻¹ is likely a C=O bond of an ester, which accounts for one of the degrees of unsaturation.

In the proton NMR spectrum, there are three signals. By comparing the height of the S-curves, we can see that the *relative* integration values are 2:2:3. With a total of 14 protons (as seen in the molecular formula), these relative integration values must correspond to 4:4:6. This indicates a high level of symmetry in the compound.

The signal just above 4 ppm has an integration of 4, indicating two methylene (CH_2) groups that are identical because of symmetry. This signal is a quartet, which means that each of these two CH_2 groups is neighboring a methyl group. The signal for these methyl groups should be a triplet with an integration of 6. That signal appears just above 1 ppm.

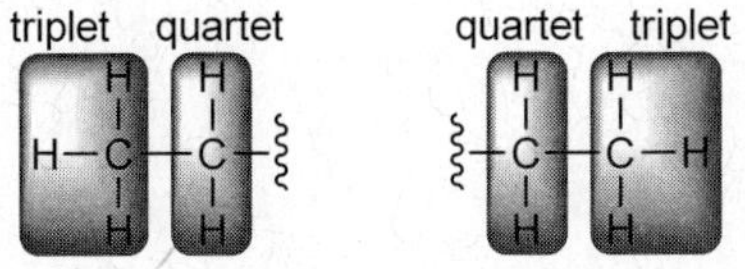

triplet   quartet   quartet   triplet

The quartet is significantly downfield (4 ppm), which indicates that it is neighboring an oxygen atom.

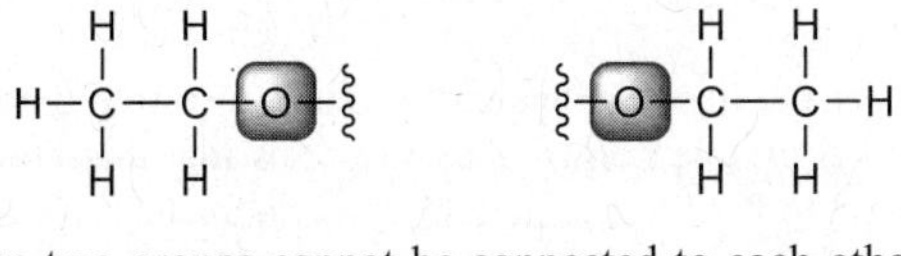

These two groups cannot be connected to each other, as that would not give us an opportunity to connect the remaining atoms in the compound.

The remaining signal, just below 3 ppm, is a singlet with an integration of 4, indicating two methylene groups that are equivalent by symmetry, with no neighboring protons:

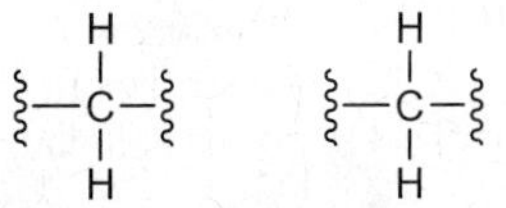

These two groups may or may not be connected to each other, but we know that they must be equivalent by symmetry, and they cannot have any neighbors.

The molecular formula indicates two degrees of unsaturation and the presence of four oxygen atoms, but so far, we have only accounted for one degree of unsaturation (C=O) and only three oxygen atoms. The fourth oxygen atom, as well as the extra degree of unsaturation can be accounted for in the following structure which has the necessary symmetry, and is consistent with all of the spectra:

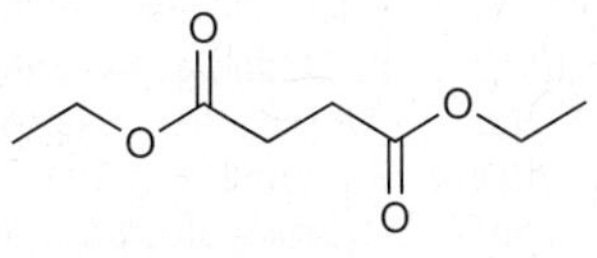

Notice that, because of symmetry, there are only four different kinds of carbon atoms in this compound, giving four signals in the carbon NMR spectrum. Two of these signals correspond to methylene (CH_2) groups, as confirmed by the presence of two upside-down signals in the DEPT-135 spectrum.

16.68. The molecular formula ($C_{12}H_8Br_2$) indicates that the compound has eight degrees of unsaturation, so the structure likely contains two aromatic rings (each of which represents four degrees of unsaturation).

In the proton NMR spectrum, there are only two signals, with the same relative integration. Since the molecule has eight protons (as seen in the molecular formula), we must conclude that each signal corresponds to four protons. Each of these signals is a doublet indicating only one neighboring proton. The following structure is consistent with this information:

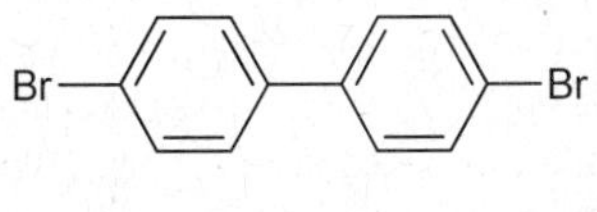

Br—⬡—⬡—Br

In this structure, there are only two different kinds of protons, labeled H_a and H_b below:

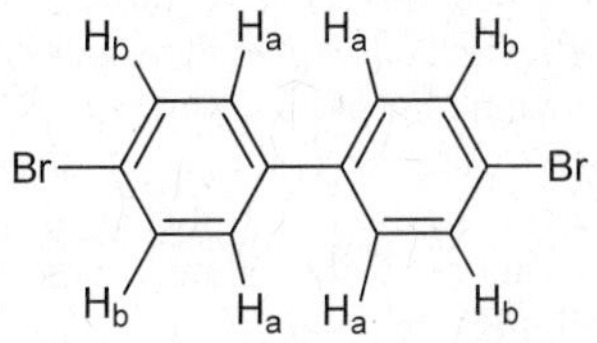

This structure is consistent with the carbon NMR spectrum, in which there are only four signals, all of which are aromatic:

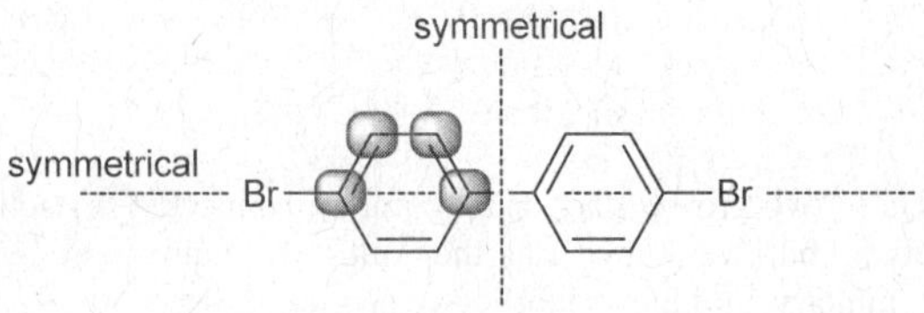

symmetrical

symmetrical

16.69.

(a) The height of the $(M+1)^{+\bullet}$ peak indicates that the compound has four carbon atoms (1.1% for each carbon atom). Since the parent ion appears at $m/z = 104$, we know that nearly half of the molecular weight is due to carbon atoms $(12 \times 4 = 48)$. The rest of the molecular weight $(104 - 48 = 56)$ must be attributed to oxygen atoms and hydrogen atoms. There is a limit to how many hydrogen atoms there can be, since there are only four carbon atoms. Even if the compound is fully saturated, it could not have more than 10 protons $(2n+2,$ where n is the number of carbon atoms). Oxygen has an atomic weight of approximately 16. Therefore, the compound must contain at least three oxygen atoms (otherwise we could not account for the rest of the compound using hydrogen atoms alone). But the compound cannot contain *more* than three oxygen atoms, because four oxygen atoms have a combined atomic weight of 64, which already blows our budget, even before we place any hydrogen atoms (remember that our budget for O and H atoms is a total mass of 56). So, we conclude that the compound must have exactly three oxygen atoms. The remaining weight is accounted for with hydrogen atoms, giving the following molecular formula: $C_4H_8O_3$.

(b) The molecular formula $(C_4H_8O_3)$ indicates that the compound has one degree of unsaturation, so the structure must either have a ring or a double bond.

In the IR spectrum, the broad signal between 3200 and 3600 cm^{-1} is characteristic of an O-H bond, and the signal at 1742 cm^{-1} is likely a C=O group of an ester.

In the proton NMR spectrum, there are four signals. By comparing the height of the S-curves, we can see that the relative integration values are 2:2:1:3. With a total of eight protons (as seen in the molecular formula), these relative integration values correspond precisely to the number of protons giving rise to each peak. The signal just above 1 ppm corresponds to three protons, and is therefore a methyl group. Since this signal is a triplet, it must be next to a methylene (CH_2) group. The signal for that methylene group appears as a quartet, just as expected (since it is next to the methyl group) above 4 ppm. The location of this signal indicates that the methylene group is likely next to an oxygen atom (since it is shifted downfield). So far, we have the following fragment:

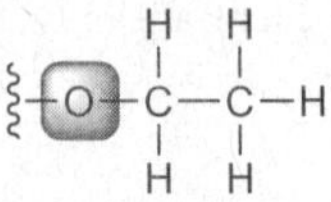

Now let's explore the remaining two signals in the proton NMR spectrum. The signal at approximately 3.6

ppm (with an integration of 1) vanishes in D_2O, indicating that it is the proton of the OH group (confirming our analysis of the IR spectrum). The singlet just above 4 ppm has an integration of 2, and therefore corresponds with an isolated methylene group (no neighbors). This methylene group is shifted significantly downfield, and our structure will have to take this into account.

In summary, we have the following fragments, which must be assembled.

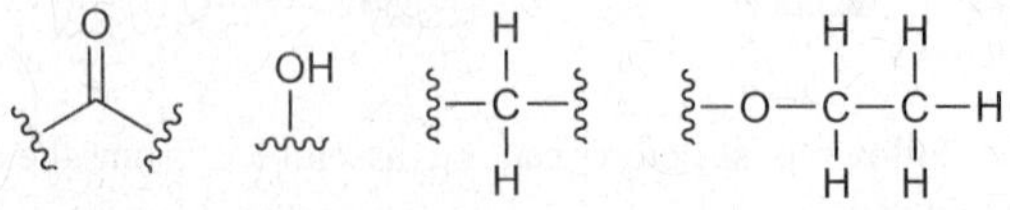

Since the C=O bond is likely part of an ester (based on the IR spectrum), we can redraw the following three fragments:

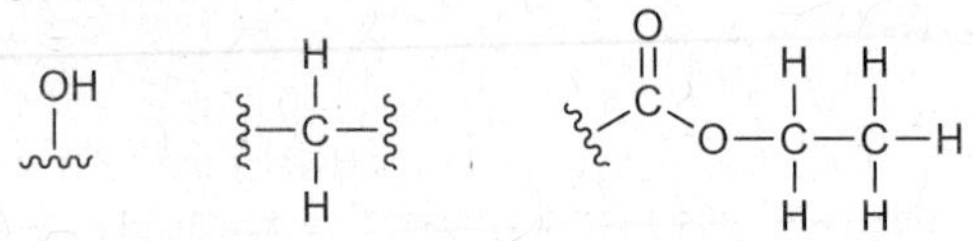

These fragments can only be assembled in one way:

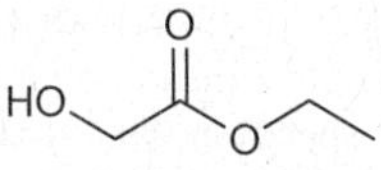

This structure is consistent with the carbon NMR spectrum, in which there is one signal for the C=O unit, two signals between 50 and 100 ppm (both of which must be methylene groups, based on the DEPT spectrum), and one signal between 0 and 50 ppm (representing the methyl group).

16.70.

N,N-dimethylformamide (DMF) has three resonance structures:

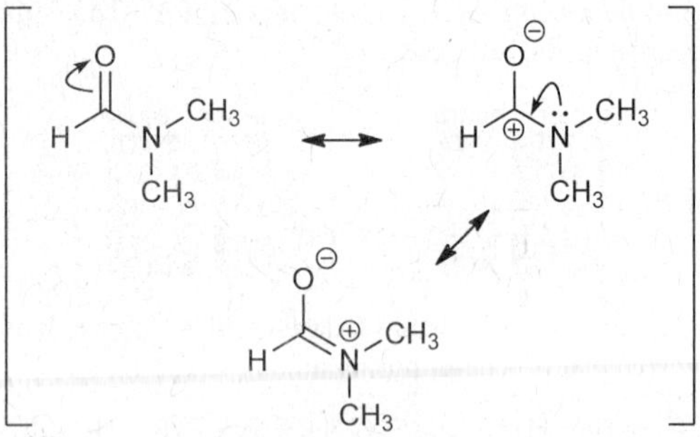

Consider the third resonance structure, in which the C-N bond is a double bond. This indicates that this bond is expected to have some double bond character. As such, there is an energy barrier associated with rotation about this bond, such that rotation of this bond occurs at a rate that it slower than the timescale of the NMR spectrometer. Therefore, the two methyl groups will appear as distinct signals in a ^1H NMR spectrum. At high temperature, more molecules will have the requisite energy to undergo free rotation about the C-N bond, so the process can occur on a timescale that is faster than the timescale of the NMR spectrometer. For this reason, the signals are expected to collapse into one signal at high temperature.

16.71. The methyl group on the right side is located in the shielding region of the π bond, so the signal for this proton is moved upfield to 0.8 ppm.

16.72. Brevianamide S has a high degree of symmetry. In fact, there is a rotational axis of symmetry that runs right through the molecule. Therefore, only half of the molecule will need to be analyzed.

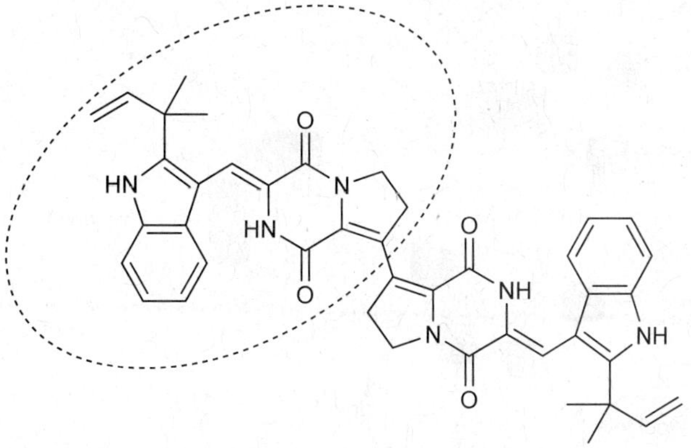

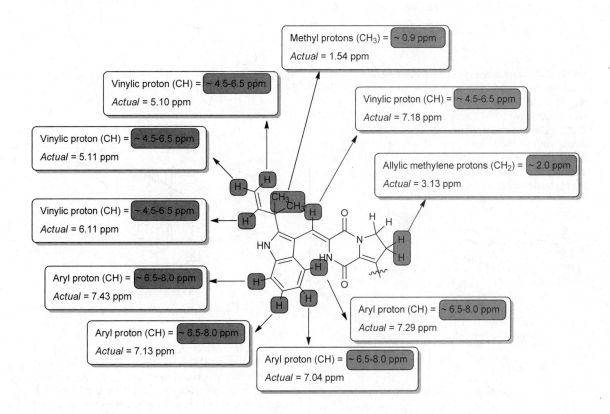

The compound lacks a chirality center and is achiral. As a result, the two protons for any methylene (CH₂) group will be equivalent to each other. There are two unique methylene groups, giving rise to two unique signals. In addition, the methyl groups will all be equivalent, giving rise to one signal. The predicted chemical shift for all signals is shown below:

Methyl protons (CH₃) = ~ 0.9 ppm
Actual = 1.54 ppm

Vinylic proton (CH) = ~ 4.5-6.5 ppm
Actual = 5.10 ppm

Vinylic proton (CH) = ~ 4.5-6.5 ppm
Actual = 7.18 ppm

Vinylic proton (CH) = ~ 4.5-6.5 ppm
Actual = 5.11 ppm

Allylic methylene protons (CH₂) = ~ 2.0 ppm
Actual = 3.13 ppm

Vinylic proton (CH) = ~ 4.5-6.5 ppm
Actual = 6.11 ppm

Aryl proton (CH) = ~ 6.5-8.0 ppm
Actual = 7.43 ppm

Aryl proton (CH) = ~ 6.5-8.0 ppm
Actual = 7.29 ppm

Aryl proton (CH) = ~ 6.5-8.0 ppm
Actual = 7.13 ppm

Aryl proton (CH) = ~ 6.5-8.0 ppm
Actual = 7.04 ppm

The multiplicity of each proton is determined with the n+1 rule, as shown below. Note: complex splitting is observed for two of the protons on the aromatic ring, as well as all three protons of the terminal olefin.

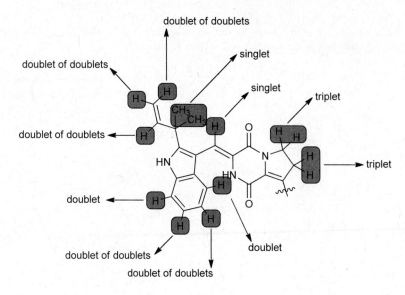

16.73. Let's first determine which diastereomer has the *R* configuration and which has the *S* configuration. When we place the hydrogen on a dash, the three substituents will be arranged such that the sequence of priorities (1–2–3) is clockwise; therefore compound **3** (having the *R* configuration) will have the H on a dash. Since compound **4** has the *S* configuration, we would simply need to invert the configuration at that center, meaning the H should be placed on a wedge.

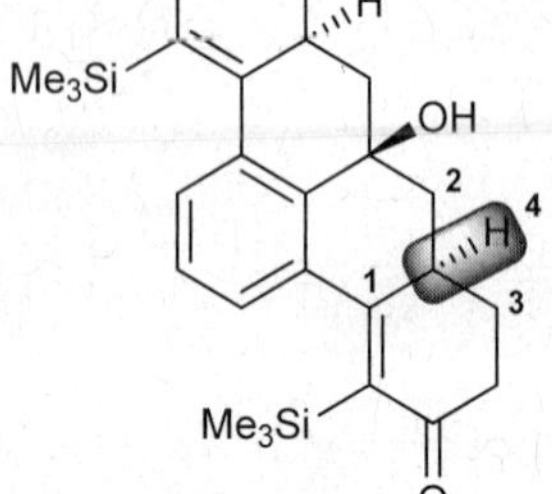

3 - (*R*) 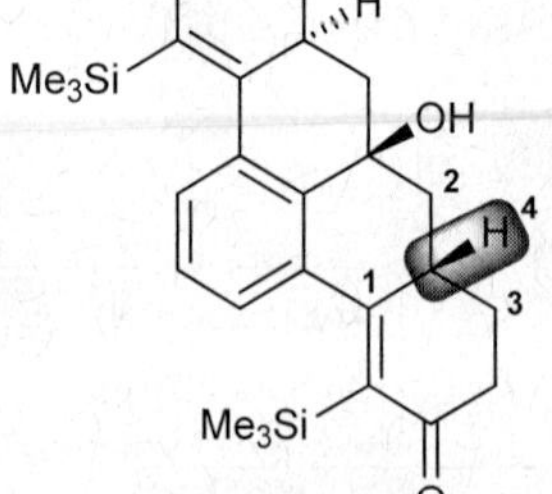   **4 - (*S*)**

Upon close inspection, compound **3** has a plane of symmetry that compound **4** lacks. In fact, compound **3** is *meso*. As a result, compound **3** should exhibit only thirteen signals in its ^{13}C NMR spectrum, while compound **4** should have twenty-three signals in its ^{13}C

NMR spectrum. This would be an easy way to distinguish between these compounds.

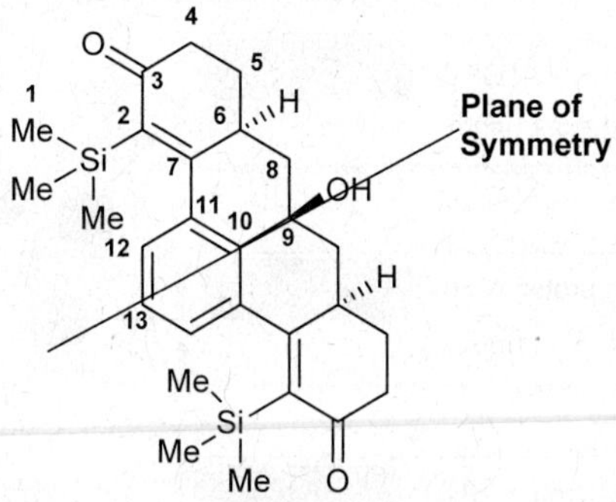

3 - (*R*)

16.74.
(a) The two most acidic protons are at C9 and C5. Deprotonation at C9 leads to a resonance-stabilized anion in which the negative charge is delocalized over three carbon atoms, while deprotonation at C5 leads to a resonance-stabilized anion in which the negative charge is delocalized over *four* carbon atoms. The latter anion is more stable than the former (since the negative charge is more highly delocalized). As such, the proton at C5 is

the most acidic proton. Deprotonation at C5 yields the following anion, with four resonance structures:

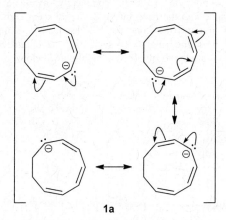

1a

(b) In an attempt to capture the nature of the resonance hybrid for each anion, **1a** and **2a** can be drawn in the following way:

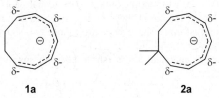

1a   **2a**

When viewed in this way, we can see that anion **1a** possesses symmetry, rendering positions C8 and C9 equivalent. Similarly, C1 and C7 are equivalent. Therefore, we expect only five signals in the proton NMR spectrum of **1a**.

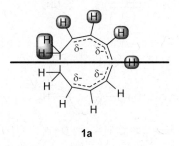

1a

In contrast, anion **2a** does not possess the same symmetry, as a result of the presence of the methyl groups. That is, each and every position is unique (for example, C1 is not equivalent to C7). Therefore, we expect nine signals for anion **2a**, shown below:

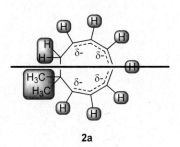

2a

Note that the two methyl groups of **2a** are identical and give one signal.

(c) Once again, we can use the resonance hybrid to explain this effect:

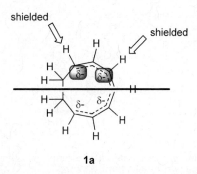

1a

The protons at positions C1 and C3 are shielded by the electron density that is distributed over those positions via resonance. The protons at positions C2 and C4 are less shielded, and they produce signals farther downfield.

(d) Inspecting the resonance hybrid, we expect four signals in the range of 3 – 4 ppm, and three signals in the range 5 – 6 ppm, as shown below:

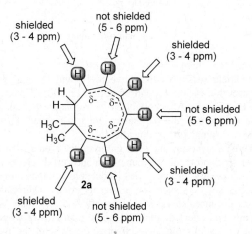

2a

Below is the actual data from the proton NMR spectrum of **2a**, which supports our prediction.

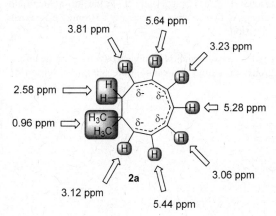

2a

16.75.

(a) The nitro group is a powerful electron-withdrawing group (primarily due to resonance), which renders the *ortho* and *para* positions electron-poor (highlighted below). This effect should cause H_a to be deshielded:

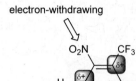

The amino group is a powerful electron-donating group (via resonance), which renders the *ortho* and *para* positions electron-rich (highlighted below). This effect should cause H_b and H_c to be shielded:

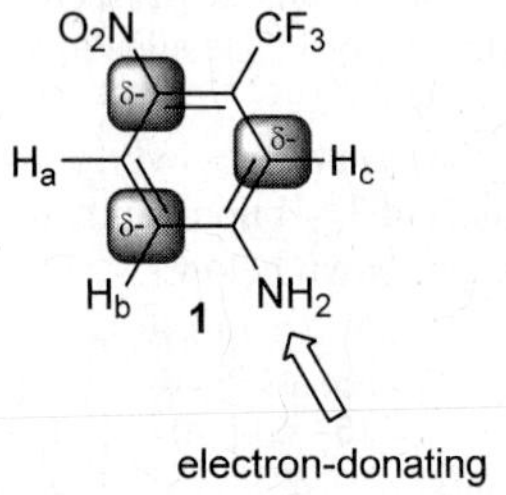

electron-donating

Based on these effects alone, we expect H_a to be downfield (deshielded) relative to H_b and H_c, and we expect that H_b and H_c will be upfield (shielded) relative to H_a. That is, based solely on the resonance effects of the nitro group and the amino group, we expect H_a to give the signal farthest downfield.

There is, however, one other group on the aromatic ring. The trifluoromethyl group is a powerful electron-withdrawing group, via induction (rather than resonance). As such, we expect its effect to diminish with distance, so H_a should be affected less than H_c. While this effect deshields H_c more than H_a, we would not expect this inductive effect to overwhelm the two resonance effects that suggest that H_a gives the most downfield signal. After all, there are two different resonance effects suggesting H_a is the most downfield signal, AND resonance is generally a stronger effect than induction.

(b) This transformation involves conversion of the amino group (a strong activator) into an amide group (a moderate activator). This can be rationalized by considering the lone pair on the nitrogen atom of the amine/amide; in **1** it is completely available for resonance into the ring, but in **2**, the adjacent carbonyl group involves this lone pair in resonance, pulling it away from the ring. As such, the *ortho* and *para* positions are the most affected, highlighted below:

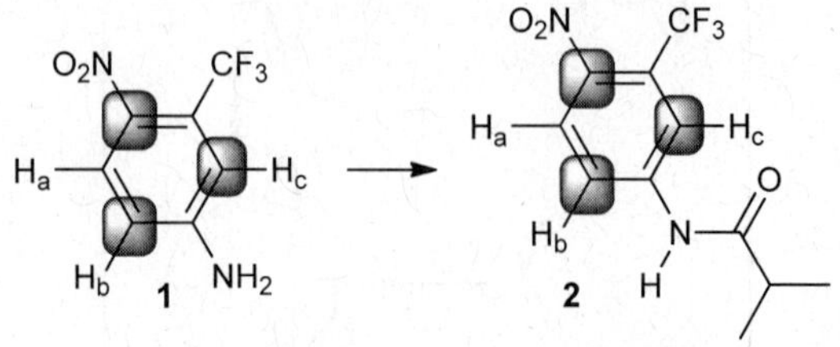

In both compound **1** and compound **2**, these three positions are shielded. But the shielding effect is greater in compound **1** than in compound **2**. In other words, these three positions become less shielded. As a result, we expect the signals for H_b and H_c to be more affected by this transformation than the signal for H_a. Specifically, H_b and H_c become less shielded as a result of this transformation, so we expect those two signals to move farther downfield.

The actual chemical shifts (shown below) support our predictions.

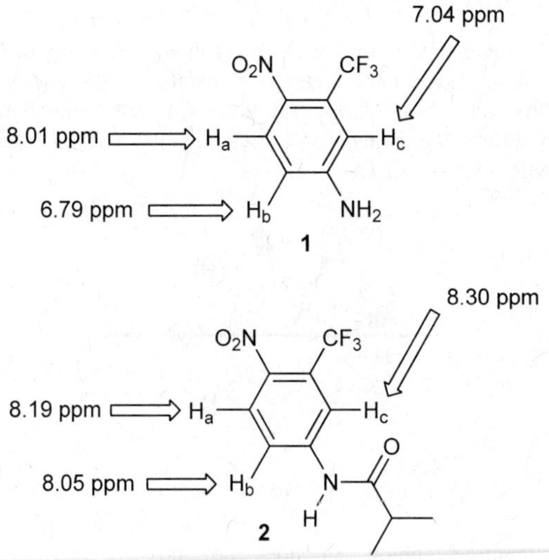

16.76.

(a) We are looking for the proton that is expected to be the most deshielded. In other words, we must identify the proton that resides in the most electron-deficient environment. Let's begin by considering resonance effects of the diethyl amino group:

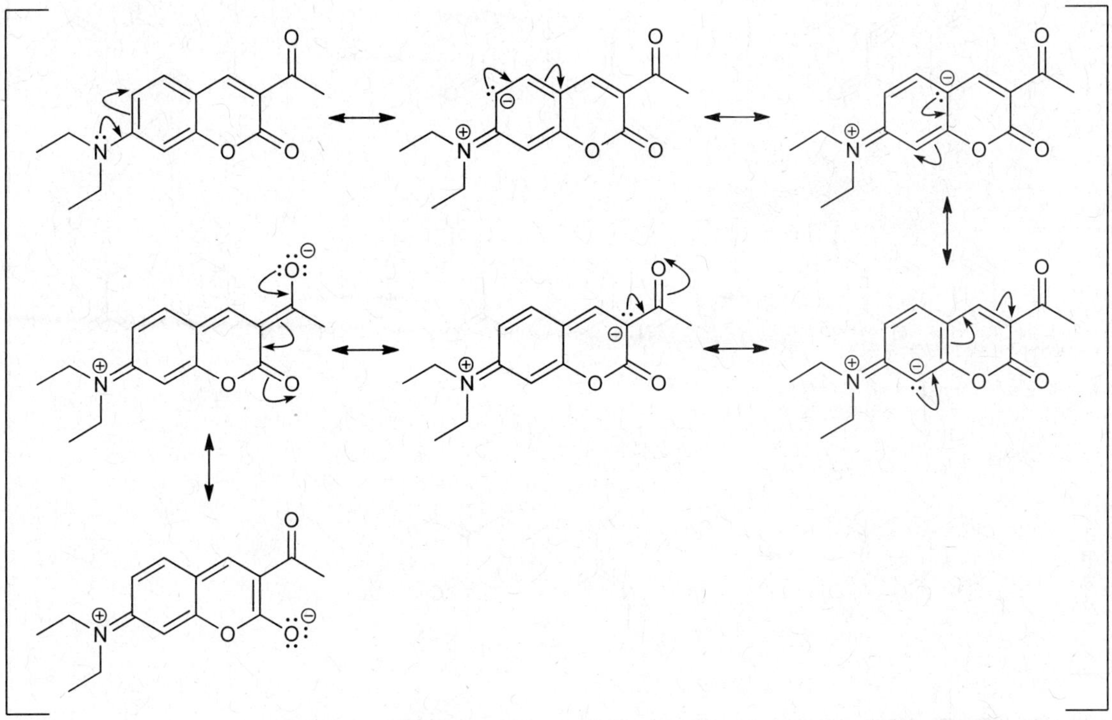

These resonance structures indicate that the diethylamino group donates electron density, so it has the opposite effect of what we are looking for; that is, the diethylamino group renders the following positions electron-rich:

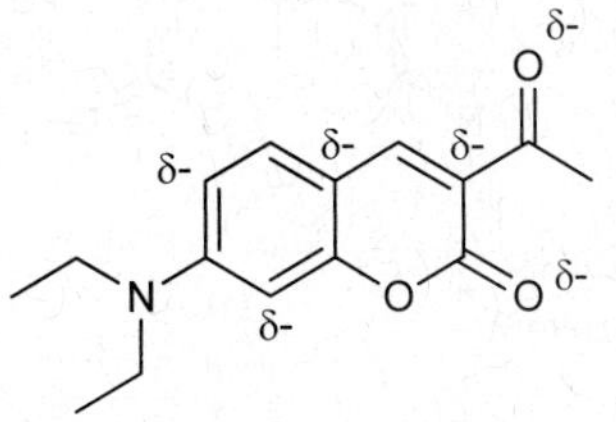

The oxygen atom incorporated in the ring has a similar effect on the same positions:

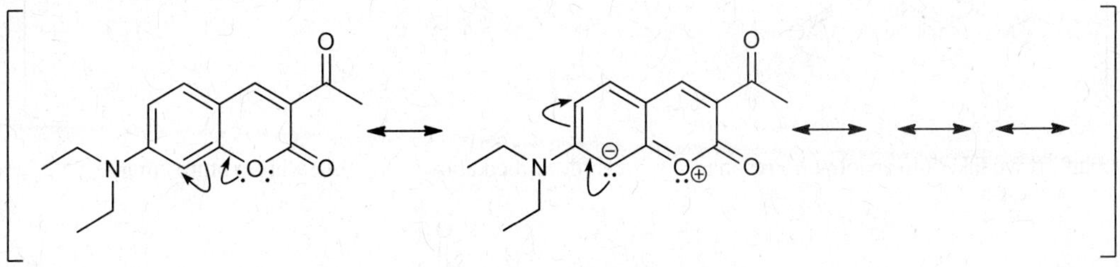

So we explore the effect of the ketone group, and we draw the following resonance structures:

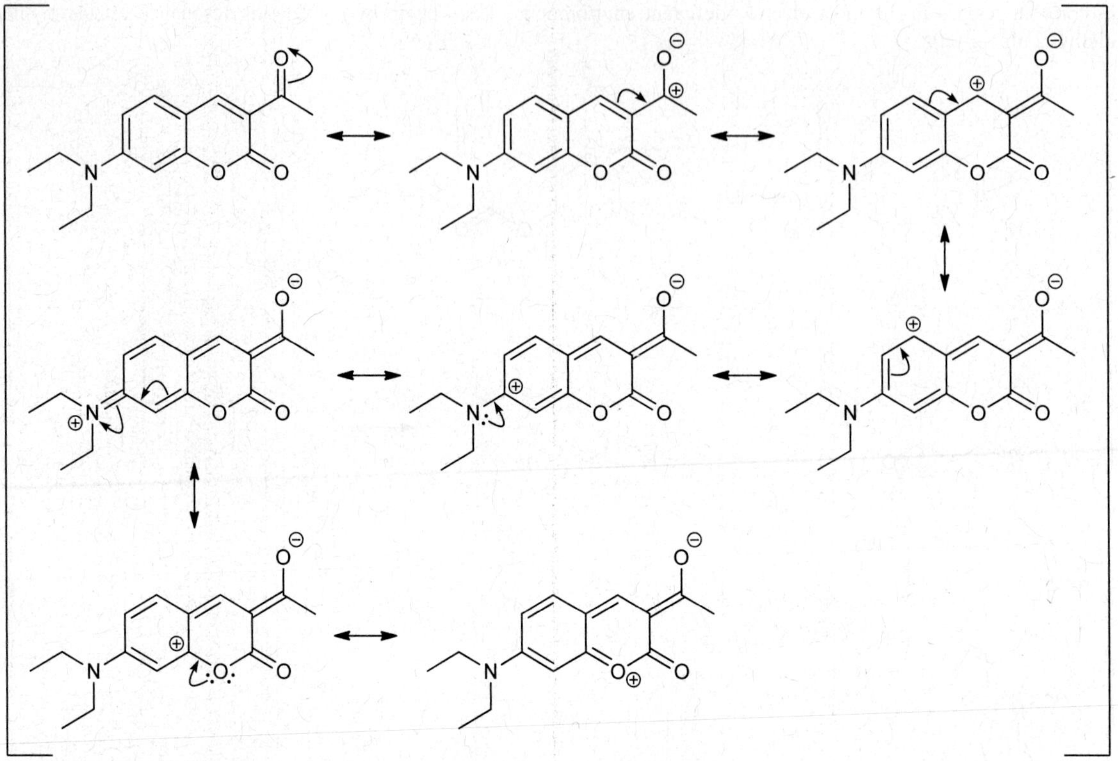

Notice that these resonance structures indicate that several positions are electron-deficient.

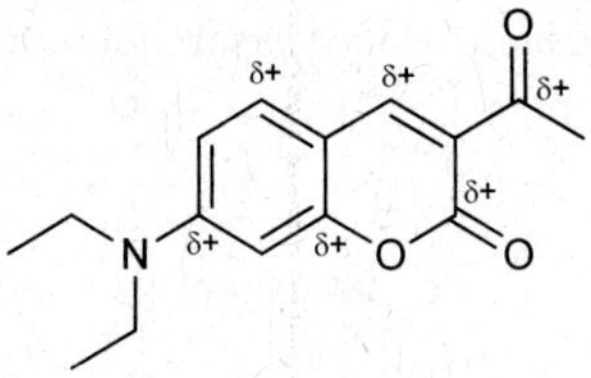

A similar effect is expected from the ester group:

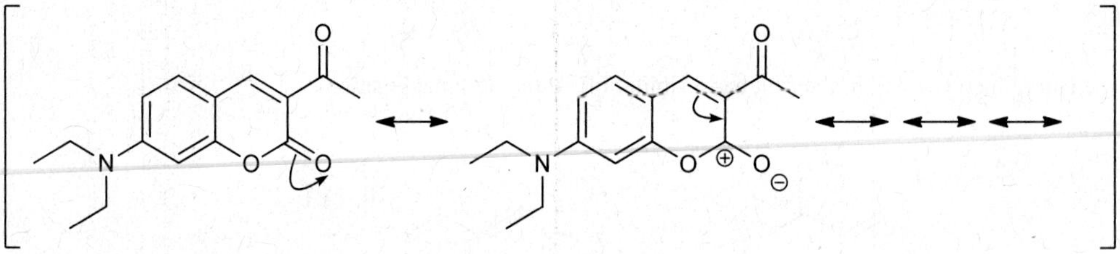

In summary, if we take into account *all* resonance effects (described above), the following picture emerges:

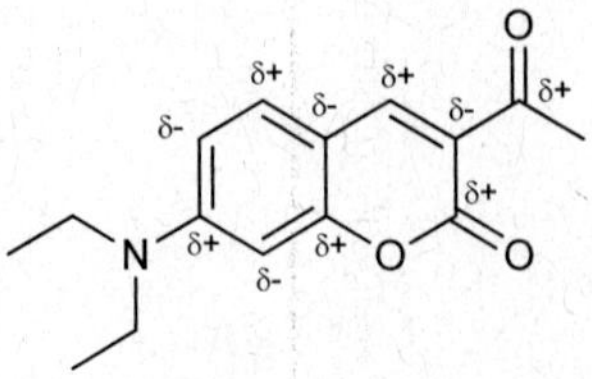

Notice that only two protons are directly attached to electron-deficient centers:

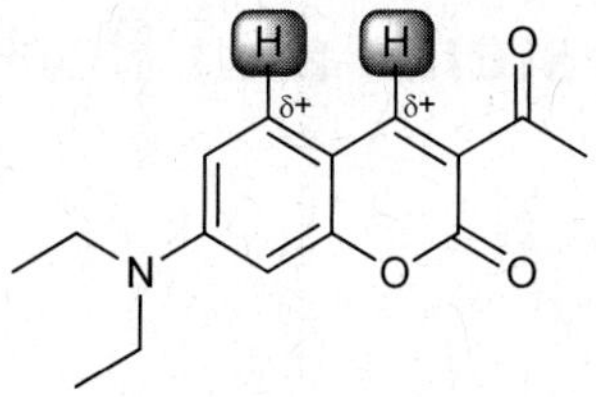

These are the two protons that are expected to produce the two signals farthest downfield in the proton NMR spectrum.

(b) In compound **1**, the C4 position is electron-deficient, as seen in the third resonance structure below:

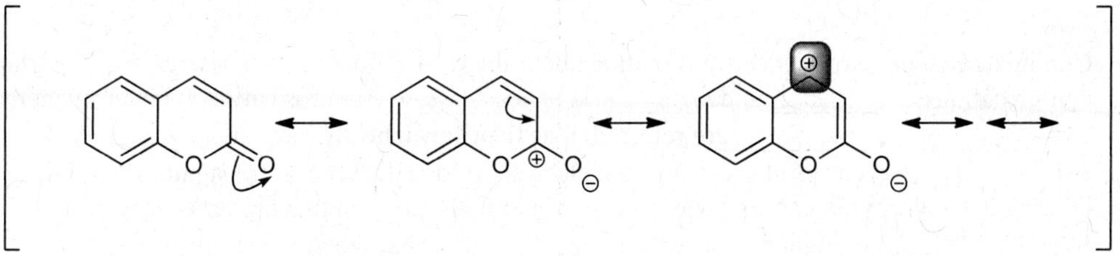

and the C3 position is electron-rich, as highlighted below:

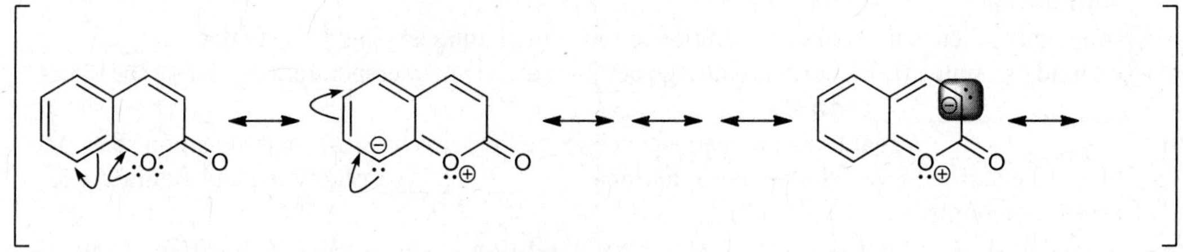

As a result, the C3-C4 π bond is highly polarized and is expected to produce a strong signal.

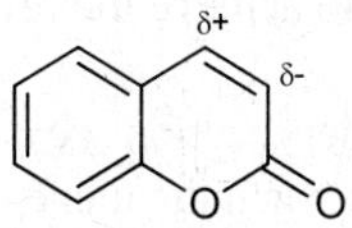

A similar effect is expected in compound **2**, but the effect should be much stronger, because the presence of the diethylamino group renders C3 even more electron-rich:

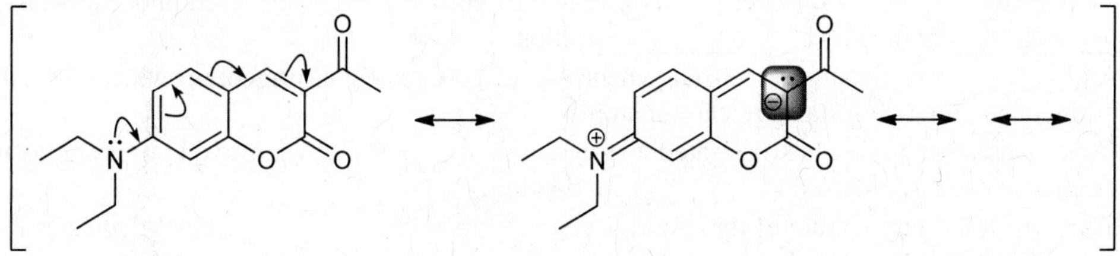

And the presence of the ketone renders C4 even more electron-deficient, as highlighted below:

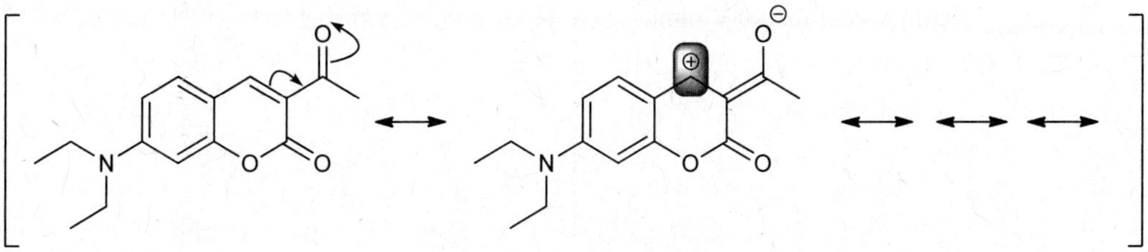

Therefore, the C3-C4 bond in compound **2** is expected to be even more polarized than the C3-C4 bond in compound **1**. So the former is expected to produce a stronger (more intense) signal.

Chapter 17
Conjugated Pi Systems and Pericyclic Reactions

Review of Concepts
Fill in the blanks below. To verify that your answers are correct, look in your textbook at the end of Chapter 17. Each of the sentences below appears verbatim in the section entitled *Review of Concepts and Vocabulary*.

- Conjugated dienes experience free-rotation about the C2-C3 bond, giving rise to two important conformations: *s*-_____ and *s*-_____. The _____ conformation is lower in energy.
- The _____ and _____ are referred to as **frontier orbitals**.
- An _____ **state** is produced when a π electron in the HOMO absorbs a photon of light bearing the appropriate energy necessary to promote the electron to a higher energy orbital.
- Reactions induced by light are called _____ **reactions**.
- When butadiene is treated with HBr, two major products are observed, resulting from _____-**addition** and _____-**addition**.
- Conjugated dienes that undergo addition at low temperature are said to be under _____ **control.** Conjugated dienes that undergo addition at elevated temperature are said to be under _____ **control.**
- _____ **reactions** proceed via a concerted process with a cyclic transition state, and they are classified as **cycloaddition reactions**, _____ **reactions**, and **sigmatropic rearrangements.**
- The Diels–Alder reaction is a **[_____] cycloaddition** in which two C-C bonds are formed simultaneously.
- High temperatures can often be used to achieve the reverse of a Diels–Alder reaction, called a _____ **Diels–Alder.**
- The starting materials for a Diels–Alder reaction are a diene, and a _____.
- The Diels–Alder reaction only occurs when the diene adopts an _____ conformation.
- When cyclopentadiene is used as the starting diene, a bridged bicyclic compound is obtained, and the _____ cycloadduct is favored over the _____ cycloadduct.
- Conservation of orbital symmetry determines whether an electrocyclic reaction occurs in a _____ fashion or a _____ fashion.
- A [_____] sigmatropic rearrangement is called a **Cope rearrangement** when all six atoms of the cyclic transition state are carbon atoms.
- Compounds that possess a conjugated π system will absorb UV or visible light to promote an electronic excitation called a _____ transition.
- The most important feature of the absorption spectrum is the _____, which indicates the wavelength of maximum absorption.
- When a compound exhibits a λ_{max} between 400 and 700 nm, the compound will absorb _____ light, rather than UV light.

Review of Skills

Fill in the blanks and empty boxes below. To verify that your answers are correct, look in your textbook at the end of Chapter 17. The answers appear in the section entitled *SkillBuilder Review*.

17.1   Proposing the Mechanism and Predicting the Products of Electrophilic Addition to Conjugated Dienes

IN THE SPACE PROVIDED, DRAW THE MECHANISM OF THE REACTION THAT IS EXPECTED TO OCCUR WHEN THE FOLLOWING COMPOUND IS TREATED WITH HBr. MAKE SURE TO DRAW ALL POSSIBLE PRODUCTS.

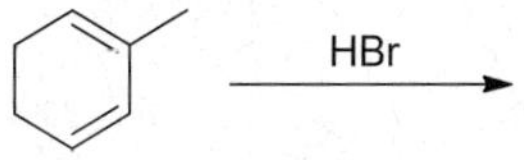

HBr

17.2   Predicting the Major Product of an Electrophilic Addition to Conjugated Dienes

DRAW THE MAJOR PRODUCT OF THE FOLLOWING REACTION:

HBr
0°C

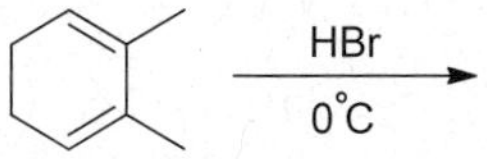

17.3   Predicting the Product of a Diels–Alder Reaction

DRAW THE MAJOR PRODUCT OF THE FOLLOWING REACTION:

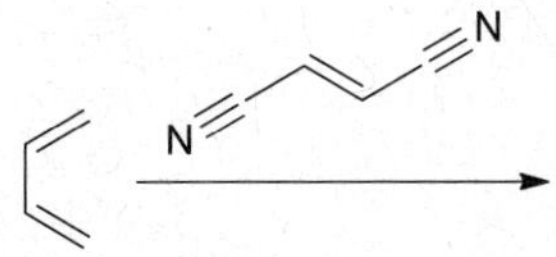

17.4   Predicting the Product of an Electrocyclic Reaction

DRAW THE MAJOR PRODUCT OF THE FOLLOWING REACTION:

hν

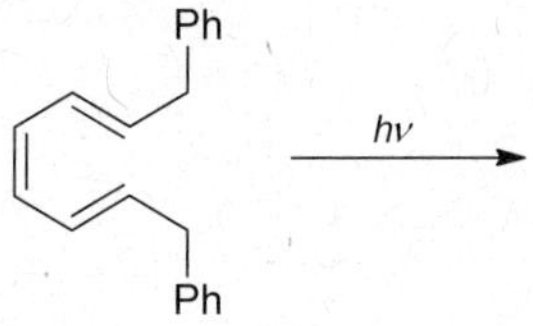

17.5   Using Woodward–Fieser Rules to Estimate λ_{max}

USE WOODWARD-FIESER RULES
TO ESTIMATE λ_{max} FOR THE
FOLLOWING COMPOUND:

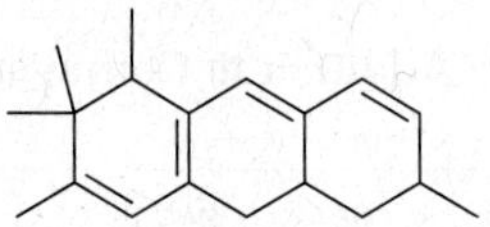

BASE VALUE =	
ADDITIONAL DOUBLE BONDS =	
AUXOCHROMIC ALKYL GROUPS =	
EXOCYCLIC DOUBLE BOND =	
HOMOANNULAR DIENE =	
TOTAL =	

<u>Review of Reactions</u>

Predict the Products for each of the following transformations. To verify that your answers are correct, look in your textbook at the end of Chapter 17. The answers appear in the section entitled *Review of Reactions*.

Preparation of Dienes

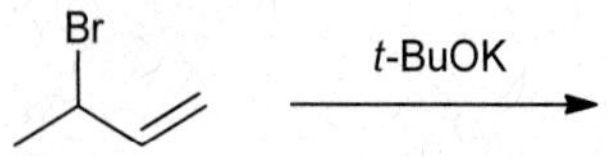

Br

$\xrightarrow{\textit{t}-BuOK}$

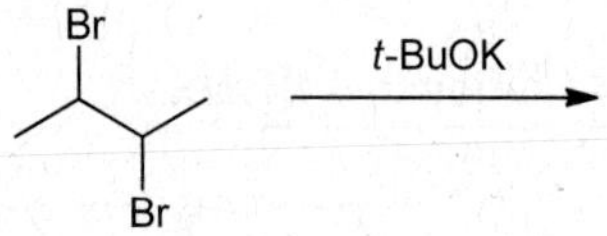

Br

Br

$\xrightarrow{\textit{t}-BuOK}$

Electrophilic Addition

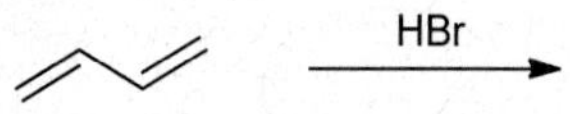

$\xrightarrow{HBr}$

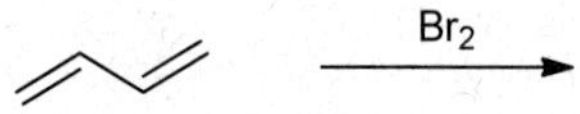

$\xrightarrow{Br_2}$

Diels–Alder Reaction

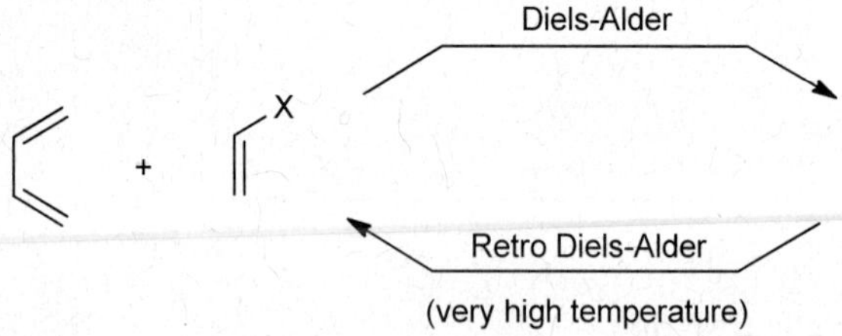

Diels-Alder

+   X

Retro Diels-Alder
(very high temperature)

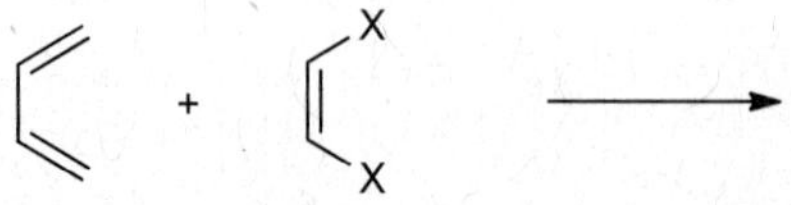

+   X
        X

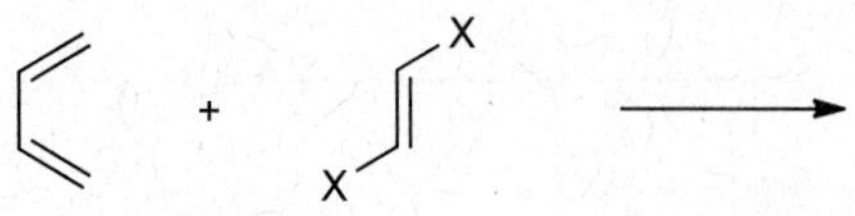

+   X
        X

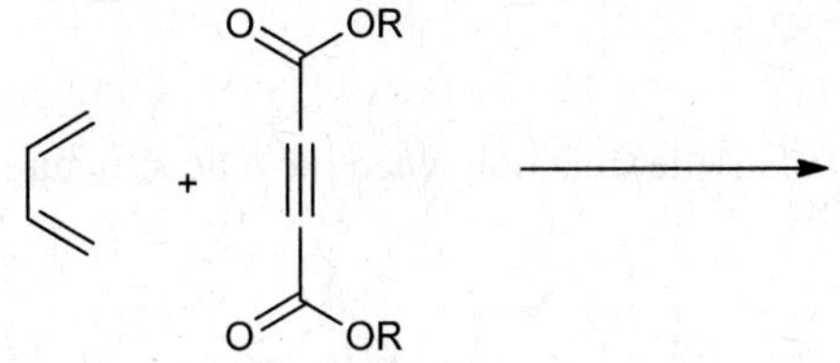

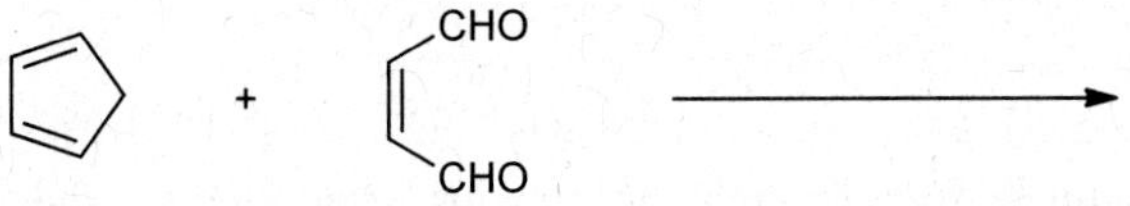

Electrocyclic Reactions

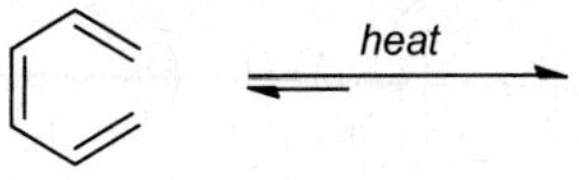

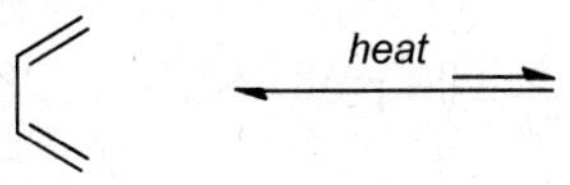

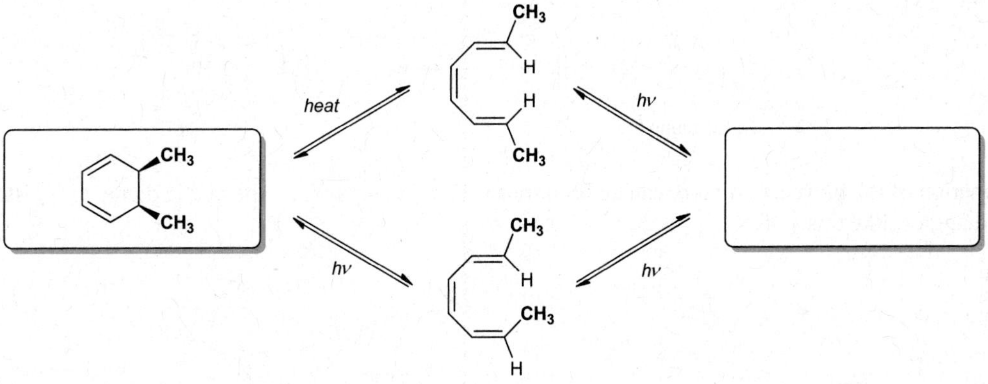

Sigmatropic Rearrangements

Cope Rearrangement

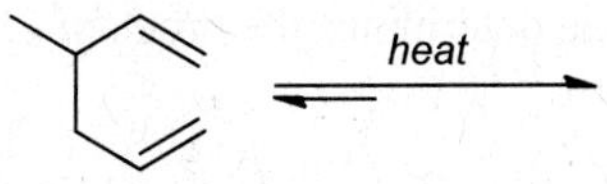

Claisen Rearrangement

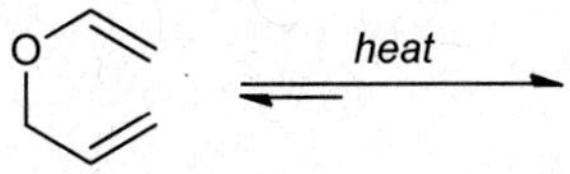

Mistakes to Avoid

Many students have trouble drawing the product(s) of a Diels-Alder reaction when bicyclic structures are involved:

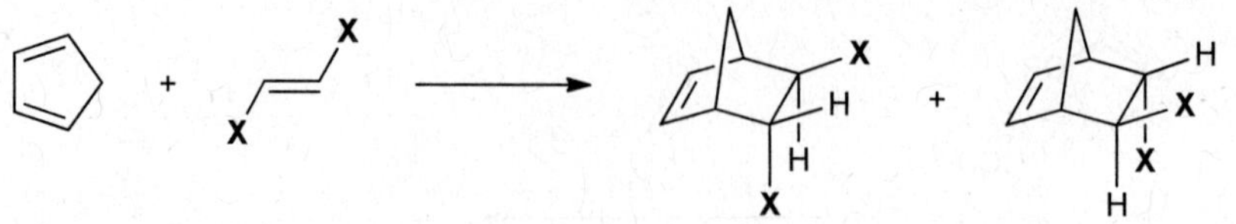

You might have trouble visualizing these structures, drawing them, or knowing where to place the substituents. The following are a few guidelines that might help you avoid making mistakes.
In a Diels-Alder reaction, an acyclic diene will give a product with a cyclohexene ring:

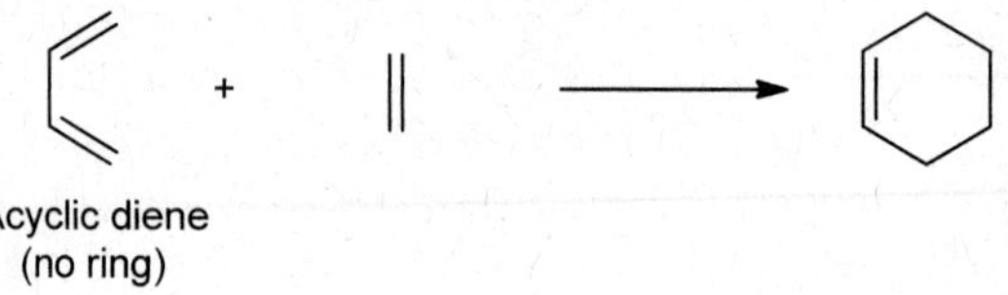

Acyclic diene
(no ring)

However, if the diene is cyclic (both π bonds are contained in the ring), then the product is a bicyclic structure:

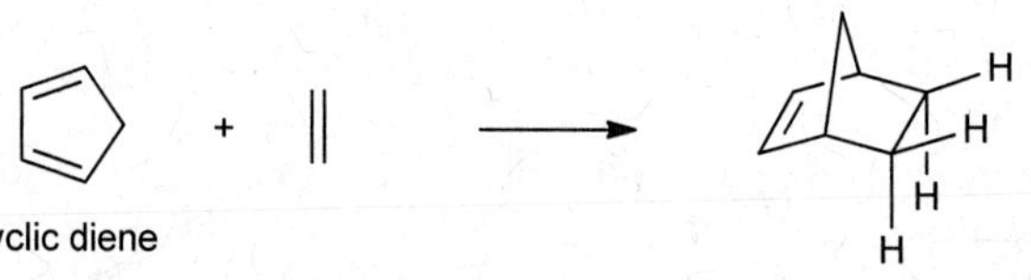

Cyclic diene

Formation of the bicyclic structure can be seen more clearly if we redraw the cyclic diene with distorted bond angles, like this:

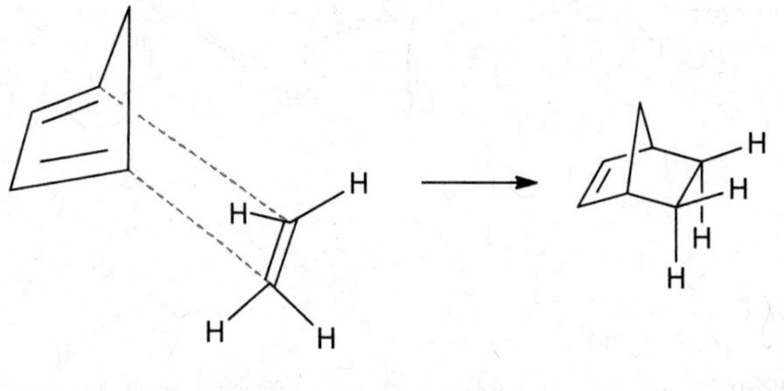

The dotted lines indicate the locations where σ bonds are forming as a result of the reaction. When the dienophile is monosubstituted or *cis*-disubstituted, the *endo* rule determines the product(s) obtained:

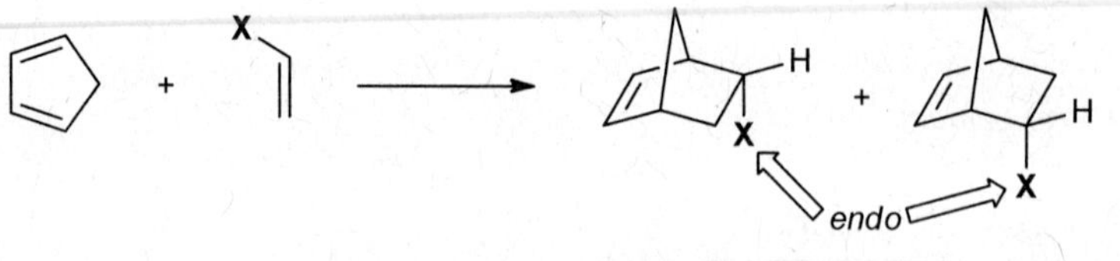

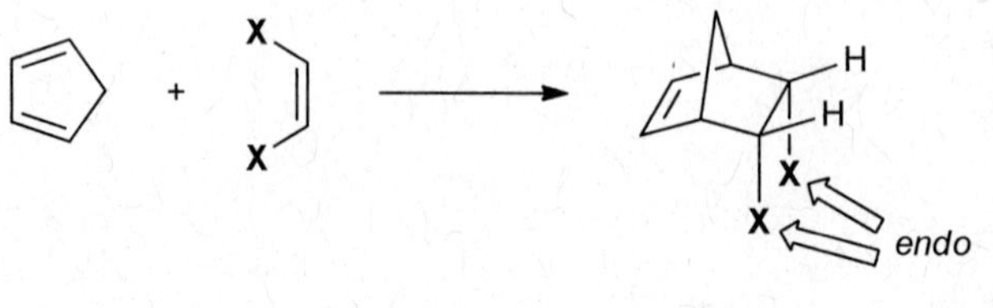

However, when a *trans*-disubstituted dienophile is used, the *endo* rule is not relevant. Two stereoisomeric products are obtained (in this case, enantiomers), and in each product, one group occupies an *endo* position while the other group occupies an *exo* position (because the *trans* configuration of the dienophile is preserved during a Diels-Alder reaction):

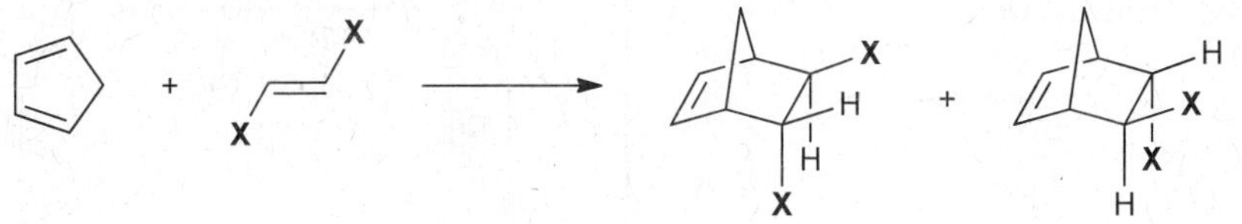

Useful reagents

The following is a list of reagents used in this chapter:

Reagents	Function
t-BuOK	A strong, sterically hindered base, used to convert a dibromide into a diene.
HBr	Will add across a conjugated π system to give two products: a 1,2 adduct and a 1,4 adduct.
Br$_2$	Will add across a conjugated π system to give two products: a 1,2 adduct and a 1,4 adduct.
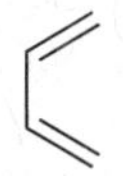	1,3-Butadiene. Can serve as a diene in a Diels-Alder reaction.
	1,3-Cyclopentadiene. Can serve as a diene in a Diels-Alder reaction.
	Can serve as a dienophile in a Diels-Alder reaction, if the substituents (X) are electron-withdrawing groups. The *cis* configuration of the dienophile is preserved in the product.
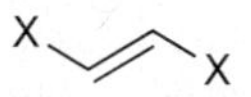	Can serve as a dienophile in a Diels-Alder reaction, if the substituents (X) are electron-withdrawing groups. The *trans* configuration of the dienophile is preserved in the product.
	Can serve as a dienophile in a Diels-Alder reaction, if the substituents (X) are electron-withdrawing groups.
heat	When you see "heat" without any other reagents indicated, consider the possibility of a pericyclic reaction (cycloaddition, electrocyclic reaction, or a sigmatropic rearrangement).
hv	When you see this term, or "light", without any other reagents indicated, consider the possibility of an electrocyclic reaction.

Solutions

17.1.
(a) The C=C bond in this compound is conjugated to two of the carboxylic acid groups:

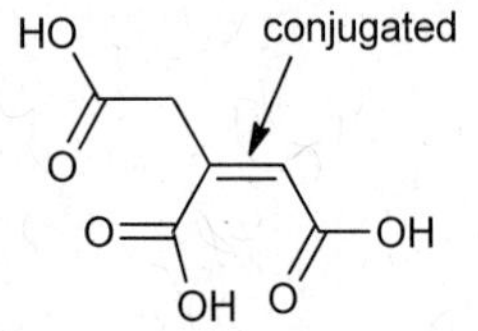

(b) The C=C bond on the left is isolated, while the other two C=C bonds are conjugated to each other.

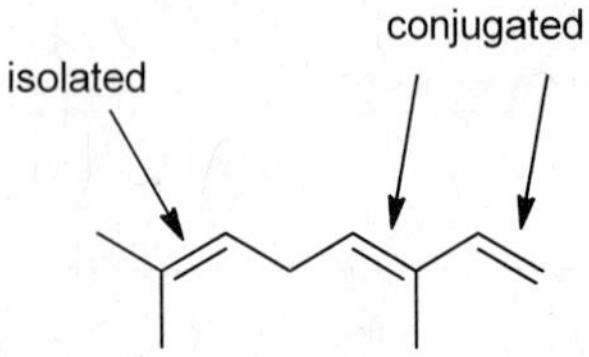

(c) One of C=C bonds is conjugated to the carbonyl group, and the other C=C bond is isolated:

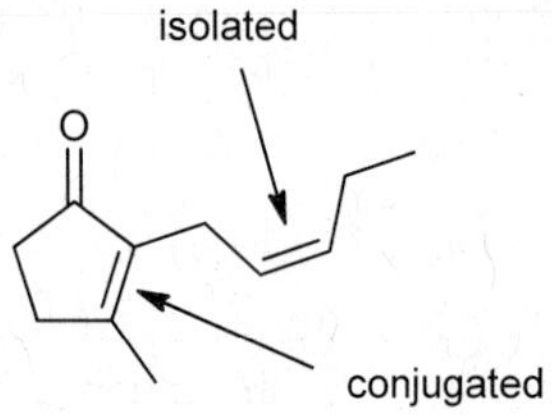

(d) One of C=C bonds is conjugated to the carbonyl group, and the other C=C bond is isolated:

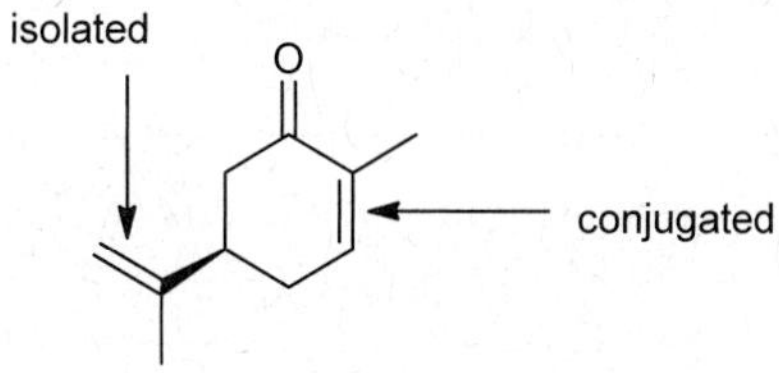

17.2. Radical bromination can be employed to install a functional group. The resulting alkyl halide can then be treated with a strong base to give an alkene. Bromination of the alkene will give a dibromide, which can then be converted into the product upon treatment with a strong, sterically hindered base, such as *t*-butoxide. This final step involves two elimination reactions, giving the desired diene:

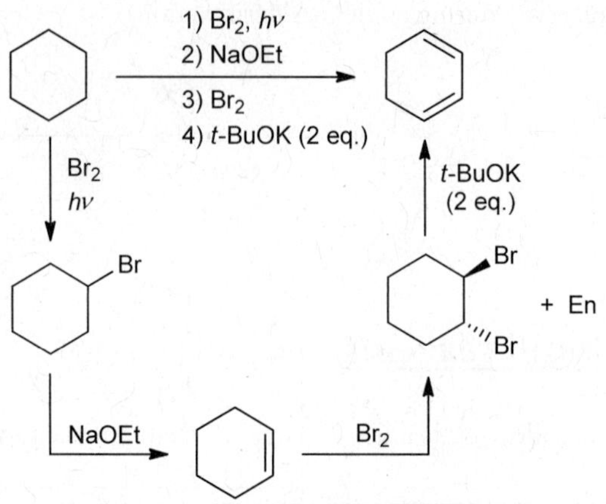

17.3. The C1–C2 bond length is expected to be the shortest because it is a double bond (comprised of both a σ bond and a π bond):

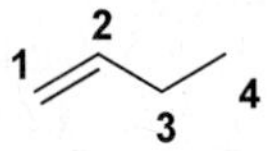

Among the other two bonds, C2–C3 is expected to be shorter than C3–C4 because the former is a C_{sp^2}–C_{sp^3} bond, while the latter is a C_{sp^3}–C_{sp^3} bond (sp^2 hybridized orbitals are closer to the nucleus than sp^3 hybridized orbitals, and therefore form shorter bonds).

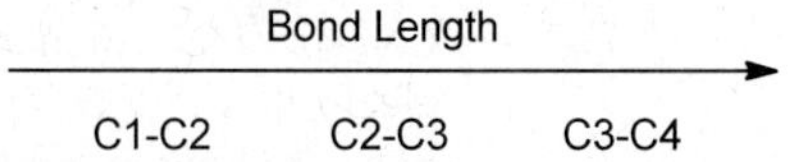

17.4.
(a) All three of these compounds will yield the same product (ethylcyclohexane) upon hydrogenation with two moles of hydrogen gas. Yet only one of these compounds is a conjugated diene, shown below.

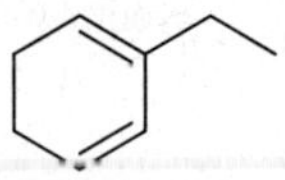

The other two compounds exhibit isolated π bonds. The conjugated diene will liberate the least heat because it is the most stable of the three compounds (lowest in energy).

(b) The following compound is expected to liberate the most heat upon hydrogenation with two moles of hydrogen gas:

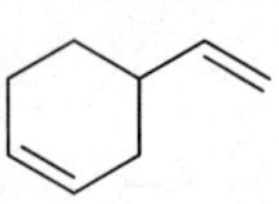

This isolated diene will liberate more heat than the other isolated diene, because the π bonds in this compound are not as highly substituted (one π bond is monosubstituted and the other is disubstituted). In the other isolated diene, the π bonds are disubstituted and trisubstituted (and therefore more stable).

17.5. In the following compound, all three π bonds are conjugated to each other, giving one extended conjugated system.

This compound is therefore more stable than the other two compounds, each of which exhibits an isolated π bond.

17.6. This compound is comprised of eight, consecutive, overlapping, *p* orbitals, giving rise to eight molecular orbitals:

There are eight π electrons, which occupy the four lower energy MO's (the bonding MO's). In the ground state, ψ_4 is the HOMO and ψ_5 is the LUMO, as shown. Photochemical excitation causes one electron to be promoted from ψ_4 to ψ_5. In the excited state, the HOMO is ψ_5 and the LUMO is ψ_6:

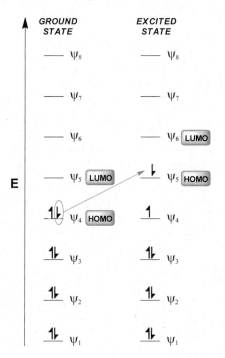

17.7.
(a) We first identify the locations where protonation can occur. There are four unique positions where protonation can occur, labeled C1 through C4:

Among these four positions, only protonation at C1 or at C4 will generate a resonance-stabilized carbocation. So, we must explore protonation at each of these positions. Let's begin with protonation at C1, which leads to a resonance–stabilized intermediate. When we draw both resonance structures, we can see that two positions are electrophilic. Nucleophilic attack can occur at either of these locations, although the product is the same in either case, leading to only one product:

Notice that the product possesses a chirality center and is therefore produced as a racemic mixture (because the chloride ion can attack either face of the allylic carbocation with equal likelihood).

Now let's consider protonation at C4. Once again, protonation leads to a resonance–stabilized intermediate. When we draw both resonance structures, we can see that two positions are electrophilic. Nucleophilic attack can occur at either of these locations, giving two possible products, as shown:

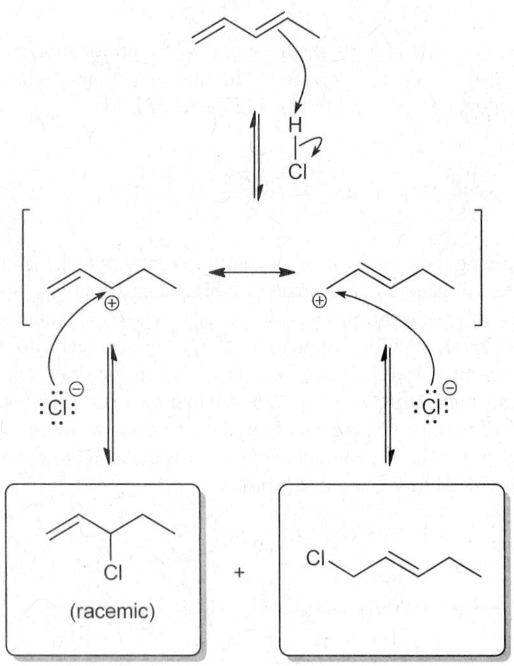

In summary, we expect the following possible products:

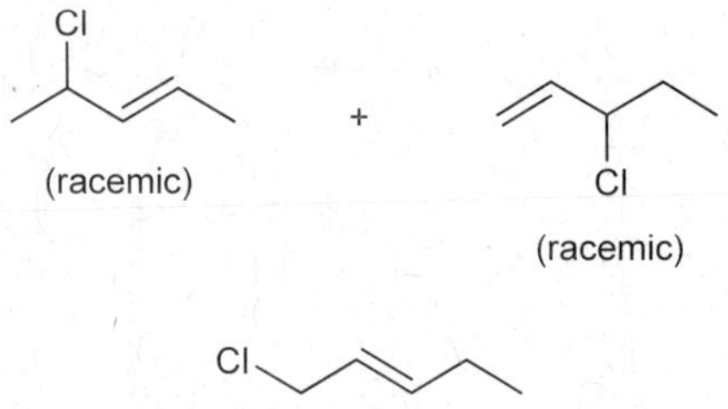

(b) We first identify the locations where protonation can occur. There are four unique positions where protonation can occur, labeled C2 through C5:

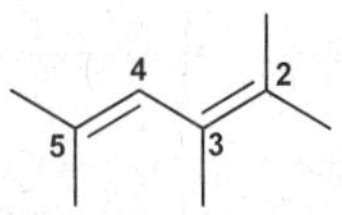

Among these four positions, only protonation at C2 or at C5 will generate a resonance-stabilized carbocation. So, we must explore protonation at each of these positions. Let's begin with protonation at C5, which leads to a resonance-stabilized intermediate. When we draw both resonance structures, we can see that two positions are electrophilic. Nucleophilic attack can occur at either of these locations, giving rise to two products:

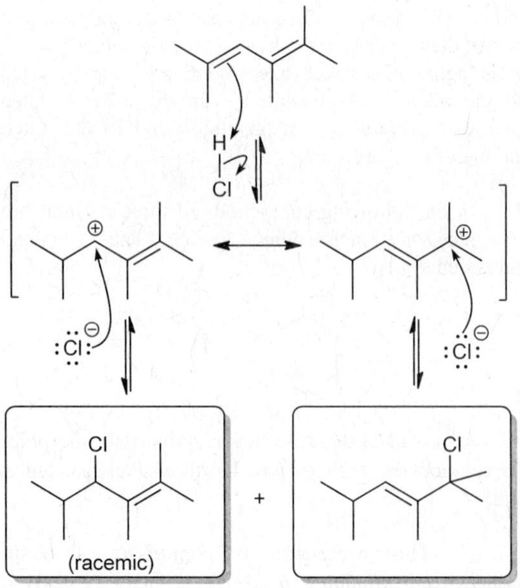

Notice that one of the products possesses a chirality center and is therefore produced as a racemic mixture (because the chloride ion can attack either face of the allylic carbocation with equal likelihood).

Now let's consider protonation at C2. Once again, protonation leads to a resonance-stabilized intermediate. When we draw both resonance structures, we can see that two positions are electrophilic. Nucleophilic attack can occur at either of these locations, giving two possible products, as shown:

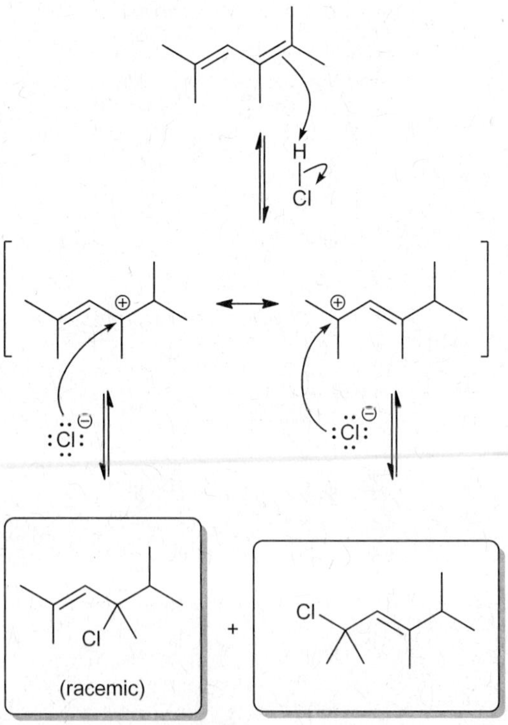

In summary, we expect the following possible products:

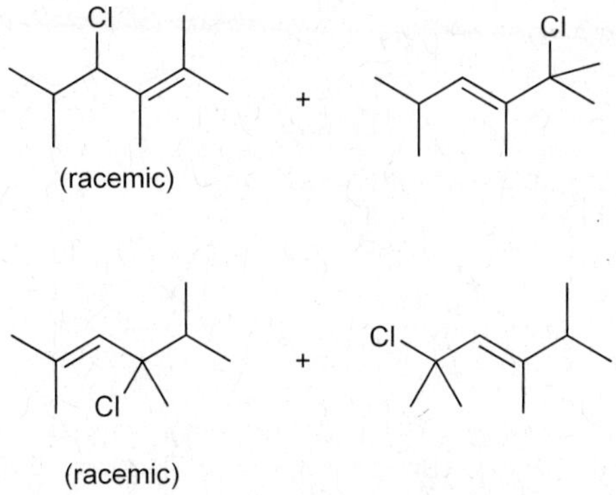

(racemic)

(racemic)

(c) The diene is symmetrical, so protonation at one end of the conjugated system is the same as protonation at the other end of the conjugated system.

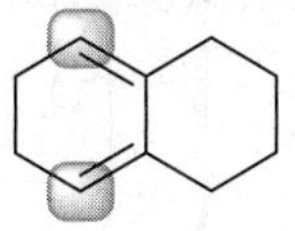

Therefore, we only need to consider protonation at one of these locations, giving rise to two possible products, as shown:

(racemic)

+

(racemic)

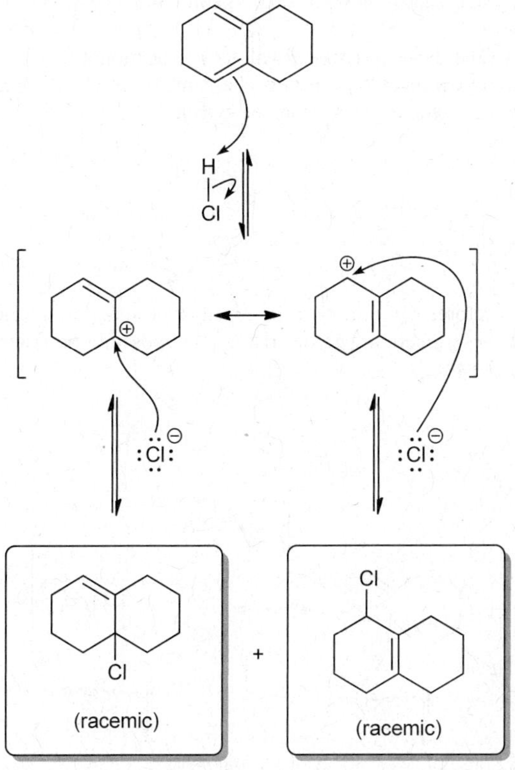

(d) We first identify the locations where protonation can occur. There are four unique positions where protonation can occur, although a resonance-stabilized intermediate can only be obtained upon protonation of one of the ends of the conjugated π system, highlighted below:

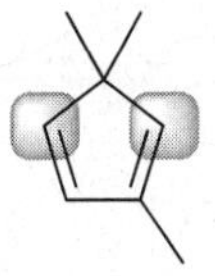

We must explore protonation at each of these positions. Let's begin with protonation of the position on the left, which leads to a resonance–stabilized intermediate. When we draw both resonance structures, we can see that two positions are electrophilic. Nucleophilic attack can occur at either of these locations, giving rise to two products:

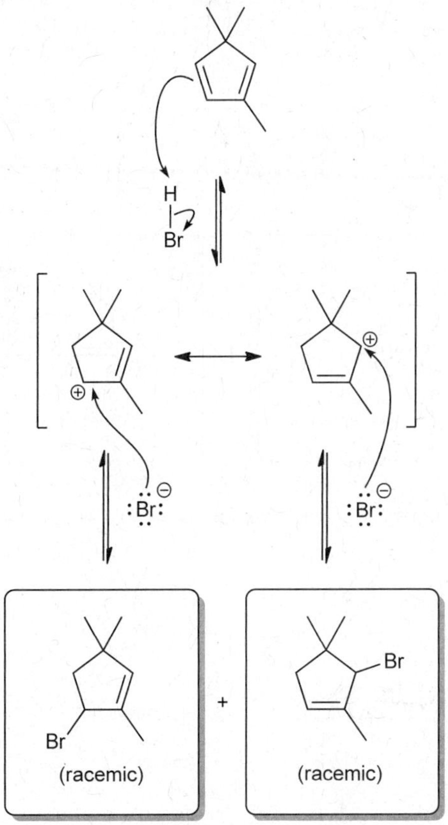

(racemic)

+

(racemic)

Notice that each of the products possesses a chirality center and is therefore produced as a racemic mixture (because the bromide ion can attack either face of the allylic carbocation with equal likelihood).

Now let's consider protonation at the position on the right. Once again, protonation leads to a resonance–stabilized intermediate. When we draw both resonance structures, we can see that two positions are electrophilic. Nucleophilic attack can occur at either of these locations, giving two possible products, as shown:

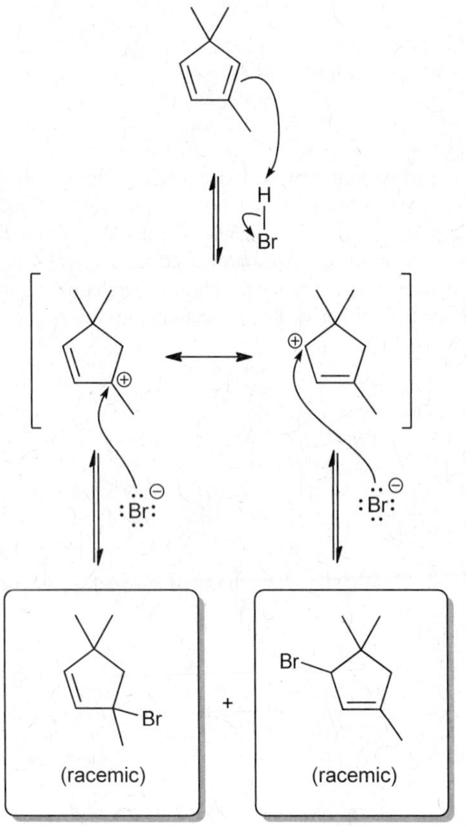

In summary, we expect the following possible products:

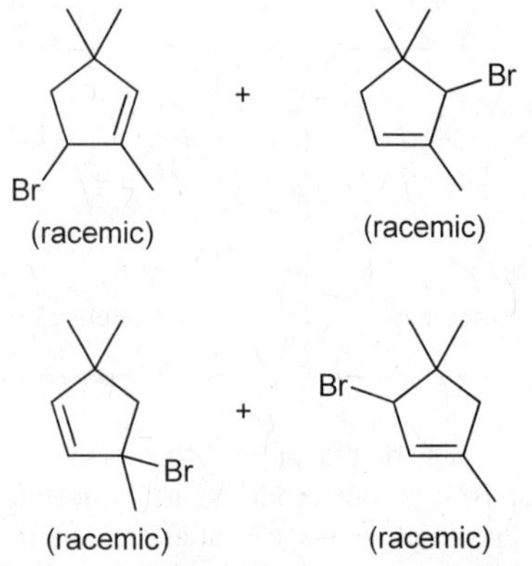

(e) The diene is symmetrical, so protonation at one end of the conjugated system is the same as protonation at the other end of the conjugated system.

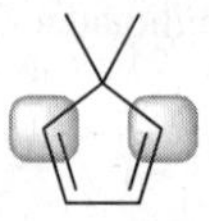

Therefore, we only need to consider protonation at one of these locations, giving rise to two possible products, as shown:

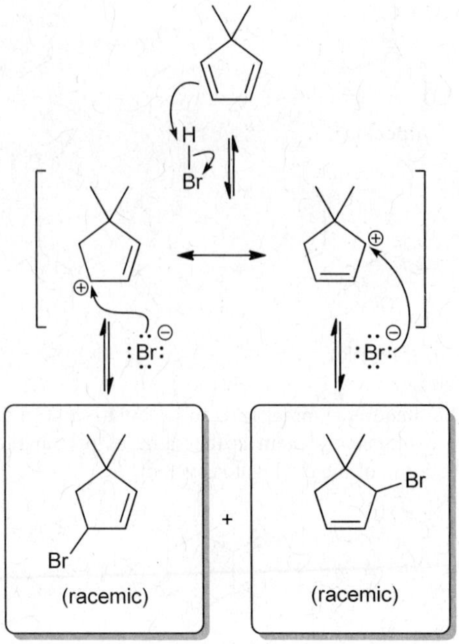

Notice that each of the products possesses a chirality center and is therefore produced as a racemic mixture (because the bromide ion can attack either face of the allylic carbocation with equal likelihood).

(f) The diene is symmetrical, so protonation at one end of the conjugated system is the same as protonation at the other end of the conjugated system.

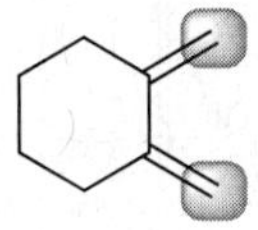

Therefore, we only need to consider protonation at one of these locations, giving rise to two possible products, as shown:

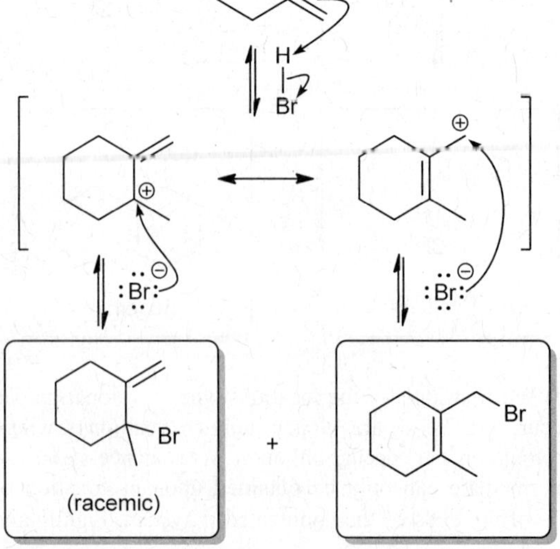

Notice that one of the products possesses a chirality center and is therefore produced as a racemic mixture (because the bromide ion can attack either face of the allylic carbocation with equal likelihood).

17.8. The first diene can be protonated either at C1 or at C4. Each of these pathways produces a resonance-stabilized carbocation. And each of these carbocations can be attacked in two positions, giving rise to four possible products.

In contrast, the second diene yields the same carbocation regardless of whether protonation occurs at C1 or at C4. This resonance-stabilized carbocation can be attacked in two positions, giving rise to two products (not four).

17.9. Even if we start with a symmetrical diene, we would still expect two products:

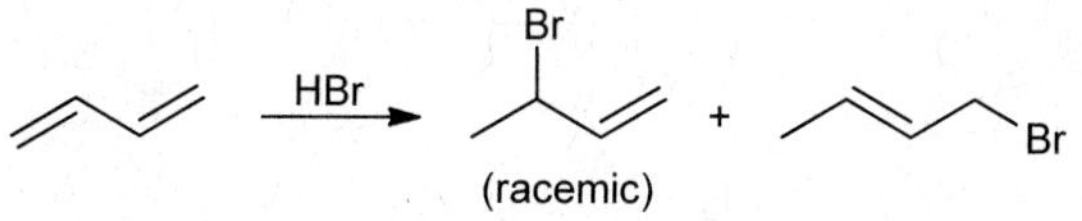

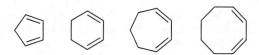

In order to obtain just one product, we must close the starting diene into a ring, so that the two products are identical.

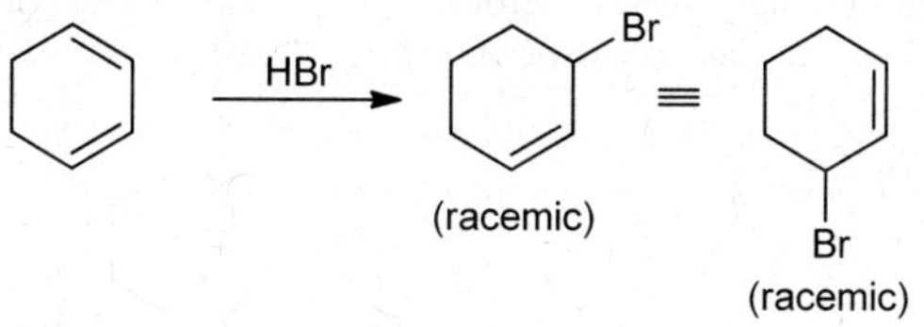

Certainly, the ring can be a different size, so there are many possible solutions to this problem.

17.10.
(a) The diene is symmetrical, so protonation at one end of the conjugated system is the same as protonation at the other end of the conjugated system:

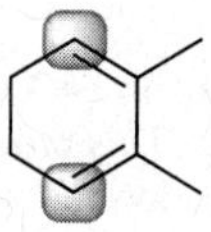

Therefore, we only need to consider protonation at one of these locations, which gives a resonance-stabilized cation:

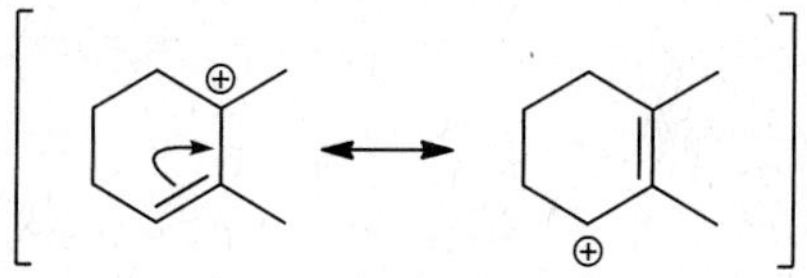

This cation has two electrophilic positions and can be attacked at either of these two positions to give the following two possible products, as shown.

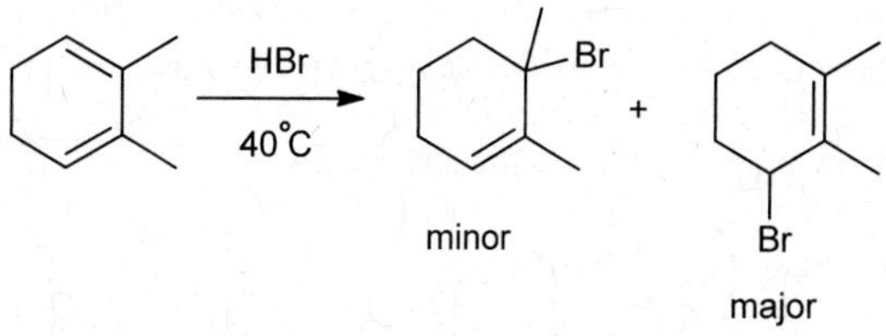

At elevated temperature, the thermodynamic product (the compound with the more substituted π bond) is expected to predominate.

(b) The diene is symmetrical, so protonation at one end of the conjugated system is the same as protonation at the other end of the conjugated system:

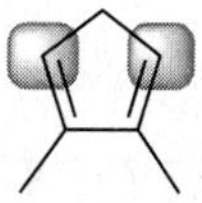

Therefore, we only need to consider protonation at one of these locations, which gives a resonance-stabilized cation:

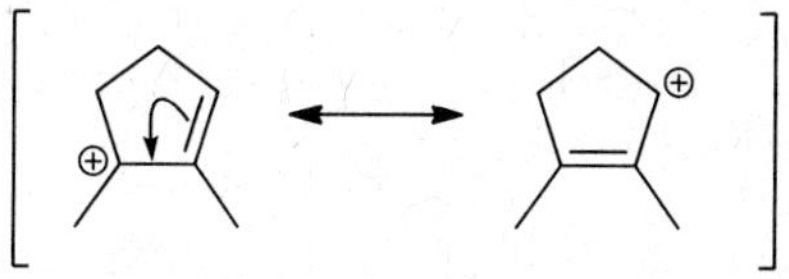

This cation has two electrophilic positions and can be attacked at either of these two positions to give the following two possible products, as shown.

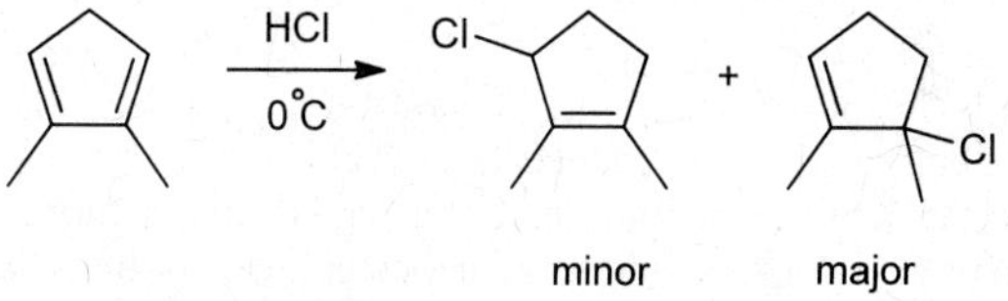

At reduced temperature, the kinetic product (resulting from 1,2-addition) is expected to predominate.

(c) The diene is symmetrical, so protonation at one end of the conjugated system is the same as protonation at the other end of the conjugated system:

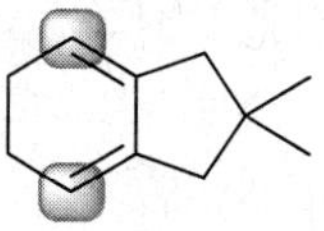

Therefore, we only need to consider protonation at one of these locations, which gives a resonance-stabilized cation:

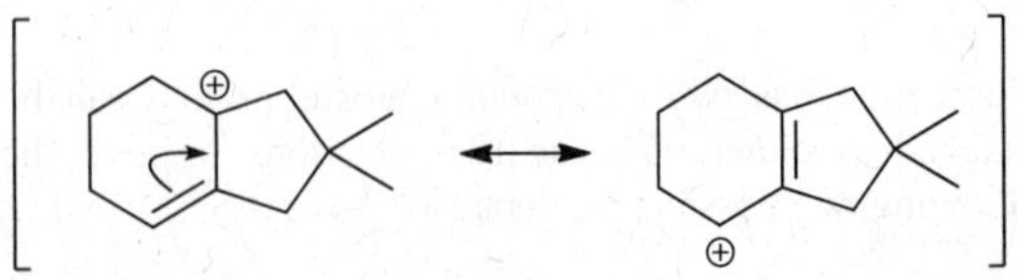

This cation has two electrophilic positions and can be attacked at either of these two positions to give the following two possible products, as shown.

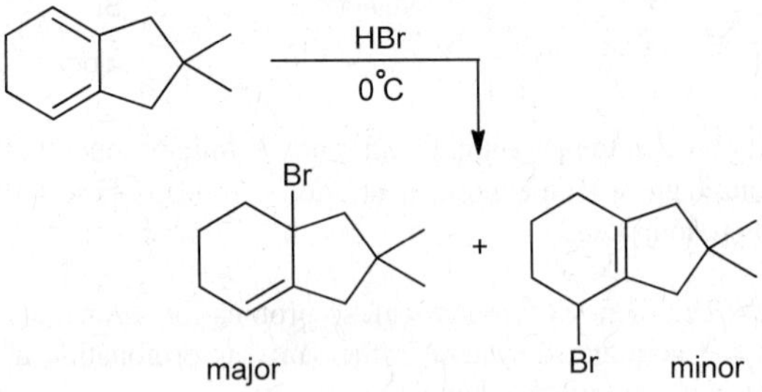

At reduced temperature, the kinetic product (resulting from 1,2-addition) is expected to predominate.

17.11. In this case, the π bond in the 1,2-adduct is more substituted than the π bond in the 1,4-adduct (trisubstituted rather than disubstituted). As a result, the 1,2-adduct predominates at either low temperature or high temperature.

17.12. In this case, 1,2-addition and 1,4-addition yield the same product.

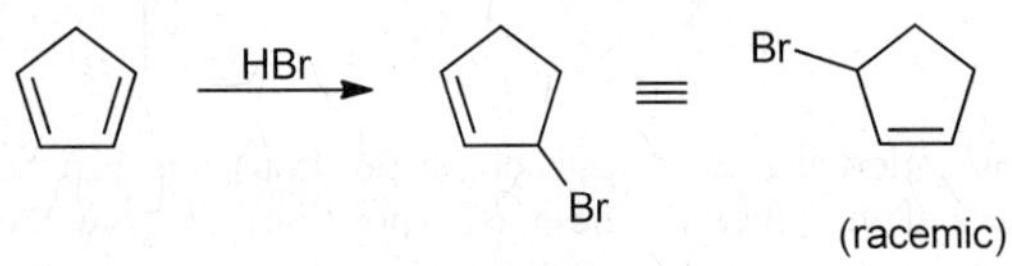

17.13.
(a) This polymer is similar in structure to neoprene, but the chloro group has been replaced with a cyano group. So the starting material should be similar in structure to chloroprene (the monomer of neoprene), except that the chloro group is replaced with a cyano group:

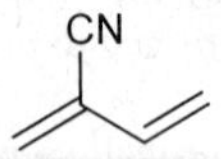

(b) This polymer is comprised of repeating units that bear two substituents (both fluoro groups). The following monomer is necessary.

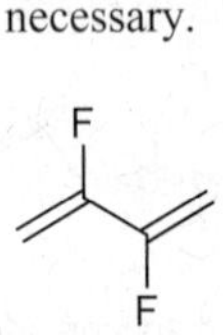

17.14.
(a) The dienophile has the *cis* configuration, which is preserved in the product. This product is superimposable on its mirror image, so it does not have an enantiomer. It is a *meso* compound.

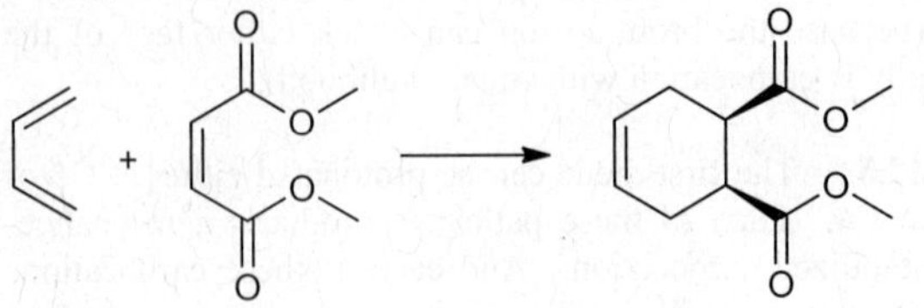

(*meso*)

(b) The dienophile has the *cis* configuration, which is preserved in the product. This product is superimposable on its mirror image, so it does not have an enantiomer. It is a *meso* compound.

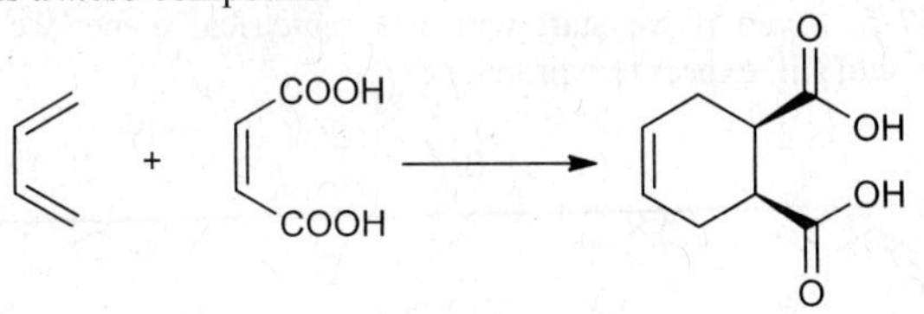

(*meso*)

(c) The dienophile has the *trans* configuration, which is preserved in the product. This product is not superimposable on its mirror image (its enantiomer), and both enantiomers are expected to be formed.

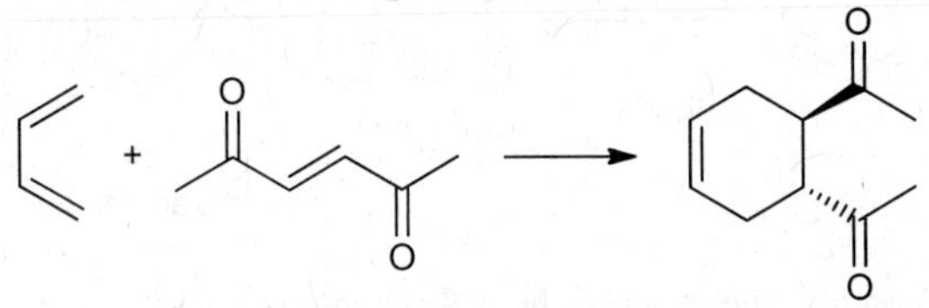

+ En

(d) The dienophile is an alkyne, and therefore, the product does not have any chirality centers (no wedges or dashes).

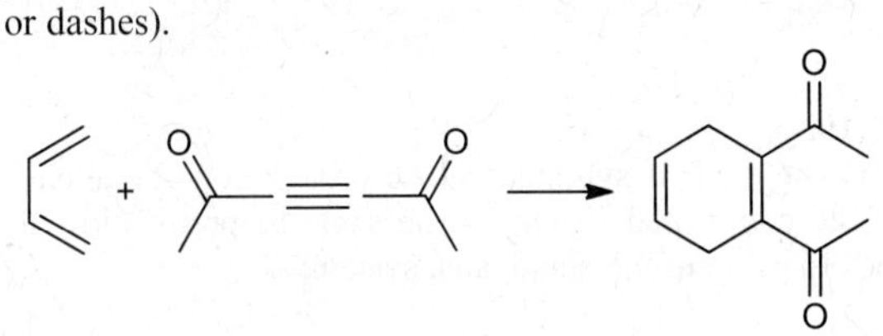

(e) The dienophile has the *cis* configuration, which is preserved in the product. This product is superimposable on its mirror image, so it does not have an enantiomer. It is a *meso* compound.

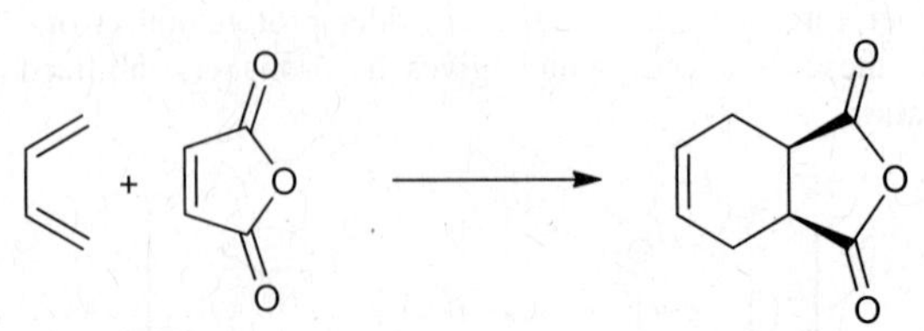

(*meso*)

(f) The dienophile is an alkyne, and therefore, the product does not have any chirality centers (no wedges or dashes).

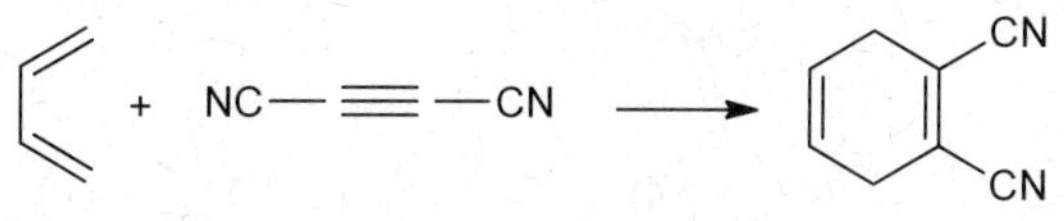

(g) The dienophile has the *cis* configuration, which is preserved in the product. This product is not superimposable on its mirror image (its enantiomer), and both enantiomers are expected to be formed.

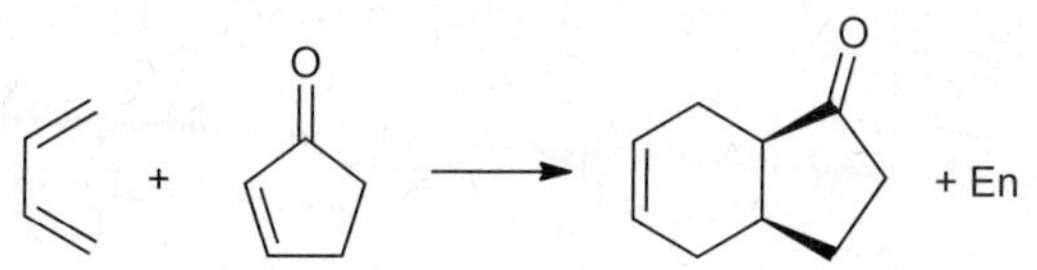

(h) The dienophile is an alkyne, and therefore, the product does not have any chirality centers (no wedges or dashes).

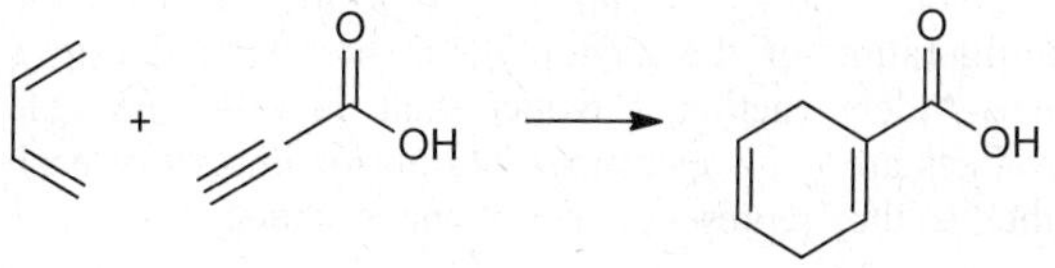

(i) The dienophile is a monosubstituted alkene, giving a cycloadduct with one chirality center. This product is not superimposable on its mirror image (its enantiomer), and both enantiomers are expected to be formed.

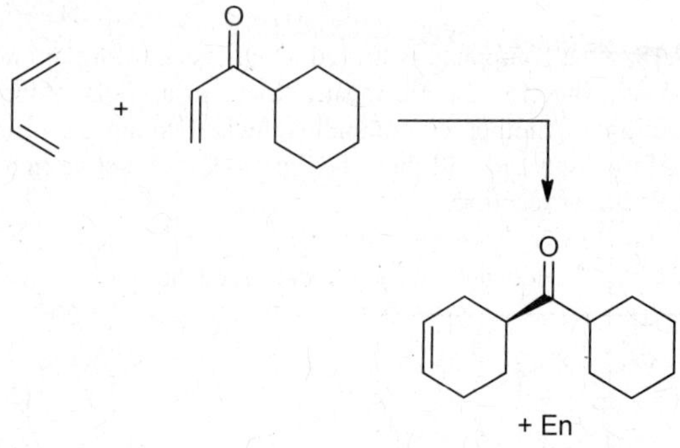

17.15. Benzoquinone has two C=C π bonds, each of which can function as a dienophile in a Diels-Alder reaction. Two successive Diels-Alder reactions will afford a tricyclic structure. The tricyclic products are diastereomers, and are formed because the second Diels-Alder reaction need not occur on the same face as the first Diels-Alder reaction.

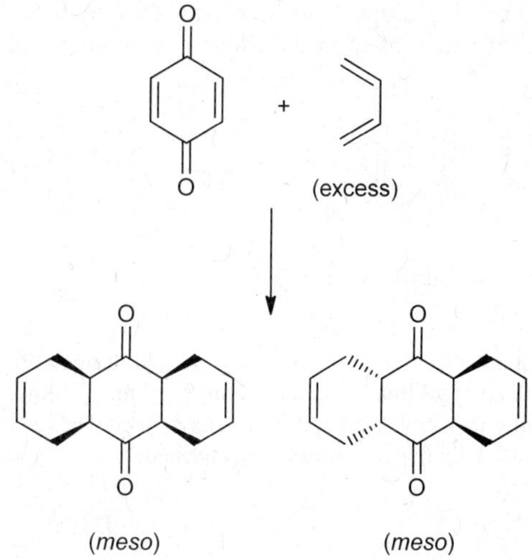

17.16. This transformation can occur via an intramolecular Diels-Alder reaction, in which a portion of the compound functions as the diene, while another portion of the compound functions as the dienophile. The result is a polycyclic compound (which is expected to be formed as a racemic mixture).

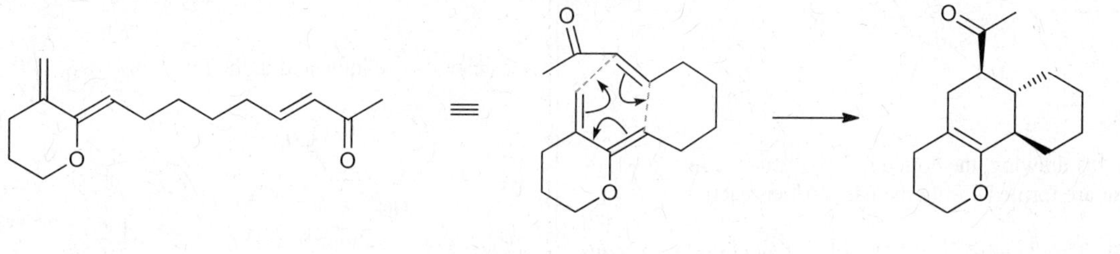

17.17. The 2*E*,4*E* isomer is expected to react more rapidly as a diene in a Diels–Alder reaction, because it can readily adopt an *s-cis* conformation.

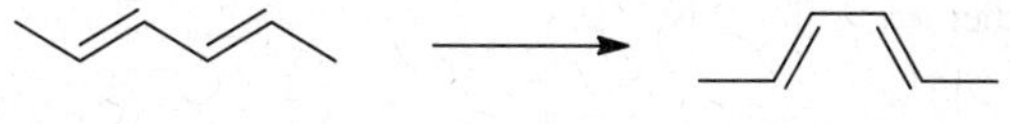

(2*E*,4*E*)-hexadiene

In contrast, the 2*Z*,4*Z* isomer is expected to react more slowly as a diene in a Diels–Alder reaction, because it cannot readily adopt an *s-cis* conformation, as a result of steric interactions.

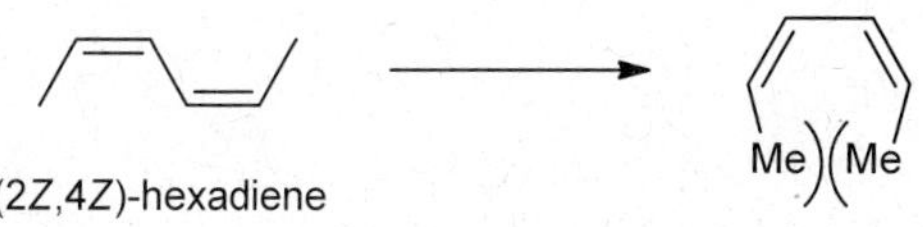

(2*Z*,4*Z*)-hexadiene

17.18. One compound is locked in an *s-cis* conformation and will therefore be the most reactive in a Diels-Alder reaction. Another compound is locked in an *s-trans* conformation and will therefore be the least reactive in a Diels-Alder reaction:

Reactivity in Diels-Alder reactions

locked in an
s-trans
conformation

locked in an *s-cis*
conformation

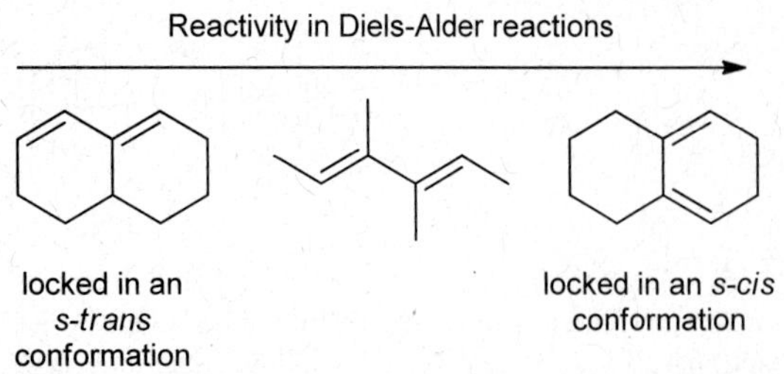

17.19.
(a) The diene is cyclic and the dienophile has a *cis* configuration, so we must consider the *endo* rule. Specifically, we draw the cycloadduct in which the cyano groups occupy *endo* positions. This product has an internal plane of symmetry and is a *meso* compound.

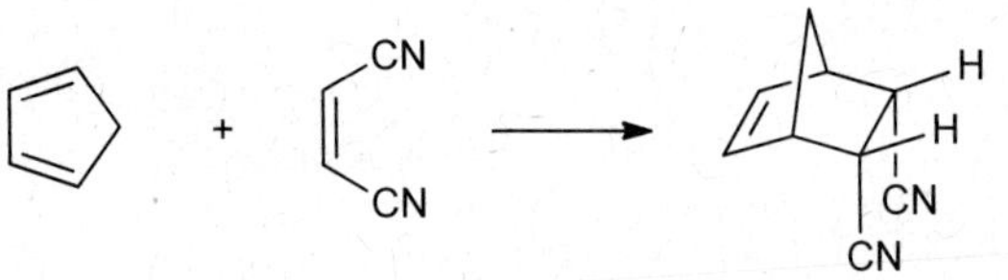

(*meso*)

If you have trouble seeing how the bicyclic framework is produced by this reaction, consider the following drawing in which the bond angles have been distorted to show how the bicyclic structure is formed:

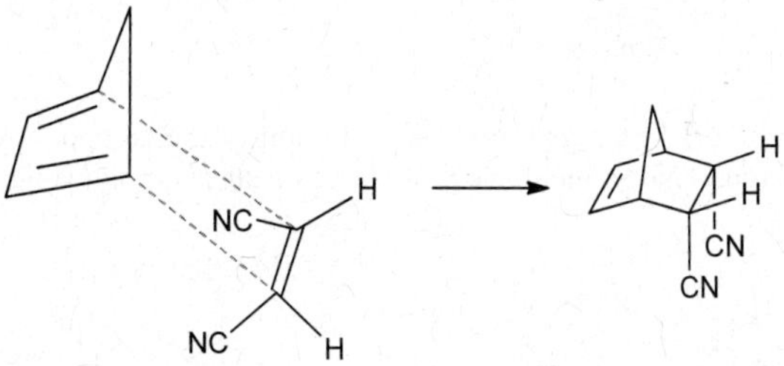

In this drawing, the dotted lines indicate the new σ bonds that are formed during the Diels-Alder reaction.

(b) The diene is cyclic, but the dienophile has a *trans* configuration. Therefore, the *endo* rule is not relevant in this case. Two products are expected, because the reaction can either occur like this,

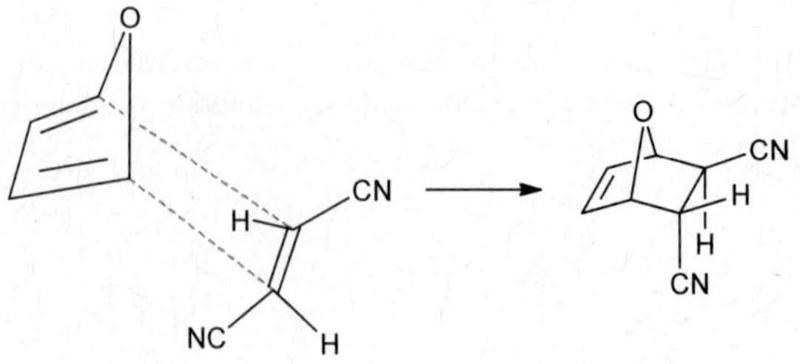

or the reaction can occur like this:

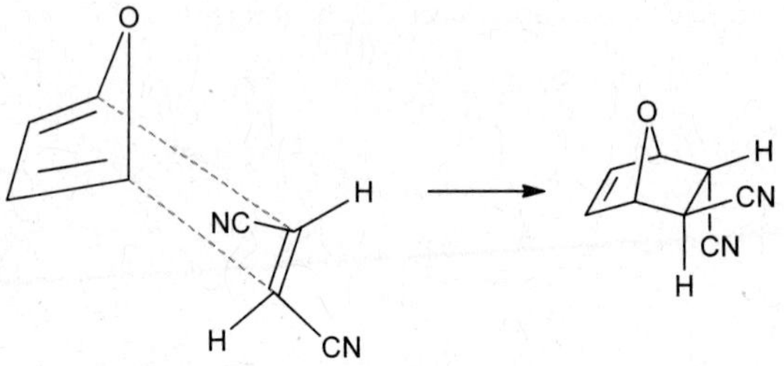

Either way, one group will occupy an *endo* position and the other group will occupy an *exo* position, because the configuration of the dienophile is preserved during a Diels-Alder reaction. Notice that the two possible products are non-superimposable mirror images of each other, so they represent a pair of enantiomers:

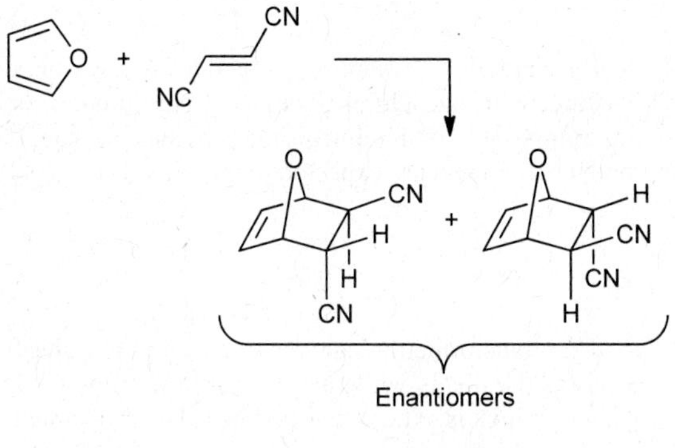

Enantiomers

This can also be indicated in the following way:

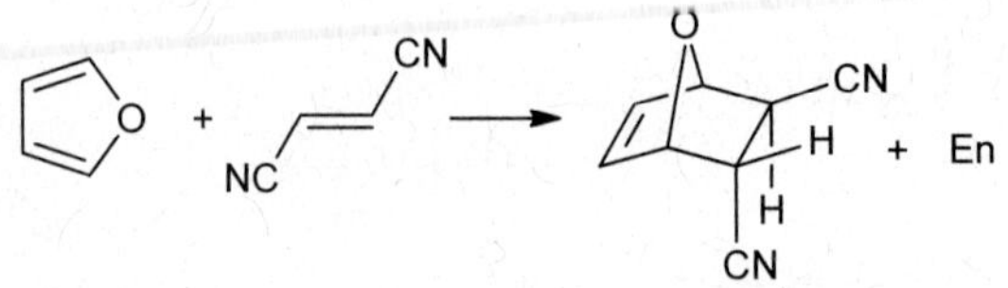

(c) The diene is cyclic, and the *endo* rule indicates that the substituent should occupy an *endo* position in the product. With this restriction in mind, the reaction can either occur like this, or the reaction can occur like this:

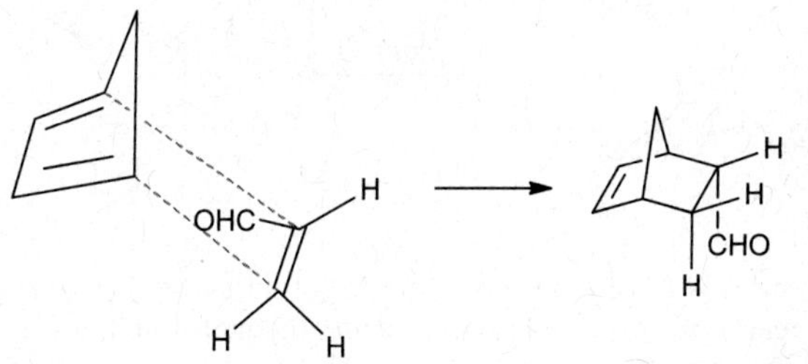

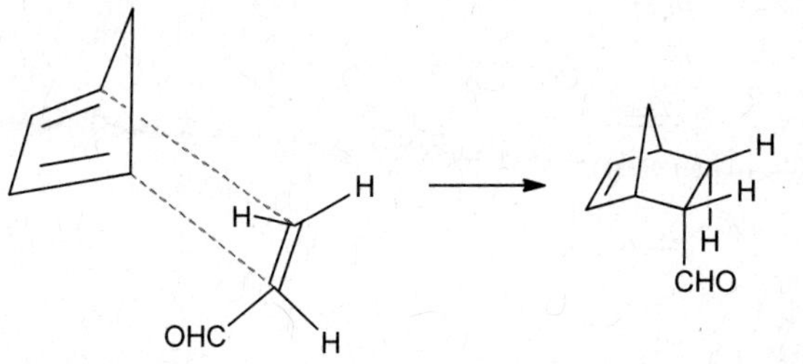

These two products represent a pair of enantiomers, and both are expected.

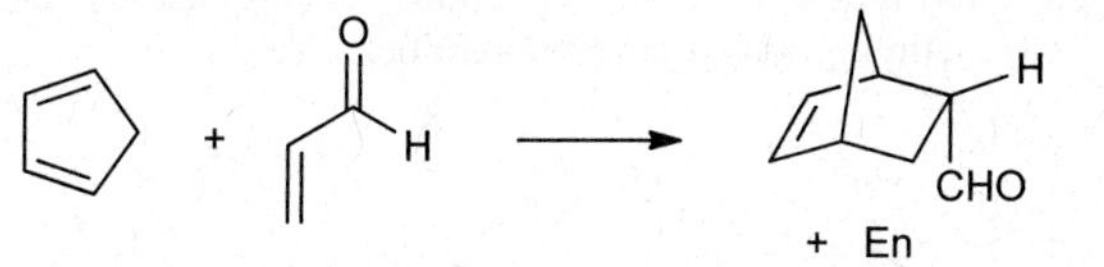

(d) The diene is cyclic, but the dienophile has a *trans* configuration. Therefore, the *endo* rule is not relevant in this case. Two products are expected, because the reaction can either occur like this,

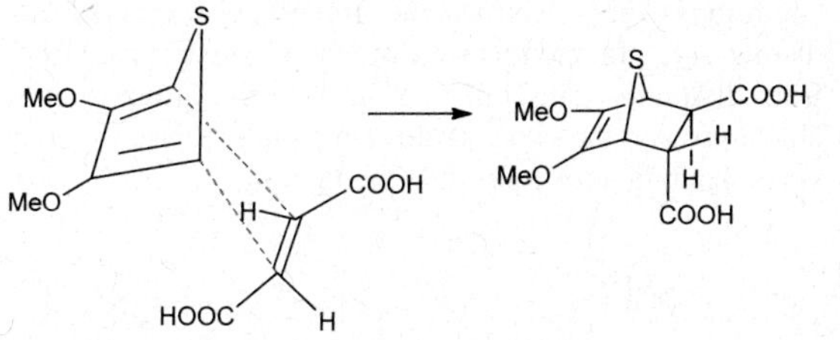

or the reaction can occur like this:

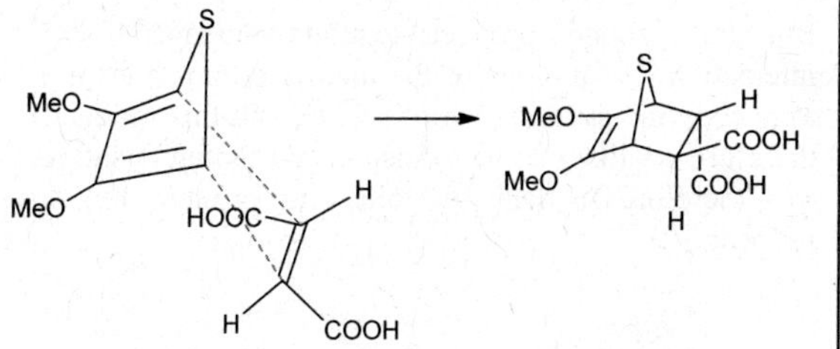

Either way, one group will occupy an *endo* position and the other group will occupy an *exo* position, because the

configuration of the dienophile is preserved during a Diels-Alder reaction. Notice that the two possible products are non-superimposable mirror images of each other, so they represent a pair of enantiomers:

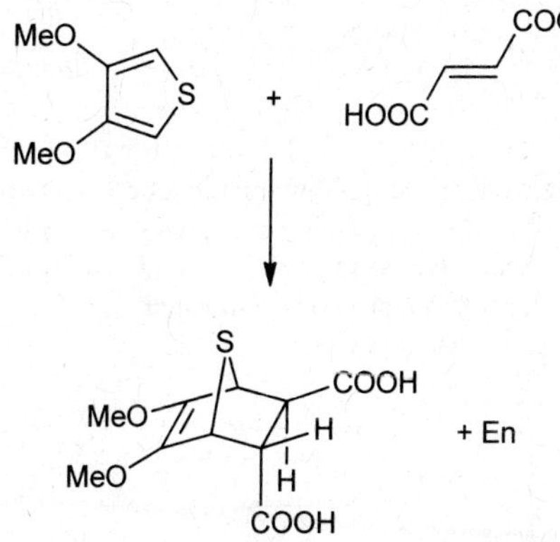

(e) The diene is cyclic, and the *endo* rule indicates that the substituent should occupy an *endo* position in the product. With this restriction in mind, the reaction can either occur like this,

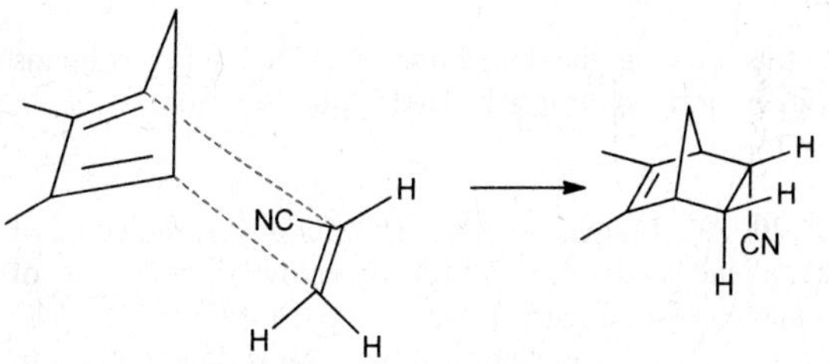

or the reaction can occur like this:

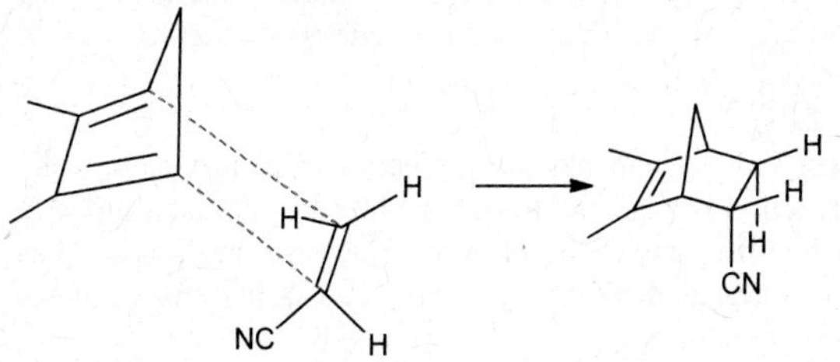

These two products represent a pair of enantiomers, and both are expected.

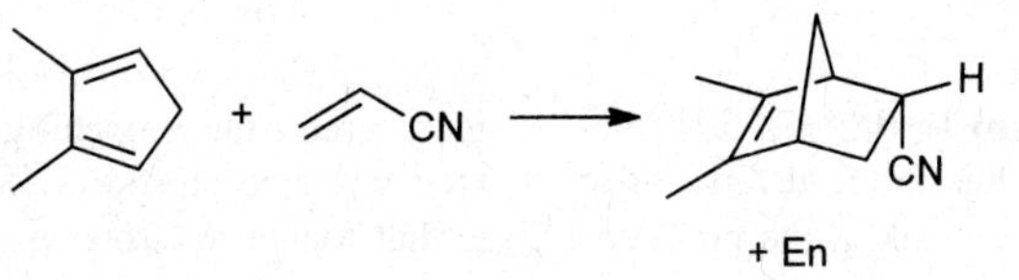

(f) The diene is cyclic and the dienophile has a *cis* configuration, so we must consider the *endo* rule. Specifically, we draw the *endo* product, which has an internal plane of symmetry in this case, so it is a *meso* compound.

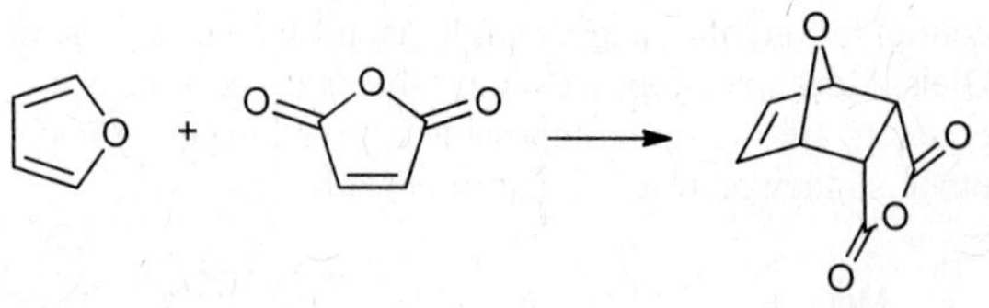

(meso)

If you have trouble seeing how the bicyclic framework is produced by this reaction, consider the following drawing in which the bond angles have been distorted to show how the bicyclic structure is formed:

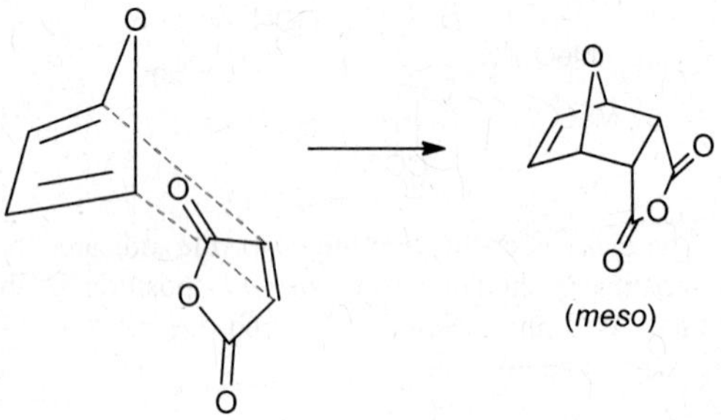

(meso)

In this drawing, the dotted lines indicate the new σ bonds that are formed during the Diels-Alder reaction.

17.20. We first consider the HOMO of one molecule of butadiene and the LUMO of another molecule of butadiene (see Figure 17.17 for the HOMO and LUMO of butadiene). The phases of these MOs do not align, so a thermal reaction is symmetry-forbidden. However, if one molecule is photochemically excited, the HOMO and LUMO of that molecule are redefined. The phases of the frontier orbitals will align under these conditions, so the reaction is expected to occur photochemically.

17.21.
(a) This system has six π electrons, so an electrocyclic reaction is expected to occur under thermal conditions to give disrotatory ring closure. The resulting product has an internal plane of symmetry and is therefore a *meso* compound.

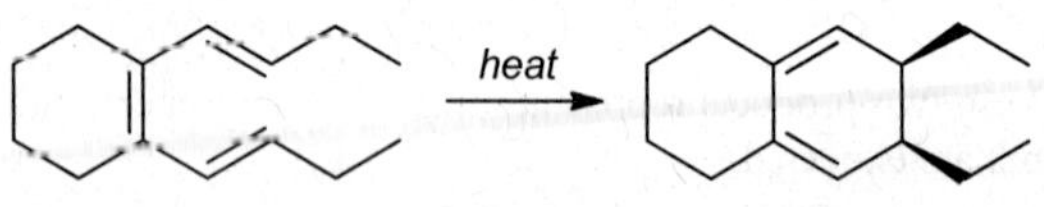

(meso)

(b) In this reaction, the four-membered ring is opening (this is the reverse of an electrocyclic ring closure). If we look at the product, we see that four π electrons are involved in the process. Under thermal conditions, this electrocyclic process is expected be conrotatory, giving the following product.

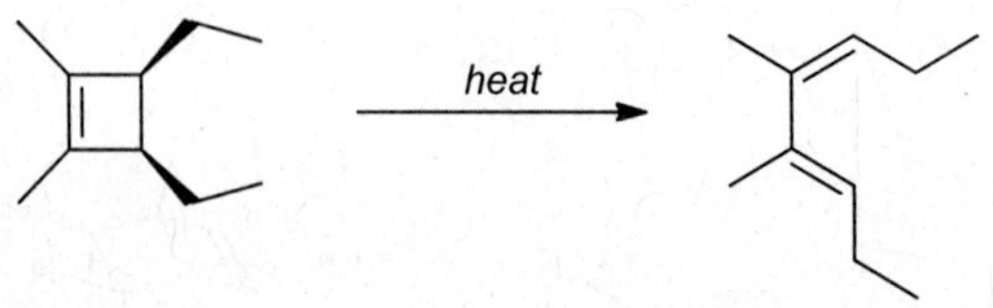

(c) This system has six π electrons, so an electrocyclic reaction is expected to occur under thermal conditions to give disrotatory ring closure. The resulting product is chiral, and both enantiomers are expected (one ethyl group rotates clockwise and the other rotates counterclockwise, or vice versa, leading to both possible enantiomers).

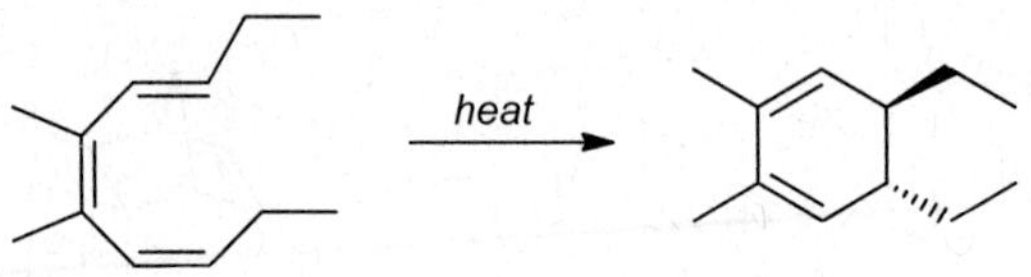

17.22.
(a) This system has six π electrons, so an electrocyclic reaction is expected to occur under photochemical conditions to give conrotatory ring closure. The resulting product is chiral, and both enantiomers are expected (either both methyl groups rotate clockwise or both methyl groups rotate counterclockwise).

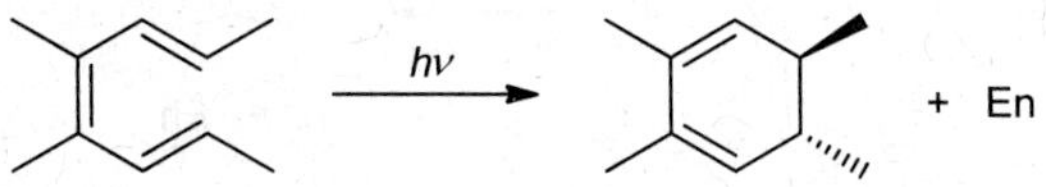

(b) In this reaction, the four-membered ring is opening (this is the reverse of an electrocyclic ring closure). If we look at the product, we see that four π electrons are involved in the process. Under photochemical conditions, this electrocyclic process is expected be disrotatory. In theory, two products can be produced from disrotatory ring-opening (one methyl group rotates clockwise and the other rotates counterclockwise, or vice versa, leading to two possible products):

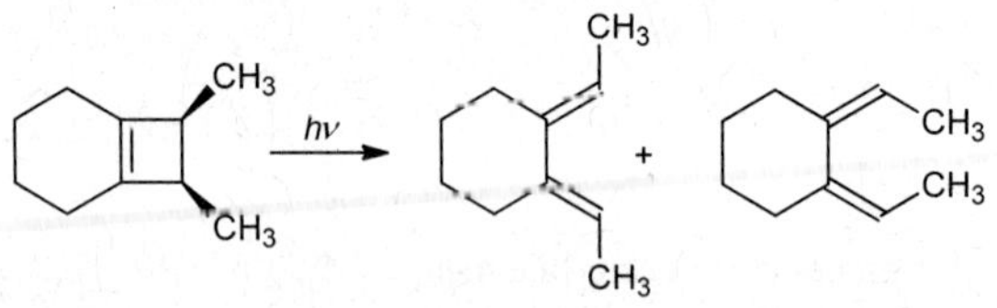

But the second product exhibits a severe steric interaction (the protons of the methyl groups are forced to occupy the same region of space), and this product is therefore not likely to be formed in substantial quantities. We therefore predict the following product for this reaction.

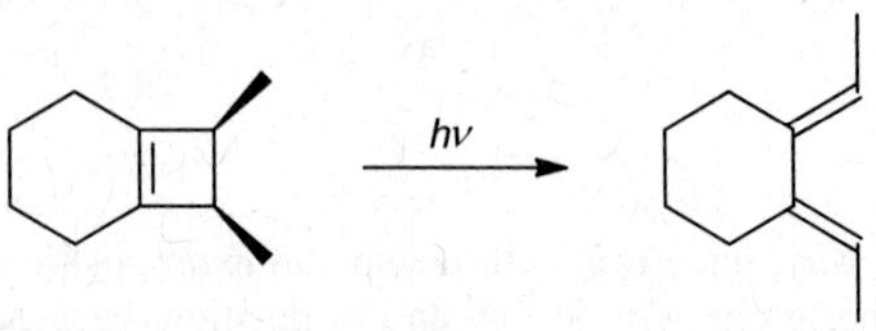

(c) This system has six π electrons, so an electrocyclic reaction is expected to occur under thermal conditions to give disrotatory ring closure. The resulting product is chiral, and both enantiomers are expected (one methyl group rotates clockwise and the other rotates counterclockwise, or vice versa, leading to both possible enantiomers).

+ En

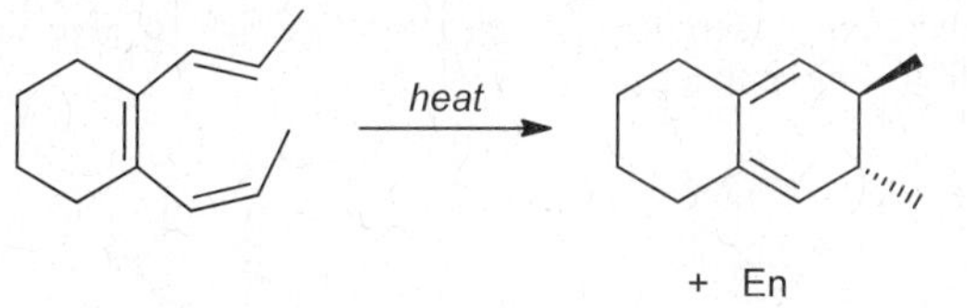

17.23. The structure of *cis*-3,4-diethylcyclobutene is shown here:

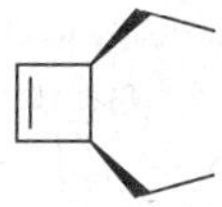

(a) Ring-opening results in a compound with four π electrons (as seen below). An electrocyclic reaction with four π electrons requires thermal conditions in order to occur in a conrotatory fashion, giving the following product:

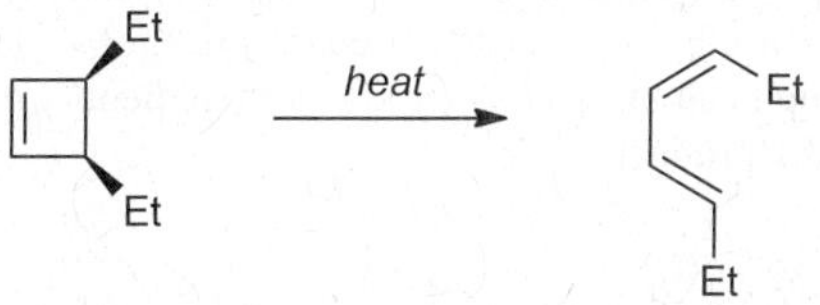

(b) Disrotatory ring-opening could potentially give two products, shown below, but the second product is not formed because it exhibits significant steric interactions (the ethyl groups are being forced too close together).

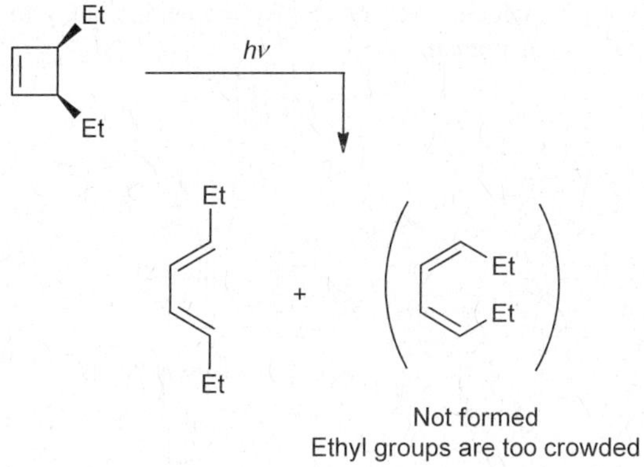

Not formed
Ethyl groups are too crowded

17.24.
(a) This system has six π electrons, so an electrocyclic reaction is expected to occur under photochemical conditions to give conrotatory ring closure. The

resulting product is a *meso* compound because it is superimposable on its mirror image:

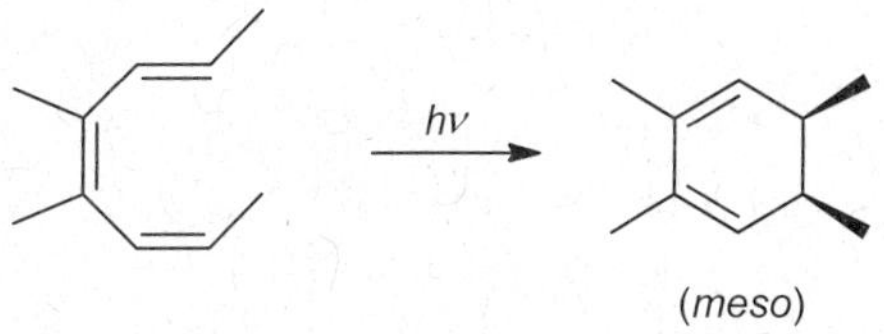

(*meso*)

This product is obtained whether both methyl groups rotate in a clockwise fashion, or whether both methyl groups rotate in a counterclockwise fashion:

(b) This system has six π electrons, so an electrocyclic reaction is expected to occur under thermal conditions to give disrotatory ring closure. The resulting product is chiral, and both enantiomers are expected (one methyl group rotates clockwise and the other rotates counterclockwise, or vice versa, leading to both possible enantiomers):

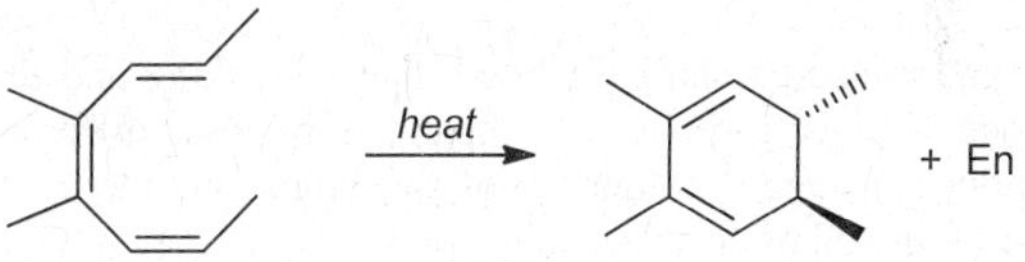

+ En

(c) This system has six π electrons, so an electrocyclic reaction is expected to occur under photochemical conditions to give conrotatory ring closure. The resulting product is chiral, and both enantiomers are expected (either both methyl groups can rotate clockwise or both can rotate counterclockwise, leading to both possible enantiomers):

+ En

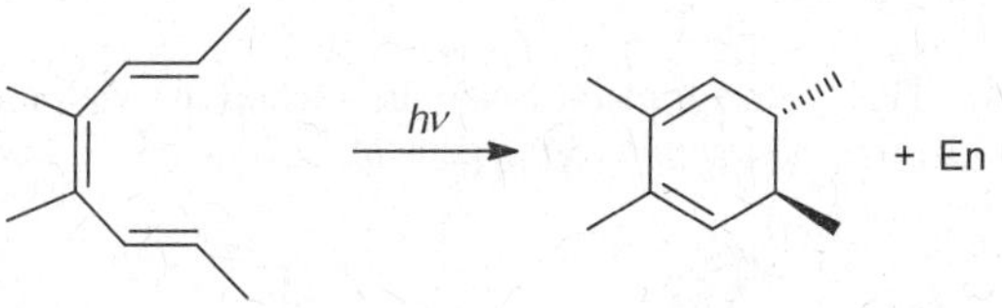

17.25.
(a) We begin by identifying the σ bond that is broken and the σ bond that is formed, highlighted here:

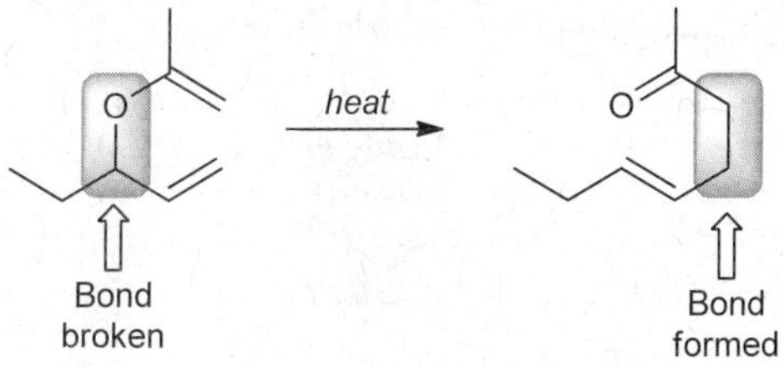

Bond broken

Bond formed

In the transition state, the bond that is breaking and the bond that is forming are separated by two different pathways, each of which is comprised of three atoms:

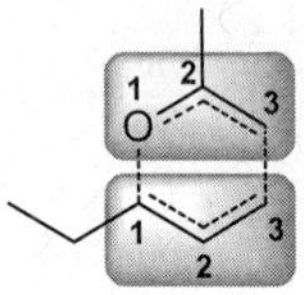

Therefore, this reaction is a [3,3] sigmatropic rearrangement.

(b) We begin by identifying the σ bond that is broken and the σ bond that is formed, highlighted here:

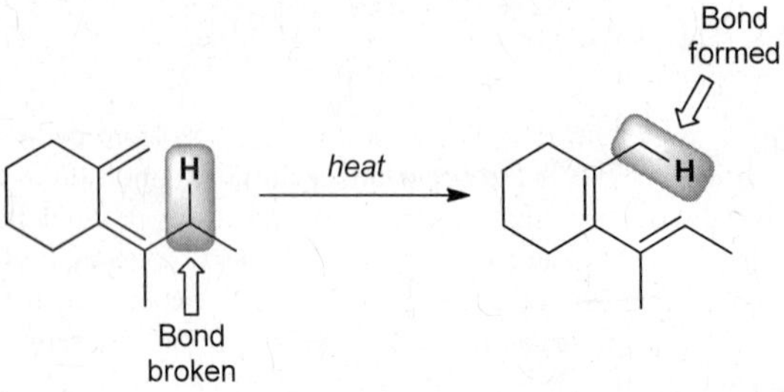

Bond
formed

heat

Bond
broken

In the transition state, the bond that is breaking and the bond that is forming are separated by two different pathways: one is comprised of five atoms and the other is comprised of only one atom:

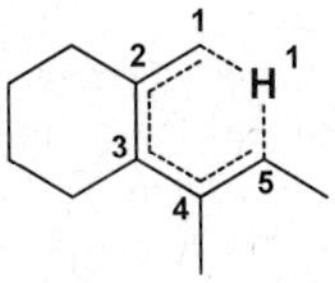

Therefore, this reaction is a [1,5] sigmatropic rearrangement.

17.26.
(a) This transformation can be achieved via the following sigmatropic rearrangement:

(b) We identify the σ bond that is broken and the σ bond that is formed, and then determine the pathways (highlighted below) that separate these bonds.

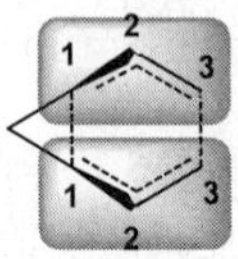

Each pathway is comprised of three atoms, so the reaction is a [3,3] sigmatropic rearrangement.

(c) The ring strain associated with the three-membered ring is alleviated. The reverse process would involve forming a high-energy, three-membered ring. The equilibrium disfavors the reverse process.

17.27.
(a) This compound is an allylic vinylic ether, and it can therefore undergo a Claisen rearrangement to give the following product:

heat

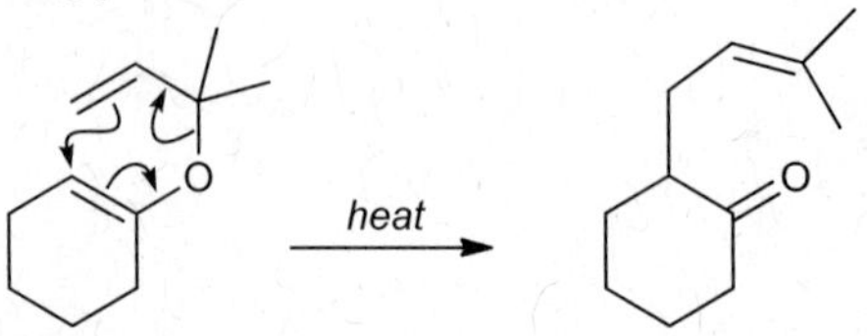

(b) This compound has two π bonds that are separated from each other by exactly three σ bonds. As such, this compound can undergo a Cope rearrangement to give the following product:

heat

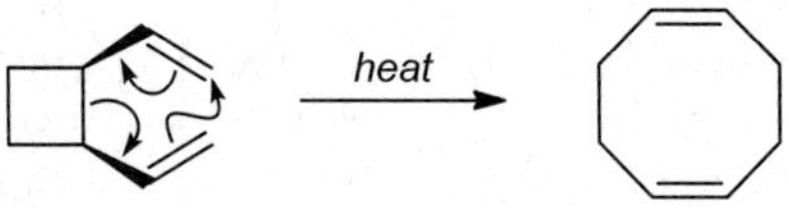

(c) This compound has two π bonds that are separated from each other by exactly three σ bonds. As such, this compound can undergo a Cope rearrangement to give the following product:

heat

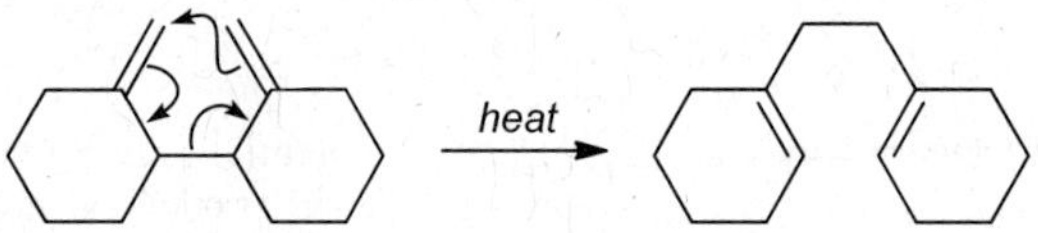

(d) This compound is an allylic vinylic ether, and it can therefore undergo a Claisen rearrangement to give a ketone, which tautomerizes to give the enol, thereby re-establishing aromaticity:

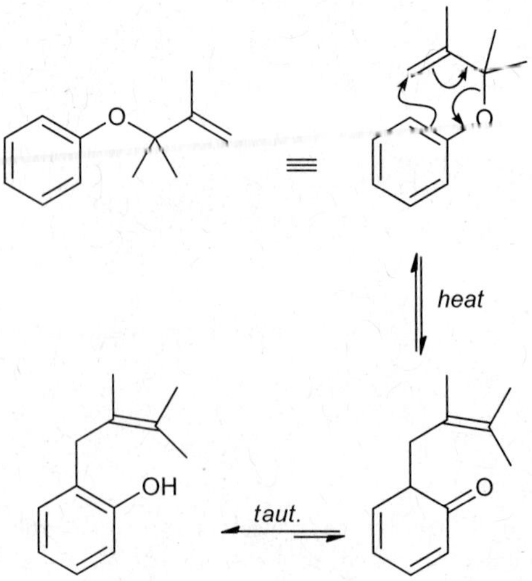

≡

heat

taut.

17.28. This compound has two π bonds that are separated from each other by exactly three σ bonds. As such, this compound can undergo a Cope rearrangement. To draw the product of this reaction, it is helpful to redraw the starting material, as shown below, so that it is easier to see the ring of electrons responsible for the transformation:

heat

17.29.
(a) This conjugated system (comprised of two π bonds) has five auxochromic alkyl groups, highlighted here:

and one exocyclic double bond:

The calculation for λ_{max} is shown here:

Base	=	217
Additional double bonds	=	0
Auxochromic alkyl groups	=	+25
Exocyclic double bond	=	+5
Homoannular diene	=	0
Total	=	247 nm

(b) This conjugated system (comprised of three π bonds) has five auxochromic alkyl groups, highlighted here:

and one exocyclic double bond:

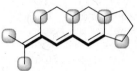

The calculation for λ_{max} is shown here:

Base	=	217
Additional double bonds	=	+30
Auxochromic alkyl groups	=	+25
Exocyclic double bond	=	+5
Homoannular diene	=	0
Total	=	277 nm

(c) This conjugated system (comprised of three π bonds) has six auxochromic alkyl groups, highlighted here:

and three exocyclic double bonds (each of the double bonds is exocyclic to a ring), shown here:

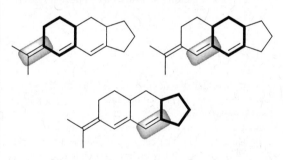

The calculation for λ_{max} is shown here:

Base	=	217
Additional double bonds	=	+30
Auxochromic alkyl groups	=	+30
Exocyclic double bonds	=	+15
Homoannular diene	=	0
Total	=	292 nm

(d) This conjugated system (comprised of three π bonds) has two of the double bonds in the same ring (homoannular):

In addition, there are seven auxochromic alkyl groups, highlighted here:

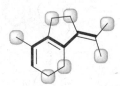

and one exocyclic double bond:

The calculation for λ_{max} is shown here:

Base	=	217
Additional double bonds	=	+30
Auxochromic alkyl groups	=	+35
Exocyclic double bonds	=	+5
Homoannular diene	=	+39
Total	=	326 nm

17.30 Only one of the compounds (shown below) has all three π bonds as part of one extended conjugated system.

Each of the other two compounds exhibits an isolated π bond. One of those compounds does exhibit a homoannular diene, which does add +39, but the compound above also has two π bonds in the same ring, so it also gets +39. The compound above will have the largest λ_{max} because it has the most extended conjugated system.

17.31.
(a) As seen in Figure 17.37 (the color wheel), the complementary color of orange is blue. Therefore, a compound that absorbs orange light will appear to be blue.
(b) As seen in Figure 17.37 (the color wheel), the complementary color of blue-green is red-orange. Therefore, a compound that absorbs blue-green light will appear to be red-orange.
(c) As seen in Figure 17.37 (the color wheel), the complementary color of orange-yellow is blue-violet. Therefore, a compound that absorbs orange-yellow light will appear to be blue-violet.

17.32.
(a) The parent ("cyclohex") indicates a six-membered ring, and the suffix ("diene") indicates the presence of two C=C bonds. The locants (1 and 4) indicate the positions of the two double bonds.

(b) The parent ("cyclohex") indicates a six-membered ring, and the suffix ("diene") indicates the presence of two C=C bonds. The locants (1 and 3) indicate the positions of the two double bonds.

(c) The parent ("pent") indicates a five-carbon chain, and the suffix ("diene") indicates the presence of two C=C bonds. The locants (1 and 3) indicate the positions of the two double bonds, and the stereodescriptor (Z) indicates the configuration of the C=C bond between C3 and C4.

(d) The parent ("hept") indicates a seven-carbon chain, and the suffix ("diene") indicates the presence of two C=C bonds. The locants (2 and 4) indicate the positions of the two double bonds, and the stereodescriptors indicate the configurations of the double bonds.

(e) The parent ("but") indicates a four-carbon chain, and the suffix ("diene") indicates the presence of two C=C bonds. The locants (1 and 3) indicate the positions of the two double bonds. There are two methyl substituents, at positions C2 and C3.

17.33. Each of the highlighted compounds possesses a conjugated π system, which is shown with darker bonds.

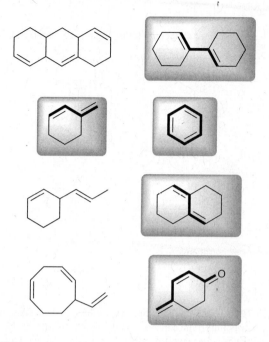

17.34.

(a) These drawings represent two different conformations of the same compound: the *s-cis* conformation and the *s-trans* conformation. These two conformations are in equilibrium at room temperature.

(b) These drawings represent two different compounds: (*Z*)-1,3,5-hexatriene and (*E*)-1,3,5-hexatriene. These compounds are diastereomers and can be isolated from one another.

(c) These drawings represent two different conformations of the same compound. In the first structure, all of the double bonds are connected in an *s-cis* fashion, while in the second compound, all of the double bonds are connected in an *s-trans* fashion. These conformations are in equilibrium at room temperature.

17.35. As seen in the solution to Problem 17.9, treatment of 1,3-cyclohexadiene with HBr produces only one product (because 1,2 addition and 1,4 addition give the same product).

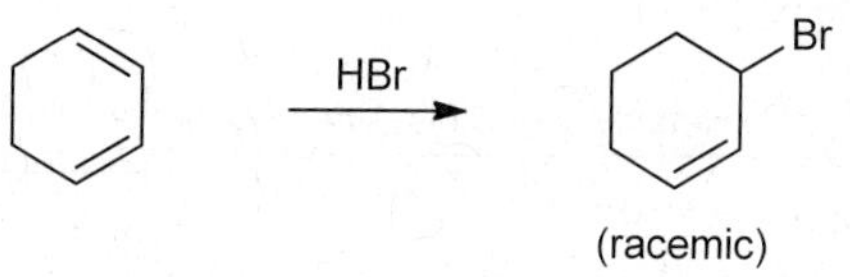

17.36. The diene is protonated to give a resonance stabilized cation, which can then be attacked in one of two locations, leading to the 1,2-adduct and the 1,4-adduct. At low temperature, the kinetic product (the 1,2-adduct) dominates.

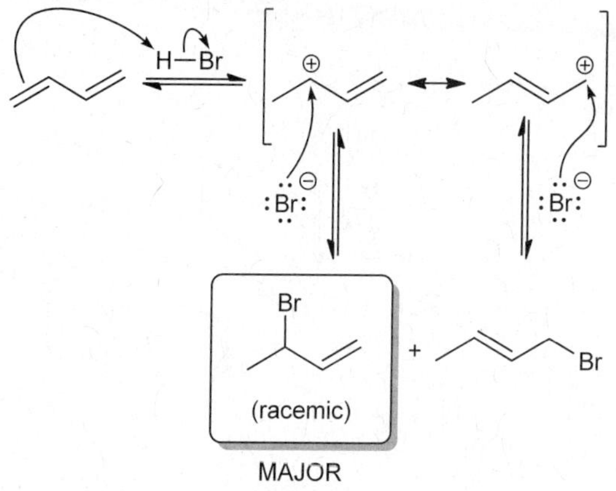

17.37. The diene is protonated to give a resonance stabilized cation, which can then be attacked in one of two locations, leading to the 1,2-adduct and the 1,4-adduct. At low temperature, the thermodynamic product (the 1,4-adduct) dominates.

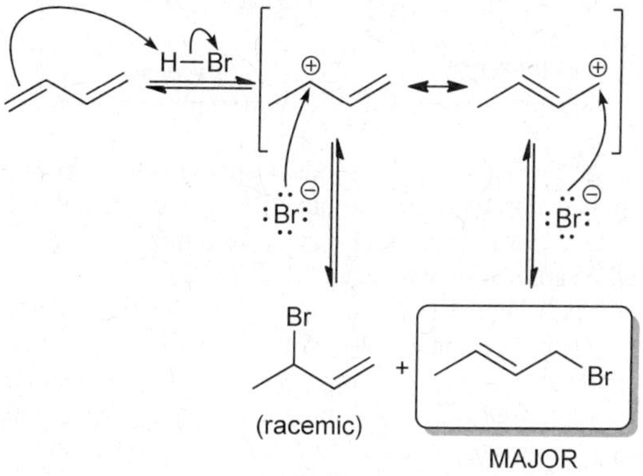

17.38. We first identify the locations where protonation can occur. There are four unique positions where protonation can occur, although a resonance-stabilized intermediate can only be obtained upon protonation of one of the ends of the conjugated π system, highlighted below:

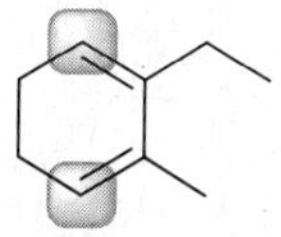

We must explore protonation at each of these positions. Let's begin with protonation of the bottom position, which leads to a resonance–stabilized intermediate. When we draw both resonance structures, we can see that two positions are electrophilic. Nucleophilic attack can occur at either of these locations, giving rise to two products:

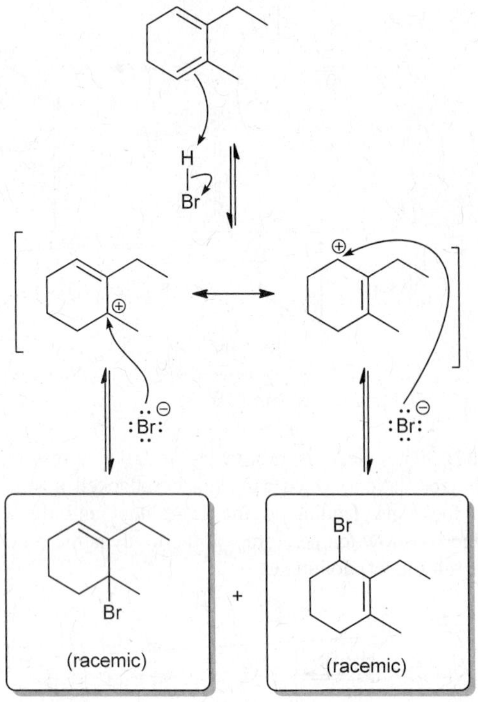

Notice that each of the products possesses a chirality center and is therefore produced as a racemic mixture (because the bromide ion can attack either face of the allylic carbocation with equal likelihood).

Now let's consider protonation at the position on top. Once again, protonation leads to a resonance–stabilized intermediate. When we draw both resonance structures, we can see that two positions are electrophilic. Nucleophilic attack can occur at either of these locations, giving two possible products, as shown:

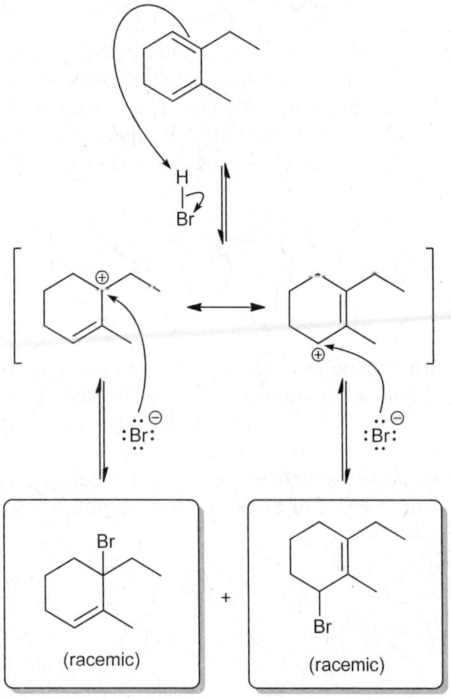

In summary, we expect the following possible products:

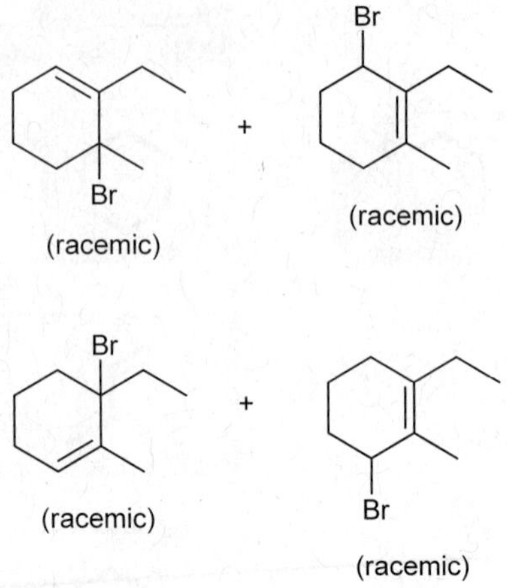

17.39. An increase in temperature allowed the system to reach equilibrium concentrations, which are determined by the relative stability of each product. Under these conditions, the 1,4-adducts predominate. Once at equilibrium, lowering the temperature will not cause a decrease in the concentration of the 1,4-adducts.

17.40.
(a) The *tert*-butyl groups provide significant steric interactions that prevent the compound from adopting an *s-cis* conformation.

(b) This diene is not conjugated.

(c) The methyl groups provide a significant steric interaction in the *s-cis* conformation that prevents the compound from adopting this conformation.

(d) This diene cannot adopt an *s-cis* conformation.

17.41. The most reactive dienophile is the one connected to two electron-withdrawing C=O groups. Then, the dienophile with only one C=O group is the next most reactive. And finally, the least reactive dienophile is the one that lacks an electron-withdrawing substituent altogether.

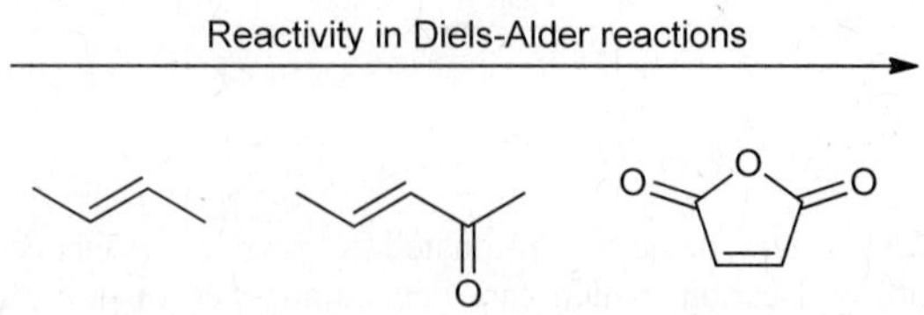

Reactivity in Diels-Alder reactions ⟶

17.42. The π bonds in 1,2-butadiene are not conjugated, and λ_{max} is therefore lower than 217 nm. In fact, it is below 200 nm, which is beyond the range used by most UV-Vis spectrometers.

17.43.

(a) In a Diels-Alder reaction, the configuration of the dienophile (in this case, *trans*) is preserved in the product, giving a pair of enantiomers:

+ En

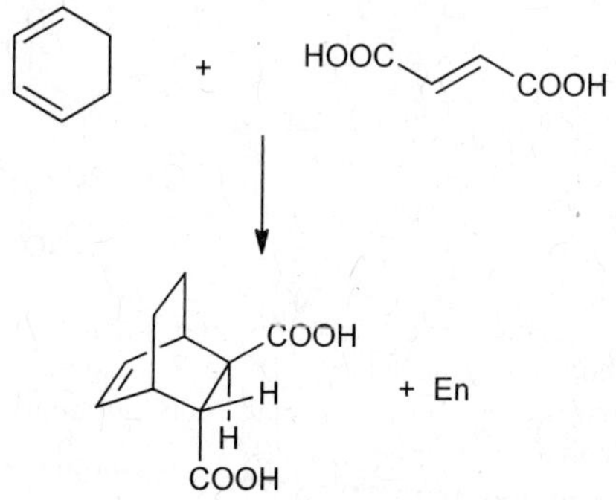

(b) In a Diels-Alder reaction, the configuration of the dienophile (in this case, *trans*) is preserved in the product, giving a pair of enantiomers. Notice that both substituents occupy *endo* (rather than *exo*) positions:

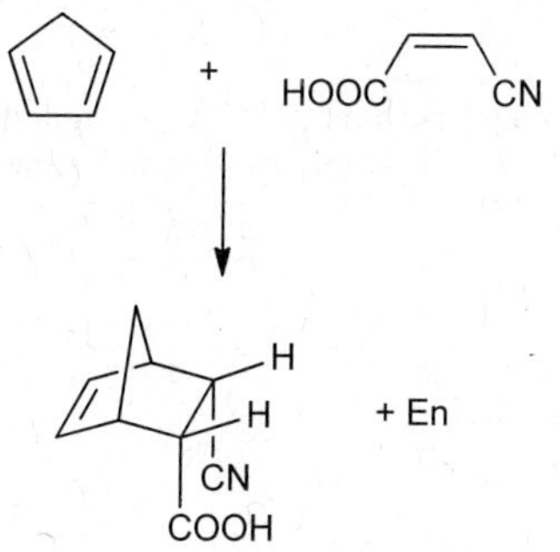

+ En

(c) In a Diels-Alder reaction, the configuration of the dienophile (in this case, *trans*) is preserved in the product, giving a pair of enantiomers:

+ En

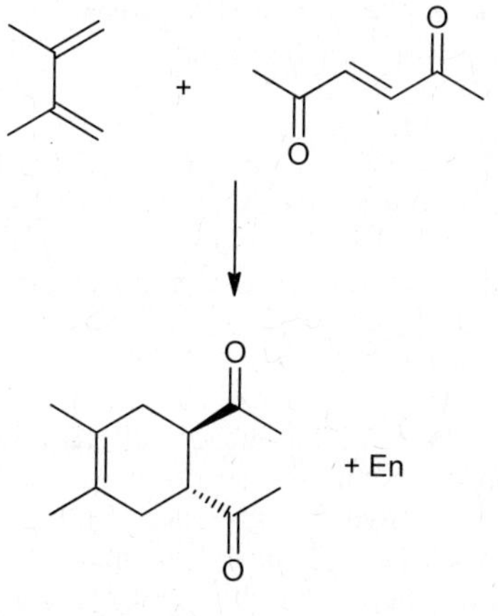

(d) The diene is cyclic and the dienophile has a *cis* configuration, so we must consider the *endo* rule. Specifically, we draw the *endo* product, which has an internal plane of symmetry in this case, so it is a *meso* compound.

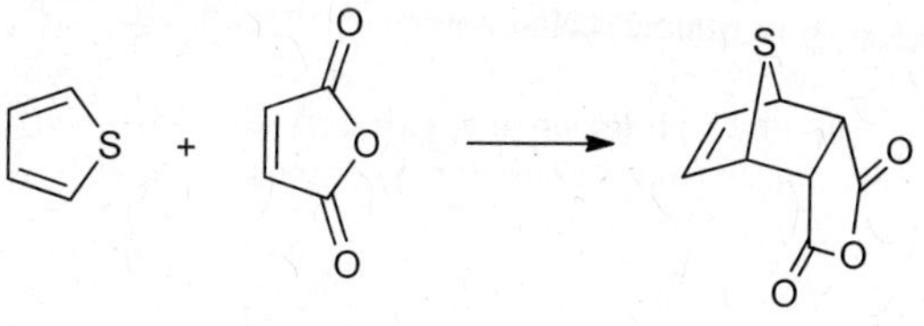

(meso)

If you have trouble seeing how the bicyclic framework is produced by this reaction, see the solution to Problem **17.19f**, which is very similar.

(e) The dienophile is an alkyne, and the product does not have any chirality centers (no wedges or dashes).

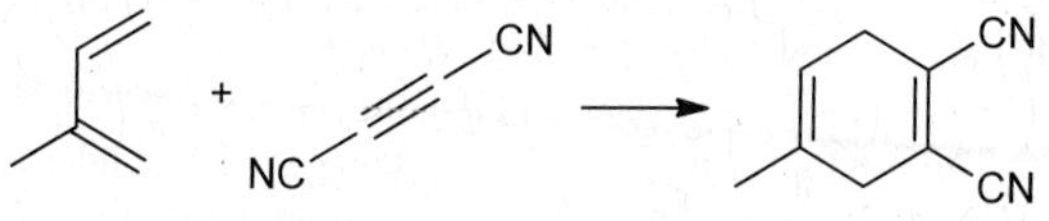

(f) The starting material is a cyclic diene, but the *endo* rule is not relevant in this case, because an alkyne is used as the dienophile, so there are no *endo* positions. The product is a *meso* compound.

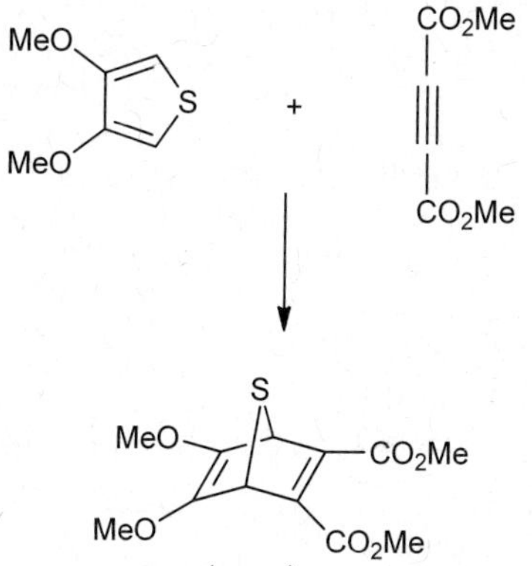

(meso)

17.44.

(a) The product exhibits a *cis* orientation of the carboxylic acid groups, so the dienophile must have the *cis* configuration as well.

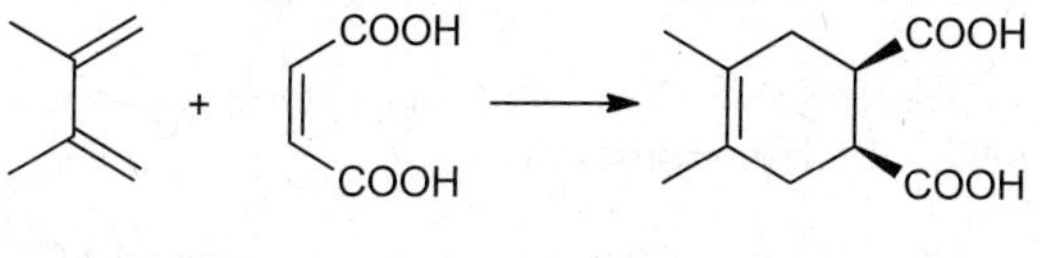

(b) The following diene and dienophile can be used to produce the desired product.

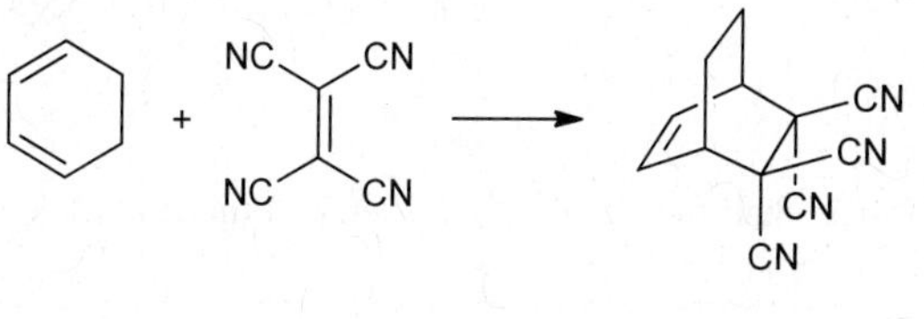

If you have trouble seeing how the bicyclic framework is produced by this reaction, consider the following drawing in which the bond angles have been distorted to show how the bicyclic structure is formed:

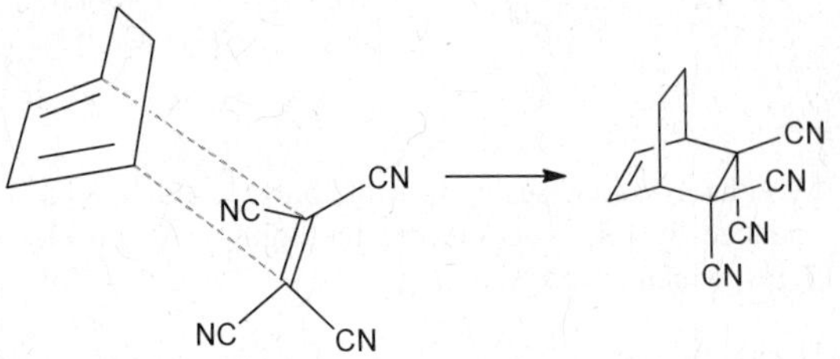

In this drawing, the dotted lines indicate the new σ bonds that are formed during the Diels-Alder reaction.

(c) The product exhibits a *cis* orientation of the aldehyde (CHO) groups, so the dienophile must have the *cis* configuration as well. Notice that the *endo* rule ensures formation of the desired product.

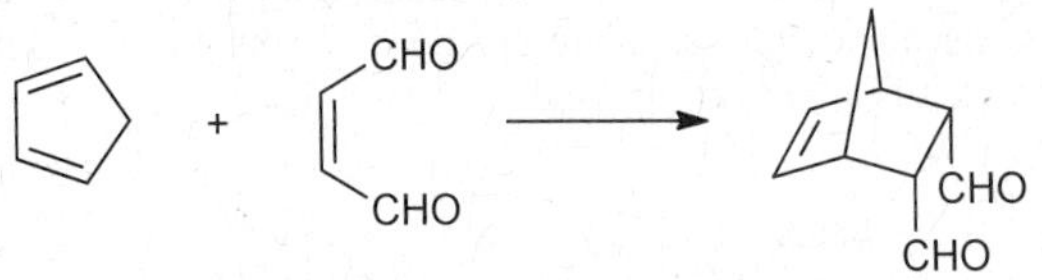

If you have trouble seeing how the bicyclic framework is produced by this reaction, consider the following drawing in which the bond angles have been distorted to show how the bicyclic structure is formed:

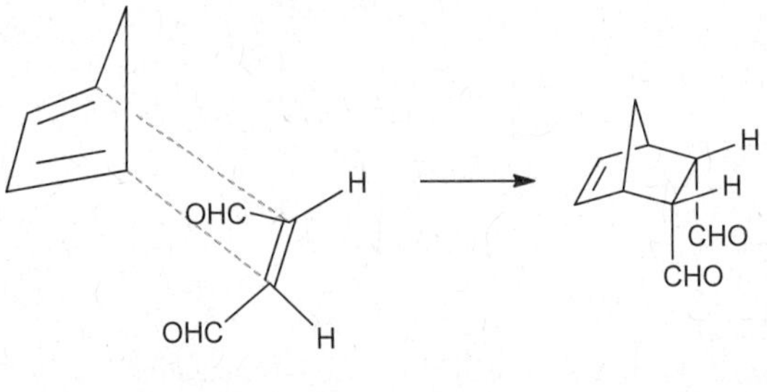

In this drawing, the dotted lines indicate the new σ bonds that are formed during the Diels-Alder reaction.

(d) The following diene and dienophile can be used to produce the desired product.

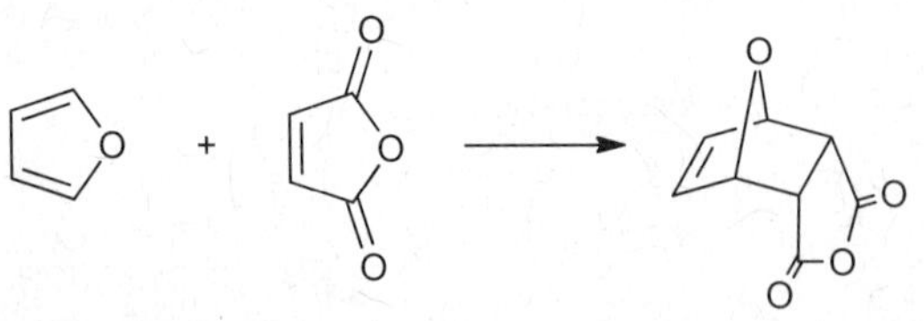

Notice that the *endo* rule ensures formation of the desired product.

(e) The product exhibits a *cis* configuration, so the dienophile must have the *cis* configuration as well (although in this case, a *trans* configuration is not possible in a six-membered ring).

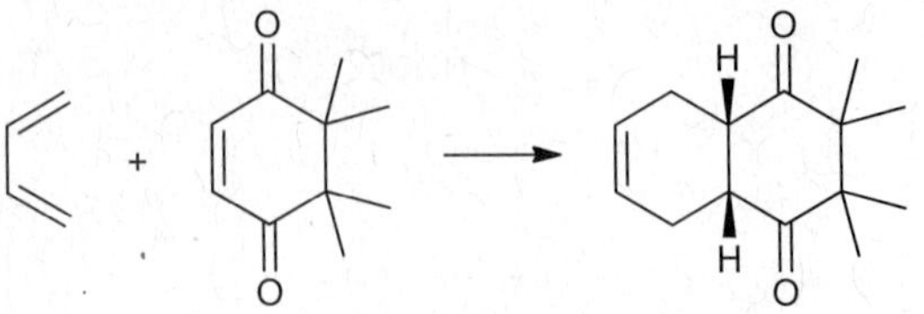

(f) The product exhibits a *cis* configuration, so the dienophile must have the *cis* configuration as well (although in this case, a *trans* configuration is not possible in a six-membered ring).

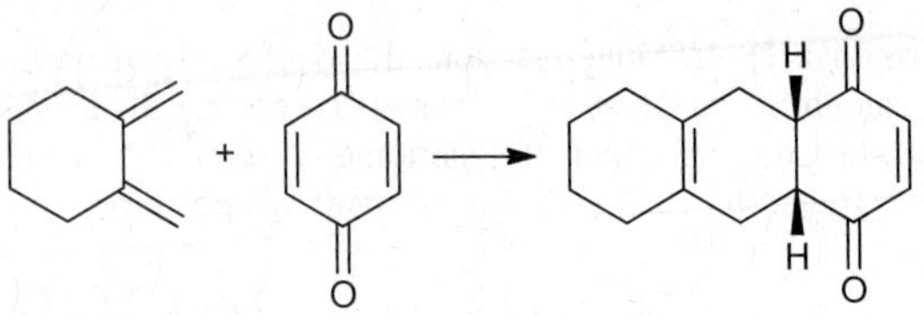

(g) The product exhibits a *trans* configuration, so the dienophile must have the *trans* configuration as well

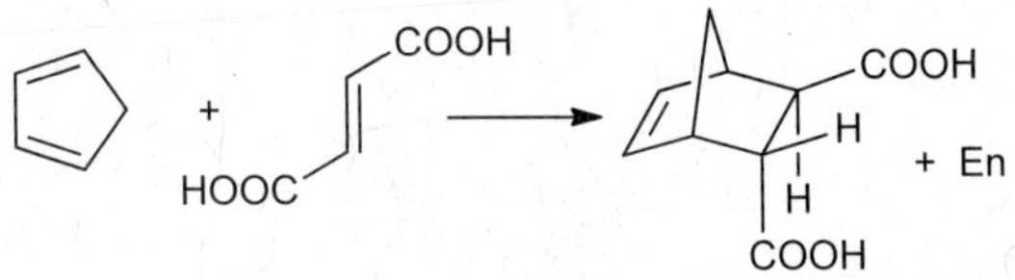

If you have trouble seeing why a pair of enantiomers is produced, see the solution to Problem **17.19b**, which is very similar to this problem (just replace the cyano groups with carboxylic acid groups)

(h) The product has a cyclohexadiene ring, so the starting dienophile must be an alkyne:

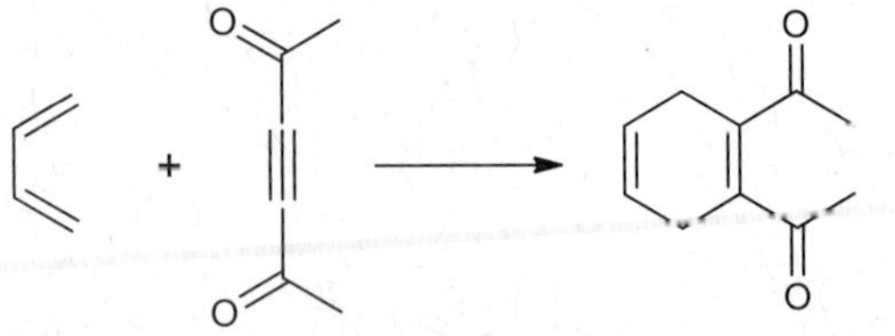

17.45. Treating 1,3-butadiene with HBr at elevated temperature gives the 1,4-adduct, which can be treated with sodium hydroxide to give an S_N2 process in which Br is replaced with a hydroxyl group. Oxidation with PCC converts the alcohol into an aldehyde, which can then be treated with another equivalent of 1,3-butadiene to give a Diels-Alder reaction that affords the product.

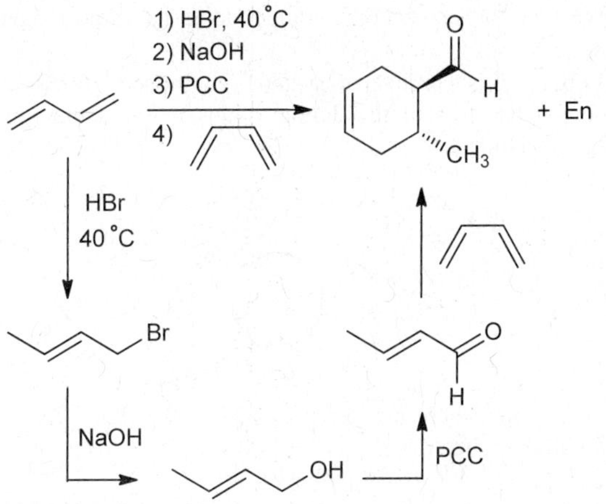

17.46. A product with molecular formula $C_{14}H_{12}O_6$ can be formed via two successive Diels-Alder reactions, as shown here:

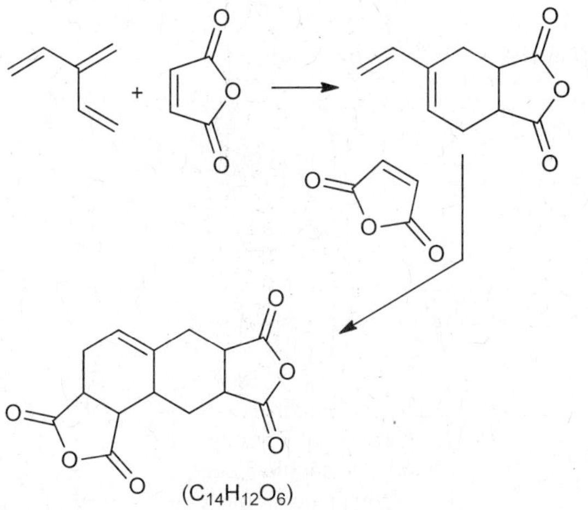

$(C_{14}H_{12}O_6)$

17.47. The following diene and dienophile would be necessary in order to produce the desired compound via a Diels-Alder reaction:

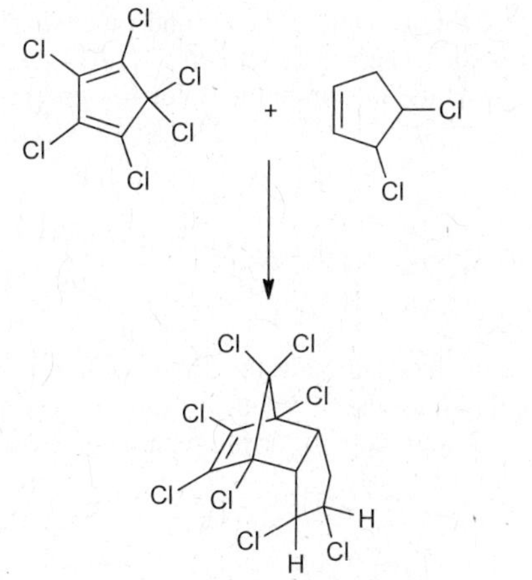

Chlordane

If you have trouble seeing how the bicyclic framework is produced by this reaction, consider the following drawing in which the bond angles have been distorted to show how the bicyclic structure is formed:

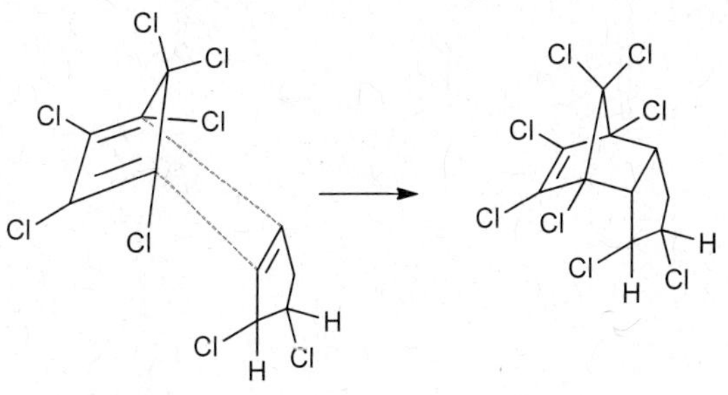

In this drawing, the dotted lines indicate the new σ bonds that are formed during the Diels-Alder reaction.

17.48. The two ends of the conjugated system are much farther apart in a seven-membered ring than they are in a five-membered ring.

17.49. The molecular formula (C_7H_{10}) indicates three degrees of unsaturation (see Section 15.16). The problem statement indicates that compound **A** will react with two equivalents of molecular hydrogen (H_2). Therefore, we can conclude that compound **A** has two π bonds, which accounts for two of the three degrees of unsaturation. The remaining degree of unsaturation must be a ring. Ozonolysis yields two products, which together account for all seven carbon atoms:

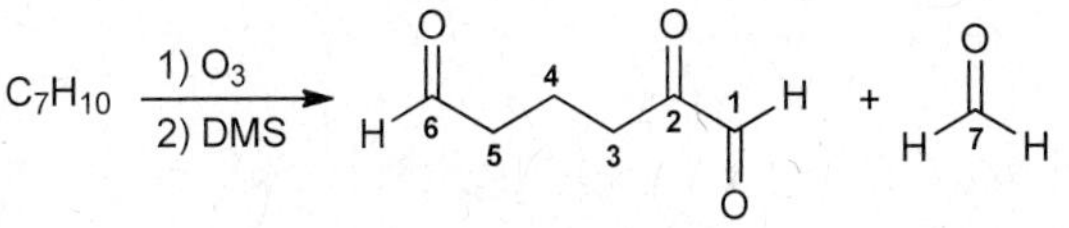

Focus on the product with three carbonyl groups. Two of them must have been connected to each other in compound **A** (as a C=C bond), and the third carbonyl group must have been connected to the carbon atom of formaldehyde (CH_2O). This gives two possibilities: Either C1 was connected to C6 (and C2 was connected to C7):

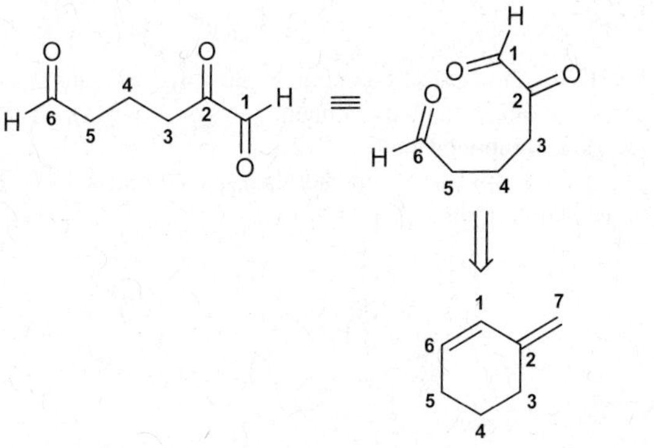

or C2 was connected to C6, and C1 was connected to C7:

In summary, we have found two possible structures for compound **A**, both of which are conjugated dienes:

17.50. One of the compounds has three π bonds in conjugation. That compound has the most extended conjugated system, so that compound is expected to have the longest λ_{max}. Of the remaining two compounds, one of them exhibits conjugation (two π bonds separated by exactly one σ bond), so it will have the next longest λ_{max}. The compound with two isolated C=C bonds will have the shortest λ_{max}:

Increasing
λ_{max}

17.51. Each of these compounds has three π bonds that comprise one extended conjugated system. However, in the first compound (shown below), two of the π bonds are in the same ring (homoannular), which adds +39 nm to the estimate for λ_{max}.

We therefore expect this compound to have the longer λ_{max}.

17.52. This conjugated system (comprised of four π bonds) has two of the double bonds in the same ring (homoannular):

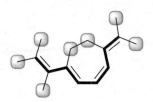

In addition, there are seven auxochromic alkyl groups, highlighted here:

and one exocyclic double bond:

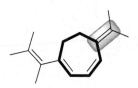

The calculation for λ_{max} is shown here:

Base	=	217
Additional double bonds	=	+60
Auxochromic alkyl groups	=	+35
Exocyclic double bonds	=	+5
Homoannular diene	=	+39
Total	=	356 nm

17.53. Notice that the carbon skeleton does not change during these reactions. It is the location of the deuteron, as well as the location of the π bonds, that changes. This is indeed characteristic of [1,5] sigmatropic rearrangements, as seen in the following general reaction mechanism:

Notice that the position of the proton changes, as well as the position of the π bonds. And we can certainly envision this process occurring with a deuteron, in place of the proton:

Much like the previous example, the position of the deuteron changes, as well as the position of the π bonds. This is exactly the type of transformation taking place in the reactions shown in the problem statement. It is therefore reasonable to explain each of these transformations with a [1,5] sigmatropic rearrangement, as shown:

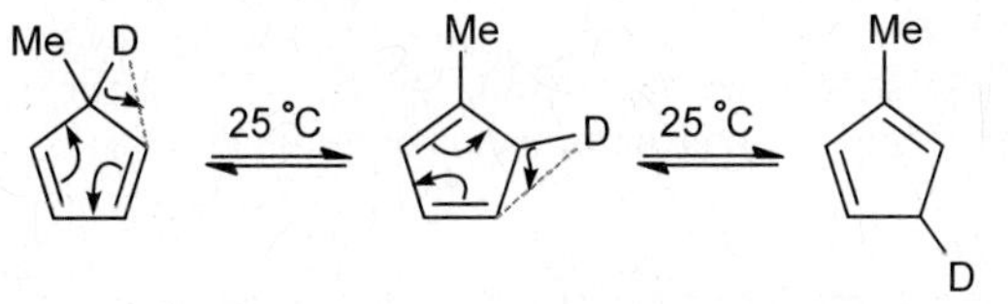

17.54. Notice that the carbon skeleton does not change during these reactions. It is the location of a deuteron, as well as the location of the π bonds, that changes. As described in the solution to Problem **17.53**, these changes are characteristic of a [1,5] sigmatropic rearrangement:

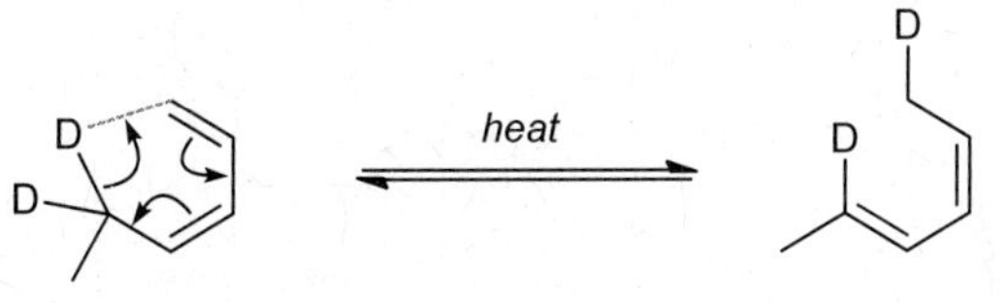

17.55. The configuration of one of the double bonds must be changed (from Z to E), although we have not seen a direct way to do this. However, we can achieve this transformation via a two-step process, in which a temporary ring is closed in a disrotatory fashion, followed by ring-opening in a conrotatory fashion, as shown here:

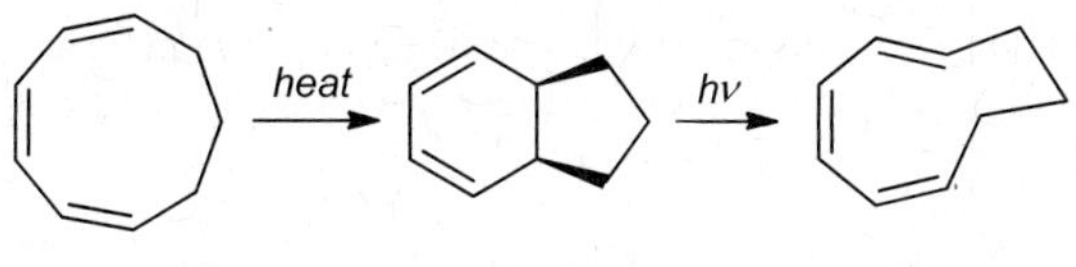

17.56.
(a) This system has six π electrons, so an electrocyclic reaction is expected to occur under thermal conditions to give disrotatory ring closure. The resulting product is a *meso* compound because it is superimposable on its mirror image:

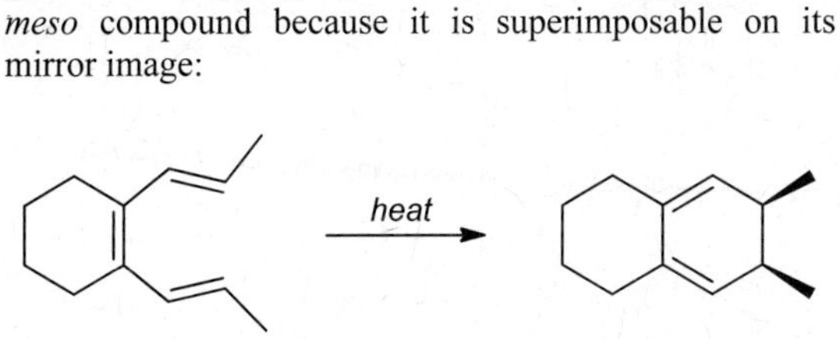

(b) This system has six π electrons, so an electrocyclic reaction is expected to occur under photochemical conditions to give conrotatory ring closure. The resulting product is chiral, and both enantiomers are expected (the methyl groups can both rotate in a

clockwise fashion, or they can both rotate in a counterclockwise fashion, giving both possible enantiomers).

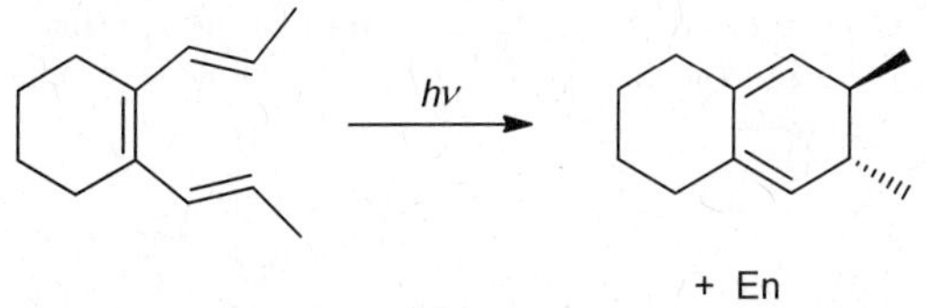

(c) This system has eight π electrons, so an electrocyclic reaction is expected to occur under thermal conditions to give conrotatory ring closure. The resulting product is chiral, and both enantiomers are expected (the methyl groups can both rotate in a clockwise fashion, or they can both rotate in a counterclockwise fashion, giving both possible enantiomers).

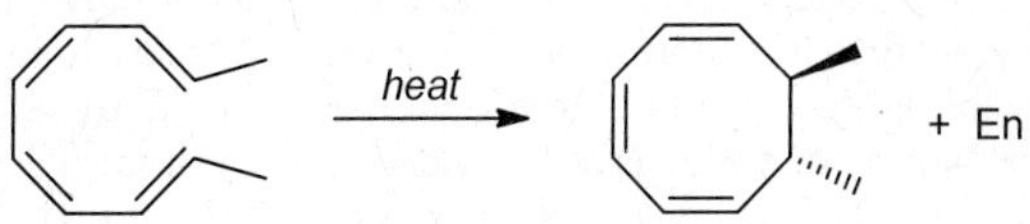

(d) This system has eight π electrons, so an electrocyclic reaction is expected to occur under photochemical conditions to give disrotatory ring closure. The resulting product is a *meso* compound because it is superimposable on its mirror image:

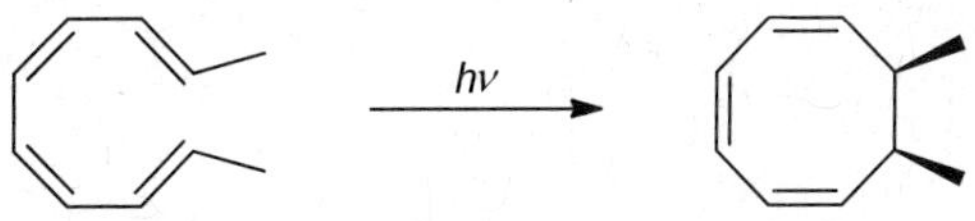

17.57. The compound on the right has a π bond in conjugation with the aromatic ring, while the compound on the left does not. Therefore, the compound on the right side of the equilibrium is expected to be more stable, and the equilibrium will favor this compound because it is lower in energy.

17.58. In this reaction, the four-membered ring is opening (this is the reverse of an electrocyclic ring closure). If we look at the product, we see that four π electrons are involved in the process. Under thermal conditions, this electrocyclic process is expected be conrotatory. In theory, two products can be produced from conrotatory ring-opening (the methyl groups can both rotate in a clockwise fashion, or they can both rotate in a counterclockwise fashion):

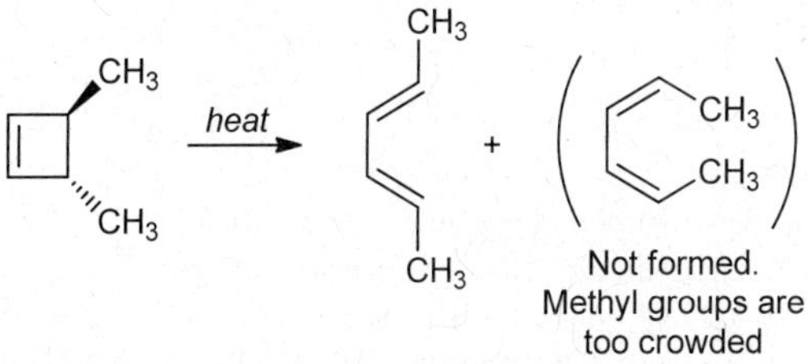

However, the second possible product exhibits a severe steric interaction (the protons of the methyl groups are forced to occupy the same region of space), and this product is therefore not likely formed in substantial quantities. We therefore predict the following product for this reaction.

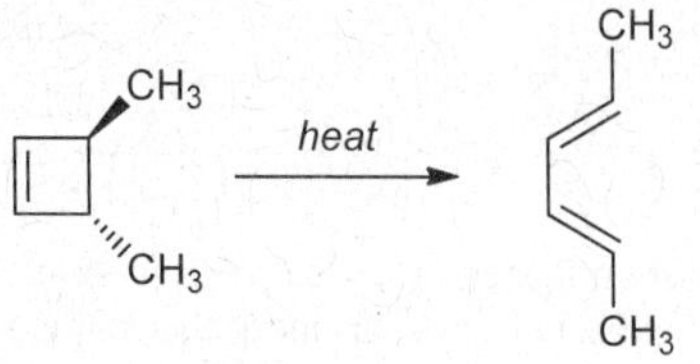

17.59.

(a) This compound is an allylic vinylic ether, and it can therefore undergo a Claisen rearrangement to give a ketone, which tautomerizes to give the enol, thereby re-establishing aromaticity. Note that it is helpful to redraw the starting material, as shown below, so that it is easier to see the ring of electrons responsible for the transformation:

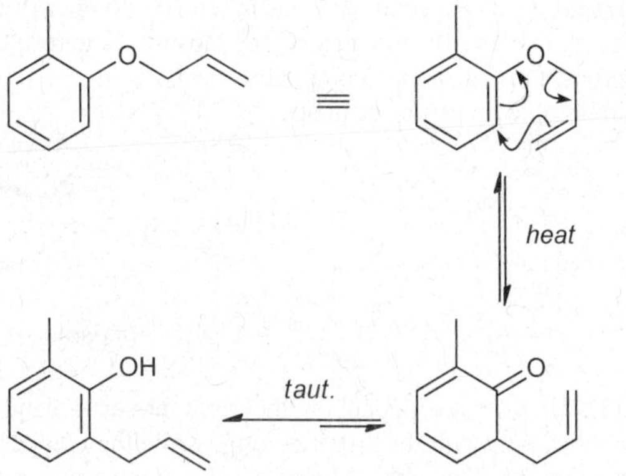

(b) This compound is an allylic vinylic ether, and it can therefore undergo a Claisen rearrangement to give an aldehyde, as shown. Note that it is helpful to redraw the starting material, as shown below, so that it is easier to see the ring of electrons responsible for the transformation:

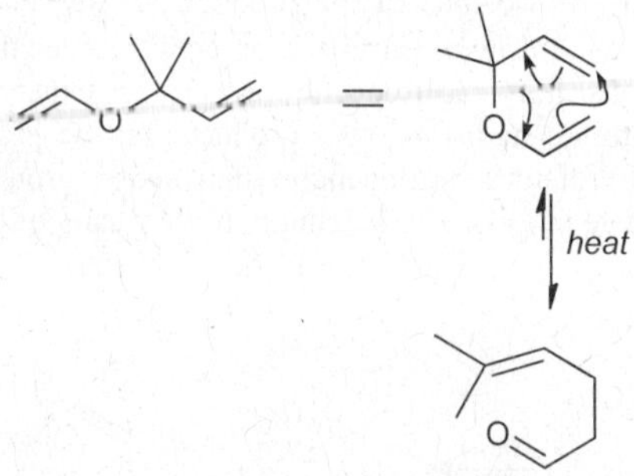

(c) This compound has two π bonds that are separated from each other by exactly three σ bonds. As such, this compound can undergo a Cope rearrangement. To draw the product of this reaction, it is helpful to redraw the

starting material, as shown below, so that it is easier to see the ring of electrons responsible for the transformation:

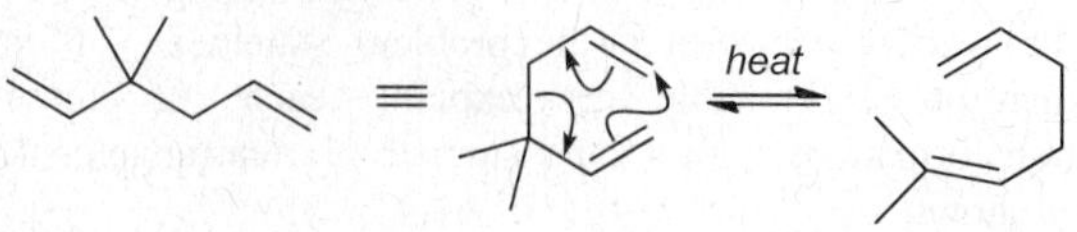

The product has a trisubstituted π bond, and is therefore more stable than the starting material. As such, the equilibrium will favor formation of the product.

17.60.

(a) α-Terpinene reacts with two equivalents of molecular hydrogen, so it must have two π bonds. These π bonds must be associated with two C=C double bonds (rather than being associated with a C≡C triple bond), because the carbon skeleton (which does not change during hydrogenation) cannot support a triple bond:

1-isopropyl-
4-methylcyclohexane

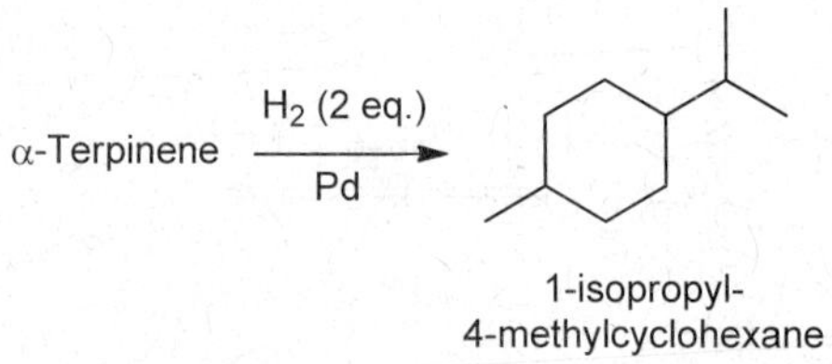

A triple bond could not have been in the ring, because a six-membered ring cannot support the linear geometry required by the *sp* hybridized carbon atoms of a triple bond. The other C-C bonds (outside the ring) can also not support a triple bond (without violating the octet rule by giving a carbon atom with five bonds). Therefore, α-terpinene must have two double bonds.

(b) The ozonolysis products indicate how the molecule must have been constructed, because ozonolysis breaks C=C bonds into C=O bonds:

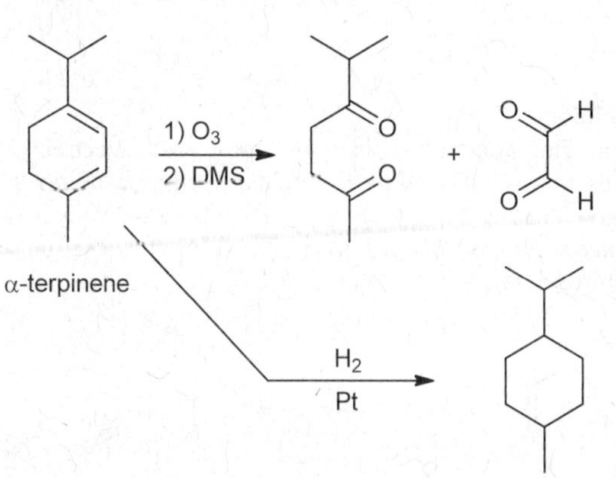

α-terpinene

(c) This conjugated system is a homoannular diene:

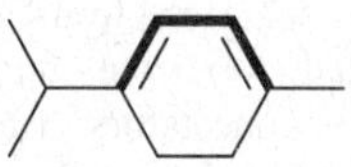

In addition, there are four auxochromic alkyl groups, highlighted here:

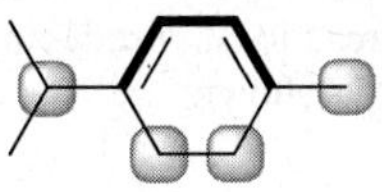

The calculation for λ_{max} is shown here:

Base	=	217
Additional double bonds	=	0
Auxochromic alkyl groups	=	+20
Exocyclic double bonds	=	0
Homoannular diene	=	+39
Total	=	276 nm

17.61. The molecular formula (C_6H_{10}) indicates two degrees of unsaturation (see Section 15.16). The problem statement indicates that all proposed structures must be conjugated dienes, which accounts for both degrees of unsaturation. That is, the proposed structures cannot have any rings. All structures must be acyclic.

To draw all possible conjugated dienes with molecular formula C_6H_{10}, we must first consider all of the different ways in which six carbon atoms can be connected to each other.

We begin with a linear chain (parent = hexane):

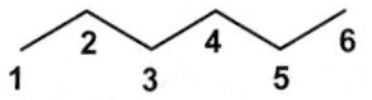

Next, we look for any skeletons where the parent is pentane (five carbon atoms). There are only two such skeletons. Specifically, we can either connect the extra CH_3 group to positions C2 or C3 of the pentane chain:

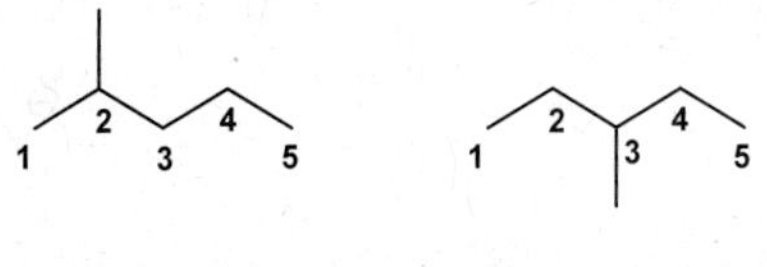

We cannot connect the CH_3 group to positions C1 or C5, as that would simply give us the linear chain (hexane), which we already drew (above). We also cannot connect the CH_3 group to position C4 as that would generate the same structure as placing the CH_3 group at the C2 position:

Next, we look for any skeletons where the parent is butane (four carbon atoms). There are only two such skeletons. Specifically, we can either connect two CH_3 groups to adjacent positions (C2 and C3) or the same position:

If we try to connect a CH_3CH_2 group to a butane chain, we end up with a pentane chain (which was has already been drawn earlier):

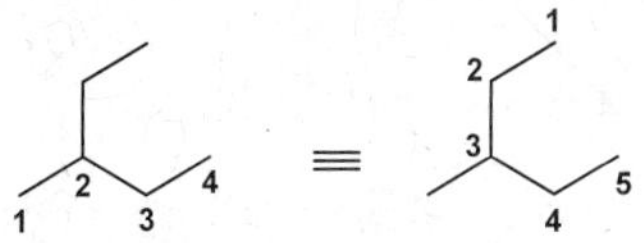

In summary, there are five different ways in which six carbon atoms can be connected:

Parent = hexane

Parent = butane

Parent = pentane

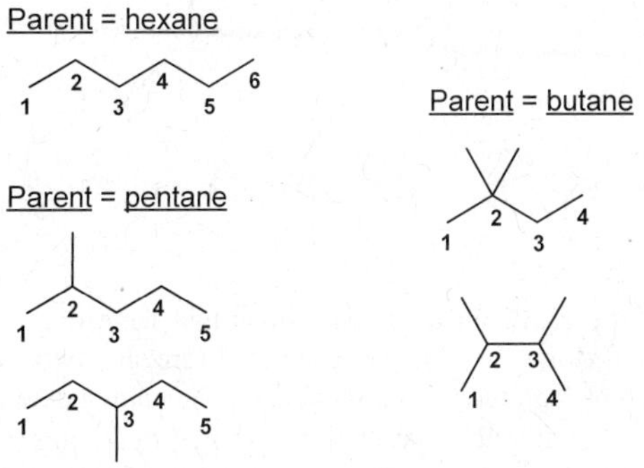

For each one of these skeletons, we must consider all of the different unique positions where the double bonds can be placed (keeping in mind that they must remain conjugated). For the first skeleton (hexane), there are many different locations where the double bonds can be placed. For example, the double bonds can be at C1 and C3 (giving two stereoisomeric options, because the double bond at C3 can have either the E configuration or the Z configuration):

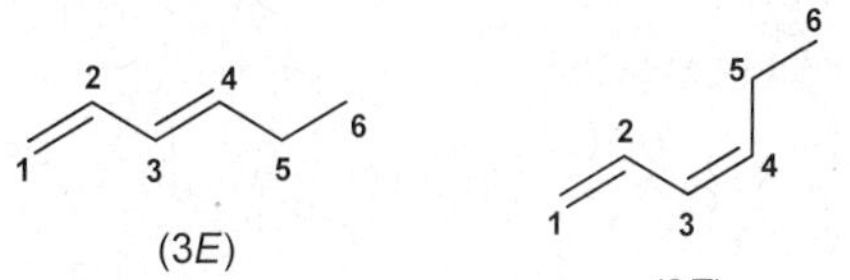

(3E)

(3Z)

or the double bonds can be placed at C2 and C4 of the hexane skeleton, in which case each of the double bonds can either have the E or Z configuration, giving three more possible structures:

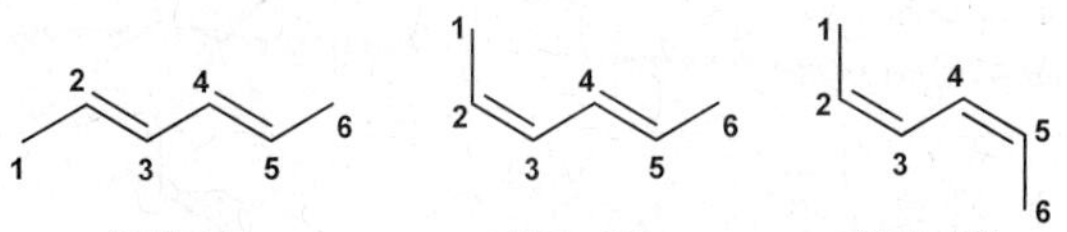

(2E, 4E)   (2Z, 4E)   (2Z, 4Z)

Next we move on to the skeletons that have only five carbon atoms in a linear chain, and for each of these skeletons, we consider all possible locations where the double bonds can be placed (including stereoisomers):

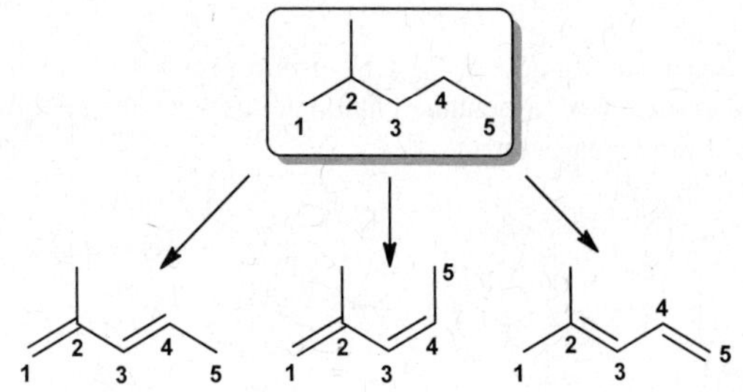

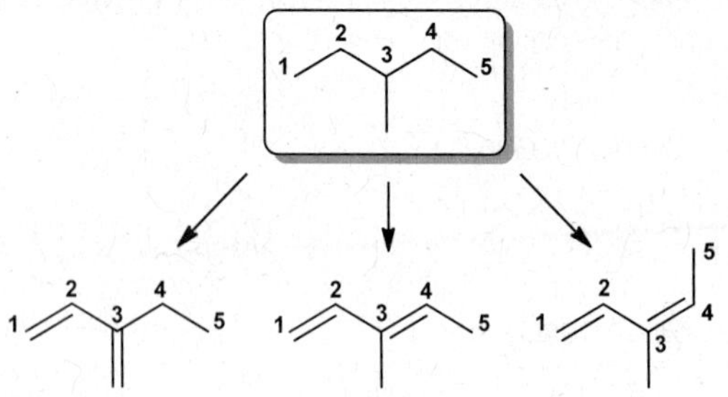

Next we move on to the skeletons that have only four carbon atoms in a linear chain. There are two such skeletons, and the first of them cannot accommodate two π bonds (without violating the octet rule by giving more the four bonds to a carbon atom). The other skeleton with only four carbon atoms in a linear chain CAN accommodate conjugated C=C bonds, but there is only such way, shown here:

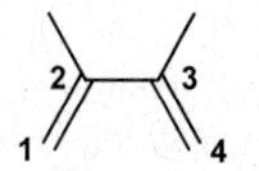

In summary, we have revealed twelve different conjugated dienes with molecular formula C_6H_{10}, shown here:

Parent = hexadiene

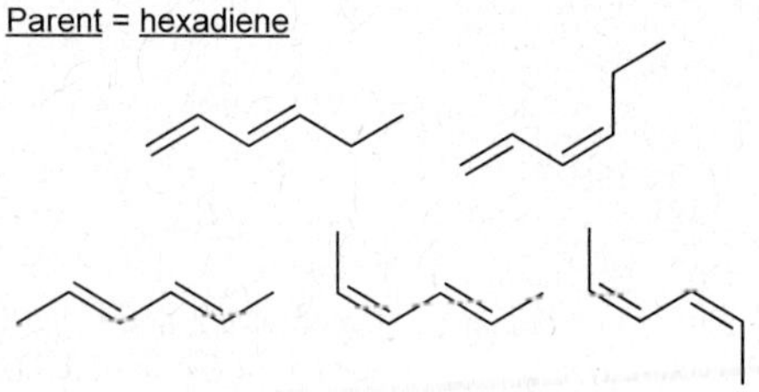

Parent = pentadiene

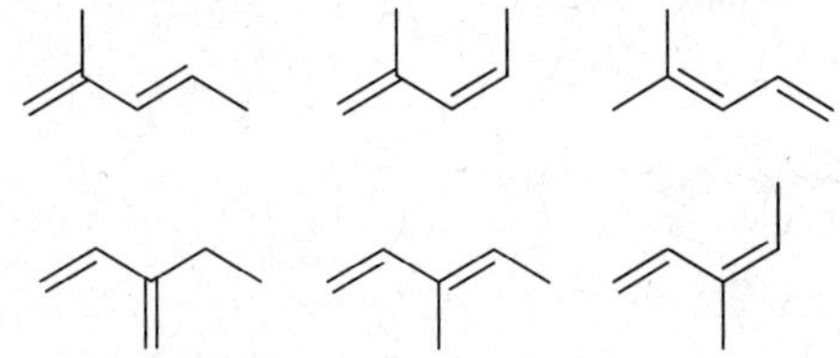

Parent = butadiene

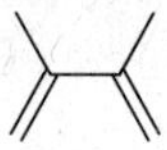

17.62. Potassium tert-butoxide is a strong, sterically hindered base, and the starting material will react with two equivalents of this base to undergo two successive elimination (E2) reactions, producing a conjugated, homoannular diene, as shown.

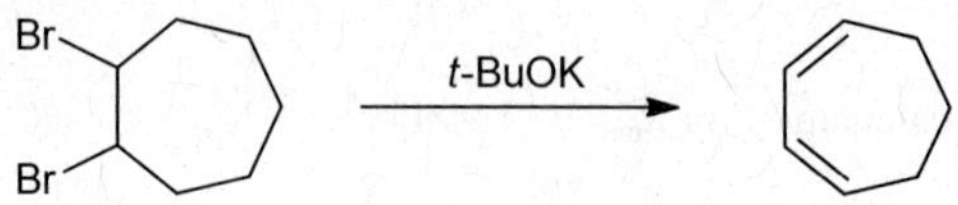

17.63.
(a) The non-conjugated isomer (shown below) will be higher in energy than the conjugated isomer.

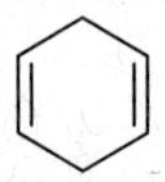

As a result, this compound will liberate more heat upon hydrogenation.

(b) The non-conjugated isomer (shown below) will be higher in energy than the conjugated isomer.

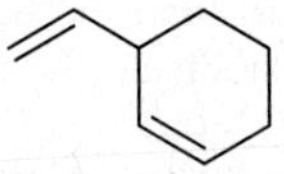

As a result, this compound will liberate more heat upon hydrogenation.

17.64. Nitroethylene should be more reactive than ethylene in a Diels–Alder reaction, because the nitro group is electron-withdrawing, via resonance:

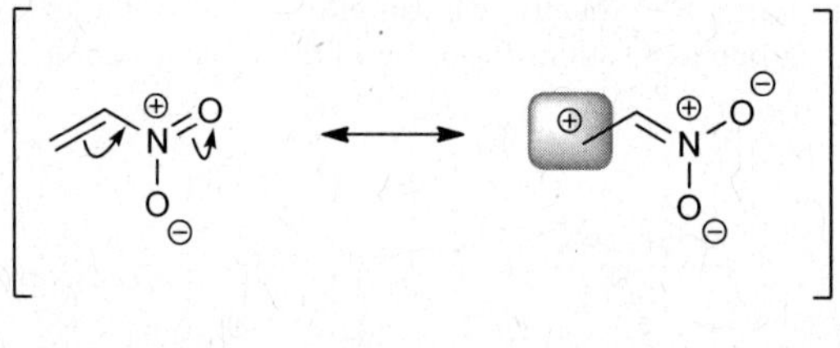

17.65. The starting material is an allylic vinylic ether, so it can undergo a Claisen rearrangement.

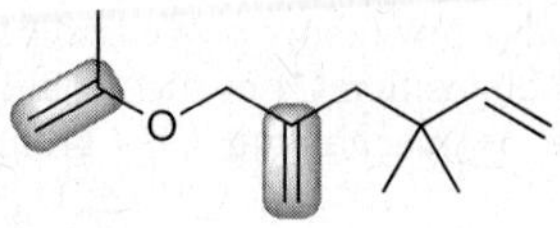

Alternatively, we also note that the starting material has two C=C bonds that are separated from each other by exactly three σ bonds, so it can undergo a Cope rearrangement:

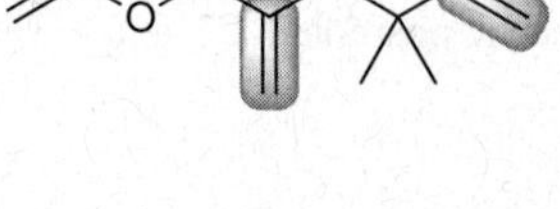

We will ultimately end up drawing both processes, and it does not matter the order in which we draw these two processes. Below, the Claisen rearrangement is drawn first, followed by the Cope rearrangement. If instead, the Cope rearrangement was drawn first, followed by the Claisen rearrangement, the same product would be obtained.

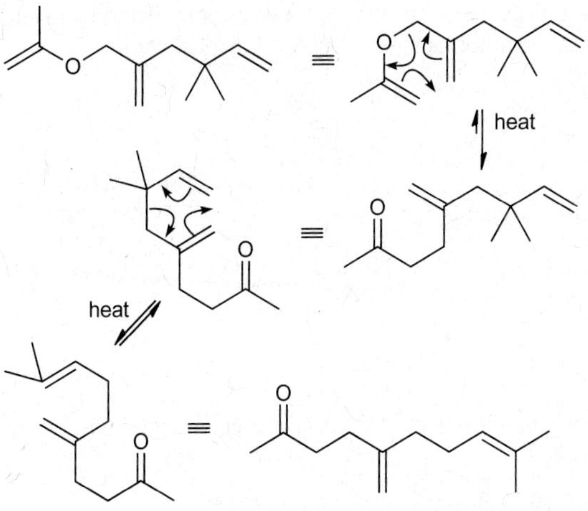

Note that for each of the sigmatropic processes above (the Claisen rearrangement and the Cope rearrangement), the compound is redrawn in such a way that enables us to clearly see the motion of the electrons that cause the reaction. You are likely to make a mistake if you try to draw the curved arrows without first redrawing the structure. That is, avoid doing this:

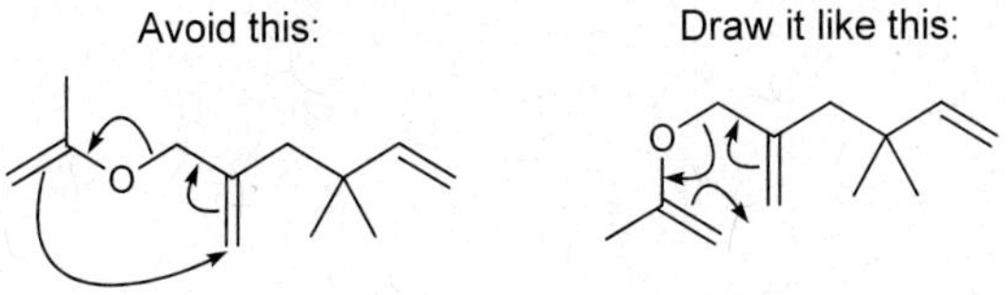

Avoid this: Draw it like this:

17.66. The diene is electron-rich in one specific location, as seen in the second resonance structure below:

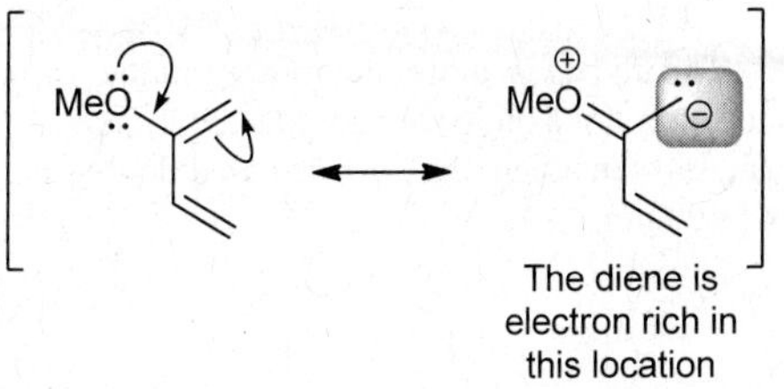

The diene is electron rich in this location

The dienophile is electron-poor in one specific location, as seen in the third resonance structure below:

The dienophile is electron poor in this location

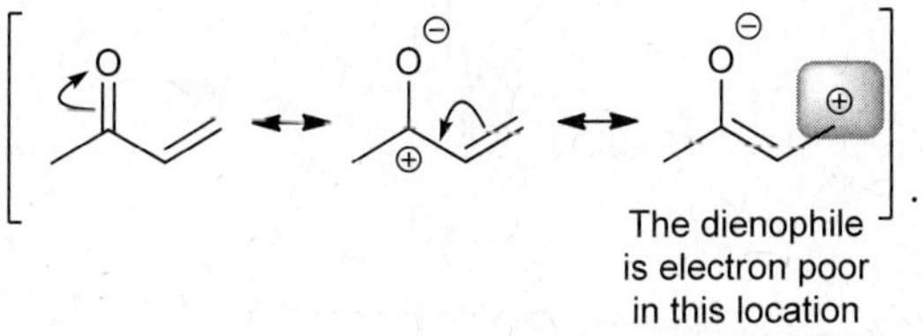

These two compounds will join in such a way that the electron-poor center lines up with the electron-rich center:

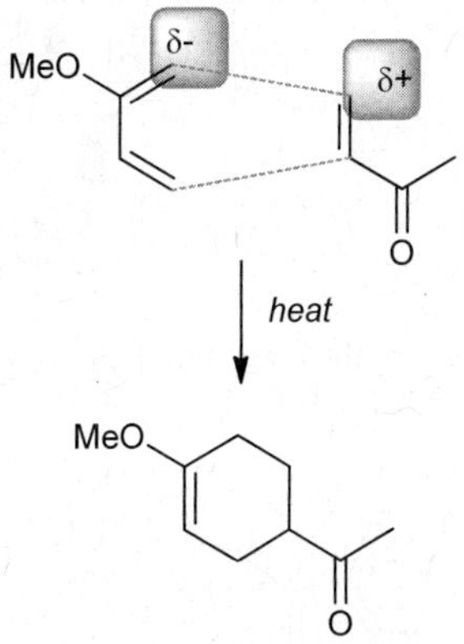

17.67. This transformation can be achieved via a retro Diels-Alder reaction (shown below), which requires elevated temperature, as described in Section 17.7.

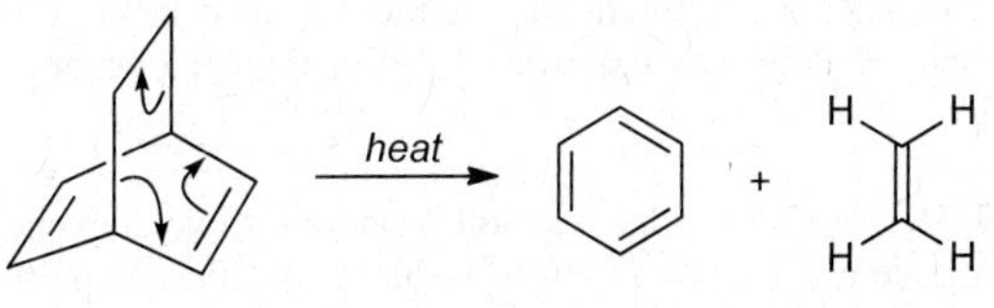

17.68. A Diels-Alder reaction, followed by a retro Diels-Alder reaction, can account for formation of the aromatic product, as shown here:

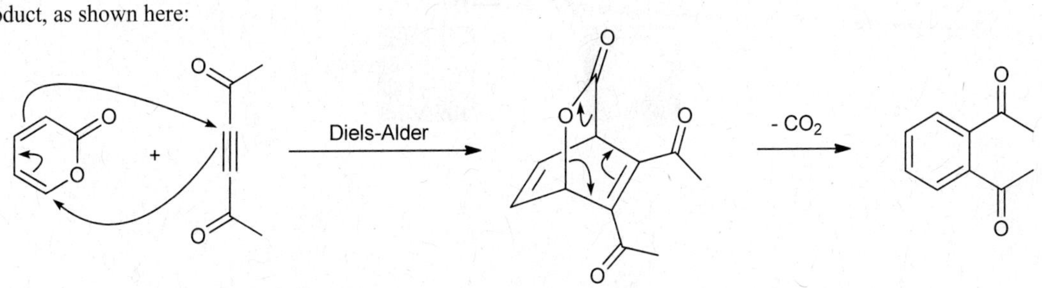

17.69. The nitrogen atom in divinyl amine is *sp²* hybridized. The lone pair is delocalized, and joins the two neighboring π bonds into one conjugated system. As such, the compound absorbs light above 200 nm (UV light). In contrast, 1,4-pentadiene has two isolated double bonds and therefore does not absorb UV light in the region between 200 and 400 nm.

17.70. The divinylcyclopropane unit is converted to a cycloheptadiene in this 3,3-sigmatropic rearrangement. Interestingly, this rearrangement also opens one ring (cyclopropane) and forms a new ring (cycloheptadiene).

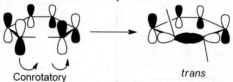

17.71. Compound **1** undergoes an electrocyclic reaction involving 6π-electrons. The reaction occurs with light, rather than heat, so we expect ring-closure to occur in a conrotatory fashion:

We therefore expect the two methyl groups to be *trans* to each other in the product:

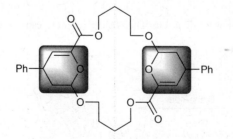

Compound **2** has rotational symmetry, but it lacks reflectional symmetry (see Section 5.6). As such, it is chiral, in much the same way that *trans*-1,2-dimethylcyclohexane is chiral.

17.72. A Diels-Alder reaction produces a cyclohexene adduct. It follows that a hetero-Diels-Alder reaction should produce a cyclohexene-type product in which one or more of the carbon atoms of the cyclohexene ring are replaced with nitrogen or oxygen. Analysis of the macrocyclic product reveals two such moieties, highlighted below.

A retrosynthetic analysis, shown below, demonstrates that this product can be produced from two equivalents of the acyclic reactant, arranged head-to-tail, where the four bonds with the wavy lines are produced in two hetero-Diels-Alder reactions.

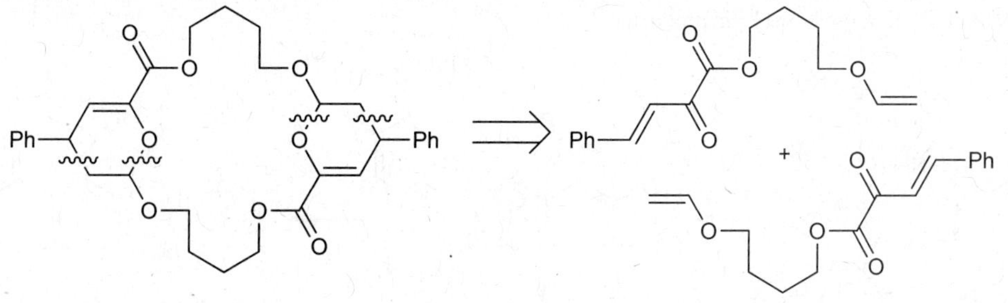

A mechanism consistent with this reaction is presented below. In the first step, the terminal alkene group of the bottom molecule serves as the "dienophile". On the top molecule, the alkene group next to the phenyl group and the adjacent, conjugated ketone serve the role of the "diene" (although in this case it is not formally a diene since one of the carbon atoms has been replaced by oxygen). This hetero-Diels-Alder reaction produces the first six-membered ring as shown. A second hetero-Diels-Alder reaction then proceeds at the other terminus of the molecule, thus producing the macrocyclic product.

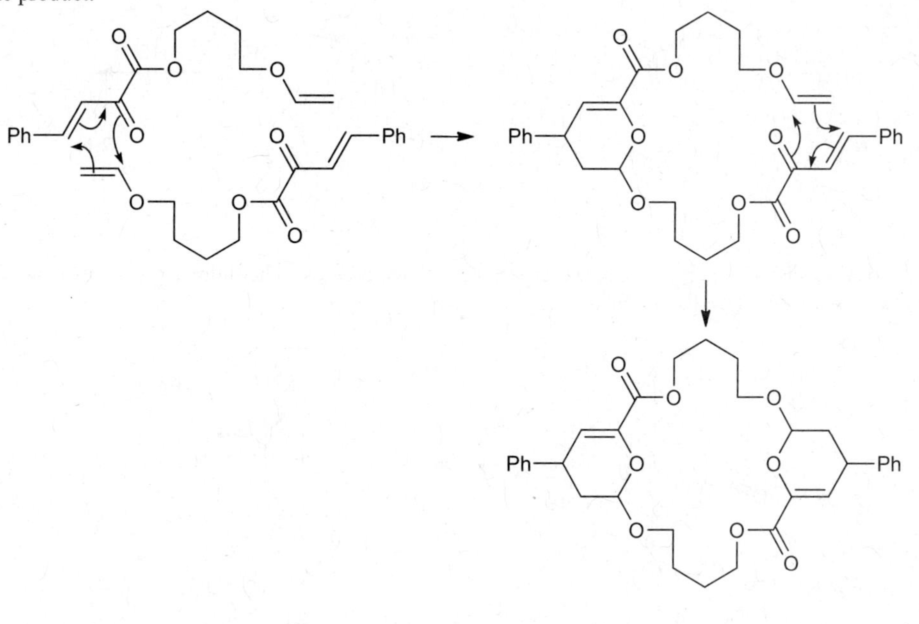

17.73. Oligofurans are highly conjugated materials. When treated with maleimide, only one product results – the [4+2] addition to the terminal furan. Notice that this product is still conjugated – there are two furans that share resonance stabilization.

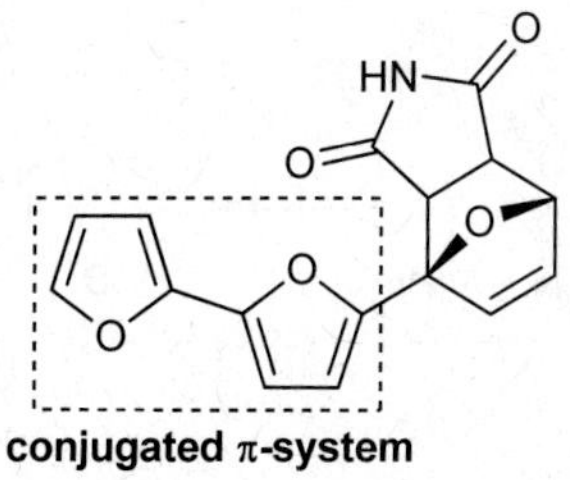

conjugated π-system

If the internal furan were to react, the conjugation between furans would be broken, leaving behind two isolated furans. This molecule would be much higher in energy because of the loss of conjugation. Since the product distribution of a Diels-Alder reaction is determined by thermodynamic considerations, the higher energy product is not obtained.

17.74. Notice that compound **1** contains a strained, four-membered ring. When this compound is heated to 120 °C, it will undergo a thermal electrocyclic reaction to form compound **2**, which possesses significantly less ring strain. The newly generated diene can then undergo a thermal, intramolecular Diels-Alder reaction with the alkyne, re-establishing aromaticity, and forming the hexacyclic product.

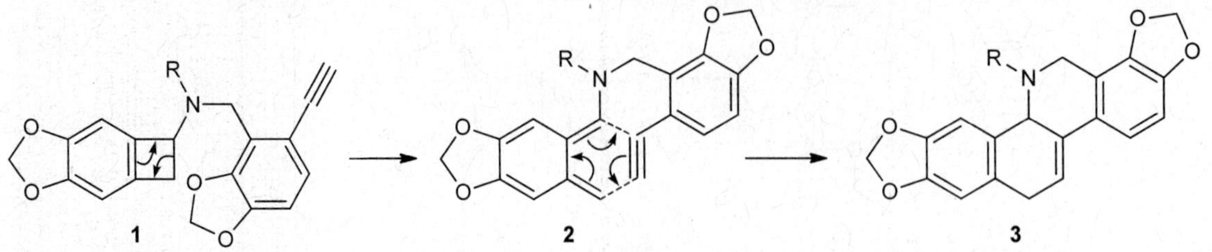

17.75. The first step involves a 6-π electrocyclic reaction that closes the first ring to form a cyclohexadiene. When this intermediate is redrawn, we can clearly see that the pendant alkene is in close enough proximity with the newly generated diene to induce an intramolecular [4+2] Diels-Alder cycloaddition, which will result in the formation of two new rings.

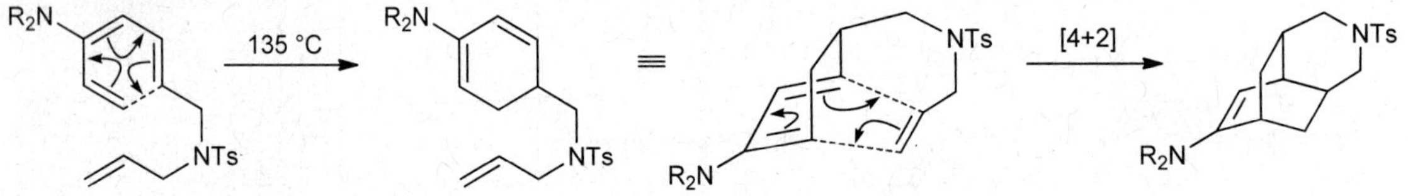

17.76.
(a) The reaction is a cycloaddition process, so we expect a concerted process. The following curved arrows represent a concerted process that would give the product:

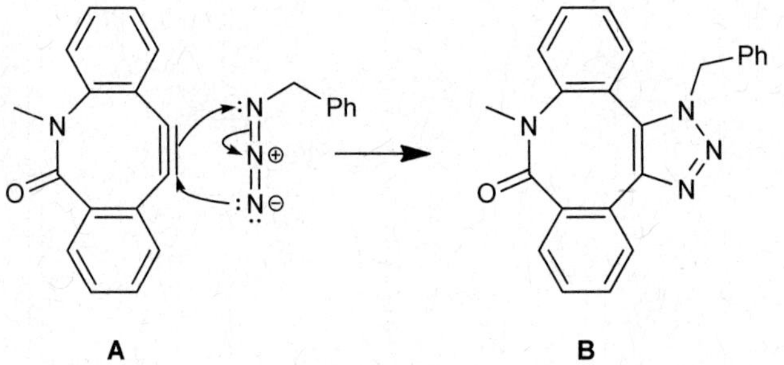

(b) Depending on the relative orientation of the azide and alkyne during the reaction, the following compound can also be formed.

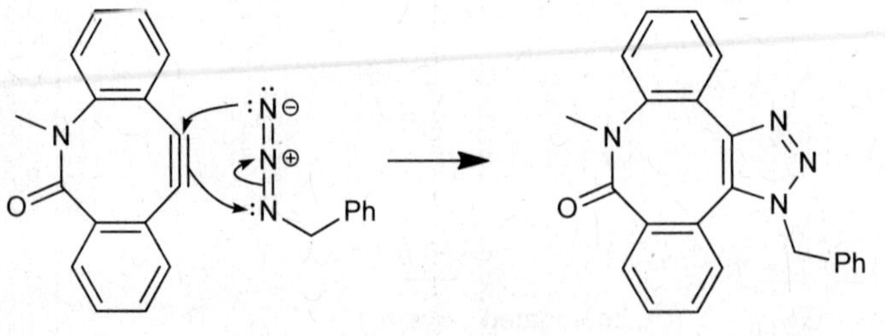

(c) Recall from Chapter 10 that the smallest isolatable cycloalkyne is cyclooctyne, which experiences significant angle strain due to the incorporation of the two adjacent *sp* hybridized carbon atoms (that should have linear geometry) into an 8-membered ring. As such, compound **A** has significant angle strain. We can infer that this strain plays an important role in the click reaction, because alkyne **D** (which is free of this strain) is unreactive under these conditions.   This angle strain increases the energy of the starting alkyne, thus decreasing the activation energy of the reaction (since it is now closer in energy to the transition state). Considering the geometry of the atoms involved in the reaction, the angle strain forces the alkyne to have bond angles closer to the angles required in the transition state leading to the sp^2 hybridized carbon atoms in the product. In other words, there is a higher activation energy associated with distorting an unstrained alkyne (180°) to an alkene (120°), compared to the analogous conversion of a strained alkyne (<180°) to an alkene (120°).

Alkyne **E** is less reactive because the non-alkyne atoms in the ring are all sp^3 hybridized; this eases some of the angle strain due to their relative conformational flexibility. The multiple sp^2 hybridized atoms in the ring of alkyne **A** increase the strain on the alkyne, thus increasing the reaction rate.

Alkyne **F** is less reactive because one of the methyl groups (on the aromatic ring) sterically hinders the approach of the benzyl azide.

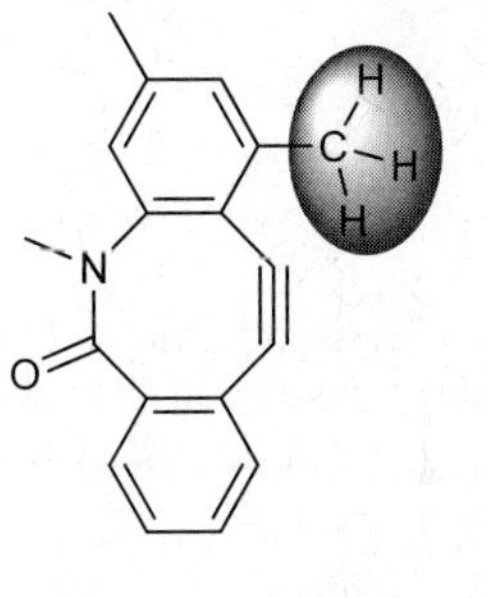

17.77.

(a) The following electrocyclic reaction will produce an eight-membered ring. Since the reaction involves eight π electrons, under thermal conditions, we expect conrotatory ring-closure, producing the *trans*-configuration seen in compound **2**. This ring-closing reaction can occur if either: (1) both ends of the forming ring rotate clockwise, or 2) both ends rotate counter-clockwise. These two possibilities can occur with equal likelihood, resulting in a racemic mixture of compound **2**.

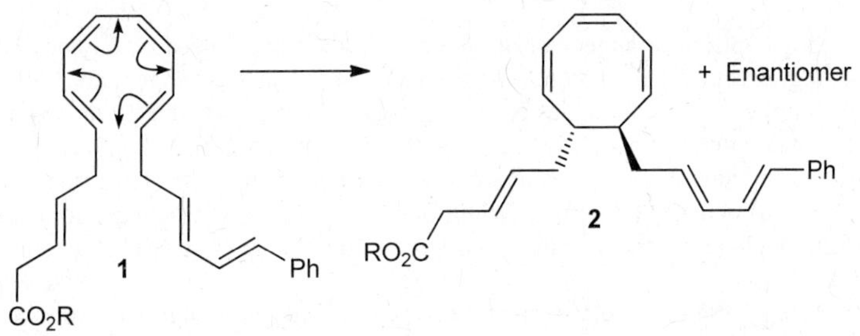

(b) The conversion of compound **2** to endiandric acid G occurs via an electrocyclic reaction involving six π electrons.

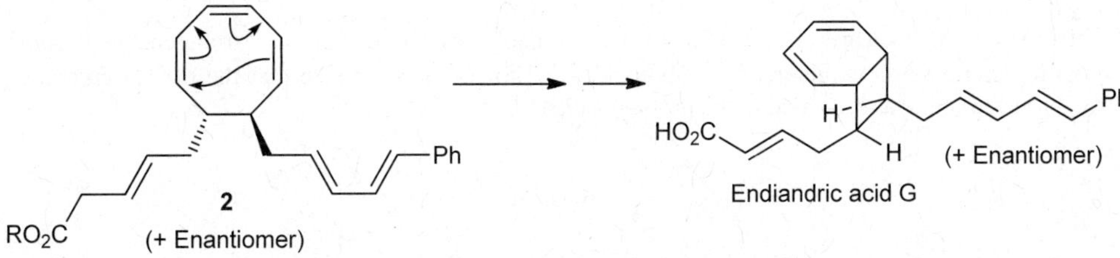

(c) Endiandric acid G contains two diene groups, but only one of them is in close proximity with the dienophile. An intramolecular Diels-Alder reaction, as shown below, gives endiandric acid C:

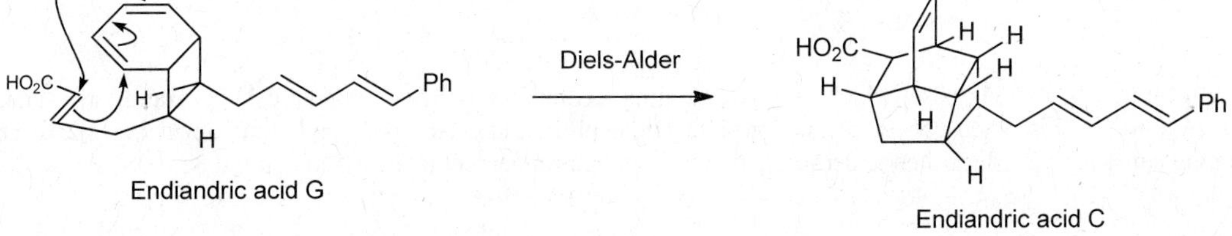

17.78. Both compounds exhibit intramolecular hydrogen bonding interactions, as shown below:

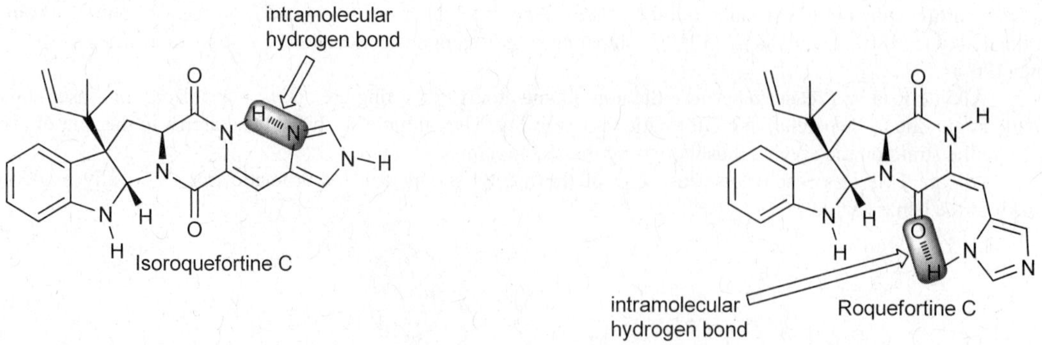

However, the hydrogen bonding interaction in isoroquefortine C involves a "ring" comprised of six atoms, while the hydrogen bonding interaction in roquefortine C involves a "ring" comprised of seven atoms.

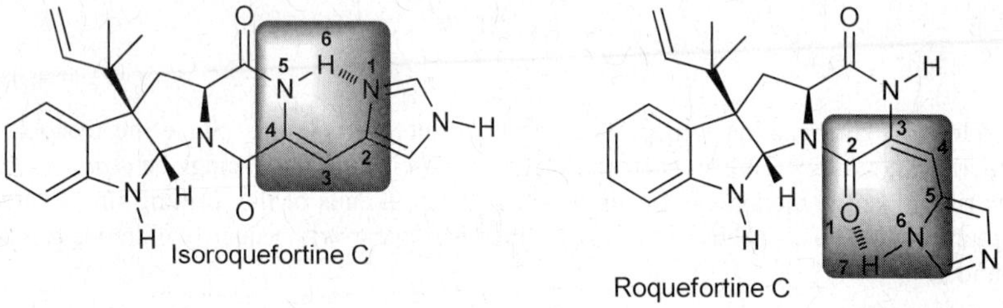

Both are expected to prefer a coplanar arrangement of all atoms in the "ring," because any deviations from coplanarity will diminish the stabilizing effect of conjugation. If confined to a coplanar arrangement, roquefortine C will have a more difficult time adopting the ideal bond angles for sp^2 hybridized centers (120°), because there are seven atoms in the "ring" rather than six. In contrast, isoroquefortine C (with its six atoms in the "ring") will not experience as much strain as it attempts to adopt the ideal bond angles in a coplanar arrangement. Deviation from the ideal bond angles or from coplanarity will cause an increase in energy. While both compounds will deviate (to a certain extent) from the ideal bond angles and from coplanarity, the deviation is expected to be more significant for roquefortine C. Therefore, isoroquefortine C is lower in energy.

This explanation is verified by the experimentally determined bond angles for roquefortine C (largest bond angle is 134°) and isoroquefortine C (largest bond angle is 127°).

17.79.
(a) The problem statement indicates that compound **3** is formed from a thio-Claisen rearrangement of compound **2**. This allows us to determine the structure of compound **2**, as shown below (notice that the carbon-carbon π bond is shown in the *Z* configuration, as indicated in the problem statement):

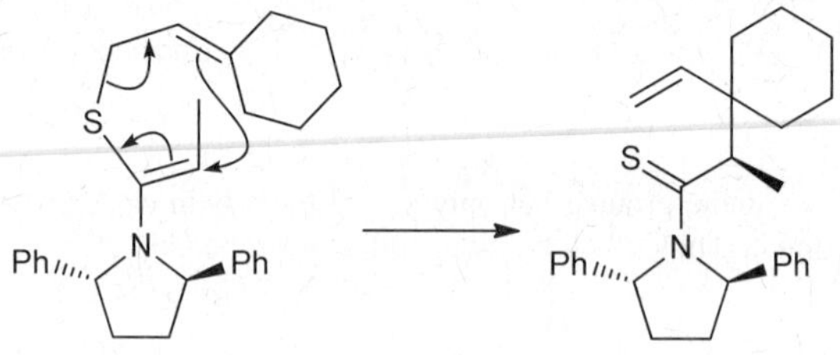

(b) Butyl lithium is a strong base, which removes the most acidic proton from compound **1**, generating a resonance-stabilized anion. This anion can then function as a nucleophile and attack the alkyl bromide in an S_N2 reaction, generating compound **2**, which then undergoes a thio-Claisen rearrangement to afford compound **3**:

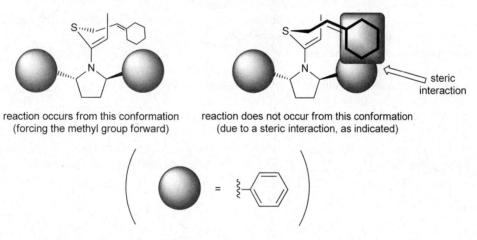

(c) The existing chirality centers have an influence on the stereochemical outcome of this reaction. Specifically, one of the phenyl groups provides steric hindrance that forces the reaction to occur in a stereospecific manner, so that the phenyl group and the cyclohexyl group remain far apart from one other:

reaction occurs from this conformation
(forcing the methyl group forward)

reaction does not occur from this conformation
(due to a steric interaction, as indicated)

steric
interaction

The reaction will proceed via the lowest energy transition state, which is most likely achieved from the first conformation above, rather than from the second conformation, in which there is a large steric interaction (highlighted).

Chapter 18
Aromatic Compounds

Review of Concepts

Fill in the blanks below. To verify that your answers are correct, look in your textbook at the end of Chapter 18. Each of the sentences below appears verbatim in the section entitled *Review of Concepts and Vocabulary*.

- When a benzene ring is a substituent, it is called a _____ **group**.
- Disubstituted derivatives of benzene can be differentiated by the use of the descriptors _____, **meta** and _____, or by the use of locants.
- Benzene is comprised of a ring of six identical C-C bonds, each of which has a bond order of _____.
- The **stabilization energy** of benzene can be measured by comparing _____ of hydrogenation.
- The stability of benzene can be explained with MO theory. The six π electrons all occupy _____ MOs.
- The presence of a fully conjugated ring of π electrons is not the sole requirement for aromaticity. The requirement for an odd number of electron pairs is called _____ **rule**.
- **Frost circles** accurately predict the relative energy levels of the _____ in a conjugated ring system.
- A compound is aromatic if it contains a ring comprised of _____ _____ and if it has a _____ number of π electrons in the ring.
- Compounds that fail the first criterion are called _____.
- Compounds that satisfy the first criterion, but have 4n electrons (rather than $4n+2$) are _____.
- Cyclic compounds containing heteroatoms, such as S, N, O, are called _____.
- Any carbon atom attached directly to a benzene ring is called a _____ **position**.
- Alkyl benzenes are oxidized at the benzylic position by _____ or _____.
- In a **Birch reduction**, the aromatic moiety is reduced to give a nonconjugated diene. The carbon atom connected to _____ is not reduced, while the carbon atom connected to _____ is reduced.

Review of Skills

Fill in the blanks and empty boxes below. To verify that your answers are correct, look in your textbook at the end of Chapter 18. The answers appear in the section entitled *SkillBuilder Review*.

18.1 Naming a Polysubstituted Benzene

PROVIDE A SYSTEMATIC NAME FOR THE FOLLOWING COMPOUND

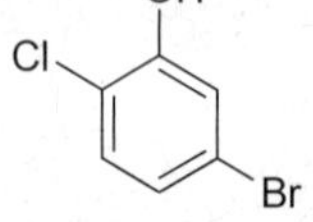

1) IDENTIFY THE PARENT
2) IDENTIFY AND NAME SUBSTITUENTS
3) ASSIGN LOCANTS TO EACH SUBSTITUENT
4) ALPHABETIZE

18.2   Determining Whether a Structure is Aromatic, Nonaromatic, or Antiaromatic

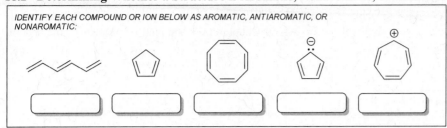

18.3   Determining Whether a Lone Pair Participates in Aromaticity

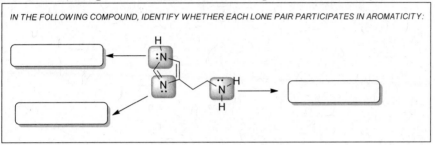

18.4   Manipulating the Side Chain of an Aromatic Compound

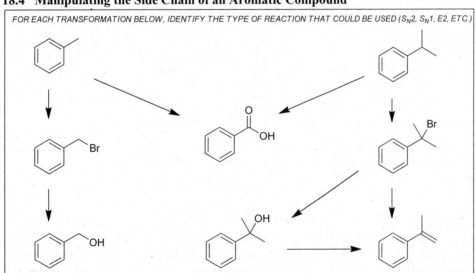

18.5   Predicting the Product of a Birch Reduction

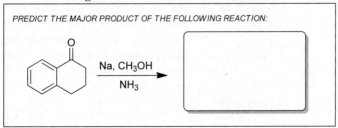

Review of Reactions

Identify the reagents necessary to achieve each of the following transformations. To verify that your answers are correct, look in your textbook at the end of Chapter 18. The answers appear in the section entitled *Review of Reactions*.

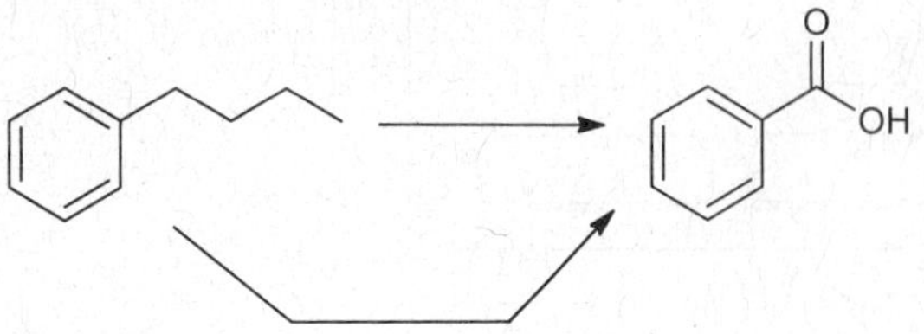

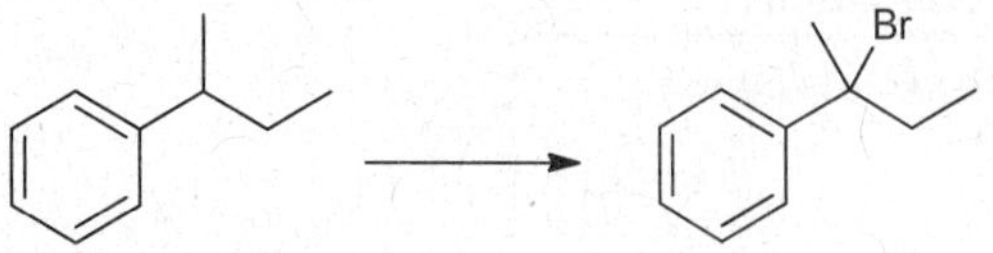

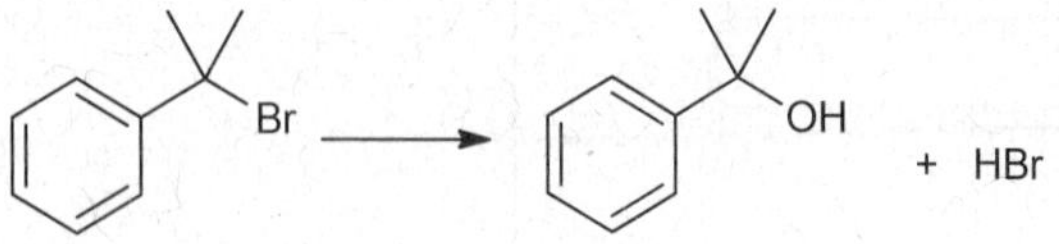

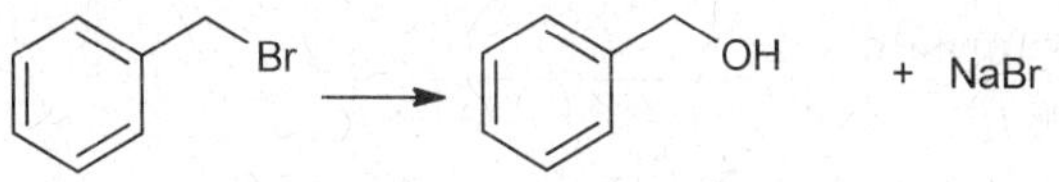

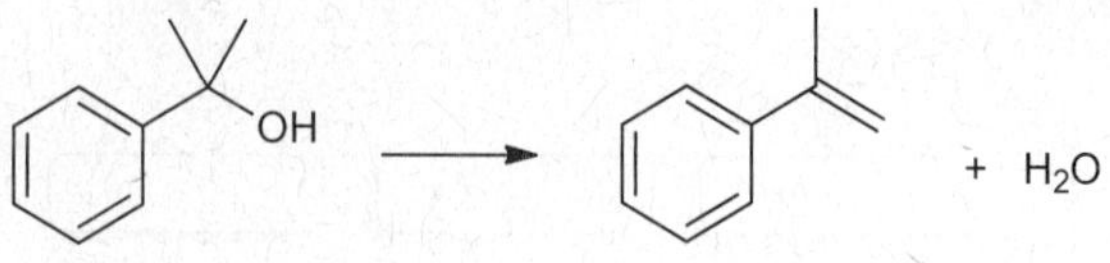

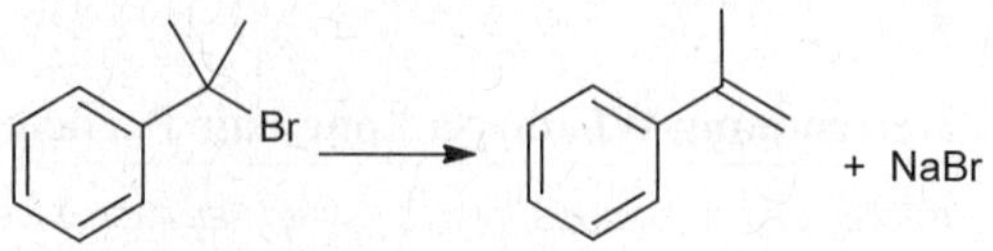

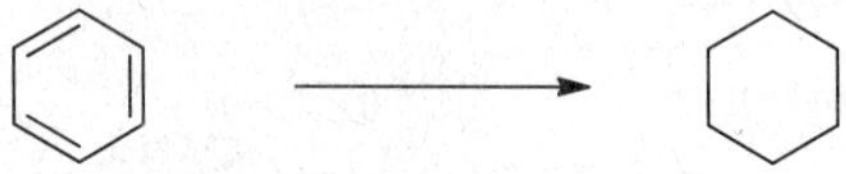

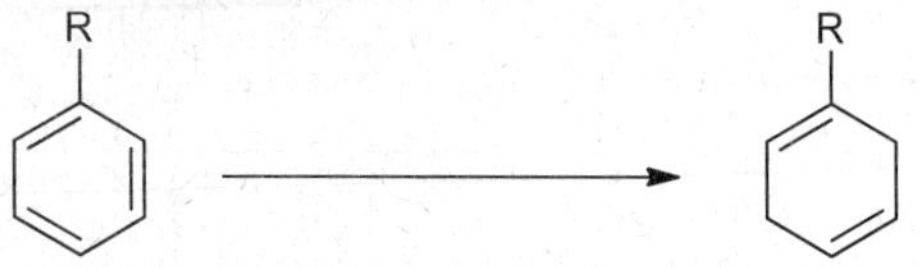

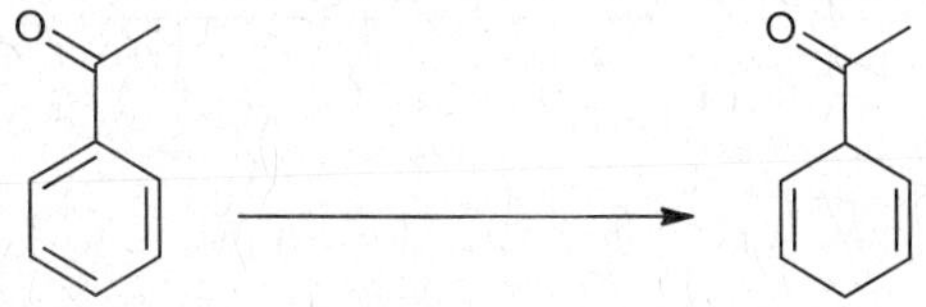

Mistakes to Avoid

When analyzing an NMR spectrum (^{1}H or ^{13}C), we must always consider the role of symmetry in affecting the number of signals in the spectrum. For example, consider the number of signals in the ^{1}H NMR spectrum of each of the following constitutional isomers:

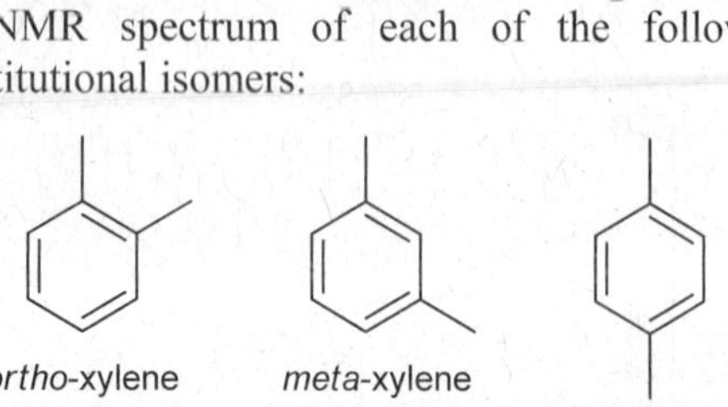

ortho-xylene   meta-xylene

para-xylene

Each of these compounds exhibits a different number of signals in its ^{1}H NMR spectrum (3 signals, 4 signals, and 2 signals, respectively). Similarly, each of these compounds exhibits a different number of signals in its ^{13}C NMR spectrum (4 signals, 5 signals, and 3 signals, respectively).

When analyzing an NMR spectrum of an aromatic compound, always make sure that the substitution pattern (i.e. monosubstituted, *ortho*-disubstituted, etc.) is fully consistent with all signals in the spectrum (taking symmetry into account). When doing so, avoid being distracted by the positions of the double bonds. After all, the π electrons are delocalized, and there are resonance structures in which the double bonds are drawn in different locations. The best way to avoid being distracted is to draw a circle to represent the aromatic ring, like this:

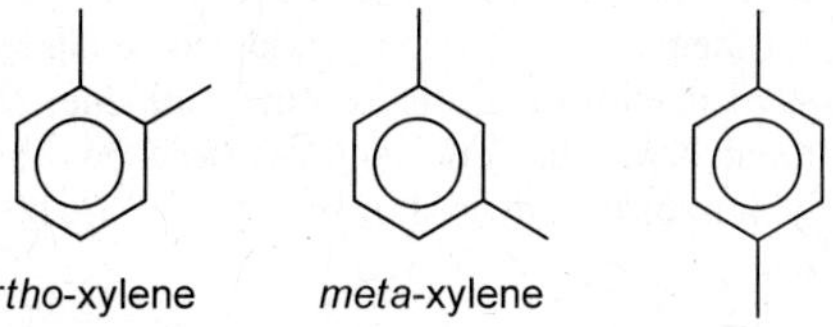

ortho-xylene   meta-xylene

para-xylene

To see the value of this approach, consider the following structure and determine how many signals are expected in the ^{1}H NMR and ^{13}C NMR spectra of this compound:

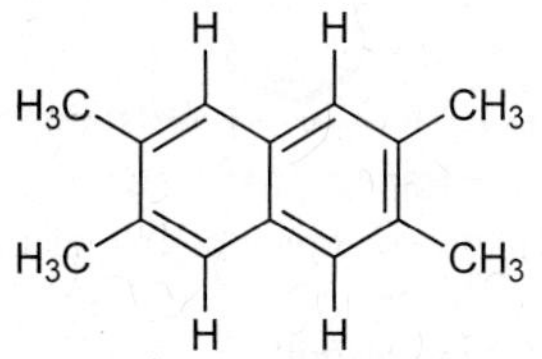

By redrawing the structure with a circle to represent aromaticity (rather than alternating single and double bonds), it becomes easier to see that

there are only two unique kinds of protons in this structure (highlighted below):

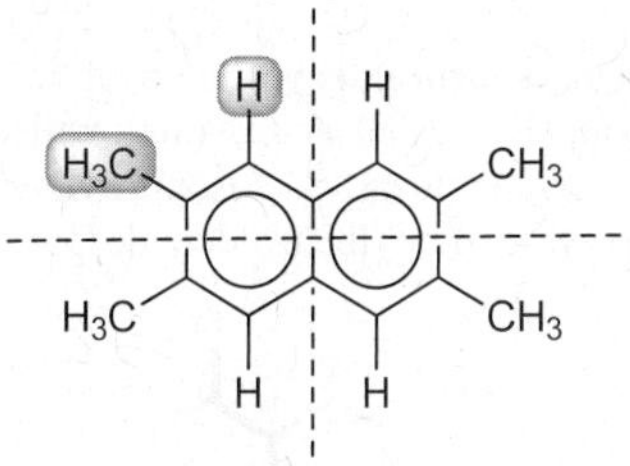

This compound will have only two signals in its ^{1}H NMR spectrum.

Similarly, the compound will have only four signals in its ^{13}C NMR spectrum:

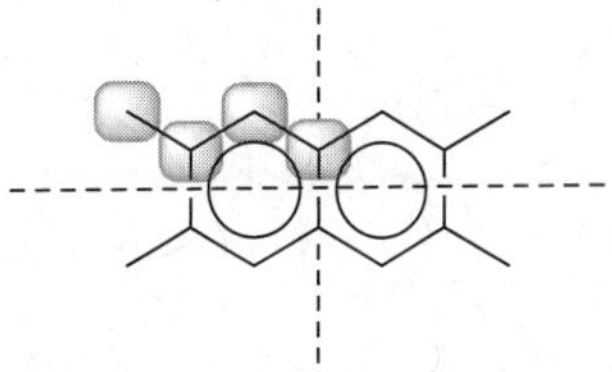

Throughout the solutions presented in this chapter, we will often represent aromatic rings with circles (particularly when symmetry is relevant to the discussion).

Useful reagents

The following is a list of reagents encountered in this chapter, as well as their specific function in the context of this chapter:

Reagents	Description
$Na_2Cr_2O_7$, H_2SO_4, H_2O	Sodium dichromate and sulfuric acid give chromic acid, which is a strong oxidizing agent that can be used to oxidize a benzylic position, provided that the benzylic position has at least one benzylic proton. The alkyl group (connected to the aromatic ring) is converted into a carboxylic acid group.
1) $KMnO_4$, H_2O, heat 2) H_3O^+	Potassium permanganate. A strong oxidizing agent that can be used to oxidize a benzylic position, provided that the benzylic position has at least one benzylic proton. The alkyl group (connected to the aromatic ring) is converted into a carboxylic acid group.
NBS, heat	N-Bromosuccinimide. A reagent that is used for radical bromination at the benzylic position.
H_2O	Water is a weak nucleophile that can be used in an S_N1 reaction with a benzylic halide.
NaOH	Hydroxide is a strong nucleophile that can be used in an S_N2 reaction with a primary benzylic halide.
conc. H_2SO_4	A strong acid that can be used to achieve acid-catalyzed dehydration to give an alkene.
NaOEt	Sodium ethoxide is a strong base that can be used to convert secondary or tertiary halides into alkenes (via an E2 process)
$3H_2$, 100 atm, 150 °C	Conditions for complete hydrogenation of benzene to give cyclohexane.
Na, CH_3OH, NH_3	Reagents for a Birch reduction, which reduces a benzene ring to give a 1,4-cyclohexadiene ring.

Solutions

18.1.

(a) The parent is benzaldehyde, shown in bold. An isopropyl group is located at C3 (also called the *meta* position), so the compound is 3-isopropylbenzaldehyde, (also called *meta*-isopropylbenzaldehyde):

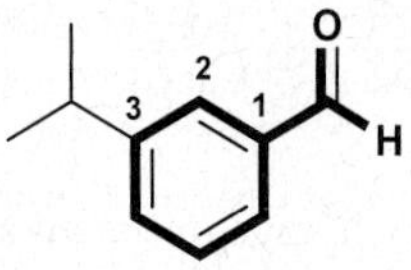

(b) The parent is toluene, shown in bold. A bromo group is located at C2 (also called the *ortho* position), so the compound is 2-bromotoluene (also called *ortho*-bromotoluene):

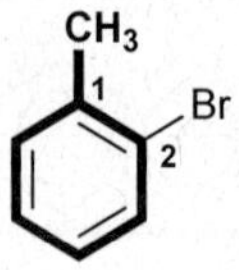

(c) The parent is phenol, shown in bold. Two nitro groups are located at C2 and C4, so the compound is called 2,4-dinitrophenol:

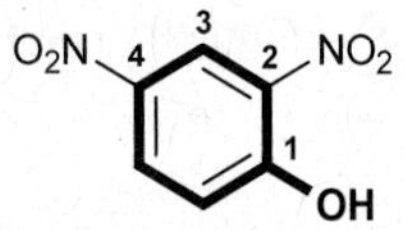

(d) The parent is benzene, shown in bold. There are three substituents (two isopropyl groups and one ethyl group). The assignment of locants, shown below, achieves the lowest numbers for all three substituents:

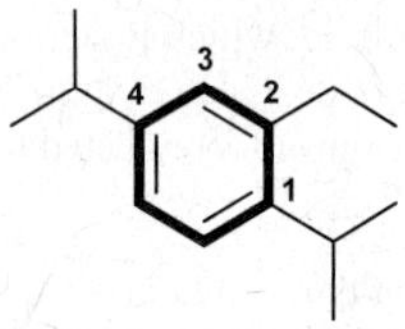

Therefore, the name of the compound is:
2-ethyl-1,4-diisopropylbenzene

(e) The parent is phenol, shown in bold. The substituents are each named and assigned a locant, as shown below. The locants start at the carbon atom connected to the OH group, and are then assigned counter-clockwise. This way, the ethyl group ("e") is assigned a lower locant than the isopropyl group ("i"), since there are no other differentiating factors that would allow us to decide which way to assign the locants.

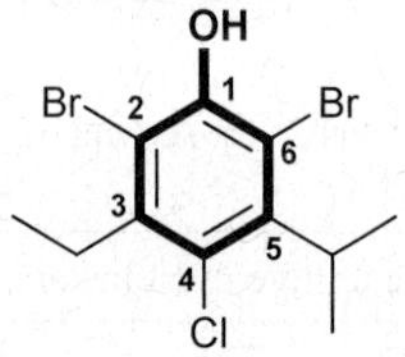

2,6-dibromo-4-chloro-3-ethyl-5-isopropylphenol

18.2. When naming this compound, there are three choices for the parent. The compound can be named as a disubstituted phenol, as a disubstituted toluene, or as a trisubstituted benzene. For each of these possibilities, the assignment of locants is shown:

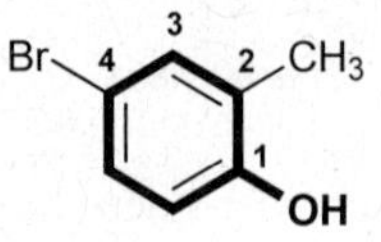

Parent = phenol 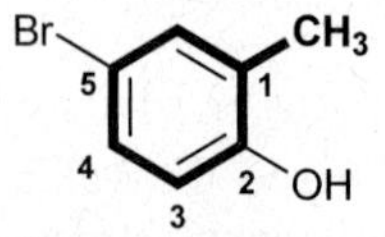   Parent = toluene

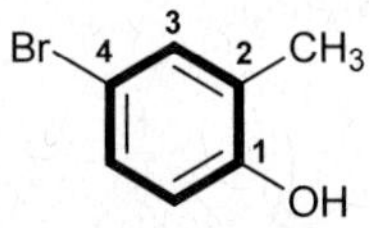

Parent = benzene

This gives the following three possible names:

(a) 4-bromo-2-methylphenol
(b) 5-bromo-2-hydroxytoluene
(c) 4-bromo-1-hydroxy-2-methylbenzene

18.3.

(a) The parent is anisole (methoxybenzene), and there are three other substituents (Br, Br, and Cl), in the following locations:

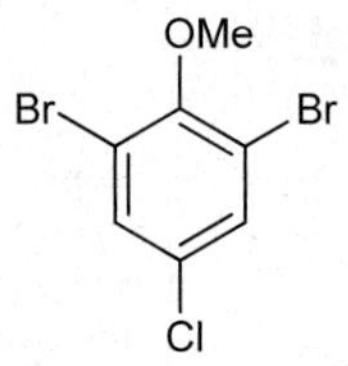

(b) The parent is phenol (hydroxybenzene), and there is one nitro group, located at C3 (the *meta* position):

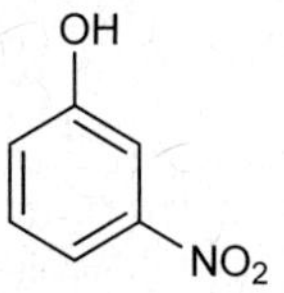

18.4. When naming this compound, there are three choices for the parent. The compound can be named as a disubstituted benzene, as a monosubstituted toluene, or as xylene. This gives the following possible names, all of which are acceptable IUPAC names:

(a) *meta*-xylene
(b) 1,3-dimethylbenzene
(c) *meta*-dimethylbenzene
(d) 3-methyltoluene
(e) *meta*-methyltoluene

18.5.

(a) Benzoic acid and peroxybenzoic acid have the following structures:

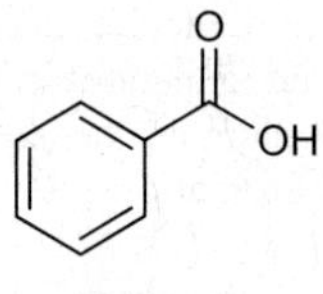

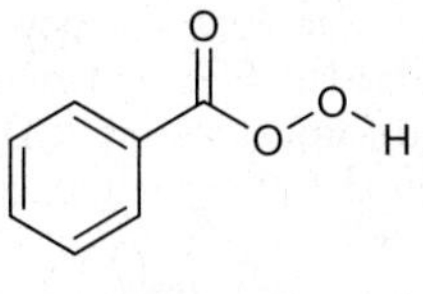

Benzoic acid   Peroxybenzoic acid

So *meta*-chloroperoxybenzoic acid must have a chlorine atom located at C3 (the *meta* position) on the aromatic ring, as shown here:

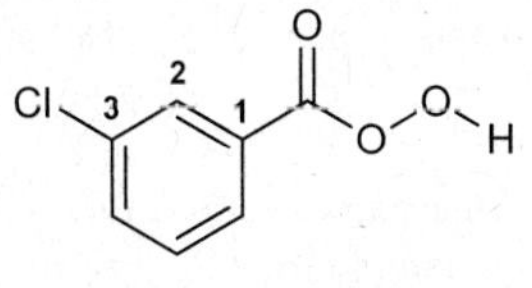

meta-chloroperoxybenzoic acid

(b) The chlorine atom has been replaced with a methyl group, so there is now a methyl group in the *meta* position. This compound must be called *meta*-methylperoxybenzoic acid (or 3-methylperoxybenzoic acid).

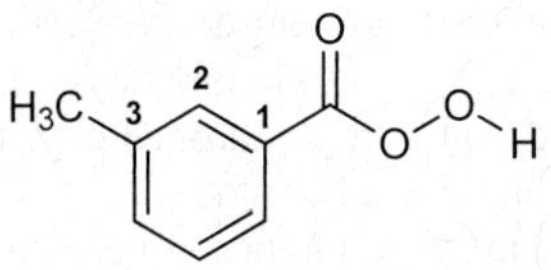

meta-methylperoxybenzoic acid
or
3-methylperoxybenzoic acid

18.6. The molecular formula (C_8H_8) indicates five degrees of unsaturation (see Section 15.16), four of which are accounted for by the aromatic ring, so the structure must also contain either a ring or a double bond (in addition to the aromatic ring). If compound **A** undergoes bromination to produce a dibromide, then it must contain a C=C double bond (outside of the aromatic ring, which does not react with Br_2). This C=C double bond accounts for the last degree of unsaturation, described above. The structure of compound **A,** as well as the structure of the resulting dibromide (compound **B**) are shown here:

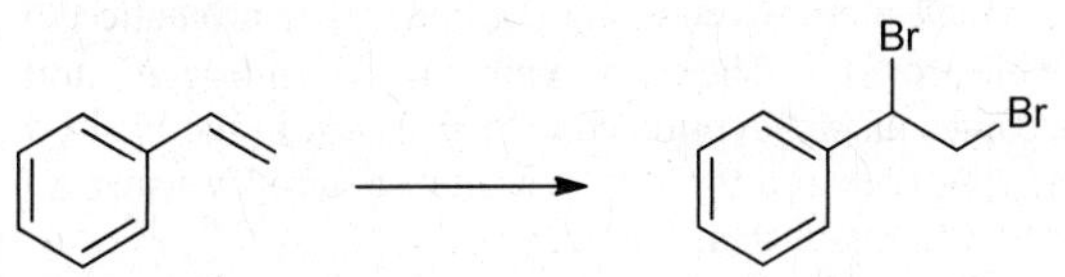

Compound A   Compound B
(C_8H_8)   ($C_8H_8Br_2$)

18.7.
(a) According to Figure 18.1, the conversion from cyclohexane to cyclohexene is uphill in energy (ΔH has a positive value).

(b) According to Figure 18.1, the conversion from cyclohexene to benzene is uphill in energy (ΔH has a positive value).
(c) According to Figure 18.1, the conversion from cyclohexadiene to benzene is downhill in energy (ΔH has a negative value).

18.8.
(a) This compound has six π bonds, for a total of twelve π electrons. Twelve is not a Hückel number, so the compound is not expected to be aromatic.
(b) This compound has seven π bonds, for a total of fourteen π electrons. Fourteen is a Hückel number. In Section 18.5, we saw that [14]annulene is somewhat destabilized by a steric interaction between the hydrogen atoms positioned inside the ring. Although [14]annulene is nonplanar, it does indeed exhibit aromatic stabilization, because the deviation from planarity is not too great.
(c) This compound has eight π bonds, for a total of sixteen π electrons. Sixteen is not a Hückel number, so the compound is not expected to be aromatic.

18.9. We draw a circle and inscribe a triangle inside the circle, with one of the connecting points of the triangle at the bottom of the circle. Each location where the triangle touches the circle represents an energy level. The energy level on the bottom of the circle is a bonding MO, and the two energy levels on top are antibonding MOs. The structure has two π electrons, which both occupy the bonding MO. This Frost circle indicates that there is only one bonding MO and it is filled, while the antibonding MOs are empty. This structure is expected to exhibit aromatic stabilization.

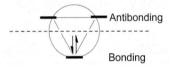

18.10. This structure exhibits a ring of continuously overlapping p orbitals, and there are 22 π electrons (a Hückel number). Therefore, the compound will be aromatic

18.11.
(a) This structure exhibits a ring of continuously overlapping p orbitals, but there are four π electrons, rendering the structure antiaromatic. In order to avoid the increase in energy associated with antiaromaticity, the lone pair is more likely to occupy an sp^3 hybridized orbital (rather than a p orbital). Without a continuous ring of overlapping p orbitals, the instability associated with antiaromaticity can be partially alleviated. Although, this change in hybridization destroys the resonance stabilization of the anion. If the lone pair does not occupy a p orbital, then it cannot participate in resonance, rendering it localized (rather than delocalized). In any event, this anion is expected to be very high in energy (very unstable).

(b) This structure exhibits a ring of continuously overlapping *p* orbitals (C+ represents a carbon atom with an empty *p* orbital), and there are two π electrons, rendering the structure aromatic.

(c) This structure exhibits a ring of continuously overlapping *p* orbitals, but there are four π electrons, rendering the structure antiaromatic. In order to avoid the increase in energy associated with antiaromaticity, the structure is likely to pucker (so that the empty *p* orbital is not in the same plane as the *p* orbitals associated with the π bonds). This change in geometry is unlikely to meaningfully compensate for the increase in energy associated with antiaromaticity. In addition, this geometric change would also decrease the stability of the positive charge by localizing it (resonance-stabilization would be reduced because the *p* orbitals would not effectively overlap).

(d) This structure exhibits a ring of continuously overlapping *p* orbitals (the lone pair occupies a *p* orbital because it is resonance stabilized). There are ten π electrons, rendering the structure aromatic.

18.12. Cyclopentadiene is more acidic because its conjugate base is highly stabilized. Deprotonation of cyclopentadiene generates an anion that is aromatic, because it is a continuous system of overlapping *p* orbitals containing six π electrons. In contrast, deprotonation of cycloheptatriene gives an anion with eight π electrons, expected to be highly unstable (antiaromatic).

more stable

18.13. The first step of an S$_N$1 process is loss of a leaving group, forming a carbocation, so we compare the carbocations that would be formed upon loss of a leaving group:

The second carbocation is more stable, because it is aromatic, and is therefore lower in energy than the first carbocation. The transition state leading to the second carbocation will be lower in energy than the transition state leading to the first carbocation, and therefore, the

second carbocation will be formed more rapidly than the first.

18.14. The first compound is more acidic because deprotonation of the first compound generates a new (second) aromatic ring. Deprotonation of the second compound does not introduce a new aromatic ring:

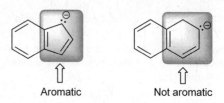

Aromatic   Not aromatic

18.15.
(a) One of the lone pairs occupies a *p* orbital, thereby rendering the compound aromatic (six π electrons). The other lone pair will occupy an *sp*^{2} hybridized orbital, in the plane of the ring, extending away from the ring.
(b) One of the lone pairs occupies a *p* orbital, thereby rendering the compound aromatic (six π electrons). The other lone pair will occupy an *sp*^{2} hybridized orbital, in the plane of the ring, extending away from the ring.
(c) If the lone pair were to occupy a *p* orbital, there would be a continuous system of overlapping *p* orbitals with eight π electrons (antiaromatic). To avoid the instability associated with antiaromaticity, the lone pair occupies an *sp*^{3} hybridized orbital.
(d) One of the lone pairs on the sulfur atom occupies a *p* orbital, thereby rendering the compound aromatic (six π electrons). The lone pair on the nitrogen atom occupies an *sp*^{2} hybridized orbital that is in the plane of the ring (because the nitrogen atom is already using a *p* orbital to establish aromaticity).
(e) There is only one lone pair (on oxygen) and it is not participating in aromaticity. That oxygen atom is already using a *p* orbital to establish aromaticity.
(f) Each nitrogen atom has one lone pair, and neither is participating in aromaticity. In each case, the nitrogen atom is already using a *p* orbital to establish aromaticity.
(g) The compound is not aromatic. In order to achieve a continuous system of overlapping *p* orbitals, each oxygen atom would need to contribute a lone pair in a *p* orbital, and that would give 8 π electrons (not a Hückel number).
(h) One of the lone pairs on the oxygen atom occupies a *p* orbital, thereby rendering the compound aromatic (six π electrons). The lone pair on the nitrogen atom occupies an *sp*^{2} hybridized orbital that is in the plane of the ring (because the nitrogen atom is already using a *p* orbital to establish aromaticity).

18.16. In addition to the aromatic rings indicated in the structures of the eight best-selling drugs (Section 18.1), each of the highlighted rings also has a continuous system of overlapping *p* orbitals, with six π electrons, and is therefore aromatic:

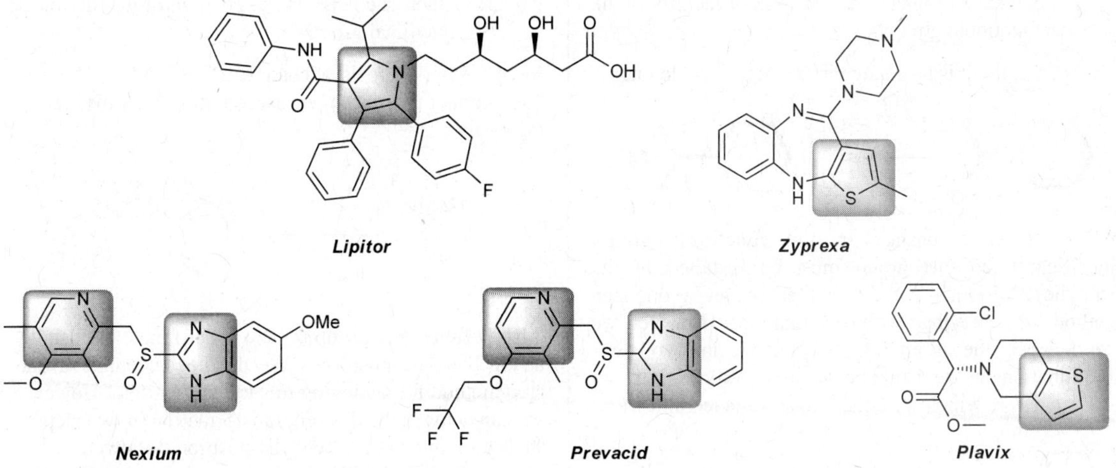

Lipitor

Zyprexa

Nexium

Prevacid

Plavix

18.17.   The first compound is expected to be more acidic (has a lower pK_a), because deprotonation restores aromaticity to the ring.

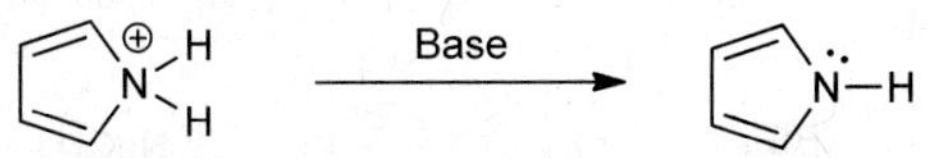

not aromatic   aromatic

The second compound is already aromatic, even before deprotonation.

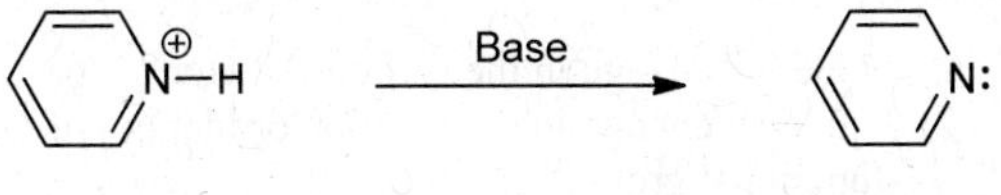

aromatic   aromatic

In this case, the driving force for losing the proton is not as great.

18.18.
(a) Yes, it has the required pharmacophore (two aromatic rings separated by one carbon atom, and a tertiary amine).

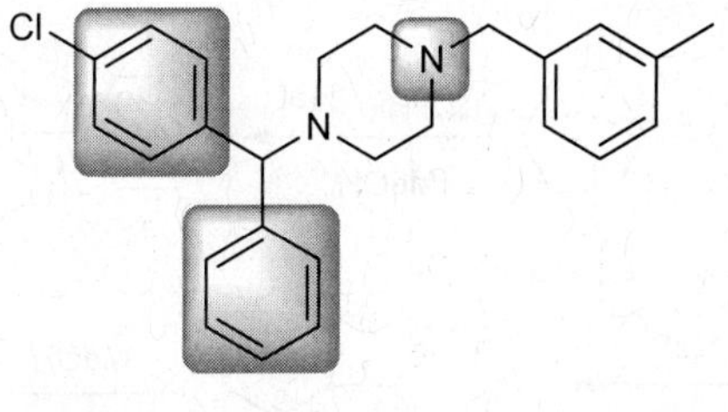

(b) Meclizine is mostly nonpolar, so it crosses the blood-brain barrier and binds with receptors in the central nervous system, causing sedation.

(c) Introduce polar functional groups that reduce the ability of the compound to cross the blood-brain barrier.

18.19.
(a) The benzylic position undergoes oxidation, giving a carboxylic acid:

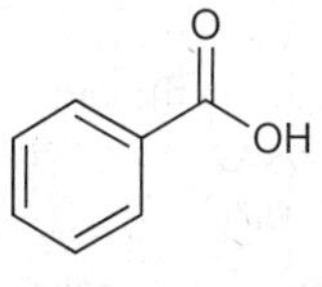

(b) Each of the benzylic positions undergoes oxidation. Notice that oxidation of the ethyl group involves cleavage of a C-C bond to give a diacid with one less carbon atom than the starting material.

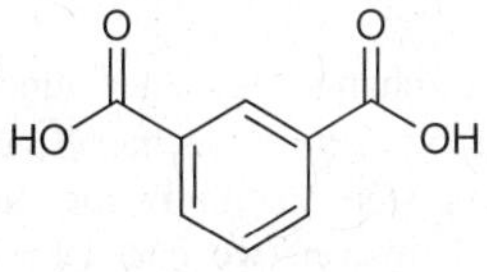

(c) There are three benzylic positions, but one of these positions (the *tert*-butyl group) lacks a benzylic proton, and as such, that position does not undergo oxidation. The other two benzylic positions undergo oxidation, giving the following diacid.

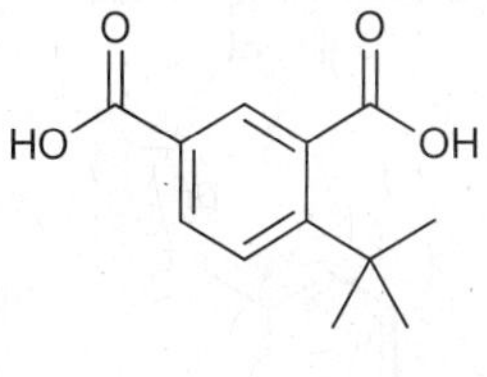

18.20.

(a) There are two questions to ask when approaching a synthesis problem.

1) Is there a change in the carbon skeleton?
2) Is there a change in the position or identity of the functional groups?

In this case, there is no change in the carbon skeleton:

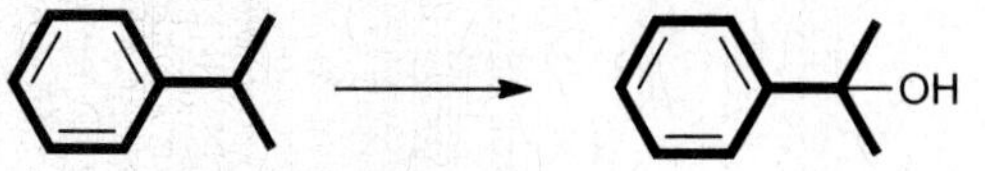

but there is a change in the functional group. Specifically, an OH group must be installed in the benzylic position. We did not encounter a one-step method for achieving this transformation. However, we can achieve the desired transformation in two steps (bromination at the benzylic position, followed by an S_N1 process in which water is used as a nucleophile).

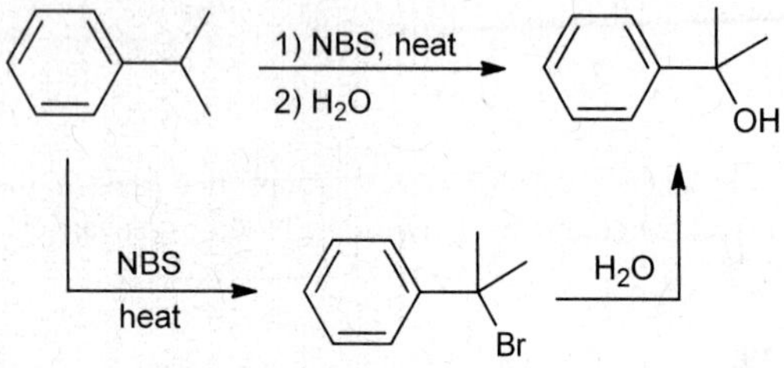

(b) There are two questions to ask when approaching a synthesis problem.

1) Is there a change in the carbon skeleton?
2) Is there a change in the position or identity of the functional groups?

In this case, there is no change in the carbon skeleton.

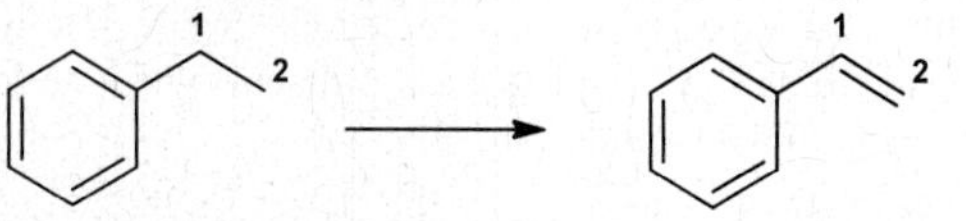

but there is a change in the functional group. Specifically, a π bond must be installed. We did not encounter a one-step method for achieving this transformation. However, we can achieve the desired transformation in two steps:
1) bromination at the benzylic position, followed by
2) elimination (E2) upon treatment with a strong base, such as sodium ethoxide:

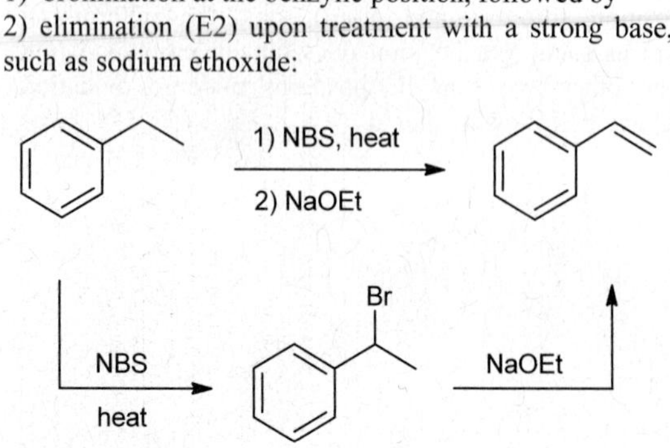

(c) There are two questions to ask when approaching a synthesis problem.

1) Is there a change in the carbon skeleton?
2) Is there a change in the position or identity of the functional groups?

In this case, there is a change in the carbon skeleton. The product has one more carbon atom than the starting material:

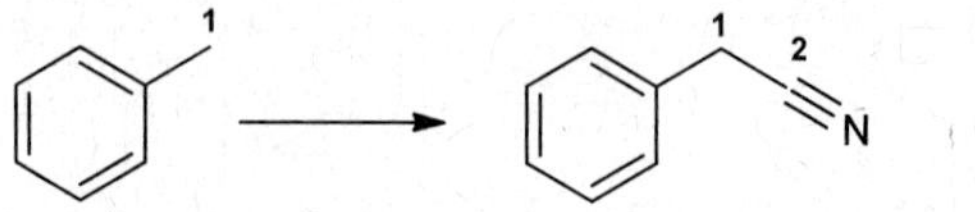

Also, a functional group (cyano group) must be installed at the benzylic position. We did not encounter a one-step method for achieving this transformation. However, we can achieve the desired transformation in two steps:
1) bromination at the benzylic position, followed by
2) substitution (S_N2), in which a cyanide ion functions as the nucleophile:

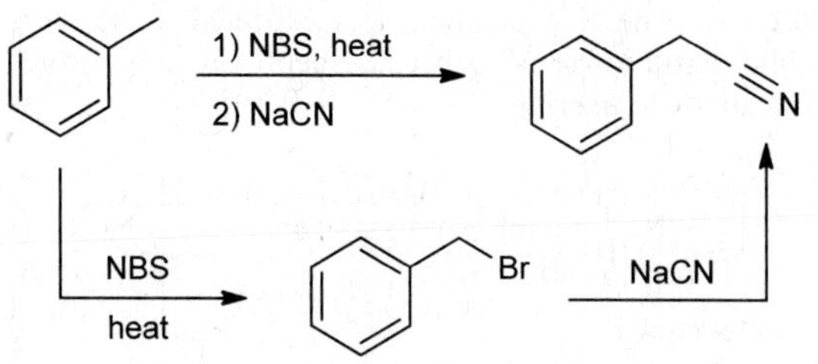

(d) There are two questions to ask when approaching a synthesis problem.

1) Is there a change in the carbon skeleton?
2) Is there a change in the position or identity of the functional groups?

In this case, there is no change in the carbon skeleton, but there is a change in the functional group. Specifically, a methoxy group (OMe) must be installed in the benzylic position. We did not encounter a one-step method for achieving this transformation. However, we can achieve the desired transformation in two steps:
1) bromination at the benzylic position, followed by
2) substitution (S_N1), in which methanol functions as the nucleophile:

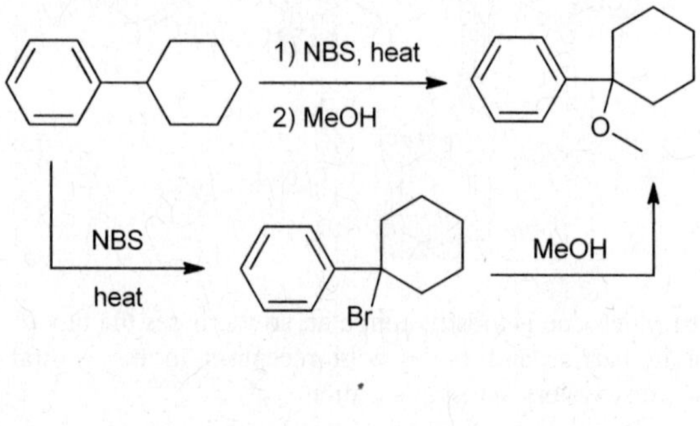

18.21. There are two questions to ask when approaching a synthesis problem.

1) Is there a change in the carbon skeleton?
2) Is there a change in the position or identity of the functional groups?

In this case, there is a change in the carbon skeleton. The product has one more carbon atom than the starting material:

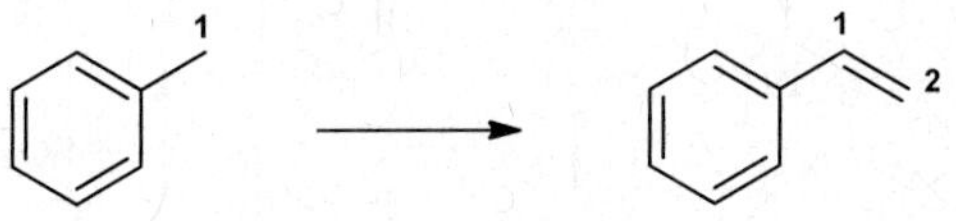

Also, a functional group (a π bond) must be installed. Let's begin with the change in the carbon skeleton. In order to increase the size of the carbon skeleton, we must form a C–C bond. A Grignard reaction will allow us to form the critical C–C bond, if we had the following aldehyde:

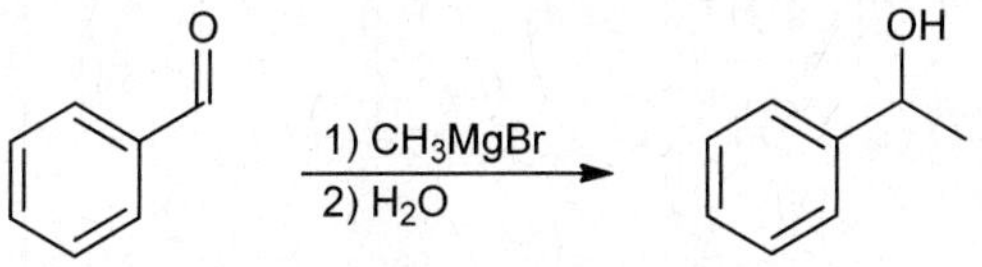

In order to use this approach, we must first convert the starting material into the necessary aldehyde. This can be achieved via benzylic bromination, followed by substitution, followed by oxidation with PCC (as shown in the following figure). Once the aldehyde is prepared, the Grignard reaction can be performed. And finally, the product of the Grignard reaction (an alcohol) can be treated with concentrated sulfuric acid to give an elimination reaction, affording the desired product:

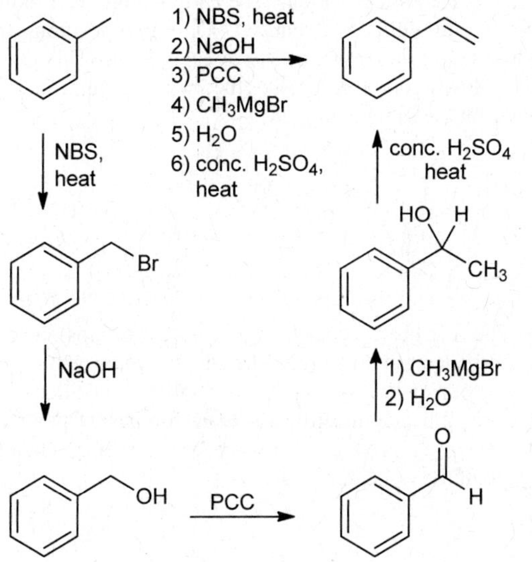

18.22. There are two questions to ask when approaching a synthesis problem.

1) Is there a change in the carbon skeleton?
2) Is there a change in the position or identity of the functional groups?

In this case, there is a change in the carbon skeleton. The product has one less carbon atom than the starting material:

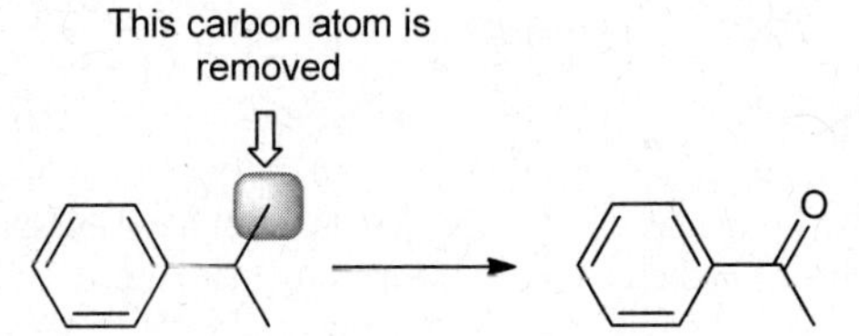

This carbon atom is removed

Also, a functional group (a C=O bond) must be installed. Let's begin with the change in the carbon skeleton. We have only learned one way (ozonolysis) to remove a carbon atom from the carbon skeleton. Ozonolysis would indeed install the necessary functional group in the correct location:

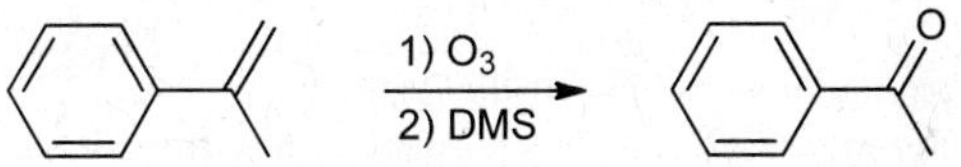

This strategy requires that we first form the alkene above, which can be accomplished via bromination of the benzylic position, followed by elimination (E2) upon treatment with a strong base (such as sodium ethoxide):

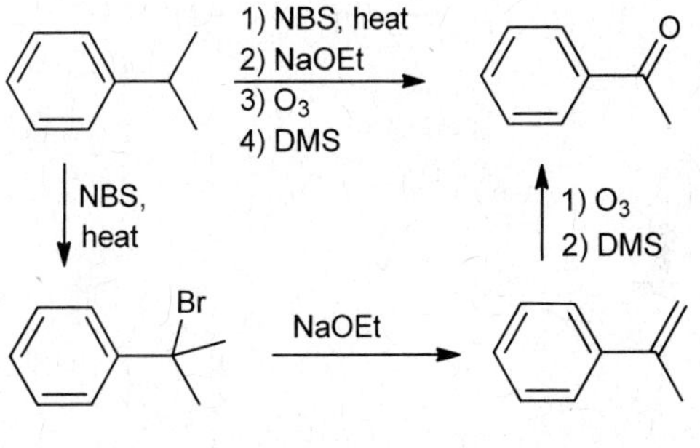

18.23. There are two questions to ask when approaching a synthesis problem.

1) Is there a change in the carbon skeleton?
2) Is there a change in the position or identity of the functional groups?

In this case, there is a change in the carbon skeleton. The product has two additional carbon atoms that are absent in the starting material:

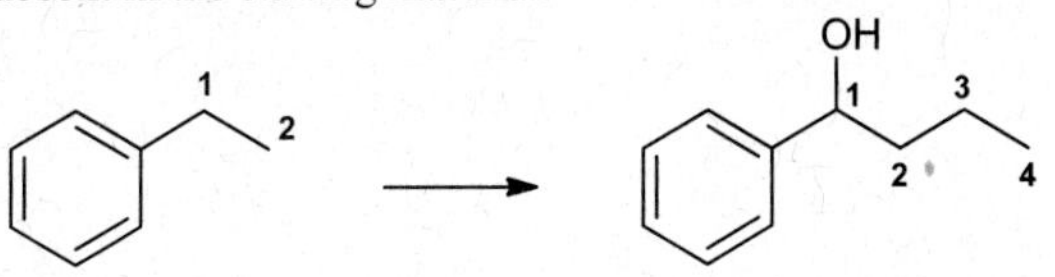

Also, a functional group (an OH group) must be installed at the benzylic position. Both of these goals (the change in carbon skeleton and installation of the functional group) can be accomplished via a Grignard reaction. That is, the following Grignard reaction can create the necessary C–C bond, while leaving an OH group in the desired location.

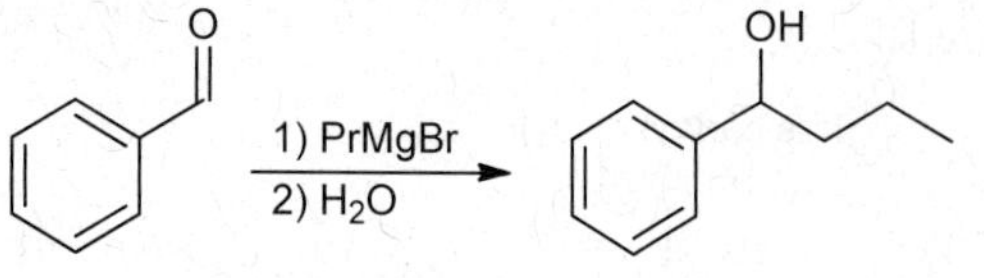

However, the necessary aldehyde does not have the same number of carbon atoms as the starting material:

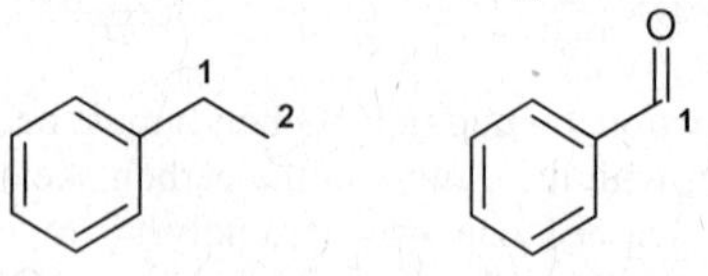

Nevertheless, we can easily convert the starting material into the desired aldehyde via the following three step process:

1) bromination at the benzylic position,
2) elimination (E2) upon treatment with a strong base,
3) ozonolysis

This strategy is shown here:

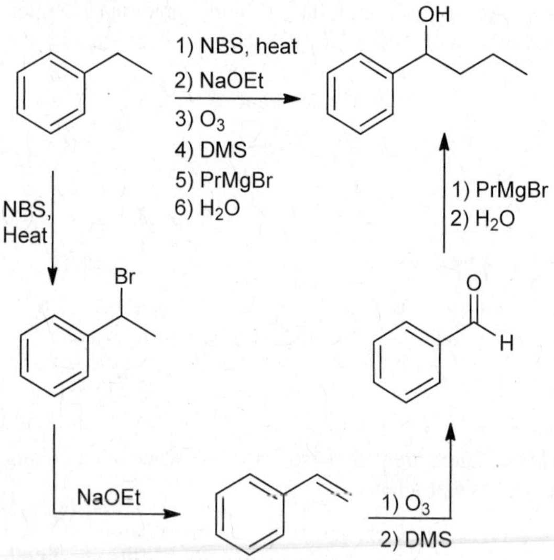

There are certainly other acceptable solutions to this problem. The solution above represents just one approach. For example, the following is another acceptable approach:

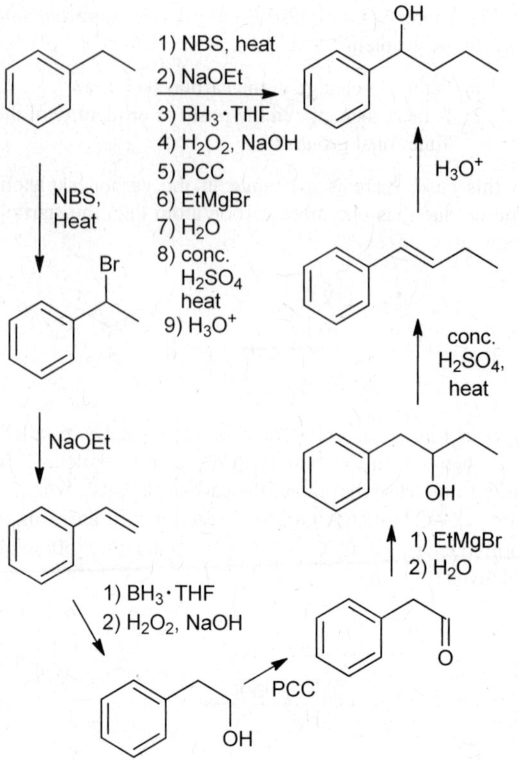

In this alternative synthetic route, the regiochemical outcome of the final step is controlled by the preference for formation of a benzylic carbocation intermediate, and the regiochemical outcome of the penultimate step is controlled by the preference for formation of a conjugated π bond.

18.24.
(a) The ring bears one substituent (an alkyl group), which is electron donating. The carbon atom next to the alkyl group will not be reduced. Only two positions are reduced, and they must be 1,4 to each other. This gives the following product (note: the two structures below represent the same compound):

(b) The ring bears two substituents (alkyl groups), which are both electron donating. The carbon atoms next to the alkyl groups will not be reduced. Only two positions are reduced, and they must be 1,4 to each other. This gives the following product (note: the two structures below represent the same compound):

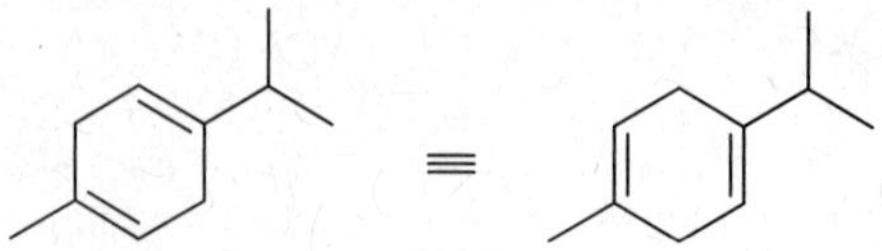

(c) The ring bears two substituents (alkyl groups), which are both electron donating. The carbon atoms next to the alkyl groups will not be reduced. Only two positions are reduced, and they must be 1,4 to each other. This gives the following product:

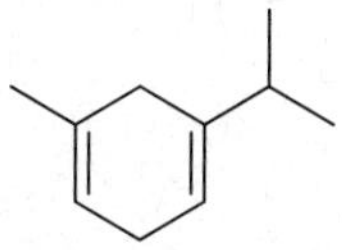

(d) The ring bears two substituents (alkyl groups), which are both electron donating. The carbon atoms next to the alkyl groups will not be reduced. Only two positions are reduced, and they must be 1,4 to each other. This gives the following product:

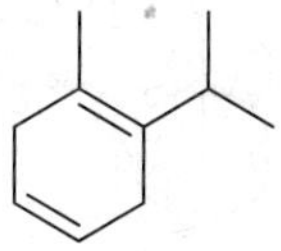

(e) The ring bears one substituent (a carboxylic acid group), which is electron withdrawing. The carbon atom next to the carboxylic acid group will be reduced. Only two positions are reduced, and they must be 1,4 to each other. This gives the following product:

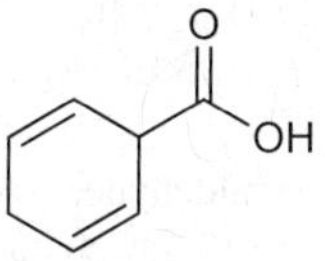

(f) The ring bears two substituents. The alkyl group is electron donating, so the carbon atom next to the alkyl group is not reduced. The carboxylic acid group is electron withdrawing, so the carbon atom next to the carboxylic acid group is reduced. Only two positions are reduced, and they must be 1,4 to each other. This gives the following product:

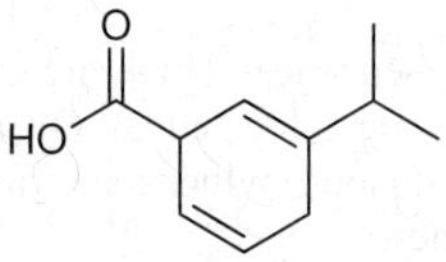

18.25.
(a) The compound exhibits a lone pair next to a π bond, so we draw two curved arrows associated with that characteristic pattern (see Chapter 2). The resulting resonance structure also exhibits a lone pair next to a π bond, so once again we draw the resulting resonance structure. This pattern is continued, until the fifth and final resonance structure is drawn. Note that the final resonance structure differs from the first resonance structure only in the position of the double bonds (that is in fact one of the other characteristic patterns seen in Chapter 2 – conjugated π bonds enclosed in a ring).

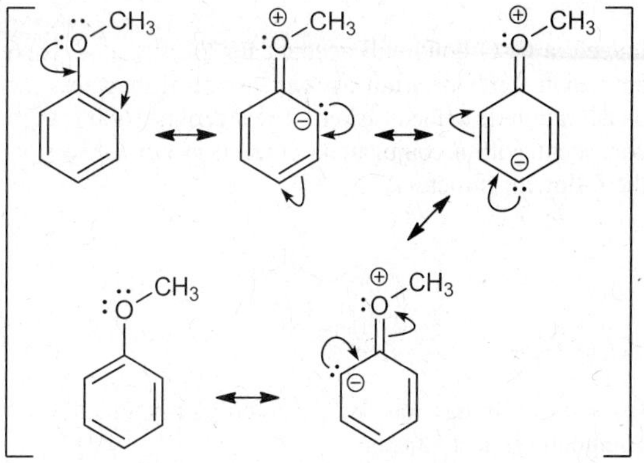

(b) The methoxy group is inductively electron withdrawing because of the electronegativity of the oxygen atom. However, the methoxy group is also electron donating by resonance (see resonance structures above, three of which exhibit a negative charge on the ring). Since resonance effects are expected to dominate over inductive effects, as described in the problem statement, we expect the methoxy group to be overall electron donating. Therefore, when this compound undergoes a Birch reduction, we expect that the carbon atom next to the methoxy group will not be reduced. Only two positions are reduced, and they must be 1,4 to each other. This gives the following product (note: the two structures below represent the same compound):

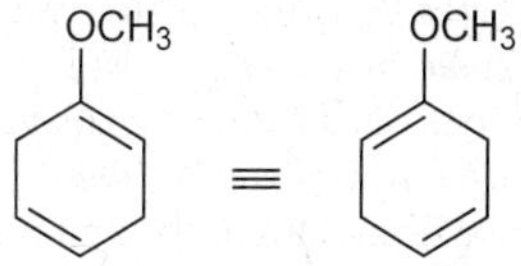

18.26.
(a) The molecular formula (C_8H_8O) indicates five degrees of unsaturation (see Section 15.16), which is highly suggestive of an aromatic ring, in addition to either one double bond or one ring. The aromatic ring is confirmed by the presence of a signal just above 3000 cm^{-1} in the IR spectrum, as well a signal at 1646 cm^{-1}. In the ^1H NMR spectrum, the multiplet at 7.5 ppm also confirms the aromatic ring. Notice that this multiplet has an integration of 5, which indicates that the aromatic ring is monosubstituted:

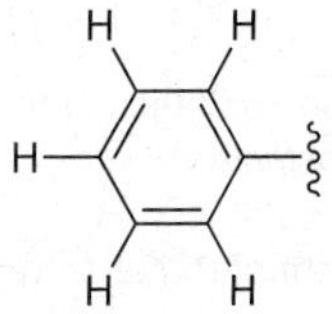

The singlet with an integration of 3 indicates a methyl group, and its chemical shift suggests it might be near a C=O bond:

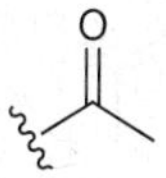

Indeed, a C=O bond will account for the signal at 1686 cm^{-1} in the IR spectrum as well as the fifth degree of unsaturation described earlier. The signal at 1686 cm^{-1} is consistent with a conjugated carbonyl group, as seen in the following structure:

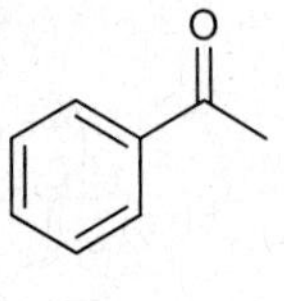

(b) As seen in Section 18.2, the common name for this compound is acetophenone.

(c) The reagents indicate a Birch reduction. The ring bears one substituent (a carbonyl group), which is electron withdrawing by resonance. The carbon atom next to the carbonyl group will be reduced. Only two positions are reduced, and they must be 1,4 to each other. This gives the following product:

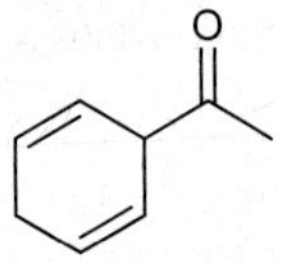

18.27.
(a) The molecular formula (C$_8$H$_{10}$) indicates four degrees of unsaturation (see Section 15.16), which is highly suggestive of an aromatic ring. The presence of an aromatic ring is confirmed by the signals just above 3000 cm^{-1} in the IR spectrum, as well as the signal near 1600 cm^{-1}. In the ^{1}H NMR spectrum, the multiplet at 7.1 ppm further confirms the presence of an aromatic ring. The integration of this signal is 4, indicating that the aromatic ring is disubstituted. The aromatic ring accounts for six of the eight carbon atoms in the compound (C$_8$H$_{10}$), so there must be two methyl groups attached to the ring (dimethyl benzene). To determine the substitution pattern (*ortho*, *meta* or *para*), we note that the ^{13}C NMR spectrum exhibits three signals in the region of 100-150 ppm, which indicates the *ortho* substitution pattern:

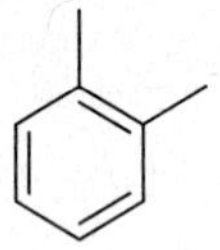

(b) As seen in Section 18.2, the common name for *ortho*-dimethylbenzene is *ortho*-xylene.

(c) When treated with chromic acid (a strong oxidizing agent), both benzylic positions are oxidized, giving the following diacid:

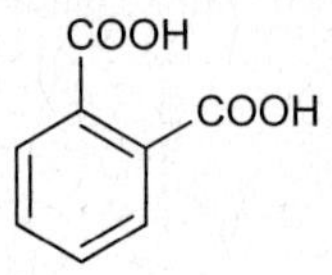

18.28.
(a) The parent is benzoic acid. An ethyl group is located at C4 (also called the *para* position), so the compound is 4-ethylbenzoic acid, (also called *para*-ethylbenzoic acid):

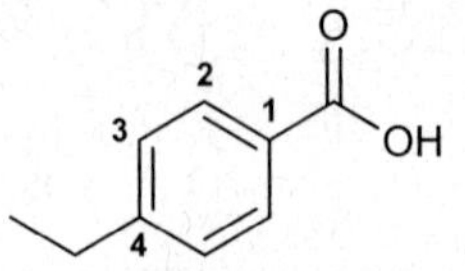

(b) The parent is phenol. A bromine atom is located at C2 (also called the *ortho* position), so the compound is 2-bromophenol, (also called *ortho*-bromophenol):

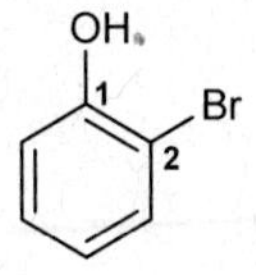

(c) The parent is phenol. A chlorine atom is located at C2 and a nitro group is located at C4, so the compound is 2-chloro-4-nitrophenol:

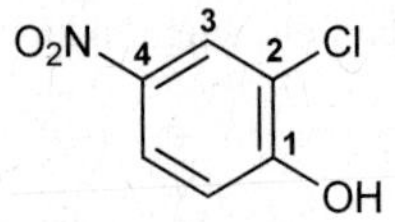

(d) The parent is benzaldehyde. A bromine atom is located at C2 and a nitro group is located at C5 (as shown), so the compound is 2-bromo-5-nitrobenzaldehyde:

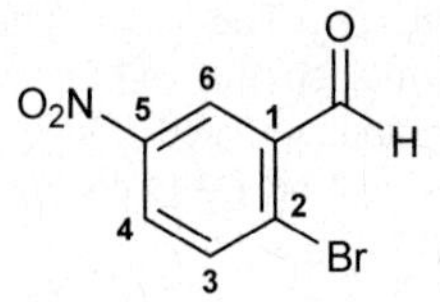

(e) The parent is benzene. There are two substituents, both isopropyl groups, located at C1 and C4, so the compound is 1,4-diisopropylbenzene, (also called *para*-diisopropylbenzene):

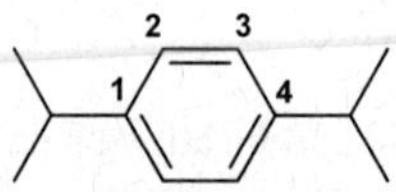

18.29.
(a) The parent is benzene, and there are two chlorine atoms that are *ortho* to each other (C1 and C2).

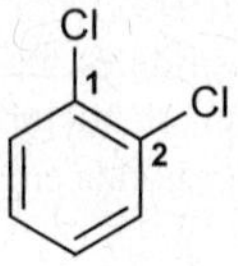

(b) As seen in Section 18.2, anisole is the common name for methoxybenzene:

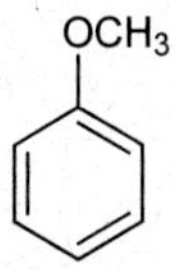

OCH₃

(c) The parent is toluene (methylbenzene), and there is a nitro group in the *meta* position (C3).

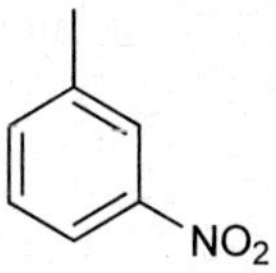

NO₂

(d) As seen in Section 18.2, aniline is the common name for aminobenzene:

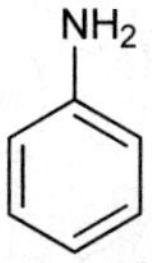

NH₂

(e) The parent is phenol (hydroxybenzene), and there are three bromine atoms, connected to positions C2, C4, and C6.

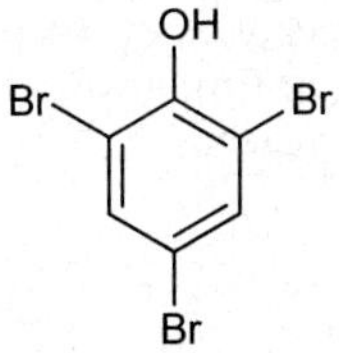

OH
Br   Br

Br

(f) As seen in Section 18.2, *para*-xylene is the common name for *para*-dimethylbenzene:

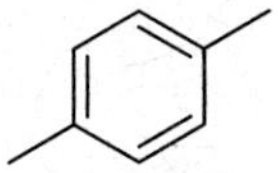

18.30. The molecular formula (C_9H_{12}) indicates four degrees of unsaturation (see Section 15.16), and the benzene ring accounts for all four degrees of unsaturation. The ring contains six carbon atoms, so the remaining three carbon atoms must be attached to the aromatic ring without any additional π bonds or rings. There are several ways to attach three carbon atoms to an aromatic ring without introducing any additional π bonds or rings. The carbon atoms can be attached as three individual methyl groups, for which there are three unique constitutional isomers:

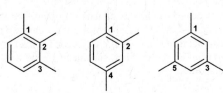

or there can be one ethyl group and one methyl group, for which there are also three constitutional isomers:

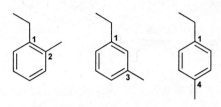

Finally, all three carbon atoms can be connected to the ring as a single substituent, and there are two ways to accomplish that, as shown here:

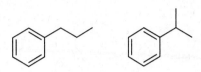

18.31. The molecular formula (C_8H_{10}) indicates four degrees of unsaturation (see Section 15.16), and the aromatic ring accounts for all four degrees of unsaturation. The ring contains six carbon atoms, so the remaining two carbon atoms must be attached to the aromatic ring without any additional π bonds or rings. There are several ways to attach two carbon atoms to an aromatic ring without introducing any additional π bonds or rings. The carbon atoms can be attached as two individual methyl groups, for which there are three unique constitutional isomers:

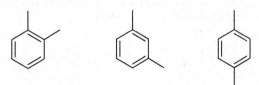

or there can simply be one ethyl group:

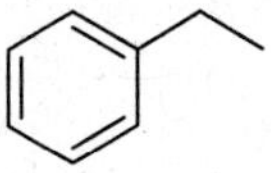

In total, we have seen four constitutional isomers.

18.32. The molecular formula (C_8H_9Cl) is similar to the molecular formula in the previous problem (C_8H_{10}), but one of the hydrogen atoms has been replaced with a chlorine atom. So, we begin our analysis by redrawing all four skeletons shown in the solution to the previous problem. In each case, we will draw a circle for the aromatic ring, rather than drawing π bonds, as it will be easier to identify symmetry without being distracted by the arbitrary locations of the double bonds:

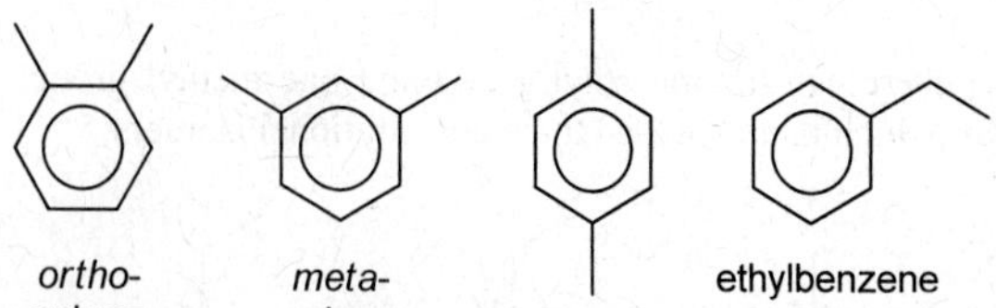

ortho-xylene   meta-xylene   para-xylene   ethylbenzene

For each one of these skeletons, we must consider all of the unique locations where a chlorine atom can be placed. Let's begin with the first skeleton (*ortho*-xylene), for which there are three unique locations where the chlorine atom can be placed:

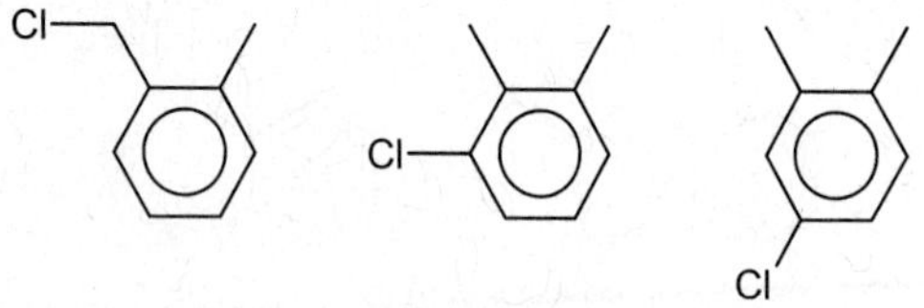

Placing the chlorine atom in any other location on this skeleton will result in one of the compounds above. For example:

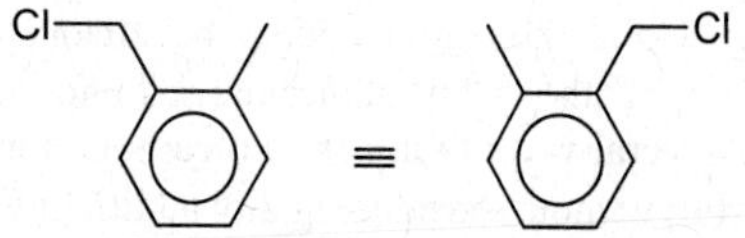

Next, we consider the second skeleton (*meta*-xylene), for which there are four unique locations where the chlorine atom can be placed:

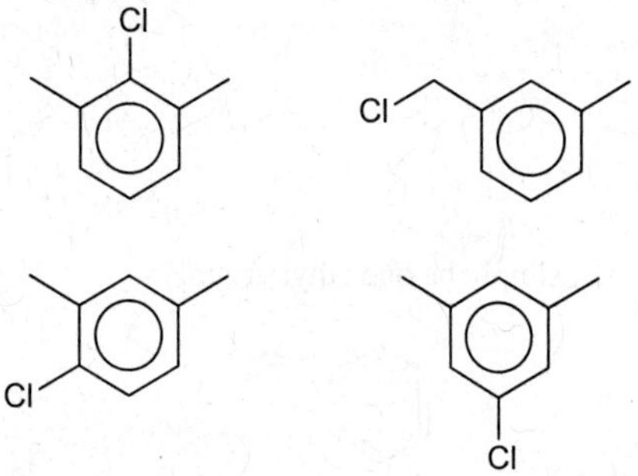

Next, we consider the third skeleton (*para*-xylene), for which there are only two unique locations where the chlorine atom can be placed:

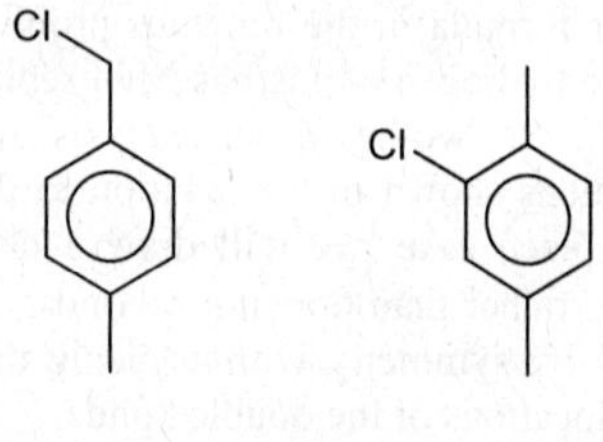

And finally, we consider the last skeleton (ethylbenzene), for which there are five unique locations where the chlorine atom can be placed:

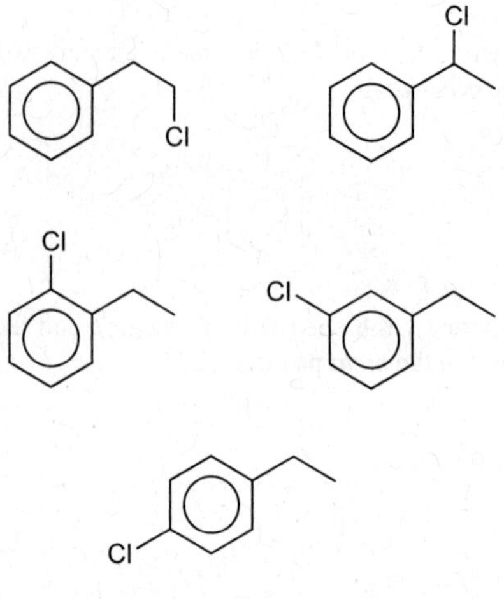

In total, we have seen fourteen constitutional isomers.

18.33. We begin by placing the first two nitro groups at positions C2 and C3:

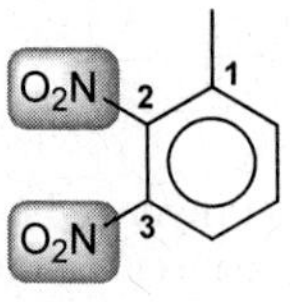

There are now three choices of where to place the third nitro group (C4, C5, or C6), giving rise to the following three constitutional isomers:

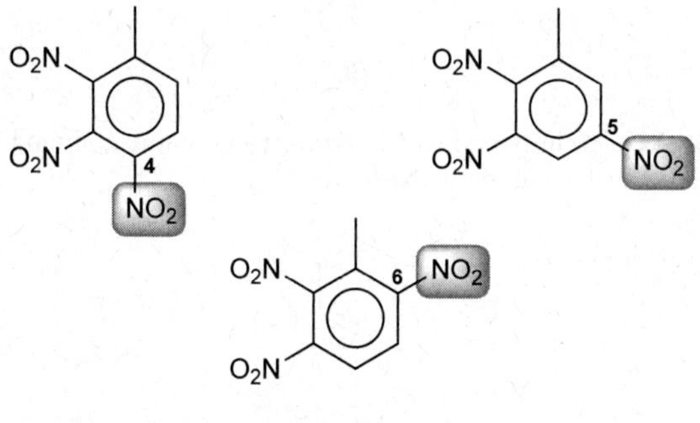

We have now exhausted all possibilities in which the first two nitro groups occupy positions C2 and C3. Next, we place the first two nitro groups at C2 and C4:

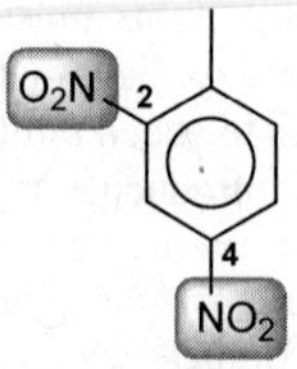

There are now two choices of where to place the third nitro group (C5 or C6), giving rise to the following two constitutional isomers, although the second isomer is the one described in the problem statement (so we will not count that one):

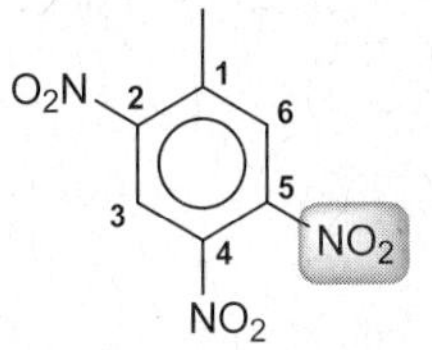

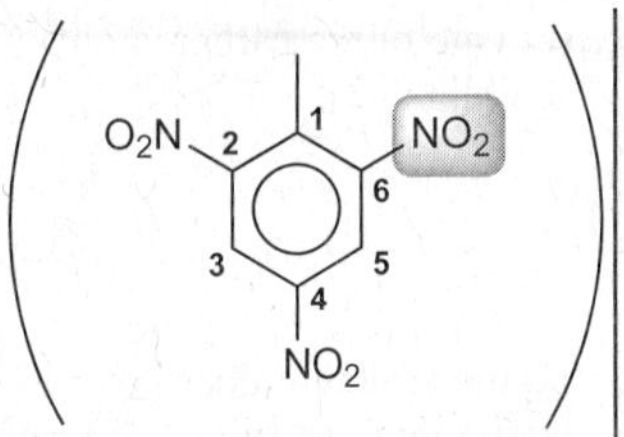

We do not count this isomer,
as per the instructions
in the problem statement

We have now exhausted all possibilities in which the first two nitro groups occupy positions C2 and C4. Next, we place the first two nitro groups at C2 and C5:

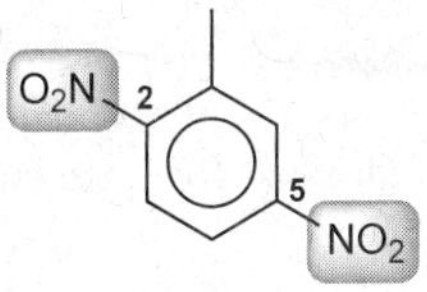

This leaves only one option for the third nitro group (C6), and this isomer has already been drawn earlier:

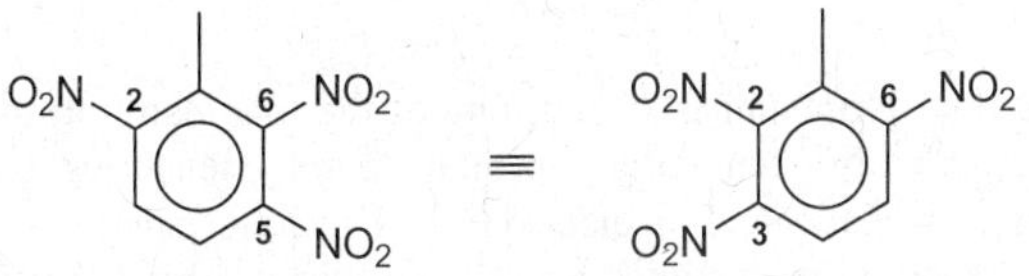

We have now considered all options for which the first nitro group is placed at C2. Now we continue by placing the first nitro group at C3. If we place the remaining two nitro groups at C4 and C5, then we obtain the following isomer:

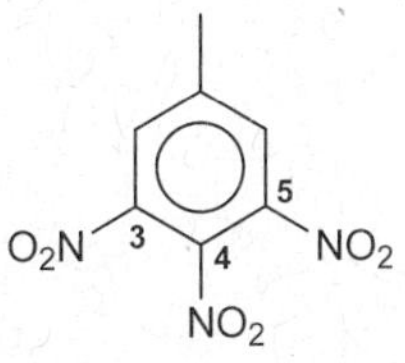

Any other arrangement of nitro groups will be identical to one of the arrangements already drawn. For example:

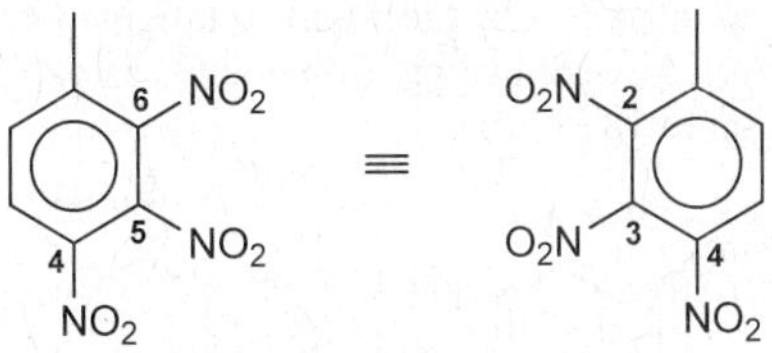

In summary, the following five isomers are consistent with the requirements described in the problem statement:

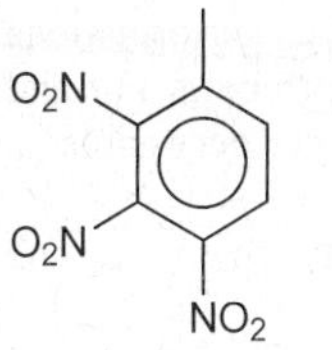

2,3,4-
trinitrotoluene

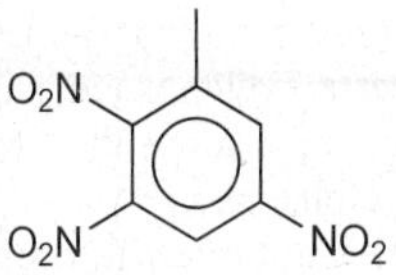

2,3,5-
trinitrotoluene

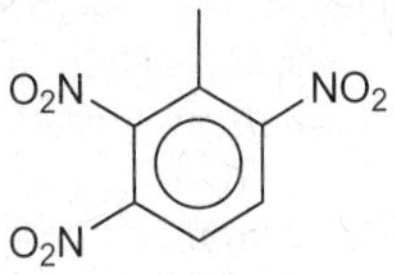

2,3,6-
trinitrotoluene

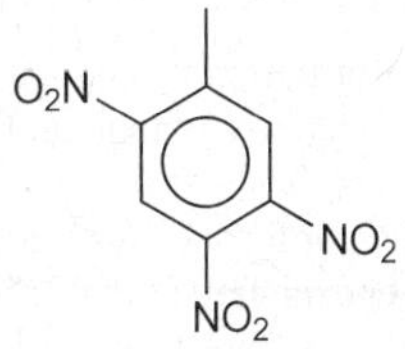

2,4,5-
trinitrotoluene

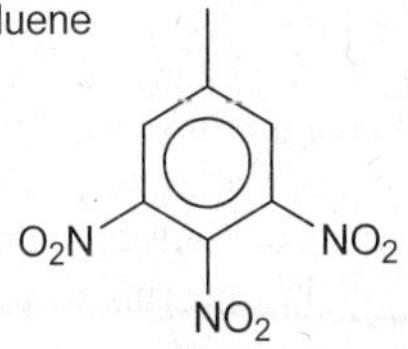

3,4,5-
trinitrotoluene

18.34.

(a) Each π bond counts as two π electrons, and the lone pair also counts as two π electrons, giving a total of ten π electrons.

(b) Each π bond counts as two π electrons. In addition, one of the lone pairs on the sulfur atom is participating in resonance (it occupies a p orbital and is delocalized) and therefore contributes two more π electrons. This gives a total of six π electrons.

(c) Each π bond counts as two π electrons, giving a total of ten π electrons.

(d) Each π bond counts as two π electrons, giving a total of four π electrons. The lone pair on the nitrogen atom does not occupy a p orbital, because the nitrogen atom is already using a p orbital to form a π bond. The lone pair occupies an sp^2 hybridized orbital and does not contribute any π electrons.

(e) Each π bond counts as two π electrons, giving a total of six π electrons. The carbocation represents an empty p orbital and does not contribute any π electrons.

18.35. The terms used in reference to cyclohexane were discussed in Chapter 4, while the terms used in reference to benzene were discussed in Chapter 18.
(a) The term *meta* refers to a 1,3-disubstituted benzene ring.
(b) A Frost circle is used to draw an energy diagram showing the relative energy levels of the MOs associated with ring comprised of a continuous conjugated π system, such as benzene.
(c) All six carbon atoms in benzene are sp^2 hybridized, with trigonal planar geometry.

(d) A chair conformation is one of the conformations that a cyclohexane ring can adopt. Benzene is flat and does not adopt the conformations that are accessible to cyclohexane.

(e) The term *ortho* refers to a 1,2-disubstituted benzene ring.

(f) All six carbon atoms in cyclohexane are sp^3 hybridized, with tetrahedral geometry.

(g) Benzene is resonance stabilized (while cyclohexane possesses no π electrons, and therefore has no resonance structures).

(h) Benzene has π electrons, while cyclohexane does not.

(i) The term *para* refers to a 1,4-disubstituted benzene ring.

(j) Cyclohexane undergoes a conformational change called ring flipping. In contrast, benzene is planar and does not undergo ring flipping.

(k) A boat conformation is one of the conformations that a cyclohexane ring can adopt. Benzene is flat and does not adopt the conformations that are accessible to cyclohexane.

18.36.
(a) Each of the rings is comprised of a continuous system of overlapping p orbitals with a total of six π electrons. As such, each ring is aromatic. The compound would be aromatic if either ring alone were aromatic. This compound is most certainly aromatic. Notice that, in determining aromaticity, each ring is considered individually.

(b) As described in Section 18.5, the hydrogen atoms positioned inside the ring experience a steric interaction that forces this compound, called [10]annulene, out of planarity. Since the molecule cannot adopt a planar conformation, the p orbitals cannot continuously overlap with each other to form one system, and as a result, [10]annulene does not meet the criteria for aromaticity. It is not aromatic.

(c) In order for this five-membered ring to exhibit a continuous system of overlapping p orbitals, each oxygen atom would have to be sp^2 hybridized (thereby placing a lone pair in a p orbital). If this were in fact to be the case, the ring would have eight π electrons (two π electrons for the π bond and two π electrons for each lone pair that occupies a p orbital). This would make the ring antiaromatic. Therefore, not all of the oxygen atoms will adopt sp^2 hybridization. At least one of them will adopt sp^3 hybridization to avoid antiaromaticity. As such, the compound does not have a continuous system of p orbitals, and is therefore not aromatic.

(d) One of the lone pairs on the oxygen atom occupies a p orbital, thereby rendering the compound aromatic (six π electrons). The lone pair on the nitrogen atom occupies an sp^2 hybridized orbital that is in the plane of the ring (because the nitrogen atom is already using a p orbital to establish aromaticity).

(e) One of the carbon atoms in the ring is sp^3 hybridized (highlighted):

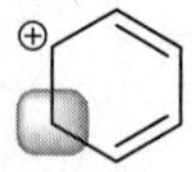

So the structure does not possess a continuous system of overlapping p orbitals. Therefore it is not aromatic.

18.37.
(a) The six-membered ring is aromatic because it is comprised of a continuous system of overlapping p orbitals with six π electrons:

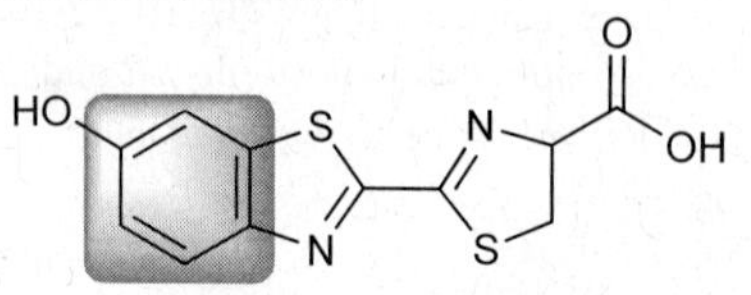

In addition, the adjacent five-membered ring is also aromatic:

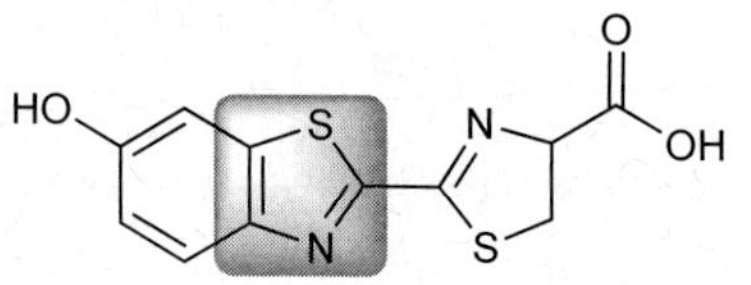

In this five-membered ring, one of the lone pairs on the sulfur atom occupies a p orbital, thereby rendering the ring aromatic (six π electrons). The lone pair on the nitrogen atom occupies an sp^2 hybridized orbital that is in the plane of the ring (because the nitrogen atom is already using a p orbital to establish aromaticity).

The other five-membered ring (not highlighted above) is not aromatic because the ring contains sp^3 hybridized atoms and is therefore not a continuous system of overlapping p orbitals:

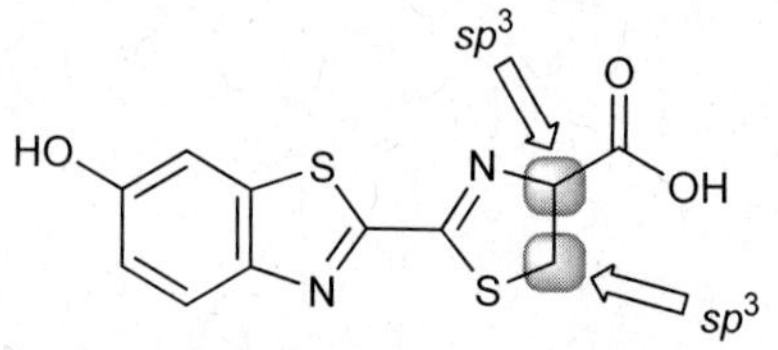

(b) Many of the lone pairs are delocalized by resonance (one lone pair on each of the sulfur atoms, as well as one lone pair on each of the OH groups). However, only one of these lone pairs (on the atom highlighted below) is participating in establishing aromaticity, as explained in the solution to 18.37(a):

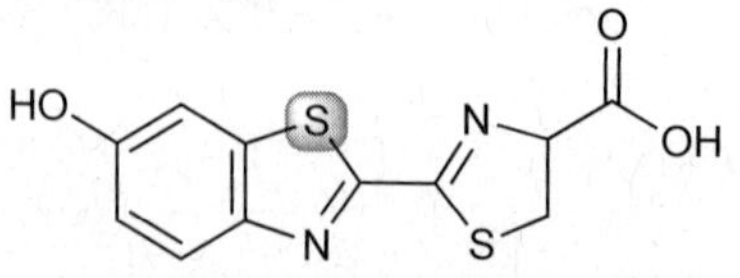

18.38.
(a) If the oxygen atom adopts sp^2 hybridization so that the lone can occupy a p orbital (to give a continuous system of overlapping p orbitals), then there would be a total of eight π electrons (six π electrons from the double

(a) A monosubstituted aromatic ring is expected to produce four signals in the region between 100 and 150 ppm in the ^{13}C NMR spectrum. In addition, the isopropyl group is expected to produce two signals (because the methyl groups are equivalent) in the region between 0 and 50 ppm. In total, we expect six signals in the ^{13}C NMR spectrum.

(b) A *para*-disubstituted aromatic ring (where the two substituents are different from each other) is expected to produce four signals in the region between 100 and 150 ppm in the ^{13}C NMR spectrum. In addition, the methyl group is expected to produce one more signal in the region between 0 and 50 ppm. In total, we expect five signals in the ^{13}C NMR spectrum.

(c) The aromatic ring is 1,3,5-trisubstituted, and all three substituents are identical (methyl groups). Because of symmetry, there are only two unique kinds of aromatic

carbon atoms, giving rise to two signals in the region between 100 and 150 ppm in the ^{13}C NMR spectrum. In addition, the three methyl groups are all equivalent and will collectively give rise to one signal in the region between 0 and 50 ppm. In total, we expect three signals in the ^{13}C NMR spectrum.

(d) The aromatic ring is 1,2,4-trisubstituted, and it lacks the symmetry necessary to render any of the carbon atoms identical. All six carbon atoms of the ring occupy unique environments, and similarly, all three methyl groups also occupy unique environments. We therefore expect six signals in the region between 100 and 150 ppm, and we expect three more signals in the region between 0 and 50 ppm, for a total of nine signals in the ^{13}C NMR spectrum.

18.49. The reagents indicate a Birch reduction, which is believed to occur via a four-step mechanism (as seen in Mechanism 18.1). In the first step, an electron is transferred to the aromatic ring, generating a radical anion, which is then protonated (with methanol functioning as the proton source). Another electron transfer process, followed by a proton transfer, results in the formation of 1,2,4,5-tetramethyl-1,4-cyclohexadiene, as shown. Notice that the four positions bearing the electron-donating alkyl groups are not reduced. The two positions without electron-donating substituents are reduced.

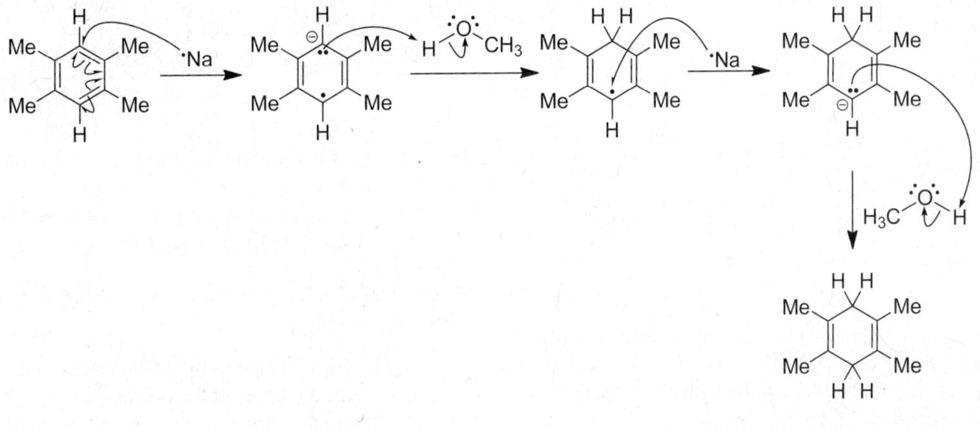

18.50. The reagents indicate a Birch reduction, and *meta*-xylene is expected to undergo a Birch reduction to produce a compound that will exhibit five signals in its ^{13}C NMR spectrum. The five unique carbon atoms are highlighted in the product:

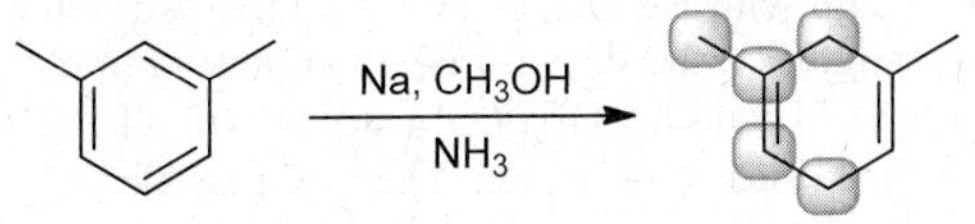

The answer cannot be *ortho*-xylene, because a Birch reduction of *ortho*-xylene would produce a compound that exhibits only four signals in its ^{13}C NMR spectrum:

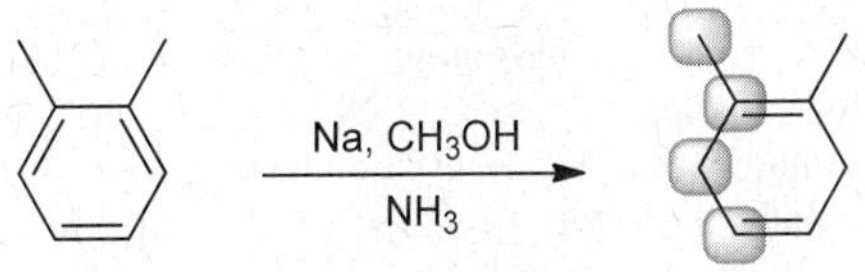

The answer cannot be *para*-xylene, because a Birch reduction of *para*-xylene produces a compound that also exhibits only four signals in its ^{13}C NMR spectrum:

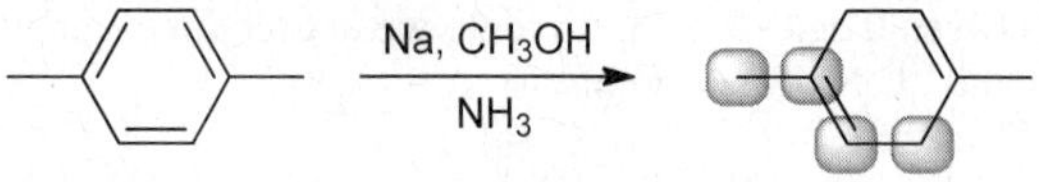

18.51.
(a) The first compound would lack C-H stretching signals just above 3000 cm^{-1}, while the second compound will have C-H stretching signals just above 3000 cm^{-1}.
(b) The ^{1}H NMR spectrum of the first compound will have only one signal, while the ^{1}H NMR spectrum of the second compound will have two signals.
(c) The ^{13}C NMR spectrum of the first compound will have only two signals, while the ^{13}C NMR spectrum of the second compound will have three signals.

18.52.   When either compound is deprotonated, an aromatic anion is generated, which can be drawn with five resonance structures. The resulting anion is the same in either case:

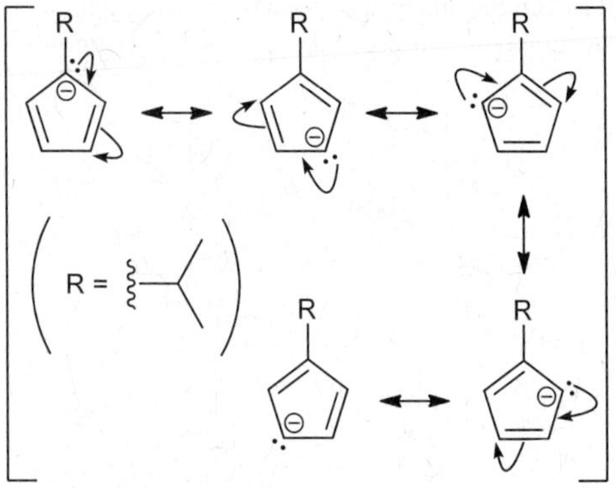

18.53.   In cycloheptatrienone, the significant resonance structures with C$^+$ and O$^-$ contribute significant character to the overall resonance hybrid, because these forms are aromatic:

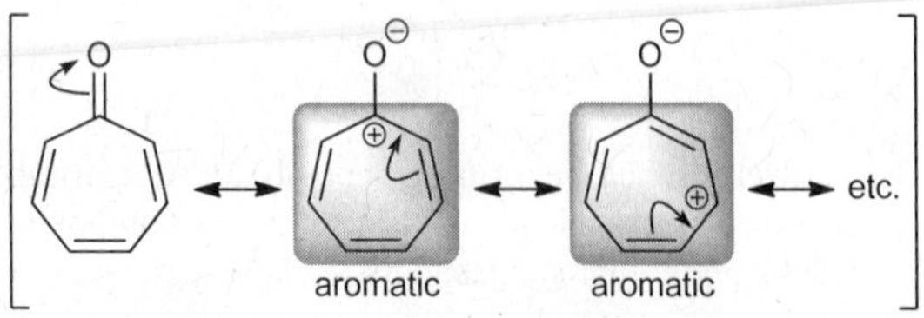

Therefore, the oxygen atom of this C=O bond is particularly electron rich. A similar analysis of cyclopentadienone reveals resonance structures with antiaromatic character:

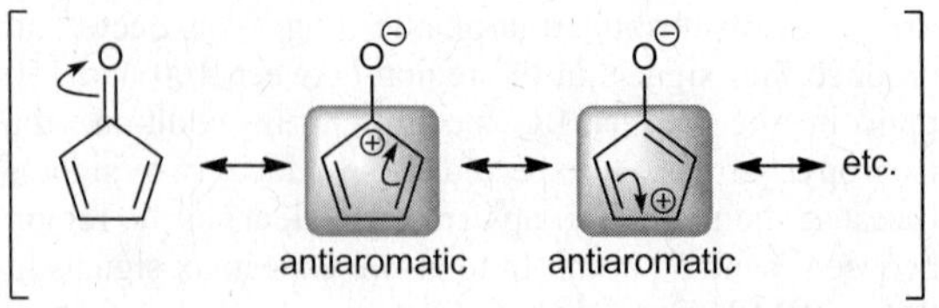

These resonance structures contribute very little character to the overall resonance hybrid, and as a result, the oxygen atom of this C=O bond is less electron rich as compared with most C=O bonds.

18.54.
(a) Each of the rings in the following resonance structure is aromatic.

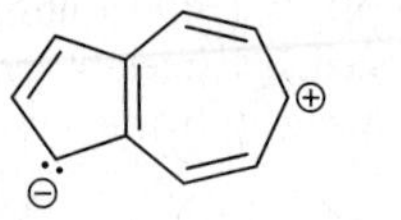

Therefore, this resonance structure contributes significant character to the overall resonance hybrid, which gives the azulene a considerable dipole moment.

(b) The following compound has resonance structures in which both rings are aromatic, as shown here:

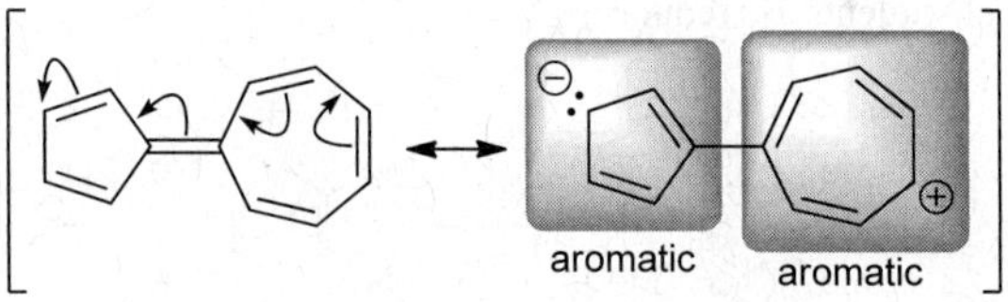

Therefore, this compound will have a significant dipole moment.
In contrast, the other compound does not have resonance structures in which both rings are aromatic.

18.55. There are two questions to ask when approaching a synthesis problem.

  1) Is there a change in the carbon skeleton?
  2) Is there a change in the position or identity of the functional groups?

In this case, there is a change in the carbon skeleton. The product has two more carbon atoms than the starting material:

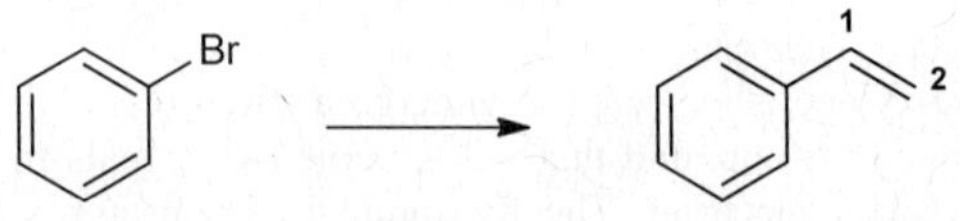

Also, a functional group (a π bond) must be installed. Let's begin with the change in the carbon skeleton. In order to increase the size of the carbon skeleton, we must form a C–C bond. A Grignard reaction will allow us to form the critical C–C bond. The resulting alcohol can then be converted into the desired alkene via an E1 process, as shown here.

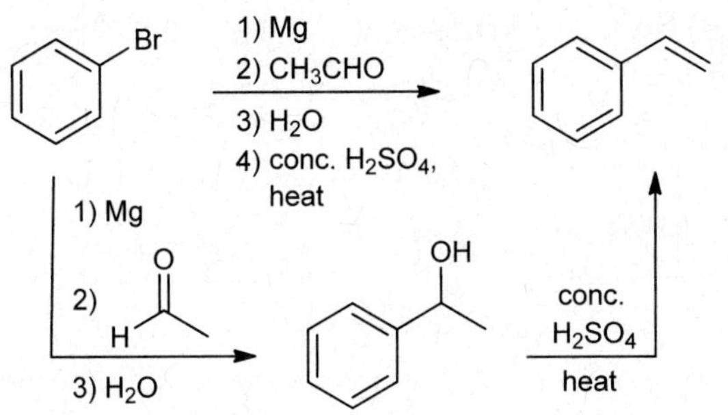

18.56. In acidic conditions, the OH group is protonated, thereby generating a better leaving group (H_2O). Loss of the leaving group generates a tertiary carbocation, which can rearrange via a hydride shift to generate a resonance-stabilized benzylic carbocation. This benzylic carbocation is then captured by a chloride ion (generated in the first step of the mechanism) to give the product, as shown.

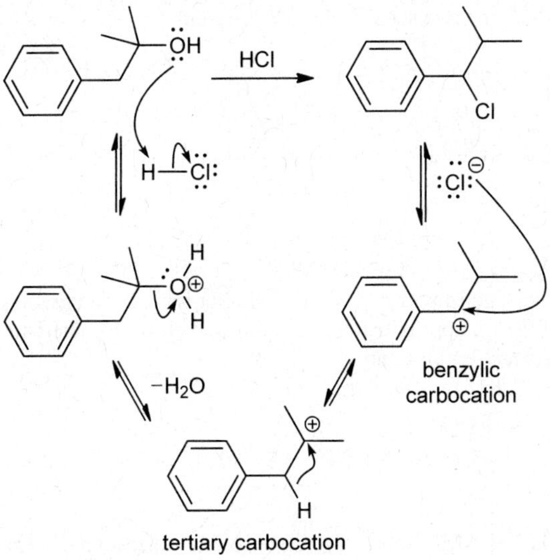

18.57. The molecular formula ($C_9H_{10}O_2$) indicates five degrees of unsaturation (see Section 15.16), which is strongly suggestive of an aromatic ring, as well as one additional double bond or ring. The signal just above 3000 cm^{-1} in the IR spectrum confirms the aromatic ring, as does the signal just above 1600 cm^{-1}. The ^{1}H NMR spectrum exhibits two doublets between 6.9 and 7.9 ppm, each with an integration of 2. This is the characteristic pattern of a disubstituted aromatic ring, in which the two substituents are different from each other:

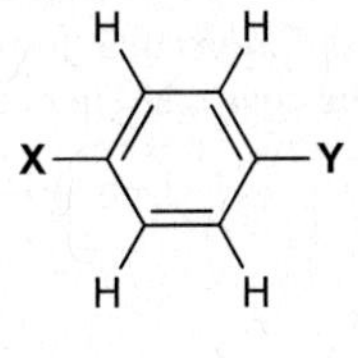

The singlet at 3.9 ppm (with an integration of 3) represents a methyl group. The chemical shift is downfield from the expected benchmark value of 0.9 ppm for a methyl group, indicating that it is likely next to an oxygen atom:

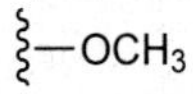

The singlet at 2.6 ppm (with an integration of 3) represents an isolated methyl group. The chemical shift of this signal suggests that the methyl group is neighboring a carbonyl group:

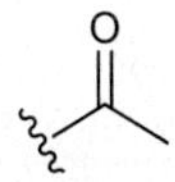

The carbonyl group accounts for one degree of unsaturation, and together with the aromatic ring, this would account for all five degrees of unsaturation. The presence of a carbonyl group is also confirmed by the signal at 196.6 ppm in the ^{13}C NMR spectrum. We have uncovered three pieces, which can only be connected in one way, as shown:

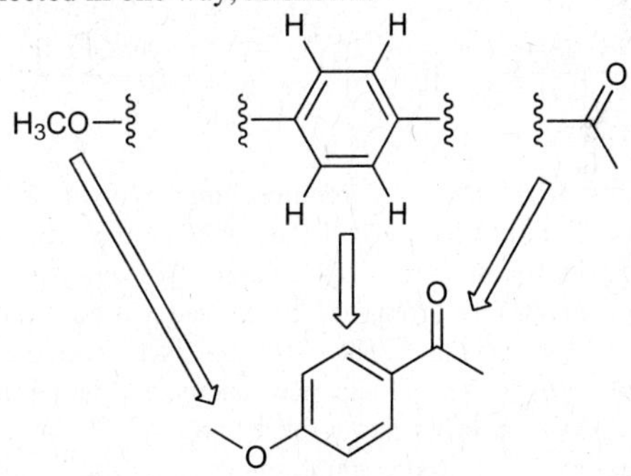

This structure is consistent with the ^{13}C NMR data: four signals for the sp^2 hybridized carbon atoms of the aromatic ring, and two signals for the sp^3 hybridized carbon atoms (one of which is above 50 ppm because it is next to an oxygen atom).

Also notice that the carbonyl group is conjugated to the aromatic ring, which explains why the signal for the C=O bond in the IR spectrum appears at 1676 cm^{-1}, rather than 1720 cm^{-1}.

18.58.
(a) This transformation does not involve a change in the carbon skeleton, but there is a change in the functional group. Specifically, an aldehyde group must be installed at the benzylic position. We have not seen a direct way to achieve the desired transformation (in one reaction flask), so we must first functionalize the benzylic position, and then convert the installed functional group into an aldehyde group. This can be achieved via bromination of the benzylic position, followed by conversion of the resulting bromide into an alcohol (via an S_N2 process), followed by oxidation with PCC to give the product.

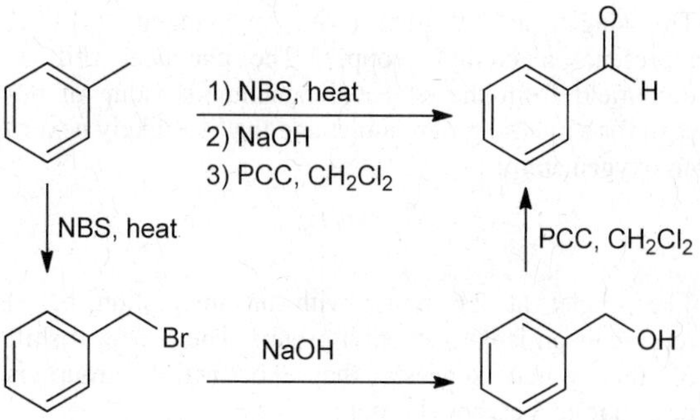

(b) If we functionalize the benzylic position (via bromination), the product can be obtained in just one more step, via a Williamson ether synthesis, as shown:

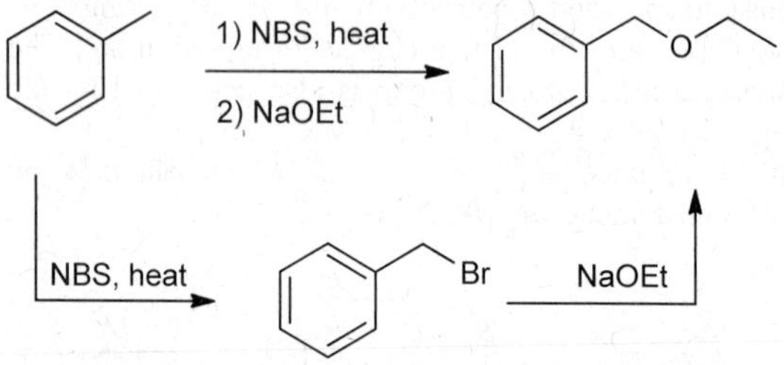

(c) This transformation requires that we move the position of the OH group. This was a strategy that was covered in Chapter 12. Specifically, we saw that the location of an OH group can be moved via elimination followed by addition. The first step is to convert the alcohol into an alkene via acid-catalyzed dehydration (E1). Then, hydroboration-oxidation will convert the alkene into the desired alcohol, via an *anti*-Markovnikov addition of H and OH across the π bond:

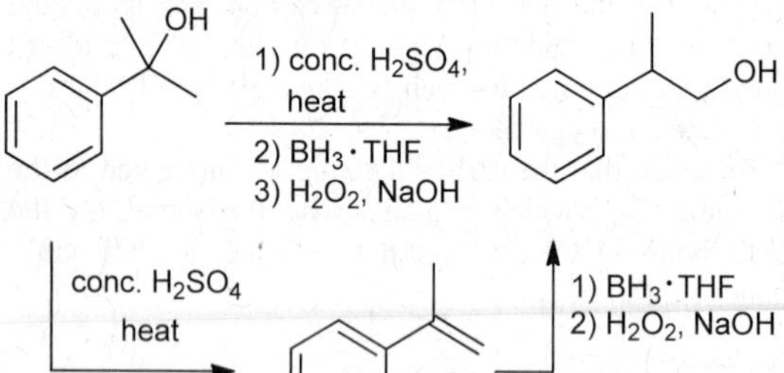

(d) The carbon skeleton is getting larger, so we must form a C–C bond. In addition, we must install a triple bond. Both of these goals can be achieved via bromination at the benzylic position, followed by an S_N2 reaction in which an acetylide ion is used as the nucleophile, as shown:

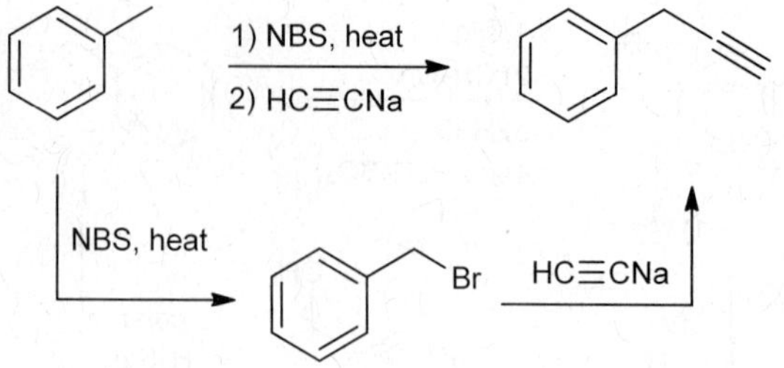

18.59. The molecular formula ($C_{11}H_{14}O_2$) indicates five degrees of unsaturation (see Section 15.16), which is highly suggestive of an aromatic ring, as well as either one double bond or one ring.

In the ^{1}H NMR spectrum, the signals near 7 ppm are likely a result of aromatic protons. Notice that the combined integration of these two signals is 4H. This, together with the distinctive splitting pattern (a pair of doublets), suggests a 1,4-disubstituted aromatic ring, in which the two substituents are different:

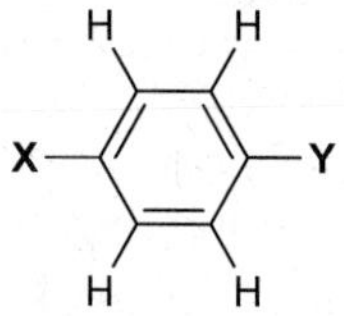

The spectrum also exhibits a singlet with an integration of 9H (at approximately 1.4 ppm) which is characteristic of a *tert*-butyl group.

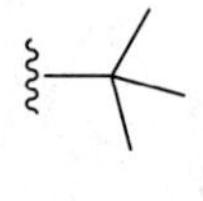

The ^{1}H NMR spectrum has only one other signal, very far downfield, characteristic of a carboxylic acid:

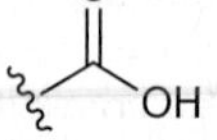

The presence of a carboxylic acid group is confirmed by the IR spectrum, which has a broad signal between 2200 and 3600 cm^{-1}. The carboxylic acid group also accounts for one degree of unsaturation, and together with the aromatic ring, they collectively account for all five degrees of unsaturation.

We have uncovered three fragments, which account for all of the atoms in the molecular formula. These three fragments can only be connected in one way, as shown:

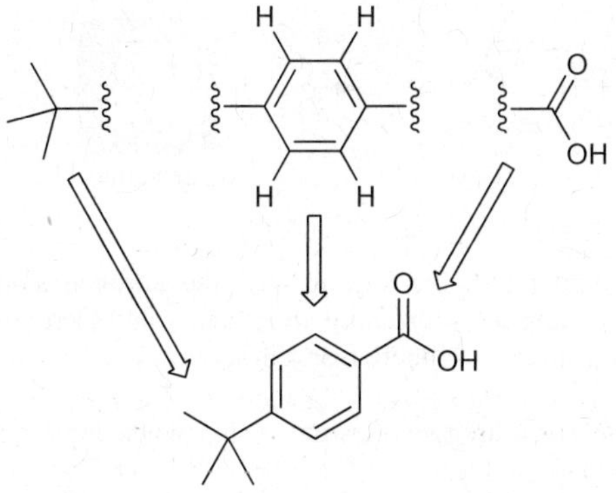

The ^{13}C NMR spectrum confirms this structure. The signal at 172.6 ppm corresponds to the carbon atom of the carbonyl group. The next four signals on the spectrum (above 100 ppm) are characteristic of a disubstituted aromatic ring, and finally, there are two signals between 0 and 50 ppm for the *tert*-butyl group (all three methyl groups are equivalent and give one signal, in addition to the signal from the benzylic carbon atom).

18.60. The molecular formula $(C_9H_{10}O)$ indicates five degrees of unsaturation (see Section 15.16), which is highly suggestive of an aromatic ring, as well as either one double bond or one ring.

In the ^1H NMR spectrum, the multiplet near 7 ppm is likely a result of aromatic protons. The integration of this multiplet is 5H, indicating a monosubstituted aromatic ring:

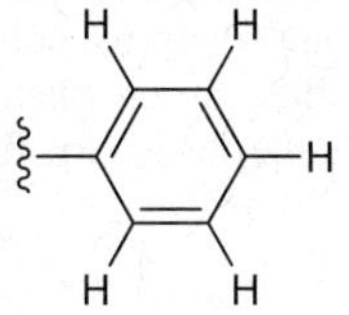

The spectrum also exhibits two triplets (near 3 ppm), indicating two methylene groups connected to each other:

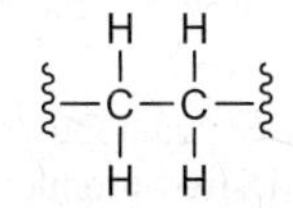

Each of these signals appears more downfield than we might expect for a methylene group (1.2 ppm), so each of these methylene groups must be connected to a group that causes a deshielding effect. This must be taken into account in our final structure.

The singlet near 10 ppm is characteristic of an aldehyde group:

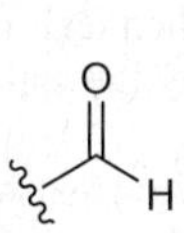

We have uncovered three fragments, which account for all of the atoms in the molecular formula. These three fragments can only be connected in one way, as shown:

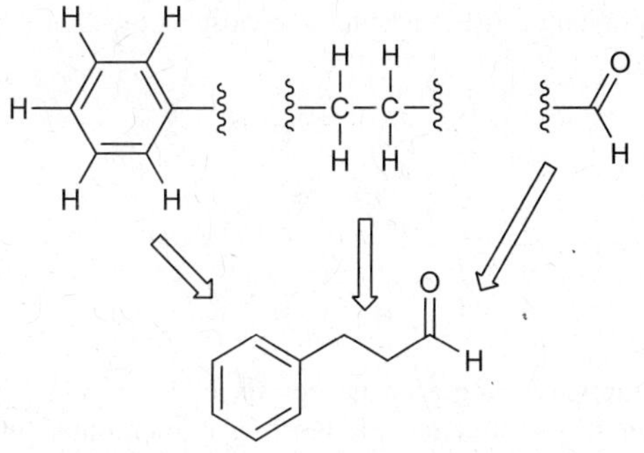

Notice that the structure explains why the two triplets appear downfield. One is shifted downfield primarily as a result of its proximity to the carbonyl group, while the other is shifted primarily as a result of its proximity to the aromatic ring.

The ^{13}C NMR spectrum confirms the structure above. The signal near 200 ppm corresponds to the carbon atom of the carbonyl group. The next four signals on the spectrum (above 100 ppm) are characteristic of a monosubstituted aromatic ring, and finally, there are two signals between 0 and 50 ppm for the methylene groups. In addition, the IR spectrum also confirms this structure. The strong signal just above 1700 cm^{-1} corresponds with the C=O bond, and the signals between 3000 and 3100 cm^{-1} correspond with the aromatic ring. Curiously, there is a signal between 3200 and 3600 which is reminiscent of an O-H signal, although a bit weaker than normal (the signal for an OH group is typically much stronger). This is perhaps due to some concentration of the tautomeric enol that is present at equilibrium (which exhibits an OH group).

18.61.
(a) The second compound holds greater promise as a potential antihistamine, because it possesses two planar aromatic rings separated from each other by one carbon atom. The first compound has only one aromatic ring. The ring with the oxygen atom is not aromatic and not planar.
(b) Yes, because it lacks polar functional groups that would prevent it from crossing the blood-brain barrier, as described in the Medically Speaking application (that appears at the end of Section 18.5).

18.62. No, this compound possesses an allene group (C=C=C). The *p* orbitals of one C=C bond of the allene group do not overlap with the *p* orbitals of the other C=C bond. This prevents the compound from having one continuous system of overlapping *p* orbitals.

18.63. The molecular formula (C_8H_{10}) indicates four degrees of unsaturation (see Section 15.16), consistent with an aromatic ring. The ^1H NMR spectrum of compound **A** has a multiplet near 7 ppm with an integration of 5H, indicating a monosubstituted aromatic ring:

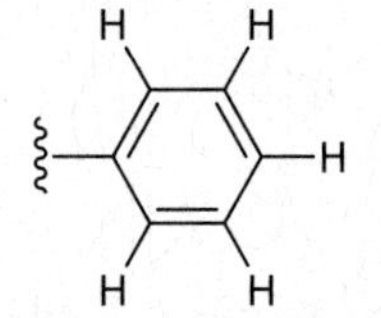

The aromatic ring accounts for six of the eight carbon atoms in the structure. If the ring is monosubstituted, then the other two carbon atoms must be an ethyl group, which is consistent with the two upfield signals in the ^{1}H NMR spectrum:

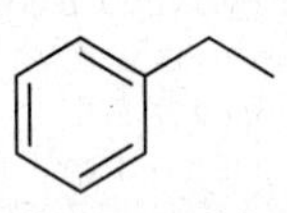

Compound **A**

Compound **B** is a constitutional isomer of **A** and exhibits four signals in its ^{13}C NMR spectrum, which is consistent with *ortho*-xylene (*meta*-xylene would produce five signals, while *para*-xylene would produce three signals in its ^{13}C NMR spectrum):

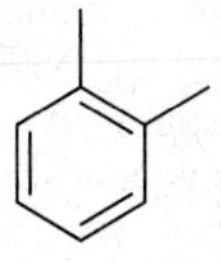

Compound **B**

Compound **C** is a constitutional isomer of **A** and **B**, and it exhibits three signals in its ^{13}C NMR spectrum, which is consistent with *para*-xylene:

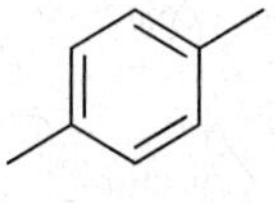

Compound **C**

Compound **D** is a constitutional isomer of **A**, **B**, and **C**, and is likely *meta*-xylene:

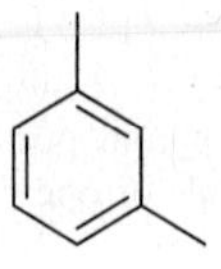

Compound **D**

18.64. The lone pair on the nitrogen atom in compound **A** is localized and is not participating in resonance. It is free to function as a base. The lone pair on the nitrogen atom in compound **B** is delocalized, allowing the compound to achieve aromatic stabilization via resonance:

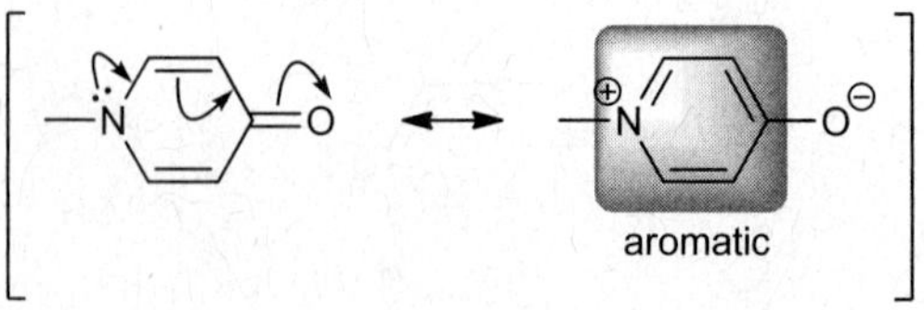

In compound **B**, the lone pair on the nitrogen atom is participating in establishing aromaticity, and therefore, it is not available to function as a base.

18.65. There are two questions to ask when approaching a synthesis problem.

1) Is there a change in the carbon skeleton?
2) Is there a change in the position or identity of the functional groups?

In this case, there is a change in the carbon skeleton. The product has one less carbon atom than the starting material:

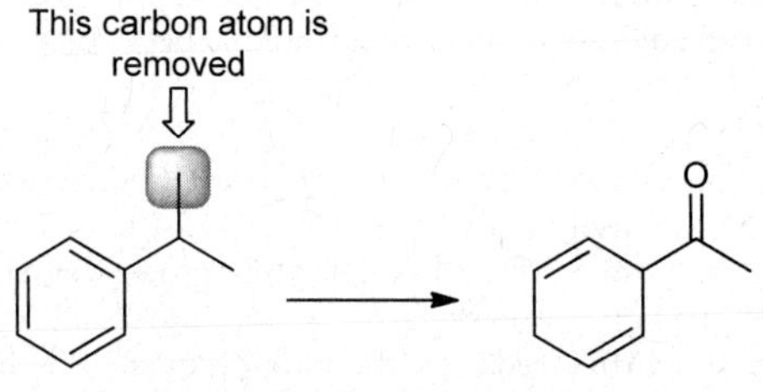

Also, a functional group (a C=O bond) must be installed, and the aromatic ring must be reduced to a diene.
Let's begin with the change in the carbon skeleton. We have only learned one way (ozonolysis) to remove a carbon atom from the carbon skeleton. Ozonolysis would indeed install the necessary functional group in the correct location:

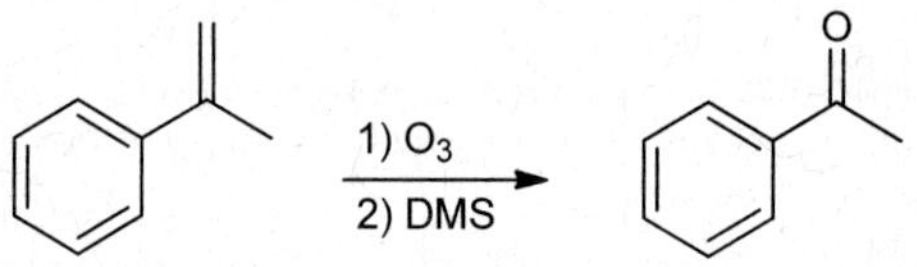

And the ketone above can be treated with Birch conditions to give the desired product:

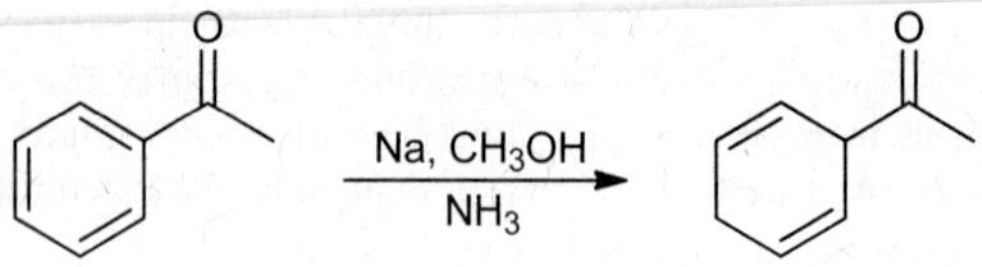

Notice that the regiochemical outcome of the Birch reduction is controlled by the carbonyl group, giving the desired product.
The strategy above requires that we first form the alkene from the starting material:

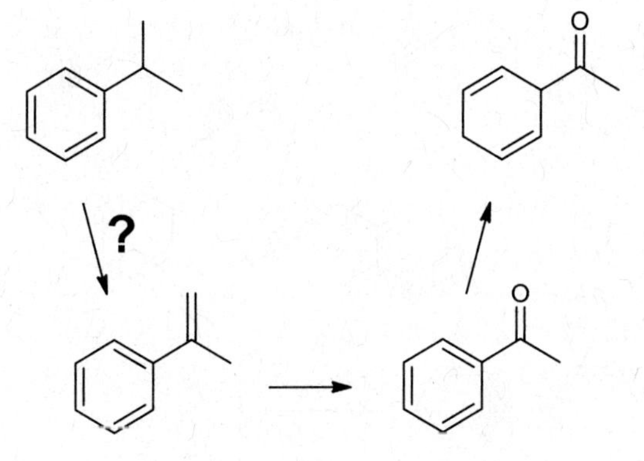

This can be accomplished via bromination of the benzylic position, followed by elimination (E2) upon treatment with a strong base (such as sodium ethoxide). The entire synthesis is summarized here:

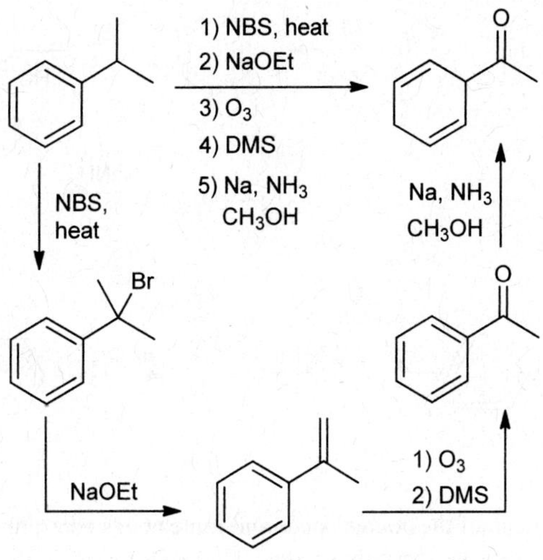

18.66. The following retrosynthetic analysis (and subsequent explanation) represents one strategy for preparing the desired compound from the given starting materials (acetylene and toluene):

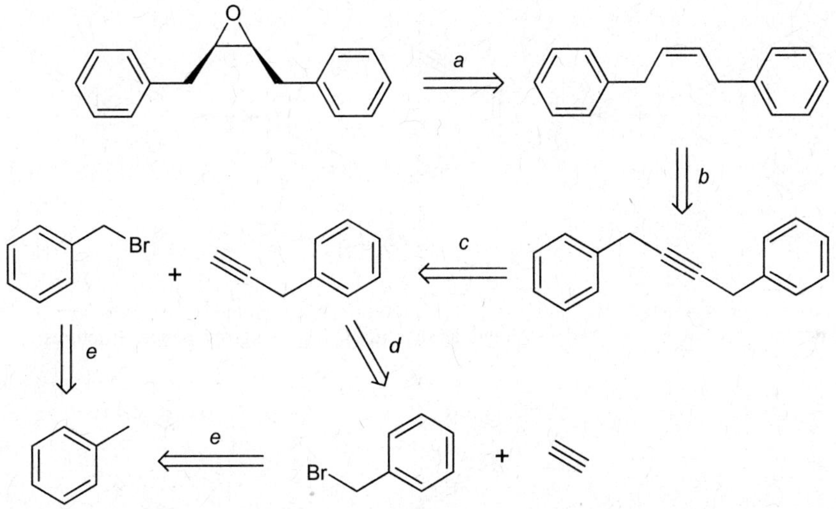

a. The *cis* epoxide can be made from the corresponding *cis* alkene, via epoxidation with MCPBA.
b. The *cis* alkene can be made from the corresponding alkyne, via a *syn* addition of molecular hydrogen (H_2) in the presence of a poisoned catalyst, such as Lindlar's catalyst.
c. The internal alkyne can be made from a terminal alkyne and benzyl bromide. The terminal alkyne is first deprotonated to give an alkynide ion, which is then treated with benzyl bromide (in an S_N2 reaction).
d. The terminal alkyne can be made from acetylene and benzyl bromide. Acetylene is first deprotonated to give an acetylide ion, which is then treated with benzyl bromide (in an S_N2 reaction).
e. Benzyl bromide can be made from toluene via radical bromination at the benzylic position.

The forward process is shown here:

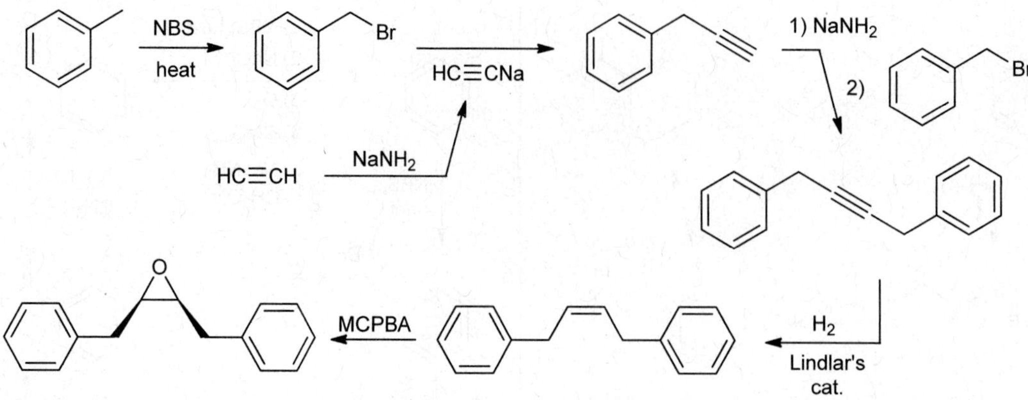

18.67.
(a) Each of the double bonds and triple bonds contributes two π electrons, and each of the unpaired electrons counts as one electron. This gives a total of 14 π electrons, so the large ring in **B** is aromatic.
(b) Each of the negative charges is delocalized over three positions, as seen in the following three resonance structures:

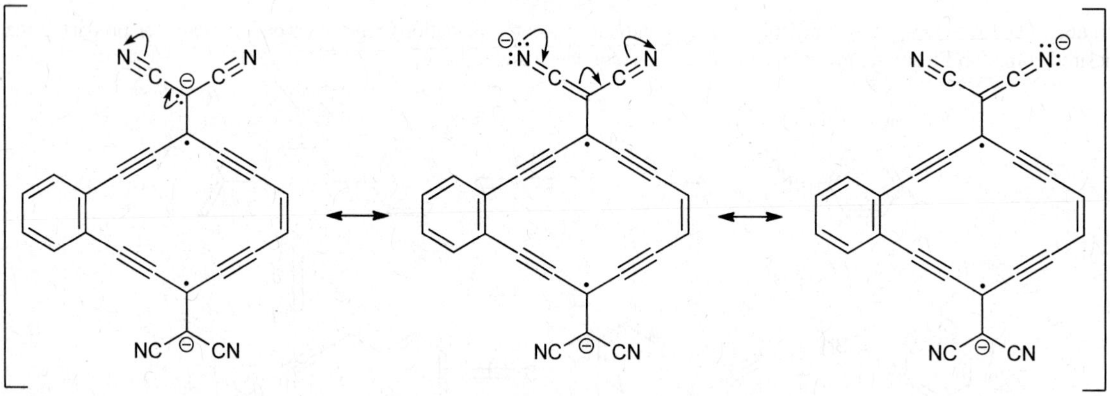

(c) Each of the unpaired electrons is highly delocalized, as seen in the following resonance structures:

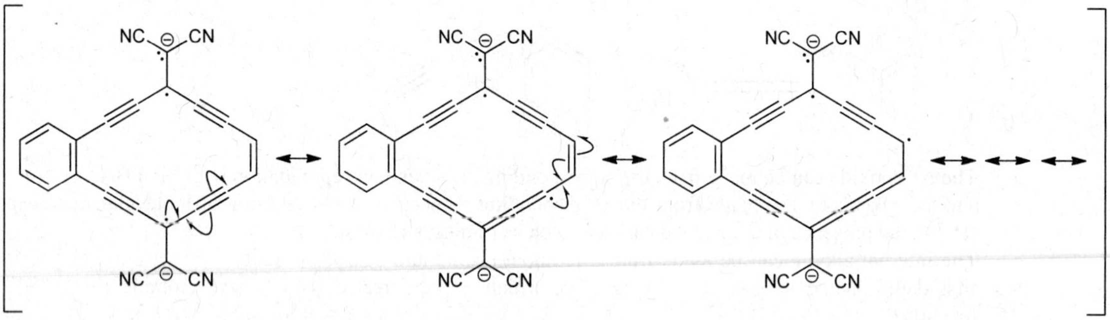

18.68. The nitrogen atom in compound **2** (highlighted below) possesses a double bond and is clearly sp^2 hybridized.

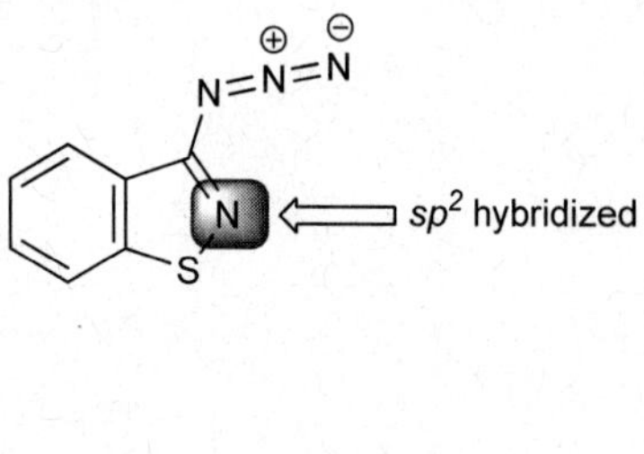

So, we must determine the hybridization state of the same nitrogen atom in compound **1**. This requires that we determine whether the tetrazole ring is aromatic. That is, the lone pair on the nitrogen atom will occupy a p orbital (rendering the nitrogen atom sp^2 hybridized) if doing so establishes aromaticity. However, if doing so establishes antiaromaticity, then the nitrogen atom would adopt sp^3 hybridization to avoid antiaromaticity.
To determine if the tetrazole ring is aromatic, we count the π electrons. The atoms highlighted below contribute a total of four π electrons (two for each π bond).

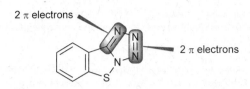

The lone pairs on those atoms must reside in sp^2 hybridized orbitals (rather than p orbitals) in the plane of the ring, and cannot contribute to aromaticity:

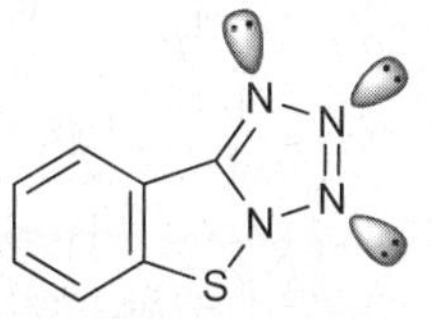

The lone pair on the remaining nitrogen atom therefore occupies a p orbital, giving a total of 6 π electrons.

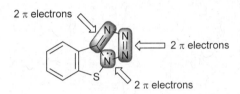

With a Hückel number of π electrons, the tetrazole ring is indeed aromatic, and as such, this nitrogen atom is sp^2 hybridized. Therefore, this nitrogen atom does not undergo a change in hybridization state during the isomerization process. It is sp^2 hybridized in both constitutional isomers.

18.69.
(a) If planar, the cyclooctatetraene ring would be antiaromatic (4n π electrons, where n = 2). It instead adopts a more stable tub conformation that is non-planar, thus avoiding the energetic cost of antiaromaticity. The tub conformation allows all four cyano groups to be oriented approximately in the same direction such that they can coordinate to a single Mg^{++} ion as shown.

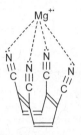

(b) The four nitriles on 1,2,4,5-tetracyanobenzene are oriented away from each other due to benzene's planar (aromatic) structure. Due to this, the four cyano groups cannot be simultaneously coordinated to the same Mg^{++} ion.

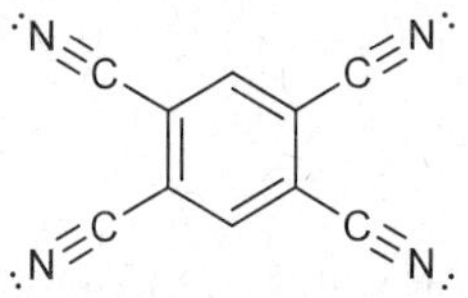

18.70.
(a) Consider the three resonance structures for naphthalene, and focus in particular on the C1-C2 bond and the C2-C3 bond:

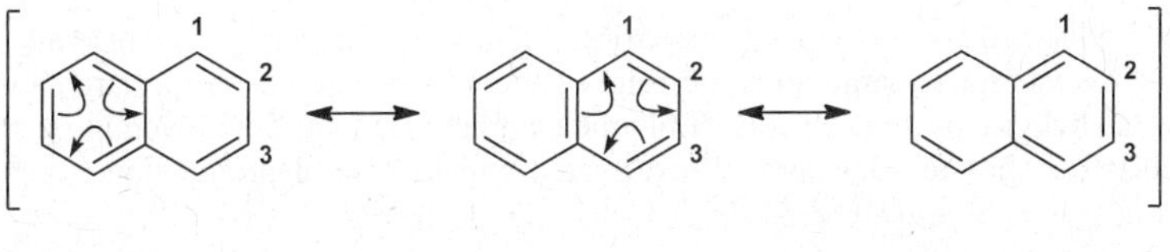

Notice that the C1-C2 bond is a double bond in two of the three resonance structures, while the C2-C3 bond is a double bond in only one of the resonance structures. As such, the former has more double-bond character than the latter, so the C1-C2 bond will have a shorter bond length. Recall that double bonds are shorter than single bonds (see Table 1.2).

(b) Begin by drawing all of the resonance structures:

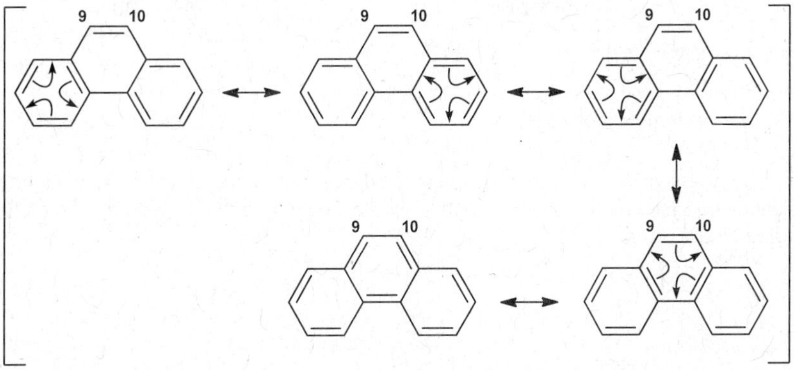

Notice that the C9-C10 bond is a double bond in four of the five resonance structures. No other bond in phenanthrene has this feature. As such, the C9-C10 bond is expected to have the most double-bond character, and consequently, the shortest bond length.

(c) In part (b), we determined that the C9-C10 bond has the most double-bond character, and in fact, this bond functions very much like a regular π bond. It will undergo addition with Br_2 to give a dibromide which retains aromaticity on two of the rings:

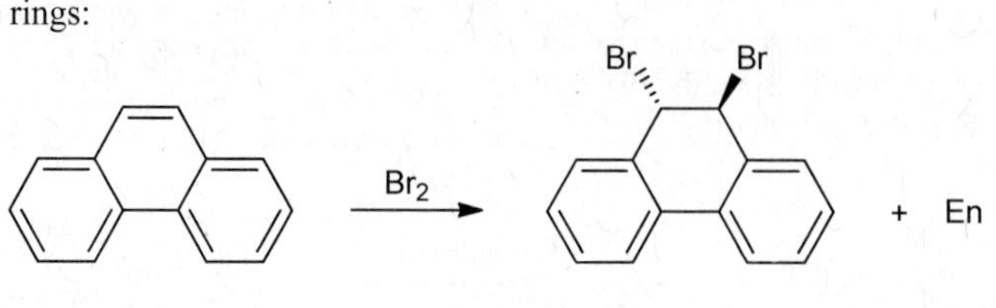

18.71.
(a) We know that *tert*-butyl groups are sterically bulky, so it is reasonable to suggest that the presence of three *tert*-butyl groups provides enough steric hindrance to slow the rate of the Diels-Alder process substantially, allowing the compound to be stable at room temperature.
(b) With these particular substituents, it is possible to draw resonance structures in which the double bonds are not positioned inside the ring, thereby diminishing the destabilizing effect of antiaromaticity.

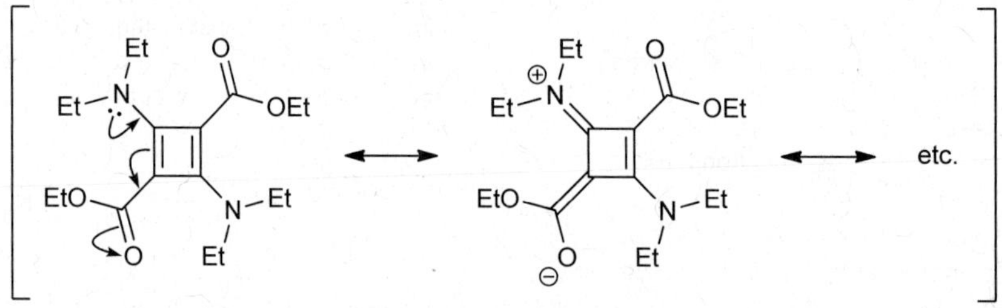

(c) Unlike systems with $4n$ π electrons, which exhibit destabilization associated with antiaromaticity, systems with $4n+2$ π electrons exhibit aromatic stabilization. For these systems, the presence of resonance structures that diminish the aromatic stabilization will have a destabilizing effect.

18.72. In Chapter 16, we learned about magnetic anisotropy, and how aromatic ring currents have a shielding effect on substituents located at the interior of large aromatic rings. The starting material has an aromatic [14]annulene ring, which causes the two methyl groups to be shielded from the external magnetic field (even more so than the protons in TMS), hence their chemical shift of -3.9 ppm. When treated with a base, deprotonation occurs to generate a new aromatic system - notice that the five-membered ring now has 6 π electrons:

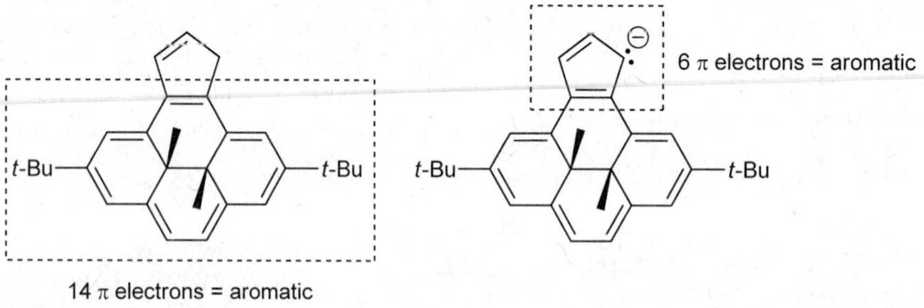

The aromatic stabilization associated with the five membered ring will cause the following highlighted π electrons (fused to the [14]annulene system) to be less available to the larger ring for conjugation, therefore resulting in a diminished ring current:

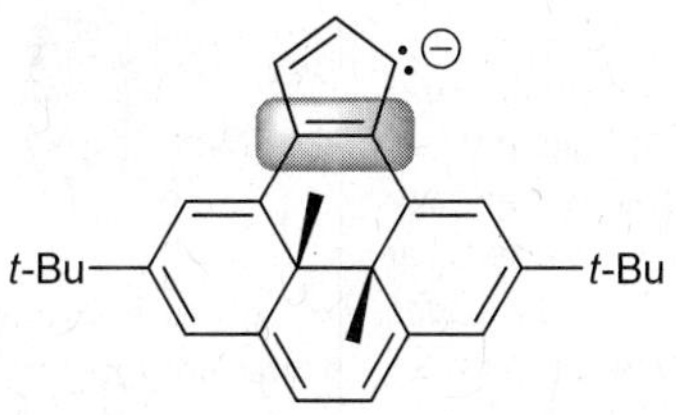

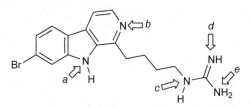

As a result, the local magnetic field established by the [14]annulene system opposes the external magnetic less significantly, causing the methyl groups to become less shielded. Being less shielded is equivalent to being more deshielded, so the signals for the methyl protons are shifted slightly downfield to –1.8 ppm.

18.73. First, let's consider the relative basicity of each nitrogen atom in the structure.

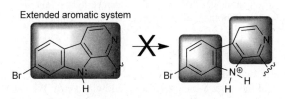

(a) The lone pair on this nitrogen atom occupies a *p*-orbital, thereby including the central ring in the aromatic stabilization of this compound. Protonation of this nitrogen is thus unlikely, as it would disrupt this extended aromatic system. Each of the two six-membered rings of the product below is still aromatic, but the extended aromatic system would be lost.

(b) The lone pair on this nitrogen atom is in an sp^2 hybridized orbital that is orthogonal to the conjugated π system of the ring. Protonation of this nitrogen is thus reasonable, as the extended aromaticity of the resulting conjugate acid remains intact.

(c) The lone pair on this nitrogen atom is delocalized, as can be demonstrated by the following resonance structure. Thus, protonation of this nitrogen atom is not likely.

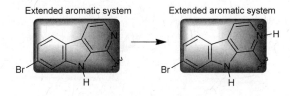

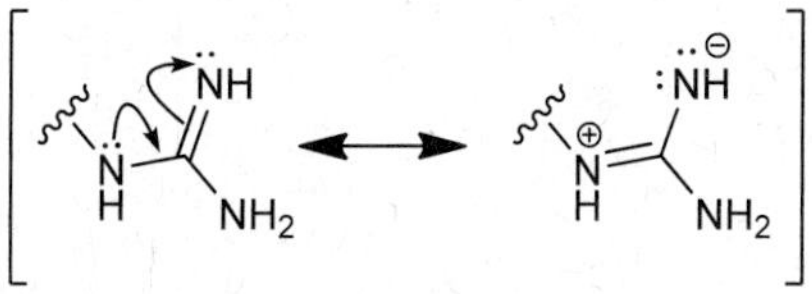

(d) The lone pair on this nitrogen atom is localized, and not involved in any resonance structures. Protonation of this site is therefore reasonable.

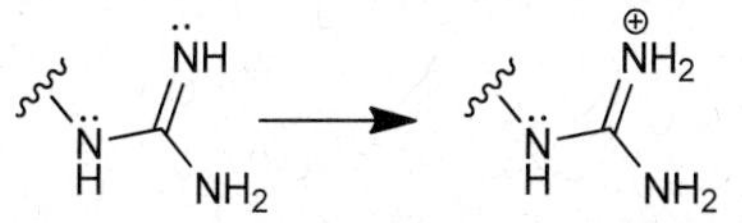

(e) The lone pair on this nitrogen atom is delocalized in a fashion analogous to location *c*. Protonation of this site is thus unlikely.

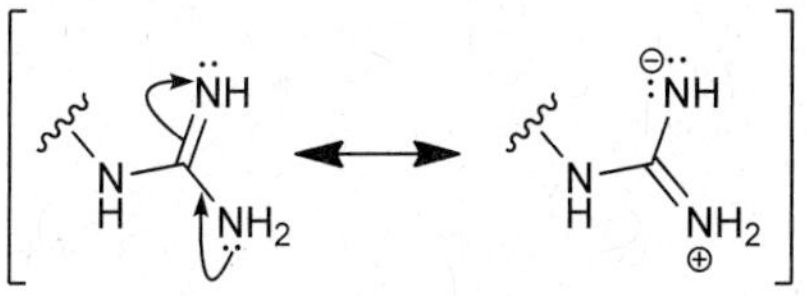

Based on the above analysis, treatment of the natural product with two equivalents of HCl is expected to afford the following structure.

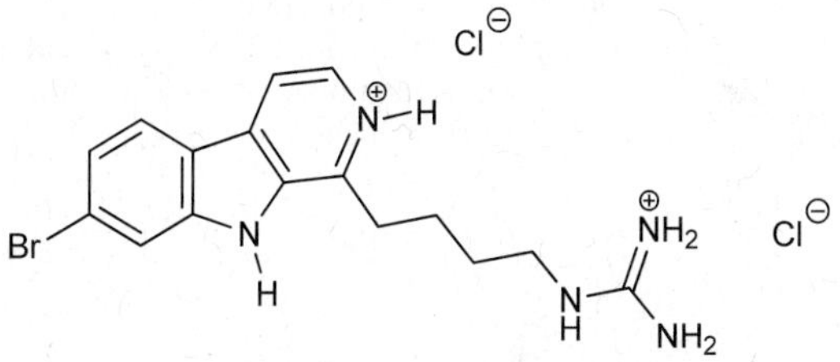

18.74. In order to propose a structure for the fragment at m/z = 39, we should consider all of the possible combinations of atoms that will provide a mass-to-charge ratio of 39. Considering the atoms present (C, H, N) in each of the analytes, there are two possible ionic formulas that fit: C_2NH^+ or $C_3H_3^+$. Considering the fact that the fragment is very common (found in 17 of 20 samples), and relatively abundant (5-84% of the base peak), it is reasonable to assume that this fragment is a relatively stable cation. While there is no obvious stable cation with a formula of C_2NH^+, we can draw an aromatic cation with a formula of $C_3H_3^+$, called a cyclopropenium cation.

Now we must consider why the compounds in the top row produce this fragment, while those in the bottom row do not. Each of the compounds in the top row contains at least three contiguous carbon atoms with at least three hydrogen atoms among them (these six atoms are necessary for formation of the cyclopropenium cation), as highlighted below.

This is a structural feature that the three compounds in the bottom row do not possess. Thus, it is reasonable to propose that the cyclopropenium fragment results from rearrangements of the fragments from the molecules in the top row.

18.75.
(a) Many resonance structures can be drawn for dication **2**. The following are just a few of them:

2a
(4 aromatic rings)   **2b**   **2c**   **2d**   **2e**

Notice that only resonance structure **2a** has four complete aromatic rings. The extra stabilization associated with this structure means that this resonance structure will contribute more character to the overall resonance hybrid, and as a result, the two neighboring central carbon atoms in dication **2** will both exhibit a large degree of carbocationic character.

(b) Recall the concept of hyperconjugation from Chapter 6. In the twisted state, a σ bond from one of the 5-membered rings can donate electron density into the adjacent carbocation of the other 5-membered ring and vice versa. We call this phenomenon cross-hyperconjugation. It is a stabilizing effect that causes the nonplanar conformation of dication **2** to be lower in energy than the planar conformation.

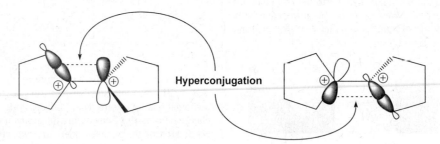

Hyperconjugation

note: benzene rings omitted for clarity

18.76.

(a) Sodium methoxide is a strong base, which removes the most acidic proton in the starting material. The resulting resonance-stabilized anion can then function as a nucleophile and attack 3-bromo-1-propyne in an S_N2 process, giving **A**.

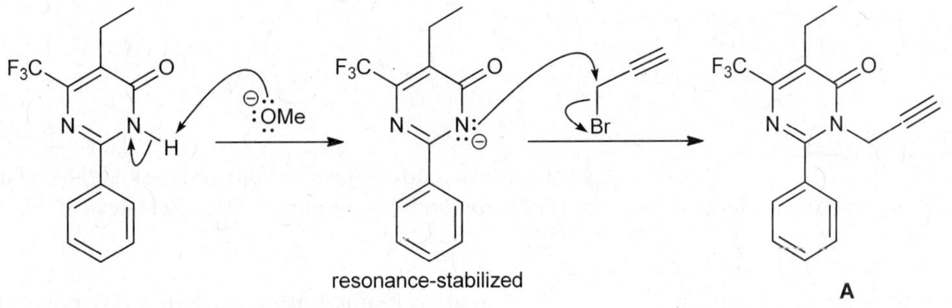

resonance-stabilized

A

(b) As mentioned in the solution to part (a), the anionic intermediate is resonance-stabilized, so we draw the resonance structures:

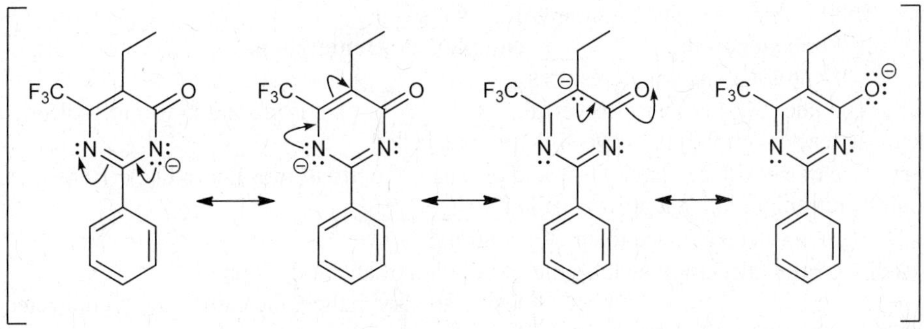

Among these resonance structures, the last one is expected to contribute the most character to the overall resonance hybrid, for two reasons: 1) the negative charge is placed on an oxygen atom (rather than on a nitrogen atom or on a carbon atom), and 2) this resonance structure exhibits a benzene-like ring. In fact, this anionic intermediate is indeed aromatic, and we expect the oxygen atom to bear the majority of the delocalized negative charge. Based on this, we can propose the following structure and mechanism of formation for **B**:

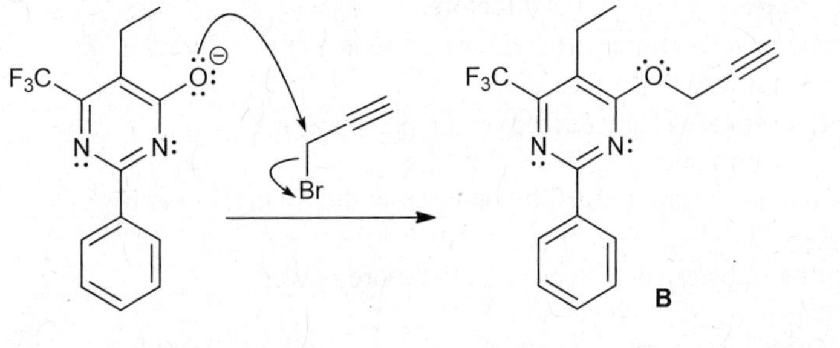

B

(c) Compound **B** has aromatic stabilization that is absent in **A**. As such, **B** is expected to be lower in energy, so the product mixture will have more of **B** than **A**.

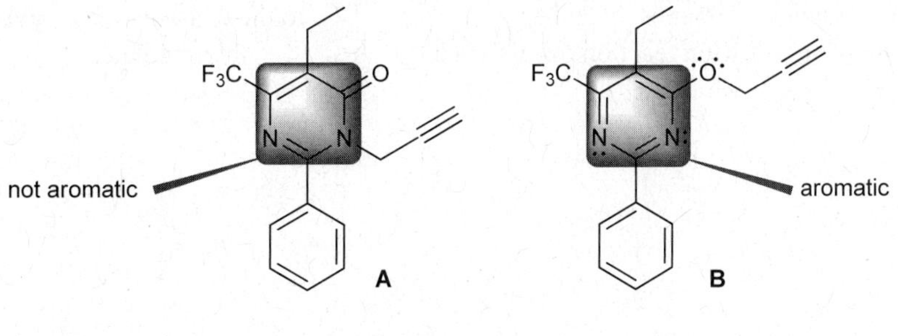

not aromatic   **A**   **B**   aromatic

Chapter 19
Aromatic Substitution Reactions

Review of Concepts

Fill in the blanks below. To verify that your answers are correct, look in your textbook at the end of Chapter 19. Each of the sentences below appears verbatim in the section entitled *Review of Concepts and Vocabulary*.

- In the presence of iron, an _____ **aromatic substitution** reaction is observed between benzene and bromine.
- Iron tribromide is a _____ acid that interacts with Br_2 and generates Br^+, which is sufficiently electrophilic to be attacked by benzene.
- Electrophilic aromatic substitution involves two steps:
    - Formation of the _____ **complex**, or **arenium ion.**
    - Deprotonation, which restores _____.
- Sulfur trioxide (SO_3) is a very powerful _____ that is present in fuming sulfuric acid. Benzene reacts with SO_3 in a reversible process called _____.
- A mixture of sulfuric acid and nitric acid produces the **nitronium ion** (NO_2^+). Benzene reacts with the nitronium ion in a process called _____.
- A nitro group can be reduced to an _____ group.
- **Friedel–Crafts alkylation** enables the installation of an alkyl group on _____. When choosing an alkyl halide, the carbon atom connected to the halogen must be _____ hybridized.
- When treated with a Lewis acid, an acyl chloride will generate an _____ **ion**, which is resonance stabilized and not susceptible to _____ rearrangements.
- When a Friedel–Crafts acylation is followed by a **Clemmensen reduction**, the net result is the installation of an _____ group.
- A methyl group is said to **activate** an aromatic ring and is an _____-_____ **director**.
- All activators are _____-_____ directors.
- A nitro group **deactivates** an aromatic ring and is a _____director.
- Most deactivators are _____directors.
- **Strong activators** are characterized by the presence of a _____ immediately adjacent to the aromatic ring.
- **Strong deactivators** are powerfully electron withdrawing, either by _____ or _____.
- When multiple substituents are present, the more powerful _____ dominates the directing effects.
- In a **nucleophilic aromatic substitution** reaction, the aromatic ring is attacked by a _____. This reaction has three requirements: 1) the ring must contain a powerful electron-withdrawing group (typically a _____ group) 2) the ring must contain a _____, and 3) the leaving group must be either _____ or _____ to the electron-withdrawing group.
- An **elimination-addition** reaction occurs via a _____ intermediate.

Review of Skills

Fill in the blanks and empty boxes below. To verify that your answers are correct, look in your textbook at the end of Chapter 19. The answers appear in the section entitled *SkillBuilder Review*.

19.1 Identifying the Effects of a Substituent

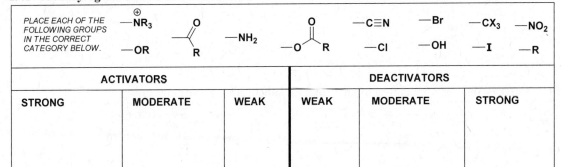

	ACTIVATORS			DEACTIVATORS	
STRONG	MODERATE	WEAK	WEAK	MODERATE	STRONG

19.2 Identifying Directing Effects for Disubstituted and Polysubstituted Benzene Rings

IN THE FOLLOWING COMPOUND, IDENTIFY THE POSITION THAT IS MOST REACTIVE TOWARDS ELECTROPHILIC AROMATIC SUBSTITUTION.

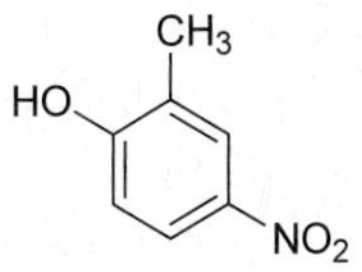

19.3 Identifying Steric Effects for Disubstituted and Polysubstituted Aromatic Benzene Rings

IN THE FOLLOWING COMPOUND, IDENTIFY THE POSITION THAT IS MOST REACTIVE TOWARDS ELECTROPHILIC AROMATIC SUBSTITUTION.

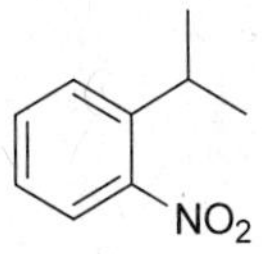

19.4 Using Blocking Groups to Control the Regiochemical Outcome of an Electrophilic Aromatic Substitution Reaction

IDENTIFY REAGENTS THAT WILL ACHIEVE THE FOLLOWING TRANSFORMATION:

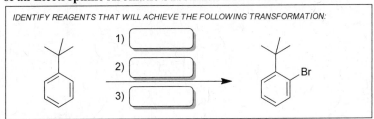

1)
2)
3)

19.5 Proposing a Synthesis for a Disubstituted Benzene Ring

IDENTIFY A THREE-STEP PROCESS FOR ACHIEVING THE FOLLOWING TRANSFORMATION:

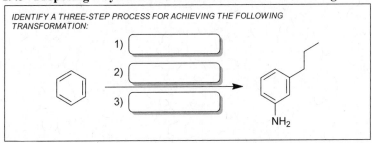

1)
2)
3)

19.6   Proposing a Synthesis for a Polysubstituted Benzene Ring

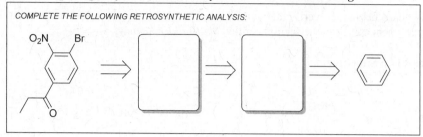

19.7   Determining the Mechanism of an Aromatic Substitution Reaction

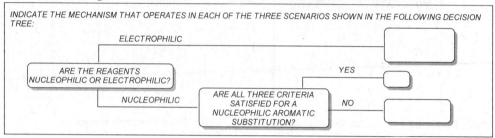

Review of Reactions

Identify the reagents necessary to achieve each of the following transformations. To verify that your answers are correct, look in your textbook at the end of Chapter 19. The answers appear in the section entitled *Review of Reactions*.

Electrophilic Aromatic Substitution

Nucleophilic Aromatic Substitution

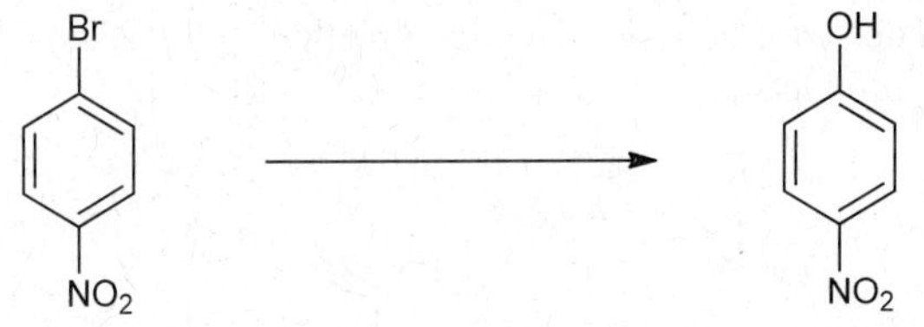

Elimination-Addition

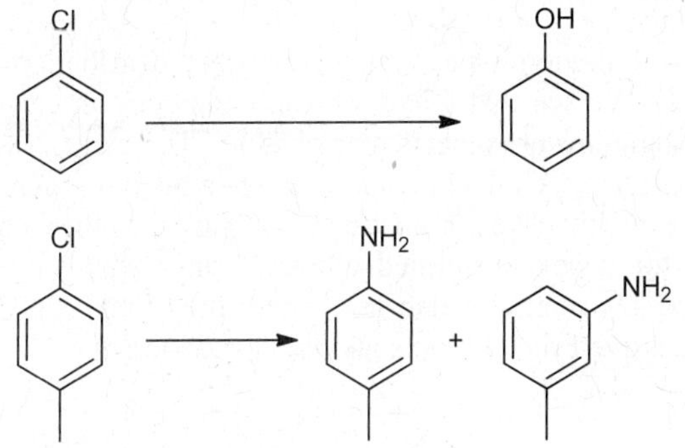

Common Mistakes to Avoid

For most *ortho-para* directors, we can rely on steric effects that will favor substitution at the *para* position rather than the *ortho* position:

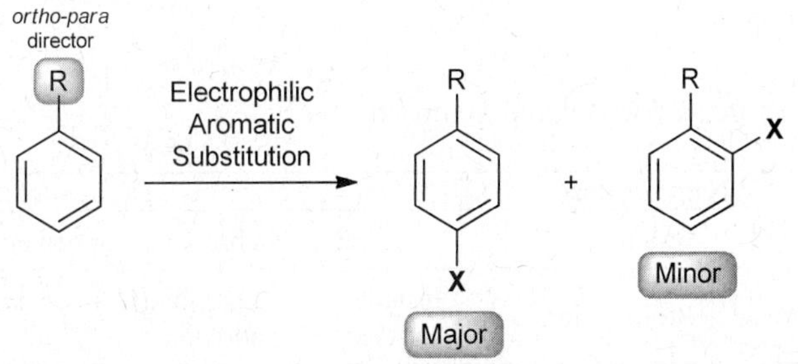

The major product results from *para* attack, and the minor product results from *ortho* attack. If we want to achieve substitution at the *ortho* position, then we must employ a blocking group:

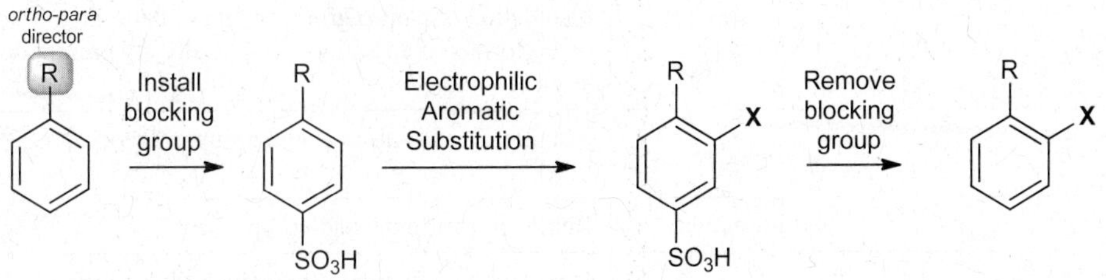

With this technique, we use the *para*-directing effects of the substituent to block (temporarily) the *para* position. This allows us to perform the desired reaction at the *ortho* position, followed by removal of the blocking group.

When designing a synthesis that involves the directing effects of an *ortho-para* director, you can generally rely on a preference for *para* attack (at the expense of *ortho* attack). A notable exception is toluene, which exhibits a very small group (a methyl group). The steric bulk provided by a methyl group is small enough to give a nearly equal mixture of *ortho* and *para* products:

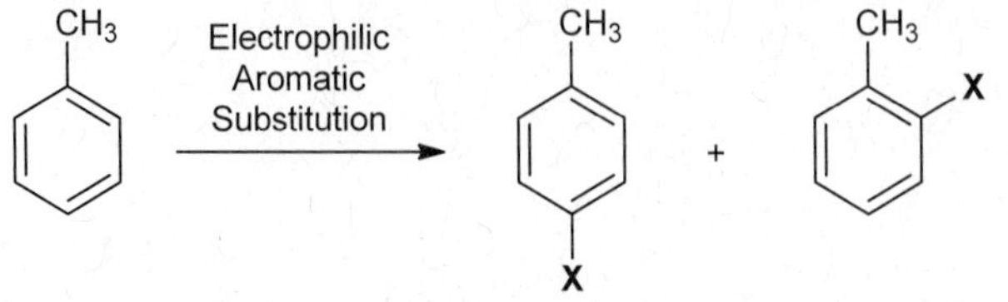

It is often difficult to predict which product will predominate, and the product distribution will be somewhat dependent on the conditions employed. We see that when toluene undergoes an electrophilic aromatic substitution reaction, a mixture of *ortho* and *para* products is unavoidable. Therefore, it would be inefficient to rely on the directing effects (to favor *para* attack over *ortho* attack) of a methyl group. Avoid making this mistake, as there are many ways that this mistake can manifest itself in a synthesis problem. For example, if you are designing a synthesis for a polysubstituted aromatic ring (where one of the substituents is a methyl group), consider installing the methyl group last, rather than first. Consider the following transformation, which requires two successive Friedel-Crafts alkylation reactions:

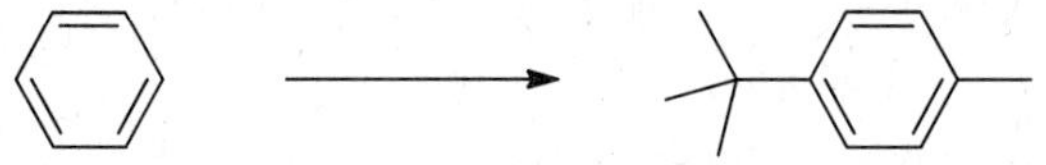

This transformation is best achieved if the *tert*-butyl group is installed prior to the methyl group. After the *tert*-butyl group is installed, subsequent methylation is forced to occur almost exclusively at the *para* position (because of the steric bulk of the *tert*-butyl group). In contrast, if methylation is performed first, then installation of the *tert*-butyl group will not proceed with the same degree of regioselectivity.

Useful reagents

The following is a list of reagents encountered in this chapter:

Reagents	Name of Reaction	Description
Br$_2$, AlBr$_3$ or FeBr$_3$	Bromination	Installation of a bromine atom on an aromatic ring.
Br$_2$	Bromination	Installation of a bromine atom on a moderately or strongly activated aromatic ring.
Cl$_2$, AlCl$_3$ or FeCl$_3$	Chlorination	Installation of a chlorine atom on an aromatic ring.
Cl$_2$	Chlorination	Installation of a chlorine atom on a moderately or strongly activated aromatic ring.
HNO$_3$, H$_2$SO$_4$	Nitration	Installation of a nitro group on an aromatic ring. Cannot be performed if an amino group is already present on the aromatic ring.
Fuming H$_2$SO$_4$	Sulfonation	Installation of a sulfonic acid group, often used as a blocking group.
Dilute H$_2$SO$_4$	Desulfonation	Removes a sulfonic acid group.
RCl, AlCl$_3$	Friedel-Crafts alkylation	Installation of an alkyl group on an aromatic ring. This process is limited to R groups that are not susceptible to carbocation rearrangements. Unless otherwise indicated, assume conditions will favor monoalkylation. This reaction cannot be performed on a moderately or strongly deactivated aromatic ring.

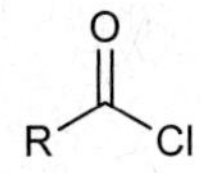 R C(=O) Cl , AlCl₃	Friedel-Crafts acylation	Installation of an acyl group on an aromatic ring. Cannot be performed on a moderately or strongly deactivated aromatic ring.
Fe or Zn, HCl	Reduction	Reduction of a nitro group to give an amino group.
Excess NBS	Benzylic bromination	Exhaustive bromination of the benzylic position.
1) KMnO₄, NaOH, heat 2) H₃O⁺	Benzylic oxidation	Oxidation of the benzylic position to give a carboxylic acid.
Zn(Hg), HCl, heat	Clemmensen reduction	Reduction of a carbonyl group to give a methylene (CH₂) group.
NaOH	S$_N$Ar or elimination-addition	A strong nucleophile used in nucleophilic aromatic substitution reactions as well as elimination-addition reactions.
NaNH₂	elimination-addition	A very strong base (and strong nucleophile) used in elimination-addition reactions.

Solutions

19.1. In the first step, the aromatic ring functions as a nucleophile and attacks the electrophile (I⁺), giving a resonance-stabilized intermediate (called a sigma complex). The sigma complex then loses a proton to restore aromaticity, affording the product:

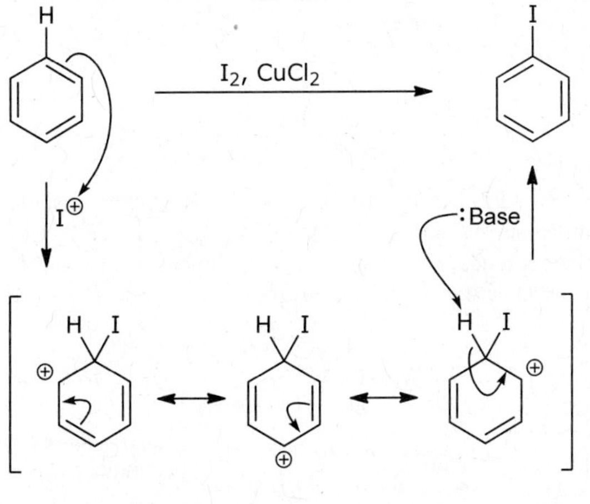

the last step of sulfonation). The aromatic ring is then protonated to give a resonance-stabilized intermediate (sigma complex), which then loses SO₃ to regenerate aromaticity.

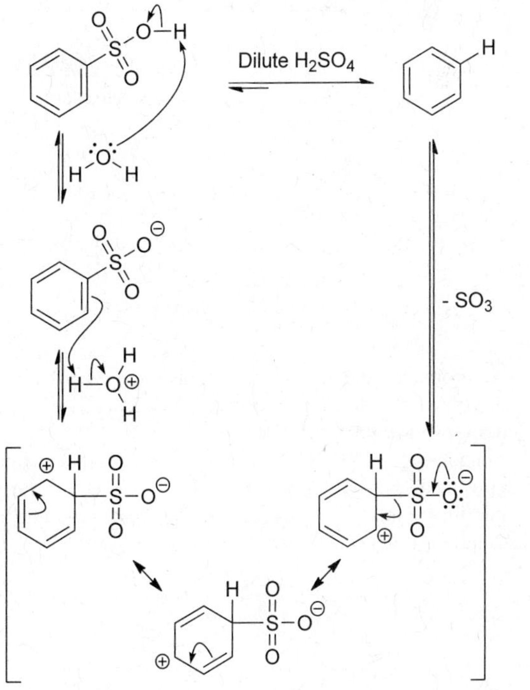

The problem statement has not provided enough information for us to determine the identity of the base that removes the proton in the last step of the process. This type of situation is extremely rare. In nearly all cases that you encounter throughout this course, you should be able to determine the identity of the base.

19.2. If we consider the fact that desulfonation is the reverse process of sulfonation, then the first step of desulfonation should be a proton transfer (since that was

19.3. The aromatic ring functions as a base and removes a deuteron from D_2SO_4, giving an intermediate sigma complex, which then loses a proton to restore aromaticity.

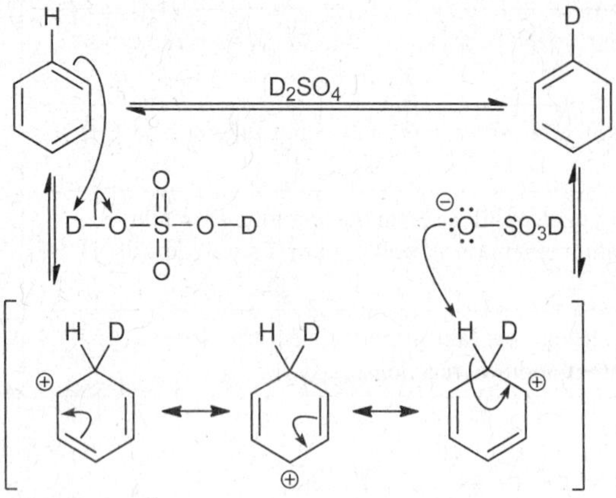

19.4.
In the presence of sulfuric acid, nitric acid can be protonated, followed by loss of water to give a nitronium ion (highlighted):

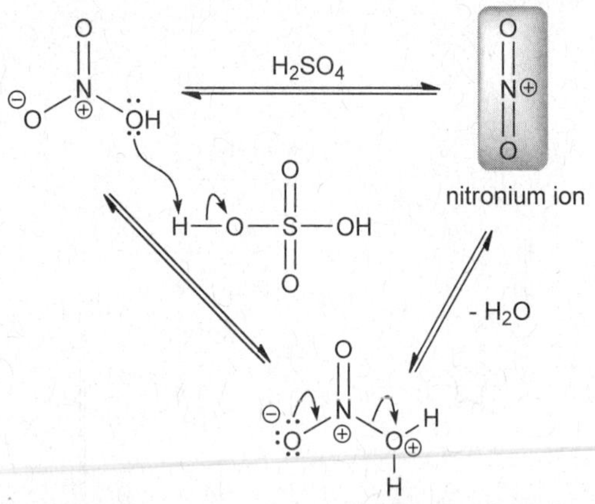

The aromatic ring then functions as a nucleophile and attacks the nitronium ion to give an intermediate sigma complex, which then loses a proton to restore aromaticity. Water is the likely base for the final deprotonation step, since water is present in large quantities (and there are no strong bases present in strongly acidic conditions).

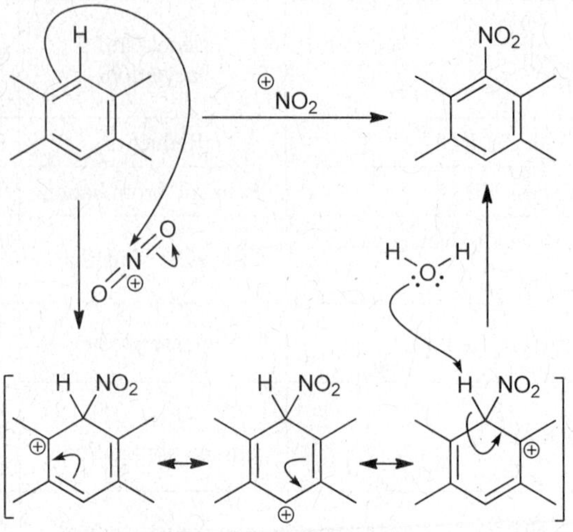

19.5.
(a) In a Friedel–Crafts alkylation, an R group is installed on an aromatic ring when the ring is treated with R–Cl in the presence of a Lewis acid. In this case, the R group is a cyclohexyl group, so we expect a cyclohexyl group to be installed on the aromatic ring, giving the following product:

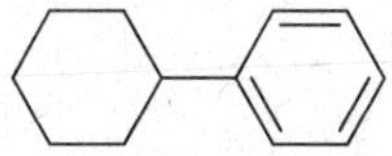

(b) In a Friedel–Crafts alkylation, an R group is installed on an aromatic ring when the ring is treated with R–Cl in the presence of a Lewis acid. In this case, the R group is a *tert*-butyl group, so we expect a *tert*-butyl group to be installed on the aromatic ring. No rearrangement is expected, since the carbocation formed upon chloride departure is tertiary.

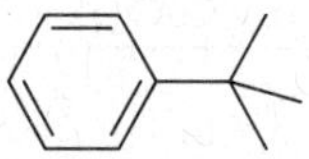

(c) In a Friedel–Crafts alkylation, an R group is installed on an aromatic ring when the ring is treated with R–Cl in the presence of a Lewis acid. In this case, the initially formed carbocation can rearrange:

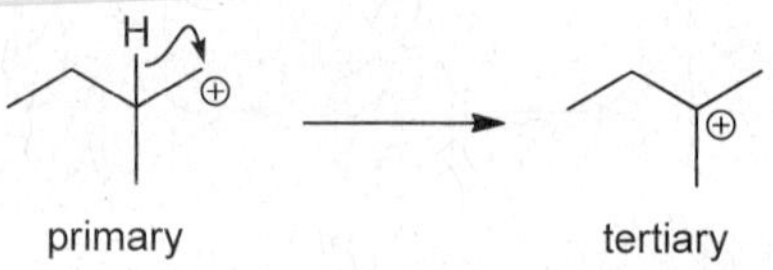

As such, there are two carbocations that can be captured by the aromatic ring in an electrophilic aromatic substitution reaction, giving rise to the following two products:

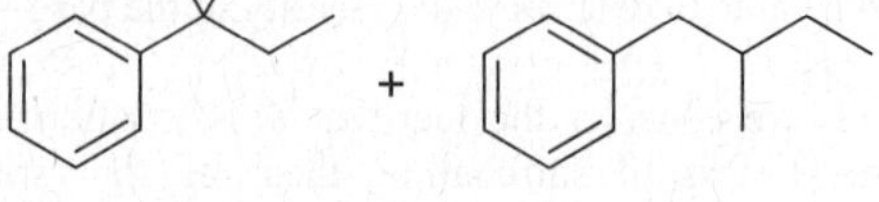

19.6. An interaction between a lone pair (on one of the chlorine atoms) and the Lewis acid generates a complex that can lose a leaving group (AlCl$_4^-$) to give a tertiary carbocation. This carbocation then functions as an electrophile in an electrophilic aromatic substitution reaction. That is, benzene attacks the carbocation to give an intermediate sigma complex, which then loses a proton to restore aromaticity. All of the steps above are then repeated once again to close the ring, except this time, the electrophilic aromatic substitution reaction occurs in an intramolecular fashion. A lone pair on the chlorine atom interacts with the Lewis acid to generate a complex that can lose a leaving group (AlCl$_4^-$) to give a tertiary carbocation. This carbocation then functions as an electrophile in an *intramolecular* electrophilic aromatic substitution reaction. That is, the aromatic ring attacks the carbocation to give an intermediate sigma complex, which then loses a proton to restore aromaticity:

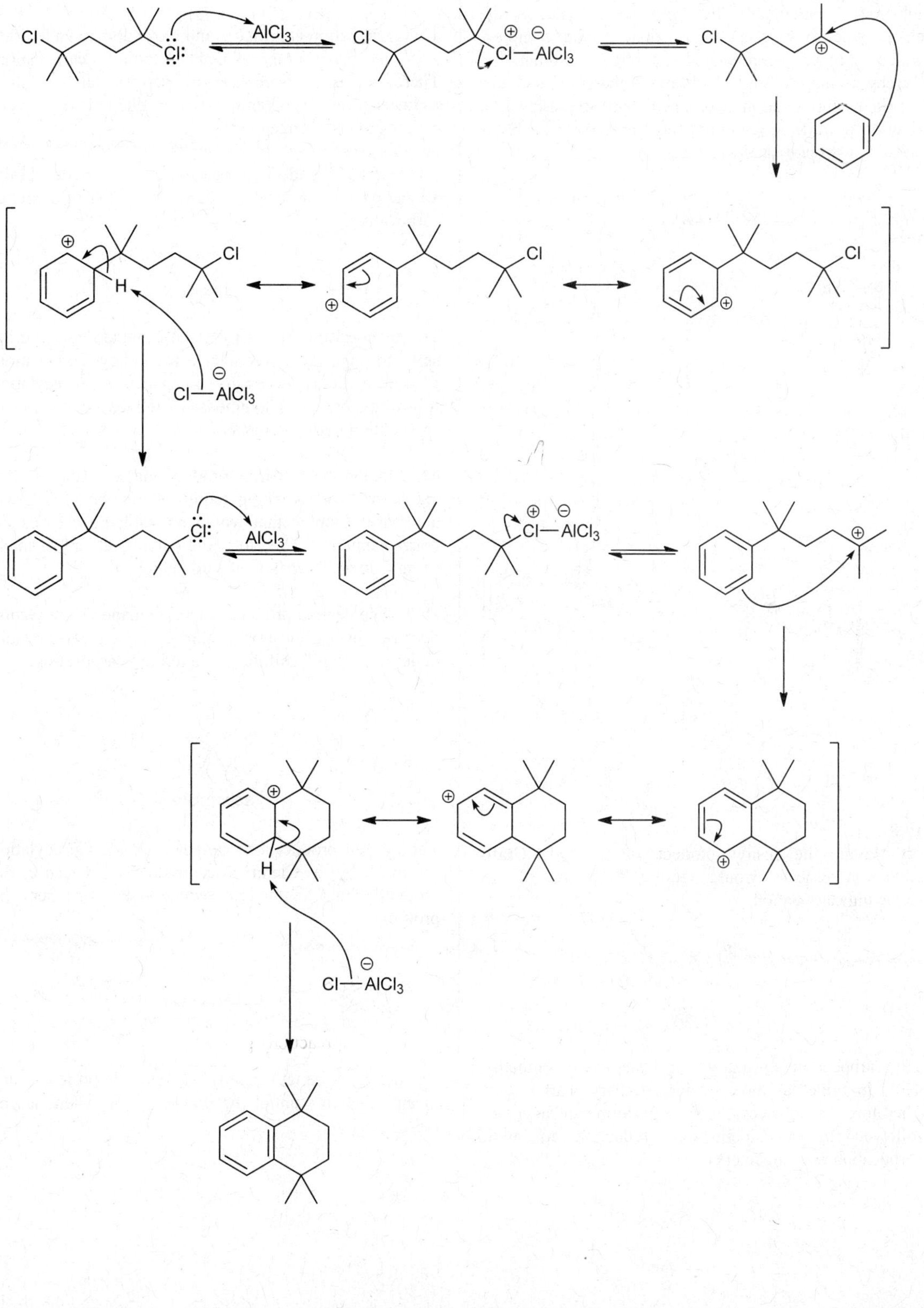

19.7. Sulfuric acid is used as an aqueous solution, so we must indicate H_3O^+ as the proton source in our mechanism, rather than H_2SO_4 (because H_2SO_4 is a stronger acid than H_3O^+, so mixing H_2SO_4 and water causes the protons to be transferred from H_2SO_4 to H_2O, giving H_3O^+).   In these acidic conditions, the alkene is protonated to give a carbocation. This carbocation then serves as an electrophile in an electrophilic aromatic substitution reaction.   The aromatic ring attacks the electrophile to give an intermediate sigma complex, which is then deprotonated to restore aromaticity, thereby giving the product. Water is the likely base for the final deprotonation step, since water is present in a solution of sulfuric acid (and there are no strong bases present in strongly acidic conditions).

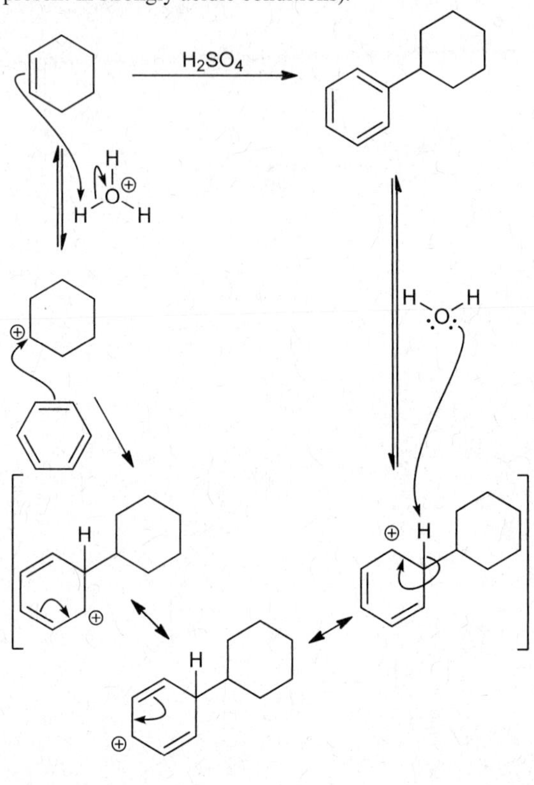

19.8.
(a) Making the desired product via a Friedel–Crafts alkylation process would require formation of the following carbocation:

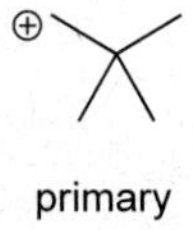

primary

This carbocation is primary, and it can undergo a methyl shift to give a more stable, tertiary carbocation. Therefore, it is necessary to perform an acylation followed by a Clemmensen reduction to avoid carbocation rearrangements.

(b) Making the desired product via a Friedel–Crafts alkylation process would require formation of the following carbocation:

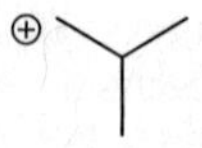

primary

This carbocation is primary, and it can undergo a hydride shift to give a more stable, tertiary carbocation. Therefore, it is necessary to perform an acylation followed by a Clemmensen reduction to avoid carbocation rearrangements.

(c) Making the desired product via a Friedel–Crafts alkylation process would require formation of a propyl carbocation.

primary

This carbocation is primary, and it can undergo a hydride shift to give a more stable, secondary carbocation. Therefore, it is necessary to perform an acylation followed by a Clemmensen reduction to avoid carbocation rearrangements.

(d) Making the desired product via a Friedel–Crafts alkylation process would require formation of a Lewis acid/base complex that would not undergo carbocation rearrangement.   Therefore, the compound can be made using a direct Friedel–Crafts alkylation.

19.9. The desired product cannot be made via alkylation because the carbocation required (shown here) would undergo a methyl shift to give a tertiary carbocation.

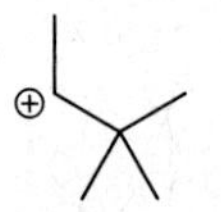

secondary

The desired product also cannot be made via acylation followed by a Clemmensen reduction, because the product of a Clemmensen reduction has two benzylic protons:

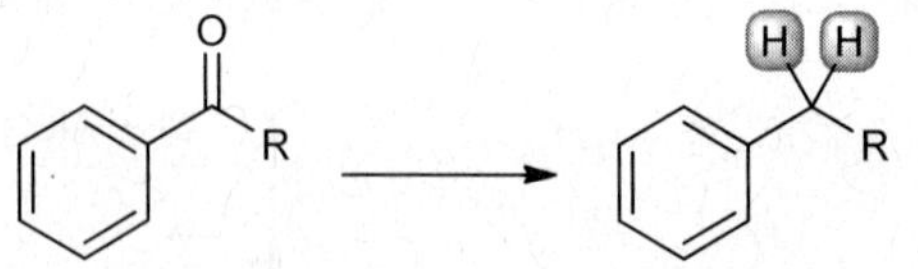

The desired product has only one benzylic proton, which means that it cannot be made via a Clemmensen reduction.

19.10. The anhydride interacts with the Lewis acid to give a complex which loses a leaving group to produce an acylium ion. This acylium ion then serves as an electrophile in an electrophilic aromatic substitution reaction. The aromatic ring attacks the electrophile to give an intermediate sigma complex, which is then deprotonated to restore aromaticity, thereby giving the product. An acetate ion is the likely base for the final deprotonation step, since acetate is a by-product of the reaction. Acetate is formed when the leaving group (of the second step of mechanism) breaks apart into $AlCl_3$ and acetate (as shown).

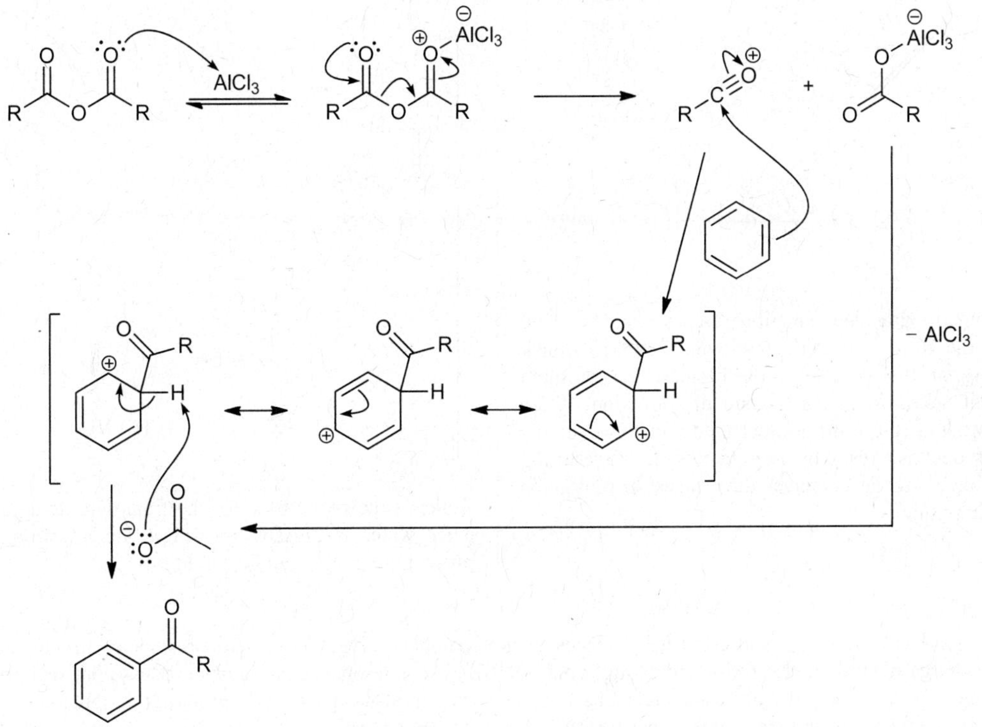

19.11. The methyl group is an activator, which directs the incoming electrophile (Br^+) to the *ortho* and *para* positions:

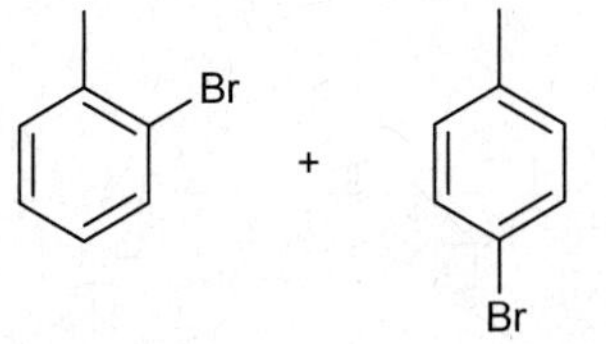

19.12.
(a) The ethoxy group is an activator, which directs the incoming electrophile (NO_2^+) to the *ortho* and *para* positions:

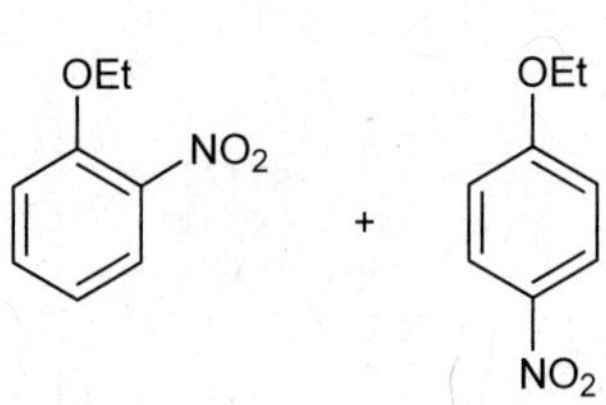

(b) In the presence of sulfuric acid, nitric acid can be protonated, followed by loss of water to give a nitronium ion:

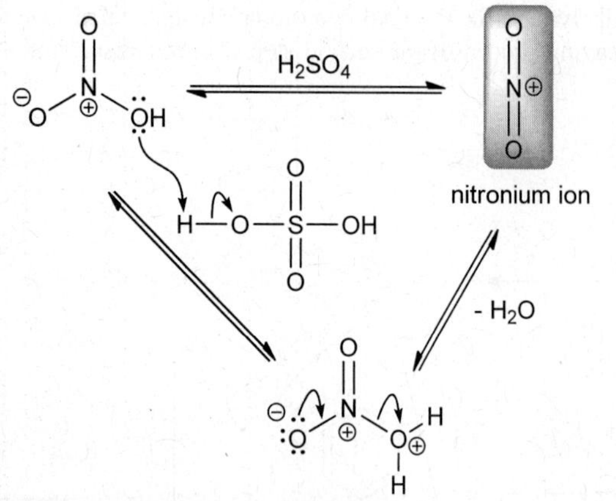

nitronium ion

The aromatic ring then functions as a nucleophile and attacks the nitronium ion. We will draw the attack occurring at the *para* position because the problem statement asked for a mechanism of formation of the major product (the *para* product predominates over the *ortho* product as a result of steric factors). The resulting intermediate sigma complex then loses a proton to restore aromaticity.

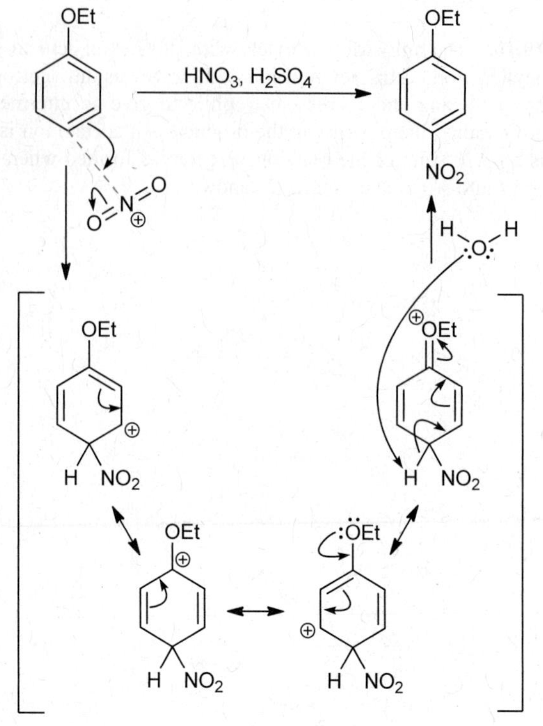

Water is the likely base for the final deprotonation step, since water is present (and there are no strong bases present in strongly acidic conditions).

19.13. As shown below, attack at C4 or C6 produces a sigma complex in which two of the resonance structures have a positive charge next to an electron-withdrawing group (NO_2). These resonance structures are less contributing to the resonance hybrid, thereby destabilizing the sigma complex. In contrast, attack at C5 produces a sigma complex for which none of the resonance structures have a positive charge next to a nitro group.

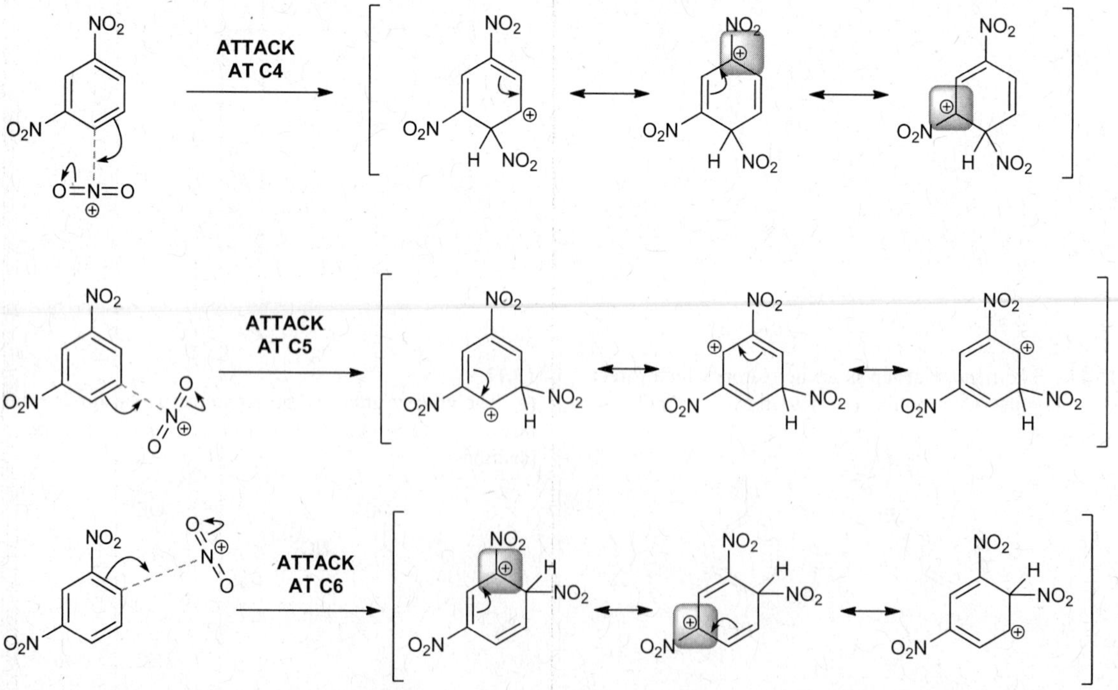

Attack at C2 is identical to attack at C4 because the starting material has symmetry that renders C2 and C4 identical.

19.14. The chlorine atom in chlorobenzene deactivates the ring relative to benzene. If benzene requires a Lewis acid for chlorination, than chlorobenzene should certainly require a Lewis acid for chlorination.

19.15. *Ortho* attack and *para* attack are preferred because each of these pathways involves a sigma complex with four resonance structures (shown below). Attack at the *meta* position involves formation of a sigma complex with only three resonance structures, which is not as stable as a sigma complex with four resonance structures. The reaction will proceed more rapidly via the lower energy sigma complex, so attack takes place at the *ortho* and *para* positions in preference to the *meta* position.

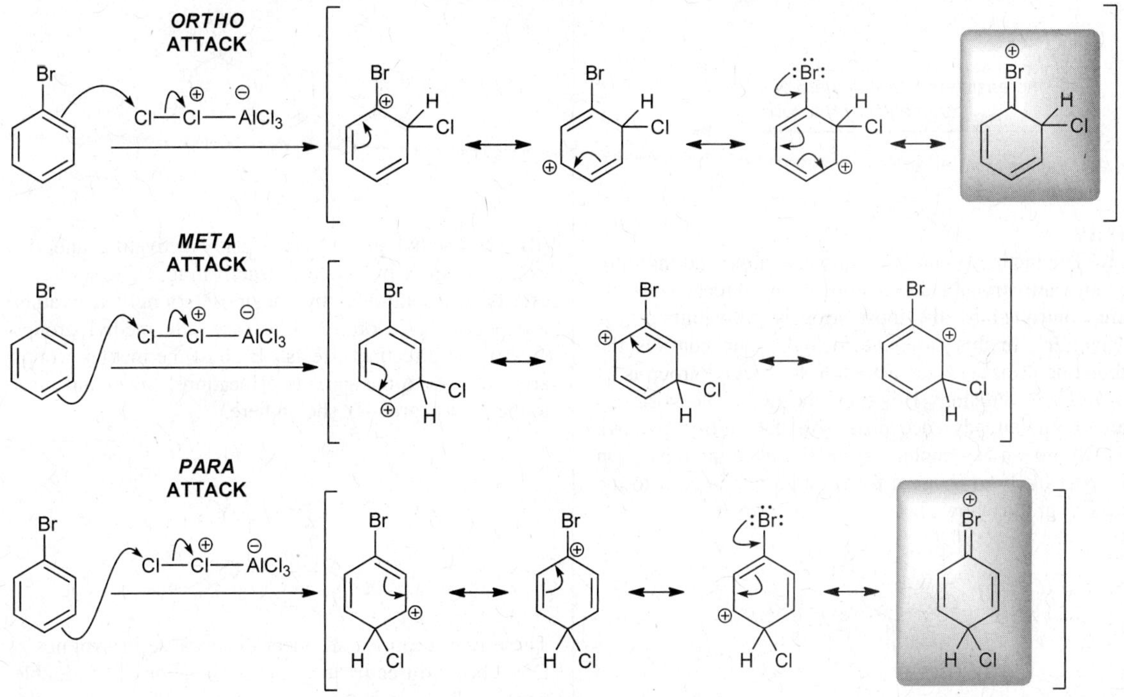

19.16.
(a) As seen in Table 19.1, the nitro is strongly deactivating and *meta*-directing.
(b) As seen in Table 19.1, an acyl group is moderately deactivating and *meta*-directing.
(c) As seen in Table 19.1, a bromine atom is weakly deactivating and *ortho, para*-directing.
(d) This aromatic ring exhibits a C=N bond (a π bond to a heteroatom) conjugated to the ring and is therefore similar to a C=O bond that is conjugated to the ring. This group is moderately deactivating and *meta*-directing.
(e) In this compound the carbonyl group of the ester is connected directly to the ring. As seen in Table 19.1, a carbonyl group is moderately deactivating and *meta*-directing.
(f) As seen in Table 19.1, the oxygen atom of an ester (connected directly to an aromatic ring) is moderately activating and *ortho, para*-directing.

19.17. One aromatic ring is connected to a carbonyl group (C=O), while the other aromatic ring is connected to an oxygen atom. The former is moderately deactivated, while the latter is moderately activated (as seen in Table 19.1).

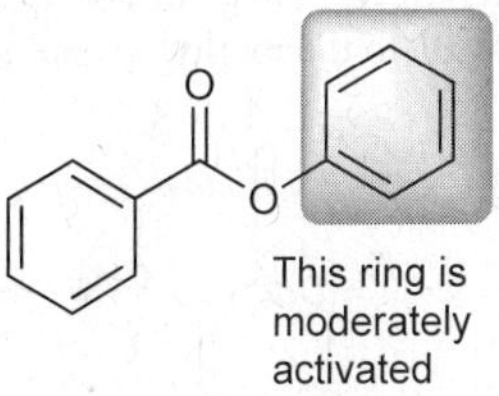

This ring is moderately activated

19.18. The are four aromatic rings, labeled **A – D**. Ring **A** is the most activated, because it exhibits a group that is a strong activator (an amino group). Ring **D** is moderately activated because it is connected to the oxygen atom of an ester. Ring **B** is weakly activated because it is connected to an alkyl group. And finally, ring **C** is deactivated because it is connected to a carbonyl group (a moderate deactivator).

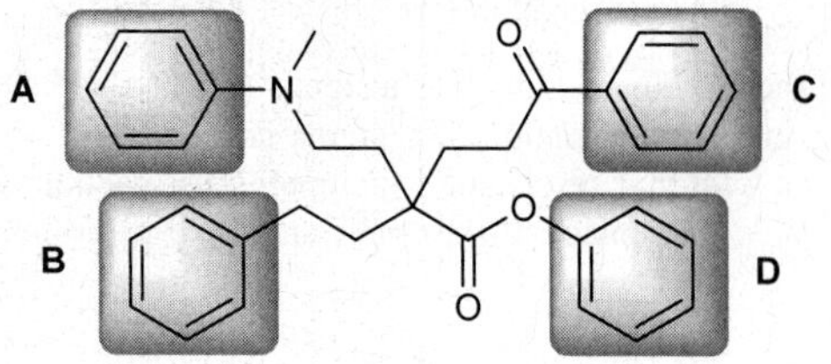

Increasing reactivity toward electrophilic aromatic substitution

C   B   D   A

19.19.

(a) The methyl group is weakly activating and the nitro groups are strongly deactivating. The directing effects are controlled by the most strongly activating group. Therefore, in this case, the methyl group controls the directing effects. As an activator, the methyl group is an *ortho-para* director. However, the two *ortho* positions are both already occupied (by the nitro groups). Therefore, an electrophilic aromatic substitution reaction is most likely to occur at the position that is *para* to the methyl group, shown here:

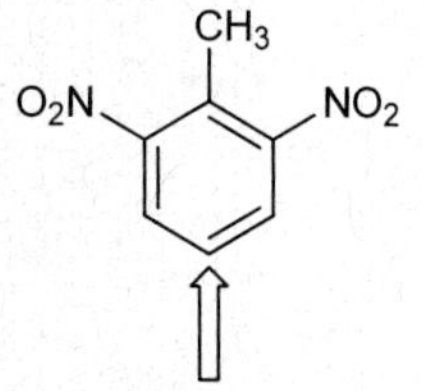

(b) The methyl group is weakly activating and the nitro groups are strongly deactivating. The directing effects are controlled by the most strongly activating group. Therefore, in this case, the methyl group controls the directing effects. As an activator, the methyl group is an *ortho-para* director. However, the *para* position and one of the *ortho* positions are already occupied (by the nitro groups). Therefore, an electrophilic aromatic substitution reaction is most likely to occur at the position that is *ortho* to the methyl group, shown here:

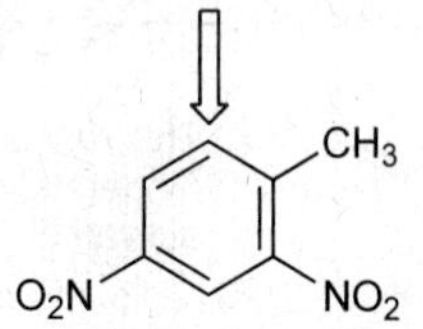

(c) This aromatic ring has three substituents. The methyl group is weakly activating. The other two groups are both esters, but they differ in the way they are connected to the ring. One group is moderately activating (the oxygen atom connected to the ring) and the other is moderately deactivating (the carbonyl group connected to the ring). The directing effects are controlled by the most strongly activating group. Therefore, in this case, the ester group (connected to the ring by its oxygen atom) controls the directing effects. As an activator, this group is an *ortho-para* director. However, the *para* position and one of the *ortho* positions are already occupied (by the other substituents). Therefore, an electrophilic aromatic substitution reaction is most likely to occur at the position that is *ortho* to the moderately activating group, shown here:

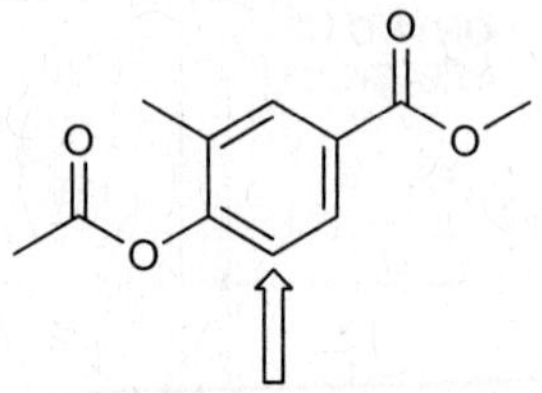

(d) The methyl groups are weakly activating and the cyano group is moderately deactivating. The directing effects are controlled by the most strongly activating group(s). Therefore, in this case, the methyl groups control the directing effects. Both of the methyl groups are directing to the same two locations (*ortho* and *para* to the methyl groups), shown here.

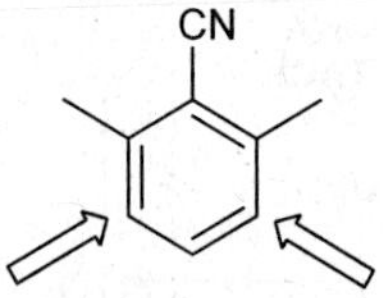

These two locations are identical because of symmetry. Don't be confused by the arbitrary location of the double bonds. Perhaps it is less distracting if we draw the structure like this:

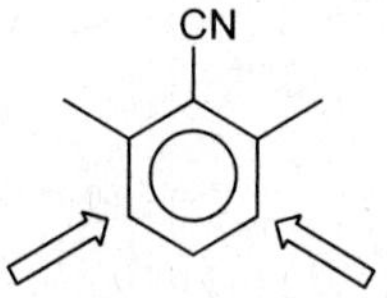

(e) This aromatic ring has three substituents. The methyl group is weakly activating. The carbonyl group (C=O) is moderately deactivating, and the amide group is moderately activating (see Table 19.1). The directing effects are controlled by the most strongly activating group. Therefore, in this case, the amide group controls the directing effects. As an activator, this group is an *ortho-para* director. However, one of the *ortho* positions is already occupied (by the methyl group). Therefore, an electrophilic aromatic substitution reaction is most likely to occur at the following positions, which are *ortho* and *para* to the amide group:

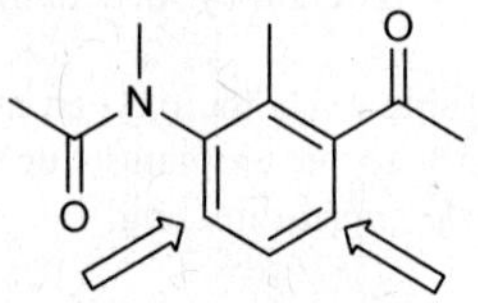

(f) The directing effects are controlled by the most strongly activating group. Therefore, in this case, the OH group (a strong activator) controls the directing effects. As an activator, this group is an *ortho-para* director. However, the *para* position and one of the *ortho* positions are already occupied (by the other substituents). Therefore, an electrophilic aromatic substitution reaction is most likely to occur at the position that is *ortho* to the OH group, shown here:

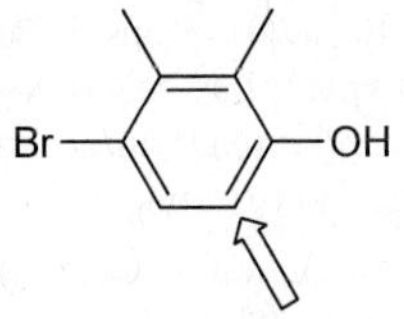

(g) The methyl groups are weakly activating and the bromo group is weakly deactivating. The directing effects are controlled by the most strongly activating group(s). Therefore, in this case, the methyl groups control the directing effects. Both of the methyl groups are directing to the same two locations (*ortho* and *para* to the methyl groups), shown here.

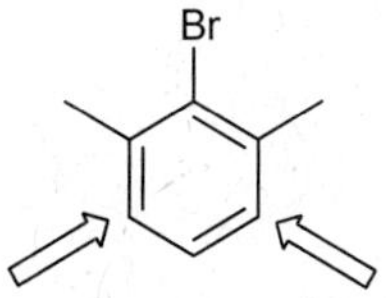

These two locations are identical because of symmetry. Don't be confused by the arbitrary location of the double bonds. Perhaps it is less distracting if we draw the structure like this:

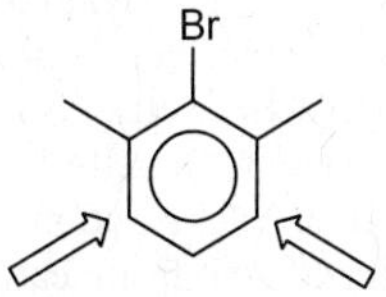

(h) The directing effects are controlled by the most strongly activating group. Therefore, in this case, the OH group (a strong activator) controls the directing effects (rather than the methoxy group, which is only a moderate activator). As an activator, the OH group is an *ortho-para* director. However, the *para* position and one of the *ortho* positions are already occupied (by the other substituents). Therefore, an electrophilic aromatic substitution reaction is most likely to occur at the position that is *ortho* to the OH group, shown here:

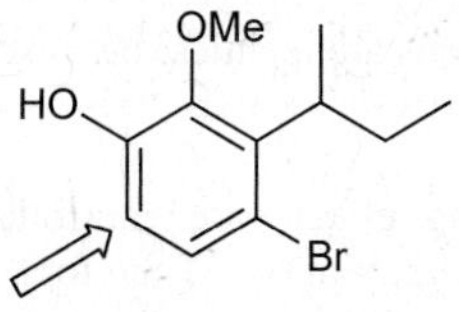

(i) This ring has four substituents (two electron-donating alkyl groups and two electron-withdrawing carbonyl

groups). The directing effects are controlled by the most strongly activating group(s). Therefore, in this case, the alkyl substituents control the directing effects. Both of the alkyl substituents are directing to the same location (*ortho* to the alkyl substituents), shown here.

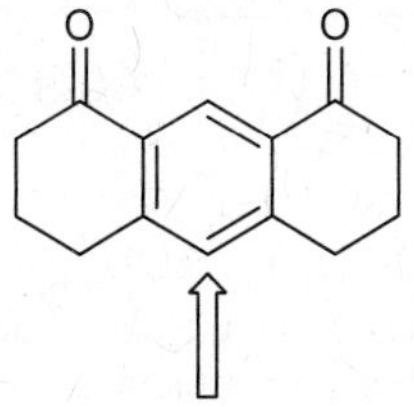

19.20.
(a) The reagents indicate a nitration reaction, so we must decide where the nitro group will be installed. The directing effects are controlled by the most strongly activating group. Therefore, in this case, the OH group (a strong activator) controls the directing effects (rather than the carbonyl groups, which are deactivators). As an activator, the OH group is an *ortho-para* director. However, the two *ortho* positions are already occupied (by the other substituents). Therefore, nitration is expected to occur at the position that is *para* to the OH group:

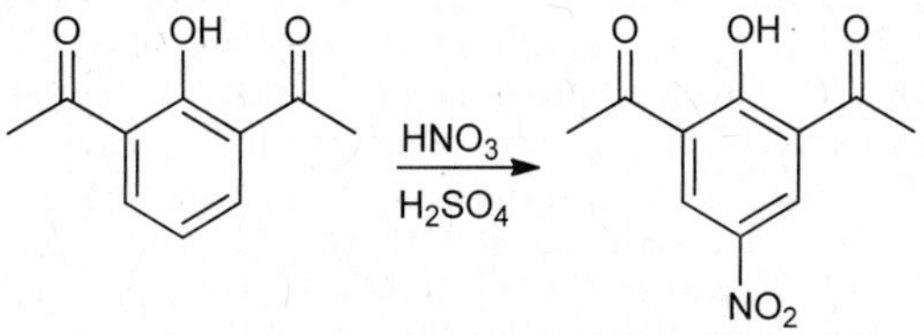

(b) The reagents indicate a bromination reaction, so we must decide where the bromine atom will be installed. The directing effects are controlled by the most strongly activating group. Therefore, in this case, the methoxy group (a moderate activator) controls the directing effects (the methyl group is a weak activator, and the other two groups are deactivators). As an activator, the methoxy group is an *ortho-para* director. However, the two *ortho* positions are already occupied (by the other substituents). Therefore, bromination is expected to occur at the position that is *para* to the methoxy group:

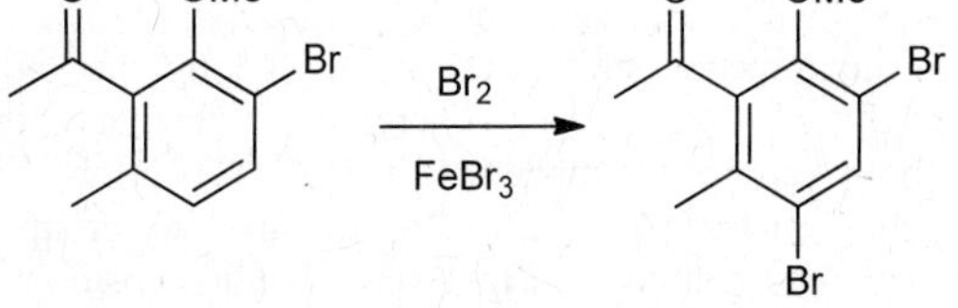

(c) The reagents indicate a sulfonation reaction, so we must decide where the sulfonic acid group will be installed. The aromatic ring has three substituents, only one of which is an activator (the methyl group). As an activator, the alkyl group is an *ortho-para* director. However, the two *ortho* positions are already occupied (by the other substituents). Therefore, sulfonation is expected to occur at the position that is *para* to the alkyl group:

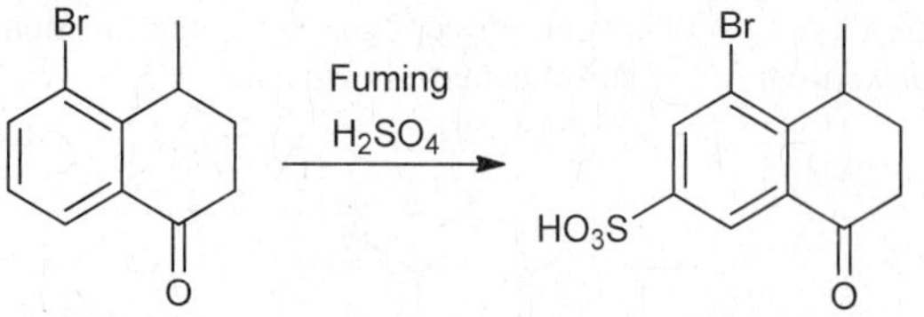

19.21. The reagents indicate a bromination reaction, so we must decide where the bromine atom will be installed. The directing effects are controlled by the most strongly activating group(s). The methyl groups are weakly activating and the bromo groups are weakly deactivating. Therefore, in this case, the methyl groups control the directing effects. Both of the methyl groups are directing to the same location (*ortho* and *para* to the methyl groups):

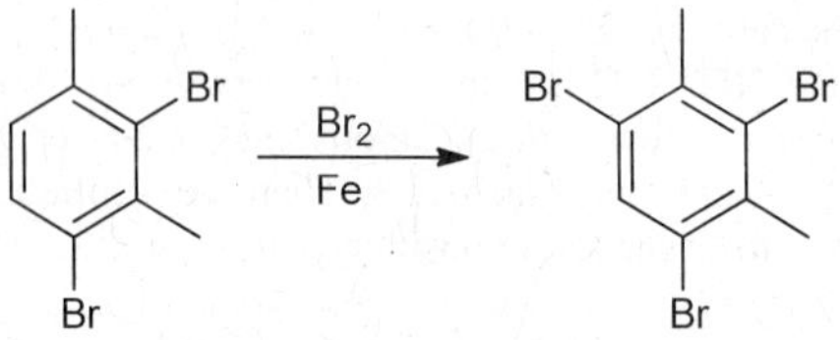

19.22.
(a) The directing effects are controlled by the most strongly activating group. Therefore, in this case, the OH group (a strong activator) controls the directing effects. As an activator, the OH group is an *ortho-para* director. However, the *para* position is already occupied (by the methyl group). Therefore, an electrophilic aromatic substitution reaction should occur at an *ortho* position. Among the two *ortho* positions, one of them (top) is sterically hindered and we do not expect the reaction to occur at that location. Therefore, the reaction is most likely to occur at the following position:

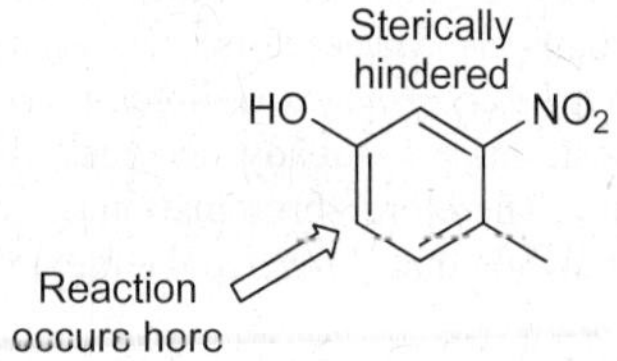

(b) The directing effects are controlled by the most strongly activating group. Therefore, in this case, the methoxy group (a moderate activator) controls the directing effects. As an activator, the methoxy group is an *ortho-para* director. However, the *para* position is already occupied (by the bromine atom). Therefore, an electrophilic aromatic substitution reaction should occur at an *ortho* position. Among the two *ortho* positions, one of them (top) is sterically hindered and we do not expect the reaction to occur at that location. Therefore, the reaction is most likely to occur at the following position:

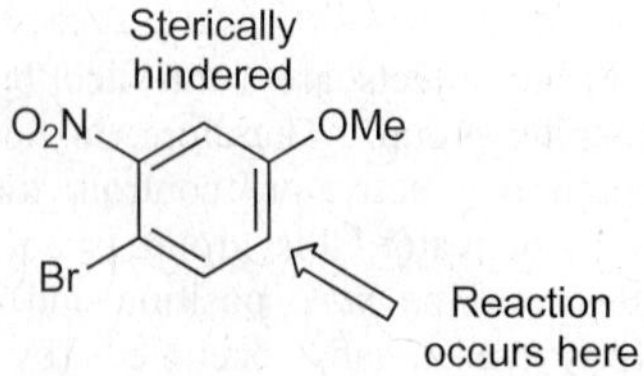

(c) The *tert*-butyl groups are weakly activating and are therefore *ortho,para* directors. They direct to the same three locations (*ortho* and *para* to themselves). However, one of these positions is sterically hindered and we do not expect the reaction to occur at that location. Therefore, the reaction is most likely to occur at one of the following positions:

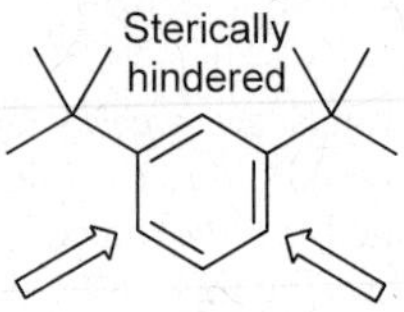

These two locations are identical because of symmetry. Don't be confused by the arbitrary location of the double bonds. Perhaps it is less distracting if we draw the structure like this:

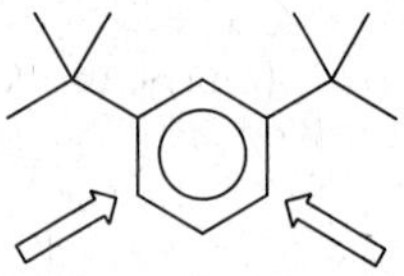

(d) This aromatic ring lacks an activating substituent. Both substituents are deactivators, and they are competing with each other (each group directs to the positions that are *meta* to itself). So electronically, the four aromatic positions are equally likely to undergo an electrophilic aromatic substitution reaction. The positions are differentiated from each other when we consider steric effects. The larger group (left) is more sterically bulky, and therefore blocks the positions closest to it. The reaction therefore occurs at one of the following locations.

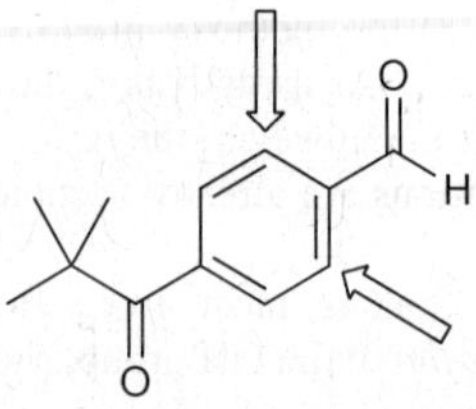

These two locations are identical because of symmetry.

(e) The directing effects are controlled by the most strongly activating group(s). Therefore, in this case, the OH groups (strong activators) control the directing effects. However, each OH group directs to different locations (*ortho* and *para* to itself), so all three available positions are activated. To differentiate between them, we note that the following position is too sterically hindered for a reaction to occur at this location:

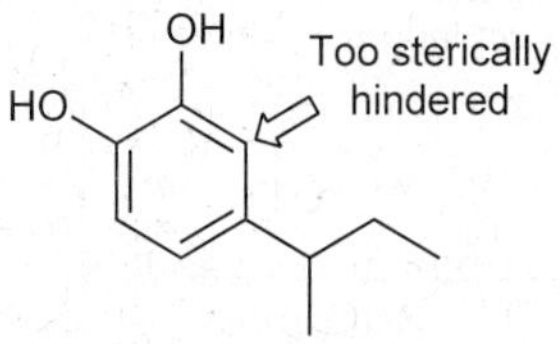

Among the other two locations, one of them is more sterically accessible than the other, so the reaction is expected to occur at this location:

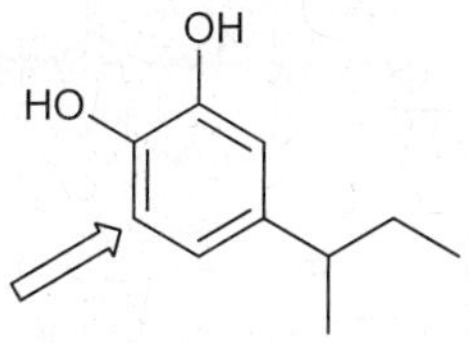

19.23. All three available positions are sterically hindered.

19.24. The reagents indicate a bromination reaction, so we must decide where the bromine atom will be installed. The aromatic ring has two substituents: an alkyl group and a carbonyl group. The former is a weak activator and the latter is a moderate deactivator. The directing effects are controlled by the more strongly activating group. Therefore, in this case, the alkyl group controls the directing effects. As an activator, the alkyl group is an *ortho-para* director. However, one of the *ortho* positions is already occupied (by the other substituent). Therefore, there are only two locations where bromination can occur (*ortho* or *para* to the alkyl group):

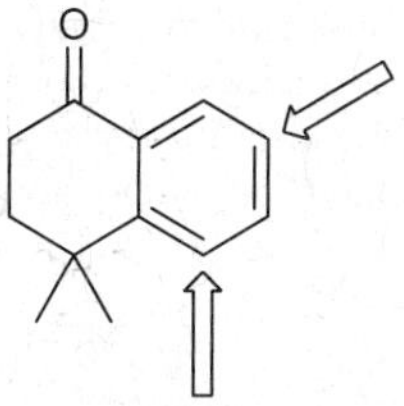

Among these two locations, one of them is more sterically hindered than the other. The reaction is expected to occur at the location that is more sterically accessible:

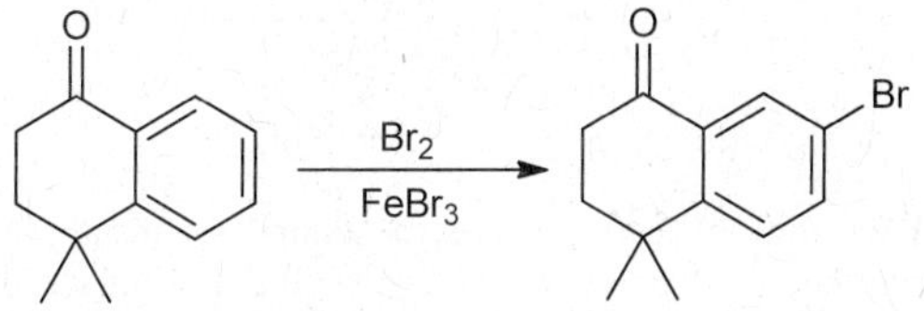

19.25.
(a) This aromatic ring has only one substituent (an alkyl group). Alkyl groups are weak activators and therefore *ortho-para* directors. An isopropyl group is sterically bulky, and we therefore expect a nitration reaction to

occur predominantly at the *para* position. In order to install a nitro group at the *ortho* position, a blocking group will be required.

(b) This aromatic ring has only one substituent (an alkyl group). Alkyl groups are weak activators and therefore *ortho-para* directors. An isopropyl group is sterically bulky, and we therefore expect a bromination reaction to occur predominantly at the *para* position. Therefore, a blocking group is not required. Direct bromination will give the desired product.

(c) The aromatic ring has two substituents: an amino group and a carbonyl group. The former is a strong activator and the latter is a moderate deactivator. The directing effects are controlled by the more strongly activating group. Therefore, in this case, the amino group controls the directing effects. As an activator, the amino group is an *ortho-para* director. However, one of the *ortho* positions is already occupied (by the other substituent). Therefore, there are only two locations where bromination can occur (*ortho* or *para* to the amino group):

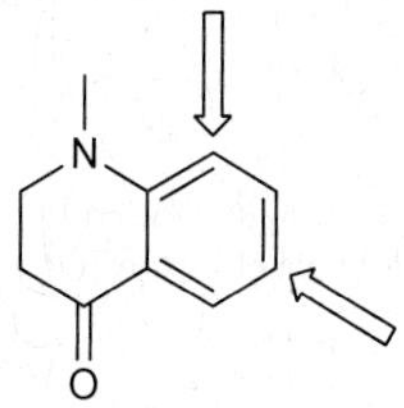

Among these two locations, one of them (top) is more sterically hindered than the other. Bromination is therefore expected to occur at the location that is more sterically accessible.

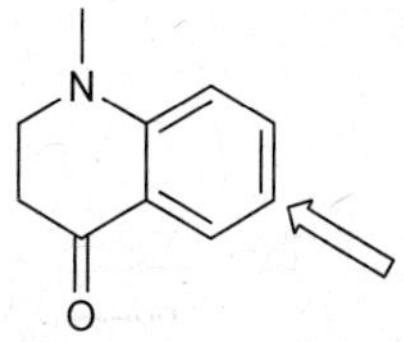

In order to install the bromine atom at the other location (*ortho* rather than *para*), a blocking group will be required.

(d) The directing effects for the desired nitration process are controlled by the most strongly activating group. Therefore, in this case, the OH group (a strong activator) controls the directing effects. As an activator, this group is an *ortho-para* director. However, one of the *ortho* positions is already occupied (by a methyl group). This leaves only two locations where the reaction can occur:

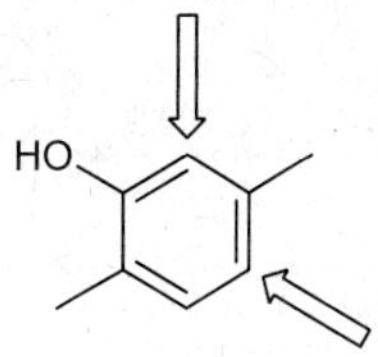

One of these locations (bottom right) is more accessible than the other, and this is exactly the position where a nitro group must be installed. Therefore, a blocking group will not be required.

19.26. The reaction conditions (dilute H_2SO_4) will remove each of the sulfonic acid groups, giving toluene as the major product:

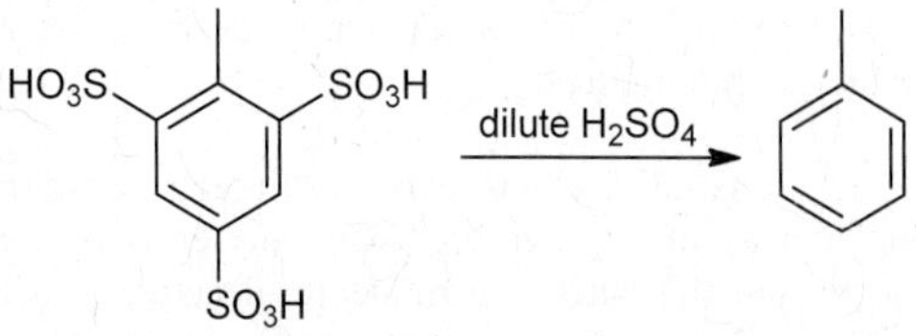

19.27.
(a) The nitro group must be installed in a position that is *meta* to each of the OH groups. Even with a blocking group, *meta* attack cannot be achieved on a highly activated ring.
(b) The position that must undergo bromination is too sterically hindered because of the presence of the *tert*-butyl groups.

19.28.
(a) The aromatic ring will undergo chlorination upon treatment with molecular chlorine (Cl_2) and a Lewis acid ($AlCl_3$).

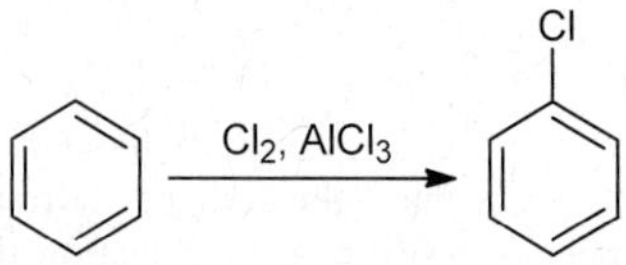

(b) The aromatic ring will undergo nitration upon treatment with a mixture of nitric acid and sulfuric acid.

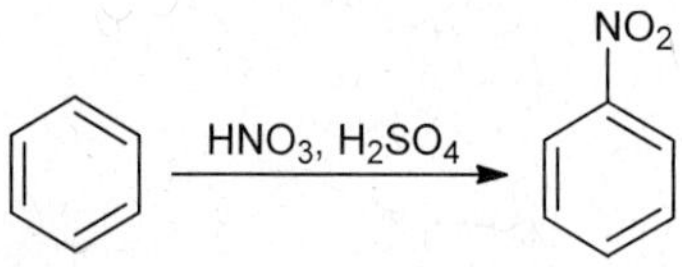

(c) The aromatic ring will undergo bromination upon treatment with molecular bromine (Br_2) and a Lewis acid ($AlBr_3$ or $FeBr_3$).

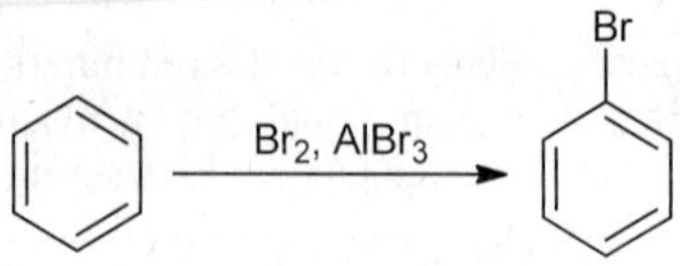

(d) The aromatic ring will undergo a Friedel-Crafts alkylation upon treatment with ethyl chloride and a Lewis acid ($AlCl_3$), to give ethylbenzene.

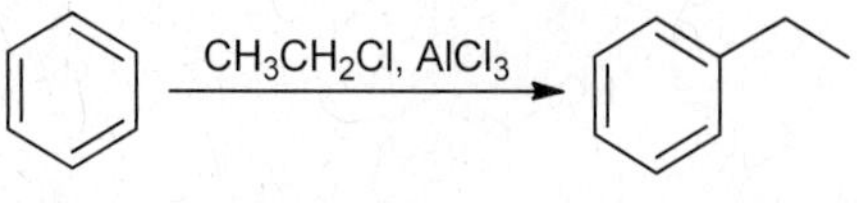

(e) A propyl group cannot be installed via a Friedel-Crafts alkylation, as a carbocation rearrangement is likely to occur. In order to install a propyl group (without rearrangement), we must first perform a Friedel-Crafts acylation, followed by a Clemmensen reduction, as shown:

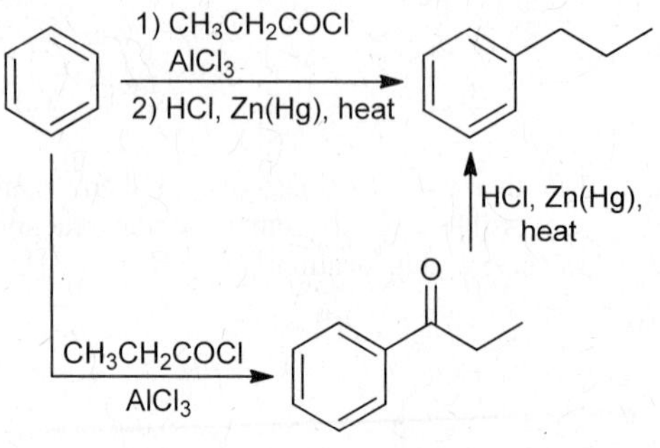

(f) An isopropyl group can be installed via a Friedel-Crafts alkylation in which isopropyl chloride is used as the alkyl halide:

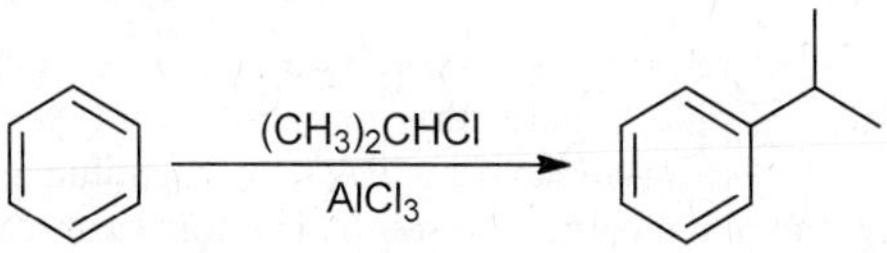

Notice that we use isopropyl chloride to install an isopropyl group, rather than using propyl chloride and relying on a carbocation rearrangement (which would be inefficient because it would likely yield a mixture of products).

(g) We did not learn a one-step method for installing an amino group on an aromatic ring, but we did learn a two-step method for achieving that transformation. Specifically, we first install a nitro group, and then we reduce the nitro group to an amino group, as shown here:

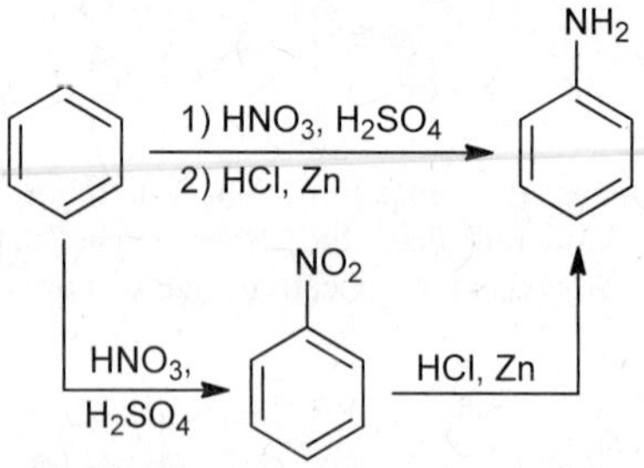

(h) We did not learn a one-step method for installing a carboxylic acid group on an aromatic ring, but we did learn a multi-step method for achieving that transformation. Specifically, we first install a methyl group, and then we oxidize the methyl group to give the desired product:

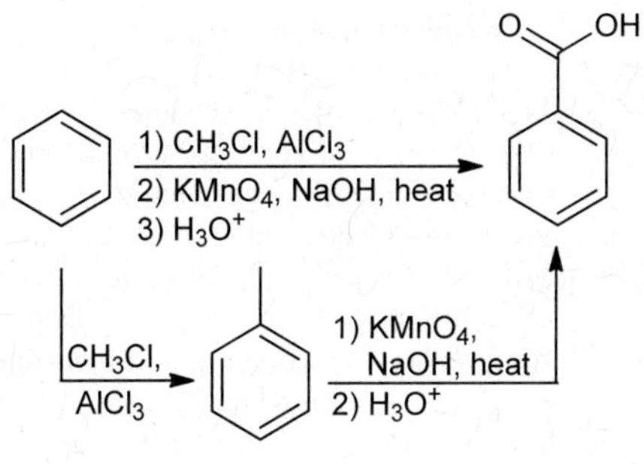

(i) The aromatic ring will undergo a Friedel-Crafts alkylation upon treatment with methyl chloride and a Lewis acid (AlCl$_3$), to give toluene.

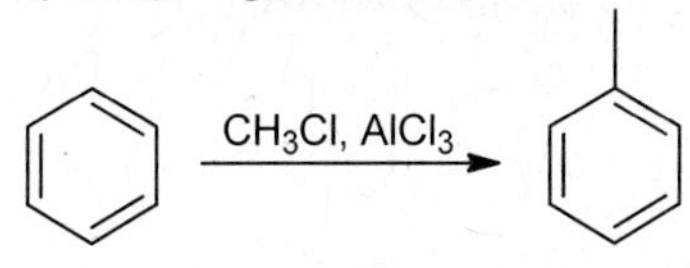

19.29.
(a) These reagents will install a sulfonic acid group on the aromatic ring (a process called sulfonation), giving benzenesulfonic acid:

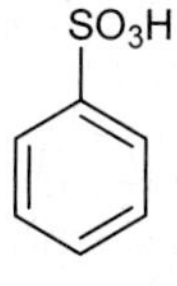

(b) These reagents will install a nitro group on the aromatic ring (a process called nitration), giving nitrobenzene:

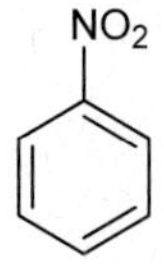

(c) These reagents will install a chlorine atom on the aromatic ring (a process called chlorination), giving chlorobenzene:

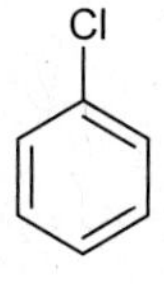

(d) These reagents will install an ethyl group on the aromatic ring (via a Friedel-Crafts alkylation), giving ethylbenzene:

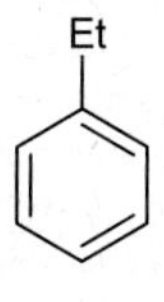

(e) These reagents will install a bromine atom on the aromatic ring (a process called bromination), giving bromobenzene:

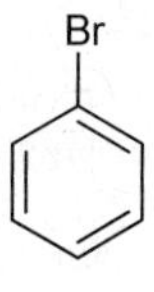

(f) These reagents will achieve the installation of a nitro group on the aromatic ring, followed by reduction of the nitro group to give aminobenzene (also called aniline):

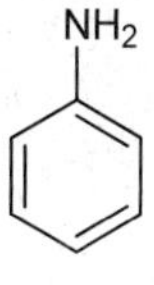

19.30.
(a) Installation of the amino group requires a two-step process (nitration, followed by reduction), while installation of the bromine atom can be achieved in just one step. Therefore, our synthesis must have at least those three steps.

Now let's consider the order of events. These two groups must be installed in an *ortho* fashion. Both groups are *ortho-para* directing, so we could theoretically install either one first, and then use a blocking technique to install the second group in the correct location (*ortho*, rather than *para*). However, if we install the amino group first, there is a concern of polybromination, since aminobenzene (also called aniline) is highly activated. To avoid this problem, we first install the bromine atom (which does not activate the ring). Then, we perform a sulfonation to block the *para* position. Nitration then installs the nitro group in the desired location (*ortho*). Then, after removing the blocking group, the nitro group is reduced to give the desired product.

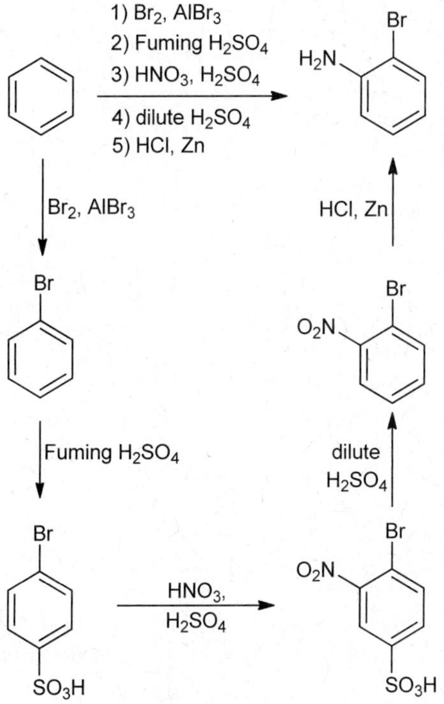

(b) Installation of the nitro group can be achieved in just one step (nitration), and installation of the chlorine atom can also be achieved in just one step.

Now let's consider the order of events. These two groups must be installed in a *meta* fashion. Only the nitro group is a *meta* director, so it must be installed first, followed by chlorination in the presence of a Lewis acid.

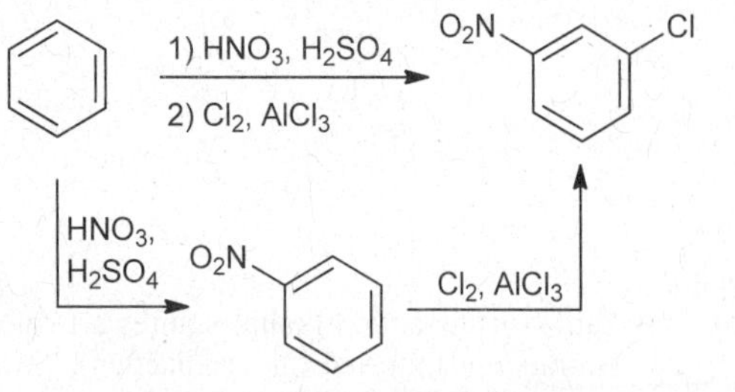

(c) Installation of the amino group requires a two-step process (nitration, followed by reduction), and installation of the propyl group also requires a two-step process (Friedel-Crafts acylation, followed by Clemmensen reduction).

Now let's consider the order of events. These two groups must be installed in a *para* fashion. Both groups are *ortho-para* directors, so we might think that either group could be installed first. However, to favor the *para* product over the *ortho* product, we must capitalize on the steric bulk of the propyl group. That is, we install the propyl group first, followed by installation of the amino group. Also, the amino group is too highly activating and cannot be installed first.

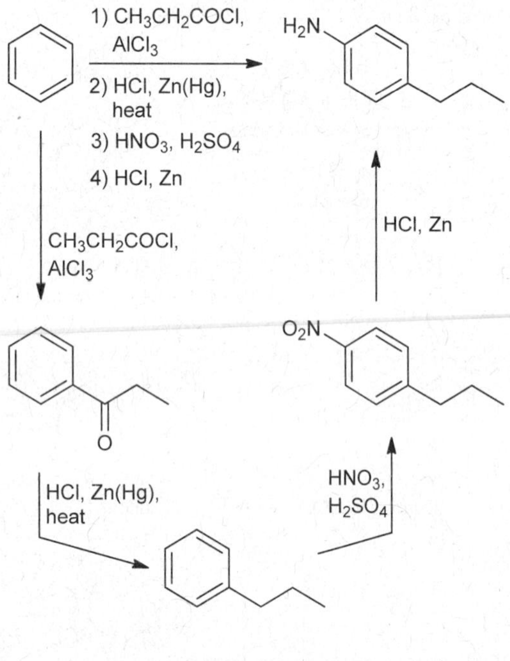

(d) Installation of the amino group requires a two-step process (nitration, followed by reduction), while

installation of the chlorine atom can be achieved in just one step.

Now let's consider the order of events. These two groups must be installed in a *meta* fashion, but both groups are *ortho-para* directors. Installation of these two groups (in either order) does not appear to give the desired *meta* substitution pattern. However, recall that installation of the amino group requires two steps. The first step is nitration, and a nitro group is a *meta*-director. So, we can achieve the desired transformation by performing the chlorination process after the nitration process but before the reduction process, as shown here:

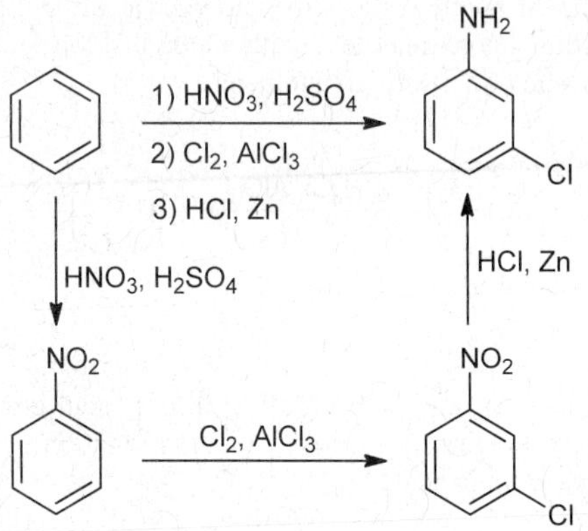

(e) Installation of the tribromomethyl group requires a two-step process (Friedel-Crafts alkylation, followed by benzylic bromination), while installation of the bromine atom can be achieved in just one step.

Now let's consider the order of events. These two groups must be installed in a *meta* fashion. Only the tribromomethyl group is a *meta* director, so it must be installed first (via the required two-step process), followed by bromination in the presence of a Lewis acid:

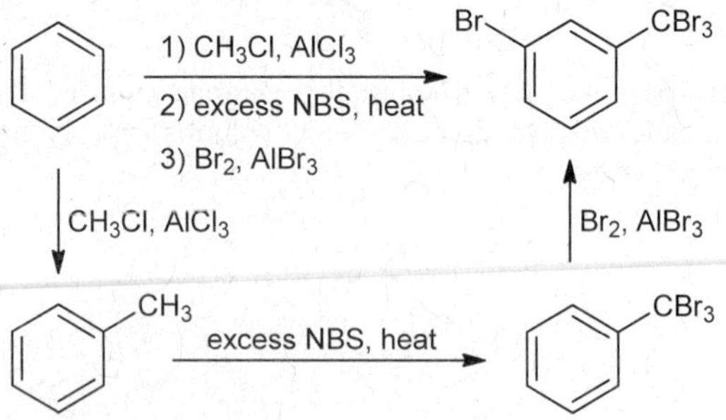

(f) Installation of the tribromomethyl group requires a two-step process (Friedel-Crafts alkylation, followed by benzylic bromination), while installation of the bromine atom can be achieved in just one step.

Now let's consider the order of events. These two groups must be installed in a *para* fashion. Only the bromo group is an *ortho-para* director, so we install that group first, followed by installation of the tribromomethyl group (which requires two steps), as shown:

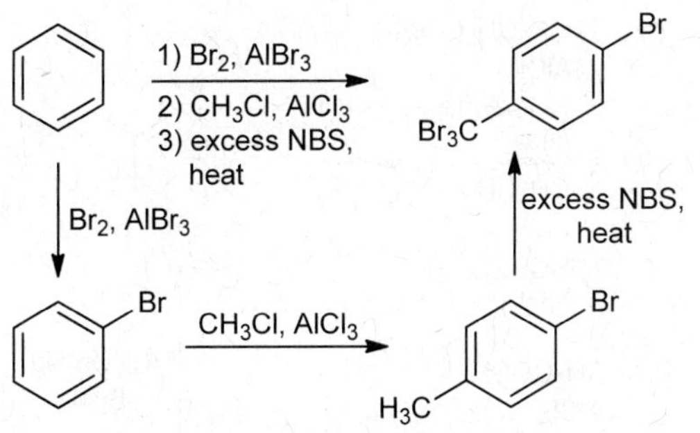

(g) This aromatic ring has two substituents, both of which are alkyl groups, although only the methyl group can be installed via a Friedel-Crafts alkylation process. The other alkyl group must be installed via a two-step process (Friedel-Crafts acylation, followed by Clemmensen reduction), because direct alkylation would result in a mixture of products (due to carbocation rearrangements).

Now let's consider the order of events. These two groups must be installed in an *ortho* fashion. Both groups are *ortho-para* directing, so we could theoretically install either one first, and then use a blocking technique to install the second group in the desired location (*ortho*, rather than *para*). However, if we install the methyl group first, then subsequent sulfonation is likely to produce a mixture of *ortho* and *para* products, since the methyl group is not very sterically demanding. To circumvent this lack of regiochemical control, it is best to install the bulky alkyl group first (using a Friedel-Crafts acylation, followed by a Clemmensen reduction), and then continue with a sulfonation step. This provides for much better selectivity during the sulfonation step, and therefore, a better yield of the final product.

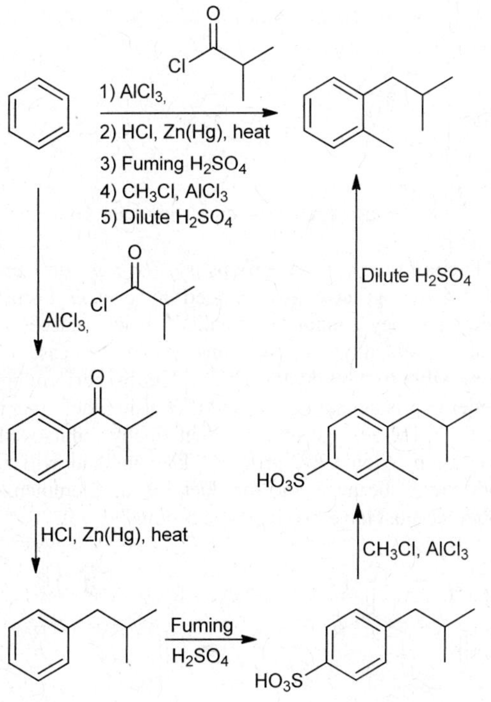

(h) Installation of the nitro group can be achieved in just one step (nitration), and installation of the isopropyl group can also be achieved in just one step (Friedel-Crafts alkylation with isopropyl chloride and aluminum trichloride).

Now let's consider the order of events. These two groups must be installed in an *ortho* fashion. Only the alkyl group is an *ortho-para* director, so it must be installed first. Then, in order to install a nitro group in the *ortho* position (rather than the *para* position), a blocking group must be used. After the blocking group is installed, nitration will install a nitro group in the desired location. Desulfonation then affords the desired product.

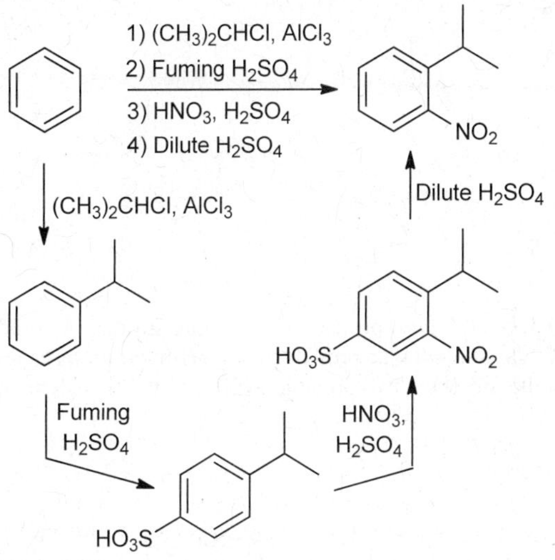

(i) Installation of the propyl group requires a two-step process (Friedel-Crafts acylation, followed by Clemmensen reduction), while installation of the chlorine atom can be achieved in just one step.

Now let's consider the order of events. These two groups must be installed in a *meta* fashion, but both groups are *ortho-para* directors. Installation of these two groups (in either order) does not appear to give the desired *meta* substitution pattern. However, as mentioned above, installation of the propyl group requires two steps. The first step is acylation, and the resulting acyl group is a *meta*-director. So, we can achieve the desired transformation by performing the chlorination process after the acylation process but before the reduction process, as shown here:

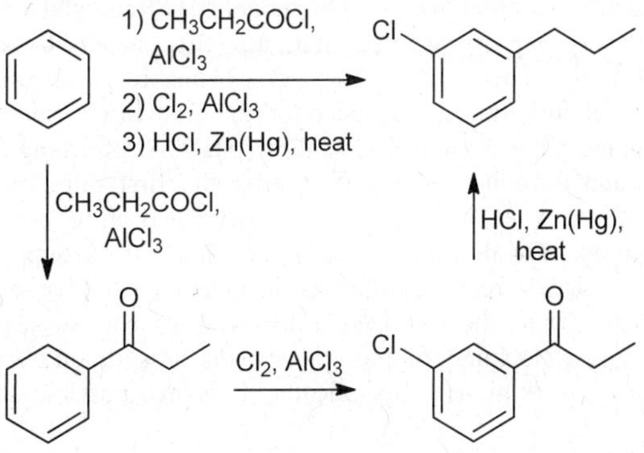

(j) Installation of each propyl group requires a two-step process (Friedel-Crafts acylation, followed by Clemmensen reduction). These two groups must be installed in a *para* fashion. Propyl groups are *ortho-para* directors, and their steric bulk is expected to cause a preference for *para*-substitution. So we simply install one propyl group (via Friedel-Crafts acylation followed by reduction), followed by the second propyl group, as shown here:

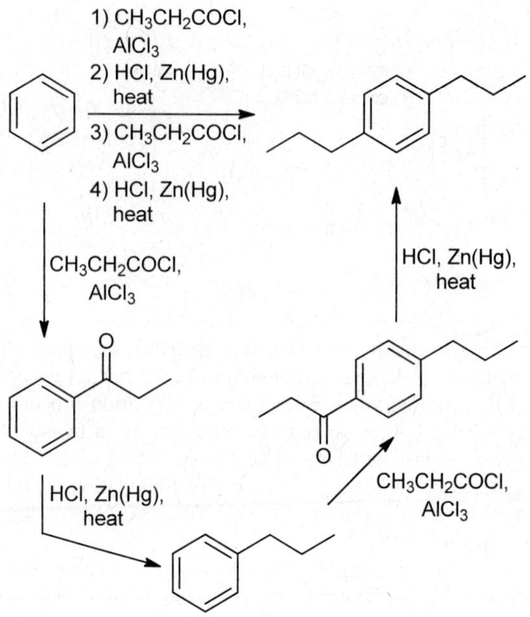

1) CH_3CH_2COCl, $AlCl_3$
2) HCl, Zn(Hg), heat
3) CH_3CH_2COCl, $AlCl_3$
4) HCl, Zn(Hg), heat

CH_3CH_2COCl, $AlCl_3$

HCl, Zn(Hg), heat

HCl, Zn(Hg), heat

CH_3CH_2COCl, $AlCl_3$

19.31. The *para* product will be more strongly favored over the *ortho* product if the *tert*-butyl group is installed first. The steric hindrance provided by a *tert*-butyl group is greater than the steric hindrance provided by an isopropyl group. Of the following two possible pathways, the first should provide a greater yield of the desired product.

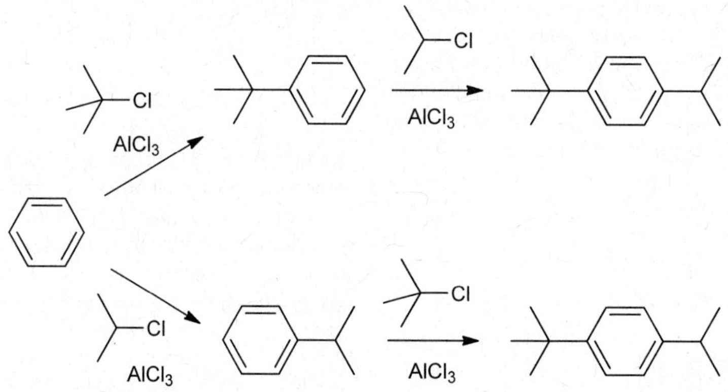

19.32.
(a) This aromatic ring has two groups (an amino group and a nitro group) arranged in an *ortho* fashion. The nitro group can be installed in one step (nitration), while installation of the amino group requires a two-step process (nitration, followed by reduction). Regardless of which group is installed first, our first step will be nitration. Then, we must decide whether to perform a second nitration or reduction of the first nitro group. Neither pathway will lead to the product. Performing a second nitration (immediately after the first nitration) will result in the groups being arranged in a *meta* fashion. On the other hand, if the first nitro group is immediately reduced after its installation, to give an amino group, then we have a different problem, even if we use a blocking group (to favor *ortho* substitution over *para* substitution). Specifically, it is not possible to perform a nitration reaction on a ring that contains an amino group.

(b) Each of the two alkyl groups is *ortho-para* directing, but the two groups are arranged in a *meta* fashion. Therefore, they cannot be installed via two successive Friedel-Crafts alkylation reactions. And in this case, it is not possible to circumvent the problem by performing a Friedel-Crafts acylation first and then using the directing effects of the acyl group (and then finally reducing the acyl group to an alkyl group). That method will not work here, because the product of a Clemmensen reduction must have two benzylic protons:

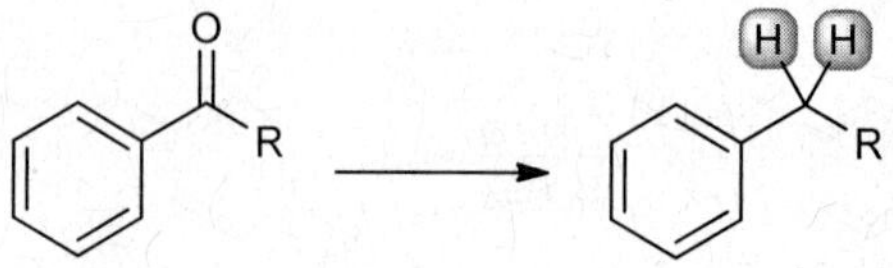

The desired product has only one benzylic proton (at each isopropyl group), which means that it cannot be made via a Clemmensen reduction.

19.33.

(a) The following retrosynthetic analysis represents one strategy for preparing the desired compound. An explanation of each of the steps (*a-c*) follows.

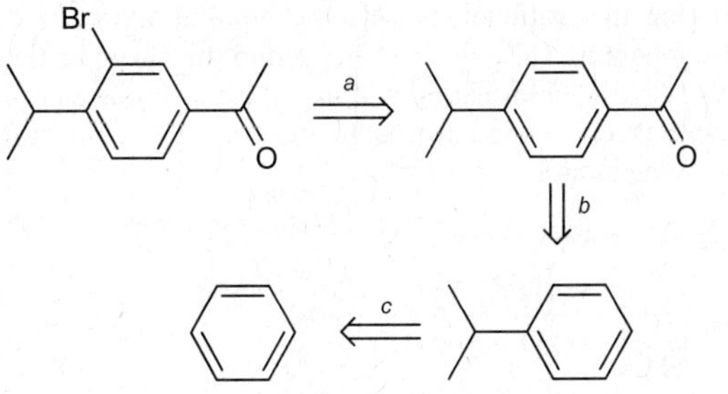

a. The product can be prepared via bromination of the disubstituted ring shown.
b. The disubstituted ring can be prepared via a Friedel-Crafts acylation.
c. The monosubstituted ring (isopropylbenzene) can be prepared via a Friedel-Crafts alkylation.

Now, let's draw the forward scheme. First, an isopropyl group is installed via a Friedel-Crafts alkylation. Then, an acyl group is installed via a Friedel-Crafts acylation. During this acylation step, the isopropyl group directs the incoming acyl group into the *para* position. And finally, bromination will install the bromine atom in the correct location (*ortho* to the isopropyl group).

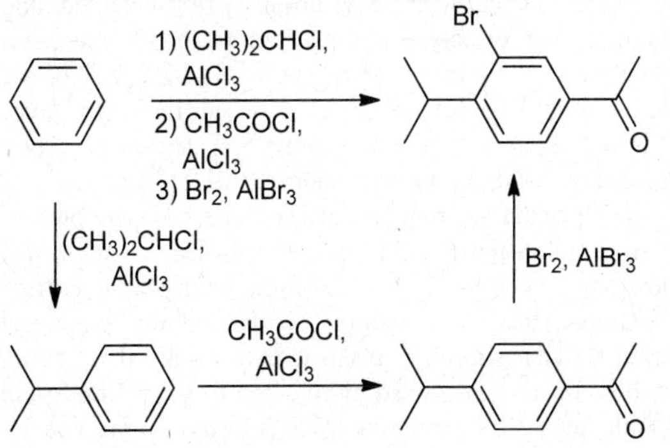

The strategy above represents just one method for making the desired the compound. There are often other acceptable solutions. For example, after installation of the isopropyl group, the acyl group and the bromo group can be installed in the opposite order (first the bromo group and then the acyl group), although this would require a blocking group (to help direct the incoming bromo group into the *ortho* position). The first method (shown above) avoids the need for a blocking group.

(b) The following retrosynthetic analysis represents one strategy for preparing the desired compound. An explanation of each of the steps (*a-d*) follows.

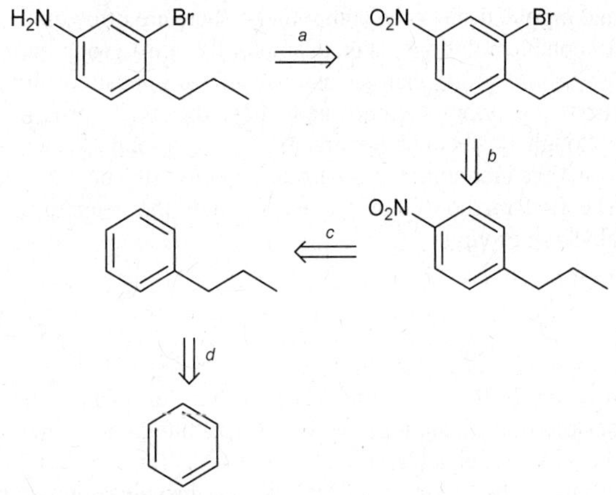

a. The product can be prepared via reduction of the nitro group.
b. The bromo group can be installed in the correct location based on the directing effects of the propyl group.
c. The nitro group is installed in the *para* position via nitration.
d. A propyl group is installed onto the ring via Friedel-Crafts acylation, followed by reduction.

Now, let's draw the forward scheme. First, a propyl group is installed via a Friedel-Crafts acylation, followed by a Clemmensen reduction. This transformation could not have been achieved via direct alkylation (due to the problem of carbocation rearrangements). Then, a nitro group is installed in the *para* position because of the directing effects (and steric effects) of the propyl group. Bromination then installs the bromine atom in the correct location (*ortho* to the activating group), and finally, reduction of the nitro group gives the product.

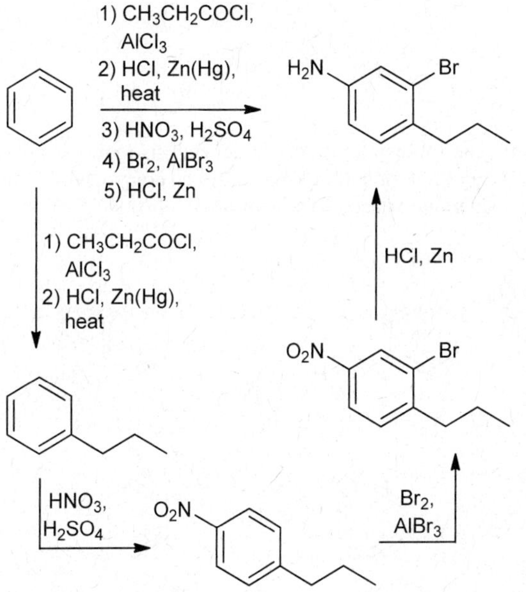

The strategy above represents just one method for making the desired the compound. There are often other acceptable solutions. For example, the nitro group and the bromo group can be installed in the opposite order (first the bromo group and then the nitro group), although this would require a blocking group (to help direct the incoming bromo group into the *ortho* position). The first method (shown above) avoids the need for a blocking group.

(c) While there is more than one way to achieve the desired transformation, the following strategy is perhaps the most efficient, as it avoids the need for a blocking group. The *tert*-butyl group is first installed, and its directing effects are exploited to install an amino group in the *para* position (installation of the amino group requires a two-step process – nitration, followed by reduction). Finally, chlorination in the presence of excess Cl_2 gives the desired product ($AlCl_3$ is not used because the ring is highly activated). The regiochemical outcome of the final step is controlled by the electronic effects of the amino group (in concert with the steric effects of the *tert*-butyl group).

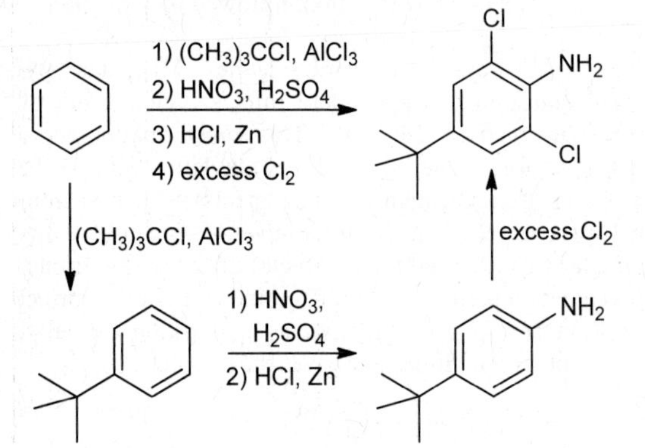

(d) The following retrosynthetic analysis represents one strategy for preparing the desired compound. An explanation of each of the steps (*a-c*) follows.

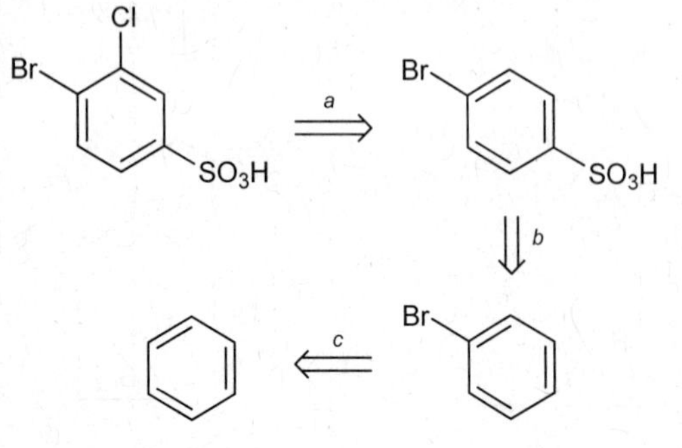

a. The product can be prepared via chlorination of the disubstituted ring shown.
b. The disubstituted ring can be prepared via sulfonation.
c. The monosubstituted ring (bromobenzene) can be prepared from benzene.

Now, let's draw the forward scheme. First, a bromine atom is installed via bromination in the presence of a Lewis acid. Then, a sulfonic acid group is installed. During this sulfonation step, the bromine atom directs the incoming sulfonic acid group into the *para* position. And finally, chlorination will install the bromine atom in the correct location (*ortho* to the *ortho-para* directing bromine atom).

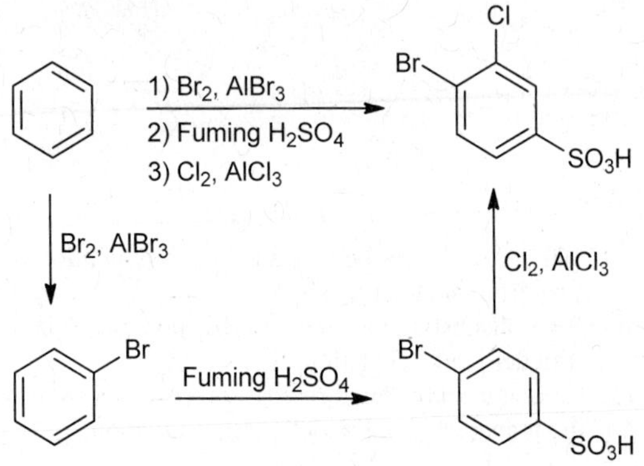

19.34.
(a) While there is more than one way to achieve the desired transformation, the following strategy is perhaps the most efficient. An alkyl group is first installed, and its directing effects are exploited to block the *para* position with a sulfonation reaction. As such, we choose an isopropyl group to provide steric bulk (to favor sulfonation at the *para* position). A *tert*-butyl group cannot be used, as we will soon oxidize this group, and benzylic oxidation requires at least one benzylic proton. With the sulfonic acid group installed in the *para* positon, we can install both bromine atoms in the correct locations (the two positions *ortho* to the isopropyl group). The isopropyl group is then oxidized to give a carboxylic acid, followed by removal of the sulfonic acid group, at which point the two chlorine atoms can be installed based on the directing effects of the bromine substituents (*ortho-para* directors):

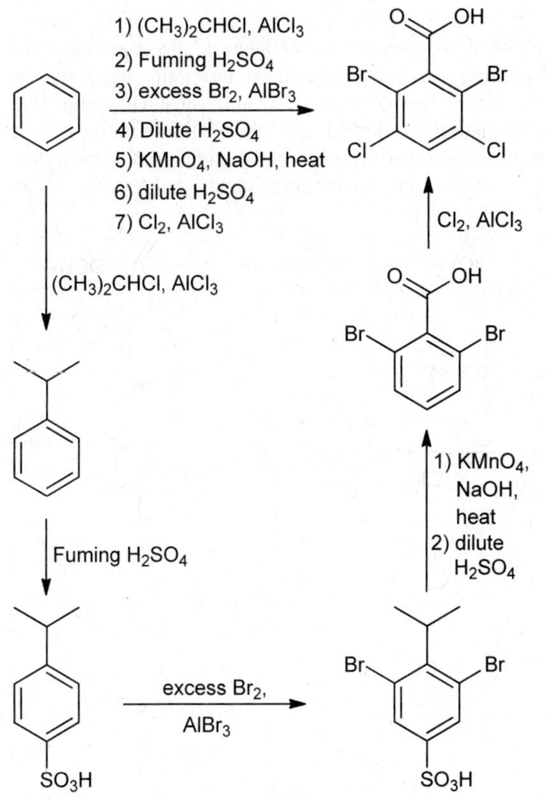
1) $(CH_3)_2CHCl$, $AlCl_3$
2) Fuming H_2SO_4
3) excess Br_2, $AlBr_3$
4) Dilute H_2SO_4
5) $KMnO_4$, NaOH, heat
6) dilute H_2SO_4
7) Cl_2, $AlCl_3$
$(CH_3)_2CHCl$, $AlCl_3$
Fuming H_2SO_4
excess Br_2, $AlBr_3$
1) $KMnO_4$, NaOH, heat
2) dilute H_2SO_4
Cl_2, $AlCl_3$
SO_3H
OH
Br
Cl

(b) The sixth position is sterically hindered by the presence of the Cl atoms.
(c) The ring is deactivated (relative to benzene) because all five groups are deactivators.

19.35. The ring contains a nitro group, as well as two leaving groups (chloride and bromide), and the reagent is a strong base (methoxide). As such, a nucleophilic aromatic substitution reaction can occur, but only if the leaving group is *ortho* or *para* to the nitro group. The bromine atom is *meta* to the nitro group, so the reaction does not occur there. Instead, chloride functions as the leaving group, because it is *para* to the nitro group.

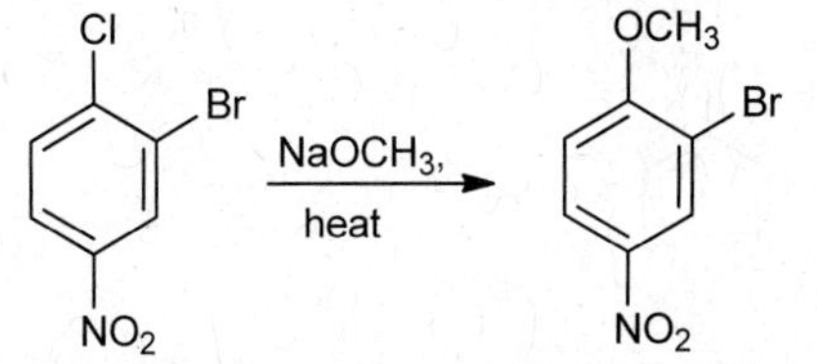
Cl
Br
$NaOCH_3$,
heat
NO_2
OCH_3

19.36. This transformation involves the installation of two groups: NH_2 and OH. Thus far, we have only seen one way (S_NAr) to install an OH group on an aromatic ring (we will see one other way in Section 19.14, and yet another method in Chapter 23). In order to install an OH group via an S_NAr reaction, the ring must exhibit both a nitro group and a leaving group:

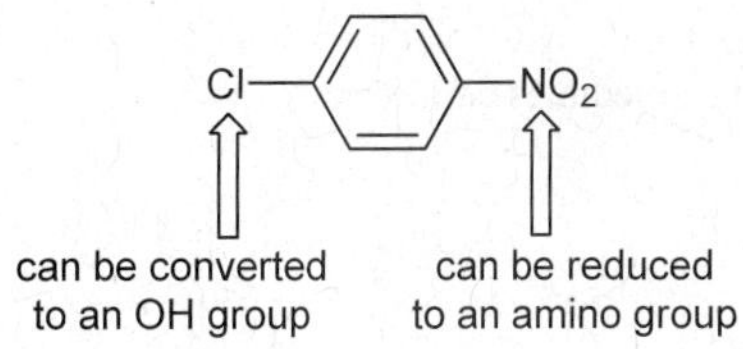
Cl
NO_2
can be converted to an OH group
can be reduced to an amino group

The chlorine atom can be replaced with an OH group, and the nitro group can be reduced to an amino group, giving the product. And the key intermediate above can be made in just two steps (chlorination, followed by nitration)
The entire synthesis is shown below. Benzene is first treated with chlorine in the presence of a Lewis acid to give chlorobenzene, which is then treated with sulfuric acid and nitric acid, to give the key intermediate. The chlorine atom is then replaced with an OH group upon treatment with hydroxide, followed by aqueous acid work-up. Reduction of the nitro group affords the desired product.

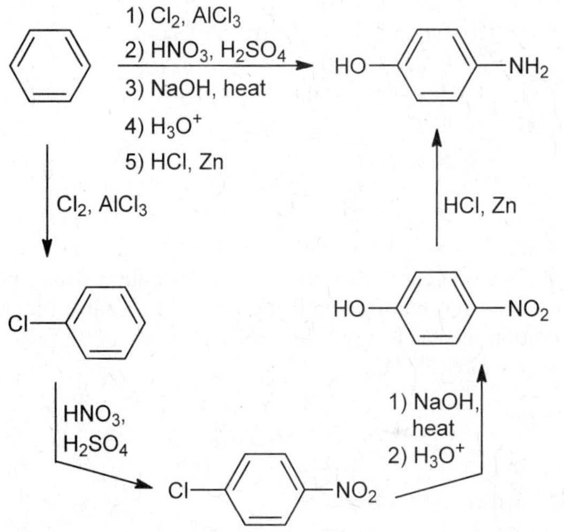
1) Cl_2, $AlCl_3$
2) HNO_3, H_2SO_4
3) NaOH, heat
4) H_3O^+
5) HCl, Zn
HO
NH_2
Cl_2, $AlCl_3$
HCl, Zn
Cl
NO_2
HNO_3, H_2SO_4
1) NaOH, heat
2) H_3O^+

19.37.
(a) Each additional nitro group serves as a reservoir of electron density and provides for an additional resonance structure in the Meisenheimer complex, thereby stabilizing the Meisenheimer complex and lowering the energy of activation for the reaction.
(b) No, a fourth nitro group would not be *ortho* or *para* to the leaving group, and would not provide resonance stabilization for the Meisenheimer complex. Therefore, we do not expect the temperature requirement to be significantly lowered.

19.38. The starting material is 4-chloro-2-methyltoluene, shown here:

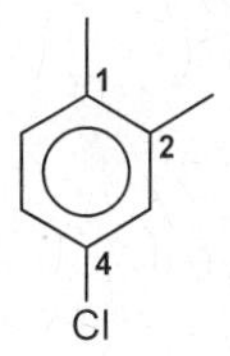
1
2
4
Cl

Upon treatment with sodium amide, two possible benzyne intermediates can form:

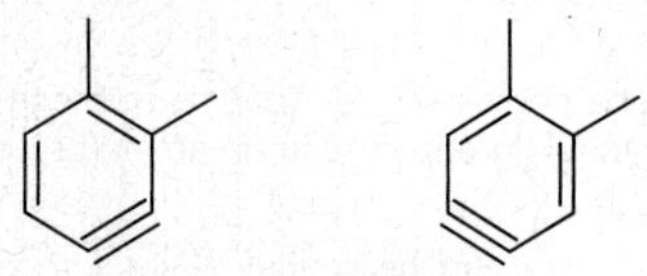

Then, for each of these benzyne intermediates, the amide ion can attack either side of the triple bond, which means that the NH$_2$ group is ultimately positioned at either C3, C4, or C5. Substitution at C4 yields the same product as substitution at C5 (because of symmetry), giving rise to only two products (rather than three):

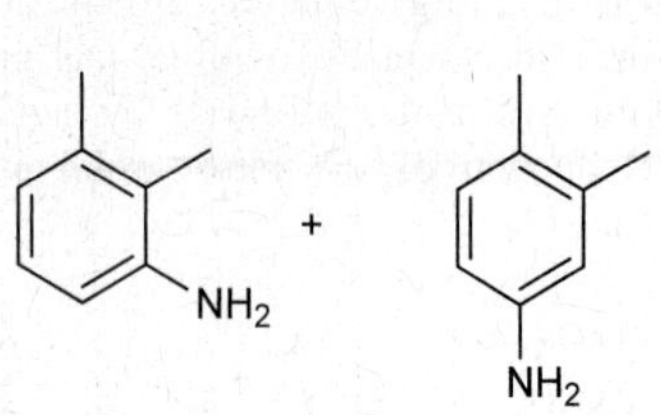

19.39. We have not yet seen a direct method for installing a methoxy group on an aromatic ring, however, we have seen a way to install an OH group (via chlorination, followed by elimination-addition). A Williamson ether synthesis can then be used to methylate the oxygen atom, giving the product.

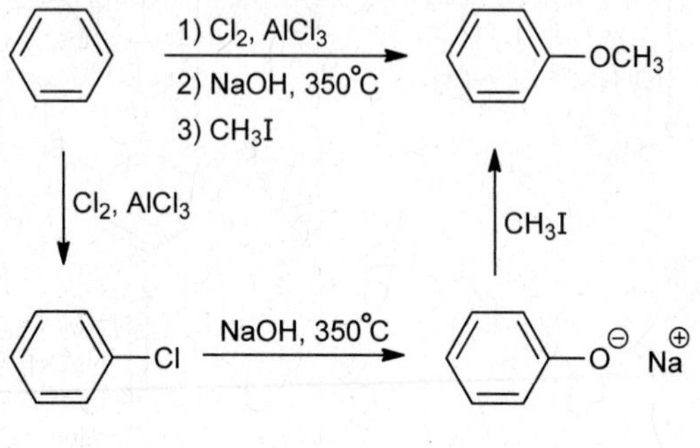

19.40.
(a) The reagent is hydroxide, which is both a strong nucleophile and a strong base. However, the three criteria for S$_N$Ar are not met (the aromatic ring lacks a nitro group), and therefore, the reaction must occur via an elimination-addition mechanism (as seen in Mechanism 19.9). The elimination-addition mechanism is also indicated by the high temperature (350°C).

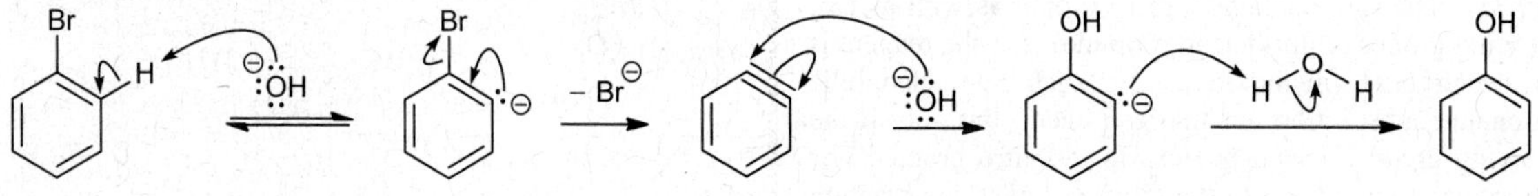

Under the basic conditions employed, the resulting product (phenol) is deprotonated to give a phenolate ion:

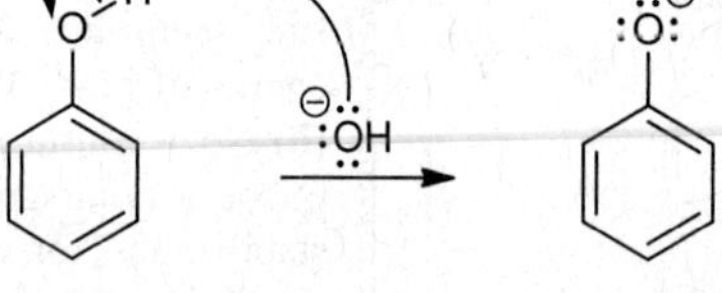

This is why aqueous acid must be introduced into the reaction flask after the reaction is complete, in order to serve as a proton source to regenerate phenol:

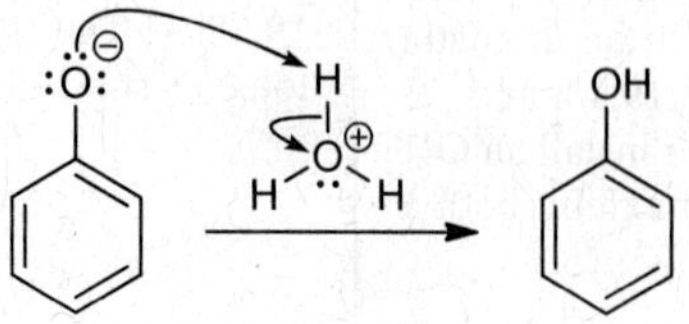

(b) These reagents provide a powerful electrophile, so the reaction must be an electrophilic aromatic substitution reaction. The first step of an electrophilic aromatic substitution reaction is formation of the electrophile that will react with the aromatic ring. This occurs when molecular bromine interacts with the Lewis acid to form a complex:

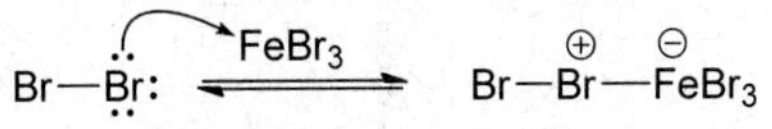

Then, one of the rings attacks the complex to give a resonance stabilized intermediate (a sigma complex), which then loses a proton to restore aromaticity (electrophilic aromatic substitution):

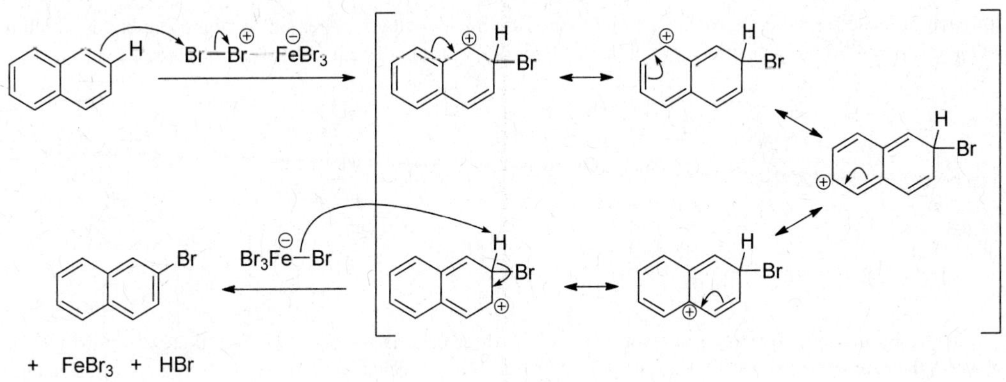

(c) The reagent (sodium amide) is a powerful nucleophile, and the three criteria for S_NAr are met (a nitro group and a leaving group that are *para* to each other), so we expect an S_NAr mechanism. In an S_NAr reaction, the first step involves the nucleophile attacking the aromatic ring to generate a resonance stabilized intermediate (a Meisenheimer complex). Loss of a leaving group then gives the product, as shown:

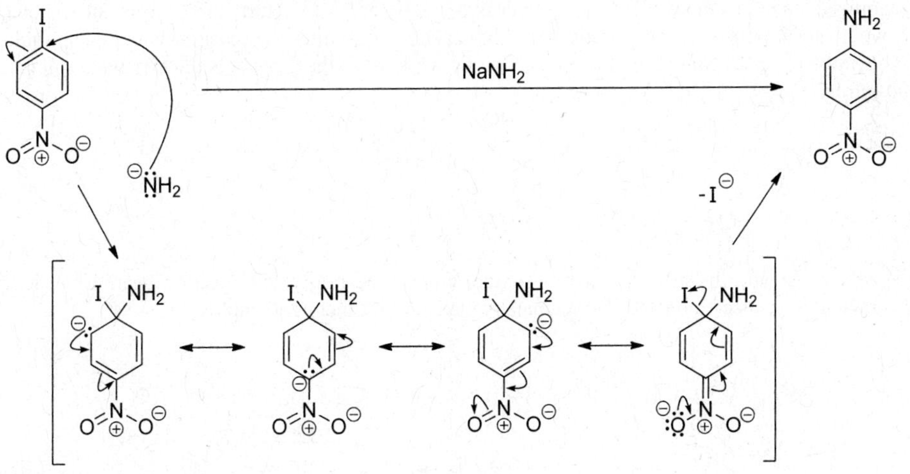

19.41. The reagent (hydroxide) is a strong nucleophile and a strong base, and the three criteria for S_NAr are not met (the aromatic ring does not have a nitro group), so we expect an elimination-addition reaction. Upon treatment with sodium hydroxide, two possible benzyne intermediates are initially formed:

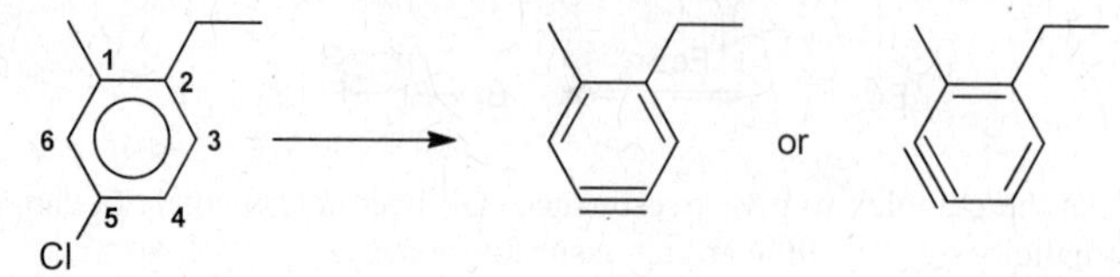

Then, for each of these benzyne intermediates, the hydroxide ion can attack either side of the triple bond, which means that the OH group is ultimately positioned at either C4, C5, or C6, giving rise to the following three products:

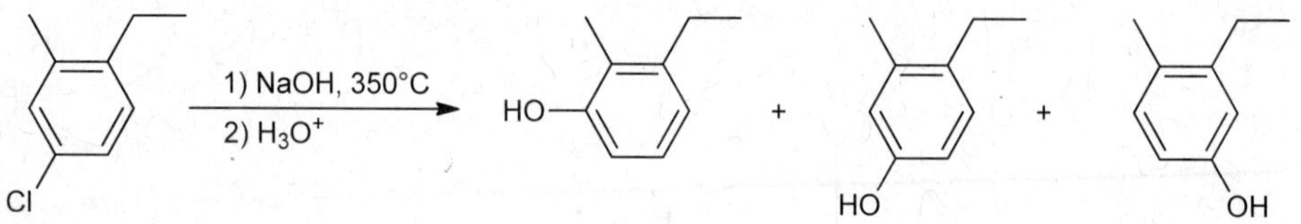

19.42.
(a) The reagent (hydroxide) is a strong nucleophile, and the three criteria for S_NAr are met (there is a nitro group and a leaving group that are *ortho* to each other), so we expect an S_NAr mechanism. In an S_NAr reaction, the nucleophile (hydroxide) replaces the leaving group (bromide) and maintains the same location:

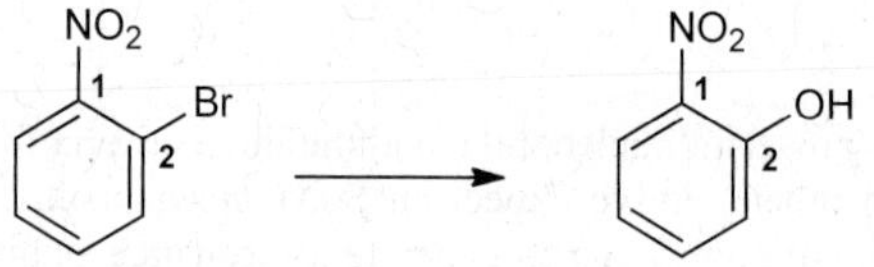

Under these basic conditions, the product (*ortho*-nitrophenol) is deprotonated upon its formation (because the resulting resonance stabilized anion is more stable than hydroxide). In order to return the proton and regenerate *ortho*-nitrophenol, we must introduce a proton source into the reaction flask after the reaction is complete (aqueous acidic work-up). The problem statement did not mention an acidic workup, so the product is the following salt (the conjugate base of *ortho*-nitrophenol):

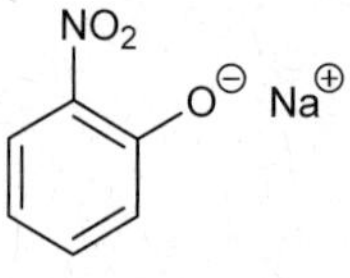

(b) In an S_NAr mechanism, the nucleophile (hydroxide) attacks the ring at the position bearing the leaving group, giving the following resonance-stabilized intermediate (called a Meisenheimer complex):

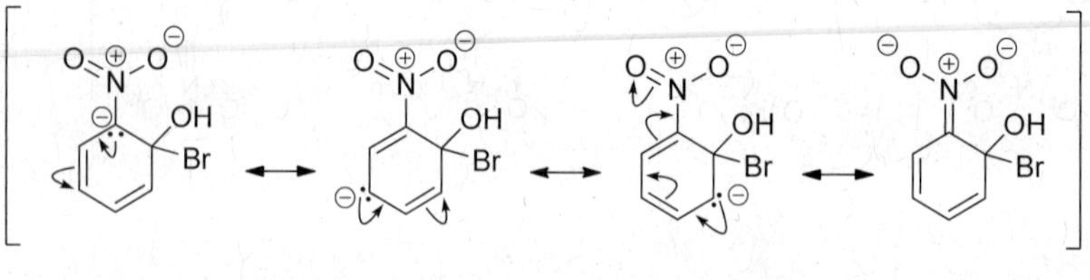

(c) No. In order for an S_NAr reaction to occur, the leaving group must be *ortho* or *para* to the nitro group. If the leaving group is *meta* to the nitro group, then the nitro group cannot function as a reservoir for electron density to stabilize the intermediate Meisenheimer complex. The Meisenheimer complex is therefore too high in energy to form at an appreciable rate.

(d) Yes. This compound exhibits a nitro group and a leaving group (bromide) that are *para* to each other.

19.43. The reagents for these reactions can be found in the Review of Reactions at the end of the chapter. They are shown again here:

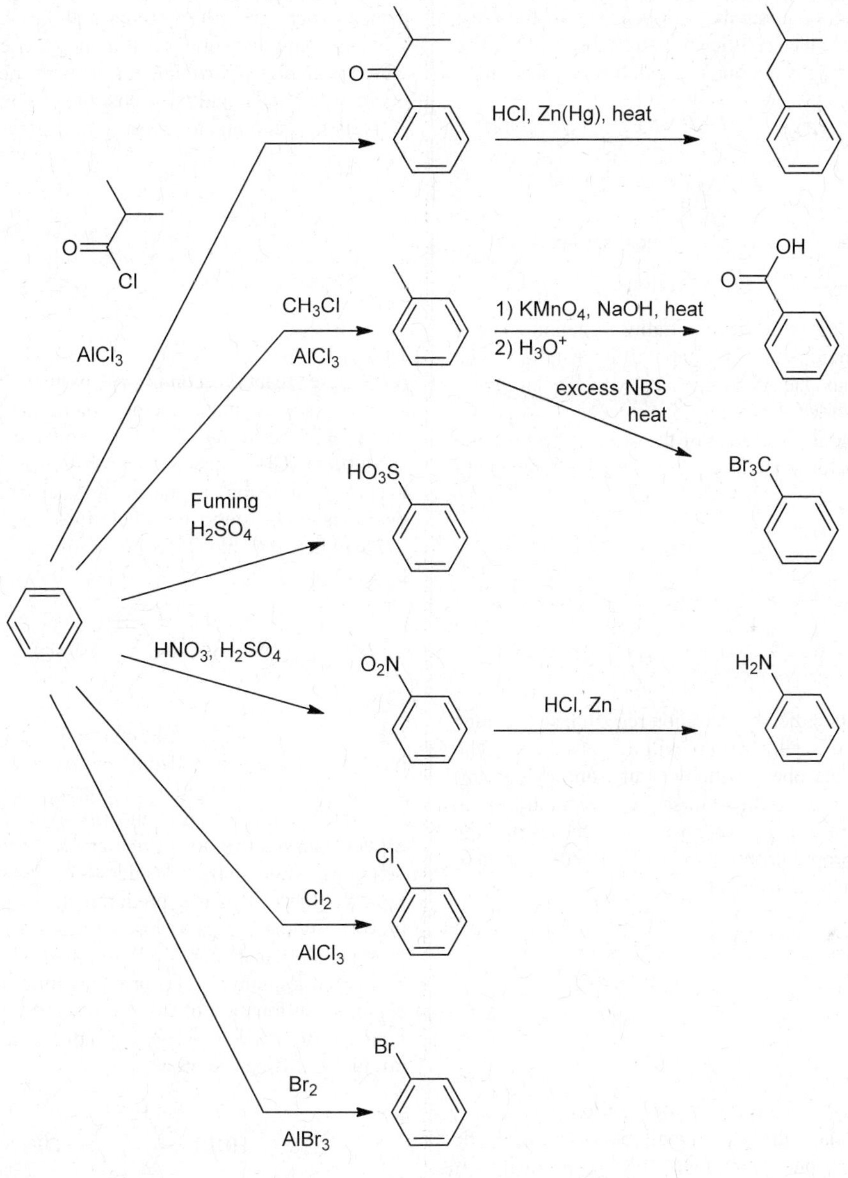

O
Cl
AlCl3
CH3Cl
AlCl3
Fuming
H2SO4
HNO3, H2SO4
Cl2
AlCl3
Br2
AlBr3
O
HCl, Zn(Hg), heat
1) KMnO4, NaOH, heat
2) H3O+
excess NBS
heat
OH
O
Br3C
HO3S
O2N
HCl, Zn
H2N
Cl
Br

19.44. Alkyl groups are activating while halogens are deactivating toward electrophilic aromatic substitution. As such, the compound with two alkyl groups will be the most activated toward electrophilic aromatic substitution, while the compound with two halogens will be the least activated:

Increasing Reactivity toward Electrophilic Aromatic Substitution
Br
Br
Br

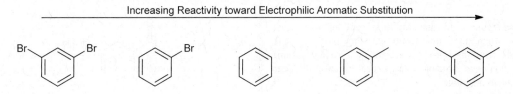

19.45. A nitro group is a strong deactivator. Therefore, among the four compounds, nitrobenzene is the least activated toward electrophilic aromatic substitution. The most activated ring is the one that exhibits two activators (OH and OMe).

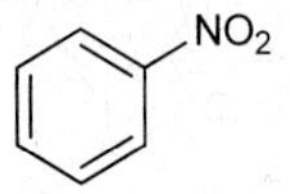

least activated 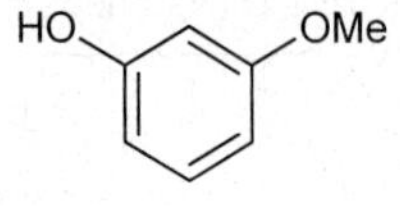   most activated

19.46.
(a) The reagents indicate a nitration reaction, so we must decide where the nitro group will be installed. The aromatic ring has one substituent (Br) which is an *ortho-para* director, so we expect nitration to occur at the *ortho* and *para* positions. Because of the size of the bromine atom, the *para* product is expected to predominate (a steric effect).

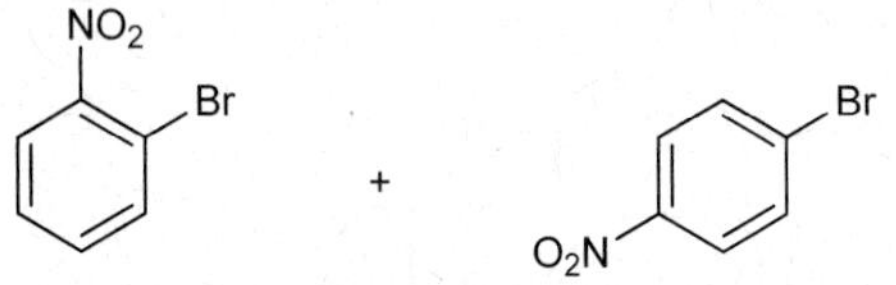

(b) The reagents indicate a nitration reaction, so we must decide where the nitro group will be installed. The aromatic ring has one substituent (an isopropyl group) which is an *ortho-para* director, so we expect nitration to occur at the *ortho* and *para* positions. Because of the size of the isopropyl group, the *para* product is expected to predominate (a steric effect).

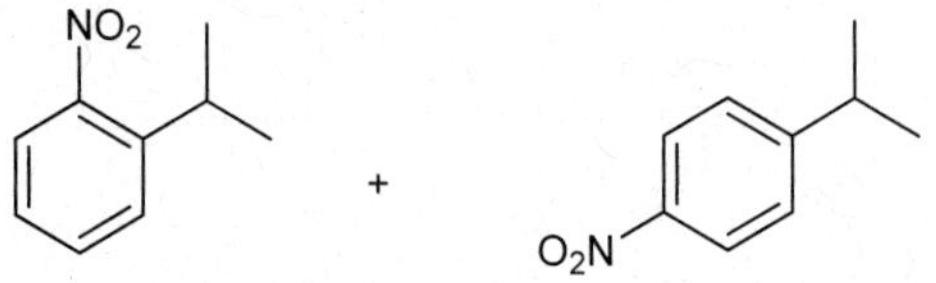

(c) The reagents indicate a nitration reaction, so we must decide where the nitro group will be installed. The aromatic ring has one substituent (a nitro group) which is a *meta* director, so we expect nitration to occur at the *meta* position.

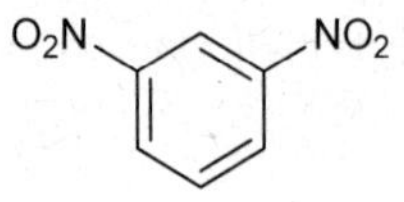

(d) The reagents indicate a nitration reaction, so we must decide where the nitro group will be installed. The aromatic ring has one substituent (a carbonyl group) which is a *meta* director, so we expect nitration to occur at the *meta* position.

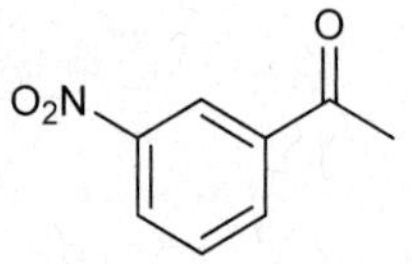

(e) The reagents indicate a nitration reaction, so we must decide where the nitro group will be installed. The aromatic ring has one substituent (a methoxy group) which is an *ortho-para* director, so we expect nitration to occur at the *ortho* and *para* positions. The *para* product is likely to predominate (because of a steric effect).

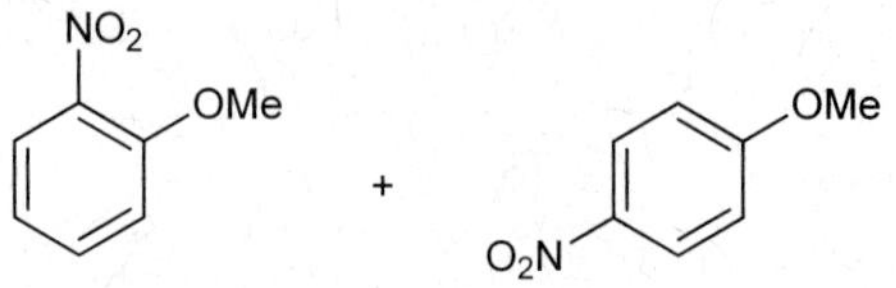

19.47.
(a) These reaction conditions indicate a sulfonation reaction, so we must decide where the sulfonic acid group will be installed. The aromatic ring has one substituent (Cl) which is an *ortho-para* director, so we expect sulfonation to occur at the *ortho* and *para* positions. For steric considerations, the *para* product (shown here) is likely to predominate over the *ortho* product.

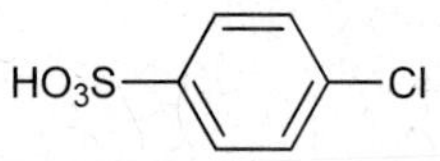

(b) These reaction conditions indicate a sulfonation reaction, so we must decide where the sulfonic acid group will be installed. The aromatic ring has one substituent (OH) which is an *ortho-para* director, so we expect sulfonation to occur at the *ortho* and *para* positions. For steric considerations, perhaps the *para* product (shown here) predominates over the *ortho* product, although it is a close call in this case. The OH group is similar in steric bulk to a methyl group, and we know that a methyl group provides little steric bulk (for example, sulfonation of toluene is expected to produce a mixture of *ortho* and *para* products, and the major product is difficult to predict).

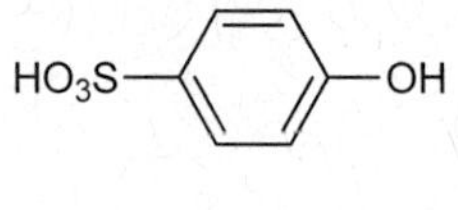

(c) These reaction conditions indicate a sulfonation reaction, so we must decide where the sulfonic acid group will be installed. The aromatic ring has one substituent (a carbonyl group) which is a *meta* director, so we expect sulfonation to occur at the *meta* position.

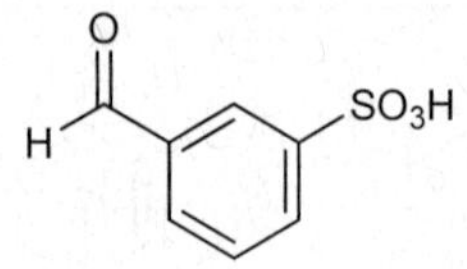

(d) These reaction conditions indicate a sulfonation reaction, so we must decide where the sulfonic acid group will be installed. The aromatic ring has two substituents (OH and NO₂). The directing effects are controlled by the more highly activating group, which is the OH group. As an activator, this group is an *ortho-para* director, so we expect sulfonation to occur at the *ortho* and *para* positions. For steric considerations, the *para* product (shown here) is likely to predominate over the *ortho* product (because formation of the *ortho* product is somewhat sterically encumbered).

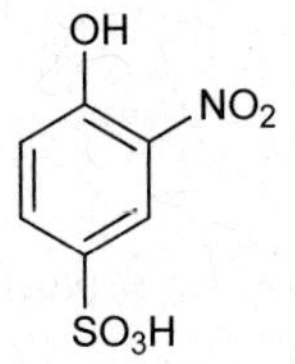

(e) These reaction conditions indicate a sulfonation reaction, so we must decide where the sulfonic acid group will be installed. The aromatic ring has two substituents (Br and Me). The directing effects are controlled by the more highly activating group, which is the methyl group. As an activator, this group is an *ortho-para* director, so we expect sulfonation to occur at the *ortho* and *para* positions. However, the *para* position is already occupied (by the bromine atom). Therefore, sulfonation can only take place at the *ortho* positions.

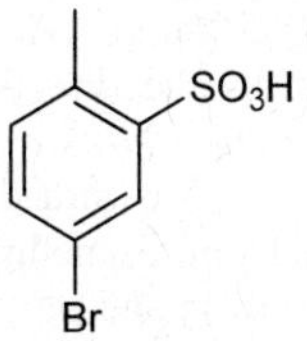

In this case, sulfonation at either *ortho* position leads to the same product. The two *ortho* positions are identical because of symmetry. Don't be confused by the arbitrary location of the double bonds. Perhaps it is less distracting if we draw the structure like this:

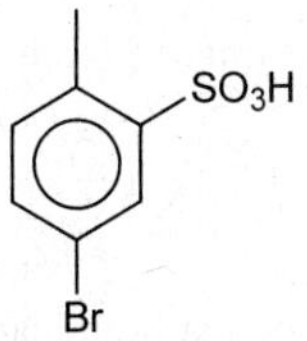

(f) These reaction conditions indicate a sulfonation reaction, so we must decide where the sulfonic acid group will be installed. The aromatic ring has one substituent (a carboxylic acid group) which is a *meta* director, so we expect sulfonation to occur at the *meta* position.

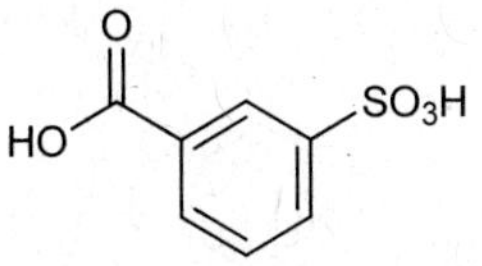

(g) These reaction conditions indicate a sulfonation reaction, so we must decide where the sulfonic acid group will be installed. The aromatic ring has two substituents (methyl and ethyl). Both groups are weak activators, so all unoccupied aromatic positions are equally activated:

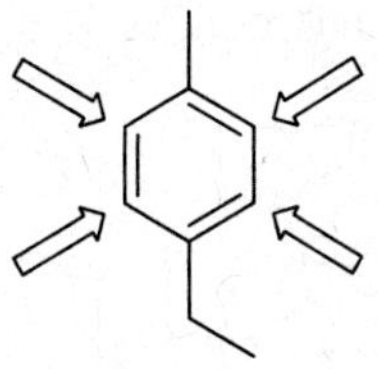

These positions are differentiated from each because of steric factors. The ethyl group is larger than the methyl group, so the reaction is expected to occur *ortho* to the methyl group, rather than *ortho* to the more sterically bulky ethyl group.

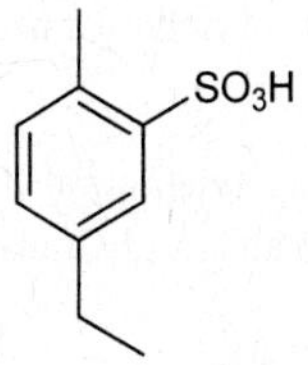

In this case, sulfonation at either *ortho* position leads to the same product. The two *ortho* positions are identical because of symmetry. Don't be confused by the arbitrary location of the double bonds. Perhaps it is less distracting if we draw the structure like this:

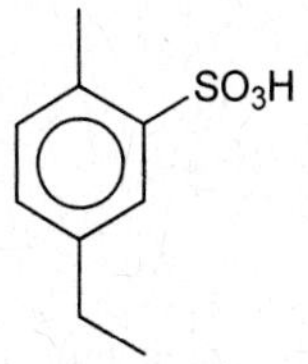

(h) These reaction conditions indicate a sulfonation reaction, so a sulfonic acid group is installed.

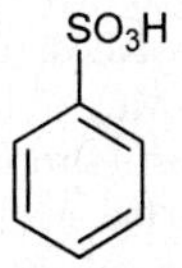

19.48. Each of these substituents appears in Table 19.1.
(a) This group is an activator and an *ortho-para* director.
(b) This group is an activator and an *ortho-para* director.
(c) This group is an activator and an *ortho-para* director.
(d) This group is a deactivator and an *ortho-para* director.
(e) This group is a deactivator and a *meta*-director.
(f) This group is a deactivator and a *meta*-director.
(g) This group is a deactivator and a *meta*-director.
(h) This group is a deactivator and a *meta*-director.
(i) This group is a deactivator and an *ortho-para* director.
(j) This group is a deactivator and a *meta*-director.

19.49.

(a) The reagents indicate a Friedel-Crafts alkylation reaction (methylation), so we must decide where the methyl group will be installed. The aromatic ring has one substituent (Cl) which is an *ortho-para* director, so we expect methylation to occur at the *ortho* and *para* positions.

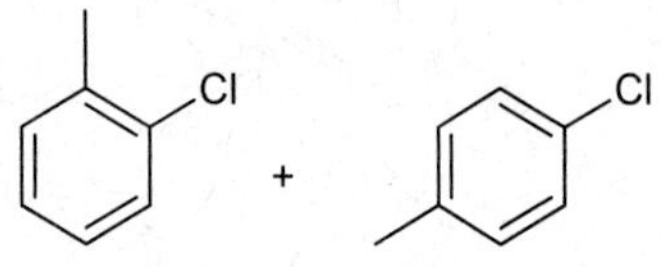

The latter is likely the major product because of steric considerations.

(b) This aromatic ring is moderately deactivated and will therefore not undergo a Friedel-Crafts alkylation.

(c) This aromatic ring is strongly deactivated and will therefore not undergo a Friedel-Crafts alkylation.

(d) The reagents indicate a Friedel-Crafts alkylation reaction (methylation), so we must decide where the methyl group will be installed. The aromatic ring has one substituent (Et) which is an *ortho-para* director, so we expect methylation to occur at the *ortho* and *para* positions.

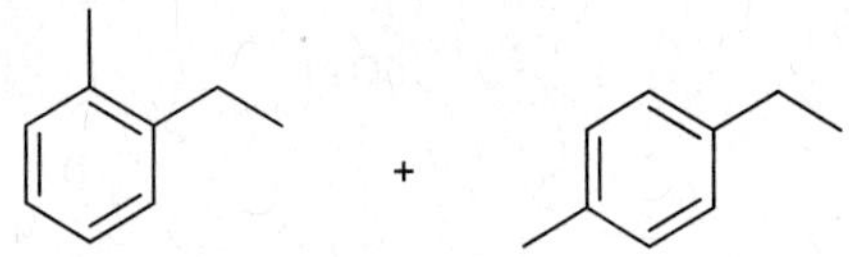

The latter is likely the major product because of steric considerations.

(e) The reagents indicate a Friedel-Crafts alkylation reaction (methylation), so we must decide where the methyl group will be installed. The aromatic ring has two substituents (I and Me). The directing effects are controlled by the most strongly activating group. Therefore, in this case, the methyl group controls the directing effects. As an activator, this group is an *ortho-para* director. However, one of the *ortho* positions is already occupied (by the other substituent). Therefore, methylation will occur at the position that is *ortho* or *para* to the methyl group, shown here:

The former is likely the major product because of steric considerations.

(f) The reagents indicate a Friedel-Crafts alkylation reaction (methylation), so we must decide where the methyl group will be installed. The aromatic ring has two substituents (propyl and methyl). Both groups are weak activators, and *ortho-para* directors. Therefore, we expect the following two products:

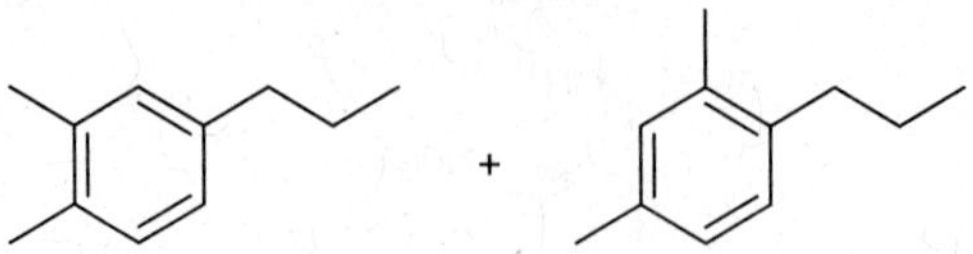

The former is likely the major product because of steric considerations.

(g) This aromatic ring has two substituents, and both are moderately deactivating. As such, the ring is too deactivated to undergo a Friedel-Crafts alkylation.

(h) The reagents indicate a Friedel-Crafts alkylation reaction (methylation), so we must decide where the methyl group will be installed. The aromatic ring has two substituents (the oxygen atom of an ester, and an alkyl group). The directing effects are controlled by the most strongly activating group. Therefore, in this case, the ester group controls the directing effects. As an activator, this group is an *ortho-para* director. However, one of the *ortho* positions is already occupied (by the other substituent). Therefore, methylation will occur at the positions that are *ortho* and *para* to the ester group, shown here:

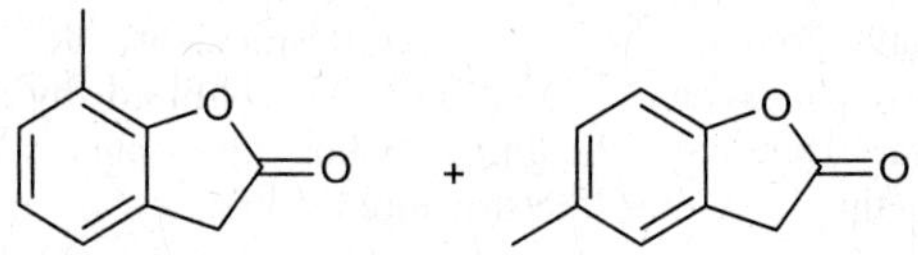

The latter is likely the major product because of steric considerations.

19.50.

(a) The reagents indicate a bromination reaction, so we must decide where the bromine atom will be installed. The aromatic ring has one substituent (Br) which is an *ortho-para* director, so we expect bromination to occur at the *ortho* and *para* positions. For steric considerations, the *para* product (shown here) is likely to predominate over the *ortho* product.

(b) The reagents indicate a bromination reaction, so we must decide where the bromine atom will be installed. The aromatic ring has one substituent (NO_2) which is a *meta* director, so we expect bromination to occur at the *meta* position:

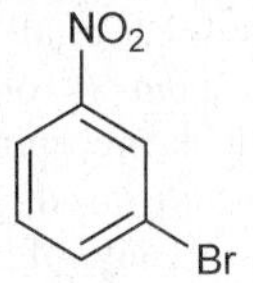

(c) The reagents indicate a bromination reaction, so we must decide where the bromine atom will be installed. The aromatic ring has two substituents (both methyl groups). Both groups are *ortho-para* directors, so all four unoccupied positions are activated:

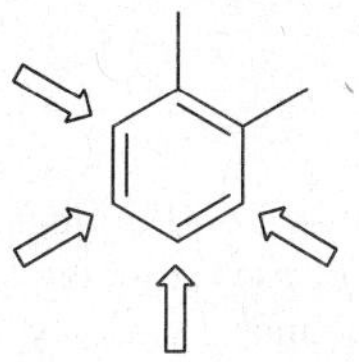

Because of symmetry, there are only two unique locations where bromination can occur:

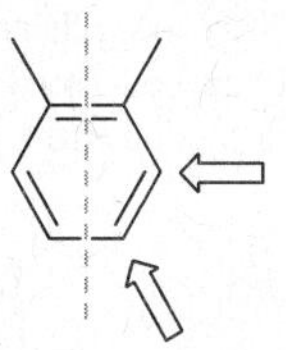

For steric considerations, the following product is expected to predominate.

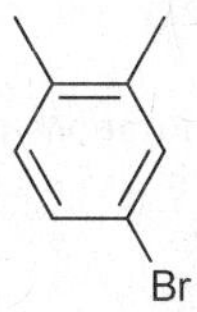

(d) The reagents indicate a bromination reaction, so we must decide where the bromine atom will be installed. The aromatic ring has one substituent (a *tert*-butyl group) which is an *ortho-para* director. The *tert*-butyl group is sterically bulky, so we expect the *para* product (shown here) to predominate over the *ortho* product.

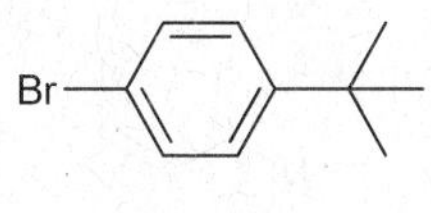

(e) The reagents indicate a bromination reaction, so we must decide where the bromine atom will be installed. The aromatic ring has one substituent (a sulfonic acid group) which is a *meta* director, so we expect bromination to occur at the *meta* position:

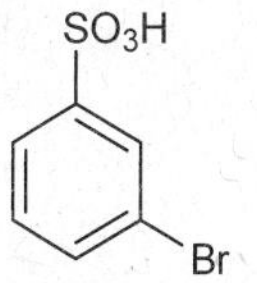

(f) The reagents indicate a bromination reaction, so we must decide where the bromine atom will be installed. The aromatic ring has one substituent (a carboxylic acid group) which is a *meta* director, so we expect bromination to occur at the *meta* position:

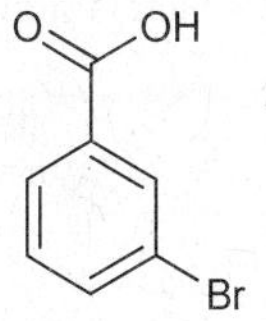

(g) The reagents indicate a bromination reaction, so we must decide where the bromine atom will be installed. The aromatic ring has one substituent (a carbonyl group) which is a *meta* director, so we expect bromination to occur at the *meta* position:

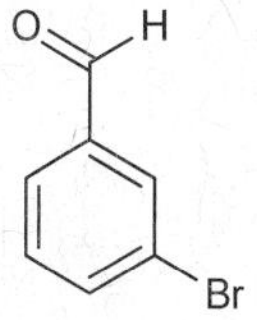

(h) The reagents indicate a bromination reaction, so we must decide where the bromine atom will be installed. The aromatic ring has two substituents (two bromine atoms). Both groups are *ortho-para* directors, so we expect these groups to direct to all four positions:

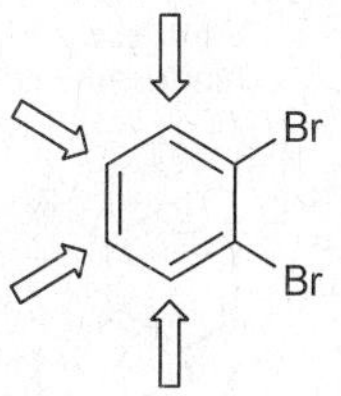

Because of symmetry, there are only two unique locations where bromination can occur:

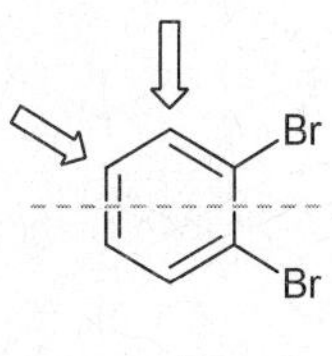

For steric considerations, the following product is expected to predominate.

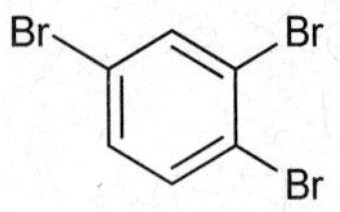

(i) The reagents indicate a bromination reaction, so we must decide where the bromine atom will be installed. The aromatic ring has two substituents: a methyl group (activator) and a nitro group (deactivator). The directing effects are controlled by the most strongly activating group. Therefore, in this case, the methyl group controls the directing effects. As an activator, this group is an *ortho-para* director, thereby activating the following three locations, one of which is sterically hindered:

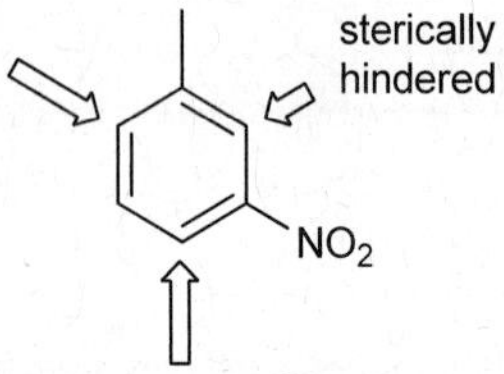

Among the remaining two positions, the position next to the nitro group is probably more hindered than the position next to the methyl group. So, we predict that bromination will occur near the methyl group:

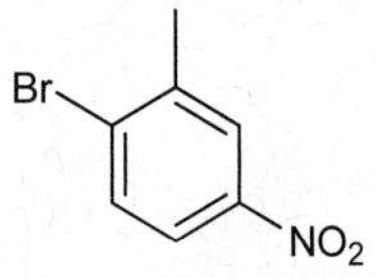

(j) The reagents indicate a bromination reaction, so we must decide where the bromine atom will be installed. The aromatic ring has two substituents (two bromine atoms). Both groups are *ortho-para* directors, so we expect these groups to direct to the following three positions, one of which is too sterically hindered:

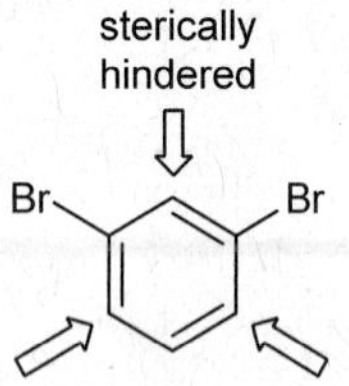

The remaining two positions are identical by symmetry. That is, bromination at either position will generate the same product:

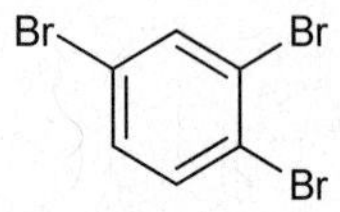

(k) The reagents indicate a bromination reaction, so we must decide where the bromine atom will be installed. The aromatic ring has two substituents (two bromine atoms), both of which are *ortho-para* directors, so we expect these groups to direct to all four positions (all of which are *ortho* to one of the halogens). Indeed, all four positions are equivalent by symmetry anyway. Bromination will install a third bromine atom which can be drawn in any of the four unoccupied positions (all are identical):

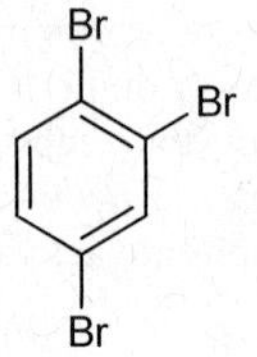

19.51. There are two aromatic rings, and each ring has two substituents. For each ring, the directing effects are controlled by the most strongly activating group. Therefore, in this case, the OR group (a moderate activator) controls the directing effects for each ring (rather than the *tert*-butyl groups, which are weak activators). As an activator, the OR group is an *ortho-para* director. However, in each ring, one of the *ortho* positions is already occupied (by the *tert*-butyl group). Therefore, nitration can occur at the following locations:

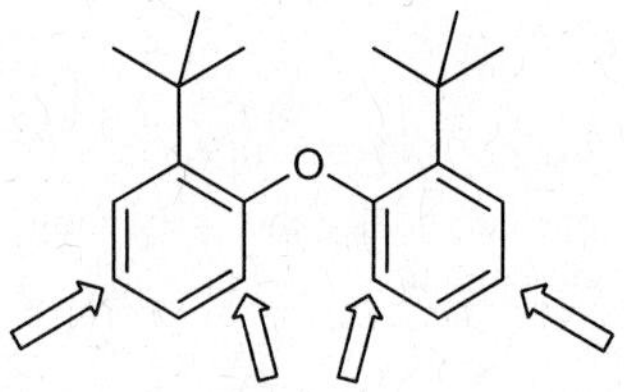

However, two of these locations are sterically hindered, so we only expect nitration at the remaining two (unencumbered) positions:

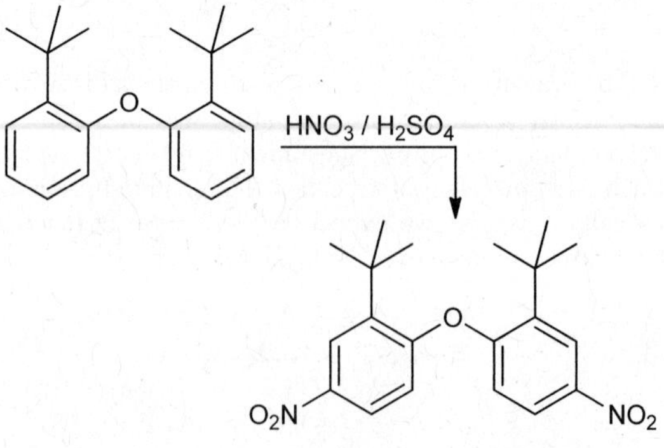

19.52. In acidic conditions, 2-methylpropene is protonated to give a tertiary carbocation, which can function as an electrophile in an electrophilic aromatic substitution reaction. The aromatic ring attacks the carbocation to give a resonance stabilized intermediate (sigma complex), followed by deprotonation which restores aromaticity. In aqueous acidic conditions, the proton source is H_3O^+ (in the beginning of the mechanism), and the base is H_2O (at the end of the mechanism).

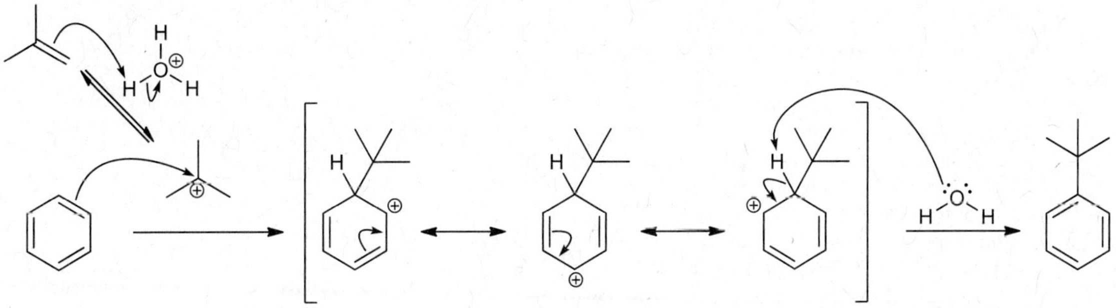

19.53.
(a) This is an electrophilic aromatic substitution reaction, and before we can draw the two steps of the process, we must begin by drawing formation of the electrophilic complex that will react with the aromatic ring. This occurs when molecular chlorine (Cl_2) attacks the Lewis acid, forming an electrophilic complex.

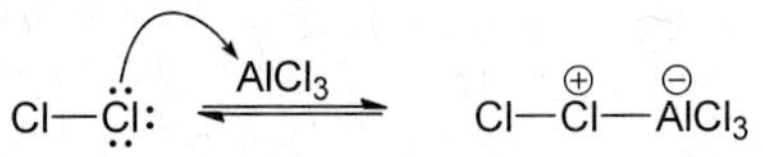

This complex serves as a delivery agent of Cl^+ (a powerful electrophile). The aromatic ring then functions as a nucleophile and attacks the electrophilic complex. The resulting intermediate sigma complex then loses a proton to restore aromaticity.

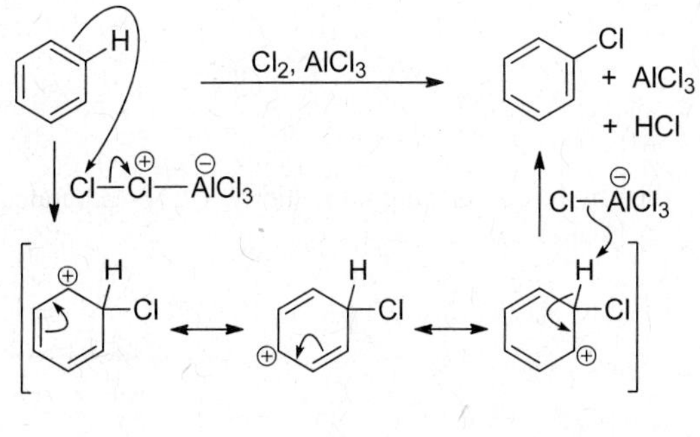

The base for the final step is $AlCl_4^-$. When it removes a proton, thereby re-establishing aromaticity, HCl is generated as a byproduct and the Lewis acid ($AlCl_3$) is regenerated (its function is catalytic).

(b) In the presence of sulfuric acid, nitric acid can be protonated, followed by loss of water to give a nitronium ion:

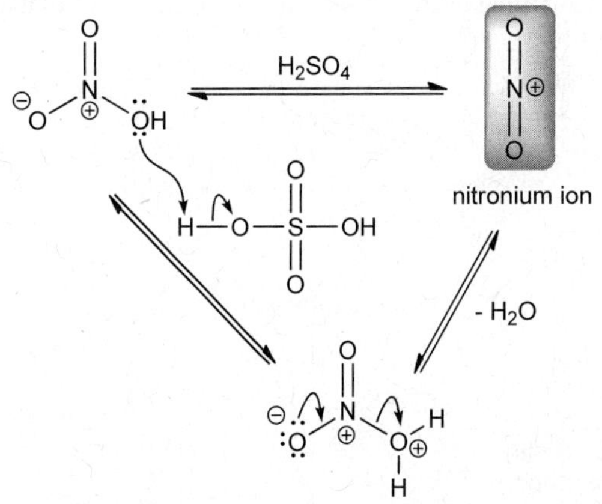

The aromatic ring then functions as a nucleophile and attacks the nitronium ion. The resulting intermediate sigma complex then loses a proton to restore aromaticity. Water is the likely base for the final deprotonation step, since water is present (and there are no strong bases present in strongly acidic conditions).

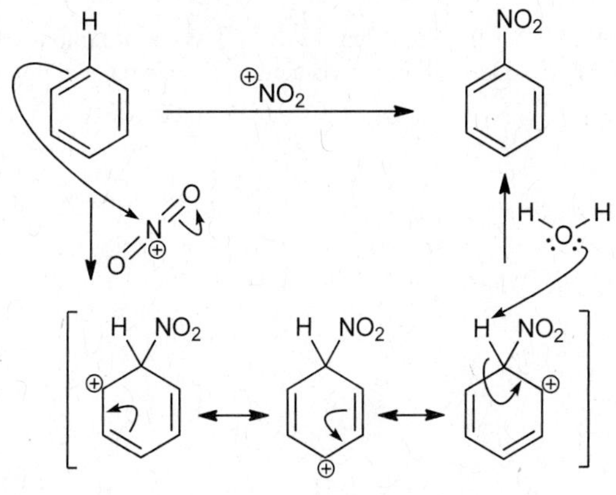

(c) This is an electrophilic aromatic substitution reaction (called sulfonation). As for all electrophilic aromatic substitution reactions, there are two key steps. First, the aromatic ring functions as a nucleophile and attacks the electrophilic species (SO_3). Then, the resulting intermediate sigma complex loses a proton to restore aromaticity. Notice that the product is an anion, which is protonated under these acidic conditions, as shown:

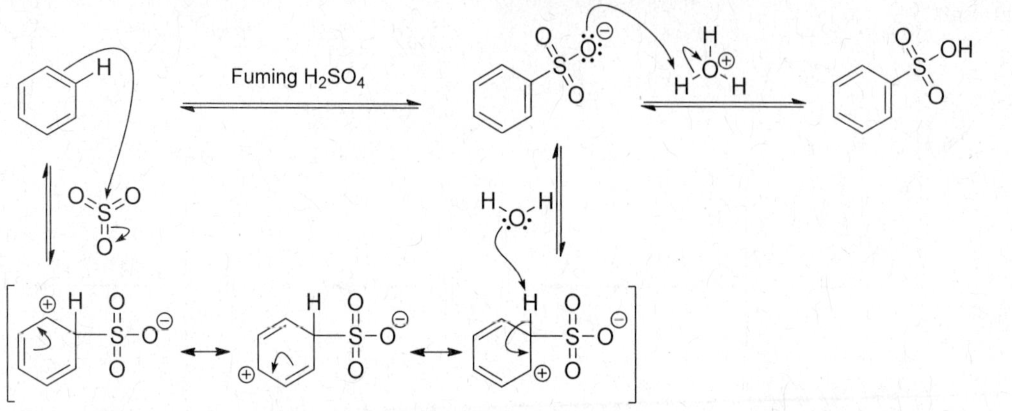

(d) This is a Friedel-Crafts alkylation process, which begins with the formation of the electrophilic complex that will react with the aromatic ring. Methyl chloride attacks the Lewis acid, forming an electrophilic complex. This complex serves as a delivery agent of H_3C^+. The aromatic ring functions as a nucleophile and attacks the electrophilic complex. The resulting intermediate sigma complex then loses a proton to restore aromaticity, as shown:

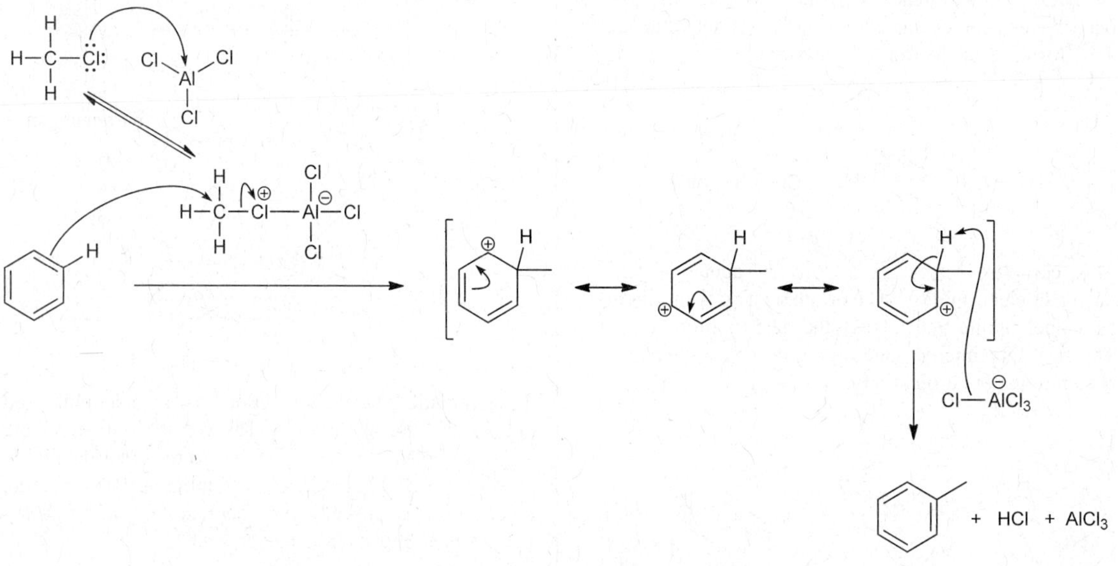

The base for the final step is $AlCl_4^-$. When it removes a proton, thereby re-establishing aromaticity, HCl is generated as a byproduct and the Lewis acid $(AlCl_3)$ is regenerated (its function is catalytic).

(e) This is an electrophilic aromatic substitution reaction, and before we can draw the two steps of the process, we must begin by drawing formation of the electrophilic complex that will react with the aromatic ring. This occurs when molecular bromine (Br_2) attacks the Lewis acid, forming an electrophilic complex.

This complex serves as a delivery agent of Br^+ (a powerful electrophile). The aromatic ring then functions as a nucleophile and attacks the electrophilic complex. The resulting intermediate sigma complex then loses a proton to restore aromaticity.

The base for the final step is $FeBr_4^-$. When it removes a proton, thereby re-establishing aromaticity, HBr is generated as a byproduct and the Lewis acid ($FeBr_3$) is regenerated (its function is catalytic).

19.54.

(a) This is an electrophilic aromatic substitution reaction, and before we can draw the two steps of the process, we must begin by drawing formation of the electrophilic complex that will react with the aromatic ring. This occurs when I–Cl attacks the Lewis acid, forming an electrophilic complex.

This complex serves as a delivery agent of I^+. The aromatic ring then functions as a nucleophile and attacks the electrophilic complex. The resulting intermediate sigma complex then loses a proton to restore aromaticity.

The base for the final step is $AlCl_4^-$. When it removes a proton, thereby re-establishing aromaticity, HCl is generated as a byproduct and the Lewis acid ($AlCl_3$) is regenerated (its function is catalytic).

(b) This transformation occurs via two successive electrophilic aromatic substitution reactions. The process begins with formation of the electrophilic complex that will react with the first aromatic ring. Methylene chloride (CH_2Cl_2) attacks the Lewis acid, forming an electrophilic complex. This complex serves as a delivery agent of ClH_2C^+. The aromatic ring functions as a nucleophile and attacks the electrophilic complex. The resulting intermediate sigma complex then loses a proton to restore aromaticity, giving benzyl chloride. Then, all of the previous steps (formation of the electrophilic complex, followed by the two steps of an electrophilic aromatic substitution reaction) are repeated again, as shown, giving the product (diphenylmethane). Notice that, for each deprotonation step, the base is $AlCl_4^-$. When it removes a proton, HCl is generated as a byproduct and the Lewis acid ($AlCl_3$) is regenerated (its function is catalytic):

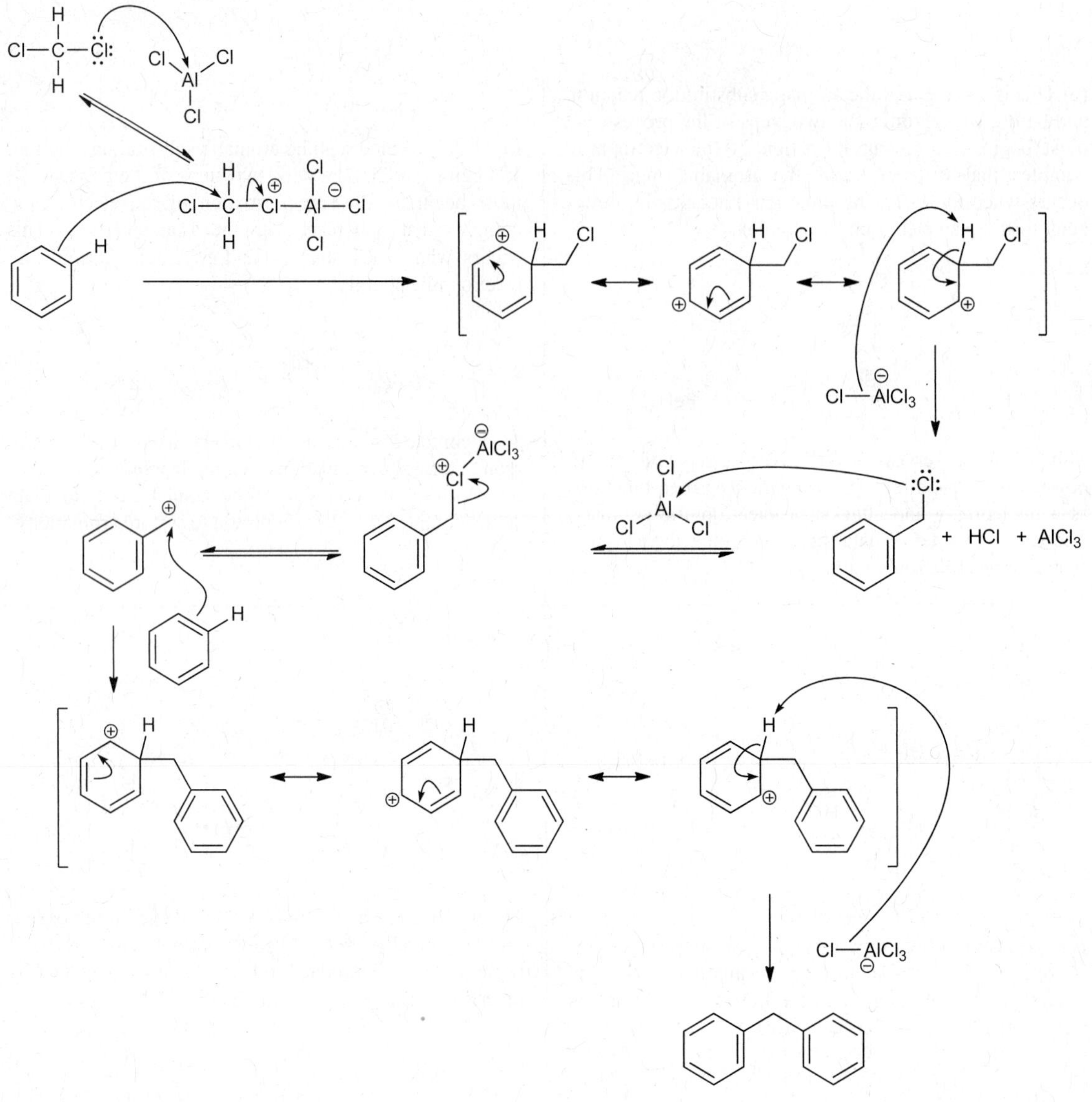

19.55. The substituent (Cl) is replaced with a different substituent (OMe). Therefore, this is an aromatic substitution reaction. The reagent (methoxide) is nucleophilic, so we must determine whether this reaction occurs via an S_NAr mechanism or via an elimination-addition mechanism. All three criteria for an S_NAr mechanism are met: 1) there is a leaving group (chloride), and 2) there is an electron-withdrawing group (the carbonyl group is electron withdrawing via resonance), and 3) the leaving group is *para* to the electron-withdrawing group. Therefore, we draw an S_NAr mechanism. In the first step, methoxide functions as a nucleophile and attacks the aromatic ring at the position bearing the leaving group, thereby forming an intermediate Meisenheimer complex. Then, a leaving group (chloride) is ejected, restoring aromaticity, and giving the product, as shown.

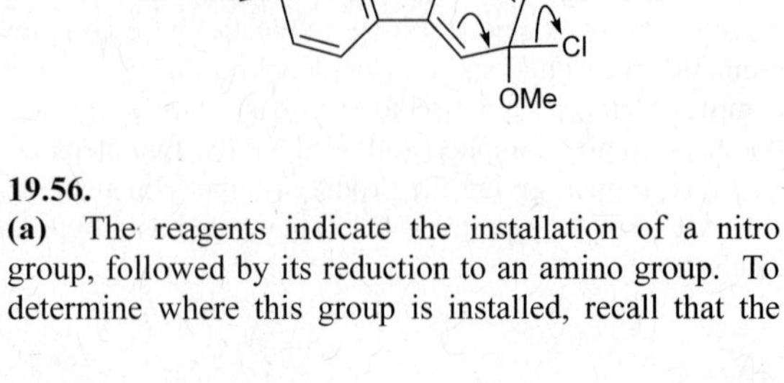

19.56.
(a) The reagents indicate the installation of a nitro group, followed by its reduction to an amino group. To determine where this group is installed, recall that the

directing effects are controlled by the most strongly activating group. In this case, the methyl group is the only activating group (albeit weakly activating), so it controls the directing effects. As an activator, this group is an *ortho-para* director. However, the *para* position is already occupied (by the other substituent). Therefore, the group is installed in a position that is *ortho* to the methyl group.

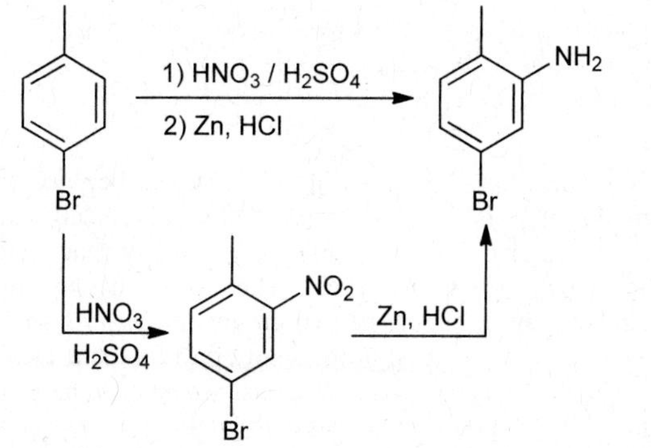

In this case, both *ortho* positions are equivalent because of symmetry, so the group can be drawn in either ortho position (don't be distracted by the arbitrary placement of the double bonds in the aromatic ring):

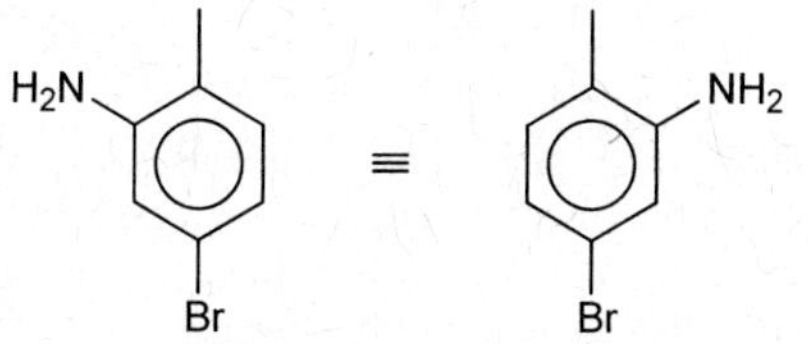

(b) The starting material is *para*-xylene, in which the following four positions are identical (because of symmetry):

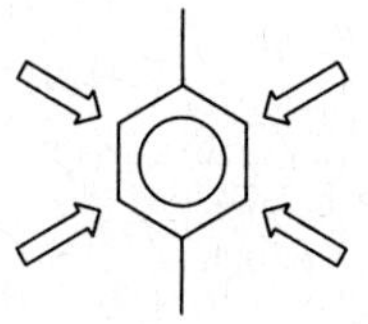

Therefore, installation of a third group leads to only one regiochemical outcome. Each group weakly activates the ring, so this compound will readily undergo a Friedel-Crafts acylation reaction, thereby installing an acyl group on the ring. Then, in the second step, the carbonyl group is reduced, giving the product shown:

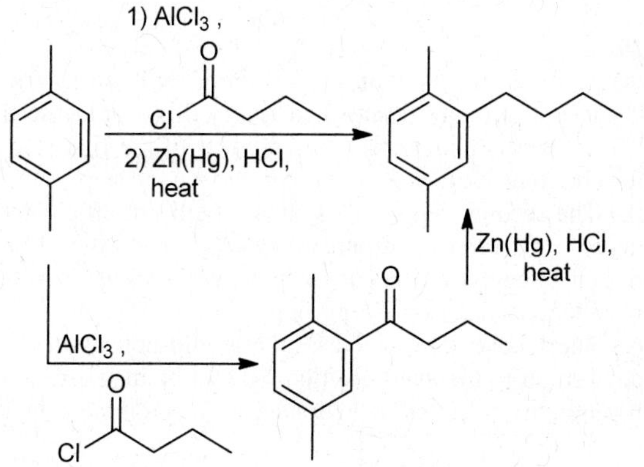

(c) The aromatic ring has only one substituent (an isopropyl group), which is an activator and therefore an *ortho-para* director. The *ortho* positions are sterically hindered (by their proximity to the bulky isopropyl group), so an electrophilic aromatic substitution reaction is expected to occur predominantly at the *para* position. In the first step, a Friedel-Crafts alkylation is used to install a methyl group in the *para* position. Then, both alkyl groups are oxidized to give a diacid, as shown:

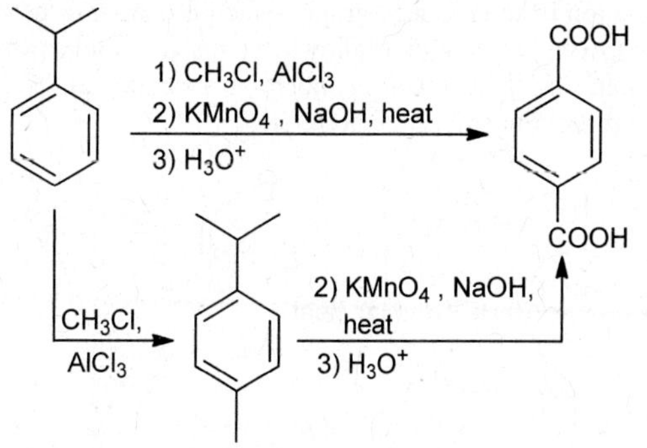

(d) The aromatic ring has only substituent (a *tert*-butyl group), which is an activator and therefore an *ortho-para* director. The *ortho* positions are very sterically hindered (by their proximity to the bulky *tert*-butyl group), so an electrophilic aromatic substitution reaction is expected to occur almost exclusively at the *para* position. In the first step, a Friedel-Crafts alkylation is used to install a methyl group in the *para* position. Then, in the next step, the benzylic protons are replaced with bromine atoms. The *tert*-butyl group does not have any benzylic protons, so no bromine atoms are installed in that location. Only the methyl group undergoes benzylic bromination (exhaustively), as shown:

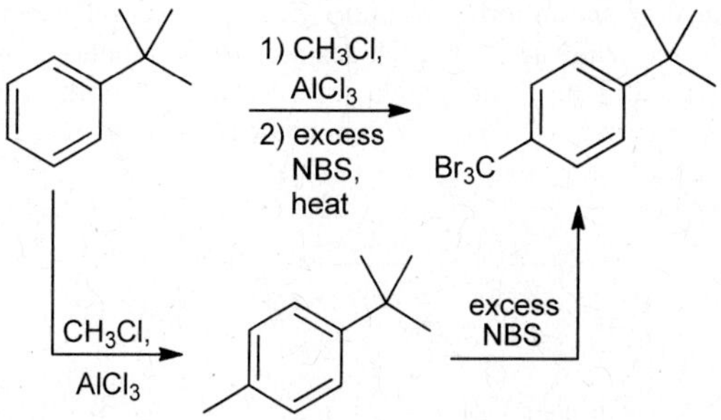

19.57.
(a) Installation of the bromine atom requires one step (bromination), and installation of the ethyl group can also be achieved in just one step (via a Friedel-Crafts alkylation). However, both of these substituents are *ortho-para* directors, and this creates a problem. If either substituent is installed first, the other substituent will not be directed to the correct location. This problem can be circumvented by installing the ethyl group via a

two-step process (Friedel-Crafts acylation, followed by reduction:

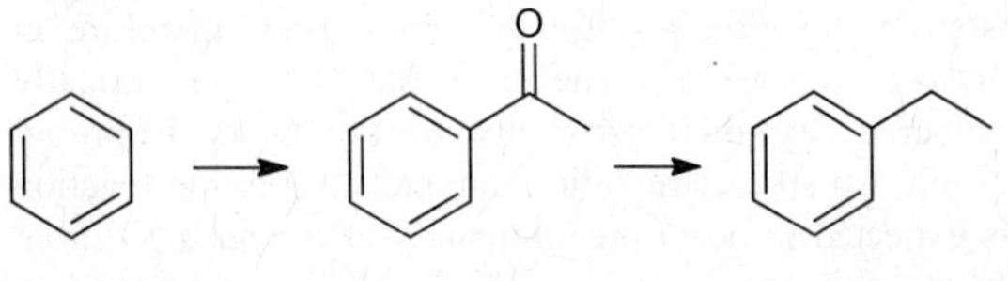

By installing the ethyl group in this way, we can capitalize on the directing effects of the carbonyl group, before reducing it. As seen below, a Friedel-Crafts acylation installs an acyl group, which is a *meta*-director. Subsequent bromination allows for the installation of a bromine atom in the *meta* position. Reduction of the acyl group then gives the product.

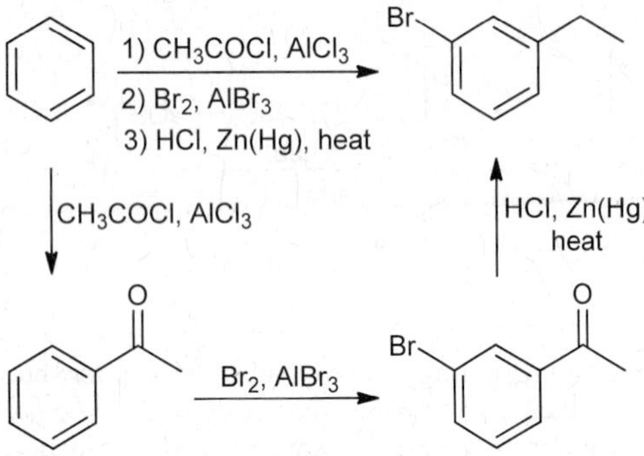

(b) Installation of the amino group requires a two-step process (nitration, followed by reduction), while installation of the bromine atom can be achieved in just one step.

Now let's consider the order of events. These two groups must be installed in a *meta* fashion, but both groups are *ortho-para* directors. Installation of these two groups (in either order) does not appear to give the desired *meta* substitution pattern. However, recall that installation of the amino group requires two steps. The first step is nitration, and a nitro group is a *meta*-director. So, we can achieve the desired transformation by performing the bromination process after the nitration process but before the reduction process, as shown here:

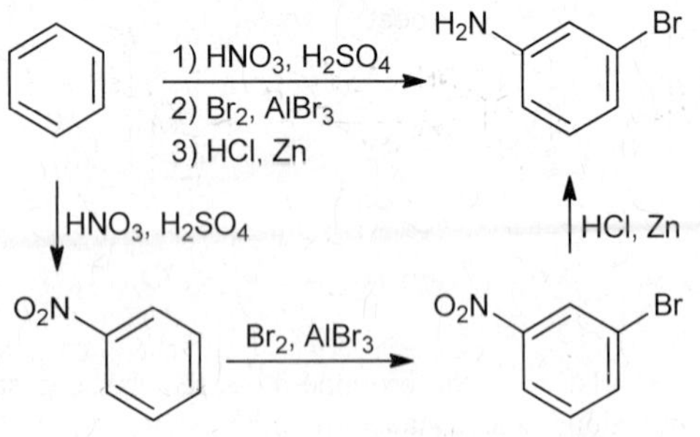

(c) Installation of the amino group requires two steps (nitration, followed by reduction), while installation of the ethyl group can be achieved in just one step (via a Friedel-Crafts alkylation). The substituents must be installed in a *meta* fashion, yet both substituents are *ortho-para* directors. To circumvent the problem, we

consider utilizing the *meta*-directing effects of the nitro group to achieve the desired regiochemical outcome:

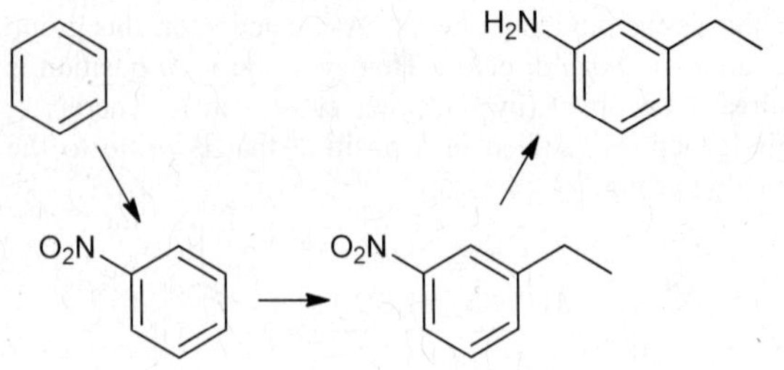

However, this strategy will not succeed because the middle step is flawed. Specifically, a Friedel-Crafts alkylation cannot be performed on a strongly deactivated ring (such as nitrobenzene). Therefore, we must find another way to install the two groups in a *meta* fashion. The trick that we used in the solution to Problem 19.57a can be used here again. That is, we install the ethyl group via a two-step process (Friedel-Crafts acylation, followed by reduction):

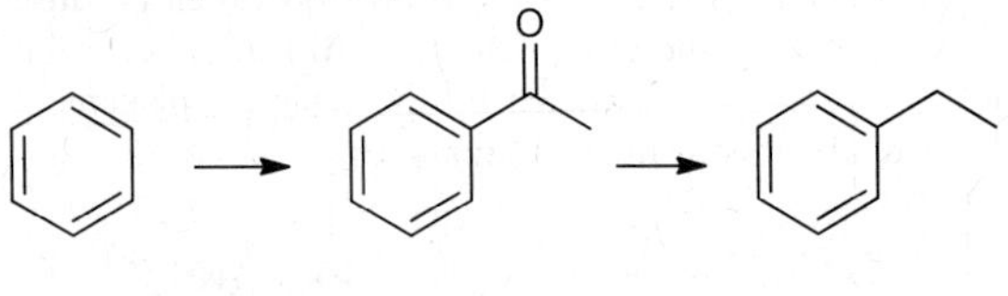

By installing the ethyl group in this way, we can capitalize on the directing effects of the carbonyl group, before reducing it. As seen below, a Friedel-Crafts acylation installs an acyl group, which is a *meta*-director. Subsequent nitration allows for the installation of a nitro group in the *meta* position. And finally, reduction (of both groups) gives the product, as shown.

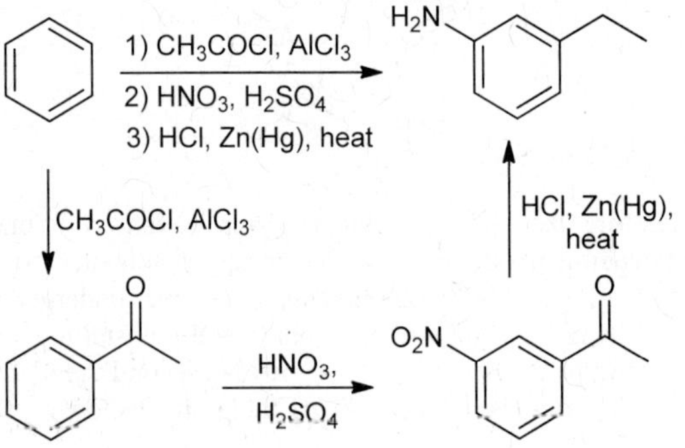

19.58.
(a) The second step of the synthesis will not work, because a strongly deactivated ring will not undergo a Friedel-Crafts alkylation. The product of the first step, nitrobenzene, will be unreactive in the second step.
(b) The second step of the synthesis will not efficiently install a propyl group, because a carbocation rearrangement can occur, which will result in the installation of an isopropyl group.
(c) The second step of the synthesis will not install the acyl group in the *meta* position. It will be installed in a position that is either *ortho* or *para* to the bromine atom.

(d) The second step of the synthesis will not install the bromine atom in the *ortho* position, because of steric hindrance from the *tert*-butyl group. Bromination will occur primarily at the *para* position.

19.59.
(a) This compound has two aromatic rings, each of which is monosubstituted. One aromatic ring (left) is connected directly to a carbonyl group and is therefore deactivated. The other aromatic ring (right) is connected to a methylene group (CH_2) and is therefore activated (the substituent is treated like an alkyl group). So, we expect monobromination to occur on the activated ring. Since the substituent is an activator, it must also be an *ortho-para* director. The *ortho* positions are sterically hindered (by the large substituent), so the reaction is expected to occur predominantly at the *para* position, indicated here:

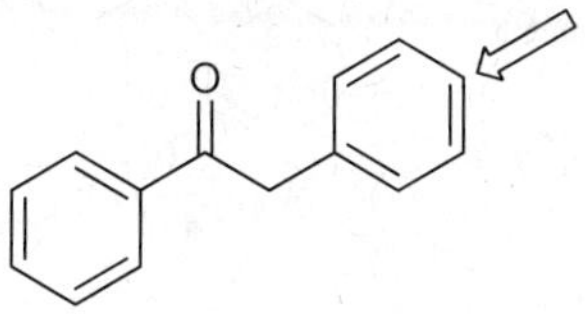

(b) This compound has two aromatic rings, each of which is monosubstituted. One aromatic ring (left) is connected directly to a carbonyl group and is therefore deactivated. The other aromatic ring (right) is connected to a nitrogen atom and is therefore activated. So, we

expect monobromination to occur on the activated ring. Since the substituent is an activator, it must also be an *ortho-para* director. The *ortho* positions are sterically hindered (by the large substituent), so the reaction is expected to occur predominantly at the *para* position, indicated here:

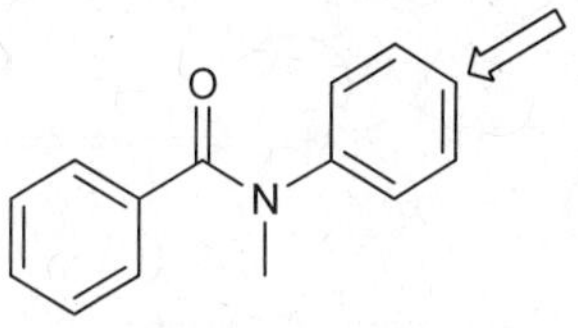

(c) This compound has two aromatic rings. The ring on the left has one only substituent, which is an activator. The ring on the right has two substituents: an activator and a deactivator. As such, we expect monobromination to occur on the left ring, which is more activated. Since the substituent (on the left ring) is an activator, it must also be an *ortho-para* director. The *ortho* positions are sterically hindered (by the large substituent), so the reaction is expected to occur predominantly at the *para* position, indicated here:

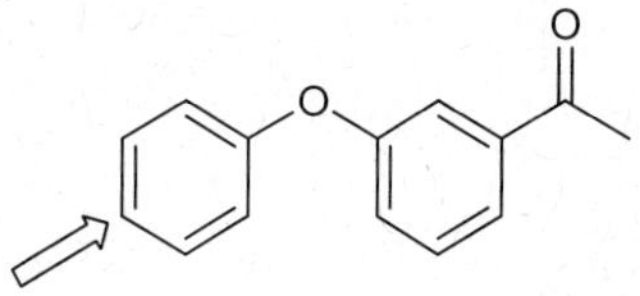

19.60. The reagent is sodium amide, which is strongly nucleophilic and strongly basic. The three criteria for S_NAr are not met (the aromatic ring lacks a nitro group), and therefore, the reaction must occur via an elimination-addition mechanism (as seen in Mechanism 19.9), shown here. The benzyne intermediate can be attacked on either side of the triple bond, leading to the two products shown:

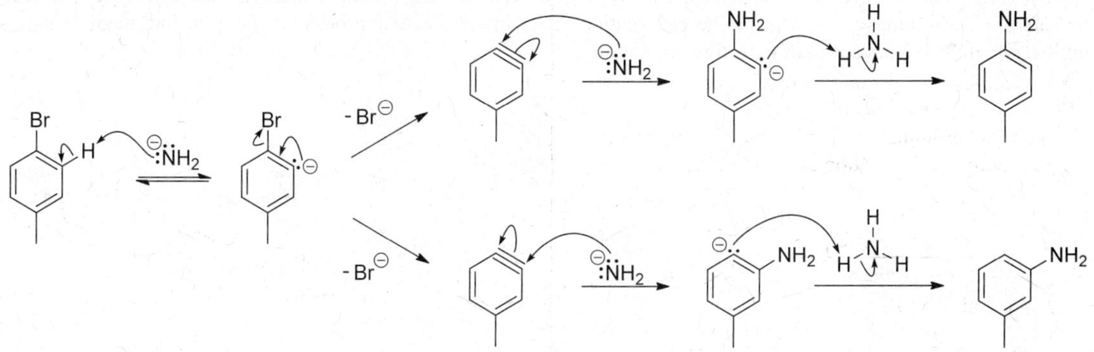

19.61. Phenol has only one substituent (OH) connected to the aromatic ring. This group is a strong activator, and is therefore an *ortho-para* director. If conditions are controlled such that three nitro groups are installed, then we expect the three nitro groups to be installed at the *ortho* and *para* positions, as shown:

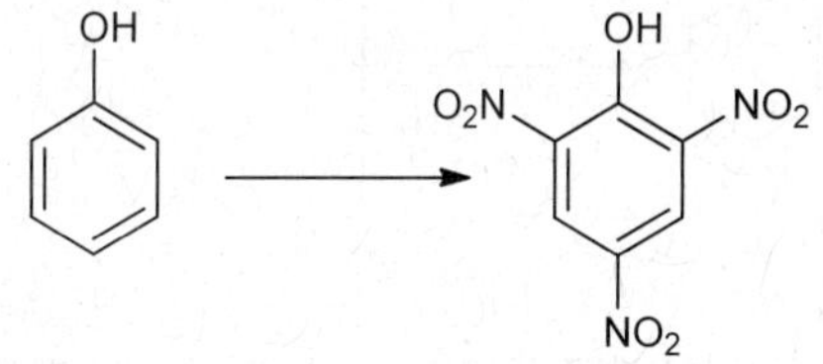

2,4,6-trinitrophenol
(picric acid)

19.62. The starting material is benzene, and the product exhibits a disubstituted aromatic ring. This will require two successive aromatic substitution reactions. The reagent (the diol) is not a very strong nucleophile, nor is it a very strong electrophile. However, in acidic conditions, the diol exists in equilibrium with a very powerful electrophile (a carbocation), as shown:

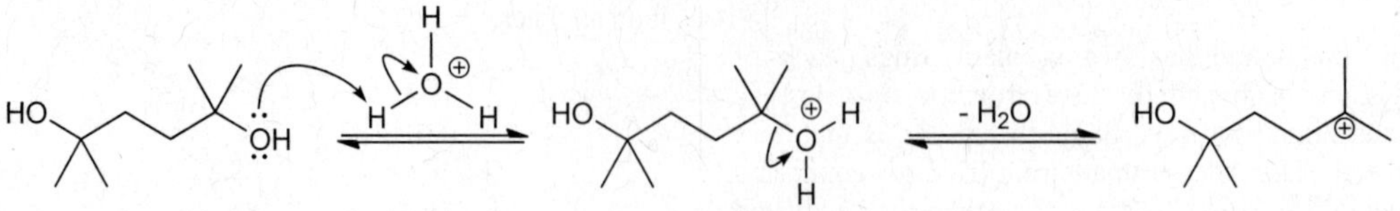

As seen above, one of the OH groups can be protonated, giving an excellent leaving group, which can leave to give a tertiary carbocation. Since the reagent is electrophilic, we expect that the reaction will proceed via electrophilic aromatic substitution. Accordingly, the aromatic ring attacks the carbocation to give an intermediate sigma complex, which then loses a proton to restore aromaticity:

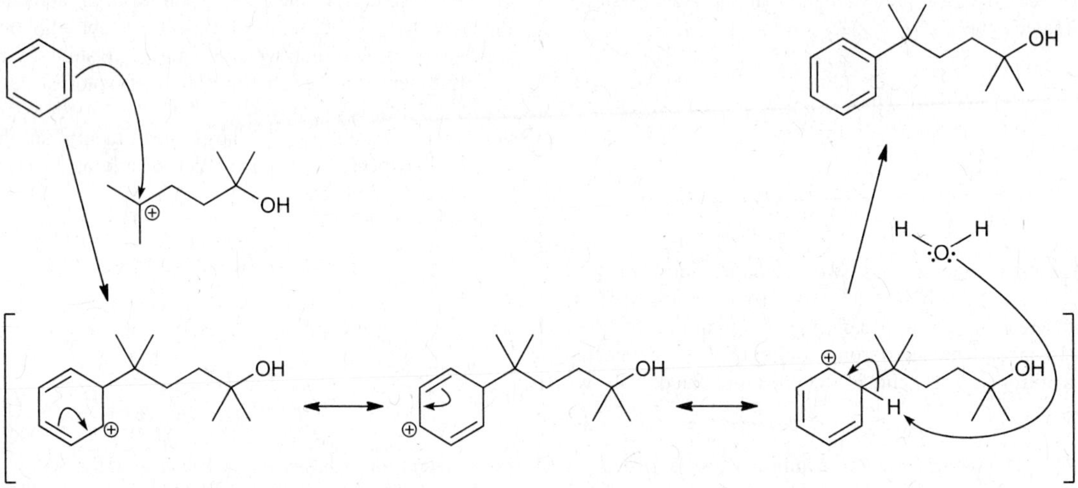

The entire process is then repeated again, except this time, the reaction will proceed in an *intramolecular* fashion, closing a ring. The OH group is protonated to give an excellent leaving group, which can leave to give a tertiary carbocation. The aromatic ring attacks the carbocation in an intramolecular process to give an intermediate sigma complex, which then loses a proton to restore aromaticity:

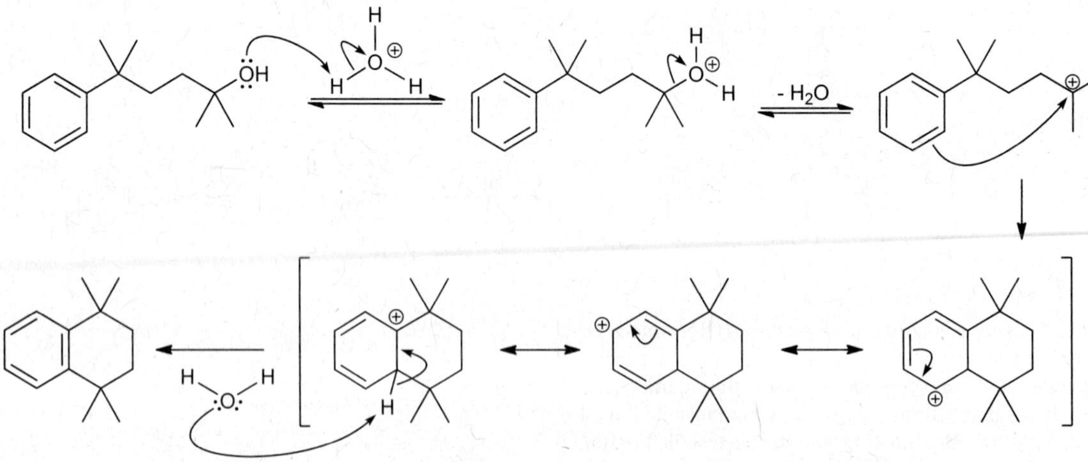

Since aqueous acidic conditions are employed (H_2SO_4 is generally dissolved in water), the proton source throughout the mechanism has been H_3O^+, and the base has always been H_2O, to be consistent with acidic conditions.

19.63. These reagents will achieve the installation of an isopropyl group on the aromatic ring. The problem statement indicates that conditions will favor dialkylation, which means that two isopropyl groups are installed. After installation of the first isopropyl group, the second isopropyl is expected to be installed predominantly in the *para* position (because alkyl groups are *ortho-para* directors, and the *ortho* positions are sterically hindered by their proximity to the isopropyl group):

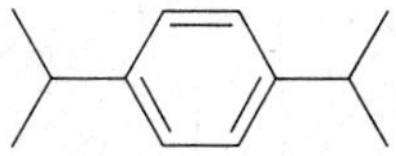

19.64.

(a) Attack at the *ortho* position gives the following resonance-stabilized intermediate. Notice that there are four resonance structures. The fourth resonance structure has the positive charge on the nitrogen atom, rather than on a carbon atom:

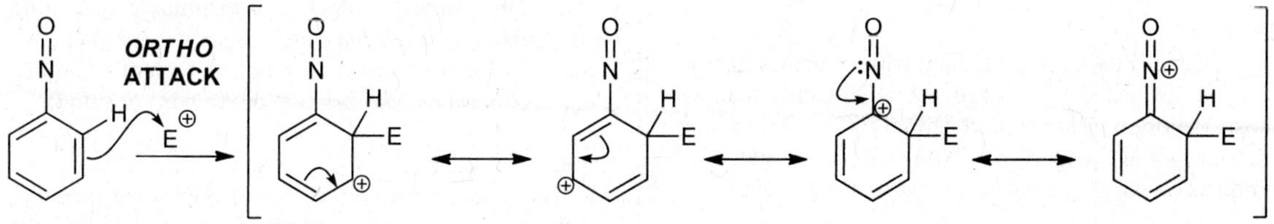

(b) Attack at the *meta* position gives the following resonance-stabilized intermediate. Notice that there are only three resonance structures. There is no resonance structure in which the positive charge is placed on the nitrogen atom:

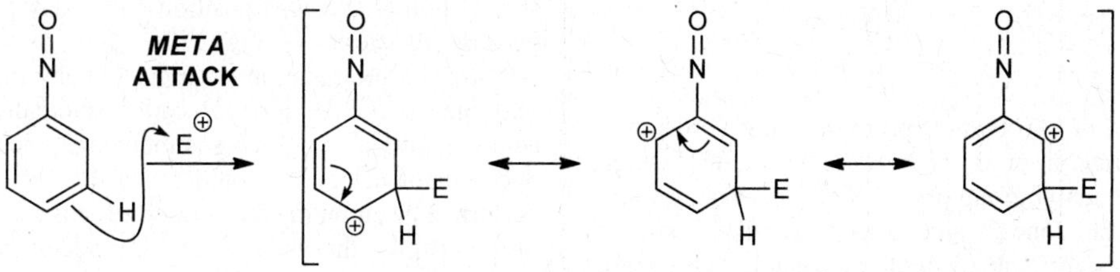

(c) Attack at the *para* position gives the following resonance-stabilized intermediate. Notice that there are four resonance structures. The fourth resonance structure has the positive charge on the nitrogen atom, rather than on a carbon atom:

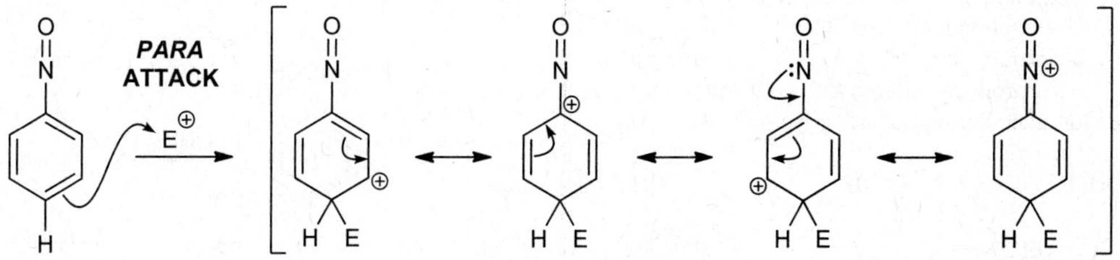

(d) The nitroso group should be *ortho-para* directing, because attack at the *ortho* or *para* position generates a sigma complex with an additional resonance structure.

(e) The nitroso group is a deactivator, yet it is an *ortho-para* director. This is unusual because nearly all deactivators are *meta* directors. We have only seen one exception. Specifically, halogens are deactivators, yet they are *ortho-para* directors. In this way, the nitroso group is similar to a chloro group.

19.65. A chlorine atom is placed at the *para* position, and a resonance-stabilized positive charge is formed, as shown:

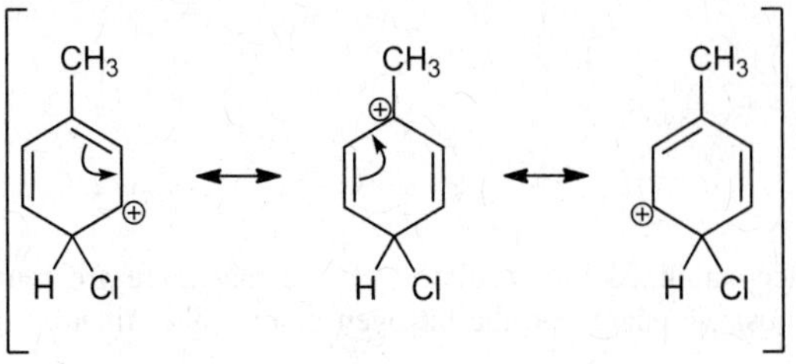

19.66.

(a) Toluene is the only compound with an activated ring, and it is expected to undergo a Friedel-Crafts reaction most rapidly. The methyl group is an *ortho-para* director, so we expect a mixture of *ortho* and *para* products (*ortho*-ethyltoluene and *para*-ethyltoluene):

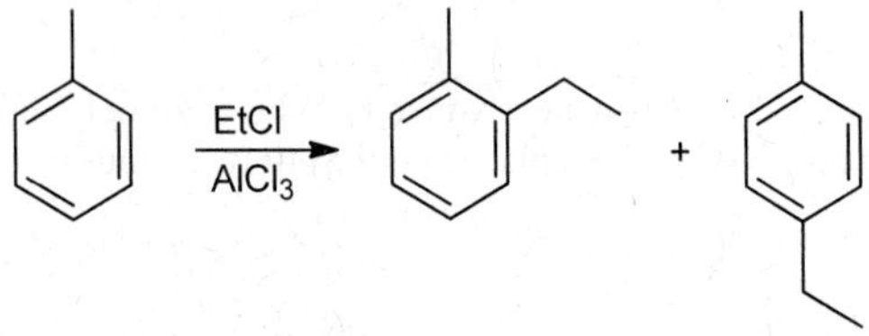

Generally, *para* substitution products are favored over *ortho* substitution products because of steric factors, However, a methyl group provides only a small amount of steric hindrance, and therefore, the product distribution is difficult to predict. That is, we cannot predict which product will predominate, and it will likely be sensitive to the conditions employed.

(b) Anisole (methoxybenzene) exhibits the most activated ring, and it is therefore expected to undergo a Friedel-Crafts reaction most rapidly. The methoxy group is an *ortho-para* director, so we expect a mixture of *ortho* and *para* products, although the *para* product will likely predominate because of steric factors:

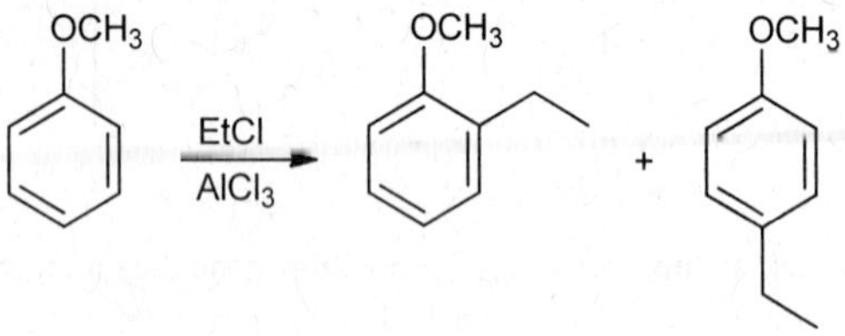

19.67. The molecular formula indicates five degrees of unsaturation, which accounts for the aromatic ring (four degrees) and the ester group (1 degree). The aromatic ring accounts for six of the eight carbon atoms, so the ester group can only have two carbon atoms. The following two structures fit this description:

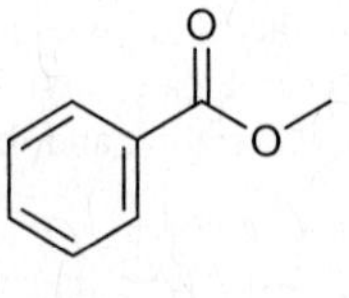

Compound A 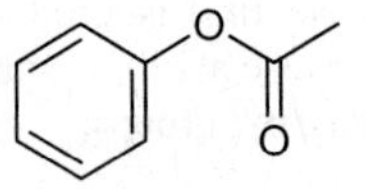   Compound B

Compound **A** must be the compound in which the carbonyl group is connected directly to the aromatic ring, because bromination of compound **A** leads to only one product. Recall that the carbonyl group is a *meta* director, and both *meta* positions are identical, which explains why there is only one monobromination product. In contrast, compound **B** exhibits an activating group, which is an *ortho-para* director. As such, we expect two monobromination products (bromination can occur at either the *ortho* position or the *para* position).

19.68.

(a) The starting material is methoxybenzene (anisole) and the product has two additional substituents, located at the *ortho* and *para* positions. The bromine atom is installed in just one step (bromination), and the nitro group can also be installed in just one step (nitration). The methoxy group is an *ortho-para* director, although substitution at the *para* position will likely predominate because of steric factors. Therefore, we must first perform the bromination reaction, thereby installing the bromine atom in the correct location. Note that no Lewis acid is required, as the ring is moderately activated; just one equivalent of Br_2 would be used. And finally, we perform a nitration reaction which installs the nitro group in the position that is *ortho* to the moderately activating methoxy group.

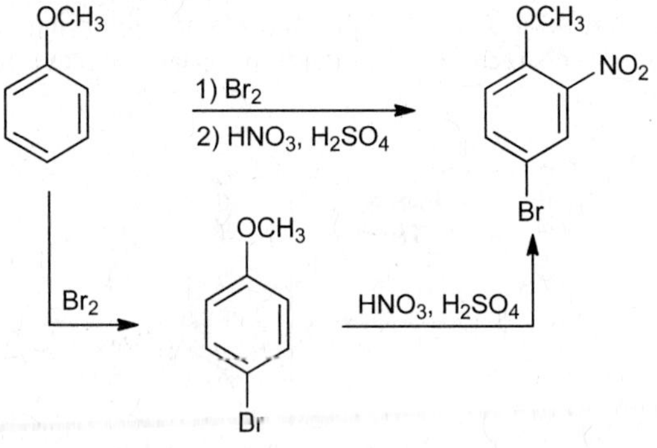

(b) The starting material is methoxybenzene (anisole) and the product has two additional substituents, located at the *ortho* and *para* positions. The bromine atom is installed in just one step (bromination), and the nitro group can also be installed in just one step (nitration). The methoxy group is an *ortho-para* director, although substitution at the *para* position will likely predominate because of steric factors. Therefore, we must first perform the nitration reaction, thereby installing the nitro group in the correct location. And then, we perform a bromination reaction which installs the bromine atom in the position that is *ortho* to the moderately activating methoxy group.

(c) The starting material exhibits an aromatic ring with only one substituent (an isopropyl group). In the product, the isopropyl group has been converted into a carboxylic acid, which can be achieved via benzylic oxidation with a strong oxidizing agent. In addition, a nitro group must be installed. Now we must determine the order of events. Both substituents (in the product) are *meta* directors, yet they are *ortho* to each other. Therefore, we must install the nitro group before the isopropyl group is oxidized, thereby relying on the *ortho-para* directing effects of an isopropyl group. A blocking group must be used because the isopropyl group provides steric bulk that would otherwise favor nitration at the *para* position, rather than the *ortho* position. By blocking the *para* position, we can perform a nitration reaction, thereby installing the nitro group in the desired location. Removing the blocking group, followed by oxidation of the benzylic position, gives the desired product.

(d) The starting material exhibits an aromatic ring with only one substituent (an isopropyl group). In the product, the isopropyl group has been converted into a carboxylic acid, which can be achieved via benzylic oxidation with a strong oxidizing agent. In addition, a nitro group must be installed. Now we must determine the order of events. Both substituents (in the product) are *meta* directors, and they are indeed *meta* to each other. Therefore, we simply oxidize the benzylic position, and then perform a nitration reaction to give the desired product.

Note that an isopropyl group is an *ortho-para* director, but the desired nitration reaction must occur at the *meta* position, so nitration cannot be the first step of our synthesis. We must first convert the alkyl group (which is an *ortho-para* director) into a carboxylic acid group (which is a *meta* director).

19.69.
(a) The reagents indicate a chlorination reaction, so we must decide where the chlorine atom will be installed. The directing effects are controlled by the most strongly activating group. In this case, none of the groups are activators, but the halogen is at least an *ortho-para* director, and therefore controls the directing effects (rather than the ester groups, which are *meta*-directing). The two *ortho* positions are already occupied (by the other substituents). Therefore, chlorination is expected to occur at the position that is *para* to the bromine atom. This is consistent with the directing effects of the ester groups as well.

(b) The reagents indicate a nitration reaction, so we must decide where the nitro group will be installed. The directing effects are controlled by the most strongly activating group. Therefore, in this case, the methoxy group controls the directing effects. As an activator, this group is an *ortho-para* director. However, one of the *ortho* positions is already occupied (by a methyl substituent). That leaves two possible locations: the remaining *ortho* position and the *para* position. The former is too sterically hindered and the reaction is not likely to occur there. Therefore, we expect nitration to occur in the *para* position, giving the following product.

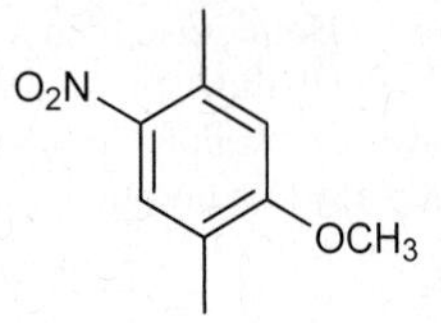

(c) The reagents indicate a sulfonation reaction, so we must decide where the sulfonic acid group will be installed. The aromatic ring has two substituents (*tert*-butyl and methyl). Both groups are weak activators, and *ortho-para* directors. Because of steric factors, we expect the sulfonic acid group to be installed *ortho* to the methyl group, rather than *ortho* to the large *tert*-butyl group.

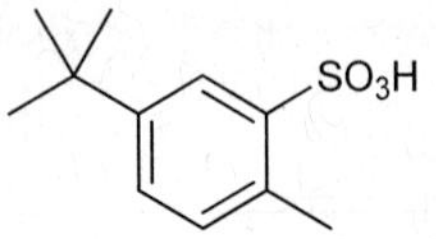

(d) The reagents indicate a nitration reaction, so we must decide where the nitro group will be installed. The directing effects are controlled by the most strongly activating group. Therefore, in this case, the OH group controls the directing effects. As an activator, this group is an *ortho-para* director. However, one of the *ortho* positions and the *para* position are already occupied (by the other substituents). Therefore, nitration can only occur in the remaining position that is *ortho* to the OH group, giving the following product.

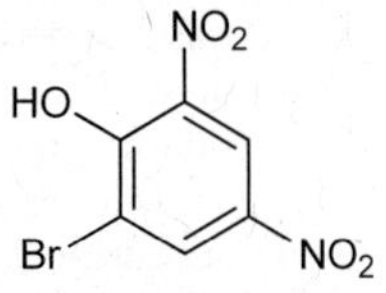

19.70.
(a) There are two possible pathways to consider, based on which C–C bond is to be formed:

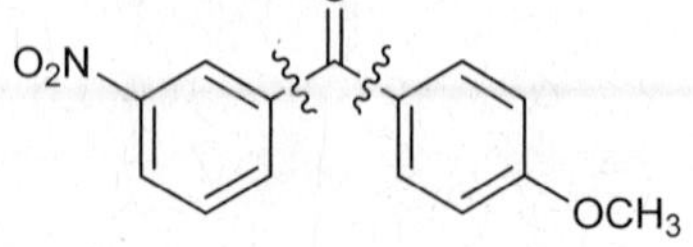

Let's consider one of these pathways (forming the bond on the left). Forming this bond via a Friedel-Crafts acylation would require the following starting materials:

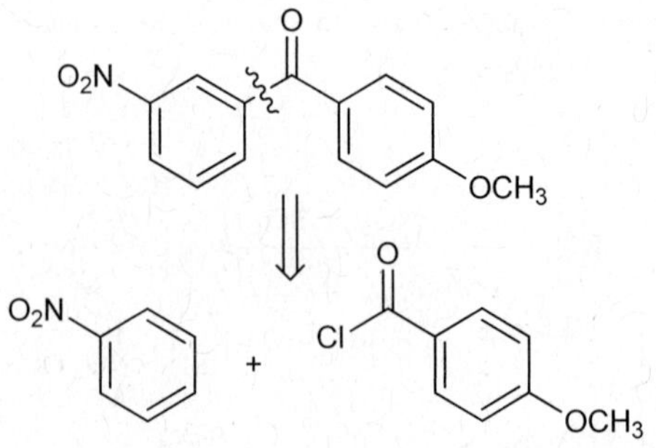

Indeed, the nitro group of nitrobenzene is *meta*-directing, so the regiochemistry is correct. However, we must remember that a Friedel-Crafts acylation cannot be performed on an aromatic ring that is strongly deactivated. Since nitrobenzene is strongly deactivated, this pathway is not feasible. So, we must explore the other possible pathway, forming this bond:

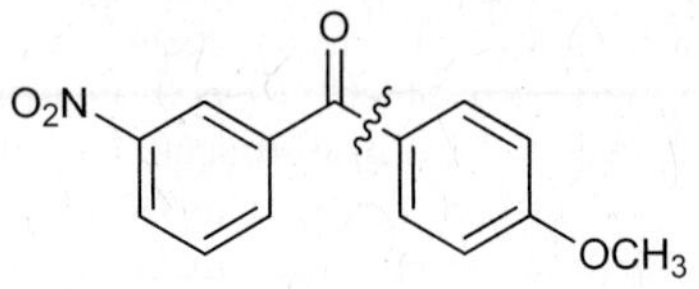

Forming this bond via a Friedel-Crafts acylation would require the following starting materials:

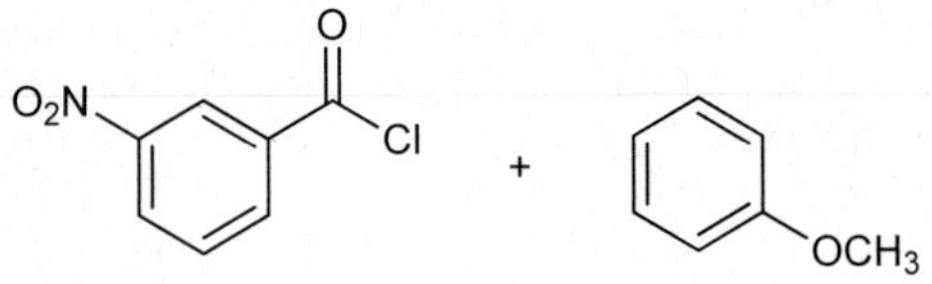

The methoxy group is an *ortho-para* director, and *para* substitution is expected to predominate (because of steric factors), giving rise to the desired product.
(b) There are two possible pathways to consider, based on which C–C bond is to be formed:

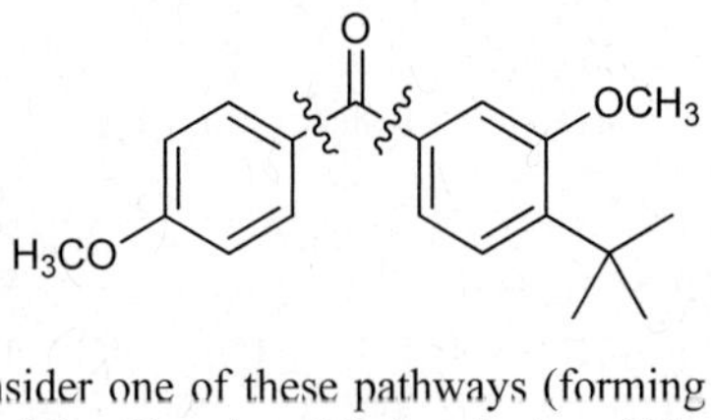

Let's consider one of these pathways (forming the bond on the right). Forming this bond via a Friedel-Crafts acylation would require the following starting materials:

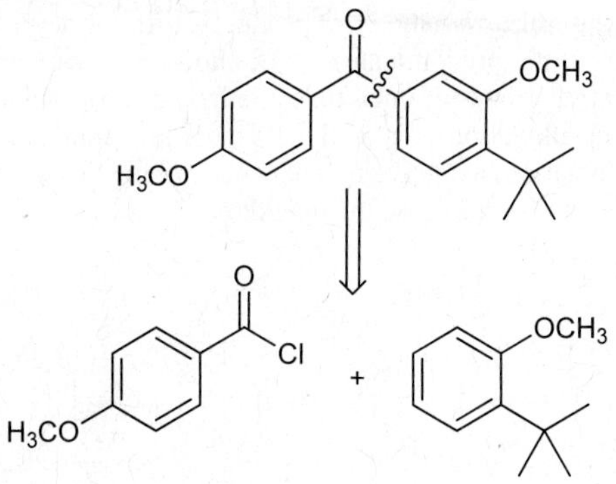

Let's consider the regiochemical outcome of a Friedel-Crafts acylation with these starting materials. The methoxy group controls the directing effects, so substitution is expected to occur at the position that is *para* to the methoxy group (not *meta* to the methoxy group).

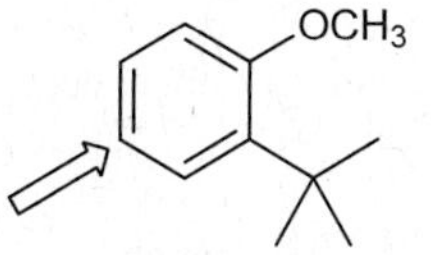

Therefore, this pathway is not feasible, because it would not give the correct product. So we must explore the other possible pathway, forming this bond:

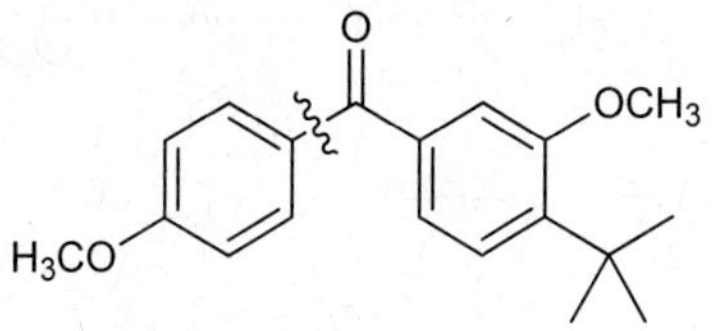

Forming this bond via a Friedel-Crafts acylation would require the following starting materials:

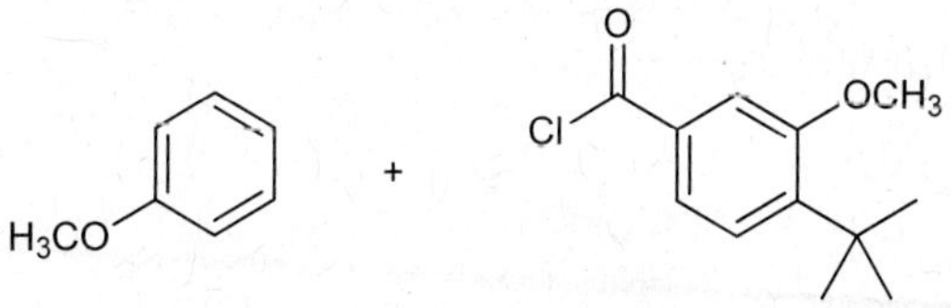

The methoxy group is an *ortho-para* director, and *para* substitution is expected to predominate (because of steric factors), giving rise to the desired product.

19.71. Attack at the C2 position proceeds via an intermediate with three resonance structures:

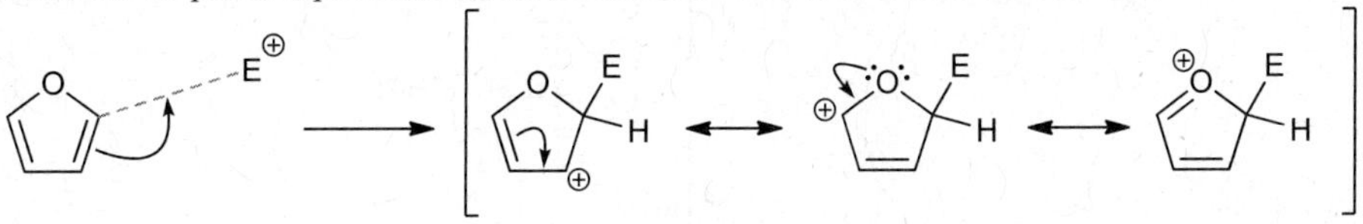

In contrast, attack at the C3 position proceeds via an intermediate with only two resonance structures:

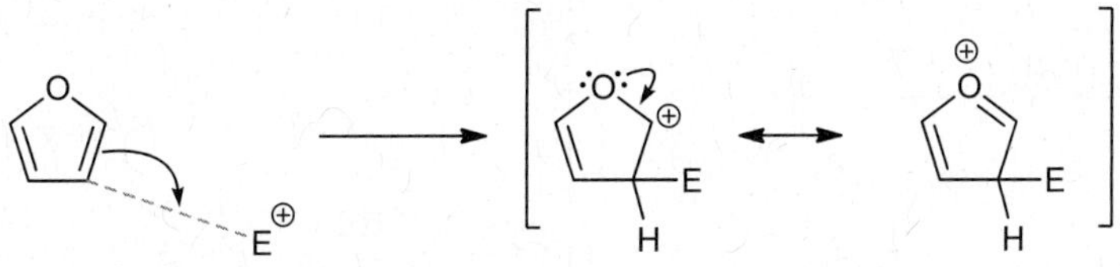

The intermediate for C2 attack is lower in energy than the intermediate for C3 attack. The transition state leading to the intermediate of C2 attack will therefore be lower in energy than the transition state leading to the intermediate of C3 attack. As a result, C2 attack occurs more rapidly.

19.72.
(a) The product has four substituents. One of them is a methoxy group, which is an *ortho-para* director. The directing effects of this substituent, if installed first, can be used to install the other groups in the correct locations:

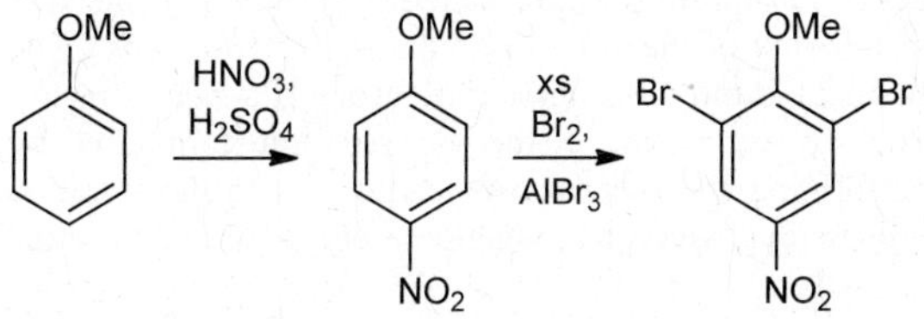

So the key to solving this problem is to convert benzene into methoxybenzene (anisole):

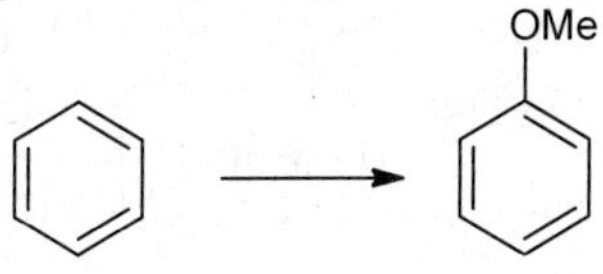

This can be achieved via the following three-step process.

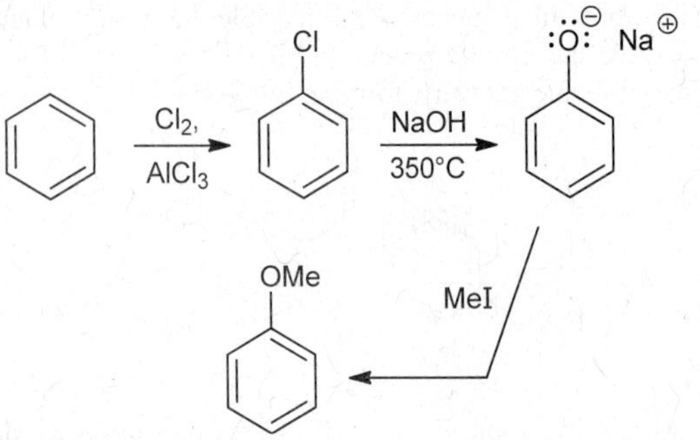

There are certainly other acceptable methods for achieving the desired transformation. Once such alternative is shown here:

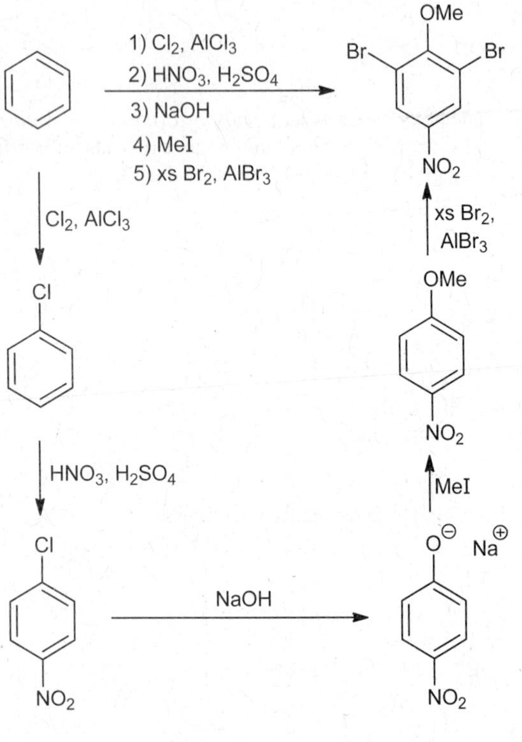

(b) The product has three substituents (a nitro group, and two carboxylic acid groups). Each of the carboxylic acid groups must be installed via a two-step method (Friedel-Crafts alkylation, followed by oxidation of the benzylic position). We could certainly install two methyl groups and then oxidize both of them:

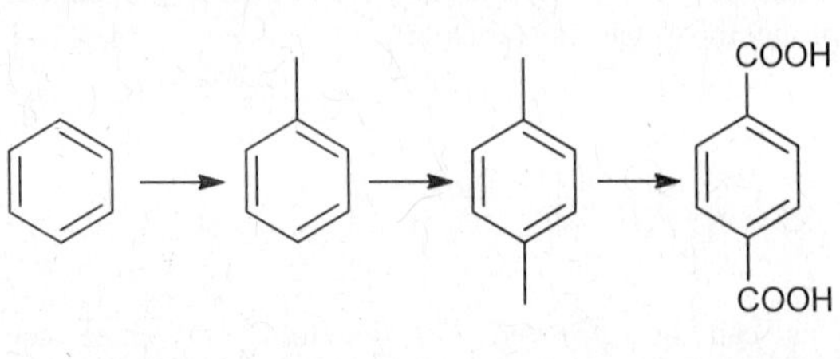

However, the second alkylation step would not proceed with good regiochemical control. Methyl groups do not provide significant steric hindrance, so we would expect

a mixture of *ortho* and *para* products (difficult to predict which will predominate). If, however, we install isopropyl groups (rather than methyl groups), then the second alkylation step will occur predominantly at the *para* position, as desired. Then, both benzylic positions can be oxidized to give the diacid:

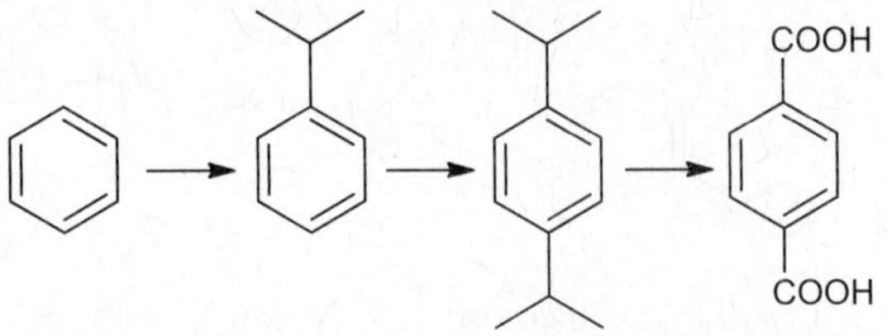

In the synthesis shown below, the nitration step is performed prior to oxidation of the benzylic groups, because nitration of an activated ring is likely to proceed more readily than nitration of a deactivated ring. Nitration of *para*-diisopropylbenzene gives only one nitration product, because all unoccupied positions on the ring are identical.

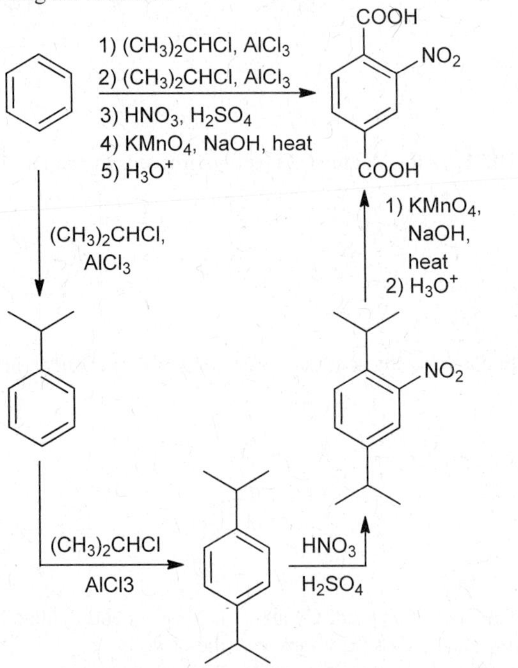

(c) Installation of the amino group requires a two-step process (nitration, followed by reduction), while installation of the chlorine atom can be achieved in just one step.

Now let's consider the order of events. These two groups must be installed in a *meta* fashion, but both groups are *ortho-para* directors. Installation of these two groups (in either order) does not appear to give the desired *meta* substitution pattern. However, recall that installation of the amino group requires two steps. The first step is nitration, and a nitro group is a *meta*-director. So, we can achieve the desired transformation by performing the chlorination process after the nitration process but before the reduction process, as shown here:

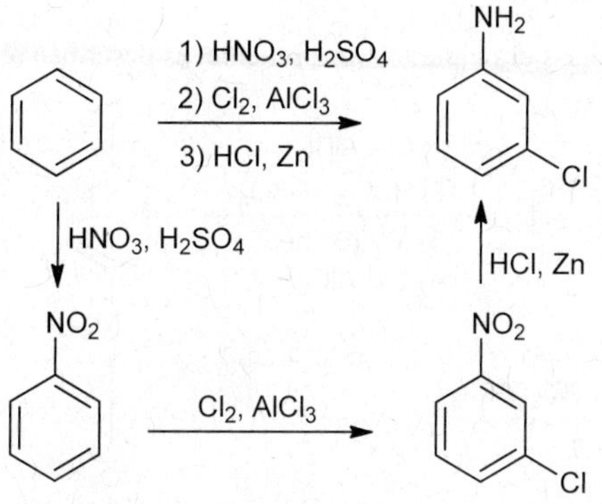

(d) If we install the *tert*-butyl group first, we can use its directing effects to install all of the other groups in the correct locations. The *tert*-butyl group is installed via a Friedel-Crafts alkylation. The directing effects of the *tert*-butyl group cause the next group to be installed in the *para* position, so the next step of our synthesis must be a nitration step. Subsequent halogenation steps are forced to proceed at the positions that are *ortho* to the *tert*-butyl group. These positions are indeed sterically hindered, but the position that is *para* to the *tert*-butyl group is already occupied (by the nitro group). So any subsequent reactions must take place at the *ortho* positions.

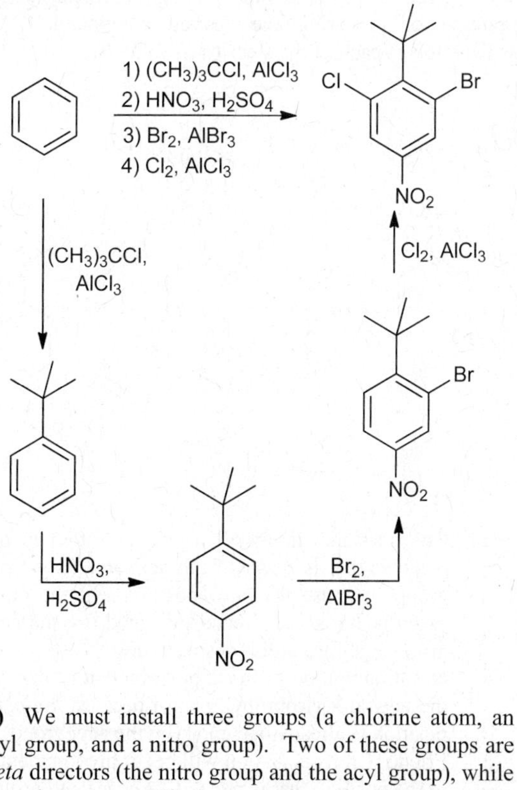

(e) We must install three groups (a chlorine atom, an acyl group, and a nitro group). Two of these groups are *meta* directors (the nitro group and the acyl group), while the other group (Cl) is an *ortho-para* director. The groups are all *meta* to each other in the product, so the Cl group must be installed last.

The first two steps of the synthesis must install the acyl group (via Friedel-Crafts acylation) and the nitro group (via nitration). These two steps cannot be performed in either order. The nitro group cannot be installed first,

because nitrobenzene is too deactivated to undergo a Friedel-Crafts alkylation. Therefore, the order of events must be: 1) Friedel-Crafts alkylation, 2) nitration, and 3) chlorination. This synthesis is shown here:

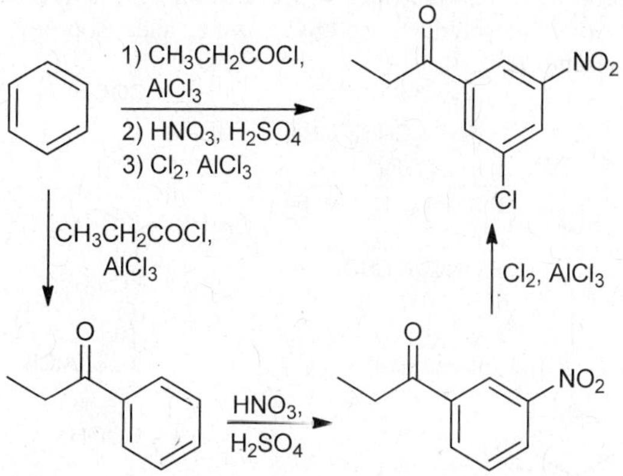

(f) There is likely more than one way to achieve the desired transformation. The following retrosynthetic analysis represents one strategy for preparing the desired compound. An explanation of each of the steps (*a-e*) follows.

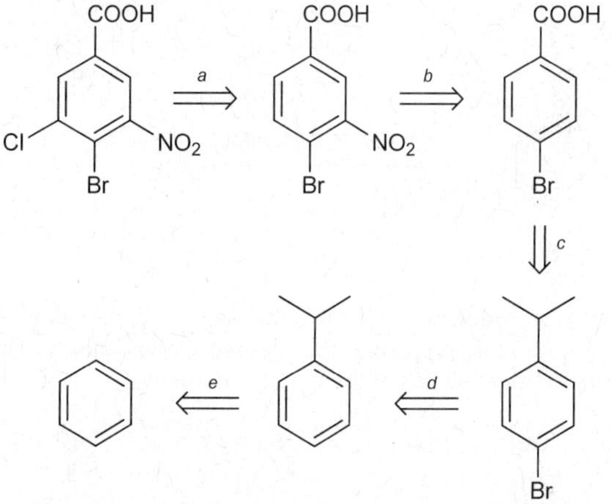

a. The product can be prepared by installing the Cl group last. None of the three groups (on the ring) are activators, but all three groups direct to the same location.

b. Installation of the nitro group would occur in the position that is *ortho* to the *ortho-para* directing group (and *meta* to the *meta*-directing group).

c. The carboxylic acid group can be made via oxidation of the benzylic position. A methyl group can certainly be oxidized to a carboxylic acid group, however, we choose a larger, sterically hindered alkyl group, as will soon be explained.

d. Bromination will occur predominantly in the *para* position, because the *ortho* positions are sterically hindered. This is indeed the reason why we chose an isopropyl group rather than a methyl group. With a methyl group, we would expect a mixture of *ortho* and *para* products.

e. Installation of an isopropyl group can be achieved via a Friedel-Crafts alkylation.

Now, let's draw the forward scheme. Note that in the first step (Friedel-Crafts alkylation), it would be inefficient to use propyl chloride and then rely on a carbocation rearrangement, as that would likely give a mixture of products (propylbenzene and isopropyl-benzene).

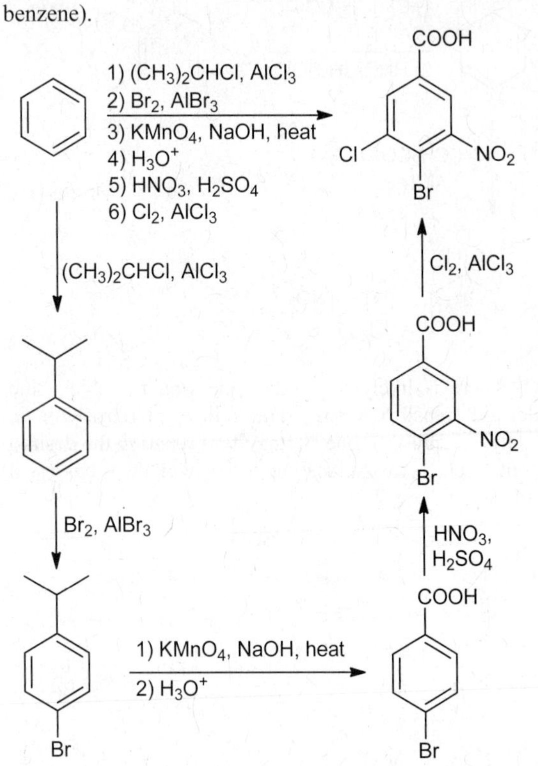

(g) The following retrosynthetic analysis represents one strategy for preparing the desired compound. An explanation of each of the steps (*a-d*) follows.

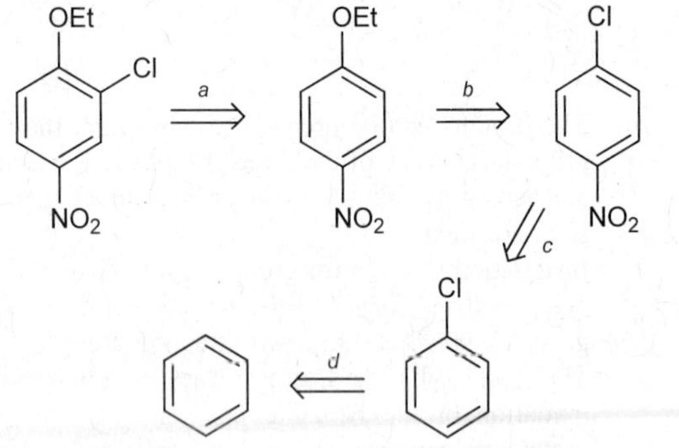

a. Chlorination will install a chlorine atom in the position that is *ortho* to the activator (OEt) because the *para* position is already occupied (by the nitro group).
b. An S_NAr reaction will replace the chlorine atom with an ethoxy group.
c. Installation of the nitro group would occur predominantly in the position that is *para* to the *ortho-para* directing Cl group.
d. Chlorination gives chlorobenzene.

Now, let's draw the forward scheme, as described above:

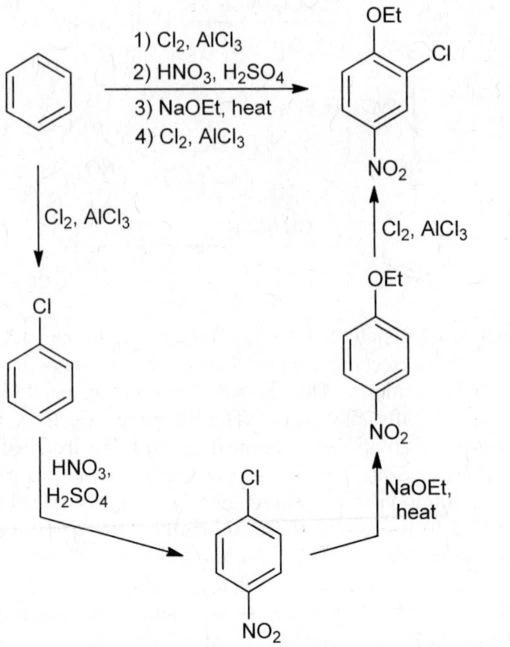

(h) The following retrosynthetic analysis represents one strategy for preparing the desired compound. An explanation of each of the steps (*a-d*) follows.

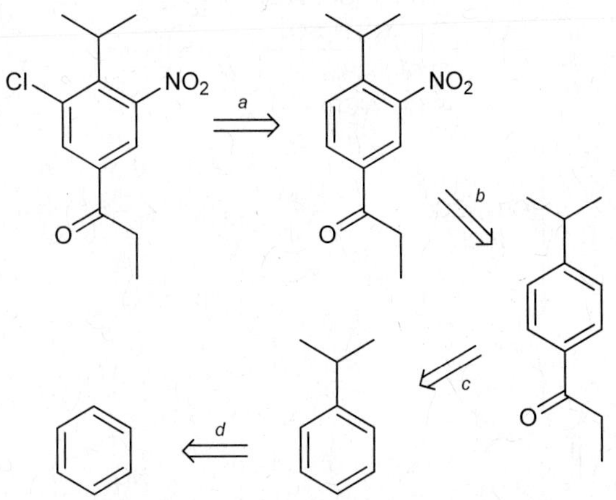

a. Chlorination will install a chlorine atom in the position that is *ortho* to the activator (isopropyl group) because the *para* position and the other *ortho* position are already occupied (by the acyl group and nitro group, respectively).
b. Nitration will occur in the position that is *ortho* to the activator (isopropyl group) because the *para* position is already occupied (by the acyl group).
c. Friedel-Crafts acylation will occur predominantly in the position that is *para* to the isopropyl group, because the isopropyl group is an *ortho-para* director, and it is sterically bulky.
d. Friedel-Crafts alkylation can be used to install an isopropyl group.

Now, let's draw the forward scheme. Note that in the first step (Friedel-Crafts alkylation), it would be inefficient to use propyl chloride and then rely on a carbocation rearrangement, as that would likely give a mixture of products (propylbenzene and isopropyl-benzene).

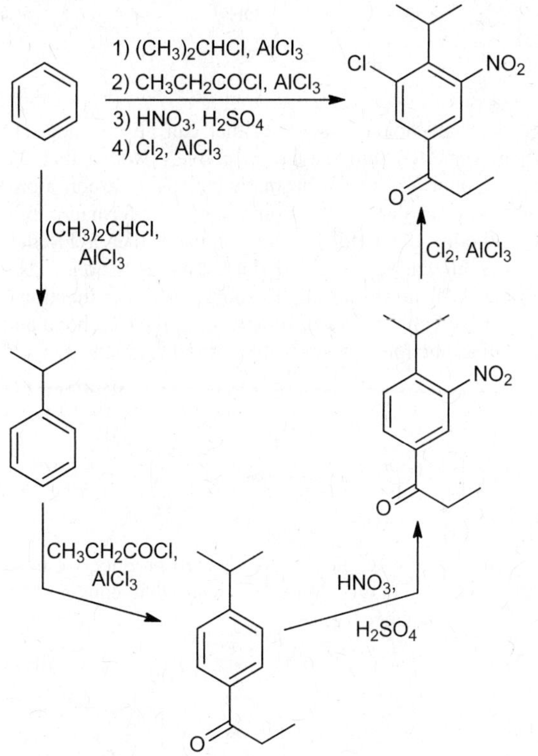

(i) The following retrosynthetic analysis represents one strategy for preparing the desired compound. An explanation of each of the steps (*a-c*) follows.

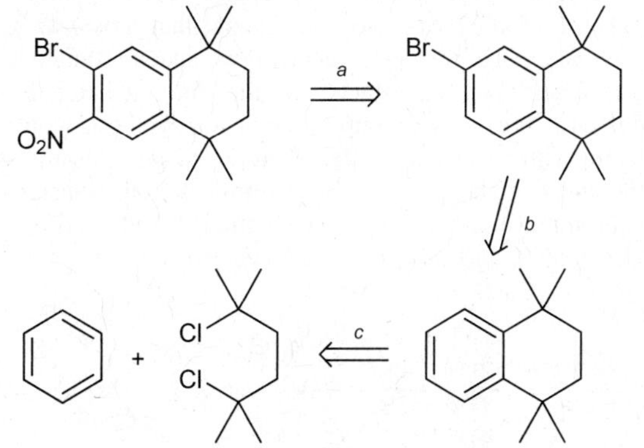

  a. Nitration will occur in the correct location because of steric factors.
  b. Bromination will install a bromine atom in the correct location since the branching on the non-aromatic ring provides significant steric bulk.
  c. Two successive Friedel-Crafts alkylation reactions (the first intermolecular and the second intramolecular) will form the desired additional ring.

Now, let's draw the forward scheme, as described above:

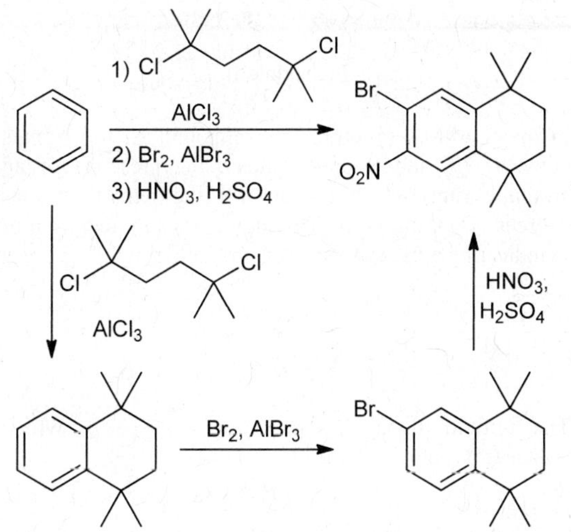

19.73. These reagents will install a methyl group, and the problem statement indicates that three methyl groups are installed. Installation of the first methyl group gives toluene. Methylation of toluene gives a mixture of *ortho*-xylene and *para*-xylene.

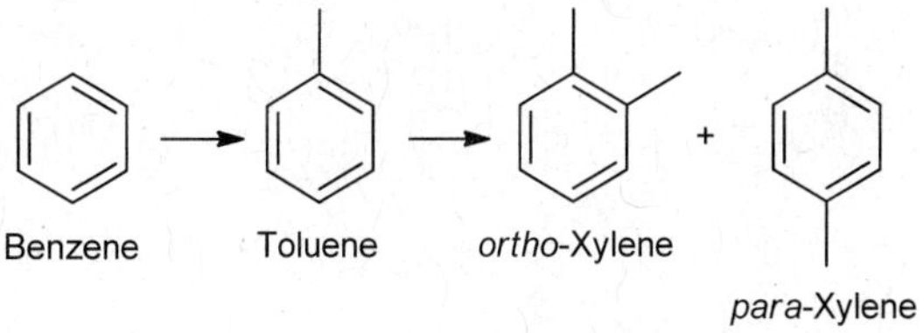

ortho-Xylene undergoes methylation at C4 because of steric factors, while *para*-xylene undergoes methylation at C2 (because all four positions are equivalent). Either way, the product is the same: 1,2,4-trimethylbenzene

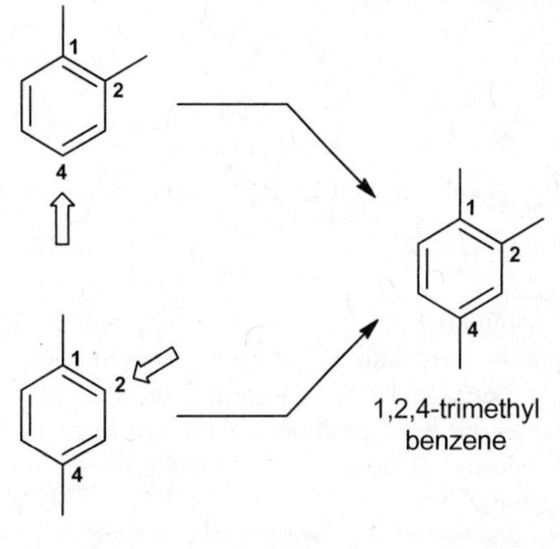

19.74. The molecular formula indicates five degrees of unsaturation. One of these degrees of unsaturation must account for the signal in the double bond region of the IR spectrum (1680 cm^{-1}). The wavenumber of this signal indicates that it is likely a C=O bond that is conjugated:

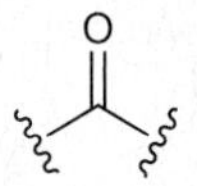

(conjugated)

In the ^{1}H NMR spectrum of compound **A**, the signals between 7.5 and 8 ppm indicate the presence of an aromatic ring, which account for the remaining four degrees of unsaturation. This signal has an integration of 5, indicating a monosubstituted aromatic ring:

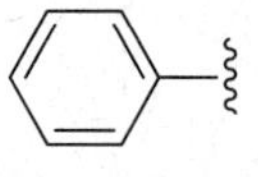

The upfield signal has an integration of 3, which indicates a methyl group:

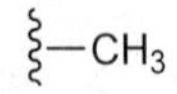

These three fragments can only be connected in one way:

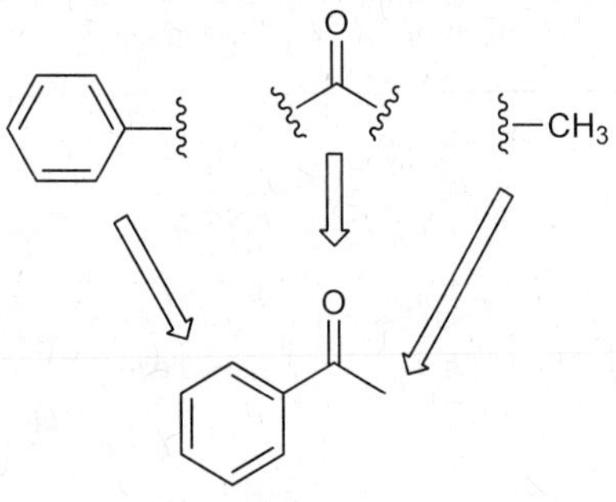

Compound **A**

Indeed, the C=O bond is conjugated in this structure. A Clemmensen reduction of compound **A** gives ethyl benzene (compound **B**):

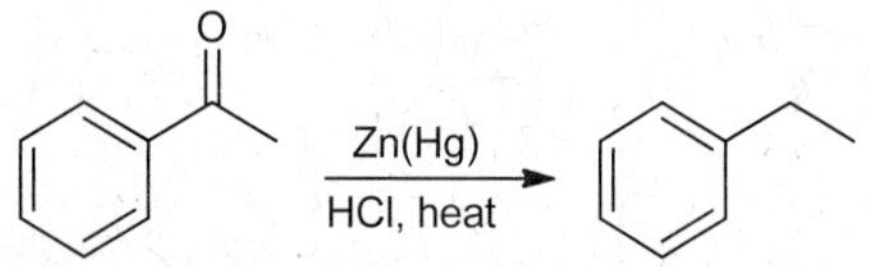

Compound **A**   Compound **B**

When compound **B** is treated with Br$_2$ and AlBr$_3$, an electrophilic aromatic substitution reaction occurs, in which a bromine atom is installed on the ring. The reaction at the *para* position is expected to occur more rapidly because *ortho* attack is sterically hindered by the ethyl group:

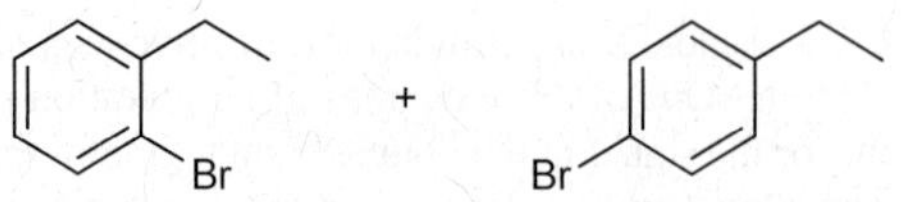

(major)

19.75.
(a) We did not learn a direct method for installing the following group on an aromatic ring:

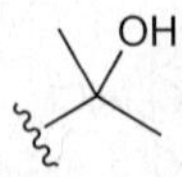

So we must identify a group that can be installed and then converted into the group above. Notice that the atom connected to the aromatic ring is a carbon atom, which means that the reaction requires the formation of a C-C bond. Therefore, we must use either a Friedel-Crafts alkylation or a Friedel-Crafts acylation. The former will install an alkyl group, *without* a functional group. In contrast, acylation will form the C-C bond and install a functional group in the desired location:

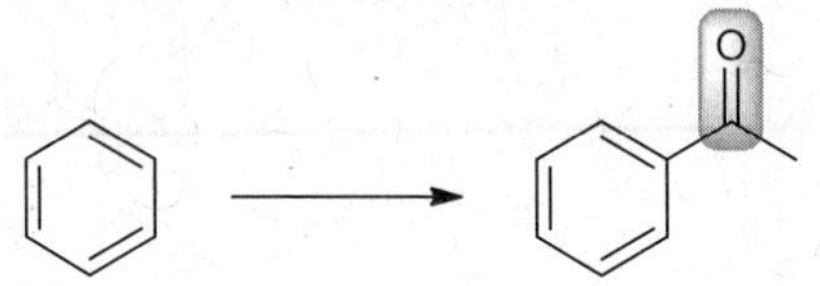

To convert this ketone into the desired product, we must introduce a methyl group in a way that converts the ketone into an alcohol:

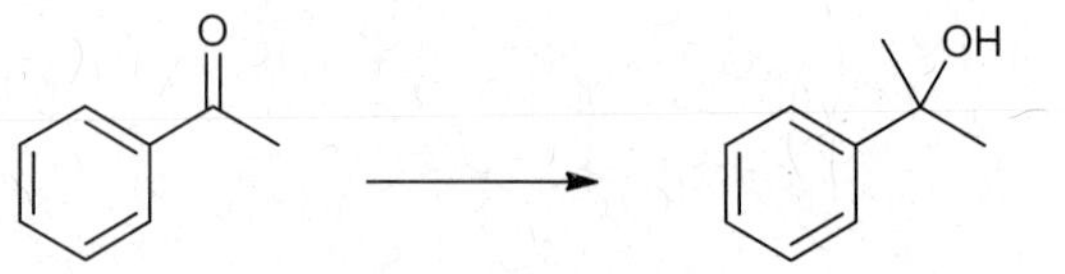

This can be achieved with a Grignard reaction, using methyl magnesium bromide.

Finally, we must consider the order of events. Notice that the product has two substituents that are *meta* to each other. This can be achieved if we take advantage of the *meta*-directing effects of the intermediate that possesses a carbonyl group. That is, our synthesis would begin with a Friedel-Crafts acylation, thereby installing the *meta*-directing acyl group. Bromination then installs a bromine atom in the *meta* position. And finally, a Grignard reaction affords the desired product:

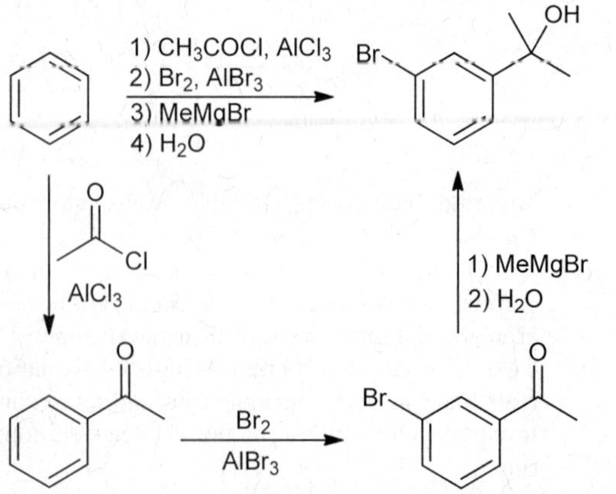

(b) We did not learn a direct method for installing the following group on an aromatic ring:

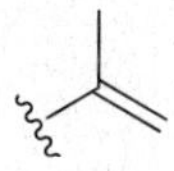

However, in the solution to Problem 19.75a, we saw how to install the following functional group:

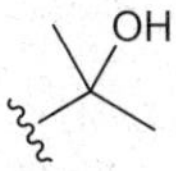

This functional group can be converted into the desired functional group in just one step:

This functional group can be converted into the desired functional group in just one step:

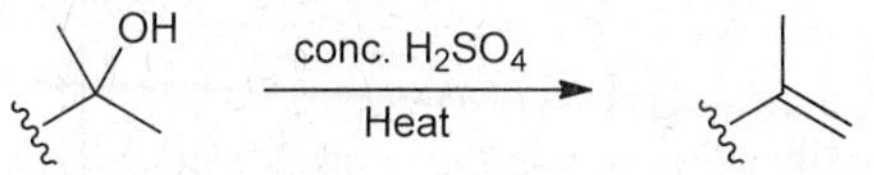

Therefore, we can perform a synthesis similar to the one described in the solution to Problem 19.75a, followed by this reaction:

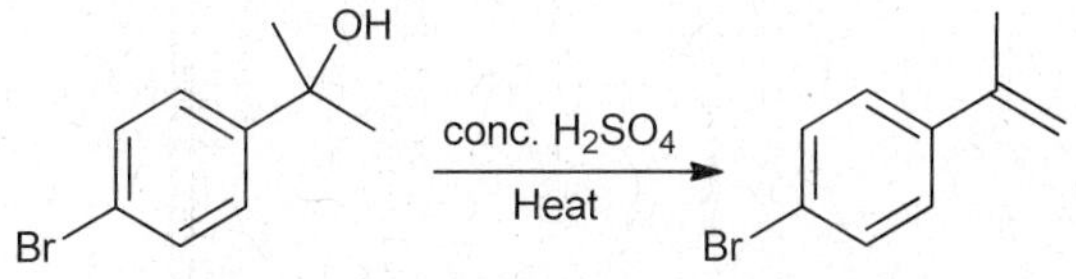

The complete synthesis is shown here. Notice that bromination is performed prior to acylation, in order to install the groups in a *para* fashion:

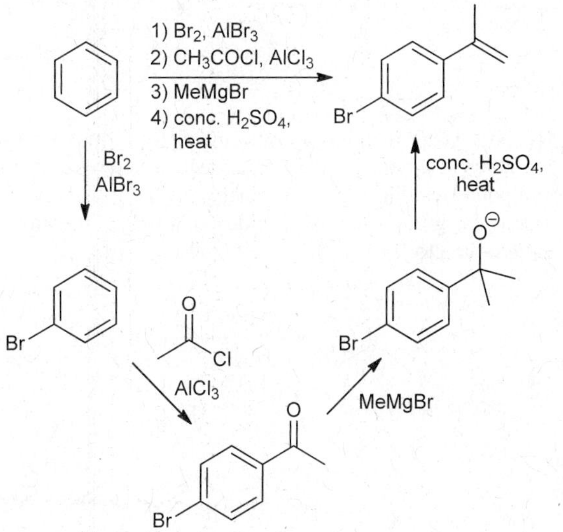

Alternatively, we can install the desired group in an entirely different way. Rather than performing a Friedel-Crafts acylation, we can perform a Friedel-Crafts alkylation to install an isopropyl group, which can then be functionalized via benzylic bromination, followed by elimination:

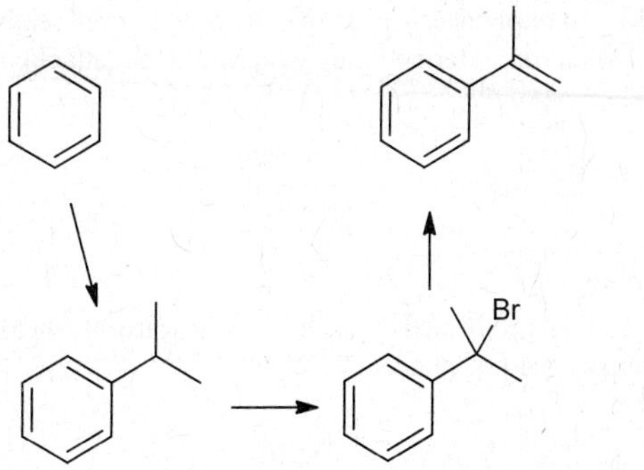

With this strategy, bromination of the aromatic ring is performed after installation of the isopropyl group, which allows us to employ the directing effects of the isopropyl group to install the bromine atom in the *para* position.

This alternative synthesis is summarized here:

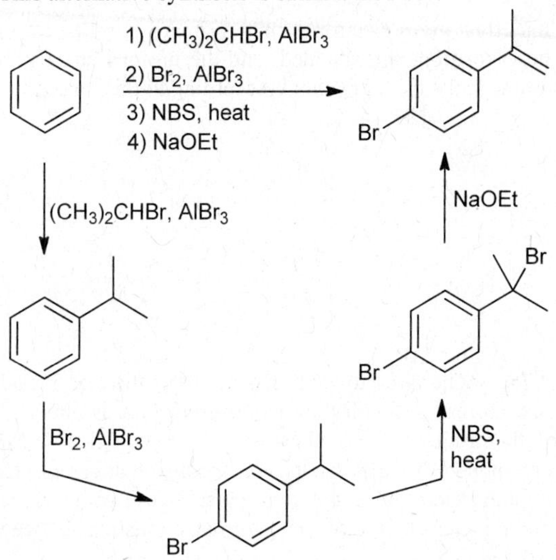

1) $(CH_3)_2CHBr$, $AlBr_3$
2) Br_2, $AlBr_3$
3) NBS, heat
4) NaOEt

19.76.
(a) The alkyl halide can interact with the Lewis acid to produce a carbocation:

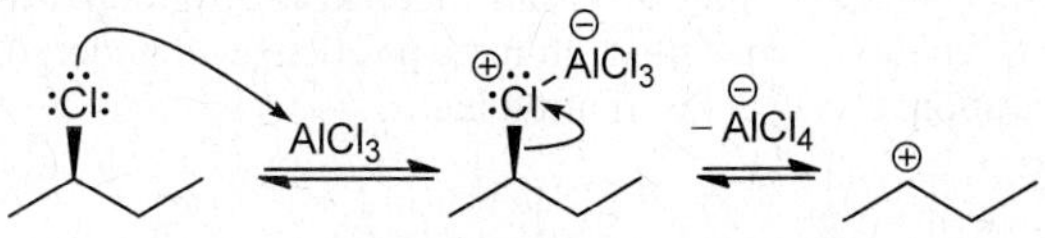

The carbocation is planar, so the aromatic ring can attack either face of the carbocation, with equal likelihood. As a result, a Friedel-Crafts alkylation with this optically active alkyl halide is expected to produce a pair of enantiomeric products:

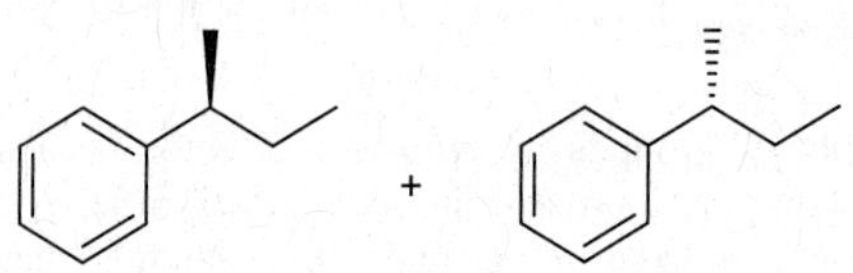

(b) As explained in part (a), the reaction proceeds via a carbocation intermediate, which can be attacked from either face, leading to a racemic mixture. Therefore, the product mixture is optically inactive.

19.77.
The OH group activates the ring toward electrophilic aromatic substitution because the OH group donates electron density via resonance.

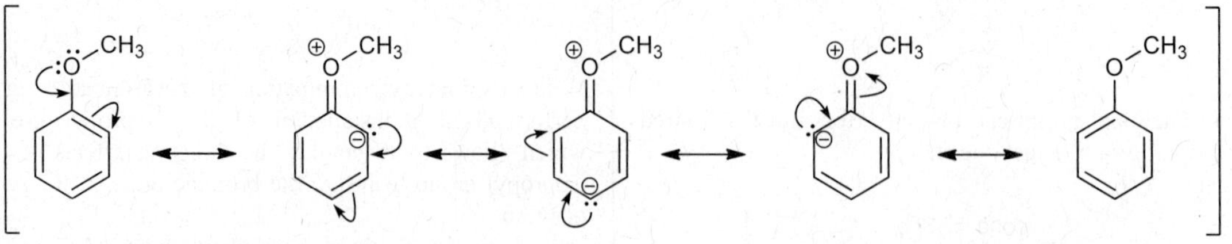

This effect gives electron density primarily to the *ortho* and *para* positions, as seen in the resonance structures above. These positions are shielded, and the protons at these positions are expected to produce signals farther upfield than protons at the *meta* position. According to this reasoning, the *meta* protons correspond with the signal at 7.2 ppm.

19.78. The upfield signal has an integration of 3, and must correspond with the methyl group that is attached to the ring. The other signal (downfield) must correspond with aromatic protons. Since that signal has an integration of 2, the ring must bear two protons, which means that three nitro groups must have been installed.

The methyl group of toluene is an *ortho-para* director, and as such, nitration is expected to occur at the *ortho* and *para* positions. Each successive nitro group deactivates the ring, thereby requiring a higher temperature in order to install the next nitro group. At high enough temperature, all three positions will undergo nitration, giving 2,4,6-trinitrotoluene:

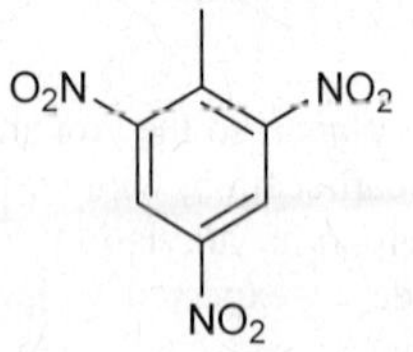

2,4,6-trinitrotoluene

19.79.
(a) A phenyl group is an *ortho-para* director, because the sigma complex formed from *ortho* attack or *para* attack is highly stabilized by resonance (the positive charge is spread over both rings). The *ortho* position is sterically hindered while the *para* position is not, so we expect nitration to occur predominantly at the *para* position:

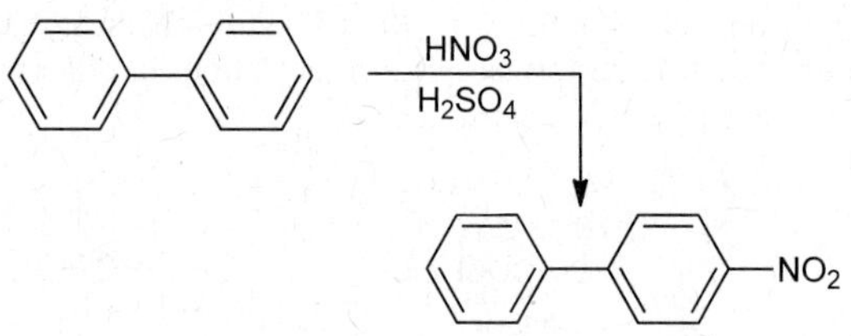

(b) This group withdraws electron density form the ring via resonance (many of the resonance structures of this compound exhibit a positive charge in the ring). As a result, this group is a moderate deactivator, and therefore a *meta*-director:

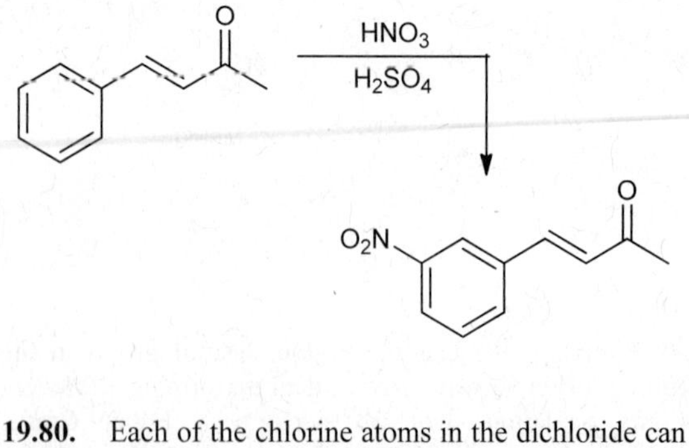

19.80. Each of the chlorine atoms in the dichloride can interact with AlCl$_3$ and initiate a Friedel-Crafts alkylation process. During the first alkylation process, a tertiary benzylic carbocation can form via a hydride shift (Note: for a description of how a carbocation rearrangement can occur at the same time as loss of a leaving group, see Figure 7.22 and corresponding text):

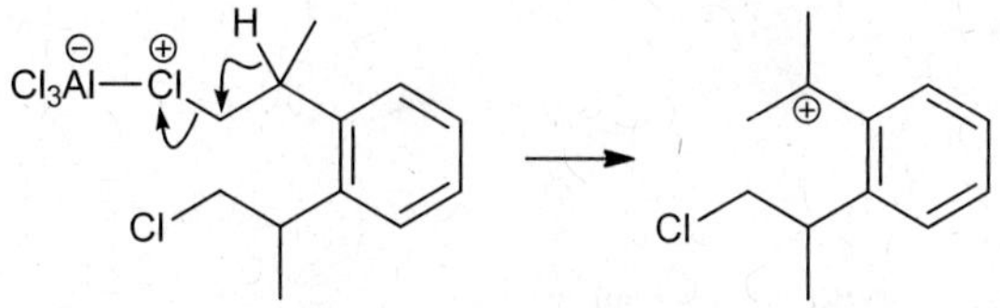

This carbocation is then attacked by benzene in an electrophilic aromatic substitution reaction, giving the following:

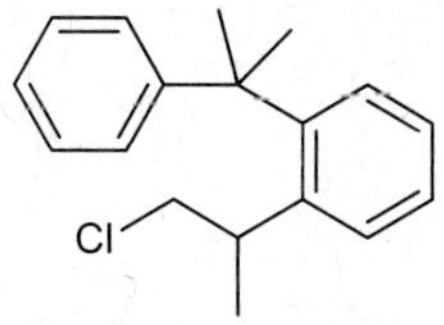

This compound still has one more chlorine atom, allowing for the second alkylation process. Once again, the second alkylation occurs in a similar fashion (via a tertiary, benzylic carbocation), but this time, the electrophilic aromatic substitution reaction proceeds in an intramolecular fashion:

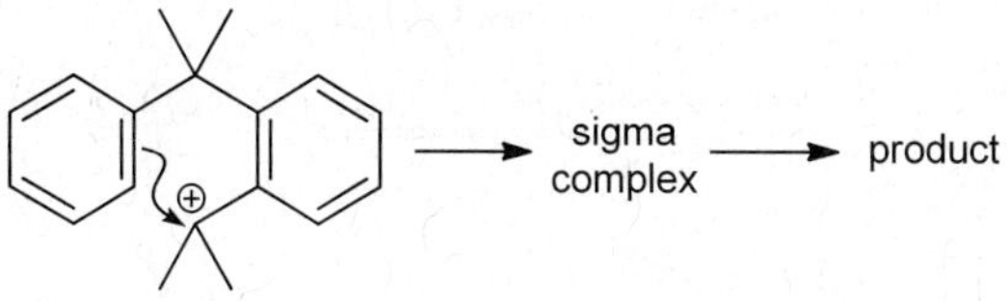

The following product is therefore expected:

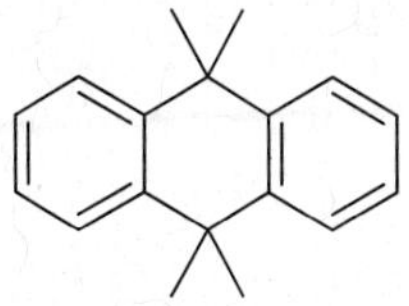

19.81.   Formaldehyde is protonated to give a resonance-stabilized cation that can serve as an electrophile in an electrophilic aromatic substitution reaction. Phenol functions as the nucleophile and attacks the electrophile, giving a resonance-stabilized intermediate (sigma complex). Water can then serve as a base and remove a proton from the sigma complex, thereby restoring aromaticity:

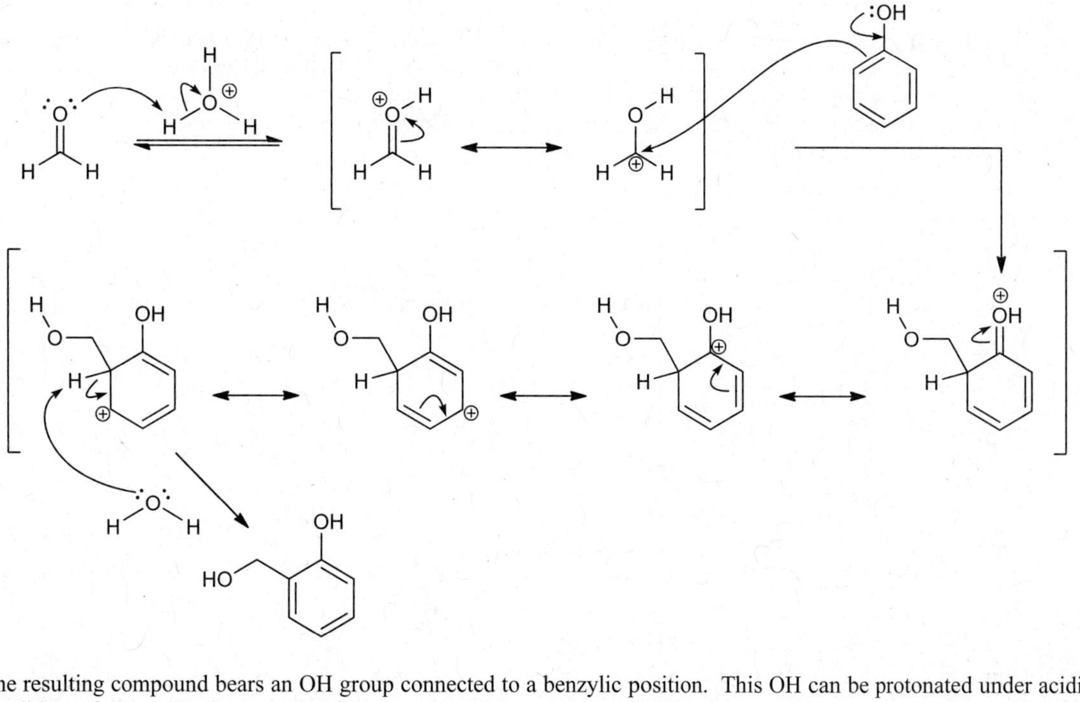

The resulting compound bears an OH group connected to a benzylic position. This OH can be protonated under acidic conditions, followed by loss of a leaving group to give a resonance-stabilized, benzylic carbocation. This carbocation can then serve as an electrophile in an electrophilic aromatic substitution reaction. Phenol functions as the nucleophile and attacks the electrophile, giving a resonance-stabilized intermediate (sigma complex). Water can then serve as a base and remove a proton from the sigma complex, thereby restoring aromaticity. This process continues, ultimately giving the structure of Bakelite:

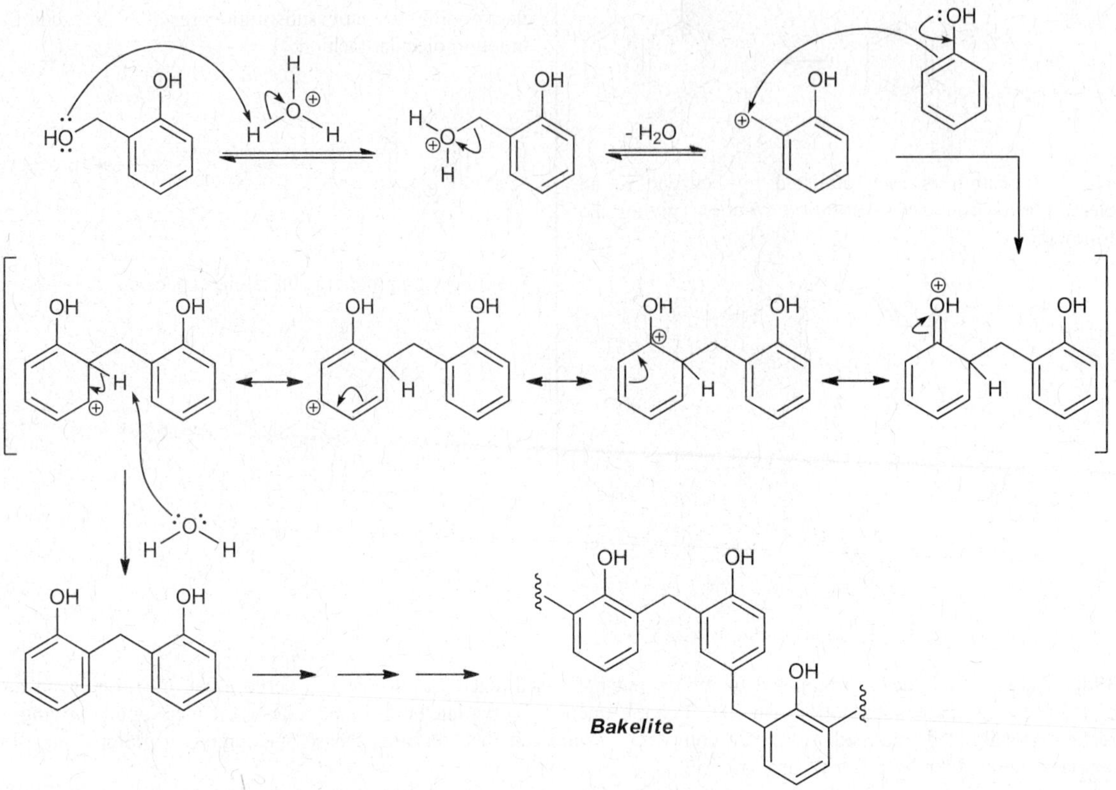

Bakelite

19.82. A *tert*-butyl group is removed from the ring and replaced with a proton, in what appears to be the reverse of an electrophilic aromatic substitution reaction. As such, we expect the reaction to proceed via an intermediate sigma complex. The aromatic ring is protonated to give a sigma complex. Water then functions as a base and removes a proton in an elimination-type reaction, where the leaving group is an aromatic ring. This step restores aromaticity, and produces isobutylene as a byproduct:

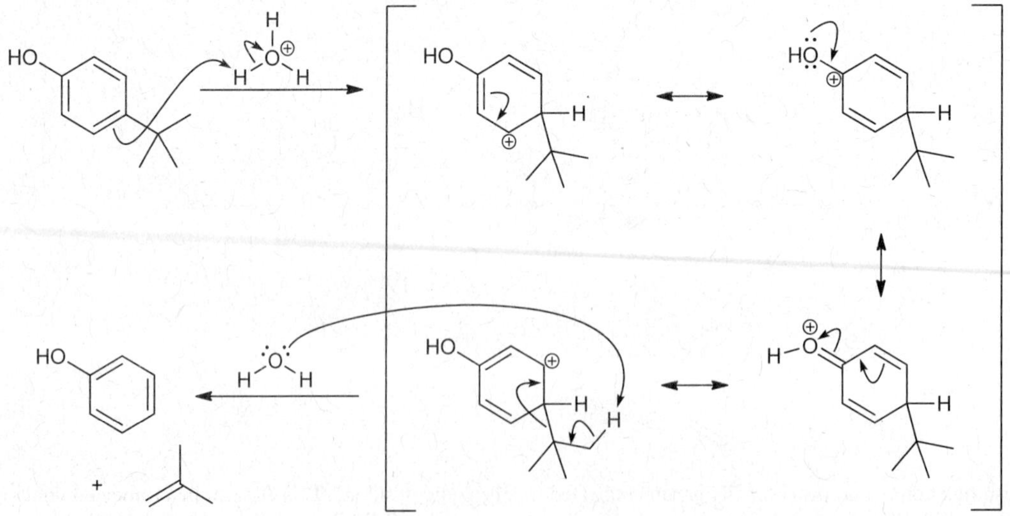

19.83. The amino group in *N,N*-dimethylaniline is a strong activator, and therefore, an *ortho-para* director. For this reason, bromination occurs at the *ortho* and *para* positions. However, in acidic conditions, the amino group is protonated to give an ammonium ion:

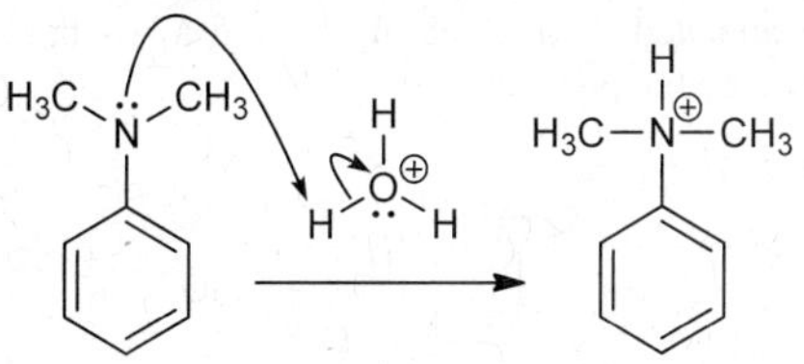

Unlike the amino group, an ammonium ion is a strong deactivator and a *meta* director. Under these conditions, nitration occurs primarily at the *meta* position.

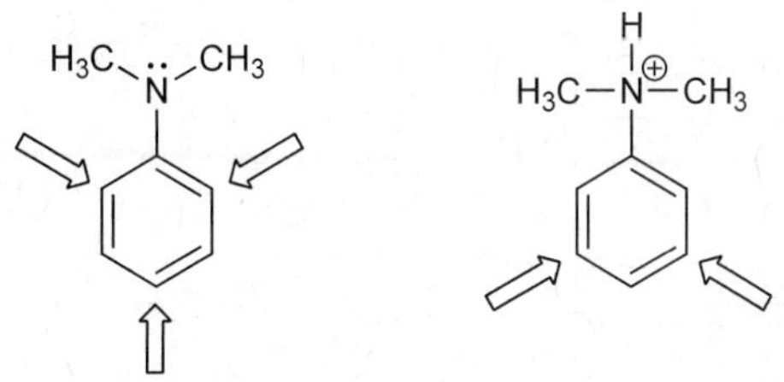

ortho-para director   *meta*-director

19.84. In all cases, the X substituent is a methoxy group. It is the Y substituent that varies in identity. It is difficult to imagine that there could be a steric effect here, since the Y substituent is far away from the alkene π bond. The trend is more easily explained when we consider the electronic effect of the Y substituent. Specifically, the methoxy group (a powerful activator) can donate electron density via resonance into the alkene π bond, making it the most nucleophilic π bond among the disubstituted stilbenes. A methyl group is only weakly activating, so its effect should be smaller than the effect of a methoxy group (giving a smaller rate constant). A chloro group is a weak deactivator, which means that the net effect of the chloro group is to withdraw electron density from the aromatic ring and the conjugated alkene π bond. As a result, when Y = Cl, the rate constant is smaller than the rate constant when Y = Me. Finally, the nitro group is powerfully electron-withdrawing, via resonance, and it will significantly reduce the nucleophilicity of the alkene π bond, rendering it the least reactive double bond towards molecular bromine.

19.85. When the acyl chloride is treated with $AlCl_3$, it might be expected that an intramolecular Friedel-Crafts acylation would occur, but that would yield a four-membered ring which is not favorable. Instead, the highly electrophilic acylium ion (generated when the acyl halide is treated with $AlCl_3$) can react with a different nucleophile that is present in the reaction mixture. Specifically, ethylene can function as a nucleophile to attack the acylium ion to give an intermediate carbocation that quickly undergoes a Friedel-Crafts alkylation, followed by re-aromatization to give the product:

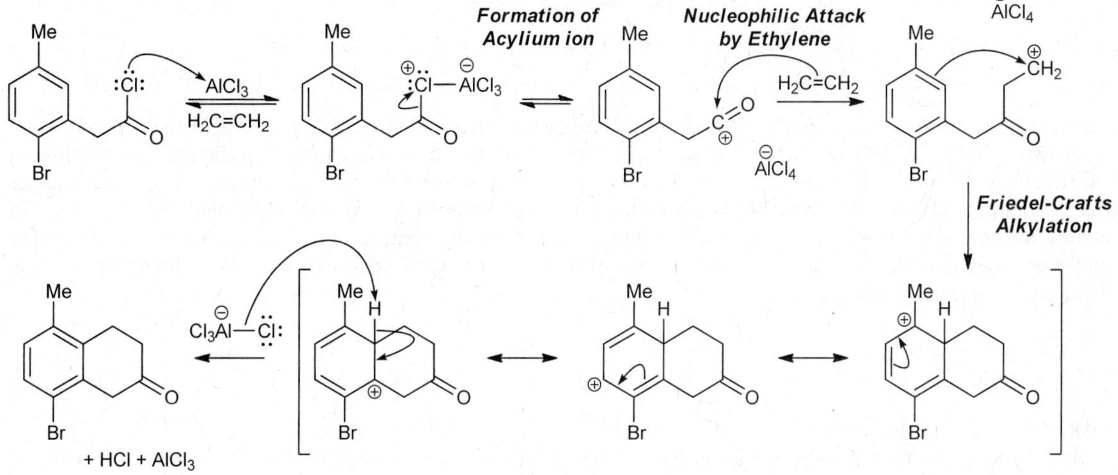

19.86.

(a) This is a reaction between a strong nucleophile (the terminal alkynide ion) and an electron-poor arene, from which a halide is expelled. This satisfies the criteria for a nucleophilic aromatic substitution mechanism, as shown below. The alkynide ion attacks the carbon atom attached to fluorine, producing a resonance-stabilized Meisenheimer complex, which loses a fluoride leaving group to give the product.

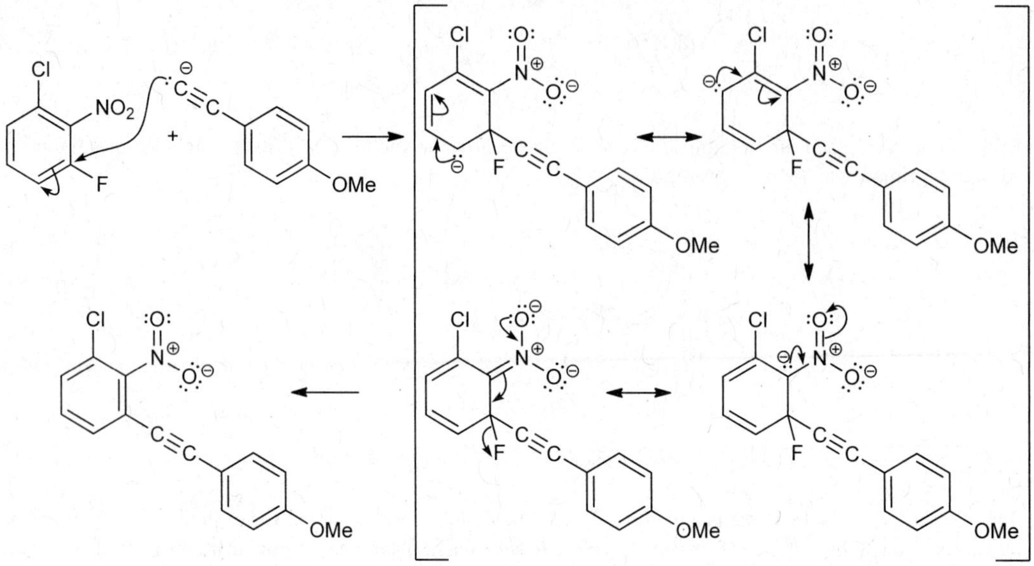

(b) The observed regiochemical outcome (that the fluoride is substituted and not the chloride, even though chloride is a better leaving group) indicates that the departure of the leaving group does not occur in the rate determining step. The leaving group leaves in step 2, so step 1 must be rate determining. In other words, the poor ability of fluoride to serve as a leaving group does not affect the rate of the reaction, since this occurs in the step that is not rate determining.

In the first step, the nucleophile prefers to attack the carbon atom attached to fluorine because of its greater electrophilicity (due to fluorine's greater electronegativity) relative to the carbon next to chlorine.

19.87.   Each of the three nitro groups is conjugated with the aromatic ring, and as a result, each of the C-N bonds has some double-bond character. The following resonance structure illustrates this for one of the nitro groups:

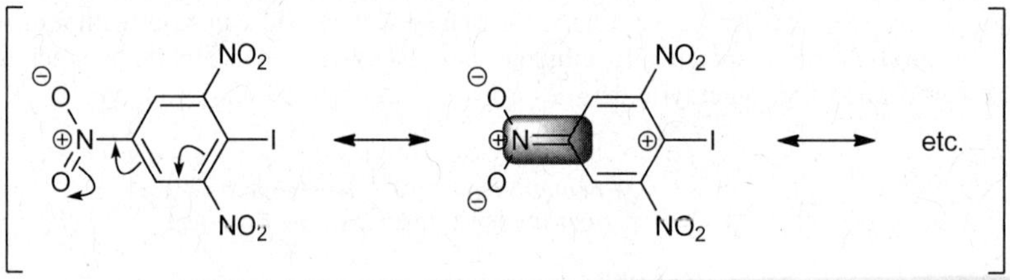

There are three nitro groups. Two of them are *ortho* to the very large iodo group. The size of the electron cloud associated with the iodo group offers steric strain that forces these two nitro groups out of planarity with the ring, in order to alleviate some of the strain. This results in reduced overlap of the *p* orbitals, and consequently, the resonance effect is diminished. As such, the C-N bonds of these two nitro groups will have less double-bond character. In contrast, the nitro group that is *para* to the iodo group is coplanar with the ring, and its C-N bond has significant double bond character – see the resonance structure above. This additional C=N character leads to a shorter bond, as is characteristic of double bonds.

19.88.

(a) In aqueous acidic conditions, compounds **1** and **2** lose tritium via an electrophilic aromatic substitution reaction. First, the aromatic ring is protonated to give a resonance-stabilized sigma complex, which then undergoes loss of T$^+$

(rather than H⁺) to re-establish aromaticity.   Notice that the methoxy groups render the sigma complex particularly stable, because the positive charge is spread over the oxygen atoms as well (via resonance).

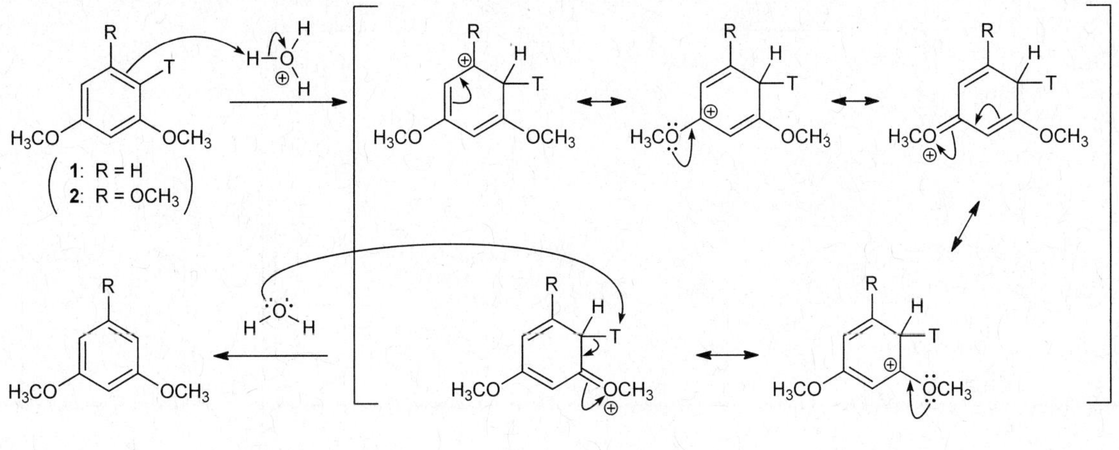

(b) Compound **2** has an extra methoxy group that is absent in compound **1**.   This group further stabilizes the intermediate sigma complex by introducing one more resonance structure (highlighted below).   That is, the positive charge is further delocalized:

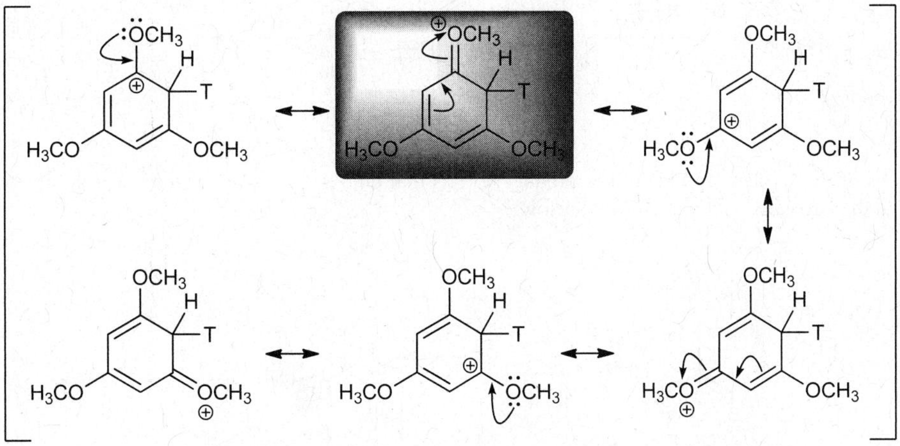

When compound **2** undergoes an electrophilic aromatic substitution reaction, the intermediate sigma complex exhibits a positive charge that is spread over three carbon atoms and three oxygen atoms.   This intermediate is more stable than the intermediate formed when compound **1** undergoes an electrophilic aromatic substitution reaction.   It is therefore expected that compound **2** will lose tritium at a faster rate than compound **1**.

19.89.   Sodium amide (NaNH₂) is a strong base, and under these conditions, a proton is abstracted from the nitrogen atom of **1** (pK_a ~ 30) to generate anion **3**. This proton will be re-introduced during the reprotonation event (work-up with water). It might seem unnecessary to draw this deprotonation step (only to protonate this location at the end of the mechanism), but this step must be drawn to indicate that you understand that this proton is sufficiently acidic that it will not survive under these conditions (because anion **3** is resonance-stabilized).   Next, deprotonation will produce **4**, which loses a bromide ion to form benzyne **5**   This intermediate is then trapped by the nearby aromatic ring to form anion **6**. Proton transfer will produce imine **7**, which can be rearomatized by the loss of a proton to form anion **8**. In the final step, protonation of **8** will form compound **2**.

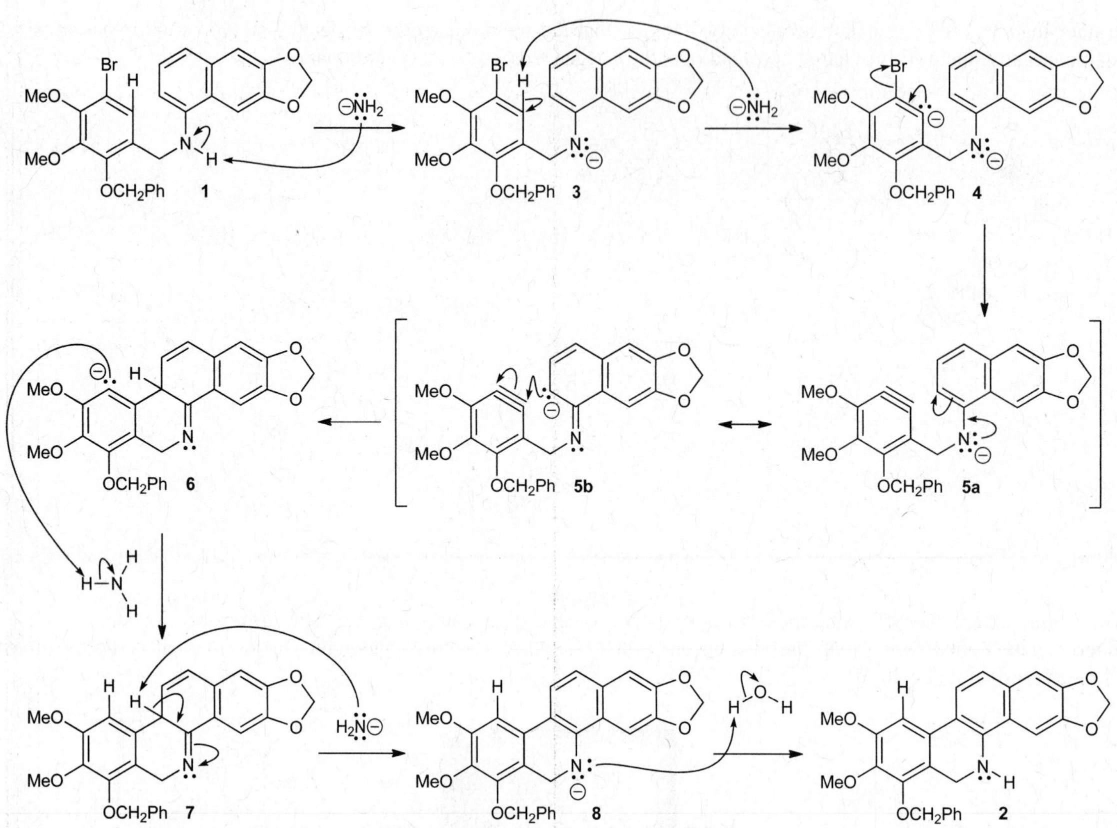

19.90.

(a) An analysis of the reactant and the minor product reveals that during the course of the reaction, a new C-C bond forms between the two carbon atoms indicated below.

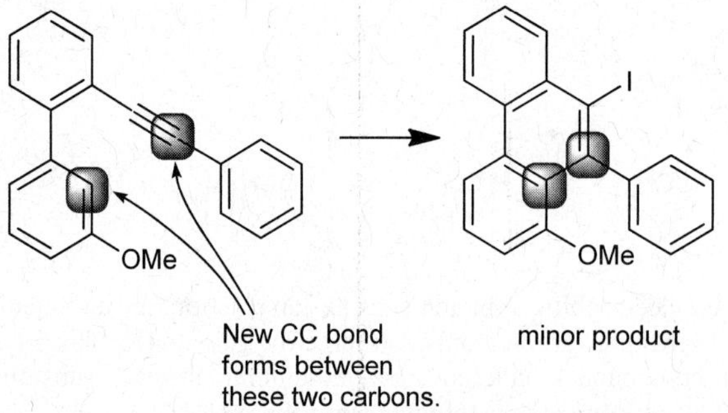

New CC bond
forms between
these two carbons.

minor product

Likewise, in the formation of the major product, an analogous new C-C bond forms after rotation around the C-C bond indicated below.

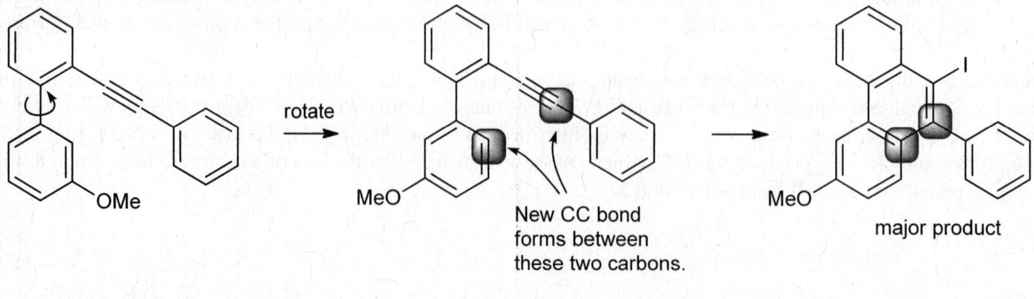

rotate

New CC bond
forms between
these two carbons.

major product

A mechanism for the formation of the minor product is presented below. The alkyne attacks ICl, resulting in the formation of a new C-I bond, and a vinyl carbocation (see below for a justification for formation of a vinyl carbocation in this case). This carbocation then serves as an electrophile in an electrophilic aromatic substitution. The π electrons from the methoxyarene attack this carbocation, resulting in the formation of a new C-C bond, and a resonance-stabilized sigma complex. This intermediate is highly conjugated and has at least seven reasonable resonance structures in addition to the four shown. Deprotonation of the sigma complex restores aromaticity, thus yielding the product.

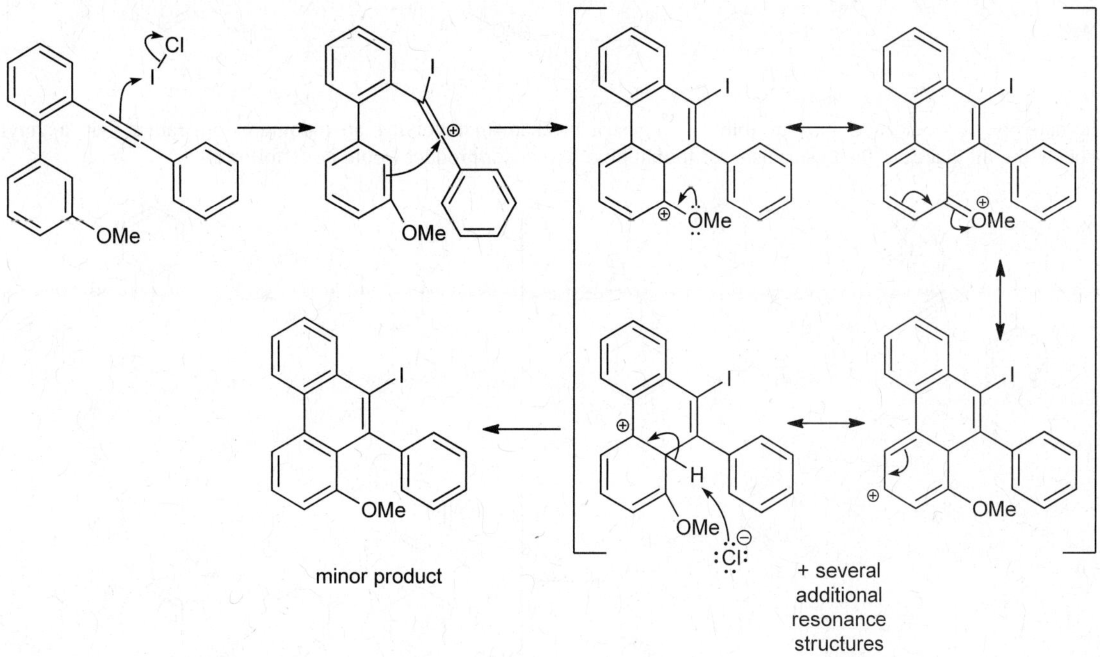

minor product

+ several
additional
resonance
structures

A mechanism to account for the formation of the major product is below. Each step is analogous to the mechanism presented above.

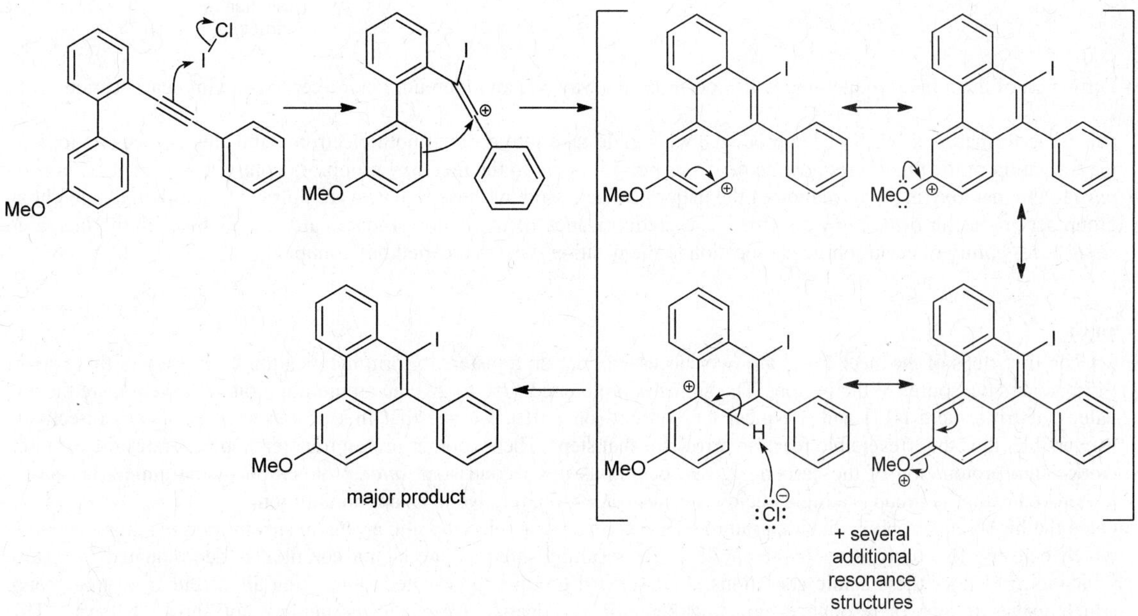

major product

+ several
additional
resonance
structures

In Chapter 10, we noted that vinyl carbocations are not very stable, and thus we should consider whether or not our first intermediate is a reasonable one. In this case, the vinyl carbocation is not an unreasonable intermediate, because it is stabilized by the adjacent aromatic ring, as demonstrated by the following resonance structures.

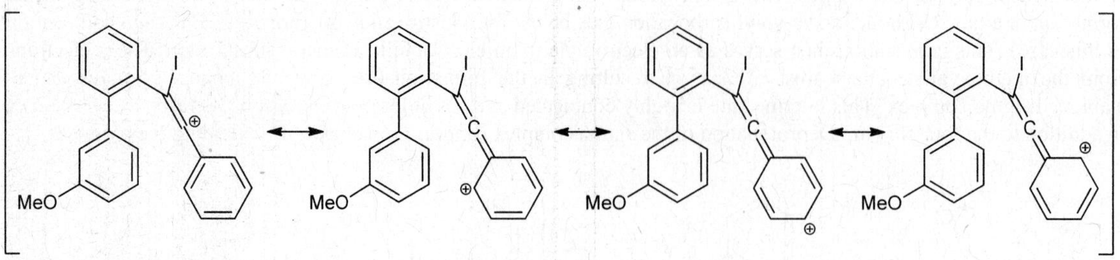

Alternatively, it would also be plausible to suggest a mechanism involving an iodonium ion, rather than a vinyl carbocation. In that case, the mechanism for formation of the major product would be as follows:

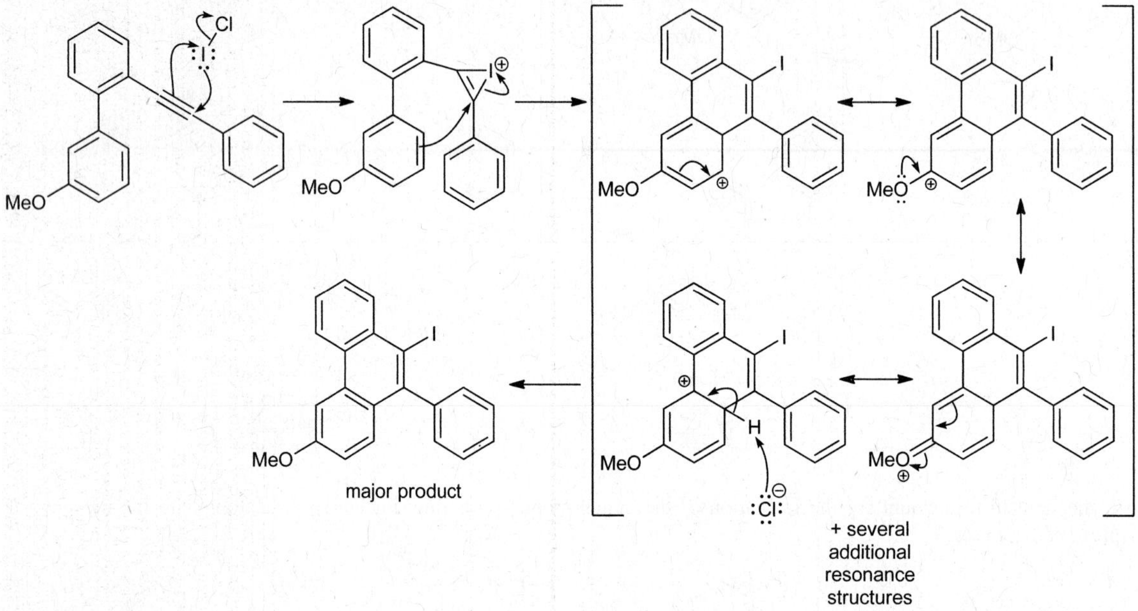

major product

+ several
additional
resonance
structures

Formation of the minor product might also occur in this way, via an iodonium ion, rather than a vinyl carbocation.

(b) In the formation of each of the products, we should take into account both electronic and steric considerations. In the formation of the minor product, the new C-C bond is *ortho* to the methoxy group. In contrast, the new C-C bond is *para* to the methoxy in the formation of the major product. Both of these products are thus consistent with the methoxy group serving as an *ortho-para* director. The predominance of the major product can be explained by the increased steric accessibility of electrophilic carbocation to the position *para* to the methoxy group.

19.91.
(a) The first steps of the mechanism are two successive, proton transfers to form the dication **2**. In each of these steps, triflic acid is the source of the proton. The amine is protonated first, to give an ammonium ion. If we compare the pK_a values of triflic acid (-14) and a typical ammonium ion (~10), we see that this proton transfer step is effectively irreversible, thus the irreversible reaction arrow for that step. Then, another proton transfer step generates the dication. Notice that protonation of the carbonyl group occurs as the second step, *after* protonation of the amine, because a protonated carbonyl group is a highly acidic species ($pK_a \approx$ -7), relative to an ammonium ion.
Once the highly electrophilic dication intermediate is formed, an electrophilic aromatic substitution reaction occurs, in which benzene functions as the nucleophile. The resulting intermediate sigma complex is deprotonated to restore aromaticity. Under these acidic conditions, the hydroxyl group is protonated, generating an excellent leaving group, which leaves to give a resonance-stabilized benzylic carbocation (resonance structures not shown below). This benzylic carbocation then functions as an electrophile in another electrophilic aromatic substitution reaction. The resulting sigma complex is then deprotonated to restore aromaticity, thereby producing ammonium ion **3**.

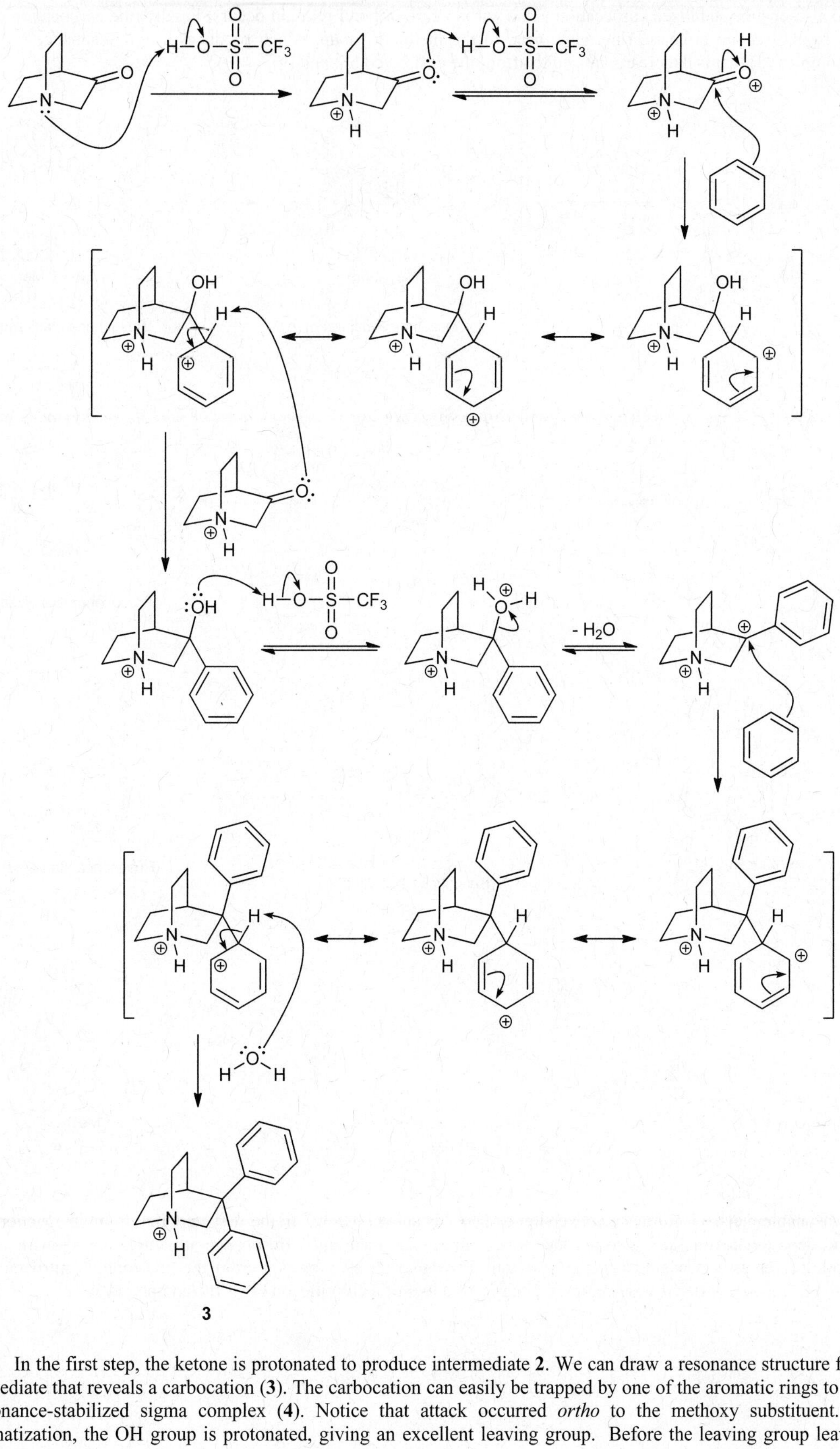

19.92. In the first step, the ketone is protonated to produce intermediate **2**. We can draw a resonance structure for this intermediate that reveals a carbocation (**3**). The carbocation can easily be trapped by one of the aromatic rings to afford a resonance-stabilized sigma complex (**4**). Notice that attack occurred *ortho* to the methoxy substituent. After rearomatization, the OH group is protonated, giving an excellent leaving group. Before the leaving group leaves, to produce a resonance-stabilized carbocation (**7**), a carbon-carbon bond rotation occurs. Finally, the carbocation of **7** is trapped by the second aromatic ring, also *ortho* to the methoxy group, which produces a new resonance-stabilized sigma complex (**8**) that is then rearomatized to afford the desired compound (**9**).

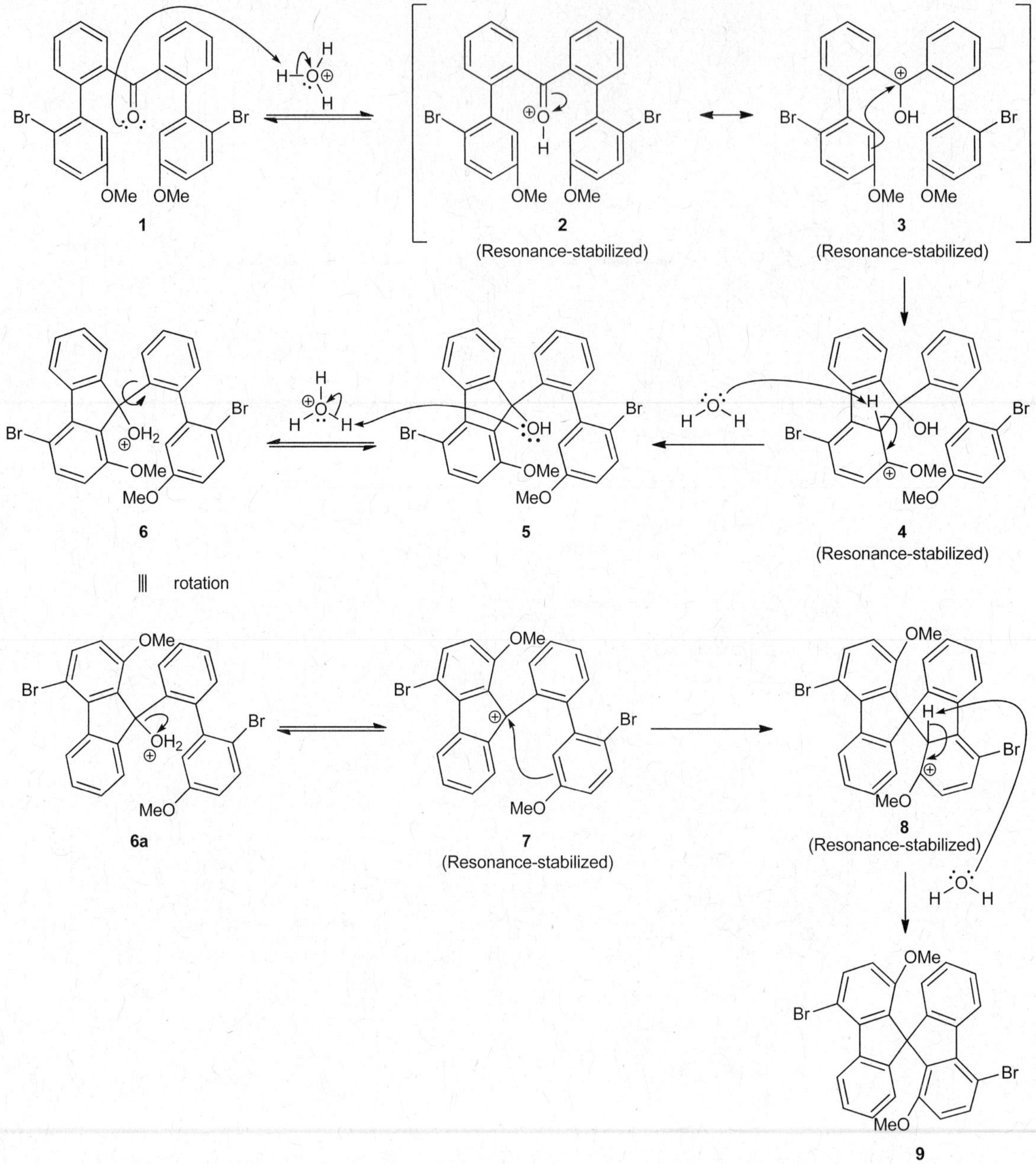

19.93.

(a) A reasonable mechanism for the conversion of **1** to **2** is shown below. In the first step, butyllithium functions as a base and deprotonates the amine to produce a resonance-stabilized anion (resonance structures not shown). Boron trichloride serves as a Lewis acid in the next step, accepting an electron pair from the nucleophilic nitrogen atom, giving the Lewis acid – Lewis base complex. Subsequent loss of a chloride ion gives the compound **2**.

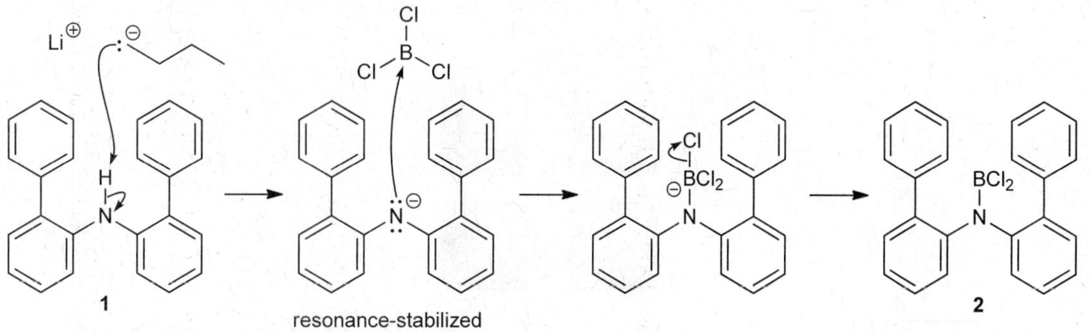

A mechanism for the final step is shown below. Overall, this step constitutes two subsequent electrophilic aromatic substitution reactions, resulting in the formation of two new B-C bonds. Addition of aluminum trichloride, a Lewis acid, serves to enhance the electrophilicity of the boron by forming a complex with one of the chlorine atoms. The boron is then attacked by one of the pendant arenes, with subsequent loss of $AlCl_4^-$ and formation of a resonance-stabilized sigma complex (several, but not all, of the many resonance structures are shown). Deprotonation with $AlCl_4^-$ restores aromaticity, giving a neutral intermediate. An identical sequence of mechanistic steps (complexation with $AlCl_3$, attack of the arene, loss of $AlCl_4^-$ and deprotonation) results in the formation of the second B-C bond, thus producing compound **3**.

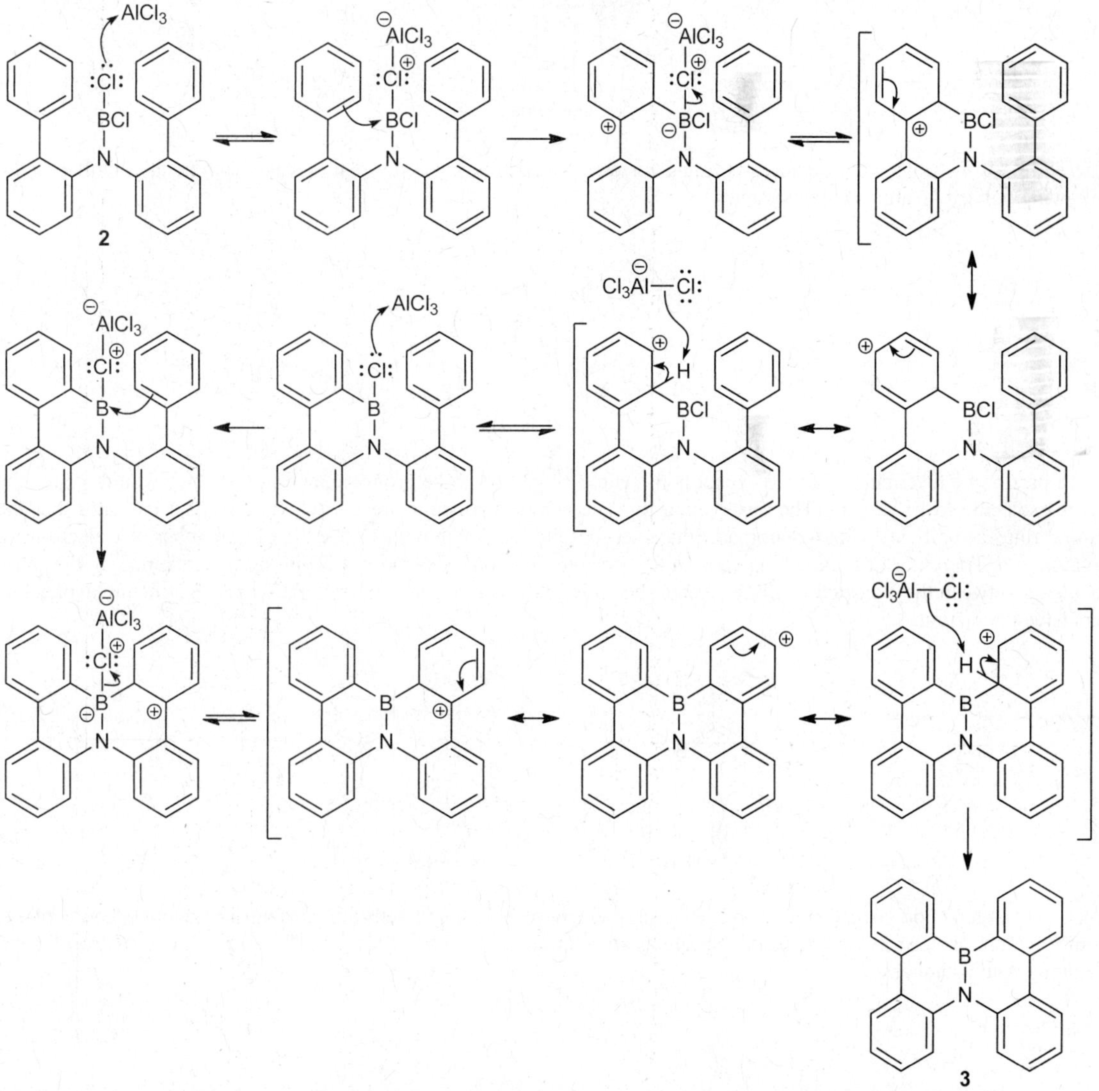

(b) Each of the four peripheral rings is clearly aromatic.

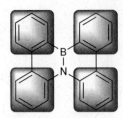

For the two central rings, the boron and nitrogen are both expected to be sp^2-hybridized, with boron having an empty p-orbital and nitrogen holding its lone pair in a p-orbital. As such, each central ring exhibits a continuous system of overlapping p orbitals with a total of 6 π electrons, thus fulfilling the criteria for aromaticity:

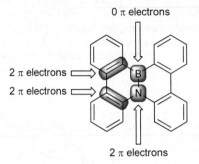

Therefore, all six rings are aromatic, providing for one extended aromatic system that is electronically similar to the following polycyclic aromatic compound:

(c) In part b, we determined that we expect both central rings in **3** to be aromatic, and as such, we would expect the entire molecule to be planar. The observation that the central two rings are twisted (non-planar) indicates that the central rings actually have limited aromatic character, despite the fact that they have the right number of π electrons to be aromatic. There must be something causing the rings to twist out of planarity. This can be explained by the steric repulsion between the indicated hydrogen atoms shown in the structures below. A slightly twisted structure avoids this repulsive interaction.

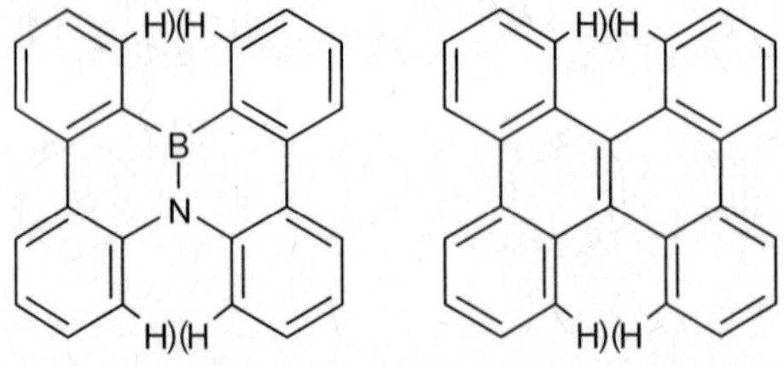

This steric interaction is very similar to the steric interaction in diphenylmethane, causing it to adopt a non-coplanar conformation, as seen in the Medically Speaking application at the end of section 18.5 (The Development of Non-Sedating Antihistamines).

Chapter 20
Ketones and Aldehydes

Review of Concepts

Fill in the blanks below. To verify that your answers are correct, look in your textbook at the end of Chapter 20. Each of the sentences below appears verbatim in the section entitled *Review of Concepts and Vocabulary*.

- The suffix "_____" indicates an aldehydic group, and the suffix "_____" is used for ketones.
- The electrophilicity of a carbonyl group derives from _____ effects, as well as _____ effects.
- A general mechanism for nucleophilic addition under basic conditions involves two steps
    1) nucleophilic attack
    2) _____
- The position of equilibrium is dependent on the ability of the nucleophile to function as a _____.
- In acidic conditions, an aldehyde or ketone will react with two molecules of alcohol to form an _____.
- The reversibility of acetal formation enables acetals to function as _____ groups for ketones or aldehydes. Acetals are stable under strongly _____ conditions.
- In acidic conditions, an aldehyde or ketone will react with a primary amine to form an _____.
- In acidic conditions, an aldehyde or ketone will react with a secondary amine to form an _____.
- In the **Wolff-Kishner reduction**, a hydrazone is reduced to an _____ under strongly basic conditions.
- _____ of acetals, imines, and enamines under acidic conditions produces ketones or aldehydes.
- In acidic conditions, an aldehyde or ketone will react with two equivalents of a thiol to form a

    _____.
- When treated with Raney nickel, thioacetals undergo **desulfurization** to yield a _____ group.
- When treated with a hydride reducing agent, such as lithium aluminum hydride (LAH) or sodium borohydride (NaBH$_4$), aldehydes and ketones are reduced to _____.
- The reduction of a carbonyl group with LAH or NaBH$_4$ is not a reversible process, because hydride does not function as a _____.
- When treated with a Grignard reagent, aldehydes and ketones are converted into alcohols, accompanied by the formation of a new _____ bond.
- Grignard reactions are not reversible, because carbanions do not function as _____.
- When treated with hydrogen cyanide (HCN), aldehydes and ketones are converted into _____. For most aldehydes and unhindered ketones, the equilibrium favors formation of the _____.
- The **Wittig reaction** can be used to convert a ketone to an _____.
- A **Baeyer-Villiger oxidation** converts a ketone to an _____ by inserting _____ next to the carbonyl group. Cyclic ketones produce cyclic esters called _____.

Review of Skills

Fill in the blanks and empty boxes below. To verify that your answers are correct, look in your textbook at the end of Chapter 20. The answers appear in the section entitled *SkillBuilder Review*.

20.1: Naming Aldehydes and Ketones

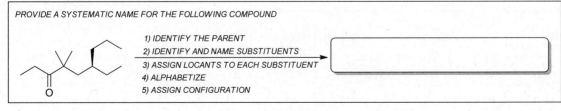

PROVIDE A SYSTEMATIC NAME FOR THE FOLLOWING COMPOUND

1) IDENTIFY THE PARENT
2) IDENTIFY AND NAME SUBSTITUENTS
3) ASSIGN LOCANTS TO EACH SUBSTITUENT
4) ALPHABETIZE
5) ASSIGN CONFIGURATION

20.2: Drawing the Mechanism of Acetal Formation

DRAW A MECHANISM FOR THE ACID-CATALYZED CONVERSION OF A KETONE TO A HEMIACETAL.
MAKE SURE TO DRAW ALL CURVED ARROWS AND INTERMEDIATES.

DRAW A MECHANISM FOR THE ACID-CATALYZED CONVERSION OF A HEMIACETAL TO AN ACETAL.
MAKE SURE TO DRAW ALL CURVED ARROWS AND INTERMEDIATES.

20.3: Drawing the Mechanism of Imine Formation

DRAW A MECHANISM FOR THE ACID-CATALYZED CONVERSION OF A KETONE TO A CARBINOLAMINE.
MAKE SURE TO DRAW ALL CURVED ARROWS AND INTERMEDIATES.

DRAW A MECHANISM FOR THE ACID-CATALYZED CONVERSION OF A CARBINOLAMINE TO AN IMINE.
MAKE SURE TO DRAW ALL CURVED ARROWS AND INTERMEDIATES.

20.4: Drawing the Mechanism of Enamine Formation

DRAW A MECHANISM FOR THE ACID-CATALYZED CONVERSION OF A KETONE TO A CARBINOLAMINE.
MAKE SURE TO DRAW ALL CURVED ARROWS AND INTERMEDIATES.

DRAW A MECHANISM FOR THE ACID-CATALYZED CONVERSION OF A CARBINOLAMINE TO AN ENAMINE.
MAKE SURE TO DRAW ALL CURVED ARROWS AND INTERMEDIATES.

20.5: Drawing the Products of a Hydrolysis Reaction

DRAW THE EXPECTED PRODUCTS WHEN THE FOLLOWING COMPOUND IS TREATED WITH AQUEOUS ACID	STEP 1 - IDENTIFY THE BOND(S) EXPECTED TO UNDERGO CLEAVAGE	STEP 2 - IDENTIFY THE CARBON ATOM THAT WILL BECOME A CARBONYL GROUP	STEP 3 - DETERMINE THE IDENTITY OF THE OTHER FRAGMENT(S)

20.6: Planning an Alkene Synthesis with a Wittig Reaction

IDENTIFY THE REACTANTS YOU WOULD USE TO PREPARE THE FOLLOWING COMPOUND VIA A WITTIG REACTION:

20.7: Proposing a Synthesis

BEGIN BY ASKING THE FOLLOWING TWO QUESTIONS:
 	
1) IS THERE A CHANGE IN THE _____?
 	
2) IS THERE A CHANGE IN THE _____?
	IF THERE IS A CHANGE IN THE CARBON SKELETON, CONSIDER ALL OF THE C-C BOND FORMING REACTIONS AND ALL OF THE C-C BOND BREAKING REACTIONS THAT YOU HAVE LEARNED SO FAR.
	
	C-C BOND-FORMING REACTIONS IN THIS CHAPTER:
	   - _____
	   - _____
	   - _____
	C-C BOND-BREAKING REACTIONS IN THIS CHAPTER:
	   - _____

Review of Reactions

Identify the reagents necessary to achieve each of the following transformations. To verify that your answers are correct, look in your textbook at the end of Chapter 20. The answers appear in the section entitled *Review of Reactions*.

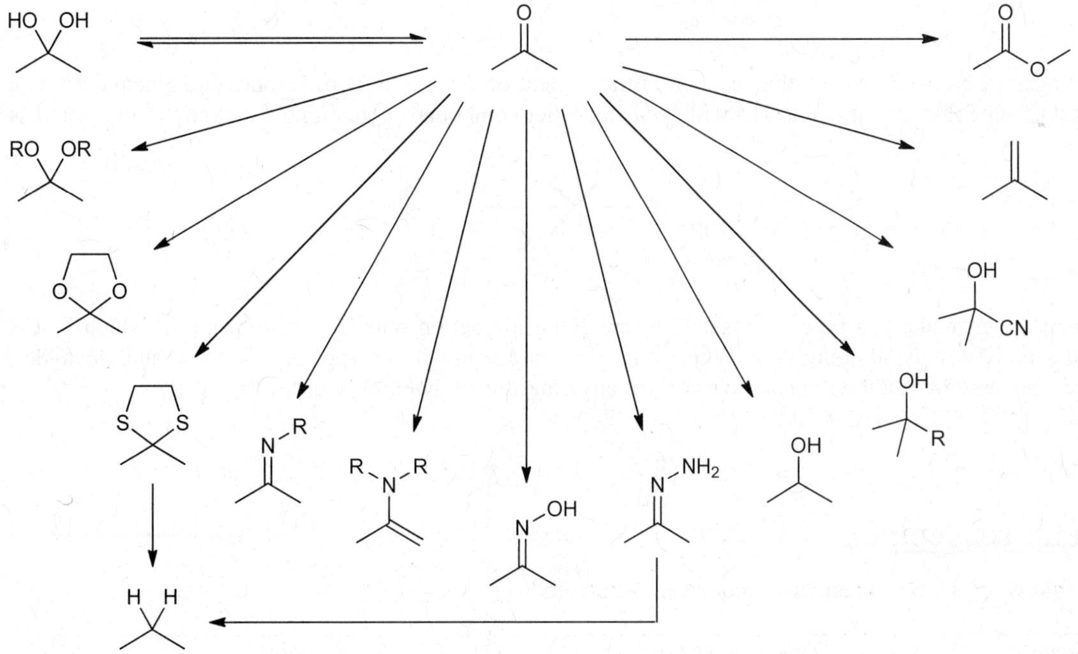

Common Mistakes to Avoid

This chapter covers many reactions. One of them is acetal formation:

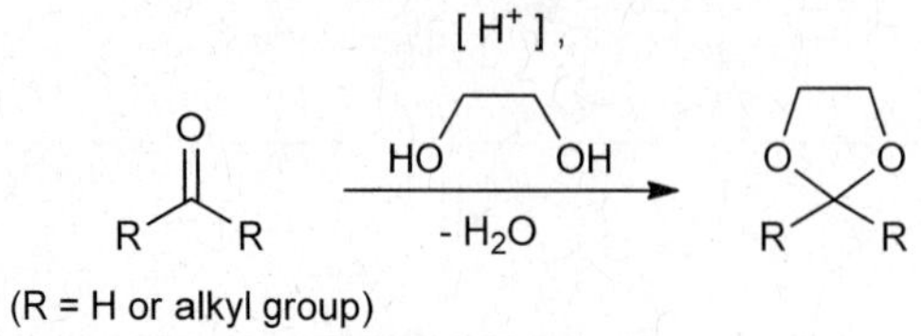

(R = H or alkyl group)

This reaction is covered in the context of the reactivity of ketones and aldehydes. The same reaction does not occur for other compounds containing a carbonyl group, such as esters:

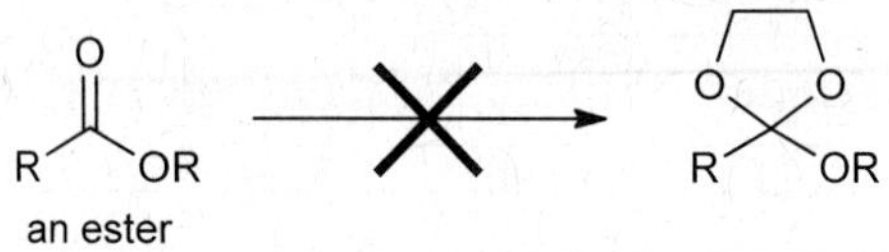

an ester

Students often make this mistake, assuming that acetal formation will work for esters, just as it does for ketones. It does not. This is a common mistake, and it should be avoided.

As another example of a reaction that is often applied in the wrong context, consider the Wittig reaction.

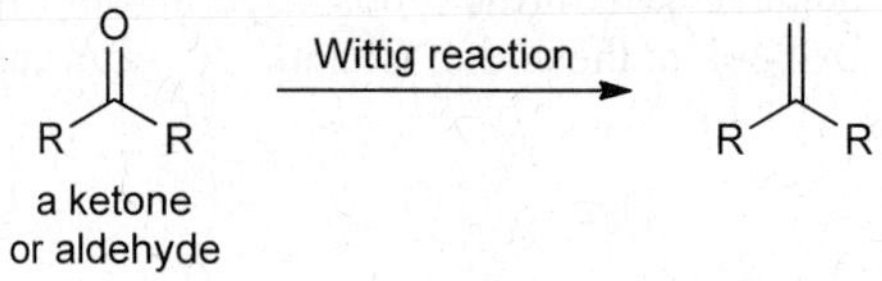

a ketone
or aldehyde

Once again, Chapter 20 covers this reaction in the context of the reactivity of ketones and aldehydes. You cannot assume that the same reaction will occur for other compounds containing a carbonyl group, such as esters:

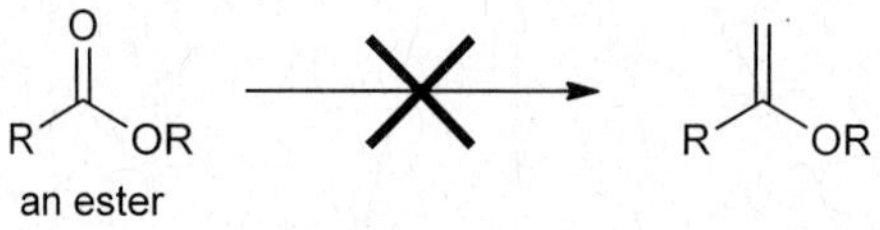

an ester

Students often make this type of mistake, by applying a reaction outside of the scope in which it was discussed. Try to avoid doing this. Whenever we cover a reaction that applies to ketones and aldehydes, you cannot assume that it will apply to esters (or any other functional group for that matter).

Useful reagents

The following is a list of reagents encountered in this chapter:

Reagents	Type of Reaction	Description
H_3O^+	Hydrate formation	Conversion of a carbonyl group to a hydrate. This process is only efficient for formaldehyde and some simple aldehydes. For ketones, the equilibrium generally does not favor formation of the hydrate.

[H$^+$], 2 ROH, (– H$_2$O)	Acetal formation	Conversion of an aldehyde or ketone into an acetal. The acetal group can be used to protect aldehydes and ketones. The acetal group is stable to basic conditions, but is removed when subjected to aqueous acidic conditions to regenerate the carbonyl group (a process called hydrolysis).
[H$^+$], HOCH$_2$CH$_2$OH, – H$_2$O	Cyclic acetal formation	Ethylene glycol can be used to convert an aldehyde or ketone into an acetal. The acetal group can be used to protect aldehydes and ketones. The acetal group is stable to basic conditions, but is removed when subjected to aqueous acidic conditions to regenerate the carbonyl group (a process called hydrolysis).
[H$^+$], HSCH$_2$CH$_2$SH, (– H$_2$O)	Cyclic thioacetal formation	Ethylene thioglycol can be used to convert an aldehyde or ketone into a cyclic thioacetal.
Raney Nickel	Desulfurization	Converts a thioacetal (or cyclic thioacetal) to an alkane.
[H$^+$], RNH$_2$, (– H$_2$O)	Imine formation	Converts an aldehyde or ketone into an imine. The imine group is removed when subjected to aqueous acidic conditions to regenerate the carbonyl group (a process called hydrolysis).
[H$^+$], R$_2$NH, (– H$_2$O)	Enamine formation	Converts an aldehyde or ketone into an enamine. The enamine group is removed when subjected to aqueous acidic conditions to regenerate the carbonyl group (a process called hydrolysis).
[H$^+$], NH$_2$OH, (– H$_2$O)	Oxime formation	Converts an aldehyde or ketone into an oxime.
[H$^+$], NH$_2$NH$_2$, (– H$_2$O)	Hydrazone formation	Converts an aldehyde or ketone into a hydrazone.
NaOH, H$_2$O, heat	Wolff-Kishner reduction	Reduces a hydrazone to an alkane.
1) LAH 2) H$_2$O	Reduction	Reduces an aldehyde or ketone to an alcohol.
1) RMgBr 2) H$_2$O	Grignard reaction	When an aldehyde or ketone is treated with a Grignard reagent (followed by water work-up), a carbon-carbon bond-forming reaction occurs, giving an alcohol that exhibits the newly formed C-C bond.
HCN, KCN	Cyanohydrin formation	Converts an aldehyde or ketone into a cyanohydrin.
H$_2$C=PPh$_3$	Wittig reaction	When an aldehyde or ketone is treated with a Wittig reagent, a carbon-carbon bond-forming reaction occurs, giving an alkene that exhibits the newly formed C=C bond in the location of the former carbonyl group.
RCO$_3$H	Baeyer-Villiger oxidation	Converts a ketone into an ester (via insertion of an oxygen atom). Converts an aldehyde into a carboxylic acid (via insertion of an oxygen atom).

Solutions

20.1.
(a) We begin by identifying the longest chain that includes the carbon atom of the carbonyl group. In this case, the parent is a chain of six carbon atoms, with the carbonyl group at C1, so the parent is hexanal. Next, we identify the substituents. There are four: two methyl groups (both at C2) and two bromine atoms (both at C5). Finally, we assemble the substituents alphabetically, giving the following name:

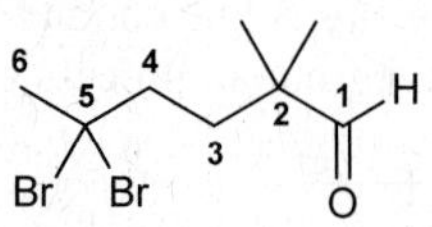

5,5-dibromo-2,2-dimethylhexanal

(b) We begin by identifying the longest chain that includes the carbon atom of the carbonyl group. In this case, the parent is a chain of six carbon atoms, with the carbonyl group at C2, so the parent is 2-hexanone. Next, we identify the substituents. There are three, all of which are methyl groups, located at C3, C4 and C5. We then assign a configuration to each of the chirality centers, giving the following name:

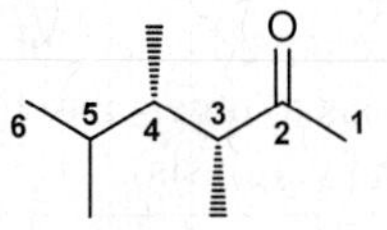

(3R,4S)-3,4,5-trimethyl-2-hexanone

(c) We begin by identifying the parent. In this case, the carbonyl group is part of a five-membered ring, so the parent is cyclopentanone. In a cyclic ketone, the carbonyl group is at C1 (by definition). Next, we identify the substituents. There are four, all of which are methyl groups (two at C2 and two at C5), giving the following name:

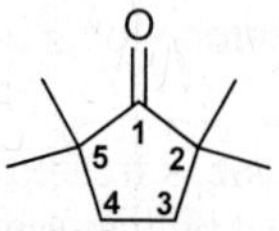

2,2,5,5-tetramethylcyclopentanone

(d) We begin by identifying the longest chain that includes the carbon atom of the carbonyl group. In this case, the parent is a chain of five carbon atoms, with the carbonyl group at C1, so the parent is pentanal. Next, we identify the substituents. There is only one substituent: a propyl group at C2, giving the following name:

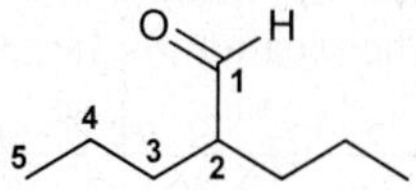

2-propylpentanal

(e) A cyclic compound containing an aldehyde adjacent to the ring is named as a cycloalkane carbaldehyde. In this case, the aldehyde group is connected to a four-membered ring (cyclobutane), giving the following name:

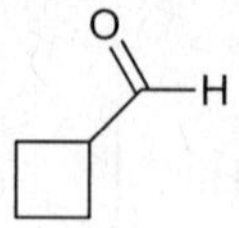

cyclobutanecarbaldehyde

20.2.
(a) The name indicates that the parent (cyclohexanone) is a six-membered ring, with a carbonyl group incorporated in the ring (a cyclic ketone). There are three substituents: two bromine atoms (both at C3) and an ethyl group (at C4). The chirality center at C4 has the S configuration:

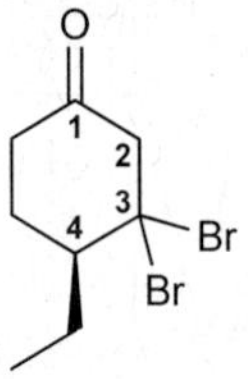

(b) The name indicates that the parent (3-pentanone) is a chain of five carbon atoms, with the carbonyl group at C3. There are two substituents (both methyl groups), located at C2 and C4, as shown:

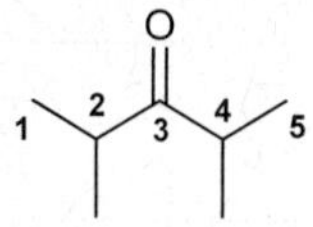

(c) The name indicates that the parent (butanal) is a chain of four carbon atoms, with the carbonyl group at C1 (by definition). There is only one substituent (Br), located at C3, which exhibits the R configuration, as shown:

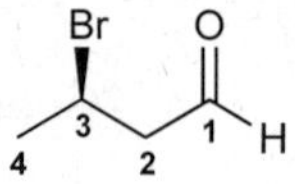

20.3. We begin by identifying the parent chain, which is a bicyclic structure. Using the skills covered in SkillBuilder 4.5, we assign the parent name for this structure and we assign a locant for the position of the carbonyl group (C2). The configurations of the two chirality centers (the bridgeheads, C1 and C4) are assigned and included in the name, as shown:

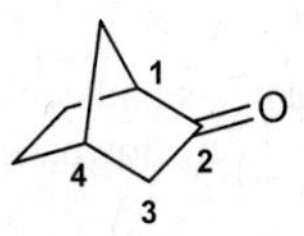

(*1S,4R*)bicyclo[2.2.1]heptan-2-one

20.4.

(a) The parent is a six-membered ring with two carbonyl groups (a dione). The locants indicate the relative positions of the carbonyl groups (C1 and C3):

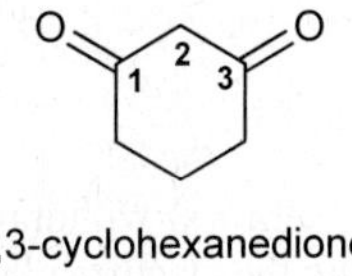

1,3-cyclohexanedione

(b) The parent is a six-membered ring with two carbonyl groups (a dione). The locants indicate the relative positions of the carbonyl groups (C1 and C4):

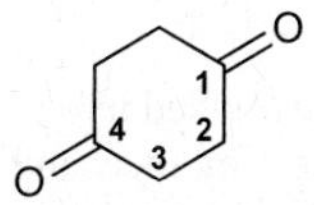

1,4-cyclohexanedione

(c) The parent is a chain of nine carbon atoms, for which three of the carbon atoms are carbonyl groups (a trione). The locants indicate the relative positions of the carbonyl groups (C2, C5, and C8):

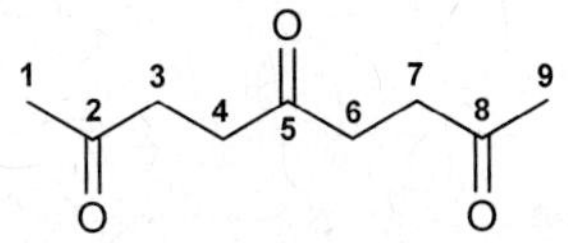

2,5,8-nonanetrione

20.5.

(a) Oxidation of a secondary alcohol (to give a ketone) can be achieved with chromic acid (H_2CrO_4), which is prepared by mixing sodium dichromate ($Na_2Cr_2O_7$) and aqueous sulfuric acid. Alternatively, PCC can affect this transformation.

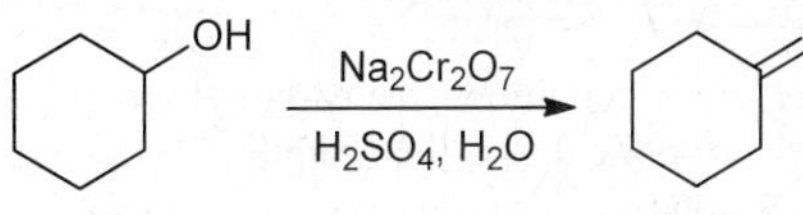

(b) Oxidation of a primary alcohol (to give an aldehyde) can be achieved upon treatment with PCC.

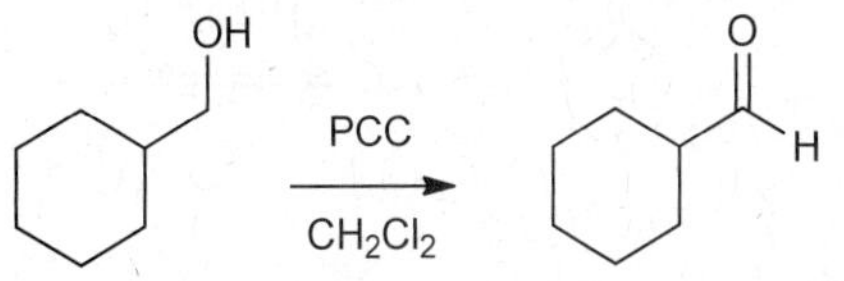

(c) A terminal alkyne can be converted into a methyl ketone upon acid-catalyzed hydration in the presence of mercuric sulfate:

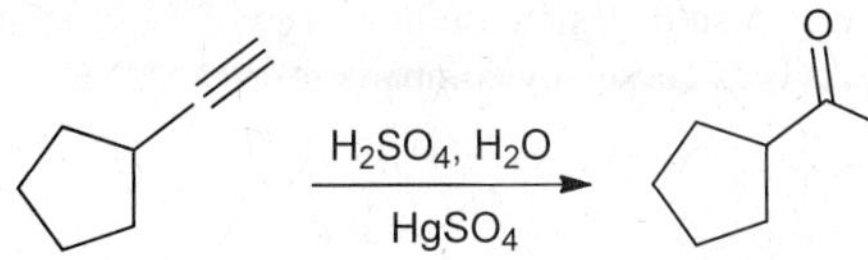

(d) A terminal alkyne can be converted into an aldehyde via hydroboration-oxidation:

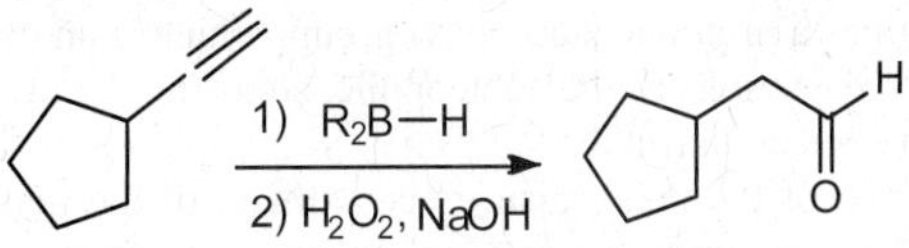

(e) A C=C bond can be cleaved into two carbonyl groups via ozonolysis. In this case, the ring is opened to give an acyclic product.

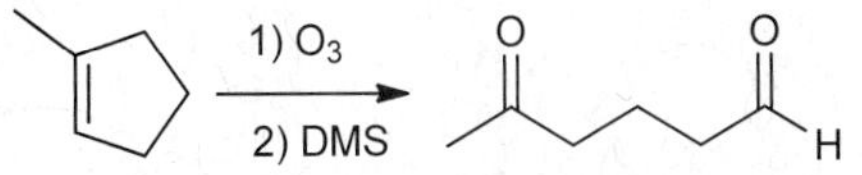

(f) An acyl group can be installed via a Friedel-Crafts acylation, using the appropriate acyl halide, as shown:

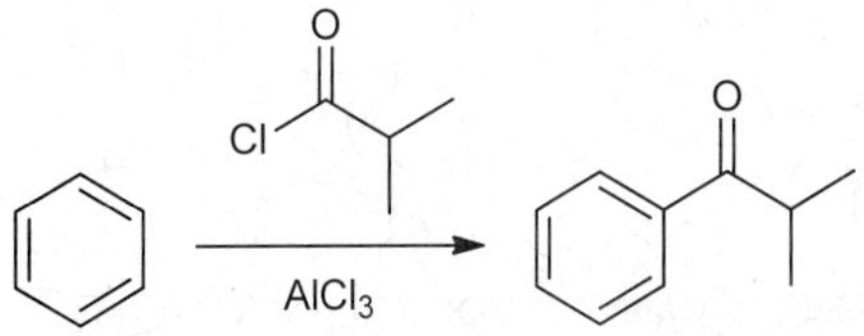

20.6.

(a) As seen in Section 13.6, a Grignard reagent is a strong nucleophile that will attack a carbonyl group to give an alkoxide ion. After the reaction is complete, water is introduced into the reaction flask, thereby serving as a proton source to protonate the alkoxide ion, giving an alcohol:

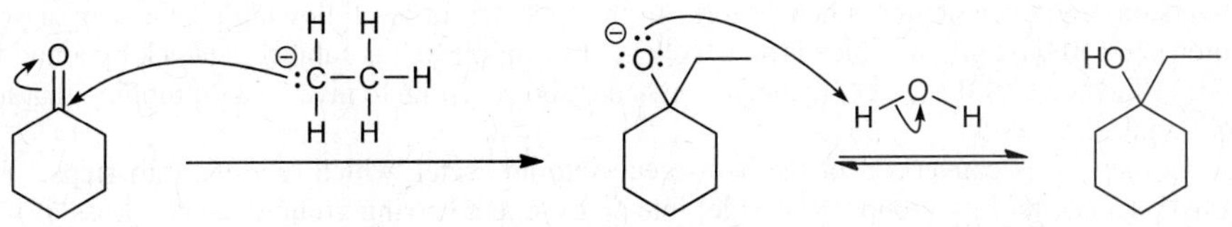

(b) Under acidic conditions, the carbonyl group is first protonated to give a resonance-stabilized cation. This cation is a strong electrophile that is then captured by a chloride ion to give the addition product shown. The equilibrium likely favors the ketone.

20.7.   The carbonyl group in hexafluoroacetone is flanked by two very powerful electron-withdrawing groups (CF_3). These groups withdraw electron density from the carbonyl group, thereby increasing the electrophilicity of the carbonyl group. The resulting increase in energy of the reactant causes the equilibrium to favor the product (the hydrate, which is now lower in energy by comparison).

20.8.

(a) Acid-catalyzed acetal formation proceeds via a seven-step process, which can be divided into two parts:
Part one is conversion of the ketone into a hemiacetal. In the first step of the following mechanism, a proton transfer generates a protonated carbonyl group, which is an excellent electrophile and is subject to attack by a nucleophile. The alcohol functions as the nucleophile, attacking the protonated carbonyl group to give an oxonium ion that loses a proton to give the hemiacetal.
Part two of the mechanism is conversion of the hemiacetal into an acetal, which requires four steps. First a proton transfer step converts a bad leaving group (hydroxide) into an excellent leaving group (water). Loss of water gives an intermediate that can be attacked by another molecule of the alcohol. The resulting oxonium ion is then deprotonated to give an acetal. Note that in all steps of the mechanism, the proton source is a protonated alcohol, and the base is the alcohol (MeOH). No strong bases are employed or formed at any point during the mechanism.

(b) Acid-catalyzed acetal formation proceeds via a seven-step process, which can be divided into two parts:
Part one is conversion of the ketone into a hemiacetal. In the first step of the following mechanism, a proton transfer generates a protonated carbonyl group, which is an excellent electrophile and is subject to attack by a nucleophile. The alcohol functions as the nucleophile, attacking the protonated carbonyl group to give an oxonium ion that loses a proton to give the hemiacetal.
Part two of the mechanism is conversion of the hemiacetal into an acetal, which requires four steps. First a proton transfer step converts a bad leaving group (hydroxide) into an excellent leaving group (water). Loss of water gives an intermediate that can be attacked by another molecule of the alcohol. The resulting oxonium ion is then deprotonated

to give an acetal. Note that in all steps of the mechanism, the proton source is a protonated alcohol, and the base is the alcohol (EtOH). No strong bases are employed or formed at any point during the mechanism.

Hemiacetal

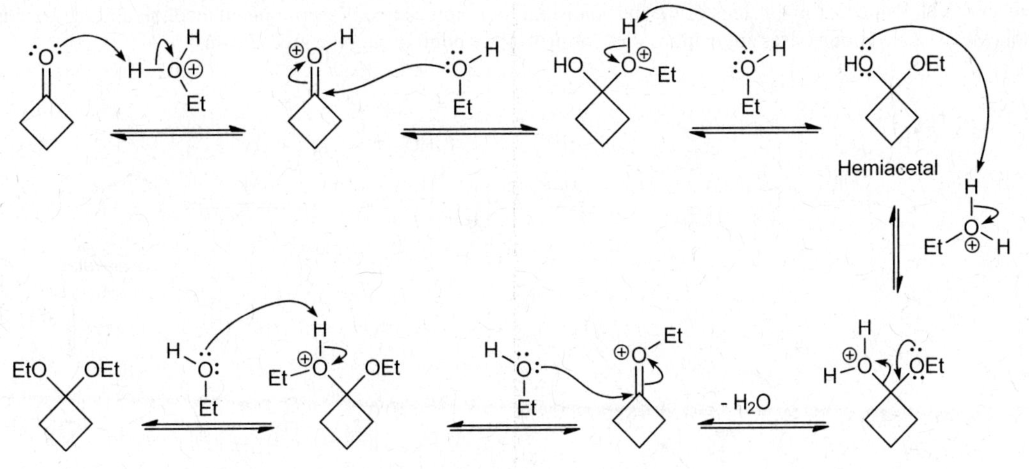

(c) Acid-catalyzed acetal formation proceeds via a seven-step process, which can be divided into two parts:
Part one is conversion of the ketone into a hemiacetal. In the first step of the following mechanism, a proton transfer generates a protonated carbonyl group, which is an excellent electrophile and is subject to attack by a nucleophile. The alcohol functions as the nucleophile, attacking the protonated carbonyl group to give an oxonium ion that loses a proton to give the hemiacetal.
Part two of the mechanism is conversion of the hemiacetal into an acetal, which requires four steps. First a proton transfer step converts a bad leaving group (hydroxide) into an excellent leaving group (water). Loss of water gives an intermediate that can be attacked by another molecule of the alcohol. The resulting oxonium ion is then deprotonated to give an acetal. Note that in all steps of the mechanism, the proton source is a protonated alcohol, and the base is the alcohol (EtOH). No strong bases are employed or formed at any point during the mechanism.

Hemiacetal

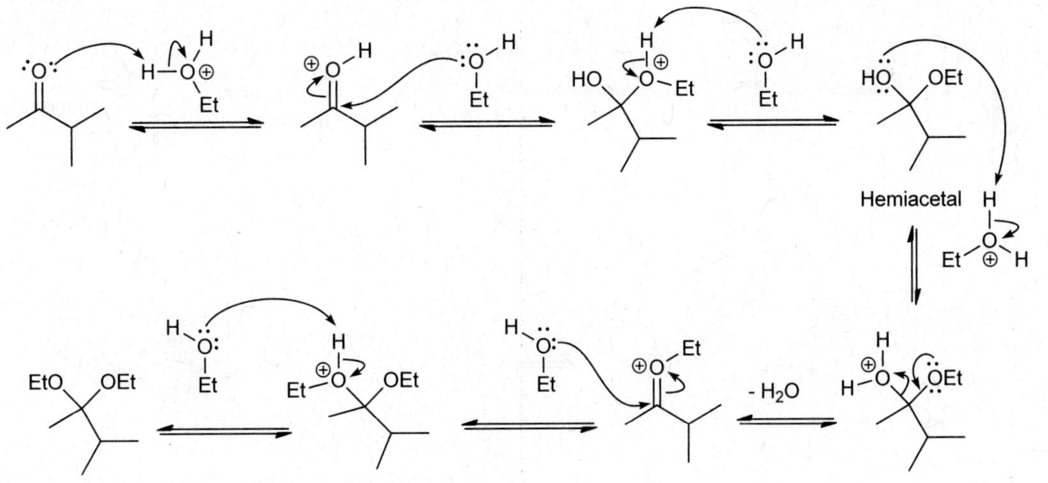

(d) Acid-catalyzed acetal formation proceeds via a seven-step process, which can be divided into two parts:
Part one is conversion of the ketone into a hemiacetal. In the first step of the following mechanism, a proton transfer generates a protonated carbonyl group, which is an excellent electrophile and is subject to attack by a nucleophile. The alcohol functions as the nucleophile, attacking the protonated carbonyl group to give an oxonium ion that loses a proton to give the hemiacetal.

Part two of the mechanism is conversion of the hemiacetal into an acetal, which requires four steps. First a proton transfer step converts a bad leaving group (hydroxide) into an excellent leaving group (water). Loss of water gives an intermediate that can be attacked by another molecule of the alcohol. The resulting oxonium ion is then deprotonated to give an acetal. Note that in all steps of the mechanism, the proton source is a protonated alcohol, and the base is the alcohol (MeOH). No strong bases are employed or formed at any point during the mechanism.

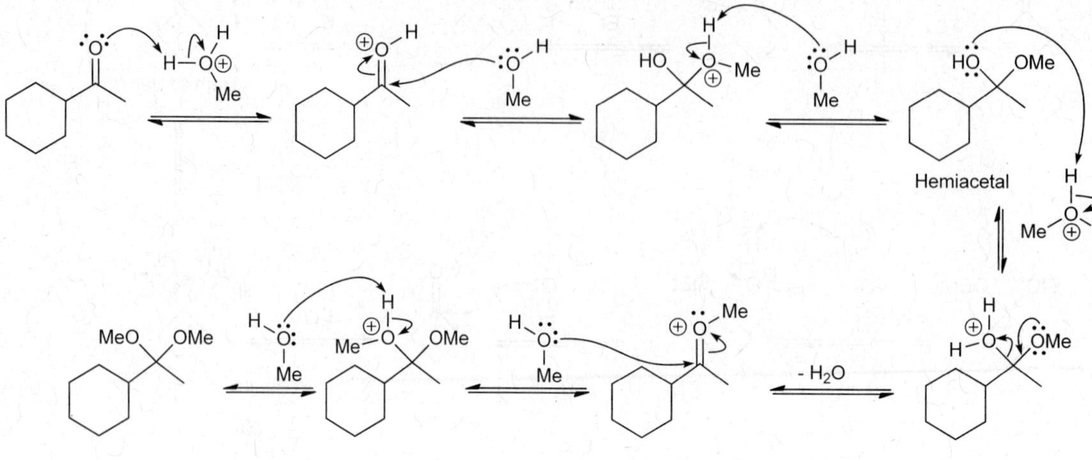

20.9.

(a) We will draw a very similar mechanism to the mechanisms drawn in the solutions to Problems 20.8a-d. Specifically, conversion of the ketone to an acetal occurs via a seven-step process. First the ketone is converted to a hemiacetal via the following three steps: 1) proton transfer, 2) nucleophilic attack, and then 3) proton transfer. And then the hemiacetal is converted into an acetal via the following four steps: 4) proton transfer, 5) loss of a leaving group, 6) nucleophilic attack, and 7) proton transfer. The difference between this mechanism and the mechanisms in problem 20.8 can be seen in step 6 (the second nucleophilic attack). In this case, step 6 occurs in an *intramolecular* fashion, giving a cyclic acetal:

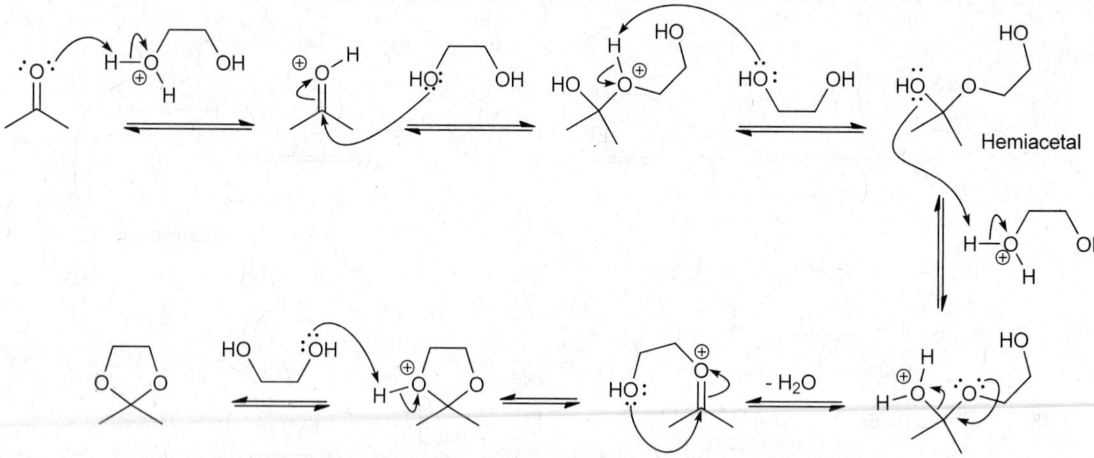

b) Acid-catalyzed acetal formation proceeds via a seven-step process, which can be divided into two parts:
Part one is conversion of the ketone into a hemiacetal. In the first step of the following mechanism, a proton transfer generates a protonated carbonyl group, which is an excellent electrophile and is subject to attack by a nucleophile. In this case, the nucleophilic attack occurs in an *intramolecular* fashion to give a five-membered ring. This oxonium ion then loses a proton to give the hemiacetal.
Part two of the mechanism is conversion of the hemiacetal into an acetal, which requires four steps. First, a proton transfer step converts a bad leaving group (hydroxide) into an excellent leaving group (water). Loss of water gives an intermediate that can be attacked by another nucleophile. Once again, an intramolecular attack generates another five-membered ring. This oxonium ion is then deprotonated to give an acetal. Note that in all steps of the mechanism, the

proton source is H_3O^+, and the base is water (H_2O). No strong bases are employed or formed at any point during the mechanism.

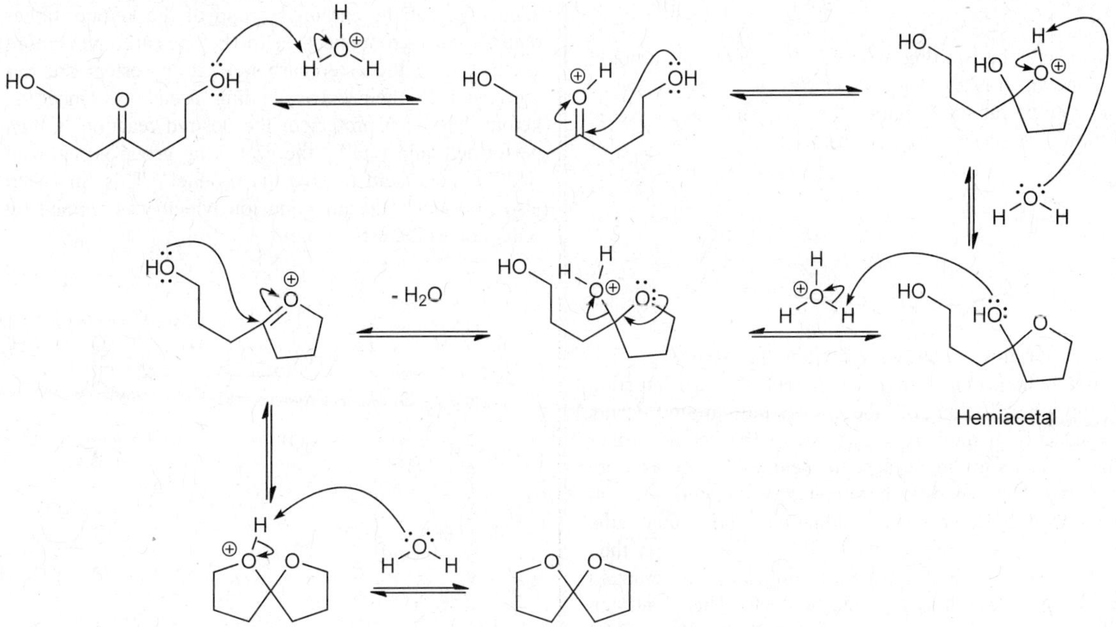

Hemiacetal

20.10.

(a) The ketone is converted into an acetal. The alcohol is MeOH so we expect two OMe groups to replace the carbonyl group, as shown:

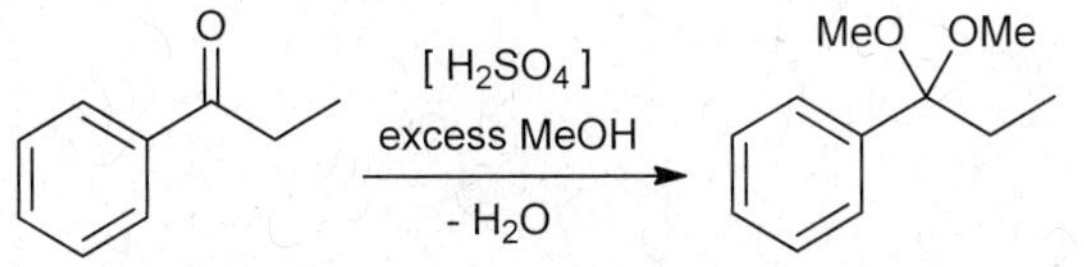

$$[\,H_2SO_4\,]$$
excess MeOH
$$-\,H_2O$$

(b) The ketone is converted into an acetal. A diol is used so we expect a cyclic acetal to be produced (step 6 of the mechanism occurs in an intramolecular fashion, much like we saw in the solution to Problem 20.9a, although there is one extra carbon atom in this case):

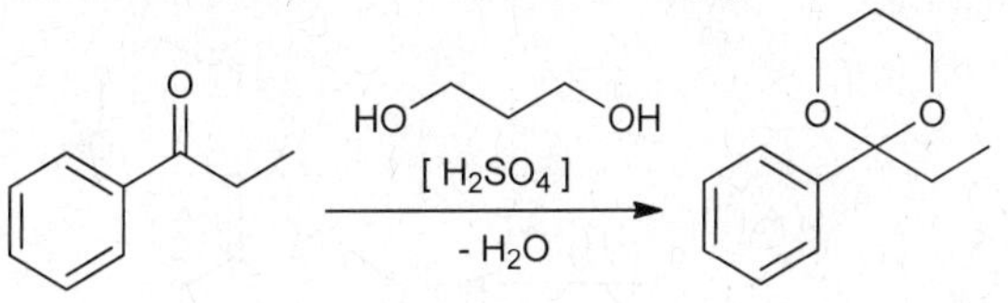

$$[\,H_2SO_4\,]$$
$$-\,H_2O$$

20.11.

(a) This type of transformation can be achieved via alkylation of a terminal alkyne, as seen in Section 10.10:

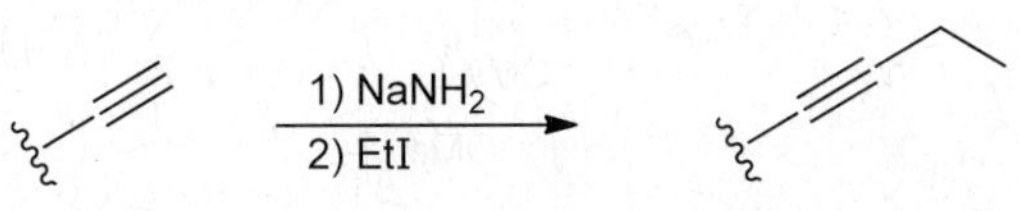

1) NaNH₂
2) EtI

However, the starting material has a carbonyl group, which is a good electrophile. In the presence of the

nucleophilic alkynide ion, the following undesired reaction can occur:

To avoid this undesired reaction, the carbonyl group is first converted into an acetal, which serves as a protecting group. The desired reaction is then performed, and finally, the protecting group (the acetal group) is removed with aqueous acid to give the product:

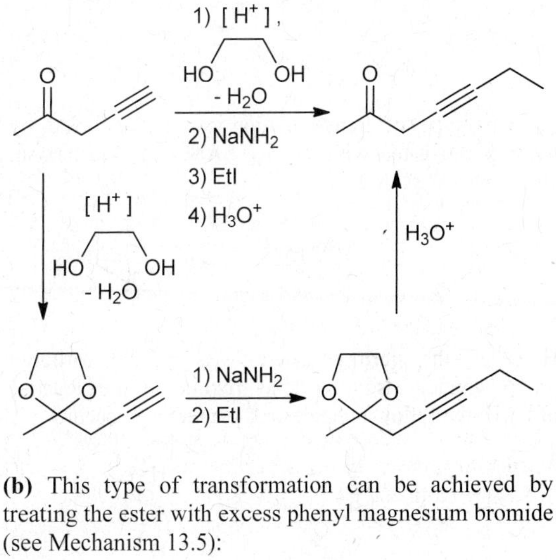

1) [H⁺],
   HO   OH
   - H₂O
2) NaNH₂
3) EtI
4) H₃O⁺

[H⁺]
HO   OH
- H₂O

H₃O⁺

1) NaNH₂
2) EtI

(b) This type of transformation can be achieved by treating the ester with excess phenyl magnesium bromide (see Mechanism 13.5):

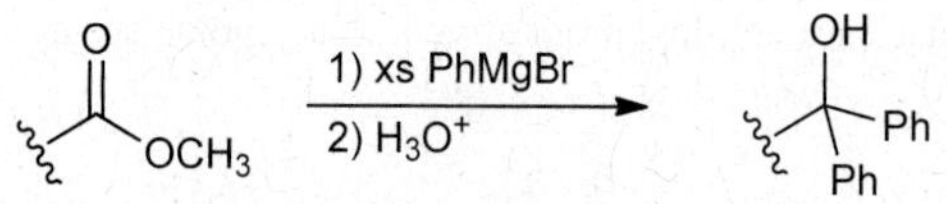

However, the starting material has another carbonyl group, which is also a good electrophile. In the presence of a powerful nucleophile (the Grignard reagent), the following undesired reaction can occur:

To avoid this undesired reaction, the ketone is first converted into an acetal, which serves as a protecting group. Installation of the protecting group occurs exclusively at the carbonyl group of the ketone, rather than the ester group, because the ketone carbonyl is more reactive than the ester carbonyl group (esters are *not* converted to their corresponding acetals). Once the ketone has been protected, the desired reaction is then performed, and finally, the protecting group is removed with aqueous acid to give the product. This final step also protonates the alkoxide ion which was formed by the Grignard reaction:

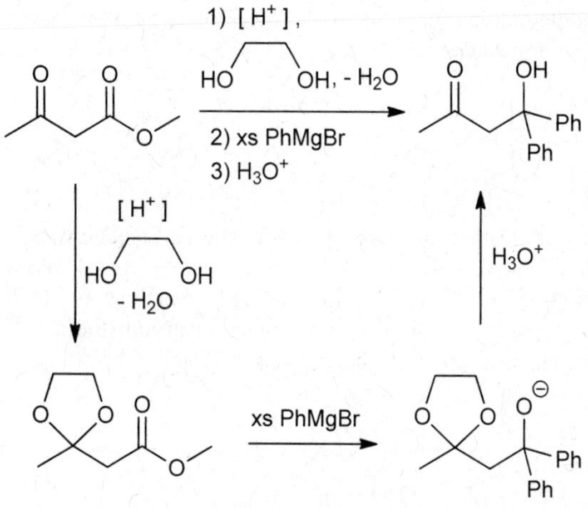

(c) This type of transformation can be achieved by treating the ester with excess LAH (see Mechanism 13.3):

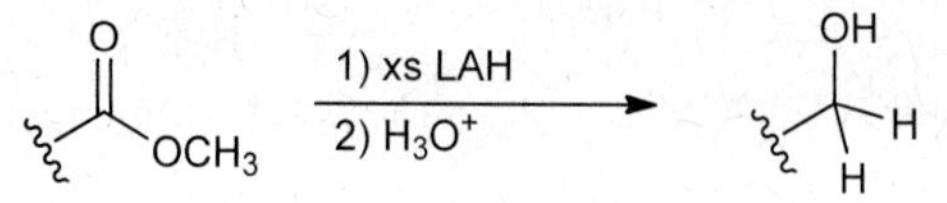

However, the starting material has another carbonyl group, which is also a good electrophile. In the presence of LAH, the following undesired reaction can occur:

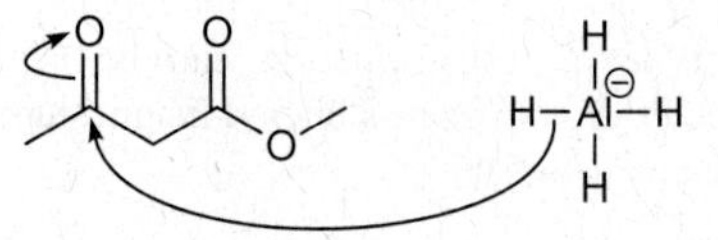

To avoid this undesired reaction, the ketone is first converted into an acetal, which serves as a protecting group. Installation of the protecting group occurs exclusively at the carbonyl group of the ketone, rather than the ester group, because the ketone carbonyl is more reactive than the ester carbonyl group (esters are *not* converted to their corresponding acetals). Once the ketone has been protected, the desired reaction is then performed, and finally, the protecting group is removed with aqueous acid to give the product. This final step also protonates the alkoxide ion which was formed by reduction of the ester group:

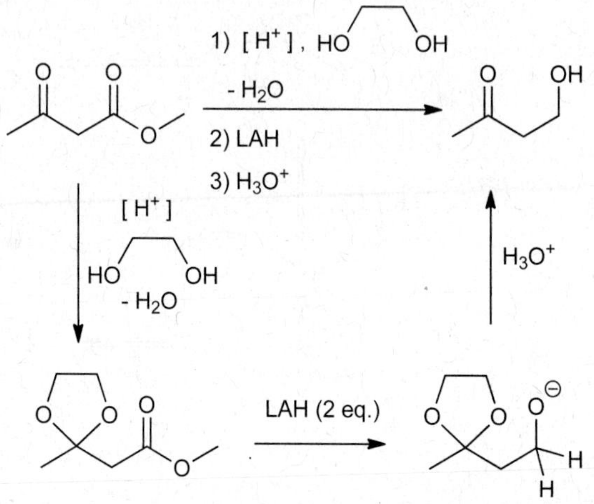

20.12.
(a) Begin by identifying the carbon atom that is connected to two oxygen atoms (highlighted):

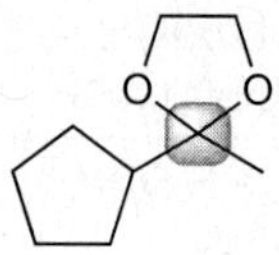

This carbon atom bears an acetal group, and will ultimately be converted into a carbonyl group in the product. The C-O bonds are cleaved, giving a ketone. Since these C-O bonds are contained in a ring, a diol is released.

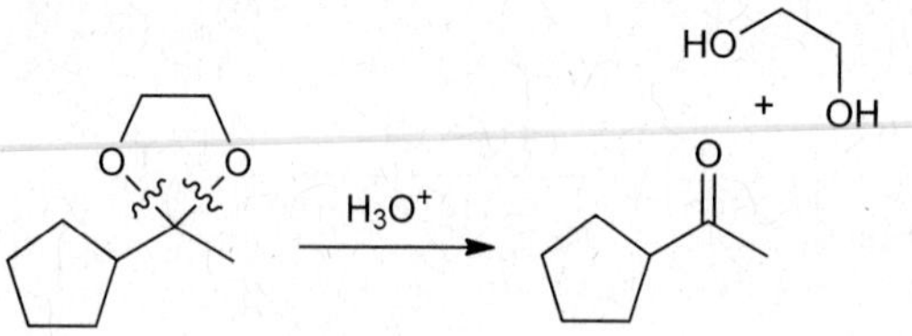

(b) Begin by identifying the carbon atom that is connected to two oxygen atoms (highlighted):

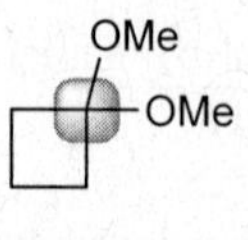

This carbon atom bears an acetal group, and will ultimately be converted into a carbonyl group in the product. The C-O bonds are cleaved, giving a ketone. Since these C-O bonds are not contained in a ring, two molecules of alcohol (methanol) are released.

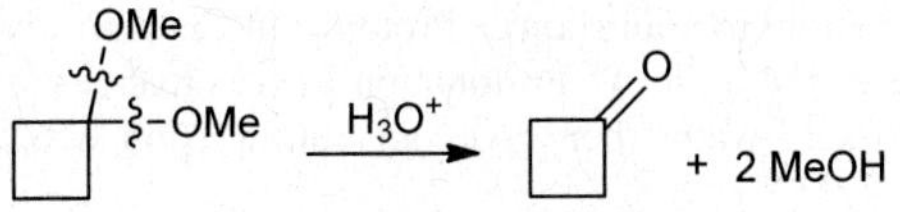

(c) Begin by identifying the carbon atom that is connected to two oxygen atoms (highlighted):

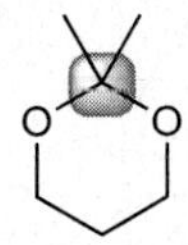

This carbon atom bears an acetal group, and will ultimately be converted into a carbonyl group in the product. The C-O bonds are cleaved, giving a ketone. Since these C-O bonds are contained in a ring, a diol is released.

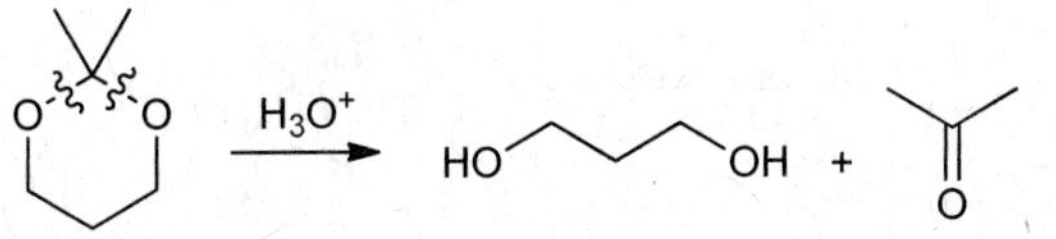

(d) Begin by identifying the carbon atom that is connected to two oxygen atoms (highlighted):

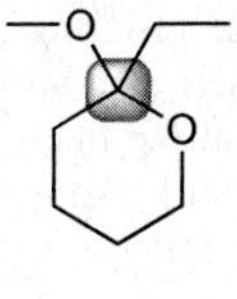

This carbon atom bears an acetal group, and will ultimately be converted into a carbonyl group in the product. The C-O bonds of the acetal group are cleaved, giving a ketone and two alcohol groups. Note that one of the alcohol groups is tethered to the ketone group (because one of the C-O bonds of the acetal group is contained in the ring, and the other C-O bond of the acetal group is not contained in the ring).

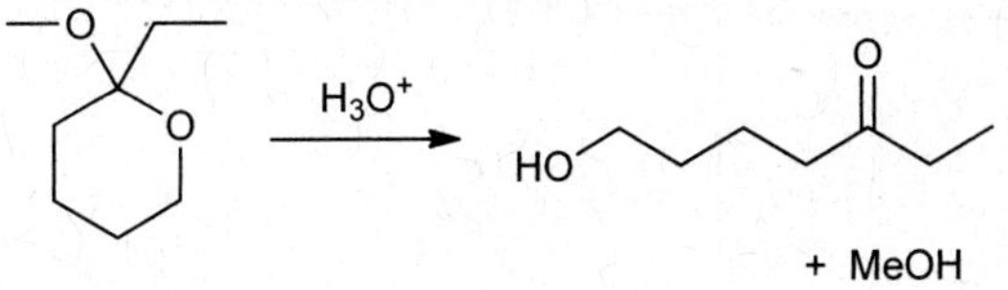

20.13. The starting material contains both a carbonyl group and an OH group, and the product is a hemiacetal. As seen in Mechanism 20.5, formation of a hemiacetal requires three steps. First, the carbonyl is protonated, rendering it more electrophilic. Then, the alcohol group attacks the protonated carbonyl group (in an intramolecular fashion), giving an oxonium ion, which is deprotonated to give the hemiacetal, as shown:

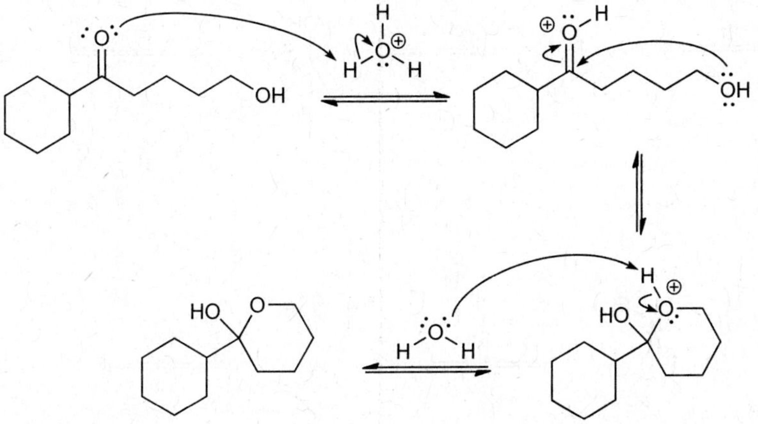

20.14. The molecular formula indicates two degrees of unsaturation. Since the product is a hemiacetal, the starting material must contain both a carbonyl group and an OH group. The carbon atom of the acetal group must have been the location of the carbonyl group. The OH group (that attacks the carbonyl group to form the hemiacetal) must be contained in the structure of compound A, which explains why a cyclic hemiacetal is formed (much like we saw in Problem **20.13**).

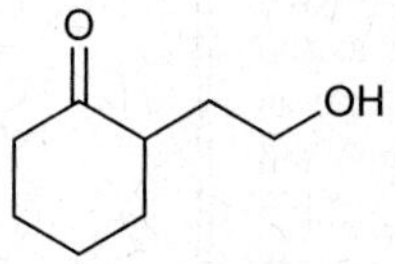

Compound A

20.15.

(a) As seen in Mechanism 20.6, imine formation proceeds via six steps. First, the amine functions as a nucleophile and attacks the ketone. The resulting intermediate is then protonated to remove the negative charge (note that the likely proton source under these conditions is a protonated amine, called an ammonium ion). Proton transfer then gives a carbinolamine (note that the likely base for this step is a molecule of the amine). Protonation then converts the bad leaving group (hydroxide) into a good leaving group (water). Loss of water then gives an iminium ion, which is deprotonated to give an imine.

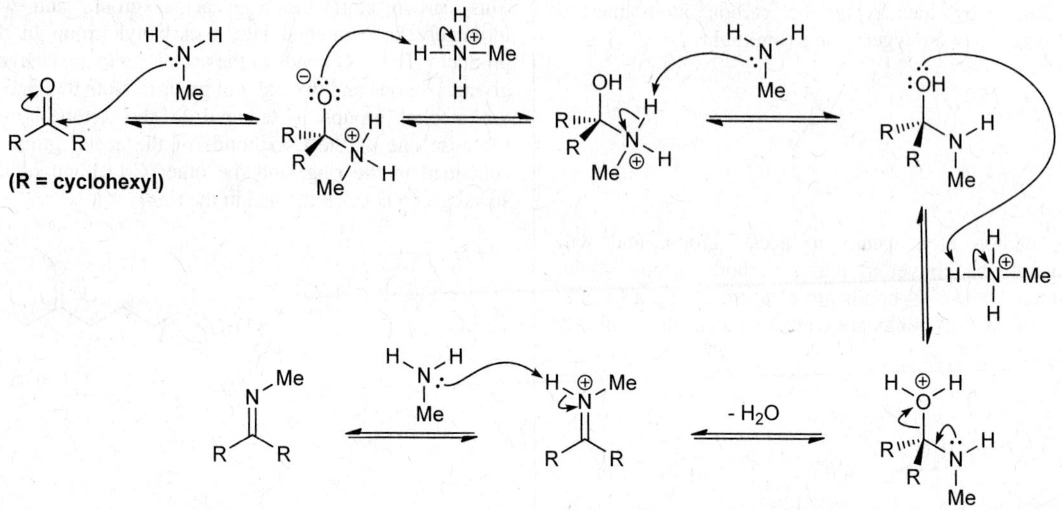

(b) As seen in the solution to Problem **20.15a**, imine formation proceeds via six steps. The same six steps are drawn again here: 1) nucleophilic attack, 2) proton transfer, 3) proton transfer, 4) proton transfer, 5) loss of a leaving group, and 6) proton transfer.

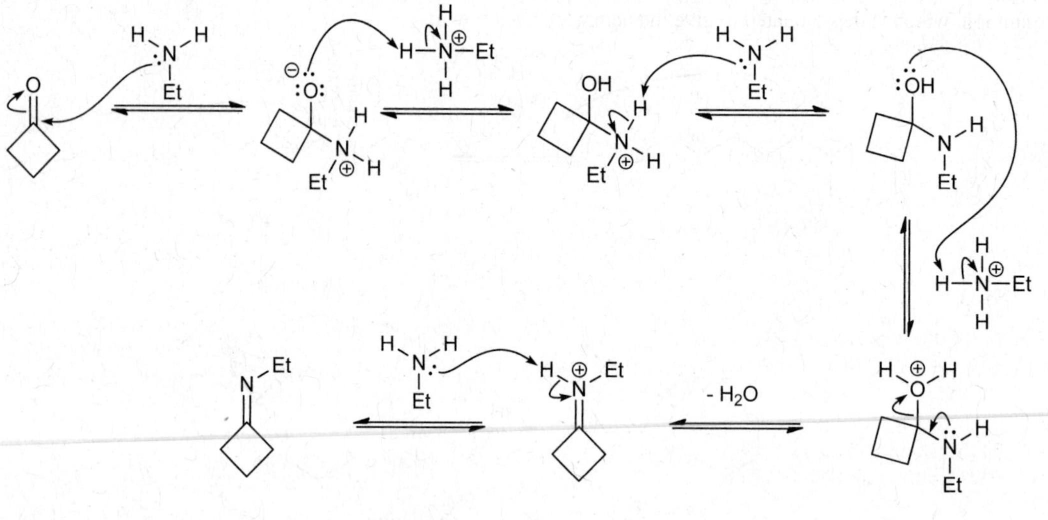

20.16.

(a) When treated with ammonia under acid-catalyzed conditions (with removal of water), a ketone is converted to an imine. In this case, ammonia (NH_3) is used as the nucleophile (rather than a primary amine, RNH_2), and as a result, the product will not have an R group connected to the nitrogen atom of the imine group. Instead, there will be a proton in that location (two of the protons from NH_3 are removed during imine formation, while the third proton remains connected to the nitrogen atom in the product).

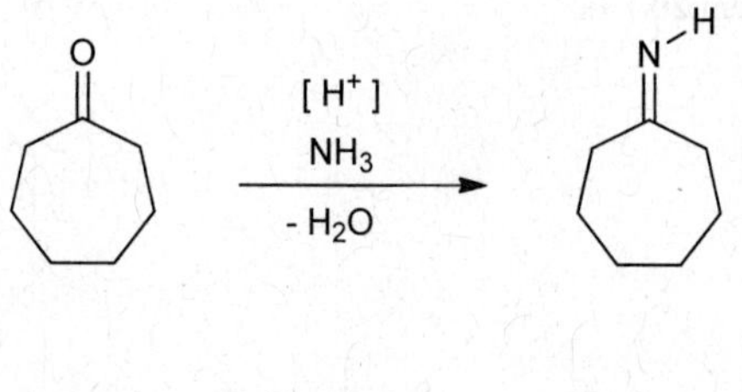

CHAPTER 20   721

(b) Upon treatment with a primary amine under acid-catalyzed conditions (with removal of water), a ketone is converted to an imine. The carbon atom of the carbonyl group ends up being connected to the nitrogen atom (with a double bond). The oxygen atom is removed, together with the two protons connected to the nitrogen atom of the amine (thus, loss of water), to give the following product:

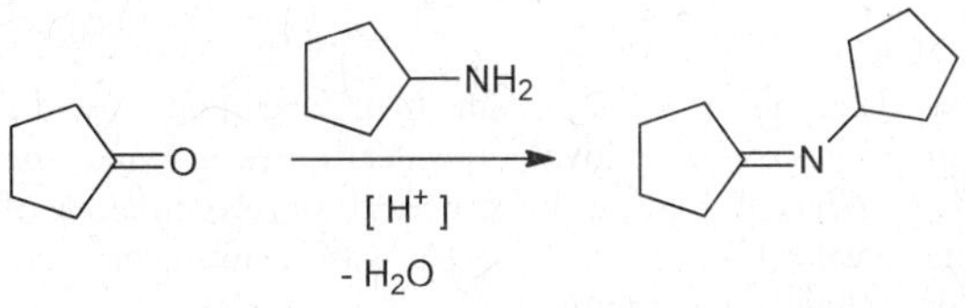

20.17.
(a) Under acid-catalyzed conditions (with removal of water), an aldehyde will react with a primary amine to give an imine. In this case, the aldehyde group and the amino group are tethered to each other (in the same molecule), and the reaction therefore occurs in an intramolecular fashion. The carbon atom of the carbonyl group ends up being connected to the nitrogen atom (with a double bond), thereby closing a five-membered ring (four carbon atoms and one nitrogen atom). The oxygen atom is removed, together with the two protons connected to the nitrogen atom of the amino group (thus, loss of water), to give the product shown. Note that locants have been assigned to help draw the product. This problem illustrates how useful locants can be when drawing the products of a reaction, especially when the reaction involves formation of a ring.

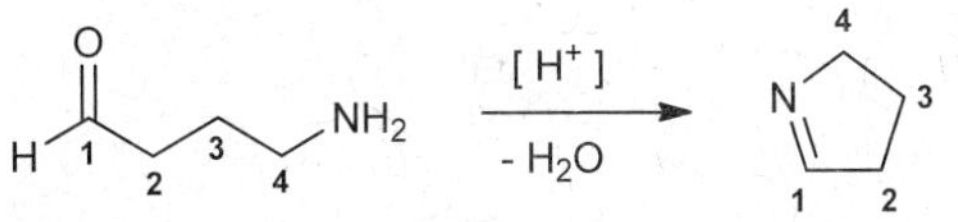

(b) This problem is very similar to the previous problem (**20.17a**), although there is a ketone group (rather than an aldehyde group), and there are two other methyl groups connected to the middle of the chain. To ensure that all substituents are drawn in the correct locations, we assign locants, much as we did in the solution to **20.17a**. This allows us to see that the methyl groups are placed at C4 in the product. Remember that the carbon atom of the carbonyl group ends up being connected to the nitrogen atom (with a double bond).

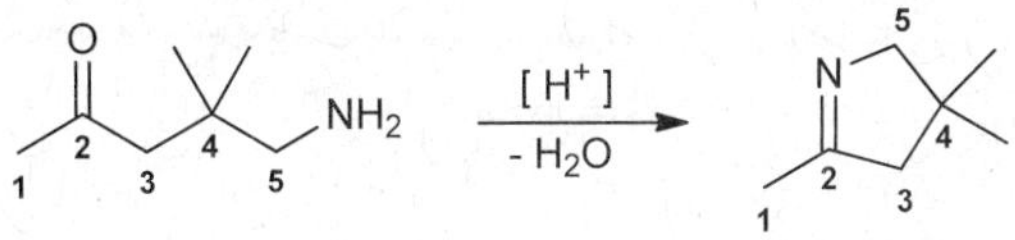

20.18.
(a) An imine can be prepared from a ketone (or aldehyde) and a primary amine. To determine the starting ketone and starting amine that must be used, we consider the location of the C=N bond in the imine. The carbon atom of this bond must have been a carbonyl group in the starting ketone, and the nitrogen atom must have been connected to two protons in the starting amine. So, the following starting materials should give the desired imine (under acid-catalyzed conditions with removal of water):

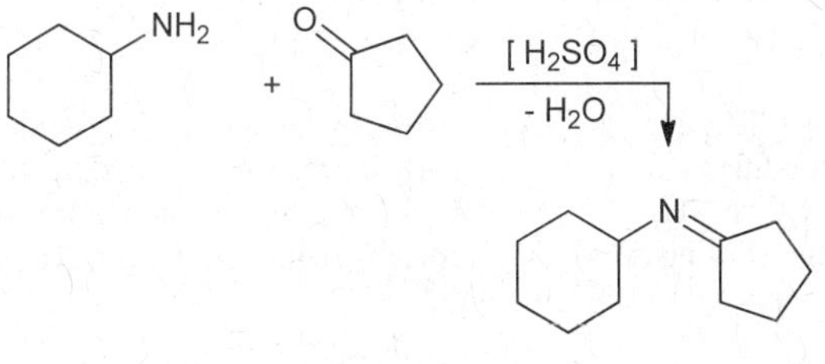

(b) An imine can be prepared from a ketone (or aldehyde) and a primary amine. To determine the starting ketone and starting amine that must be used, we consider the location of the C=N bond in the imine. The carbon atom of this bond must have been a carbonyl group in the starting ketone, and the nitrogen atom must have been connected to two protons in the starting amine. So, the following starting materials should give the desired imine (under acid-catalyzed conditions with removal of water):

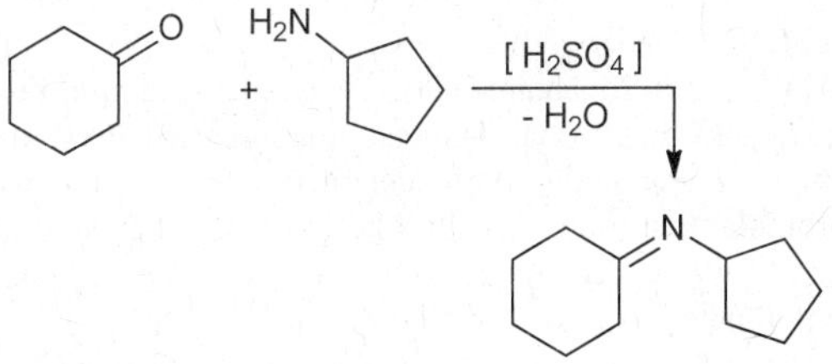

(c) An imine can be prepared from a ketone (or aldehyde) and a primary amine. To determine the starting ketone and starting amine that must be used, we consider the location of the C=N bond in the imine. The carbon atom of this bond must have been a carbonyl group in the starting ketone, and the nitrogen atom must have been connected to two protons in the starting amine. In this case, the C=N bond is incorporated into a ring, which means that the starting ketone and the starting amine must have been tethered to each other in the starting material (thereby closing into a cyclic imine). That is, imine formation must occur in an intramolecular fashion (under acid-catalyzed conditions with removal of water):

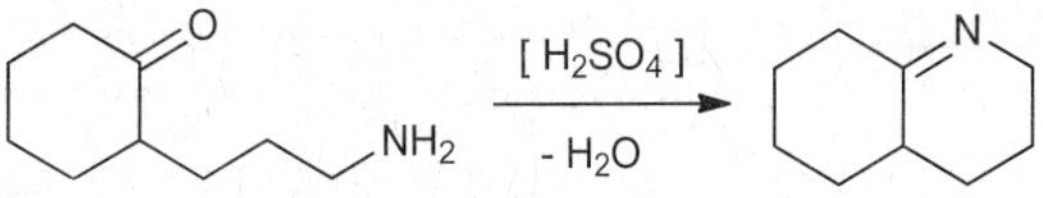

20.19.
(a) Treating a ketone with hydroxylamine under acidic conditions (with removal of water) gives an oxime ($R_2C=NOH$). During this process, the oxygen atom of

the starting ketone is replaced with a nitrogen atom connected to an OH group:

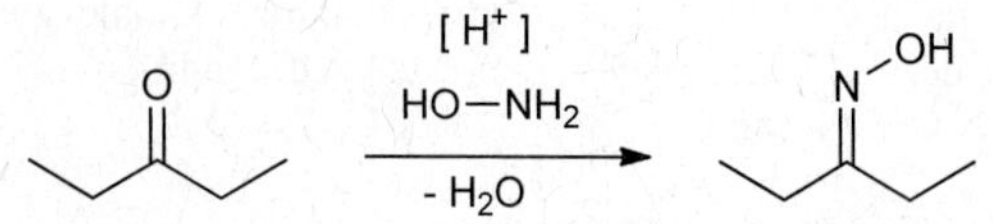

(b) Treating a ketone with hydrazine under acidic conditions (with removal of water) gives a hydrazone ($R_2C=NNH_2$). During this process, the oxygen atom of the starting ketone is replaced with a nitrogen atom connected to an NH_2 group:

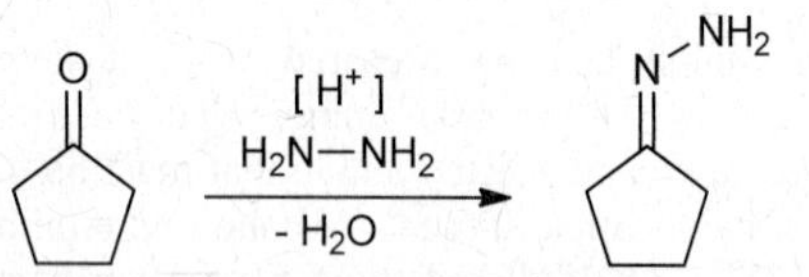

20.20.
(a) An oxime can be made by treating the corresponding ketone with hydroxyl amine, in acid catalyzed conditions (with removal of water), as shown.

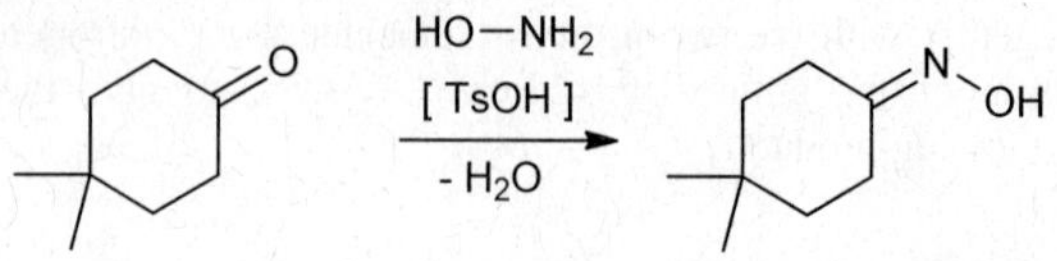

(b) Treating a ketone with hydrazine under acidic conditions (with removal of water) gives a hydrazone ($R_2C=NNH_2$). During this process, the oxygen atom of the starting ketone is replaced with a nitrogen atom connected to an NH_2 group:

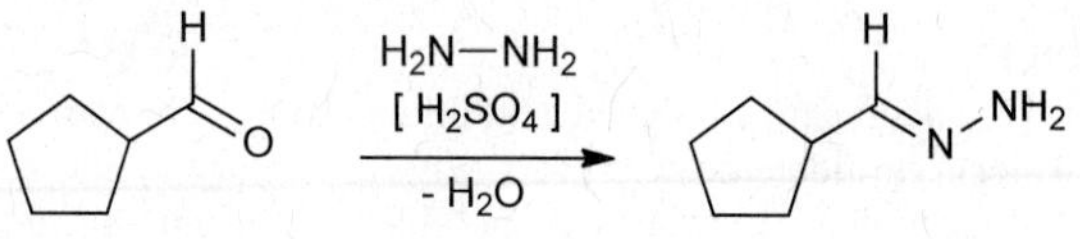

20.21.
(a) As seen in Mechanism 20.7, enamine formation proceeds via six steps. First, the amine functions as a nucleophile and attacks the ketone. The resulting intermediate is then protonated to remove the negative charge (note that the likely proton source under these conditions is a protonated amine, or ammonium ion). Proton transfer then gives a carbinolamine (note that the likely base for this step is a molecule of the amine). Protonation then converts the bad leaving group (hydroxide) into a good leaving group (water). Loss of water then gives an iminium ion, which is deprotonated to give an enamine.

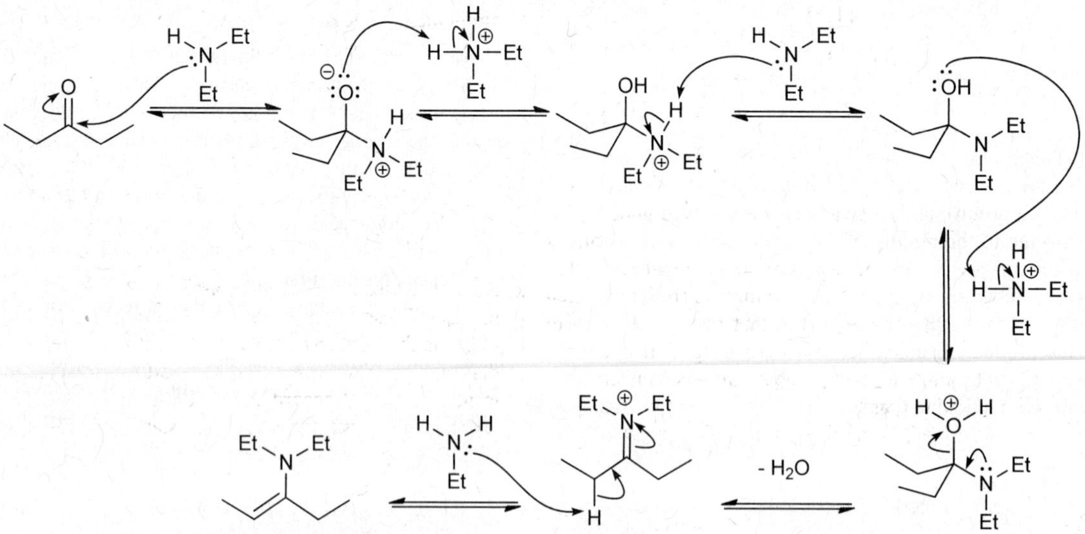

(b) As seen in Mechanism 20.7, enamine formation proceeds via six steps. First, the amine functions as a nucleophile and attacks the ketone. The resulting intermediate is then protonated to remove the negative charge (note that the likely proton source under these conditions is a protonated amine, or ammonium ion). Proton transfer then gives a carbinolamine (note that the likely base for this step is a molecule of the amine). Protonation then converts the bad leaving group (hydroxide) into a good leaving group (water). Loss of water then gives an iminium ion, which is deprotonated to give an enamine.

20.22.

(a) Upon treatment with a secondary amine under acid-catalyzed conditions (with removal of water), a ketone is converted to an enamine. The carbon atom of the carbonyl group ends up being connected to the nitrogen atom (with a single bond), and that same carbon atom ends up having a double bond to a neighboring carbon atom, as shown. In the process, the oxygen atom is removed altogether, as well as two protons (one from the nitrogen atom and the other from a carbon atom) to give loss of water.

(b) Upon treatment with a secondary amine under acid-catalyzed conditions (with removal of water), a ketone is converted to an enamine. The carbon atom of the carbonyl group ends up being connected to the nitrogen atom (with a single bond), and that same carbon atom ends up having a double bond to a neighboring carbon atom, as shown. In the process, the oxygen atom is removed altogether, as well as two protons (one from the nitrogen atom and the other from a carbon atom) to give loss of water.

20.23.

(a) Under acid-catalyzed conditions (with removal of water), a ketone will react with a secondary amine to give an enamine. In this case, the carbonyl group and the amino group are tethered to each other (in the same molecule), and therefore, the reaction occurs in an intramolecular fashion, forming a five-membered ring (four carbon atoms and one nitrogen atom). The carbon atom of the carbonyl group ends up being connected to the nitrogen atom (with a single bond). That same carbon atom ends up having a double bond to a neighboring carbon atom. The oxygen atom is removed altogether, as well as two protons (one from the nitrogen atom and the other from a carbon atom) to give loss of water.

Note that locants have been assigned to help draw the product. These locants do not have to conform to IUPAC rules, as they are just helpful tools that were are using to draw the product correctly. The starting material above is not numbered according to IUPAC rules but that is OK, as long as we don't name the starting material using these incorrect locants.

(b) Under acid-catalyzed conditions (with removal of water), a ketone will react with a secondary amine to give an enamine. In this case, the carbonyl group and the amino group are tethered to each other (in the same molecule), and therefore, the reaction occurs in an intramolecular fashion, forming a five-membered ring (four carbon atoms and one nitrogen atom). The carbon atom of the carbonyl group ends up being connected to the nitrogen atom (with a single bond). That same carbon atom ends up having a double bond to a neighboring carbon atom. The oxygen atom is removed altogether, as well as two protons (one from the nitrogen atom and the other from a carbon atom) to give loss of water. Locants have been assigned to help draw the product.

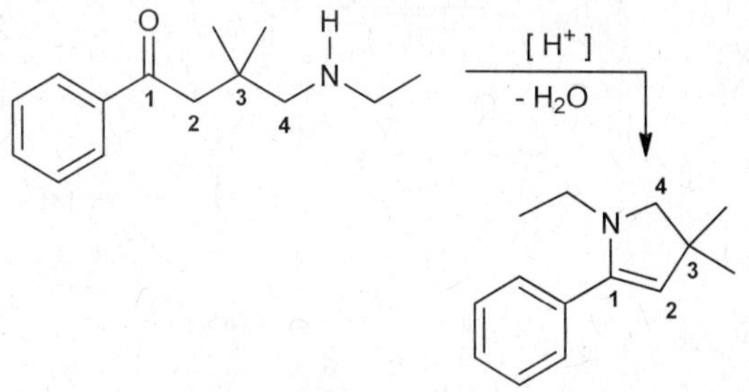

20.24.
(a) An enamine can be prepared from a ketone (or aldehyde) and a secondary amine. To determine the starting ketone and starting amine that must be used, we consider the location of the vinylic carbon atom (highlighted) connected directly to nitrogen:

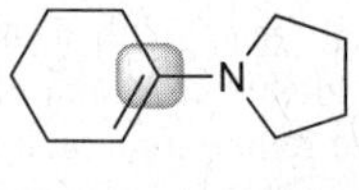

This carbon atom must have been a carbonyl group in the starting ketone, and the nitrogen atom must have been connected to one proton in the starting amine. So, the following starting materials should give the desired enamine (under acid-catalyzed conditions with removal of water):

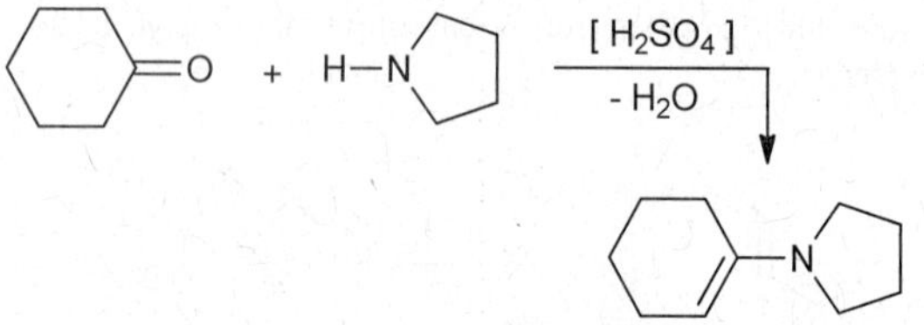

(b) An enamine can be prepared from a ketone (or aldehyde) and a secondary amine. To determine the starting ketone and starting amine that must be used, we consider the location of the vinylic carbon atom (highlighted) connected directly to nitrogen:

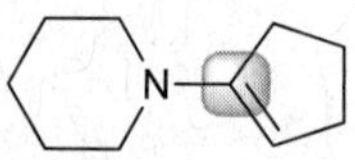

This carbon atom must have been a carbonyl group in the starting ketone, and the nitrogen atom must have been connected to one proton in the starting amine. So, the following starting materials should give the desired enamine (under acid-catalyzed conditions with removal of water):

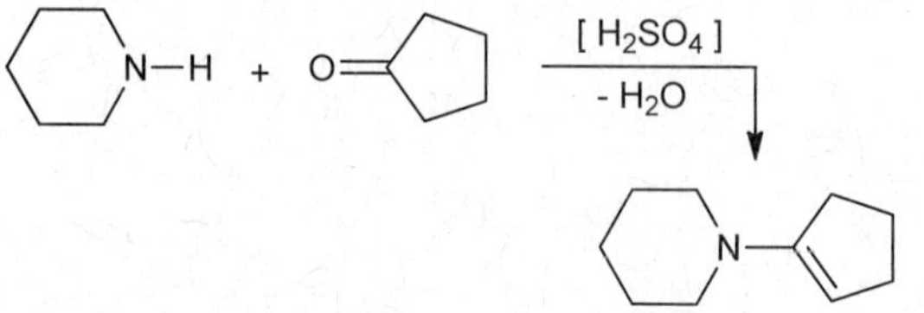

(c) An enamine can be prepared from a ketone (or aldehyde) and a secondary amine. To determine the starting ketone and starting amine that must be used, we consider the location of the vinylic carbon atom (highlighted) connected directly to nitrogen:

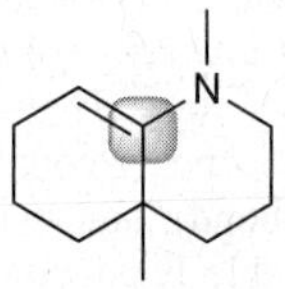

This carbon atom must have been a carbonyl group in the starting ketone, and the nitrogen atom must have been connected to one proton in the starting amine. In this case, the C-N bond of the enamine group is incorporated into a ring, which means that the starting ketone and the starting amine must have been tethered to each other in the starting material (thereby closing into a cyclic enamine). That is, enamine formation must occur in an intramolecular fashion (under acid-catalyzed conditions with removal of water):

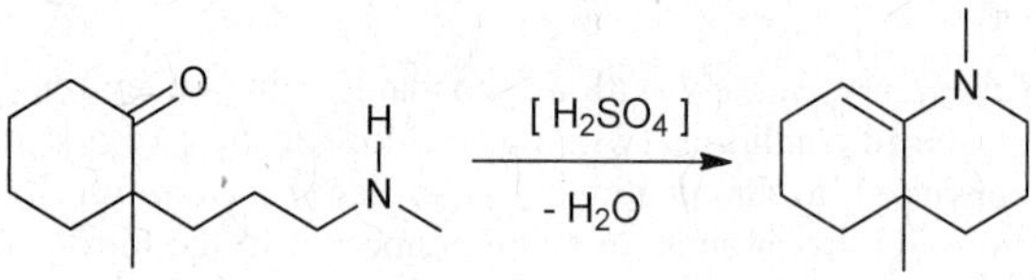

20.25. The starting material is a ketone, and the reagents indicate formation of a hydrazone, followed by a Wolff-Kishner reduction. The net result is complete reduction of the carbonyl group to a methylene group (CH_2):

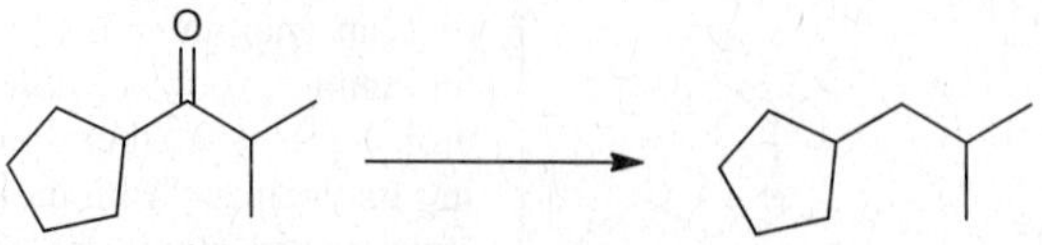

Formation of the hydrazone occurs via a six-step process (Mechanism 20.6). First, hydrazine functions as a nucleophile and attacks the ketone. The resulting intermediate is then protonated to remove the negative charge (note that the likely proton source under these conditions is a protonated hydrazine). Proton transfer then gives a carbinolamine (note that the likely base for this step is hydrazine). Protonation then converts the bad leaving group (hydroxide) into a good leaving group (water). Loss of water then gives an iminium ion, which is deprotonated to give the hydrazone.

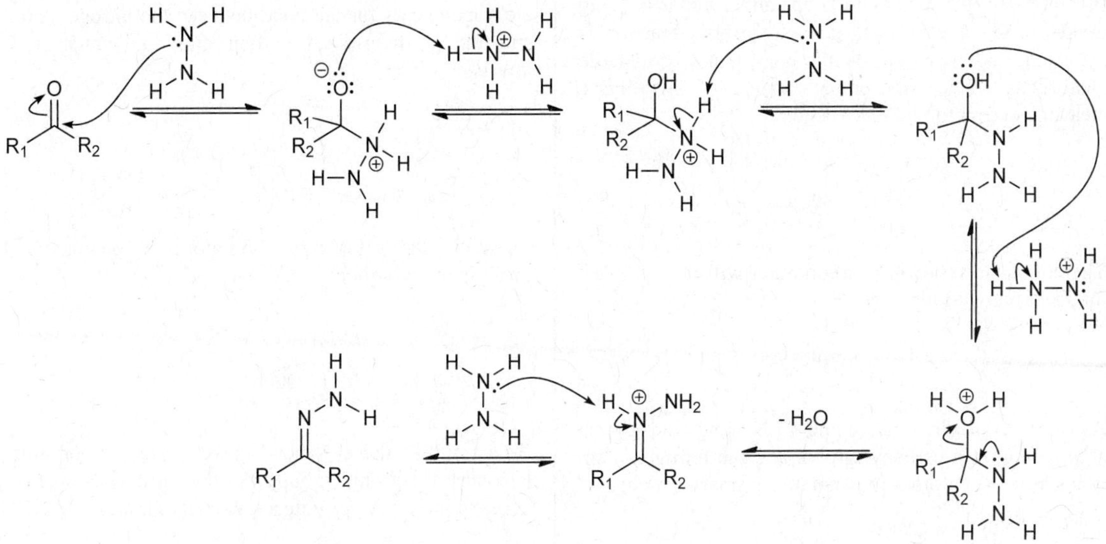

And then, under basic conditions, the hydrazone is reduced via the process shown here. As seen in Mechanism 20.8, three successive proton transfer steps, followed by loss of nitrogen gas, gives a carbanion, which is protonated by water to give the product:

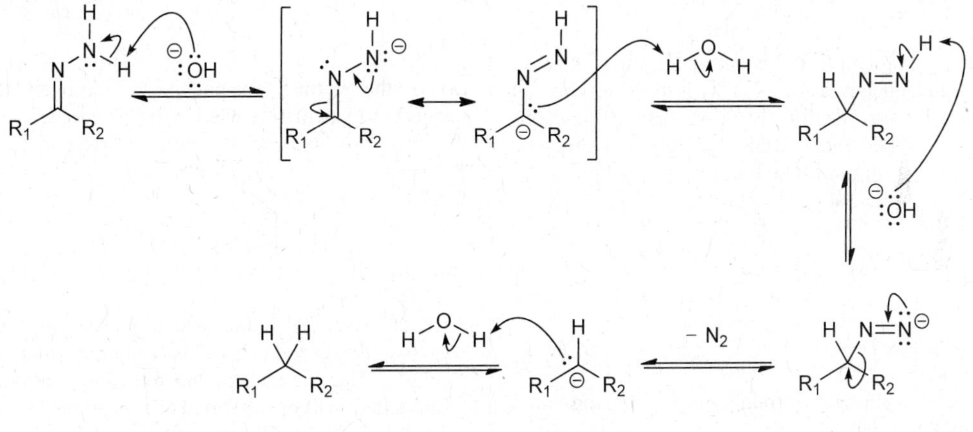

20.26.

(a) The starting compound is an acetal, and it is being treated with aqueous acid, so a hydrolysis reaction is expected. We first identify the bonds that will undergo cleavage. When an acetal undergoes hydrolysis, cleavage occurs for the C-O bonds of the acetal group:

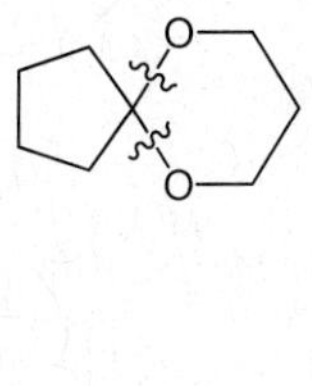

Therefore, the following carbon atom will be converted into a carbonyl group:

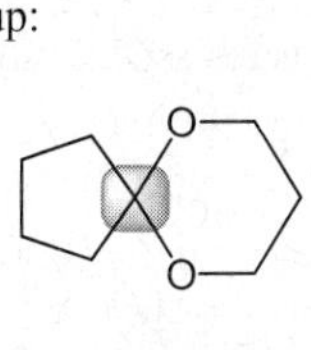

In the process, a diol is released, as shown:

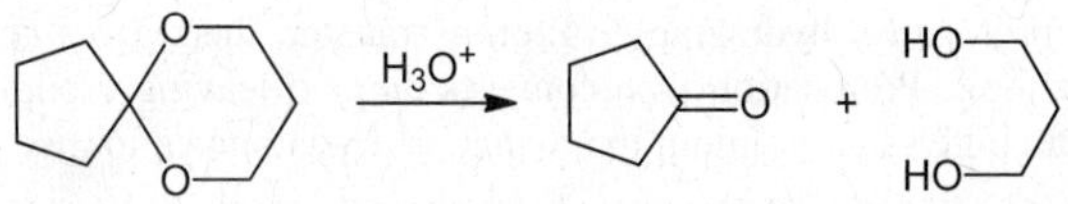

(b) The starting compound is an imine, and it is being treated with aqueous acid, so a hydrolysis reaction is expected. We first identify the bond that will undergo cleavage. When an imine undergoes hydrolysis, cleavage occurs for the C=N bond:

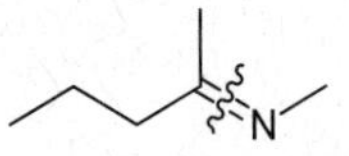

Therefore, the following carbon atom will be converted into a carbonyl group:

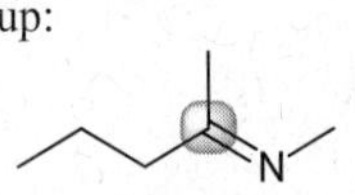

As a result of cleavage of the C=N bond, the carbon atom becomes a carbonyl group, and the nitrogen atom will accept two protons to generate a primary amine:

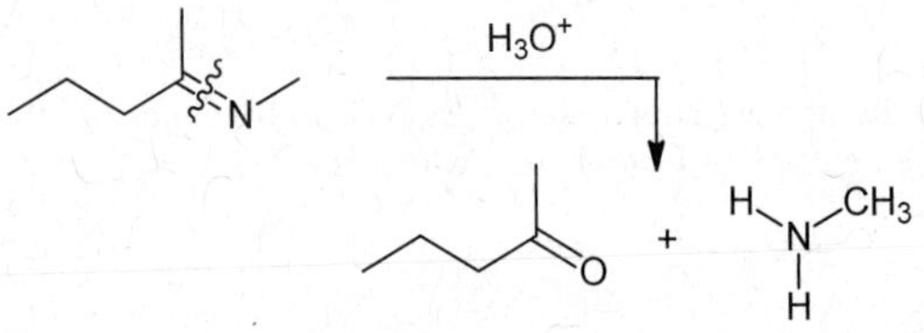

(c) The starting compound is an acetal, and it is being treated with aqueous acid, so a hydrolysis reaction is expected. We first identify the bonds that will undergo cleavage. When an acetal undergoes hydrolysis, cleavage occurs for the C-O bonds of the acetal group:

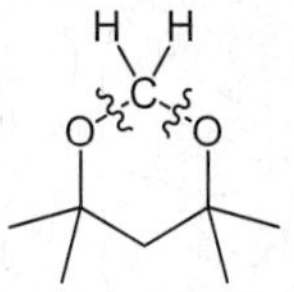

Therefore, the following carbon atom will be converted into a carbonyl group:

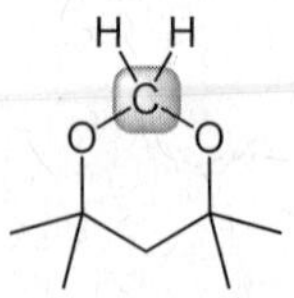

In the process, a diol is released, as shown:

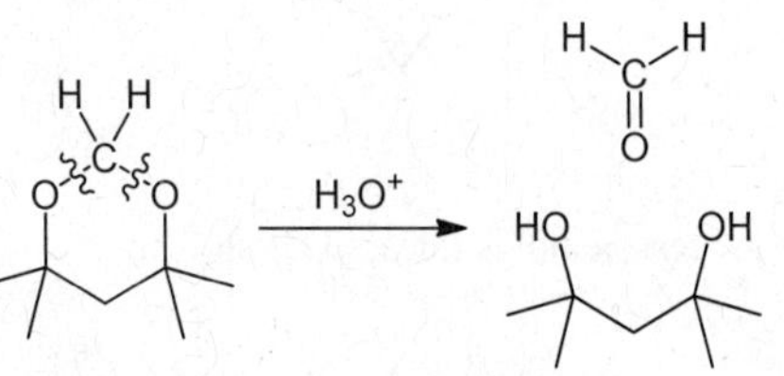

(d) The starting compound is an enamine, and it is being treated with aqueous acid, so a hydrolysis reaction is expected. We first identify the bond that will undergo cleavage. When an enamine undergoes hydrolysis, cleavage occurs for the bond between the nitrogen atom and the sp^2-hybridized carbon atom to which it is attached:

Therefore, the following carbon atom will be converted into a carbonyl group:

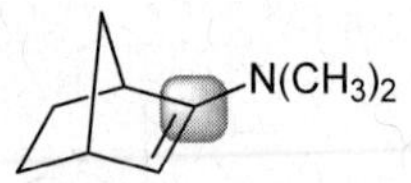

As a result of the C-N bond cleavage, the carbon atom becomes a carbonyl group, and the nitrogen atom will accept a proton to generate a secondary amine:

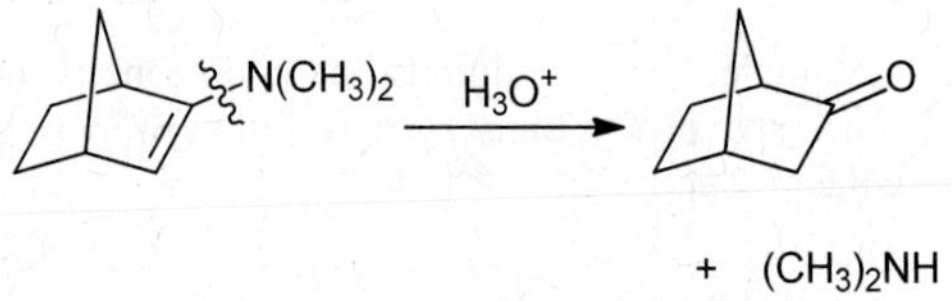

20.27.
(a) In the starting compound, the nitrogen atom is connected to two different vinylic positions. As such, there are two C-N bonds that will undergo hydrolysis:

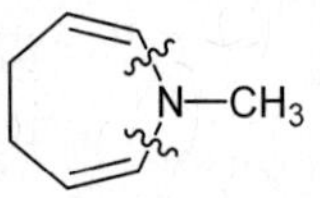

Each of these C-N bonds will be cleaved in the same way that the C-N bond of an enamine group can be cleaved. That is, each of the following carbon atoms (highlighted) will be converted into a carbonyl group:

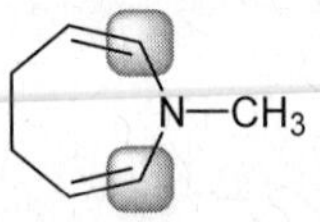

In the process, an amine is released, as shown:

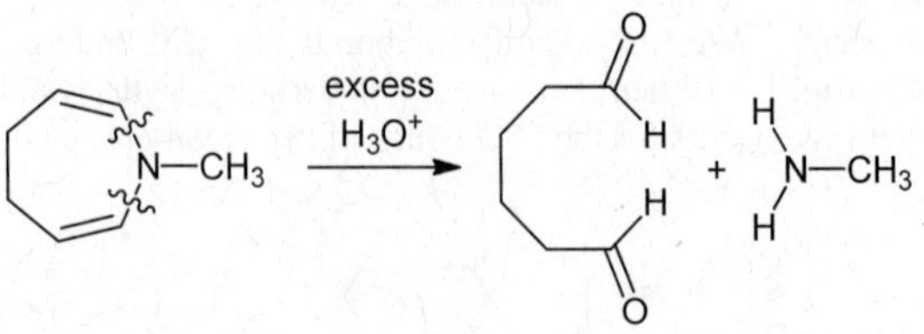

(b) The starting compound exhibits a carbon atom that is connected to two oxygen atoms, as shown, and is therefore an acetal:

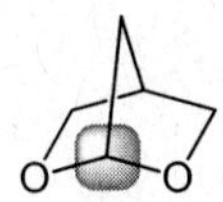

When an acetal is treated with aqueous acid, it is expected to undergo hydrolysis. We first identify the bonds that will undergo cleavage. When an acetal undergoes hydrolysis, cleavage occurs for the C-O bonds of the acetal group:

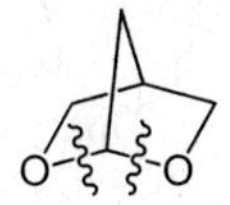

Each of these bonds is broken, thereby converting the carbon atom of the acetal group into a carbonyl group. In the process, each of the oxygen atoms will accept a proton to become an OH group, giving the following product, which exhibits a carbonyl group, as well as two OH groups:

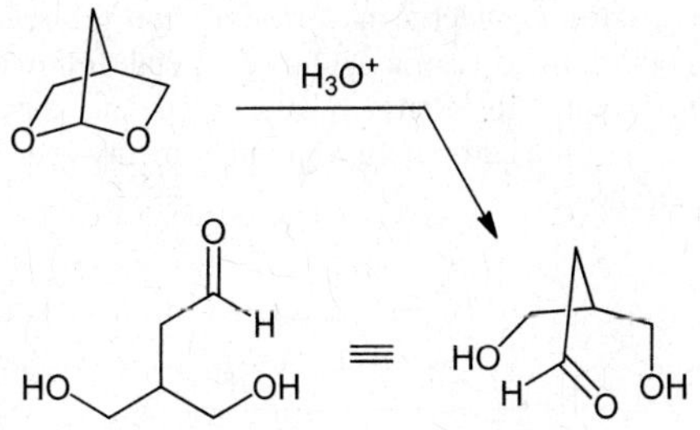

20.28. In aqueous acidic conditions, one of the nitrogen atoms is protonated to give an ammonium ion. Loss of a leaving group then generates an iminium ion, which is attacked by water to give an oxonium ion. Loss of a proton then gives an alcohol (the likely base is water). Protonation of the neighboring nitrogen atom gives a leaving group, which leaves, thereby expelling a fragment with one carbon atom. A proton transfer provides the first equivalent of formaldehyde. The other equivalents are formed in a similar way.

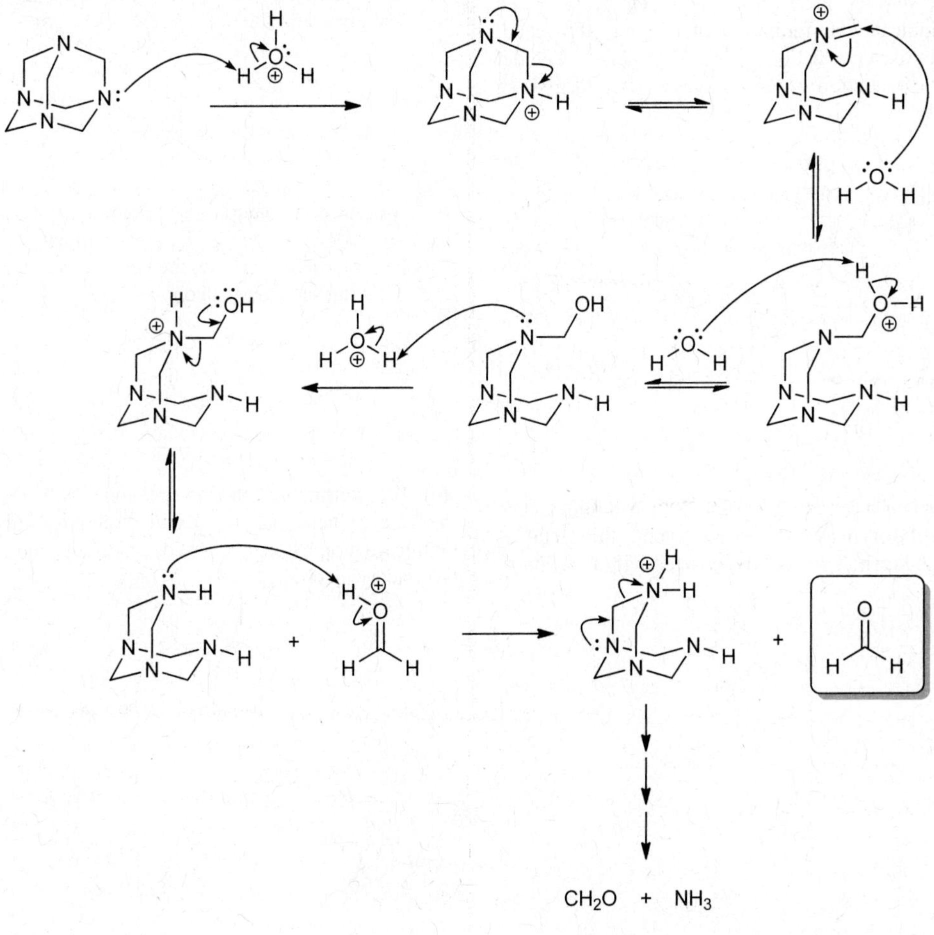

$$CH_2O \quad + \quad NH_3$$

20.29.

(a) The starting material is a ketone, and the reagents indicate the formation of a cyclic thioacetal, followed by desulfurization with Raney nickel. The net result is conversion of the carbonyl group into a methylene (CH$_2$) group, as shown.

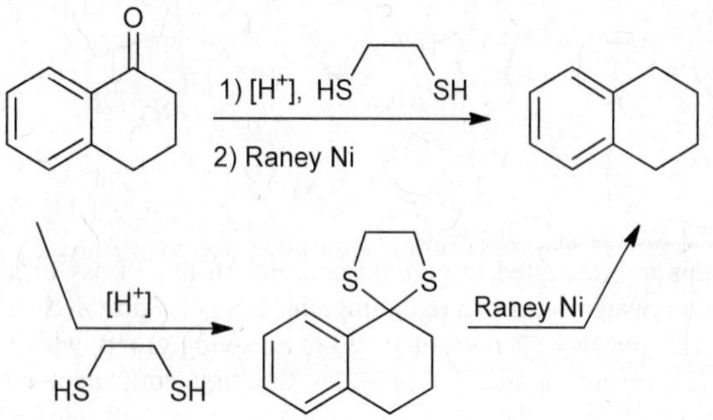

(b) The starting material is an aldehyde, and the reagents indicate the formation of a cyclic thioacetal, followed by desulfurization with Raney nickel. The net result is the reduction of the carbonyl group to give an alkane.

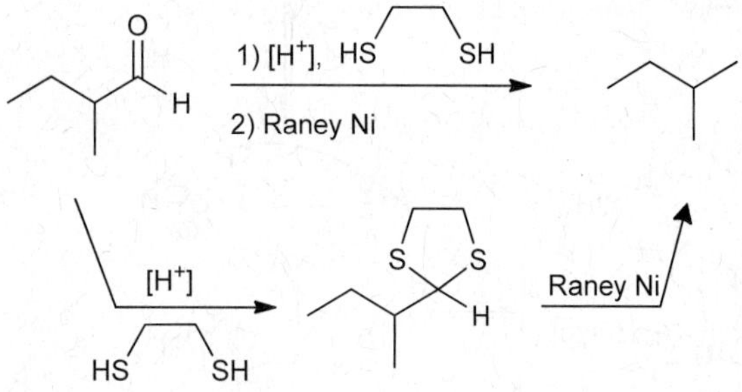

20.30. When a ketone is treated with a dithiol in the presence of an acid catalyst, a cyclic thioacetal is formed. The carbonyl group is replaced with C-S bonds, as shown:

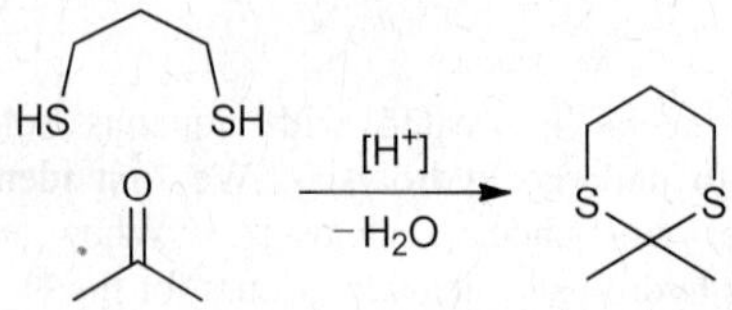

20.31.

(a) The starting material is a ketone, and LAH is a hydride reducing agent. Upon treatment with LAH, followed by water work-up, the ketone is reduced to give the following secondary alcohol:

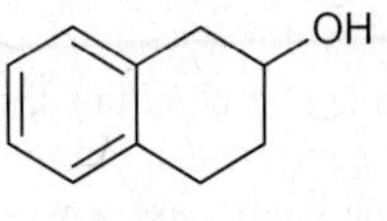

(b) The starting material is an aldehyde, and NaBH$_4$ is a hydride reducing agent. Upon treatment with NaBH$_4$ and methanol, the aldehyde is reduced to give the following primary alcohol:

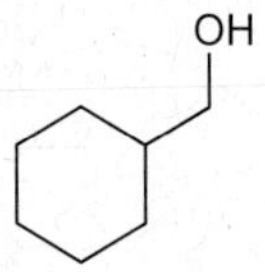

(c) The starting material is a ketone, and LAH is a hydride reducing agent. Upon treatment with LAH, followed by water work-up, the ketone is reduced to give the following secondary alcohol:

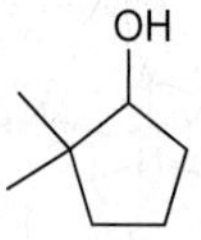

(d) The starting material is a ketone, and NaBH$_4$ is a hydride reducing agent. Upon treatment with NaBH$_4$ and methanol, the ketone is reduced to give the following secondary alcohol:

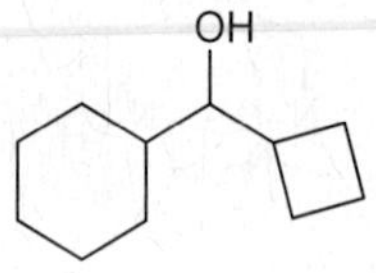

20.32.

(a) The following mechanism is consistent with the description given in the problem statement. After a hydroxide ion attacks one molecule of benzaldehyde, the resulting intermediate functions as a hydride delivery agent to attack another molecule of benzaldehyde, giving a carboxylic acid and an alkoxide ion. The alkoxide ion then deprotonates the carboxylic acid, generating a more stable carboxylate ion. When the reaction is complete, aqueous acid is added to the reaction flask in order to protonate the carboxylate ion, giving benzoic acid.

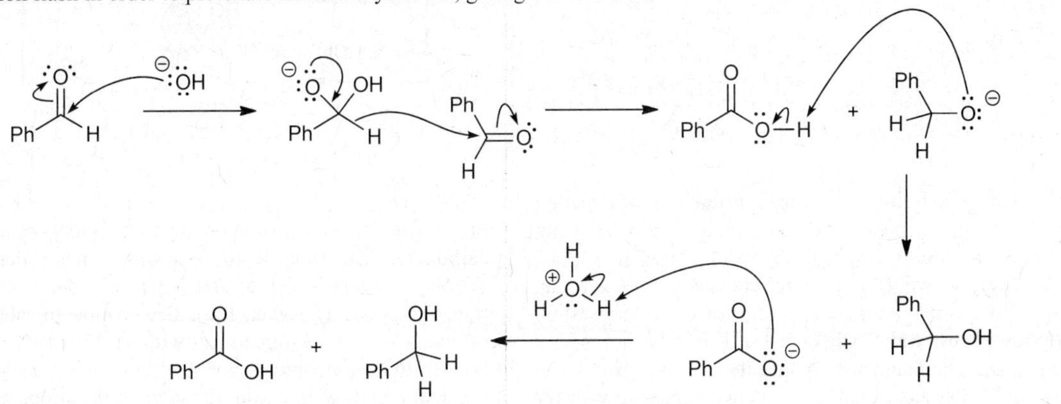

(b) The function of H_3O^+ in the second step is to serve as a proton source to protonate the carboxylate ion, giving benzoic acid.

(c) Water is only a weak acid ($pK_a = 15.7$), and is not sufficiently strong to serve as a proton source for a carboxylate ion (pK_a of PhCOOH is 4.2). See Section 3.5 for a discussion of this topic.

20.33.

(a) The starting material is a ketone, and ethyl magnesium bromide is a Grignard reagent (a strong nucleophile). Upon treatment with a Grignard reagent, followed by water work-up, the ketone is converted to a tertiary alcohol (with installation of an ethyl group):

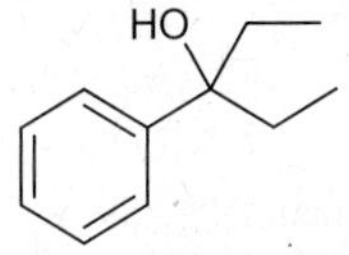

(b) The starting material is an aldehyde, and phenyl magnesium bromide is a Grignard reagent (a strong nucleophile). Upon treatment with a Grignard reagent, followed by water work-up, the aldehyde is converted to a secondary alcohol (with installation of a phenyl group):

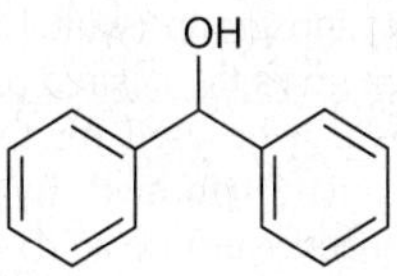

(c) The starting material is a ketone, and phenyl magnesium bromide is a Grignard reagent (a strong nucleophile). Upon treatment with a Grignard reagent, followed by water work-up, the ketone is attacked, giving the alkoxide ion shown. Treatment with aqueous acid then protonates the alkoxide ion (giving a tertiary

alcohol) and also hydrolyzes the acetal group, giving a ketone group:

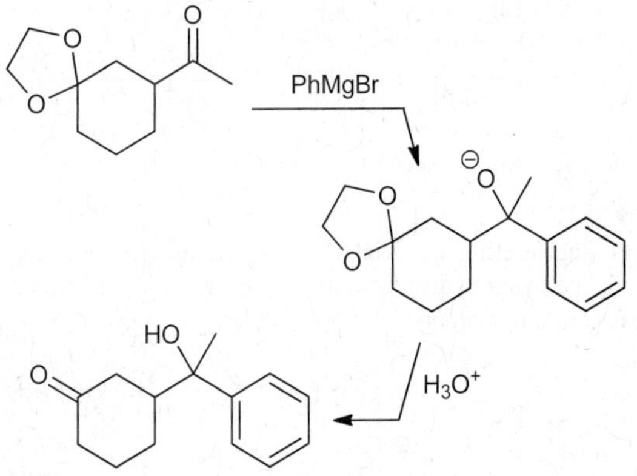

20.34.

(a) This transformation requires installation of a methyl group at the α position of an alcohol. We have not learned a direct way to achieve this transformation. However, if we first oxidize the alcohol to give a ketone, then treatment with methyl magnesium (followed by water work-up) will install the methyl group and simultaneously reduce the ketone back to an alcohol. The net result is the conversion of a secondary alcohol into a tertiary alcohol, as shown. A variety of oxidizing agents can be used for the oxidation step, including chromic acid (formed by mixing sodium dichromate with aqueous sulfuric acid).

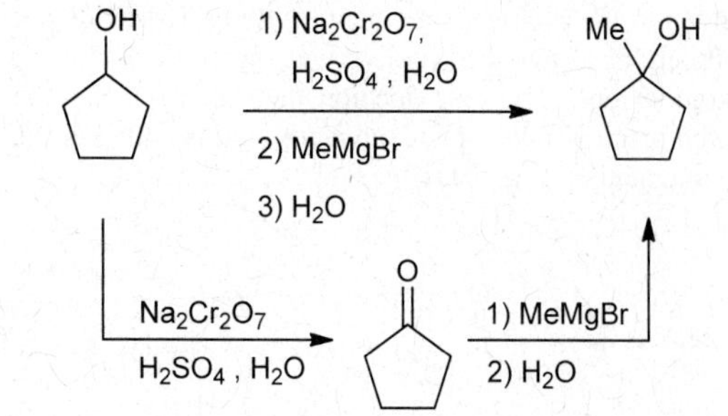

(b) This transformation requires installation of a methyl group at the α position of an alcohol. We have not learned a direct way to achieve this transformation. However, if we first oxidize the alcohol to give an aldehyde, then treatment with methyl magnesium (followed by water work-up) will install the methyl group and simultaneously reduce the aldehyde back to an alcohol. The net result is the conversion of a primary alcohol into a secondary alcohol, as shown. The oxidation step must be performed with PCC (rather than chromic acid), to give an aldehyde. The use of chromic acid would result in formation of a carboxylic acid, rather than an aldehyde.

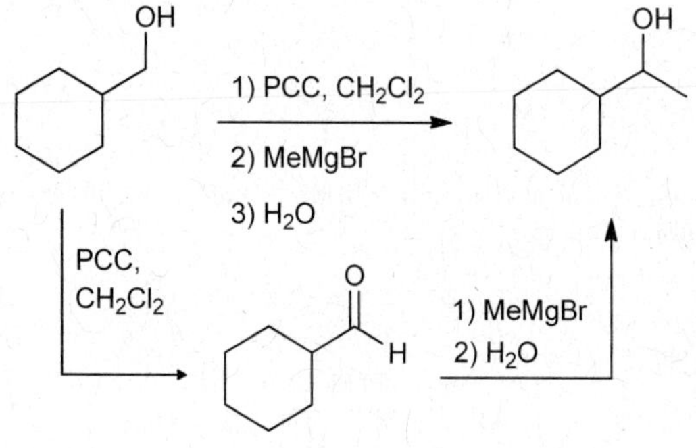

20.35.
(a) The starting material is a ketone, and the reagents indicate cyanohydrin formation, followed by reduction to give an amino alcohol.

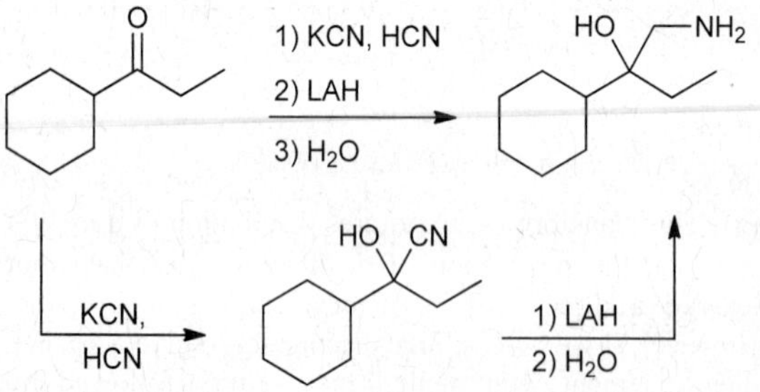

(b) The starting material is an aldehyde, and the reagents indicate cyanohydrin formation, followed by hydrolysis of the cyano group to give a carboxylic acid group.

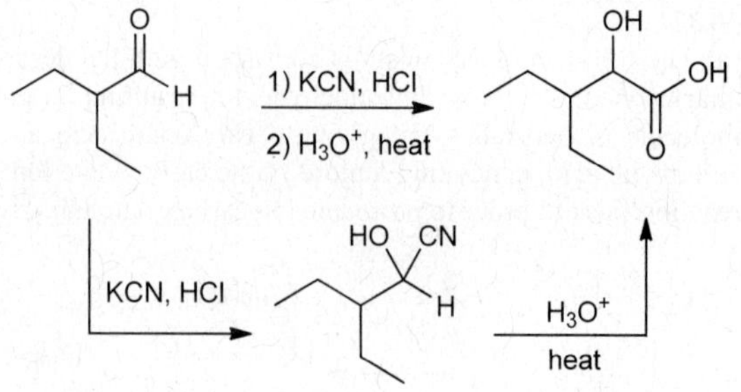

20.36.
(a) This transformation requires installation of a carboxylic acid group at the α position of an alcohol. We have not learned a direct way to achieve this transformation. However, if we first oxidize the alcohol to give a ketone, then treatment with KCN and HCl will convert the ketone into a cyanohydrin (thereby installing the extra carbon atom in the correct location, while simultaneously converting the carbonyl group back into an alcohol). Hydrolysis of the cyano group then gives the desired product. A variety of oxidizing agents can be used for the oxidation step, including chromic acid (formed by mixing sodium dichromate with aqueous sulfuric acid).

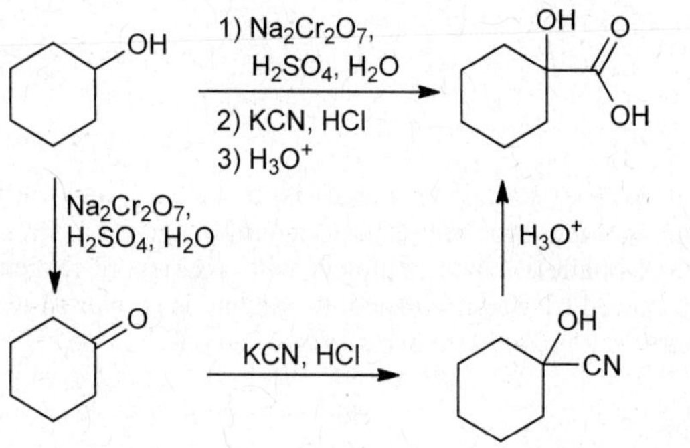

(b) This transformation requires installation of a CH$_2$NH$_2$ group at the α position of an alcohol. We have not learned a direct way to achieve this transformation. However, if we first oxidize the alcohol to give a ketone, then treatment with KCN and HCl will convert the ketone into a cyanohydrin (thereby installing the extra carbon atom in the correct location, while simultaneously converting the carbonyl group back into an alcohol). Reduction of the cyano group (with LAH, followed by water work-up) then gives the desired product. A variety of oxidizing agents can be used for the oxidation step, including chromic acid (formed by mixing sodium dichromate with aqueous sulfuric acid).

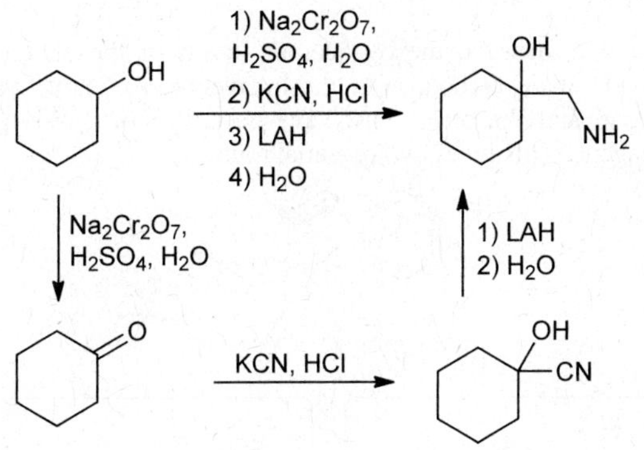

20.37.
(a) We focus on the two carbon atoms of the double bond. One carbon atom must have been a carbonyl group and the other must have been part of a Wittig reagent. This gives two potential routes:

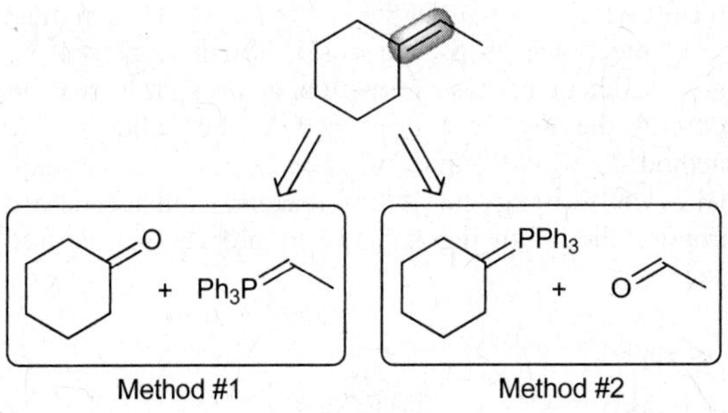

In each case, we focus on the Wittig reagent (which must be prepared via an S_N2 process). Method #1 will be more efficient because formation of the Wittig reagent requires the use of a primary alkyl halide, while method #2 would require the use of a secondary alkyl halide in forming the Wittig reagent. Therefore, we propose the following strategy to prepare the desired alkene:

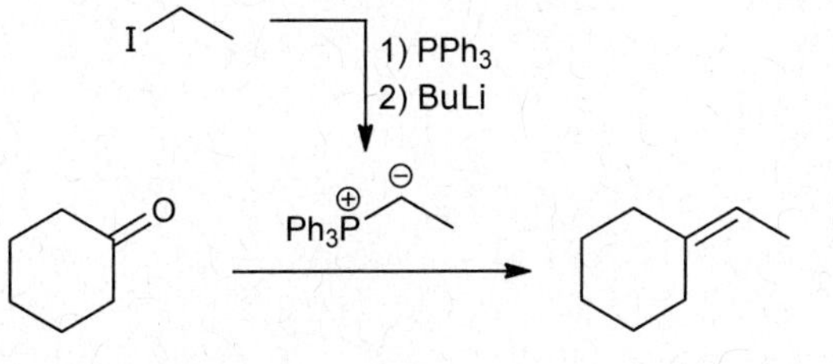

(b) We focus on the two carbon atoms of the double bond. One carbon atom must have been a carbonyl group and the other must have been part of a Wittig reagent. This gives two potential routes:

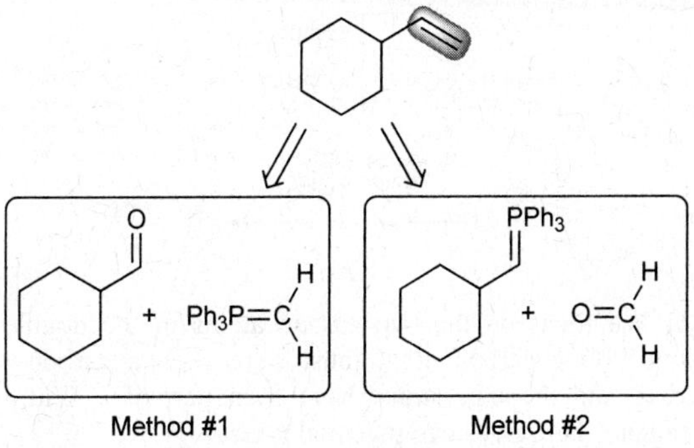

In each case, we focus on the Wittig reagent (which must be prepared via an S_N2 process). Method #1 will be more efficient because formation of the Wittig reagent requires the use of a methyl halide, while method #2 would require the use of a somewhat hindered primary alkyl halide in forming the Wittig reagent. Therefore, we propose the following strategy to prepare the desired alkene:

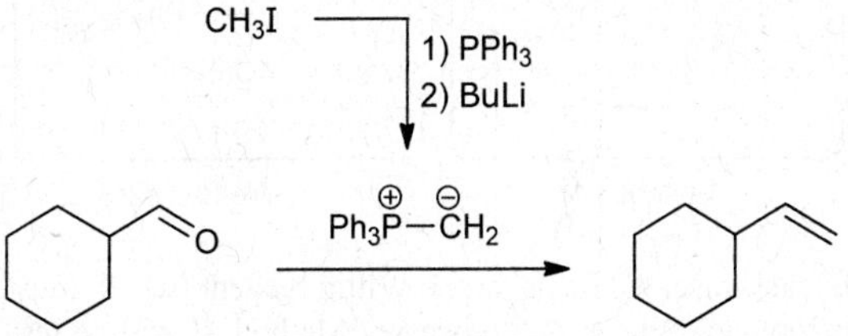

(c) We focus on the two carbon atoms of the double bond. One carbon atom must have been a carbonyl group and the other must have been part of a Wittig reagent. This gives two potential routes:

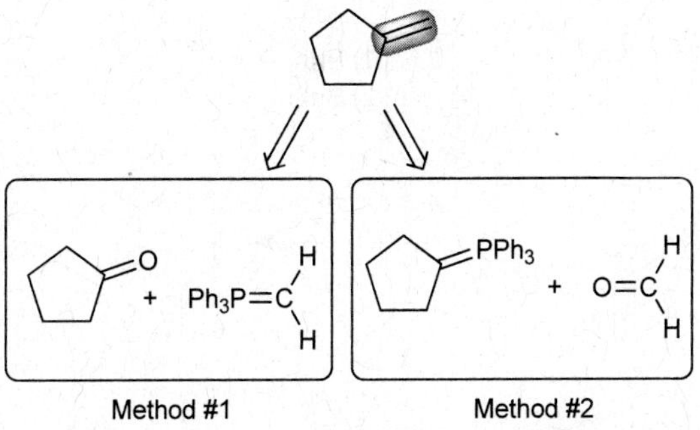

In each case, we focus on the Wittig reagent (which must be prepared via an S_N2 process). Method #1 will be more efficient because formation of the Wittig reagent requires the use of a methyl halide, while method #2 would require the use of a secondary alkyl halide in forming the Wittig reagent. Therefore, we propose the following strategy to prepare the desired alkene:

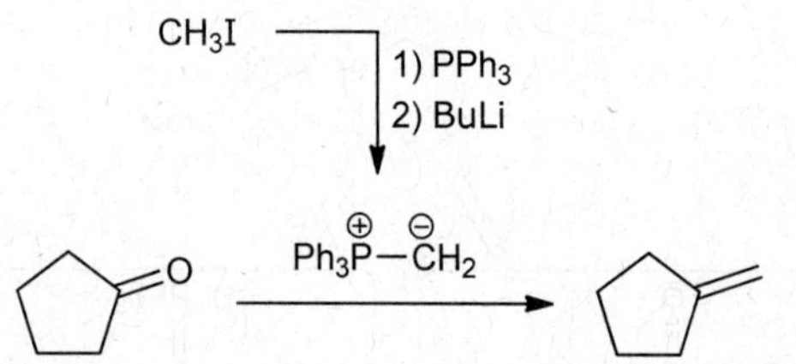

(d) We focus on the two carbon atoms of the double bond. One carbon atom must have been a carbonyl group and the other must have been part of a Wittig reagent. This gives two potential routes:

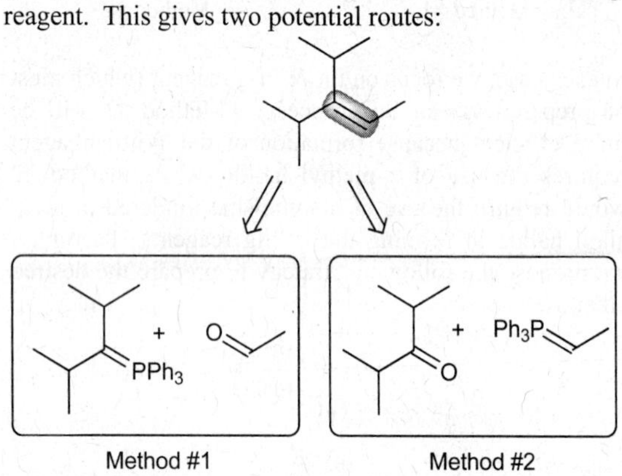

Method #1   Method #2

In each case, we focus on the Wittig reagent (which must be prepared via an S_N2 process). Method #1 will be less efficient because formation of the Wittig reagent requires the use of a secondary alkyl halide, while method #2 would require the use of a primary alkyl halide in forming the Wittig reagent. Therefore, we propose the following strategy to prepare the desired alkene:

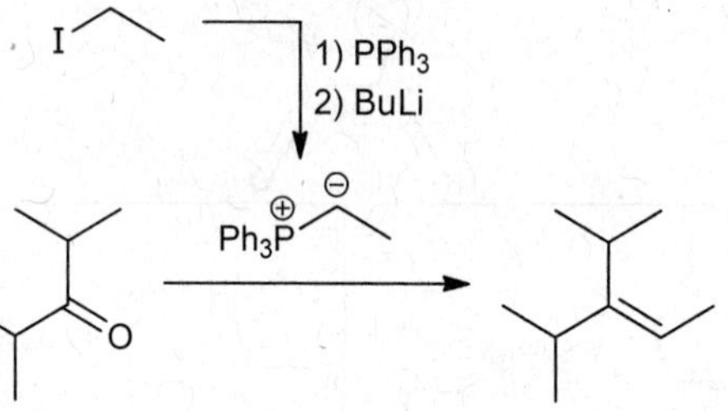

(e) We focus on the two carbon atoms of the double bond. One carbon atom must have been a carbonyl group and the other must have been part of a Wittig reagent. This gives two potential routes:

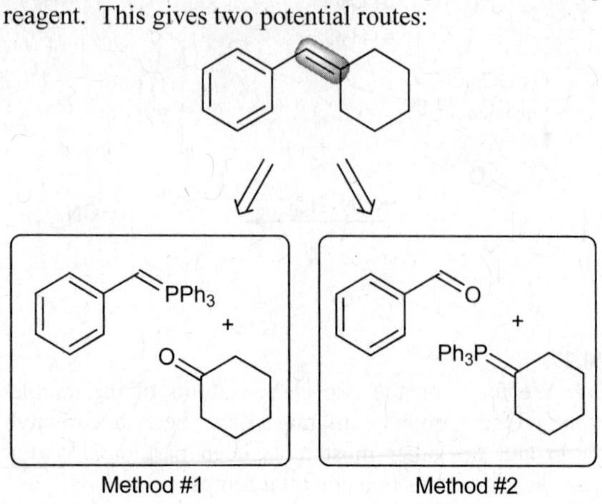

Method #1   Method #2

In each case, we focus on the Wittig reagent (which must be prepared via an S_N2 process). Method #1 will be more efficient because formation of the Wittig reagent requires the use of a primary benzylic halide, while method #2 would require the use of a secondary alkyl halide in forming the Wittig reagent. Therefore, we propose the following strategy to prepare the desired alkene:

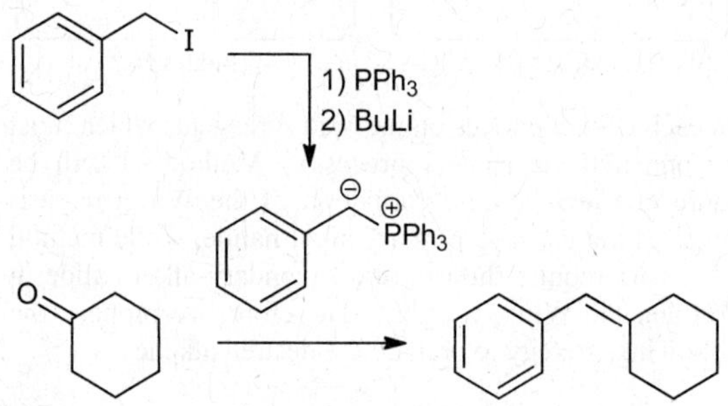

20.38. The product has 40 carbon atoms, while the starting material has 20 carbon atoms. Clearly, we will need to join two equivalents of the starting material to give one equivalent of the product. To make the necessary double bond via a Wittig reaction, an aldehyde must be treated with a Wittig reagent. As shown, the necessary Wittig reagent is prepared by treating the starting material with triphenylphosphine, followed by butyllithium. The aldehyde is prepared by treating the starting material with hydroxide (in an S$_N$2 reaction), followed by oxidation with PCC, as shown:

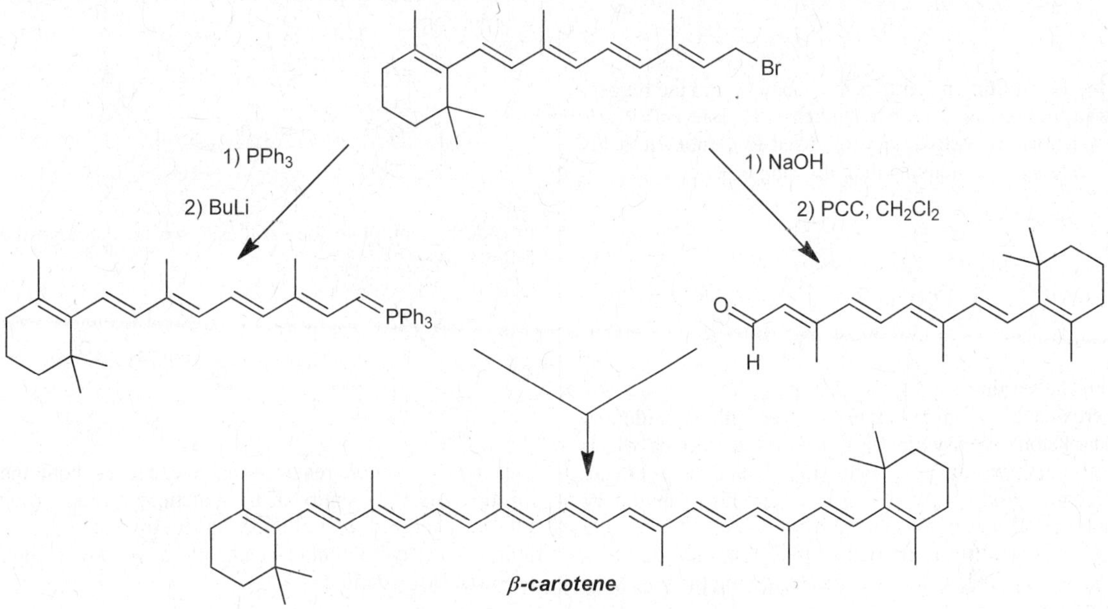

β-carotene

20.39.
(a) The product has one more carbon atom than the starting material. Also, the product has a double bond that must be installed. The extra carbon atom and the double bond can both be installed simultaneously with a Wittig reaction:

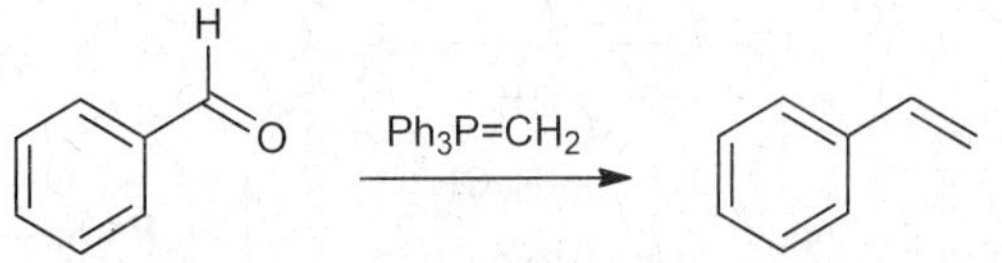

To prepare the necessary aldehyde, we simply oxidize the starting alcohol with PCC. This gives the following synthesis:

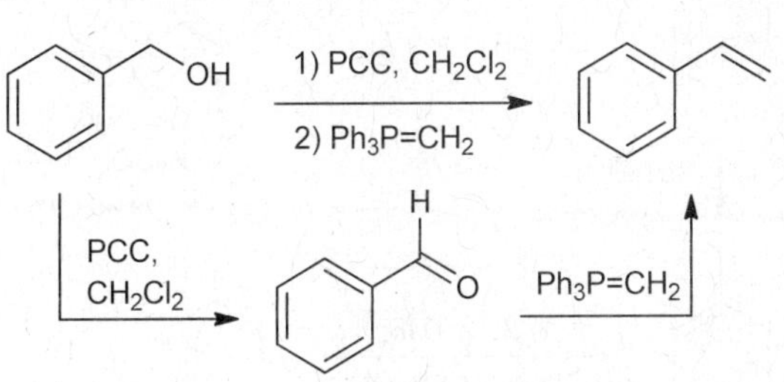

(b) The product has one more carbon atom than the starting material. Also, the position of the double bond must be moved. The extra carbon atom and the double bond can both be placed in the correct location if the last step of our synthesis is a Wittig reaction:

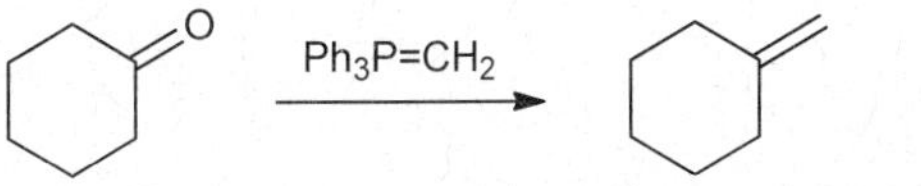

This strategy requires converting the starting alkene (cyclohexene) into the ketone shown above (cyclohexanone), which can be achieved in two steps (via acid-catalyzed hydration, followed by oxidation with chromic acid):

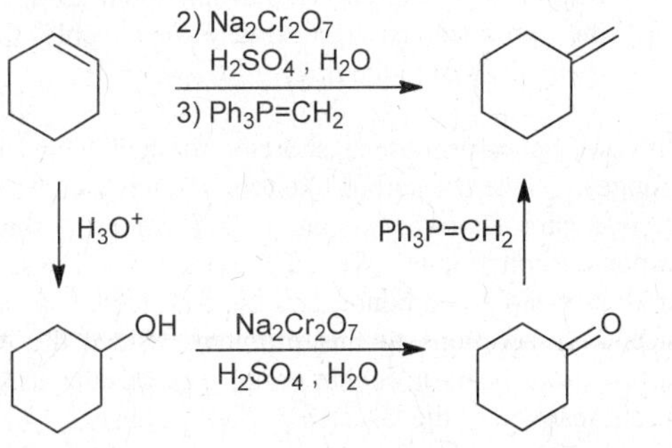

20.40.
(a) The starting material is a ketone, and the reagent is a peroxy acid, which indicates a Baeyer-Villiger oxidation. The ketone is unsymmetrical, so we must decide where to insert the oxygen atom (i.e., which side of the carbonyl group). According to the trends of migratory

aptitude that we encountered, we would expect the oxygen atom to be inserted on the left side (secondary) rather than the right side (primary), giving the following product:

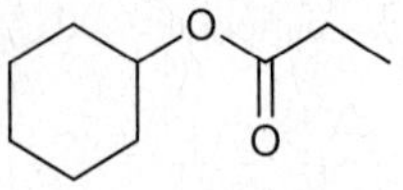

(b) The starting material is an aldehyde, and the reagent is a peroxy acid, which indicates a Baeyer-Villiger oxidation. Aldehydes are oxidized to carboxylic acids with this process, giving the following product:

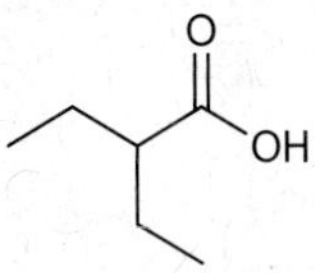

(c) The starting material is a ketone, and the reagent is a peroxy acid, which indicates a Baeyer-Villiger oxidation. The ketone is unsymmetrical, so we must decide where to insert the oxygen atom (i.e., which side of the carbonyl group). According to the trends of migratory aptitude that we encountered, we would expect the oxygen atom to be inserted on the right side (tertiary) rather than the left side (secondary), giving the following product:

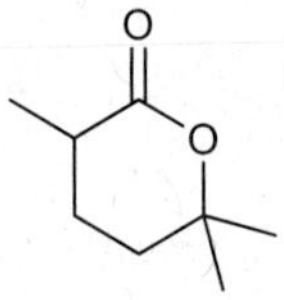

20.41.
(a) We begin by asking the following two questions:

1) *Is there a change in the carbon skeleton?* Yes, the carbon skeleton is increasing in size by one carbon atom.
2) *Is there a change in the functional groups?* Yes, the starting material lacks a functional group, and the product has an exocyclic double bond (extending from the ring).

Now we must propose a strategy for achieving these changes. Since the carbon skeleton is increasing in size by one carbon atom, we know that we must form a carbon-carbon bond. We have seen several ways to make carbon-carbon bonds, and there are certainly many acceptable solutions to this problem. Since the extra carbon atom bears a double bond, we consider using a Wittig reaction as the last step of our synthesis:

This step will install the extra carbon atom while simultaneously installing a double bond in the desired location. In order for this strategy to work, we must find a way to convert the starting material into the ketone above. Since the starting material lacks a functional group, the first step our synthesis must be a radical bromination:

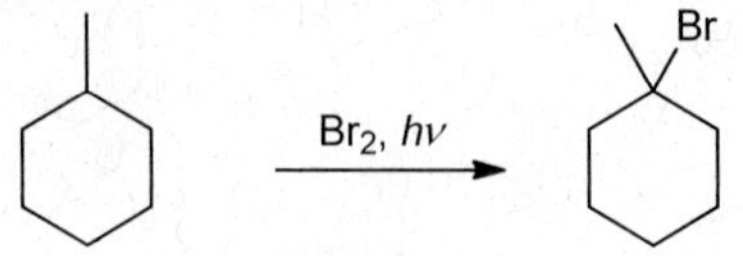

In order to complete the synthesis, we must bridge the following gap:

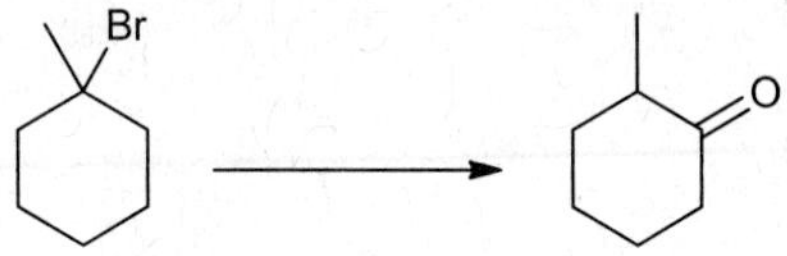

This transformation requires that we change both the location and the identity of the functional group. This can be achieved via elimination to give an alkene, followed by hydroboration-oxidation to give an alcohol, followed by oxidation:

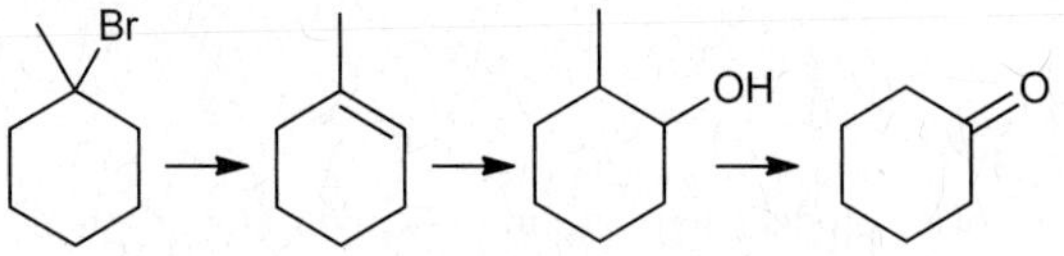

The entire proposed synthesis is summarized here:

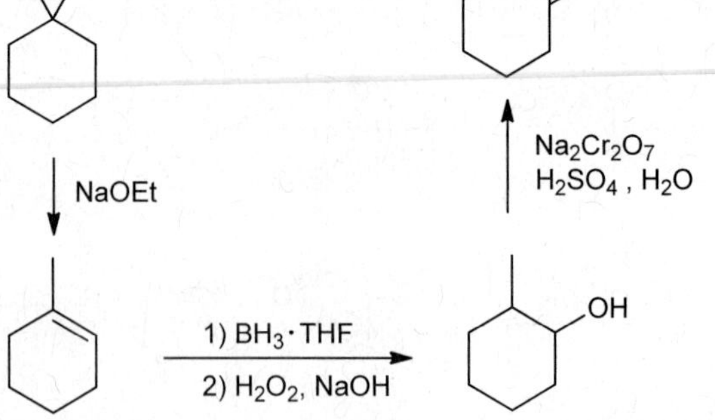

(b) We begin by asking the following two questions:

1) *Is there a change in the carbon skeleton?* Yes, the carbon skeleton is increasing in size by one carbon atom.
2) *Is there a change in the functional groups?* Yes, the starting material has a triple bond, and the product is an alcohol.

Now we must propose a strategy for achieving these changes. There are certainly many acceptable solutions to this problem. One such solution derives from the following retrosynthetic analysis. An explanation of each of the steps (*a-c*) follows.

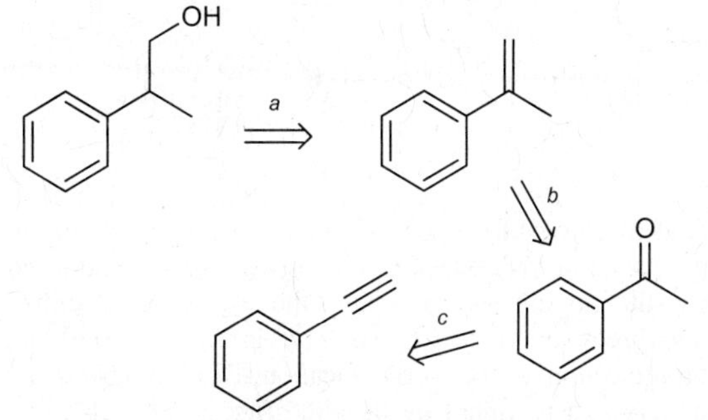

a. The desired product is a primary alcohol, which can be made via hydroboration-oxidation of an alkene.
b. The alkene can be made from a ketone via a Wittig reaction.
c. The ketone can be made from the starting alkyne (via acid-catalyzed hydration).

Now let's draw the forward scheme. The starting alkyne is treated with aqueous acid in the presence of mercuric sulfate to give a hydration reaction (the initially formed enol rapidly tautomerizes to give a methyl ketone). This ketone is then treated with a Wittig reagent, thereby forming the crucial carbon-carbon bond. The resulting alkene is then expected to undergo hydroboration-oxidation to give *anti*-Markovnikov addition of H and OH across the double bond, affording the desired alcohol.

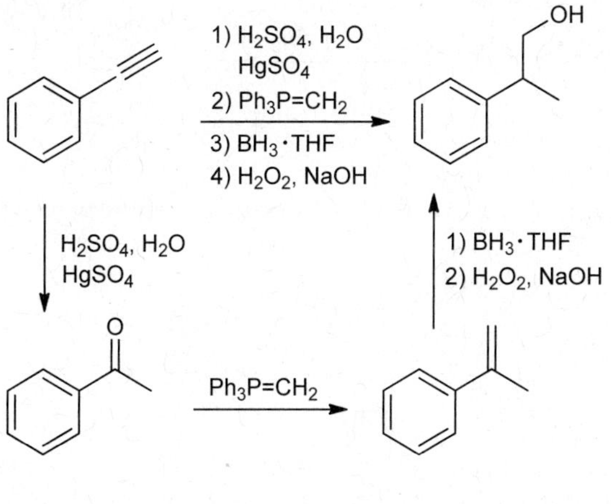

(c) There are certainly many acceptable solutions to this problem. One such solution is virtually identical to the solution presented for Problem **20.41a**, except that the last step has been replaced with a Grignard reaction:

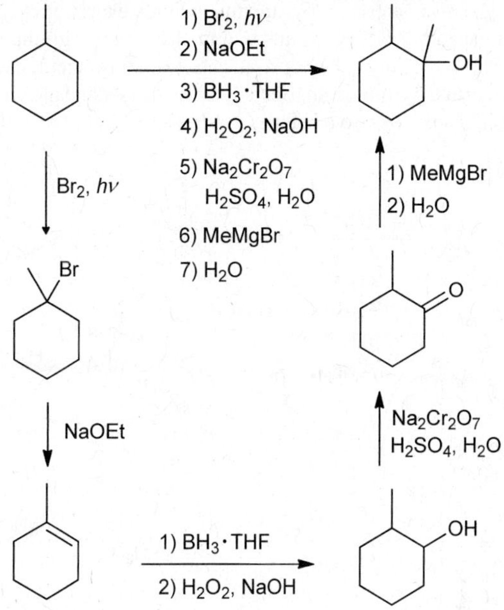

(d) We begin by asking the following two questions:

1) *Is there a change in the carbon skeleton?* Yes, the starting material is cyclic and the product is acyclic. In addition, the starting material has seven carbon atoms, while the product has nine. Two carbon atoms must be installed and the ring must be opened.
2) *Is there a change in the functional groups?* Yes, the starting material lacks a functional group, and the product has two functional groups (C=O and OH).

Now we must propose a strategy for achieving these changes. The ring can be opened via ozonolysis, but only if we first install a double bond in the ring:

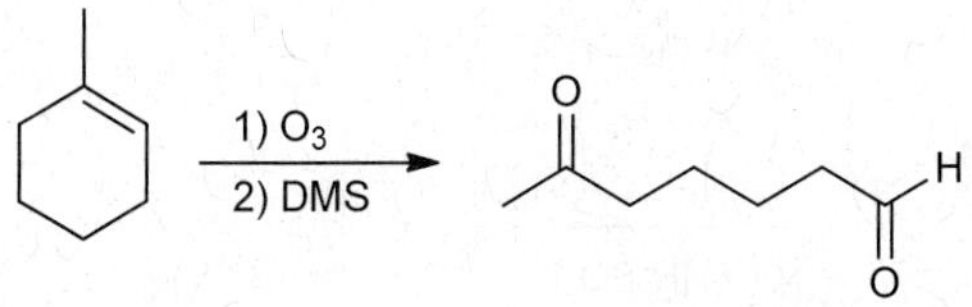

The necessary cycloalkene can be generated from the starting material in just two steps: 1) radical bromination installs a bromine atom in the tertiary position, and 2) subsequent treatment of the resulting tertiary alkyl halide with a strong base (such as sodium ethoxide) gives the cycloalkene above:

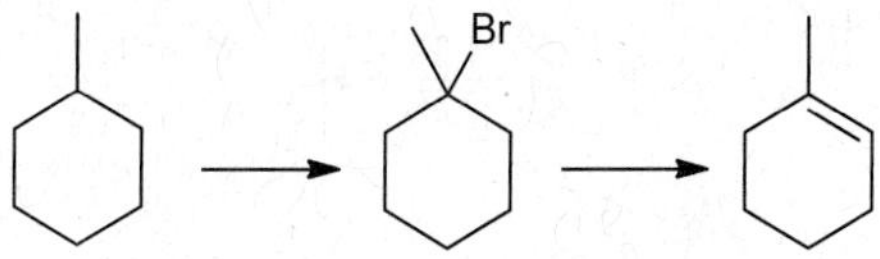

This cycloalkene then undergoes ozonolysis to open the ring. After ozonolysis has been performed, the resulting dicarbonyl compound can be converted into the product in just two steps. First, treatment with excess methyl magnesium bromide installs two methyl groups, giving a diol. When treated with a strong oxidizing agent, the tertiary OH group is unaffected, while the secondary OH group is oxidized to a ketone.

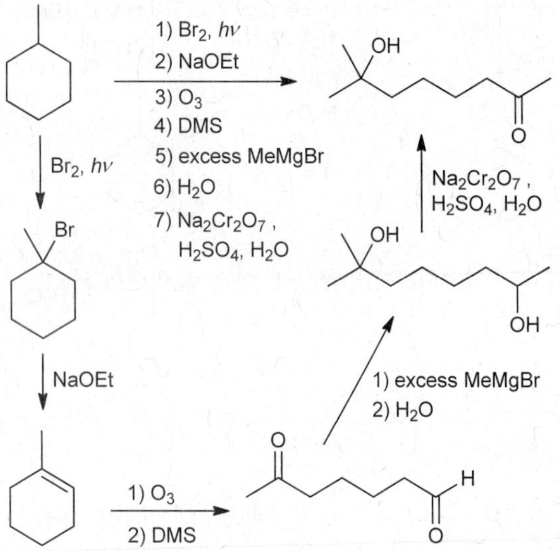

(e) The product is an imine, which can be made from the corresponding aldehyde and primary amine:

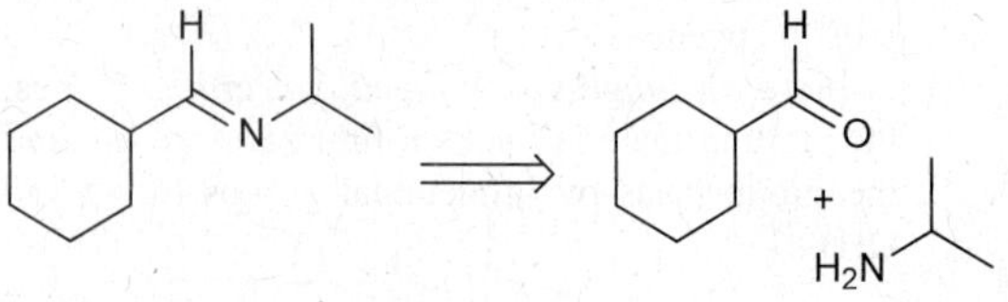

This aldehyde can be made from the starting material via hydroboration-oxidation, followed by oxidation with PCC, as shown:

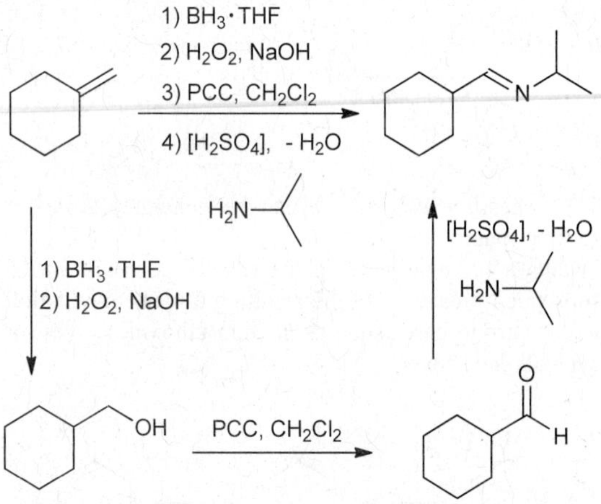

(f) While we certainly learned a way to install a bromine atom on a ring (bromination) we did not learn a way to install the other substituent (highlighted):

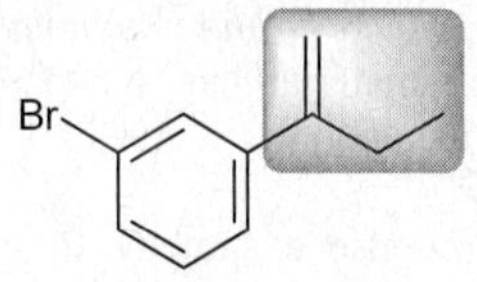

However, this substituent can be made from the following ketone, using a Wittig reaction:

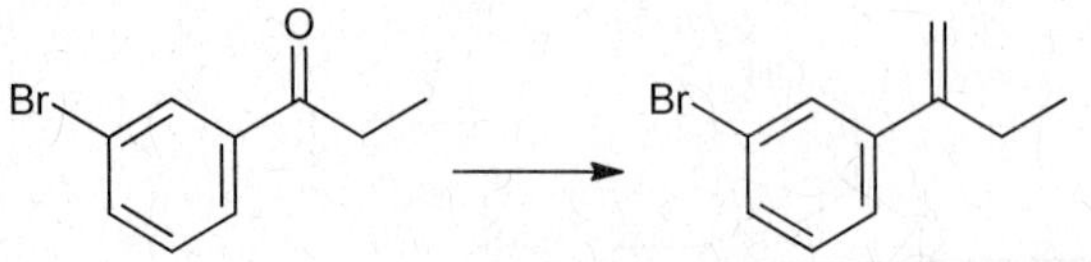

And we have indeed seen a way to install an acyl group on a ring (Friedel-Crafts acylation). The two substituents are *meta* to each other, so we must utilize the directing effects of the acyl group to install the bromine atom in the correct location (Br is an *ortho-para* director, so it cannot be installed first). After the acyl group has been installed, the product can be obtained via bromination (which installs a bromine atom in the *meta* position), followed by a Wittig reaction, as shown:

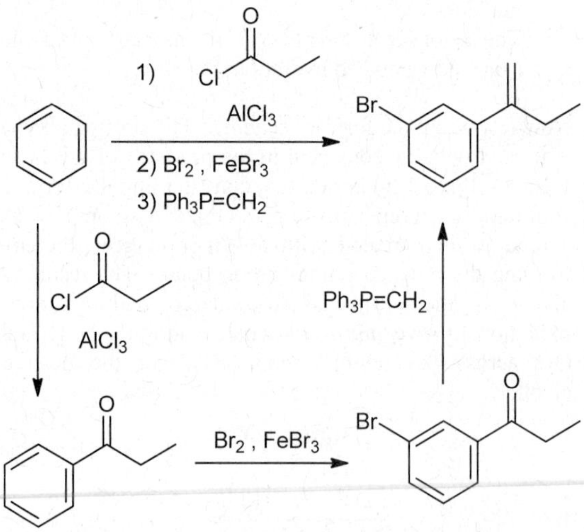

(g) The desired product has a carbon atom (highlighted) connected to two oxygen atoms, so this compound is an acetal:

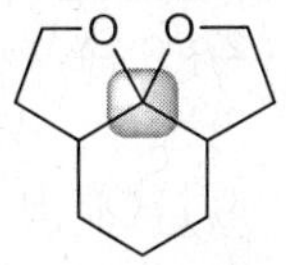

As such, the last step of our synthesis might be formation of the acetal group from the following precursor:

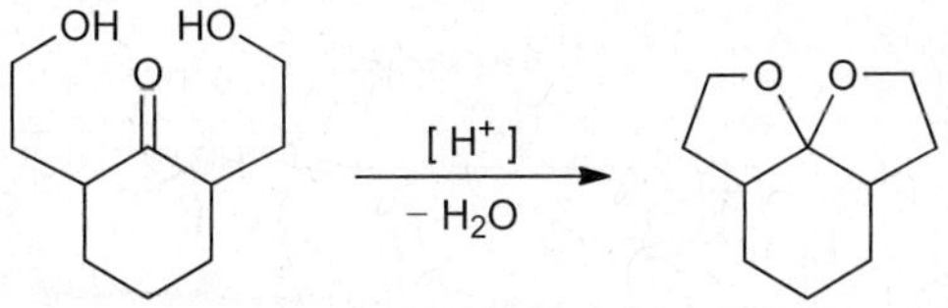

Notice that the reaction occurs in an intramolecular fashion because the two OH groups and the carbonyl group are tethered together.

Now we must determine how to make the precursor above from the starting material:

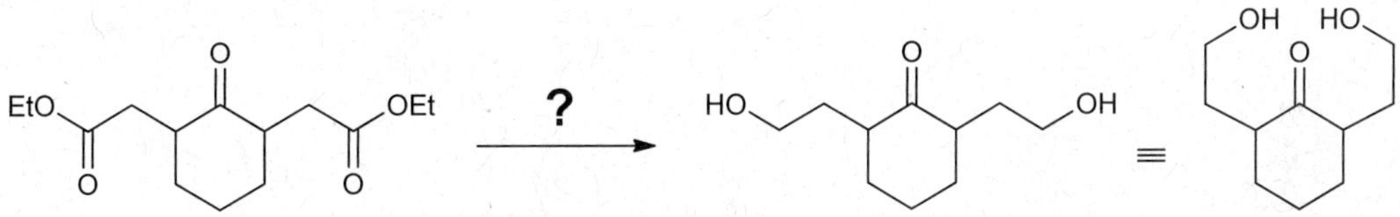

It might be tempting to treat the starting material with a reducing agent, such as LAH (xs), so it will reduce the two ester groups. Unfortunately, excess LAH will also reduce the ketone:

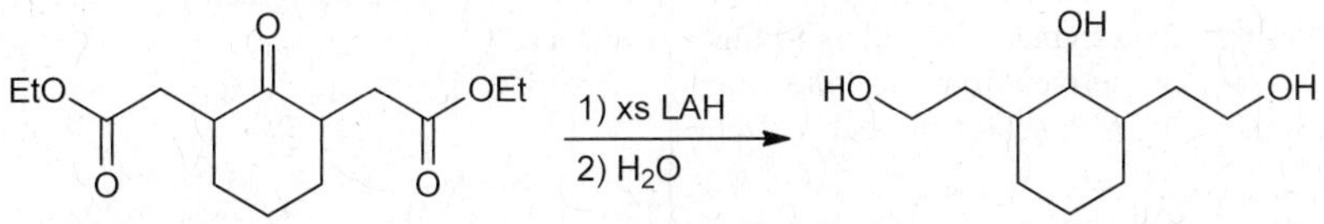

And once all three functional groups are reduced to a triol, we are stuck, because we have not learned a way to selectively oxidize the secondary OH group. This problem can be circumvented by using a protecting group first. That is, the ketone can be converted into an acetal group, thereby protecting it prior to the reduction step:

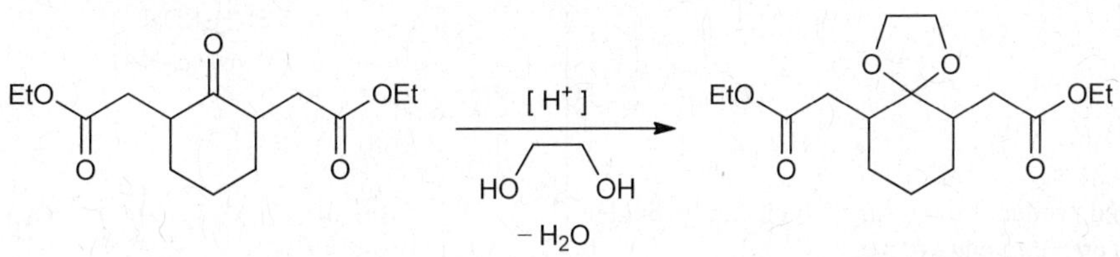

Notice that the ester groups are NOT converted into acetals. Only the ketone group is converted to an acetal. Subsequent reaction with xs LAH causes reduction of both ester groups (the acetal group is stable under these conditions). Treatment with aqueous acid will then protonate the dianion (each ester group is reduced to an alkoxide when treated with xs LAH). In addition, aqueous acid will also remove the acetal group, converting it back into a ketone. This allows us to perform the final step of the synthesis (intramolecular acetal formation) to give the product. The complete synthesis is shown here:

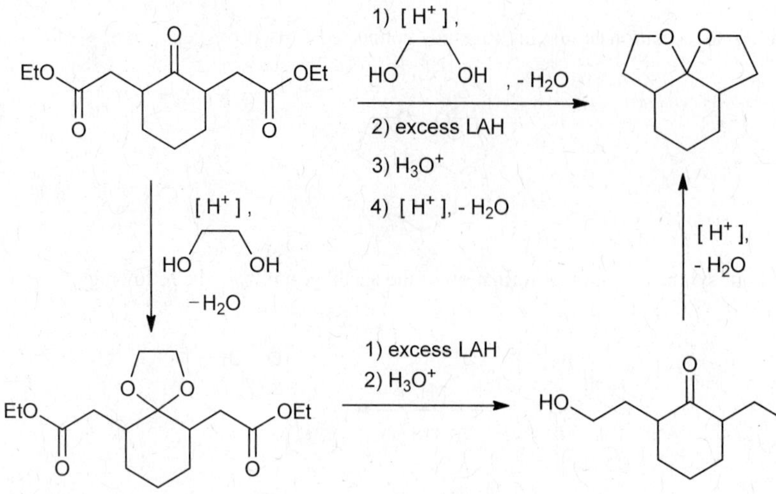

20.42.

(a) There are certainly many acceptable solutions to this problem. One such solution derives from the following retrosynthetic analysis. An explanation of each of the steps (*a-c*) follows.

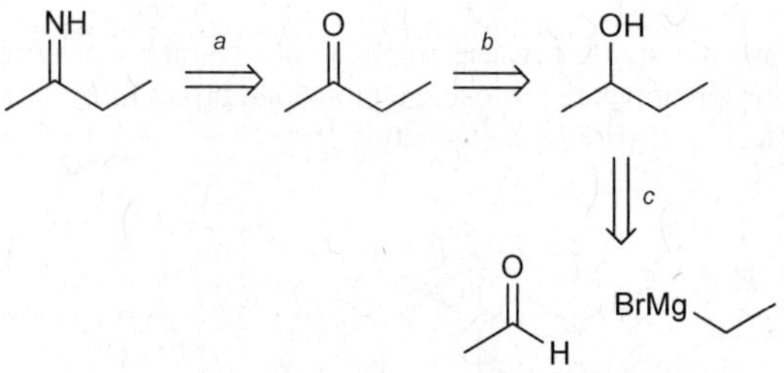

a. The desired product is an imine, which can be made from the corresponding ketone.

b. The ketone can be made via oxidation of the corresponding secondary alcohol.

c. The secondary alcohol can be made via a Grignard reaction from compounds containing no more than two carbon atoms.

Now let's draw the forward scheme. Ethyl bromide is converted into ethyl magnesium bromide (a Grignard reagent), which is then treated with acetaldehyde (to give a Grignard reaction), followed by water work-up, to give 2-butanol. Oxidation gives 2-butanone, which can then be converted into the desired imine upon treatment with

ammonia in acid-catalyzed conditions (with removal of water):

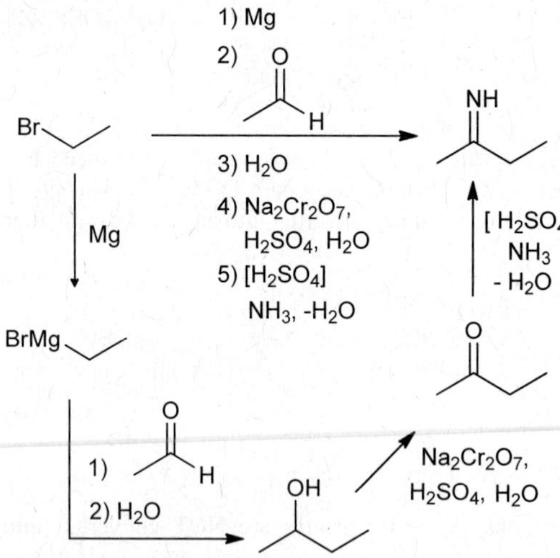

(b) There are certainly many acceptable solutions to this problem. One such solution derives from the following retrosynthetic analysis. An explanation of each of the steps (*a-e*) follows.

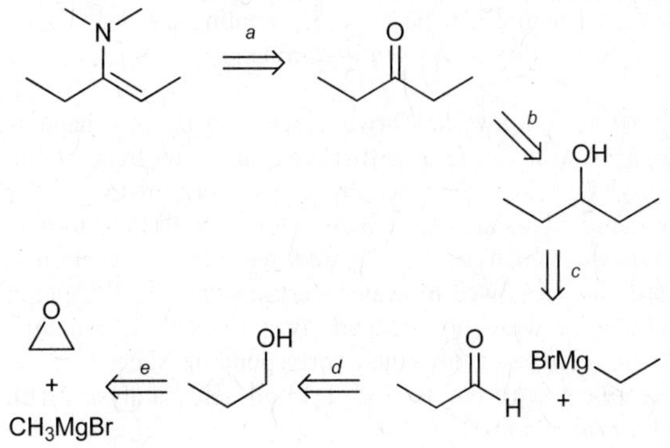

a. The desired product is an enamine, which can be made from the corresponding ketone (3-pentanone).
b. The ketone can be made via oxidation of the corresponding secondary alcohol (3-pentanol).
c. The secondary alcohol can be made via a Grignard reaction between ethyl magnesium bromide and an aldehyde (propanal).
d. Propanal can be made via oxidation of the corresponding alcohol (1-propanol).
e. 1-Propanol can be made by treating ethylene oxide with methyl magnesium bromide.

Now let's draw the forward scheme. Ethylene oxide is treated with methyl magnesium bromide, followed by water work-up to give 1-propanol. This alcohol is then oxidized to an aldehyde with PCC, and the resulting aldehyde is then treated with ethyl magnesium bromide, followed by water work-up, to give 3-pentanol. Oxidation with chromic acid gives 3-pentanone, which is then converted to the corresponding enamine upon treatment with dimethylamine and acid catalysis (with removal of water):

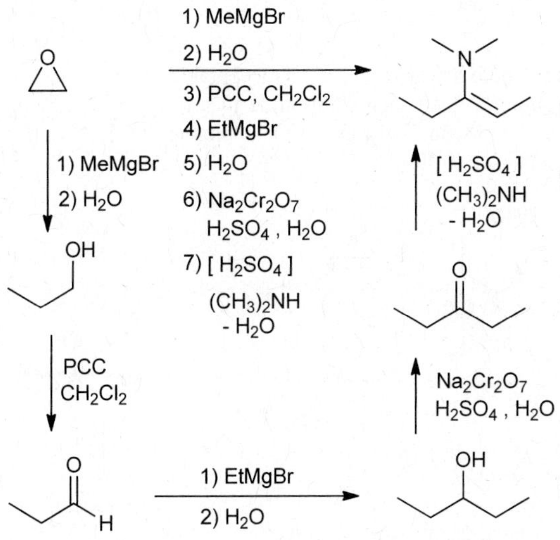

(c) The product is a cyclic acetal, which can be made from the corresponding ketone:

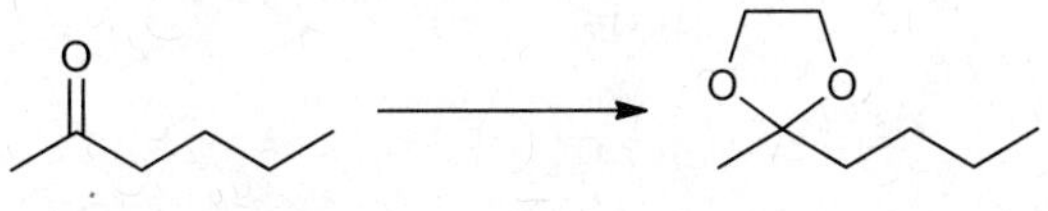

This ketone can be prepared in a variety of ways, so there are certainly many acceptable solutions to this problem. One such solution derives from the following retrosynthetic analysis. An explanation of each of the steps (*a-e*) follows.

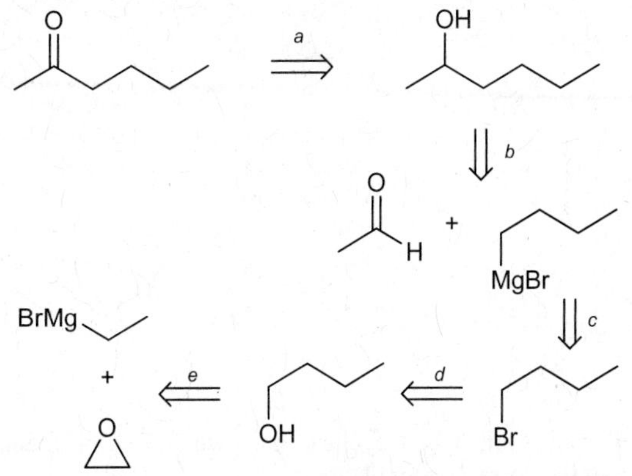

a. The ketone can be made via oxidation of the corresponding secondary alcohol (2-hexanol).
b. 2-Hexanol can be made via a Grignard reaction between acetaldehyde and butyl magnesium bromide.
c. Butyl magnesium bromide can be made from 1-bromobutane.
d. 1-Bromobutane can be made from 1-butanol via a substitution process.
e. 1-Butanol can be made by treating ethylene oxide with ethyl magnesium bromide.

Now let's draw the forward scheme. Bromoethane is converted into a Grignard reagent and then treated with ethylene oxide, followed by water work-up to give 1-butanol. This alcohol is then converted to the corresponding alkyl bromide upon treatment with PBr$_3$. The alkyl bromide is then converted into a Grignard reagent (via insertion of Mg), and then treated with acetaldehyde, followed by water work-up to give 2-hexanol. Oxidation with chromic acid gives 2-hexanone, which is then converted to the corresponding acetal upon treatment with ethylene glycol and acid catalysis (with removal of water):

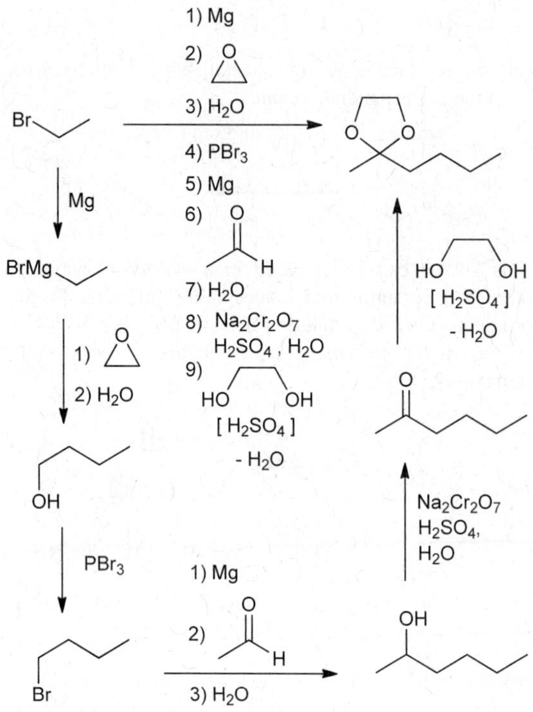

(d) The product is a cyclic acetal, which can be made from the corresponding ketone:

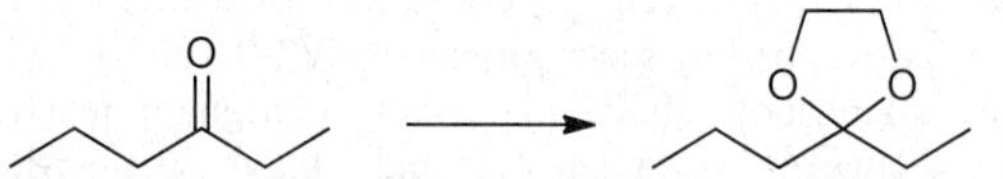

This ketone can be prepared in a variety of ways, so there are certainly many acceptable solutions to this problem. One such solution derives from the following retrosynthetic analysis. An explanation of each of the steps (*a-d*) follows.

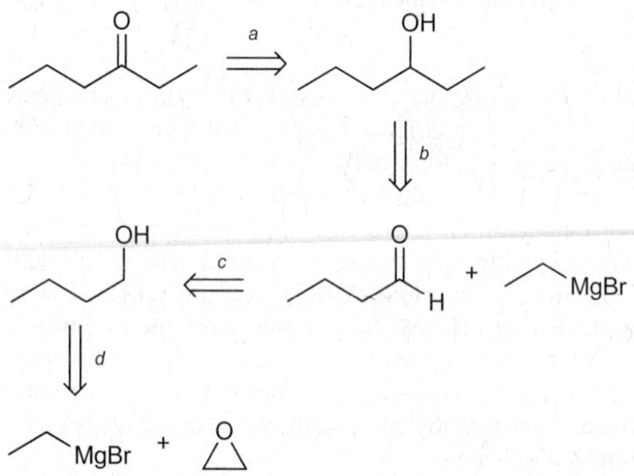

a. The ketone can be made via oxidation of the corresponding secondary alcohol (3-hexanol).
b. 3-Hexanol can be made via a Grignard reaction between butanal and ethyl magnesium bromide.
c. Butanal can be via oxidation of 1-butanol.

d. 1-Butanol can be made by treating ethylene oxide with ethyl magnesium bromide.

Now let's draw the forward scheme. Bromoethane is converted into a Grignard reagent and then treated with ethylene oxide, followed by water work-up to give 1-butanol. This alcohol is then oxidized with PCC to give butanal, which is then treated with ethyl magnesium bromide (followed by water work-up) to give 3-hexanol. Oxidation with chromic acid gives 3-hexanone, which is then converted to the corresponding acetal upon treatment with ethylene glycol and acid catalysis (with removal of water):

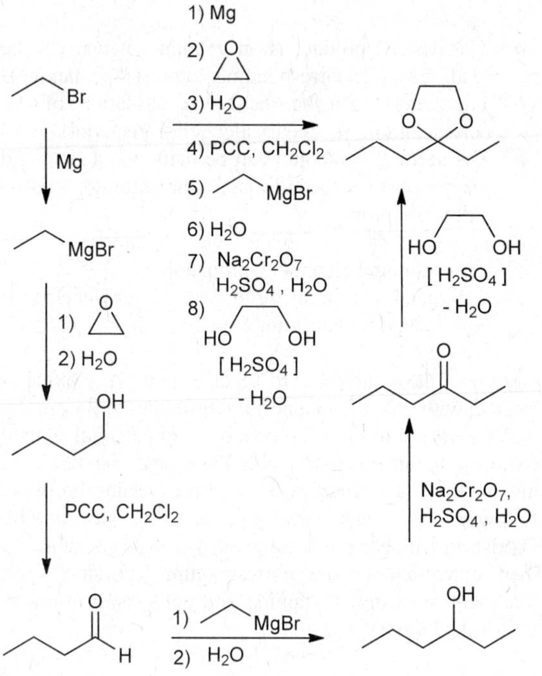

(e) There are certainly many acceptable solutions to this problem. One such solution derives from the following retrosynthetic analysis. An explanation of each of the steps (*a-e*) follows.

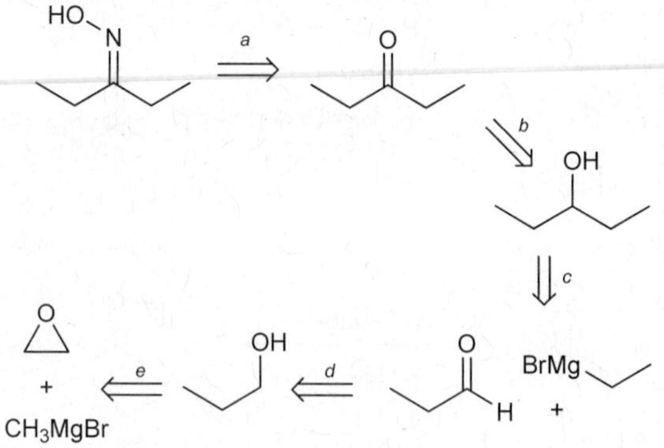

a. The desired product is an oxime, which can be made from the corresponding ketone (3-pentanone).
b. The ketone can be made via oxidation of the corresponding secondary alcohol (3-pentanol).
c. The secondary alcohol can be made via a Grignard reaction between ethyl magnesium bromide and an aldehyde (propanal).
d. Propanal can be made via oxidation of the corresponding alcohol (1-propanol).
e. 1-Propanol can be made by treating ethylene oxide with methyl magnesium bromide.

Now let's draw the forward scheme. Ethylene oxide is treated with methyl magnesium bromide, followed by water work-up to give 1-propanol. This alcohol is then oxidized to an aldehyde with PCC, and the resulting aldehyde is then treated with ethyl magnesium bromide, followed by water work-up, to give 3-pentanol. Oxidation with chromic acid gives 3-pentanone, which is then converted to the corresponding oxime upon treatment with hydroxylamine and acid catalysis (with removal of water):

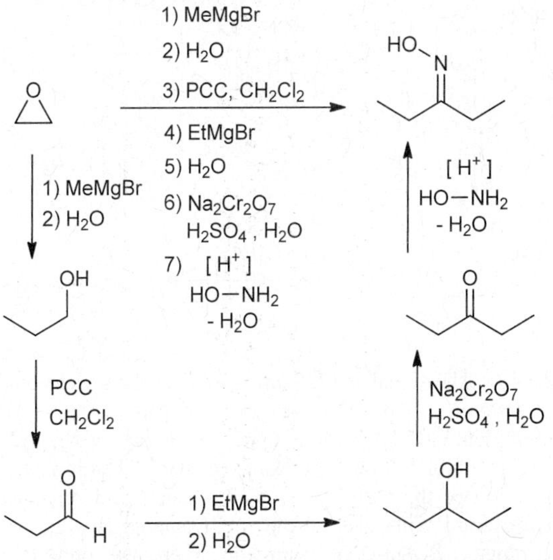

(f) There are certainly many acceptable solutions to this problem. One such solution derives from the following retrosynthetic analysis. An explanation of each of the steps (*a-e*) follows.

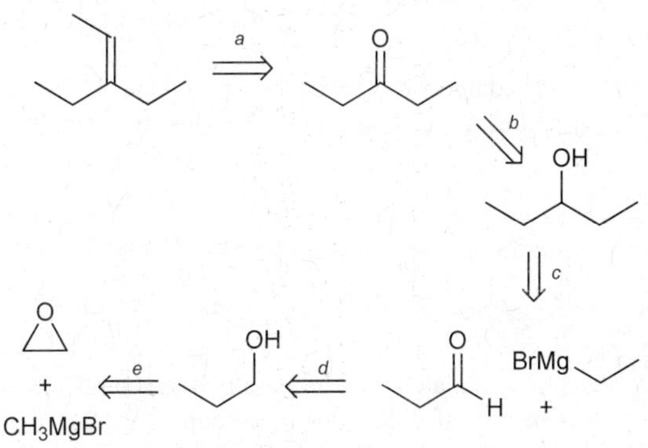

a. The desired product is an alkene, which can be made from a ketone (3-pentanone) via a Wittig reaction.
b. The ketone can be made via oxidation of the corresponding secondary alcohol (3-pentanol).
c. The secondary alcohol can be made via a Grignard reaction between ethyl magnesium bromide and an aldehyde (propanal).
d. Propanal can be made via oxidation of the corresponding alcohol (1-propanol).
e. 1-Propanol can be made by treating ethylene oxide with methyl magnesium bromide.

Now let's draw the forward scheme. Ethylene oxide is treated with methyl magnesium bromide, followed by water work-up to give 1-propanol. This alcohol is then oxidized to an aldehyde with PCC, and the resulting aldehyde is then treated with ethyl magnesium bromide, followed by water work-up, to give 3-pentanol. Oxidation with chromic acid gives 3-pentanone, which is then converted to the product upon treatment with the appropriate Wittig reagent, thereby installing two carbon atoms and a double bond in the correct location:

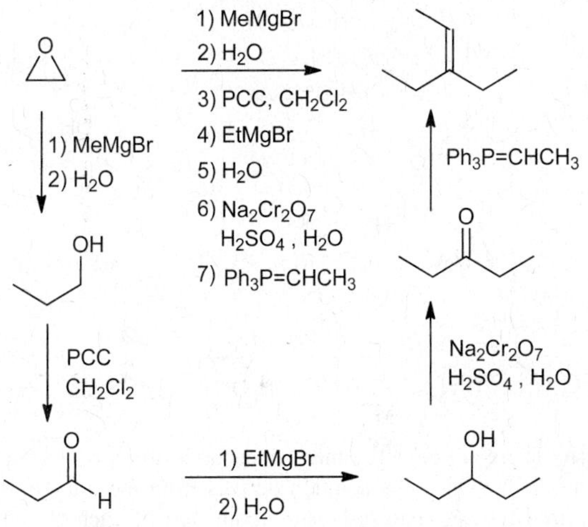

(g) There are certainly many acceptable solutions to this problem. One such solution derives from the following retrosynthetic analysis. An explanation of each of the steps (*a-d*) follows.

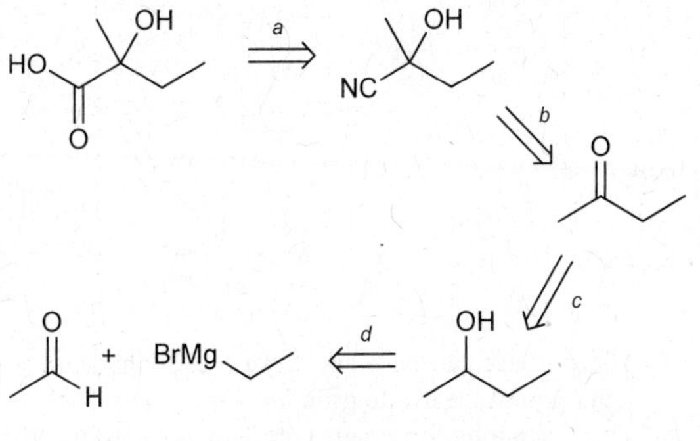

a. The product can be made from a cyanohydrin, via hydrolysis of the cyano group.

b. The cyanohydrin can be made from the corresponding ketone (2-butanone).

c. 2-Butanone can be made via oxidation of the corresponding secondary alcohol (2-butanol).

d. 2-Butanol can be made via a Grignard reaction between acetaldehyde and ethyl magnesium bromide.

Now let's draw the forward scheme. Ethyl bromide is converted into ethyl magnesium bromide, which is then treated with acetaldehyde (to give a Grignard reaction), followed by water work-up, to give 2-butanol. Oxidation of 2-butanol with chromic acid gives 2-butanone, which can then be converted into a cyanohydrin upon treatment with KCN and HCl. And finally, hydrolysis of the cyano group gives the desired product:

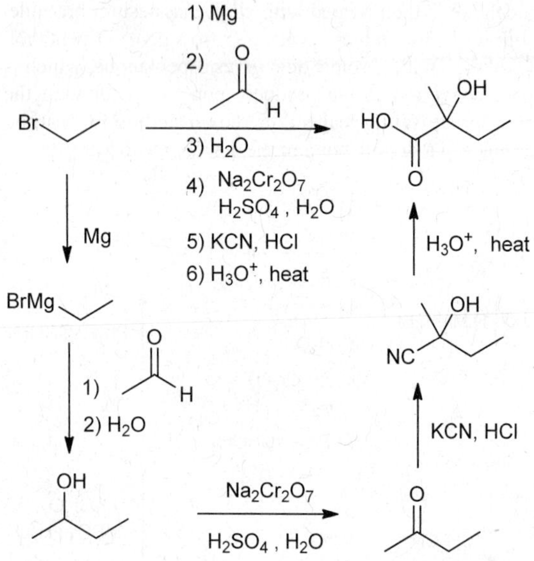

(h) There are certainly many acceptable solutions to this problem. One such solution derives from the following retrosynthetic analysis. An explanation of each of the steps (*a-d*) follows.

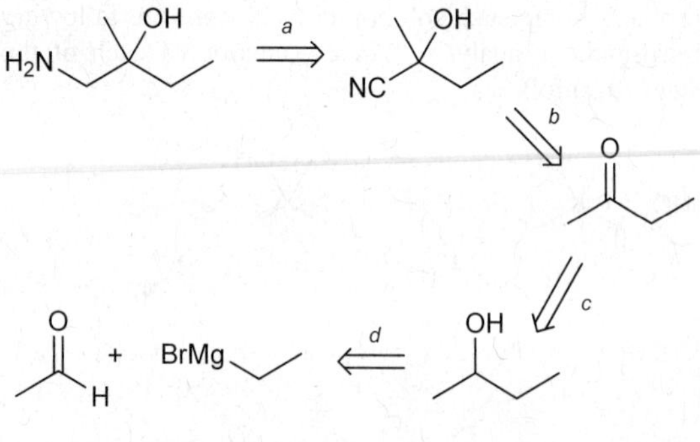

a. The product can be made from a cyanohydrin, via reduction of the cyano group.

b. The cyanohydrin can be made from the corresponding ketone (2-butanone).

c. 2-Butanone can be made via oxidation of the corresponding secondary alcohol (2-butanol).

d. 2-Butanol can be made via a Grignard reaction between acetaldehyde and ethyl magnesium bromide.

Now let's draw the forward scheme. Ethyl bromide is converted into ethyl magnesium bromide, which is then treated with acetaldehyde (to give a Grignard reaction), followed by water work-up, to give 2-butanol. Oxidation of 2-butanol with chromic acid gives 2-butanone, which can then be converted into a cyanohydrin upon treatment with KCN and HCl. And finally, reduction of the cyano group gives the desired product:

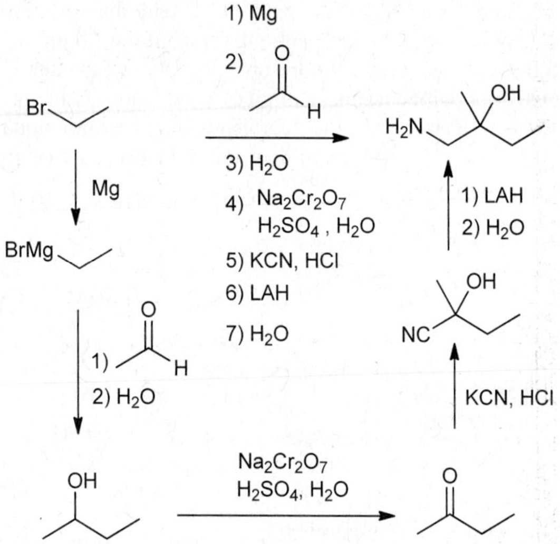

20.43. The signal at 1720 cm^{-1} indicates the presence of a carbonyl group. That carbonyl group is reduced to a methylene (CH$_2$) group upon treatment with 1,2-ethanedithiol followed by Raney nickel. Therefore, the position of the carbonyl group (in compound **A**) must correspond with one of the methylene groups in the product. Because of symmetry, there are only two unique methylene positions in the product.

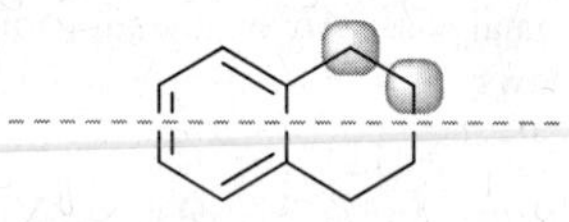

Therefore, compound **A** must be one of the following two structures:

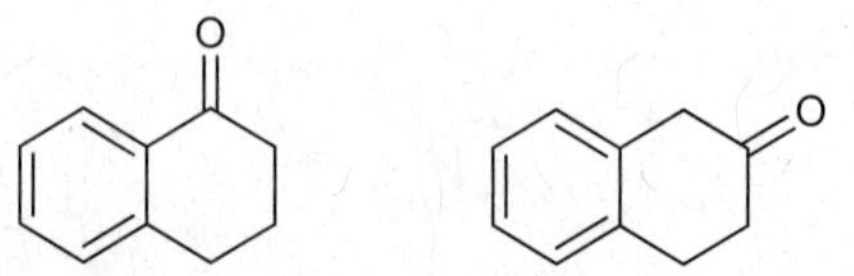

We can differentiate between these two structures based on the nature of the carbonyl group. In the first structure, the carbonyl group is conjugated to the aromatic ring. In the second structure, the carbonyl group is isolated. The signal at 1720 cm^{-1} indicates an isolated carbonyl group (a conjugated carbonyl group would produce a signal near 1680 cm^{-1}). Therefore, compound **A** must be the structure with the isolated carbonyl group.

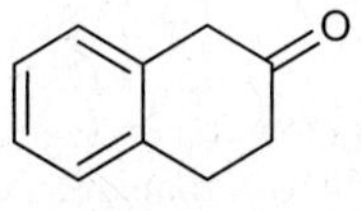

Compound **A**

20.44.
(a) We begin by identifying the parent. In this case, the carbonyl group is part of a five-membered ring, so the parent is cyclopentanone. In a cyclic ketone, the carbonyl group is at C1 (by definition). Next, we identify the substituents. There are two (a propyl group at C2 and a methyl group at C3), which are organized alphabetically in the name (methyl precedes propyl). We then assign a configuration to each of the chirality centers, giving the following name:

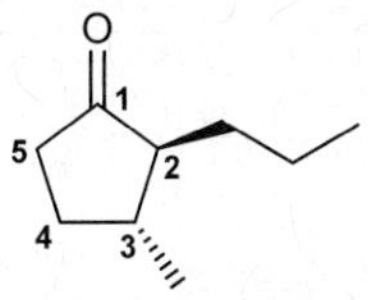

(2S,3R)-3-methyl-2-propylcyclopentanone

(b) A cyclic compound containing an aldehyde adjacent to the ring is named as a cycloalkane carbaldehyde. In this case, the aldehyde group is connected to a six-membered ring (cyclohexane), giving the following name:

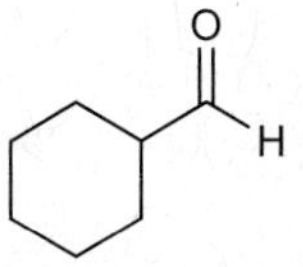

cyclohexanecarbaldehyde

(c) This compound is an aldehyde with a parent of four carbon atoms, so we might think the parent should be butanal. But this parent also contains a double bond, which is indicated by changing "an" to "en" (as in propane vs. propene). Therefore, in this case, it is 2-butenal. Notice that the location of the double bond (between C2 and C3) is indicated with a single locant (2). The location of the aldehyde group is at C1 (by definition) and need not be specified. The methyl group at C3 must be indicated the name. The name does not have a stereodescriptor (*E* or *Z*) because the C=C bond is not stereoisomeric (one of the vinylic positions is connected to two identical groups).

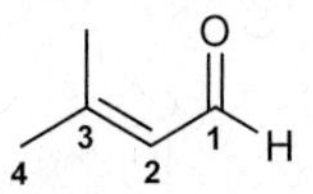

3-methyl-2-butenal

(d) The parent is a chain of six carbon atoms, with the carbonyl group at C3, so the parent is 3-hexanone. Next, we identify the substituents. There is only one: a methyl group at C4. We then assign a configuration to the chirality center, giving the following name:

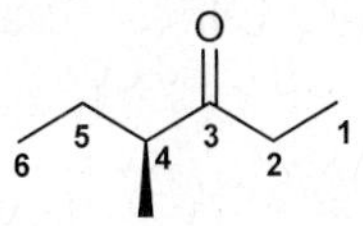

(*S*)-4-methyl-3-hexanone

20.45.
(a) The parent indicates three carbon atoms, and the suffix "dial" indicates two aldehyde groups at either end of the parent chain:

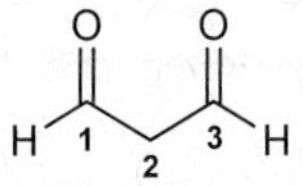

(b) The parent ("butanal") indicates four carbon atoms, and the suffix "al" indicates an aldehyde group (which is at C1, by definition). A phenyl substituent is located at C4:

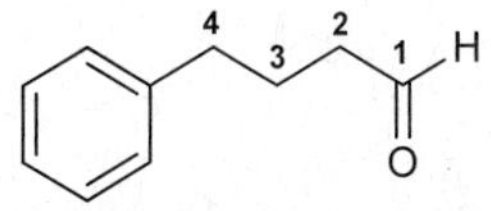

(c) The parent ("butanal") indicates four carbon atoms, and the suffix "al" indicates an aldehyde group (which is at C1, by definition). A phenyl substituent is located at C3, which is a chirality center with the *S* configuration:

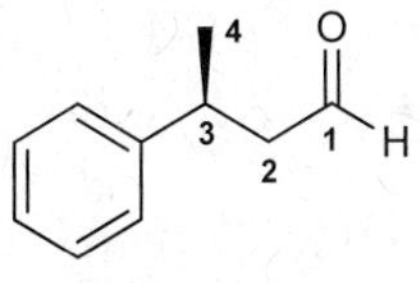

(d) The parent ("heptanone") indicates seven carbon atoms, and the suffix "one" indicates a ketone. The carbonyl group is located at C4, and there are four methyl groups (two at C3 and two at C5):

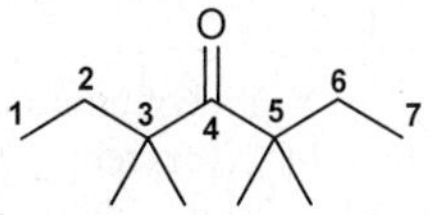

(e) The parent ("pentanal") indicates five carbon atoms, and the suffix "al" indicates an aldehyde group (which is at C1, by definition). An OH group is located at C3, which is a chirality center with the *R* configuration:

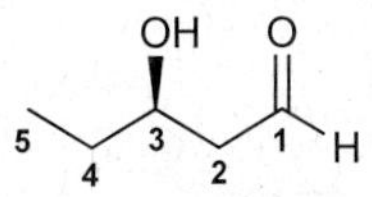

(f) The parent is acetophenone, and there is an OH group in the *meta* position (C3 of the ring):

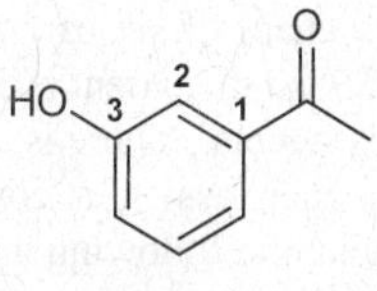

744   **CHAPTER 20**

(g) The parent is benzaldehyde, and there are three nitro groups, located at C2, C4, and C6:

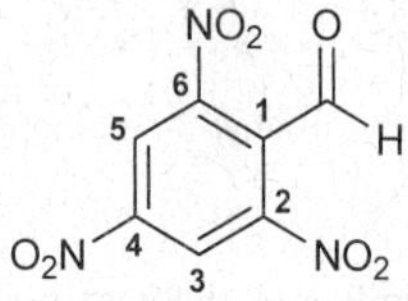

(h) The parent is acetaldehyde (which is the common name for ethanal), and there are three bromine atoms. Since C1 must be connected to an H (in order to be an aldehyde), there are only three positions where the bromine atoms can be placed, and they occupy all three of those positions. This is why locants were not necessary in the name.

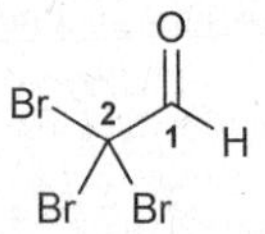

(i) The parent ("pentanone") indicates five carbon atoms, and the suffix "one" indicates a ketone. The carbonyl group is located at C2, and there are two OH groups, located at C3 and C4, both of which are chirality centers with the *R* configuration:

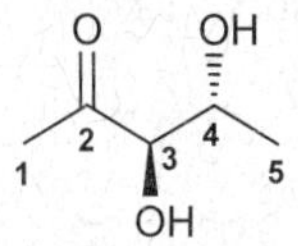

20.46. The molecular formula (C₄H₈O) indicates one degree of unsaturation (see Section 15.16), which accounts for the carbonyl group of an aldehyde:

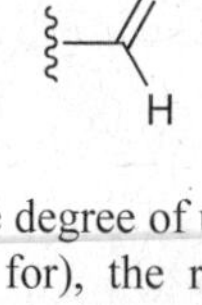

Since there is only one degree of unsaturation (which has now been accounted for), the remaining three carbon atoms do not comprise a ring and do not possess any π bonds. There are only two ways to connect three carbon atoms (without a ring or a π bond), shown here:

Therefore, there are only two aldehydes with molecular formula C₄H₈O, shown here:

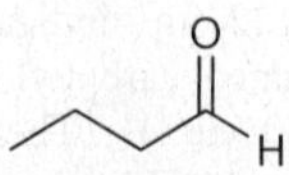

butanal 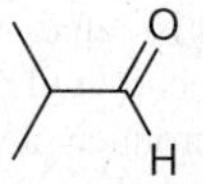   2-methylpropanal

20.47. The molecular formula (C₅H₁₀O) indicates one degree of unsaturation (see Section 15.16), which accounts for the carbonyl group of an aldehyde:

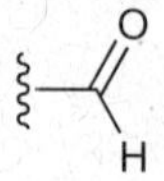

Since there is only one degree of unsaturation (which has now been accounted for), the remaining four carbon atoms do not comprise a ring and do not possess any π bonds. There are only four ways to connect four carbon atoms (without a ring or a π bond), shown here:

Linear Chain of 4

Linear Chain of 3

Linear Chain of 2

Therefore, there are only four aldehydes with molecular formula C₅H₁₀O, shown here. One of these aldehydes exhibits a chirality center (highlighted):

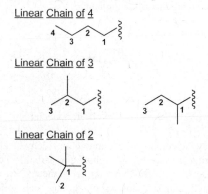

pentanal   2-methylbutanal

3-methylbutanal   2,2-dimethylpropanal

20.48. The molecular formula (C₆H₁₂O) indicates one degree of unsaturation (see Section 15.16), which accounts for the carbonyl group of a ketone:

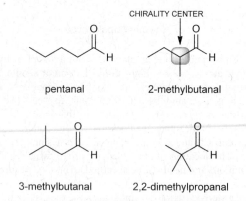

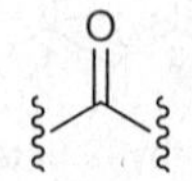

Since there is only one degree of unsaturation (which has now been accounted for), the remaining five carbon atoms do not comprise a ring and do not possess any π bonds. Now let's consider all of the different unique ways of connecting five carbon atoms around a carbonyl group. We can immediately rule out any isomers for which we place all five carbon atoms on one side of the carbonyl group, as that would generate an aldehyde, not a ketone. So, either there are four carbon atoms on one side of the carbonyl group and one carbon atom on the other side, OR, there are two carbon atoms on one side and three carbon atoms on the other side. We must explore each of these possibilities.

If we first consider having two carbon atoms on one side and three carbon atoms on the other side, there are only two such isomers, shown here:

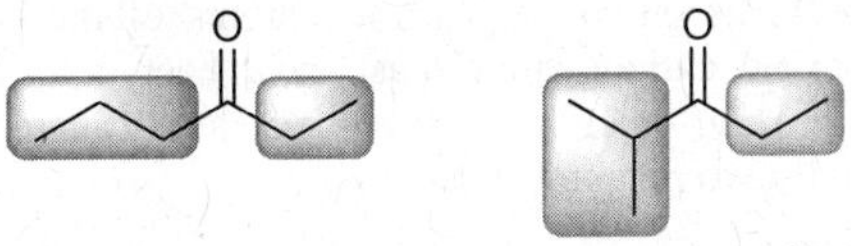

Now we consider having four carbon atoms on one side of the carbonyl group and one carbon atom on the other side. The side with four carbon atoms can be arranged in one of four possible ways, shown here, giving rise to four more isomers:

Linear Chain of 4   Linear Chain of 3

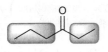

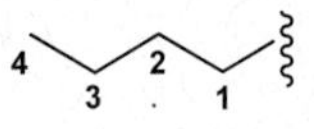

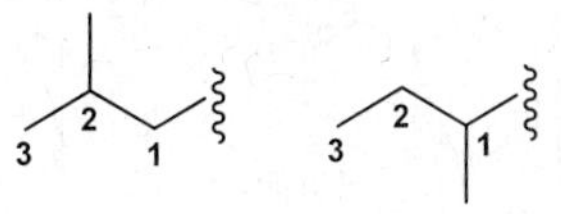

Linear Chain of 2

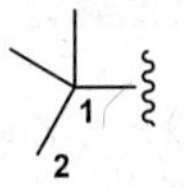

In total, there are six different isomeric ketones with molecular formula $C_6H_{12}O$, shown here. One of these ketones exhibits a chirality center (highlighted):

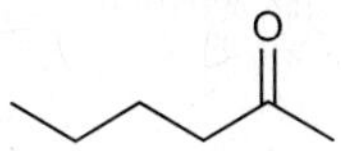

2-hexanone   3-hexanone

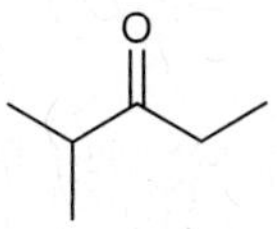

2-methyl-3-pentanone

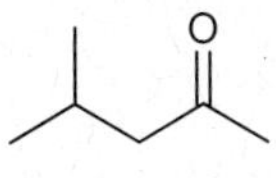

4-methyl-2-pentanone

CHIRALITY CENTER

3-methyl-2-pentanone 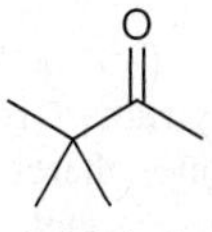   3,3-dimethyl-2-butanone

20.49. The carbonyl group of a ketone is unlikely to appear at C-1 because if it did, the compound would likely be called an aldehyde rather than a ketone. There are exceptions, such as 1-phenylketones (one example is 1-phenyl-1-butanone).

20.50.

(a) The following compound is expected to react with a nucleophile more rapidly, because aldehydes are more reactive than ketones toward nucleophiles, as a result of electronic effects (aldehydes are more electrophilic than ketones) as well as steric effects (See Section 20.4).

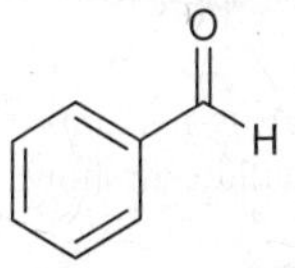

(b) The following compound is expected to react with a nucleophile more rapidly, because each of the trifluoromethyl groups is very powerfully electron-withdrawing, thereby rendering the carbonyl group even more electrophilic.

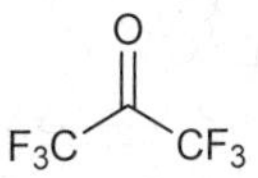

20.51

(a) The Wittig reagent attacks the carbonyl group, thereby forming a C=C bond between the carbon atom of the carbonyl group and the carbon atom attached to phosphorus (in the Wittig reagent). The resulting alkene could have either the E or Z configuration, as shown, although the Z isomer is likely to be a minor (or very minor) product in this case, as a result of steric considerations.

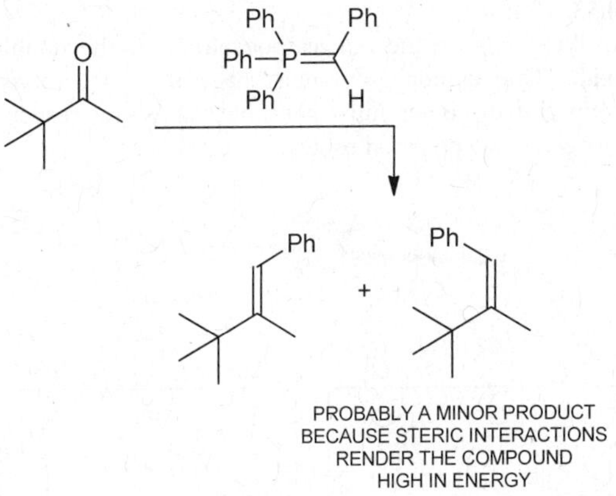

PROBABLY A MINOR PRODUCT BECAUSE STERIC INTERACTIONS RENDER THE COMPOUND HIGH IN ENERGY

(b) The Wittig reagent attacks the carbonyl group, thereby forming a C=C bond between the carbon atom of the carbonyl group and the carbon atom attached to phosphorus (in the Wittig reagent). The resulting alkene could have either the *E* or *Z* configuration, as shown.

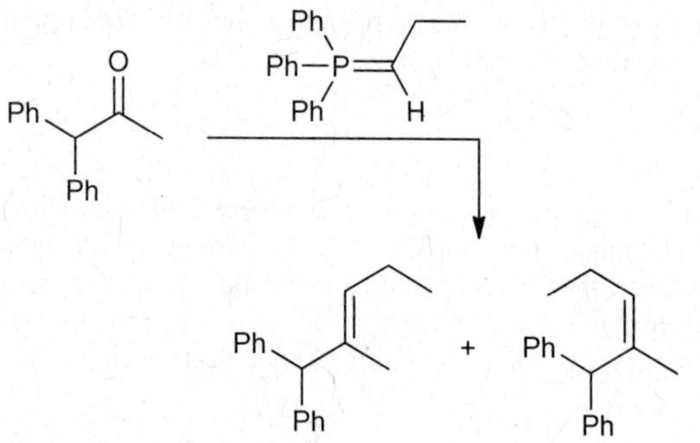

20.52. Each of the Wittig reagents can be made from the corresponding alkyl halide, as shown:

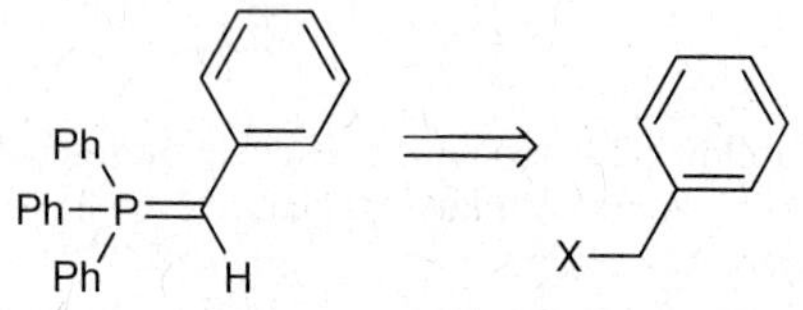

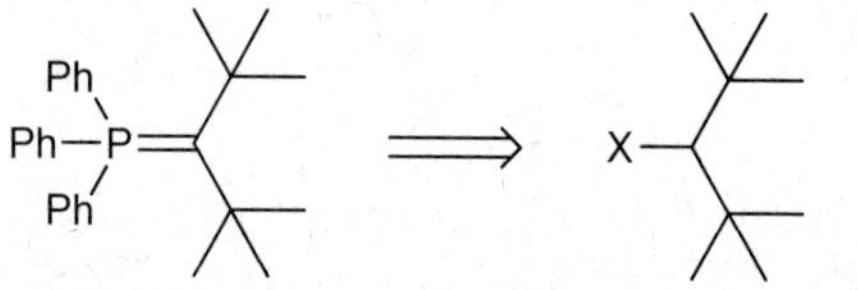

In the first case, the alkyl halide is primary, while in the second case, the alkyl halide is secondary (although this secondary alkyl halide is likely to behave much like a tertiary alkyl halide in that it exhibits significant steric hindrance). The latter alkyl halide will be more difficult to convert into a Wittig reagent, because it is too sterically hindered to undergo an S_N2 attack.

20.53.
(a) We focus on the two carbon atoms of the double bond. One carbon atom must have been a carbonyl group and the other must have been a Wittig reagent. This gives two potential routes:

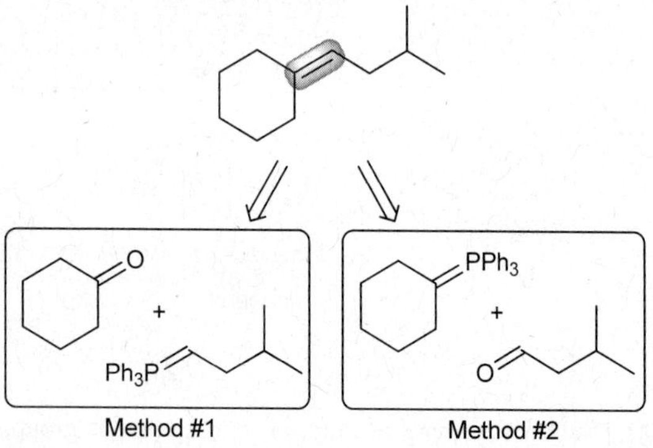

In each case, we focus on the Wittig reagent (which must be prepared via an S_N2 process). Method #1 will be

more efficient because formation of the Wittig reagent requires the use of a primary alkyl halide, while method #2 would require the use of a secondary alkyl halide in forming the Wittig reagent. Therefore, we propose the following strategy to prepare the desired alkene:

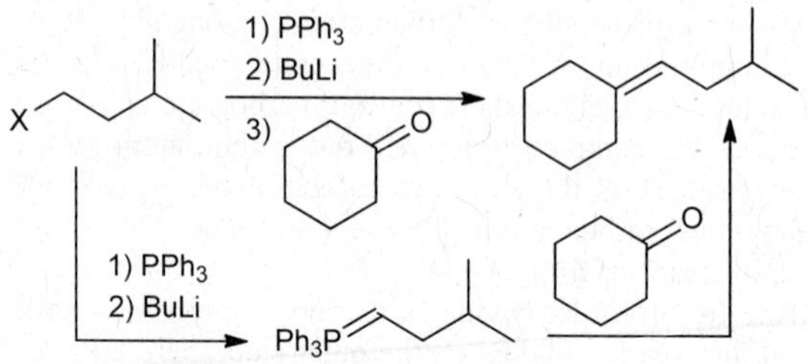

(b) We focus on the two carbon atoms of the double bond. One carbon atom must have been a carbonyl group and the other must have been a Wittig reagent. This gives two potential routes:

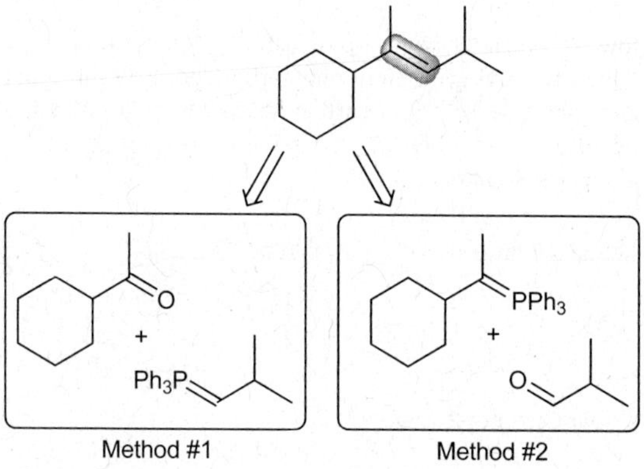

In each case, we focus on the Wittig reagent (which must be prepared via an S_N2 process). Method #1 will be more efficient because formation of the Wittig reagent requires the use of a primary alkyl halide, while method #2 would require the use of a secondary alkyl halide in forming the Wittig reagent. Therefore, we propose the following strategy to prepare the desired alkene:

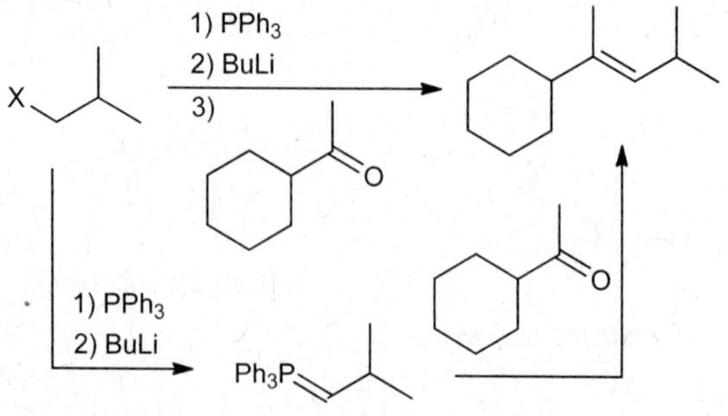

20.54.
(a) A Grignard reagent will attack the carbonyl group of a ketone, thereby forming a C-C bond and giving a tertiary alcohol. In this case, there are two possible routes that can be used to form the product via a

Grignard reaction (the Grignard reaction can be used to install either a methyl group or an ethyl group), as shown:

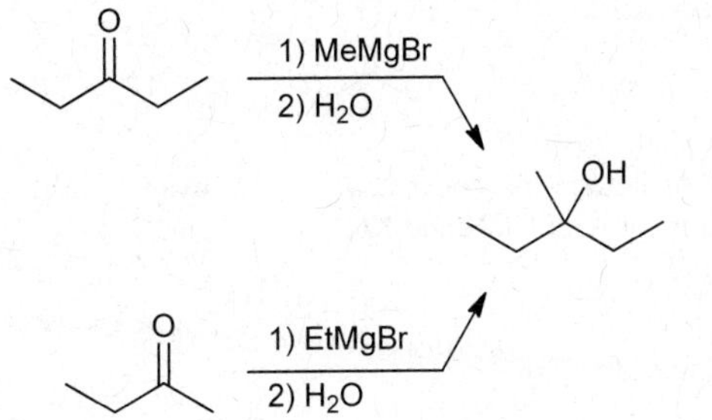

(b) A Grignard reagent will attack the carbonyl group of a ketone, thereby forming a C-C bond and giving an alcohol. In this case, treating cyclohexanone with ethyl magnesium bromide (followed by water work-up) gives the desired product:

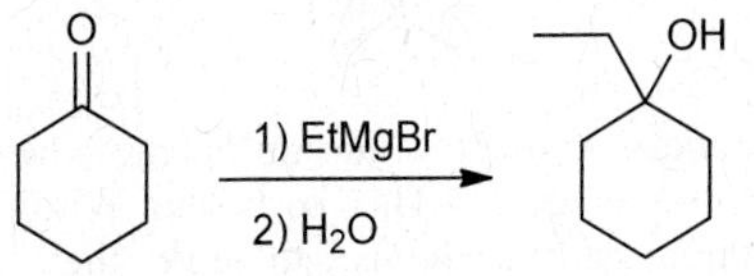

(c) Treating benzophenone with phenyl magnesium bromide (followed by water work-up) will afford the desired product via a Grignard reaction:

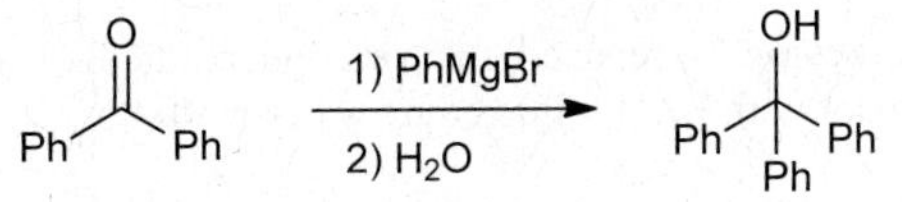

(d) A Grignard reagent will attack the carbonyl group of a ketone, thereby forming a C-C bond and giving an alcohol. In this case, there are two possible routes that can be used to form the product via a Grignard reaction (the Grignard reaction can be used to install either a phenyl group or a butyl group), as shown:

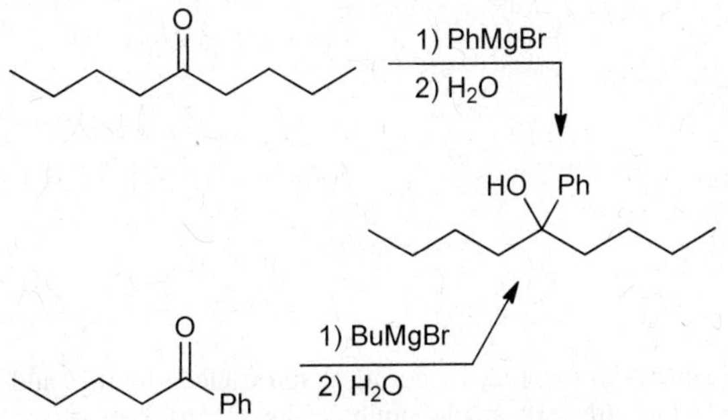

20.55. The following retrosynthetic analysis relies on a different strategy for opening the ring (rather than ozonolysis). Specifically, the final step might involve opening a cyclic ester (called a lactone) into the desired diol via reduction with LAH (as seen in Section 13.4). This ester can be made from the starting material, as shown. An explanation of each of the steps (*a-d*) follows.

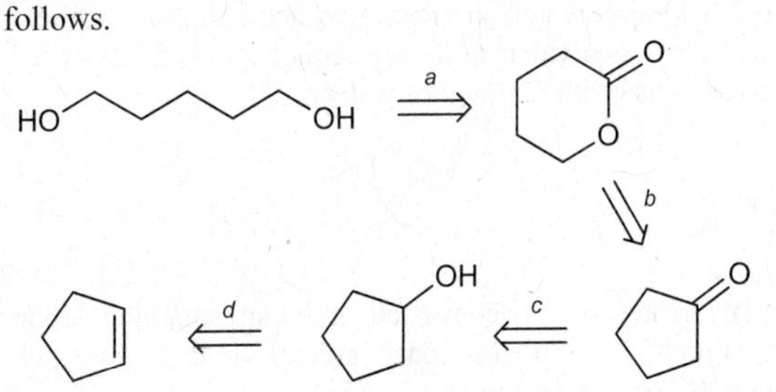

a. The diol can be made from the lactone shown, upon reduction with xs LAH.
b. The lactone can be made from the corresponding ketone (cyclopentanone).
c. Cyclopentanone can be made via oxidation of the corresponding secondary alcohol (cyclopentanol).
d. Cyclopentanol can be made from cyclopentene via acid-catalyzed hydration.

Now let's draw the forward scheme. Cyclopentene is treated with aqueous acid to give cyclopentanol, which is then oxidized with chromic acid to give cyclopentanone. A Baeyer-Villiger oxidation converts the ketone into a lactone (a cyclic ester), which can then be converted into the product upon treatment with excess LAH followed by water work-up:

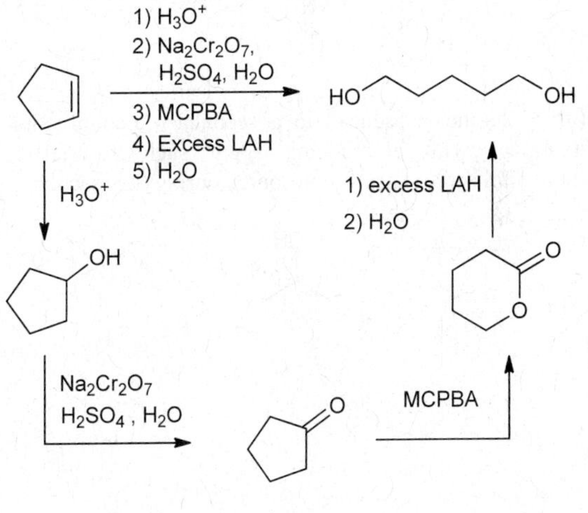

20.56.
(a) A ketone is converted into an imine upon treatment with ammonia in acid catalyzed conditions (with removal of water).

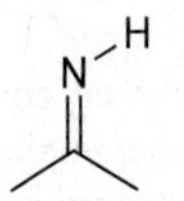

(b) A ketone is converted into an imine upon treatment with a primary amine in acid catalyzed conditions (with removal of water).

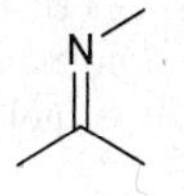

(c) A ketone is converted into an acetal upon treatment with two equivalents of an alcohol in acid catalyzed conditions (with removal of water).

EtO   OEt

(d) A ketone is converted into an enamine upon treatment with a secondary amine in acid catalyzed conditions (with removal of water).

(e) A ketone is converted into a hydrazone upon treatment with hydrazine in acid catalyzed conditions (with removal of water).

N   NH$_2$

(f) A ketone is converted into an oxime upon treatment with hydroxylamine in acid catalyzed conditions (with removal of water).

N   OH

(g) A ketone is reduced to a secondary alcohol upon treatment with a reducing agent such as NaBH$_4$ (methanol serves as a proton source during the reaction).

OH

(h) A ketone is converted into an ester upon treatment with a peroxyacid. This process is called a Baeyer-Villiger oxidation.

O

O

(i) A ketone is converted into a cyanohydrin upon treatment with HCN and KCN.

HO   CN

(j) Ethyl magnesium bromide is a Grignard reagent (a very strong nucleophile) and it will attack a ketone to give a tertiary alcohol (with installation of the ethyl group at the position that is α to the OH group):

HO

(k) A ketone is converted into an alkene when treated with a Wittig reagent. This particular Wittig reagent installs three carbon atoms, to give the following product:

(l) A ketone is reduced to a secondary alcohol upon treatment with LAH followed by water work-up.

OH

20.57. Under acid-catalyzed conditions, the carbonyl group can be protonated, rendering it more electrophilic, and subject to attack by a nucleophile. There are two options for the attacking nucleophile in the second step of the mechanism: 1) the protonated carbonyl group is tethered to an OH group that can function as a nucleophile and attack the protonated carbonyl group in an intramolecular fashion, or 2) the reagent (ethanol) can serve as a nucleophile and attack the protonated carbonyl group. Indeed, both of these processes will occur, as the product is an acetal. But we must decide which one is likely to occur first. As a general rule, intramolecular reactions will occur more rapidly than intermolecular reactions. As such, the second step of the mechanism will show the OH group attacking the protonated carbonyl group in an intramolecular fashion to give an oxonium ion, which then loses a proton to give a hemiacetal (the base for this step is likely a molecule of ethanol). Protonation of the hemiacetal gives a good leaving group (water) which leaves to give a resonance-stabilized cation that is attacked by ethanol. The resulting oxonium ion is then deprotonated to afford the product.

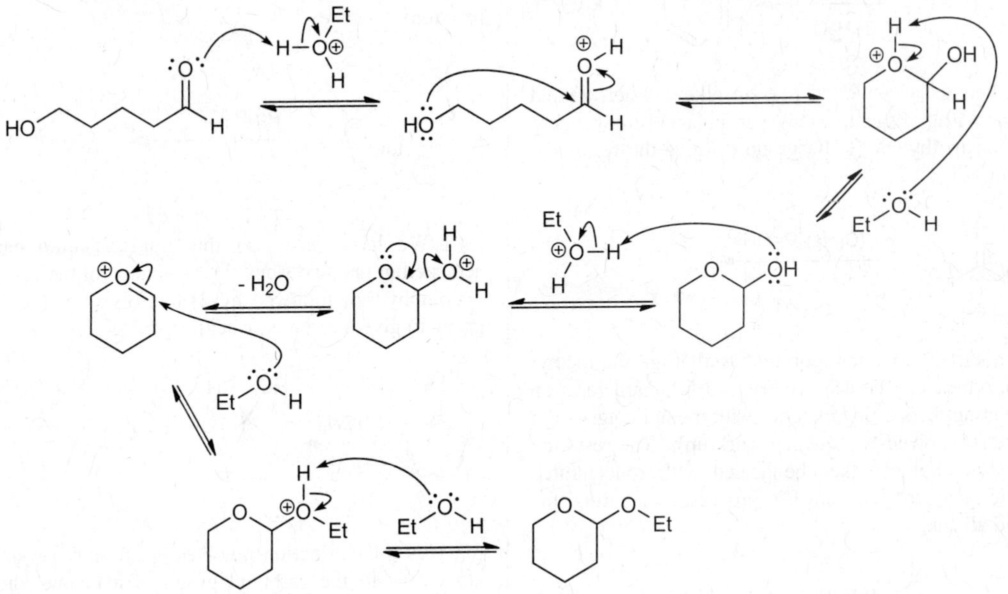

20.58. This transformation requires that we perform a Grignard reaction selectively at the carbonyl group of the ketone and not at the carbonyl group of the aldehyde. Since the aldehyde is more reactive than the ketone, we will need a method to protect the aldehyde group. This can be achieved via acetal formation. If we treat the starting material with an alcohol *without* using special distillation techniques to remove water (as it is formed), then the equilibrium will favor conversion of the aldehyde into an acetal, while conversion of the ketone into an acetal will be disfavored. As such, we can use this method to selectively convert the aldehyde into an acetal. Once protected, the desired Grignard reaction is performed, followed by deprotection, as shown:

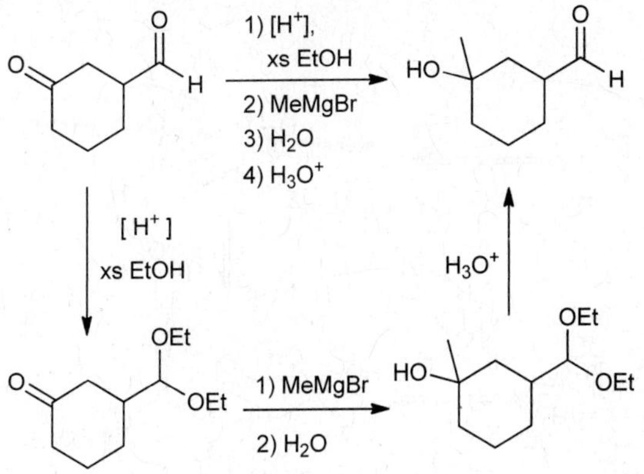

20.59. When ethylene glycol is treated with a ketone or aldehyde in acidic conditions, a cyclic acetal is formed:

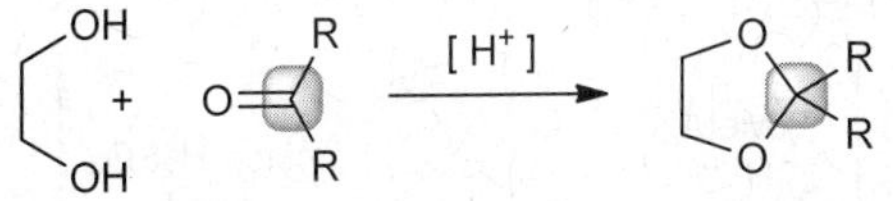

Catechol also has two OH groups (like ethylene glycol). So, when treated with a ketone or aldehyde in acidic conditions, catechol can behave much like ethylene glycol, giving formation of a cyclic acetal:

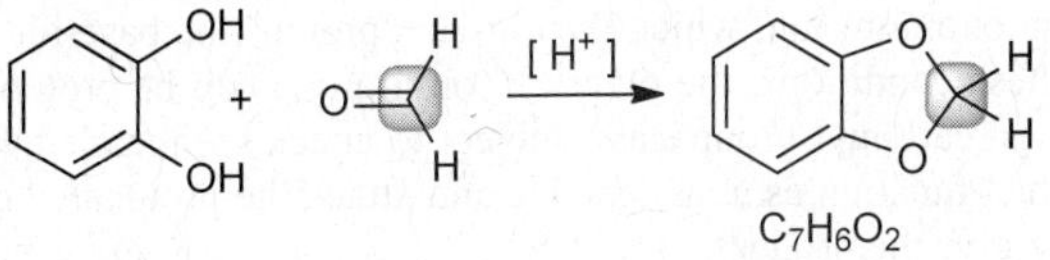

$C_7H_6O_2$

20.60.
(a) A ketone is reduced to a secondary alcohol upon treatment with LAH followed by water work-up.

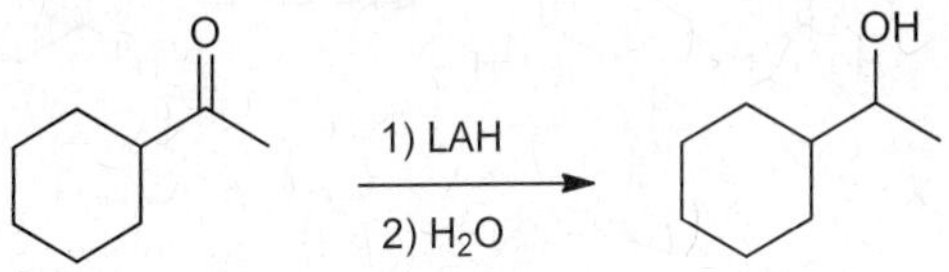

(b) When treated with phenyl magnesium bromide, a ketone will undergo a Grignard reaction, resulting in a tertiary alcohol (with installation of a phenyl group), as shown:

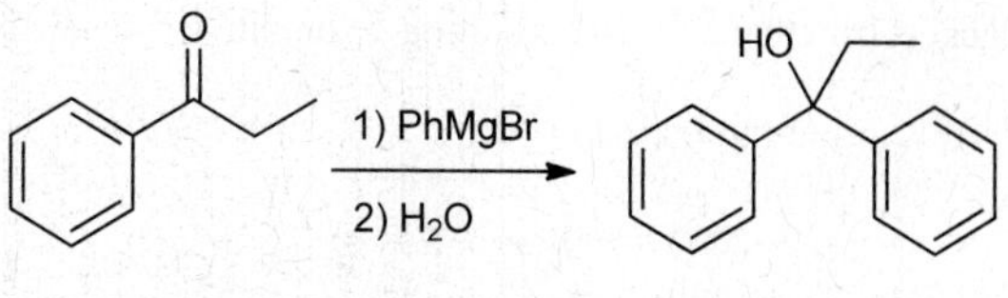

(c) A ketone is converted into an alkene when treated with a Wittig reagent. This particular Wittig reagent installs a methylene (CH_2) group, to give the following product:

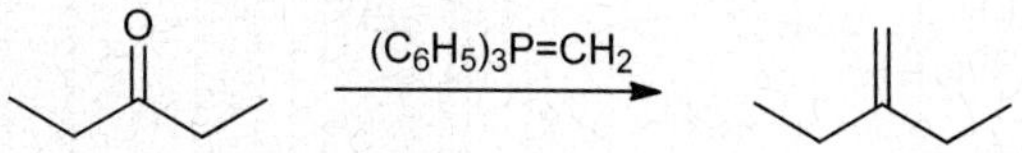

20.61.

(a) This transformation requires installation of a methyl group, which can be achieved with a Grignard reaction (by treating the starting material with methyl magnesium bromide, followed by aqueous work-up). The resulting tertiary alcohol can then be heated with concentrated sulfuric acid to give an E1 process, generating the desired alkene.

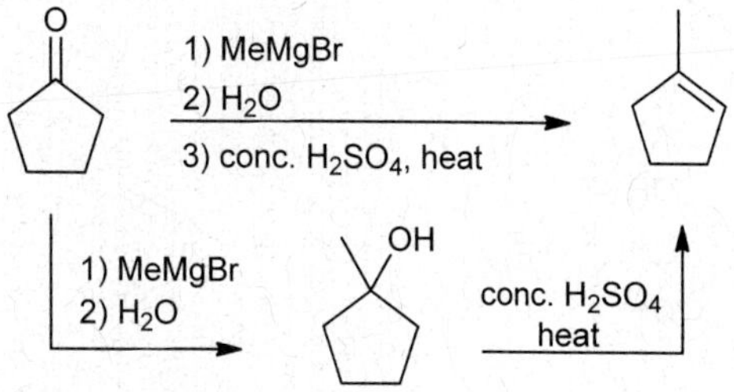

(b) This transformation requires installation of a carbon atom, while converting the carbonyl group into a C=C bond. This can be achieved in just one step, via a Wittig reaction:

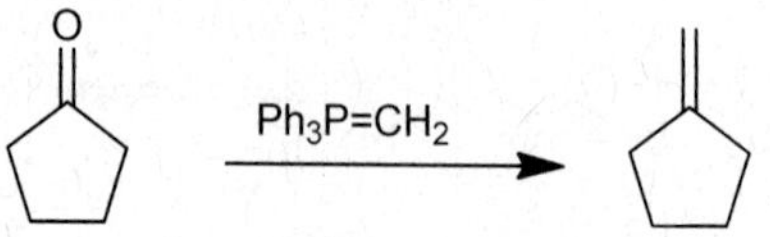

(c) We have seen that this transformation can be achieved in just two steps: 1) conversion of the ketone to a cyanohydrin, followed by 2) hydrolysis of the cyano group to give a carboxylic acid:

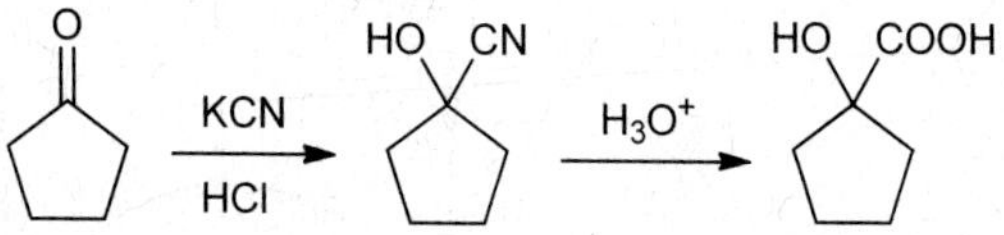

(d) This transformation involves insertion of an oxygen atom next to the carbonyl group of a ketone, thereby converting the ketone into a cyclic ester. This can be accomplished with a Baeyer-Villiger oxidation, which requires the use of a peroxyacid (such as MCPBA).

20.62. Under acid-catalyzed conditions, one of the carbonyl groups can be protonated, rendering it more electrophilic, and subject to attack by a nucleophile. Water serves as a nucleophile, attacking the protonated carbonyl group to give an oxonium ion, which then loses a proton (the base for this deprotonation step is likely a molecule of water). Under these conditions, the other carbonyl group can be protonated, rendering it more electrophilic (just as we saw with the first carbonyl group), and subject to attack by a nucleophile. One of the OH groups (present in the intermediate), can then function as a nucleophile and attack the protonated carbonyl group, giving an oxonium ion, which is deprotonated to give the product.

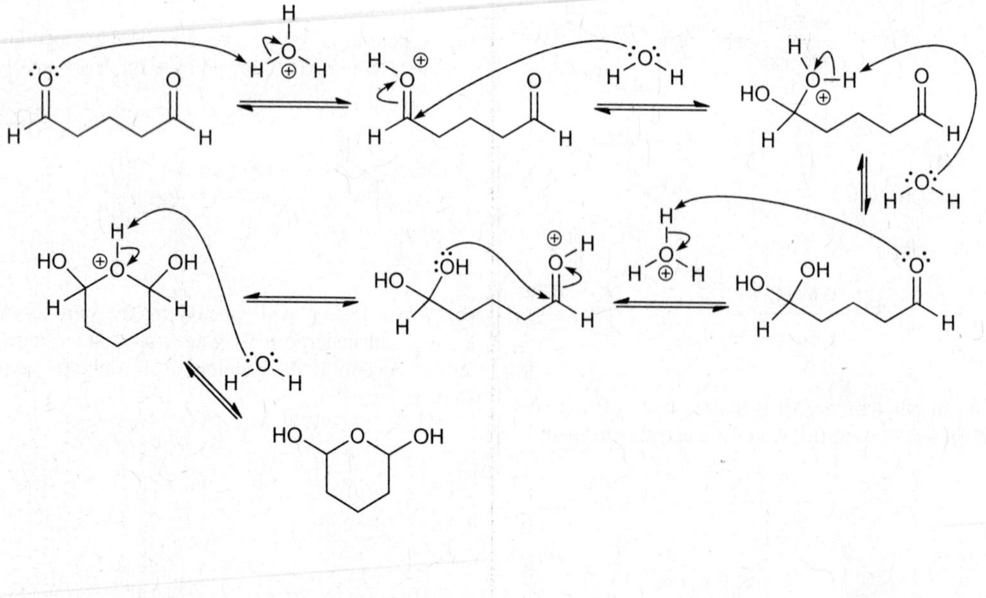

20.63.

(a) To determine the products of hydrolysis of an enamine, we must first identify the bond that will undergo cleavage. Bond cleavage is expected for the bond between the nitrogen atom and the sp^2-hybridized carbon atom to which it is attached:

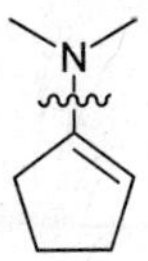

Therefore, the following carbon atom will be converted into a carbonyl group:

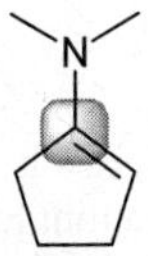

As a result of the C-N bond cleavage, the carbon atom becomes a carbonyl group, and the nitrogen atom will accept a proton to generate a secondary amine (dimethyl amine):

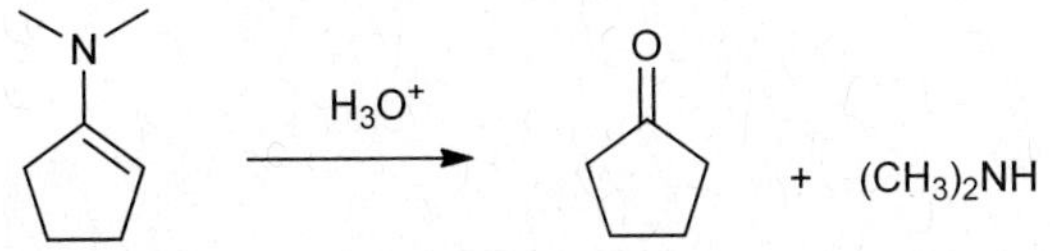

$$ \xrightarrow{H_3O^+} \qquad + \quad (CH_3)_2NH $$

(b) To determine the products of hydrolysis of an imine, we must first identify the bond that will undergo cleavage. Bond cleavage is expected for the C=N bond:

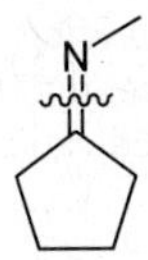

Therefore, the following carbon atom will be converted into a carbonyl group:

As a result of cleavage of the C=N bond, the carbon atom becomes a carbonyl group, and the nitrogen atom will accept two protons to generate a primary amine (methyl amine):

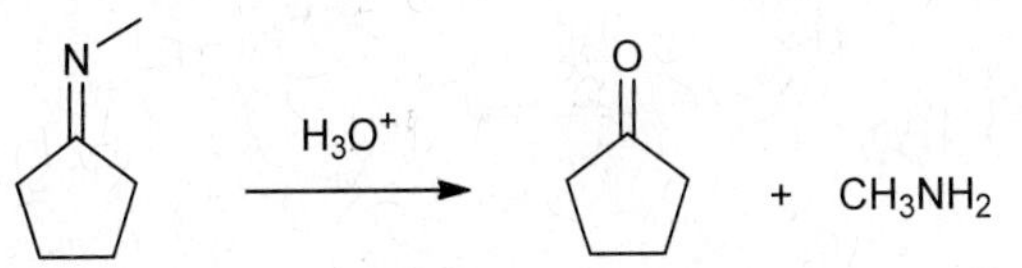

$$ \xrightarrow{H_3O^+} \qquad + \quad CH_3NH_2 $$

(c) To determine the products of hydrolysis of an acetal, we must first identify the bonds that will undergo cleavage. Bond cleavage is expected for the C-O bonds of the acetal group:

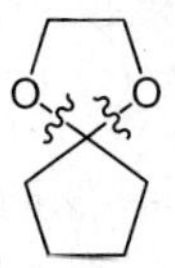

Therefore, the following carbon atom will be converted into a carbonyl group:

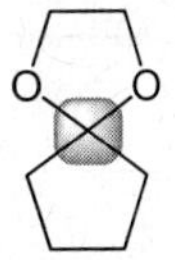

In the process, a diol is released, as shown:

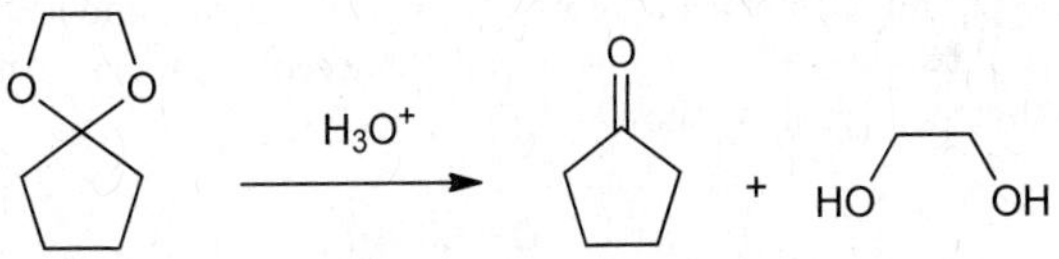

$$ \xrightarrow{H_3O^+} \qquad + \quad HO\text{——}OH $$

(d) The starting compound exhibits a carbon atom that is connected to two oxygen atoms, as shown, and is therefore an acetal:

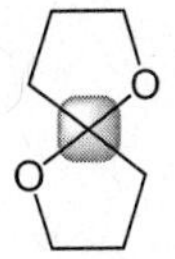

When an acetal is treated with aqueous acid, it will undergo hydrolysis. During acetal hydrolysis, cleavage occurs for the C-O bonds of the acetal group:

Each of these bonds is broken, thereby converting the carbon atom of the acetal group into a carbonyl group. In the process, each of the oxygen atoms will accept a proton to become an OH group, giving the following product, which exhibits a carbonyl group, as well as two OH groups:

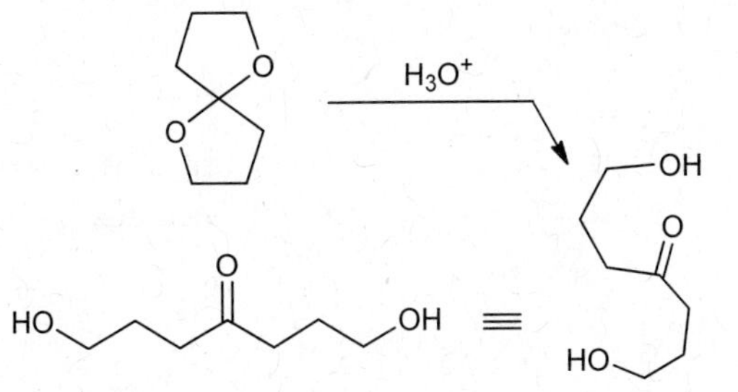

$$ \xrightarrow{H_3O^+} $$

(e) The starting compound exhibits a carbon atom that is connected to two oxygen atoms, as shown, and is therefore an acetal:

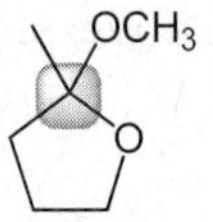

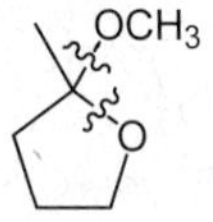

Each of these bonds is broken, thereby converting the carbon atom of the acetal group into a carbonyl group. In the process, each of the oxygen atoms will accept a proton to become an OH group, giving a hydroxy-ketone and methanol:

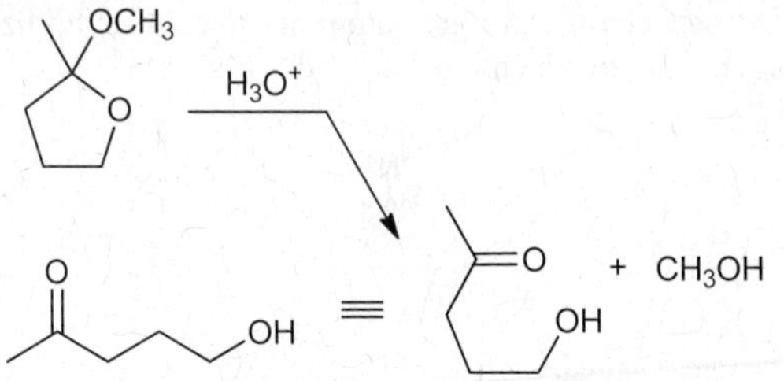

When an acetal is treated with aqueous acid, it will undergo hydrolysis. During acetal hydrolysis, cleavage occurs for the C-O bonds of the acetal group:

20.64. This compound contains three different functional groups that will each undergo hydrolysis. The enamine (upper left corner) is hydrolyzed to give a ketone and a secondary amine. The imine (bottom left) is hydrolyzed to give an aldehyde and an amino group (tethered together). And finally, the cyclic acetal is hydrolyzed to give a ketone and ethylene glycol, as shown:

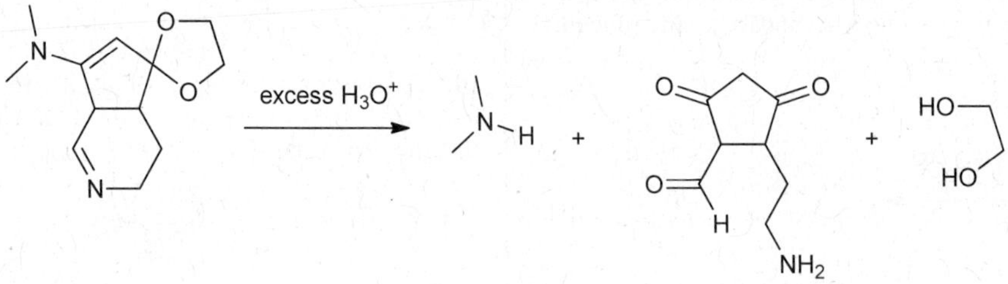

20.65.
(a) This transformation involves hydrolysis of an enamine under acidic conditions, and therefore, we propose a mechanism that is the reverse of the mechanism for enamine formation (all of the same intermediates, but in reverse order). Under acid-catalyzed conditions, the π bond of the enamine can be protonated, giving an iminium ion which is then attacked by water to give an oxonium ion. Loss of a proton gives a carbinolamine intermediate. Protonation of the amino group gives a good leaving group, which leaves to give a resonance-stabilized cation that is deprotonated to give the product.

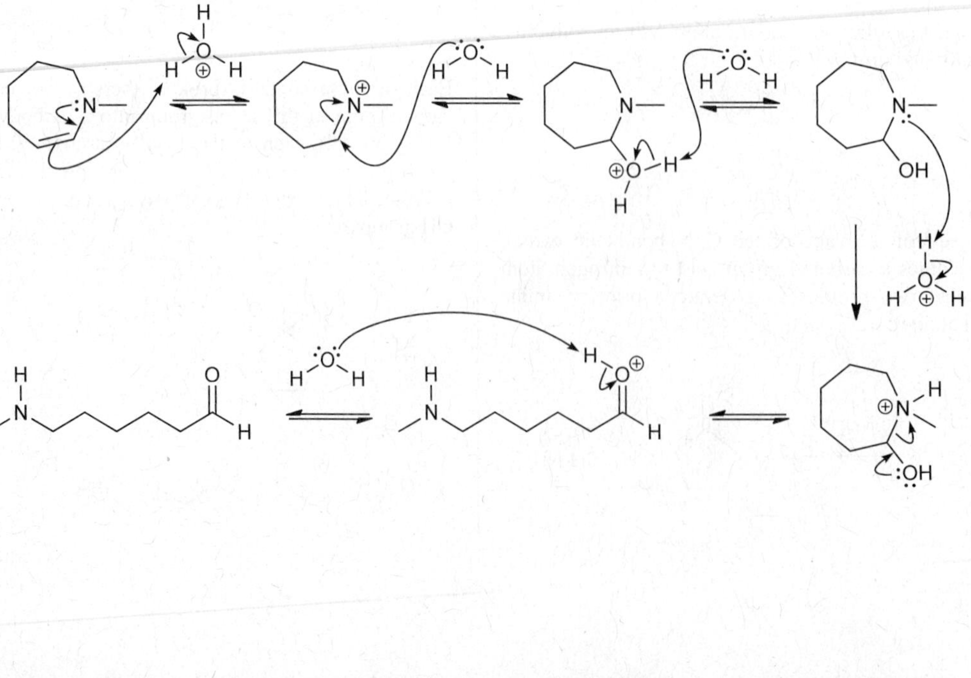

(b) This transformation involves hydrolysis of an imine under acidic conditions, and therefore, we propose a mechanism that is the reverse of the mechanism for imine formation (all of the same intermediates, but in reverse order). Under acid-catalyzed conditions, the imine is protonated, giving an iminium ion which is then attacked by water to give an oxonium ion. Loss of a proton gives a carbinolamine intermediate. Protonation of the amino group gives a good leaving group, which leaves to give a resonance-stabilized cation that is deprotonated to give the product.

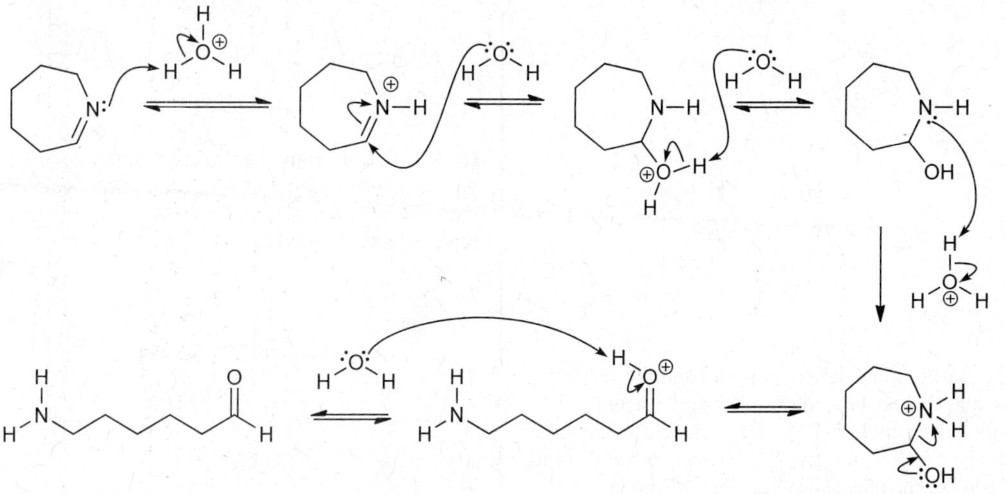

(c) This transformation involves hydrolysis of an acetal under acidic conditions, and therefore, we propose a mechanism that is the reverse of the mechanism for acetal formation (all of the same intermediates, but in reverse order). Under acid-catalyzed conditions, one of the oxygen atoms can be protonated (we can protonate either oxygen atom, and the end result will be the same), giving an oxonium ion. Loss of leaving group gives a resonance-stabilized cation that can be attacked by water to give another oxonium ion. Deprotonation gives a hemiacetal (water likely functions as the base for this step). Protonation, followed by loss of a leaving group gives a protonated aldehyde, which is deprotonated to give the product.

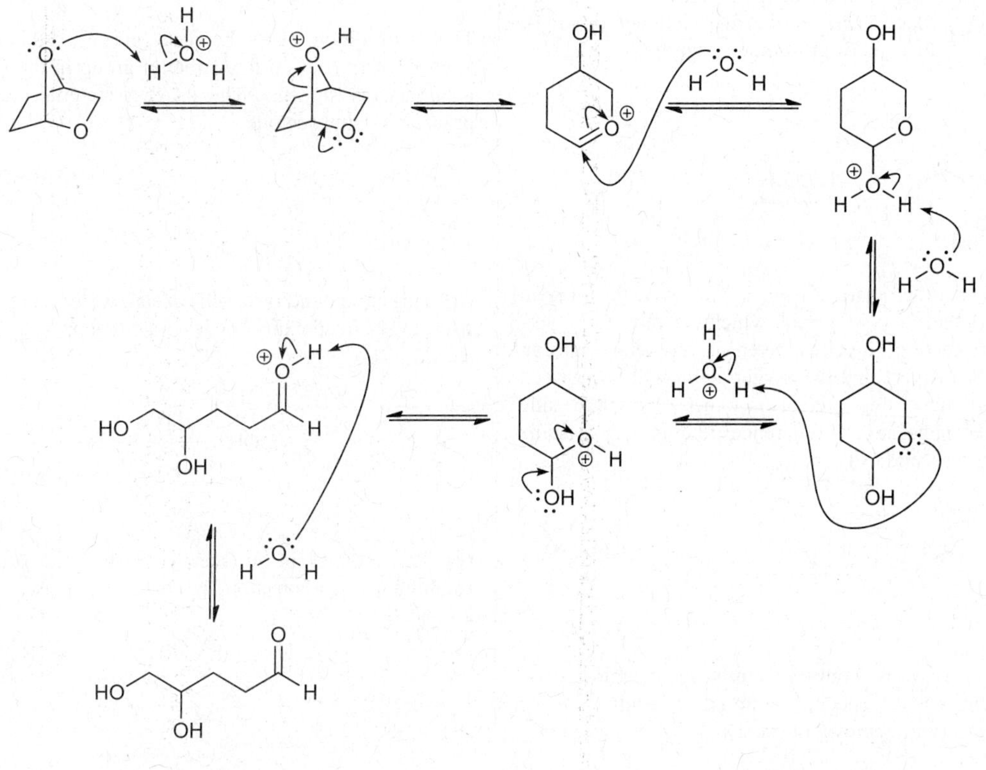

20.66.
(a) The starting material is a ketone, and the reagent is a substituted derivative of hydrazine. We therefore expect the following substituted hydrazone as a product:

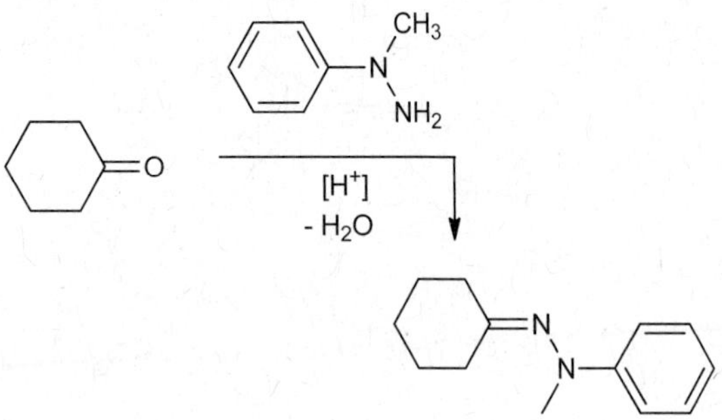

(b) The starting material is a ketone, and the reagent is phenyl magnesium bromide (a Grignard reagent, which is a very strong nucleophile). The Grignard reagent will attack the ketone to give a tertiary alcohol (with installation of the phenyl group at the α position):

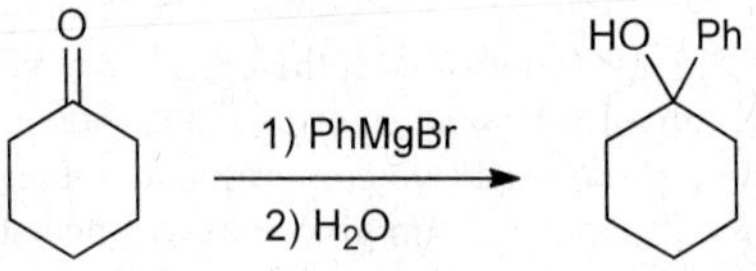

(c) The starting material is a ketone, and the reagent (CH$_3$CO$_3$H) is a peroxy acid, which indicates a Baeyer-Villiger reaction, thereby converting the ketone into an ester. We expect that the oxygen atom will be inserted on the right side (tertiary) rather than left side (phenyl), because of differences in migratory aptitude (tertiary > phenyl).

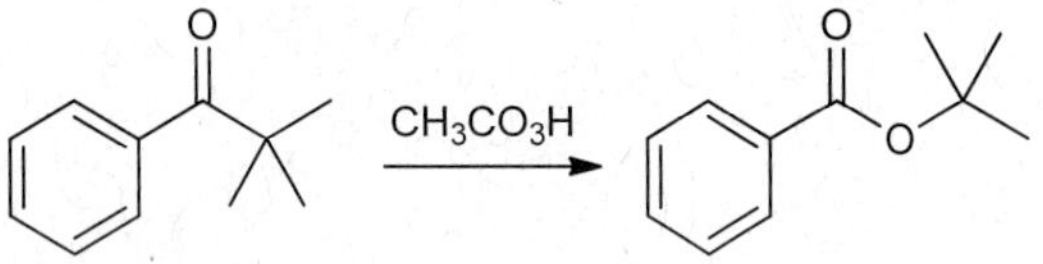

(d) The starting material is a ketone, and the reagent (CH$_3$CO$_3$H) is a peroxy acid, which indicates a Baeyer-Villiger reaction, thereby converting the ketone into an ester. We expect that the oxygen atom will be inserted on the right side (tertiary) rather than left side (secondary), because of differences in migratory aptitude (tertiary > secondary).

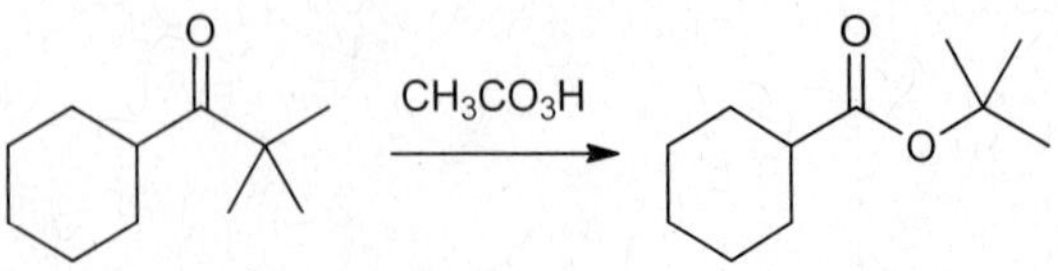

(e) A ketone is converted into an enamine upon treatment with a secondary amine in acid catalyzed conditions (with removal of water).

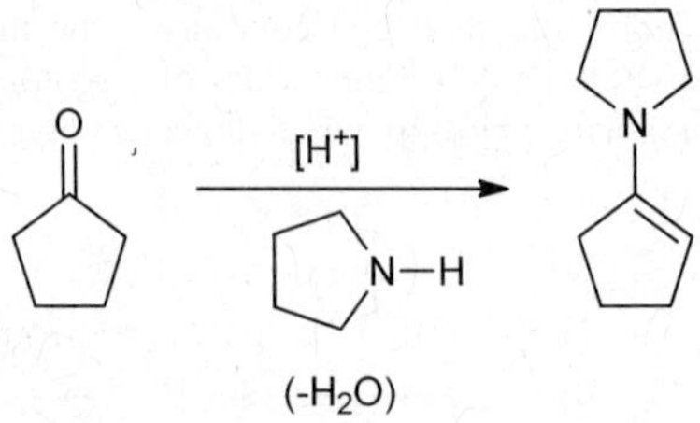

(f) A ketone is converted into an imine upon treatment with a primary amine in acid catalyzed conditions (with removal of water). In the process, the C=O bond is replaced a C=N bond.

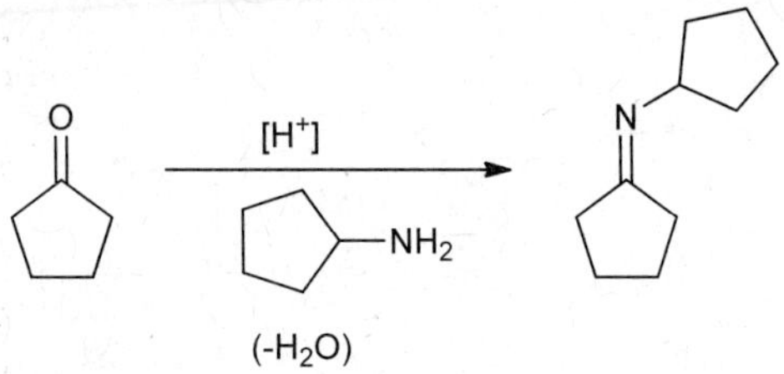

20.67.
(a) The desired product is a cyclic acetal, which will require a diol and the appropriate ketone or aldehyde. To determine the identity of the starting ketone or aldehyde, we find the carbon atom that is connected to two oxygen atoms (highlighted):

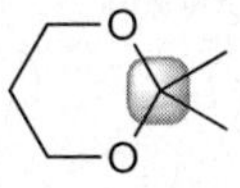

This carbon atom bears the acetal group, so this carbon atom must have been the carbonyl group in the starting materials, as shown, The starting materials are 1,3-propanediol and acetone.

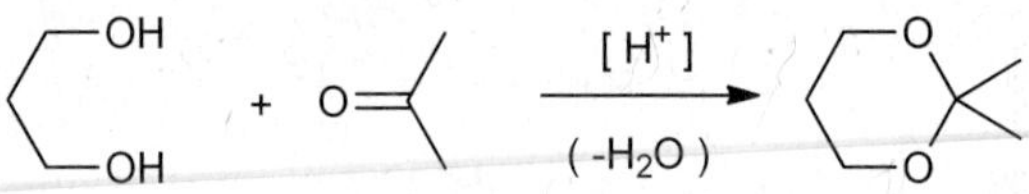

(b) This compound is a cyclic acetal, which can be made from the corresponding hydroxy-ketone and ethanol, as shown:

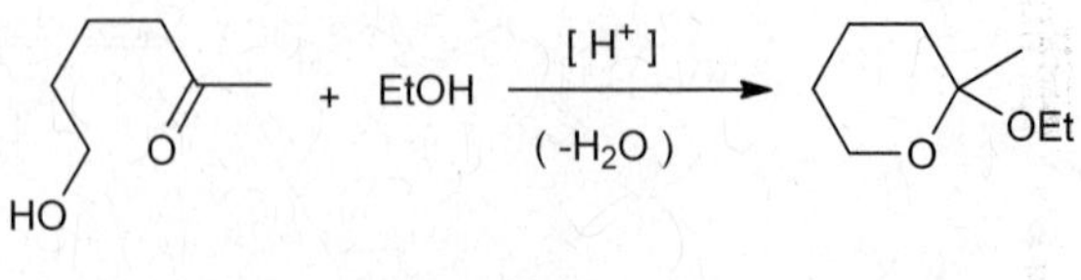

(c) The product is an acetal, because the following (highlighted) carbon atom is connected to two oxygen atoms:

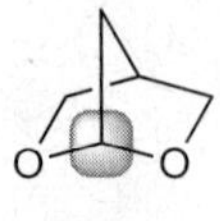

This carbon atom must have been the carbonyl group in the starting material, as shown:

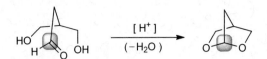

The starting material exhibits a carbonyl group, as well as two OH groups, and can be redrawn like this:

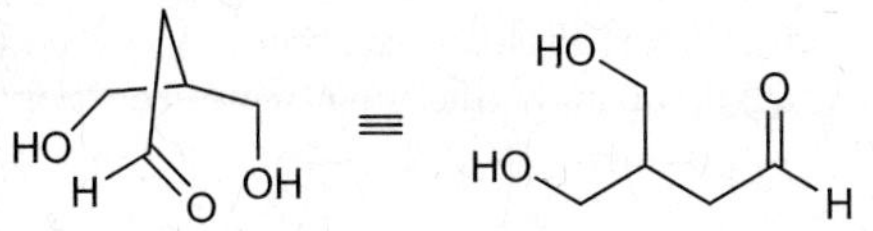

20.68. The desired product is an acetal, because the following (highlighted) carbon atom is connected to two oxygen atoms:

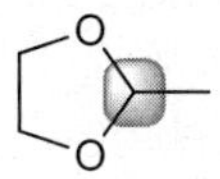

This carbon atom must have been the carbonyl group in the starting material, as shown:

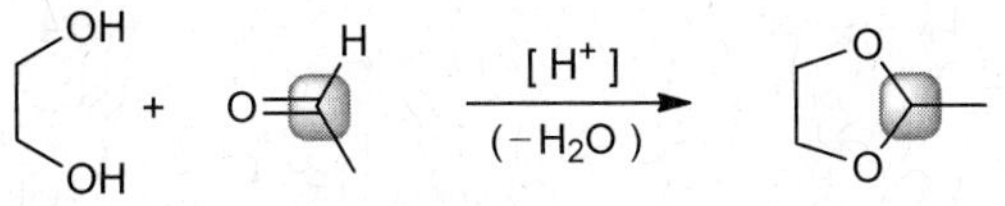

The diol and aldehyde shown above (ethylene glycol and acetaldehyde) can both be prepared from ethanol, as shown in the following retrosynthetic analysis. An explanation of each of the steps (a-d) follows.

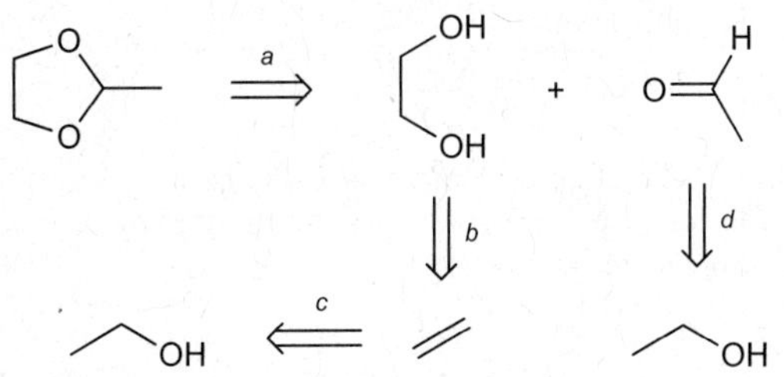

a. The acetal can be made from ethylene glycol and acetaldehyde.
b. Ethylene glycol can be made from ethylene via dihydroxylation.
c. Ethylene can be made from ethanol via acid-catalyzed dehydration.
d. Acetaldehyde can be prepared from ethanol via oxidation (with PCC)

Now let's draw the forward scheme. Ethanol is heated with concentrated sulfuric acid to give ethylene. Subsequent treatment with potassium permanganate (or osmium tetroxide) gives ethylene glycol.
Another equivalent of ethanol is oxidized with PCC to give an aldehyde, which is then treated with ethylene glycol in acid-catalyzed conditions (with removal of water) to give the desired acetal, as shown.

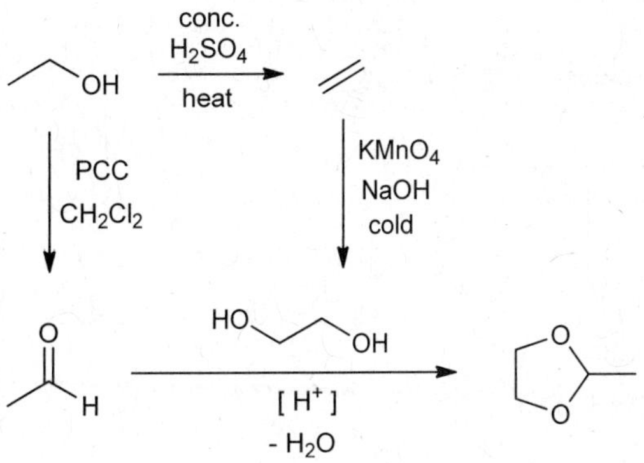

20.69.
(a) The desired product is an acetal, because the following (highlighted) carbon atom is connected to two oxygen atoms:

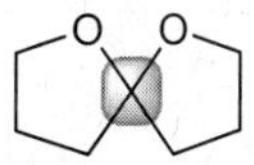

Therefore, this acetal can be made (via acetal formation) from the following dihydroxyketone:

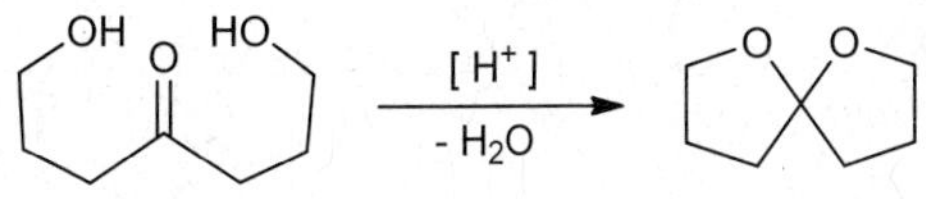

Now we must find a way to convert the starting material into the dihydroxyketone above. It might be tempting to perform ozonolysis on the starting material, followed by reduction:

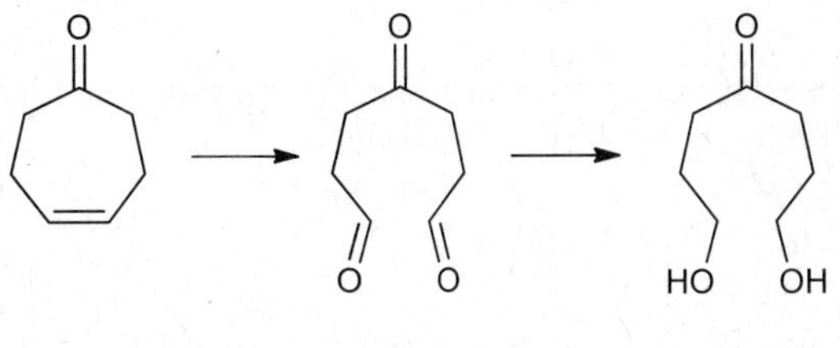

However, the reduction step is problematic, because we don't know a way to selectively reduce the aldehyde groups in the presence of a ketone. Therefore, treatment with a reducing agent (such as LAH) would result in a triol:

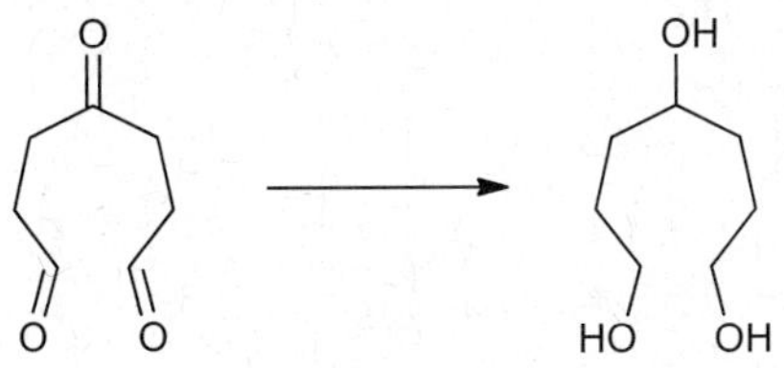

To circumvent this problem, we must first protect the ketone, before opening the ring with ozonolysis.
The entire synthesis is summarized here:

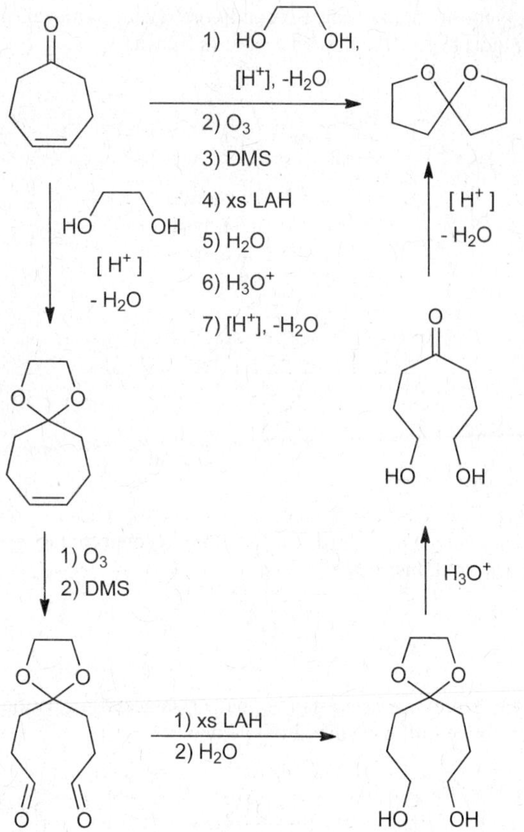

1) HO‿OH,
   [H⁺], -H₂O
2) O₃
3) DMS
4) xs LAH
5) H₂O
6) H₃O⁺
7) [H⁺], -H₂O

[H⁺]
- H₂O

HO‿OH
[H⁺]
- H₂O

1) O₃
2) DMS

1) xs LAH
2) H₂O

H₃O⁺

(b) The desired product is an acetal, because the following (highlighted) carbon atom is connected to two oxygen atoms:

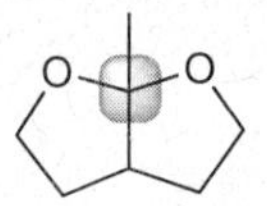

Therefore, this acetal can be made (via acetal formation) from the following dihydroxyketone:

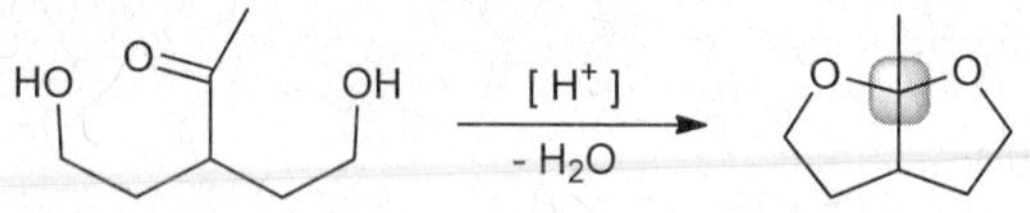

[H⁺]
- H₂O

Now we must find a way to convert the starting material into the dihydroxyketone above. It might be tempting to simply reduce the cyclic ester with xs LAH:

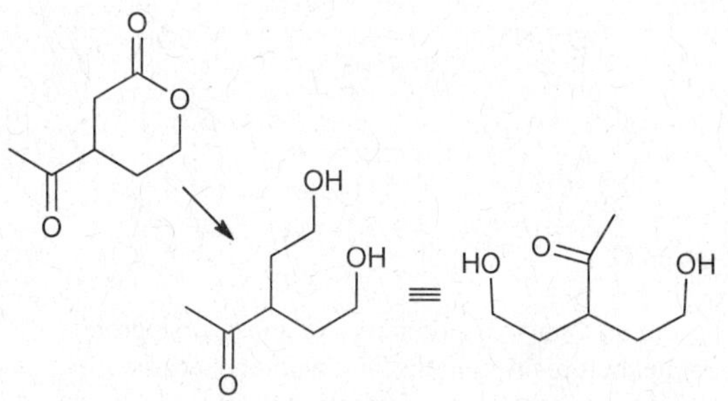

However, this reduction step is problematic, because treatment with a reducing agent (such as LAH) would result in a triol:

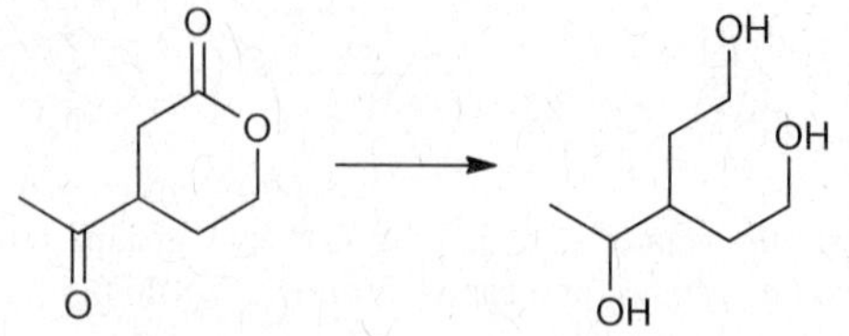

To circumvent this problem, we must first protect the ketone, before opening the ring with reduction. The entire synthesis is summarized here:

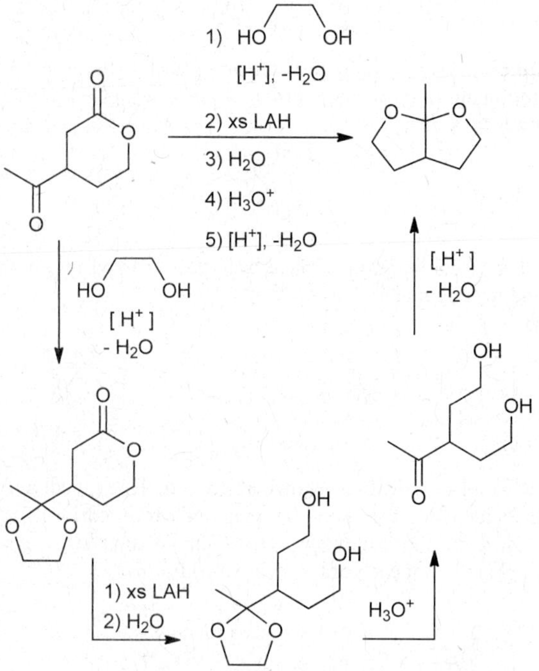

1) HO‿OH
   [H⁺], -H₂O
2) xs LAH
3) H₂O
4) H₃O⁺
5) [H⁺], -H₂O

HO‿OH
[H⁺]
- H₂O

[H⁺]
- H₂O

1) xs LAH
2) H₂O

H₃O⁺

20.70. The starting compound exhibits a carbon atom that is connected to two oxygen atoms, as shown, and is therefore an acetal:

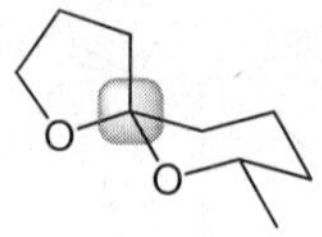

When an acetal is treated with aqueous acid, it is expected to undergo hydrolysis. We first identify the bonds that will undergo cleavage. When an acetal undergoes hydrolysis, cleavage occurs for the C-O bonds of the acetal group:

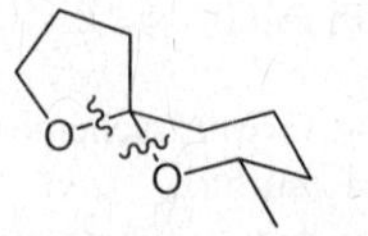

Each of these bonds is broken, thereby converting the carbon atom of the acetal group into a carbonyl group.

In the process, each of the oxygen atoms will accept a proton to become an OH group, giving the following product, which exhibits a carbonyl group, as well as two OH groups:

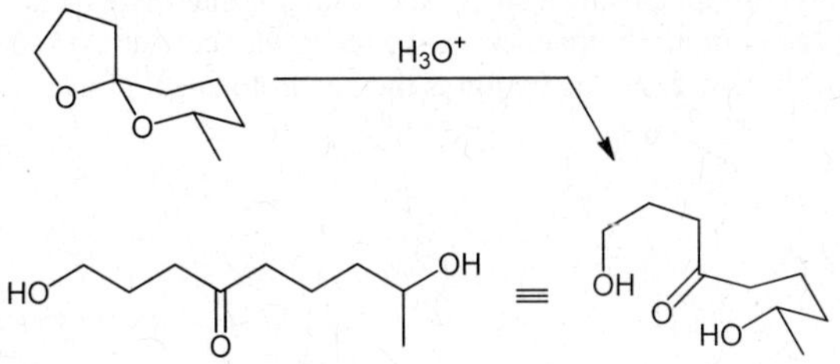

20.71.

(a) The product has three more carbon atoms than the starting material, which requires a C-C bond-forming reaction. Also, the position of the double bond must be moved. The extra three carbon atoms and the double bond can both be installed in the correct location if the last step of our synthesis is a Wittig reaction:

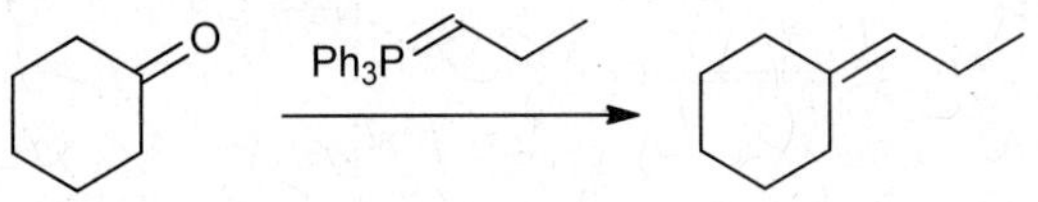

This strategy requires converting the starting alkene (cyclohexene) into the ketone shown above (cyclohexanone), which can be achieved in two steps (via acid-catalyzed hydration, followed by oxidation with chromic acid). A Wittig reaction then gives the final product:

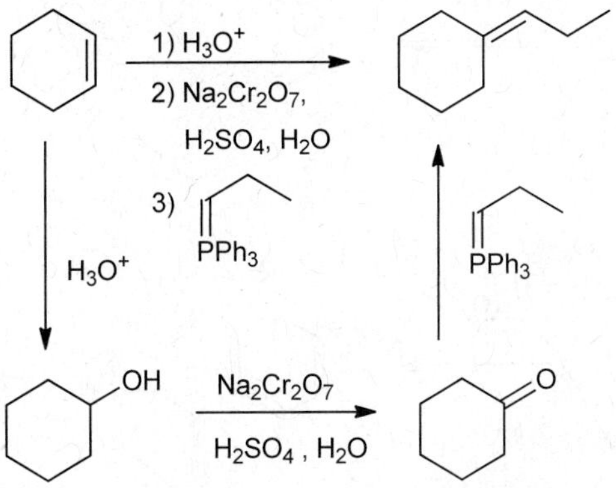

(b) The product is an acetal, which can be made from the corresponding ketone:

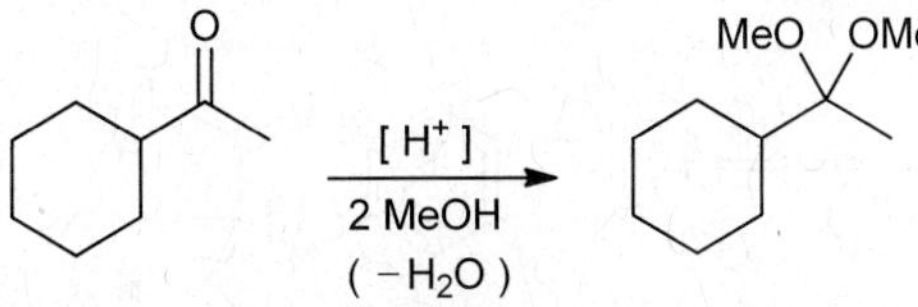

This ketone can be made from the starting material in just two steps. First, the dibromide is converted to an alkyne upon treatment with excess sodium amide (via two successive E2 reactions), followed by water work-up (to protonate the resulting alkynide ion). The terminal alkyne is then treated with aqueous acid, in the presence of mercuric sulfate, giving a hydration reaction. The initially formed enol will rapidly tautomerize to give a ketone. Conversion of the ketone to an acetal then gives the desired product.

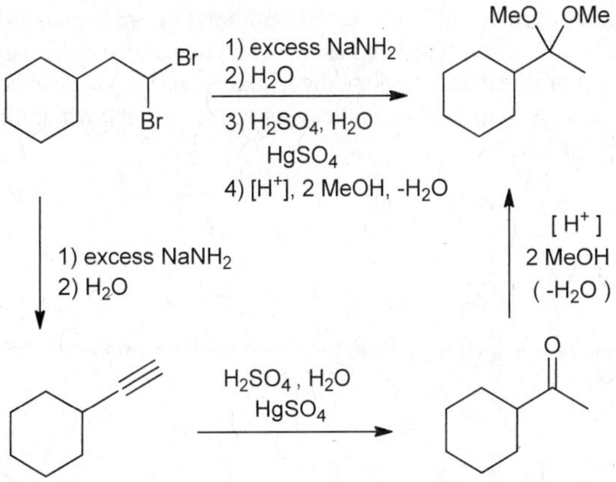

(c) The desired product is an acetal, because the following (highlighted) carbon atom is connected to two oxygen atoms:

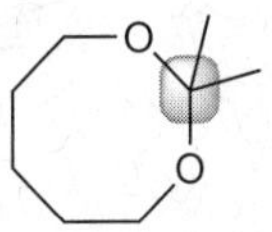

Therefore, this acetal can be made (via acetal formation) from the following ketone and diol:

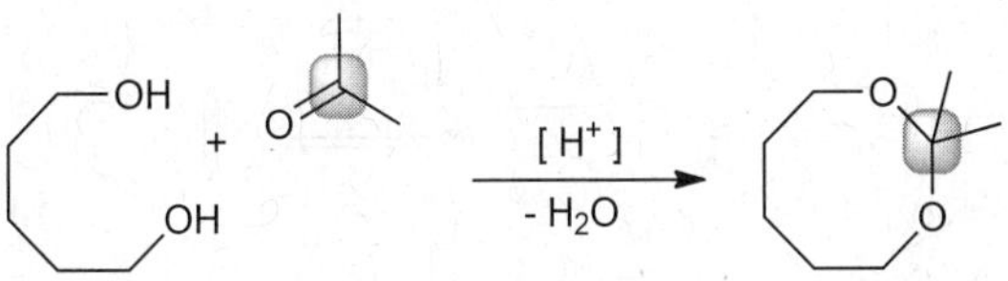

Now we must find a way to convert the starting cycloalkene into the diol shown above. This can be achieved in just two steps. First, the ring is opened with an ozonolysis reaction to produce a dialdehyde. Then, the dialdehyde is reduced to a diol upon treatment with two equivalents of a reducing agent. Acetal formation then gives the final product.

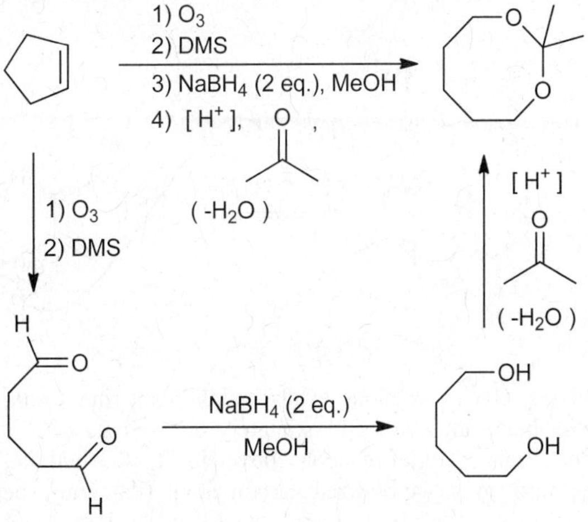

20.72. A lone pair on the nitrogen atom attacks one of the carbonyl groups, giving an intermediate that is protonated (under acidic conditions). The resulting cation is then deprotonated, followed by protonation of the OH group, thereby converting a bad leaving group into a good leaving group (water). Loss of the leaving group gives a cation, which is then deprotonated. The second carbonyl group is then attacked by a nucleophile, this time in an intramolecular fashion. The NH$_2$ group functions as a nucleophile and attacks the carbonyl group, closing a ring. The resulting intermediate is then protonated to remove the negative charge, followed by deprotonation. Protonation of the OH group converts a bad leaving group into a good leaving group (water), which then leaves. The resulting cation is then deprotonated to give the product.

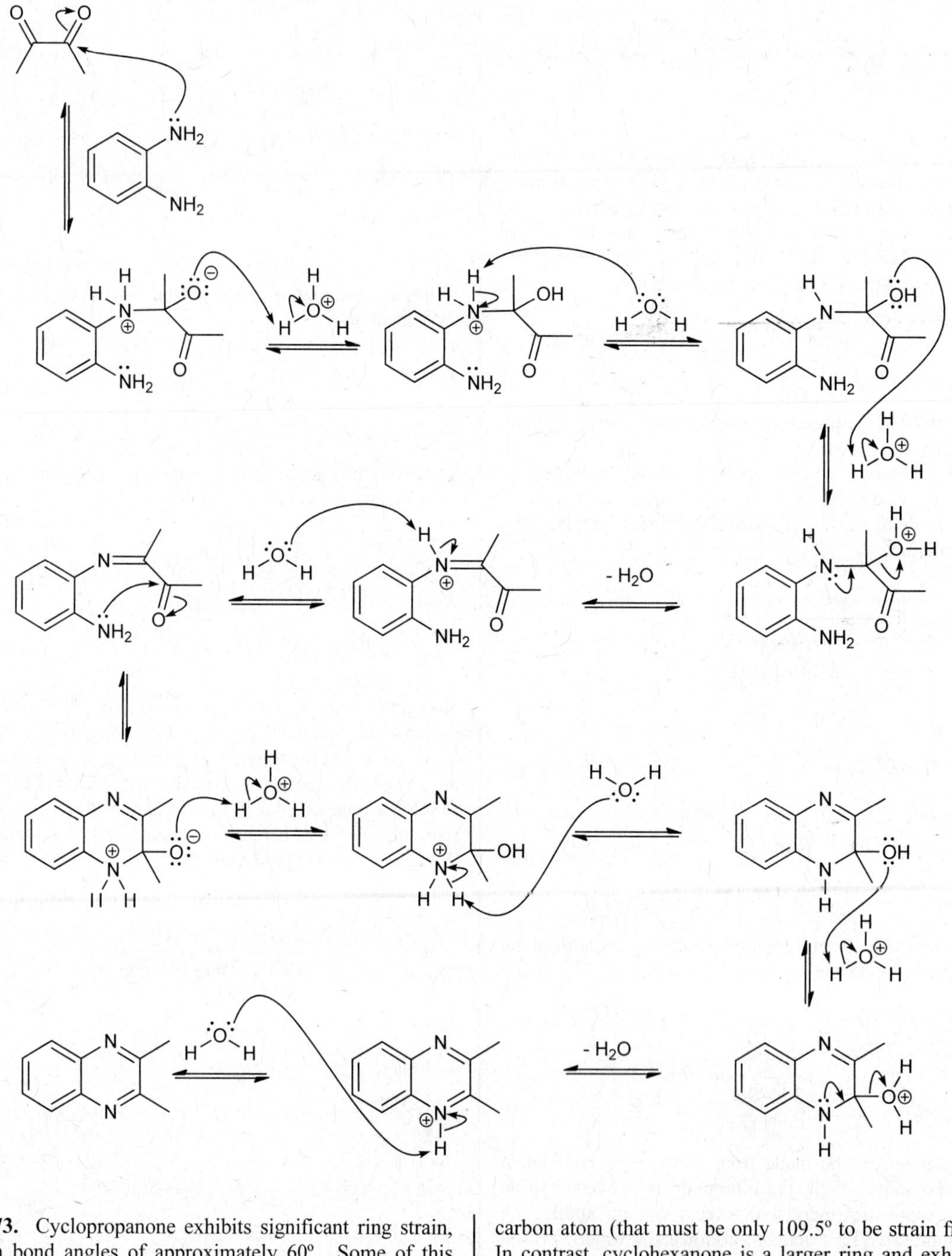

20.73. Cyclopropanone exhibits significant ring strain, with bond angles of approximately 60°. Some of this ring strain is relieved upon conversion to the hydrate, because an sp^2-hybridized carbon atom (that must be 120° to be strain free) is replaced by an sp^3-hybridized carbon atom (that must be only 109.5° to be strain free). In contrast, cyclohexanone is a larger ring and exhibits only minimal ring strain. Conversion of cyclohexanone to its corresponding hydrate does not alleviate a significant amount of ring strain.

20.74. 1,2-dioxane has two adjacent oxygen atoms and is therefore a peroxide. Like other peroxides, it is extremely unstable and potentially explosive.
1,3-dioxane has two oxygen atoms separated by one carbon atom. This compound is therefore an acetal. Like other acetals, it is only stable under basic conditions, but undergoes hydrolysis under mildly acidic conditions.
1,4-dioxane is neither a peroxide nor an acetal. It is therefore stable under basic conditions as well as mildly acidic conditions, and is used as a common solvent because of its inert behavior.

20.75.
(a) The desired product is an acetal, because the following (highlighted) carbon atom is connected to two oxygen atoms:

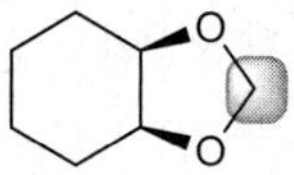

Therefore, this acetal can be made (via acetal formation) from the following diol and formaldehyde:

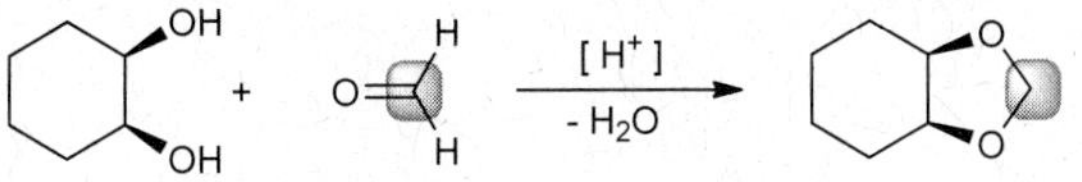

The diol can be made from the starting alkene in just one step, giving the following two-step synthesis:

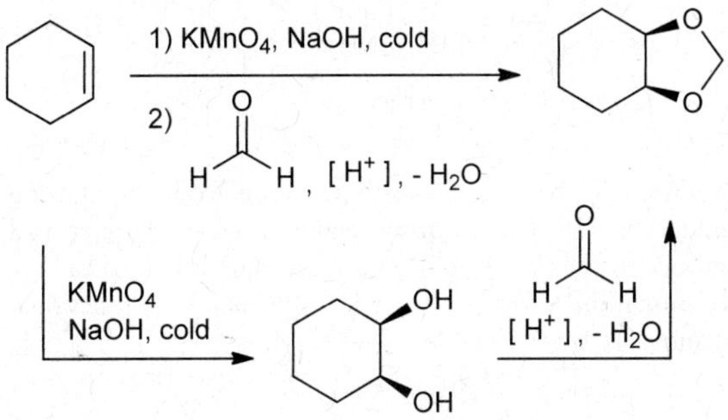

(b) There are certainly many acceptable solutions to this problem. One such solution derives from the following retrosynthetic analysis. An explanation of each of the steps (*a-c*) follows.

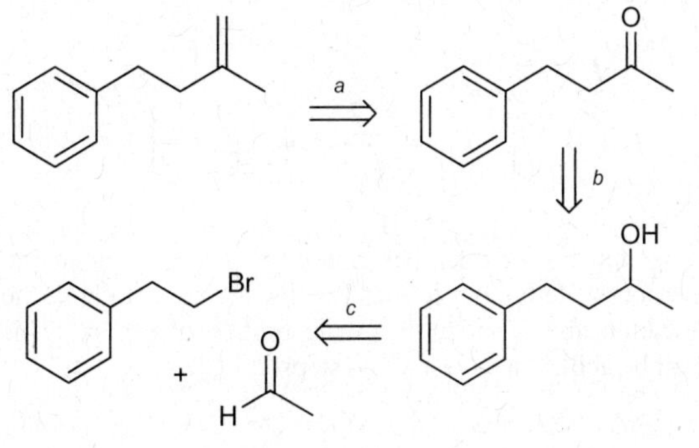

a. The desired alkene can be made from a ketone via a Wittig reaction.
b. The ketone can be made via oxidation of the corresponding secondary alcohol.
c. The alcohol can be made from the starting material via a Grignard reaction.

Now let's draw the forward scheme. Treating the starting material with magnesium gives a Grignard reagent, which is then treated with acetaldehyde (to give a Grignard reaction), followed by water work-up, to give the alcohol. Oxidation of the alcohol with chromic acid gives a ketone, which can then be converted into the desired product upon treatment with a Wittig reagent:

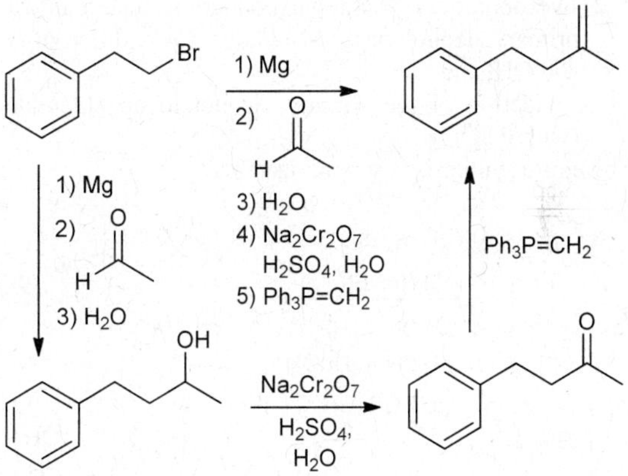

(c) The desired product is an acetal, because the following (highlighted) carbon atom is connected to two oxygen atoms:

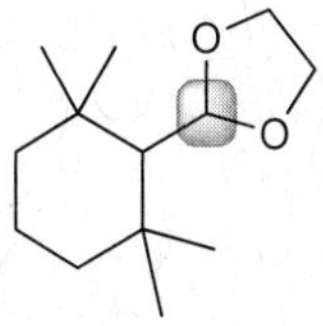

Therefore, this acetal can be made (via acetal formation) from the following aldehyde:

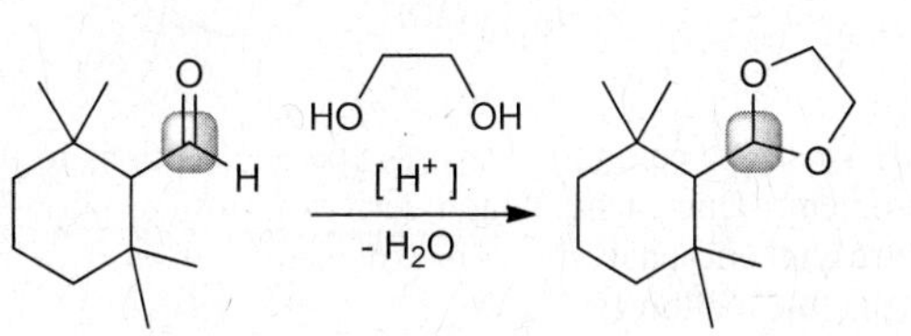

So we will need to make this aldehyde from the starting material. But the starting material lacks a functional group, so the first step of our synthesis must be a radical bromination process in order to install a functional group:

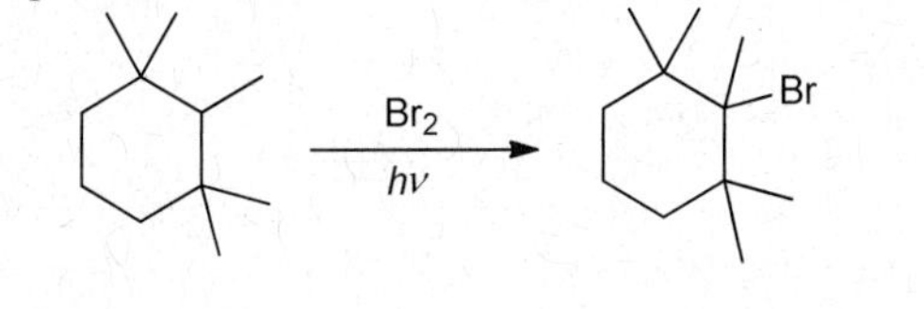

With the functional group installed, we must now bridge the gap between the first and last steps of the synthesis:

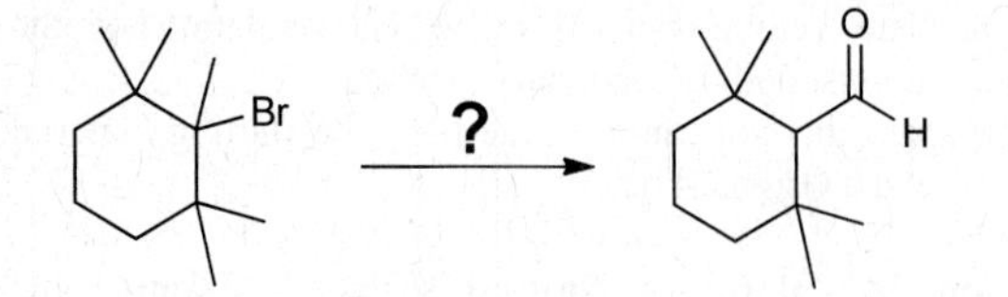

This transformation does not involve a change in the carbon skeleton, but it does involve a change in both the location and the identity of the functional group. This can be achieved in just a few steps:

1) elimination to give an alkene (upon treatment with a strong base),
2) hydroboration-oxidation to convert the alkene into a primary alcohol via *anti*-Markovnikov addition of H and OH, and
3) oxidation of the primary alcohol to an aldehyde (with PCC).

The entire synthesis is summarized here:

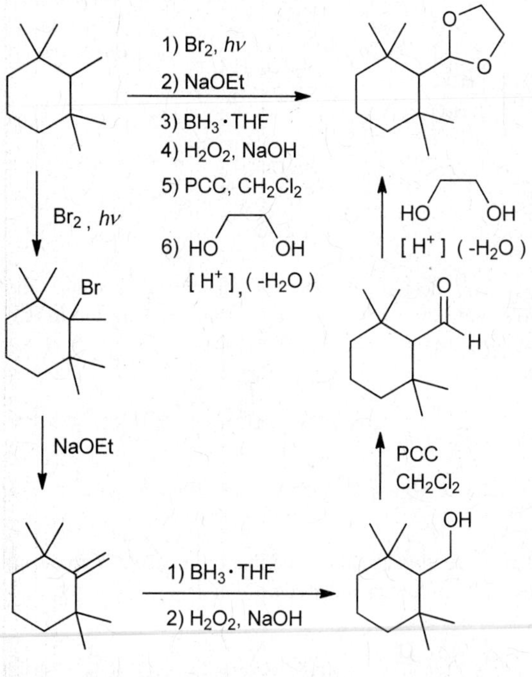

(d) There are certainly many acceptable solutions to this problem. One such solution derives from the following retrosynthetic analysis. An explanation of each of the steps (*a-c*) follows.

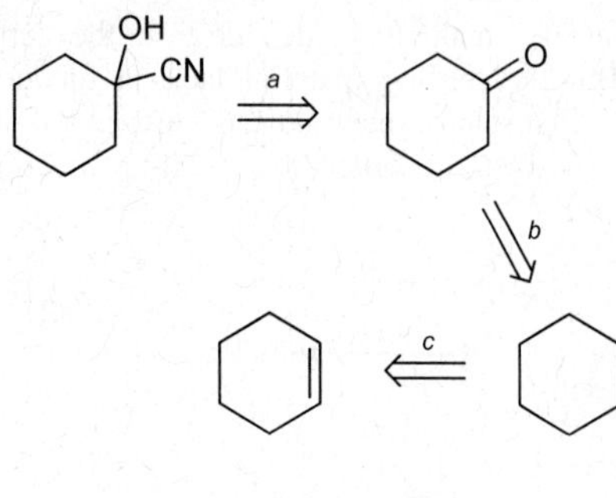

a. The product is a cyanohydrin, which can be made from the corresponding ketone.
b. The ketone can be made via oxidation of the corresponding secondary alcohol.
c. The alcohol can be made from the starting alkene via acid-catalyzed hydration.

Now let's draw the forward scheme. The starting alkene is converted to an alcohol upon treatment with aqueous acid. Oxidation of the alcohol with chromic acid gives a ketone, which can then be converted into the desired cyanohydrin upon treatment with KCN and HCl:

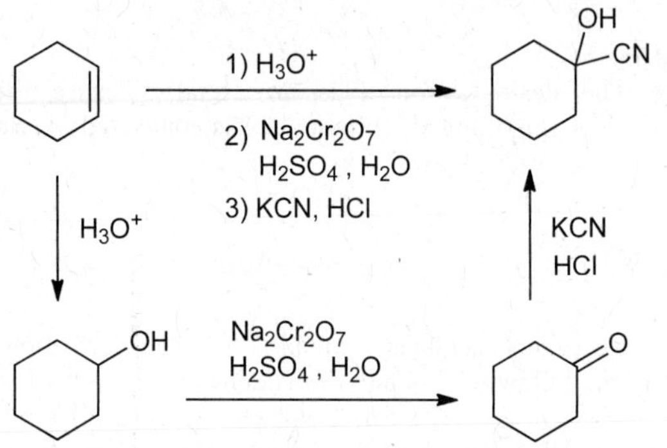

(e) The desired product is an imine, which can be made from the corresponding ketone:

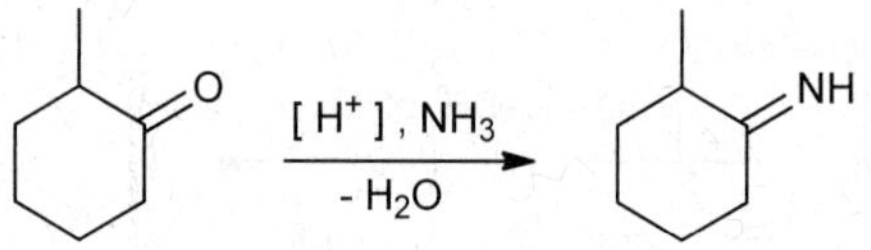

So we will need to make this ketone from the starting material. But the starting material lacks a functional group, so the first step of our synthesis must be a radical bromination process in order to install a functional group:

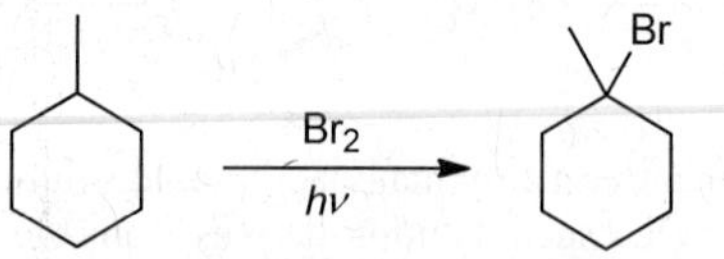

With the functional group installed, we must now bridge the gap between the first and last steps of the synthesis:

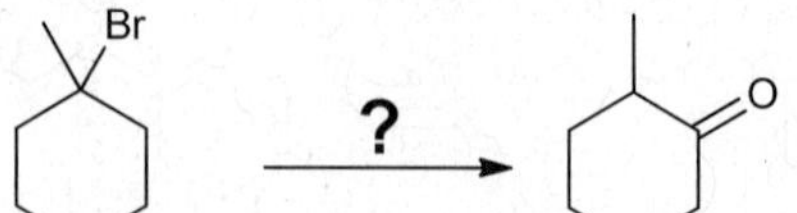

This transformation does not involve a change in the carbon skeleton, but it does involve a change in both the location and the identity of the functional group. This can be achieved in just a few steps:

1) elimination to give an alkene (upon treatment with a strong base),
2) hydroboration-oxidation to convert the alkene into a secondary alcohol via *anti*-Markovnikov addition of H and OH, and
3) oxidation of the secondary alcohol to a ketone (with chromic acid).

The entire synthesis is summarized here:

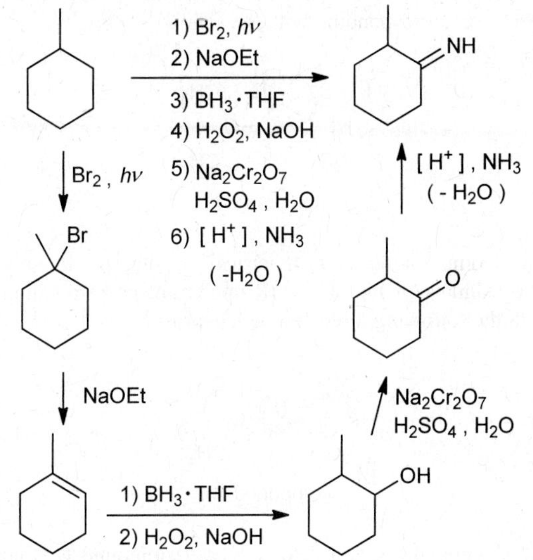

(f) This transformation does not involve a change in the carbon skeleton, but it does involve a change in both the location and the identity of the functional group:

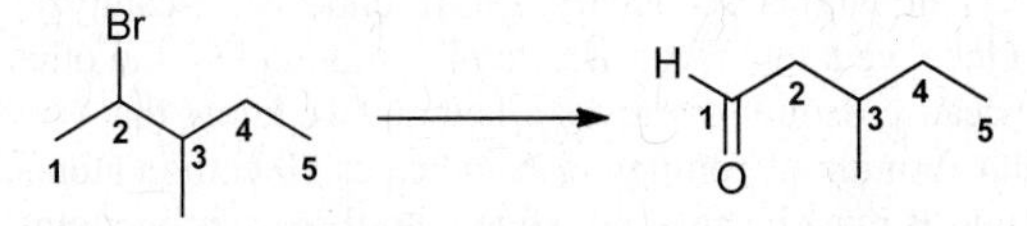

This can be achieved in just a few steps:
1) elimination with a strong, sterically hindered base to give the less substituted alkene,
2) hydroboration-oxidation to convert the alkene into a primary alcohol via *anti*-Markovnikov addition of H and OH, and
3) oxidation of the primary alcohol to an aldehyde (with PCC):

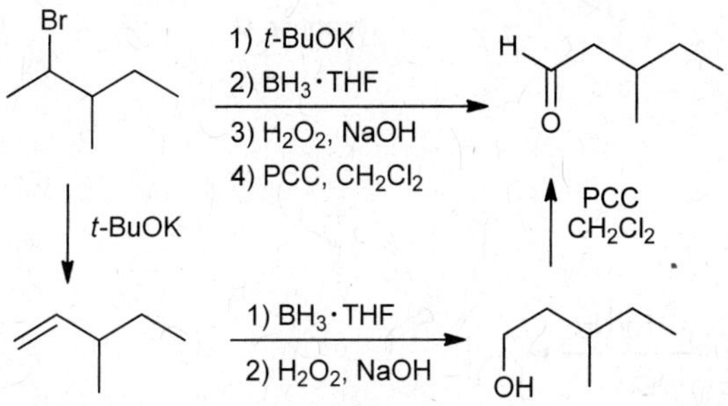

(g) The product is an enamine, which can be prepared from the corresponding ketone:

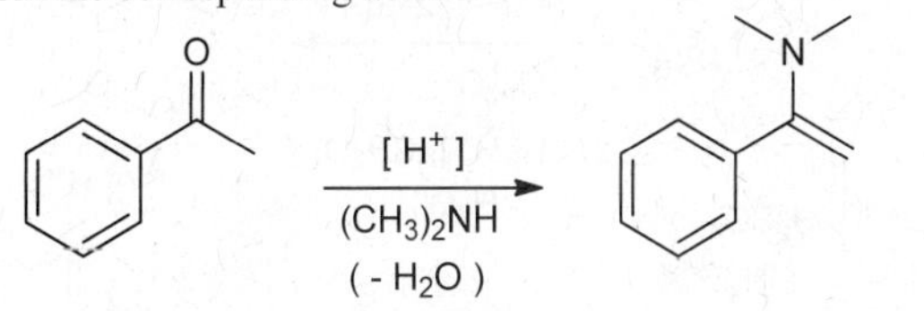

This ketone can be made from benzene via a Friedel-Crafts acylation:

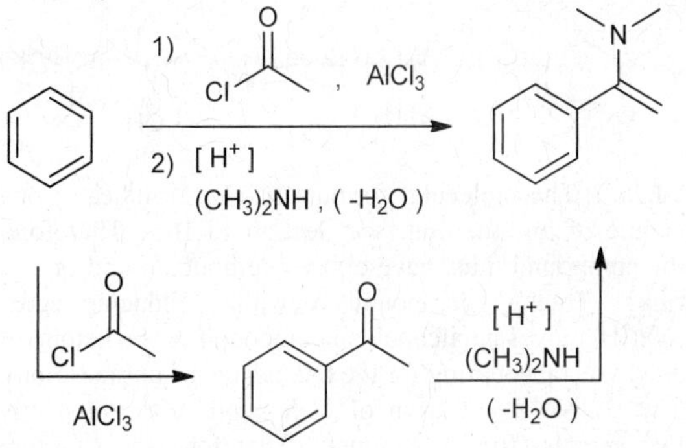

(h) The desired product is an acetal, because the following (highlighted) carbon atom is connected to two oxygen atoms:

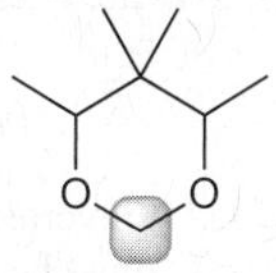

Therefore, this acetal can be made (via acetal formation) from the following diol and formaldehyde, as shown:

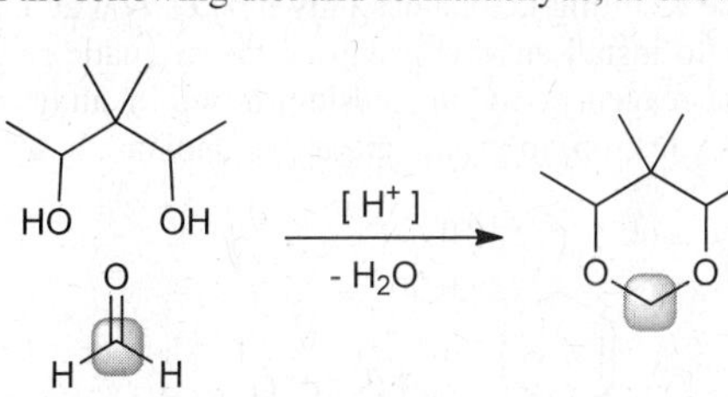

So we will need to make this diol from the starting material. This can be achieved in just two steps. First, the starting diyne is treated with sulfuric acid, in the presence of mercuric sulfate, to give a dione. Then, the dione can be reduced with two equivalents of LAH, followed by water work-up, to give the diol, which is then converted into the desired acetal.

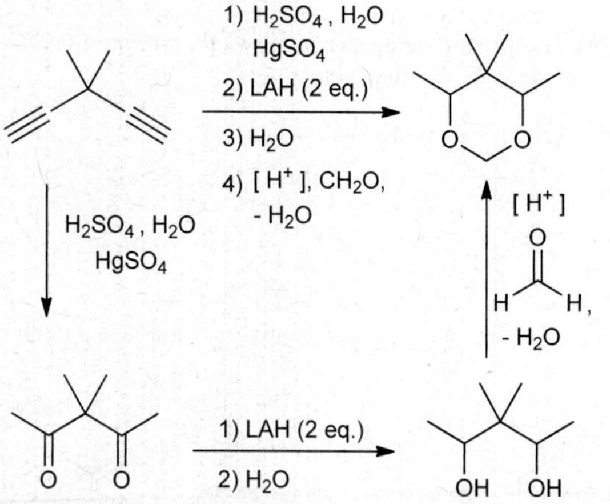

20.76. The molecular formula ($C_7H_{14}O$) indicates one degree of unsaturation (see Section 15.16). Therefore, the compound must have either one double bond or one ring. Treating compound **A** with a reducing agent ($NaBH_4$) gives an alcohol, so compound **A** is a ketone or aldehyde (accounting for the one degree of unsaturation). The ^1H NMR spectrum of compound **A** exhibits only two signals (for 14 protons). Therefore, the structure must be symmetrical. The two signals in the ^1H NMR spectrum are characteristic of an isopropyl group, which means that compound **A** must be diisopropyl ketone (or 2,4-dimethyl-3-pentanone). Conversion of compound **A** into a thioacetal, followed by desulfurization with Raney nickel, gives compound **B**. Notice that the carbonyl group has been reduced to a methylene (CH_2) group.

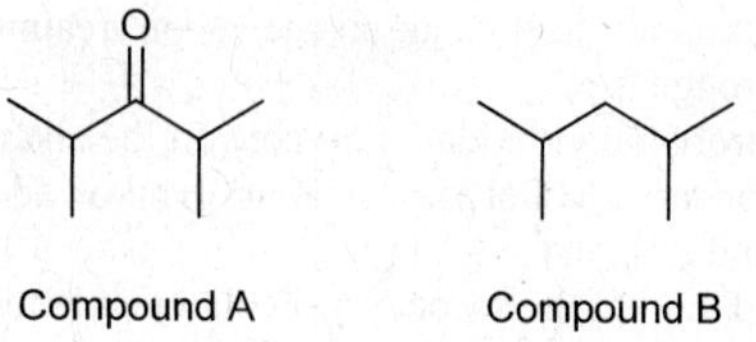

Compound A   Compound B

(a) Compound **B** is symmetrical, much like compound **A**, giving rise to only three signals in its ^1H NMR spectrum, corresponding with the following protons:

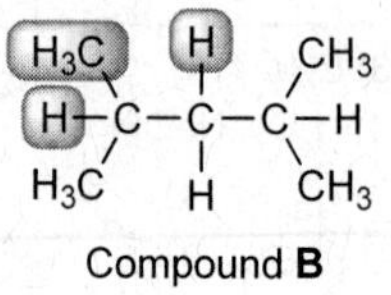

Compound **B**

(b) Compound **B** is symmetrical, giving rise to only three signals in its ^{13}C NMR spectrum, corresponding with the following three unique locations:

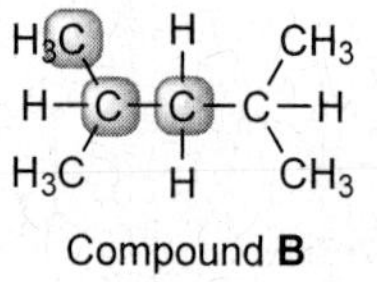

Compound **B**

(c) Compound **A** is a ketone, while compound **B** is an alkane. Therefore, compound **A** will exhibit a strong signal near 1715 cm^{-1}, while compound **B** will not exhibit a signal in the same region.

20.77. Compound **C** is converted to an enamine upon treatment with a secondary amine under acid-catalyzed conditions, so compound **C** must be the corresponding ketone. Once we know the structure of compound **C**, the other structures can be identified. Compound **A** must be an alkene, because ozonolysis gives a ketone (we know that only one carbon atom is lost during this process, because the molecular formula of compound **A** indicates 10 carbon atoms, while the resulting ketone has only nine carbon atoms). Compound **B** must be an acyl halide with three carbon atoms, in order to install an acyl group on the aromatic ring via a Friedel-Crafts acylation. Compound **D** is formed when a Grignard reagent (ethyl magnesium bromide) attacks the ketone to give an alkoxide ion which is then protonated (via aqueous work-up) to give a tertiary alcohol, as shown:

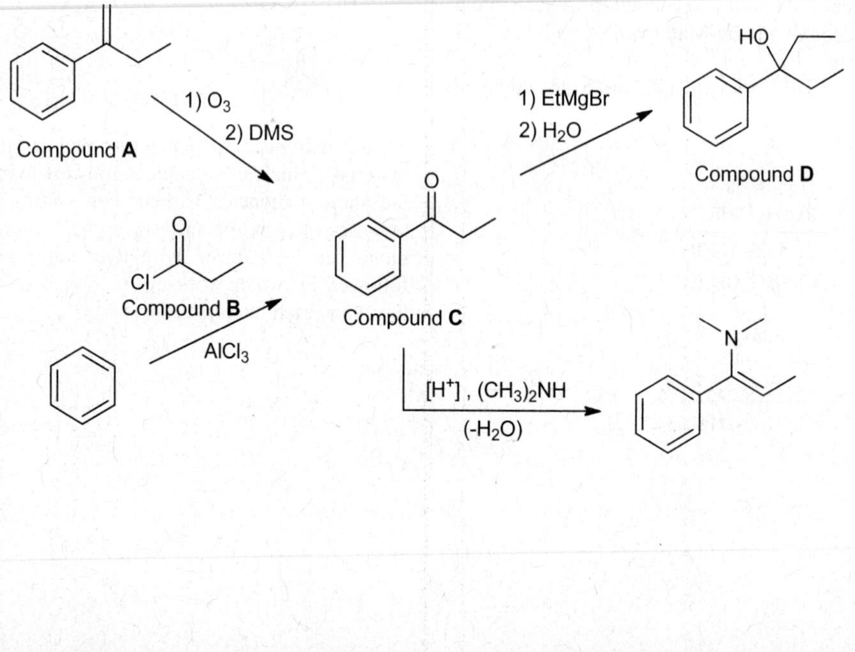

20.78. Cyclohexene is converted to cyclohexanol upon treatment with aqueous acid (acid-catalyzed hydration). Cyclohexanol is oxidized to cyclohexanone upon treatment with a strong oxidizing agent. Upon treatment with hydrazine in acid-catalyzed conditions, cyclohexanone is converted into the corresponding hydrazone. A Wolff-Kishner reduction then gives cyclohexane.

The conversion of cyclohexene to cyclohexane can be achieved more directly, in one step, via hydrogenation:

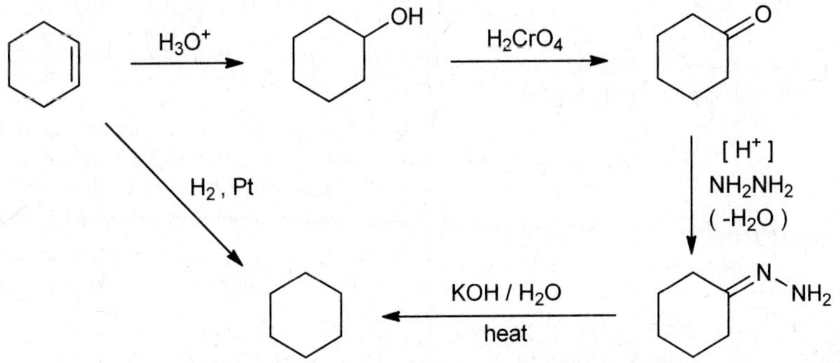

20.79. Benzene undergoes an electrophilic aromatic substitution reaction when treated with Br_2 and a Lewis acid, such as $FeBr_3$, to give bromobenzene. Subsequent treatment with magnesium gives a Grignard reagent (phenyl magnesium bromide), which reacts with formaldehyde to give benzyl alcohol (after aqueous work-up). This alcohol is oxidized by PCC to give benzaldehyde, which is then converted into an acetal upon treatment with ethylene glycol under acid-catalyzed conditions.

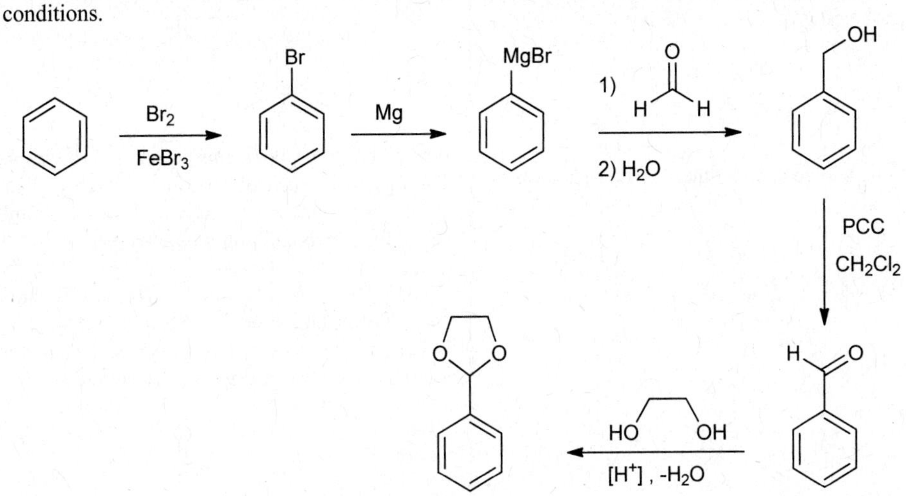

20.80.

(a) The problem statement indicates that the compound is an aldehyde. The molecular formula (C_4H_6O) indicates two degrees of unsaturation (see Section 15.16), but the aldehyde group only accounts for one of the degrees of unsaturation. The signal at 1715 cm^{-1} (stretching of the carbonyl group) indicates that the carbonyl group is NOT conjugated. If it were conjugated, the signal would be expected to appear at lower wavenumber (below 1700 cm^{-1}).

The following two structures are consistent with the requirements described above. The first aldehyde is acyclic and exhibits a double bond that is not conjugated to the carbonyl group, while the second aldehyde is cyclic:

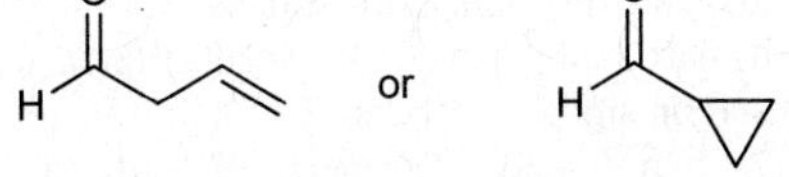

(b) The acyclic aldehyde would exhibit four signals in its ^{13}C NMR spectrum, while cyclopropyl carbaldehyde would exhibit only three signals in its ^{13}C NMR spectrum (two of the carbon atoms of the cyclopropyl group are identical).

20.81. The molecular formula ($C_9H_{10}O$) indicates five degrees of unsaturation (see Section 15.16), which is highly suggestive of an aromatic ring, in addition to either one double bond or one ring.

In the ^1H NMR spectrum, the signals near 7 ppm are likely a result of aromatic protons. Notice that the combined integration of these signals is 5H, indicating a monosubstituted aromatic ring:

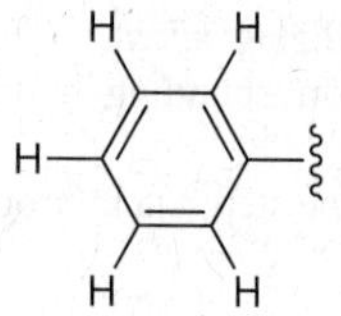

The spectrum also exhibits the characteristic pattern of an ethyl group (a quartet with an integration of 2 and a triplet with an integration of 3):

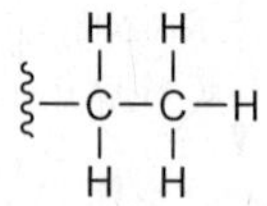

If we inspect the two fragments that we have determined thus far (the monosubstituted aromatic ring and the ethyl group), we will find that these two fragments account for nearly all of the atoms in the molecular formula ($C_9H_{10}O$). We only need to account for one more carbon atom and one oxygen atom. And let's not forget that our structure still needs one more degree of unsaturation, suggesting a carbonyl group:

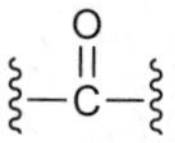

There is only one way to connect these three fragments.

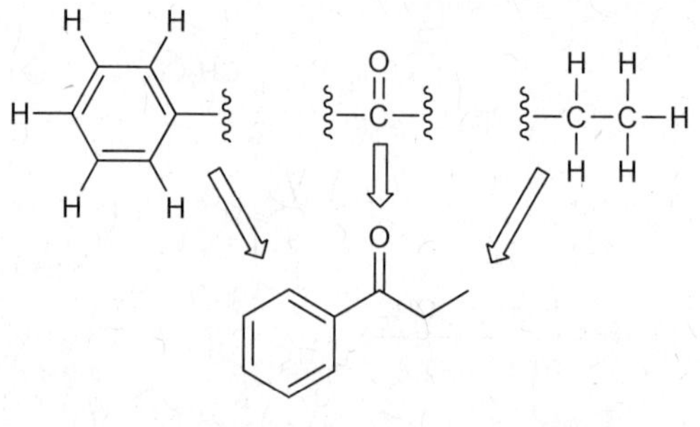

This structure is consistent with the ^{13}C NMR spectrum. The signal near 200 ppm corresponds with the carbon atom of the carbonyl group. A monosubstituted aromatic ring gives four signals between 100 and 150 ppm, and there are two signals between 0 and 50 ppm, corresponding to the carbon atoms of the ethyl group. The signal in the IR spectrum (at 1687 cm^{-1}) is consistent with a conjugated carbonyl group.

20.82. The molecular formula ($C_{13}H_{10}O$) indicates nine degrees of unsaturation (see Section 15.16), which is highly suggestive of two aromatic rings, in addition to either one double bond or one ring. The ^{13}C NMR spectrum exhibits only five signals, which must account for all thirteen carbon atoms in the compound. Therefore, many of the carbon atoms are identical, as a result of symmetry. There are four signals between 100 and 150 ppm, indicating a monosubstituted aromatic ring. To account for so many degrees of unsaturation, as

well as the symmetry that must be present, we propose two monosubstituted aromatic rings, rather than just one:

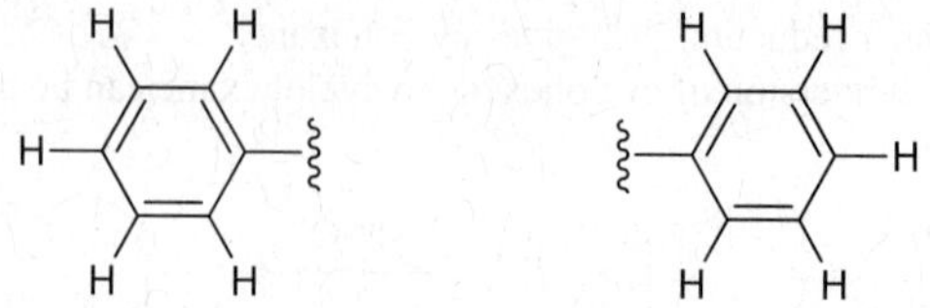

If we inspect these two fragments, we will find that they account for nearly all of the atoms in the molecular formula ($C_{13}H_{10}O$). We only need to account for one more carbon atom and one oxygen atom. And let's not forget that our structure still needs one more degree of unsaturation, suggesting a carbonyl group:

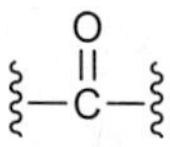

There is only one way to connect these fragments, as shown:

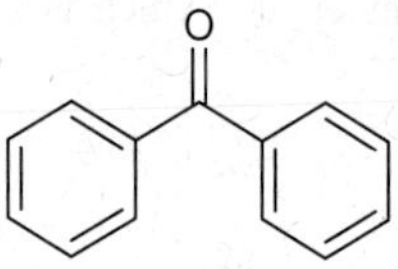

The carbonyl group in this compound (benzophenone) is conjugated to each of the rings, which explains why it produces a signal at a relatively low wavenumber (1660 cm^{-1}) for a carbonyl group of a ketone.

20.83. The molecular formula ($C_9H_{18}O$) indicates one degree of unsaturation (see Section 15.16). The problem statement indicates that the compound is a ketone, which accounts for the one degree of unsaturation:

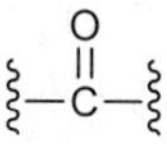

With only one signal in the ^1H NMR spectrum, the structure must have a high degree of symmetry, such that all eighteen protons are equivalent. This can be achieved with two tert-butyl groups:

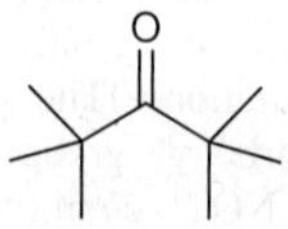

This compound is a ketone with a parent chain of five carbon atoms. The carbonyl group is located at C3, and there are four methyl groups (two at C2 and two at C4):

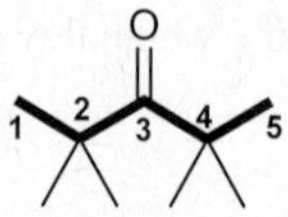

2,2,4,4-Tetramethyl-3-pentanone

20.84.

(a) In acid-catalyzed conditions, the starting material is protonated. There are two locations where protonation can occur (the lone pair of the nitrogen atom, or a lone pair of the oxygen atom). The nitrogen atom is more likely protonated first, because it is a stronger base (a protonated amine, called an ammonium ion, is a much weaker acid than a protonated ether, called an oxonium ion, as seen in the pK_a table on the inside cover of the textbook). Loss of a leaving group (dimethyl amine) gives a resonance-stabilized cation, which is then attacked by water. The resulting oxonium ion is then deprotonated to give a cyclic hemiacetal. Protonation, followed by loss of a leaving group gives a protonated carbonyl group, which then loses a proton to give the product. Notice that water functions as the base in each deprotonation step.

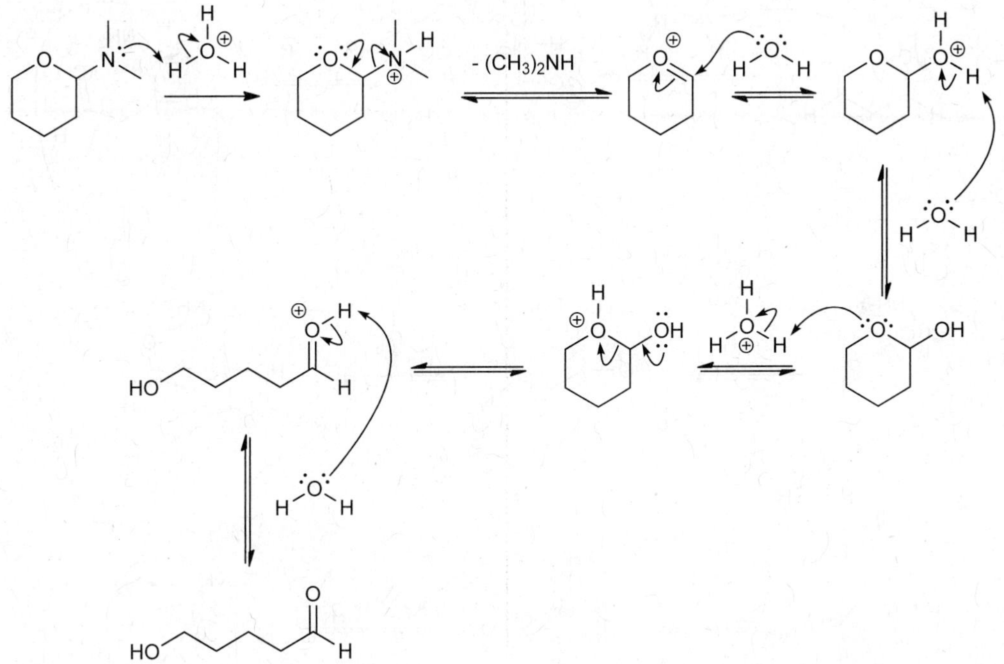

(b) The starting material is a vinyl ether, and it is being subjected to aqueous acidic conditions. This indicates that protonation is likely the first step of the mechanism. There are two locations to consider for protonation: the oxygen atom or the π bond. Protonation of the oxygen atom does not result in a resonance-stabilized cation, while protonation of the π bond does indeed result in a resonance-stabilized cation. As such, the first step is protonation of the π bond. The resulting intermediate is then attacked by water to give an oxonium ion, which is then deprotonated to give a hemiacetal. Protonation, followed by loss of a leaving group gives a protonated carbonyl group, which then loses a proton to give the product. Notice that water functions as the base in each deprotonation step.

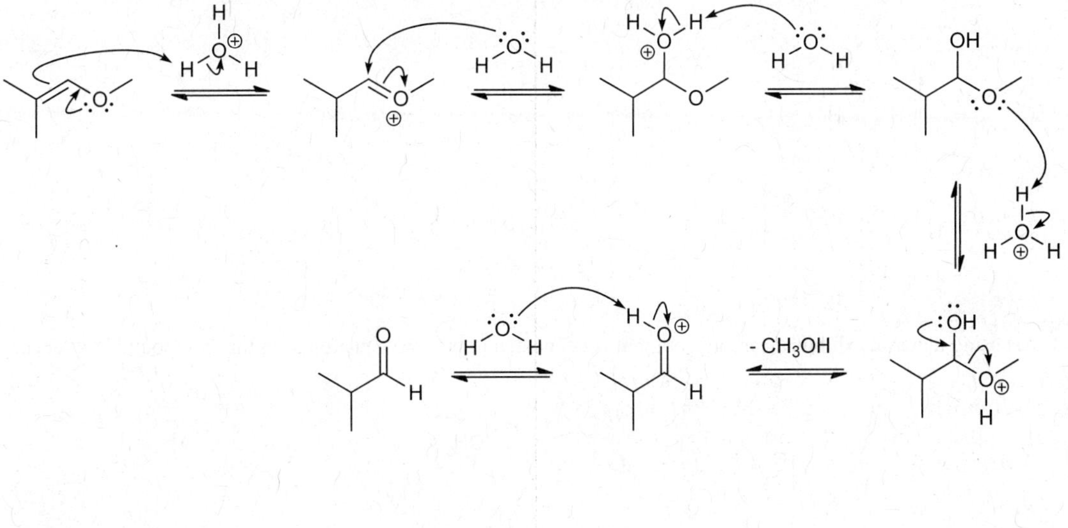

(c) Hydrazine is sufficiently nucleophilic to attack a carbonyl group directly (without prior activation of the carbonyl group via protonation). The resulting intermediate undergoes two successive proton transfer steps, giving an intermediate that is free of formal charges. Protonation of the OH group converts a bad leaving group into a good one (water). Loss of the leaving group gives a resonance-stabilized cation, which then loses a proton to give a hydrazone. An intramolecular nucleophilic attack gives an intermediate that undergoes two proton transfer steps to give an intermediate free of formal charges. Protonation of the OH group converts a bad leaving group into a good one (water). Loss of the leaving group gives a resonance-stabilized cation, which then loses a proton to give the product, which is aromatic.

(d) The starting material exhibits a carbon atom that is connected to two oxygen atoms, so this compound is an acetal:

Protonation of the acetal (specifically at the oxygen atom in the bottom right corner of the structure) results in an oxonium ion that can lose a leaving group to give a resonance-stabilized cation. This intermediate has two OH groups. If the more distant OH group attacks the C=O bond, the resulting oxonium ion can lose a proton to give the product. Notice that the base for the deprotonation step is water.

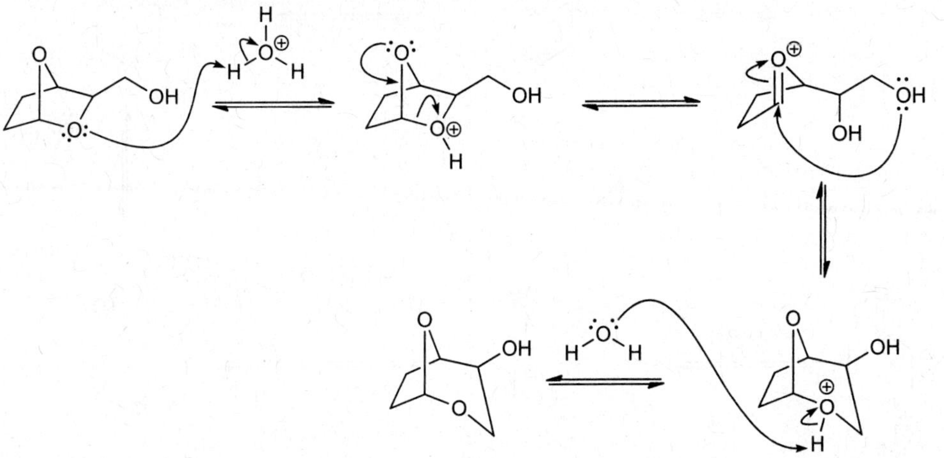

(e) The starting material exhibits a carbon atom that is connected to two oxygen atoms, so this compound is an acetal. Protonation of the acetal (specifically at the oxygen atom on the left) results in an oxonium ion that can lose a leaving group to give a resonance-stabilized cation. This intermediate has two OH groups. If the one on the right side attacks the C=O bond, the resulting oxonium ion can lose a proton to give the product. Notice that the base for the deprotonation step is water.

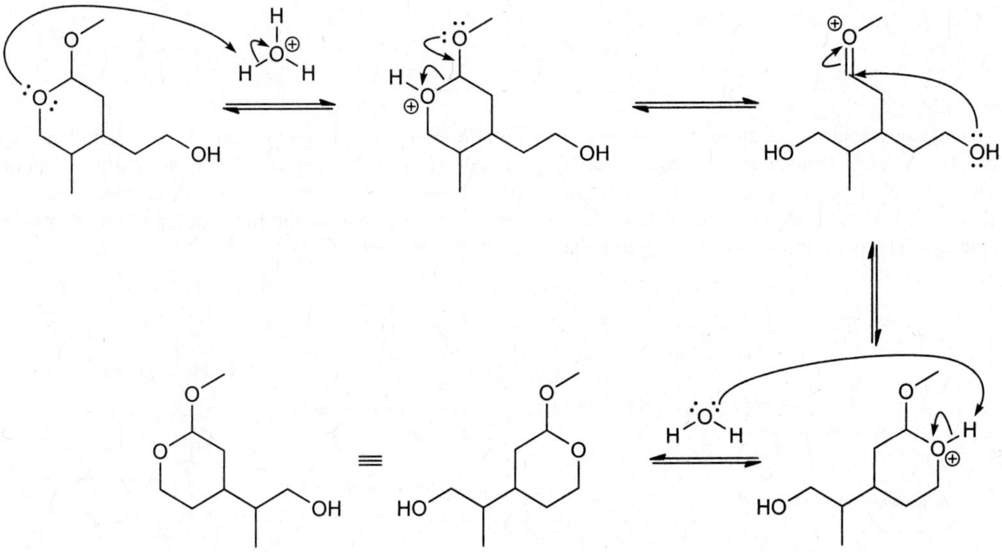

(f) The starting material is a vinyl ether, and it is being subjected to acidic conditions. This indicates that protonation is likely the first step of the mechanism. There are two locations to consider for protonation: the oxygen atom or the π bond. Protonation of the oxygen atom does not result in a resonance-stabilized cation, while protonation of the π bond does indeed result in a resonance-stabilized cation. As such, the first step is protonation of the π bond. The likely proton source is the conjugate acid of ethylene glycol, which received its proton from the acid (TsOH). Protonation results in a resonance-stabilized cation, which can then attacked by ethylene glycol to give an oxonium ion, followed by deprotonation to give a hemiacetal. Protonation of the methoxy group, followed by loss of a leaving group (methanol) gives a protonated carbonyl group, which then functions as an electrophile in an intramolecular nucleophilic attack. Deprotonation then gives the product.

Notice that ethylene glycol functions as the base in each deprotonation step.

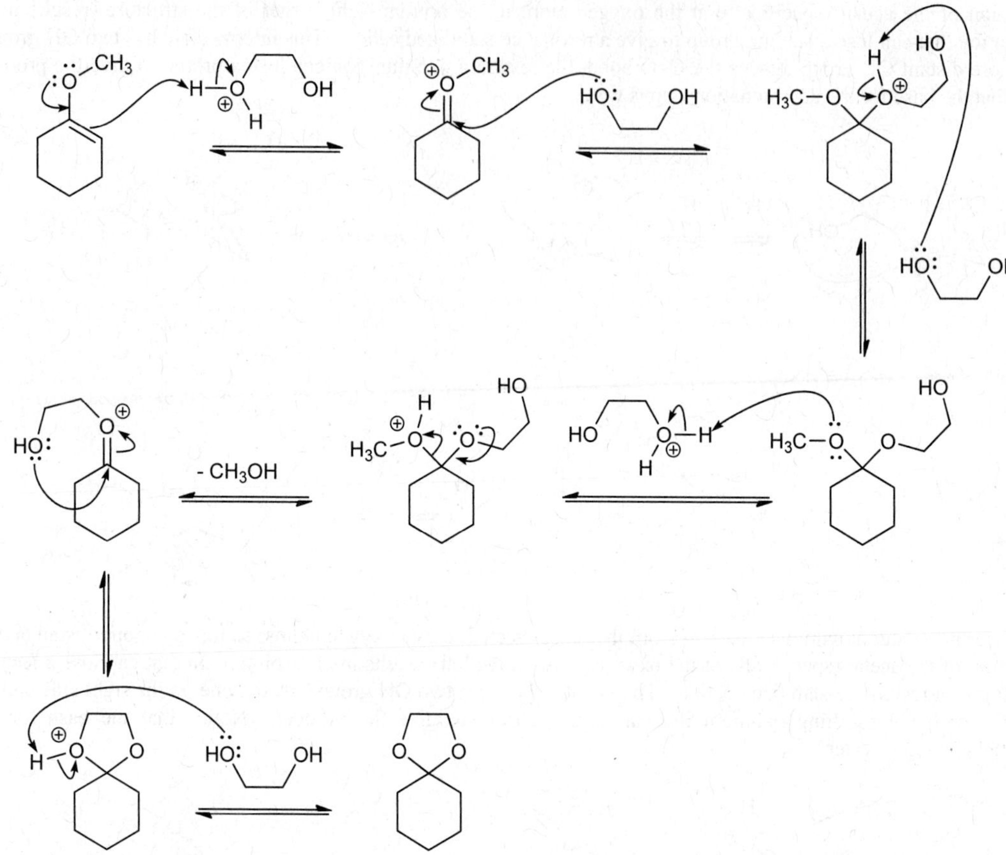

20.85. In aqueous acidic conditions, the carbonyl group of formaldehyde is protonated, thereby rendering it more electrophilic. Another molecule of formaldehyde (that has not been protonated) can function as a nucleophile and attack the protonated carbonyl group. The resulting resonance-stabilized cation functions as an electrophile and is attacked by another molecule of formaldehyde, giving yet another resonance-stabilized cation. An intramolecular attack gives an oxonium ion, which is then deprotonated to give the product.

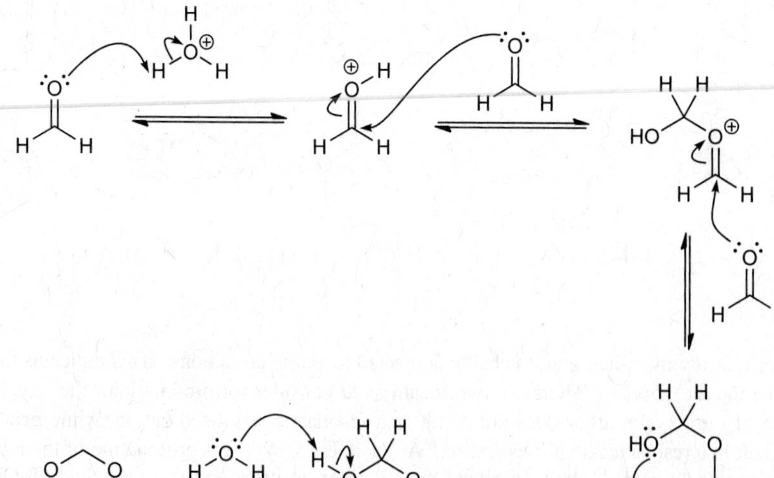

metaformaldehyde

20.86. The ketone group (the more reactive carbonyl) must first be protected by converting it into an acetal. Then, the ester can be reduced with xs LAH to give an alcohol (if the ketone had not been protected, it would have also been reduced by LAH). Mild oxidation of the alcohol with PCC gives an aldehyde, which is then converted to the corresponding thioacetal.

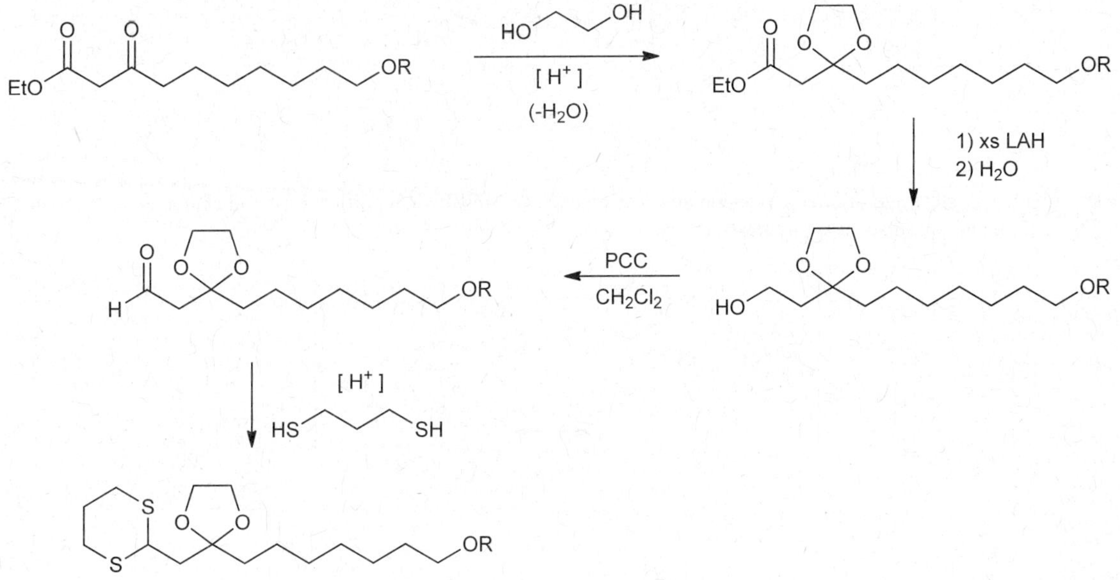

20.87. For this problem, it is best to first work in the retrosynthetic direction. Cyclic ethers such as **1** can be generated from **2** via the Williamson ether synthesis. The substrate for this reaction can arise from a diol, such as **3**, which in turn can come from reduction of a cyclic ester such as **4**. Finally, we know that we can prepare lactone **4** from ketone **5** using the Baeyer-Villiger reaction.

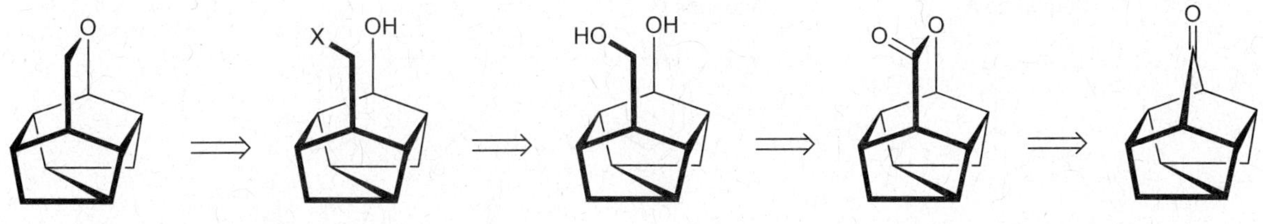

In the forward direction, treatment of **5** with a peroxy acid generates a lactone (**4**), which is reduced with LAH to give diol **3**. When treated with one equivalent of TsCl in pyridine, the primary alcohol can be selectively converted to a tosylate (**2, X = OTs**). One can also form a leaving group at this site by using one equivalent of SOCl₂ (X = Cl). Finally, treatment of compound **2** with NaH will result in deprotonation of the alcohol and ring closure to form the desired ether.

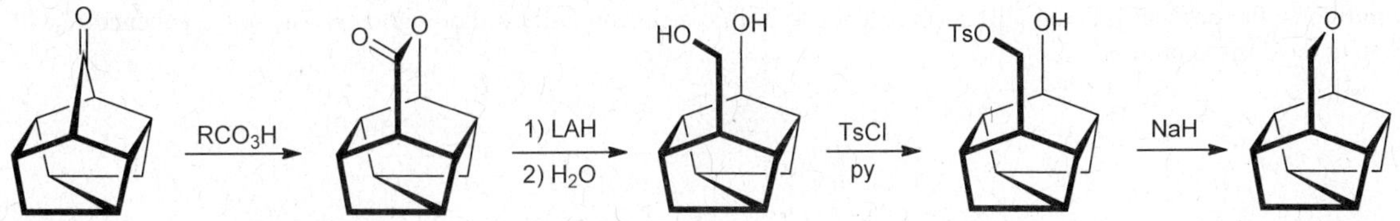

The starting ketone has rotational symmetry, which the product lacks. As such, we expect the product to have many more signals in its proton NMR spectrum than the starting ketone.

20.88. The two conformational isomers are the structures shown here:

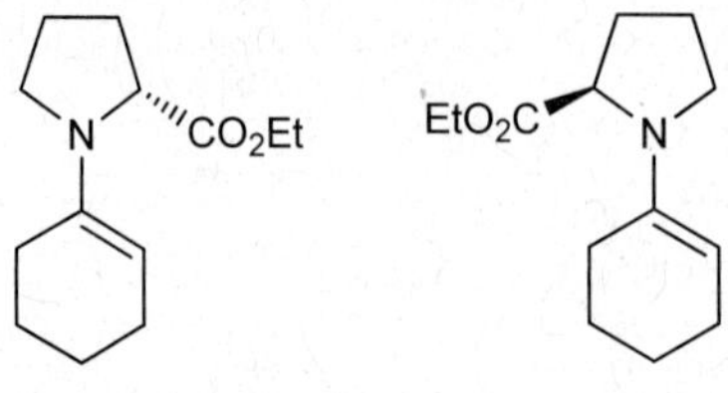

They have a very slow rate of interconversion due to resonance that results in significant π-bond character between the nitrogen atom and the adjacent sp^2 hybridized carbon atom.

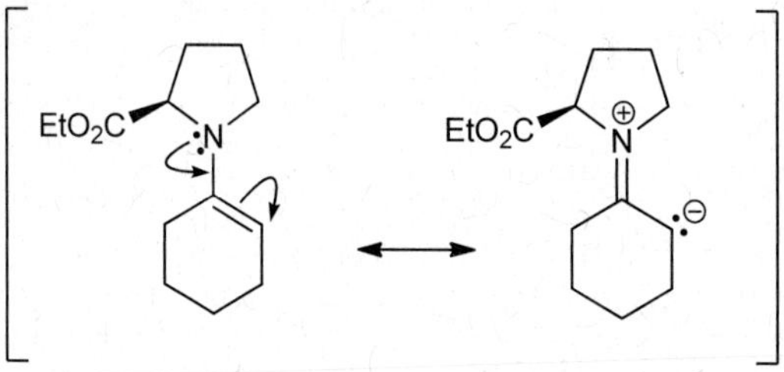

20.89. In all three reaction sequences, the first two steps involve the addition of a Grignard reagent to the ketone, followed by alcohol dehydration using concentrated sulfuric acid and heat. The products of each of these operations are shown below:

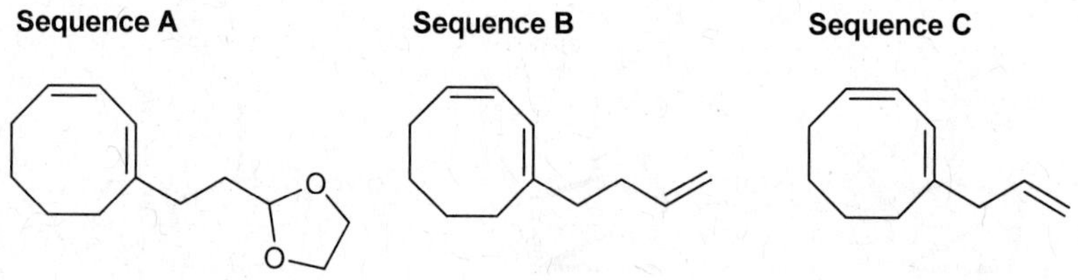

Sequence A   **Sequence B**   **Sequence C**

In sequence A, the final step is acetal deprotection using acetic acid and water, which will produce the desired product, aldehyde **2**. Under these conditions no undesired side reactions are expected to occur. In sequence B, ozonolysis of the terminal alkene will certainly produce an aldehyde, however the molecule also contains two additional alkenes within the 8-membered ring. Since ozone is a non-selective oxidant, if this compound were subjected to O_3/DMS, all three alkene groups would react! Therefore, the major product of the reaction will not be aldehyde **2**, but a compound with three different carbonyl groups (as well as two other fragments). In sequence C, hydroboration/oxidation will convert the terminal alkene to a primary alcohol, which will then be transformed into the desired product via oxidation with PCC. However, just like we saw in sequence B, the two other π bonds will react under the hydroboration/oxidation conditions; the product isolated will not be aldehyde **2**. In conclusion, after a thorough analysis, only sequence A will lead to the desired product.

20.90. The synthesis of aldehyde **2** from aromatic aldehyde **1** can be accomplished via the following two steps:
1) A Wittig reaction can be performed with phosphonium salt **3** in the presence of a strong base (the investigators chose *t*-BuOK) to afford alkene **4** as a mixture of E/Z isomers.
2) Heating compound **4** in a suitable solvent then affords the desired aldehyde via a Claisen rearrangement:

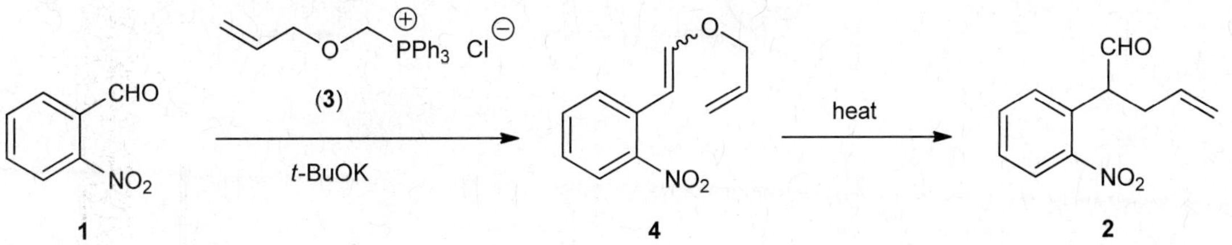

20.91. Both the Wolff-Kishner reduction (mechanism 20.8) and the Bamford-Stevens-Shapiro reaction involve the base-catalyzed decomposition of a hydrazone via the loss of nitrogen gas (N_2) and generation of a carbanion. The difference is that the former generates an sp^3-hybridized carbanion that is quickly protonated in the aqueous reaction environment to form a methylene group whereas the latter generates an sp^2-hybridized carbanion that can be used as a nucleophile in its anhydrous reaction environment. The first step of the Bamford-Stevens-Shapiro reaction, just like we saw with the Wolff-Kishner reduction, involves deprotonation of the hydrazone using base. In this example, the base is *tert*-butyl lithium. The resulting intermediate (**4**) is then deprotonated again using a second equivalent of *tert*-butyl lithium. This second deprotonation occurs at the protons next to the hydrazone, producing intermediate **5**. As this carbanion forms, a fragmentation occurs whereby an alkene is formed via the loss of the sulfonyl leaving group. Once intermediate **6** is generated, nitrogen gas is expelled along with the formation of an sp^2-hybridized carbanion (**2**). In this example, carbanion **2** is then allowed to react with aldehyde **3** as an electrophile; after several additional steps, frondosin B is formed.

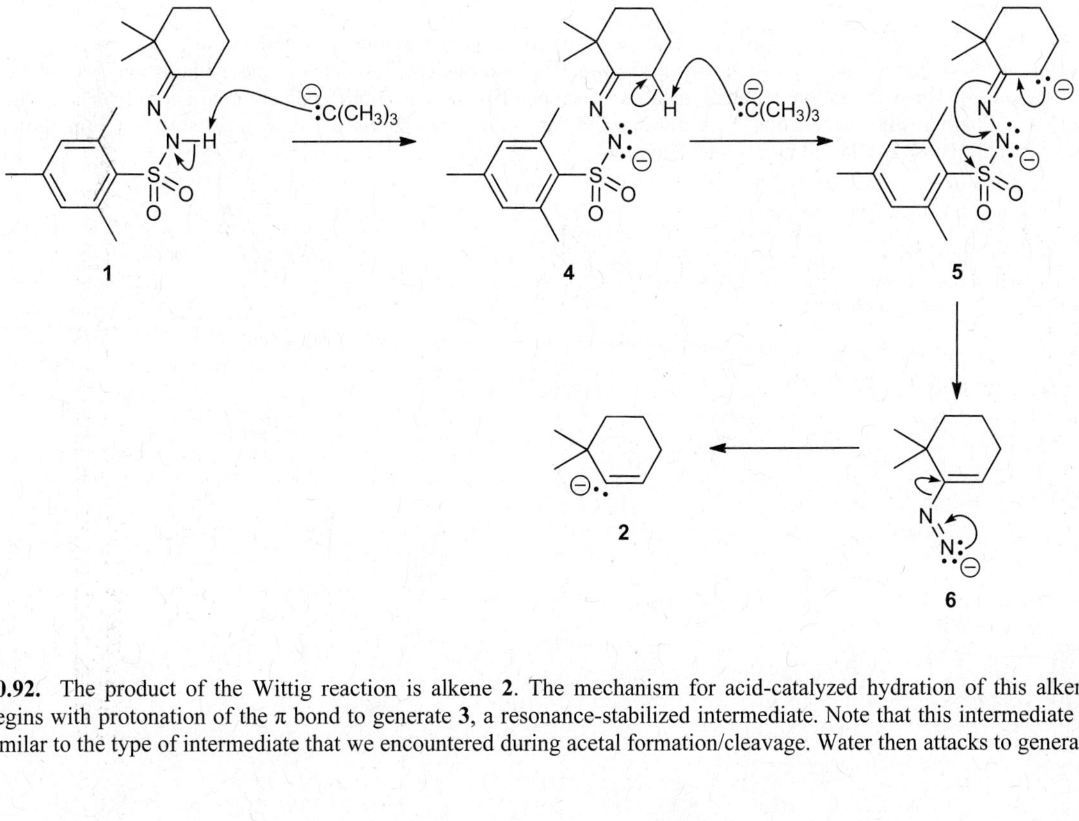

20.92. The product of the Wittig reaction is alkene **2**. The mechanism for acid-catalyzed hydration of this alkene begins with protonation of the π bond to generate **3**, a resonance-stabilized intermediate. Note that this intermediate is similar to the type of intermediate that we encountered during acetal formation/cleavage. Water then attacks to generate a tetrahedral intermediate (**4**) which can be deprotonated to form hemiacetal **5**. Protonation, followed by regeneration of the carbonyl group via loss of methanol, will produce protonated aldehyde **7**, which is deprotonated in the final step to afford aldehyde **8**.

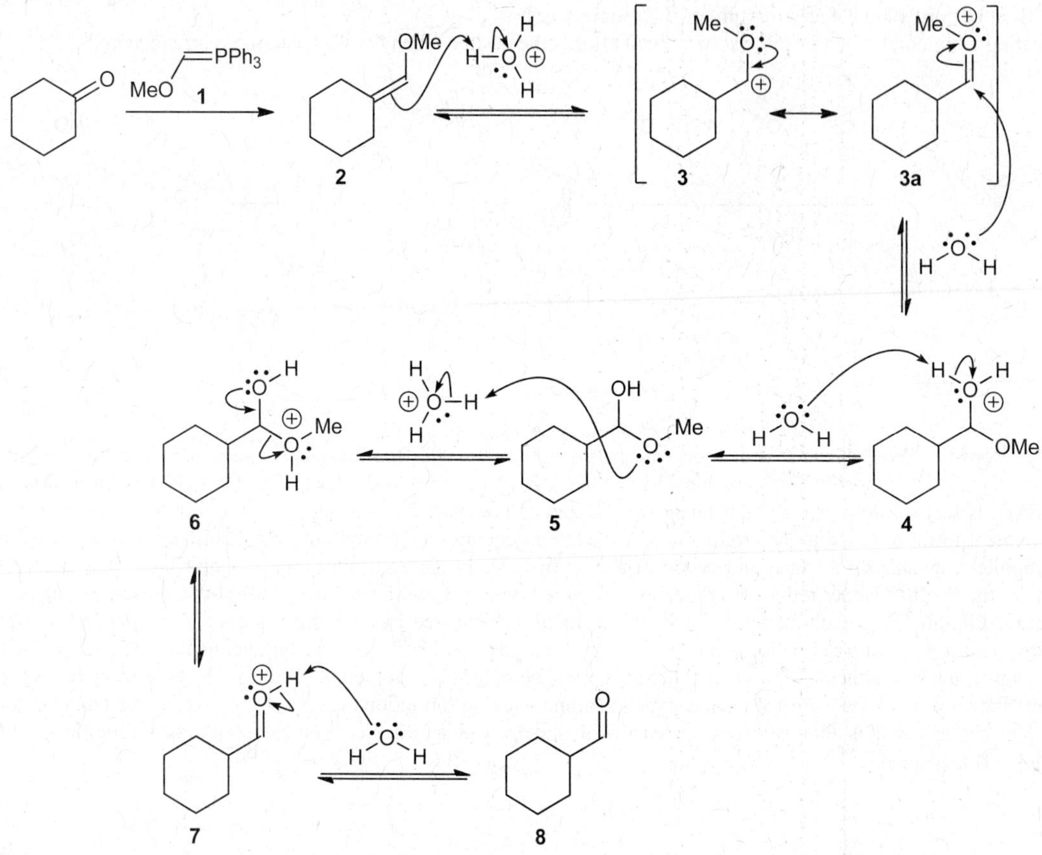

20.93. In the first step, the acid chloride reacts with AlCl₃ to form a resonance-stabilized acylium ion (**4**). In the absence of an aromatic ring, the C=C π bond will function as a nucleophile and trap the acylium ion to produce a carbocation (**5**). This carbocation is transformed into compound **1** if AlCl₄⁻ transfers a chloride ion to the carbocation (path **A**). Alternatively, carbocation **5** is transformed into compound **2** via a 1,2-hydride shift to form tertiary carbocation **6** (path **B**), followed by deprotonation.

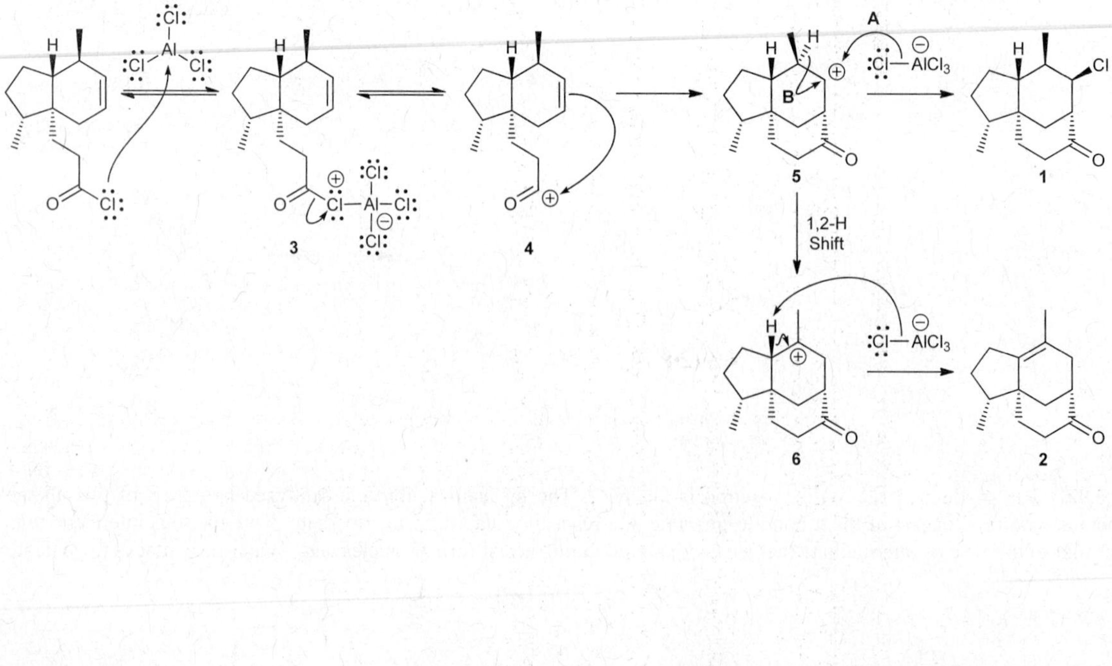

20.94. In the first step, one of the carbonyl groups of compound **1** is protonated. Next, a carbon-carbon bond rotation (**3** to **4**) will bring the protonated carbonyl into close proximity to the phenolic OH group. This OH groups attacks the protonated carbonyl to form tetrahedral intermediate **5**, which is deprotonated by acetone to form hemiacetal **6**. The OH group of the hemiacetal is then protonated, turning it into an excellent leaving group. Water leaves, regenerating the carbonyl group to form compound **8**. Next, the OH group at the top of the molecule attacks a protonated molecular of acetone to form tetrahedral intermediate **9**. After a proton transfer, we can draw a resonance structure of **10**, which exhibits a carbocation (**10a**). The alcohol then traps the carbocation, followed by a final proton transfer, which generates compound **2**.

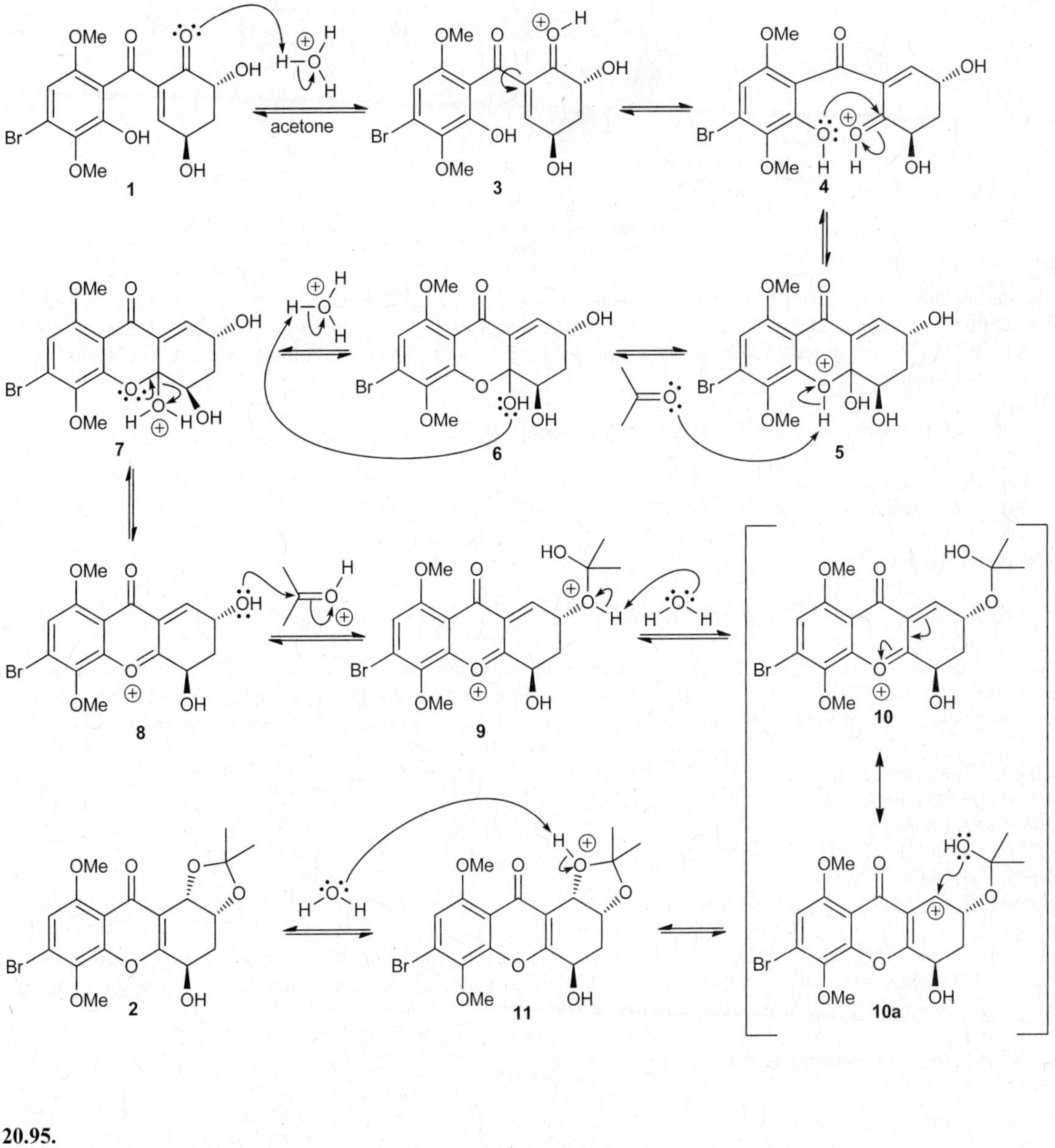

20.95.
(a) We begin by determining the HDI for compound **B**, as shown here:

$$\text{HDI} = \tfrac{1}{2}(2C + 2 + N - H - X) = \tfrac{1}{2}(2 \cdot 8 + 2 + 0 - 14 - 0) = \tfrac{1}{2}(4) = 2$$

Therefore, compound **B** has two degrees of unsaturation. Now let's consider the IR data. Only two peaks are given, but they are important bands: a broad band around 3300 cm^{-1} is typical for an alcohol group, and 2117 cm^{-1} is diagnostic for a triple bond – either an alkyne (C≡C) or a nitrile (C≡N). Since nitrogen is absent from the molecular formula, the band at 2117 cm^{-1} must correspond with a C≡C triple bond. This accounts for both degrees of unsaturation.

At this stage consider the ^{1}H NMR data. Also, take into consideration the structure of the starting material; we know that the product should have at least some degree of similarity. The ^{1}H NMR data shows six types of protons; the easiest to assign is the broad singlet at 1.56 δ which is likely a proton of an alcohol; this is further confirmed by the IR band at 3305 cm^{-1}.

To assign the remaining signals, let's look at the starting material and consider what its ^{1}H NMR must look like.

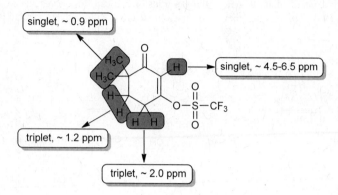

The starting material has no chirality centers, and only 4 unique protons. The two methyl groups will appear as a singlet near 0.9 ppm, the allylic methylene group is expected to appear as a triplet near 2.0 ppm, the methylene next to the gem-dimethyl group is expected to be a triplet near 1.2 ppm, and finally, the vinyl proton should be a singlet in the range 4.5-6.5 ppm.

Now let's take another look at the ^{1}H NMR data for compound B.

0.89 δ (6H, singlet)
~~1.49 δ (1H, broad singlet)~~
1.56 δ (2H, triplet)
1.95 δ (1H, singlet)
2.19 δ (2H, triplet)
3.35 δ (2H, singlet)

We've already assigned the 1.49 ppm broad singlet as an OH proton, so we can cross it off. We can also make the assumption that the 0.89 ppm singlet that integrates for 6 protons must be the gem-dimethyl group, and the two triplets at 1.56 ppm and 2.19 ppm are the two methylene groups from the starting material. Let's cross them out as well.

~~0.89 δ (6H, singlet)~~
~~1.49 δ (1H, broad singlet)~~
~~1.56 δ (2H, triplet)~~
1.95 δ (1H, singlet)
~~2.19 δ (2H, triplet)~~
3.35 δ (2H, singlet)

The vinyl proton of the starting material (4.5 – 6.5 ppm) is absent in Compound **B**. In its place, there are two signals: a singlet at 3.35 ppm, which (because of chemical shift and integration) is likely a methylene group alpha to an OH group, and the singlet at 1.95 which integrates to 1 proton.

Let's take a look at the starting material again:

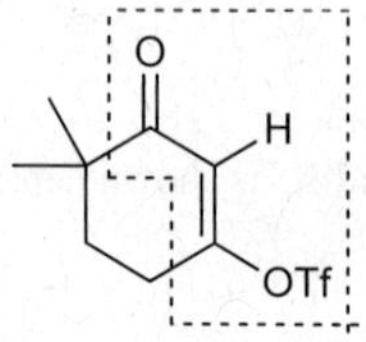

It seems that the left half is represented in compound **B**, however the right half has changed dramatically. From the molecular formula we know that there is no longer any sulfur in the molecule, so the triflate group must be gone. We also know that Compound B must exhibit an OH group and a C≡C triple bond (from the IR spectral data). If we were to take an eraser and eliminate the triflate, form an alkyne where the alkene once was, and reduce the ketone to an alcohol we would arrive at this structure:

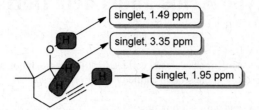

Does this structure fit the data left over for compound B? The methylene group alpha to the OH group should be a singlet with an integration of 2. The chemical shift of 3.35 ppm is also in the acceptable range for a proton next to oxygen. And the proton of the alkyne should also be a singlet, with an integration of 1. The chemical shift is slightly less than what the table in chapter 16 shows (~2.5 ppm), but it is certainly within a margin of error. The molecular formula also matches.

(b) In the first step, LAH delivers hydride to the carbonyl group to form tetrahedral intermediate **2**. Next, the carbonyl group is regenerated which induces carbon-carbon bond cleavage to simultaneously generate the alkyne via the expulsion of the triflate leaving group, producing aldehyde **3**. The *anti* orientation of the electrons in the carbon-carbon single bond and the triflate leaving group facilitated this E2-like process. In the presence of LAH, another hydride ion can be delivered once more, to attack the aldehyde. This forms tetrahedral intermediate **4**, which is protonated upon workup to form compound **B**.

Chapter 21
Carboxylic Acids and Their Derivatives

Review of Concepts

Fill in the blanks below. To verify that your answers are correct, look in your textbook at the end of Chapter 21. Each of the sentences below appears verbatim in the section entitled *Review of Concepts and Vocabulary*.

- Treatment of a carboxylic acid with a strong base yields a _____ salt.
- The pK_a of most carboxylic acids is between ____ and _____.
- Using the **Henderson-Hasselbalch equation**, it can be shown that carboxylic acids exist primarily as _____ at **physiological pH**.
- Electron-_____ substituents can increase the acidity of a carboxylic acid.
- When treated with aqueous acid, a nitrile will undergo _____, yielding a carboxylic acid.
- Carboxylic acids are reduced to _____ upon treatment with lithium aluminum hydride or borane.
- Carboxylic acid derivatives exhibit the same _____ state as carboxylic acids.
- Carboxylic acid derivatives differ in reactivity, with _____ being the most reactive and _____ the least reactive.
- When drawing a mechanism, avoid formation of a strong _____ in acidic conditions, and avoid formation of a strong _____ in basic conditions.
- When a nucleophile attacks a carbonyl group to form a tetrahedral intermediate, always reform the carbonyl group if possible, but avoid expelling _____ or _____.
- When treated with an alcohol, acid chlorides are converted into _____.
- When treated with ammonia, acid chlorides are converted into _____.
- When treated with a _____ reagent, acid chlorides are converted into alcohols with the introduction of two alkyl groups.
- The reactions of anhydrides are the same as the reactions of _____ except for the identity of the leaving group.
- When treated with a strong base followed by an alkyl halide, carboxylic acids are converted into _____.
- In a process called the **Fischer esterification**, carboxylic acids are converted into esters when treated with an _____ in the presence of _____.
- Esters can be hydrolyzed to yield carboxylic acids upon treatment with either aqueous base or aqueous _____. Hydrolysis under basic conditions is also called _____.
- When treated with lithium aluminum hydride, esters are reduced to yield _____. If the desired product is an aldehyde, then _____ is used as a reducing agent instead of LAH.
- When treated with a _____ reagent, esters are reduced to yield alcohols, with the introduction of two alkyl groups.
- When treated with excess LAH, amides are converted into _____.
- Nitriles are converted to amines when treated with _____.

Review of Skills

Fill in the blanks and empty boxes below. To verify that your answers are correct, look in your textbook at the end of Chapter 21. The answers appear in the section entitled *SkillBuilder Review*.

21.1 Drawing the Mechanism of a Nucleophilic Acyl Substitution Reaction

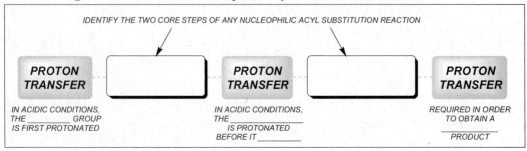

21.2   Interconverting Functional Groups

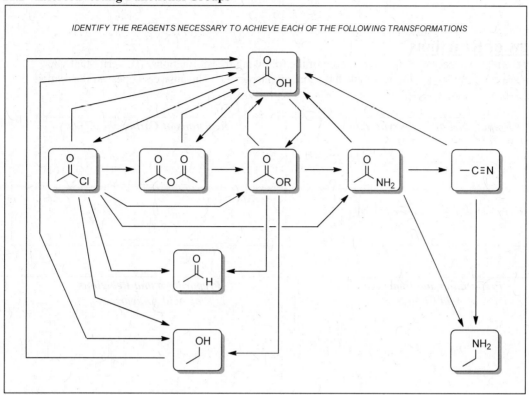

IDENTIFY THE REAGENTS NECESSARY TO ACHIEVE EACH OF THE FOLLOWING TRANSFORMATIONS

21.3   Choosing the Most Efficient C-C Bond-Forming Reaction

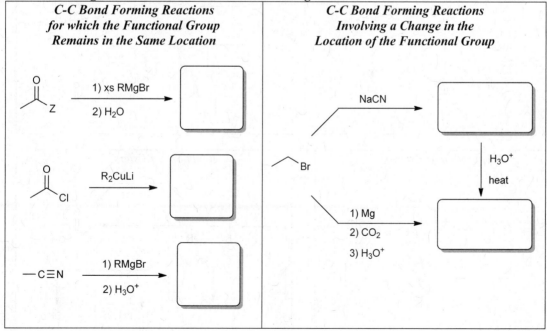

Review of Reactions

Identify the reagents necessary to achieve each of the following transformations. To verify that your answers are correct, look in your textbook at the end of Chapter 21. The answers appear in the section entitled *Review of Reactions*.

Preparation of Carboxylic acids

Reactions of Carboxylic Acids

Preparation and Reactions of Acid Chlorides

Preparation and Reactions of Acid Anhydrides

Preparation of Esters

Reactions of Esters

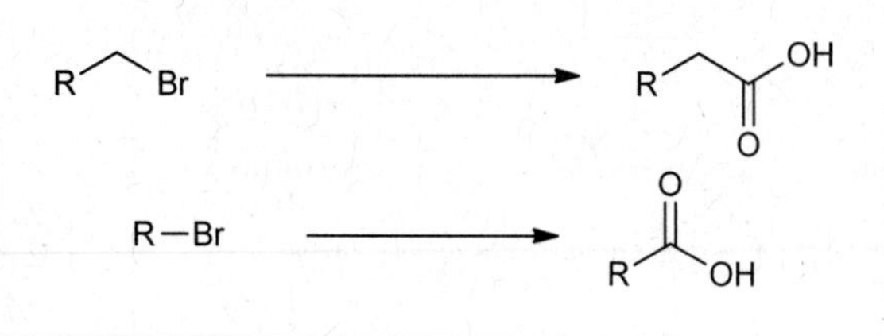

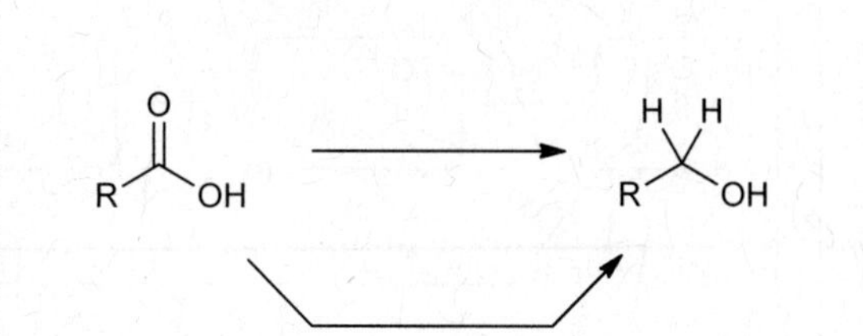

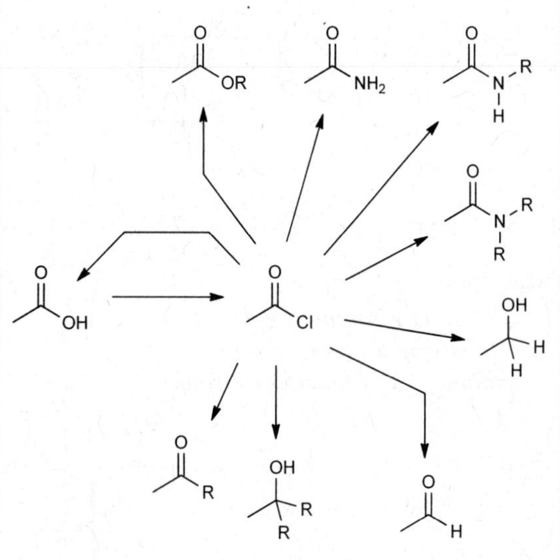

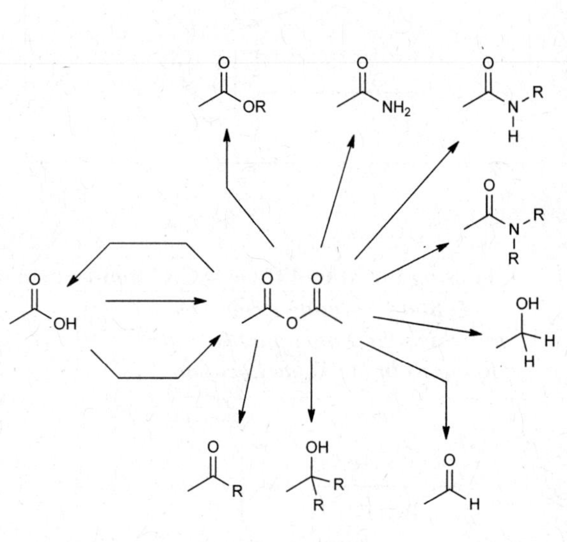

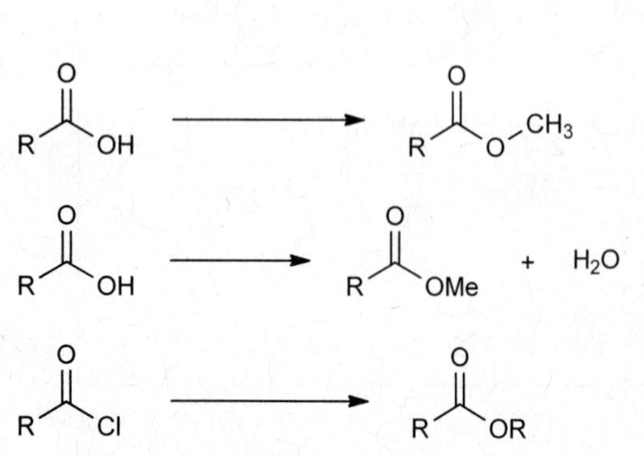

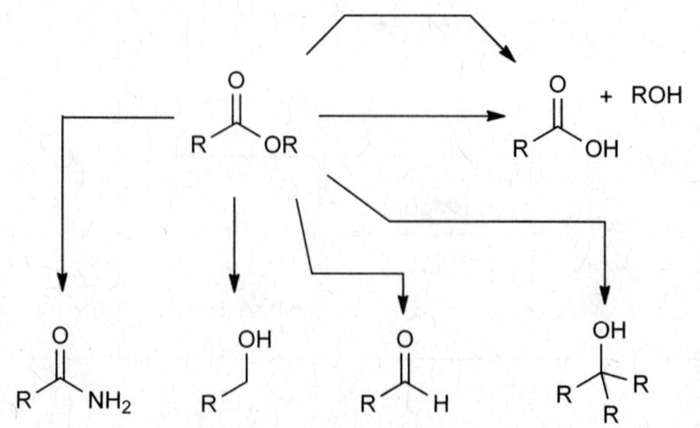

Preparation of Amides	Reactions of Amides

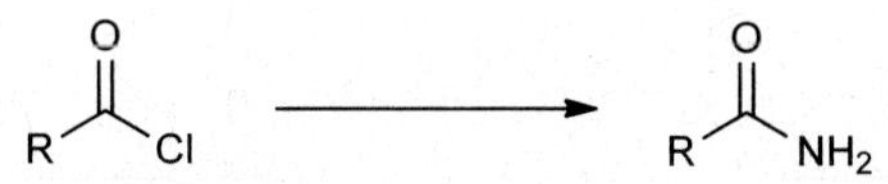

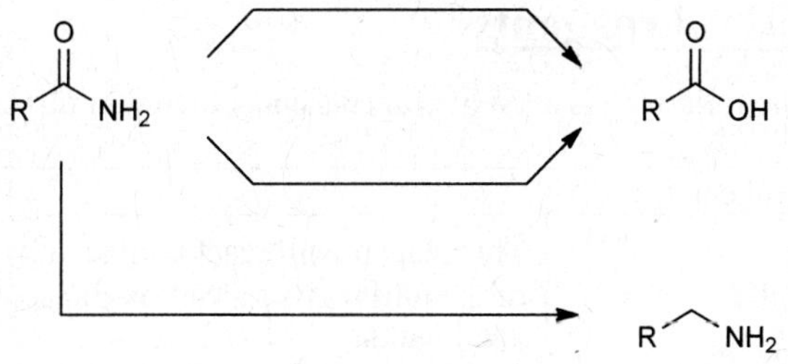

Preparation of Nitriles	Reactions of Nitriles

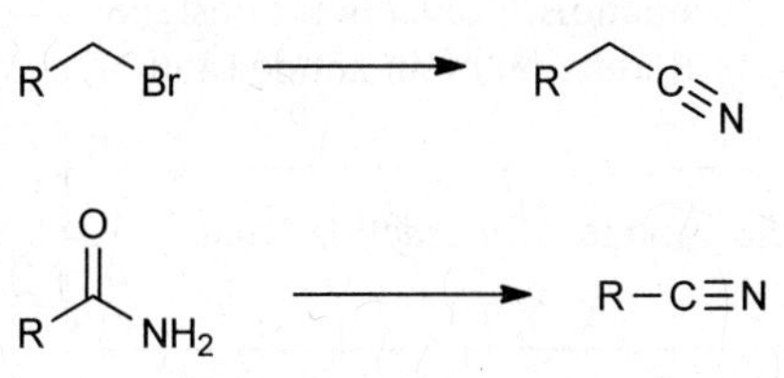

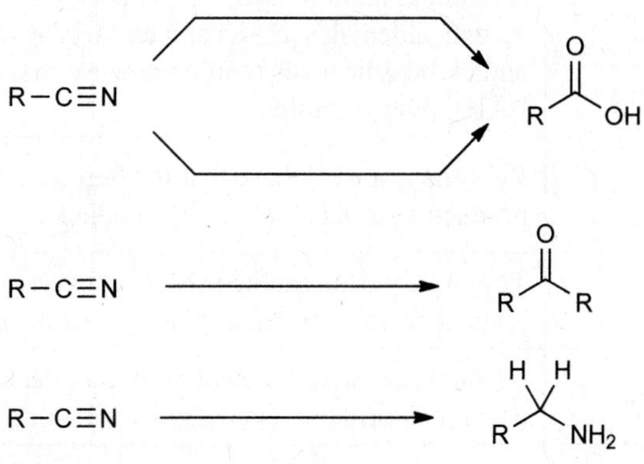

Common Mistake to Avoid

This chapter covers many reactions. One of these reactions is between an acid chloride and a lithium dialkyl cuprate, giving a ketone as the product:

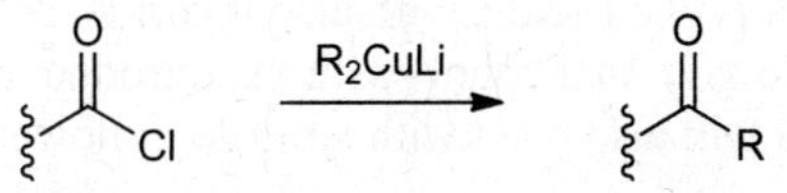

The resulting ketone is not further attacked by the lithium dialkyl cuprate (unlike a Grignard reagent, which would attack the ketone). For some reason, students commonly propose a similar reaction between an ester and a lithium dialkyl cuprate:

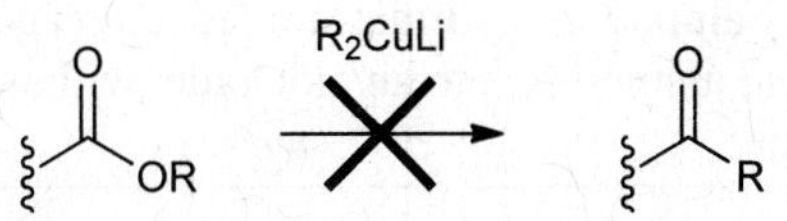

This reaction will not work. If a lithium dialkyl cuprate will not attack a ketone, then it certainly won't attack an ester (which is less electrophilic than a ketone). Students often make this type of mistake, by applying a reaction outside of the scope in which it was discussed. Try to avoid doing this. Whenever we cover a reaction that applies to a particular functional group, you cannot assume that it will apply to other, less reactive, functional groups as well.

Useful reagents

The following is a list of reagents encountered in this chapter:

Reagents	Description
NaCN	This reagent will react with an alkyl halide to give a nitrile. Subsequent hydrolysis of the nitrile gives a carboxylic acid, with one more carbon atom than the starting alkyl halide.
1) Mg 2) CO_2 3) H_3O^+	These reagents can be used to convert an alkyl halide into a carboxylic acid, with the introduction of one carbon atom. Insertion of magnesium gives a Grignard reagent, which then attacks carbon dioxide to give a carboxylate ion, which is the protonated upon acid work-up.
1) LAH 2) H_2O	Lithium aluminum hydride is a powerful hydride reducing agent. It will reduce ketone, aldehydes, esters and carboxylic acids to give alcohols. Reduction of esters and carboxylic acids requires the use of excess LAH. Reduction of an amide (with LAH) gives an amine.
	Pyridine is a weak base that is often used as an "acid sponge," for reactions that produce a strong acid as a by-product.
$SOCl_2$	Thionyl chloride can be used to convert a carboxylic acid into an acid halide. This reagent can also be used to dehydrate an amide to give a nitrile.
ROH	Alcohols are weak nucleophiles and weak bases. An alcohol can be used to convert an acid chloride or an acid anhydride into an ester.
NH_3	Ammonia is both a base and a nucleophile. Excess ammonia can be used to convert an acid chloride or an acid anhydride into an amide.
RNH_2	Primary amines are bases and nucleophiles. Excess amine can be used to convert an acid chloride or an acid anhydride into an amide.
R_2NH	Secondary amines are bases and nucleophiles. Excess amine can be used to convert an acid chloride or an acid anhydride into an amide.
1) xs RMgBr 2) H_2O	A Grignard reagent is a strong nucleophile. Two equivalents of a Grignard reagent will react with an acid chloride, with an anhydride, or with an ester, followed by water work-up, to give an alcohol (with the introduction of two R groups). A Grignard reagent will also react with a nitrile, followed by water work-up, to give a ketone.
R_2CuLi	A lithium dialkyl cuprate is a weak nucleophile. It will react with an acid chloride to give a ketone, but it will not react with ketones or esters.
1) $LiAl(OR)_3H$ 2) H_2O	Lithium trialkoxy aluminum hydrides are reducing agents that will convert an acid chloride or an acid anhydride into an aldehyde, without subsequent reduction of the resulting aldehyde.
1) DIBAH 2) H_2O	Diisobutyl aluminum hydride is a hydride reducing agent that will convert an ester into an aldehyde.
H_3O^+	Aqueous acid will cause hydrolysis of an acid chloride, an anhydride, an ester, an amide, or a nitrile to give a carboxylic acid.
$[H^+]$, ROH	Under acidic conditions, an alcohol will react with a carboxylic acid via a Fischer esterification, giving an ester.

Solutions

21.1.

(a) In this molecule, the longest chain that contains both carboxylic acid groups is comprised of five carbon atoms, so the parent name (pentane) is given the suffix "–dioic acid", resulting in the IUPAC name pentanedioic acid. The common name is glutaric acid.

IUPAC name = pentanedioic acid
Common name = glutaric acid

(b) In this molecule, the longest chain that contains the carboxylic acid group is comprised of four carbon atoms, so the "e" in the parent name (butane) is replaced with the suffix "-oic acid", resulting in the IUPAC name butanoic acid. The common name is butyric acid.

IUPAC name = butanoic acid
Common name = butyric acid

(c) The carboxylic acid group is attached to a ring, so the IUPAC name for this compound uses the name of the ring followed by the suffix "-carboxylic acid"; thus this compound is benzenecarboxylic acid. The common name is benzoic acid.

IUPAC name = benzenecarboxylic acid
Common name = benzoic acid

(d) In this molecule, the longest chain that contains both carboxylic acid groups is comprised of four carbon atoms, so the parent name (butane) is given the suffix "–dioic acid", resulting in the IUPAC name butanedioic acid. The common name is succinic acid.

IUPAC name = butanedioic acid
Common name = succinic acid

(e) In this molecule, the parent chain is comprised of two carbon atoms, and the "e" in the parent name (ethane) is replaced with the suffix "-oic acid", resulting in the IUPAC name ethanoic acid. The common name is acetic acid.

IUPAC name = ethanoic acid
Common name = acetic acid

(f) In this molecule, which contains only one carbon atom, the "e" in the parent name (methane) is replaced with the suffix "-oic acid", resulting in the IUPAC name methanoic acid. The common name is formic acid.

IUPAC name = methanoic acid
Common name = formic acid

21.2.

(a) This molecule has a four-membered ring (cyclobutane) connected to a carboxylic acid group.

(b) The parent (butyric acid) has a four-carbon chain with a carboxylic acid at one terminus. The name indicates that there are two substituents (both chlorine atoms) at the C3 position.

(c) The parent (glutaric acid) has a five-carbon chain with a carboxylic acid at each terminus. The name indicates that there are two substituents (both methyl groups) at the C3 position.

21.3.

(a) We begin by identifying the parent. The carbon atom of the carboxylic acid group must be included in the parent. The longest chain that includes this carbon atom is six carbon atoms in length, so the parent is hexanoic acid. There are four substituents (highlighted), all of which are methyl groups. Notice that the parent chain is numbered starting from the carbonyl carbon (defined as C1). According to this numbering scheme, two methyl groups are at C3, and two are at C4. We use the prefix "tetra" to indicate four methyl groups.

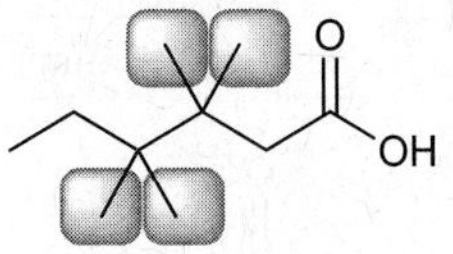

3,3,4,4-tetramethylhexanoic acid

(b) We begin by identifying the parent. The carbon atom of the carboxylic acid must be included in the parent. The longest chain that includes this carbon atom is five carbon atoms in length, so the parent is pentanoic acid. There is one substituent: a propyl group (highlighted). Notice that the parent chain is numbered starting from the carbonyl carbon (defined as C1). According to this numbering scheme, the propyl group is at C2.

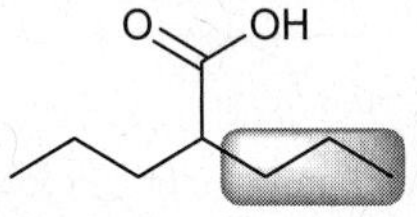

2-propylpentanoic acid

(c) We begin by identifying the parent. The carbon atom of the carboxylic acid must be included in the parent. The longest chain that includes this carbon atom is three carbon atoms in length, so the parent is propanoic acid.

There are two substituents: an amino group and a phenyl group. Notice that the parent chain is numbered starting from the carbonyl carbon (defined as C1). According to this numbering scheme, the amino group is at C2, and the phenyl group is at C3. The substituents are listed alphabetically. Finally, we assign a configuration to the chirality center.

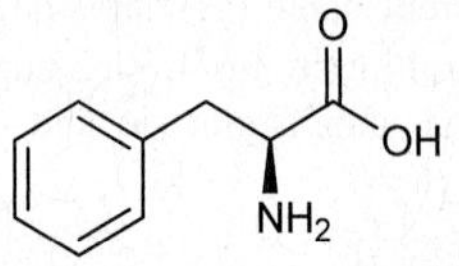

(*S*)-2-amino-3-phenylpropanoic acid

21.4. The compound on the right is more acidic because its conjugate base is resonance-stabilized:

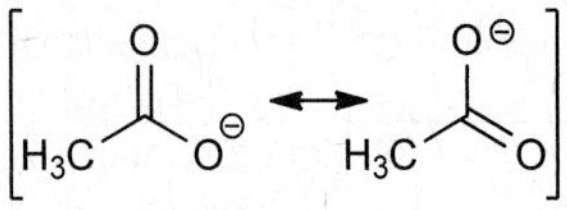

The conjugate base of the other compound is not resonance stabilized.

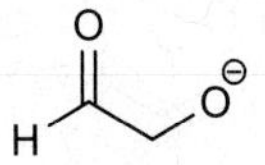

21.5. The conjugate base is resonance stabilized, with the negative charge spread over two oxygen atoms (much like the conjugate base of a carboxylic acid), as shown below. There are three additional resonance structures that have not been drawn (all of which exhibit the negative charge on a carbon atom).

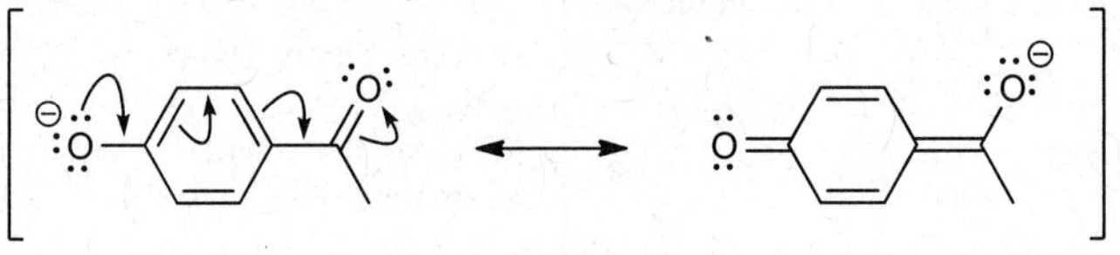

21.6. The first step is to draw the conjugate base of each molecule, including all relevant resonance structures. *meta*-Hydroxyacetophenone is expected to be less acidic than *para*-hydroxyacetophenone, because in the conjugate base of the former, the negative charge is spread over only one oxygen atom (and three carbon atoms). In contrast, the conjugate base of *para*-hydroxyacetophenone has the negative charge spread over two oxygen atoms (and three carbon atoms). The additional resonance structure of the latter conjugate base renders it more stable.

21.7. The first step is to identify the acid (formic acid) and base (hydroxide). The mechanism for this acid-base reaction requires two curved arrows, as shown below. The tail of the first arrow is placed on a lone pair of the hydroxide oxygen atom and the head is placed on the acidic hydrogen atom of formic acid. The tail of the second arrow is placed on the O-H bond and the head is placed on the oxygen atom. The resulting carboxylate salt is named starting with the inorganic cation (potassium) followed by replacing the "-ic acid" suffix from formic acid to "-ate" giving the name potassium formate.

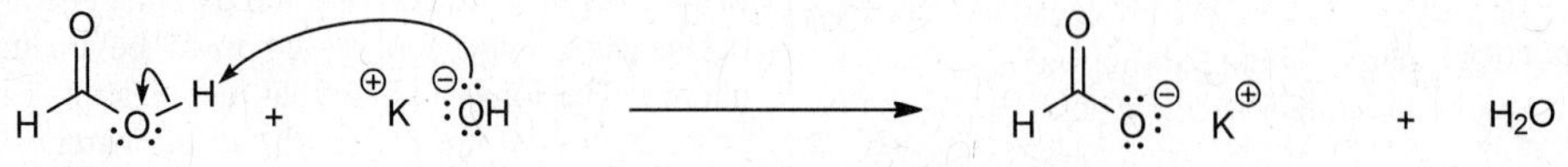

formic acid   potassium formate

21.8. To determine the relative amounts of acetic acid and its conjugate base (acetate), we plug the pK_a of acetic acid (4.76) and the pH (5.76) into the rearranged Henderson-Hasselbach equation as shown below:

$$\frac{[\text{conjugate base}]}{[\text{acid}]} = 10^{(pH - pK_a)} = 10^{(5.76 - 4.76)} = 10^1 = 10$$

The result shows that the conjugate base predominates under these conditions (at a ratio of 10:1).

21.9. For each set of acids, we need to assess the relative stability of each of the conjugate bases. A more stable conjugate base means that the corresponding acid is more acidic. When comparing the structures, we need to consider the nature of the substituents and their position relative to the carboxylic acid functional group. Electron-withdrawing groups stabilize the conjugate base while electron-donating groups destabilize the conjugate base. Substituents that are closer to the carboxylic acid have a greater affect on acidity.

(a) Each of the three acids has the same four-carbon parent: butyric acid. The first two molecules have two electron-withdrawing chlorine substituents each, rendering them more acidic than the third molecule, which has two electron-donating methyl groups. The difference between the first two molecules is in the relative positions of the two chlorine atoms. The compound with the two chlorine atoms on positions C2 and C3 is more acidic than the compound with the chlorine atoms on positions C2 and C4. The chlorine atom on C3 has a greater effect than the chlorine atom on C4. The correct order, in increasing acidity, is thus: 3,4-dimethylbutyric acid < 2,4-dichlorobutyric acid < 2,3-dichlorobutyric acid.

(b) Each of the three acids has the same three-carbon parent: propionic acid. The first molecule has an electron-withdrawing bromine substituent on C3 and the other two molecules have two bromine substituents each. The compound with two bromine atoms on C2 is the most acidic because the two bromine atoms are closest to the carboxylic acid group. The compound with two bromine atoms on C3 is more acidic than the compound with one carbon atom on C3 due to the additive affect of the two bromine atoms. The correct order, in increasing acidity, is thus: 3-bromopropionic acid < 3,3-dibromopropionic acid < 2,2-dibromo-propionic acid.

21.10.
(a) The starting material is an alcohol (ethanol) and the product is the corresponding carboxylic acid with two carbon atoms (acetic acid). Reagents that accomplish this oxidation reaction are: $Na_2Cr_2O_7$, H_2SO_4, H_2O.

(b) The conversion of toluene to benzoic acid requires oxidation of the methyl group to give a carboxylic acid group. Recall that an alkyl group attached to an aromatic ring is oxidized to a carboxylic acid by a strong oxidizing agent, as long as the alkyl group has at least one benzylic hydrogen atom. This conversion can thus be accomplished using: $Na_2Cr_2O_7$, H_2SO_4, H_2O.

(c) The conversion of benzene to benzoic acid requires the installation of a carbon atom on the carbon skeleton of the starting material. One approach to accomplish this transformation is to install a methyl group using a Friedel-Crafts alkylation (CH_3Cl, $AlCl_3$) followed by an oxidation of the methyl group using strongly oxidizing conditions, as shown.

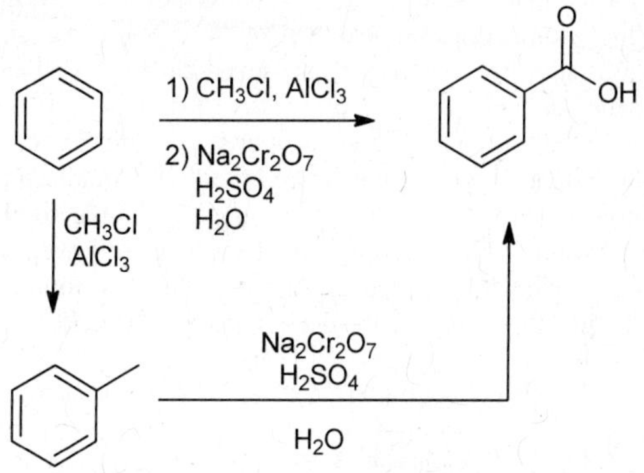

There are certainly other acceptable solutions. For example, we can perform a bromination reaction to give bromobenzene, followed by treatment with magnesium to give phenyl magnesium bromide. This Grignard reagent can then be treated with CO_2, followed by acid work-up, to give benzoic acid. This alternate solution illustrates an important point. For most of the synthesis problems that you will encounter, there is rarely only one correct approach. Most often, there are multiple correct ways to approach the problem.

(d) The starting material (1-bromobutane) has four carbon atoms, and the product (pentanoic acid) has five carbon atoms, so we need to propose a synthesis that involves installation of the fifth carbon atom. We can use NaCN to convert the four-carbon starting material to a five-carbon synthetic intermediate (a nitrile) via an S_N2 process. Hydrolysis of this intermediate (H_3O^+, heat) converts the nitrile to a carboxylic acid, giving the desired product. An alternate approach is to convert the starting alkyl halide to the corresponding Grignard reagent using Mg, followed by reaction with CO_2 to produce a carboxylate. Protonation with aqueous acid gives the desired carboxylic acid.

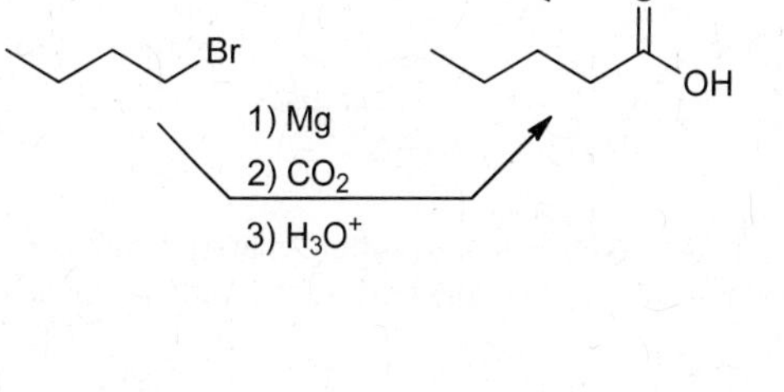

(e) The conversion of ethylbenzene to benzoic acid requires an oxidation of the benzylic carbon atom of the starting material to form a carboxylic acid group. Recall that an alkyl group attached to an aromatic ring is oxidized to a carboxylic acid by a strong oxidizing agent, as long as the alkyl group has at least one benzylic hydrogen atom. This conversion can thus be accomplished using: $Na_2Cr_2O_7$, H_2SO_4, H_2O.

(f) The starting material (bromocyclohexane) has six carbon atoms, and the product (cyclohexanecarboxylic acid) has seven carbon atoms, so we need to propose a synthesis that includes the installation of the seventh carbon atom. We can use NaCN to convert the six-carbon starting material to a seven-carbon synthetic intermediate via an S_N2 process. Hydrolysis of this intermediate (H_3O^+, heat) converts the nitrile to a carboxylic acid, giving the desired product. An alternate approach is to convert the starting alkyl halide to the corresponding Grignard reagent using Mg, followed by reaction with CO_2 to produce a carboxylate. Protonation with aqueous acid gives the desired carboxylic acid.

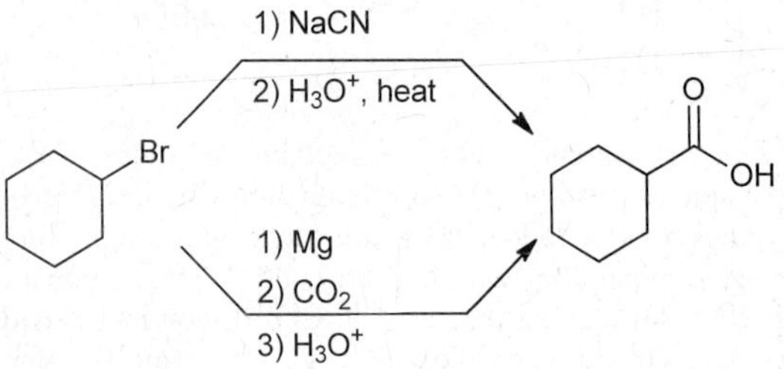

21.11.
(a) This synthesis requires the conversion of a six-carbon starting material (bromobenzene) to a seven-carbon product (benzyl alcohol), so we must include a reaction to form a new C-C bond. There are certainly multiple solutions to this problem. One such solution involves conversion of the starting material into a Grignard reagent, which can then be treated with CO_2 followed by acidic workup (H_3O^+) to produce benzoic acid. Reduction of the carboxylic acid using lithium aluminum hydride, followed by protonation with water, produces the desired product (benzyl alcohol) as shown.

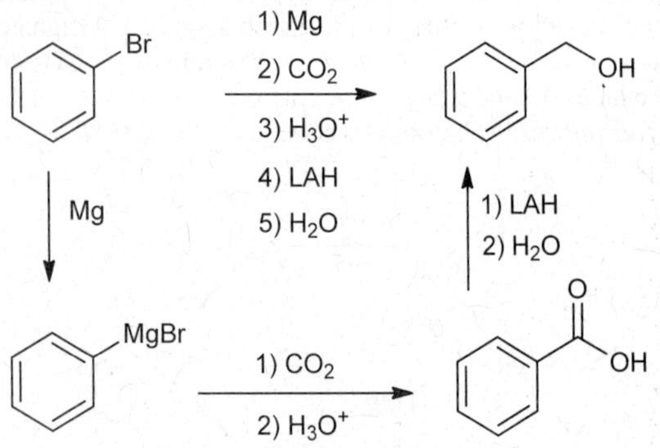

(b) In this synthesis the starting material (toluene) and product (benzyl alcohol) have seven carbon atoms each.

Oxidation of the starting material to benzoic acid can be accomplished with a suitable oxidizing agent to give benzoic acid. Reduction of the resulting carboxylic acid using lithium aluminum hydride followed by protonation with water produces the desired product (benzyl alcohol) as shown. Alternatively, a bromine atom can be installed in the benzylic position using NBS/heat. Subsequent reaction with NaOH produces the desired product via an S_N2 reaction, as shown:

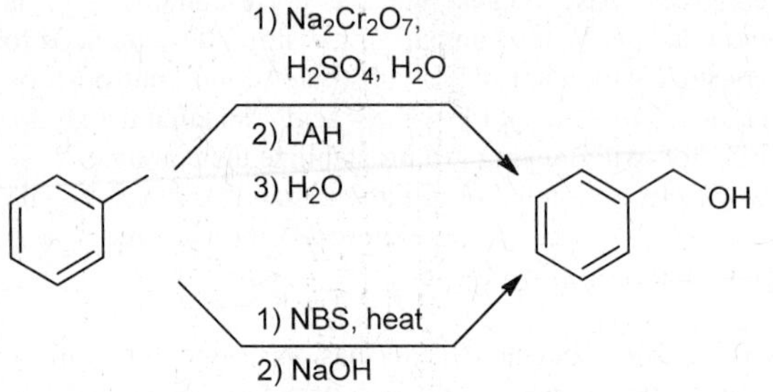

21.12.
(a) This symmetric anhydride is named by replacing "acid" from the corresponding carboxylic acid (propionic acid) with the suffix "anhydride", giving the name propionic anhydride.

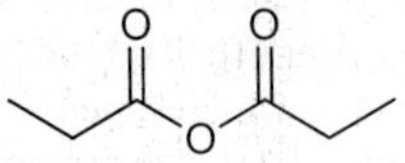

propionic anhydride

(b) This amide is named as a derivative of the carboxylic acid "propionic acid" by replacing the "-ic acid" suffix with "-amide". The two phenyl groups attached to the nitrogen atom are listed as substituents. Their position is indicated by the locant "N" thus giving the name N,N-diphenylpropionamide.

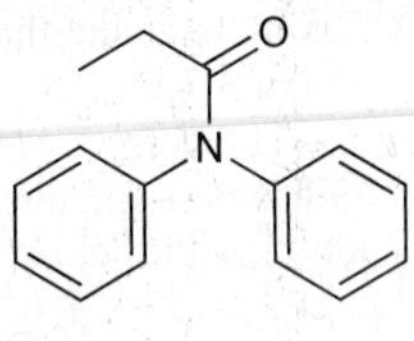

N,N-diphenylpropionamide

(c) This diester is named as a derivative of the parent dicarboxylic acid "succinic acid" by replacing the "-ic acid" suffix with "-ate". The two methyl groups attached to the oxygen atoms are indicated at the beginning, thus giving the name dimethyl succinate.

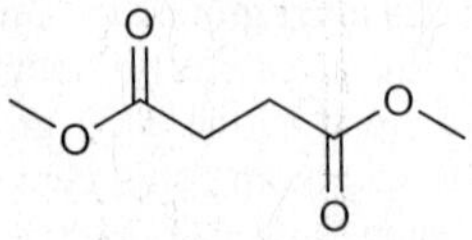

dimethyl succinate

(d) This amide is named as a derivative of the carboxylic acid "cyclobutanecarboxylic acid" by replacing the "-carboxylic acid" suffix with "-carboxamide". The two alkyl groups attached to the nitrogen atom (ethyl and methyl) are listed in alphabetical order as substituents. Their position is indicated with the locant "*N*" thus giving the name *N*-ethyl-*N*-methylcyclobutanecarbox-amide.

N-ethyl-*N*-methylcyclobutanecarboxamide

(e) This nitrile is named as a derivative of the carboxylic acid "butyric acid" by replacing the "-ic acid" suffix with "-onitrile" giving the name butyronitrile.

butyronitrile

(f) This ester is named as a derivative of the parent carboxylic acid "butyric acid" by replacing the "-ic acid" suffix with "-ate". The propyl group attached to the oxygen atom is indicated at the beginning, thus giving the name propyl butyrate.

propyl butyrate

(g) This cyclic anhydride is named as a derivative of the parent dicarboxylic acid "succinic acid" by replacing "acid" with "anhydride", giving the name succinic anhydride.

succinic anhydride

(h) This ester is named as a derivative of the parent carboxylic acid "benzoic acid" by replacing the "-oic acid" suffix with "-ate". The methyl group attached to the oxygen atom is indicated at the beginning, thus giving the name methyl benzoate.

methyl benzoate

(i) This ester is named as a derivative of the parent carboxylic acid "acetic acid" by replacing the "-ic acid" suffix with "-ate". The phenyl group attached to the oxygen atom is indicated at the beginning, thus giving the name phenyl acetate.

phenyl acetate

21.13.

(a) The parent (oxalic acid) is a dicarboxylic acid with two carbon atoms. The suffix "-ic acid" is replaced with "-ate", indicating that this is a diester. The name indicates that the two alkyl groups attached to the oxygen atoms are methyl groups.

(b) The parent (cyclopentanecarboxylic acid) is a ring with five carbon atoms attached to a carboxylic acid. The suffix "-ic acid" is replaced with "-ate", indicating that this is an ester. The name indicates that the group attached to the oxygen atom is a phenyl group.

(c) The parent (propionic acid) is a carboxylic acid with three carbon atoms. The suffix "-ic acid" is replaced with "-amide", indicating that this compound is an amide. The name indicates that there is one methyl group attached to the nitrogen atom.

(d) The parent (propionic acid) is a carboxylic acid with three carbon atoms. The suffix "-ic acid" is replaced with "-yl chloride", indicating that this compound is an acid chloride.

21.14.

(a) This mechanism has three steps: 1) nucleophilic attack, 2) loss of a leaving group and 3) proton transfer. The first step (nucleophilic attack) requires two curved arrows, which show the amine functioning as a nucleophile and attacking the electrophilic carbonyl group, resulting in a tetrahedral intermediate. In step two (loss of a leaving group), the carbonyl group is reformed and the chloride ion leaves, shown with the two curved arrows. In the third step (proton transfer), a second equivalent of the amine serves as a base, deprotonating the cationic intermediate, resulting in the formation of an amide, as shown.

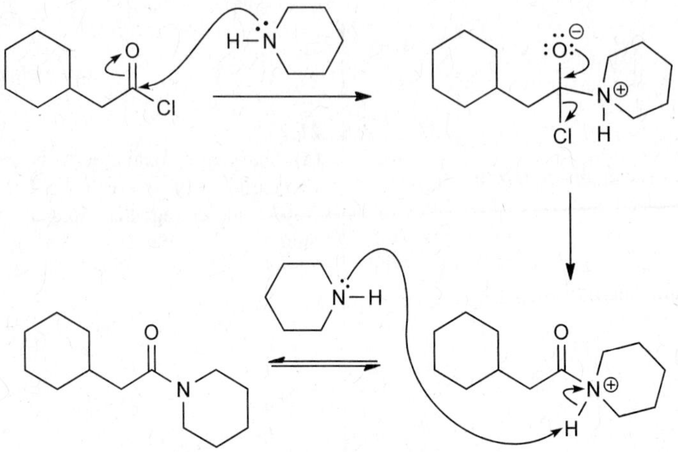

(b) This mechanism has three steps: 1) nucleophilic attack, 2) loss of a leaving group and 3) proton transfer. The first step (nucleophilic attack) requires two curved arrows, which show methanol functioning as a nucleophile and attacking the electrophilic carbonyl group, resulting in a tetrahedral intermediate. In step two (loss of a leaving group), the carbonyl group is reformed and the chloride ion leaves, as shown with two curved arrows. In the third step (proton transfer), a second equivalent of methanol serves as a base, deprotonating the cationic intermediate, resulting in the formation of an ester, as shown:

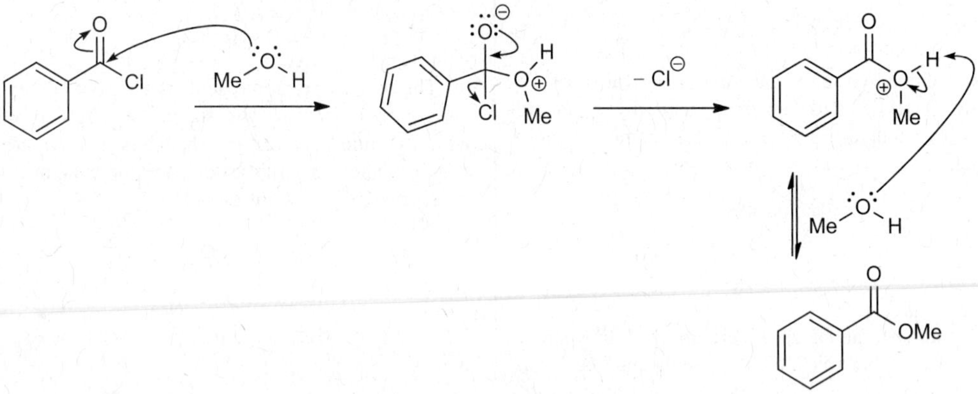

(c) This mechanism has two steps: 1) nucleophilic attack, and 2) loss of a leaving group. The first step (nucleophilic attack), requires two curved arrows, which show the carboxylate ion functioning as a nucleophile and attacking the electrophilic carbonyl group, resulting in a tetrahedral intermediate. In step two (loss of a leaving group), the carbonyl group is reformed and chloride leaves, as shown with two curved arrows, resulting in the formation of an anhydride, as shown.

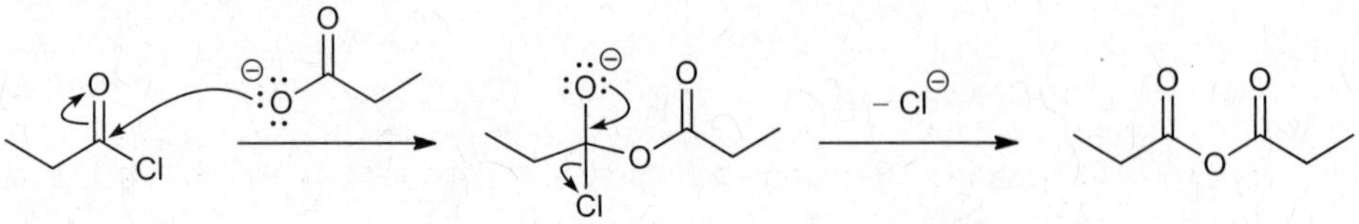

(d) This mechanism has three steps: 1) nucleophilic attack, 2) loss of a leaving group and 3) proton transfer. The first step (nucleophilic attack) requires two curved arrows, which show ammonia functioning as a nucleophile and attacking the electrophilic carbonyl group, resulting in a tetrahedral intermediate. In step two (loss of a leaving group), the carbonyl group is reformed and chloride leaves, as shown with two curved arrows. In the third step (proton transfer), a second equivalent of ammonia serves as a base, deprotonating the cationic intermediate, resulting in the formation of an amide, as shown.

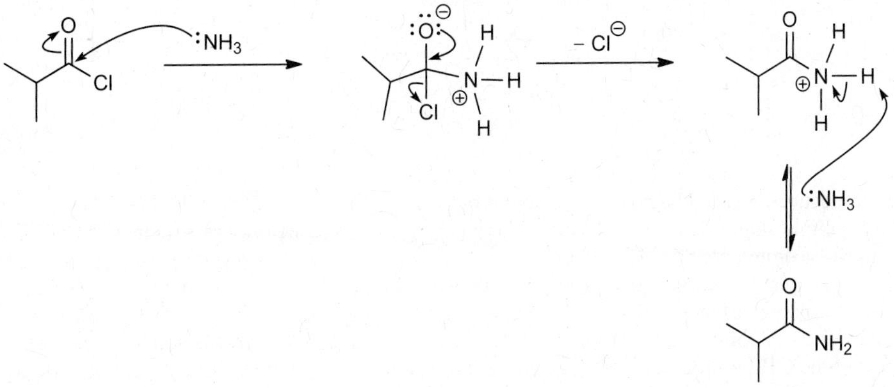

(e) This reaction occurs under acidic conditions, so we must avoid formation of a strong base. Thus, proton transfers are required at multiple stages in the mechanism. The mechanism shown below has six steps: 1) proton transfer, 2) nucleophilic attack, 3) proton transfer, 4) proton transfer, 5) loss of a leaving group and 6) proton transfer. The first step (proton transfer) requires two curved arrows to show protonation of the carbonyl group, resulting in the formation of an activated electrophile. In step two (nucleophilic attack), water serves as a nucleophile, attacking the activated electrophilic carbonyl group, producing a cationic tetrahedral intermediate. We cannot immediately expel the NH_2 group at this stage, as this would result in the formation of a strong base (NH_2^-). This must be avoided in acidic conditions. In step 3 (proton transfer), water serves as a base, resulting in a neutral intermediate. Subsequently, in step 4 (proton transfer), a proton is transferred from H_3O^+ to the nitrogen atom, as shown. In step 5 (loss of leaving group), the carbonyl group is reformed and ammonia serves as the leaving group. In step 6 (proton transfer), ammonia serves as a base which deprotonates the cationic intermediate, resulting in the formation of the two products shown.

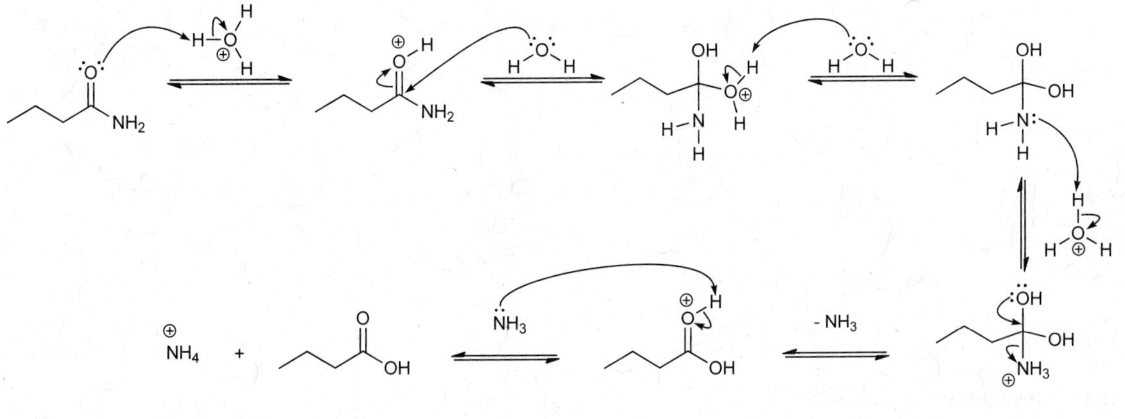

(f) This process occurs via three-step mechanism: 1) nucleophilic attack, 2) loss of a leaving group and 3) proton transfer. The first step (nucleophilic attack) requires two curved arrows, which show hydroxide functioning as a nucleophile and attacking the electrophilic carbonyl group, resulting in an anionic tetrahedral intermediate. In step two (loss of a leaving group), the carbonyl group is reformed and methoxide leaves, resulting in the formation of a carboxylic acid. In the third step (proton transfer), methoxide serves as a base, deprotonating the carboxylic acid intermediate, resulting in the formation of two products: a carboxylate ion and methanol. This final step (in which a strong base deprotonates the carboxylic acid) is the driving force for the reaction (formation of a resonance-stabilized anion).

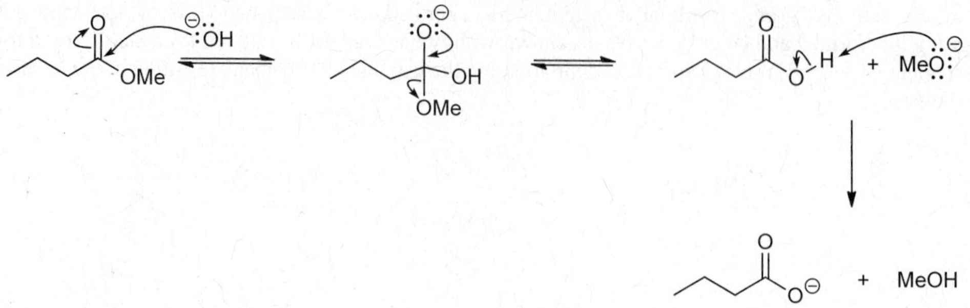

(g) This reaction occurs under acidic conditions, so we must avoid formation of a strong base. Thus, proton transfers are required at multiple stages in the mechanism. The mechanism shown below has six steps: 1) proton transfer, 2) nucleophilic attack, 3) proton transfer, 4) proton transfer, 5) loss of a leaving group and 6) proton transfer. The first step (proton transfer) requires two curved arrows to show the transfer of a proton from $MeOH_2^+$ to the carbonyl group, resulting in the formation of an activated electrophile. In step two (nucleophilic attack), methanol serves as a nucleophile attacking the protonated carbonyl group, producing a cationic tetrahedral intermediate. We cannot immediately expel the OH group at this stage, as this would result in the formation of a strong base (hydroxide). This must be avoided in acidic conditions. In step 3 (proton transfer), methanol serves as a base, resulting in a neutral intermediate. Subsequently, in step 4 (proton transfer) a proton is transferred from $MeOH_2^+$ to the uncharged oxygen atom, as shown. In step 5 (loss of leaving group), the carbonyl group is reformed and water serves as the leaving group. In step 6 (proton transfer), methanol serves as a base which deprotonates the cationic intermediate, resulting in formation of the ester.

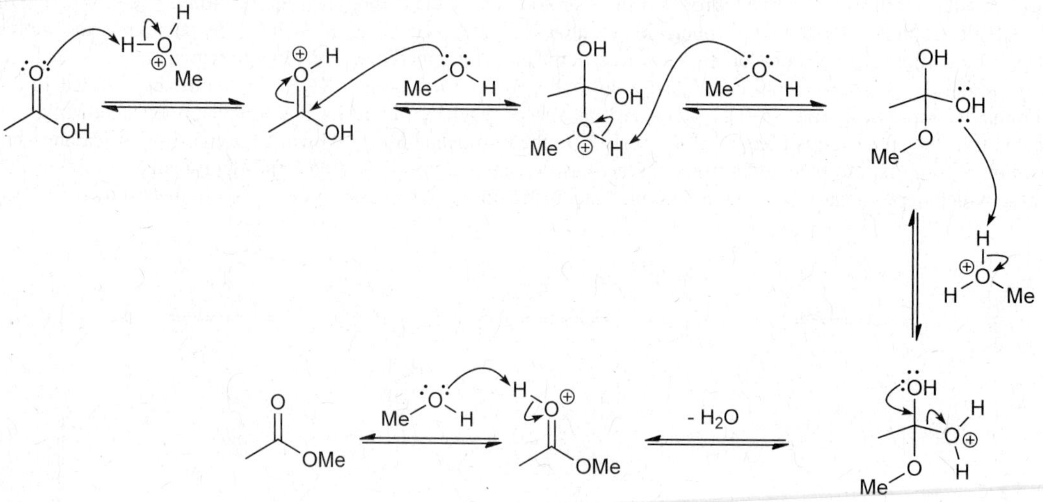

21.15. This mechanism has three steps: 1) nucleophilic attack, 2) loss of a leaving group and 3) proton transfer. The first step (nucleophilic attack) requires two curved arrows, which shows the alcohol functioning as a nucleophile and attacking the electrophilic carbonyl group, resulting in a tetrahedral intermediate. In step two (loss of a leaving group), the carbonyl group is reformed and chloride leaves, shown with two curved arrows. In the third step (proton transfer), a second equivalent of the alcohol serves as a base, deprotonating the cationic intermediate, resulting in formation of the ester.

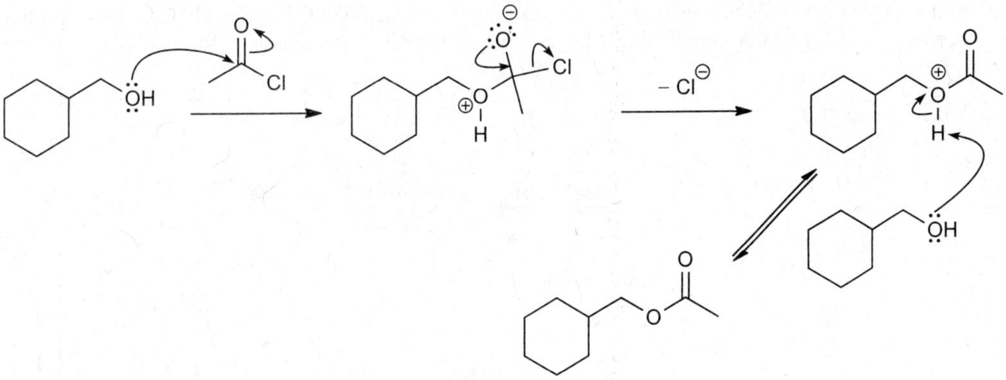

21.16. This reaction occurs under acidic conditions, so we must avoid formation of a strong base. Thus, proton transfers are required at multiple stages in the mechanism. The mechanism shown below has six steps: 1) proton transfer, 2) nucleophilic attack, 3) proton transfer, 4) proton transfer, 5) loss of a leaving group and 6) proton transfer. The first step (proton transfer), requires two curved arrows to show the transfer of a proton from H_3O^+ to the carbonyl group, resulting in formation of an activated electrophile. In step two (nucleophilic attack), the tethered OH group serves as a nucleophile attacking the protonated carbonyl group, producing a cyclic, cationic tetrahedral intermediate. We cannot immediately expel the OH group at this stage, as this would result in the formation of a strong base (hydroxide). This must be avoided in acidic conditions. This OH group must first be protonated. However, protonation at this stage would result in an intermediate with two positive charges, which should be avoided, if possible. Therefore, in the next step (step 3), water serves as a base, resulting in a neutral intermediate. Subsequently, in step 4 (proton transfer), a proton is transferred from H_3O^+ to the OH group, as shown. In step 5 (loss of leaving group), the carbonyl group is reformed and water serves as the leaving group. In step 6 (proton transfer), water serves as a base which deprotonates the cationic intermediate, resulting in formation of a cyclic ester.

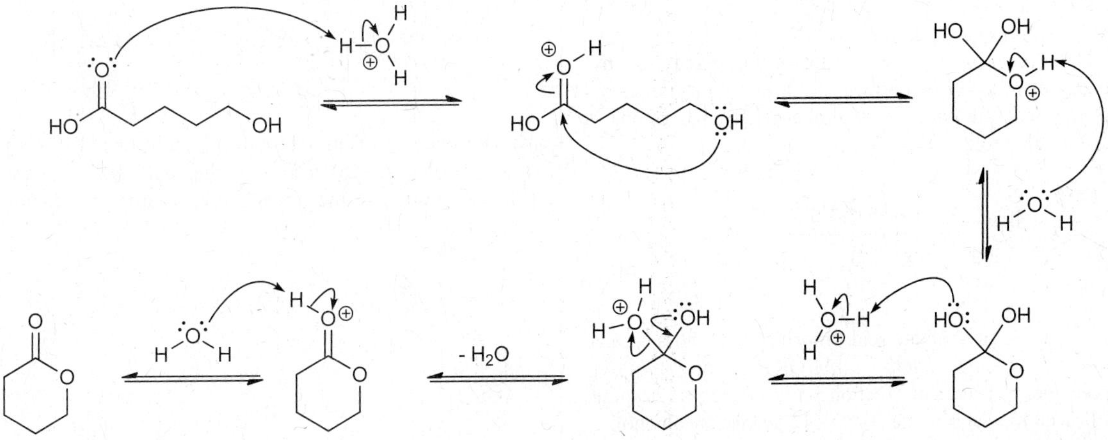

21.17. This reaction occurs under acidic conditions, so we must avoid formation of a strong base. Thus, proton transfers are required at multiple stages in the mechanism. The mechanism shown below has six steps: 1) proton transfer, 2) nucleophilic attack, 3) proton transfer, 4) proton transfer, 5) loss of a leaving group and 6) proton transfer. The first step (proton transfer), requires two curved arrows to show the transfer of a proton from $MeOH_2^+$ to the carbonyl group, resulting in formation of an activated electrophile. In step two (nucleophilic attack), methanol serves as a nucleophile attacking the protonated carbonyl group, producing a cationic tetrahedral intermediate. We cannot immediately expel the amine group at this stage, as this would result in the formation of a strong base. This must be avoided in acidic conditions. This nitrogen atom must first be protonated. However, protonation at this stage would result in an intermediate with two positive charges, which should be avoided, if possible. Therefore, in the next step (step 3), methanol serves as a base, resulting in a neutral intermediate. Subsequently, in step 4 (proton transfer) a proton is transferred from $MeOH_2^+$ to the nitrogen atom, as shown. In step 5 (loss of leaving group), the carbonyl group is reformed and the amine serves as the leaving group, resulting in the opening of the ring. In step 6 (proton transfer), methanol serves as a base which deprotonates the cationic intermediate, resulting in formation of a bifunctional product.

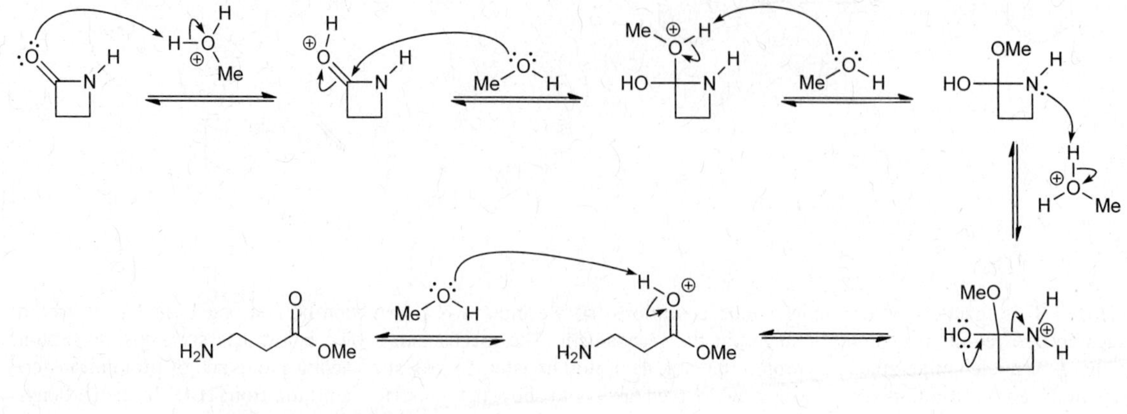

21.18.

(a) The reaction of an acid chloride with excess LAH, followed by water work-up, results in the formation of the corresponding alcohol shown below.

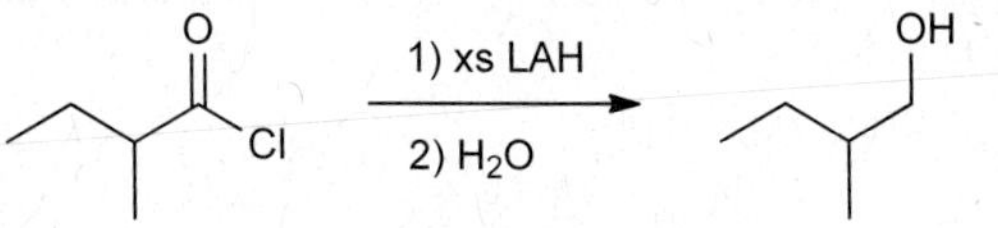

(b) The reaction of an acid chloride with excess phenyl magnesium bromide, followed by water work-up, results in the incorporation of two phenyl groups, giving a tertiary alcohol.

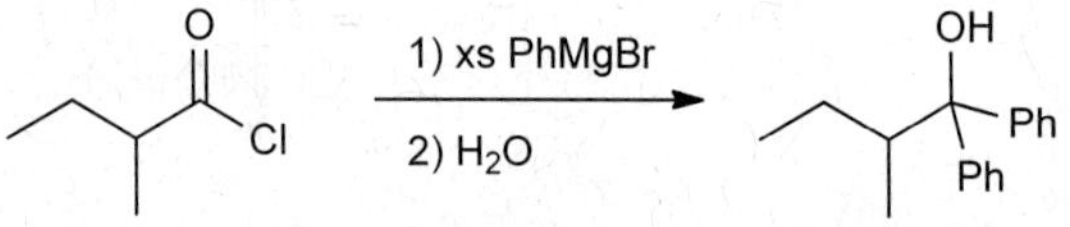

(c) The reaction of an acid chloride with the selective hydride-reducing agent, LiAl(OR)$_3$H, produces an aldehyde. Subsequent reaction with a Grignard reagent, followed by water work-up, gives a secondary alcohol.

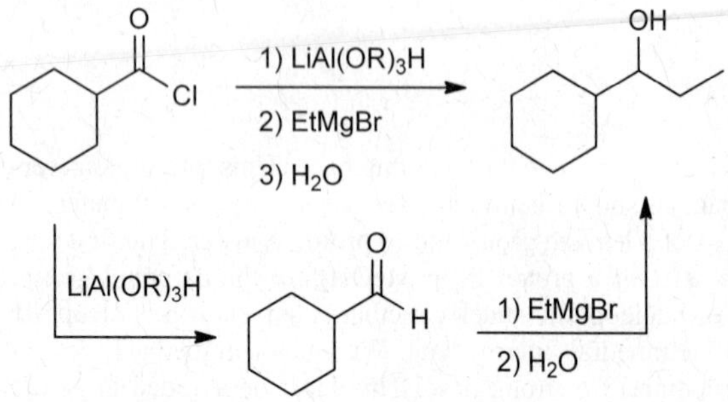

(d) The reaction of an acid chloride with a lithium dialkyl cuprate (a selective carbon nucleophile) produces a ketone. Subsequent reaction with LAH, followed by water work-up, gives a secondary alcohol.

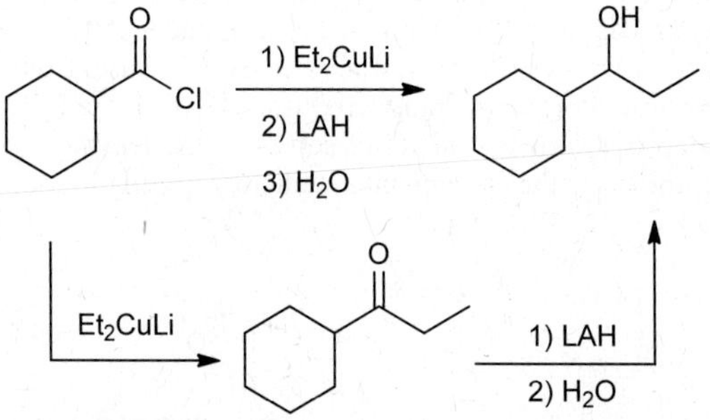

(e) The reaction of an acid chloride with phenol (in the presence of pyridine) results in the replacement of the chlorine atom with the phenol oxygen atom, producing an ester, as shown.

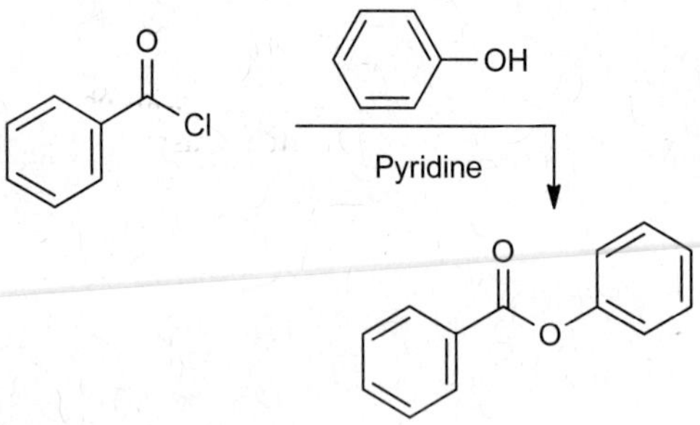

(f) The reaction of an acid chloride with two equivalents of an amine results in the replacement of the chlorine atom with the amine nitrogen atom, producing an amide, as shown.

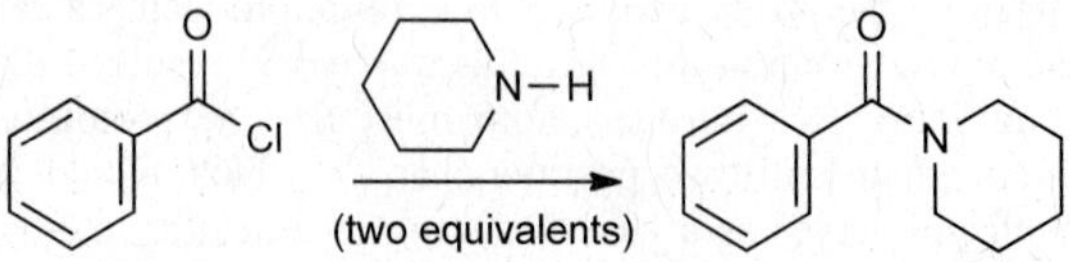

21.19. The conversion of benzyl alcohol to benzoyl chloride requires oxidation of the benzylic carbon atom. Subsequent reaction with thionyl chloride results in the conversion of the carboxylic acid to the desired acid chloride.

21.20. The reaction between an acid chloride and a Grignard reagent occurs via the following mechanism. In the first step (nucleophilic attack), the anionic carbon atom of the Grignard reagent serves as a nucleophile and attacks the electrophilic carbonyl group, resulting in a tetrahedral intermediate. In step two (loss of a leaving group), the carbonyl group is reformed and chloride leaves, shown with two curved arrows. In step three (nucleophilic attack), a second equivalent of the Grignard reagent attacks the carbonyl group of the intermediate ketone, resulting in the formation of another tetrahedral intermediate.

Once this reaction is complete, concentrated acid is added to the reaction flask. H_3O^+ serves as an acid, protonating the alkoxide ion and producing an alcohol. Under these strongly acidic conditions, the alcohol is further protonated, giving an oxonium ion. Loss of water generates a tertiary carbocation, which can then be deprotonated (an E1 process). Note that removal of this proton results in the more substituted alkene (the Zaitsev product).

21.21.
(a) The reaction of an acid anhydride with phenol results in replacement of the carboxylate leaving group with the phenol oxygen atom. After a proton transfer, the following ester and carboxylic acid are produced.

(b) The reaction of an acid anhydride with diethylamine results in replacement of the carboxylate leaving group with the amine nitrogen atom. After a proton transfer, the following amide and a carboxylate ion are produced. You might be wondering why a carboxylate ion is drawn rather than a carboxylic acid. This will be discussed in Chapter 23, but here is a preview: In the presence of excess diethylamine, the resulting carboxylic acid is deprotonated to give a carboxylate ion (compare the pK_a values of an ammonium ion and a carboxylic acid, which can be found in the pK_a table on the inside cover of the textbook).

(c) The reaction of a phenol derivative with acetic anhydride results in the replacement of the carboxylate leaving group with the phenol oxygen atom. After a proton transfer, the following ester and carboxylic acid are produced.

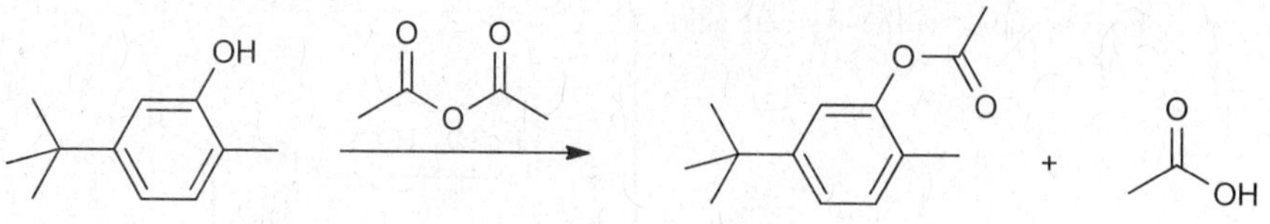

(d) The reaction of an acid anhydride with a cyclic secondary amine results in the replacement of the carboxylate leaving group with the amine nitrogen atom. After a proton transfer, the following amide and carboxylic acid are produced. You might be wondering why a carboxylate ion is drawn rather than a carboxylic acid. This will be discussed in Chapter 23, but here is a preview: In the presence of excess amine, the resulting carboxylic acid is deprotonated to give a carboxylate ion (compare the pK_a values of an ammonium ion and a carboxylic acid, which can be found in the pK_a table on the inside cover of the textbook).

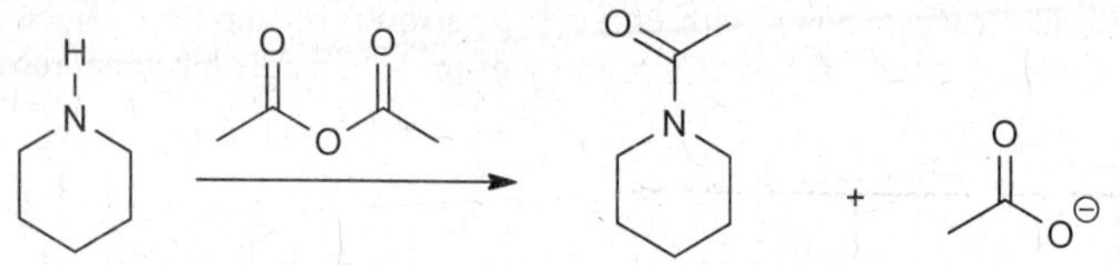

21.22. Three methods of converting benzoic acid to ethyl benzoate are shown below. In the first method, benzoic acid is deprotonated by NaOH. The intermediate salt (sodium benzoate) serves as a nucleophile in a subsequent S_N2 reaction with ethyl iodide. The second method is a Fischer esterification process, in which ethanol serves as both the solvent and a weak nucleophile. In the third method, benzoic acid is first converted to benzoyl chloride, and subsequently treated with ethanol (in the presence of pyridine) to produce the desired ester.

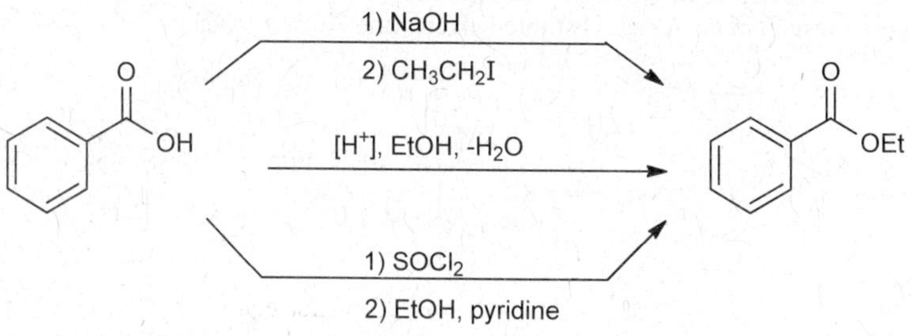

21.23.
(a) Oxidation of benzyl alcohol to benzoic acid is accomplished using strongly oxidizing conditions, as shown below. Subsequent conversion to the corresponding ethyl ester can be accomplished by any of the three methods shown, as described above in the solution to problem 21.22.

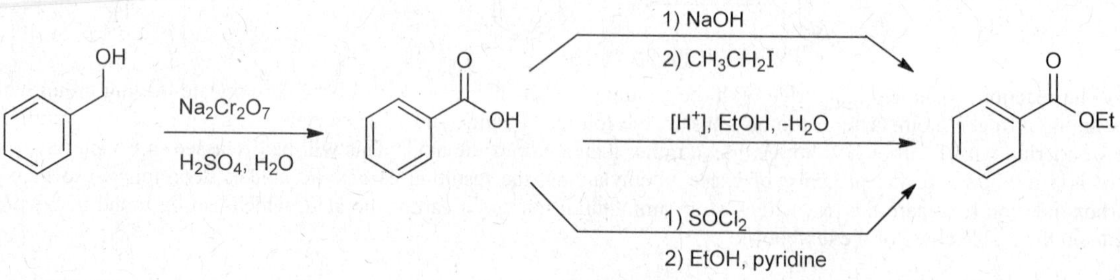

(b) Oxidation of styrene to benzoic acid is accomplished using strongly oxidizing conditions, as shown below. Subsequent conversion to the corresponding ethyl ester can be accomplished by any of the three methods shown, as described above in the solution to problem 21.22.

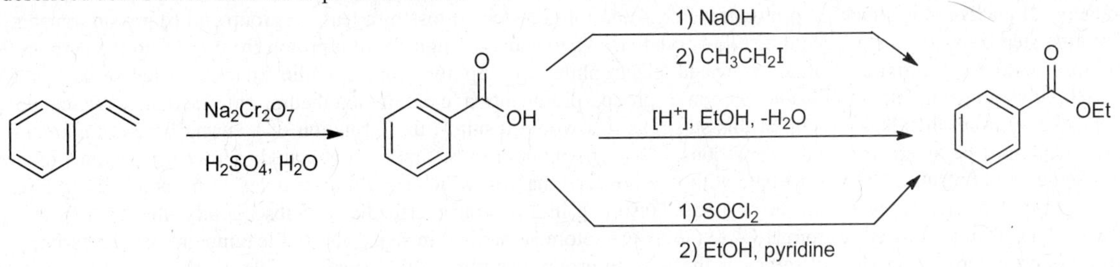

21.24.

(a) The first equivalent of lithium aluminum hydride reduces the ester to an aldehyde in two mechanistic steps (nucleophilic attack of LAH, then loss of methoxide). A second equivalent further reduces the aldehyde to the corresponding alkoxide, which is subsequently protonated by water. Overall, LAH supplies two equivalents of hydride that are incorporated into the product.

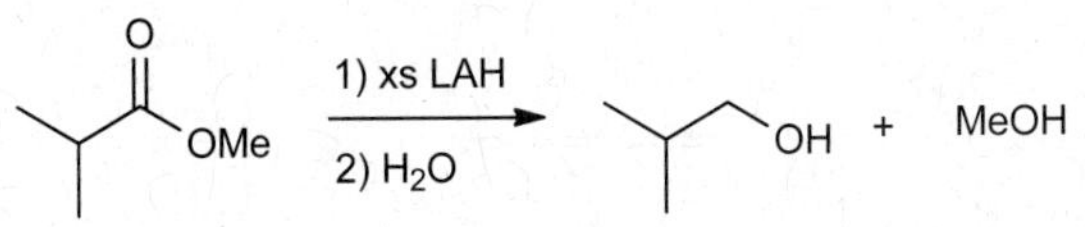

(b) The first equivalent of the Grignard reagent attacks the carbonyl group, thereby converting the ester into a ketone in two mechanistic steps (nucleophilic attack of EtMgBr, followed by loss of methoxide). A second equivalent of the Grignard reagent then attacks the carbonyl group of the ketone intermediate to produce an alkoxide ion, which is subsequently protonated by water. Overall, two ethyl substituents are incorporated into the product.

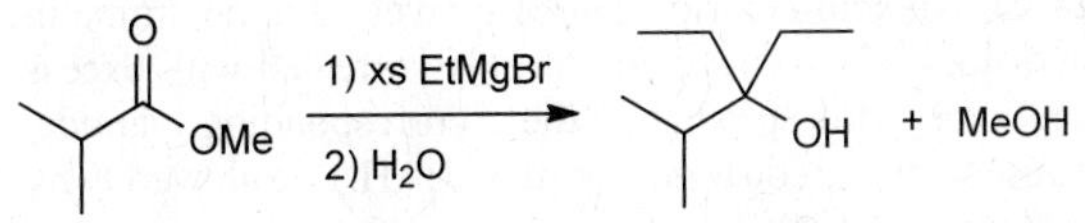

(c) The first equivalent of lithium aluminum hydride reduces the ester to an aldehyde in two mechanistic steps (nucleophilic attack of LAH, followed by loss of a leaving group, which remains tethered to the aldehyde group via the alkyl chain). A second equivalent of lithium aluminum hydride further reduces the aldehyde to the corresponding alkoxide. The resulting dianion is protonated upon treatment with water. Overall, LAH supplies two equivalents of hydride that are incorporated into the product.

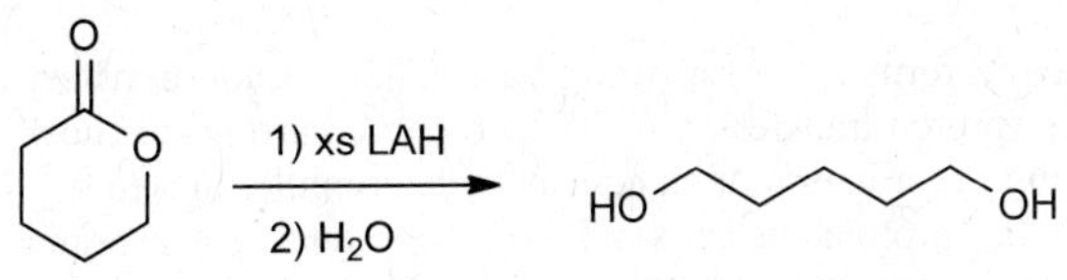

(d) Upon treatment with aqueous acid, an ethyl ester is hydrolyzed to the corresponding carboxylic acid and ethanol. The reaction occurs under acid-catalyzed conditions, in which an OH group ultimately replaces the OEt group of the ester.

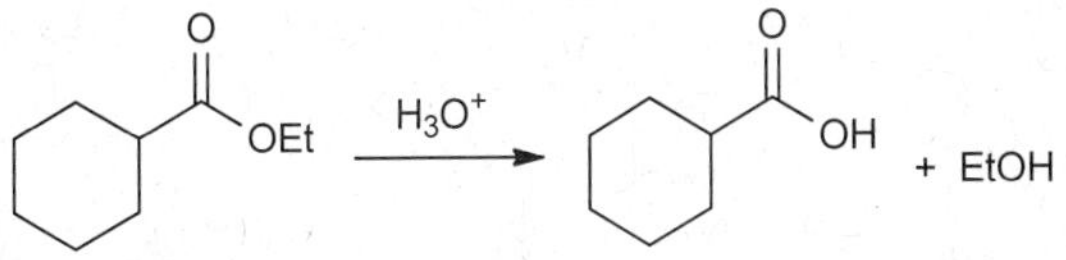

(e) In step 1, benzoic acid is deprotonated by NaOH. The intermediate salt (sodium benzoate) serves as a nucleophile in a subsequent S_N2 reaction with ethyl iodide, giving an ester, as shown.

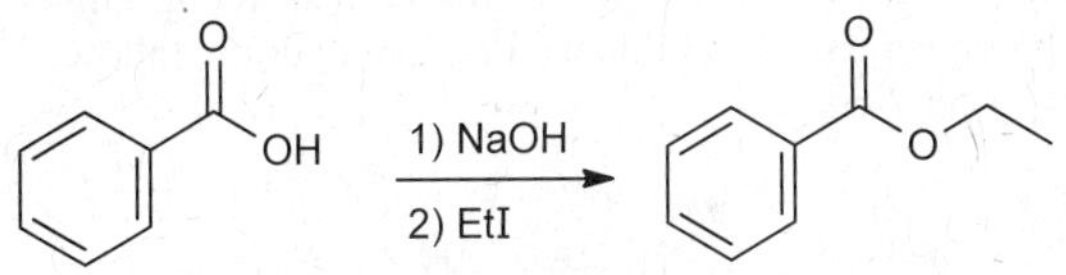

(f) The first equivalent of the Grignard reagent attacks the carbonyl group, thereby converting the cyclic ester into an acyclic ketone in two mechanistic steps (nucleophilic attack of EtMgBr, followed by loss of the leaving group, which remains tethered to the molecule via the alkyl chain attached to the aromatic ring). A second equivalent of the Grignard reagent attacks the carbonyl group of the ketone intermediate to produce an alkoxide ion, which is subsequently protonated by water. Overall, two ethyl substituents are incorporated into the product.

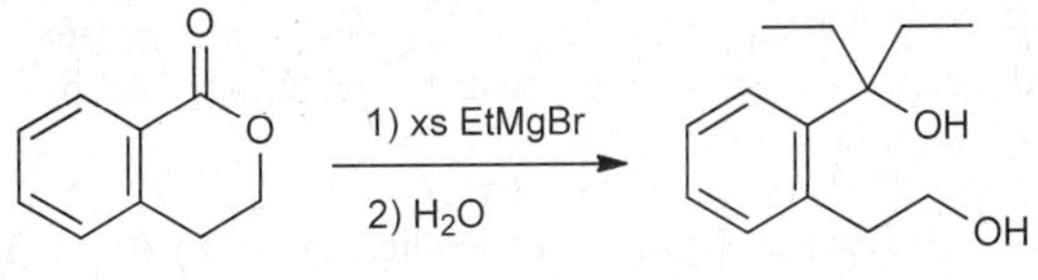

21.25. This reaction occurs under acidic conditions, so we must avoid formation of a strong base. Thus, proton transfers are required at multiple stages in the mechanism. The mechanism shown below has six steps: 1) proton transfer, 2) nucleophilic attack, 3) proton transfer, 4) proton transfer, 5) loss of a leaving group and 6) proton transfer. The first step (proton transfer), requires two curved arrows to show the transfer of a proton from H_3O^+ to the carbonyl group, resulting in formation of an activated electrophile. In step two (nucleophilic attack), water serves as a nucleophile, attacking the protonated carbonyl group, producing a cationic tetrahedral intermediate. We cannot immediately expel the alkoxy group at this stage, as this would result in the formation of a strong base (an alkoxide ion). This must be avoided in acidic conditions. This oxygen atom must first be protonated. However, protonation at this stage would result in an intermediate with two positive charges, which should be avoided, if possible. Therefore, in the next step (step 3), water serves as a base, resulting in a neutral intermediate. Subsequently, in step 4 (proton transfer) a proton is transferred from H_3O^+ to the oxygen atom, as shown. In step 5 (loss of leaving group), the carbonyl group is reformed and an alcohol serves as the leaving group, resulting in the opening of the ring. In step 6 (proton transfer), water serves as a base which deprotonates the cationic intermediate, resulting in formation of a bifunctional product.

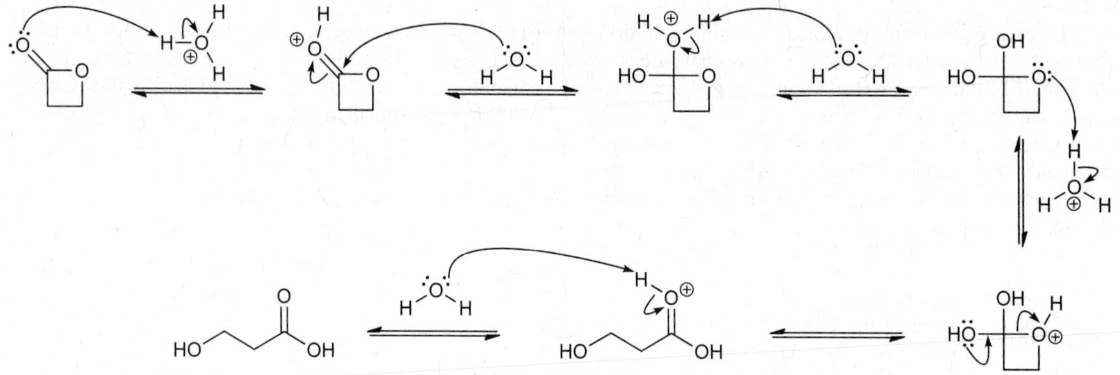

21.26.

(a) An amide is converted to the corresponding amine upon treatment with lithium aluminum hydride, followed by water work-up.

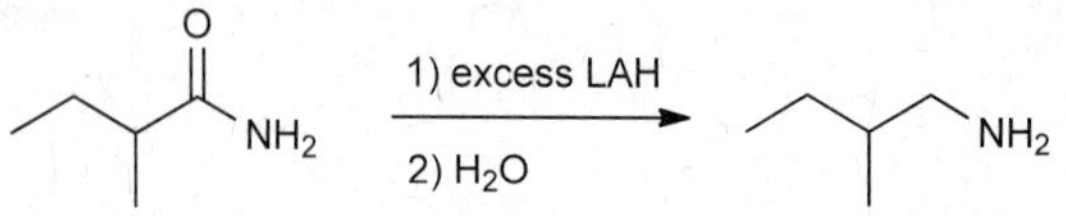

(b) The reaction of an acid chloride with excess ammonia results in replacement of the chloride leaving group with an NH_2 group.

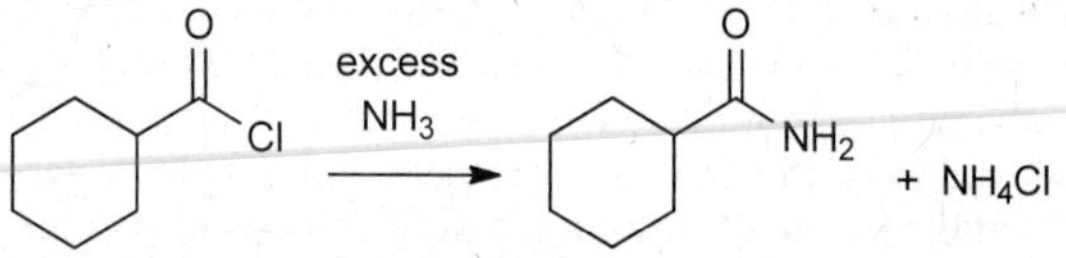

(c) An amide is hydrolyzed to give a carboxylic acid upon treatment with aqueous acid at elevated temperature. Under these acidic conditions, the by-product (ammonia) is protonated, resulting in formation of an ammonium ion.

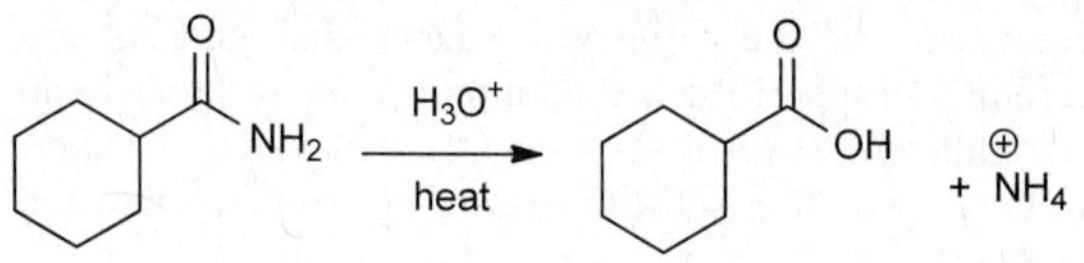

21.27. An acid chloride can be converted to an amine in two steps. First, the acid chloride is treated with excess ammonia to produce the corresponding amide. Subsequent reduction with LAH, followed by protonation with water, produces benzyl amine, as shown.

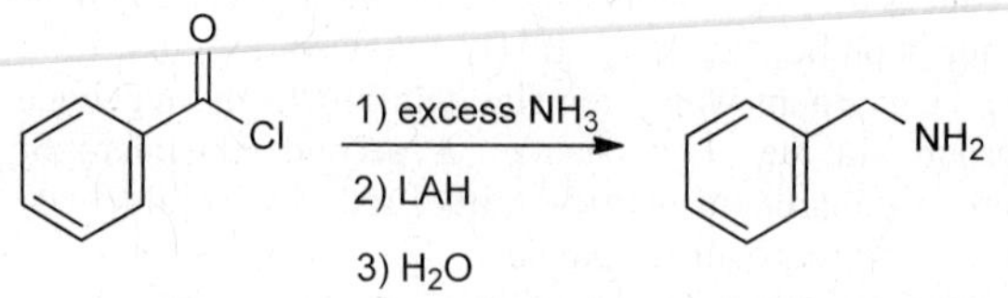

21.28.

(a) This reaction occurs under acidic conditions, so we must avoid formation of a strong base. Thus, proton transfers are required at multiple stages in the mechanism. The first step (proton transfer), requires two curved arrows to show the transfer of a proton from H_3O^+ to the carbonyl group, resulting in formation of an activated electrophile. In step two (nucleophilic attack), water serves as a nucleophile attacking the protonated carbonyl group, producing a cationic tetrahedral intermediate. We cannot immediately expel the amine group at this stage, as this would result in the formation of a strong base. This must be avoided in acidic conditions. This nitrogen atom must first be protonated.

However, protonation at this stage would result in an intermediate with two positive charges, which should be avoided, if possible. Therefore, in the next step (step 3), water serves as a base, resulting in a neutral intermediate. Subsequently, in step 4 (proton transfer), a proton is transferred from H_3O^+ to the nitrogen atom, as shown. In step 5 (loss of leaving group), the carbonyl group is reformed and the amine serves as the leaving group, resulting in the opening of the ring. In step 6 (proton transfer), water serves as a base which deprotonates the cationic intermediate, resulting in formation of a bifunctional product. Under these acidic conditions, the amino group in the product is protonated to give an ammonium ion.

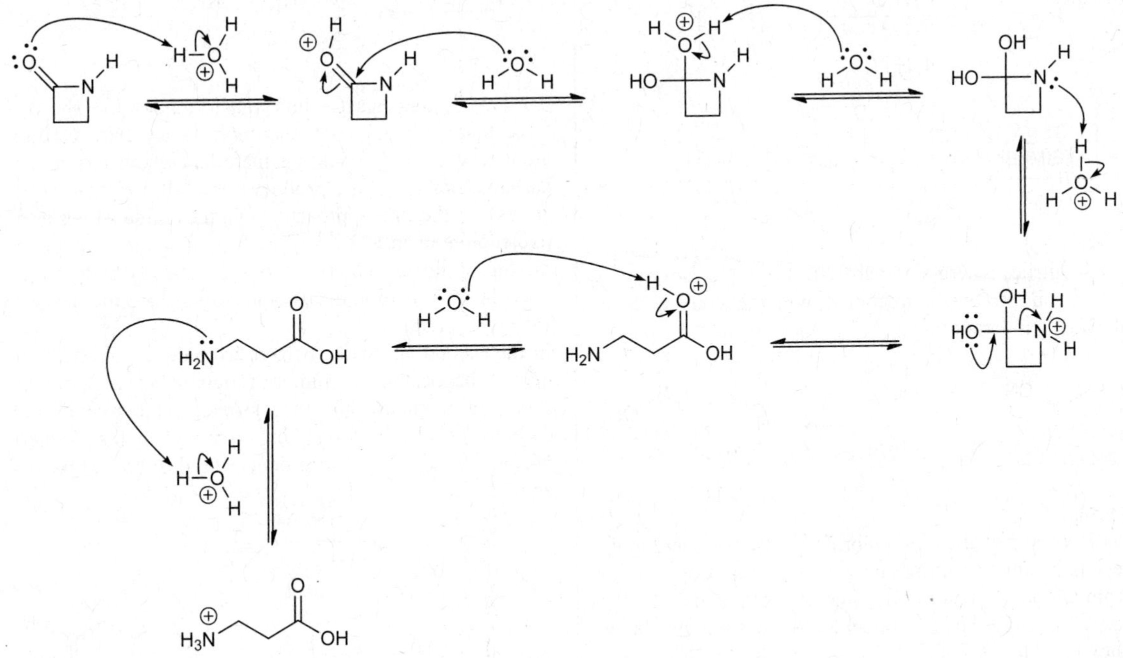

(b) This mechanism has three steps: 1) nucleophilic attack, 2) loss of a leaving group and 3) proton transfer. The first step (nucleophilic attack) requires two curved arrows, which show hydroxide functioning as a nucleophile and attacking the electrophilic carbonyl group, resulting in an anionic tetrahedral intermediate. In step two (loss of a leaving group), the carbonyl group is reformed as a result of cleavage of the carbon-nitrogen bond, thereby opening up the ring and resulting in formation of a bifunctional anionic intermediate containing a carboxylic acid group and a deprotonated amine. In the third step (proton transfer) the anionic nitrogen atom serves as a base, deprotonating the carboxylic acid group in an intramolecular process, resulting in the formation of a carboxylate ion.

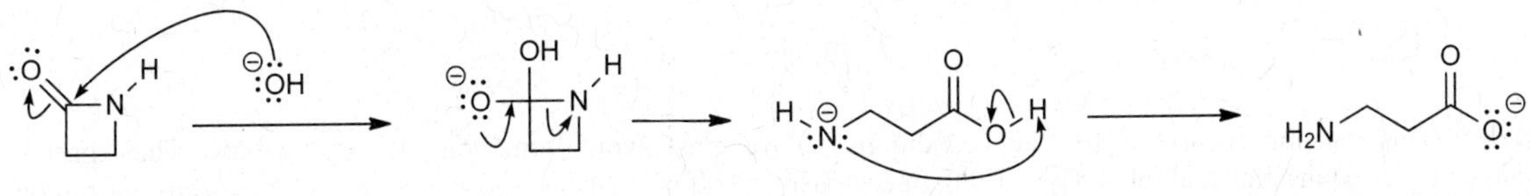

21.29.
(a) LAH reduces the C≡N triple bond to a single bond via incorporation of two equivalents of hydride, producing the primary amine shown.

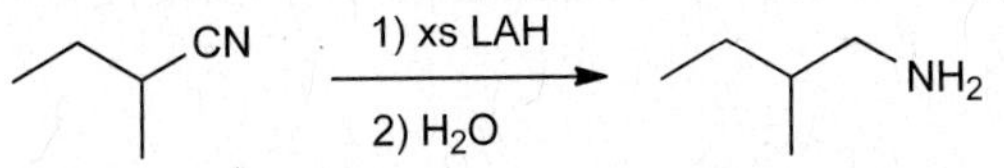

(b) Treating benzyl bromide with sodium cyanide results in the formation of an intermediate nitrile (shown) via an S_N2 reaction. Subsequent attack by a Grignard reagent produces an anionic intermediate which is then protonated and hydrolyzed with H_3O^+ to form the corresponding ketone.

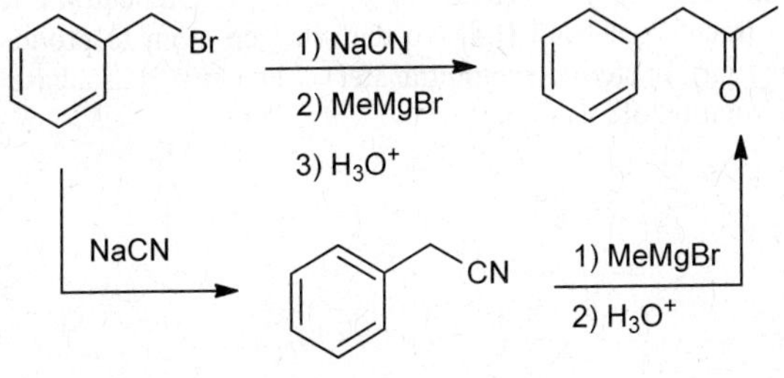

(c) Reaction of a nitrile with a Grignard reagent produces an anionic intermediate which is subsequently protonated and hydrolyzed with H_3O^+ to form a ketone, as shown. Reduction with LAH, followed by water, converts the ketone to the corresponding secondary alcohol.

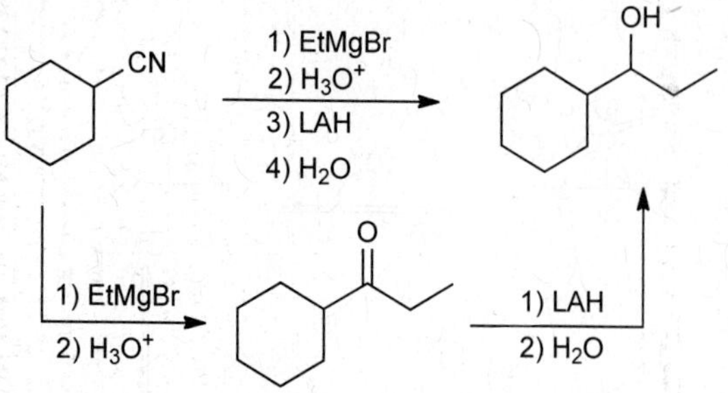

(d) A nitrile undergoes hydrolysis to give a carboxylic acid upon prolonged treatment with aqueous acid at elevated temperature.

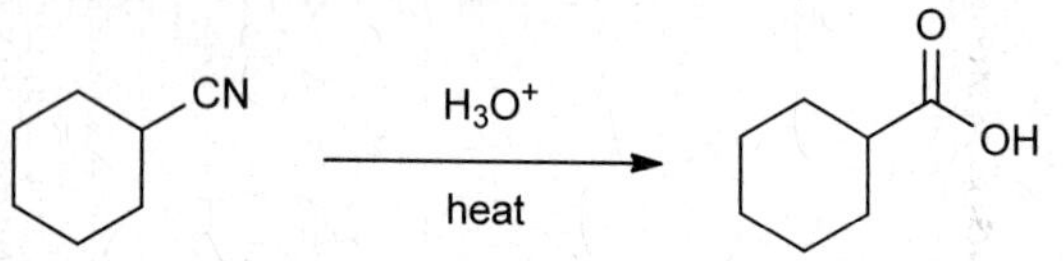

21.30.
(a) Reaction of the acid chloride with excess ammonia results in substitution of chloride with the nitrogen atom from ammonia. Thionyl chloride serves to dehydrate the resulting amide thus producing the desired nitrile as shown.

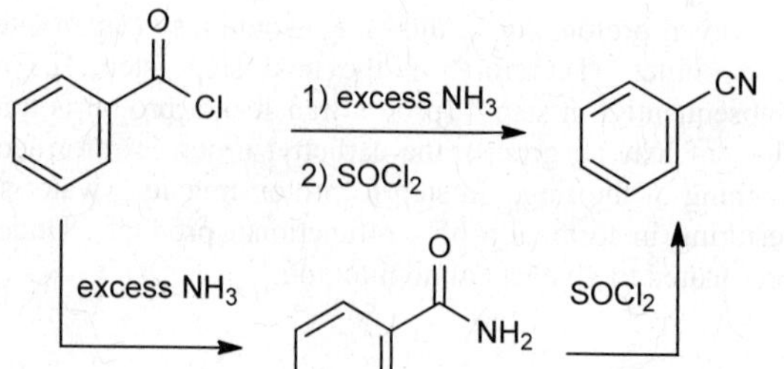

(b) Two approaches for this transformation are shown. Note that each approach incorporates an extra carbon atom to convert the starting material (which has seven carbon atoms) into the product (which has eight carbon atoms). In the first approach, sodium cyanide serves as a nucleophile in an S_N2 reaction, displacing the bromide to produce a nitrile. The nitrile is subsequently hydrolyzed upon heating with aqueous acid, to produce the desired carboxylic acid.

In the second approach, benzyl bromide is converted to benzyl magnesium bromide (a Grignard reagent), which serves as a nucleophile in a subsequent reaction with carbon dioxide. The resulting carboxylate ion is then protonated with H_3O^+ to produce the desired carboxylic acid.

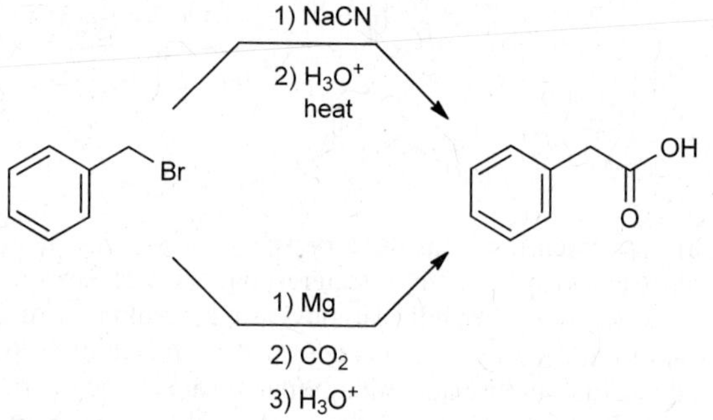

21.31. This reaction occurs under acidic conditions, so we must avoid formation of a strong base. Thus, proton transfers are required at multiple stages in this mechanism. The mechanism shown below has five steps: 1) proton transfer, 2) nucleophilic attack, 3) proton transfer, 4) proton transfer and 5) proton transfer. The first step (proton transfer), requires two curved arrows to show the transfer of a proton from H_3O^+ to the nitrogen atom, resulting in the formation of an activated electrophile. In step two (nucleophilic attack), water serves as a nucleophile attacking the activated electrophilic carbon atom of the protonated nitrile, producing a cationic intermediate. In step 3 (proton transfer), water serves as a base, resulting in a neutral intermediate. Subsequently, in step 4 (proton transfer), a proton is transferred from H_3O^+ to the nitrogen atom to produce a cationic intermediate (two key resonance structures are shown). In step 5 (proton transfer), water serves as a base which deprotonates the cationic intermediate, resulting in the formation of an amide.

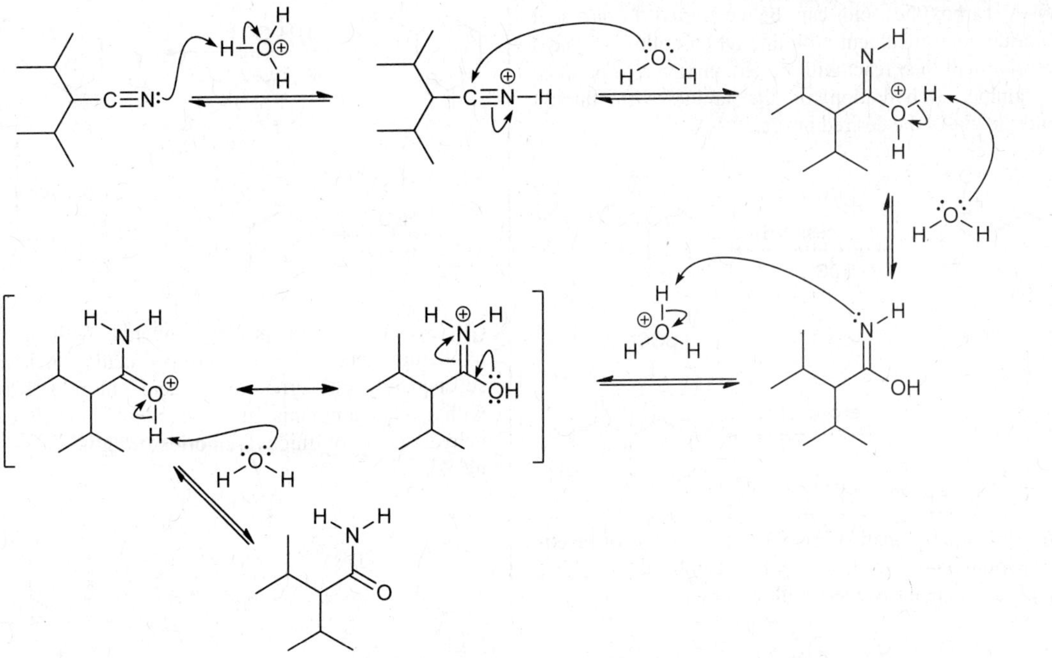

21.32.

(a) Acid catalyzed hydrolysis of the ester produces the carboxylic acid, which can then be converted to the acid chloride upon treatment with thionyl chloride.

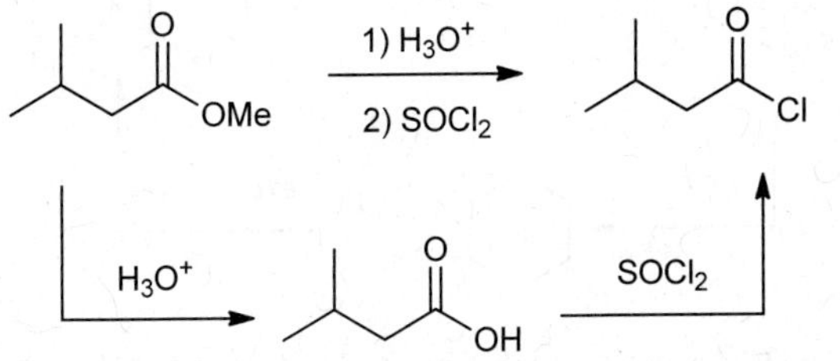

(b) Upon treatment with thionyl chloride, a carboxylic acid is converted to the corresponding acid chloride, which can then be treated with excess ammonia to produce an amide. Reduction of the amide with LAH, followed by protonation with water, yields the desired primary amine.

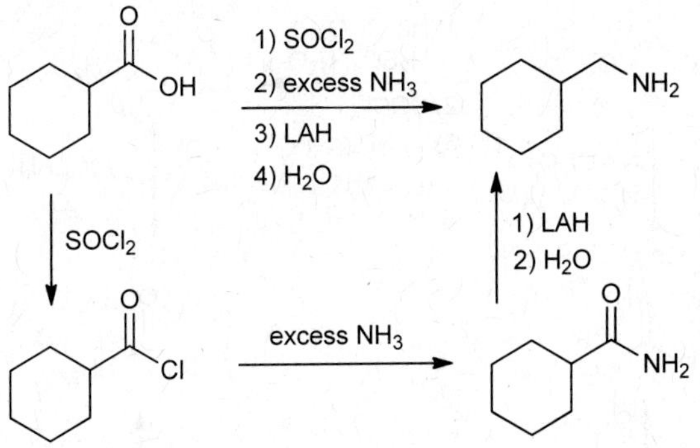

(c) Hydrolysis of acetic anhydride with water produces acetic acid, which can be converted to the desired product (acetyl chloride) upon treatment with thionyl chloride.

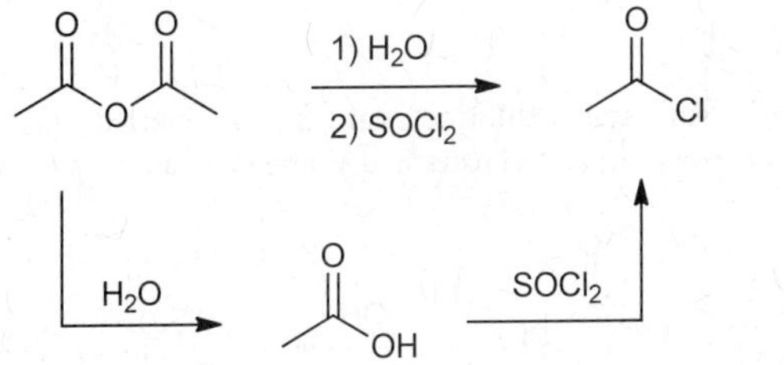

(d) The transformation requires conversion of an ester to an amine (with no change in the carbon skeleton). One way to achieve this transformation involves initial hydrolysis of the ester. The resulting carboxylic acid is converted to the acid chloride with thionyl chloride, which subsequently reacts with excess ammonia to produce the amide. Reduction of the amide with LAH, followed by water work-up, yields the desired primary amine.

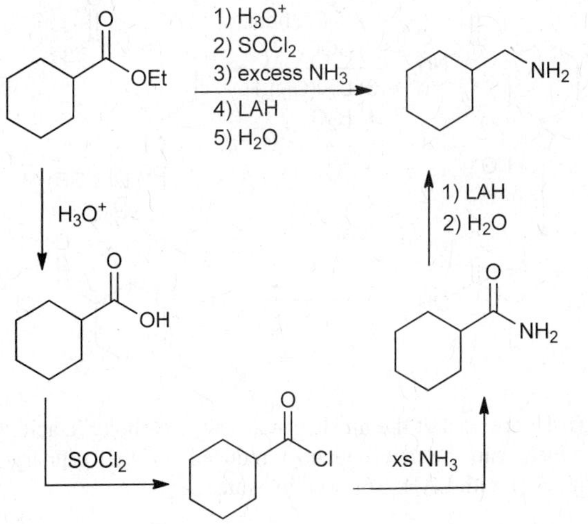

(e) A carboxylic acid can be converted to an acid chloride upon treatment with thionyl chloride. The acid chloride will then react with excess ammonia to produce an amide. Dehydration of the amide with thionyl chloride yields the desired nitrile.

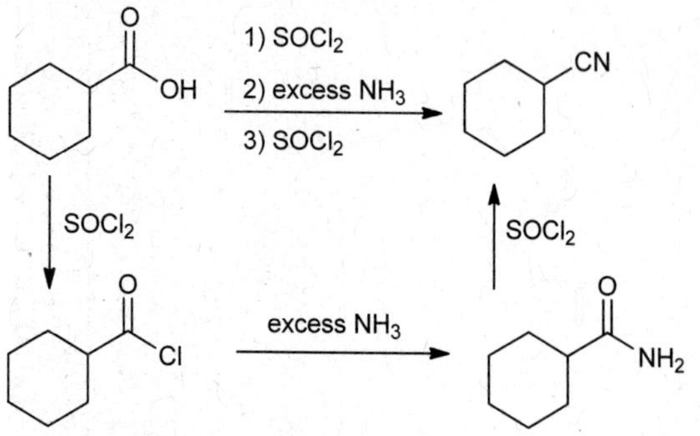

(f) This transformation involves hydrolysis of acetic anhydride to give acetic acid, which can be achieved in a single step, upon treatment with water.

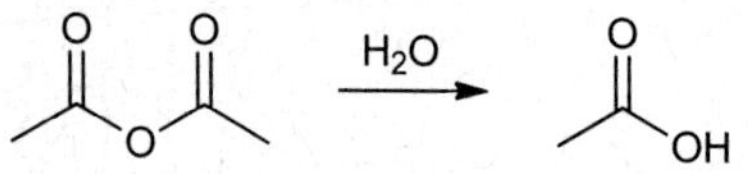

(g) An acid chloride can be converted to the corresponding ethyl ester in a single step, as shown.

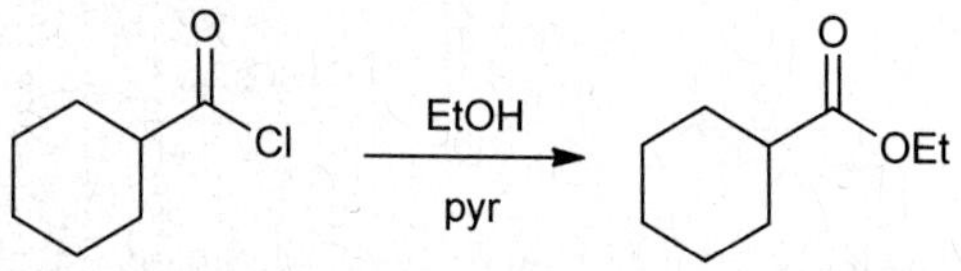

(h) Hydrolysis of the amide with H_3O^+ (and heat) produces a carboxylic acid, which can be converted to an acid chloride upon treatment with thionyl chloride. Subsequent reaction with a sterically hindered lithium trialkoxy aluminum hydride reagent, followed by water work-up, produces the desired aldehyde.

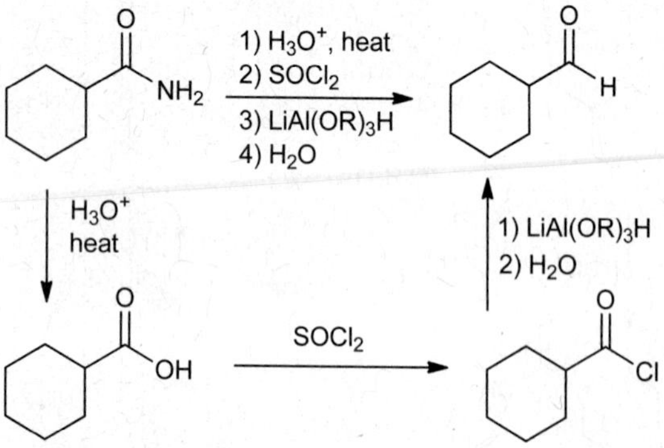

(i) Hydrolysis of the nitrile produces the carboxylic acid, which can be subsequently reduced to the primary alcohol with LAH, followed by water.

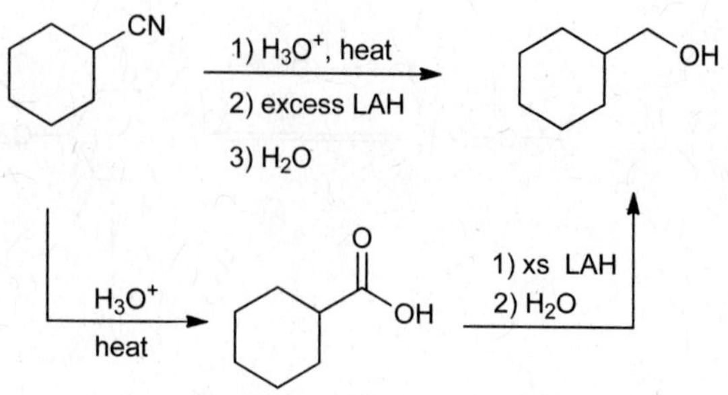

(j) Oxidation of a primary alcohol with a strong oxidizing agent yields a carboxylic acid, which can be subsequently converted to an acid chloride. Reaction with excess ammonia gives an amide, which can be dehydrated with thionyl chloride to give the desired nitrile.

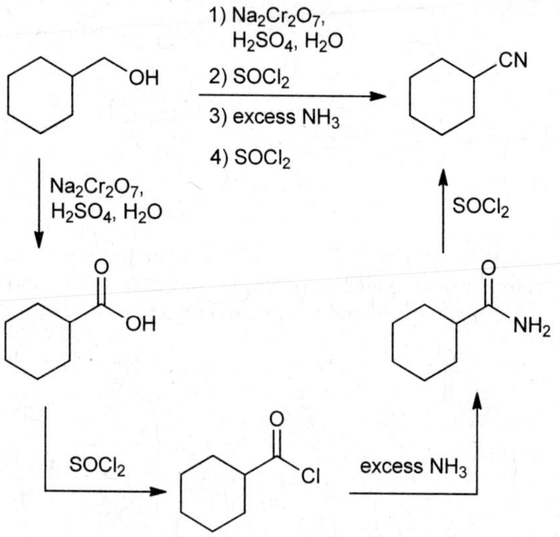

21.33. According to Figure 21.11, this transformation will require a minimum of four synthetic steps (if we are limited to reactions covered in this chapter). First, the alcohol is oxidized to a carboxylic acid, followed by conversion to an acid halide. The acid halide is then converted into an amide, followed by reduction to give an amine, as shown:

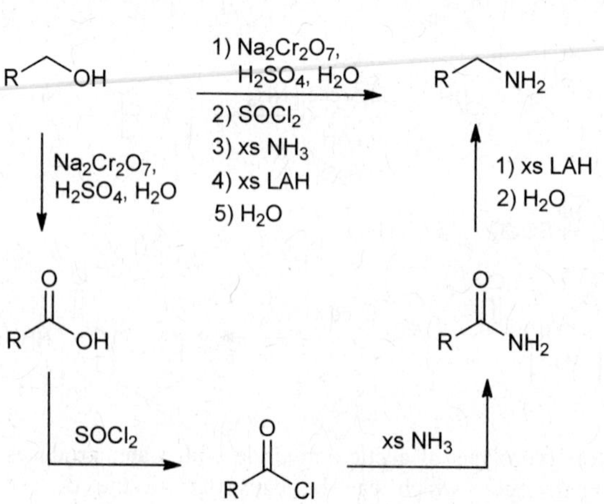

21.34. The starting material (1-hexene) has a functional group located between C1 and C2 (a double bond), while the product has a functional group located only at C1. As such, we will need to perform an addition reaction in an *anti*-Markovnikov fashion. We have seen two such processes (radical addition of HBr or hydroboration-oxidation). The synthesis shown here is based on the latter approach. *Anti*-Markovnikov addition of water (via hydroboration-oxidation) converts 1-hexene to 1-hexanol. Subsequent oxidation produces the carboxylic acid, which is then converted to the acid chloride upon treatment with thionyl chloride.

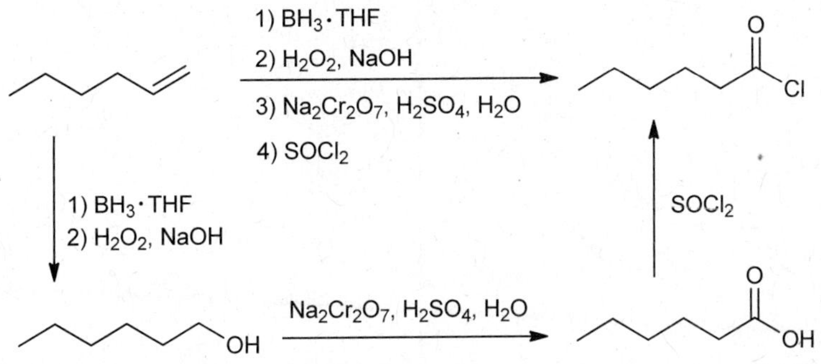

Alternatively, the alkene can be converted into a primary alkyl halide upon treatment with HBr in the presence of peroxides at elevated temperature. The primary alkyl halide is then converted to hexanol upon treatment with hydroxide (S_N2). Oxidation followed by treatment with thionyl chloride (just as above) gives the product.

21.35.

(a) The starting material has seven carbon atoms, and the product has nine carbon atoms. This requires installation of an ethyl group via a carbon-carbon bond-forming reaction. There are certainly several ways to achieve the installation of a single ethyl group. Let's first consider one way that will NOT work. Specifically, we cannot install the ethyl group via the reaction between an acid chloride and a Grignard reagent, as that would install two ethyl groups:

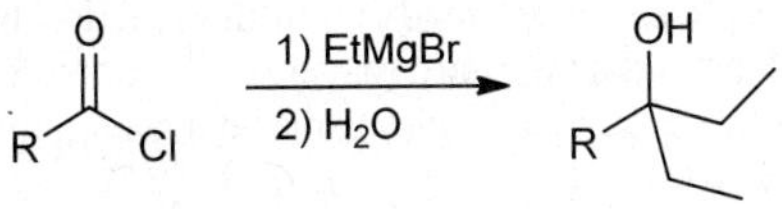

This reaction cannot be controlled to install a single ethyl group. However, a lithium dialkyl cuprate will attack an acid chloride just once, installing just one ethyl group:

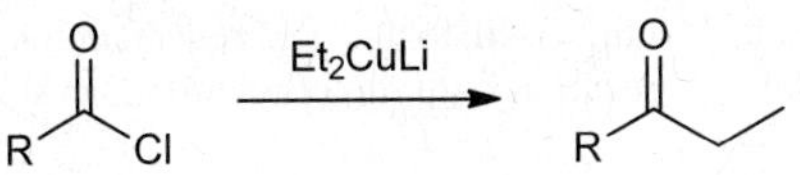

In order to use this method to install an ethyl group, we must first convert the starting material into an acid halide (which can be accomplished by treating the acid with thionyl chloride). Then, after installation of the ethyl group, we must convert the ketone into the final product (which can be achieved via reduction):

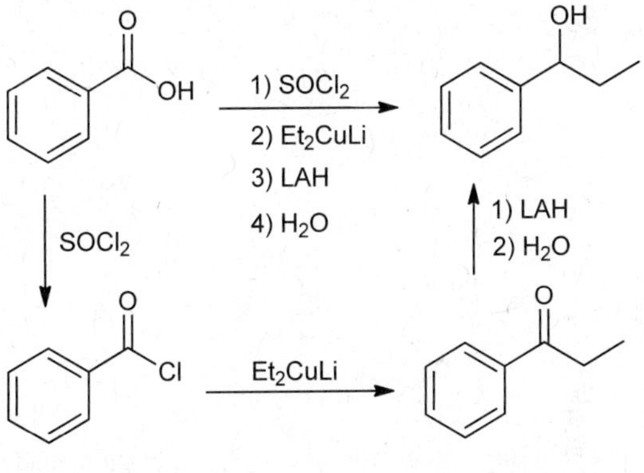

Alternatively, a single ethyl group can be installed via the reaction between an aldehyde and a Grignard reagent. This strategy gives the following synthesis: The carboxylic acid is first converted to an acid halide, followed by subsequent treatment with LiAl(OR)$_3$H to give an aldehyde. The aldehyde can then be treated with ethyl magnesium bromide, followed by aqueous work-up, to give the desired product. This alternative strategy demonstrates that there is rarely only one correct way to approach a synthesis problem. THIS IS TRUE FOR NEARLY ALL OF THE SYNTHESIS PROBLEMS THAT WE ENCOUNTER.

(b) One solution to this problem is shown here (there are certainly other acceptable solutions). Bromobenzene is converted to phenyl magnesium bromide (a Grignard reagent) and then treated with formaldehyde, followed by water work-up, to give benzyl alcohol, which serves as a nucleophile in a subsequent reaction with acetyl chloride to give the desired ester.

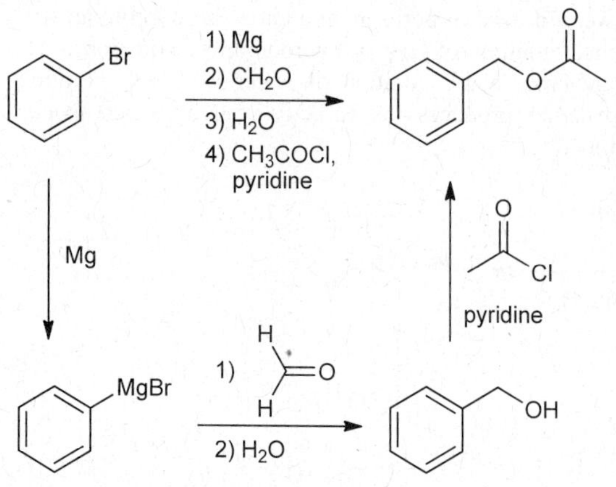

(c) There are certainly many acceptable solutions to this problem. One such solution derives from the following retrosynthetic analysis. An explanation of each of the steps (*a-d*) follows.

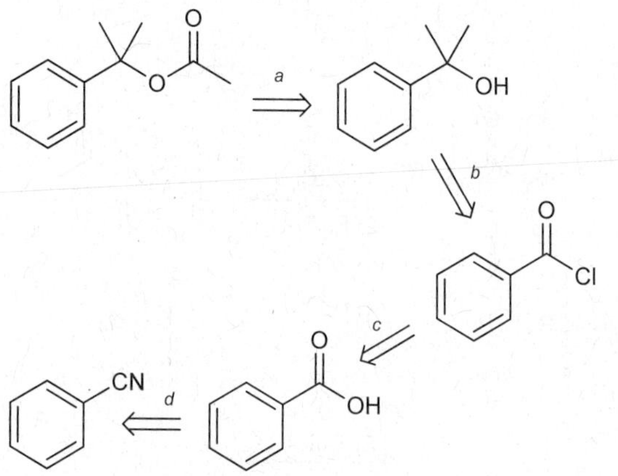

a. The desired product is an ester, which can be made via acetylation of the appropriate tertiary alcohol.

b. The tertiary alcohol can be made from an acid halide upon treatment with excess Grignard reagent.

c. The acid halide can be made from the corresponding carboxylic acid (benzoic acid).

d. Benzoic acid can be made from the starting material via hydrolysis.

Now let's draw the forward scheme. Hydrolysis of the nitrile to the carboxylic acid, followed by reaction with thionyl chloride, produces the acid chloride. Reaction with excess methyl magnesium bromide, followed by water work-up, results in the formation of a tertiary alcohol, with the incorporation of two new methyl groups. The tertiary alcohol can then serve as a nucleophile in an acetylation reaction (upon treatment with acetyl chloride) to give the desired ester.

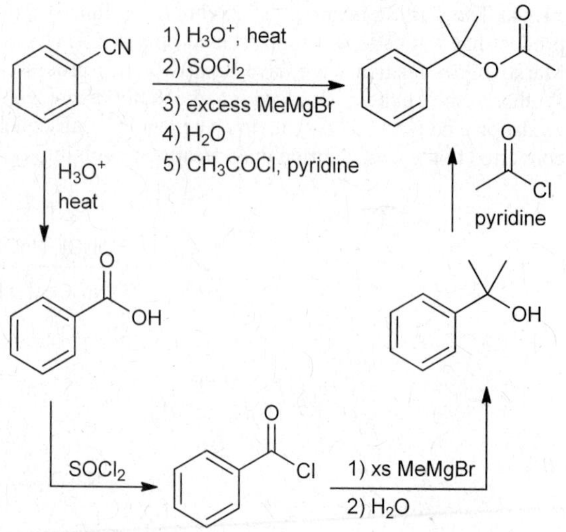

21.36. There are certainly many acceptable solutions to this problem. One such solution derives from the following retrosynthetic analysis. An explanation of each of the steps (*a-c*) follows.

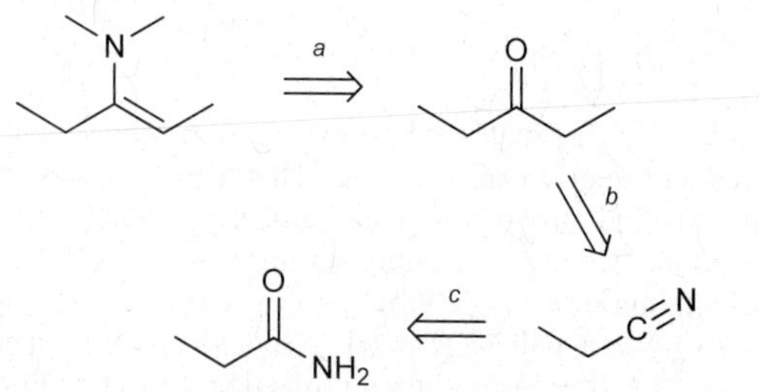

a. The desired product is an enamine, which can be made from the corresponding ketone.

b. The ketone can be made from the reaction between a nitrile and a Grignard reagent.

c. The nitrile can be made from the starting amide via dehydration.

Now let's draw the forward scheme. Dehydration of the amide with thionyl chloride produces a nitrile. Subsequent reaction with ethyl magnesium bromide, followed by aqueous acidic work-up, yields 3-pentanone, which can be converted to the desired enamine via an acid-catalyzed reaction with dimethylamine, as shown.

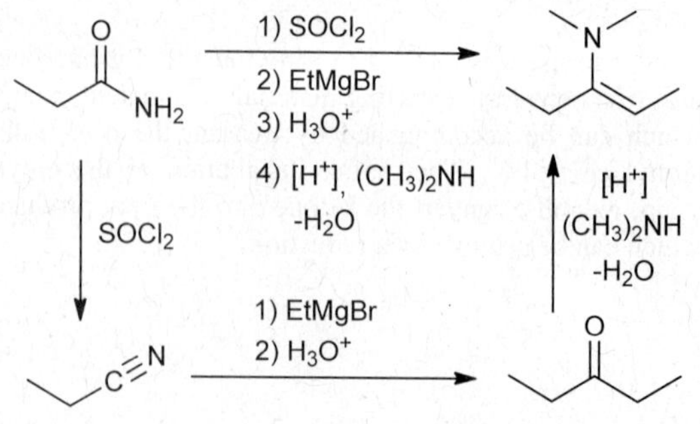

21.37.

(a) There are certainly many acceptable solutions to this problem. One such solution derives from the following retrosynthetic analysis. An explanation of each of the steps (*a-f*) follows.

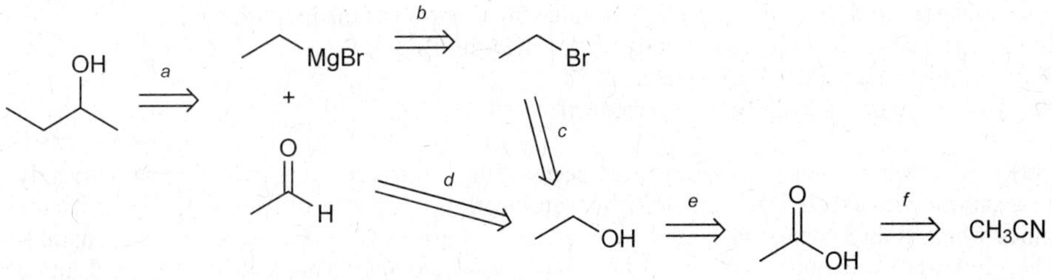

a. The desired alcohol can be made from a Grignard reaction between acetaldehyde and ethyl magnesium bromide.
b. Ethyl magnesium bromide is made from ethyl bromide, by insertion of magnesium.
c. Ethyl bromide can be made from ethanol upon treatment with PBr$_3$.
d. Acetaldehyde can be made from ethanol via oxidation with PCC.
e. Ethanol can be made via reduction of acetic acid.
f. Acetic acid can be made via hydrolysis of acetonitrile.

Now let's draw the forward scheme. Hydrolysis of acetonitrile gives acetic acid which is subsequently reduced to ethanol upon treatment with excess LAH, followed by water work-up. Upon treatment with PBr$_3$, ethanol is converted to ethyl bromide which is then converted to ethyl magnesium bromide (a Grignard reagent). A Grignard reaction with acetaldehyde (produced by PCC oxidation of ethanol, as shown), followed by water work-up, produces the desired alcohol, 2-butanol.

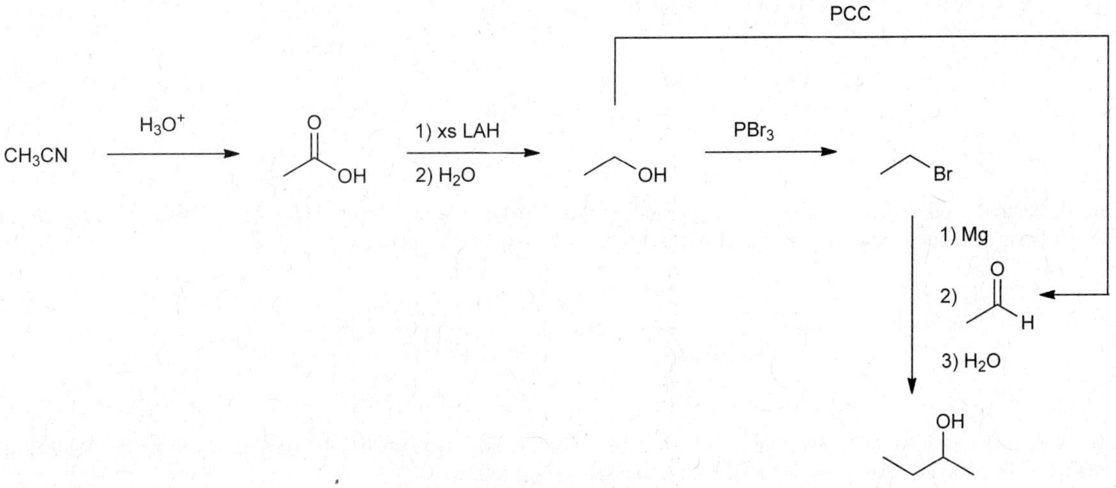

(b) There are certainly many acceptable solutions to this problem. One such solution derives from the following retrosynthetic analysis. An explanation of each of the steps (*a-g*) follows.

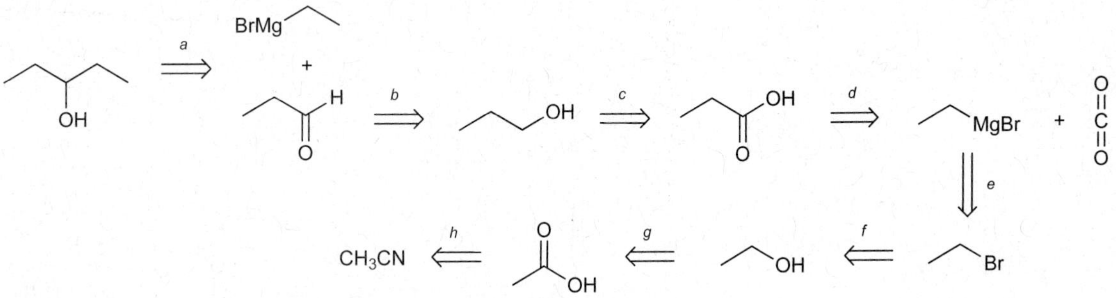

a. The desired alcohol can be made from a Grignard reaction between propanal and ethyl magnesium bromide (formed in step *d*).
b. Propanal can be made from 1-propanol, upon treatment with PCC.
c. 1-Propanol can be made from propanoic acid, upon treatment with excess LAH, followed by water work-up.
d. Propanoic acid can be made from a reaction between ethyl magnesium bromide and carbon dioxide.
e. Ethyl magnesium bromide is made from ethyl bromide, by insertion of magnesium.
f. Ethyl bromide can be made from ethanol upon treatment with PBr$_3$.
g. Ethanol can be made via reduction of acetic acid.
h. Acetic acid can be made via hydrolysis of acetonitrile.

Now let's draw the forward scheme. Hydrolysis of acetonitrile gives acetic acid which is subsequently reduced to ethanol upon treatment with excess LAH, followed by water work-up. Upon treatment with PBr$_3$, ethanol is converted to ethyl bromide which is then converted to ethyl magnesium bromide (a Grignard reagent). A Grignard reaction with carbon dioxide, followed by protonation with H$_3$O$^+$, gives propanoic acid. This acid is converted into propanal via reduction (with excess LAH) followed by oxidation with PCC. Reaction with ethyl magnesium bromide (prepared as described above), followed by water work-up, gives the desired alcohol, 3-pentanol.

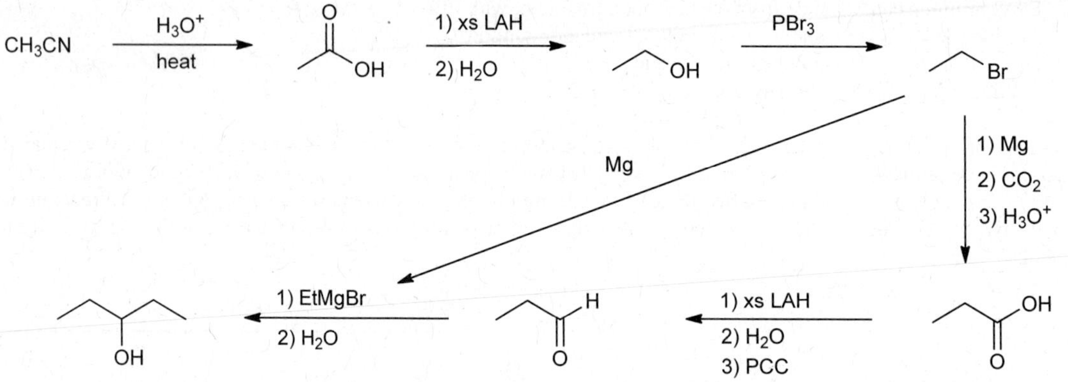

(c) The desired product can be made from the product of **21.37(a)** in just two steps. Therefore, we would first perform the synthesis described in the solution to **21.37(a)**, followed by these two reactions:

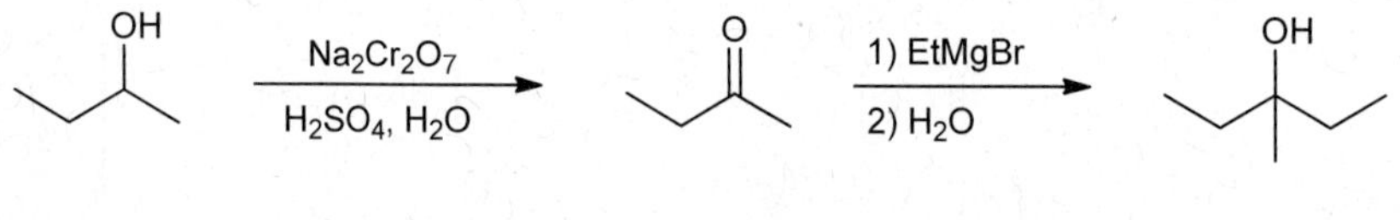

(d) There are certainly many acceptable solutions to this problem. One such solution derives from the following retrosynthetic analysis. An explanation of each of the steps (*a*-*g*) follows.

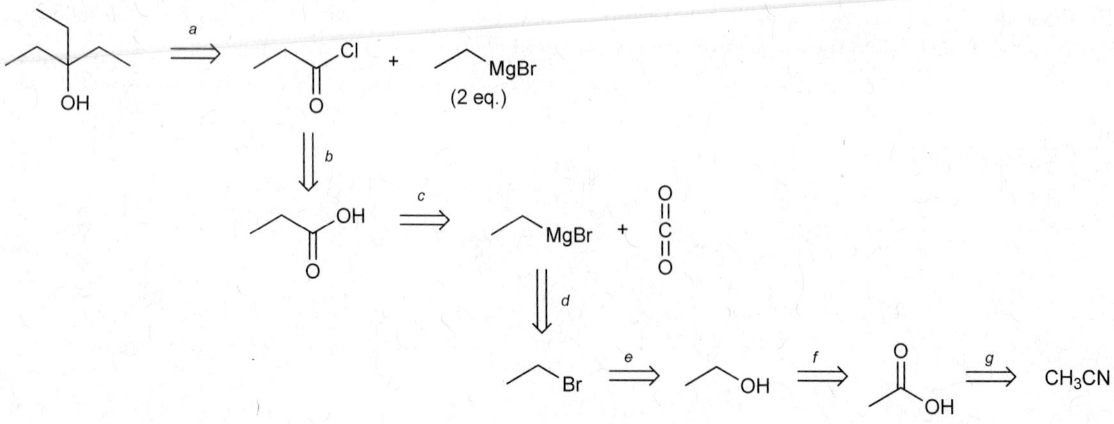

a. The desired alcohol can be made from a Grignard reaction between propanoyl chloride and two equivalents of ethyl magnesium bromide (formed in step *d*).

b. Propanoyl chloride can be made from propanoic acid, upon treatment with thionyl chloride.

c. Propanoic acid can be made from a reaction between ethyl magnesium bromide and carbon dioxide.

d. Ethyl magnesium bromide is made from ethyl bromide, by insertion of magnesium.

e. Ethyl bromide can be made from ethanol upon treatment with PBr$_3$.

f. Ethanol can be made via reduction of acetic acid.

g. Acetic acid can be made via hydrolysis of acetonitrile.

Now let's draw the forward scheme. Hydrolysis of acetonitrile gives acetic acid which is subsequently reduced to ethanol upon treatment with excess LAH, followed by water work-up. Upon treatment with PBr$_3$, ethanol is converted to ethyl bromide which is then converted to ethyl magnesium bromide (a Grignard reagent). A Grignard reaction with carbon dioxide followed by protonation with H$_3$O$^+$ gives propanoic acid. Conversion to the acid chloride, followed by reaction with excess ethyl magnesium bromide (prepared as described above) produces the desired tertiary alcohol, 3-ethyl-3-pentanol.

21.38. The signal at 1740 cm^{-1} indicates the presence of a carbonyl group (likely of an ester group) that is not conjugated with the aromatic ring (it would be at a lower wavenumber if it was conjugated). The cyclic ester (lactone) below fits the description provided and would indeed result in the diol shown upon reduction with two equivalents of LAH, followed by water work-up.

21.39.

(a) Each of these acids is a *para*-substituted benzoic acid. Relative acid strength depends on the electron withdrawing (or electron donating) capacity of the substituent. Electron withdrawing groups pull electron density away from the ring, thus stabilizing the anionic charge on the conjugate base, thus giving a stronger acid. Electron donating groups have the opposite effect: they donate electron density into the ring, thus destabilizing the anionic charge on the conjugate base, resulting in a weaker acid. Accordingly, the acids below are arranged in order of increasing acid strength. As seen in Table 19.1, a methoxy group is strongly electron-donating; a methyl group is weakly electron-donating; a bromine atom is weakly electron-withdrawing; a carbonyl group is a moderate electron-withdrawing group; a nitro group is a strong electron-withdrawing group.

(b) When comparing the acids below, the difference in acidity is related to the proximity of the electron-withdrawing bromine atom to the carboxylic acid group. The closer the bromine atom is to the carboxylic acid, the more it stabilizes the anionic charge on the conjugate base. Accordingly, the strongest acid in this series has the bromine atom *alpha* to the carboxylic acid, followed by the isomer with the bromine atom on a *beta* position, and then the isomer with the bromine atom on a *gamma* position, as shown below.

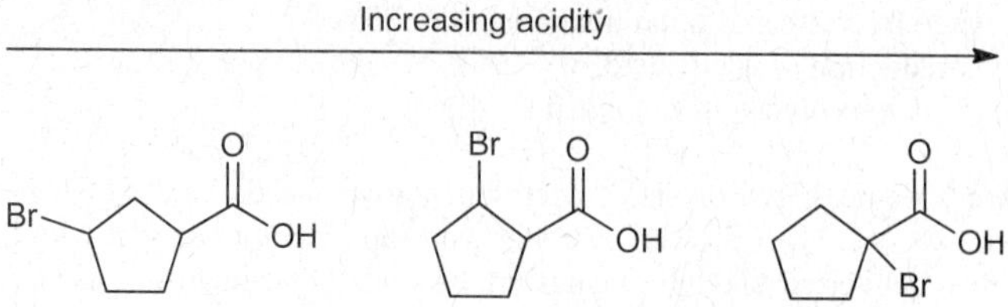

Increasing acidity

21.40.

(a) The second carboxylic acid group is electron withdrawing, and stabilizes the conjugate base that is formed when the first proton is removed.

(b) The carboxylate ion is electron rich and it destabilizes the conjugate base that is formed when the second proton is removed.

(c) Since both pK_1 and pK_2 are lower than 7.3, they are both expected to be largely deprotonated at physiological pH, resulting in the dianion shown below.

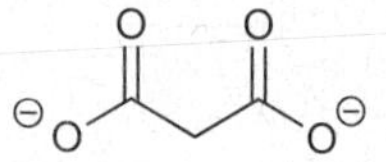

(d) The number of methylene (CH_2) groups separating the carboxylic acid groups is greater in succinic acid than in malonic acid. Therefore, the inductive effects described above in parts (a) and (b) are not as strong.

21.41.

(a) When a carboxylic acid group is attached to a ring, it is named as an alkanecarboxylic acid. This compound (with a ring composed of five carbon atoms) is thus cyclopentanecarboxylic acid.

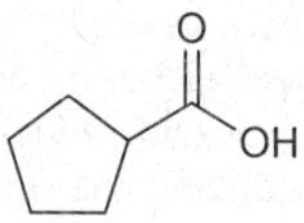

cyclopentanecarboxylic acid

(b) An amide is named by replacing the suffix "ic acid" or "oic acid" with "amide". The corresponding carboxylic acid is named cyclopentanecarboxylic acid. Replacement of "ic acid" with "amide" produces the name cyclopentanecarboxamide.

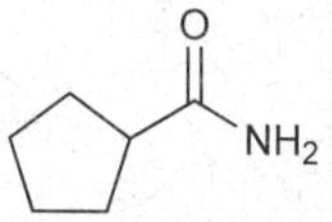

cyclopentanecarboxamide

(c) This acid chloride is named by replacing the "ic acid" from the parent (benzoic acid) with "yl chloride" to produce benzoyl chloride.

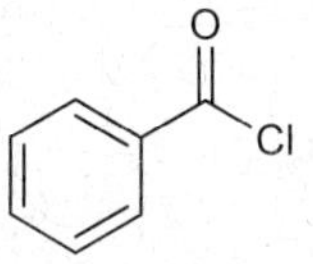

benzoyl chloride

(d) An ester is named by indicating the alkyl group (in this case, ethyl) attached to the oxygen atom of the ester, followed by the parent name of the corresponding carboxylic acid in which the "ic acid" is replaced by "ate". That is, the parent carboxylic acid (acetic acid) becomes acetate, giving the name ethyl acetate.

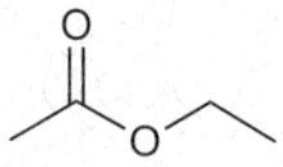

ethyl acetate

(e) The chain that contains the carboxylic acid group is comprised of six carbon atoms, so the "e" in the parent name (hexane) is replaced with the suffix "-oic acid", resulting in the IUPAC name hexanoic acid.

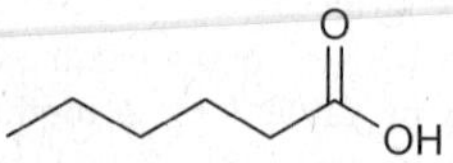

hexanoic acid

(f) The chain that contains the acid chloride group is comprised of five carbon atoms, so the "e" in the parent name (pentane) is replaced with the suffix "-oyl chloride", resulting in the IUPAC name pentanoyl chloride.

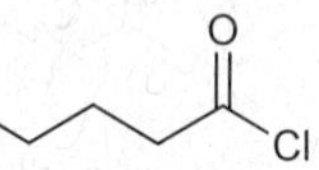

pentanoyl chloride

(g) The chain that contains the amide group is comprised of six carbon atoms, so the "e" in the parent name (hexane) is replaced with the suffix "-amide", resulting in the IUPAC name hexanamide.

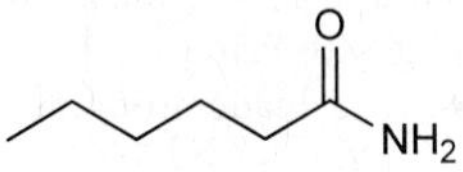

hexanamide

21.42. The common name for each molecule is shown below:

(a)

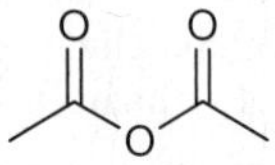

acetic anhydride

(b)

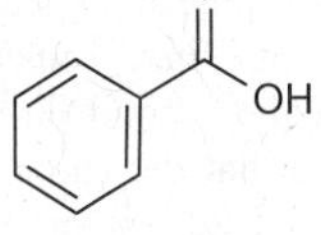

benzoic acid

(c)

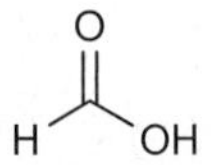

formic acid

(d)

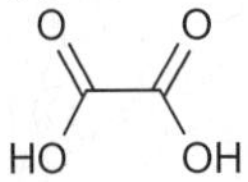

oxalic acid

21.43. A molecular formula of $C_6H_{12}O_2$ corresponds with one degree of unsaturation (see Section 15.16), which accounts for the carboxylic acid group. Since there are no other degrees of unsaturation, all of the isomers must be acyclic, saturated carboxylic acids. There are eight isomers that fit this description, shown below. These isomers are identified by methodically considering each possible parent chain. There is only one isomer with a parent chain of six carbon atoms (the first isomer shown). Then, there are three isomers that have a parent chain of five carbon atoms (with one methyl substituent). The methyl group can be located at C2, C3 or C4 (it cannot be at C5, because that would simply generate a parent chain of six carbon atoms, and we have already accounted for that isomer). Then, there are several isomers with a parent chain of only four carbons (with either two methyl groups or with one ethyl group). Once again, these isomers are drawn methodically. The two methyl groups can both be at C2, or they can be at C2 and C3, or both can be at C3. And finally, there can be an ethyl group at C2. Notice that, for a four-carbon chain, an ethyl group cannot be placed at C3, as that would generate a structure with a parent of five carbon atoms (and we have already accounted for that isomer).

Each carboxylic acid is named by identifying the longest chain containing the carboxylic acid group and replacing the "e" at the end of the alkane name with "oic acid". Each chain is numbered with the carboxylic acid carbon atom being C1, and the substituents are identified accordingly. Three of the isomers exhibit chirality centers (highlighted).

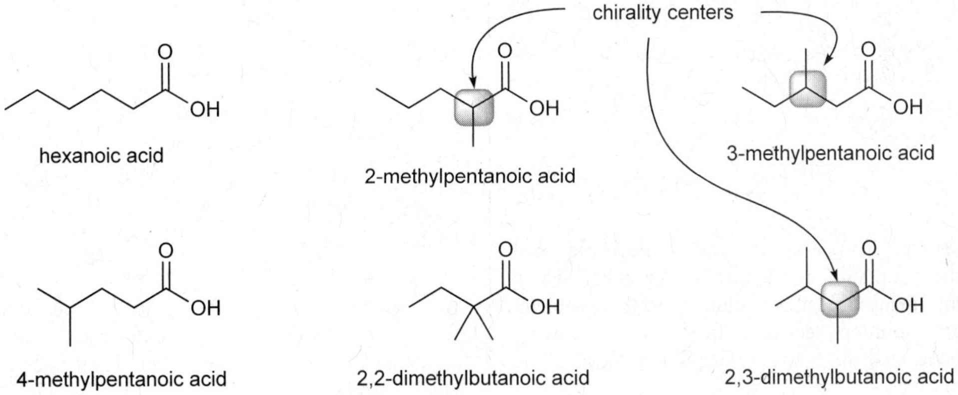

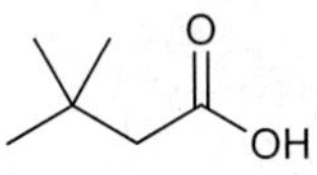

chirality centers

hexanoic acid

2-methylpentanoic acid

3-methylpentanoic acid

4-methylpentanoic acid

2,2-dimethylbutanoic acid

2,3-dimethylbutanoic acid

3,3-dimethylbutanoic acid

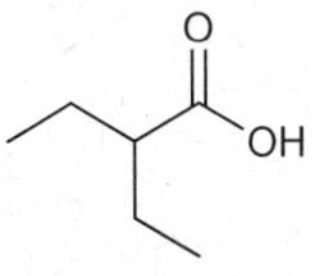

2-ethylbutanoic acid

21.44. There are only two constitutional isomers, shown below. Each one is named by identifying the longest chain containing the acid chloride group and replacing the "e" in the parent name with the suffix "-oyl chloride". The first isomer is thus named butanoyl chloride. In the second (branched) isomer, the chain is numbered so that the carbonyl group is C1, which puts the methyl substituent on C2, resulting in the IUPAC name 2-methylpropanoyl chloride.

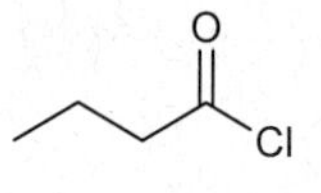

butanoyl chloride

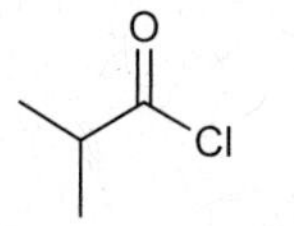

2-methylpropanoyl chloride

21.45.

(a) Pentanoic acid is converted to 1-pentanol using a strong reducing agent (LAH), followed by water work-up, as shown.

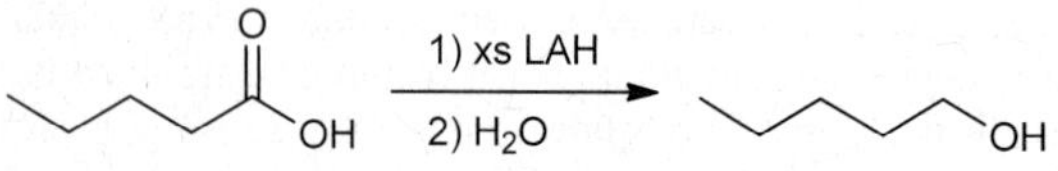

(b) Pentanoic acid is initially converted to 1-pentanol using a strong reducing agent (LAH), followed by water work-up, as shown. Conversion to the tosylate followed by reaction with a strong, sterically hindered base produces 1-pentene via an E2 reaction. Note that a sterically hindered base is required in the last step because a non-sterically hindered base (*i.e.*, NaOEt) would result in the formation of the S_N2 product as the major product.

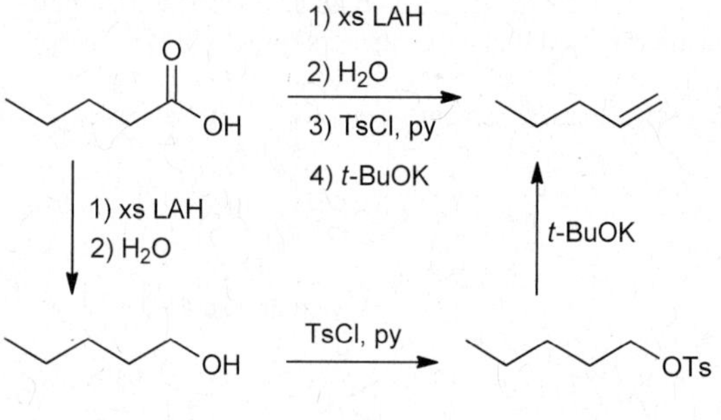

(c) Conversion of pentanoic acid to hexanoic acid requires the installation of an extra carbon atom. There are certainly many acceptable solutions to this problem. One such solution derives from the following retrosynthetic analysis. An explanation of each of the steps (*a-d*) follows.

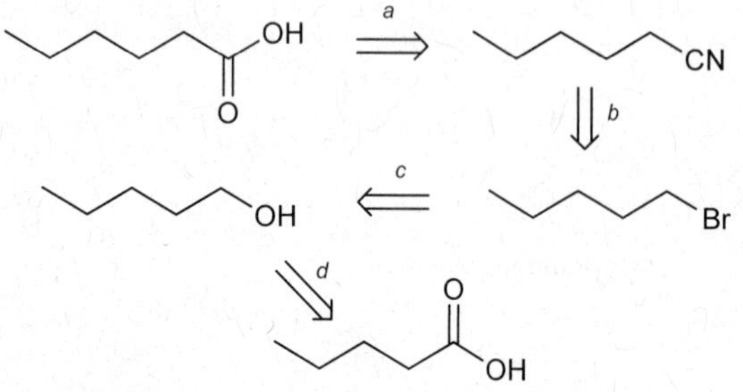

a. The desired product can be made via hydrolysis of a nitrile.

b. The nitrile can be made from 1-bromopentane (via an S_N2 process).

c. 1-Bromopentane can be made from 1-pentanol (upon treatment with PBr_3).

d. 1-Pentanol can be made from pentanoic acid via reduction.

Now let's draw the forward scheme. Reduction of pentanoic acid gives 1-pentanol. Treating this alcohol with PBr_3 produces 1-bromopentane. Subsequent S_N2 substitution with sodium cyanide installs an extra carbon atom, producing a nitrile, which can be hydrolyzed to the desired product under acidic conditions:

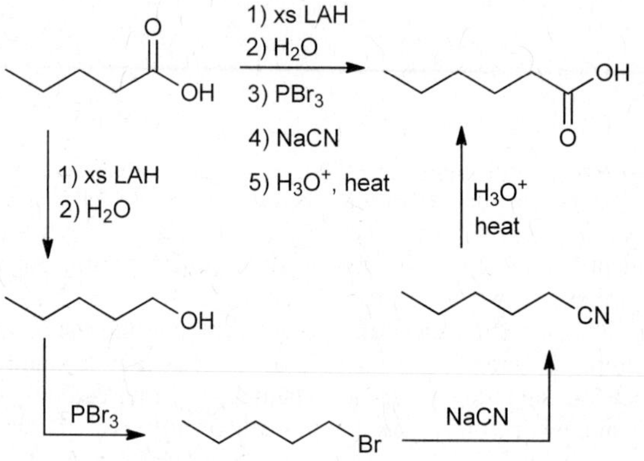

21.46.

(a) *Anti*-Markovnikov addition of water (via hydroboration / oxidation) produces 1-pentanol, which is subsequently oxidized to pentanoic acid using a strong oxidizing agent.

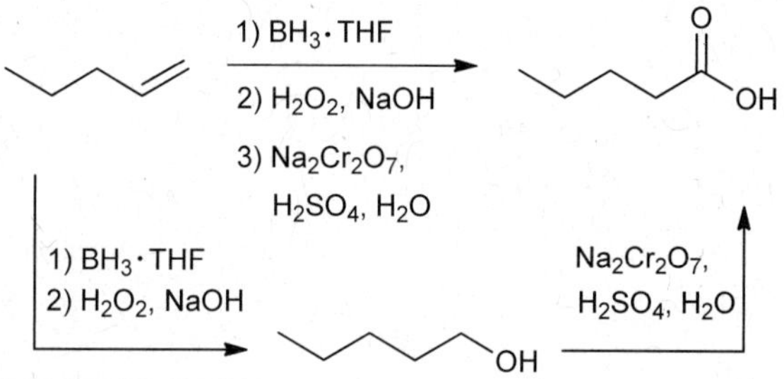

(b) Conversion of 1-bromobutane to pentanoic acid requires the installation of an extra carbon atom. This extra carbon atom can be installed via an S_N2 process in which bromide is replaced with cyanide, thereby converting the alkyl halide into a nitrile. Subsequent acid-catalyzed hydrolysis produces the desired carboxylic acid, pentanoic acid.

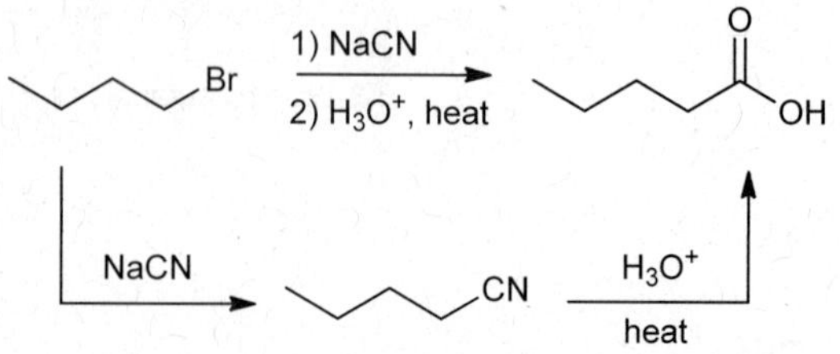

21.47. As discussed in Chapter 19, the methoxy group is electron donating via resonance, but electron withdrawing via induction. The resonance donation effect is stronger, but only significantly affects the acidity when the methoxy group is in an *ortho* or *para* position relative to the carboxylic acid. Note that in the third resonance structure for the conjugate base of the *para* derivative below, there is a negative charge next to the carboxylate group (a destabilizing effect). In contrast, none of the resonance structures of the corresponding *meta* derivative have this destabilizing feature, and as such, the conjugate base of the *meta* derivative is more stable than the conjugate base of the *para* derivative. In fact, the conjugate base of the *meta* derivative is even more stable than the conjugate base of benzoic acid, because of the electron withdrawing effect of the methoxy group (in the absence of strong resonance effects that are present in the *ortho* or *para* derivative).

21.48.
(a) Reaction of hexanoyl chloride with an excess of ethyl amine produces the corresponding amide, where the chloride leaving group has been replaced with the nitrogen atom of the amine, as shown.

(b) Reaction of hexanoyl chloride with an excess of LAH, followed by water work-up, reduces the carboxylic acid group to the corresponding primary alcohol.

(c) Reaction of hexanoyl chloride with ethanol and pyridine produces the corresponding ester, where the chloride leaving group has been replaced with an ethoxy group.

(d) Reaction of hexanoyl chloride with water (in the presence of pyridine) produces the parent carboxylic acid, where the chloride leaving group has been replaced with an OH group.

(e) Reaction of hexanoyl chloride with sodium benzoate produces the corresponding anhydride, where the chloride leaving group has been replaced with an oxygen atom of sodium benzoate, as shown.

(f) Reaction of hexanoyl chloride with excess ammonia produces the corresponding amide, where the chloride leaving group has been replaced with an NH_2 group.

(g) Reaction of hexanoyl chloride with lithium diethyl cuprate produces a ketone, where the chloride leaving group has been replaced with the ethyl group from the diethyl cuprate, as shown.

(h) Reaction of hexanoyl chloride with excess ethyl magnesium bromide, followed by water work-up, produces a tertiary alcohol, where two new ethyl groups have been incorporated into the product. The first equivalent of ethyl magnesium bromide attacks the carbonyl group of the acid chloride to give a ketone (nucleophilic attack of EtMgBr, followed by loss of chloride). A second equivalent of ethyl magnesium bromide then attacks the carbonyl group of the ketone intermediate to produce an alkoxide ion, which is subsequently protonated during work-up to give the following tertiary alcohol.

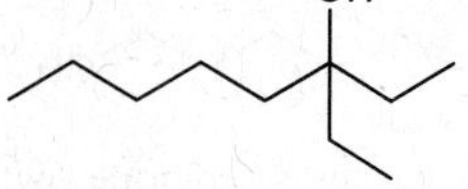

21.49.
(a) Reaction with thionyl chloride converts a carboxylic acid into the corresponding acid chloride, shown here.

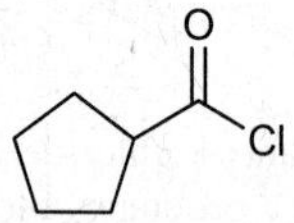

(b) A carboxylic acid is reduced to the corresponding alcohol upon treatment with LAH, followed by water work-up.

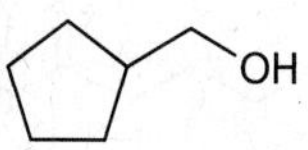

(c) Reaction with sodium hydroxide deprotonates the carboxylic acid to yield the sodium carboxylate.

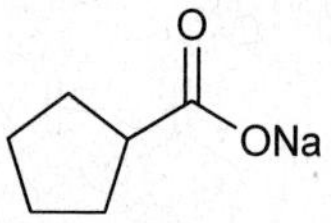

(d) Reaction with ethanol and catalytic acid converts the carboxylic acid to the ethyl ester shown, via a Fischer esterification.

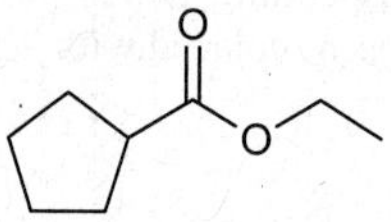

21.50.
(a) A carboxylic acid is reduced to the corresponding alcohol upon treatment with LAH, followed by water work-up.

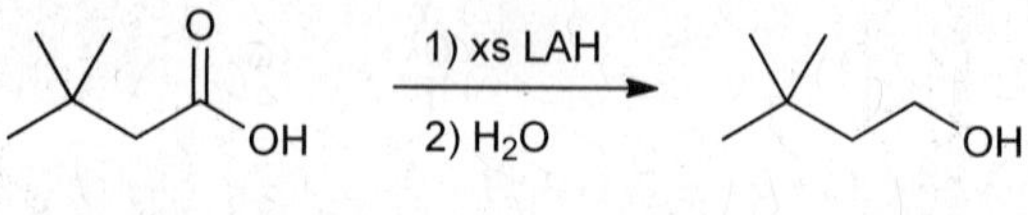

(b) Reaction with thionyl chloride converts the carboxylic acid to the corresponding acid chloride. Subsequent reaction with dimethylamine (in the presence of pyridine) converts the acid chloride to the corresponding dimethylamide.

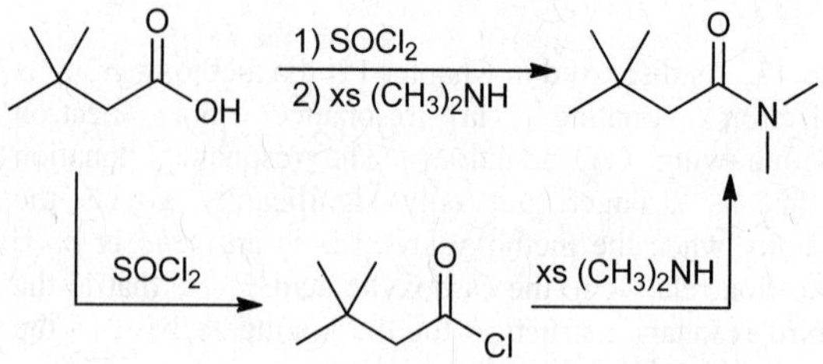

(c) An amide is converted into the corresponding nitrile upon treatment with thionyl chloride.

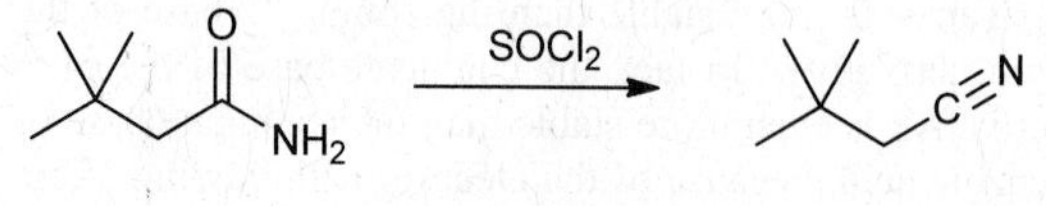

(d) Acid catalyzed hydrolysis of an ester gives a carboxylic acid. Subsequent reaction with acetyl chloride and pyridine causes an acetylation reaction that gives the following anhydride as the product.

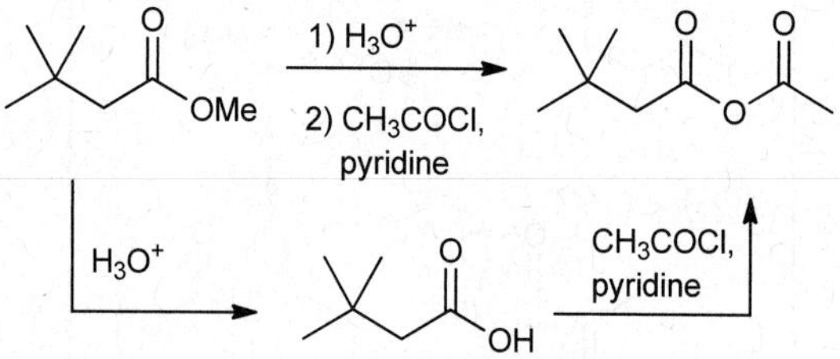

(e) An ester is converted into an aldehyde upon treatment with DIBAH, followed by water work-up.

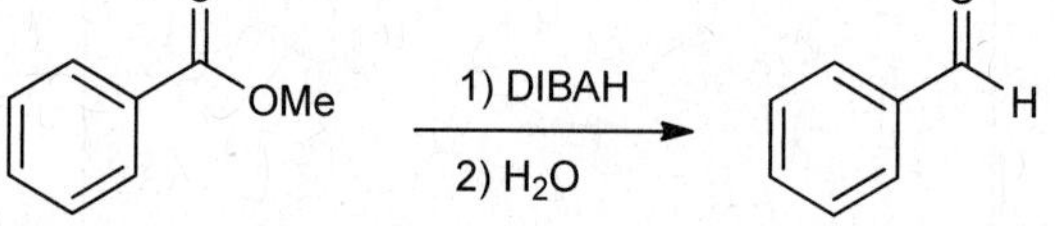

(f) Phenol serves as a nucleophile in this reaction. Phenol replaces the acetate leaving group on acetic anhydride resulting in the formation of a phenyl ester and acetic acid, as shown. This process is called an acetylation reaction.

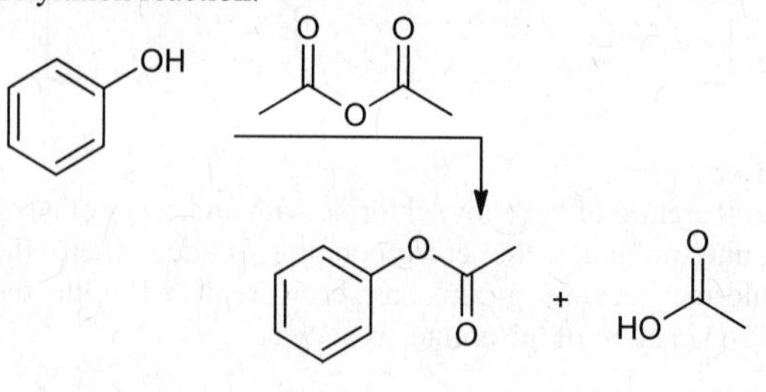

(g) Diphenylamine serves as a nucleophile in this reaction. Diphenylamine replaces the chloride leaving group on acetyl chloride resulting in the formation of the diphenylamide shown here:

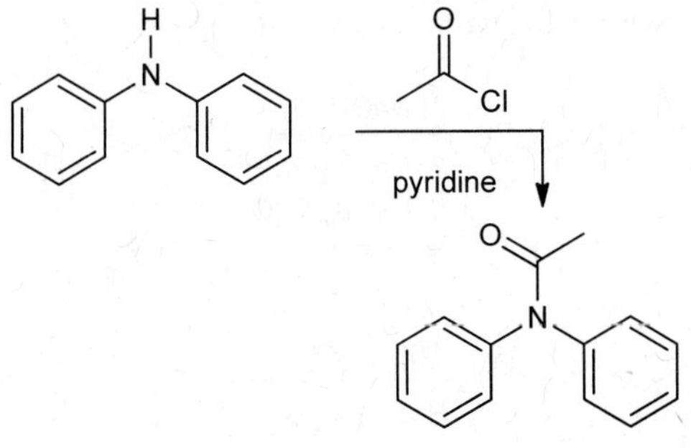

(h) The first equivalent of lithium aluminum hydride reduces the cyclic ester to an acyclic aldehyde (via nucleophilic attack of LAH, followed by loss of an alkoxide leaving group, which remains tethered to the aldehyde group via the alkyl chain). A second equivalent of LAH further reduces the aldehyde to give a dianion, which is subsequently protonated during water work-up to give a diol.

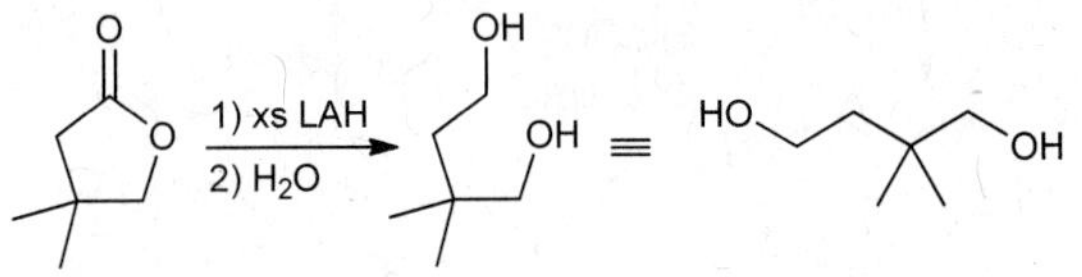

(i) Acid-catalyzed hydration of the cyclic amide results in formation of a bifunctional molecule containing both a carboxylic acid and an amine. The amine nitrogen atom is further protonated under the acidic conditions of the reaction to produce the ammonium group.

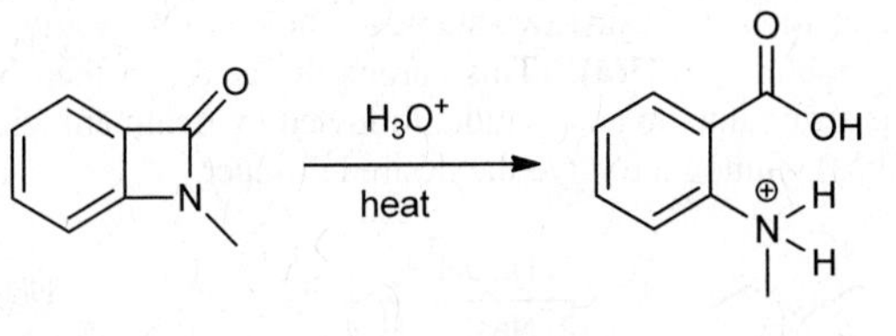

(j) Acid-catalyzed hydration of the cyclic ester (lactone) results in formation of a bifunctional molecule containing both a carboxylic acid group and a phenol group.

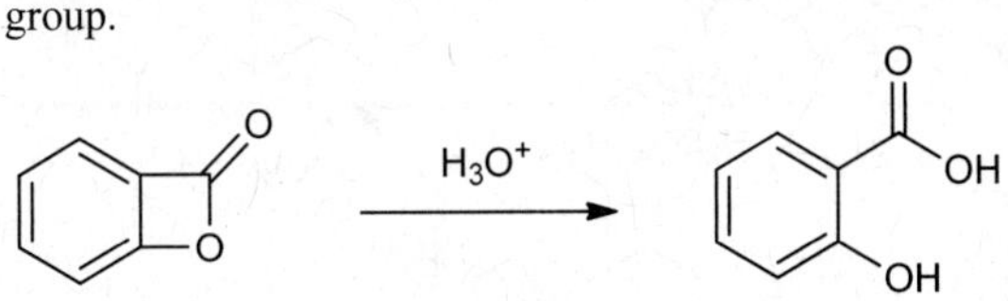

21.51.
(a) The new bond that forms as a result of a Fischer esterification is the σ bond between the carbonyl group and the oxygen atom connected to it, as shown. Making this disconnection, it becomes evident that benzoic acid and phenol would produce the desired ester under acidic conditions.

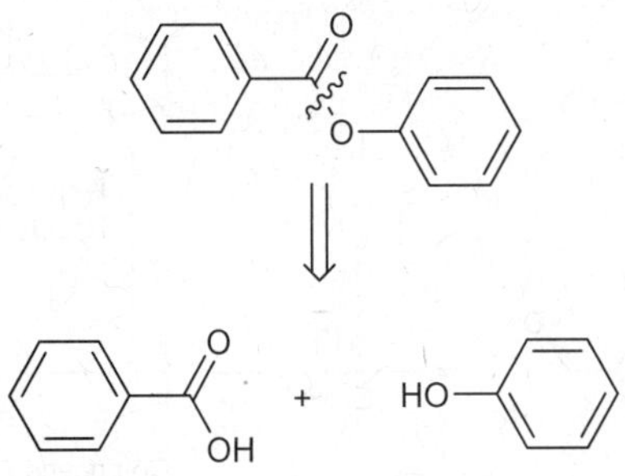

(b) The new bond that forms as a result of a Fischer esterification is the σ bond between the carbonyl group and the oxygen atom connected to it, as shown. Making this disconnection, it becomes evident that butyric acid and isopropanol would produce the desired ester under acidic conditions.

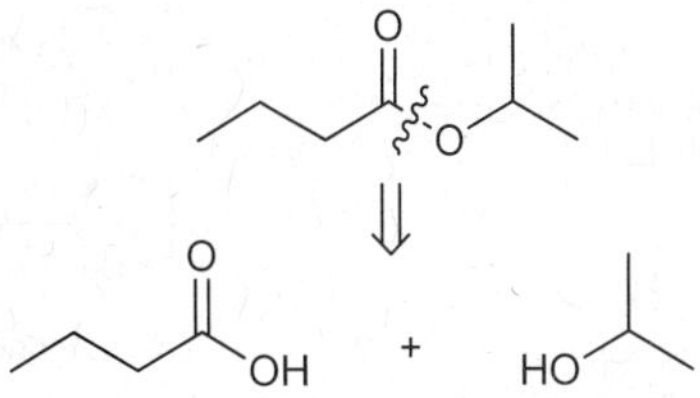

(c) The new bond that forms as a result of a Fischer esterification is the σ bond between the carbonyl group and the oxygen atom connected to it, as shown. Making this disconnection, it becomes evident that propanoic acid and *tert*-butanol would produce the desired ester under acidic conditions.

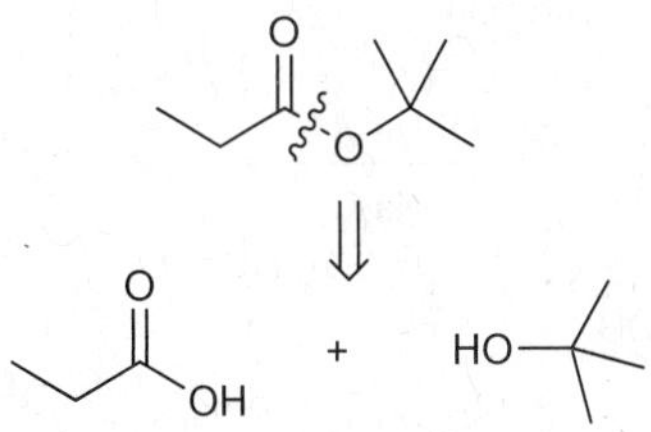

21.52. Oxidation of the primary alcohol gives the corresponding carboxylic acid (**A**). Reaction of **A** with thionyl chloride converts the carboxylic acid to the acid chloride (**B**). Reaction of **B** with excess ammonia produces the amide (**C**). Carboxylic acid (**A**) undergoes Fischer esterification upon reaction with ethanol and catalytic acid to produce the ethyl ester (**D**). Reaction of acid chloride **B** with a lithium trialkoxyaluminum hydride produces aldehyde **F**, which can also be made from ester **D** by reaction with DIBAH (**E**).

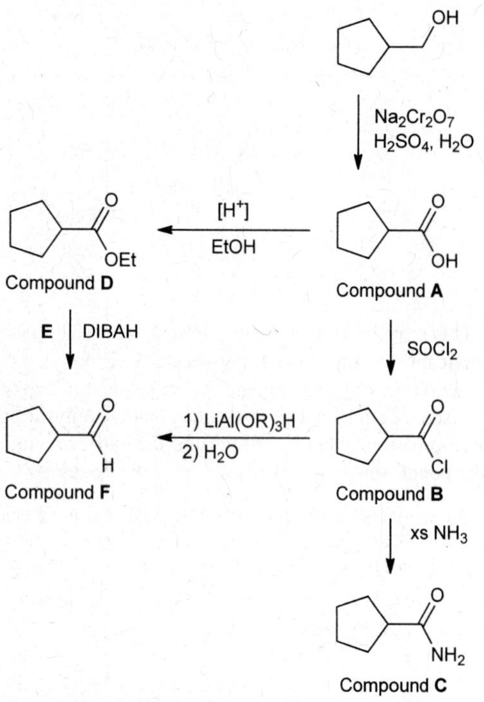

Compound D

E | DIBAH

Compound F

Compound A

Compound B

Compound C

21.53.
(a) This transformation does not involve a change in the carbon skeleton. Only the identity of the functional group must be changed. To accomplish this transformation, the starting material can be treated with NaOH to give 1-pentanol via an S_N2 reaction. Subsequent oxidation of the primary alcohol gives the desired carboxylic acid.

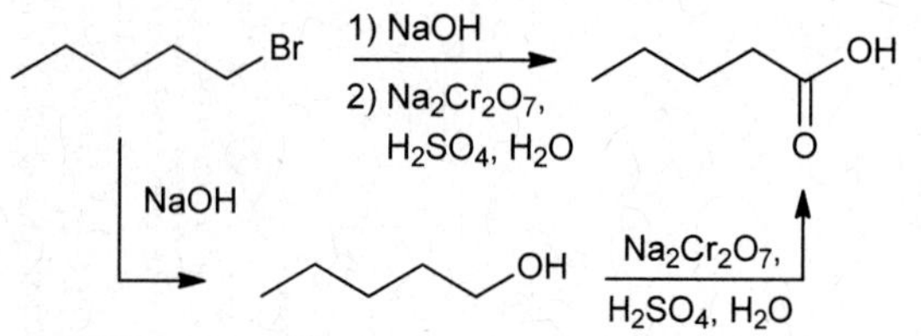

(b) This transformation involves a change in the carbon skeleton (one extra carbon atom must be inserted), as well as a change in the identity and location of the functional group. To accomplish this transformation, the starting material can be treated with NaCN, thereby converting 1-bromopentane into hexanenitrile via an S_N2 reaction. Subsequent acid catalyzed hydrolysis of the nitrile (upon treatment with aqueous acid) gives hexanoic acid.

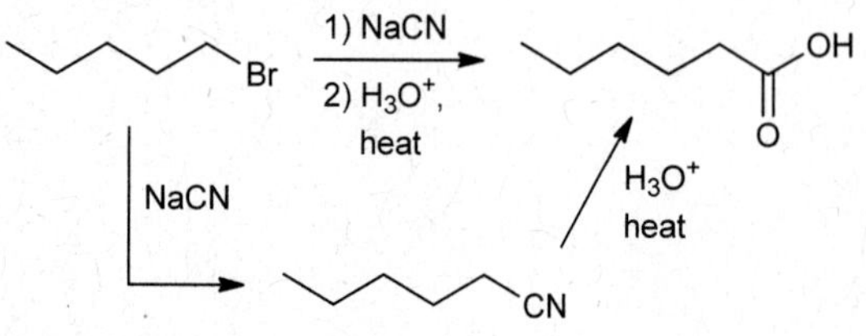

(c) First, 1-bromopentane can be converted into pentanoic acid in just two steps, as shown in the solution to Problem **21.53(a)**. Then, treating this carboxylic acid with $SOCl_2$ will give the desired acid chloride.

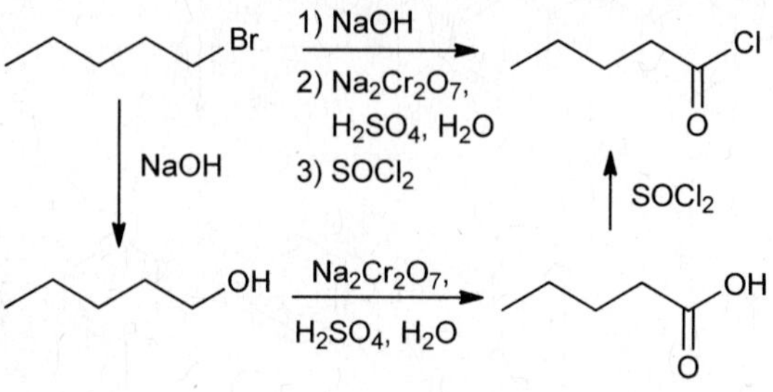

(d) First, 1-bromopentane can be converted into hexanoic acid in just two steps, as shown in the solution to Problem **21.53(b)**. Then, this carboxylic acid can be converted into an acid halide, followed by treatment with excess ammonia to give the desired product:

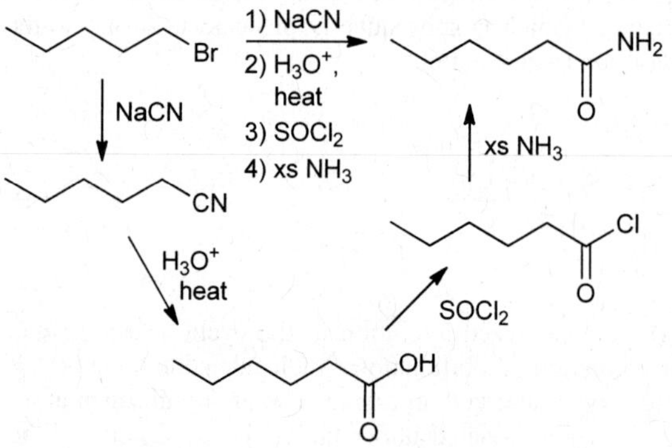

(e) First, 1-bromopentane can be converted into pentanoic acid in just two steps, as shown in the solution to Problem **21.53(a)**. This carboxylic acid can then be converted into an acid halide, followed by treatment with excess ammonia to give the desired product:

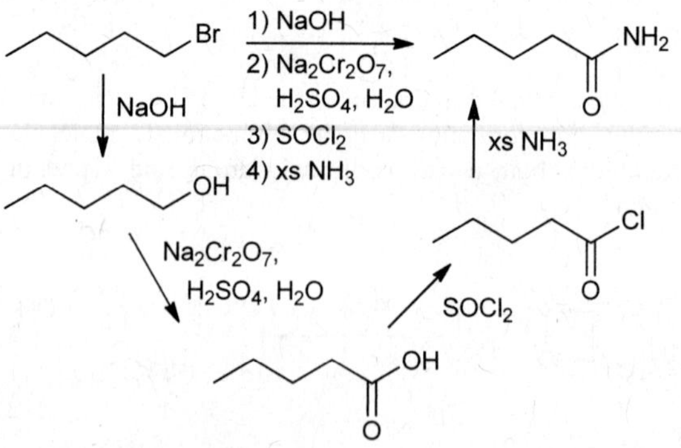

(f) First, 1-bromopentane can be converted into hexanoic acid in just two steps, as shown in the solution to Problem **21.53(b)**. Upon treatment with ethanol in acid-catalyzed conditions, the carboxylic acid can be converted into an ester via a Fischer esterification process:

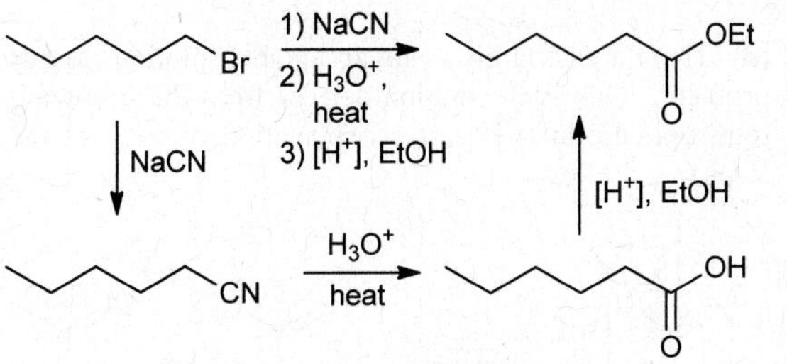

21.54.

(a) There are certainly many acceptable solutions to this problem. One such solution derives from the following retrosynthetic analysis. An explanation of each of the steps (*a-c*) follows.

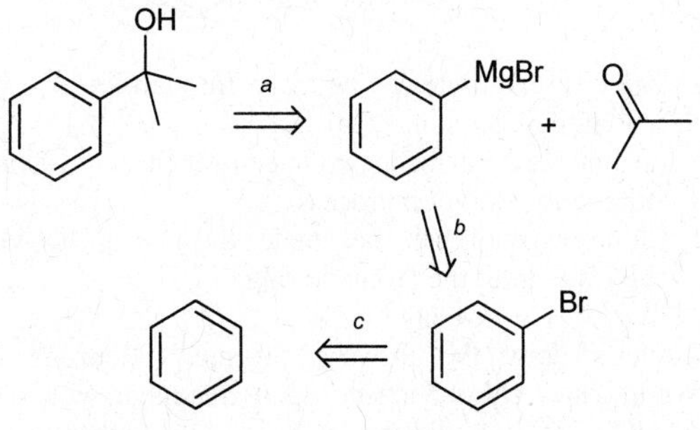

a. The desired product can be made via a Grignard reaction between phenyl magnesium bromide and acetone.
b. Phenyl magnesium bromide can be made from bromobenzene via insertion of magnesium.
c. Bromobenzene can be made from benzene via bromination.

Now let's draw the forward scheme. Benzene is converted into bromobenzene upon treatment with Br_2 and $AlBr_3$ (via an electrophilic aromatic substitution reaction). Bromobenzene is subsequently converted into phenyl magnesium bromide, which is then treated with acetone (in a Grignard reaction), followed by water work-up, to give the desired tertiary alcohol.

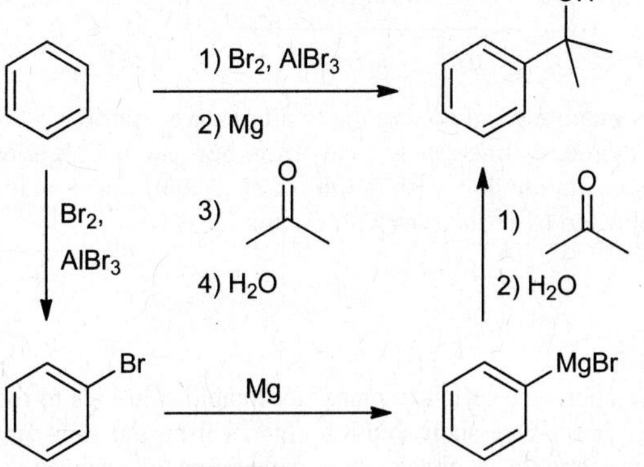

(b) There are certainly many acceptable solutions to this problem. One such solution derives from the following retrosynthetic analysis. An explanation of each of the steps (*a-e*) follows.

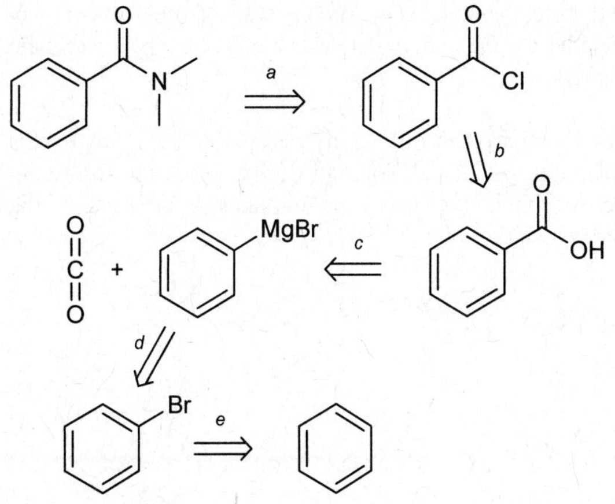

a. The desired amide can be made from the corresponding acid chloride (benzoyl chloride).
b. Benzoyl chloride can be made from benzoic acid upon treatment with thionyl chloride.
c. Benzoic acid can be made from the reaction between phenyl magnesium bromide and carbon dioxide.
d. Phenyl magnesium bromide can be made from bromobenzene via insertion of magnesium.
e. Bromobenzene can be made from benzene via bromination.

Now let's draw the forward scheme. Benzene is converted into bromobenzene upon treatment with Br_2 and $AlBr_3$ (via an electrophilic aromatic substitution reaction). Bromobenzene is subsequently converted into phenyl magnesium bromide, which is then treated with carbon dioxide, followed by an acidic workup, to give benzoic acid. Conversion to the acid chloride, followed by reaction with dimethylamine, yields the desired amide.

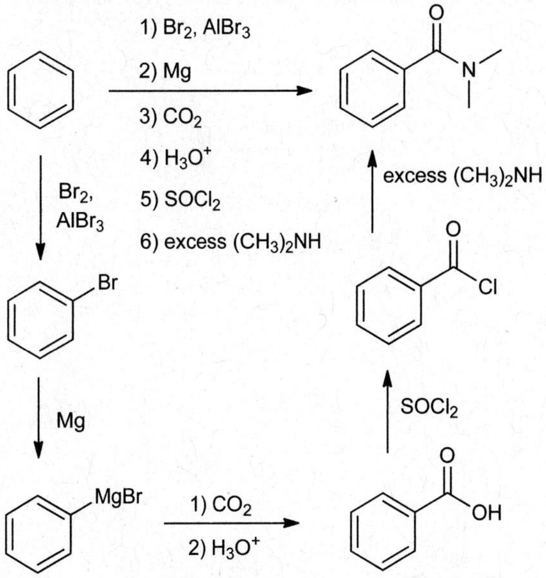

As mentioned, there are many alternative solutions. For example, benzoic acid can be made from benzene via Friedel-Crafts methylation, followed by benzylic oxidation.

(c) There are certainly many acceptable solutions to this problem. One such solution derives from the following retrosynthetic analysis. An explanation of each of the steps (*a-d*) follows.

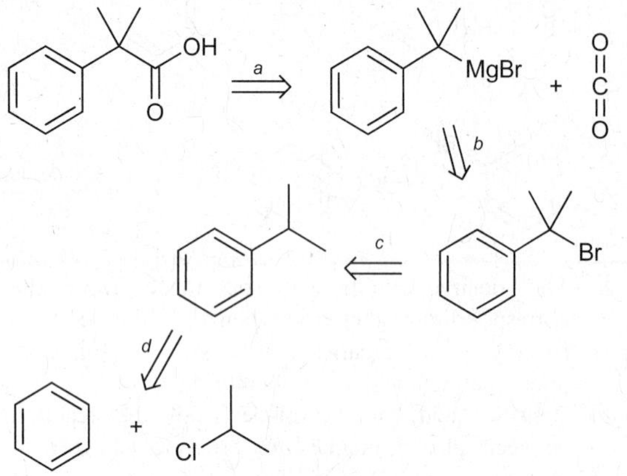

a. The desired carboxylic acid can be made from the reaction between a Grignard reagent and CO_2.
b. The Grignard reagent can be made from the corresponding tertiary benzylic halide.
c. The benzylic halide can be made from isopropyl benzene via benzylic bromination.
d. Isopropylbenzene can be made from benzene via a Friedel-Crafts alkylation.

Now let's draw the forward scheme. Benzene is converted into isopropyl benzene upon treatment with 2-chloropropane and a Lewis acid (via a Friedel-Crafts alkylation). Benzylic bromination replaces the benzylic hydrogen atom with a bromine atom. Conversion to a Grignard reagent, followed by reaction with carbon dioxide and subsequent acidification, gives the desired carboxylic acid.

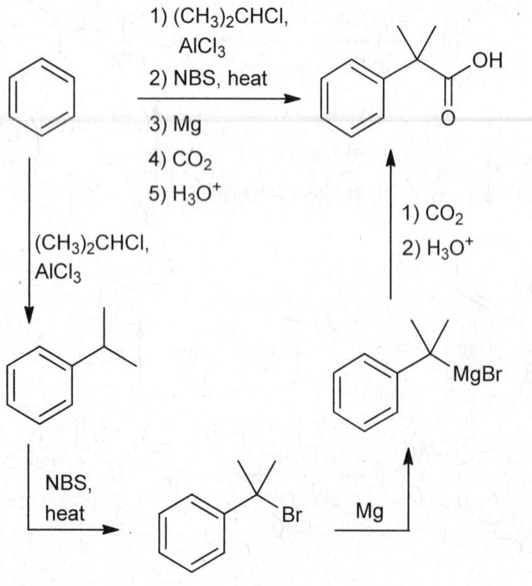

(d) There are certainly many acceptable solutions to this problem. One such solution derives from the following retrosynthetic analysis. An explanation of each of the steps (*a-c*) follows.

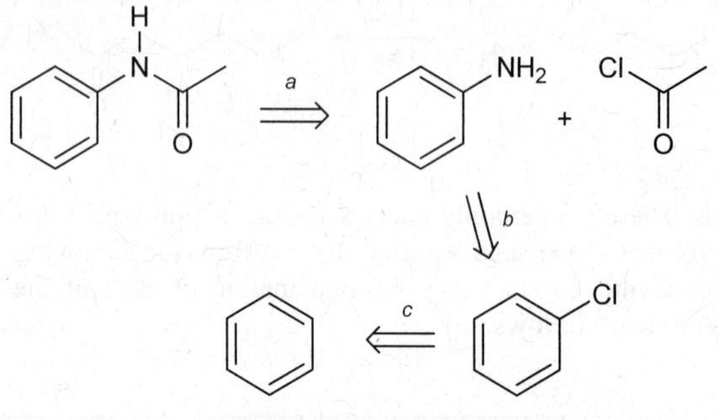

a. The desired amide can be made from aniline via an acetylation reaction.
b. Aniline can be made from chlorobenzene via an elimination-addition process.
c. Chlorobenzene can be made from benzene via chlorination of the aromatic ring.

Now let's draw the forward scheme. Benzene is converted into chlorobenzene upon treatment with Cl_2 and $AlCl_3$ (via an electrophilic aromatic substitution reaction). Chlorobenzene is then converted to aniline via an elimination-addition reaction. Reaction with acetyl chloride (in the presence of pyridine) converts aniline to the desired amide.

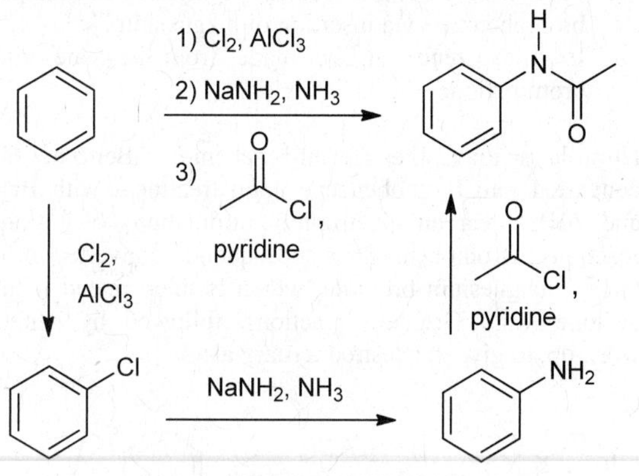

As mentioned, there are many alternative solutions. For example, aniline can be made from benzene via nitration (upon treatment with sulfuric acid and nitric acid), followed by reduction (with Zn and HCl).

21.55.
(a) There are certainly many acceptable solutions to this problem. One such solution derives from the following retrosynthetic analysis. An explanation of each of the steps (*a-c*) follows.

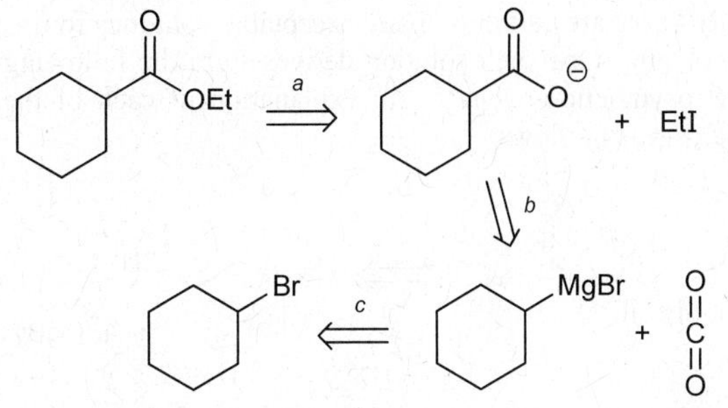

c. The carboxylic acid can be made from the corresponding primary alcohol via oxidation.

Now let's draw the forward scheme. The primary alcohol is oxidized to the carboxylic acid, and subsequently converted to the acid chloride upon treatment with thionyl chloride. Reaction with lithium diethyl cuprate then produces the desired ketone.

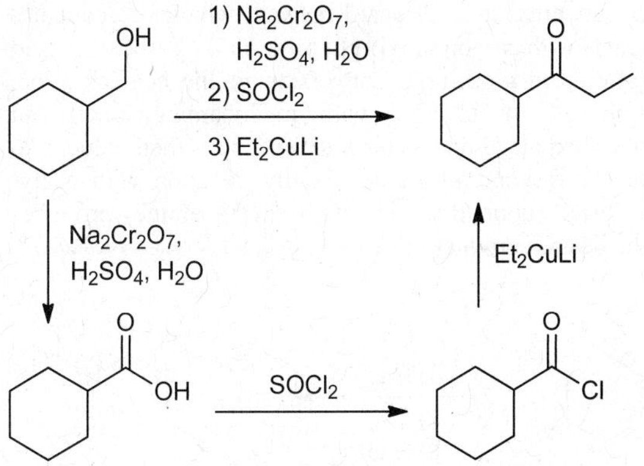

a. The desired ester can be made from a reaction between the corresponding carboxylate ion and ethyl iodide (S_N2).
b. The carboxylate ion can be made from the reaction between a Grignard reagent and CO_2.
c. The Grignard reagent can be made from the corresponding secondary bromide.

Now let's draw the forward scheme. Bromocyclohexane is converted to a Grignard reagent, which subsequently reacts with carbon dioxide to produce a carboxylate ion. This anion then serves as a nucleophile in an S_N2 reaction with iodoethane, giving the desired product.

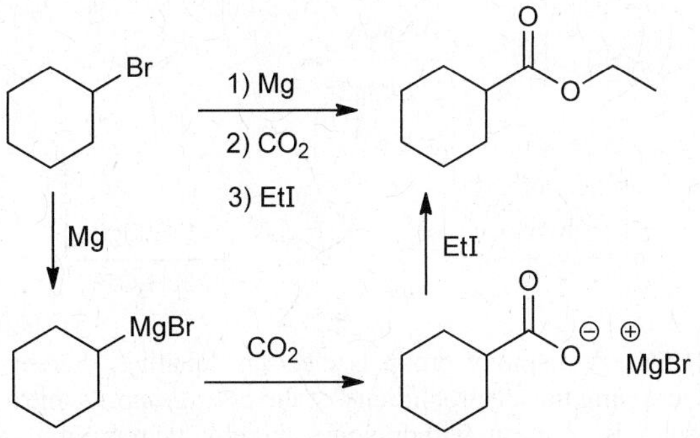

(b) There are certainly many acceptable solutions to this problem. One such solution derives from the following retrosynthetic analysis. An explanation of each of the steps (*a-c*) follows.

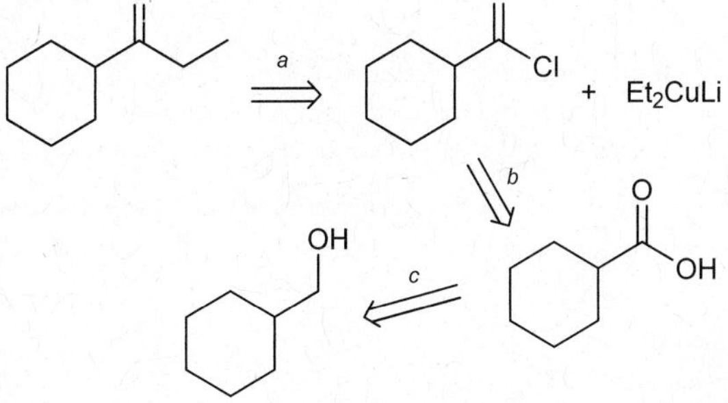

a. The desired ketone can be made from a reaction between an acid chloride and lithium diethyl cuprate.
b. The acid chloride can be made from the corresponding carboxylic acid.

(c) There are certainly many acceptable solutions to this problem. One such solution derives from the following retrosynthetic analysis. An explanation of each of the steps (*a-f*) follows.

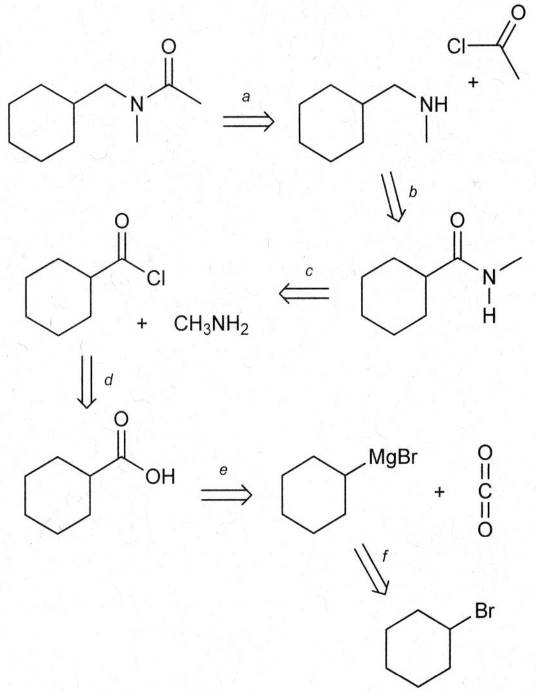

a. The desired amide can be made via acetylation of the corresponding secondary amine.
b. The secondary amine can be made via reduction of the corresponding amide.
c. The amide can be made from the corresponding acid halide (upon treatment with excess methyl amine).

d. The acid halide can be made from the corresponding carboxylic acid, upon treatment with thionyl chloride.

e. The carboxylic acid can be made from the reaction between a Grignard reagent and CO_2.

f. The Grignard reagent can be made from the corresponding secondary alkyl bromide.

Now let's draw the forward scheme. Bromocyclohexane is converted to a Grignard reagent, which subsequently reacts with carbon dioxide to produce the carboxylic acid (after acid work-up). Conversion to the acid chloride, followed by reaction with methylamine, yields an intermediate amide. This intermediate is then reduced to the corresponding amine. Finally, reaction with acetyl chloride (upon treatment with excess amine) produces the desired product.

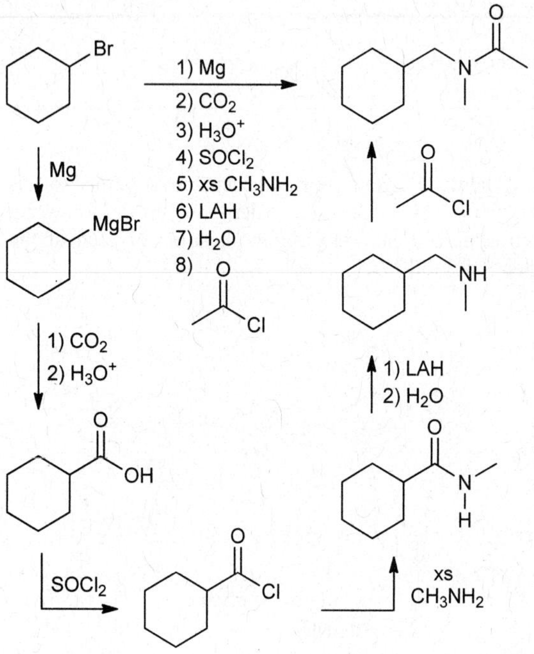

(d) There are certainly many acceptable solutions to this problem. One such solution derives from the following retrosynthetic analysis. An explanation of each of the steps (*a-b*) follows.

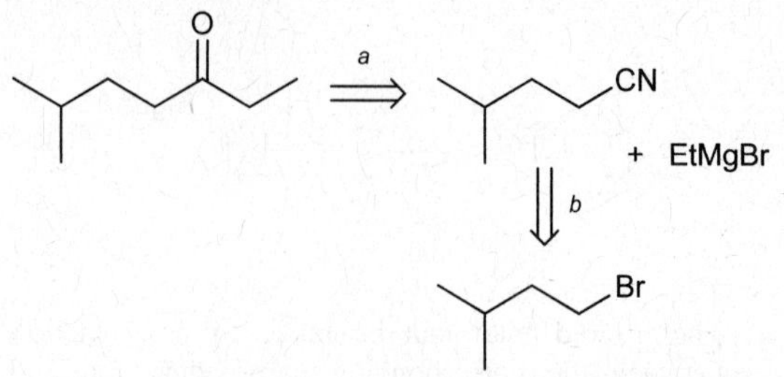

a. The desired ketone can be made from the reaction between a Grignard reagent and a nitrile.

b. The nitrile can be made from the corresponding primary bromide via an S_N2 process.

Now let's draw the forward scheme. Reaction of 1-bromo-3-methylbutane with sodium cyanide produces the nitrile, which is subsequently converted to the desired ketone upon treatment with ethyl magnesium bromide, followed by aqueous acid, as shown.

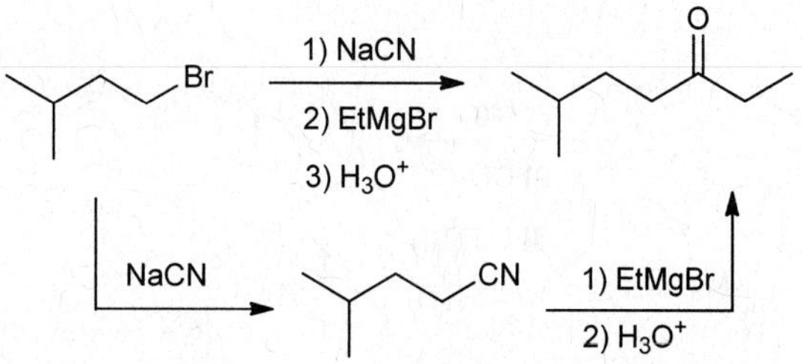

21.56. A methoxy group is electron donating, thereby decreasing the electrophilicity of the ester group. A nitro group is electron withdrawing, thereby increasing the electrophilicity of the ester group.

21.57. The reagents for each of these transformations can be found in Figure 21.11.

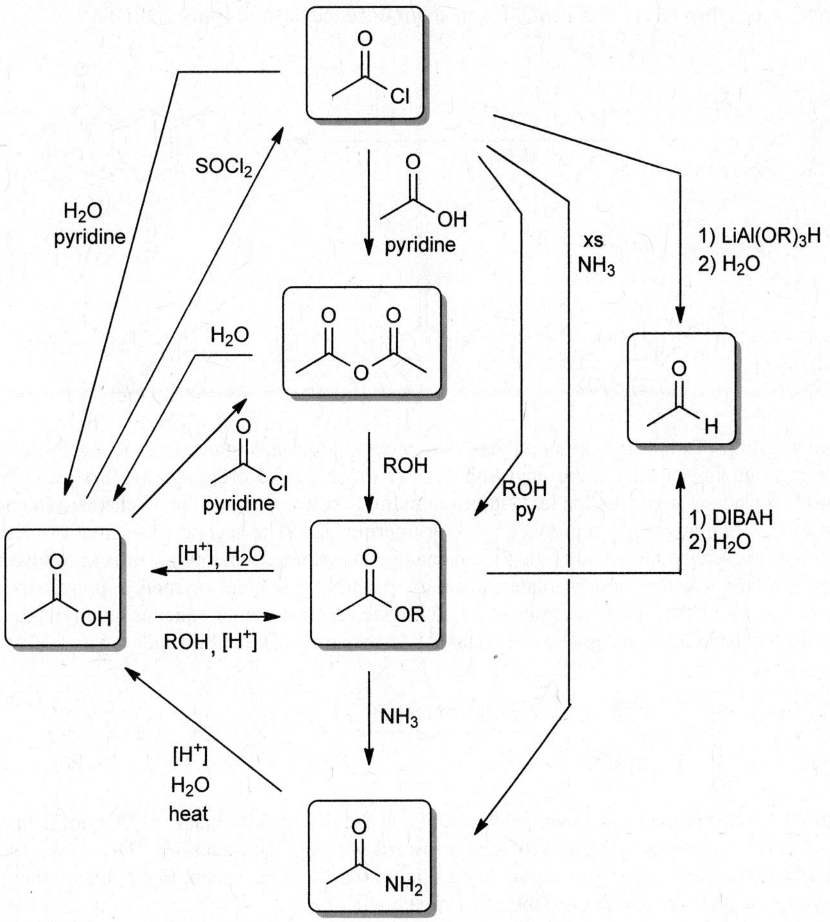

21.58. There are certainly many acceptable solutions to this problem. One such solution derives from the following retrosynthetic analysis. An explanation of each of the steps (*a-d*) follows.

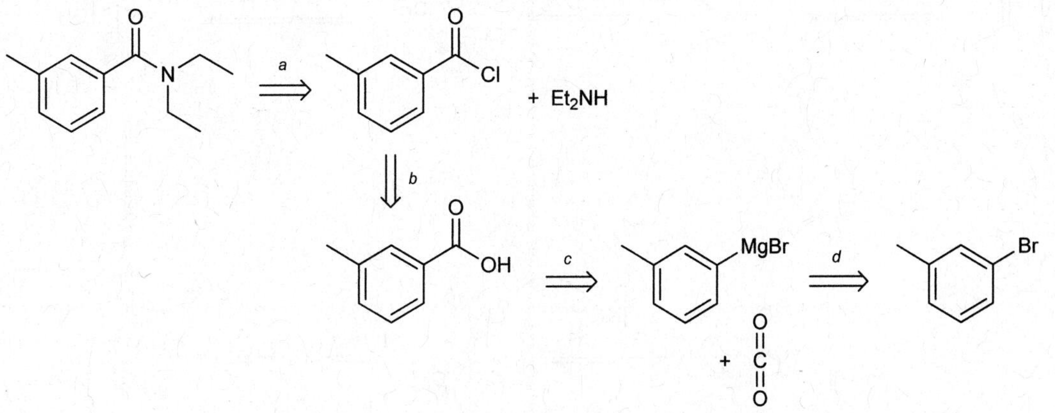

a. The desired amide can be made from the corresponding acid halide (upon treatment with excess diethylamine).
b. The acid halide can be made from the corresponding carboxylic acid, upon treatment with thionyl chloride.
c. The carboxylic acid can be made from the reaction between a Grignard reagent and CO_2.
d. The Grignard reagent can be made from the starting material, via insertion of magnesium.

Now let's draw the forward scheme. Conversion of *meta*-bromotoluene to a Grignard reagent, followed by reaction with carbon dioxide and subsequent acidification, produces *meta*-methylbenzoic acid. The carboxylic acid is converted to the acid chloride, which then reacts with diethylamine to produce the desired amide, DEET.

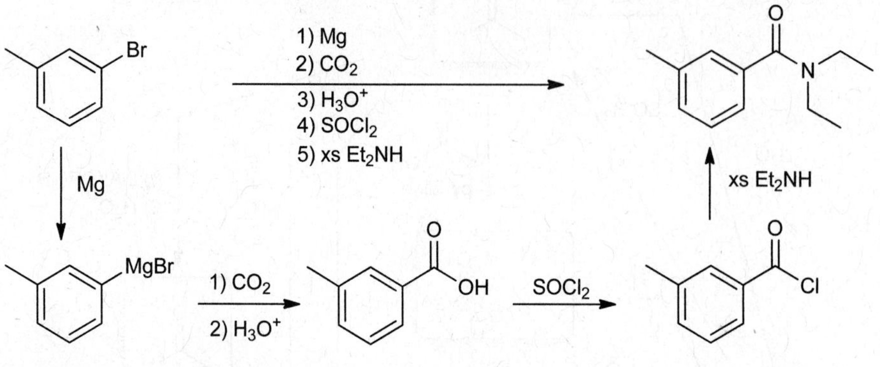

21.59. The carbonyl group of diphenyl carbonate has two phenoxide groups attached to it. Each of these groups can serve as a leaving group in a nucleophilic acyl substitution reaction. Accordingly, the first equivalent of methyl magnesium bromide replaces one phenoxide leaving group in the first two steps of the mechanism (nucleophilic attack, followed by loss of the leaving group) to produce an ester intermediate. The second equivalent of methyl magnesium bromide then replaces the second phenoxide group in an analogous manner (nucleophilic attack, followed by loss of the leaving group) to produce a ketone intermediate (acetone). A third equivalent of methyl magnesium bromide then attacks the carbonyl group of the ketone to produce *tert*-butoxide. Work-up with aqueous acid (H_3O^+) protonates *tert*-butoxide, as well as the phenoxide ions, giving *tert*-butanol and two equivalents of phenol.

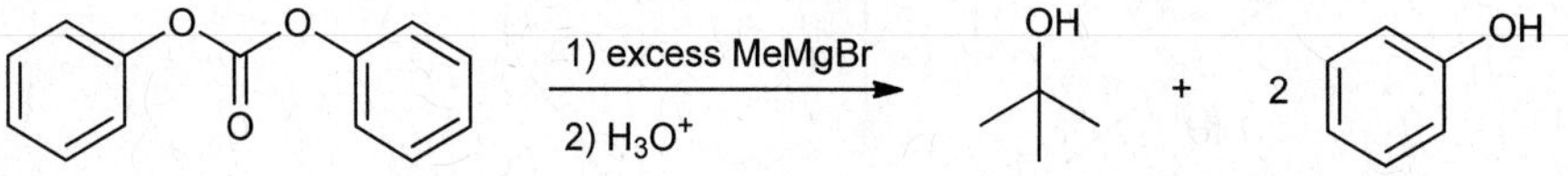

21.60. A mechanism for this reaction is shown below, in which the isotopically labeled ^{18}O atom is highlighted with a gray box. Protonation of the carbonyl group activates it toward nucleophilic attack by $^{18}OH_2$. Two successive proton transfers, followed by expulsion of the non-labeled oxygen atom (as a leaving group, H_2O) and another proton transfer, result in the formation of acetic acid with one labeled oxygen atom.

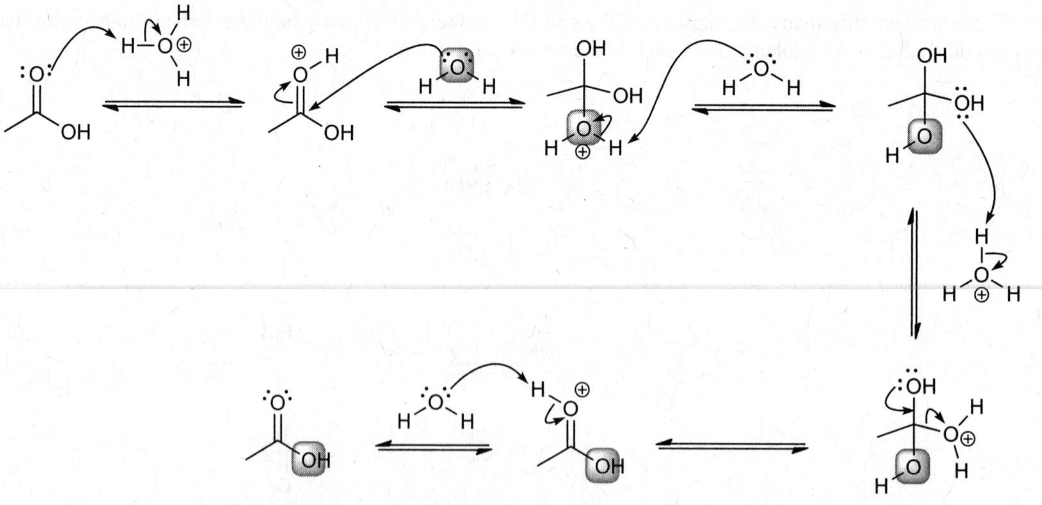

Two successive proton transfer steps (protonation, followed by deprotonation) give a molecule of acetic acid in which the labeled oxygen atom is incorporated in the carbonyl group, as shown.

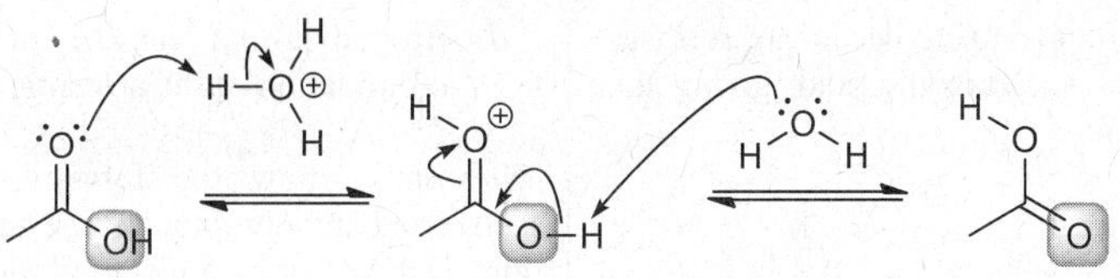

21.61.

(a) An alcohol (ROH) is used as a representative nucleophile in the mechanism below, which has six steps: 1) nucleophilic attack, 2) loss of a leaving group, 3) proton transfer, 4) nucleophilic attack, 5) loss of a leaving group and 6) proton transfer. ROH attacks the carbonyl group of phosgene to form a tetrahedral intermediate which subsequently expels a chloride ion. A proton transfer produces the neutral, monochlorinated intermediate shown below. A second round of these three steps (nucleophilic attack by ROH, loss of chloride and proton transfer) results in the overall substitution of the second chloride ion with the oxygen atom from the alcohol, thereby producing a carbonate ester, as shown.

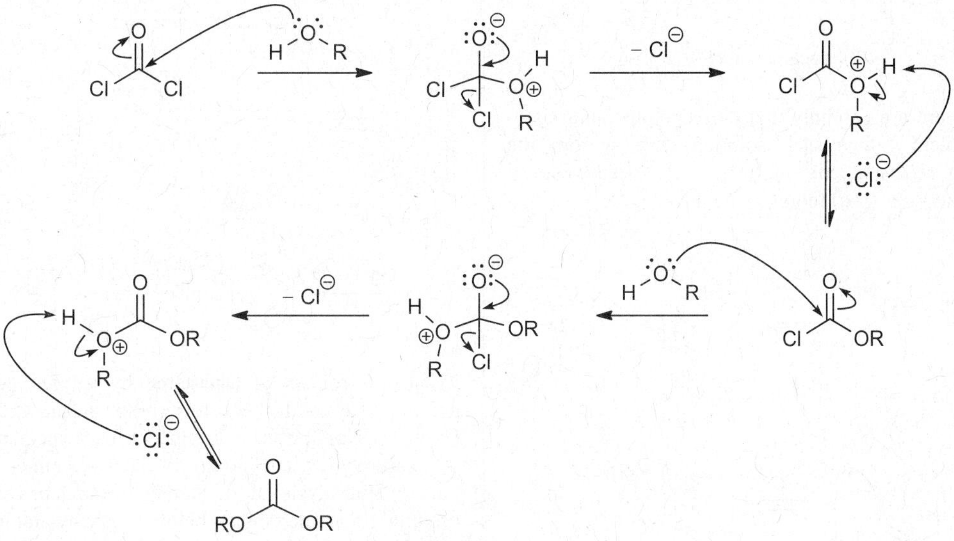

(b) The molecule below is produced from the reaction between phosgene and ethylene glycol via a mechanism analogous to the one in part (a) above. In this case, the second nucleophilic attack (step 4) is an intramolecular reaction, leading to the cyclic product shown here.

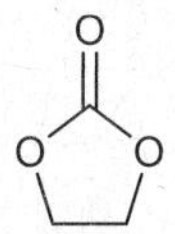

(c) Excess phenyl magnesium bromide reacts with phosgene to produce a tertiary alcohol (shown below), which results from the incorporation of three molar equivalents of the Grignard reagent. The first equivalent attacks the carbonyl group to produce a tetrahedral anionic intermediate; subsequent expulsion of a chloride ion gives an acid chloride intermediate. Likewise, a second equivalent of the Grignard reagent attacks the carbonyl group, producing a second anionic tetrahedral

intermediate. Subsequent expulsion of a chloride leaving group gives a ketone intermediate, which is attacked by a third equivalent of phenyl magnesium bromide. Protonation of the resulting alkoxide ion (by H_2O) gives the tertiary alcohol, shown here.

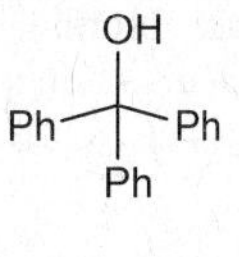

21.62.
(a) Hydrolysis of an ester group produces an alcohol and a carboxylic acid.

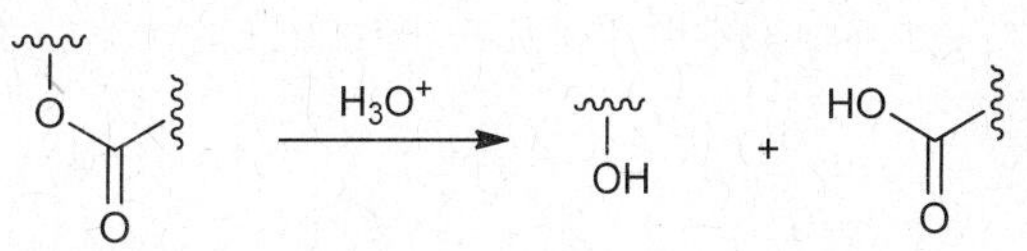

As such, hydrolysis of fluphenazine decanoate releases the hydrophobic chain as a carboxylic acid, giving the following primary alcohol:

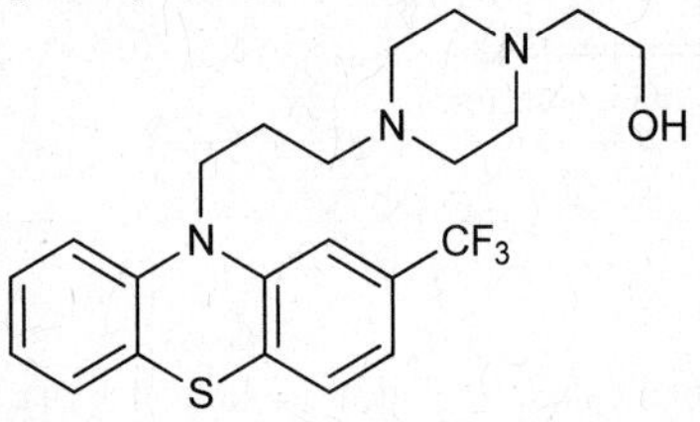

(b) The by-product of the reaction is a carboxylic acid containing ten carbon atoms. The "e" at the end of the parent alkane name (decane) is replaced with the suffix "-oic acid" to give the IUPAC name decanoic acid.

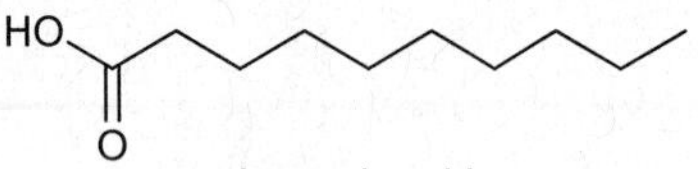

decanoic acid

21.63. There are certainly many acceptable solutions to this problem. One such solution derives from the following retrosynthetic analysis. An explanation of each of the steps (*a-d*) follows.

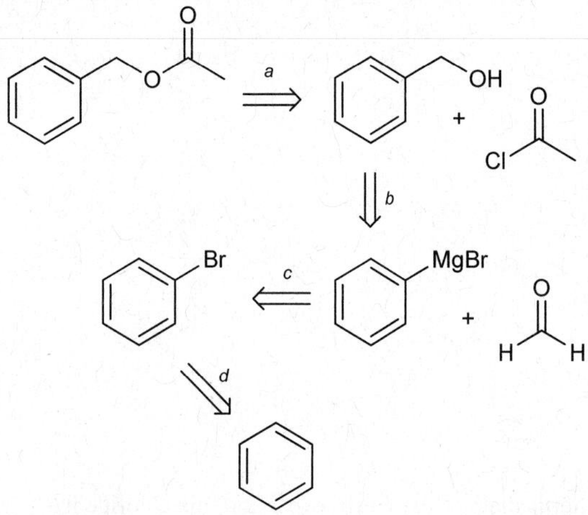

a. The ester can be made via acetylation of the corresponding alcohol (benzyl alcohol).
b. The alcohol can be made from the reaction between phenyl magnesium bromide and formaldehyde.
c. Phenyl magnesium bromide can be made from bromobenzene, via insertion of magnesium.

d. Bromobenzene can be made from benzene via bromination of the aromatic ring.

Now let's draw the forward scheme. Benzene is converted into bromobenzene upon treatment with Br_2 and $AlBr_3$ (via an electrophilic aromatic substitution reaction). Bromobenzene is then converted to phenyl magnesium bromide (a Grignard reagent), which is then treated with formaldehyde, followed by water work-up, to give benzyl alcohol. This alcohol then serves as a nucleophile in a subsequent acyl substitution reaction with acetyl chloride and pyridine to produce benzyl acetate (a process called acetylation).

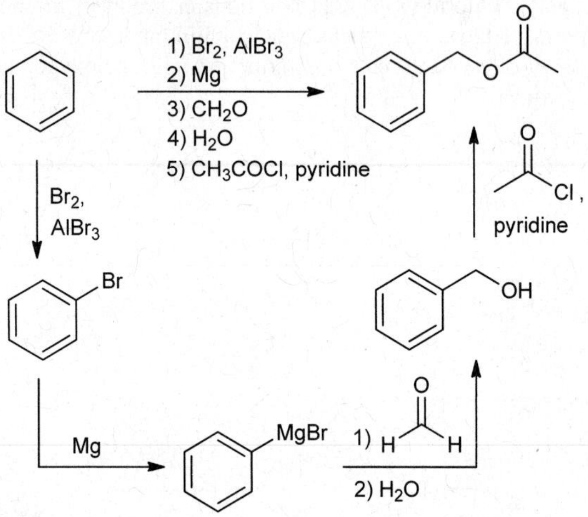

21.64. Hydrolysis of aspartame hydrolyzes both the amide group and the ester group in the molecule. Hydrolysis of the ester group produces methanol and the carboxylic acid group in phenylalanine, shown below. Hydrolysis of the amide group converts this group to an amine (shown below on phenylalanine) and a carboxylic acid (on the left side of aspartic acid below). Note that the stereochemistry at both chirality centers is retained because none of the bonds to the chirality centers are broken in this transformation.

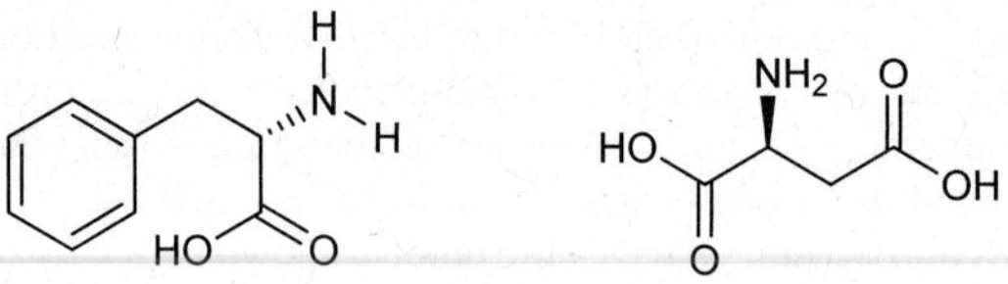

phenylalanine   aspartic acid

21.65.

(a) The mechanism shown below has three steps: 1) nucleophilic attack, 2) loss of a leaving group and 3) proton transfer. Phenol serves as a nucleophile in the first step, attacking the carbonyl group of the acid chloride to produce a tetrahedral intermediate. Reformation of the carbonyl group and expulsion of the leaving group (chloride) produces a cationic intermediate which is subsequently deprotonated by pyridine to give an ester.

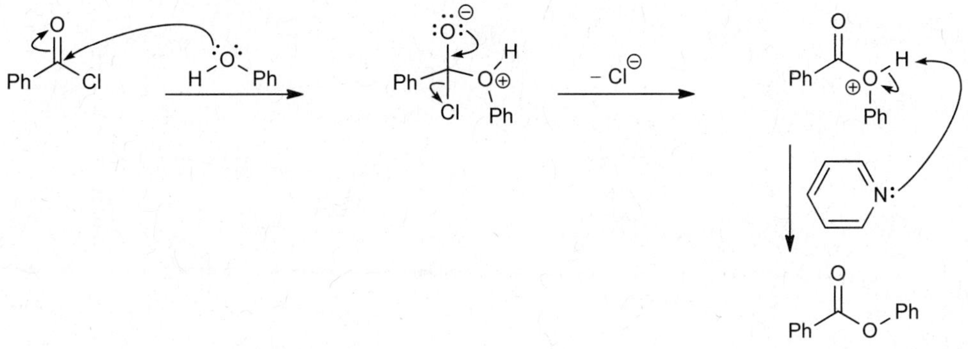

(b) The first part of this reaction (saponification) has three mechanistic steps: 1) nucleophilic attack, 2) loss of a leaving group and 3) proton transfer. The first step (nucleophilic attack) requires two curved arrows, which show hydroxide functioning as a nucleophile and attacking the electrophilic carbonyl group, resulting in an anionic tetrahedral intermediate. In step two (loss of a leaving group), the carbonyl group is reformed with loss of an alkoxide ion, resulting in opening of the ring and formation of a carboxylic acid. In the third step (proton transfer), the tethered alkoxide ion serves as a base, deprotonating the carboxylic acid intermediate, resulting in the formation of a carboxylate ion. This third step (in which a strong base deprotonates the carboxylic acid) is the driving force of the reaction. Subsequent work-up with aqueous acid protonates the carboxylate ion, regenerating the carboxylic acid.

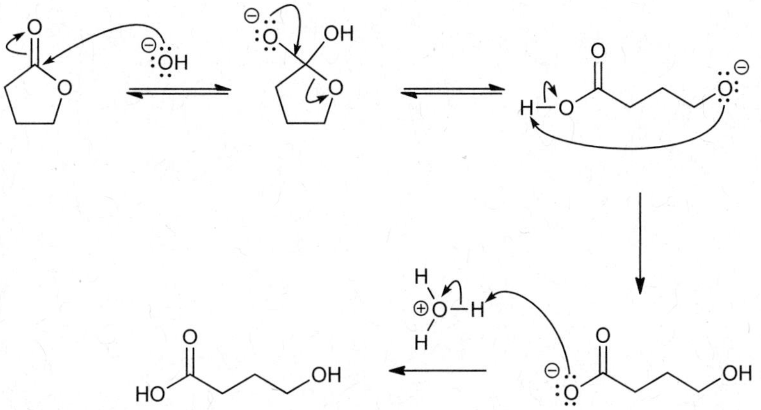

(c) The first part of this reaction (saponification) has three mechanistic steps: 1) nucleophilic attack, 2) loss of a leaving group and 3) proton transfer. The first step (nucleophilic attack) requires two curved arrows, which show hydroxide functioning as a nucleophile and attacking the electrophilic carbonyl group, resulting in an anionic tetrahedral intermediate. In step two (loss of a leaving group), the carbonyl group is reformed with loss of an alkoxide ion, resulting in opening of the ring and formation of a carboxylic acid. In the third step (proton transfer), the tethered alkoxide ion serves as a base, deprotonating the carboxylic acid intermediate, resulting in the formation of a carboxylate ion. This third step (in which a strong base deprotonates the carboxylic acid) is the driving force of the reaction. Subsequent work-up with aqueous acid protonates the carboxylate ion, regenerating the carboxylic acid.

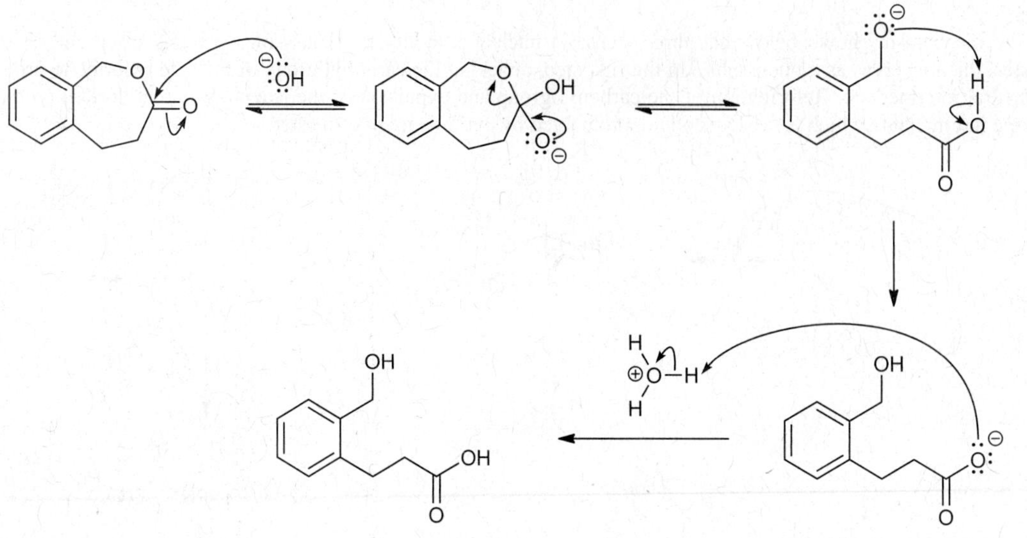

(d) The mechanism shown below has six steps: 1) nucleophilic attack, 2) loss of a leaving group, 3) proton transfer, 4) nucleophilic attack, 5) loss of a leaving group and 6) proton transfer. Hydrazine (NH_2NH_2) has two nucleophilic centers (each nitrogen atom has a lone pair), each of which can attack an acid chloride group. In the first step, hydrazine attacks one of the acid chloride groups to form a tetrahedral intermediate which subsequently expels a leaving group (chloride). Pyridine then functions as a base and removes a proton, giving an intermediate that bears no formal charges. Subsequent intramolecular nucleophilic attack by the second nitrogen atom on the second acid chloride group, followed by loss of chloride, results in an intermediate that is deprotonated to give the product. Once again, pyridine functions as the base for deprotonation.

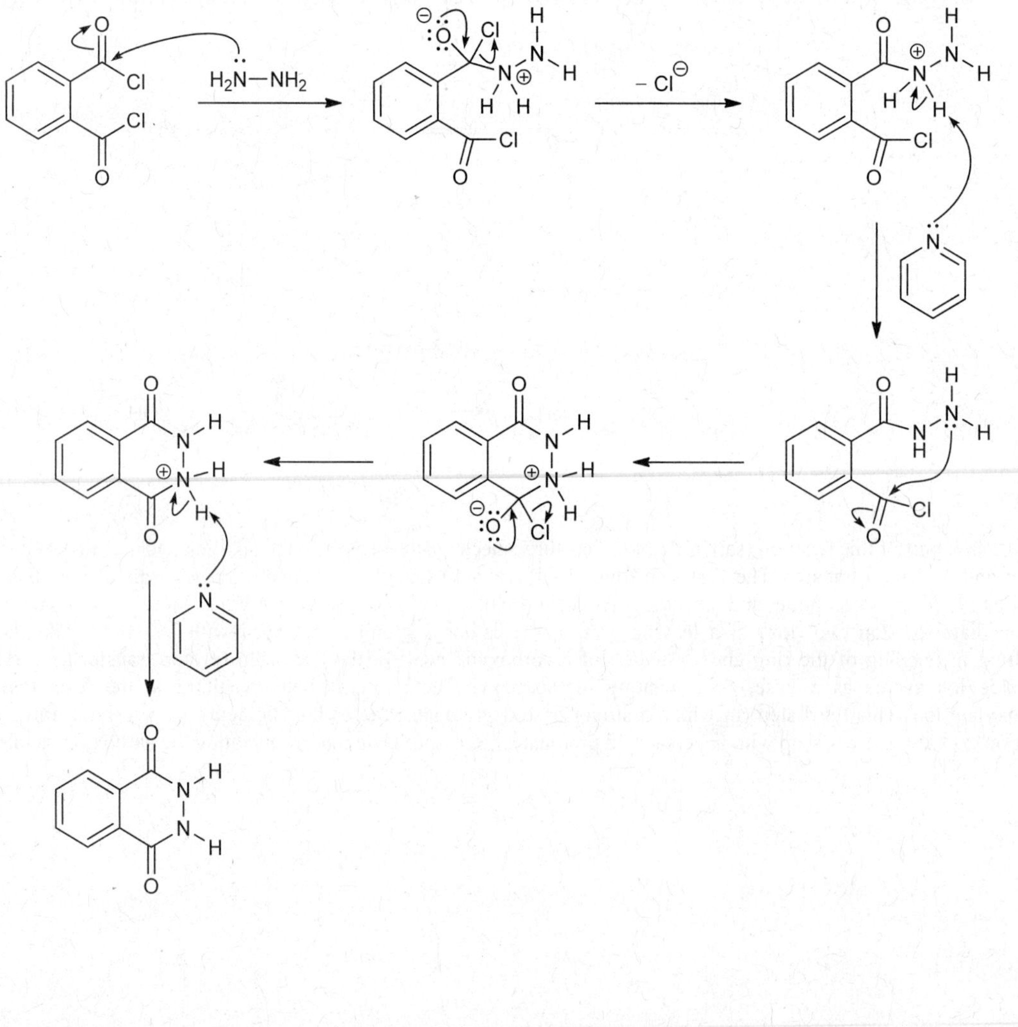

(e) The mechanism shown below has five steps: 1) nucleophilic attack, 2) loss of a leaving group, 3) nucleophilic attack, 4) proton transfer and 5) proton transfer. In the first step (nucleophilic attack), the anionic carbon atom of ethyl magnesium bromide serves as a nucleophile and attacks the electrophilic carbonyl group, resulting in a tetrahedral intermediate. In step two (loss of a leaving group), the carbonyl group is reformed with loss of an alkoxide ion, resulting in opening of the ring. In step three (nucleophilic attack) a second equivalent of ethyl magnesium bromide attacks the carbonyl group of the intermediate ketone, resulting in the formation of another tetrahedral intermediate. After the reaction is complete, a proton source (water) is introduced into the reaction flask, thereby protonating the dianion. Each anion is protonated separately, so two separate steps are required.

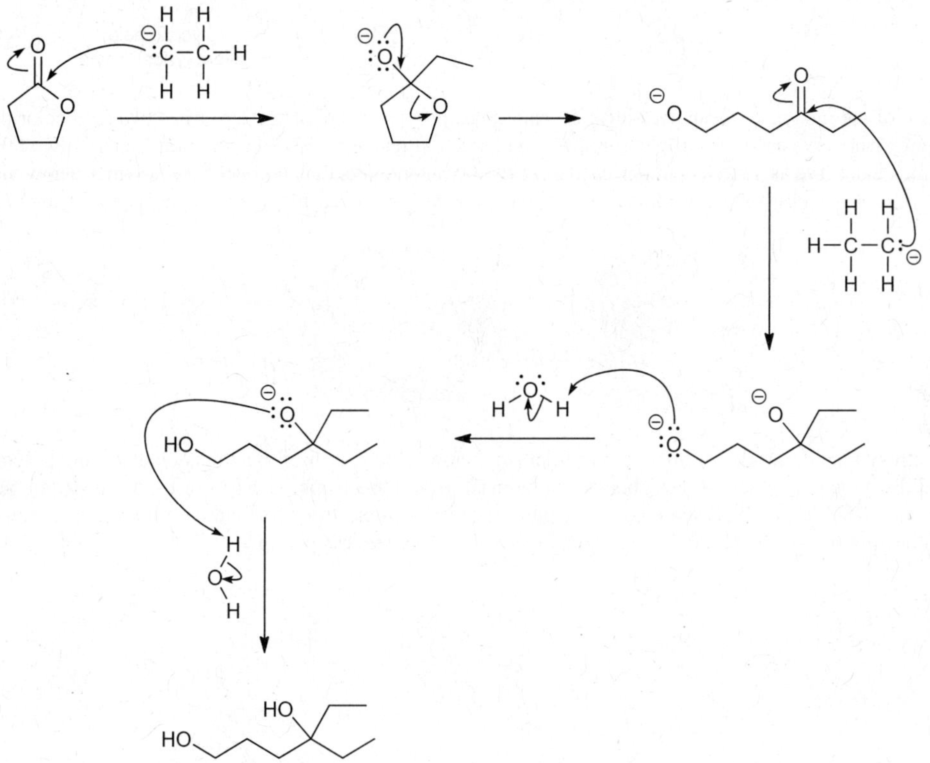

21.66. The three chlorine atoms withdraw electron density via induction. This effect renders the carbonyl group more electrophilic, and thus more reactive toward hydrolysis.

21.67. When treated with aqueous acid, each of the C-O bonds (on either side of the carbonyl group) is expected to undergo cleavage via an acid-catalyzed nucleophilic acyl substitution reaction. This produces a diol, shown below:

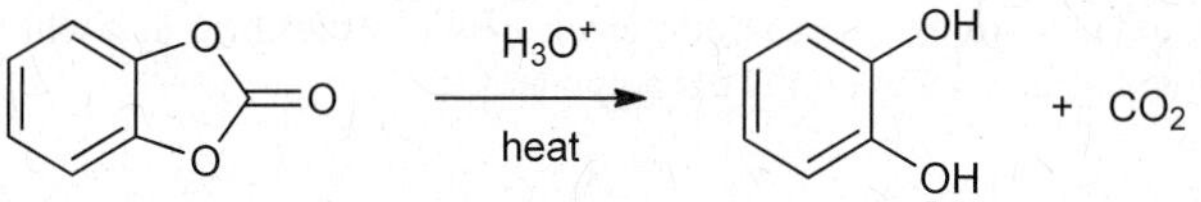

21.68
(a) When treated with aqueous acid, the ester group is hydrolyzed via a nucleophlic acyl substitution reaction in which water functions as a nucleophile, giving the active drug shown below. Note that the configuration of each chirality center is conserved, because the bonds to those chirality centers were not involved in the reaction.

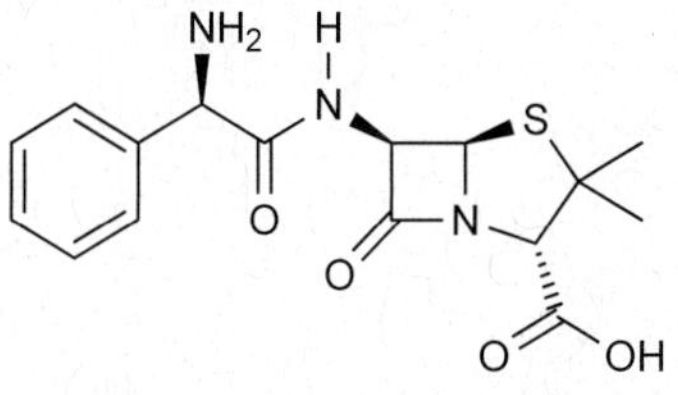

(b) The active drug is ampicillin, as indicated in the Medically Speaking box.

21.69. Hydrolysis of the ester groups in Dexon™ results in cleavage of the bonds indicated by the arrows in the figure below. This results in the formation of glycolic acid. The IUPAC name is based on a parent carboxylic acid called acetic acid, with indication of an alcohol (hydroxy group) at C2. Since acetic acid only has one location that can bear a substituent, a locant is not required to indicate the position of the OH group. The IUPAC name is thus hydroxyacetic acid.

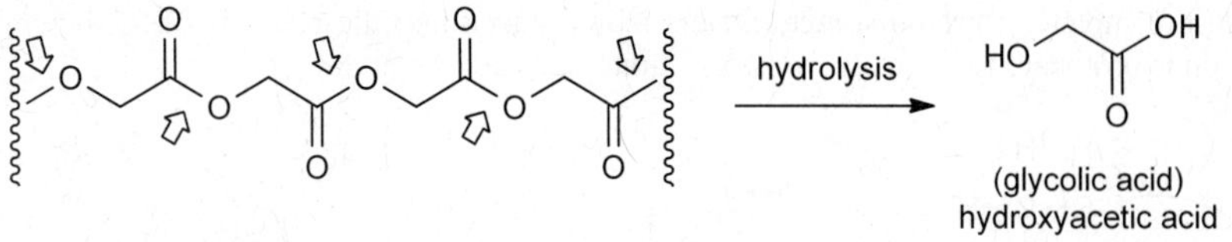

(glycolic acid)
hydroxyacetic acid

21.70. Each of the monomers bears two identical functional groups. The first monomer has two acid chloride groups, each of which can serve as an electrophile in the polymerization reaction. The second monomer has two OH groups, each of which can serve as a nucleophile. Reaction of the two monomers thus produces the polymer below via a series of nucleophilic acyl substitution reactions. The monomeric origins of each section of the polymer are highlighted below.

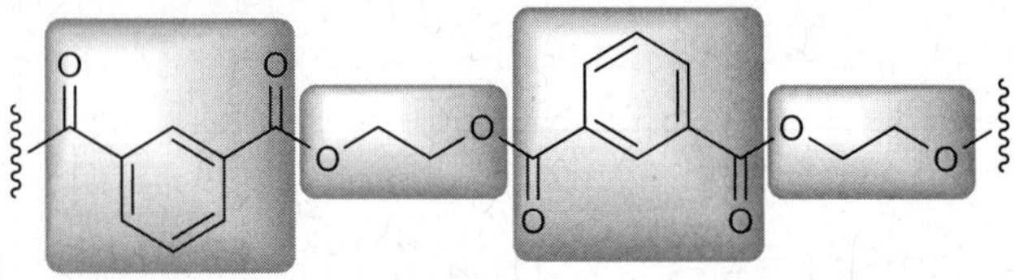

21.71. A retrosynthetic analysis of the polymer is shown below, where each of the bonds made during polymerization is indicated by an arrow. Each of these bonds can be made from a nucleophilic acyl substitution reaction between an amine and an acid chloride. Reaction of one monomer (bearing two electrophilic acid chloride groups) with the other monomer (bearing two nucleophilic amino groups) will produce the desired polymer.

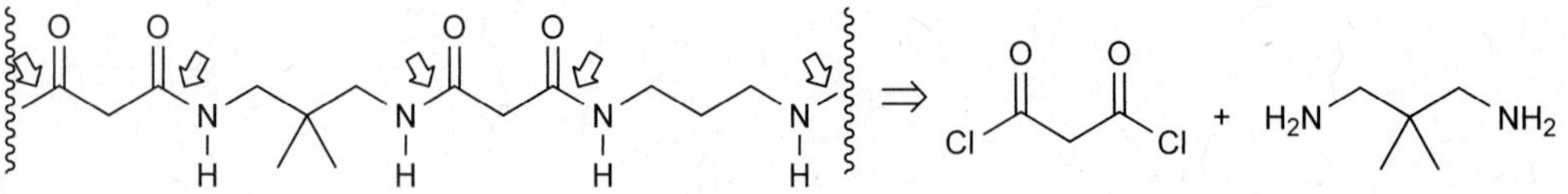

21.72. *meta*-Hydroxybenzoyl chloride (structure below) has a nucleophilic center (the OH group) as well as a strong electrophilic center (the acid chloride group) in a single molecule, thus making it susceptible to facile polymerization via the mechanism below. Each nucleophilic acyl substitution reaction has three steps: 1) nucleophilic attack, 2) loss of a leaving group and 3) proton transfer. In the first step (nucleophilic attack), the phenol oxygen atom of one molecule attacks the electrophilic carbonyl group on a second molecule resulting in the formation of a tetrahedral intermediate. A chloride leaving group is expelled in step 2 (loss of a leaving group), along with reformation of the carbonyl group. A proton transfer gives an intermediate that does not bear any formal charges. This intermediate (like the reactant) has a nucleophilic center (phenol oxygen atom) and a strong electrophilic center (the acid chloride group), thus allowing further reactions via an analogous pathway to produce a polymer.

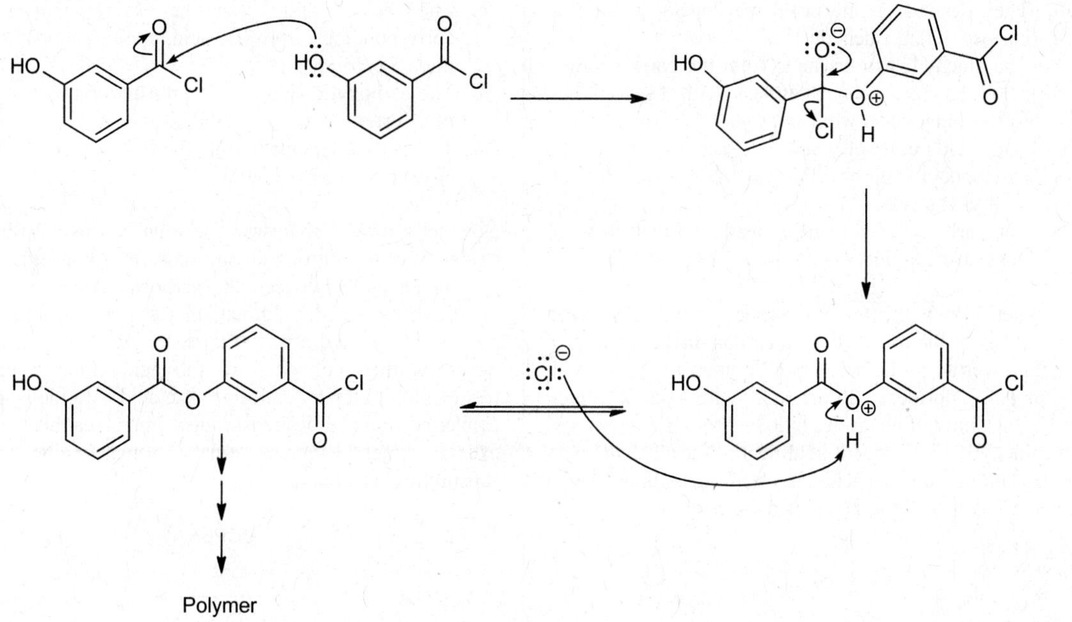

Polymer

21.73.
(a) There are certainly many acceptable solutions to this problem. One such solution derives from the following retrosynthetic analysis. An explanation of each of the steps (*a-c*) follows.

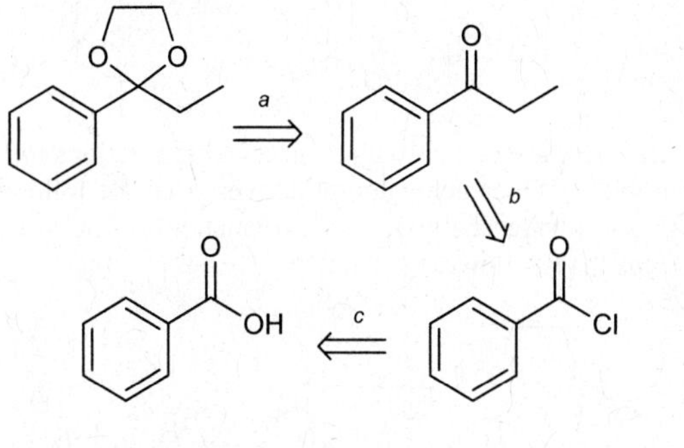

a. The desired cyclic acetal can be made from the corresponding ketone (via acetal formation).
b. The ketone can be made from benzoyl chloride, upon treatment with lithium diethyl cuprate.
c. The acid halide can be prepared from benzoic acid.

Now let's draw the forward scheme. Benzoic acid is converted to benzoyl chloride upon treatment with thionyl chloride. Subsequent reaction with lithium diethyl cuprate installs an ethyl group, giving a ketone. An acid-catalyzed reaction with ethylene glycol (with removal of water) produces the desired cyclic acetal.

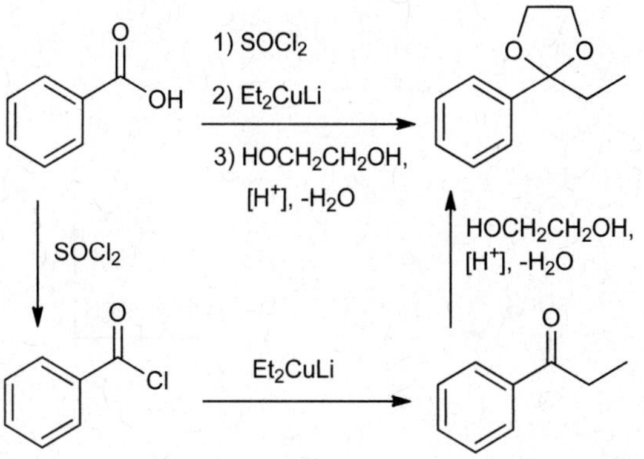

(b) There are certainly many acceptable solutions to this problem. One such solution derives from the following retrosynthetic analysis. An explanation of each of the steps (*a-d*) follows.

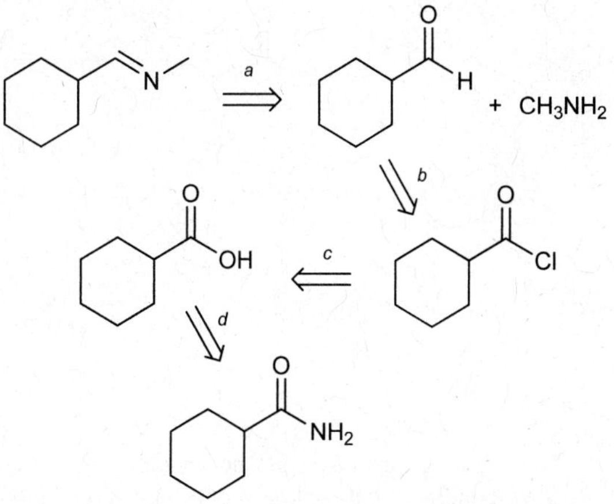

a. The desired imine can be made from the corresponding aldehyde.

b. The aldehyde can be made from the corresponding acid chloride, upon treatment with LiAl(OR)$_3$H, followed by water work-up.

c. The acid chloride can be prepared from the corresponding carboxylic acid, upon treatment with thionyl chloride.

d. The carboxylic acid can be made via hydrolysis of the starting amide.

Now let's draw the forward scheme. Acid catalyzed hydrolysis of the amide gives a carboxylic acid which is then converted to the acid chloride upon treatment with thionyl chloride. Reaction with a lithium trialkoxyaluminum hydride, followed by water, produces the aldehyde. Subsequent treatment of the aldehyde with methylamine under acid-catalyzed conditions (with removal of water) gives the desired imine.

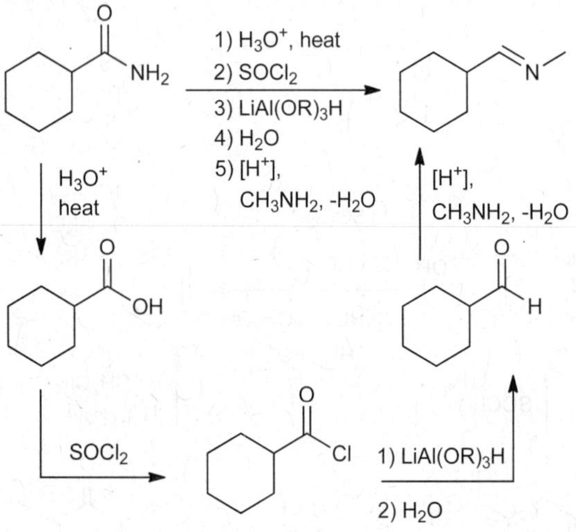

(c) There are certainly many acceptable solutions to this problem. One such solution derives from the following retrosynthetic analysis. An explanation of each of the steps (*a-d*) follows.

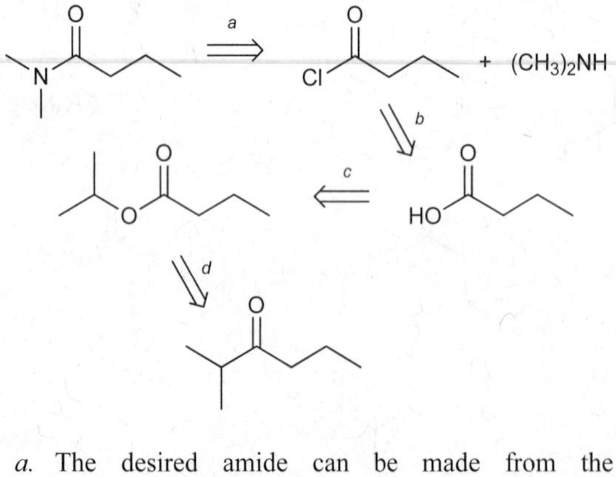

a. The desired amide can be made from the corresponding acid halide upon treatment with excess dimethylamine.

b. The acid halide can be made from the corresponding carboxylic acid, upon treatment with thionyl chloride.

c. The carboxylic acid can be prepared via hydrolysis of an ester.

d. The ester can be made from the starting ketone via a Baeyer-Villiger oxidation.

Now let's draw the forward scheme. Baeyer-Villiger oxidation of the starting ketone achieves the insertion of an oxygen atom between the carbonyl group and the more substituted alkyl group, thereby giving an ester. Acid-catalyzed hydrolysis of the ester gives butyric acid, which is then converted to the acid chloride upon treatment with thionyl chloride. The acid chloride is then converted into the desired product upon treatment with excess dimethyl amine (via a nucleophilic acyl substitution reaction).

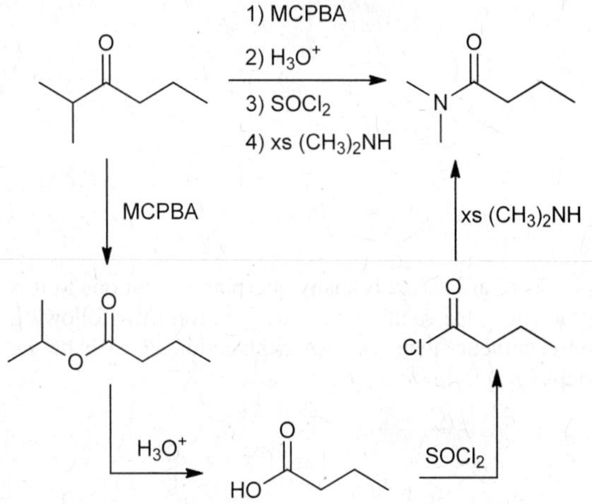

(d) There are certainly many acceptable solutions to this problem. One such solution derives from the following retrosynthetic analysis. An explanation of each of the steps (*a-d*) follows.

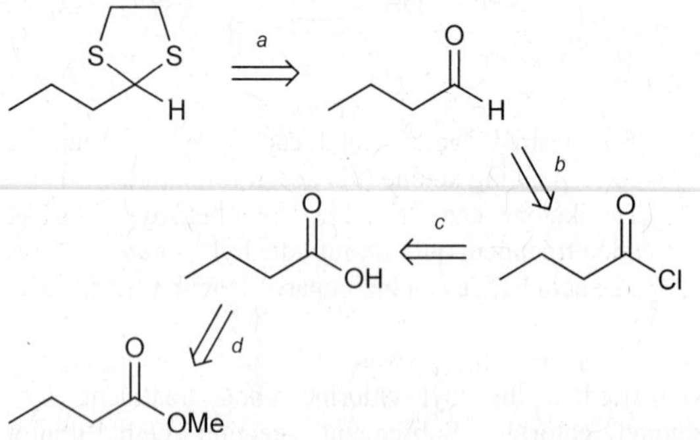

a. The cyclic thioacetal can be made from the corresponding aldehyde.

b. The aldehyde can be made from the corresponding acid halide, upon treatment with LiAl(OR)$_3$H, followed by water work-up.

c. The acid halide can be made from the corresponding carboxylic acid, upon treatment with thionyl chloride.

d. The carboxylic acid can be made from the starting ester via hydrolysis.

Now let's draw the forward scheme. The ester undergoes hydrolysis upon treatment with aqueous acid, giving butyric acid, which is subsequently converted to an acid chloride upon treatment with thionyl chloride. Reaction with a lithium trialkoxyaluminum hydride produces an aldehyde, which is then converted into the desired cyclic thioacetal.

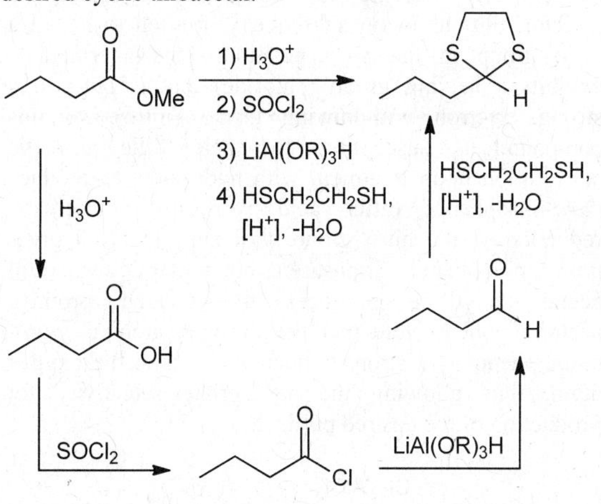

(e) There are certainly many acceptable solutions to this problem. One such solution derives from the following retrosynthetic analysis. An explanation of each of the steps (a-c) follows.

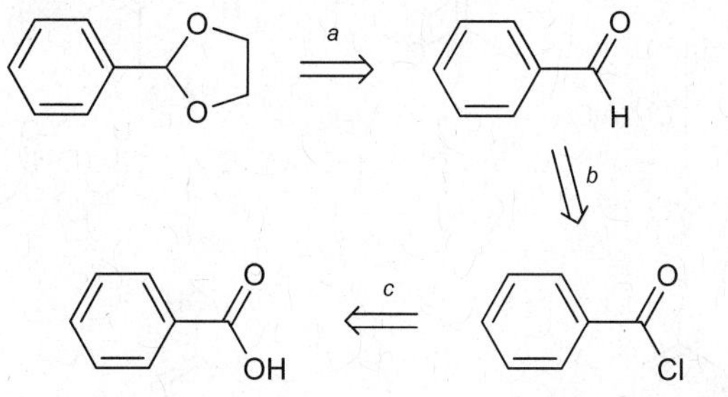

a. The cyclic acetal can be made from the corresponding aldehyde (benzaldehyde) via acetal formation.

b. Benzaldehyde can be made from the corresponding acid halide (benzoyl chloride).

c. Benzoyl chloride can be made from benzoic acid.

Now let's draw the forward scheme. Benzoic acid is converted to benzoyl chloride upon treatment with thionyl chloride. Benzoyl chloride is then converted into an aldehyde upon treatment with a lithium trialkoxyaluminum hydride. The aldehyde is then treated with ethylene glycol under acid-catalyzed conditions (with removal of water), giving the desired cyclic acetal.

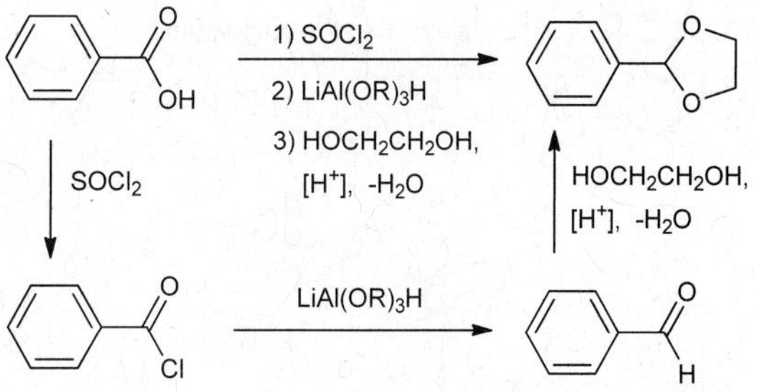

(f) There are certainly many acceptable solutions to this problem. One such solution derives from the following retrosynthetic analysis. An explanation of each of the steps (a-c) follows.

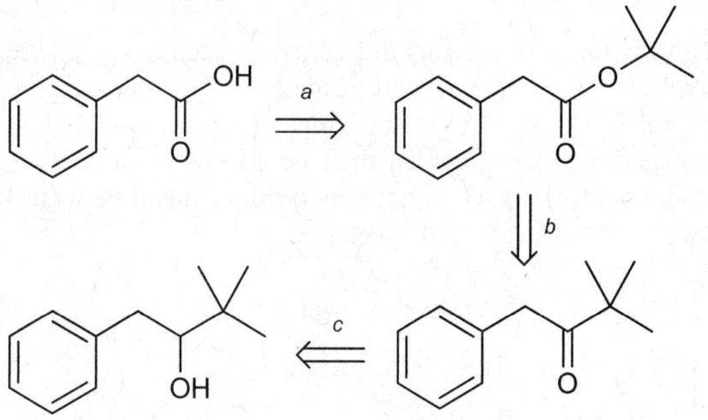

a. The carboxylic acid can be made from an ester, via hydrolysis.

b. The ester can be made from a ketone via a Baeyer-Villiger oxidation.

c. The ketone can be made from the starting material via oxidation of the starting alcohol.

Now let's draw the forward scheme. Oxidation of a secondary alcohol gives a ketone, which is then converted into an ester via a Baeyer-Villiger oxidation (this process inserts an oxygen atom between the carbonyl group and the more substituted alkyl group). Finally, acid-catalyzed hydrolysis of the ester gives the desired carboxylic acid.

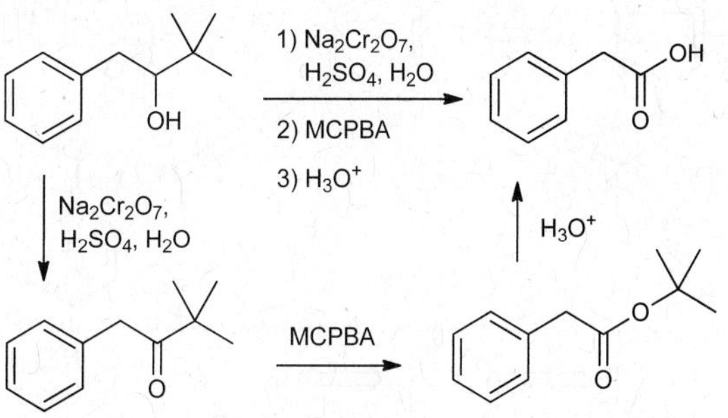

(g) There are certainly many acceptable solutions to this problem. One such solution derives from the following retrosynthetic analysis. An explanation of both steps follows.

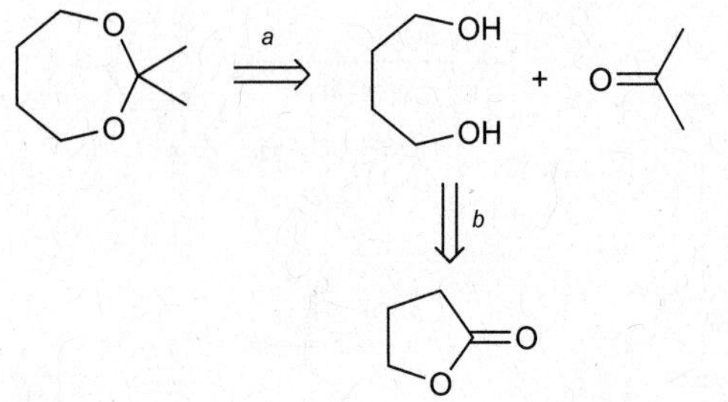

a. The product is a cyclic acetal, which can be made from a diol and a ketone (via acetal formation).

b. The diol can be made from the starting material via reduction with excess LAH.

Now let's draw the forward scheme. Reduction of the cyclic ester (lactone) with excess LAH, followed by water work-up, causes the ring to open, giving 1,4-butanediol. This diol can then be treated with acetone under acid-catalyzed conditions (with removal of water) to give the desired cyclic acetal.

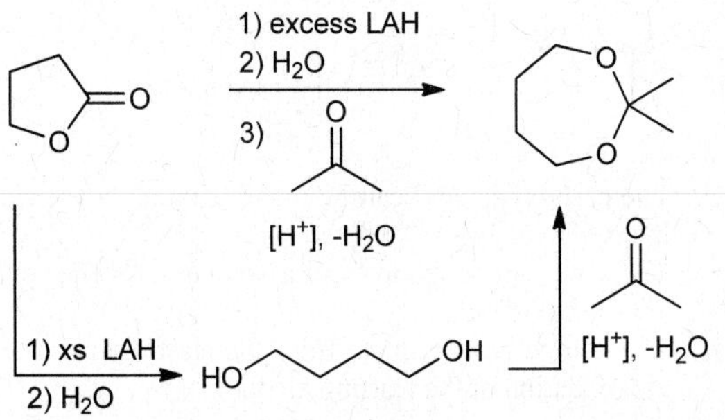

21.74. There are certainly many acceptable solutions to this problem. One such solution derives from the following retrosynthetic analysis. An explanation of each of the steps (*a-d*) follows.

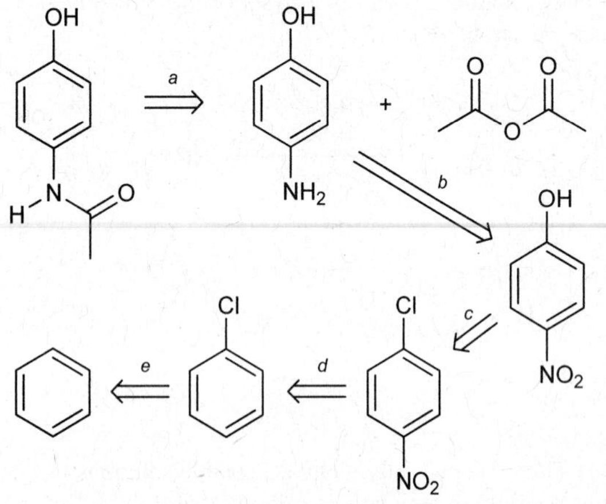

a. The product can be made via acetylation of *para*-aminophenol.

b. *para*-Aminophenol can be made from *para*-nitrophenol via reduction of the nitro group.

c. *para*-Nitrophenol can be made from *para*-chloronitrobenzene via an S_NAr process.

d. *para*-Chloronitrobenzene can be made via the nitration of chlorobenzene.

e. Chlorobenzene can be made from benzene via chlorination of the aromatic ring.

Now let's draw the forward scheme. Benzene is converted into chlorobenzene via an electrophilic aromatic substitution (upon treatment with Cl_2 and $AlCl_3$). The chlorine substituent is an *ortho-para* director, thus allowing subsequent nitration to install a nitro group in the *para* position. This intermediate exhibits a leaving group (chloride) that is *para* to a strong electron withdrawing group (nitro), so this compound is susceptible to nucleophilic aromatic substitution upon treatment with hydroxide, to produce *para*-nitrophenol (after acid work-up). Subsequent reduction of the nitro group with zinc and HCl gives *para*-aminophenol. Exposure to one molar equivalent of acetic anhydride produces the desired product, acetominophen. Note that the nitrogen atom of *para*-aminophenol is a stronger nucleophile than the oxygen atom, thus allowing the appropriate selectivity for production of the desired product.

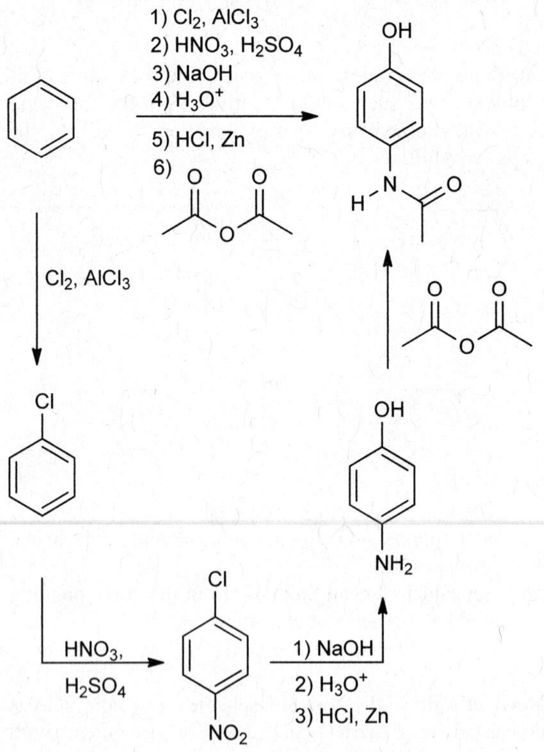

21.75. The mechanism shown below has 12 steps: 1) proton transfer, 2) loss of a leaving group, 3) nucleophilic attack, 4) proton transfer, 5) proton transfer, 6) loss of a leaving group, 7) proton transfer, 8) nucleophilic attack, 9) proton transfer, 10) proton transfer, 11) loss of a leaving group and 12) proton transfer. In step 1 (proton transfer), one of the acetal oxygen atoms is protonated by H_3O^+, activating it as a leaving group (either oxygen atom can be protonated, which will ultimately lead to the same product). In step 2, an alcohol group leaves as a leaving group. Water then attacks the activated carbonyl group in step 3 (nucleophilic attack). In step 4 (proton transfer), water deprotonates the cationic oxygen atom, and in step 5 (proton transfer), the other oxygen atom in the hemiacetal is protonated by H_3O^+, producing a cationic intermediate. In step 6, an alcohol serves as a leaving group. Protonation of the carbonyl group (step 7, proton transfer) activates this carbonyl group for intramolecular attack by one of the tethered alcohol groups (step 8, nucleophilic attack), thus forming a ring (a five-membered ring is more likely formed than a more strained, four-membered ring). Deprotonation by water (step 9, proton transfer) followed by protonation by H_3O^+ (step 10, proton transfer) produces a cationic intermediate with an activated leaving group (water). In step 11, water serves as a leaving group. In step 12 (proton transfer), water serves as a weak base to deprotonate the cationic oxygen atom, resulting in the formation of the final product (a lactone).

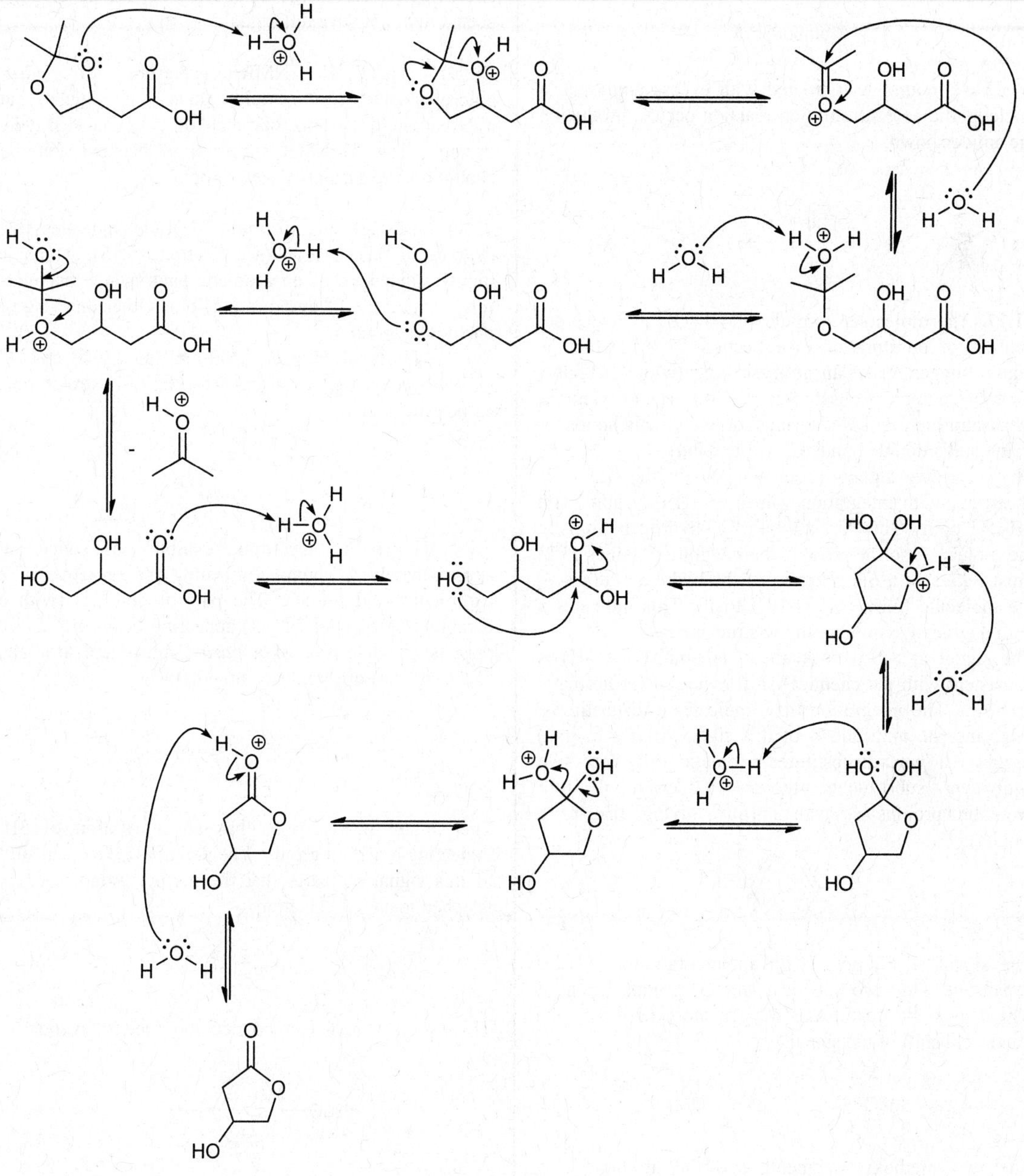

21.76. The carboxylic acid has a molecular formula of $C_5H_{12}O_2$. Thionyl chloride replaces the OH group on the carboxylic acid with a chlorine atom, thus the molecular formula of the resulting acid chloride is $C_5H_{11}ClO$. Considering the possible acid chloride isomers with this molecular formula, only one (compound **A**) gives a single signal in its ^1H NMR spectrum

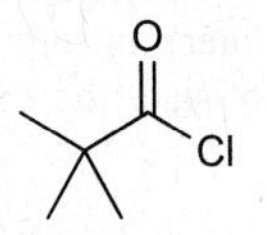

Compound **A**

When compound **A** is treated with excess ammonia, a nucleophilic acyl substitution reaction occurs, producing the amide shown.

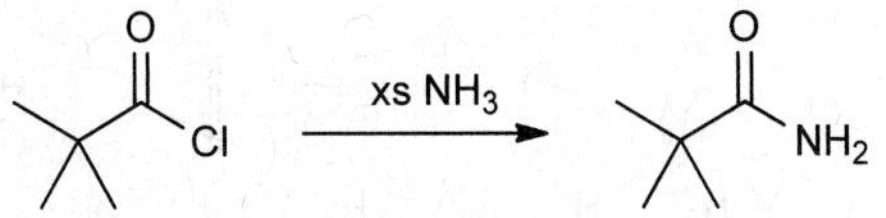

21.77. The molecular formula ($C_{10}H_{10}O_4$) indicates six degrees of unsaturation (see Section 15.16), which is highly suggestive of an aromatic ring (which accounts for four degrees of unsaturation) plus two more degrees of unsaturation (either two rings, or two double bonds, or a ring and a double bond, or a triple bond).

There are two signals (both singlets) in the ^1H NMR spectrum, with integration values of 3H (4.0 ppm) and 2H (8.1 ppm). There are a total of 10 hydrogen atoms in the molecule, so the ratio of the integration values (3:2) must represent a 6H:4H ratio of the hydrogen atoms in the molecule (6H + 4H = 10H total). This indicates a high degree of symmetry in the structure.

The signal at 8.1 ppm (with an integration of 4H) is consistent with the chemical shift expected for aromatic protons. The integration (4H) indicates a disubstituted ring, and the multiplicity of this signal (it is a singlet) suggests a *para*-disubstituted aromatic ring with two equivalent substituents, thereby rendering all four aromatic protons equivalent (thus giving rise to a singlet):

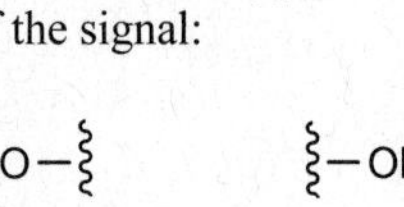

The signal at 4.0 ppm (with an integration of 6H) is consistent with two identical methyl groups, each of which must be next to an oxygen atom to justify the downfield shift of the signal:

$$MeO-\xi \qquad \xi-OMe$$

The two methoxy fragments and the aromatic ring account for all of the atoms in the molecular formula except for two carbon atoms and two oxygen atoms.

Since we still need to account for two more degrees of unsaturation, and since the proposed structure must retain its high degree of symmetry, we propose the following structure:

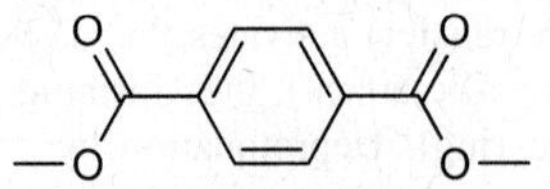

21.78. An IR spectrum of butyric acid should have a broad signal between 2200 and 3600 cm^{-1} due to the O-H stretch of the carboxylic acid. An IR spectrum of ethyl acetate will not have this signal.

21.79. The ^1H NMR spectrum of *para*-chlorobenzaldehyde should have a signal at approximately 10 ppm corresponding to the aldehydic proton. The ^1H NMR spectrum of benzoyl chloride should not have a signal near 10 ppm.

21.80. The molecular formula ($C_8H_8O_3$) indicates five degrees of unsaturation (see Section 15.16), which is highly suggestive of an aromatic ring (which accounts for four degrees of unsaturation) plus either one ring or one double bond.

The broad signal between 2200 cm^{-1} and 3600 cm^{-1} in the IR spectrum is consistent with the O-H stretch of a carboxylic acid.

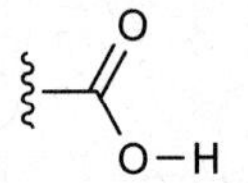

The ^1H NMR spectrum exhibits a signal at approximately 12 ppm, confirming the presence of a carboxylic acid group. The pair of doublets (with a combined integration of 4H) appearing between 7 and 8 ppm is characteristic of a *para*-disubstituted aromatic ring (with two different substituents):

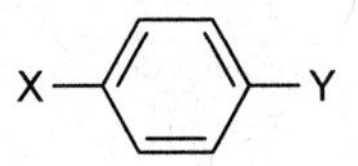

$$X-\langle\rangle-Y$$

The singlet near 4 ppm has an integration of 3H, indicating a methyl group. The downfield chemical shift of this signal indicates that the methyl group is likely attached to an oxygen atom:

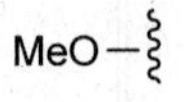

$$MeO-\xi$$

There is only one way to connect the three fragments:

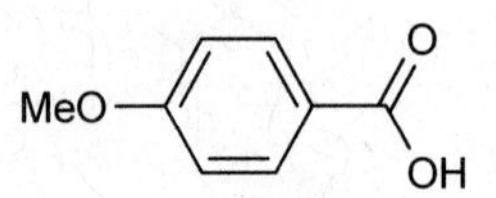

The ^{13}C NMR spectrum is consistent with this structure. The most downfield signal (172.8 ppm) is consistent with the carbonyl group. A disubstituted aromatic ring (bearing two different substituents) is expected to produce four signals between 100 and 150 ppm. We do in fact see four signals, although one of them is above 150 ppm, which likely corresponds with the carbon atom connected to the methoxy group (an oxygen atom is electron-withdrawing, causing a deshielding effect). Finally, the signal between 50 and 100 ppm is consistent with the carbon atom of the methoxy group (an sp^3 hybridized carbon atom attached to an electronegative atom).

21.81. The mechanism shown below has 6 steps: 1) proton transfer, 2) nucleophilic attack, 3) proton transfer, 4) proton transfer, 5) loss of a leaving group and 6) proton transfer. In step 1 (proton transfer), the carbonyl group is protonated by H_3O^+, activating it as an electrophile. In step 2 (nucleophilic attack), the tethered alcohol group serves as a nucleophile in an intramolecular nucleophilic attack. In step 3 (proton transfer), water deprotonates the oxonium ion, and in step 4 (proton transfer), a different oxygen atom is protonated by H_3O^+, activating it as a leaving group. In step 5 (loss of a leaving group), an alcohol serves as a leaving group. Deprotonation (step 6) gives the final, rearranged product. If the oxygen atom of the OH group in the starting material is an isotopic label (as indicated by the highlighted boxes), then we would expect the label to be incorporated into the ring of the product, as shown.

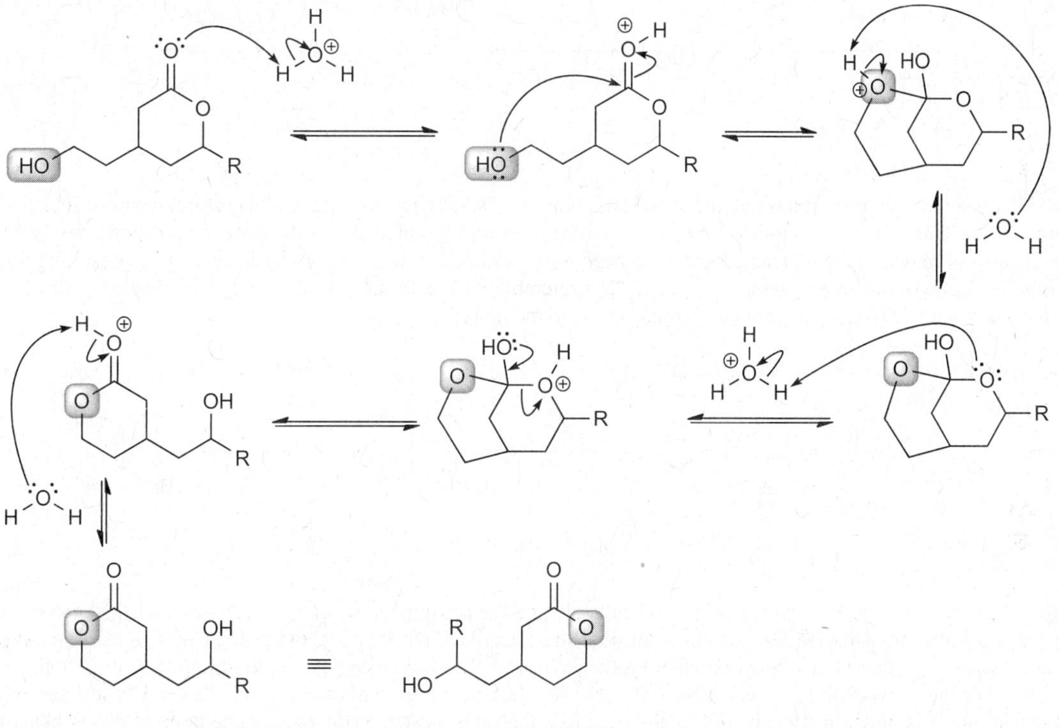

21.82. The lone pair of the nitrogen atom (of the amide group) is participating in aromaticity and is therefore unavailable to donate electron density into the carbonyl group. As a result, the carbonyl group is more electrophilic than the carbonyl group of a regular amide (where the lone pair contributes significant electron density to the carbonyl group via resonance). Also, when this compound functions as an electrophile in a nucleophilic acyl substitution reaction, the leaving group is particularly stable because it is an aromatic anion in which the negative charge is spread over all five atoms of the aromatic ring. With such a good leaving group, this compound more closely resembles the reactivity of an acid halide than an amide.

21.83.
(a) DMF, like most amides, exhibits restricted rotation about the bond between the carbonyl group and the nitrogen atom, due to the significant contribution of the resonance form with a C=N double bond. This restricted rotation causes the methyl groups to be in different electronic environments. They are not chemically equivalent, and will therefore produce two different signals (in addition to the signal from the other proton in the compound). Upon treatment with excess LAH, followed by water work-up, DMF is reduced to an amine:

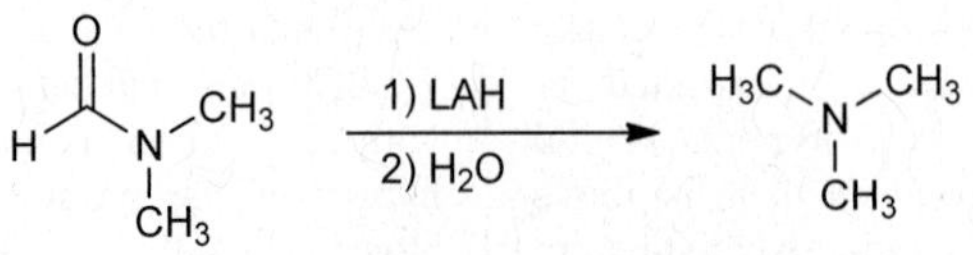

This amine does not exhibit restricted rotation. As such, all of the methyl groups are now chemically equivalent and will together produce only one signal.

(b) Restricted rotation causes the methyl groups to be in different electronic environments. As a result, the ^{13}C NMR spectrum of DMF should have three signals.

21.84. The first step of the synthesis involves deprotonation of the alcohol group in compound **1** using NaH, generating an alkoxide ion. This alkoxide ion is then treated with the chiral 2-bromo ethyl ester, to give an S_N2 reaction (note the inversion of configuration of the chirality center bearing the methyl group). Reduction of the ester with DIBAH provides an aldehyde, which is transformed into the terminal olefin (compound **2**) using a Wittig reaction.

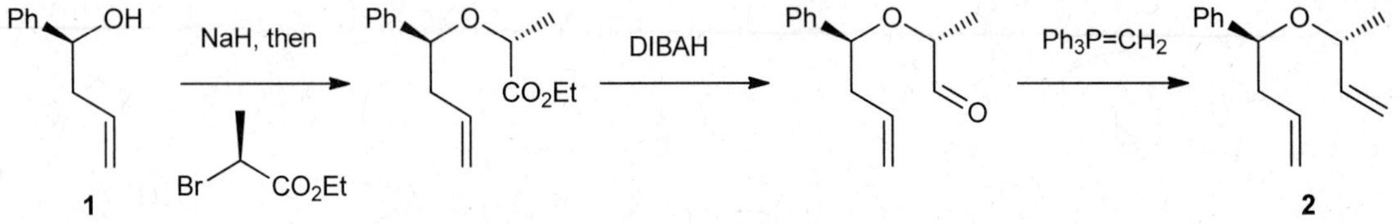

21.85. Notice that the product has one more carbon atom than the starting material, and therefore, we must introduce a carbon atom. This can be accomplished by first installing a leaving group at one of the benzylic positions, followed by an S_N2 process in which the leaving group is replaced with cyanide. Acid-catalyzed hydrolysis of the resulting nitrile affords the desired carboxylic acid. This route is preferable to the formation of a Grignard reagent followed by condensation with CO_2, as there are two bromine atoms in the molecule.

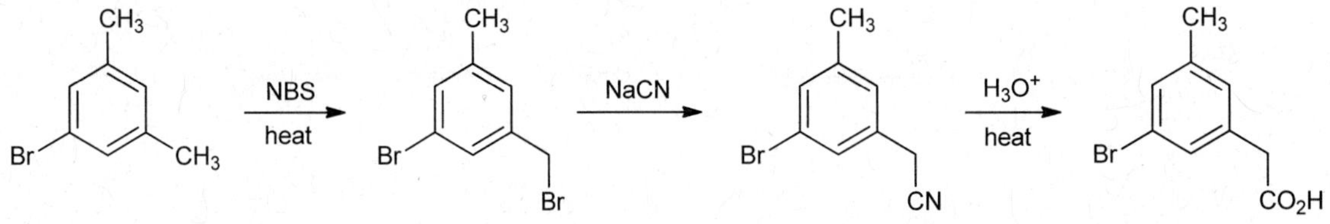

21.86. A possible synthesis is shown below. When analyzing the starting material, **1**, and the product, **2**, it can be seen that the alcohol is transformed into an ether and that the ester is converted to a different ester. So first determine a method to make an ether; the Williamson ether synthesis (base + RX) is a convenient method for ether formation.

For the conversion of the ester to a different ester, notice that the carbonyl carbon of ester **1** is still part of the side chain, but it is now a methylene (CH_2) and therefore the ester must be reduced at some point (LAH is known to reduce esters). Reduction of the ester affords a 1° alcohol, which can then be converted to the new ester:

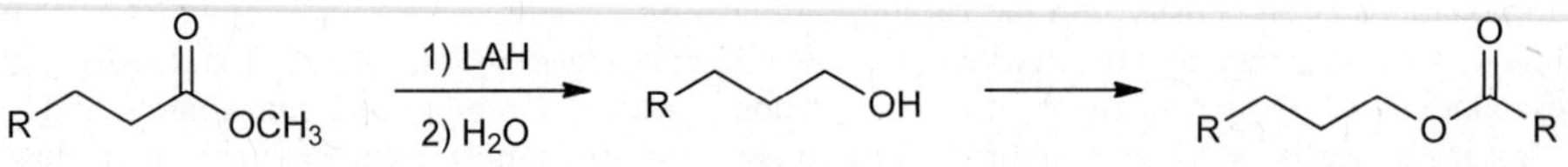

Conversion of the alcohol into the ester could be achieved either via an acid-catalyzed Fischer esterification ($ROH + RCO_2H$) or the via addition of an alcohol to an acid chloride ($ROH + RCOCl$). The latter method is preferable, because acidic conditions (employed by the first method) could produce undesired side reactions with other functional groups present.

Finally, the order of addition of these reagents is important so that you don't get an undesired product. See the correct order below, which shows the formation of ether **3** first (Williamson ether synthesis), followed by reduction of the ester to give alcohol **4**, and finally, conversion of alcohol **4** to the desired ester **2** using the corresponding acid chloride.

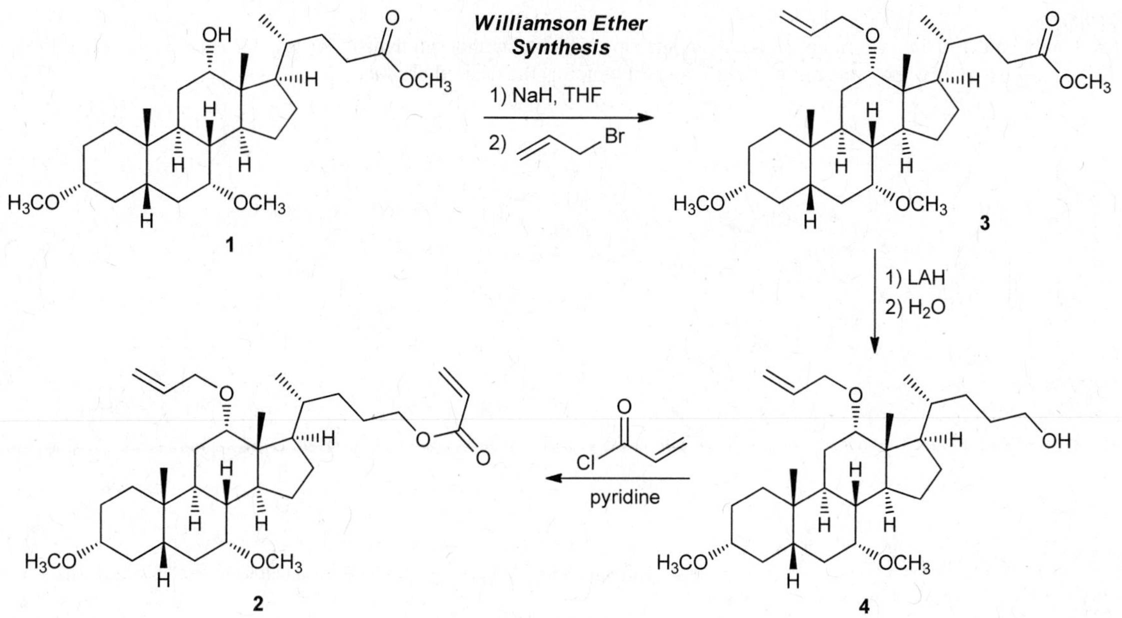

Williamson Ether Synthesis

1) NaH, THF
2) allyl bromide

3

1) LAH
2) H₂O

acryloyl chloride, pyridine

2

4

21.87.
The desired transformation can be achieved via reduction of the carboxylic acid, followed by substitution. Direct conversion of the resultant alcohol may be accomplished using PBr₃, or one can utilize a two-step method involving: 1) tosylate formation using TsCl and pyridine followed by, 2) S_N2 displacement using sodium bromide in DMSO:

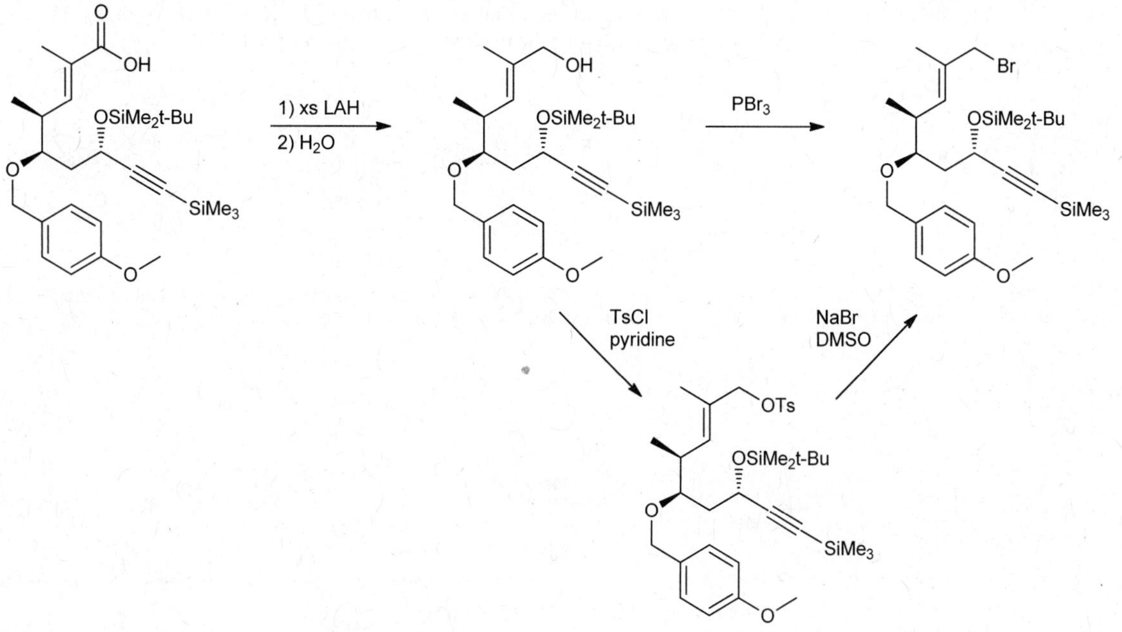

1) xs LAH
2) H₂O

PBr₃

TsCl pyridine

NaBr DMSO

21.88.

(a) Compound 1 is a nitrile (it contains a cyano group), which can undergo hydrolysis under these reaction conditions. This gives rise to three possible products, only one of which is the desired product:

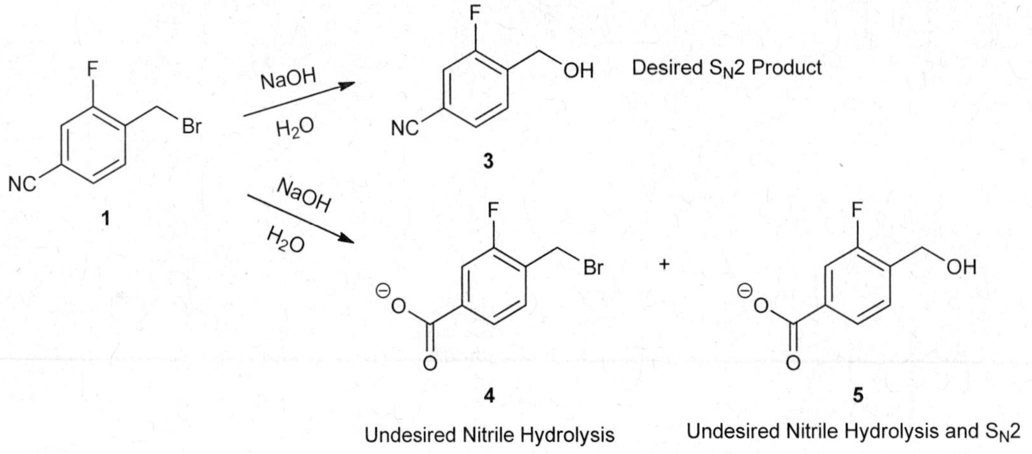

(b) A possible mechanism for the reaction sequence is illustrated below. The conversion of compound **1** to compound **2** is an S_N2 displacement of the bromide with acetate ($CH_3CO_2^-$). The reaction of the ester with NaOMe/MeOH is expected to proceed via a mechanism in which methoxide (CH_3O^-) attacks the carbonyl carbon of the ester (just as hydroxide would during hydrolysis). This generates a tetrahedral intermediate that collapses back down to give a new ester and alkoxide, which is protonated by the solvent to give the desired alcohol, compound **3**.

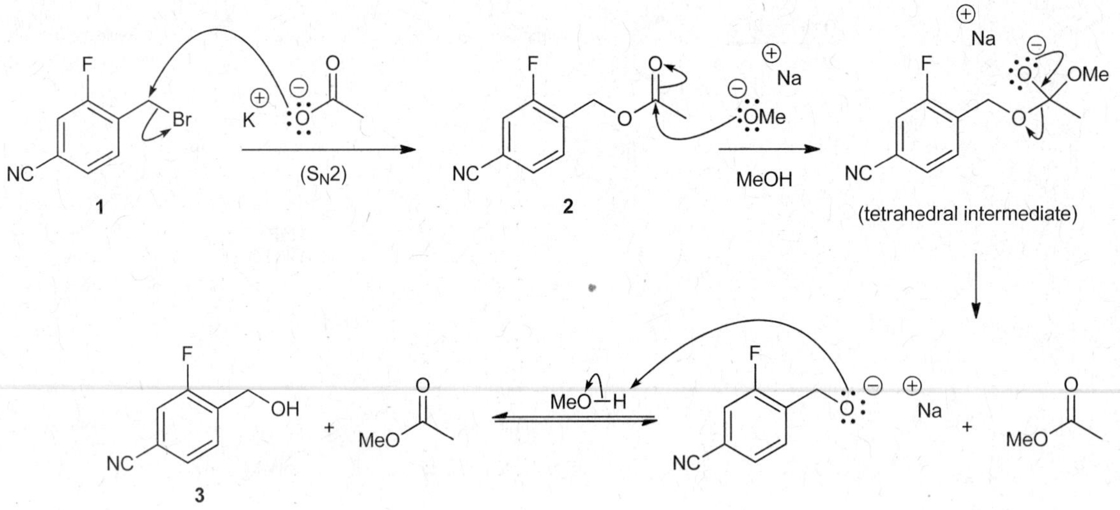

21.89. Note that methyl benzoate is the reference compound among the series of compounds examined. So, benzoates with rate constants larger in value than 1.7 $M^{-1}min^{-1}$ are more reactive while those with lower values are less reactive than the reference compound. The aromatic ring of methyl *p*-nitrobenzoate is considerably lower in electron density due to resonance interaction between the nitro group and the ring. However, in one of the resonance forms (structure A below), the electron deficient carbon atom of the ring is adjacent to the carbonyl carbon atom. Thus, the electrophilicity at the carbonyl functional group is the largest, making its reactivity towards nucleophilic attack by hydroxide anion the largest in this compound.

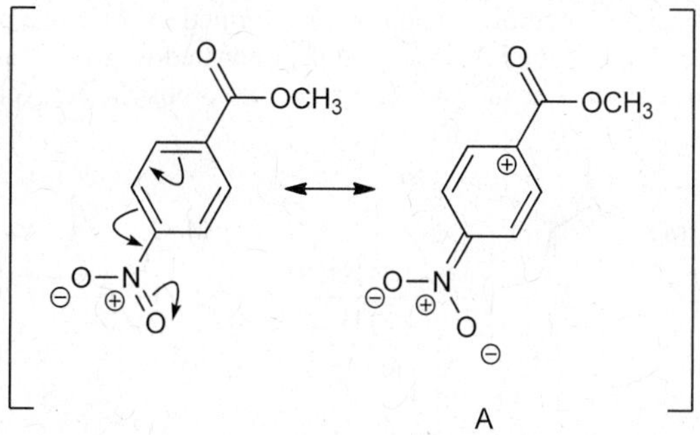

A

While the aromatic ring of methyl *m*-nitrobenzoate is also expected to be just as deficient in electron density as that of the *para* isomer, the positive charge on the ring never occupies the carbon atom adjacent to the carbonyl carbon atom (structure B). So, the electron deficiency (electrophilicity) at the carbonyl carbon atom is not as large as that of the *para* isomer. Thus, the reaction rate is not as large as that of the *para* isomer.

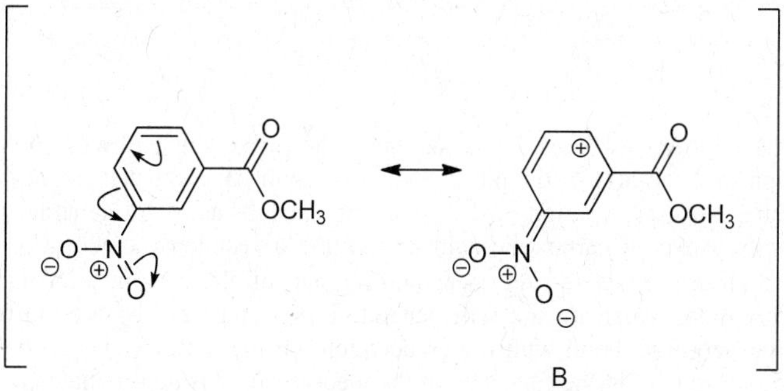

B

The main electronic interaction between the halogen atoms and the aromatic ring is induction by which both chlorine and bromine are expected to withdraw electron density from the ring. Since chlorine is more electronegative than bromine, the *m*-chlorophenyl ring is more electron deficient than the *m*-bromophenyl ring, and thus more reactive.

The remaining three benzoates are less reactive than the reference compound. This suggests that the substituents found in these compounds are all electron-donating so as to increase the electron density at the carbonyl group. The increased electron density lowers the electrophilicity at the carbonyl group and thus lowers their reactivity towards hydroxide ion. The methyl group is only weakly electron-donating, so it raises the electron density the least. According to electronegativity trends, since a nitrogen atom is more electron-donating than oxygen, the amino group is the best electron donor via resonance and thus lowers the reactivity of the carbonyl carbon the most (structure C).

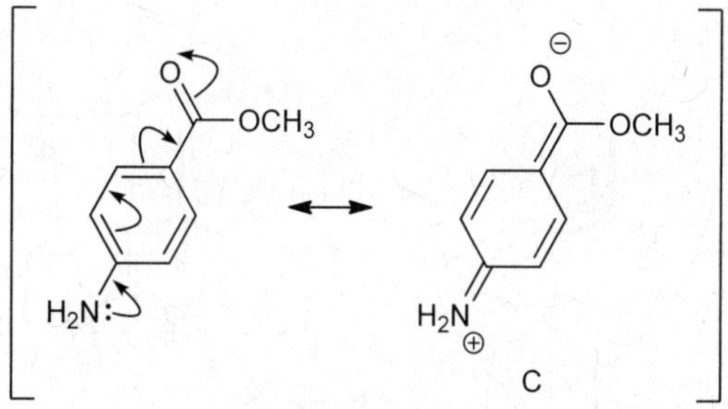

C

In this problem, we see that the influence of the resonance effects dominates those of induction. We also learn that the precise resonance forms involved can make a difference in the overall reactivity (*p*- vs *m*-nitrobenzoates).

21.90. Compound **1** is a ketone, so treatment with a peroxy acid, such as MCPBA, will give the corresponding lactone (cyclic ester). Notice that the oxygen atom is inserted between the carbonyl group and the bridgehead position (because that position is more substituted than the other side of the carbonyl group). Hydrolysis of the lactone gives compound **3**. In Corey's synthesis, this hydrolysis step was performed under basic conditions (saponification), and under those conditions, the product (compound **3**) would be deprotonated to give a carboxylate ion. Acid workup is necessary in order to protonate the carboxylate ion and regenerate the carboxylic acid (compound **3**).

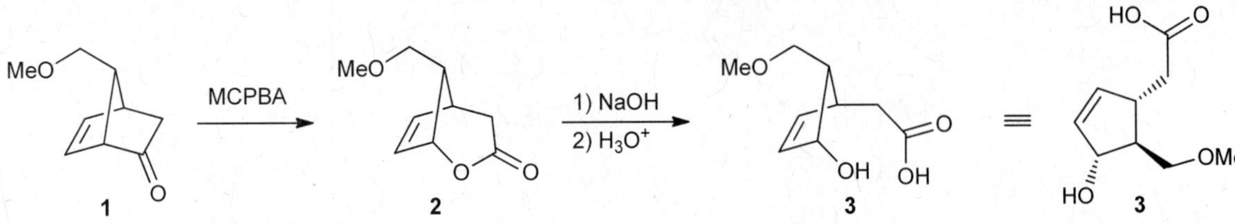

21.91. The first step of the synthesis involves deprotonation of the carboxylic acid with sodium hydride, followed by intramolecular esterification of **2** via loss of the mesylate group, giving **3**. Note that the newly formed ring is on the bottom face of the molecule. Next, protonation of the ester with aqueous acid will generate an activated carbonyl (**4**) that spontaneously leaves as a neutral carboxylic acid, generating a secondary carbocation (**5**). Next, a 1,2-carbon migration occurs so that the four-membered ring opens and the pair of electrons in the C-C σ-bond moves over one carbon to satisfy the carbocation, which in turn will generate a new, tertiary carbocation (**6**). In the next step, the carboxylic acid will form a new C-O bond with the carbocation, on the bottom face of the molecule, to generate a resonance-stabilized intermediate (**7**). In the final step of the mechanism, **7** is deprotonated to produce the desired ring system (**8**).

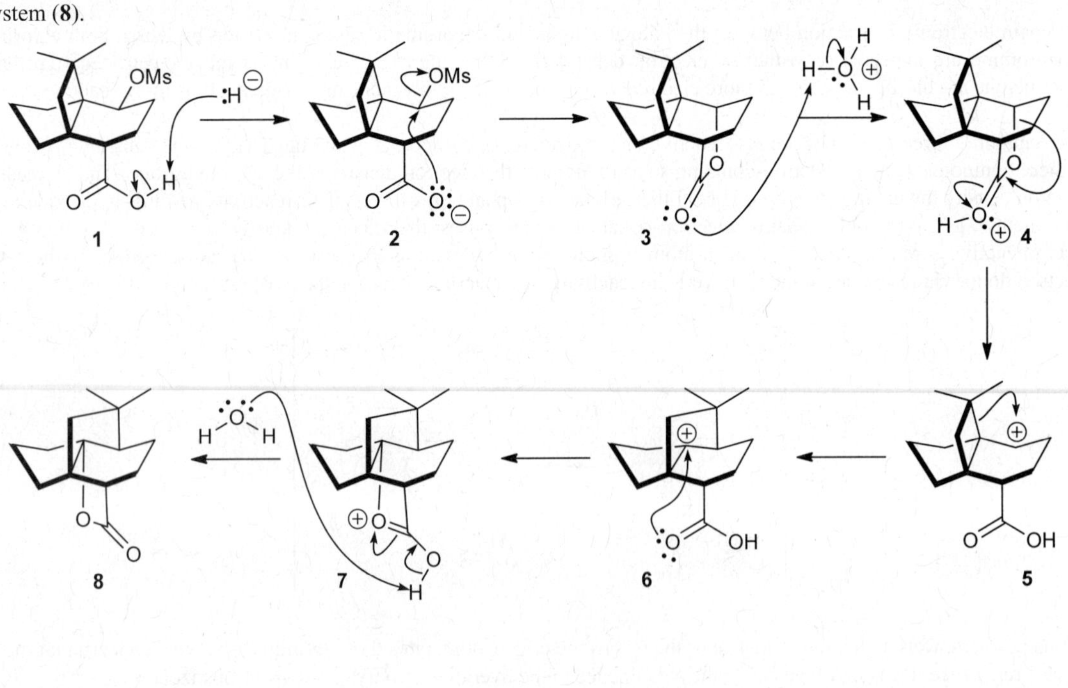

21.92. Deprotonation of alcohol **1** with sodium hydride will produce an alkoxide (**3**) that can easily react at the lactone carbonyl to form intermediate **4** (note: the bond angles in **4** have been exaggerated). Once formed, the unstable tetrahedral intermediate quickly decomposes to generate a new 5-membered lactone via the expulsion of an alkoxide leaving group. While it may be difficult to see in its current form, intermediate **5** can be redrawn as **5a**, which clearly shows the desired fused 5,5-ring system of the product with the proper stereochemistry at the ring junction. In the final step, the alkoxide in **5a** reacts at the carbon-bromine bond in an S_N2 fashion to produce epoxide **2**.

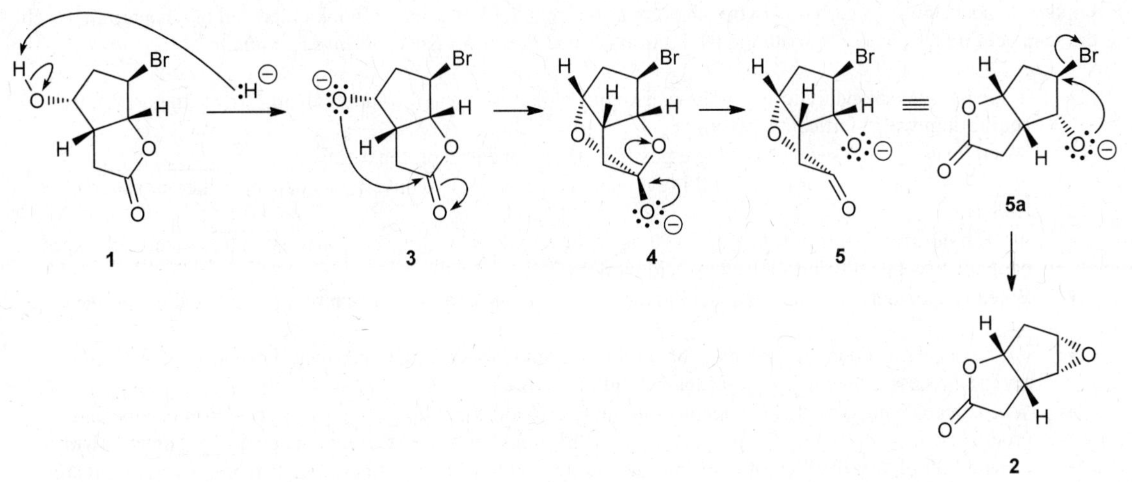

21.93. In the first step, methoxide attacks the lactone carbonyl to produce tetrahedral intermediate **2**. In the second step, the carbonyl is reformed via the loss of a leaving group, which in this case is the oxygen atom that was part of the 6-membered ring, leading to alkoxide **3**. Alpha to this newly formed alkoxide is an epoxide; generation of a new carbonyl via alkoxide opening of the epoxide will produce aldehyde **4**. Finally, the alkoxide that was generated from the epoxide opening can now attack the ester carbonyl to form tetrahedral intermediate **5**, which quickly loses methoxide to form lactone **6**.

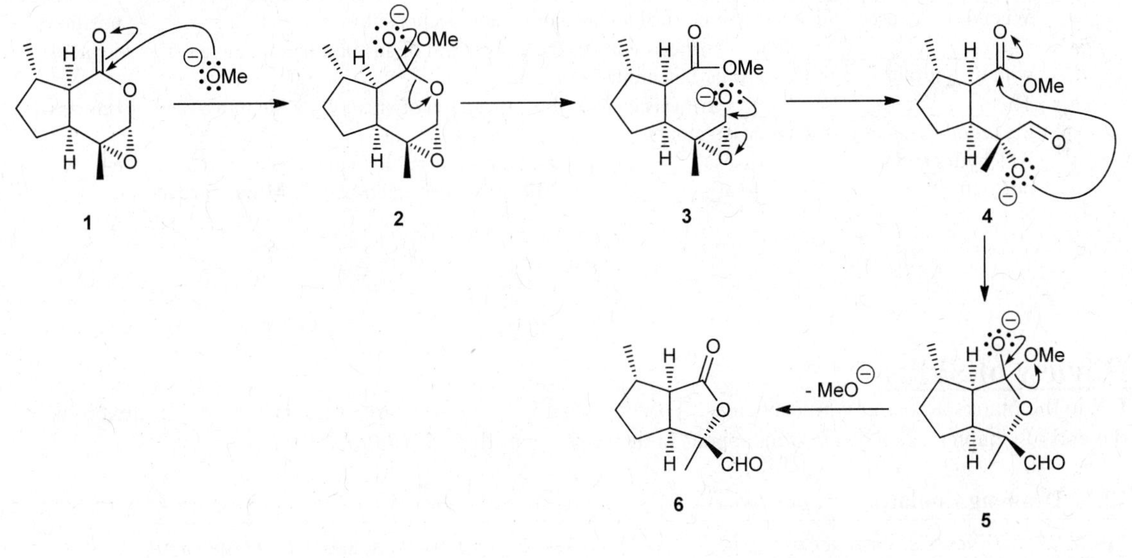

Chapter 22
Alpha Carbon Chemistry: Enols and Enolates

Review of Concepts

Fill in the blanks below. To verify that your answers are correct, look in your textbook at the end of Chapter 22. Each of the sentences below appears verbatim in the section entitled *Review of Concepts and Vocabulary*.

- In the presence of catalytic acid or base, a ketone will exist in equilibrium with an _____. In general, the equilibrium will significantly favor the _____.
- When treated with a strong base, the α position of a ketone is deprotonated to give an _____.
- _____ or _____ will irreversibly and completely convert an aldehyde or ketone into an enolate.
- In the **haloform reaction**, a _____ ketone is converted into a carboxylic acid upon treatment with excess base and excess halogen followed by acid workup.
- When an aldehyde is treated with sodium hydroxide, an **aldol addition reaction** occurs, and the product is a _____.
- For most simple aldehydes, the position of equilibrium favors the aldol product. For most ketones, the reverse process, called a _____-**aldol reaction** is favored.
- When an aldehyde is heated in aqueous sodium hydroxide, an **aldol** _____ **reaction** occurs, and the product is an _____. Elimination of water occurs via an _____ **mechanism**.
- **Crossed aldol**, or **mixed aldol reactions** are aldol reactions that occur between different partners and are only efficient if one partner lacks _____ or if a **directed aldol addition** is performed.
- Intramolecular aldol reactions show a preference for formation of _____ and ____-membered rings.
- When an ester is treated with an alkoxide base, a **Claisen condensation reaction** occurs, and the product is a _____.
- The α position of a ketone can be alkylated by forming an enolate and treating it with an _____.
- For unsymmetrical ketones, reactions with _____ at low temperature favor formation of the kinetic enolate, while reactions with _____ at room temperature favor the thermodynamic enolate.
- When LDA is used with an unsymmetrical ketone, alkylation occurs at the _____ position.
- The _____ **synthesis** enables the conversion of an alkyl halide into a carboxylic acid with the introduction of two new carbon atoms.
- The _____ **synthesis** enables the conversion of an alkyl halide into a methyl ketone with the introduction of two new carbon atoms.
- Aldehydes and ketones that possess _____-unsaturation are susceptible to nucleophilic attack at the β position. This reaction is called a _____ **addition,** or *1,4-addition,* or a **Michael reaction**.

Review of Skills

Fill in the blanks and empty boxes below. To verify that your answers are correct, look in your textbook at the end of Chapter 22. The answers appear in the section entitled *SkillBuilder Review*.

22.1 Drawing Enolates

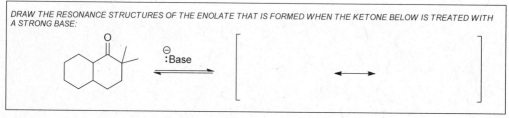

22.2 Predicting the Products of an Aldol Addition Reaction

PREDICT THE PRODUCT OF THE ALDOL ADDITION REACTION THAT OCCURS WHEN THE FOLLOWING ALDEHYDE IS TREATED WITH SODIUM HYDROXIDE:

NaOH

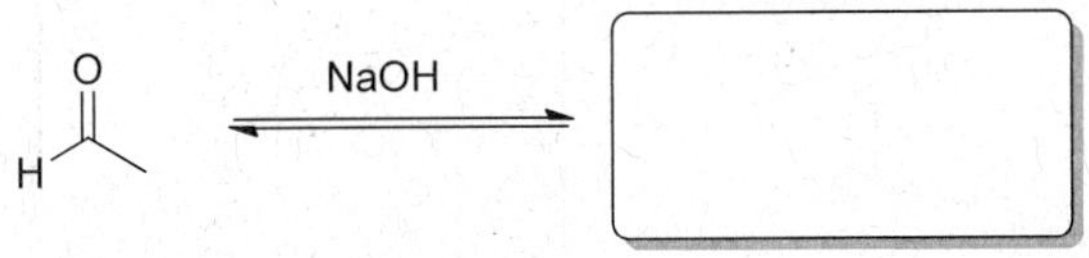

22.3 Drawing the Product of an Aldol Condensation

DRAW THE PRODUCT OF THE ALDOL CONDENSATION REACTION THAT OCCURS WHEN THE FOLLOWING COMPOUND IS HEATED WITH SODIUM HYDROXIDE:

NaOH, heat

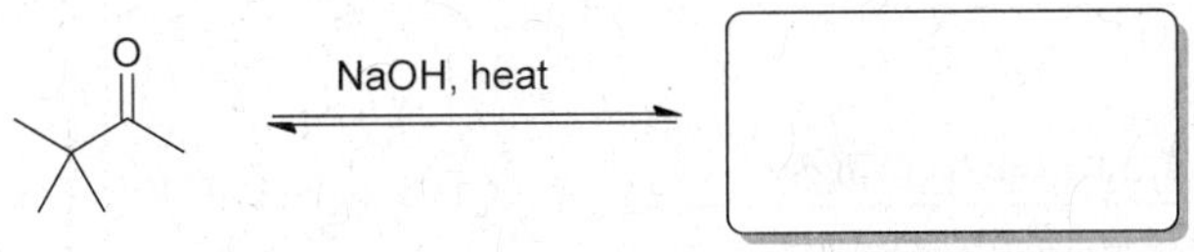

22.4 Identifying the Reagents Necessary for a Crossed Aldol Reaction

IDENTIFY REAGENTS THAT WILL ACHIEVE THE FOLLOWING TRANSFORMATION:

1)

2)

3)

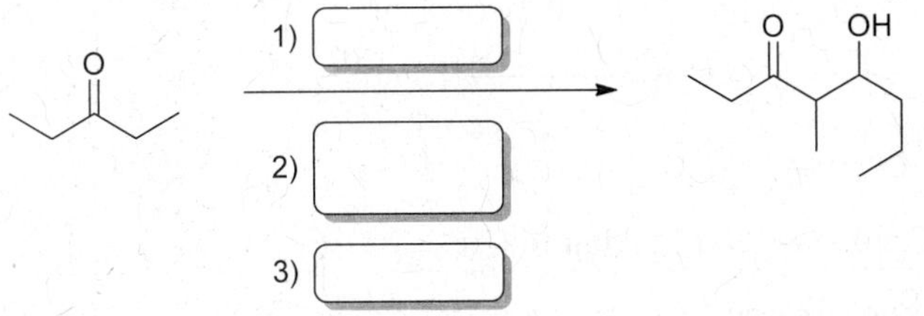

22.5 Using the Malonic Ester Synthesis

IDENTIFY REAGENTS THAT WILL ACHIEVE THE FOLLOWING TRANSFORMATION:

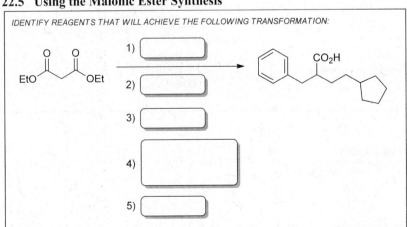

1)

2)

3)

4)

5)

22.6   Using the Acetoacetic Ester Synthesis

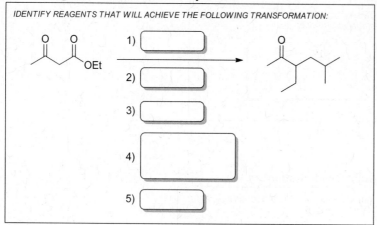

22.7   Determining When to Use a Stork Enamine Synthesis

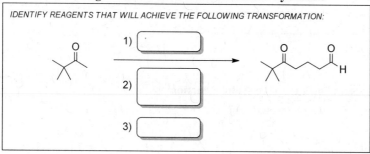

22.8   Determining which Addition or Condensation Reaction to Use

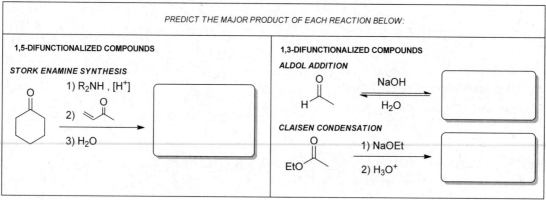

22.9   Alkylating the α and β Positions

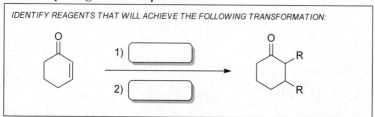

Review of Reactions

Identify the reagents necessary to achieve each of the following transformations. To verify that your answers are correct, look in your textbook at the end of Chapter 22. The answers appear in the section entitled *Review of Reactions*.

Alpha Halogenation

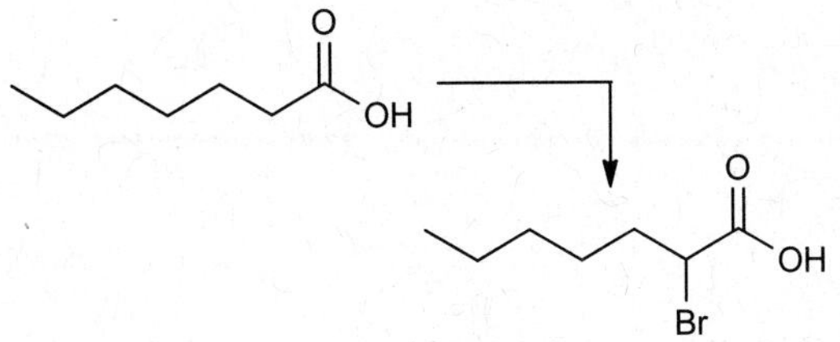

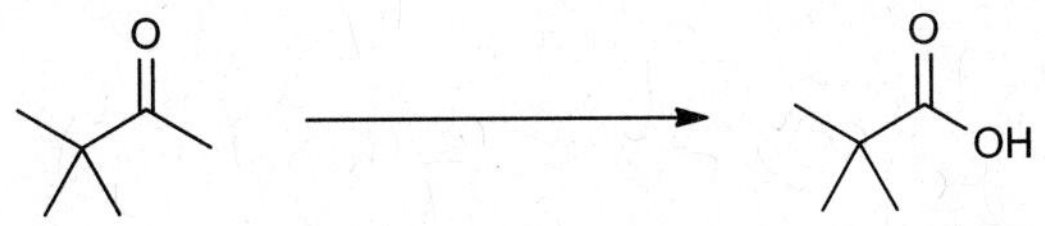

Aldol Reactions

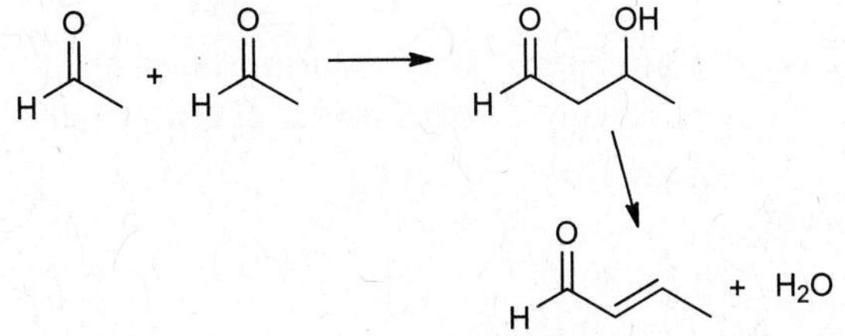

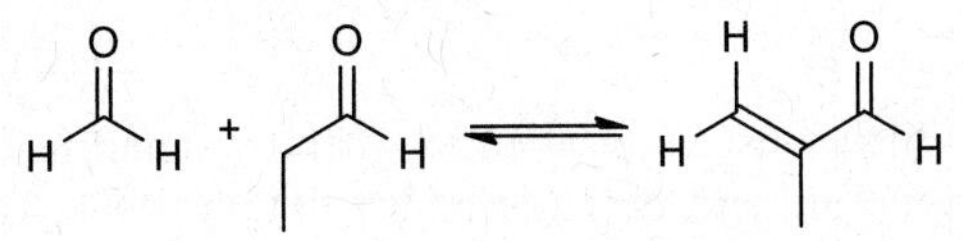

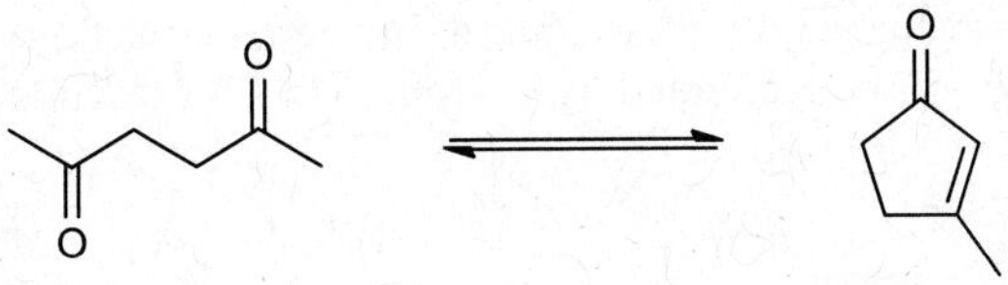

Claisen Condensation

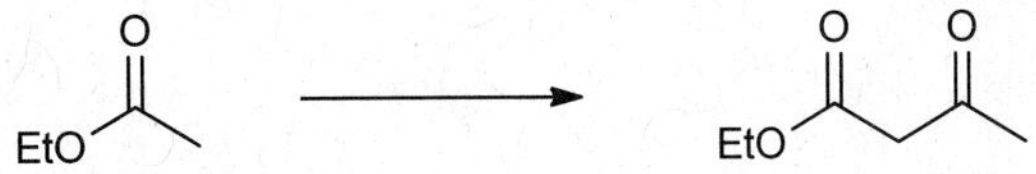

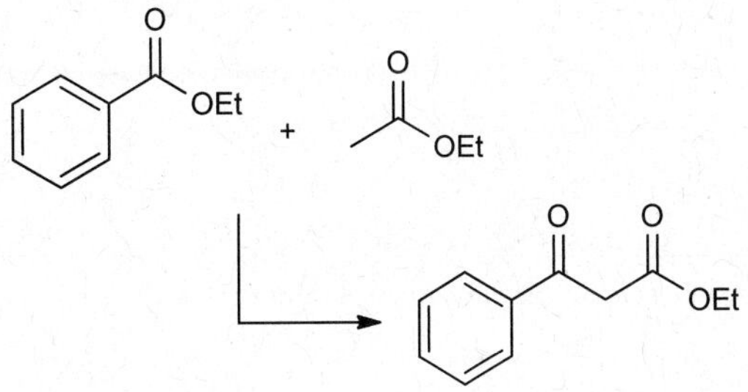

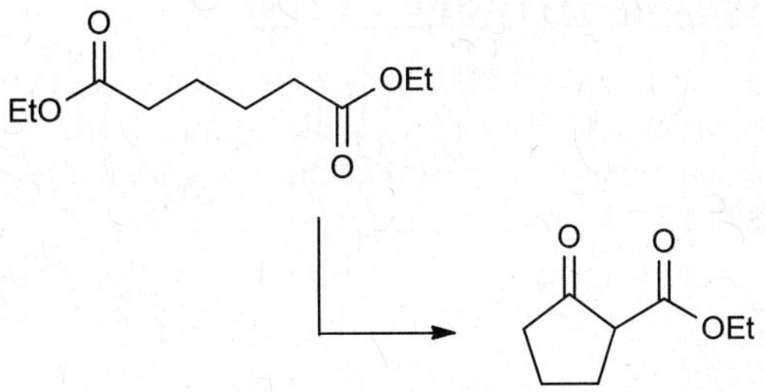

Alkylation

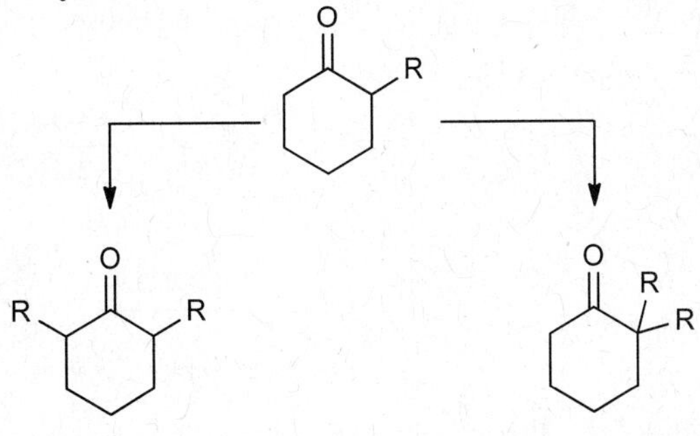

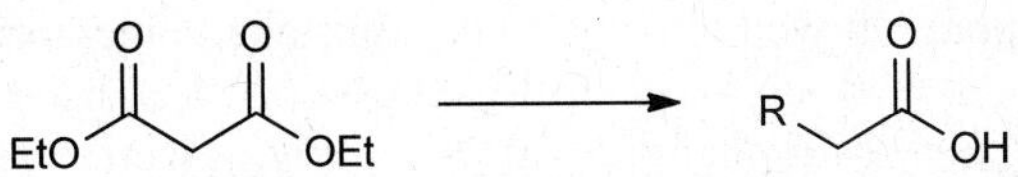

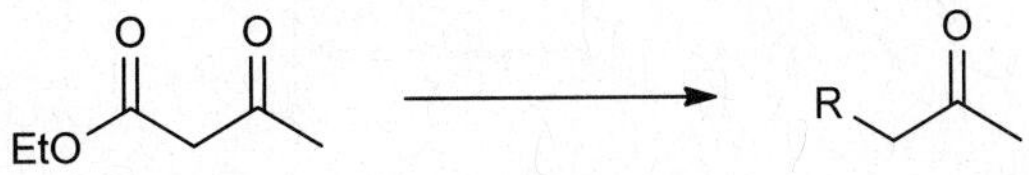

Michael Additions

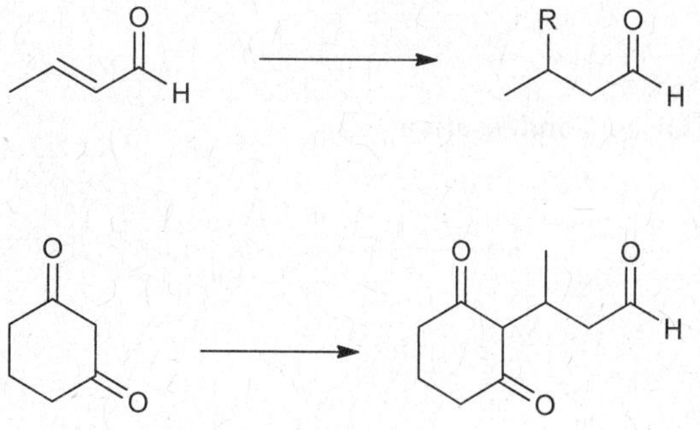

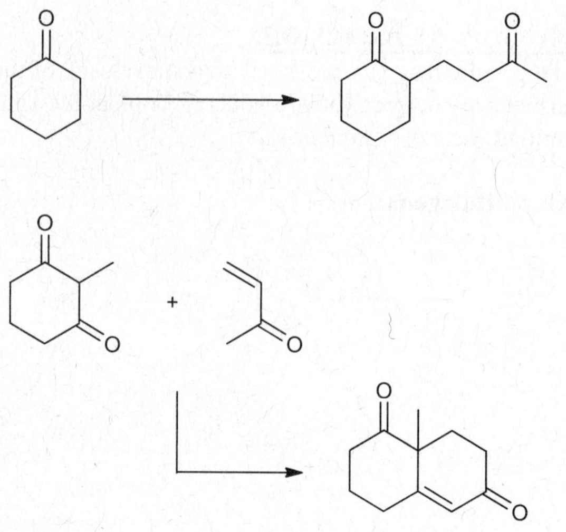

Common Mistake to Avoid

This chapter covers many reactions. Some of them require aqueous work-up, some require aqueous acidic work-up, while others require no work-up at all. Students often confuse the appropriate work-up conditions by associating the work-up conditions with the reagents, rather than the reaction occurring. For example, consider the following reaction:

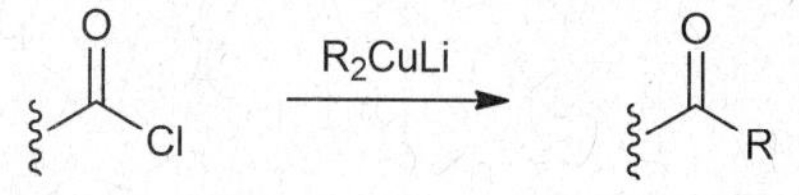

This reaction (covered in Section 21.8) involves a lithium dialkyl cuprate being used as a nucleophile, giving a ketone as a product. In contrast, consider the following reaction (from Chapter 22), which also employs a lithium dialkyl cuprate as a nucleophile:

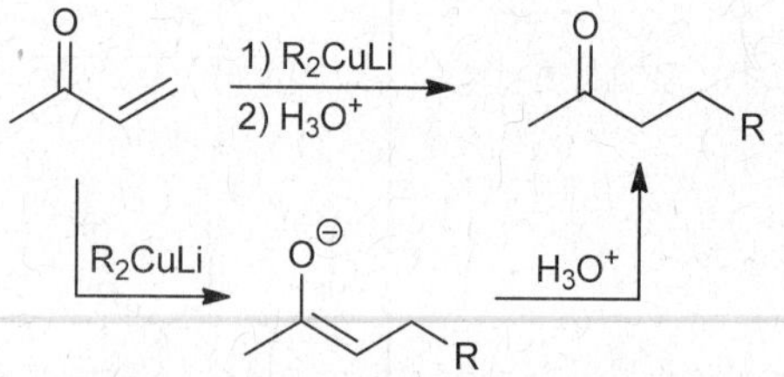

In this case, the initial product of the reaction is an enolate, which must be protonated, thereby requiring acidic work-up. This was not the case for the reaction between an acid chloride and a lithium dialkyl cuprate. It would therefore be a mistake to memorize that lithium dialkyl cuprates always require aqueous acidic work-up (or to memorize the opposite), because it depends on the situation. Rather than memorizing arbitrary rules that don't always apply, it would be wiser to focus on understanding *why* certain reactions require work-up while others do not. Your understanding will be facilitated by a strong focus on reaction mechanisms.

Useful reagents

The following is a list of reagents encountered in this chapter:

Reagents	Type of Reaction	Description
$[H_3O^+]$, Br_2	α-Bromination	These reagents can be used to install a bromine atom at the α position of a ketone (or aldehyde). Subsequent treatment of the resulting α-bromoketone with pyridine gives an α,β-unsaturated ketone. This two-step process can be used to introduce α,β-unsaturation into a ketone or aldehyde.
1) Br_2, PBr_3 2) H_2O	Hell-Volhard-Zelinsky reaction	These reagents can be used to install a bromine atom at the α position of a carboxylic acid.
1) NaOH, Br_2 2) H_3O^+	Haloform reaction	These reagents can be used to convert a methyl ketone into a carboxylic acid. This process is most efficient when the other α position (of the starting ketone) bears no protons.
NaOH, H_2O	Aldol addition reaction	Aqueous sodium hydroxide will cause an aldol addition reaction between two equivalents of an aldehyde or ketone to give a β-hydroxyaldehyde (or a β-hydroxyketone).
NaOH, H_2O, heat	Aldol condensation	Aqueous sodium hydroxide and heat will cause an aldol condensation between two equivalents of an aldehyde or ketone to give an α,β-unsaturated aldehyde (or an α,β-unsaturated ketone).
1) NaOEt 2) H_3O^+	Claisen condensation	These reagents will cause two equivalents of an ester to undergo a condensation reaction, giving a β-ketoester.
1) LDA, -78°C 2) RX	Alkylation	These conditions can be used to install an alkyl group at the less-substituted α position of an unsymmetrical ketone (via the kinetic enolate).
1) NaH, 25°C 2) RX	Alkylation	These conditions can be used to install an alkyl group at the more-substituted α position of an unsymmetrical ketone (via the thermodynamic enolate).
	Malonic ester synthesis	Diethyl malonate can be converted into a substituted carboxylic acid upon treatment with ethoxide, followed by an alkyl halide, followed by aqueous acid.
	Acetoacetic ester synthesis	Ethyl acetoacetate can be converted into a derivative of acetone upon treatment with ethoxide, followed by an alkyl halide, followed by aqueous acid.
1) R_2CuLi 2) H_3O^+	Michael reaction	A lithium dialkyl cuprate is a weak nucleophile and can serve as a Michael donor. It will react with a suitable Michael acceptor (see Table 22.2).

Solutions

22.1. Under acid-catalyzed conditions, the carbonyl group is first protonated, generating a resonance-stabilized intermediate, which is then deprotonated at the α position to give an enol. Notice that the acid for the protonation step is a hydronium ion, and the base for the deprotonation step is water, consistent with acidic conditions.

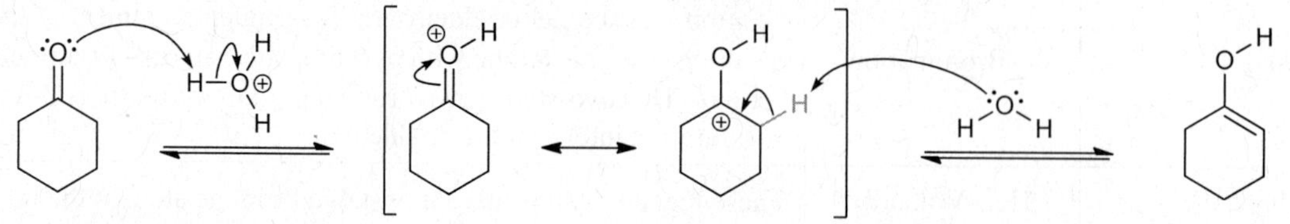

22.2. If we carefully inspect the solution to the previous problem, we find that the final step of the mechanism is deprotonation of the α position, thereby converting the resonance-stabilized cationic intermediate into an enol. Therefore, the reverse of this process must begin with protonation of the α position, thereby converting the enol into a resonance-stabilized cationic intermediate. Subsequent deprotonation of this intermediate gives the ketone. Notice that the acid for the protonation step is a hydronium ion, and the base for the deprotonation step is water, consistent with acidic conditions.

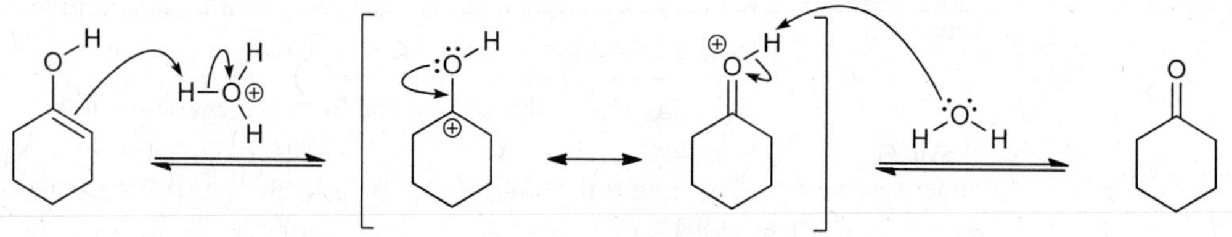

22.3. Under base-catalyzed conditions, the α position is first deprotonated, generating a resonance-stabilized anionic intermediate. The oxygen atom in this intermediate is then protonated to give an enol. Since the ketone is unsymmetrical, the two α positions are not equivalent. Therefore, the enol can be formed at either α position, as shown below. Notice that, in each case, the base for the deprotonation step is a hydroxide ion, and the acid for the protonation step is water, consistent with basic conditions.

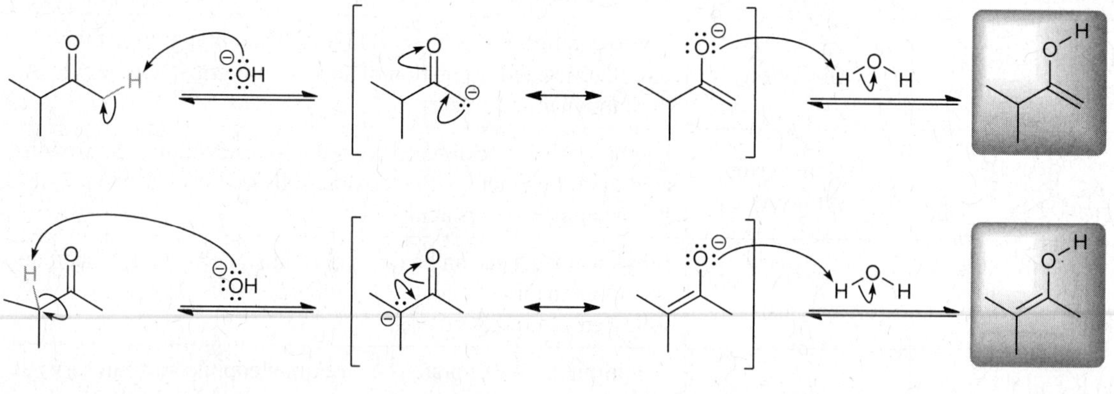

22.4.
(a) This compound has two α positions, although they are identical because the ketone is symmetrical. Deprotonation at either location will lead to the same enolate ion, which has the following resonance structures:

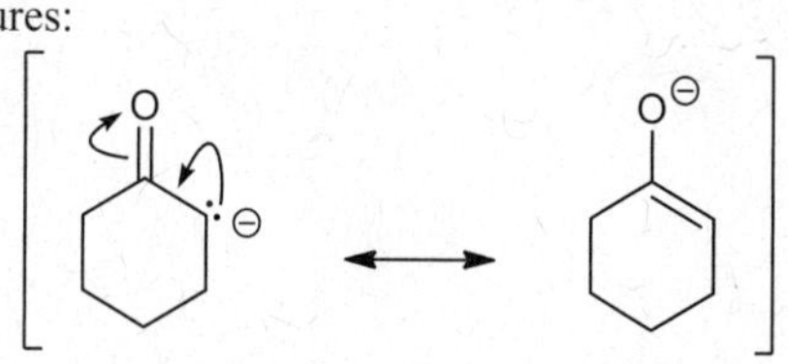

(b) This compound has two α positions, although only one of these positions bears protons. Deprotonation at that location will lead to an enolate ion with the following resonance structures:

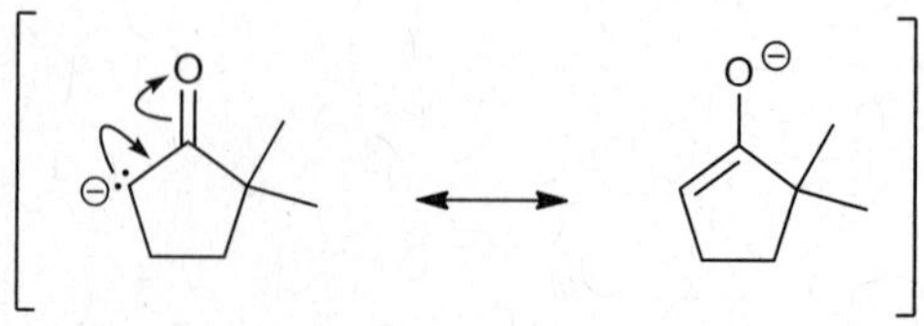

(c) This compound has two α positions, although only one of these positions bears protons. Deprotonation at that location will lead to an enolate ion with the following resonance structures:

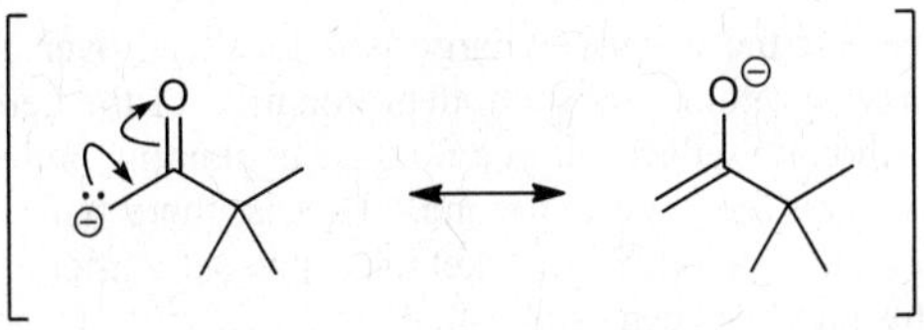

(d) This compound has two α positions, although only one of these positions bears protons. Deprotonation at that location will lead to an enolate ion with the following resonance structures:

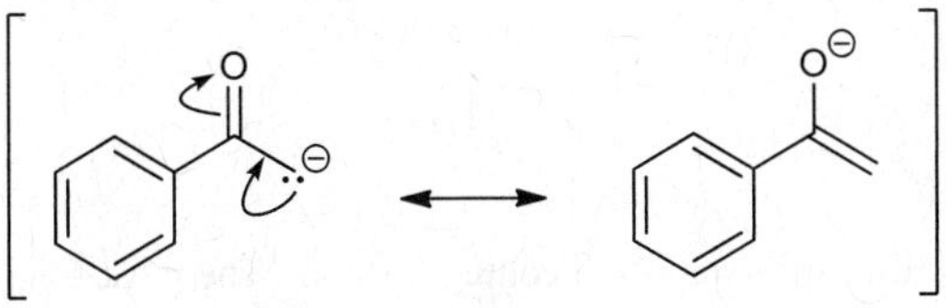

(e) This compound is an aldehyde and therefore has only one α position. Deprotonation at that location will lead to an enolate ion with the following resonance structures:

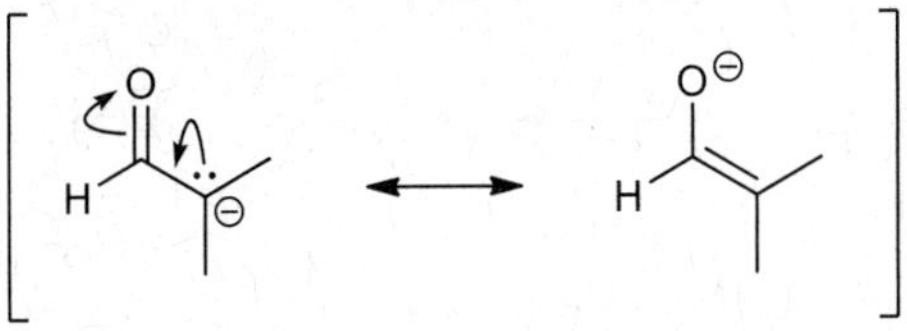

22.5. We begin by considering the structure of 2-methylcyclohexanone. This structure has two α positions (highlighted):

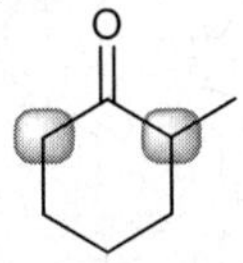

These two positions are not identical, because the ketone is unsymmetrical. Therefore, either of these positions can be deprotonated, giving rise to the following two possible enolate ions (each of which is resonance stabilized):

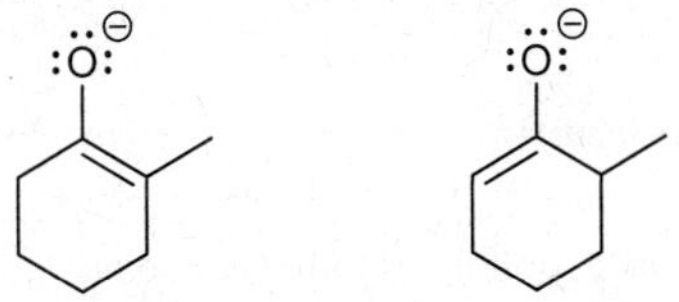

22.6.
(a) This compound has three α positions:

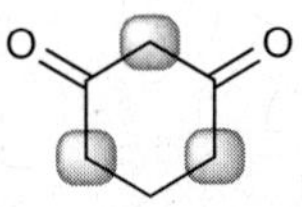

Among these positions, the central position is α to both carbonyl groups, and therefore, deprotonation occurs at this location. The resulting anion is a doubly stabilized enolate ion, which is particularly stable:

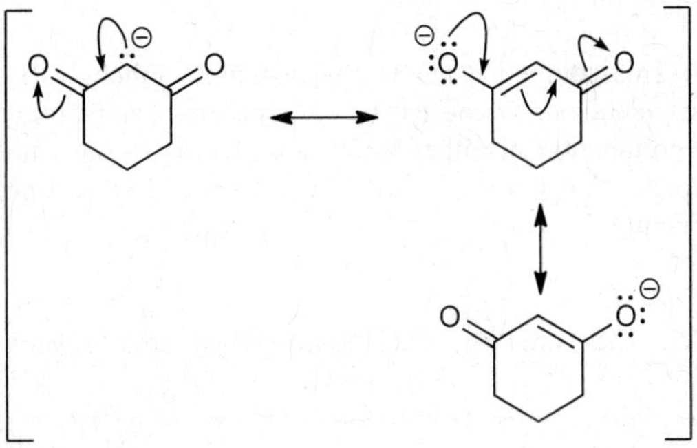

If ethoxide is used as the base to form the enolate, then enolate formation can be treated as nearly complete. That is, there will not be a substantial amount of diketone present after the equilibrium has been established.

(b) This compound has four α positions:

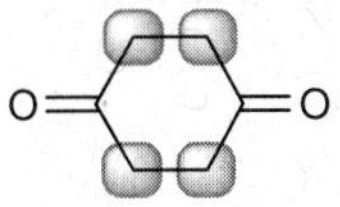

All four of these positions are identical because of symmetry (the structure has been rotated to make the symmetry more apparent). Therefore, deprotonation at any one of these positions results in the same enolate, which has the following two resonance structures:

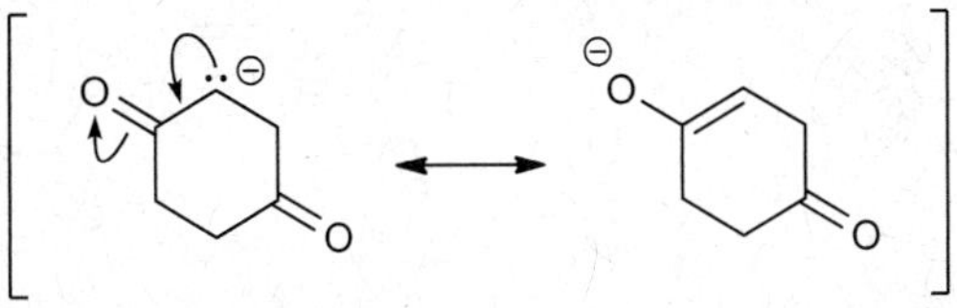

Notice that the negative charge is delocalized over only one oxygen atom (not two). As such, deprotonation of the diketone with ethoxide will result in a mixture containing both the enolate and the starting diketone. That is, there will be a substantial amount of diketone present after the equilibrium has been established.

(c) This compound has two α positions, although only one of these positions bears protons. Deprotonation at that location will lead to an enolate ion with the following resonance structures:

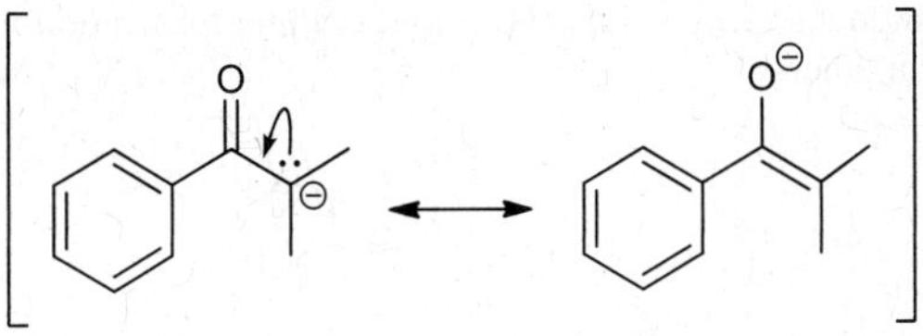

Notice that the negative charge is delocalized over only one oxygen atom. As such, deprotonation of the ketone with ethoxide will result in a mixture containing both the enolate and the starting ketone. That is, there will be a substantial amount of ketone present after the equilibrium has been established.

(d) This compound has two α positions, although they are identical because the ketone is symmetrical. Deprotonation at either location will lead to the same enolate ion, which has the following resonance structures:

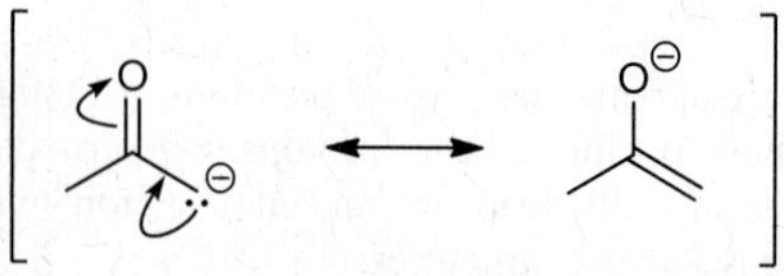

Notice that the negative charge is delocalized over only one oxygen atom. As such, deprotonation of the ketone with ethoxide will result in a mixture containing both the enolate and the starting ketone. That is, there will be a substantial amount of ketone present after the equilibrium has been established.

22.7. There are three α positions that bear protons, shown here:

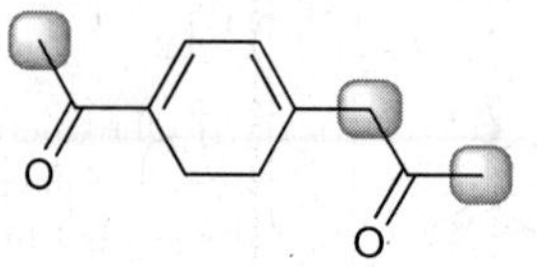

Deprotonation at any one of these three locations will result in a resonance-stabilized conjugate base. The most acidic proton is the one that leads to the most stable conjugate base, shown below. Notice that the negative charge is spread over two oxygen atoms (just like a doubly stabilized enolate), as well as three carbon atoms. This anion is therefore more stable than an ethoxide ion, so the equilibrium will favor formation of this resonance-stabilized anion:

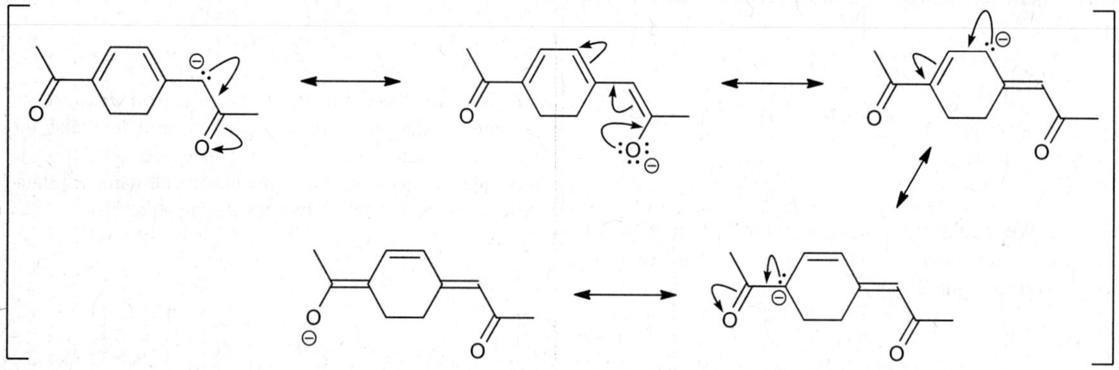

22.8.
(a) 2,4-Dimethyl-3,5-heptanedione is more acidic because its conjugate base is a doubly stabilized enolate. The other compound (4,4-dimethyl-3,5-heptanedione) cannot form a doubly stabilized enolate because there are no protons connected to the carbon atom that is in between both carbonyl groups.

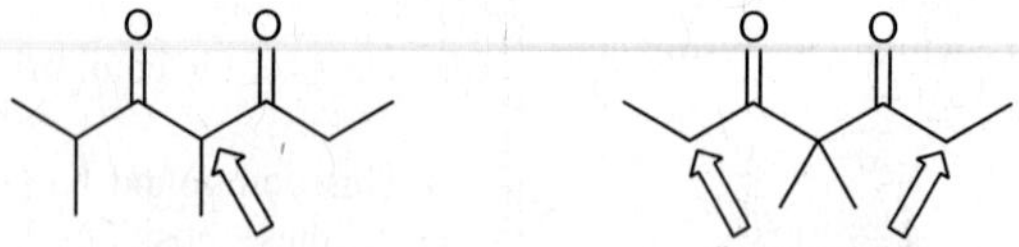

(more acidic)   (less acidic)

(b) 1,3-Cyclopentanedione is more acidic because its conjugate base is a doubly stabilized enolate. The other compound (1,2-cyclopentanedione) cannot form a doubly stabilized enolate because the carbonyl groups are adjacent to each other.

(more acidic)   (less acidic)

(c) Acetophenone is more acidic than benzaldehyde because the former has α protons and the latter does not.

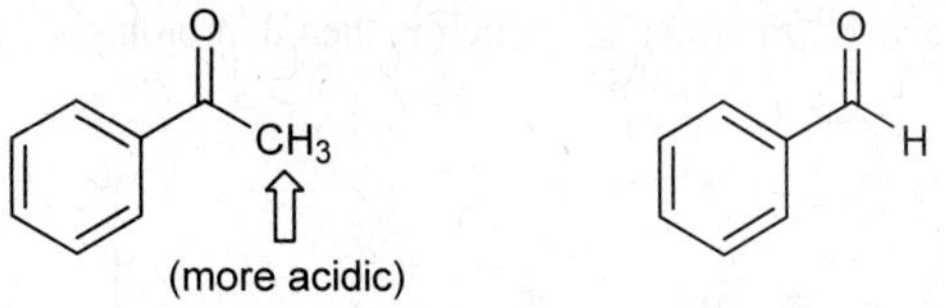

(more acidic)

22.9.

(a) These reagents indicate bromination at the α position, followed by elimination to give an α,β-unsaturated ketone:

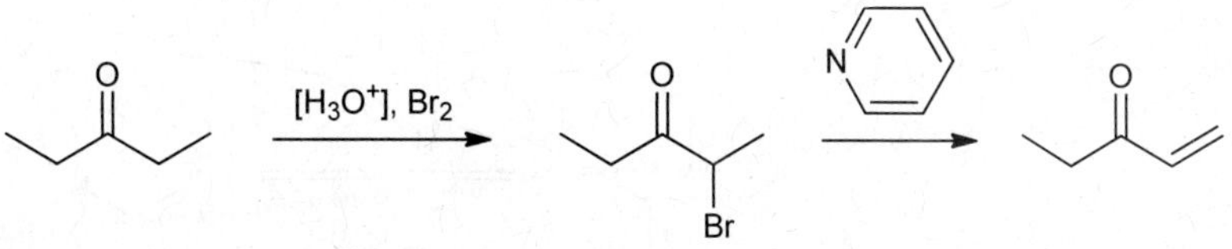

Below is a mechanism accounting for the entire transformation. Under acid-catalyzed conditions, the carbonyl group is protonated, giving a resonance-stabilized intermediate, which can then be deprotonated to give an enol. There is only a small amount of enol present at equilibrium, but its steady presence is responsible for the bromination process (as the enol is consumed by reacting with Br$_2$, the equilibrium is adjusted to replenish the small concentration of enol). The enol is a nucleophile and can attack molecular bromine (Br$_2$) to give an intermediate which is then deprotonated by water to give the product of the first reaction.

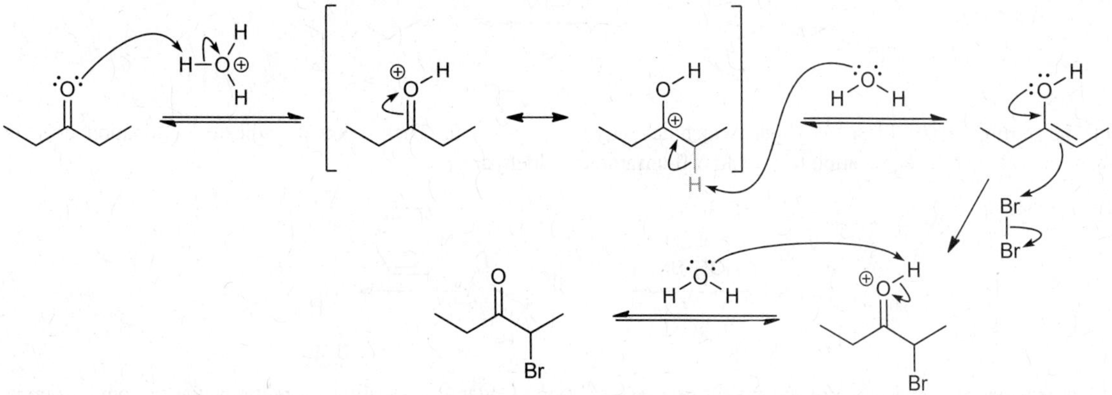

When this α-bromoketone is subsequently treated with pyridine, an E2 reaction gives the product:

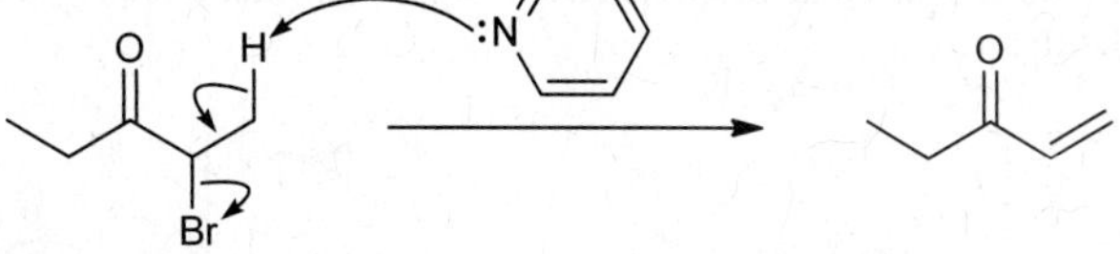

(b) These reagents indicate bromination at the more substituted α position, followed by elimination to give an α,β-unsaturated ketone:

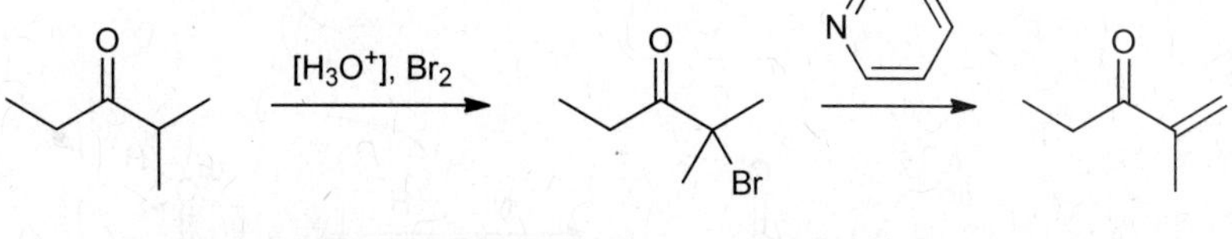

Below is a mechanism accounting for the entire transformation. Under acid-catalyzed conditions, the carbonyl group is protonated, giving a resonance-stabilized intermediate, which can then be deprotonated to give an enol (the more substituted enol is favored over the less substituted enol at equilibrium). There is only a small amount of enol present at equilibrium, but its steady presence is responsible for the bromination process (as the enol is consumed by reacting with Br$_2$, the equilibrium is adjusted to replenish the small concentration of enol). The enol is a nucleophile and can attack molecular bromine (Br$_2$) to give an intermediate which is then deprotonated by water to give the product of the first reaction.

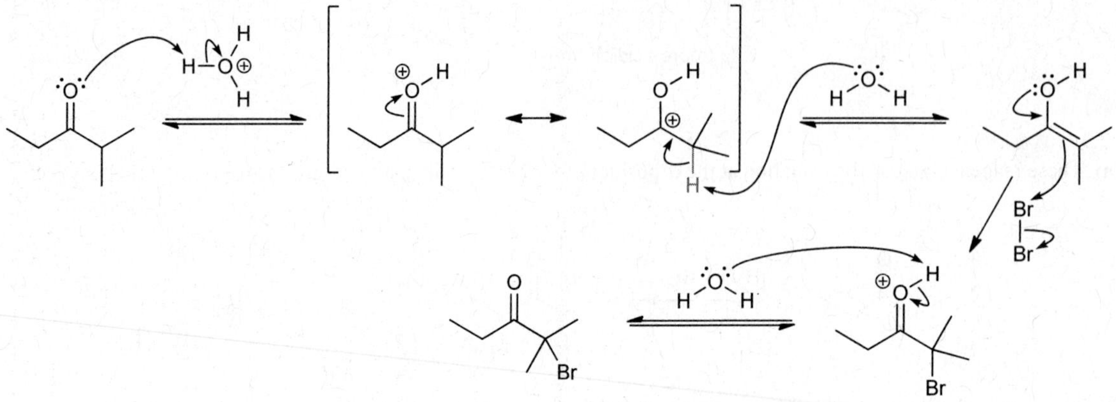

When this α-bromoketone is subsequently treated with pyridine, an E2 reaction gives the product:

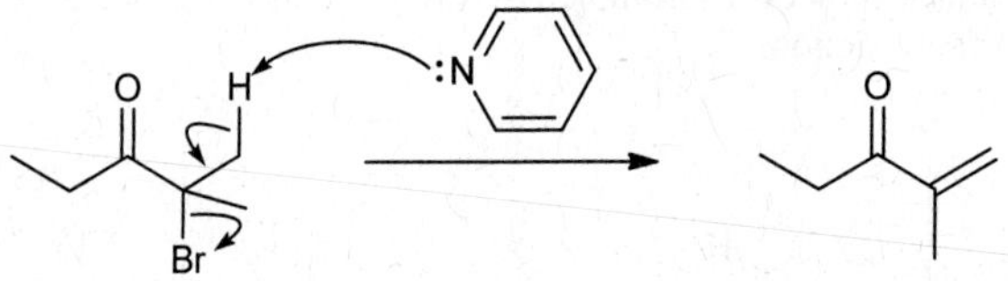

(c) The starting material is an aldehyde, which has only one α position. These reagents indicate bromination at the α position, followed by elimination to give an α,β-unsaturated aldehyde:

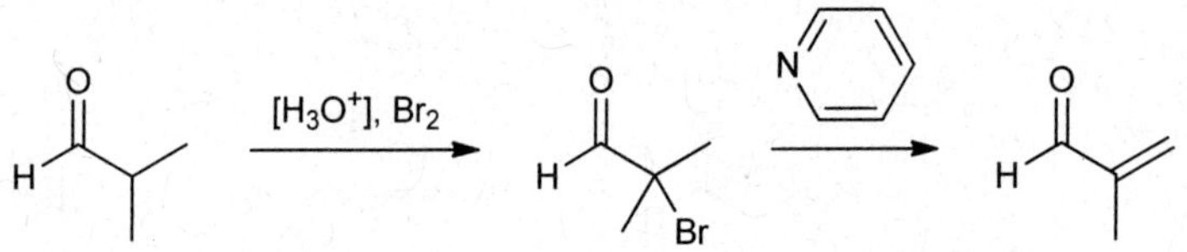

Below is a mechanism accounting for the entire transformation. Under acid-catalyzed conditions, the carbonyl group is protonated, giving a resonance-stabilized intermediate, which can then be deprotonated to give an enol. There is only a small amount of enol present at equilibrium, but its steady presence is responsible for the bromination process (as the enol is consumed by reacting with Br$_2$, the equilibrium is adjusted to replenish the small concentration of enol). The enol is a nucleophile and can attack molecular bromine (Br$_2$) to give an intermediate which is then deprotonated by water to give the product of the first reaction.

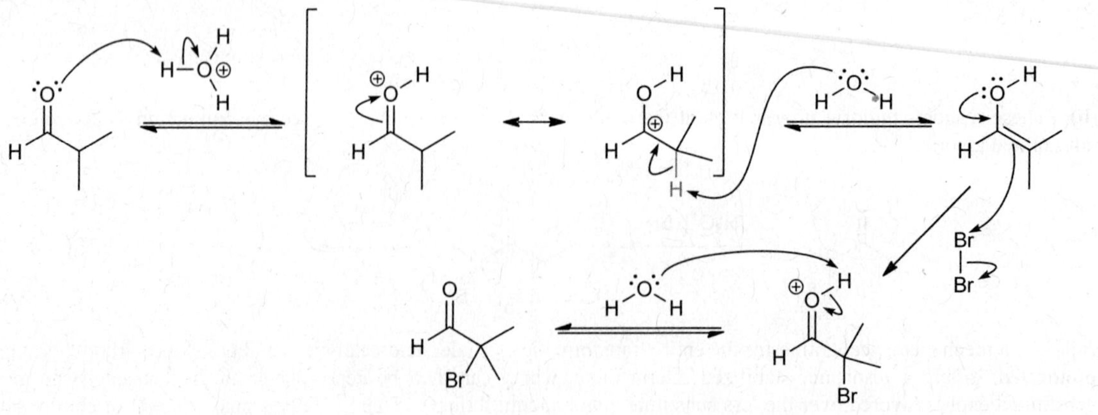

When this α-bromoaldehyde is subsequently treated with pyridine, an E2 reaction gives the product:

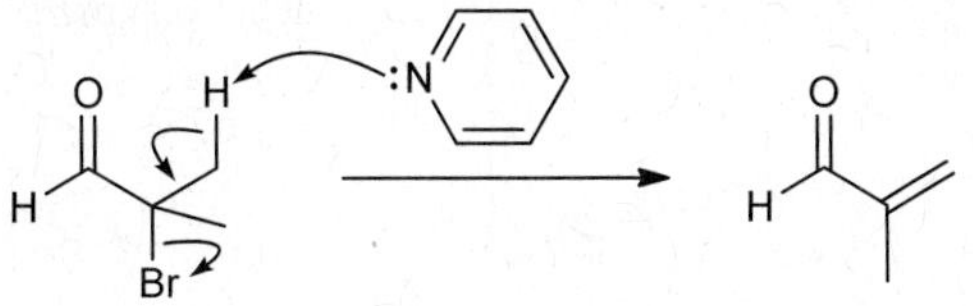

22.10.

(a) The product is an α,β-unsaturated ketone, which can be prepared from the corresponding saturated ketone (via acid-catalyzed halogenation followed by elimination):

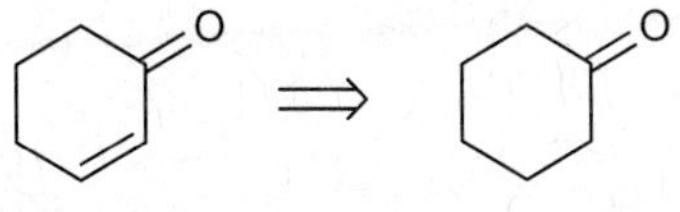

The saturated ketone can be prepared from the starting secondary alcohol via oxidation. The complete synthesis is shown here. The first step employs chromic acid as the oxidizing agent. Alternatively, PCC can be used to affect the same transformation.

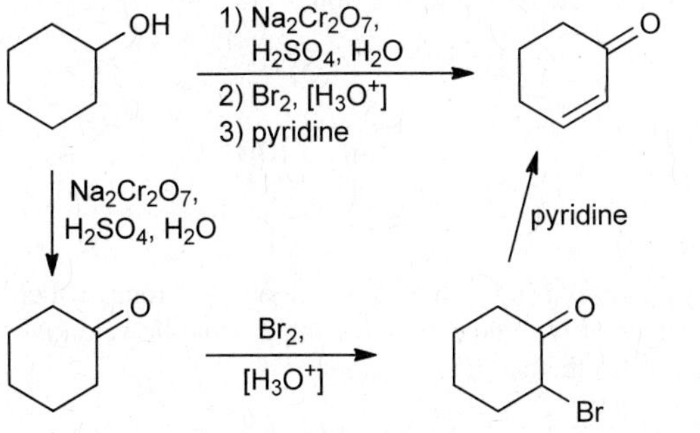

(b) The product is an α,β-unsaturated aldehyde, which can be prepared from the corresponding saturated aldehyde (via acid-catalyzed halogenation followed by elimination):

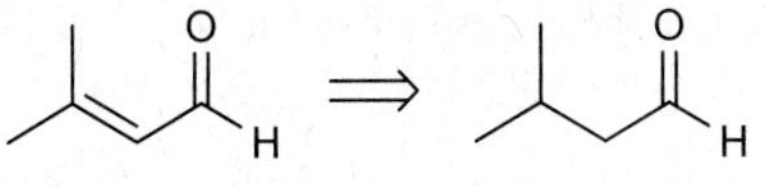

The saturated aldehyde can be prepared from the starting primary alcohol via oxidation with PCC.

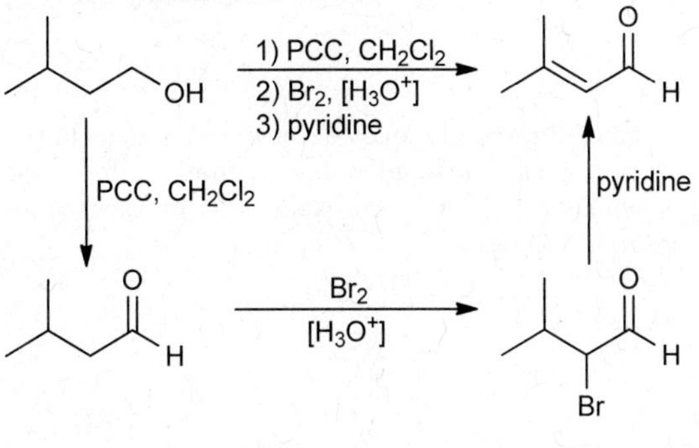

22.11.

(a) The starting material is a carboxylic acid, and the reagents indicate a Hell-Volhard-Zelinsky reaction. This process installs a bromine atom at the α position, as shown:

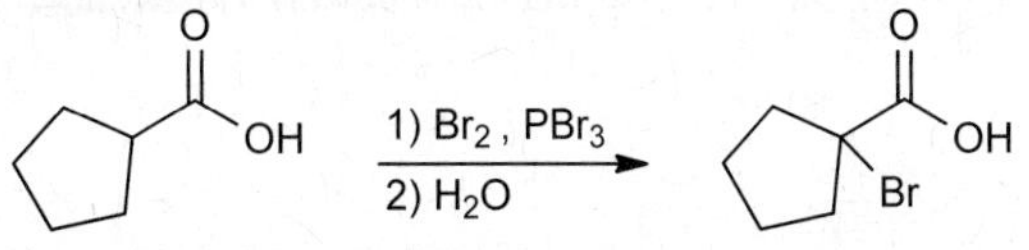

(b) The starting material is a carboxylic acid, and the reagents indicate a Hell-Volhard-Zelinsky reaction. This process installs a bromine atom at the α position, as shown:

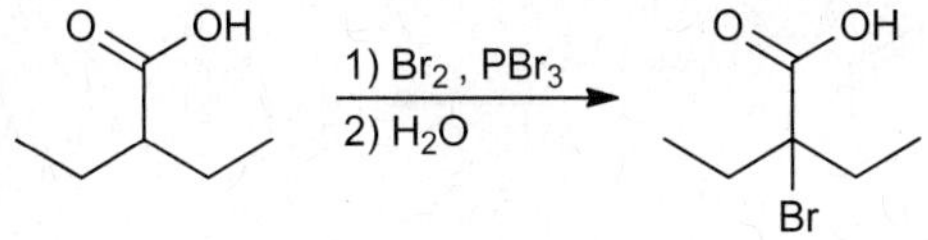

22.12.

(a) There are certainly many acceptable solutions to this problem. One such solution derives from the following retrosynthetic analysis. An explanation of each of the steps (*a-c*) follows.

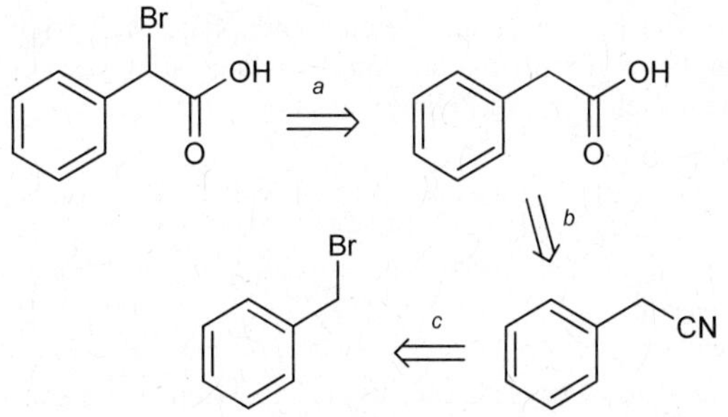

a. The product can be made from the corresponding carboxylic acid via bromination at the α position.

b. The carboxylic acid can be prepared via hydrolysis of the corresponding nitrile.

c. The nitrile can be made from the starting material (benzyl bromide) via an S$_N$2 process in which cyanide is used as a nucleophile.

Now let's draw the forward scheme. Benzyl bromide is treated with sodium cyanide, giving an S$_N$2 reaction that results in formation of a nitrile. Upon treatment with aqueous acid, the nitrile is hydrolyzed to give a carboxylic acid. Bromination at the α position then gives the product, as shown.

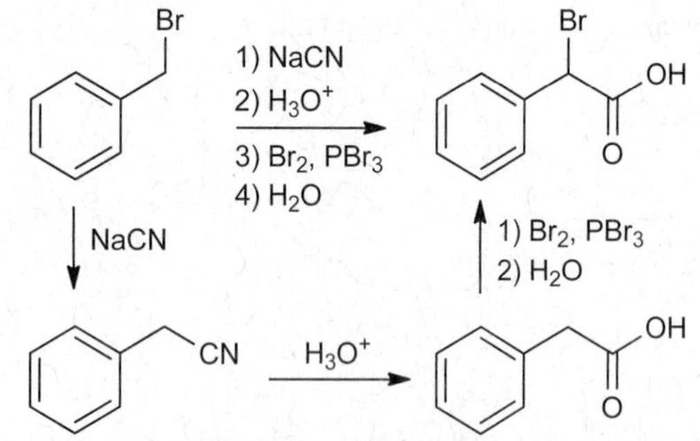

Alternatively, the carboxylic acid can be prepared via a Grignard reaction between benzyl magnesium bromide and carbon dioxide, followed by acid work-up, as shown:

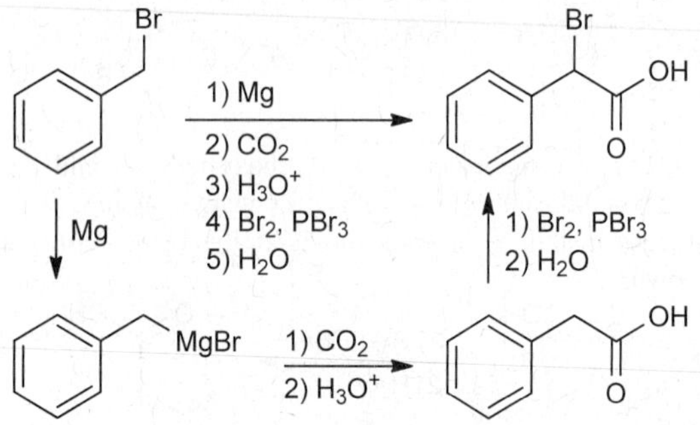

This alternate synthesis demonstrates an important point that has been stressed several times throughout this solutions manual. Specifically, synthesis problems will generally have multiple correct solutions. There is rarely only one correct solution to a synthesis problem.

(b) The product is an α-bromo carboxylic acid, which can be prepared from the corresponding carboxylic acid (via a Hell-Volhard-Zelinsky reaction):

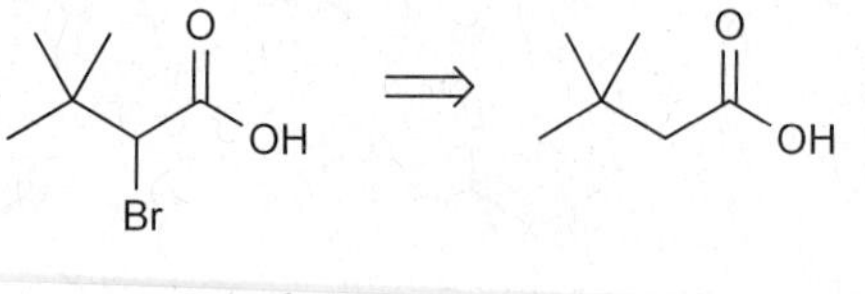

This carboxylic acid can be prepared from the starting primary alcohol via oxidation with chromic acid:

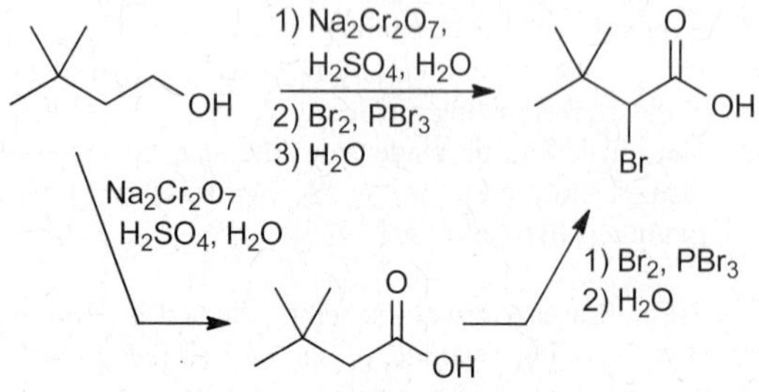

(c) The product is an α-bromo carboxylic acid, which can be prepared from the corresponding carboxylic acid (via a Hell-Volhard-Zelinsky reaction):

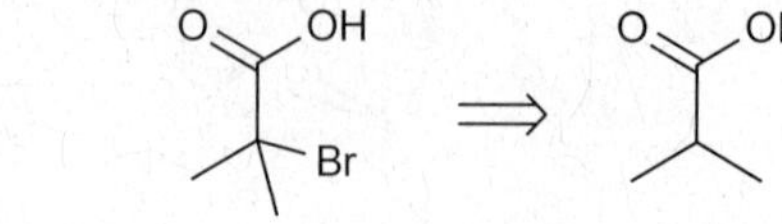

This carboxylic acid can be prepared from the starting nitrile via hydrolysis with aqueous acid, as shown:

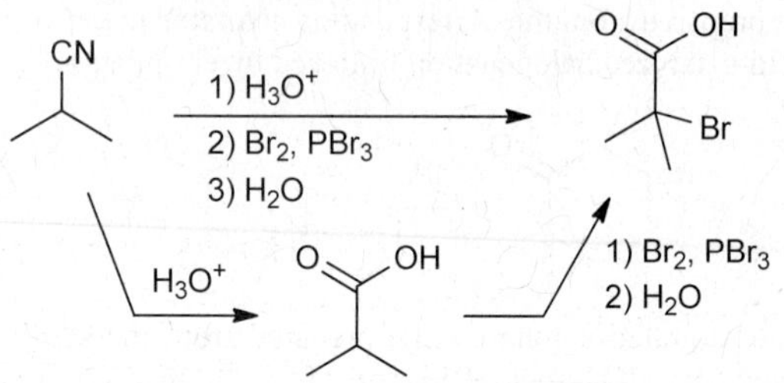

22.13.
(a) The starting material is a methyl ketone, which is converted into the corresponding carboxylic acid (shown here) via the haloform reaction:

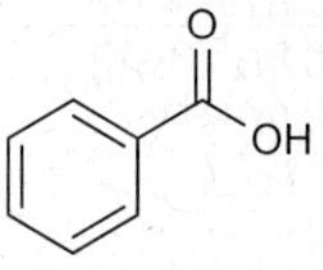

(b) The starting material is a methyl ketone, which is converted into the corresponding carboxylic acid (shown here) via the haloform reaction:

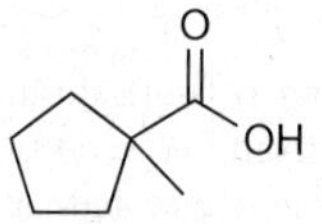

(c) The starting material is a methyl ketone, which is converted into the corresponding carboxylic acid (shown here) via the haloform reaction:

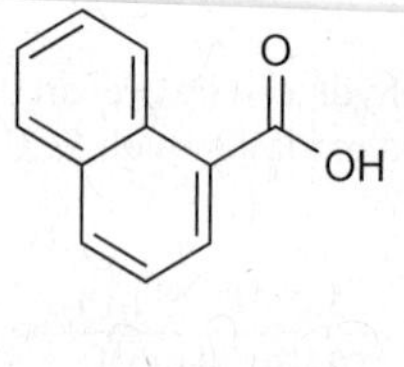

22.14.
(a) There are certainly many acceptable solutions to this problem. One such solution derives from the following retrosynthetic analysis. An explanation of each of the steps (*a-c*) follows.

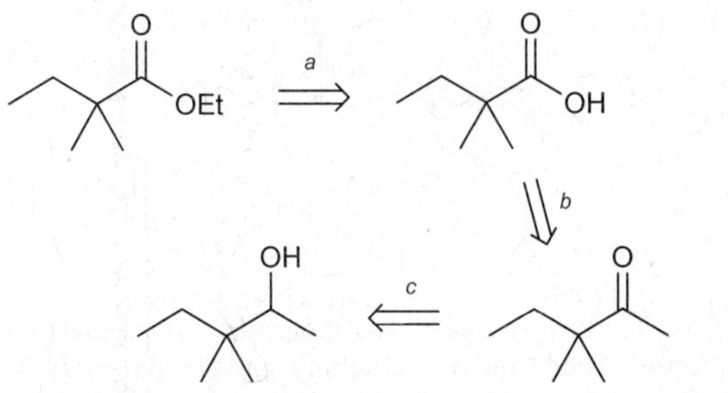

a. The product is an ester, so it can be made from the corresponding carboxylic acid via a Fischer esterification.
b. The carboxylic acid can be prepared from the corresponding methyl ketone via a haloform reaction.
c. The ketone can be made via oxidation of the corresponding secondary alcohol.

Now let's draw the forward scheme. The starting alcohol is oxidized upon treatment with chromic acid (alternatively, PCC can be used for this step). The resulting ketone is then treated with molecular bromine (Br_2) and sodium hydroxide, followed by aqueous acid, to give a carboxylic acid (via a haloform reaction). Finally, the carboxylic acid is treated with ethanol in the presence of an acid catalyst, giving the desired ester (via a Fischer esterification).

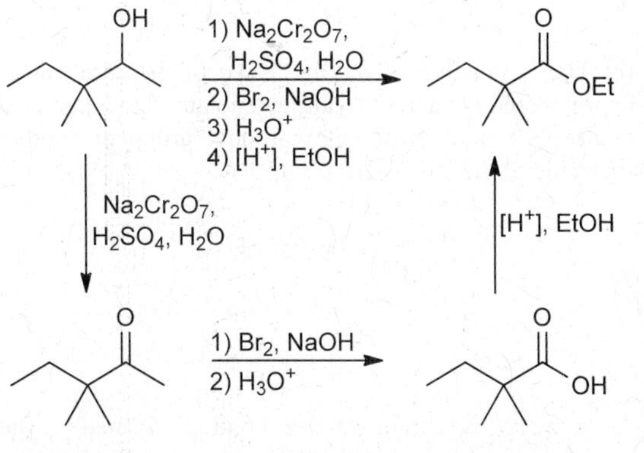

(b) There are certainly many acceptable solutions to this problem. One such solution derives from the following retrosynthetic analysis. An explanation of each of the steps (a-c) follows.

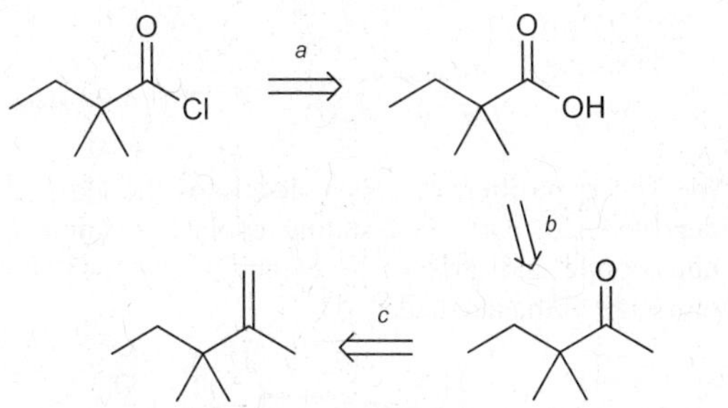

a. The product is an acid chloride which can be made from the corresponding carboxylic acid.

b. The carboxylic acid can be prepared from the corresponding methyl ketone via a haloform reaction.
c. The ketone can be made from the starting alkene via ozonolysis.

Now let's draw the forward scheme. Ozonolysis converts the starting alkene into a ketone (with loss of a carbon atom). The resulting ketone is then treated with molecular bromine (Br_2) and sodium hydroxide, followed by aqueous acid, to give a carboxylic acid (via a haloform reaction). Finally, the carboxylic acid is converted into an acid chloride upon treatment with thionyl chloride.

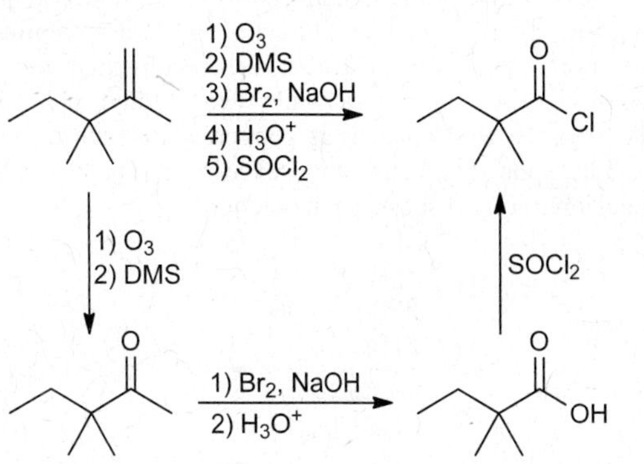

(c) The starting material is an acetal. Upon treatment with aqueous acid, the acetal is converted to a ketone, which can then be converted into the desired carboxylic acid via a haloform reaction.

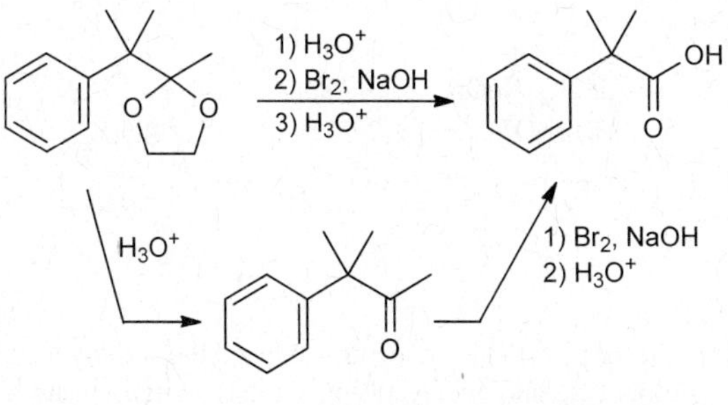

(d) There are certainly many acceptable solutions to this problem. One such solution derives from the following retrosynthetic analysis. An explanation of each of the steps (a-d) follows.

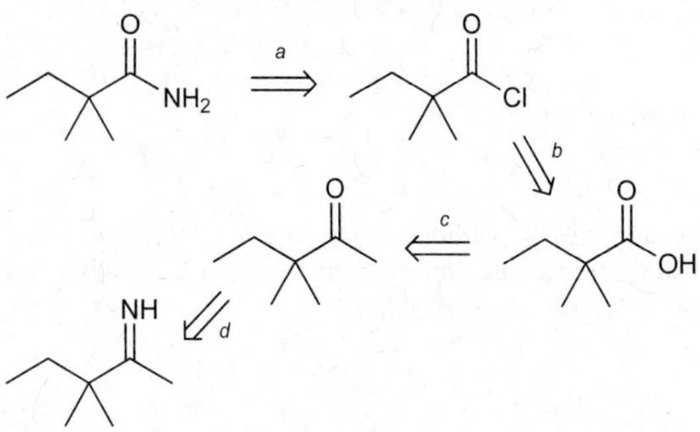

a. The product is an amide, which can be made from the corresponding acid chloride.

b. The acid chloride can be made from the corresponding carboxylic acid upon treatment with thionyl chloride.

c. The carboxylic acid can be prepared from the corresponding methyl ketone via a haloform reaction.

d. The ketone can be made via hydrolysis of the starting imine.

Now let's draw the forward scheme. The starting imine is hydrolyzed upon treatment with aqueous acid to give a ketone. The ketone is then treated with molecular bromine (Br$_2$) and sodium hydroxide, followed by aqueous acid, to give a carboxylic acid (via a haloform reaction). The carboxylic acid is then converted into an acid chloride upon treatment with thionyl chloride. Finally, the acid chloride is converted into the desired amide upon treatment with excess ammonia (via a nucleophilic acyl substitution reaction).

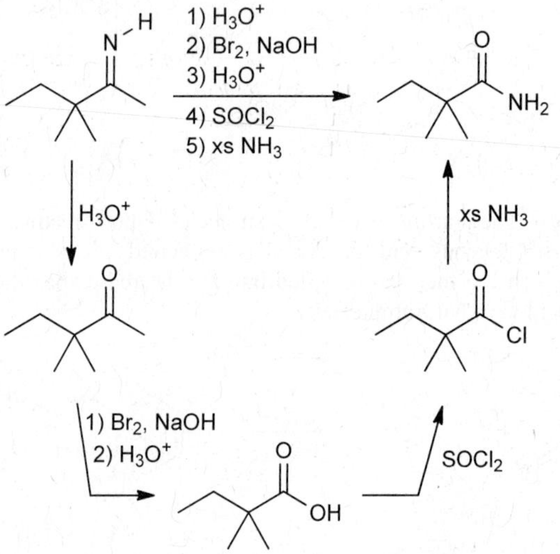

22.15.
(a) The α position of one molecule of the aldehyde is deprotonated, and the resulting enolate functions as a nucleophile and attacks the carbonyl group of another molecule of the aldehyde:

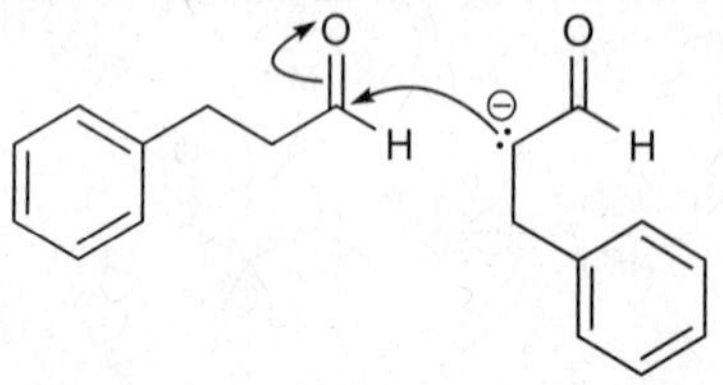

As a result, a carbon-carbon bond is formed. The resulting alkoxide ion is then protonated to give the following β-hydroxy aldehyde:

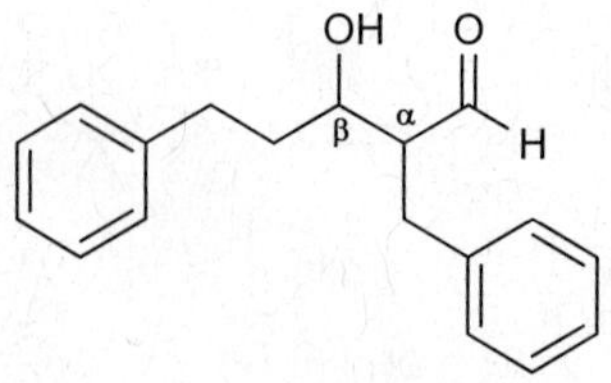

(b) The α position of one molecule of the aldehyde is deprotonated, and the resulting enolate functions as a nucleophile and attacks the carbonyl group of another molecule of the aldehyde:

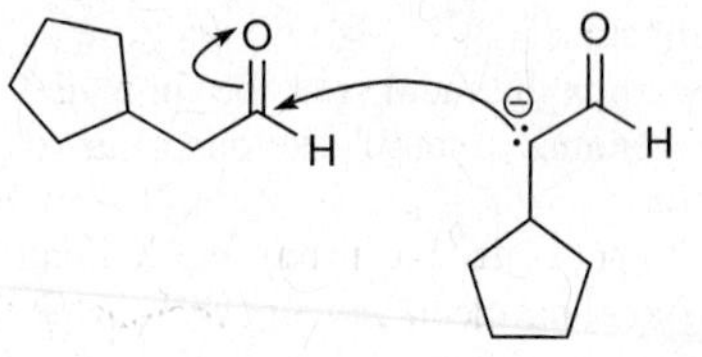

As a result, a carbon-carbon bond is formed. The resulting alkoxide ion is then protonated to give the following β-hydroxy aldehyde:

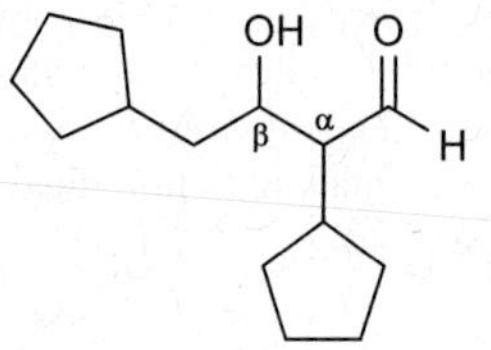

(c) The α position of one molecule of the aldehyde is deprotonated, and the resulting enolate functions as a nucleophile and attacks the carbonyl group of another molecule of the aldehyde:

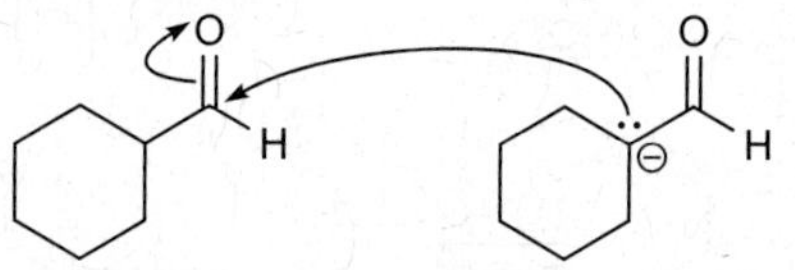

As a result, a carbon-carbon bond is formed. The resulting alkoxide ion is then protonated to give the following β-hydroxy aldehyde:

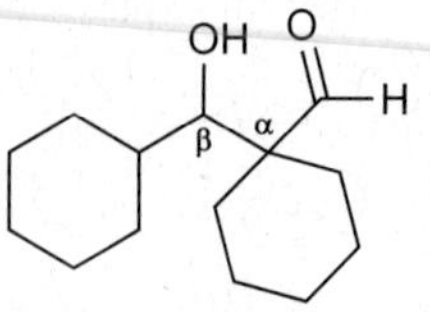

(d) The α position of one molecule of the aldehyde is deprotonated, and the resulting enolate functions as a nucleophile and attacks the carbonyl group of another molecule of the aldehyde:

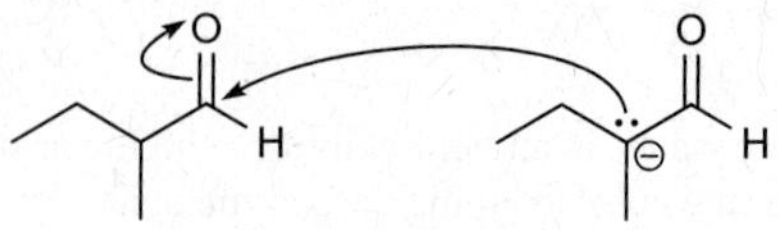

As a result, a carbon-carbon bond is formed. The resulting alkoxide ion is then protonated to give the following β-hydroxy aldehyde:

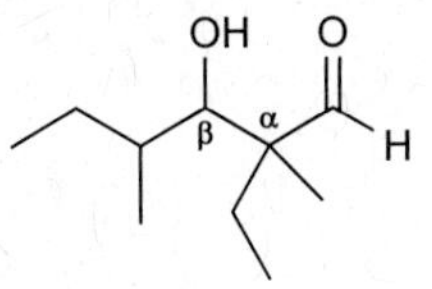

22.16.

(a) This compound has two α positions, although they are identical because the ketone is symmetrical. That is, deprotonation at either location will lead to the same enolate. This enolate can then function as a nucleophile and attack the carbonyl group of another molecule of the ketone. As a result, a carbon-carbon bond is formed. The resulting alkoxide ion is then protonated to give a β-hydroxy ketone, as shown:

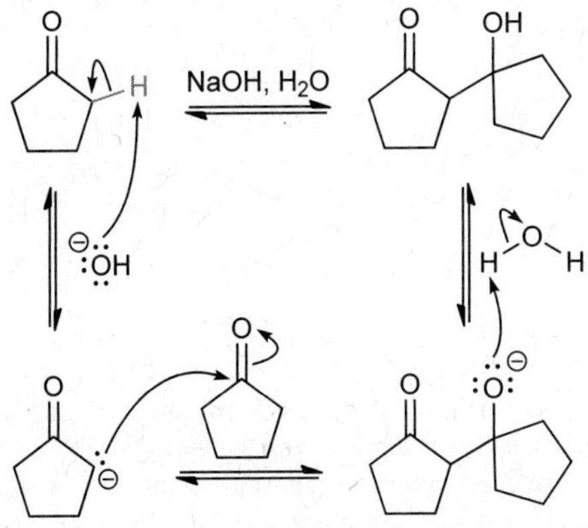

(b) This compound has two α positions, although only one of these positions bears protons. Deprotonation at that location will lead to an enolate ion that can function as a nucleophile and attack the carbonyl group of another molecule of the ketone. As a result, a carbon-carbon bond is formed. The resulting alkoxide ion is then protonated to give a β-hydroxy ketone, as shown:

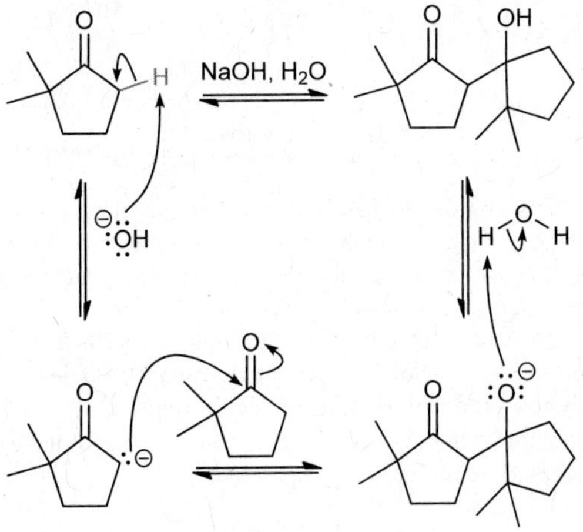

(c) This compound has two α positions, although they are identical because the ketone is symmetrical. That is, deprotonation at either location will lead to the same enolate. This enolate can then function as a nucleophile and attack the carbonyl group of another molecule of the ketone. As a result, a carbon-carbon bond is formed. The resulting alkoxide ion is then protonated to give a β-hydroxy ketone, as shown:

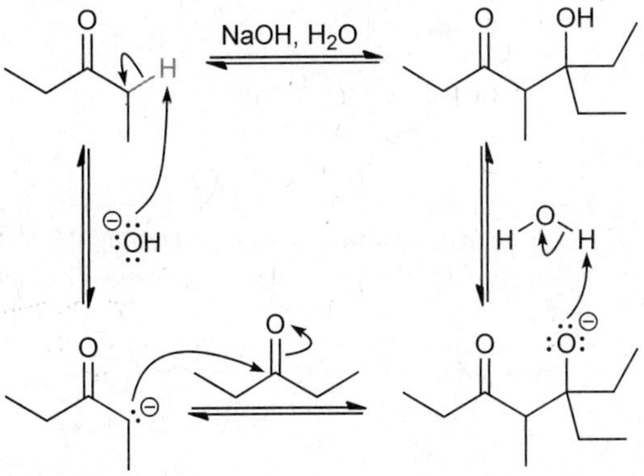

(d) This compound has two α positions, although only one of these positions bears protons. Deprotonation at that location will lead to an enolate ion that can function as a nucleophile and attack the carbonyl group of another molecule of the ketone. As a result, a carbon-carbon bond is formed. The resulting alkoxide ion is then protonated to give a β-hydroxy ketone, as shown:

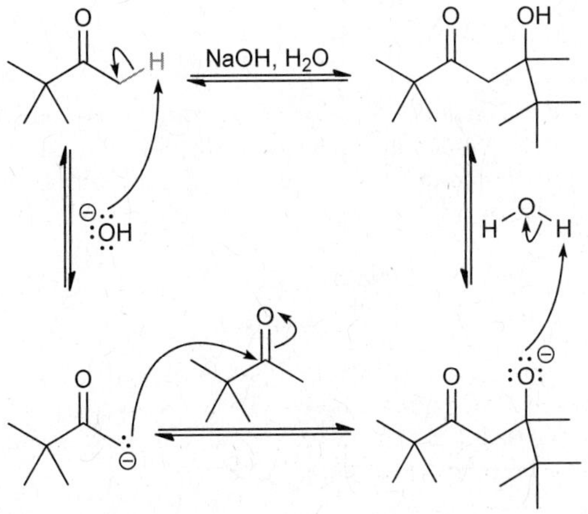

22.17. The first step of an aldol addition reaction is deprotonation at the α position, but this compound has no α protons.

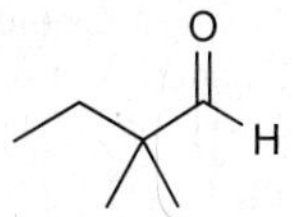

22.18. The problem statement indicates that oxidation of the alcohol with PCC gives an aldehyde, which indicates that the alcohol must be primary. There are only two primary alcohols with the molecular formula $C_4H_{10}O$:

Oxidation of the first compound (1-butanol) gives 1-butanal, which is expected to produce four signals in its ^{1}H NMR spectrum. In contrast, oxidation of the second compound (2-methyl-1-propanol) is expected to produce an aldehyde that exhibits only three signals in its ^{1}H NMR spectrum:

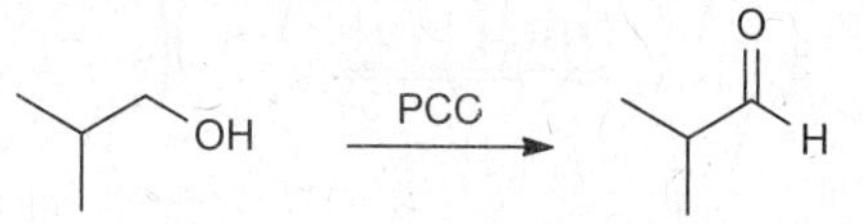

Upon treatment with aqueous sodium hydroxide, this aldehyde is deprotonated to give an enolate ion that can function as a nucleophile and attack the carbonyl group of another molecule of the aldehyde:

As a result, a carbon-carbon bond is formed. The resulting alkoxide ion is then protonated to give a β-hydroxy aldehyde, as shown:

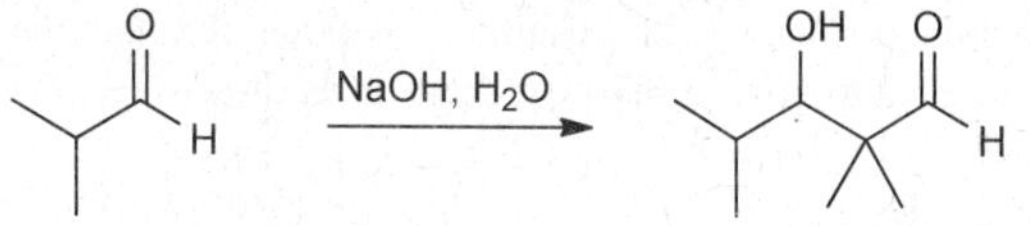

22.19. Treatment of acetaldehyde with aqueous sodium hydroxide results in a β-hydroxy aldehyde that can be converted into the desired diol via reduction with LAH, followed by water work-up, as shown:

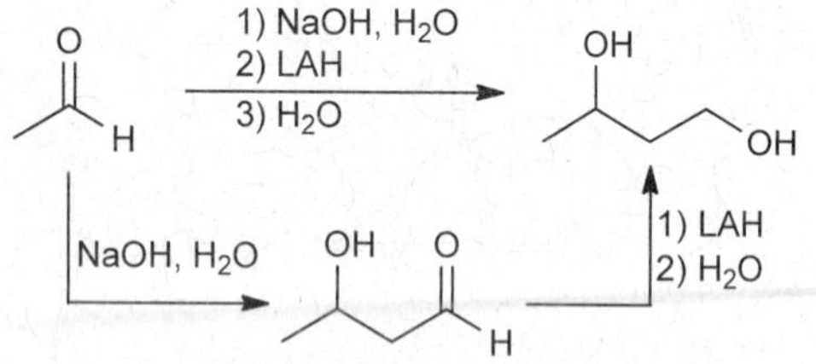

22.20.

(a) Two molecules of the aldehyde are redrawn such that two α protons of one molecule are directly facing the carbonyl group of another molecule (highlighted). When drawn in this way, it is easier to predict the product without having to draw the entire mechanism. We simply remove the two α protons and the oxygen atom, and we replace them with a double bond. In this case, two stereoisomers are possible, so we draw the product that is likely to have fewer steric interactions:

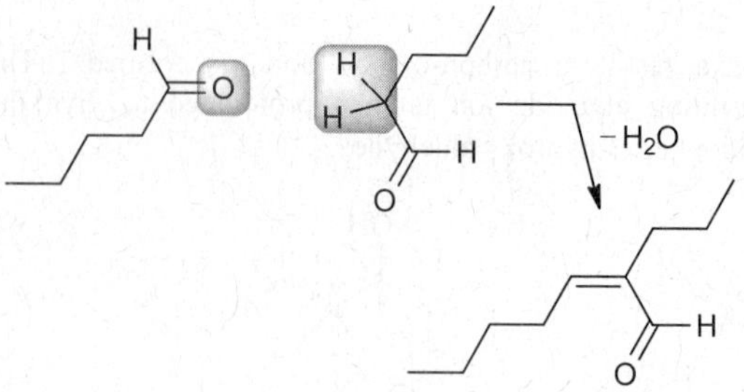

(b) Two molecules of the aldehyde are redrawn such that two α protons of one molecule are directly facing the carbonyl group of another molecule (highlighted). We then remove the two α protons and the oxygen atom, and we replace them with a double bond. In this case, two stereoisomers are possible, so we draw the product that is likely to have fewer steric interactions:

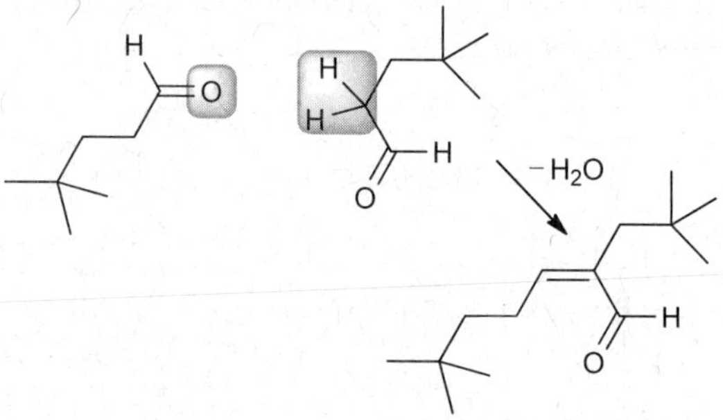

(c) Two molecules of the aldehyde are redrawn such that two α protons of one molecule are directly facing the carbonyl group of another molecule (highlighted). We then remove the two α protons and the oxygen atom, and we replace them with a double bond. In this case, two stereoisomers are possible, so we draw the product that is likely to have fewer steric interactions:

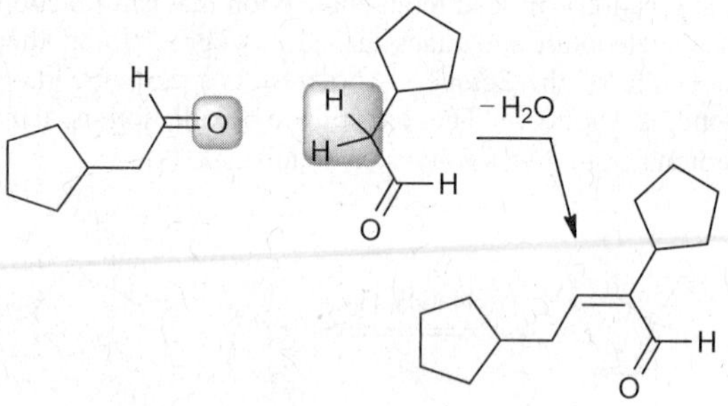

(d) This compound has two α positions, although they are identical because the ketone is symmetrical, so we only need to consider the reaction occurring at one of these locations.

Two molecules of the ketone are redrawn such that two α protons of one molecule are directly facing the carbonyl group of another molecule (highlighted). We then remove the two α protons and the oxygen atom, and we replace them with a double bond:

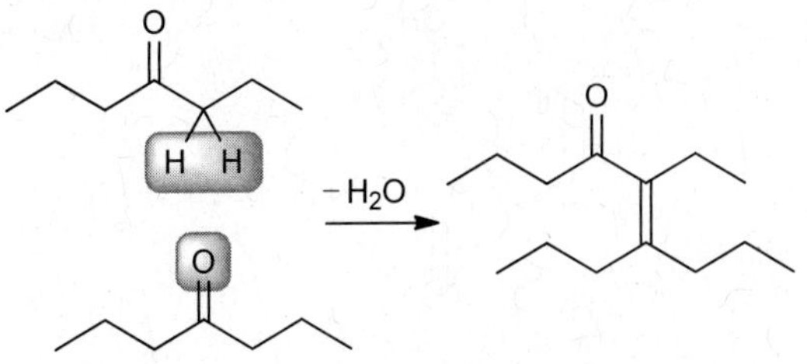

(e) This compound has two α positions, although they are identical because the ketone is symmetrical, so we only need to consider the reaction occurring at one of these locations.

Two molecules of the ketone are redrawn such that two α protons of one molecule are directly facing the carbonyl group of another molecule (highlighted). We then remove the two α protons and the oxygen atom, and we replace them with a double bond:

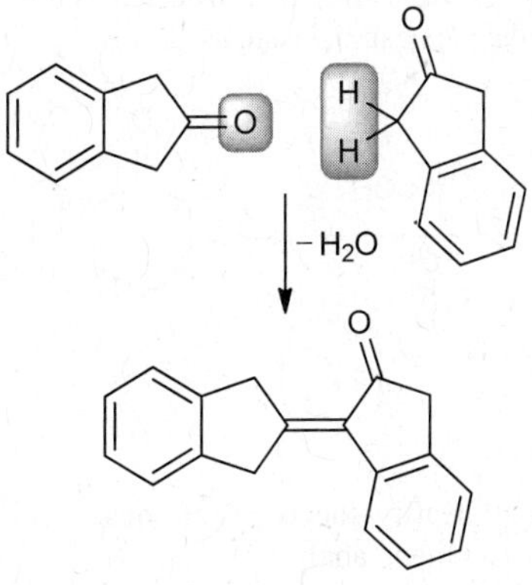

(f) This compound has two α positions, although they are identical because the ketone is symmetrical, so we only need to consider the reaction occurring at one of these locations.

Two molecules of the ketone are redrawn such that two α protons of one molecule are directly facing the carbonyl group of another molecule (highlighted). We then remove the two α protons and the oxygen atom, and we replace them with a double bond:

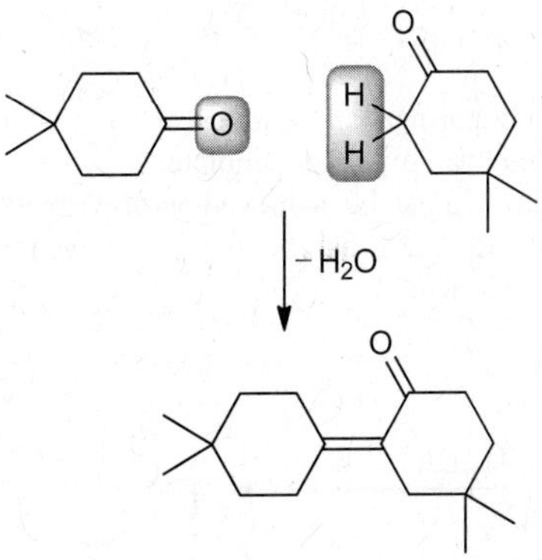

22.21.
(a) First identify the carbon-carbon bond (indicated with a wavy line below) that is formed as a result of an aldol

condensation. This double bond must have been formed via the loss of two hydrogen atoms and an oxygen atom. The β position of the condensation product must have originally been a carbonyl group in the starting material:

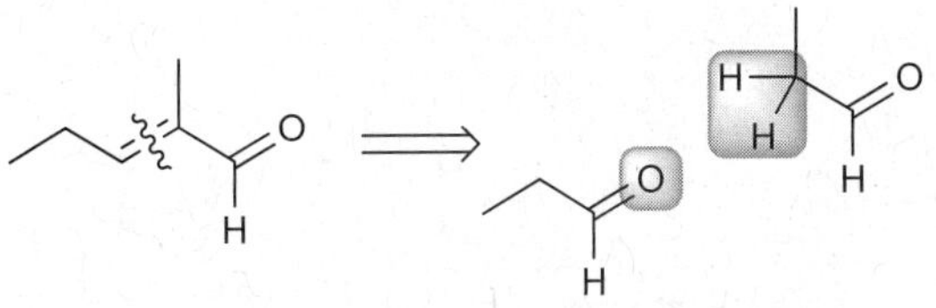

(b) First identify the carbon-carbon bond (indicated with a wavy line below) that is formed as a result of an aldol condensation. This double bond must have been formed via the loss of two hydrogen atoms and an oxygen atom. The β position of the condensation product must have originally been a carbonyl group in the starting material:

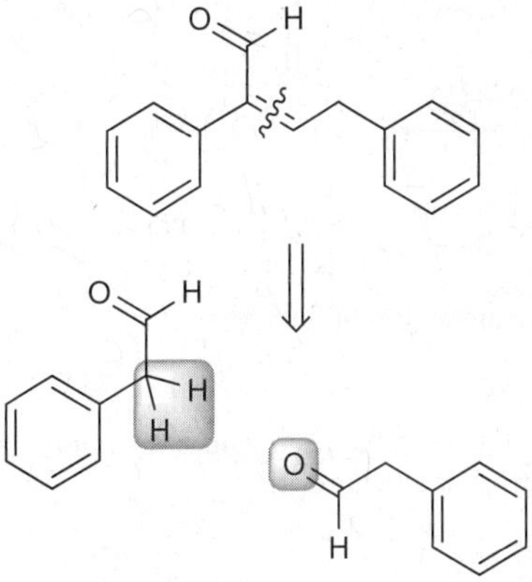

(c) First identify the carbon-carbon bond (indicated with a wavy line below) that is formed as a result of an aldol condensation. This double bond must have been formed via the loss of two hydrogen atoms and an oxygen atom. The β position of the condensation product must have originally been a carbonyl group in the starting material:

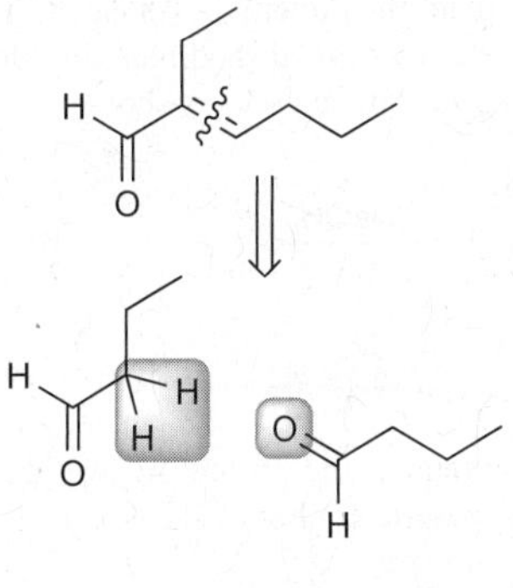

22.22.   We begin by considering the structure of 2-butanone, which has two α positions (highlighted):

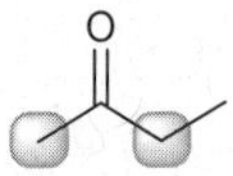

854   **CHAPTER 22**

These two positions are not identical, because the ketone is unsymmetrical. Therefore, an aldol reaction can occur at either of these locations. A reaction at the more-substituted α position gives the following two diastereomers:

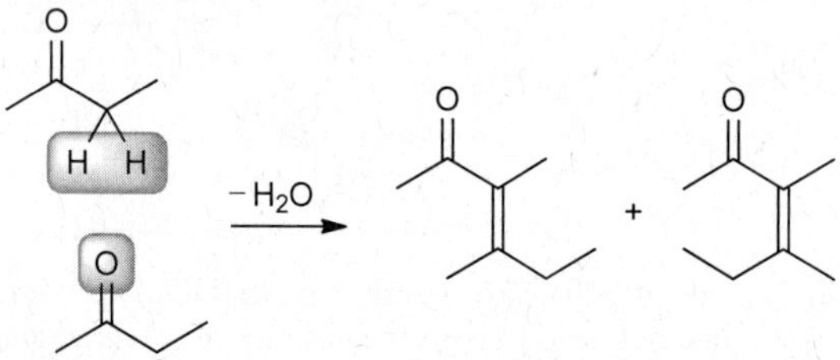

Similarly, a reaction at the less substituted α position also gives rise to two diastereomeric products:

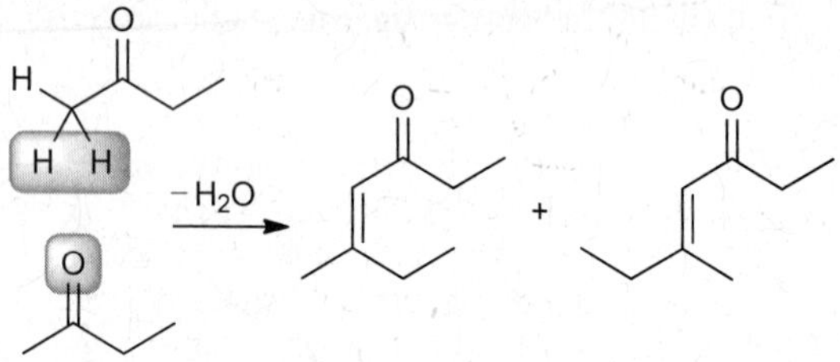

Thus, there are a total of four products.

22.23.
(a) We first identify the α and β positions, and then apply a retrosynthetic analysis:

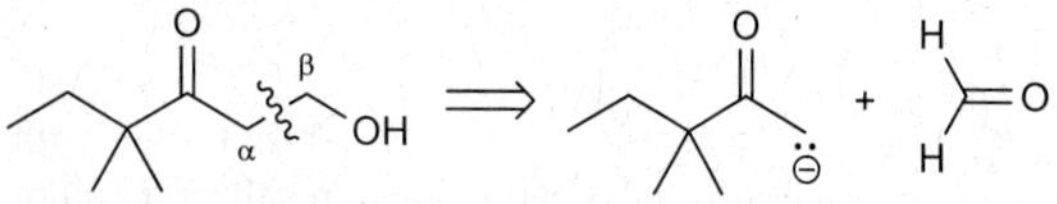

This transformation can be achieved with an aldol reaction between two different partners. One of the partners (the ketone) only has one α position that can be deprotonated (giving only one possible enolate), and the other partner (formaldehyde) lacks α protons and is more electrophilic than the ketone. As such, LDA is not required as a base for this directed aldol addition. Sodium hydroxide can be used, as shown:

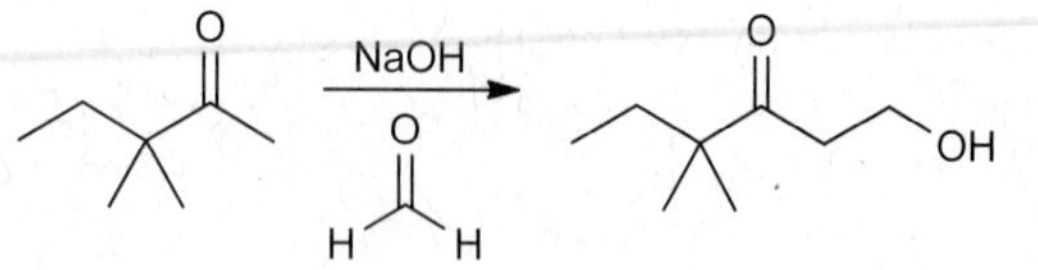

(b) We first identify the α and β positions, and then apply a retrosynthetic analysis:

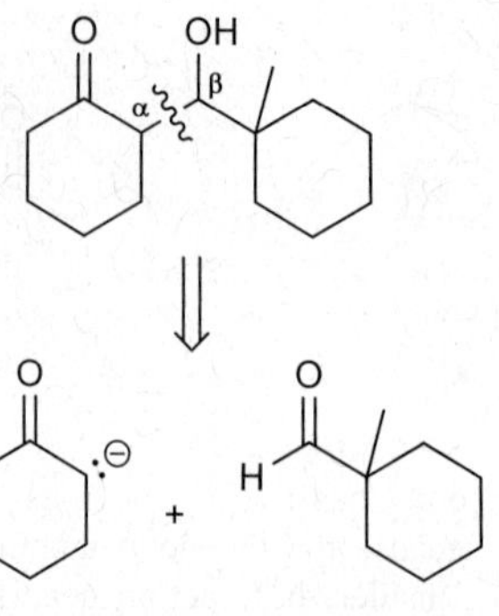

This transformation can be achieved with an aldol reaction between two different partners. One of the partners (the ketone) only has one α position that can be deprotonated (giving only one possible enolate), and the other partner (the aldehyde) lacks α protons and is more electrophilic than the ketone. As such, LDA is not required as a base for this directed aldol addition. Sodium hydroxide can be used, as shown:

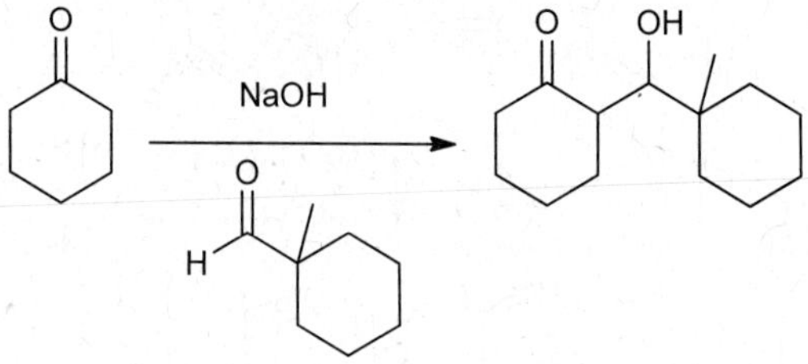

(c) We first identify the α and β positions, and then apply a retrosynthetic analysis:

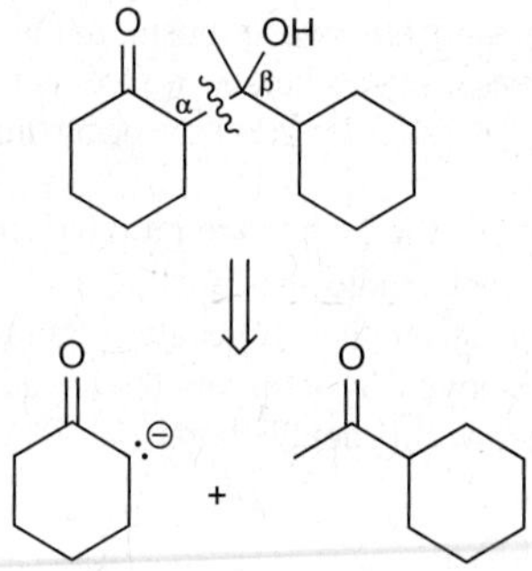

Two different partners are required, each of which can be deprotonated to give an enolate. Therefore, this transformation must be achieved with a directed aldol reaction, using LDA as the base, followed by water work-up, as shown:

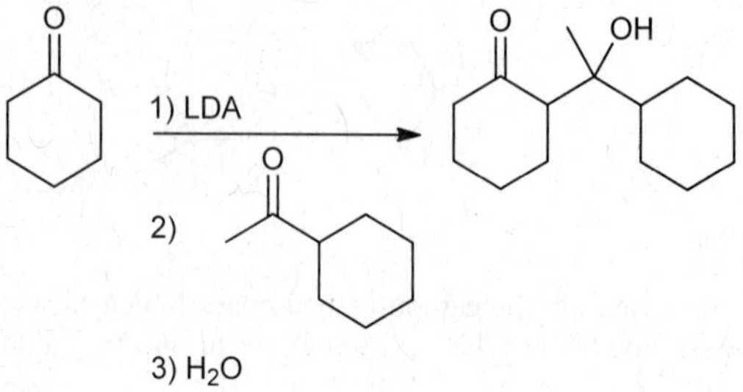

(d) We first identify the α and β positions, and then apply a retrosynthetic analysis to identify the starting materials:

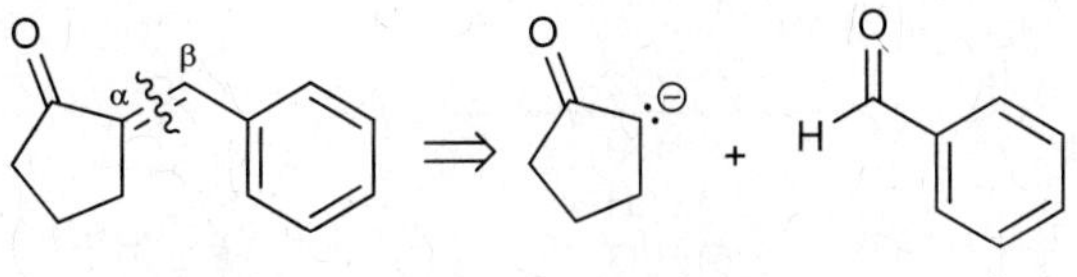

This transformation can be achieved with an aldol reaction between two different partners. One of the partners (the ketone) only has one α position that can be deprotonated (giving only one possible enolate), and the other partner (the aldehyde) lacks α protons and is more electrophilic than the ketone. As such, LDA is not required as a base for this directed aldol addition. Sodium hydroxide can be used, as shown:

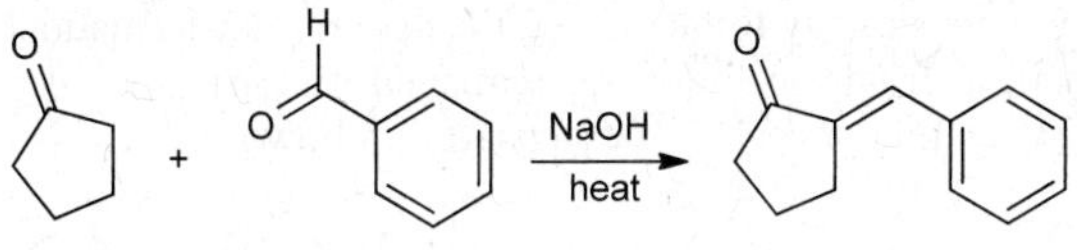

(e) We first identify the α and β positions, and then apply a retrosynthetic analysis:

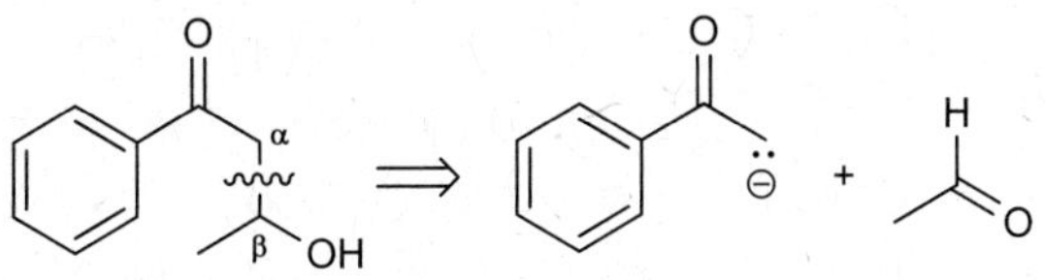

Two different partners are required, each of which can be deprotonated to give an enolate. Therefore, this transformation must be achieved with a directed aldol reaction, using LDA as the base, followed by water work-up, as shown:

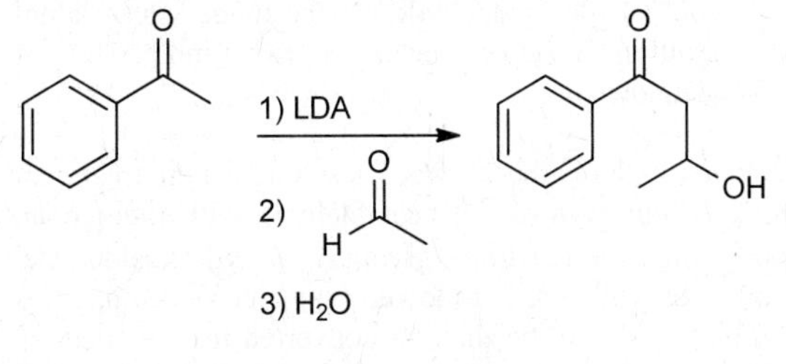

22.24.
(a) One strategy for achieving the desired transformation derives from the following retrosynthetic analysis. An explanation of each of the steps (*a-c*) follows.

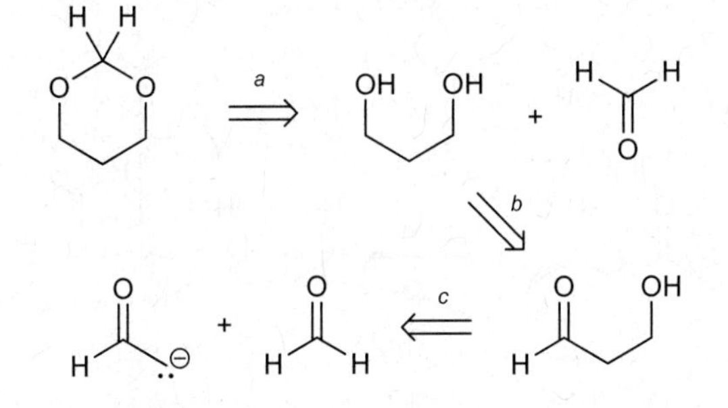

a. The product is a cyclic acetal, which can be prepared from a diol and formaldehyde.
b. The diol can be made via reduction of a β-hydroxyaldehyde.
c. The β-hydroxyaldehyde can be made via a directed aldol addition reaction between acetaldehyde and formaldehyde.

Now let's draw the forward scheme. We begin the synthesis with a directed aldol addition because the reaction involves two different partners (acetaldehyde and formaldehyde), which would otherwise give two possible addition products. That is, the enolate of acetaldehyde would be able to attack either acetaldehyde or formaldehyde, if both were present. It is true that formaldehyde is a better electrophile than acetaldehyde, so the desired product would be expected to predominate. Nevertheless, the minor product can be avoided altogether by using a directed aldol addition reaction. LDA is used as the base to deprotonate acetaldehyde irreversibly, with enolate formation proceeding to completion. This enolate can then be treated with formaldehyde, followed by water work-up, to give the aldol addition product (a β-hydroxyaldehyde). Reduction with LAH, followed by water work-up, gives a diol, which can be then be converted into the desired acetal upon treatment with formaldehyde in acid-catalyzed conditions (with removal of water).

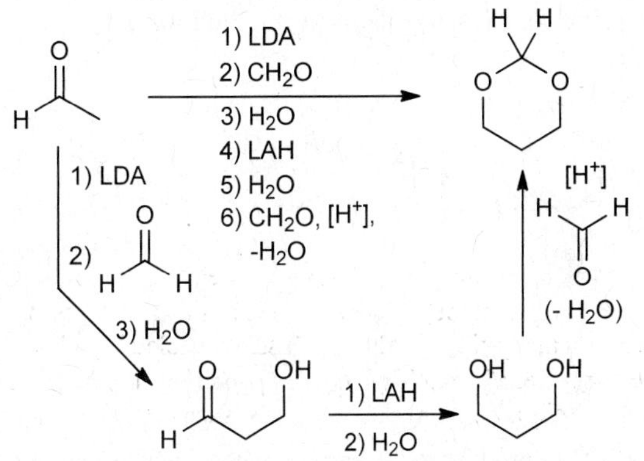

(b) One strategy for achieving the desired transformation derives from the following retrosynthetic analysis. An explanation of each of the steps (*a-c*) follows.

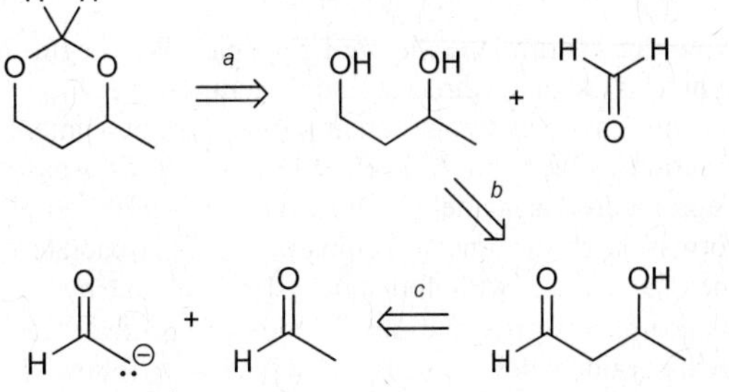

a. The product is a cyclic acetal, which can be prepared from a diol and formaldehyde.

b. The diol can be made via reduction of a β-hydroxyaldehyde.

c. The β-hydroxyaldehyde can be made via an aldol addition reaction between two molecules of acetaldehyde.

Now let's draw the forward scheme. Upon treatment with sodium hydroxide, acetaldehyde will undergo an aldol addition reaction, giving a β-hydroxyaldehyde. Reduction with LAH, followed by water work-up, gives a diol, which can be then be converted into the desired acetal upon treatment with formaldehyde in acid-catalyzed conditions (with removal of water).

(c) One strategy for achieving the desired transformation derives from the following retrosynthetic analysis. An explanation of each of the steps (*a-c*) follows.

a. The product is a cyclic acetal, which can be prepared from a diol and acetaldehyde.

b. The diol can be made via reduction of a β-hydroxyaldehyde.

c. The β-hydroxyaldehyde can be made via a directed aldol addition reaction between acetaldehyde and formaldehyde.

Now let's draw the forward scheme. We begin the synthesis with a directed aldol addition because the reaction involves two different partners (see comment in solution to Problem **22.24a**). LDA is used as a base to deprotonate acetaldehyde irreversibly, with enolate formation proceeding to completion. This enolate can then be treated with formaldehyde, followed by water work-up, to give the aldol addition product (a β-hydroxyaldehyde). Reduction with LAH, followed by water work-up, gives a diol, which can be then be

converted into the desired acetal upon treatment with acetaldehyde in acid-catalyzed conditions (with removal of water).

(d) One strategy for achieving the desired transformation derives from the following retrosynthetic analysis. An explanation of each of the steps (*a-c*) follows.

a. The product is a cyclic acetal, which can be prepared from a diol and acetaldehyde.

b. The diol can be made via reduction of a β-hydroxyaldehyde.

c. The β-hydroxyaldehyde can be made via an aldol addition reaction between two molecules of acetaldehyde.

Now let's draw the forward scheme. Upon treatment with sodium hydroxide, acetaldehyde will undergo an aldol addition reaction, giving a β-hydroxyaldehyde. Reduction with LAH, followed by water work-up, gives a diol, which can be then be converted into the desired acetal upon treatment with acetaldehyde in acid-catalyzed conditions (with removal of water).

22.25. This process is an intramolecular aldol condensation reaction. As such, we draw a mechanism with the same mechanistic steps found in Mechanism 22.6. First, hydroxide functions as a base and deprotonates the starting dione to give an enolate. This enolate ion is a nucleophilic center, and it will attack the carbonyl group present in the same structure (an intramolecular process), thereby closing a five-membered ring. The resulting alkoxide ion is then protonated to give a β-hydroxy ketone. Hydroxide then functions as a base again, deprotonating the α position. The resulting enolate then ejects a hydroxide ion, giving the condensation product, as shown:

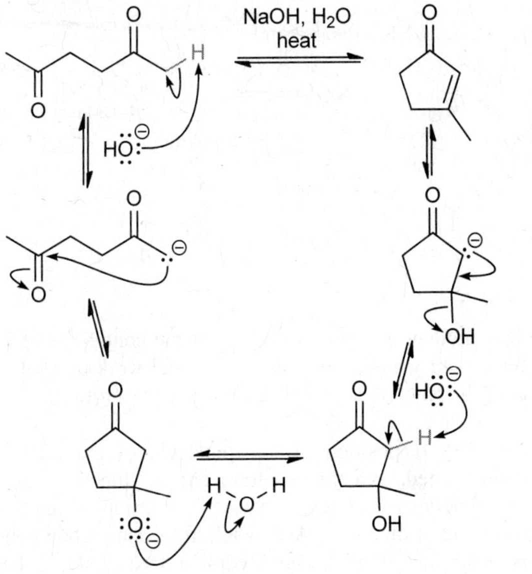

22.26. The following process is the reverse of the mechanism shown in the previous problem. As such, all of the intermediates are identical to the intermediates in the previous problem, but they appear in reverse order. First, hydroxide attacks the β position of the α,β-unsaturated ketone, giving an enolate. The enolate is then protonated to give a β-hydroxyketone, which is subsequently deprotonated to give an alkoxide ion. The carbonyl group is then reformed, with loss of an enolate as a leaving group (this is a retro-aldol process). The resulting enolate is then protonated to give the dione. Notice that each of the protonation steps employs water as an acid, consistent with basic conditions (strong acids are not measurably present under these conditions).

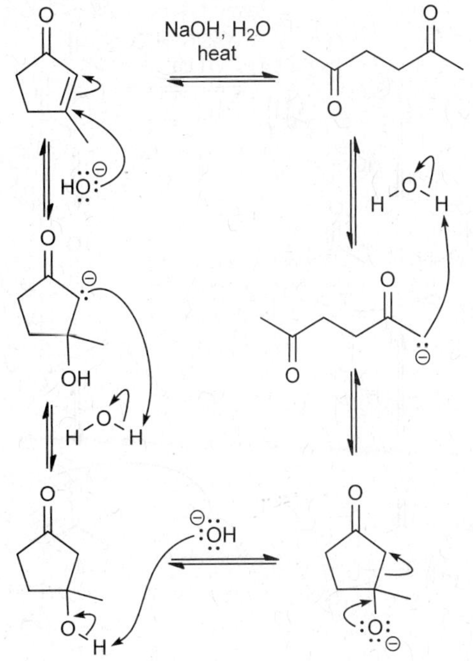

22.27. This process is an intramolecular aldol condensation reaction, and is similar to Problem **22.25**, with one additional methylene (CH_2) group in between the two carbonyl groups. As such, a six-membered ring is formed, rather than a five-membered ring. Other than this small difference, this mechanism is identical to the mechanism shown in the solution to Problem **22.25**.

First, hydroxide functions as a base and deprotonates the starting dione to give an enolate. This enolate ion is a nucleophilic center, and it will attack the carbonyl group present in the same structure (an intramolecular process), thereby closing a six-membered ring. The resulting alkoxide ion is then protonated to give a β-hydroxy ketone. Hydroxide then functions as a base again, deprotonating the α position. The resulting enolate then ejects a hydroxide ion, giving the condensation product, as shown:

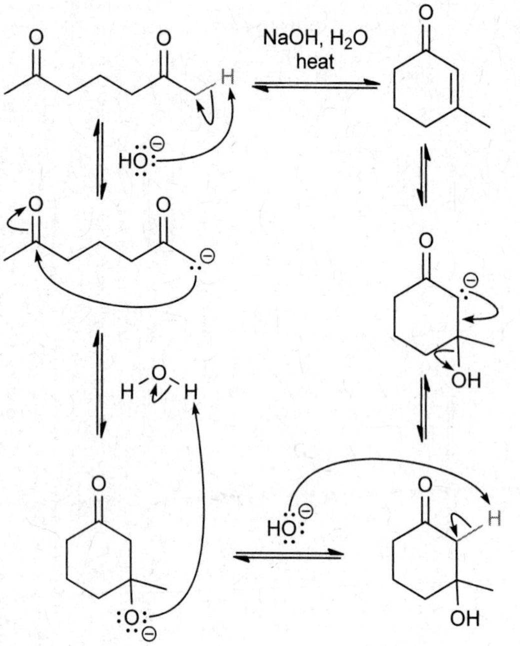

22.28.

(a) The starting material is an ester, and the product is a β-ketoester. Therefore, this process is a Claisen condensation. Since the OR group of the ester is an ethoxy group, we must use sodium ethoxide as the base, in order to avoid transesterification.

(b) The starting material is an ester, and the product is a β-ketoester. Therefore, this process is a Claisen condensation. Since the OR group of the ester is a *tert*-butoxy group, we must use potassium *tert*-butoxide as the base, in order to avoid transesterification.

22.29.

(a) The α position of one molecule of the ester is deprotonated, and the resulting enolate functions as a nucleophile and attacks the carbonyl group of another molecule of the ester. As a result, a carbon-carbon bond is formed, giving a tetrahedral intermediate. The carbonyl group is then reformed via loss of an ethoxide ion, affording a β-ketoester, as shown:

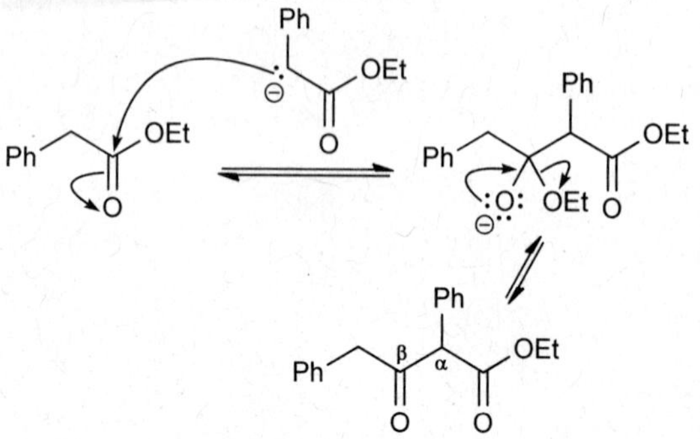

This product is deprotonated under the conditions of its formation, which is the reason for acid work-up (H_3O^+) after the reaction is complete (to return the proton).

(b) The α position of one molecule of the ester is deprotonated, and the resulting enolate functions as a nucleophile and attacks the carbonyl group of another molecule of the ester. As a result, a carbon-carbon bond is formed, giving a tetrahedral intermediate. The carbonyl group is then reformed via loss of a methoxide ion, affording a β-ketoester, as shown:

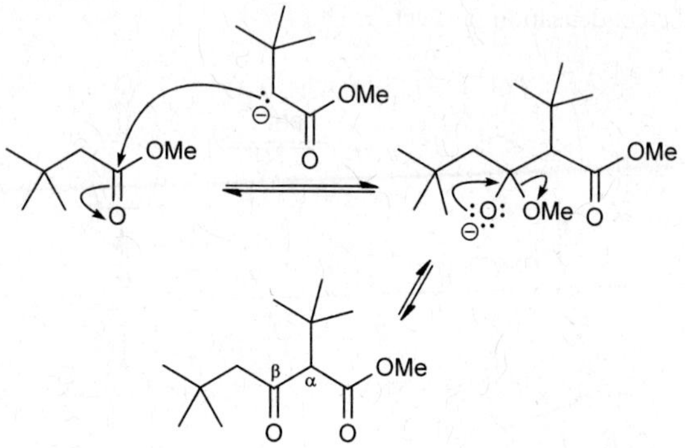

This product is deprotonated under the conditions of its formation, which is the reason for acid work-up (H_3O^+) after the reaction is complete (to return the proton).

(c) The α position of one molecule of the ester is deprotonated, and the resulting enolate functions as a nucleophile and attacks the carbonyl group of another molecule of the ester. As a result, a carbon-carbon bond is formed, giving a tetrahedral intermediate. The carbonyl group is then reformed via loss of an ethoxide ion, affording a β-ketoester, as shown:

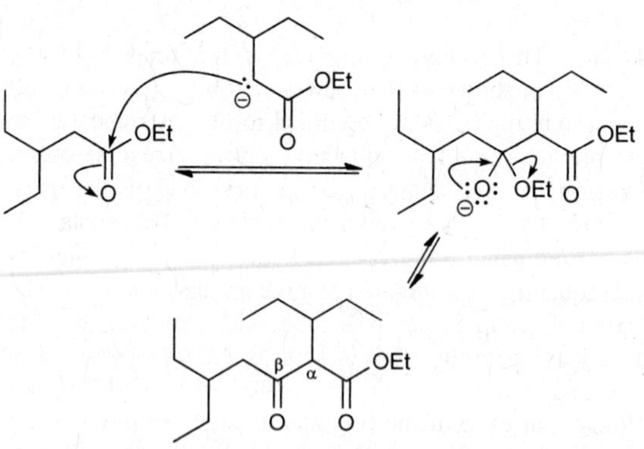

This product is deprotonated under the conditions of its formation, which is the reason for acid work-up (H_3O^+) after the reaction is complete (to return the proton).

22.30.

(a) We first identify the α and β positions, and then apply a retrosynthetic analysis. The α position is the location between the two carbonyl groups, and the β position bears the keto group:

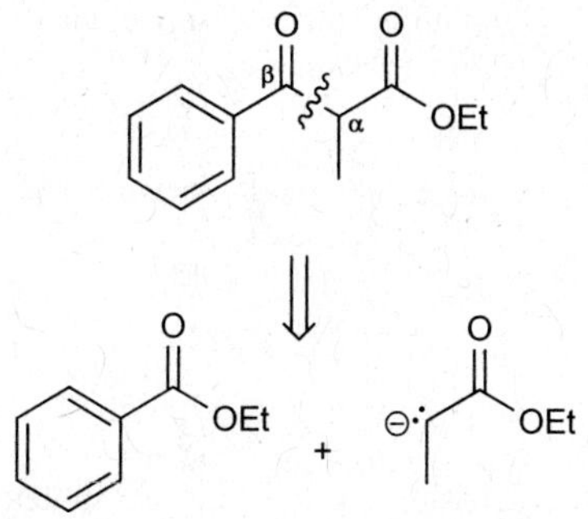

Since the two partners are different, we use a crossed Claisen condensation. LDA is used as the base in the first step, and the final step of the process is aqueous acidic work-up, as shown:

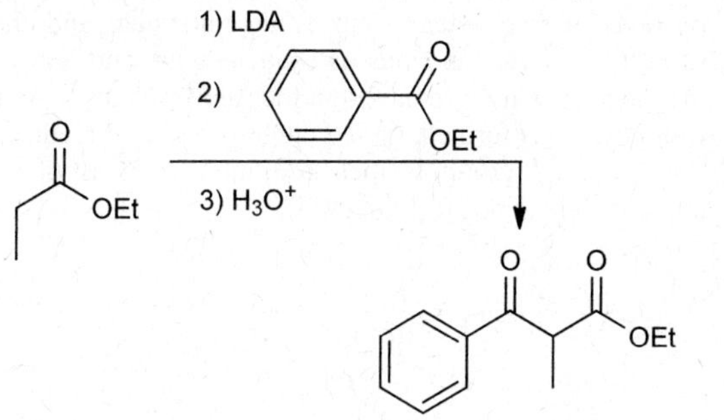

(b) We first identify the α and β positions, and then apply a retrosynthetic analysis. The α position is the location between the two carbonyl groups, and the β position bears the keto group:

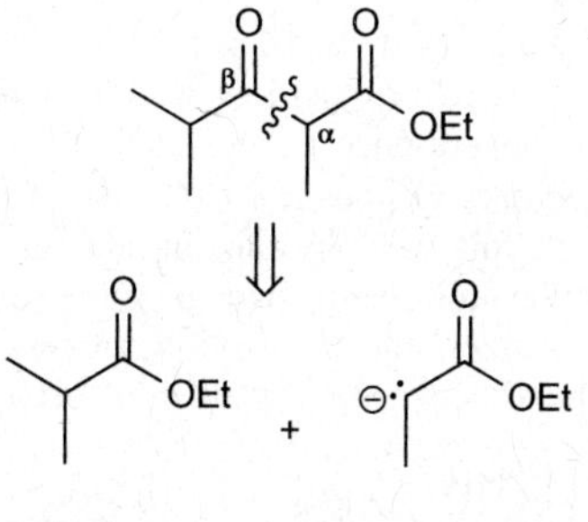

Since the two partners are different, we use a crossed Claisen condensation. LDA is used as the base in the first step, and the final step of the process is aqueous acidic work-up, as shown:

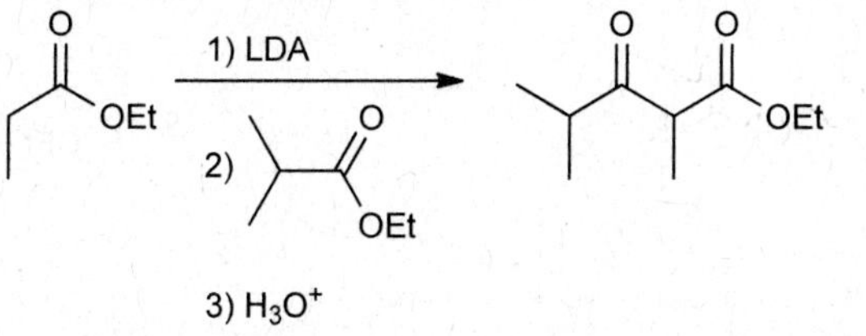

(c) We first identify the α and β positions, and then apply a retrosynthetic analysis. The α position is the

location between the two carbonyl groups, and the β position bears the keto group:

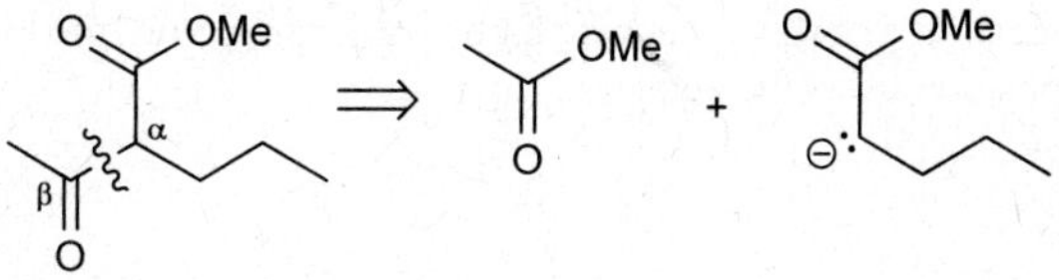

Since the two partners are different, we use a crossed Claisen condensation. LDA is used as the base in the first step, and the final step of the process is aqueous acidic work-up, as shown:

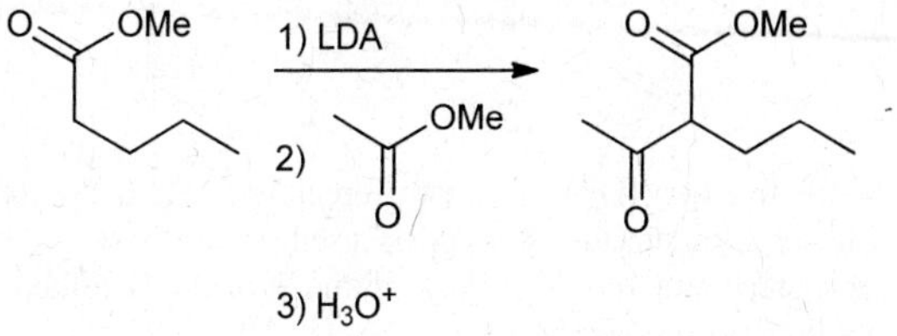

(d) We first identify the α and β positions, and then apply a retrosynthetic analysis. The α position is the location between the two carbonyl groups, and the β position bears the keto group:

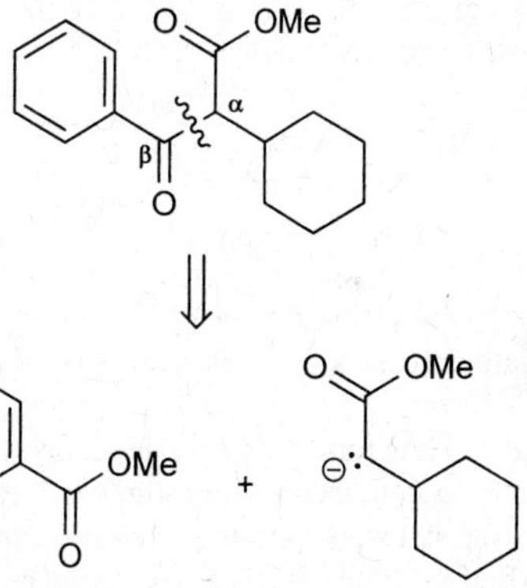

Since the two partners are different, we use a crossed Claisen condensation. LDA is used as the base in the first step, and the final step of the process is aqueous acidic work-up, as shown:

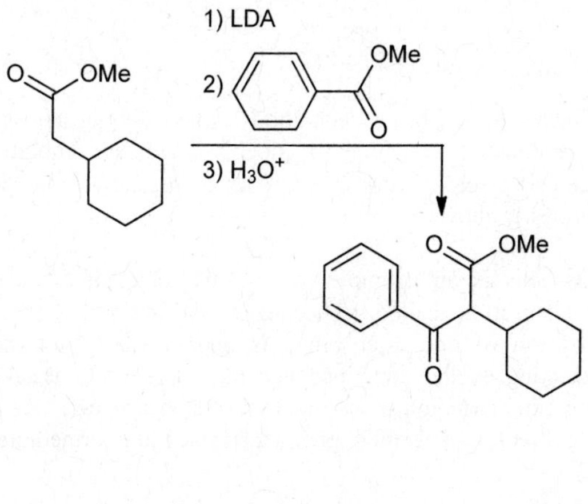

(e) We first identify the α and β positions, and then apply a retrosynthetic analysis. The α position is the location between the two carbonyl groups, and the β position bears the keto group:

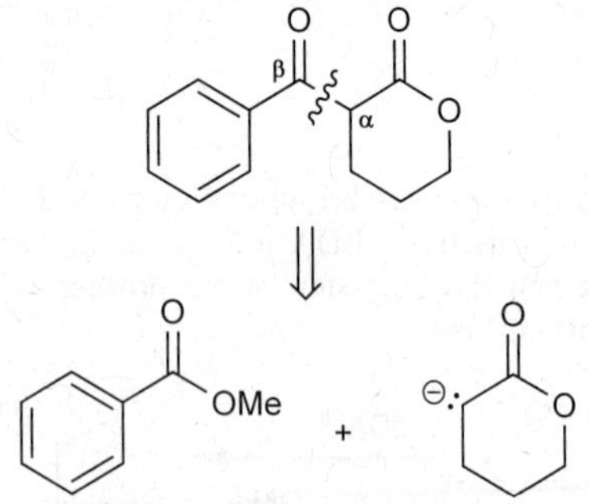

Since the two partners are different, we use a crossed Claisen condensation. LDA is used as the base in the first step, and the final step of the process is aqueous acidic work-up, as shown:

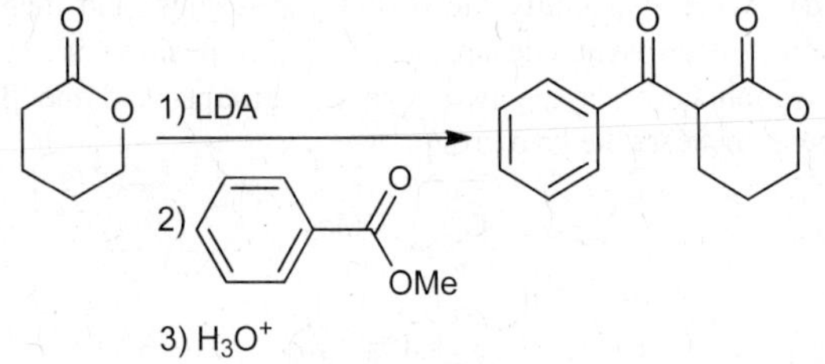

22.31.

(a) This is an example of an intramolecular Claisen condensation (called a Dieckmann cyclization). The α position of one ester group is deprotonated, and the resulting enolate functions as a nucleophile and attacks the other carbonyl group within the same structure. As a result, a ring is formed, giving a tetrahedral intermediate. The carbonyl group is then reformed via loss of an ethoxide ion, giving a β-ketoester:

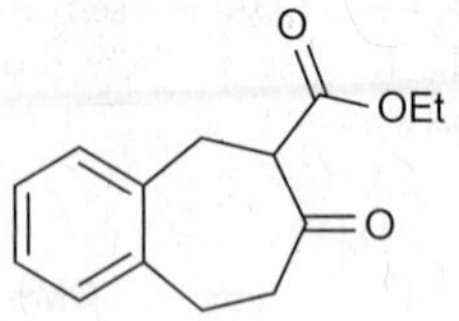

Under these basic conditions, the β-ketoester is deprotonated to give a doubly-stabilized enolate, requiring acidic work-up in order to regenerate the β-ketoester above.

(b) This is an example of an intramolecular Claisen condensation (called a Dieckmann cyclization). The α position of one ester group is deprotonated, and the resulting enolate functions as a nucleophile and attacks the other carbonyl group within the same structure. As a result, a ring is formed, giving a tetrahedral intermediate.

The carbonyl group is then reformed via loss of an ethoxide ion, giving a β-ketoester:

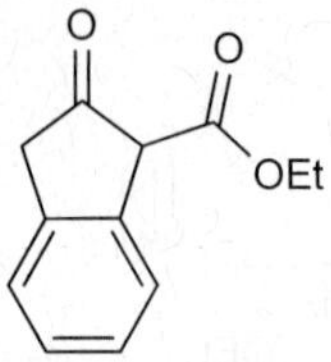

Under these basic conditions, the β-ketoester is deprotonated to give a doubly-stabilized enolate, requiring acidic work-up in order to regenerate the β-ketoester above.

(c) This is an example of an intramolecular Claisen condensation (called a Dieckmann cyclization). The α position of one ester group is deprotonated, and the resulting enolate functions as a nucleophile and attacks the other carbonyl group within the same structure. As a result, a ring is formed, giving a tetrahedral intermediate. The carbonyl group is then reformed via loss of an ethoxide ion, giving a β-ketoester:

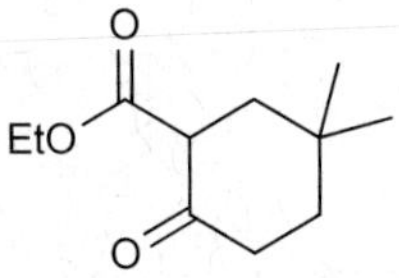

Under these basic conditions, the β-ketoester is deprotonated to give a doubly-stabilized enolate, requiring acidic work-up in order to regenerate the β-ketoester above.

22.32. There are two α positions which are not identical (because of the presence of the methyl group at C3). Therefore, either α position (C2 or C6) can be deprotonated, followed by an intramolecular attack, leading to the following two possible condensation products. That is, the cyclization process can either result in a bond between C2 and C7 or between C6 and C1:

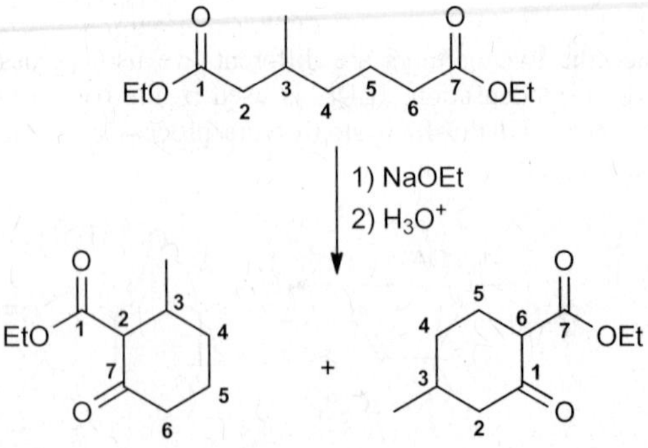

22.33.

(a) The starting material is a ketone, which has two α positions. With LDA as the base (at low temperature), we expect deprotonation to occur at the less substituted site, giving the kinetic enolate. This enolate is then treated with methyl iodide to give an S$_N$2 reaction, thereby installing the methyl group at the less substituted α position, as shown.

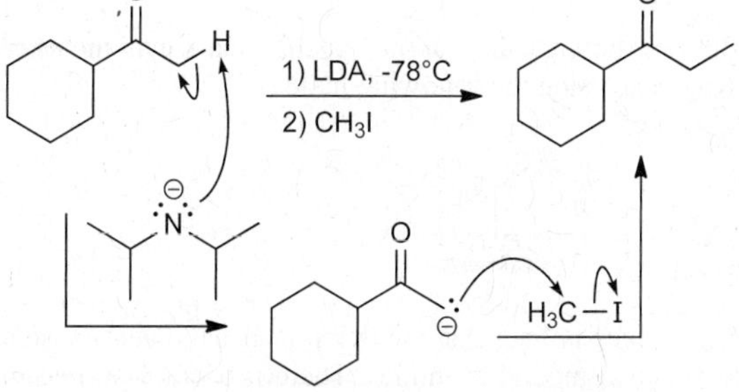

(b) The starting material is a ketone, which has two α positions. With NaH as the base, we expect deprotonation to occur at the more substituted site, giving the thermodynamic enolate. This enolate is then treated with benzyl bromide to give an S$_N$2 reaction, thereby installing the benzyl group at the more substituted α position, as shown.

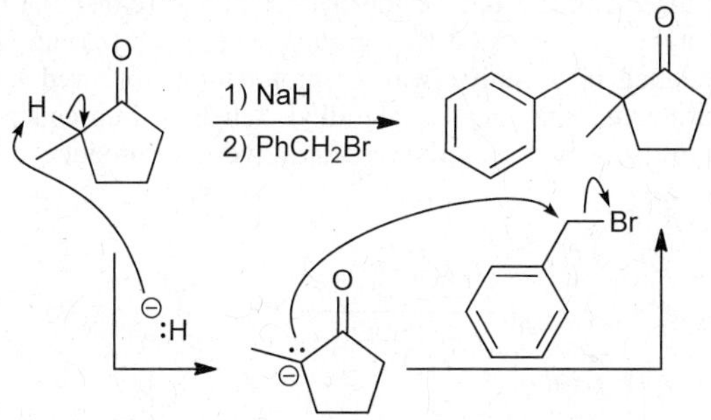

(c) The starting material is a ketone, which has two α positions, although they are identical because the ketone is symmetrical. Deprotonation at either location will lead to the same enolate ion. In the first step, LDA functions as a base and deprotonates the ketone to give an enolate. This enolate is then treated with ethyl iodide to give an S$_N$2 reaction, thereby installing an ethyl group. Subsequent treatment with LDA (at low tempereature), followed by methyl iodide, installs a methyl group at the other (less substituted) α position via the kinetic enolate. The net result is the installation of an ethyl group at one α position and the installation of a methyl group at the other α position:

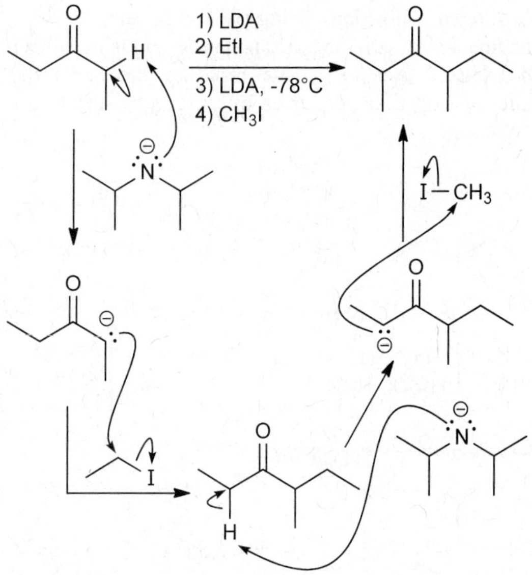

22.34. This transformation does not involve a change in the identity or location of the functional group (a hydroxyl group), but it does involve a change in the carbon skeleton:

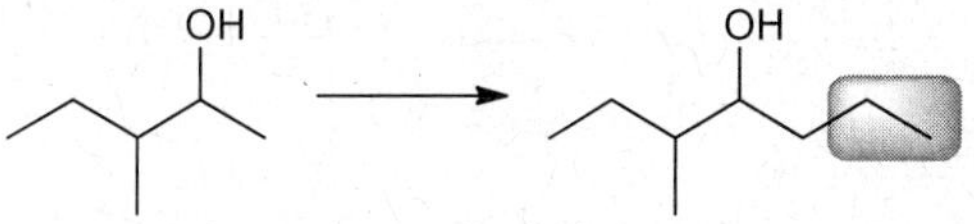

An ethyl group must be installed, although we have not learned a way to do this in one step. A multi-step strategy is necessary. One strategy for achieving this transformation derives from the following retrosynthetic analysis. An explanation of each of the steps (*a-c*) follows.

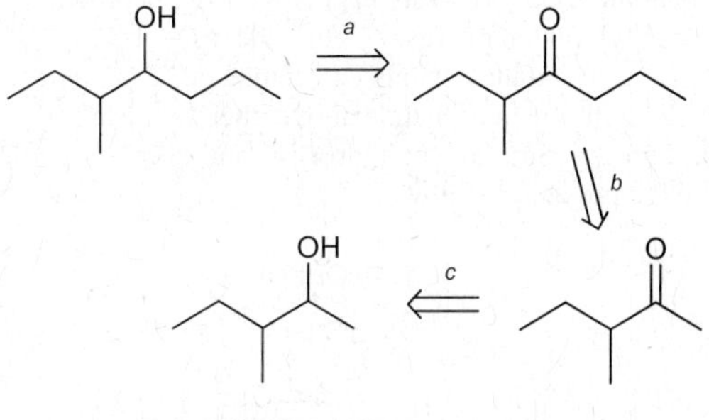

a. The alcohol can be made via reduction of the corresponding ketone.
b. The ethyl group can be installed through alkylation of 3-methyl-2-pentanone (via the kinetic enolate).
c. 3-Methyl-2-pentanone can be made from the starting alcohol via oxidation.

Now let's draw the forward scheme. The starting alcohol is oxidized with chromic acid to give 3-methyl-2-pentanone. Alternatively, PCC can be used to affect the same transformation. The ketone is then treated with LDA at low temperature to deprotonate the less substituted α position, giving the kinetic enolate. Upon treatment with ethyl iodide, an ethyl group is installed in the desired location. The resulting ketone is then reduced with LAH, followed by water work-up, to give the product.

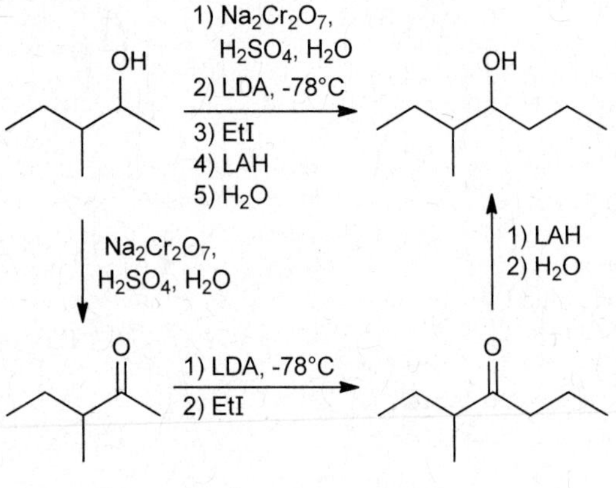

1) Na$_2$Cr$_2$O$_7$, H$_2$SO$_4$, H$_2$O
2) LDA, -78°C
3) EtI
4) LAH
5) H$_2$O

Na$_2$Cr$_2$O$_7$, H$_2$SO$_4$, H$_2$O

1) LAH
2) H$_2$O

1) LDA, -78°C
2) EtI

22.35.
(a) The product is a carboxylic acid that has the following (highlighted) group connected to the α position:

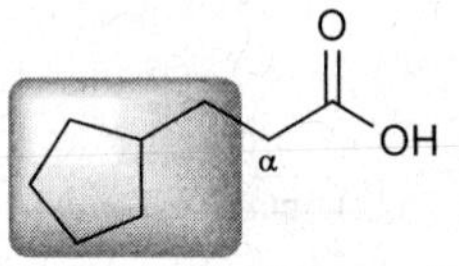

This group can be installed via a malonic ester synthesis, using the following alkyl halide:

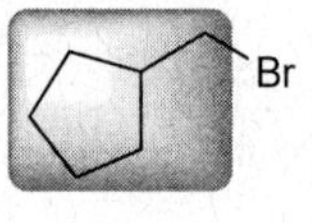

A malonic ester synthesis begins with the deprotonation of diethyl malonate (using ethoxide as a base). The resulting resonance-stabilized conjugate base is then treated with the alkyl halide above, thereby installing the alkyl group. Subsequent hydrolysis and decarboxylation give the product, as shown:

1) NaOEt
2) [cyclopentylmethyl bromide]
3) H$_3$O$^+$, heat

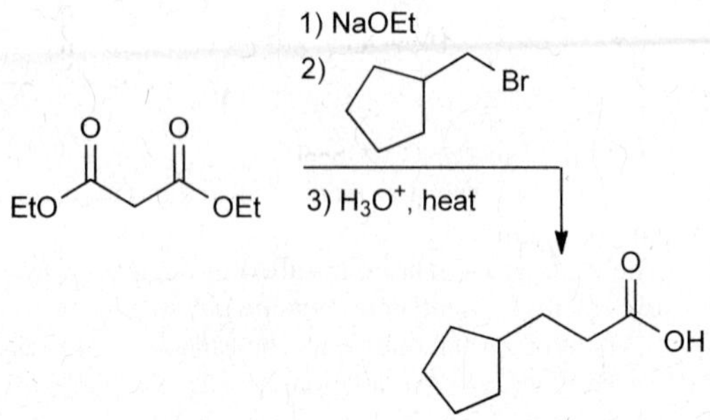

(b) The product is a carboxylic acid that has the following two (highlighted) groups connected to the α position:

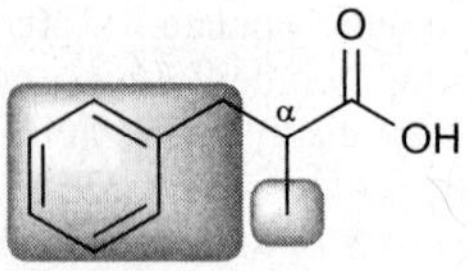

Both of these groups can be installed via a malonic ester synthesis, using the following halides:

You might notice that methyl iodide has been chosen, rather than methyl bromide. There is a practical reason for this choice (methyl iodide is a liquid at room temperature, while methyl bromide is a gas, rendering the latter more difficult to work with). This fact does not appear anywhere in the textbook, so there is no way that you could have known this (unless your instructor informed you of this fact).

A malonic ester synthesis begins with the deprotonation of diethyl malonate (using ethoxide as a base). The resulting resonance-stabilized conjugate base is then treated with the one of the halides above (either one), thereby installing the first group. The second group is installed in a similar way (deprotonation, followed by treatment with the second alkyl halide). Subsequent hydrolysis and decarboxylation give the product, as shown:

1) NaOEt
2) PhCH$_2$Br
3) NaOEt
4) CH$_3$I
5) H$_3$O$^+$, heat

1) NaOEt
2) PhCH$_2$Br

H$_3$O$^+$, heat

1) NaOEt
2) CH$_3$I

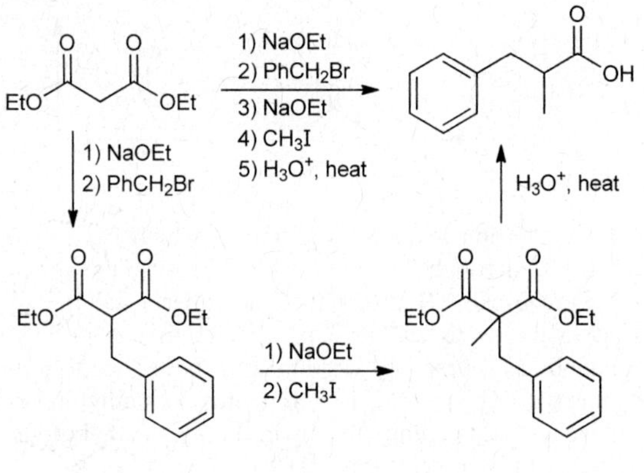

(c) The product is a carboxylic acid that has two propyl groups (highlighted) connected to the α position:

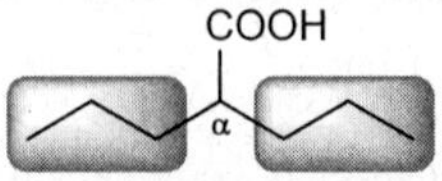

COOH

Both of these groups can be installed via a malonic ester synthesis, using propyl iodide. A malonic ester synthesis begins with the deprotonation of diethyl malonate (using ethoxide as a base). The resulting resonance-stabilized conjugate base is then treated with propyl iodide, thereby installing the first propyl group. The second propyl group is installed in a similar way (deprotonation, followed by treatment with propyl iodide). Subsequent hydrolysis and decarboxylation give the product, as shown:

1) NaOEt
2) ⌐⌐⌐I
3) NaOEt
4) ⌐⌐I
5) H₃O⁺, heat

1) NaOEt
2) ⌐⌐⌐I

1) NaOEt
2) ⌐⌐⌐I

H₃O⁺, heat

(d) The product is a carboxylic acid that has the following two (highlighted) groups connected to the α position:

Both of these groups can be installed via a malonic ester synthesis, using propyl bromide and methyl iodide, respectively. The reason for using methyl iodide (rather than methyl bromide) was discussed in the solution to Problem **22.35b**.

A malonic ester synthesis begins with the deprotonation of diethyl malonate (using ethoxide as a base). The resulting resonance-stabilized conjugate base is then treated with the one of the alkyl halides above (either one), thereby installing the first alkyl group. The second alkyl group is installed in a similar way (deprotonation, followed by treatment with the second alkyl halide). Subsequent hydrolysis and decarboxylation give the product, as shown:

1) NaOEt
2) CH₃I
3) NaOEt
4) ⌐⌐Br
5) H₃O⁺, heat

1) NaOEt
2) CH₃I

1) NaOEt
2) ⌐⌐Br

H₃O⁺, heat

(e) The product is a carboxylic acid that has the following two (highlighted) groups connected to the α position:

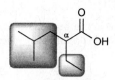

Both of these groups can be installed via a malonic ester synthesis, using ethyl iodide and isobutyl iodide, respectively.

A malonic ester synthesis begins with the deprotonation of diethyl malonate (using ethoxide as a base). The resulting resonance-stabilized conjugate base is then treated with the one of the alkyl halides above (either one), thereby installing the first alkyl group. The second alkyl group is installed in a similar way (deprotonation, followed by treatment with the second alkyl halide). Subsequent hydrolysis and decarboxylation give the product, as shown:

1) NaOEt
2) EtI
3) NaOEt
4) ⌐⌐I
5) H₃O⁺, heat

1) NaOEt
2) EtI

1) NaOEt
2) ⌐⌐I

H₃O⁺, heat

22.36.
(a) The desired product is an ester, which can be made from the corresponding carboxylic acid via a Fischer esterification:

The carboxylic acid can be made with a malonic ester synthesis, as shown in the following scheme:

1) NaOEt
2) ⌐⌐Br
3) H₃O⁺, heat
4) [H⁺], EtOH, (-H₂O)

1) NaOEt
2) ⌐⌐Br
3) H₃O⁺, heat

[H⁺], EtOH, (-H₂O)

(b) The desired product is a primary alcohol, which can be made from the corresponding carboxylic acid via reduction:

This carboxylic acid can be made with a malonic ester synthesis, as shown in the following scheme:

EtO
OEt
1) NaOEt
2) PhCH2Br
3) H3O+, heat
4) xs LAH
5) H2O
Ph
OH
1) NaOEt
2) PhCH2Br
3) H3O+, heat
1) xs LAH
2) H2O
Ph
OH

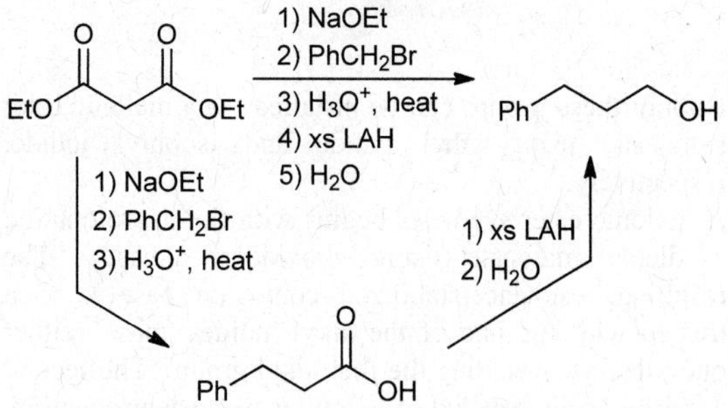

(c) The desired product is an amide, which can be made from the corresponding carboxylic acid (via an acid halide), as shown:

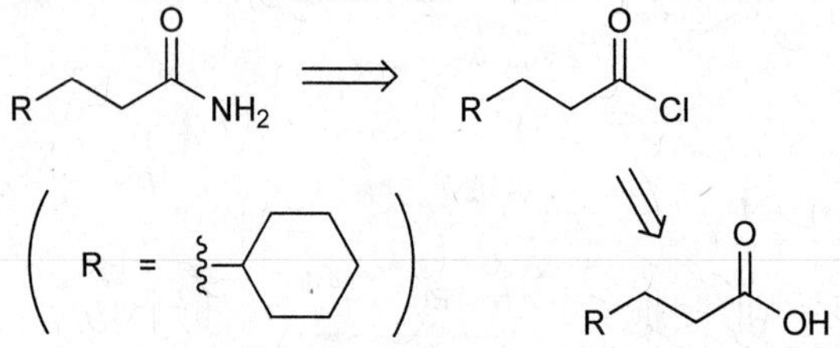

R
NH2
R
Cl
R =
R
OH

This carboxylic acid can be made with a malonic ester synthesis, as shown in the following scheme:

EtO
OEt
1) NaOEt
2) R Br
3) H3O+, heat
4) SOCl2
5) excess NH3
R
NH2
1) NaOEt
2) R Br
3) H3O+, heat
xs NH3
R
OH
SOCl2
R
Cl

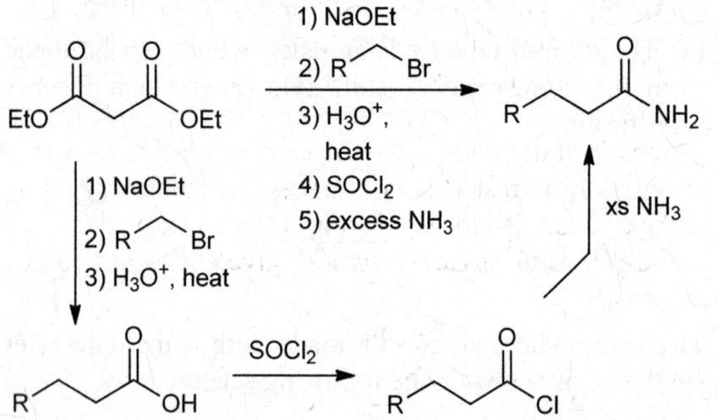

22.37. Preparation of the desired compound requires the installation of three alkyl groups at the α position.

α
OH

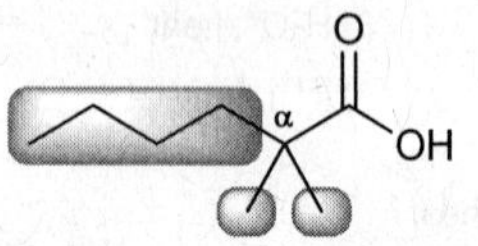

The malonic ester synthesis can only be used to install two alkyl groups because the starting material (diethyl malonate) has only two α protons.

α
EtO
OEt
H
H

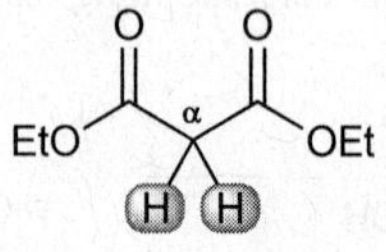

22.38. When a dibromide is used (rather than two separate alkyl halides), a cyclic product is expected, as shown:

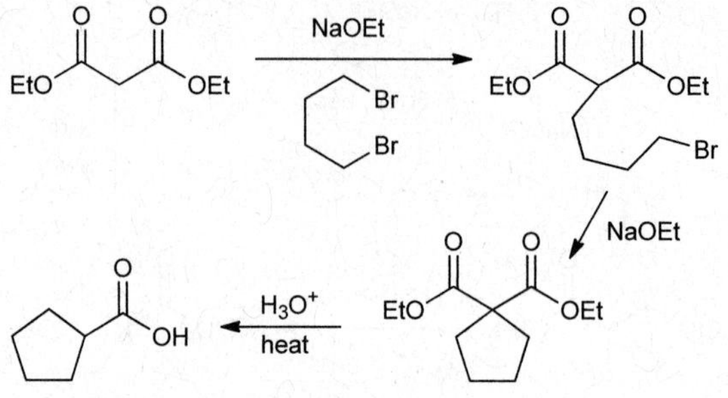

EtO
OEt
NaOEt
Br
Br
EtO
OEt
Br
NaOEt
EtO
OEt
H3O+
heat
OH

22.39.

(a) The product is a methyl ketone that has the following (highlighted) group connected to the α position:

α

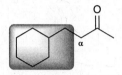

This group can be installed via an acetoacetic ester synthesis, using the following alkyl halide:

I

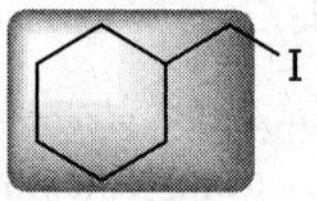

An acetoacetic ester synthesis begins with the deprotonation of ethyl acetoacetate (using ethoxide as a base). The resulting resonance-stabilized conjugate base is then treated with the alkyl halide above, thereby installing the alkyl group. Subsequent hydrolysis and decarboxylation give the product, as shown:

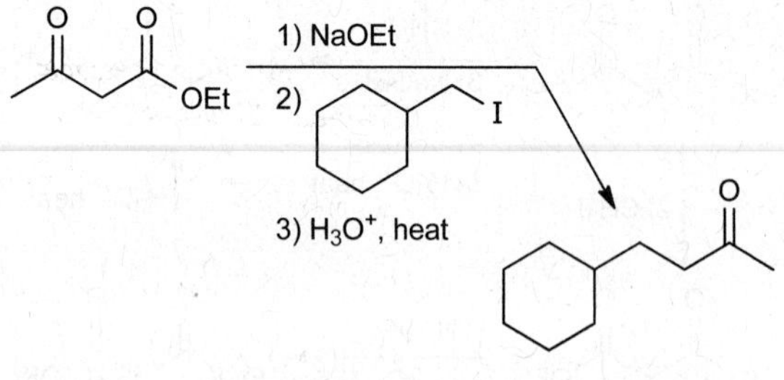

OEt
1) NaOEt
2)
I
3) H3O+, heat

(b) The product is a methyl ketone that has the following two (highlighted) groups connected to the α position:

α

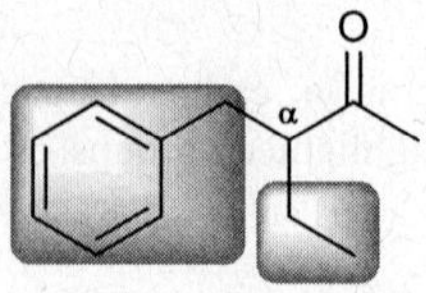

Each of these groups can be installed via an acetoacetic ester synthesis, using the following halides:

An acetoacetic ester synthesis begins with the deprotonation of ethyl acetoacetate (using ethoxide as a base). The resulting resonance-stabilized conjugate base is then treated with one of the halides above, to install one of the two groups. The other group is installed in a similar way (deprotonation with a base, followed by alkylation). Subsequent hydrolysis and decarboxylation give the product, as shown:

1) NaOEt
2) EtI
3) NaOEt
4) PhCH₂Br
5) H₃O⁺, heat

1) NaOEt
2) EtI

1) NaOEt
2) PhCH₂Br

H₃O⁺, heat

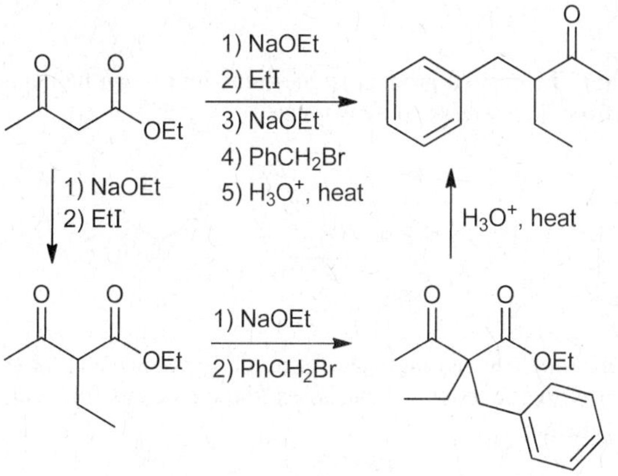

(c) The product is a methyl ketone that has the following two (highlighted) groups connected to the α position:

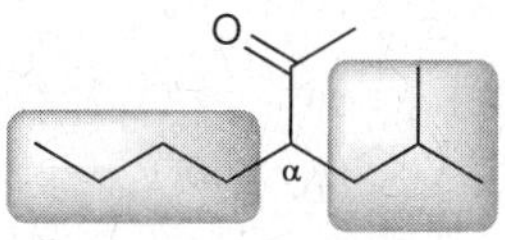

Each of these groups can be installed via an acetoacetic ester synthesis, using the following alkyl halides:

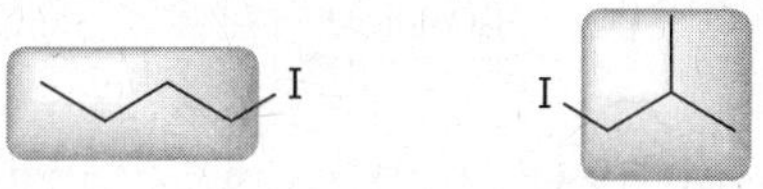

An acetoacetic ester synthesis begins with the deprotonation of ethyl acetoacetate (using ethoxide as a base). The resulting resonance-stabilized conjugate base is then treated with one of the alkyl halides above, thereby installing one of the two alkyl groups. The other alkyl group is installed in a similar way (deprotonation with a base, followed by alkylation). Subsequent hydrolysis and decarboxylation give the product, as shown:

1) NaOEt
2)
3) NaOEt
4) I
5) H₃O⁺, heat

1) NaOEt
2) I

3) NaOEt
4) I

H₃O⁺, heat

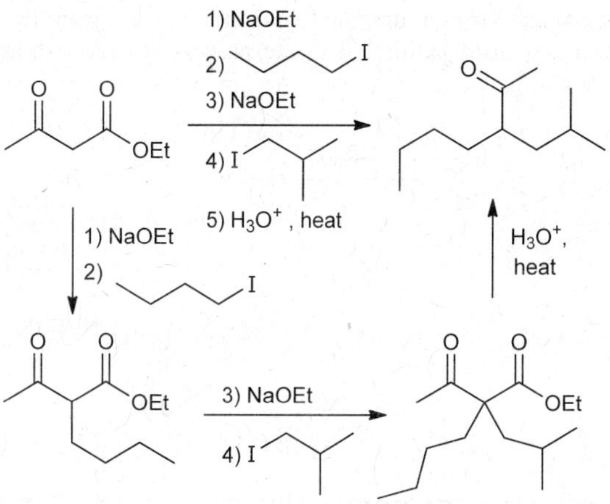

(d) The product is a methyl ketone that has the following two (highlighted) groups connected to the α position:

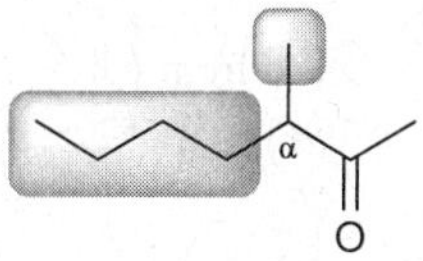

Each of these groups can be installed via an acetoacetic ester synthesis, using methyl iodide and butyl iodide, respectively.

An acetoacetic ester synthesis begins with the deprotonation of ethyl acetoacetate (using ethoxide as a base). The resulting resonance-stabilized conjugate base is then treated with one of the alkyl halides above, thereby installing one of the two alkyl groups. The other alkyl group is installed in a similar way (deprotonation with a base, followed by alkylation). Subsequent hydrolysis and decarboxylation give the product, as shown:

1) NaOEt
2) CH₃I
3) NaOEt
4) I
5) H₃O⁺, heat

1) NaOEt
2) CH₃I

3) NaOEt
4) I

H₃O⁺, heat

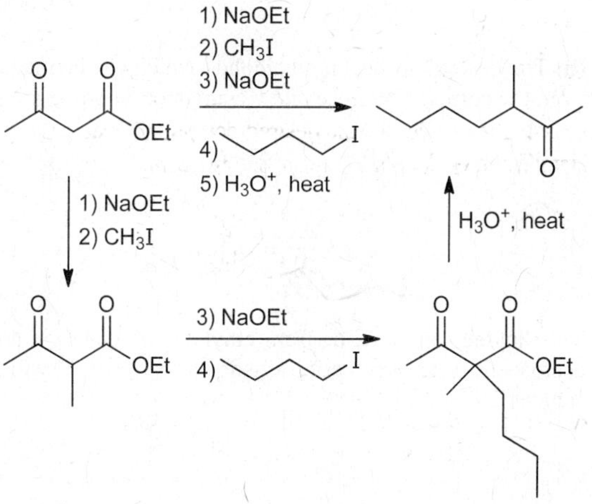

22.40. When a dibromide is used (rather than two separate alkyl halides), a cyclic product is expected, as shown:

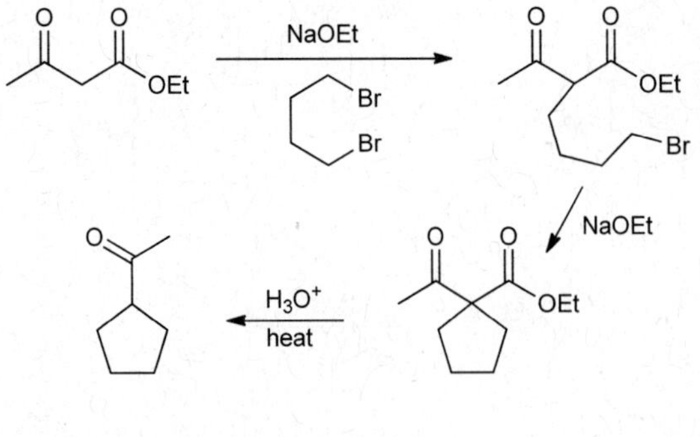

22.41.
(a) The desired product is an acetal, which can be made from the corresponding ketone:

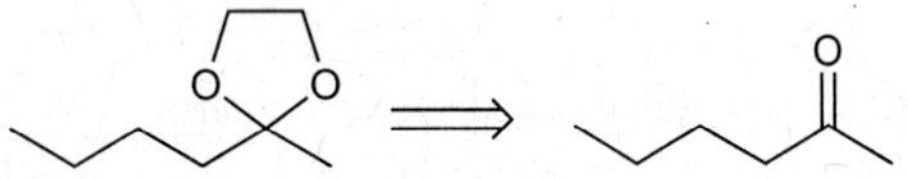

This ketone can be made from ethyl acetoacetate via an acetoacetic ester synthesis, as shown in the following scheme:

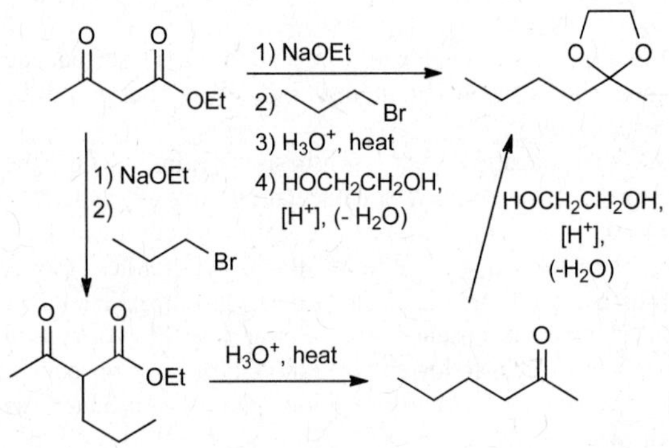

(b) The desired product is an alcohol, which can be made from the corresponding ketone via a reduction process:

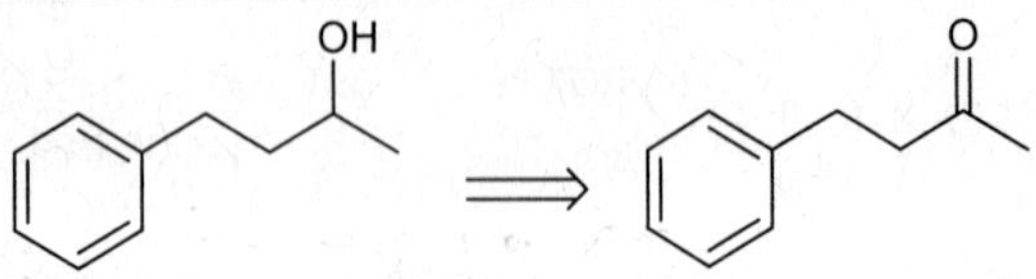

This ketone can be made from ethyl acetoacetate via an acetoacetic ester synthesis, as shown in the following scheme:

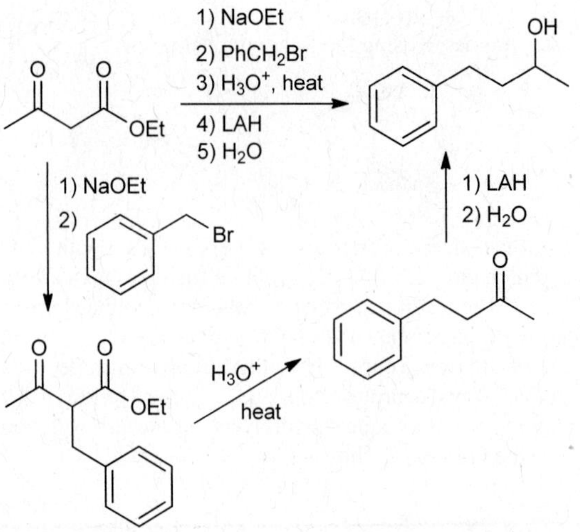

(c) The desired product is an imine, which can be made from the corresponding ketone:

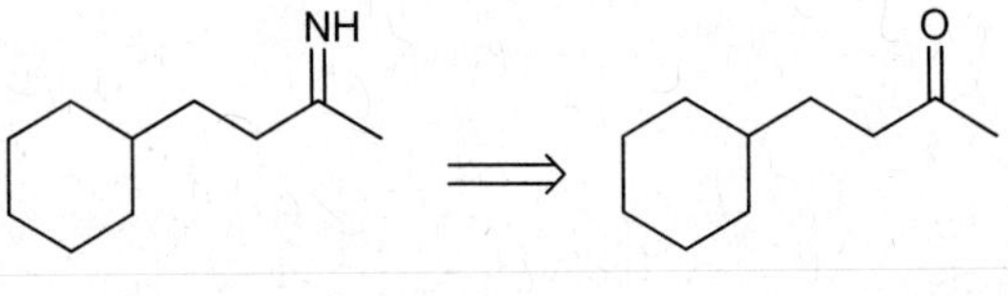

This ketone can be made from ethyl acetoacetate via an acetoacetic ester synthesis, as shown in the following scheme:

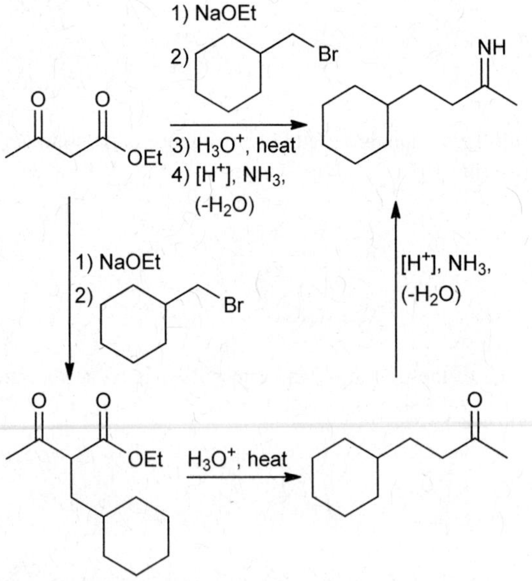

22.42. Preparation of the desired compound requires the installation of three alkyl groups at the α position.

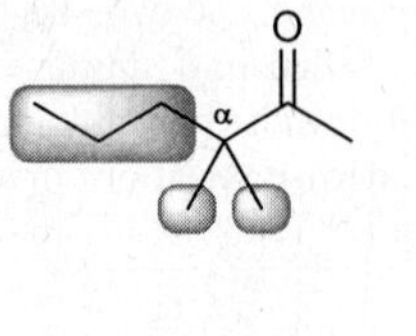

The acetoacetic ester synthesis can only be used to install two alkyl groups because the starting material (ethyl acetoacetate) has only two α protons.

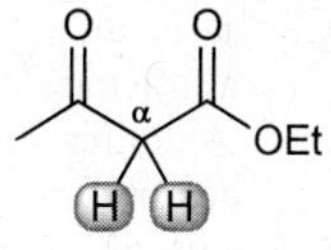

22.43. Using the approach described in the problem statement, the product can be made from an acyclic diester, as shown in the following retrosynthetic analysis:

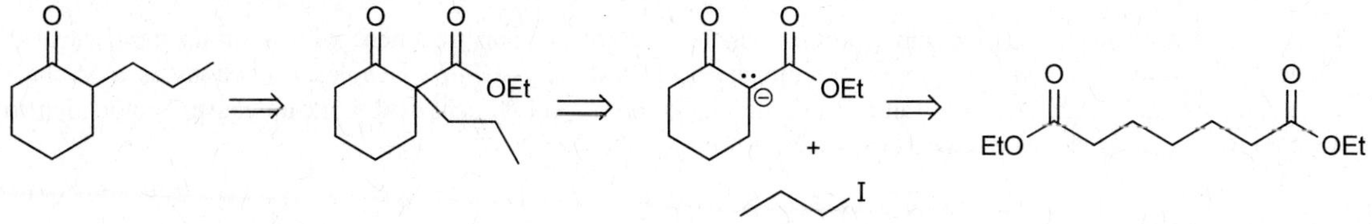

The diester can be made from the corresponding diacid (via Fischer esterification of both carboxylic acid groups), and the diacid can be made via oxidation of the starting diol:

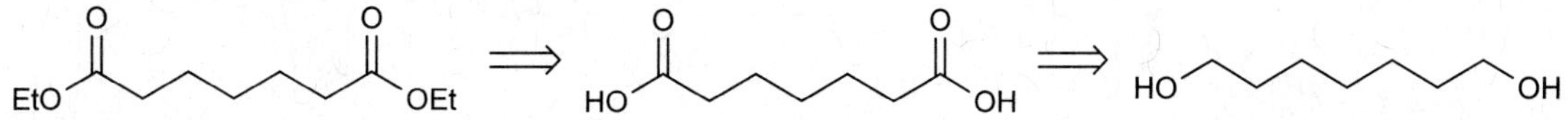

Now let's draw the forward scheme. The starting diol is converted to a diacid upon treatment with chromic acid. This diacid is then converted to a diester upon treatment with ethanol and an acid catalyst (with removal of water). The resulting diester will undergo a Dieckmann cyclization upon treatment with sodium ethoxide, followed by aqueous acid work-up. Alkylation, followed by hydrolysis and decarboxylation, gives the product, as shown.

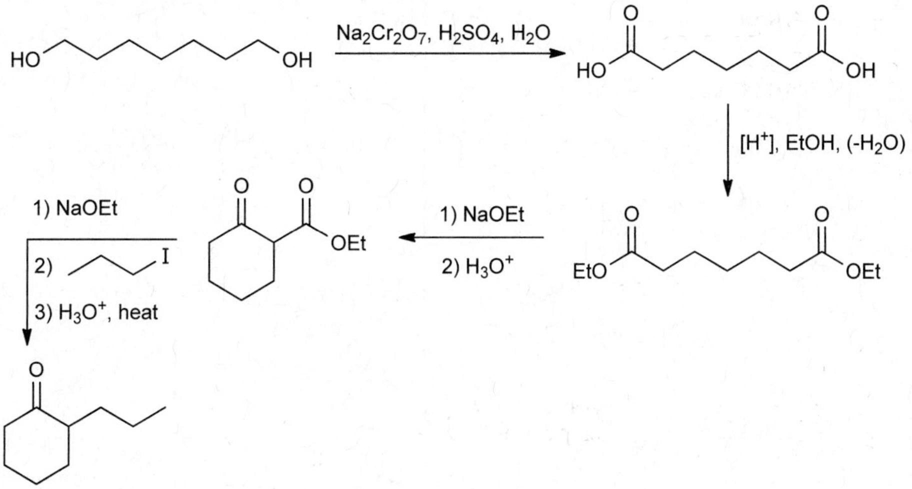

22.44.

(a) The starting material is an α,β-unsaturated ketone, which can function as a Michael acceptor (see Table 22.2), and the reagent (lithium diethyl cuprate) can function as a Michael donor. As such, we expect a Michael reaction, thereby installing an ethyl group at the β position:

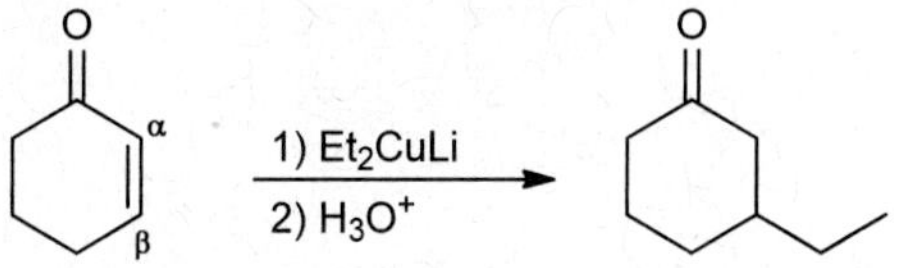

(b) The starting material is an α,β-unsaturated nitrile, which can function as a Michael acceptor (see Table 22.2), and the reagent (lithium diethyl cuprate) can function as a Michael donor. As such, we expect a Michael reaction, thereby installing an ethyl group at the β position:

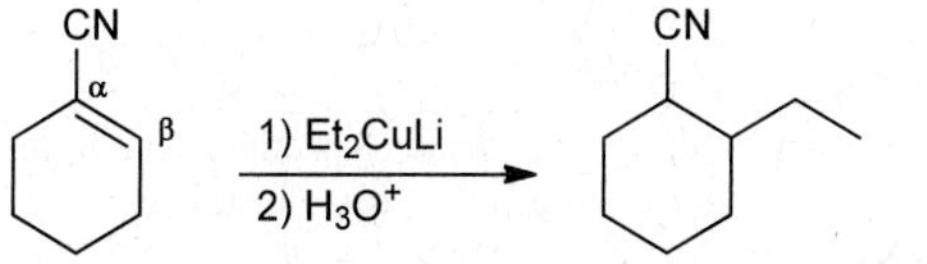

(c) The starting material is an α,β-unsaturated ester, which can function as a Michael acceptor (see Table 22.2), and the reagent (lithium diethyl cuprate) can function as a Michael donor. As such, we expect a Michael reaction, thereby installing an ethyl group at the β position:

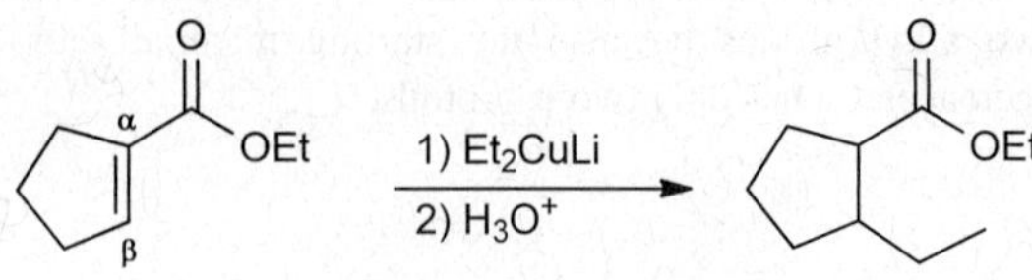

22.45. The starting material has an acidic proton, which is removed upon treatment with a strong base, such as hydroxide. The resulting resonance-stabilized conjugate base (a doubly-stabilized enolate) functions as a Michael donor and attacks the Michael acceptor (an α,β-unsaturated ketone). Subsequent acid work-up causes protonation to give an enol, which tautomerizes to give a ketone, as shown.

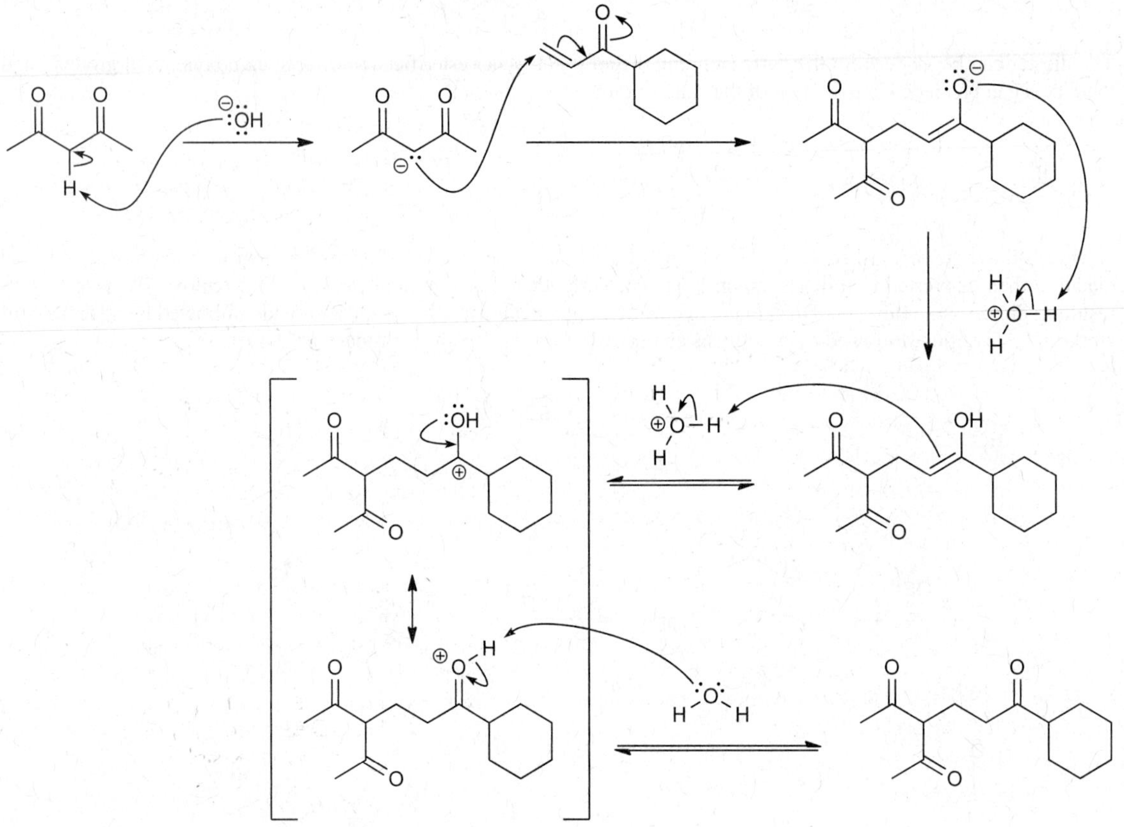

22.46.
(a) Recall that the malonic ester synthesis is useful for creating carboxylic acids that possess either one or two alkyl groups at the α position:

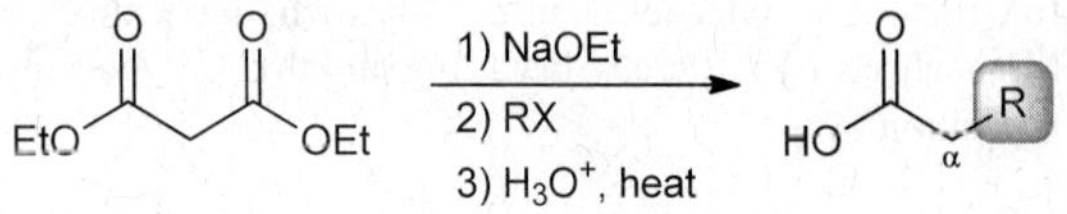

Therefore, making the desired product via a malonic ester synthesis would require installation of the following highlighted group:

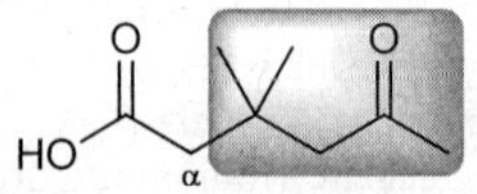

Installation of this group via alkylation (as seen in Section 22.5) would require a tertiary alkyl halide:

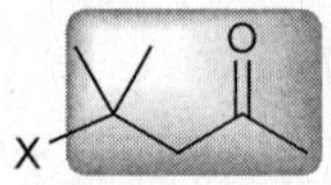

However, a tertiary alkyl halide will not undergo an S_N2 reaction because it is too sterically hindered. So this method will not work. The problem statement indicates that a Michael reaction can be used to achieve the desired transformation. That is, we would use an electrophile with the same carbon skeleton as the alkyl halide above, but the electrophilic position will be the β position of an α,β-unsaturated ketone:

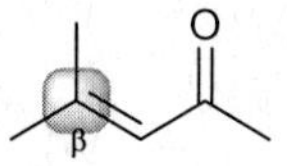

This compound can function as a Michael acceptor, thereby allowing the desired transformation, as seen in the following synthesis:

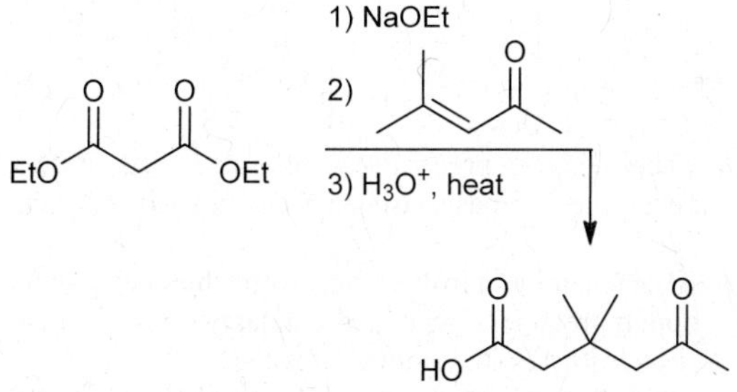

(b) As described in the solution to Problem **22.46a**, the conjugate base of diethyl malonate can function as a Michael donor. So we must identify the appropriate Michael acceptor, shown here:

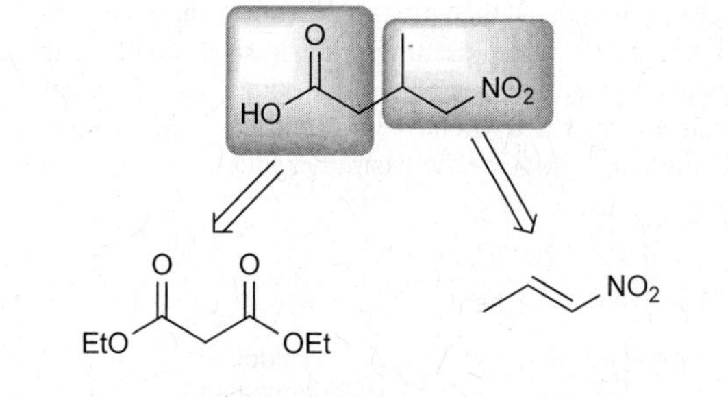

The forward scheme is shown here:

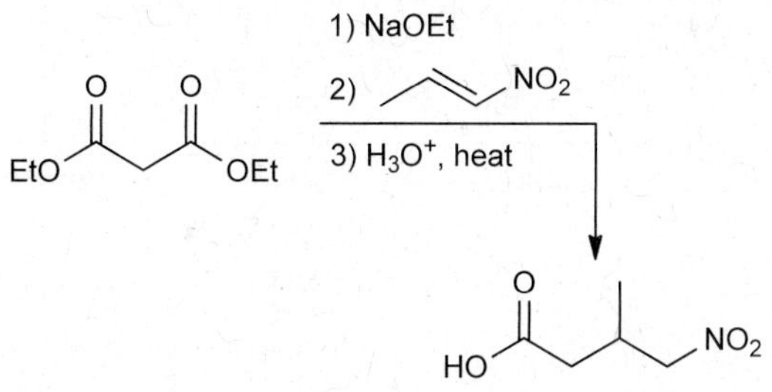

22.47.
(a) With the following retrosynthetic analysis, we can identify the starting reagents necessary to prepare this product via a Stork enamine synthesis:

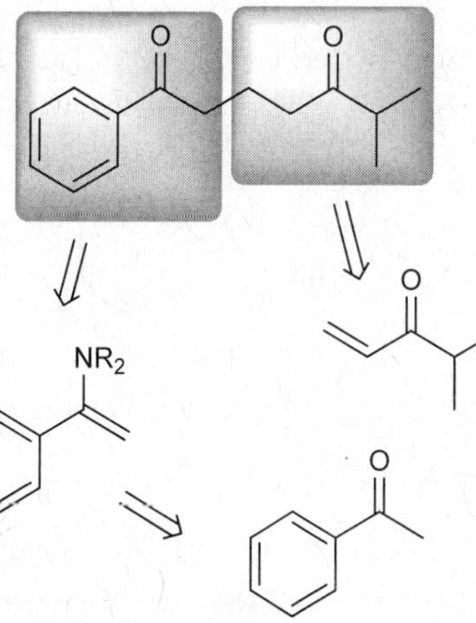

The forward scheme is shown here. The starting ketone is first treated with a secondary amine in acidic conditions (with removal of water) to give an enamine. This enamine is then used as a Michael donor in a Michael reaction with an α,β-unsaturated ketone. Aqueous acidic work-up gives the desired product.

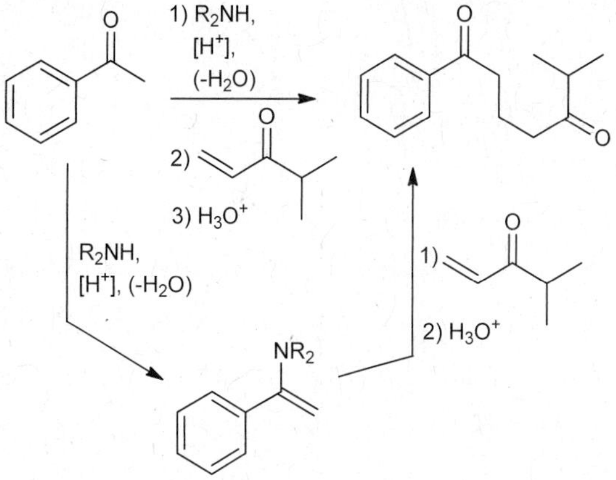

(b) With the following retrosynthetic analysis, we can identify the starting reagents necessary to prepare this product via a Stork enamine synthesis:

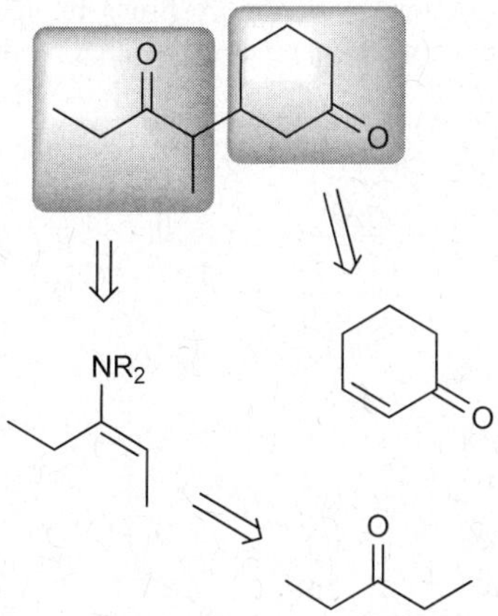

The forward scheme is shown here. The starting ketone is first treated with a secondary amine in acidic conditions (with removal of water) to give an enamine. This enamine is then used as a Michael donor in a Michael reaction with an α,β-unsaturated ketone. Aqueous acidic work-up gives the desired product.

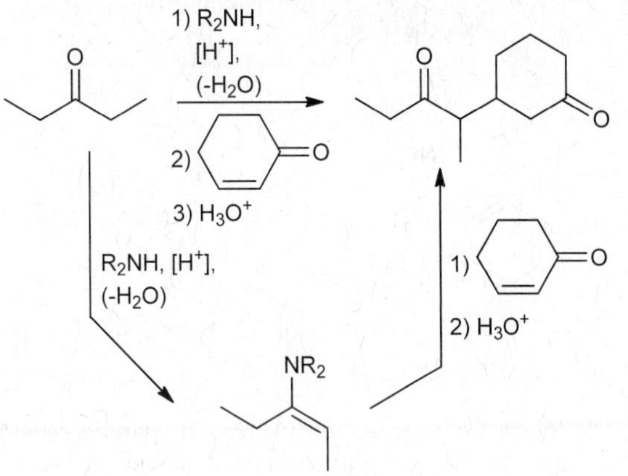

(c) With the following retrosynthetic analysis, we can identify the starting reagents necessary to prepare this product via a Stork enamine synthesis:

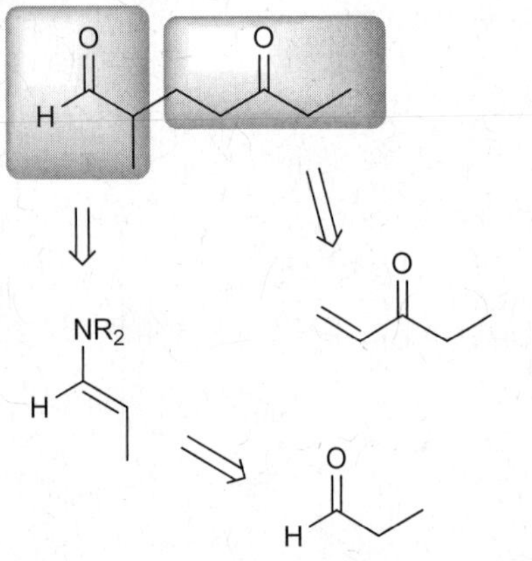

The forward scheme is shown here. The starting ketone is first treated with a secondary amine in acidic conditions (with removal of water) to give an enamine. This enamine is then used as a Michael donor in a Michael reaction with an α,β-unsaturated ketone. Aqueous acidic work-up gives the desired product.

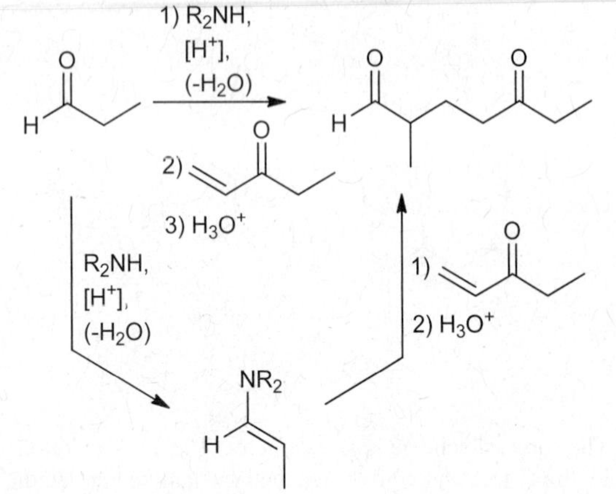

22.48. With the following retrosynthetic analysis, we can identify the starting reagents necessary to prepare this product via a Stork enamine synthesis:

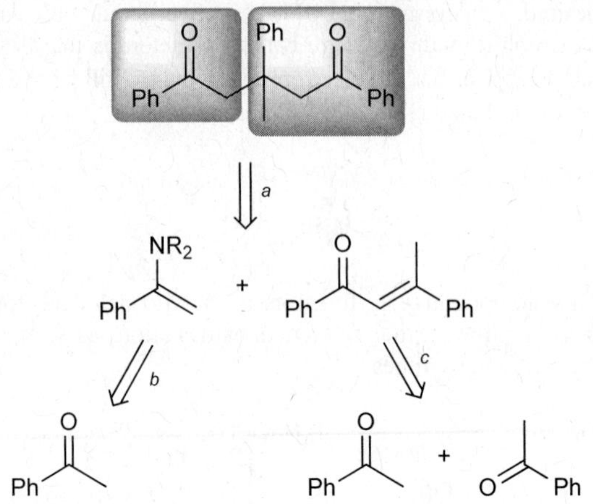

a. The desired product can be made via a Stork enamine synthesis using an enamine as a Michael donor.

b. The enamine can be made from the acetophenone upon treatment with a secondary amine in acidic conditions (with removal of water).

c. The desired product can be made via an aldol condensation between two equivalents of acetophenone.

The forward scheme is shown here. Two equivalents of acetophenone are used to create an α,β-unsaturated ketone, which is then treated with an enamine, which was formed upon treatment of acetophenone with a secondary amine in acidic conditions (with removal of water). After the Michael reaction is complete, aqueous acidic work-up gives the desired product.

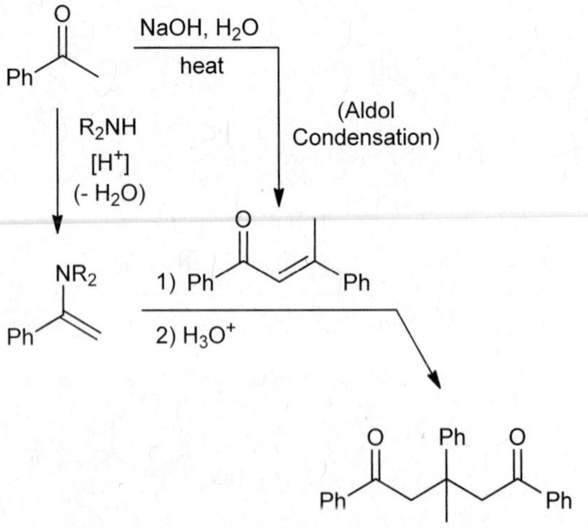

22.49. In the presence of a strong base, the β-dicarbonyl compound is deprotonated, giving a resonance-stabilized intermediate (doubly stabilized enolate) which then functions as a Michael donor, attacking the β position of the α,β-unsaturated ketone in a Michael reaction. The resulting enolate is then protonated, and then subsequently deprotonated to give a different enolate (these two steps represent equilibration of the enolates). The new enolate then attacks one of the carbonyl groups to initiate an aldol condensation. The resulting alkoxide ion is then protonated by water. Subsequent deprotonation and loss of hydroxide gives the product. Notice that water functions as the acid for all protonation steps, consistent with basic conditions (strong acids are not measurably present under these conditions).

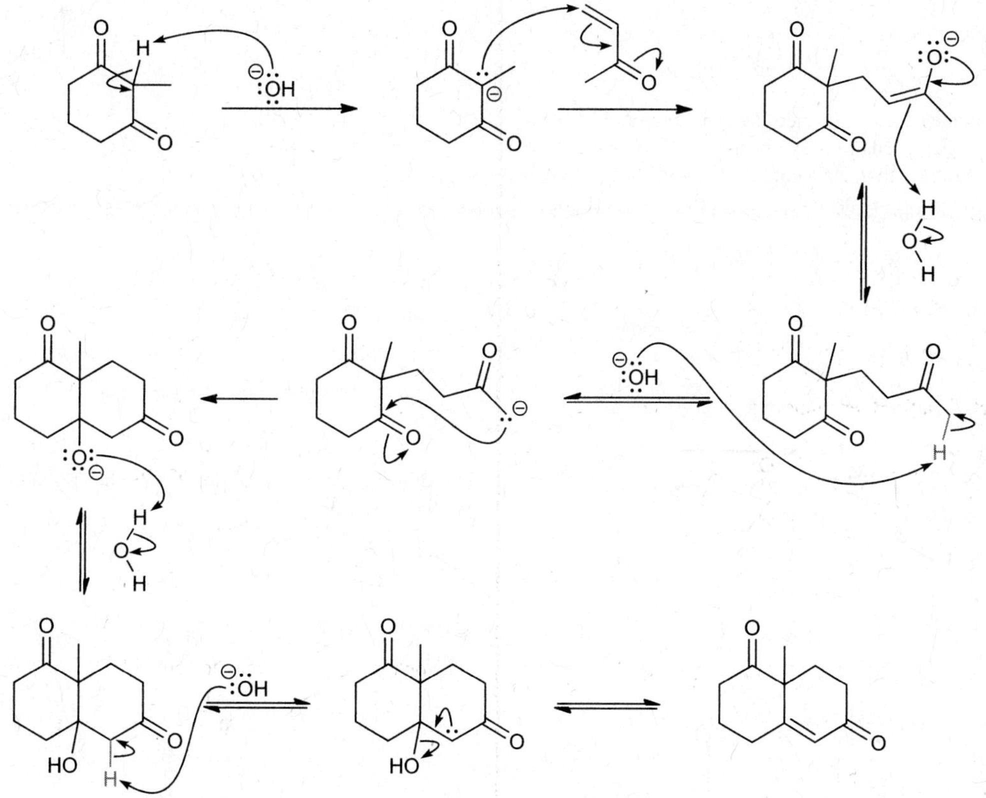

22.50. A Robinson annulation is comprised of a Michael reaction, followed by an intramolecular aldol condensation. To determine the starting materials necessary to prepare the desired product via a Robinson annulation, we draw the following retrosynthetic analysis:

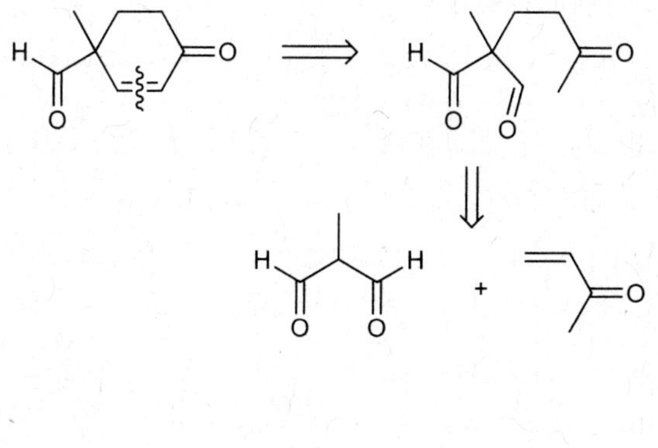

These two steps do not represent two separate reactions. A Robinson annulation can be performed in one reaction flask, as shown in the following forward scheme:

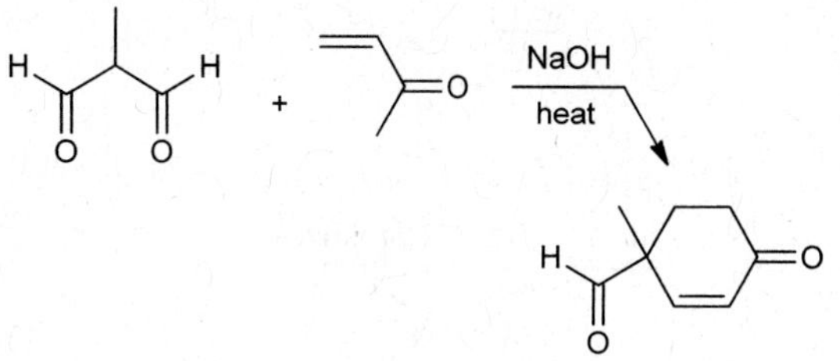

22.51.
(a) The product is 1,5-difunctionalized:

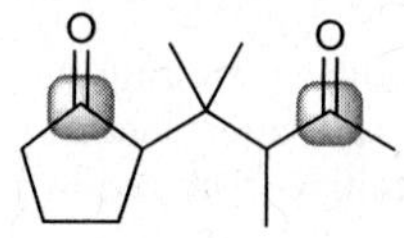

Therefore, we consider preparing the product via a Michael reaction, which would require the following starting materials:

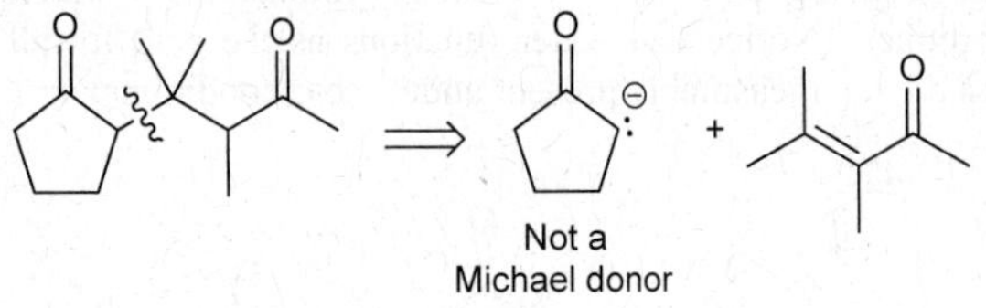

Not a
Michael donor

Since enolates are not efficient Michael donors, we must consider a Stork enamine synthesis (in which we use an enamine, rather than an enolate, as a Michael donor). The enamine can be made directly from cyclopentanone.

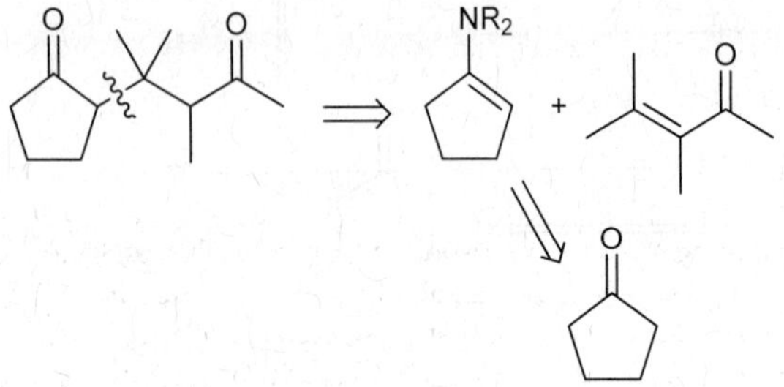

The forward scheme is shown here:

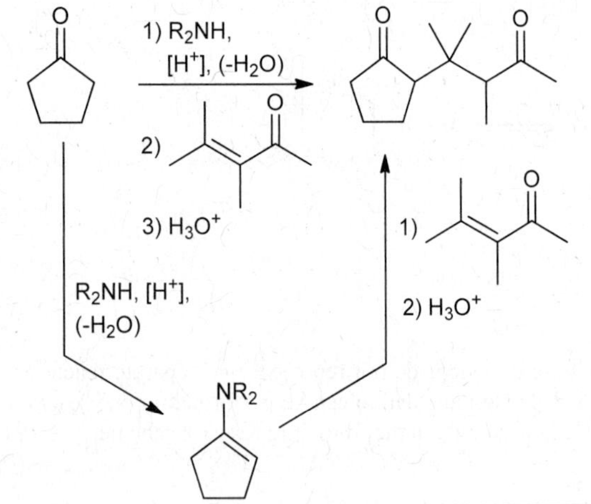

1) R₂NH,
   [H⁺], (-H₂O)

2)

3) H₃O⁺

R₂NH, [H⁺],
(-H₂O)

1)

2) H₃O⁺

NR₂

(b) The product is 1,3-difunctionalized:

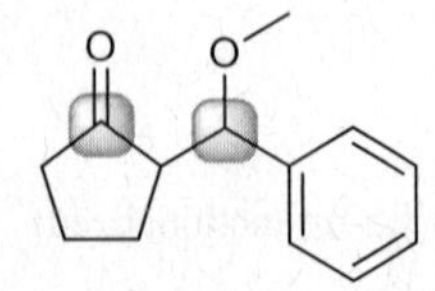

Therefore, we consider preparing the product via either an aldol reaction or a Claisen condensation. In this case, a directed aldol addition, followed by methylation of the

resulting alkoxide ion (in a Williamson ether synthesis), gives the desired product:

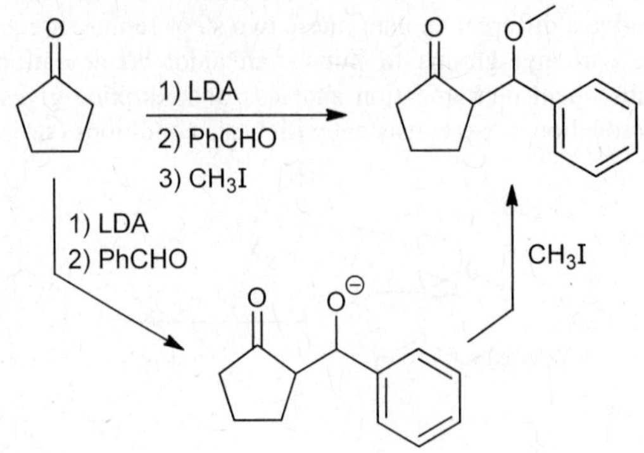

1) LDA

2) PhCHO

3) CH₃I

1) LDA

2) PhCHO

CH₃I

(c) The product is 1,5-difunctionalized:

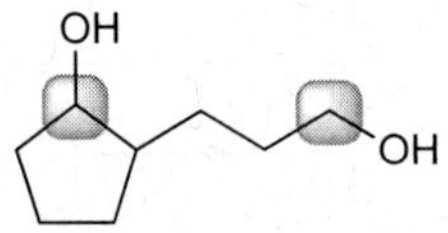

Therefore, we consider preparing the product via a Michael reaction:

Not a
Michael donor

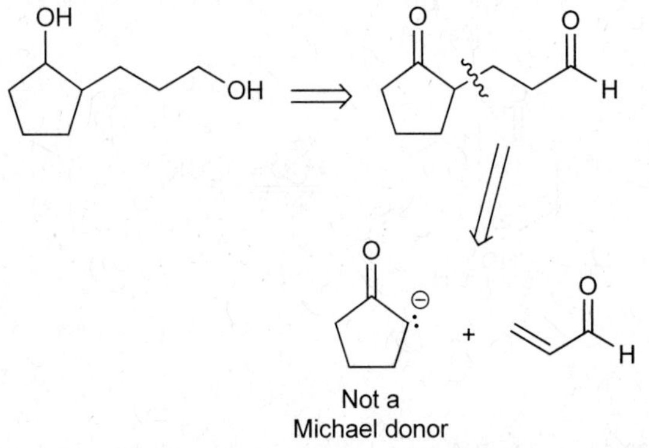

This strategy will not work, because it involves the use of an enolate, which is not an efficient Michael donor. Therefore, we consider a Stork enamine synthesis (in which we use an enamine, rather than an enolate, as a Michael donor). The enamine can be made directly from cyclopentanone.

NR₂

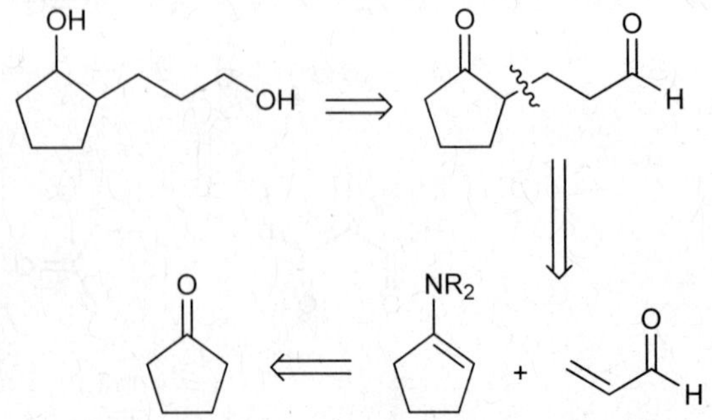

The forward scheme is shown here:

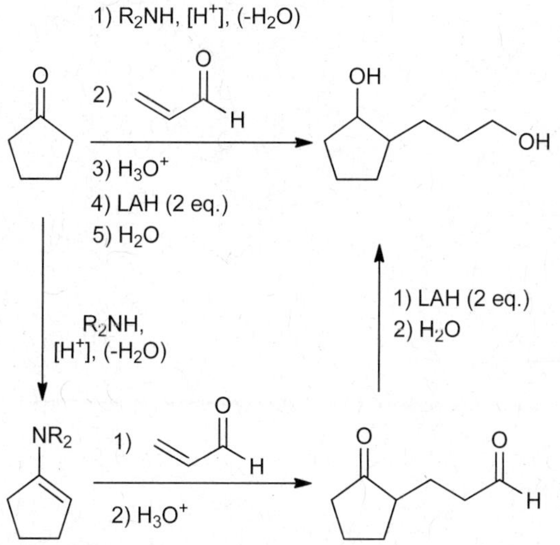

22.52.

(a) The product is 1,3-difunctionalized:

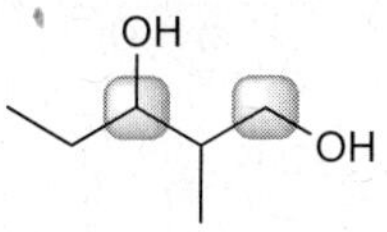

Therefore, we consider preparing the product via either an aldol reaction or a Claisen condensation. In this case, an aldol addition reaction can be employed, as shown in the following retrosynthetic analysis.

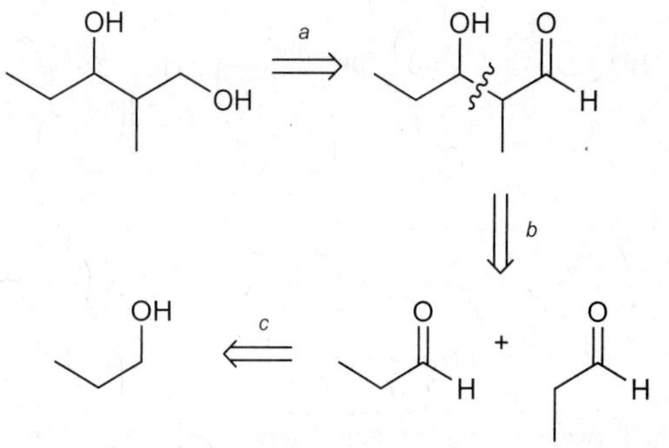

a. The product can be made via reduction of a β-hydroxyaldehyde.

b. The β-hydroxyaldehyde can be made via an aldol addition reaction between two molecules of propanal.

c. Propanal can be made via oxidation of 1-propanol with PCC.

Now let's draw the forward scheme. Upon treatment with PCC, 1-propanol is oxidized to give propanal. Treating propanal with sodium hydroxide then gives a β-hydroxyaldehyde (via an aldol addition reaction between two molecules of propanal). Reduction with LAH, followed by water work-up, gives the product.

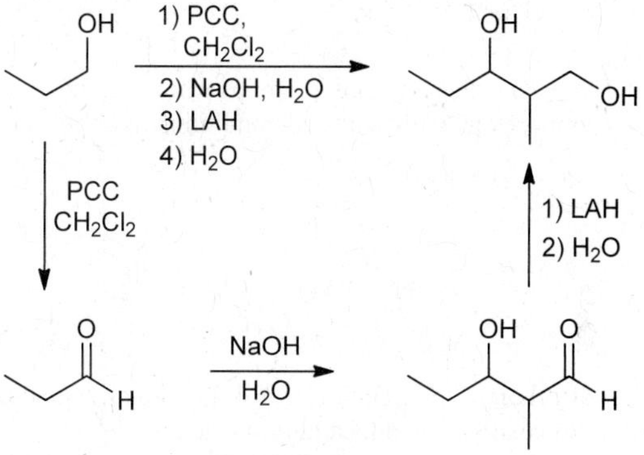

(b) The product is 1,5-difunctionalized:

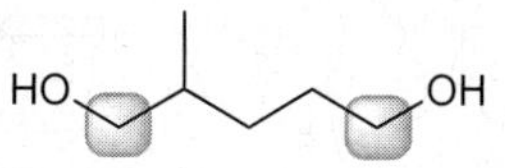

Therefore, we consider preparing the product via a Michael reaction:

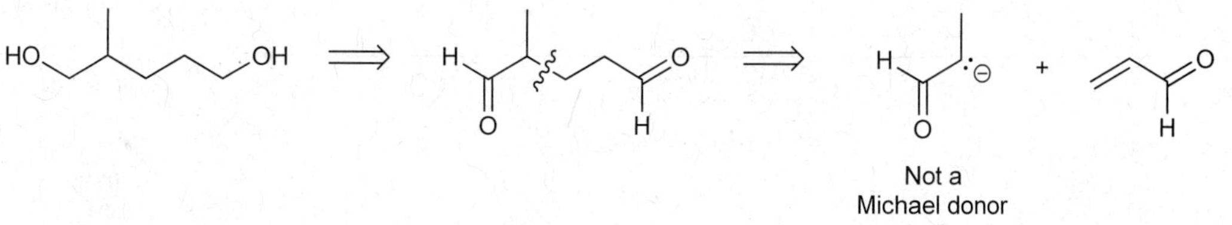

This strategy will not work, because it involves the use of an enolate, which is not an efficient Michael donor. Therefore, we consider a Stork enamine synthesis (in which we use an enamine, rather than an enolate, as a Michael donor). Both the Michael donor and the Michael acceptor can be made from propanal, which can be made from 1-propanol via oxidation with PCC:

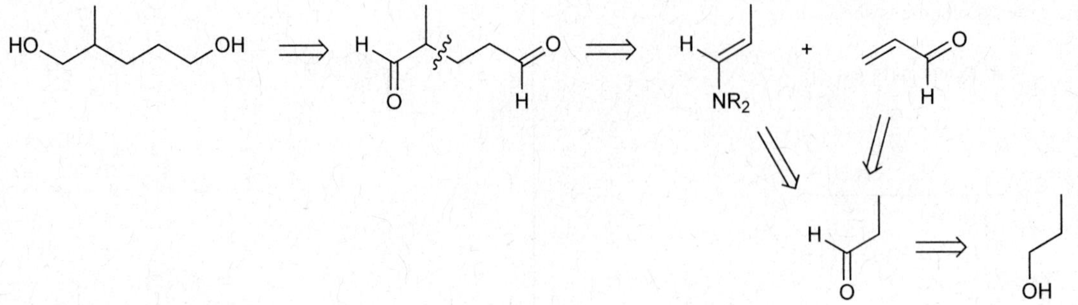

The forward scheme is shown here:

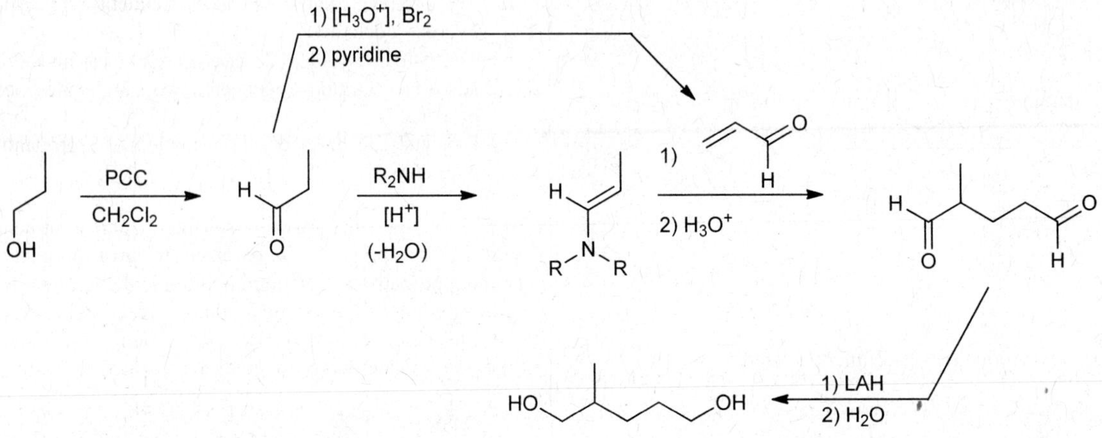

(c) The product has two imine groups which can be made from the corresponding dicarbonyl compound upon treatment with ammonia in acid-catalyzed conditions (with removal of water):

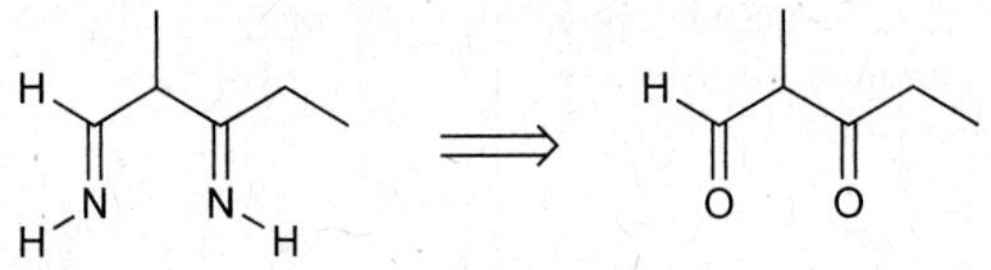

This dicarbonyl compound is 1,3-difunctionalized and can be made from a β-hydroxyaldehyde, which can be made from propanal via an aldol addition reaction:

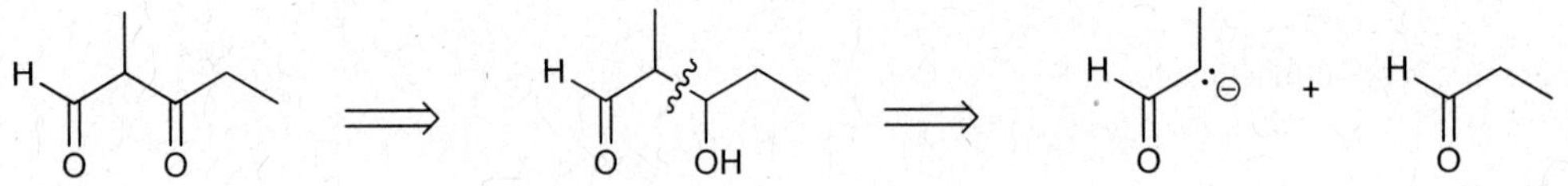

And propanal can be made from 1-propanol via oxidation with PCC. The forward scheme is shown here. Notice that the third step of this synthesis employs PCC, rather than chromic acid, to avoid oxidation of the aldehyde group.

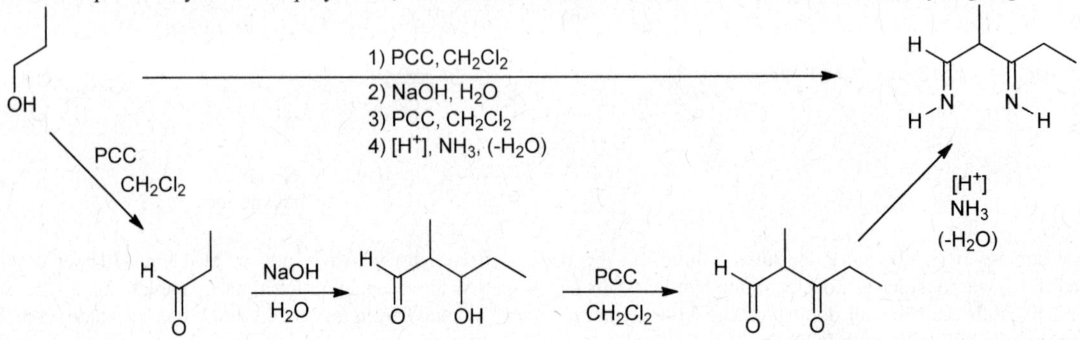

22.53.

(a) The compound possesses three functional groups, and can be assembled in a variety of ways. One method capitalizes on the 1,5-arrangement of two of the functional groups (highlighted):

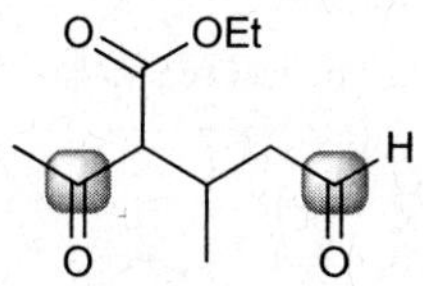

Therefore, we consider preparing the product via a Michael reaction. The following retrosynthetic analysis is based on assembly of the carbon skeleton via a Michael reaction, as well as an aldol condensation reaction to prepare the Michael acceptor, and a Claisen condensation to prepare the Michael donor:

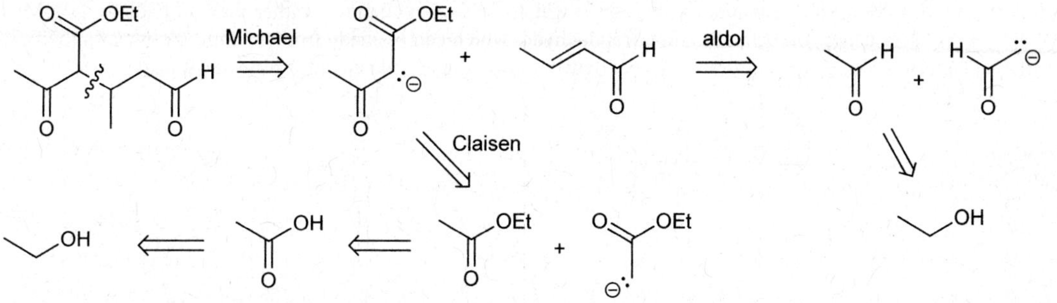

The forward scheme for this strategy is shown below. One equivalent of ethanol is oxidized with PCC to give acetaldehyde, which is then heated with aqueous sodium hydroxide to give an α,β-unsaturated aldehyde (via an aldol condensation reaction). Another equivalent of ethanol is oxidized with chromic acid to give a carboxylic acid, which is then treated with ethanol under acidic conditions to give an ester (via Fischer esterification). The ester is then converted into a β-ketoester (via a Claisen condensation). The β-ketoester is then deprotonated with ethoxide to give a doubly stabilized enolate which then attacks the α,β-unsaturated aldehyde to give a Michael reaction:

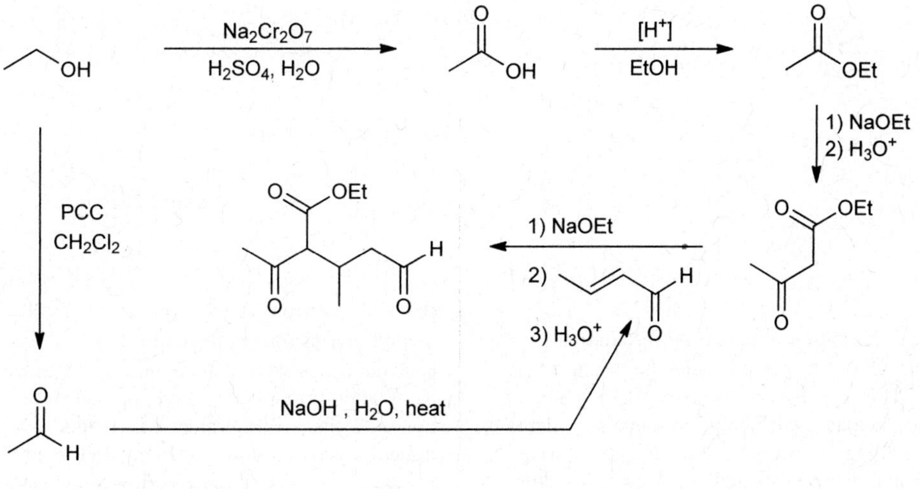

(b) The product is 1,5-difunctionalized:

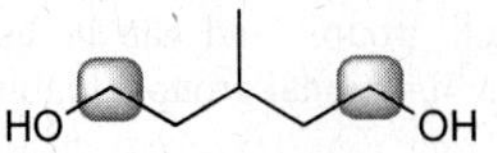

Therefore, we consider preparing the product via a Michael reaction:

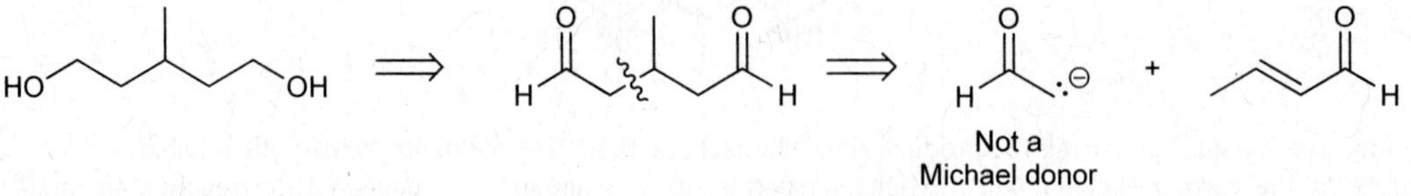

Not a
Michael donor

This strategy will not work, because it involves the use of an enolate, which is not an efficient Michael donor. Therefore, we consider a Stork enamine synthesis (in which we use an enamine, rather than an enolate, as a Michael donor). The Michael donor can be made from acetaldehyde, which can be made from ethanol via oxidation with PCC. The Michael acceptor can be made from two equivalents of acetaldehyde via an aldol condensation:

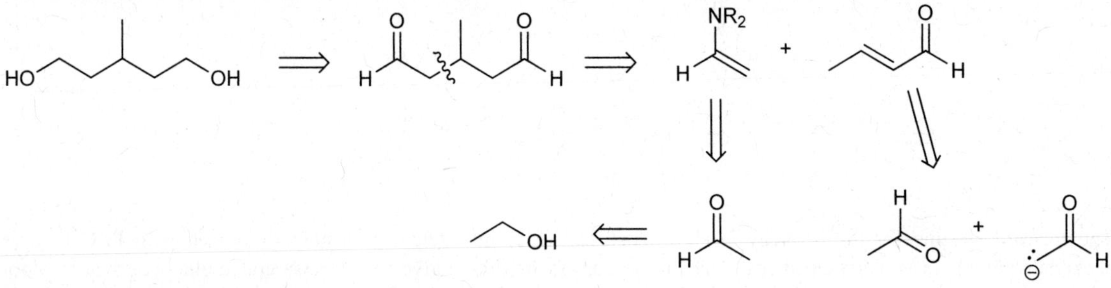

The forward scheme is shown here:

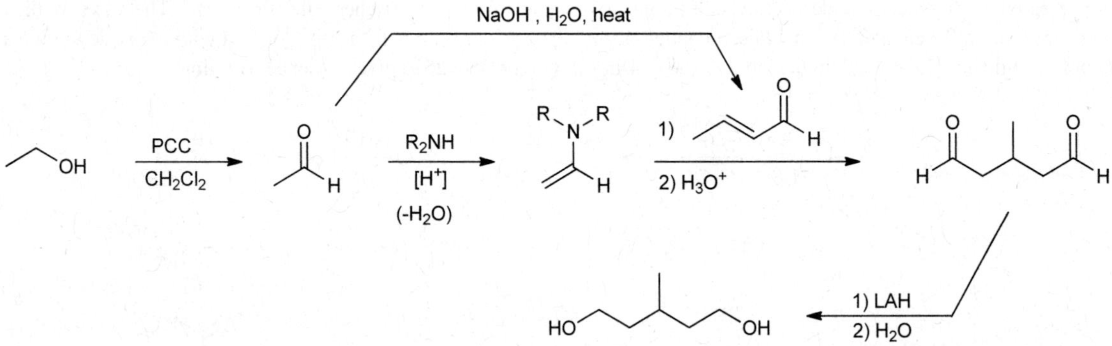

NaOH , H₂O, heat

PCC
CH₂Cl₂

R₂NH
[H⁺]
(-H₂O)

1)
2) H₃O⁺

1) LAH
2) H₂O

22.54.
(a) This transformation requires the installation of two groups (one at the α position and the other at the β position). This can be achieved by treating the α,β-unsaturated ketone with lithium dimethyl cuprate, thereby installing a methyl group and generating an enolate, which is then treated with benzyl iodide to install a benzyl group:

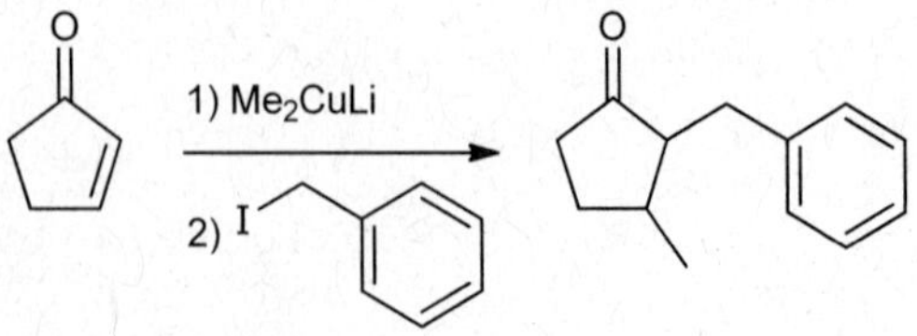

1) Me₂CuLi

2) I

(b) This transformation requires the installation of two methyl groups (one at the α position and the other at the β position), as well as reduction of the aldehyde group. Installation of the two methyl groups can be achieved by treating the α,β-unsaturated ketone with lithium dimethyl cuprate, followed by methyl iodide. Reduction is then achieved with LAH, followed by water work-up.

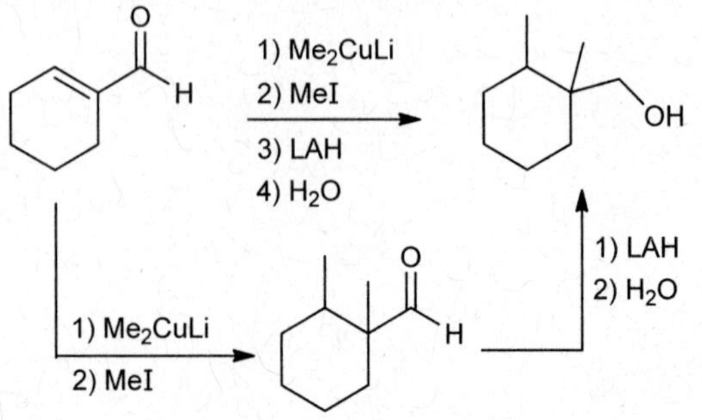

1) Me₂CuLi
2) MeI
3) LAH
4) H₂O

1) Me₂CuLi
2) MeI

1) LAH
2) H₂O

(c) This transformation requires the installation of two methyl groups (one at the α position and the other at the β position), as well as conversion of the aldehyde group into an acid halide. Installation of the two methyl groups can be achieved by treating the α,β-unsaturated aldehyde with lithium dimethyl cuprate, followed by methyl iodide. The aldehyde is then converted into an acid halide via oxidation with chromic acid (to give a carboxylic acid) followed by treatment with thionyl chloride.

Note than the acid chloride group is formed at the end of the synthesis, because if it were formed in the beginning of the synthesis, then lithium dimethyl cuprate would react with the acid chloride group.

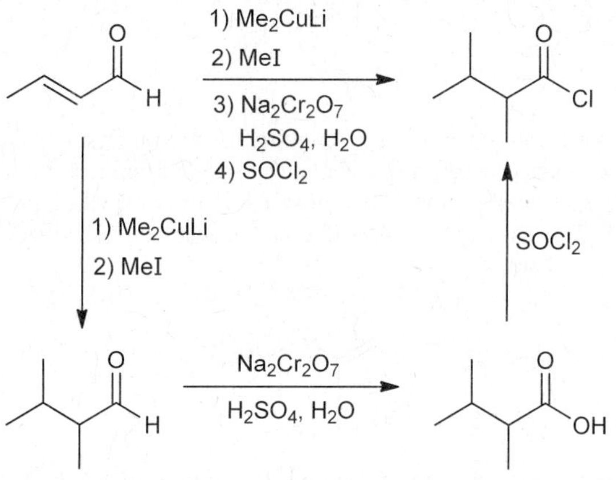

(d) If the acetal is hydrolyzed with aqueous acid, the resulting α,β-unsaturated ketone can be treated with lithium diethyl cuprate, followed by ethyl iodide, to install the two ethyl groups in the correct locations. The ketone can then be converted back into an acetal upon treatment with ethylene glycol under acidic conditions (with removal of water).

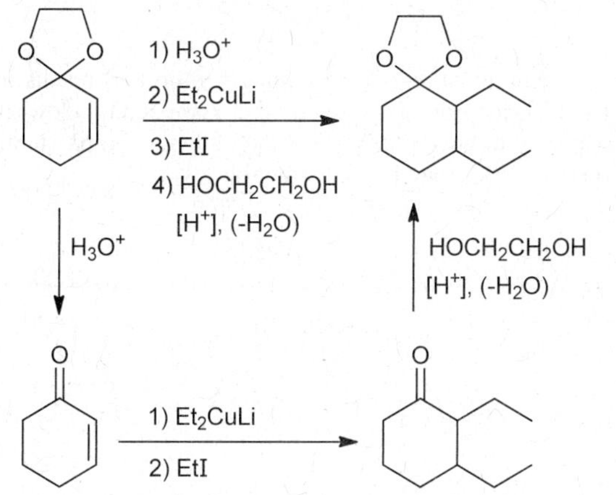

(e) The imine can be prepared from the corresponding aldehyde:

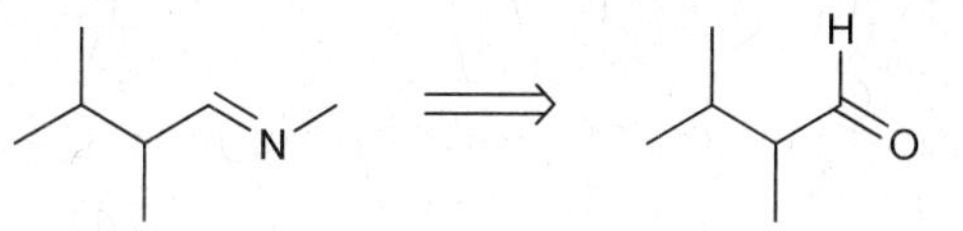

This aldehyde can be made from an α,β-unsaturated aldehyde, with installation of two methyl groups (one at the α position and the other at the β position):

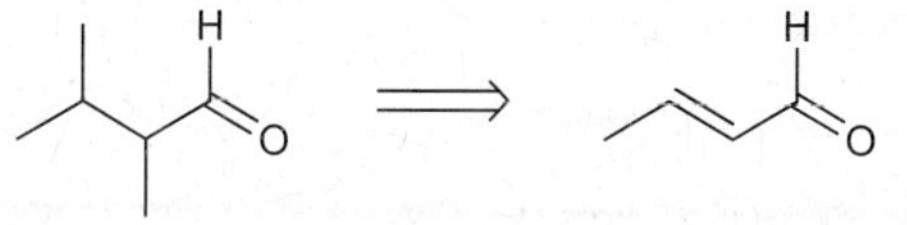

The α,β-unsaturated aldehyde can be made from the starting material via oxidation with PCC. The entire synthesis is summarized here:

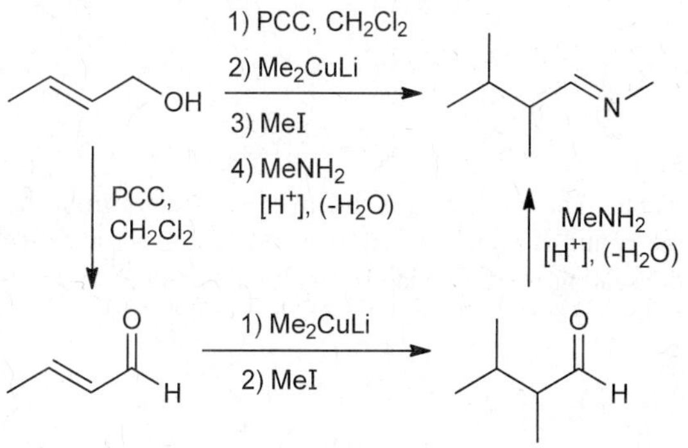

(f) This problem is very similar to the previous problem, although the final step is reduction of the aldehyde with a Clemmensen reduction to give an alkane.

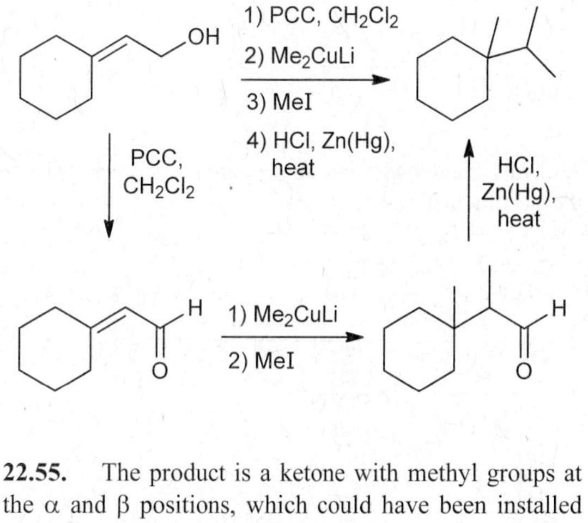

22.55. The product is a ketone with methyl groups at the α and β positions, which could have been installed by treating the following α,β-unsaturated ketone with lithium dimethyl cuprate, followed by methyl iodide:

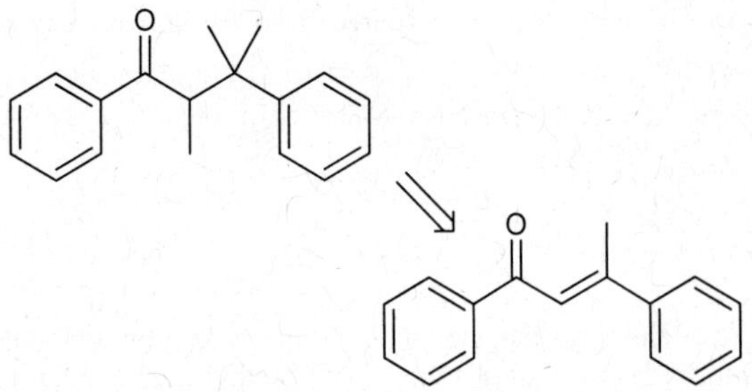

This α,β-unsaturated ketone can be prepared via an aldol condensation, as shown:

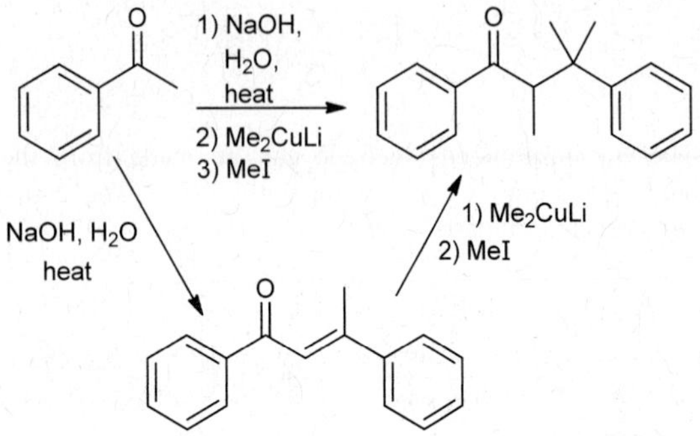

22.56. The product is an aldehyde with alkyl groups at the α and β positions, which could have been installed by treating the following α,β-unsaturated aldehyde with lithium diethyl cuprate, followed by methyl iodide:

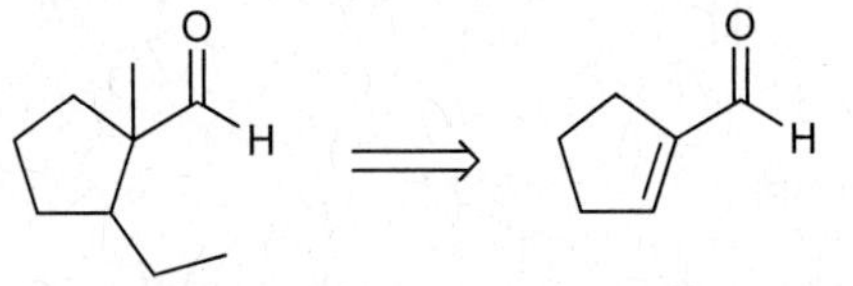

This α,β-unsaturated aldehyde can be prepared via an intramolecular aldol condensation, as shown:

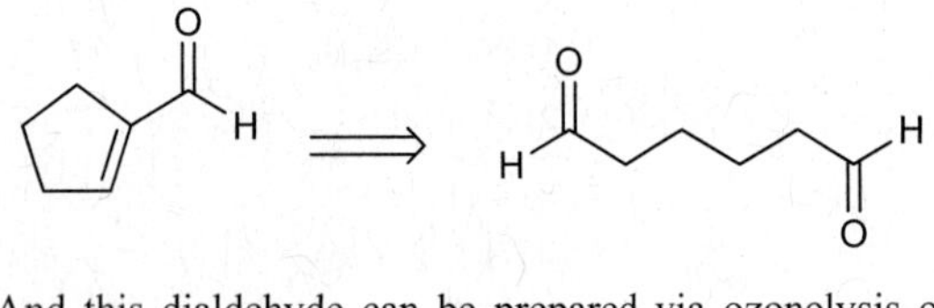

And this dialdehyde can be prepared via ozonolysis of the starting material, as shown:

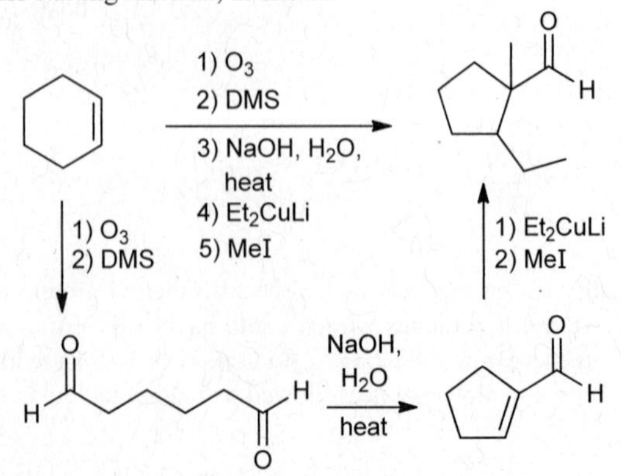

22.57.
(a) Deprotonation (of the highlighted proton) results in a resonance-stabilized enolate ion. Therefore, the highlighted proton is the most acidic proton (with a pK_a just below 20), because its removal leads to a stabilized conjugate base.

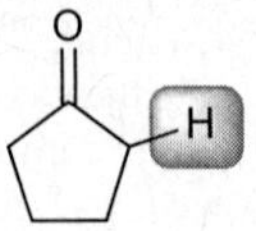

(b) This compound does not have an acidic proton, and is expected to have a pK_a above 20.

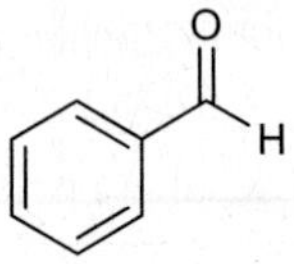

(c) Deprotonation (of the highlighted proton) results in a resonance stabilized enolate ion. Therefore, the highlighted proton is the most acidic proton (with a pK_a just below 20), because its removal leads to a stabilized conjugate base.

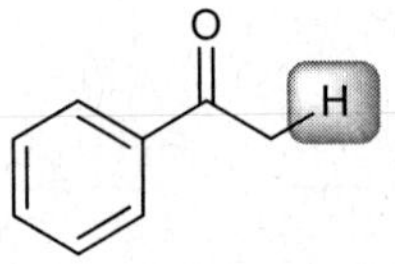

(d) Deprotonation (of the highlighted proton) results in a resonance stabilized conjugate base (a doubly-stabilized enolate). Therefore, the highlighted proton is the most acidic proton (with a pK_a just below 20), because its removal leads to the most stable conjugate base possible.

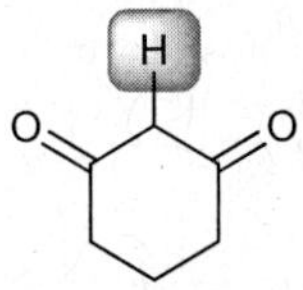

(e) Deprotonation (of the highlighted proton) results in an alkoxide ion. As such, the compound below is expected to have a pK_a lower than 20 (see the pK_a table on the inside cover of the textbook).

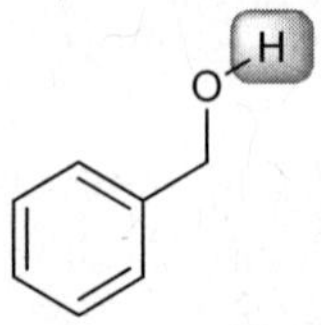

22.58. The most acidic proton is connected to the position that is α to two carbonyl groups (in between both carbonyl groups). Deprotonation at this location leads to a resonance stabilized conjugate base in which the negative charge is spread over two oxygen atoms and one carbon atom).

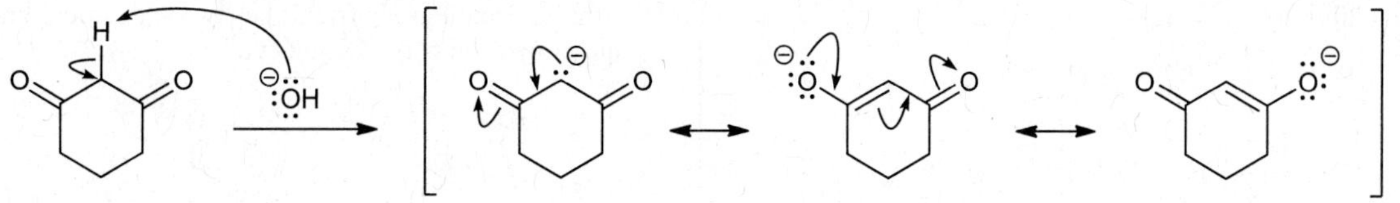

22.59.

(a) The most acidic proton is connected to the position that is α to both carbonyl groups (in between both carbonyl groups). Deprotonation at this location leads to a resonance stabilized conjugate base in which the negative charge is spread over two oxygen atoms and one carbon atom).

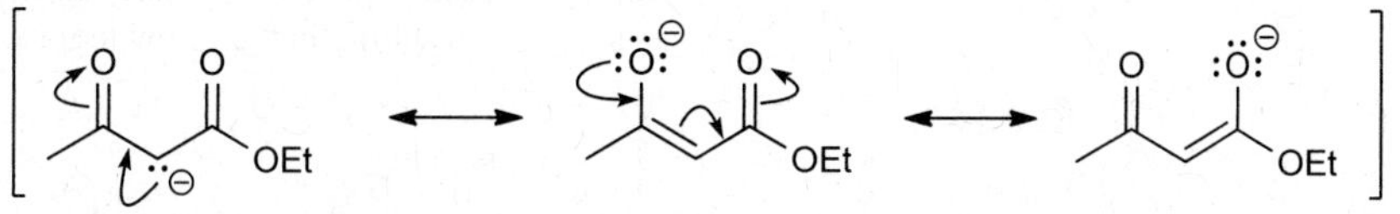

(b) The most acidic proton is connected to the position that is α to both carbonyl groups (in between both carbonyl groups). Deprotonation at this location leads to a resonance stabilized conjugate base in which the negative charge is spread over two oxygen atoms and one carbon atom).

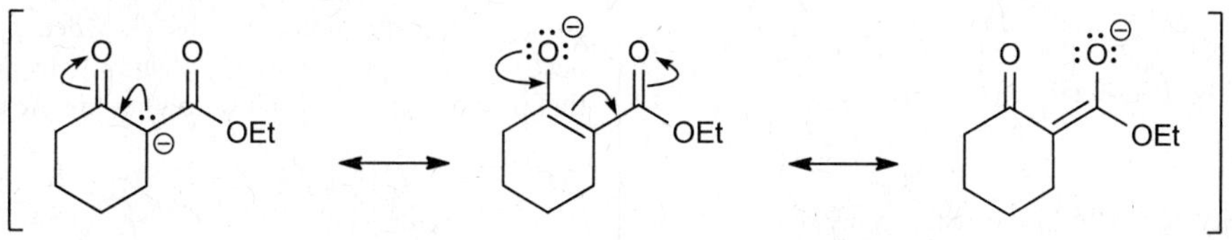

(c) The most acidic proton is connected to the position that is α to the carbonyl group as well as the cyano group (in between both groups). Deprotonation at this location leads to a resonance stabilized conjugate base in which the negative charge is spread over an oxygen atom, a nitrogen atom and a carbon atom).

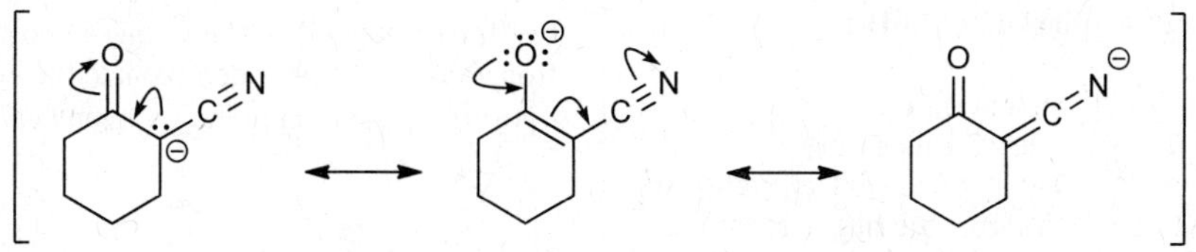

22.60. The most acidic compound is the one that exhibits a position that is α to three carbonyl groups (deprotonation of this compound gives a conjugate base in which the negative charge is spread over three oxygen atoms and one carbon atom). The next most acidic compound is the one that exhibits a position that is α to two carbonyl groups (deprotonation of this compound gives a conjugate base in which the negative charge is spread over two oxygen atoms and one carbon atom). Of the remaining two compounds, an alcohol is generally more acidic than a ketone (see the pK_a table on the inside cover of the textbook).

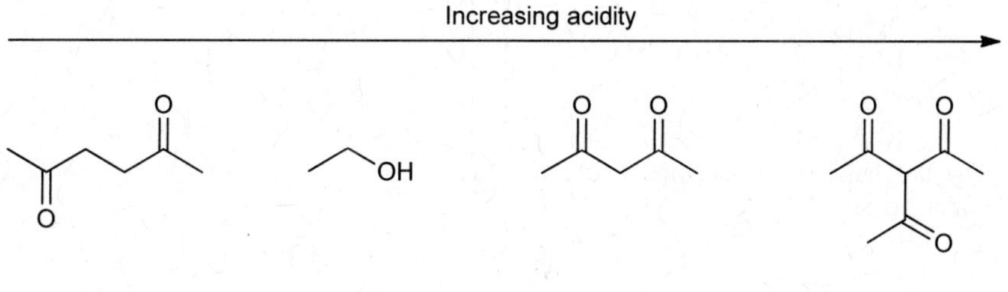

Increasing acidity

22.61.

(a) This enol does not exhibit a significant presence at equilibrium. Ketones are generally favored at equilibrium.

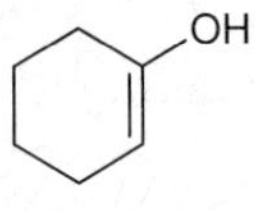

(b) The following enol does exhibit a significant presence at equilibrium because it exhibits conjugation as well as intramolecular hydrogen bonding (between the oxygen atom of the carbonyl group and the proton of the OH group):

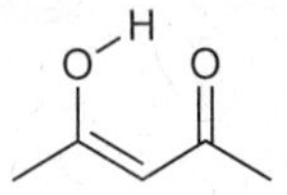

(c) This compound has a position that is α to both carbonyl groups, but that position does not bear any protons. As such, we must evaluate the following enol:

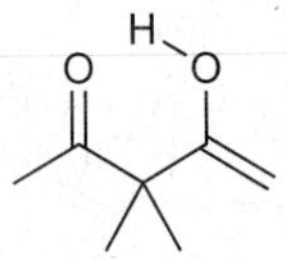

This enol does not exhibit the stability associated with conjugation, as seen in the solution to the previous problem (**22.61b**). However, it does exhibit intramolecular hydrogen bonding. As such, there is one factor that stabilizes the enol, so it is expected to exhibit a significant presence at equilibrium, (albeit less significant than the enol shown in **22.61b**).

(d) The following enol does indeed exhibit a significant presence at equilibrium, because it is aromatic. In fact, in this case, the ketone does not exhibit a significant presence at equilibrium. The aromatic ring is so strongly favored, that we cannot detect the ketone present in the mixture.

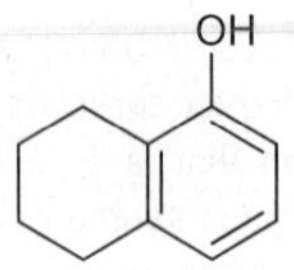

22.62. Ethyl acetoacetate has two carbonyl groups:

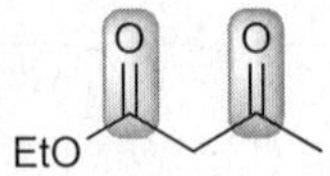

One of them (left) has only one α position and can therefore only form one enol:

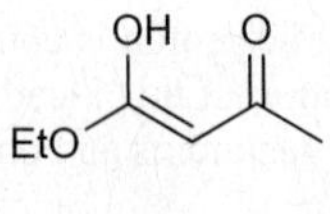

The other carbonyl group (right) has two α positions and can therefore form two different enols:

In total, there are three enol isomers.

22.63.

(a) This compound has only one α position. LDA is a strong base and it will deprotonate the compound (at the α position), resulting in the following enolate.

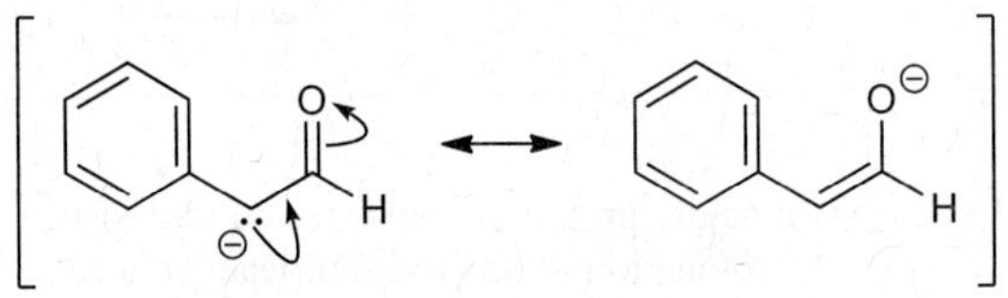

(b) This compound has two α positions. LDA is a strong, sterically hindered base, so deprotonation will occur at the less substituted position. Deprotonation at that position results in the following kinetic enolate:

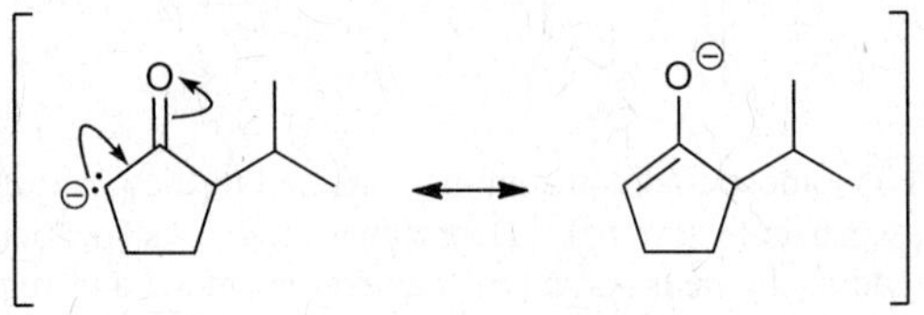

(c) This compound has only one α position. LDA is a strong base and it will deprotonate the compound (at the α position), resulting in the following enolate:

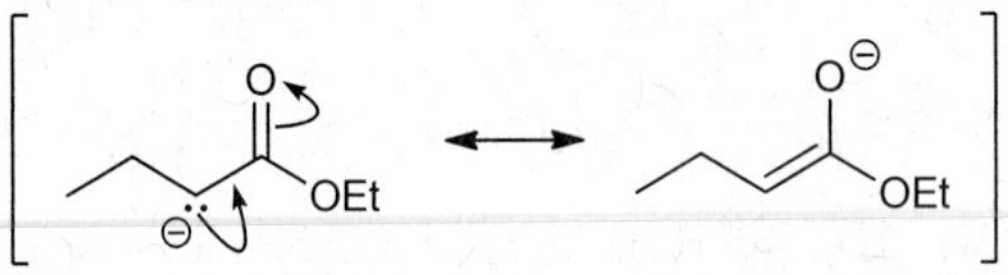

(d) This compound has two α positions. LDA is a strong, sterically hindered base, so deprotonation will occur at the less substituted position. Deprotonation at that position results in the following kinetic enolate:

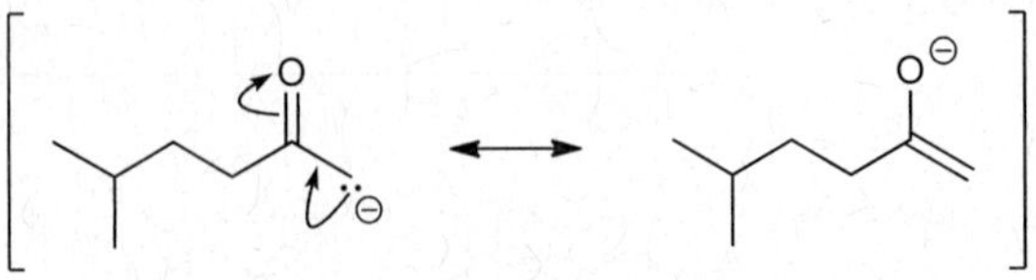

22.64. Deprotonation at the highlighted γ position results in an anion that has three resonance structures. The negative charge is spread over one oxygen atom and two carbon atoms:

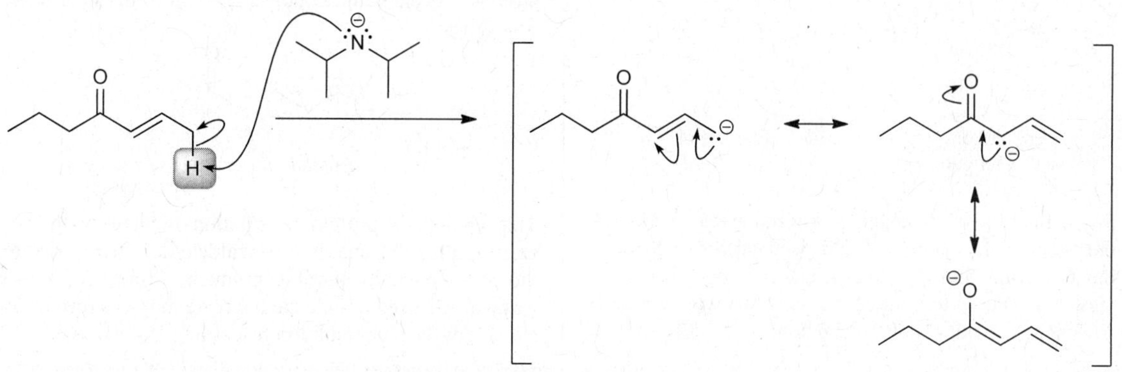

22.65. Deprotonation at the α carbon changes the hybridization state of the α carbon from sp^3 (tetrahedral) to sp^2 (planar). When the α position is protonated once again, the proton can be placed on either side of the planar α carbon, resulting in racemization:

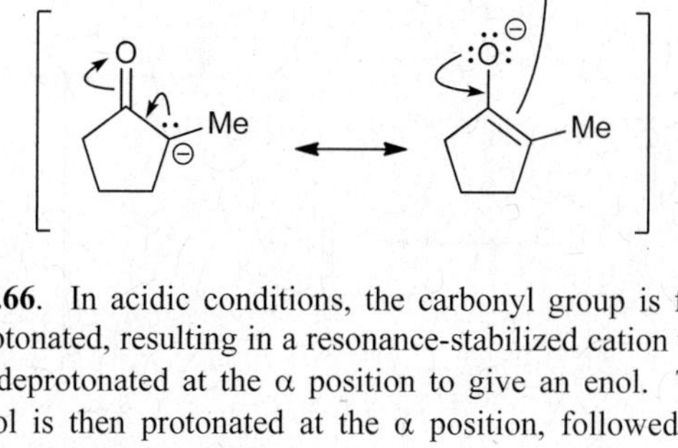

22.66. In acidic conditions, the carbonyl group is first protonated, resulting in a resonance-stabilized cation that is deprotonated at the α position to give an enol. The enol is then protonated at the α position, followed by deprotonation to give the product:

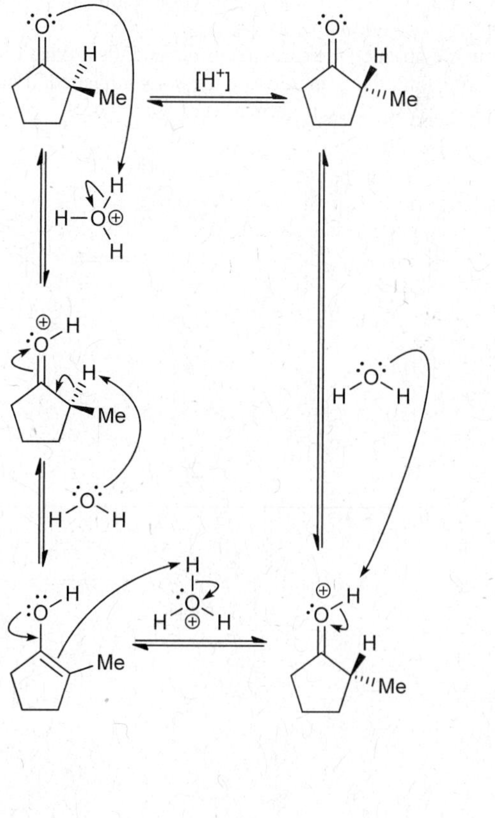

22.67. Each of the starting aldehydes has an α position that can be deprotonated, giving two possible enolates:

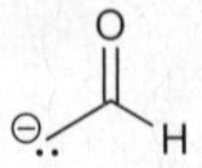

Enolate of
acetaldehyde

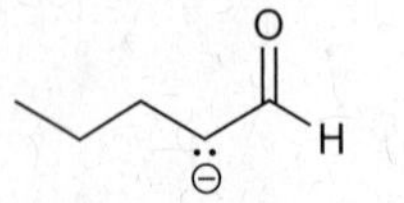

Enolate of
pentanal

So, there are two nucleophiles in solution, as well as two electrophiles (acetaldehyde and pentanal), giving rise to the following four possible products. In each case, a wavy line is used to indicate the bond that was formed as a result of the aldol addition reaction:

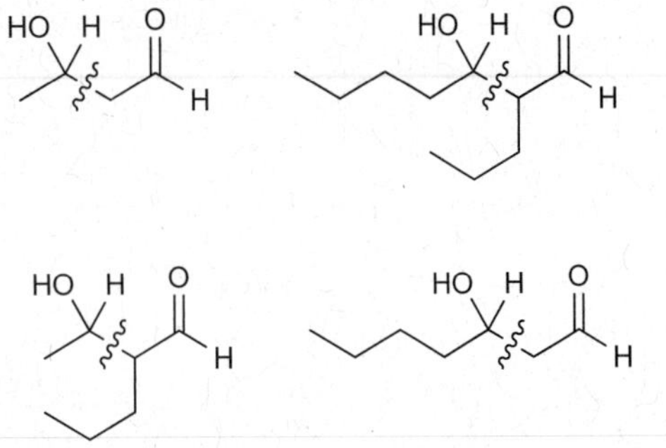

22.68. Hexanal has an α position that bears protons, but the α position of benzaldehyde does not bear any protons. As such, only one enolate can form under these conditions:

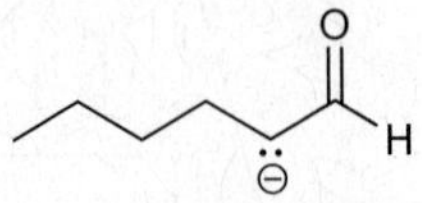

Enolate of hexanal

This enolate is present in solution together with two electrophiles (hexanal and benzaldehyde), giving rise to the following two possible products. In each case, a wavy line is used to indicate the bond that was formed as a result of the aldol addition reaction:

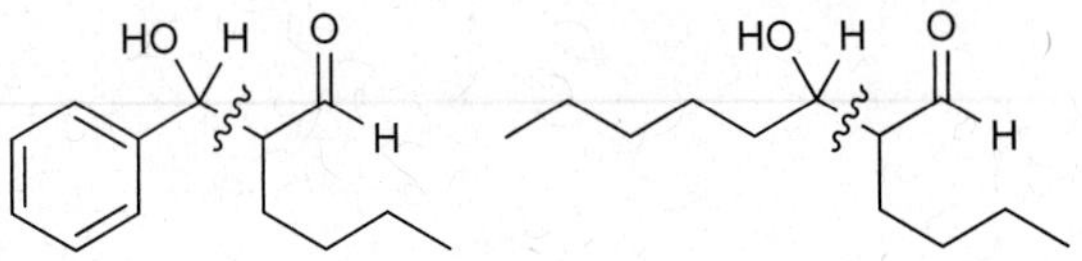

22.69. The carbonyl group is protonated, giving a resonance-stabilized intermediate that is then deprotonated to give an enol. Protonation of the enol results in a resonance-stabilized, benzylic carbocation intermediate that is then deprotonated to give the product.

In the product, the carbonyl group and the aromatic ring are conjugated. However, in the starting material, the carbonyl group and the aromatic ring are *not* conjugated. Formation of conjugation results in a decrease in energy which serves as a driving force in formation of the product.

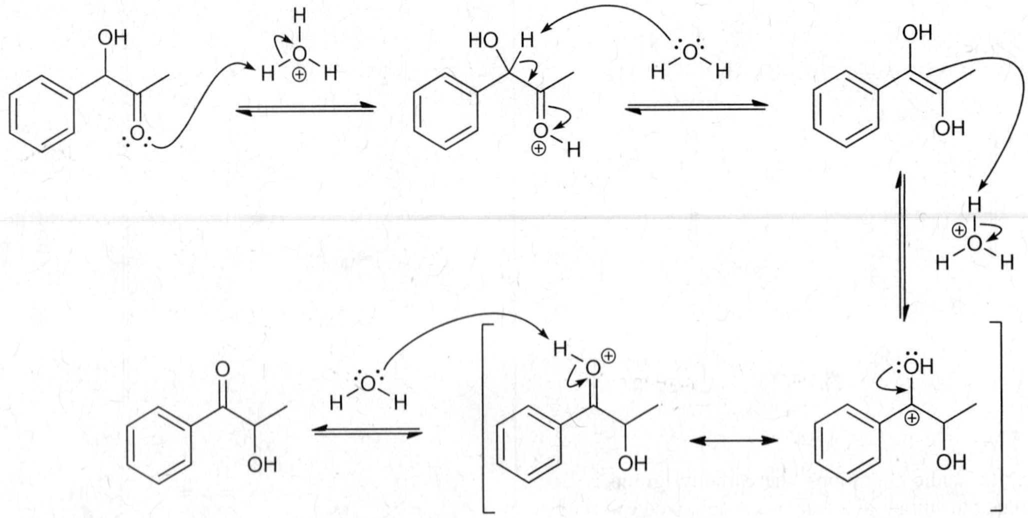

22.70. Hydroxide functions as a base and deprotonates the α position, giving a resonance-stabilized enolate (only the more significant resonance structure is drawn below) that is protonated to give an enol. The enol is then deprotonated to give another enolate ion, which is then protonated to give the product.

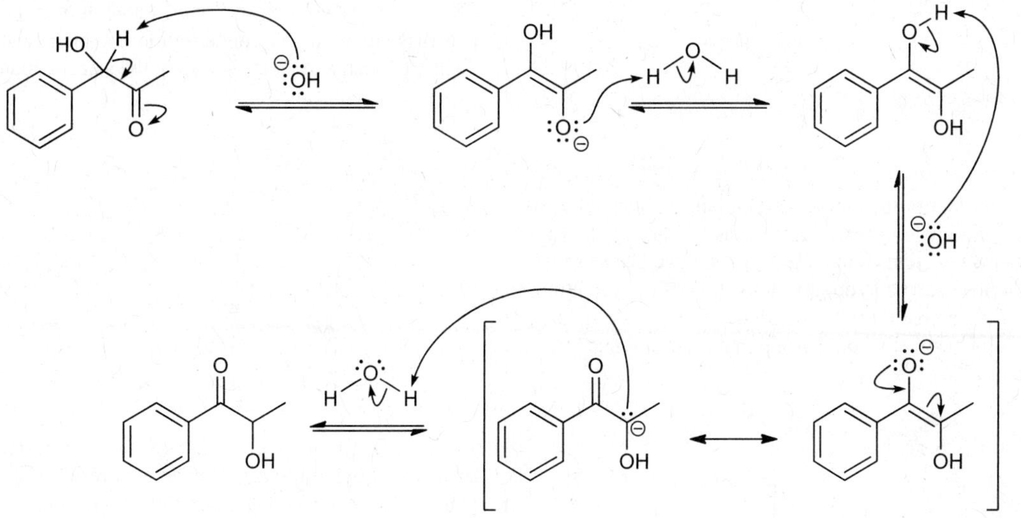

22.71.

(a) This compound (acetophenone) has only one α position that bears protons. Two molecules of the ketone are redrawn such that two α protons of one molecule are directly facing the carbonyl group of another molecule (highlighted). We then remove the two α protons and the oxygen atom, and we replace them with a double bond. In this case, two stereoisomeric products are possible, and we draw the *E* isomer, rather than the *Z* isomer, because the former has fewer steric interactions.

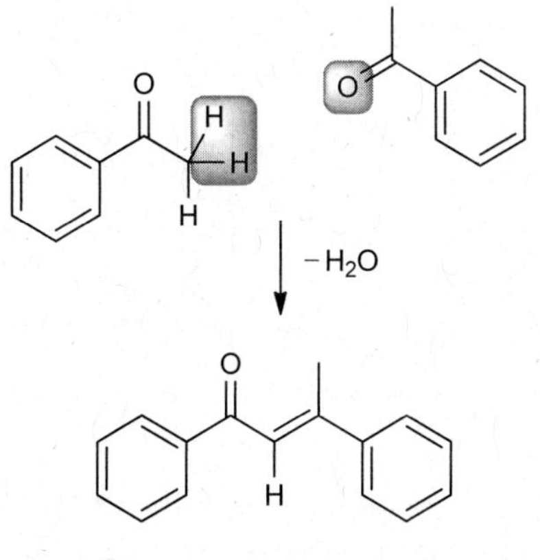

(b) This compound is an aldehyde, so there is only one α position. Two molecules of the aldehyde are redrawn such that two α protons of one molecule are directly facing the carbonyl group of another molecule (highlighted). We then remove the two α protons and the oxygen atom, and we replace them with a double bond. In this case, two stereoisomeric products are possible, and we draw the isomer that exhibits fewer steric interactions.

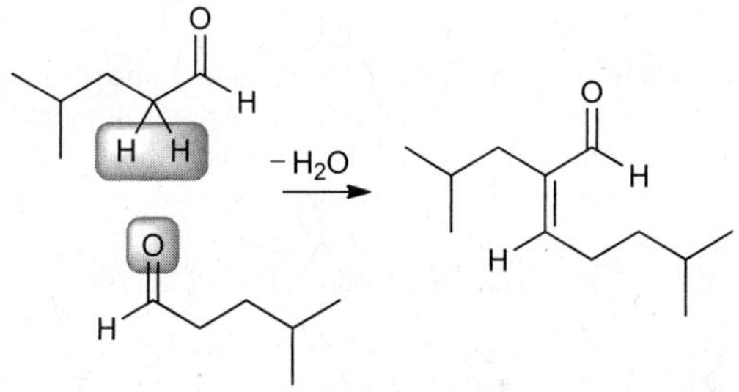

(c) This compound is an aldehyde, so there is only one α position. Two molecules of the aldehyde are redrawn such that two α protons of one molecule are directly facing the carbonyl group of another molecule (highlighted). We then remove the two α protons and the oxygen atom, and we replace them with a double bond. In this case, two stereoisomeric products are possible, and we draw the isomer that exhibits fewer steric interactions.

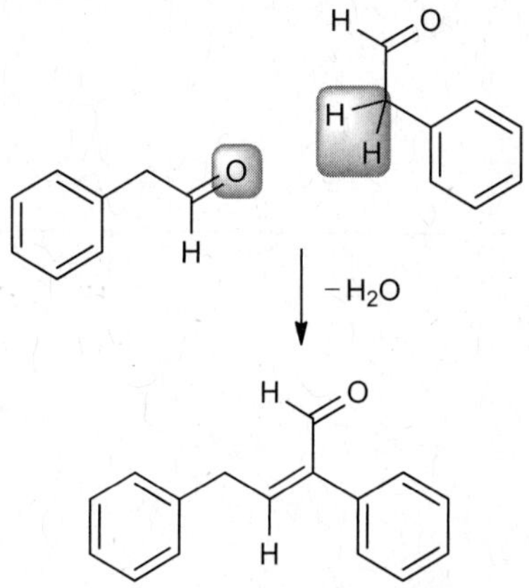

22.72. Trimethylacetaldehyde does not have any α protons, and therefore, a base cannot deprotonate the α position (the first step of an aldol reaction). As such, this compound cannot undergo an aldol reaction.

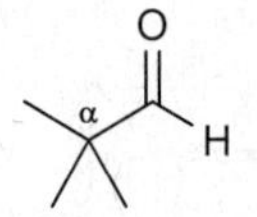

22.73.
(a) First identify the carbon-carbon bond (indicated with a wavy line below) that is formed as a result of an aldol condensation. This double bond must have been formed via the loss of two hydrogen atoms and an oxygen atom. The β position of the condensation product must have originally been a carbonyl group in the starting material:

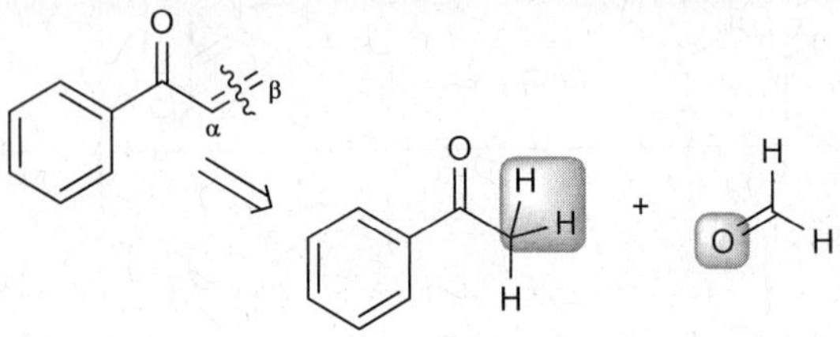

(b) First identify the carbon-carbon bond (indicated with a wavy line below) that is formed as a result of an aldol condensation. This double bond must have been formed via the loss of two hydrogen atoms and an oxygen atom. The β position of the condensation product must have originally been a carbonyl group in the starting material:

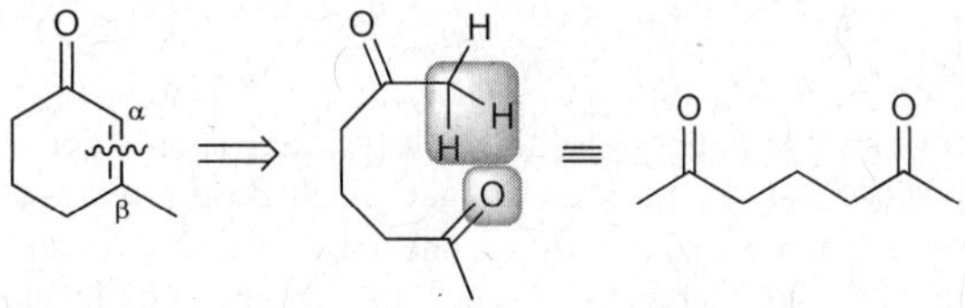

(c) First identify the carbon-carbon bond (indicated with a wavy line below) that is formed as a result of an aldol condensation. This double bond must have been formed via the loss of two hydrogen atoms and an oxygen atom. The β position of the condensation product must have originally been a carbonyl group in the starting material:

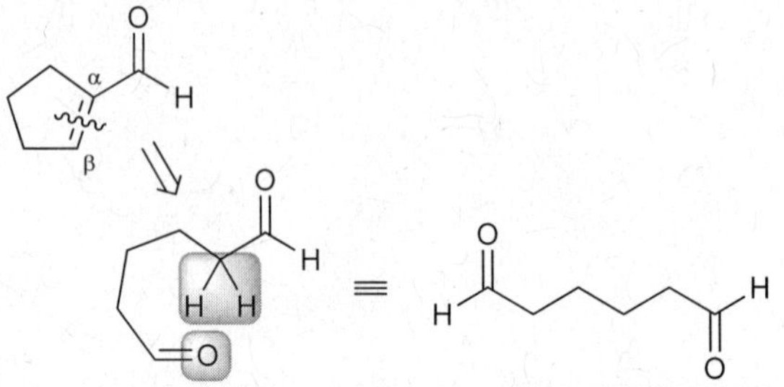

(d) First identify the carbon-carbon bond (indicated with a wavy line below) that is formed as a result of an aldol condensation. This double bond must have been formed via the loss of two hydrogen atoms and an oxygen atom. The β position of the condensation product must have originally been a carbonyl group in the starting material:

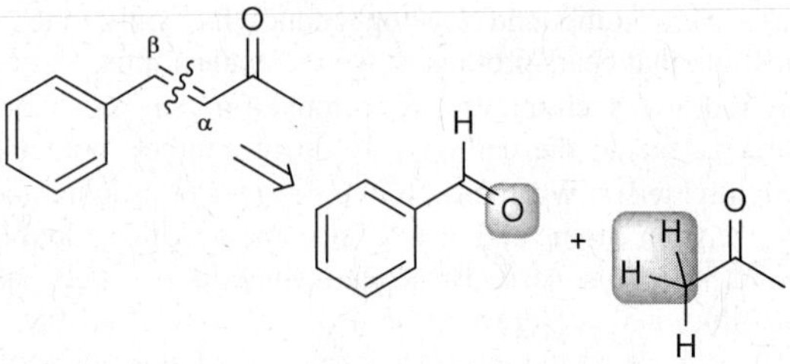

22.74. In acidic conditions, the nucleophilic agent must be an enol, rather than an enolate, because enolate ions are fairly basic and are therefore incompatible with acidic conditions. In the first step, the carbonyl group is protonated, giving a resonance-stabilized intermediate, which is then deprotonated at the α position to give an enol. The enol then functions as a nucleophile and attacks another protonated carbonyl group. The resulting resonance-stabilized cation is then deprotonated to give the aldol addition product.

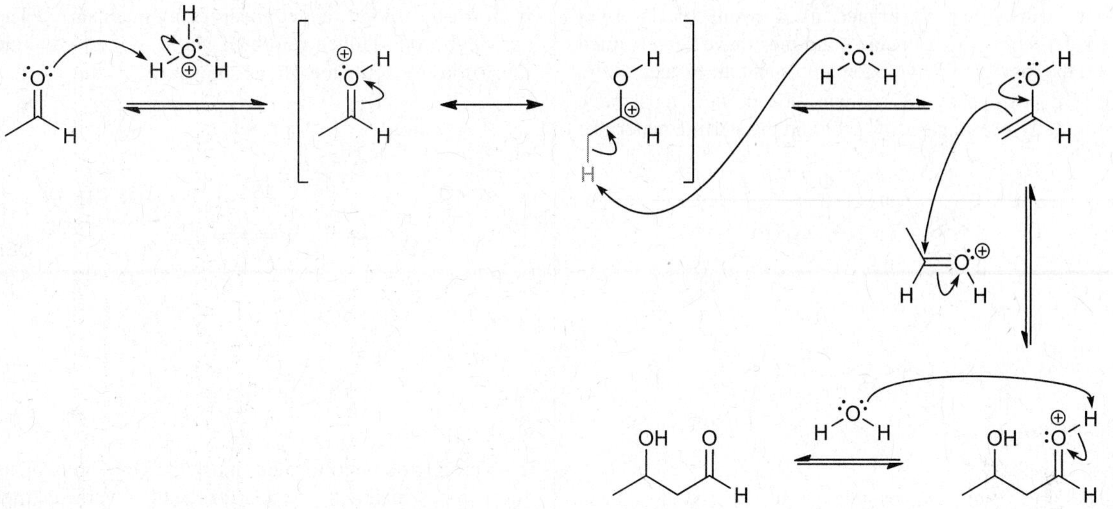

22.75.
(a) In acidic conditions, the nucleophilic agent must be an enol, rather than an enolate, because enolate ions are fairly basic and are therefore incompatible with acidic conditions. The reaction still occurs at the α position, installing a bromine atom at that position (α-halogenation), giving the following product.

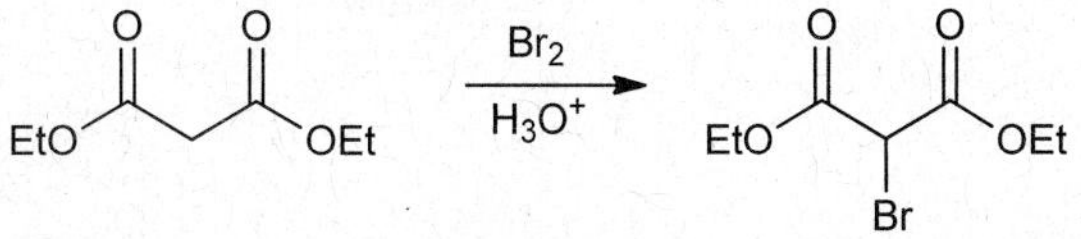

(b) In the first step, the carbonyl group is protonated, giving a resonance-stabilized intermediate, which is then deprotonated at the α position to give an enol. The enol then functions as a nucleophile and attacks molecular bromine (Br$_2$). The resulting resonance-stabilized cation is then deprotonated to give the product.

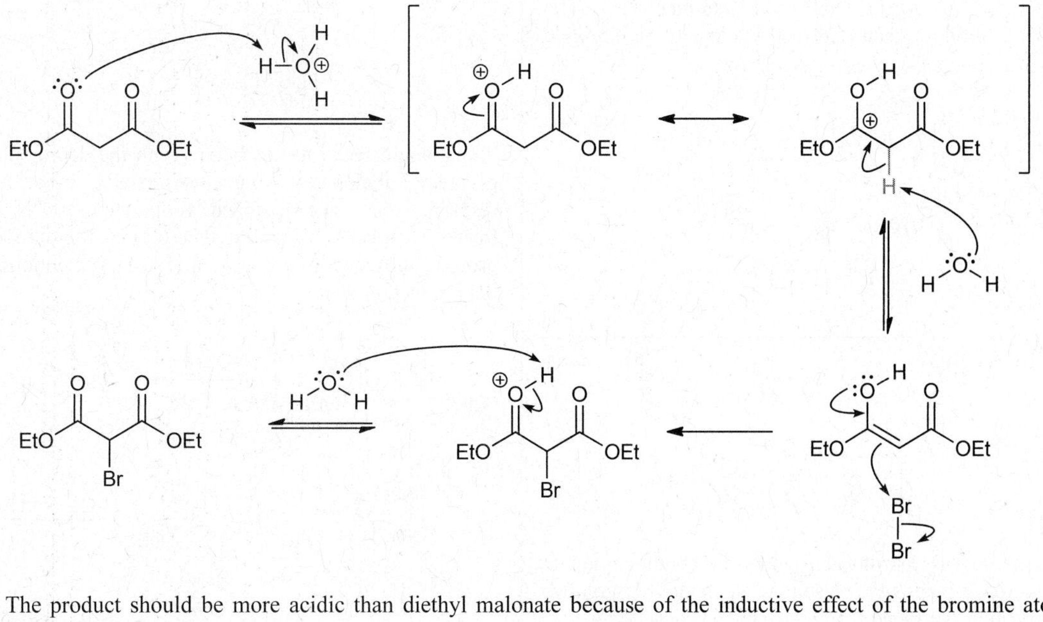

(c) The product should be more acidic than diethyl malonate because of the inductive effect of the bromine atom, which stabilizes the conjugate base.

22.76. Cinnamaldehyde is an an α,β-unsaturated aldehyde, so it can be made via an aldol condensation. To determine the starting materials necessary, first identify the carbon-carbon bond (indicated with a wavy line below) that is formed as a result of an aldol condensation. This double bond must have been formed via the loss of two hydrogen atoms and an oxygen atom. The β position of the condensation product must have originally been a carbonyl group in the starting material:

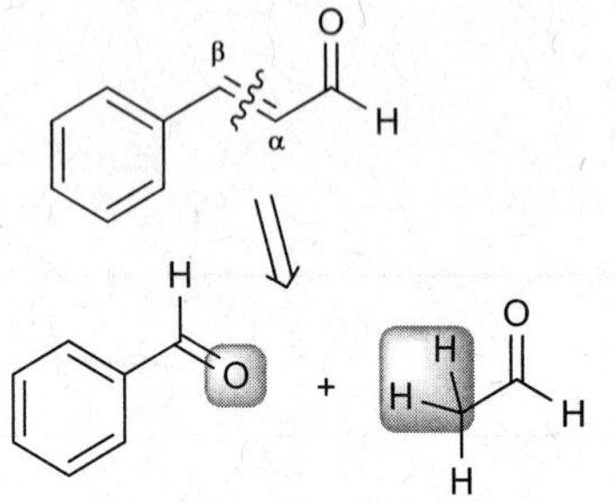

Therefore, cinnamaldehyde can be made from benzaldehyde and acetaldehyde, as shown:

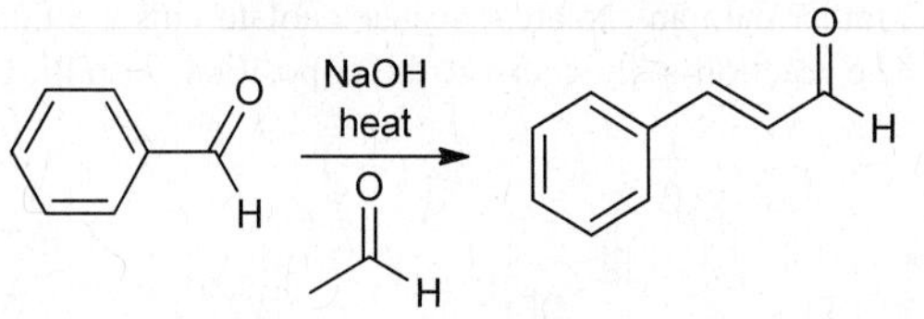

22.77.

(a) The α position of one molecule of the ester is deprotonated, and the resulting enolate functions as a nucleophile and attacks the carbonyl group of another molecule of the ester. As a result, a carbon-carbon bond is formed, giving a tetrahedral intermediate. The carbonyl group is then reformed via loss of an ethoxide ion, affording a β-ketoester, as shown:

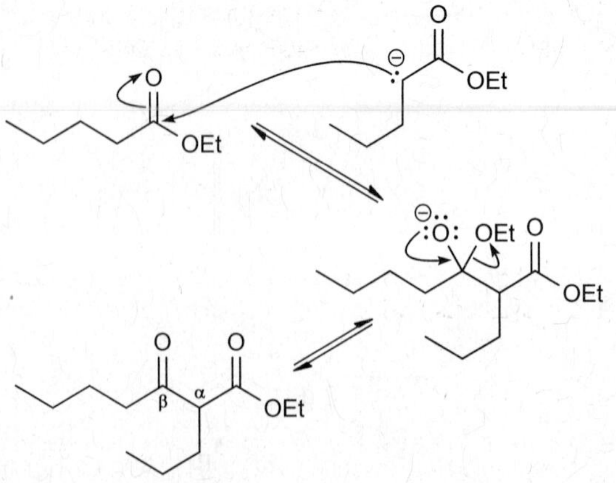

This product is deprotonated under the conditions of its formation, which is the reason for the acid work-up after the reaction is complete (to return the proton).

(b) The α position of one molecule of the ester is deprotonated, and the resulting enolate functions as a nucleophile and attacks the carbonyl group of another molecule of the ester. As a result, a carbon-carbon bond is formed, giving a tetrahedral intermediate. The carbonyl group is then reformed via loss of an ethoxide ion, affording a β-ketoester, as shown:

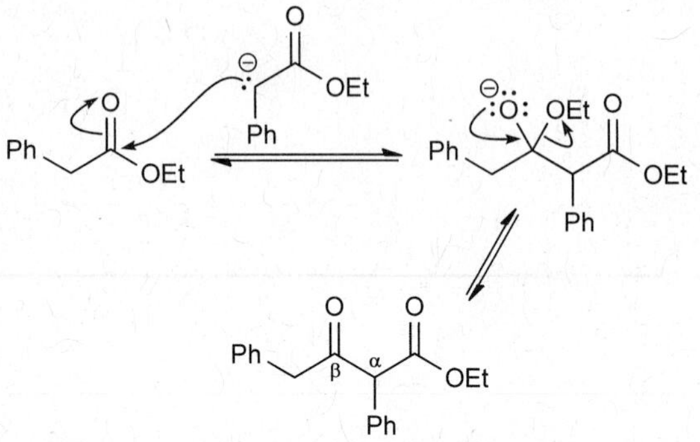

This product is deprotonated under the conditions of its formation, which is the reason for the acid work-up after the reaction is complete (to return the proton).

22.78.

(a) The product is a carboxylic acid that has the following (highlighted) group connected to the α position:

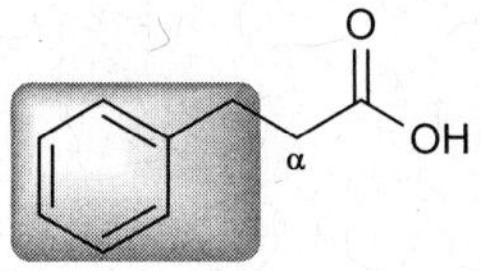

This group can be installed via a malonic ester synthesis, using benzyl bromide:

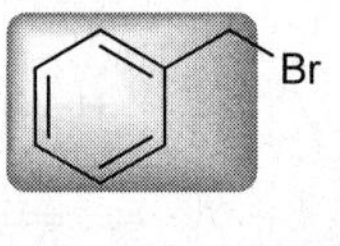

A malonic ester synthesis begins with the deprotonation of diethyl malonate (using ethoxide as a base). The resulting resonance-stabilized conjugate base is then treated with benzyl bromide, thereby installing the alkyl group. Subsequent hydrolysis and decarboxylation give the product, as shown:

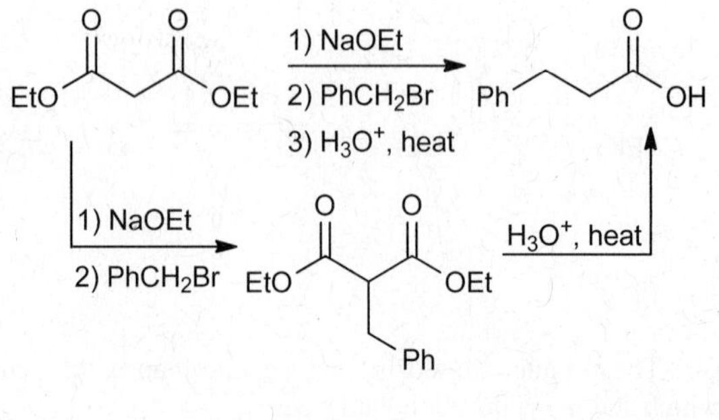

(b) The product is a carboxylic acid that has two methyl groups (highlighted) connected to the α position:

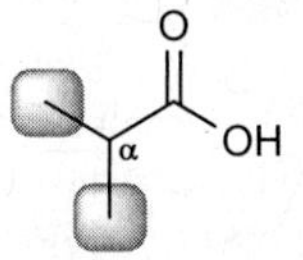

Each of these groups can be installed via a malonic ester synthesis, using methyl iodide. The reason for using methyl iodide (rather than methyl bromide) was discussed in the solution to Problem **22.35b**.

A malonic ester synthesis begins with the deprotonation of diethyl malonate (using ethoxide as a base). The resulting resonance-stabilized conjugate base is then treated with methyl iodide, thereby installing the first methyl group. The second methyl group is installed in a similar way (deprotonation, followed by treatment with methyl iodide). Subsequent hydrolysis and decarboxylation give the product, as shown:

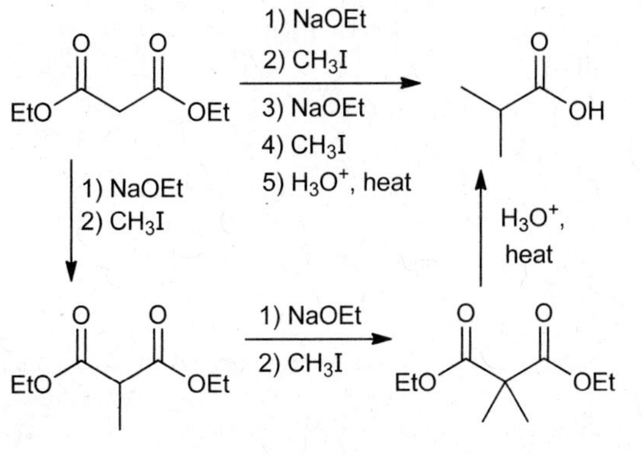

(c) The product is a carboxylic acid that has two benzyl groups (highlighted) connected to the α position:

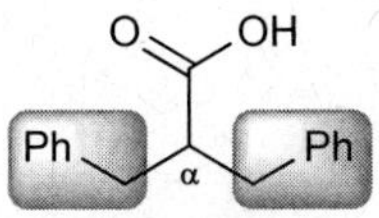

Each of these groups can be installed via a malonic ester synthesis, using benzyl bromide.

A malonic ester synthesis begins with the deprotonation of diethyl malonate (using ethoxide as a base). The resulting resonance-stabilized conjugate base is then treated with benzyl bromide, thereby installing the first benzyl group. The second benzyl group is installed in a similar way (deprotonation, followed by treatment with benzyl bromide). Subsequent hydrolysis and decarboxylation give the product, as shown:

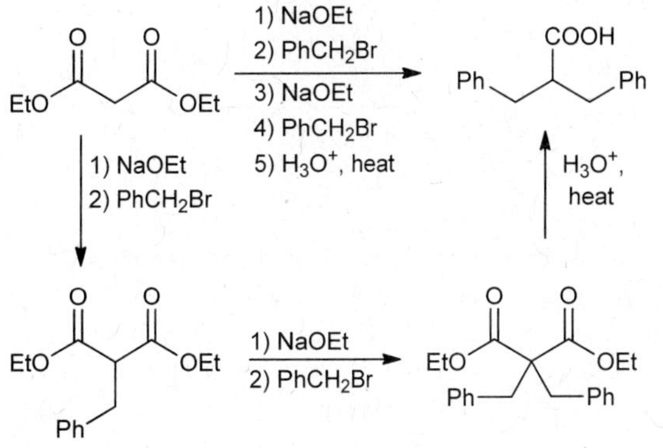

22.79.
(a) The product is a methyl ketone that has a benzyl group (highlighted) connected to the α position:

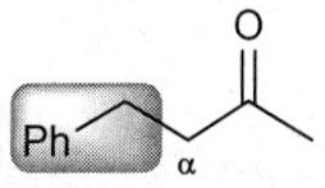

This group can be installed via an acetoacetic ester synthesis, using benzyl bromide.

An acetoacetic ester synthesis begins with the deprotonation of ethyl acetoacetate (using ethoxide as a base). The resulting resonance-stabilized conjugate base is then treated with benzyl bromide, thereby installing a benzyl group. Subsequent hydrolysis and decarboxylation give the product, as shown:

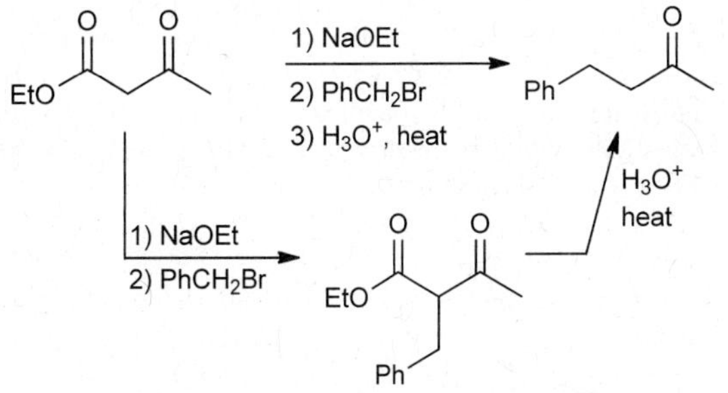

(b) The product is a methyl ketone that has two methyl groups (highlighted) connected to the α position:

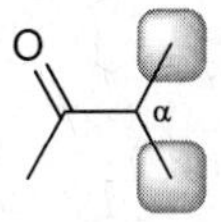

Each of these groups can be installed via an acetoacetic ester synthesis, using methyl iodide to install each methyl group.

An acetoacetic ester synthesis begins with the deprotonation of ethyl acetoacetate (using ethoxide as a base). The resulting resonance-stabilized conjugate base is then treated with methyl iodide, thereby installing the first methyl group. The second methyl group is installed in a similar way (deprotonation with a base, followed by alkylation). Subsequent hydrolysis and decarboxylation give the product, as shown:

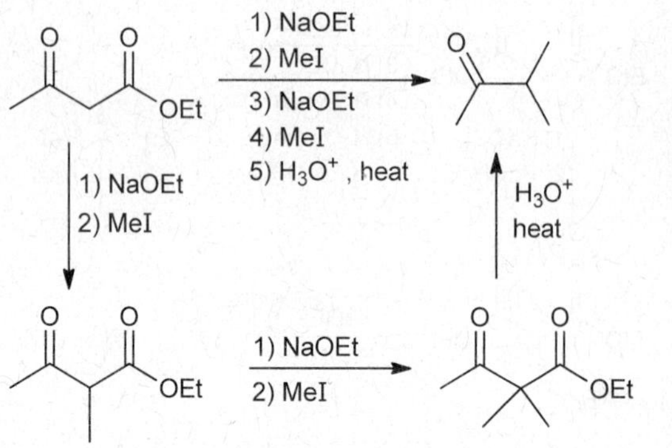

(c) The product is a methyl ketone that has the following two (highlighted) groups connected to the α position:

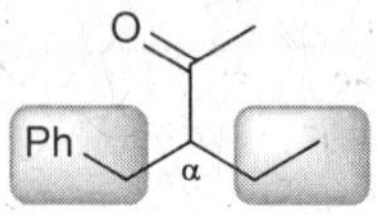

Each of these groups can be installed via an acetoacetic ester synthesis, using the following halides:

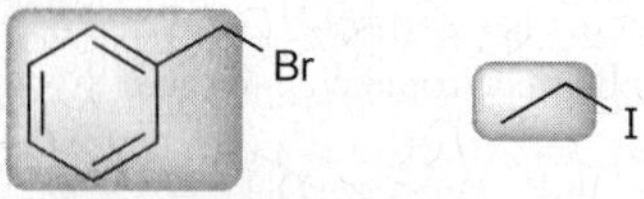

An acetoacetic ester synthesis begins with the deprotonation of ethyl acetoacetate (using ethoxide as a base). The resulting resonance-stabilized conjugate base is then treated with one of the halides above, to install one of the two groups. The other group is installed in a similar way (deprotonation with a base, followed by alkylation). Subsequent hydrolysis and decarboxylation give the product, as shown:

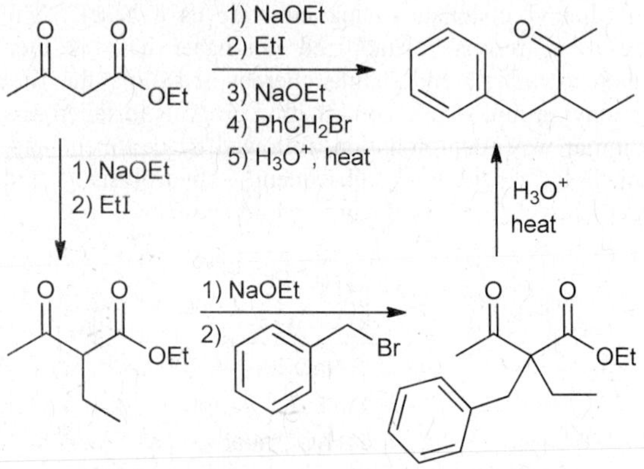

22.80. Protonation of the isolated π bond gives a tertiary carbocation, which is then deprotonated to give a fully conjugated system. Protonation of the carbonyl group then gives a resonance stabilized cation (there are two additional significant resonance structures that have not been drawn). This intermediate is then deprotonated to generate aromaticity, which is the driving force for this process.

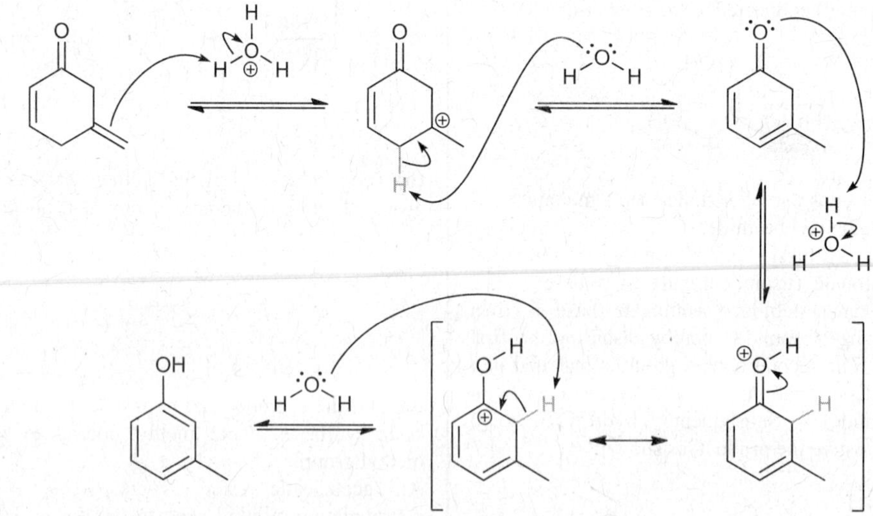

22.81. The reaction conditions suggest an aldol condensation. This compound has three α positions. During an aldol condensation, one of the α positions must be deprotonated to give an enolate, which will attack the other carbonyl group (in an intramolecular process). But we must decide which of the three possible enolates gives rise to the product, as all three possible enolates are expected to be present at equilibrium. If either of the interior α positions is deprotonated to give an enolate ion, the resulting intramolecular attack would generate a four-membered ring:

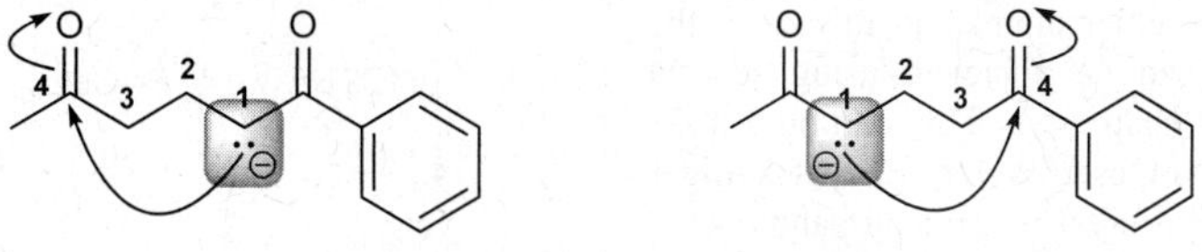

However, the third enolate can participate in an intramolecular attack that gives a six-membered ring:

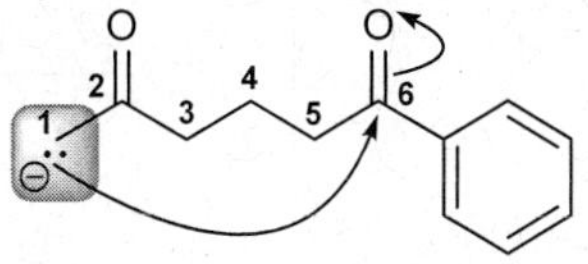

Formation of a six-membered ring is favored over formation of a four-membered ring, because the former is relatively strain-free, while the latter is not (and therefore higher in energy).

Now that we have identified which carbon-carbon bond will be formed during an intramolecular aldol condensation, we can draw the product by removing the following two highlighted α protons and the oxygen atom, and we replace them with a double bond, giving a product with the molecular formula $C_{12}H_{12}O$:

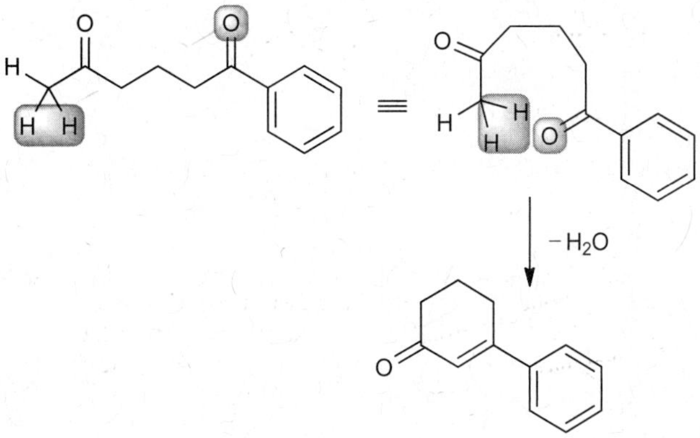

22.82. This transformation requires the installation of a methyl group at a β position:

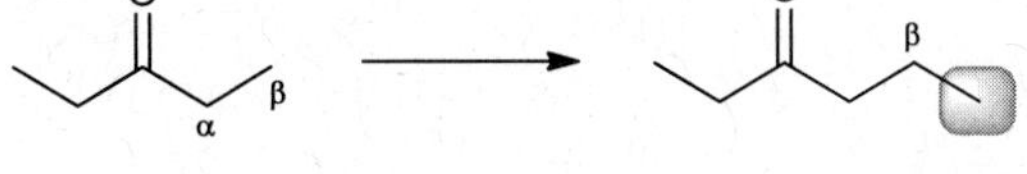

We did not learn a way to install a group at the β position of a saturated ketone, however, we did learn a way to install a methyl group at the β position of an α,β-unsaturated ketone:

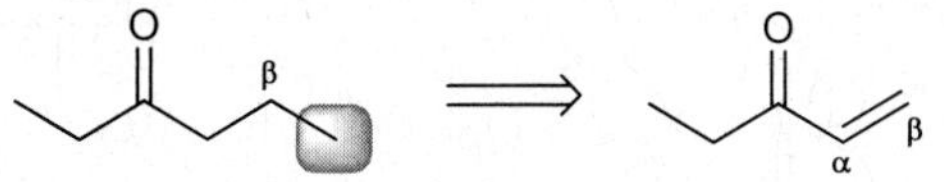

This α,β-unsaturated ketone can be prepared from the starting material via bromination at the α position under acidic conditions, thereby installing a leaving group, which can then be removed in an elimination process upon treatment with a base (pyridine):

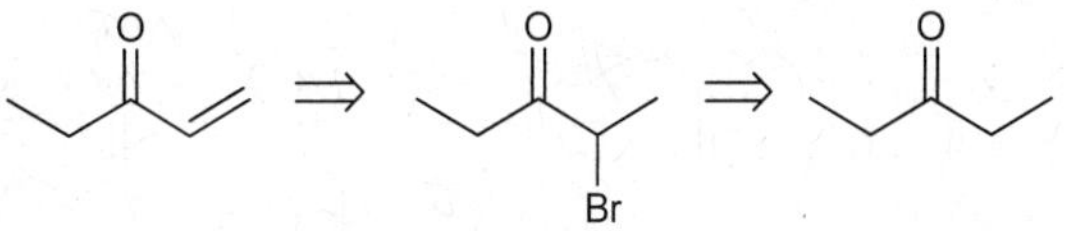

The forward scheme is shown here:

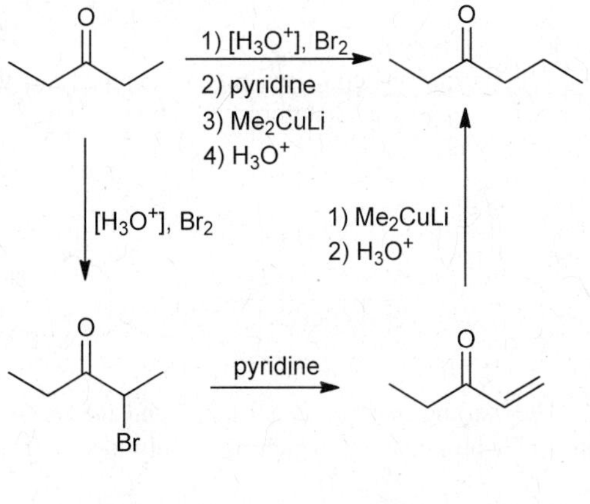

As mentioned so many times throughout this entire course, there are almost always multiple correct solutions to a synthesis problem. For example, in this case, the desired transformation can be achieved using several reactions from previous chapters. The starting ketone can be converted into an ester via a Baeyer-Villiger oxidation, followed by hydrolysis to give propanoic acid. This acid can then be converted to an acid chloride upon treatment with thionyl chloride, followed by conversion to the product upon treatment with lithium dipropyl cuprate:

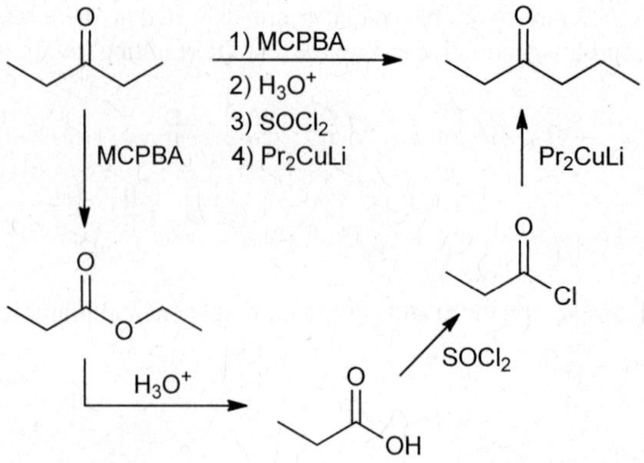

22.83. The ester is converted into a β-ketoester via a crossed Claisen condensation. Upon treatment with aqueous acid at elevated temperature, the ester group is hydrolyzed and the resulting β-ketoacid (not drawn below) undergoes decarboxylation to give acetophenone. A crossed aldol condensation (with formaldehyde) gives an α,β-unsaturated ketone. Treating the α,β-unsaturated ketone with lithium diethyl cuprate, followed by water work-up, installs an ethyl group at the β position, and removes the unsaturation between the α and β positions. Another crossed aldol condensation (with formaldehyde again) gives the final product.

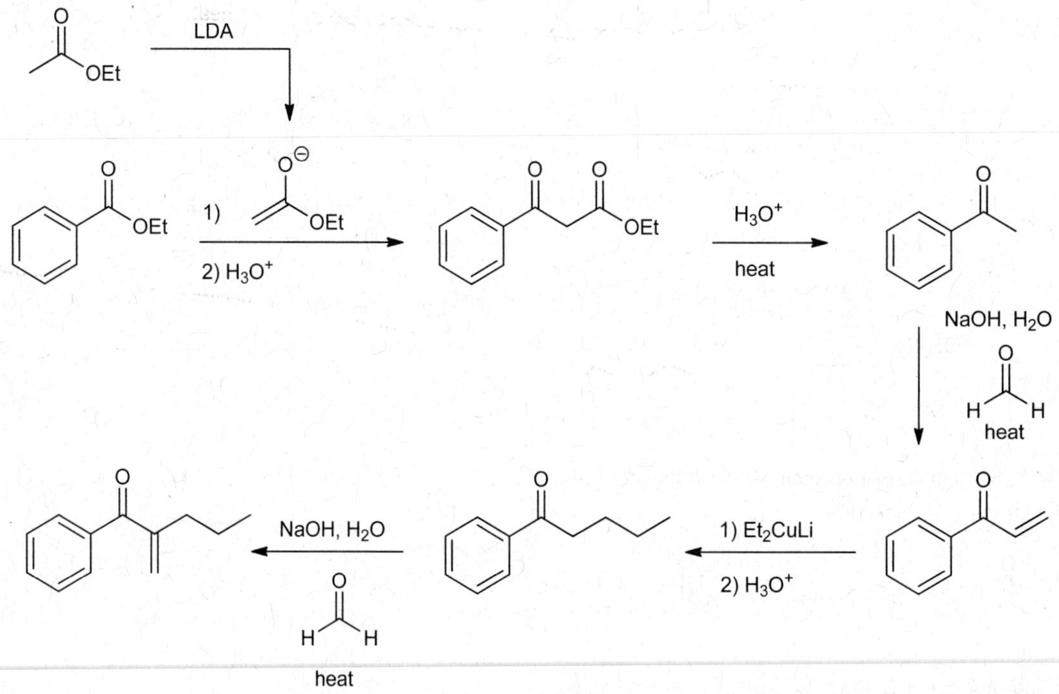

22.84.
(a) The starting material is a methyl ketone, and the reagents indicate a haloform reaction, giving a carboxylic acid:

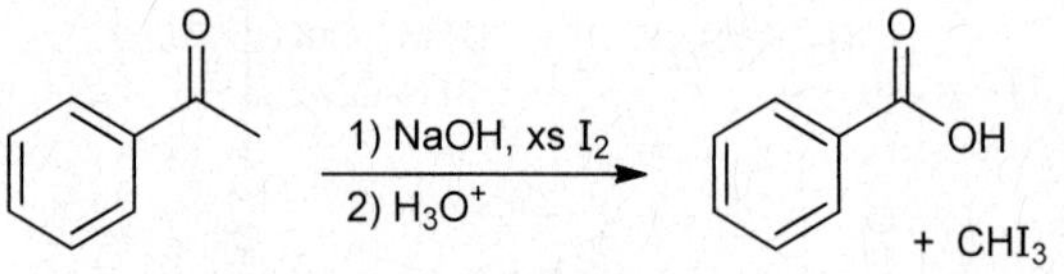

(b) The starting material is a ketone, and the reagents indicate α-bromination, which will only occur at an α position that bears protons. In this case, there is only one such position. Under these conditions, a bromine atom is installed at this α position, to give the following product:

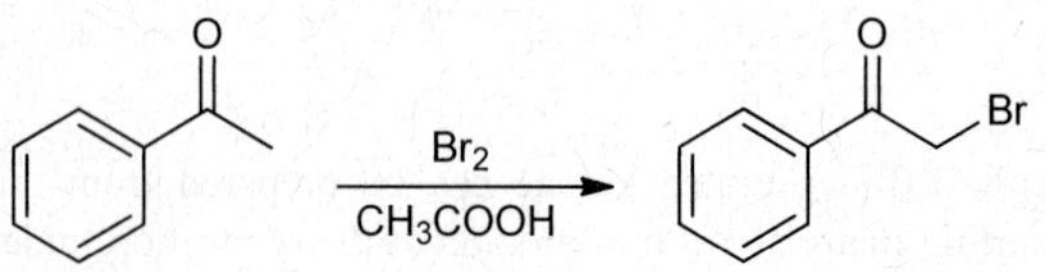

(c) The starting material is a ketone, and the reagents indicate an aldol condensation. The starting ketone has only one α position that bears protons, and the reaction occurs at this location. To draw the product, two molecules of the ketone are redrawn such that two α protons of one molecule are directly facing the carbonyl group of another molecule (highlighted). We then remove the two α protons and the oxygen atom, and we replace them with a double bond. In this case, two stereoisomers are possible, so we draw the product with fewer steric interactions:

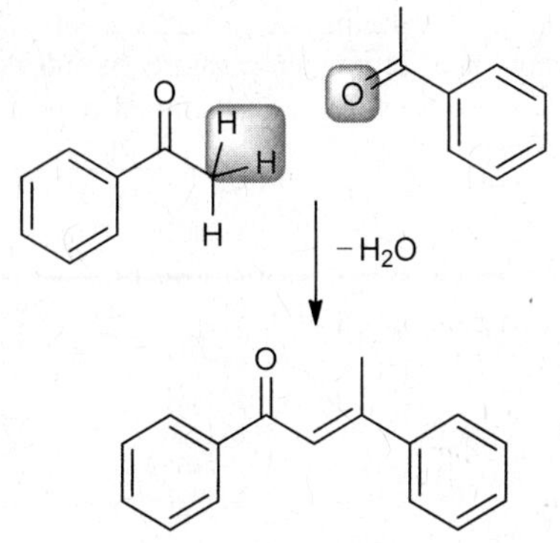

22.85. This transformation represents a retro-aldol reaction, which occurs via a mechanism that is the reverse of an aldol condensation (all the same intermediates, but in reverse order). First a hydroxide ion functions as a nucleophile and attacks the electrophilic β position of the α,β-unsaturated ketone. The resulting enolate is then protonated to give a β-hydroxyketone. Deprotonation gives an alkoxide ion, which then reforms a carbonyl group by expelling an enolate ion as a leaving group. This enolate ion is then protonated to give cyclohexanone. Notice that water is the proton source for the protonation steps, consistent with basic conditions (strong acids, such as hydronium ions, are not measurably present under these conditions).

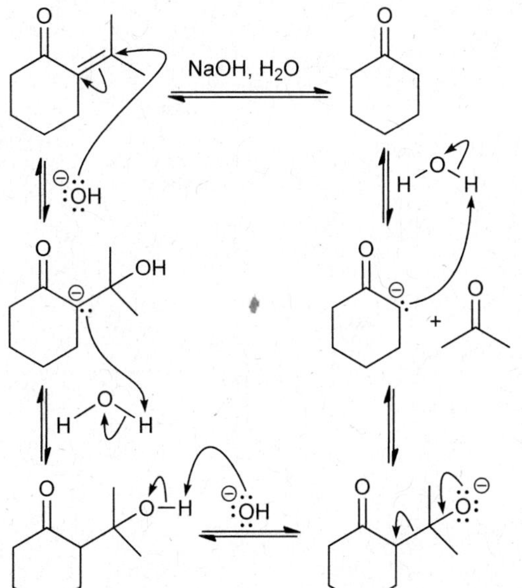

22.86.

(a) Upon treatment with aqueous acid, each of the four ester groups is hydrolyzed, giving a compound with four carboxylic acid groups, shown here:

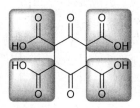

Each of these four carboxylic acid groups is β to a carbonyl group, and will therefore undergo decarboxylation upon heating. This gives the dione shown below, as well as four equivalents of ethanol (from hydrolysis) and four equivalents of carbon dioxide (from decarboxylation):

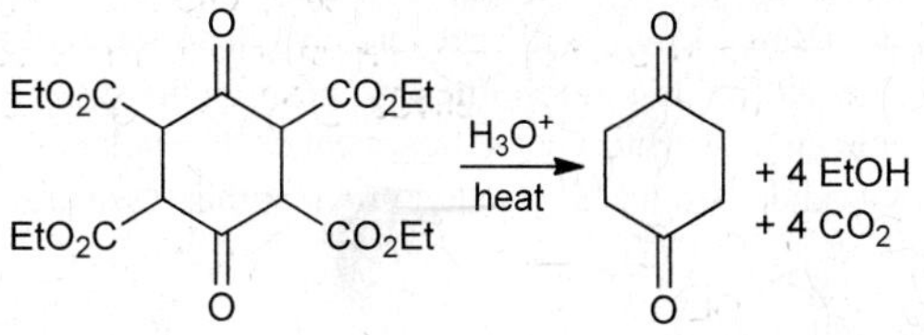

(b) The starting material is a diester. Upon treatment with ethoxide, an intramolecular Claisen condensation (followed by acid work-up) gives a β-ketoester via formation of a ring. Upon heating with aqueous acid, the β-ketoester is hydrolyzed to a β-ketoacid (not shown), which then undergoes decarboxylation (under the conditions of its formation) to give the ketone shown below. Notice that the configuration of each chirality center remains unchanged because the chirality centers are not involved in the reaction.

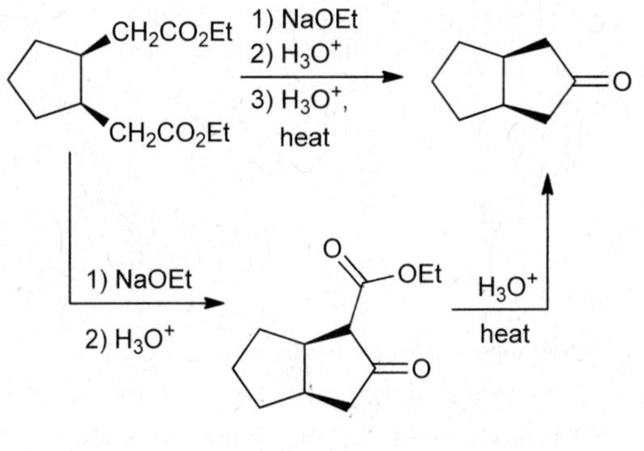

(c) The starting material is an unsymmetrical ketone. Treatment with bromine in aqueous acidic conditions gives α-bromination, which is expected to occur at the more substituted position. Subsequent treatment of the resulting α-bromoketone with pyridine (a base) gives an elimination reaction to afford an α,β-unsaturated ketone. Treatment of the α,β-unsaturated ketone with lithium diethyl cuprate, followed by methyl iodide, achieves the installation of an ethyl group at the β position and a methyl group at the α position:

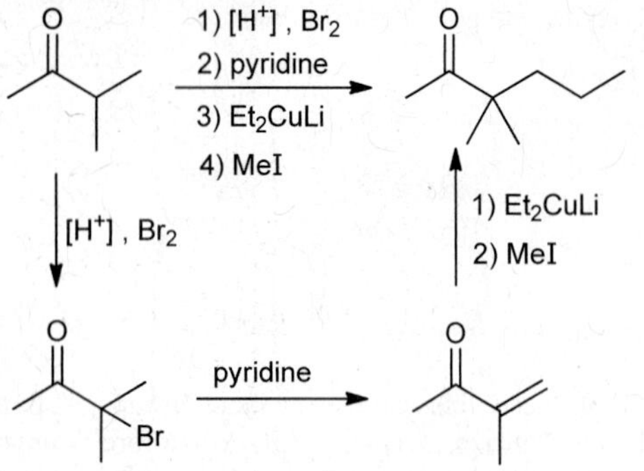

(d) The starting material is an unsymmetrical ketone, and LDA is a strong, sterically hindered base. At low temperature, LDA will irreversibly deprotonate the ketone at the less substituted α position to give the kinetic enolate. Subsequent treatment of the enolate with ethyl iodide will install an ethyl group at this α position:

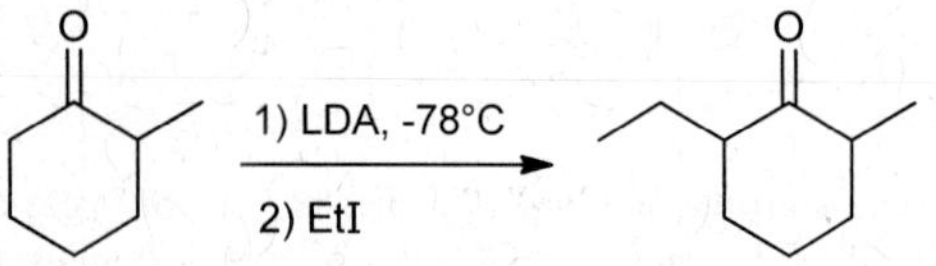

22.87.
(a) One strategy for achieving the desired transformation derives from the following retrosynthetic analysis. An explanation of each of the steps (a-d) follows.

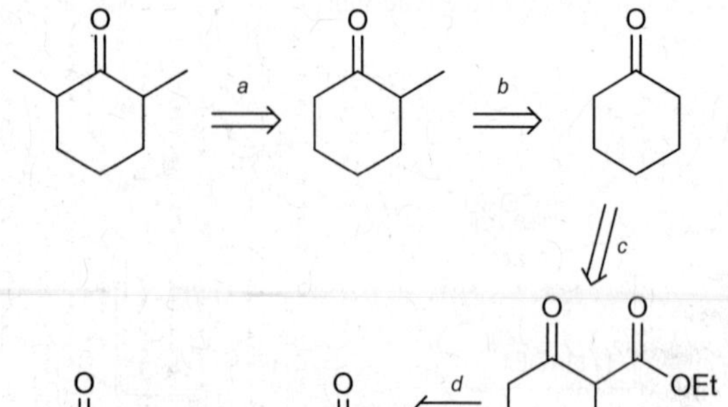

a. The product can be made from 2-methylcyclo-hexanone via alkylation of an α position (with LDA as the base, to control the regiochemical outcome).
b. 2-Methylcyclohexanone can be made from cyclohexanone via alkylation of an α position.
c. Cyclohexanone can be made from a β-ketoester via hydrolysis and subsequent decarboxylation.
d. The β-ketoester can be made from the starting material via a Dieckmann cyclization.

Now let's draw the forward scheme. Treating the starting diester with ethoxide, followed by acid work-up, gives a β-ketoester via a Dieckmann cyclization. Upon treatment with aqueous acid and heat, the β-ketoester is hydrolyzed to give a β-keto acid, which then undergoes decarboxylation to give cyclohexanone. Two subsequent alkylation processes will install the two methyl groups. The choice of base in the first alkylation is not so critical, because cyclohexanone is symmetrical (both α positions are identical). But during the second alkylation process, LDA must be used at low temperature, in order to install the methyl group at the less substituted α position (via the kinetic enolate).

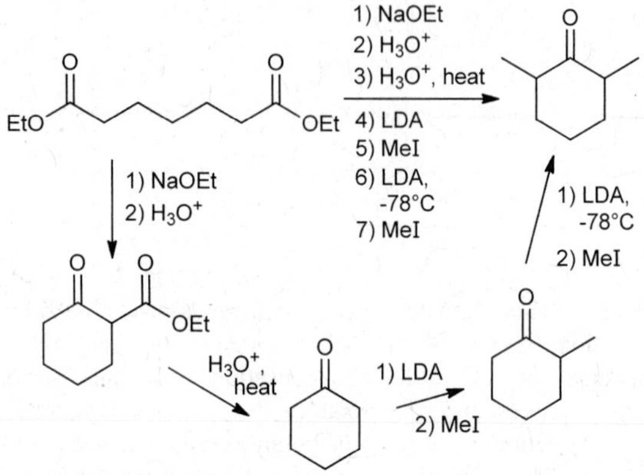

Alternatively, and perhaps more efficiently, installation of the first methyl group can be performed immediately after the Dieckmann cyclization (before hydrolysis and decarboxylation). In this way, the anionic product of the Dieckmann cyclization (without acid work-up) is used as a nucleophile to attack methyl iodide in an S_N2 process, thereby installing the first methyl group, as shown.

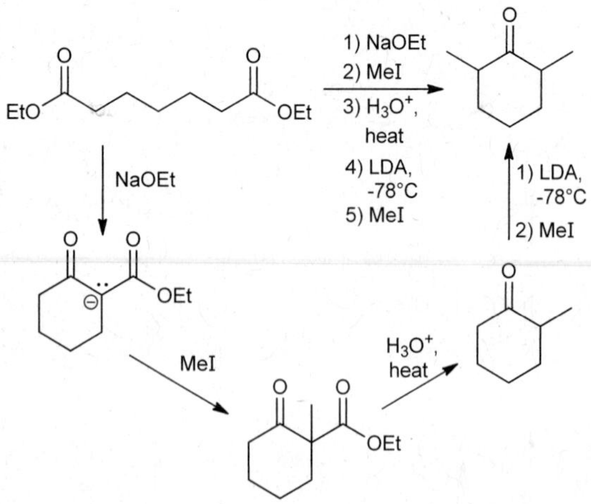

(b) The product can be made via two successive aldol condensation reactions, one of which is intramolecular and the other is intermolecular:

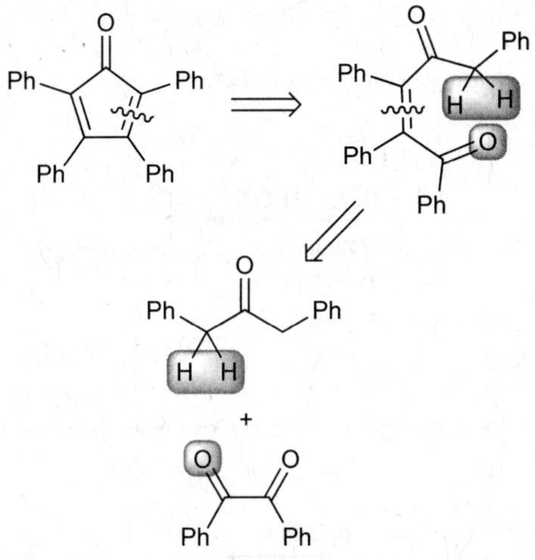

The entire transformation can be achieved in one reaction flask, by treating the starting material with the dione above in basic conditions.

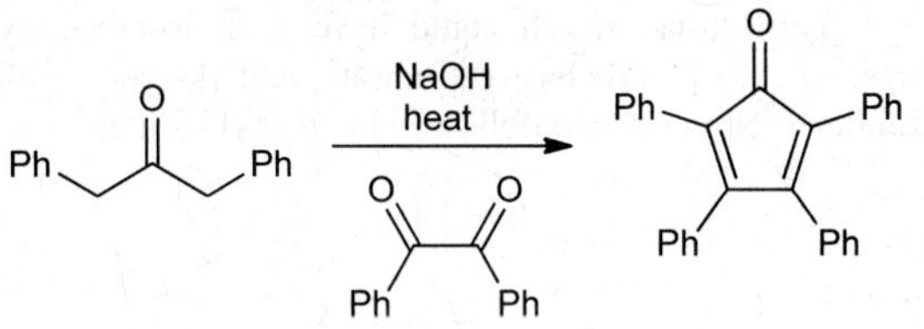

(c) The product is 1,5-difunctionalized:

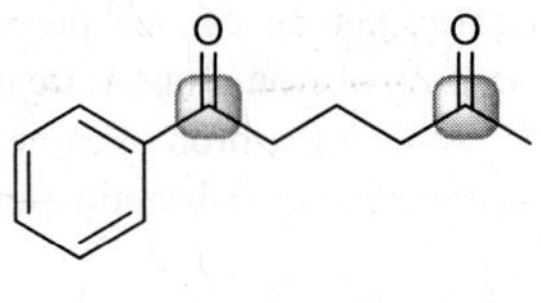

Therefore, we consider preparing the product via a Michael reaction:

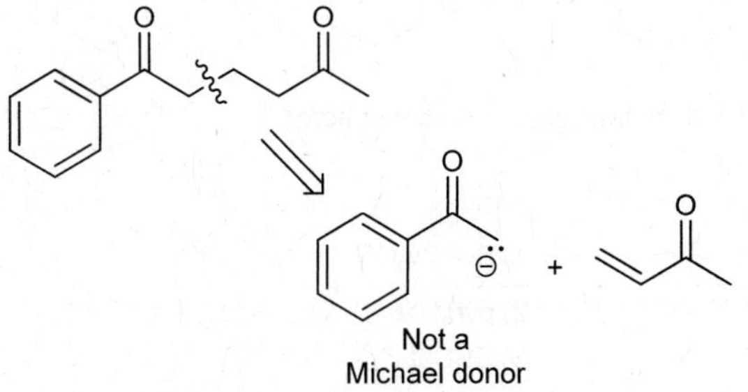

Not a
Michael donor

This strategy will not work, because it involves the use of an enolate, which is not an efficient Michael donor. Therefore, we consider a Stork enamine synthesis (in which we use an enamine, rather than an enolate, as a Michael donor). The enamine can be made directly from the starting material (acetophenone).

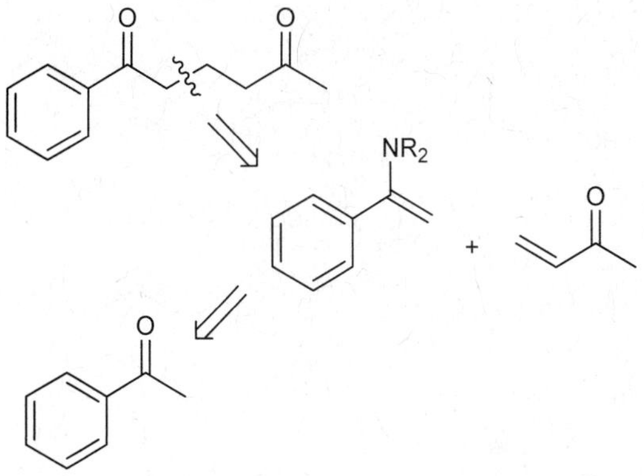

The forward scheme is shown here:

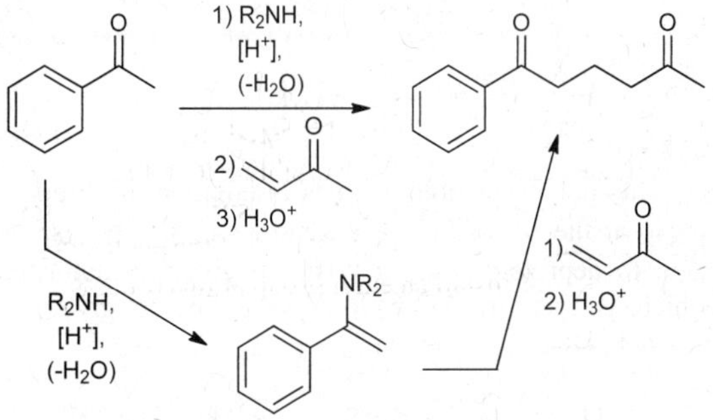

22.88. There are certainly many ways to achieve this transformation, which involves a change in the carbon skeleton. We begin by considering a directed aldol condensation, followed by reduction:

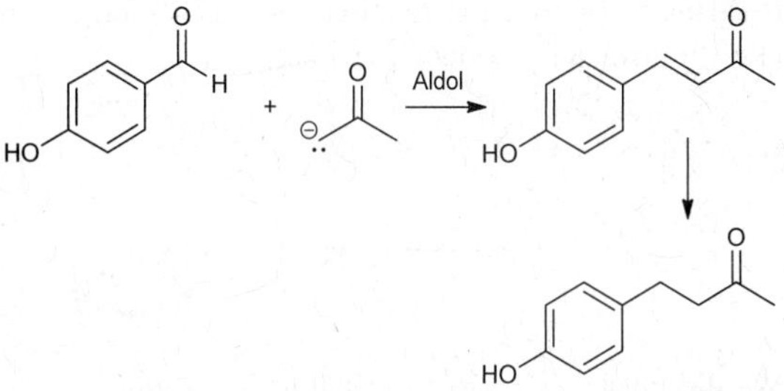

This strategy suffers from a fatal flaw. The phenolic OH group is more acidic than acetone. Therefore, it is not possible to form the enolate of acetone without first deprotonating the phenolic OH group. And deprotonation in that location would generate a resonance stabilized anion, in which the negative charge is spread over several positions including the oxygen atom of the carbonyl group, thereby deactivating the aldehyde group as an electrophile. This obstacle can be circumvented by protecting the OH group before the desired transformation is performed, and then deprotecting with TBAF at the end of the synthesis (see Section 13.7).

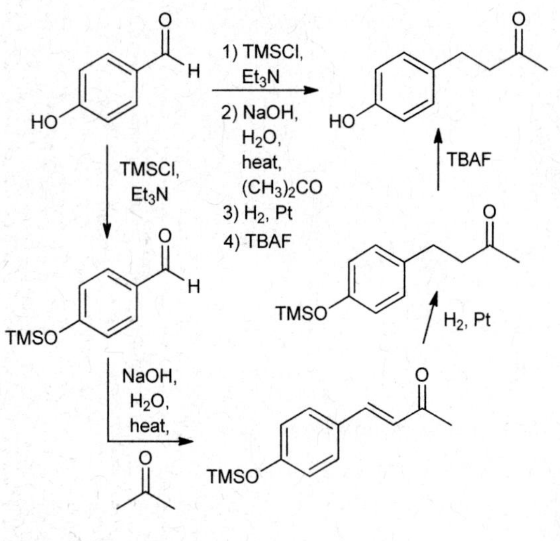

22.89.

(a) This transformation involves installation of an ethyl group at the α position of a ketone. A strong base is used to deprotonate the α position, giving an enolate, which is then treated with ethyl iodide to give the product via an S$_N$2 reaction.

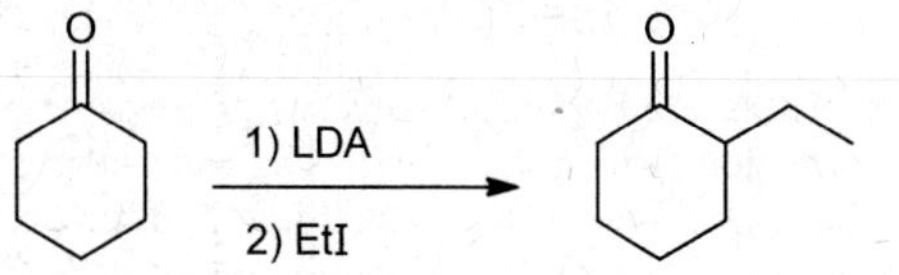

(b) This transformation requires the installation of an ethyl group at a β position:

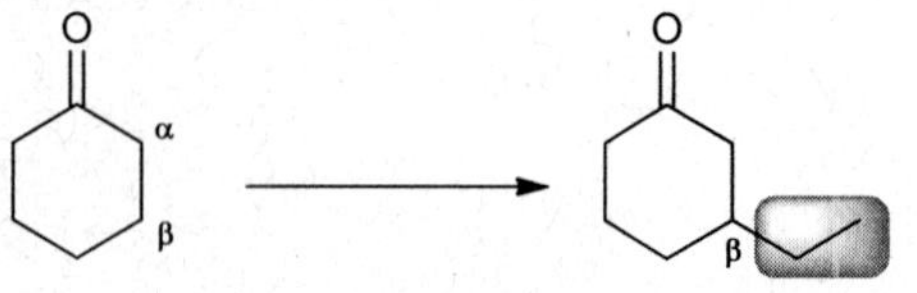

We did not learn a way to install an alkyl group at the β position of a saturated ketone, however, we did learn a way to install an alkyl group at the β position of an α,β-unsaturated ketone:

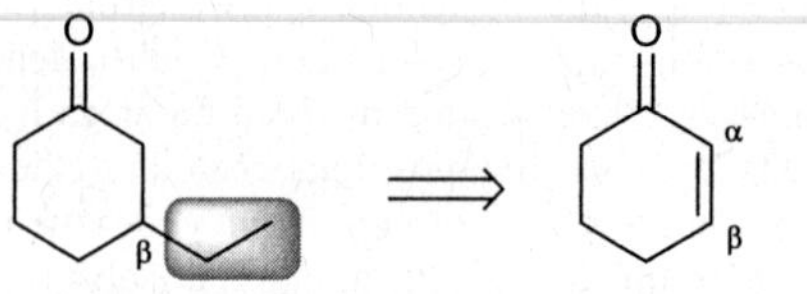

This α,β-unsaturated ketone can be prepared from the corresponding α-bromoketone upon treatment with a base (pyridine). And the α-bromoketone can be made from the starting material via α-bromination under acidic conditions:

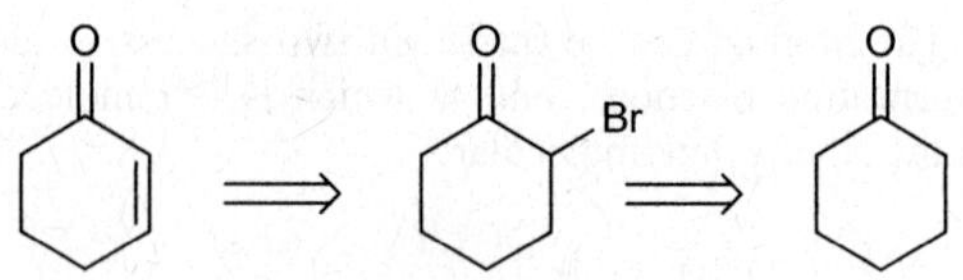

The forward scheme is shown here:

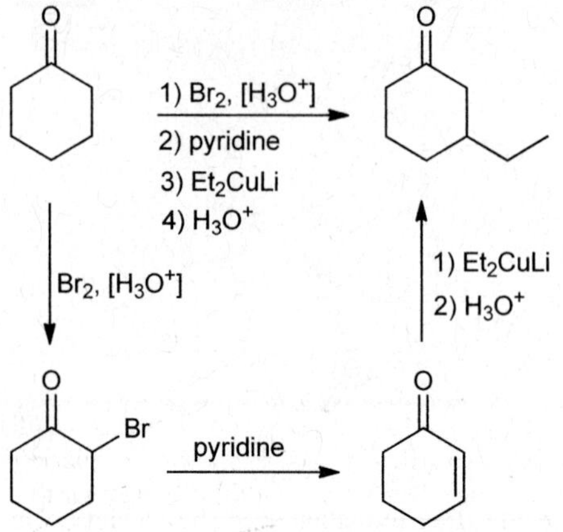

(c) The product is a ketone with alkyl groups at the α and β positions, which could have been installed by treating the following α,β-unsaturated ketone with lithium diethyl cuprate, followed by methyl iodide:

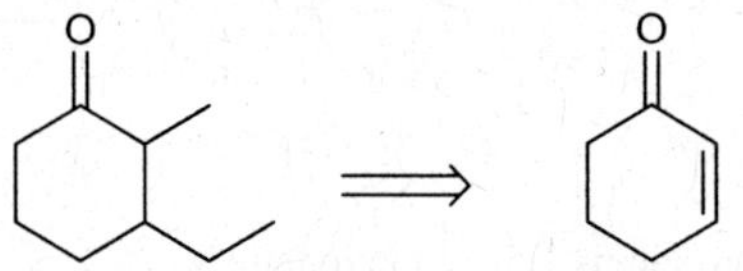

This α,β-unsaturated ketone can be prepared from the corresponding α-bromoketone upon treatment with a base (pyridine). And the α-bromoketone can be made from the starting material via α-bromination under acidic conditions:

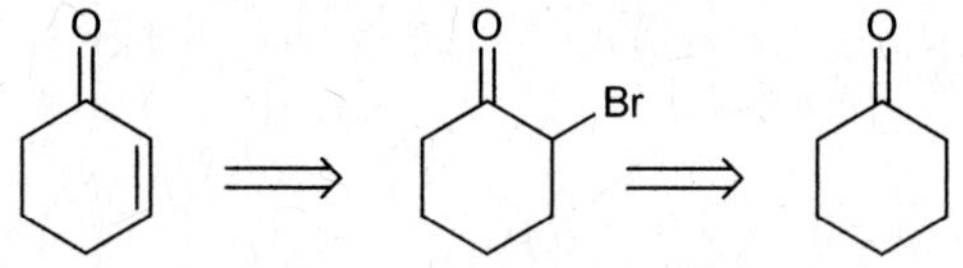

The forward scheme is shown here:

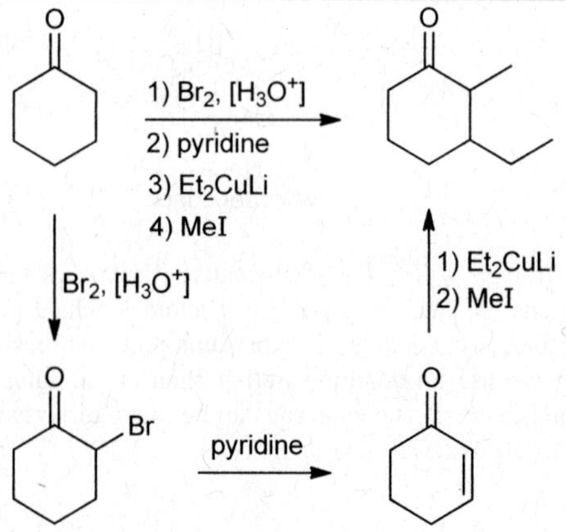

(d) The product is 1,5-difunctionalized:

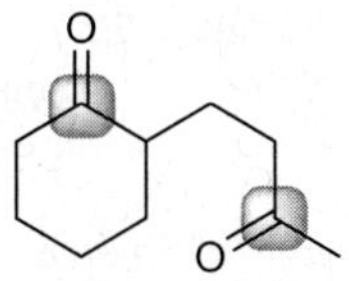

Therefore, we consider preparing the product via a Michael reaction:

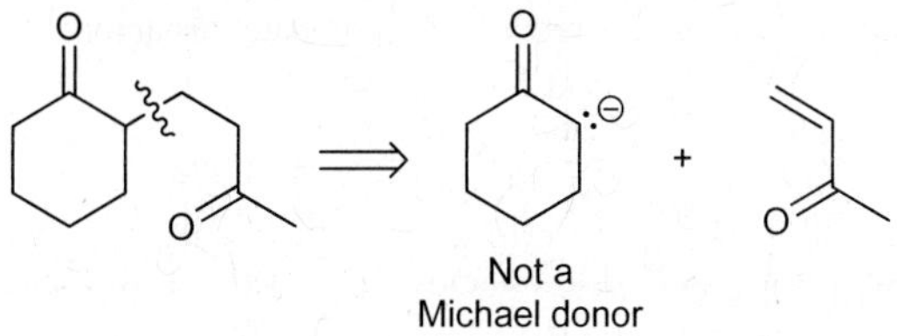

Not a
Michael donor

This strategy will not work, because it involves the use of an enolate, which is not an efficient Michael donor. Therefore, we consider a Stork enamine synthesis (in which we use an enamine, rather than an enolate, as a Michael donor). The enamine can be made from the starting ketone upon treatment with a secondary amine under acid-catalyzed conditions (with removal of water):

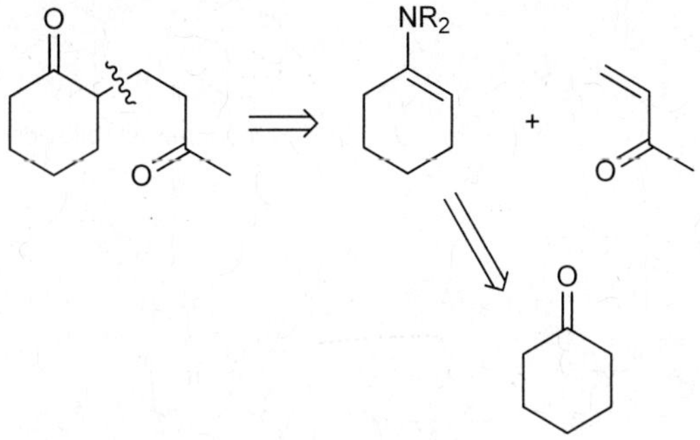

The forward scheme is shown here:

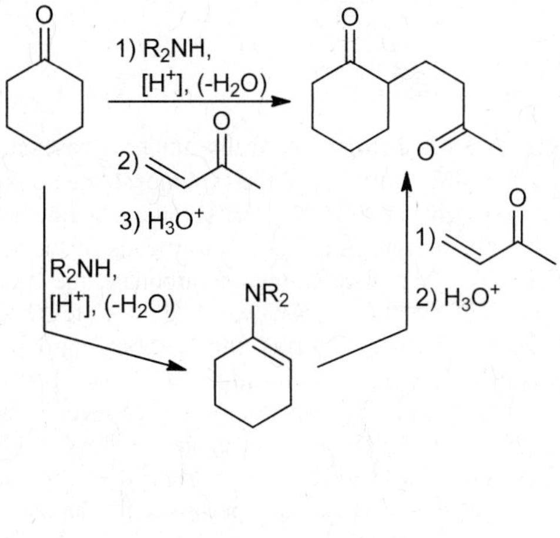

(e) The product is a ketone that exhibits α,β-unsaturation on either side of the carbonyl group. This suggests two

aldol condensation reactions, one at each α position of the starting ketone (cyclohexanone):

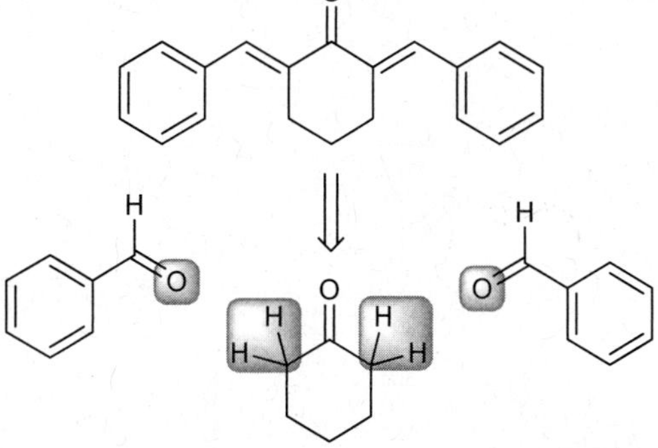

Both aldol condensation reactions can be performed in a single reaction flask, by treating cyclohexanone with excess benzaldehyde under basic conditions:

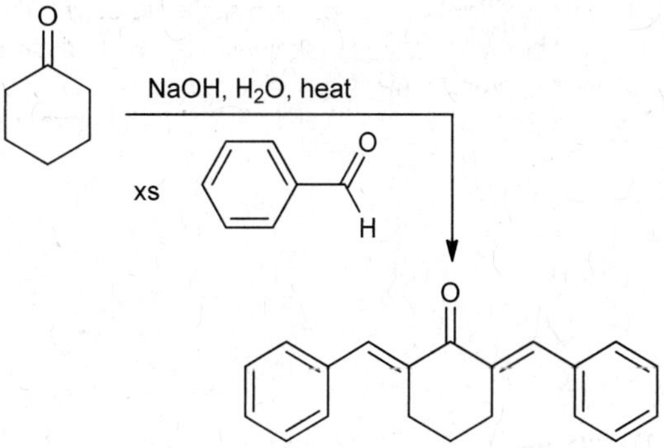

(f) The product is 1,5-difunctionalized:

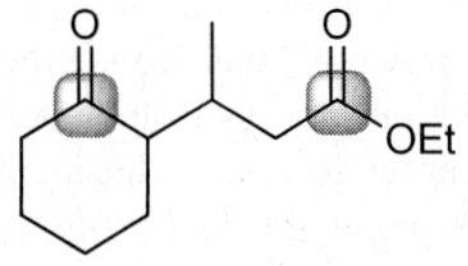

Therefore, we consider preparing the product via a Michael reaction:

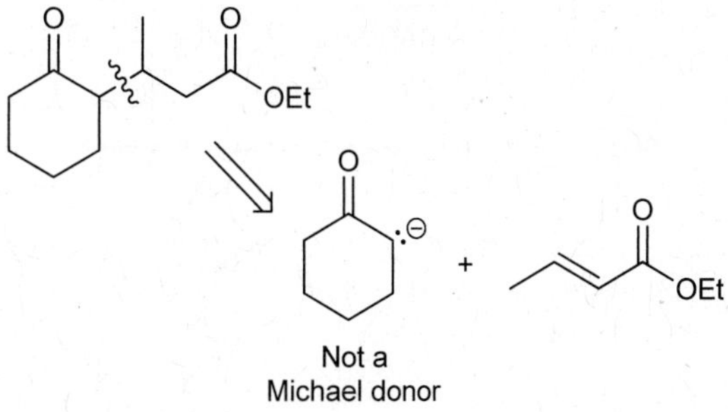

Not a
Michael donor

This strategy will not work, because it involves the use of an enolate, which is not an efficient Michael donor. Therefore, we consider a Stork enamine synthesis (in which we use an enamine, rather than an enolate, as a Michael donor). The enamine can be made from the starting ketone upon treatment with a secondary amine under acid-catalyzed conditions (with removal of water):

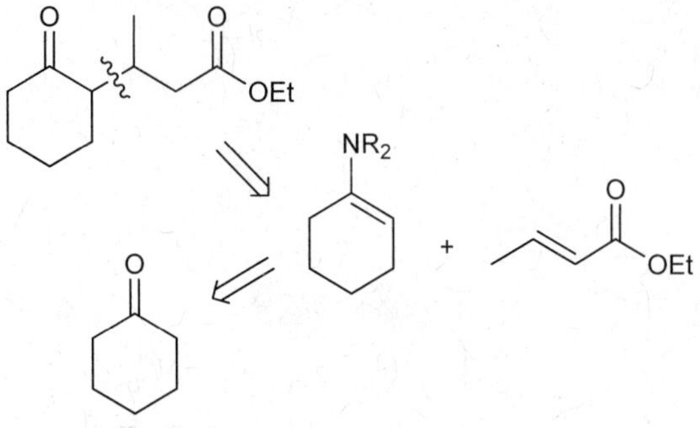

The forward scheme is shown here:

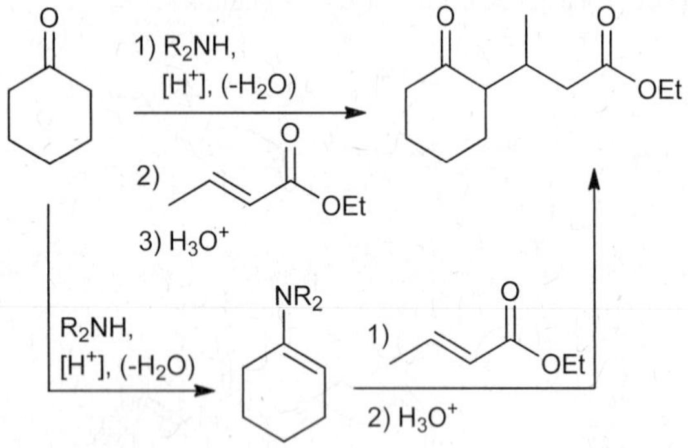

(g) Two subsequent alkylation processes will install the two methyl groups. The choice of base in the first alkylation is not so critical, because cyclohexanone is symmetrical (both α positions are identical). But during the second alkylation process, LDA must be used at low temperature, in order to install the methyl group at the less substituted α position (via the kinetic enolate).

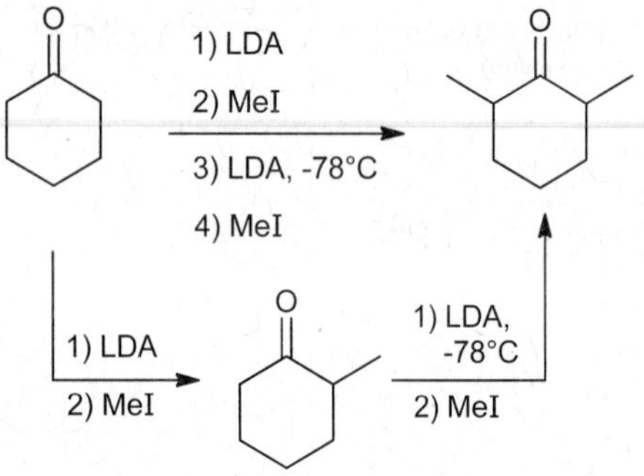

22.90. LDA is a strong, sterically hindered base, and it will irreversibly deprotonate the cyclohexanone (at the α position) to give an enolate. The ketone is symmetrical, so deprotonation at either α position leads to the same enolate ion. When treated with an ester, the enolate ion will attack the ester, to give a tetrahedral intermediate, which reforms the carbonyl group by expelling ethoxide. The resulting β-dicarbonyl compound is then deprotonated to give a doubly-stabilized enolate. Indeed, the formation of this resonance-stabilized anion is a driving force for this reaction. After the reaction is complete, an acid is required to protonate this anion.

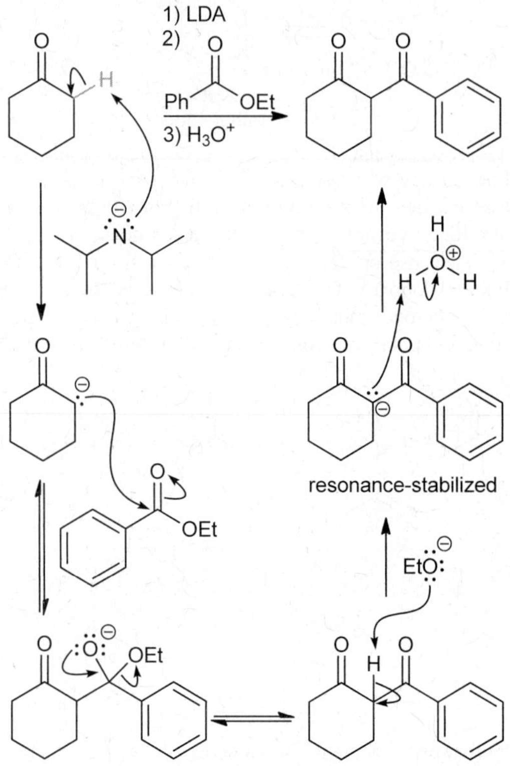

22.91. LDA is a strong, sterically hindered base, and it will irreversibly deprotonate the cyclohexanone (at the α position) to give an enolate. The ketone is symmetrical, so deprotonation at either α position leads to the same enolate ion. When treated with a carbonate, the enolate ion will attack the carbonate, to give a tetrahedral intermediate, which reforms the carbonyl group by expelling ethoxide. The resulting β-ketoester is then deprotonated to give a doubly-stabilized enolate. Indeed, the formation of this resonance-stabilized anion is a driving force for this reaction. After the reaction is complete, an acid is required to protonate this anion.

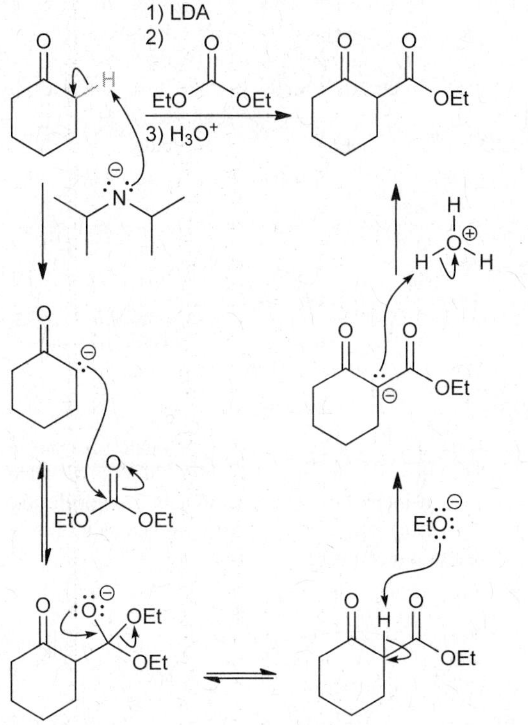

22.92. LDA is a strong, sterically hindered base, and it will irreversibly deprotonate the ester (at the α position) to give an ester enolate. When treated with a ketone, the enolate ion will attack the ketone, to give an alkoxide ion. After the reaction is complete, an acid is required to protonate this alkoxide ion.

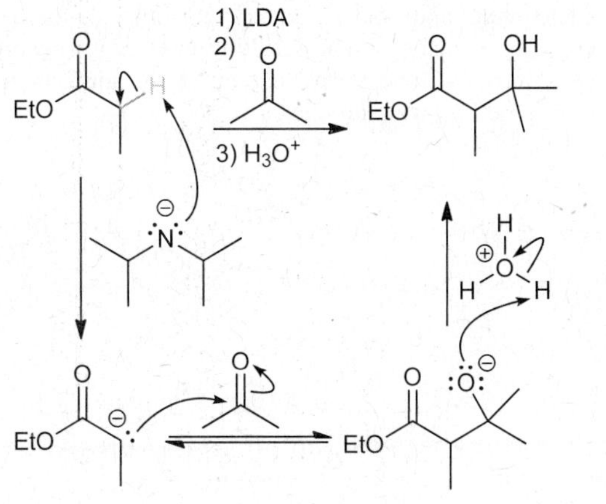

22.93.
(a) The following mechanism is consistent with the description in the problem statement.

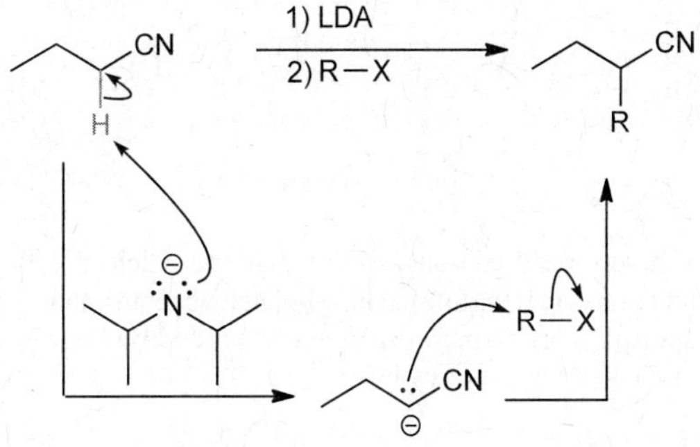

(b) Benzyl bromide is converted into a nitrile (via an S$_N$2 reaction in which cyanide functions as a nucleophile). This nitrile can then undergo two successive alkylation processes, installing two methyl groups at the α position. Hydrolysis of the nitrile then gives a carboxylic acid.

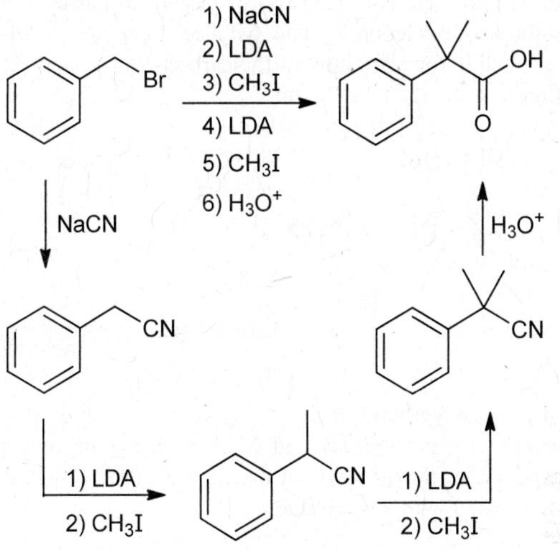

22.94.
(a) First, we look for the 1,5-difunctionalization which is the hallmark of a Michael reaction. A retrosynthetic analysis reveals the Michael donor (stabilized nucleophile) and Michael acceptor that are responsible for formation of the carbon-carbon bond that is indicated with a wavy line:

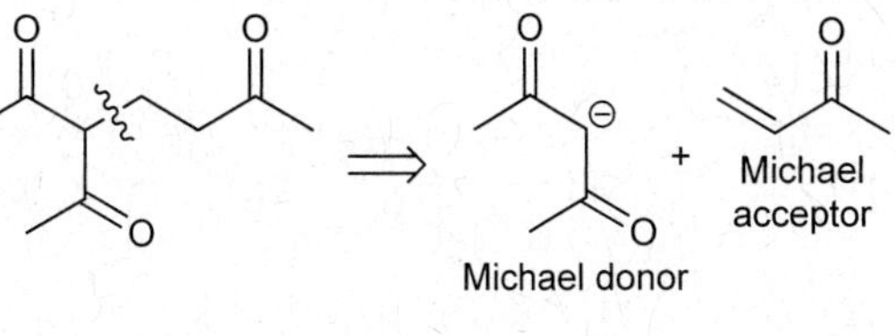

(b) A retrosynthetic analysis reveals the Michael donor (stabilized nucleophile) and Michael acceptor that are responsible for formation of the carbon-carbon bond that is indicated with a wavy line:

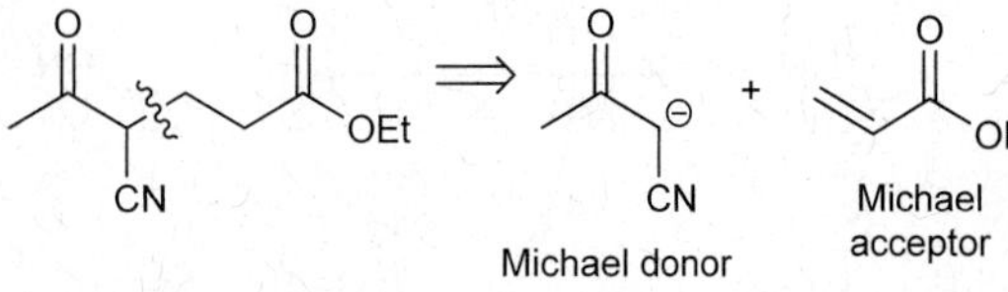

(c) A retrosynthetic analysis reveals the Michael donor (stabilized nucleophile) and Michael acceptor that are responsible for formation of the carbon-carbon bond that is indicated with a wavy line:

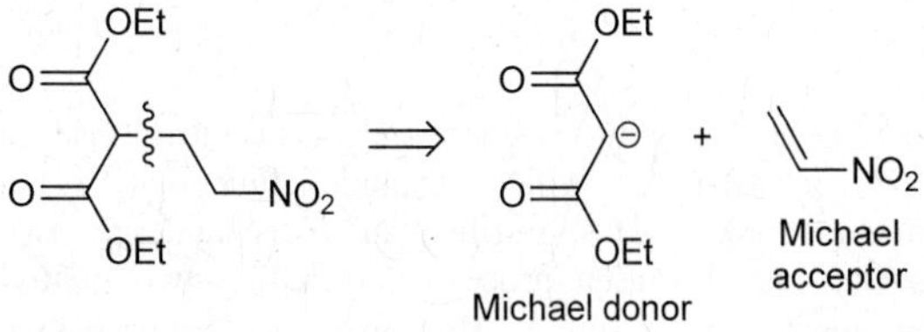

(d) A retrosynthetic analysis reveals the Michael donor (stabilized nucleophile) and Michael acceptor that are responsible for formation of the carbon-carbon bond that is indicated with a wavy line:

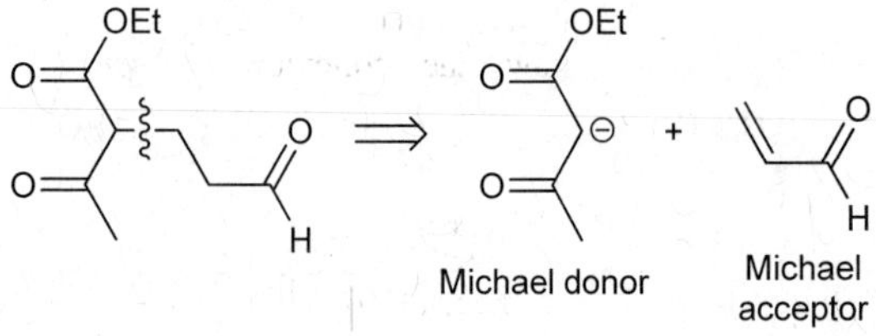

(e) A retrosynthetic analysis reveals the Michael donor (stabilized nucleophile) and Michael acceptor that are responsible for formation of the carbon-carbon bond that is indicated with a wavy line:

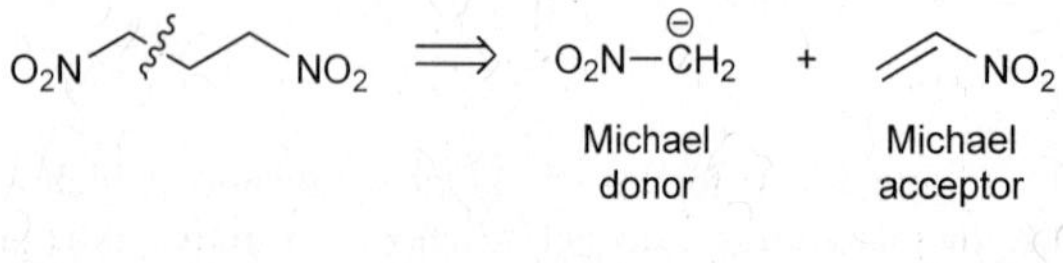

22.95.
(a) The conjugate base of diethyl malonate functions as a nucleophile and attacks propyl bromide in an S_N2 process, expelling bromide as a leaving group, and giving the following product:

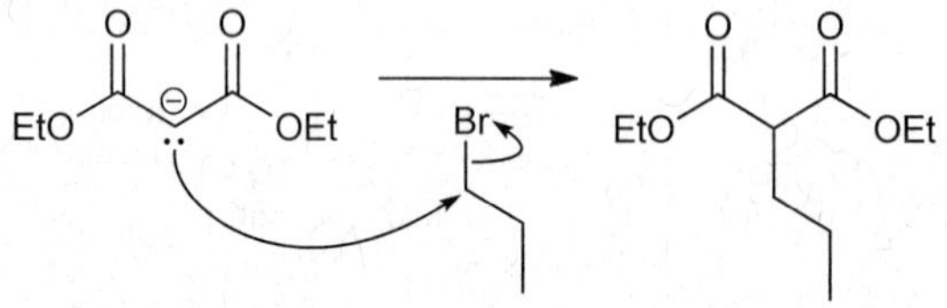

(b) The conjugate base of diethyl malonate functions as a nucleophile and attacks the epoxide at the less substituted (more accessible) position, thereby opening the epoxide and forming an alkoxide ion. Acid work-up converts the alkoxide ion into an alcohol, as shown:

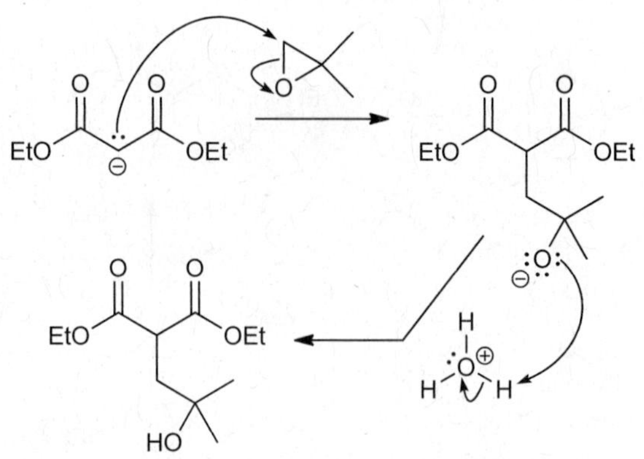

(c) The conjugate base of diethyl malonate functions as a nucleophile and attacks the acid chloride to give a tetrahedral intermediate, which expels a chloride ion to reform a carbonyl group (via a nucleophilic acyl substitution reaction):

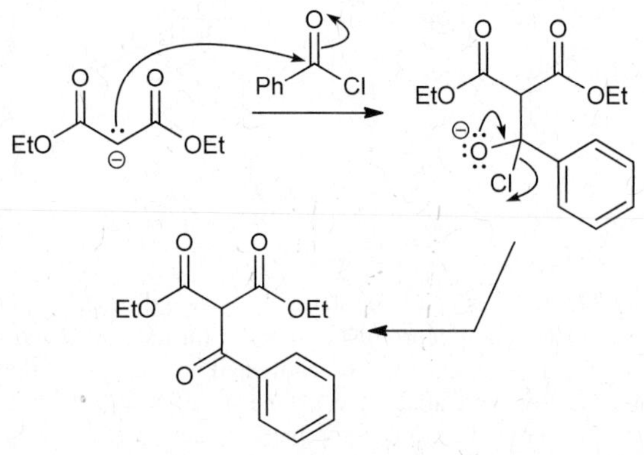

(d) The conjugate base of diethyl malonate functions as a nucleophile and attacks the β position of the α,β-unsaturated ketone. The resulting enolate is converted back into a ketone upon treatment with aqueous acid, giving the following product:

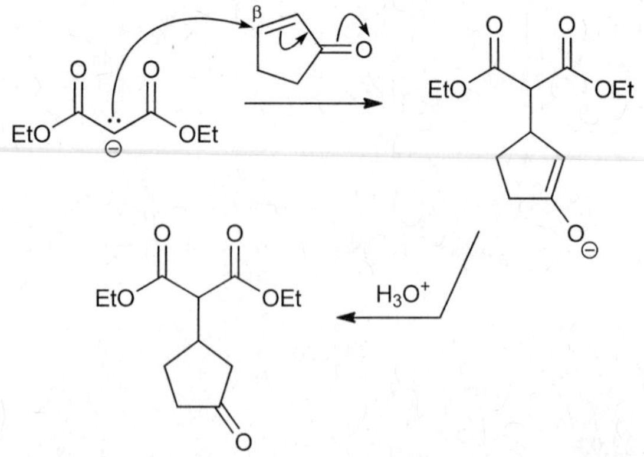

(e) The conjugate base of diethyl malonate functions as a nucleophile and attacks benzyl iodide in an S_N2 process, expelling iodide as a leaving group, and giving the following product:

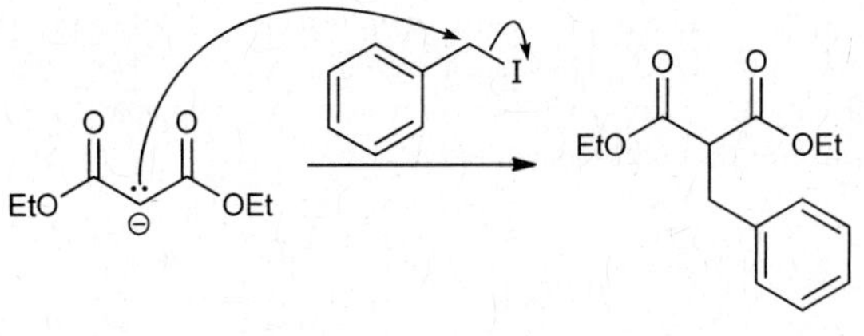

(f) The conjugate base of diethyl malonate functions as a nucleophile and attacks the β position of the α,β-unsaturated nitrile. The resulting intermediate is converted back into a nitrile upon acid work-up, giving the following product:

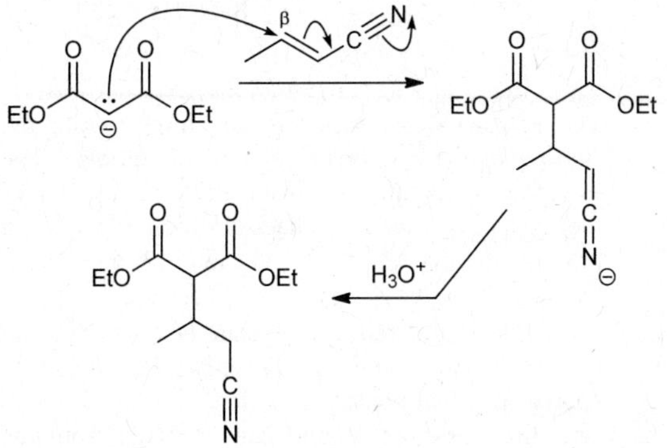

(g) The conjugate base of diethyl malonate functions as a nucleophile and attacks the acid anhydride to give a tetrahedral intermediate, which expels an acetate ion to reform a carbonyl group (via a nucleophilic acyl substitution reaction):

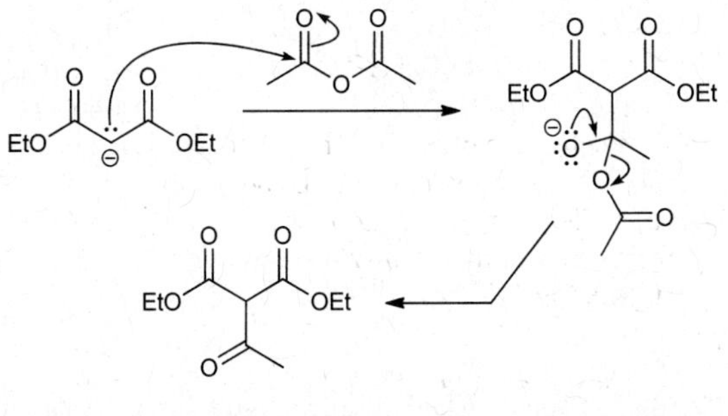

(h) The conjugate base of diethyl malonate functions as a nucleophile and attacks the β position of the α,β-unsaturated nitro compound. The resulting intermediate is converted back into a nitro compound upon acid work-up, giving the following product:

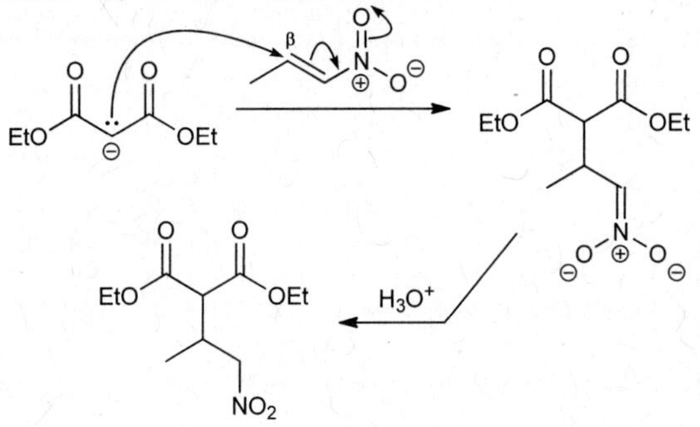

22.96. A Robinson annulation is comprised of a Michael addition, followed by an intramolecular aldol condensation, as shown:

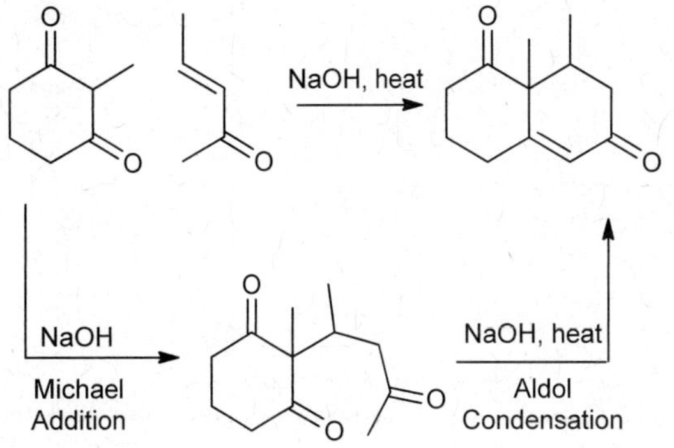

22.97. A Robinson annulation is comprised of a Michael reaction, followed by an intramolecular aldol condensation. To determine the starting materials necessary to prepare the desired product via a Robinson annulation, we draw the following retrosynthetic analysis:

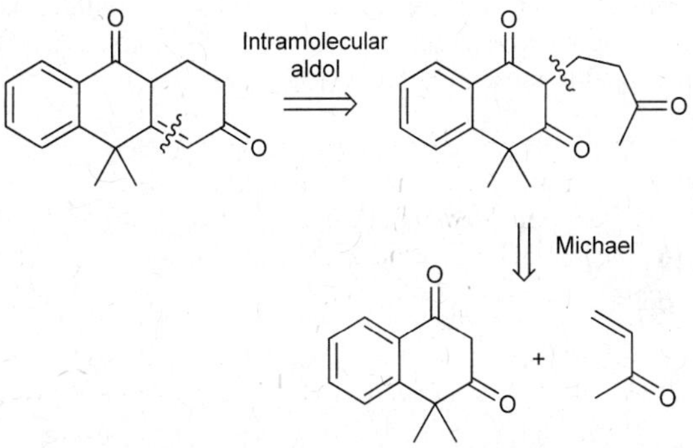

These two steps do not represent two separate reactions. A Robinson annulation can be performed in one reaction flask, as shown in the following forward scheme:

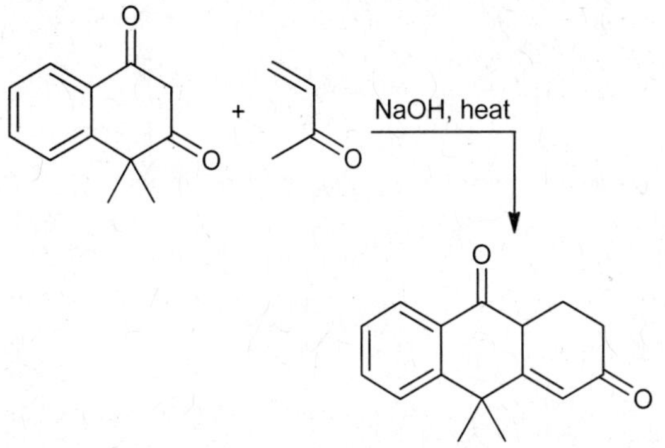

22.98. Notice the similarity between this transformation and an aldol condensation:

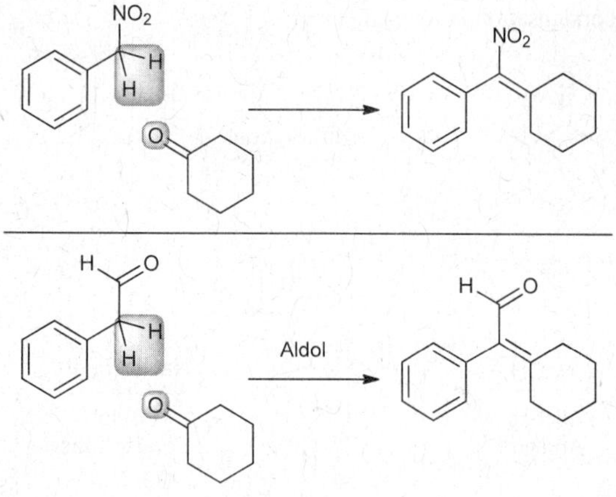

Indeed, we will draw a mechanism (below) that is extremely similar to the mechanism of an aldol condensation. In the first step, hydroxide functions as a base and deprotonates the position adjacent to the nitro group, giving a resonance-stabilized conjugate base (much like an enolate). This conjugate base can function as a nucleophile and attack cyclohexanone, giving an alkoxide ion. Protonation of the alkoxide ion gives an alcohol. Deprotonation, followed by loss of hydroxide, gives the product. Notice that the protonation step employs water as the proton source, consistent with basic conditions (strong acids, such as hydronium, are not measurably present):

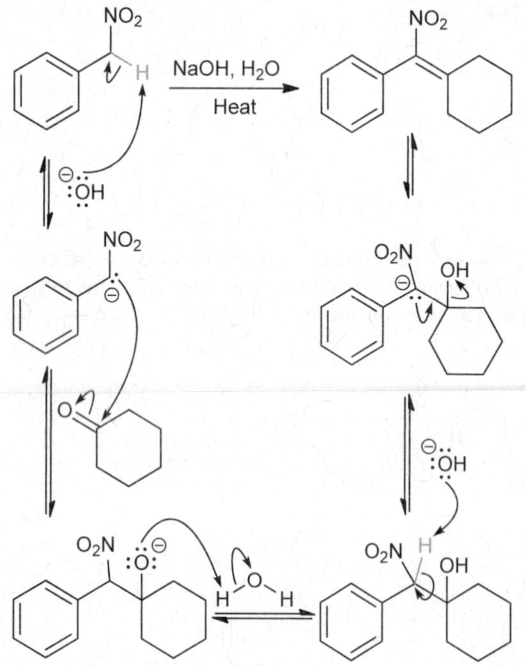

22.99. There are certainly many acceptable solutions to this problem. One such solution derives from the following retrosynthetic analysis. An explanation of each of the steps (*a-c*) follows.

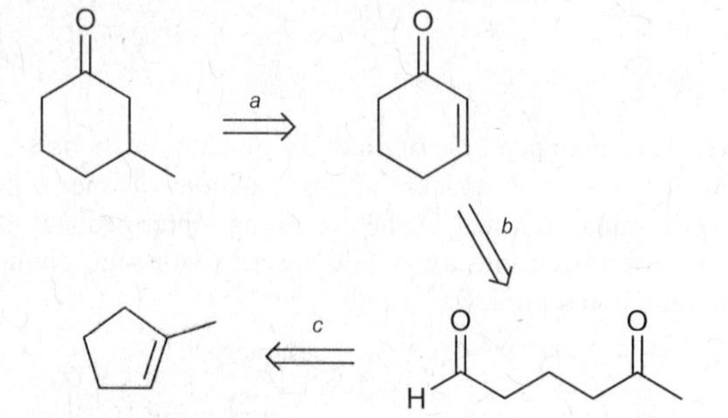

a. The product has seven carbon atoms, while the starting material only has six carbon atoms. The extra carbon atom can be installed in the last step of the synthesis, via a Michael addition (between a lithium dialkyl cuprate and an α,β-unsaturated ketone).

b. The α,β-unsaturated ketone can be made via an intramolecular aldol condensation, starting the appropriate dicarbonyl compound.

c. The dicarbonyl compound can be made from the starting material via ozonolysis.

Now let's draw the forward scheme:

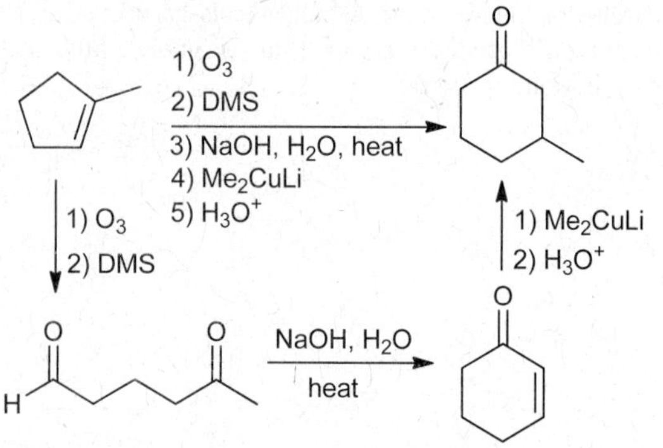

22.100. A ketone generally produces a strong signal at approximately 1720 cm^{-1} (C=O stretching), while an alcohol produces a broad signal between 3200 and 3600 cm^{-1} (O-H stretching). These regions of an IR spectrum can be inspected to determine whether the ketone or the enol predominates.

22.101. Under acidic conditions, one of the OH groups is protonated. If the middle OH group is protonated, the resulting leaving group (water) can leave to give a secondary carbocation. A hydride shift then gives a resonance-stabilized cation that is deprotonated to give a hydroxyaldehyde. Protonation of the OH group (to give a good leaving group), followed by an E2 process, gives the product. In this last step, an E2 process is more likely than an E1 process, because the latter would involve formation of a primary carbocation. Notice that water is the base for the deprotonation steps, consistent with acidic conditions (strong bases, such as hydroxide, are not measurably present under these conditions).

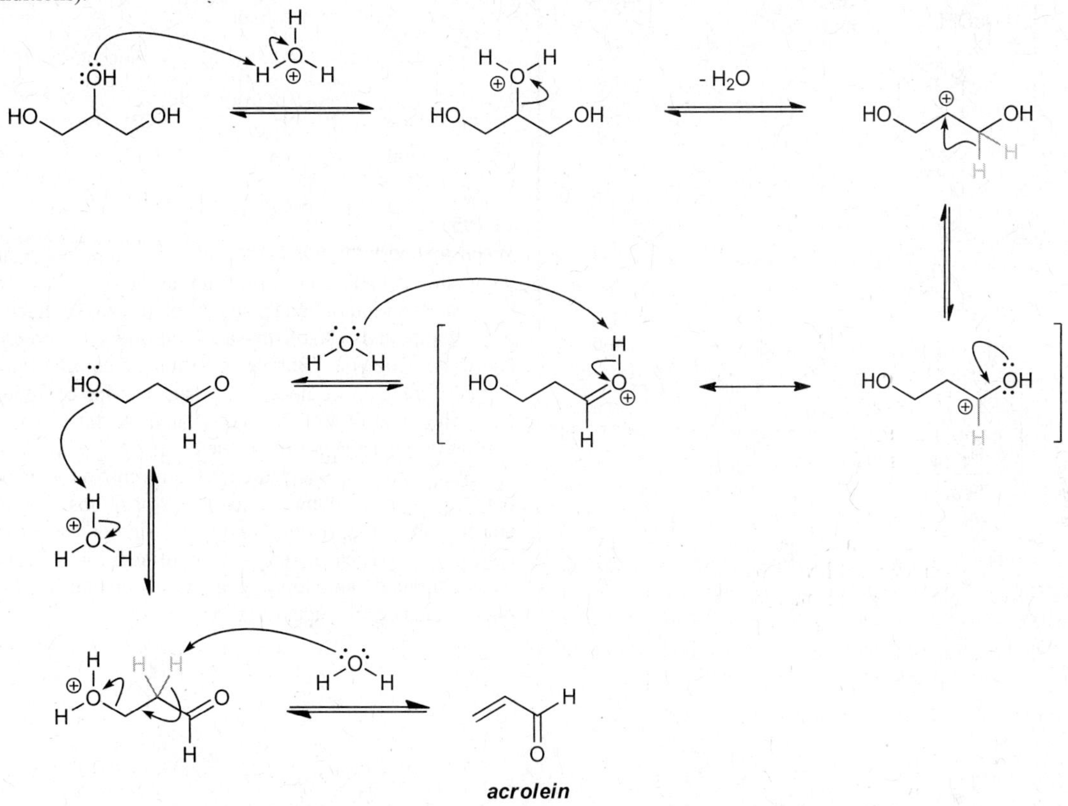

acrolein

22.102. Upon treatment with aqueous acid, the nitrile is hydrolyzed to give a β-ketoacid, which undergoes decarboxylation at elevated temperature to give the following ketone:

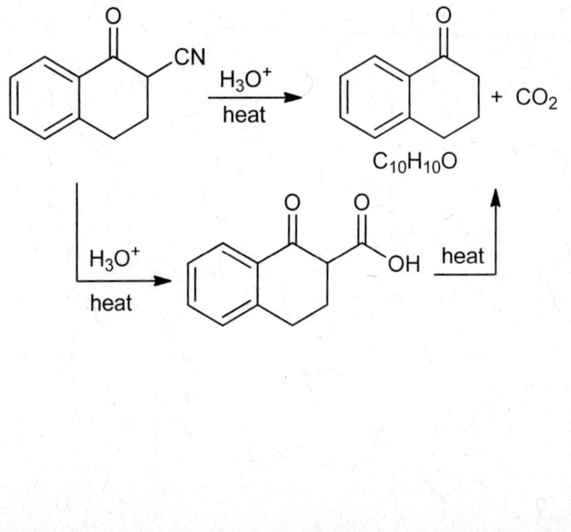

22.103. Using the strategy described in the problem statement, the desired lactone can be made if we use the following epoxide, instead of ethylene oxide:

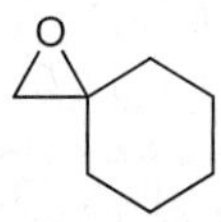

The synthesis, as described in the problem statement, is shown here (follow the location of the cyclohexyl group):

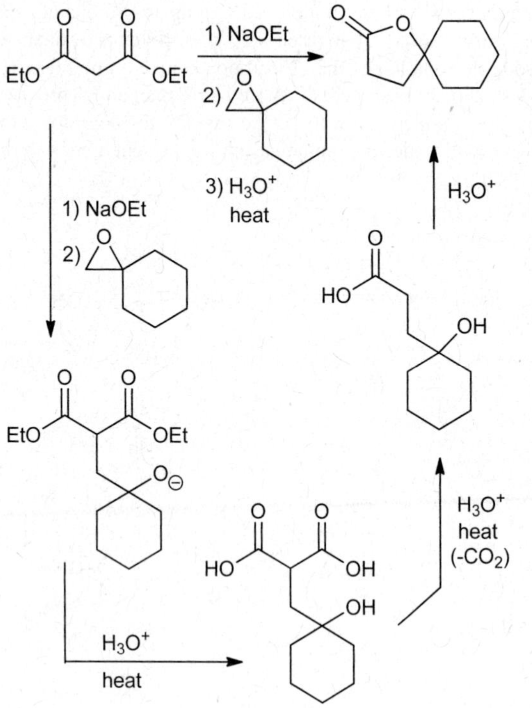

22.104. Upon treatment with aqueous acid at elevated temperature, the ester group is hydrolyzed to a carboxylic acid group, and the acetal is hydrolyzed to a ketone. Under these conditions, the resulting β-keto acid undergoes decarboxylation to give the ketone shown below:

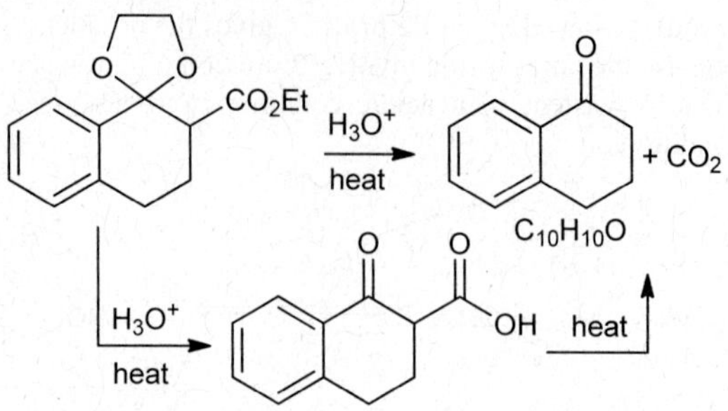

22.105.
When treated with aqueous acid, both compound **A** and compound **B** undergo racemization at the α position (via the enol as an intermediate, see Problem **22.65**). Each of these compounds establishes an equilibrium between *cis* and *trans* isomers. But the position of equilibrium is very different for compound **A** than it is for compound **B**. The equilibrium for compound **A** favors a *cis* configuration, because that is the configuration for which the compound can adopt a chair conformation in which both groups occupy equatorial positions. The equilibrium for compound **B** favors a *trans* configuration, because that is the configuration for which that compound can adopt a chair conformation in which both groups occupy equatorial positions.

22.106. Protonation of the carbonyl group gives a resonance-stabilized cation, which is then deprotonated to give the product. This process is an example of tautomerization:

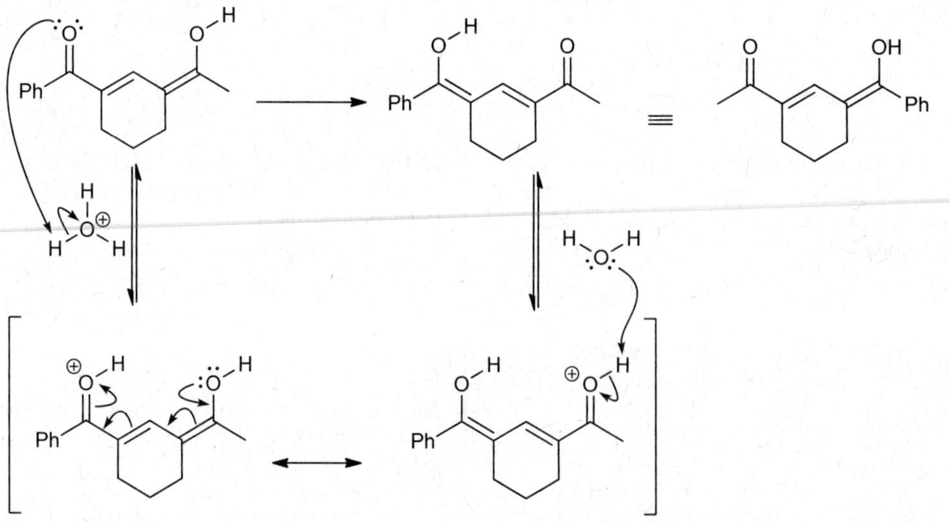

22.107. A retro-aldol reaction opens the ring into an acyclic diketone, which then closes up again via an intramolecular aldol condensation:

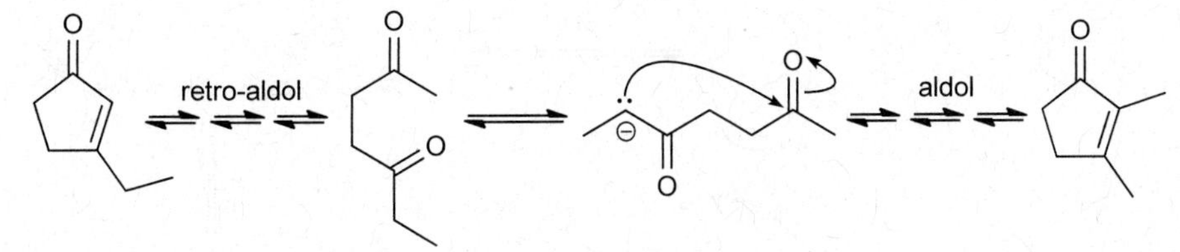

The retro-aldol process proceeds via a mechanism similar to the mechanism seen in the solution to Problem **22.26**. In the first step, hydroxide attacks the β position of the α,β-unsaturated ketone, giving an enolate. The enolate is then protonated to give a β-hydroxyketone, which is subsequently deprotonated to give an alkoxide ion. The carbonyl group is then formed, with loss of an enolate as a leaving group. The resulting enolate is then protonated to give the diketone. Then, an intramolecular aldol reaction occurs, thereby closing a six-membered ring. First, hydroxide functions as a base, giving a new enolate. This enolate ion is a nucleophilic center, and it will attack the carbonyl group present in the same structure (an intramolecular process), thereby closing a five-membered ring. The resulting alkoxide ion is then protonated to give a β-hydroxy ketone. Hydroxide then functions as a base again, deprotonating the α position. The resulting enolate then ejects a hydroxide ion, giving the condensation product, as shown. Notice that each of the protonation steps employs water as the proton source, consistent with basic conditions (strong acids are not measurably present under these conditions).

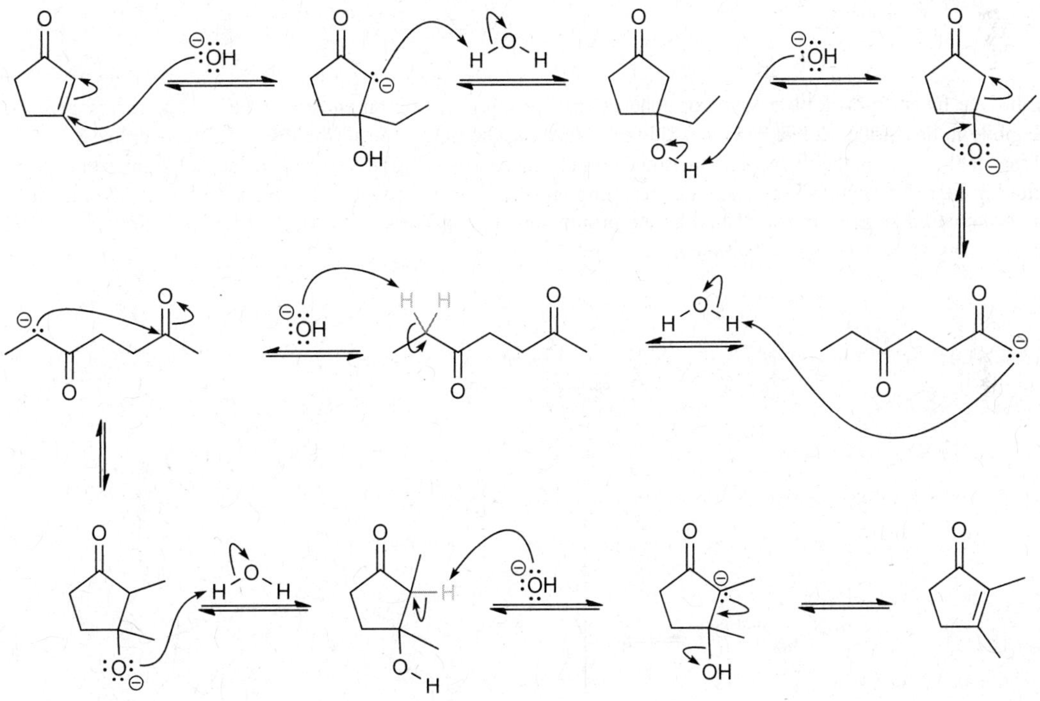

22.108.
(a) Hydroxide functions as a base and deprotonates the α position of the ketone, giving an enolate. The enolate then functions as a nucleophile in an intramolecular Michael addition, attacking the β position of the α,β-unsaturated ketone. The resulting enolate ion is then protonated to give a ketone, which is then further deprotonated to give a new enolate (all possible enolates are present at equilibrium). This enolate then attacks the other carbonyl group in an intramolecular attack, giving an alkoxide ion, which is then protonated to give the product. Notice that each of the protonation steps employs water as the proton source, consistent with basic conditions (strong acids are not measurably present under these conditions).

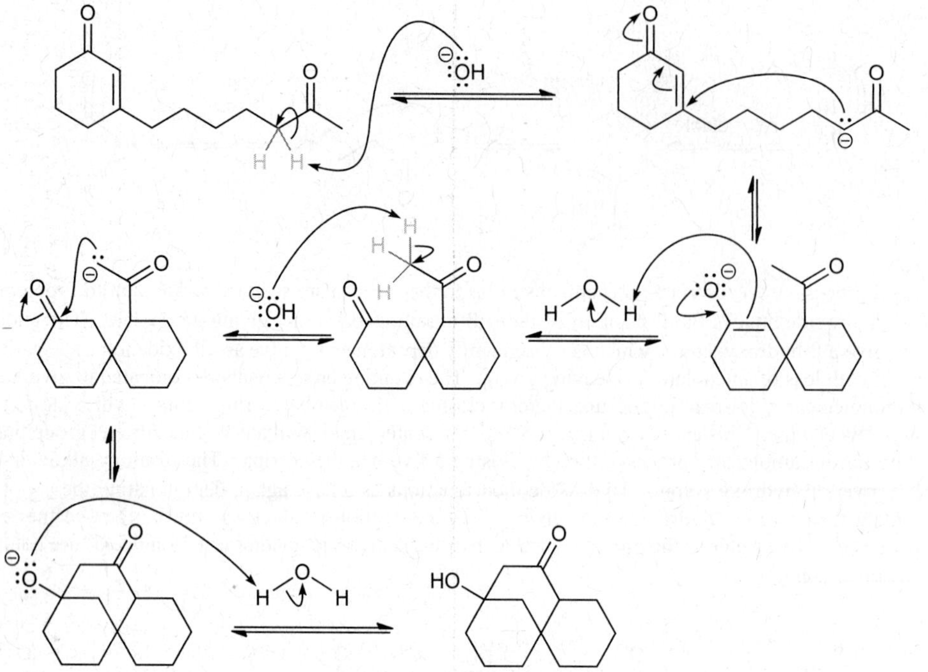

(b) Ethoxide functions as a base and deprotonates an α position, giving an enolate. The enolate then functions as a nucleophile in an intramolecular Michael addition, attacking the β position of the other α,β-unsaturated ketone. The resulting enolate ion is then functions as a nucleophile in another intramolecular Michael addition, attacking the β position of the α,β-unsaturated ketone. The resulting enolate is then protonated to give the product. Notice that each of the protonation steps employs ethanol as the proton source, consistent with basic conditions (strong acids are not measurably present under these conditions).

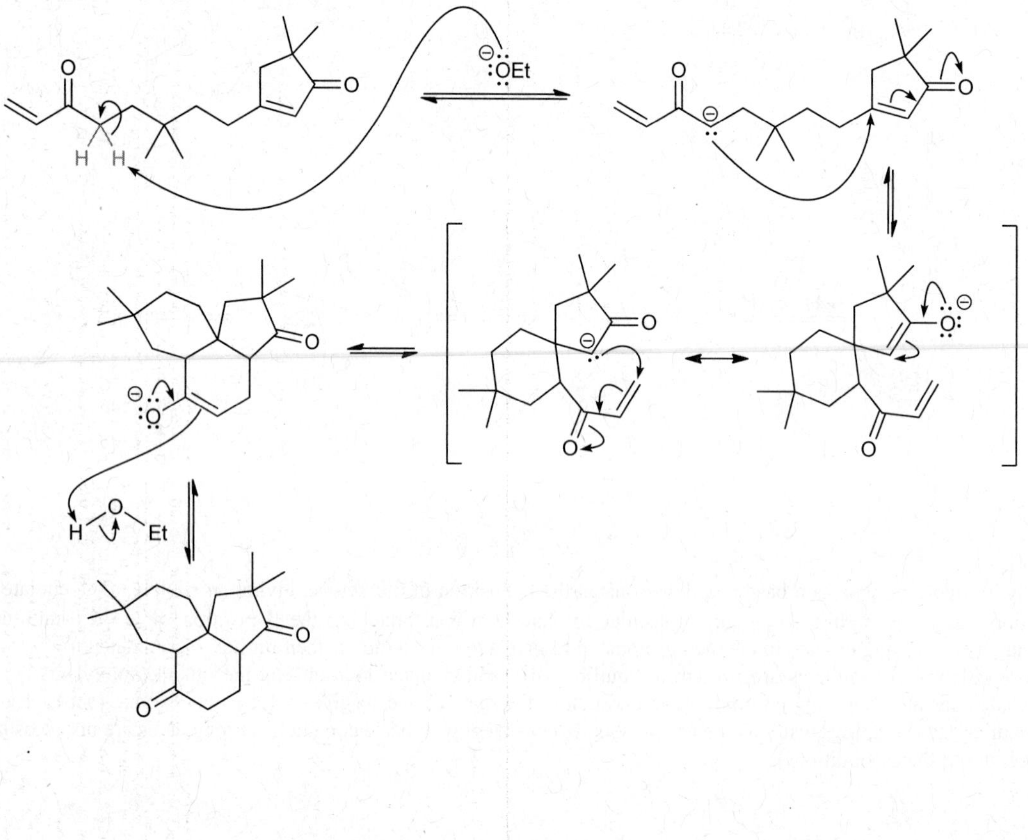

22.109. Direct alkylation would require performing an S_N2 reaction on a tertiary substrate, which will not occur. Instead the enolate would function as a base and E2 elimination would be observed instead of S_N2. The desired transformation can be achieved via a directed aldol condensation, followed by a Michael addition, as shown:

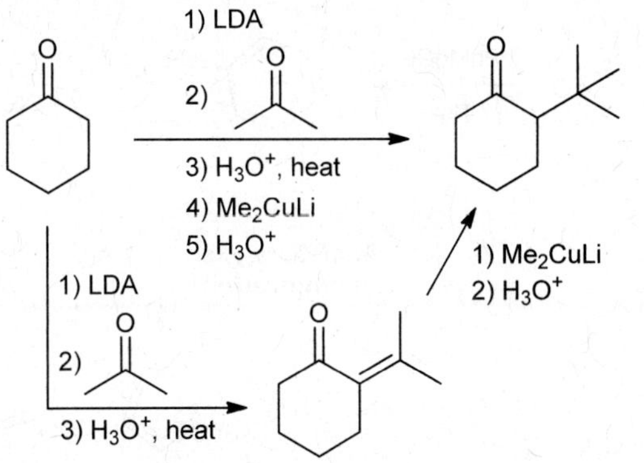

22.110. Intermediate **A** is the corresponding enamine, while the alkylation product **B** is 2-methylcyclohexanone:

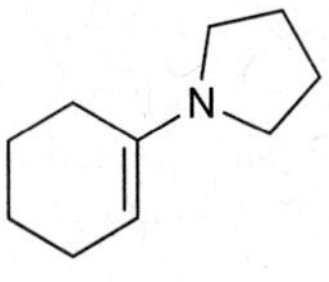

Product **B** is a result of the hydrolysis of the iminium ion shown below, which forms after alkylation of enamine **A** by methyl iodide:

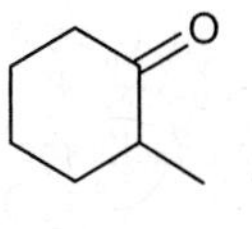

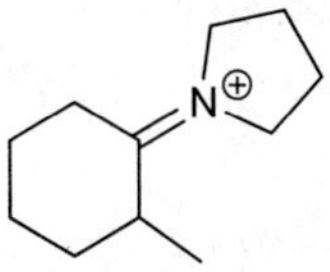

22.111.
(a) Only one isomer of the enamine forms due to rotational symmetry possessed by the secondary amine. Due to this symmetry property, the two possible conformational isomers for this enamine are equivalent.

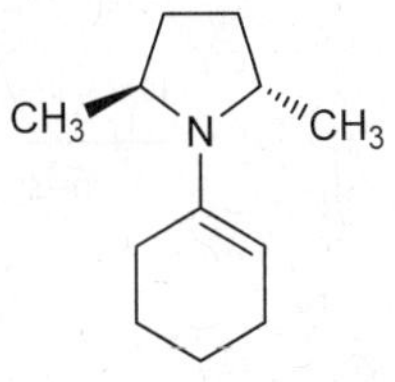

(b) When the enamine in part (a) reacts with methyl iodide, the methyl group closer to the α carbon atom of the enamine is oriented below the plane of the page. So, approach by the electrophile from that direction is hindered, resulting in (R)-2-methylcyclohexanone being the minor product. In contrast, since the approach of methyl iodide from the top face of the page is unimpeded by this substituent, (S)-2-methylcyclohexanone is the major product. And since this is the only isomer of the enamine undergoing alkylation, the S-enantiomer is the major product formed. This means that the rotational symmetry of the starting secondary amine is responsible for the % ee achieved.

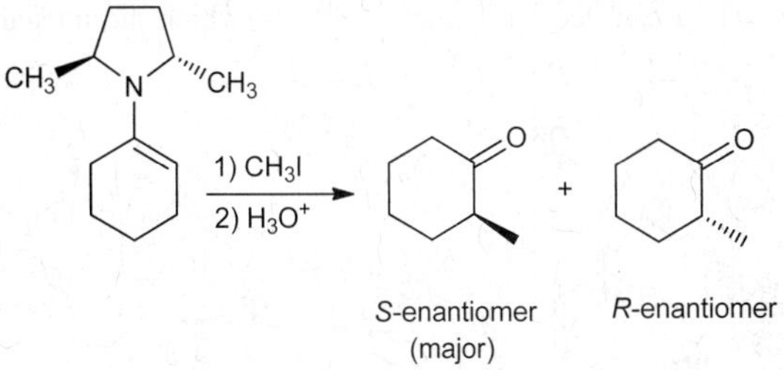

22.112. While there are three alpha positions in the molecule, deprotonation at only one of them (at the terminal methyl) will lead to the formation of a 5-membered ring. The first reaction is a base-catalyzed, intramolecular aldol condensation reaction, affording an α,β-unsaturated ketone. Addition of the Grignard reagent phenylmagnesium chloride gives an alkoxide ion. Treatment with sulfuric acid causes protonation of the alkoxide ion, followed by acid-catalyzed dehydration to afford the highly conjugated diene shown below.

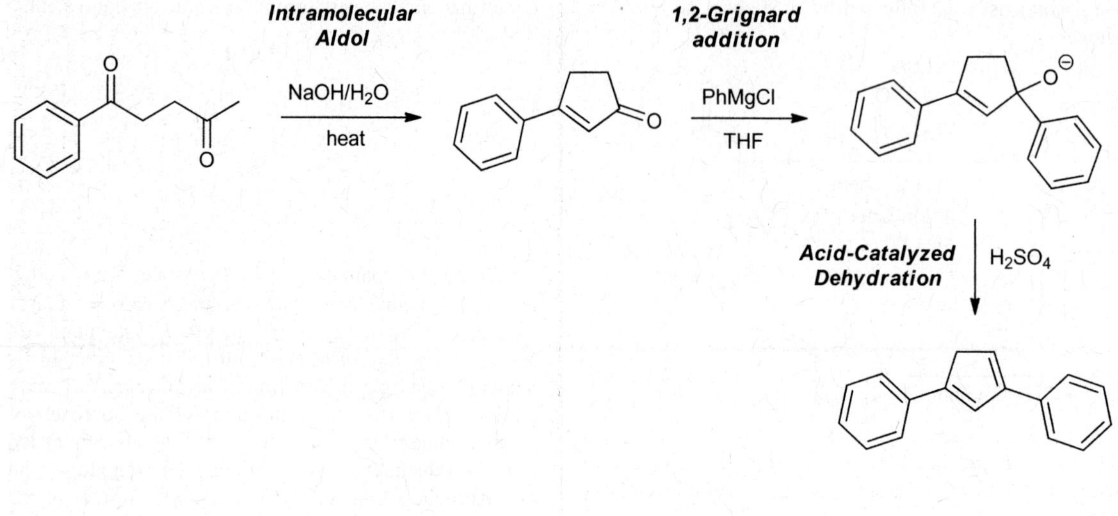

22.113.
(a) Under strongly basic conditions, an enolate is formed. Then, under these conditions, an elimination-addition reaction can occur (Section 19.14), in which the nucleophilic enolate attacks the tethered benzyne in a ring-forming reaction, followed by protonation.

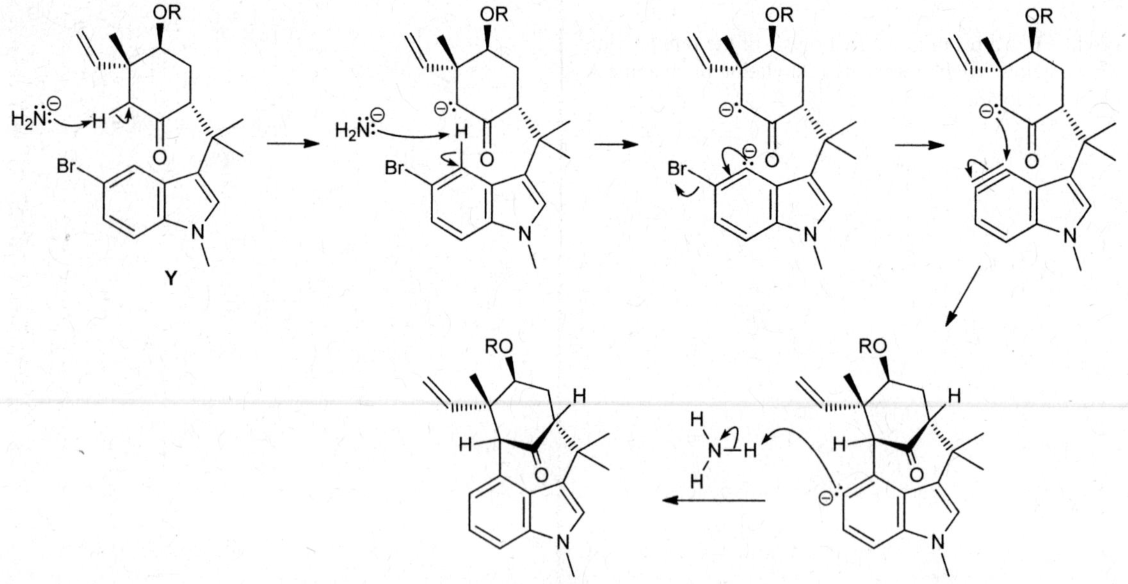

(b) The enolate ion is an ambident nucleophile, which means that it can attack from the oxygen, as well as from the alpha carbon atom. If the oxygen atom of the enolate functions as the nucleophile and attacks the benzyne unit, the following side product is formed:

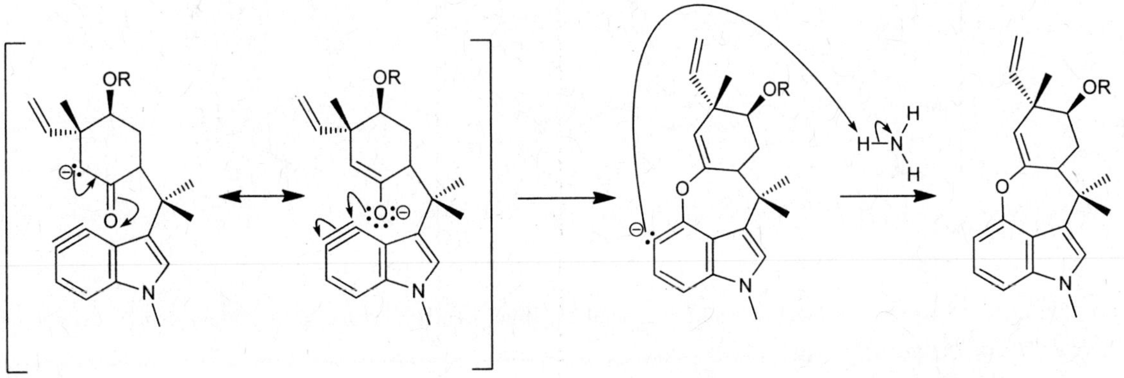

22.114. When the enamine below reacts with methyl acrylate, the ester group is oriented below the plane of the page. So, approach by the electrophile from the back face of the ring is hindered, resulting in the *S* enantiomer being the minor product. In contrast, since the approach of methyl acrylate from the top face of the ring is unimpeded by this substituent, the *R* enantiomer is the major product.

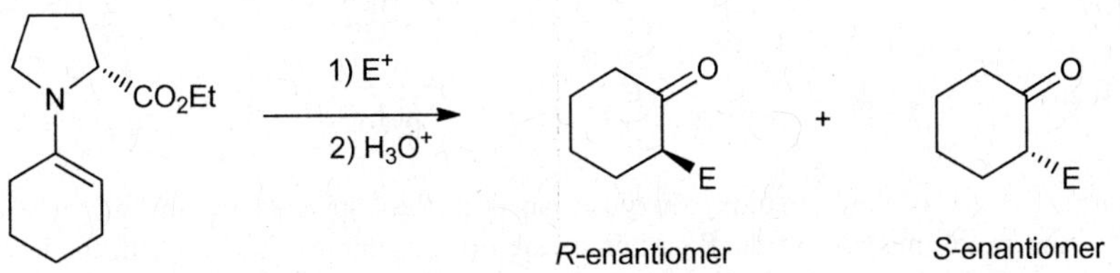

On the other hand, in the case of the enamine below, the ester group is oriented sufficiently far away from the reaction center (the alpha carbon atom of the enamine) such that it imparts much less stereoselectivity in the Michael reaction, making approach by the electrophile from the top or bottom side of the page to be roughly equally accessible. Thus, the two enantiomers form in about 1:1 ratio.

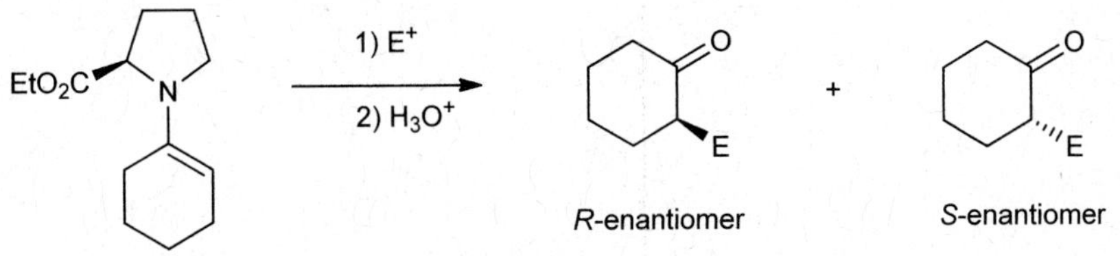

So, there are a total of four possible reaction pathways and only one pathway is hindered by the substituent attached to the chirality center. This means that there are two pathways that result in the formation of the *R* isomer, while only one pathway that leads to the *S* isomer. As such, we expect an enantiomeric excess that favors the *R* configuration.

22.115.
(a) Cyclic hemiacetal **A** is in equilibrium with its open chain form, which has an aldehyde group and a hydroxyl group. The aldehyde group can undergo a Wittig reaction when treated with a stabilized ylide ($Ph_3P=CHCO_2Et$), giving an α,β-unsaturated ester. Then, the hydroxyl group is converted to a tosylate group upon treatment with tosyl chloride:

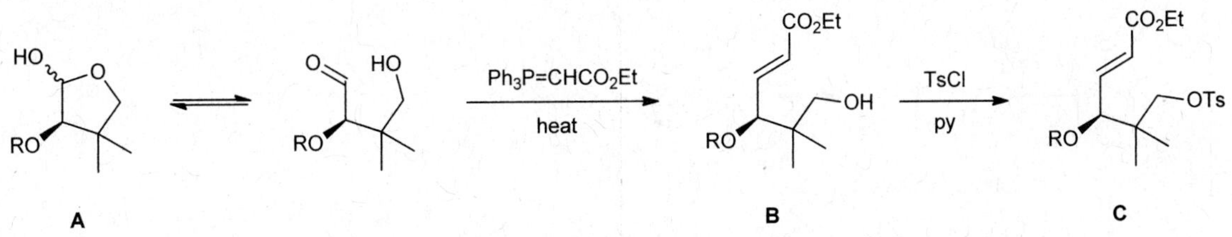

(b) The conversion of compound **C** to **D** begins with a Michael addition reaction, in which Me₂CuLi functions as a Michael donor and attacks the α,β-unsaturated ester (the Michael acceptor) to afford an enolate. This enolate can then function as a nucleophile in an intramolecular S_N2-type process, in which the enolate undergoes α-alkylation to close the cyclopentyl ring system, giving compound **D**:

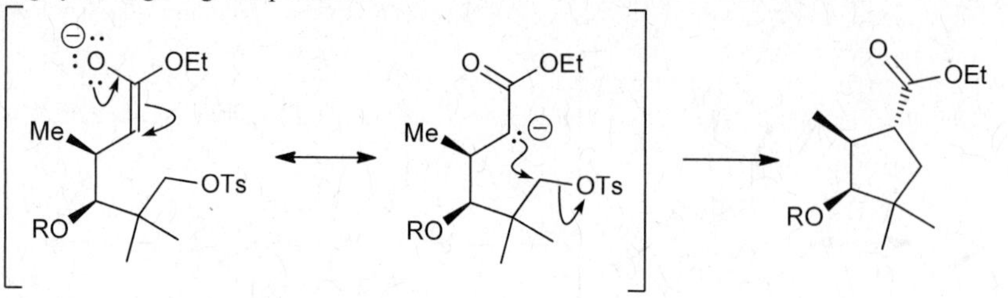

D

(c) The ester is first reduced to an alcohol (compound **E**), which can be acylated, using either acetic anhydride or acetyl chloride and pyridine:

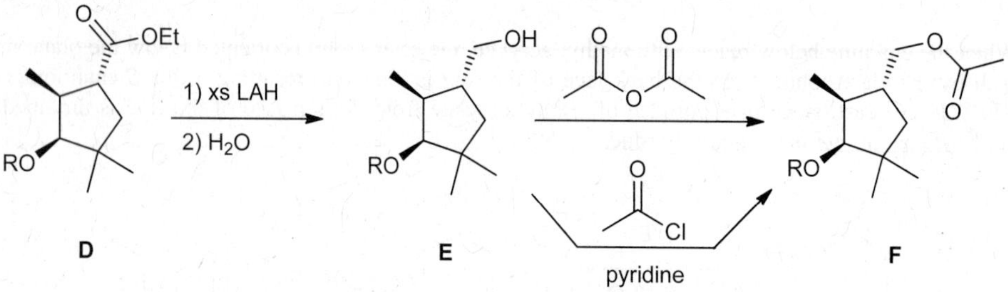

1) xs LAH

2) H₂O

pyridine

D   **E**   **F**

22.116. In the starting material, the dimethoxybenzyl group is in the "up" position on the chiral carbon, from the perspective drawn. This group thus attacks the β carbon of the α,β-unsaturated ketone from the top face, pushing the other aromatic ring "down", so that it ends up *cis* to the hydrogen on the adjacent chirality center.

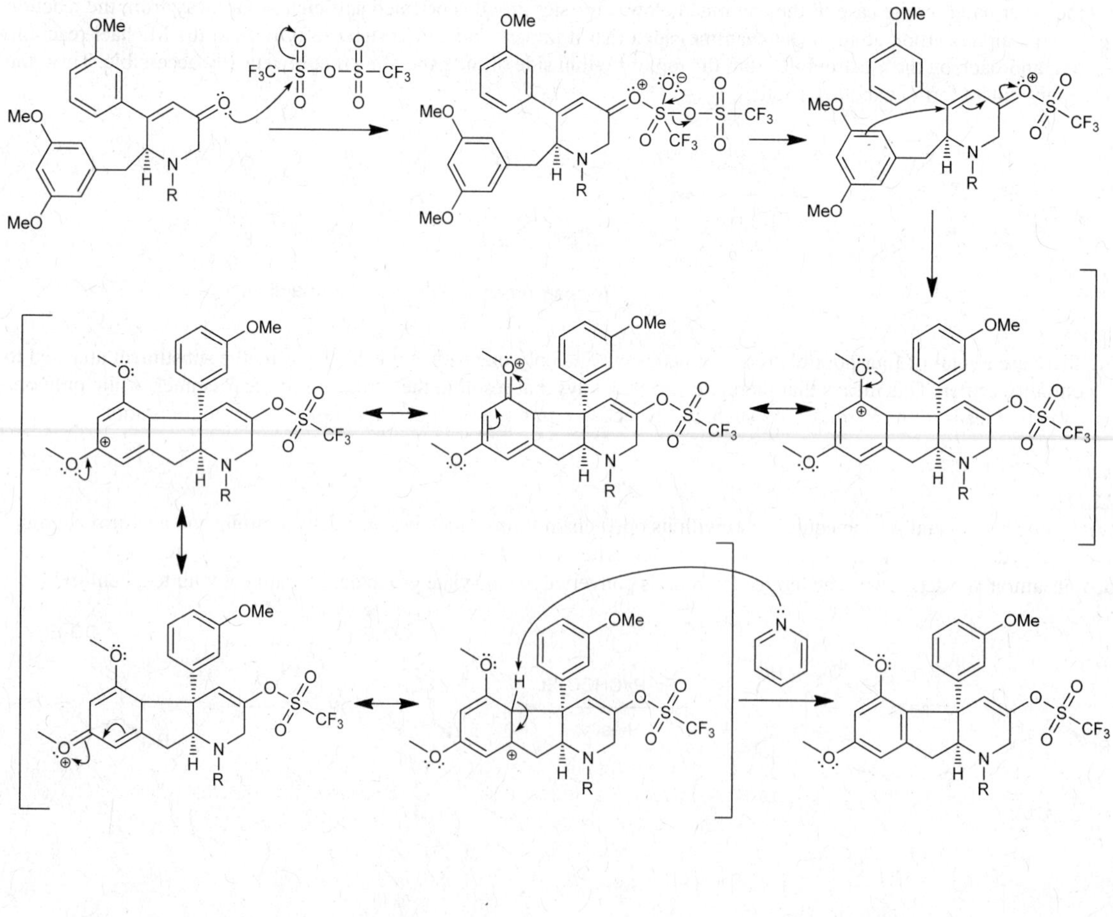

22.117. The mechanism for this reaction involves two successive Michael addition reactions. The sequence begins with the deprotonation of diethylmalonate with potassium carbonate to give a stabilized anion (Michael donor) that attacks the α,β-unsaturated ketone **1** (Michael acceptor) in a Michael fashion to afford enolate **3**. After protonation of enolate **3** to give compound **4**, the β-diester moiety is then deprotonated to give the resonance-stabilized anion **5**. At this point a second Michael addition occurs as the anion (Michael donor) attacks the carbon containing the two thio ethers (Michael acceptor) in a conjugate fashion to give enolate **6**. Finally, the electrons of the enolate come back down to make the ketone and the α,β-unsaturation by expelling the ethanethiolate anion as a leaving group to afford the final product **2**.

Alternatively, the reverse order of these two Michael additions will likewise afford the product.

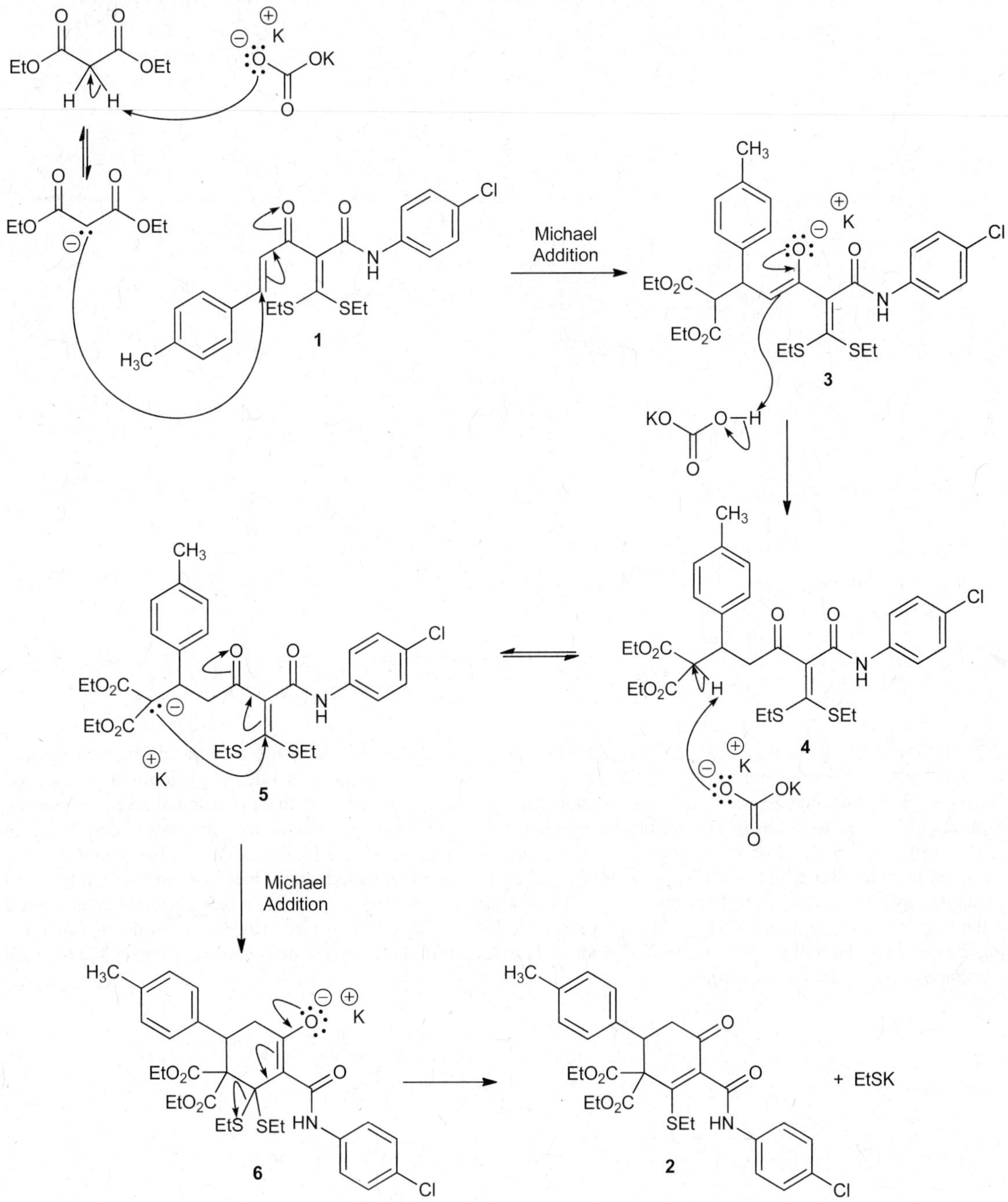

22.118. Methoxide functions as a base and deprotonates the most acidic position, leading to a doubly-stabilized enolate. This enolate then functions as a nucleophile and attacks the ester in an intramolecular nucleophilic acyl substitution reaction. The resulting tetrahedral intermediate loses methoxide to reform the carbonyl group. The OH group is then deprotonated to give an alkoxide ion, which then attacks the newly formed carbonyl group to give another alkoxide ion, which is protonated to give the product.

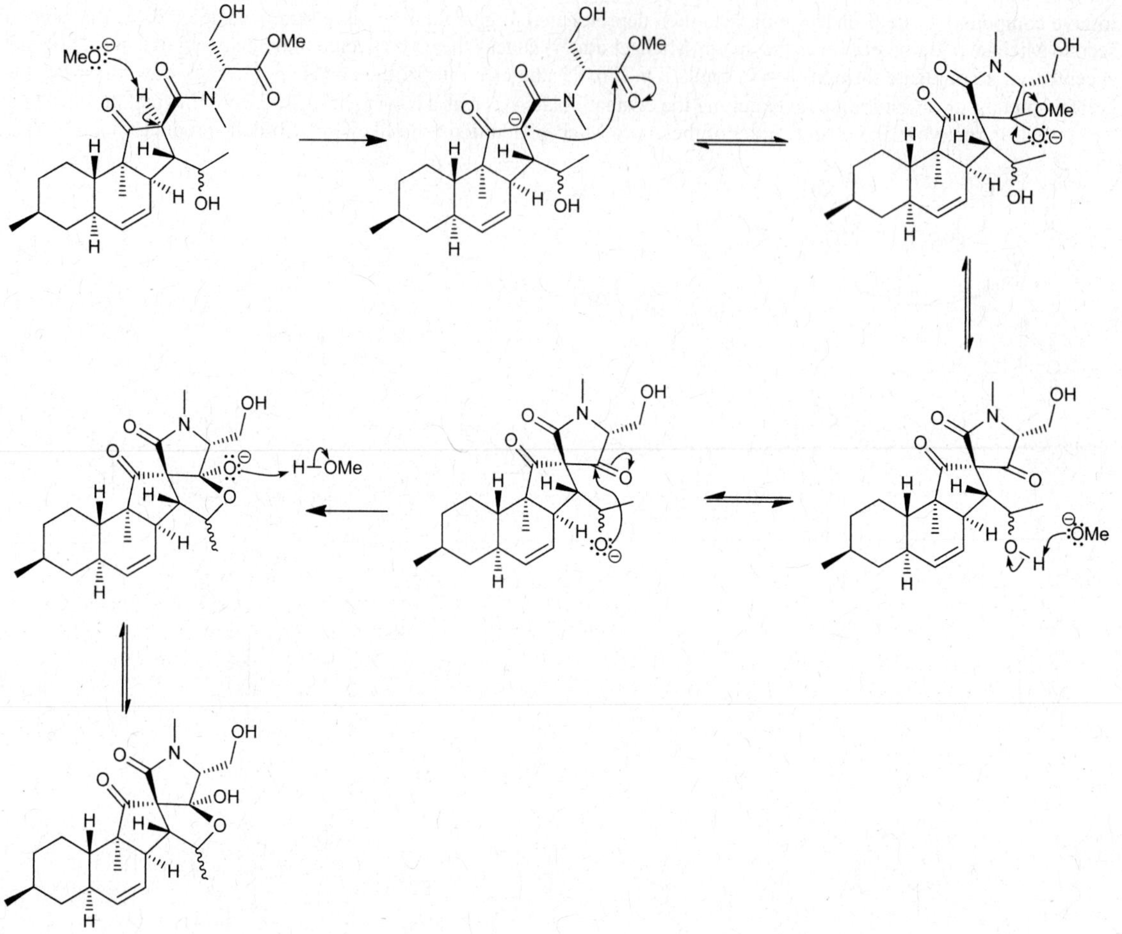

22.119. Ketone **1** is flanked on one side by a phenyl group and on the other side by a 6-membered ring bearing only one alpha proton. When it is treated with NaHMDS, only one enolate can form – alpha deprotonation will produce **3a**, which can be redrawn in a pseudo-chair transition state (**3b**). In this orientation, note that the terminal alkene is pseudo-axial; because of the locations of the two other alkenes, there are no 1,3-diaxial interactions that would destabilize an axial substituent. Upon a close inspection of **3b**, it becomes clear that there is a [3,3]-relationship between the newly generated enolate alkene and the terminal alkene attached to the 6-membered ring. Once formed, the enolate will spontaneously undergo a [3,3]-sigmatropic reaction to form alkoxide **4**. Note the observed facial selectivity as a result of the position of the terminal alkene. This will be responsible for the newly formed stereocenter in the 8-membered ring. Protonation of **4** will produce the desired product. This variant of the Claisen rearrangement (Chapter 17) is called an anionic oxy-Claisen rearrangement.

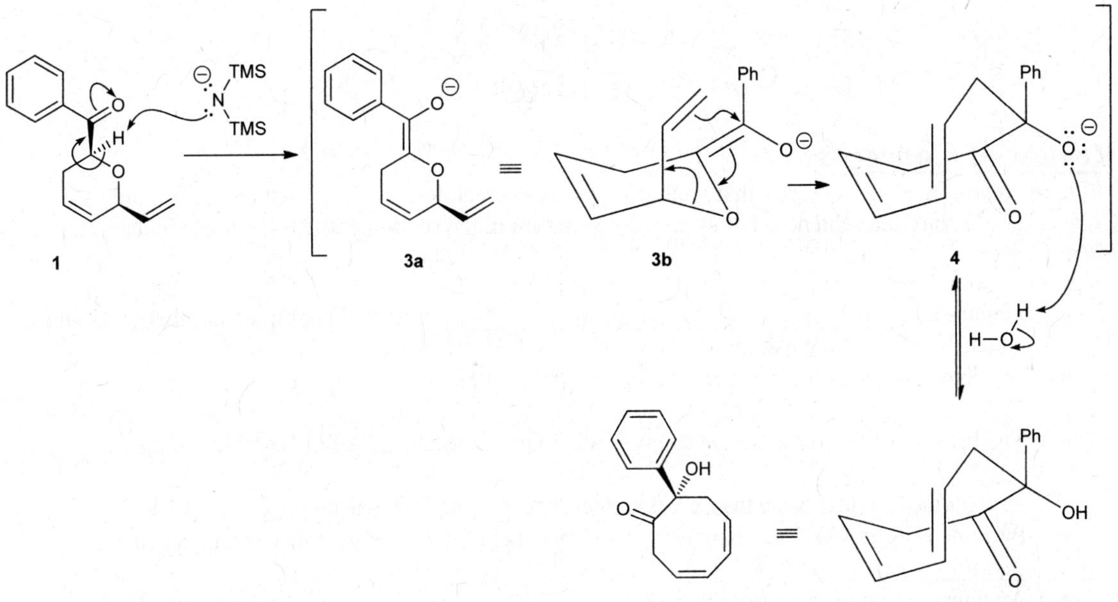

1   3a   3b   4

2

22.120. Although this reaction looks difficult, it is less intimidating if you can see the two atoms that are being connected (alpha carbon of the lactone with the isotopically labeled ^{14}C aldehyde). Before we can draw the mechanism of this aldol process, we must recognize that the compound contains a carboxylic acid group, which will likely be deprotonated under these basic conditions. So the first step of our mechanism is deprotonation of the carboxylic acid group (the last step in our mechanism will be to give this proton back during the HCl work-up):

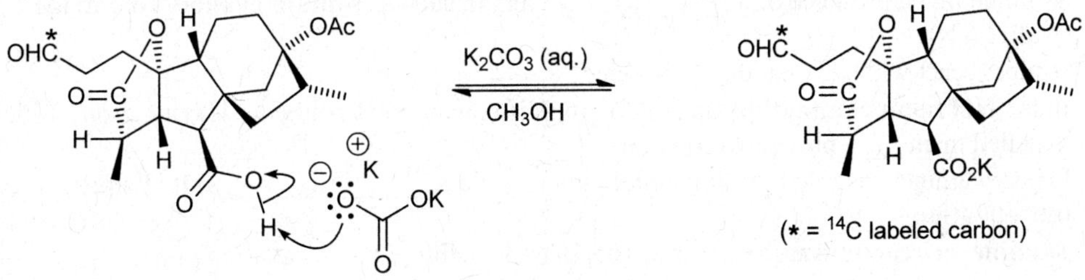

K$_2$CO$_3$ (aq.)

CH$_3$OH

(* = ^{14}C labeled carbon)

In the second step of the mechanism, an enolate is formed (alpha to the ester group), which then undergoes an aldol-type reaction by attacking the pendant ^{14}C-labeled aldehyde to afford an alkoxide:

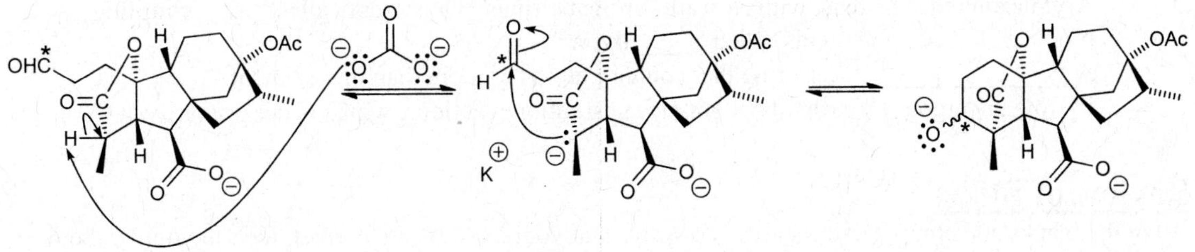

The work-up for this reaction is completed by the addition of aqueous hydrochloric acid, protonating the alkoxide as well as the carboxylate, which gives the desired radiolabled aldol-type product. In drawing the mechanism for these protonation steps, we use H$_3$O$^+$ as the acid (remember that H$_3$O$^+$ is the strongest acid that can be present in an aqueous solution):

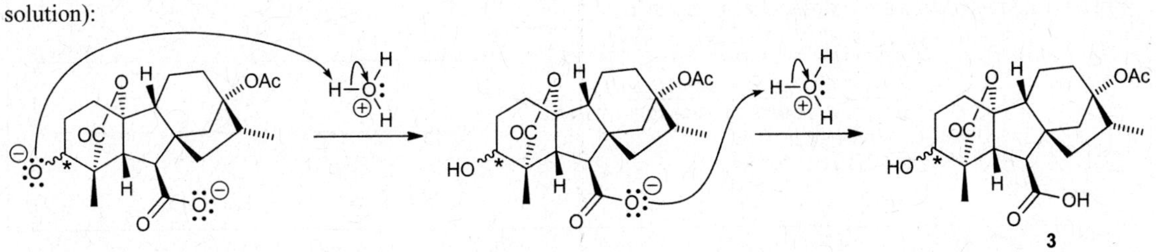

3

Chapter 23
Amines

Review of Concepts

Fill in the blanks below. To verify that your answers are correct, look in your textbook at the end of Chapter 23. Each of the sentences below appears verbatim in the section entitled *Review of Concepts and Vocabulary*.

- Amines are _____, _____, or _____, depending on the number of groups attached to the nitrogen atom.
- The lone pair on the nitrogen atom of an amine can function as a _____ or _____.
- The basicity of an amine can be quantified by measuring the pK_a of the corresponding _____.
- Aryl amines are less basic than alkyl amines, because the lone pair is _____.
- Pyridine is a stronger base than pyrrole, because the lone pair in pyrrole participates in _____.
- An amine group exists primarily as _____ at physiological pH.
- The **azide synthesis** involves treating an _____ with sodium azide, followed by _____.
- The _____ **synthesis** generates primary amines upon treatment of potassium phthalimide with an alkyl halide, followed by hydrolysis or reaction with N_2H_4.
- Amines can be prepared via **reductive amination**, in which a ketone or aldehyde is converted into an imine in the presence of a _____ agent, such as **sodium cyanoborohydride** (NaBH$_3$CN).
- Amines react with acyl halides to produce _____.
- In the **Hofmann elimination**, an amino group is converted into a better leaving group which is expelled in an _____ process to form an _____.
- Primary amines react with a nitrosonium ion to yield a _____ **salt** in a process called **diazotization**.
- **Sandmeyer reactions** utilize copper salts (CuX), enabling the installation of a halogen or a _____ group.
- In the **Schiemann reaction**, an aryl diazonium salt is converted into a fluorobenzene by treatment with _____.
- Aryldiazonium salts react with activated aromatic rings in a process called _____ **coupling**, to produce colored compounds called _____ **dyes**.
- A _____ is a ring that contains atoms of more than one element.
- Pyrrole undergoes electrophilic aromatic substitution reactions, which occur primarily at C__.

Review of Skills

Fill in the blanks and empty boxes below. To verify that your answers are correct, look in your textbook at the end of Chapter 23. The answers appear in the section entitled *SkillBuilder Review*.

23.1 Naming an Amine

PROVIDE A SYSTEMATIC NAME FOR THE FOLLOWING COMPOUND:

1) IDENTIFY THE PARENT
2) IDENTIFY AND NAME SUBSTITUENTS
3) ASSIGN LOCANTS TO EACH SUBSTITUENT
4) ALPHABETIZE
5) ASSIGN CONFIGURATION

23.2 Preparing a Primary Amine via the Gabriel Reaction

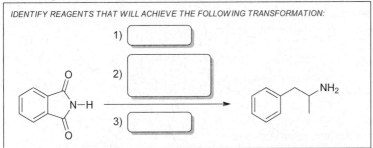

23.3 Preparing an Amine via a Reductive Amination

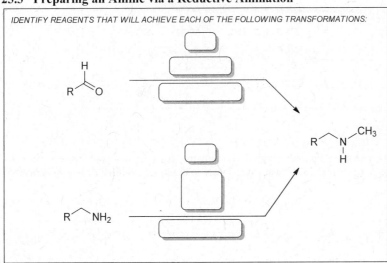

23.4 Synthesis Strategies

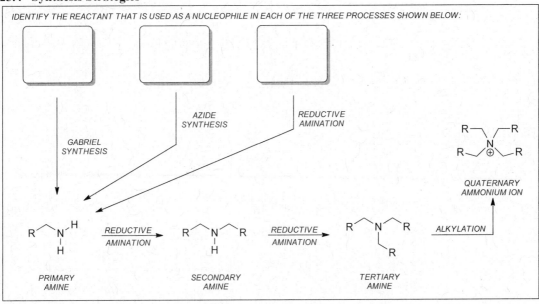

23.5   Predicting the Product of a Hofmann Elimination

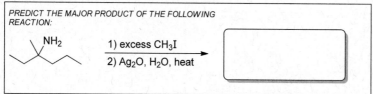

PREDICT THE MAJOR PRODUCT OF THE FOLLOWING
REACTION:

1) excess CH₃I
2) Ag₂O, H₂O, heat

23.6   Determining the Reactants for Preparing an Azo Dye

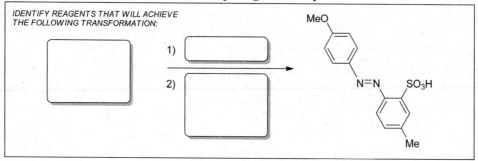

IDENTIFY REAGENTS THAT WILL ACHIEVE
THE FOLLOWING TRANSFORMATION:

1)

2)

Review of Reactions

Identify the reagents necessary to achieve each of the following transformations. To verify that your answers are
correct, look in your textbook at the end of Chapter 23. The answers appear in the section entitled *Review of Reactions*.

Preparation of Amines

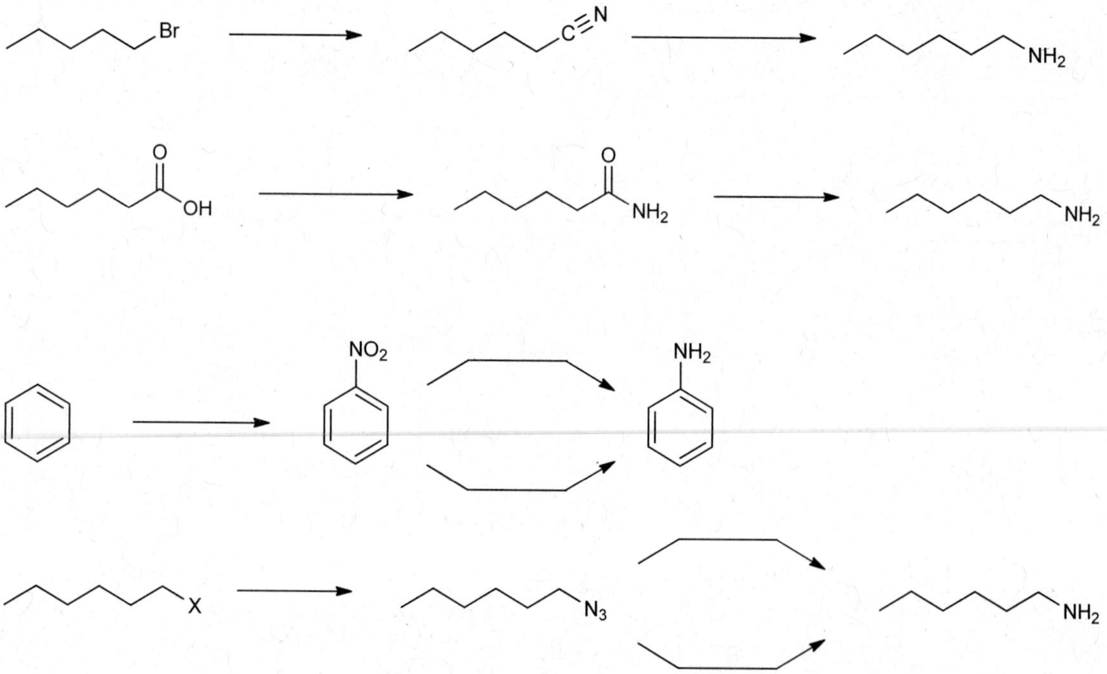

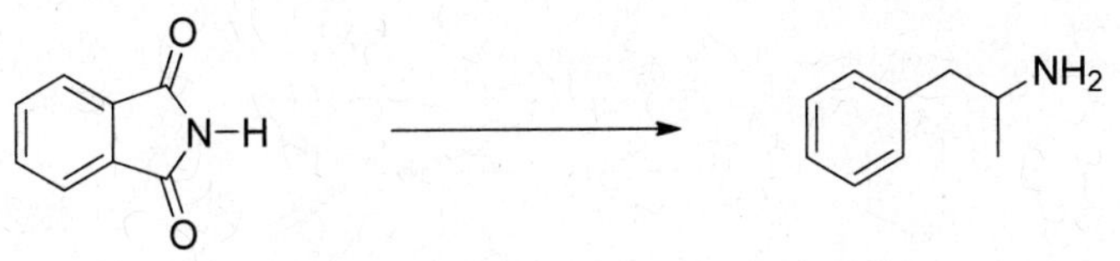

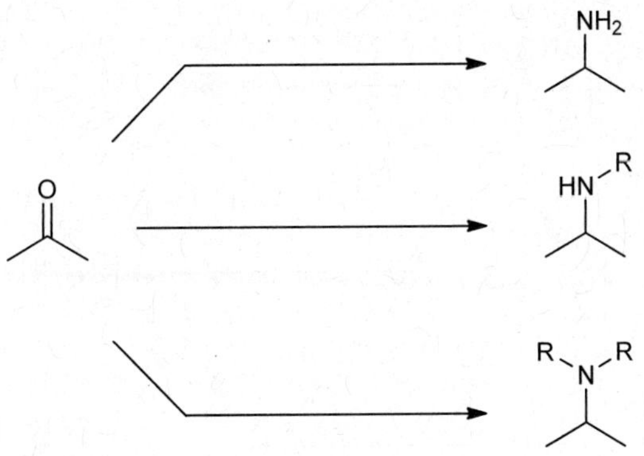

Reactions of Amines

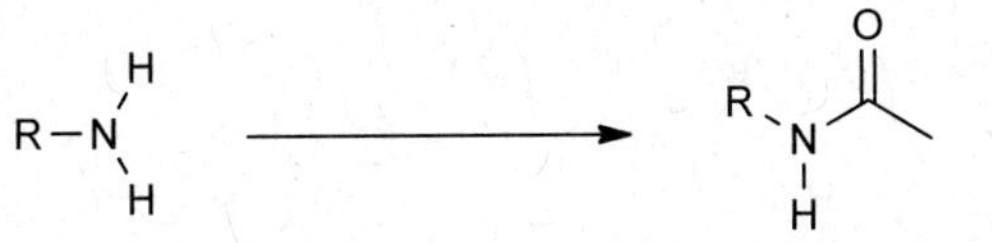

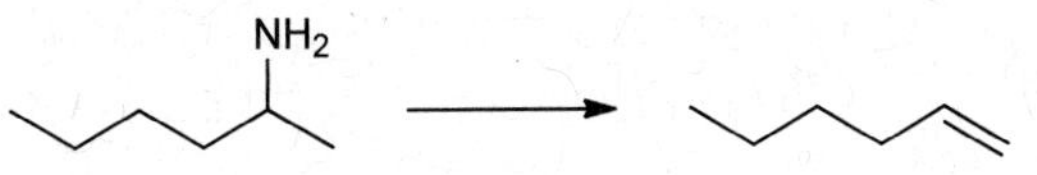

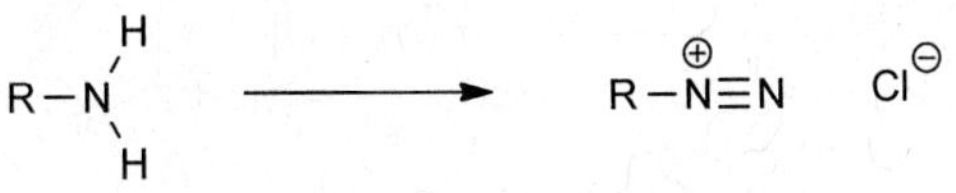

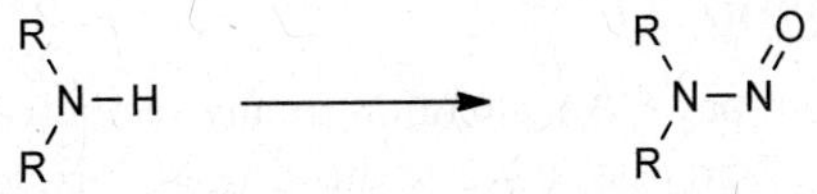

Reactions of Aryldiazonium Salts

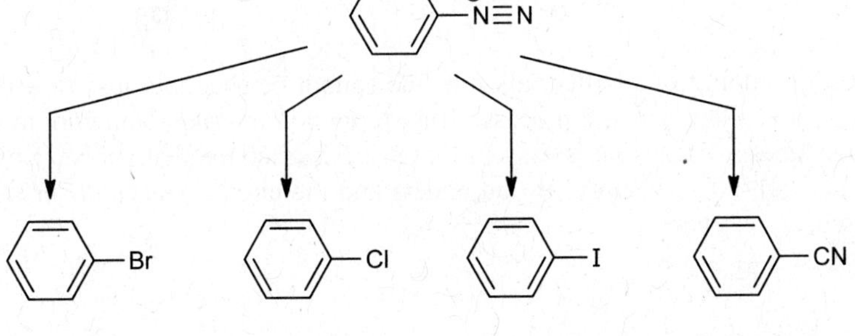

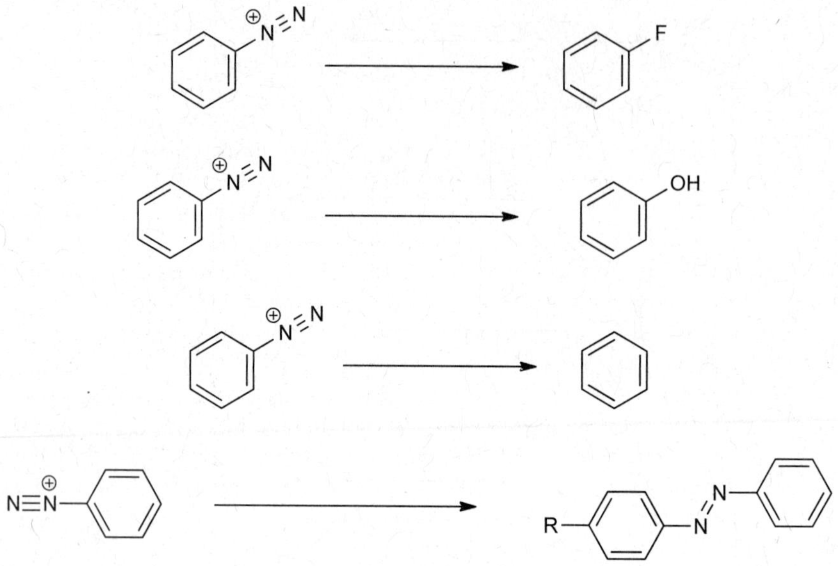

Reactions of Nitrogen Heterocycles

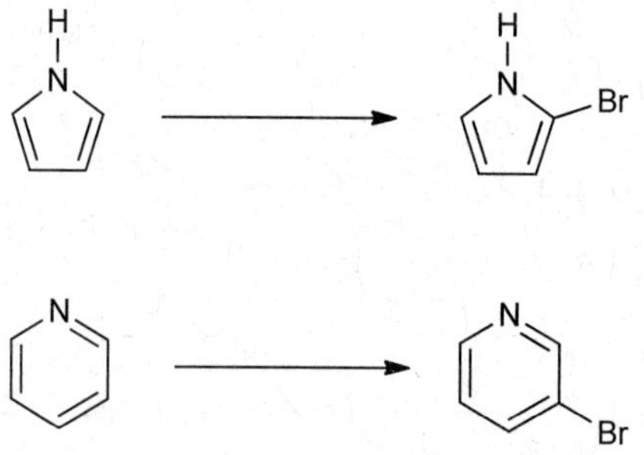

Common Mistake to Avoid

Whenever you learn a new reaction, pay close attention to any restrictions that may apply. For example, the Gabriel synthesis employs an S_N2 process to create the critical C–N bond of a primary amine:

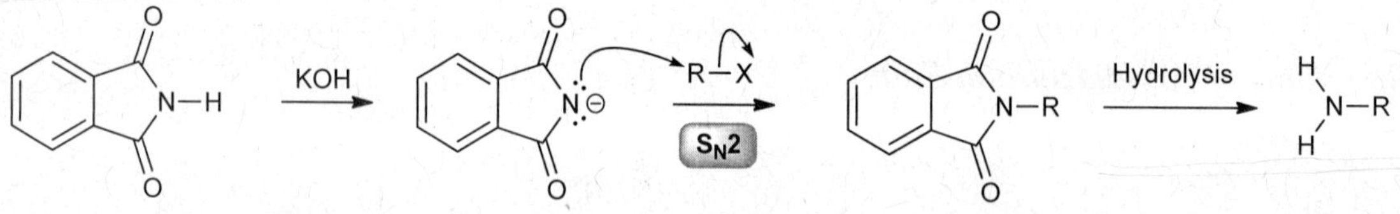

Since an S_N2 process is employed, a tertiary alkyl halide cannot be used, because tertiary alkyl halides are too sterically hindered to undergo an S_N2 process. It is a common mistake to attempt to use a tertiary alkyl halide in a Gabriel synthesis, because it is easy to forget the restrictions that apply. Keep this in mind for all reactions that you study. Make sure that you understand the circumstances under which each reaction can or cannot be used.

Useful reagents

The following is a list of reagents encountered in this chapter:

Reagents	Type of Reaction	Description
1) NaCN 2) xs LAH 3) H_2O	Preparation of an amine (from an alkyl halide)	These reagents can be used to convert an alkyl halide into an amine with the introduction of one carbon atom (from the cyano group).
1) $SOCl_2$ 2) xs NH_3 3) xs LAH 4) H_2O	Preparation of an amine (from a carboxylic acid)	These reagents can be used to convert a carboxylic acid into an amine, without a change in the carbon skeleton.
1) Fe, H_3O^+ 2) NaOH	Reduction	These reagents can be used to reduce an aryl nitro group into an amino group. The first step employs acidic conditions, so the amine is protonated (under the conditions of its formation) to give an ammonium ion. The ammonium ion is then deprotonated upon basic work-up, giving the amine.
1) NaN_3 2) LAH 3) H_2O	Azide synthesis	These reagents can be used to convert an alkyl halide into an amine, without a change in the carbon skeleton. The last two steps (reduction and water work-up) can be replaced with hydrogenation in the presence of a metal catalyst (H_2, Pt)
	Gabriel synthesis	Phthalimide is the starting material for the Gabriel synthesis, which can be used to prepare primary amines. Phthalimide is treated with KOH to give potassium phthalimide, which is then treated with an alkyl halide, giving an S_N2 reaction. The product of the S_N2 process is then hydrolyzed (upon treatment with hydrazine or aqueous acid) to release the amine.
$NaBH_3CN$	Reductive amination	In the presence of an acid catalyst, sodium cyanoborohydride can be used to achieve a reductive amination. The reaction occurs between a ketone (or aldehyde) and an amine (or ammonia). This process can be used to convert a primary amine into a secondary amine. Similarly, a secondary amine is converted into a tertiary amine.
	Acetylation	An amine will undergo acetylation (giving an amide) when treated with acetyl chloride.
1) Excess CH_3I 2) Ag_2O, H_2O, heat	Hofmann elimination	These reagents can be used to achieve elimination of H and NH_2 to give an alkene. When there are two possible regiochemical outcomes for the elimination process, the less substituted alkene predominates.
$NaNO_2$, HCl	Reactions with nitrous acid	A mixture of sodium nitrite and HCl will convert a primary amine into a diazonium salt. Under the same conditions, a secondary amine is converted into an N-nitrosamine.
CuBr	Sandmeyer reaction	When an aryldiazonium salt is treated with CuBr, the diazonium group is replaced with a bromine atom.
CuCl	Sandmeyer reaction	When an aryldiazonium salt is treated with CuCl, the diazonium group is replaced with a chlorine atom.
CuI	Sandmeyer reaction	When an aryldiazonium salt is treated with CuI, the diazonium group is replaced with an iodine atom.
CuCN	Sandmeyer reaction	When an aryldiazonium salt is treated with CuCN, the diazonium group is replaced with a cyano group.
HBF_4	Fluorination (Schiemann reaction)	When an aryldiazonium salt is treated with HBF_4, the diazonium group is replaced with a fluorine atom.
H_2O, heat	Preparation of phenol	When an aryldiazonium salt is treated with water and heat, the diazonium group is replaced with an OH group.
H_3PO_2	Reduction	When an aryldiazonium salt is treated with H_3PO_2, the diazonium group is replaced with a hydrogen atom.

Solutions

23.1.

(a) This compound is an amine that has only one alkyl group connected to the nitrogen atom. Since this alkyl group is complex, we must name the compound as an alkanamine (rather than an alkyl amine). The parent is comprised of four carbon atoms (thus, butanamine), and the amino group is located at C1. There are two methyl groups, both located at C3.

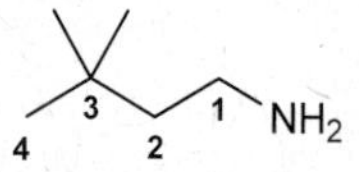

3,3-dimethyl-1-butanamine

(b) This compound is an amine that has only one simple alkyl group (a cyclopentyl group) connected to the nitrogen atom, so we can name this compound as an alkyl amine, rather than an alkanamine. Therefore, this compound is cyclopentylamine.

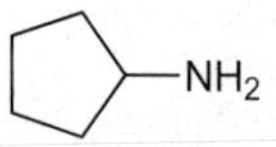

cyclopentylamine

(c) This compound is an amine that has three simple alkyl groups (two methyl groups and a cyclopentyl group) connected to the nitrogen atom, so we can name this compound as a trialkyl amine, rather than an alkanamine. The alkyl groups are listed in alphabetical order:

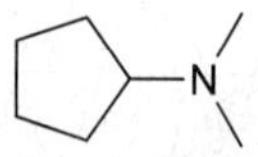

cyclopentyldimethylamine

(d) This compound is an amine that has three simple alkyl groups (all ethyl groups) connected to the nitrogen atom, so we can name this compound as a trialkyl amine, rather than an alkanamine:

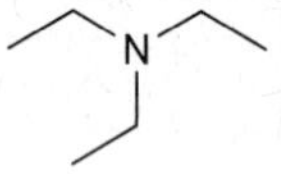

triethylamine

(e) This compound is an amine that has only one alkyl group connected to the nitrogen atom. Since this alkyl group is complex, we must name the compound as an alkanamine (rather than an alkyl amine). The parent is a six-membered ring (thus, cyclohexanamine), and there is an isopropyl group located at C3. There are two chirality centers, and the configuration of each is listed at the beginning of the name:

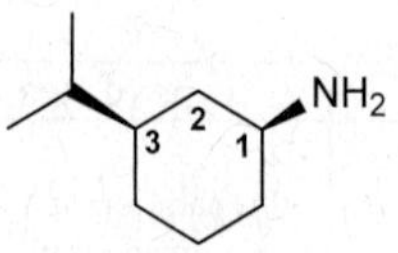

(1S,3R)-3-isopropylcyclohexanamine

(f) This compound has two functional groups (an OH group and an NH_2 group). The OH group takes priority, so the compound is named as an alcohol (cyclohexanol), with the amino group listed as a substituent, located at C3. There are two chirality centers, and the configuration of each is listed at the beginning of the name:

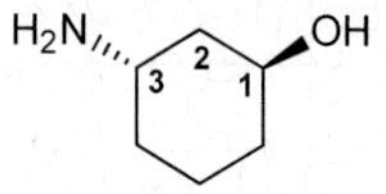

(1S,3S)-3-aminocyclohexanol

23.2.

(a) The name indicates a dialkyl amine, in which both alkyl groups are simple groups (a cyclohexyl group and a methyl group):

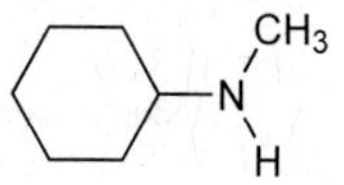

(b) The name indicates a trialkyl amine, in which all three alkyl groups are cyclobutyl groups:

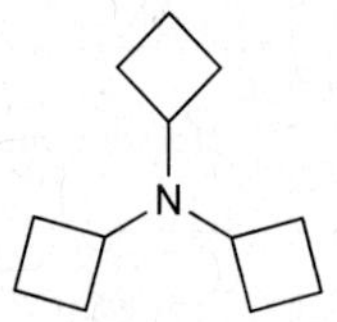

(c) The parent is aniline (or aminobenzene), and there are two ethyl groups (one at C2 and the other at C4):

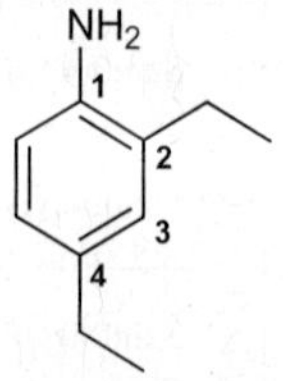

(d) The parent is a six-membered ring that bears an amino group (thus, cyclohexanamine). There is a methyl group at C2, and the configuration of each chirality center (C1 and C2) is indicated in the name:

(e) The parent is benzaldehyde, and there is an amino substituent in the *ortho* position.

23.3. The molecular formula (C_3H_9N) indicates no degrees of unsaturation (see Section 15.16), so all of the isomers must be acyclic amines. There is only one isomer in which the nitrogen atom is connected to three alkyl groups, all of which are methyl groups, thus the name trimethylamine:

trimethylamine

Likewise, there is also only one isomer in which the nitrogen atom is connected to two alkyl groups (which must be ethyl and methyl groups):

ethyl methyl amine

And finally, there are two isomers in which the nitrogen atom is connected to only one alkyl group (either a propyl group or an isopropyl group):

propylamine   isopropylamine

In total, there are four constitutional isomers with the molecular formula C_3H_9N.

23.4. The primary amine has two N-H bonds and is expected to exhibit the highest extent of hydrogen bonding, and therefore, the highest boiling point. The tertiary amine lacks N-H bonds, and is therefore expected to have the lowest boiling point.

Increasing boiling point

23.5.
(a) This amine has more than five carbon atoms per amino group (there are eight carbon atoms and only one amino group). Therefore, this compound is not expected to be water soluble.
(b) This amine has fewer than five carbon atoms per amino group (there are only three carbon atoms and one amino group). Therefore, this compound is expected to be water soluble.
(c) This diamine has fewer than five carbon atoms per amino group (there are six carbon atoms and two amino groups). Therefore, this compound is expected to be water soluble.

23.6.
(a) The following compound is expected to be a stronger base because the lone pair is localized, and therefore more available to function as a base.

The other compound exhibits a delocalized lone pair, and is therefore a weaker base.

(b) The following compound is expected to be a stronger base because the lone pair is not participating in aromaticity. It is available to function as a base.

In contrast, the other compound is aromatic, and the lone pair is delocalized (in order to establish aromaticity), so it is unavailable to function as base.

(c) The following compound is expected to be a stronger base because the nitrogen atom has a lone pair that is localized, and therefore more available to function as a base.

The other compound exhibits a nitrogen atom with a delocalized lone pair, and is therefore a weaker base.

(d) The following compound is expected to be a stronger base because the lone pair is not participating in aromaticity. The lone pair occupies an sp^2 hybridized orbital (directed away from the ring, in the plane of the ring) and is available to function as a base.

In contrast, the other compound (shown in the problem statement) exhibits a nitrogen atom with a lone pair that is highly delocalized (in order to establish aromaticity in the five-membered ring), so it is unavailable to function as base.

23.7. In all of these compounds, the lone pair (on the nitrogen atom) is delocalized throughout two aromatic rings:

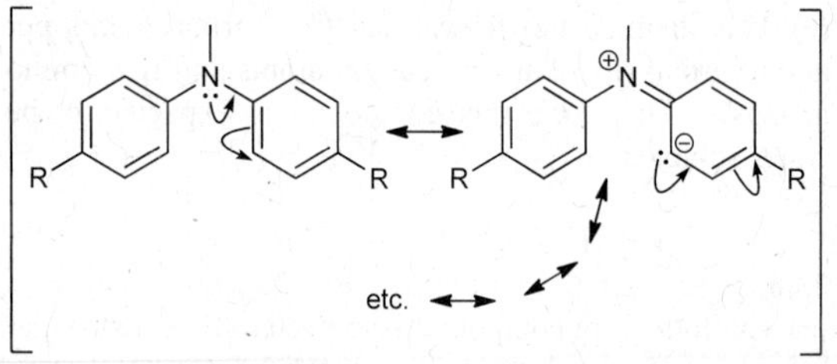

One of the three compounds (shown in the problem statement) has two methyl groups (electron-donating), which destabilize the delocalized charge:

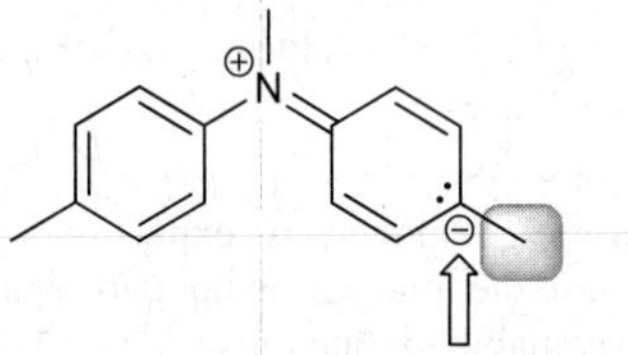

Charge is destabilized by methyl group

This compound is the strongest base, because the delocalization effect is diminished by the effect of the alkyl groups.

In contrast, the following compound has an aldehyde group. This group is electron-withdrawing, and a resonance structure can be drawn in which the nitrogen atom bears a positive charge, and the oxygen atom bears a negative charge:

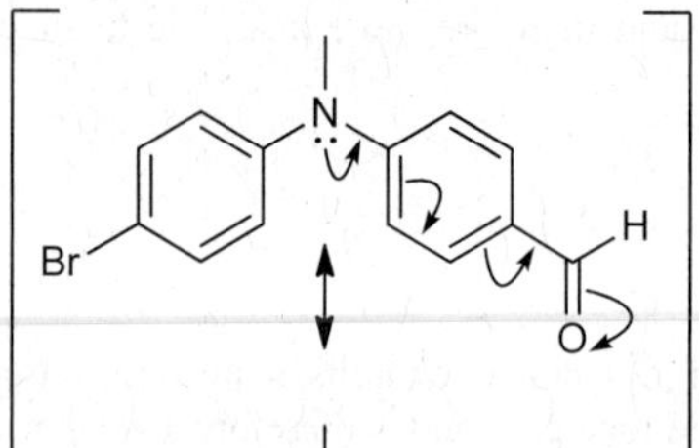

This resonance contributor is significant, and it renders the lone pair highly delocalized, and therefore a very poor base. In summary, we predict the following order of base strength:

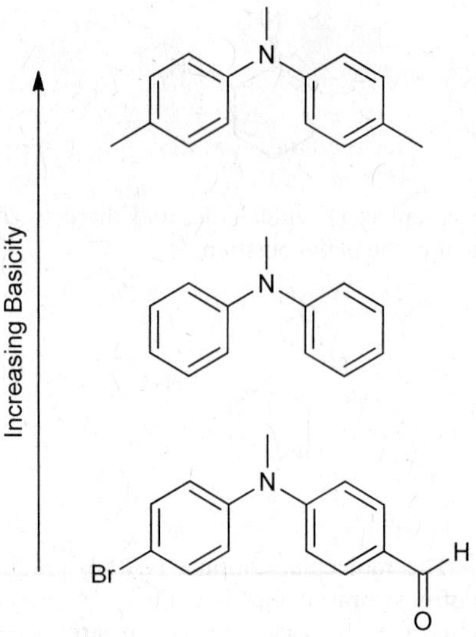

23.8. In the reactant, the lone pair of the amino group is delocalized via resonance. In the product, the lone pair of the amino group is localized, and is therefore more available to function as a base.

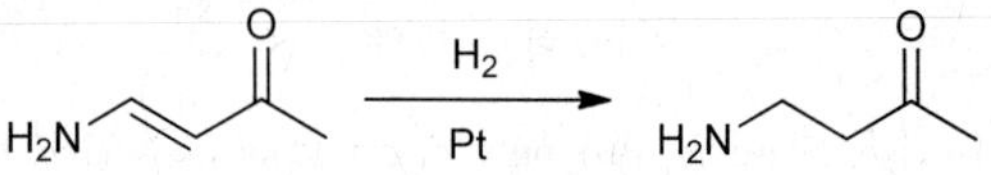

23.9.
(a) At physiological pH, the amino group exists primarily as a charged ammonium ion:

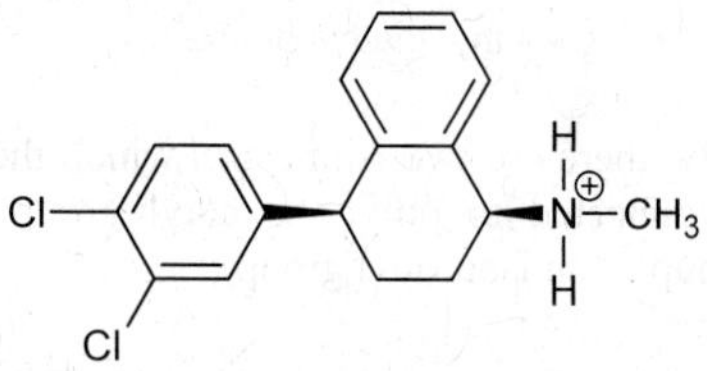

(b) At physiological pH, the amino group exists primarily as a charged ammonium ion:

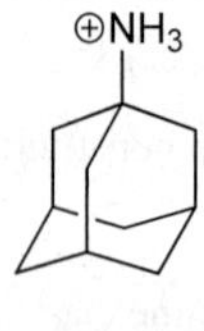

(c) At physiological pH, the amino group exists primarily as a charged ammonium ion:

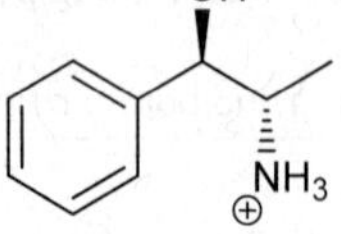

23.10.

(a) Butylamine can be made from 1-bromopropane, as shown. Treatment with sodium cyanide gives a nitrile (via an S_N2 reaction). Reduction of the nitrile with excess lithium aluminum hydride, followed by water work-up, gives the product:

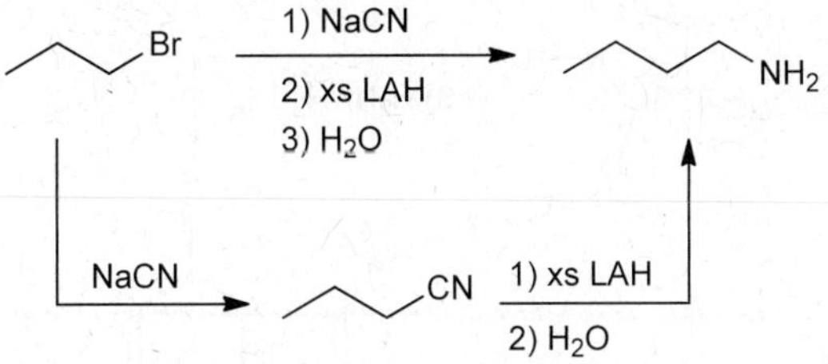

Alternatively, butylamine can be made from butanoic acid, as shown. Treatment with thionyl chloride gives an acid chloride, which can be treated with excess ammonia to give an amide. The amide is then reduced with excess lithium aluminum hydride, followed by water work-up, to give the product:

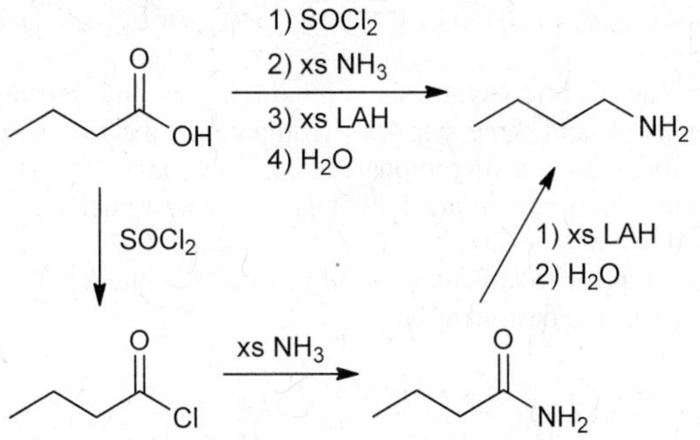

(b) The desired amine can be made from benzyl bromide, as shown. Treatment with sodium cyanide gives a nitrile (via an S_N2 reaction). Reduction of the nitrile with excess lithium aluminum hydride, followed by water work-up, gives the product:

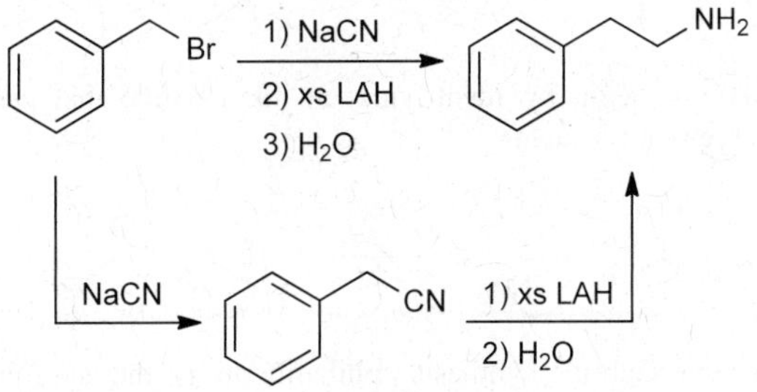

Alternatively, the desired amine can be made from the corresponding carboxylic acid, as shown below. Treatment with thionyl chloride gives an acid chloride, which can then be treated with excess ammonia to give an amide. The amide is then reduced with excess lithium aluminum hydride, followed by water work-up, to give the product:

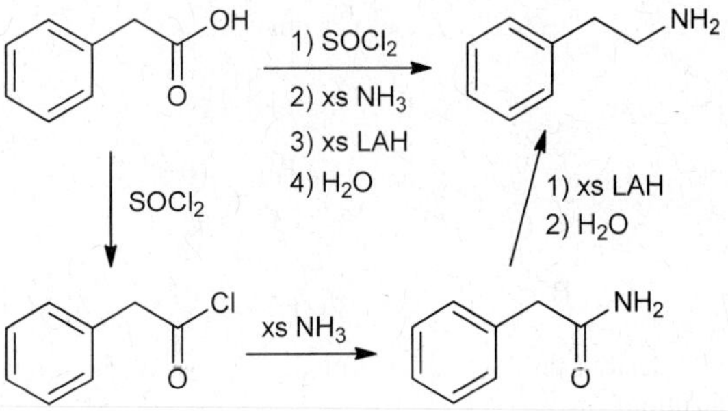

(c) The desired amine can be made from bromocyclohexane, as shown. Treatment with sodium cyanide gives a nitrile (via an S_N2 reaction). Reduction of the nitrile with excess lithium aluminum hydride, followed by water work-up, gives the product:

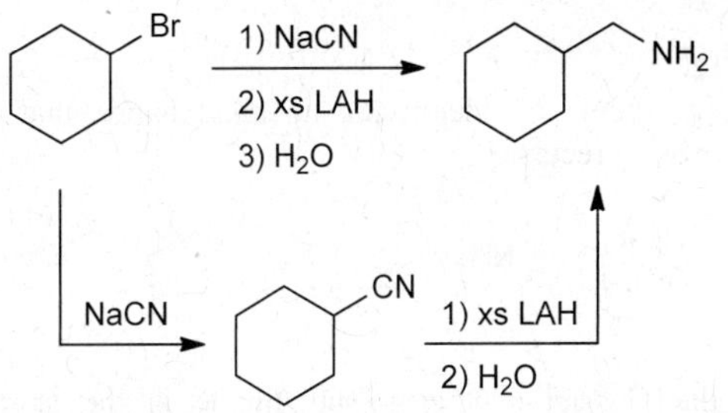

Alternatively, the desired amine can be made from the corresponding carboxylic acid, as shown below. Treatment with thionyl chloride gives an acid chloride, which can then be treated with excess ammonia to give an amide. The amide is then reduced with excess lithium aluminum hydride, followed by water work-up, to give the product:

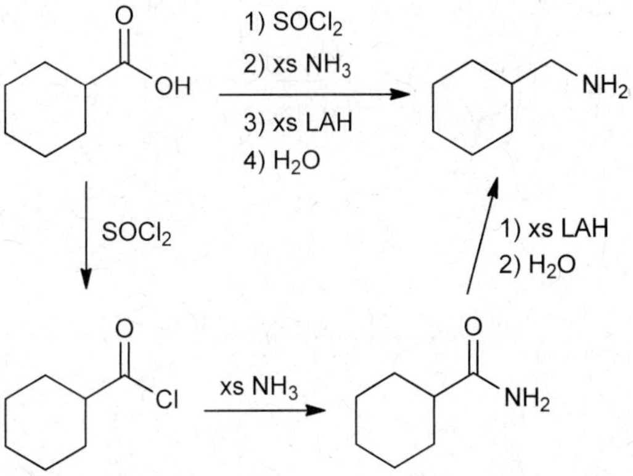

23.11. This compound cannot be prepared from an alkyl halide or a carboxylic acid, using the methods described in this section, because both methods produce an amine with two alpha protons:

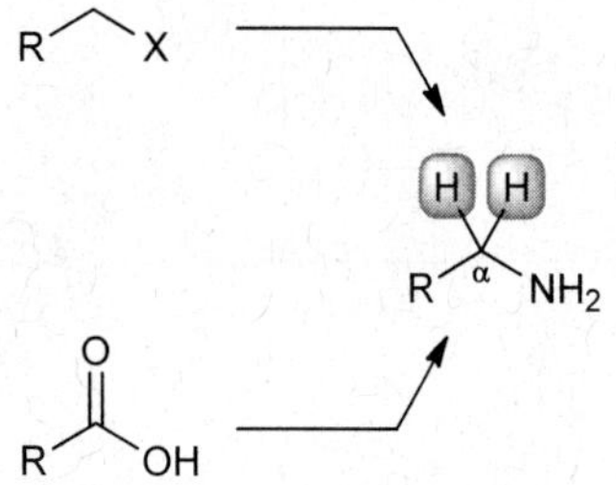

The desired product has two methyl groups at the alpha position:

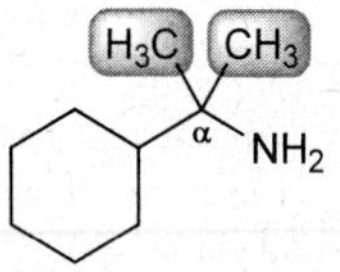

So this product cannot be made with either of the synthetic methods above.

23.12.
(a) We begin by identifying an alkyl halide that can serve as a precursor:

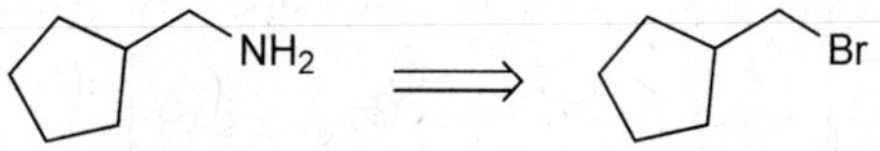

In the Gabriel synthesis, phthalimide is the starting material, and three steps are required. In the first step, phthalimide is deprotonated by hydroxide to give potassium phthalimide, which can serve as a nucleophile and attack the alkyl halide above in an S_N2 process. Subsequent treatment with hydrazine (or aqueous acid) releases the desired amine:

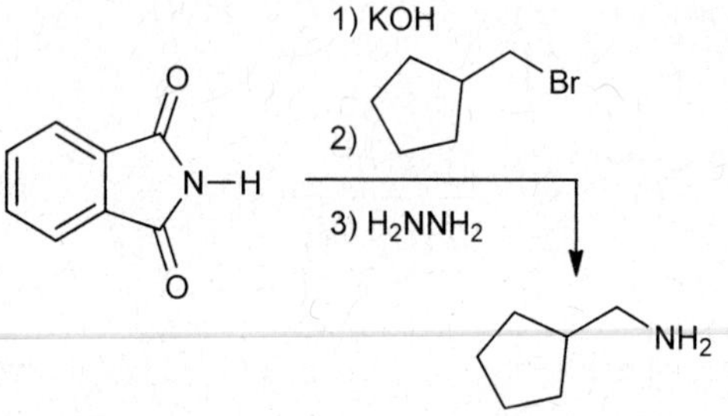

(b) We begin by identifying a halide that can serve as a precursor:

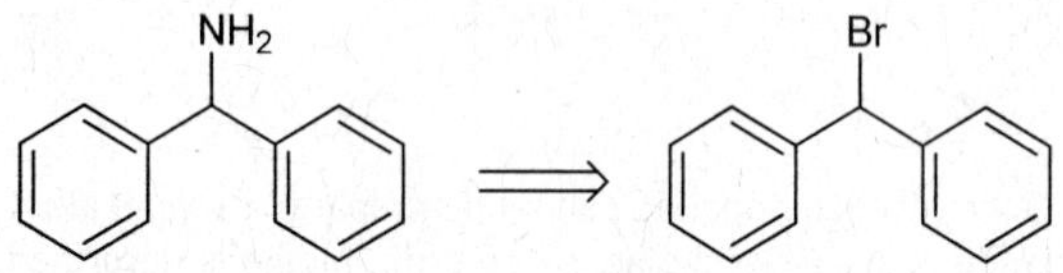

In the Gabriel synthesis, phthalimide is the starting material, and three steps are required. In the first step, phthalimide is deprotonated by hydroxide to give potassium phthalimide, which can serve as a nucleophile and attack the halide above in an S_N2 process.

Subsequent treatment with hydrazine (or aqueous acid) releases the desired amine:

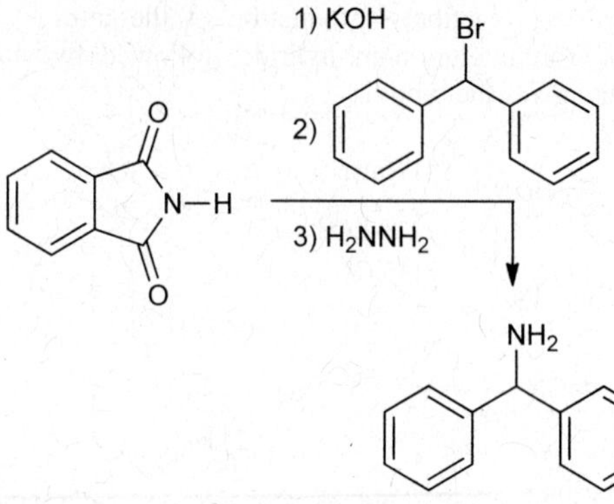

(c) We begin by identifying an alkyl halide that can serve as a precursor:

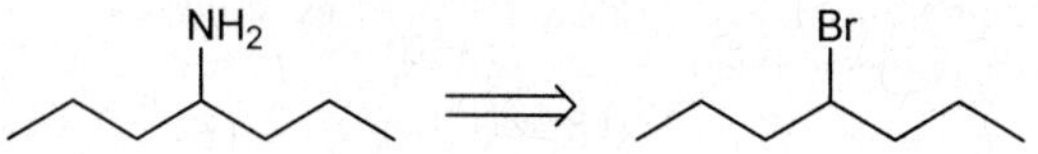

In the Gabriel synthesis, phthalimide is the starting material, and three steps are required. In the first step, phthalimide is deprotonated by hydroxide to give potassium phthalimide, which can serve as a nucleophile and attack the alkyl halide above in an S_N2 process. Subsequent treatment with hydrazine (or aqueous acid) releases the desired amine:

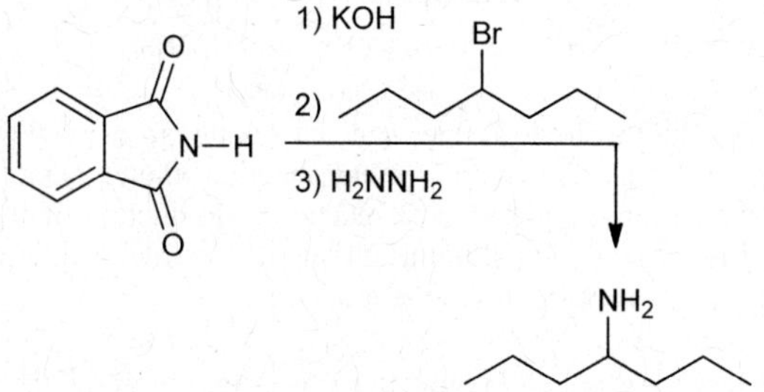

(d) We begin by identifying an alkyl halide that can serve as a precursor:

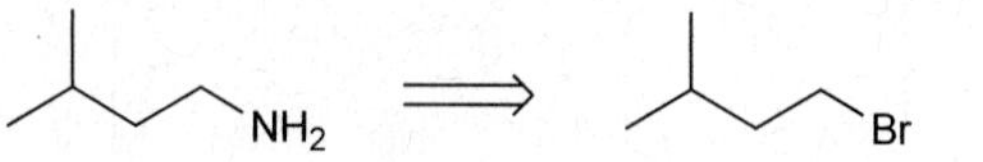

In the Gabriel synthesis, phthalimide is the starting material, and three steps are required. In the first step, phthalimide is deprotonated by hydroxide to give potassium phthalimide, which can serve as a nucleophile and attack the alkyl halide above in an S_N2 process. Subsequent treatment with hydrazine (or aqueous acid) releases the desired amine:

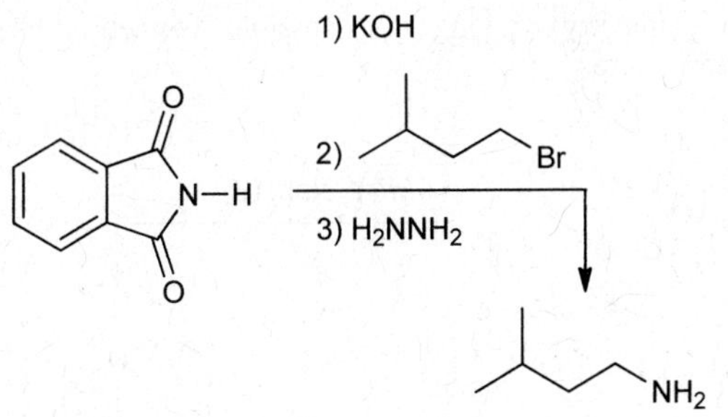

1) KOH

2) [structure: isobutyl bromide]

3) H_2NNH_2

[reaction scheme: phthalimide to amine with NH_2]

23.13.

(a) We begin by identifying an alkyl halide that can serve as a precursor:

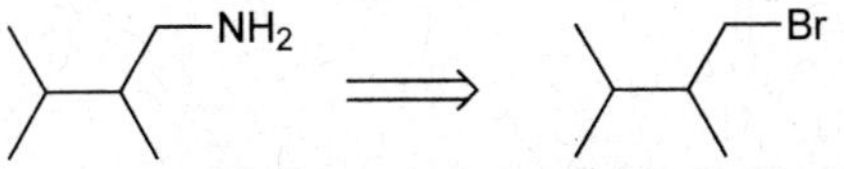

[structure: amine $\Longrightarrow$ alkyl bromide]

This alkyl halide can be made from the starting material via the following reaction sequence. Radical bromination installs a functional group (Br) at a tertiary position. The resulting tertiary alkyl bromide will then undergo elimination upon treatment with a strong, sterically hindered base to give the less substituted alkene. Radical addition of HBr, in the presence of peroxides, gives the necessary alkyl halide:

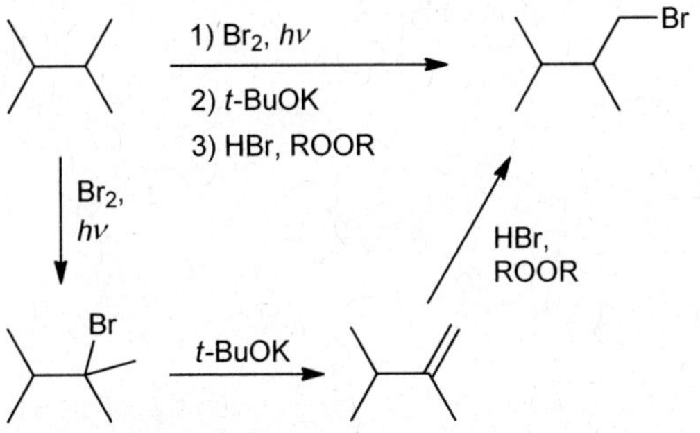

1) Br_2, hv
2) t-BuOK
3) HBr, ROOR

Br_2, hv

HBr, ROOR

t-BuOK

[reaction structures]

Once the necessary alkyl halide has been prepared, it can be treated with potassium phthalimide, followed by hydrazine (or aqueous acid), to give the desired amine, as shown:

1) [potassium phthalimide structure] N⁻ K⁺

2) H_2NNH_2

[structure: alkyl bromide to NH_2]

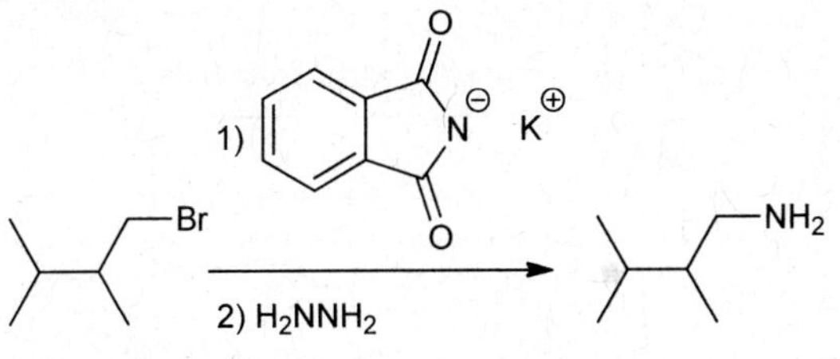

(b) We begin by identifying an alkyl halide that can serve as a precursor:

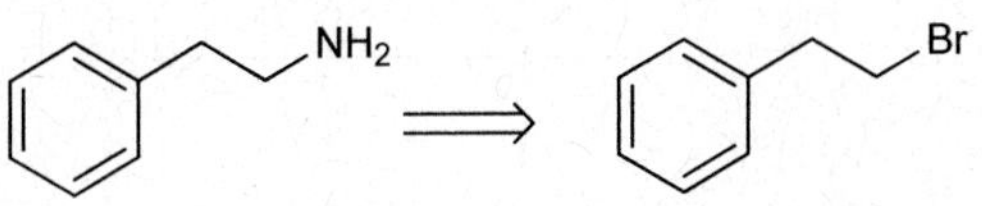

[structure: phenethylamine NH_2 $\Longrightarrow$ alkyl bromide Br]

This alkyl halide can be made from the starting carboxylic acid via the following reaction sequence. Reduction with excess lithium aluminum hydride, followed by water work-up, gives an alcohol. Treatment with PBr_3 then converts the alcohol into the desired alkyl halide:

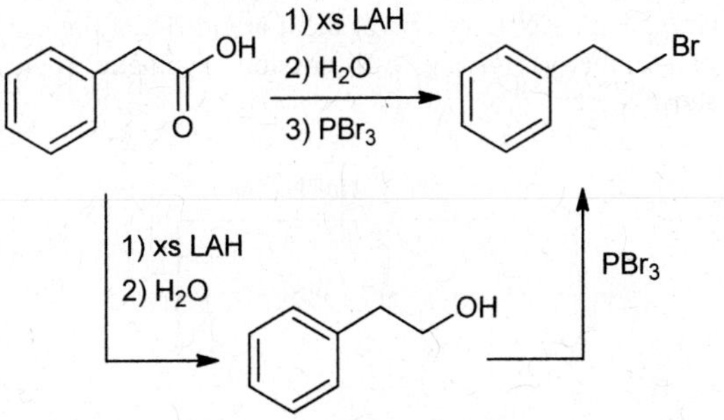

1) xs LAH
2) H_2O
3) PBr_3

[phenylacetic acid to alkyl bromide]

1) xs LAH
2) H_2O

PBr_3

[phenethyl alcohol OH]

Once the necessary alkyl halide has been prepared, it can be treated with potassium phthalimide, followed by hydrazine (or aqueous acid), to give the desired amine, as shown:

1) [potassium phthalimide structure] N⁻ K⁺

2) H_2NNH_2

[structure: phenethyl bromide Br to NH_2]

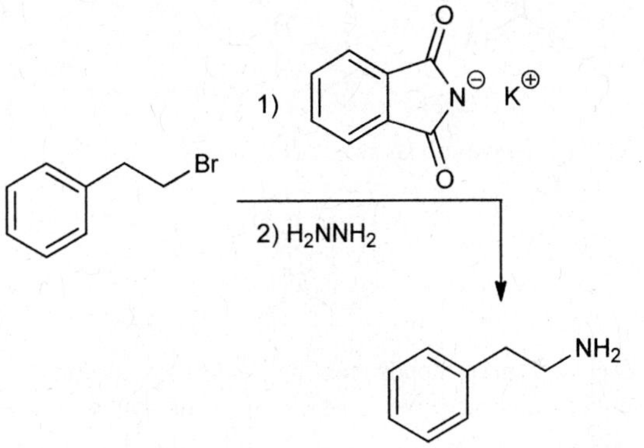

23.14.

(a) The compound has two C-N bonds:

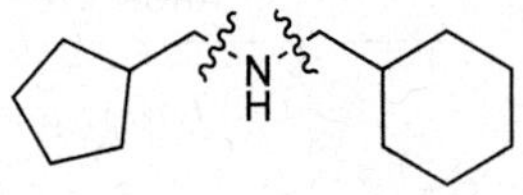

[structure: cyclopentylmethyl-N(H)-cyclohexylmethyl]

Each of these bonds can be made via a reductive amination, giving two possible synthetic routes, shown here:

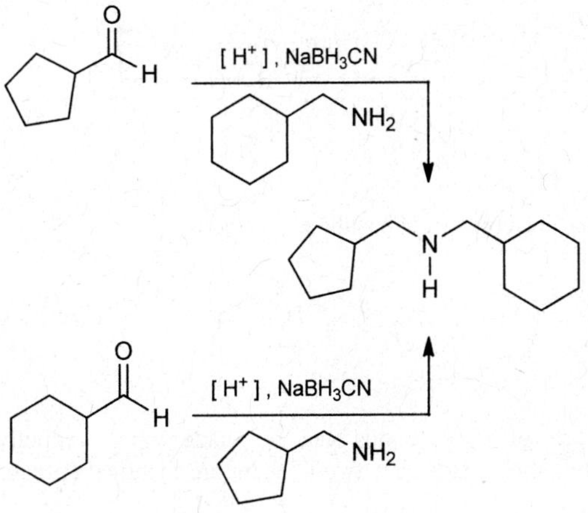

[H⁺], $NaBH_3CN$

[cyclopentanecarbaldehyde + cyclohexylmethylamine NH_2]

[product: secondary amine N-H]

[H⁺], $NaBH_3CN$

[cyclohexanecarbaldehyde + cyclopentylmethylamine NH_2]

(b) The compound has three C-N bonds:

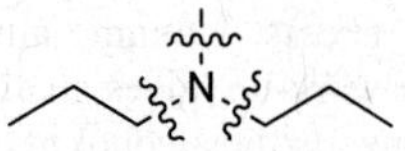

Each of these bonds can be made via a reductive amination. However two of them are identical (because of symmetry), giving two possible synthetic routes, shown here:

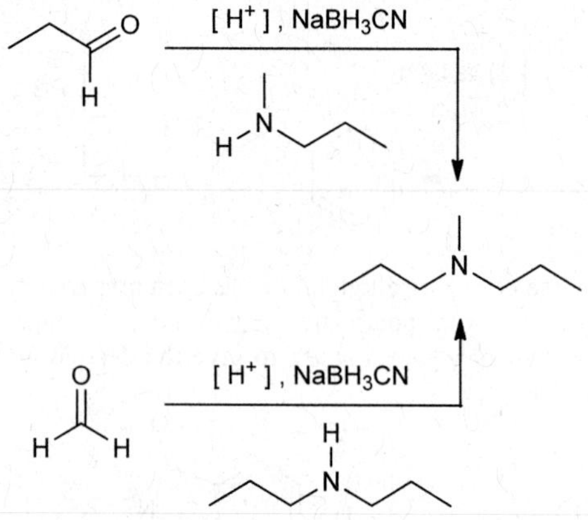

(c) The compound has two C-N bonds:

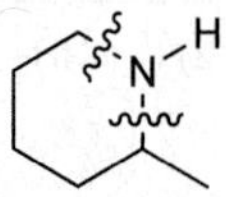

Each of these bonds can be made via a reductive amination, giving two possible synthetic routes, shown here:

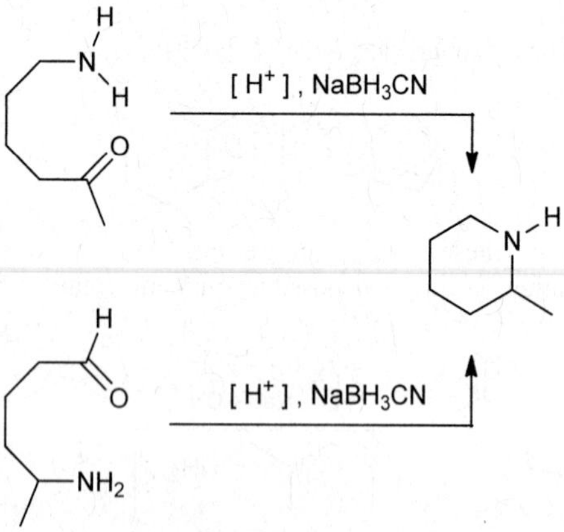

(d) The compound has three C-N bonds:

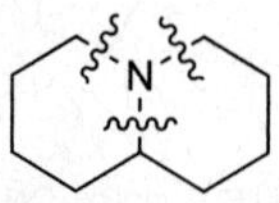

Each of these bonds can be made via a reductive amination. However two of them are identical (because

of symmetry), giving two possible synthetic routes, shown here:

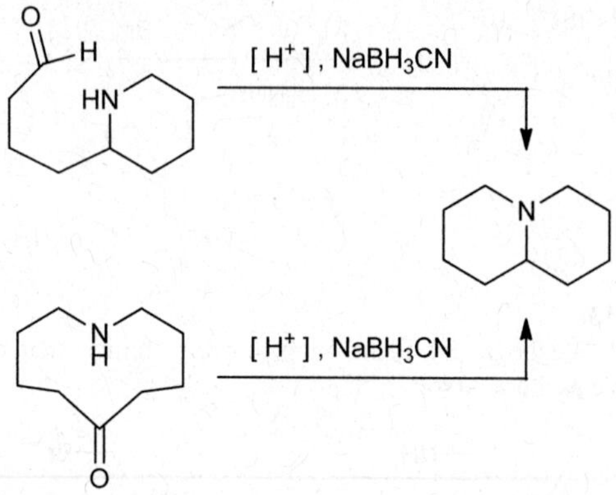

(e) The compound has three C-N bonds:

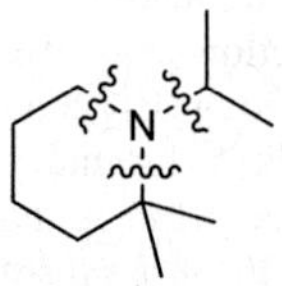

However, one of these bonds cannot be made via a reductive amination, because the starting material cannot have a pentavalent carbon atom:

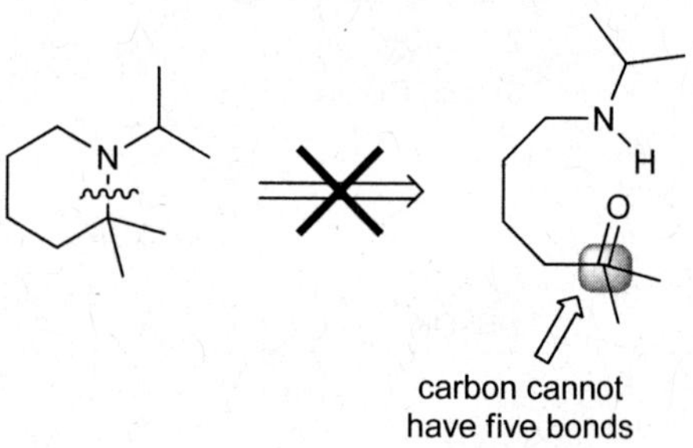

carbon cannot have five bonds

The other two C-N bonds can be made via reductive amination, giving two possible synthetic routes, shown here:

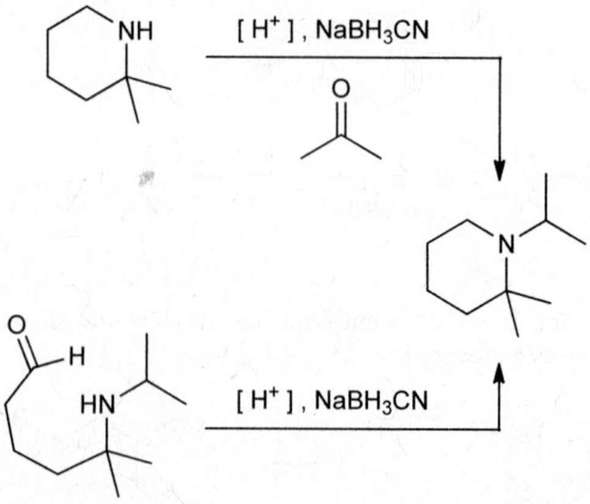

23.15. Phenylacetone is expected to give a secondary amine upon treatment with methyl amine in the presence of sodium cyanoborohydride and an acid catalyst, as shown:

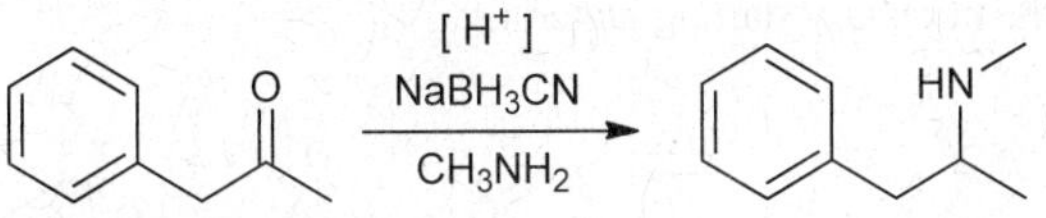

Phenylacetone

23.16. The last step of reductive amination is the reduction of a C=N bond. That step introduces a hydrogen atom at the alpha position (the carbon atom that is connected to the nitrogen atom in the product):

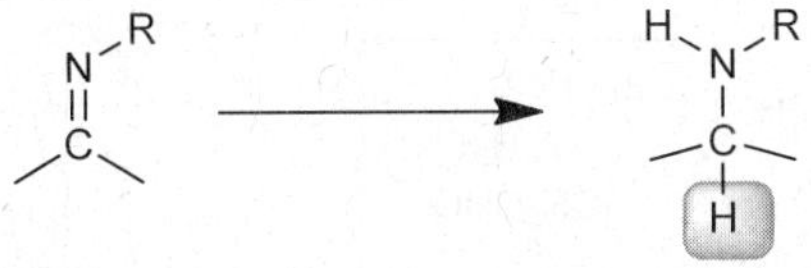

As a result, the product of a reductive amination must have at least one proton at the alpha position. In the case of tri-*tert*-butyl amine, there are three alpha positions, and none of them bears a proton. Each of the alpha positions has three alkyl groups and no protons.

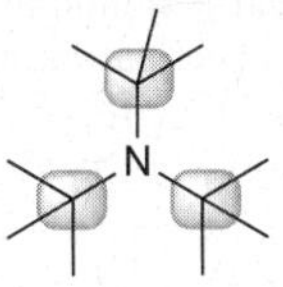

Therefore, this compound cannot be made with a reductive amination.

23.17. There are certainly many acceptable solutions to this problem. One such solution derives from the following retrosynthetic analysis. An explanation of each of the steps (*a-d*) follows.

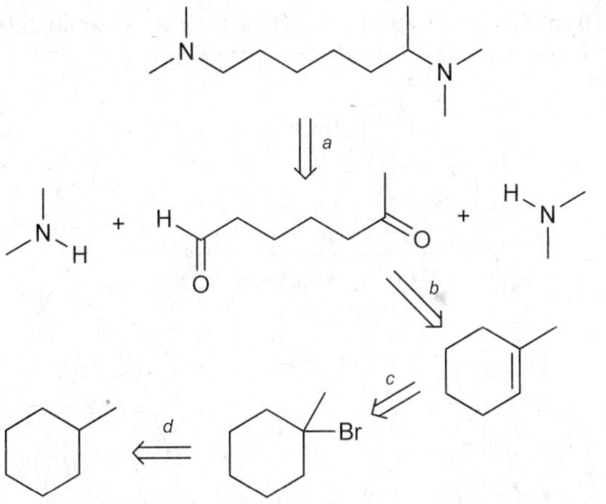

a. The product can be made from the corresponding dicarbonyl compound via reductive amination with excess dimethyl amine (thereby converting each carbonyl group into a dimethyl amino group).

b. The dicarbonyl compound can be made via ozonolysis of 1-methylcyclohexene.

c. 1-Methylcyclohexene can be made from 1-bromo-1-methylcyclohexane via elimination with a strong base.

d. 1-Bromo-1-methylcyclohexane can be made from the starting material via radical bromination at the tertiary position.

Now let's draw the forward scheme. Radical bromination of the starting cycloalkane gives a tertiary alkyl bromide, which is then converted into an alkene upon treatment with a strong base, such as ethoxide. Ozonolysis causes cleavage of the C=C bond, thereby opening the ring and giving a dicarbonyl compound, which can then be converted into the product via reductive amination, upon treatment with excess dimethylamine and sodium cyanoborohydride with acid catalysis.

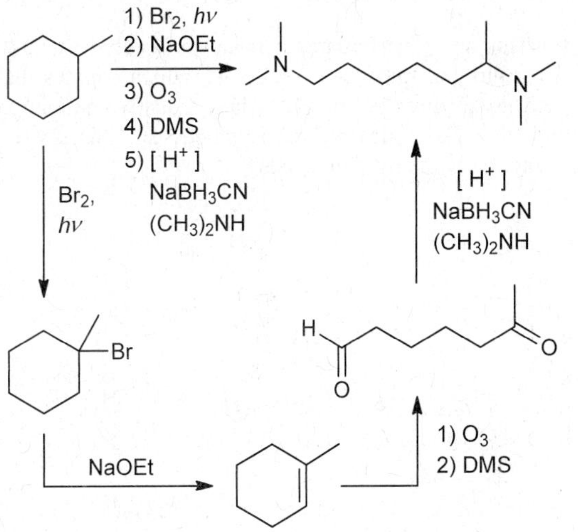

23.18.
(a) This amine is secondary (it bears two alkyl groups). The source of nitrogen is ammonia, which dictates that each group must be installed via a reductive amination process. The following retrosynthetic analysis reveals the necessary starting materials:

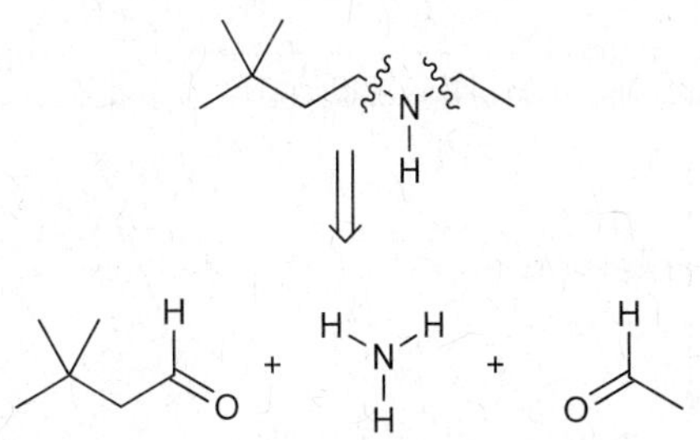

Each C-N bond can be formed via a reductive amination, as shown in the following forward scheme:

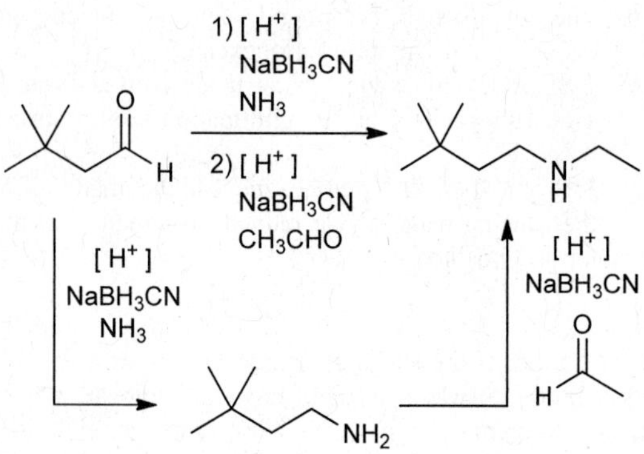

(b) Cyclopentyl amine can be made from cyclopentanone and ammonia, via a reductive amination, as shown:

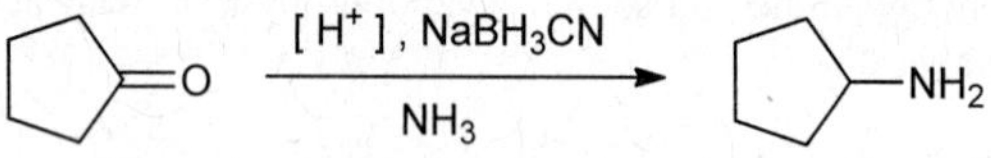

(c) This amine is tertiary (it bears three alkyl groups). The source of nitrogen is ammonia, which dictates that each group must be installed via a reductive amination process. The following retrosynthetic analysis reveals the necessary starting materials:

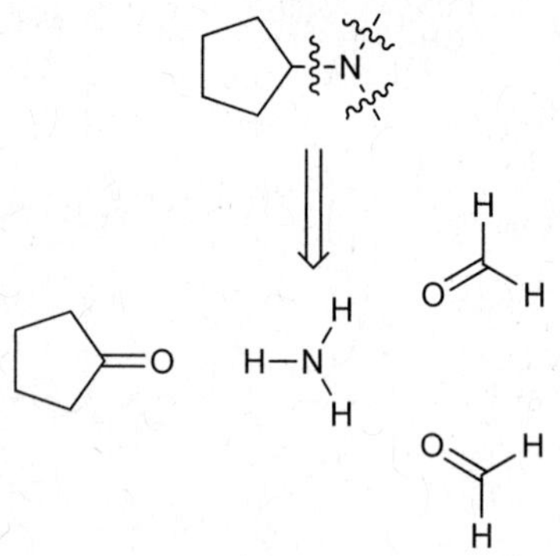

Each C-N bond can be formed via a reductive amination, as shown in the following forward scheme:

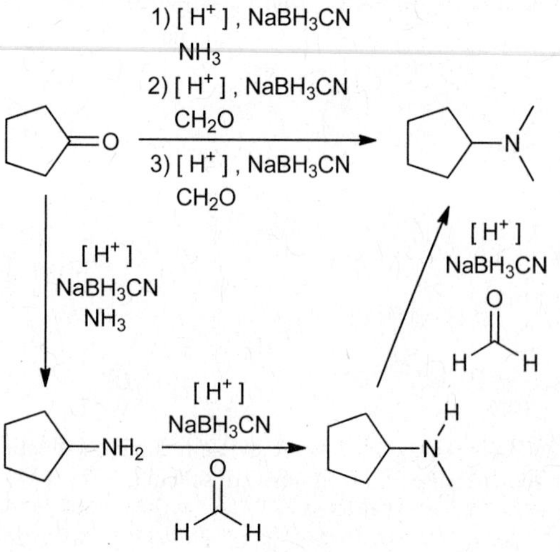

(d) This amine is secondary (it bears two ethyl groups). The source of nitrogen is ammonia, which dictates that each group must be installed via a reductive amination process. The following retrosynthetic analysis reveals the necessary starting materials:

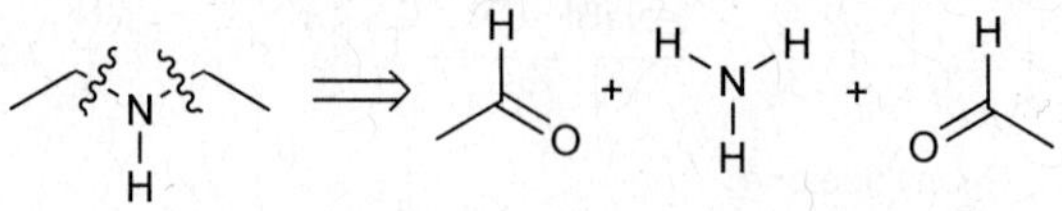

Each C-N bond can be formed via a reductive amination, as shown in the following forward scheme:

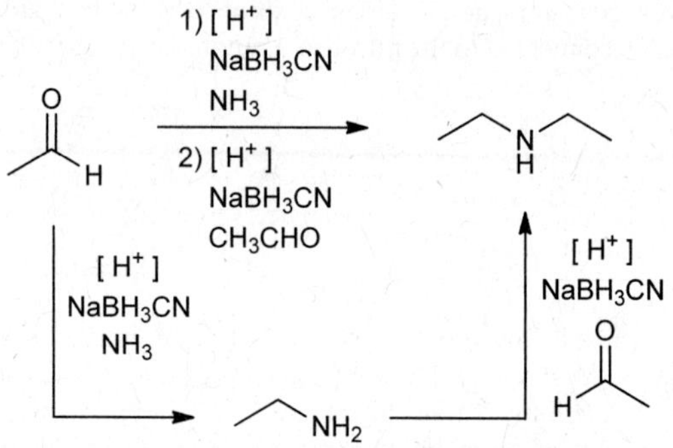

(e) This amine is tertiary (it bears three ethyl groups). The source of nitrogen is ammonia, which dictates that each group must be installed via a reductive amination process. The following retrosynthetic analysis reveals the necessary starting materials:

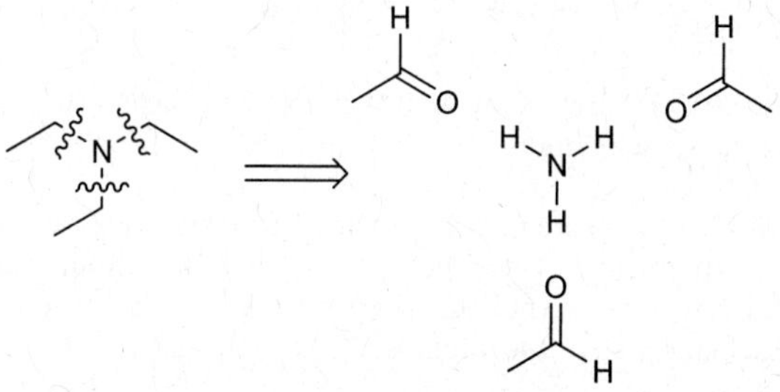

Each C-N bond can be formed via a reductive amination, as shown in the following forward scheme:

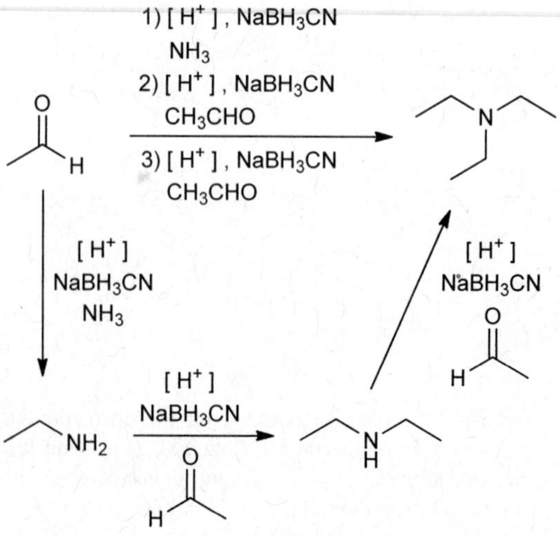

(f) This amine is tertiary (it bears three alkyl groups). The source of nitrogen is ammonia, which dictates that each group must be installed via a reductive amination process. The following retrosynthetic analysis reveals the necessary starting materials:

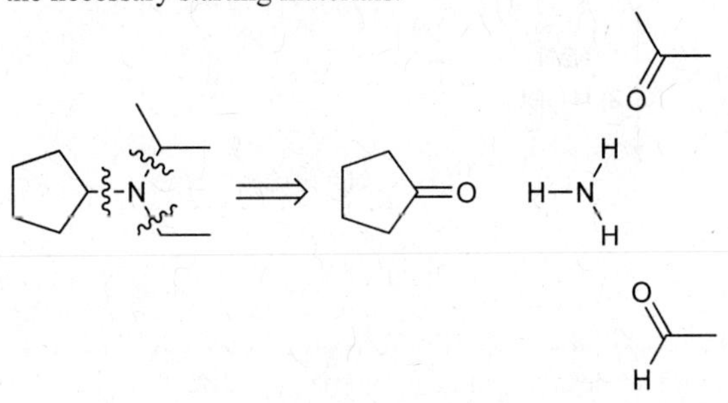

Each C-N bond can be formed via a reductive amination, as shown in the following forward scheme:

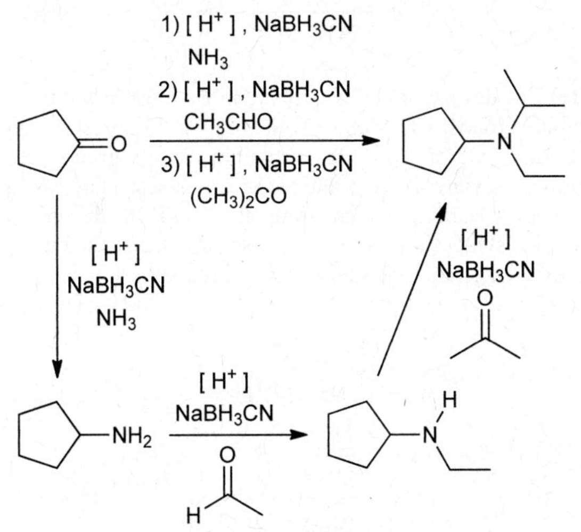

23.19.

(a) The desired amine is secondary, so we must install two alkyl groups. The first alkyl group is installed via a Gabriel synthesis, and the remaining alkyl group is installed via a reductive amination process. There is a choice regarding which group to install via the initial Gabriel synthesis, so we choose the least sterically hindered group (the group whose installation involves the least hindered alkyl halide):

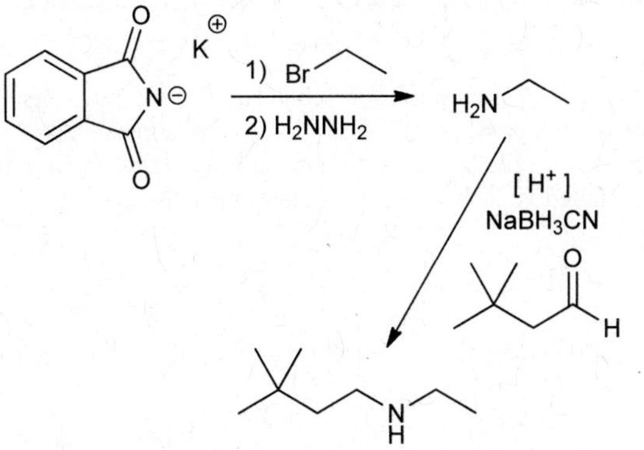

(b) The desired amine is primary, so we only need to install one alkyl group, which can be achieved with a Gabriel synthesis, as shown:

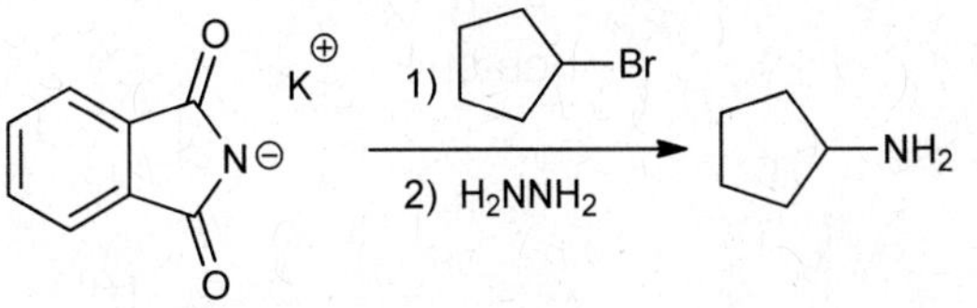

(c) The desired amine is tertiary, so we must install three alkyl groups. The first alkyl group is installed via a Gabriel synthesis, and the remaining alkyl groups are installed via reductive amination processes. There is a choice regarding which group to install via the initial Gabriel synthesis, so we choose the least sterically hindered group (one of the methyl groups):

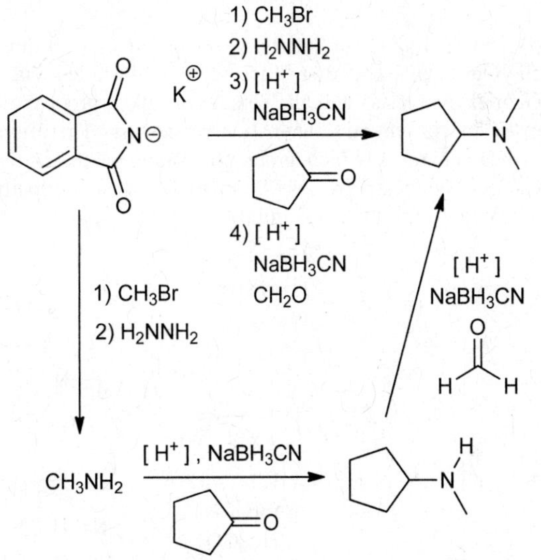

(d) The desired amine is secondary, so we must install two alkyl groups. The first ethyl group is installed via a Gabriel synthesis, and the remaining ethyl group is installed via a reductive amination process:

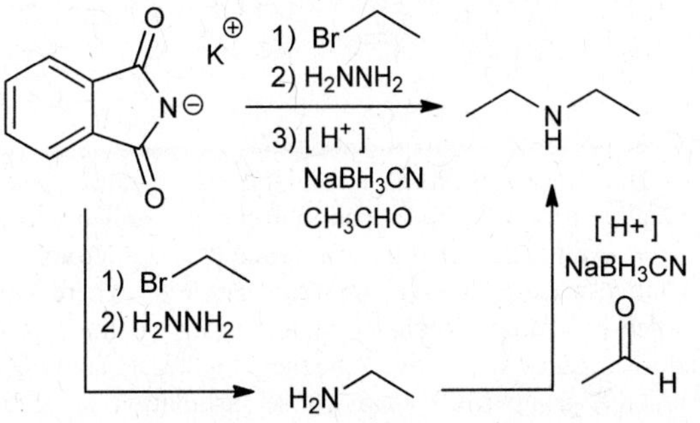

(e) The desired amine is tertiary, so we must install three alkyl groups (all ethyl groups). The first ethyl group is installed via a Gabriel synthesis, and the remaining ethyl groups are installed via reductive amination processes.

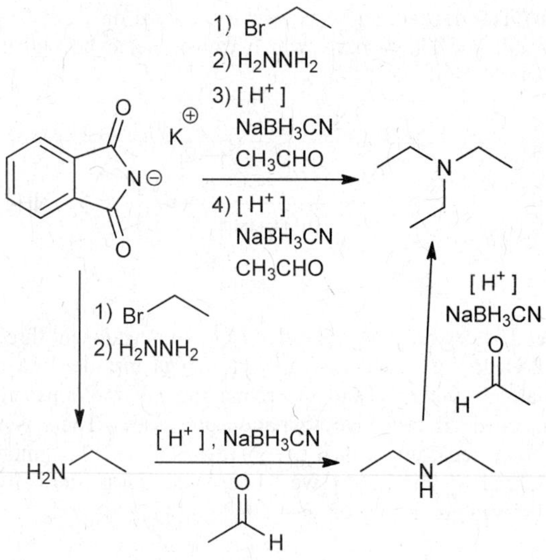

(f) The desired amine is tertiary, so we must install three alkyl groups. The first alkyl group is installed via a Gabriel synthesis, and the remaining alkyl groups are installed via reductive amination processes. There is a choice regarding which group to install via the initial Gabriel synthesis, so we choose the least sterically hindered group (the ethyl group):

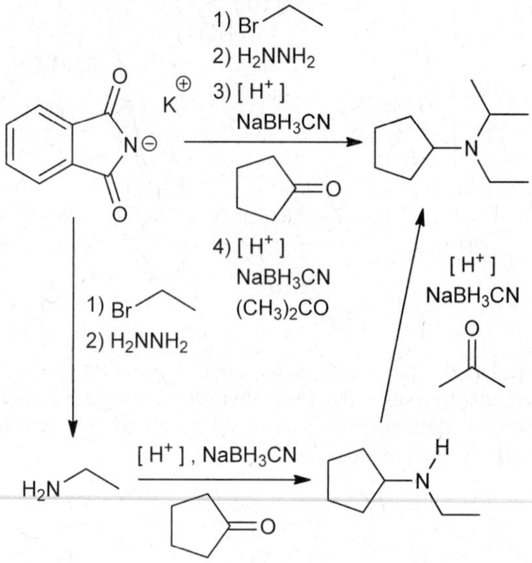

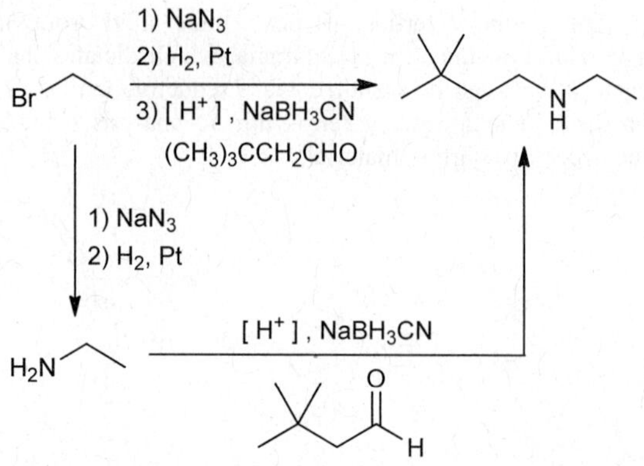

(b) The desired amine is primary, so we only need to install one alkyl group, which can be achieved via an azide synthesis, as shown:

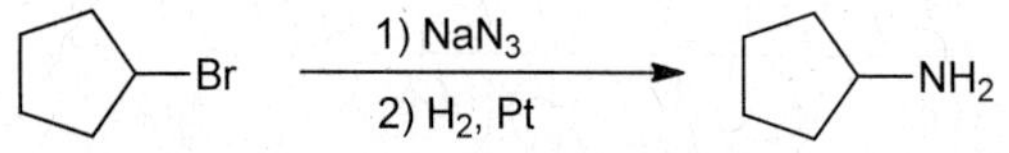

(c) The desired amine is tertiary, so we must install three alkyl groups. The first alkyl group is installed via an azide synthesis, and the remaining alkyl groups are installed via reductive amination processes. There is a choice regarding which group to install via the initial azide synthesis, so we choose the least sterically hindered group (one of the methyl groups):

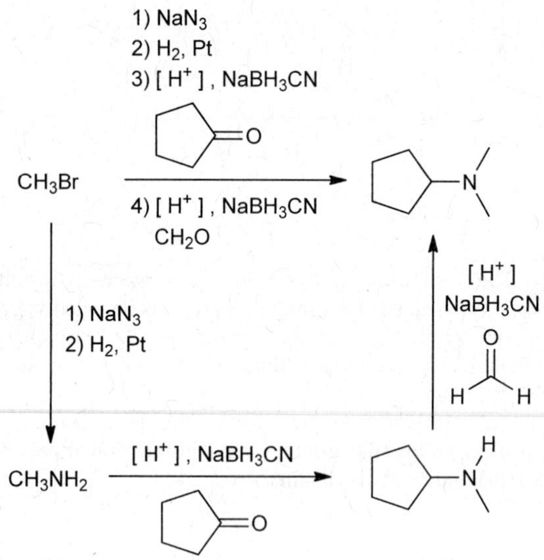

23.20.
(a) The desired amine is secondary, so we must install two alkyl groups. The first alkyl group is installed via an azide synthesis, and the remaining alkyl group is installed via a reductive amination process. There is a choice regarding which group to install via the initial azide synthesis, so we choose the least sterically hindered group (the group whose installation involves the least hindered alkyl halide):

(d) The desired amine is secondary, so we must install two alkyl groups. The first ethyl group is installed via an azide synthesis, and the remaining ethyl group is installed via a reductive amination process:

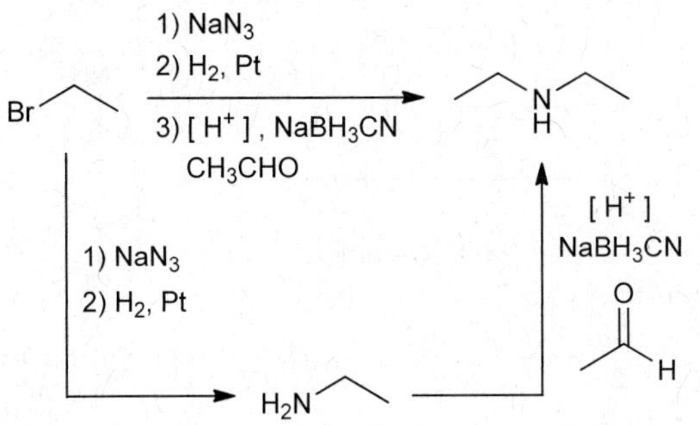

(e) The desired amine is tertiary, so we must install three alkyl groups (all ethyl groups). The first ethyl group is installed via an azide synthesis, and the remaining ethyl groups are installed via reductive amination processes.

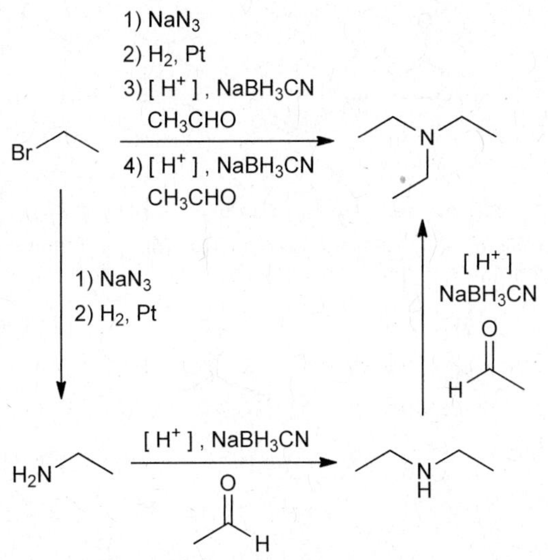

(f) The desired amine is tertiary, so we must install three alkyl groups. The first alkyl group is installed via an azide synthesis, and the remaining alkyl groups are installed via reductive amination processes. There is a choice regarding which group to install via the initial azide synthesis, so we choose the least sterically hindered group (the ethyl group):

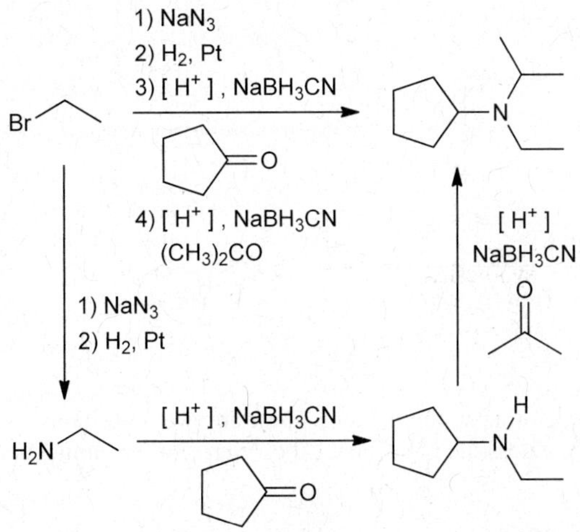

23.21. This amine is tertiary, and each C-N bond can be made via a reductive amination process, as shown in the following retrosynthetic analysis:

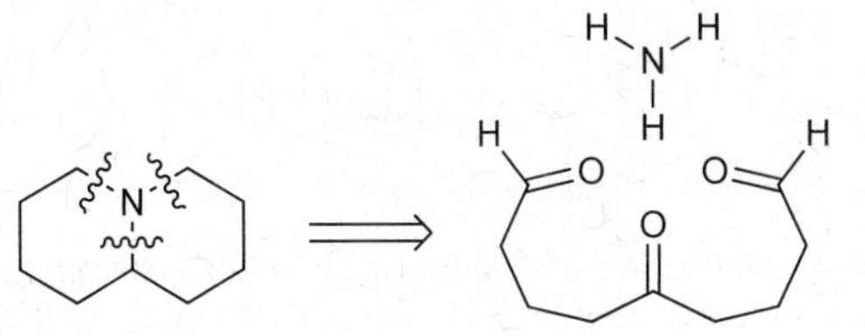

The necessary tricarbonyl compound can be made from the starting material via ozonolysis, as shown in the following forward scheme:

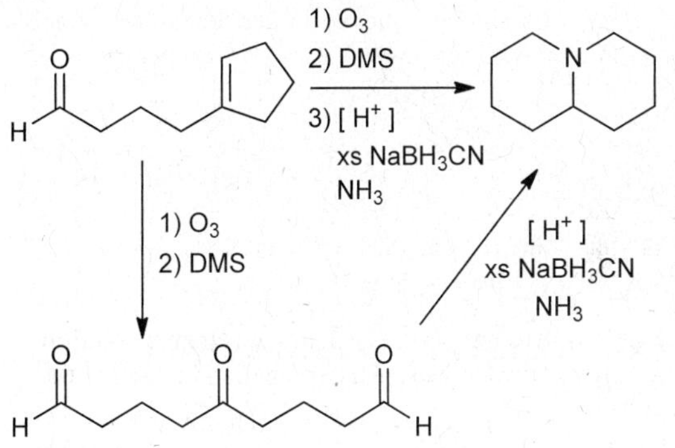

23.22. Acetylation of the amino group allows for direct nitration of the ring (in the *para* position). After nitration is complete, the acetyl group can be removed in aqueous basic or acidic conditions:

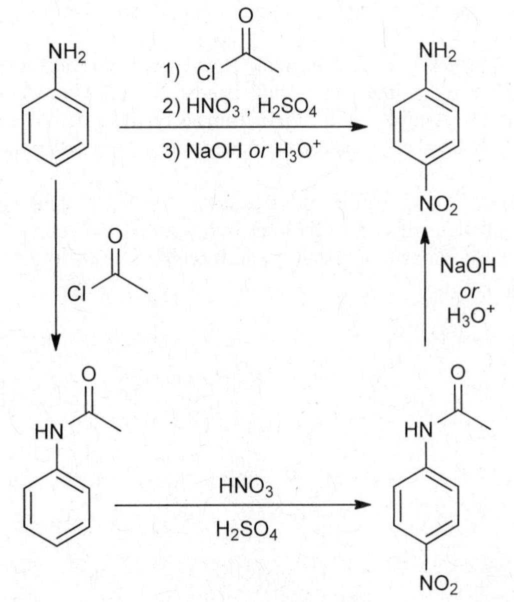

23.23. Direct chlorination of nitrobenzene would result in a *meta*-disubstituted product (because the nitro group is *meta*-directing). So we must first reduce the nitro group into an amino group, thereby converting a *meta* director into an *ortho-para* director:

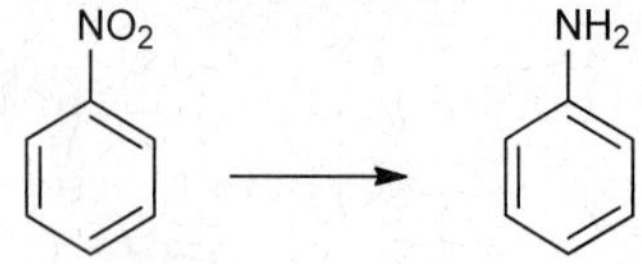

Monochlorination of aniline (in the *para* position) will then give the product. Unfortunately, aniline will not efficiently undergo monochlorination (the ring is too highly activated to install just one chlorine atom). However, the strongly activating effect of the amino group can be temporarily diminished via acetylation. Then, after monochlorination has been performed, the acetyl group can be removed in aqueous basic or acidic conditions. The entire synthesis is shown here:

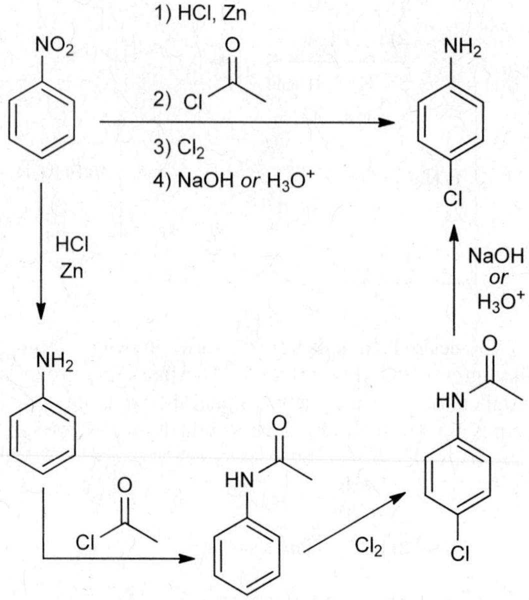

23.24. There are certainly many acceptable solutions to this problem. One such solution derives from the following retrosynthetic analysis. An explanation of each of the steps (*a-d*) follows.

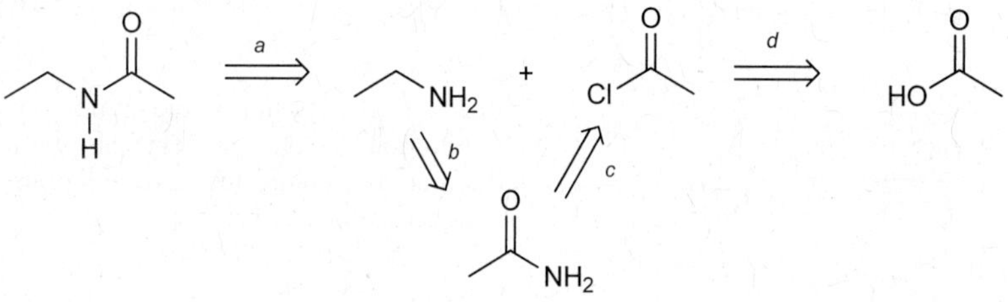

a. The product can be made from ethyl amine and acetyl chloride, via a nucleophilic acyl substitution reaction.
b. Ethyl amine can be made from acetamide via reduction with LAH, followed by water work-up.
c. Acetamide can be made from acetyl chloride, via a nucleophilic acyl substitution reaction.
d. Acetyl chloride can be made from acetic acid upon treatment with thionyl chloride.

Now let's draw the forward scheme. Acetic acid is treated with thionyl chloride to give acetyl chloride. One equivalent of acetyl chloride is converted into ethyl amine (via aminolysis, followed by reduction), which is then treated with another equivalent of acetyl chloride to give the product:

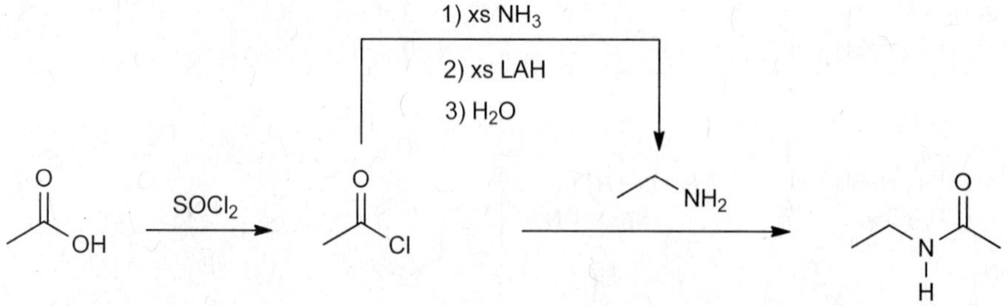

23.25.
(a) The starting material is an amine, and the reagents indicate a Hofmann elimination. There are two β positions, but they are identical because of symmetry. As such, there is only one possible regiochemical outcome for the elimination process:

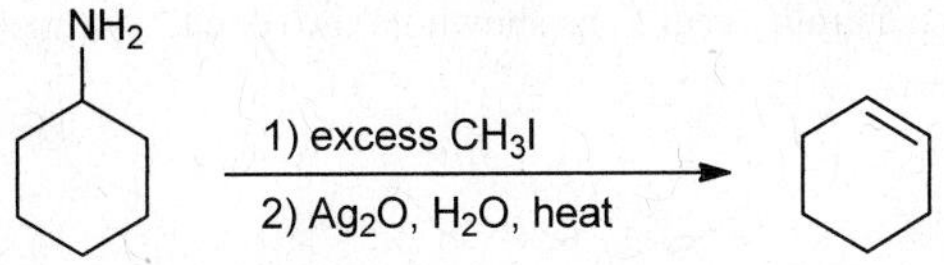

(b) The starting material is an amine, and the reagents indicate a Hofmann elimination. There are two β positions, and we expect elimination to occur at the β position that leads to the less substituted alkene:

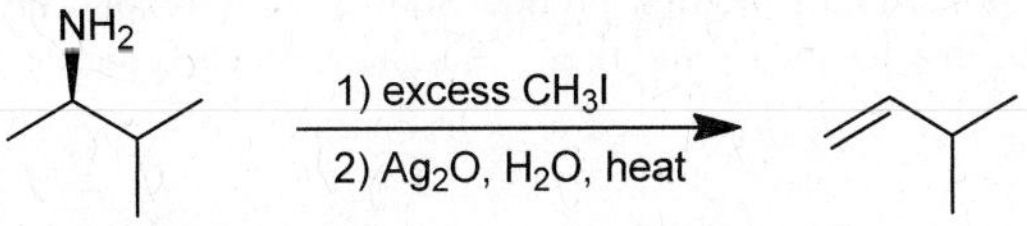

(c) The starting material is an amine, and the reagents indicate a Hofmann elimination. There are two β positions, and we expect elimination to occur at the β position that leads to the less substituted alkene:

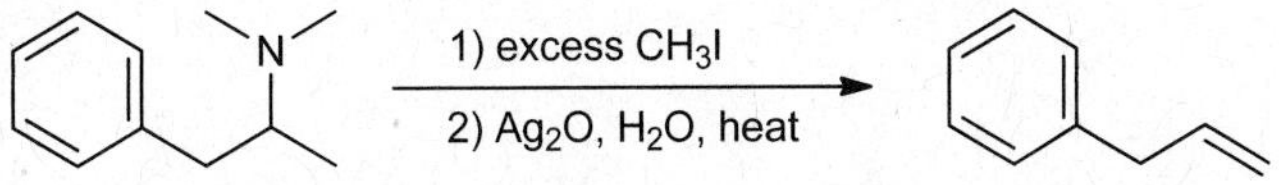

23.26. The starting material has nine carbon atoms, while the product has ten. The identity of the functional group has also changed, so we must propose a synthesis that introduces the tenth carbon atom and installs a π bond in the appropriate location. There are certainly many ways to achieve the desired transformation. One method involves introduction of the tenth carbon atom via conversion of the starting alkyl halide into a nitrile upon treatment with cyanide (an S_N2 reaction). Reduction of the nitrile with excess LAH, followed by water work-up, gives an amine, which can then be converted into the desired alkene via a Hofmann elimination, as shown:

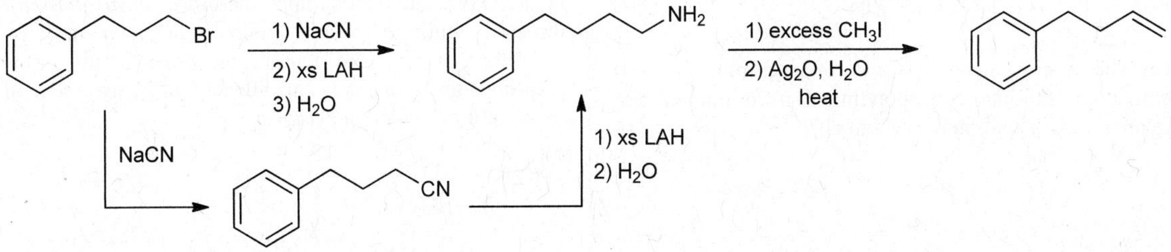

As an alternate approach, the starting alkyl halide can be treated with NaOH to give an alcohol, which can be oxidized (with PCC) to give an aldehyde. This aldehyde can be then be converted directly into the product with a Wittig reaction.

23.27. The structure of the intermediate alkene can be determined from the products of ozonolysis (butanal and pentanal). Based on the ozonolysis products alone, we cannot determine the configuration of the alkene (E or Z):

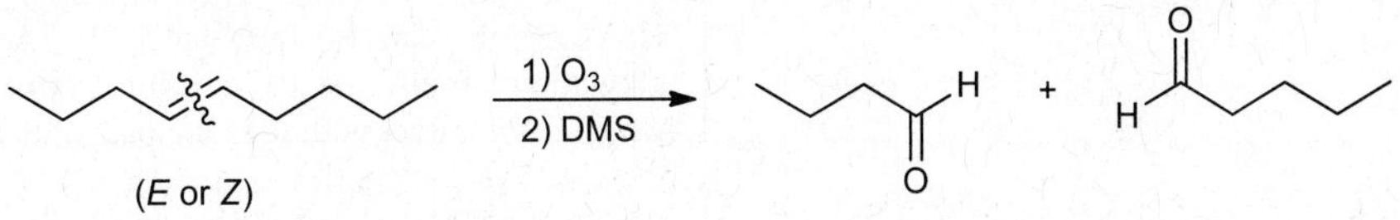

The *E* alkene can be made via a Hofmann elimination from two possible amines:

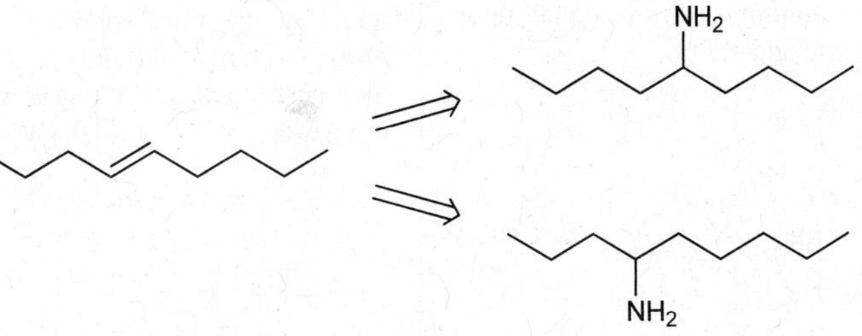

But only one of these amines lacks a chirality center, as shown below (due to symmetry):

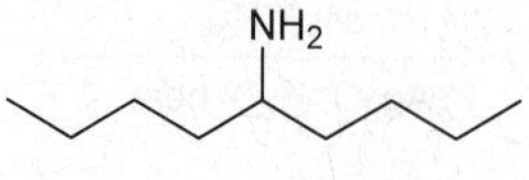

Compound **A**

23.28. The third product is perhaps the most revealing. It indicates that the structure of PCP must contain an aromatic ring for which the benzylic position is connected to a nitrogen atom, and the same benzylic position is also part of a cyclohexyl ring. This justifies formation of the first product shown. The second product indicates that the nitrogen atom in PCP must be incorporated in a six-membered ring. This ring is opened during formation of the third product.

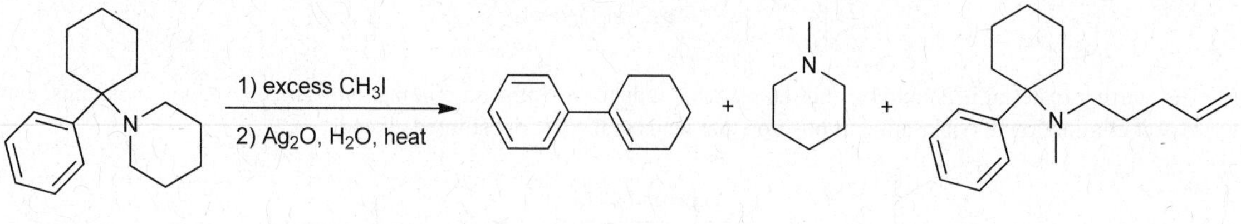

23.29.

(a) The starting material is a primary amine, so it is converted into the corresponding diazonium salt upon treatment with sodium nitrite and HCl:

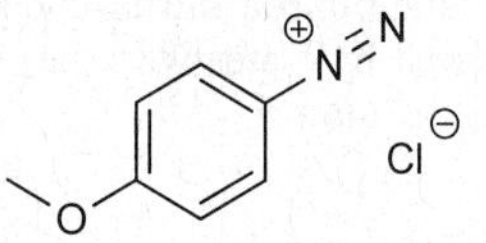

(b) The starting material is a secondary amine, so it is converted into the corresponding N-nitrosamine upon treatment with sodium nitrite and HCl:

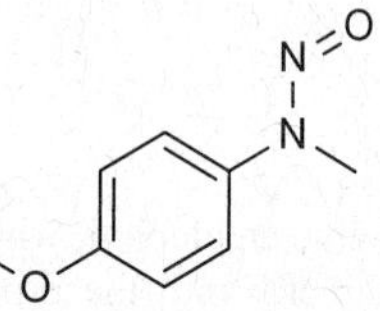

(c) The starting material is a secondary amine, so it is converted into the corresponding N-nitrosamine upon treatment with sodium nitrite and HCl:

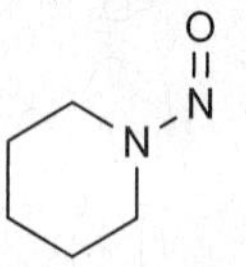

(d) The starting material is a secondary amine, so it is converted into the corresponding N-nitrosamine upon treatment with sodium nitrite and HCl:

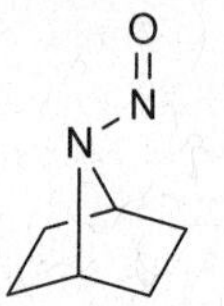

23.30.

(a) The desired transformation requires installation of an isopropyl group in the *para* position, as well as conversion of the NH_2 group into a cyano group. The former can be achieved via a Friedel-Crafts alkylation, while the latter can be achieved with a Sandmeyer reaction (via a diazonium ion).

Now let's consider the order of events. An amino group is a strong activator, and therefore an *ortho-para* director, while a cyano group is a *meta* director. Therefore, in order to achieve *para*-disubstitution, the isopropyl group must be installed before conversion of the amino group into a cyano group:

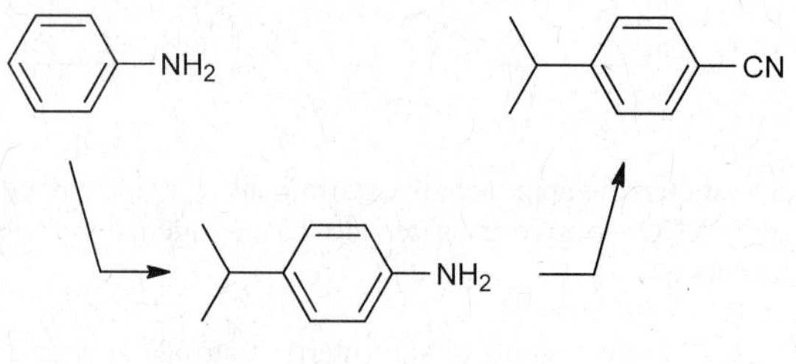

However, this strategy has one flaw. A Friedel-Crafts alkylation requires the use of a Lewis acid ($AlCl_3$), which can interact with the lone pair of the amino group, thereby converting the activating amino group into a deactivating group (see Section 23.8). As such, a Friedel-Crafts alkylation will not work. This issue can be avoided by acetylating the amino group first, thereby reducing the nucleophilicity of the lone pair on the nitrogen atom. The desired Friedel-Crafts alkylation is then performed (installing an isopropyl group), followed by hydrolysis to restore the amino group:

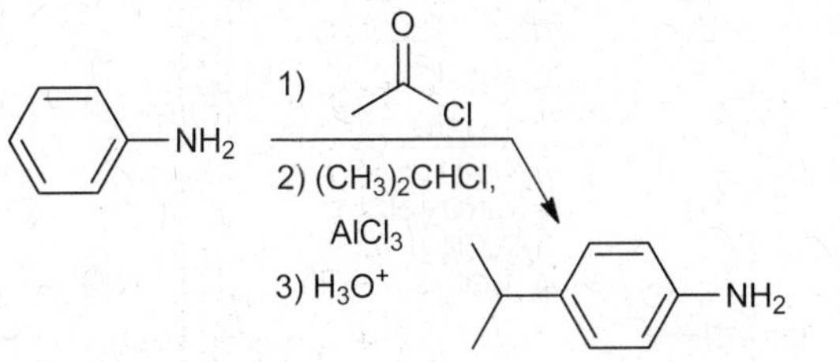

And finally, conversion of the amino group into a cyano group is achieved in two steps. First, *para*-isopropylaniline is treated with sodium nitrite and HCl, giving an aromatic diazonium ion, which is then treated with CuCN to give a Sandmeyer reaction that affords the desired product:

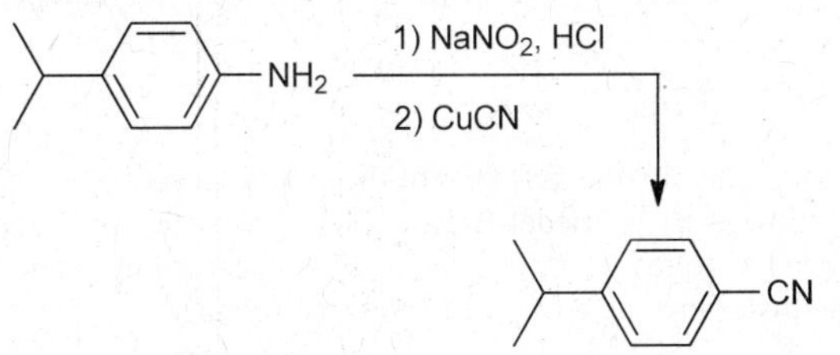

(b) The *meta*-directing effect of the nitro group enables installation of a bromine atom in the correct location (the *meta* position):

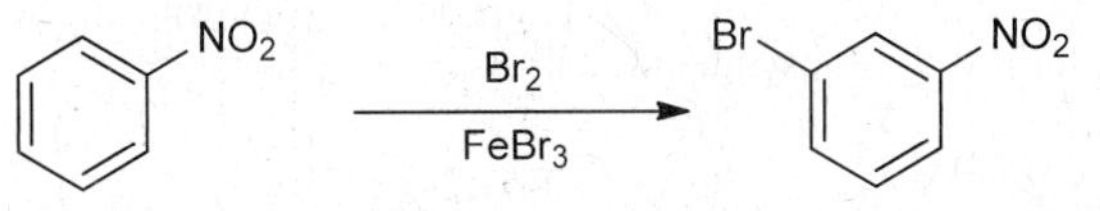

Now we must replace the nitro group with a bromine atom. One method for accomplishing this transformation involves the use of a Sandmeyer reaction, as seen in the following retrosynthetic analysis:

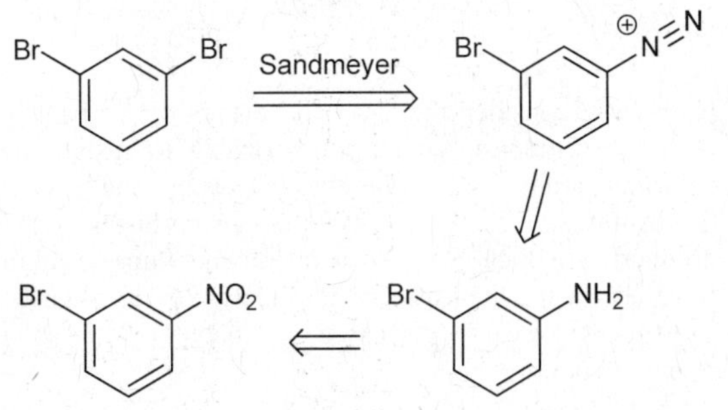

Now let's draw the forward scheme. The starting material is treated with Br$_2$ and a Lewis acid, thereby installing a bromine atom in the *meta* position. Reduction of the nitro group gives *meta*-bromoaniline, which is then converted into an aromatic diazonium ion upon treatment with sodium nitrite and HCl. The aromatic diazonium ion is then treated with CuBr to give a Sandmeyer reaction that affords the desired product, as shown:

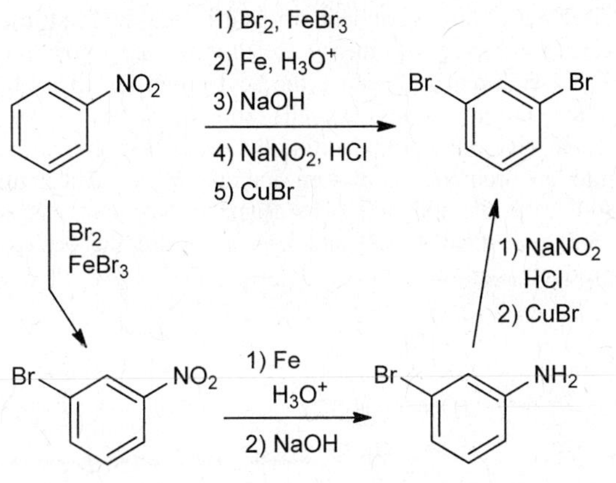

(c) The starting material is benzene and the product is disubstituted. Specifically, we must install a propyl group and a hydroxyl group. Installation of a propyl group can be achieved with a Friedel-Crafts acylation (followed by reduction). Installation of a hydroxyl group can be achieved via a diazonium ion, as shown in the following retrosynthetic analysis:

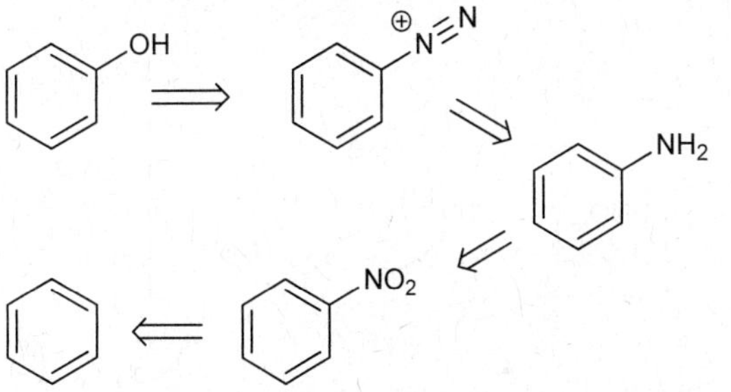

Now let's consider the order of events. In order to achieve *meta*-disubstitution, we must capitalize either on the *meta*-directing effects of a nitro group, or on the *meta*-directing effects of an acyl group:

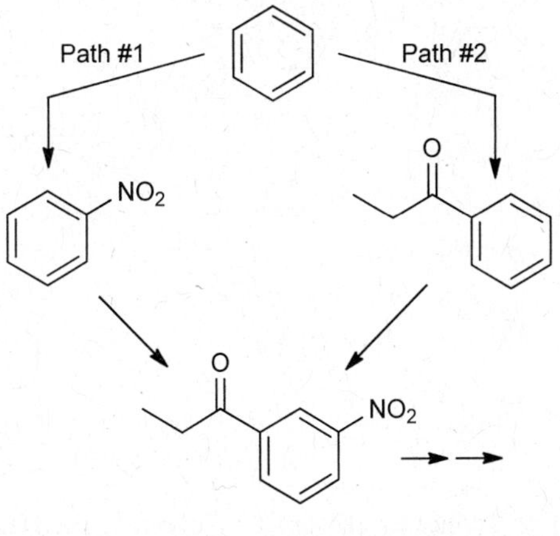

The first path is flawed, because it involves a Friedel-Crafts acylation process on a strongly deactivated ring, which will not occur. Therefore, only the second pathway is viable.

Now let's draw the forward scheme. A Friedel-Crafts acylation will install an acyl group, which is *meta*-

directing. Upon treatment with nitric acid and sulfuric acid, a nitro group is then installed in the *meta* position. The conditions that reduce the acyl group are likely to reduce the nitro group as well, followed by basic work-up, to give *meta*-propylaniline, which is then converted into an aromatic diazonium ion upon treatment with sodium nitrite and HCl. The aromatic diazonium ion is then treated with H$_2$O and heat to afford the desired product, as shown:

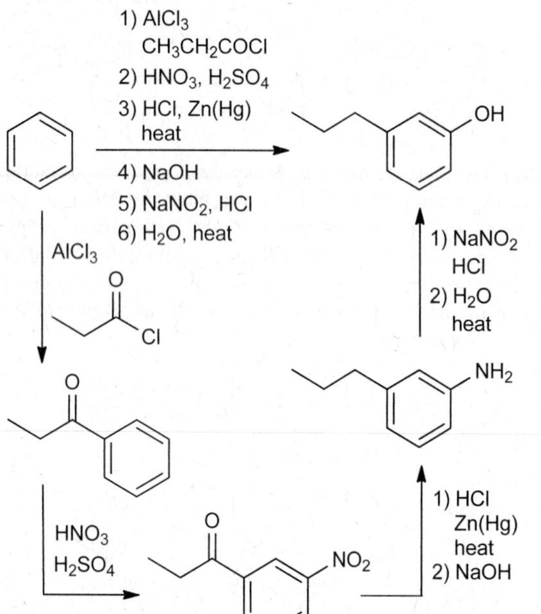

(d) Installation of a *tert*-butyl group can be achieved with a Friedel-Crafts alkylation. Installation of a carboxylic acid group can be achieved via a diazonium ion, as shown in the following retrosynthetic analysis:

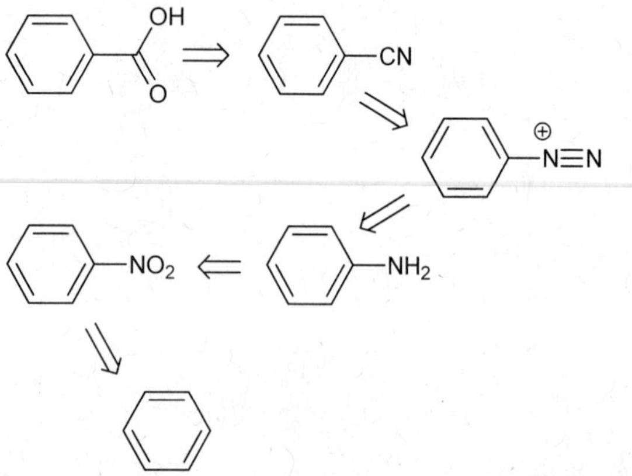

Now let's consider the order of events. The desired product is *para*-disubstituted, which can be achieved by installing the *tert*-butyl group first. This group is very large and will favor nitration at the *para* position, as seen in the following forward scheme:

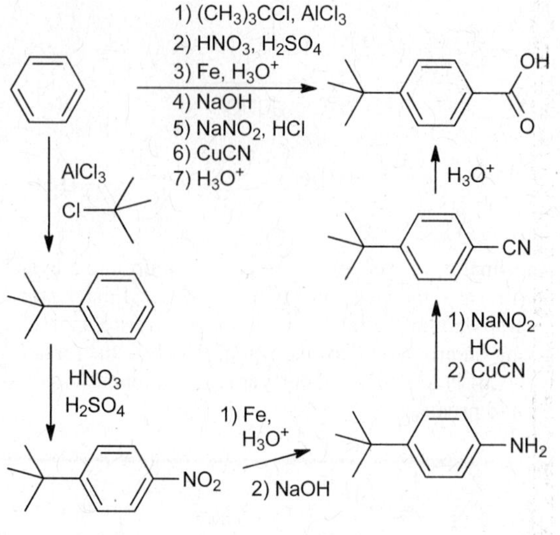

Alternatively, the first step of the synthesis above can be followed by a Friedel-Crafts alkylation (with MeCl and AlCl$_3$), followed by oxidation of the benzylic position with chromic acid to give the desired product.

(e) Installation of a chlorine atom can be achieved by treating benzene with Cl$_2$ and AlCl$_3$ (via an electrophilic aromatic substitution reaction). Installation of a fluorine atom cannot be achieved via a similar process, but it can be achieved via a diazonium ion, as shown in the following retrosynthetic analysis:

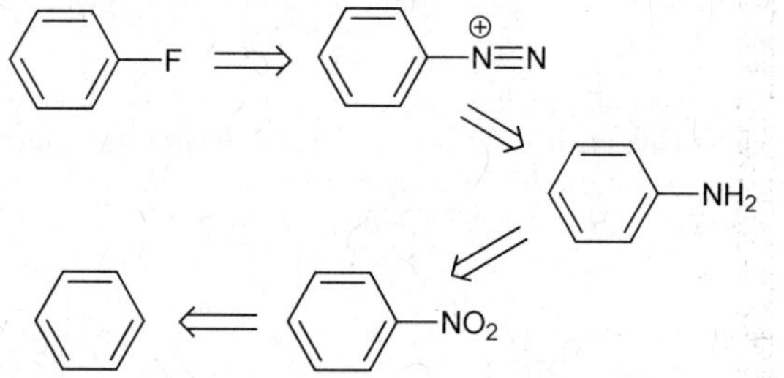

Now let's consider the order of events. Chlorine is larger than fluorine, so it is reasonable to install the chlorine atom first (thereby favoring the *para*-disubstituted product over the *ortho*-disubstituted product). In fact, we cannot install the fluorine atom first, because the directing effects of a fluorine substituent were not discussed in Chapter 19 (beyond scope of course).

Now let's draw the forward scheme. The starting material is treated with Cl$_2$ and a Lewis acid, thereby installing a chlorine atom. A nitro group is then installed in the *para* position, upon treatment with nitric acid and sulfuric acid. Reduction of the nitro group gives *para*-chloroaniline, which is then converted into an aromatic diazonium ion upon treatment with sodium nitrite and HCl. The aromatic diazonium ion is then treated with HBF$_4$ to give the desired product (a Schiemann reaction), as shown:

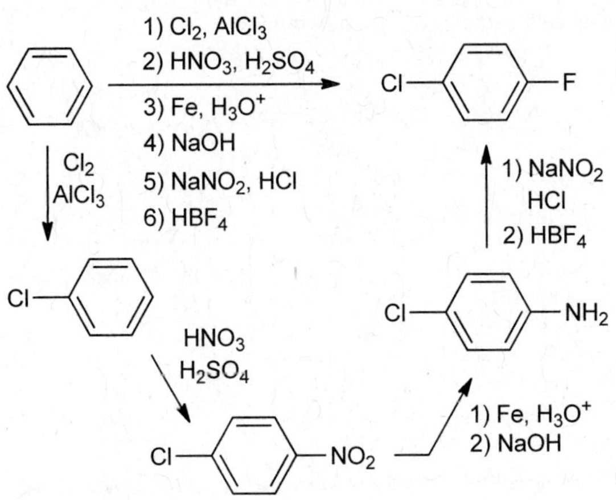

(f) This transformation can be achieved by replacing the chlorine atom with an amino group (via elimination-addition), followed by conversion of aniline into a diazonium ion, followed by subsequent treatment with CuBr (via a Sandmeyer reaction):

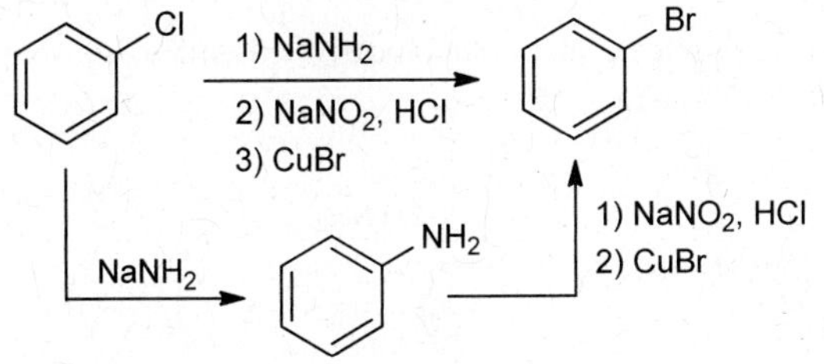

23.31.
(a) The starting material has two aromatic rings. The ring bearing the amino group is more highly activated. During the azo coupling process, the activated ring functions as the nucleophile, and the other ring must function as the diazonium ion, as shown in the following retrosynthetic analysis:

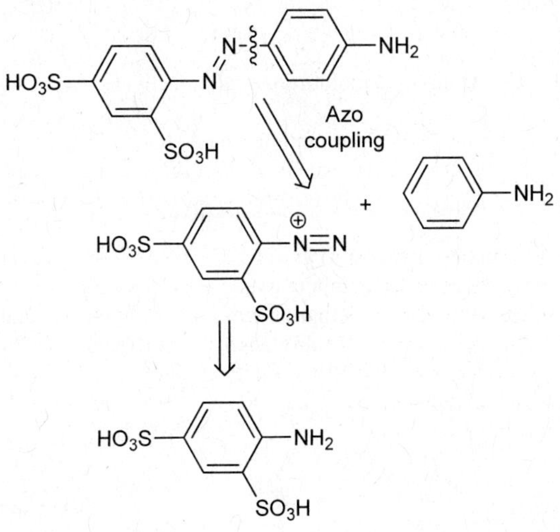

The forward scheme is shown here:

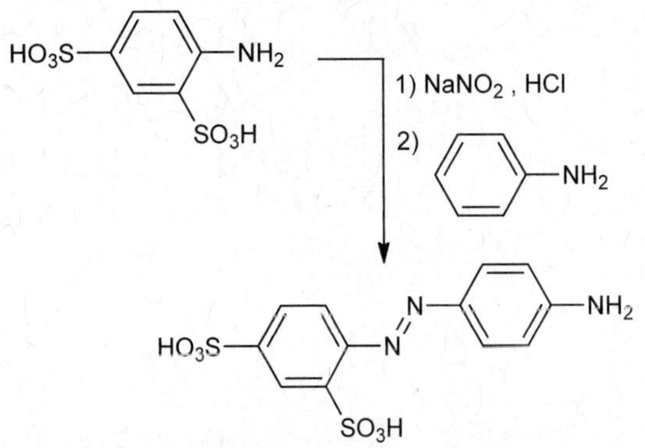

(b) The ring bearing the hydroxyl group is more highly activated ring. During the azo coupling process, the activated ring functions as the nucleophile, and the other ring must function as the diazonium ion, as shown in the following retrosynthetic analysis:

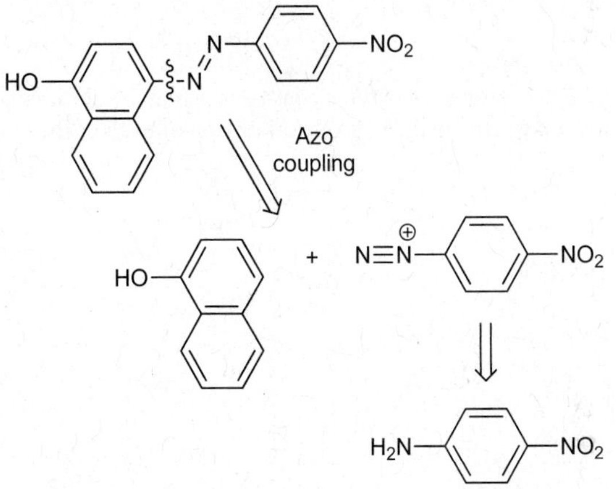

The forward scheme is shown here:

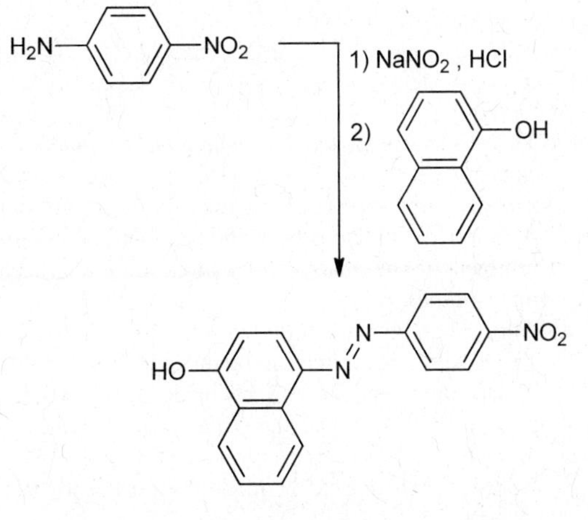

(c) The ring bearing the dimethylamino group is the more highly activated ring. During the azo coupling process, the activated ring functions as the nucleophile, and the other ring must function as the diazonium ion, as shown in the following retrosynthetic analysis:

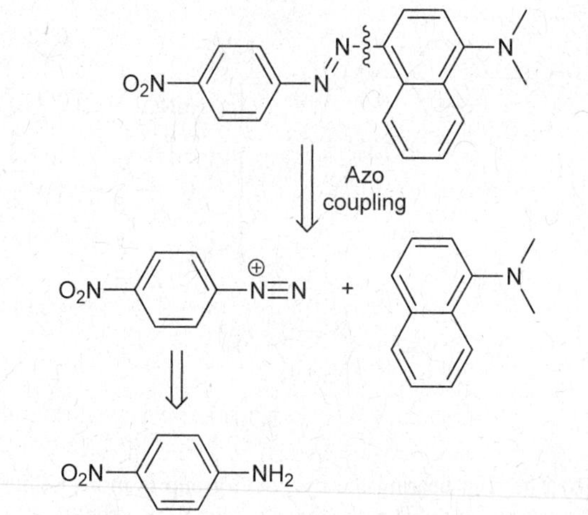

The forward scheme is shown here:

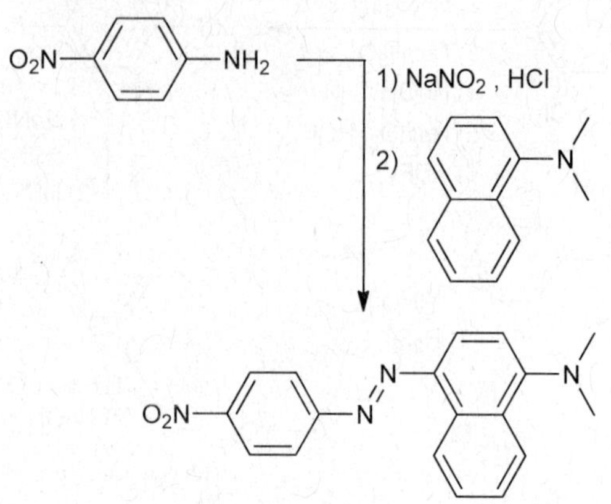

Azo
coupling

23.32. There are certainly many acceptable solutions to this problem. One such solution derives from the following retrosynthetic analysis. An explanation of each of the steps (*a-g*) follows.

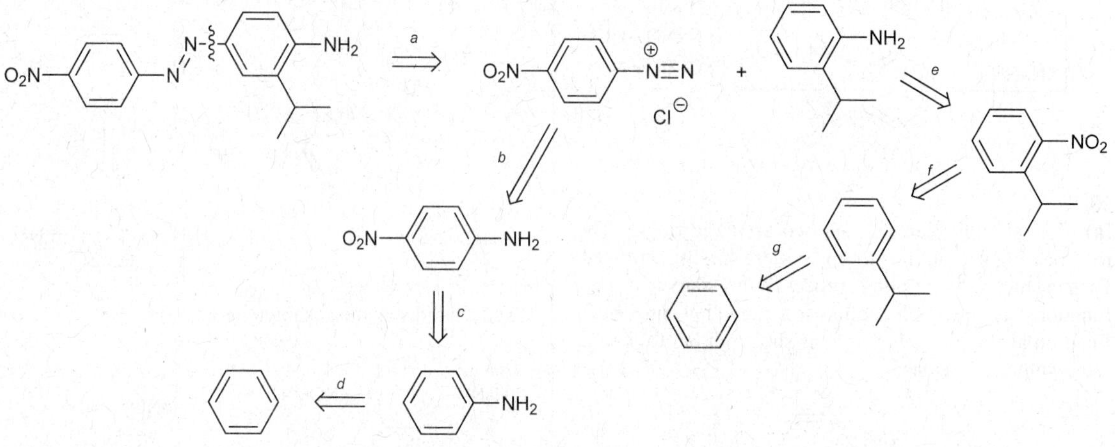

a. The product is an azo dye, which can be made via azo coupling from a diazonium ion and an activated aromatic ring.
b. The diazonium ion can be made from *para*-nitroaniline, upon treatment with sodium nitrite and HCl.
c. *para*-Nitroaniline can be made from aniline via nitration. This process requires acetylation prior to nitration, and removal of the acetyl group after nitration (because aniline will not directly undergo nitration to give *para*-nitroaniline).
d. Aniline can be made from benzene via nitration followed by reduction of the nitro group.
e. The substituted aniline can be made via reduction of the corresponding nitro compound.
f. The nitro compound can be made via nitration. This process requires sulfonation prior to nitration, so that nitration will occur at the *ortho* position (rather than the *para* position). Desulfonation is then required after nitration (to remove the sulfonic acid group).
g. Isopropyl benzene can be made from benzene via Friedel-Crafts alkylation.

The forward scheme is shown here:

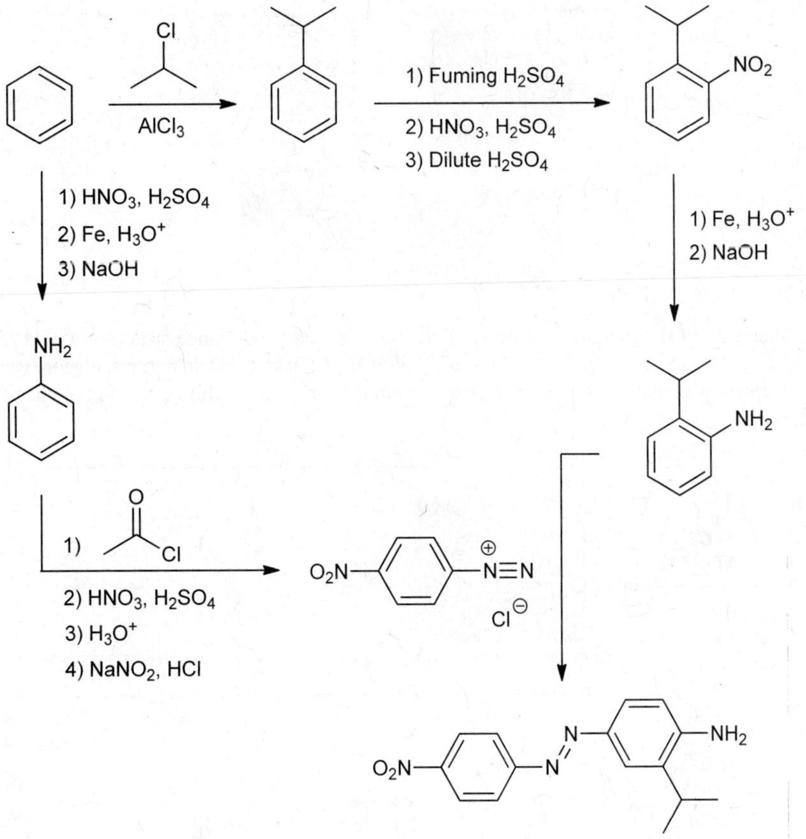

23.33.

(a) An azo coupling reaction will give the following product:

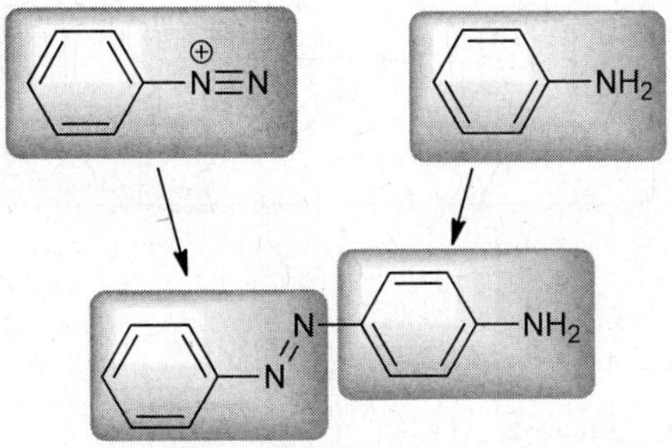

(b) An azo coupling reaction will give the following product:

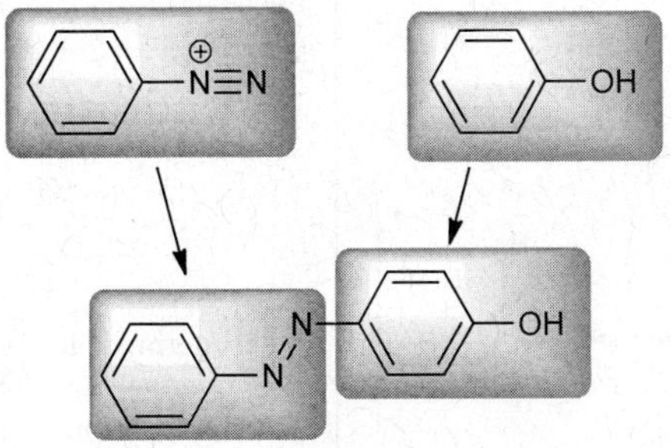

(c) An azo coupling reaction will give the following product:

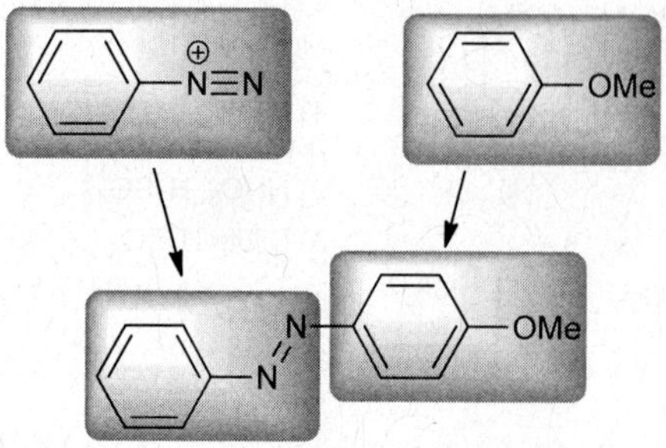

23.34. Attack at either C2 or C4 generates an intermediate that exhibits a resonance structure with a nitrogen atom that bears a positive charge and lacks an octet (highlighted below). Attack at C3 generates a more stable intermediate (carbon is more electropositive than nitrogen so a carbon atom can be better stabilize a positive charge associated with an unfilled octet):

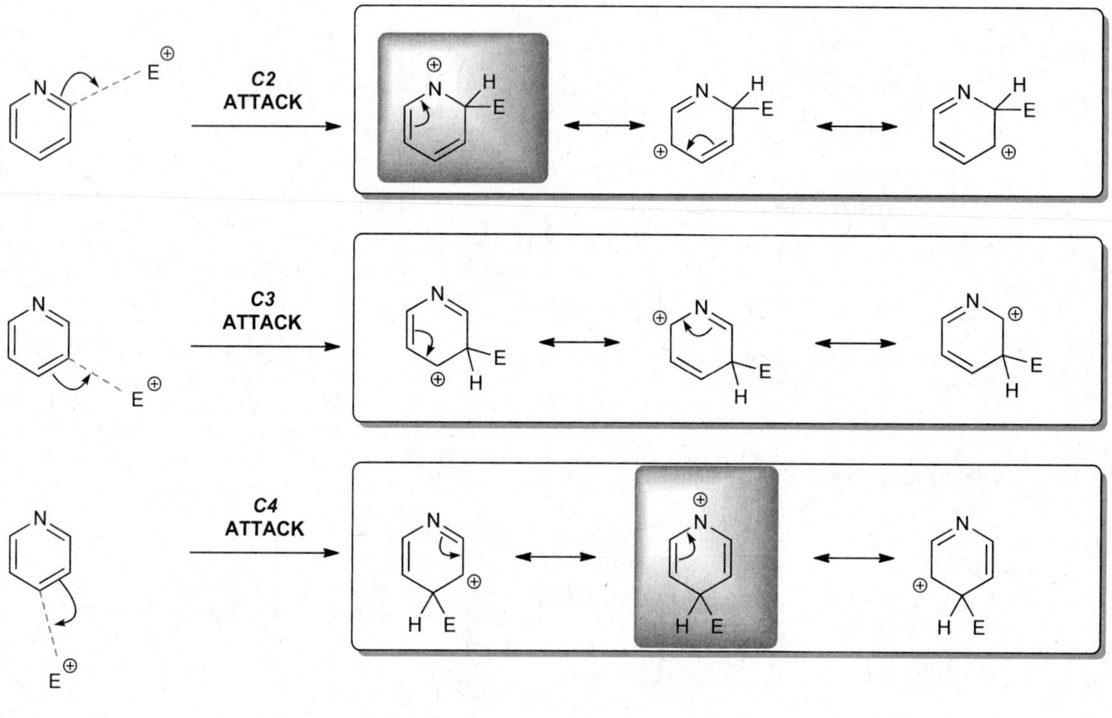

23.35. Attack at the C2 position proceeds via an intermediate with three resonance structures:

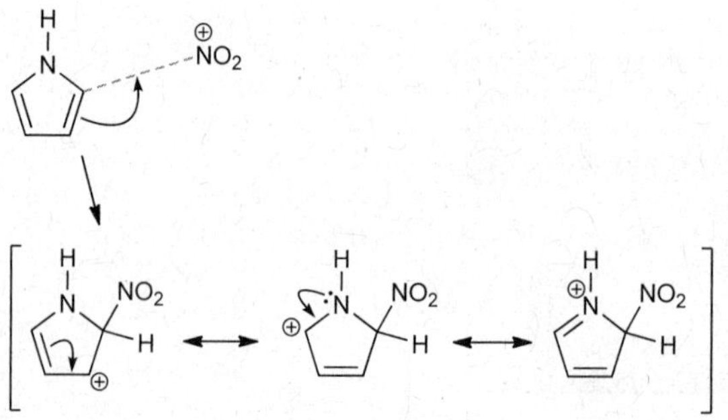

In contrast, attack at the C3 position proceeds via an intermediate with only two resonance structures:

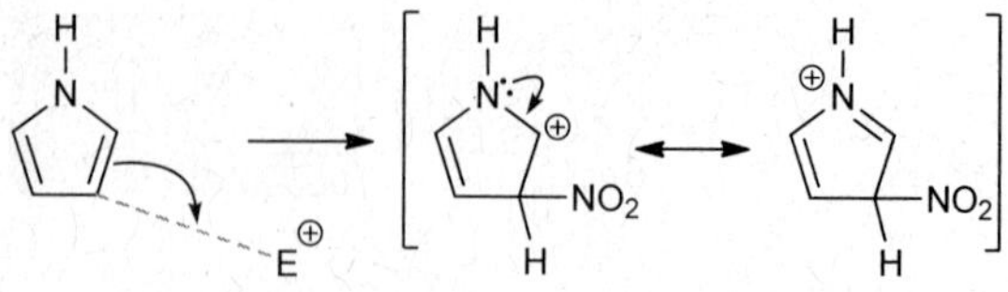

The intermediate for C2 attack is lower in energy than the intermediate for C3 attack. The transition state leading to the intermediate of C2 attack will therefore be lower in energy than the transition state leading to the intermediate of C3 attack. As a result, C2 attack occurs more rapidly, giving the following product:

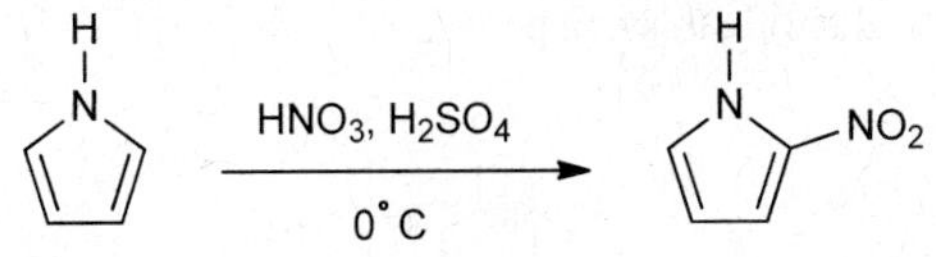

23.36.
(a) The second compound will have an N-H stretching signal between 3300 and 3500 cm^{-1}. The first compound will not have such a signal.
(b) When treated with HCl, the first compound will be protonated to form an ammonium salt that will produce an IR signal between 2200 and 3000 cm^{-1}. The second compound is not an amine and will not exhibit the same behavior.

23.37.
(a) The ^{1}H NMR spectrum of the first compound will have a singlet resulting from the N-methyl group. The ^{1}H NMR spectrum of the second compound is not expected to exhibit any singlets.
(b) The ^{1}H NMR spectrum of the first compound will have six signals, while the ^{1}H NMR spectrum of the second compound will have only three signals.

23.38. The designation "primary" indicates that two hydrogen atoms are attached to the nitrogen atom, while the designation "secondary" indicates that one hydrogen atom is attached to the nitrogen atom:

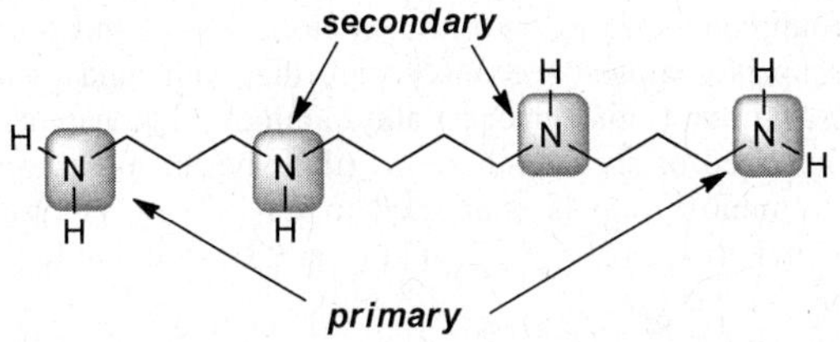

23.39.
(a) The lone pair that is farther away from the rings is the most basic, because that lone pair is localized. The lone pair of the other nitrogen atom is delocalized via resonance.
(b) The dimethylamino group exhibits a localized lone pair, and as such, it is expected to exist primarily as a charged ammonium ion (p$K_a \sim 10$; see pK_a table on inside cover of textbook) at physiological pH. In contrast, the other nitrogen atom exhibits a delocalized lone pair, and it is not expected to be protonated at physiological pH (see discussion of the Henderson-Hasselbalch equation in Section 21.3).

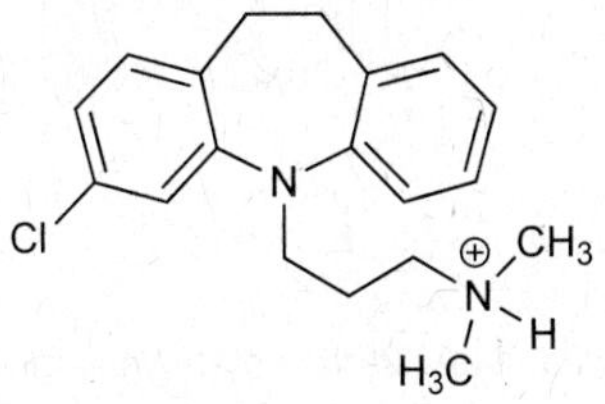

23.40. The nitrogen atom of the amide group exhibits a delocalized lone pair, so this lone pair will certainly not be the most basic. Indeed, amides do not function as bases. Each of the remaining two nitrogen atoms exhibits a localized lone pair. The nitrogen atom of the aromatic system has the localized lone pair in an sp^2 hybridized orbital (in the plane of the ring, and going away from the ring). In contrast, the other nitrogen atom (highlighted) has the localized lone pair in an sp^3 hybridized orbital. The sp^3 hybridized nitrogen atom (highlighted) is expected to be a better base than the sp^2 hybridized nitrogen atom, because the former has a lone pair that is farther away from the nucleus (held less tightly) and is therefore more available to function as a base. Also, a comparison of pK_a values (see inside cover of textbook) indicates that pyridine is a weaker base than triethyl amine (compare the pK_a values of their conjugate acids).

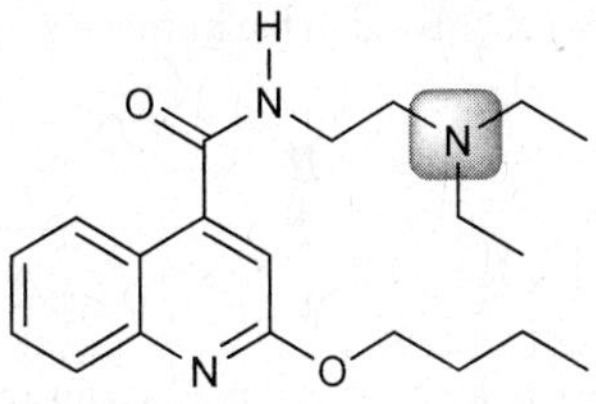

23.41.
(a) Pyridine is a weaker base than trimethylamine because the lone pair of pyridine occupies an sp^2 hybridized orbital, rather than an sp^3 hybridized orbital. By occupying an sp^2 hybridized orbital, the electrons of the lone pair have more s character and are therefore closer to the positively charged nucleus, rendering them less basic. As such, trimethylamine is a stronger base than pyridine:

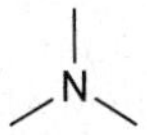

(b) The nitrogen atom of an amide group exhibits a lone pair that is highly delocalized and is therefore not expected to function as a base. Pyridine is a stronger base because the lone pair is localized (the lone pair occupies an sp^2 hybridized orbital):

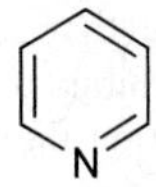

(c) Each compound exhibits a nitrogen atom incorporated in a five-membered aromatic ring. The lone pair on each of these nitrogen atoms is participating in aromatic stabilization, and will not function as a base. The difference between these two compounds is in the location of the other nitrogen atom. In the first compound, the nitrogen atom is adjacent to the aromatic ring, and as such, its lone pair is delocalized into the ring. The second compound (shown below) is a stronger base, because it exhibits a lone pair that is localized (highlighted):

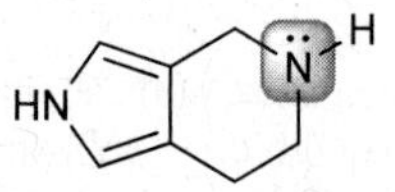

23.42.

(a) The parent is aniline (aminobenzene), and there are two alkyl groups connected to the nitrogen atom (an ethyl group and an isopropyl group):

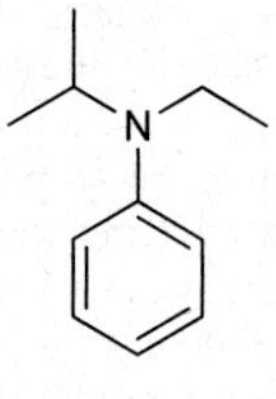

(b) The name indicates a three-membered ring connected to a nitrogen atom, as well as two substituents (both methyl groups) connected to the nitrogen atom.

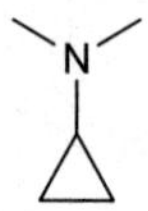

(c) The parent is a five-membered chain (pentane) that bears an amino group at C2. In addition, there is a dimethyl amino group located at C3. The configuration of each chirality center (C2 and C3) is indicated in the name:

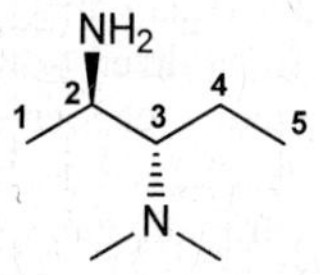

(d) The name indicates a primary amine in which the nitrogen atom is connected to a benzyl (PhCH$_2$–) group:

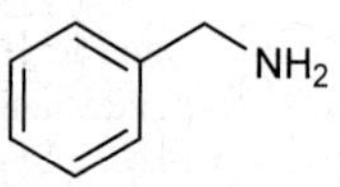

23.43. Only one of the nitrogen atoms (highlighted) has a localized lone pair. As such, this nitrogen atom is significantly more basic than the other two nitrogen atoms, each of which exhibits a highly delocalized lone pair. The nitrogen atom on the left is part of an amide group, and is not expected to function as base, while the nitrogen atom on the right is using its lone pair to establish aromaticity. So that lone pair is also unavailable to serve as a base.

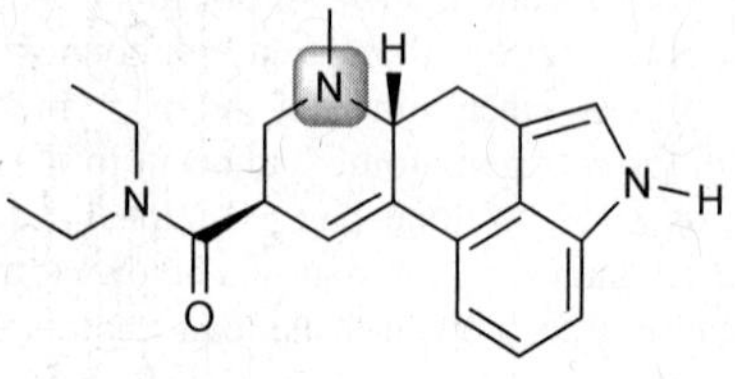

23.44.

(a) Recall that an atom bearing four different groups is a chirality center. There are two chirality centers (highlighted) in this compound:

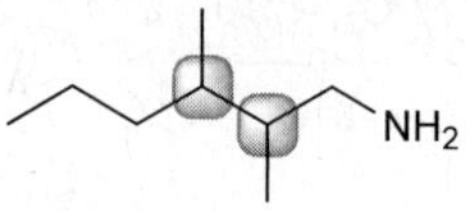

(b) Recall that an atom bearing four different groups is a chirality center. There are two chirality centers (highlighted) in this compound. Notice that, in this case, the nitrogen atom is a chirality center because it is connected to four different groups (one of which is a lone pair).

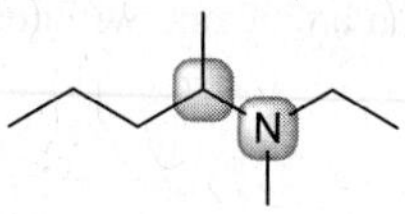

(c) Recall that an atom bearing four different groups is a chirality center. There is only one chirality center (highlighted) in this compound. Notice that, in this case, the nitrogen atom is not a chirality center because it is connected to two identical groups (two propyl groups).

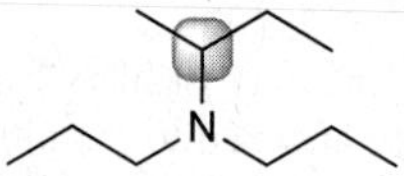

23.45.

(a) This compound is an amine that has only one alkyl group connected to the nitrogen atom. Since this alkyl group is complex, we must name the compound as an alkanamine (rather than an alkyl amine). The parent is comprised of six carbon atoms (thus, hexanamine), and the amino group is connected to C1. There are four methyl groups (two at C2 and two at C3), resulting in the following name:

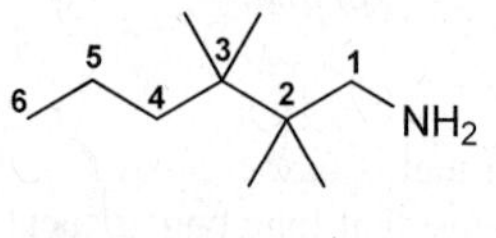

2,2,3,3-tetramethyl-1-hexanamine

(b) This compound has two functional groups (a carbonyl group and an NH$_2$ group). The carbonyl group takes priority, so the compound is named as a ketone (cyclohexanone), with the amino group listed as a substituent, located at C4. In addition, there are two methyl groups, both located at C2. The configuration of the chirality center is listed at the beginning of the name:

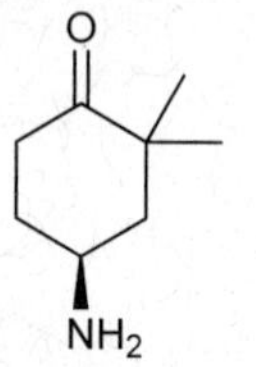

(S)-4-amino-2,2-dimethylcyclohexanone

(c) This compound is an amine in which the nitrogen atom is connected to three alkyl groups (two cyclobutyl groups and a methyl group) connected to the nitrogen atom, so we can name this compound as a trialkyl amine, rather than an alkanamine. The alkyl groups are listed in alphabetical order:

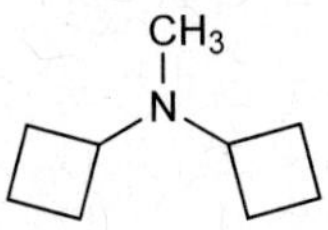

dicyclobutylmethylamine

(d) The parent is aniline, and there are two methyl groups (one at C2 and the other at C6), as well as one bromine atom located at C3:

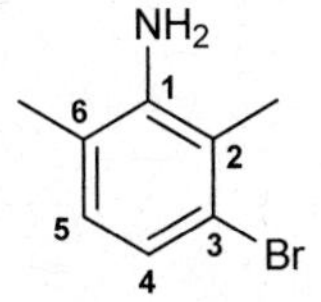

3-bromo-2,6-dimethylaniline

(e) The parent is aniline, and there are three substituents: a propyl group at C3 and two methyl groups (both connected to the nitrogen atom).

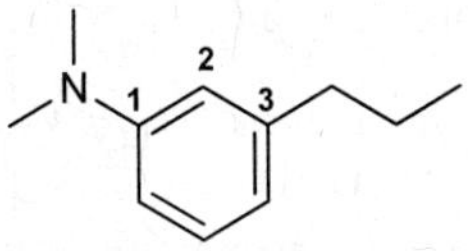

N,N-dimethyl-3-propylaniline

(f) The parent is pyrrole, and there are three substituents: a methyl group connected to the nitrogen atom, and two ethyl groups at C2 and C5.

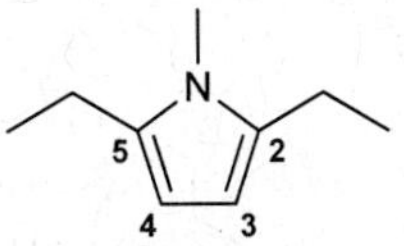

2,5-diethyl-N-methyl pyrrole

23.46. The molecular formula ($C_4H_{11}N$) indicates no degrees of unsaturation (see Section 15.16), so all of the isomers must be acyclic amines. There is only one isomer that is a tertiary amine:

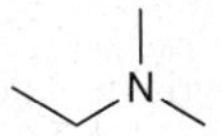

ethyldimethylamine

And there are three isomers that are all secondary amines:

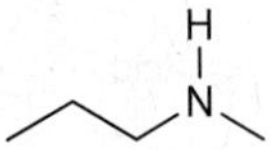

methylpropylamine 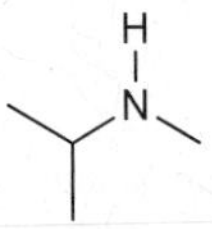   isopropylmethylamine 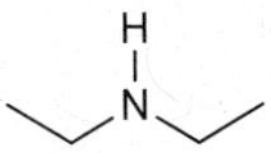   diethylamine

And finally, there are four isomers that are all primary amines:

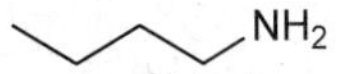

1-butanamine

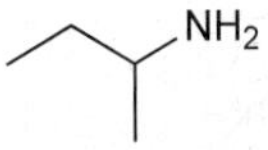

2-butanamine

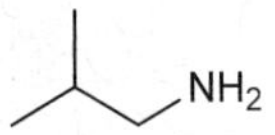

2-methyl-1-propanamine

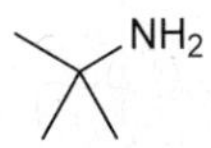

2-methyl-2-propanamine

In total, there are eight constitutional isomers with the molecular formula $C_4H_{11}N$.

23.47. The molecular formula ($C_5H_{13}N$) indicates no degrees of unsaturation (see Section 15.16), so all of the isomers must be acyclic amines. The following three isomers are all tertiary amines (acyclic and fully saturated), and none of them have a chirality center:

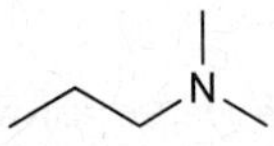

dimethylpropylamine 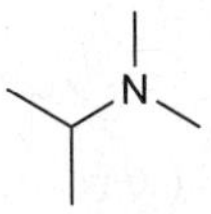   isopropyldimethylamine

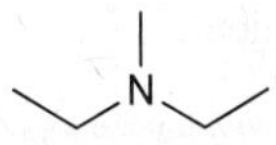

diethylmethylamine

23.48.

(a) The lone pair on pyridine functions as a base and deprotonates acetic acid, giving a pyridinium ion and an acetate ion, as shown. Two curved arrows must be drawn. The first curved arrow shows the base attacking the proton, and the second curved arrow shows heterolytic cleavage of the O-H bond:.

Base   Acid

(b) The tertiary amine functions as a base and deprotonates the carboxylic acid (benzoic acid), giving an ammonium ion and a benzoate ion, as shown. Two curved arrows must be drawn. The first curved arrow shows the base attacking the proton, and the second curved arrow shows heterolytic cleavage of the O-H bond:

Base   Acid

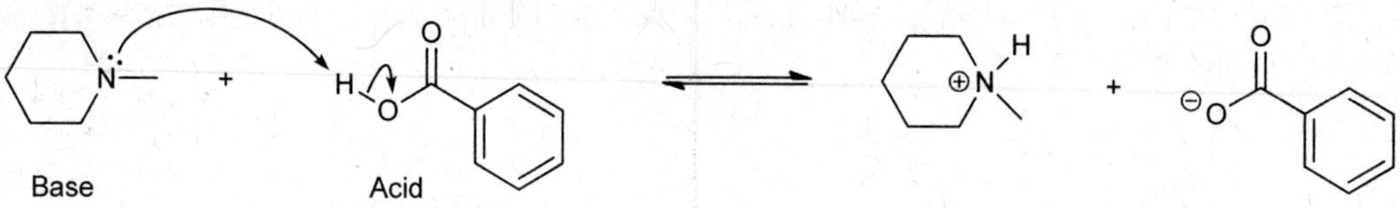

23.49.

(a) The amino group of aniline is an *ortho-para* director and it strongly activates the ring toward electrophilic aromatic substitution. When treated with excess Br_2, we expect bromination to occur in the two *ortho* positions and the *para* position, giving 2,4,6-tribromoaniline:

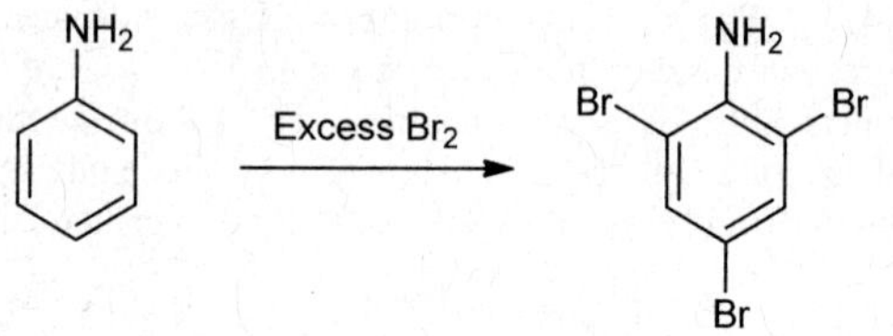

(b) Aniline is a strong nucleophile. When treated with an acid chloride, the amino group undergoes acylation, giving the following product. Pyridine functions as an acid sponge to neutralize the HCl that is produced as a by-product of the reaction.

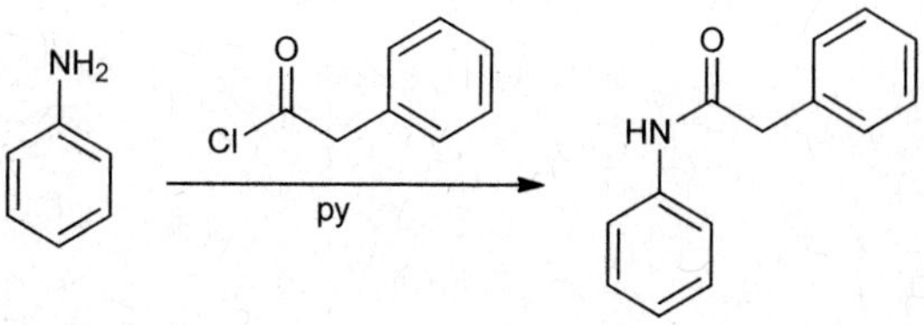

(c) When treated with excess methyl iodide, the amino group of aniline undergoes exhaustive alkylation to give a quaternary ammonium salt:

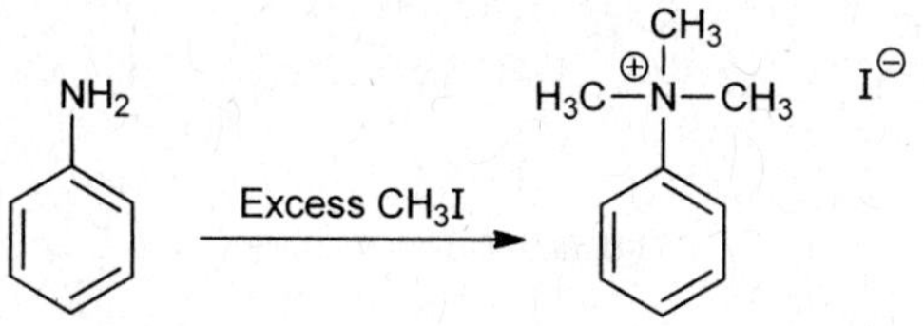

(d) Treating aniline with sodium nitrite and HCl gives a diazonium ion, which is then converted into benzene upon treatment with H_3PO_2:

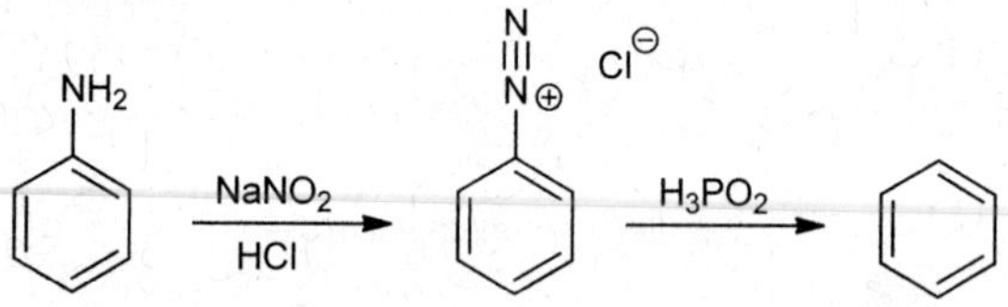

(e) Treating aniline with sodium nitrite and HCl gives a diazonium ion, which is then converted into benzonitrile via a Sandmeyer reaction (upon treatment with CuCN):

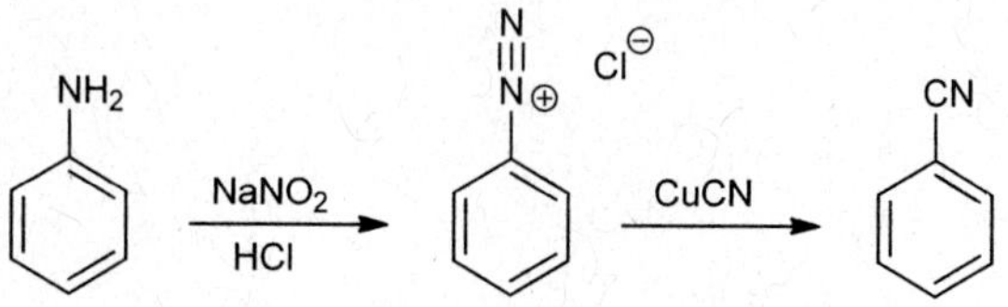

23.50.

(a) This transformation does not involve a change in the carbon skeleton. Only the identity of the functional group must be changed. This can be achieved by converting the alcohol into an alkyl halide (upon treatment with PBr$_3$), followed by an azide synthesis, as shown below. Alternatively, the alkyl halide can be converted into the desired amine via a Gabriel synthesis.

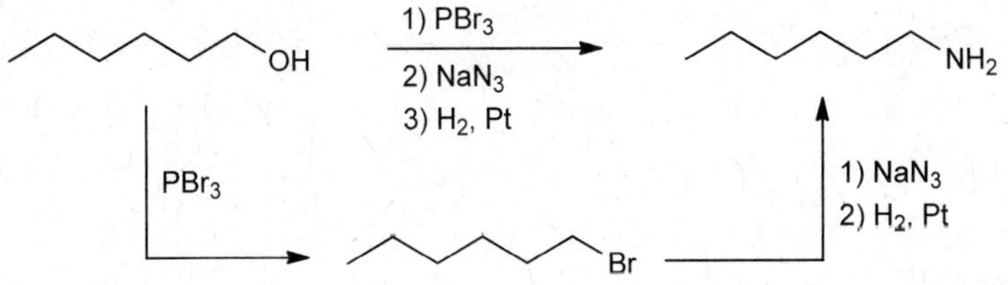

(b) This transformation involves a change in the carbon skeleton, as well as a change in the identity and location of the functional group. There are certainly many ways to install the extra carbon atom and manipulate the functional group as necessary. One method involves converting the alcohol into an alkyl halide (upon treatment with PBr$_3$), followed by an S$_N$2 reaction with cyanide as the nucleophile, thereby giving a nitrile. Reduction of the nitrile with excess lithium aluminum hydride, followed by water work-up, gives the product:

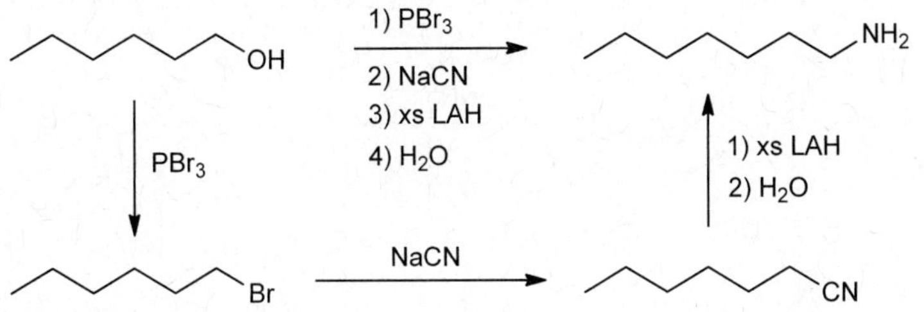

(c) The starting material has six carbon atoms, while the product has only five carbon atoms. In order to remove a carbon atom, a carbon-carbon bond must be broken, which can be accomplished via ozonolysis:

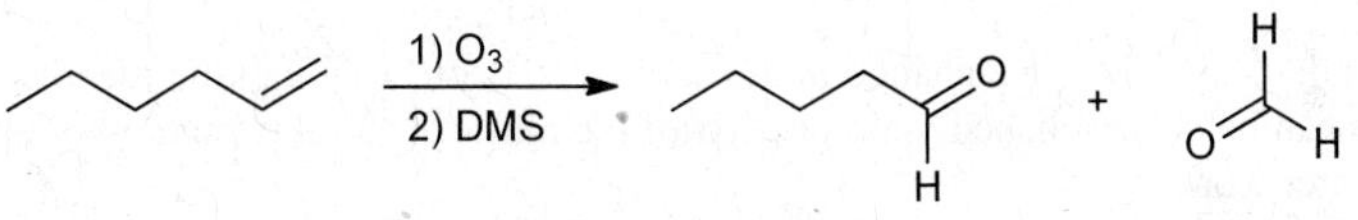

This strategy requires that we first convert the starting alcohol into the alkene above, which can be achieved by treating the alcohol with PBr$_3$, giving an alkyl halide, followed by elimination with a strong, sterically hindered base (such as *tert*-butoxide) to give an alkene. Ozonolysis of the alkene gives pentanal, which can be converted into the product via reductive amination, as shown:

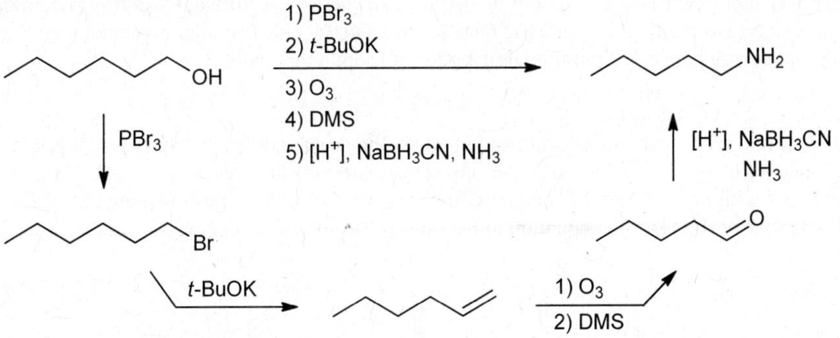

23.51.

(a) This transformation does not involve a change in the carbon skeleton. Only the identity of the functional group must be changed. This can be achieved via an azide synthesis, as shown below. Alternatively, the alkyl halide can be converted into the desired amine via a Gabriel synthesis.

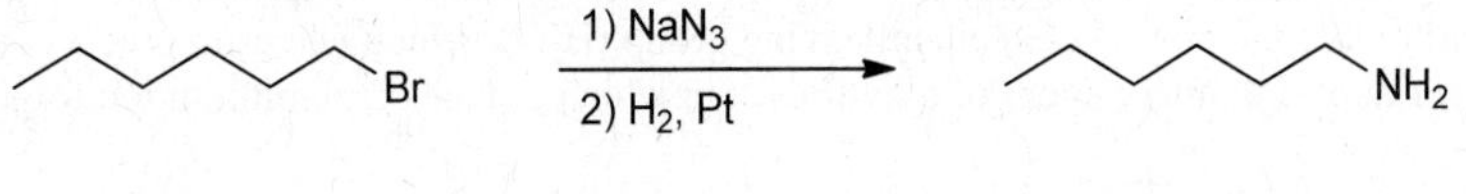

(b) This transformation involves a change in the carbon skeleton, as well as a change in the identity and location of the functional group. There are certainly many ways to install the extra carbon atom and manipulate the functional group as necessary. One method involves an S_N2 reaction with cyanide as the nucleophile, thereby converting the alkyl bromide into a nitrile. Reduction of the nitrile with excess lithium aluminum hydride, followed by water work-up, gives the product:

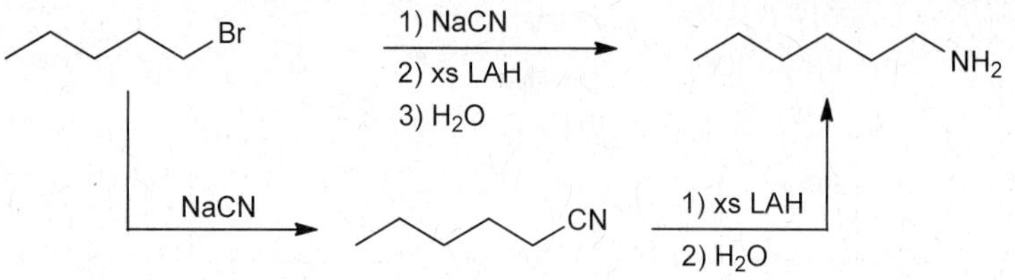

(c) This transformation does not involve a change in the carbon skeleton, although the identity of the functional group must be changed. There are certainly many ways to change the identity of the functional group. One method involves conversion of the carboxylic acid to the corresponding amide (upon treatment with thionyl chloride to give an acid chloride, followed by treatment with excess NH_3). Reduction of the amide with excess lithium aluminum hydride, followed by water work-up, gives the product:

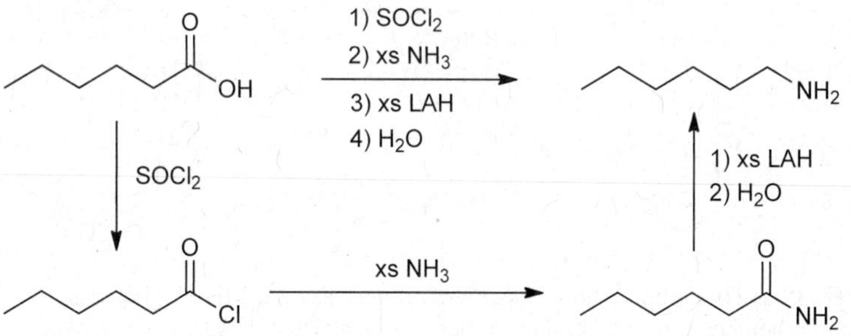

(d) This transformation does not involve a change in the carbon skeleton, although the identity of the functional group must be changed. This can be accomplished via reduction of the nitrile (upon treatment with excess lithium aluminum hydride, followed by water work-up):

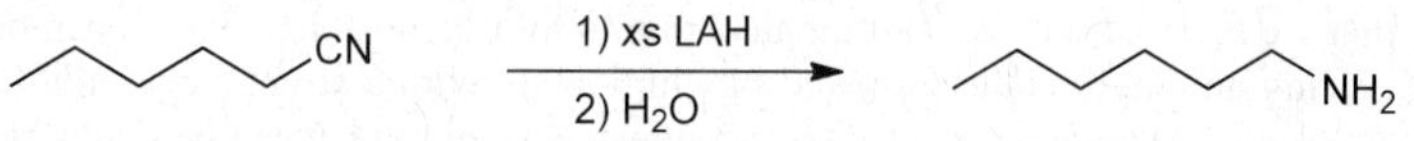

23.52. Aziridine has significant ring strain, and this strain will increase significantly during pyramidal inversion (as the bond angle must increase during the geometric change associated with pyramidal inversion (see Figure 23.3). This provides a significant energy barrier for pyramidal inversion at room temperature.

23.53. The diethylamino group exhibits a localized lone pair, and as such, it is expected to exist primarily as a charged ammonium ion ($pK_a \sim 10$; see pK_a table on inside cover of textbook) at physiological pH. In contrast, the other nitrogen atom exhibits a highly delocalized lone pair, and it is not expected to be protonated at physiological pH (see discussion of the Henderson-Hasselbalch equation in Section 21.3).

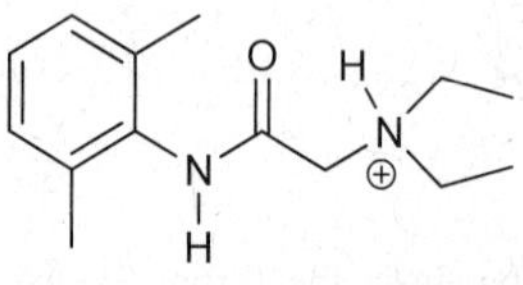

23.54. The following mechanism is based on Mechanism 20.6 (imine formation), although the final step is reduction, rather than a proton transfer step. In the first step, ammonia is a strong nucleophile and will attack the aldehyde directly. The resulting intermediate is then protonated, followed by subsequent deprotonation to give a carbinolamine. Protonation of the carbinolamine gives an excellent leaving group (H_2O), which leaves to give an iminium ion. Finally, sodium cyanoborohydride is a delivery agent of a hydride ion, which reduces the iminium ion to give methyl amine, as shown.

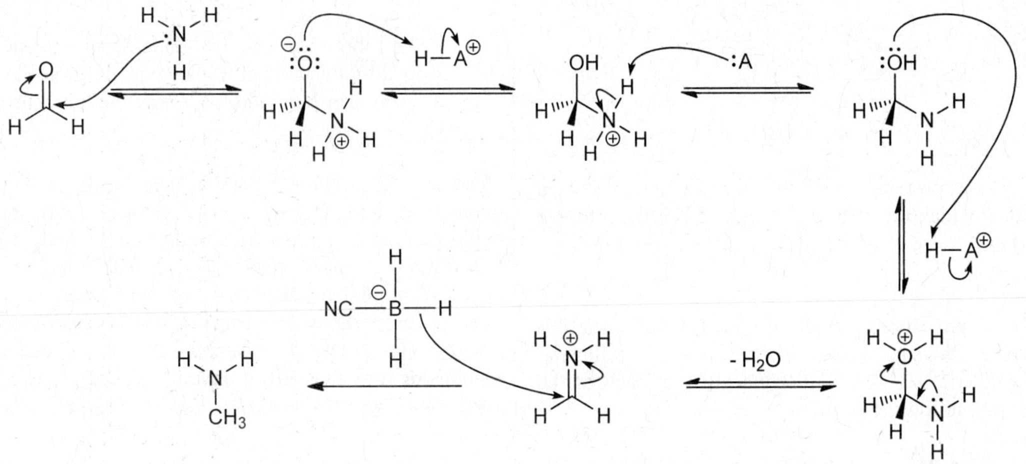

23.55. In acidic conditions, the amino group is protonated to give an ammonium ion. The ammonium group is a powerful deactivator and a *meta*-director.

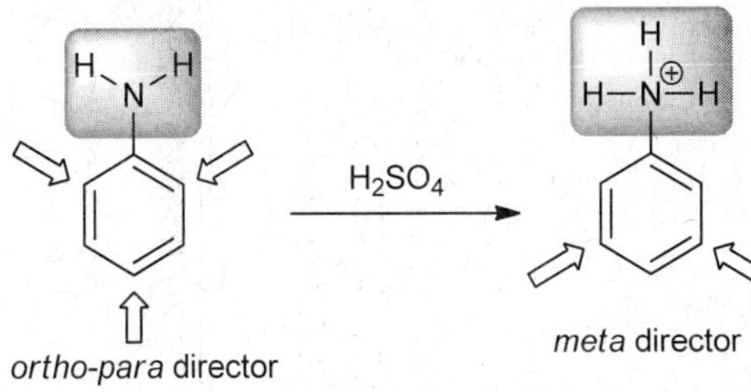

H₂SO₄

ortho-para director → *meta* director

23.56.

(a) In each compound, the lone pair of the amino group is delocalized because it is adjacent to the aromatic ring. However, the lone pair is more strongly delocalized for the compound that exhibits a nitro group in the *para* position (rather than the *meta* position). This extra delocalization is a result of the following additional resonance structure, in which electron density is delocalized onto the nitro group:

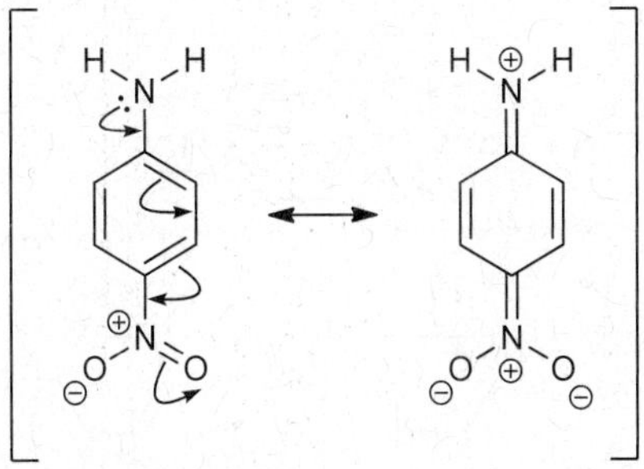

A similar resonance structure (in which electron density is delocalized onto the nitro group) cannot be drawn for *meta*-nitroaniline. As such, the lone pair in *meta*-nitroaniline is less delocalized and is therefore a stronger base. This explains why *para*-nitroaniline is a weaker base than *meta*-nitroaniline.

(b) The basicity of *ortho*-nitroaniline should be closer in value to *para*-nitroaniline, because a resonance structure can be drawn in which the lone pair is delocalized onto the nitro group:

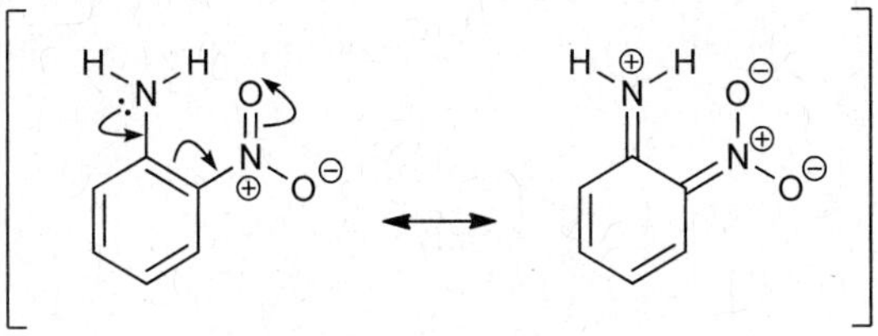

23.57. The starting amine exhibits two positions that are β to the dimethyl amino group:

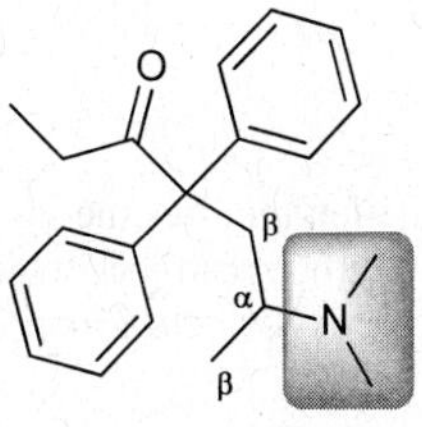

During a Hofmann elimination, the amino group is removed, and a double bond is formed between the α position and the less substituted β position, giving the following product:

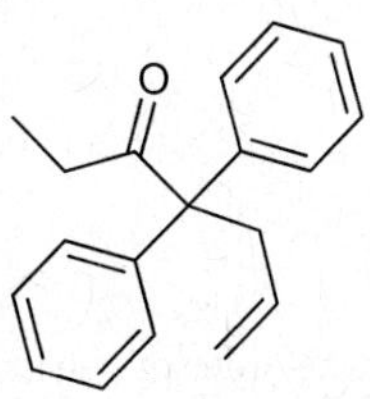

23.58. Protonation of the oxygen atom gives a cation in which the positive charge is delocalized over three locations:

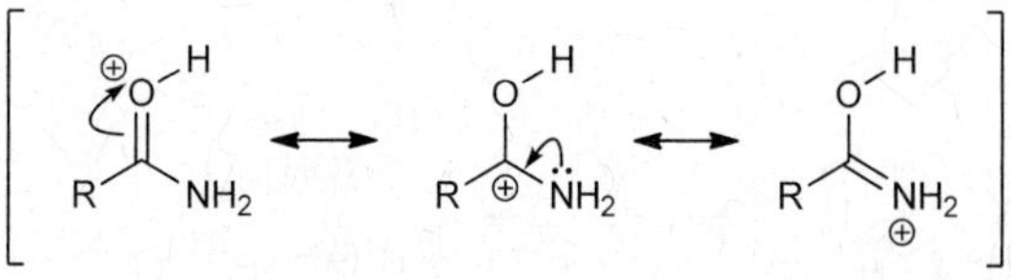

In contrast, protonation of the nitrogen atom gives a cation in which the charge is localized (on the nitrogen atom).

23.59.
(a) There are certainly many acceptable solutions to this problem. One such solution derives from the following retrosynthetic analysis. An explanation of each of the steps (*a-e*) follows.

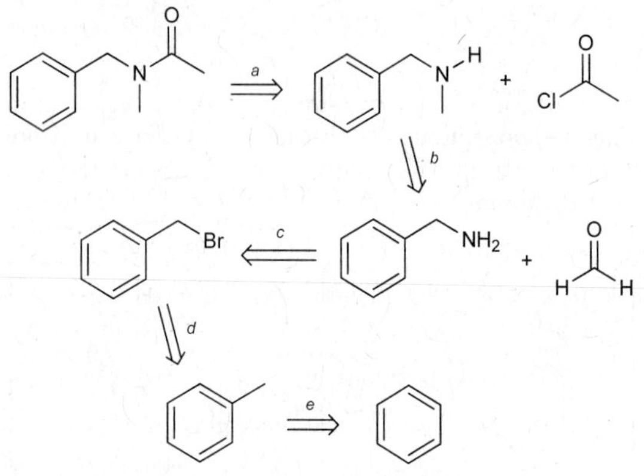

a. The product can be made via acetylation of the corresponding secondary amine.
b. The secondary amine can be made from the corresponding primary amine (benzyl amine) via a reductive amination.

c. Benzyl amine can be made from benzyl bromide via an azide synthesis.
d. Benzyl bromide can be made from toluene via radical bromination at the benzylic position.
e. Toluene can be made from benzene via a Friedel-Crafts alkylation.

The forward scheme is shown here. Benzene is first converted into toluene upon treatment with methyl chloride and aluminum trichloride (via a Friedel-Crafts alkylation). Upon treatment with NBS and heat, a bromine atom is installed at the benzylic position, giving benzyl bromide. An azide synthesis then converts benzyl bromide into benzyl amine. A reductive amination then installs a methyl group, followed by acetylation with acetyl chloride to give the product:

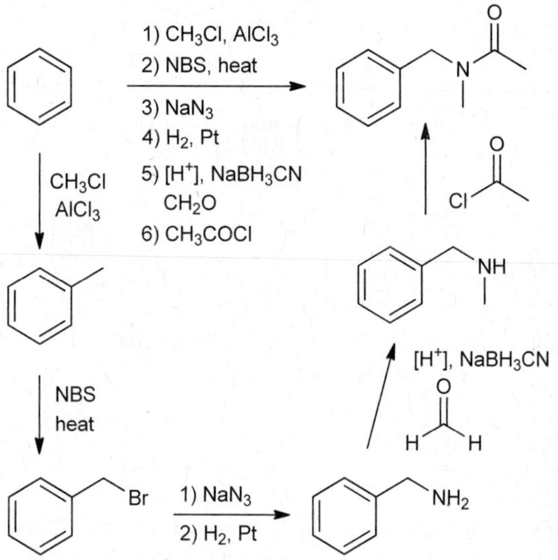

(b) This transformation requires the installation of a cyclohexyl group as well as a carboxylic acid group. Installation of the cyclohexyl group can be achieved with a Friedel-Crafts alkylation, using chlorocyclohexane and $AlCl_3$. Installation of a carboxylic acid group can be achieved via a diazonium ion, as shown in the following retrosynthetic analysis:

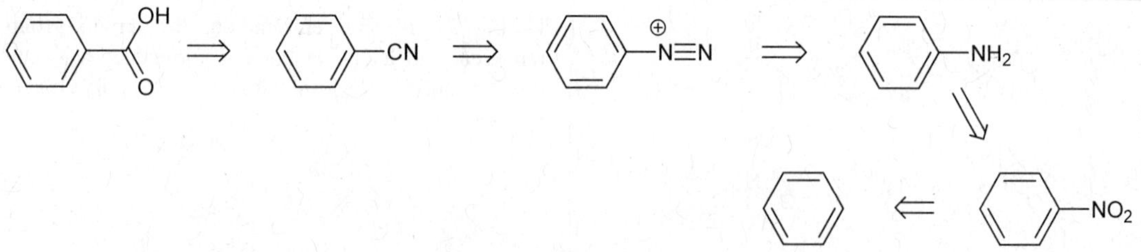

Now let's consider the order of events. The desired product is *para*-disubstituted, which can be achieved by installing the cyclohexyl group first. This group is very large and will favor nitration at the *para* position, as seen in the following forward scheme:

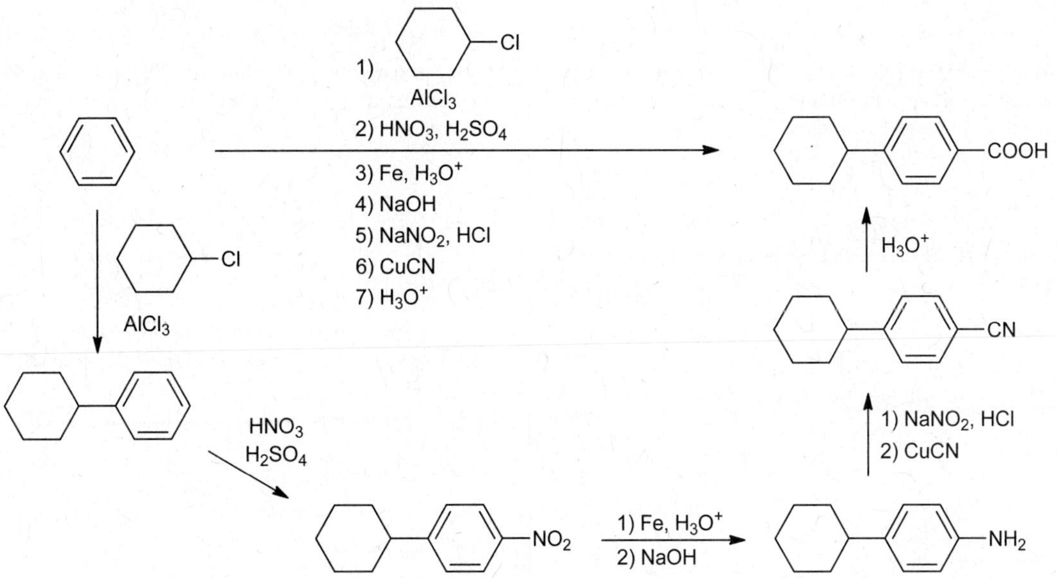

1)

AlCl₃

2) HNO₃, H₂SO₄

3) Fe, H₃O⁺
4) NaOH
5) NaNO₂, HCl
6) CuCN
7) H₃O⁺

COOH

H₃O⁺

CN

1) NaNO₂, HCl
2) CuCN

AlCl₃

HNO₃
H₂SO₄

NO₂

1) Fe, H₃O⁺
2) NaOH

NH₂

23.60. In Chapter 21, we learned that an amide linkage can be hydrolyzed under aqueous basic conditions. In this case, the compound has two amide linkages, each of which is hydrolyzed via a nucleophilic acyl substitution process. Hydroxide functions as a nucleophile and attacks one of the carbonyl groups to give a tetrahedral intermediate that reforms the carbonyl group by expelling a negatively charge nitrogen atom as a leaving group. The resulting anion then undergoes an intramolecular proton transfer step, giving a more stable carboxylate ion. Then, the remaining amide group undergoes hydrolysis in a similar way. Specifically, hydroxide attacks the carbonyl group to give a tetrahedral intermediate that reforms the carbonyl group by expelling a negatively charged nitrogen atom as a leaving group. Under these conditions, the carboxylic acid group is deprotonated to give a carboxylate ion, and the amide ion is protonated (by water) to give an amine:

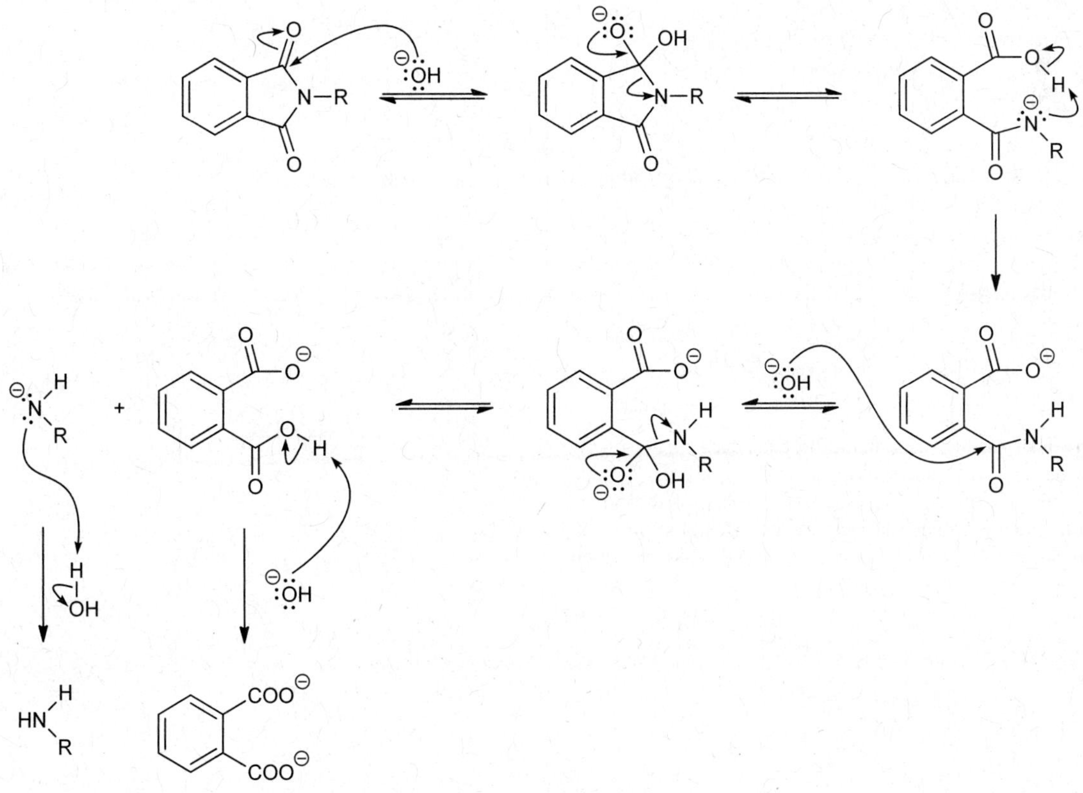

23.61. Hydrazine releases the amine via two successive nucleophilic acyl substitution reactions (as shown), giving the following by-product:

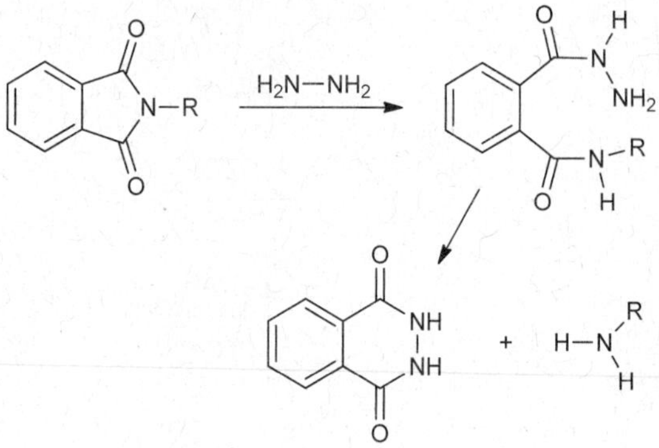

23.62.

(a) The starting material is a secondary amine and the reagents indicate a Hofmann elimination. In the first step, two methyl groups are installed, converting the secondary amine into a quaternary ammonium ion. Then, treatment with aqueous silver oxide and heat results in cleavage of a C-N bond (thereby opening the ring) and formation of a double bond, as shown:

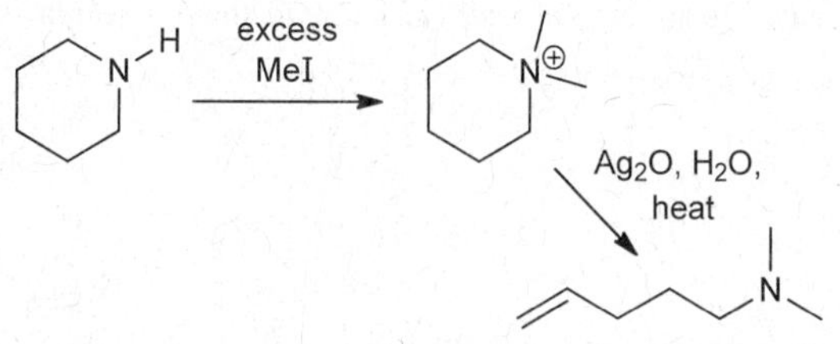

(b) This is a Gabriel synthesis. Since the alkyl halide is ethyl bromide, the product is ethylamine:

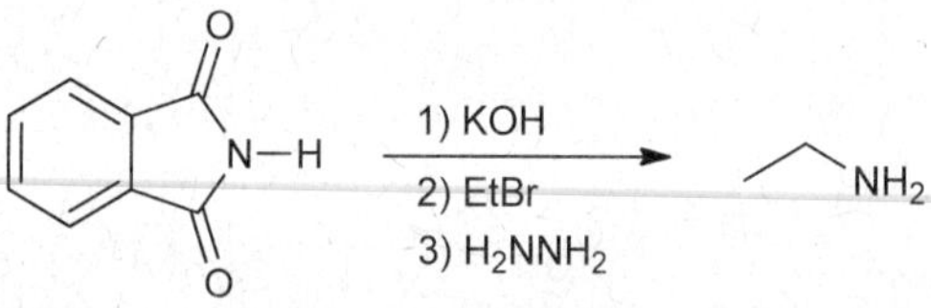

(c) The starting alkyl halide is converted into a nitrile upon treatment with sodium cyanide in a polar aprotic solvent (via an S_N2 process). Hydrolysis of the nitrile with aqueous acid and heat gives a carboxylic acid, which is then converted into an acid chloride upon

treatment with thionyl chloride. The acid chloride is then converted into an amide upon treatment with excess ammonia, via a nucleophilic acyl substitution reaction.

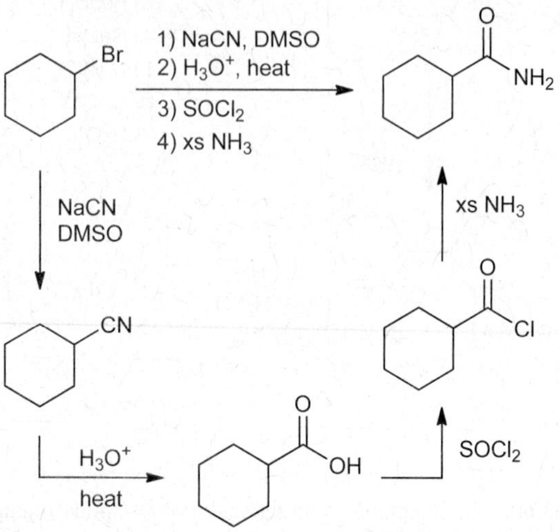

(d) Treating benzene with a mixture of nitric and sulfuric acid results in nitration of the aromatic ring, giving nitrobenzene. Subsequent treatment of nitrobenzene with iron in aqueous acid (followed by basic work-up) results in reduction of the nitro group, giving aniline. Aniline is converted into a diazonium ion upon treatment with sodium nitrite and HCl, and the diazonium ion is converted into benzonitrile via a Sandmeyer reaction (with CuCN):

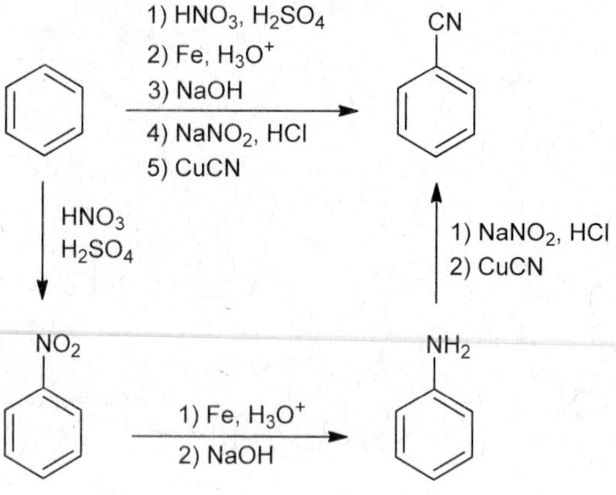

23.63. The carbonyl group is converted into a dimethyl amino group via a reductive amination process. A Hofmann elimination then gives an alkene. The double bond cannot be formed at a bridgehead position (Bredt's rule), so there is only one possible regiochemical outcome for the elimination process. Ozonolysis of the alkene gives a dialdehyde (in a *cis* configuration), which is then converted into a diamine via reductive amination of each carbonyl group.

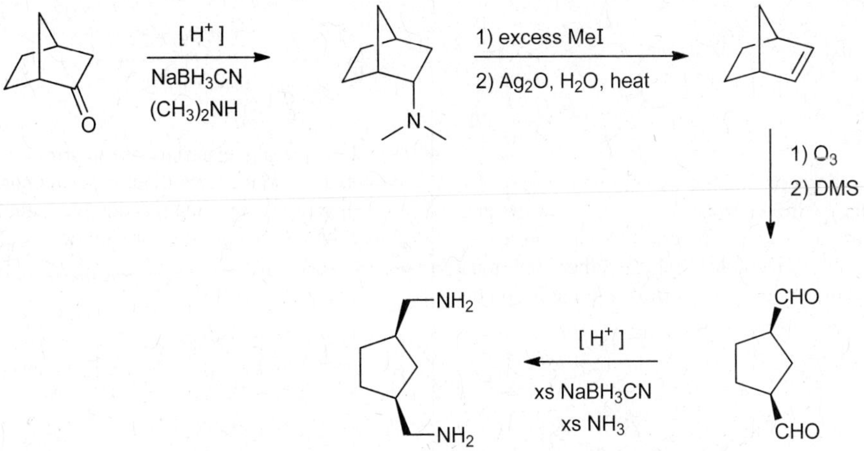

23.64. The conjugate base of pyrrole is highly stabilized because it is an aromatic anion and it is resonance stabilized, spreading the negative charge over all five atoms of the ring, as shown below. Pyrrole is relatively acidic (compared with other amines) because its conjugate base is so highly stabilized.

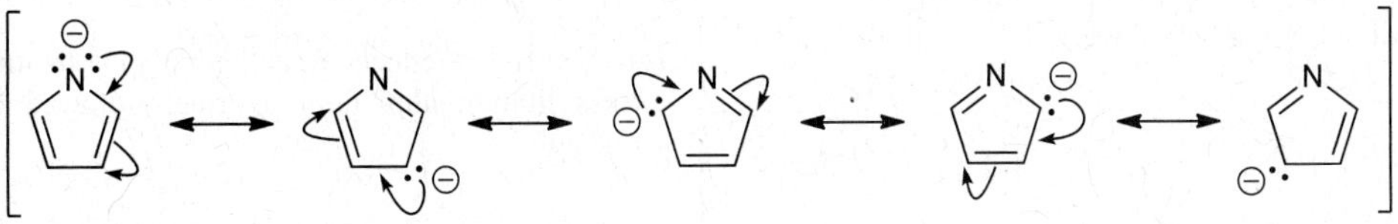

23.65. The desired amine is primary (it has only one C–N bond), and it can be made from the following ketone and ammonia via a reductive amination:

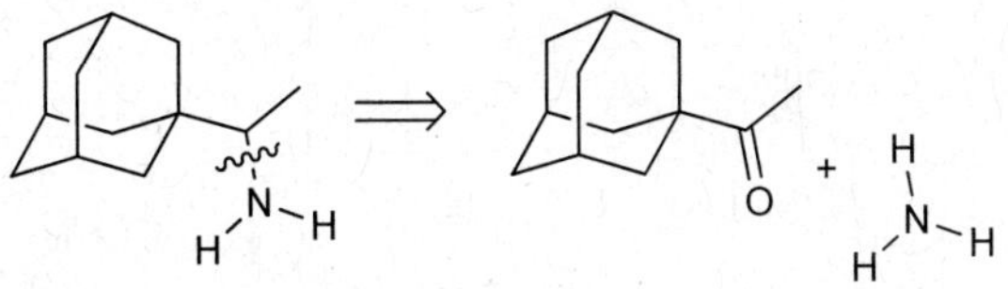

23.66.
(a) The compound has three C-N bonds:

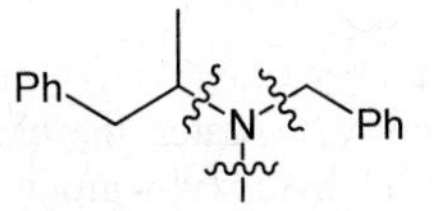

Each of these bonds can be made via a reductive amination, giving three possible synthetic routes. Two are shown below (the third route begins with formaldehyde):

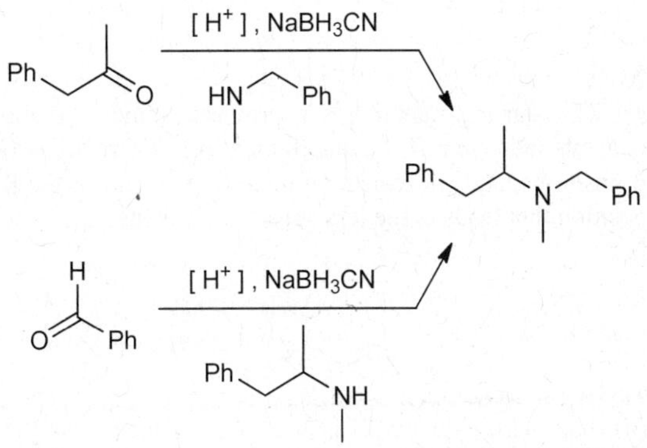

23.67.
(a) The starting amine is primary, so it is converted into the following diazonium salt upon treatment with sodium nitrite and HCl:

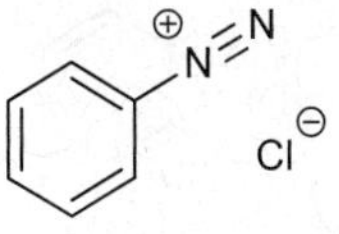

(b) The starting amine is secondary, so it is converted into the following *N*-nitrosamine upon treatment with sodium nitrite and HCl:

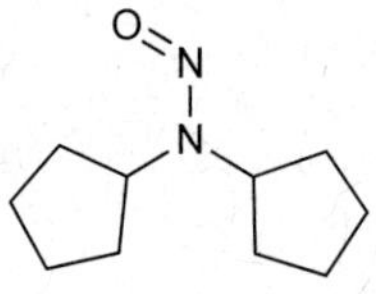

23.68. The starting material has two aromatic rings. The ring bearing the hydroxyl group is more highly activated. During the azo coupling process, the activated ring functions as the nucleophile, and the other ring must function as the diazonium ion, as shown in the following retrosynthetic analysis:

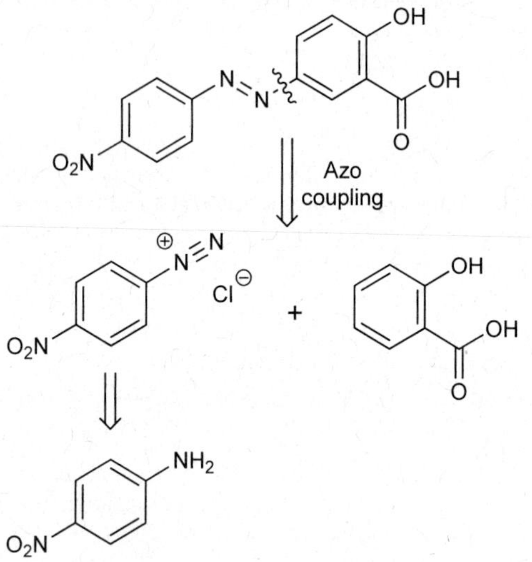

23.69.
(a) The starting material is a primary amine, and the reagents indicate a Hofmann elimination. There are two β positions, and we expect elimination to occur at the β position that leads to the less substituted alkene:

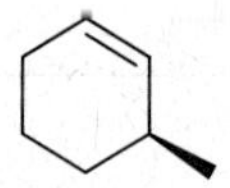

(b) The starting material is a primary amine, and the reagents indicate a Hofmann elimination. There are two β positions, but only one of these positions bears a proton (which is necessary for elimination to occur). So, there is only one possible regiochemical outcome for this Hofmann elimination, giving the following alkene:

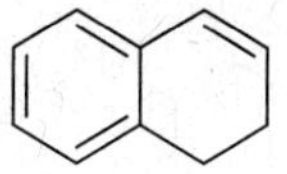

23.70.
(a) Reduction of the nitro group, followed by basic work-up, gives *meta*-bromoaniline:

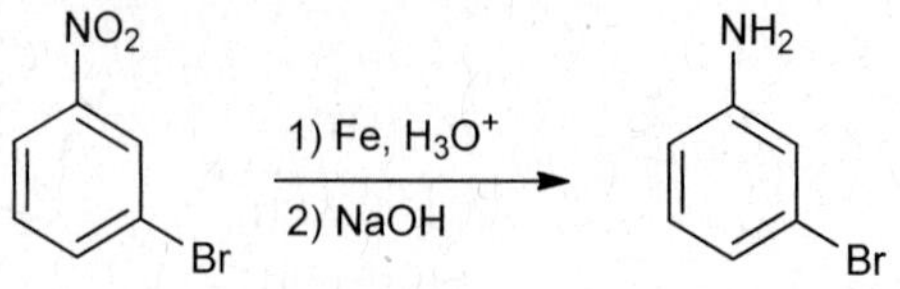

(b) The starting materials are a primary amine and a ketone, and the reagents (sodium cyanoborohydride and an acid catalyst) indicate a reductive amination process, giving the following secondary amine:

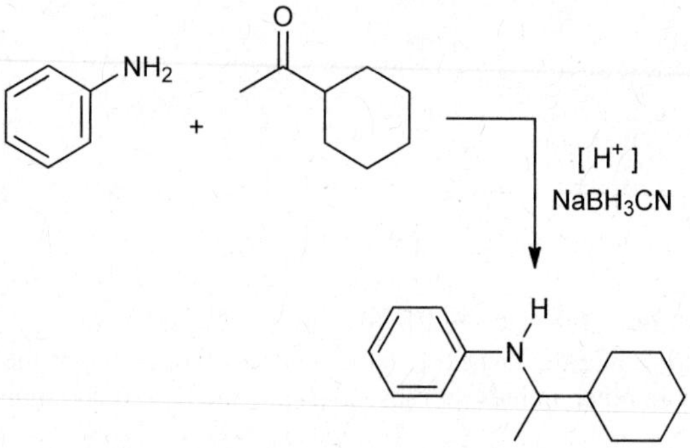

(c) A nitrile is reduced to an amine upon treatment with excess lithium aluminum hydride, followed by water work-up:

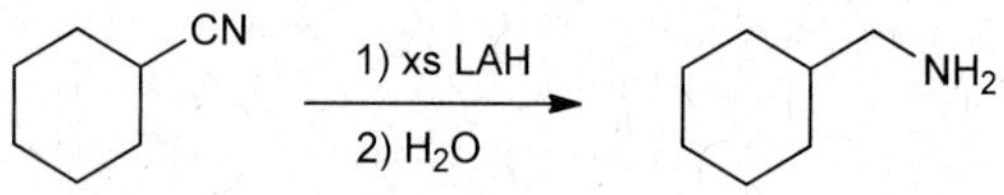

(d) An amide is reduced to an amine upon treatment with excess lithium aluminum hydride, followed by water work-up:

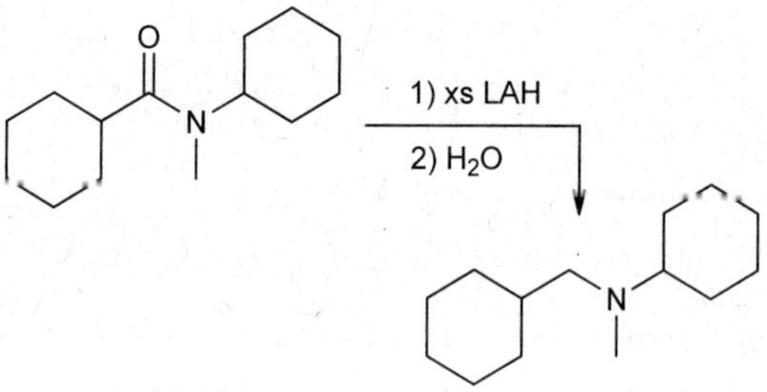

23.71.
(a) Upon treatment with water, the diazonium group is replaced with a hydroxyl group, giving *meta*-bromophenol:

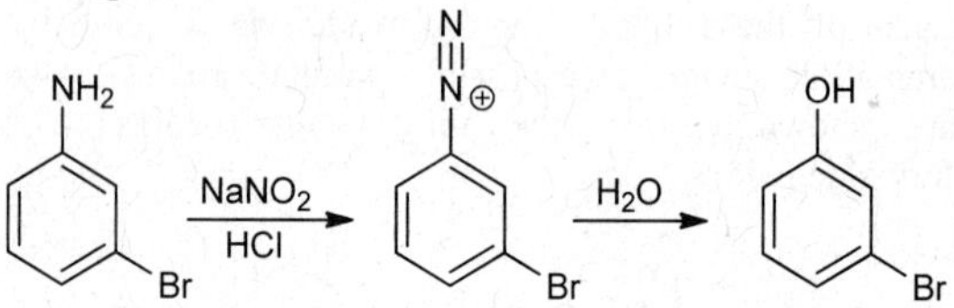

(b) Upon treatment with HBF₄, the diazonium group is replaced with a fluorine atom (via a Schiemann reaction), as shown:

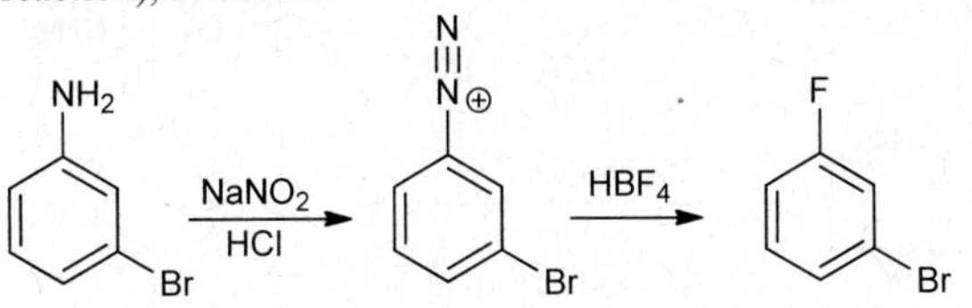

(c) Upon treatment with CuCN, the diazonium group is replaced with a cyano group via a Sandmeyer reaction:

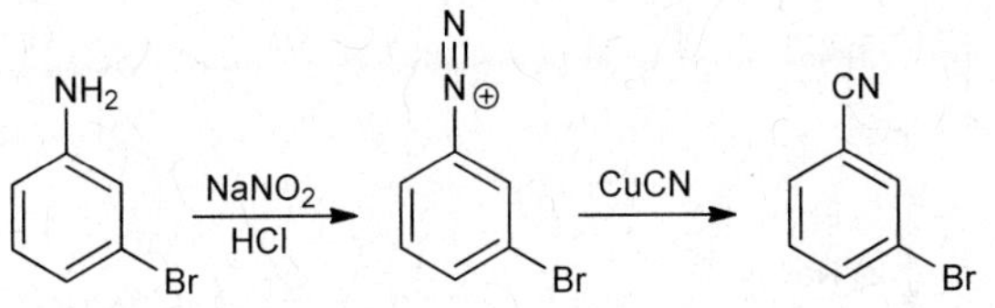

(d) Upon treatment with H₃PO₂, the diazonium group is replaced with a hydrogen atom, giving bromobenzene:

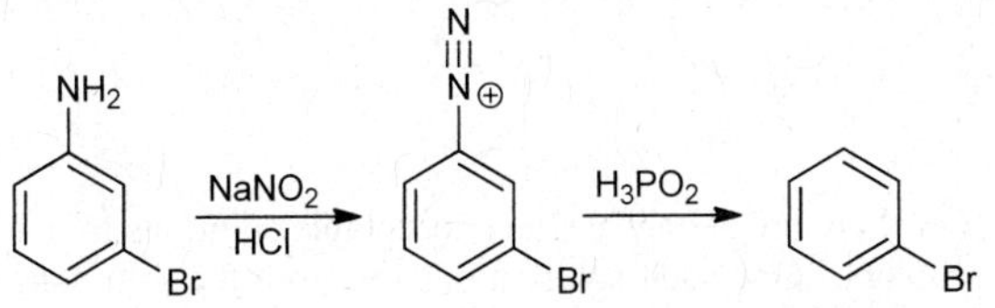

(e) Upon treatment with CuBr, the diazonium group is replaced with a bromine atom via a Sandmeyer reaction:

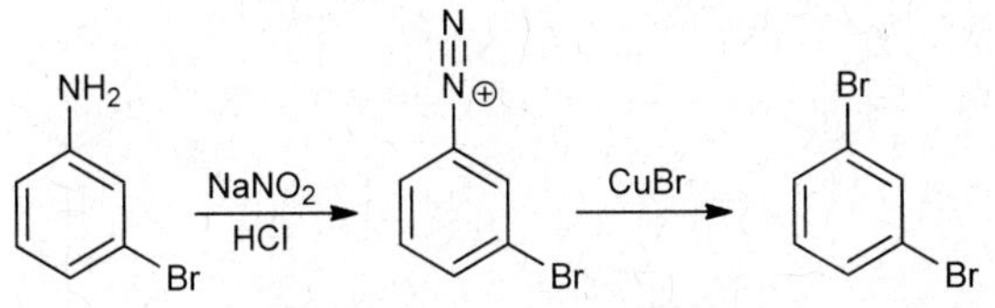

23.72. In this case, the carbonyl group and the amino group are tethered together (both functional groups are present in one compound), so we expect an *intramolecular* reductive amination to occur, thereby forming a new ring to give a bicyclic product, as shown:

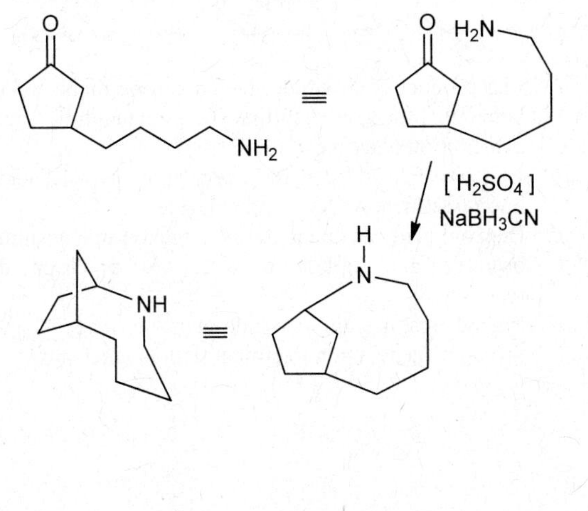

23.73.

(a) There are certainly many acceptable solutions to this problem. One such solution derives from the following retrosynthetic analysis. An explanation of each of the steps (*a-f*) follows.

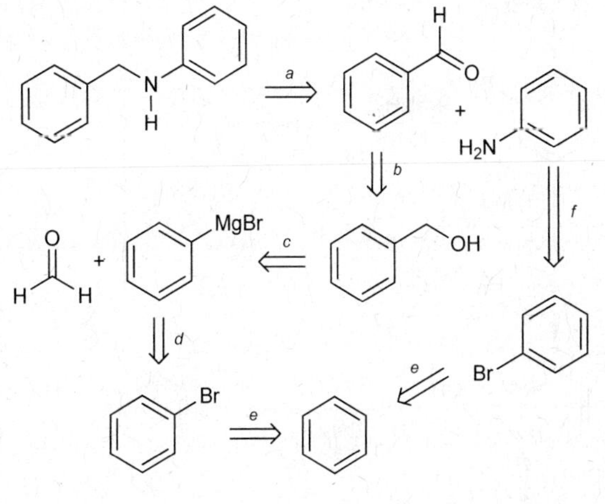

a. The product is a secondary amine, which can be made from benzaldehyde and aniline via a reductive amination.

b. Benzaldehyde can be made from benzyl alcohol via oxidation (with PCC).

c. Benzyl alcohol can be made from phenyl magnesium bromide and formaldehyde, via a Grignard reaction.

d. Phenyl magnesium bromide can be made from bromobenzene, upon treatment with magnesium.

e. Bromobenzene can be made from benzene via an electrophilic aromatic substitution reaction.

f. Aniline can be made from bromobenzene via elimination-addition.

The forward scheme is shown here. Benzene is first converted into bromobenzene upon treatment with bromine in the presence of a Lewis acid (AlBr₃). Treating bromobenzene with magnesium gives a Grignard reagent, which can be further treated with formaldehyde, followed by water work-up, to give benzyl alcohol (via Grignard reaction). Oxidation with PCC gives benzaldehyde, which is then treated with aniline in a reductive amination process to give the product. Aniline can be made from bromobenzene upon treatment with sodium amide in ammonia (via elimination-addition):

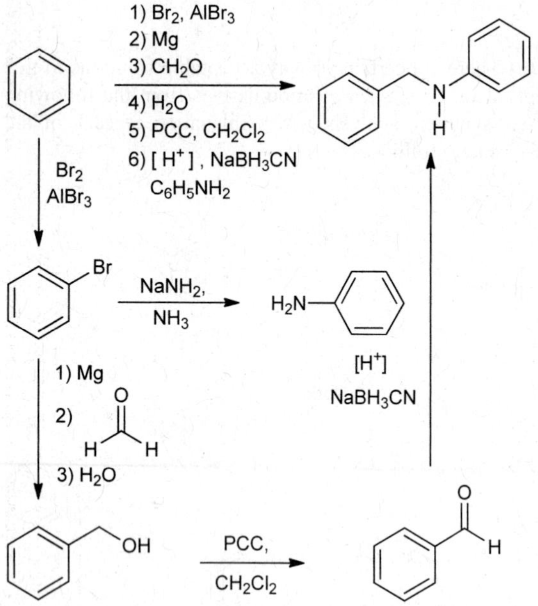

1) Br$_2$, AlBr$_3$
2) Mg
3) CH$_2$O
4) H$_2$O
5) PCC, CH$_2$Cl$_2$
6) [H$^+$] , NaBH$_3$CN
   C$_6$H$_5$NH$_2$

Br$_2$
AlBr$_3$

NaNH$_2$,
NH$_3$

[H$^+$]
NaBH$_3$CN

1) Mg
2) (formaldehyde)
3) H$_2$O

PCC,
CH$_2$Cl$_2$

(b) This transformation requires the installation of an amide group and three chlorine atoms (in the *ortho* and *para* positions). Installation of the amide group can be achieved via a diazonium ion, as shown here:

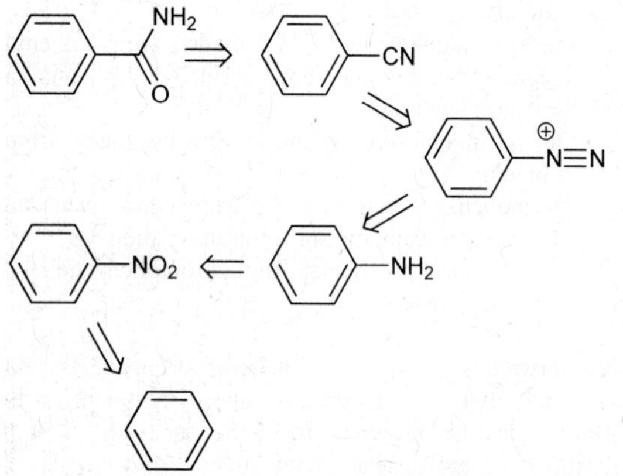

In order to install three chlorine atoms (in the positions that are *ortho* and *para* to the amide group), we must perform chlorination (with excess chlorine) during a stage in the process when the ring is highly activated (thereby giving trichlorination). This can be accomplished immediately prior to making the diazonium ion, because aniline is strongly activated toward electrophilic aromatic substitution, giving trichlorination, as desired.

The forward scheme is shown here. Benzene is first treated with a mixture of sulfuric acid and nitric acid, giving nitrobenzene. Reduction, followed by basic work-up, gives aniline, which will undergo trichlorination when treated with excess chlorine to give 2,4,6-trichloroaniline. Treatment with sodium nitrite and HCl converts the substituted aniline into a diazonium ion, which can then be treated with CuCN to give a

nitrile (via a Sandmeyer reaction). Treating the nitrile with aqueous acid then gives the product.

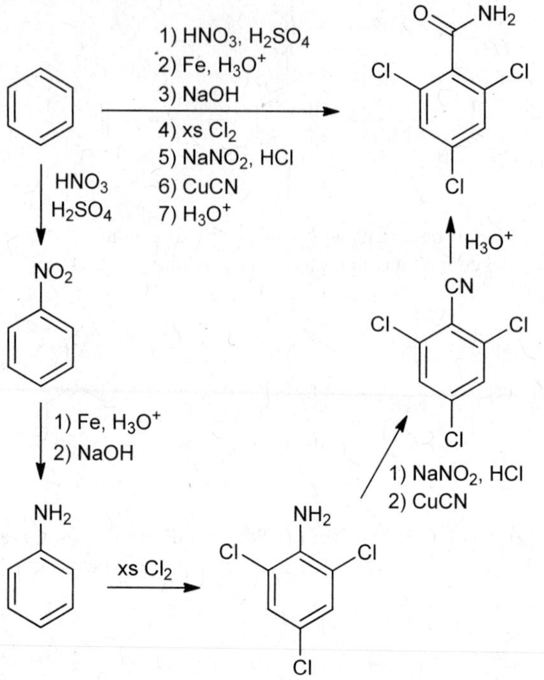

1) HNO$_3$, H$_2$SO$_4$
2) Fe, H$_3$O$^+$
3) NaOH
4) xs Cl$_2$
5) NaNO$_2$, HCl
6) CuCN
7) H$_3$O$^+$

HNO$_3$
H$_2$SO$_4$

NO$_2$

1) Fe, H$_3$O$^+$
2) NaOH

NH$_2$

xs Cl$_2$

H$_3$O$^+$

CN

1) NaNO$_2$, HCl
2) CuCN

(c) There are certainly many acceptable solutions to this problem. One such solution derives from the following retrosynthetic analysis. An explanation of each of the steps (*a-f*) follows.

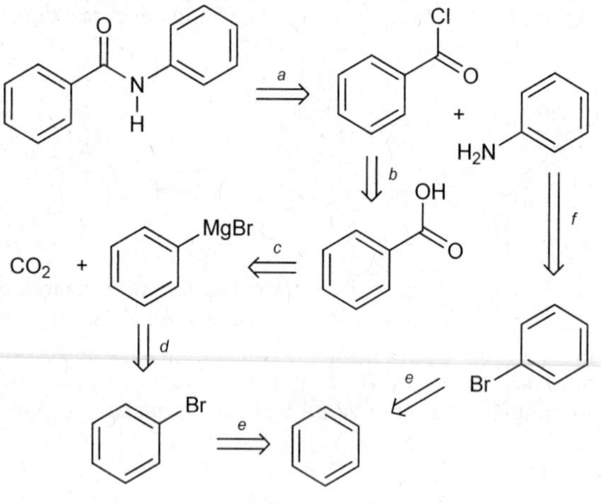

CO$_2$ +

a. The product is an amide, which can be made from benzoyl chloride and aniline via a nucleophilic acyl substitution reaction.

b. Benzoyl chloride can be made from benzoic acid upon treatment with thionyl chloride.

c. Benzoic acid can be made from phenyl magnesium bromide and carbon dioxide, via a Grignard reaction.

d. Phenyl magnesium bromide can be made from bromobenzene, upon treatment with magnesium.

e. Bromobenzene can be made from benzene via an electrophilic aromatic substitution reaction.
f. Aniline can be made from bromobenzene via elimination-addition.

The forward scheme is shown here. Benzene is first converted into bromobenzene upon treatment with bromine in the presence of a Lewis acid ($AlBr_3$). Treating bromobenzene with magnesium gives a Grignard reagent, which can be further treated with carbon dioxide, followed by acidic work-up, to give benzoic acid. Treating benzoic acid with thionyl chloride gives benzoyl chloride, which is then treated with aniline in a nucleophilic acyl substitution to give the product. Aniline can be made from bromobenzene upon treatment with sodium amide in ammonia (via elimination-addition):

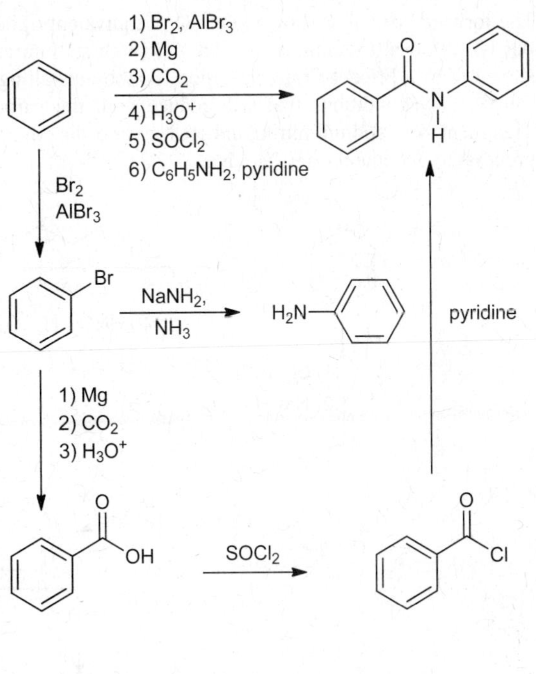
1) Br2, AlBr3
2) Mg
3) CO2
4) H3O+
5) SOCl2
6) C6H5NH2, pyridine
Br2
AlBr3
Br
NaNH2,
NH3
H2N
pyridine
1) Mg
2) CO2
3) H3O+
OH
SOCl2
Cl

(d) There are certainly many acceptable solutions to this problem. One such solution derives from the following retrosynthetic analysis. An explanation of each of the steps (*a-f*) follows.

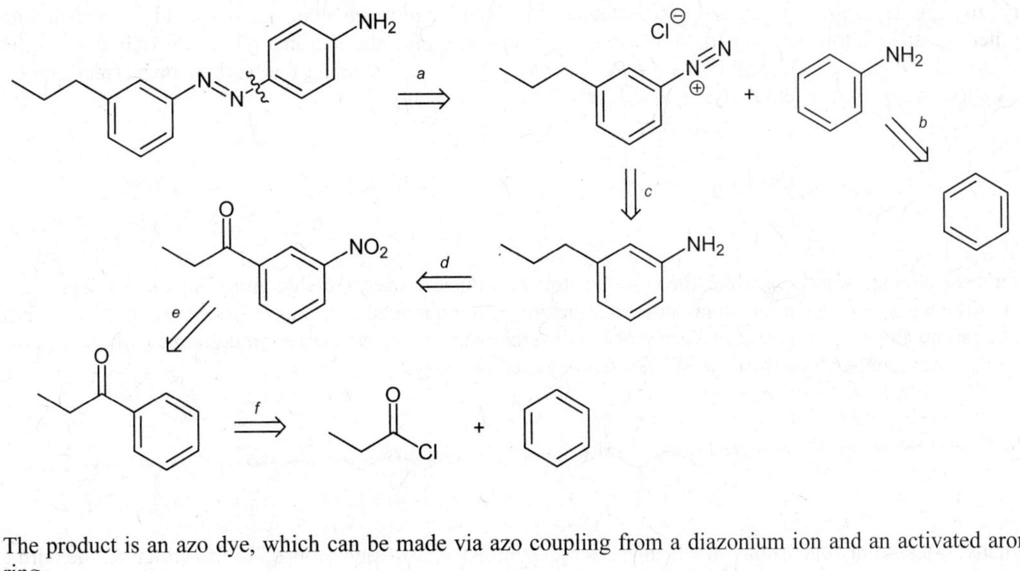
NH2
a
Cl
N≡N
+
b
c
NO2
d
NH2
e
f
Cl
+

a. The product is an azo dye, which can be made via azo coupling from a diazonium ion and an activated aromatic ring.
b. Aniline can be made from benzene via chlorination followed by treatment with sodium amide in liquid ammonia (elimination-addition).
c. The diazonium ion can be made from *meta*-propylaniline, upon treatment with sodium nitrite and HCl.
d. *meta*-Propylaniline can be made via reduction of the disubstituted ring shown.
e. The disubstituted ring can be prepared via nitration of an aromatic ketone (giving nitration at the *meta* position).
f. The aromatic ketone can be made via a Friedel-Crafts acylation.

The forward scheme is shown here. One equivalent of benzene is converted into aniline via chlorination (with Cl_2 and $AlCl_3$) followed by elimination-addition (with sodium amide in liquid ammonia). Another equivalent of benzene is subjected to a Friedel-Crafts acylation (thereby installing an acyl group), followed by nitration (in the *meta* position), followed by conditions that will reduce both the carbonyl group and the nitro group, giving *meta*-propylaniline. Treatment with sodium nitrite and HCl gives a diazonium ion, which is then treated with aniline (in an azo compling process) to give the desired azo dye.

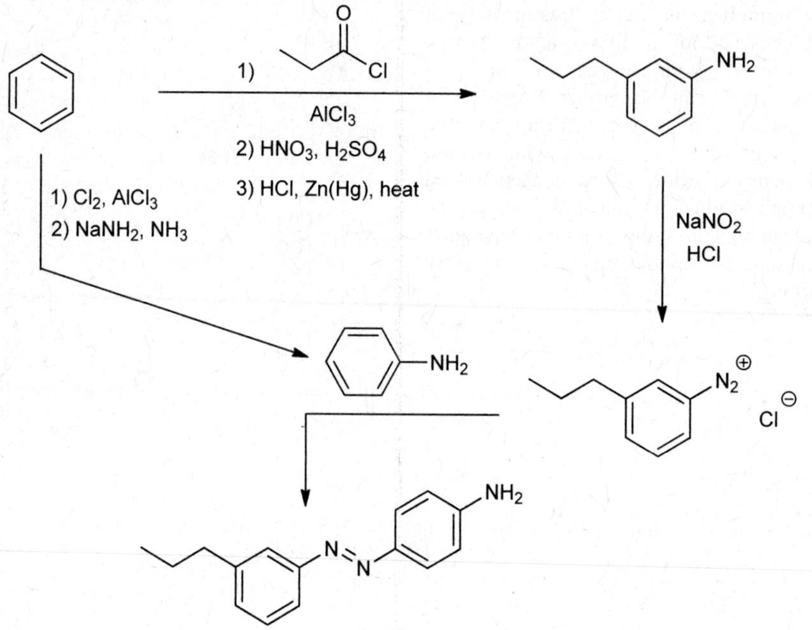

23.74. The molecular formula ($C_6H_{15}N$) indicates no degrees of unsaturation (see Section 15.16), so all of the isomers must be saturated, acyclic amines. The IR data indicates that we are looking for structures that lack an N-H bond (i.e., tertiary amines). Let's first consider all tertiary amines in which the nitrogen atom is connected to two methyl groups. Since there must be a total of six carbon atoms in each structure (and the two methyl groups only account for two carbon atoms), we must consider all of the different ways in which the remaining four carbon atoms can be connected. There are four ways, shown here (butyl, isobutyl, *sec*-butyl, and *tert*-butyl):

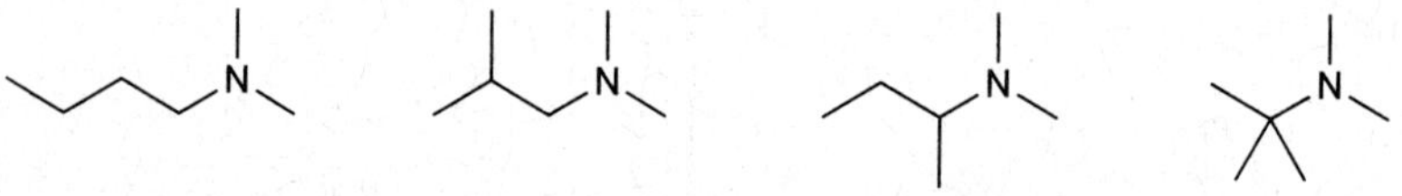

Now let's consider all isomers in which the nitrogen atom is connected to one methyl group and one ethyl group. Since there must be a total of six carbon atoms in each structure (while one methyl group and one ethyl group only account for three carbon atoms), we must consider all of the different ways in which the remaining three carbon atoms can be connected. There are only two ways, shown here (propyl and isopropyl):

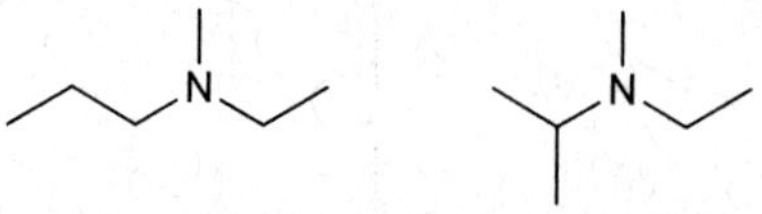

And finally, there is only one isomer in which the nitrogen atom has two ethyl groups. In this structure, the third group is also an ethyl group, giving triethylamine:

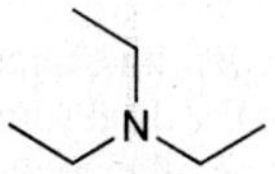

In total, we have seen seven different tertiary amines with the molecular formula $C_6H_{15}N$.

23.75. The compound has two nitrogen atoms. One of the nitrogen atoms (adjacent to the aromatic ring) exhibits a delocalized lone pair, while the other nitrogen atom (of the NH_2 group) exhibits a localized lone pair. The localized lone pair is more nucleophilic than the delocalized lone pair, so only the NH_2 group is converted into a quaternary ammonium ion, as shown here:

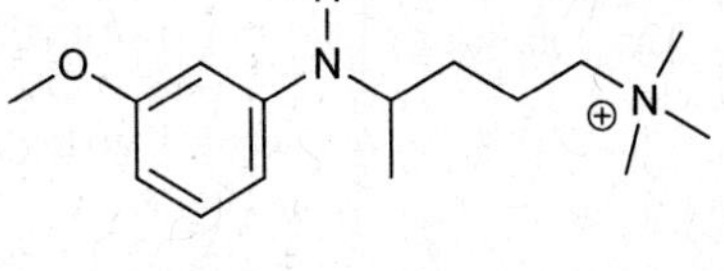

23.76. Pyrrole functions as a nucleophile (preferentially at C2, as discussed in Section 23.12) and attacks acetyl chloride, giving a tetrahedral intermediate that can then expel a chloride ion (as a leaving group), thereby regenerating a carbonyl group. The resulting cation is resonance-stabilized, much like a sigma complex. Pyridine then functions as a base and removes a proton, thereby restoring aromaticity and generating the product:

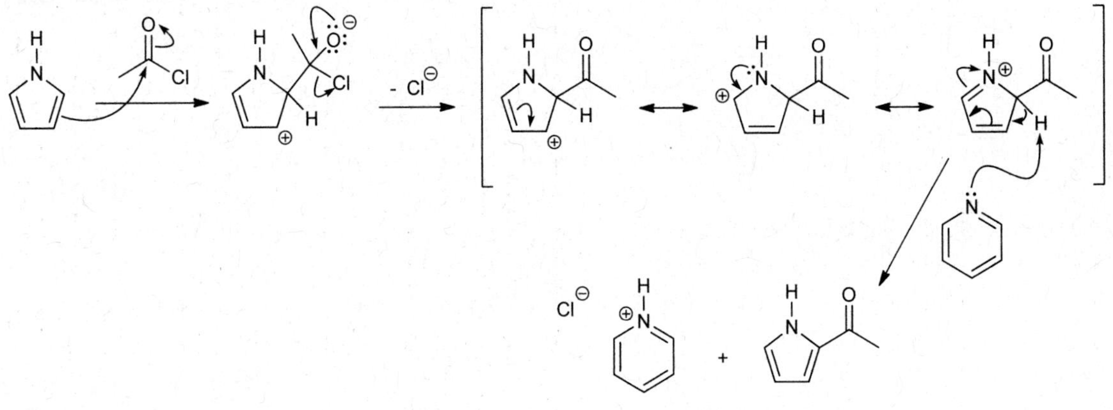

23.77. The molecular formula ($C_5H_{13}N$) indicates no degrees of unsaturation (see Section 15.16), so all of the isomers must be saturated, acyclic amines. The IR data indicates that we are looking for structures that lack an N-H bond (i.e., tertiary amines). As seen in the solution to Problem **23.47**, there are three isomers that fit this description:

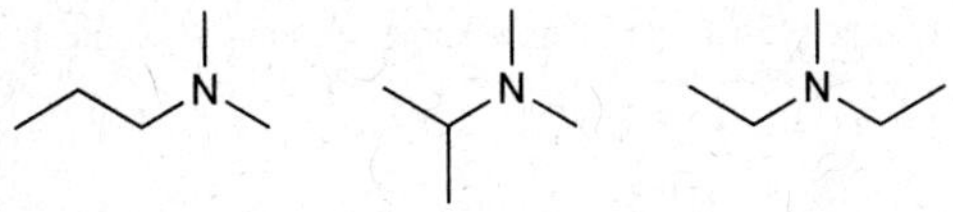

The first compound above is expected to exhibit four signals in its ^1H NMR spectrum. Only the latter two isomers are expected to produce three signals in their ^1H NMR spectra:

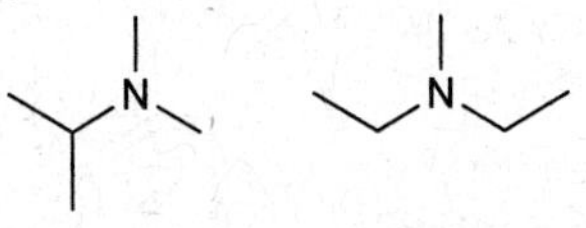

23.78. The molecular formula ($C_8H_{17}N$) indicates one degree of unsaturation (see Section 15.16), so the structure must contain either one double bond or one ring (but not both). We are given the product obtained when coniine is subjected to a Hofmann elimination, which

allows us to determine the structure of coniine, as shown in the following retrosynthetic analysis:

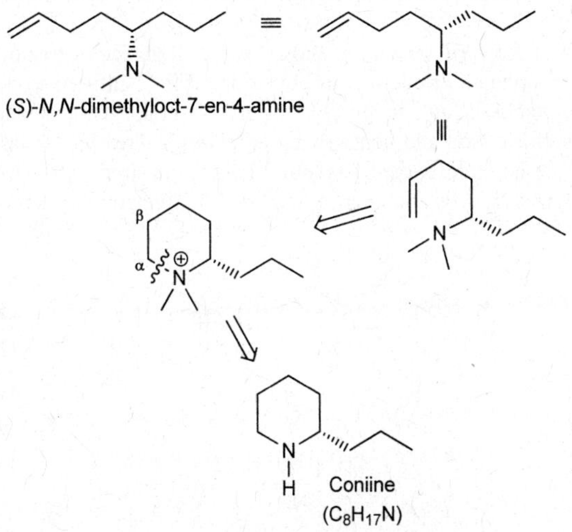

(S)-N,N-dimethyloct-7-en-4-amine

Coniine
($C_8H_{17}N$)

The N-H bond in coniine is responsible for the one peak above 3000 cm^{-1} in the IR spectrum.

23.79. The molecular formula ($C_4H_{10}N_2$) indicates one degree of unsaturation (see Section 15.16), so the structure must contain either one double bond or one ring (but not both). The ^{1}H NMR spectrum has only two signals, indicating a high degree of symmetry. One of these signals vanishes in D$_2$O, indicating a labile proton, consistent with an N-H bond. The other signal must account for all of the other protons. The following structure accounts for all of the observations:

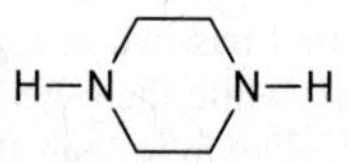

The two N-H protons are identical (because of symmetry), and they produce one signal in the ^{1}H NMR spectrum. The four methylene (CH$_2$) groups are all identical, giving rise to the second signal in the ^{1}H NMR spectrum.

23.80.   There are certainly many acceptable solutions to this problem. One such solution derives from the following retrosynthetic analysis, in which the final step of the synthesis employs the strategy described in the problem statement (opening an epoxide with an amine functioning as the nucleophile). An explanation of each of the steps (*a-f*) follows.

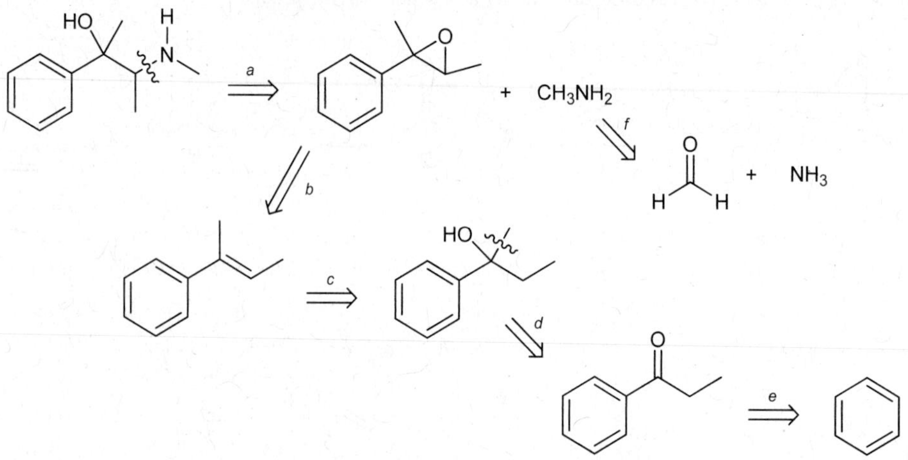

a. The product can be made by treating the appropriate epoxide with methyl amine, as described in the problem statement.

b. The epoxide can be made from the corresponding alkene, upon treatment with a peroxy acid, such as MCPBA.

c. The alkene can be made from an alcohol, via acid-catalyzed dehydration.

d. The alcohol can be made from a ketone, via a Grignard reaction (with methyl magnesium bromide, followed by water work-up).

e. The ketone can be made from benzene via a Friedel-Crafts acylation.

f. Methylamine can be made from formaldehyde and ammonia via a reductive amination.

The forward scheme is shown here. Benzene is treated with an acyl chloride and AlCl$_3$, thereby installing an acyl group via a Friedel-Crafts acylation. The resulting ketone is then treated with methyl magnesium bromide, followed by water work-up, to give a tertiary alcohol. This alcohol undergoes dehydration upon treatment with concentrated sulfuric acid and heat, giving an alkene. Treating the alkene with a peroxy acid, such as MCPBA, gives an epoxide. The epoxide is then converted into the desired product upon treatment with methyl amine (which can be made from formaldehyde and ammonia via a reductive amination process).

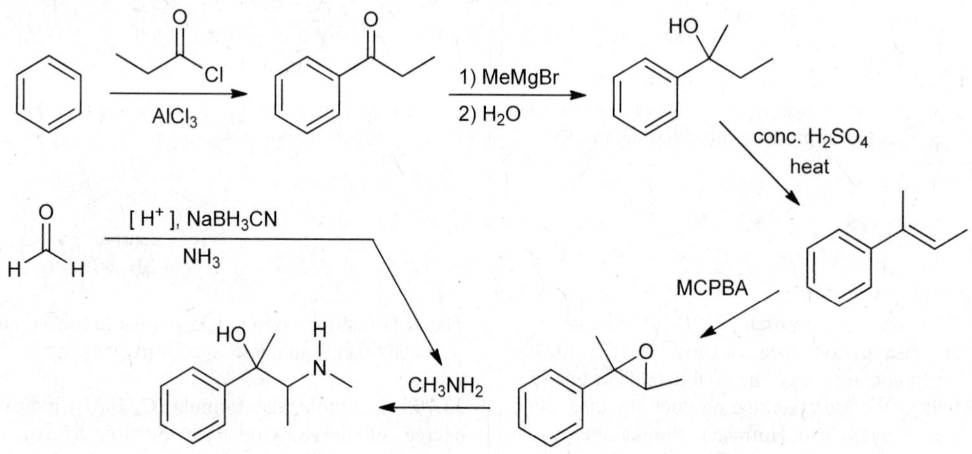

23.81. There are certainly many acceptable solutions to this problem. One such solution derives from the following retrosynthetic analysis. An explanation of each of the steps (*a-e*) follows.

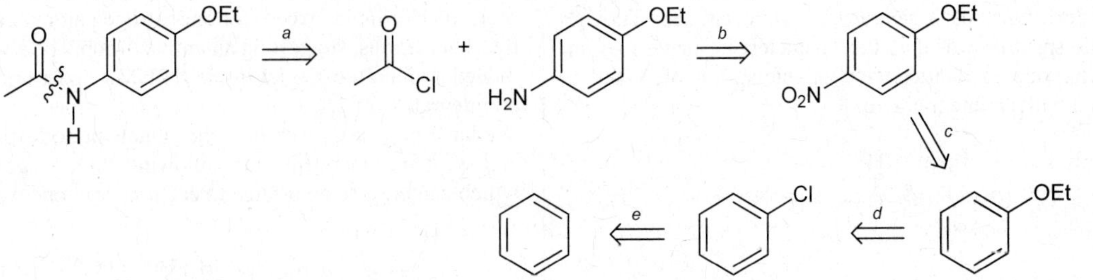
OEt
a
b
c
d
e
+
Cl
H₂N
O₂N
N
H
O

a. The product has an amide group, which can be made via acetylation of the amino group in 4-ethoxyaniline.
b. 4-Ethoxyaniline can be made from the corresponding nitro compound via reduction.
c. The nitro compound can be made from ethoxybenzene, via nitration of the aromatic ring (in the *para* position).
d. Ethoxybenzene can be made from chlorobenzene, via elimination-addition (upon treatment with hydroxide at high temperature), followed by a Williamson ether synthesis.
e. Chlorobenzene can be made from benzene via an electrophilic aromatic substitution reaction.

The forward scheme is shown here. Benzene is treated with Cl_2 and $AlCl_3$, thereby installing a chlorine atom. Heating chlorobenzene (at 350°C) in the presence of hydroxide gives an elimination-addition process that gives a phenolate ion as the product. Rather than protonating the phenolate ion to give phenol, we can treat the phenolate ion with ethyl iodide, giving ethoxybenzene via an S_N2 process. When ethoxybenzene is treated with a mixture of sulfuric acid and nitric acid, nitration occurs at the *para* position (the ethoxy group is an *ortho-para* director, and the *para* position is favored because of steric factors). Reduction of the nitro group, followed by acetylation of the resulting amino group, gives the desired product.

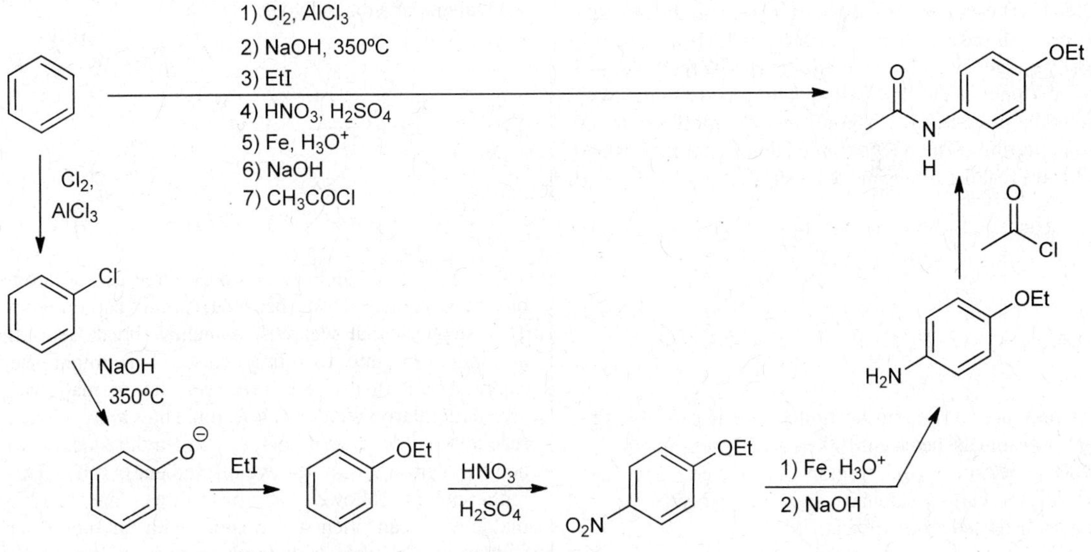
1) Cl₂, AlCl₃
2) NaOH, 350°C
3) EtI
4) HNO₃, H₂SO₄
5) Fe, H₃O⁺
6) NaOH
7) CH₃COCl
Cl₂,
AlCl₃
Cl
NaOH
350°C
EtI
HNO₃
H₂SO₄
1) Fe, H₃O⁺
2) NaOH
OEt
O₂N
H₂N
N
H
O
Cl

23.82. The starting material can lose both carbon dioxide and nitrogen gas, as shown below, to give a very reactive benzyne intermediate. This intermediate can then react with furan in a Diels-Alder reaction, as first described in Section 19.14, to give the cycloadduct shown.

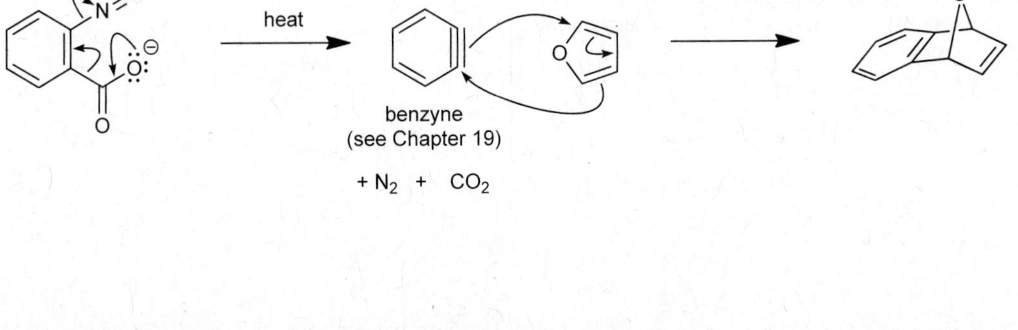
heat
benzyne
(see Chapter 19)
+ N₂ + CO₂

23.83. The molecular formula ($C_6H_{15}N$) indicates no degrees of unsaturation (see Section 15.16), so the structure does not contain a π bond or a ring. That is, the structure must be a saturated, acyclic amine. The ^1H NMR spectrum exhibits the characteristic pattern of an ethyl group (a quartet with an integration of 2 and a triplet with an integration of 3).

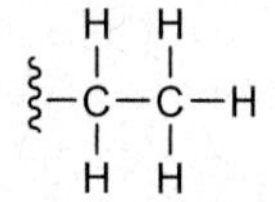

There are no other signals in the ^1H NMR spectrum, indicating a high degree of symmetry. That is, these two signals must account for all fifteen protons in the compound, indicating that there are three equivalent ethyl groups. The compound is therefore triethylamine:

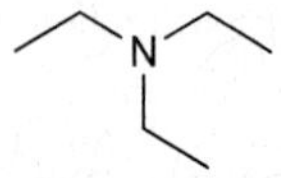

This analysis is confirmed by the ^{13}C NMR spectrum, which has only two signals (one signal for the three equivalent methyl groups and another signal for the three equivalent methylene groups).

23.84. The molecular formula ($C_8H_{11}N$) indicates four degrees of unsaturation (see Section 15.16), which is highly suggestive of an aromatic ring. The multiplet just above 7 ppm in the ^1H NMR spectrum corresponds with aromatic protons, which confirms the presence of an aromatic ring. The integration of this signal is 5, which indicates that the ring is monosubstituted:

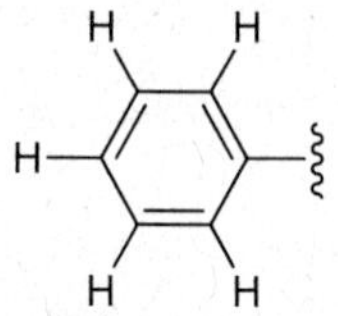

The presence of a monosubstituted ring is confirmed by the four signals between 100 and 150 ppm in the ^{13}C NMR spectrum (the region associated with sp^2 hybridized carbon atoms), just as expected for a monosubstituted aromatic ring:

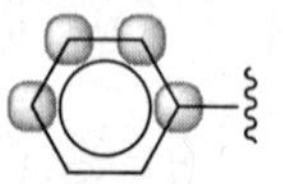

In the ^1H NMR spectrum, the pair of triplets (each with an integration of 2) indicates a pair of neighboring methylene groups:

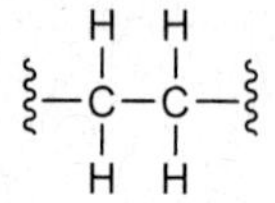

These methylene groups account for the two upfield signals in the ^{13}C NMR spectrum.

Thus far, we have accounted for all of the atoms in the molecular formula, except for one nitrogen atom and two hydrogen atoms, suggesting an amino group. This would indeed explain the singlet in the ^1H NMR spectrum with an integration of 2.

We have now analyzed all of the signals in both spectra, and we have uncovered the following three fragments, which can only be connected to each other in one way:

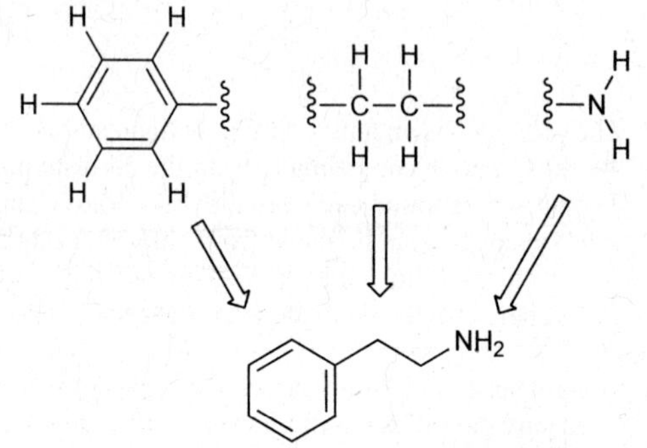

23.85. The product has four C-N bonds, each of which can be prepared via a reductive amination process. As such, the product can be made from the following starting materials (two equivalents of ammonia and two equivalents of a dialdehyde):

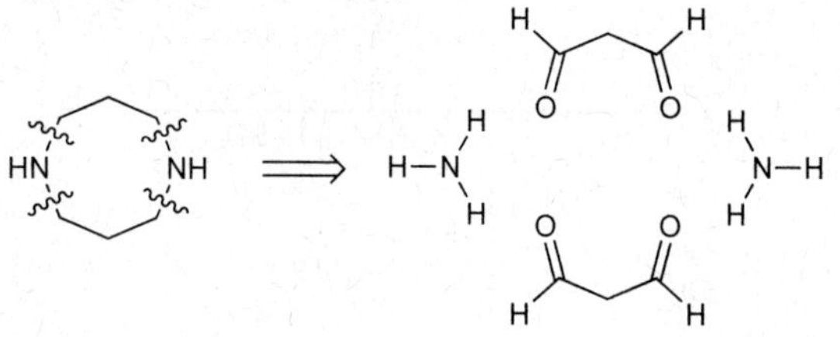

The necessary dialdehyde has only three carbon atoms, but the starting material (benzene) has six carbon atoms. This suggests that we must somehow break apart the aromatic ring into two fragments. This might seem impossible at first, as we have seen that aromatic rings are particularly stable. We did, however, cover a reaction that destroys aromaticity (a Birch reduction will convert benzene into 1,4-cyclohexadiene). If a Birch reduction is followed by ozonolysis, the resulting dialdehyde can then be treated with ammonia and sodium cyanoborohydride (with an acid catalyst) to give the product:

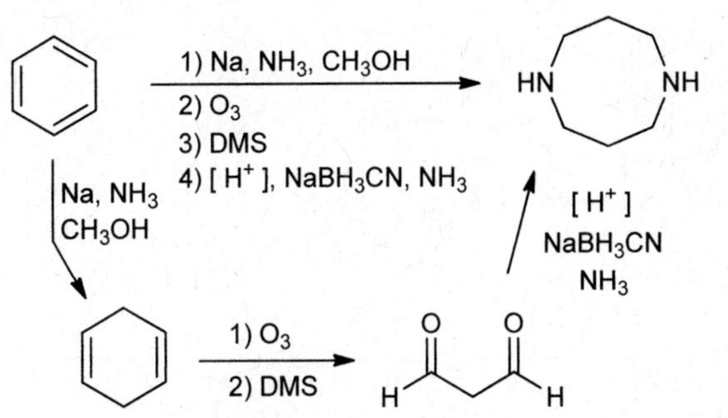

23.86. The starting material exhibits a five-membered ring, while the product exhibits a six-membered ring that contains a nitrogen atom. Since we have not learned a way to insert a nitrogen atom into an existing ring, we must consider opening the ring, and then closing it back up again (in a way that incorporates the nitrogen atom into the ring). There are certainly many acceptable synthetic routes. One such route derives from the following retrosynthetic analysis. An explanation of each of the steps (*a-d*) follows.

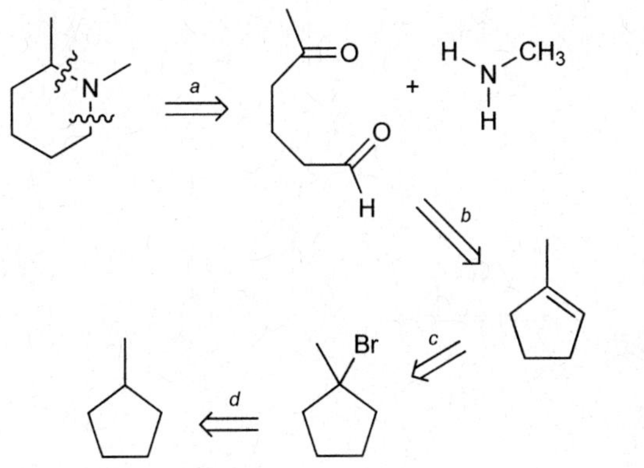

a. The product is a tertiary amine, and it can be made from a dicarbonyl compound and methyl amine, via two reductive amination processes.

b. The dicarbonyl compound can be made via ozonolysis of 1-methylcyclopentene.

c. 1-Methylcyclopentene can be made from 1-bromo-1-methylcyclopentane via an elimination reaction.

d. 1-Bromo-1-methylcyclopentane can be made from methylcyclopentane via radical bromination.

The forward scheme is shown here. Methylcyclopentane will undergo radical bromination selectively at the tertiary position, giving a tertiary alkyl bromide. This alkyl bromide will undergo an elimination reaction upon treatment with a strong base, such as sodium ethoxide. Ozonolysis of the resulting alkene gives a dicarbonyl compound, which can then be converted into the product upon treatment with methyl amine and sodium cyanoborohydride (with acid catalysis):

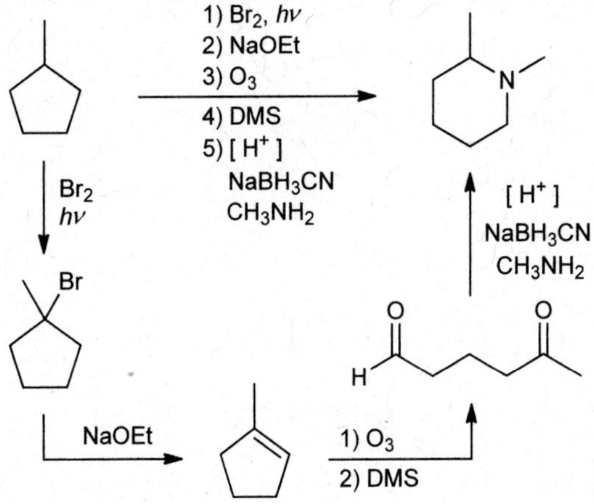

23.87. Sodium nitrite is protonated upon treatment with HCl, giving nitrous acid (HONO), which can be further protonated under these acidic conditions. The resulting cation can lose water (an excellent leaving group), giving a nitrosonium ion, as shown:

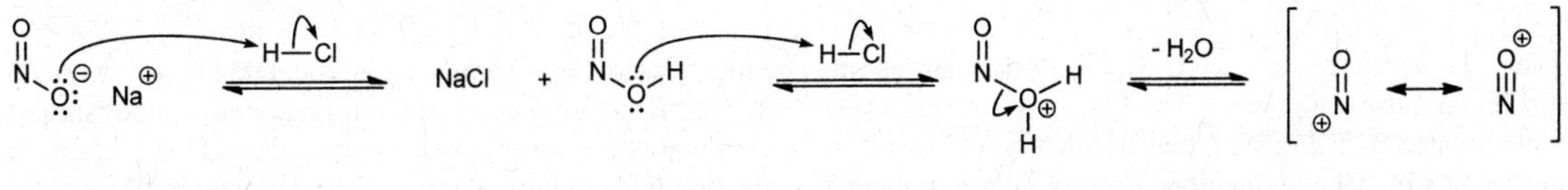

The amino group attacks the nitrosonium ion, giving a cation, which then loses a proton to give an intermediate *N*-nitrosamine. Protonation, followed by deprotonation, gives a tautomer of the *N*-nitrosamine. Protonation of this tautomer, followed by loss of a leaving group (water) gives a diazonium ion. Loss of the diazonium group (as N_2 gas) would generate a primary carbocation, which is unlikely to occur because of the high energy cost associated with primary carbocations. However, a hydride shift can occur at the same time as the leaving group leaves (see the discussion at the very end of Section 7.6), giving a secondary carbocation. A subsequent methyl shift generates a more stable, tertiary benzylic carbocation. Finally, deprotonation gives the product. Notice that in acidic conditions, water functions as the base for all deprotonation steps, rather than hydroxide (which is not measurably present in acidic conditions).

23.88. Protonation of the highlighted nitrogen atom results in a cation that is highly resonance stabilized. Protonation of either of the other nitrogen atoms would not result in a resonance-stabilized cation.

23.89. Two steps are required. The secondary amine must be methylated to give a tertiary amine, and the halogen (Cl) must be replaced with azide. The first step can be achieved via a reductive amination (the nitrogen atom cannot simply be methylated by using MeI, because that would result in over-alkylation, giving the quaternary salt, R_4N^+ I^-). Then, in the second step, Cl can function as a leaving group in an S_N2 reaction with sodium azide, to afford compound **2**.

23.90. We begin by identifying the bond that must be made:

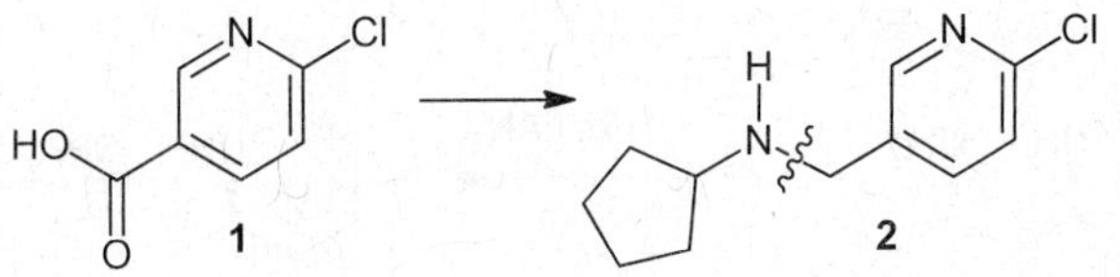

Forming this C-N bond will require the use of cyclopentyl amine:

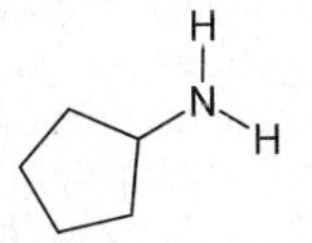

In order to form the desired C-N bond, the carboxylic acid can be reduced to an alcohol (which can be achieved with LAH), but then we must decide what to do with the alcohol.

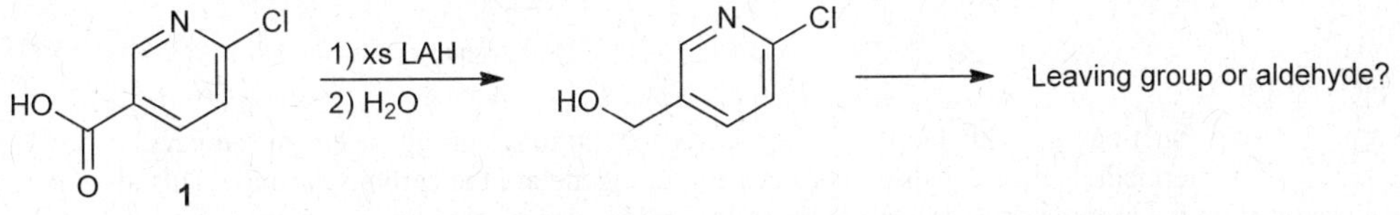

1) xs LAH
2) H$_2$O

Leaving group or aldehyde?

We can either convert the alcohol into a leaving group (OTs, Cl, Br) and then treat it with cyclopentyl amine in an S$_N$2 reaction, or we can convert the alcohol into an aldehyde and then perform a reductive amination with cyclopentyl amine:

S$_N$2

reductive
amination

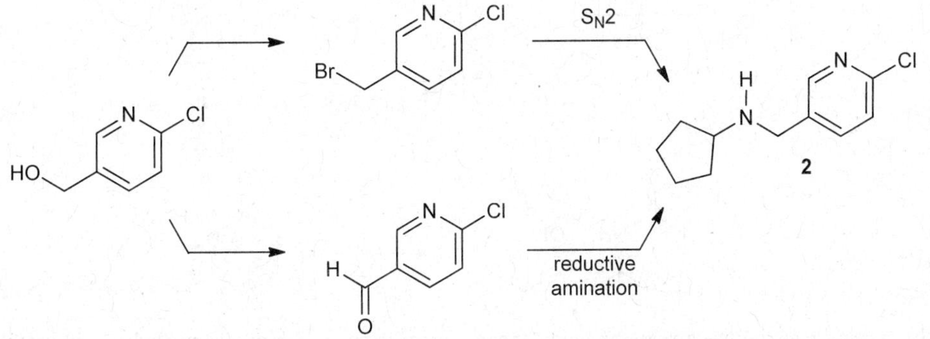

The first path is expected to be inefficient, because it will be difficult to achieve monoalkylation. More likely, polyalkylation will occur, especially with the activated benzylic leaving group. The second path is expected to be more efficient, which gives the following synthesis:

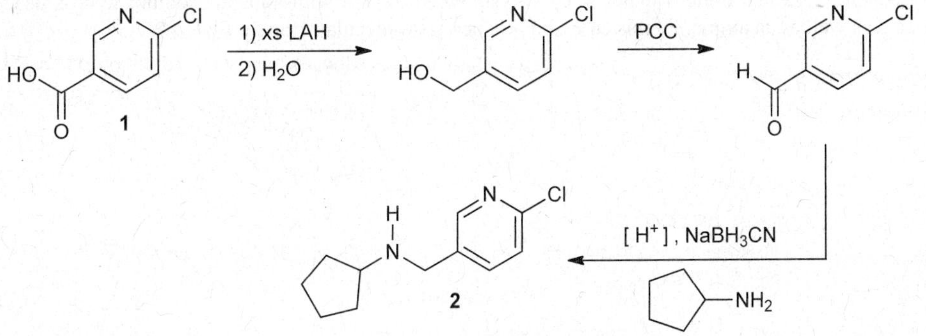

1) xs LAH
2) H$_2$O

PCC

[H$^+$], NaBH$_3$CN

23.91.

(a) An amide is converted to an amine upon treatment with xs LAH:

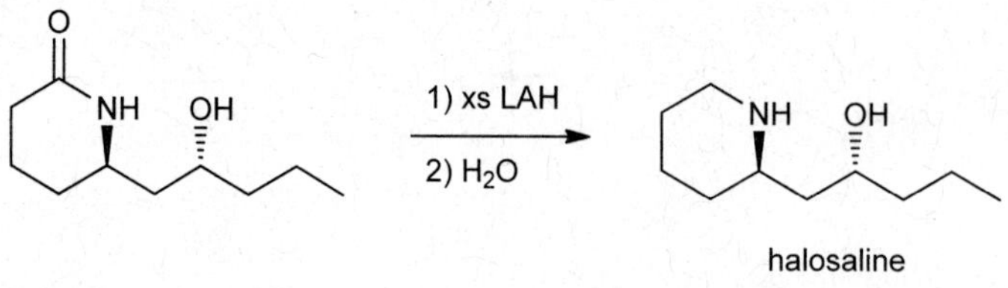

halosaline

(b) In Section 23.5, we learned that LAH is able to reduce an azide to a primary amine. However, in addition to an azide, the starting material also contains a lactone (cyclic ester) functional group. Since LAH is a nonselective reducing agent, it will also reduce the lactone to a diol under these reaction conditions:

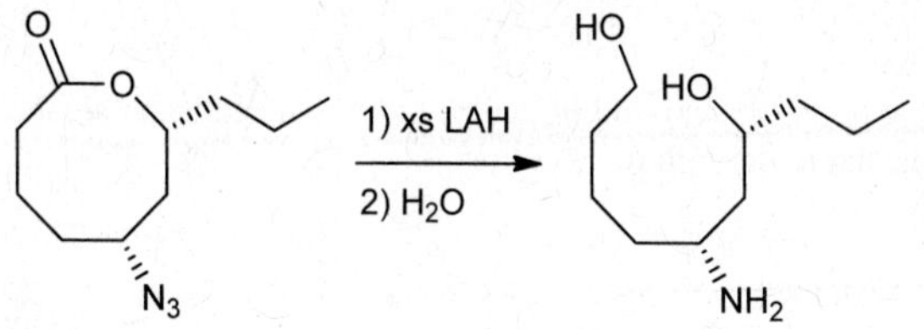

(c) The amino group functions as a nucleophile and attacks the carbonyl group in an intramolecular process. The resulting tetrahedral intermediate then expels an alkoxide ion to regenerate the carbonyl group. This alkoxide ion then removes a proton from the protonated amide group, in an intramolecular process, giving the product.

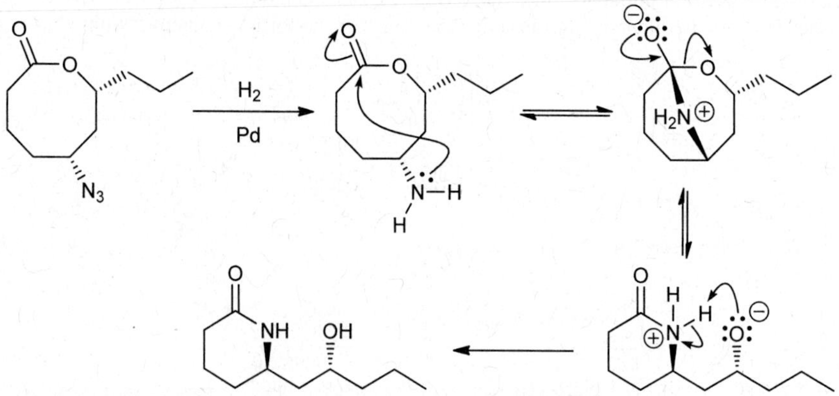

23.92.   The starting material has a primary amino group. When treated with excess methyl iodide, it will undergo exhaustive methylation to produce a quaternary ammonium salt. When this salt is treated with NaOH, first an anion exchange occurs, followed by an E2 elimination to produce an alkene (Hofmann elimination). At this stage it would be useful to take inventory of the other functional groups in the molecule. Compound **5** also contains an ethyl ester. We saw in Section 21.11 that esters will undergo saponification when treated with aqueous sodium hydroxide. We also learned in Section 21.12 that amides undergo hydrolysis when treated with aqueous sodium hydroxide, to afford a carboxylic acid as well as an amine. In this case, the product has molecular formula $C_{10}H_{17}NO_2$.

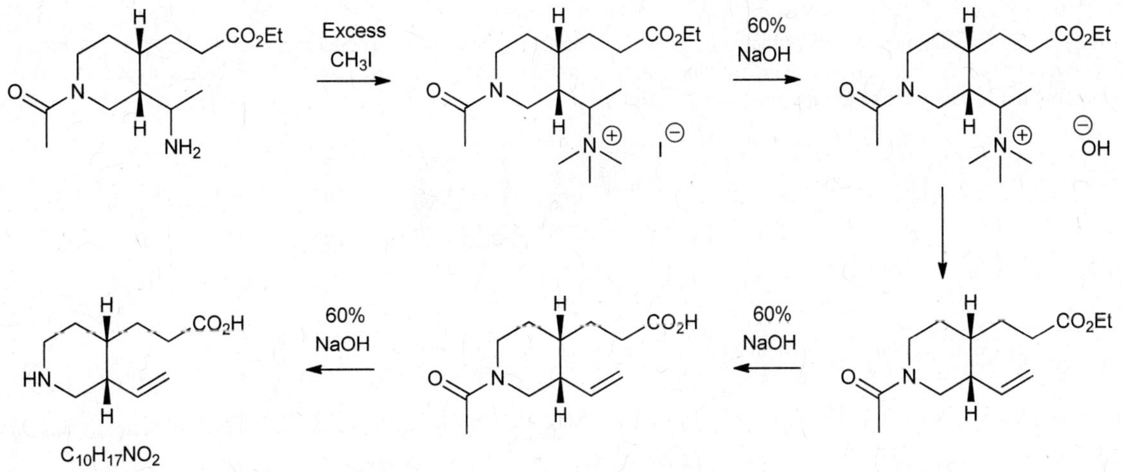

$C_{10}H_{17}NO_2$

23.93.

(a) The ketone group of compound **2** undergoes reductive amination (the ester group is unreactive under these conditions) to give compound **3**, as shown below:

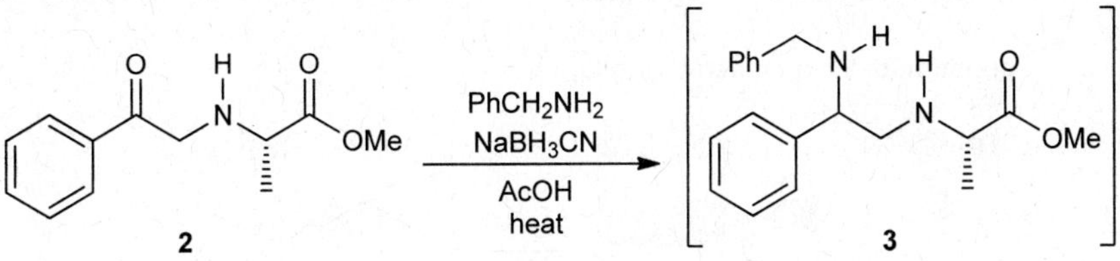

Notice that compound **2** has only one chirality center (highlighted below), and its configuration is not affected during the conversion of **2** to **3**. Compound **3** has two chirality centers (highlighted below), because a new chirality center is created during the conversion of **2** to **3**:

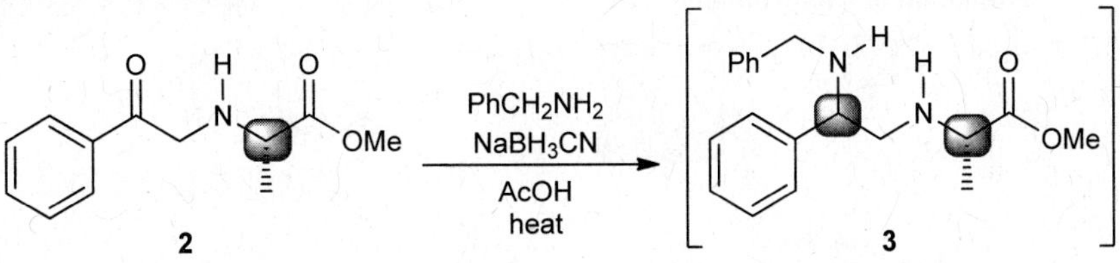

This new chirality center is formed when an iminium ion (formed from compound **2**) is reduced by the hydride reducing agent (NaBH₃CN) to give **3**:

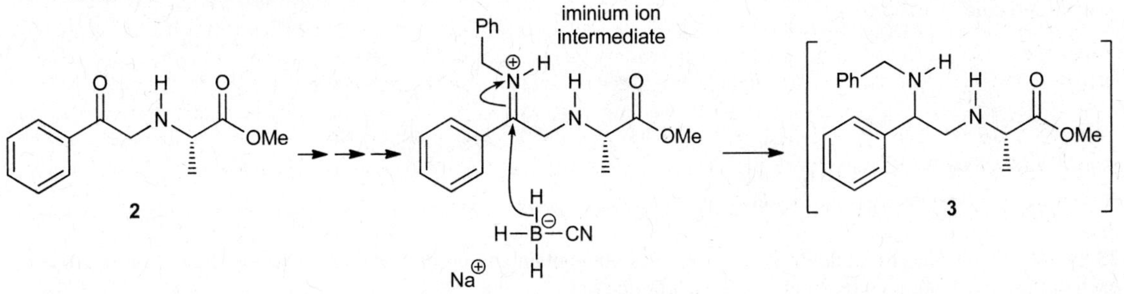

Focus carefully on the reduction step above, in which the new chirality center is created. Since the reducing agent can approach from either face of the C=N π bond, we would expect the following two diastereomeric products:

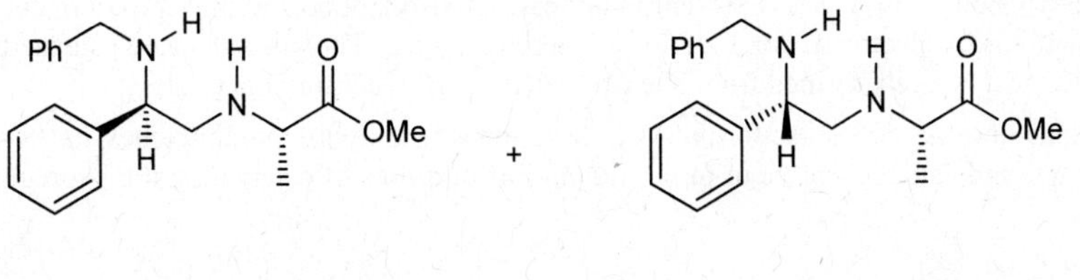

(b) The problem statement indicates that compound **3** undergoes a reaction that produces a cyclic amide. So we inspect the structure of **3** to determine which functional groups can react with each other to produce a cyclic amide. Compound **3** has two amine groups and one ester group:

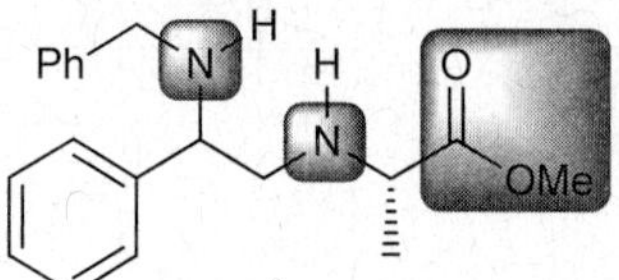

3

And we learned in Section 21.11 that an amide can be formed from an amine and an ester:

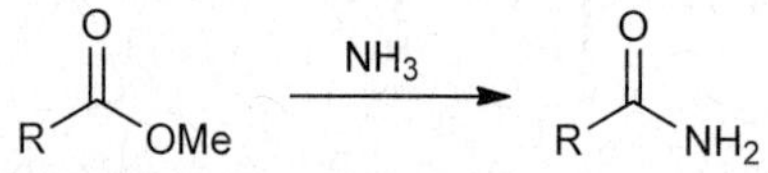

This type of reaction is generally slow and inefficient, but in our case, the reaction is intramolecular, so it can occur more rapidly. The reaction can occur in two possible ways, either forming a 3-membered ring or a 6-membered ring, depending on which amine group reacts with the ester group:

Formation of a 3-membered ring

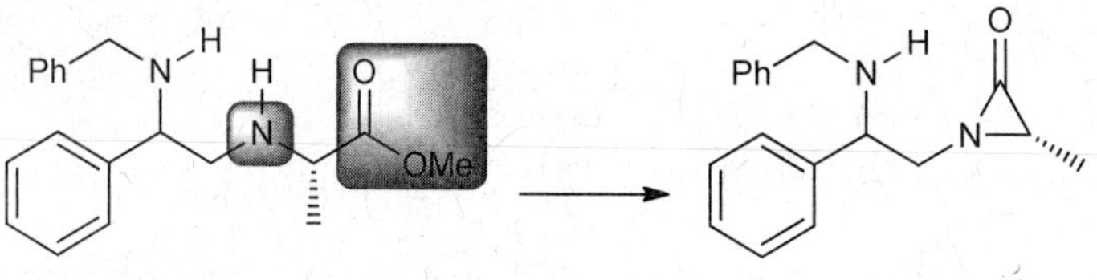

Formation of a 6-membered ring

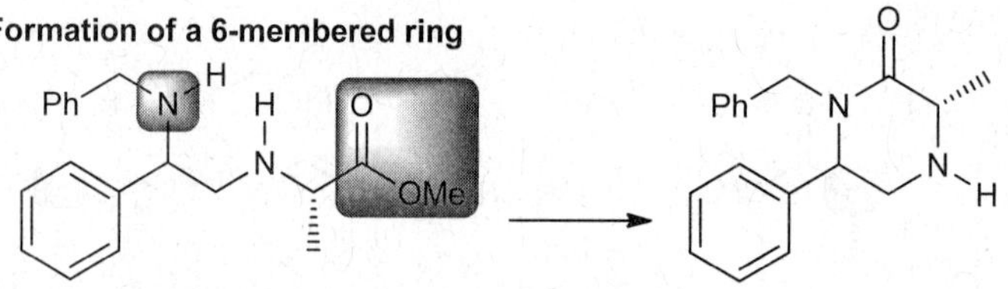

The 3-membered ring will be higher in energy than the 6-membered ring as a result of significant ring strain which is present in the former and absent in the latter. As such, the 6-membered ring will be formed as the product.

Since compound **3** was formed as a diastereomeric mixture, we expect that compound **4** will also be produced as a diastereomeric mixture:

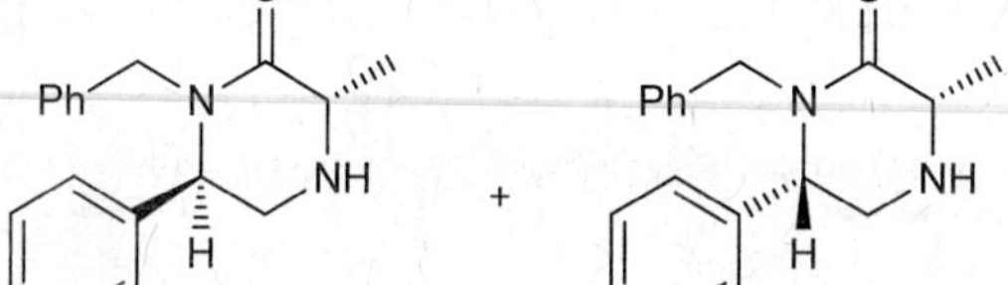

23.94. The molecular formula for the product does not contain any bromine atoms. Therefore, the leaving group for each of the two substitution reactions will be a bromide ion.

We should not ignore the valuable information contained in the molecular formula ($C_{29}H_{50}N_6OSi_2$). First of all, both silicon atoms are still present in the product, so the silyl protecting groups are not removed (that should be expected, since we learned in Chapter 13 that silyl protecting groups are removed either with acidic conditions or with a source of fluoride, such as TBAF, neither of which is present in our case). Also, notice that the starting materials have 22 carbon atoms and 7 carbon atoms, respectively, for a total of 29 carbon atoms. The product also has exactly 29 carbon atoms, indicating that the product is likely formed from one equivalent of each starting material.

We are not done analyzing the molecular formula. Calculation of the hydrogen deficiency index (as seen in Chapter 15) for the product will indicate the degree of unsaturation. For purposes of calculating the hydrogen deficiency index, we will treat silicon in the same way that we treat carbon, because Si and C are in the same group (same column) of the periodic table. The hydrogen deficiency index of the product is 10, which indicates 10 degrees of unsaturation. Let's compare this with the degrees of unsaturation in the starting materials. The first compound has seven degrees of unsaturation (three rings and four π bonds), while the second compound contains two degrees of unsaturation (the C-C π bond, and the C-O π bond of the ester group). Therefore, the two starting materials have a combined hydrogen deficiency index of 9. When we compare this to the product (HDI = 10), we conclude that formation of the product must involve either the formation of a new ring or a new π bond.

Since we know that each bromide must be ejected during a substitution process, let's start by determining which bromide is expected to participate more readily in an S_N2 process. Comparison of the location for each Br indicates that one is a 1° bromide, while the other is a 2° bromide. We expect the 1° bromide to be more reactive towards a nucleophile, because it is less hindered than the 2° bromide.

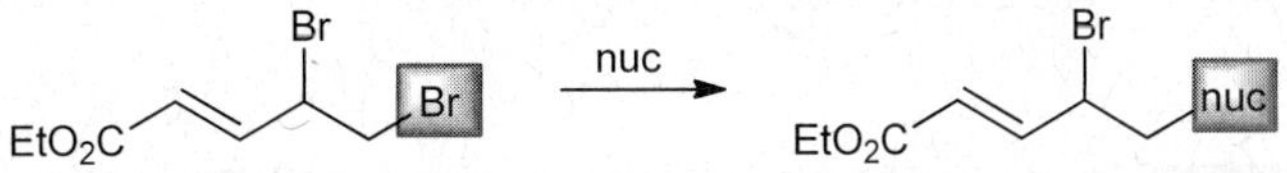

Now let's search for a nucleophilic center in the other starting material. There are two amino groups in the other starting material; however, one of them exhibits a lone pair that is delocalized (via resonance) into the aromatic ring. Since the lone pair of this amino group is delocalized, it is expected to be less reactive as a nucleophile. The other amino group has a localized lone pair, and is a much better nucleophile.

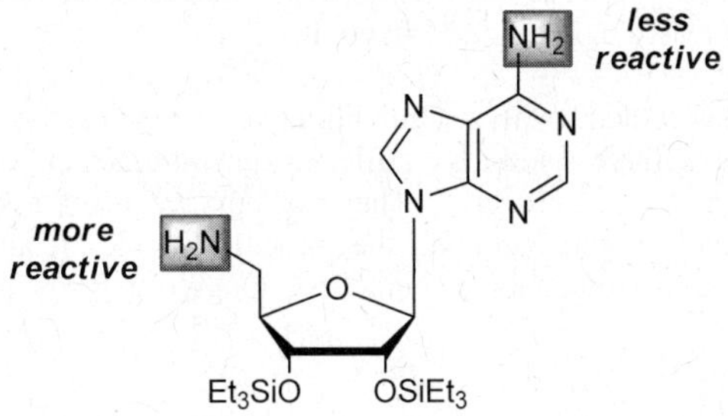

Now that we have identified the nucleophilic center in one starting material and the electrophilic center in the other starting material, we are ready to draw the product of the S_N2 reaction that can occur between them:

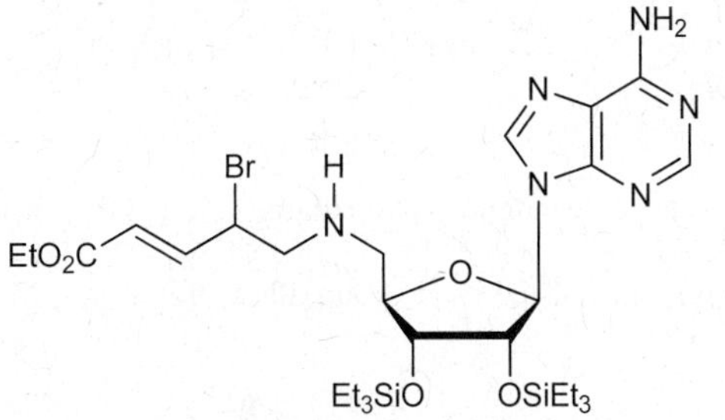

Finally, loss of the other bromide can occur via an intramolecular S_N2-type process, in which the localized lone pair on the nitrogen atom (closest to Br) functions as a nucleophile (again) to give a three-membered ring called an aziridine.

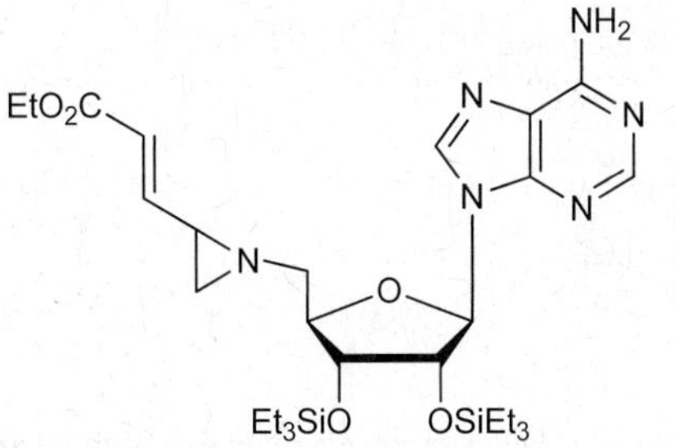

This displacement is similar to the conversion of a halohydrin to an epoxide under basic conditions. In this case, the base is DBU.

23.95.

(a) Hydrogenation of an azide is expected to give an amine.

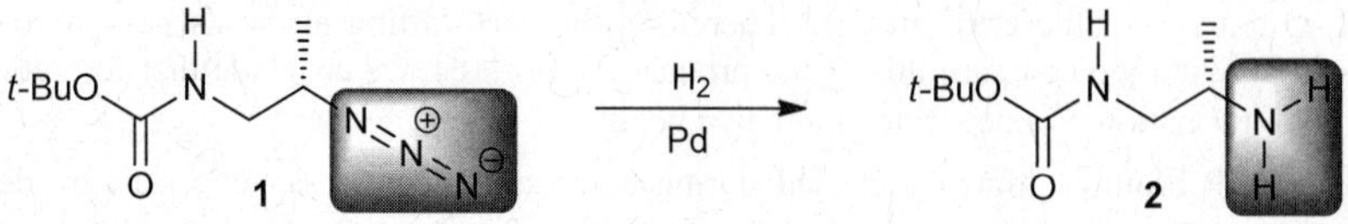

In the reaction, compound **1** (with *t*-BuO carbamate, N—H, and azide N=N⁺=N⁻ groups) is treated with H₂ / Pd to give compound **2**.

Compound **2** is a primary amine, and the problem statement indicates that it reacts with methyl acetoacetate to give an imine. We saw in Chapter 20 that an imine is the product generated when a primary amine is treated with a ketone. Methyl acetoacetate does have a ketone group, so we expect that compound **3** will have the following structure:

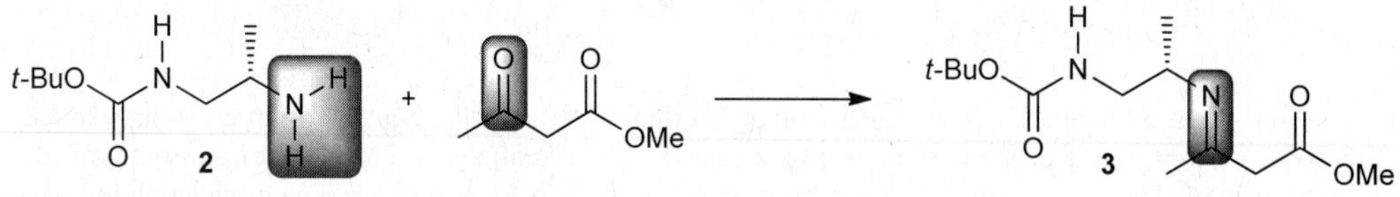

Compound **2** + methyl acetoacetate (with two C=O groups and OMe) → compound **3** (imine with OMe ester).

In Chapter 20, we saw that the process was catalyzed by acid, but in this case, the problem statement indicates that acid catalyst was not employed, and the reaction proceeded without it.

(b) Tautomerization can occur either in acid-catalyzed conditions or in base-catalyzed conditions. In fact, even if we do not introduce either acid or base, there should be sufficient quantities of either (adsorbed to the surface of the glassware) to catalyze the tautomerization process. The problem statement asks us to draw an acid-catalyzed mechanism. Much like keto-enol tautomerization, the process should require two steps (protonation and deprotonation). In acidic conditions, protonation occurs first, to give a resonance stabilized cation, which then undergoes deprotonation:

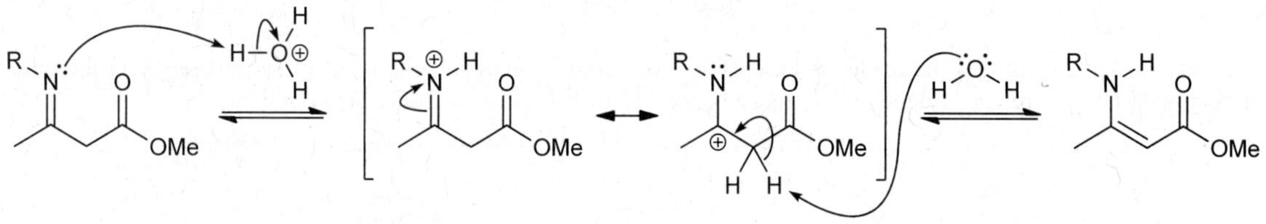

(c) The following are three reasons why the enamine is particularly stable in this case:

(1)   The presence and proximity of the ester group allows for conjugation with the C=C bond, which is a stabilizing factor:

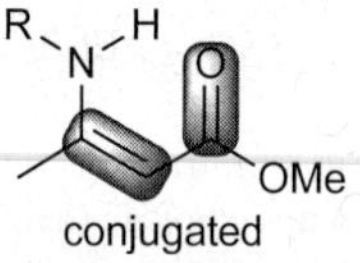

conjugated

(2)   The lone pair of this enamine is particularly delocalized, as a result of the resonance structure shown below. This delocalization contributes to the stability of the enamine (and therefore of the entire molecule).

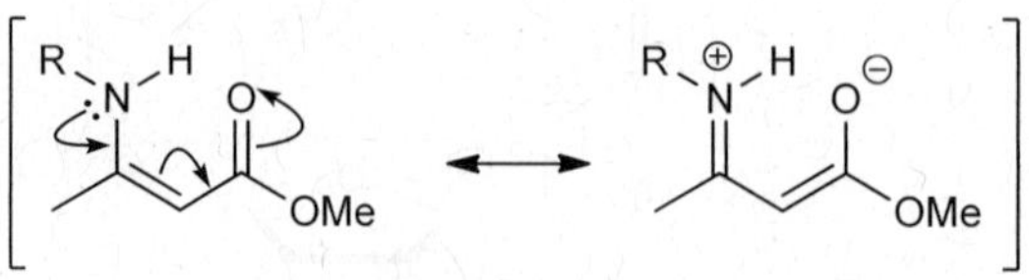

(3)   The presence and proximity of the ester group enables intramolecular hydrogen bonding, which is also a stabilizing factor.

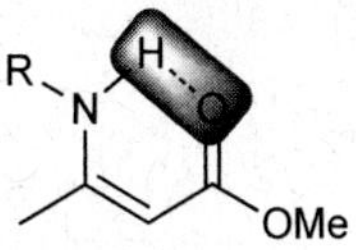

All three factors (above) contribute to the enhanced stability of the enamine in this particular case, and as such, the equilibrium favors formation of the enamine.

23.96. The amino group of compound **1** attacks the carbonyl group of the aldehyde to produce intermediate **3**. After three successive proton transfer steps, water is lost, producing iminium ion **7**. In section 23.6, we saw that iminium ions are sufficiently activated (electrophilic) to be reduced by sodium cyanoborohydride. In the absence of a hydride source, they can also be trapped by a nucleophile. In this example, the nucleophile is the attached aromatic ring. In Chapter 19, we learned about electrophilic aromatic substitution reactions (such as the Friedel-Crafts acylation); we know that the two methoxy groups in **7** sufficiently activate the aromatic ring, enabling the ring to function as a nucleophile and attack an electrophile. Nucleophilic attack by the aromatic ring will produce intermediate **8**, in which the electrophilic group has been added *para* to one of the methoxy groups. Next, rearomatization will produce compound **9**. This process is known as the Pictet-Spengler condensation reaction. In the final stage of the synthesis, the lone pair on the nitrogen atom will function as a nucleophile and attack, via S_N2, the 1° alkyl chloride to form the 5-membered ring. A final proton transfer produces desired product, compound **2**.

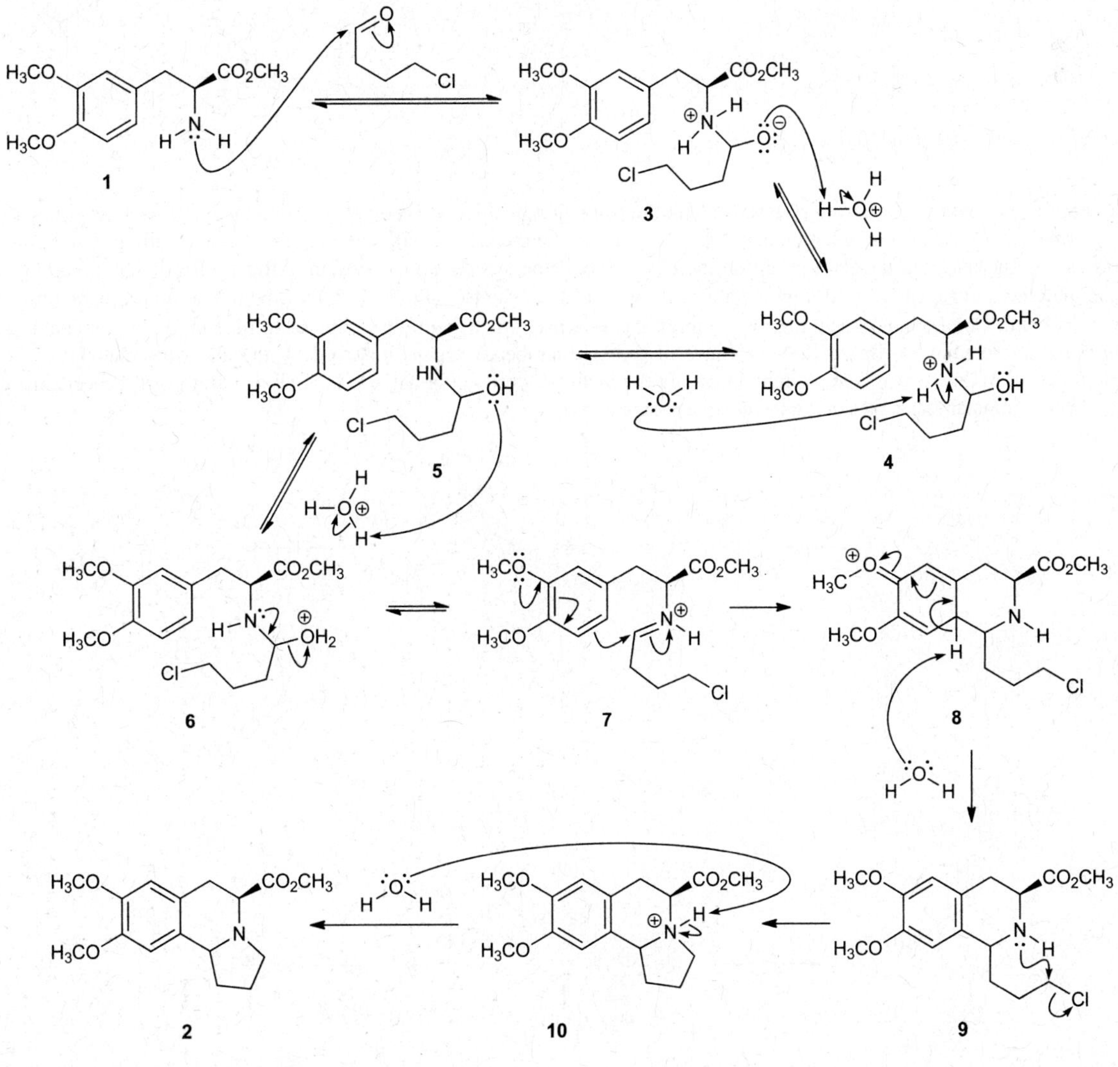

23.97. Trifluoroacetic acid is a proton source, which can supply a proton to the hydroxyl group of **1**, thereby converting it into a good leaving group. The lone pair on nitrogen can then initiate a fragmentation process that will generate a C-N π-bond, via the breakage of a C-C σ-bond and the generation of a C-C π-bond. This process is known as a Grob fragmentation. Notice that the intermediate is an iminium ion (**3**); once generated, it can be easily reduced by sodium cyanoborohydride to generate the product (**4**).

23.98. If we redraw the starting material so that the pendant π bond is in close proximity to the diene, we can envision a thermal [4+2] Diels Alder reaction occurring. Notice that the dienophile will approach the diene, resulting in the oxo-bridge on one face and the ester group on the other face of the newly formed system. After cyclization, compound **2** can undergo a nitrogen-induced fragmentation that will result in the formation of a C-N π-bond via the breakage of the C-O σ-bond. Compound **3** has a highly electrophilic α,β-unsaturated iminium ion that can be trapped by methanol to produce intermediate **4**. Because the top face of the six-membered ring is hindered by the alkoxide, methanol will approach from the bottom face, thereby installing the methoxy group on a dash. The final two steps of the mechanism are proton transfer steps, producing the desired compound (**6**).

Chapter 24
Carbohydrates

Review of Concepts

Fill in the blanks below. To verify that your answers are correct, look in your textbook at the end of Chapter 24. Each of the sentences below appears verbatim in the section entitled *Review of Concepts and Vocabulary*.

- **Carbohydrates** are polyhydroxy _____ or ketones.
- Simple sugars are called _____ and are generally classified as **aldoses** and _____.
- For all **D sugars**, the chirality center farthest from the carbonyl group has the ___ configuration.
- Aldohexoses can form cyclic hemi_____ that exhibit a **pyranose** ring.
- Cyclization produces two stereoisomeric hemiacetals, called _____. The newly created chirality center is called the _____ **carbon.**
- In the **α anomer**, the hydroxyl group at the anomeric position is _____ to the CH$_2$OH group, while in the **β anomer**, the hydroxyl group is _____ to the CH$_2$OH group.
- Anomers equilibrate by a process called _____, which is catalyzed by either _____ or _____.
- Some carbohydrates, such as D-fructose, can also form five-membered rings, called _____ rings.
- Monosaccharides are converted into their ester derivatives when treated with excess _____.
- Monosaccharides are converted into their ether derivatives when treated with excess _____ and silver oxide.
- When treated with an alcohol under acid-catalyzed conditions, monosaccharides are converted into acetals, called _____. Both anomers are formed.
- Upon treatment with sodium borohydride an aldose or ketose can be reduced to yield an _____.
- When treated with a suitable oxidizing agent, an aldose can be oxidized to yield an _____.
- When treated with HNO$_3$, an aldose is oxidized to give a dicarboxylic acid called an _____.
- D-Glucose and D-mannose are **epimers** and are interconverted under strongly _____ conditions.
- The **Kiliani-Fischer synthesis** can be used to lengthen the chain of an _____.
- The **Wohl degradation** can be used to shorten the chain of an _____.
- _____ are comprised of two monosaccharide units, joined together via a glycosidic linkage.
- **Polysaccharides** are polymers consisting of repeating monosaccharide units linked by _____ bonds.
- When treated with an _____ in the presence of an acid catalyst, monosaccharides are converted into their corresponding **N-glycosides**.

Review of Skills

Fill in the blanks and empty boxes below. To verify that your answers are correct, look in your textbook at the end of Chapter 24. The answers appear in the section entitled *SkillBuilder Review*.

24.1   Drawing the Cyclic Hemiacetal of a Hydroxyaldehyde

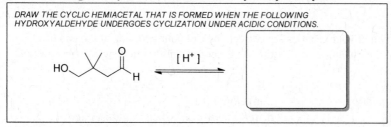

24.2: Drawing a Haworth Projection of an Aldohexose

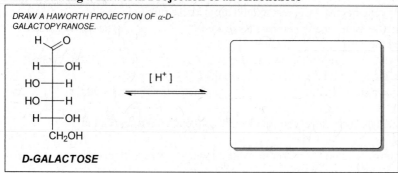

24.3: Drawing the More Stable Chair Conformation of a Pyranose Ring

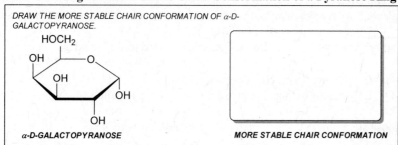

24.4   Identifying a Reducing Sugar

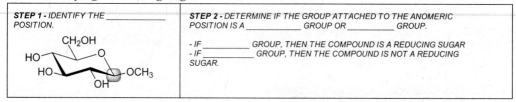

24.5   Determining Whether a Disaccharide Is a Reducing Sugar

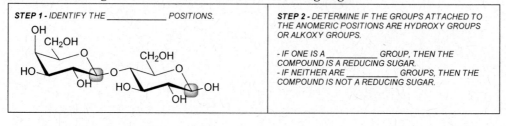

Review of Reactions

Identify the reagents necessary to achieve each of the following transformations. To verify that your answers are correct, look in your textbook at the end of Chapter 24. The answers appear in the section entitled *Review of Reactions*.

Hemiacetal Formation

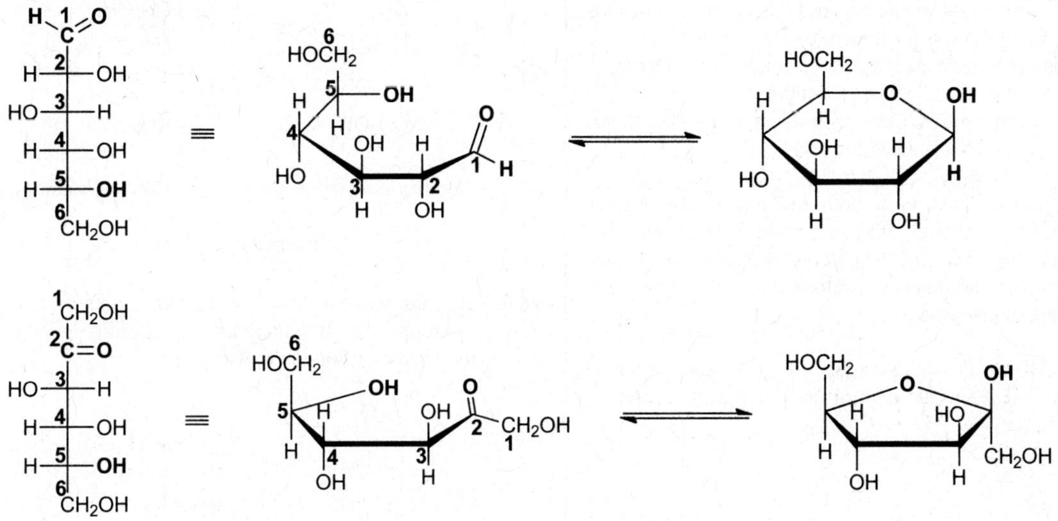

Chain Lengthening and Chain Shortening

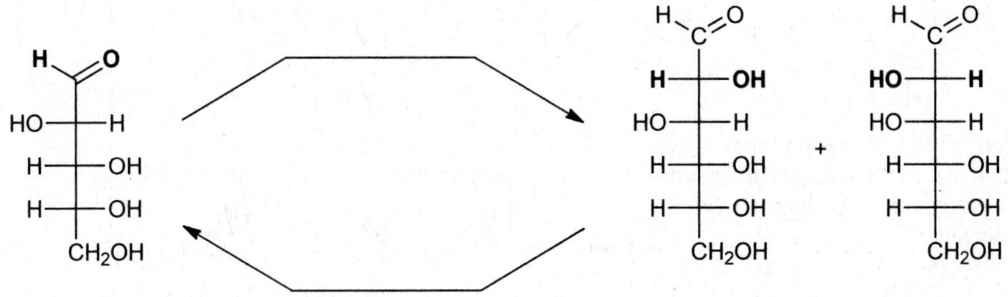

Reactions of Monosaccharides

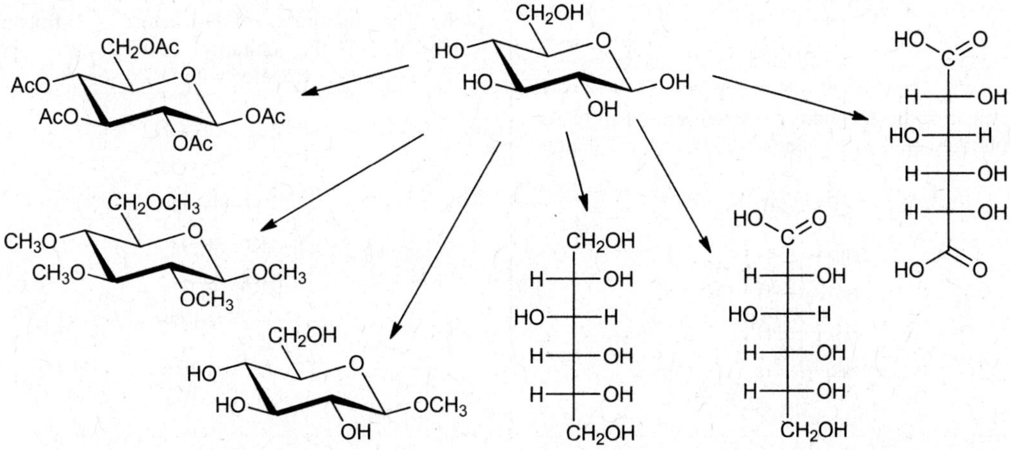

Solutions

24.1.
(a) This compound is an aldehyde and it is comprised of six carbon atoms, so it is an aldohexose.
(b) This compound is an aldehyde and it is comprised of five carbon atoms, so it is an aldopentose.
(c) This compound is a ketone and it is comprised of five carbon atoms, so it is a ketopentose.
(d) This compound is an aldehyde and it is comprised of four carbon atoms, so it is an aldotetrose.
(e) This compound is a ketone and it is comprised of six carbon atoms, so it is a ketohexose.

24.2. Both are hexoses so both have molecular formula ($C_6H_{12}O_6$). Although they have the same molecular formula, they have different constitution – one is an aldehyde and the other is a ketone. Therefore, they are constitutional isomers.

24.3. All are D sugars except for (b), which is an L sugar. The configuration of each chirality center is shown here:
(a) 2S, 3S, 4R, 5R
(b) 2R, 3S, 4S
(c) 3R, 4R
(d) 2S, 3R
(e) 3S, 4S, 5R
Pay special attention to the following trend: The configuration of each chirality center is *R* when the OH group is on the right side of the Fischer projection, and the configuration is *S* when the OH group is on the left side.

24.4.
(a) As seen in the solution to the previous problem, the *R* configuration is characterized by an OH group on the right side of the Fischer projection. D-Allose has all of the OH groups on the right side:

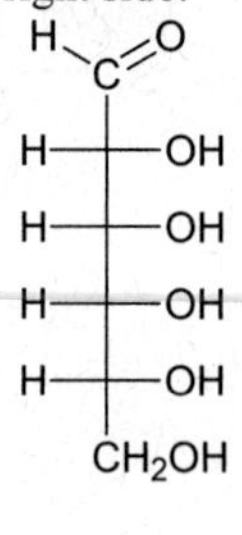

(b) L-Allose is the enantiomer of D-allose, so all of the OH groups are on the right side of the Fischer projection:

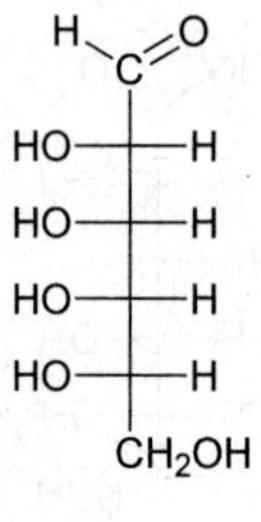

24.5. A ketotetrose has four carbon atoms and a ketone group, leaving only one chirality center, giving the following two enantiomers:

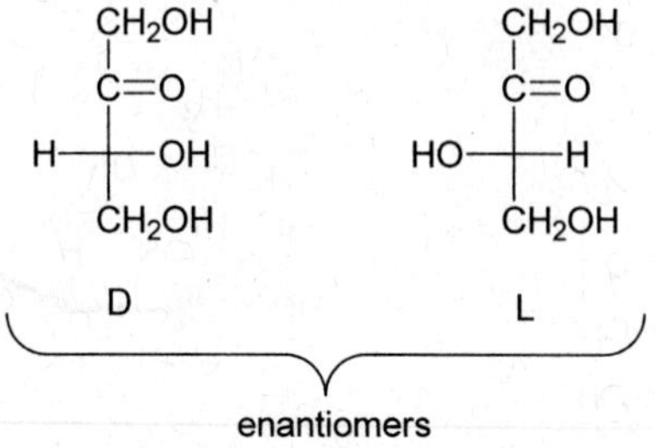

enantiomers

24.6. An aldotetraose has four carbon atoms and an aldehyde group. As such, there are two chirality centers, giving the following four stereoisomers:

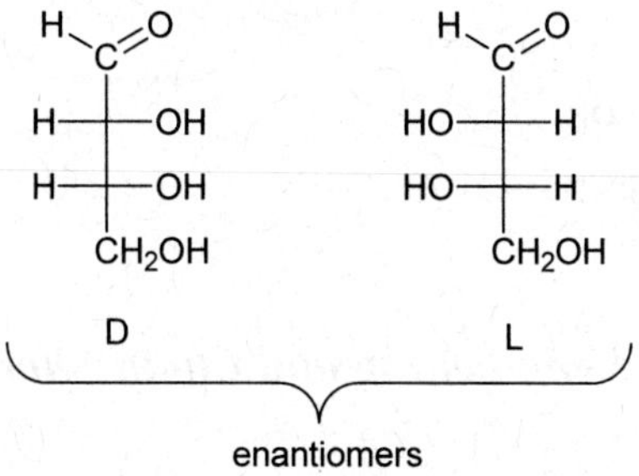

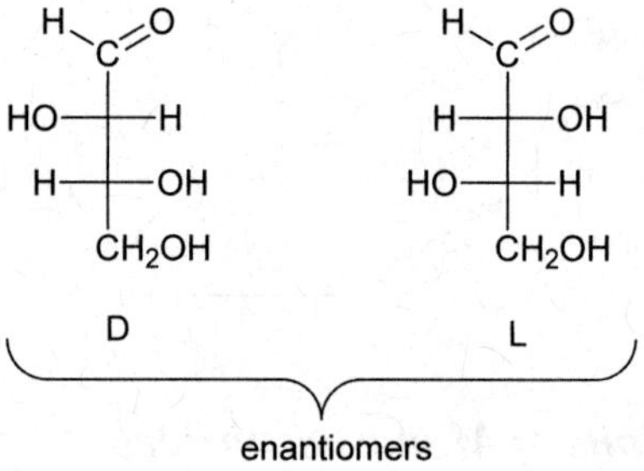

enantiomers

enantiomers

24.7. The enantiomer of D-fructose is L-fructose, in which all of the chirality centers have opposite configuration (as compared with D-fructose):

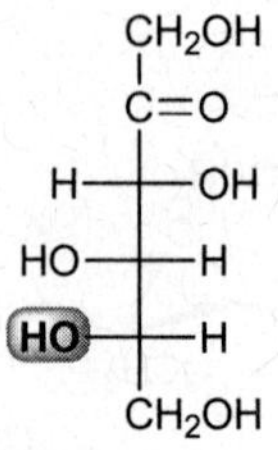

L-Fructose

24.8. Both D-fructose and D-glucose have the molecular formula ($C_6H_{12}O_6$). However, they have different constitution – one is a ketone, and the other is an aldehyde. Therefore, they are constitutional isomers.

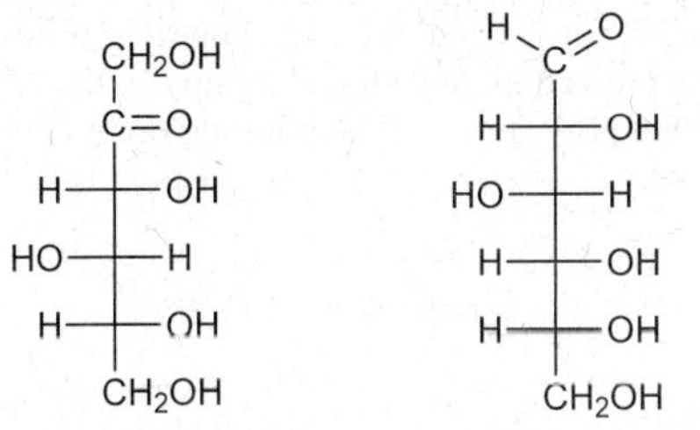

D-Fructose   **D-Glucose**

24.9.
(a) We use a numbering system to determine the size of the ring that is formed. Four carbon atoms and one oxygen atom are incorporated into a five-membered ring. There are two methyl groups at C4, which must be drawn in the product:

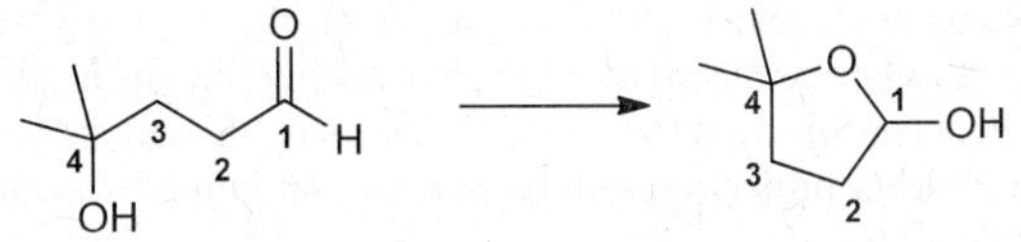

(b) We use a numbering system to determine the size of the ring that is formed. Five carbon atoms and one oxygen atom are incorporated into a six-membered ring, as shown. There are two methyl groups at C5, which must be drawn in the product:

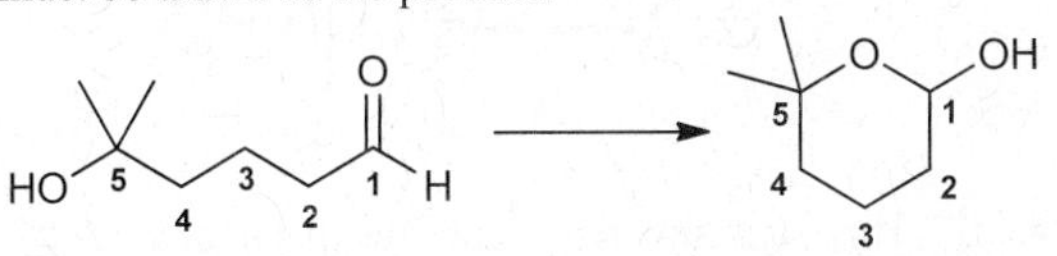

(c) We use a numbering system to determine the size of the ring that is formed. Six carbon atoms and one oxygen atom are incorporated into a seven-membered ring, as shown. There are two methyl groups at C3, which must be drawn in the product:

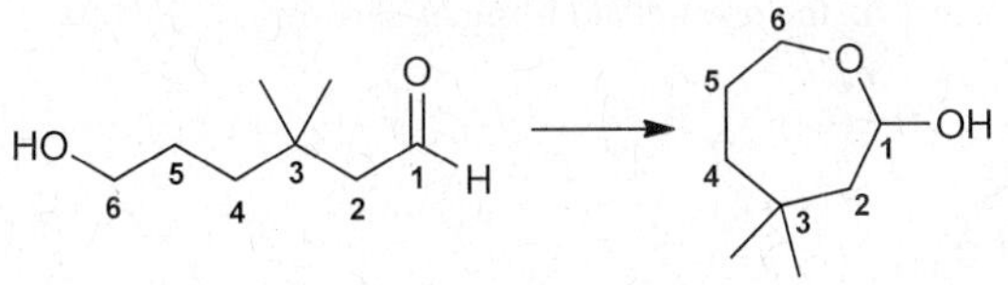

(d) We use a numbering system to determine the size of the ring that is formed. Four carbon atoms and one oxygen atom are incorporated into a five-membered ring, as shown. There are two methyl groups at C4 of the ring and one methyl group at C1 of the ring, which must be drawn in the product:

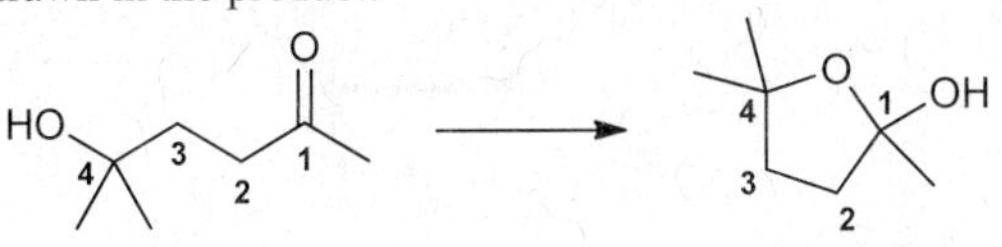

24.10. We must use a numbering system, just as we did in the previous problem. There are two methyl groups at C4, which must be drawn in the starting material:

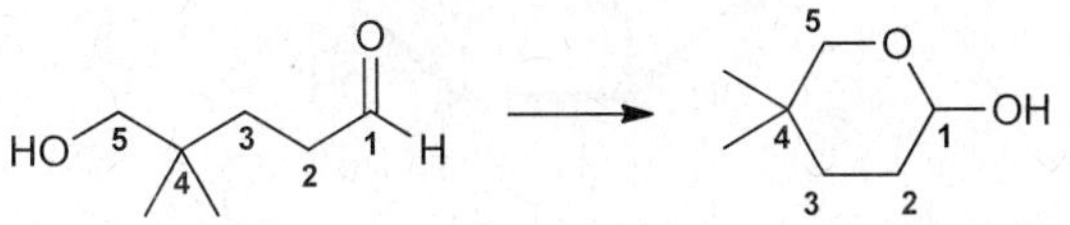

24.11.
(a) The carbonyl group can be attacked by the OH group that is connected to C4, giving a five-membered ring;

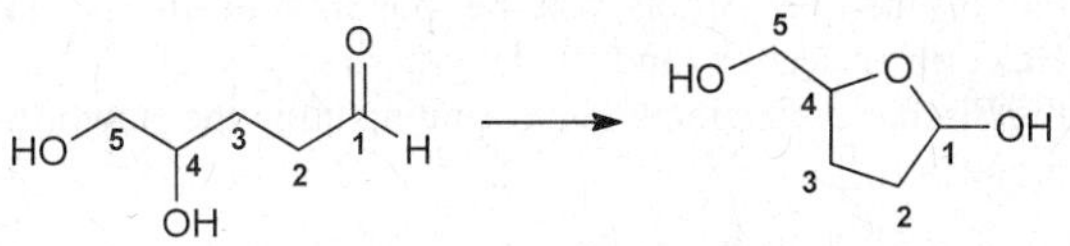

or the carbonyl group can be attacked by the OH group that is connected to C5, giving a six-membered ring:

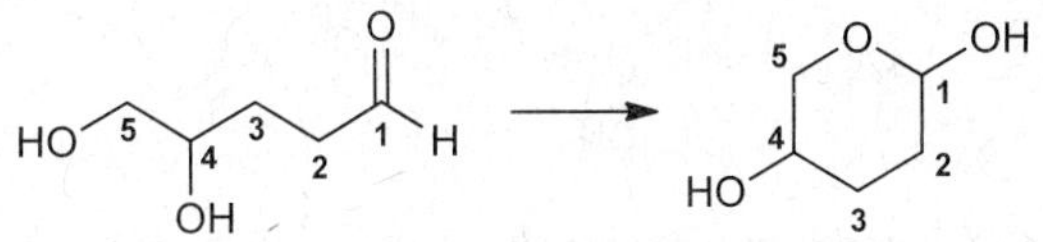

(b) The six-membered ring is expected to predominate because it has less ring strain than a five-membered ring.

24.12.
(a) We begin by drawing the skeleton of the Haworth projection. Then, we draw the CH_2OH group (at C6) pointing up. The anomeric OH group is then drawn pointing up (the β anomer). Finally, the remaining groups are drawn). All OH groups on the right side of the Fischer projection will be pointing down in the Haworth projection (and all OH groups on the left side of the Fischer projection will be pointing up in the Haworth projection).

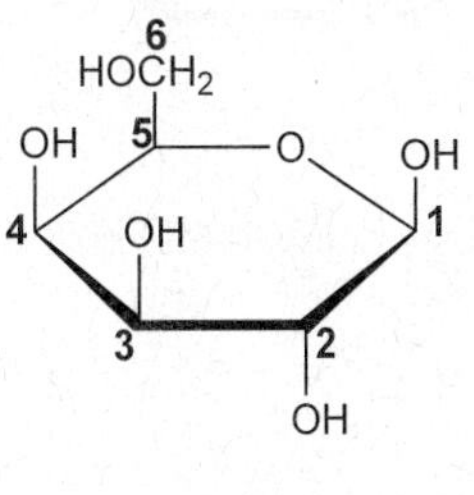

(b) We begin by drawing the skeleton of the Haworth projection. Then, we draw the CH_2OH group (at C6) pointing up. The anomeric OH group is then drawn pointing down (the α anomer). Finally, the remaining groups are drawn. All OH groups on the right side of the Fischer projection will be pointing down in the Haworth projection (and all OH groups on the left side of the Fischer projection will be pointing up in the Haworth projection).

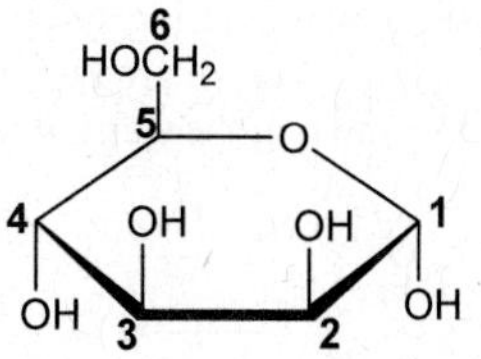

(c) We begin by drawing the skeleton of the Haworth projection. Then, we draw the CH$_2$OH group (at C6) pointing up. The anomeric OH group is then drawn pointing down (the α anomer). Finally, the remaining groups are drawn). All OH groups on the right side of the Fischer projection will be pointing down in the Haworth projection (and all OH groups on the left side of the Fischer projection will be pointing up in the Haworth projection).

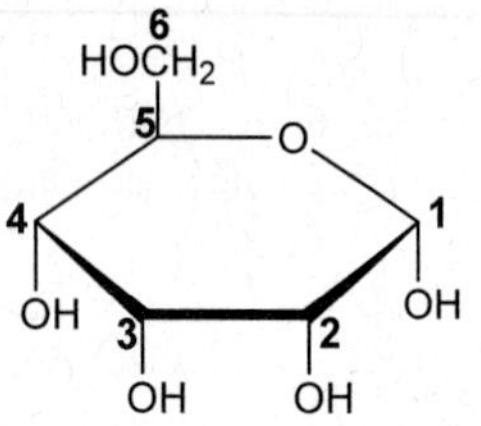

(d) We begin by drawing the skeleton of the Haworth projection. Then, we draw the CH$_2$OH group (at C6) pointing up. The anomeric OH group is then drawn pointing up (the β anomer). Finally, the remaining groups are drawn). All OH groups on the right side of the Fischer projection will be pointing down in the Haworth projection (and all OH groups on the left side of the Fischer projection will be pointing up in the Haworth projection).

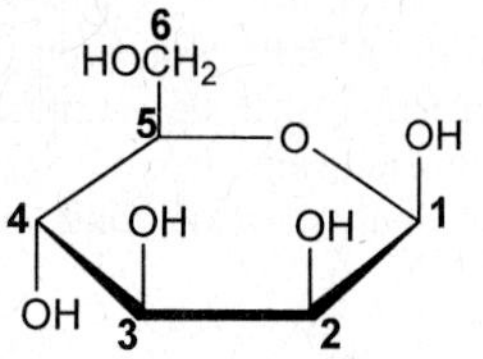

(e) We begin by drawing the skeleton of the Haworth projection. Then, we draw the CH$_2$OH group (at C6) pointing up. The anomeric OH group is then drawn pointing up (the β anomer). Finally, the remaining groups are drawn). All OH groups on the right side of the Fischer projection will be pointing down in the Haworth projection (and all OH groups on the left side of the Fischer projection will be pointing up in the Haworth projection).

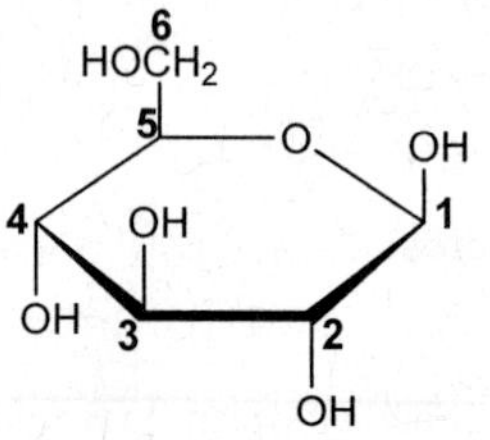

(f) We begin by drawing the skeleton of the Haworth projection. Then, we draw the CH$_2$OH group (at C6) pointing up. The anomeric OH group is then drawn pointing down (the α anomer). Finally, the remaining groups are drawn). All OH groups on the right side of the Fischer projection will be pointing down in the Haworth projection (and all OH groups on the left side of the Fischer projection will be pointing up in the Haworth projection).

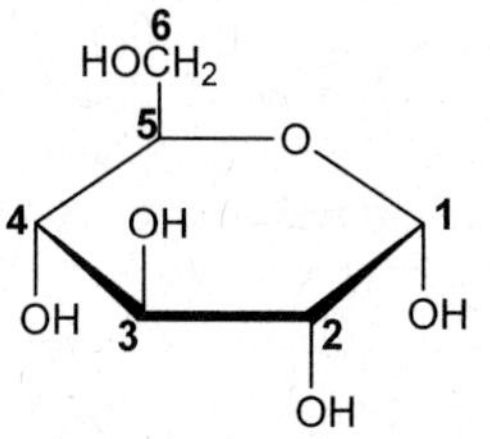

24.13. This structure represents the β anomer of the cyclic form of D-galactose and is therefore called β-D-galactopyranose.

24.14. The two pyranose forms are in equilibrium with each other, via the open-chain form, as shown:

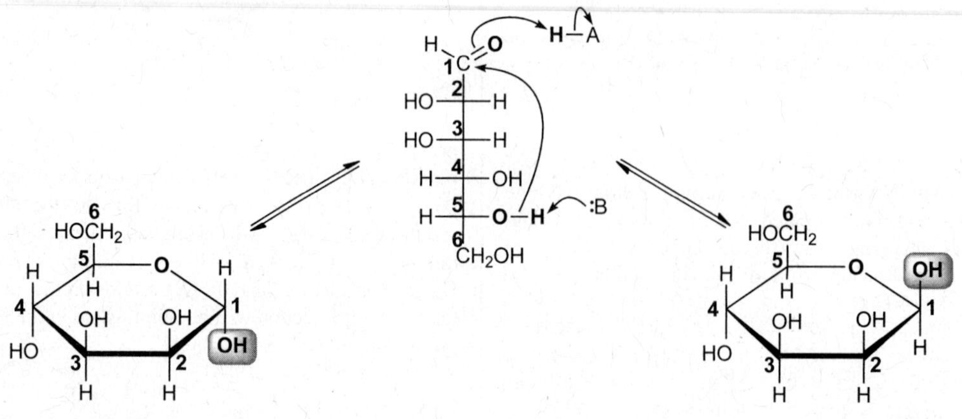

α-D-Mannopyranose   *β-D-Mannopyranose*

24.15.   The two pyranose forms are in equilibrium with each other, via the open-chain form, as shown:

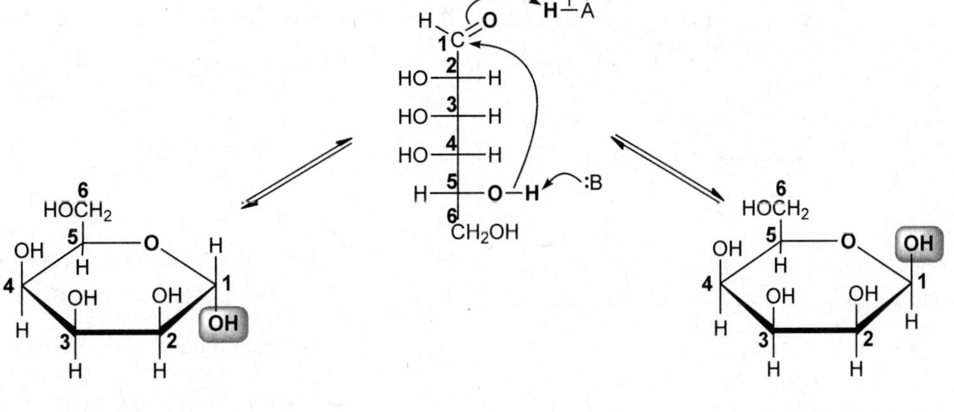

α-D-Talopyranose   *β-D-Talopyranose*

24.16.
(a) The skeleton of the chair is drawn with an oxygen atom in the upper back-right corner. Each substituent is then labeled as "up" or "down" and placed on the chair accordingly. The anomeric OH group is drawn pointing up, indicating the β anomer. In this chair conformation, most of the substituents occupy equatorial positions, so this is the more stable chair conformation of β-D-galactopyranose.

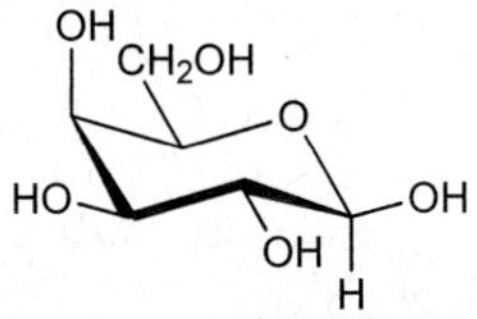

(b) The skeleton of the chair is drawn with an oxygen atom in the upper back-right corner. Each substituent is then labeled as "up" or "down" and placed on the chair accordingly. The anomeric OH group is drawn pointing down, indicating the α anomer. In this chair conformation, most of the substituents occupy equatorial positions, so this is the more stable chair conformation of α-D-glucopyranose.

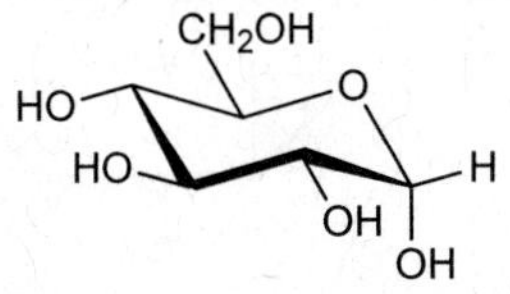

(c) The skeleton of the chair is drawn with an oxygen atom in the upper back-right corner. Each substituent is then labeled as "up" or "down" and placed on the chair accordingly. The anomeric OH group is drawn pointing up, indicating the β anomer. In this chair conformation, most of the substituents occupy equatorial positions, so

this is the more stable chair conformation of β-D-glucopyranose.

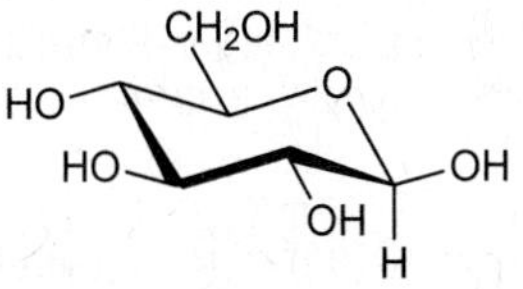

24.17.   The anomeric position becomes an aldehyde group in the open-chain form. The OH group at C5 (of the open-chain form) must be pointing to the right, because this is a D sugar (CH_2OH is "up" in the chair conformation). The remaining three groups (at C2, C3 and C4) are all pointing "down", so they must be on the right side of the Fischer projection of the open-chain form. This structure represents D-allose.

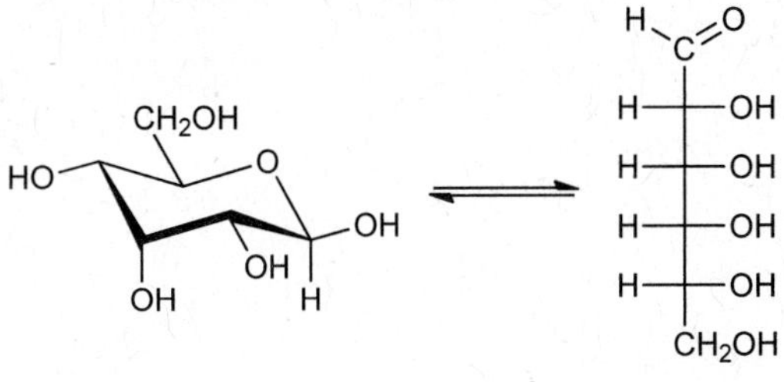

D-Allose

24.18.   Both chair conformations of β-D-glucopyranose are shown below. The less stable conformation is the one in which all substituents occupy axial positions.

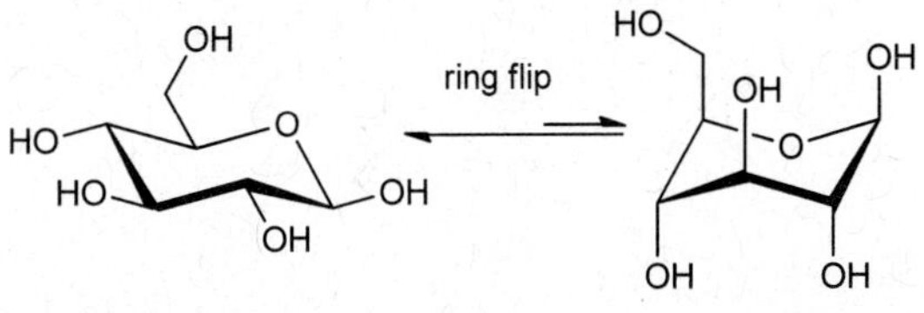

more stable   less stable

24.19.

(a) The anomeric OH group is drawn pointing down (thus, the α anomer). The remaining two OH groups (at C2 and C3) are both on the right side of the Fischer projection of D-erythrose, so both OH groups are both drawn pointing down in the Haworth projection.

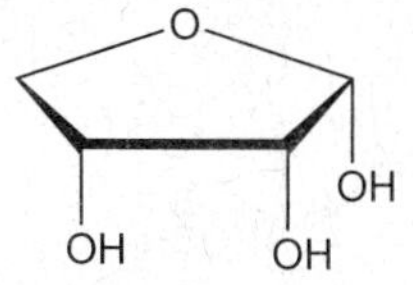

(b) The anomeric OH group is drawn pointing up (thus, the β anomer). The remaining two OH groups (at C2 and C3) are both on the right side of the Fischer projection of D-erythrose, so both OH groups are both drawn pointing down in the Haworth projection.

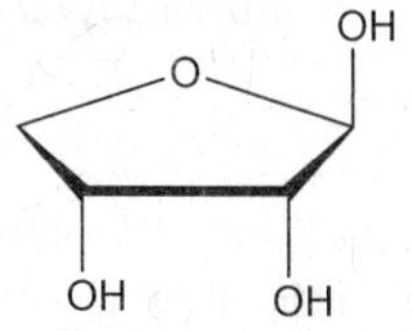

(c) The anomeric OH group is drawn pointing down (thus, the α anomer). The OH group on the left side of

the Fischer projection (at C2 of D-threose) is drawn pointing up in the Haworth projection (at C2). Similarly, the OH group on the right side of the Fischer projection (at C3 of D-threose) is drawn pointing down in the Haworth projection (at C3).

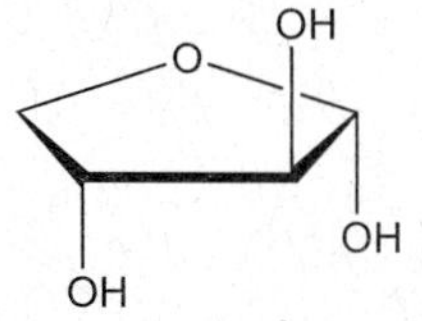

(d) The anomeric OH group is drawn pointing up (thus, the β anomer). The OH group on the left side of the Fischer projection (at C2 of D-threose) is drawn pointing up in the Haworth projection (at C2). Similarly, the OH group on the right side of the Fischer projection (at C3 of D-threose) is drawn pointing down in the Haworth projection (at C3).

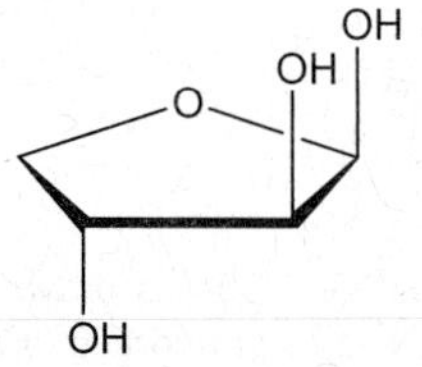

24.20. The carbonyl group is first protonated, thereby rendering it more electrophilic and more susceptible to nucleophilic attack by one of the OH groups. The OH group at C4 will attack the protonated carbonyl group to generate a furanose (five-membered) ring. Deprotonation (with water functioning as a base) gives the product, as shown.

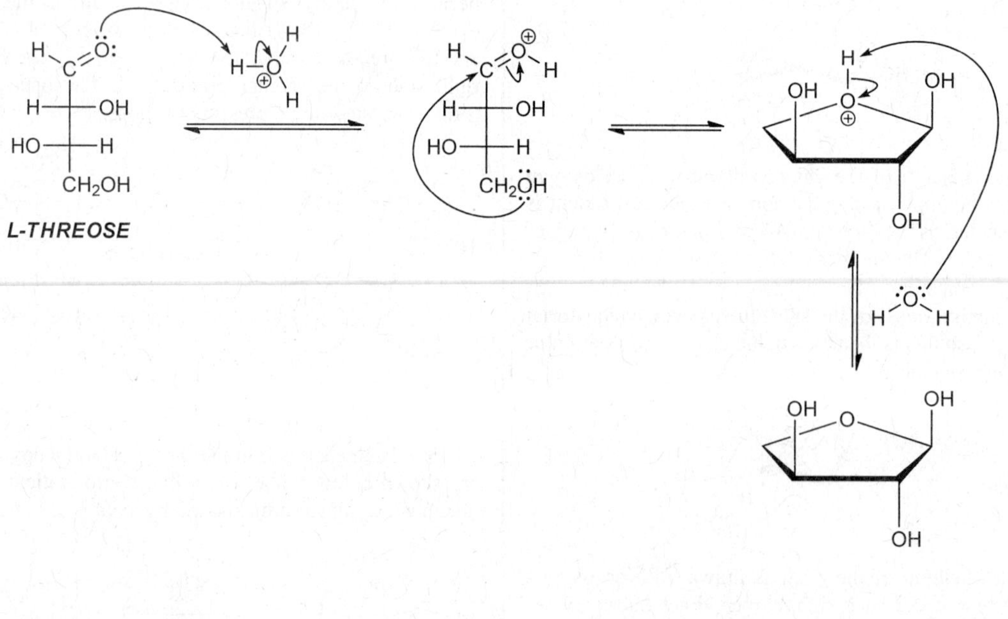

24.21. The carbonyl group is first protonated, thereby rendering it more electrophilic and more susceptible to nucleophilic attack by one of the OH groups. The OH group at C5 will attack the protonated carbonyl group to generate a furanose (five-membered) ring. Deprotonation (with water functioning as a base) gives the product, as shown.

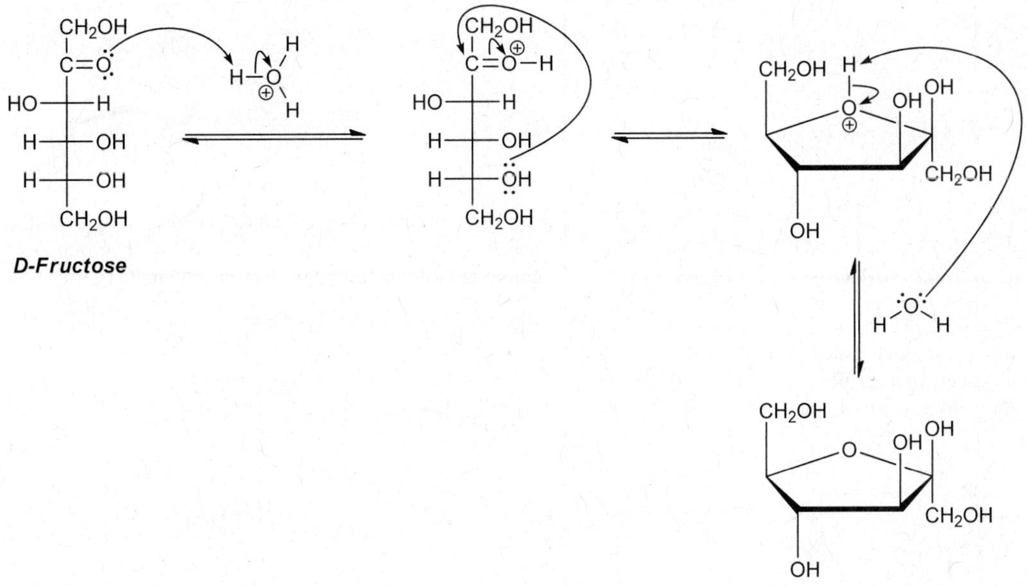

D-Fructose

24.22. The name (α-D-fructopyranose) indicates that the open chain form is D-fructose, shown here:

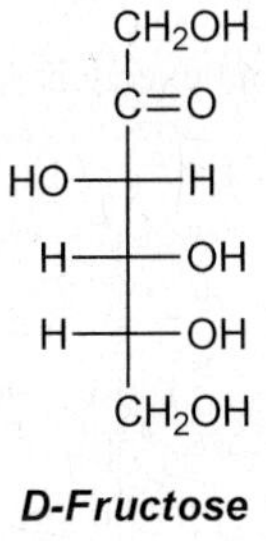

D-Fructose

24.23.

(a) Upon treatment with excess acetic anhydride and pyridine, all of the OH groups undergo acetylation, as shown:

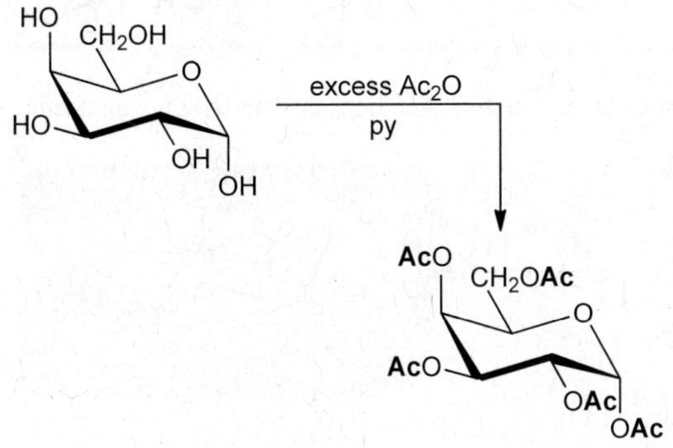

(b) Upon treatment with excess acetic anhydride and pyridine, all of the OH groups undergo acetylation, as shown:

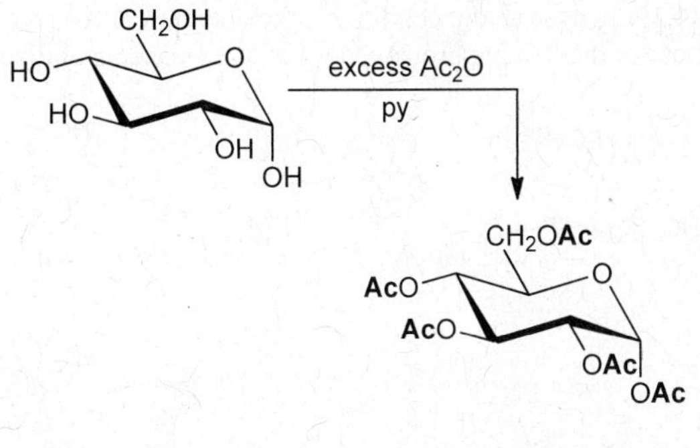

(c) Upon treatment with excess acetic anhydride and pyridine, all of the OH groups undergo acetylation, as shown:

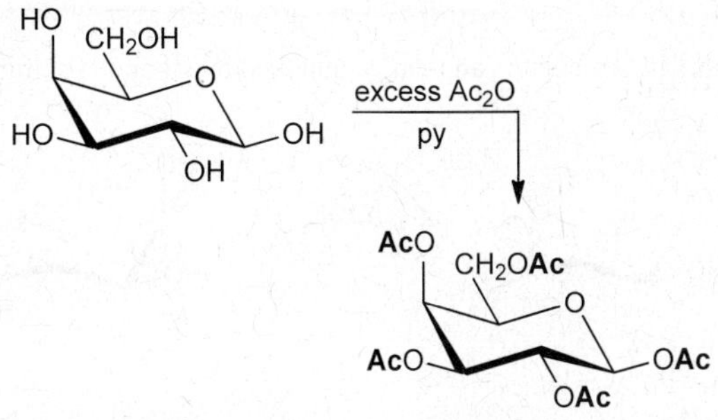

978   **CHAPTER 24**

24.24.
(a) Upon treatment with excess methyl iodide in the presence of silver oxide, all of the OH groups are converted into methoxy groups, as shown:

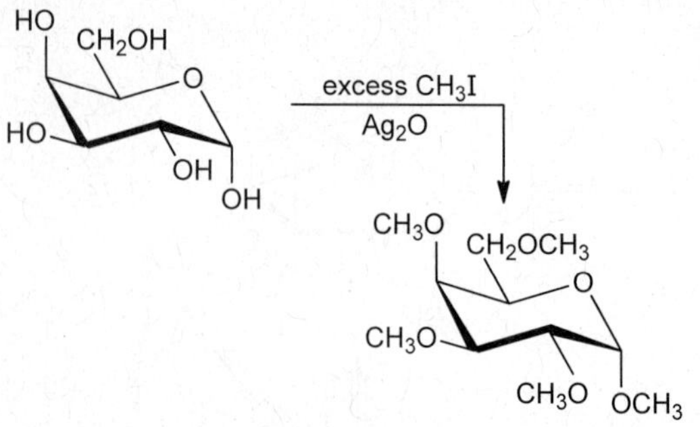

(b) Upon treatment with excess methyl iodide in the presence of silver oxide, all of the OH groups are converted into methoxy groups, as shown:

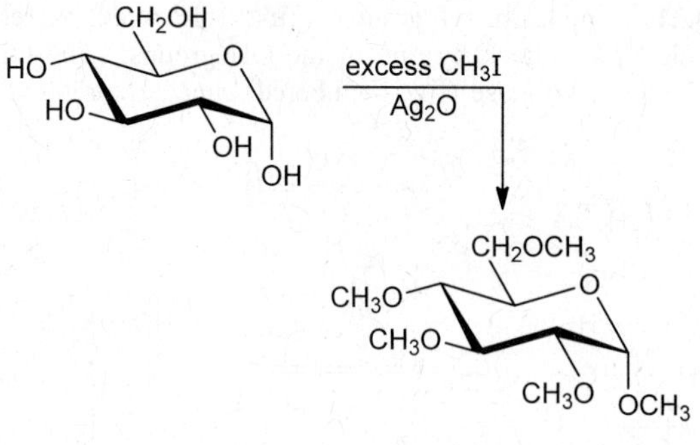

(c) Upon treatment with excess methyl iodide in the presence of silver oxide, all of the OH groups are converted into methoxy groups, as shown:

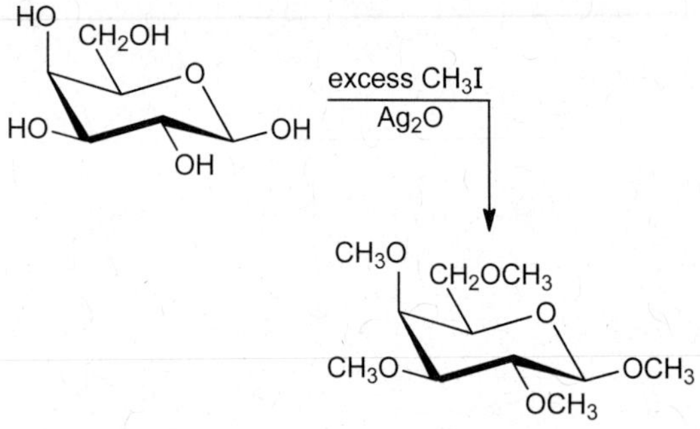

24.25. Under acidic conditions, the anomeric OH group can be protonated, giving an excellent leaving group (water). Loss of the leaving group generates a resonance-stabilized cation intermediate.

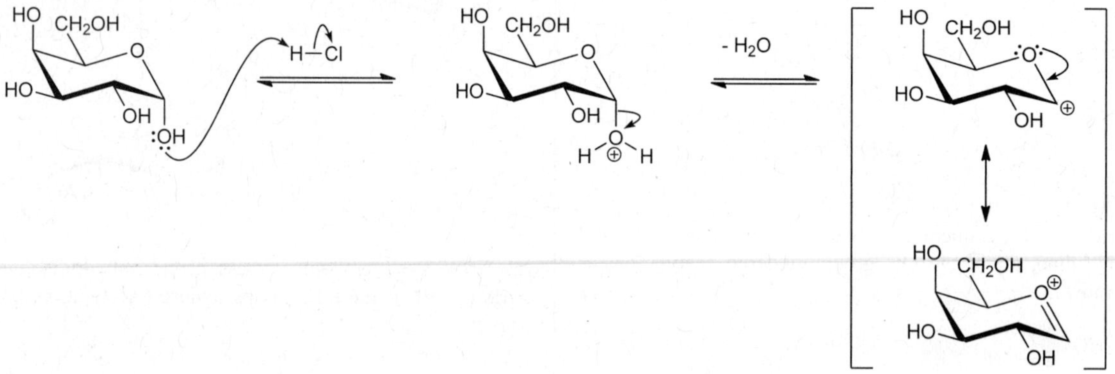

This intermediate can then be attacked by ethanol, giving an oxonium ion, which is then deprotonated to give an acetal:

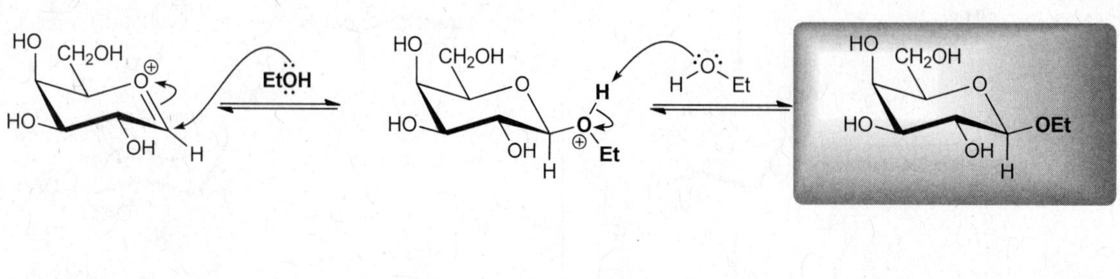

Notice that ethanol is shown to attack from above, but it can also attack from below, giving the following acetal:

24.26. Under acidic conditions, the anomeric methoxy group can be protonated, giving an excellent leaving group (methanol). Loss of the leaving group generates a resonance-stabilized cation intermediate (resonance structures not shown) which can then be attacked by methanol. This attack can occur from either above or from below (as seen in the previous problem) giving a mixture of both anomers. In the last step of the mechanism, the oxonium ion is then deprotonated:

24.27.

(a) D-Mannose is epimeric with D-glucose at C2, as shown:

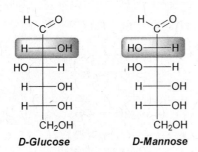

(b) D-Allose is epimeric with D-glucose at C3, as shown:

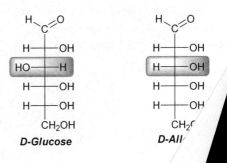

(c) D-Galactose is epimeric with D-glucose at C4, as shown:

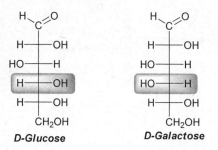

D-Glucose   **D-Galactose**

24.28. Reduction of the carbonyl group generates the same product in each case. This can be seen by rotating one of the products by 180°, as shown:

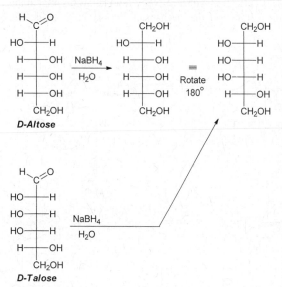

D-Altose

D-Talose

24.29. Reduction of the carbonyl group generates the same *meso* product in each case. This can be seen by rotating one of the products by 180°, as shown:

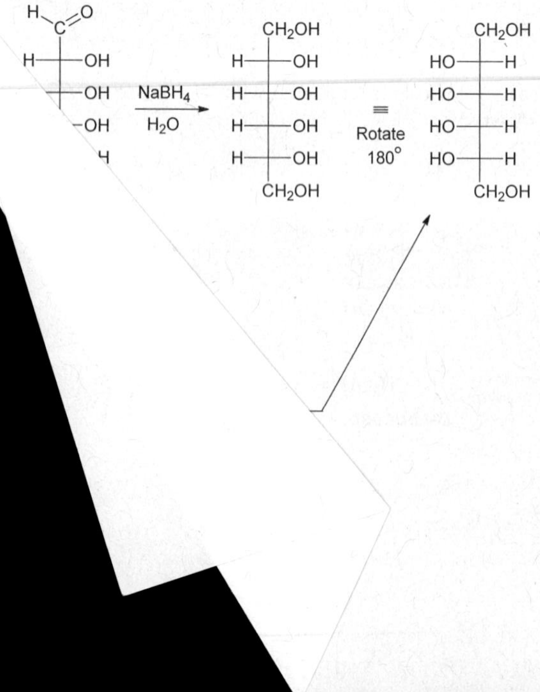

24.30. Reduction of either D-allose or D-galactose will produce a *meso* alditol. *Meso* compounds are optically inactive:

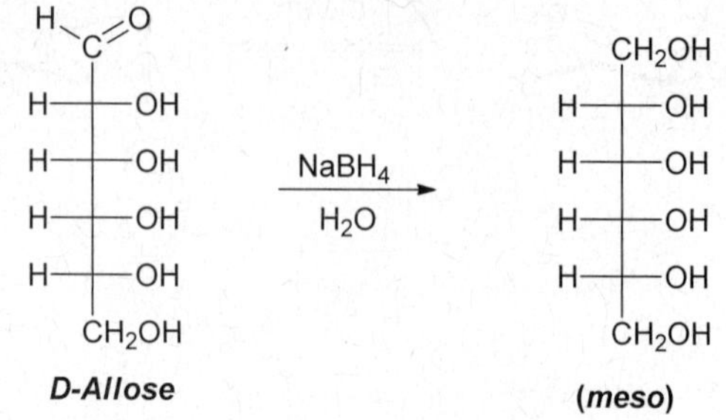

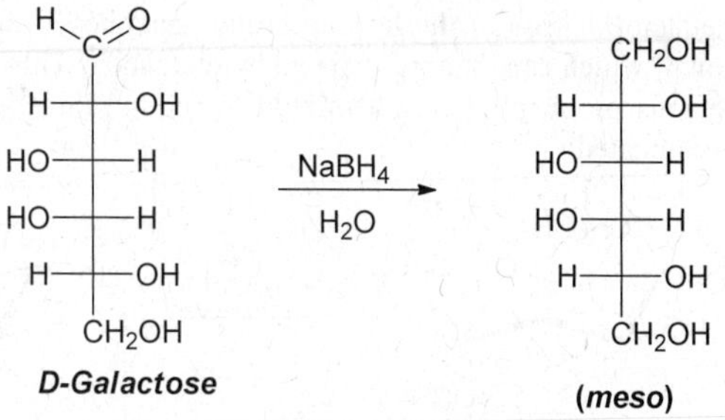

D-Allose   **(meso)**

D-Galactose   **(meso)**

24.31.
(a) The anomeric position is occupied by a methoxy group. Therefore, this compound is an acetal and is not a reducing sugar.

(b) The anomeric position is occupied by an OH group. Therefore, this compound is a reducing sugar.

(c) The anomeric position is occupied by an OH group. Therefore, this compound is a reducing sugar.

24.32.
(a) The open chain form of this compound is D-galactose, which is oxidized under these conditions to give the following aldonic acid:

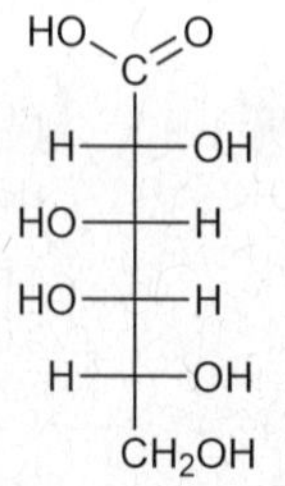

D-Galactonic acid

(b) The open chain form of this compound is D-galactose, which is oxidized under these conditions to give the following aldonic acid:

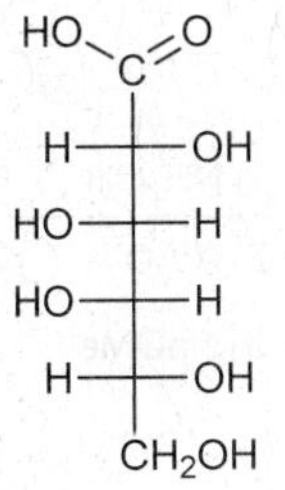

D-Galactonic acid

(c) The open chain form of this compound is D-glucose, which is oxidized under these conditions to give the following aldonic acid:

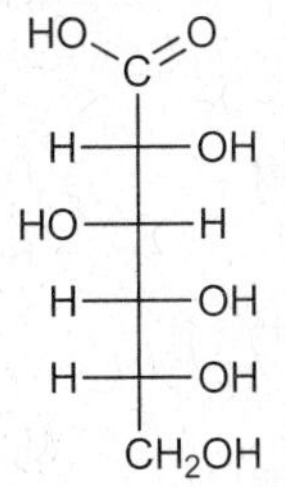

D-Gluconic acid

(d) The open chain form of this compound is D-glucose, which is oxidized under these conditions to give the following aldonic acid:

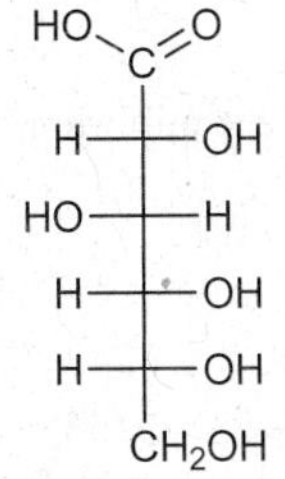

D-Gluconic acid

24.33. This compound will not be a reducing sugar because the anomeric position is an acetal group.

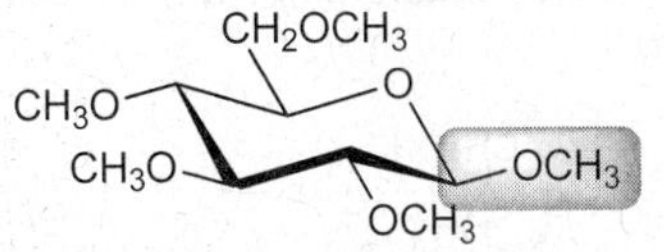

β-D-Glucopyranose pentamethyl ether

24.34.
(a) In a Kiliani-Fischer synthesis, the chain is lengthened, with C1 becoming C2 in the product. Both possible configurations of the C2 position are obtained, giving the following epimers:

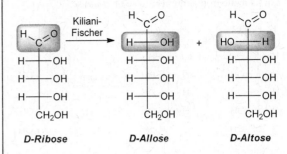

D-Ribose   **D-Allose**   **D-Altose**

(b) In a Kiliani-Fischer synthesis, the chain is lengthened, with C1 becoming C2 in the product. Both possible configurations of the C2 position are obtained, giving the following epimers:

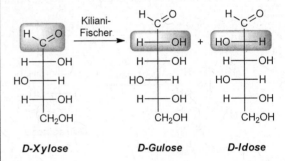

D-Xylose   **D-Gulose**   **D-Idose**

(c) In a Kiliani-Fischer synthesis, the chain is lengthened, with C1 becoming C2 in the product. Both possible configurations of the C2 position are obtained, giving the following epimers:

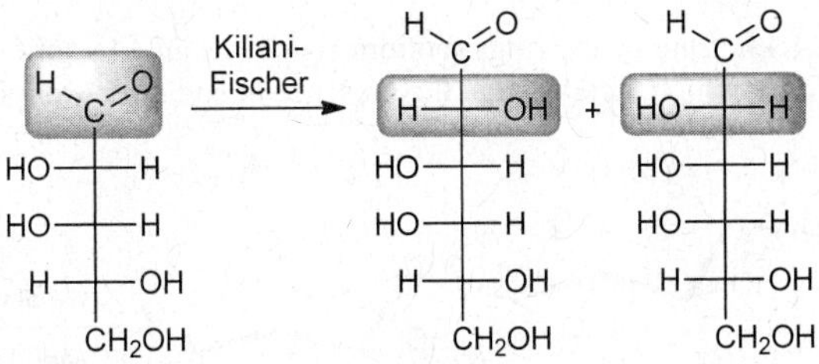

D-Lyxose   **D-Galactose**   **D-Talose**

24.35. Conversion of D-erythrose (which has four carbon atoms) to D-ribose (which has five carbon atoms) requires a chain-lengthening process. This process will produce D-ribose together with its C2 epimer, D-arabinose, as shown:

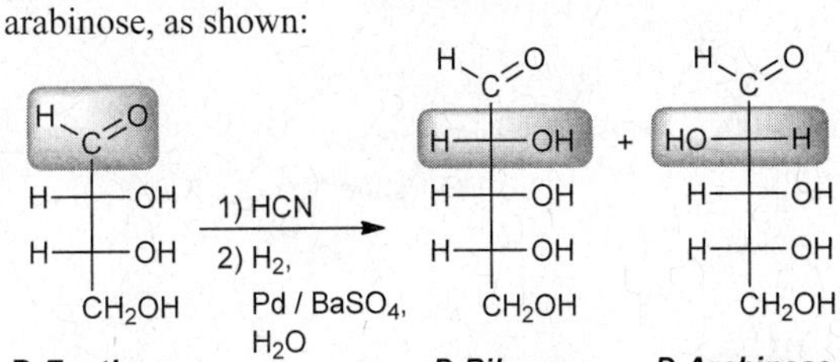

D-Erythrose   **D-Ribose**   **D-Arabinose**

24.36. A Wohl degradation involves the removal of a carbon atom from an aldose. As shown below, D-ribose can be made from either D-allose or D-altose:

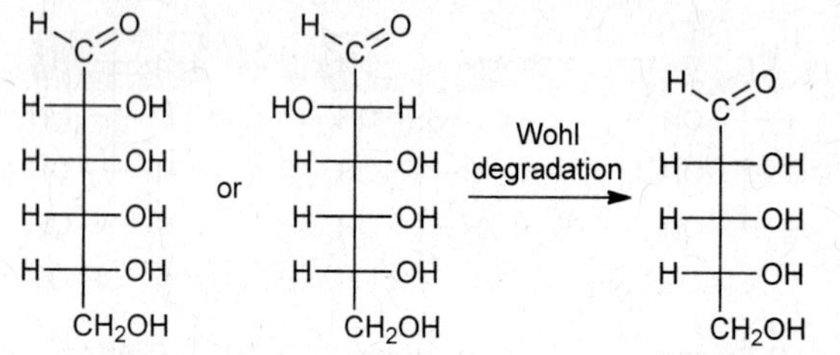

D-Allose   **D-Altose**   **D-Ribose**

24.37. A Wohl degradation will remove a carbon atom from an aldose. This method can be used to convert D-ribose into D-erythrose, as shown:

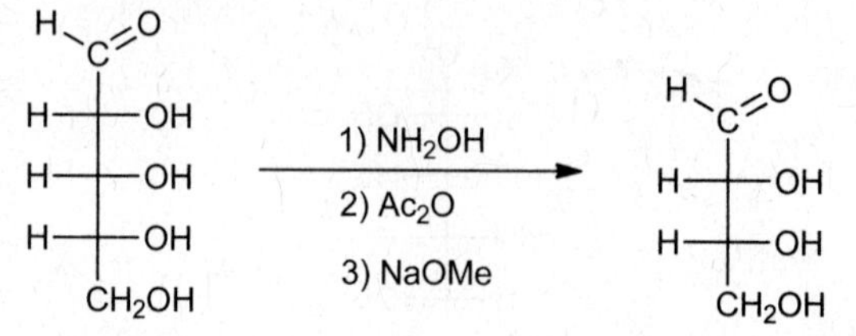

D-Ribose   **D-Erythrose**

24.38. A Wohl degradation will remove a carbon atom from D-glucose. This carbon atom is then restored with a Kiliani-Fischer synthesis, giving D-glucose and its C2 epimer, D-mannose.

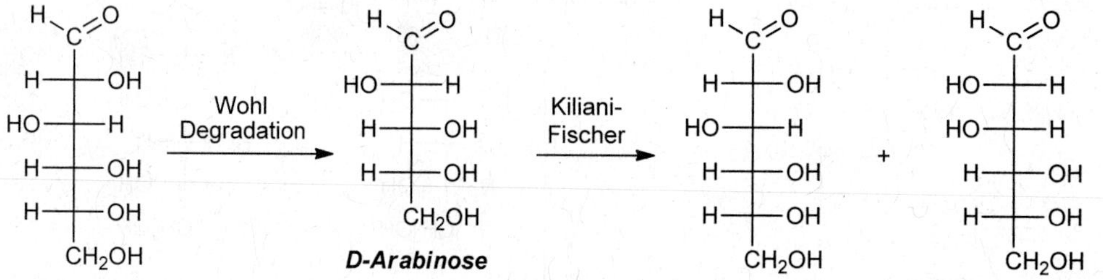

D-Glucose   **D-Arabinose**   **D-Glucose**   **D-Mannose**

24.39.
(a) One of the anomeric positions (bottom right) bears an OH group. Therefore, this disaccharide is a reducing sugar.
(b) Both anomeric positions bear acetal groups, so this disaccharide is not a reducing sugar.
(c) Both anomeric positions bear acetal groups, so this disaccharide is not a reducing sugar.

24.40. One of the rings (bottom right) has an anomeric OH group. As such, it is in equilibrium with the open chain form, which is reduced in the presence of sodium borohydride, as shown:

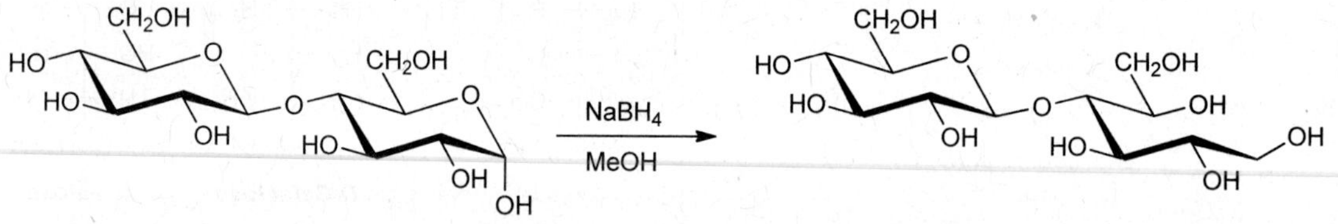

24.41.
(a) One of the rings (bottom right) has an anomeric OH group. As such, it is in equilibrium with the open chain form, which is reduced in the presence of sodium borohydride, as shown:

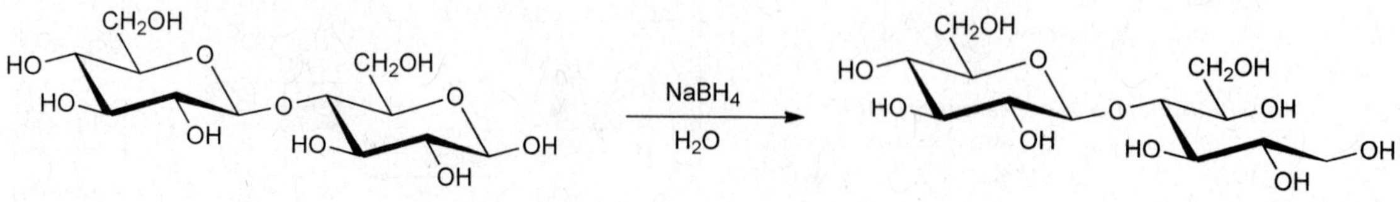

(b) One of the rings (bottom right) has an anomeric OH group, and is therefore in equilibrium with the open chain form, which is oxidized in the presence of Br_2 and H_2O (at pH = 6), as shown:

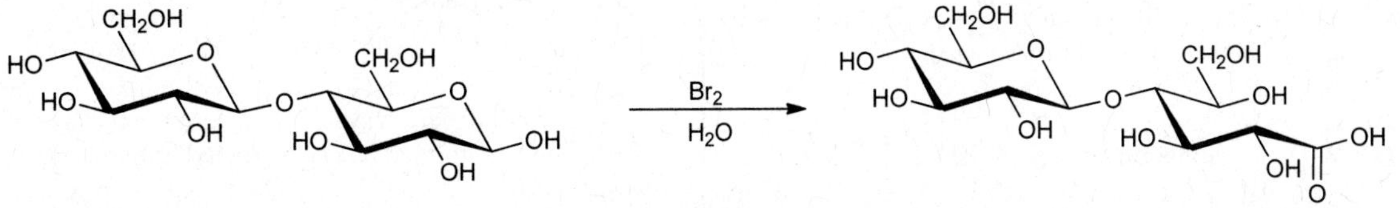

(c) One of the rings (bottom right) has an anomeric OH group. Upon treatment with methanol in acidic conditions, this OH group undergoes methylation, giving a methoxy group, as shown:

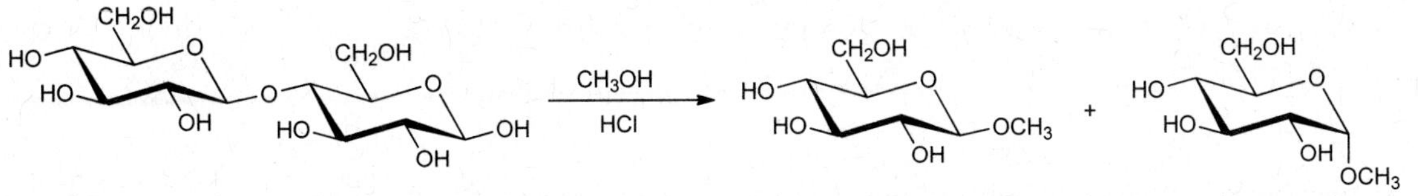

(d) Upon treatment with excess acetic anhydride and pyridine, each of the OH groups undergoes acetylation, as shown:

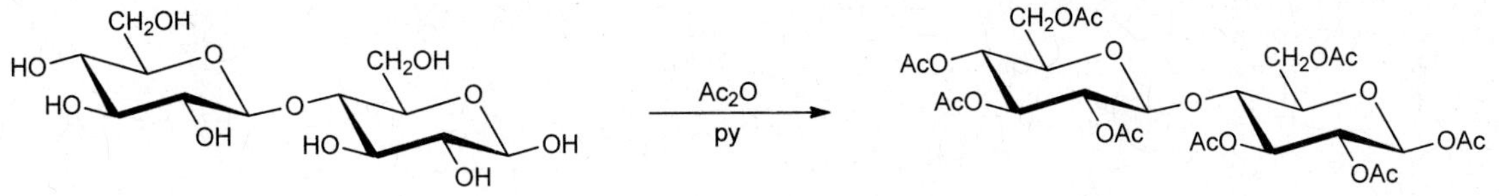

24.42.

(a) The OH group connected to C3 is pointing to the right, so this is a D-sugar. The functional group at C1 is an aldehyde group, so the compound is an aldose. And finally, the compound has four carbon atoms, so it is a tetrose. In summary, this compound is a D-aldotetrose.

(b) The OH group connected to C4 is pointing to the left, so this is an L-sugar. The functional group at C1 is an aldehyde group, so the compound is an aldose. And finally, the compound has five carbon atoms, so it is a pentose. In summary, this compound is an L-aldopentose.

(c) The OH group connected to C4 is pointing to the right, so this is a D-sugar. The functional group at C1 is an aldehyde group, so the compound is an aldose. And finally, the compound has five carbon atoms, so it is a pentose. In summary, this compound is a D-aldopentose.

(d) The OH group connected to C5 is pointing to the right, so this is a D-sugar. The functional group at C1 is an aldehyde group, so the compound is an aldose. And finally, the compound has six carbon atoms, so it is a hexose. In summary, this compound is a D-aldohexose.

(e) The OH group connected to C4 is pointing to the right, so this is a D-sugar. The functional group at C2 is a ketone group, so the compound is a ketose. And finally, the compound has five carbon atoms, so it is a pentose. In summary, this compound is a D-ketopentose

24.43. D-Glyceraldehyde has the *R* configuration, while L-glyceraldehyde has the *S* configuration. Therefore:
(a) The chirality center has the *R* configuration. This compound is D-glyceraldehyde.
(b) The chirality center has the *S* configuration. This compound is L-glyceraldehyde.
(c) The chirality center has the *R* configuration. This compound is D-glyceraldehyde.
(d) The chirality center has the *S* configuration. This compound is L-glyceraldehyde.

24.44.
(a) This compound is D-Glucose (see Figure 24.5).
(b) This compound is D-Mannose (see Figure 24.5).
(c) This compound is D- Galactose (see Figure 24.5).
(d) This compound is L-Glucose (see Figure 24.5).

24.45.
(a) D-Ribose is epimeric with D-arabinose at C2.
(b) D-Arabinose is epimeric with D-lyxose at C2.
(c) The enantiomer of D-ribose has the opposite configuration (*S*, rather than *R*) for all three chirality centers:

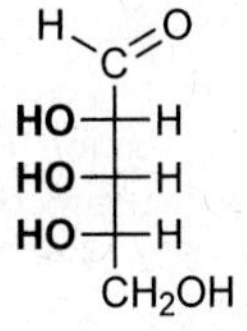

L-Ribose

(d) They are the same compound. That is, the enantiomer of D-arabinose is L-arabinose, which is also the C2 epimer of L-ribose.

(e) They are diastereomers because they are stereoisomers that are not mirror images.

24.46.

(a) We use a numbering system to determine the size of the ring that is formed. The carbonyl group can be attacked by the OH group that is connected to C4, giving a five-membered ring:

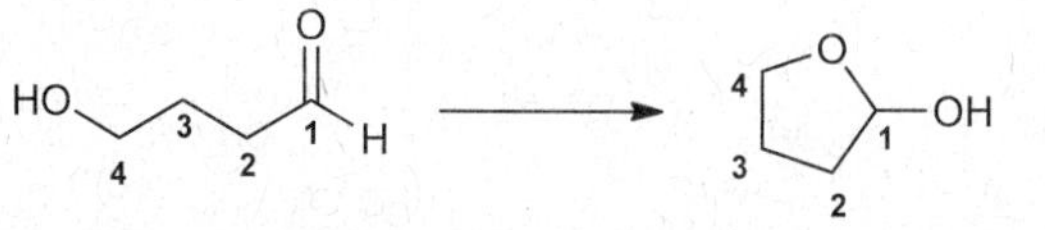

(b) We use a numbering system to determine the size of the ring that is formed. The carbonyl group can be attacked by the OH group that is connected to C5, giving a six-membered ring. Notice that there is a methyl group connected to C1 of the ring, as well as a methyl group connected to C5 of the ring:

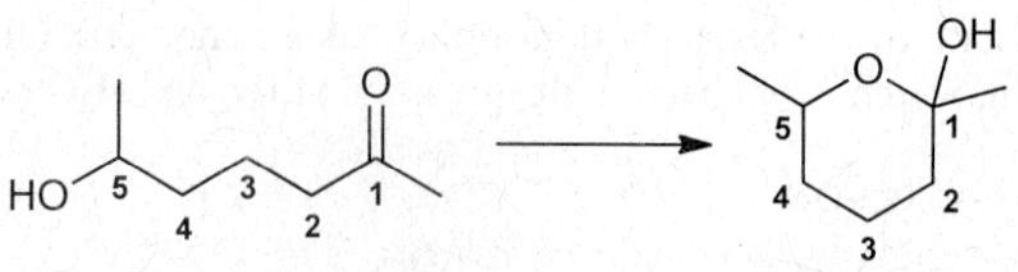

(c) We use a numbering system to determine the size of the ring that is formed. The carbonyl group can be attacked by the OH group that is connected to C6, giving a seven-membered ring. Notice that there is a methyl group connected to C6 of the ring:

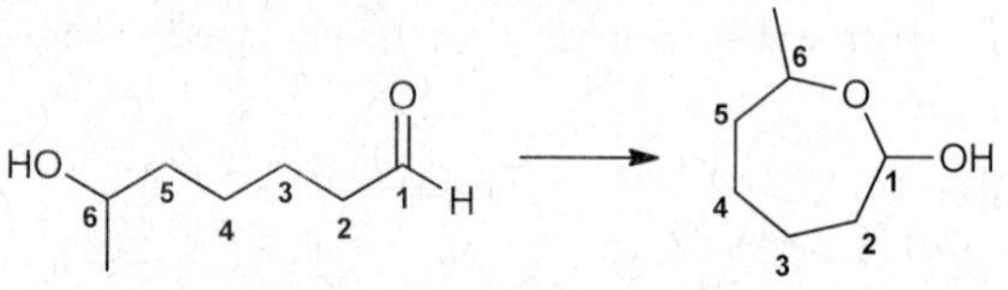

24.47. We must use a numbering system, just as we did in the previous problem. There are two methyl groups at C3, which must be drawn in the starting material:

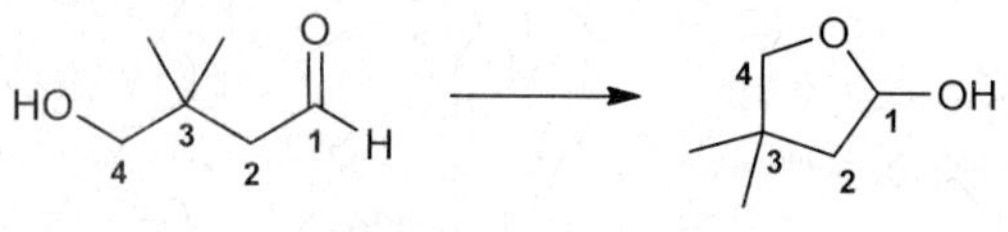

24.48.

(a) The following are the two pyranose forms (α and β anomers) of D-ribose:

D-ribose   **α pyranose ring**   **β pyranose ring**

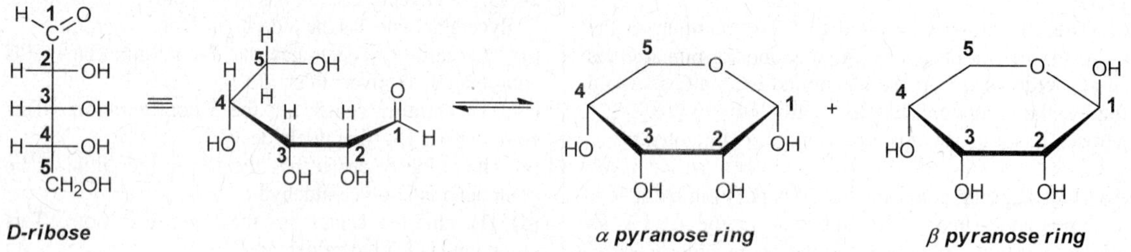

(b) The following are the two furanose forms (α and β anomers) of D-ribose:

D-ribose   **α furanose ring**   **β furanose ring**

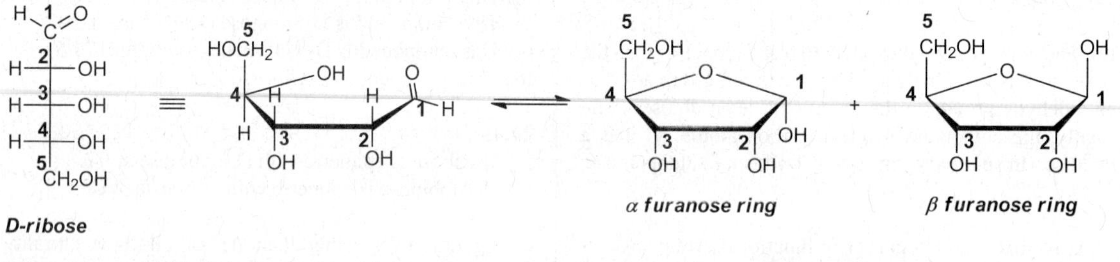

24.49.

(a) These compounds are diastereomers that differ from each other in the configuration of only one chirality center. Therefore, they are epimers.

(b) These compounds are stereoisomers that are not mirror images of one another. Therefore, they are diastereomers.

(c) These compounds are non-superimposable mirror images of one another. Therefore, they are enantiomers.

(d) These structures are two different representations of the same compound (β-D-glucopyranose).

24.50. Upon treatment with aqueous acid, the anomeric methoxy group is replaced with an anomeric hydroxy group. The open chain form of the resulting cyclic hemiacetal is D-glucose.

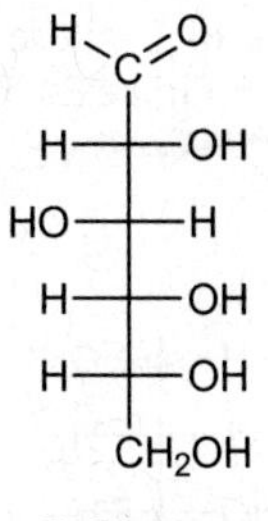

D-Glucose

24.51. To assign the configuration of each chirality center, we can use the rule of thumb that was pointed out in the solution to Problem **24.3**. Specifically, an OH on the right side of the Fischer projection indicates the *R* configuration, while an OH on the left side of the Fischer projection indicates the *S* configuration:

(a)

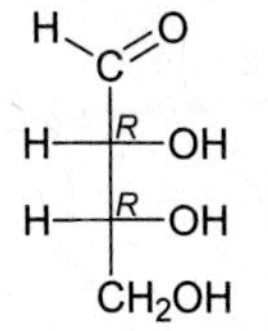

(b)

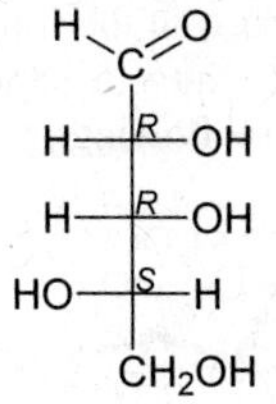

(c)

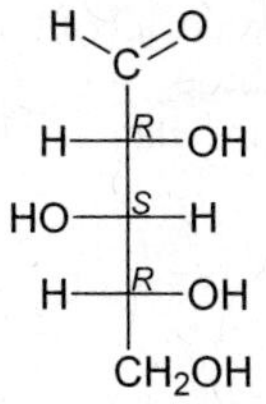

(d)

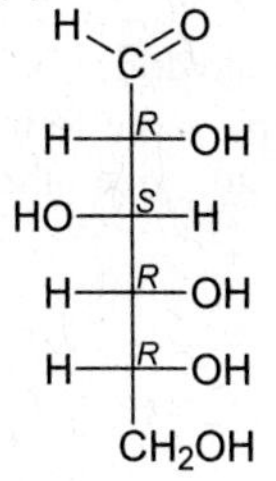

(e)

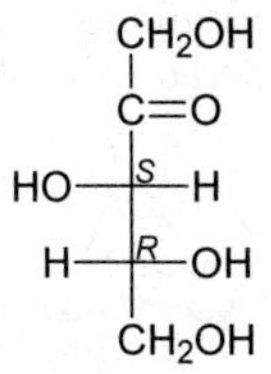

24.52. The structures below can be found in Figure 24.5:

(a)

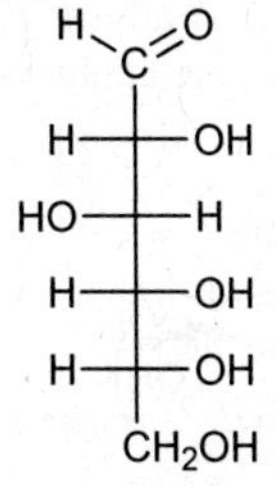

D-Glucose

(b)

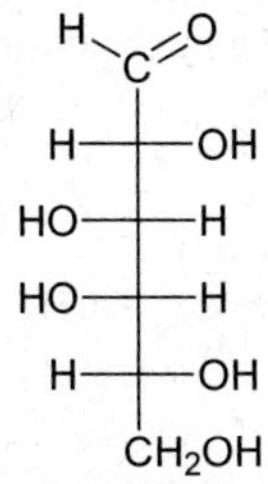

D-Galactose

(c)

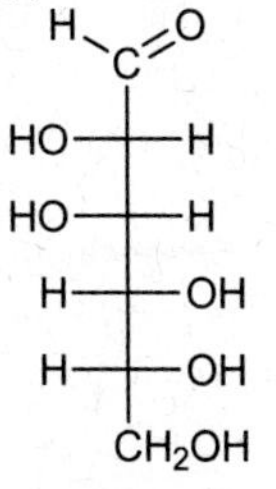

D-Mannose

(d)

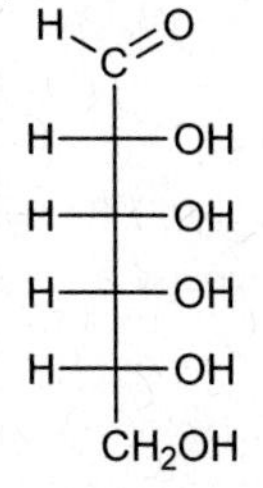

D-Allose

24.53.

(a) We begin by drawing the open chain form of D-fructose (see Figure 24.6). This compound is closed into a furanose form, so we draw a Haworth projection of a furanose skeleton (a five-membered ring with the oxygen atom in the back). We use a numbering system, shown below, to assist us. Next, we draw the CH_2OH group (connected to C5) pointing up, because this is a D sugar. The anomeric OH group is then drawn pointing up (the β anomer). Finally, the remaining OH groups are drawn. Any OH groups on the right side of the Fischer projection will be pointing down in the Haworth projection, while any OH groups on the left side of the Fischer projection will be pointing up in the Haworth projection):

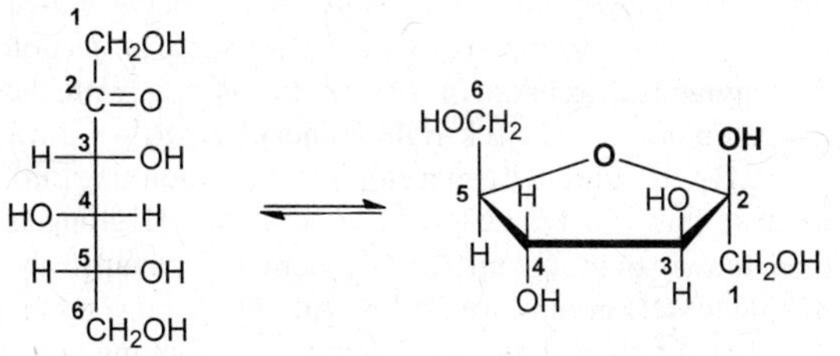

D-Fructose

(b) We begin by considering the open chain form of D-galactose (see Figure 24.5). This compound is closed into a pyranose form, so we draw a Haworth projection of a pyranose skeleton (a six-membered ring with the oxygen atom in the back right corner). Next, we draw the CH_2OH group (connected to C5) pointing up, because this is a D sugar. The anomeric OH group is then drawn pointing up (the β anomer). Finally, the remaining OH groups are drawn. Any OH groups on the right side of the Fischer projection will be pointing down in the Haworth projection, while any OH groups on the left side of the Fischer projection will be pointing up in the Haworth projection):

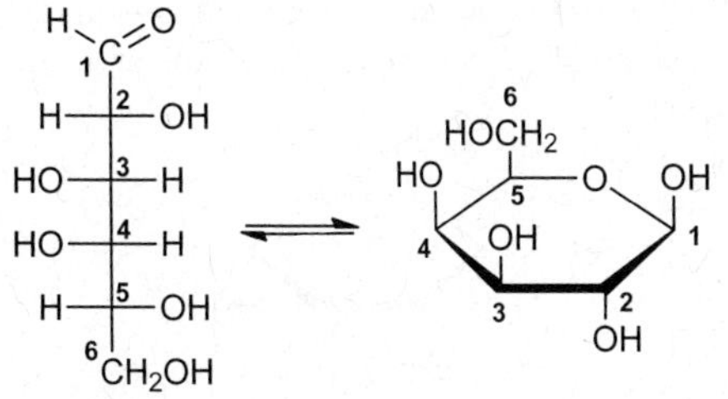

D-Galactose

(c) We begin by considering the open chain form of D-glucose (see Figure 24.5). This compound is closed into a pyranose form, so we draw a Haworth projection of a pyranose skeleton (a six-membered ring with the oxygen atom in the back right corner). Next, we draw the CH_2OH group (connected to C5) pointing up, because this is a D sugar. The anomeric OH group is then drawn pointing up (the β anomer). Finally, the remaining OH groups are drawn. Any OH groups on the right side of the Fischer projection will be pointing down in the Haworth projection, while any OH groups on the left side of the Fischer projection will be pointing up in the Haworth projection):

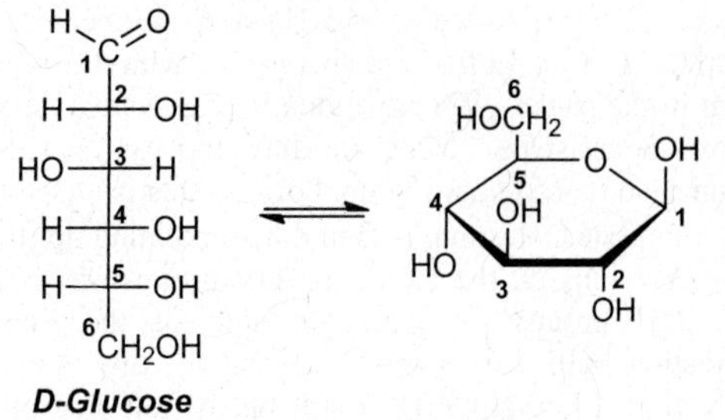

D-Glucose

(d) We begin by considering the open chain form of D-mannose (see Figure 24.5). This compound is closed into a pyranose form, so we draw a Haworth projection of a pyranose skeleton (a six-membered ring with the oxygen atom in the back right corner). Next, we draw the CH$_2$OH group (connected to C5) pointing up, because this is a D sugar. The anomeric OH group is then drawn pointing up (the β anomer). Finally, the remaining OH groups are drawn. Any OH groups on the right side of the Fischer projection will be pointing down in the Haworth projection, while any OH groups on the left side of the Fischer projection will be pointing up in the Haworth projection):

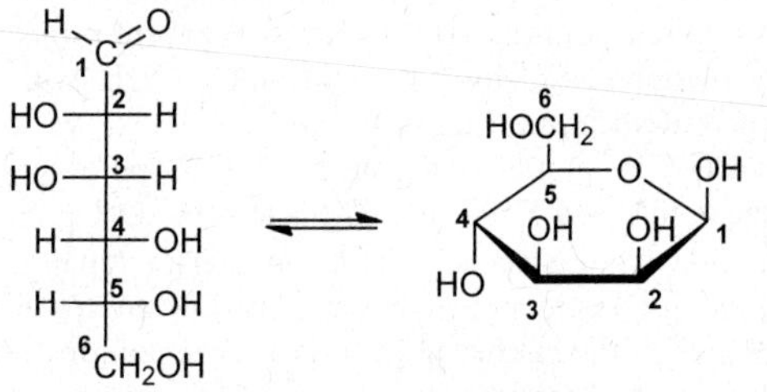

D-Mannose

24.54. D-allose is the aldohexose that is epimeric with D-glucose at C3. The α-pyranose form of D-allose is shown here:

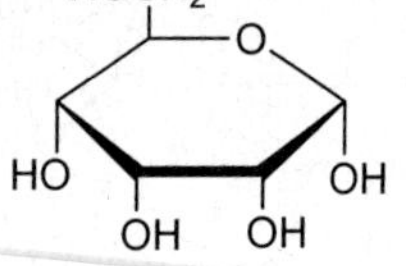

24.55.
(a) This structure represents the α anomer of the pyranose form of D-allose and is therefore called α-D-allopyranose.
(b) This structure represents the β anomer of the pyranose form of D-galactose and is therefore called β-D-galactopyranose.
(c) This structure represents an acetal of the β anomer of the pyranose form of D-glucose and is called methyl β-D-glucopyranoside .

24.56. The structures of D-allose, D-galactose and D-glucose can be found in Figure 24.5.

(a)   **(b)**   **(c)**

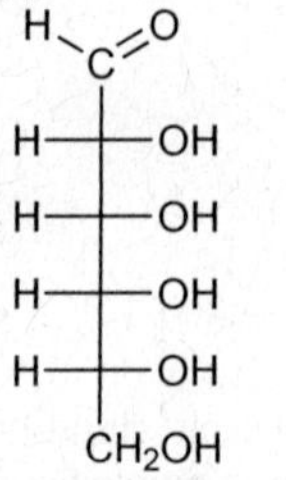

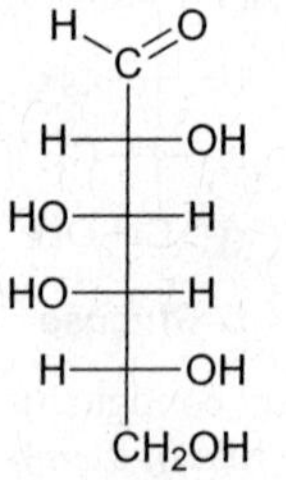

D-Allose   D-Galactose   D-Glucose

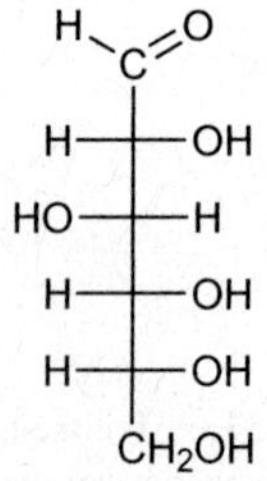

24.57.
(a) When treated with excess methyl iodide and silver oxide, all of the OH groups in the β-pyranose form of D-allose are converted into methoxy groups (via methylation), giving the following compound:

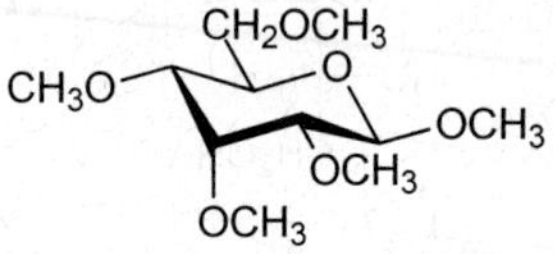

(b) When treated with excess acetic anhydride and pyridine, each of the OH groups in the β-pyranose form of D-allose will undergo acetylation to give the following compound:

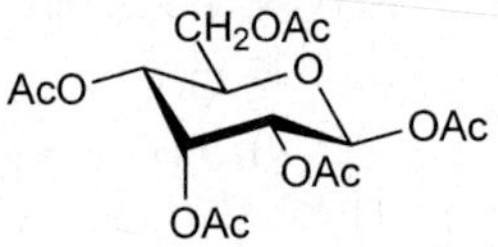

(c) When treated with methanol and HCl, the anomeric OH group in the β-pyranose form of D-allose is converted into a methoxy group. Both possible anomers are formed:

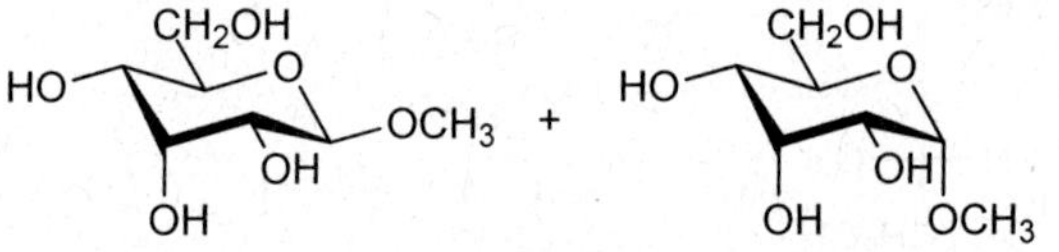

24.58. The product, shown below, is optically inactive because it is a *meso* compound:

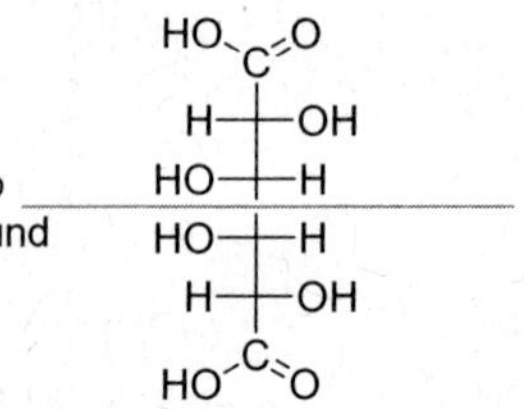

24.59. Upon treatment with nitric acid, D-allose undergoes oxidation to give an aldaric acid that is optically inactive because it is a *meso* compound:

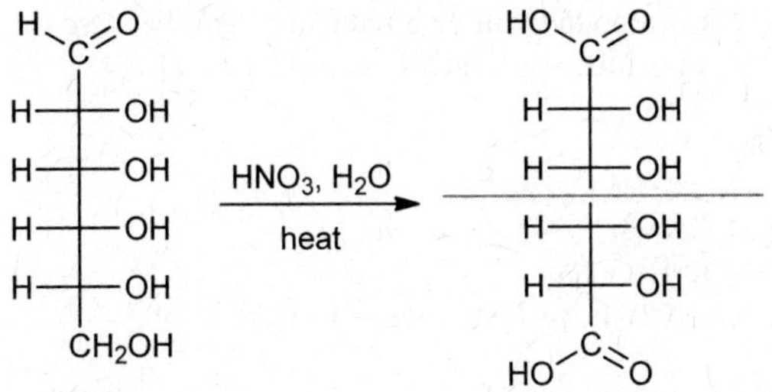

D-Allose

optically inactive

(*meso* compound)

24.60. The skeleton of the chair is drawn with an oxygen atom in the upper back-right corner. Each substituent is then labeled as "up" or "down" and placed on the chair accordingly. The anomeric OH group is drawn pointing down, indicating the α anomer. In this chair conformation, the largest substituent (CH$_2$OH) occupies an equatorial position, so this is the more stable chair conformation of α-D-altropyranose.

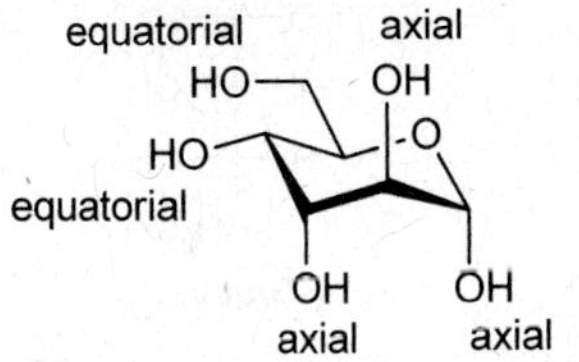

24.61. Upon treatment with excess methyl iodide in the presence of silver oxide, all of the OH groups are converted into methoxy groups, as shown. Upon treatment with aqueous acid, the acetal is hydrolyzed, giving both anomers of D-galactopyranose:

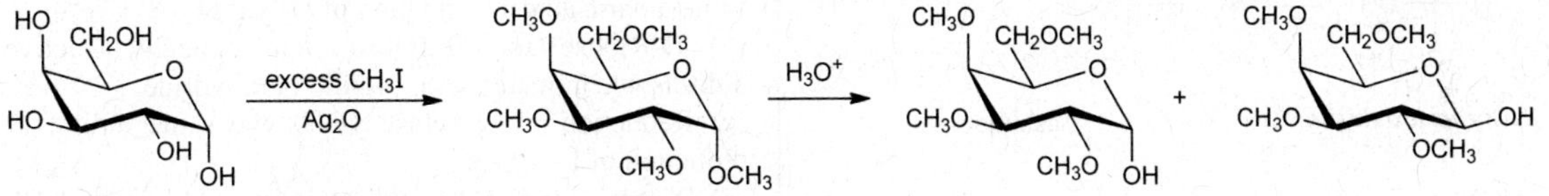

24.62.
(a) These compounds (see Figure 24.5) are stereoisomers that are not mirror images of each other, so they are diastereomers.
(b) D-Ribose and D-arabinose are C2 epimers. Therefore, removing the OH group from C2 of either compound will result in the same structure. That is, 2-deoxy-D-ribose and 2-deoxy-D-arabinose are the same compound.

24.63. 2-Ketohexoses have three chirality centers. The configuration of one of these chirality centers (at C5) is fixed because the problem statement asks only for D sugars. That leaves two other chirality centers (C3 and C4), giving rise to the following four stereoisomers:

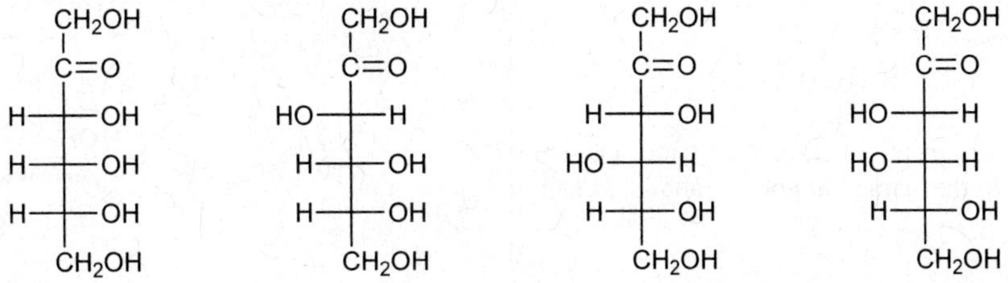

24.64. A Wohl degradation involves the removal of a carbon atom from an aldose. As shown below, D-ribose can be made from either D-allose or D-altose:

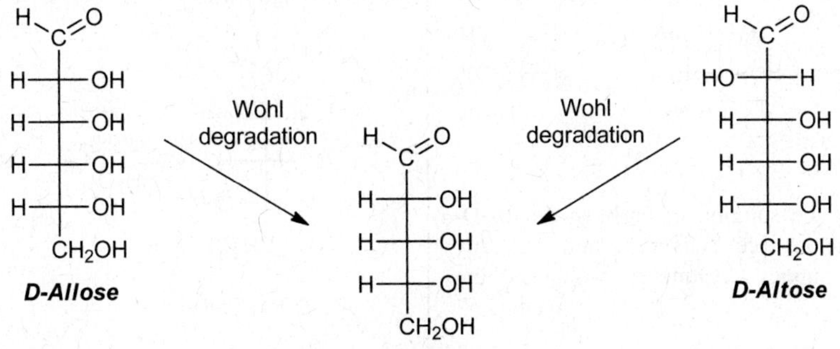

D-Allose   D-Ribose   D-Altose

24.65. In a Kiliani-Fischer synthesis, the chain is lengthened, with C1 becoming C2 in the product. Both possible configurations of the C2 position are obtained, giving the following epimers:

D-Arabinose ***D-Glucose*** ***D-Mannose***

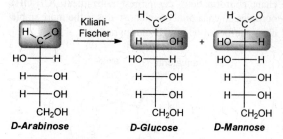

24.66. The aldehyde group is converted into a cyanohydrin. The newly installed chirality center can have either R or S configuration, giving the following diastereomers:

D-Glyceraldehyde Diastereomers

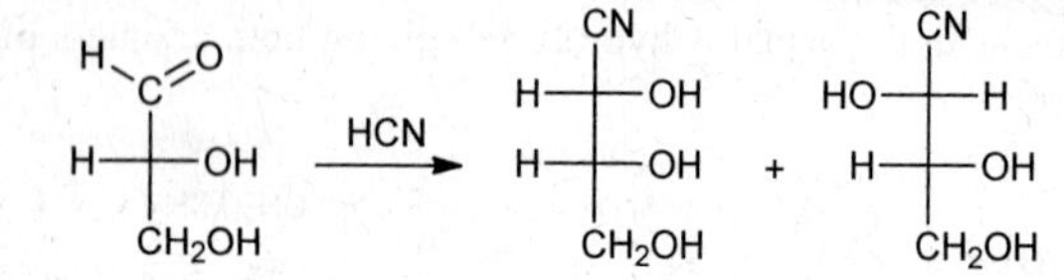

24.67.
(a) Upon treatment with sodium borohydride, the aldehyde group of D-glucose is reduced to an alcohol, giving the following alditol:

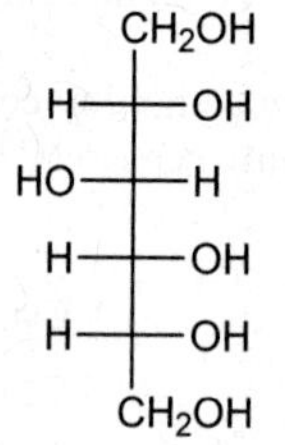

(b) As shown, treatment of L-gulose with sodium borohydride gives the same alditol as above (when rotated 180°).

L-Gulose

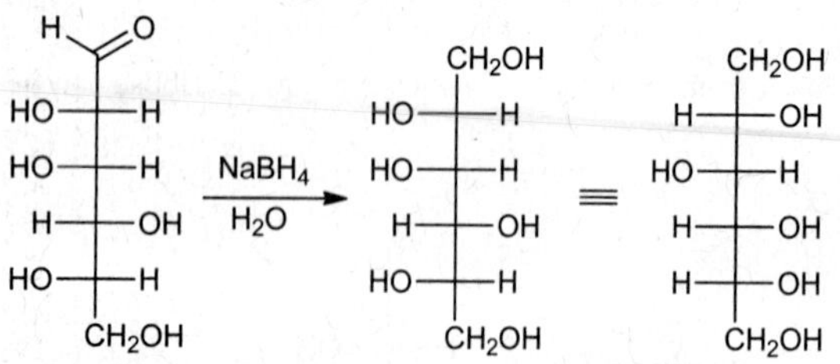

24.68. As seen in the solution to Problem **24.30**, D-allose and D-galactose are converted into optically inactive alditols upon treatment with sodium borohydride.

24.69.
(a) This compound will not be a reducing sugar because the anomeric position is an acetal group.
(b) This compound will be a reducing sugar because the anomeric position bears an OH group.

24.70.
(a) CH_3OH, HCl
(b) CH_3OH, HCl
(c) HNO_3, H_2O, heat
(d) excess CH_3I, Ag_2O followed by H_3O^+

24.71.
(a) The methoxy group is replaced with an OH group. Under these conditions, both anomers are formed, giving α-D-glucopyranose and β-D-glucopyranose.
(b) The ethoxy group is replaced with an OH group. Under these conditions, both anomers are formed, giving α-D-galactopyranose and β-D-galactopyranose.

24.72.
(a) Oxidation of D-Arabinose with nitric acid gives the same aldaric acid as oxidation of D-lyxose.
(b) D-Ribose and D-xylose yield optically inactive alditols when treated with sodium borohydride.
(c) Reduction of D-xylose yields the same alditol as reduction of L-xylose.
(d) D-xylose can close into a β-pyranose form in which all substituents are equatorial.

24.73. Trehalose is a disaccharide assembled from two equivalents of the α-pyranose form of D-glucose. Trehalose is not a reducing sugar, which means that the two rings must be fused at the anomeric positions (so there is no anomeric OH group). The disaccharide is assembled from the α-pyranose form of each equivalent of D-glucose:

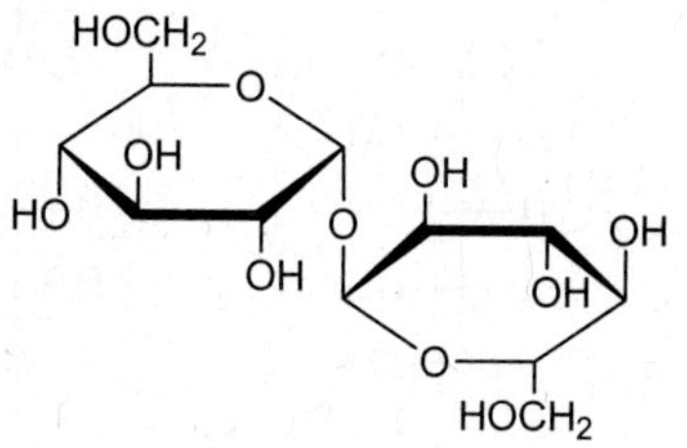

24.74. Reduction of D-xylose gives the following structure (D-xylitol):

D-Xylose ***D-Xylitol***

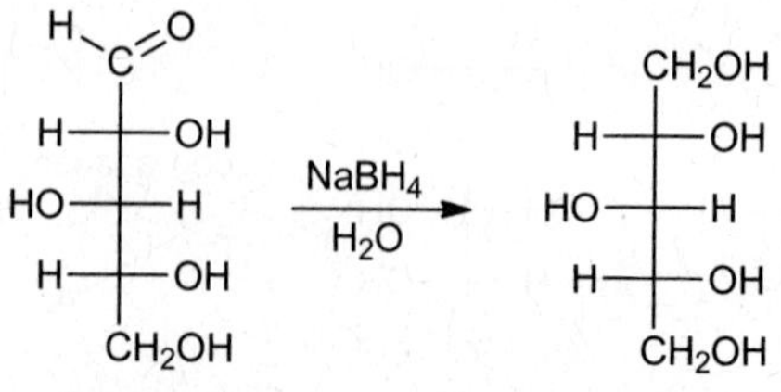

24.75. The 1→6-α-glycoside linkage of isomaltose is illustrated below:

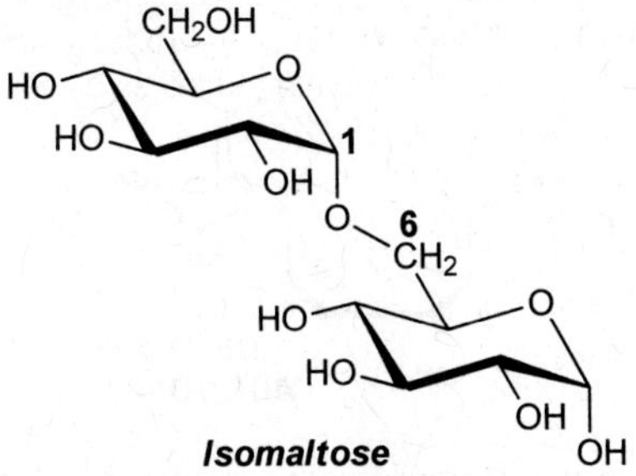

**Isomaltose
(a 1→ 6-α-glycoside)**

24.76.
(a) No, it is not a reducing sugar because the anomeric position has an acetal group.
(b) The acetal group is hydrolyzed, giving both anomers of the cyclic hemiacetal:

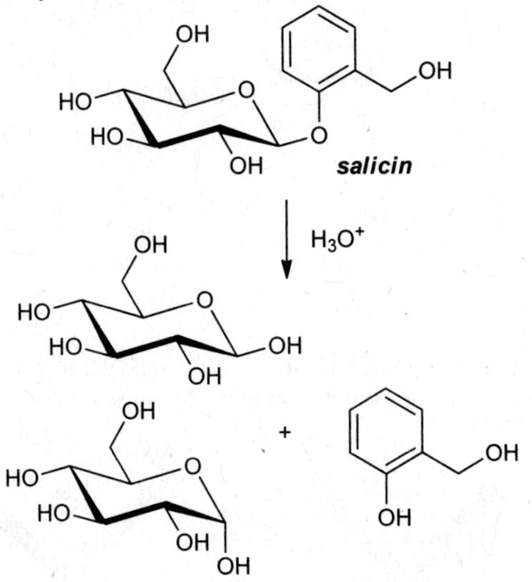

salicin

(c) Salicin is a β-glycoside.

(d) Upon treatment with acetic anhydride and pyridine, all of the OH groups undergo acetylation, giving the product shown:

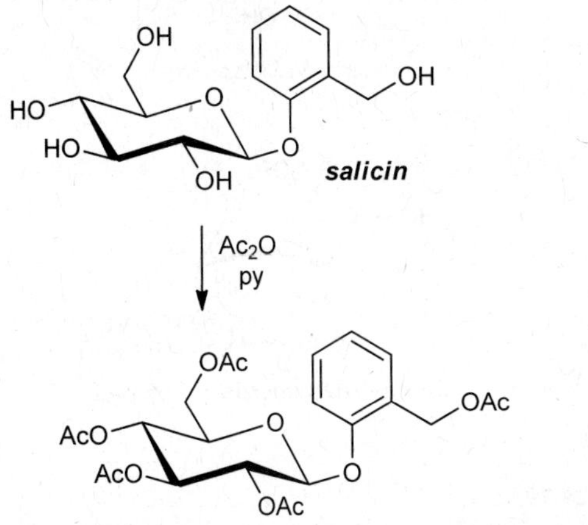

salicin

(e) No. In the absence of acid catalysis, the acetal group is not readily hydrolyzed.

24.77. The anomeric OH group is protonated under acidic conditions, giving an excellent leaving group (water). Loss of the leaving group gives a resonance-stabilized cation (resonance structure not shown). This cation is then attacked by phenol, giving an oxonium ion that is deprotonated to give the product:

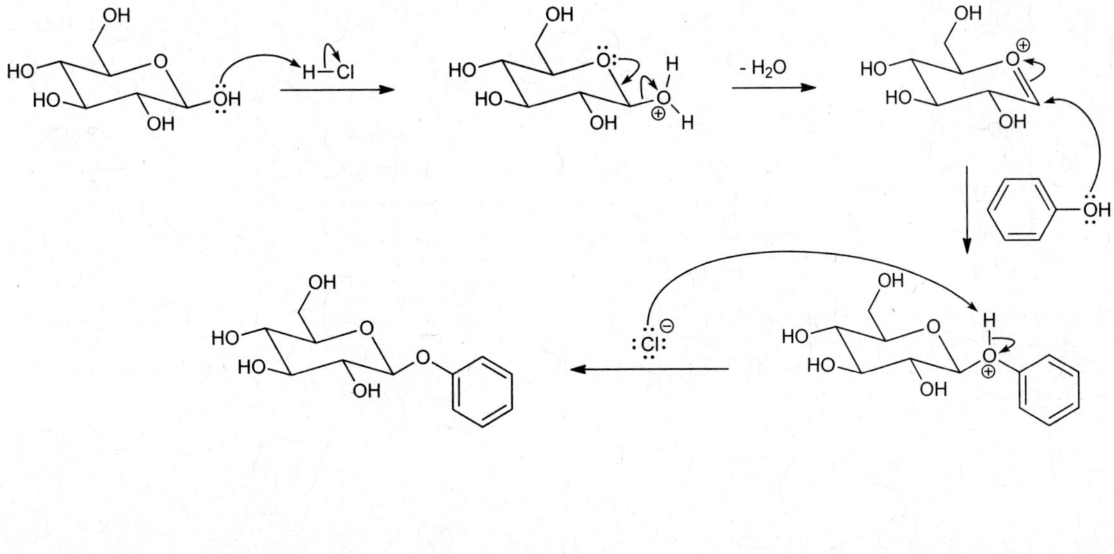

24.78. The following α and β anomers are obtained when D-glucose is treated with aniline:

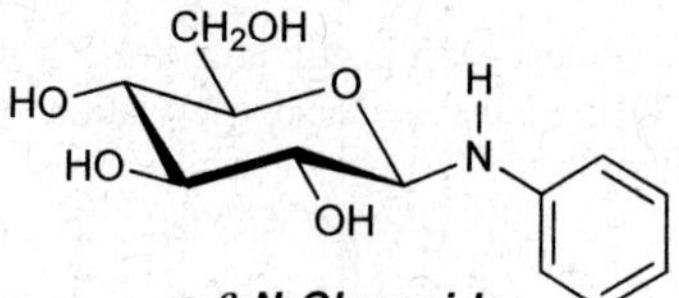

a β-N-Glycoside

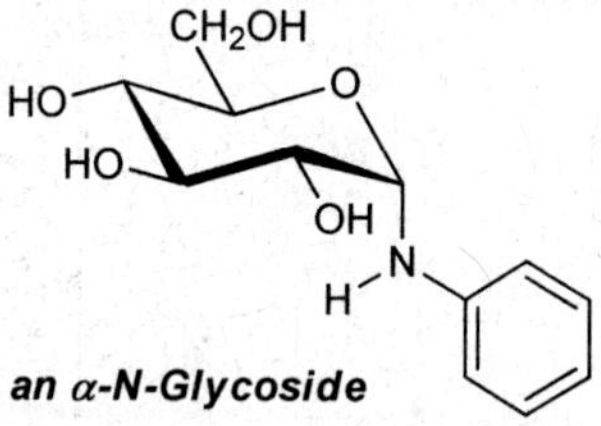

an α-N-Glycoside

24.79.

(a) The following nucleoside is formed from 2-deoxy-D-ribose and adenine. Notice that this structure differs from adenosine (see Figure 24.13) only at the C2 position. Specifically, the C2 position in this structure does not bear an OH group (as compared with adenosine):

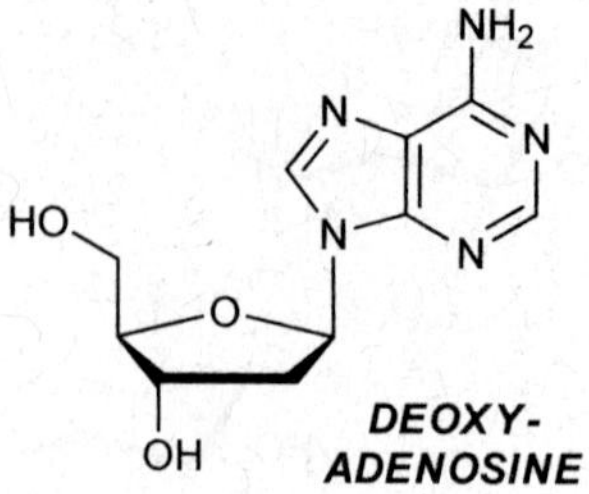

DEOXY-ADENOSINE

(b) The following nucleoside, called guanosine (see Figure 24.13), is formed from D-ribose and guanine:

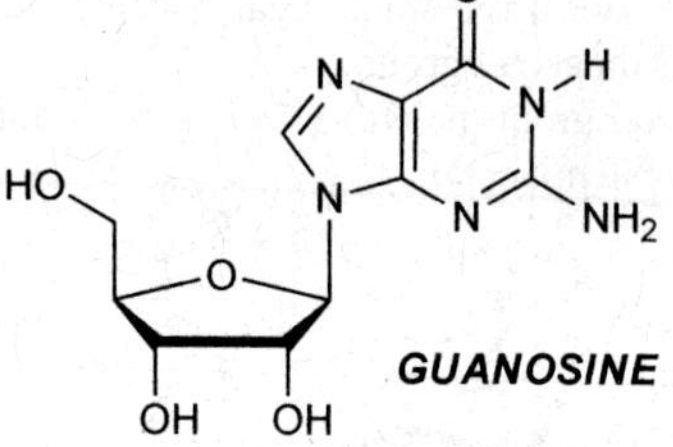

GUANOSINE

24.80. There are only four D-aldopentoses, all which are shown below. Only the first two are reduced to give an optically active alditol, as shown. The latter two are reduced to give *meso* compounds (not optically active):

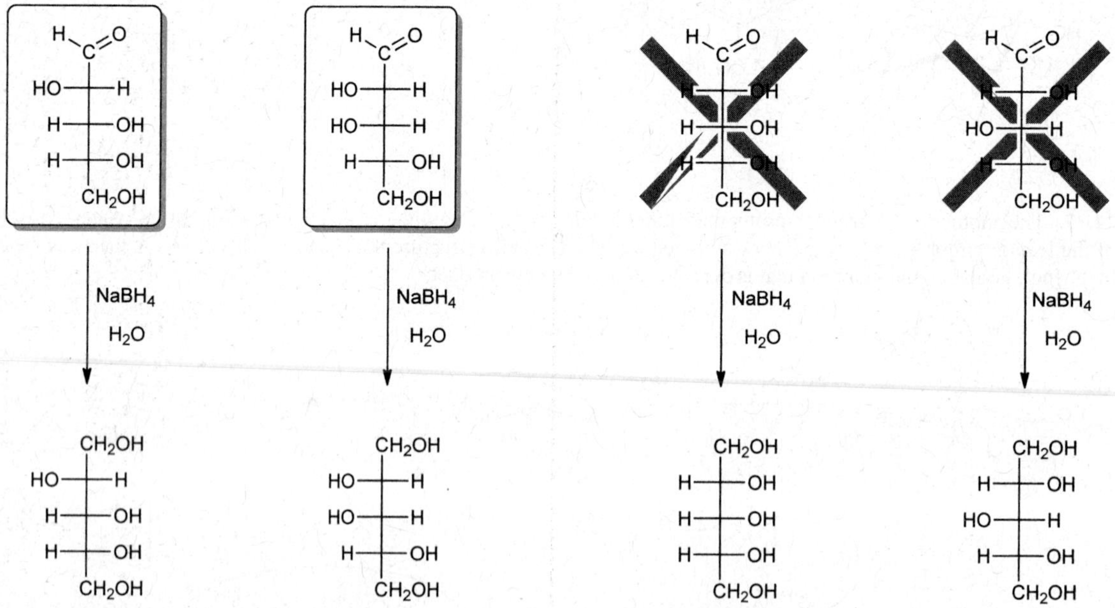

(optically active) (optically active) (optically inactive) (optically inactive)

24.81.

(a) D-Gluconic acid is formed when the C1 position of D-glucose undergoes oxidation to give a carboxylic acid, shown here:

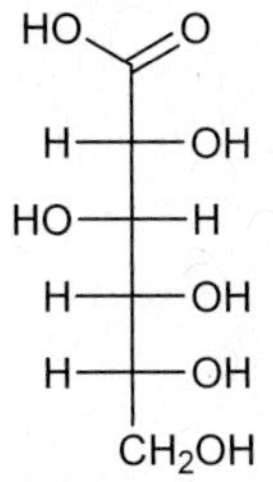

(b) A numbering system is used to help draw the product. The CH₂OH group (connected to C5) is drawn pointing up, because the starting material is a D sugar. Any OH groups on the right side of the Fischer projection will be pointing down in the Haworth projection, while any OH groups on the left side of the Fischer projection will be pointing up in the Haworth projection):

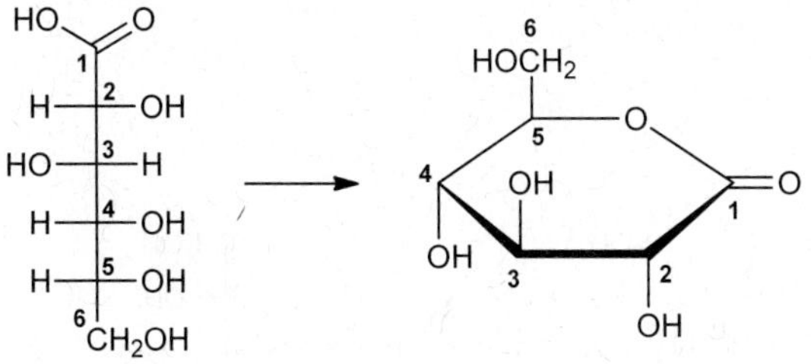

(c) Yes. The compound has chirality centers, and it is not a *meso* compound. Therefore, it will be optically active.

(d) The gluconic acid is a carboxylic acid and its IR spectrum is expected to have a broad signal between 2200 and 3600 cm^{-1}. The IR spectrum of the lactone will not have this broad signal.

24.82. In order for the CH₂OH group to occupy an equatorial position, all of the OH groups on the ring must occupy axial positions. The total energy cost associated with the steric interactions of the axial OH groups is more than the energy cost associated with one (albeit larger) CH₂OH group in an axial position. Therefore, the equilibrium will favor the form in which the CH₂OH group occupies an axial position. The structure of L-idose is shown here:

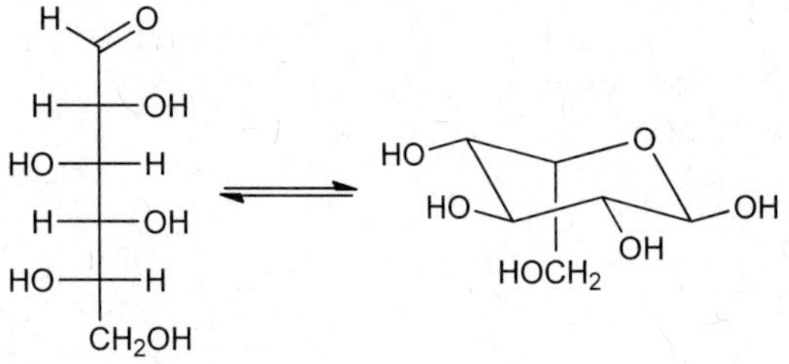

L-Idose

24.83. The molecular formula ($C_6H_{12}O_6$) indicates that compound **A** is a hexose (carbohydrate with six carbon atoms). Compound **A** is a reducing sugar, so it must be an aldohexose (rather than a ketohexose). Two successive Wohl degradations of compound **A** gives D-erythrose, which indicates the configurations of C4 and C5 of the aldohexose (both positions have the *R* configuration, just as in D-erythrose). Compound **A** is epimeric with glucose at C3, which indicates the *R* configuration at C3 (D-glucose has the *S* configuration at C3). Finally, C2 has the *R* configuration, giving the structure shown below (D-allose). The β-pyranose form of D-allose has also been drawn below. When treated with excess ethyl iodide in the presence of silver oxide, all of the OH groups in the β-pyranose form of compound **A** undergo alkylation, thereby converting them into ethoxy groups, as shown:

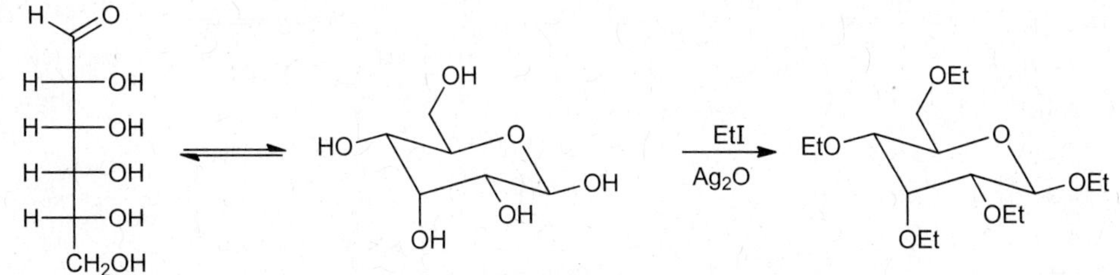

D-Allose
(Compound A)

β-pyranose form

24.84. Glucose can adopt a chair conformation in which all of the substituents on the ring occupy equatorial positions. Therefore, D-glucose can achieve a lower energy conformation than any of the other D-aldohexoses.

24.85. The following mechanism consists entirely of deprotonation and protonation steps. Notice that for each deprotonation step, hydroxide is used as the base, while water is used as the proton source for each protonation step (consistent with basic conditions):

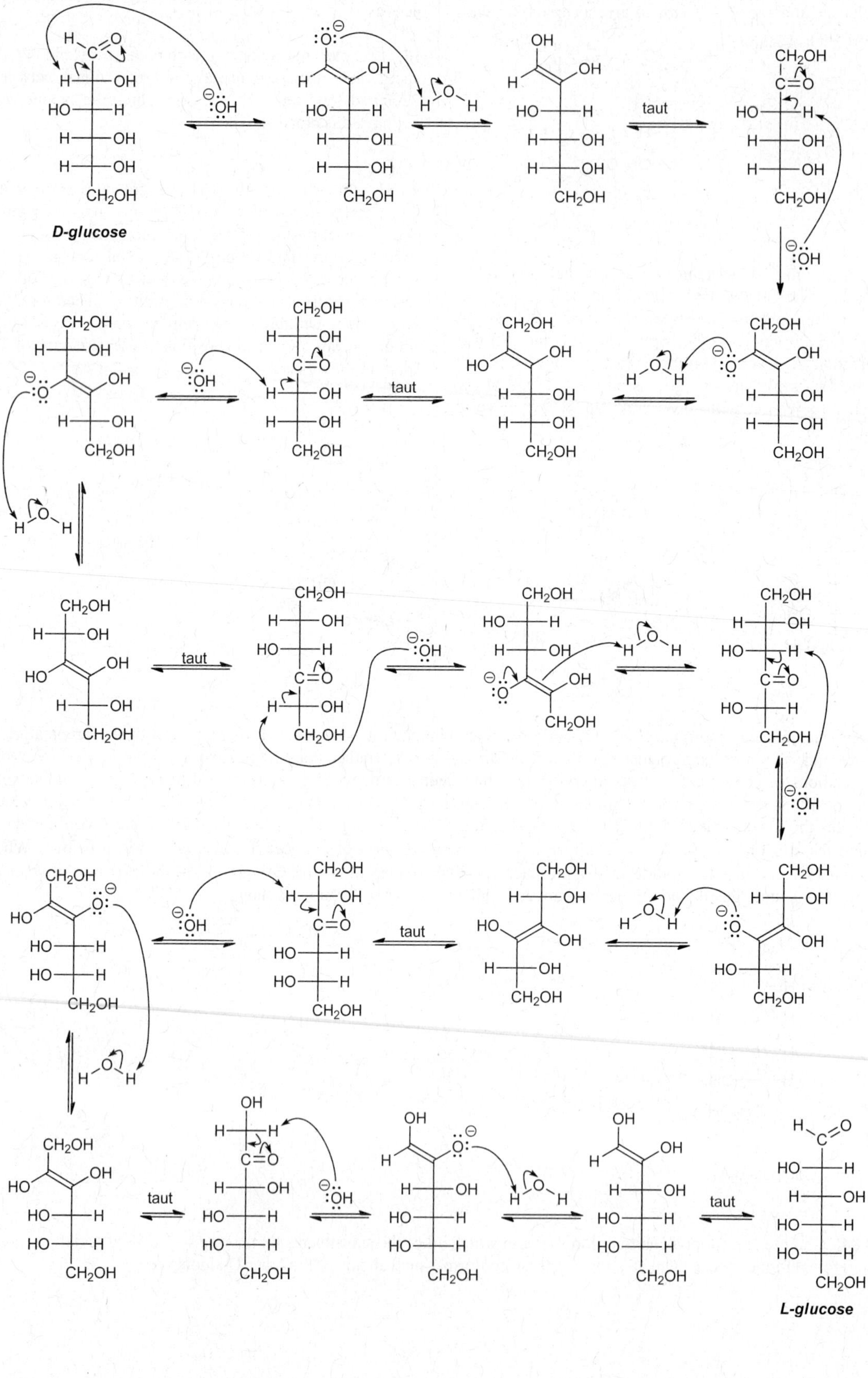

D-glucose

L-glucose

24.86. Compound **X** is a D-aldohexose that can adopt a β-pyranose form with only one axial substituent. Recall that D-glucose has all substituents in equatorial positions, so compound **X** must be epimeric with D-glucose either at C2 (D-mannose), C3 (D-allose), or C4 (D-galactose).

Compound **X** undergoes a Wohl degradation to produce an aldopentose, which is converted into an optically active alditol when treated with sodium borohydride. Therefore, compound **X** cannot be D-allose, because a Wohl degradation of D-allose followed by reduction produces an optically *inactive* alditol.

We conclude that compound **X** must be either D-mannose or D-galactose.

The identity of compound **X** can be determined by treating compound **X** with sodium borohohydride. Reduction of D-mannose should give an optically active alditol, while reduction of D-galactose gives an optically inactive alditol.

24.87.

(a) Compound **A** is a D-aldopentose. Therefore, there are four possible structures to consider (Figure 24.4).

When treated with sodium borohydride, compound **A** is converted into an alditol that exhibits three signals in its ^{13}C NMR spectrum. Therefore, compound **A** must be D-ribose or D-xylose both of which are reduced to give symmetrical alditols (thus, three signals for five carbon atoms).

When compound **A** undergoes a Kiliani-Fischer synthesis, both products can be treated with nitric acid to give optically active aldaric acids. Therefore, compound **A** cannot be D-ribose, because when D-ribose undergoes a Kiliani-Fischer synthesis, one of the products is D-allose, which is oxidized to give an optically inactive aldaric acid. We conclude that the structure of compound **A** must be D-xylose.

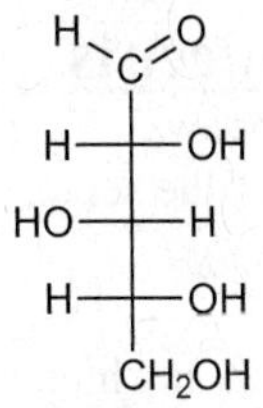

D-Xylose

(b) Compound D is expected have six signals in its ^{13}C NMR spectrum, while compound E is expected to have only three signals in its ^{13}C NMR spectrum.

Compound D   Compound E

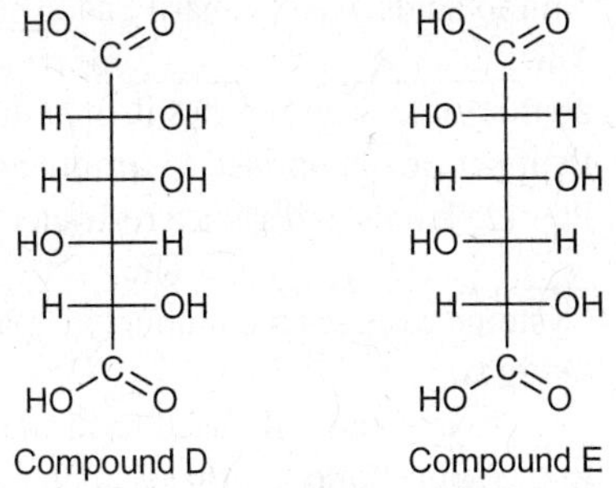

Chapter 25
Amino Acids, Peptides, and Proteins

Review of Concepts

Fill in the blanks below. To verify that your answers are correct, look in your textbook at the end of Chapter 25. Each of the sentences below appears verbatim in the section entitled *Review of Concepts and Vocabulary*.

- Amino acids in which the two functional groups are separated by exactly one carbon atom are called _____ **amino acids.**
- Amino acids are coupled together by amide linkages called _____ **bonds.**
- Relatively short chains of amino acids are called _____ .
- Only twenty amino acids are abundantly found in proteins, all of which are ___ **amino acids,** except for _____ which lacks a chirality center.
- Amino acids exist primarily as _____ at physiological pH
- The _____ of an amino acid is the pH at which the concentration of the zwitterionic form reaches its maximum value.
- Peptides are comprised of **amino acid** _____ joined by peptide bonds.
- Peptide bonds experience restricted rotation, giving rise to two possible conformations, called _____ and _____. The _____ conformation is generally more stable.
- Cysteine residues are uniquely capable of being joined to one another via _____ **bridges.**
- _____ is commonly used to form peptide bonds.
- In the **Merrifield synthesis**, a peptide chain is assembled while tethered to _____ .
- The **primary structure** of a protein is the sequence of _____ .
- The **secondary structure** of a protein refers to the _____ _____ of localized regions of the protein. Two particularly stable arrangements are the ___ **helix** and ____ **pleated sheet.**
- The **tertiary structure** of a protein refers to its _____ .
- Under conditions of mild heating, a protein can unfold, a process called _____ .
- **Quaternary structure** arises when a protein consists of two or more folded polypeptide chains, called _____ , that aggregate to form one protein complex.

Review of Skills

Fill in the blanks and empty boxes below. To verify that your answers are correct, look in your textbook at the end of Chapter 25. The answers appear in the section entitled *SkillBuilder Review*.

25.1 Determining the Predominant Form of an Amino Acid at a Specific pH

CONSIDER THE FOLLOWING AMINO ACID, AND DRAW THE FORM THAT PREDOMINATES AT PHYSIOLOGICAL pH.

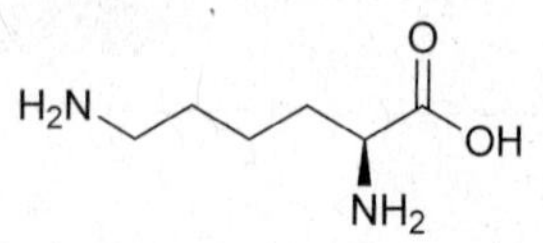

25.2 Using the Amidomalonate Synthesis

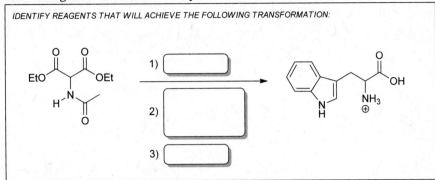

IDENTIFY REAGENTS THAT WILL ACHIEVE THE FOLLOWING TRANSFORMATION:

1)

2)

3)

25.3 Drawing a Peptide

DRAW A BOND-LINE STRUCTURE FOR THE TRIPEPTIDE Phe-Val-Trp.

25.4 Sequencing a Peptide via Enzymatic Cleavage

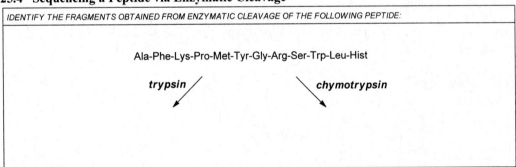

IDENTIFY THE FRAGMENTS OBTAINED FROM ENZYMATIC CLEAVAGE OF THE FOLLOWING PEPTIDE:

Ala-Phe-Lys-Pro-Met-Tyr-Gly-Arg-Ser-Trp-Leu-Hist

trypsin   *chymotrypsin*

25.5 Planning the Synthesis of a Dipeptide

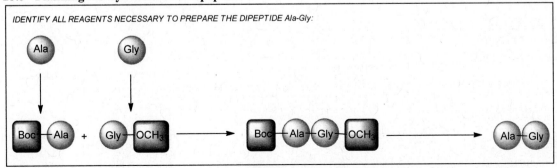

IDENTIFY ALL REAGENTS NECESSARY TO PREPARE THE DIPEPTIDE Ala-Gly:

Ala   Gly

Boc–Ala + Gly–OCH₃ ⟶ Boc–Ala–Gly–OCH₃ ⟶ Ala–Gly

25.6   Preparing a Peptide using the Merrifield Synthesis

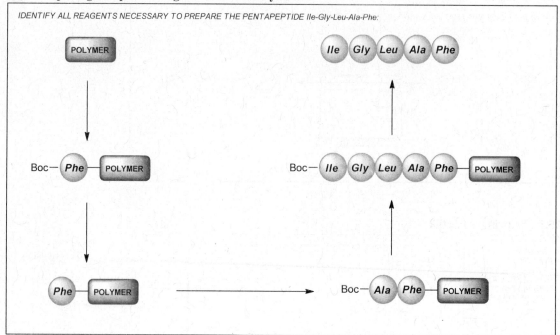

IDENTIFY ALL REAGENTS NECESSARY TO PREPARE THE PENTAPEPTIDE Ile-Gly-Leu-Ala-Phe:

Review of Reactions

Identify the reagents necessary to achieve each of the following transformations. To verify that your answers are correct, look in your textbook at the end of Chapter 25. The answers appear in the section entitled *Review of Reactions*.

Analysis of Amino Acids

Synthesis of Amino Acids

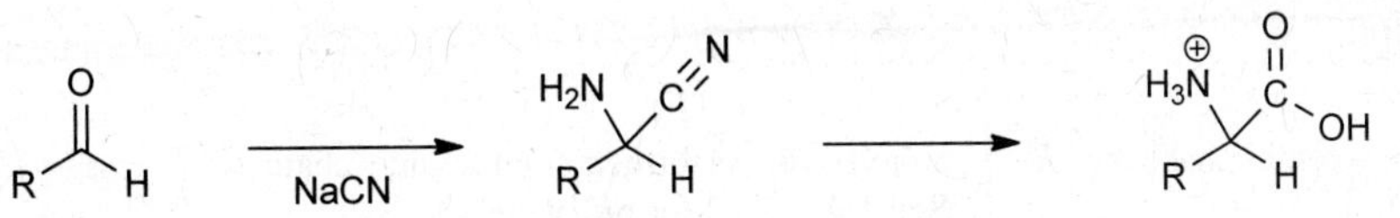

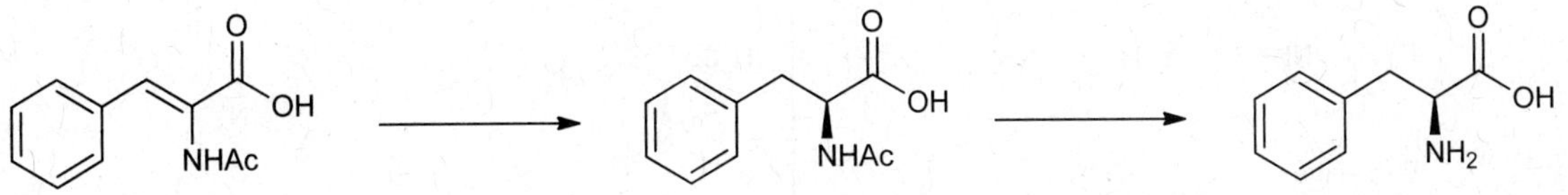

Analysis of Amino Acids

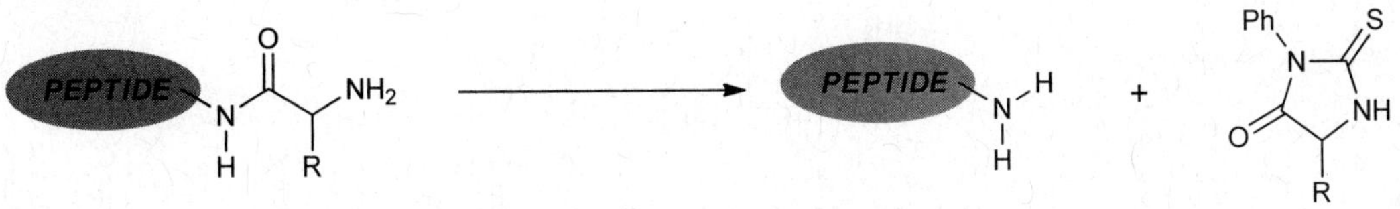

Synthesis of Peptides

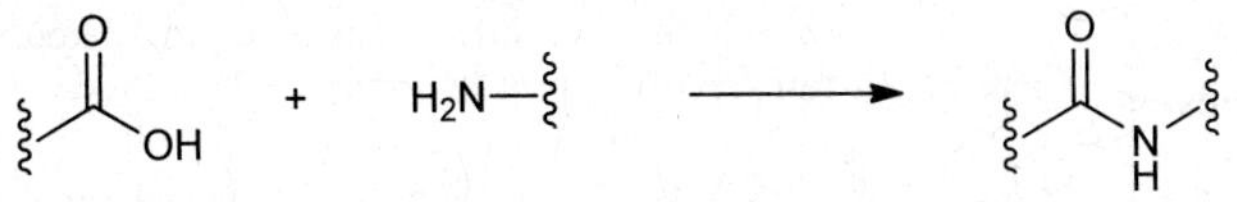

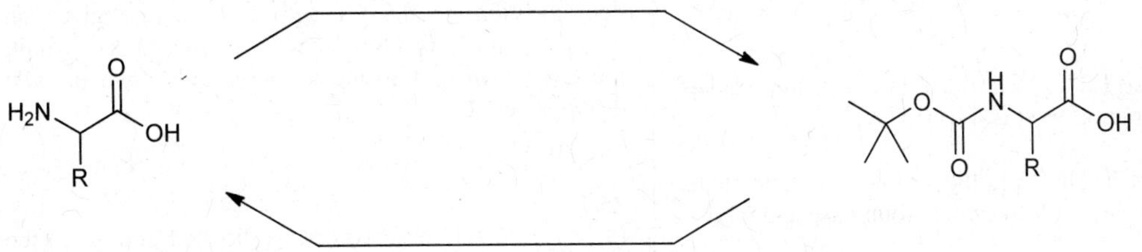

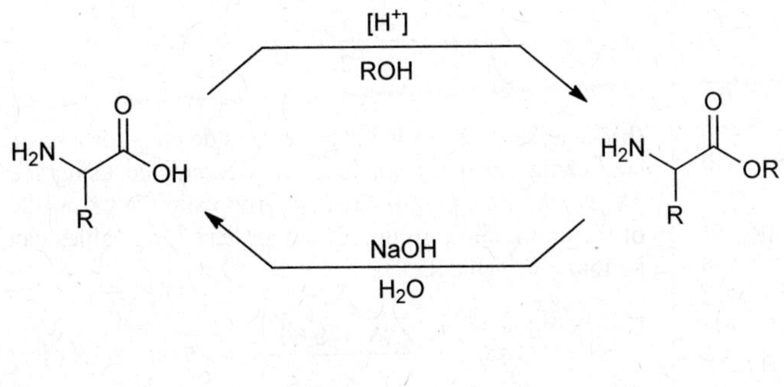

Solutions

25.1. In each case, the chirality center has the *R* configuration (see SkillBuilder 5.9).

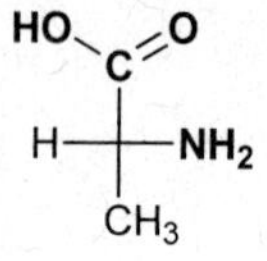

D-Alanine 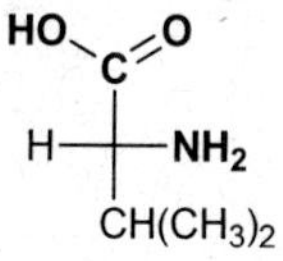   D-Valine

25.2. The structure of each of the following amino acids can be found in Table 25.1.

(a)

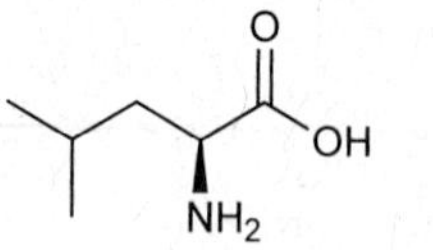

(b)

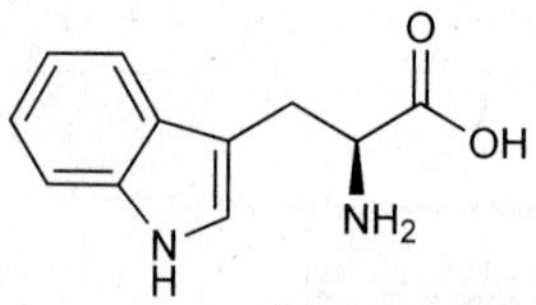

(c)

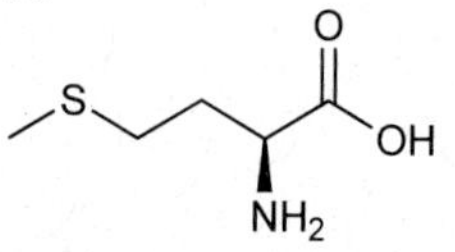

(d)

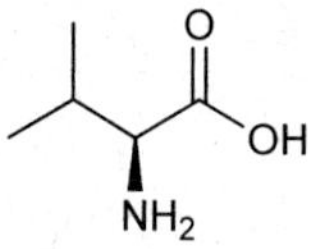

analysis is performed for the side chain (if necessary). See Table 25.2 for pK_a values.

(a)

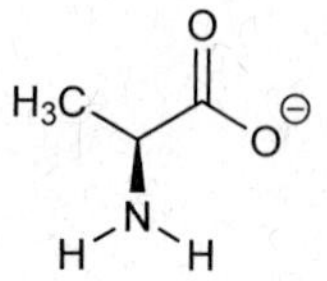

(b)

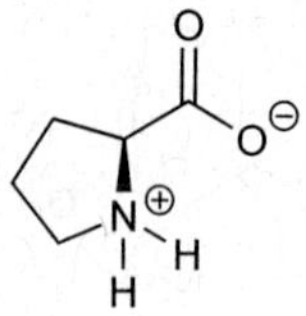

(c)

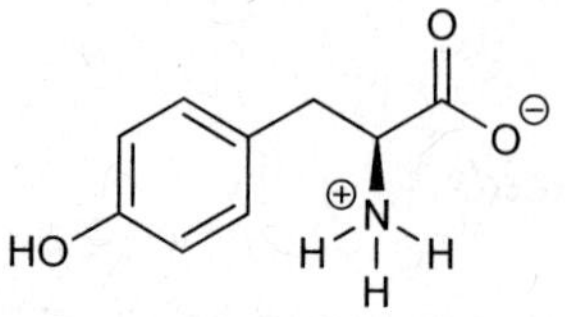

(d)

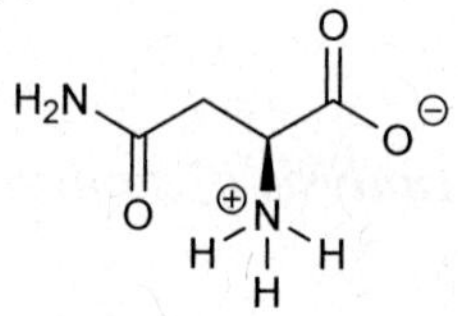

(e)

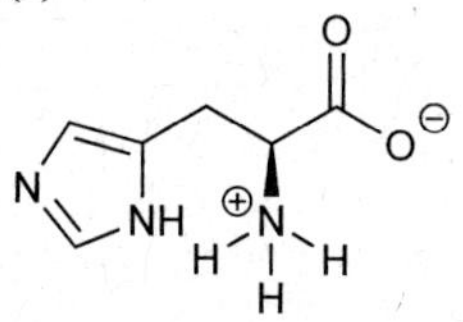

(f)

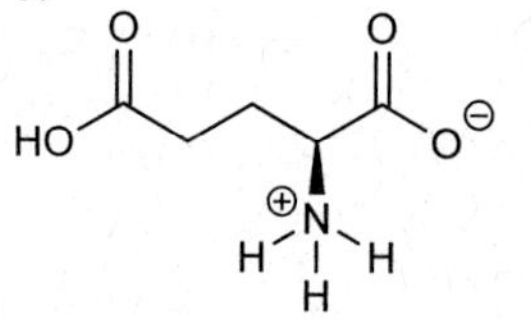

25.3.
(a) As seen in Table 25.1, the following amino acids exhibit a cyclic structure: Pro, Phe, Trp, Tyr, and His.

(b) As seen in Table 25.1, the following amino acids exhibit an aromatic side chain: Phe, Trp, Tyr, and His.

(c) As seen in Table 25.1, the following amino acids exhibit a side chain with a basic group: Arg, His, and Lys.

(d) As seen in Table 25.1, the following amino acids exhibit a sulfur atom: Met and Cys.

(e) As seen in Table 25.1, the following amino acids exhibit a side chain with an acidic group: Asp and Glu.

(f) As seen in Table 25.1, the following amino acids exhibit a side chain containing a proton that will likely participate in hydrogen bonding: Pro, Trp, Asn, Gln, Ser, Thr, Tyr, Cys, Asp, Glu, Arg, His, and Lys.

25.4. In each case, we first identify the pK_a of the carboxylic acid group and determine which form predominates. The protonated form (RCOOH) will predominate if pH < pK_a, while the carboxylate ion will predominate if pH > pK_a. Next, we identify the pK_a of the α-amino group and determine which form predominates. The protonated form (RNH$_3^+$) will predominate if pH < pK_a, while the uncharged form (RNH$_2$) will predominate if pH > pK_a. Finally, a similar

25.5. Arginine has a basic side chain, while asparagine does not. At a pH of 11, arginine exists predominantly in a form in which the side chain is protonated. Therefore, it can serve as a proton donor.

25.6. Tyrosine possesses a phenolic proton which is more readily deprotonated because deprotonation forms a resonance-stabilized phenolate ion. In contrast, deprotonation of the OH group of serine gives an alkoxide ion that is not resonance-stabilized. As a result, the OH group of tyrosine is more acidic than the OH group of serine.

25.7.
(a) Aspartic acid has two carboxylic acid groups, so the pI of aspartic acid is calculated using the pK_a values of the two carboxylic acid groups, as shown here (pK_a values can be found in Table 25.2):

$$pI = \frac{1.88 + 3.65}{2} = 2.77$$

(b) Leucine does not have an acidic side chain or a basic side chain, so the pI of leucine is calculated using the pK_a value of the carboxylic acid group and the pK_a value of the ammonium group, as shown here (pK_a values can be found in Table 25.2):

$$pI = \frac{2.36 + 9.60}{2} = 5.98$$

(c) Lysine has two ammonium groups, so the pI of lysine is calculated using the pK_a values of the two ammonium groups, as shown here (pK_a values can be found in Table 25.2):

$$pI = \frac{8.95 + 10.93}{2} = 9.74$$

(d) Proline does not have an acidic side chain or a basic side chain, so the pI of proline is calculated using the pK_a value of the carboxylic acid group and the pK_a value of the ammonium group, as shown here (pK_a values can be found in Table 25.2):

$$pI = \frac{1.99 + 10.60}{2} = 6.30$$

25.8.
(a) Aspartic acid has two carboxylic acid groups, so it is expected to have the lowest pI.
(b) Glutamic acid has two carboxylic acid groups, so it is expected to have the lowest pI.

25.9. Leucine and isoleucine both exhibit the same pK_a value for the carboxylic acid group. Similarly, both leucine and isoleucine exhibit the same pK_a value for the amino group. As such, the pI value of leucine is expected to be the same as the pI value of isoleucine.

25.10. The pI of Phe = 5.48, the pI of Trp = 6.11, and the pI of Leu = 6.00. Using these values, we make the following predictions:
(a) At pH = 6.0, Phe will travel the farthest distance.
(b) At pH = 5.0, Trp will travel the farthest distance.

25.11. The following aldehyde is expected when L-leucine is treated with ninhydrin:

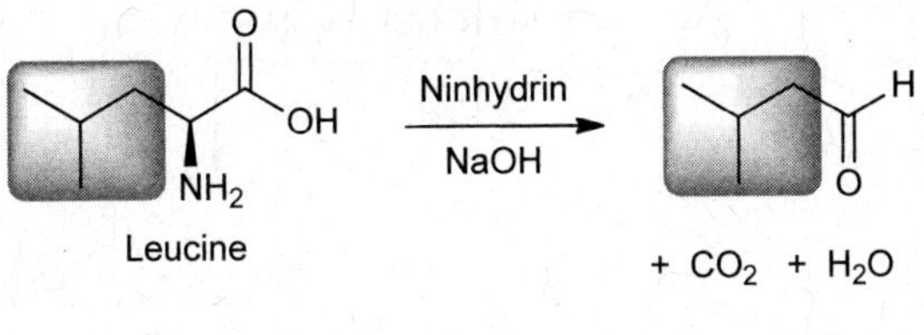

25.12.
(a) Racemic leucine can be made using a Hell-Volhard-Zelinsky reaction, as shown:

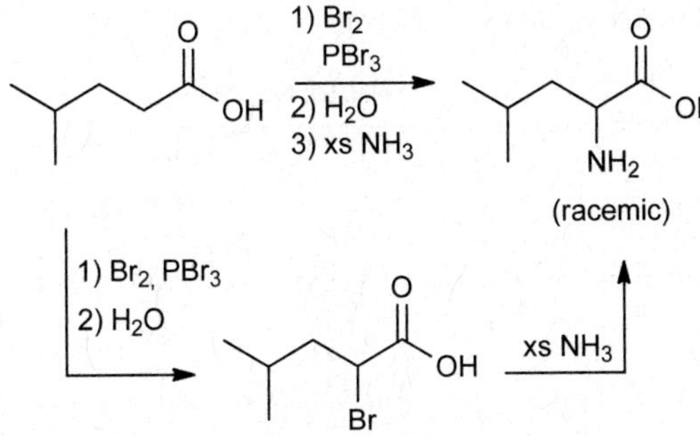

(b) Racemic alanine can be made using a Hell-Volhard-Zelinsky reaction, as shown:

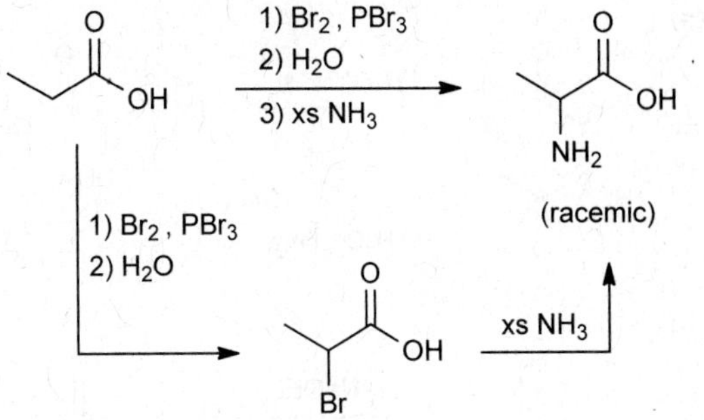

(c) Racemic valine can be made using a Hell-Volhard-Zelinsky reaction, as shown:

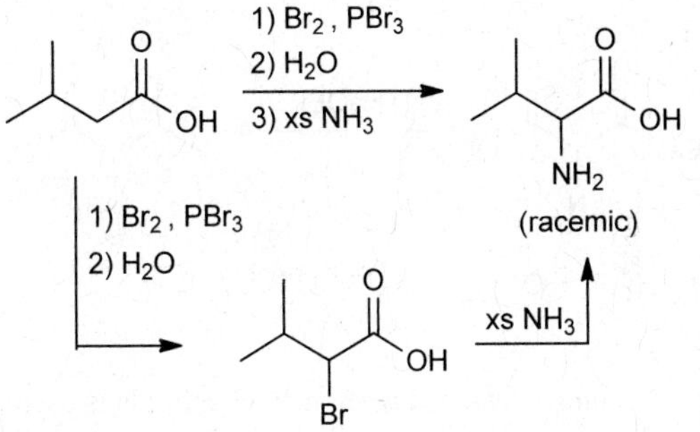

25.13. In each case, the process is a Hell-Volhard-Zelinsky reaction, which will give the following amino acids:

(a)

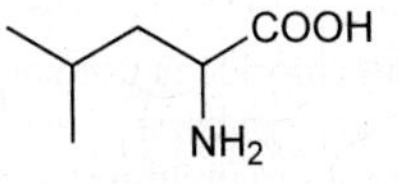

Leucine

(b)

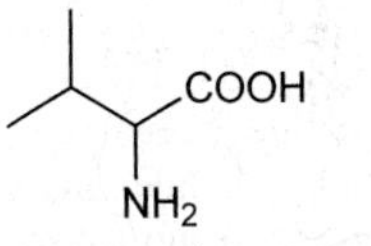

Valine

(c)

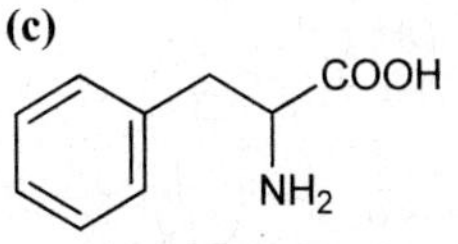

Phenylalanine

(d)

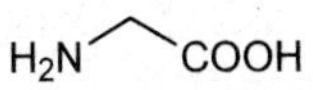

Glycine

25.14. In each case, we begin by identifying the side chain connected to the α position. Then, we identify the necessary alkyl halide and ensure that it is not tertiary (because a tertiary alkyl halide will not undergo an S_N2 reaction). An amidomalonate synthesis is performed using acetamidomalonate as the starting material, which is first treated with sodium ethoxide. The resulting conjugate base (a doubly stabilized enolate) is then treated with the alkyl halide, followed by hydrolysis with aqueous acid and heat, to give the desired amino acid. Note that the final step employs acidic conditions, so the amino group of the resulting amino acid is protonated:

(a)

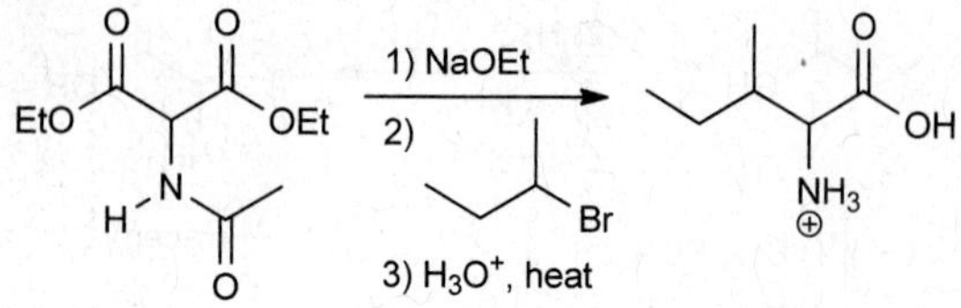

(b)

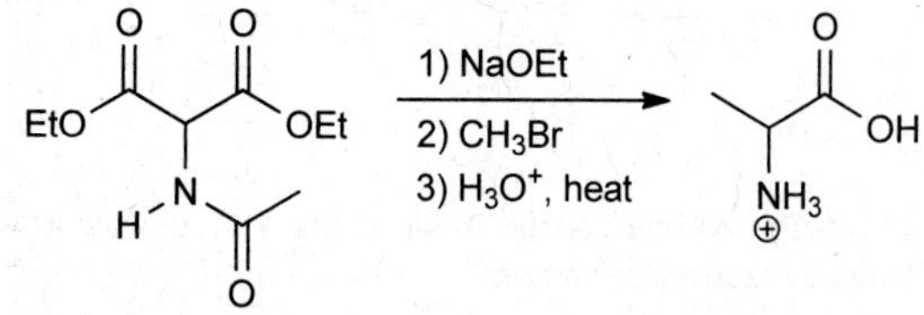

(c)

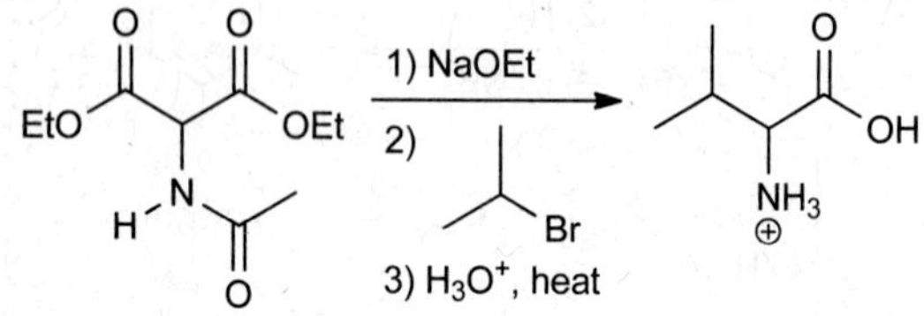

25.15.
(a) Alanine is obtained when methyl chloride is used as the alkyl halide in an amidomalonate synthesis. The methyl group (from methyl chloride) is highlighted in the product:

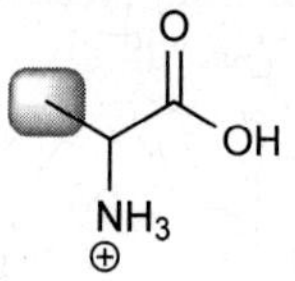

(b) Valine is obtained when isopropyl chloride is used as the alkyl halide in an amidomalonate synthesis. The isopropyl group (from isopropyl chloride) is highlighted in the product:

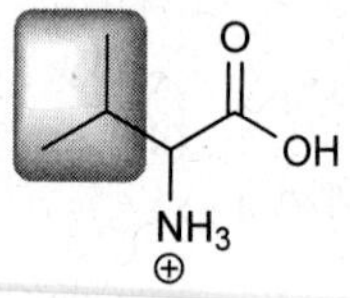

(c) Leucine is obtained when 2-methyl-1-chloropropane is used as the alkyl halide in an amidomalonate synthesis. The alkyl group (from the alkyl chloride) is highlighted in the product:

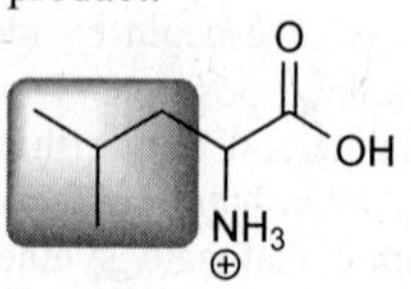

25.16.   Leucine can be prepared via the amidomalonate synthesis with higher yields than isoleucine, because the former requires an S_N2 reaction with a primary alkyl halide, while the latter requires an S_N2 reaction with a secondary (more hindered) alkyl halide.

25.17.
(a)   Methionine can be prepared from the aldehyde below via a Strecker synthesis, as shown:

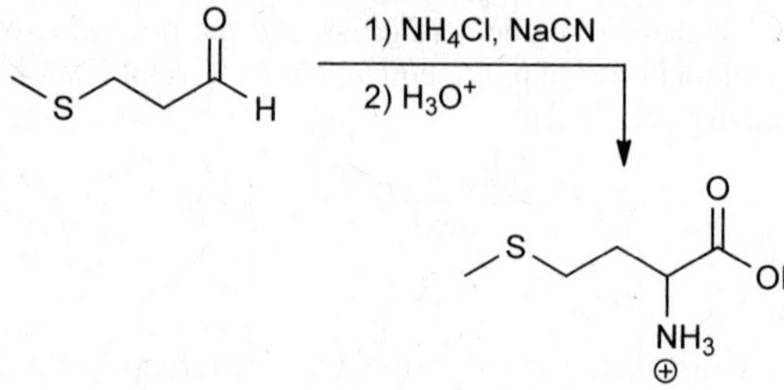

(b) Histidine can be prepared from the aldehyde below via a Strecker synthesis, as shown:

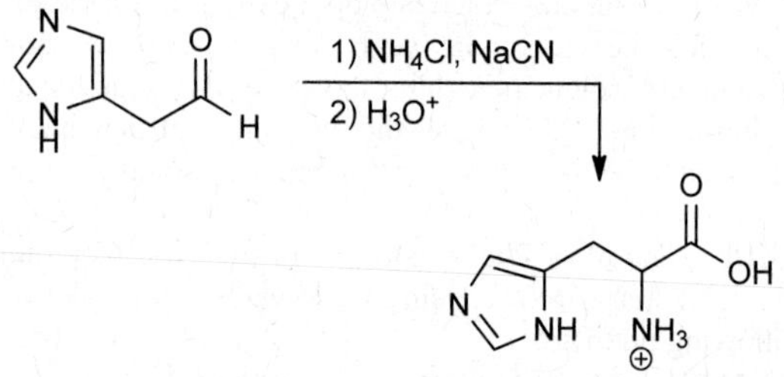

(c) Phenylalanine can be prepared from the aldehyde below via a Strecker synthesis, as shown:

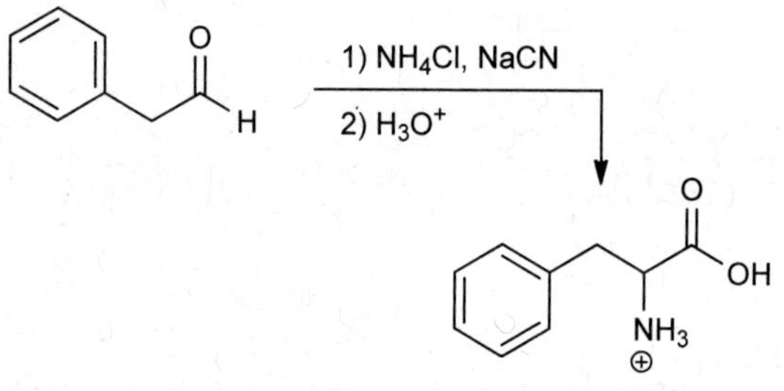

(d) Leucine can be prepared from the aldehyde below via a Strecker synthesis, as shown:

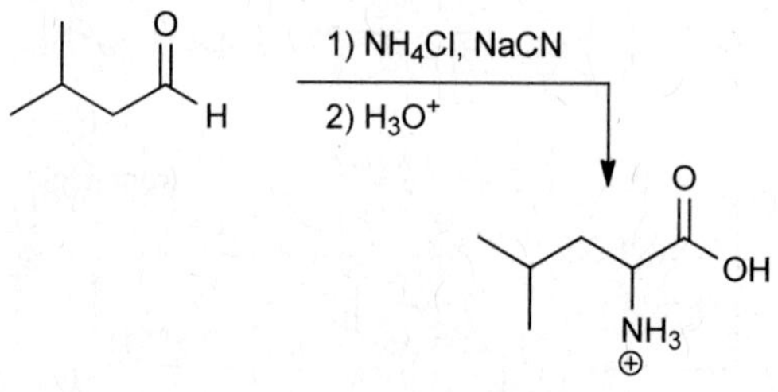

25.18.
(a) Acetaldehyde is converted into a racemic mixture of alanine (via a Strecker synthesis), as shown:

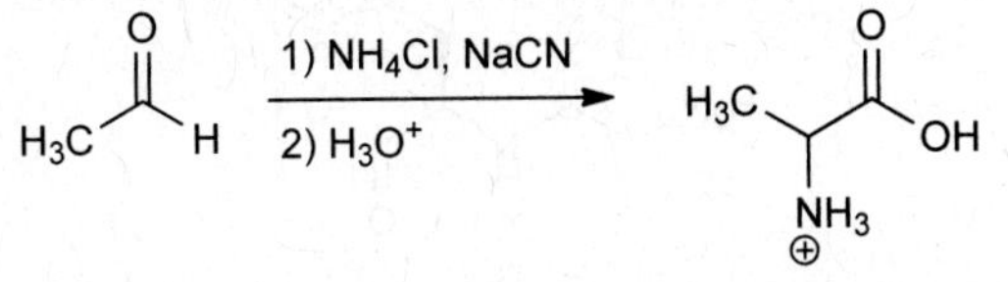

(b) 3-Methylbutanal is converted into a racemic mixture of leucine (via a Strecker synthesis), as shown:

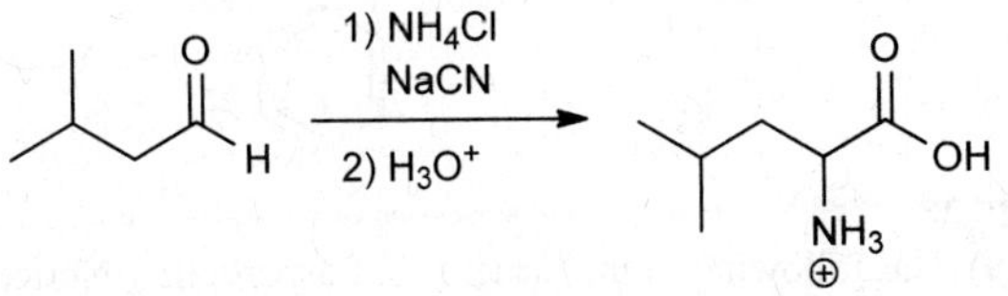

(c) 2-Methylpropanal is converted into a racemic mixture of valine (via a Strecker synthesis), as shown:

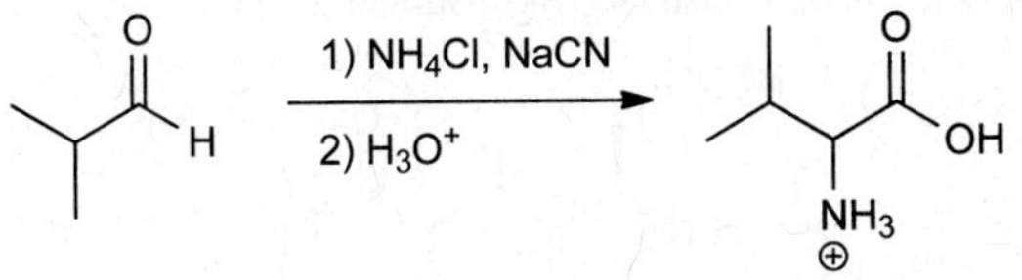

25.19.
(a) L-Alanine can be prepared from the compound below via an asymmetric catalytic hydrogenation:

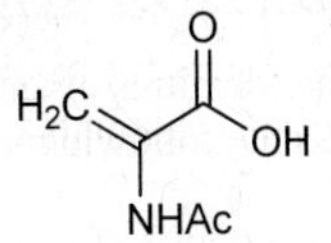

(b) L-Valine can be prepared from the compound below via an asymmetric catalytic hydrogenation:

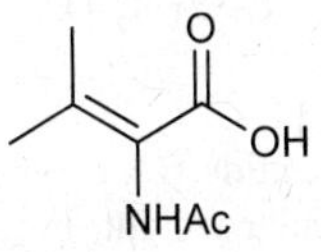

(c) L-Leucine can be prepared from the compound below via an asymmetric catalytic hydrogenation:

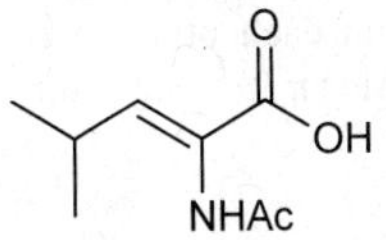

(d) L-Tyrosine can be prepared from the compound below via an asymmetric catalytic hydrogenation:

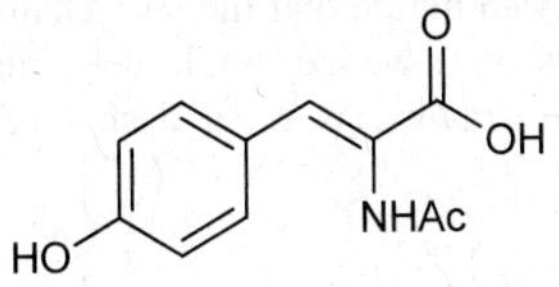

25.20. Glycine does not possess a chirality center, so the use of a chiral catalyst is unnecessary. Also, there is no alkene that would lead to glycine upon hydrogenation.

25.21. For each of the following peptides, the N terminus is drawn on the left and the C terminus on the right. Side chains at the top of the drawing are on wedges, while side chains on the bottom of the drawing are on dashes. The identity of each side chain can be found in Table 25.1.
(a)   **(b)**

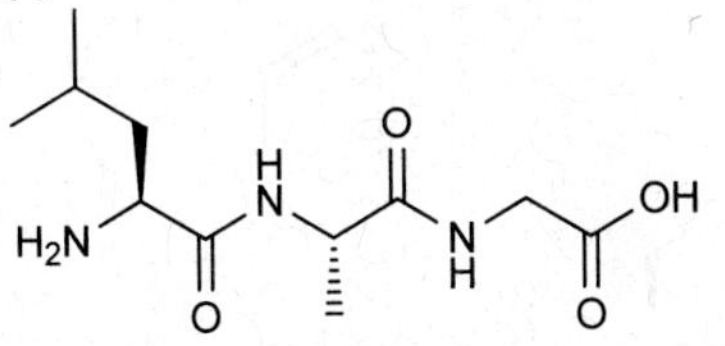

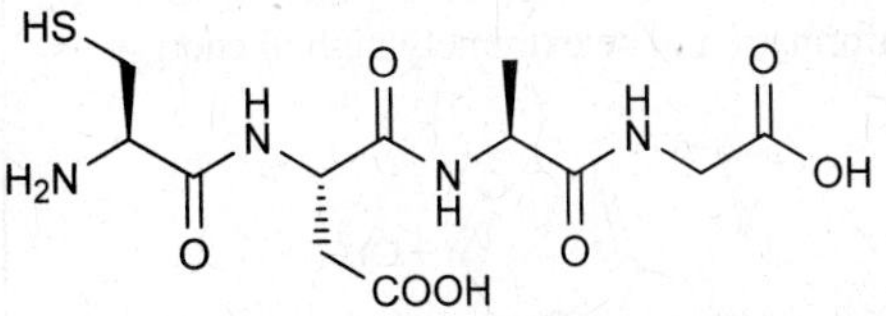

(c)

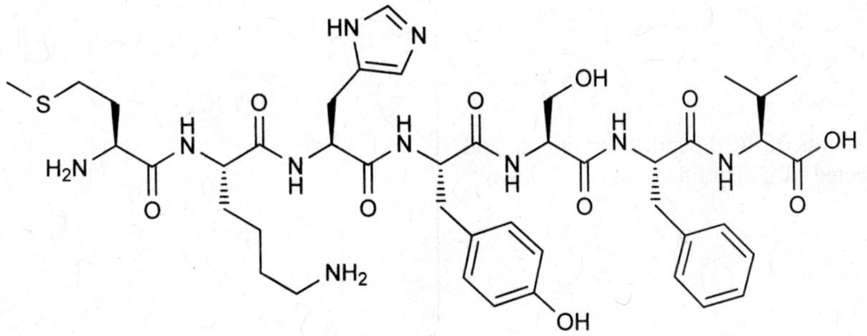

25.22. Based on the identities of side chains (see Table 25.1), this peptide has the following sequence:

Leu-Ala-Phe-Cys-Asp

This sequence can be summarized with the following one-letter abbreviations:

L-A-F-C-D.

25.23. The first peptide (Cys-Tyr-Leu) is expected to have a higher molecular weight, because the amino acid residues have larger side chains (see Table 25.1).

25.24. These peptides have the same molecular formula but they differ from each other in their connectivity of atoms (or constitution). As such, they are constitutional isomers.

25.25. The following is the *s-trans* conformation of the dipeptide Phe-Phe. Notice that the N-terminus is on the left, while the C-terminus is on the right, as per accepted convention. Also notice that the side chain at the top of the drawing is on a wedge, while the side chain at the bottom of the drawing is on a dash (see SkillBuilder 25.3).

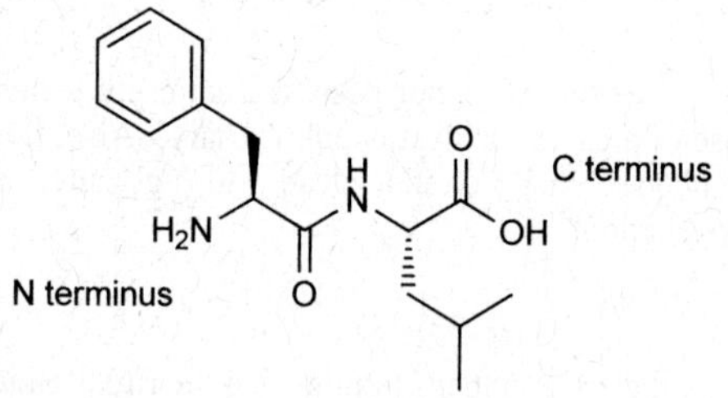

25.26. In the *s-cis* conformation, the phenyl groups will experience a severe steric interaction, thereby causing the *s-cis* conformation to be extremely high in energy:

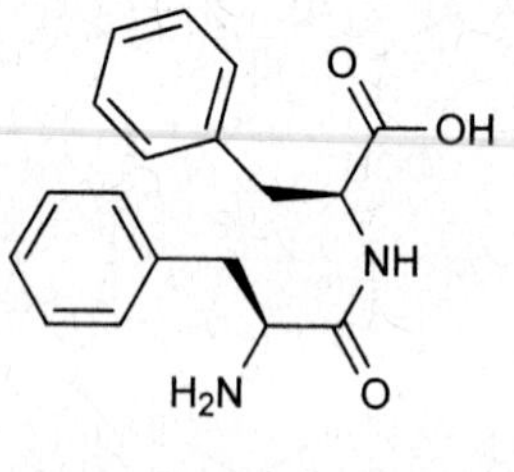

25.27. Two equivalents of the dipeptide are drawn, and they are then connected by a disulfide bridge, as shown:

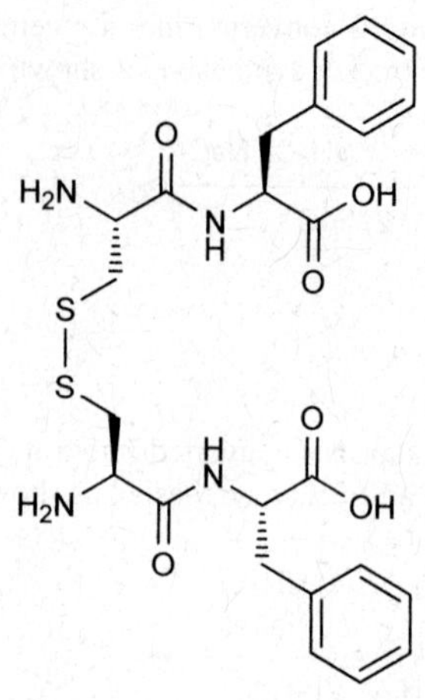

25.28.
(a) The following is the structure of aspartame. Notice that the N-terminus is on the left, while the C-terminus is on the right, as per accepted convention. In this case, the C-terminus is an ester (rather than a carboxylic acid). Also notice that the side chain at the top of the drawing is on a wedge, while the side chain at the bottom of the drawing is on a dash (see SkillBuilder 25.3).

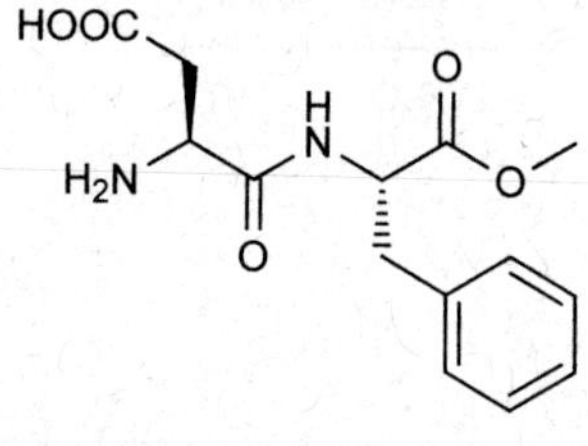

(b) The compound above has two chirality centers, giving rise to a total of four possible stereoisomers. The structure above represents one of these stereoisomers. The other three isomers are shown here:

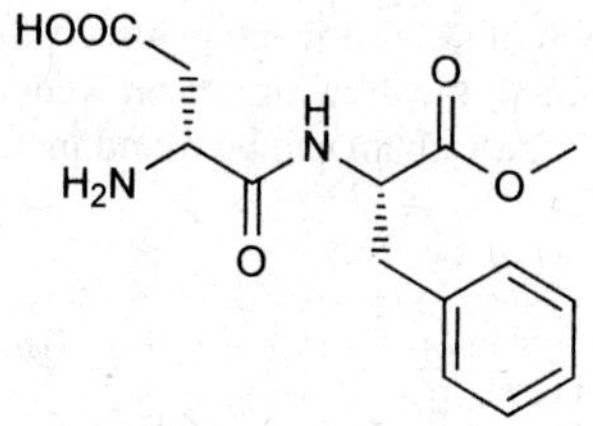

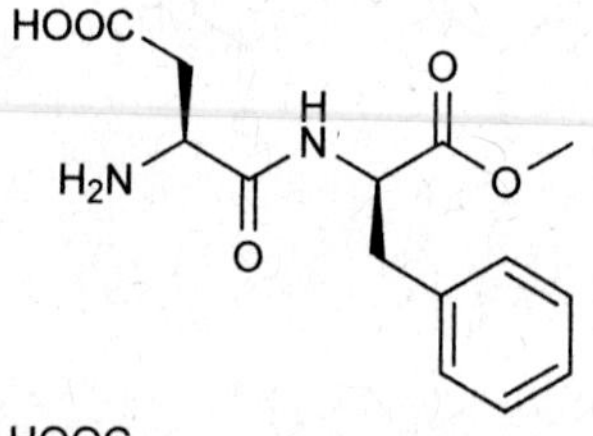

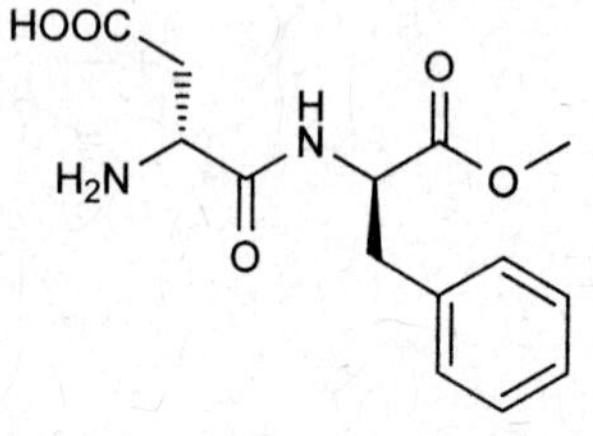

25.29. In the following structure, each of the amino acid residues has been highlighted and labeled:

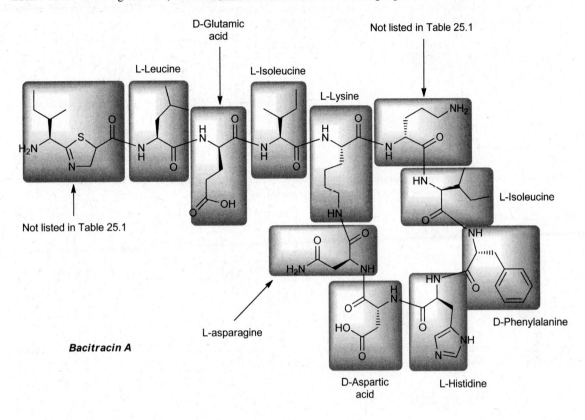

25.30. An Edman degradation will remove the amino acid residue at the N terminus, and Ala is the N terminus in Ala-Phe-Val. Therefore, alanine is removed, giving the following PTH derivative:

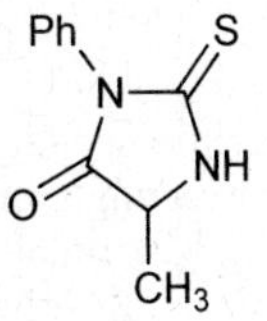

25.31. Only one of the trypsin fragments has a C terminus that is not arginine or lysine. This fragment, which ends with valine, must be the last fragment in the peptide sequence. The remaining three trypsin fragments can be placed in the proper order by analyzing the chymotrypsin fragments. The correct peptide sequence is:

Ala-ValMet-Phe-Val-Ala-Tyr-Lys-Pro-Val-Ile-Leu-Arg-Trp-His-Phe-Met-Cys-Arg-Gly-Pro-Phe-Ala-Val

25.32. The following tetrapeptide will be cleaved by chymotrypsin to give Ala-Phe and Val-Lys:

Ala-Phe-Val-Lys

25.33. Cleavage with trypsin will produce Phe-Arg, while cleavage with chymotrypsin will produce Arg-Phe. These dipeptides are not the same. They are constitutional isomers.

25.34.

(a) We begin by installing the appropriate protecting groups. Then, upon treatment with DCC, the protected amino acids are coupled. And finally, the protecting groups are removed, as shown:

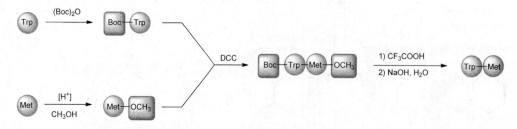

(b) We begin by installing the appropriate protecting groups. Then, upon treatment with DCC, the protected amino acids are coupled. And finally, the protecting groups are removed, as shown:

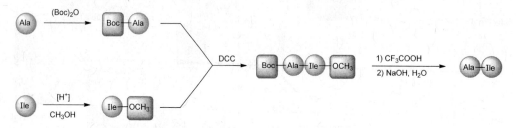

(c) We begin by installing the appropriate protecting groups. Then, upon treatment with DCC, the protected amino acids are coupled. And finally, the protecting groups are removed, as shown:

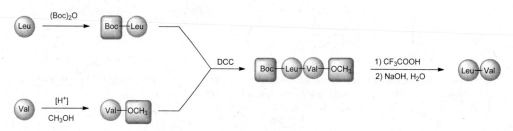

25.35. The first two amino acid residues (in the desired peptide sequence) are Ile and Phe. So we must begin with those amino acids. We first install the appropriate protecting groups. Then, upon treatment with DCC, the protected amino acids are coupled. The protecting group at the C-terminus is then removed and the resulting unprotected C-terminus is coupled with the appropriate protected amino acid (glycine, protected at the C-terminus), using DCC. And finally, the protecting groups are removed, as shown:

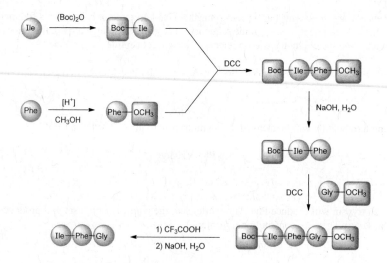

25.36.   The first two amino acid residues (in the desired peptide sequence) are Leu and Val. So we must begin with those amino acids. We first install the appropriate protecting groups. Then, upon treatment with DCC, the protected amino acids are coupled. The protecting group at the C-terminus is then removed and the resulting unprotected C-terminus is coupled with the appropriate protected amino acid, using DCC. Each additional residue is installed via deprotection of the C-terminus followed by coupling with the appropriate protected amino acid, as shown. Finally, the protecting groups are removed, giving the desired pentapeptide:

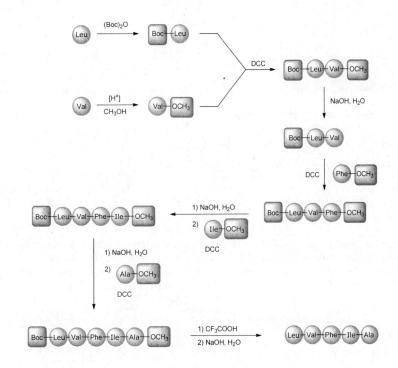

25.37.
(a) First, we attach the appropriate Boc-protected residue to the polymer:

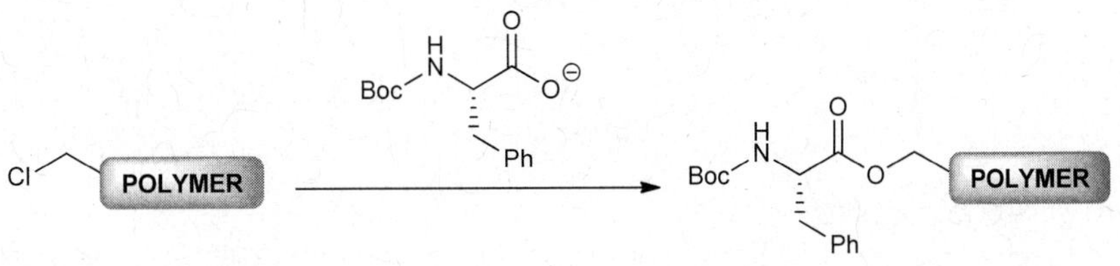

Then, the Boc protecting group is removed and a new peptide bond is formed with a Boc-protected amino acid, using DCC. This two-step process (removal of the Boc protecting group, followed by peptide bond formation) is then repeated to install each additional residue, until the desired sequence has been assembled. Finally, the Boc protecting group is removed and the desired peptide is detached from the polymer, as shown:

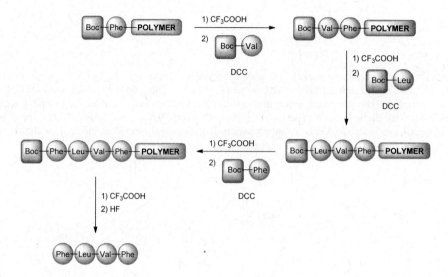

(b) First, we attach the appropriate Boc-protected residue to the polymer:

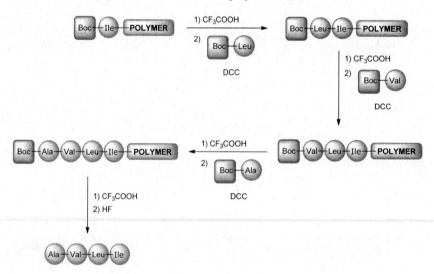

Then, the Boc protecting group is removed and a new peptide bond is formed with a Boc-protected amino acid, using DCC. This two-step process (removal of the Boc protecting group, followed by peptide bond formation) is then repeated to install each additional residue, until the desired sequence has been assembled. Finally, the Boc protecting group is removed and the desired peptide is detached from the polymer, as shown:

25.38. First, a protected valine residue is connected to the polymer. After deprotection, a protected alanine residue is installed. Then, after deprotection again, a protected phenylalanine residue is installed. Deprotection, followed by detachment from the polymer, gives the following tripeptide:

(N terminus)   Val-Ala-Phe   (C terminus)

25.39. The regions that contain repeating glycine and/or alanine units are the most likely regions to form β sheets:

Trp-His-Pro-Ala-Gly-Gly-Ala-Val-His-Cyst-Asp-Ser-Arg-Arg-Ala-Gly-Ala-Phe

25.40. In each case, the carboxylic acid group is drawn in its deprotonated form (as a carboxylate ion), and the amino group is drawn in its protonated form (as an ammonium ion):

(a)

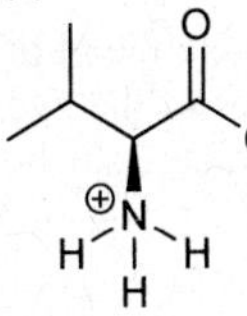

(b)

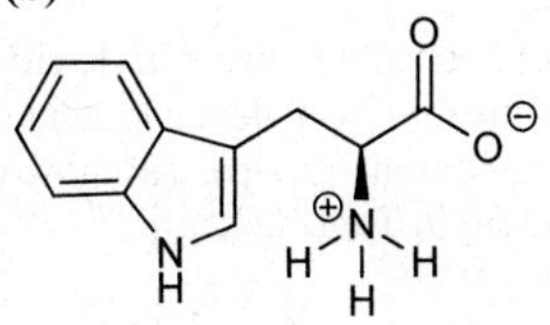

(c)

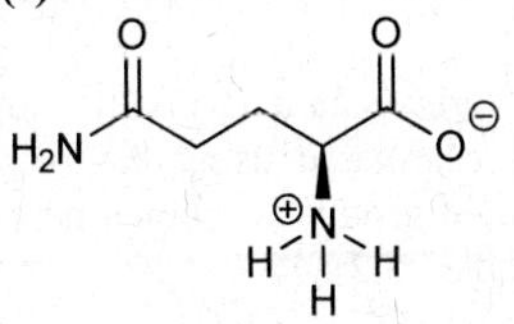

(d)

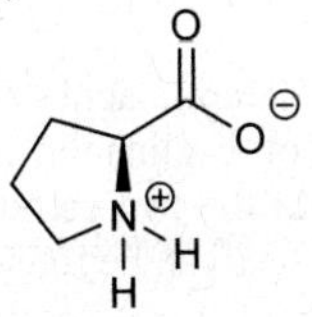

25.41. When applying the Cahn-Ingold-Prelog convention for assigning the configuration of a chirality center, the amino group generally receives the highest priority (1), followed by the carboxylic acid group (2), followed by the side chain (3), and finally the H (4). Accordingly, the *S* configuration is assigned to L amino acids. Cysteine is the one exception because the side chain has a higher priority than the carboxylic acid group. As a result, the *R* configuration is assigned.

25.42.
(a) L-threonine has two chirality centers (see Table 25.1). The configuration of each of these chirality centers is shown in the following Fischer projection:

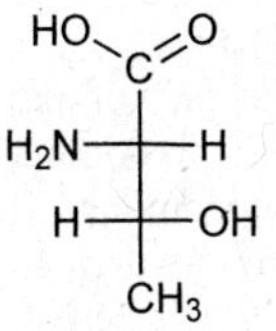

(b) L-Serine has only one chirality center (see Table 25.1). The configuration of this chirality center is shown in the following Fischer projection:

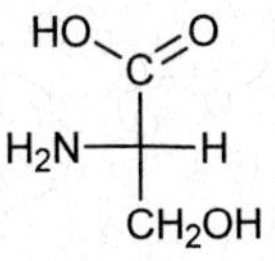

(c) L-Phenylalanine has only one chirality center (see Table 25.1). The configuration of this chirality center is shown in the following Fischer projection:

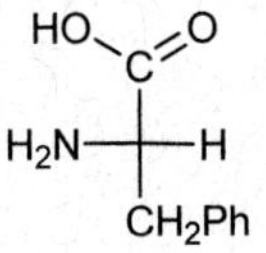

(d) L-Asparagine has only one chirality center (see Table 25.1). The configuration of this chirality center is shown in the following Fischer projection:

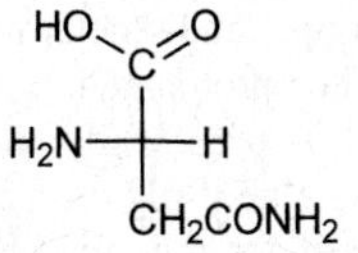

25.43.
(a) Isoleucine and threonine each have two chirality centers (see Table 25.1).

(b) The Cahn-Ingold Prelog convention (SkillBuilder 5.4) gives the following configurations:

Isoleucine = 2*S*, 3*S*

Threonine = 2*S*, 3*R*

25.44. Isoleucine has two chirality centers, so we expect four possible stereoisomers, shown here. The configuration of each chirality center is shown.

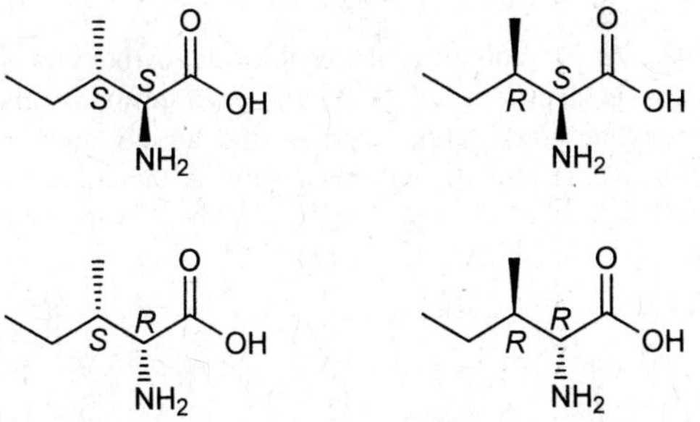

25.45. Protonation of the highlighted nitrogen atom gives a conjugate acid that is highly stabilized by resonance (the positive charge is highly delocalized).

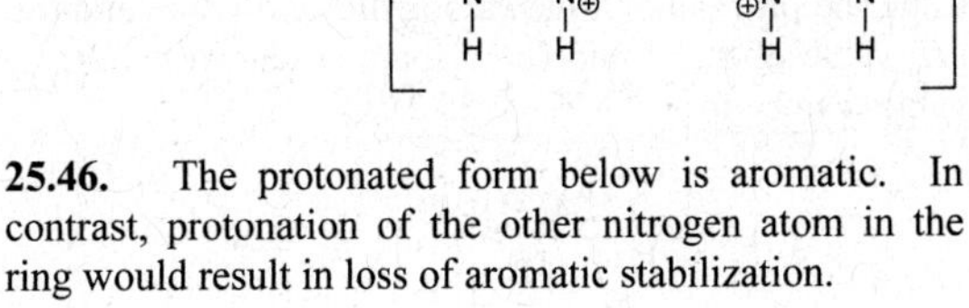

25.46. The protonated form below is aromatic. In contrast, protonation of the other nitrogen atom in the ring would result in loss of aromatic stabilization.

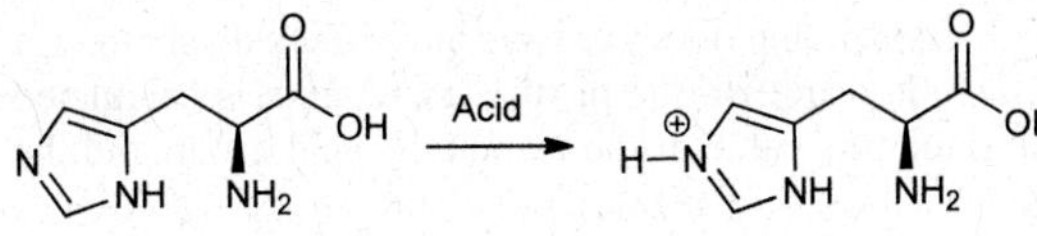

25.47. In each case, we first identify the pK_a of the carboxylic acid group and determine which form predominates. The protonated form (RCOOH) will predominate if pH < pK_a, while the carboxylate ion will predominate if pH > pK_a. Next, we identify the pK_a of the α-amino group and determine which form predominates. The protonated form (RNH$_3^+$) will predominate if pH < pK_a, while the uncharged form (RNH$_2$) will predominate if pH > pK_a. Finally, a similar analysis is performed for the side chain (if necessary). See Table 25.2 for pK_a values.

(a)   **(b)**

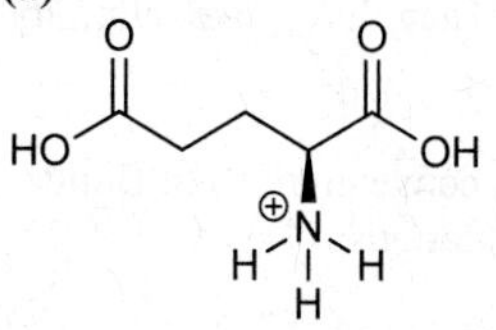

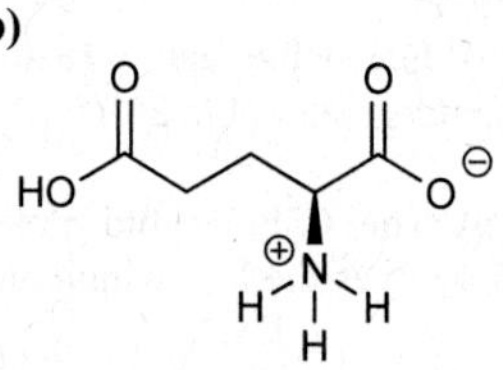

(c) 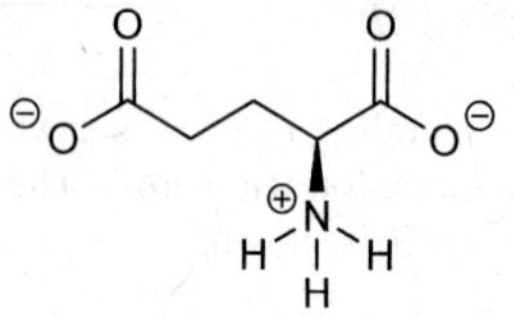   **(d)**

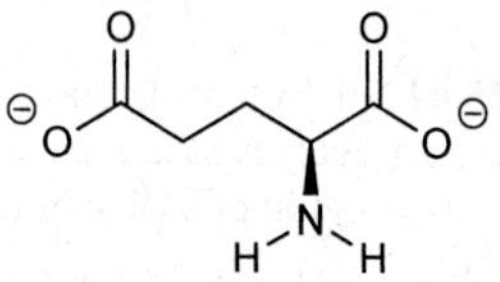

25.48. At physiological pH, each of the carboxylic acid groups is deprotonated (and will exist primarily as a carboxylate ion), while each of the amino groups is protonated (and will exist primarily as an ammonium ion):

(a)   **(b)**

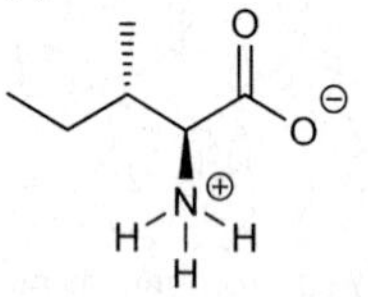

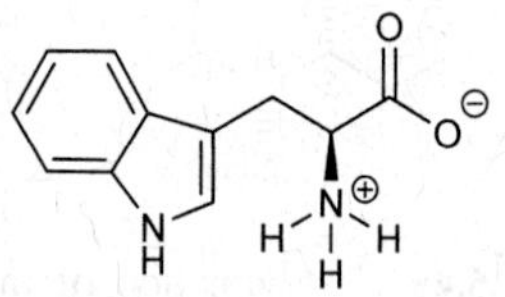

(c) 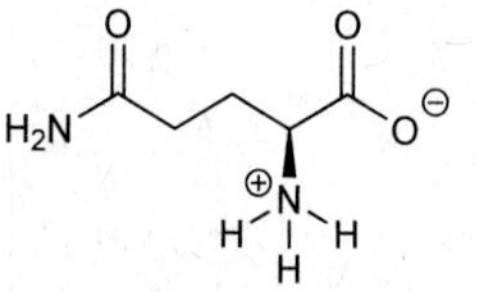   **(d)**

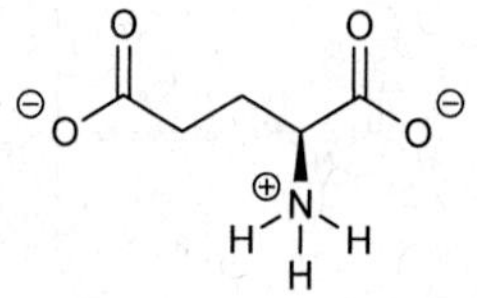

25.49.
(a) L-Alanine does not have an acidic side chain or a basic side chain, so the pI of L-alanine is calculated using the pK_a value of the carboxylic acid group and the pK_a value of the ammonium group, as shown here (pK_a values can be found in Table 25.2):

$$pI = \frac{2.34 + 9.69}{2} = 6.02$$

(b) L-Asparagine does not have an acidic side chain or a basic side chain, so the pI of L-asparagine is calculated using the pK_a value of the carboxylic acid group and the

pK_a value of the ammonium group, as shown here (pK_a values can be found in Table 25.2):

$$pI = \frac{2.02 + 8.80}{2} = 5.41$$

(c) L-Histidine has a basic side chain. As such, the pI of L-histidine is calculated using the pK_a values of the two ammonium groups, as shown here (pK_a values can be found in Table 25.2):

$$pI = \frac{9.17 + 6.00}{2} = 7.58$$

(d) L-Glutamic acid has two carboxylic acid groups, so the pI of L-glutamic acid is calculated using the pK_a values of the two carboxylic acid groups, as shown here (pK_a values can be found in Table 25.2):

$$pI = \frac{2.19 + 4.25}{2} = 3.22$$

25.50. Lysozyme is likely to be comprised primarily of amino acid residues that contain basic side chains (arginine, histidine, and lysine), while pepsin is comprised primarily of amino acid residues that contain acidic side chains (aspartic acid and glutamic acid).

25.51. First, we must calculate the pI for each amino acid (using the procedure shown in the solution to Problem 25.49). Next, we identify the pK_a of the carboxylic acid group and determine which form predominates. The protonated form (RCOOH) will predominate if pI < pK_a, while the carboxylate ion will predominate if pI > pK_a.

Then, we identify the pK_a of the α-amino group and determine which form predominates. The protonated form (RNH$_3^+$) will predominate if pI < pK_a, while the uncharged form (RNH$_2$) will predominate if pI > pK_a. Finally, a similar analysis is performed for the side chain (if necessary). See Table 25.2 for pK_a values.

(a)   **(b)**

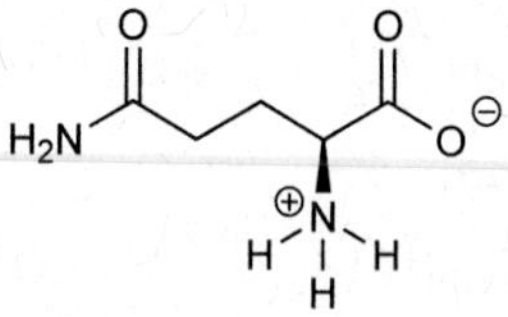

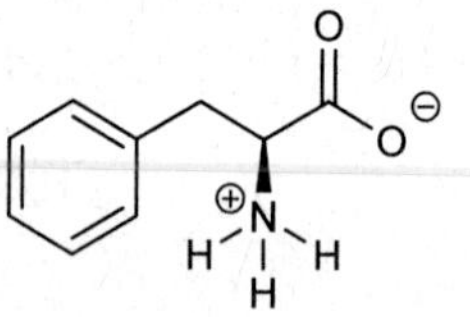

(c) 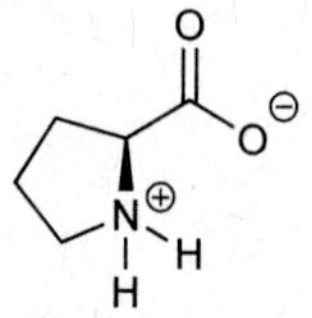   **(d)**

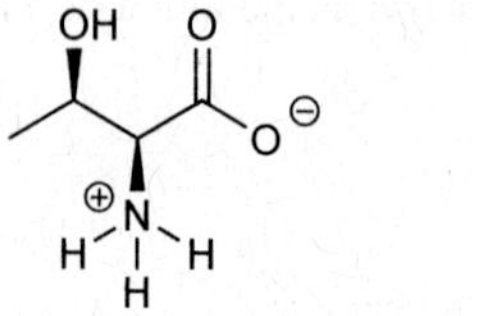

25.52. Under strongly basic conditions (NaOH), the carboxylic acid group exists as a carboxylate ion.

Under these conditions, the α position can be deprotonated, giving a dianion (resonance-stabilized). This dianion can be protonated by water (at the α position), thereby regenerating the carboxylate ion. In the process, racemization occurs at the α position because the α position is sp^2 hybridized (trigonal planar) in the dianion intermediate. Protonation of the dianion can occur on either face of the plane (with equal likelihood), giving a racemic mixture.

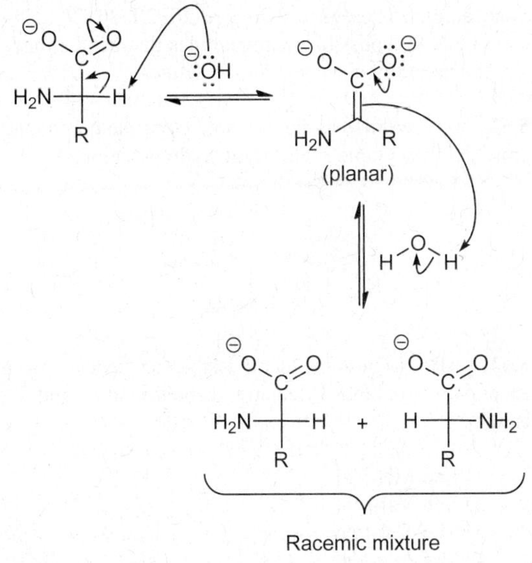

Racemic mixture

25.53. The pI of Gly = 5.97, the pI of Gln = 5.65, and the pI of Asn = 5.41. Using these values, we make the following predictions:
(a) At pH = 6.0, Asn will travel the farthest distance.
(b) At pH = 5.0, Gly will travel the farthest distance.

25.54. When treated with ninhydrin, the carboxylic acid group (connected to the α position) and the amino group (connected to the α position) are both removed, and the α position becomes an aldehyde group, giving the following products:
(a)   **(b)**

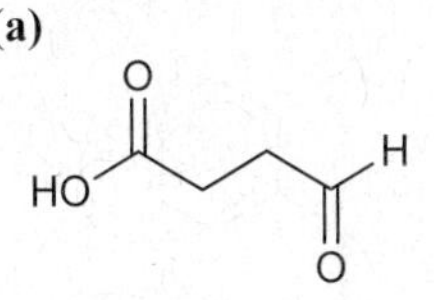

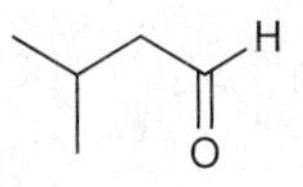

(c)

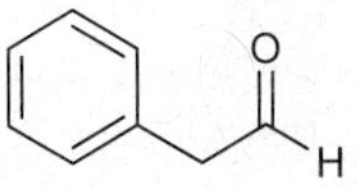

(d) Ninhydrin does not react with proline because the amino group is not primary.

25.55.
(a) When treated with ninhydrin, each of the amino acids (except proline) is converted into an aldehyde with complete removal of the carboxylic acid group and the amino group (that are connected to the α position). Therefore, the identity of each aldehyde indicates the side chain of the corresponding amino acid from which it was made (see Table 25.1). This analysis reveals that the starting mixture must have contained methionine, valine, and glycine.

(b) The following purple product is obtained whenever ninhydrin reacts with an amino acid (except for proline).

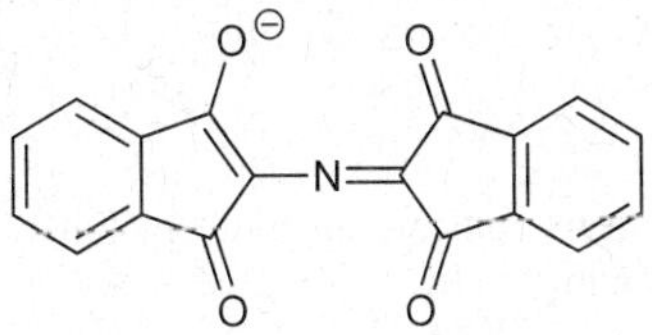

(c) The compound is highly conjugated and has a λ_{max} that is greater than 400 nm (see Section 17.12)

25.56. Valine can be made from the following aldehyde (via a Strecker synthesis), as shown:

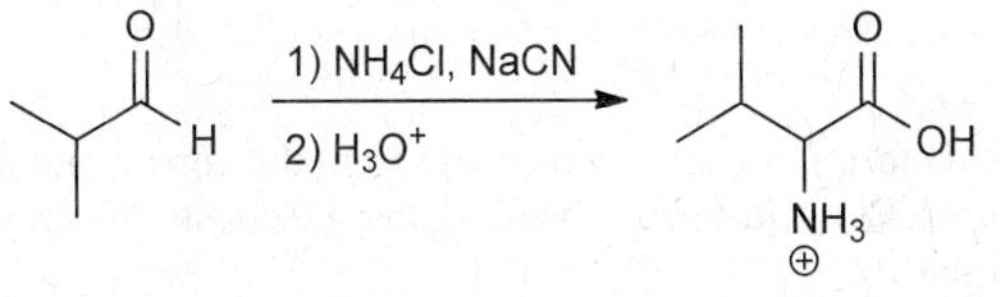

25.57. Alanine can be prepared via the amidomalonate synthesis with higher yields than valine, because the former requires an S_N2 reaction with a primary alkyl halide, while the latter requires an S_N2 reaction with a secondary (more hindered) alkyl halide.

25.58. The side chain (R) of glycine is a hydrogen atom (H). Therefore, no alkyl group needs to be installed at the α position.

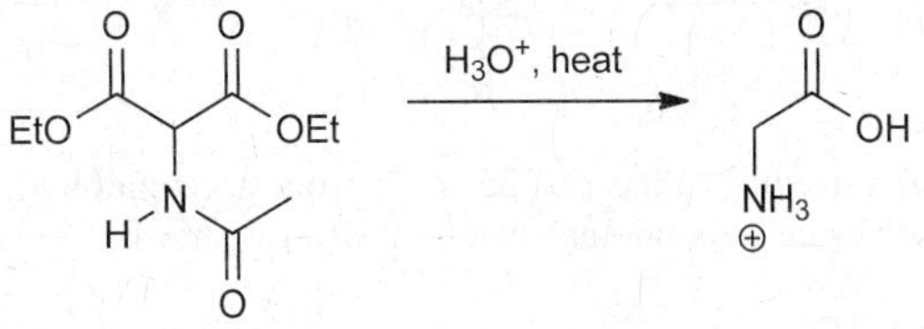

25.59.
(a) The reagents indicate a Hell-Volhard-Zelinsky reaction (thereby installing a bromine atom at the α position), followed by an S_N2 reaction (thereby replacing the bromine atom with an amino group). The product is an amino acid (phenylalanine), as shown:

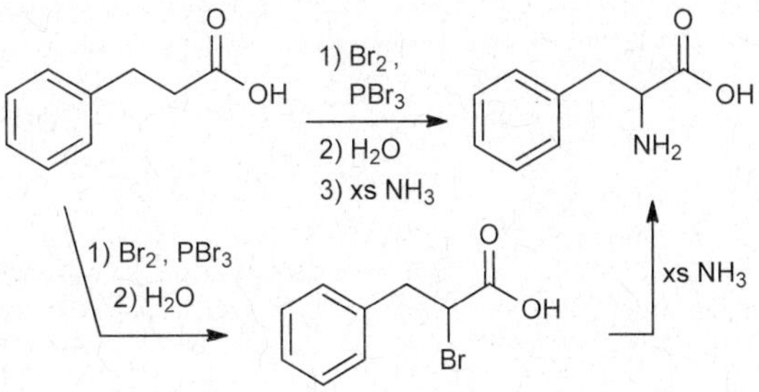

(b) The reagents indicate a Strecker synthesis, giving an amino acid (phenylalanine), as shown:

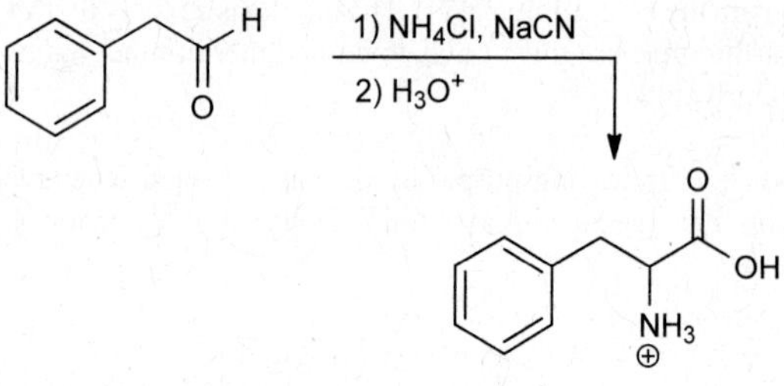

1) NH_4Cl, NaCN
2) H_3O^+

(c) The reagents indicate an amidomalonate synthesis, giving an amino acid (alanine), as shown:

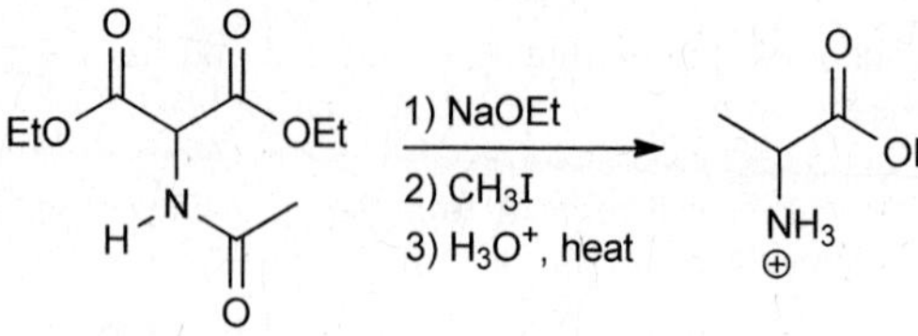

1) NaOEt
2) CH_3I
3) H_3O^+, heat

25.60.
(a) Racemic valine can be made from the corresponding carboxylic acid using a Hell-Volhard-Zelinsky reaction, as shown:

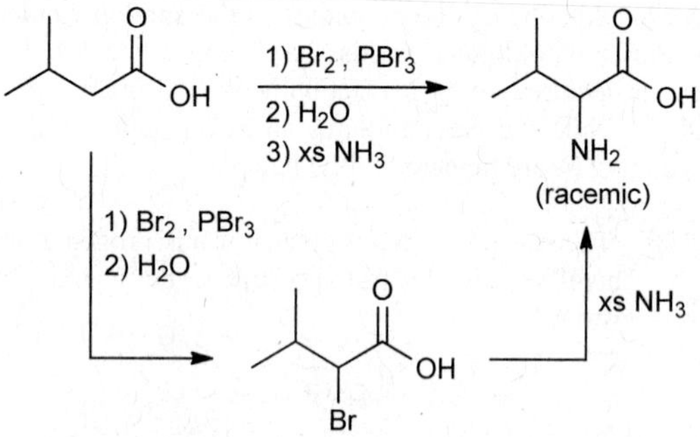

1) Br_2, PBr_3
2) H_2O
3) xs NH_3

(racemic)

1) Br_2, PBr_3
2) H_2O

xs NH_3

(b) Racemic valine can be made from acetamidomalonate using an amidomalonate synthesis, as shown:

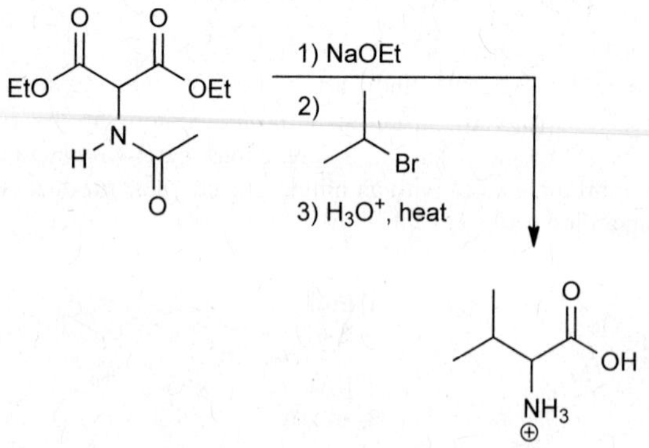

1) NaOEt
2)
3) H_3O^+, heat

(c) Racemic valine can be made from 2-methylpropanal using a Strecker synthesis, as shown:

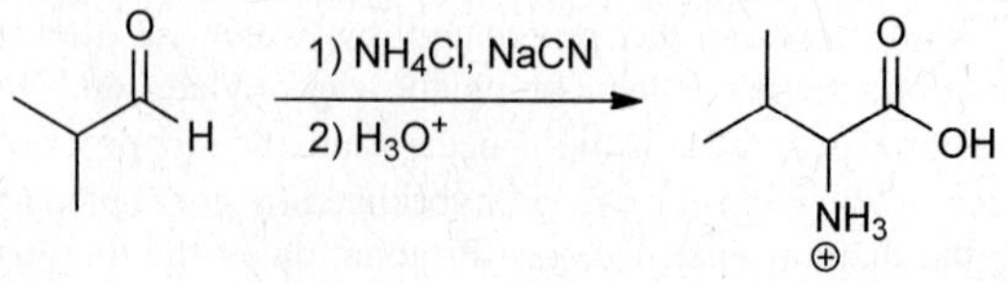

1) NH_4Cl, NaCN
2) H_3O^+

25.61. A pentapeptide has five amino acid residues, each of which can be any of the 20 naturally occurring amino acids. Therefore, there are $20 \times 20 \times 20 \times 20 \times 20 = 20^5 = 3,200,000$ possible pentapeptides that can be made from the naturally occurring amino acids.

25.62. L-Histidine can be prepared from the compound below via an asymmetric catalytic hydrogenation:

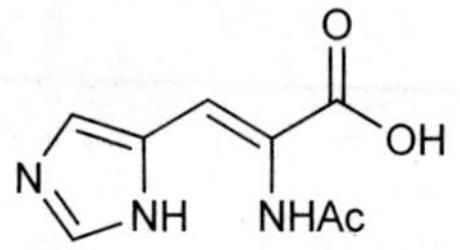

25.63. Below are the six possible sequences for a tripeptide containing L-leucine, L-methionine, and L-histidine:

1) Leu-Met-Val
2) Leu-Val-Met
3) Met-Val-Leu,
4) Met-Leu-Val
5) Val-Met-Leu
6) Val-Leu-Met

25.64. The N terminus of this tripeptide is drawn on the left and the C terminus on the right. Side chains at the top of the drawing are on wedges, while side chains on the bottom of the drawing are on dashes. The identity of each side chain can be found in Table 25.1.

At physiological pH, the amino groups exist primarily in their protonated form (ammonium ions) while the carboxylic acid groups exist primarily in their deprotonated form (carboxylate ions):

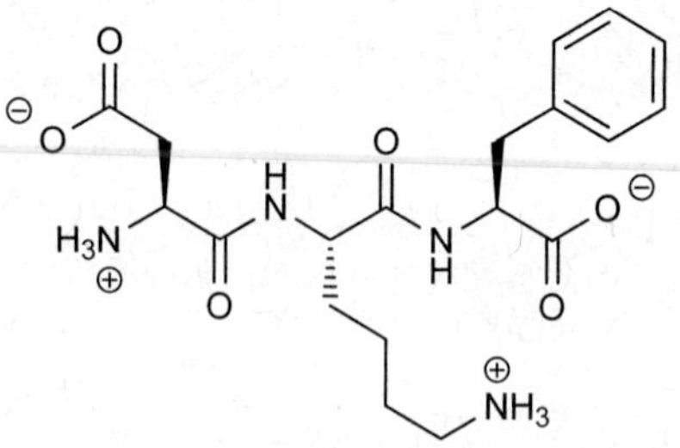

25.65. The N terminus is drawn on the left and the C terminus on the right. Side chains at the top of the drawing are on wedges, while side chains on the bottom of the drawing are on dashes. The identity of each side chain can be found in Table 25.1.

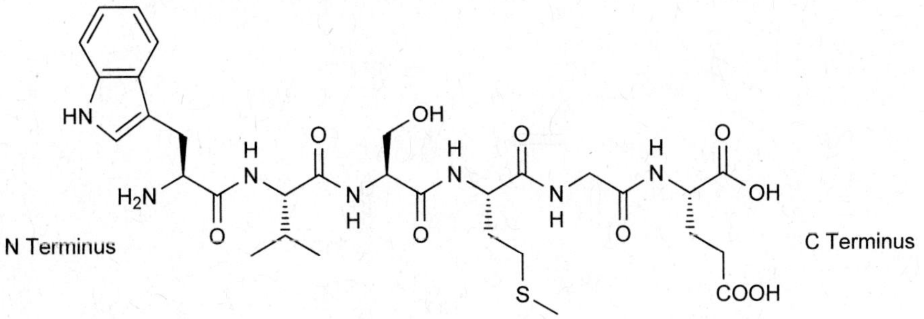

25.66. The N terminus is drawn on the left and the C terminus on the right. Side chains at the top of the drawing are on wedges, while side chains on the bottom of the drawing are on dashes. The identity of each side chain can be found in Table 25.1.

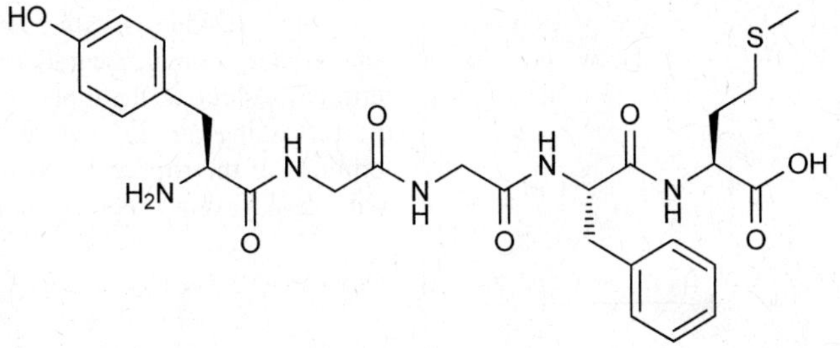

25.67. The structure of aspartame is shown in the solution to Problem 25.28. At physiological pH, the carboxylic acid group is deprotonated (and will exist primarily as a carboxylate ion), while the amino group is protonated (and will exist primarily as an ammonium ion):

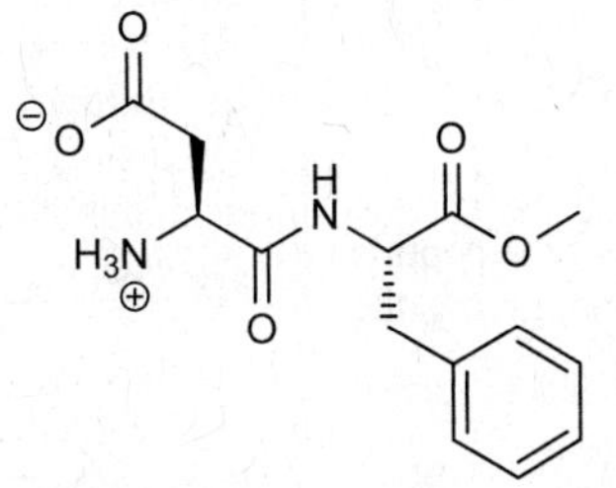

25.68. The following retrosynthetic analysis reveals the two amino acids (cysteine and valine) that are most likely utilized during the biosynthesis of penicillin antibiotics:

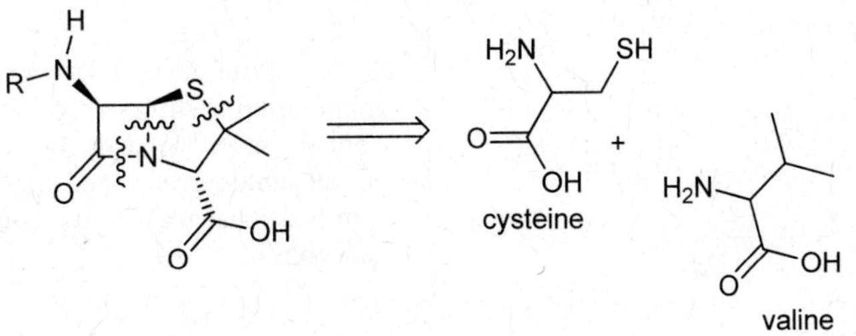

25.69. The following retrosynthetic analysis reveals the three amino acids (tyrosine, serine, and glycine) that are necessary for biosynthesis of the fluorophore:

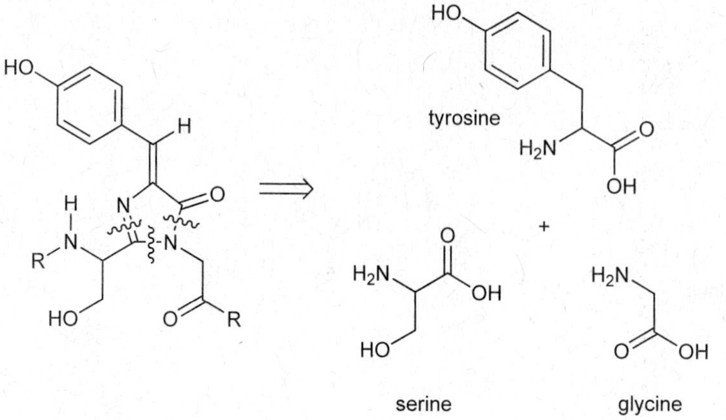

25.70. If a tripeptide does not react with phenyl isothiocyanate, then it must not have a free N terminus. It must be a cyclic tripeptide. Below are the two possible cyclic tripeptides:

25.71.
(a) Trypsin catalyzes the hydrolysis of the peptide bond at the carboxyl side of arginine, giving the following two fragments:

Arg   +   Pro-Pro-Gly-Phe-Ser-Pro-Phe-Arg

(b) Chymotrypsin catalyzes the hydrolysis of the peptide bonds at the carboxyl side of phenylalanine, giving the following three fragments:

Arg-Pro-Pro-Gly-Phe   +   Ser-Pro-Phe   +   Arg

25.72. The R group (highlighted) in the PTH derivative indicates the identity of the N-terminal residue. Since this R group is a benzylic group ($-CH_2Ph$), the N-terminal residue must be phenylalanine.

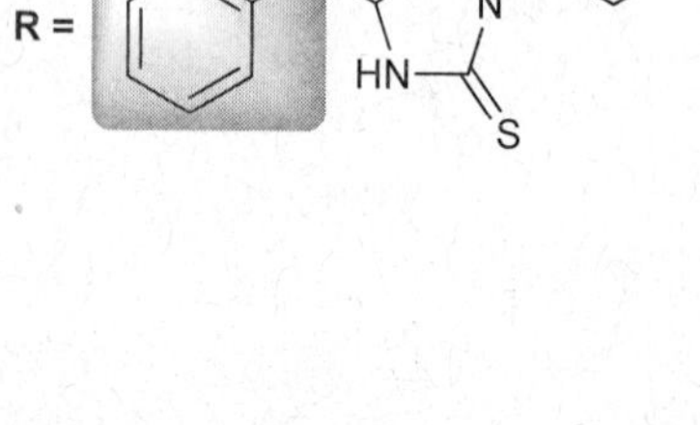

25.73. The first Edman degradation indicates that the N-terminal residue is valine (the R group is isopropyl). The second Edman degradation indicates that the N-terminal residue of the dipeptide is alanine. And finally, the remaining amino acid (glycine) must be at the C-terminus of the tripeptide. In summary, the tripeptide is Val-Ala-Gly, drawn here:

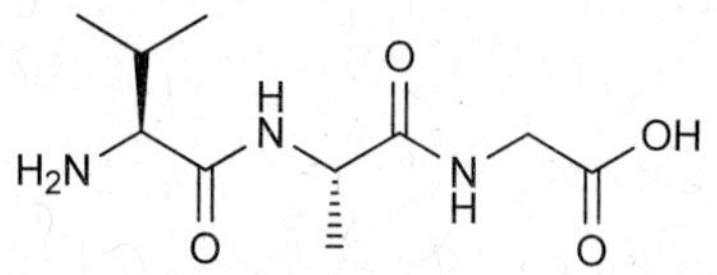

25.74. Only one of the trypsin fragments has a C terminus that is not arginine or lysine. This fragment (which ends with threonine), must be the last fragment in the peptide sequence. The remaining three trypsin fragments can be placed in the proper order by analyzing the chymotrypsin fragments. The correct peptide sequence is:

His-Ser-Gln-Gly-Thr-Phe-Thr-Ser-Asp-Tyr-Ser-Lys-Tyr-Leu-Asp-Ser-Arg-Arg-Ala-Gln-Asp-Phe-Val-Gln-Trp-Leu-Met-Asn-Thr

There cannot be any disulfide bridges in this peptide, because it has no cysteine residues, and only cysteine residues form disulfide bridges.

25.75. Prior to acetylation, the nitrogen atom of the amino group is sufficiently nucleophilic to attack phenyl isothiocyanate. Acetylation converts the amino group into an amide group, and the lone pair of the nitrogen atom is delocalized via resonance, rendering it much less nucleophilic.

25.76.

(a) When treated with acid and methanol, the carboxylic acid group is converted into a methyl ester (via a Fischer esterification process), and the amino group is protonated, giving the following compound:

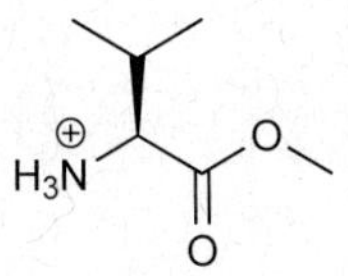

(b) When treated with di-*tert*-butyl dicarbonate, the amino group is protected with a Boc protecting group, giving the following product:

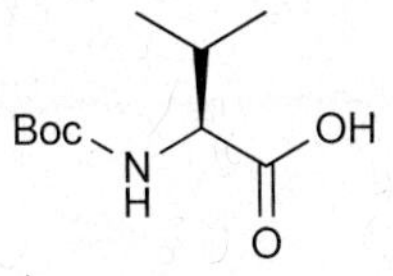

(c) Under basic conditions, the amino group is not protonated, and the carboxylic acid group is deprotonated, giving a carboxylate ion:

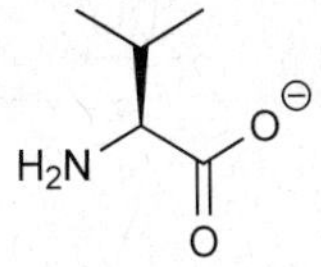

(d) Under acidic conditions, the amino group is protonated to give an ammonium ion, and the carboxylic acid group will be in its protonated form:

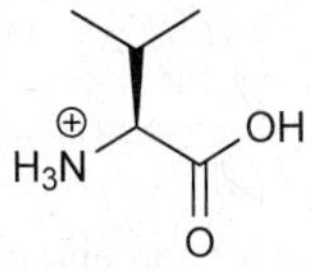

25.77.   We begin by installing the appropriate protecting groups. Then, upon treatment with DCC, the protected amino acids are coupled. And finally, the protecting groups are removed, as shown:

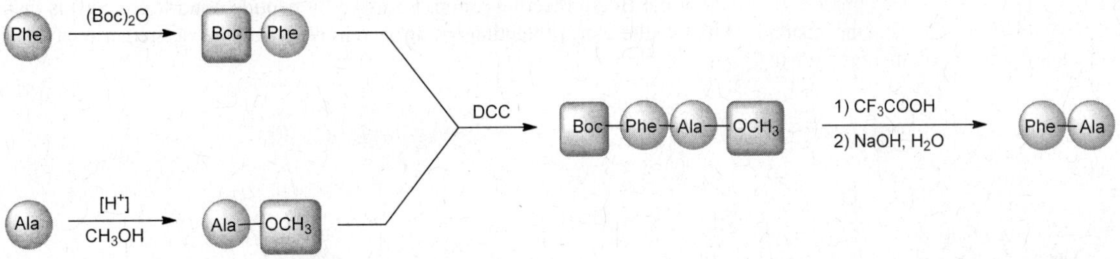

25.78.   When a mixture of L-phenylalanine and L-alanine is treated with DCC, there are four possible dipeptides: 1) Phe-Ala, or 2) Ala-Phe, or 3) Phe-Phe, or 4) Ala-Ala. These four possibilities are drawn below:

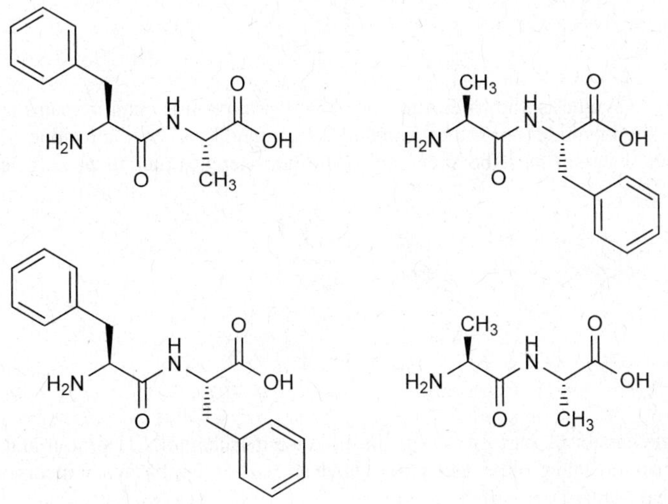

25.79. We begin by installing the appropriate protecting groups. Then, upon treatment with DCC, the protected amino acids are coupled. And finally, the protecting groups are removed, as shown:

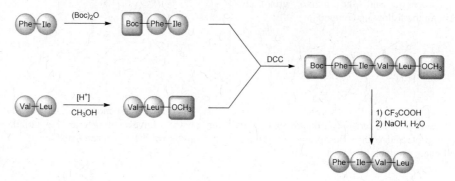

25.80. First, we attach the appropriate Boc-protected residue to the polymer:

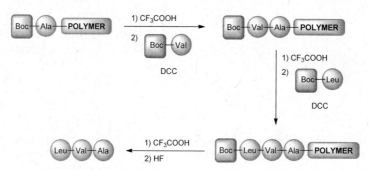

Then, the Boc protecting group is removed and a new peptide bond is formed with a Boc-protected amino acid (valine), using DCC. This two-step process (removal of the Boc protecting group, followed by peptide bond formation) is then repeated to install the leucine residue. Finally, the Boc protecting group is removed and the desired tripeptide is detached from the polymer, as shown:

25.81. During a Merrifield synthesis, the C-terminus of the growing peptide chain remains anchored to the polymer. The C-terminus of the desired peptide (leucine enkephalin) is occupied by a leucine residue. Therefore, the following Boc-protected amino acid (leucine) must be anchored to the polymer in order to prepare leucine enkephalin via a Merrifield synthesis:

25.82. A proline residue cannot be part of an α helix, because it lacks an N-H proton and does not participate in hydrogen bonding. (The amino acid proline does indeed have an N-H group, but when incorporated into a peptide, the proline residue does not have an N-H group)

25.83. The amino group attacks one of the carbonyl groups of DCC, giving a tetrahedral intermediate. The carbonyl group is then reformed upon expulsion of a resonance-stabilized leaving group. The resulting cation is then deprotonated to give the product:

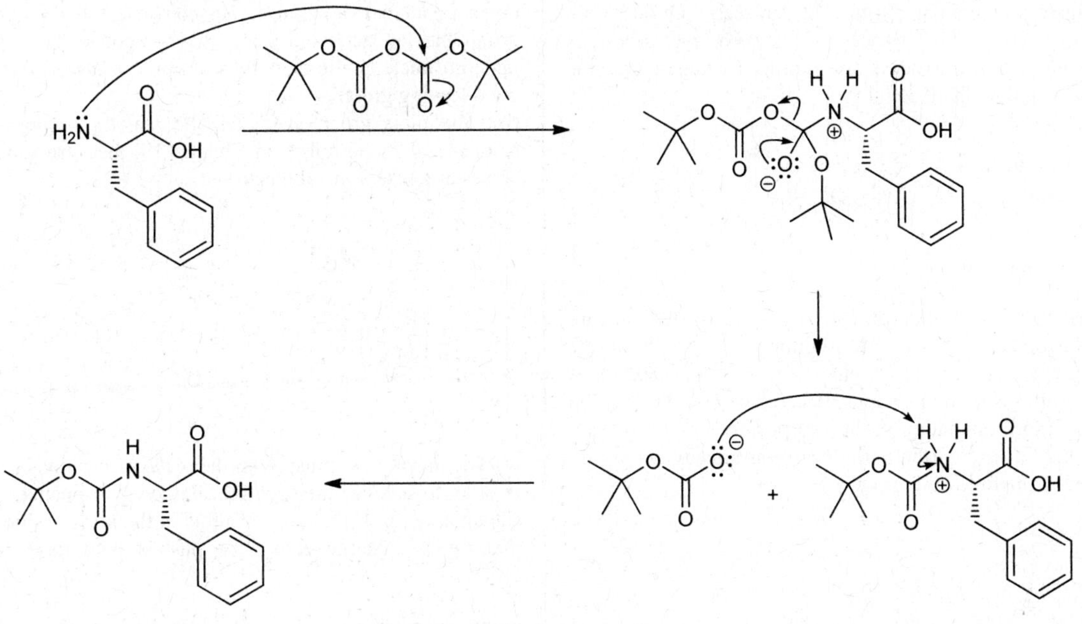

25.84. The following alkyl halide would be necessary in order to prepare tyrosine via an amidomalonate synthesis. This starting material possesses both a nucleophilic center (the OH group) as well as an electrophilic center (it is a primary benzylic bromide). As such, the molecules can react with each other via an S_N2 process, thereby forming the polymer shown:

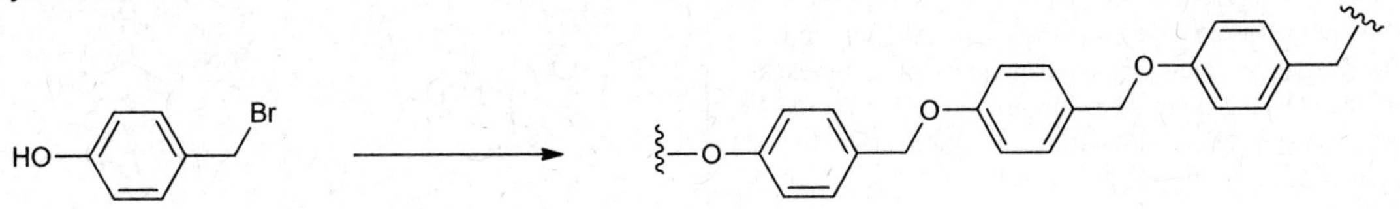

25.85. The stabilized enolate ion (formed in the first step) can function as a base, rather than a nucleophile, giving an E2 reaction:

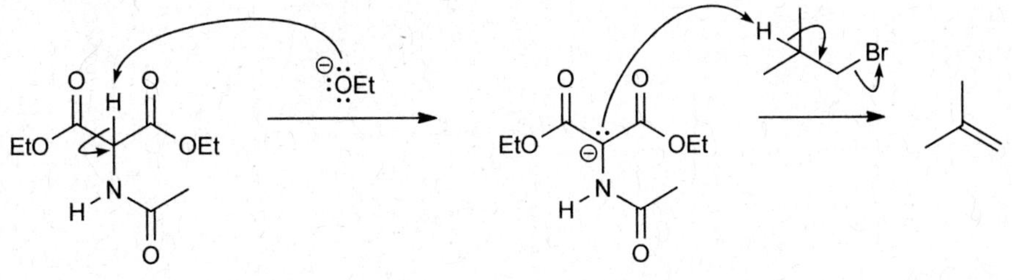

25.86. The lone pair on that nitrogen atom is highly delocalized via resonance and is participating in aromaticity. Accordingly, the lone pair is not available to function as a base.

25.87.

(a) The Hell-Volhard-Zelinsky reaction will install a bromine atom at the α position of a carboxylic acid. This bromine atom can then be replaced with an amino group via an S_N2 process. As such, the following carboxylic acid is necessary in order to prepare tyrosine via a Hell-Volhard-Zelinsky reaction:

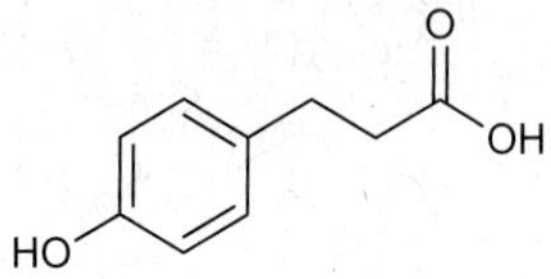

(b) The aromatic ring is highly activated toward electrophilic aromatic substitution, as a result of the presence of the OH group, which is an *ortho-para* director (as seen in Chapter 19). Therefore, the ring can undergo bromination in the two positions that are *ortho* to the OH group (the *para* position is already occupied), giving the following product:

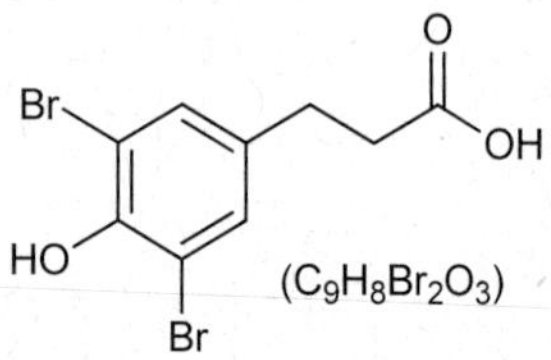

$(C_9H_8Br_2O_3)$

25.88. At low temperature, the barrier to rotation keeps the two methyl groups in different electronic environments (one is *cis* to the C=O bond and the other is *trans* to the C=O bond), and as a result, they give rise to separate signals. At high temperature, there is sufficient energy to overcome the energy barrier, and the protons change electronic environments on a timescale that is faster than the timescale of the NMR spectrometer. The result is an averaging effect which gives rise to only one signal.

25.89.

(a) The COOH group does not readily undergo nucleophilic acyl substitution because the OH group is not a good leaving group. By converting the COOH group into an activated ester, the compound can now undergo nucleophilic acyl substitution because it has a good leaving group.

(b) The nitro group stabilizes the leaving group via resonance. As described in Chapter 19, the nitro group serves as a reservoir for electron density:

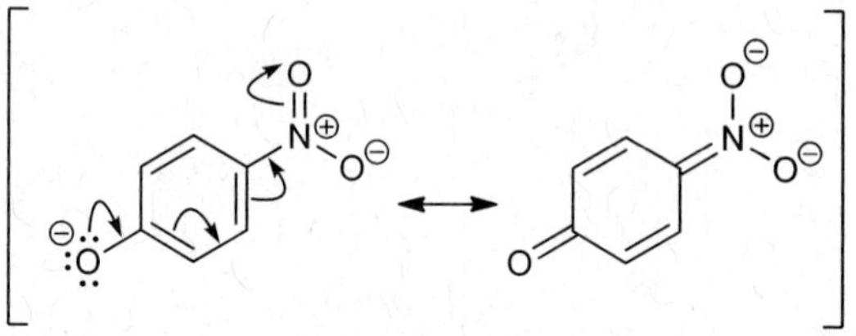

(c) The nitro group must be in the *ortho* or *para* position in order to stabilize the negative charge via resonance (as shown above). If the nitro group is in the *meta* position, the negative charge cannot be pushed onto the nitro group.

25.90. Hydrolysis of the ester group gives threonine, as shown here:

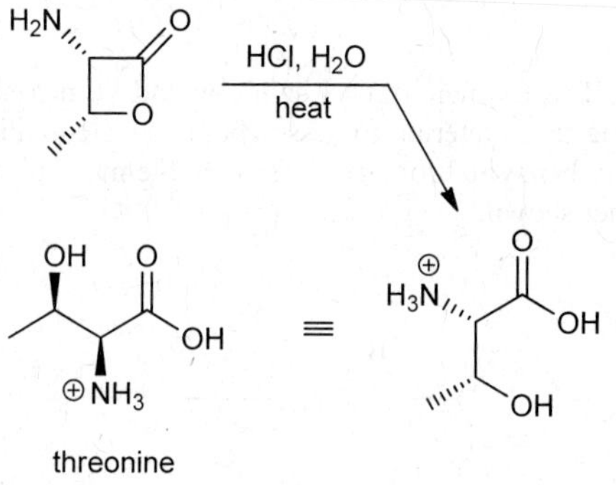

threonine

Chapter 26
Lipids

Review of Concepts

Fill in the blanks below. To verify that your answers are correct, look in your textbook at the end of Chapter 26. Each of the sentences below appears verbatim in the section entitled *Review of Concepts and Vocabulary*.

- **Lipids** are naturally occurring compounds that are extracted from cells using _____ solvents.
- **Complex lipids** readily undergo _____, while **simple lipids** do not.
- _____ are high molecular weight esters that are constructed from carboxylic acids and alcohols.
- _____ are the triesters formed from glycerol and three long-chain carboxylic acids, called **fatty acids**. The resulting triglyceride is said to contain three fatty acid _____.
- For saturated fatty acids, the melting point increases with increasing _____ _____. The presence of a _____ double bond causes a decrease in the melting point.
- Triglycerides that are solids at room temperature are called _____, while those that are liquids at room temperature are called _____.
- Triglycerides containing unsaturated fatty acid residues will undergo hydrogenation. During the hydrogenation process, some of the double bonds can isomerizes to give _____ π bonds
- In the presence of molecular oxygen, triglycerides are particularly susceptible to oxidation at the _____ position to produce hydroperoxides.
- Transesterification of triglycerides can be achieved either via _____ catalysis or _____ catalysis to produce biodiesel.
- _____ are similar in structure to triglycerides except that one of the three fatty acid residues is replaced by a phosphoester group.
- The structures of **steroids** are based on a tetracyclic ring system, involving three six-membered rings and one _____-membered ring.
- The ring fusions are all _____ in most steroids, giving steroids their rigid geometry.
- All steroids, including cholesterol, are biosynthesized from _____.
- Prostaglandins contain twenty carbon atoms and are characterized by a _____-membered ring with two side chains.
- **Terpenes** are a class of naturally occurring compounds that can be thought of as being assembled from _____ units.
- A terpene with 10 carbon atoms is called a _____, while a terpene with 20 carbon atoms is called a _____.

Review of Skills

Fill in the blanks and empty boxes below. To verify that your answers are correct, look in your textbook at the end of Chapter 26. The answers appear in the section entitled *SkillBuilder Review*.

26.1 Comparing Molecular Properties of Triglycerides

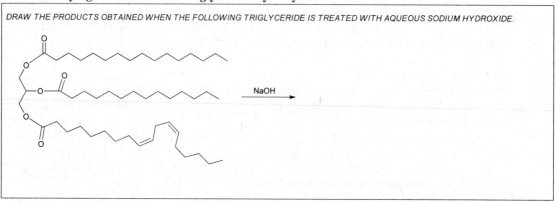

CIRCLE THE TRIGLYCERIDE BELOW THAT IS EXPECTECD TO HAVE A HIGHER MELTING POINT.

26.2 Identifying the Products of Triglyceride Hydrolysis

DRAW THE PRODUCTS OBTAINED WHEN THE FOLLOWING TRIGLYCERIDE IS TREATED WITH AQUEOUS SODIUM HYDROXIDE.

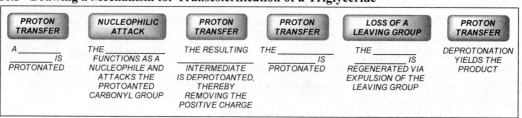

NaOH

26.3 Drawing a Mechanism for Transesterification of a Triglyceride

PROTON TRANSFER	NUCLEOPHILIC ATTACK	PROTON TRANSFER	PROTON TRANSFER	LOSS OF A LEAVING GROUP	PROTON TRANSFER
A _____ IS PROTONATED	THE _____ FUNCTIONS AS A NUCLEOPHILE AND ATTACKS THE PROTOANTED CARBONYL GROUP	THE RESULTING _____ INTERMEDIATE IS DEPROTOANTED, THEREBY REMOVING THE POSITIVE CHARGE	THE _____ IS PROTONATED	THE _____ IS REGENERATED VIA EXPULSION OF THE LEAVING GROUP	DEPROTONATION YIELDS THE PRODUCT

26.4 Identifying Isoprene Units in a Terpene

IDENTIFY THE ISOPRENE UNITS IN THE FOLLOWING TERPENE

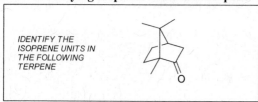

Review of Reactions

Identify the reagents necessary to achieve each of the following transformations. To verify that your answers are correct, look in your textbook at the end of Chapter 26. The answers appear in the section entitled *Review of Reactions*.

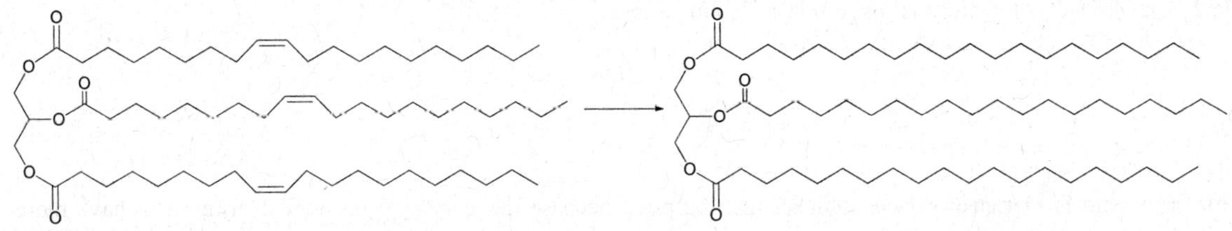

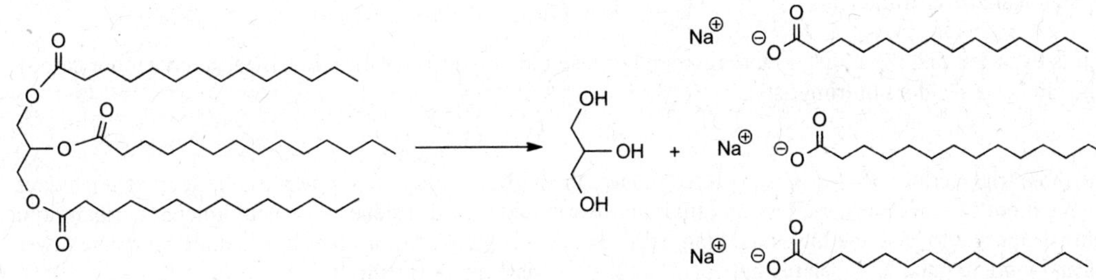

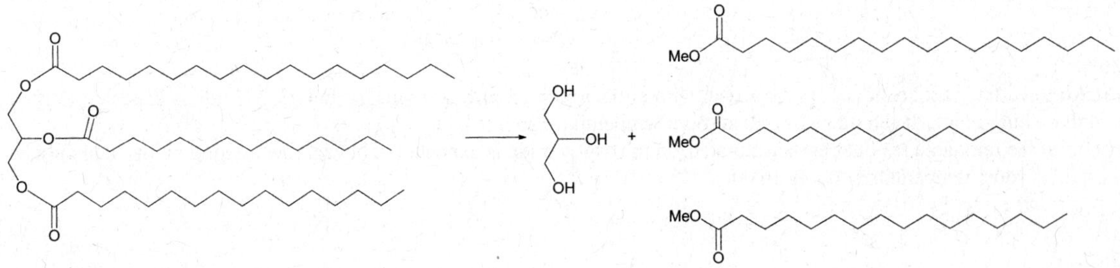

Solutions

26.1   Hydrolysis of triacontyl hexadecanoate gives a carboxylic acid with sixteen carbon atoms and an alcohol with thirty carbon atoms, as shown:

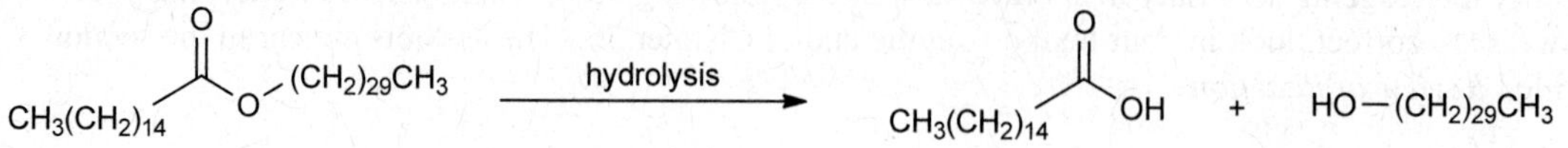

26.2. The products of hydrolysis suggest the following ester:

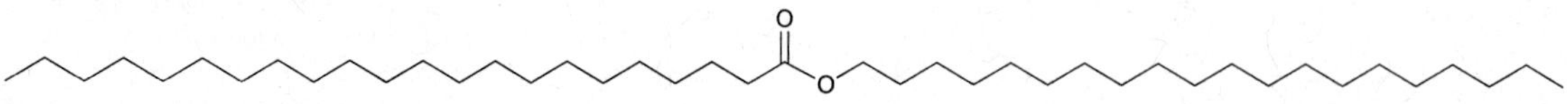

26.3.
(a) Trimyristin is expected to have a higher melting point because the fatty acid residues in trimyristin have more carbon atoms than the fatty acid residues of trilaurin.

(b) Triarachidin is expected to have a higher melting point because the fatty acid residues in triarachidin have more carbon atoms and less unsaturation than the fatty acid residues of trilinolein.

(c) Triolein is expected to have a higher melting point because the fatty acid residues in triolein have less unsaturation than the fatty acid residues of trilinolein.

(d) Tristearin is expected to have a higher melting point because the fatty acid residues in tristearin have more carbon atoms than the fatty acid residues of trimyristin.

26.4.   Of the three triglycerides, tristearin is expected to have the highest melting point because the fatty acid residues in tristearin have more carbon atoms and less unsaturation than the fatty acid residues of tripalmitolein or tripalmitin. Tripalmitolein is expected to have the lowest melting point because the fatty acid residues in tripalmitolein have fewer carbon atoms and more unsaturation than the fatty acid residues of tristearin or tripalmitin.

26.5.   The fatty acid residues in triarachidin have more carbon atoms than the fatty acid residues in tristearin. Therefore, triarachadin is expected to have a higher melting point. It should be a solid at room temperature, and should therefore be classified as a fat, rather than an oil. Therefore, triglycerides made from lauric acid will also have a low melting point.

26.6.
(a) All three fatty acid residues are saturated, with either 16 or 18 carbon atoms, so the triglyceride is expected to have a high melting point. It should be a solid at room temperature, so it is a fat.
(b) All three fatty acid residues are unsaturated, so the triglyceride is expected to have a low melting point. It should be a liquid at room temperature, so it is an oil.

26.7.
(a) The C=C bond in each oleic acid residue undergoes hydrogenation, giving the following saturated triglyceride:

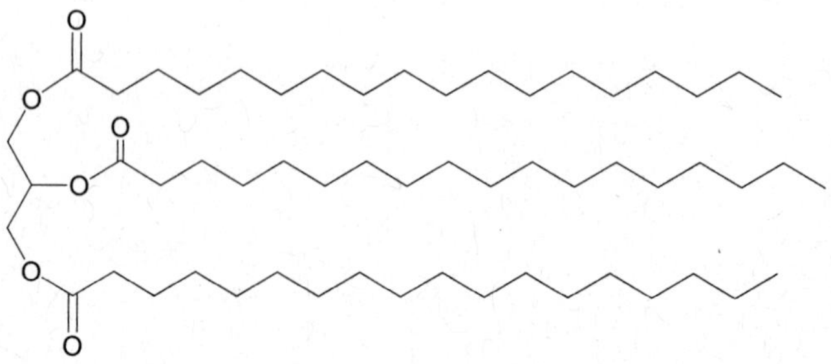

(b) The triglyceride obtained from hydrogenation (shown above) has three stearic acid residues, so it is called tristearin.
(c) Tristearin is expected to have a higher melting point because the fatty acid residues in tristearin are saturated, while the fatty acid residues of triolein are unsaturated.
(d) Upon treatment with aqueous base, each of the ester groups will undergo hydrolysis, giving three equivalents of stearic acid.

26.8. There are three fatty acid residues. Partial hydrogenation indicates that either one or two of these residues has a double bond (in the *trans* configuration). There are two isomers that exhibit one C=C bond, and there are two isomers that exhibit two C=C bonds, as shown here:

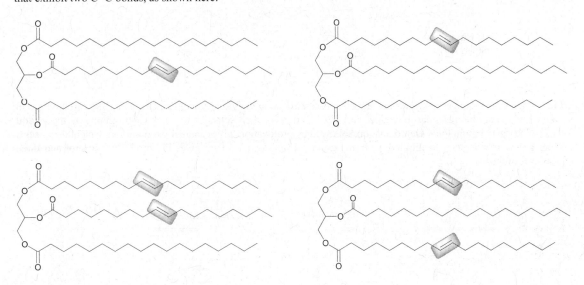

26.9. When a triglyceride is treated with aqueous base, each of the ester groups is hydrolyzed, giving glycerol and three carboxylate ions, as shown:

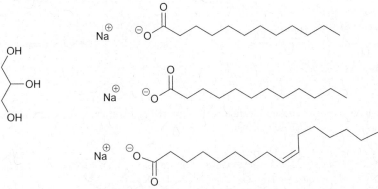

26.10. The products of hydrolysis indicate that the starting triglyceride has three lauric acid residues, as shown:

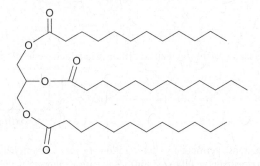

26.11. The products of hydrolysis indicate that the starting triglyceride has two lauric acid residues and one palmitic acid residue. In order to be optically inactive, the palmitic acid residue must be connected to C2 of the glycerol backbone, as shown below. Otherwise, the highlighted position (C2 of the glycerol backbone) would be a chirality center:

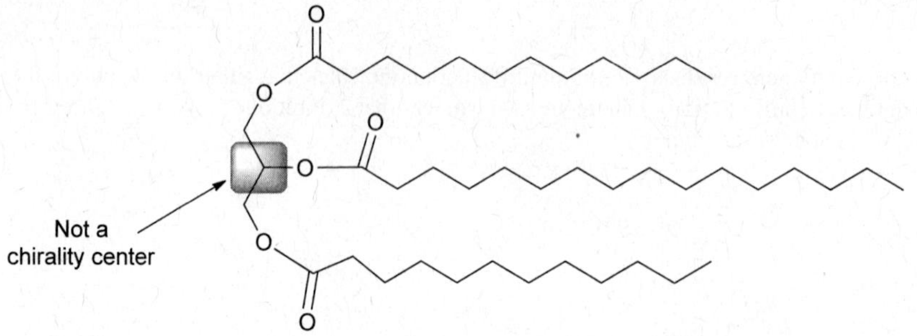

Not a
chirality center

26.12. Each of the three ester groups undergoes transesterification via the following mechanism. The carbonyl group is protonated and the resulting resonance-stabilized cation is then attacked by methanol, giving a tetrahedral intermediate (an oxonium ion). Deprotonation, followed by protonation, gives another oxonium ion, which then loses a leaving group, thereby regenerating the carbonyl group. Deprotonation then gives the products (glycerol and three equivalents of the ester):

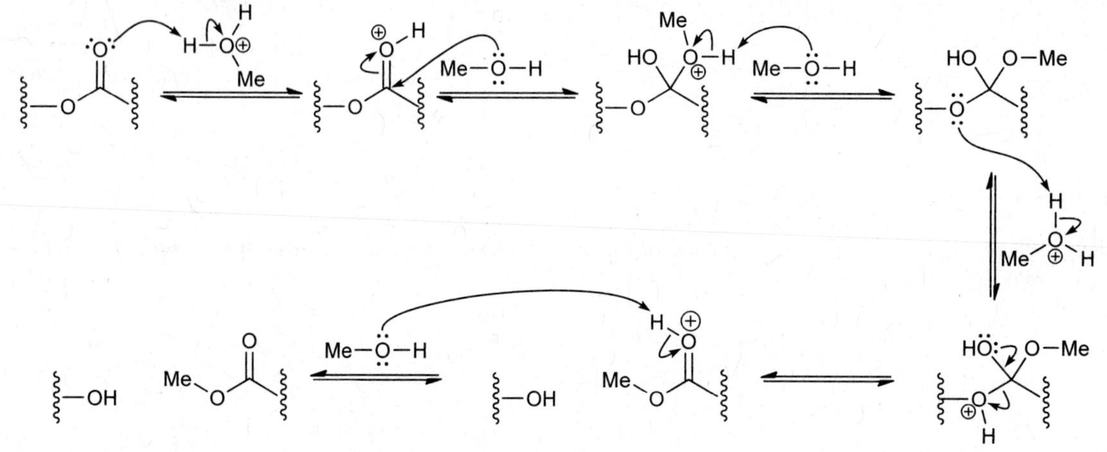

26.13. When triolein undergoes transesterification with isopropyl alcohol, the glycerol backbone is released, along with three equivalents of the isopropyl ester shown:

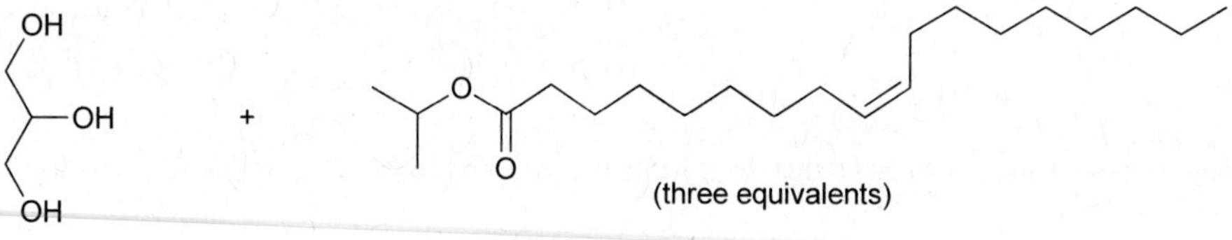

(three equivalents)

26.14.
(a) Hydroxide functions as a catalyst by establishing an equilibrium in which some ethoxide ions are present.

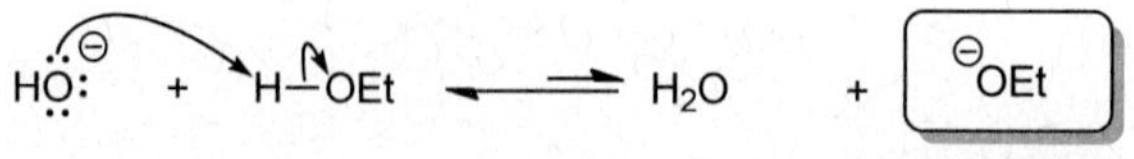

HO:⁻ + H—OEt ⇌ H₂O + ⁻OEt

ethoxide

Then, each ester group undergoes transesterification via the following mechanism. The carbonyl group is attacked by ethoxide, giving a tetrahedral intermediate (an alkoxide ion). The carbonyl group is then reformed via expulsion of an alkoxide leaving group, which is then protonated by water, giving glycerol and three equivalents of an ethyl ester, as shown.

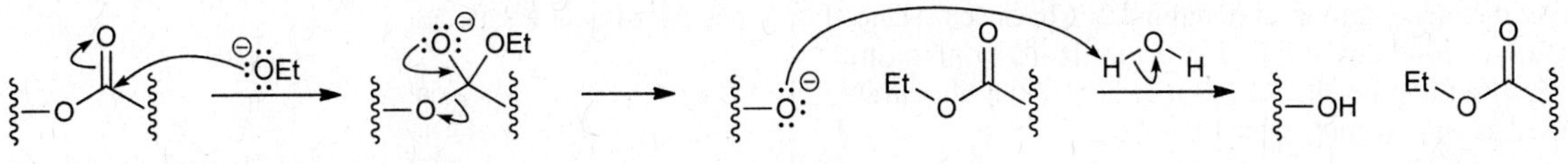

(b) Hydroxide can function as a nucleophile and attack each ester group directly, giving hydrolysis rather than transesterification.

26.15.
(a) The following lecithin has two myristic acid residues:

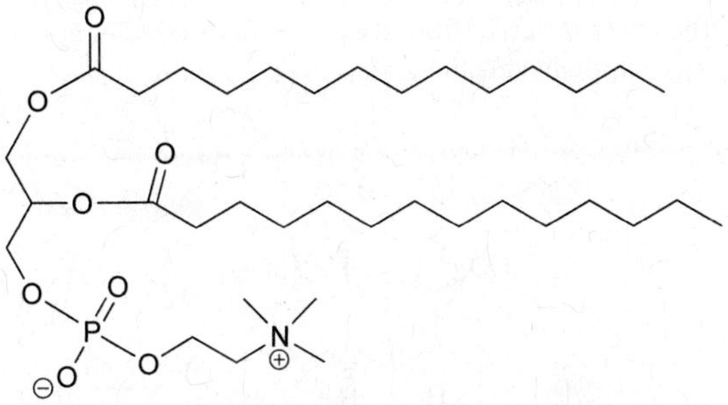

(b) The C2 position (of a lecithin) will be a chirality center and will generally have the *R* configuration, as shown:

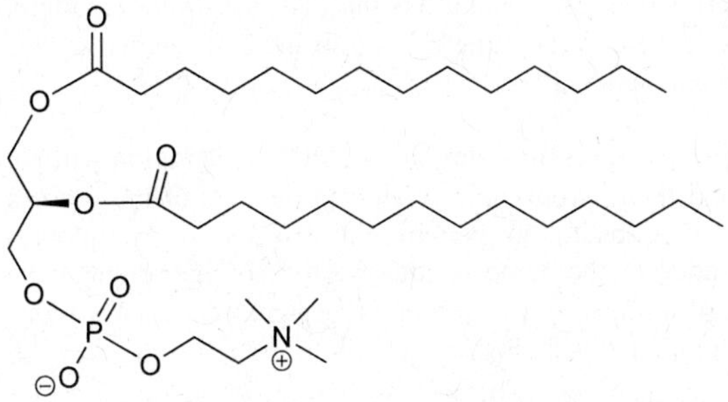

(c) No. The C2 position would no longer be a chirality center, because it would be connected to two identical groups.

26.16.
(a) Each of the following two cephalins has one palmitic acid residue and one oleic acid residue. In each case, the C2 position is a chirality center with the *R* configuration.

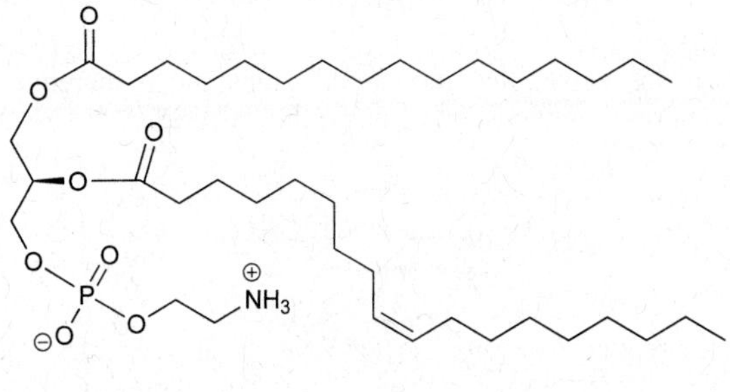

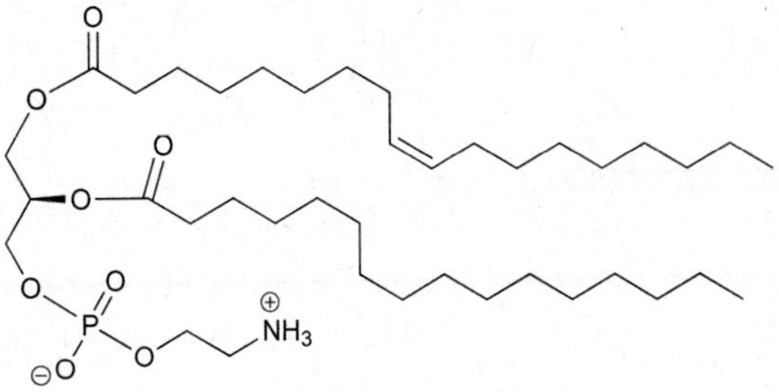

(b) Yes. The C2 position would still be connected to four different groups so it would still be a chirality center.

26.17. The phosphate group has three resonance structures, as shown:

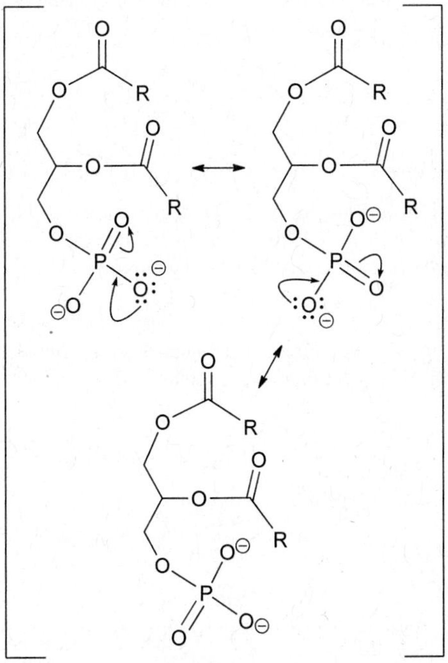

26.18. Octanol has a longer hydrophobic tail than hexanol and is therefore more efficient at crossing the nonpolar environment of the cell membrane.

26.19. No. Glycerol has three OH groups (hydrophilic) and no hydrophobic tail. It cannot cross the nonpolar environment of the cell membrane.

26.20. One of these compounds is extremely similar in structure to norethindrone (an oral contraceptive), except that the methyl group of norethindrone has been replaced with an ethyl group. This structure is likely norgestrel, because the problem statement indicates that nogestrel is used as an oral contraceptive.

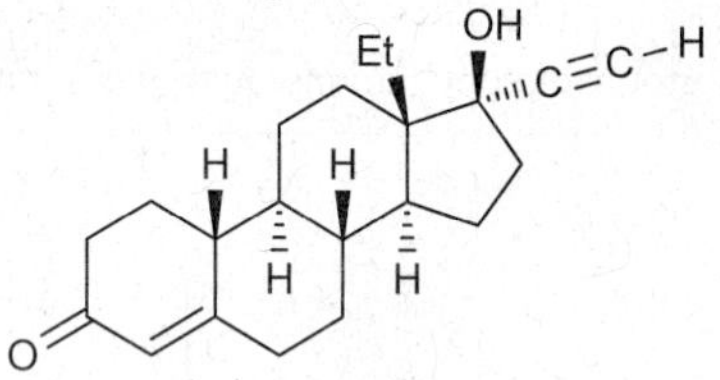

norgestrel

Oxymetholone is an anabolic steroid and it is similar in structure to nandrolone (a synthetic androgen analogue):

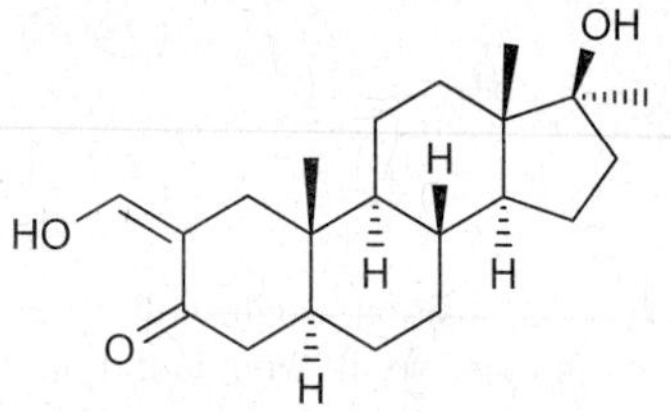

oxymetholone

26.21. A ring-flip is not possible for *trans*-decalin because one of the rings would have to achieve a geometry that resembles a six-membered ring with a *trans*-alkene, which is not possible.

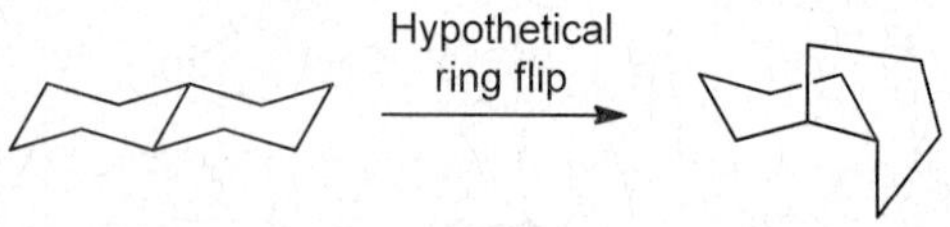

Hypothetical ring flip

The ring fusions of cholesterol all resemble the ring fusion in *trans*-decalin, so none of the rings in cholesterol are free to undergo ring-flipping.

26.22.
(a) The two rings are fused together in a *trans*-decalin system. The position of each substituent is labeled:

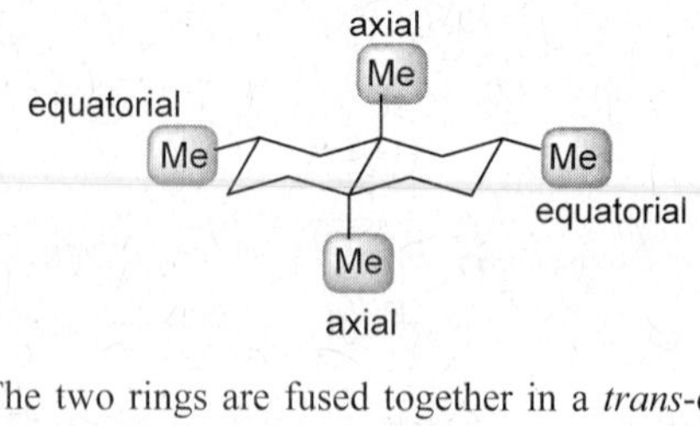

(b) The two rings are fused together in a *trans*-decalin system. The position of each substituent is labeled:

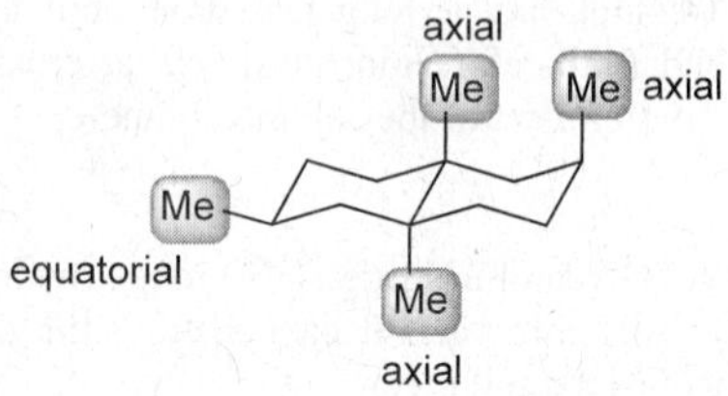

(c) Each set of neighboring rings are fused together, much like in a *trans*-decalin system. The position of each substituent is labeled:

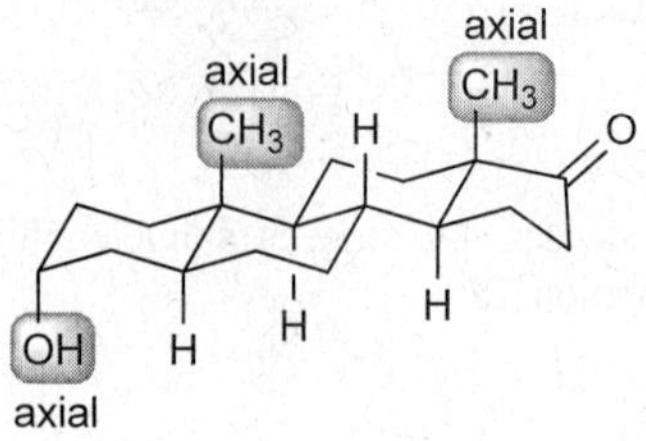

26.23. The structure of prednisolone acetate is shown below. As described in the problem statement, this structure is different from the structure of cortisol in two ways, highlighted below:

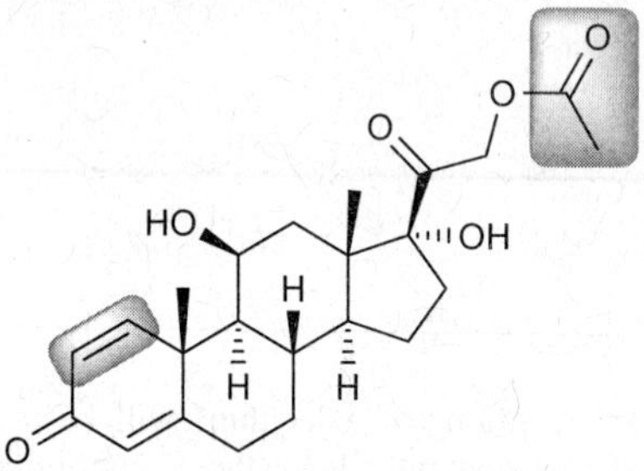

26.24.
(a) This prostaglandin has the PGE substitution pattern, and there is only one π bond in the side chains, so this compound is PGE_1 .

(b) This prostaglandin has the PGF substitution pattern, and there is only one π bond in the side chains. For a PGF substitution pattern, an additional descriptor is added to the name to indicate the configuration of the OH groups. A *cis* diol is designated as α, so this compound is $PGF_{1\alpha}$.

26.25.
(a) This terpene has ten carbon atoms and is therefore comprised of two isoprene units, shown here:

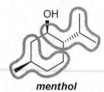

menthol

(b) This terpene has ten carbon atoms and is therefore comprised of two isoprene units, shown here:

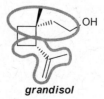

grandisol

(c) This terpene has ten carbon atoms and is therefore comprised of two isoprene units, shown here:

carvone

26.26.
(a) Yes, this compound has 10 carbon atoms and is comprised of two isoprene units.
(b) No, this compound is not a terpene, because it has 11 carbon atoms. In order to be a terpene, the number of carbon atoms must be divisible by 5.
(c) No, this compound is not a terpene, because it has 11 carbon atoms. In order to be a terpene, the number of carbon atoms must be divisible by 5.
(d) This compound has 10 carbon atoms, but the branching pattern cannot be achieved by joining two isoprene units, so this compound is not a terpene.

26.27. The pyrophosphate leaving group is expelled to give a resonance-stabilized (allylic) carbocation. The π bond of isopentyl phosphate then functions as a nucleophile and attacks the carbocation. Finally, a basic amino acid residue of the enzyme removes a proton to give the product:

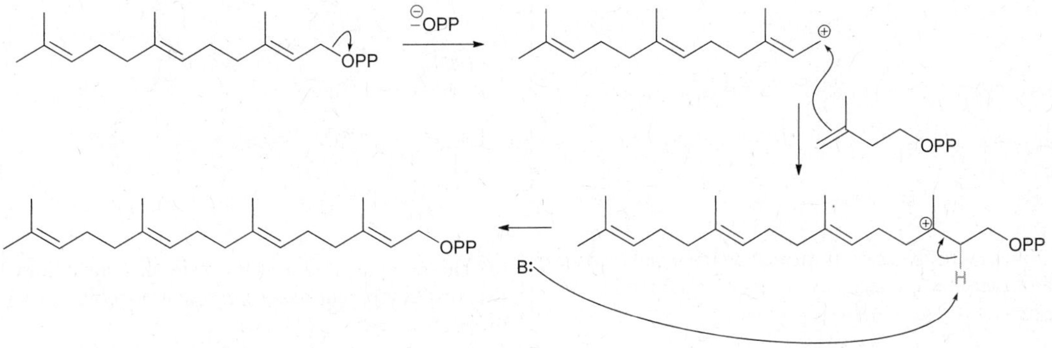

26.28. The pyrophosphate leaving group is expelled to give a resonance-stabilized (allylic) carbocation. The π bond of isopentyl phosphate then functions as a nucleophile and attacks the carbocation. A basic amino acid residue of the enzyme then removes a proton to give geranyl pyrophosphate. The previous three steps are then repeated to give farnesyl pyrophosphate, followed by an elimination process to give α-farnesene, as shown:

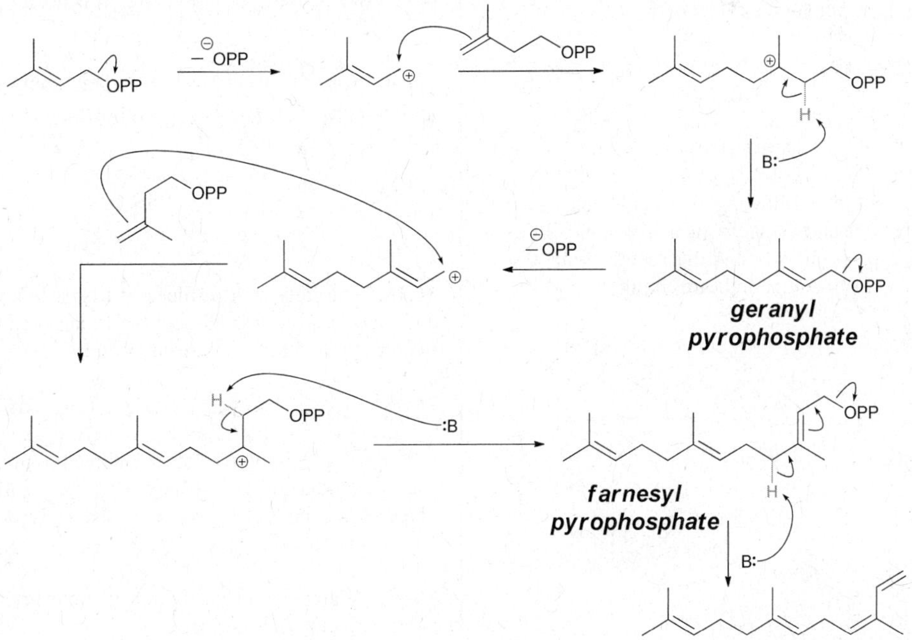

geranyl pyrophosphate

farnesyl pyrophosphate

α-farnesene

26.29.
(a) As seen in Section 26.6, stanozolol is a steroid.
(b) As seen in Section 26.8, lycopene is a terpene.
(c) As seen in Section 26.3, tristearin is a triglyceride.
(d) As seen in Section 26.5, lecithins are phospholipids.
(e) As seen in Section 26.7, PGF$_2$ is a prostaglandin.
(f) As seen in Section 26.2, pentadecyl octadecanoate is a wax.

26.30.
(a) When treated with excess molecular hydrogen and a metal catalyst (Ni), the C=C bond in each palmitoleic acid residue undergoes hydrogenation, giving the following saturated triglyceride:

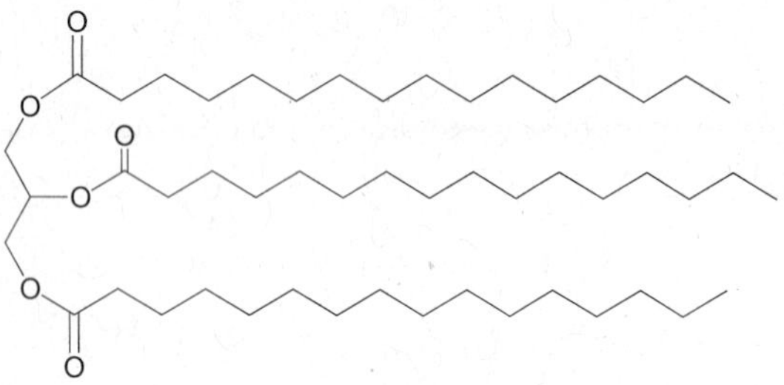

(b) When a triglyceride is treated with aqueous base, each of the ester groups is hydrolyzed, thereby releasing glycerol:

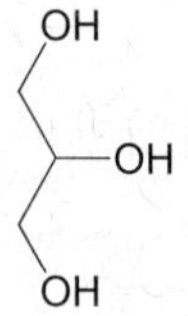

and three carboxylate ions, as shown:

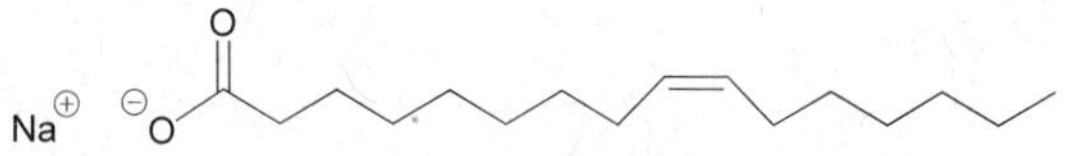

(three equivalents)

26.31. Each of the following two cephalins has one lauric acid residue and one myristic acid residue. In each case, the C2 position is a chirality center with the *R* configuration. Both compounds are chiral:

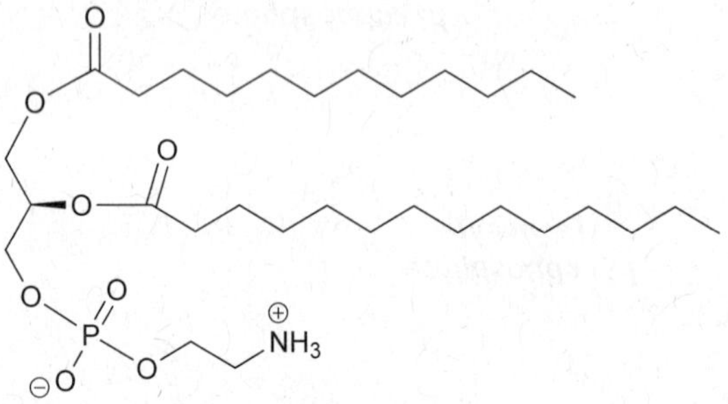

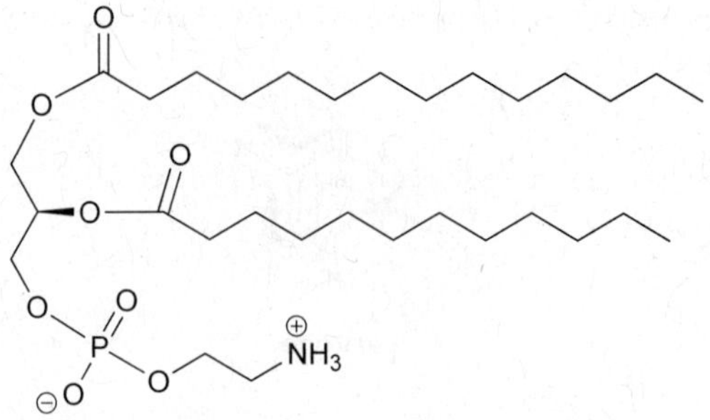

26.32. The fatty acid residues in this triglyceride are saturated, and will not react with molecular hydrogen.

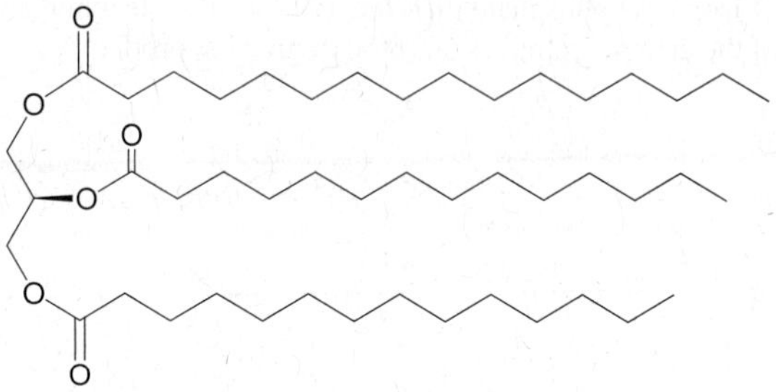

26.33.
(a) This compound is an amino acid. It is not a lipid.
(b) This compound has a large hydrophobic tail and is therefore a lipid.
(c) Lycopene is terpene, which is a type of lipid.
(d) Trimyristin is a triglyceride, which is a type of lipid.
(e) Palmitic acid has a large hydrophobic tail and is therefore a lipid.
(f) D-Glucose is a carbohydrate. It is not a lipid.
(g) Testosterone is a steroid, which is a type of lipid.
(h) D-Mannose is a carbohydrate. It is not a lipid.

26.34. *trans*-Oleic acid has 18 carbon atoms and a *trans* π bond between C9 and C10, as shown:

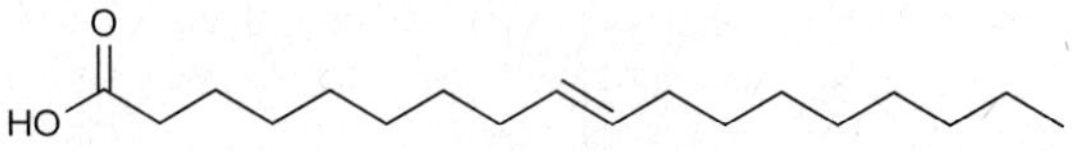

26.35. The fatty acid residues of tristearin are saturated and are therefore less susceptible to auto-oxidation than the unsaturated fatty acid residues in triolein.

26.36. A monoglyceride exhibits two OH groups (of the glycerol backbone) and is therefore expected to be the most water-soluble of the three compounds. A triglyceride has no OH groups (all three positions of the glycerol backbone are occupied), so a triglyceride will be the least water-soluble.

26.37. Water would not be appropriate because it is a polar solvent, and terpenes are nonpolar compounds. Hexane is a nonpolar solvent and would be suitable.

26.38.
(a) As seen in Table 26.1, palmitic acid is a saturated fatty acid.
(b) As seen in Table 26.1, myristic acid is a saturated fatty acid.
(c) As seen in Table 26.1, oleic acid is an unsaturated fatty acid.
(d) As seen in Table 26.1, lauric acid is a saturated fatty acid.
(e) As seen in Table 26.1, linoleic acid is an unsaturated fatty acid.
(f) As seen in Table 26.1, arachidonic acid is an unsaturated fatty acid.

26.39. As seen in Table 26.1, arachidonic acid has four carbon-carbon double bonds.

26.40.
(a) No. It is an oil.
(b) No. The fatty acid residues in triolein are unsaturated, so triolein is reactive towards molecular hydrogen in the presence of Ni.
(c) Yes. It undergoes hydrolysis to produce unsaturated fatty acids.
(d) Yes. It is a complex lipid because it undergoes hydrolysis.
(e) No. It is not an ester with a high molecular weight. It is not a wax.
(f) No. It does not have a phosphate group.

26.41.
(a) Yes. It is a fat.
(b) Yes. The fatty acid residues in tristearin are saturated, so tristearin is not reactive towards molecular hydrogen in the presence of Ni.
(c) No. It undergoes hydrolysis to produce fatty acids that are saturated.
(d) Yes. It is a complex lipid because it undergoes hydrolysis.

(e) No. It is not an ester with a high molecular weight. It is not a wax.
(f) No. It does not have a phosphate group.

26.42. The products of hydrolysis suggest the following ester:

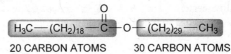

26.43. Trimyristin is expected to have a lower melting point than tripalmitin because the former is comprised of fatty acid residues with fewer carbon atoms (14 instead of 16).

trimyristin

tripalmitin

26.44. Each of the three ester groups undergoes transesterification via the following mechanism. The carbonyl group is protonated and the resulting resonance-stabilized cation is then attacked by isopropanol, giving a tetrahedral intermediate (an oxonium ion). Deprotonation, followed by protonation, gives another oxonium ion, which then loses a leaving group, thereby regenerating the carbonyl group. Deprotonation then gives the products (glycerol and three equivalents of the isopropyl ester):

26.45. See the solution to Problem 26.14.

26.46. In order to be optically inactive, the palmitic acid residue must be connected to C2 of the glycerol backbone (shown below). Otherwise, C2 would be a chirality center. This way, C2 is connected to two identical groups, so it is not a chirality center.

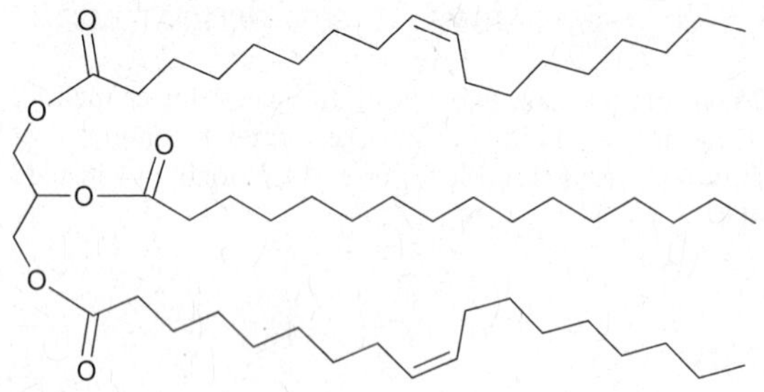

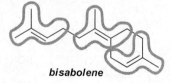

26.47. In order for the triglyceride to be optically active, the palmitic acid residue cannot be connected to C2 of the glycerol backbone (as explained in the previous problem).

26.48. The carbon skeleton of cholesterol is redrawn, but all wedges are replaced with dashes, and all dashes are replaced with wedges, giving the following structure (the enantiomer of cholesterol):

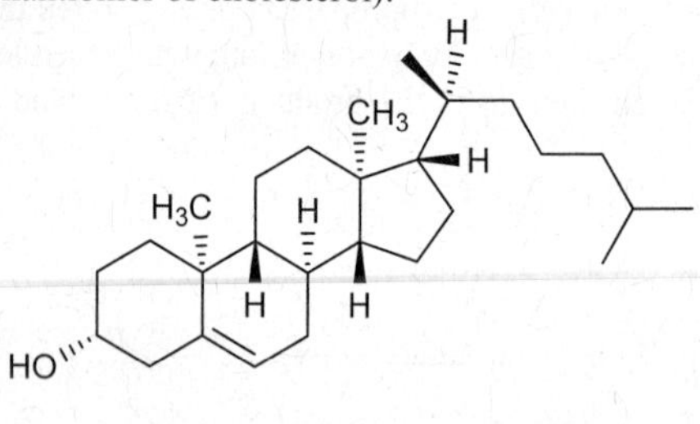

26.49.
(a) This terpene has fifteen carbon atoms and is therefore comprised of three isoprene units, shown here:

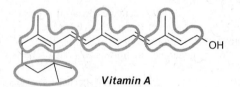

bisabolene

(b) This terpene has twenty carbon atoms and is therefore comprised of four isoprene units, shown here:

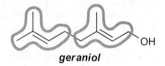

flexibilene

(c) This terpene has fifteen carbon atoms and is therefore comprised of three isoprene units, shown here:

humulene

(d) This terpene has twenty carbon atoms and is therefore comprised of four isoprene units, shown here:

Vitamin A

(e) This terpene has ten carbon atoms and is therefore comprised of two isoprene units, shown here:

geraniol

(f) This terpene has ten carbon atoms and is therefore comprised of two isoprene units, shown here:

sabinene

26.50.

(a) The polar head and the two hydrophobic tails are labeled in the following structure:

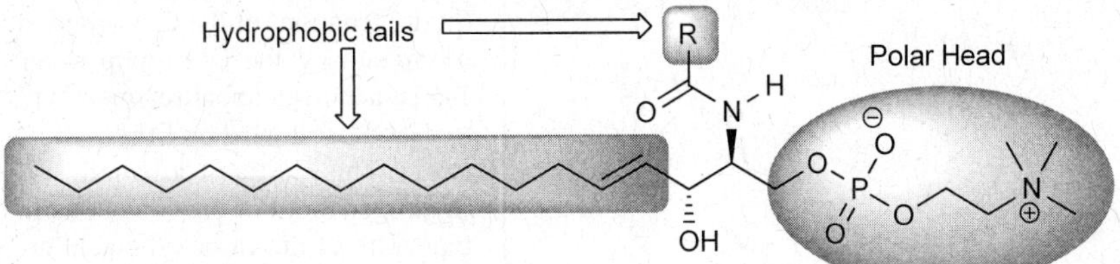

(b) Yes, they have one polar head and two hydrophobic tails. See Figure 26.6.

26.51.

(a) The epoxide can be formed on the top face of the π bond, as shown:

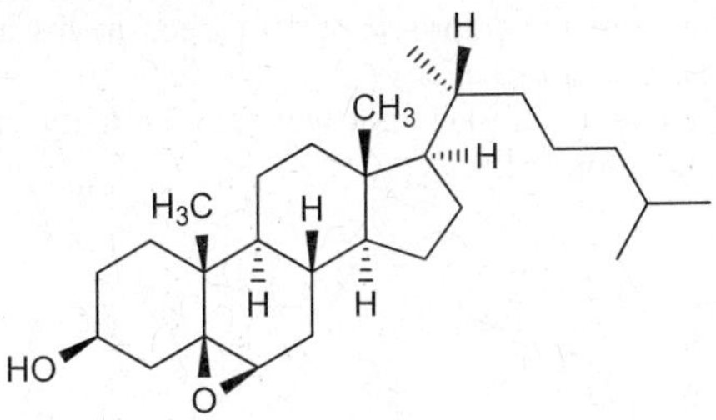

or the epoxide can be formed on the bottom face of the π bond, as shown:

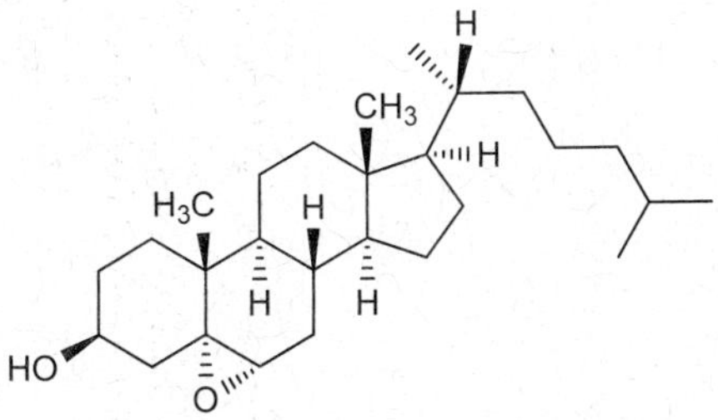

(b) The methyl group (C19) provides steric hindrance that blocks one side of the π bond, and only the following epoxide is obtained:

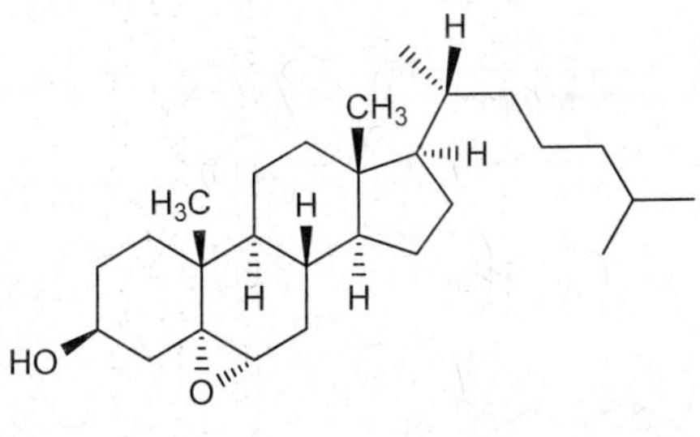

26.52.

(a) Estradiol has an aromatic ring that bears an OH group. As such, the ring is strongly activated toward electrophilic aromatic substitution. Upon treatment with excess Br$_2$, bromination occurs at the two positions that are *ortho* to the OH group, which is an *ortho-para* director.

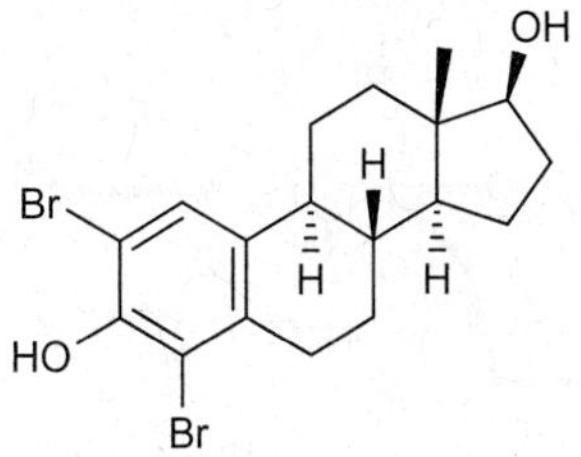

The *para* position is already occupied so bromination does not occur at that location.

(b) Upon treatment with PCC, the secondary alcohol is oxidized to give the following ketone:

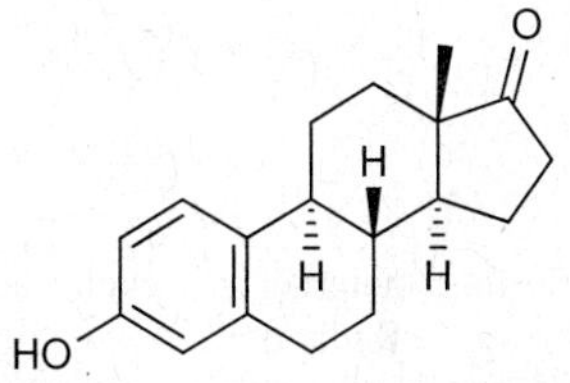

(c) Upon treatment with a strong base, followed by excess ethyl iodide, each of the OH group undergoes alkylation, thereby converting the OH groups into ethoxy groups, as shown:

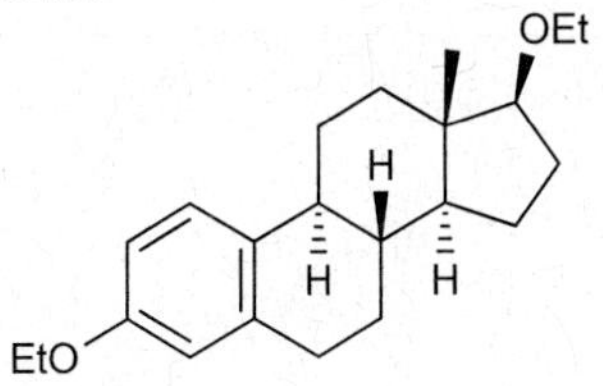

(d) Upon treatment with excess acetyl chloride in the presence of pyridine, each of the OH group undergoes acetylation, giving the following product:

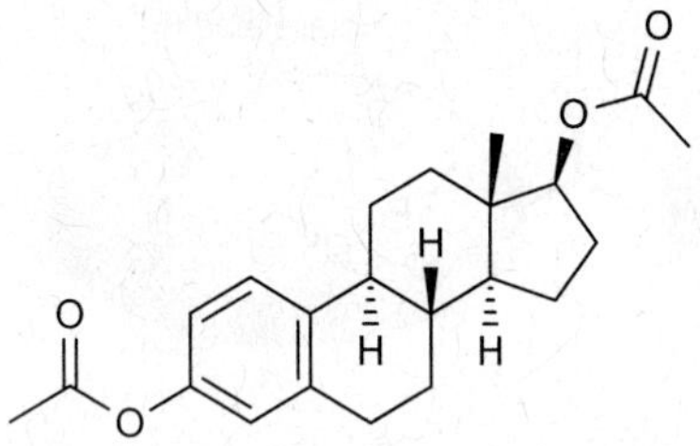

26.53. Every one of the OH groups in sucrose (see Section 24.7) is converted into an ester group, as shown below, where R is used to represent the hydrophobic tail of each lauric acid residue. This compound is not superimposable on its mirror image, so it is chiral (much like sucrose).

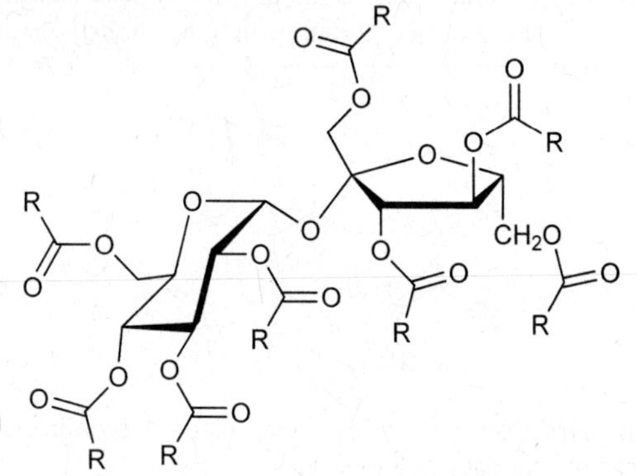

26.54.
(a) This transformation requires reduction (hydrogenation) of the C=C bond in oleic acid, which can be achieved upon treatment with H_2 in the presence of Ni.

(b) This transformation requires reduction (hydrogenation) of the C=C bond in oleic acid, as well as conversion of the OH group to an ethoxy group. This can be achieved upon treatment with H_2 and Ni, followed by NaOH, followed by EtI.

(c) This transformation requires reduction (hydrogenation) of the C=C bond in oleic acid, as well as reduction of the carboxylic acid group to give a primary alcohol. This can be achieved upon treatment with H_2 and Ni, followed by LAH, followed by water work-up.

(d) Ozonolysis (O_3, followed by DMS) followed by oxidation with $Na_2Cr_2O_7$ and H_2SO_4 will generate the desired dicarboxylic acid.

(e) This transformation requires reduction (hydrogenation) of the C=C bond in oleic acid, as well as installation of a bromine atom at the α position. This can be achieved upon treatment with H_2 and Ni, followed by PBr_3 and Br_2, followed by H_2O.

26.55.
(a) Limonene is comprised of 10 carbon atoms and is therefore a monoterpene.
(b) The compound does not have any chirality centers and is, therefore, achiral:

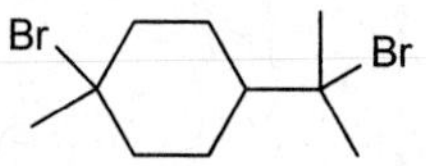

(c) Ozonolysis of limonene causes cleavage of each C=C bond, giving a tricarbonyl compound and formaldehyde, as shown:

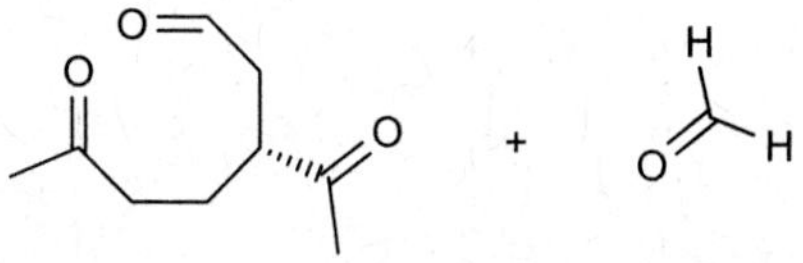

26.56. The starting material is a cyclic acetal. Upon treatment with aqueous acid, the acetal is opened to give a dihydroxyaldehyde. Reduction of the aldehyde group gives glycerol, which is then converted into the desired triglyceride upon treatment with an excess of the acyl halide in the presence of pyridine:

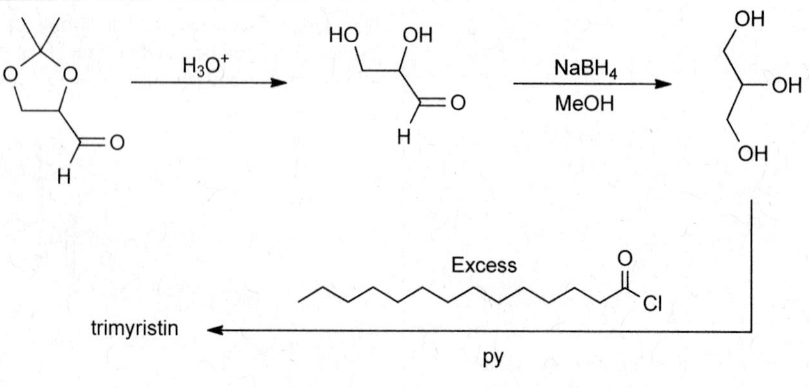

26.57.

(a) Fats and oils have a glycerol backbone connected to three fatty acid residues. Plasmalogens also have a glycerol backbone, but it is only connected to two fatty acid residues. The third group is not a fatty acid residue.

(b) Each of the ester groups is hydrolyzed upon treatment with aqueous base, giving the following products:

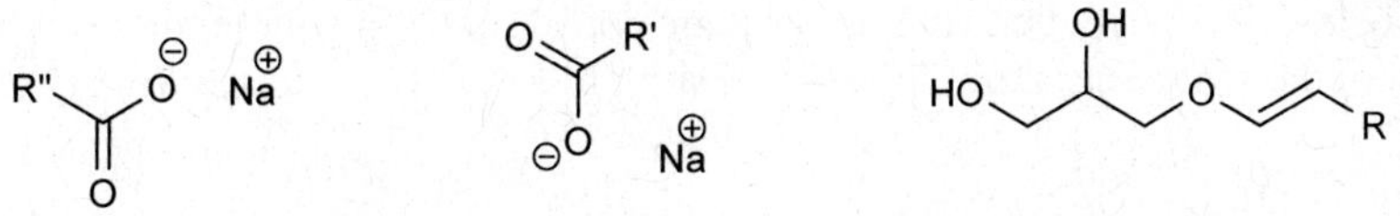

(c) In aqueous acid, the two ester groups undergo hydrolysis, just as we saw in basic conditions. Under these conditions, the ether also undergoes acidic cleavage, thereby freeing glycerol and an enol. Upon its formation, the enol rapidly tautomerizes to give an aldehyde. In summary, we expect the following products:

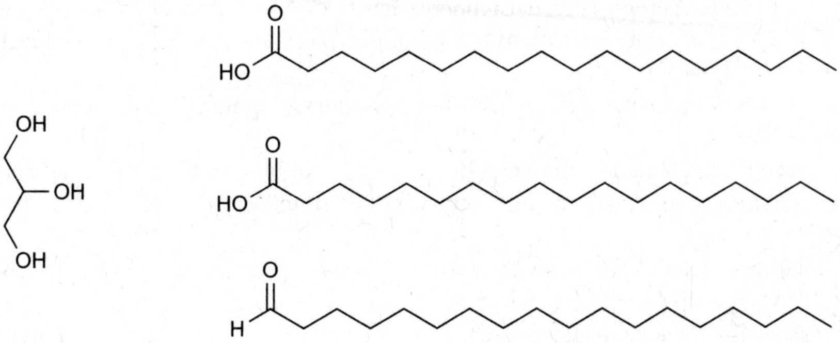

Chapter 27
Synthetic Polymers

Review of Concepts

Fill in the blanks below. To verify that your answers are correct, look in your textbook at the end of Chapter 27. Each of the sentences below appears verbatim in the section entitled *Review of Concepts and Vocabulary*.

- Polymers are comprised of repeating units that are constructed by joining _____ together.
- A _____ is a polymer made up of a single type of monomer. Polymers made from two or more different types of monomers are called _____.
- In a _____ **copolymer**, different homopolymer subunits are connected together in one chain. In a _____ **copolymer**, sections of one homopolymer have been grafted onto a chain of another homopolymer.
- Monomers can join together to form **addition polymers** by cationic, anionic, or _____ addition.
- Most derivatives of ethylene will undergo _____ polymerization under suitable conditions.
- Cationic addition is only efficient with derivatives of ethylene that contain an electron-_____ group.
- Anionic addition is only efficient with derivatives of ethylene that contain an electron-_____ group.
- Polymers generated via condensation reactions are called _____ **polymers**.
- _____-**growth polymers** are formed under conditions in which each monomer is added to the growing chain one at a time. The monomers do not react directly with each other. ,
- _____-**growth polymers** are formed under conditions in which the individual monomers react with each other to form _____, which are then joined together to form polymers.
- **Crossed-linked polymers** contain _____ bridges or branches that connect neighboring chains.
- **Thermoplastics** are polymers that are _____ at room temperature but _____ when heated. They are often prepared in the presence _____ to prevent the polymer from being brittle.
- _____ are polymers that return to their original shape after being stretched.
- _____ **polymers** can be broken down by enzymes produced by microorganisms in the soil.

Review of Skills

Fill in the blanks and empty boxes below. To verify that your answers are correct, look in your textbook at the end of Chapter 27. The answers appear in the section entitled *SkillBuilder Review*.

27.1 Determining Which Polymerization Technique is More Efficient

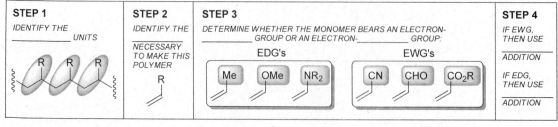

27.2   Identifying the Monomers Required to Produce a Desired Condensation Polymer

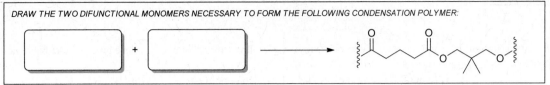

Review of Reactions

Identify the reagents necessary to achieve each of the following transformations. To verify that your answers are correct, look in your textbook at the end of Chapter 27. The answers appear in the section entitled *Review of Reactions*.

Reactions for Formation of Chain-Growth Polymers

$$\text{R} \quad \xrightarrow[\text{heat or light}]{\text{ROOR}} \quad \text{R} \quad \text{R} \quad \text{R}$$

$$\text{EDG} \quad \xrightarrow{\text{BF}_3 \,,\, \text{H}_2\text{O}} \quad \text{EDG} \quad \text{EDG} \quad \text{EDG}$$

$$\text{EWG} \quad \xrightarrow[\text{2) H}_2\text{O or CO}_2]{\text{1) BuLi}} \quad \text{EWG} \quad \text{EWG} \quad \text{EWG}$$

Reactions for Formation of Step-Growth Polymers

$$\underset{\text{R}}{\overset{\text{O}}{\|}}\text{OH} \quad + \quad \text{H—O—R} \longrightarrow$$

$$\underset{\text{R}}{\overset{\text{O}}{\|}}\text{O}^{\text{R}} \quad + \quad \text{H}_2\text{O}$$

$$\underset{\text{Cl}}{\overset{\text{O}}{\|}}\text{Cl} \quad \longrightarrow \quad \underset{\text{RO}}{\overset{\text{O}}{\|}}\text{OR}$$

$$\underset{\text{R}}{\overset{\text{O}}{\|}}\text{OH} \quad \longrightarrow \quad \underset{\text{R}}{\overset{\text{O}}{\|}}\text{NHR}$$

$$\text{R}_{\text{N}}{=}\text{C}{=}\text{O} \quad + \quad \text{ROH} \quad \longrightarrow \quad \text{R}_{\text{N}}\text{C}_{\text{OR}}$$

Solutions

27.1.
(a) Polymerization of vinyl acetate gives poly(vinyl acetate):

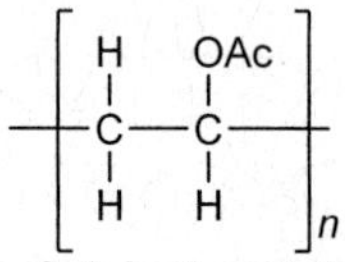

poly(vinyl acetate)

(b) Polymerization of vinyl bromide gives poly(vinyl bromide):

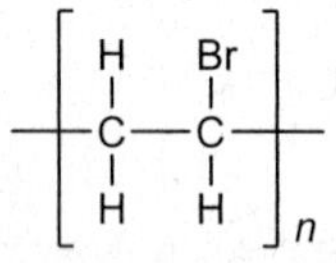

poly(vinyl bromide)

(c) Polymerization of α-butylene gives poly-α-butylene:

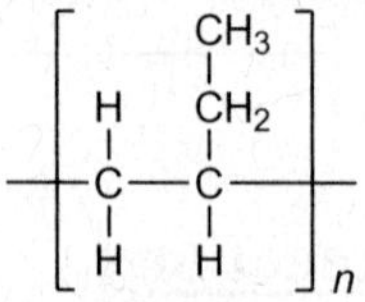

poly-α-butylene

27.2. Poly(methyl acrylate) can be made from methyl acrylate, shown here:

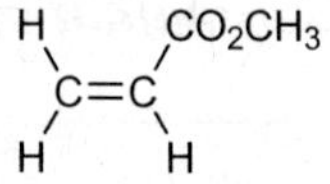

methyl acrylate

27.3. The following structure represents an alternating copolymer constructed from styrene and ethylene. The styrene and ethylene units are highlighted:

27.4. The following structure represents a block copolymer constructed from propylene and vinyl chloride. The propylene and vinyl chloride units are highlighted:

27.5. This copolymer can be made from isobutylene and styrene, as shown:

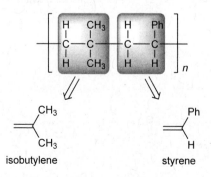

isobutylene   styrene

27.6. In each case, we identify the nature of the vinylic group, which determines the conditions to use. Anionic conditions are used if the vinylic group is electron-withdrawing, while cationic conditions are used for an an electron-donating group:

(a) A cyano group is an electron-withdrawing substituent (see Table 19.1 and associated discussion), so preparation of this compound would be best achieved via anionic addition.

(b) A methoxy group is an electron-donating substituent (see Table 19.1 and associated discussion), so preparation of this compound would be best achieved via cationic addition.

(c) Methyl groups are electron-donating substituents (see Table 19.1 and associated discussion), so preparation of this compound would be best achieved via cationic addition.

(d) An acetate group is an electron-donating substituent (see Table 19.1 and associated discussion), so preparation of this compound would be best achieved via cationic addition.

(e) A nitro group is an electron-withdrawing substituent (see Table 19.1 and associated discussion), so preparation of this compound would be best achieved via anionic addition.

(f) A trichloromethyl group is an electron-withdrawing substituent (see Table 19.1 and associated discussion), so preparation of this compound would be best achieved via anionic addition.

27.7. An acetate group is more powerfully electron donating (via resonance) than a methyl group (via hyperconjugation), so vinyl acetate is expected be the most reactive toward cationic polymerization. A nitro group is electron withdrawing, so nitroethylene is

expected to be the least reactive toward cationic polymerization.

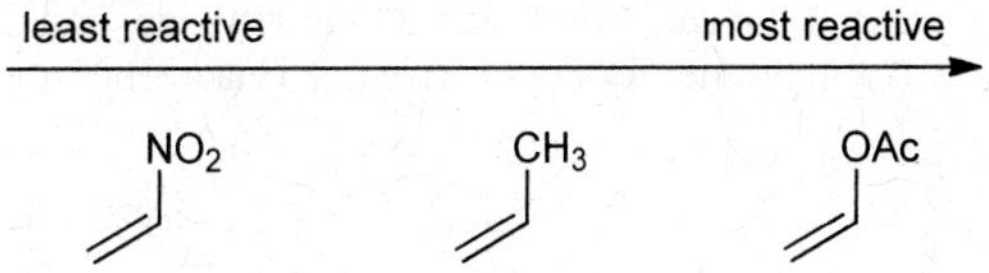

27.8. A nitro group is a very powerful electron withdrawing group (see Table 19.1 and associated discussion), so nitroethylene is expected to be the most reactive toward anionic polymerization. A chlorine atom is only weakly electron-withdrawing, as compared with a nitro group or a carbonyl group (see Table 19.1 and associated discussion), so vinyl chloride is expected to be the least reactive toward anionic polymerization.

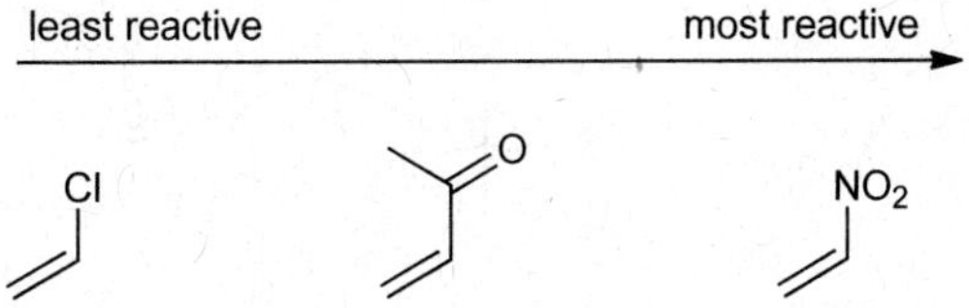

27.9. A benzylic anion, cation or radical will be stabilized by resonance.

27.10. In the initiation step, water attacks one molecule of the monomer, giving a carbanion. This carbanion then attacks another molecule of the monomer in a propagation step. This propagation step repeats itself, thereby growing the polymer chain. A termination step can occur if the carbanion is protonated by water, as shown:

Initiation

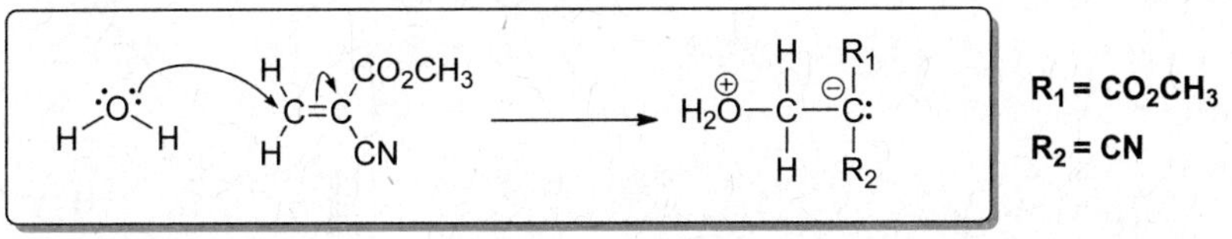

$R_1 = CO_2CH_3$

$R_2 = CN$

Propagation

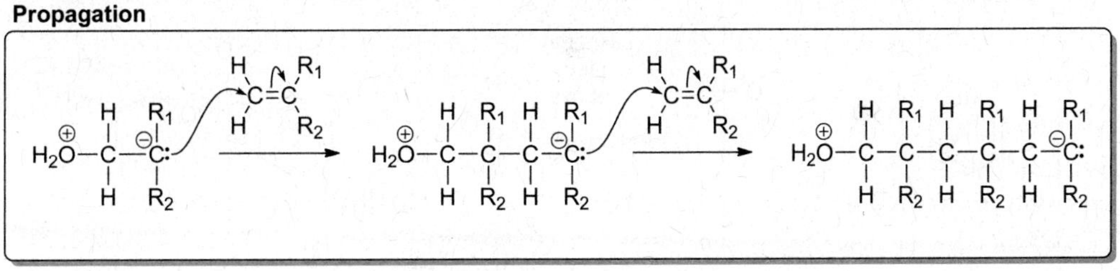

Termination

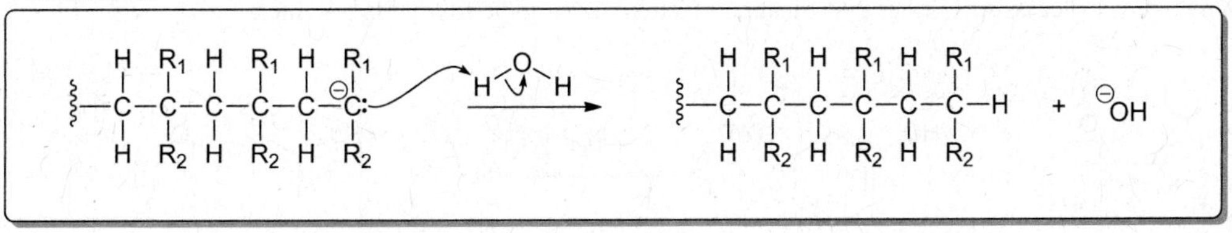

27.11. Protonation of one of the carbonyl groups renders it even more electrophilic, and it is then attacked by ethylene glycol to give a tetrahedral intermediate (an oxonium ion). Two successive proton transfer steps convert the oxonium ion into another oxonium ion, which can lose water to regenerate the C=O bond. Deprotonation generates an ester. This ester has a carbonyl group on the left side, and an OH group on the right side. As a result, this compound can serve as a monomer for polymerization, (via a repetition of the steps described above).

27.12. Oxalic acid bears two carboxylic acid groups, while resorcinol bears two OH groups. These two compounds can polymerize via successive Fischer esterification reactions, giving the following polymer:

oxalic acid   *resorcinol*

27.13. Each of the amide groups can be made via the reaction between a carboxylic acid group and an amino group. Therefore, Kevlar can be made from the following dicarboxylic acid and the following diamine:

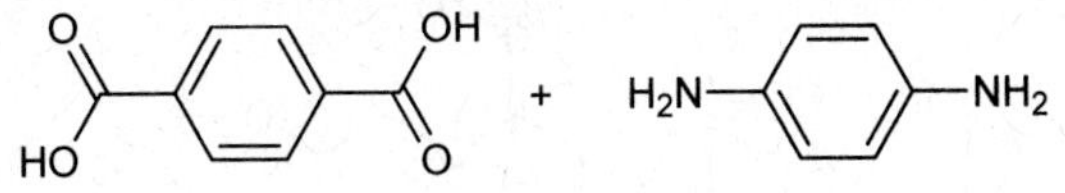

27.14.
(a) Each of the ester groups can be made from the reaction between a carboxylic acid group and an alcohol (via a Fischer esterification reaction). Therefore, the desired polymer can be made from the following diol and the following dicarboxylic acid:

(b) Each carbonate group can be made from the reaction between phosgene and two alcohols. Therefore, the desired polymer can be made from the following diol and phosgene:

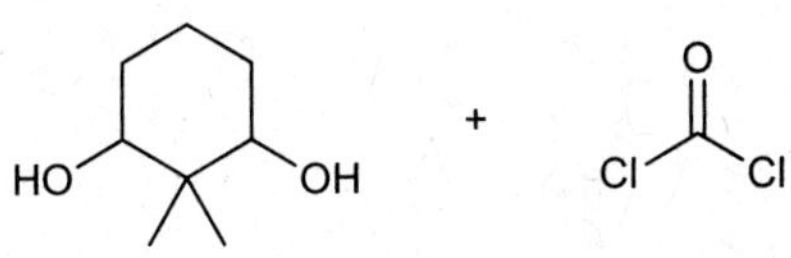

27.15. Each of the OH groups in 1,4-butanediol can attack phosgene, giving the following polymer:

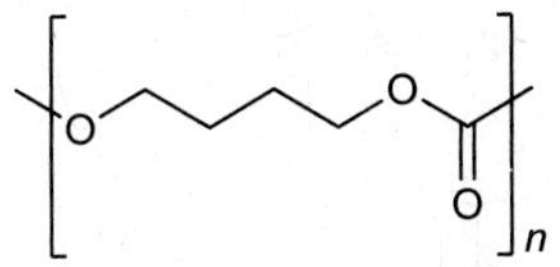

27.16.
(a) ε-Aminocaproic acid has both a carboxylic acid group and an amino group (which can react with each, in an intermolecular fashion, to give an amide linkage). As such, this compound will polymerize to from the following polymer:

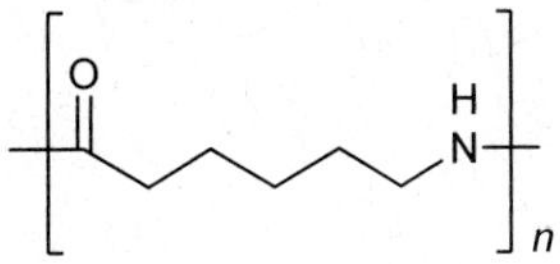

(b) Nylon 6 exhibits a smaller repeating unit than Nylon 6,6.

27.17.
(a) Each monomer has two growth points, so we expect that polymerization will generate a step-growth polymer.

(b) When these monomers react are used to form a copolymer, the growing polymer chain has only one growth point, so we expect that polymerization will generate a chain-growth polymer.

27.18. This polymer exhibits repeating carbonate groups, so it can be made from phosgene and the appropriate diol. Since the diol has two growth points (and since the growing oligomers also have two growth points), this polymer can be classified as a step-growth polymer.

27.19. Polyisobutylene does not have any chirality centers.

27.20. LDPE is used to make Ziploc bags (a flexible product, like trash bags) and HDPE is used to make folding tables (an inflexible product, like Tupperware).

27.21. Protonation of the ester renders it even more electrophilic, and it is then attacked by water to give a tetrahedral intermediate (an oxonium ion). Two successive proton transfer steps convert the oxonium ion into another oxonium ion, which can lose a leaving group to regenerate the C=O bond. Deprotonation generates a carboxylic acid (and an alcohol). These steps are then repeated to give terephthalic acid and ethylene glycol, as shown:

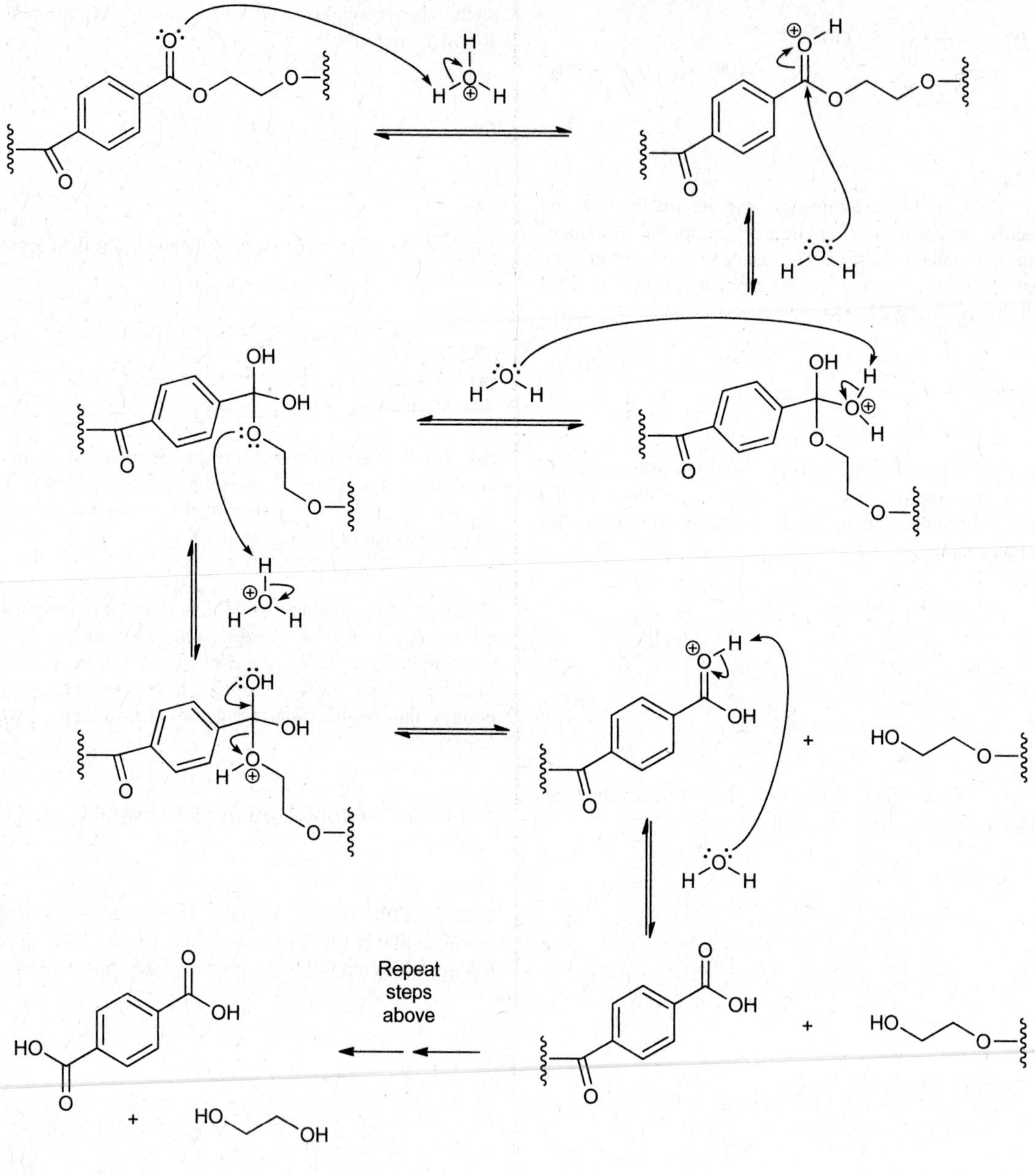

27.22.

(a) Polymerization of nitroethylene gives polynitro-ethylene:

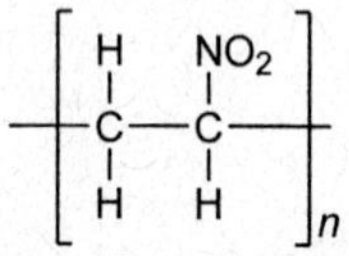

polynitroethylene

(b) Polymerization of acrylonitrile gives polyacrylo-nitrile:

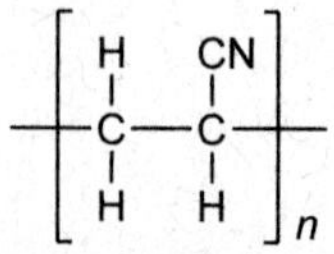

polyacrylonitrile

(c) Polymerization of vinylidene fluoride gives poly(vinylidene fluoride):

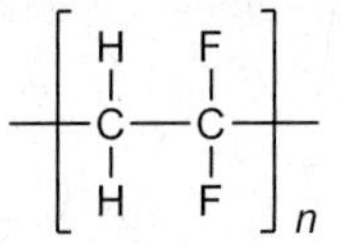

poly(vinylidene fluoride)

27.23.
(a) Each of the ester groups can be made from the reaction between a carboxylic acid group and an alcohol (via a Fischer esterification reaction). Therefore, the desired polymer can be made from the following dicarboxylic acid and the following diol:

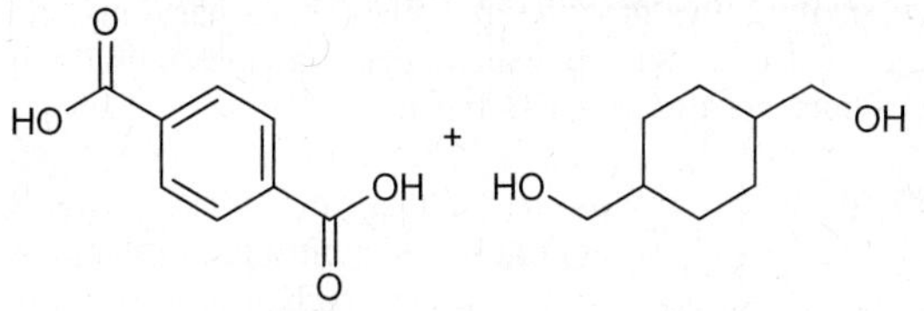

(b) A Fischer esterification requires acidic conditions.

27.24. This copolymer can be made from the following monomers:

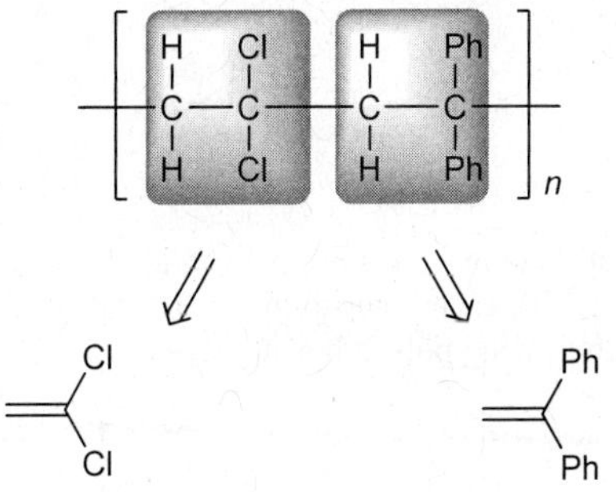

27.25. The following structure represents a block copolymer constructed from isobutylene and styrene. The isobutylene and styrene units are highlighted:

27.26. The following structure represents an alternating copolymer constructed from vinyl chloride and ethylene. The vinyl chloride and ethylene units are highlighted:

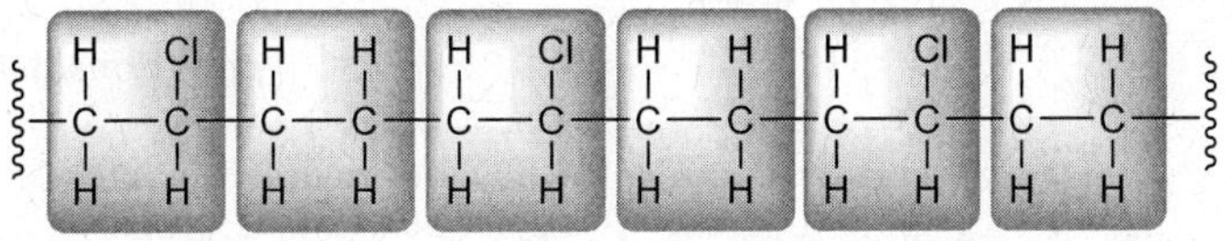

27.27. An acetate group is an electron-donating substituent (via resonance), while the other two groups (CN and Cl) are electron-withdrawing substituents. Therefore, vinyl acetate is expected be the most reactive toward cationic polymerization.

27.28. A cyano group is an electron-withdrawing substituent (via resonance), while the other two groups (acetate and methyl) are both electron-donating substituents. Therefore, the compound bearing the cyano group is expected be the most reactive toward anionic polymerization.

27.29. All three polymers are step-growth polymers, because in each case, the growing oligomers have two growth points.
(a) The starting materials are a diacid and a diamine, which can be linked together via amide groups, giving the following polymer:

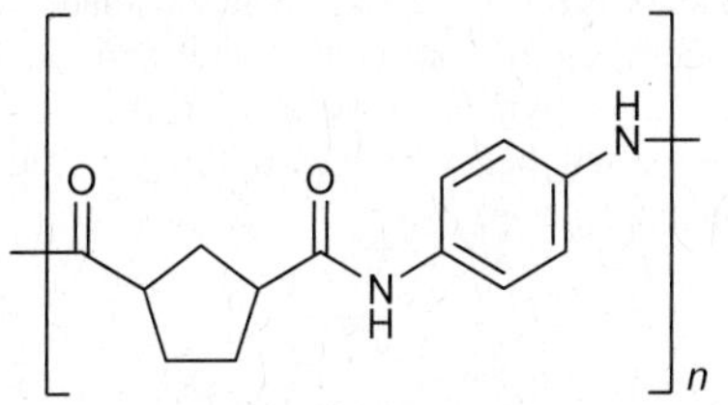

(b) The starting materials are a diol and a diisocyanate, which can be linked together as carbamate groups, giving the following polyurethane:

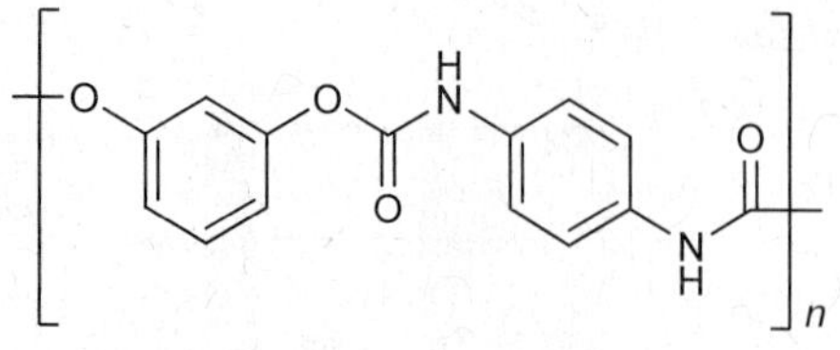

(c) The starting materials are a diol and phosgene, which will react with each other to give carbonate groups, and thus the following polycarbonate:

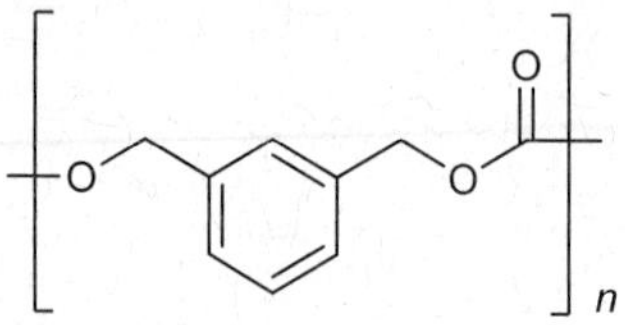

27.30.
(a) The starting materials are a diacid and a diamine, which can be linked together via amide groups, giving the following polymer:

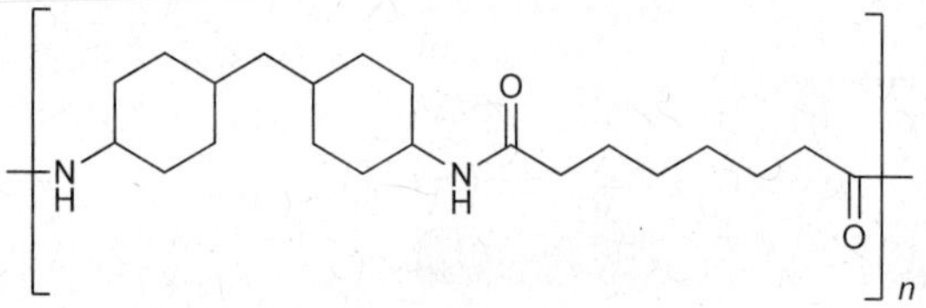

(b) Quiana is a polyamide.
(c) Quiana is a step-growth polymer, because each of the growing oligomers has two growth points.
(d) Quiana is a condensation polymer because it is made via a condensation process (between carboxylic acid and amino groups).

27.31.
(a) Each of the amide groups can be made from the reaction between a carboxylic acid and an amino group. Therefore, this polymer can be made from the following monomer, which bears both the amino group and the carboxylic acid group:

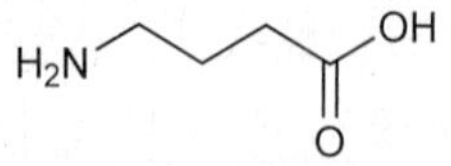

(b) Each of the ester groups can be made from the reaction between a carboxylic acid and an alcohol. Therefore, this polymer can be made from the following monomer, which bears both a hydroxyl group and a carboxylic acid group:

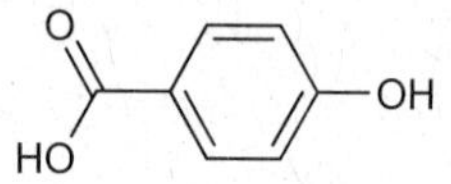

27.32. The starting materials are a diol and phosgene, which will react with each other to give carbonate groups, and thus the following polycarbonate:

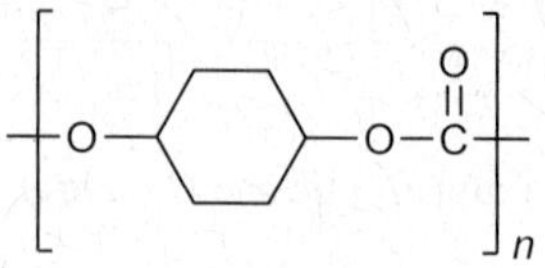

27.33.
(a) Each monomer has two growth points, so we expect that polymerization will generate a step-growth polymer.

(b) When these monomers react are used to form a copolymer, the growing polymer chain has only one growth point, so we expect that polymerization will generate a chain-growth polymer.

27.34. Nitro groups are among the most powerful electron-withdrawing groups, and a nitro group stabilizes a negative charge on an adjacent carbon atom, thereby facilitating anionic polymerization.

27.35. Shower curtains are made from PVC, which is a thermoplastic polymer. To prevent the polymer from being brittle, the polymer is prepared in the presence of plasticizers which become trapped between the polymer chains where they function as lubricants. Over time, the plasticizers evaporate, and the polymer becomes brittle.

27.36.
(a) Polyformaldehyde is a polymer that is assembled from repeating formaldehyde (CH_2O) units, as shown:

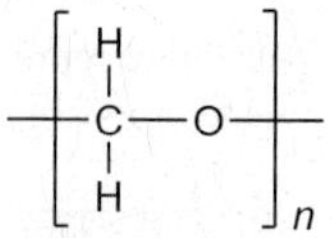

polyformaldehyde

(b) Polyformaldehyde has repeating ether groups, so it is a polyether.
(c) The growing polymer chain has only one growth point, so polyformaldehyde is classified as a chain-growth polymer.
(d) Polyformaldehyde is an addition polymer, because it is formed via successive addition reactions (involving the π bond in each molecule of formaldehyde).

27.37. It bears an electron-withdrawing group (CN) that can stabilize a negative charge via resonance, but it also bears an electron-donating group (OMe) that can stabilize a positive charge via resonance.

27.38. The nitro group serves as a reservoir of electron density that stabilizes a negative charge via resonance (see Chapter 19).

27.39. The methoxy group is an electron donating group that stabilizes a positive charge via resonance (see Chapter 19).

27.40. A methoxy group can only donate electron density via resonance if it is located in an *ortho* or *para* position. It cannot stabilize the developing carbocation if it is located in a *meta* position (see Chapter 19).

27.41.
(a) In a syndiotactic polymer, the chirality centers exhibit alternating configuration, as shown:

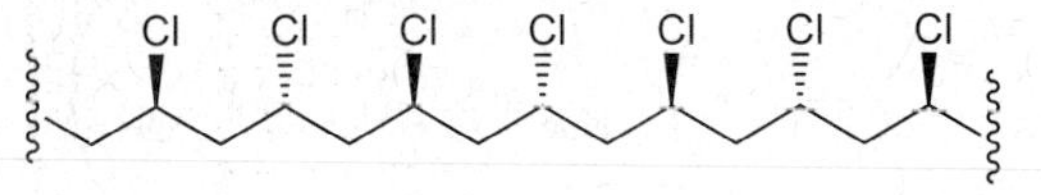

(b) In an isotactic polymer, the chirality centers all exhibit the same configuration, as shown:

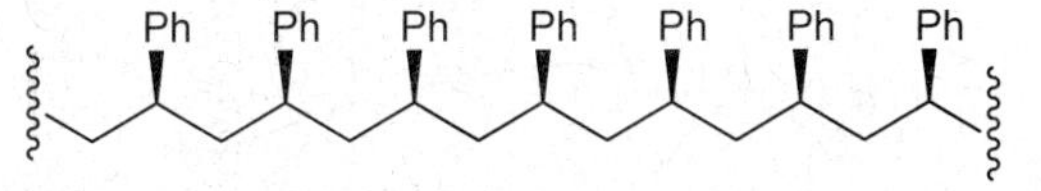

27.42.
(a) The desired polymer is a polyurethane, which can be prepared from the following diisocyanate and the following diol:

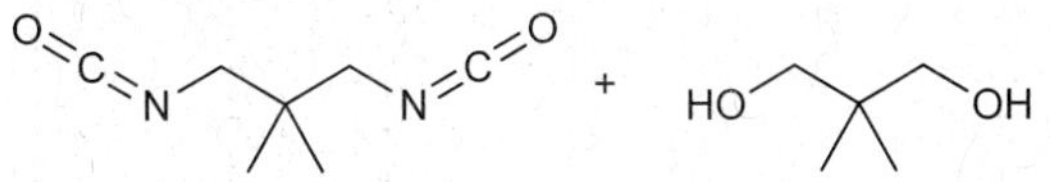

(b) Each monomer has two growth points, so we expect that polymerization will generate a step-growth polymer.
(c) Polyurethanes are classified as addition polymers (see end of Section 27.5), because they are formed via successive addition reactions.

27.43. A ketone will react with a primary amine (under acid-catalyzed conditions) to give an imine (see Section 20.6). The starting materials are a dione and a diamine, so we expect formation of the following polyimine:

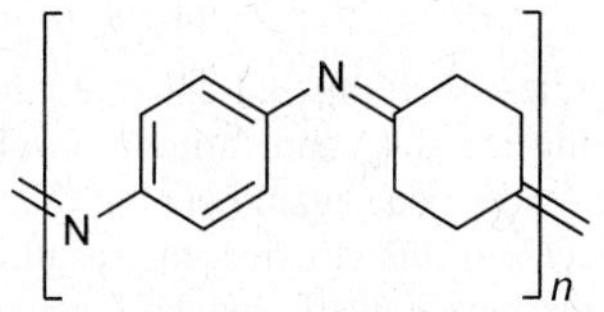

27.44. Vinyl alcohol is an enol, which is not stable. If it is prepared, it undergoes rapid tautomerization to give an aldehyde, which will not produce the desired product upon polymerization.

27.45. The ester groups undergo hydrolysis in basic conditions, which breaks down the polymer into monomers.

27.46.
(a) The carbocation that is initially formed is a secondary carbocation, and it can undergo a carbocation rearrangement to give a more stable, tertiary carbocation. In some cases, the secondary carbocation will be added to the growing polymer chain before it has a chance to rearrange. In other cases, the secondary carbocation will rearrange first and then be added to the growing polymer chain. The result is the incorporation of two different repeating units in the growing polymer chain.

(b) The following structure represents a segment of the random copolymer described in the solution to **27.46(a)**. The repeating units are highlighted:

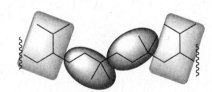

(c) Yes, because a secondary carbocation is formed when 3,3-dimethyl-1-butene is protonated, and a methyl shift can occur that converts the secondary carbocation into a tertiary carbocation.

27.47.
(a) As described in the problem statement, the epoxide ring is opened with a strong nucleophile to form an alkoxide ion, which then functions as a nucleophile and attacks another molecule of ethylene oxide. This process repeats itself, thereby forming poly(ethylene oxide), as shown:

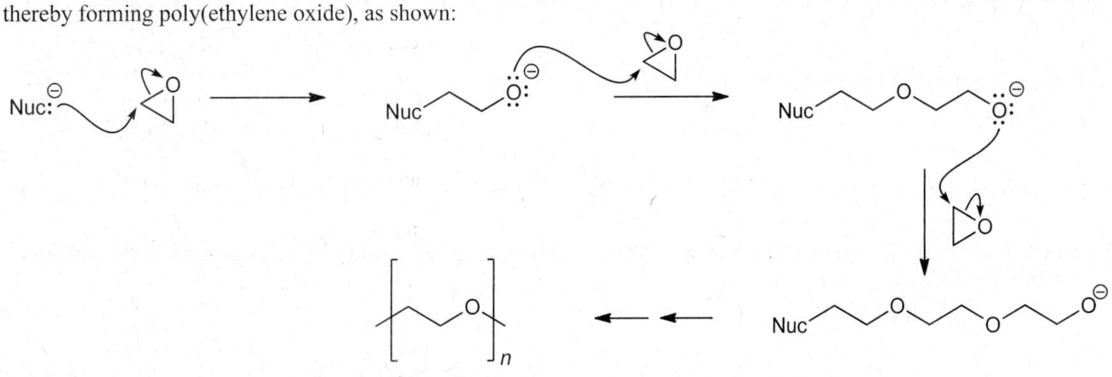

poly(ethylene oxide)

(b) Under acidic conditions, an epoxide can be protonated. A nucleophile can then attack the protonated epoxide, thereby opening the ring, and forming an alcohol. An alcohol is a weak nucleophile and it can then attack another protonated epoxide, once again opening the ring. The resulting oxonium ion is then deprotonated. If the base for this step is a molecule of the epoxide, the resulting protonated epoxide can then serve as the electrophile for the next step. This process can repeat itself, thereby forming poly(ethylene oxide), as shown:

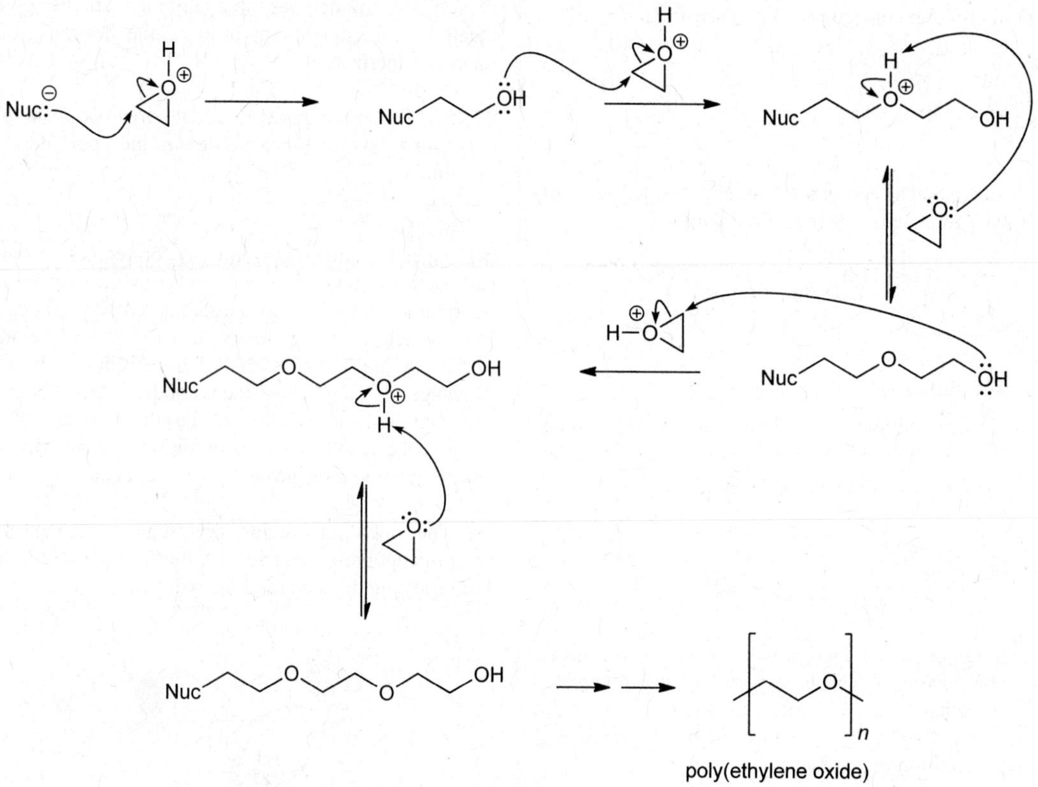

poly(ethylene oxide)

(c) The desired polymer is similar in structure to poly(ethylene oxide), but there is a *gem*-dimethyl group present in the repeating unit. This polymer can be made if the starting epoxide also bears a *gem*-dimethyl group:

(d) Preparation of this polymer would require the following epoxide:

Acidic conditions will be required, because the epoxide is too sterically hindered to be attacked under basic conditions (see Section 14.10).

27.48. Each of the highlighted positions represents an acetal group (see Section 20.5):

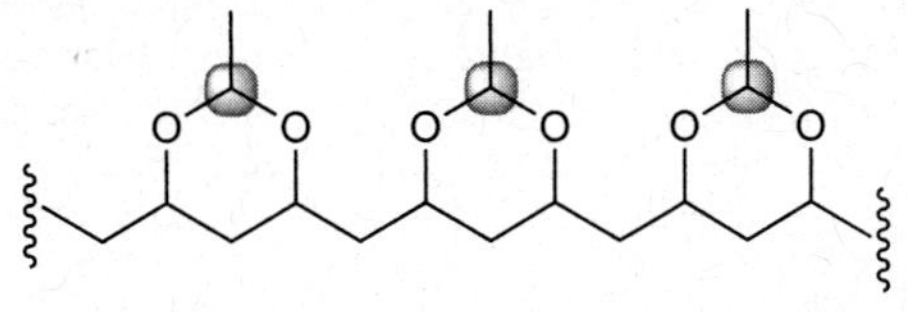

Therefore, this polymer can be made via acetal formation, from poly(vinyl alcohol) and acetaldehyde:

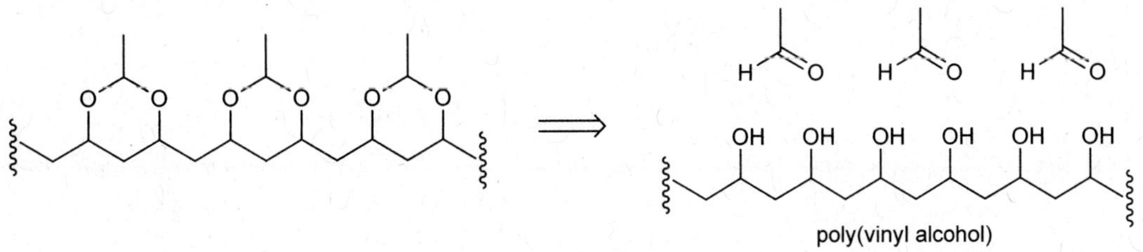

poly(vinyl alcohol)

Poly(vinyl alcohol) can be made from vinyl acetate in just two steps (as seen in Problem 27.44). Acetaldehyde can also be made from vinyl acetate (upon treatment with aqueous acid). Under these conditions, the acetate group is hydrolyzed, giving an enol, which tautomerizes to give acetaldehyde:

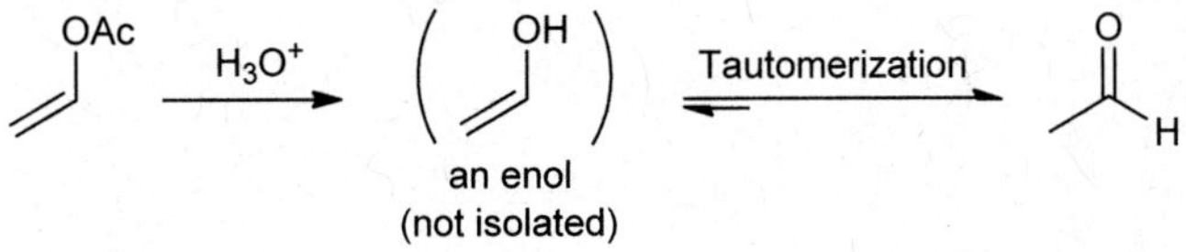

an enol
(not isolated)

The forward scheme is shown here. Polymerization of vinyl acetate gives poly(vinyl acetate), which can be treated with aqueous acid to give poly(vinyl alcohol), as seen in Problem 27.44. This polymer can then be treated with acetaldehyde (formed by treating vinyl acetate with aqueous acid) to give the desired polymer:

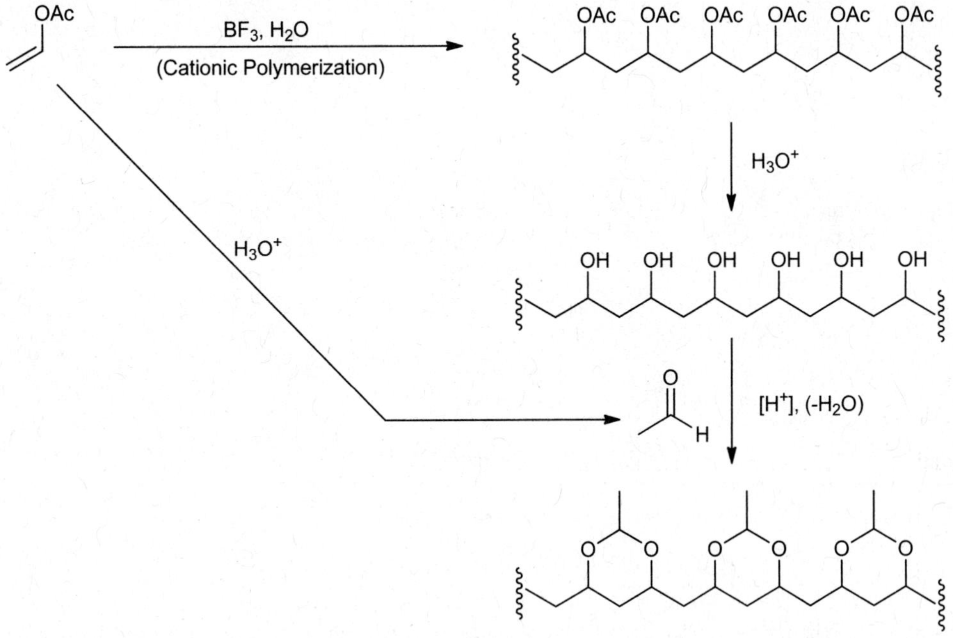